AF814345

The Most Accepted
Crash Course Programme

JEE Main in 40 DAYS

Mathematics

The Most Accepted
Crash Course Programme

JEE Main in 40 DAYS

Mathematics

Amit Kumar,
Vivak

arihant
SUCCESS AND NOTHING LESS

arihant
SUCCESS AND NOTHING LESS

ARIHANT PRAKASHAN (Series), MEERUT

卐 **Administrative & Production Offices**

Regd. Office

'Ramchhaya' 4577/15, Agarwal Road, Darya Ganj, New Delhi -110002
Tele: 011- 47630600, 43518550

卐 **Head Office**

Kalindi, TP Nagar, Meerut (UP) - 250002
Tel: 0121-7156203, 7156204

卐 **Sales & Support Offices**

Agra, Ahmedabad, Bengaluru, Bareilly, Chennai, Delhi, Guwahati, Hyderabad, Jaipur, Jhansi, Kolkata, Lucknow, Nagpur & Pune.

卐 **PRICE** ₹525.00

卐 **PO No :** TXT-XX-XXXXXXX-X-XX

Published by Arihant Publications (India) Ltd.

For further information about the books published by Arihant, log on to www.arihantbooks.com or e-mail at info@arihantbooks.com

Follow us on

PREFACE

It is a fact that nearly 10 lacs students would be in the race with you in JEE Main, the gateway to some of the prestigious engineering and technology institutions in the country, requires that you take it seriously and head-on. A slight underestimation or wrong guidance will ruin all your prospects. You have to earmark the topics in the syllabus and have to master them in concept-driven-problem-solving ways, considering the thrust of the questions being asked in JEE Main.

The book **40 Days JEE Main Mathematics** serves the above cited purpose in perfect manner. At whatever level of preparation you are before the exam, this book gives you an accelerated way to master the whole JEE Main Mathematics Syllabus. It has been conceived keeping in mind the latest trend of questions, and the level of different types of students.

The whole syllabus of Mathematics has been divided into day-wise-learning modules with clear groundings into concepts and sufficient practice with solved and unsolved questions on that day. After every few days you get a Unit Test based upon the topics covered before that day. On last three days you get three full-length Mock Tests, making you ready to face the test. It is not necessary that you start working with this book in 40 days just before the exam. You may start and finish your preparation of JEE Main much in advance before the exam date. This will only keep you in good frame of mind and relaxed, vital for success at this level.

Salient Features

- Concepts discussed clearly and directly without being superfluous. **Only the required material** for JEE Main being described comprehensively to keep the students focussed.
- Exercises for each day give you the collection of **only the Best Questions** of the concept, giving you the perfect practice in less time.
- Each day has two Exercises; **Foundation Questions Exercise** having Topically Arranged Questions & **Progressive Question Exercise** having higher Difficulty Level Questions.
- **All types of Objective Questions** included in Daily Exercises (Single Option Correct, Assertion & Reason, etc).
- Along with Daywise Exercises, there above also the **Unit Tests** & **Full Length Mock Tests**.
- At the end, there are all Online Solved Papers of JEE Main 2023; Jan & April attempts.

We are sure that 40 Days Mathematics for JEE Main will give you a fast way to prepare for Mathematics without any other support or guidance.

Publisher

CONTENTS

Preparing JEE Main 2024 Mathematics in 40 Days !

SYLLABUS

MATHEMATICS

UNIT 1 Sets, Relations and Functions

Sets and their representation; Union, intersection and complement of sets and their algebraic properties; Power set; Relation, Types of relations, equivalence relations, functions;. one-one, into and onto functions, composition of functions.

UNIT 2 Complex Numbers and Quadratic Equations

Complex numbers as ordered pairs of reals, Representation of complex numbers in the form *a+ib* and their representation in a plane, Argand diagram, algebra of complex numbers, modulus and argument
(or amplitude) of a complex number, square root of a complex number, triangle inequality, Quadratic equations in real and complex number system and their solutions. Relation between roots and
co-efficients, nature of roots, formation of quadratic equations with given roots.

UNIT 3 Matrices and Determinants

Matrices, algebra of matrices, types of matrices, determinants and matrices of order two and three. Properties of determinants, evaluation of deter-minants, area of triangles using determinants. Adjoint and evaluation of inverse of a square matrix using determinants and elementary transformations, Test of consistency and solution of simultaneous linear equations in two or three variables using determinants and matrices.

UNIT 4 Permutations and Combinations

Fundamental principle of counting, permutation as an arrangement and combination as selection, Meaning of $P(n,r)$ and $C(n,r)$, simple applications.

UNIT 5 Mathematical Induction

Principle of Mathematical Induction and its simple applications.

UNIT 6 Binomial Theorem and its Simple Applications

Binomial theorem for a positive integral index, general term and middle term, properties of Binomial coefficients and simple applications.

UNIT 7 Sequences and Series

Arithmetic and Geometric progressions, insertion of arithmetic, geometric means between two given numbers. Relation between AM and GM Sum upto n terms of special series: $\sum n, \sum n^2, \sum n^3$. Arithmetico - Geometric progression.

UNIT 8 Limit, Continuity and Differentiability

Real valued functions, algebra of functions, polynomials, rational, trigonometric, logarithmic and exponential functions, inverse functions. Graphs of simple functions. Limits, continuity and differentiability.

Differentiation of the sum, difference, product and quotient of two functions. Differentiation of trigonometric, inverse trigonometric, logarithmic exponential, composite and implicit functions derivatives of order upto two. Rolle's and Lagrange's Mean Value Theorems. Applications of derivatives: Rate of change of quantities, monotonic - increasing and decreasing functions, Maxima and minima of functions of one variable, tangents and normals.

UNIT 9 Integral Calculus

Integral as an anti - derivative. Fundamental integrals involving algebraic, trigonometric, exponential and

logarithmic functions. Integration by substitution, by parts and by partial fractions. Integration using trigonometric identities.

Evaluation of simple integrals of the type

$$\int \frac{dx}{x^2 \pm a^2}, \quad \int \frac{dx}{\sqrt{x^2 \pm a^2}}, \quad \int \frac{dx}{a^2 - x^2}, \quad \int \frac{dx}{\sqrt{a^2 - x^2}},$$

$$\int \frac{dx}{ax^2 + bx + c}, \quad \int \frac{dx}{\sqrt{ax^2 + bx + c}}, \quad \int \frac{(px + q)\,dx}{ax^2 + bx + c},$$

$$\int \frac{(px + q)\,dx}{\sqrt{ax^2 + bx + c}}, \quad \int \sqrt{a^2 \pm x^2}\,dx \quad \text{and} \quad \int \sqrt{x^2 - a^2}\,dx$$

Integral as limit of a sum. Fundamental Theorem of Calculus. Properties of definite integrals. Evaluation of definite integrals, determining areas of the regions bounded by simple curves in standard form.

UNIT 10 Differential Equations

Ordinary differential equations, their order and degree. Formation of differential equations. Solution of differential equations by the method of separation of variables, solution of homogeneous and linear differential equations of the type $\frac{dy}{dx} + p(x)\,y = q(x)$

UNIT 11 Coordinate Geometry

Cartesian system of rectangular coordinates in a plane, distance formula, section formula, locus and its equation, translation of axes, slope of a line, parallel and perpendicular lines, intercepts of a line on the coordinate axes.

- **Straight Lines**

  Various forms of equations of a line, intersection of lines, angles between two lines, conditions for concurrence of three lines, distance of a point from a line, equations of internal and external bisectors of angles between two lines, coordinates of centroid, orthocentre and circumcentre of a triangle, equation of family of lines passing through the point of intersection of two lines.

- **Circles, Conic Sections**

  Standard form of equation of a circle, general form of the equation of a circle, its radius and centre, equation of a circle when the end points of a diameter are given, points of intersection of a line and a circle with the centre at the origin and condition for a line to be tangent to a circle, equation of the tangent. Sections of cones, equations of conic sections (parabola, ellipse and hyperbola) in standard forms, condition for $y = mx + c$ to be a tangent and point (s) of tangency.

UNIT 12 Three Dimensional Geometry

Coordinates of a point in space, distance between two points, section formula, direction ratios and direction cosines, angle between two intersecting lines. Skew lines, the shortest distance between them and its equation. Equations of a line and a plane in different forms, intersection of a line and a plane, coplanar lines.

UNIT 13 Vector Algebra

Vectors and scalars, addition of vectors, components of a vector in two dimensions and three dimensional space, scalar and vector products, scalar and vector triple product.

UNIT 14 Statistics and Probability

Measures of Dispersion Calculation of mean, median, mode of grouped and ungrouped data. Calculation of standard deviation, variance and mean deviation for grouped and ungrouped data.

Probability Probability of an event, addition and multiplication theorems of probability, Baye's theorem, probability distribution of a random variate, Bernoulli trials and Binomial distribution.

UNIT 15 Trigonometry

Trigonometrical identities and equations. Trigonometrical functions. Inverse trigonometrical functions and their properties. Heights and Distances.

UNIT 16 Mathematical Reasoning

Statements, logical operations and implies, implied by, if and only if. Understanding of tautology, contradiction, converse and contra positive.

HOW THIS BOOK IS
USEFUL FOR YOU ?

As the name suggest, this is the perfect book for your recapitulation of the whole syllabus, as it provides you a capsule course on the subject covering the syllabi of JEE Main, with the smartest possible tactics as outlined below:

1. REVISION PLAN

The book provides you with a practical and sound revision plan.

The chapters of the book have been designed day-wise to guide the students in a planned manner through day-by-day, during those precious 35-40 days. Every day you complete a chapter/a topic, also take an exercise on the chapter. So that you can check & correct your mistakes, answers with hints & solutions also have been provided. By 37th day from the date you start using this book, entire syllabus gets revisited.

Again, as per your convenience/preparation strategy, you can also divide the available 30-35 days into two time frames, first time slot of 3 weeks and last slot of 1 & 1/2 week. Utilize first time slot for studies and last one for revising the formulas and important points. Now fill the time slots with subjects/topics and set key milestones. Keep all the formulas, key points on a couple of A4 size sheets as ready-reckner on your table and go over them time and again. If you are done with notes, prepare more detailed inside notes and go over them once again. Study all the 3 subjects every day. Concentrate on the topics that have more weightage in the exam that you are targeting.

2. MOCK TESTS

Once you finish your revision on 37th day, the book provides you with full length mock tests for day 38th, 39th, & 40th, thereby ensures your total & full proof preparation for the final show.

The importance of solving previous years' papers and 10-15 mock tests cannot be overemphasized. Identify your weaknesses and strengths. Work towards your strengths i.e., devote more time to your strengths to be 100% sure and confident. In the last time frame of 1 & 1/2 week, don't take-up anything new, just revise what you have studied before. Be exam-ready with quality mock tests in between to implement your winning strategy.

3. FOCUS TOPICS

Based on past years question paper trends, there are few topics in each subject which have more questions in exam than other. So far Mathematics is concerned it may be summed up as below:

Calculus, Trigonometry, Algebra, Coordinate Geometry & Vector 3D.

More than 80% of questions are normally asked from these topics.

However, be prepared to find a completely changed pattern for the exam than noted above as examiners keep trying to weed out 'learn by rot practice'. One should not panic by witnessing a new pattern , rather should be tension free as no one will have any upper hand in the exam.

4. IMPROVES STRIKE RATE AND ACCURACY

The book even helps to improve your strike rate & accuracy. When solving practice tests or mock tests, try to analyze where you are making mistakes-where are you wasting your time; which section you are doing best. Whatever mistakes you make in the first mock test, try to improve that in second. In this way, you can make the optimum use of the book for giving perfection to your preparation.

What most students do is that they revise whole of the syllabus but never attempt a mock and thus they always make mistake in main exam and lose the track.

5. LOG OF LESSONS

During your preparations, make a log of Lesson's Learnt. It is specific to each individual as to where the person is being most efficient and least efficient. Three things are important - what is working, what's not working and how would you like to do in your next mock test.

6. TIME MANAGEMENT

Most candidates who don't make it to good medical colleges are not good in one area- Time Management. And, probably here lies the most important value addition that's the book provides in an aspirant's preparation. Once the students go through the content of the book precisely as given/directed, he/she learns the tactics of time management in the exam.

Realization and strengthening of what you are good at is very helpful, rather than what one doesn't know. Your greatest motto in the exam should be, how to maximize your scoring with the given level of preparation. You have to get about 200 plus marks out of a total of about 400 marks for admission to a good NIT (though for a good branch one needs to do much better than that). Remember that one would be doomed if s/he tries to score 400 in about 3 hours.

7. ART OF PROBLEM SOLVING

The book also let you to master the art of problem solving. The key to problem solving does not lie in understanding the solution to the problem but to find out what clues in the problem leads you to the right solution. And, that's the reason Hints & Solutions are provided with the exercises after each chapter of the book. Try to find out the reason by analyzing the level of problem & practice similar kind of problems so that you can master the tricks involved. Remember that directly going though the solutions is not going to help you at all.

8. POSITIVE PERCEPTION

The book put forth for its readers a 'Simple and Straightforward' concept of studies, which is the best possible, time-tested perception for 11th hour revision / preparation.

The content of the book has been presented in such a lucid way so that you can enjoy what you are reading, keeping a note of your already stressed mind & time span.

Cracking JEE Main is not a matter of life and death. Do not allow panic and pressure to create confusion. Do some yoga and prayers. Enjoy this time with studies as it will never come back.

DAY ONE

Sets, Relations and Functions

Learning & Revision for the Day

- Sets
- Venn Diagram
- Operations on Sets
- Law of Algebra of Sets
- Cartesian Product of Sets
- Relations
- Composition of Relations
- Functions or Mapping
- Composition of Functions

Sets

- A **set** is a well-defined class or collection of the objects.
- Sets are usually denoted by the symbol $A, B, C, \dots$ and its elements are denoted by a, b, c, ... etc.
- If a is an element of a set A, then we write $a \in A$ and if not then we write $a \notin A$.

Representations of Sets

There are two methods of representing a set :

- In **roster method**, a set is described by listing elements, separated by commas, within curly braces$\{\neq\}$. e.g. A set of vowels of English alphabet may be described as $\{a, e, i, o, u\}$.
- In **set-builder method**, a set is described by a property $P(x)$, which is possessed by all its elements x. In such a case the set is written as $\{x : P(x) \text{ holds}\}$ or $\{x \mid P(x) \text{ holds}\}$, which is read as the set of all x such that $P(x)$ holds. e.g. The set $P = \{0, 1, 4, 9, 16, \dots\}$ can be written as $P = \{x^2 \mid x \in Z\}$.

Types of Sets

- The set which contains no element at all is called the **null set** (empty set or void set) and it is denoted by the symbol 'ϕ' or '$\{\}$' and if it contains a single element, then it is called **singleton set**.
- A set in which the process of counting of elements definitely comes to an end, is called a **finite set,** otherwise it is an **infinite set.**
- Two sets A and B are said to be **equal set** iff every element of A is an element of B and also every element of B is an element of A. i.e. $A = B$, if $x \in A \Leftrightarrow x \in B$.

PREP MIRROR
Your Personal Preparation Indicator

- No. of Questions in Exercises (x)—
- No. of Questions Attempted (y)—
- No. of Correct Questions (z)—
  (Without referring Explanations)

- Accuracy Level ($z/y \times 100$)—
- Prep Level ($z/x \times 100$)—

In order to expect good rank in JEE, your Accuracy Level should be above 85 & Prep Level should be above 75.

- **Equivalent sets** have the same number of elements but not exactly the same elements.
- A set that contains all sets in a given context is called **universal set** (U).
- Let A and B be two sets. If every element of A is an element of B, then A is called a **subset** of B, i.e. $A \subseteq B$.
- If A is a subset of B and $A \neq B$, then A is a **proper subset** of B. i.e. $A \subset B$.
- The null set ϕ is a subset of every set and every set is a subset of itself i.e. $\phi \subset A$ and $A \subseteq A$ for every set A. They are called **improper subsets** of A.
- If S is any set, then the set of all the subsets of S is called the **power set** of S and it is denoted by $P(S)$. Power set of a given set is always non-empty. If A has n elements, then $P(A)$ has 2^n elements.

NOTE
- The set $\{\phi\}$ is not a null set. It is a set containing one element ϕ.
- Whenever we have to show that two sets A and B are equal show that $A \subseteq B$ and $B \subseteq A$.
- If a set A has m elements, then the number m is called **cardinal number** of set A and it is denoted by $n(A)$. Thus, $n(A) = m$.

Venn Diagram

The combination of rectangles and circles is called **Venn Euler diagram** or Venn diagram. In Venn diagram, the universal set is represented by a rectangular region and a set is represented by circle on some closed geometrical figure. Where, A is the set and U is the universal set.

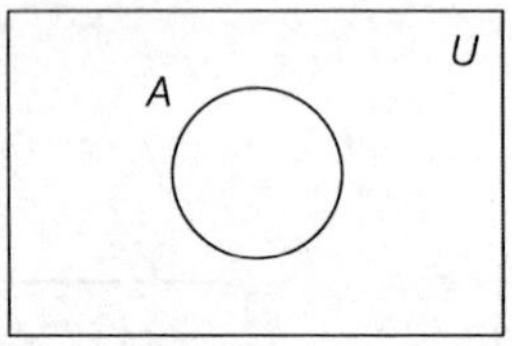

Operations on Sets

- The **union** of sets A and B is the set of all elements which are in set A or in B or in both A and B.
  i.e. $A \cup B = \{x : x \in A \text{ or } x \in B\}$

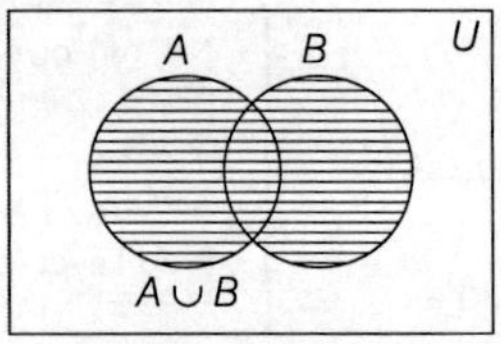

- The **intersection** of A and B is the set of all those elements that belong to both A and B.
  i.e. $A \cap B = \{x : x \in A \text{ and } x \in B\}$.

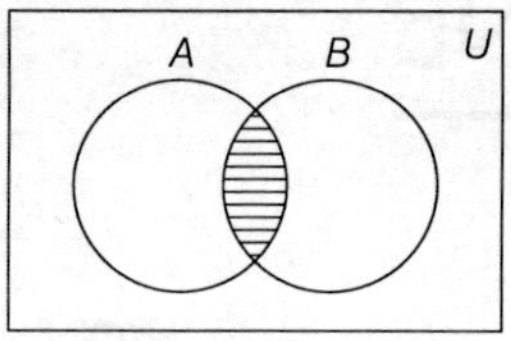

$A \cap B$

- If $A \cap B = \phi$, then A and B are called **disjoint sets.**
- Let U be an universal set and A be a set such that $A \subset U$. Then, **complement of A** with respect to U is denoted by A' or A^c or $\overline{A}$ or $U - A$. It is defined as the set of all those elements of U which are not in A.

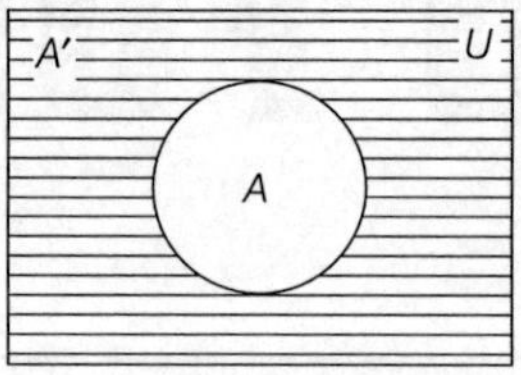

- The **difference** $A - B$ is the set of all those elements of A which does not belong to B.
  i.e. $A - B = \{x : x \in A \text{ and } x \notin B\}$
  and $B - A = \{x : x \in B \text{ and } x \notin A\}$.

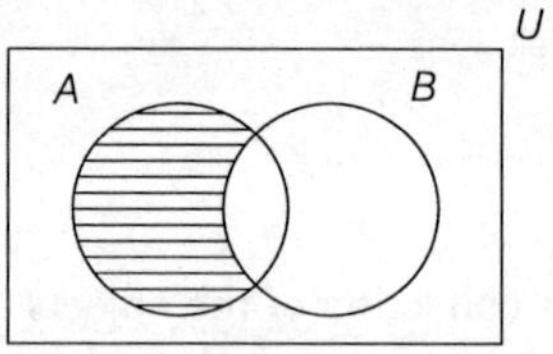

$A - B$

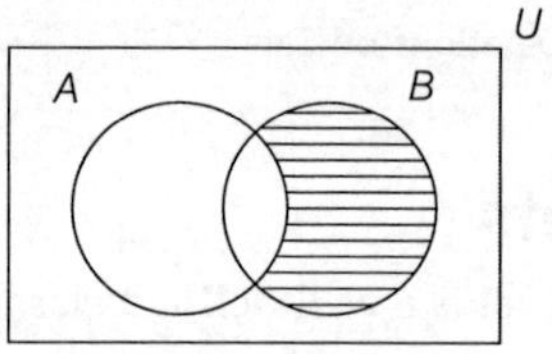

$B - A$

- The symmetric difference of sets A and B is the set $(A - B) \cup (B - A)$ and is denoted by $A \Delta B$.
  i.e. $A \Delta B = (A - B) \cup (B - A)$

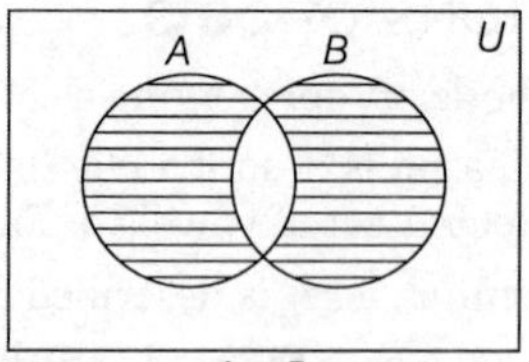

$A \Delta B$

Law of Algebra of Sets

If A, B and C are any three sets, then

1. **Idempotent Laws**
   (i) $A \cup A = A$ (ii) $A \cap A = A$
2. **Identity Laws**
   (i) $A \cup \phi = A$ (ii) $A \cap U = A$
3. **Distributive Laws**
   (i) $A \cup (B \cap C) = (A \cup B) \cap (A \cup C)$
   (ii) $A \cap (B \cup C) = (A \cap B) \cup (A \cap C)$

4. **De-Morgan's Laws**
   (i) $(A \cup B)' = A' \cap B'$
   (ii) $(A \cap B)' = A' \cup B'$
   (iii) $A - (B \cap C) = (A - B) \cup (A - C)$
   (iv) $A - (B \cup C) = (A - B) \cap (A - C)$
5. **Associative Laws**
   (i) $(A \cup B) \cup C = A \cup (B \cup C)$
   (ii) $A \cap (B \cap C) = (A \cap B) \cap C$
6. **Commutative Laws**
   (i) $A \cup B = B \cup A$ (ii) $A \cap B = B \cap A$
   (iii) $A \Delta B = B \Delta A$

Important Results on Operation of Sets

1. $A - B = A \cap B'$
2. $B - A = B \cap A'$
3. $A - B = A \Leftrightarrow A \cap B = \phi$
4. $(A - B) \cup B = A \cup B$
5. $(A - B) \cap B = \phi$
6. $A \subseteq B \Leftrightarrow B' \subseteq A'$
7. $(A - B) \cup (B - A) = (A \cup B) - (A \cap B)$
8. $n(A \cup B) = n(A) + n(B) - n(A \cap B)$
9. $n(A \cup B) = n(A) + n(B)$
   $\Leftrightarrow$ A and B are disjoint sets.
10. $n(A - B) = n(A) - n(A \cap B)$
11. $n(A \Delta B) = n(A) + n(B) - 2n(A \cap B)$
12. $n(A \cup B \cup C) = n(A) + n(B) + n(C) - n(A \cap B) - n(B \cap C) - n(A \cap C) + n(A \cap B \cap C)$
13. $n(A' \cup B') = n(A \cap B)' = n(U) - n(A \cap B)$
14. $n(A' \cap B') = n(A \cup B)' = n(U) - n(A \cup B)$

Cartesian Product of Sets

Let A and B be any two non-empty sets. Then the cartesian product $A \times B$, is defined as set of all ordered pairs (a, b) such that $a \in A$ and $b \in B$.
i.e.

- $A \times B = \{(a, b) : a \in A \text{ and } b \in B\}$
- $B \times A = \{(b, a) : b \in B \text{ and } a \in A\}$
  and $A \times A = \{(a, b) : a, b \in A\}$.
- $A \times B = \phi$, if either A or B is an empty set.
- If $n(A) = p$ and $n(B) = q$, then
  $n(A \times B) = n(A) \cdot n(B) = pq$.
- $A \times (B \cup C) = (A \times B) \cup (A \times C)$
- $A \times (B \cap C) = (A \times B) \cap (A \times C)$
- $A \times (B - C) = (A \times B) - (A \times C)$
- $(A \times B) \cap (C \times D) = (A \cap C) \times (B \cap D)$.

Relations

- Let A and B be two non-empty sets, then **relation** R from A to B is a subset of $A \times B$, i.e. $R \subseteq A \times B$.
- If $(a, b) \in R$, then we say a is related to b by the relation R and we write it as aRb.
- Domain of $R = \{a : (a, b) \in R\}$ and range of $R = \{b : (a, b) \in R\}$.
- If $n(A) = p$ and $n(B) = q$, then the total number of relations from A to B is 2^{pq}.

Types of Relations

- Let A be any non-empty set and R be a relation on A. Then,
  (i) R is said to be **reflexive** iff $(a, a) \in R, \forall a \in A$.
  (ii) R is said to be **symmetric** iff
  $(a, b) \in R$
  $\Rightarrow$ $(b, a) \in R, \forall a, b \in A$
  (iii) R is said to be a **transitive** iff $(a, b) \in R$ and $(b, c) \in R$
  $\Rightarrow$ $(a, c) \in R, \forall a, b, c \in A$
  i.e. aRb and bRc $\Rightarrow$ $aRc, \forall a, b, c \in A$.
- The relation $I_A = \{(a, a) : a \in A\}$ on A is called the **identity relation** on A.
- R is said to be an **equivalence relation** iff
  (i) it is reflexive i.e. $(a, a) \in R, \forall a \in A$.
  (ii) it is symmetric i.e. $(a, b) \in R \Rightarrow (b, a) \in R, \forall a, b \in A$
  (iii) it is transitive
  i.e. $(a, b) \in R$ and $(b, c) \in R$
  $\Rightarrow$ $(a, c) \in R, \forall a, b, c \in A$

Inverse Relation

Let R be a relation from set A to set B, then the **inverse of R**, denoted by R^{-1}, is defined by
$R^{-1} = \{(b, a) : (a, b) \in R\}$. Clearly, $(a, b) \in R \Leftrightarrow (b, a) \in R^{-1}$.

NOTE
- The intersection of two equivalence relations on a set is an equivalence relation on the set.
- The union of two equivalence relations on a set is not necessarily an equivalence relation on the set.
- If R is an equivalence relation on a set A, then R^{-1} is also an equivalence relation A.

Composition of Relations

Let R and S be two relations from set A to B and B to C respectively, then we can define a relation SoR from A to C such that $(a, c) \in SoR \Leftrightarrow \exists\, b \in B$ such that $(a, b) \in R$ and $(b, c) \in S$. This relation is called the **composition of R and S.**

$$RoS \neq SoR$$

Functions or Mapping

- If A and B are two non-empty sets, then a rule f which associates each $x \in A$, to a unique member $y \in B$, is called a function from A to B and it is denoted by $f : A \to B$.
- The set A is called the **domain** of $f(D_f)$ and set B is called the **codomain** of $f(C_f)$.
- The set consisting of all the f-images of the elements of the domain A, called the range of $f(R_f)$.

NOTE
- A relation will be a function, if no two distinct ordered pairs have the same first element.
- Every function is a relation but every relation is not necessarily a function.
- The number of functions from a finite set A into finite set B is $\{n(B)\}^{n(A)}$.

Different Types of Functions

Let f be a function from A to B, i.e. $f : A \to B$. Then,

f is said to be **one-one function** or injective function, if different elements of A have different images in B.

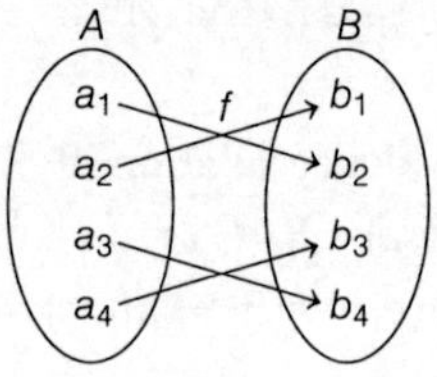

Methods to Check One-One Function

Method I If $f(x) = f(y) \Rightarrow x = y$, then f is one-one.

Method II A function is one-one iff no line parallel to X-axis meets the graph of function at more than one point.

- The number of one-one function that can be defined from a finite set A into finite set B is $\begin{cases} {}^{n(B)}P_{n(A)}, & \text{if } n(B) \ge n(A) \\ 0, & \text{otherwise} \end{cases}$.
- f is said to be a **many-one function**, if two or more elements of set A have the same image in B.

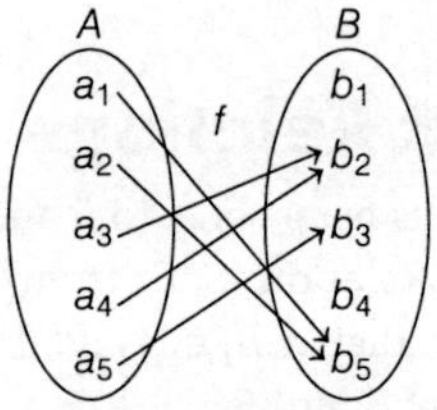

i.e. $f : A \to B$ is a many-one function, if it is not a one-one function.

- f is said to be **onto function** or **surjective function**, if each element of B has its pre-image in A.

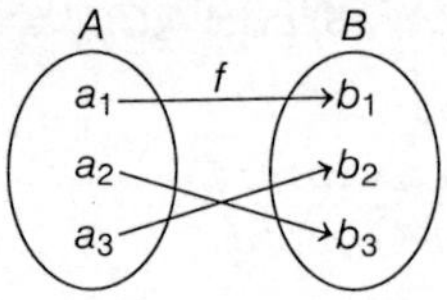

Method to Check Onto Function

Find the range of $f(x)$ and show that range of $f(x) =$ codomain of $f(x)$.

- Any polynomial function of odd degree is always onto.
- The number of onto functions that can be defined from a finite set A containing n elements onto a finite set B containing 2 elements $= 2^n - 2$.
- If $n(A) \ge n(B)$, then number of onto function is 0.
- If A has m elements and B has n elements, where $m < n$, then number of onto functions from A to B is $n^m - {}^nC_1 (n-1)^m + {}^nC_2 (n-2)^m - \dots, m < n$.
- f is said to be an **into function**, if there exists atleast one element in B having no pre-image in A. i.e. $f : A \to B$ is an into function, if it is not an onto function.

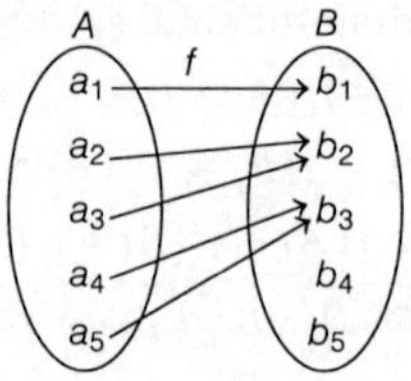

- f is said to be a **bijective function**, if it is one-one as well as onto.

NOTE
- If $f : A \to B$ is a bijective, then A and B have the same number of elements.
- If $n(A) = n(B) = m$, then number of bijective map from A to B is $m!$.

Composition of Functions

Let $f : A \to B$ and $g : B \to C$ are two functions. Then, the composition of f and g, denoted by

$gof : A \to C$, is defined as,

$gof\,(x) = g[f(x)], \forall\ x \in A$.

NOTE
- gof is defined only if $f(x)$ is an element of domain of g.
- Generally, $gof \ne fog$.

DAY PRACTICE SESSION 1

FOUNDATION QUESTIONS EXERCISE

1 If $Q=\left\{x: x=\frac{1}{y}, \text{ where } y \in N\right\}$, then

(a) $0 \in Q$ (b) $1 \in Q$ (c) $2 \in Q$ (d) $\frac{2}{3} \in Q$

2 If $P(A)$ denotes the power set of A and A is the void set, then what is number of elements in $P\{P\{P\{P(A)\}\}\}$?

(a) 0 (b) 1 (c) 4 (d) 16

3 If $X=\{4^n-3n-1: n \in N\}$ and $Y=\{9(n-1): n \in N\}$; where N is the set of natural numbers,then $X \cup Y$ is equal to → JEE Mains 2014

(a) N (b) Y-X (c) X (d) Y

4 If A, B and C are three sets such that $A \cap B=A \cap C$ and $A \cup B=A \cup C$, then

(a) $A=C$ (b) $B=C$ (c) $A \cap B=\phi$ (d) $A=B$

5 Suppose $A_1, A_2, \ldots, A_{30}$ are thirty sets each having 5 elements and $B_1, B_2, \ldots, B_n$ are n sets each having 3 elements. Let $\bigcup_{i=1}^{30} A_i=\bigcup_{j=1}^{n} B_j=S$ and each element of S belongs to exactly 10 of A_i's and exactly 9 of B_j' s. The value of n is equal to → NCERT Exemplar

(a) 15 (b) 3
(c) 45 (d) None of these

6 If A and B are two sets and $A \cup B \cup C=U$. Then, $\{(A-B) \cup(B-C) \cup(C-A)\}'$ is equal to

(a) $A \cup B \cup C$ (b) $A \cup(B \cap C)$
(c) $A \cap B \cap C$ (d) $A \cap(B \cup C)$

7 Let X be the universal set for sets A and B, if $n(A)=200$, $n(B)=300$ and $n(A \cap B)=100$, then $n(A' \cap B')$ is equal to 300 provided $n(X)$ is equal to

(a) 600 (b) 700 (c) 800 (d) 900

8 If $n(A)=1000$, $n(B)=500$, $n(A \cap B) \geq 1$ and $n(A \cup B)=P$, then

(a) $500 \leq P \leq 1000$ (b) $1001 \leq P \leq 1498$
(c) $1000 \leq P \leq 1498$ (d) $1000 \leq P \leq 1499$

9 If $n(A)=4$, $n(B)=3$, $n(A \times B \times C)=24$, then $n(C)$ is equal to

(a) 2 (b) 288 (c) 12 (d) 1

10 If $R=\{(3,3),(6,6),(9,9),(12,12),(6,12),(3,9),(3,12),(3,6)\}$ is a relation on the set $A=\{3,6,9,12\}$. The relation is

(a) an equivalence relation
(b) reflexive and symmetric
(c) reflexive and transitive
(d) only reflexive

11 Let $R=\{(x, y): x, y \in N$ and $x^2-4xy+3y^2=0\}$, where N is the set of all natural numbers. Then, the relation R is → JEE Mains 2013

(a) reflexive but neither symmetric nor transitive
(b) symmetric and transitive
(c) reflexive and symmetric
(d) reflexive and transitive

12 If $g(x)=1+\sqrt{x}$ and $f\{g(x)\}=3+2\sqrt{x}+x$, then $f(x)$ is equal to

(a) $1+2x^2$ (b) $2+x^2$
(c) $1+x$ (d) $2+x$

13 Let $f(x)=ax+b$ and $g(x)=cx+d$, $a \neq 0$, $c \neq 0$. Assume $a=1$, $b=2$, if $(fog)(x)=(gof)(x)$ for all x. What can you say about c and d?

(a) c and d both arbitrary (b) $c=1$ and d is arbitrary
(c) c is arbitrary and $d=1$ (d) $c=1$, $d=1$

14 If R is relation from $\{11,12,13\}$ to $\{8,10,12\}$ defined by $y=x-3$. Then, R^{-1} is

(a) $\{(8,11),(10,13)\}$ (b) $\{(11,18),(13,10)\}$
(c) $\{(10,13),(8,11)\}$ (d) None of these

15 Let R be a relation defined by $R=\{(4,5),(1,4),(4,6),(7,6),(3,7)\}$, then R^{-1} OR is

(a) $\{(1,1),(4,4),(4,7),(7,4),(7,7),(3,3)\}$
(b) $\{(1,1),(4,4),(7,7),(3,3)\}$
(c) $\{(1,5),(1,6),(3,6)\}$
(d) None of the above

16 Let A be a non-empty set of real numbers and $f: A \rightarrow A$ be such that $f(f(x))=x$, $\forall x \in R$. Then, $f(x)$ is

(a) a bijection (b) one-one but not onto
(c) onto but not one-one (d) neither one-one nor onto

17 The function f satisfies the functional equation $3f(x)+2f\left(\frac{x+59}{x-1}\right)=10x+30$ for all real $x \neq 1$. The value of $f(7)$ is

(a) 8 (b) 4 (c) -8 (d) 11

18 The number of onto mapping from the set $A=\{1,2, \ldots 100\}$ to set $B=\{1,2\}$ is

(a) $2^{100}-2$ (b) 2^{100} (c) $2^{99}-2$ (d) 2^{99}

19 Let $f: R-\{n\} \rightarrow R$ be a function defined by $f(x)=\frac{x-m}{x-n}$, where $m \neq n$. Then,

(a) f is one-one onto (b) f is one-one into
(c) f is many-one onto (d) f is many-one into

20 A function f from the set of natural numbers to integers defined by $f(n)=\begin{cases}\frac{n-1}{2}, & \text{when } n \text{ is odd} \\ -\frac{n}{2}, & \text{when } n \text{ is even}\end{cases}$ is

(a) one-one but not onto (b) onto but not one-one
(c) both one-one and onto (d) neither one-one nor onto

21 Let $f : N \to N$ defined by $f(x) = x^2 + x + 1$, $x \in N$, then f is

(a) one-one onto (b) many-one onto
(c) one-one but not onto (d) None of these

22 Let R be the real line. Consider the following subsets of the plane $R \times R$.

$$S = \{(x, y): y = x + 1 \text{ and } 0 < x < 2\}$$
and $$T = \{(x, y): x - y \text{ is an integer}\}$$

Which one of the following is true?

(a) T is an equivalence relation on R but S is not
(b) Neither S nor T is an equivalence relation on R
(c) Both S and T are equivalence relations on R
(d) S is an equivalence relation on R but T is not

23 Consider the following relations

$R = \{(x, y) \mid x$ and y are real numbers and $x = wy$ for some rational number $w\}$;

$S = \left\{\left(\frac{m}{n}, \frac{p}{q}\right) \middle| m, n, p \text{ and } q \text{ are integers such that } n, q \neq 0 \text{ and } qm = pn\right\}$. Then,

(a) R is an equivalence relation but S is not an equivalence relation
(b) neither R nor S is an equivalence relation
(c) S is an equivalence relation but R is not an equivalence relation
(d) R and S both are equivalence relations

24 If $f(x) = \begin{cases} 2 + x, & x \geq 0 \\ 4 - x, & x < 0 \end{cases}$, then $f(f(x))$ is given by

(a) $f(f(x)) = \begin{cases} 4 + x, & x \geq 0 \\ 6 - x, & x < 0 \end{cases}$ (b) $f(f(x)) = \begin{cases} 4 + x, & x \geq 0 \\ x, & x < 0 \end{cases}$

(c) $f(f(x)) = \begin{cases} 4 - x, & x \geq 0 \\ x, & x < 0 \end{cases}$ (d) $f(f(x)) = \begin{cases} 4 - 2x, & x \geq 0 \\ 4 + 2x, & x < 0 \end{cases}$

25 **Statement I** A relation is defined by

$f(x) = \begin{cases} x^2, & 0 \leq x \leq 3 \\ 2x, & 3 \leq x \leq 9 \end{cases}$ is a function.

Statement II In a function, every member must have a unique image.

(a) Statement I is true, Statement II is true; Statement II is a correct explanation for Statement I
(b) Statement I is true, Statement II is true; Statement II is not a correct explanation for Statement I
(c) Statement I is true; Statement II is false
(d) Statement I is false; Statement II is true

DAY PRACTICE SESSION 2

PROGRESSIVE QUESTIONS EXERCISE

1 If $f(x) + 2f\left(\frac{1}{x}\right) = 3x$, $x \neq 0$ and $S = \{x \in R : f(x) = f(-x)\}$; then S → JEE Mains 2016

(a) is an empty set
(b) contains exactly one element
(c) contains exactly two elements
(d) contains more than two elements

2 $\left\{x \in R : \frac{2x - 1}{x^3 + 4x^2 + 3x} \in R\right\}$ is equal to

(a) $R - \{0\}$ (b) $R - \{0, 1, 3\}$
(c) $R - \{0, -1, -3\}$ (d) $R - \left\{0, -1, -3, \frac{1}{2}\right\}$

3 Given the relation $R = \{(1, 2)\ (2, 3)\}$ on the set $A = \{1, 2, 3\}$, the minimum number of ordered pairs which when added to R make it an equivalence relation is

(a) 5 (b) 7 (c) 6 (d) 8

4 The set $(A \cup B \cup C) \cap (A \cap B' \cap C')' \cap C'$ is equal to → NCERT Exemplar

(a) $B \cap C'$ (b) $A \cap C$
(c) $B' \cap C'$ (d) None of these

5 Let $A = \{1, 2, 3, 4\}$, $B = \{2, 4, 6\}$. Then the number of sets C such that $A \cap B \subseteq C \subseteq A \cup B$ is

(a) 6 (b) 9 (c) 8 (d) 10

6 On the set N of all natural numbers define the relation R by aRb iff the g.c.d. of a and b is 2, then R is

(a) reflexive but not symmetric (b) symmetric only
(c) reflexive and transitive (d) equivalence relation

7 Suppose f is a function satisfying $f(x + f(x)) = 4f(x)$ and $f(1) = 4$. The value of $f(21)$ is

(a) 16 (b) 64 (c) 4 (d) 44

8 Let $f\left(x + \frac{1}{x}\right) = x^2 + \frac{1}{x^2}$, $x \neq 0$, then $f(x)$ is equal to

(a) x^2 (b) $x^2 - 1$
(c) $x^2 - 2$ (d) $x^2 + 1$

9 Let $f(x) = \frac{x}{\sqrt{1 + x^2}}$, the $\underbrace{fofofo \ldots of(x)}_{x \text{ times}}$ is

(a) $\frac{x}{\sqrt{1 + \left(\sum_{r=1}^{n} r\right) x^2}}$ (b) $\frac{x}{\sqrt{1 + \left(\sum_{r=1}^{n} 1\right) x^2}}$

(c) $\left(\frac{x}{\sqrt{1 + x^2}}\right)^x$ (d) $\frac{nx}{\sqrt{1 + nx^2}}$

10 If two sets A and B are having 99 elements in common, then the number of elements common to each of the sets $A \times B$ and $B \times A$ are

(a) 2^{99} (b) 99^2 (c) 100 (d) 18

ANSWERS

SESSION 1	**1.** (b)	**2.** (d)	**3.** (d)	**4.** (b)	**5.** (c)	**6.** (c)	**7.** (b)	**8.** (d)	**9.** (a)	**10.** (c)
	11. (a)	**12.** (b)	**13.** (b)	**14.** (a)	**15.** (a)	**16.** (a)	**17.** (b)	**18.** (a)	**19.** (b)	**20.** (c)
	21. (c)	**22.** (a)	**23.** (c)	**24.** (a)	**25.** (d)					
SESSION 2	**1.** (c)	**2.** (c)	**3.** (b)	**4.** (a)	**5.** (c)	**6.** (b)	**7.** (b)	**8.** (c)	**9.** (b)	**10.** (b)

Hints and Explanations

SESSION 1

1 Clearly, $\frac{1}{y} \neq 0, 2$ and $\frac{2}{3}$ $\quad [\because y \in N]$

$\therefore \frac{1}{y}$ can be 1.

$\Rightarrow \quad x = 1 \in Q$

2 The number of elements in power set of A is 1.

$\therefore \quad P\{P(A)\} = 2^1 = 2$

$\Rightarrow \quad P\{P\{P(A)\}\} = 2^2 = 4$

$\Rightarrow \quad P\{P\{P\{P(A)\}\}\} = 2^4 = 16$

3 We have,

$X = \{4^n - 3n - 1 : n \in N\}$

$X = \{0, 9, 54, 243, \ldots\}$ [put $n = 1, 2, 3, \ldots$]

$Y = \{9(n-1) : n \in N\}$

$Y = \{0, 9, 18, 27, \ldots\}$ [put $n = 1, 2, 3, \ldots$]

It is clear that $X \subset Y$.

$\therefore \quad X \cup Y = Y$

4 Clearly, $A \cap B = A \cap C$ and $A \cup B = A \cup C$ possible if

$B = C$

5 Number of elements in $A_1 \cup A_2 \cup A_3 \cup \ldots \cup A_{30}$ is 30×5 but each element is used 10 times, so

$n(S) = \frac{30 \times 5}{10} = 15 \quad \ldots(i)$

Similarly, number of elements in $B_1 \cup B_2 \ldots \cup B_n$ is $3n$ but each element is repeated 9 times, so

$n(S) = \frac{3n}{9}$

$\Rightarrow \quad 15 = \frac{3n}{9}$ [from Eq. (i)]

$\Rightarrow \quad n = 45$

6 From Venn Euler's diagram,

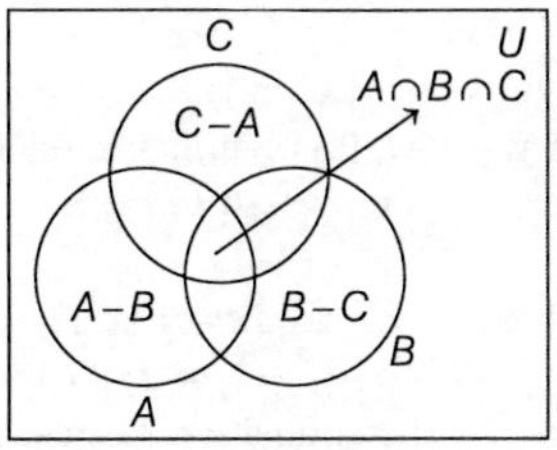

It is clear that,

$\{(A - B) \cup (B - C) \cup (C - A)\}'$
$= A \cap B \cap C$

7 $\because n(A \cup B) = n(A) + n(B) - n(A \cap B)$

$\therefore n(A \cup B) = 200 + 300 - 100 = 400$

$\therefore n(A' \cap B') = n(A \cup B)' = n(X) - n(A \cup B)$

$\Rightarrow \quad 300 = n(X) - 400$

$\Rightarrow \quad n(X) = 700$

8 We know,

$n(A \cup B) = n(A) + n(B) - n(A \cap B)$

$\therefore \quad P = 1500 - n(A \cap B)$

$\Rightarrow \quad n(A \cap B) = 1500 - P$

Clearly, $1 \leq n(A \cap B) \leq 500$

[$\because$ maximum number of elements common in A and $B = 500$]

$\Rightarrow \quad 1 \leq 1500 - P \leq 500$

$\Rightarrow \quad -1499 \leq -P \leq -1000$

$\Rightarrow \quad 1000 \leq P \leq 1499$

9 We know,

$n(A \times B \times C) = n(A) \times n(B) \times n(C)$

$\therefore \quad n(C) = \frac{24}{4 \times 3} = 2$

10 Since for each $a \in A$, $(a, a) \in R$. R is reflexive relation.

Now, $(6, 12) \in R$ but $(12, 6) \notin R$. So, it is not a symmetric relation.

Also, $(3, 6), (6, 12) \in R \Rightarrow (3, 12) \in R$

$\Rightarrow R$ is transitive.

11 $\because a^2 - 4a \cdot a + 3a^2 = 4a^2 - 4a^2 = 0$

$\therefore \; (a, a) \in R, \forall a \in N \Rightarrow R$ is reflexive.

Now, as $a^2 - 4ab + 3b^2 = 0$

but $b^2 - 4ba + 3a^2 \neq 0$

$\therefore R$ is not symmetric.

Also, $(a, b) \in R$ and $(b, c) \in R$

$\not\Rightarrow \; (a, c) \in N$

So, R is not transitive.

12 Given, $g(x) = 1 + \sqrt{x}$

and $f\{g(x)\} = 3 + 2\sqrt{x} + x \quad \ldots(i)$

$\Rightarrow f(1 + \sqrt{x}) = 3 + 2\sqrt{x} + x$

Put $1 + \sqrt{x} = y \Rightarrow x = (y-1)^2$

$\therefore \quad f(y) = 3 + 2(y-1) + (y-1)^2$

$= 2 + y^2$

$\therefore \quad f(x) = 2 + x^2$

13 Here, $(fog)(x) = f\{g(x)\} = a(cx + d) + b$

and $(gof)(x) = g\{f(x)\} = c(ax + b) + d$

Since, $cx + d + 2 = cx + 2c + d$

[$\because a = 1, b = 2$]

Hence, $c = 1$ and d is arbitrary.

14 R is a relation from $\{11, 12, 13\}$ to $\{8, 10, 12\}$ defined by

$y = x - 3 \Rightarrow x - y = 3$

$\therefore \quad R = \{(11, 8), (13, 10)\}$

Hence, $R^{-1} = \{(8, 11), (10, 13)\}$

15 Clearly, $R^{-1} = \{(5, 4), (4, 1), (6, 4), (6, 7), (7, 3)\}$

Now, as $(4, 5) \in R$ and $(5, 4) \in R^{-1}$,

therefore $(4, 4) \in R^{-1}OR$

Similarly, $(1, 4) \in R$ and $(4, 1) \in R^{-1}$

$\Rightarrow \quad (1, 1) \in R^{-1}OR$

$(4, 6) \in R$ and $(6, 7) \in R^{-1}$

$\Rightarrow \quad (4, 7) \in R^{-1}OR$

$(7, 6) \in R$ and $(6, 7) \in R^{-1}$

$\Rightarrow \quad (7, 7) \in R^{-1}OR$

$(7, 6) \in R$ and $(6, 4) \in R^{-1}$

$\Rightarrow$ $(7, 4) \in R^{-1}OR$
and $(3, 7) \in R$ and $(7, 3) \in R^{-1}$
$\Rightarrow$ $(3, 3) \in R^{-1}OR$
Hence, $R^{-1}OR = \{(1, 1), (4, 4), (4, 7), (7, 7), (7, 4), (3, 3)\}$

16 Let $x, y \in A$ such that $f(x) = f(y)$, then
$$f(f(x)) = f(f(y))$$
$\Rightarrow$ $x = y$
$\Rightarrow f$ is one-one.
Also, for any $a \in A$, we have
$$f(f(a)) = a$$
$\Rightarrow f(b) = a$, where $b = f(a) \in A$
Thus, for each $a \in A$ (codomain) there exists $b = f(a) \in A$ such that $f(b) = a$
$\therefore$ f is onto.
Hence f is a bijective function.

17 We have, $3f(x) + 2f\left(\frac{x+59}{x-1}\right) = 10x + 30$ ... (i)
On replacing x by $\frac{x+59}{x-1}$, we get
$$3f\left(\frac{x+59}{x-1}\right) + 2f(x) = \frac{40x+560}{x-1} \quad \text{...(ii)}$$
On solving Eqs. (i) and (ii), we get
$$f(x) = \frac{6x^2 - 4x - 242}{x-1}$$
$$\therefore \quad f(7) = \frac{6 \times 49 - 4 \times 7 - 242}{6} = 4$$

18 We know that if $n(A) = n$ and $n(B) = 2$, the number of onto relations from A to $B = 2^n - 2$
$\therefore$ Required number of relations
$= 2^{100} - 2$

19 Suppose for any $x, y \in R$,
$$f(x) = f(y)$$
$$\Rightarrow \quad \frac{x-m}{x-n} = \frac{y-m}{y-n}$$
$\Rightarrow$ $x = y$
So, f is one-one.
Let $\alpha \in R$ be such that $f(x) = \alpha$
$$\therefore \quad \frac{x-m}{x-n} = \alpha \Rightarrow x = \frac{m - n\alpha}{1-\alpha}$$
Clearly, $x \notin R$ for $\alpha = 1$
So, f is not onto.

20 Let $x, y \in N$ and both be even.
Then, $f(x) = f(y) \Rightarrow -\frac{x}{2} = -\frac{y}{2}$
$\Rightarrow$ $x = y$
Again, $x, y \in N$ and both are odd.
Then, $f(x) = f(y) \Rightarrow x = y$
So, f is one-one
Since, each negative integer is an image of even natural number and positive integer is an image of odd natural number. So, f is onto.

21 Let $x, y \in N$ such that $f(x) = f(y)$
$\Rightarrow$ $x^2 + x + 1 = y^2 + y + 1$
$\Rightarrow$ $(x^2 - y^2) = y - x$
$\Rightarrow$ $(x - y)(x + y + 1) = 0$
$\Rightarrow$ $x = y$ or $x = -y - 1 \notin N$
$\Rightarrow$ $x = y$
$\Rightarrow f$ is one-one.
But f is not onto, as $1 \in N$ does not have any pre-image.
$\therefore$ f is one-one but not onto.

22 Since, $(1, 2) \in S$ but $(2, 1) \notin S$
Thus S is not symmetric.
Hence, S is not an equivalence relation.
Given, $T = \{(x, y) : (x - y) \in I\}$
Now, $x - x = 0 \in I$, it is reflexive relation.
Again, now $(x - y) \in I$
$\Rightarrow y - x \in I$, it is symmetric relation.
Let $x - y = I_1$
and $y - z = I_2$
Then, $x - z = (x - y) + (y - z)$
$= I_1 + I_2 \in I$
So, T is also transitive. Hence, T is an equivalence relation.

23 Since, the relation R is defined as
$R = \{(x, y) \mid x, y$ are real numbers and $x = wy$ for some rational number $w\}$.
(a) **Reflexive** xRx as $x = 1x$
Here, $w = 1 \in$ Rational number
So, the relation R is reflexive.
(b) **Symmetric** $xRy \not\Rightarrow yRx$ as $0R1$ but $1 \not R 0$
So, the relation R is not symmetric.
Thus, R is not equivalence relation.
Now, for the relation S, defined as,
$S = \left\{\left(\frac{m}{n}, \frac{p}{q}\right) \middle| m, n, p \text{ and } q \in \text{integers}\right.$ such that $n, q \neq 0$ and $qm = pn\}$
(a) **Reflexive** $\frac{m}{n} S \frac{m}{n} \Rightarrow mn = mn$ [true]
Hence, the relation S is reflexive.
(b) **Symmetric** $\frac{m}{n} S \frac{p}{q} \Rightarrow mq = np$
$\Rightarrow np = mq \Rightarrow \frac{p}{q} S \frac{m}{n}$
Hence, the relation S is symmetric.
(c) **Transitive** $\frac{m}{n} S \frac{p}{q}$ and $\frac{p}{q} S \frac{r}{s}$
$\Rightarrow$ $mq = np$ and $ps = rq$
$\Rightarrow mq \cdot ps = np \cdot rq \Rightarrow ms = nr$
$\Rightarrow$ $\frac{m}{n} = \frac{r}{s}$
$\Rightarrow$ $\frac{m}{n} S \frac{r}{s}$
So, the relation S is transitive.
Hence, the relation S is equivalence relation.

24 Clearly,
$$f(f(x)) = \begin{cases} 2 + f(x), & f(x) \geq 0 \\ 4 - f(x), & f(x) < 0 \end{cases}$$
$$= \begin{cases} 2 + (2 + x), & x \geq 0 \\ 2 + (4 - x), & x < 0 \end{cases}$$
$$= \begin{cases} 4 + x, & x \geq 0 \\ 6 - x, & x < 0 \end{cases}$$

25 **Statement I** $f(x) = \begin{cases} x^2, & 0 \leq x \leq 3 \\ 2x, & 3 \leq x \leq 9 \end{cases}$
Now, $f(3) = 9$
Also, $f(3) = 2 \times 3 = 6$
Here, we see that for one value of x, there are two different values of $f(x)$.
Hence, it is not a function but Statement II is true.

SESSION 2

1 We have, $f(x) + 2f\left(\frac{1}{x}\right) = 3x$, $x \neq 0$ ... (i)
$$\therefore \quad f\left(\frac{1}{x}\right) + 2f(x) = \frac{3}{x} \quad \text{... (ii)}$$
$\left[\text{replacing } x \text{ by } \frac{1}{x}\right]$
On multiplying Eq. (ii) by 2 and then subtracting it from Eq. (i), we get
$$-3f(x) = 3x - \frac{6}{x}$$
$$\Rightarrow \quad f(x) = \frac{2}{x} - x$$
Now, consider $f(x) = f(-x)$
$$\Rightarrow \quad \frac{2}{x} - x = -\frac{2}{x} + x \Rightarrow \frac{4}{x} = 2x$$
$$\Rightarrow \quad x^2 = 2 \Rightarrow x = \pm\sqrt{2}$$
Thus, x contains exactly two elements.

2 Clearly, $\frac{2x-1}{x^3 + 4x^2 + 3x} \in R$ only when $x^3 + 4x^2 + 3x \neq 0$
Consider $x^3 + 4x^2 + 3x = 0$
$\Rightarrow$ $x(x^2 + 4x + 3) = 0$
$\Rightarrow$ $x(x + 1)(x + 3) = 0$
$\Rightarrow$ $x = 0, -1, -3$
$$\therefore \quad \left\{x \in R : \frac{2x-1}{x^3 + 4x^2 + 3x} \in R\right\} = R - \{0, -1, -3\}$$

3. For R to be an equivalence relation, R must be reflexive, symmetric and transitive.
R will be reflexive if it contains (1, 1), (2, 2) and (3, 3)
R will be symmetric if it contains (2, 1) and (3, 2)
R will be transitive if it contains (1, 3) and (3, 1)
Hence, minimum number of ordered pairs = 7

4 $(A \cup B \cup C) \cap (A \cap B' \cap C')' \cap C'$
$= (A \cup B \cup C) \cap (A' \cup B \cup C) \cap C'$
$= (\phi \cup B \cup C) \cap C'$
$= (B \cup C) \cap C'$
$= (B \cap C') \cup \phi = B \cap C'$

5 Here, $A \cap B = \{2, 4\}$
and $A \cup B = \{1, 2, 3, 4, 6\}$
$\because A \cap B \subseteq C \subseteq A \cup B$
$\therefore$ C can be {2, 4}, {1, 2, 4}, {3, 2, 4}, {6, 2, 4}, {1, 6, 2, 4}, {6, 3, 2, 4}, {1, 3, 2, 4}, {1, 2, 3, 4, 6}
Thus, number of set C which satisfy the given condition is 8.

6 Clearly, g.c.d $(a, a) = a$, $\forall\ a \in N$
$\therefore$ R is not reflexive.
If g.c.d $(a, b) = 2$, then g.c.d (b, a) is also 2.
Thus, $aRb \Rightarrow bRa$
Hence, R is symmetric.
According to given option, R is symmetric only.

7 We have,
$f(x + f(x)) = 4\,f(x)$ and $f(1) = 4$
On putting $x = 1$, we get
$$f(1 + f(1)) = 4f(1)$$
$$\Rightarrow f(1 + f(1)) = 16$$
$$\Rightarrow f(1 + 4) = 16$$
$$\Rightarrow f(5) = 16$$
On putting, $x = 5$, we get
$$f(5 + f(5)) = 4f(5)$$
$$\Rightarrow f(5 + 16) = 4 \times 16$$
$$\Rightarrow f(21) = 64$$

8 We have,
$$f\left(x + \frac{1}{x}\right) = x^2 + \frac{1}{x^2}$$
$$= \left(x + \frac{1}{x}\right)^2 - 2$$
$$\Rightarrow f(x) = x^2 - 2$$

9 We have, $f(x) = \dfrac{x}{\sqrt{1 + x^2}}$
$$\Rightarrow f(f(x)) = \frac{f(x)}{\sqrt{1 + (f(x))^2}}$$
$$= \frac{\dfrac{x}{\sqrt{1 + x^2}}}{\sqrt{1 + \dfrac{x^2}{1 + x^2}}}$$
$$= \frac{x}{\sqrt{1 + 2x^2}}$$
Similarly, $f(f(f(x))) = \dfrac{x}{\sqrt{1 + 3x^2}}$
$$\vdots \qquad \vdots$$
$$\underbrace{fofo \ldots of of(x)}_{n \text{ times}} = \frac{x}{\sqrt{1 + nx^2}}$$
$$= \frac{x}{\sqrt{1 + \left(\sum_{r=1}^{n} 1\right)x^2}}$$

10 We know,
$$(A \times B) \cap (C \times D) = (A \cap C) \times (B \cap D)$$
$$\therefore (A \times B) \cap (B \times A) = (A \cap B) \times (B \cap A)$$
Thus, number of elements common to $A \times B$ and $B \times A$
$= n((A \times B) \cap (B \times A))$
$= n((A \cap B) \times (B \cap A))$
$= n(A \cap B) \times n(B \cap A)$
$= 99 \times 99 = 99^2$

DAY TWO

Complex Numbers

Learning & Revision for the Day

- Complex Numbers and Its Representation
- Algebra and Equality of Complex Numbers
- Conjugate and Modulus of a Complex Number
- Argument or Amplitude of a Complex Number
- Different forms of a Complex Number
- Concept of Rotation
- Square Root of a Complex Number
- De-Moivre's Theorem
- Cube Roots of Unity
- nth Roots of Unity
- Applications of Complex Numbers in Geometry

Complex Numbers and Its Representation

- A number in the form of $z = x + iy$, where $x, y \in R$ and $i = \sqrt{-1}$, is called a **complex number**. The real numbers x and y are respectively called **real** and **imaginary** parts of complex number z.
  i.e. $x = \text{Re}(z)$, $y = \text{Im}(z)$ and the symbol i is called ***iota***.
- A complex number $z = x + iy$ is said to be purely real if $y = 0$ and purely imaginary if $x = 0$.
- **Integral power of iota** (i)
  (i) $i = \sqrt{-1}$, $i^2 = -1$, $i^3 = -i$ and $i^4 = 1$
  (ii) If n is an integer, then $i^{4n} = 1$, $i^{4n+1} = i$, $i^{4n+2} = -1$ and $i^{4n+3} = -i$
  (iii) $i^n + i^{n+1} + i^{n+2} + i^{n+3} = 0$

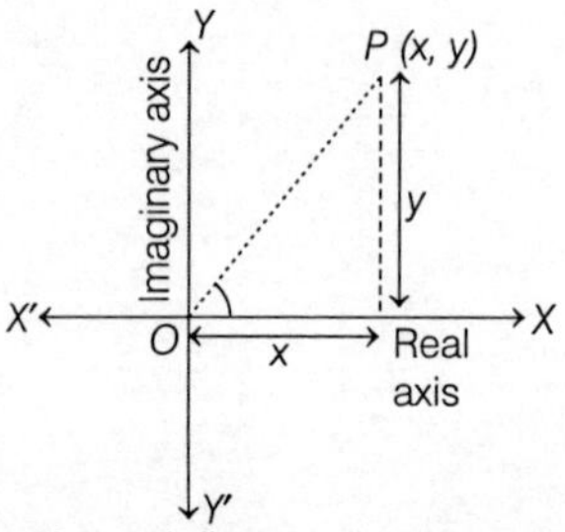

Argand diagram

- The complex number $z = x + iy$ can be represented by a point P in a plane called **argand plane** or **Gaussian plane** or **complex plane**. The coordinates of P are referred to the rectangular axes XOX' and YOY' which are called **real** and **imaginary axes**, respectively.

PREP MIRROR
Your Personal Preparation Indicator

- No. of Questions in Exercises (x)—
- No. of Questions Attempted (y)—
- No. of Correct Questions (z)—
  (Without referring Explanations)

- Accuracy Level ($z/y \times 100$)—
- Prep Level ($z/x \times 100$)—

In order to expect good rank in JEE, your Accuracy Level should be above 85 & Prep Level should be above 75.

Algebra and Equality of Complex Numbers

If $z_1 = x_1 + i y_1$ and $z_2 = x_2 + i y_2$ are two complex numbers, then

(i) $z_1 + z_2 = (x_1 + x_2) + i(y_1 + y_2)$

(ii) $z_1 - z_2 = (x_1 - x_2) + i(y_1 - y_2)$

(iii) $z_1 z_2 = (x_1 x_2 - y_1 y_2) + i(x_1 y_2 + x_2 y_1)$

(iv) $\dfrac{z_1}{z_2} = \dfrac{(x_1 x_2 + y_1 y_2) + i(x_2 y_1 - x_1 y_2)}{x_2^2 + y_2^2}$

(v) z_1 and z_2 are said to be equal if $x_1 = x_2$ and $y_1 = y_2$.

NOTE • Complex numbers does not possess any inequality, e.g. $3 + 2i > 1 + 2i$ does not make any sense.

Conjugate and Modulus of a Complex Number

- If $z = x + iy$ is a complex number, then **conjugate** of z is denoted by $\bar{z}$ and is obtained by replacing i by $-i$.
  i.e. $\bar{z} = x - iy$
- If $z = x + iy$, then **modulus** or **magnitude** of z is denoted by $|z|$ and is given by $|z| = \sqrt{x^2 + y^2}$

Results on Conjugate and Modulus

(i) $\overline{(\bar{z})} = z$

(ii) $z + \bar{z} = 2\,\text{Re}(z),\ z - \bar{z} = 2i\,\text{Im}(z)$

(iii) $z = \bar{z} \Leftrightarrow z$ is purely real.

(iv) $z + \bar{z} = 0 \Leftrightarrow z$ is purely imaginary.

(v) $\overline{z_1 \pm z_2} = \bar{z}_1 \pm \bar{z}_2$

(vi) $\overline{z_1 z_2} = \bar{z}_1 \bar{z}_2$

(vii) $\overline{\left(\dfrac{z_1}{z_2}\right)} = \dfrac{\bar{z}_1}{\bar{z}_2}$, if $z_2 \neq 0$

(viii) If $z = \begin{vmatrix} a_1 & a_2 & a_3 \\ b_1 & b_2 & b_3 \\ c_1 & c_2 & c_3 \end{vmatrix}$, then $\bar{z} = \begin{vmatrix} \bar{a}_1 & \bar{a}_2 & \bar{a}_3 \\ \bar{b}_1 & \bar{b}_2 & \bar{b}_3 \\ \bar{c}_1 & \bar{c}_2 & \bar{c}_3 \end{vmatrix}$

where $a_i, b_i, c_i; (i = 1, 2, 3)$ are complex numbers.

(ix) $|z| = 0 \Leftrightarrow z = 0$

(x) $|z| = |\bar{z}| = |-z| = |-\bar{z}|$

(xi) $-|z| \le \text{Re}(z), \text{Im}(z) \le |z|$

(xii) $|z_1 z_2| = |z_1||z_2|$

(xiii) $\left|\dfrac{z_1}{z_2}\right| = \dfrac{|z_1|}{|z_2|}$, if $|z_2| \neq 0$

(xiv) $|z_1 \pm z_2|^2 = |z_1|^2 + |z_2|^2 \pm \bar{z}_1 z_2 \pm z_1 \bar{z}_2$
$= |z_1|^2 + |z_2|^2 \pm 2\,\text{Re}(z_1 \bar{z}_2)$

(xv) $|z^n| = |z|^n, n \in N$

(xvi) **Reciprocal of a complex number** For non-zero complex number $z = x + iy$, the reciprocal is given by $z^{-1} = \dfrac{1}{z} = \dfrac{\bar{z}}{|z|^2}$.

(xvii) **Triangle Inequality**
(a) $|z_1 + z_2| \le |z_1| + |z_2|$ (b) $|z_1 + z_2| \ge ||z_1| - |z_2||$
(c) $|z_1 - z_2| \le |z_1| + |z_2|$ (d) $|z_1 - z_2| \ge ||z_1| - |z_2||$

Argument or Amplitude of a Complex Number

Let $z = x + iy$ be a complex number, represented by a point $P(x, y)$ in the argand plane. Then, the angle θ which OP makes with the positive direction of Real axis (X-axis) is called the argument or amplitude of z and it is denoted by **arg (z)** or **amp (z)**.

The argument of z, is given by $\theta = \tan^{-1}\left(\dfrac{y}{x}\right)$

- The value of argument θ which satisfies the inequality $-\pi < \theta \le \pi$, is called principal value of argument.
- The principal value of arg(z) is θ, $\pi - \theta$, $-\pi + \theta$ or $-\theta$ according as z lies in the 1st, 2nd, 3rd or 4th quadrants respectively, where $\theta = \tan^{-1}\left|\dfrac{y}{x}\right|$.

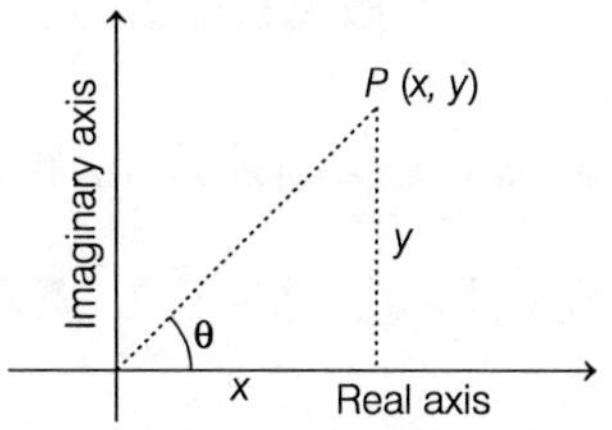

- Argument of z is not unique. General value of argument of z is $2n\pi + \theta$.

Results on Argument

If z, z_1 and z_2 are complex numbers, then

(i) $\arg(\bar{z}) = -\arg(z)$

(ii) $\arg(z_1 z_2) = \arg(z_1) + \arg(z_2)$

(iii) $\arg\left(\dfrac{z_1}{z_2}\right) = \arg(z_1) - \arg(z_2)$

(iv) The general value of $\arg(\bar{z})$ is $2n\pi - \arg(z)$.

(v) If z is purely imaginary then $\arg(z) = \pm\dfrac{\pi}{2}$.

(vi) If z is purely real then $\arg(z) = 0$ or π.

(vii) If $|z_1 + z_2| = |z_1 - z_2|$, then
$\arg\left(\dfrac{z_1}{z_2}\right)$ or $\arg(z_1) - \arg(z_2) = \dfrac{\pi}{2}$

(viii) If $|z_1 + z_2| = |z_1| + |z_2|$, then $\arg(z_1) = \arg(z_2)$

Different forms of a Complex Number

- **Polar or Trigonometrical Form** of $z = x + iy$ is $z = r(\cos\theta + i\sin\theta)$, where $r = |z|$ and $\theta = \arg(z)$.
  If we use the general value of the argument θ, then the polar form of z is $z = r[\cos(2n\pi + \theta) + i\sin(2n\pi + \theta)]$, where n is an integer.
- **Euler's form** of $z = x + iy$ is $z = re^{i\theta}$, where $r = |z|, \theta = \arg(z)$ and $e^{i\theta} = \cos\theta + i\sin\theta$.

Concept of Rotation

Let z_1, z_2, z_3 be the vertices of ΔABC as shown in figure, then

$$\alpha = \arg\left(\frac{z_3 - z_1}{z_2 - z_1}\right) \text{ and } \frac{z_3 - z_1}{z_2 - z_1} = \frac{|z_3 - z_1|}{|z_2 - z_1|} e^{i\alpha}$$

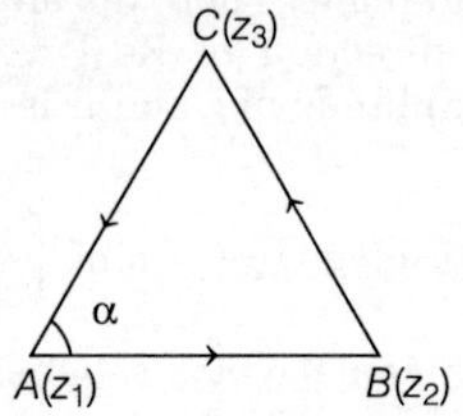

NOTE • Always mark the direction of arrow in anti-clockwise sense and keep that complex number in the numerator on which the arrow goes.

Square Root of a Complex Number

- If $z = a + ib$, then
$$\sqrt{z} = \sqrt{a+ib} = \pm \frac{1}{\sqrt{2}}[\sqrt{|z|+a} + i\sqrt{|z|-a}]$$
- If $z = a - ib$, then $\sqrt{z} = \sqrt{a-ib} = \pm \frac{1}{\sqrt{2}}[\sqrt{|z|+a} - i\sqrt{|z|-a}]$

De-Moivre's Theorem

- If n is any integer, then $(\cos\theta + i\sin\theta)^n = \cos n\theta + i\sin n\theta$
- If n is any rational number, then one of the values of $(\cos\theta + i\sin\theta)^n$ is $\cos n\theta + i\sin n\theta$.
- If n is any positive integer, then
$$(\cos\theta + i\sin\theta)^{1/n} = \cos\left(\frac{2k\pi + \theta}{n}\right) + i\sin\left(\frac{2k\pi + \theta}{n}\right)$$
where, $k = 0, 1, 2, \ldots n-1$

Cube Root of Unity

Cube roots of unity are $1, \omega, \omega^2$

where, $\omega = \dfrac{-1+\sqrt{3}i}{2}$ and $\omega^2 = \dfrac{-1-\sqrt{3}i}{2}$

Properties of Cube Roots of Unity

(i) $1 + \omega + \omega^2 = 0$

(ii) $\omega^3 = 1$

(iii) $1 + \omega^n + \omega^{2n} = \begin{cases} 0 & \text{if } n \neq 3m, \quad m \in N \\ 3 & \text{if } n = 3m, \quad m \in N \end{cases}$

*n*th Roots of Unity

By nth root of unity we mean any complex number z which satisfies the equation $z^n = 1$.

(i) The nth roots of unity are $1, \alpha, \alpha^2, \ldots, \alpha^{n-1}$, where $\alpha = e^{\frac{i2\pi}{n}}$

(ii) $1 + \alpha + \alpha^2 + \alpha^3 + \ldots + \alpha^{n-1} = 0$

(iii) $1 \cdot \alpha \cdot \alpha^2 \ldots \alpha^{n-1} = [-1]^{n-1}$

Applications of Complex Numbers in Geometry

1. Distance between $A(z_1)$ and $B(z_2)$ is given by $AB = |z_2 - z_1|$.
2. Let point $P(z)$ divides the line segment joining $A(z_1)$ and $B(z_2)$ in the ratio $m : n$. Then,

   (i) for internal division, $z = \dfrac{mz_2 + nz_1}{m+n}$

   (ii) for external division, $z = \dfrac{mz_2 - nz_1}{m-n}$
3. Let ABC be a triangle with vertices $A(z_1)$, $B(z_2)$ and $C(z_3)$, then centroid $G(z)$ of the ΔABC is given by $z = \frac{1}{3}(z_1 + z_2 + z_3)$

   Area of ΔABC is given by $\Delta = \dfrac{1}{2}\begin{vmatrix} z_1 & \bar{z}_1 & 1 \\ z_2 & \bar{z}_2 & 1 \\ z_3 & \bar{z}_3 & 1 \end{vmatrix}$
4. For an equilateral triangle ABC with vertices $A(z_1)$, $B(z_2)$ and $C(z_3)$, $z_1^2 + z_2^2 + z_3^2 = z_2z_3 + z_3z_1 + z_1z_2$
5. The general equation of a straight line is $\bar{a}z + a\bar{z} + b = 0$, where a is a complex number and b is a real number.
6. (i) An equation of the circle with centre at z_0 and radius r, is $|z - z_0| = r$

   (ii) $|z - z_0| < r$ represents the interior of circle and $|z - z_0| > r$ represents the exterior of circle.

   (iii) General equation of a circle is $z\bar{z} + a\bar{z} + \bar{a}z + b = 0$, where b is real number, with centre is $-a$ and radius is $\sqrt{a\bar{a} - b}$.
7. If z_1 and z_2 are two fixed points and $k > 0$, $k \neq 1$ is a real number, then $\dfrac{|z - z_1|}{|z - z_2|} = k$ represents a circle.

   For $k = 1$, it represents perpendicular bisector of the segment joining $A(z_1)$ and $B(z_2)$.
8. If end points of diameter of a circle are $A(z_1)$ and $B(z_2)$ and $P(z)$ be any point on the circle, then equation of circle in diameter form is
$$(z - z_1)(\bar{z} - \bar{z}_2) + (z - z_2)(\bar{z} - \bar{z}_1) = 0$$

DAY PRACTICE SESSION 1

FOUNDATION QUESTIONS EXERCISE

1 Real part of $\dfrac{1}{1-\cos\theta+i\sin\theta}$ is

(a) $-\dfrac{1}{2}$ (b) $\dfrac{1}{2}$ (c) $\dfrac{1}{2}\tan\theta/2$ (d) 2

2 A value of θ, for which $\dfrac{2+3i\sin\theta}{1-2i\sin\theta}$ is purely imaginary, is

(a) $\dfrac{\pi}{3}$ (b) $\dfrac{\pi}{6}$ (c) $\sin^{-1}\dfrac{\sqrt{3}}{4}$ (d) $\sin^{-1}\left(\dfrac{1}{\sqrt{3}}\right)$

3 $\sum_{n=1}^{13}(i^n+i^{n+1})$ is equal to

(a) i (b) $i-1$ (c) $-i$ (d) 0

4 If $\dfrac{z-1}{z+1}$ is a purely imaginary number (where, $z\neq -1$), then the value of $|z|$ is

(a) -1 (b) 1 (c) 2 (d) -2

5 If $z_1\neq 0$ and z_2 are two complex numbers such that $\dfrac{z_2}{z_1}$ is a purely imaginary number, then $\left|\dfrac{2z_1+3z_2}{2z_1-3z_2}\right|$ is equal to

→ JEE Mains 2013

(a) 2 (b) 5 (c) 3 (d) 1

6 If $f(z)=\dfrac{7-z}{1-z^2}$, where $z=1+2i$, then $|f(z)|$ is equal to

(a) $\dfrac{|z|}{2}$ (b) $|z|$

(c) $2|z|$ (d) None of these

7 If $8iz^3+12z^2-18z+27i=0$, then the value of $|z|$ is

(a) $3/2$ (b) $2/3$ (c) 1 (d) $3/4$

8 If a complex number z satisfies the equation $z+\sqrt{2}|z+1|+i=0$, then $|z|$ is equal to → JEE Mains 2013

(a) 2 (b) $\sqrt{3}$ (c) $\sqrt{5}$ (d) 1

9 If α and β are two different complex numbers such that $|\alpha|=1, |\beta|=1$, then the expression $\left|\dfrac{\beta-\alpha}{1-\bar{\alpha}\beta}\right|$ is equal to

(a) $\dfrac{1}{2}$ (b) 1

(c) 2 (d) None of these

10 If $|z|=1$ and $\omega=\dfrac{z-1}{z+1}$ (where $z\neq -1$), then $\text{Re}(\omega)$ is

(a) 0 (b) $-\dfrac{1}{|z+1|^2}$

(c) $\dfrac{\sqrt{2}}{|z+1|^2}$ (d) None of these

11 If $\left|z-\dfrac{4}{z}\right|=2$, then the maximum value of $|z|$ is → AIEEE 2009

(a) $\sqrt{3}+1$ (b) $\sqrt{5}+1$ (c) 2 (d) $2+\sqrt{2}$

12 If z is a complex number such that $|z|\geq 2$, then the minimum value of $\left|z+\dfrac{1}{2}\right|$ → JEE Mains 2014

(a) is equal to 5/2
(b) lies in the interval (1, 2)
(c) is strictly greater than 5/2
(d) is strictly greater than 3/2 but less than 5/2

13 If $|z_1|=2$, $|z_2|=3$ then $|z_1+z_2+5+12i|$ is less than or equal to

(a) 8 (b) 18 (c) 10 (d) 5

14 If $|z|<\sqrt{3}-1$, then $|z^2+2z\cos\alpha|$ is

(a) less than 2 (b) $\sqrt{3}+1$
(c) $\sqrt{3}-1$ (d) None of these

15 The number of complex numbers z such that $|z-1|=|z+1|=|z-i|$, is

(a) 0 (b) 1 (c) 2 (d) ∞

16 Number of solutions of the equation $|z|^2+7\bar{z}=0$ is/are

(a) 1 (b) 2 (c) 4 (d) 6

17 If $z\bar{z}+(3-4i)z+(3+4i)\bar{z}=0$ represent a circle, the area of the circle in square units is

(a) 5π (b) 10π (c) $25\pi^2$ (d) 25π

18 If $z=1+\cos\left(\dfrac{\pi}{5}\right)+i\sin\left(\dfrac{\pi}{5}\right)$, then $\{\sin(\arg(z))\}$ is equal to

(a) $\sqrt{\dfrac{10-2\sqrt{5}}{4}}$ (b) $\dfrac{\sqrt{5}-1}{4}$

(c) $\dfrac{\sqrt{5}+1}{4}$ (d) None of these

19 If z is a complex number of unit modulus and argument θ, then $\arg\left(\dfrac{1+z}{1+\bar{z}}\right)$ equals to → JEE Mains 2013

(a) $-\theta$ (b) $\dfrac{\pi}{2}-\theta$ (c) θ (d) $\pi-\theta$

20 Let z and ω are two non-zero complex numbers such that $|z|=|\omega|$ and $\arg z+\arg\omega=\pi$, then z equals

(a) $\bar{\omega}$ (b) ω
(c) $-\bar{\omega}$ (d) $-\omega$

21 If $|z-1|=1$, then $\arg(z)$ is equal to

(a) $\dfrac{1}{2}\arg(z)$ (b) $\dfrac{1}{3}\arg(z+1)$

(c) $\dfrac{1}{2}\arg(z-1)$ (d) None of these

22 Let $z = \cos\theta + i\sin\theta$. Then the value of $\sum_{m=1}^{15} \text{Im}(z^{2m-1})$ at, $\theta = 2°$, is

(a) $\frac{1}{\sin 2°}$ (b) $\frac{1}{3\sin 2°}$ (c) $\frac{1}{2\sin 2°}$ (d) $\frac{1}{4\sin 2°}$

23 If $z = (i)^{(i)^{(i)}}$, where $i = \sqrt{-1}$, then $|z|$ is equal to

(a) 1 (b) $e^{-\pi/2}$ (c) 0 (d) $e^{\pi/2}$

24 $\left(\frac{1 + i\sin\frac{\pi}{8} + \cos\frac{\pi}{8}}{1 - i\sin\frac{\pi}{8} + \cos\frac{\pi}{8}}\right)^8$ equals to

(a) 2^8 (b) 0 (c) -1 (d) 1

25 If $1, \alpha_1, \alpha_2, \ldots, \alpha_{n-1}$ are the nth roots of unity, then $(2-\alpha_1)(2-\alpha_2)\ldots(2-\alpha_{n-1})$ is equal to

(a) n (b) 2^n (c) $2^n + 1$ (d) $2^n - 1$

26 If $\omega(\neq 1)$ is a cube root of unity and $(1+\omega)^7 = A + B\omega$. Then, (A, B) is equal to

(a) (1,1) (b) (1, 0) (c) (−1, 1) (d) (0, 1)

27 If $\alpha, \beta \in C$ are the distinct roots of the equation $x^2 - x + 1 = 0$, then $\alpha^{101} + \beta^{107}$ is equal to → JEE Mains 2018

(a) −1 (b) 0 (c) 1 (d) 2

28 If $x^2 + x + 1 = 0$, then $\sum_{r=1}^{25}\left(x^r + \frac{1}{x^r}\right)^2$ is equal to

(a) 25 (b) 25ω
(c) $25\omega^2$ (d) None of these

29 Let ω be a complex number such that $2\omega + 1 = z$, where $z = \sqrt{-3}$. If $\begin{vmatrix} 1 & 1 & 1 \\ 1 & -\omega^2 - 1 & \omega^2 \\ 1 & \omega^2 & \omega^7 \end{vmatrix} = 3k$, then k is equal to → JEE Mains 2017

(a) $-z$ (b) z (c) -1 (d) 1

30 The value $\begin{vmatrix} 1+\omega & \omega^2 & 1+\omega^2 \\ -\omega & -(1+\omega^2) & (1+\omega) \\ -1 & -(1+\omega^2) & 1+\omega \end{vmatrix}$, where ω is cube root of unity, is equal to

(a) 2ω (b) $3\omega^2$ (c) $-3\omega^2$ (d) 3ω

31 If a, b and c are integers not all equal and ω is a cube root of unity (where, $\omega \neq 1$), then minimum value of $|a + b\omega + c\omega^2|$ is equal to

(a) 0 (b) 1 (c) $\frac{\sqrt{3}}{2}$ (d) $\frac{1}{2}$

32 Let $\omega = e^{i\pi/3}$, and a, b, c, x, y, z be non-zero complex numbers such that:

$$a + b + c = x;\ a + b\omega + c\omega^2 = y;\ a + b\omega^2 + c\omega = z$$

Then the value of $\frac{|x|^2 + |y|^2 + |z|^2}{|a|^2 + |b|^2 + |c|^2}$ is:

(a) 1 (b) 2 (c) 3 (d) 4

33 If $\text{Re}\left(\frac{1}{z}\right) = 3$, then z lies on

(a) circle with centre on Y-axis
(b) circle with centre on X-axis not passing through origin
(c) circle with centre on X-axis passing through origin
(d) None of the above

34 If the imaginary part of $(2z+1)/(iz+1)$ is −2, then the locus of the point representing z in the complex plane is

(a) a circle (b) a straight line
(c) a parabola (d) None of these

35 If $|z| = 1$ and $z \neq \pm 1$, then all the values of $\frac{z}{1-z^2}$ lie on

(a) a line not passing through the origin
(b) $|z| = \sqrt{2}$
(c) the X-axis
(d) the Y-axis

36 If $\omega = \frac{z}{z - \frac{i}{3}}$ and $|\omega| = 1$, then z lies on

(a) a circle (b) an ellipse
(c) a parabola (d) a straight line

37 If z_1 and z_2 are two complex numbers such that $\frac{z_1}{z_2} + \frac{z_2}{z_1} = 1$, then

(a) z_1, z_2 are collinear
(b) z_1, z_2 and the origin form a right angled triangle
(c) z_1, z_2 and the origin form an equilateral triangle
(d) None of the above

38 A complex number z is said to be unimodular, if $|z| = 1$. Suppose z_1 and z_2 are complex numbers such that $\frac{z_1 - 2z_2}{2 - z_1\bar{z}_2}$ is unimodular and z_2 is not unimodular.

Then, the point z_1 lies on a → JEE Mains 2015

(a) straight line parallel to X-axis
(b) straight line parallel to Y-axis
(c) circle of radius 2
(d) circle of radius $\sqrt{2}$

39 If $|z^2 - 1| = |z|^2 + 1$, then z lies on

(a) a real axis (b) an ellipse
(c) a circle (d) imaginary axis

40 Let z satisfy $|z| = 1$ and $z = 1 - \bar{z}$ → JEE Mains 2013

Statement I z is a real number.

Statement II Principal argument of z is $\pi/3$.

(a) Statement I is true, Statement II is true; Statement II is a correct explanation for statement I
(b) Statement I is true, Statement II is true; Statement II is not a correct explanation for Statement I
(c) Statement I is true, Statement II is false
(d) Statement I is false, Statement II is true

DAY PRACTICE SESSION 2

PROGRESSIVE QUESTIONS EXERCISE

1 For positive integers n_1 and n_2 the value of the expression $(1+i)^{n_1}+(1+i^3)^{n_1}+(1+i^5)^{n_2}+(1+i^7)^{n_2}$ where $i=\sqrt{-1}$, is a real number iff

(a) $n_1=n_2$ (b) $n_2=n_2-1$ (c) $n_1=n_2+1$ (d) $\forall n_1$ and n_2

2 If $z\neq 1$ and $\dfrac{z^2}{z-1}$ is real, then the point represented by the complex number z lies

(a) on the imaginary axis
(b) either on the real axis or on a circle passing through the origin
(c) on a circle with centre at the origin
(d) either on the real axis or on a circle not passing through the origin

3 Let ω be the complex number $\cos\dfrac{2\pi}{3}+i\sin\dfrac{2\pi}{3}$. Then the number of distinct complex numbers z satisfying $\begin{vmatrix} z+1 & \omega & \omega^2 \\ \omega & z+\omega^2 & 1 \\ \omega^2 & 1 & z+\omega \end{vmatrix}=0$ is equal to

(a) 0 (b) 1 (c) 2 (d) 4

4 The locus of $z=x+iy$ which satisfying the inequality $\log_{1/2}|z-1|>\log_{1/2}|z-i|$ is given by

(a) $x+y<0$ (b) $x-y>0$ (c) $x-y<0$ (d) $x+y>0$

5 Let $z_1=10+6i$, $z_2=4+6i$. If z is any complex number such that $\arg(z-z_1)/(z-z_2)=\pi/4$, then $|z-7-9i|$ is equal to

(a) 18 (b) $3\sqrt{2}$ (c) $3/\sqrt{2}$ (d) None of these

6 Let $z=x+iy$ be a complex number where x and y are integers. Then the area of the rectangle whose vertices are the roots of the equation $z\bar{z}^3+\bar{z}z^3=350$ is

(a) 48 (b) 32 (c) 40 (d) 80

7 If $\alpha+i\beta=\cot^{-1}(z)$, where $z=x+iy$ and α is a constant, then the locus of z is

(a) $x^2+y^2-x\cot 2\alpha-1=0$
(b) $x^2+y^2-2x\cot\alpha-1=0$
(c) $x^2+y^2-2x\cot 2\alpha+1=0$
(d) $x^2+y^2-2x\cot 2\alpha-1=0$

8 If a complex number z lies in the interior or on the boundary of a circle of radius 3 and centre at $(-4, 0)$, then the greatest and least value of $|z+1|$ are

(a) 5, 0 (b) 6, 1 (c) 6, 0 (d) None of these

9 If z is any complex number satisfying $|z-3-2i|\leq 2$, then the minimum value of $|2z-6+5i|$ is

(a) 2 (b) 3 (c) 5 (d) 6

10 A man walks a distance of 3 units from the origin towards the North-East (N 45° E) direction. From there, he walks a distance of 4 units towards the North-West (N 45° W) direction to reach a point P. Then the position of P in the Argand plane is

(a) $3e^{i\pi/4}+4i$ (b) $(3-4i)e^{i\pi/4}$
(c) $(4+3i)e^{i\pi/4}$ (d) $(3+4i)e^{i\pi/4}$

11 If $1, a_1, a_2 \dots a_{n-1}$ are n^{th} roots of unity, then $\dfrac{1}{1-a_1}+\dfrac{1}{1-a_2}+\dots+\dfrac{1}{1-a_{n-1}}$ equals to

(a) $\dfrac{2^n-1}{n}$ (b) $\dfrac{n-1}{2}$ (c) $\dfrac{n}{n-1}$ (d) None of these

12 For $z,\omega\in C$, if $|z|^2\omega-|\omega|^2 z=z-\omega$, then z is equal to

(a) ω or $\bar{\omega}$ (b) ω or $\omega/|\omega|^2$
(c) $\bar{\omega}$ or $\omega/|\omega|^2$ (d) None of these

13 The value of $\displaystyle\sum_{k=1}^{10}\left(\sin\frac{2k\pi}{11}+i\cos\frac{2k\pi}{11}\right)$ is

(a) 1 (b) -1 (c) $-i$ (d) i

14 Let z_1 and z_2 be roots of the equation $z^2+pz+q=0$, $p,q\in C$. Let A and B represents z_1 and z_2 in the complex plane. If $\angle AOB=\alpha\neq 0$ and $OA=OB$; O is the origin, then $p^2/4q$ is equal to

(a) $\sin^2(\alpha/2)$ (b) $\tan^2(\alpha/2)$ (c) $\cos^2(\alpha/2)$ (d) None of these

15 If $1,\omega$ and ω^2 are the three cube roots of unity α,β,γ are the cube roots of $p, q<0$, then for any x, y, z the expression $\left(\dfrac{x\alpha+y\beta+z\gamma}{x\beta+y\gamma+z\alpha}\right)$ is equal to

(a) 1 (b) ω (c) ω^2 (d) None of these

ANSWERS

SESSION 1

1. (b)	**2.** (d)	**3.** (b)	**4.** (b)	**5.** (d)	**6.** (a)	**7.** (a)	**8.** (c)	**9.** (b)	**10.** (a)
11. (b)	**12.** (b)	**13.** (b)	**14.** (a)	**15.** (b)	**16.** (b)	**17.** (d)	**18.** (b)	**19.** (c)	**20.** (c)
21. (c)	**22.** (d)	**23.** (a)	**24.** (c)	**25.** (d)	**26.** (a)	**27.** (c)	**28.** (d)	**29.** (a)	**30.** (c)
31. (b)	**32.** (c)	**33.** (c)	**34.** (b)	**35.** (d)	**36.** (d)	**37.** (c)	**38.** (c)	**39.** (d)	**40.** (d)

SESSION 2

1. (d)	**2.** (b)	**3.** (b)	**4.** (b)	**5.** (b)	**6.** (a)	**7.** (d)	**8.** (c)	**9.** (c)	**10.** (d)
11. (b)	**12.** (b)	**13.** (c)	**14.** (c)	**15.** (c)					

Hints and Explanations

SESSION 1

1 Let $z = \dfrac{1}{1-\cos\theta + i\sin\theta}$

$$= \frac{1}{2\sin^2(\theta/2) + 2i\sin(\theta/2)\cos(\theta/2)}$$

$$= \frac{1}{2i\sin(\theta/2)}\frac{1}{[\cos(\theta/2) - i\sin(\theta/2)]}$$

$$= \frac{\cos(\theta/2) + i\sin(\theta/2)}{2i\sin(\theta/2)} = \frac{1}{2} + \frac{1}{2i}\cot(\theta/2)$$

$$= \frac{1}{2} - i\cdot\frac{1}{2}\cot\theta/2$$

2 Let $z = \dfrac{2+3i\sin\theta}{1-2i\sin\theta}$ is purely imaginary

then we have

$\text{Re}(z) = 0$

Consider, $z = \dfrac{2+3i\sin\theta}{1-2i\sin\theta}$

$$= \frac{(2+3i\sin\theta)(1+2i\sin\theta)}{(1-2i\sin\theta)(1+2i\sin\theta)}$$

$$= \frac{(2-6\sin^2\theta) + (4\sin\theta + 3\sin\theta)i}{1+4\sin^2\theta}$$

$\because \quad \text{Re}(z) = 0$

$\therefore \quad \dfrac{2-6\sin^2\theta}{1+4\sin^2\theta} = 0$

$\Rightarrow \quad \sin^2\theta = \dfrac{1}{3} \Rightarrow \sin\theta = \pm\dfrac{1}{\sqrt{3}}$

$\Rightarrow \quad \theta = \pm\sin^{-1}\left(\dfrac{1}{\sqrt{3}}\right)$

3 $\displaystyle\sum_{n=1}^{13}(i^n + i^{n+1}) = (1+i)\sum_{n=1}^{13} i^n$

$$= (1+i)\frac{i(1-i^{13})}{1-i}$$

$$= i - 1 \quad [\because i^{13} = i,\ i^2 = -1]$$

4 Let $z = x + iy$

$$\frac{z-1}{z+1} = \frac{x+iy-1}{x+iy+1} = \frac{(x-1)+iy}{(x+1)+iy} \times \frac{(x+1)-iy}{(x+1)-iy}$$

$$= \frac{(x-1)(x+1) - iy(x-1) + iy(x+1) - i^2y^2}{(x+1)^2 - i^2y^2}$$

$$= \frac{x^2 - 1 + iy(x+1-x+1) + y^2}{(x+1)^2 + y^2}$$

$$\Rightarrow \frac{z-1}{z+1} = \frac{(x^2+y^2-1)}{(x+1)^2+y^2} + \frac{i(2y)}{(x+1)^2+y^2}$$

Since, $\dfrac{z-1}{z+1}$ is purely imaginary.

$\therefore \quad \text{Re}\left(\dfrac{z-1}{z+1}\right) = 0$

$\Rightarrow \quad \dfrac{x^2+y^2-1}{(x+1)^2+y^2} = 0$

$\Rightarrow \quad x^2 + y^2 - 1 = 0$

$\Rightarrow \quad x^2 + y^2 = 1$

$\Rightarrow \quad |z|^2 = 1 \Rightarrow |z| = 1$

5 Given, $\dfrac{z_2}{z_1}$ is a purely imaginary

Let $z = ni$. Then,

$$\left|\frac{2z_1 + 3z_2}{2z_1 - 3z_2}\right| = \left|\frac{2 + 3\cdot\frac{z_2}{z_1}}{2 - 3\cdot\frac{z_2}{z_1}}\right| = \frac{|2+3ni|}{|2-3ni|}$$

$$= \frac{\sqrt{4+9n^2}}{\sqrt{4+9n^2}} = 1$$

6 Given, $f(z) = \dfrac{7-z}{1-z^2}$ and $z = 1+2i$

$\therefore \quad f(z) = \dfrac{7-(1+2i)}{1-(1+2i)^2}$

$$= \frac{6-2i}{1-(1-4+4i)} = \frac{6-2i}{4-4i}$$

$$= \frac{6-2i}{4(1-i)} \times \frac{1+i}{(1+i)} = \frac{6+4i+2}{4(1^2-i^2)}$$

$$= \frac{8+4i}{4(2)} = \frac{1}{2}(2+i)$$

Now, $|f(z)| = \dfrac{\sqrt{4+1}}{2} = \dfrac{\sqrt{5}}{2} = \dfrac{|z|}{2}$

$[\because z = 1+2i, \text{ given} \Rightarrow |z| = \sqrt{5}]$

7 Given, $8iz^3 + 12z^2 - 18z + 27i = 0$

$\Rightarrow \quad 4z^2(2iz+3) + 9i(2iz+3) = 0$

$\Rightarrow \quad (2iz+3)(4z^2+9i) = 0$

$\Rightarrow \quad 2iz + 3 = 0 \text{ or } 4z^2 + 9i = 0$

$\therefore \quad |z| = \dfrac{3}{2}$

8 We have, $(x+iy) + \sqrt{2}|x+iy+1| + i = 0$

[put $z = x+iy$]

$\Rightarrow \quad (x+iy) + \sqrt{2}\sqrt{(x+1)^2 + y^2} + i = 0$

$\Rightarrow \quad x + \sqrt{2}\sqrt{(x+1)^2+y^2} = 0$

and $\quad y + 1 = 0$

$\Rightarrow \quad x + \sqrt{2}\sqrt{(x+1)^2 + (-1)^2} = 0$

and $\quad y = -1$

$\Rightarrow \quad x^2 = 2[(x+1)^2 + 1]$

$\Rightarrow \quad x^2 = 2x^2 + 4x + 4$

$\Rightarrow \quad x^2 + 4x + 4 = 0 \Rightarrow (x+2)^2 = 0$

$\Rightarrow \quad x = -2$

$\therefore \quad z = -2 - i \Rightarrow |z| = \sqrt{4+1} = \sqrt{5}$

9 $\left|\dfrac{\beta-\alpha}{1-\bar{\alpha}\beta}\right| = \left|\dfrac{\beta-\alpha}{\beta\cdot\bar{\beta} - \bar{\alpha}\beta}\right|$

$\left[\begin{array}{l}\because \ |\beta| = 1 \\ \text{and } |\beta|^2 = \beta\bar{\beta} = 1\end{array}\right]$

$$= \left|\frac{\beta-\alpha}{\beta(\bar{\beta}-\bar{\alpha})}\right| = \frac{1}{|\beta|}\frac{|\beta-\alpha|}{|\overline{\beta-\alpha}|} = \frac{|\beta-\alpha|}{|\beta-\alpha|} = 1$$

$[\because |z| = |\bar{z}|]$

10 Given, $|z| = 1$

$\Rightarrow \quad z\bar{z} = 1$

Now, $2\text{Re}(\omega) = \omega + \bar{\omega} = \dfrac{z-1}{z+1} + \dfrac{\bar{z}-1}{\bar{z}+1}$

$$= \frac{(z-1)(\bar{z}+1) + (\bar{z}-1)(z+1)}{|z+1|^2}$$

$$= \frac{2z\bar{z} - 2}{|z+1|^2} = 0 \quad [\because z\bar{z} = 1]$$

$\therefore \quad \text{Re}(\omega) = 0.$

11 $|z| = \left|\left(z - \dfrac{4}{z}\right) + \dfrac{4}{z}\right|$

$\Rightarrow \quad |z| \le \left|z - \dfrac{4}{z}\right| + \dfrac{4}{|z|}$

$\Rightarrow \quad |z| \le 2 + \dfrac{4}{|z|}$

$\Rightarrow \quad \dfrac{|z|^2 - 2|z| - 4}{|z|} \le 0$

Since, $|z| > 0$

$\Rightarrow \quad |z|^2 - 2|z| - 4 \le 0$

$\Rightarrow [|z| - (\sqrt{5}+1)][|z| - (1-\sqrt{5})] \le 0$

$\Rightarrow \quad 1 - \sqrt{5} \le |z| \le \sqrt{5} + 1$

12 $|z| \ge 2$ is the region on or outside circle whose centre is (0,0) and radius is 2.

Minimum $\left|z + \dfrac{1}{2}\right|$ is distance of z, which lie on circle $|z| = 2$ from $\left(-\dfrac{1}{2}, 0\right)$.

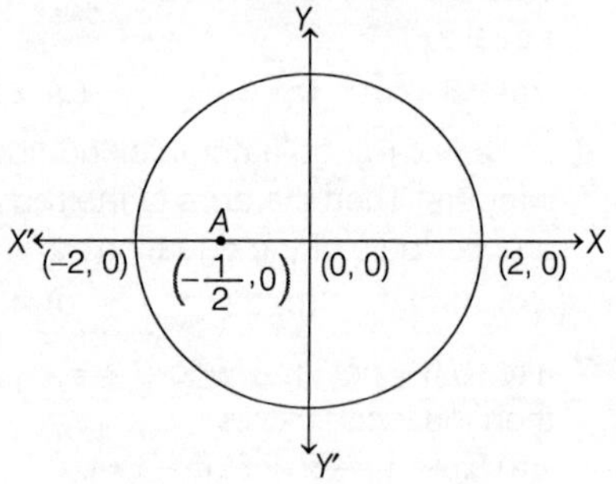

$\therefore$ Minimum $\left|z + \dfrac{1}{2}\right|$

$= \text{Distance of } \left(-\dfrac{1}{2}, 0\right) \text{ from } (-2, 0)$

$$= \sqrt{\left(-2 + \frac{1}{2}\right)^2 + 0} = \frac{3}{2}$$

Alternate Method

We know, $|z_1 + z_2| \ge ||z_1| - |z_2||$

$\therefore \quad \left|z + \dfrac{1}{2}\right| \ge \left||z| - \left|\dfrac{1}{2}\right|\right| = \left||z| - \dfrac{1}{2}\right|$

$$\ge \left|z - \frac{1}{2}\right| = \frac{3}{2}$$

$\therefore \quad \left|z+\frac{1}{2}\right| \geq \frac{3}{2}$

$\therefore$ Minimum value of $\left|z+\frac{1}{2}\right|$ is $\frac{3}{2}$.

13 **Fact:** $|z_1 + z_2 + \ldots + z_n| \leq |z_1| + |z_2| + \ldots + |z_n|$

$\therefore \quad |z_1 + z_2 + (5+12i)| \leq |z_1| + |z_2| + |5+12i| = 2+3+13 = 18$

14 Consider $|z^2 + 2z\cos\alpha| \leq |z|^2 + 2|z|$

$|\cos\alpha| \leq |z|^2 + 2|z|$

$< (\sqrt{3}-1)^2 + 2(\sqrt{3}-1)$

$= 3+1-2\sqrt{3}+2\sqrt{3}-2 = 2$

$\therefore \quad |z^2 + 2z\cos\alpha| < 2$

15 Let $z = x + iy$

$|z-1| = |z+1|$

$\text{Re } z = 0 \Rightarrow x = 0$

$|z-1| = |z-i| \Rightarrow x = y$

$|z+1| = |z-i| \Rightarrow y = -x$

Since, only (0, 0) will satisfy all conditions.

$\therefore$ Number of complex number $z = 1$.

16 **Given** $|z|^2 + 7\bar{z} = 0$

$\Rightarrow \quad z\bar{z} + 7\bar{z} = 0 \Rightarrow \bar{z}(z+7) = 0$

Case (i) : $\bar{z} = 0, \quad \therefore z = 0 = 0 + i0$

Case (ii) : $z = -7 \quad \therefore z = -7 + 0i$

Hence, there is only two solutions.

$z = 0$ and $z = -7$

17 Given $z\bar{z} + (3-4i)z + (3+4i)\bar{z} = 0$

Let $z = x + iy$

Then, $z\bar{z} = x^2 + y^2$

$\therefore \quad x^2 + y^2 + (3-4i)(x+iy) + (3+4i)(x-iy) = 0$

$\Rightarrow \quad x^2 + y^2 + 6x + 8y = 0$

$\Rightarrow \quad (x^2 + 6x) + (y^2 + 8y) = 0$

$\Rightarrow \quad (x+3)^2 + (y+4)^2 = 3^2 + 4^2$

$\Rightarrow \quad [x-(-3)]^2 + [y-(-4)]^2 = 5^2$

So, area of circle be $\pi R^2 = 25\pi$

$[\because R = \text{radius} = 5]$

18 If $z = 1 + \cos\theta + i\sin\theta$, then $\arg(z) = \frac{\theta}{2}$

$\therefore \quad \arg(z) = \frac{\pi/5}{2} = \frac{\pi}{10}$

$\Rightarrow \sin(\arg z)$

$= \sin\left(\frac{\pi}{10}\right) = \sin 18° = \frac{\sqrt{5}-1}{4}$

19 Given, $|z| = 1$ and $\arg z = \theta$

$\therefore \quad z = e^{i\theta}$ and $\bar{z} = \frac{1}{z}$

Now, $\arg\left(\frac{1+z}{1+\bar{z}}\right) = \arg\left(\frac{1+z}{1+\frac{1}{z}}\right)$

$= \arg(z) = \theta$

20 Let $|z| = |\omega| = r$ and let $\arg\omega = \theta$

Then, $\omega = r(\cos\theta + i\sin\theta) = re^{i\theta}$

and $\arg z = \pi - \theta$

Hence, $z = r(\cos(\pi-\theta) + i\sin(\pi-\theta))$

$= r(-\cos\theta + i\sin\theta)$

$= -r(\cos\theta - i\sin\theta)$

$z = -\bar{\omega}$

21 Given, $|z-1| = 1 \Rightarrow z - 1 = e^{i\theta}$,

where $\arg(z-1) = \theta \quad \ldots\text{(i)}$

$\Rightarrow \quad z = e^{i\theta} + 1$

$\Rightarrow \quad z = 1 + \cos\theta + i\sin\theta$

$[\because e^{i\theta} = \cos\theta + i\sin\theta]$

$= 2\cos^2\frac{\theta}{2} + 2i\sin\frac{\theta}{2}\cdot\cos\frac{\theta}{2}$

$\Rightarrow \arg(z) = \frac{\theta}{2} = \frac{1}{2}\arg(z-1)$ [from Eq. (i)]

22 Given that $z = \cos\theta + i\sin\theta = e^{i\theta}$

$\therefore \quad \sum_{m=1}^{15} \text{Im}(z^{2m-1}) = \sum_{m=1}^{15} \text{Im}(e^{i\theta})^{2m-1}$

$= \sum_{m=1}^{15} \text{Im}\, e^{i(2m-1)\theta}$

$= \sin\theta + \sin 3\theta + \sin 5\theta + \ldots + \sin 29\theta$

$= \frac{\sin\left(\theta + \frac{14\cdot 2\theta}{2}\right)\sin\left(\frac{15\cdot 2\theta}{2}\right)}{\sin\left(\frac{2\theta}{2}\right)}$

$= \frac{\sin(15\theta)\sin(15\theta)}{\sin\theta} = \frac{1}{4\sin 2°} \quad [\because \theta = 2°]$

23 Clearly, $i = \cos\frac{\pi}{2} + i\sin\frac{\pi}{2} = e^{i\pi/2}$

$\therefore \quad (i)^i = (e^{i\pi/2})^i = e^{i^2\cdot\frac{\pi}{2}} = e^{-\pi/2}$

Now, $(i)^{(i)^{(i)}} = (i)^{e^{-\pi/2}} \Rightarrow z = (i)^{e^{-\pi/2}}$

$\Rightarrow \quad |z| = |i|^{e^{-\pi/2}} = 1$

24 Let $z = \cos\frac{\pi}{8} + i\sin\frac{\pi}{8}$

Then, $\frac{1}{z} = \cos\frac{\pi}{8} - i\sin\frac{\pi}{8}$

Now, $\left(\frac{1+\cos\frac{\pi}{8}+i\sin\frac{\pi}{8}}{1+\cos\frac{\pi}{8}-i\sin\frac{\pi}{8}}\right)^8 = \left(\frac{1+z}{1+z^{-1}}\right)^8$

$= \left(\frac{(1+z)z}{(1+z)}\right)^8 = z^8 = \left(\cos\frac{\pi}{8} + i\sin\frac{\pi}{8}\right)^8$

$= \cos 8\cdot\frac{\pi}{8} + i\sin 8\cdot\frac{\pi}{8}$

[using De-moivre's theorem]

$= \cos\pi = -1 \; (\because \sin\pi = 0)$

25 Clearly, $(x-1)(x-\alpha_1)(x-\alpha_2)\ldots(x-\alpha_{n-1}) = x^n - 1$

Putting $x = 2$, we get

$(2-\alpha_1)(2-\alpha_2)\ldots(2-\alpha_{n-1}) = 2^n - 1$

26 We have, $(1+\omega)^7 = A + B\omega$

We know that $1 + \omega + \omega^2 = 0$

$\therefore \quad 1 + \omega = -\omega^2$

$\Rightarrow \quad (-\omega^2)^7 = A + B\omega$

$\Rightarrow \quad -\omega^{14} = A + B\omega$

$\Rightarrow -\omega^2 = A + B\omega \Rightarrow 1 + \omega = A + B\omega$

$[\because \omega^{14} = \omega^{12}\cdot\omega^2 = \omega^2]$

On comparing both sides, we get

$A = 1, B = 1$

27 α, β are the roots of $x^2 - x + 1 = 0$

$\because$ Roots of $x^2 - x + 1 = 0$ are $-\omega, -\omega^2$

$\therefore \quad$ Let $\alpha = -\omega$ and $\beta = -\omega^2$

$\Rightarrow \quad \alpha^{101} + \beta^{107} = (-\omega)^{101} + (-\omega^2)^{107}$

$= -(\omega^{101} + \omega^{214}) = -(\omega^2 + \omega)$

$[\because \omega^{3n+2} = \omega^2 \text{ and } \omega^{3n+1} = \omega]$

$= -(-1) = 1 \quad [1 + \omega + \omega^2 = 0]$

28 $x^2 + x + 1 = 0$

$\Rightarrow \quad x = \omega, \omega^2$

So, $x^r + \frac{1}{x^r} = \omega^r + \frac{1}{\omega^r} = -1$

or 2 according as r is not divisible by 3 or divisible by 3.

$\therefore$ Required sum

$= 17(-1)^2 + 8\cdot 2^2 = 49$

29 Given, $z = 2\omega + 1$

$\Rightarrow \omega = \frac{-1+z}{2} \Rightarrow \omega = \frac{-1+\sqrt{3}i}{2}$

$[\because z = \sqrt{-3}]$

$\Rightarrow \omega$ is complex cube root of unity

Now, $\begin{vmatrix} 1 & 1 & 1 \\ 1 & -\omega^2 - 1 & \omega^2 \\ 1 & \omega^2 & \omega^7 \end{vmatrix} = 3k$

$\Rightarrow \quad \begin{vmatrix} 1 & 1 & 1 \\ 1 & \omega & \omega^2 \\ 1 & \omega^2 & \omega \end{vmatrix} = 3k$

$\left[\begin{array}{r} \because 1 + \omega + \omega^2 = 0 \\ \omega^7 = \omega \end{array}\right]$

Applying $R_1 \to R_1 + R_2 + R_3$, we get

$\begin{vmatrix} 3 & 1+\omega+\omega^2 & 1+\omega+\omega^2 \\ 1 & \omega & \omega^2 \\ 1 & \omega^2 & \omega \end{vmatrix} = 3k$

$\Rightarrow \begin{vmatrix} 3 & 0 & 0 \\ 1 & \omega & \omega^2 \\ 1 & \omega^2 & \omega \end{vmatrix} = 3k$

$\Rightarrow \quad 3(\omega^2 - \omega^4) = 3k$

$\Rightarrow \quad k = \omega^2 - \omega \Rightarrow k = -1 - 2\omega$

$\Rightarrow \quad k = -(1 + 2\omega) \Rightarrow k = -z$

30 Using $1 + \omega + \omega^2 = 0$, we get

$\Delta = \begin{vmatrix} 1+\omega & \omega^2 & -\omega \\ 1+\omega^2 & \omega & -\omega^2 \\ \omega^2+\omega & \omega & -\omega^2 \end{vmatrix}$

Applying $C_1 \to C_1 + C_2$,

$\Delta = \begin{vmatrix} 0 & \omega^2 & -\omega \\ 0 & \omega & -\omega^2 \\ \omega^2 + 2\omega & \omega & -\omega^2 \end{vmatrix}$

$= (\omega^2 + 2\omega)(-\omega + \omega^2) = -3\omega^2$

31 $|a + b\omega + c\omega^2|^2 = (a + b\omega + c\omega^2)(a + b\bar{\omega} + c\bar{\omega}^2)$

$= (a + b\omega + c\omega^2)(a + b\omega^2 + c\omega)$ [$\because \bar{\omega} = \omega^2$ and $\bar{\omega}^2 = \omega$]

$= a^2 + b^2 + c^2 - ab - bc - ca$

$= \frac{1}{2}[(a-b)^2 + (b-c)^2 + (c-a)^2]$

So, it has minimum value 1 for $a = b = 1$ and $c = 2$.

32 Clearly, $|x|^2 + |y|^2 + |z|^2 = x\bar{x} + y\bar{y} + z\bar{z}$

$= (a+b+c)(\bar{a}+\bar{b}+\bar{c}) + (a + b\omega + c\omega^2)(\bar{a} + \bar{b}\bar{\omega} + \bar{c}\bar{\omega}^2) + (a + b\omega^2 + c\omega)(\bar{a} + \bar{b}\bar{\omega}^2 + \bar{c}\bar{\omega})$

$= 3(|a|^2 + |b|^2 + |c|^2)$

$\Rightarrow \dfrac{|x|^2 + |y|^2 + |z|^2}{|a|^2 + |b|^2 + |c|^2} = 3$

33 Given, $\text{Re}\left(\frac{1}{z}\right) = 3 \Rightarrow \text{Re}\left(\frac{\bar{z}}{|z|^2}\right) = 3$ $\left[\because \frac{1}{z} = \frac{\bar{z}}{|z|^2}\right]$

$\Rightarrow \frac{x}{x^2 + y^2} = 3 \Rightarrow 3x^2 + 3y^2 - x = 0$

So, it is a circle whose centre is on X-axis and passes through the origin.

34 $\dfrac{2z+1}{iz+1} = \dfrac{(2x+1) + 2iy}{(1-y) + ix}$

$= \dfrac{[(2x+1) + 2iy]\cdot[(1-y) - ix]}{(1-y)^2 - i^2x^2}$

$= \dfrac{(2x - y + 1) - (2x^2 + 2y^2 + x - 2y)i}{1 + x^2 + y^2 - 2y}$

$\therefore$ Imaginary part

$= \dfrac{-(2x^2 + 2y^2 + x - 2y)}{1 + x^2 + y^2 - 2y} = -2$

$\Rightarrow x + 2y - 2 = 0$, which is a straight line.

35 Clearly, $\dfrac{z}{1 - z^2} = \dfrac{z}{z\bar{z} - z^2} = \dfrac{1}{\bar{z} - z}$, which is always imaginary.

36 $|\omega| = 1 \Rightarrow |z| = \left|z - \frac{i}{3}\right|$

It is the perpendicular bisector of the line segment joining $(0, 0)$ to $\left(0, \frac{1}{3}\right)$ i.e. the line $y = \frac{1}{6}$.

37 Given, $\dfrac{z_1}{z_2} + \dfrac{z_2}{z_1} = 1 \Rightarrow z_1^2 + z_2^2 = z_1z_2$

$\Rightarrow z_1^2 + z_2^2 + z_3^2 = z_1z_2 + z_1z_3 + z_2z_3$, where $z_3 = 0$

So, z_1, z_2 and the origin form an equilateral triangle.

38 Given, z_2 is not unimodular i.e. $|z_2| \neq 1$ and $\dfrac{z_1 - 2z_2}{2 - z_1\bar{z}_2}$ is unimodular.

$\Rightarrow \left|\dfrac{z_1 - 2z_2}{2 - z_1\bar{z}_2}\right| = 1$

$\Rightarrow |z_1 - 2z_2|^2 = |2 - z_1\bar{z}_2|^2$

$\Rightarrow (z_1 - 2z_2)(\bar{z}_1 - 2\bar{z}_2) = (2 - z_1\bar{z}_2)(2 - \bar{z}_1z_2)$ $[\because z\bar{z} = |z|^2]$

$\Rightarrow |z_1|^2 + 4|z_2|^2 - 2\bar{z}_1z_2 - 2z_1\bar{z}_2 = 4 + |z_1|^2|z_2|^2 - 2\bar{z}_1z_2 - 2z_1\bar{z}_2$

$\Rightarrow (|z_2|^2 - 1)(|z_1|^2 - 4) = 0$

$\because |z_2| \neq 1$

$\therefore |z_1| = 2$

Let $z_1 = x + iy \Rightarrow x^2 + y^2 = (2)^2$

Point z_1 lies on a circle of radius 2.

39 Let $z = re^{i\theta}$

Then, $|r^2e^{2i\theta} - 1| = r^2 + 1$

$\Rightarrow (r^2\cos 2\theta - 1)^2 + (r^2\sin 2\theta)^2 = (r^2 + 1)^2$

$\Rightarrow r^4 - 2r^2\cos 2\theta + 1 = r^4 + 2r^2 + 1$

$\Rightarrow \cos 2\theta = -1 \Rightarrow \theta = \frac{\pi}{2}$

$\Rightarrow z = r\left(\cos\frac{\pi}{2} + i\sin\frac{\pi}{2}\right) = ir$

40 Let $z = x + iy$

Then, $x^2 + y^2 = 1$

and $x + iy = 1 - (x - iy)$

$\Rightarrow x^2 + y^2 = 1$ and $2x = 1 \Rightarrow x = \frac{1}{2}$

and $y = \pm\frac{\sqrt{3}}{2}$

$\therefore z = \frac{1}{2} \pm \frac{\sqrt{3}}{2}i$

Now, take, $z = \frac{1}{2} + \frac{\sqrt{3}}{2}i$

$\therefore \theta = \tan^{-1}\left(\frac{\sqrt{3}/2}{1/2}\right) = \frac{\pi}{3}$

SESSION 2

1 Clearly, $(1+i)^{n_1} + (1+i^3)^{n_1} + (1+i^5)^{n_2} + (1+i^7)^{n_2}$

$= (1+i)^{n_1} + (1-i)^{n_1} + (1+i)^{n_2} + (1-i)^{n_2}$

$= \left[\sqrt{2}\left(\cos\frac{\pi}{4} + i\sin\frac{\pi}{4}\right)\right]^{n_1} + \left[\sqrt{2}\left(\cos\left(-\frac{\pi}{4}\right) + i\sin\left(-\frac{\pi}{4}\right)\right)\right]^{n_1} + \left[\sqrt{2}\left(\cos\left(-\frac{\pi}{4}\right) + i\sin\left(-\frac{\pi}{4}\right)\right)\right]^{n_2} + \left[\sqrt{2}\left(\cos\frac{\pi}{4} + i\sin\frac{\pi}{4}\right)\right]^{n_2}$

$= (\sqrt{2})^{n_1}\left[\cos\frac{n_1\pi}{4} + i\sin\frac{n_1\pi}{4}\right] + (\sqrt{2})^{n_1}\left[\cos\frac{n_1\pi}{4} - i\sin i\frac{n_1\pi}{4}\right] + (\sqrt{2})^{n_2}\left[\cos\frac{n_2\pi}{4} - i\sin n_2\frac{\pi}{4}\right] + (\sqrt{2})^{n_2}\left[\cos\frac{n_2\pi}{4} + i\sin\frac{n_2\pi}{4}\right]$

$= (\sqrt{2})^{n_1}\left[2\cos\frac{n_1\pi}{4}\right] + (\sqrt{2})^{n_2}\left[2\cos\frac{n_2\pi}{4}\right]$

which is purely real $\forall\, n_1, n_2$.

2 Clearly, $\dfrac{z^2}{z-1} = \dfrac{\bar{z}^2}{\bar{z}-1}$

$\Rightarrow z\bar{z}z - z^2 = \bar{z}z\bar{z} - \bar{z}^2$

$\Rightarrow |z|^2(z - \bar{z}) - (z - \bar{z})(z + \bar{z}) = 0$

$\Rightarrow (z - \bar{z})(|z|^2 - (z + \bar{z})) = 0$

Either $z = \bar{z} \Rightarrow$ real axis

or $|z|^2 = z + \bar{z} \Rightarrow z\bar{z} - z - \bar{z} = 0$

i.e. $(x^2 + y^2 = 2x)$

represents a circle passing through origin.

3 $\omega = e^{2\pi i/3}$ = imaginary cube root of unity

$\therefore 1 + \omega + \omega^2 = 0$

Now, $\Delta = \begin{vmatrix} z+1 & \omega & \omega^2 \\ \omega & z+\omega^2 & 1 \\ \omega^2 & 1 & z+\omega \end{vmatrix}$

$= \begin{vmatrix} z & z & z \\ \omega & z+\omega^2 & 1 \\ \omega^2 & 1 & z+\omega \end{vmatrix}$

(applying $R_1 \to R_1 + R_2 + R_3$)

$= z\begin{vmatrix} 1 & 1 & 1 \\ \omega & z+\omega^2 & 1 \\ \omega^2 & 1 & z+\omega \end{vmatrix}$

$= z\{[(z+\omega^2)(z+\omega) - 1] + [\omega^2 - \omega(z+\omega)] + [\omega - \omega^2(z+\omega^2)]\}$

$= z\{z^2 + z(\omega + \omega^2) + \omega^3 - 1 - \omega z - \omega^2 z\} = z^3$

$\therefore \Delta = 0 \Rightarrow z^3 = 0 \Rightarrow z = 0$ is the only solution.

4 In the problem, base $= 1/2 \in (0, 1)$

$\therefore |z - 1| < |z - i| \Rightarrow |z-1|^2 < |z-i|^2$

$\Rightarrow (z-1)(\bar{z}-1) < (z-i)(\bar{z}+i)$ $[\because |z|^2 = z\bar{z}]$

$\Rightarrow (1+i)z + (1-i)\bar{z} > 0$

$\Rightarrow (z + \bar{z}) + i(z - \bar{z}) > 0$

$\Rightarrow \left(\frac{z+\bar{z}}{2}\right) + i\left(\frac{z-\bar{z}}{2}\right) > 0$

$\Rightarrow \left(\frac{z+\bar{z}}{2}\right) - \left(\frac{z-\bar{z}}{2i}\right) > 0$

$\Rightarrow \text{Re}(z) - \text{Im}(z) > 0 \Rightarrow x - y > 0$

5 $\arg\left(\dfrac{z - (10+6i)}{z - (4+6i)}\right) = \dfrac{\pi}{4}$

$\Rightarrow \tan^{-1}\dfrac{y-6}{x-10} - \tan^{-1}\dfrac{y-6}{x-4} = \dfrac{\pi}{4}$

[take $z = x + iy$]

$$\Rightarrow \frac{\frac{y-6}{x-10}-\frac{y-6}{x-4}}{1+\frac{(y-6)(y-6)}{(x-10)(x-4)}}=1$$

$\Rightarrow x^2+y^2-14x-18y+112=0$

$\Rightarrow (x-7)^2+(y-9)^2=18=(3\sqrt{2})^2$

$\Rightarrow |z-(7+9i)|=3\sqrt{2}$

6 We have, $z\bar{z}=(z^2+\bar{z}^2)=350$

$\Rightarrow 2(x^2+y^2)(x^2-y^2)=350$

$\Rightarrow (x^2+y^2)(x^2-y^2)=175$

Since $x, y \in I$, the only possible case which gives integral solutions, is

$x^2+y^2=25$ ...(i)

$x^2-y^2=7$ ...(ii)

From Eqs. (i) and (ii) $x^2=16;\ y^2=9$

$\Rightarrow x=\pm 4;\ y=\pm 3 \Rightarrow$ Area $=48$

7 We have, $\alpha+i\beta=\cot^{-1}(z)$

$\Rightarrow \cot(\alpha+i\beta)=x+iy$

and $\cot(\alpha-i\beta)=x-iy$

Now, consider

$$\cot 2\alpha=\cot[(\alpha+i\beta)+(\alpha-i\beta)]=\frac{\cot(\alpha+i\beta)\cdot\cot(\alpha-i\beta)-1}{\cot(\alpha+i\beta)+\cot(\alpha-i\beta)}=\frac{(x^2+y^2-1)}{2x}$$

$\therefore x^2+y^2-2x\cot 2\alpha-1=0$

8 Given, $|z+4|\le 3$

Now, $|z+1|=|z+4-3|$

$\le |z+4|+|3|\le 3+3=6$

Hence, greatest value of $|z+1|=6$

Since, least value of the modulus of a complex number is 0.

Consider, $|z+1|=0 \Rightarrow z=-1$

Now, $|z+4|=|-1+4|=3$

$\Rightarrow |z+4|\le 3$ is satisfied by $z=-1$.

$\therefore$ Least value of $|z+1|=0$

9 $|z-3-2i|\le 2$

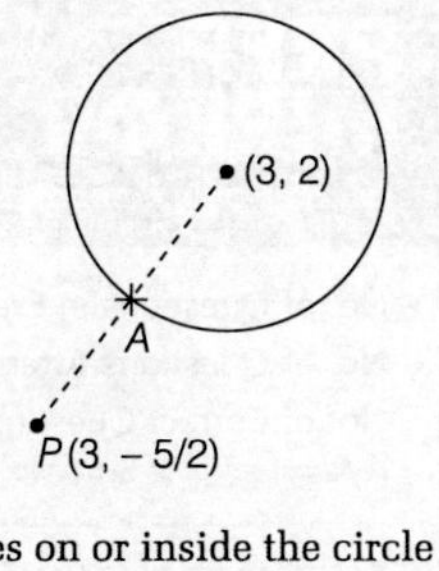

$\Rightarrow$ z lies on or inside the circle

$(x-3)^2+(y-2)^2=2^2=4$ ... (i)

$$|2z-6+5i|=2\left|z-\left(3-\frac{5}{2}i\right)\right|=2\left[\text{Distance of } z \text{ from }\left(3,-\frac{5}{2}\right)\right]$$

where z lies on circle (i).

$$\therefore \min|2z-6+5i|=2PA=2\left(\frac{9}{2}-2\right)=5$$

10 Clearly, $\frac{0-3e^{i\pi/4}}{z-3e^{i\pi/4}}=\frac{3}{4}e^{i\pi/2}$

$\therefore \frac{-3e^{-i\pi/4}}{z-3e^{i\pi/4}}=\frac{3}{4}i$

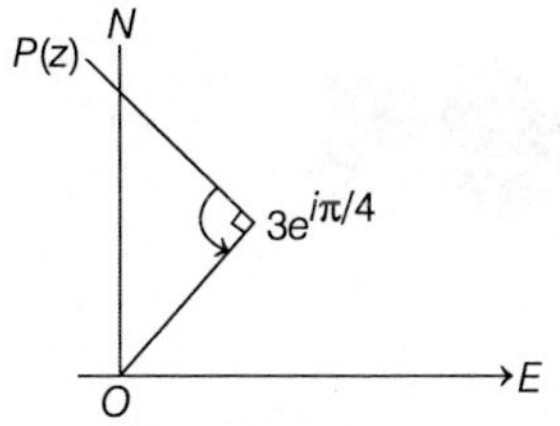

$z-3e^{i\pi/4}=4ie^{i\pi/4}$

$z=(3+4i)e^{i\pi/4}$

11 Given $z^n=1$, where

$z=1, a_1, a_2, \ldots a_{n-1}$ ...(i)

Let $\alpha=\frac{1}{1-z}$, then $z=1-\frac{1}{\alpha}$

$\therefore \left(1-\frac{1}{\alpha}\right)^n=1$ [by (i)]

$\Rightarrow (\alpha-1)^n-\alpha^n=0$

$\Rightarrow -C_1\alpha^{n-1}+C_2\alpha^{n-2}+\ldots+(-1)^n=0$

where, $\alpha=\frac{1}{1-a_1}, \frac{1}{1-a_2}, \ldots, \frac{1}{1-a_{n-1}}$

$$\Rightarrow \frac{1}{1-a_1}+\frac{1}{1-a_2}+\ldots+\frac{1}{1-a_{n-1}}=\frac{C_2}{C_1}=\frac{n(n-1)}{2/n}=\frac{(n-1)}{2}$$

12 $|z|^2\omega-|\omega|^2 z=z-\omega$ ... (i)

$\Rightarrow (|z|^2+1)\omega=(|\omega|^2+1)z$

$\Rightarrow \frac{z}{\omega}=\frac{|z|^2+1}{|\omega|^2+1}=$ real

$\Rightarrow \frac{z}{\omega}=\frac{\bar{z}}{\bar{\omega}} \Rightarrow \frac{z\bar{\omega}}{\omega}=\bar{z}\omega$ ... (ii)

Also, from Eq. (i), $z\bar{z}\omega-\omega\bar{\omega}z=z-\omega$

$\Rightarrow z\bar{z}\omega-\omega\bar{z}\omega-z+\omega=0$

$\Rightarrow (\bar{z}\omega-1)(z-\omega)=0 \Rightarrow z=\omega$ or $\bar{z}\omega=1$

i.e. $z\bar{\omega}=1$

$\Rightarrow z=\omega$ or $z=\frac{1}{\bar{\omega}}=\omega/|\omega|^2$

13 We have, $\sum_{k=1}^{10}\left(\sin\frac{2k\pi}{11}+i\cos\frac{2k\pi}{11}\right)$

$$=i\sum_{k=1}^{10}\left(\cos\frac{2k\pi}{11}-i\sin\frac{2k\pi}{11}\right)=i\sum_{k=1}^{10}e^{\frac{-i2k\pi}{11}}=i\sum_{k=1}^{10}\alpha^k$$

where $\alpha=e^{-i2\pi/11}$

$=i[\alpha+\alpha^2+\alpha^3+\ldots+\alpha^{10}]$

$$=i\frac{\alpha(1-\alpha^{10})}{1-\alpha}=i\frac{(\alpha-\alpha^{11})}{1-\alpha}$$

$$=i\frac{(\alpha-1)}{(1-\alpha)} \quad [\because \alpha^{11}=\cos 2\pi-\sin 2\pi=1]$$

$=-i$

14 Given, $z^2+pz+q=0$

$\Rightarrow z_1+z_2=-p$ and $z_1z_2=q$

$\because OA=OB$

$\Rightarrow |z_1|=|z_2|$

$\therefore \frac{z_2}{z_1}=e^{i\alpha}=\cos\alpha+i\sin\alpha$

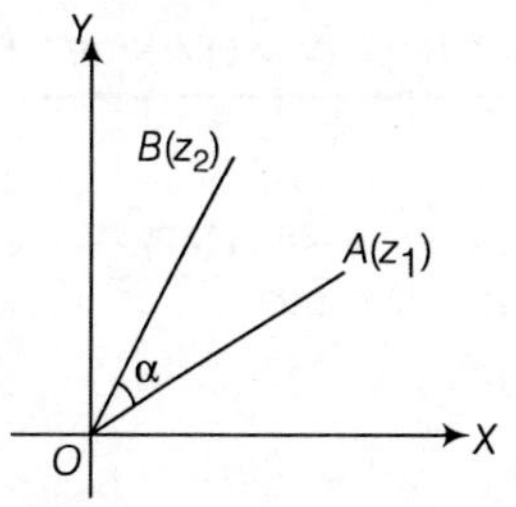

$$\Rightarrow \frac{z_1+z_2}{z_1}=1+\cos\alpha+i\sin\alpha=2\cos\frac{\alpha}{2}\left(\cos\frac{\alpha}{2}+i\sin\frac{\alpha}{2}\right)$$

$$\Rightarrow \frac{(z_1+z_2)^2}{z_1^2}=4\cos^2\frac{\alpha}{2}e^{i\alpha}=4\cos^2\alpha/2\cdot\frac{z_2}{z_1}$$

$\Rightarrow (z_1+z_2)^2=4\cos^2\frac{\alpha}{2}z_1z_2$

$\Rightarrow p^2=4q\cos^2\alpha/2$

$\therefore \frac{p^2}{4q}=\cos^2\frac{\alpha}{2}$

15 $\because p<0$, take $p=-q^3 (q>0)$

$\therefore p^{1/3}=q(-1)^{1/3}=-q, -q\omega, -q\omega^2$

Now, take $\alpha=-q, \beta=-q\omega, \gamma=-q\omega^2$

Then, given expression

$$=\frac{x+y\omega+z\omega^2}{x\omega+y\omega^2+z}=\omega^2$$

DAY THREE

Sequence and Series

Learning & Revision for the Day

- Definition
- Arithmetic Progression (AP)
- Arithmetic Mean (AM)
- Geometric Progression (GP)
- Geometric Mean (GM)
- Arithmetico-Geometric Progression (AGP)
- Sum of Special Series
- Summation of Series by the Difference Method

Definition

- By a **sequence** we mean a list of numbers, arranged according to some definite rule.

  or

  We define a sequence as a function whose domain is the set of natural numbers or some subsets of type {1, 2, 3, ... k}.
- If $a_1, a_2, a_3, \ldots, a_n, \ldots$ is a sequence, then the expression $a_1 + a_2 + a_3 + \ldots + a_n + \ldots$ is called the **series**.
- If the terms of a sequence follow a certain pattern, then it is called a **progression**.

Arithmetic Progression (AP)

- It is a sequence in which the difference between any two consecutive terms is always same.
- An AP can be represented as $a, a+d, a+2d, a+3d, \ldots$ where, a is the first term, d is the common difference.
- The nth term, $t_n = a + (n-1)d$
- Common difference $d = t_n - t_{n-1}$
- The nth term from end, $t_n = l - (n-1)d$, where l is the last term.
- Sum of first n terms, $S_n = \frac{n}{2}[2a + (n-1)d] = \frac{n}{2}[a + l]$, where l is the last term.
- If sum of n terms is S_n, then nth term is $t_n = S_n - S_{n-1}$, $t_n = \frac{1}{2}[t_{n-k} + t_{n+k}]$, where $k < n$

PREP MIRROR
Your Personal Preparation Indicator

- No. of Questions in Exercises (x)—
- No. of Questions Attempted (y)—
- No. of Correct Questions (z)—
  (Without referring Explanations)
- Accuracy Level ($z/y \times 100$)—
- Prep Level ($z/x \times 100$)—

In order to expect good rank in JEE, your Accuracy Level should be above 85 & Prep Level should be above 75.

NOTE
- Any three numbers in AP can be taken as $a-d, a, a+d$.
- Any four numbers in AP can be taken as $a-3d, a-d, a+d, a+3d$.
- Any five numbers in AP can be taken as $a-2d, a-d, a, a+d, a+2d$.
- Three numbers a, b, c are in AP iff $2b = a + c$.

An Important Result of AP

- In a finite AP, $a_1, \dots, a_n$, the sum of the terms equidistant from the beginning and end is always same and equal to the sum of first and last term i.e. $a_1 + a_n = a_k + a_{n-(k-1)}, \forall k = 1, 2, 3, \dots, n-1$.

Arithmetic Mean (AM)

- If a, A and b are in AP, then $A = \frac{a+b}{2}$ is the arithmetic mean of a and b.
- If $a, A_1, A_2, \dots, A_n, b$ are in AP, then $A_1, A_2, \dots, A_n$ are the n arithmetic means between a and b.
- The n arithmetic means, $A_1, A_2, \dots, A_n$, between a and b are given by the formula, $A_r = a + \frac{r(b-a)}{n+1} \ \forall r = 1, 2, \dots n$
- Sum of n AM's inserted between a and b is nA i.e. $A_1 + A_2 + A_3 + \dots + A_n = n\left(\frac{a+b}{2}\right)$

NOTE
- The AM of n numbers $a_1, a_2, \dots, a_n$ is given by $\text{AM} = \frac{(a_1 + a_2 + a_3 + \dots + a_n)}{n}$

Geometric Progression (GP)

- It is a sequence in which the ratio of any two consecutive terms is always same.
- A GP can be represented as $a, ar, ar^2, \dots$ where, a is the first term and r is the common ratio.
- The nth term, $t_n = ar^{n-1}$
- The nth term from end, $t_n' = \frac{l}{r^{n-1}}$, where l is the last term.
- Sum of first n terms, $S_n = \begin{cases} a\left(\frac{1-r^n}{1-r}\right), & r \neq 1 \\ na, & r = 1 \end{cases}$
- If $|r| < 1$, then the sum of infinite GP is $S_\infty = \frac{a}{1-r}$

NOTE
- Any three numbers in GP can be taken as $\frac{a}{r}, a, ar$.
- Any four numbers in GP can be taken as $\frac{a}{r^3}, \frac{a}{r}, ar, ar^3$.
- Any five numbers in GP can be taken as $\frac{a}{r^2}, \frac{a}{r}, a, ar, ar^2$.
- Three non-zero numbers a, b, c are in GP iff $b^2 = ac$.
- If a, b and c are in AP as well as GP, then $a = b = c$.
- If $a > 0$ and $r > 1$ or $a < 0$ and $0 < r < 1$, then the GP will be an increasing GP.
- If $a > 0$ and $0 < r < 1$ or $a < 0$ and $r > 1$, then the GP will be a decreasing GP.

Important Results on GP

- If $a_1, a_2, a_3, \dots, a_n$ is a GP of positive terms, then $\log a_1, \log a_2, \dots, \log a_n$ is an AP and *vice-versa*.
- In a finite GP, $a_1, a_2, \dots, a_n$, the product of the terms equidistant from the beginning and the end is always same and is equal to the product of the first and the last term. i.e. $a_1 a_n = a_k \cdot a_{n-(k-1)}, \forall k = 1, 2, 3, \dots, n-1$.

Geometric Mean (GM)

- If a, G and b are in GP, then $G = \sqrt{ab}$ is the geometric mean of a and b.
- If $a, G_1, G_2, \dots, G_n, b$ are in GP, then $G_1, G_2, \dots, G_n$ are the n geometric means between a and b.
- The n GM's, $G_1, G_2, \dots, G_n$, inserted between a and b, are given by the formula, $G_r = a\left(\frac{b}{a}\right)^{\frac{r}{n+1}}$.
- Product of n GM's, inserted between a and b, is the nth power of the single GM between a and b, i.e. $G_1 \cdot G_2 \cdot \dots \cdot G_n = G^n = (ab)^{n/2}$.

NOTE
- If a and b are of opposite signs, then their GM can not exist.
- If A and G are respectively the AM and GM between two numbers a and b, then a, b are given by $[A \pm \sqrt{(A+G)(A-G)}]$.
- If $a_1, a_2, a_3, \dots, a_n$ are positive numbers, then their GM $= (a_1 a_2 a_3 \dots a_n)^{1/n}$.

Arithmetico-Geometric Progression (AGP)

- A progression in which every term is a product of a term of AP and corresponding term of GP, is known as arithmetico-geometric progression.
- If the series of AGP be $a + (a+d)r + (a+2d)r^2 + \dots + \{a + (n-1)d\}r^{n-1} + \dots$, then

(i) $S_n = \frac{a}{1-r} + \frac{dr(1-r^{n-1})}{(1-r)^2} - \frac{\{a+(n-1)d\}r^n}{1-r}, r \neq 1$

(ii) $S_\infty = \frac{a}{1-r} + \frac{dr}{(1-r)^2}, |r| < 1$

Method to find the Sum of n-terms of Arithmetic Geometric Progression

Usually, we do not use the above formula to find the sum of n terms.

Infact we use the mechanism by which we derived the formula, shown below:

Let, $S_n = a + (a+d)r + (a+2d)r^2$

$$+ \dots + (a + (n-1)d)\,r^{n-1} \quad \dots\text{(i)}$$

Step I Multiply each term by r (Common ratio of GP) and obtain a new series

$\Rightarrow \quad r\,S_n = ar + (a+d)\,r^2 + \dots +$

$$(a + (n-2)d)\,r^{n-1} + (a + (n-1)d)\,r^n \quad \dots\text{(ii)}$$

Step II Subtract the new series from the original series by shifting the terms of new series by one term

$\Rightarrow (1-r)S_n = a + [dr + dr^2 + \dots + dr^{n-1}] - (a + (n-1)d)\,r^n$

$$\Rightarrow S_n(1-r) = a + dr\left(\frac{1-r^{n-1}}{1-r}\right) - (a + (n-1)d)\,r^n$$

$$\Rightarrow \quad S_n = \frac{a}{1-r} + dr\left(\frac{1-r^{n-1}}{(1-r)^2}\right) - \frac{(a+(n-1)\,d)}{1-r}\,r^n$$

Sum of Special Series

- Sum of first n natural numbers,

$$1 + 2 + \dots + n = \Sigma n = \frac{n(n+1)}{2}$$

- Sum of squares of first n natural numbers,

$$1^2 + 2^2 + \dots + n^2 = \Sigma n^2 = \frac{n(n+1)(2n+1)}{6}$$

- Sum of cubes of first n natural numbers,

$$1^3 + 2^3 + 3^3 + \dots + n^3 = \Sigma n^3 = \left[\frac{n(n+1)}{2}\right]^2$$

(i) Sum of first n even natural numbers

$$2 + 4 + 6 + \dots + 2n = n(n+1)$$

(ii) Sum of first n odd natural numbers

$$1 + 3 + 5 + \dots + (2n-1) = n^2$$

Summation of Series by the Difference Method

If nth term of a series cannot be determined by the methods discussed so far. Then, nth term can be determined by the method of difference, if the difference between successive terms of series are either in AP or in GP, as shown below:

Let $T_1 + T_2 + T_3 + \dots$ be a given infinite series.

If $T_2 - T_1, T_3 - T_2, \dots$ are in AP or GP, then T_n can be found by following procedure.

Clearly, $S_n = T_1 + T_2 + T_3 + \dots + T_n \quad \dots\text{(i)}$

Again, $S_n = T_1 + T_2 + \dots + T_{n-1} + T_n \quad \dots\text{(ii)}$

$\therefore \quad S_n - S_n = T_1 + (T_2 - T_1) + (T_3 - T_2) + \dots + (T_n - T_{n-1}) - T_n$

$\Rightarrow \quad T_n = T_1 + (T_2 - T_1) + (T_3 - T_2) + \dots + (T_n - T_{n-1})$

$\Rightarrow \quad T_n = T_1 + t_1 + t_2 + t_3 + t_{n-1}$

where, $t_1, t_2, t_3, \dots$ are terms of the new series $\Rightarrow S_n = \sum_{r=1}^{n} T_r$

DAY PRACTICE SESSION 1

FOUNDATION QUESTIONS EXERCISE

1 If $\log_3 2, \log_3(2^x - 5)$ and $\log_3\left(2^x - \frac{7}{2}\right)$ are in AP, then x is equal to

(a) 2 (b) 3 (c) 4 (d) 2, 3

2 The number of numbers lying between 100 and 500 that are divisible by 7 but not by 21 is

(a) 57 (b) 19 (c) 38 (d) None of these

3 If 100 times the 100th term of an AP with non-zero common difference equals the 50 times its 50th term, then the 150th term of this AP is

(a) – 150 (b) 150 times its 50th term
(c) 150 (d) zero

4 If $a_1, a_2, \dots, a_{n+1}$ are in AP, then $\frac{1}{a_1a_2} + \frac{1}{a_2a_3} + \dots + \frac{1}{a_na_{n+1}}$ is

(a) $\frac{n-1}{a_1\,a_{n+1}}$ (b) $\frac{1}{a_1\,a_{n+1}}$ (c) $\frac{n+1}{a_1\,a_{n+1}}$ (d) $\frac{n}{a_1\,a_{n+1}}$

5 A man arranges to pay off a debt of ₹ 3600 by 40 annual instalments which are in AP. When 30 of the instalments are paid, he dies leaving one-third of the debt unpaid. The value of the 8th instalment is

(a) ₹ 35 (b) ₹ 50
(c) ₹ 65 (d) None of these

6 Let $a_1, a_2, a_3, \dots$ be an AP, such that $\frac{a_1 + a_2 + \dots + a_p}{a_1 + a_2 + a_3 + \dots + a_q} = \frac{p^3}{q^3}; p \neq q$, then $\frac{a_6}{a_{21}}$ is equal to

→ JEE Mains 2013

(a) $\frac{41}{11}$ (b) $\frac{121}{1681}$
(c) $\frac{11}{41}$ (d) $\frac{121}{1861}$

7 A person is to count 4500 currency notes. Let a_n denotes the number of notes he counts in the nth minute. If $a_1 = a_2 = \ldots = a_{10} = 150$ and $a_{10}, a_{11}, \ldots$ are in AP with common difference – 2, then the time taken by him to count all notes, is

(a) 24 min (b) 34 min (c) 125 min (d) 135 min

8 If $\log_{\sqrt{3}} a^2 + \log_{(3)^{1/3}} a^2 + \log_{3^{1/4}} a^2 + \ldots$ upto 8th term $= 44$, then the value of a is

(a) $\pm\sqrt{3}$ (b) $2\sqrt{2}$ (c) $\frac{1}{\sqrt{2}}$ (d) None of these

9 n arithmetic means are inserted between 7 and 49 and their sum is found to be 364, then n is

(a) 11 (b) 12 (c) 13 (d) 14

10 If $x = 111\ldots1$ (20 digits), $y = 333\ldots3$ (10 digits) and $z = 222\ldots2$ (10 digits), then $\frac{x - y^2}{z}$ is equal to

(a) 1 (b) 2 (c) 1/2 (d) 3

11 If the 2nd, 5th and 9th terms of a non-constant AP are in GP, then the common ratio of this GP is → JEE Mains 2016

(a) $\frac{8}{5}$ (b) $\frac{4}{3}$ (c) 1 (d) $\frac{7}{4}$

12 Three positive numbers form an increasing GP. If the middle term in this GP is doubled,then new numbers are in AP. Then, the common ratio of the GP is → JEE Mains 2014

(a) $\sqrt{2}+\sqrt{3}$ (b) $3+\sqrt{2}$ (c) $2-\sqrt{3}$ (d) $2+\sqrt{3}$

13 A GP consists of an even number of terms. If the sum of all the terms is 5 times the sum of terms occupying odd places,then its common ratio is

(a) 2 (b) 3 (c) 4 (d) 5

14 The sum of first 20 terms of the sequence 0.7, 0.77, 0.777, ..., is → JEE Mains 2013

(a) $\frac{7}{81}[179 - 10^{20}]$ (b) $\frac{7}{9}[99 - 10^{-20}]$

(c) $\frac{7}{81}[179 + 10^{-20}]$ (d) $\frac{7}{9}[99 + 10^{-20}]$

15 If x, y and z are distinct prime numbers, then

(a) x, y and z may be in AP but not in GP
(b) x, y and z may be in GP but not in AP
(c) x, y and z can neither be in AP nor in GP
(d) None of the above

16 Let $n\,(>1)$ be a positive integer, then the largest integer m such that $(n^m + 1)$ divides $(1 + n + n^2 + \ldots + n^{127})$, is

(a) 32 (b) 8 (c) 64 (d) 16

17 An infinite GP has first term x and sum 5, then x belongs

(a) $x < -10$ (b) $-10 < x < 0$ (c) $0 < x < 10$ (d) $x > 10$

18 The length of a side of a square is a metre. A second square is formed by joining the mid-points of these squares. Then, a third square is formed by joining the mid-points of the second square and so on. Then, sum of the area of the squares which carried upto infinity is

(a) a^2m^2 (b) $2a^2m^2$ (c) $3a^2m^2$ (d) $4a^2m^2$

19 If $|a| < 1$ and $|b| < 1$, then the sum of the series $1 + (1 + a)b + (1 + a + a^2)b^2 + (1 + a + a^2 + a^3)\, b^3 + \ldots$ is

(a) $\frac{1}{(1-a)(1-b)}$ (b) $\frac{1}{(1-a)(1-ab)}$

(c) $\frac{1}{(1-b)(1-ab)}$ (d) $\frac{1}{(1-a)(1-b)(1-ab)}$

20 A man saves ₹ 200 in each of the first three months of his service. In each of the subsequent months his saving increases by ₹ 40 more than the saving of immediately previous month. His total saving from the start of service will be ₹ 11040 after → AIEEE 2011

(a) 19 months (b) 20 months
(c) 21 months (d) 18 months

21 If one GM, g and two AM's, p and q are inserted between two numbers a and b, then $(2p - q)(p - 2q)$ is equal to

(a) g^2 (b) $-g^2$ (c) $2g$ (d) $3g^2$

22 If five GM's are inserted between 486 and $\frac{2}{3}$, then fourth GM will be

(a) 4 (b) 6 (c) 12 (d) – 6

23 The sum to 50 terms of the series

$$1 + 2\left(1 + \frac{1}{50}\right) + 3\left(1 + \frac{1}{50}\right)^2 + \ldots, \text{ is given by}$$

(a) 2500 (b) 2550
(c) 2450 (d) None of these

24 If $(10)^9 + 2(11)^1(10)^8 + 3(11)^2(10)^7 + \ldots + 10(11)^9 = k(10)^9$, then k is equal to → JEE Mains 2014

(a) $\frac{121}{10}$ (b) $\frac{441}{100}$ (c) 100 (d) 110

25 The sum of the infinity of the series

$$1 + \frac{2}{3} + \frac{6}{3^2} + \frac{10}{3^3} + \frac{14}{3^4} + \ldots \text{ is}$$

(a) 3 (b) 4 (c) 6 (d) 2

26 The sum of the series $1^3 + 3^3 + 5^3 + \ldots$ upto 20 terms is

(a) 319600 (b) 321760
(c) 306000 (d) 347500

27 Let $a_1, a_2, a_3, \ldots, a_{49}$ be in AP such that $\sum_{k=0}^{12} a_{4k+1} = 416$ and $a_9 + a_{43} = 66$. If $a_1^2 + a_2^2 + \ldots + a_{17}^2 = 140\,m$, then m is equal to → JEE Mains 2018

(a) 66 (b) 68 (c) 34 (d) 33

28 Let $a, b, c \in R$. if $f(x) = ax^2 + bx + c$ be such that $a + b + c = 3$ and $f(x + y) = f(x) + f(y) + xy,\ \forall\, x, y \in R$, then $\sum_{n=1}^{10} f(n)$ is equal to → JEE Mains 2017

(a) 330 (b) 165 (c) 190 (d) 255

29 The sum of the series $(2)^2 + 2(4)^2 + 3(6)^2 + \ldots$ upto 10 terms is → JEE Mains 2013

(a) 11300 (b) 11200 (c) 12100 (d) 12300

30 Let A be the sum of the first 20 terms and B be the sum of the first 40 terms of the series

$1^2 + 2\cdot 2^2 + 3^2 + 2\cdot 4^2 + 5^2 + 2\cdot 6^2 + \ldots$

If $B - 2A = 100\lambda$, then λ is equal to → JEE Mains 2018

(a) 232 (b) 248 (c) 464 (d) 496

31 The sum of first 9 terms of the series $\frac{1^3}{1} + \frac{1^3+2^3}{1+3} + \frac{1^3+2^3+3^3}{1+3+5} + \ldots$ is → JEE Mains 2015

(a) 71 (b) 96 (c) 142 (d) 192

32 The sum $\frac{3}{1^2} + \frac{5}{1^2+2^2} + \frac{7}{1^2+2^2+3^2} + \ldots$ upto 11 terms is → JEE Mains 2013

(a) $\frac{7}{2}$ (b) $\frac{11}{4}$ (c) $\frac{11}{2}$ (d) $\frac{60}{11}$

33 The sum of the series $1 + \frac{1}{1+2} + \frac{1}{1+2+3} + \ldots$ upto 10 terms is → JEE Mains 2013

(a) $\frac{18}{11}$ (b) $\frac{22}{13}$ (c) $\frac{20}{11}$ (d) $\frac{16}{9}$

DAY PRACTICE SESSION 2

PROGRESSIVE QUESTIONS EXERCISE

1 The value of $1^2 + 3^2 + 5^2 + \ldots + 25^2$ is → JEE Mains 2013

(a) 2925 (b) 1469 (c) 1728 (d) 1456

2 If the function f satisfies the relation $f(x+y) = f(x)\cdot f(y)$ for all natural numbers $x, y, f(1) = 2$ and $\sum_{r=1}^{n} f(a+r) = 16(2^n - 1)$, then the natural number a is

(a) 2 (b) 3 (c) 4 (d) 5

3 If the sum of an infinite GP is $\frac{7}{2}$ and sum of the squares of its terms is $\frac{147}{16}$, then the sum of the cubes of its terms is

(a) $\frac{315}{19}$ (b) $\frac{700}{39}$ (c) $\frac{985}{13}$ (d) $\frac{1029}{38}$

4 The sum of the infinite series $\frac{5}{13} + \frac{55}{13^2} + \frac{555}{13^3} + \ldots$ is

(a) $\frac{31}{18}$ (b) $\frac{65}{32}$ (c) $\frac{65}{36}$ (d) $\frac{75}{36}$

5 Given sum of the first n terms of an AP is $2n + 3n^2$. Another AP is formed with the same first term and double of the common difference, the sum of n terms of the new AP is

(a) $n + 4n^2$ (b) $6n^2 - n$ (c) $n^2 + 4n$ (d) $3n + 2n^2$

6 For $0 < \theta < \frac{\pi}{2}$, if $x = \sum_{n=0}^{\infty} \cos^{2n}\theta$, $y = \sum_{n=0}^{\infty} \sin^{2n}\theta$ and $z = \sum_{n=0}^{\infty} \cos^{2n}\theta \sin^{2n}\theta$, then xyz is equal to

(a) $xz + y$ (b) $x + y + z$ (c) $yz + x$ (d) $x + y - z$

7 If $\frac{1}{1^4} + \frac{1}{2^4} + \frac{1}{3^4} + \ldots$ to $\infty = \frac{\pi^4}{90}$, then $\frac{1}{1^4} + \frac{1}{3^4} + \frac{1}{5^4} + \ldots$ to ∞ is equal to

(a) $\frac{\pi^4}{96}$ (b) $\frac{\pi^4}{45}$ (c) $\frac{89\pi^4}{90}$ (d) $\frac{\pi^4}{90}$

8 If S_n is the sum of first n terms of a GP : $\{a_n\}$ and S_n' is the sum of another GP : $\{1/a_n\}$, then S_n equals

(a) $\frac{S_n'}{a_1 a_n}$ (b) $a_1 a_n S_n'$ (c) $\frac{a_1}{a_n} S_n'$ (d) $\frac{a_n}{a_1} S_n'$

9 If the sum of the first ten terms of the series

$\left(1\frac{3}{5}\right)^2 + \left(2\frac{2}{5}\right)^2 + \left(3\frac{1}{5}\right)^2 + 4^2 + \left(4\frac{4}{5}\right)^2 + \ldots$, is $\frac{16}{5}m$,

then m is equal to → JEE Mains 2016

(a) 102 (b) 101 (c) 100 (d) 99

10 If m is the AM of two distinct real numbers l and n $(l, n > 1)$ and G_1, G_2 and G_3 are three geometric means between l and n, then $G_1^4 + 2G_2^4 + G_3^4$ equals → JEE Mains 2015

(a) $4l^2mn$ (b) $4lm^2n$
(c) $4lmn^2$ (d) $4l^2m^2n^2$

11 The sum of the series $(\sqrt{2}+1) + 1 + (\sqrt{2}-1) + \ldots \infty$ is

(a) $\sqrt{2}$ (b) $2 + 3\sqrt{2}$ (c) $2 - 3\sqrt{2}$ (d) $\frac{4+3\sqrt{2}}{2}$

12 The largest term common to the sequences 1, 11, 21, 31, ... to 100 terms and 31, 36, 41, 46, ... to 100 terms is

(a) 531 (b) 471 (c) 281 (d) 521

13 If a, b, c are in GP and x is the AM between a and b, y the AM between b and c, then

(a) $\frac{a}{x} + \frac{c}{y} = 1$ (b) $\frac{a}{x} + \frac{c}{y} = 2$
(c) $\frac{a}{x} + \frac{c}{y} = 3$ (d) None of these

14 Suppose a, b and c are in AP and a^2, b^2 and c^2 are in GP. If $a < b < c$ and $a + b + c = \frac{3}{2}$, then the value of a is

(a) $\frac{1}{2\sqrt{2}}$ (b) $\frac{1}{2\sqrt{3}}$ (c) $\frac{1}{2} - \frac{1}{\sqrt{3}}$ (d) $\frac{1}{2} - \frac{1}{\sqrt{2}}$

15 For any three positive real numbers a, b and c, if $9(25a^2 + b^2) + 25(c^2 - 3ac) = 15b(3a + c)$, then

(a) b, c and a are in GP ➔ JEE Mains 2017
(b) b, c and a are in AP
(c) a, b and c are in AP
(d) a, b and c are in GP

16 If $S_1, S_2, S_3, \ldots$ are the sum of infinite geometric series whose first terms are 1, 2, 3,... and whose common ratios $\frac{1}{2}, \frac{1}{3}, \frac{1}{4}, \ldots$ respectively, then $S_1^2 + S_2^2 + S_3^2 + \ldots + S_{10}^2$ is equal to

(a) 485 (b) 495
(c) 500 (d) 505

17 **Statement I** The sum of the series $1 + (1 + 2 + 4) + (4 + 6 + 9) + (9 + 12 + 16) + \ldots + (361 + 380 + 400)$ is 8000.

Statement II $\sum_{k=1}^{n} [k^3 - (k-1)^3] = n^3$, for any natural number n.

(a) Statement I is true, Statement II is true; Statement II is a correct explanation for Statement I
(b) Statement I is true, Statement II is ture; Statement II is not a correct explanation for Statement I
(c) Statement I is true; Statement II is false
(d) Statement I is false; Statement II is true

ANSWERS

SESSION 1	**1** (b)	**2** (c)	**3** (d)	**4** (d)	**5** (c)	**6** (b)	**7** (b)	**8** (a)	**9** (c)	**10** (a)
	11 (b)	**12** (d)	**13** (c)	**14** (c)	**15** (a)	**16** (c)	**17** (c)	**18** (b)	**19** (c)	**20** (c)
	21 (b)	**22** (b)	**23** (a)	**24** (c)	**25** (a)	**26** (a)	**27** (c)	**28** (a)	**29** (c)	**30** (b)
	31 (b)	**32** (c)	**33** (c)							
SESSION 2	**1** (a)	**2** (b)	**3** (d)	**4** (c)	**5** (b)	**6** (b)	**7** (a)	**8** (b)	**9** (b)	**10** (b)
	11 (d)	**12** (d)	**13** (b)	**14** (d)	**15** (b)	**16** (d)	**17** (a)			

Hints and Explanations

SESSION 1

1 $\because 2\log_3(2^x - 5) = \log_3 2 + \log_3\left(2^x - \frac{7}{2}\right)$

$\Rightarrow \quad (2^x - 5)^2 = 2\left(2^x - \frac{7}{2}\right)$

$\Rightarrow \quad t^2 + 25 - 10t = 2t - 7$ [put $2^x = t$]

$\Rightarrow \quad t^2 - 12t + 32 = 0$

$\Rightarrow \quad (t - 8)(t - 4) = 0$

$\Rightarrow \quad 2^x = 8$ or $2^x = 4$

$\therefore \quad x = 3$ or $x = 2$

At, $x = 2$, $\log_3 (2^x - 5)$ is not defined.

Hence, $x = 3$ is the only solution.

2 The numbers between 100 and 500 that are divisible by 7 are 105, 112, 119, 126, ..., 490, 497.

Let such numbers be n.

$\therefore \quad t_n = a_n + (n-1)d$

$\Rightarrow \quad 497 = 105 + (n-1) \times 7$

$\Rightarrow \quad n - 1 = 56$

$\Rightarrow \quad n = 57$

The numbers between 100 and 500 that are divisible by 21 are 105, 126, 147, ..., 483.

Let such numbers be m.

$\therefore \quad 483 = 105 + (m-1) \times 21$

$\Rightarrow \quad 18 = m - 1 \Rightarrow m = 19$

$\therefore$ Required number $= n - m = 57 - 19 = 38$

3 Let a be the first term and d $(d \neq 0)$ be the common difference of a given AP, then

$T_{100} = a + (100-1)d = a + 99d$

$T_{50} = a + (50-1)d = a + 49d$

$T_{150} = a + (150-1)d = a + 149d$

Now, according to the given condition,

$$100 \times T_{100} = 50 \times T_{50}$$

$\Rightarrow \quad 100(a + 99d) = 50(a + 49d)$

$\Rightarrow \quad 2(a + 99d) = (a + 49d)$

$\Rightarrow \quad 2a + 198d = a + 49d$

$\Rightarrow \quad a + 149d = 0$

$\therefore \quad T_{150} = 0$

4 Let d be the common difference of given AP and let

$S = \frac{1}{a_1 a_2} + \frac{1}{a_2 a_3} + \ldots + \frac{1}{a_n a_{n+1}}$. Then,

$$S = \frac{1}{d}\left[\frac{d}{a_1 a_2} + \frac{d}{a_2 a_3} + \ldots + \frac{d}{a_n a_{n+1}}\right]$$

$$= \frac{1}{d}\left[\frac{a_2 - a_1}{a_1 a_2} + \frac{a_3 - a_2}{a_2 a_3} + \ldots + \frac{a_{n+1} - a_n}{a_n a_{n+1}}\right]$$

$$= \frac{1}{d}\left[\left(\frac{1}{a_1} - \frac{1}{a_2}\right) + \left(\frac{1}{a_2} - \frac{1}{a_3}\right) + \ldots + \left(\frac{1}{a_n} - \frac{1}{a_{n+1}}\right)\right]$$

$$= \frac{1}{d}\left[\frac{1}{a_1} - \frac{1}{a_{n+1}}\right] = \frac{1}{d}\left[\frac{a_{n+1} - a_1}{a_1 a_{n+1}}\right]$$

$$= \frac{1}{d}\left[\frac{a_1 + nd - a_1}{a_1 a_{n+1}}\right] = \frac{n}{a_1 a_{n+1}}$$

5 Given, $3600 = \frac{40}{2}[2a + (40-1)d]$

$\Rightarrow \quad 3600 = 20(2a + 39d)$

$\Rightarrow \quad 180 = 2a + 39d \qquad \dots(i)$

After 30 instalments one-third of the debt is unpaid.

Hence, $\frac{3600}{3} = 1200$ is unpaid and 2400 is paid.

Now, $2400 = \frac{30}{2}\{2a + (30-1)d\}$

$\therefore \quad 160 = 2a + 29d \qquad \dots(ii)$

On solving Eqs. (i) and (ii), we get

$a = 51, d = 2$

Now, the value of 8th instalment

$= a + (8-1)d$

$= 51 + 7\cdot 2 = ₹\,65$

6 Given that, $\frac{a_1 + a_2 + \dots + a_p}{a_1 + a_2 + \dots + a_q} = \frac{p^3}{q^3}$

$\Rightarrow \quad \frac{\frac{p}{2}[2a_1 + (p-1)d]}{\frac{q}{2}[2a_2 + (q-1)d]} = \frac{p^3}{q^3}$

where, d is a common difference of an AP.

$\Rightarrow \quad \frac{2a_1 + (p-1)d}{2a_2 + (q-1)d} = \frac{p^2}{q^2}$

$\Rightarrow \quad \frac{a_1 + (p-1)\frac{d}{2}}{a_2 + (q-1)\frac{d}{2}} = \frac{p^2}{q^2}$

On putting $p = 11$ and $q = 41$, we get

$\frac{a_1 + (11-1)\frac{d}{2}}{a_2 + (41-1)\frac{d}{2}} = \frac{(11)^2}{(41)^2}$

$\Rightarrow \quad \frac{a_1 + 5d}{a_2 + 20d} = \frac{121}{1681}$

$\Rightarrow \quad \frac{a_6}{a_{21}} = \frac{121}{1681}$

7 Number of notes that the person counts in 10 min

$= 10 \times 150 = 1500$

Since, $a_{10}, a_{11}, a_{12}, \dots$ are in AP with common difference -2.

Let n be the time taken to count remaining 3000 notes.

Then, $\frac{n}{2}[2 \times 148 + (n-1) \times -2] = 3000$

$\Rightarrow \quad n^2 - 149n + 3000 = 0$

$\Rightarrow \quad (n-24)(n-125) = 0$

$\therefore \quad n = 24$ and 125

Then, the total time taken by the person to count all notes

$= 10 + 24$

$= 34$ min

8 $S_n = \log a^2 \left[\frac{1}{\frac{1}{2}\log 3} + \frac{1}{\frac{1}{3}\log 3} + \frac{1}{\frac{1}{4}\log 3} + \dots \text{ upto 8th term}\right]$

$\Rightarrow \quad \frac{\log a^2}{\log 3}[2 + 3 + 4 + \dots + 9] = 44$ [given]

$\Rightarrow \quad 44 \log a^2 = 44 \log 3$

$\therefore \quad a = \pm\sqrt{3}$

9 We know that,

$A_1 + A_2 + \dots + A_n = nA$, where

$A = \frac{a+b}{2}$

$\therefore \quad 364 = \left(\frac{7 + 49}{2}\right)n$

$\Rightarrow \quad n = \frac{364 \times 2}{56} = 13$

10 Given, $x = \frac{1}{9}(999\dots9) = \frac{1}{9}(10^{20} - 1)$

$y = \frac{1}{3}(999\dots9) = \frac{1}{3}(10^{10} - 1)$

and $z = \frac{2}{9}(999\dots9) = \frac{2}{9}(10^{10} - 1)$

$\therefore \quad \frac{x - y^2}{z} = \frac{10^{20} - 1 - (10^{10} - 1)^2}{2(10^{10} - 1)}$

$= \frac{10^{10} + 1 - (10^{10} - 1)}{2} = 1$

11 Let a be the first term and d be the common difference.

Then, we have $a + d, a + 4d, a + 8d$ in GP,

i.e. $(a + 4d)^2 = (a + d)(a + 8d)$

$\Rightarrow a^2 + 16d^2 + 8ad = a^2 + 8ad + ad + 8d^2$

$\Rightarrow \quad 8d^2 = ad$

$\Rightarrow \quad 8d = a \qquad [\because d \neq 0]$

Now, common ratio,

$r = \frac{a + 4d}{a + d} = \frac{8d + 4d}{8d + d} = \frac{12d}{9d} = \frac{4}{3}$

12 Let a, ar, ar^2 be in GP (where, $r > 1$).

On multiplying middle term by 2, we get that $a, 2ar, ar^2$ are in AP.

$\Rightarrow 4ar = a + ar^2 \Rightarrow r^2 - 4r + 1 = 0$

$\Rightarrow \quad r = \frac{4 \pm \sqrt{16-4}}{2} = 2 \pm \sqrt{3}$

$\therefore \quad r = 2 + \sqrt{3}$ [$\because$ AP is increasing]

13 Let the GP be $a, ar, ar^2, ar^3, ar^{2n-2}, ar^{2n-1}$.

where, $a, ar^2, ar^4, ar^6, \dots$ occupy odd places and $ar, ar^3, ar^5, ar^7, \dots$ occupy even places.

Given, sum of all terms $= 5 \times$ sum of terms occupying odd places, i.e.

$a + ar + ar^2 + \dots + ar^{2n-1}$

$= 5 \times (a + ar^2 + ar^4 + \dots + ar^{2n-2})$

$\Rightarrow \quad \frac{a(r^{2n} - 1)}{r - 1} = \frac{5a[(r^2)^n - 1]}{r^2 - 1}$

$\left[\because S_n = \frac{a(r^n - 1)}{r - 1}\right]$

$\Rightarrow \quad \frac{r^{2n} - 1}{r - 1} = \frac{5(r^{2n} - 1)}{(r-1)(r+1)}$

$\Rightarrow \quad 1 = \frac{5}{r+1} \Rightarrow r + 1 = 5 \Rightarrow r = 4$

14 Let $S = 0.7 + 0.77 + 0.777 + \dots$ upto 20 terms

$= \frac{7}{10} + \frac{77}{10^2} + \frac{777}{10^3} + \dots$ upto 20 terms

$= 7\left[\frac{1}{10} + \frac{11}{10^2} + \frac{111}{10^3} + \dots \text{ upto 20 terms}\right]$

$= \frac{7}{9}\left[\frac{9}{10} + \frac{99}{100} + \frac{999}{1000} + \dots \text{ upto 20 terms}\right]$

$= \frac{7}{9}\left[\left(1 - \frac{1}{10}\right) + \left(1 - \frac{1}{10^2}\right) + \left(1 - \frac{1}{10^3}\right) + \dots \text{ upto 20 terms}\right]$

$= \frac{7}{9}\left[(1 + 1 + \dots \text{ upto 20 terms}) - \left(\frac{1}{10} + \frac{1}{10^2} + \frac{1}{10^3} + \dots \text{ upto 20 terms}\right)\right]$

$= \frac{7}{9}\left[20 - \frac{\frac{1}{10}\left\{1 - \left(\frac{1}{10}\right)^{20}\right\}}{1 - \frac{1}{10}}\right]$

$\left[\because \text{sum of n terms of GP, } S_n = \frac{a(1 - r^n)}{1 - r}, \text{ where } r < 1\right]$

$= \frac{7}{9}\left[20 - \frac{1}{9}\left\{1 - \left(\frac{1}{10}\right)^{20}\right\}\right]$

$= \frac{7}{9}\left[\frac{179}{9} + \frac{1}{9}\left(\frac{1}{10}\right)^{20}\right] = \frac{7}{81}[179 + 10^{-20}]$

15 x, y, z are in GP

$\Leftrightarrow \quad y^2 = xz$

$\Leftrightarrow \quad x$ is factor of y. Which is not possible, as y is a prime number.

If $x = 3$, $y = 5$ and $z = 7$, then they are in AP.

Thus, x, y and z may be in AP but not in GP.

16 Clearly,

$$1 + n + n^2 + \dots + n^{127} = \frac{n^{128} - 1}{n - 1} \qquad \left[\because S_n = \frac{a(r^n - 1)}{r - 1}\right]$$

$$= \frac{(n^{64} - 1)(n^{64} + 1)}{n - 1}$$

$$= (1 + n + n^2 + \dots + n^{63})(n^{64} + 1)$$

Thus, the largest value of m for which $n^m + 1$ divides $1 + n + n^2 + \dots + n^{127}$ is 64.

17 Since, $S_\infty = \frac{x}{1 - r} = 5 \Rightarrow r = \frac{5 - x}{5}$

For infinite GP, $|r| < 1$

$\Rightarrow -1 < \frac{5 - x}{5} < 1 \Rightarrow -10 < -x < 0$

$\therefore \quad 0 < x < 10$

18 Sum of the area of the squares which carried upto infinity

$$= a^2 + \frac{a^2}{2} + \frac{a^2}{4} + \dots$$

$$= \frac{a^2}{1 - \frac{1}{2}} = 2a^2m^2$$

19 Clearly, $1 + (1 + a)b + (1 + a + a^2)b^2 + (1 + a + a^2 + a^3)b^3 + \dots \infty$

$$= \sum_{n=1}^{\infty} (1 + a + a^2 + \dots + a^{n-1})b^{n-1}$$

$$= \sum_{n=1}^{\infty} \left(\frac{1 - a^n}{1 - a}\right) b^{n-1}$$

$$= \frac{1}{1 - a}\left[\sum_{n=1}^{\infty} b^{n-1} - \sum_{n=1}^{\infty} a^n b^{n-1}\right]$$

$$= \frac{1}{1 - a}\left[\sum_{n=1}^{\infty} b^{n-1} - a\sum_{n=1}^{\infty} (ab)^{n-1}\right]$$

$$= \frac{1}{1 - a}[1 + b + b^2 + \dots \infty] - \frac{a}{1 - a}[1 + ab + (ab)^2 + \dots]$$

$$= \frac{1}{1 - a}\cdot\frac{1}{1 - b} - \frac{a}{1 - a}\cdot\frac{1}{1 - ab} \qquad [\because |b| < 1 \text{ and } |ab| = |a||b| < 1]$$

$$= \frac{1 - ab - a(1 - b)}{(1 - a)(1 - b)(1 - ab)}$$

$$= \frac{1 - ab - a + ab}{(1 - a)(1 - b)(1 - ab)}$$

$$= \frac{1}{(1 - b)(1 - ab)}$$

20 Let the time taken to save ₹ 11040 be $(n+3)$ months.

For first 3 months he saves ₹ 200 each month.

In $(n+3)$ months, $3 \times 200 + \frac{n}{2}\{2(240) + (n - 1) \times 40\} = 11040$

$$\Rightarrow 600 + \frac{n}{2}\{40(12 + n - 1)\} = 11040$$

$$\Rightarrow 600 + 20n(n + 11) = 11040$$

$$\Rightarrow 30 + n^2 + 11n = 552$$

$$\Rightarrow n^2 + 11n - 522 = 0$$

$$\Rightarrow n^2 + 29n - 18n - 522 = 0$$

$$\Rightarrow n(n + 29) - 18(n + 29) = 0$$

$$\Rightarrow (n - 18)(n + 29) = 0$$

$\therefore \ n = 18$ [neglecting $n = -29$]

$\therefore$ Total time $= (n + 3) = 21$ months

21 Since, $g = \sqrt{ab}$. Also, a, p, q and b are in AP.

So, common difference d is $\frac{b - a}{3}$.

$$\therefore \quad p = a + d = a + \frac{b - a}{3} = \frac{2a + b}{3}$$

$$q = b - d = b - \frac{b - a}{3} = \frac{a + 2b}{3}$$

Now, $(2p - q)(p - 2q)$

$$= \frac{(4a + 2b - a - 2b)}{3}\cdot\frac{(2a + b - 2a - 4b)}{3}$$

$$= -ab = -g^2$$

22 Here, $a = 486$ and $b = \frac{2}{3}$

We know that, $G_r = a\left(\frac{b}{a}\right)^{\frac{r}{n+1}}$

$$\therefore G_4 = 486\left(\frac{2}{3}\cdot\frac{1}{486}\right)^{4/6} \qquad [\because \text{here, } n = 5]$$

$$= 486\left(\frac{1}{3\cdot 243}\right)^{4/6}$$

$$= 486\left(\frac{1}{729}\right)^{4/6} = 486\cdot\frac{1}{3^4} = 6$$

23 Let $x = 1 + \frac{1}{50}$ and S_{50} be the sum of first 50 terms of the given series.

Then, $S_{50} = 1 + 2x + 3x^2 + \dots + 50x^{49}$ ...(i)

$\Rightarrow xS_{50} = x + 2x^2 + \dots + 49x^{49} + 50x^{50}$ ...(ii)

$\Rightarrow (1 - x)S_{50} = 1 + x + x^2 + x^3 + \dots + x^{49} - 50x^{50}$

[subtracting Eq. (ii) from Eq. (i)]

$$\Rightarrow S_{50}(1 - x) = \frac{1 - x^{50}}{1 - x} - 50x^{50}$$

$$\Rightarrow S_{50}\left(\frac{-1}{50}\right) = \frac{1 - x^{50}}{\left(\frac{-1}{50}\right)} - 50x^{50} \qquad \left[\because x = 1 + \frac{1}{50}\right]$$

$$\Rightarrow S_{50}\left(\frac{-1}{50}\right) = -50 + 50x^{50} - 50x^{50}$$

$$\Rightarrow S_{50} = 2500.$$

24 Given, $k \cdot 10^9 = 10^9 + 2(11)^1(10)^8 + 3(11)^2(10)^7 + \dots + 10(11)^9$

$$k = 1 + 2\left(\frac{11}{10}\right) + 3\left(\frac{11}{10}\right)^2 + \dots + 10\left(\frac{11}{10}\right)^9 \quad \text{...(i)}$$

$$\left(\frac{11}{10}\right)k = 1\left(\frac{11}{10}\right) + 2\left(\frac{11}{10}\right)^2 + \dots + 9\left(\frac{11}{10}\right)^9 + 10\left(\frac{11}{10}\right)^{10} \quad \text{...(ii)}$$

On subtracting Eq.(ii) from Eq.(i), we get

$$k\left(1 - \frac{11}{10}\right) = 1 + \frac{11}{10} + \left(\frac{11}{10}\right)^2 + \dots + 9\left(\frac{11}{10}\right)^9 - 10\left(\frac{11}{10}\right)^{10}$$

$$\Rightarrow k\left(\frac{10 - 11}{10}\right) = \frac{1\left[\left(\frac{11}{10}\right)^{10} - 1\right]}{\left(\frac{11}{10} - 1\right)} - 10\left(\frac{11}{10}\right)^{10}$$

$$\left[\because \text{in GP, sum of } n \text{ terms} = \frac{a(r^n - 1)}{r - 1}, \text{ when } r > 1\right]$$

$$\Rightarrow -k = 10\left[10\left(\frac{11}{10}\right)^{10} - 10 - 10\left(\frac{11}{10}\right)^{10}\right]$$

$\therefore \quad k = 100$

25 Let $S = 1 + \frac{2}{3} + \frac{6}{3^2} + \frac{10}{3^3} + \frac{14}{3^4} + \dots$

$$= 1 + \frac{2}{3}\left[1 + \frac{3}{3} + \frac{5}{3^2} + \frac{7}{3^3} + \dots\right]$$

$$= 1 + \frac{2}{3}\left[\frac{1}{1 - 1/3} + \frac{2\cdot 1/3}{(1 - 1/3)^2}\right]$$

$$\left[\because \text{sum of infinite AGP, is } S_\infty = \frac{a}{1 - r} + \frac{dr}{(1 - r)^2}\right]$$

$$= 1 + \frac{2}{3}\left[\frac{3}{2} + \frac{2}{3}\cdot\frac{9}{4}\right] = 1 + \frac{2}{3}\cdot 2\cdot\frac{3}{2} = 3$$

26 $1^3 + 3^3 + \dots + 39^3 = 1^3 + 2^3 + 3^3 + \dots + 40^3 - (2^3 + 4^3 + 6^3 + \dots + 40^3)$

$$= \left(\frac{40 \times 41}{2}\right)^2 - 8(1^3 + 2^3 + 3^3 + \dots + 20^3)$$

$$= (20 \times 41)^2 - 8\left(\frac{20 \times 21}{2}\right)^2$$

$$= 20^2[41^2 - 2(21)^2]$$

$$= 319600$$

27 Let $a_1 = a$ and d = common difference

$\because a_1 + a_5 + a_9 + \cdots + a_{49} = 416$

$\therefore a + (a + 4d) + (a + 8d) + \ldots(a + 48d) = 416$

$\Rightarrow \frac{13}{2}(2a + 48d) = 416$

$\Rightarrow a + 24d = 32$ ...(i)

Also, we have $a_9 + a_{43} = 66$

$\therefore a + 8d + a + 42d = 66$

$\Rightarrow 2a + 50d = 66$

$\Rightarrow a + 25d = 33$ ...(ii)

Solving Eqs. (i) and (ii), we get

$a = 8$ and $d = 1$

Now, $a_1^2 + a_2^2 + a_3^2 + \cdots + a_{17}^2 = 140m$

$8^2 + 9^2 + 10^2 + \ldots + 24^2 = 140m$

$\Rightarrow (1^2 + 2^2 + 3^2 + \ldots + 24^2) - (1^2 + 2^2 + 3^2 + \ldots + 7^2) = 140m$

$\Rightarrow \frac{24 \times 25 \times 49}{6} - \frac{7 \times 8 \times 15}{6} = 140m$

$\Rightarrow \frac{3 \times 7 \times 8 \times 5}{6}(7 \times 5 - 1) = 140m$

$\Rightarrow 7 \times 4 \times 5 \times 34 = 140m$

$\Rightarrow 140 \times 34 = 140m$

$\Rightarrow m = 34$

28 We have, $f(x) = ax^2 + bx + c$

Now, $f(x + y) = f(x) + f(y) + xy$

Put $y = 0 \Rightarrow f(x) = f(x) + f(0) + 0$

$\Rightarrow f(0) = 0$

$\Rightarrow c = 0$

Again, put $y = -x$

$\therefore f(0) = f(x) + f(-x) - x^2$

$\Rightarrow 0 = ax^2 + bx + ax^2 - bx - x^2$

$\Rightarrow 2ax^2 - x^2 = 0 \Rightarrow a = \frac{1}{2}$

Also, $a + b + c = 3$

$\Rightarrow \frac{1}{2} + b + 0 = 3 \Rightarrow b = \frac{5}{2}$

$\therefore f(x) = \frac{x^2 + 5x}{2}$

Now, $f(n) = \frac{n^2 + 5n}{2} = \frac{1}{2}n^2 + \frac{5}{2}n$

$\therefore \sum_{n=1}^{10} f(n) = \frac{1}{2}\sum_{n=1}^{10} n^2 + \frac{5}{2}\sum_{n=1}^{10} n$

$= \frac{1}{2} \cdot \frac{10 \times 11 \times 21}{6} + \frac{5}{2} \times \frac{10 \times 11}{2}$

$= \frac{385}{2} + \frac{275}{2} = \frac{660}{2} = 330$

29 Series $(2)^2 + 2(4)^2 + 3(6)^2 + \ldots$

$= 4\{1 \cdot 1^2 + 2 \cdot 2^2 + 3 \cdot 3^2 + \ldots\}$

$\therefore T_n = 4n \cdot n^2$

and $S_n = \Sigma T_n = 4\Sigma n^3 = 4\left[\frac{n(n + 1)}{2}\right]^2$

Now, $S_{10} = [10 \cdot (10 + 1)]^2$

$= (110)^2 = 12100$

30 We have,

$1^2 + 2 \cdot 2^2 + 3^2 + 2 \cdot 4^2 + 5^2 + 2 \cdot 6^2 + \ldots$

A = sum of first 20 terms

B = sum of first 40 terms

$\therefore A = 1^2 + 2 \cdot 2^2 + 3^2 + 2 \cdot 4^2 + 5^2 + 2 \cdot 6^2 + \ldots + 2 \cdot 20^2$

$A = (1^2 + 2^2 + 3^2 + \ldots + 20^2) + (2^2 + 4^2 + 6^2 + \ldots + 20^2)$

$A = (1^2 + 2^2 + 3^2 + \ldots + 20^2) + 4(1^2 + 2^2 + 3^2 + \ldots + 10^2)$

$A = \frac{20 \times 21 \times 41}{6} + \frac{4 \times 10 \times 11 \times 21}{6}$

$A = \frac{20 \times 21}{6}(41 + 22) = \frac{20 \times 21 \times 63}{6}$

Similarly,

$B = (1^2 + 2^2 + 3^2 + \ldots + 40^2) + 4(1^2 + 2^2 + \ldots + 20^2)$

$B = \frac{40 \times 41 \times 81}{6} + \frac{4 \times 20 \times 21 \times 41}{6}$

$B = \frac{40 \times 41}{6}(81 + 42) = \frac{40 \times 41 \times 123}{6}$

Now, $B - 2A = 100\lambda$

$\therefore \frac{40 \times 41 \times 123}{6} - \frac{2 \times 20 \times 21 \times 63}{6} = 100\lambda$

$\Rightarrow \frac{40}{6}(5043 - 1323) = 100\lambda$

$\Rightarrow \frac{40}{6} \times 3720 = 100\lambda$

$\Rightarrow 40 \times 620 = 100\lambda$

$\Rightarrow \lambda = \frac{40 \times 620}{100} = 248$

31 Write the nth term of the given series and simplify it to get its lowest form. Then, apply, $S_n = \Sigma T_n$.

Given series is

$\frac{1^3}{1} + \frac{1^3 + 2^3}{1 + 3} + \frac{1^3 + 2^3 + 3^3}{1 + 3 + 5} + \ldots\infty$

Let T_n be the nth term of the given series.

$\therefore T_n = \frac{1^3 + 2^3 + 3^3 + \ldots + n^3}{1 + 3 + 5 + \ldots \text{upto } n \text{ terms}}$

$= \frac{\left\{\frac{n(n + 1)}{2}\right\}^2}{n^2} = \frac{(n + 1)^2}{4}$

Now, $S_9 = \sum_{n=1}^{9} \frac{(n + 1)^2}{4} = \frac{1}{4}[(2^2 + 3^2 + \ldots + 10^2) + 1^2 - 1^2]$

$= \frac{1}{4}\left[\frac{10(10 + 1)(20 + 1)}{6} - 1\right]$

$= \frac{384}{4} = 96$

32 $T_n = \frac{2n + 1}{(1^2 + 2^2 + \ldots + n^2)}$

$= \frac{2n + 1}{\frac{n(n + 1)(2n + 1)}{6}} = \frac{6}{n(n + 1)}$

$= 6\left(\frac{1}{n} - \frac{1}{(n + 1)}\right)$

$T_1 = 6\left(\frac{1}{1} - \frac{1}{2}\right), T_2 = 6\left[\frac{1}{2} - \frac{1}{3}\right], \ldots$

$T_{11} = 6\left[\frac{1}{11} - \frac{1}{12}\right]$

$\therefore S = 6\left[\frac{1}{1} - \frac{1}{12}\right] = \frac{6 \times 11}{12} = \frac{11}{2}$

33 nth term of the series is

$T_n = \frac{1}{\frac{n(n + 1)}{2}} = \frac{2}{n(n + 1)}$

$\Rightarrow T_n = 2\left\{\frac{1}{n} - \frac{1}{n + 1}\right\}$

$\Rightarrow T_1 = 2\left(\frac{1}{1} - \frac{1}{2}\right), T_2 = 2\left(\frac{1}{2} - \frac{1}{3}\right),$

$T_3 = 2\left(\frac{1}{3} - \frac{1}{4}\right), \ldots, T_{10} = 2\left(\frac{1}{10} - \frac{1}{11}\right)$

$\therefore S_{10} = T_1 + T_2 + \ldots + T_{20}$

$= 2\left[1 - \frac{1}{2} + \frac{1}{2} - \frac{1}{3} + \frac{1}{3} - \frac{1}{4} + \ldots + \frac{1}{10} - \frac{1}{11}\right]$

$= 2\left(1 - \frac{1}{11}\right)$

$= 2 \cdot \frac{10}{11} = \frac{20}{11}$

SESSION 2

1 Let $S = 1^2 + 3^2 + 5^2 + \ldots + 25^2$

$= (1^2 + 2^2 + 3^2 + 4^2 + \ldots + 25^2) - (2^2 + 4^2 + 6^2 + \ldots + 24^2)$

$= (1^2 + 2^2 + 3^2 + 4^2 + \ldots + 25^2) - 2^2(1 + 2^2 + 3^2 + \ldots + 12^2)$

$= \frac{25(25 + 1)(2 \times 25 + 1)}{6} - 4 \times \frac{12(12 + 1)(2 \times 12 + 1)}{6}$

$= \frac{25 \times 26 \times 51}{6} - \frac{4 \times 12 \times 13 \times 25}{6}$

$= 25 \times 13 \times 17 - 4 \times 2 \times 13 \times 25$

$= 5525 - 2600 = 2925$

2 Now, $f(2) = f(1 + 1)$

$= f(1) \cdot f(1) = 2^2$ and $f(3) = 2^3$

Similarly, $f(n) = 2^n$

$\therefore 16(2^n - 1) = \sum_{r=1}^{n} f(a + r) = \sum_{r=1}^{n} 2^{a+r}$

$= 2^a(2 + 2^2 + \ldots + 2^n)$

$$= 2^a \cdot 2\left(\frac{2^n - 1}{2 - 1}\right) \quad \text{[GP series]}$$
$$= 2^{a+1}(2^n - 1)$$
$$\Rightarrow \quad 2^{a+1} = 16 = 2^4$$
$$\therefore \quad a = 3$$

3 Let GP be $a, ar, ar^2, \ldots, |r| < 1$.
According to the question,
$$\frac{a}{1-r} = \frac{7}{2}, \frac{a^2}{1-r^2} = \frac{147}{16}$$
On eliminating a, we get
$$\frac{147}{16}(1 - r^2) = \left(\frac{7}{2}\right)^2 (1 - r)^2$$
$$\Rightarrow 3(1 + r) = 4(1 - r) \Rightarrow r = \frac{1}{7}, a = 3$$
$\therefore$ Sum of cubes
$$= \frac{a^3}{1 - r^3} = \frac{(3)^3}{1 - \left(\frac{1}{7}\right)^3} = \frac{1029}{38}$$

4 Let $S = \frac{5}{13} + \frac{55}{13^2} + \frac{555}{13^3} + \ldots \quad \ldots\text{(i)}$

and $\frac{S}{13} = \frac{5}{13^2} + \frac{55}{13^3} + \ldots \quad \ldots\text{(ii)}$

On subtracting Eq. (ii) from Eq. (i), we get
$$\frac{12}{13}S = \frac{5}{13} + \frac{50}{13^2} + \frac{500}{13^3} + \ldots$$
which is a GP with common ratio $\frac{10}{13}$.
$$\therefore \quad S = \frac{13}{12} \times \left[\frac{5}{13} \div \left(1 - \frac{10}{13}\right)\right] = \frac{65}{36}$$
$$\left[\because S_\infty = \frac{a}{1-r}\right]$$

5 Here, $T_1 = S_1 = 2(1) + 3(1)^2 = 5$
$$T_2 = S_2 - S_1 = 16 - 5 = 11$$
$$[\because S_2 = 2(2) + 3(2)^2 = 16]$$
$$T_3 = S_3 - S_2 = 33 - 16 = 17$$
$$[\because S_3 = 2(3) + 3(3)^2 = 33]$$
Hence, sequence is 5, 11, 17.
$\therefore$ $a = 5$ and $d = 6$
For new AP, $A = 5, D = 2 \times 6 = 12$
$$\therefore \quad S'_n = \frac{n}{2}[2 \times 5 + (n-1)12]$$
$$= 6n^2 - n$$

6 Sum of three infinite GP's are
$$x = \frac{1}{1 - \cos^2\theta} = \frac{1}{\sin^2\theta},$$
Similarly,
$$y = \frac{1}{\cos^2\theta} \text{ and } z = \frac{1}{1 - \cos^2\theta \sin^2\theta}$$
Now, $\frac{1}{x} + \frac{1}{y} = 1 \quad [\because \sin^2\theta + \cos^2\theta = 1]$
$$\Rightarrow \quad x + y = xy$$
and
$$\frac{1}{z} = 1 - \cos^2\theta \sin^2\theta$$
$$= 1 - \frac{1}{xy} = \frac{xy - 1}{xy}$$
$$\Rightarrow \quad xy = xyz - z$$
$$\therefore \quad xyz = xy + z = x + y + z$$

7 Let $S = \frac{1}{1^4} + \frac{1}{3^4} + \frac{1}{5^4} + \ldots + \text{to } \infty$

Since, $\frac{1}{1^4} + \frac{1}{2^4} + \frac{1}{3^4} + \ldots \text{ to } \infty = \frac{\pi^4}{90}$
$$\therefore \left(\frac{1}{1^4} + \frac{1}{3^4} + \frac{1}{5^4} + \ldots \text{ to } \infty\right) + \left(\frac{1}{2^4} + \frac{1}{4^4} + \frac{1}{6^4} + \ldots \text{ to } \infty\right) = \frac{\pi^4}{90}$$
$$\Rightarrow \left(\frac{1}{1^4} + \frac{1}{3^4} + \frac{1}{5^4} + \ldots \text{ to } \infty\right) + \frac{1}{2^4}\left(\frac{1}{1^4} + \frac{1}{2^4} + \frac{1}{3^4} + \ldots + \text{ to } \infty\right) = \frac{\pi^4}{90}$$
$$\Rightarrow S + \frac{1}{16} \cdot \frac{\pi^4}{90} = \frac{\pi^4}{90}$$
$$\left(\because \frac{1}{1^4} + \frac{1}{2^4} + \frac{1}{3^4} + \ldots \text{ to } \infty = \frac{\pi^4}{90}\right)$$
$$\Rightarrow \quad S = \frac{\pi^4}{90}\left(1 - \frac{1}{16}\right) = \frac{15\pi^4}{16 \times 90} = \frac{\pi^4}{96}$$

8 Let $a_n = ar^{n-1}$.

Then, $S_n = \frac{a(1 - r^n)}{1 - r}$

and $S'_n = \frac{\left(\frac{1}{a}\right)\left[1 - \left(\frac{1}{r}\right)^n\right]}{1 - \frac{1}{r}}$
$$\left[\because \text{first term of } \left\{\frac{1}{a_n}\right\} \text{ is } \frac{1}{a} \text{ and common ratio is } \frac{1}{r}\right]$$
$$= \frac{\left(\frac{1}{a}\right)(r^n - 1)}{r^n(r - 1)} \cdot r$$
$$= \frac{1 - r^n}{1 - r} \cdot \frac{1}{a \cdot r^{n-1}}$$
$$= \frac{1 - r^n}{1 - r} \cdot \frac{1}{a_n} = \frac{a(1 - r^n)}{1 - r} \cdot \frac{1}{aa_n}$$
$$= S_n \cdot \frac{1}{a_1 a_n}$$
$$\Rightarrow \quad S_n = a_1 a_n S'_n$$

9 Let S_{10} be the sum of first ten terms of the series.
Then, we have
$$S_{10} = \left(1\frac{3}{5}\right)^2 + \left(2\frac{2}{5}\right)^2 + \left(3\frac{1}{5}\right)^2 + 4^2 + \left(4\frac{4}{5}\right)^2 + \ldots \text{ to 10 terms}$$
$$= \left(\frac{8}{5}\right)^2 + \left(\frac{12}{5}\right)^2 + \left(\frac{16}{5}\right)^2 + 4^2 + \left(\frac{24}{5}\right)^2 + \ldots \text{ to 10 terms}$$
$$= \frac{1}{5^2}(8^2 + 12^2 + 16^2 + 20^2 + 24^2 + \ldots \text{ to 10 terms})$$
$$= \frac{4^2}{5^2}(2^2 + 3^2 + 4^2 + 5^2 + \ldots \text{ to 10 terms})$$
$$= \frac{4^2}{5^2}(2^2 + 3^2 + 4^2 + 5^2 + \ldots + 11^2)$$
$$= \frac{16}{25}((1^2 + 2^2 + \ldots + 11^2) - 1^2)$$
$$= \frac{16}{25}\left(\frac{11 \cdot (11 + 1)(2 \cdot 11 + 1)}{6} - 1\right)$$
$$= \frac{16}{25}(506 - 1) = \frac{16}{25} \times 505$$
$$\Rightarrow \quad \frac{16}{5} m = \frac{16}{25} \times 505$$
$$\Rightarrow \quad m = 101$$

10 Given, m is the AM of l and n.
$$\therefore \quad l + n = 2m$$
and G_1, G_2, G_3 are geometric means between l and n.
So, l, G_1, G_2, G_3, n are in GP.
Let r be the common ratio of this GP.
$$\therefore G_1 = lr, G_2 = lr^2, G_3 = lr^3;$$
$$n = lr^4 \Rightarrow r = \left(\frac{n}{l}\right)^{1/4}$$
Now, $G_1^4 + 2G_2^4 + G_3^4 = (lr)^4 + 2(lr^2)^4 + (lr^3)^4$
$$= l^4 \times r^4(1 + 2r^4 + r^8) = l^4 \times r^4(r^4 + 1)^2$$
$$= l^4 \times \frac{n}{l}\left(\frac{n + l}{l}\right)^2 = ln \times 4m^2 = 4lm^2n$$
$$[\because n + l = 2m]$$

11 Given series is a geometric series with $a = \sqrt{2} + 1$ and $r = \sqrt{2} - 1$.
$\therefore$ Required sum
$$= \frac{a}{1 - r} = \frac{\sqrt{2} + 1}{1 - (\sqrt{2} - 1)} = \frac{\sqrt{2} + 1}{2 - \sqrt{2}}$$
$$= \frac{(\sqrt{2} + 1)(2 + \sqrt{2})}{(2 - \sqrt{2})(2 + \sqrt{2})}$$
$$= \frac{2\sqrt{2} + 2 + 2 + \sqrt{2}}{4 - 2} = \frac{4 + 3\sqrt{2}}{2}$$

12 Clearly, the common terms of the given sequences are
31, 41, 51, ...
Now, 100th term of 1, 11, 21, 31, ... is
$1 + 99 \times 10 = 991$
and 100th term of 31, 36, 41, 46, ... is
$31 + 99 \times 5 = 526$.
Let the largest common term be 526.
Then, $526 = 31 + (n - 1)10$
$$\Rightarrow \quad (n - 1)10 = 495$$

$\Rightarrow \quad n - 1 = 49.5$

$\Rightarrow \quad n = 50.5$

But n is an integer, $n = 50$.

Hence, the largest common term is $31 + (50 - 1)10 = 521$.

13 Since, a, b, c are in GP.

$\therefore \quad b^2 = ac \quad \ldots(i)$

Also, as x is A between a and b

$\therefore \quad x = \frac{a+b}{2} \quad \ldots(ii)$

Similarly, $y = \frac{b+c}{2} \quad \ldots(iii)$

Now, consider $\frac{a}{x} + \frac{c}{y} = \frac{2a}{a+b} + \frac{2c}{b+c}$ [using Eqs. (ii) and (iii)]

$= 2\left[\frac{ab + ac + ac + bc}{ab + ac + b^2 + bc}\right]$

$= 2\left[\frac{ab + bc + 2ac}{ab + bc + 2ac}\right]$

$= 2$ [using Eq. (i)]

14 Since, a, b, c are in AP

$\therefore \quad 2b = a + c \quad \ldots(i)$

Also, as a^2, b^2 and c^2 are in GP

$\therefore \quad b^4 = a^2c^2 \quad \ldots(ii)$

$\because \quad a + b + c = \frac{3}{2}$

$\therefore \quad 3b = \frac{3}{2}$ [using Eq. (i)]

$\Rightarrow \quad b = \frac{1}{2}$

$\Rightarrow \quad a + c = 1$ [using Eq. (i)]

and $ac = \frac{1}{4}$ or $-\frac{1}{4}$ [using Eq. (ii)]

Case I When $a + c = 1$ and $ac = \frac{1}{4}$

In this case,

$(a - c)^2 = (a + c)^2 - 4ac = 0$

$\Rightarrow \quad a = c$

But $a \neq c$, as $a < c$.

Case II When $a + c = 1$ and $ac = -\frac{1}{4}$

In this case, $(a - c)^2 = 1 + 1 = 2$

$\Rightarrow \quad a - c = \pm\sqrt{2}$

But $a < c$, $a - c = -\sqrt{2}$

On solving $a + c = 1$ and $a - c = -\sqrt{2}$, we get

$a = \frac{1}{2} - \frac{1}{\sqrt{2}}$.

15 We have, $225a^2 + 9b^2 + 25c^2 - 75ac - 45ab - 15bc = 0$

$\Rightarrow (15a)^2 + (3b)^2 + (5c)^2 - (15a)(5c) - (15a)(3b) - (3b)(5c) = 0$

$\Rightarrow \frac{1}{2}[(15a - 3b)^2 + (3b - 5c)^2 + (5c - 15a)^2] = 0$

$\Rightarrow \quad 15a = 3b, 3b = 5c$ and $5c = 15a$

$\therefore \quad 15a = 3b = 5c$

$\Rightarrow \quad \frac{a}{1} = \frac{b}{5} = \frac{c}{3} = \lambda$ (say)

$\Rightarrow \quad a = \lambda, b = 5\lambda, c = 3\lambda$

Hence, b, c and a are in AP.

16 Here, S_r is sum of an infinite GP, r is first term and $\frac{1}{r+1}$ is common ratio

$$S_r = \frac{r}{1 - \frac{1}{r+1}} = r + 1$$

$\Rightarrow \sum_{r=1}^{10} S_r^2 = 2^2 + 3^2 + \ldots + 11^2$

$= 1^2 + 2^2 + 3^2 + \ldots + 11^2 - 1$

$= \frac{11 \times 12 \times 23}{6} - 1 = 505$

17 Statement I

Let $S = (1) + (1+2+4) + (4+6+9) + (9+12+16) + \ldots + (361+380+400)$

$= (0+0+1) + (1+2+4) + (4+6+9) + (9+12+16) + \ldots + (361+380+400)$

Now, we can clearly observe the elements in each bracket.

The general term of the series is

$T_r = (r-1)^2 + (r-1)r + (r^2)$

Now, the sum to n terms of the series is

$$S_n = \sum_{r=1}^{n}[(r-1)^2 + (r-1)r + (r)^2]$$

$$= \sum_{r=1}^{n}\left[\frac{r^3 - (r-1)^3}{r - (r-1)}\right]$$

$[\because (a^3 - b^3) = (a-b)(a^2 + ab + b^2)]$

$$= \sum_{r=1}^{n}[r^3 - (r-1)^3]$$

$= (1^3 - 0^3) + (2^3 - 1^3) + (3^3 - 2^3) + \ldots + [n^3 - (n-1)^3]$

Rearranging the terms, we get

$S_n = -0^3 + (1^3 - 1^3) + (2^3 - 2^3) + (3^3 - 3^3) + \ldots + [(n-1)^3 - (n-1)^3] + n^3$

$= n^3$

$\Rightarrow S_{20} = 8000$

Hence, Statement I is correct.

Statement II We have, already proved in the Statement I that

$S_n = \sum_{r=1}^{n}(r^3 - (r-1)^3) = n^3$

Hence, Statement II is also correct and it is a correct explanation for Statement I.

DAY FOUR

Quadratic Equation and Inequalities

Learning & Revision for the Day

- Quadratic Equation
- Relation between Roots and Coefficients
- Formation of an Equation
- Transformation of Equations
- Maximum and Minimum Value of $ax^2 + bx + c$
- Sign of Quadratic Expression
- Position of Roots
- Inequalities
- Arithmetic-Geometric-Harmonic Mean Inequality
- Logarithm Inequality

Quadratic Equation

- An equation of the form $ax^2 + bx + c = 0$, where $a \neq 0$, a, b and c, $x \in R$, is called a **real quadratic equation**. Here a, b and c are called the **coefficients** of the equation.
- The quantity $D = b^2 - 4ac$ is known as the **discriminant** of the equation $ax^2 + bx + c = 0$ and its roots are given by $x = \dfrac{-b \pm \sqrt{D}}{2a}$
- An equation of the form $az^2 + bz + c = 0$, where $a \neq 0$, a, b and c, $z \in C$ (complex) is called a **complex quadratic equation** and its roots are given by $z = \dfrac{-b \pm \sqrt{D}}{2a}$.

Nature of Roots of Quadratic Equation

Let $a, b, c \in R$ and $a \neq 0$, then the equation $ax^2 + bx + c = 0$

(i) has real and distinct roots if and only if $D > 0$.

(ii) has real and equal roots if and only if $D = 0$.

(iii) has complex roots with non-zero imaginary parts if and only if $D < 0$.

PREP MIRROR

Your Personal Preparation Indicator

- No. of Questions in Exercises (x)—
- No. of Questions Attempted (y)—
- No. of Correct Questions (z)—
  (Without referring Explanations)

- Accuracy Level $(z/y \times 100)$—
- Prep Level $(z/x \times 100)$—

In order to expect good rank in JEE, your Accuracy Level should be above 85 & Prep Level should be above 75.

Some Important Results

(i) If $p + iq$ (where, $p, q \in R, q \neq 0$) is one root of $ax^2 + bx + c = 0$, then second root will be $p - iq$

(ii) If $a, b, c \in Q$ and $p + \sqrt{q}$ is an **irrational root** of $ax^2 + bx + c = 0$, then other root will be $p - \sqrt{q}$.

(iii) If $a, b, c \in Q$ and D is a perfect square, then $ax^2 + bx + c = 0$ has **rational roots.**

(iv) If $a = 1$, b, $c \in I$ and roots of $ax^2 + bx + c = 0$ are rational numbers, then these roots must be integers.

(v) If the roots of $ax^2 + bx + c = 0$ are both positive, then the signs of a and c should be like and opposite to the sign of b.

(vi) If the roots of $ax^2 + bx + c = 0$ are both negative, then signs of a, b and c should be like.

(vii) If the roots of $ax^2 + bx + c = 0$ are reciprocal to each other, then $c = a$.

(viii) In the equation $ax^2 + bx + c = 0$ $(a, b, c \in R)$, if $a + b + c = 0$, then the roots are $1, \frac{c}{a}$ and if $a - b + c = 0$, then the roots are -1 and $-\frac{c}{a}$.

Relation between Roots and Coefficients

Quadratic Roots

If α and β are the roots of quadratic equation $ax^2 + bx + c = 0$; $a \neq 0$, then sum of roots $= \alpha + \beta = -\frac{b}{a}$

and product of roots $= \alpha\beta = \frac{c}{a}$.

And, also $ax^2 + bx + c = a(x - \alpha)(x - \beta)$

Cubic Roots

If α, β and γ are the roots of cubic equation $ax^3 + bx^2 + cx + d = 0; a \neq 0$, then $\alpha + \beta + \gamma = -\frac{b}{a}$

$$\beta\gamma + \gamma\alpha + \alpha\beta = \frac{c}{a}$$

and $$\alpha\beta\gamma = -\frac{d}{a}$$

Common Roots (Conditions)

Suppose that the quadratic equations are $ax^2 + bx + c = 0$ and $a'x^2 + b'x + c' = 0$.

(i) When **one root** is common, then the condition is $(a'c - ac')^2 = (bc' - b'c)(ab' - a'b)$.

(ii) When **both roots** are common, then the condition is $\frac{a}{a'} = \frac{b}{b'} = \frac{c}{c'}$.

Formation of an Equation

Quadratic Equation

If the roots of a quadratic equation are α and β, then the equation will be of the form $x^2 - (\alpha + \beta)x + \alpha\beta = 0$.

Cubic Equation

If α, β and γ are the roots of the cubic equation, then the equation will be form of

$$x^3 - (\alpha + \beta + \gamma)x^2 + (\alpha\beta + \beta\gamma + \gamma\alpha)x - \alpha\beta\gamma = 0.$$

Transformation of Equations

Let the given equation be

$$a_0x^n + a_1x^{n-1} + \ldots + a_{n-1}x + a_n = 0 \quad \ldots(A)$$

Then, the equation

(i) whose roots are k $(\neq 0)$ times roots of the Eq. (A), is obtained by replacing x by $\frac{x}{k}$ in Eq. (A).

(ii) whose roots are the negatives of the roots of Eq. (A), is obtained by replacing x by $-x$ in Eq. (A).

(iii) whose roots are k more than the roots of Eq. (A), is obtained by replacing x by $(x - k)$ in Eq. (A).

(iv) whose roots are reciprocals of the roots of Eq. (A), is obtained by replacing x by $1/x$ in Eq. (A) and then multiply both the sides by x^n.

Maximum and Minimum Value of $ax^2 + bx + c$

(i) When $a > 0$, then minimum value of $ax^2 + bx + c$ is $\frac{-D}{4a}$ or $\frac{4ac - b^2}{4a}$ at $x = \frac{-b}{2a}$

(ii) When $a < 0$, then maximum value of $ax^2 + bx + c$ is $\frac{-D}{4a}$ or $\frac{4ac - b^2}{4a}$ at $x = \frac{-b}{2a}$

Sign of Quadratic Expression

Let $f(x) = ax^2 + bx + c$, where a, b and $c \in R$ and $a \neq 0$.

(i) If $a > 0$ and $D < 0$, then $f(x) > 0, \forall x \in R$.

(ii) If $a < 0$ and $D < 0$, then $f(x) < 0, \forall x \in R$.

(iii) If $a > 0$ and $D = 0$, then $f(x) \geq 0, \forall x \in R$.

(iv) If $a < 0$ and $D = 0$, then $f(x) \leq 0, \forall x \in R$.

(v) If $a > 0$, $D > 0$ and $f(x) = 0$ have two real roots α and β, where $(\alpha < \beta)$, then $f(x) > 0$, $\forall x \in (-\infty, \alpha) \cup (\beta, \infty)$ and $f(x) < 0$, $\forall x \in (\alpha, \beta)$.

(vi) If $a < 0, D > 0$ and $f(x) = 0$ have two real roots α and β, where $(\alpha < \beta)$, then $f(x) < 0$, $\forall x \in (-\infty, \alpha) \cup (\beta, \infty)$ and $f(x) > 0$, $\forall x \in (\alpha, \beta)$.

Position of Roots

Let $ax^2 + bx + c = 0$ has roots α and β. Then, we have the following conditions:

(i) with respect to one real number (k).

Situation	Required conditions
$\alpha < \beta < k$	$D \ge 0,\ af(k) > 0,\ k > -b/2a$
$k < \alpha < \beta$	$D \ge 0,\ af(k) > 0,\ k < -b/2a$
$\alpha < k < \beta$	$D > 0,\ af(k) < 0$

(ii) with respect to two real numbers k_1 and k_2.

Situation	Required conditions
$k_1 < \alpha < \beta < k_2$	$D \ge 0,\ af(k_1) > 0,$ $af(k_2) > 0,\ k_1 < -b/2a < k_2$
$\alpha < k_1 < k_2 < \beta$	$D > 0,\ af(k_1) < 0,\ af(k_2) < 0$
$k_1 < \alpha < k_2 < \beta$	$D > 0,\ f(k_1)f(k_2) < 0$

Inequalities

Let a and b be two real numbers. If $a - b$ is negative, we say that a is less than b $(a < b)$ and if $a - b$ is positive, then a is greater than b $(a > b)$. This shows the inequalities concept.

Important Results on Inequalities

(i) If $a > b$, then $a \pm c > b \pm c, \forall c \in R$.

(ii) If $a > b$, then

(a) for $m > 0$, $am > bm, \frac{a}{m} > \frac{b}{m}$

(b) for $m < 0$, $am < bm, \frac{a}{m} < \frac{b}{m}$

(iii) (a) If $a > b > 0$, then

- $a^2 > b^2$ • $|a| > |b|$ • $\frac{1}{a} < \frac{1}{b}$

(b) If $a < b < 0$, then

- $a^2 > b^2$ • $|a| > |b|$ • $\frac{1}{a} > \frac{1}{b}$

(iv) If $a < 0 < b$, then

(a) $a^2 > b^2$, if $|a| > |b|$ (b) $a^2 < b^2$, if $|a| < |b|$

(v) If $a < x < b$ and a, b are positive real numbers, then $a^2 < x^2 < b^2$.

(vi) If $a < x < b$ and a is negative number and b is positive number, then

(a) $0 \le x^2 < b^2$, if $|b| > |a|$ (b) $0 \le x^2 < a^2$, if $|a| > |b|$

(vii) If $a_i > b_i > 0$, where $i = 1, 2, 3, \dots, n$, then

$$a_1 a_2 a_3 \dots a_n > b_1 b_2 b_3 \dots b_n.$$

(viii) If $a_i > b_i$, where $i = 1, 2, 3, \dots, n$, then

$$a_1 + a_2 + a_3 + \dots + a_n > b_1 + b_2 + \dots + b_n.$$

(ix) If $|x| < a$, then

(a) for $a > 0$, $-a < x < a$.

(b) for $a < 0$, $x \in \phi$.

Arithmetic-Geometric-Harmonic Mean Inequality

The Arithmetic-Geometric-Harmonic Mean of positive real numbers is defined as follows

Arithmetic Mean $\ge$ Geometric Mean $\ge$ Harmonic Mean

(i) If $a, b > 0$ then $\frac{a+b}{2} \ge \sqrt{ab} \ge \frac{2}{\frac{1}{a} + \frac{1}{b}}$

(ii) If $a_i > 0$, where $i = 1, 2, 3, \dots, n$, then

$$\frac{a_1 + a_2 + \dots + a_n}{n} \ge (a_1 \cdot a_2 \dots a_n)^{1/n} \ge \frac{n}{\frac{1}{a_1} + \frac{1}{a_2} + \dots + \frac{1}{a_n}}$$

Logarithm Inequality

If a is a positive real number other than 1 and $a^x = m$, then x is called the logarithm of m to the base a, written as $\log_a m$. In $\log_a m$, m should always be positive.

(i) If $m \le 0$, then $\log_a m$ will be meaningless.

(ii) $\log_a m$ exists, if $m, a > 0$ and $a \ne 1$.

Important Results on Logarithm

(i) $a^{\log_a x} = x; a > 0, \ne 1, x > 0$

(ii) $a^{\log_b x} = x^{\log_b a}; a, b > 0, \ne 1, x > 0$

(iii) $\log_a a = 1, a > 0, \ne 1$

(iv) $\log_a x = \frac{1}{\log_x a}; x, a > 0, \ne 1$

(v) $\log_a x = \log_a b \ \log_b x = \frac{\log_b x}{\log_b a}; a, b > 0, \ne 1, x > 0$

(vi) For $x > 0; a > 0, \ne 1$

(a) $\log_{a^n}(x) = \frac{1}{n}\log_a x$

(b) $\log_{a^n} x^m = \left(\frac{m}{n}\right)\log_a x$

(vii) For $x > y > 0$

(a) $\log_a x > \log_a y$, if $a > 1$

(b) $\log_a x < \log_a y$, if $0 < a < 1$

(viii) If $a > 1$ and $x > 0$, then

(a) $\log_a x > p \Rightarrow x > a^p$

(b) $0 < \log_a x < p \Rightarrow 0 < x < a^p$

(ix) If $0 < a < 1$, then

(a) $\log_a x > p \Rightarrow 0 < x < a^p$

(b) $0 < \log_a x < p \Rightarrow a^p < x < 1$

DAY PRACTICE SESSION 1

FOUNDATION QUESTIONS EXERCISE

1 If $\sqrt{3x^2-7x-30}+\sqrt{2x^2-7x-5}=x+5$, then x is equal to

(a) 2 (b) 3 (c) 6 (d) 5

2 The number of solutions for equation $x^2-5|x|+6=0$ is

(a) 4 (b) 3 (c) 2 (d) 1

3 The roots of the equation $|2x-1|^2-3|2x-1|+2=0$ are

(a) $\left\{-\frac{1}{2},0,\frac{1}{2}\right\}$ (b) $\left\{-\frac{1}{2},0,\frac{3}{2}\right\}$ (c) $\left\{-\frac{3}{2},\frac{1}{2},0,1\right\}$ (d) $\left\{-\frac{1}{2},0,1,\frac{3}{2}\right\}$

4 The product of all the values of x satisfying the equation $(5+2\sqrt{6})^{x^2-3}+(5-2\sqrt{6})^{x^2-3}=10$ is

(a) 4 (b) 6 (c) 8 (d) 19

5 The root of the equation $2(1+i)x^2-4(2-i)x-5-3i=0$, where $i=\sqrt{-1}$, which has greater modulus, is

(a) $\frac{3-5i}{2}$ (b) $\frac{5-3i}{2}$ (c) $\frac{3+i}{2}$ (d) $\frac{3i+1}{2}$

6 $x^2+x+1+2k(x^2-x-1)=0$ is perfect square for how many value of k

(a) 2 (b) 0 (c) 1 (d) 3

7 If the roots of $(a^2+b^2)x^2-2(bc+ad)x+c^2+d^2=0$ are equal, then

(a) $\frac{a}{b}=\frac{c}{d}$ (b) $\frac{a}{c}+\frac{b}{d}=0$ (c) $\frac{a}{d}=\frac{b}{c}$ (d) $a+b=c+d$

8 The least value of $|\alpha|$ for which $\tan\theta$ and $\cot\theta$ are roots of the equation $x^2+ax+1=0$, is

(a) 2 (b) 1 (c) 1/2 (d) 0

9 If one root of the equation $x^2-\lambda x+12=0$ is even prime while $x^2+\lambda x+\mu=0$ has equal roots, then μ is equal to

(a) 8 (b) 16 (c) 24 (d) 32

10 If $a+b+c=0$, then the roots of the equation $4ax^2+3bx+2c=0$, where $a,b,c\in R$ are

(a) real and distinct (b) imaginary
(c) real and equal (d) infinite

11 The equation $(\cos\beta-1)x^2+(\cos\beta)x+\sin\beta=0$ in the variable x has real roots, then β lies in the interval

(a) $(0,2\pi)$ (b) $(-\pi,0)$ (c) $\left(-\frac{\pi}{2},\frac{\pi}{2}\right)$ (d) $(0,\pi)$

12 If $ax^2+2bx-3c=0$ has no real root and $\frac{3c}{4}<a+b$, then the range of c is

(a) $(-1,1)$ (b) $(0,1)$
(c) $(0,\infty)$ (d) $(-\infty,0)$

13 If a, b and c are real numbers in AP, then the roots of $ax^2+bx+c=0$ are real for

(a) all a and c (b) no a and c
(c) $\left|\frac{c}{a}-7\right|\geq 4\sqrt{3}$ (d) $\left|\frac{a}{c}+7\right|\geq 2\sqrt{3}$

14 If $P(x)=ax^2+bx+c$ and $Q(x)=-ax^2+dx+c=0$; $ac\neq 0$, then the equation $P(x)\cdot Q(x)=0$ has

(a) four real roots (b) exactly two real roots
(c) either two or four real roots (d) atmost two real roots

15 The rational values of a in $ax^2+bx+1=0$ if $\frac{1}{4+\sqrt{3}}$ is a root, are

(a) $a=13, b=-8$ (b) $a=-13, b=8$
(c) $a=13, b=8$ (d) $a=-13, b=-8$

16 If $1-i$, is a root of the equation $x^2+ax+b=0$, where $a,b\in R$, then the values of a and b are

(a) 1,–1 (b) 2,–2
(c) 3,–3 (d) None of these

17 The values of p for which one root of the equation $x^2-30x+p=0$ is the square of the other, is/are

(a) Only 125 (b) 125 and – 216
(c) 125 and 215 (d) Only 216

18 If the roots of the quadratic equation $\frac{x-m}{mx+1}=\frac{x+n}{nx+1}$ are reciprocal to each other, then

(a) $n=0$ (b) $m=n$ (c) $m+n=1$ (d) $m^2+n^2=1$

19 Let α and α^2 be the roots of $x^2+x+1=0$, then the equation whose roots are α^{31} and α^{62}, is

(a) $x^2-x+1=0$ (b) $x^2+x-1=0$
(c) $x^2+x+1=0$ (d) $x^{60}+x^{30}+1=0$

20 If α and β are the roots of $x^2-a(x-1)+b=0$, then the value of $\frac{1}{\alpha^2-a\alpha}+\frac{1}{\beta^2-a\beta}+\frac{2}{a+b}$ is

(a) $\frac{4}{a+b}$ (b) $\frac{1}{a+b}$ (c) 0 (d) –1

21 The value of a for which the sum of the squares of the roots of the equation $x^2-(a-2)x-a-1=0$ assume the least value is → AIEEE 2005

(a) 2 (b) 3 (c) 0 (d) 1

22 If α and β be the roots of the equation $2x^2+2(a+b)x+a^2+b^2=0$, then the equation whose roots are $(\alpha+\beta)^2$ and $(\alpha-\beta)^2$, is

(a) $x^2-2abx-(a^2-b^2)^2=0$ (b) $x^2-4abx-(a^2-b^2)^2=0$
(c) $x^2-4abx+(a^2-b^2)^2=0$ (d) None of these

23 Let α,β be the roots of $x^2-2x\cos\phi+1=0$, then the equation whose roots are α^n and β^n, is

(a) $x^2-2x\cos n\phi-1=0$ (b) $x^2-2x\cos n\phi+1=0$
(c) $x^2-2x\sin n\phi+1=0$ (d) $x^2+2x\sin n\phi-1=0$

24 The harmonic mean of the roots of the equation $(5+\sqrt{2})x^2-(4+\sqrt{5})x+8+2\sqrt{5}=0$ is

(a) 2 (b) 4 (c) 6 (d) 8

25 If the ratio of the roots of $\lambda x^2 + \mu x + \nu = 0$ is equal to the ratio of the roots of $x^2 + x + 1 = 0$, then λ, μ and ν are in

(a) AP (b) GP
(c) HP (d) None of these

26 If the roots of the equation $\frac{1}{x+p} + \frac{1}{x+q} = \frac{1}{r}$ are equal in magnitude but opposite in sign, then the product of the roots is

(a) $-2(p^2+q^2)$ (b) $-(p^2+q^2)$
(c) $\frac{-(p^2+q^2)}{2}$ (d) $-pq$

27 If the roots of the equation $ax^2 + bx + c = 0$ of the form $\frac{k+1}{k}$ and $\frac{k+2}{k+1}$, then $(a+b+c)^2$ is equal to

(a) $2b^2 - ac$ (b) Σa^2 (c) $b^2 - 4ac$ (d) $b^2 - 2ac$

28 If α and β are the roots of the equation $ax^2 + bx + c = 0$ such that $\beta < \alpha < 0$, then the quadratic equation whose roots are $|\alpha|, |\beta|$, is given by

(a) $|a|x^2 + |b|x + |c| = 0$ (b) $ax^2 - |b|x + c = 0$
(c) $|a|x^2 - |b|x + |c| = 0$ (d) $a|x|^2 + b|x + c = 0$

29 If α and β be the roots of $x^2 + px + q = 0$, then $\frac{(\omega\alpha + \omega^2\beta)(\omega^2\alpha + \omega\beta)}{\frac{\alpha^2}{\beta} + \frac{\beta^2}{\alpha}}$ is equal to

(a) $-\frac{q}{p}$ (b) $\alpha\beta$ (c) $-\frac{p}{q}$ (d) ω

30 If α and β are roots of the equation $x^2 + px + 3.\frac{p}{4} = 0$, such that $|\alpha - \beta| = \sqrt{10}$, then p belongs **→ JEE Mains 2013**

(a) $\{2, -5\}$ (b) $\{-3, 2\}$ (c) $\{-2, 5\}$ (d) $\{3, -5\}$

31 Sachin and Rahul attemped to solve a quadratic equation. Sachin made a mistake in writing down the constant term and ended up in roots (4, 3). Rahul made a mistake in writing down coefficient of x to get roots (3, 2). The correct roots of equation are **→ AIEEE 2011**

(a) −4, −3 (b) 6, 1 (c) 4, 3 (d) −6, −1

32 If α and β are the roots of $x^2 + x + 2 = 0$ and γ, δ are the roots of $x^2 + 3x + 4 = 0$, then $(\alpha+\gamma)(\alpha+\delta)(\beta+\gamma)(\beta+\delta)$ is equal to

(a) −18 (b) 18 (c) 24 (d) 44

33 In $\Delta PQR, R = \frac{\pi}{2}$. If $\tan\frac{P}{2}$ and $\tan\frac{Q}{2}$ are the roots of the equation $ax^2 + bx + c = 0$, then

(a) $a = b + c$ (b) $b = c + a$
(c) $c = a + b$ (d) $b = c$

34 If the equation $k(6x^2+3) + rx + 2x^2 - 1 = 0$ and $6k(2x^2+1) + px + 4x^2 - 2 = 0$ have both roots common, then $2r - p$ is equal to

(a) 2 (b) 1 (c) 0 (d) k

35 If the equations $x^2 + 2x + 3 = 0$ and $ax^2 + bx + c = 0$, $a, b, c \in R$, have a common root, then $a:b:c$ is **→ AIEEE 2012**

(a) 1 : 2 : 3 (b) 3 : 2 : 1 (c) 1 : 3 : 2 (d) 3 : 1 : 2

36 The equation formed by decreasing each root of $ax^2 + bx + c = 0$ by 1 is $2x^2 + 8x + 2 = 0$, then

(a) $a = -b$ (b) $b = -c$ (c) $c = -a$ (d) $b = a + c$

37 If $f(x) = x^2 + 2bx + 2c^2$ and $g(x) = -x^2 - 2cx + b^2$ such that min $f(x) >$ max $g(x)$, then the relation between b and c is

(a) $|c| < |b|\sqrt{2}$ (b) $0 < c < b\sqrt{2}$
(c) $|c| < |b| 2$ (d) $|c| > |b|\sqrt{2}$

38 If $a \in R$ and $a_1, a_2, a_3 \dots, a_n \in R$, then $(x - a_1)^2 + (x - a_2)^2 + \dots + (x - a_n)^2$ assumes its least value at $x =$

(a) $a_1 + a_2 + \dots + a_n$ (b) $2(a_1 + a_2 + a_3 + \dots + a_n)$
(c) $n(a_1 + a_2 + \dots + a_n)$ (d) None of these

39 If the roots of the equation $bx^2 + cx + a = 0$ is imaginary, then for all real values of x, the expression $3b^2x^2 + 6bcx + 2c^2$ is **→ AIEEE 2009**

(a) greater than $4ab$ (b) less than $4ab$
(c) greater than $-4ab$ (d) less than $-4ab$

40 If $x^2 + 2ax + 10 - 3a > 0$ for all $x \in R$, then

(a) $-5 < a < 2$ (b) $a < -5$
(c) $a > 5$ (d) $2 < a < 5$

41 If the expression $\left(ax - 1 + \frac{1}{x}\right)$ is non-negative for all positive real x, then the minimum value of a must be

(a) 0 (b) $\frac{1}{2}$
(c) $\frac{1}{4}$ (d) None of these

42 The number of real solutions of the equation $\left(\frac{9}{10}\right)^x = -3 + x - x^2$ is

(a) 0 (b) 1
(c) 2 (d) None of these

43 If α and β be the roots of the quadratic equation $ax^2 + bx + c = 0$ and k be a real number, then the condition, so that $\alpha < k < \beta$ is given by

(a) $ac > 0$ (b) $ak^2 + bk + c = 0$
(c) $ac < 0$ (d) $a^2k^2 + abk + ac < 0$

44 $|2x - 3| < |x + 5|$, then x belongs to

(a) (−3,5) (b) (5, 9) (c) $\left(-\frac{2}{3}, 8\right)$ (d) $\left(-8, \frac{2}{3}\right)$

45 The least integral value α of x such that $\frac{x-5}{x^2+5x-14} > 0$, satisfies **→ JEE Mains 2013**

(a) $\alpha^2 + 3\alpha - 4 = 0$ (b) $\alpha^2 - 5\alpha + 4 = 0$
(c) $\alpha^2 - 7\alpha + 6 = 0$ (d) $\alpha^2 + 5\alpha - 6 = 0$

46 The solution set of $\frac{|x-2|-1}{|x-2|-2} \le 0$ is

(a) $[0, 1] \cup (3, 4)$ (b) $[0, 1] \cup [3, 4]$
(c) $[-1, 1) \cup (3, 4]$ (d) None of these

47 Number of integral solutions of $\frac{x+2}{x^2+1} > \frac{1}{2}$ is

(a) 0 (b) 1 (c) 2 (d) 3

48 If the product of n positive numbers is 1, then their sum is

(a) a positive integer (b) divisible by n
(c) equation to $n + \frac{1}{n}$ (d) greater than or equal to n

49 The minimum value of $P = bcx + cay + abz$, when $xyz = abc$, is

(a) $3abc$ (b) $6abc$
(c) abc (d) $4abc$

50 If a, b and c are distinct three positive real numbers, then $(a+b+c)\left(\frac{1}{a}+\frac{1}{b}+\frac{1}{c}\right)$ is

(a) > 1 (b) > 9
(c) < 9 (d) None of these

51 If a, b, c are positive real numbers such that $a + b + c = 1$, then the greatest value of $(1-a)(1-b)(1-c)$, is

(a) $\frac{1}{27}$ (b) $\frac{8}{27}$ (c) $\frac{4}{27}$ (d) 9

52 If a, b, c, d are positive real numbers such that $a + b + c + d = 2$, then $M = (a+b)(c+d)$ satisfies the relation

(a) $0 \le M \le 1$ (b) $1 \le M \le 2$ (c) $2 \le M \le 3$ (d) $3 \le M \le 4$

53 $\log_2(x^2 - 3x + 18) < 4$, then x belongs to

(a) (1, 2) (b) (2, 16)
(c) (1, 16) (d) None of these

54 If $\log_{0.3}(x-1) > \log_{0.09}(x-1)$, then x lies in

(a) (1, 2) (b) $(-\infty, 1)$
(c) $(2, \infty)$ (d) None of these

55 What is the solution set of the following inequality?

$$\log_x\left(\frac{x+5}{1-3x}\right) > 0$$

(a) $0 < x < \frac{1}{3}$ (b) $x \ge 3$
(c) $\frac{1}{3} < x < 1$ (d) None of these

DAY PRACTICE SESSION 2

PROGRESSIVE QUESTIONS EXERCISE

1 Let $S = \{x \in R : x \ge 0$ and $2\left|\sqrt{x} - 3\right| + \sqrt{x}(\sqrt{x} - 6) + 6 = 0$

Then, S → JEE Mains 2018

(a) is an empty set
(b) contains exactly one element
(c) contains exactly two elements
(d) contains exactly four elements

2 The roots of the equation $2^{x+2} \cdot 3^{3x/(x-1)} = 9$ are given by

(a) $1 - \log_2 3, 2$ (b) $\log_2\left(\frac{2}{3}\right), 1$
(c) $2, -2$ (d) $-2, 1 - \frac{(\log 3)}{(\log 2)}$

3 Let α and β be the roots equation $x^2 - 6x - 2 = 0$. If $a_n = \alpha^n - \beta^n$ for $n \ge 1$, then the value of $\frac{a_{10} - 2a_8}{2a_9}$ is equal to → JEE Mains 2015

(a) 6 (b) -6 (c) 3 (d) -3

4 If $x^2 - 5x + 1 = 0$, then $\frac{x^{10}+1}{x^5}$ is equal to

(a) 2424 (b) 3232
(c) 2525 (d) None of these

5 If $a < b < c < d$, then the roots of the equation $(x-a)(x-c) + 2(x-b)(x-d) = 0$ are

(a) real and distinct (b) real and equal
(c) imaginary (d) None of these

6 Let α and β be the roots of equation $px^2 + qx + r = 0$, $p \ne 0$. If p, q and r are in AP and $\frac{1}{\alpha} + \frac{1}{\beta} = 4$, then the value of $|\alpha - \beta|$ is → JEE Mains 2014

(a) $\frac{\sqrt{61}}{9}$ (b) $\frac{2\sqrt{17}}{9}$ (c) $\frac{\sqrt{34}}{9}$ (d) $\frac{2\sqrt{13}}{9}$

7 If α and β are the roots of the equation $ax^2 + bx + c = 0$ ($a \ne 0, a, b, c$ being different), then $(1 + \alpha + \alpha^2)(1 + \beta + \beta^2)$ is

(a) zero (b) positive
(c) negative (d) None of these

8 The minimum value of the sum of real numbers $a^{-5}, a^{-4}, 3a^{-3}, 1, a^8$ and a^{10} $a > 0$ is

(a) 9 (b) 8 (c) 2 (d) 1

9 For $a > 0$, the roots of the equation $\log_{ax} a + \log_x a^2 + \log_{a^2x} a^3 = 0$ is given by

(a) $a^{-4/3}$ (b) $a^{-3/4}$ (c) $a^{1/2}$ (d) a^{-1}

10 If a, b and c are in AP and if the equations $(b-c)x^2+(c-a)x+(a-b)=0$ and $2(c+a)x^2+(b+c)x=0$ have a common root, then

(a) a^2, b^2 and c^2 are in AP (b) a^2, c^2 and b^2 are in AP
(c) c^2, a^2 and b^2 are in AP (d) None of these

11 If the equations $x^2+ax+12=0$, $x^2+bx+15=0$ and $x^2+(a+b)x+36=0$ have a common positive root, then the ordered pair (a, b) is

(a) $(-6,-7)$ (b) $(-7,-8)$
(c) $(-6,-8)$ (d) $(-8,-7)$

12 If x is real, then the maximum and minimum value of the expression $\frac{x^2-3x+4}{x^2+3x+4}$ will be

(a) 2, 1 (b) $5, \frac{1}{5}$
(c) $7, \frac{1}{7}$ (d) None of these

13 If $a \in R$ and the equation $-3(x-[x])^2+2(x-[x])+a^2=0$ (where, $[x]$ denotes the greatest integer $\leq x$) has no integral solution, then all possible values of a lie in the interval → JEE Mains 2014

(a) $(-1,0)\cup(0,1)$ (b) $(1,2)$
(c) $(-2,-1)$ (d) $(-\infty,-2)\cup(2,\infty)$

14 If $a=\cos\frac{2\pi}{7}+i\sin\frac{2\pi}{7}$, then the quadratic equation whose roots are $\alpha=a+a^2+a^4$ and $\beta=a^3+a^5+a^6$, is

(a) $x^2-x+2=0$ (b) $x^2+2x+2=0$
(c) $x^2+x+2=0$ (d) $x^2+x-2=0$

15 If α and β are roots of $375x^2-25x-2=0$ and $S_n=\alpha^n+\beta^n$, then $\lim_{n\to\infty}\sum_{r=1}^{n} S_r$ is equal to

(a) $\frac{7}{116}$ (b) $\frac{1}{12}$
(c) $\frac{29}{358}$ (d) None of these

16 If $S=\{a\in N, 1\leq a\leq 100\}$ and $[\tan^2 x]-\tan x-a=0$ has real roots, where [.] denotes the greatest integer function, then number of elements in set S equals

(a) 2 (b) 5 (c) 6 (d) 9

17 The sum of all real values of x satisfying the equation $(x^2-5x+5)^{x^2+4x-60}=1$ is → JEE Mains 2016

(a) 3 (b) -4 (c) 6 (d) 5

18 If λ is an integer and α, β are the roots of $4x^2-16x+\frac{\lambda}{4}=0$ such that $1<\alpha<2$ and $2<\beta<3$, then the possible values of λ are

(a) {60, 64, 68} (b) {61, 62, 63}
(c) {49, 50,..., 62, 63} (d) {62, 65, 68, 71, 75}

19 If α and β are the roots of the equation $ax^2+bx+c=0$, then the quadratic equation whose roots are $\frac{\alpha}{1+\alpha}$ and $\frac{\beta}{1+\beta}$ is

(a) $ax^2-b(1-x)+c(1-x)^2=0$
(b) $ax^2-b(x-1)+c(x-1)^2=0$
(c) $ax^2+b(1-x)+c(1-x)^2=0$
(d) $ax^2+b(x+1)+c(1+x)^2=0$

20 If both the roots of the quadratic equation $x^2-2kx+k^2+k-5=0$ are less than 5, then k lies in the interval → AIEEE 2005

(a) $[4,5]$ (b) $(-\infty,4)$ (c) $(6,\infty)$ (d) $(5,6)$

21 All the values of m for which both roots of the equation $x^2-2mx+m^2-1=0$ are greater than -2 but less than 4 lie in the interval

(a) $m>3$ (b) $-1<m<3$
(c) $1<m<4$ (d) $-2<m<0$

22 Let α and β be real and z be a complex number. If $z^2+\alpha z+\beta=0$ has two distinct roots on the line $\text{Re}(z)=1$, then it is necessary that → AIEEE 2011

(a) $\beta\in(-1,0)$ (b) $|\beta|=1$ (c) $\beta\in[1,\infty)$ (d) $\beta\in(0,1)$

23 The equation $e^{\sin x}-e^{-\sin x}-4=0$ has → AIEEE 2012

(a) infinite number of real roots
(b) no real root
(c) exactly one real root
(d) exactly four real roots

24 If a, b, c, d are positive real numbers such that $a+\frac{1}{b}=4$, $b+\frac{1}{c}=1$, $c+\frac{1}{d}=4$ and $d+\frac{1}{a}=1$, then

(a) $a=c$ and $b=d$ (b) $b=d$ but $a\neq c$
(c) $ab=1$ and $cd\neq 1$ (d) $cd=1$ and $ab\neq 1$

(ω and ω^2 are complex cube roots of unity)

25 Let $f: R\to R$ be a continuous function defined by $f(x)=\frac{1}{e^x+2e^{-x}}$

Statement I $f(c)=\frac{1}{3}$, for some $c\in R$.

Statement II $0<f(x)\leq\frac{1}{2\sqrt{2}}, \forall x\in R$. → AIEEE 2010

(a) Statement I is false, Statement II is true
(b) Statement I is true, Statement II is true. Statement II is a correct explanation of Statement I
(c) Statement I is true, Statement II is true; Statement II is not a correct explanation for Statement I
(d) Statement I is true, Statement II is false

ANSWERS

SESSION 1

1. (c)	**2.** (a)	**3.** (d)	**4.** (c)	**5.** (a)	**6.** (a)	**7.** (c)	**8.** (a)	**9.** (b)	**10.** (a)
11. (d)	**12.** (d)	**13.** (c)	**14.** (c)	**15.** (a)	**16.** (d)	**17.** (b)	**18.** (a)	**19.** (c)	**20.** (c)
21. (d)	**22.** (b)	**23.** (b)	**24.** (b)	**25.** (b)	**26.** (c)	**27.** (c)	**28.** (c)	**29.** (a)	**30.** (c)
31. (b)	**32.** (d)	**33.** (c)	**34.** (c)	**35.** (a)	**36.** (b)	**37.** (d)	**38.** (d)	**39.** (c)	**40.** (a)
41. (c)	**42.** (a)	**43.** (d)	**44.** (c)	**45.** (d)	**46.** (b)	**47.** (d)	**48.** (d)	**49.** (a)	**50.** (b)
51. (b)	**52.** (a)	**53.** (a)	**54.** (a)	**55.** (d)					

SESSION 2

1. (c)	**2.** (d)	**3.** (c)	**4.** (c)	**5.** (a)	**6.** (d)	**7.** (b)	**8.** (b)	**9.** (a)	**10.** (b)
11. (b)	**12.** (c)	**13.** (a)	**14.** (c)	**15.** (b)	**16.** (d)	**17.** (a)	**18.** (c)	**19.** (c)	**20.** (b)
21. (b)	**22.** (c)	**23.** (b)	**24.** (a)	**25.** (b)					

Hints and Explanations

SESSION 1

1 We have,

$\sqrt{3x^2-7x-30}+\sqrt{2x^2-7x-5}=x+5$

$\Rightarrow \sqrt{3x^2-7x-30}=(x+5)-\sqrt{2x^2-7x-5}$

On squaring both sides, we get

$$3x^2-7x-30 = x^2+25+10x+(2x^2-7x-5) - 2(x+5)\sqrt{2x^2-7x-5}$$

$\Rightarrow -10x-50=-2(x+5)\sqrt{2x^2-7x-5}$

$x+5\neq 0, \sqrt{2x^2-7x-5}=5$

[$\because x=-5$ does not satisfy the given equation]

$\Rightarrow 2x^2-7x-30=0$

$\therefore \quad x=6$

2 Given equation is $x^2-5|x|+6=0$

When $x\geq 0$, $x^2-5x+6=0$

and when $x<0$, $x^2+5x+6=0$

$\Rightarrow \quad x^2-3x-2x+6=0;\ x\geq 0$

and $\quad x^2+3x+2x+6=0;\ x<0$

$\Rightarrow \quad (x-3)(x-2)=0,\ x\geq 0$

and $\quad (x+3)\cdot(x+2)=0,\ x<0$

$\therefore \quad x=3,\ x=2$ and $x=-3,\ x=-2$

There are four solutions of this equation.

3 Given equation is

$$|2x-1|^2-3|2x-1|+2=0$$

Let $|2x-1|=t$, then

$$t^2-3t+2=0$$

$\Rightarrow \quad (t-1)(t-2)=0 \Rightarrow t=1,2$

$\Rightarrow \quad |2x-1|=1$ and $|2x-1|=2$

$\Rightarrow \quad 2x-1=\pm1$ and $2x-1=\pm2$

$\Rightarrow \quad x=1,0$ and $x=\frac{3}{2},-\frac{1}{2}$

4 $\because\ 5-2\sqrt{6}=\frac{1}{5+2\sqrt{6}}$

$\therefore \quad t+\frac{1}{t}=10,$

where $\quad t=(5+2\sqrt{6})^{x^2-3}$ ...(i)

$\Rightarrow \quad t^2-10t+1=0$

$\Rightarrow \quad t=5\pm 2\sqrt{6}$

or $\quad t=(5+2\sqrt{6})^{\pm1}$ ...(ii)

From Eqs. (i) and (ii),

$x^2-3=\pm 1$

$\Rightarrow \quad x^2=2, 4$

$\Rightarrow \quad x=-\sqrt{2}, \sqrt{2}, -2, 2$

$\therefore$ Required product = 8

5 The given equation is

$$2(1+i)x^2-4(2-i)x-5-3i=0$$

$$\Rightarrow x=\frac{4(2-i)\pm\sqrt{16(2-i)^2+8(1+i)(5+3i)}}{4(1+i)}$$

$$=-\frac{i}{1+i} \text{ or } \frac{4-i}{1+i}=\frac{-1-i}{2} \text{ or } \frac{3-5i}{2}$$

Now, $\left|\frac{-1-i}{2}\right|=\sqrt{\frac{1}{4}+\frac{1}{4}}=\sqrt{\frac{1}{2}}$

and $\left|\frac{3-5i}{2}\right|=\sqrt{\frac{9}{4}+\frac{25}{4}}=\sqrt{\frac{17}{2}}$

Also, $\sqrt{\frac{17}{2}}>\sqrt{\frac{1}{2}}$

Hence, required root is $\frac{3-5i}{2}$.

6 Given equation

$(1+2k)x^2+(1-2k)x+(1-2k)=0$

If equation is a perfect square then root are equal

i.e. $(1-2k)^2-4(1+2k)(1-2k)=0$

i.e. $\quad k=\frac{1}{2},\frac{-3}{10}$.

Hence, total number of values = 2.

7 Since, roots are real.

$\therefore \{2(bc+ad)\}^2=4(a^2+b^2)(c^2+d^2)$

$\Rightarrow 4b^2c^2+4a^2d^2+8abcd=4a^2c^2+4a^2d^2+4b^2c^2+4b^2d^2$

$\Rightarrow \quad 4a^2c^2+4b^2d^2-8abcd=0$

$\Rightarrow \quad 4(ac-bd)^2=0$

$\Rightarrow \quad ac=bd$

$\Rightarrow \quad \frac{a}{d}=\frac{b}{c}$

8 Given equation is $x^2+ax+1=0$

Since, roots are real

$\therefore \quad a^2-4\geq 0\Rightarrow |a|\geq 2$

Thus, the least value of $|a|$ is 2.

9 We know that only even prime is 2,

$\therefore\ (2)^2-\lambda(2)+12=0.$

$\Rightarrow \quad \lambda=8$ ...(i)

$\because x^2+\lambda x+\mu=0$ has equal roots.

$\therefore \quad \lambda^2-4\mu=0 \quad [\because D=0]$

$\Rightarrow \quad (8)^2-4\mu=0\Rightarrow \mu=16$

10 Here, $D=(3b)^2-4(4a)(2c)$

$=9b^2-32ac=9(-a-c)^2-32ac$

$=9a^2-14ac+9c^2$

$$=9c^2\left[\left(\frac{a}{c}\right)^2-\frac{14}{9}\cdot\frac{a}{c}+1\right]$$

$$=9c^2\left[\left(\frac{a}{c}-\frac{7}{9}\right)^2-\frac{49}{81}+1\right]>0$$

Hence, the roots are real and distinct.

11 For real roots, discriminant,

$$D = b^2 - 4ac \ge 0$$
$$= \cos^2\beta - 4(\cos\beta - 1)\sin\beta \ge 0$$
$$= \cos^2\beta + 4(1-\cos\beta)\sin\beta \ge 0$$

So, $\sin\beta$ should be > 0. $[\because \cos^2\beta \ge 0, 1-\cos\beta \ge 0]$

$\Rightarrow \quad \beta \in (0, \pi)$

12 Here, $D = 4b^2 + 12ca < 0$

$\Rightarrow \quad b^2 + 3ca < 0 \quad \ldots(i)$

$\Rightarrow \quad ca < 0$

If $c > 0$, then $a < 0$

Also, $\dfrac{3c}{4} < a + b$

$\Rightarrow \quad 3ca > 4a^2 + 4ab$

$\Rightarrow b^2 + 3ca > 4a^2 + 4ab + b^2$

$= (2a+b)^2 \ge 0 \quad \ldots(ii)$

From (i) and (ii), $c > 0$, is not true.

$\therefore \quad c < 0$

13 Since, $D \ge 0$

$\therefore \quad b^2 - 4ac \ge 0$

$\Rightarrow \left(\dfrac{c+a}{2}\right)^2 - 4ac \ge 0 \quad [\because 2b = a + c]$

$\Rightarrow \quad c^2 - 14ca + a^2 \ge 0$

$\Rightarrow \left(\dfrac{c}{a}\right)^2 - 14\left(\dfrac{c}{a}\right) + 1 \ge 0$

$\Rightarrow \quad \left(\dfrac{c}{a} - 7\right)^2 \ge 48$

$\Rightarrow \quad \left|\dfrac{c}{a} - 7\right| \ge 4\sqrt{3}$

14 Let D_1 and D_2 be the discriminants of given equation, respectively. Then

$$D_1 + D_2 = b^2 - 4ac + d^2 + 4ac$$
$$= b^2 + d^2 > 0$$

So, either D_1 and D_2 are positive or atleast one D's is positive.

Therefore, $P(x) \cdot Q(x) = 0$ has either two or four real roots.

15 One root $= \dfrac{1}{4+\sqrt{3}} \times \dfrac{4-\sqrt{3}}{4-\sqrt{3}} = \dfrac{4-\sqrt{3}}{13}$

$\therefore$ Other root $= \dfrac{4+\sqrt{3}}{13}$

$\therefore$ The quadratic equation is

$$x^2 - \left(\frac{4+\sqrt{3}}{13} + \frac{4-\sqrt{3}}{13}\right)x + \frac{4+\sqrt{3}}{13} \cdot \frac{4-\sqrt{3}}{13} = 0$$

or $\quad 13x^2 - 8x + 1 = 0$

This equation must be identical with $ax^2 + bx + 1 = 0$;

$\therefore \quad a = 13$ and $b = -8$.

16 Sum of roots

$$\frac{-a}{1} = (1-i) + (1+i) \Rightarrow a = -2.$$

[since, non-real complex roots occur in conjugate pairs]

Product of roots,

$$\frac{b}{1} = (1-i)(1+i) \Rightarrow b = 2$$

17 Let roots be α and α^2.

Then, $\alpha + \alpha^2 = 30$ and $\alpha^3 = p$

$\Rightarrow \quad \alpha^2 + \alpha - 30 = 0$

$\Rightarrow \quad (\alpha + 6)(\alpha - 5) = 0$

$\therefore \quad \alpha = -6, 5$

$\Rightarrow \quad p = \alpha^3 = (-6)^3 = -216$

and $\quad p = (5)^3 = 125$

$\therefore \quad p = 125$ and -216

18 Given, $\dfrac{x-m}{mx+1} = \dfrac{x+n}{nx+1}$

$\Rightarrow x^2(m-n) + 2mnx + (m+n) = 0$

Roots are $\alpha, \dfrac{1}{\alpha}$ respectively, then

$$\alpha \cdot \frac{1}{\alpha} = \frac{m+n}{m-n}$$

$\Rightarrow m - n = m + n \Rightarrow n = 0.$

19 Since, α, α^2 be the roots of the equation

$$x^2 + x + 1 = 0$$

$\therefore \quad \alpha + \alpha^2 = -1 \quad \ldots(i)$

and $\alpha^3 = 1 \quad \ldots(ii)$

Now, $\quad \alpha^{31} + \alpha^{62} = \alpha^{31}(1 + \alpha^{31})$

$\Rightarrow \quad \alpha^{31} + \alpha^{62} = \alpha^{30} \cdot \alpha(1 + \alpha^{30} \cdot \alpha)$

$\Rightarrow \quad \alpha^{31} + \alpha^{62} = (\alpha^3)^{10} \cdot \alpha\{1 + (\alpha^3)^{10} \cdot \alpha\}$

$\Rightarrow \quad \alpha^{31} + \alpha^{62} = \alpha(1+\alpha)$ [from Eq. (ii)]

$\Rightarrow \quad \alpha^{31} + \alpha^{62} = -1$ [from Eq. (i)]

Again, $\quad \alpha^{31} \cdot \alpha^{62} = \alpha^{93}$

$\Rightarrow \quad \alpha^{31} \cdot \alpha^{62} = [\alpha^3]^{31} = 1$

$\therefore$ Required equation is

$$x^2 - (\alpha^{31} + \alpha^{62})x + \alpha^{31} \cdot \alpha^{62} = 0$$

$\Rightarrow \quad x^2 + x + 1 = 0$

20 Since, α and β are the roots of $x^2 - ax + a + b = 0$, then

$$\alpha + \beta = a$$

and $\quad \alpha\beta = a + b$

$\Rightarrow \quad \alpha^2 + \alpha\beta = a\alpha$

$\Rightarrow \quad \alpha^2 - a\alpha = -(a+b)$

and $\quad \alpha\beta + \beta^2 = a\beta$

$\Rightarrow \quad \beta^2 - a\beta = -(a+b)$

$\therefore \quad \dfrac{1}{\alpha^2 - a\alpha} + \dfrac{1}{\beta^2 - a\beta} + \dfrac{2}{a+b}$

$$= \frac{1}{-(a+b)} + \frac{1}{-(a+b)} + \frac{2}{(a+b)} = 0$$

21 Let α and β be the roots of equation

$$x^2 - (a-2)x - a - 1 = 0$$

Then, $\quad \alpha + \beta = a - 2$ and $\alpha\beta = -a - 1$

Now, $\alpha^2 + \beta^2 = (\alpha+\beta)^2 - 2\alpha\beta$

$\Rightarrow \quad \alpha^2 + \beta^2 = (a-2)^2 + 2(a+1)$

$\Rightarrow \quad \alpha^2 + \beta^2 = a^2 - 2a + 6$

$\Rightarrow \quad \alpha^2 + \beta^2 = (a-1)^2 + 5$

The value of $\alpha^2 + \beta^2$ will be least, if

$$a - 1 = 0$$

$\Rightarrow \quad a = 1$

22 Since, α and β are the roots of the equation

$$2x^2 + 2(a+b)x + a^2 + b^2 = 0$$

$\therefore \quad (\alpha+\beta)^2 = (a+b)^2$ and $\alpha\beta = \dfrac{a^2+b^2}{2}$

Now, $(\alpha-\beta)^2 = (\alpha+\beta)^2 - 4\alpha\beta$

$$= (a+b)^2 - 4\left(\frac{a^2+b^2}{2}\right)$$
$$= -(a-b)^2$$

Now, the required equation whose roots are $(\alpha+\beta)^2$ and $(\alpha-\beta)^2$ is

$$x^2 - \{(\alpha+\beta)^2 + (\alpha-\beta)^2\}x + (\alpha+\beta)^2(\alpha-\beta)^2 = 0$$

$\Rightarrow \quad x^2 - \{(a+b)^2 - (a-b)^2\}x - (a+b)^2(a-b)^2 = 0$

$\Rightarrow \quad x^2 - 4abx - (a^2 - b^2)^2 = 0$

23 The given equation is $x^2 - 2x\cos\phi + 1 = 0$

$\therefore \quad x = \dfrac{2\cos\phi \pm \sqrt{4\cos^2\phi - 4}}{2}$

$$= \cos\phi \pm i\sin\phi$$

Let $\alpha = \cos\phi + i\sin\phi$, then

$$\beta = \cos\phi - i\sin\phi$$

$\therefore \alpha^n + \beta^n = (\cos\phi + i\sin\phi)^n + (\cos\phi - i\sin\phi)^n$

$$= 2\cos n\phi$$

and $\alpha^n\beta^n = (\cos n\phi + i\sin n\phi) \cdot (\cos n\phi - i\sin n\phi)$

$$= \cos^2 n\phi + \sin^2 n\phi = 1$$

$\therefore$ Required equation is

$$x^2 - 2x\cos n\phi + 1 = 0$$

24 Given equation is

$$(5+\sqrt{2})x^2 - (4+\sqrt{5})x + 8 + 2\sqrt{5} = 0$$

Let x_1 and x_2 are the roots of the equation, then

$x_1 + x_2 = \dfrac{4+\sqrt{5}}{5+\sqrt{2}} \quad \ldots(i)$

and $\quad x_1x_2 = \dfrac{8+2\sqrt{5}}{5+\sqrt{2}} = \dfrac{2(4+\sqrt{5})}{5+\sqrt{2}}$

$= 2(x_1 + x_2) \quad \ldots(ii)$

Clearly, harmonic mean

$$= \frac{2x_1x_2}{x_1+x_2} = \frac{4(x_1+x_2)}{(x_1+x_2)} = 4$$

[from Eq. (ii)]

25 Let α, β and α', β' be the roots of the given equations, respectively.

$\therefore \quad \alpha + \beta = -\dfrac{\mu}{\lambda}, \alpha\beta = \dfrac{\nu}{\lambda} \quad \ldots(i)$

and $\quad \alpha' = \omega$ and $\beta' = \omega^2$

$\because \quad \dfrac{\alpha}{\beta} = \dfrac{\alpha'}{\beta'}$ [given]

$\Rightarrow \quad \dfrac{\alpha}{\beta} = \dfrac{\omega}{\omega^2} \Rightarrow \beta = \alpha\omega$

From Eq. (i),

$\alpha + \alpha\omega = -\frac{\mu}{\lambda}, \alpha^2\omega = \frac{\nu}{\lambda}$

$\Rightarrow \quad -\alpha\omega^2 = -\frac{\mu}{\lambda}, \alpha^2\omega = \frac{\nu}{\lambda}$

$[\because -\omega^2 = 1 + \omega]$

$\Rightarrow \quad \frac{\mu^2}{\lambda^2} = \frac{\nu}{\lambda} \Rightarrow \mu^2 = \lambda\nu$

26 Simplified form of given equation is

$(2x + p + q)r = (x + p)(x + q)$

$\Rightarrow x^2 + (p + q - 2r)x - (p + q)r + pq = 0$

Since, sum of roots = 0

$\Rightarrow \quad -(p + q - 2r) = 0$

$\Rightarrow \quad r = \frac{p+q}{2}$

and product of roots

$= -(p + q)r + pq$

$= -\frac{(p+q)^2}{2} + pq$

$= -\frac{1}{2}(p^2 + q^2)$

27 Clearly, sum of roots,

$\frac{k+1}{k} + \frac{k+2}{k+1} = -\frac{b}{a}$ ... (i)

and product of roots,

$\frac{k+1}{k} \times \frac{k+2}{k+1} = \frac{c}{a}$

$\Rightarrow \quad \frac{k+2}{k} = \frac{c}{a}$

$\Rightarrow \quad \frac{2}{k} = \frac{c}{a} - 1 = \frac{c-a}{a} \Rightarrow k = \frac{2a}{c-a}$

On putting the value of k in Eq. (i), we get

$\frac{c+a}{2a} + \frac{2c}{c+a} = -\frac{b}{a}$

$\Rightarrow \quad (c + a)^2 + 4ac = -2b(a + c)$

$\Rightarrow \quad (a + c)^2 + 2b(a + c) = -4ac$

$\Rightarrow (a + c)^2 + 2b(a + c) + b^2 = b^2 - 4ac$

$\Rightarrow \quad (a + b + c)^2 = b^2 - 4ac$

28 Since, α and β are the roots of the equation $ax^2 + bx + c = 0$

$\therefore \quad \alpha + \beta = -\frac{b}{a}$ and $\alpha\beta = \frac{c}{a}$

Now, sum of roots $= |\alpha| + |\beta|$

$= -\alpha - \beta \quad (\because \beta < \alpha < 0)$

$= -\left(-\frac{b}{a}\right) = \left|\frac{b}{a}\right| \quad (\because |\alpha| + |\beta| > 0)$

and product of roots $= |\alpha||\beta| = \left|\frac{c}{a}\right|$

Hence, required eqution is

$x^2 - \left|\frac{b}{a}\right|x + \left|\frac{c}{a}\right| = 0$

$\Rightarrow \quad |a|x^2 - |b|x + |c| = 0$

29 Since, α and β are the roots of the equation $x^2 + px + q = 0$ therefore $\alpha + \beta = -p$ and $\alpha\beta = q$

Now, $(\omega\alpha + \omega^2\beta)(\omega^2\alpha + \omega\beta)$

$= \alpha^2 + \beta^2 + (\omega^4 + \omega^2)\alpha\beta \quad (\because \omega^3 = 1)$

$= \alpha^2 + \beta^2 - \alpha\beta \quad (\because \omega + \omega^2 = -1)$

$= (\alpha + \beta)^2 - 3\alpha\beta = p^2 - 3q$

Also, $\frac{\alpha^2}{\beta} + \frac{\beta^2}{\alpha} = \frac{\alpha^3 + \beta^3}{\alpha\beta}$

$= \frac{(\alpha + \beta)^3 - 3\alpha\beta(\alpha + \beta)}{\alpha\beta} = \frac{p(3q - p^2)}{q}$

$\therefore$ The given expression

$= \frac{(p^2 - 3q)}{\frac{p(3q - p^2)}{q}} = -\frac{q}{p}$

30 Clearly, $\alpha + \beta = -p$ and $\alpha\beta = \frac{3p}{4}$

Also, $\quad (\alpha - \beta)^2 = 10$

$\Rightarrow \quad (\alpha + \beta)^2 - 4\alpha\beta = 10$

$\Rightarrow \quad p^2 - 3p = 10$

$\Rightarrow \quad (p + 2)(p - 5) = 0$

$\therefore \quad p = -2, 5$

31 Let the quadratic equation be $ax^2 + bx + c = 0$ and its roots are α and β.

Sachin made a mistake in writing down constant term.

$\therefore$ Sum of roots is correct i.e. $\alpha + \beta = 7$

Rahul made mistake in writing down coefficient of x.

$\therefore$ Product of roots is correct.

i.e. $\alpha \cdot \beta = 6$

$\Rightarrow$ Correct quadratic equation is

$x^2 - (\alpha + \beta)x + \alpha\beta = 0.$

$\Rightarrow x^2 - 7x + 6 = 0$ having roots 1 and 6.

32 Since, $\alpha + \beta = -1, \alpha\beta = 2,$

$\gamma + \delta = -3,$ and $\gamma\delta = 4$

$\therefore (\alpha + \gamma)(\alpha + \delta)(\beta + \gamma)(\beta + \delta)$

$= (\alpha^2 - 3\alpha + 4)(\beta^2 - 3\beta + 4)$

$= 4 - 3\alpha\beta^2 + 4\beta^2 - 3\alpha^2\beta + 9\alpha\beta - 12\beta + 4\alpha^2 - 12\alpha + 16$

$= 4 - 3(2)\beta + 4\beta^2 + 4\alpha^2 - 3(2)\alpha + 9(2) - 12(\beta + \alpha) + 16$

$= 4 - 6\beta + 4(\alpha^2 + \beta^2) - 6\alpha + 18 + 12 + 16$

$= 50 + 6 + 4[(\alpha + \beta)^2 - 2\alpha\beta]$

$= 56 - 12 = 44$

33 Given, $R = \frac{\pi}{2} \Rightarrow P + Q = \frac{\pi}{2}$

$\Rightarrow \frac{P}{2} + \frac{Q}{2} = \frac{\pi}{4}$

$\Rightarrow \quad 1 = \tan\frac{\pi}{4} = \frac{\tan\frac{P}{2} + \tan\frac{Q}{2}}{1 - \tan\frac{P}{2}\tan\frac{Q}{2}}$

$\Rightarrow \quad 1 = \frac{-\frac{b}{a}}{1 - \frac{c}{a}} = \frac{b}{c-a} \Rightarrow a + b = c$

34 Given, $(6k + 2)x^2 + rx + 3k - 1 = 0$

and $(12k + 4)x^2 + px + 6k - 2 = 0$

For both common roots,

$\frac{6k+2}{12k+4} = \frac{r}{p} = \frac{3k-1}{6k-2}$

$\Rightarrow \quad \frac{r}{p} = \frac{1}{2} \Rightarrow 2r - p = 0$

35 Given equations are

$x^2 + 2x + 3 = 0$ ... (i)

and $\quad ax^2 + bx + c = 0$ ... (ii)

Since, Eq. (i) has imaginary roots.

So, Eq. (ii) will also have both roots same as Eq. (i)

Thus, $\frac{a}{1} = \frac{b}{2} = \frac{c}{3}$

Hence, $a:b:c$ is 1:2:3.

36 α, β be the roots of $ax^2 + bx + c = 0$

$\therefore \quad \alpha + \beta = -b/a, \alpha\beta = c/a$

Now roots are $\alpha - 1, \beta - 1$

Their sum, $\alpha + \beta - 2 = (-b/a) - 2$

$= -8/2 = -4$

Their product,

$(\alpha - 1)(\beta - 1) = \alpha\beta - (\alpha + \beta) + 1$

$= c/a + b/a + 1 = 1$

$\therefore b/a = 2$ i.e. $b = 2a$

also $c + b = 0 \Rightarrow b = -c.$

37 $\min f(x) = -\frac{D}{4a} = -\frac{4b^2 - 8c^2}{4}$

$= -(b^2 - 2c^2)$

(upward parabola)

$\max g(x) = -\frac{D}{4a} = \frac{4c^2 + 4b^2}{4}$

$= b^2 + c^2$

(downward parabola)

Now, $2c^2 - b^2 > b^2 + c^2$

$\Rightarrow \quad c^2 > 2b^2 \Rightarrow |c| > \sqrt{2}\,|b|$

38 We have,

$(x - a_1)^2 + (x - a_2)^2 + \ldots + (x - a_n)^2$

$= n x^2 - 2x(a_1 + a_2 + \ldots + a_n) + (a_1^2 + a_2^2 + \ldots + a_n^2)$

So, it attains its minimum value at

$x = \frac{2(a_1 + a_2 + \ldots + a_n)}{2n}$

$= \frac{a_1 + a_2 + \ldots + a_n}{n} \quad \left[\text{using} : x = \frac{-b}{2a}\right]$

39 Given $bx^2 + cx + a = 0$ has imaginary roots.

$\Rightarrow c^2 - 4ab < 0 \Rightarrow c^2 < 4ab$ ...(i)

Let $f(x) = 3b^2x^2 + 6bcx + 2c^2$

Here, $3b^2 > 0$

So, the given expression has a minimum value.

$\therefore$ Minimum value $= \frac{-D}{4a}$

$= \frac{4ac - b^2}{4a} = \frac{4(3b^2)(2c^2) - 36b^2c^2}{4(3b^2)}$

$= -\dfrac{12b^2c^2}{12b^2} = -c^2 > -4ab$

[from Eq. (i)]

40 According to given condition,

$4a^2 - 4(10 - 3a) < 0$

$\Rightarrow a^2 + 3a - 10 < 0$

$\Rightarrow (a+5)(a-2) < 0$

$\Rightarrow -5 < a < 2.$

41 We have, $ax - 1 + \dfrac{1}{x} \geq 0$

$\Rightarrow \dfrac{ax^2 - x + 1}{x} \geq 0$

$\Rightarrow ax^2 - x + 1 \leq 0$ as $x > 0$

It will hold if $a > 0$ and $D \leq 0$

$a > 0$ and $1 - 4a \leq 0 \Rightarrow a \geq \dfrac{1}{4}$

Therefore, the minimum value of a is $\dfrac{1}{4}$.

42 Let $f(x) = -3 + x - x^2$.

Then, $f(x) < 0$ for all x, because coeff. of $x^2 < 0$ and disc. < 0.

Thus, LHS of the given equation is always positive whereas the RHS is always less than zero. Hence, the given equation has no solution.

43 Let $f(x) = ax^2 + bx + c$.

Then, k lies between α and β, if $a\,f(k) < 0$

$\Rightarrow a(ak^2 + bk + c) < 0$

$\Rightarrow a^2k^2 + abk + ac < 0.$

44 We have, $|2x-3| < |x+5|$

$\Rightarrow |2x-3| - |x+5| < 0$

$$\Rightarrow \begin{cases} 3-2x+x+5 < 0, & x \leq -5 \\ 3-2x-x-5 < 0, & -5 < x \leq \dfrac{3}{2} \\ 2x-3-x-5 < 0, & x > \dfrac{3}{2} \end{cases}$$

$$\Rightarrow \begin{cases} x > 8, & x \leq -5 \\ x > -\dfrac{2}{3}, & -5 < x \leq \dfrac{3}{2} \\ x < 8, & x > \dfrac{3}{2} \end{cases}$$

$\Rightarrow x \in \left(-\dfrac{2}{3}, \dfrac{3}{2}\right] \cup \left(\dfrac{3}{2}, 8\right)$

$\Rightarrow x \in \left(-\dfrac{2}{3}, 8\right)$

45 $\dfrac{x-5}{x^2+5x-14} > 0$

$\Rightarrow \dfrac{(x+7)(x-2)(x-5)}{(x-2)^2(x+7)^2} > 0$

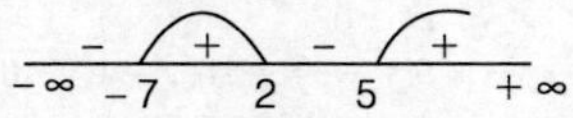

$\Rightarrow x \in (-7, 2) \cup (5, \infty)$

So, the least integral value α of x is -6, which satisfy the equation

$\alpha^2 + 5\alpha - 6 = 0$

46 Given, $\dfrac{|x-2|-1}{|x-2|-2} \leq 0$

Let $|x-2| = k$

Then, given equation,

$\dfrac{k-1}{k-2} \leq 0 \Rightarrow \dfrac{(k-1)(k-2)}{(k-2)^2} \leq 0$

$\Rightarrow (k-1)(k-2) \leq 0 \Rightarrow 1 \leq k \leq 2$

$\Rightarrow 1 \leq |x-2| \leq 2$

***Case* I** When $1 \leq |x-2|$

$\Rightarrow |x-2| \geq 1$

$\Rightarrow x-2 \geq 1$ or $x-2 \leq -1$

$\Rightarrow x \geq 3$ and $x \leq 1$ ...(i)

***Case* II** When $|x-2| \leq 2$

$\Rightarrow -2 \leq x-2 \leq 2$

$\Rightarrow -2+2 \leq x \leq 2+2$

$\Rightarrow 0 \leq x \leq 4$ ...(ii)

From (i) and (ii), $x \in [0, 1] \cup [3, 4]$

47 $\dfrac{x+2}{x^2+1} - \dfrac{1}{2} > 0$

$\Rightarrow \dfrac{-x^2 - 1 + 2x + 4}{2(x^2+1)} > 0$

$\Rightarrow \dfrac{3 + 2x - x^2}{2(x^2+1)} > 0$

Since, denominator is positive

$\therefore 3 + 2x - x^2 > 0$

$\Rightarrow -1 < x < 3$

$\Rightarrow x = 0, 1, 2$

48 Let $a_1, a_2, \ldots, a_n$ be n positive integers such that $a_1 a_2 \ldots a_n = 1$.

Using AM $\geq$ GM, we have

$\Rightarrow \dfrac{a_1 + a_2 + \ldots + a_n}{n} \geq (a_1 a_2 \ldots a_n)^{1/n}$

$\Rightarrow a_1 + a_2 + \ldots a_n \geq n.$

49 Since, $AM \geq GM$

$\therefore \dfrac{bcx + cay + abz}{3} \geq (a^2b^2c^2 \cdot xyz)^{1/3}$

$\Rightarrow bcx + cay + abz \geq 3abc$

$[\because xyz = abc]$

50 We know that, $AM > GM$

$\therefore \dfrac{a+b+c}{3} > (abc)^{1/3}$ ... (i)

and $\dfrac{\frac{1}{a} + \frac{1}{b} + \frac{1}{c}}{3} > \left(\dfrac{1}{abc}\right)^{1/3}$

$\Rightarrow \left(\dfrac{1}{a} + \dfrac{1}{b} + \dfrac{1}{c}\right) > \dfrac{3}{(abc)^{1/3}}$ ... (ii)

From (i) and (ii), we get

$\left(\dfrac{a+b+c}{3}\right)\left(\dfrac{1}{a} + \dfrac{1}{b} + \dfrac{1}{c}\right) > \dfrac{3}{(abc)^{1/3}} \cdot (abc)^{1/3}$

$\Rightarrow (a+b+c)\left(\dfrac{1}{a} + \dfrac{1}{b} + \dfrac{1}{c}\right) > 9$

51 Using GM $\leq$ AM, we have

$\{(1-a)(1-b)(1-c)^{1/3}\} \leq \dfrac{1-a+1-b+1-c}{3}$

$\Rightarrow (1-a)(1-b)(1-c) \leq \dfrac{8}{27}$

Hence, the greatest value is $\dfrac{8}{27}$.

52 Using AM $\geq$ GM, we have

$\dfrac{(a+b)+(c+d)}{2} \geq \{(a+b)(c+d)\}^{1/2}$

$\Rightarrow \dfrac{2}{2} \geq M^{1/2} \Rightarrow M \leq 1$

As $a, b, c, d > 0$.

Therefore, $M = (a+b)(c+d) > 0$.

Hence, $0 \leq M \leq 1$.

53 $\log_2(x^2 - 3x + 18) < 4$

$\Rightarrow x^2 - 3x + 18 < 16$

$(\because \log_a b < c \Leftrightarrow b < a^c$, if $a > 1)$

$\Rightarrow x^2 - 3x + 2 < 0$

$\Rightarrow (x-1)(x-2) < 0 \Rightarrow x \in (1, 2)$

54. Clearly, $x - 1 > 0 \Rightarrow x > 1$

and $\log_{0.3}(x-1) > \log_{(0.3)^2}(x-1)$

$\Rightarrow \log_{0.3}(x-1) > \dfrac{1}{2}\log_{0.3}(x-1)$

$\Rightarrow \log_{0.3}(x-1) > 0 \Rightarrow x < 2$

Hence, $x \in (1, 2)$

55 By definition of $\log x$, $x > 0$ and

$\left(\dfrac{x+5}{1-3x}\right) > 0$

$\Rightarrow \dfrac{(x+5)(1-3x)}{(1-3x)^2} > 0$

$\Rightarrow (x+5)(1-3x) > 0$

$\Rightarrow (x+5)(3x-1) < 0$

$\Rightarrow -5 < x < 1/3$

As $x > 0,\ 0 < x < 1/3$

$\therefore \dfrac{x+5}{1-3x} < 1 \Rightarrow x < -1$

This does not satisfy $0 < x < 1/3$.

Hence, there is no solution.

SESSION 2

1 We have

$2|\sqrt{x} - 3| + \sqrt{x}(\sqrt{x} - 6) + 6 = 0$

Let $\sqrt{x} - 3 = y$

$\Rightarrow \sqrt{x} = y + 3$

$\therefore 2|y| + (y+3)(y-3) + 6 = 0$

$\Rightarrow 2|y| + y^2 - 3 = 0$

$\Rightarrow |y|^2 + 2|y| - 3 = 0$

$\Rightarrow (|y| + 3)(|y| - 1) = 0$

$\Rightarrow |y| \neq -3 \Rightarrow |y| = 1$

$\Rightarrow \quad y = \pm 1$

$\Rightarrow \quad \sqrt{x} - 3 = \pm 1$

$\Rightarrow \quad \sqrt{x} = 4, 2$

$\Rightarrow \quad x = 16, 4$

2 We have, $2^{x+2} \cdot 3^{3x/(x-1)} = 9$

Taking log on both sides, we get

$$(x+2)\log 2 + \frac{3x}{(x-1)}\log 3 = 2\log 3$$

$$\Rightarrow (x^2 + x - 2)\log 2 + 3x\log 3 = 2(x-1)\log 3$$

$$\Rightarrow x^2\log 2 + (\log 2 + \log 3)x - 2\log 2 + 2\log 3 = 0$$

$$\therefore x = \frac{-(\log 2 + \log 3) \pm \sqrt{\{(\log 2 + \log 3)^2 - 4\log 2(-2\log 2 + 2\log 3)\}}}{2\log 2}$$

$$= \frac{-(\log 2 + \log 3) \pm \sqrt{\{(3\log 2)^2 - 6\log 2\log 3 + (\log 3)^2\}}}{2\log 2}$$

$$= \frac{-(\log 2 + \log 3) \pm (3\log 2 - \log 3)}{2\log 2}$$

$$\therefore x = -2, 1 - \frac{\log 3}{\log 2}$$

3 Given, α and β are the roots of the equation $x^2 - 6x - 2 = 0$

$\because a_n = \alpha^n - \beta^n, n \geq 1 \Rightarrow a_{10} = \alpha^{10} - \beta^{10}$

$a_8 = \alpha^8 - \beta^8$ and $a_9 = \alpha^9 - \beta^9$

Now, consider

$\dfrac{a_{10} - 2a_8}{2a_9}$	[$\because \alpha$ and β are the roots of
$= \dfrac{\alpha^{10} - \beta^{10} - 2(\alpha^8 - \beta^8)}{2(\alpha^9 - \beta^9)}$	$x^2 - 6x - 2 = 0$ or $x^2 = 6x + 2$
$= \dfrac{\alpha^8(\alpha^2 - 2) - \beta^8(\beta^2 - 2)}{2(\alpha^9 - \beta^9)}$	$\Rightarrow \alpha^2 = 6\alpha + 2$ $\Rightarrow \alpha^2 - 2 = 6\alpha$
$= \dfrac{\alpha^8 \cdot 6\alpha - \beta^8 \cdot 6\beta}{2(\alpha^9 - \beta^9)}$	and $\beta^2 = 6\beta + 2$ $\Rightarrow \beta^2 - 2 = 6\beta$]
$= \dfrac{6\alpha^9 - 6\beta^9}{2(\alpha^9 - \beta^9)} = \dfrac{6}{2} = 3$	

4 We have, $x^2 - 5x + 1 = 0$

$$x + \frac{1}{x} = 5 \Rightarrow x^2 + \frac{1}{x^2} = 5^2 - 2 = 23$$

and $$x^3 + \frac{1}{x^3} = 5^3 - 3 \times 5 = 110$$

Now, $$\left(x^2 + \frac{1}{x^2}\right)\left(x^3 + \frac{1}{x^3}\right) = 23 \times 110 = 2530$$

$$\Rightarrow x^5 + \frac{1}{x^5} = 2530 - \left(x + \frac{1}{x}\right) = 2525$$

5 Given equation can be rewritten as

$3x^2 - (a + c + 2b + 2d)x + ac + 2bd = 0$

Now, discriminant, D

$= (a + c + 2b + 2d)^2 - 4 \cdot 3(ac + 2bd)$

$= \{(a + 2d) + (c + 2b)\}^2 - 12(ac + 2bd)$

$= \{(a + 2d) - (c + 2b)\}^2 + 4(a + 2d)(c + 2b) - 12(ac + 2bd)$

$= \{(a + 2d) - (c + 2b)\}^2 - 8ac + 8ab + 8dc - 8bd$

$= \{(a + 2d) - (c + 2b)\}^2 + 8(c - b)(d - a)$

Which is +ve, since $a < b < c < d$.

Hence, roots are real and distinct.

6 Since, α and β are roots of

$px^2 + qx + r = 0, p \neq 0$

$$\therefore \quad \alpha + \beta = \frac{-q}{p}, \alpha\beta = \frac{r}{p}$$

Since, p, q and r are in AP.

$\therefore \quad 2q = p + r$

Also, $$\frac{1}{\alpha} + \frac{1}{\beta} = 4 \Rightarrow \frac{\alpha + \beta}{\alpha\beta} = 4$$

$$\Rightarrow \quad \alpha + \beta = 4\alpha\beta \Rightarrow \frac{-q}{p} = \frac{4r}{p}$$

$\because \quad 2q = p + r$

$\Rightarrow \quad 2(-4r) = p + r \Rightarrow p = -9r$

$$\because \quad \alpha + \beta = \frac{-q}{p} = \frac{4r}{p} = \frac{4r}{-9r} = -\frac{4}{9}$$

and $$\alpha\beta = \frac{r}{p} = \frac{r}{-9r} = \frac{1}{-9}$$

Now, consider

$$(\alpha - \beta)^2 = (\alpha + \beta)^2 - 4\alpha\beta = \frac{16}{81} + \frac{4}{9} = \frac{16 + 36}{81}$$

$$\Rightarrow (\alpha - \beta)^2 = \frac{52}{81}$$

$$\Rightarrow |\alpha - \beta| = \frac{2}{9}\sqrt{13}$$

7 $\alpha + \beta = -b/a$ and $\alpha\beta = c/a$

Now, $(1 + \alpha + \alpha^2)(1 + \beta + \beta^2)$

$= 1 + (\alpha + \beta) + (\alpha^2 + \beta^2) + \alpha\beta + \alpha\beta(\alpha + \beta) + (\alpha\beta)^2$

$= 1 + (\alpha + \beta) + (\alpha + \beta)^2 - \alpha\beta + \alpha\beta(\alpha + \beta) + (\alpha\beta)^2$

$$= 1 - \frac{b}{a} + \frac{b^2}{a^2} - \frac{c}{a} + \left(\frac{c}{a}\right)\left(-\frac{b}{a}\right) + \frac{c^2}{a^2}$$

$$= \frac{(a^2 + b^2 + c^2 - ab - bc - ca)}{a^2}$$

$$= [(a - b)^2 + (b - c)^2 + (c - a)^2]/2a^2$$

which is positive.

8 Using AM $\geq$ GM

$$\therefore \frac{a^{-5} + a^{-4} + a^{-3} + a^{-3} + a^{-3} + 1 + a^8 + a^{10}}{8}$$

$$\geq (a^{-5} \cdot a^{-4} \cdot a^{-3} \cdot a^{-3} \cdot a^{-3} \cdot 1 \cdot a^8 \cdot a^{10})^{1/8}$$

$$\Rightarrow a^{-5} + a^{-4} + 3a^{-3} + 1 + a^8 + a^{10} \geq 8 \cdot 1$$

Hence, minimum value is 8.

9 We have,

$$\frac{\log_a a}{\log_a a + \log_a x} + \frac{2\log_a a}{\log_a x} + \frac{3\log_a a}{2\log_a a + \log_a x} = 0$$

$$\Rightarrow \quad \frac{1}{1 + t} + \frac{2}{t} + \frac{3}{2 + t} = 0$$

(let $\log_a x = t$)

$$\Rightarrow \quad \frac{2t + t^2 + 2t^2 + 6t + 4 + 3t^2 + 3t}{t(1 + t)(2 + t)} = 0$$

$\Rightarrow \quad 6t^2 + 11t + 4 = 0$

$\Rightarrow \quad 6t^2 + 8t + 3t + 4 = 0$

$\Rightarrow \quad (2t + 1)(3t + 4) = 0$

$$\Rightarrow \quad t = -\frac{1}{2}, -\frac{4}{3}$$

$$\Rightarrow \quad \log_a x = -\frac{1}{2}, -\frac{4}{3}$$

$$\therefore \quad x = a^{-1/2}, a^{-4/3}$$

10 Since, $x = 1$ is a root of first equation. If α is another root of first equation, then

$$\alpha \times 1 = \alpha = \frac{a - b}{b - c}$$

(Product of roots)

$$= \frac{2a - 2b}{2b - 2c} = \frac{2a - (a + c)}{a + c - 2c} = 1$$

So, the roots of first equation are 1 and 1.

Since, the equations have a common root, 1 is also a root of second equation.

$\Rightarrow \quad 2(c + a) + b + c = 0$

$\Rightarrow \quad 2(2b) + b + c = 0$

[since, a, b and c are in AP]

$\Rightarrow \quad c = -5b$

Also, $\quad a + c = 2b \Rightarrow a = 2b - c = 2b + 5b = 7b$

$\therefore \ a^2 = 49b^2, c^2 = 25b^2$

Hence, a^2, c^2 and b^2 are in AP.

11 On adding first and second equations, we get

$$2x^2 + (a + b)x + 27 = 0$$

On subtracting above equation from given third equation, we get

$$x^2 - 9 = 0 \Rightarrow x = 3, -3$$

Thus, common positive root is 3.

Now, $(3)^2 + 3a + 12 = 0 \Rightarrow a = -7$

and $9 + 3b + 15 = 0 \Rightarrow b = -8$

Hence, the order pair (a, b) is $(-7, -8)$.

12 Let $$y = \frac{x^2 - 3x + 4}{x^2 + 3x + 4}$$

$\Rightarrow \ (y - 1)x^2 + 3(y + 1)x + 4(y - 1) = 0$

$\because x$ is real.

$\therefore \quad D \geq 0$

$\Rightarrow \quad 9(y + 1)^2 - 16(y - 1)^2 \geq 0$

$\Rightarrow \quad -7y^2 + 50y - 7 \geq 0$

$\Rightarrow \quad 7y^2 - 50y + 7 \leq 0$

$\Rightarrow \quad (y - 7)(7y - 1) \leq 0$

$\Rightarrow \quad y \le 7$ and $y \ge \dfrac{1}{7}$

$\Rightarrow \quad \dfrac{1}{7} \le y \le 7$

Hence, maximum value is 7 and minimum value is $\dfrac{1}{7}$.

13 Here, $a \in R$ and equation is

$-3\{x-[x]\}^2 + 2\{x-[x]\} + a^2 = 0$

Let $t = x - [x]$, then

$-3t^2 + 2t + a^2 = 0$

$\Rightarrow \quad t = \dfrac{1 \pm \sqrt{1+3a^2}}{3}$

$\because \quad t = x - [x] = \{x\}$ [fractional part]

$\therefore \quad 0 \le t \le 1$

$\Rightarrow \quad 0 \le \dfrac{1 \pm \sqrt{1+3a^2}}{3} < 1$

$[\because 0 \le \{x\} < 1]$

But $1 - \sqrt{1+3a^2} < 0$ therefore

$0 \le \dfrac{1+\sqrt{1+3a^2}}{3} < 1$

$\Rightarrow \quad \sqrt{1+3a^2} < 2$

$\Rightarrow \quad 1 + 3a^2 < 4 \Rightarrow a^2 - 1 < 0$

$\Rightarrow \quad (a+1)(a-1) < 0$

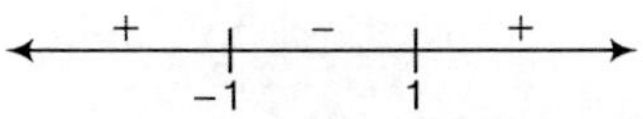

$\Rightarrow \quad a \in (-1, 1)$

For no integral solution we consider the interval $(-1, 0) \cup (0, 1)$.

14 Given, $a = \cos \dfrac{2\pi}{7} + i \sin \dfrac{2\pi}{7}$

$\therefore \quad a^7 = \cos 2\pi + i \sin 2\pi = 1$

$[\because e^{i\theta} = \cos\theta + i\sin\theta]$

Also, $\alpha = a + a^2 + a^4$,

$\beta = a^3 + a^5 + a^6$

Then, the sum of roots,

$S = \alpha + \beta = a + a^2 + a^3 + a^4 + a^5 + a^6$

$\Rightarrow \quad S = \dfrac{a(1-a^6)}{1-a} = \dfrac{a-a^7}{1-a}$

$= \dfrac{a-1}{1-a} = -1 \quad [\because a^7 = 1]$

and product of the roots,

$P = \alpha\beta = (a + a^2 + a^4)(a^3 + a^5 + a^6)$

$= a^4 + a^5 + 1 + a^6 + 1 + a^2 + 1 + a + a^3 \quad [\because a^7 = 1]$

$= 3 + (a + a^2 + a^3 + a^4 + a^5 + a^6)$

$= 3 - 1 = 2$

Hence, the required quadratic equation is $x^2 + x + 2 = 0$

15 Since, α and β are the roots of $375x^2 - 25x - 2 = 0$.

$\therefore \quad \alpha + \beta = \dfrac{25}{375} = \dfrac{1}{15}$

and $\quad \alpha\beta = -\dfrac{2}{375}$

Now, consider

$$\lim_{n\to\infty} \sum_{r=1}^{n} S_r = \lim_{n\to\infty} \sum_{r=1}^{n} (\alpha^r + \beta^r)$$

$$= (\alpha + \alpha^2 + \alpha^3 + \dots \infty) + (\beta + \beta^2 + \beta^3 + \dots \infty)$$

$$= \frac{\alpha}{1-\alpha} + \frac{\beta}{1-\beta} = \frac{\alpha - \alpha\beta + \beta - \alpha\beta}{(1-\alpha)(1-\beta)}$$

$$= \frac{\alpha + \beta - 2\alpha\beta}{1 - (\alpha+\beta) + \alpha\beta}$$

$$= \frac{\frac{1}{15} + \frac{4}{375}}{1 - \frac{1}{15} - \frac{2}{375}}$$

$$= \frac{25 + 4}{375 - 25 - 2}$$

$$= \frac{29}{348} = \frac{1}{12}$$

16 Here, $[\tan^2 x]$ = integer

and a = integer

So, $\tan x$ is also an integer.

Then, $\tan^2 x - \tan x - a = 0$

$\Rightarrow \quad a = \tan x(\tan x - 1) = I(I - 1)$

= Product of two consecutive integers

$\therefore \quad a = 2, 6, 12, 20, 30, 42, 56, 72, 90$

Hence, set S has 9 elements.

17 Given, $(x^2 - 5x + 5)^{x^2+4x-60} = 1$

Clearly, this is possible when

I. $x^2 + 4x - 60 = 0$ and $x^2 - 5x + 5 \ne 0$

or

II. $x^2 - 5x + 5 = 1$

or

III. $x^2 - 5x + 5 = -1$ and $x^2 + 4x - 60$ = Even integer

Case I When $x^2 + 4x - 60 = 0$

$\Rightarrow \quad x^2 + 10x - 6x - 60 = 0$

$\Rightarrow \quad x(x+10) - 6(x+10) = 0$

$\Rightarrow \quad (x+10)(x-6) = 0$

$\Rightarrow \quad x = -10$ or $x = 6$

Note that, for these two values of x, $x^2 - 5x + 5 \ne 0$

Case II When $x^2 - 5x + 5 = 1$

$\Rightarrow \quad x^2 - 5x + 4 = 0$

$\Rightarrow \quad x^2 - 4x - x + 4 = 0$

$\Rightarrow \quad (x-4)(x-1) = 0$

$\Rightarrow \quad x = 4$ or $x = 1$

Case III When $x^2 - 5x + 5 = -1$

$\Rightarrow \quad x^2 - 5x + 6 = 0$

$\Rightarrow \quad x^2 - 2x - 3x + 6 = 0$

$\Rightarrow \quad x(x-2) - 3(x-2) = 0$

$\Rightarrow \quad (x-2)(x-3) = 0$

$\Rightarrow \quad x = 2$ or $x = 3$

Now, when $x = 2$,

$x^2 + 4x - 60 = 4 + 8 - 60 = -48$, which is an even integer.

When $x = 3$,

$x^2 + 4x - 60 = 9 + 12 - 60 = -39$, which is not an even integer.

Thus, in this case, we get $x = 2$

Hence, the sum of all real values of

$x = -10 + 6 + 4 + 1 + 2 = 3$

18 Given, $4x^2 - 16x + \dfrac{\lambda}{4} = 0$

$\therefore \quad x = \dfrac{16 \pm \sqrt{(256 - 4\lambda)}}{8}$

$= \dfrac{8 \pm \sqrt{(64-\lambda)}}{4}$

$\Rightarrow \quad \alpha, \beta = 2 \pm \dfrac{\sqrt{(64-\lambda)}}{4}$

Here, $64 - \lambda > 0$

$\therefore \lambda < 64$

Also, $1 < \alpha < 2$ and $2 < \beta < 3$

$\therefore \quad 1 < 2 - \dfrac{\sqrt{64-\lambda}}{4} < 2$

and $\quad 2 < 2 + \dfrac{\sqrt{64-\lambda}}{4} < 3$

$\Rightarrow \quad -1 < -\dfrac{\sqrt{64-\lambda}}{4} < 0$

and $\quad 0 < \dfrac{\sqrt{(64-\lambda)}}{4} < 1$

$\Rightarrow \quad 1 > \dfrac{\sqrt{(64-\lambda)}}{4} > 0$

and $\quad 0 < \dfrac{\sqrt{(64-\lambda)}}{4} < 1$

i.e. $\quad 0 < \dfrac{\sqrt{(64-\lambda)}}{4} < 1$

$\Rightarrow \quad 0 < \sqrt{(64-\lambda)} < 4$

$\Rightarrow \quad 0 < 64 - \lambda < 16 \Rightarrow \lambda > 48$

or $\quad 48 < \lambda < 64$

$\therefore \quad \lambda = \{49, 50, 51, 52, \dots, 63\}$

19 Since, roots of $ax^2 + bx + c = 0$ are α and β. Hence, roots of $cx^2 + bx + a = 0$, will be $\dfrac{1}{\alpha}$ and $\dfrac{1}{\beta}$. Now, if we replace x by $x - 1$, then roots of $c(x-1)^2 + b(x-1) + a = 0$ will be $1 + \dfrac{1}{\alpha}$ and $1 + \dfrac{1}{\beta}$. Now, again replace x by $\dfrac{1}{x}$, we will get $c(1-x)^2 + b(1-x) + ax^2 = 0$, whose roots are $\dfrac{\alpha}{1+\alpha}$ and $\dfrac{\beta}{1+\beta}$.

20 Let $f(x) = x^2 - 2kx + k^2 + k - 5$

Since, both roots are less than 5.

$\therefore \quad D \ge 0, -\dfrac{b}{2a} < 5$ and $f(5) > 0$

Here, $\quad D = 4k^2 - 4(k^2 + k - 5)$

$= -4k + 20 \ge 0$

$\Rightarrow \quad k \le 5 \quad \dots \text{(i)}$

$-\frac{b}{2a} < 5 \Rightarrow k < 5$ ... (ii)

and $f(5) > 0$

$\Rightarrow$ $25 - 10k + k^2 + k - 5 > 0$

$\Rightarrow$ $k^2 - 9k + 20 > 0$

$\Rightarrow$ $(k-5)(k-4) > 0$

$\Rightarrow$ $k < 4$ and $k > 5$ ...(iii)

From (i), (ii) and (iii), we get

$k < 4$

21 Since, both roots of equation $x^2 - 2mx + m^2 - 1 = 0$ are greater than -2 but less than 4.

$\therefore$ $D \geq 0, -2 < -\frac{b}{2a} < 4,$

$f(4) > 0$ and $f(-2) > 0$

Now, $D \geq 0$

$\Rightarrow$ $4m^2 - 4m^2 + 4 \geq 0$

$\Rightarrow$ $4 > 0 \Rightarrow m \in R$ ... (i)

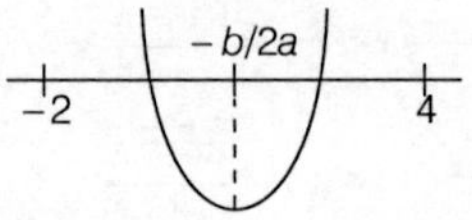

$-2 < -\frac{b}{2a} < 4$

$\Rightarrow$ $-2 < \left(\frac{2m}{2 \cdot 1}\right) < 4$

$\Rightarrow$ $-2 < m < 4$ ... (ii)

$f(4) > 0$

$\Rightarrow$ $16 - 8m + m^2 - 1 > 0$

$\Rightarrow$ $m^2 - 8m + 15 > 0$

$\Rightarrow$ $(m-3)(m-5) > 0$

$\Rightarrow$ $-\infty < m < 3$ and $5 < m < \infty$ ... (iii)

and $f(-2) > 0$

$\Rightarrow$ $4 + 4m + m^2 - 1 > 0$

$\Rightarrow$ $m^2 + 4m + 3 > 0$

$(m+3)(m+1) > 0$

$\Rightarrow$ $-\infty < m < -3$ and

$-1 < m < \infty$ ... (iv)

From (i), (ii), (iii) and (iv), we get m lie between -1 and 3.

22 Let $z = x + iy$, given $\text{Re}(z) = 1$

$\therefore$ $x = 1 \Rightarrow z = 1 + iy$

Since, the complex roots are conjugate to each other.

So, $z = 1 + iy$ and $1 - iy$ are two roots of $z^2 + \alpha z + \beta = 0$.

$\because$ Product of roots $= \beta$

$\Rightarrow$ $(1 + iy)(1 - iy) = \beta$

$\therefore$ $\beta = 1 + y^2 \geq 1 \Rightarrow \beta \in [1, \infty)$

23 Given equation is

$$e^{\sin x} - e^{-\sin x} = 4 \Rightarrow e^{\sin x} - \frac{1}{e^{\sin x}} = 4$$

Let $e^{\sin x} = t$, then $t - \frac{1}{t} = 4$

$\Rightarrow$ $t^2 - 1 - 4t = 0 \Rightarrow t^2 - 4t - 1 = 0$

$\Rightarrow$ $t = \frac{4 \pm \sqrt{16 + 4}}{2}$

$t = 2 \pm \sqrt{5} \Rightarrow e^{\sin x} = 2 \pm \sqrt{5}$

But $-1 \leq \sin x \leq 1 \Rightarrow e^{-1} \leq e^{\sin x} \leq e^1$

$\Rightarrow$ $e^{\sin x} \in \left[\frac{1}{e}, e\right]$

Also, $0 < e < 2 + \sqrt{5}$

Hence, given equation has no solution.

24 Using AM > GM, we have

$a + \frac{1}{b} > 2\sqrt{\frac{a}{b}}$, $b + \frac{1}{c} > 2\sqrt{\frac{b}{c}}$

$c + \frac{1}{d} > 2\sqrt{\frac{c}{d}}$ and $d + \frac{1}{a} > 2\sqrt{\frac{d}{a}}$

$$\left(a + \frac{1}{b}\right)\left(b + \frac{1}{c}\right)\left(c + \frac{1}{d}\right)\left(d + \frac{1}{a}\right) > 16$$

But, $\left(a + \frac{1}{b}\right)\left(b + \frac{1}{c}\right)\left(c + \frac{1}{d}\right)$

$\left(d + \frac{1}{a}\right) = 4 \times 1 \times 4 \times 1 = 16$

$\therefore a = \frac{1}{b}, b = \frac{1}{c}, c \ \frac{1}{d}$ and $d = \frac{1}{a}$

$\Rightarrow a = \frac{1}{b} = 2,\ b = \frac{1}{c} = \frac{1}{2},\ c = \frac{1}{d} = 2$

and $d = \frac{1}{a} = \frac{1}{2}$

$\Rightarrow a = 2, b = \frac{1}{2}, c = 2$ and $d = \frac{1}{2}$

$\Rightarrow a = c$ and $b = d$

25 We have, $f(x) = \dfrac{1}{e^x + \dfrac{2}{e^x}}$

Using $AM \geq GM$, we get

$$\frac{e^x + \frac{2}{e^x}}{2} \geq \left(e^x \cdot \frac{2}{e^x}\right)^{1/2}, \text{ as } e^x > 0$$

$\Rightarrow$ $e^x + \frac{2}{e^x} \geq 2\sqrt{2}$

$\Rightarrow$ $0 < \dfrac{1}{e^x + \dfrac{2}{e^x}} \leq \dfrac{1}{2\sqrt{2}}$

$\therefore$ $0 < f(x) \leq \frac{1}{2\sqrt{2}}, \forall x \in R$

Statement II is true and Statement I is also true as for some 'c'.

$\Rightarrow$ $f(c) = \frac{1}{3}$ [for $c = 0$]

which lies betwen 0 and $\frac{1}{2\sqrt{2}}$.

So, Statement II is correct explanation of Statement I.

DAY FIVE

Matrices

Learning & Revision for the Day

- Matrix
- Types of Matrices
- Equality of Matrices
- Algebra of Matrices
- Transpose of a Matrix
- Some Special Matrices
- Trace of a Matrix
- Equivalent Matrices
- Invertible Matrices

Matrix

- A **matrix** is an arrangement of numbers in rows and columns.
- A matrix having m rows and n columns is called a matrix of order $m \times n$ and the number of elements in this matrix will be mn.
- A matrix of order $m \times n$ is of the form $A = \begin{bmatrix} a_{11} & a_{12} & a_{13} & \cdots & a_{1n} \\ a_{21} & a_{22} & a_{23} & \cdots & a_{2n} \\ \cdots & \cdots & \cdots & \cdots & \cdots \\ a_{m1} & a_{m2} & a_{m3} & \cdots & a_{mn} \end{bmatrix}$

Some important terms related to matrices

- The element in the ith row and jth column is denoted by a_{ij}.
- The elements $a_{11}, a_{22}, a_{33}, \ldots\ldots$ are called diagonal elements.
- The line along which the diagonal elements lie is called the principal diagonal or simply the diagonal of the matrix.

Types of Matrices

- If all elements of a matrix are zero, then it is called a **null** or **zero matrix** and it is denoted by O.
- A matrix which has only one row and any number of columns is called a **row matrix** and if it has only one column and any number of rows, then it is called a **column matrix**.
- If in a matrix, the number of rows and columns are equal, then it is called a **square matrix**. If $A = [a_{ij}]_{n \times n}$, then it is known as square matrix of order n.
- If in a matrix, the number of rows is less/greater than the number of columns, then it is called **rectangular matrix**.
- If in a square matrix, all the non-diagonal elements are zero, it is called a **diagonal matrix**.

PREP MIRROR
Your Personal Preparation Indicator

- No. of Questions in Exercises (x)—
- No. of Questions Attempted (y)—
- No. of Correct Questions (z)—
  (Without referring Explanations)

- Accuracy Level ($z/y \times 100$)—
- Prep Level ($z/x \times 100$)—

In order to expect good rank in JEE, your Accuracy Level should be above 85 & Prep Level should be above 75.

- If in a square matrix, all non-diagonal elements are zero and diagonal elements are equal, then it is called a **scalar matrix.**
- If in a square matrix, all non-diagonal elements are zero and diagonal elements are unity, then it is called an **unit** (identity) **matrix**. We denote the identity matrix of order n by I_n and when order is clear from context then we simply write it as I.
- In a square matrix, if $a_{ij} = 0$, $\forall\ i > j$, then it is called an **upper triangular matrix** and if $a_{ij} = 0, \forall i < j$, then it is called a **lower triangular matrix.**

NOTE • The diagonal elements of diagonal matrix may or may not be zero.

Equality of Matrices

Two matrices A and B are said to be equal, if they are of same order and all the corresponding elements are equal.

Algebra of Matrices

- If $A = [a_{ij}]_{m \times n}$ and $B = [b_{ij}]_{m \times n}$ be two matrices of same order, then $A + B = [a_{ij} + b_{ij}]_{m \times n}$ and $A - B = [a_{ij} - b_{ij}]_{m \times n}$, where $i = 1, 2, \ldots, m$, $j = 1, 2, \ldots, n$.
- If $A = [a_{ij}]$ be an $m \times n$ matrix and k be any scalar, then, $kA = [ka_{ij}]_{m \times n}$.
- If $A = [a_{ij}]_{m \times n}$ and $B = [b_{ij}]_{n \times p}$ be any two matrices such that number of columns of A is equal to the number of rows of B, then the product matrix $AB = [c_{ij}]$, of order $m \times p$, where $c_{ij} = \sum_{k=1}^{n} a_{ik}\, b_{kj}$.

Some Important Properties

- $A + B = B + A$ (Commutativity of addition)
- $(A + B) + C = A + (B + C)$ (Associativity of addition)
- $\alpha (A + B) = \alpha A + \alpha B$, where α is any scalar.
- $(\alpha + \beta) A = \alpha A + \beta A$, where α and β are any scalars.
- $\alpha (\beta A) = (\alpha\beta) A$, where α and β are any scalars.
- $(AB)C = A(BC)$ (Associativity of multiplication)
- $AI = A = IA$
- $A(B + C) = AB + AC$ (Distributive property)

NOTE
- $A^2 = A \cdot A$, $A^3 = A \cdot A \cdot A = A^2 \cdot A^1, \ldots$
- If the product AB is possible, then it is not necessary that the product BA is also possible. Also, it is not necessary that $AB = BA$.
- The product of two non-zero matrices can be a zero matrix.

Transpose of a Matrix

Let A be $m \times n$ matrix, then the matrix obtained by interchanging the rows and columns of A is called the transpose of A and is denoted by A' or A^C or A^T.

If A be $m \times n$ matrix, A' will be $n \times m$ matrix.

Important Results

(i) If A and B are two matrices of order $m \times n$, then $(A \pm B)' = A' \pm B'$

(ii) If k is a scalar, then $(k A)' = k A'$

(iii) $(A')' = A$

(iv) $(AB)' = B' A'$

(v) $(A^n)' = (A')^n$

Some Special Matrices

- A square matrix A is called an **idempotent matrix,** if it satisfies the relation $A^2 = A$.
- A square matrix A is called **nilpotent matrix** of order k, if it satisfies the relation $A^k = O$, for some $k \in N$.
- The least value of k is called the index of the nilpotent matrix A.
- A square matrix A is called an **involutary matrix,** if it satisfies the relation $A^2 = I$.
- A square matrix A is called an **orthogonal matrix,** if it satisfies the relation $AA' = I$ or $A' A = I$.
- A square matrix A is called **symmetric matrix,** if it satisfies the relation $A' = A$.
- A square matrix A is called **skew-symmetric matrix,** if it satisfies the relation $A' = -A$.

NOTE
- If A and B are idempotent matrices, then $A + B$ is idempotent iff $AB = -BA$.
- If $A = \begin{bmatrix} a_1 & a_2 & a_3 \\ b_1 & b_2 & b_3 \\ c_1 & c_2 & c_3 \end{bmatrix}$ is orthogonal, then
  $\Sigma a_i^2 = \Sigma b_i^2 = \Sigma c_i^2 = 1$ and $\Sigma a_i b_i = \Sigma b_i c_i = \Sigma a_i c_i = 0$
- If A and B are symmetric matrices of the same order, then
  (i) AB is symmetric if and only if $AB = BA$.
  (ii) $A \pm B$, $AB + BA$ are also symmetric matrices.
- If A and B are two skew-symmetric matrices, then
  (i) $A \pm B$, $AB - BA$ are skew-symmetric matrices.
  (ii) $AB + BA$ is a symmetric matrix.
- Every square matrix can be uniquely expressed as the sum of symmetric and skew-symmetric matrices.
  i.e. $A = \frac{1}{2}(A + A') + \frac{1}{2}(A - A')$, where $\frac{1}{2}(A + A')$ and $\frac{1}{2}(A - A')$ are symmetric and skew-symmetric respectively.

Trace of a Matrix

The sum of the diagonal elements of a square matrix A is called the trace of A and is denoted by $tr(A)$.

(i) $tr(\lambda A) = \lambda\, tr(A)$ (ii) $tr(A) = tr(A')$
(iii) $tr(AB) = tr(BA)$

Equivalent Matrices

Two matrices A and B are said to be **equivalent**, if one is obtained from the other by one or more elementary operations and we write $A \sim B$.

Following types of operations are called ***elementary operations.***

(i) Interchanging any two rows (columns).
This transformation is indicated by

$$R_i \leftrightarrow R_j \ (C_i \leftrightarrow C_j)$$

(ii) Multiplication of the elements of any row (column) by a non-zero scalar quantity, indicated as

$$R_i \to kR_i \ (C_i \to kC_i)$$

(iii) Addition of constant multiple of the elements of any row (column) to the corresponding elements of any other row (column), indicated as

$$R_i \to R_i + kR_j \ (C_i \to C_i + kC_j).$$

Invertible Matrices

- A square matrix A of order n is said to be **invertible** if there exists another square matrix B of order n such that $AB = BA = I$.
- The matrix B is called the inverse of matrix A and it is denoted by A^{-1}.

Some Important Results

- Inverse of a square matrix, if it exists, is unique.
- $AA^{-1} = I = A^{-1}A$
- If A and B are invertible, then $(AB)^{-1} = B^{-1}A^{-1}$
- $(A^{-1})^T = (A^T)^{-1}$
- If A is symmetric, then A^{-1} will also be symmetric matrix.
- Every orthogonal matrix is invertible.

DAY PRACTICE SESSION 1

FOUNDATION QUESTIONS EXERCISE

1 If $A = \begin{bmatrix} 0 & 1 \\ 1 & 1 \end{bmatrix}$ and $B = \begin{bmatrix} 0 & -1 \\ 1 & 0 \end{bmatrix}$, then which of the following is correct? → NCERT Exemplar

(a) $(A+B)\cdot(A-B) = A^2 + B^2$ (b) $(A+B)\cdot(A-B) = A^2 - B^2$
(c) $(A+B)\cdot(A-B) = I$ (d) None of these

2 If p, q, r are 3 real numbers satisfying the matrix equation, $[p\ q\ r]\begin{bmatrix} 3 & 4 & 1 \\ 3 & 2 & 3 \\ 2 & 0 & 2 \end{bmatrix} = [3\ 0\ 1]$, then $2p + q - r$ is equal to → JEE Mains 2013

(a) -3 (b) -1 (c) 4 (d) 2

3 In a upper triangular matrix $n \times n$, minimum number of zeroes is

(a) $\frac{n(n-1)}{2}$ (b) $\frac{n(n+1)}{2}$
(c) $\frac{2n(n-1)}{2}$ (d) None of these

4 Let $A = \begin{bmatrix} 1 & 2 \\ 3 & 4 \end{bmatrix}$ and $B = \begin{bmatrix} a & 0 \\ 0 & b \end{bmatrix}$; $a, b \in N$. Then,

(a) there exists more than one but finite number of B's such that $AB = BA$
(b) there exists exactly one B such that $AB = BA$
(c) there exist infinitely many B's such that $AB = BA$
(d) there cannot exist any B such that $AB = BA$

5 If $A = \begin{bmatrix} 1 & -2 & 3 \\ -4 & 2 & 5 \end{bmatrix}$ and $B = \begin{bmatrix} 2 & 3 \\ 4 & 5 \\ 2 & 1 \end{bmatrix}$, then

(a) AB, BA exist and are equal
(b) AB, BA exist and are not equal
(c) AB exists and BA does not exist
(d) AB does not exist and BA exists

6 If $\omega \neq 1$ is the complex cube root of unity and matrix $H = \begin{bmatrix} \omega & 0 \\ 0 & \omega \end{bmatrix}$, then H^{70} is equal to

(a) H (b) 0 (c) $-H$ (d) H^2

7 If A and B are 3×3 matrices such that $AB = A$ and $BA = B$, then

(a) $A^2 = A$ and $B^2 \neq B$ (b) $A^2 \neq A$ and $B^2 = B$
(c) $A^2 = A$ and $B^2 = B$ (d) $A^2 \neq A$ and $B^2 \neq B$

8 For each real number x such that $-1 < x < 1$, let

$$A(x) = \begin{bmatrix} \frac{1}{1-x} & \frac{-x}{1-x} \\ \frac{-x}{1-x} & \frac{1}{1-x} \end{bmatrix} \text{ and } z = \frac{x+y}{1+xy}. \text{ Then,}$$

(a) $A(z) = A(x) + A(y)$
(b) $A(z) = A(x)\,[A(y)]^{-1}$
(c) $A(z) = A(x)\cdot A(y)$
(d) $A(z) = A(x) - A(y)$

9 If $A(\alpha)=\begin{bmatrix}\cos\alpha & -\sin\alpha & 0\\ \sin\alpha & \cos\alpha & 0\\ 0 & 0 & 1\end{bmatrix}$, then $A(\alpha)\,A(\beta)$ is equal to

(a) $A(\alpha\beta)$ (b) $A(\alpha+\beta)$ (c) $A(\alpha-\beta)$ (d) None

10 If A is 3×4 matrix and B is a matrix such that $A'B$ and BA' are both defined, then B is of the type

(a) 4×3 (b) 3×4 (c) 3×3 (d) 4×4

11 If $A=\begin{bmatrix}1 & 2 & 2\\ 2 & 1 & -2\\ a & 2 & b\end{bmatrix}$ is a matrix satisfying the equation $AA^T=9I$, where I is 3×3 identity matrix, then the ordered pair (a,b) is equal to → JEE Mains 2015

(a) $(2,-1)$ (b) $(-2,1)$ (c) $(2,1)$ (d) $(-2,-1)$

12 If $E=\begin{bmatrix}0 & 0 & 1\\ 0 & 0 & 1\\ 0 & 0 & 0\end{bmatrix}$ and $F=\begin{bmatrix}1 & 0 & 0\\ 0 & 1 & 0\\ 0 & 0 & 1\end{bmatrix}$, then E^2F+F^2E

(a) F (b) E (c) 0 (d) None

13 If A and B are two invertible matrices and both are symmetric and commute each other, then

(a) both $A^{-1}B$ and $A^{-1}B^{-1}$ are symmetric
(b) neither $A^{-1}B$ nor $A^{-1}B^{-1}$ are symmetric
(c) $A^{-1}B$ is symmetric but $A^{-1}B^{-1}$ is not symmetric
(d) $A^{-1}B^{-1}$ is symmetric but $A^{-1}B$ is not symmetric

14 If neither α nor β are multiples of $\pi/2$ and the product AB of matrices

$$A=\begin{bmatrix}\cos^2\alpha & \sin\alpha\cos\alpha\\ \cos\alpha\sin\alpha & \sin^2\alpha\end{bmatrix}$$

and $$B=\begin{bmatrix}\cos^2\beta & \cos\beta\sin\beta\\ \cos\beta\sin\beta & \sin^2\beta\end{bmatrix}$$

is null matrix, then $\alpha-\beta$ is

(a) 0 (b) multiple of π
(c) an odd multiple of $\pi/2$ (d) None of these

15 The matrix $\begin{bmatrix}1 & 2 & 3\\ 1 & 2 & 3\\ -1 & -2 & -3\end{bmatrix}$ is

(a) idempotent (b) nilpotent
(c) involutary (d) orthogonal

16 If $A=\begin{bmatrix}\cos\theta & -\sin\theta\\ \sin\theta & \cos\theta\end{bmatrix}$, then

(a) A is skew-symmetric (b) symmetric
(c) idempotent (d) orthogonal

17 If $A=\begin{bmatrix}a & a^2-1 & -2\\ a+1 & 1 & a^2+4\\ -2 & 4a & 5\end{bmatrix}$ is symmetric, then a is

(a) -2 (b) 2 (c) -1 (d) None

18 If $A=\begin{bmatrix}1 & 2 & 2\\ 2 & 1 & -2\\ x & 2 & y\end{bmatrix}$ and $A^TA=AA^T=I$, then xy is equal to

(a) -1 (b) 1 (c) 2 (d) -2

19 If A and B are symmetric matrices of the same order and $X=AB+BA$ and $Y=AB-BA$, then $(XY)^T$ is equal to

(a) XY (b) YX
(c) $-YX$ (d) None of these

20 Let $A=\begin{bmatrix}1 & 0 & 0\\ 0 & 1 & 1\\ 0 & -2 & 4\end{bmatrix}$, $I=\begin{bmatrix}1 & 0 & 0\\ 0 & 1 & 0\\ 0 & 0 & 1\end{bmatrix}$ and $A^{-1}=\left[\frac{1}{6}(A^2+cA+dI)\right]$. The values of c and d are

(a) $(-6,-11)$ (b) $(6,11)$
(c) $(-6,11)$ (d) $(6,-11)$

21 Elements of a matrix A of order 9×9 are defined as $a_{ij}=\omega^{i+j}$ (where ω is cube root of unity), then trace (A) of the matrix is

(a) 0 (b) 1 (c) ω (d) ω^2

22 If $A=\begin{bmatrix}1 & -1 & 1\\ 2 & 1 & -3\\ 1 & 1 & 1\end{bmatrix}$ and $A^{-1}=\frac{1}{10}\begin{bmatrix}4 & 2 & 2\\ -5 & 0 & \alpha\\ 1 & -2 & 3\end{bmatrix}$, then α is equal to

(a) -2 (b) 5 (c) 2 (d) -1

23 If A is skew-symmetric and $B=(I-A)^{-1}(I+A)$, then B is

(a) symmetric
(b) skew-symmetric
(c) orthogonal
(d) None of the above

24 Let A be a square matrix satisfying $A^2+5A+5I=O$. The inverse of $A+2I$ is equal to

(a) $A-2I$ (b) $A+3I$
(c) $A-3I$ (d) does not exist

25 Let $A=\begin{bmatrix}1 & 0\\ 1/3 & 1\end{bmatrix}$. Then A^{48} is

(a) $\begin{bmatrix}1 & 0\\ (1/3)^{48} & 1\end{bmatrix}$ (b) $\begin{bmatrix}1 & 0\\ \frac{3}{2}\left[1-\frac{1}{3^{48}}\right] & 1\end{bmatrix}$

(c) $\begin{bmatrix}1 & 0\\ 16 & 1\end{bmatrix}$ (d) None of these

26 If X is any matrix of order $n\times p$ and I is an identity matrix of order $n\times n$, then the matrix $M=I-X(X'X)^{-1}X'$ is

I. Idempotent matrix
II. $MX=O$

(a) Only I is correct (b) Only II is correct
(c) Both I and II are correct (d) None of them is correct

27 Let A and B be two symmetric matrices of order 3.

Statement I $A(BA)$ and $(AB)A$ are symmetric matrices.

Statement II AB is symmetric matrix, if matrix multiplication of A with B is commutative.

(a) Statement I is true, Statement II is true; Statement II is a correct explanation for Statement I
(b) Statement I is true, Statement II is true; Statement II is not a correct explanation for Statement I
(c) Statement I is true; Statement II is false
(d) Statement I is false; Statement II is true

28 Consider the following relation R on the set of real square matrices of order 3.

$R = \{(A, B) : A = P^{-1}BP$ for some invertible matrix $P\}$

Statement I R is an equivalence relation.

Statement II For any two invertible 3×3 matrices M and N, $(MN)^{-1} = N^{-1}M^{-1}$.

(a) Statement I is false, Statement II is true
(b) Statement I is true, Statement II is true; Statement II is correct explanation of Statement I
(c) Statement I is true, Statement II is true; Statement II is not a correct explanation of Statement I
(d) Statement I is true, Statement II is false

DAY PRACTICE SESSION 2

PROGRESSIVE QUESTIONS EXERCISE

1 If $A = \begin{bmatrix} 2 & 1 \\ -4 & -2 \end{bmatrix}$, then $I + 2A + 3A^2 + \ldots \infty$ is equal to

(a) $\begin{bmatrix} 4 & 1 \\ -4 & 0 \end{bmatrix}$ (b) $\begin{bmatrix} 3 & 1 \\ -4 & -1 \end{bmatrix}$ (c) $\begin{bmatrix} 5 & 2 \\ -8 & -3 \end{bmatrix}$ (d) $\begin{bmatrix} 5 & 2 \\ -3 & -8 \end{bmatrix}$

2 The matrix A that commute with the matrix $\begin{pmatrix} 1 & 2 \\ 3 & 4 \end{pmatrix}$ is

(a) $A = \frac{1}{2}\begin{pmatrix} 2a & 2b \\ 3b & 2a+3b \end{pmatrix}$ (b) $A = \frac{1}{2}\begin{pmatrix} 2b & 2a \\ 3a & 2a+3b \end{pmatrix}$
(c) $A = \frac{1}{3}\begin{pmatrix} 2a+3b & 2a \\ 3a & 2a+3b \end{pmatrix}$ (d) None of these

3 The total number of matrices that can be formed using 5 different letters such that no letter is repeated in any matrix, is

(a) 5! (b) 2×5^5
(c) $2 \times (5!)$ (d) None of these

4 If A is symmetric and B is a skew-symmetric matrix, then for $n \in N$, which of the following is not correct?

(a) A^n is symmetric
(b) B^n is symmetric if n is even
(c) A^n is symmetric if n is odd only
(d) B^n is skew-symmetric if n is odd

5 Consider three matrices $X = \begin{bmatrix} 2 & 1 \\ 4 & 1 \end{bmatrix}$, $Y = \begin{bmatrix} 5 & 4 \\ 6 & 5 \end{bmatrix}$ and $Z = \begin{bmatrix} 5 & -4 \\ -6 & 5 \end{bmatrix}$. Then, the value of the sum

$$tr(X) + tr\left(\frac{XYZ}{2}\right) + tr\left(\frac{X(YZ)^2}{4}\right) + tr\left(\frac{X(YZ)^3}{8}\right) + \ldots \text{ to } \infty \text{ is}$$

(a) 6 (b) 9
(c) 12 (d) None of these

6 If both $A - \frac{1}{2}I$ and $A + \frac{1}{2}I$ are orthogonal matrices, then

(a) A is orthogonal
(b) A is skew-symmetric matrix
(c) A is symmetric matrix
(d) None of the above

7 If $A = \begin{bmatrix} \frac{-1+i\sqrt{3}}{2i} & \frac{-1-i\sqrt{3}}{2i} \\ \frac{1+i\sqrt{3}}{2i} & \frac{1-i\sqrt{3}}{2i} \end{bmatrix}$, $i = \sqrt{-1}$ and $f(x) = x^2 + 2$, then $f(A)$ is equal to

(a) $\left(\frac{5-i\sqrt{3}}{2}\right)\begin{bmatrix} 1 & 0 \\ 0 & 1 \end{bmatrix}$ (b) $\left(\frac{3-i\sqrt{3}}{2}\right)\begin{bmatrix} 1 & 0 \\ 0 & 1 \end{bmatrix}$
(c) $\begin{bmatrix} 1 & 0 \\ 0 & 1 \end{bmatrix}$ (d) $(2+i\sqrt{3})\begin{bmatrix} 1 & 0 \\ 0 & 1 \end{bmatrix}$

8 If $A = \begin{bmatrix} 1 & 0 \\ 1 & 1 \end{bmatrix}$, then A^n is equal to

(a) $2^{n-1}A - (n-1)I$ (b) $nA - (n-1)I$
(c) $2^{n-1}A + (n-1)I$ (d) $nA + (n-1)I$

9 Let $A = \begin{bmatrix} \cos\theta & \sin\theta \\ -\sin\theta & \cos\theta \end{bmatrix}$. Let $A^n = [b_{ij}]_{2\times2}$. Define $\lim_{n\to\infty} A^n = \lim_{n\to\infty} [b_{ij}]_{2\times2}$. Then $\lim_{n\to\infty}\left(\frac{A^n}{n}\right)$ is

(a) zero matrix (b) unit matrix
(c) $\begin{bmatrix} 0 & 1 \\ -1 & 0 \end{bmatrix}$ (d) limit does not exist

10 If B is skew-symmetric matrix of order n and A is $n \times 1$ column matrix and $A^T BA = [p]$, then

(a) $p < 0$ (b) $p = 0$
(c) $p > 0$ (d) Nothing can be said

11 If A, B and $A+B$ are idempotent matrices, then AB is equal to

(a) BA (b) $-BA$ (c) I (d) O

12 If $P = \begin{bmatrix} \frac{\sqrt{3}}{2} & \frac{1}{2} \\ -\frac{1}{2} & \frac{\sqrt{3}}{2} \end{bmatrix}, A = \begin{bmatrix} 1 & 1 \\ 0 & 1 \end{bmatrix}$ and $Q = PAP^T$, then $P^T Q^{2019} P$ is equal to

(a) $\begin{bmatrix} 1 & 2019 \\ 0 & 1 \end{bmatrix}$

(b) $\begin{bmatrix} 4+2019\sqrt{3} & 6057 \\ 2019 & 4-2019\sqrt{3} \end{bmatrix}$

(c) $\frac{1}{4}\begin{bmatrix} 2+\sqrt{3} & 1 \\ -1 & 2-\sqrt{3} \end{bmatrix}$

(d) $\frac{1}{4}\begin{bmatrix} 2019 & 2-\sqrt{3} \\ 2+\sqrt{3} & 2019 \end{bmatrix}$

13 Which of the following is an orthogonal matrix?

(a) $\frac{1}{7}\begin{bmatrix} 6 & 2 & -3 \\ 2 & 3 & 6 \\ 3 & -6 & 2 \end{bmatrix}$ (b) $\frac{1}{7}\begin{bmatrix} 6 & 2 & 3 \\ 2 & -3 & 6 \\ 3 & 6 & -2 \end{bmatrix}$

(c) $\frac{1}{7}\begin{bmatrix} -6 & -2 & -3 \\ 2 & 3 & 6 \\ -3 & 6 & 2 \end{bmatrix}$ (d) $\frac{1}{7}\begin{bmatrix} 6 & -2 & 3 \\ 2 & 2 & -3 \\ -6 & 2 & 3 \end{bmatrix}$

14 If $A_1, A_3, \ldots, A_{2n-1}$ are n skew-symmetric matrices of same order, then $B = \sum_{r=1}^{n} (2r-1)(A_{2r-1})^{2r-1}$ will be

(a) symmetric
(b) skew-symmetric
(c) neither symmetric nor skew-symmetric
(d) data not adequate

15 Let matrix $A = \begin{bmatrix} a & b & c \\ b & c & a \\ c & a & b \end{bmatrix}$, where a, b, c are real positive numbers with $abc = 1$. If $A^T A = I$, then $a^3 + b^3 + c^3$ is

(a) 3 (b) 4
(c) 2 (d) None of these

16 If A is an 3×3 non-singular matrix such that $AA' = A'A$ and $B = A^{-1}A'$, then BB' equals → JEE Mains 2014

(a) $(B^{-1})'$ (b) $I + B$
(c) I (d) B^{-1}

17 A is a 3×3 matrix with entries from the set $\{-1, 0, 1\}$. The probability that A is neither symmetric nor skew-symmetric is

(a) $\frac{3^9 - 3^6 - 3^3 + 1}{3^9}$ (b) $\frac{3^9 - 3^6 - 3^3}{3^9}$

(c) $\frac{3^9 - 3^6 + 1}{3^9}$ (d) $\frac{3^9 - 3^3 + 1}{3^9}$

ANSWERS

SESSION 1

1. (d)	**2.** (a)	**3.** (a)	**4.** (c)	**5.** (b)	**6.** (a)	**7.** (c)	**8.** (c)	**9.** (b)	**10.** (b)
11. (d)	**12.** (b)	**13.** (a)	**14.** (c)	**15.** (b)	**16.** (d)	**17.** (b)	**18.** (c)	**19.** (c)	**20.** (c)
21. (a)	**22.** (b)	**23.** (c)	**24.** (b)	**25.** (c)	**26.** (c)	**27.** (b)	**28.** (c)		

SESSION 2

1. (c)	**2.** (a)	**3.** (c)	**4.** (c)	**5.** (a)	**6.** (b)	**7.** (d)	**8.** (b)	**9.** (a)	**10.** (b)
11. (b)	**12.** (a)	**13.** (a)	**14.** (b)	**15.** (d)	**16.** (c)	**17.** (a)			

Hints and Explanations

SESSION 1

1 Here,

$$A+B=\begin{bmatrix}0&1\\1&1\end{bmatrix}+\begin{bmatrix}0&-1\\1&0\end{bmatrix}=\begin{bmatrix}0&0\\2&1\end{bmatrix}$$

$$A-B=\begin{bmatrix}0&1\\1&1\end{bmatrix}-\begin{bmatrix}0&-1\\1&0\end{bmatrix}=\begin{bmatrix}0&2\\0&1\end{bmatrix}$$

$$A^2=A\cdot A=\begin{bmatrix}0&1\\1&1\end{bmatrix}\begin{bmatrix}0&1\\1&1\end{bmatrix}$$

$$=\begin{bmatrix}0+1&0+1\\0+1&1+1\end{bmatrix}=\begin{bmatrix}1&1\\1&2\end{bmatrix}$$

and $B^2=B\cdot B=\begin{bmatrix}0&-1\\1&0\end{bmatrix}\cdot\begin{bmatrix}0&-1\\1&0\end{bmatrix}$

$$=\begin{bmatrix}-1&0\\0&-1\end{bmatrix}$$

$$\therefore\ A^2+B^2=\begin{bmatrix}1&1\\1&2\end{bmatrix}+\begin{bmatrix}-1&0\\0&-1\end{bmatrix}=\begin{bmatrix}0&1\\1&1\end{bmatrix}$$

$$A^2-B^2=\begin{bmatrix}1&1\\1&2\end{bmatrix}-\begin{bmatrix}-1&0\\0&-1\end{bmatrix}=\begin{bmatrix}2&1\\1&3\end{bmatrix}$$

and $(A+B)(A-B)=\begin{bmatrix}0&0\\2&1\end{bmatrix}\begin{bmatrix}0&2\\0&1\end{bmatrix}$

$$=\begin{bmatrix}0+0&0+0\\0+0&4+1\end{bmatrix}=\begin{bmatrix}0&0\\0&5\end{bmatrix}$$

Clearly, $(A+B)(A-B)\neq A^2-B^2$

$$\neq A^2+B^2\neq I.$$

2 $[3p+3q+2r,\ 4p+2q+0,$

$$p+3q+2r]=[3\ 0\ 1]$$

$\Rightarrow 3p+3q+2r=3,\ 4p+2q=0,$

$$p+3q+2r=1$$

$\Rightarrow p=1, q=-2, r=3$

$\therefore\ 2p+q-r=2-2-3=-3$

3 We know that, a square matrix $A=[a_{ij}]$ is said to be an upper triangular matrix if $a_{ij}=0,\ \forall\ i>j$.

Consider, an upper triangular matrix

$$A=\begin{bmatrix}1&2&3\\0&5&6\\0&0&7\end{bmatrix}_{3\times3}$$

Here, number of zeroes $=3=\dfrac{3(3-1)}{2}$

$\therefore$ Minimum number of zeroes

$$=\frac{n(n-1)}{2}$$

4 Clearly, $AB=\begin{bmatrix}1&2\\3&4\end{bmatrix}\begin{bmatrix}a&0\\0&b\end{bmatrix}=\begin{bmatrix}a&2b\\3a&4b\end{bmatrix}$

and $BA=\begin{bmatrix}a&0\\0&b\end{bmatrix}\begin{bmatrix}1&2\\3&4\end{bmatrix}=\begin{bmatrix}a&2a\\3b&4b\end{bmatrix}$

If $AB=BA$, then $a=b$.

Hence, $AB=BA$ is possible for infinitely many values of B's.

5 Here, A is 2×3 matrix and B is 3×2 matrix.

$\therefore$ Both AB and BA exist, and AB is a 2×2 matrix, while BA is 3×3 matrix.

$\therefore$ $AB\neq BA.$

6 Clearly,

$$H^2=\begin{bmatrix}\omega&0\\0&\omega\end{bmatrix}\begin{bmatrix}\omega&0\\0&\omega\end{bmatrix}=\begin{bmatrix}\omega^2&0\\0&\omega^2\end{bmatrix}$$

$$H^3=\begin{bmatrix}\omega^2&0\\0&\omega^2\end{bmatrix}\begin{bmatrix}\omega&0\\0&\omega\end{bmatrix}=\begin{bmatrix}\omega^3&0\\0&\omega^3\end{bmatrix}$$

$$\therefore\ H^{70}=\begin{bmatrix}\omega^{70}&0\\0&\omega^{70}\end{bmatrix}=\begin{bmatrix}\omega^{69}\cdot\omega&0\\0&\omega^{69}\cdot\omega\end{bmatrix}$$

$$=\begin{bmatrix}(\omega^3)^{23}\cdot\omega&0\\0&(\omega^3)^{23}\cdot\omega\end{bmatrix}$$

$$=\begin{bmatrix}\omega&0\\0&\omega\end{bmatrix}=H \qquad [\because \omega^3=1]$$

7 Since, $AB=A$

$\therefore$ $B=I\ \Rightarrow\ B^2=B$

Similarly, $BA=B$

$\Rightarrow$ $A=I$

$\Rightarrow$ $A^2=A$

Hence, $A^2=A$ and $B^2=B$

8 We have,

$$A(x)=\frac{1}{1-x}\begin{bmatrix}1&-x\\-x&1\end{bmatrix} \quad \text{...(i)}$$

$$\therefore\ A(y)=\frac{1}{1-y}\begin{bmatrix}1&-y\\-y&1\end{bmatrix} \quad \text{...(ii)}$$

and $A(z)=\dfrac{1}{1-\dfrac{(x+y)}{1+xy}}$

$$\begin{bmatrix}1&-\dfrac{(x+y)}{1+xy}\\-\dfrac{(x+y)}{1+xy}&1\end{bmatrix}$$

$$=\frac{1+xy}{1+xy-x-y}\begin{bmatrix}1&-\dfrac{(x+y)}{1+xy}\\-\dfrac{(x+y)}{1+xy}&1\end{bmatrix}$$

$$=\frac{1+xy}{(1-x)(1-y)}\begin{bmatrix}1&-\dfrac{(x+y)}{1+xy}\\-\dfrac{(x+y)}{1+xy}&1\end{bmatrix}$$

$$=\frac{1}{(1-x)(1-y)}$$

$$\begin{bmatrix}1+xy&-(x+y)\\-(x+y)&1+xy\end{bmatrix} \quad \text{...(iii)}$$

Now, consider

$$A(x)\cdot A(y)=\frac{1}{(1-x)(1-y)}\cdot$$

$$\begin{bmatrix}1&-x\\-x&1\end{bmatrix}\begin{bmatrix}1&-y\\-y&1\end{bmatrix}$$

$$=\frac{1}{(1-x)(1-y)}\cdot$$

$$\begin{bmatrix}1+xy&-(x+y)\\-(x+y)&1+xy\end{bmatrix} \quad \text{...(iv)}$$

From Eqs. (iii) and (iv), we get

$$A(z)=A(x)\cdot A(y).$$

9 $A(\alpha)\,A(\beta)=\begin{bmatrix}\cos\alpha&-\sin\alpha&0\\\sin\alpha&\cos\alpha&0\\0&0&1\end{bmatrix}$

$$\times\begin{bmatrix}\cos\beta&-\sin\beta&0\\\sin\beta&\cos\beta&0\\0&0&1\end{bmatrix}$$

$$=\begin{bmatrix}\cos(\alpha+\beta)&-\sin(\alpha+\beta)&0\\\sin(\alpha+\beta)&\cos(\alpha+\beta)&0\\0&0&1\end{bmatrix}$$

$$=A(\alpha+\beta)$$

10 Clearly, order of A' is 4×3.

Now, for $A'B$ to be defined, order of B should be $3\times m$ and for BA' to be defined, order of B should be $n\times4$.

Thus, for both $A'B$ and BA' to be defined, order of B should be 3×4.

11 Given, $A=\begin{bmatrix}1&2&2\\2&1&-2\\a&2&b\end{bmatrix}$

$$\Rightarrow\quad A^T=\begin{bmatrix}1&2&a\\2&1&2\\2&-2&b\end{bmatrix}$$

Now, $AA^T=\begin{bmatrix}1&2&2\\2&1&-2\\a&2&b\end{bmatrix}\begin{bmatrix}1&2&a\\2&1&2\\2&-2&b\end{bmatrix}$

$$=\begin{bmatrix}9&0&a+4+2b\\0&9&2a+2-2b\\a+4+2b&2a+2-2b&a^2+4+b^2\end{bmatrix}$$

It is given that, $AA^T=9I$

$$\Rightarrow\begin{bmatrix}9&0&a+4+2b\\0&9&2a+2-2b\\a+4+2b&2a+2-2b&a^2+4+b^2\end{bmatrix}$$

$$=\begin{bmatrix}9&0&0\\0&9&0\\0&0&9\end{bmatrix}$$

On comparing, we get

$$a+4+2b=0$$

$$\Rightarrow\quad a+2b=-4 \quad \text{...(i)}$$

$$2a+2-2b=0$$

$\Rightarrow \quad a-b=-1 \quad \ldots(\text{ii})$

and $\quad a^2+4+b^2=9 \quad \ldots(\text{iii})$

On solving Eqs. (i) and (ii), we get

$a=-2, b=-1$

This satisfies Eq. (iii) also.

Hence, $(a,b)\equiv(-2,-1)$

12 F is unit matrix $\Rightarrow F^2 = F$

and $\quad E^2F + F^2E = E^2 + E$

Also, $E^2 = \begin{bmatrix} 0 & 0 & 1 \\ 0 & 0 & 1 \\ 0 & 0 & 0 \end{bmatrix} \times \begin{bmatrix} 0 & 0 & 1 \\ 0 & 0 & 1 \\ 0 & 0 & 0 \end{bmatrix}$

$= \begin{bmatrix} 0 & 0 & 0 \\ 0 & 0 & 0 \\ 0 & 0 & 0 \end{bmatrix}$

$\therefore E^2 + E = E.$

13 Consider, $(A^{-1}B)^T = B^T(A^{-1})^T$

$= B^T(A^T)^{-1} = B A^{-1}$

$[\because A^T = A \text{ and } B^T = B]$

$= A^{-1}B$

$[\because AB = BA \Rightarrow A^{-1}(AB)A^{-1}$

$= A^{-1}(BA)A^{-1} \Rightarrow BA^{-1} = A^{-1}B]$

$\Rightarrow A^{-1}B$ is symmetric.

Now, consider

$(A^{-1}B^{-1})^T = ((BA)^{-1})^T$

$= ((AB)^{-1})^T \quad [\because AB = BA]$

$= (B^{-1}A^{-1})^T = (A^{-1})^T(B^{-1})^T$

$= (A^T)^{-1}(B^T)^{-1} = A^{-1}B^{-1}$

$\Rightarrow A^{-1}B^{-1}$ is also symmetric.

14 $AB = \begin{bmatrix} \cos^2\alpha & \cos\alpha\sin\alpha \\ \cos\alpha\sin\alpha & \sin^2\alpha \end{bmatrix}$

$\times \begin{bmatrix} \cos^2\beta & \cos\beta\sin\beta \\ \cos\beta\sin\beta & \sin^2\beta \end{bmatrix}$

$= \begin{bmatrix} \cos\alpha\cos\beta\cos(\alpha-\beta) & \cos\alpha\sin\beta\cos(\alpha-\beta) \\ \sin\alpha\cos\beta\cos(\alpha-\beta) & \sin\alpha\sin\beta\cos(\alpha-\beta) \end{bmatrix}$

$= \begin{bmatrix} 0 & 0 \\ 0 & 0 \end{bmatrix}$

$\Rightarrow \quad \cos(\alpha-\beta) = 0$

$\Rightarrow \quad \alpha - \beta = (2n+1)\pi/2$

15 Let $\quad A = \begin{bmatrix} 1 & 2 & 3 \\ 1 & 2 & 3 \\ -1 & -2 & -3 \end{bmatrix}$

Then, $A^2 = \begin{bmatrix} 0 & 0 & 0 \\ 0 & 0 & 0 \\ 0 & 0 & 0 \end{bmatrix}$

Hence, A is nilpotent matrix of index 2.

16 $A' = \begin{bmatrix} \cos\theta & \sin\theta \\ -\sin\theta & \cos\theta \end{bmatrix} \neq A$ or $-A$.

$A A' = \begin{bmatrix} \cos\theta & -\sin\theta \\ \sin\theta & \cos\theta \end{bmatrix}$

$\begin{bmatrix} \cos\theta & \sin\theta \\ -\sin\theta & \cos\theta \end{bmatrix} = \begin{bmatrix} 1 & 0 \\ 0 & 1 \end{bmatrix} = I$

$\therefore A$ is orthogonal.

17 A is symmetric

$\Rightarrow a^2 - 1 = a + 1, a^2 + 4 = 4a$

$\Rightarrow a^2 - a - 2 = 0, a^2 - 4a + 4 = 0$

$\Rightarrow \quad a = 2.$

18 Since, A is orthogonal, each row is orthogonal to the other rows.

$\Rightarrow \quad R_1 \cdot R_3 = 0$

$\Rightarrow \quad x + 4 + 2y = 0$

Also, $\quad R_2 \cdot R_3 = 0$

$\Rightarrow \quad 2x + 2 - 2y = 0$

On solving, we get $x = -2$, $y = -1$

$\therefore \quad xy = 2$

19 Since, A and B are symmetric matrices

$\therefore \quad X = AB + BA$

will be a symmetric matrix and

$Y = AB - BA$ will be a skew-symmetric matrix.

Thus, we get $X^T = X$ and $Y^T = -Y$

Now, consider $(XY)^T = Y^TX^T$

$= (-Y)(X) = -YX$

20 Clearly, $6A^{-1} = A^2 + cA + dI$

$\Rightarrow \quad (6A^{-1})A = (A^2 + cA + dI)A$

$[\because$ Post multiply both sides by $A]$

$\Rightarrow \quad 6(A^{-1}A) = A^3 + cA^2 + dIA$

$\Rightarrow \quad 6I = A^3 + cA^2 + dA$

$[\because A^{-1}A = I \text{ and } IA = A]$

$\Rightarrow \quad A^3 + cA^2 + dA - 6I = O \quad \ldots(\text{i})$

Here, $A^2 = A \cdot A = \begin{bmatrix} 1 & 0 & 0 \\ 0 & 1 & 1 \\ 0 & -2 & 4 \end{bmatrix} \times$

$\begin{bmatrix} 1 & 0 & 0 \\ 0 & 1 & 1 \\ 0 & -2 & 4 \end{bmatrix} = \begin{bmatrix} 1 & 0 & 0 \\ 0 & -1 & 5 \\ 0 & -10 & 14 \end{bmatrix}$

and $A^3 = A^2 \cdot A = \begin{bmatrix} 1 & 0 & 0 \\ 0 & -1 & 5 \\ 0 & -10 & 14 \end{bmatrix} \times$

$\begin{bmatrix} 1 & 0 & 0 \\ 0 & 1 & 1 \\ 0 & -2 & 4 \end{bmatrix} = \begin{bmatrix} 1 & 0 & 0 \\ 0 & -11 & 19 \\ 0 & -38 & 46 \end{bmatrix}$

Now, from Eq. (i), we get

$\begin{bmatrix} 1 & 0 & 0 \\ 0 & -11 & 19 \\ 0 & -38 & 46 \end{bmatrix} + c\begin{bmatrix} 1 & 0 & 0 \\ 0 & -1 & 5 \\ 0 & -10 & 14 \end{bmatrix}$

$+ d\begin{bmatrix} 1 & 0 & 0 \\ 0 & 1 & 1 \\ 0 & -2 & 4 \end{bmatrix}$

$- 6\begin{bmatrix} 1 & 0 & 0 \\ 0 & 1 & 0 \\ 0 & 0 & 1 \end{bmatrix} = \begin{bmatrix} 0 & 0 & 0 \\ 0 & 0 & 0 \\ 0 & 0 & 0 \end{bmatrix}$

$\Rightarrow \begin{bmatrix} 1 + c + d - 6 & 0 & 0 \\ 0 & -11 - c + d - 6 & 19 + 5c + d \\ 0 & -38 - 10c - 2d & 46 + 14c + 4d - 6 \end{bmatrix}$

$= \begin{bmatrix} 0 & 0 & 0 \\ 0 & 0 & 0 \\ 0 & 0 & 0 \end{bmatrix}$

$\Rightarrow \quad 1 + c + d - 6 = 0;$

$-11 - c + d - 6 = 0$

$\Rightarrow \quad c + d = 5; -c + d = 17$

On solving, we get $c = -6, d = 11.$

These value also satisfy other equations.

21 Clearly, $tr(A) = a_{11} + a_{22} + a_{33} + a_{44}$

$+ a_{55} + a_{66} + a_{77} + a_{88} + a_{99}$

$= \omega^2 + \omega^4 + \omega^6 + \omega^8 + \omega^{10} + \omega^{12}$

$+ \omega^{14} + \omega^{16} + \omega^{18}$

$= (\omega^2 + \omega + 1) + (\omega^2 + \omega + 1)$

$+ (\omega^2 + \omega + 1) [\because \omega^{3n} = 1, n \in N]$

$= 0 + 0 + 0 \quad [\because 1 + \omega + \omega^2 = 0]$

$= 0$

22 Clearly, $AA^{-1} = I$

Now, if R_1 of A is multiplied by C_3 of A^{-1}, we get $2 - \alpha + 3 = 0 \Rightarrow \alpha = 5$

23 Consider,

$BB^T = (I - A)^{-1}(I + A)(I + A)^T[(I - A)^{-1}]^T$

$= (I - A)^{-1}(I + A)(I - A)(I + A)^{-1}$

$= (I - A)^{-1}(I - A)(I + A)(I + A)^{-1}$

$= I \cdot I = I$

Hence, B is an orthogonal matrix.

24 We have, $A^2 + 5A + 5I = O$

$\Rightarrow \quad A^2 + 5A + 6I = I$

$\Rightarrow \quad (A + 2I)(A + 3I) = I$

$\Rightarrow \quad A + 2I$ and $A + 3I$ are inverse of each other.

25 If $A = \begin{bmatrix} 1 & 0 \\ a & 1 \end{bmatrix}$, then $A^2 = \begin{bmatrix} 1 & 0 \\ 2a & 1 \end{bmatrix}$

$A^3 = \begin{bmatrix} 1 & 0 \\ 3a & 1 \end{bmatrix}, \ldots, A^n = \begin{bmatrix} 1 & 0 \\ na & 1 \end{bmatrix}$

Here, $a = 1/3$,

$\therefore \quad A^{48} = \begin{bmatrix} 1 & 0 \\ 16 & 1 \end{bmatrix}$

26 We have, $M = I - X(X'X)^{-1}X'$

$= I - X(X^{-1}(X')^{-1})X'$

$[\because (AB)^{-1} = B^{-1}A^{-1}]$

$= I - (XX^{-1})((X')^{-1}X')$

[by associative property]

$= I - I \times I \quad [\because AA^{-1} = I = A^{-1}A]$

$= I - I \quad [\because I^2 = I]$

$= O$

Clearly, $M^2 = O = M$

So, M is an idempotent matrix. Also, $MX = O$.

27 Given, $A^T = A$ and $B^T = B$

Statement I $[A(BA)]^T = (BA)^T \cdot A^T$
$= (A^T B^T) A^T$
$= (AB) A = A (BA)$

So, $A(BA)$ is symmetric matrix. Similarly, $(AB) A$ is symmetric matrix. Hence, Statement I is true. Also, Statement II is true but not a correct explanation of Statement I.

28 Given, $R = \{(A, B) : A = P^{-1} BP$ for some invertible matrix $P\}$

For Statement I

(i) **Reflexive** ARA

$\Rightarrow \quad A = P^{-1} AP$

which is true only, if $P = I$. Thus, $A = P^{-1}AP$ for some invertible matrix P. So, R is Reflexive.

(ii) **Symmetric**

$ARB \Rightarrow A = P^{-1}BP$

$\Rightarrow \; PAP^{-1} = P(P^{-1}BP)P^{-1}$

$\Rightarrow \; PAP^{-1} = (PP^{-1})B(PP^{-1})$

$\therefore \quad B = PAP^{-1}$

Now, let $Q = P^{-1}$

Then, $B = Q^{-1} AQ \Rightarrow BRA$

$\Rightarrow R$ is symmetric.

(iii) **Transitive** ARB and BRC

$\Rightarrow \quad A = P^{-1}BP$

and $B = Q^{-1}CQ$

$\Rightarrow \quad A = P^{-1}(Q^{-1}CQ)P$
$= (P^{-1}Q^{-1})C(QP)$
$= (QP)^{-1}C(QP)$

So, ARC.

$\Rightarrow R$ is transitive

So, R is an equivalence relation.

For Statement II It is always true that $(MN)^{-1} = N^{-1}M^{-1}$

Hence, both statements are true but second is not the correct explanation of first.

SESSION 2

1 Clearly, $A^2 = \begin{bmatrix} 2 & 1 \\ -4 & -2 \end{bmatrix}\begin{bmatrix} 2 & 1 \\ -4 & -2 \end{bmatrix}$

$= \begin{bmatrix} 0 & 0 \\ 0 & 0 \end{bmatrix} = O$

$\therefore \; I + 2A + 3A^2 + \ldots = I + 2A$

$= \begin{bmatrix} 1 & 0 \\ 0 & 1 \end{bmatrix} + \begin{bmatrix} 4 & 2 \\ -8 & -4 \end{bmatrix} = \begin{bmatrix} 5 & 2 \\ -8 & -3 \end{bmatrix}$

2 Let $A = \begin{pmatrix} a & b \\ c & d \end{pmatrix}$ be a matrix that commute with $\begin{pmatrix} 1 & 2 \\ 3 & 4 \end{pmatrix}$. Then,

$$\begin{pmatrix} a & b \\ c & d \end{pmatrix}\begin{pmatrix} 1 & 2 \\ 3 & 4 \end{pmatrix} = \begin{pmatrix} 1 & 2 \\ 3 & 4 \end{pmatrix}\begin{pmatrix} a & b \\ c & d \end{pmatrix}$$

$$\Rightarrow \begin{pmatrix} a + 3b & 2a + 4b \\ c + 3d & 2c + 4d \end{pmatrix} = \begin{pmatrix} a + 2c & b + 2d \\ 3a + 4c & 3b + 4d \end{pmatrix}$$

On equating the corresponding elements, we get

$a + 3b = a + 2c \Rightarrow 3b = 2c$ ...(i)

$2a + 4b = b + 2d \Rightarrow 2a + 3b = 2d$ ...(ii)

$c + 3d = 3a + 4c \Rightarrow a + c = d$ ...(iii)

$2c + 4d = 3b + 4d \Rightarrow 3b = 2c$ ...(iv)

Thus, A can be taken as

$$\begin{pmatrix} a & b \\ \frac{3b}{2} & a + \frac{3}{2}b \end{pmatrix} = \frac{1}{2}\begin{pmatrix} 2a & 2b \\ 3b & 2a + 3b \end{pmatrix}$$

3 Clearly, matrix having five elements is of order 5×1 or 1×5.

$\therefore$ Total number of such matrices $= 2 \times 5!$.

4 $(A^n)' = (A\,A \cdots A)' = (A'A' \cdots A')$
$= (A')^n = A^n$ for all n

$\therefore A^n$ is symmetric for all $n \in N$.

Also, B is skew-symmetric

$\Rightarrow \quad B' = -B$.

$\therefore \; (B^n)' = (B\,B \cdots B)' = (B'B' \cdots B')$
$= (B')^n$
$= (-B)^n = (-1)^n B^n$.

$\Rightarrow B^n$ is symmetric if n is even and is skew-symmetric if n is odd.

5 Here, $YZ = \begin{bmatrix} 5 & 4 \\ 6 & 5 \end{bmatrix}\begin{bmatrix} 5 & -4 \\ -6 & 5 \end{bmatrix} = \begin{bmatrix} 1 & 0 \\ 0 & 1 \end{bmatrix}$

$$\therefore \; tr(X) + tr\left(\frac{XYZ}{2}\right) + tr\left(\frac{X(YZ)^2}{4}\right) + tr\left(\frac{X(YZ)^3}{8}\right) + \ldots$$

$= tr(X) + tr\left(\frac{X}{2}\right) + tr\left(\frac{X}{4}\right) + \ldots$

$= tr(X) + \frac{1}{2}tr(X) + \frac{1}{4}tr(X) + \ldots$

$= tr(X)\left[1 + \frac{1}{2} + \frac{1}{2^2} + \ldots\right]$

$= tr(X)\dfrac{1}{1 - \frac{1}{2}}$

$= 2\,tr(X) = 2(2 + 1) = 6$

6 Since, both $A - \frac{1}{2}I$ and $A + \frac{1}{2}I$ are orthogonal, therefore, we have

$$\left(A - \frac{1}{2}I\right)'\left(A - \frac{1}{2}I\right) = I$$

$$\Rightarrow \left(A' - \frac{1}{2}I\right)\left(A - \frac{1}{2}I\right) = I \quad \ldots\text{(i)}$$

and $\left(A + \frac{1}{2}I\right)'\left(A + \frac{1}{2}I\right) = I$

$$\Rightarrow \left(A' + \frac{1}{2}I\right)\left(A + \frac{1}{2}I\right) = I \quad \ldots\text{(ii)}$$

From Eq. (i), we get

$$A'A - \frac{1}{2}IA' - \frac{1}{2}IA + \frac{1}{4}I = I$$

$$\Rightarrow A'A - \frac{1}{2}A' - \frac{1}{2}A + \frac{1}{4}I = I \quad \ldots\text{(iii)}$$

Similarly, from Eq. (ii), we get

$$A'A + \frac{1}{2}A' + \frac{1}{2}A + \frac{1}{4}I = I \quad \ldots\text{(iv)}$$

On subtracting Eq. (iii) from Eq. (iv), we get

$A + A' = O$

or $\quad A' = -A$

Hence, A is a skew-symmetric matrix.

7 We have, $A = \begin{bmatrix} \frac{\omega}{i} & \frac{\omega^2}{i} \\ -\frac{\omega^2}{i} & \frac{-\omega}{i} \end{bmatrix} = \frac{\omega}{i}\begin{bmatrix} 1 & \omega \\ -\omega & -1 \end{bmatrix}$

$\therefore \quad A^2 = -\omega^2 \begin{bmatrix} 1 - \omega^2 & 0 \\ 0 & 1 - \omega^2 \end{bmatrix}$

$= \begin{bmatrix} -\omega^2 + \omega^4 & 0 \\ 0 & -\omega^2 + \omega^4 \end{bmatrix}$

$= \begin{bmatrix} -\omega^2 + \omega & 0 \\ 0 & -\omega^2 + \omega \end{bmatrix}$

$\because \quad f(x) = x^2 + 2$ [given]

$\therefore \quad f(A) = A^2 + 2I$

$= \begin{bmatrix} -\omega^2 + \omega & 0 \\ 0 & -\omega^2 + \omega \end{bmatrix} + \begin{bmatrix} 2 & 0 \\ 0 & 2 \end{bmatrix}$

$= (-\omega^2 + \omega + 2)\begin{bmatrix} 1 & 0 \\ 0 & 1 \end{bmatrix}$

$= (3 + 2\omega)\begin{bmatrix} 1 & 0 \\ 0 & 1 \end{bmatrix}$

$= (2 + i\sqrt{3})\begin{bmatrix} 1 & 0 \\ 0 & 1 \end{bmatrix}$

8 $A^2 = \begin{bmatrix} 1 & 0 \\ 1 & 1 \end{bmatrix}\begin{bmatrix} 1 & 0 \\ 1 & 1 \end{bmatrix} = \begin{bmatrix} 1 & 0 \\ 2 & 1 \end{bmatrix}$

$A^3 = \begin{bmatrix} 1 & 0 \\ 2 & 1 \end{bmatrix}\begin{bmatrix} 1 & 0 \\ 1 & 1 \end{bmatrix} = \begin{bmatrix} 1 & 0 \\ 3 & 1 \end{bmatrix}$

....

....

$A^n = \begin{bmatrix} 1 & 0 \\ n & 1 \end{bmatrix}$

$= \begin{bmatrix} n & 0 \\ n & n \end{bmatrix} - \begin{bmatrix} n - 1 & 0 \\ 0 & n - 1 \end{bmatrix}$

$= nA - (n - 1)I$

9 $A^2 = \begin{bmatrix} \cos\theta & \sin\theta \\ -\sin\theta & \cos\theta \end{bmatrix}\begin{bmatrix} \cos\theta & \sin\theta \\ -\sin\theta & \cos\theta \end{bmatrix}$

$= \begin{bmatrix} \cos 2\theta & \sin 2\theta \\ -\sin 2\theta & \cos 2\theta \end{bmatrix}$

Similarly, $A^3 = \begin{bmatrix} \cos 3\theta & \sin 3\theta \\ -\sin 3\theta & \cos 3\theta \end{bmatrix}$ etc

$\therefore \quad A^n = \begin{bmatrix} \cos n\theta & \sin n\theta \\ -\sin n\theta & \cos n\theta \end{bmatrix}$

$= \begin{bmatrix} b_{11} & b_{12} \\ b_{21} & b_{22} \end{bmatrix}$

Now, $\lim_{n\to\infty} \frac{A^n}{n} = \begin{bmatrix} 0 & 0 \\ 0 & 0 \end{bmatrix}$ as $\lim_{n\to\infty} \frac{b_{ij}}{n} = 0$

10 $A^T B A = [p] \Rightarrow (A^TBA)^T = [p]^T = [p]$

$\Rightarrow \quad A^TB^TA = A^T(-B)A = [p]$

$\Rightarrow \quad [-p] = [p] \Rightarrow p = 0.$

11 Since, A, B and $A + B$ are idempotent matrix

$\therefore A^2 = A; B^2 = B$ and $(A + B)^2 = A + B$

Now, consider $(A + B)^2 = A + B$

$\Rightarrow \quad A^2 + B^2 + AB + BA = A + B$

$\Rightarrow \quad A + B + AB + BA = A + B$

$\Rightarrow \quad AB = -BA$

12 P is orthogonal matrix as $P^TP = I$

$Q^{2019} = (PAP^T)(PAP^T)$

$...(PAP^T) = PA^{2019}P^T$

$\therefore \; P^TQ^{2019}P = P^T \cdot PA^{2019}P^T \cdot P = A^{2019}$

Now, $A^2 = \begin{bmatrix} 1 & 1 \\ 0 & 1 \end{bmatrix}\begin{bmatrix} 1 & 1 \\ 0 & 1 \end{bmatrix} = \begin{bmatrix} 1 & 2 \\ 0 & 1 \end{bmatrix}$

$A^3 = \begin{bmatrix} 1 & 2 \\ 0 & 1 \end{bmatrix}\begin{bmatrix} 1 & 1 \\ 0 & 1 \end{bmatrix} = \begin{bmatrix} 1 & 3 \\ 0 & 1 \end{bmatrix}$

$\Rightarrow \quad A^{2019} = \begin{bmatrix} 1 & 2019 \\ 0 & 1 \end{bmatrix}$

13 We know that a matrix

$A = \begin{bmatrix} a_1 & a_2 & a_3 \\ b_1 & b_2 & b_3 \\ c_1 & c_2 & c_3 \end{bmatrix}$ will be orthogonal if

$AA' = I$, which implies

$\Sigma a_i^2 = \Sigma b_i^2 = \Sigma c_i^2 = 1$

and $\quad \Sigma a_i b_i = \Sigma b_i c_i = \Sigma c_i a_i = 0$

Now, from the given options, only

$\frac{1}{7}\begin{bmatrix} 6 & 2 & -3 \\ 2 & 3 & 6 \\ 3 & -6 & 2 \end{bmatrix}$ satisfies these conditions.

Hence, $\frac{1}{7}\begin{bmatrix} 6 & 2 & -3 \\ 2 & 3 & 6 \\ 3 & -6 & 2 \end{bmatrix}$ is an orthogonal matrix.

14 We have,

$B = A_1 + 3A_3^3 + \ldots + (2n-1)A_{2n-1}^{2n-1}$

Now, $B^T = (A_1 + 3A_3^3$

$+ \ldots + (2n-1)A_{2n-1}^{2n-1})^T$

$= A_1^T + (3A_3^3)^T + \ldots + ((2n-1)A_{2n-1}^{2n-1})^T$

$= A_1^T + 3(A_3^T)^3$

$+ \ldots + (2n-1)(A_{2n-1}^T)^{2n-1}$

$= -A - 3A_3^3 - \ldots - (2n-1)A_{2n-1}^{2n-1}$

[$\because A_1, A_3, \ldots, A_{2n-1}$ are skew-symmetric matrices

$\therefore (A_i)^T = -A_i \; \forall \, i = 1, 3, 5, \ldots, 2n-1$]

$= -[A + 3A_3^3 + \ldots + (2n-1)A_{2n-1}^{2n-1}]$

$= -B$

Hence, B is a skew-symmetric matrix.

15 $A^TA = \begin{bmatrix} a & b & c \\ b & c & a \\ c & a & b \end{bmatrix} \times \begin{bmatrix} a & b & c \\ b & c & a \\ c & a & b \end{bmatrix}$

$= \begin{bmatrix} a^2+b^2+c^2 & ab+bc+ca & ac+ab+bc \\ ab+bc+ca & b^2+c^2+a^2 & ab+bc+ca \\ ab+bc+ca & ab+bc+ca & a^2+b^2+c^2 \end{bmatrix}$

$A^TA = I \Rightarrow a^2 + b^2 + c^2 = 1$

and $\quad ab + bc + ca = 0$

Since $\quad a, b, c > 0,$

$\therefore ab + bc + ca \neq 0$ and hence no real value of $a^3 + b^3 + c^3$ exists.

16 $AA' = A'A, B = A^{-1}A'$.

$BB' = (A^{-1}A')(A^{-1} \cdot A')'$

$= (A^{-1}A')[(A')'(A^{-1})']$

$= (A^{-1}A')[A(A')^{-1}]$

$[\because (A^{-1})' = (A')^{-1}]$

$= A^{-1}(A'A)(A')^{-1}$

$= A^{-1}(AA')(A')^{-1} \quad [\because A'A = AA']$

$= (A^{-1}A)[A'(A')^{-1}]$

$= I \cdot I = I$

17 Total number of matrices $= 3^9$. A is symmetric, then $a_{ij} = a_{ji}$.

Now, 6 places (3 diagonal, 3 non-diagonal), can be filled from any of $-1, 0, 1$ in 3^6 ways. A is skew-symmetric, then diagonal entries are '0' and a_{12}, a_{13}, a_{23} can be filled from any of $-1, 0, 1$ in 3^3 ways. Zero matrix is common.

$\therefore$ Favourable matrices are $3^9 - 3^6 - 3^3 + 1$.

Hence, required probability

$= \frac{3^9 - 3^6 - 3^3 + 1}{3^9}$

DAY SIX

Determinants

Learning & Revision for the Day

- Determinants
- Properties of Determinants
- Cyclic Determinants
- Area of Triangle by using Determinants
- Minors and Cofactors
- Adjoint of a Matrix
- Inverse of a Matrix
- Solution of System of Linear Equations in Two and Three Variables

Determinants

Every square matrix A can be associated with a number or an expression which is called its determinant and it is denoted by det (A) or $|A|$ or Δ.

If $A = \begin{bmatrix} a_{11} & a_{12} & \cdots & a_{1n} \\ a_{21} & a_{22} & \cdots & a_{2n} \\ \vdots & \vdots & \vdots & \vdots \\ a_{n1} & a_{n2} & \cdots & a_{nn} \end{bmatrix}$, then det $(A) = \begin{vmatrix} a_{11} & a_{12} & \cdots & a_{1n} \\ a_{21} & a_{22} & \cdots & a_{2n} \\ \vdots & \vdots & \vdots & \vdots \\ a_{n1} & a_{n2} & \cdots & a_{nn} \end{vmatrix}$

- If $A = \begin{bmatrix} a & b \\ c & d \end{bmatrix}$, then $|A| = \begin{vmatrix} a & b \\ c & d \end{vmatrix} = ad - bc$
- If $A = \begin{bmatrix} a & b & c \\ p & q & r \\ u & v & w \end{bmatrix}$, then $|A| = \begin{vmatrix} a & b & c \\ p & q & r \\ u & v & w \end{vmatrix}$

$$= a\begin{vmatrix} q & r \\ v & w \end{vmatrix} - b\begin{vmatrix} p & r \\ u & w \end{vmatrix} + c\begin{vmatrix} p & q \\ u & v \end{vmatrix} \quad \text{[expanding along } R_1]$$

$$= a(qw - vr) - b(pw - ur) + c(pv - uq)$$

There are six ways of expanding a determinant of order 3 corresponding to each of three rows (R_1, R_2, R_3) and three columns (C_1, C_2, C_3).

NOTE
- Rule to put + or – sign in the expansion of determinant of order 3. $\begin{vmatrix} + & - & + \\ - & + & - \\ + & - & + \end{vmatrix}$
- A square matrix A is said to be singular, if $|A| = 0$ and non-singular, if $|A| \neq 0$.

PREP MIRROR
Your Personal Preparation Indicator

- No. of Questions in Exercises (x)—
- No. of Questions Attempted (y)—
- No. of Correct Questions (z)—
  (Without referring Explanations)
- Accuracy Level $(z/y \times 100)$—
- Prep Level $(z/x \times 100)$—

In order to expect good rank in JEE, your Accuracy Level should be above 85 & Prep Level should be above 75.

Properties of Determinants

(i) If each element of a row (column) is zero, then $\Delta = 0$.

(ii) If two rows (columns) are proportional, then $\Delta = 0$.

(iii) $|A^T| = |A|$, where A^T is a transpose of a matrix.

(iv) If any two rows (columns) are interchanged, then Δ becomes $-\Delta$.

(v) If each element of a row (column) of a determinant is multiplied by a constant k, then the value of the new determinant is k times the value of the original determinant

(vi) $\det(kA) = k^n \det(A)$, if A is of order $n \times n$.

(vii) If each element of a row (column) of a determinant is written as the sum of two or more terms, then the determinant can be written as the sum of two or more determinants i.e.

$$\begin{vmatrix} a_1 + a_2 & b & c \\ p_1 + p_2 & q & r \\ u_1 + u_2 & v & w \end{vmatrix} = \begin{vmatrix} a_1 & b & c \\ p_1 & q & r \\ u_1 & v & w \end{vmatrix} + \begin{vmatrix} a_2 & b & c \\ p_2 & q & r \\ u_2 & v & w \end{vmatrix}$$

(viii) If a scalar multiple of any row (column) is added to another row (column), then Δ is unchanged

i.e. $\begin{vmatrix} a & b & c \\ p & q & r \\ u & v & w \end{vmatrix} = \begin{vmatrix} a & b & c \\ p+ka & q+kb & r+kc \\ u & v & w \end{vmatrix}$, which is obtained by the operation $R_2 \to R_2 + kR_1$

Product of Determinants

If $|A| = \begin{vmatrix} a_1 & b_1 & c_1 \\ a_2 & b_2 & c_2 \\ a_3 & b_3 & c_3 \end{vmatrix}$ and $|B| = \begin{vmatrix} \alpha_1 & \beta_1 & \gamma_1 \\ \alpha_2 & \beta_2 & \gamma_2 \\ \alpha_3 & \beta_3 & \gamma_3 \end{vmatrix}$, then

$$|A| \times |B| = \begin{vmatrix} a_1\alpha_1 + b_1\beta_1 + c_1\gamma_1 & a_1\alpha_2 + b_1\beta_2 + c_1\gamma_2 & a_1\alpha_3 + b_1\beta_3 + c_1\gamma_3 \\ a_2\alpha_1 + b_2\beta_1 + c_2\gamma_1 & a_2\alpha_2 + b_2\beta_2 + c_2\gamma_2 & a_2\alpha_3 + b_2\beta_3 + c_2\gamma_3 \\ a_3\alpha_1 + b_3\beta_1 + c_3\gamma_1 & a_3\alpha_2 + b_3\beta_2 + c_3\gamma_2 & a_3\alpha_3 + b_3\beta_3 + c_3\gamma_3 \end{vmatrix} = |AB|$$

[multiplying row by row]

We can multiply rows by columns or columns by rows or columns by columns

NOTE
- $|AB| = |A|\ |B| = |BA| = |A^TB| = |AB^T| = |A^TB^T|$
- $|A^n| = |A|^n, n \in Z^+$

Cyclic Determinants

In a cyclic determinant, the elements of row (or column) are arranged in a systematic order and the value of a determinant is also in systematic order.

(i) $\begin{vmatrix} 1 & x & x^2 \\ 1 & y & y^2 \\ 1 & z & z^2 \end{vmatrix} = (x-y)(y-z)(z-x)$

(ii) $\begin{vmatrix} 1 & x & x^3 \\ 1 & y & y^3 \\ 1 & z & z^3 \end{vmatrix} = (x-y)(y-z)(z-x)(x+y+z)$

(iii) $\begin{vmatrix} 1 & x^2 & x^3 \\ 1 & y^2 & y^3 \\ 1 & z^2 & z^3 \end{vmatrix} = (x-y)(y-z)(z-x)(xy+yz+zx)$

(iv) $\begin{vmatrix} a & b & c \\ b & c & a \\ c & a & b \end{vmatrix} = -(a+b+c)(a^2+b^2+c^2-ab-bc-ca)$
$= -(a^3+b^3+c^3-3abc)$

(v) $\begin{vmatrix} a & bc & abc \\ b & ca & abc \\ c & ab & abc \end{vmatrix} = \begin{vmatrix} a & a^2 & a^3 \\ b & b^2 & b^3 \\ c & c^2 & c^3 \end{vmatrix} = abc(a-b)(b-c)(c-a)$

Area of Triangle by using Determinants

If $A(x_1, y_1), B(x_2, y_2)$ and $C(x_3, y_3)$ are vertices of ΔABC, then

$$\text{Area of } \Delta ABC = \frac{1}{2}\left\|\begin{matrix} x_1 & y_1 & 1 \\ x_2 & y_2 & 1 \\ x_3 & y_3 & 1 \end{matrix}\right\|$$

$$= \frac{1}{2}\,|[x_1(y_2 - y_3) + x_2(y_3 - y_1) + x_3(y_1 - y_2)]|$$

If these three points are collinear, then $\begin{vmatrix} x_1 & y_1 & 1 \\ x_2 & y_2 & 1 \\ x_3 & y_3 & 1 \end{vmatrix} = 0$ and *vice-versa*.

Minors and Cofactors

The **minor** M_{ij} of the element a_{ij} is the determinant obtained by deleting the ith row and jth column of Δ.

If $\Delta = \begin{vmatrix} a_{11} & a_{12} & a_{13} \\ a_{21} & a_{22} & a_{23} \\ a_{31} & a_{32} & a_{33} \end{vmatrix}$,

then $M_{11} = \begin{vmatrix} a_{22} & a_{23} \\ a_{32} & a_{33} \end{vmatrix}, M_{12} = \begin{vmatrix} a_{21} & a_{23} \\ a_{31} & a_{33} \end{vmatrix}$ etc.

The **cofactor** C_{ij} of the element a_{ij} is $(-1)^{i+j}M_{ij}$.

If $\Delta = \begin{vmatrix} a_{11} & a_{12} & a_{13} \\ a_{21} & a_{22} & a_{23} \\ a_{31} & a_{32} & a_{33} \end{vmatrix}$, then $C_{11} = \begin{vmatrix} a_{22} & a_{23} \\ a_{32} & a_{33} \end{vmatrix}, C_{12} = -\begin{vmatrix} a_{21} & a_{23} \\ a_{31} & a_{33} \end{vmatrix}$ etc.

The sum of product of the elements of any row (or column) with their corresponding cofactors is equal to the value of determinant.

i.e. $\Delta = a_{11}C_{11} + a_{12}C_{12} + a_{13}C_{13}$
$= a_{21}C_{21} + a_{22}C_{22} + a_{23}C_{23}$
$= a_{31}C_{31} + a_{32}C_{32} + a_{33}C_{33}$

But if elements of a row (or column) are multiplied with cofactors of any other row (or column), then their sum is zero.

Adjoint of a Matrix

If $A = [a_{ij}]_{nxn}$, then adjoint of A, denoted by adj (A), is defined as $[C_{ij}]^T_{nxn}$, where C_{ij} is the cofactor of a_{ij}.

If $A = \begin{bmatrix} a_{11} & a_{12} & a_{13} \\ a_{21} & a_{22} & a_{23} \\ a_{31} & a_{32} & a_{33} \end{bmatrix}$, then adj $(A) = \begin{bmatrix} C_{11} & C_{12} & C_{13} \\ C_{21} & C_{22} & C_{23} \\ C_{31} & C_{32} & C_{33} \end{bmatrix}^T$

NOTE • If $A = \begin{bmatrix} a & b \\ c & d \end{bmatrix}$, then adj $(A) = \begin{bmatrix} d & -b \\ -c & a \end{bmatrix}$.

Properties of Adjoint of a Matrix

Let A be a square matrix of order n, then

(i) $(\text{adj } A)A = A(\text{adj } A) = |A| \cdot I_n$

(ii) $|\text{adj } A| = |A|^{n-1}$, if $|A| \neq 0$

(iii) $\text{adj }(AB) = (\text{adj } B)(\text{adj } A)$

(iv) $\text{adj}(A^T) = (\text{adj } A)^T$

(v) $\text{adj}(\text{adj } A) = |A|^{n-2} A$, if $|A| \neq 0$

(vi) $|\text{adj}(\text{adj } A)| = |A|^{(n-1)^2}$, if $|A| \neq 0$

Inverse of a Matrix

Let A be any non-singular (i.e. $|A| \neq 0$) square matrix, then inverse of A can be obtained by following two ways.

1. Using determinants

In this, $$A^{-1} = \frac{1}{|A|} \text{adj }(A)$$

2. Using Elementary operations

In this, first write $A = IA$ (for applying row operations) or $A = AI$ (for applying column operations) and then reduce A of LHS to I, by applying elementary operations simultaneously on A of LHS and I of RHS. If it reduces to $I = PA$ or $I = AP$, then $P = A^{-1}$.

Properties of Inverse of a Matrix

(i) A square matrix is invertible if and only if it is non-singular.

(ii) If $A = \text{diag}(\lambda_1, \lambda_2, \ldots, \lambda_n)$, then $A^{-1} = \text{diag}(\lambda_1^{-1}, \lambda_2^{-1}, \ldots, \lambda_n^{-1})$ provided $\lambda_i \neq 0 \forall, i = 1, 2, \ldots n$.

Solution of System of Linear Equations in Two and Three Variables

Let system of linear equations in three variables be
$$a_1x + b_1y + c_1z = d_1,\ a_2x + b_2y + c_2z = d_2$$
and $$a_3x + b_3y + c_3z = d_3.$$

Now, we have two methods to solve these equations.

1. Matrix Method

In this method we first write the above system of equations in matrix form as shown below.

$$\begin{bmatrix} a_1 & b_1 & c_1 \\ a_2 & b_2 & c_2 \\ a_3 & b_3 & c_3 \end{bmatrix} \begin{bmatrix} x \\ y \\ z \end{bmatrix} = \begin{bmatrix} d_1 \\ d_2 \\ d_3 \end{bmatrix} \text{ or } AX = B$$

where, $A = \begin{bmatrix} a_1 & b_1 & c_1 \\ a_2 & b_2 & c_2 \\ a_3 & b_3 & c_3 \end{bmatrix}, X = \begin{bmatrix} x \\ y \\ z \end{bmatrix}$ and $B = \begin{bmatrix} d_1 \\ d_2 \\ d_3 \end{bmatrix}$

***Case* I When system of equations is non-homogeneous (i.e. when $B \neq 0$).**

- If $|A| \neq 0$, then the system of equations is consistent and has a unique solution given by $X = A^{-1}B$.
- If $|A| = 0$ and (adj A), $B \neq 0$, then the system of equations is inconsistent and has no solution.
- If $|A| = 0$ and $(\text{adj } A) \cdot B = O$, then the system of equations may be either consistent or inconsistent according as the system have infinitely many solutions or no solution.

***Case* II When system of equations is homogeneous (i.e. when $B = 0$).**

- If $|A| \neq 0$, then system of equations has only trivial solution, namely $x = 0, y = 0$ and $z = 0$.
- If $|A| = 0$, then system of equations has non-trivial solution, which will be infinite in numbers.

2. Cramer's Rule Method

In this method we first determine

$$D = \begin{vmatrix} a_1 & b_1 & c_1 \\ a_2 & b_2 & c_2 \\ a_3 & b_3 & c_3 \end{vmatrix},\ D_1 = \begin{vmatrix} d_1 & b_1 & c_1 \\ d_2 & b_2 & c_2 \\ d_3 & b_3 & c_3 \end{vmatrix},$$

$$D_2 = \begin{vmatrix} a_1 & d_1 & c_1 \\ a_2 & d_2 & c_2 \\ a_3 & d_3 & c_3 \end{vmatrix} \text{ and } D_3 = \begin{vmatrix} a_1 & b_1 & d_1 \\ a_2 & b_2 & d_2 \\ a_3 & b_3 & d_3 \end{vmatrix}$$

***Case* I When system of equations is non-homogeneous**

- If $D \neq 0$, then it is consistent with unique solution given by $x = \frac{D_1}{D}, y = \frac{D_2}{D}, z = \frac{D_3}{D}$.
- If $D = 0$ and atleast one of D_1, D_2 and D_3 is non-zero, then it is inconsistent (no solution).
- If $D = D_1 = D_2 = D_3 = 0$, then it may be consistent or inconsistent according as the system have infinitely many solutions or no solution.

Case II When system of equations is homogeneous

- If $D \neq 0$, then $x = y = z = 0$ is the only solution, i.e. the trivial solution.
- If $D = 0$, then it has infinitely many solutions.

Above methods can be used, in a similar way, for the solution of system of linear equations in two variables.

DAY PRACTICE SESSION 1

FOUNDATION QUESTIONS EXERCISE

1 If $x = cy + bz$, $y = az + cx$ and $z = bx + ay$, where x, y and z are not all zero, then $a^2 + b^2 + c^2$ is equal to

(a) $1 + 2abc$ (b) $1 - 2abc$
(c) $1 + abc$ (d) $abc - 1$

2 Consider the set A of all determinants of order 3 with entries 0 or 1 only. Let B be the subset of A consisting of all determinants with value 1 and C be the subset of A consisting of all determinants with value -1. Then,

(a) C is empty
(b) B and C have the same number of elements
(c) $A = B \cup C$
(d) B has twice as many elements as C

3 If x, y and z are positive, then $\begin{vmatrix} 1 & \log_x y & \log_x z \\ \log_y x & 1 & \log_y z \\ \log_z x & \log_z y & 1 \end{vmatrix}$ is equal to

(a) 0 (b) 1 (c) -1 (d) None of these

4 If a, b and c are cube roots of unity, then
$\begin{vmatrix} e^a & e^{2a} & e^{3a}-1 \\ e^b & e^{2b} & e^{3b}-1 \\ e^c & e^{2c} & e^{3c}-1 \end{vmatrix}$ is equal to

(a) 0 (b) e (c) e^2 (d) e^3

5 If $px^4 + qx^3 + rx^2 + sx + t$
$= \begin{vmatrix} x^2+3x & x-1 & x+3 \\ x+1 & -2x & x-4 \\ x-3 & x+4 & 3x \end{vmatrix}$, where p, q, r, s and t are constants, then t is equal to

(a) 0 (b) 1 (c) 2 (d) -1

6 If $f(x) = \begin{vmatrix} 1 & x & x+1 \\ 2x & x(x-1) & (x+1)x \\ 3x(x-1) & x(x-1)(x-2) & (x+1)x(x-1) \end{vmatrix}$, then $f(50)$ is equal to

(a) 0 (b) 50 (c) 1 (d) -50

7 If α, β and γ are the roots of the equation $x^3 + px + q = 0$, then the value of the determinant $\begin{vmatrix} \alpha & \beta & \gamma \\ \beta & \gamma & \alpha \\ \gamma & \alpha & \beta \end{vmatrix}$ is

(a) 0 (b) -2 (c) 2 (d) 4

8 If ω is a cube root of unity, then a root of the following
$\begin{vmatrix} x-\omega-\omega^2 & \omega & \omega^2 \\ \omega & x-\omega-1 & 1 \\ \omega^2 & 1 & x-1-\omega^2 \end{vmatrix} = 0$ is

(a) $x = 0$ (b) $x = -1$
(c) $x = \omega$ (d) None of these

9 If ω is an imaginary cube root of unity, then the value of
$\begin{vmatrix} a & b\omega^2 & a\omega \\ b\omega & c & b\omega^2 \\ c\omega^2 & a\omega & c \end{vmatrix}$ is

(a) $a^3 + b^3 + c^2 - 3abc$ (b) $a^2b - b^2c$
(c) 0 (d) $a^2 + b^2 + c^2$

10 If $\begin{vmatrix} x-4 & 2x & 2x \\ 2x & x-4 & 2x \\ 2x & 2x & x-4 \end{vmatrix} = (A + Bx)(x - A)^2$, then the ordered pair (A, B) is equal to → JEE Mains 2018

(a) $(-4,-5)$ (b) $(-4,3)$
(c) $(-4,5)$ (d) $(4,5)$

11 If x, y, z are non-zero real numbers and
$\begin{vmatrix} 1+x & 1 & 1 \\ 1+y & 1+2y & 1 \\ 1+z & 1+z & 1+3z \end{vmatrix} = 0$,
then $x^{-1} + y^{-1} + z^{-1}$ is equal to

(a) 0 (b) -1
(c) -3 (d) -6

12 If $\begin{vmatrix} b+c & c+a & a+b \\ c+a & a+b & b+c \\ a+b & b+c & c+a \end{vmatrix} = k \begin{vmatrix} a & b & c \\ b & c & a \\ c & a & b \end{vmatrix}$, then k is equal to

(a) 0 (b) 1 (c) 2 (d) 3

13 Let a, b and c be such that $(b + c) \neq 0$. If
$\begin{vmatrix} a & a+1 & a-1 \\ -b & b+1 & b-1 \\ c & c-1 & c+1 \end{vmatrix} + \begin{vmatrix} a+1 & b+1 & c-1 \\ a-1 & b-1 & c+1 \\ (-1)^{n+2}a & (-1)^{n+1}b & (-1)^n c \end{vmatrix} = 0$,
then the value of n is

(a) zero (b) an even integer
(c) an odd integer (d) an integer

14 If one of the roots of the equation
$\begin{vmatrix} 7 & 6 & x^2-13 \\ 2 & x^2-13 & 2 \\ x^2-13 & 3 & 7 \end{vmatrix} = 0$ is $x = 2$, then sum of all other five roots is

(a) -2 (b) 0
(c) $2\sqrt{5}$ (d) $\sqrt{15}$

15 If a, b and c are sides of a scalene triangle, then the value of $\begin{vmatrix} a & b & c \\ b & c & a \\ c & a & b \end{vmatrix}$ is → JEE Mains 2013

(a) non-negative (b) negative
(c) positive (d) non-positive

16 If the system of linear equations

$$x + ky + 3z = 0,\ 3x + ky - 2z = 0$$

and $\quad 2x + 4y - 3z = 0$

has a non-zero solution (x, y, z), then $\frac{xz}{y^2}$ is equal to

→ JEE Mains 2018

(a) −10 (b) 10
(c) −30 (d) 30

17 If $A = [a_{ij}]_{n \times n}$ and $a_{ij} = (i^2 + j^2 - ij)(j - i)$, n is odd, then which of the following is not the value of tr(A).

(a) 0 (b) $|A|$
(c) $2|A|$ (d) None of these

18 If the coordinates of the vertices of an equilateral triangle with sides of length a are $(x_1, y_1), (x_2, y_2)$ and (x_3, y_3), then $\begin{vmatrix} x_1 & y_1 & 1 \\ x_2 & y_2 & 1 \\ x_3 & y_3 & 1 \end{vmatrix}^2$ is equal to

→ NCERT Exemplar

(a) $\frac{a^4}{4}$ (b) $\frac{3a^2}{4}$ (c) $\frac{5a^4}{4}$ (d) $\frac{3a^4}{4}$

19 The points $A(a, b + c), B(b, c + a)$ and $C(c, a + b)$ are

(a) vertices of an isoscele triangle
(b) vertices of an equilateral triangle
(c) collinear
(d) None of the above

20 If A_1, B_1, C_1, ... are respectively the co-factors of the elements $x_1, y_1, z_1, \ldots$ of the determinant

$$\Delta = \begin{vmatrix} x_1 & y_1 & z_1 \\ x_2 & y_2 & z_2 \\ x_3 & y_3 & z_3 \end{vmatrix}, \text{ then } \begin{vmatrix} B_2 & C_2 \\ B_3 & C_3 \end{vmatrix} \text{ equals to}$$

(a) $x_1\Delta$ (b) $x_1x_3\Delta$
(c) $(x_1 + y_1)\Delta$ (d) None of these

21 If $A = \begin{bmatrix} 2 & -3 \\ -4 & 1 \end{bmatrix}$, then adj $(3A^2 + 12A)$ is equal to

→ JEE Mains 2017

(a) $\begin{bmatrix} 72 & -84 \\ -63 & 51 \end{bmatrix}$ (b) $\begin{bmatrix} 51 & 63 \\ 84 & 72 \end{bmatrix}$
(c) $\begin{bmatrix} 51 & 84 \\ 63 & 72 \end{bmatrix}$ (d) $\begin{bmatrix} 72 & -63 \\ -84 & 51 \end{bmatrix}$

22 Which of the following is/are incorrect?

(i) Adjoint of a symmetric matrix is symmetric
(ii) Adjoint of a unit matrix is a unit matrix
(iii) $A(adj\ A) = (adj\ A) = |A| I$
(iv) Adjoint of a diagonal matrix is a diagonal matrix .

(a) (i) (b) (ii)
(c) (iii) and (iv) (d) None of these

23 If $P = \begin{bmatrix} 1 & \alpha & 3 \\ 1 & 3 & 3 \\ 2 & 4 & 4 \end{bmatrix}$ is the adjoint of a 3×3 matrix A and $|A| = 4$, then α is equal to

→ JEE Mains 2013

(a) 4 (b) 11 (c) 5 (d) 0

24 If $A = \begin{bmatrix} 5a & -b \\ 3 & 2 \end{bmatrix}$ and $A\ adj\ A = AA^T$, then $5a + b$ is equal to

→ JEE Mains 2016

(a) − 1 (b) 5 (c) 4 (d) 13

25 If $A = \begin{bmatrix} a & b & c \\ x & y & z \\ p & q & r \end{bmatrix}, B = \begin{bmatrix} q & -b & y \\ -p & a & -x \\ r & -c & z \end{bmatrix}$ and if A is invertible, then which of the following is not true?

(a) $|A| = -|B|$ and $|adj\ A| \neq |adj\ B|$
(b) $|A| = -|B|$ and $|adj\ A| = |adj\ B|$
(c) A is invertible iff B is invertible
(b) None of the above

26 If A an 3×3 non-singular matrix such that $AA' = A'A$ and $B = A^{-1}A'$, then BB' is equal to

→ JEE Mains 2014

(a) $I + B$ (b) I (c) B^{-1} (d) $(B^{-1})'$

27 If $\begin{bmatrix} 1 & -\tan\theta \\ \tan\theta & 1 \end{bmatrix} \begin{bmatrix} 1 & \tan\theta \\ -\tan\theta & 1 \end{bmatrix}^{-1} = \begin{bmatrix} a & -b \\ b & a \end{bmatrix}$, then

(a) $a = 1, b = 1$
(b) $a = \sin 2\theta, b = \cos 2\theta$
(c) $a = \cos 2\theta, b = \sin 2\theta$
(d) None of the above

28 If $A = \begin{bmatrix} 1 & 0 & 0 \\ 2 & 1 & 0 \\ 3 & 2 & 1 \end{bmatrix}$ and u_1, u_2 are column matrices such that $Au_1 = \begin{bmatrix} 1 \\ 0 \\ 0 \end{bmatrix}$ and $Au_2 = \begin{bmatrix} 0 \\ 1 \\ 0 \end{bmatrix}$, then $u_1 + u_2$ is equal to

(a) $\begin{bmatrix} -1 \\ 1 \\ 0 \end{bmatrix}$ (b) $\begin{bmatrix} -1 \\ 1 \\ -1 \end{bmatrix}$ (c) $\begin{bmatrix} -1 \\ -1 \\ 0 \end{bmatrix}$ (d) $\begin{bmatrix} 1 \\ -1 \\ -1 \end{bmatrix}$

29 The number of values of k, for which the system of equations $(k+1)x + 8y = 4k$ and $kx + (k+3)y = 3k - 1$ has no solution, is

→ JEE Mains 2013

(a) infinite (b) 1
(c) 2 (d) 3

30 The system of equations

$$\alpha x + y + z = \alpha - 1,\ x + \alpha y + z = \alpha - 1$$

and $\quad x + y + \alpha z = \alpha - 1$

has no solution, if α is

(a) 1 (b) not −2
(c) either −2 or 1 (d) −2

31 If the system of linear equations

→ JEE Mains 2013

$$x_1 + 2x_2 + 3x_3 = 6,\ x_1 + 3x_2 + 5x_3 = 9$$
$$2x_1 + 5x_2 + ax_3 = b$$

is consistent and has infinite number of solutions, then

(a) $a = 8$, b can be any real number
(b) $b = 15$, a can be any real number
(c) $a \in R - \{8\}$ and $b \in R - \{15\}$
(d) $a = 8, b = 15$

32 If the trivial solution is the only solution of the system of equations

$$x - ky + z = 0,\ kx + 3y - kz = 0$$
and $$3x + y - z = 0$$

Then, the set of all values of k is

(a) $\{2, -3\}$ (b) $R - \{2, -3\}$ (c) $R - \{2\}$ (d) $R - \{-3\}$

33 Let A, other than I or $-I$, be a 2×2 real matrix such that $A^2 = I$, I being the unit matrix. Let tr (A) be the sum of diagonal elements of A. → JEE Mains 2013

Statement I tr $(A) = 0$

Statement II det $(A) = -1$

(a) Statement I is true, Statement II is true; Statement II is a correct explanation for Statement I
(b) Statement I is true, Statement II is true; Statement II is not a correct explanation for Statement I
(c) Statement I is true; Statement II is false
(d) Statement I is false; Statement II is true

34 The set of all values of λ for which the system of linear equations $2x_1 - 2x_2 + x_3 = \lambda x_1$, $2x_1 - 3x_2 + 2x_3 = \lambda x_2$ and $-x_1 + 2x_2 = \lambda x_3$ has a non-trivial solution.

(a) is an empty set → JEE Mains 2015
(b) is a singleton set
(c) contains two elements
(d) contains more than two elements

35 **Statement I** Determinant of a skew-symmetric matrix of order 3 is zero.

Statement II For any matrix A, $\det(A^T) = \det(A)$ and $\det(-A) = -\det(A)$.

Where, $\det(A)$ denotes the determinant of matrix A. Then, → JEE Mains 2013

(a) Statement I is true and Statement II is false
(b) Both statements are true
(c) Both statements are false
(d) Statement I is false and Statement II is true

DAY PRACTICE SESSION 2

PROGRESSIVE QUESTIONS EXERCISE

1 If $a^2 + b^2 + c^2 = -2$ and

$$f(x) = \begin{vmatrix} 1+a^2x & (1+b^2)x & (1+c^2)x \\ (1+a^2)x & 1+b^2x & (1+c^2)x \\ (1+a^2)x & (1+b^2)x & 1+c^2x \end{vmatrix},$$

then $f(x)$ is a polynomial of degree.

(a) 2 (b) 3 (c) 0 (d) 1

2 If A is a square matrix of order 3 such that $|A| = 2$, then $|(\text{adj } A^{-1})^{-1}|$ is

(a) 1 (b) 2 (c) 4 (d) 8

3 The equations $(k-1)x + (3k+1)y + 2kz = 0$,
$$(k-1)x + (4k-2)y + (k+3)z = 0$$
and $$2x + (3k+1)y + 3(k-1)z = 0$$
gives non-trivial solution for some values of k, then the ratio $x : y : z$ when k has the smallest of these values.

(a) 3:2:1 (b) 3:3:2 (c) 1:3:1 (d) 1:1:1

4 If $x = 1 + 2 + 4 + \ldots$ upto k terms, $y = 1 + 3 + 9 + \ldots$ upto k terms and $c = 1 + 5 + 25 + \ldots$ upto k terms. Then,

$$\Delta = \begin{vmatrix} x & 2y & 4z \\ 3 & 3 & 3 \\ 2^k & 3^k & 5^k \end{vmatrix} \text{ equals to}$$

(a) $(20)^k$ (b) 5^k (c) 0 (d) $2^k + 3^k + 5^k$

5 Product of roots of equation $\begin{vmatrix} 1+2x & 1 & 1-x \\ 2-x & 2+x & 3+x \\ x & 1+x & 1-x^2 \end{vmatrix} = 0$

(a) $\frac{1}{2}$ (b) $\frac{3}{4}$ (c) $\frac{4}{3}$ (d) $\frac{1}{4}$

6 If the equations $a(y+z) = x, b(z+x) = y, c(x+y) = z$ have non-trivial solution, then $\frac{1}{1+a} + \frac{1}{1+b} + \frac{1}{1+c}$ is equal to

(a) 1 (b) 2 (c) -1 (d) -2

7 Let a, b and c be positive real numbers. The following system of equations in x, y and z.

$$\frac{x^2}{a^2} + \frac{y^2}{b^2} - \frac{z^2}{c^2} = 1, \frac{x^2}{a^2} - \frac{y^2}{b^2} + \frac{z^2}{c^2} = 1 \text{ and } -\frac{x^2}{a^2} + \frac{y^2}{b^2} + \frac{z^2}{c^2} = 1$$

has

(a) no solution (b) unique solution
(c) infinitely many solutions (d) finitely many solutions

8 If S is the set of distinct values of b for which the following system of linear equations

$$x + y + z = 1,\ x + ay + z = 1 \text{ and } ax + by + z = 0$$

has no solution, then S is → JEE Mains 2017

(a) an infinite set
(b) a finite set containing two or more elements
(c) singleton set
(d) a empty set

9 If M is a 3×3 matrix, where $M^T M = I$ and $\det(M) = 1$, then the value of det $(M - I)$ is

(a) -1 (b) 1
(c) 0 (d) None of these

10 If $a_1, a_2, \ldots, a_n, \ldots$ are in GP, then the determinant

$$\Delta = \begin{vmatrix} \log a_n & \log a_{n+1} & \log a_{n+2} \\ \log a_{n+3} & \log a_{n+4} & \log a_{n+5} \\ \log a_{n+6} & \log a_{n+7} & \log a_{n+8} \end{vmatrix} \text{ is equal to}$$

(a) 2 (b) 4 (c) 0 (d) 1

11 Let A be a square matrix of order 2 with $|A| \neq 0$ such that $|A + |A| \text{adj}(A)| = 0$, then the value of $|A - |A| \text{adj}(A)|$ is

(a) 1 (b) 2
(c) 3 (d) 4

12 Let P and Q be 3×3 matrices $P \neq Q$. If $P^3 = Q^3$ and $P^2Q = Q^2P$, then the determinant of $(P^2 + Q^2)$ is

(a) -2 (b) 1
(c) 0 (d) -1

13 If $\alpha, \beta \neq 0$, $f(n) = \alpha^n + \beta^n$ and $\begin{vmatrix} 3 & 1+f(1) & 1+f(2) \\ 1+f(1) & 1+f(2) & 1+f(3) \\ 1+f(2) & 1+f(3) & 1+f(4) \end{vmatrix}$

$= K(1-\alpha)^2 (1-\beta)^2 (\alpha - \beta)^2$, then K is equal to

→ JEE Mains 2014

(a) $\alpha\beta$ (b) $\frac{1}{\alpha\beta}$ (c) 1 (d) -1

14 Area of triangle whose vertices are $(a, a^2), (b, b^2), (c, c^2)$ is $\frac{1}{2}$, and the area of triangle whose vertices are $(p, p^2), (q, q^2)$ and (r, r^2) is 4, then the value of

$$\begin{vmatrix} (1+ap)^2 & (1+bp)^2 & (1+cp)^2 \\ (1+aq)^2 & (1+bq)^2 & (1+cq)^2 \\ (1+ar)^2 & (1+br)^2 & (1+cr)^2 \end{vmatrix} \text{ is}$$

(a) 2 (b) 4 (c) 8 (d) 16

15 Let $\det A = \begin{vmatrix} l & m & n \\ p & q & r \\ 1 & 1 & 1 \end{vmatrix}$ and if $(l-m)^2 + (p-q)^2 = 9$, $(m-n)^2 + (q-r)^2 = 16$, $(n-l)^2 + (r-p)^2 = 25$, then the value of $(\det A)^2$ equals to

(a) 36 (b) 100 (c) 144 (b) 169

ANSWERS

SESSION 1	**1** (b)	**2** (b)	**3** (a)	**4** (a)	**5** (a)	**6** (a)	**7** (a)	**8** (a)	**9** (c)	**10** (c)
	11 (c)	**12** (c)	**13** (c)	**14** (a)	**15** (b)	**16** (b)	**17** (d)	**18** (d)	**19** (c)	**20** (a)
	21 (b)	**22** (d)	**23** (b)	**24** (b)	**25** (a)	**26** (b)	**27** (c)	**28** (d)	**29** (d)	**30** (d)
	31 (d)	**32** (b)	**33** (b)	**34** (c)	**35** (a)					
SESSION 2	**1** (a)	**2** (c)	**3** (d)	**4** (c)	**5** (a)	**6** (b)	**7** (d)	**8** (c)	**9** (c)	**10** (c)
	11 (d)	**12** (c)	**13** (c)	**14** (d)	**15** (c)					

Hints and Explanations

SESSION 1

1 Given equation can be rewritteen as

$$x - cy - bz = 0$$
$$cx - y + az = 0$$
$$bx + ay - z = 0$$

Since, x, y and z are not all zero

$\therefore$ The above system have non-trivial solution.

$$\therefore \begin{vmatrix} 1 & -c & -b \\ c & -1 & a \\ b & a & -1 \end{vmatrix} = 0$$

$$\Rightarrow 1(1 - a^2) + c(-c - ab) - b(ac + b) = 0$$
$$\Rightarrow \quad 1 - a^2 - c^2 - abc - abc - b^2 = 0$$
$$\Rightarrow \quad a^2 + b^2 + c^2 = 1 - 2abc$$

2 If we interchange any two rows of a determinant in the set B, its value becomes -1 and hence it is in C. Likewise, for every determinant in C, there is corresponding determinant in B. So, B and C have the same number of elements.

3 Let $\Delta = \begin{vmatrix} 1 & \log_x y & \log_x z \\ \log_y x & 1 & \log_y z \\ \log_z x & \log_z y & 1 \end{vmatrix}$

$$= \begin{vmatrix} \frac{\log x}{\log x} & \frac{\log y}{\log x} & \frac{\log z}{\log x} \\ \frac{\log x}{\log y} & \frac{\log y}{\log y} & \frac{\log z}{\log y} \\ \frac{\log x}{\log z} & \frac{\log y}{\log z} & \frac{\log z}{\log z} \end{vmatrix}$$

By taking common factors from the rows, we get

$$\Delta = \frac{1}{\log x \cdot \log y \cdot \log z} \begin{vmatrix} \log x & \log y & \log z \\ \log x & \log y & \log z \\ \log x & \log y & \log z \end{vmatrix}$$

Now, by taking common factor from the columns, we get $\begin{vmatrix} 1 & 1 & 1 \\ 1 & 1 & 1 \\ 1 & 1 & 1 \end{vmatrix} = 0$

4 Let $\Delta = \begin{vmatrix} e^a & e^{2a} & e^{3a} \\ e^b & e^{2b} & e^{3b} \\ e^c & e^{2c} & e^{3c} \end{vmatrix} - \begin{vmatrix} e^a & e^{2a} & 1 \\ e^b & e^{2b} & 1 \\ e^c & e^{2c} & 1 \end{vmatrix}$

$$= e^a \cdot e^b \cdot e^c \begin{vmatrix} 1 & e^a & e^{2a} \\ 1 & e^b & e^{2b} \\ 1 & e^c & e^{2c} \end{vmatrix} + \begin{vmatrix} e^a & 1 & e^{2a} \\ e^b & 1 & e^{2b} \\ e^c & 1 & e^{2c} \end{vmatrix}$$

$$= \begin{vmatrix} 1 & e^a & e^{2a} \\ 1 & e^b & e^{2b} \\ 1 & e^c & e^{2c} \end{vmatrix} - \begin{vmatrix} 1 & e^a & e^{2a} \\ 1 & e^b & e^{2b} \\ 1 & e^c & e^{2c} \end{vmatrix}$$

$$= 0 \quad [\because a + b + c = 0 \Rightarrow e^{a+b+c} = 1]$$

5 Put $x = 0$ in the given equation, we get

$$t = \begin{vmatrix} 0 & -1 & 3 \\ 1 & 0 & -4 \\ -3 & 4 & 0 \end{vmatrix} = -12 + 12 = 0$$

6 On taking common factors x from C_2, $(x + 1)$ from C_3 and $(x - 1)$ from R_3, we get

$$f(x) = x(x^2 - 1) \begin{vmatrix} 1 & 1 & 1 \\ 2x & x - 1 & x \\ 3x & x - 2 & x \end{vmatrix}$$

$$= x(x^2 - 1) \begin{vmatrix} 1 & 0 & 0 \\ 2x & -(x + 1) & 1 \\ 3x & -2(x + 1) & 2 \end{vmatrix}$$

$$= 0 \quad \begin{bmatrix} C_3 \to C_3 - C_2, \\ C_2 \to C_2 - C_1 \end{bmatrix}$$

$$\therefore \quad f(50) = 0$$

7 Clearly, $\alpha + \beta + \gamma = 0$

On applying $C_1 \to C_1 + C_2 + C_3$ in the given determinant, we get

$$\therefore \Delta = \begin{vmatrix} \alpha + \beta + \gamma & \beta & \gamma \\ \alpha + \beta + \gamma & \gamma & \alpha \\ \alpha + \beta + \gamma & \alpha & \beta \end{vmatrix}$$

$$= \begin{vmatrix} 0 & \beta & \gamma \\ 0 & \gamma & \alpha \\ 0 & \alpha & \beta \end{vmatrix} = 0$$

8 On applying $C_1 \to C_1 + C_2 + C_3$, we get

$$\begin{vmatrix} x & \omega & \omega^2 \\ x & x + \omega^2 & 1 \\ x & 1 & x + \omega \end{vmatrix} = 0$$

$$[\because 1 + \omega + \omega^2 = 0]$$

$$\Rightarrow \quad x \begin{vmatrix} 1 & \omega & \omega^2 \\ 1 & x + \omega^2 & 1 \\ 1 & 1 & x + \omega \end{vmatrix} = 0$$

$$\therefore \quad x = 0$$

9 $\Delta = \begin{vmatrix} a(1 + \omega) & b\omega^2 & a\omega \\ b(\omega + \omega^2) & c & b\omega^2 \\ c(\omega^2 + 1) & a\omega & c \end{vmatrix}$

$$[\because C_1 \to C_1 + C_3]$$

$$= \begin{vmatrix} -a\omega^2 & b\omega^2 & a\omega \\ -b & c & b\omega^2 \\ -c\omega & a\omega & c \end{vmatrix}$$

$$= \omega^2 \cdot \omega \begin{vmatrix} -a & b & a\omega^2 \\ -b & c & b\omega^2 \\ -c & a & c\omega^2 \end{vmatrix}$$

$$= -\omega^5 \begin{vmatrix} a & b & a \\ b & c & b \\ c & a & c \end{vmatrix} = 0$$

10 Given, $\begin{vmatrix} x - 4 & 2x & 2x \\ 2x & x - 4 & 2x \\ 2x & 2x & x - 4 \end{vmatrix}$

$$= (A + Bx)(x - A)^2$$

On applying $C_1 \to C_1 + C_2 + C_3$, we get

$$\begin{vmatrix} 5x - 4 & 2x & 2x \\ 5x - 4 & x - 4 & 2x \\ 5x - 4 & 2x & x - 4 \end{vmatrix}$$

$$= (A + Bx)(x - A)^2$$

On taking common $(5x - 4)$ from C_1, we get

$$(5x - 4) \begin{vmatrix} 1 & 2x & 2x \\ 1 & x - 4 & 2x \\ 1 & 2x & x - 4 \end{vmatrix}$$

$$= (A + Bx)(x - A)^2$$

On applying $R_2 \to R_2 - R_1$ and $R_3 \to R_3 - R_1$, we get

$$\therefore (5x - 4) \begin{vmatrix} 1 & 2x & 2x \\ 0 & -x - 4 & 0 \\ 0 & 0 & -x - 4 \end{vmatrix}$$

$$= (A + Bx)(x - A)^2$$

On expanding along C_1, we get

$$(5x - 4)(x + 4)^2 = (A + Bx)(x - A)^2$$

On comparing, we get

$$A = -4 \text{ and } B = 5$$

11 Let $\Delta = \begin{vmatrix} 1 + x & 1 & 1 \\ 1 + y & 1 + 2y & 1 \\ 1 + z & 1 + z & 1 + 3z \end{vmatrix} = 0$

On applying $C_1 \to C_1 - C_3$ and $C_2 \to C_2 - C_3$, we get

$$\Delta = \begin{vmatrix} x & 0 & 1 \\ y & 2y & 1 \\ -2z & -2z & 1 + 3z \end{vmatrix}$$

On expanding along R_1, we get

$$\Delta = x[2y(1 + 3z) + 2z] + 1[-2yz + 4yz] = 0$$

$$\Rightarrow \quad 2[xy + 3xyz + xz] + 2yz = 0$$

$$\Rightarrow \quad xy + yz + zx + 3xyz = 0$$

$$\Rightarrow \quad \frac{1}{x} + \frac{1}{y} + \frac{1}{z} = -3$$

12 Let $\Delta = \begin{vmatrix} b+c & c+a & a+b \\ c+a & a+b & b+c \\ a+b & b+c & c+a \end{vmatrix}$

$$= \begin{vmatrix} 2(a+b+c) & c+a & a+b \\ 2(a+b+c) & a+b & b+c \\ 2(a+b+c) & b+c & c+a \end{vmatrix}$$

[using $C_1 \to C_1 + C_2 + C_3$]

$$= 2\begin{vmatrix} a+b+c & c+a & a+b \\ a+b+c & a+b & b+c \\ a+b+c & b+c & c+a \end{vmatrix}$$

[taking common 2 from C_1]

$$= 2\begin{vmatrix} a+b+c & -b & -c \\ a+b+c & -c & -a \\ a+b+c & -a & -b \end{vmatrix}$$

[using $C_2 \to C_2 - C_1, C_3 \to C_3 - C_1$]

$$= 2\begin{vmatrix} a & b & c \\ b & c & a \\ c & a & b \end{vmatrix}$$

[using $C_1 \to C_1 + C_2 + C_3$]

On comparing, $k = 2$

13 $\begin{vmatrix} a & a+1 & a-1 \\ -b & b+1 & b-1 \\ c & c-1 & c+1 \end{vmatrix} + (-1)^n \begin{vmatrix} a+1 & b+1 & c-1 \\ a-1 & b-1 & c+1 \\ a & -b & c \end{vmatrix}$

$$= \begin{vmatrix} a & a+1 & a-1 \\ -b & b+1 & b-1 \\ c & c-1 & c+1 \end{vmatrix} + (-1)^n \begin{vmatrix} a+1 & a-1 & a \\ b+1 & b-1 & -b \\ c-1 & c+1 & c \end{vmatrix}$$

$[\because |A'| = |A|]$

$$= \begin{vmatrix} a & a+1 & a-1 \\ -b & b+1 & b-1 \\ c & c-1 & c+1 \end{vmatrix} + (-1)^{n+1} \begin{vmatrix} a+1 & a & a-1 \\ b+1 & -b & b-1 \\ c-1 & c & c+1 \end{vmatrix}$$

$$= [1 + (-1)^{n+2}] \begin{vmatrix} a & a+1 & a-1 \\ -b & b+1 & b-1 \\ c & c-1 & c+1 \end{vmatrix}$$

This is equal to zero only, if $n+2$ is an odd i.e. n is an odd integer.

14 $\begin{vmatrix} 7 & 6 & x^2-13 \\ 2 & x^2-13 & 2 \\ x^2-13 & 3 & 7 \end{vmatrix} = 0$

On applying $R_1 \to R_1 + R_2 + R_3$, we get

$$\begin{vmatrix} x^2-4 & x^2-4 & x^2-4 \\ 2 & x^2-13 & 2 \\ x^2-13 & 3 & 7 \end{vmatrix} = 0$$

$$\Rightarrow (x^2-4)\begin{vmatrix} 1 & 1 & 1 \\ 2 & x^2-13 & 2 \\ x^2-13 & 3 & 7 \end{vmatrix} = 0$$

Now, on applying $C_2 \to C_2 - C_1$ and $C_3 \to C_3 - C_1$, we get

$$(x^2-4)\begin{vmatrix} 1 & 0 & 0 \\ 2 & x^2-15 & 0 \\ x^2-13 & 16-x^2 & 20-x^2 \end{vmatrix} = 0$$

On expanding along R_1, we get

$$(x^2-4)(x^2-15)(x^2-20) = 0$$

Thus, other five roots are

$$-2, \pm\sqrt{15}, \pm 2\sqrt{5}$$

Hence, sum of other five roots is -2.

15 Let $\Delta = \begin{vmatrix} a & b & c \\ b & c & a \\ c & a & b \end{vmatrix}$

$= a(bc - a^2) - b(b^2 - ac) + c(ab - c^2)$

$= abc - a^3 - b^3 + abc + abc - c^3$

$= -(a^3 + b^3 + c^3 - 3abc)$

$= -(a+b+c)(a^2 + b^2 + c^2 - ab - bc - ca)$

$= -\frac{1}{2}(a+b+c)\{(a-b)^2 + (b-c)^2 + (c-a)^2\}$

< 0, where $a \neq b \neq c$

16 We have,

$x + ky + 3z = 0$; $3x + ky - 2z = 0$;
$2x + 4y - 3z = 0$

System of equation has non-zero solution, if

$$\begin{vmatrix} 1 & k & 3 \\ 3 & k & -2 \\ 2 & 4 & -3 \end{vmatrix} = 0$$

$\Rightarrow (-3k + 8) - k(-9 + 4) + 3(12 - 2k) = 0$

$\Rightarrow \quad -3k + 8 + 9k - 4k + 36 - 6k = 0$

$\Rightarrow \quad -4k + 44 = 0 \Rightarrow k = 11$

Let $z = \lambda$, then we get

$x + 11y + 3\lambda = 0$ ...(i)

$3x + 11y - 2\lambda = 0$ ...(ii)

and $2x + 4y - 3\lambda = 0$ ...(iii)

On solving Eqs. (i) and (ii), we get

$$x = \frac{5\lambda}{2}, y = \frac{-\lambda}{2}, z = \lambda$$

$$\Rightarrow \quad \frac{xz}{y^2} = \frac{5\lambda^2}{2 \times \left(-\frac{\lambda}{2}\right)^2} = 10$$

17 We have,

$$a_{ij} = (i^2 + j^2 - ij)(j - i)$$

$\therefore \quad a_{ji} = (i^2 + j^2 - ij)(i - j)$

$= -(i^2 + j^2 - ij)(j - i) = -a_{ij}$

$\Rightarrow$ A is a skew-symmetric matrix of odd order.

$\therefore \quad \text{tr}(A) = 0$ and $|A| = 0$

18 If $(x_1, y_1), (x_2, y_2)$ and (x_3, y_3) are the vertices of a triangle, then Area

$$= \frac{1}{2}\begin{vmatrix} x_1 & y_1 & 1 \\ x_2 & y_2 & 1 \\ x_3 & y_3 & 1 \end{vmatrix} \quad \text{...(i)}$$

Also, we know that, if a is the length of an equilateral triangle, then

$$\text{Area} = \frac{\sqrt{3}}{4}a^2 \quad \text{...(ii)}$$

From Eqs. (i) and (ii), we get

$$\frac{\sqrt{3}}{4}a^2 = \frac{1}{2}\begin{vmatrix} x_1 & y_1 & 1 \\ x_2 & y_2 & 1 \\ x_3 & y_3 & 1 \end{vmatrix}$$

$$\Rightarrow \quad \frac{\sqrt{3}}{2}a^2 = \begin{vmatrix} x_1 & y_1 & 1 \\ x_2 & y_2 & 1 \\ x_3 & y_3 & 1 \end{vmatrix}$$

On squaring both sides, we get

$$\frac{3}{4}a^4 = \begin{vmatrix} x_1 & y_1 & 1 \\ x_2 & y_2 & 1 \\ x_3 & y_3 & 1 \end{vmatrix}^2$$

19 Clearly, area of

$$(\Delta ABC) = \frac{1}{2}\begin{vmatrix} a & b+c & 1 \\ b & c+a & 1 \\ c & a+b & 1 \end{vmatrix}$$

$$= \frac{1}{2}(a+b+c)\begin{vmatrix} a & 1 & 1 \\ b & 1 & 1 \\ c & 1 & 1 \end{vmatrix} = 0$$

$[\because C_2 \to C_2 + C_1]$

20 Clearly, $B_2 = \begin{vmatrix} x_1 & z_1 \\ x_3 & z_3 \end{vmatrix} = x_1 z_3 - x_3 z_1$

$C_2 = -\begin{vmatrix} x_1 & y_1 \\ x_3 & y_3 \end{vmatrix} = -(x_1 y_3 - x_3 y_1)$

$B_3 = -\begin{vmatrix} x_1 & z_1 \\ x_2 & z_2 \end{vmatrix} = -(x_1 z_2 - x_2 z_1)$

and $C_3 = \begin{vmatrix} x_1 & y_1 \\ x_2 & y_2 \end{vmatrix} = x_1 y_2 - y_1 x_2$

$$\therefore \begin{vmatrix} B_2 & C_2 \\ B_3 & C_3 \end{vmatrix}$$

$$= \begin{vmatrix} x_1 z_3 - x_3 z_1 & -(x_1 y_3 - y_1 x_3) \\ -(x_1 z_2 - x_2 z_1) & x_1 y_2 - x_2 y_1 \end{vmatrix}$$

$$= \begin{vmatrix} x_1 z_3 & -x_1 y_3 \\ -x_1 z_2 + x_2 z_1 & x_1 y_2 - x_2 y_1 \end{vmatrix} + \begin{vmatrix} -x_3 z_1 & y_1 x_3 \\ -x_1 z_2 + x_2 z_1 & x_1 y_2 - x_2 y_1 \end{vmatrix}$$

$$= \begin{vmatrix} x_1 z_3 & -x_1 y_3 \\ -x_1 z_2 & x_1 y_2 \end{vmatrix} + \begin{vmatrix} x_1 z_3 & -x_1 y_3 \\ x_2 z_1 & -x_2 y_1 \end{vmatrix} + \begin{vmatrix} -x_3 z_1 & y_1 x_3 \\ -x_1 z_2 & x_1 y_2 \end{vmatrix} + \begin{vmatrix} -x_3 z_1 & y_1 x_3 \\ x_2 z_1 & -x_2 y_1 \end{vmatrix}$$

$= x_1^2(z_3 y_2 - z_2 y_3) - x_1 x_2(z_3 y_1 - z_1 y_3)$
$- x_1 x_3(z_1 y_2 - z_2 y_1) + x_2 x_3(z_1 y_1 - z_1 y_1)$

$= x_1[x_1(z_3 y_2 - z_2 y_3) - x_2(z_3 y_1 - z_1 y_3) + x_3(z_2 y_1 - z_1 y_2)]$

$= x_1 \, \Delta$

21 We have, $A=\begin{bmatrix}2 & -3\\ -4 & 1\end{bmatrix}$

$\therefore\ A^2 = A\cdot A$

$=\begin{bmatrix}2 & -3\\ -4 & 1\end{bmatrix}\begin{bmatrix}2 & -3\\ -4 & 1\end{bmatrix}$

$=\begin{bmatrix}4+12 & -6-3\\ -8-4 & 12+1\end{bmatrix}=\begin{bmatrix}16 & -9\\ -12 & 13\end{bmatrix}$

Now, $3A^2+12A = 3\begin{bmatrix}16 & -9\\ -12 & 13\end{bmatrix} + 12\begin{bmatrix}2 & -3\\ -4 & 1\end{bmatrix}$

$=\begin{bmatrix}48 & -27\\ -36 & 39\end{bmatrix}+\begin{bmatrix}24 & -36\\ -48 & 12\end{bmatrix}$

$=\begin{bmatrix}72 & -63\\ -84 & 51\end{bmatrix}$

$\therefore \text{adj}(3A^2+12A)=\begin{bmatrix}51 & 63\\ 84 & 72\end{bmatrix}$

22 All the given statements are true.

23 Given, adj $A = P = \begin{bmatrix}1 & \alpha & 3\\ 1 & 3 & 3\\ 2 & 4 & 4\end{bmatrix}$

Find the determinant of P and apply the formula

$|\text{adj}\,A| = |A|^{n-1}$

24 Given, $A=\begin{bmatrix}5a & -b\\ 3 & 2\end{bmatrix}$

and A adj $A = AA^T$

Clearly, $A(\text{adj}\,A) = |A|\,I_2$

[$\because$ if A is square matrix of order n, then $A(\text{adj}\,A) = (\text{adj}\,A)\cdot A = |A|\,I_n$]

$=\begin{vmatrix}5a & -b\\ 3 & 2\end{vmatrix} I_2$

$=(10a+3b)\,I_2$

$=(10a+3b)\begin{bmatrix}1 & 0\\ 0 & 1\end{bmatrix}$

$=\begin{bmatrix}10a+3b & 0\\ 0 & 10a+3b\end{bmatrix}$ ...(i)

and $AA^T = \begin{bmatrix}5a & -b\\ 3 & 2\end{bmatrix}\begin{bmatrix}5a & 3\\ -b & 2\end{bmatrix}$

$=\begin{bmatrix}25a^2+b^2 & 15a-2b\\ 15a-2b & 13\end{bmatrix}$ ...(ii)

$\because\ A(\text{adj}\,A) = AA^T$

$\therefore \begin{bmatrix}10a+3b & 0\\ 0 & 10a+3b\end{bmatrix}=\begin{bmatrix}25a^2+b^2 & 15a-2b\\ 15a-2b & 13\end{bmatrix}$

[using Eqs. (i) and (ii)]

$\Rightarrow\ 15a-2b=0$

$\Rightarrow\ a=\dfrac{2b}{15}$ ...(iii)

and $10a+3b=13$ ...(iv)

On substituting the value of 'a' from Eq. (iii) in Eq. (iv), we get

$10\left(\dfrac{2b}{15}\right)+3b=13$

$\Rightarrow\ \dfrac{20b+45b}{15}=13$

$\Rightarrow\ \dfrac{65b}{15}=13 \Rightarrow b=3$

Now, substituting the value of b in Eq. (iii), we get $5a=2$

Hence, $5a+b=2+3=5$

25 Clearly, $|B|=\begin{vmatrix}q & -b & y\\ -p & a & -x\\ r & -c & z\end{vmatrix}$

$=-\begin{vmatrix}q & -b & y\\ p & -a & x\\ r & -c & z\end{vmatrix}$

(taken (-1) common from R_2)

$=(+)\begin{vmatrix}q & b & y\\ p & a & x\\ r & c & z\end{vmatrix}$

(taken (-1) common from C_2)

$=-\begin{vmatrix}p & a & x\\ q & b & y\\ r & c & z\end{vmatrix}$ $(\because R_1 \leftrightarrow R_2)$

$=-\begin{vmatrix}a & x & p\\ b & y & q\\ c & z & r\end{vmatrix}$

$(\because C_1\leftrightarrow C_2$ and then $C_2\leftrightarrow C_3)$

$=-|A^T| = -|A|$

Thus, $|A| = -|B|$

Hence, $|A|\neq 0$ iff $|B|\neq 0$

$\therefore A$ is invertible iff B is invertible

Also, $|\text{adj}\,A| = |A|^2 = |B|^2 = |\text{adj}\,B|$

26 Given, A is non-singular matrix

$\Rightarrow |A|\neq 0.$

Also we have, $AA^T = A^TA$ and $B = A^{-1}A^T$

Now, $BB' = (A^{-1}A^T)(A^{-1}A^T)^T$

$= A^{-1}A^T\,A\,(A^{-1})^T$ $\quad[\because (A')' = A]$

$= A^{-1}AA^T\,(A^{-1})^T [\because AA^T = A^TA]$

$= IA^T\,(A^{-1})^T = A^T\,(A^{-1})^T$

$= (A^{-1}A)^T\ [\because (AB)^T = B^TA^T]$

$= I^T = I$

27 We have,

$\begin{bmatrix}1 & -\tan\theta\\ \tan\theta & 1\end{bmatrix}\begin{bmatrix}1 & \tan\theta\\ -\tan\theta & 1\end{bmatrix}^{-1}=\begin{bmatrix}a & -b\\ b & a\end{bmatrix}$

$\Rightarrow \begin{bmatrix}1 & -\tan\theta\\ \tan\theta & 1\end{bmatrix}\cdot\dfrac{1}{1+\tan^2\theta}\begin{bmatrix}1 & -\tan\theta\\ \tan\theta & 1\end{bmatrix}=\begin{bmatrix}a & -b\\ b & a\end{bmatrix}$

$\Rightarrow \dfrac{1}{1+\tan^2\theta}\begin{bmatrix}1-\tan^2\theta & -2\tan\theta\\ 2\tan\theta & 1-\tan^2\theta\end{bmatrix}=\begin{bmatrix}a & -b\\ b & a\end{bmatrix}$

$\Rightarrow \begin{bmatrix}\dfrac{1-\tan^2\theta}{1+\tan^2\theta} & \dfrac{-2\tan\theta}{1+\tan^2\theta}\\ \dfrac{2\tan\theta}{1+\tan^2\theta} & \dfrac{1-\tan^2\theta}{1+\tan^2\theta}\end{bmatrix}=\begin{bmatrix}a & -b\\ b & a\end{bmatrix}$

$\Rightarrow \begin{bmatrix}\cos 2\theta & -\sin 2\theta\\ \sin 2\theta & \cos 2\theta\end{bmatrix}=\begin{bmatrix}a & -b\\ b & a\end{bmatrix}$

$\Rightarrow a=\cos 2\theta$ and $b=\sin 2\theta$

28 On adding Au_1 and Au_2, we get

$Au_1 + Au_2 = \begin{bmatrix}1\\0\\0\end{bmatrix}+\begin{bmatrix}0\\1\\0\end{bmatrix}=\begin{bmatrix}1+0\\0+1\\0+0\end{bmatrix}$

$\Rightarrow\ A(u_1+u_2)=\begin{bmatrix}1\\1\\0\end{bmatrix}$

Since, A is non-singular matrix, i.e. $|A|\neq 0$, on multiplying both sides by A^{-1}, we get

$A^{-1}A(u_1+u_2) = A^{-1}\begin{bmatrix}1\\1\\0\end{bmatrix}$

$\Rightarrow u_1+u_2 = \begin{bmatrix}1 & 0 & 0\\ 2 & 1 & 0\\ 3 & 2 & 1\end{bmatrix}^{-1}\times\begin{bmatrix}1\\1\\0\end{bmatrix}$ ...(i)

Now, $|A| = \begin{vmatrix}1 & 0 & 0\\ 2 & 1 & 0\\ 3 & 2 & 1\end{vmatrix}$

$=1(1-0)-0+0=1$

$\therefore$ adj of A

$=\begin{bmatrix}1 & -2 & 1\\ 0 & 1 & -2\\ 0 & 0 & 1\end{bmatrix}^T=\begin{bmatrix}1 & 0 & 0\\ -2 & 1 & 0\\ 1 & -2 & 1\end{bmatrix}$

$\therefore\ A^{-1}=\begin{bmatrix}1 & 0 & 0\\ -2 & 1 & 0\\ 1 & -2 & 1\end{bmatrix}$ $\quad[\because |A|=1]$

From Eq. (i),

$u_1+u_2=\begin{bmatrix}1 & 0 & 0\\ -2 & 1 & 0\\ 1 & -2 & 1\end{bmatrix}\times\begin{bmatrix}1\\1\\0\end{bmatrix}$

$\Rightarrow u_1+u_2=\begin{bmatrix}1+0+0\\ -2+1+0\\ 1-2+0\end{bmatrix}=\begin{bmatrix}1\\-1\\-1\end{bmatrix}$

29 Given equations can be written in matrix form as $AX=B$

where, $A=\begin{bmatrix}k+1 & 8\\ k & k+3\end{bmatrix}, X=\begin{bmatrix}x\\y\end{bmatrix}$

and $B=\begin{bmatrix}4k\\ 3k-1\end{bmatrix}$

For no solution, $|A|=0$ and (adj A) $B\neq 0$

Now, $|A| = \begin{vmatrix} k+1 & 8 \\ k & k+3 \end{vmatrix} = 0$

$\Rightarrow \quad (k+1)(k+3) - 8k = 0$

$\Rightarrow \quad k^2 + 4k + 3 - 8k = 0$

$\Rightarrow \quad k^2 - 4k + 3 = 0$

$\Rightarrow \quad (k-1)(k-3) = 0$

$\therefore \quad k = 1, k = 3$

Also, $\text{adj } A = \begin{bmatrix} k+3 & -8 \\ -k & k+1 \end{bmatrix}$

$\therefore (\text{adj } A) B = \begin{bmatrix} k+3 & -8 \\ -k & k+1 \end{bmatrix} \begin{bmatrix} 4k \\ 3k-1 \end{bmatrix}$

$= \begin{bmatrix} (k+3)(4k) - 8(3k-1) \\ -4k^2 + (k+1)(3k-1) \end{bmatrix}$

$= \begin{bmatrix} 4k^2 - 12k + 8 \\ -k^2 + 2k - 1 \end{bmatrix}$

Put $k = 1$, then

$(\text{adj } A) B = \begin{bmatrix} 4 - 12 + 8 \\ -1 + 2 - 1 \end{bmatrix} = \begin{bmatrix} 0 \\ 0 \end{bmatrix}$, not true.

Put $k = 3$, then $(\text{adj } A)$

$B = \begin{bmatrix} 36 - 36 + 8 \\ -9 + 6 - 1 \end{bmatrix} = \begin{bmatrix} 8 \\ -4 \end{bmatrix} \neq 0$, true.

Hence, the required value of k is 3.

Alternate Method Condition for the system of equations has no solution is

$\frac{a_1}{a_2} = \frac{b_1}{b_2} \neq \frac{c_1}{c_2} \Rightarrow \frac{k+1}{k} = \frac{8}{k+3} \neq \frac{4k}{3k-1}$

Take $\frac{k+1}{k} = \frac{8}{k+3}$

$\Rightarrow \quad k^2 + 4k + 3 = 8k$

$\Rightarrow \quad k^2 - 4k + 3 = 0$

$\Rightarrow \quad (k-1)(k-3) = 0$

$\therefore \quad k = 1, 3$

If $k = 1$, then $\frac{8}{1+3} \neq \frac{4 \cdot 1}{2}$, [false]

and $k = 3$, then $\frac{8}{6} \neq \frac{4 \cdot 3}{9-1}$, [true]

$\therefore \quad k = 3$

Hence, only one value of k exists.

30 The system of given equations has no solution, if

$$\begin{vmatrix} \alpha & 1 & 1 \\ 1 & \alpha & 1 \\ 1 & 1 & \alpha \end{vmatrix} = 0$$

On applying $C_1 \to C_1 + C_2 + C_3$, we get

$$\begin{vmatrix} \alpha+2 & 1 & 1 \\ \alpha+2 & \alpha & 1 \\ \alpha+2 & 1 & \alpha \end{vmatrix} = 0$$

On applying $R_2 \to R_2 - R_1$, $R_3 \to R_3 - R_1$,

$$(\alpha+2)\begin{vmatrix} 1 & 1 & 1 \\ 0 & \alpha-1 & 0 \\ 0 & 0 & \alpha-1 \end{vmatrix} = 0$$

$\Rightarrow \quad (\alpha+2)(\alpha-1)^2 = 0$

$\therefore \quad \alpha = 1, -2$

But $\alpha = 1$ makes given three equations same. So, the system of equations has infinite solution. Hence, answer is $\alpha = -2$ for which the system of equations has no solution.

31 For consistency $|A| = 0$ and $(\text{adj } A) B = O$

$$\begin{vmatrix} 1 & 2 & 3 \\ 1 & 3 & 5 \\ 2 & 5 & a \end{vmatrix} = 0$$

$\Rightarrow 1(3a - 25) - 2(a - 10) + 3(5 - 6) = 0$

$\Rightarrow \quad a = 8$

On solving, $(\text{adj } A)B = O$, we get

$b = 15$

32 Since, $x - ky + z = 0$,

$kx + 3y - kz = 0$ and

$3x + y - z = 0$ has trivial solution.

$$\therefore \begin{vmatrix} 1 & -k & 1 \\ k & 3 & -k \\ 3 & 1 & -1 \end{vmatrix} \neq 0$$

$\Rightarrow \quad 1(-3 + k) + k(-k + 3k) + 1(k - 9) \neq 0$

$\Rightarrow \quad k - 3 + 2k^2 + k - 9 \neq 0$

$\Rightarrow \quad 2k^2 + 2k - 12 \neq 0$

$\Rightarrow \quad k^2 + k - 6 \neq 0$

$\Rightarrow \quad (k+3)(k-2) \neq 0$

$\therefore \quad k \neq 2, -3$

$k \in R - \{2, -3\}$

33 Let $A = \begin{bmatrix} a & b \\ c & d \end{bmatrix}$ such that $A \neq I, -I$

and $\begin{bmatrix} a & b \\ c & d \end{bmatrix}^2 = \begin{bmatrix} 1 & 0 \\ 0 & 1 \end{bmatrix}$

$\Rightarrow \begin{bmatrix} a^2 + bc & ab + bd \\ ac + cd & bc + d^2 \end{bmatrix} = \begin{bmatrix} 1 & 0 \\ 0 & 1 \end{bmatrix}$

$\Rightarrow \quad b(a + d) = 0, c(a + d) = 0$

and $\quad a^2 + bc = 1, bc + d^2 = 1$

$\Rightarrow \quad a = 1, d = -1, b = c = 0$ or

$a = -1, d = 1, b = c = 0$

$A = \begin{bmatrix} 1 & 0 \\ 0 & -1 \end{bmatrix}$, then

$\det(A) = -1$ and $\text{tr}(A) = 1 - 1 = 0$

34 Given system of linear equations are

$2x_1 - 2x_2 + x_3 = \lambda x_1$

$\Rightarrow \quad (2 - \lambda)x_1 - 2x_2 + x_3 = 0 \quad \text{...(i)}$

$2x_1 - 3x_2 + 2x_3 = \lambda x_2$

$\Rightarrow \quad 2x_1 - (3 + \lambda)x_2 + 2x_3 = 0 \quad \text{...(ii)}$

and $\quad -x_1 + 2x_2 = \lambda x_3$

$\Rightarrow \quad -x_1 + 2x_2 - \lambda x_3 = 0 \quad \text{...(iii)}$

Since, the system has non-trivial solution.

$$\therefore \begin{vmatrix} 2-\lambda & -2 & 1 \\ 2 & -(3+\lambda) & 2 \\ -1 & 2 & -\lambda \end{vmatrix} = 0$$

$\Rightarrow (2 - \lambda)(3\lambda + \lambda^2 - 4) + 2(-2\lambda + 2) + 1(4 - 3 - \lambda) = 0$

$\Rightarrow \quad (2 - \lambda)(\lambda^2 + 3\lambda - 4) + 4(1 - \lambda) + (1 - \lambda) = 0$

$\Rightarrow (2 - \lambda)(\lambda + 4)(\lambda - 1) + 5(1 - \lambda) = 0$

$\Rightarrow \quad (\lambda - 1)[(2 - \lambda)(\lambda + 4) - 5] = 0$

$\Rightarrow \quad (\lambda - 1)(\lambda^2 + 2\lambda - 3) = 0$

$\Rightarrow \quad (\lambda - 1)[(\lambda - 1)(\lambda + 3)] = 0$

$\Rightarrow \quad (\lambda - 1)^2(\lambda + 3) = 0$

$\Rightarrow \quad \lambda = 1, 1, -3$

35 Determinant of skew-symmetric matrix of odd order is zero and of even order is perfect square.

Also, $\quad \det(A^T) = \det(A)$

and $\quad \det(-A) = (-1)^n \det(A)$

Hence, Statement II is false.

SESSION 2

1 Given that,

$$f(x) = \begin{vmatrix} 1 + a^2x & (1 + b^2)x & (1 + c^2)x \\ (1 + a^2)x & 1 + b^2x & (1 + c^2)x \\ (1 + a^2)x & (1 + b^2)x & 1 + c^2x \end{vmatrix}$$

On applying $C_1 \to C_1 + C_2 + C_3$, we get

$$f(x) = \begin{vmatrix} 1 + a^2x + x + b^2x + x + c^2x & (1 + b^2)x & (1 + c^2)x \\ x + a^2x + 1 + b^2x + x + c^2x & 1 + b^2x & (1 + c^2)x \\ x + a^2x + x + b^2x + 1 + c^2x & (1 + b^2)x & 1 + c^2x \end{vmatrix}$$

$$= \begin{vmatrix} 1 + (a^2 + b^2 + c^2 + 2)x & (1 + b^2)x & (1 + c^2)x \\ 1 + (a^2 + b^2 + c^2 + 2)x & 1 + b^2x & (1 + c^2)x \\ 1 + (a^2 + b^2 + c^2 + 2)x & (1 + b^2)x & 1 + c^2x \end{vmatrix}$$

$$= \begin{vmatrix} 1 & (1 + b^2)x & (1 + c^2)x \\ 1 & 1 + b^2x & (1 + c^2)x \\ 1 & (1 + b^2)x & 1 + c^2x \end{vmatrix}$$

$[\because a^2 + b^2 + c^2 = -2$, given$]$

On applying

$R_1 \to R_1 - R_3$, $R_2 \to R_2 - R_3$, we get

$$= \begin{vmatrix} 0 & 0 & x - 1 \\ 0 & 1 - x & x - 1 \\ 1 & (1 + b^2)x & 1 + c^2x \end{vmatrix}$$

$$= \begin{vmatrix} 0 & x - 1 \\ 1 - x & x - 1 \end{vmatrix} = (x - 1)^2$$

Hence, $f(x)$ is of degree 2.

2 Clearly, $\left|\text{adj} A^{-1}\right| = |A^{-1}|^2 = \dfrac{1}{|A|^2}$

and $|(\text{adj} A^{-1})^{-1}| = \dfrac{1}{|(\text{adj} A^{-1})|}$

$= |A|^2 = 2^2 = 4$

3 For non - trivial solution,

$$\begin{vmatrix} k-1 & 3k+1 & 2k \\ k-1 & 4k-2 & k+3 \\ 2 & 3k+1 & 3(k-1) \end{vmatrix} = 0$$

$\Rightarrow \quad k = 0$ or 3

Now, when $k = 0$, then the equation becomes

$-x + y = 0 \quad$ ...(i)

$-x - 2y + 3z = 0 \quad$ ...(ii)

and $\quad 2x + y - 3z = 0 \quad$ (iii)

From (i), we get $x = y = \lambda$ (say), Then, from Eq. (ii), we get

$-\lambda - 2\lambda + 3z = 0$

$\Rightarrow \quad 3z = 3\lambda$

$\Rightarrow \quad z = \lambda$

$\therefore \quad x:y:z = 1:1:1$

4 Clearly, $x = 2^k - 1, y = \dfrac{3^k - 1}{2}$

and $z = \dfrac{5^k - 1}{4}$

$\left[\because \text{sum to } n \text{ terms of } a \text{ GP is given by } \dfrac{a(r^n - 1)}{r - 1}\right]$

Now, $\Delta = \begin{vmatrix} 2^k - 1 & 3^k - 1 & 5^k - 1 \\ 3 & 3 & 3 \\ 2^k & 3^k & 5^k \end{vmatrix}$

$= \begin{vmatrix} 2^k & 3^k & 5^k \\ 3 & 3 & 3 \\ 2^k & 3^k & 5^k \end{vmatrix} - 3\begin{vmatrix} 1 & 1 & 1 \\ 1 & 1 & 1 \\ 2^k & 3^k & 5^k \end{vmatrix}$

$= 0 - 3 \times 0 = 0$

5 Let $p(x) = \begin{vmatrix} 1+2x & 1 & 1-x \\ 2-x & 2+x & 3+x \\ x & 1+x & 1-x^2 \end{vmatrix} = 0$

Clearly, product of roots $= \dfrac{\text{Constant term}}{\text{Coefficient of } x^4}$

Here, constant term is given by

$P(0) = \begin{vmatrix} 1 & 1 & 1 \\ 2 & 2 & 3 \\ 0 & 1 & 1 \end{vmatrix} = -1$

and coefficient of x^4 is -2

$\therefore$ Product of roots is $\dfrac{1}{2}$.

6 Here, $\begin{vmatrix} -1 & a & a \\ b & -1 & b \\ c & c & -1 \end{vmatrix} = 0$

On applying, $C_2 \to C_2 - C_1$ and $C_3 \to C_3 - C_1$, we get

$\begin{vmatrix} -1 & a+1 & a+1 \\ b & -(b+1) & 0 \\ c & 0 & -(1+c) \end{vmatrix} = 0$

On applying $R_1 \to \dfrac{R_1}{a+1}, R_2 \to \dfrac{R_2}{b+1}$

and $R_3 \to \dfrac{R_3}{c+1}$, we get

$\begin{vmatrix} -\dfrac{1}{a+1} & 1 & 1 \\ \dfrac{b}{b+1} & -1 & 0 \\ \dfrac{c}{c+1} & 0 & -1 \end{vmatrix} = 0$

$\Rightarrow \quad -\dfrac{1}{a+1} + \dfrac{b}{b+1} + \dfrac{c}{c+1} = 0$

$\therefore -\dfrac{1}{a+1} + 1 - \dfrac{1}{b+1} + 1 - \dfrac{c}{c+1} = 0$

$\Rightarrow \quad \dfrac{1}{a+1} + \dfrac{1}{b+1} + \dfrac{1}{c+1} = 2$

7 Let $\dfrac{x^2}{a^2} = X, \dfrac{y^2}{b^2} = Y, \dfrac{z^2}{c^2} = Z$

Then, $X + Y - Z = 1, X - Y + Z = 1,$
$-X + Y + Z = 1$

Now, coefficient matrix is

$A = \begin{bmatrix} 1 & 1 & -1 \\ 1 & -1 & 1 \\ -1 & 1 & 1 \end{bmatrix}$

Here, $|A| \neq 0$

$\therefore$ It has unique solution, *viz.*, $X = 1$, $Y = 1$ and $Z = 1$

Hence , $x = \pm a; y = \pm b$ and $z = \pm c$.

8 Here, $\Delta = \begin{vmatrix} 1 & 1 & 1 \\ 1 & a & 1 \\ a & b & 1 \end{vmatrix}$

$= 1(a - b) - 1(1 - a) + 1(b - a^2)$

$= -(a-1)^2$

$\Delta_1 = \begin{vmatrix} 1 & 1 & 1 \\ 1 & a & 1 \\ 0 & b & 1 \end{vmatrix}$

$= 1(a - b) - 1(1) + 1(b) = (a - 1)$

$\Delta_2 = \begin{vmatrix} 1 & 1 & 1 \\ 1 & 1 & 1 \\ a & 0 & 1 \end{vmatrix}$

$= 1(1) - 1(1 - a) + 1(0 - a) = 0$

and $\Delta_3 = \begin{vmatrix} 1 & 1 & 1 \\ 1 & a & 1 \\ a & b & 0 \end{vmatrix}$

$= 1(-b) - 1(-a) + 1(b - a^2)$

$= -a(a-1)$

For $a = 1$

$\Delta = \Delta_1 = \Delta_2 = \Delta_3 = 0$

For $b = 1$ only

$x + y + z = 1, x + y + z = 1$

and $\quad x + y + z = 0$

i.e. no solution $\quad$ ($\because$ RHS is not equal)

Hence, for no solution, $b = 1$ only.

9 Clearly,

$\det(M - I) = \det(M - I) \cdot \det(M)$

$[\because \det(M) = 1]$

$= \det(M - I)\det(M^T)$

$[\because \det(A^T) = \det(A)]$

$= \det(MM^T - M^T)$

$= \det(I - M^T) \quad [\because MM^T = I]$

$= -\det(M^T - I)$

$= -\det(M^T - I)^T$

$= -\det(M - I)$

$\Rightarrow \quad 2\det(M - I) = 0$

$\Rightarrow \quad \det(M - I) = 0$

10 Since, $a_1, a_2, \ldots, a_n, \ldots$ are in GP, then, $\log a_n, \log a_{n+1}, \log a_{n+2}, \ldots,$ $\log a_{n+8}, \ldots$are in AP.

Given that,

$\Delta = \begin{vmatrix} \log a_n & \log a_{n+1} & \log a_{n+2} \\ \log a_{n+3} & \log a_{n+4} & \log a_{n+5} \\ \log a_{n+6} & \log a_{n+7} & \log a_{n+8} \end{vmatrix}$

$\therefore \Delta = \begin{vmatrix} a & a+d & a+2d \\ a+3d & a+4d & a+5d \\ a+6d & a+7d & a+8d \end{vmatrix}$

where a and d are the first term and common difference of AP.

On applying $R_2 \to 2R_2$, we get

$\Delta = \dfrac{1}{2}\begin{vmatrix} a & a+d & a+2d \\ 2a+6d & 2a+8d & 2a+10d \\ a+6d & a+7d & a+8d \end{vmatrix}$

On applying $R_2 \to R_2 - R_3$, we get

$\Delta = \dfrac{1}{2}\begin{vmatrix} a & a+d & a+2d \\ a & a+d & a+2d \\ a+6d & a+7d & a+8d \end{vmatrix} = 0$

11 Let $A = \begin{bmatrix} a & b \\ c & d \end{bmatrix}$. Then,

$|A| = ad - bc = k$ (say)

and $\quad \text{adj}(A) = \begin{bmatrix} d & -b \\ -c & a \end{bmatrix}$

Now, $|A + |A| \text{adj}(A)| = 0$

$\Rightarrow \quad \begin{vmatrix} a+kd & (1-k)b \\ (1-k)c & d+ka \end{vmatrix} = 0$

$\Rightarrow \quad (a + kd)(d + ka) - (1-k)^2 bc$

$\Rightarrow \quad ad + a^2k + kd^2 + k^2ad$
$-(1 + k^2 - 2k)bc$

$\Rightarrow (ad - bc) + (ad - bc)k^2 + k(a^2 + d^2 + 2bc) = 0$

$\Rightarrow (ad - bc) + (ad - bc)k^2 + k(a^2 + d^2) + 2(ad - k)) = 0$ $[\because bc = ad - k]$

$\Rightarrow (ad - bc) + (ad - bc)k^2 + k(a + d)^2 - 2k^2 = 0$

$\Rightarrow (k + k^3 - 2k^2) + k(a + d)^2 = 0$

$\Rightarrow k[(k-1)^2 + (a+d)^2] = 0$

$\Rightarrow k = 1$ and $a + d = 0$

Now, $|A - |A| \text{adj} A|$

$$= \begin{vmatrix} a-kd & (1+k)b \\ (1+k)c & d-ak \end{vmatrix} = \begin{vmatrix} a-d & 2b \\ 2c & d-a \end{vmatrix}$$

$= -(a-d)^2 - 4bc$

$= -((a+d)^2 - 4ad) - 4bc$

$= 4ad - 4bc = 4k = 4$

12 On subtracting the given equation, we get

$P^3 - P^2Q = Q^3 - Q^2P$

$\Rightarrow P^2(P - Q) = Q^2(Q - P)$

$\Rightarrow (P - Q)(P^2 + Q^2) = 0$ ...(i)

Now, if $|P^2 + Q^2| \neq 0$

$(P^2 + Q^2)$ is invertible.

On post multiply both sides by $(P^2 + Q^2)^{-1}$, we get

$P - Q = 0$, which is a contradiction.

Hence, $|P^2 + Q^2| = 0$

13 Let $\Delta = \begin{vmatrix} 3 & 1 + f(1) & 1 + f(2) \\ 1 + f(1) & 1 + f(2) & 1 + f(3) \\ 1 + f(2) & 1 + f(3) & 1 + f(4) \end{vmatrix}$

$$\Rightarrow \Delta = \begin{vmatrix} 3 & 1 + \alpha + \beta & 1 + \alpha^2 + \beta^2 \\ 1 + \alpha + \beta & 1 + \alpha^2 + \beta^2 & 1 + \alpha^3 + \beta^3 \\ 1 + \alpha^2 + \beta^2 & 1 + \alpha^3 + \beta^3 & 1 + \alpha^4 + \beta^4 \end{vmatrix}$$

$$= \begin{vmatrix} 1\cdot1 + 1\cdot1 + 1\cdot1 & 1\cdot1 + 1\cdot\alpha + 1\cdot\beta & 1\cdot1 + 1\cdot\alpha^2 + 1\cdot\beta^2 \\ 1\cdot1 + \alpha\cdot1 + \beta\cdot1 & 1\cdot1 + \alpha\cdot\alpha + \alpha\cdot\beta & 1\cdot1 + \alpha\cdot\alpha^2 + \beta\cdot\beta^2 \\ 1\cdot1 + 1\cdot\alpha^2 + 1\cdot\beta^2 & 1\cdot1 + \alpha^2\cdot\alpha + \beta^2\cdot\beta & 1\cdot1 + \alpha^2\cdot\alpha^2 + \beta^2\cdot\beta^2 \end{vmatrix}$$

$$= \begin{vmatrix} 1 & 1 & 1 \\ 1 & \alpha & \beta \\ 1 & \alpha^2 & \beta^2 \end{vmatrix} \begin{vmatrix} 1 & 1 & 1 \\ 1 & \alpha & \beta \\ 1 & \alpha^2 & \beta^2 \end{vmatrix} = \begin{vmatrix} 1 & 1 & 1 \\ 1 & \alpha & \beta \\ 1 & \alpha^2 & \beta^2 \end{vmatrix}^2$$

On expanding, we get

$\Delta = (1 - \alpha)^2 (1 - \beta)^2 (\alpha - \beta^2)$

Hence, $K(1 - \alpha)^2 (1 - \beta)^2 (\alpha - \beta)^2$

$= (1 - \alpha)^2 (1 - \beta)^2 (\alpha - \beta)^2$

$\therefore \quad K = 1$

14 Let $\Delta = \begin{vmatrix} (1 + ap)^2 & (1 + bp)^2 & (1 + cp)^2 \\ (1 + aq)^2 & (1 + bq)^2 & (1 + cq)^2 \\ (1 + ar)^2 & (1 + br)^2 & (1 + cr)^2 \end{vmatrix}$

$$= \begin{vmatrix} 1 + 2ap + a^2p^2 & 1 + 2bp + b^2p^2 & 1 + 2cp + c^2p^2 \\ 1 + 2aq + a^2q^2 & 1 + 2bq + b^2q^2 & 1 + 2cq + c^2q^2 \\ 1 + 2ar + a^2r^2 & 1 + 2br + b^2r^2 & 1 + 2cr + c^2r^2 \end{vmatrix}$$

$$= \begin{vmatrix} 1 & p & p^2 \\ 1 & q & q^2 \\ 1 & r & r^2 \end{vmatrix} \times \begin{vmatrix} 1 & 2a & a^2 \\ 1 & 2b & b^2 \\ 1 & 2c & c^2 \end{vmatrix}$$

$$= 2\left(\begin{vmatrix} 1 & p & p^2 \\ 1 & q & q^2 \\ 1 & r & r^2 \end{vmatrix} \times \begin{vmatrix} 1 & a & a^2 \\ 1 & b & b^2 \\ 1 & c & c^2 \end{vmatrix} \right)$$

$= 2(2A_1 \times 2A_2) = 2(8 \times 1) = 16$

15 According to given conditions we get a right angled triangle whose vertices are $(n, r), (m, q)$ and (l, p).

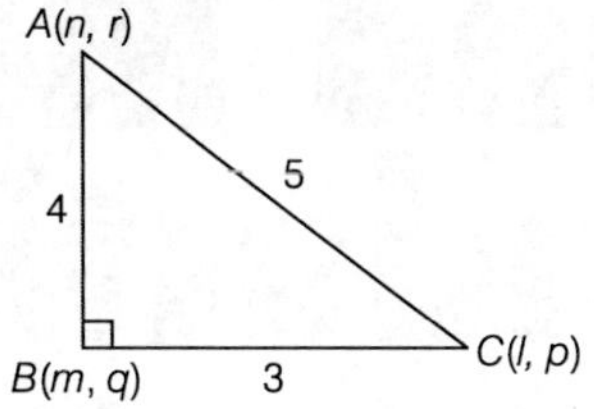

Also, we have, $|A| = \begin{vmatrix} l & m & n \\ p & q & r \\ 1 & 1 & 1 \end{vmatrix}$

$$\Rightarrow |A|^2 = \begin{vmatrix} l & p & 1 \\ m & q & 1 \\ n & r & 1 \end{vmatrix}^2 = [2ar(\Delta ABC)]^2$$

$$= \left[2 \times \frac{1}{2} \times 3 \times 4\right]^2 = 144$$

DAY SEVEN

Binomial Theorem and Mathematical Induction

Learning & Revision for the Day

- Binomial Theorem
- Binomial Theorem for Positive Index
- Properties of Binomial Coefficient
- Applications of Binomial Theorem
- Binomial Theorem for Negative/Rational Index
- Principle of Mathematical Induction

Binomial Theorem

Binomial theorem describes the algebraic expansion of powers of a binomial. According to this theorem, it is possible to expand $(x + y)^n$ into a sum involving terms of the form $ax^b y^c$, where the exponents b and c are non-negative integers with $b + c = n$. The coefficient a of each term is a specific positive integer depending on n and b, is known as the binomial coefficient $\binom{n}{b}$.

Binomial Theorem for Positive Index

An algebraic expression consisting of two terms with (+) ve or (–)ve sign between them, is called binomial expression.

If n is any positive integer,

then $(x + a)^n = {}^nC_0 x^n + {}^nC_1 x^{n-1} a + \cdots + {}^nC_n a^n$

$$= \sum_{r=0}^{n} {}^nC_r \cdot x^{n-r} a^r,$$ where x and a are **real** (complex) **numbers**.

(i) The coefficient of terms equidistant from the beginning and the end, are equal.

(ii) $(x - a)^n = {}^nC_0 x^n - {}^nC_1 x^{n-1} a + \cdots + (-1)^n \, {}^nC_n a^n$

(iii) $(1 + x)^n = {}^nC_0 + {}^nC_1 x + {}^nC_2 x^2 + \cdots + {}^nC_n x^n$

(iv) Total number of terms in the expansion $(x + a)^n$ is $(n + 1)$.

(v) If n is a positive integer, then the number of terms in $(x + y + z)^n$ is $\frac{(n+1)(n+2)}{2}$.

PREP MIRROR
Your Personal Preparation Indicator

- No. of Questions in Exercises (x)—
- No. of Questions Attempted (y)—
- No. of Correct Questions (z)—
  (Without referring Explanations)

- Accuracy Level $(z/y \times 100)$—
- Prep Level $(z/x \times 100)$—

In order to expect good rank in JEE, your Accuracy Level should be above 85 & Prep Level should be above 75.

(vi) The number of terms in the expansion of

$$(x+a)^n + (x-a)^n = \begin{cases} \dfrac{n+2}{2}, & \text{if } n \text{ is even} \\ \dfrac{n+1}{2}, & \text{if } n \text{ is odd} \end{cases}$$

(vii) The number of terms in the expansion of

$$(x+a)^n - (x-a)^n = \begin{cases} \dfrac{n}{2}, & \text{if } n \text{ is even} \\ \dfrac{n+1}{2}, & \text{if } n \text{ is odd} \end{cases}$$

General Term and Middle Term

(i) Let $(r+1)$th term be the **general term** in the expansion of $(x+a)^n$.

$$T_{r+1} = {}^nC_r x^{n-r} a^r$$

(ii) If expansion is $(x-a)^n$, then the **general term** is

$$(-1)^r \cdot {}^nC_r\, x^{n-r} a^r.$$

(iii) The **middle term** in the expansion of $(a+x)^n$.

(a) ***Case I*** If n is even, then $\left(\dfrac{n}{2}+1\right)$ th term is middle term.

(b) ***Case II*** If n is odd, then $\dfrac{(n+1)}{2}$th term and $\dfrac{(n+3)}{2}$th terms are middle terms.

(iv) $(p+1)^{\text{th}}$ term from end $=(n-p+1)^{\text{th}}$ term from beginning.

(v) For making a term independent of x we put $r=n$ in general term of $(x+a)^n$, so we get ${}^nC_n a^n$, that is independent of x.

NOTE If the coefficients of rth, $(r+1)$th, $(r+2)$th term of $(1+x)^n$ are in AP, then $n^2-(4r+1)n+4r^2=2$

Greatest Term

If T_r and T_{r+1} be the rth and $(r+1)$th terms in the expansion of $(1+x)^n$, then

$$\frac{T_{r+1}}{T_r} = \frac{{}^nC_r \cdot x^r}{{}^nC_{r-1} \cdot x^{r-1}} = \frac{n-r+1}{r} \cdot x$$

Let numerically, T_{r+1} be the greatest term in the above expansion. Then, $T_{r+1} \ge T_r$ or $\dfrac{T_{r+1}}{T_r} \ge 1$.

$$\therefore \quad \frac{n-r+1}{r}|x| \ge 1 \text{ or } r \le \frac{(n+1)}{(1+|x|)}|x| \quad \text{...(i)}$$

(i) Now, substituting values of n and x in Eq. (i), we get $r \le m+f$ or $r \le m$, where m is a positive integer and f is a fraction such that $0<f<1$.

(ii) When $r \le m+f$, T_{m+1} is the greatest term, when $r \le m$, T_m and T_{m+1} are the greatest terms and both are equal.

(iii) The coefficients of the middle terms in the expansion of $(a+x)^n$ are called **greatest coefficients.**

Properties of Binomial Coefficients

In the binomial expansion of $(1+x)^n$,

$(1+x)^n = {}^nC_0 + {}^nC_1 \cdot x + {}^nC_2 \cdot x^2 + \ldots + {}^nC_r \cdot x^r + \ldots {}^nC_n \cdot x^n$,

where, ${}^nC_0, {}^nC_1, \ldots, {}^nC_n$ are the coefficients of various powers of x are called **binomial coefficients** and it is also written as $C_0, C_1, \ldots, C_n$ or $\binom{n}{0}, \binom{n}{1}, \ldots, \binom{n}{n}$

- ${}^nC_r = {}^nC_{n-r}$
- ${}^nC_{r_1} = {}^nC_{r_2} \Rightarrow r_1 = r_2$ or $r_1 + r_2 = n$
- ${}^nC_r + {}^nC_{r-1} = {}^{n+1}C_r$
- $\dfrac{{}^nC_r}{{}^nC_{r-1}} = \dfrac{n-r+1}{r}$
- $r \cdot {}^nC_r = n \cdot {}^{n-1}C_{r-1}$
- $\dfrac{{}^nC_r}{r+1} = \dfrac{{}^{n+1}C_{r+1}}{n+1}$
- $C_0 + C_1 + C_2 + \ldots + C_n = 2^n$
- $C_0 + C_2 + C_4 + \ldots = C_1 + C_3 + C_5 + \ldots = 2^{n-1}$
- $C_0 - C_1 + C_2 - C_3 + \ldots + (-1)^n \cdot C_n = 0$
- $C_0^2 + C_1^2 + C_2^2 + \ldots + C_n^2 = {}^{2n}C_n = \dfrac{(2n)!}{(n!)^2}$
- $C_0^2 - C_1^2 + C_2^2 - C_3^2 + \ldots = \begin{cases} (-1)^{n/2} \cdot {}^nC_{n/2}, & \text{if } n \text{ is even} \\ 0, & \text{if } n \text{ is odd} \end{cases}$
- $C_0 \cdot C_r + C_1 \cdot C_{r+1} + \ldots + C_{n-r} \cdot C_n = {}^{2n}C_{n-r} = \dfrac{(2n)!}{(n-r)!\,(n+r)!}$
- $C_1 - 2C_2 + 3C_3 - \ldots = 0$
- $C_0 + 2C_1 + 3C_2 + \ldots + (n+1) \cdot C_n = (n+2)\,2^{n-1}$
- $C_0 - C_2 + C_4 - C_6 + \ldots = \sqrt{2^n} \cdot \cos \dfrac{n\pi}{4}$
- $C_1 - C_3 + C_5 - C_7 + \ldots = \sqrt{2^n} \cdot \sin \dfrac{n\pi}{4}$

Applications of Binomial Theorem

1. R-f Factor Relation

Here, we are going to discuss problems involving $(\sqrt{A}+B)^n = I + f$, where I and n are positive integers $0 \le f \le 1$, $|A - B^2| = k$ and $|\sqrt{A} - B| < 1$.

2. Divisibility Problem

In the expansion, $(1+\alpha)^n$. We can conclude that, $(1+\alpha)^n - 1$ is divisible by α, i.e. it is a multiple of α.

3. Differentiability Problem

Sometimes to generalise the result we use the differentiation.

$$(1+x)^n = {}^nC_0 + {}^nC_1 x + {}^nC_2 x^2 + \ldots + {}^nC_n x^n$$

On differentiating w.r.t. x, we get

$n(1+x)^{n-1} = 0 + {}^nC_1 + 2\cdot x \cdot {}^nC_2 + \ldots + n \cdot {}^nC_n \cdot x^{n-1}$

Put $x = 1$, we get, $n2^{n-1} = {}^nC_1 + 2\,{}^nC_2 + \ldots + n\,{}^nC_n$

Binomial Theorem for Negative/Rational Index

Let n be a rational number and x be a real number such that $|x| < 1$, then $(1+x)^n = 1 + nx + \frac{n(n-1)}{2!}x^2 + \frac{n(n-1)(n-2)}{3!}x^3 + \ldots$

- If n is a positive integer, then $(1+x)^n$ contains $(n+1)$ terms i.e. a finite number of terms. When n is any negative integer or rational number, then expansion of $(1+x)^n$ contains infinitely many terms.
- When n is a positive integer, then expansion of $(1+x)^n$ is valid for all values of x. If n is any negative integer or rational number, then expansion of $(1+x)^n$ is valid for the values of x satisfying the condition $|x| < 1$.
  (i) $(1+x)^{-1} = 1 - x + x^2 - x^3 + \ldots$
  (ii) $(1-x)^{-1} = 1 + x + x^2 + x^3 + \ldots$
  (iii) $(1+x)^{-2} = 1 - 2x + 3x^2 - 4x^3 + \ldots$
  (iv) $(1-x)^{-2} = 1 + 2x + 3x^2 + 4x^3 + \ldots$

Principle of Mathematical Induction

In algebra, there are certain results that are formulated in terms of n, where n is a positive integer. Such results can be proved by a specific technique, which is known as the principle of mathematical induction.

First Principle of Mathematical Induction

It consists of the following three steps

***Step* I** Actual verification of the proposition for the starting value of i.

***Step* II** Assuming the proposition to be true for k, $k \geq i$ and proving that it is true for the value $(k+1)$ which is next higher integer.

***Step* III** To combine the above two steps. Let $p(n)$ be a statement involving the natural number n such that
(i) $p(1)$ is true i.e. $p(n)$ is true for $n = 1$.
(ii) $p(m+1)$ is true, whenever $p(m)$ is true
i.e. $p(m)$ is true $\Rightarrow p(m+1)$ is true. Then, $p(n)$ is true for all natural numbers n.
Product of r consecutive integers is divisible by $r!$

DAY PRACTICE SESSION 1

FOUNDATION QUESTIONS EXERCISE

1 If $(1+ax)^n = 1 + 8x + 24x^3 + \ldots$, then the values of a and n are

(a) 2, 4 (b) 2, 3 (c) 3, 6 (d) 1, 2

2 The coefficient of x^n in the expansion of $(1+x)^{2n}$ and $(1+x)^{2n-1}$ are in the ratio → NCERT Exemplar

(a) 1 : 2 (b) 1 : 3
(c) 3 : 1 (d) 2 : 1

3 The value of $(1.002)^{12}$ upto fourth place of decimal is

(a) 1.0242 (b) 1.0245
(c) 1.0004 (d) 1.0254

4 The coefficient of x^4 in the expansion of $(1+x+x^2+x^3)^n$ is

(a) ${}^nC_4$ (b) ${}^nC_4 + {}^nC_2$
(c) ${}^nC_4 + {}^nC_2 + {}^nC_2$ (d) ${}^nC_4 + {}^nC_2 + {}^nC_1 \cdot {}^nC_2$

5 If the middle term of $\left(\frac{1}{x} + x\sin x\right)^{10}$ is equal to $7\frac{7}{8}$, then the value of x is → NCERT Exemplar

(a) $2n\pi + \frac{\pi}{6}$ (b) $n\pi + \frac{\pi}{6}$
(c) $n\pi + (-1)^n \frac{\pi}{6}$ (d) $n\pi + (-1)\frac{\pi}{3}$

6 If the 7th term in the binomial expansion of $\left(\frac{3}{\sqrt[3]{84}} + \sqrt{3}\ln x\right)^9$, $x > 0$ is equal to 729, then x can be → JEE Mains 2013

(a) e^2 (b) e (c) $e/2$ (d) $2e$

7 If the number of terms in the expansion of $\left(1 - \frac{2}{x} + \frac{4}{x^2}\right)^n$, $x \neq 0$ is 28, then the sum of the coefficients of all the terms in this expansion, is → JEE Mains 2016

(a) 64 (b) 2187 (c) 243 (d) 729

8 In the binomial expansion of $(a-b)^n$, $n \geq 5$, the sum of 5th and 6th terms is zero, then $\frac{a}{b}$ is equal to

(a) $\frac{5}{n-4}$ (b) $\frac{6}{n-5}$
(c) $\frac{n-5}{6}$ (d) $\frac{n-4}{5}$

9 In the expansion of the following expression $1 + (1+x) + (1+x)^2 + \ldots + (1+x)^n$, the coefficient of $x^4 (0 \leq k \leq n)$ is

(a) ${}^{n+1}C_{k+1}$ (b) ${}^nC_k$
(c) ${}^nC_{n-k-1}$ (d) None of these

10 The coefficient of t^{24} in the expansion of $(1+t^2)^{12}(1+t^{12})(1+t^{24})$ is

(a) ${}^{12}C_6+2$ (b) ${}^{12}C_5$ (c) ${}^{12}C_6$ (d) ${}^{12}C_7$

11 The coefficient of x^{53} in the following expansion $\sum_{m=0}^{100} {}^{100}C_m(x-3)^{100-m}\cdot 2^m$ is

(a) ${}^{100}C_{47}$ (b) ${}^{100}C_{53}$ (c) $-{}^{100}C_{53}$ (d) $-{}^{100}C_{100}$

12 If p is a real number and if the middle term in the expansion of $\left(\frac{p}{2}+2\right)^8$ is 1120, then the value of p is → NCERT Exemplar

(a) ± 3 (b) ± 1
(c) ± 2 (d) None of these

13 The constant term in the expansion of $\left(1+x+\frac{2}{x}\right)^6$, is

(a) 479 (b) 517 (c) 569 (d) 581

14 If in the expansion of $(1+x)^m(1-x)^n$, the coefficient of x and x^2 are 3 and –6 respectively, them m is

(a) 6 (b) 9 (c) 12 (d) 24

15 If n is a positive integer, then $(\sqrt{3}+1)^{2n}-(\sqrt{3}-1)^{2n}$ is → AIEEE 2012

(a) an irrational number
(b) an odd positive integer
(c) an even positive integer
(d) a rational number other than positive integers

16 If the $(r+1)$ th term in the expansion of $\left(\sqrt[3]{\frac{a}{\sqrt{b}}}+\sqrt{\frac{b}{\sqrt[3]{a}}}\right)^{21}$ has the same power of a and b, then the value of r is

(a) 9 (b) 10 (c) 8 (d) 6

17 If x^{2k} occurs in the expansion of $\left(x+\frac{1}{x^2}\right)^{n-3}$, then

(a) $n-2k$ is a multiple of 2 (b) $n-2k$ is a multiple of 3
(c) $k=0$ (d) None of these

18 The ratio of the coefficient of x^{15} to the term independent of x in the expansion of $\left(x^2+\frac{2}{x}\right)^{15}$, is → JEE Mains 2013

(a) 7 : 16 (b) 7 : 64 (c) 1 : 4 (d) 1 : 32

19 The greatest term in the expansion of $\sqrt{3}\left(1+\frac{1}{\sqrt{3}}\right)^{20}$ is

(a) $\binom{20}{7}\frac{1}{27}$ (b) $\binom{20}{6}\frac{1}{81}$
(c) $\frac{1}{9}\binom{20}{9}$ (d) None of these

20 The largest term in the expansion of $(3+2x)^{50}$, where $x=\frac{1}{5}$ is

(a) 5th (b) 3th (c) 7th (d) 6th

21 If the sum of the coefficients in the expansion of $(x-2y+3z)^n$ is 128, then the greatest coefficient in the expansion of $(1+x)^n$ is

(a) 35 (b) 20 (c) 10 (d) None of these

22. If for positive integers $r>1$, $n>2$, the coefficient of the $(3r)$th and $(r+2)$th powers of x in the expansion of $(1+x)^{2n}$ are equal, then

(a) $n=2r$ (b) $n=3r$
(c) $n=2r+1$ (d) None of these

23 If $a_n=\sum_{r=0}^{n}\frac{1}{{}^nC_r}$, then $\sum_{r=0}^{n}\frac{r}{{}^nC_r}$ is equal to

(a) $(n-1)a_n$ (b) na_n
(c) $\frac{1}{2}na_n$ (d) None of these

24 $\sum_{r=0}^{n}(-1)^r({}^nC_r)\frac{1+rx}{1+nx}$ is equal to

(a) 1 (b) –1 (c) n (d) 0

25 $\binom{30}{0}\binom{30}{10}-\binom{30}{1}\binom{30}{11}+\ldots+\binom{30}{20}\binom{30}{30}$ is equal to

(a) ${}^{30}C_{11}$ (b) ${}^{60}C_{10}$ (c) ${}^{30}C_{10}$ (d) ${}^{65}C_{55}$

26 The value of $({}^{21}C_1-{}^{10}C_1)+({}^{21}C_2-{}^{10}C_2)+({}^{21}C_3-{}^{10}C_3)+({}^{21}C_4-{}^{10}C_4)+\ldots+({}^{21}C_{10}-{}^{10}C_{10})$ is → JEE Mains 2017

(a) $2^{21}-2^{11}$ (b) $2^{21}-2^{10}$ (c) $2^{20}-2^9$ (d) $2^{20}-2^{10}$

27 The sum of the series ${}^{20}C_0-{}^{20}C_1+{}^{20}C_2-{}^{20}C_3+\ldots+{}^{20}C_{10}$ is → AIEEE 2007

(a) $-{}^{20}C_{10}$ (b) $\frac{1}{2}{}^{20}C_{10}$ (c) 0 (d) ${}^{20}C_{10}$

28 If $(1+x)^n=C_0+C_1x+C_2x^2+\ldots+C_nX^n$, then the value of $C_0+2C_1+3C_2+\ldots+(n+1)C_n$ will be

(a) $(n+2)2^{n-1}$ (b) $(n+1)2^n$
(c) $(n+1)2^{n-1}$ (d) $(n+2)2^n$

29 If $n>(8+3\sqrt{7})^{10}, n\in N$, then the least value of n is

(a) $(8+3\sqrt{7})^{10}-(8-3\sqrt{7})^{10}$
(b) $(8+3\sqrt{7})^{10}+(8-3\sqrt{7})^{10}$
(c) $(8+3\sqrt{7})^{10}-(8-3\sqrt{7})^{10}+1$
(d) $(8+3\sqrt{7})^{10}-(8-3\sqrt{7})^{10}-1$

30 $49^n+16n-1$ is divisible by

(a) 3 (b) 19 (c) 64 (d) 29

31 If $A=1000^{1000}$ and $B=(1001)^{999}$, then

(a) $A>B$ (b) $A=B$
(c) $A<B$ (d) None of these

32 If ${}^{n-1}C_r=(k^2-3)\cdot{}^nC_{r+1}$, then k belongs to

(a) $(-\infty,-2]$ (b) $[2,\infty)$ (c) $[-\sqrt{3},\sqrt{3}]$ (d) $(\sqrt{3},2]$

33 The remainder left out when $8^{2n}-(62)^{2n+1}$ is divided by 9, is

(a) 0 (b) 2 (c) 7 (d) 8

34 If x is positive, the first negative term in the expansion of $(1+x)^{27/5}$ is → AIEEE 2003

(a) 7th term (b) 5th term (c) 8th term (d) 6th term

35 Let $P(n): n^2+n+1\ (n \in N)$ is an even integer. Therefore, $P(n)$ is true

(a) for $n>1$ (b) for all n (c) for $n>2$ (d) None of these

36 For all $n \in N, 1\times 1! + 2\times 2! + 3\times 3! + \ldots + n\times n!$ is equal to → NCERT Exemplar

(a) $(n+1)!-2$ (b) $(n+1)!$

(c) $(n+1)!-1$ (d) $(n+1)!-3$

37 For each $n \in N$, $2^{3n}-1$ is divisible by

(a) 8 (b) 16

(c) 32 (d) None of these

38 Let $S(k) = 1+3+5+\ldots+(2k-1) = 3+k^2$. Then, which of the following is true? → AIEEE 2004

(a) $S(1)$ is correct

(b) $S(k) \Rightarrow S(k+1)$

(c) $S(k) \not\Rightarrow S(k+1)$

(d) Principle of mathematical induction can be used to prove the formula

DAY PRACTICE SESSION 2

PROGRESSIVE QUESTIONS EXERCISE

1 The coefficient of x^{2m+1} in the expansion of

$$E = \frac{1}{(1+x)(1+x^2)(1+x^4)(1+x^8)\ldots(1+x^{2m})}, |x|<1 \text{ is}$$

(a) 3 (b) 2 (c) 1 (d) 0

2 $C_1 - \frac{C_2}{2} + \frac{C_3}{3} - \ldots + (-1)^{n-1}\frac{C_n}{n}$ is equal to

(a) $1-\frac{1}{2}+\frac{1}{3}-\ldots+\frac{(-1)^{n-1}}{n}$ (b) $1+\frac{1}{2}+\frac{1}{3}+\ldots+\frac{1}{n}$

(c) $1+\frac{1}{2}+\frac{1}{3}+\ldots+\frac{1}{n-1}$ (d) None of these

3 If the coefficient of x^5 in $\left[ax^2+\frac{1}{bx}\right]^{10}$ is a times and equal to the coefficient of x^{-5} in $\left[ax-\frac{1}{b^2x^2}\right]^{10}$, then the value of ab is

(a) $(b)^{-3}$ (b) $-(b)^6$ (c) $(b)^{-1}$ (d) None of these

4 The sum of coefficients of integral powers of x in the binomial expansion of $(1-2\sqrt{x})^{50}$, is → JEE Mains 2015

(a) $\frac{1}{2}(3^{50}+1)$ (b) $\frac{1}{2}(3^{50})$

(c) $\frac{1}{2}(3^{50}-1)$ (d) $\frac{1}{2}(2^{50}+1)$

5 The term independent of x in expansion of $\left(\frac{x+1}{x^{2/3}-x^{1/3}+1}-\frac{x-1}{x-x^{1/2}}\right)$ is → JEE Mains 2013

(a) 4 (b) 120 (c) 210 (d) 310

6 If $(1+x)^n = C_0 + C_1x + C_2x^2 + \ldots + C_nx^n$, then $C_0^2 + C_1^2 + C_2^2 + C_3^2 + \ldots + C_n^2$ is equal to

(a) $\frac{n!}{n!n!}$ (b) $\frac{(2n)!}{n!n!}$

(c) $\frac{(2n)!}{n!}$ (d) None of these

7 If a and d are two complex numbers, then the sum to $(n+1)$ terms of the following series

$$aC_0 - (a+d)C_1 + (a+2d)C_2 - \ldots + \ldots \text{ is}$$

(a) $\frac{a}{2^n}$ (b) na

(c) 0 (d) None of these

8 $\sum_{p=1}^{n}\sum_{m=p}^{n}\binom{n}{m}\binom{m}{p}$ is equal to

(a) 3^n (b) 2^n

(c) 3^n+2^n (d) 3^n-2^n

9 The sum of the series $\sum_{r=0}^{n}(-1)^r\,{}^nC_r\left(\frac{1}{2^r}+\frac{3^r}{2^{2r}}+\frac{7^r}{2^{3r}}+\frac{15^r}{2^{4r}}+\ldots+m \text{ terms}\right)$ is

(a) $\frac{2^{mn}-1}{2^{mn}(2^n-1)}$ (b) $\frac{2^{mn}-1}{2^n-1}$

(c) $\frac{2^{mn}+1}{2^n+1}$ (d) None of these

10 The value of x, for which the 6th term in the expansion of $\left\{2^{\log_2\sqrt{9^{x-1}+7}}+\frac{1}{2^{(1/5)\log_2(3^{x-1}+1)}}\right\}^7$ is 84, is equal to

(a) 4 (b) 3 (c) 2 (d) 5

11 If the last term in the binomial expansion of $\left(2^{1/3}-\frac{1}{\sqrt{2}}\right)^n$ is $\left(\frac{1}{3^{5/3}}\right)^{\log_3 8}$, then the 5th term from the beginning is

(a) 210 (b) 420

(c) 105 (d) None of these

12 The sum of the coefficients of all odd degree terms in the expansion of $(x+\sqrt{x^3-1})^5+(x-\sqrt{x^3-1})^5, (x>1)$ is → JEE Mains 2018

(a) −1 (b) 0 (c) 1 (d) 2

13 The greatest value of the term independent of x, as α varies over R, in the expansion of $\left(x\cos\alpha + \frac{\sin\alpha}{x}\right)^{20}$ is

(a) $^{20}C_{10}$ (b) $^{20}C_{15}$ (c) $^{20}C_{19}$ (d) None of these

14 **Statement I** For each natural number $n, (n+1)^7 - n^7 - 1$ is divisible by 7.

Statement II For each natural number n, $n^7 - n$ is divisible by 7. → AIEEE 2011

(a) Statement I is false, Statement II is true
(b) Statement I is true, Statement II is true, Statement II is correct explanation of Statement I.
(c) Statement I is true, Statement II is true; Statement II is not a correct explanation of Statement I
(d) Statement I is true, Statement II is false

15 If the ratio of the fifth term from the beginning to the fifth term from the end in the expansion of $\left(\sqrt[4]{2} + \frac{1}{\sqrt[4]{3}}\right)^n$ is $\sqrt{6}:1$, then

Statement I The value of n is 10.

Statement II $\dfrac{2^{\frac{n-4}{4}} \cdot 3^{-1}}{2 \cdot 3^{\frac{4+n}{4}}} = \sqrt{6}$ → NCERT Exemplar

(a) Statement I is true; Statement II is true; Statement II is a correct explanation for Statement I
(b) Statement I is true; Statement II is true; Statement II is not a correct explanation for Statement I
(c) Statement I is true; Statement II is false
(d) Statement I is false; Statement II is true

ANSWERS

SESSION 1

1 (a)	**2** (d)	**3** (a)	**4** (d)	**5** (c)	**6** (b)	**7** (d)	**8** (d)	**9** (a)	**10** (a)
11 (c)	**12** (c)	**13** (d)	**14** (c)	**15** (a)	**16** (a)	**17** (b)	**18** (d)	**19** (a)	**20** (c)
21 (a)	**22** (c)	**23** (c)	**24** (d)	**25** (c)	**26** (d)	**27** (b)	**28** (a)	**29** (b)	**30** (c)
31 (a)	**32** (d)	**33** (b)	**34** (c)	**35** (d)	**36** (c)	**37** (d)	**38** (b)		

SESSION 2

1 (c)	**2** (b)	**3** (b)	**4** (a)	**5** (c)	**6** (b)	**7** (c)	**8** (d)	**9** (a)	**10** (c)
11 (a)	**12** (d)	**13** (d)	**14** (b)	**15** (c)					

Hints and Explanations

1 Given that, $(1+ax)^n = 1 + 8x + 24x^2 + \ldots$

$\Rightarrow 1 + \frac{n}{1}ax + \frac{n(n-1)}{1\cdot 2}a^2x^2 + \ldots$
$= 1 + 8x + 24x^2 + \ldots$

On comparing the coefficients of x, x^2, we get

$na = 8, \frac{n(n-1)}{1\cdot 2}a^2 = 24$

$\Rightarrow na(n-1)a = 48$

$\Rightarrow 8(8-a) = 48 \Rightarrow 8 - a = 6$

$\Rightarrow a = 2 \Rightarrow n = 4$

2 Coefficient of x^n in $(1+x)^{2n} = {}^{2n}C_n$

and coefficient of x^n in $(1+x)^{2n-1} = {}^{2n-1}C_n$

∴ Required ratio

$$= \frac{^{2n}C_n}{^{2n-1}C_n} = \frac{\frac{(2n)!}{n!n!}}{\frac{(2n-1)!}{n!(n-1)!}} = 2:1$$

3 We have, $(1.002)^{12}$ or it can be rewritten as $(1+0.002)^{12}$

$\Rightarrow (1.002)^{12} = 1 + {}^{12}C_1(0.002) + {}^{12}C_2(0.002)^2 + {}^{12}C_3(0.002)^3 + \ldots$

We want the answer upto 4 decimal places and as such we have left further expansion.

$\therefore (1.002)^{12} = 1 + 12(0.002) + \frac{12\cdot 11}{1\cdot 2}(0.002)^2 + \frac{12\cdot 11\cdot 10}{1\cdot 2\cdot 3}(0.002)^3 + \ldots$

$= 1 + 0.024 + 2.64\times 10^{-4} + 1.76\times 10^{-6} + \ldots$

$= 1.0242$

4 $(1 + x + x^2 + x^3)^n = \{(1+x)^n(1+x^2)^n\}$

$= (1 + {}^nC_1x + {}^nC_2x^2 + {}^nC_3x^3 + {}^nC_4x^4 + \ldots {}^nC_nx^n)$
$(1 + {}^nC_1x^2 + {}^nC_2x^4 + \ldots + {}^nC_nx^{2n})$

Therefore, the coefficient of x^4

$= {}^nC_2 + {}^nC_2\,{}^nC_1 + {}^nC_4$
$= {}^nC_4 + {}^nC_2 + {}^nC_1\,{}^nC_2$

5 $\left(\frac{1}{x} + x\sin x\right)^{10}$

Here, $n = 10$ [even]

$\Rightarrow$ Middle term $= \left(\frac{10}{2} + 1\right)$th $=$ 6th

$T_6 = {}^{10}C_5\left(\frac{1}{x}\right)^{10-5}(x\sin x)^5$

$\Rightarrow 252(\sin x)^5 = 7\frac{7}{8} = \frac{63}{8}$

$\Rightarrow (\sin x)^5 = \frac{1}{32} \Rightarrow \sin x = \frac{1}{2}$

$\Rightarrow \sin x = \sin \pi/6$

$\therefore x = n\pi + (-1)^n\frac{\pi}{6}$

6 $T_7 = {}^9C_6\left(\frac{3}{\sqrt[3]{84}}\right)^3(\sqrt{3}\ln x)^6 = 729$

$\Rightarrow \frac{84\times 3^3}{84}\times 3^3\times(\ln x)^6 = 729$

$= (\ln x)^6 = 1$

$\Rightarrow x = e$

7 Clearly number of terms in the expansion of

$\left(1-\frac{2}{x}+\frac{4}{x^2}\right)^n$ is $\frac{(n+2)(n+1)}{2}$ or ${}^{n+2}C_2$.

[assuming $\frac{1}{x}$ and $\frac{1}{x^2}$ distinct]

$\therefore \quad \frac{(n+2)(n+1)}{2}=28$

$\Rightarrow \quad (n+2)(n+1)=56=(6+1)(6+2)$

$\Rightarrow \quad n=6$

Hence, sum of coefficients

$=(1-2+4)^6=3^6=729$

8 Since, in a binomial expansion of $(a-b)^n$, $n\geq 5$, the sum of 5th and 6th terms is equal to zero.

$\therefore \quad {}^nC_4a^{n-4}(-b)^4+{}^nC_5a^{n-5}(-b)^5=0$

$\Rightarrow \quad \frac{n!}{(n-4)!4!}a^{n-4}\cdot b^4$

$-\frac{n!}{(n-5)!5!}a^{n-5}b^5=0$

$\Rightarrow \quad \frac{n!}{(n-5)!4!}a^{n-5}\cdot b^4\left(\frac{a}{n-4}-\frac{b}{5}\right)=0$

$\Rightarrow \quad \frac{a}{b}=\frac{n-4}{5}$

9 The given expression is $1+(1+x)+(1+x)^2+\ldots+(1+x)^n$ being in GP.

Let, $S=1+(1+x)+(1+x)^2+\ldots+(1+x)^n$

$=\frac{(1+x)^{n+1}-1}{(1+x)-1}=x^{-1}[(1+x)^{n+1}-1]$

$\therefore$ The coefficient of x^k in S.

= The coefficient of x^{k+1} in $[(1+x)^{n+1}-1]$

$={}^{n+1}C_{k+1}$

10 We have, $(1+t^2)^{12}(1+t^{12})(1+t^{24})$

$=(1+{}^{12}C_1t^2+{}^{12}C_2t+\ldots+{}^{12}C_6t^{12}$

$+\ldots+{}^{12}C_{12}t^{24}+\ldots)(1+t^{12}+t^{24}+t^{36})$

$\therefore$ Coefficient of t^{24} in

$(1+t^2)^{12}(1+t^{12})(1+t^{24})$

$={}^{12}C_6+{}^{12}C_{12}+1={}^{12}C_6+2$

11 The given sigma expansion

$\sum_{m=0}^{100}{}^{100}C_m(x-3)^{100-m}\cdot 2^m$ can be written

as $[(x-3)+2]^{100}=(x-1)^{100}=(1-x)^{100}$

$\therefore$ Coefficient of x^{53} in

$(1-x)^{100}=(-1)^{53}\,{}^{100}C_{53}=-{}^{100}C_{53}$

12 Given expression is $\left(\frac{p}{2}+2\right)^8$

Here, $\quad n=8$ [even]

$\Rightarrow$ Middle term $=\left(\frac{8}{2}+1\right)$th term

= 5 th term

$T_5={}^8C_4(p/2)^{8-4}(2^4)$

$\Rightarrow \quad \frac{8\times7\times6\times5}{4\times3\times2\times1}\times\frac{p^4}{2^4}\times2^4=1120$

$\Rightarrow \quad p^4=16$

$\Rightarrow \quad p=\pm 2$

13 $\left(1+x+\frac{2}{x}\right)^6=1+\binom{6}{1}\left(x+\frac{2}{x}\right)$

$+\binom{6}{2}\left(x+\frac{2}{x}\right)^2+\ldots+\binom{6}{6}\left(x+\frac{2}{x}\right)^6$

$\therefore$ Constant term

$=1+\binom{6}{2}\binom{2}{1}2^1+\binom{6}{4}\binom{4}{2}2^2+$

$\binom{6}{6}\binom{6}{3}2^3$

$=1+60+360+160=581$

14 $(1+x)^m(1-x)^n$

$=\left\{1+mx+\frac{m(m-1)x^2}{2!}+\ldots\right\}$

$\left[1-nx+\frac{n(n-1)}{2!}x^2-\ldots\right]$

$=1+(m-n)x$

$+\left[\frac{n^2-n}{2}-mn+\frac{(m^2-m)}{2}\right]x^2+\ldots$

Given, $\quad m-n=3\Rightarrow n=m-3$

and $\quad \frac{n^2-n}{2}-mn+\frac{m^2-m}{2}=-6$

$\Rightarrow \quad \frac{(m-3)(m-4)}{2}-m(m-3)$

$+\frac{m^2-m}{2}=-6$

$\Rightarrow \quad m^2-7m+12-2m^2+6m$

$+m^2-m+12=0$

$\Rightarrow \quad -2m+24=0\Rightarrow m=12$

15 $(\sqrt{3}+1)^{2n}={}^{2n}C_0(\sqrt{3})^{2n}+{}^{2n}C_1(\sqrt{3})^{2n-1}$

$+{}^{2n}C_2(\sqrt{3})^{2n-2}+\ldots+{}^{2n}C_{2n}(\sqrt{3})^{2n-2n}$

$(\sqrt{3}-1)^{2n}={}^{2n}C_0(\sqrt{3})^{2n}(-1)^0$

$+{}^{2n}C_1(\sqrt{3})^{2n-1}(-1)^1+{}^{2n}C_2(\sqrt{3})^{2n-2}(-1)^2+\ldots$

$+{}^{2n}C_{2n}(\sqrt{3})^{2n-2n}(-1)^{2n}$

Adding both the binomial expansions above, we get

$(\sqrt{3}+1)^{2n}-(\sqrt{3}-1)^{2n}=2[{}^{2n}C_1(\sqrt{3})^{2n-1}$

$+{}^{2n}C_3(\sqrt{3})^{2n-3}+{}^{2n}C_5(\sqrt{3})^{2n-5}$

$+\ldots+{}^{2n}C_{2n-1}(\sqrt{3})^{2n-(2n-1)}]$

which is most certainly an irrational number because of odd powers of $\sqrt{3}$ in each of the terms.

16 $\therefore$ General term is

$T_{r+1}={}^{21}C_r\left(\sqrt[3]{\frac{a}{\sqrt{b}}}\right)^{21-r}\left(\sqrt{\frac{b}{\sqrt[3]{a}}}\right)^r$

$={}^{21}C_r\,a^{7-\frac{r}{2}}\cdot b^{\frac{2r}{3}-\frac{7}{2}}$

$\because$ Power of a = Power of b [given]

$\Rightarrow \quad 7-\frac{r}{2}=\frac{2}{3}r-\frac{7}{2}$

$\therefore \quad r=9$

17 The general term in the expansion of $\left(x+\frac{1}{x^2}\right)^{n-3}$ is given by

$T_{r+1}={}^{n-3}C_r(x)^{n-3-r}\left(\frac{1}{x^2}\right)^r$

$={}^{n-3}C_rx^{n-3-3r}$

As x^{2k} occurs in the expansion of $\left(x+\frac{1}{x^2}\right)^{n-3}$, we must have

$n-3-3r=2k$ for some non-negative integer r.

$\Rightarrow \quad 3(1+r)=n-2k$

$\Rightarrow \quad n-2k$ is a multiple of 3.

18 $T_{r+1}={}^{15}C_r(x^2)^{15-r}\cdot\left(\frac{2}{x}\right)^r$

$={}^{15}C_rx^{30-2r}\cdot2^r\cdot x^{-r}$

$={}^{15}C_r\cdot x^{30-3r}\cdot2^r \quad \ldots(i)$

For coefficient of x^{15}, put $30-3r=15$

$\Rightarrow 3r=15\Rightarrow r=5$

$\therefore$ Coefficient of $x^{15}={}^{15}C_5\cdot2^5$

For coefficient of independent of x i.e. x^0 put $30-3r=0$

$\Rightarrow \quad r=10$

$\therefore$ Coefficient of $x^0={}^{15}C_{10}\cdot2^{10}$

By condition $\Rightarrow \frac{\text{Coefficient of } x^{15}}{\text{Coefficient of } x^0}$

$=\frac{{}^{15}C_5\cdot2^5}{{}^{15}C_{10}\cdot2^{10}}=\frac{{}^{15}C_{10}\cdot2^5}{{}^{15}C_{10}\cdot2^{10}}=1:32$

19 Greatest term in the expansion of $(1+x)^n$ is T_{r+1}

where, $r=\left[\frac{(n+1)x}{1+x}\right]$

Here, $n=20$, $x=\frac{1}{\sqrt{3}}$

$\therefore \quad r=\left[\frac{21}{\sqrt{3}+1}\right]$

$=[10.5(\sqrt{3}-1)]=(7.69)\approx 7$

Hence, greatest term is

$\sqrt{3}\binom{20}{7}\left(\frac{1}{\sqrt{3}}\right)^7=\binom{20}{7}\frac{1}{27}$.

20 $\because \quad (3+2x)^{50}=3^{50}\left(1+\frac{2x}{3}\right)^{50}$

Here, $\quad T_{r+1}=3^{50}\,{}^{50}C_r\left(\frac{2x}{3}\right)^r$

and $\quad T_r=3^{50}\,{}^{50}C_{r-1}\left(\frac{2x}{3}\right)^{r-1}$

But $\quad x=\frac{1}{5}$ (given)

$\therefore \quad \frac{T_{r+1}}{T_r}\geq1\Rightarrow\frac{{}^{50}C_r}{{}^{50}C_{r-1}}\cdot\frac{2}{3}\cdot\frac{1}{5}\geq1$

$\Rightarrow \quad 102-2r\geq15r\Rightarrow r\leq6$

21 Sum of the coefficients in the expansion of $(x-2y+3z)^n$ is $(1-2+3)^n = 2^n$

(put $x = y = z = 1$)

$\therefore \quad 2^n = 128 \Rightarrow n = 7$

Therefore, the greatest coefficient in the expansion of $(1+x)^7$ is ${}^7C_3$ or ${}^7C_4$ because both are equal to 35.

22 In the expansion of $(1+x)^{2n}$, the general term $= {}^{2n}C_k x^k, 0 \le k \le 2n$

As given for $r > 1, n > 2$,

$${}^{2n}C_{3r} = {}^{2n}C_{r+2}$$

$\Rightarrow$ Either $3r = r + 2$

or $3r = 2n - (r+2)$

$(\because {}^nC_x = {}^nC_y \Rightarrow x + y = n \text{ or } x = y)$

$\Rightarrow \quad r = 1$ or $n = 2r + 1$

We take the relation only

$n = 2r + 1 \qquad (\because r > 1)$

23 Let $b = \sum_{r=0}^{n} \frac{r}{{}^nC_r} = \sum_{r=0}^{n} \frac{n-(n-r)}{{}^nC_r}$

$$= n\sum_{r=0}^{n} \frac{1}{{}^nC_r} - \sum_{r=0}^{n} \frac{n-r}{{}^nC_r}$$

$$= na_n - \sum_{r=0}^{n} \frac{n-r}{{}^nC_{n-r}} \quad (\because {}^nC_r = {}^nC_{n-r})$$

$$= na_n - b \Rightarrow 2b = na_n \Rightarrow b = \frac{n}{2}a_n$$

24 Let $E = \sum_{r=0}^{n} (-1)^r\, {}^nC_r \left(\frac{1+rx}{1+nx}\right)$

$$= \left(\frac{1}{1+nx}\right)\sum_{r=0}^{n} (-1)^r\, {}^nC_r (1+rx)$$

$$= \left(\frac{1}{1+nx}\right)\left\{\sum_{r=0}^{n} (-1)^r \cdot {}^nC_r + x\sum_{r=0}^{n} r(-1)^r\, {}^nC_r\right\}$$

$$= \left(\frac{1}{1+nx}\right)(0+0) = 0$$

$[\because {}^nC_0 - {}^nC_1 + {}^nC_2 - {}^nC_3 + \ldots (-1)^n\, {}^nC_n = 0]$

25 Let

$$A = \binom{30}{0}\binom{30}{10} - \binom{30}{1}\binom{30}{11} + \binom{30}{2}\binom{30}{12} - \ldots + \binom{30}{20}\binom{30}{30}$$

or $A = {}^{30}C_0 \cdot {}^{30}C_{10} - {}^{30}C_1 \cdot {}^{30}C_{11} + {}^{30}C_2 \cdot {}^{30}C_{12} - \ldots + {}^{30}C_{20} \cdot {}^{30}C_{30}$

= Coefficient of x^{20} in $(1+x)^{30}(1-x)^{30}$

= Coefficient of x^{20} in $(1-x^2)^{30}$

= Coefficient of x^{20} in $\sum_{r=0}^{30} (-1)^r\, {}^{30}C_r (x^2)^r$

$= (-1)^{10}\, {}^{30}C_{10}$

(for coefficient of x^{20}, let $r = 10$)

$= {}^{30}C_{10}$

26 $({}^{21}C_1 - {}^{10}C_1) + ({}^{21}C_2 - {}^{10}C_2) + ({}^{21}C_3 - {}^{10}C_3) + \ldots + ({}^{21}C_{10} - {}^{10}C_{10})$

$= ({}^{21}C_1 + {}^{21}C_2 + \ldots + {}^{21}C_{10}) - ({}^{10}C_1 + {}^{10}C_2 + \ldots + {}^{10}C_{10})$

$= \frac{1}{2}({}^{21}C_1 + {}^{21}C_2 + \ldots + {}^{21}C_{20}) - (2^{10} - 1)$

$= \frac{1}{2}({}^{21}C_1 + {}^{21}C_2 + \ldots + {}^{21}C_{21} - 1) - (2^{10} - 1)$

$= \frac{1}{2}(2^{21} - 2) - (2^{10} - 1) = 2^{20} - 1 - 2^{10} + 1$

$= 2^{20} - 2^{10}$

27 We know that,

$(1+x)^{20} = {}^{20}C_0 + {}^{20}C_1 x + \ldots + {}^{20}C_{10}x^{10} + \ldots + {}^{20}C_{20}x^{20}$

On putting $x = -1$ in the above expansion, we get

$0 = {}^{20}C_0 - {}^{20}C_1 + \ldots - {}^{20}C_a + {}^{20}C_{10} - {}^{20}C_{11} + \ldots + {}^{20}C_{20}$

$\Rightarrow \quad 0 = {}^{20}C_0 - {}^{20}C_1 + \ldots - {}^{20}C_9 + {}^{20}C_{10} - {}^{20}C_9 + \ldots + {}^{20}C_0$

$\Rightarrow 0 = 2({}^{20}C_0 - {}^{20}C_1 + \ldots - {}^{20}C_9) + {}^{20}C_{10}$

$\Rightarrow \quad {}^{20}C_{10} = 2({}^{20}C_0 - {}^{20}C_1 + \ldots + {}^{20}C_{10})$

$\Rightarrow \quad {}^{20}C_0 - {}^{20}C_1 + \ldots + {}^{20}C_{10} = \frac{1}{2}\,{}^{20}C_{10}$

28 Since, $x(1+x)^n = xC_0 + C_1x^2 + C_2x^3 + \ldots + C_nx^{n+1}$

On differentiating w.r.t. x, we get

$(1+x)^n + nx(1+x)^{n-1} = C_0 + 2C_1x + 3C_2x^2 + \ldots + (n+1)C_nx^n$

Put $x = 1$, we get

$C_0 + 2C_1 + 3C_2 + \ldots + (n+1)C_n = 2^n + n2^{n-1} = 2^{n-1}(n+2)$

29 Let $f = (8 - 3\sqrt{7})^{10}$, here $0 < f < 1$

$\therefore \quad (8+3\sqrt{7})^{10} + (8-3\sqrt{7})^{10}$ is an integer, hence this is the value of n.

30 We have,

$49^n + 16n - 1 = (1+48)^n + 16n - 1$

$= 1 + {}^nC_1(48) + {}^nC_2(48)^2 + \ldots + {}^nC_n(48)^n + 16n - 1$

$= (48n + 16n) + {}^nC_2(48)^2 + {}^nC_3(48)^3 + \ldots + {}^nC_n(48)^n$

$= 64n + 8^2[{}^nC_2 \cdot 6^2 + {}^nC_3 \cdot 6^3 \cdot 8 + {}^nC_4 \cdot 6^4 8^2 + \ldots + {}^nC_n \cdot 6^n \cdot 8^{n-2}]$

Hence, $49^n + 16n - 1$ is divisible by 64.

31 Since, $\left(1 + \frac{1}{n}\right)^n < 3$ for $\forall n \in N$

Now, $\frac{(1001)^{999}}{(1000)^{1000}} = \frac{1}{1001}\left(\frac{1001}{1000}\right)^{1000}$

$= \frac{1}{1001}\left(1 + \frac{1}{1000}\right)^{1000} < \frac{1}{1001} \cdot 3 < 1$

$(1001)^{999} < (1000)^{1000}$

$\therefore \quad B < A$

32 Since, ${}^{n-1}C_r = (k^2 - 3)\frac{n}{r+1}\,{}^{n-1}C_r$

$\Rightarrow \quad k^2 - 3 = \frac{r+1}{n}$

$\Rightarrow \quad 0 < k^2 - 3 \le 1$

$\left[\because n \ge r \Rightarrow \frac{r+1}{n} \le 1 \text{ and } n, r > 0\right]$

$\Rightarrow \quad 3 < k^2 \le 4$

Hence, $k \in [-2, -\sqrt{3}] \cup (\sqrt{3}, 2)$

33 $8^{2n} - (62)^{2n+1} = (1+63)^n - (63-1)^{2n+1}$

$= (1+63)^n + (1-63)^{2n+1}$

$= [1 + {}^nC_1 \cdot 63 + {}^nC_2 \cdot (63)^2 + \ldots + (63)^n] + [1 - {}^{2n-1}C_1 \cdot 63 + {}^{(2n+1)}C_2 \cdot (63)^2 - \ldots + (-1)(63)^{(2n+1)}]$

$= 2 + 63[{}^nC_1 + {}^nC_2(63) + \ldots + (63)^{n-1} - {}^{(2n+1)}C_1 + {}^{(2n+1)}C_2(63) - \ldots + (-1)(63)^{2n}]$

Hence, remainder is 2.

34 Since, $(r+1)$th term in the expansion of $(1+x)^{27/5}$

$$= \frac{\frac{27}{5}\left(\frac{27}{5} - 1\right)\ldots\left(\frac{27}{5} - r + 1\right)}{r!}x^r$$

Now, this term will be negative, if the last factor in numerator is the only one negative factor.

$\Rightarrow \quad \frac{27}{5} - r + 1 < 0 \Rightarrow \frac{32}{5} < r$

$\Rightarrow \quad 6.4 < r \Rightarrow$ least value of r is 7.

Thus, first negative term will be 8th.

35 Given, $P(n): n^2 + n + 1$

At $n = 1, P(1): 3$, which is not an even integer.

Thus, $P(1)$ is not true.

Also, $n(n+1) + 1$ is always an odd integer.

36 Let the statement $P(n)$ be defined as

$P(n): 1 \times 1! + 2 \times 2! + 3 \times 3! \ldots + n \times n! = (n+1)! - 1$

for all natural numbers n.

Note that $P(1)$ is true, since

$P(1): 1 \times 1! = 1 = 2 - 1 = 2! - 1$

Assume that $P(n)$ is true for some natural number k, i.e.

$P(k): 1 \times 1! + 2 \times 2! + 3 \times 3! + \ldots + k \times k! = (k+1)! - 1 \quad \ldots(i)$

To prove $P(k+1)$ is true, we have

$P(k+1): 1 \times 1! + 2 \times 2! + 3 \times 3! + \ldots + k \times k! + (k+1) \times (k+1)!$

$= (k+1)! - 1 + (k+1)! \times (k+1)$ [by Eq. (i)]

$= (k+1+1)(k+1)! - 1$

$= (k+2)(k+1)! - 1 = (k+2)! - 1$

Thus, $P(k+1)$ is true, whenever $P(k)$ is true. Therefore, by the principle of mathematical induction, $P(n)$ is true for all natural numbers n.

37 Now, $2^{3n}-1=(2^3)^n-1=(1+7)^n-1$

$$=1+{}^nC_1\cdot 7+{}^nC_2\cdot 7^2+\dots+{}^nC_n\cdot 7^n-1$$
$$=7[{}^nC_1+{}^nC_2 7+\dots+{}^nC_n\cdot 7^{n-1}]$$

Hence, 7 divides $2^{3n}-1$ for all $n\in N$.

38 $S(k)=1+3+5+\dots+(2k-1)=3+k^2$

Put $k=1$ in both sides, we get

LHS = 1 and RHS = 3 + 1 = 4

$\Rightarrow$ LHS $\neq$ RHS

Put $(k+1)$ in both sides in the place of k, we get

LHS $=1+3+5+\dots+(2k-1)+(2k+1)$

RHS $=3+(k+1)^2=3+k^2+2k+1$

Let LHS = RHS

$$1+3+5+\dots+(2k-1)+(2k+1)=3+k^2+2k+1$$
$$\Rightarrow 1+3+5+\dots+(2k-1)=3+k^2$$

If $S(k)$ is true, then $S(k+1)$ is also true.

Hence, $S(k)\Rightarrow S(k+1)$

SESSION 2

1 Multiplying the numerator and denominator by $1-x$, we have

$$E=\frac{1-x}{(1-x)(1+x)(1+x^2)(1+x^4)\dots(1+x^{2^m})}$$
$$=\frac{1-x}{(1-x^2)(1+x^2)(1+x^4)\dots(1+x^{2^m})}$$
$$=\frac{1-x}{(1-x^4)(1+x^4)\dots(1+x^{2^m})}$$
$$=\frac{1-x}{(1-x^{2^{m+1}})}=(1-x)(1-x^{2^{m+1}})^{-1}$$
$$=(1-x)(1+x^{2^{m+1}}+x^{2^{m+2}}+\dots)$$

$\therefore$ Coefficient of $x^{2^{m+1}}$ is 1.

2 Since, $(1-x)^n=C_0-C_1\cdot x+C_2\cdot x^2-C_3\cdot x^3+\dots$

$$\Rightarrow 1-(1-x)^n=C_1\cdot x-C_2\cdot x^2+C_3\cdot x^3-\dots$$
$$\Rightarrow \frac{1-(1-x)^n}{x}=C_1-C_2\cdot x+C_3\cdot x^2-\dots$$
$$\Rightarrow \int_0^1(C_1-C_2\cdot x+C_3\cdot x^2-\dots)\,dx=\int_0^1\frac{1-(1-x)^n}{1-(1-x)}\,dx$$
$$\Rightarrow \frac{C_1}{1}-\frac{C_2}{2}+\frac{C_3}{3}-\dots=\int_0^1\frac{1-x^n}{1-x}\,dx$$
$$\left[\because \int_0^1 f(x)\,dx=\int_0^1 f(1-x)\,dx\right]$$
$$=\int_0^1(1+x+x^2+\dots+x^{n-1})\,dx$$
$$=\left[x+\frac{x^2}{2}+\dots+\frac{x^n}{n}\right]_0^1=1+\frac{1}{2}+\frac{1}{3}+\dots+\frac{1}{n}$$

3 General term is

$$T_{r+1}={}^{10}C_r\cdot(a\cdot x^2)^{10-r}\cdot\left(\frac{1}{bx}\right)^r$$
$$={}^{10}C_r\cdot(a)^{10-r}\left(\frac{1}{b}\right)^r(x)^{20-3r}$$

Since, x^5 occurs in T_{r+1}.

$\therefore$ $20-3r=5$

$\Rightarrow 3r=15\Rightarrow r=5$

So, the coefficient of x^5 is ${}^{10}C_5(a)^5(b)^{-5}$.

Again, let x^{-5} occurs in T_{r+1} of

$$\left[a\cdot x-\frac{1}{b^2\cdot x^2}\right]^{10}\text{ is }{}^{10}C_r(ax)^{10-r}\left(-\frac{1}{b^2x^2}\right)^r$$
$$={}^{10}C_r(a)^{10-r}\left(-\frac{1}{b^2}\right)^r(x)^{10-3r}$$
$$10-3r=-5\Rightarrow 15=3r\Rightarrow r=5$$

So, the coefficient of x^{-5} is $-{}^{10}C_5\dfrac{a^5}{b^{10}}$.

According to the given condition,

$${}^{10}C_5\frac{a^5}{b^5}=-a\,{}^{10}C_5\frac{a^5}{b^{10}}$$
$$\Rightarrow -b^5=a\Rightarrow -b^6=ab$$

4 Let T_{r+1} be the general term in the expansion of $(1-2\sqrt{x})^{50}$.

$$\therefore\quad T_{r+1}={}^{50}C_r\,(1)^{50-r}(-2x^{1/2})^r$$
$$={}^{50}C_r\cdot 2^r\cdot x^{r/2}(-1)^r$$

For the integral power of x and r should be even integer.

$$\therefore\text{ Sum of coefficients}=\sum_{r=0}^{25}{}^{50}C_{2r}(2)^{2r}$$
$$=\frac{1}{2}[(1+2)^{50}+(1-2)^{50}]=\frac{1}{2}[3^{50}+1]$$

Alternate Method

We have,

$$(1-2\sqrt{x})^{50}=C_0-C_1\cdot 2\sqrt{x}+C_2(2\sqrt{x})^2+\dots+C_{50}(2\sqrt{x})^{50}\quad\dots(i)$$
$$(1+2\sqrt{x})^{50}=C_0+C_1\cdot 2\sqrt{x}+C_2(2\sqrt{x})^2+\dots+C_{50}(2\sqrt{x})^{50}\quad\dots(ii)$$

On adding Eqs. (i) and (ii), we get

$$(1-2\sqrt{x})^{50}+(1+2\sqrt{x})^{50}=2[C_0+C_2(2\sqrt{x})^2+\dots+C_{50}(2\sqrt{x})^{50}]$$
$$\Rightarrow\frac{(1-2\sqrt{x})^{50}+(1+2\sqrt{x})^{50}}{2}=C_0+C_2(2\sqrt{x})^2+\dots+C_{50}(2\sqrt{x})^{50}$$

On putting $x=1$, we get

$$\frac{(1-2\sqrt{1})^{50}+(1+2\sqrt{1})^{50}}{2}=C_0+C_2(2)^2+\dots+C_{50}(2)^{50}$$
$$\Rightarrow\frac{(-1)^{50}+(3)^{50}}{2}=C_0+C_2(2)^2+\dots+C_{50}(2)^{50}$$
$$\Rightarrow\frac{1+3^{50}}{2}=C_0+C_2(2)^2+\dots+C_{50}(2)^{50}$$

5
$$\left[\frac{x+1}{x^{2/3}-x^{1/3}+1}-\frac{(x-1)}{x-x^{1/2}}\right]^{10}$$
$$=\left[\frac{(x^{1/3})^3+1^3}{x^{2/3}-x^{1/3}+1}-\frac{\{(\sqrt{x})^2-1\}}{\sqrt{x}(\sqrt{x}-1)}\right]^{10}$$
$$=\left[\frac{(x^{1/3}+1)(x^{2/3}+1-x^{1/3})}{x^{2/3}-x^{1/3}+1}-\frac{\{(\sqrt{x})^2-1\}}{\sqrt{x}(\sqrt{x}-1)}\right]^{10}$$
$$=\left[(x^{1/3}+1)-\frac{(\sqrt{x}+1)}{\sqrt{x}}\right]^{10}=(x^{1/3}-x^{-1/2})^{10}$$

$\therefore$ The genreal term is

$$T_{r+1}={}^{10}C_r(x^{1/3})^{10-r}(-x^{-1/2})^r$$
$$={}^{10}C_r(-1)^r x^{\frac{10-r}{3}-\frac{r}{2}}$$

For independent for x, put

$$\frac{10-r}{3}-\frac{r}{2}=0\Rightarrow 20-2r-3r=0$$
$$\Rightarrow\quad 20=5r\Rightarrow r=4$$
$$\therefore\quad T_5={}^{10}C_4=\frac{10\times 9\times 8\times 7}{4\times 3\times 2\times 1}=210$$

6 We have,

$$(1+x)^n=C_0+C_1x+C_2x^2+\dots+C_nx^n\quad\dots(i)$$
$$\text{and}\left(1+\frac{1}{x}\right)^n=C_0+C_1\frac{1}{x}+C_2\left(\frac{1}{x}\right)^2+\dots+C_n\left(\frac{1}{x}\right)^n\quad\dots(ii)$$

On multiplying Eqs. (i) and (ii) and taking coefficient of constant terms in right hand side $=C_0^2+C_1^2+C_2^2+\dots+C_n^2$

In right hand side $(1+x)^n\left(1+\dfrac{1}{x}\right)^n$ or in $\dfrac{1}{x^n}(1+x)^{2n}$ or term containing x^n in $(1+x)^{2n}$. Clearly, the coefficient of x^n in $(1+x)^{2n}$ is equal to ${}^{2n}C_n=\dfrac{(2n)!}{n!n!}$.

7 We can write,

$aC_0-(a+d)C_1+(a+2d)C_2-\dots$ upto $(n+1)$ terms

$$=a(C_0-C_1+C_2-\dots)+d(-C_1+2C_2-3C_3+\dots)\quad\dots(i)$$

We know,

$$(1-x)^n=C_0-C_1x+C_2x^2-\dots+(-1)^nC_nx^n\quad\dots(ii)$$

On differentiating Eq. (ii) w.r.t. x, we get

$$-n(1-x)^{n-1}=-C_1+2C_2x-\dots+(-1)^nC_nnx^{n-1}\quad\dots(iii)$$

On putting $x=1$ in Eqs. (ii) and (iii), we get

$$C_0-C_1+C_2-\dots+(-1)^nC_n=0\quad\dots(iv)$$
$$\text{and }-C_1+2C_2-\dots+(-1)^nC_n=0\quad\dots(v)$$

From Eq. (i),

$aC_0-(a+d)C_1+(a+2d)C_2-\ldots+$ upto $(n+1)$ terms

$= a\cdot 0 + d\cdot 0 = 0$ [from Eqs. (iv) and (v)]

8 Since, $\binom{n}{m}\binom{m}{p} = \dfrac{n!}{(n-m)!\,p!\,(m-p)!}$

$= \binom{n}{p}\binom{n-p}{m-p}$

$\therefore$ Given series can be rewritten as

$\sum_{p=1}^{n}\sum_{m=p}^{n}\binom{n}{p}\binom{n-p}{m-p}$

$= \sum_{p=1}^{n}\binom{n}{p}\sum_{m=p}^{n}\binom{n-p}{m-p}$

$= \sum_{p=1}^{n}\binom{n}{p}\sum_{t=0}^{n-p}\binom{n-p}{t}$

$= \sum_{p=1}^{n}\binom{n}{p}2^{n-p}$ [put $m-p=t$]

$= 2^n\sum_{p=1}^{n}\binom{n}{p}\cdot\dfrac{1}{2^p} = 2^n\left[\left(1+\dfrac{1}{2}\right)^n-1\right]$

$= 3^n - 2^n$

9 $\sum_{r=0}^{n}(-1)^r$

${}^nC_r\left\{\dfrac{1}{2^r}+\dfrac{3^r}{2^{2r}}+\dfrac{7^r}{2^{3r}}+\ldots \text{upto } m \text{ terms}\right\}$

$= \sum_{r=0}^{n}(-1)^r\,{}^nC_r\cdot\dfrac{1}{2^r}+\sum_{r=0}^{n}(-1)^r\cdot{}^nC_r\dfrac{3^r}{2^{2r}}$

$+\sum_{r=0}^{n}(-1)^r\,{}^nC_r\dfrac{7^r}{2^{3r}}+\ldots$

$= \left(1-\dfrac{1}{2}\right)^n+\left(1-\dfrac{3}{4}\right)^n+\left(1-\dfrac{7}{8}\right)^n+\ldots$

upto m terms

$= \dfrac{1}{2^n}+\dfrac{1}{2^{2n}}+\dfrac{1}{2^{3n}}\ldots$ upto m terms

$= \dfrac{\dfrac{1}{2^n}\left\{1-\left(\dfrac{1}{2^n}\right)^m\right\}}{\left(1-\dfrac{1}{2^n}\right)} = \dfrac{2^{mn}-1}{2^{mn}(2^n-1)}$

10 We have,

$\left[2^{\log_2\sqrt{9^{x-1}+7}}+\dfrac{1}{2^{(1/5)\log_2(3^{x-1}+1)}}\right]^7$

$= \left[\sqrt{9^{x-1}+7}+\dfrac{1}{(3^{x-1}+1)^{1/5}}\right]^7$

$\therefore T_6 = {}^7C_5(\sqrt{9^{x-1}+7})^{7-5}\left[\dfrac{1}{(3^{x-1}+1)^{1/5}}\right]^5$

$= {}^7C_5(9^{x-1}+7)\dfrac{1}{(3^{x-1}+1)}$

$\Rightarrow\ 84 = {}^7C_5\dfrac{(9^{x-1}+7)}{(3^{x-1}+1)}$

$\Rightarrow\ 9^{x-1}+7 = 4(3^{x-1}+1)$

$\Rightarrow\ \dfrac{3^{2x}}{9}+7 = 4\left(\dfrac{3^x}{3}+1\right)$

$\Rightarrow\ 3^{2x}-12(3^x)+27 = 0$

$\Rightarrow\ y^2-12y+27 = 0$ (put $y = 3^x$)

$\Rightarrow\ (y-3)(y-9) = 0$

$\Rightarrow\ y = 3, 9$

$\Rightarrow\ 3^x = 3, 9$

$\Rightarrow\ x = 1, 2$

11 Last term of $\left(2^{1/3}-\dfrac{1}{\sqrt{2}}\right)^n$ is

$T_{n+1} = {}^nC_n(2^{1/3})^{n-n}\left(-\dfrac{1}{\sqrt{2}}\right)^n$

$= {}^nC_n(-1)^n\dfrac{1}{2^{n/2}} = \dfrac{(-1)^n}{2^{n/2}}$

Also, we have

$\left(\dfrac{1}{3^{5/3}}\right)^{\log_3 8} = 3^{-(5/3)\log_3 2^3} = 2^{-5}$

Thus, $\dfrac{(-1)^n}{2^{n/2}} = 2^{-5} \Rightarrow \dfrac{(-1)^n}{2^{n/2}} = \dfrac{(-1)^{10}}{2^5}$

$\Rightarrow\ \dfrac{n}{2} = 5 \Rightarrow n = 10$

Now, $T_5 = T_{4+1} = {}^{10}C_4(2^{1/3})^{10-4}\left(-\dfrac{1}{\sqrt{2}}\right)^4$

$= \dfrac{10!}{4!6!}(2^{1/3})^6(-1)^4(2^{-1/2})^4$

$= 210(2)^2(1)(2^{-2}) = 210$

12 **Key idea** $= (a+b)^n+(a-b)^n$

$= 2({}^nC_0a^n+{}^nC_2a^{n-2}b^2+{}^nC_4a^{n-4}b^4\ldots)$

We have

$(x+\sqrt{x^3-1})^5+(x-\sqrt{x^3-1})^5, x>1$

$= 2({}^5C_0x^5+{}^5C_2x^3(\sqrt{x^3-1})^2$

$+{}^5C_4x(\sqrt{x^3-1})^4)$

$= 2(x^5+10x^3(x^3-1)+5x(x^3-1)^2)$

$= 2(x^5+10x^6-10x^3+5x^7-10x^4+5x)$

Sum of coefficients of all odd degree terms is

$2(1-10+5+5) = 2$

13 The general term in the expansion of $\left(x\cos\alpha+\dfrac{\sin\alpha}{x}\right)^{20}$ is

${}^{20}C_r(x\cos\alpha)^{20-r}\left(\dfrac{\sin\alpha}{x}\right)^r$

$= {}^{20}C_rx^{20-2r}(\cos\alpha)^{20-r}(\sin\alpha)^r$

For this term to be independent of x, we get

$20-2r = 0$

$\Rightarrow\ r = 10$

Let β = Term indeoendent of x

$= {}^{20}C_{10}(\cos\alpha)^{10}(\sin\alpha)^{10}$

$= {}^{20}C_{10}(\cos\alpha\sin\alpha)^{10}$

$= {}^{20}C_{10}\left(\dfrac{\sin 2\alpha}{2}\right)^{10}$

Thus, the greatest possible value of β is ${}^{20}C_{10}\left(\dfrac{1}{2}\right)^{10}$.

14 Let $P(n) = (n)^7-n$

By mathematical induction

For $n=1$,

$P(1) = 0$, which is divisible by 7.

For $n = k$

$P(k) = k^7-k$

Let $P(k)$ be divisible by 7.

$\therefore\ k^7-k = 7\lambda$, for some $\lambda\in N$ ... (i)

For $n = k+1$,

$P(k+1) = (k+1)^7-(k+1)$

$= ({}^7C_0k^7+{}^7C_1k^6+{}^7C_2k^5+\ldots+{}^7C_6\cdot k$

$+{}^7C_7)-(k+1)$

$= (k^7-k)+7\{k^6+3k^5+\ldots+k\}$

$= 7\lambda+7\{k^6+3k^5+\ldots+k\}$ [Using Eq. (i)]

$\Rightarrow$ Divisible by 7.

So, both statements are true and Statement II is correct explanation of Statement I.

15 We know that, in the expansion of $(a+b)^n$, pth term from the end is $(n-p+2)$th term from the beginning.

So, 5th term from the end is

$= (n-5+2)$th term from the beginning

$= (n-3)$th term from the beginning

$= (n-4+1)$th term from the beginning ...(i)

$\therefore$ We have,

$\left(\sqrt[4]{2}+\dfrac{1}{\sqrt[4]{3}}\right)^n = \left(2^{1/4}+\dfrac{1}{3^{1/4}}\right)^n$

Now, 5th term from the beginning is

$T_{4+1} = {}^nC_4(2^{1/4})^{n-4}(3^{-1/4})^4$

$= {}^nC_4\,2^{\frac{n-4}{4}}\,3^{-1}$ ...(ii)

And 5th term from the end is

$T_{(n-4)+1} = {}^nC_{n-4}(2^{1/4})^{n-n+4}(3^{1/4})^{n-4}$

$= {}^nC_4\,2\cdot 3^{-\frac{n+4}{4}}$

$[\because {}^nC_r = {}^nC_{n-r}]$ ...(iii)

So, from the given condition, we have

$\dfrac{\text{Fifth term from the beginning}}{\text{Fifth term from the end}} = \dfrac{\sqrt{6}}{1}$

$\Rightarrow\ \dfrac{{}^nC_4\cdot 2^{\frac{n-4}{4}}\cdot 3^{-1}}{{}^nC_4\cdot 2\cdot 3^{\frac{-n+4}{4}}} = \dfrac{\sqrt{6}}{1}$

$\Rightarrow\ 2^{\frac{n-4}{4}-1}\cdot 3^{-1-\left(\frac{4-n}{4}\right)} = \sqrt{6}$

$\Rightarrow\ 2^{\frac{n-8}{4}}\,3^{\frac{n-8}{4}} = 6^{1/2}$

$\Rightarrow\ (2\times 3)^{\frac{n-8}{4}} = (2\cdot 3)^{1/2}$

$\Rightarrow\ \dfrac{n-8}{4} = 1/2$

$\Rightarrow\ n = 2+8 \quad \therefore n = 10$

DAY EIGHT

Permutations and Combinations

Learning & Revision for the Day

- Fundamental Principle of Counting
- Factorial Notation
- Permutations
- Circular Permutations
- Combinations
- Applications of Permutations and Combinations
- Prime Factors
- Division of Objects into Groups

Fundamental Principle of Counting

The fundamental principle of counting is a way to figure out the total number of ways in which different events can occur. If a certain work A can be done in m ways and another work B in n ways, then

(i) the number of ways of doing the work C, which is done only when either A or B is done, is $m + n$. (addition principle)

(ii) the number of ways of doing the work C, which is done only when both A and B are done, is mn. (multiplication principle)

Factorial Notation

The product of first n natural numbers is denoted by $n!$ and read as 'factorial n'.

Thus, $n! = n(n-1)(n-2)\ldots 3\cdot 2\cdot 1$

Properties of Factorial Notation

1. $0! = 1! = 1$
2. Factorials of negative integers and fractions are not defined.
3. $n! = n(n-1)! = n(n-1)(n-2)!$
4. $\dfrac{n!}{r!} = n(n-1)(n-2)\ldots(r+1)$

Permutations

- Permutation means **arrangement** of things. The number of permutations of n different things taken r at a time is ${}^nP_r$.
- ${}^nP_r = n(n-1)(n-2)\ldots(n-r+1) = \dfrac{n!}{(n-r)!}, 0 \le r \le n$

PREP MIRROR

Your Personal Preparation Indicator

- No. of Questions in Exercises (x)—
- No. of Questions Attempted (y)—
- No. of Correct Questions (z)—
  (Without referring Explanations)

- Accuracy Level ($z/y \times 100$)—
- Prep Level ($z/x \times 100$)—

In order to expect good rank in JEE, your Accuracy Level should be above 85 & Prep Level should be above 75.

Properties of ^nP_r

(i) $^nP_0 = 1, {}^nP_1 = n, {}^nP_n = n!$ (ii) $^nP_r + r \cdot {}^nP_{r-1} = {}^{n+1}P_r$

(iii) $^nP_r = n \, {}^{n-1}P_{r-1}$ (iv) $^nP_r = (n - r + 1) \, {}^nP_{r-1}$

(v) $^{n-1}P_r = (n - r) \, {}^{n-1}P_{r-1}$

Important Results on Permutations

(i) Number of permutations of n different things taken r at a time when a particular thing is to be always included in each arrangement is $r \cdot {}^{n-1}P_{r-1}$.

(ii) Number of permutations of n different things taken r at a time, when a particular thing is never taken in each arrangement is $^{n-1}P_r$.

(iii) The number of permutations of n different things taken r at a time, allowing repetitions is n^r.

(iv) The permutations of n things of which p are identical of one sort, q are identical of second sort, r are identical of third sort, is $\dfrac{n!}{p!q!r!}$, where $p + q + r = n$.

(v) **Arrangements**

(a) The number of ways in which m different things and n different things ($m + 1 \geq n$) can be arranged in a row, so that no two things of second kind come together is $m! \, {}^{(m+1)}P_n$.

(b) The number of ways in which m different things and n different things ($m \geq n$) can be arranged in a row so that all the second type of things come together is $(m + 1)! \, n!$

(vi) **Dearrangement** The number of dearrangements (No object goes to its scheduled place) of n objects, is

$$n!\left[\frac{1}{2!} - \frac{1}{3!} + \frac{1}{4!} - \dots (-1)^n \frac{1}{n!}\right]$$

(vii) **Sum of Digits**

(a) Sum of numbers formed by taking all the given n digits (excluding 0) is (sum of all the n digits) $\times (n - 1)! \times (111 \dots n$ times).

(b) Sum of the numbers formed by taking all the given n digits (including 0) is (sum of all the n digits) $\times [(n - 1)! \times (111 \dots n$ times$) - (n - 2) \times \{111 \dots (n - 1)$ times$\}]$.

Circular Permutations

- If different objects are arranged along a closed curve, then permutation is known as circular permutation.
- The number of circular permutations of n different things taken all at a time is $(n - 1)!$. If clockwise and anti-clockwise orders are taken as different.
- If clockwise and anti-clockwise circular permutations are considered to be same, then it is $\dfrac{(n-1)!}{2}$.
- Number of circular permutations of n different things, taken r at a time, when clockwise and anti-clockwise orders are taken as different, is $\dfrac{^nP_r}{r}$.
- Number of circular permutations of n different things, taken r at a time, when clockwise and anti-clockwise orders are not different, is $\dfrac{^nP_r}{2r}$.

Important Results on Circular Permutations

(i) The number of ways in which m different things and n different things (where, $m \geq n$) can be arranged in a circle, so that no two things of second kind come together is $(m - 1)! \, {}^mP_n$.

(ii) The number of ways in which m different things and n different things can be arranged in a circle, so that all the second type of things come together is $m! \, n!$.

(iii) The number of ways in which m different things and n different things (where, $m \geq n$) can be arranged in the form of garland, so that no two things of second kind come together is $(m - 1)! \, {}^mP_n/2$.

(iv) The number of ways in which m different things and n different things can be arranged in the form of garland, so that all the second type of things come together is $m! \, n!/2$.

Combinations

Combination means **selection** of things. The number of combinations of n different things taken r at a time is ^nC_r or $\dbinom{n}{r}$.

$$^nC_r = \binom{n}{r} = \frac{n(n-1)(n-2)\dots(n-r+1)}{r!} = \frac{n!}{r!(n-r)!} = \frac{^nP_r}{r!}$$

Properties of ^nC_r

(i) $^nC_r = {}^nC_{n-r}$ (ii) $^nC_x = {}^nC_y \Rightarrow x = y$ or $x + y = n$

(iii) $^nC_r + {}^nC_{r-1} = {}^{n+1}C_r$

Important Results on Combinations

(i) The number of combinations of n different things, taken r at a time, when p particular things always occur is $^{n-p}C_{r-p}$.

(ii) The number of combinations of n different things, taken r at a time, when p particular things never occur is $^{n-p}C_r$.

(iii) The number of selections of zero or more things out of n different things is $^nC_0 + {}^nC_1 + \dots + {}^nC_n = 2^n$.

(iv) The number of selections of one or more things out of n different things is $^nC_1 + {}^nC_2 + \dots + {}^nC_n = 2^n - 1$.

(v) The number of selections of zero or more things out of n identical things $= n + 1$.

(vi) The number of selections of one or more things out of n identical things $= n$.

(vii) The number of selections of one or more things from $p+q+r$ things, where p are alike of one kind, q are alike of second kind and rest are alike of third kind, is $[(p+1)(q+1)(r+1)]-1$.

(viii) The number of selections of one or more things from p identical things of one kind, q identical things of second kind, r identical things of third kind and n different things, is $(p+1)(q+1)(r+1)2^n-1$

(ix) If there are m items of one kind, n items of another kind and so on. Then, the number of ways of choosing r items out of these items = coefficient of x^r in
$$(1+x+x^2+\ldots+x^m)(1+x+x^2+\ldots+x^n)\ldots$$

(x) If there are m items of one kind, n items of another kind and so on. Then, the number of ways of choosing r items out of these items such that atleast one item of each kind is included in every selection = coefficient of x^r in
$$(x+x^2+\ldots+x^m)(x+x^2+\ldots+x^n)\ldots$$

Applications of Permutations and Combinations

Functional Applications

If the set A has m elements and B has n elements, then

(i) the number of functions from A to B is n^m.

(ii) the number of one-one functions from A to B is ${}^nP_m, m \le n$.

(iii) the number of onto functions from A to B is
$$n^m - \binom{n}{1}(n-1)^m + \binom{n}{2}(n-2)^m - \ldots, m \le n.$$

(iv) the number of bijections from A to B is $n!$, if $m=n$.

Geometrical Applications

(i) Number of triangles formed from n points, when no three points are collinear, is ${}^nC_3$.

(ii) Out of n non-concurrent and non-parallel straight lines, the points of intersection are ${}^nC_2$.

(iii) The number of diagonals in a polygon of n sides is ${}^nC_2 - n$.

(iv) The number of total triangles formed by the n points on a plane of which m are collinear, is ${}^nC_3 - {}^mC_3$.

(v) The number of total different straight lines, formed by the n points on a plane, of which m are collinear, is ${}^nC_2 - {}^mC_2 + 1$.

Prime Factors

Let $n = p_1^{\alpha_1} \cdot p_2^{\alpha_2} \cdot p_3^{\alpha_3} \ldots p_r^{\alpha_r}$, where $p_i, i = 1, 2, \ldots, r$ are distinct primes and $\alpha_i, i = 1, 2, \ldots, r$ are positive integers.

(i) Number of divisor of n is
$$(\alpha_1+1)(\alpha_2+1)(\alpha_3+1)\ldots(\alpha_r+1).$$

(ii) Sum of divisors of n is $\dfrac{(p_1^{\alpha_1+1}-1)}{(p_1-1)}\dfrac{(p_2^{\alpha_2+1}-1)}{(p_2-1)}\ldots\dfrac{(p_r^{\alpha_r+1}-1)}{(p_r-1)}$

(iii) If p is a prime such that p^r divides $n!$ but p^{r+1} does not divide $n!$.

Then, $r = \left[\dfrac{n}{p}\right] + \left[\dfrac{n}{p^2}\right] + \ldots$

Division of Objects into Groups

Objects are Different

(i) The number of ways of dividing n different objects into 3 groups of size p, q and r $(p+q+r=n)$ is

(a) $\dfrac{n!}{p!q!r!}$; p, q and r are unequal.

(b) $\dfrac{n!}{p!2!(q!)^2}$; $q=r$ (c) $\dfrac{n!}{3!(p!)^3}$; $p=q=r$

(ii) The number of ways in which n different things can be distributed into r different groups, if empty groups are allowed, is r^n.

(iii) The number of ways in which n different things can be distributed into r different groups, if empty groups are not allowed, is
$$r^n - \binom{r}{1}(r-1)^n + \binom{r}{2}(r-2)^n - \ldots + (-1)^{r-1}\,{}^rC_{r-1}\cdot 1$$

Objects are Identical

(i) The number of ways of dividing n identical objects among r persons such that each one may get atmost n objects, is $\binom{n+r-1}{r-1}$. In other words, the total number of ways of dividing n identical objects into r groups, if blank groups are allowed, is ${}^{n+r-1}C_{r-1}$.

(ii) The total number of ways of dividing n identical objects among r persons, each one of whom, receives atleast one item, is ${}^{n-1}C_{r-1}$. In other words, the number of ways in which n identical things can be divided into r groups such that blank groups are not allowed, is ${}^{n-1}C_{r-1}$.

(iii) Number of non-negative integral solutions of the equation $x_1 + x_2 + \ldots + x_r = n$ is equivalent to number of ways of distributing n identical objects into r groups if empty groups are allowed, which is ${}^{n+r-1}C_{r-1}$.

(iv) Number of positive integral solutions of the equation $x_1 + x_2 + \ldots + x_r = n$ is equivalent to the number of ways of distributing n identical objects into r groups such that no group empty, which is ${}^{n-1}C_{r-1}$.

(v) Number of integral solutions of the equation $x_1 + x_2 + \ldots + x_r = n$, where $a \le x_i \le b, \forall i = 1, 2, \ldots, r$, is given by coefficient of x^n in $(x^a + x^{a+1} + x^{a+2} + \ldots + x^b)^r$.

DAY PRACTICE SESSION 1

FOUNDATION QUESTIONS EXERCISE

1 The value of $2^n [1 \cdot 3 \cdot 5 \dots (2n-3)(2n-1)]$ is

(a) $\frac{(2n)!}{n!}$ (b) $\frac{(2n)!}{2^n}$ (c) $\frac{n!}{(2n)!}$ (d) None of these

2 The value of $1 \cdot 1! + 2 \cdot 2! + 3 \cdot 3! + \dots + n \cdot n!$, is

(a) $(n+1)!$ (b) $(n+1)!+1$
(c) $(n+1)!-1$ (d) None of these

3 How many different nine-digit numbers can be formed from the digits of the number 223355888 by rearrangement of the digits so that the odd digits occupy even places?

(a) 16 (b) 36 (c) 60 (d) 180

4 A library has a copies of one book, b copies of each of two books, c copies of each of three books and single copies of d books. The total number of ways in which these books can be arranged, is

(a) $\frac{(a+b+c+d)!}{a!\,b!\,c!}$ (b) $\frac{(a+2b+3c+d)!}{a!(b!)^2(c!)^3}$

(c) $\frac{(a+2b+3c+d)}{a!\,b!\,c!}$ (d) None of these

5 The number of words which can be formed out of the letters of the word ARTICLE, so that vowels occupy the even place is → NCERT Exemplar

(a) 1440 (b) 144 (c) 7! (d) ${}^4C_4 \times {}^3C_3$

6 In how many ways the letters of the word 'ARRANGE' can be arranged without altering the relative positions of vowels and consonants?

(a) 36 (b) 26
(c) 62 (d) None of these

7 If the letters of the word 'SACHIN' are arranged in all possible ways and these words are written in dictionary order, then the word 'SACHIN' appears at serial number

(a) 600 (b) 601 (c) 602 (d) 603

8 The number of integers greater than 6000 that can be formed, using the digits 3, 5, 6, 7 and 8 without repetition, is → JEE Mains 2015

(a) 216 (b) 192 (c) 120 (d) 72

9 The number of 5-digits telephone numbers having atleast one of their digits repeated, is → NCERT Exemplar

(a) 90000 (b) 100000 (c) 30240 (d) 69760

10 If all the words (with or without meaning) having five letters, formed using the letters of the word SMALL and arranged as in dictionary, then the position of the word SMALL is → JEE Mains 2016

(a) 46th (b) 59th
(c) 52nd (d) 58th

11 The total number of permutations of $n\,(>1)$ different things taken not more than r at a time, when each thing may be repeated any number of times is

(a) $\frac{n(n^n-1)}{n-1}$ (b) $\frac{n^r-1}{n-1}$

(c) $\frac{n(n^r-1)}{n-1}$ (d) None of these

12 If $\frac{{}^nP_{r-1}}{a} = \frac{{}^nP_r}{b} = \frac{{}^nP_{r+1}}{c}$, then

(a) $b^2 = a(b+c)$ (b) $c^2 = a(b+c)$
(c) $ab = a^2 + bc$ (d) $bc = a^3 + b^2$

13 The range of the function $f(x) = {}^{7-x}P_{x-3}$ is → AIEEE 2004

(a) {1, 2, 3} (b) {1, 2, 3, 4, 5, 6}
(c) {1, 2, 3, 4} (d) {1, 2, 3, 4, 5}

14 Find the number of different words that can be formed from the letters of the word TRIANGLE, so that no vowels are together. → NCERT Exemplar

(a) 14000 (b) 14500 (c) 14400 (d) 14402

15 In a class of 10 students, there are 3 girls. The number of ways they can be arranged in a row, so that no 2 girls are consecutive is $k \cdot 8!$, where k is equal to

(a) 12 (b) 24 (c) 36 (d) 42

16 The sum of all the 4-digit numbers that can be formed using the digits 1, 2, 3, 4 and 5 without repetition of the digits is

(a) 399960 (b) 288860 (c) 301250 (d) 420210

17 If eleven members of a committee sit at a round table so that the President and Secretary always sit together, then the number of arrangements is

(a) $10! \times 2$ (b) $10!$
(c) $9! \times 2$ (d) None of these

18 Let $f : \{1, 2, 3, 4, 5\} \to \{1, 2, 3, 4, 5\}$ that are onto and $f(x) \neq x$, is equal to

(a) 9 (b) 44
(c) 16 (d) None of these

19 There are 4 balls of different colours and 4 boxes of same colours as those of the balls. The number of ways in which the balls, one in each box, could be placed such that a ball does not go to box of its own colour is

(a) 8 (b) 9 (c) 7 (d) 1

20 How many different words can be formed by jumbling the letters in the word 'MISSISSIPPI' in which no two S are adjacent? → AIEEE 2008

(a) $7 \cdot {}^6C_4 \cdot {}^8C_4$ (b) $8 \cdot {}^6C_4 \cdot {}^7C_4$
(c) $6 \cdot 7 \cdot {}^8C_4$ (d) $6 \cdot 8 \cdot {}^7C_4$

21 The number of ways in which seven persons can be arranged at a round table, if two particular persons may not sit together is

(a) 480 (b) 120
(c) 80 (d) None of these

22 The number of ways in which 6 men and 5 women can sit at a round table, if no two women are to sit together, is given by → AIEEE 2003

(a) $6!\times 5!$ (b) 30 (c) $5!\times 4!$ (d) $7!\times 5!$

23 The value of $^{47}C_4 + \sum_{r=1}^{5} {}^{52-r}C_3$ is equal to

(a) $^{47}C_6$ (b) $^{52}C_5$
(c) $^{52}C_4$ (d) None of these

24 If $^nC_3 + {}^nC_4 > {}^{n+1}C_3$, then

(a) $n>6$ (b) $n>7$
(c) $n<6$ (d) None of these

25 The number of ways in which we can choose a committee from four men and six women so that the committee includes atleast two men and exactly twice as many women as men is → NCERT Exemplar

(a) 94 (b) 126
(c) 128 (d) None of these

26 A box contains 2 white balls, 3 black balls and 4 red balls. The number of ways of drawing 3 balls from the box, if atleast one black ball is included, is

(a) 36 (b) 42 (c) 56 (d) 64

27 A student is allowed to select atmost n books from a collection of $(2n+1)$ books. If the number of ways in which he can select atleast one book is 63, then n is equal to

(a) 3 (b) 4 (c) 6 (d) 5

28 Let A and B two sets containing 2 elements and 4 elements, respectively. The number of subsets of $A\times B$ having 3 or more elements is → JEE Mains 2013

(a) 256 (b) 220 (c) 219 (d) 211

29 In an examination of 9 papers a candidate has to pass in more papers than the number of papers in which he fails, in order to be successful. The number of ways in which he can be unsuccessful is

(a) 255 (b) 256 (c) 193 (d) 319

30 There are two urns. Urn A has 3 distinct red balls and urn B has 9 distinct blue balls. From each urn, two balls are taken out at random and then transferred to the other. The number of ways in which this can be done, is → AIEEE 2010

(a) 3 (b) 36 (c) 66 (d) 108

31 If the total number of m elements subsets of the set $A=\{a_1, a_2, a_3, \dots, a_n\}$ is λ times the number of m elements subsets containing a_4, then n is

(a) $(m-1)\lambda$ (b) $m\lambda$ (c) $(m+1)\lambda$ (d) 0

32 A guard of 12 men is formed from a group of n soldiers in all possible ways. If the number of times two particular soldiers A and B are together on guard is thrice the number of times three particular soldiers C, D, E are together on guard, then n is equal to

(a) 18 (b) 24 (c) 32 (d) 36

33 In a steamer, there are stalls for 12 animals and there are horses, cows and calves (not less than 12 each) ready to be shipped. In how many ways, can the ship load be made?

(a) $3^{12}-1$ (b) 3^{12} (c) $(12)^3-1$ (d) $(12)^3$

34 There are 10 points in a plane, out of these 6 are collinear. If N is the number of triangles formed by joining these points, then → AIEEE 2011

(a) $N>190$ (b) $N\le 100$
(c) $100<N\le 140$ (d) $140<N\le 190$

35 Let T_n be the number of all possible triangles formed by joining vertices of an n-sided regular polygon. If $T_{n+1}-T_n=10$, then the value of n is → JEE Mains 2013

(a) 7 (b) 5
(c) 10 (d) 8

36 On the sides AB, BC, CA of a ΔABC, 3, 4, 5 distinct points (excluding vertices A, B, C) are respectively chosen. The number of triangles that can be constructed using these chosen points as vertices are → JEE Mains 2013

(a) 210 (b) 205
(c) 215 (d) 220

37 The number of divisors of the number of 38808 (excluding 1 and the number itself) is

(a) 70 (b) 72
(c) 71 (d) None of these

38 If a, b, c, d, e are prime integers, then the number of divisors of ab^2c^2de excluding 1 as a factor is

(a) 94 (b) 72 (c) 36 (d) 71

39 The number of ways of distributing 8 identical balls in 3 distinct boxes, so that no box is empty, is

(a) 5 (b) $\binom{8}{3}$
(c) 3^8 (d) 21

40 If 4 dice are rolled, then the number of ways of getting the sum 10 is

(a) 56 (b) 64
(c) 72 (d) 80

DAY PRACTICE SESSION 2

PROGRESSIVE QUESTIONS EXERCISE

1 From 6 different novels and 3 different dictionaries, 4 novels and 1 dictionary are to be selected and arranged in a row on a shelf, so that the dictionary is always in the middle. The number of such arrangements is → JEE Mains 2018

(a) atleast 1000 (b) less than 500
(c) atleast 500 but less than 750
(d) atleast 750 but less than 1000

2 Sixteen men compete with one another in running, swimming and riding. How many prize lists could be made, if there were altogether 6 prizes for different values, one for running, 2 for swimming and 3 for riding?

(a) $16 \times 15 \times 14$ (b) $16^3 \times 15^2 \times 14$
(c) $16^3 \times 15 \times 14^2$ (d) $16^2 \times 15 \times 14$

3 The number of ways of dividing 52 cards amongst four players so that three players have 17 cards each and the fourth players just one card, is

(a) $\frac{52!}{(17!)^3}$ (b) 52! (c) $\frac{52!}{17!}$ (d) None of these

4 If the letters of the word MOTHER are written in all possible orders and these words are written out as in a dictionary, then the rank of the word MOTHER is

(a) 240 (b) 261 (c) 308 (d) 309

5 Let $S = \left\{\begin{pmatrix} a_{11} & a_{12} \\ a_{21} & a_{22} \end{pmatrix} : a_{ij} \in \{0, 1, 2\}, a_{11} = a_{22}\right\}$. Then, the number of non-singular matrices in the set S, is → JEE Mains 2013

(a) 27 (b) 24 (c) 10 (d) 20

6 A group of 6 is chosen from 10 men and 7 women so as to contain atleast 3 men and 2 women. The number of ways this can be done, if two particular women refuse to serve on the same group, is

(a) 8000 (b) 7800 (c) 7600 (d) 7200

7 A man X has 7 friends, 4 of them are ladies and 3 are men. His wife Y also has 7 friends, 3 of them are ladies and 4 are men. Assume X and Y have no common friends. Then, the total number of ways in which X and Y together can throw a party inviting 3 ladies and 3 men, so that 3 friends of each of X and Y are in this party, is → JEE Mains 2017

(a) 485 (b) 468 (c) 469 (d) 484

8 The number of ways in which we can select four numbers from 1 to 30 so as to exclude every selection of four consecutive numbers is

(a) 27378 (b) 27405 (c) 27399 (d) None of these

9 The number of numbers divisible by 3 that can be formed by four different even digits is

(a) 36 (b) 18 (c) 0 (d) None of these

10 The total number of integral solutions (x, y, z) such that $xyz = 24$ is

(a) 36 (b) 90 (c) 120 (d) None of these

11 Assuming the balls to be identical except for difference in colours, the number of ways in which one or more balls can be selected from 10 white, 9 green and 7 black balls is → AIEEE 2012

(a) 880 (b) 629 (c) 630 (d) 879

12 The number of ways in which an examiner can assign 30 marks to 8 questions, giving not less than 2 marks to any question, is → JEE Mains 2013

(a) ${}^{30}C_7$ (b) ${}^{21}C_8$ (c) ${}^{21}C_7$ (d) ${}^{30}C_8$

13 The number of divisors of the form $4n + 1, n \geq 0$ of the number $10^{10}\,11^{11}\,13^{13}$ is

(a) 750 (b) 840 (c) 924 (d) 1024

14 Out of 8 sailors on a boat, 3 can work only one particular side and 2 only the other side. Then, number of ways in which the sailors can be arranged on the boat is

(a) 2718 (b) 1728 (c) 7218 (d) None of these

15 In a cricket match between two teams X and Y, the team X requires 10 runs to win in the last 3 balls. If the possible runs that can be made from a ball be 0, 1, 2, 3, 4, 5 and 6. The number of sequence of runs made by the batsman is

(a) 12 (b) 18 (c) 21 (d) 36

ANSWERS

SESSION 1

1. (a)	**2.** (c)	**3.** (c)	**4.** (b)	**5.** (b)	**6.** (a)	**7.** (b)	**8.** (b)	**9.** (d)	**10.** (d)
11. (c)	**12.** (a)	**13.** (a)	**14.** (c)	**15.** (d)	**16.** (a)	**17.** (c)	**18.** (b)	**19.** (b)	**20.** (a)
21. (a)	**22.** (a)	**23.** (c)	**24.** (a)	**25.** (a)	**26.** (d)	**27.** (a)	**28.** (c)	**29.** (b)	**30.** (d)
31. (b)	**32.** (c)	**33.** (b)	**34.** (b)	**35.** (b)	**36.** (b)	**37.** (a)	**38.** (d)	**39.** (d)	**40.** (d)

SESSION 2

1. (a)	**2.** (b)	**3.** (a)	**4.** (d)	**5.** (d)	**6.** (b)	**7.** (a)	**8.** (a)	**9.** (a)	**10.** (c)
11. (d)	**12.** (c)	**13.** (c)	**14.** (b)	**15.** (d)					

Hints and Explanations

SESSION 1

1 Clearly, $[1\cdot 3\cdot 5 \ldots (2n-3)(2n-1)]2^n$

$$= \frac{1\cdot 2\cdot 3\cdot 4\cdot 5\cdot 6 \ldots (2n-1)(2n)2^n}{2\cdot 4\cdot 6 \ldots 2n}$$

$$= \frac{(2n)!\,2^n}{2^n(1\cdot 2\cdot 3 \ldots n)} = \frac{(2n)!}{n!}$$

2 Clearly, given expression $= \sum_{r=1}^{n} r\cdot r!$

$$= \sum_{r=1}^{n} ((r+1)-1)\,r! = \sum_{r=1}^{n} ((r+1)! - r!)$$

$= (2! - 1!) + (3! - 2!) + \ldots + ((n+1)! - n!)$

$= (n+1)! - 1$

3 In a nine digits number, there are four even places for the four odd digits 3, 3, 5, 5.

$\therefore$ Required number of ways $= \frac{4!}{2!\,2!} \cdot \frac{5!}{2!\,3!} = 60$

4 Total number of books $= a + 2b + 3c + d$

$\therefore$ The total number of arrangements

$$= \frac{(a + 2b + 3c + d)!}{a!(b!)^2(c!)^3}$$

5 In a word ARTICLE, vowels are A, E, I and consonants are C, L, R, T.

In a seven letter word, there are three even places in which three vowels are placed in 3! ways. In rest of the four places, four consonants are placed in 4! ways.

$\therefore$ Required number of ways

$= 3! \times 4! = 6 \times 24 = 144$

6 Clearly, the consonants in their positions can be arranged in $\frac{4!}{2!} = 12$ ways and the vowels in their positions can be arranged in $\frac{3!}{2!} = 3$ ways

$\therefore$ Total number of arrangements

$= 12 \times 3 = 36$

7 The letters of given word are A, C, H, I, N, S.

Now, the number of words starting with $A = 5!$

the number of words starting with $C = 5!$

the number of words starting with $H = 5!$

the number of words starting with $I = 5!$

and the number of words starting with $N = 5!$.

Then, next word is SACHIN.

So, the required serial number is

$= (5\cdot 5!) + 1 = 601.$

8 The integer greater than 6000 may be of 4 digits or 5 digits. So, here two cases arise.

***Case* I** When number is of 4 digit.

Four digit number can start from 6,7 or 8.

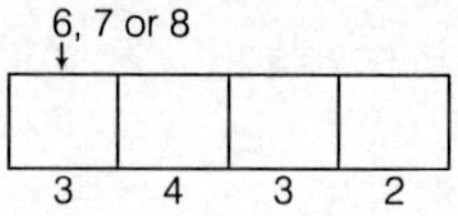

Thus, total number of 4-digit number, which are greater than

$6000 = 3 \times 4 \times 3 \times 2 = 72$

***Case* II** When number is of 5 digit.

Total number of 5-digit number, which are greater than $6000 = 5! = 120$

$\therefore$ Total number of integers

$= 72 + 120 = 192$

9 Using the digits 0, 1, 2, ..., 9 the number of five digits telephone numbers which can be formed is 10^5 (since, repetition is allowed). The number of five digits telephone numbers, which have none of the digits repeated $= {}^{10}P_5 = 30240$

$\therefore$ The required number of telephone number $= 10^5 - 30240 = 69760$

10 Clearly, number of words start with

$\text{A} = \frac{4!}{2!} = 12$

Number of words start with L = 4! = 24

Number of words start with M $= \frac{4!}{2!} = 12$

Number of words start with SA $= \frac{3!}{2!} = 3$

Number of words start with SL = 3! = 6

Note that, next word will be "SMALL".

Hence, position of word "SMALL" is 58th.

11 When we arrange things one at a time, the number of possible permutations is n. When we arrange them two at a time the number of possible permutations are $n \times n = n^2$ and so on. Thus, the total number of permutations are

$$n + n^2 + \ldots + n^r = \frac{n(n^r - 1)}{n - 1} \quad [\because n > 1]$$

12 $\because \frac{{}^nP_{r-1}}{a} = \frac{{}^nP_r}{b} = \frac{{}^nP_{r+1}}{c}$

From first two terms $\frac{b}{a} = n - r + 1$

From last two terms $\frac{c}{b} = n - r$

Hence, $\frac{b}{a} = \frac{c}{b} + 1 \Rightarrow b^2 = a(b + c)$

13 Given that, $f(x) = {}^{7-x}P_{x-3}$. The above function is defined, if $7 - x \geq 0$ and $x - 3 \geq 0$ and $7 - x \geq x - 3$.

$\Rightarrow \quad x \leq 7, x \geq 3$ and $x \leq 5$

$\therefore \quad D_f = \{3, 4, 5\}$

Now, $\quad f(3) = {}^4P_0 = 1$

$f(4) = {}^3P_1 = 3$ and $f(5) = {}^2P_2 = 2$

$\therefore \quad R_f = \{1, 2, 3\}$

14 In a word TRIANGLE, vowels are (A, E, I) and consonants are (G, L, N, R, T). First, we fix the 5 consonants in alternate position in 5! ways.

_G_L_N_R_T_

In rest of the six blank position, three vowels can be arranged in ${}^6P_3$ ways.

$\therefore$ Total number of different words

$= 5! \times {}^6P_3 = 120 \times 6 \times 5 \times 4 = 14400$

15 The 7 boys can be arranged in row in 7! ways. There will be 6 gaps between them and one place before them and one place after them. The 3 girls can be arranged in a row in ${}^8P_3 = 8\cdot 7\cdot 6$ ways

$\therefore$ Required number of ways

$= 7! \times 8\cdot 7\cdot 6 = 42\cdot 8!$

16 Required sum

= (Sum of all the n-digits)

$\times {}^{n-1}P_{r-1} \times (111 \ldots r \text{ times})$

$= (1 + 2 + 3 + 4 + 5)\,{}^4P_3 \times (1111)$

$= 15 \times 24 \times 1111 = 399960$

17 Since, out of eleven members two members sit together, the number of arrangements $= 9! \times 2$

($\because$ two members can be sit in two ways.)

18 Total number of functions

= Number of dearrangement of 5 objects

$$= 5!\left(\frac{1}{2!} - \frac{1}{3!} + \frac{1}{4!} - \frac{1}{5!}\right) = 44$$

19 Use the number of dearrangements

i.e. $n!\left\{1 - \frac{1}{1!} + \frac{1}{2!} - \frac{1}{3!} + \ldots + (-1)^n \frac{1}{n!}\right\}$

Here, $n = 4$

So, the required number of ways

$$= 4!\left\{\frac{1}{2!} - \frac{1}{3!} + \frac{1}{4!}\right\} = 12 - 4 + 1 = 9$$

20 Given word is MISSISSIPPI.

Here, I = 4 times, S = 4 times,

P = 2 times, M = 1 time

_M_I_I_I_I_P_P_

$\therefore$ Required number of words

$$= {}^8C_4 \times \frac{7!}{4!\,2!} = {}^8C_4 \times \frac{7 \times 6!}{4!\,2!} = 7\cdot {}^8C_4 \cdot {}^6C_4$$

21 Clearly, remaining 5 persons can be seated in 4! ways. Now, on five cross marked places person can sit in ${}^5P_2$ ways.

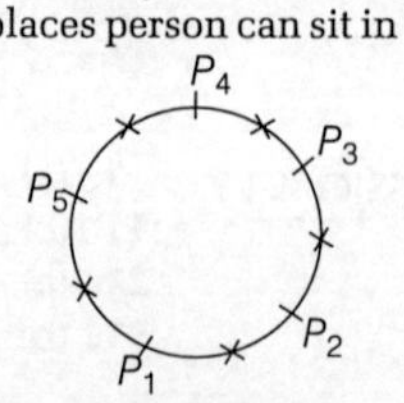

So, number of arrangements

$$= 4! \times \frac{5!}{3!} = 24 \times 20 = 480 \text{ ways}$$

22 First, we fix the position of men, number of ways in which men can sit = 5!.

Now, the number of ways in which women can sit = $^6P_5$

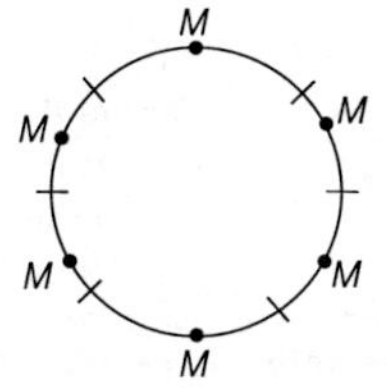

$\therefore$ Total number of ways

$= 5! \times {}^6P_5 = 5! \times 6!$

23 $^{47}C_4 + \sum_{r=1}^{5} {}^{52-r}C_3 = {}^{47}C_4 + {}^{51}C_3 + {}^{50}C_3 + {}^{49}C_3 + {}^{48}C_3 + {}^{47}C_3$

$= {}^{51}C_3 + {}^{50}C_3 + {}^{49}C_3 + {}^{48}C_3 + ({}^{47}C_3 + {}^{47}C_4)$

$= {}^{52}C_4$

24 $^nC_3 + {}^nC_4 > {}^{n+1}C_3$

$\Rightarrow {}^{n+1}C_4 > {}^{n+1}C_3 \quad ({}^nC_r + {}^nC_{r+1} = {}^{n+1}C_{r+1})$

$\Rightarrow \frac{{}^{n+1}C_4}{{}^{n+1}C_3} > 1 \Rightarrow \frac{n-2}{4} > 1 \Rightarrow n > 6$

25 The number of ways in which we can choose a committee

= Choose two men and four women

+ Choose three men and six women

$= {}^4C_2 \times {}^6C_4 + {}^4C_3 \times {}^6C_6$

$= 6 \times 15 + 4 \times 1 = 90 + 4 = 94$

26 The number of ways of drawing 1 black and 2 non-black balls is

$^3C_1 \cdot {}^6C_2 = 3 \cdot 15 = 45$

The number of ways of drawing 2 black and 1 non-black ball is

$^3C_2 \cdot {}^6C_1 = 3 \cdot 6 = 18$

The number of ways of drawing 3 black balls is $^3C_3 = 1$

$\therefore$ Number of ways = 45 + 18 + 1 = 64

27 He can select 1, 2, ... or n books.

The number of ways to select atleast one book is

$^{2n+1}C_1 + {}^{2n+1}C_2 + \dots + {}^{2n+1}C_n$

$= \frac{1}{2}({}^{2n+1}C_1 + {}^{2n+1}C_2 + \dots + {}^{2n+1}C_n + {}^{2n+1}C_{n+1} + \dots + {}^{2n+1}C_{2n})$

$= \frac{1}{2}(2^{2n+1} - {}^{2n+1}C_0 - {}^{2n+1}C_{2n+1})$

$= 2^{2n} - 1 = 63$ [given]

$\Rightarrow \quad 2^{2n} = 64 = 2^6 \Rightarrow n = 3$

28 Given, $n(A) = 2$, $n(B) = 4$.

$\therefore \; n(A \times B) = 8$

The number of subsets of $A \times B$ having 3 or more elements $= {}^8C_3 + {}^8C_4 + \dots + {}^8C_8$

$= ({}^8C_0 + {}^8C_1 + {}^8C_2 + {}^8C_3 + \dots + {}^8C_8) - ({}^8C_0 + {}^8C_1 + {}^8C_2)$

$= 2^8 - {}^8C_0 - {}^8C_1 - {}^8C_2$

$= 256 - 1 - 8 - 28 = 219$

$[\because 2^n = {}^nC_0 + {}^nC_1 + \dots + {}^nC_n]$

29 Clearly, the candidate is unsuccessful, if he fails in 9 or 8 or 7 or 6 or 5 papers.

$\therefore$ Numbers of ways to be unsuccessful

$= {}^9C_9 + {}^9C_8 + {}^9C_7 + {}^9C_6 + {}^9C_5$

$= {}^9C_0 + {}^9C_1 + {}^9C_2 + {}^9C_3 + {}^9C_4$

$= \frac{1}{2}({}^9C_0 + {}^9C_1 + \dots + {}^9C_9)$

$= \frac{1}{2}(2^9) = 2^8 = 256$

30

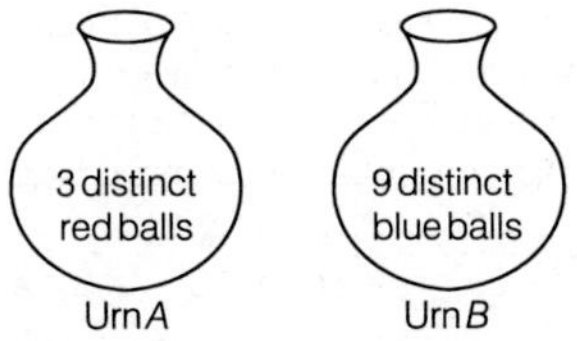

The number of ways in which 2 balls from urn A and 2 balls from urn B can be selected $= {}^3C_2 \times {}^9C_2 = 3 \times 36 = 108$

31 Total number of m elements subsets of

$A = {}^nC_m \quad \dots$(i)

and number of m elements subsets of A each containing element $a_4 = {}^{n-1}C_{m-1}$

According to the question,

$= \lambda \cdot {}^{n-1}C_{m-1}$

$\Rightarrow \quad \frac{n}{m} \cdot {}^{n-1}C_{m-1} = \lambda \cdot {}^{n-1}C_{m-1}$

$\Rightarrow \quad \lambda = \frac{n}{m} \Rightarrow n = m\lambda$

32 Number of times A and B are together on guard is $\binom{n-2}{10}$.

Number of times C, D and E are together on guard is $\binom{n-3}{9}$.

According to the question,

$\binom{n-2}{10} = 3\binom{n-3}{9}$

$\Rightarrow \quad n - 2 = 30 \Rightarrow n = 32$

33 First stall can be filled in 3 ways, second stall can be filled in 3 ways and so on.

$\therefore$ Number of ways of loading steamer

$= {}^3C_1 \times {}^3C_1 \times \dots \times {}^3C_1$ (12 times)

$= 3 \times 3 \times \dots \times 3$ (12 times) $= 3^{12}$

34 If out of n points, m are collinear, then

Number of triangles $= {}^nC_3 - {}^mC_3$

$\therefore$ Required number of triangles

$= {}^{10}C_3 - {}^6C_3 = 120 - 20$

$\Rightarrow \quad N = 100$

35 $T_n = {}^nC_3$, hence $T_{n+1} = {}^{n+1}C_3$

Now, $\quad T_{n+1} - T_n = 10$

$\Rightarrow \quad {}^{n+1}C_3 - {}^nC_3 = 10$ [given]

$\Rightarrow \quad \frac{(n+1)\,n\,(n-1)}{3!} - \frac{n\,(n-1)(n-2)}{3!} = 10$

$\Rightarrow \quad \frac{n\,(n-1)}{3!}(n + 1 - n + 2) = 10$

$\Rightarrow \quad \frac{n\,(n-1)}{3!} \times 3 = 10$

$\Rightarrow \quad n^2 - n - 20 = 0 \Rightarrow n = 5$

36 Required number of triangles that can be constructed using these chosen points as vertices $= {}^{12}C_3 - {}^3C_3 - {}^4C_3 - {}^5C_3$

Here, we subtract those cases in which points are collinear

$= 220 - 1 - 4 - 10 = 220 - 15 = 205$

37 Since, $38808 = 2^3 \times 3^2 \times 7^2 \times 11^1$

$\therefore$ Number of divisors

$= 4 \times 3 \times 3 \times 2 - 2 = 72 - 2 = 70$

38 The number of divisors of ab^2c^2de

$= (1+1)(2+1)(2+1)(1+1)(1+1) - 1$

$= 2 \cdot 3 \cdot 3 \cdot 2 \cdot 2 - 1 = 71$

39 Required number of ways is equal to the number of positive integer solutions of the equation $x + y + z = 8$ which equal to $\binom{8-1}{3-1} = \binom{7}{2} = 21$

40 Coefficient of x^{10} in $(x + x^2 + \dots + x^6)^4$

= Coefficient of x^6 in $(1 + x + \dots + x^5)^4$

$= (1 - x^6)^4 (1 - x)^{-4}$ in $(1 - 4x^6 + \dots)$

$\left[1 + \binom{4}{1}x + \dots\right]$

Hence, coefficient of x^6 is $\binom{9}{6} - 4 = 80$.

SESSION 2

1 Given, 6 different novels and 3 different dictionaries.

Number of ways of selecting 4 novels from 6 novels is $^6C_4 = \frac{6!}{2!4!} = 15$

Number of ways of selecting 1 dictionary from 3 dictionaries is

$^3C_1 = \frac{3!}{1!2!} = 3$

Number of arrangement of 4 novels and 1 dictionary where dictionary is always in the middle, is 4!

Required number of arrangement

$15 \times 3 \times 4! = 45 \times 24 = 1080$

2 Number of ways of giving one prize for running = 16

Number of ways of giving two prizes for swimming = 16×15

Number of ways of giving three prizes for riding = $16 \times 15 \times 14$

$\therefore$ Required ways of giving prizes

$= 16 \times 16 \times 15 \times 16 \times 15 \times 14$

$= 16^3 \times 15^2 \times 14$

3 For the first player, cards can be distributed in the $^{52}C_{17}$ ways. Now, out of 35 cards left 17 cards can be distributed for second player in $^{52}C_{17}$ ways.

Similarly, for third player in ${}^{18}C_{17}$ ways. One card for the last player can be distributed in ${}^{1}C_{1}$ way.
Therefore, the required number of ways for the proper distribution.
$= {}^{52}C_{17} \times {}^{35}C_{17} \times {}^{18}C_{17} \times {}^{1}C_{1}$
$= \dfrac{52!}{35!\,17!} \times \dfrac{35!}{18!17!} \times \dfrac{18!}{17!1!} \times 1! = \dfrac{52!}{(17!)^3}$

4 The number of words starting from E $= 5! = 120$
The number of words starting from H $= 5! = 120$
The number of words starting from ME $= 4! = 24$
The number of words starting from MH $= 4! = 24$
The number of words starting from MOE $= 3! = 6$
The number of words starting from MOH $= 3! = 6$
The number of words starting from MOR $= 3! = 6$
The number of words starting from MOTE $= 2! = 2$
The number of words starting from MOTHER $= 1! = 1$
Hence, rank of the word MOTHER
$= 2(120) + 2(24) + 3(6) + 2 + 1$
$= 309$

5 A matrix whose determinant is non-zero is called a non-singular matrix.
Here, we have
$$S = \left\{\begin{pmatrix} a_{11} & a_{12} \\ a_{21} & a_{22}\end{pmatrix}; a_{ij} \in \{0, 1, 2\}, a_{11} = a_{22}\right\}$$
Clearly, $n(s) = 27$
[$\because$ for $a_{11} = a_{22}$, we have 3 choices, for a_{12}, we have 3 choices and for a_{21}, we have 3 choices]
Now, $\begin{vmatrix} a_{11} & a_{12} \\ a_{21} & a_{22}\end{vmatrix} = 0$
$\Rightarrow a_{11} \cdot a_{22} - a_{12} \cdot a_{21} = 0$
$\Rightarrow (a_{11})^2 = a_{12}\, a_{21} = 0$ [$\because a_{11} = a_{22}$]
$\Rightarrow a_{12}a_{21} = 0, 1, 4$ [$\because a_{11} \in \{0, 1, 2\}$]
Consider $a_{12}\, a_{21} = 0$, this is possible in 5 cases
$a_{12}a_{21} = 1$, this is possible in only 1 case
$a_{12}\, a_{21} = 4$, this is possible in only 1 case
Thus, number of singular matrices in S are 7.
Hence, number of non-singular matrices in S are $27 - 7 = 20$.

6 Let the men be $M_1, M_2, \dots, M_{10}$ and women be $W_1, W_2, \dots, W_7$.
Let W_1 and W_2 do not want to be on the same group. The six members group can contain 4 men and 2 women or 3 men and 3 women.
The number of ways of forming $4M, 2W$ group is
$${}^{10}C_4({}^{5}C_2 + {}^{2}C_1 \cdot {}^{5}C_1) = 4200$$
where, ${}^{5}C_2$ is the number of ways without W_1 and W_2 and ${}^{5}C_1$ is the number of ways with W_1 and without W_2 or with W_2 and without W_1.
The number of ways of forming $3M, 3W$ group is ${}^{10}C_3({}^{5}C_3 + {}^{2}C_1\,{}^{5}C_2) = 3600$
where, ${}^{5}C_3$ is the number of ways without W_1 and W_2 and ${}^{5}C_2$ is the number of ways with W_1 or W_2 but not both.
$\therefore$ Number of ways $= 4200 + 3600 = 7800$

7 Given, X has 7 friends, 4 of them are ladies and 3 are men while Y has 7 friends, 3 of them are ladies and 4 are men.
$\therefore$ Total number of required ways
$= {}^{3}C_3 \times {}^{4}C_0 \times {}^{4}C_0 \times {}^{3}C_3$
$+ {}^{3}C_2 \times {}^{4}C_1 \times {}^{4}C_1 \times {}^{3}C_2$
$+ {}^{3}C_1 \times {}^{4}C_2 \times {}^{4}C_2 \times {}^{3}C_1$
$+ {}^{3}C_0 \times {}^{4}C_3 \times {}^{4}C_3 \times {}^{3}C_0$
$= 1 + 144 + 324 + 16 = 485$

8 The number of ways of selecting four numbers from 1 to 30 without any restriction is ${}^{30}C_4$. The number of ways of selecting four consecutive numbers [i.e. (1, 2, 3, 4), (2, 3, 4, 5), ..., (27, 28, 29, 30)] is 27.
Hence, the number of ways of selecting four integers which excludes selection of consecutive four numbers is
$${}^{30}C_4 - 27 = \frac{30 \times 29 \times 28 \times 27}{24} - 27$$
$$= 27378$$

9 Possible even digits are 2,4,6,8,0.
***Case* I** Number has digits 4,6,8,0.
(Here, sum of digits is 18, divisible by 3)
□ □ □ □
Number of arrangements $= 3 \times 3!$
[Ist place can be filled using 4, 6, 8]
$= 3 \times 6 = 18$
***Case* II** Number has digits 2, 4, 6, 0
(Here, sum of digits is 12, divisible by 3)
□ □ □ □
1st place cannot be filled by 0.
Number of arrangements $= 3 \times 3! = 18$
$\therefore$ Number of numbers $= 18 + 18 = 36$

10 $24 = 2 \cdot 3 \cdot 4, 2 \cdot 2 \cdot 6, 1 \cdot 6 \cdot 4, 1 \cdot 3 \cdot 8, 1 \cdot 2 \cdot 12, 1 \cdot 1 \cdot 24$
[as product of three positive integers]
$\therefore$ The total number of positive integral solutions of $xyz = 24$ is
equal to $3! + \dfrac{3!}{2!} + 3! + 3! + 3! + \dfrac{3!}{2!}$ i.e. 30.
Any two of the numbers in each factorisation may be negative. So, the number of ways to associate negative sign in each case is ${}^{3}C_2$ i.e. 3.
$\therefore$ Total number of integral solutions
$= 30 + 3 \times 30 = 120$

11 The number of ways to choose zero or more from white balls $= (10 + 1)$
[$\because$ all white balls are mutually identical]
Number of ways to choose zero or more from green balls $= (9 + 1)$
[$\because$ all green balls are mutually identical]
Number of ways to choose zero or more from black balls $= (7 + 1)$
[$\because$ all black balls are mutually identical]
Hence, number of ways to choose zero or more balls of any colour
$= (10 + 1)(9 + 1)(7 + 1)$
Also, number of ways to choose zero balls from the total $= 1$
Hence, the number of ways to choose atleast one ball
[irrespective of any colour]
$= (10 + 1)(9 + 1)(7 + 1) - 1$
$= 880 - 1 = 879$

12 Let $x_1, x_2, \dots, x_8$ denote the marks assign to 8 questions.
$\therefore x_1 + x_2 + \dots + x_8 = 30$
Also, $x_1, x_2, \dots, x_8 \geq 2$
Let, $u_1 = x_1 - 2, u_2 = x_2 - 2 \dots u_8 = x_8 - 2$
Then, $(u_1 + 2 + u_2 + 2 + \dots + u_8 + 2) = 30$
$\Rightarrow u_1 + u_2 + \dots + u_8 = 14$
$\therefore$ Total number of solutions
$= {}^{14+8-1}C_{8-1} = {}^{21}C_7$

13 $2^{10}5^{10}11^{11}13^{13}$ has a divisor of the form $2^{\alpha} \cdot 5^{\beta} \cdot 11^{\gamma} \cdot 13^{\delta}$, where
$\alpha = 0, 1, 2, \dots, 10; \beta = 0, 1, 2, \dots, 10;$
$\gamma = 0, 1, 2, \dots, 11; \delta = 0, 1, 2, \dots, 13$
It is of the form $4n + 1$, if
$\alpha = 0; \beta = 0, 1, 2, \dots, 10;$
$\gamma = 0, 2, 4, \dots, 10;$
$\delta = 0, 1, 2, \dots, 13.$
$\therefore$ Number of divisors
$= 11 \times 6 \times 14 = 924$

14 Let the particular side on which 3 particular sailors can work be named A and on the other side by B on which 2 particular sailors can work. Thus, we are left with 3 sailors only. Selection of one sailor for side $A = {}^{3}C_1 = 3$ and, then we are left with 2 sailors for the other side. Now, on each side, 4 sailors can be arranged in 4! ways.
$\therefore$ Total number of arrangements
$= 3 \times 24 \times 24 = 1728$

15 Required number is the coefficient of x^{10} in $(1 + x + x^2 + \dots + x^6)^3$
$= (1 - x^7)^3 (1 - x)^{-3} = (1 - 3x^7 + \dots)$
$$\left[1 + \binom{3}{1}x + \binom{4}{2}x^2 + \dots\right]$$
Hence, coefficient of x^{10} is
$$\binom{12}{10} - 3\binom{5}{3} = 36.$$

DAY NINE

Unit Test 1

(Algebra)

1 If $f(x)=\begin{vmatrix}\cos x & x & 1\\ 2\sin x & x^2 & 2x\\ \tan x & x & 1\end{vmatrix}$, then $\lim_{x\to 0}\frac{f'(x)}{x}$ is equal to

(a) 1 (b) −1 (c) 2 (d) −2

2 If $\log_{0.5}(x-1)<\log_{0.25}(x-1)$, then x lies in the interval

(a) $(2,\infty)$ (b) $(3,\infty)$ (c) $(-\infty,0)$ (d) $(0,3)$

3 Sum of n terms of series $12+16+24+40+\ldots$ will be

(a) $2(2^n-1)+8n$
(b) $2(2^n-1)+6n$
(c) $3(2^n-1)+8n$
(d) $4(2^n-1)+8n$

4 Let a, b and $c\in R$ and $a\neq 0$. If α is a root of $a^2x^2+bx+c=0$, β is a root of $a^2x^2-bx-c=0$ and $0<\alpha<\beta$, then the equation of $a^2x^2+2bx+2c=0$ has a root γ, that always satisfies

(a) $\gamma=\alpha$ (b) $\gamma=\beta$
(c) $\gamma=(\alpha+\beta)/2$ (d) $\alpha<\gamma<\beta$

5 Between two numbers whose sum is $2\frac{1}{6}$ an even number of arithmetic means are inserted. If the sum of these means exceeds their number by unity, then the number of means are

(a) 12 (b) 10
(c) 8 (d) None of these

6 If $A=\begin{bmatrix}1 & 0\\ 1 & 1\end{bmatrix}$, then which of the following is not true?

(a) $\lim_{n\to\infty}\frac{1}{n^2}A^{-n}=\begin{bmatrix}0 & 0\\ -1 & 0\end{bmatrix}$ (b) $\lim_{n\to\infty}\frac{1}{n}A^{-n}=\begin{bmatrix}0 & 0\\ -1 & 0\end{bmatrix}$

(c) $A^{-n}=\begin{bmatrix}1 & 0\\ -n & 1\end{bmatrix}$, $\forall\, n\in N$ (d) None of these

7 From 50 students taking examinations in Mathematics, Physics and Chemistry, 37 passed Mathematics, 24 Physics and 43 Chemistry. Atmost 19 passed Mathematics and Physics, atmost 29 passed Mathematics and Chemistry and atmost 20 passed Physics and Chemistry. The largest possible number that could have passed all three examinations is

(a) 11 (b) 12 (c) 13 (d) 14

8 The inequality $|z-4|<|z-2|$ represents the region given by

(a) $\text{Re}(z)>0$ (b) $\text{Re}(z)<0$
(c) $\text{Re}(z)>3$ (d) None of these

9 If 1, ω and ω^2 be the three cube roots of unity, then $(1+\omega)(1+\omega^2)(1+\omega^4)\ldots 2n$ factors is equal to

(a) 1 (b) −1
(c) 0 (d) None of these

10 If $a<0$, then the positive root of the equation $x^2-2a|x-a|-3a^2=0$ is

(a) $a(-1-\sqrt{6})$ (b) $a(1-\sqrt{2})$
(c) $a(1-\sqrt{6})$ (d) $a(1+\sqrt{2})$

11 The common roots of the equations $z^3+2z^2+2z+1=0$ and $z^{1985}+z^{100}+1=0$ are

(a) $-1, \omega$ (b) $-1, \omega^2$
(c) ω, ω^2 (d) None of these

12 Let z_1, z_2 and z_3 be three points on $|z|=1$. If θ_1, θ_2 and θ_3 are the arguments of z_1, z_2 and z_3 respectively, then $\cos(\theta_1-\theta_2)+\cos(\theta_2-\theta_3)+\cos(\theta_3-\theta_1)$

(a) $\geq\frac{3}{2}$ (b) $\geq-\frac{3}{2}$

(c) $\leq\frac{-3}{2}$ (d) None of these

13 If the roots of the equation
$$(a^2+b^2)x^2+2(bc+ad)x+(c^2+d^2)=0$$
are real, then a^2, bd and c^2 are in
(a) AP (b) GP
(c) HP (d) None of these

14 150 workers were engaged to finish a piece of work in a certain number of days. Four workers dropped the second day, four more workers dropped the third day and so on. It takes 8 more days to finish the work now. Then, the number of days in which the work was completed is
(a) 29 days (b) 24 days (c) 25 days (d) 26 days

15 Let R be a relation defined by $R=\{(x,x^3): x$ is a prime number $<10\}$, then which of the following is true?
(a) $R=\{(1,1),(2,8),(3,27),(4,64),(5,125),(6,216),(7,343),(8,512),(9,729)\}$
(b) $R=\{(2,8),(3,27),(5,125),(7,343)\}$
(c) $R=\{(2,8),(4,64),(6,216),(8,512)\}$
(d) None of the above

16 If $b>a$, then the equation $(x-a)(x-b)-1=0$, has
(a) both the roots in $[a,b]$
(b) both the roots in $(-\infty, a]$
(c) both the roots in (b,∞)
(d) one root in $(-\infty, a)$ and other in (b,∞)

17 The value of x satisfying $\log_2(3x-2)=\log_{1/2}x$ is
(a) $-\frac{1}{3}$ (b) 2
(c) $\frac{1}{2}$ (d) None of these

18 If $f(x,n)=\sum_{r=1}^{n}\log_x\left(\frac{r}{x}\right)$, then the value of x satisfying the equation $f(x,11)=f(x,12)$ is
(a) 10 (b) 11
(c) 12 (d) None of these

19 The three numbers a, b and c between 2 and 18 are such that their sum is 25, the numbers $2, a$ and b are consecutive terms of an AP and the numbers b, c and 18 are consecutive terms of a GP. The three numbers are
(a) 3, 8, 14 (b) 2, 9, 14
(c) 5, 8, 12 (d) None of these

20 If X is the set of all complex numbers z such that $|z|=1$, then the relation R defined on X by
$|\arg z_1-\arg z_2|=\frac{2\pi}{3}$, is
(a) reflexive (b) symmetric
(c) transitive (d) anti-symmetric

21 If α and β are the roots of the equation $ax^2-2bx+c=0$, then $\alpha^3\beta^3+\alpha^2\beta^3+\alpha^3\beta^2$ is equal to
(a) $\frac{c^2}{a^3}(c-2b)$ (b) $\frac{c^2}{a^3}(c+2b)$
(c) $\frac{bc^2}{a^3}$ (d) None of these

22 If the roots of the equation $x^2-2ax+a^2+a-3=0$ are real and less than 3, then
(a) $a<2$ (b) $2\le a\le 3$ (c) $3\le a\le 4$ (d) $a>4$

23 The integer k for which the inequality
$x^2-2(4k-1)x+15k^2-2k-7>0$ is valid for any x, is
(a) 2 (b) 3
(c) 4 (d) None of these

24 The maximum sum of the series
$20+19\frac{1}{3}+18\frac{2}{3}+18+\ldots$ is
(a) 310 (b) 290
(c) 320 (d) None of these

25 The number of common terms to the two sequences 17, 21, 25, ..., 417 and 16, 21, 26, ..., 466 is
(a) 21 (b) 19 (c) 20 (d) 91

26 If the sum of the first three terms of a GP is 21 and the sum of the next three terms is 168, then the first term and the common ratio is
(a) 3, 4 (b) 2, 4
(c) 3, 2 (d) None of these

27 The sum to n terms of the series
$$\frac{1}{1+1^2+1^4}+\frac{2}{1+2^2+2^4}+\frac{3}{1+3^2+3^4}+\ldots,\text{ is}$$
(a) $\frac{n^2+n}{2(n^2+n+1)}$ (b) $\frac{n^2-n}{2(n^2+n+1)}$
(c) $\frac{n^2+n}{2(n^2-n+1)}$ (d) None of these

28 If C is a skew-symmetric matrix of order n and X is $n\times 1$ column matrix, then $X'CX$ is a
(a) scalar matrix (b) unit matrix
(c) null matrix (d) None of these

29 Which of the following is correct?
(a) Skew-symmetric matrix of an even order is always singular
(b) Skew-symmetric matrix of an odd order is non-singular
(c) Skew-symmetric matrix of an odd order is singular
(d) None of the above

30 If $\begin{vmatrix} a_1 & b_1 & c_1 \\ a_2 & b_2 & c_2 \\ a_3 & b_3 & c_3 \end{vmatrix}=5$, then the value of
$$\Delta=\begin{vmatrix} b_2c_3-b_3c_2 & a_3c_2-a_2c_3 & a_2b_3-a_3b_2 \\ b_3c_1-b_1c_3 & a_1c_3-a_3c_1 & a_3b_1-a_1b_3 \\ b_1c_2-b_2c_1 & a_2c_1-a_1c_2 & a_1b_2-a_2b_1 \end{vmatrix}$$ is
(a) 5 (b) 25 (c) 125 (d) 0

31 The number of seven letter words that can be formed by using the letters of the word 'SUCCESS' so that the two C are together but no two S are together, is
(a) 24 (b) 36
(c) 54 (d) None of these

32 The greatest integer less than or equal to $(\sqrt{2}+1)^6$ is
(a) 196 (b) 197
(c) 198 (d) 199

33 If A and B are square matrices such that $B = -A^{-1}BA$, then
(a) $AB + BA = O$
(b) $(A+B)^2 = A^2 - B^2$
(c) $(A+B)^2 = A^2 + 2AB + B^2$
(d) $(A+B)^2 = A + B$

34 The determinant $\begin{vmatrix} xp+y & x & y \\ yp+z & y & z \\ 0 & xp+y & yp+z \end{vmatrix} = 0$, if
(a) x, y, z are in AP (b) x, y, z are in GP
(c) x, y, z are in HP (d) xy, yz, zx are in AP

35 The value of k, for which the system of equations $x + ky + 3z = 0$, $3x + ky - 2z = 0$ and $2x + 3y - 4z = 0$ possess a non-trivial solution over the set of rationals, is
(a) $-\frac{33}{2}$ (b) $\frac{33}{2}$
(c) 11 (d) None of these

36 The number of groups that can be made from 5 different green balls, 4 different blue balls and 3 different red balls, if atleast 1 green and 1 blue ball is to be included is
(a) 3700 (b) 3720
(c) 4340 (d) None of these

37 If n is an integer greater than 1, then
$a - {}^nC_1(a-1) + {}^nC_2(a-2) + \ldots + (-1)^n(a-n)$ is equal to
(a) a (b) 0 (c) a^2 (d) 2^n

38 If α, β and γ are the roots of the equation $x^2(x+e) = e(x+1)$. Then, the value of the determinant
$\begin{vmatrix} 1+\alpha & 1 & 1 \\ 1 & 1+\beta & 1 \\ 1 & 1 & 1+\gamma \end{vmatrix}$ is
(a) -1 (b) 1
(c) 0 (d) $1 + \frac{1}{\alpha} + \frac{1}{\beta} + \frac{1}{\gamma}$

39 For all natural number $n > 1$, $2^{4n} - 15n - 1$ is divisible by
(a) 225 (b) 125
(c) 325 (d) None of these

40 If x is so small that its two and higher power can be neglected and if $(1-2x)^{-1/2}(1-4x)^{-5/2} = 1 + kx$, then k is equal to
(a) -2 (b) 1
(c) 10 (d) 11

41 If $x = -5$ is a root of $\begin{vmatrix} 2x+1 & 4 & 8 \\ 2 & 2x & 2 \\ 7 & 6 & 2x \end{vmatrix} = 0$, then the other two roots are
(a) 3, 3.5 (b) 1, 3.5
(c) 3, 6 (d) 2, 6

42 The 8th term of $\left(3x + \frac{2}{3x^2}\right)^{12}$ when expanded in ascending power of x, is
(a) $\frac{228096}{x^3}$ (b) $\frac{228096}{x^9}$
(c) $\frac{328179}{x^9}$ (d) None of these

43 The greatest term in the expansion of $(3-5x)^{11}$ when $x = \frac{1}{5}$, is
(a) 55×3^9 (b) 46×3^9
(c) 55×3^6 (d) None of these

44 In an examination a candidate has to pass in each of the papers to be successful. If the total number of ways to fail is 63, how many papers are there in the examination?
(a) 6 (b) 8 (c) 10 (d) 12

45 A is a set containing n elements. A subset P of A is chosen. The set A is reconstructed by replacing the elements of P. A subset Q of A is again chosen. The number of ways of choosing P and Q, so that $P \cap Q$ contains exactly two elements is
(a) $9 \cdot {}^nC_2$ (b) $3^n - {}^nC_2$
(c) $2 \cdot {}^nC_n$ (d) None of these

46 If the sets A and B are defined as
$$A = \{(x, y) : y = \frac{1}{x}, 0 \neq x \in R\}$$
and $B = \{(x, y) : y = -x, x \in R\}$, then
(a) $A \cap B = A$ (b) $A \cap B = B$
(c) $A \cap B = \phi$ (d) None of these

47 There are 16 points in a plane no three of which are in a straight line except 8 which are all in a straight line. The number of triangles can be formed by joining them equals
(a) 1120 (b) 560 (c) 552 (d) 504

48 The value of the natural numbers n such that inequality $2^n > 2n + 1$ is valid, is
(a) for $n \geq 3$ (b) for $n < 3$ (c) for mn (d) for any n

49 Let $w = \cos\left(\frac{2\pi}{7}\right) + i\sin\left(\frac{2\pi}{7}\right)$ and $\alpha = w + w^2 + w^4$ and $\beta = w^3 + w^5 + w^6$, then $\alpha + \beta$ is equal to
(a) 0 (b) -1 (c) -2 (d) 1

50 Let $H(x) = \frac{f(x)}{g(x)}$, where $f(x) = 1 - 2\sin^2 x$ and $g(x) = \cos 2x, \forall f : R \to [-1, 1]$ and $g : R \to [-1, 1]$. Domain and range of $H(x)$ are respectively
(a) R and $\{1\}$
(b) R and $\{0, 1\}$
(c) $R \sim \{(2n+1)\frac{\pi}{4}\}$ and $\{1\}, n \in I$
(d) $R \sim \left\{(2n+1)\frac{\pi}{2}\right\}$ and $\{0, 1\}, n \in I$

51 Let T_n denote the number of triangles which can be formed using the vertices of a regular polygon of n sides. If $T_{n+1} - T_n = 21$, then n equals

(a) 4 (b) 6
(c) 7 (d) None of these

52 If r is a real number such that $|r| < 1$ and if $a = 5(1 - r)$, then

(a) $0 < a < 5$ (b) $-5 < a < 5$
(c) $0 < a < 10$ (d) $0 \le a < 10$

Directions (Q. Nos. 53-57) *Each of these questions contains two statements : Statement* I *(Assertion) and Statement* II *(Reason). Each of these questions also has four alternative choices, only one of which is the correct answer. You have to select one of the codes (a), (b), (c) and (d) given below.*

(a) Statement I is true, Statement II is true; Statement II is a correct explanation for Statement I
(b) Statement I is true, Statement II is true; Statement II is not a correct explanation for Statement I
(c) Statement I is true; Statement II is false
(d) Statement I is false; Statement II is true

53 **Statement I** The number of natural numbers which divide 10^{2009} but not 10^{2008} is 4019.

Statement II If p is a prime, then number of divisors of p^n is $p^{n+1} - 1$.

54 Suppose $A = \begin{bmatrix} 1 & 0 \\ 0 & -1 \end{bmatrix}$ and $B = \begin{bmatrix} 2 & 0 \\ 0 & 2 \end{bmatrix}$. Let X be a 2×2 matrices such that $X' AX = B$.

Statement I X is non-singular and $\det(X) = \pm 2$.

Statement II X is a singular matrix.

55 The general term in the expansion of $(a + x)^n$ is ${}^nC_r\, a^{n-r}\, x^r$.

Statement I The third term in the expansion of $\left(2x + \frac{1}{x^2}\right)^m$ does not contain x. The value of x for which that term equal to the second term in the expansion of $(1 + x^3)^{30}$ is 2.

Statement II $(a + x)^n = \sum_{r=0}^{n} {}^nC_r\, a^{n-r} x^r$.

56 Sets A and B have four and eight elements, respectively.

Statement I The minimum number of elements in $A \cup B$ is 8.

Statement II $A \cap B = 5$

57 Let $a \ne 0, p \ne 0$ and $\Delta = \begin{vmatrix} a & b & c \\ 0 & p & q \\ p & q & 0 \end{vmatrix}$

Statement I If the equations $ax^2 + bx + c = 0$ and $px + q = 0$ have a common root, then $\Delta = 0$.

Statement II If $\Delta = 0$, then the equations $ax^2 + bx + c = 0$ and $px + q = 0$ have a common root.

58 Assume X, Y, Z, W and P are matrices of order $2 \times n$, $3 \times k$, $2 \times p$, $n \times 3$ and $p \times k$, respectively. Now, consider the following statements

I. $PY + WY$ will be defined for $k = 3$ and $p = n$.

II. The order of the matrix $7X - 5Z$ is $n \times 2$ (if $p = n$).

Choose the correct option.

(a) Only I is true (b) Only II is true
(c) Both I and II are true (d) Neither I nor II is true

ANSWERS

1. (d)	**2.** (a)	**3.** (d)	**4.** (d)	**5.** (a)	**6.** (a)	**7.** (d)	**8.** (c)	**9.** (a)	**10.** (b)
11. (c)	**12.** (b)	**13.** (b)	**14.** (c)	**15.** (b)	**16.** (d)	**17.** (d)	**18.** (c)	**19.** (c)	**20.** (b)
21. (b)	**22.** (a)	**23.** (b)	**24.** (a)	**25.** (c)	**26.** (c)	**27.** (a)	**28.** (c)	**29.** (c)	**30.** (b)
31. (a)	**32.** (b)	**33.** (a)	**34.** (b)	**35.** (b)	**36.** (b)	**37.** (b)	**38.** (c)	**39.** (a)	**40.** (d)
41. (b)	**42.** (a)	**43.** (a)	**44.** (a)	**45.** (d)	**46.** (c)	**47.** (d)	**48.** (a)	**49.** (b)	**50.** (c)
51. (c)	**52.** (c)	**53.** (c)	**54.** (c)	**55.** (b)	**56.** (c)	**57.** (c)	**58.** (a)		

Hints and Explanations

1 Applying $R_1 \to R_1 - R_3$

$$f(x) = \begin{vmatrix} \cos x - \tan x & 0 & 0 \\ 2\sin x & x^2 & 2x \\ \tan x & x & 1 \end{vmatrix}$$

$$= (\cos x - \tan x)(x^2 - 2x^2)$$
$$= -x^2(\cos x - \tan x)$$

$\therefore f'(x) = -2x(\cos x - \tan x) - x^2(-\sin x - \sec^2 x)$

$\therefore \lim_{x\to 0} \frac{f'(x)}{x} = \lim_{x\to 0}[-2(\cos x - \tan x) + \lim_{x\to 0} x(\sin x + \sec^2 x)]$
$= -2 \times 1 = -2$

2 Given, $\log_{0.5}(x-1) < \log_{0.25}(x-1)$

$\Rightarrow \log_{0.5}(x-1) < \log_{(0.5)^2}(x-1)$

$\Rightarrow \log_{0.5}(x-1) < \frac{1}{2}\log_{0.5}(x-1)$

$\Rightarrow \log_{0.5}(x-1) < 0 \Rightarrow x - 1 > 1$

$\therefore x > 2$

3 Let, $S_n = 12 + 16 + 24 + \ldots + T_n$

Again, $S_n = 12 + 16 + \ldots + T_n$

$0 = (12 + 4 + 8 + 16 + \ldots$ upto n terms$) - T_n$

$\therefore T_n = 12 + \frac{4(2^{n-1}-1)}{2-1} = 2^{n+1} + 8$

On putting $n = 1, 2, 3, \ldots$, we get

$T_1 = 2^2 + 8, T_2 = 2^3 + 8, T_3 = 2^4 + 8 \ldots$

$\therefore S_n = T_1 + T_2 + \ldots + T_n$
$= (2^2 + 2^3 + \ldots$ upto n terms$)$ $+ (8 + 8 + \ldots$ upto n terms$)$
$= \frac{2^2(2^n - 1)}{2-1} + 8n$
$= 4(2^n - 1) + 8n$

4 Let $f(x) = a^2x^2 + 2bx + 2c$

$\because$ We have, $a^2\alpha^2 + b\alpha + c = 0$

and $a^2\beta^2 - b\beta - c = 0$

$\therefore f(\alpha) = a^2\alpha^2 + 2b\alpha + 2c = b\alpha + c = -a^2\alpha^2$

$f(\beta) = a^2\beta^2 + 2b\beta + 2c = 3(b\beta + c) = 3a^2\beta^2$

But $0 < \alpha < \beta \Rightarrow \alpha, \beta$ are real number.

$\therefore f(\alpha) < 0, f(\beta) > 0$

Hence, $\alpha < \gamma < \beta$.

5 Let $2n$ arithmetic means be $A_1, A_2, \ldots, A_{2n}$ between a and b.

Then, $A_1 + A_2 + \ldots + A_{2n} = \frac{a+b}{2} \times 2n$

$= \frac{13/6}{2} \times 2n = \frac{13n}{6}$

Given, $A_1 + A_2 + \ldots + A_{2n} = 2n + 1$

$\Rightarrow 2n + 1 = \frac{13n}{6}$

$\Rightarrow 12n + 6 = 13n$

$\therefore n = 6$

Hence, the number of means $= 2 \times 6 = 12$

6 $A^{-1} = \begin{bmatrix} 1 & 0 \\ -1 & 1 \end{bmatrix}$

$\Rightarrow (A^{-1})^2 = \begin{bmatrix} 1 & 0 \\ -1 & 1 \end{bmatrix}\begin{bmatrix} 1 & 0 \\ -1 & 1 \end{bmatrix} = \begin{bmatrix} 1 & 0 \\ -2 & 1 \end{bmatrix}$

Similarly, $(A^{-1})^n = \begin{bmatrix} 1 & 0 \\ -n & 1 \end{bmatrix}$

Now, $\lim_{n\to\infty} \frac{1}{n} A^{-n} = \lim_{n\to\infty} \begin{bmatrix} 1/n & 0 \\ -1 & 1/n \end{bmatrix} = \begin{bmatrix} 0 & 0 \\ -1 & 0 \end{bmatrix}$

and $\lim_{n\to\infty} \frac{1}{n^2} A^{-n} = \lim_{n\to\infty} \begin{bmatrix} 1/n^2 & 0 \\ -\frac{1}{n} & 1/n^2 \end{bmatrix} = \begin{bmatrix} 0 & 0 \\ 0 & 0 \end{bmatrix}$

7 Given, $n(M \cup P \cup C) = 50$,

$n(M) = 37, n(P) = 24, n(C) = 43$

$n(M \cap P) \le 19, n(M \cap C) \le 29, n(P \cap C) \le 20$

$\therefore n(M \cup P \cup C) = n(M) + n(P) + n(C) - n(M \cap P) - n(M \cap C) - n(P \cap C) + n(M \cap P \cap C)$

$\Rightarrow 50 = 37 + 24 + 43 - n(M \cap P) - n(M \cap C) - n(P \cap C) + n(M \cap P \cap C)$

$\Rightarrow n(M \cap P \cap C) = n(M \cap P) + n(M \cap C) + n(P \cap C) - 54$

$\therefore n(M \cap P \cap C) \le 19 + 29 + 20 - 54 = 14$

8 $|(x-4) + iy|^2 < |(x-2) + iy|^2$ [let $z = x + iy$]

$\Rightarrow (x-4)^2 + y^2 < (x-2)^2 + y^2$

$\Rightarrow -4x < -12 \Rightarrow x > 3$

$\therefore \text{Re}(z) > 3$

9 $(1+\omega)(1+\omega^2)(1+\omega^4)(1+\omega^8) \ldots$ to $2n$ factors

$= (1+\omega)(1+\omega^2)(1+\omega)(1+\omega^2) \ldots$ to $2n$ factors

$= [(1+\omega)(1+\omega) \ldots$ to n factors$]$ $[(1+\omega^2)(1+\omega^2) \ldots$ to n factors$]$

$= (1+\omega)^n(1+\omega^2)^n$

$= (1 + \omega + \omega^2 + \omega^3)^n$

$= (0 + \omega^3)^n = \omega^{3n} = 1$

10 If $x \ge a$, then $x^2 - 2a(x-a) - 3a^2 = 0$

$\Rightarrow x^2 - 2ax - a^2 = 0$

$\therefore x = \frac{2a \pm \sqrt{4a^2 + 4a^2}}{2} = a(1 \pm \sqrt{2})$

Since, $x \ge a$

$\therefore x = a(1 + \sqrt{2})$, it is impossible because $a < 0$

$\therefore x = a(1 - \sqrt{2})$

If $x < a$, then $x^2 + 2a(x-a) - 3a^2 = 0$

$\Rightarrow x^2 + 2ax - 5a^2 = 0$

$\therefore x = (-1 \pm \sqrt{6})a$

[impossible $x < a$ and $a < 0$]

11 $z^3 + 2z^2 + 2z + 1 = 0$

$\Rightarrow (z+1)(z^2 + z + 1) = 0$

$\Rightarrow z = -1, \omega, \omega^2$

But $z = -1$ does not satisfy the second equation.

Hence, common roots are ω and ω^2.

12 We have, $|z_1| = |z_2| = |z_3| = 1$

Now, $|z_1 + z_2 + z_3| \ge 0$

$\Rightarrow |z_1|^2 + |z_2|^2 + |z_3|^2 + 2\,\text{Re}(z_1\bar{z}_2 + z_2\bar{z}_3 + z_3\bar{z}_1) \ge 0$

$\Rightarrow 3 + 2[\cos(\theta_1 - \theta_2) + \cos(\theta_2 - \theta_3) + \cos(\theta_3 - \theta_1)] \ge 0$

$\Rightarrow \cos(\theta_1 - \theta_2) + \cos(\theta_2 - \theta_3) + \cos(\theta_3 - \theta_1) \ge -\frac{3}{2}$

13 Here, $D \ge 0$

$\therefore 4(bc + ad)^2 - 4(a^2 + b^2)(c^2 + d^2) \ge 0$

$\Rightarrow b^2c^2 + a^2d^2 + 2abcd - a^2c^2 - a^2d^2 - b^2c^2 - b^2d^2 \ge 0$

$\Rightarrow (ac - bd)^2 \le 0$

$\Rightarrow ac - bd = 0$

[since, square of any expression cannot be negative]

$\therefore b^2d^2 = a^2c^2$

Hence, a^2, bd and c^2 are in GP.

14 Here, $a = 150$ and $d = -4$

$$S_n = \frac{n}{2}[2 \times 150 + (n-1)(-4)]$$
$$= n(152 - 2n)$$

Had the workers not dropped, then the work would have finished is $(n - 8)$ days with 150 workers working on each day.

$\therefore \quad n(152 - 2n) = 150(n - 8)$

$\Rightarrow n^2 - n - 600 = 0$

$\Rightarrow (n - 25)(n + 24) = 0$

$\therefore \quad n = 25$

[since, n cannot be negative]

15 Given, x is a prime < 10

$\therefore \; x = \{2, 3, 5, 7\}$

Now, from $R = \{(x, x^3) : x = 2, 3, 5, 7\}$
$= \{(2, 8), (3, 27), (5, 125), (7, 343)\}$

16 Let $f(x) = (x - a)(x - b) - 1$

We observe that the coefficient of x^2 in $f(x)$ is positive and $f(a) = f(b) = -1$. Thus, the graph of $f(x)$ is as shown in figure given below

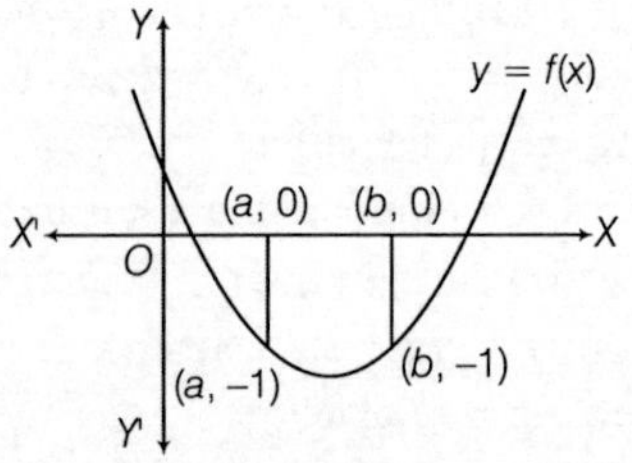

It is evident from the graph that one of the roots of $f(x) = 0$ lies in $(-\infty, a)$ and the other root lies in (b, ∞).

17 Given, $\log_2(3x - 2) = \log_{1/2} x$

$\Rightarrow \log_2(3x - 2) = -\log_2 x$

$\Rightarrow \log_2(3x - 2) = \log_2 x^{-1}$

$\Rightarrow 3x - 2 = x^{-1}$

$\Rightarrow 3x^2 - 2x - 1 = 0$

$\Rightarrow (3x + 1)(x - 1) = 0$

$\Rightarrow x = 1 \text{ or } x = -\frac{1}{3}$

$\therefore \; x = 1$

[since, negative of x cannot satisfy the given equation]

18 $f(x, n) = \sum_{r=1}^{n} (\log_x r - \log_x x)$

$= \sum_{r=1}^{n} (\log_x r - 1) = \log_x (1 \cdot 2 \ldots n) - n$

$= \log_x n! - n$

Given, $\quad f(x, 11) = f(x, 12)$

$\Rightarrow \log_x(11!) - 11 = \log_x(12!) - 12$

$\Rightarrow \log_x\left(\frac{12!}{11!}\right) = 1 \Rightarrow \log_x (12) = 1$

$\therefore \; x = 12$

19 Given, $a + b + c = 25$ ...(i)

Since, 2, a, b are in AP, therefore
$2a = 2 + b$ ...(ii)

Since, b, c, 18 are in GP, therefore
$c^2 = 18b$ ...(iii)

From Eqs. (i) and (ii), we get
$3b = 48 - 2c$

From Eq. (iii), we get
$c^2 = 6(48 - 2c) = 288 - 12c$

$\Rightarrow c^2 + 12c - 288 = 0$

$\Rightarrow (c + 24)(c - 12) = 0$

$\Rightarrow c = 12 \text{ as } c \neq -24$

$\therefore \; b = 8 \text{ and } a = 5$

20 $\because |z| = 1 \Rightarrow z = \cos\theta + i \sin\theta$

$\therefore \arg(z) = \theta$

Let, $\arg(z_1) = \theta_1$ and $\arg(z_2) = \theta_2$

Then, $z_1 R z_2 \Rightarrow |\arg z_1 - \arg z_2| = \frac{2\pi}{3}$

$\Rightarrow z_1 R z_2 \Rightarrow |\theta_1 - \theta_2| = \frac{2\pi}{3}$

$\Rightarrow |\theta_2 - \theta_1| = \frac{2\pi}{3}$

$\Rightarrow z_2 R z_1$

Hence, it is symmetric.

21 Here, $\alpha + \beta = \frac{2b}{a}$ and $\alpha\beta = \frac{c}{a}$

Now, $(\alpha\beta)^3 + \alpha^2\beta^2(\beta + \alpha)$

$= \left(\frac{c}{a}\right)^3 + \frac{c^2}{a^2}\left(\frac{2b}{a}\right)$

$= \frac{c^2(c + 2b)}{a^3}$

22 According to the question,

$D \geq 0$ and $f(3) > 0$

$\Rightarrow 4a^2 - 4(a^2 + a - 3) \geq 0$

and $\; 3^2 - 2a(3) + a^2 + a - 3 > 0$

$\Rightarrow -a + 3 \geq 0$ and $a^2 - 5a + 6 > 0$

$\Rightarrow a \leq 3$ and $a < 2$ or $a > 3$

$\therefore \; a < 2$

23 Let $f(x) = x^2 - 2(4k - 1)x + 15k^2 - 2k - 7$, then $f(x) > 0$

$\therefore \; D < 0$

$\Rightarrow 4(4k - 1)^2 - 4(15k^2 - 2k - 7) < 0$

$\Rightarrow k^2 - 6k + 8 < 0 \Rightarrow 2 < k < 4$

Hence, required integer value of k is 3.

24 Here, $a = 20, d = -\frac{2}{3}$

As the common difference is negative, the terms will become negative after some stage. So, the sum is maximum, if only positive terms are added.

Now, $T_n = 20 + (n - 1)\left(-\frac{2}{3}\right) \geq 0$

$\Rightarrow 60 - 2(n - 1) \geq 0$

$\Rightarrow 62 \geq 2n \Rightarrow 31 \geq n$

Thus, the first 31 terms are non-negative.

$\therefore$ Maximum sum,

$S_{31} = \frac{31}{2}\left[2 \times 20 + (31 - 1)\left(-\frac{2}{3}\right)\right]$

$= \frac{31}{2}(40 - 20) = 310$

25 First series has common difference 4 and second series has common difference 5.

Hence, the series with common terms has common difference is equal to the LCM of 4 and 5 i.e. 20.

Since, the first common term is 21. So, the series will be 21, 41, 61, ..., 401 which has 20 terms.

26 Given, $a_1 + a_2 + a_3 = 21$

$\Rightarrow \quad a(1 + r + r^2) = 21$

and $\quad a_4 + a_5 + a_6 = 168$

$\Rightarrow \quad ar^3(1 + r + r^2) = 168$

$\therefore \quad r^3 = 8 \Rightarrow r = 2$

and $\quad a(1 + 2 + 4) = 21$

$\therefore \quad a = 3$

27 $T_r = \frac{r}{1 + r^2 + r^4}, r = 1, 2, 3, \ldots, n$

$= \frac{r}{(r^2 + r + 1)(r^2 - r + 1)}$

$= \frac{1}{2}\left[\frac{1}{r^2 - r + 1} - \frac{1}{r^2 + r + 1}\right]$

$\therefore \sum_{r=1}^{n} T_r = \frac{1}{2}\sum_{r=1}^{n}\left[\frac{1}{r^2 - r + 1} - \frac{1}{r^2 + r + 1}\right]$

$= \frac{1}{2}\left[\left(1 - \frac{1}{3}\right) + \left(\frac{1}{3} - \frac{1}{7}\right) + \ldots + \left(\frac{1}{n^2 - n + 1} - \frac{1}{n^2 + n + 1}\right)\right]$

$= \frac{1}{2}\left[1 - \frac{1}{n^2 + n + 1}\right] = \frac{n^2 + n}{2(n^2 + n + 1)}$

28 Here, X is $n \times 1$, C is $n \times n$ and X' is $1 \times n$ order matrix. Therefore, $X' C X$ is 1×1 order matrix. Let $X' C X = K$

Then, $(X' C X)' = X' C' X''$
$= X'(-C)X = -K$

$\Rightarrow \quad 2K = O$

$\therefore \quad K = O$

29 Since, the determinant of a skew-symmetric matrix of an odd order is zero. Therefore, the matrix is singular.

30 We know that, if A is a square matrix of order n and B is the matrix of cofactors of elements of A. Then,

$|B| = |A|^{n-1}$

$\therefore \Delta = |A|^{3-1} = 5^{3-1} = 25$

31 Considering CC as single letter, U,CC, E can be arranged in 3 ! ways

Here, × U× CC× E×

Hence, the required number of ways

$= 3! \cdot {}^4C_3 = 24$

32 Let $(\sqrt{2} + 1)^6 = I + F$, where I is an integer and $0 \le F < 1$

Let $f = (\sqrt{2} - 1)^6$

Now, $\sqrt{2} - 1 = \dfrac{1}{\sqrt{2} + 1}$

$\Rightarrow 0 < \sqrt{2} - 1 < 1$

Also, $I + F + f = (\sqrt{2} + 1)^6 + (\sqrt{2} - 1)^6$

$= 2[{}^6C_0 \cdot 2^3 + {}^6C_2 \cdot 2^2 + {}^6C_4 \cdot 2 + {}^6C_6]$

$= 2(8 + 60 + 30 + 1) = 198$

Hence, $F + f = 198 - I$ is an integer.

But $0 < F + f < 2$

$\therefore F + f = 1$ and $I = 197$

33 Given, $B = -A^{-1}BA$

$\Rightarrow AB = -A(A^{-1}BA)$

$\Rightarrow AB = -I(BA)$

$\therefore AB + BA = O$

34 Applying $R_3 \to R_3 - pR_1 - R_2$

$$\Rightarrow \begin{vmatrix} xp + y & x & y \\ yp + z & y & z \\ -(xp^2 + 2yp + z) & 0 & 0 \end{vmatrix} = 0$$

$\Rightarrow -(xp^2 + 2yp + z)(xz - y^2) = 0$

Hence, x, y and z are in GP.

35 For non-trivial solution, we must have

$$\begin{vmatrix} 1 & k & 3 \\ 3 & k & -2 \\ 2 & 3 & -4 \end{vmatrix} = 0$$

Applying

$R_2 \to R_2 - 3R_1, R_3 \to R_3 - 2R_1$,

$$\begin{vmatrix} 1 & k & 3 \\ 0 & -2k & -11 \\ 0 & 3-2k & -10 \end{vmatrix} = 0$$

$\Rightarrow 20k + 11(3 - 2k) = 0$

$\Rightarrow k = \dfrac{33}{2}$

36 Atleast one green ball can be selected out of 5 green balls in $2^5 - 1$, i.e. 31 ways. Similarly, atleast one blue ball can be selected from 4 blue balls in $2^4 - 1 = 15$ ways and atleast one red or not red can be selected in $2^3 = 8$ ways.

Hence, the required number of ways $= 31 \times 15 \times 8 = 3720$

37 LHS $= a[C_0 - C_1 + C_2 - C_3 + \ldots + (-1)^n C_n]$

$+ [C_1 - 2C_2 + 3C_3 - \ldots + (-1)^{n-1} \cdot nC_n]$

$= a \cdot 0 + 0 = 0$

38 Given, equation is

$x^3 + ex^2 - ex - e = 0$

Applying $R_2 \to R_2 - R_1$

and $R_3 \to R_3 - R_2$,

$$\Delta = \begin{vmatrix} (1 + \alpha) & 1 & 1 \\ -\alpha & \beta & 0 \\ 0 & -\beta & \gamma \end{vmatrix}$$

$= (1 + \alpha)(\beta\gamma - 0) + \alpha(\gamma) + \alpha\beta$

$= \alpha\beta + \beta\gamma + \gamma\alpha + \alpha\beta\gamma$

$= -e + e = 0$

39 Now, $2^{4n} = (1 + 15)^n$

$= 1 + {}^nC_1 \cdot 15 + {}^nC_2 \cdot 15^2 + {}^nC_3 \cdot 15^3 + \ldots$

$\therefore 2^{4n} - 1 - 15n = 15^2 [{}^nC_2 + {}^nC_3 \cdot 15 + \ldots]$

$= 225k$

Hence, it is divisible by 225.

40 $(1 - 2x)^{-1/2}(1 - 4x)^{-5/2}$

$= (1 + x)(1 + 10x)$ [neglecting higher power]

$= 1 + 11x$ [neglecting higher power]

$= 1 + kx$

$\therefore k = 11$

41 Applying $R_1 \to R_1 + R_2 + R_3$, we get

$$\begin{vmatrix} 2x + 10 & 2x + 10 & 2x + 10 \\ 2 & 2x & 2 \\ 7 & 6 & 2x \end{vmatrix} = 0$$

Taking $2x + 10$ common from R_1 and applying

$C_2 \to C_2 - C_1, C_3 \to C_3 - C_1$, we get

$$2(x + 5)\begin{vmatrix} 1 & 0 & 0 \\ 2 & 2x - 2 & 0 \\ 7 & -1 & 2x - 7 \end{vmatrix} = 0$$

$\Rightarrow 2(x + 5)(2x - 2)(2x - 7) = 0$

$\therefore x = -5, 1, 3.5$

42 For ascending power of x, we take the expression

$$\left(\frac{2}{3x^2} + 3x\right)^{12}$$

$\therefore T_8$ in $\left(\dfrac{2}{3x^2} + 3x\right)^{12}$

$= {}^{12}C_7 \left(\dfrac{2}{3x^2}\right)^{12-7} (3x)^7$

$= \dfrac{12!}{7!5!}\left(\dfrac{2}{3x^2}\right)^5 (3x)^7$

$= \dfrac{12 \times 11 \times 10 \times 9 \times 8}{5 \times 4 \times 3 \times 2} \times \dfrac{2^5 \times 3^2}{x^3}$

$= \dfrac{228096}{x^3}$

43 $(3 - 5x)^{11} = 3^{11}\left(1 - \dfrac{5x}{3}\right)^{11}$

$= 3^{11}\left(1 - \dfrac{1}{3}\right)^{11}$ $\left[\because x = \dfrac{1}{5}\right]$

$\therefore$ Greatest term $= \dfrac{|x|(n + 1)}{(|x| + 1)}$

$= \dfrac{\left|-\dfrac{1}{3}\right|(11 + 1)}{\left|-\dfrac{1}{3}\right| + 1} = 3$

Now, $T_3 = 3^{11} \cdot {}^{11}C_2 \left(-\dfrac{1}{3}\right)^2$

$= 3^{11}\left(\dfrac{11 \cdot 10}{1 \cdot 2} \times \dfrac{1}{9}\right) = 55 \times 3^9$

44 Let the number of papers be n.

$\therefore$ Total number of ways to fail or pass

$= {}^nC_0 + {}^nC_1 + \ldots + {}^nC_n = 2^n$

$\therefore$ Total number of ways to fail

$= 2^n - 1$

[since, there is only one way to pass]

According to the question,

$2^n - 1 = 63 \Rightarrow 2^n = 2^6 \Rightarrow n = 6$

45 Let $A = \{a_1, a_2, \ldots, a_n\}, 1 \le i \le n$

(i) $a_i \in P, a_i \in Q$ (ii) $a_i \notin P, a_i \notin Q$

(iii) $a_i \notin P, a_i \in Q$ (iv) $a_i \in P, a_i \notin Q$

So, $P \cap Q$ contains exactly two elements, taking 2 elements in (i) and $(n - 2)$ elements in (ii), (iii) and (iv).

$\therefore$ Required number of ways

$= {}^nC_2 \times 3^{n-2}$

46 It is clear from the graph that two curves do not intersect anywhere.

$\therefore A \cap B = \phi$

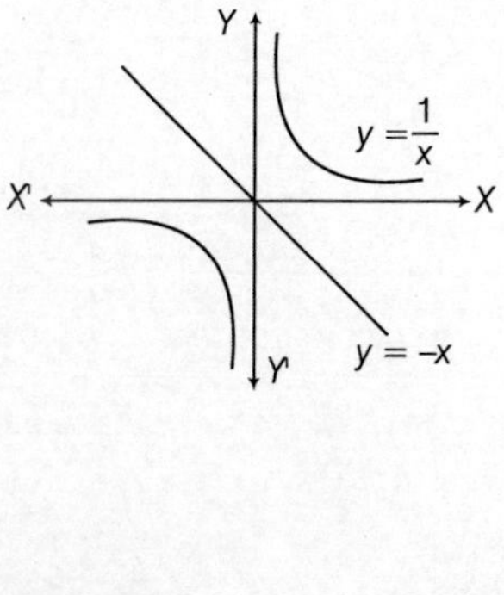

47 $\therefore$ Required number of ways
$= {}^{16}C_3 - {}^{8}C_3 = 560 - 56 = 504$

48 Check through options, the condition $2^n > 2n + 1$ is valid for $n \geq 3$.

49 Here, $\alpha + \beta = \sum_{k=1}^{6} w^k = \dfrac{w(1 - w^6)}{1 - w}$
$= -1 \quad [\because w^7 = 1]$

50 $\because H(x) = \dfrac{f(x)}{g(x)} = \dfrac{1 - 2\sin^2 x}{\cos 2x}$
$= \dfrac{\cos 2x}{\cos 2x} = 1$

But $\cos 2x \neq 0$

$\Rightarrow 2x \neq n\pi + \dfrac{\pi}{2}, n \in I$

$\therefore x \in R \sim \left\{(2n+1)\dfrac{\pi}{4}, n \in I\right\}$

and range $= \{1\}$

51 $T_n = {}^nC_3$ and $T_{n+1} - T_n = 21$

$\Rightarrow {}^{n+1}C_3 - {}^nC_3 = 21$

$\Rightarrow {}^nC_2 + {}^nC_3 - {}^nC_3 = 21$

$\Rightarrow {}^nC_2 = 21$

$\Rightarrow \dfrac{n(n-1)}{2} = 21$

$\Rightarrow n^2 - n - 42 = 0$

$\Rightarrow (n - 7)(n + 6) = 0$

$\therefore n = 7 \quad [\because n \neq -6]$

52 Since, $|r| < 1 \Rightarrow -1 < r < 1$

Also, $a = 5(1 - r) \Rightarrow 0 < a < 10$

$[\because$ at $r = -1, a = 10$ and at $r = 1, a = 0]$

53 The number of divisor of $10^m = 2^m 5^m$ is $(m+1)^2$.

$\therefore$ Number of divisors which divide 10^{2009} but not 10^{2008} is

$(2010)^2 - (2009)^2 = 4019$

54 If $X = O$, then $X'AX = O \Rightarrow B = O$, a contradiction.

Let $\det(X) = a$, then $\det(X') = a$

$\therefore \det(X'AX) = \det(B)$

$\Rightarrow a(-1)a = -4$

$[\because \det(X'AX) = \det(X')\det(A)\det(X)]$

$\therefore a = \pm 2$

As $\det(X) \neq 0$, X cannot be a singular matrix.

55 $T_3 = {}^mC_2(2x)^{m-2}\left(\dfrac{1}{x^2}\right)^2$

$= {}^mC_2(2)^{m-2} \cdot x^{m-6}$

For independent of x, put

$m - 6 = 0 \Rightarrow m = 6$

$\therefore T_3 = {}^6C_2(2)^{6-2} = 15 \times 16 = 240$

According to the question,

${}^{30}C_1 x^3 = 240$

$\Rightarrow x^3 = 8$

$\Rightarrow x = 2$

56 $n(A \cup B) = n(A) + n(B) - n(A \cap B)$

$= 4 + 8 - n(A \cap B)$

$= 12 - n(A \cap B)$

Since, maximum number of element in $n(A \cap B) = 4$.

$\therefore$ Minimum number of element is

$n(A \cup B) = 12 - 4 = 8$

57 If λ is a common root of $ax^2 + bx + c = 0$ and $px + q = 0$, then $a\lambda^2 + b\lambda + c = 0$, $p\lambda + q = 0$ and $p\lambda^2 + q\lambda = 0$

Eliminating λ, we obtained $\Delta = 0$.

For Statement II, expanding Δ along C_1, we obtain

$aq^2 + p(bq - cp) = 0$

or $a\left(-\dfrac{q}{p}\right)^2 + b\left(-\dfrac{q}{p}\right) + c = 0$

Thus, $ax^2 + bx + c = 0$ and $px + q = 0$ have a common root.

58 I. Matrices P and Y are of the orders $p \times k$ and $3 \times k$, respectively.

Therefore, matrix PY will be defined, if $k = 3$.

Consequently PY will be of the order $p \times k$. Matrices W and Y are of the orders $n \times 3$ and $3 \times k$, respectively.

Since, the number of columns in W is equal to the number of rows in Y, matrix WY is well-defined and is of the order $n \times k$.

Matrices PY and WY can be added only when their orders are same.

However, PY is of the order $p \times k$ and WY is of the order $n \times k$, therefore we must have $p = n$.

Thus, $k = 3$ and $p = n$ are the restrictions on n, k and p, so that $PY + WY$ will be defined.

II. Matrix X is of the order $2 \times n$.

Therefore, matrix $7X$ is also of the same order.

Matrix Z is of the order $2 \times p$, i.e. $2 \times n$. $[\because n = p]$

Therefore, matrix $5Z$ is also of the same order.

Now, both the matrices $7X$ and $5Z$ are of the order $2 \times n$.

Thus, matrix $7X - 5Z$ is well-defined and is of the order $2 \times n$.

DAY TEN

Real Function

Learning & Revision for the Day

- Real Valued Function and Real Function
- Domain and Range of real Function
- Algebra of Real Functions
- Inverse Function
- Basic Functions
- Nature of a Functions

Real Valued Function and Real Function

A Function $f: A \to B$ is said to be a **real valued function** if $B \subseteq R$ (the set of real numbers), if both A and B are subset of R (the set of real numbers) then f is called a **real function**.

NOTE Every real function is a real valued function but converse need not be true.

Domain and Range of Real Function

The **domain** of $y = f(x)$ is the set of all real x for which $f(x)$ is defined (real).

Range of $y = f(x)$ is collection of all distinct images corresponding to each real number in the domain.

NOTE If $f: A \to B$, then A will be domain of f and B will be codomain of f.

To find range

(i) First of all find the domain of $y = f(x)$.

(ii) If domain has finite number of points, then range is the set of f – images of these points.

(iii) If domain is R or R – {some finite points}, express x in terms of y and find the values of y for which the values of x lie in the domain.

(iv) If domain is a finite interval, find the least and the greatest values for range using monotonicity.

Algebra of Real Functions

Let $f: X \to R$ and $g: X \longrightarrow R$ be two real functions. Then,

- The sum $f + g: X \longrightarrow R$ defined as $(f + g)(x) = f(x) + g(x)$.
- The difference $f - g$: $X - R$, defined as $(f - g)(x) = f(x) - g(x)$

PREP MIRROR
Your Personal Preparation Indicator

- No. of Questions in Exercises (x)—
- No. of Questions Attempted (y)—
- No. of Correct Questions (z)— *(Without referring Explanations)*

- Accuracy Level ($z/y \times 100$)—
- Prep Level ($z/x \times 100$)—

In order to expect good rank in JEE, your Accuracy Level should be above 85 & Prep Level should be above 75.

- The **product** $fg : X \longrightarrow R$, defined as $(fg)(x) = f(x)\,g(x)$
- $f + g$ and fg are defined only, if f and g have the same domain. In case the domain of f and g are different, domain of $f + g$ or fg = Domain of $f \cap$ Domain of g.
- The product $cf : X \longrightarrow R$, defined as $(cf)(x) = cf(x)$, where c is a real number.
- The quotient $\frac{f}{g}$ is a function defined as $\frac{f}{g}(x) = \frac{f(x)}{g(x)}$, provided $g(x) \neq 0, x \in X$
- If domain of $y = f(x)$ and $y = g(x)$ are D_1 and D_2 respectively, then the domain of $f(x) \pm g(x)$ or $f(x) \cdot g(x)$ is $D_1 \cap D_2$, while domain of $\frac{f(x)}{g(x)}$ is

$$D_1 \cap D_2 - \{x : g(x) = 0\}.$$

Equal or Identical Functions

Two functions f and g are said to be equal, if

(i) the domain of f = the domain of g

(ii) the range of f = the range of g

(iii) $f(x) = g(x), \forall\, x \in$ domain

Inverse Functions

- If $f : A \to B$ is a bijective function, then the mapping $f^{-1} : B \to A$ which associate each element $b \in B$ to a unique element $a \in A$ such that $f(a) = b$, is called the **inverse function** of f.

$$f^{-1}(b) = a \Leftrightarrow f(a) = b$$

- The curves $y = f(x)$ and $y = f^{-1}(x)$ are mirror images of each other in the line mirror $y = x$.
- f is invertible iff f is one-one and onto.
- Inverse of bijective function is unique and bijective.
- The solution of $f(x) = f^{-1}(x)$ are same as the solution of $f(x) = x$.
- If $fog = I = gof$, then f and g are inverse of each other.
- $fof^{-1} = I_B, f^{-1}of = I_A$ and $(f^{-1})^{-1} = f$.
- If f and g are two bijections such that (gof) exists, then gof is also bijective function and $(gof)^{-1} = f^{-1}og^{-1}$.

Basic Functions

Basic functions can be categorised into the following categories.

1. Algebraic Functions

A function, say $f(x)$, is called an algebraic function, if it consists finite number of terms involving powers and roots of the independent variable x and the four algebraic operations $+, -, \times$ and $\div$.

Some algebraic functions are given below

(i) **Polynomial Function**

(a) The function

$$f(x) = a_0x^n + a_1x^{n-1} + a_2x^{n-2} + \ldots + a_{n-1}x + a_n$$

where, $a_0, a_1, a_2, \ldots, a_n$ are real numbers and $n \in N$ is known as **polynomial function**. If $a_0 \neq 0$, then n is the degree of polynomial function.

(b) Domain of polynomial function is R.

(c) A polynomial of odd degree has its range $(-\infty, \infty)$ but a polynomial of even degree has a range which is always subset of R.

(ii) **Constant Function** The function $f(x) = k$, where k is constant, is known as constant function. Its domain is R and range is $\{k\}$,

(iii) **Identity Function** The function $f(x) = x$, is known as **identity function**. Its domain is R and range is R.

(iv) **Rational Function** The function $f(x) = \frac{p(x)}{q(x)}$, where $p(x)$ and $q(x)$ are polynomial functions and $q(x) \neq 0$, is called rational function.

Its domain is $R - \{x \mid q(x) = 0\}$.

(v) **Irrational Function** The algebraic functions containing one or more terms having non-integral rational power of x are called irrational functions.

e.g., $y = f(x) = 2\sqrt{x} - \sqrt[3]{x} + 6$

(vi) **Reciprocal Function** The function $f(x) = \frac{1}{x}$ is called the reciprocal function of x. Its domain is $R - \{0\}$ and range is $R - \{0\}$.

2. Piecewise Functions

Piecewise functions are special type of algebraic functions.

(i) **Absolute Valued Function** (Modulus Function) The function $f(x) = |x| = \begin{cases} x, & x \geq 0 \\ -x, & x < 0 \end{cases}$ is called modulus function.

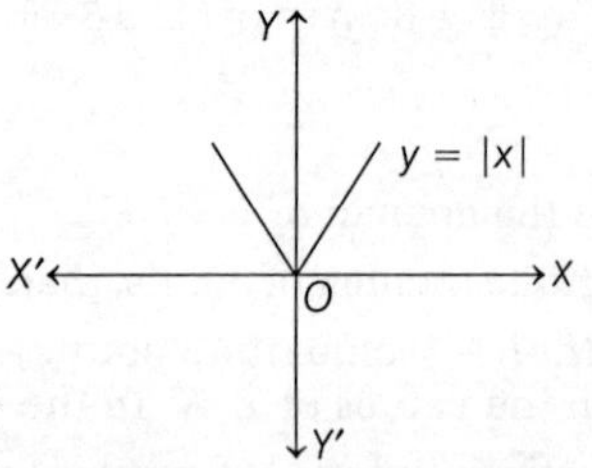

Its domain is R and range is $[0, \infty)$.

Properties of Modulus Function

(a) $|x| \leq a \Rightarrow -a \leq x \leq a \, (a > 0)$

(b) $|x| \geq a \Rightarrow x \leq -a$ or $x \geq a \, (a > 0)$

(c) $|x \pm y| \leq |x| + |y|$

(d) $|x \pm y| \geq ||x| - |y||$

(ii) **Signum Function** The function $f(x)$

$$= \operatorname{sgn}(x) = \begin{cases} \dfrac{|x|}{x} \text{ or } \dfrac{x}{|x|}, & \text{if } x \neq 0 \\ 0, & \text{if } x = 0 \end{cases} = \begin{cases} 1, & \text{if } x > 0 \\ -1, & \text{if } x < 0 \\ 0, & \text{if } x = 0 \end{cases}$$

called signum function.

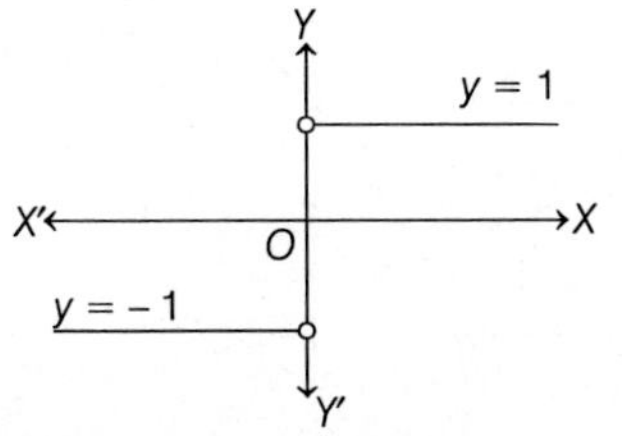

Its domain is R and range is $\{-1, 0, 1\}$.

(iii) **Greatest Integer Function** The symbol $[x]$ indicates the integral part of x which is nearest and smaller than to x. It is also known as floor of x.

The function $f(x) = [x] = \begin{cases} x \ \forall \ x \in I \\ n, \ n \le x < n+1, \ n \in I \end{cases}$ is called greatest integer function.

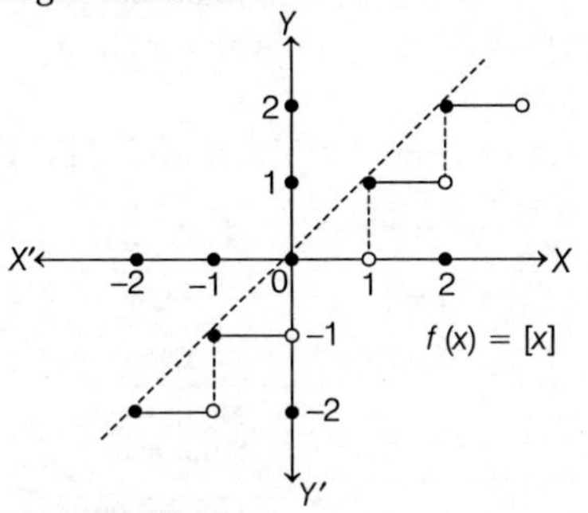

Its domain is R and range is I.

(iv) **Fractional Part Function** The symbol $\{x\}$ indicates the fractional part of x. i.e. $\{x\} = x - [x], x \in R$

$\therefore \quad y = \{x\} = x - [x]$

The function $f(x) = \{x\} = \begin{cases} 0, \ \forall \ x \in I \\ x - n, \ n \le x < n+1, \ n \in I \end{cases}$

is called the fractional part function.

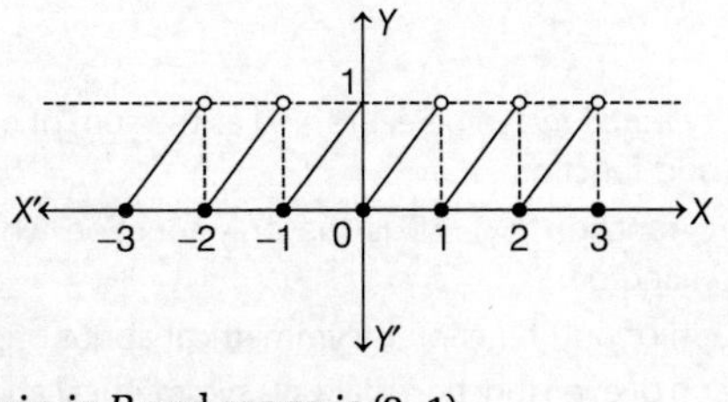

Its domain is R and range is $(0, 1)$.

(v) **Least Integer Function** The symbol (x) indicates the integer part of x which is nearest and greater than x.

The function $f(x) = (x) = \begin{cases} x, \ \forall \ x \in I \\ n+1, \ n < x \le n+1 \ \ n \in I \end{cases}$

is called least integer function.

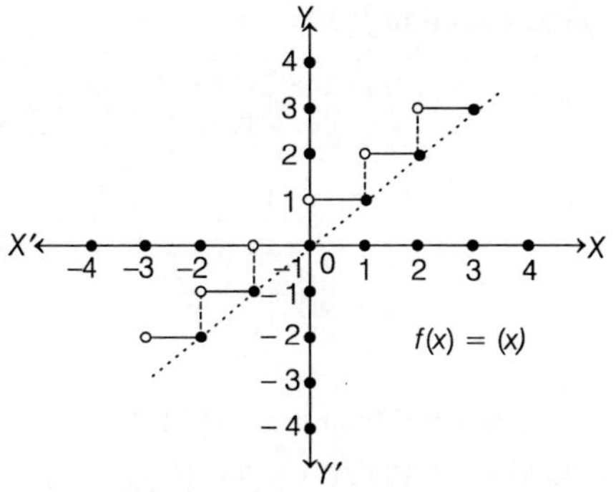

Its domain $\in R$ and range $\in I$.

(vi) **Transcendental function** The function which is not algebraic is called transcendental function.

(vii) **Exponential Function** The function $f(x) = a^x, a > 0, a \neq 1$, is called an exponential function.

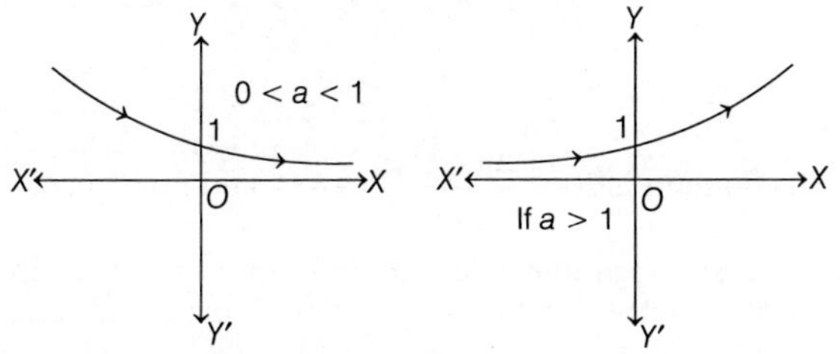

Its domain is R and range is $(0, \infty)$.

It is a one-one into function.

(viii) **Logarithmic Function** The function $f(x) = \log_a x, (x, a > 0)$ and $a \neq 1$ is called logarithmic function.

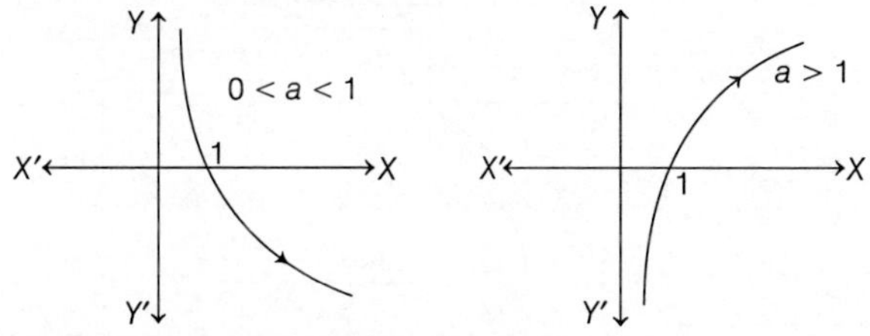

Its domain is $(0, \infty)$ and range is R.

It is a one-one into function.

(ix) **Trigonometric Functions** Some standard trigonometric functions with their domain and range, are given below

(a) **Sine Function** $f(x) = \sin x$,

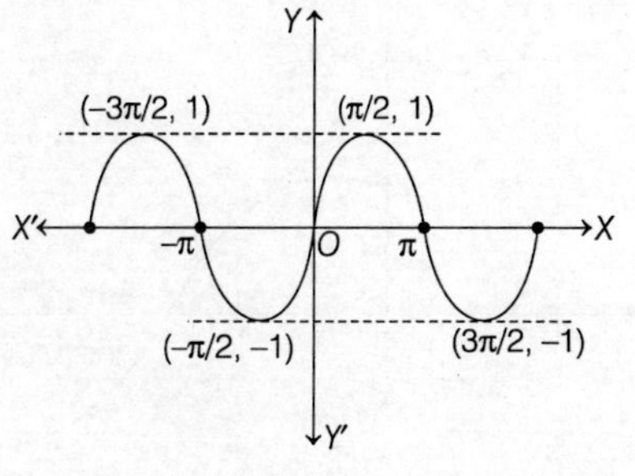

Its domain is R and the range is $[-1, 1]$.

(b) **Cosine Function** $f(x) = \cos x$,

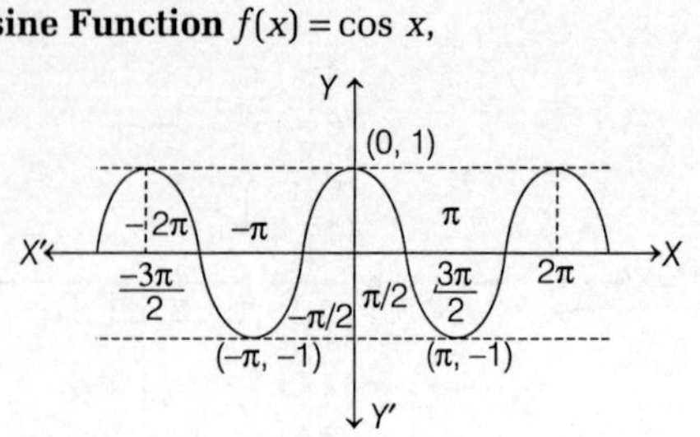

Its domain is R and the range is $[-1, 1]$.

(c) **Tangent Function** $f(x) = \tan x$,

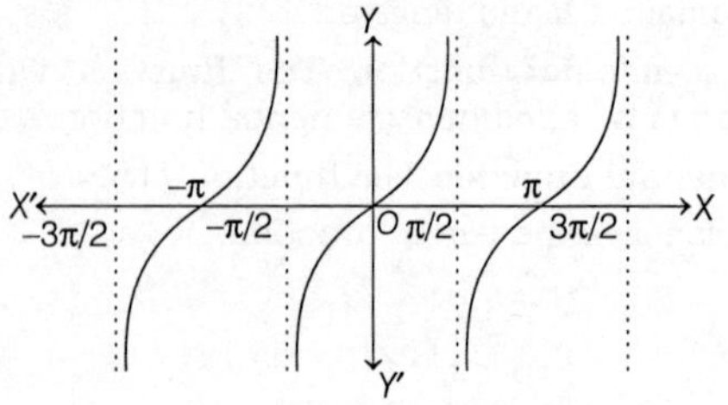

Its domain is $R - \left\{\frac{(2n+1)\pi}{2}, n \in I\right\}$ and range is R.

(x) **Inverse Trigonometric Function** Some standard inverse trigonometric functions with their domain and range, are given below.

(a) $y = f(x) = \sin^{-1} x$

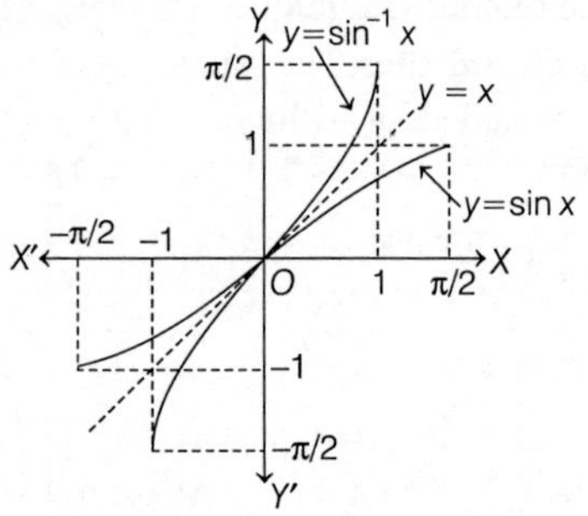

Its domain is $[-1, 1]$ and range is $\left[-\frac{\pi}{2}, \frac{\pi}{2}\right]$.

(b) $y = f(x) = \cos^{-1} x$

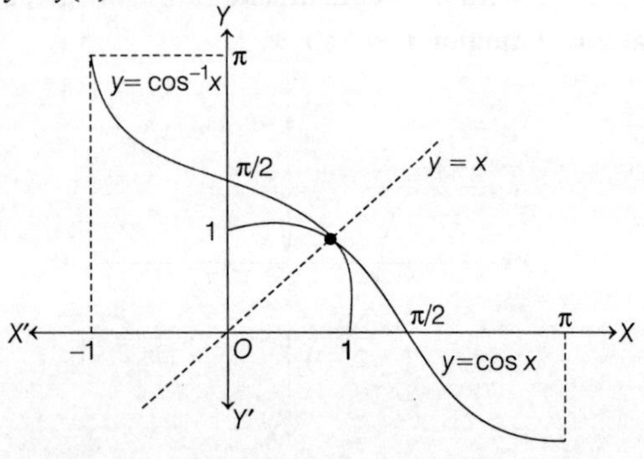

Its, domain is $[-1, 1]$ and range is $[0, \pi]$.

(c) $y = f(x) = \tan^{-1} x$

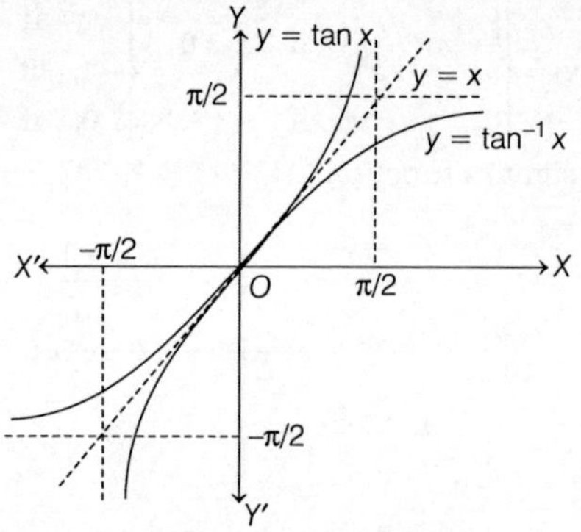

Its domain is R and range is $\left(-\frac{\pi}{2}, \frac{\pi}{2}\right)$.

Nature of a Function

A function $f(x)$ is said to be an **odd** function, if

$$f(-x) = -f(x), \forall x.$$

A function $f(x)$ is said to be an **even** function, if $f(-x) = f(x)$, $\forall x$.

Different Conditions for Even and Odd Functions

$f(x)$	$g(x)$	$f(x) + g(x)$	$f(x) - g(x)$
Odd	Odd	Odd	Odd
Even	Even	Even	Even
Odd	Even	Neither odd nor even	Neither odd nor even
Even	Odd	Neither odd nor even	Neither odd nor even
$f(x)g(x)$	$f(x)/g(x)$	$(gof)(x)$	$(fog)(x)$
Even	Even	Odd	Odd
Even	Even	Even	Even
Odd	Odd	Even	Even
Odd	Odd	Even	Even

NOTE

- Every function can be expressed as the sum of an even and an odd function.
- Zero function $f(x) = 0$ is the only function which is both even and odd.
- Graph of odd function is symmetrical about origin.
- Graph of even function is always symmetrical about Y-axis.

3. Periodic Function

- A function $f(x)$ is said to be periodic function, if there exists a positive real number T, such that $f(x + T) = f(x)$, $\forall\ x \in R$.
- The smallest value of T is called the Fundamental period of $f(x)$.

Properties of Periodic Function

(i) If $f(x)$ is periodic with period T, then $cf(x)$, $f(x+c)$ and $f(x) \pm c$ is periodic with period T.

(ii) If $f(x)$ is periodic with period T, then $kf(cx+d)$ has period $\frac{T}{|c|}$.

(iv) If $f(x)$ is periodic with period T_1 and $g(x)$ is periodic with period T_2, then $f(x)+g(x)$ is periodic with period equal to LCM of T_1 and T_2, provided there is no positive k, such that $f(k+x)=g(x)$ and $g(k+x)=f(x)$.

(iv) If $f(x)$ is a periodic function with period T and $g(x)$ is any function, such that range of $f \subseteq$ domain of g, then gof is also periodic with period T.

Periods of Some Important Functions

Function	Periods
$\sin x, \cos x, \sec x, \text{cosec } x, (\sin x)^{2n+1}, (\cos x)^{2n+1}, (\sec x)^{2n+1}, (\text{cosec } x)^{2n+1}$	2π
$\tan x, \cot x, \tan^n x, \cot^n x, (\sin x)^{2n}, (\cos x)^{2n}, (\sec x)^{2n}, (\text{cosec } x)^{2n}, \lvert\sin x\rvert, \lvert\cos x\rvert, \lvert\tan x\rvert, \lvert\cot x\rvert, \lvert\sec x\rvert, \lvert\text{cosec } x\rvert$	π
$\lvert\sin x+\cos x\rvert$, $\sin^4 x+\cos^4 x$, $\lvert\sec x\rvert+\lvert\text{cosec } x\rvert$, $\lvert\tan x\rvert+\lvert\cot x\rvert$	$\pi/2$
$x-[x]$	1
Algebraic functions like $\sqrt{x}, x^2, x^2+5c,\ldots$ etc.	Period does not exist

DAY PRACTICE SESSION 1

FOUNDATION QUESTIONS EXERCISE

1 Two sets A and B are defined as follows
$A=\{(x,y): y=e^{2x}, x\in R\}$ and
$B=\{(x,y): y=x^2, x\in R\}$, then

(a) $A\subset B$ (b) $B\subset A$
(c) $A\cup B$ (d) $A\cap B=\phi$

2 Domain of definition of the function $f(x)=\sqrt{\sin^{-1}(2x)+\frac{\pi}{6}}$ for real valued x, is

(a) $\left[-\frac{1}{4},\frac{1}{2}\right]$ (b) $\left[-\frac{1}{2},\frac{1}{2}\right]$
(c) $\left(-\frac{1}{2},\frac{1}{9}\right)$ (d) $\left[-\frac{1}{4},\frac{1}{4}\right]$

3 The domain of the function $f(x)=\frac{1}{\sqrt{|x|-x}}$ is

(a) $(0,\infty)$ (b) $(-\infty,0)$
(c) $(-\infty,\infty)-(0)$ (d) $(-\infty,\infty)$

4 Domain of definition of the function
$f(x)=\frac{3}{4-x^2}+\log_{10}(x^3-x)$, is → AIEEE 2003

(a) $(1,2)$
(b) $(-1,0)\cup(1,2)$
(c) $(1,2)\cup(2,\infty)$
(d) $(-1,0)\cup(1,2)\cup(2,\infty)$

5. If $f: R\to R$ is a function satisfying the property $f(x+1)+f(x+3)=2$ for all $x\in R$, then f is

(a) periodic with period 3
(b) periodic with period 4
(c) non- periodic
(d) periodic with period 5

6 The period of the function $f(x)=\sin^3 x+\cos^3 x$ is

(a) 2π (b) π
(c) $\frac{2\pi}{3}$ (d) None of these

7 Let $f: R\to R$ be defined by $f(x)=x^2+1$.
Then, pre-images of 17 and -3, respectively are

(a) $\phi,\{4,-4\}$ (b) $\{3,-3\},\phi$ → NCERT Exemplar
(c) $\{4,-4\},\phi$ (d) $\{4,-4\},\{2,-2\}$

8 Suppose $f(x)=(x+1)^2$ for $x\geq -1$. If $g(x)$ is the function, whose graph is reflection of the graph of $f(x)$ w.r.t. the line $y=x$, then $g(x)$ is equal to

(a) $-\sqrt{x}-1, x\geq 0$
(b) $\frac{1}{(x+1)^2}, x>-1$
(c) $\sqrt{x+1}, x\geq -1$
(d) $\sqrt{x}-1, x\geq 0$

9 The function $f: R\to\left[-\frac{1}{2},\frac{1}{2}\right]$ defined as $f(x)=\frac{x}{1+x^2}$ is → JEE Main 2017

(a) invertible
(b) injective but not surjective
(c) surjective but not injective
(d) neither injective nor surjective

10 The inverse of the function $f(x)=\frac{e^x-e^{-x}}{e^x+e^{-x}}+2$ is given by

(a) $\log_e\left(\frac{x-2}{x-1}\right)^{1/2}$ (b) $\log_e\left(\frac{x-1}{3-x}\right)^{1/2}$
(c) $\log_e\left(\frac{x}{2-x}\right)^{1/2}$ (d) $\log_e\left(\frac{x-1}{x+1}\right)^{-2}$

11 If $f(x)$ is an invertible function, and $g(x) = 2f(x) + 5$, then the value of g^{-1} is

(a) $2f^{-1}(x) - 5$ (b) $\frac{1}{2f^{-1}(x)+5}$

(c) $\frac{1}{2}f^{-1}(x) + 5$ (d) $f^{-1}\left(\frac{x-5}{2}\right)$

12 Let $f:(2,3) \to (0,1)$ be defined by $f(x) = x - [x]$, then $f^{-1}(x)$ is equal to

(a) $x - 2$ (b) $x + 1$ (c) $x - 1$ (d) $x + 2$

13 For a real number x, $[x]$ denotes the integral part of x.

The value of $\left[\frac{1}{2}\right] + \left[\frac{1}{2} + \frac{1}{100}\right] + \left[\frac{1}{2} + \frac{2}{100}\right] + \ldots + \left[\frac{1}{2} + \frac{99}{100}\right]$ is

(a) 49 (b) 50 (c) 48 (d) 51

14 If $f^2(x) \cdot f\left(\frac{1-x}{1+x}\right) = x^3$, [where, $x \neq -1, 1$ and $f(x) \neq 0$], then find $|[f(-2)]|$ (where [.] is the greatest integer function)

(a) $1/x$ (b) $1 - x$ (c) 1 (d) 2

15 If $g(x) = 1 + x - [x]$ and $f(x) = \begin{cases} -1, & x < 0 \\ 0, & x = 0 \\ 1, & x > 0 \end{cases}, \forall x$, then $f\{g(x)\}$ is equal to

(a) x (b) 1 (c) $f(x)$ (d) $g(x)$

16. The function $f(x) = \log(x + \sqrt{x^2 + 1})$, is

(a) an even function (b) an odd function
(c) a periodic function
(d) neither an even nor an odd function

17. **Statement I** $f(x) = |x-2| + |x-3| + |x-5|$ is an odd function for all values of x lie between 3 and 5.

Statement II For odd function $f(-x) = -f(x)$

(a) Statement I is true, Statement II is true; Statement II is a correct explanation for Statement I
(b) Statement I is true, Statement II is true; Statement II is not a correct explanation for Statement I
(c) Statement I is true; Statement II is false
(d) Statement I is false; Statement II is true

18. If domain of $f(x)$ and $g(x)$ are D_1 and D_2 respectively, then domain of $f(x) + g(x)$ is $D_1 \cap D_2$, then

Statement I The domain of the function $f(x) = \sin^{-1} x + \cos^{-1} x + \tan^{-1} x$ is $[-1, 1]$.

Statement II $\sin^{-1} x$ and $\cos^{-1} x$ is defined in $|x| \leq 1$ and $\tan^{-1} x$ is defined for all x.

(a) Statement I is true, Statement II is true; Statement II is a correct explanation for Statement I
(b) Statement I is true, Statement II is true; Statement II is not a correct explanation for Statement I
(c) Statement I is true; Statement II is false
(d) Statement I is false; Statement II is true

19. **Statement I** The period of $f(x) = 2\cos\frac{1}{3}(x-\pi) + 4\sin\frac{1}{3}(x-\pi)$ is 3π.

Statement II If T is the period of $f(x)$, then the period of $f(ax+b)$ is $\frac{T}{|a|}$.

(a) Statement I is true, Statement II is true; Statement II is a correct explanation for Statement I
(b) Statement I is true, Statement II is true; Statement II is not a correct explanation for Statement I
(c) Statement I is true; Statement II is false
(d) Statement I is false; Statement II is true

20. If the range of $f(x)$ is collection of all outputs $f(x)$ corresponding to each real number in the domain, then

Statement I The range of $\log\left(\frac{1}{1+x^2}\right)$ is $(-\infty, \infty)$.

Statement II When $0 < x \leq 1$, $\log x \in (-\infty, 0]$.

(a) Statement I is true, Statement II is true; Statement II is a correct explanation for Statement I
(b) Statement I is true, Statement II is true; Statement II is not a correct explanation for Statement I
(c) Statement I is true; Statement II is false
(d) Statement I is false; Statement II is true

DAY PRACTICE SESSION 2

PROGRESSIVE QUESTIONS EXERCISE

1 Domain of $f(x) = \sqrt{\frac{x-1}{x-2\{x\}}}$, where {} denotes the fractional part of x, is

(a) $(-\infty, 0) \cup (0, 2]$ (b) $[1, 2)$
(c) $(-\infty, \infty) \sim [0, 2)$ (d) $(-\infty, 0) \cup (0, 1] \cup [2, \infty)$

2 Range of $f(x) = [|\sin x| + |\cos x|]$, where $[\cdot]$ denotes the greatest integer function, is

(a) {0} (b) {0, 1} (c) {1} (d) None of these

3 If $[x^2] + x - a = 0$ has a solution, where $a \in N$ and $a \leq 20$, then total number of different values of a can be

(a) 2 (b) 3 (c) 4 (d) 6

4 Total number of solutions of $[x]^2 = x + 2\{x\}$, where $[\cdot]$ and {·} denotes the greatest integer function and fractional part respectively, is equal to

(a) 2 (b) 4
(c) 6 (d) None of these

5 If $f(x) = \sin x + \cos x$, $g(x) = x^2 - 1$, then $g\{f(x)\}$ is invertible in the domain

(a) $\left[0, \frac{\pi}{2}\right]$ (b) $\left[-\frac{\pi}{4}, \frac{\pi}{4}\right]$

(c) $\left[-\frac{\pi}{2}, \frac{\pi}{2}\right]$ (d) $[0, \pi]$

6 Let $f(x) = x^{10} + a\cdot x^8 + b\cdot x^6 + cx^4 + dx^2$ be a polynomial function with real corfficient. If $f(1) = 1$ and $f(2) = -5$, then the minimum number of distinct real zeroes of $f(x)$ is

(a) 5 (b) 6

(c) 7 (d) 8

7 If $f: R \to R, f(x) = x^3 + 3$, and $g: R \to R, g(x) = 2x + 1$, then $f^{-1}og^{-1}(23)$ equals

(a) 2 (b) 3

(c) 4 (d) 5

8 If $f(x)$ and $g(x)$ are two functions such that $f(x) = [x] + [-x]$ and $g(x) = \{x\} \forall x \in R$ and $h(x) = f(g(x))$; then which of the following is incorrect ?
([·] denotes greatest integer function and {·} denotes fractional part function).

(a) $f(x)$ and $h(x)$ are inertial functions
(b) $f(x) = g(x)$ has no solution
(c) $f(x) + h(x) > 0$ has no solution
(b) $f(x) - h(x)$ is a periodic function

9 The period of the function $f(x) = [6x + 7] + \cos \pi x - 6x$, where [·] denotes the greatest integer function, is

(a) 3 (b) 2π (c) 2 (d) None of these

10 The number of real solutions of the equation $\log_{0.5}|x| = 2|x|$ is.

(a) 1 (b) 2 (c) 0 (d) None of these

ANSWERS

SESSION 1	1 (d)	2 (a)	3 (b)	4 (d)	5 (b)	6 (a)	7 (c)	8 (d)	9 (c)	10 (b)
	11 (d)	12 (d)	13 (b)	14 (d)	15 (b)	16 (b)	17 (b)	18 (a)	19 (d)	20 (d)
SESSION 2	1 (d)	2 (c)	3 (c)	4 (b)	5 (b)	6 (a)	7 (a)	8 (b)	9 (c)	10 (b)

Hints and Explanations

SESSION 1

1 Set A represents the set of points lying on the graph of an exponential function and set B represents the set of points lying on the graph of the polynomial.
Take $e^{2x} = x^2$, then the two curves does not intersect. Hence, there is no point common between them.

2 For $f(x)$ to be defined,

$$\sin^{-1}(2x) + \frac{\pi}{6} \geq 0$$

$$\Rightarrow \quad -\frac{\pi}{6} \leq \sin^{-1}(2x) \leq \frac{\pi}{2}$$

$$\Rightarrow \quad \sin\left(-\frac{\pi}{6}\right) \leq 2x \leq \sin\left(\frac{\pi}{2}\right)$$

$$\Rightarrow \quad -\frac{1}{4} \leq x \leq \frac{1}{2}$$

$$\Rightarrow \quad x \in \left[-\frac{1}{4}, \frac{1}{2}\right]$$

3 $y = \dfrac{1}{\sqrt{|x| - x}}$

For domain, $|x| - x > 0$

$\Rightarrow \quad |x| > x$

i.e. only possible, if $x < 0$.

$\therefore \quad x \in (-\infty, 0)$

4 Given, $f(x) = \dfrac{3}{4 - x^2} + \log_{10}(x^3 - x)$

For domain of $f(x)$,

$x^3 - x > 0$

$\Rightarrow \quad x(x-1)(x+1) > 0$

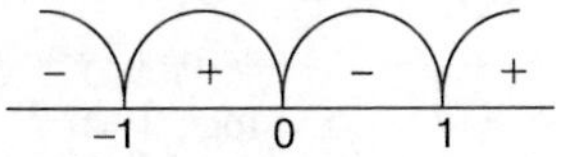

$\Rightarrow \quad x \in (-1, 0) \cup (1, \infty)$

and $\quad 4 - x^2 \neq 0$

$\Rightarrow \quad x \neq \pm 2$

$\Rightarrow \quad x \in (-\infty, -2) \cup (-2, 2) \cup (2, \infty)$

So, common region is
$(-1, 0) \cup (1, 2) \cup (2, \infty)$.

6 We have,

$$f(x+1) + f(x+3) = 2 \quad \text{...(i)}$$

On replacing x by $x + 2$, we get

$$f(x+3) + f(x+5) = 2 \quad \text{...(ii)}$$

On subtracting Eq. (ii) from Eq. (i), we get

$$f(x+1) - f(x+5) = 0$$

$$\Rightarrow \quad f(x+1) = f(x+5)$$

Now, on replacing x by $x - 1$, we get

$$f(x) = f(x+4)$$

Hence, f is periodic with period 4.

6 $f(x) = \left[\dfrac{3\sin x - \sin 3x}{4} + \dfrac{3\cos x + \cos 3x}{4}\right]$

$\therefore$ Period of $f(x)$ = LCM of period of

$\{\sin x, \cos x, \sin 3x, \cos 3x\}$

$$= \frac{\text{LCM of } \{2\pi, 2\pi\}}{\text{HCF of } \{1, 3\}} = 2\pi$$

7 Let $y = x^2 + 1$

$\Rightarrow \quad x = \pm\sqrt{y - 1}$

$\therefore \quad f^{-1}(x) = \pm\sqrt{x - 1}$

$\therefore \quad f^{-1}(17) = \pm\sqrt{17 - 1} = \pm 4$

and $f^{-1}(-3) = \pm\sqrt{-3 - 1}$

$= \pm\sqrt{-4} \notin R$

$\therefore \quad f^{-1}(-3) = \phi$

8 Let $y = (x + 1)^2$ for $x \geq -1$

$\Rightarrow \quad \pm\sqrt{y} = x + 1 \Rightarrow \quad \sqrt{y} = x + 1$

$\Rightarrow \quad y \geq 0, x + 1 \geq 0$

$\Rightarrow \quad x = \sqrt{y} - 1$

$\Rightarrow \quad f^{-1}(y) = \sqrt{y} - 1$

$\Rightarrow \quad f^{-1}(x) = \sqrt{x} - 1, x \geq 0$

9 We have, $f(x)=\dfrac{x}{1+x^2}$

$\therefore \quad f\left(\dfrac{1}{x}\right)=\dfrac{\dfrac{1}{x}}{1+\dfrac{1}{x^2}}$

$=\dfrac{x}{1+x^2}=f(x)$

$\therefore \quad f\left(\dfrac{1}{2}\right)=f(2)$

or $\quad f\left(\dfrac{1}{3}\right)=f(3)$

and so on.

So, $f(x)$ is many-one function.

Again, let $y=f(x)$

$\Rightarrow \quad y=\dfrac{x}{1+x^2}$

$\Rightarrow \quad y+x^2y=x$

$\Rightarrow \quad yx^2-x+y=0$

As, $\quad x\in R$

$\therefore \quad (-1)^2-4(y)(y)\geq 0$

$\Rightarrow \quad 1-4y^2\geq 0$

$\Rightarrow \quad y\in\left[\dfrac{-1}{2},\dfrac{1}{2}\right]$

$\therefore$ Range = Codomain $=\left[\dfrac{-1}{2},\dfrac{1}{2}\right]$

So, $f(x)$ is surjective.

Hence, $f(x)$ is surjective but not injective.

10 Given, $y=\dfrac{e^x-e^{-x}}{e^x+e^{-x}}+2$

$\Rightarrow \quad y=\dfrac{e^{2x}-1}{e^{2x}+1}+2$

$\Rightarrow \quad e^{2x}=\dfrac{1-y}{y-3}=\dfrac{y-1}{3-y}$

$\Rightarrow \quad x=\dfrac{1}{2}\log_e\left(\dfrac{y-1}{3-y}\right)$

$\Rightarrow \quad f^{-1}(y)=\log_e\left(\dfrac{y-1}{3-y}\right)^{1/2}$

$\Rightarrow \quad f^{-1}(x)=\log_e\left(\dfrac{x-1}{3-x}\right)^{1/2}$

11 We have, $g(x)=2f(x)+5$

Now, on replacing x by $g^{-1}(x)$, we get

$g(g^{-1}(x))=2f(g^{-1}(x))+5$

$\Rightarrow \quad x=2f(g^{-1}(x))+5$

$\Rightarrow \quad f(g^{-1}(x))=\dfrac{x-5}{2}$

$\Rightarrow \quad g^{-1}(x)=f^{-1}\left(\dfrac{x-5}{2}\right)$

12 $f:(2,3)\to(0,1)$ and $f(x)=x-[x]$

$\therefore \quad f(x)=y=x-2\Rightarrow x=y+2$

$\Rightarrow \quad f^{-1}(x)=x+2$

13 $\because [x]$ denotes the integral part of x.

Hence, after term $\left[\dfrac{1}{2}+\dfrac{50}{100}\right]$ each term will be one. Hence, the sum of given series will be 50.

14 $f^2(x)\cdot f\left(\dfrac{1-x}{1+x}\right)=x^3 \quad \ldots(\text{i})$

On replacing x by $\dfrac{1-x}{1+x}$, we get

$f^2\left(\dfrac{1-x}{1+x}\right)f(x)=\left(\dfrac{1-x}{1+x}\right)^3 \quad \ldots(\text{ii})$

From Eqs. (i) and (ii),

$f^3(x)=x^6\left(\dfrac{1+x}{1-x}\right)^3$

$\Rightarrow \quad f(x)=x^2\left(\dfrac{1+x}{1-x}\right)$

$\Rightarrow \quad f(-2)=\dfrac{-4}{3}\Rightarrow[f(-2)]=-2$

$\Rightarrow \quad |[f(-2)]|=2$

15 $\because \; g(x)=1+x-[x]$ [put $x=n\in Z$]

$\therefore \quad g(x)=1+x-x=1$

and $g(x)=1+n+k-n=1+k$ [put $x=n+k$]

[where, $n\in Z, 0<k<1$]

Now, $f\{g(x)\}=\begin{cases}-1, & g(x)<0\\ 0, & g(x)=0\\ 1, & g(x)>0\end{cases}$

Clearly, $\quad g(x)>0,\ \forall\ x$

So, $\quad f\{g(x)\}=1,\ \forall\ x$

16 Given that, $f(x)=\log(x+\sqrt{x^2+1})$

Now, $\quad f(-x)=\log(-x+\sqrt{x^2+1})$

$\therefore \; f(x)+f(-x)=\log(x+\sqrt{x^2+1})$

$+\log(-x+\sqrt{x^2+1})$

$=\log(1)=0$

Hence, $f(x)$ is an odd function.

17 Here, $f(x)=\begin{cases}-3x+10, & \forall\ x\leq 2\\ -x+6, & \forall\ 2<x\leq 3\\ x, & \forall\ 3<x\leq 5\\ 3x-10, & \forall\ x>5\end{cases}$

$\therefore \quad f(x)=x,\ \forall\ 3<x<5$

$\Rightarrow \quad f(-x)=-x=-f(x)$

18 Since, $\sin^{-1}x$ is defined in $[-1,1]$, $\cos^{-1}x$ is defined in $[-1,1]$ and $\tan^{-1}x$ is defined in R.

Hence, $f(x)$ is defined in $[-1,1]$.

19 Period of $2\cos\dfrac{1}{3}(x-\pi)$ and $4\sin\dfrac{1}{3}(x-\pi)$ are $\dfrac{2\pi}{1/3},\dfrac{2\pi}{1/3}$ or $6\pi, 6\pi$

$\therefore$ Period of their sum $=6\pi$

20 Range of $\dfrac{1}{1+x^2}$ is $(0,1)$ and domain R

$\therefore \quad \log\left(\dfrac{1}{1+x^2}\right)\in(-\infty,0]$

SESSION 2

1 We have, $\dfrac{x-1}{x-2\{x\}}\geq 0$, here two cases arise

***Case* I** $x\geq 1$ and $x>2\{x\}$

$\Rightarrow \quad x\geq 2$

$\therefore \quad x\in[2,\infty)$.

***Case* II** $x\leq 1$ and $x<2\{x\}$

$\Rightarrow \quad x<1$ and $x\neq 0$.

$\therefore\ x\in(-\infty,0)\cup(0,1)$.

Finally, $x=1$ is also a point of the domain.

2 $y=|\sin x|+|\cos x|$

$\Rightarrow \quad y^2=1+|\sin 2x|$

$\Rightarrow \quad 1\leq y^2\leq 2\Rightarrow \quad y\in[1,\sqrt{2}]$

$\therefore \quad f(x)=[y]=1,\ \forall\ x\in R$

3 Since, $[x^2]+x-a=0$

$\therefore\ x$ has to be an integer.

$\Rightarrow \quad a=x^2+x=x(x+1)$

Thus, a can be 2, 6, 12, 20.

4 $[x]^2=x+2\{x\}$

$\Rightarrow \quad [x]^2=[x]+3\{x\}$

$\Rightarrow \quad \{x\}=\dfrac{[x]^2-[x]}{3}$

$\Rightarrow \quad 0\leq\dfrac{[x]^2-[x]}{3}<1$

$\Rightarrow \quad [x]\in\left(\dfrac{1-\sqrt{13}}{2},0\right]\cup\left[1,\dfrac{1+\sqrt{13}}{2}\right)$

$\Rightarrow \quad [x]=-1,0,1,2$

$\Rightarrow \quad \{x\}=\dfrac{2}{3},0,0,\dfrac{2}{3}$

$\therefore \quad x=-\dfrac{1}{3},0,1,\dfrac{8}{3}$

5 $g\{f(x)\}=(\sin x+\cos x)^2-1$ is invertible.

$\Rightarrow \quad g\{f(x)\}=\sin 2x$

We know that, $\sin x$ is bijective only when $x\in\left[-\dfrac{\pi}{2},\dfrac{\pi}{2}\right]$.

Thus, $g\{f(x)\}$ is bijective, if

$-\dfrac{\pi}{2}\leq 2x\leq\dfrac{\pi}{2}$.

$\therefore \quad -\dfrac{\pi}{4}\leq x\leq\dfrac{\pi}{4}$

6 Since, $f(x)$ is an even function.

$\therefore$ Its graph is symmetrical about Y-axis

Also, we have,

$f(1) = 1$ and $f(2) = -5$

$\Rightarrow f(-1) = 1$ and $f(-2) = -5$

According to these information, we have the following graph

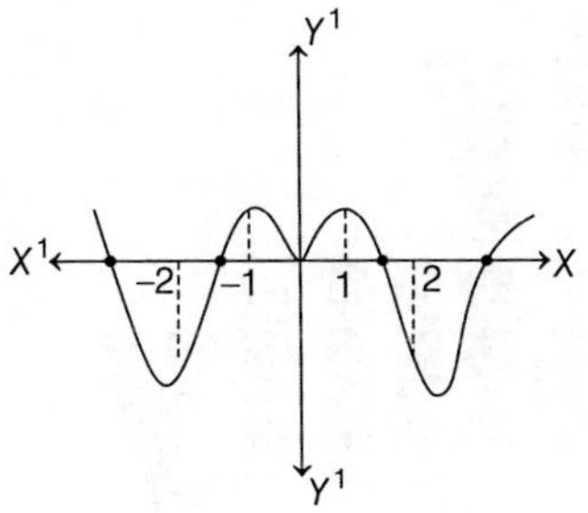

Thus, minimum number of zeroes is 5.

7 Clearly, $f^{-1}og^{-1}(23) = (gof)^{-1}(23)$

Here, $gof(x) = 2(x^3 + 3) + 1$
$= 2x^3 + 7$

Now, let $y = (gof)^{-1}(23)$

$\Rightarrow (gof)(y) = 23$

$\Rightarrow 2y^3 + 7 = 23$

$\Rightarrow 2y^3 = 16$

$\Rightarrow y^3 = 8$

$\Rightarrow y = 2$

Hence, $f^{-1}og^{-1}(23) = 2$

8 We have,

$$f(x) = [x] + [-x] = \begin{cases} 0, & \text{if } x \in I \\ -1, & \text{if } x \notin I \end{cases}$$

$$g(x) = \{x\} = \begin{cases} 0, & \text{if } x \in I \\ (x), & \text{if } x \notin I \end{cases}$$

and $h(x) = f(g(x))$

$$= f(\{x\}) = \begin{cases} f(0), & x \in I \\ f(\{x\}), & x \notin I \end{cases} = \begin{cases} 0, & x \in I \\ -1, & x \notin I \end{cases}$$

Clearly, option (b) is incorrect.

9 We have, $f(x) = [6x + 7] + \cos \pi x - 6x$

$= [6x] + 7 + \cos \pi x - 6x$

$= 7 + \cos \pi x - \{6x\}$ $[\because \{x\} = x - [x]]$

Now, as $\{6x\}$ has period $\frac{1}{6}$ and $\cos \pi x$ has the period 2, therefore the period of $f(x) = \text{LCM}\left(2, \frac{1}{6}\right)$ which is 2.

Hence, the period is 2.

10 For the solution of given equation, let us draw the graph of $y = \log_{0.5}|x|$ and $y = 2|x|$

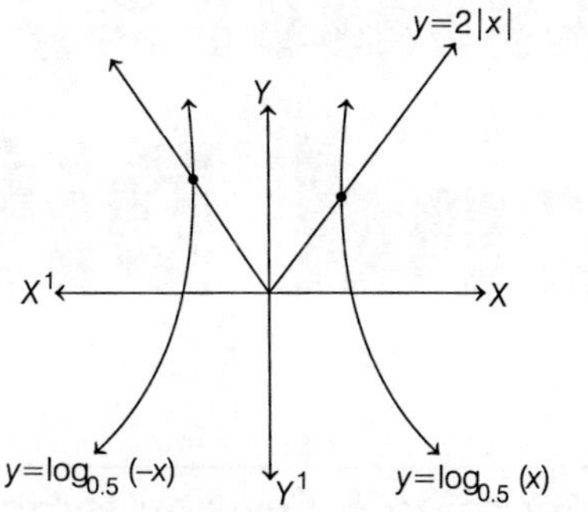

From the graph it is clear that there are two solution.

DAY ELEVEN

Limits, Continuity and Differentiability

Learning & Revision for the Day

- Limits
- Important Results on Limit
- Methods to Evaluate Limits
- Continuity
- Differentiability

Limits

Let $y = f(x)$ be a function of x. If the value of $f(x)$ tend to a definite number as x tends to a, then the number so obtained is called the **limit** of $f(x)$ at $x = a$ and we write it as $\lim_{x \to a} f(x)$.

- If $f(x)$ approaches to l_1 as x approaches to 'a' from left, then l_1 is called the **left hand limit** of $f(x)$ at $x = a$ and symbolically we write it as $f(a-0)$ or $\lim_{x \to a^-} f(x)$ or $\lim_{h \to 0} f(a-h)$
- Similarly, **right hand limit** can be expressed as
$$f(a+0) \text{ or } \lim_{x \to a^+} f(x) \text{ or } \lim_{h \to 0} f(a+h)$$
- $\lim_{x \to a} f(x)$ exists iff $\lim_{x \to a^-} f(x)$ and $\lim_{x \to a^+} f(x)$ exist and equal.

Fundamental Theorems on Limits

If $\lim_{x \to a} f(x) = l$ and $\lim_{x \to a} g(x) = m$ (where, l and m are real numbers), then

(i) $\lim_{x \to a} \{f(x) + g(x)\} = l + m$ [sum rule]

(ii) $\lim_{x \to a} \{f(x) - g(x)\} = l - m$ [difference rule]

(iii) $\lim_{x \to a} \{f(x) \cdot g(x)\} = l \cdot m$ [product rule]

(iv) $\lim_{x \to a} k \cdot f(x) = k \cdot l$ [constant multiple rule]

(v) $\lim_{x \to a} \dfrac{f(x)}{g(x)} = \dfrac{l}{m}, m \neq 0$ [quotient rule]

(vi) If $\lim_{x \to a} f(x) = +\infty$ or $-\infty$, then $\lim_{x \to a} \dfrac{1}{f(x)} = 0$

(vii) $\lim_{x \to a} |f(x)| = \left|\lim_{x \to a} f(x)\right|$

(viii) $\lim_{x \to a} \log\{f(x)\} = \log\{\lim_{x \to a} f(x)\}$, provided $\lim_{x \to a} f(x) > 0$

PREP MIRROR
Your Personal Preparation Indicator

- No. of Questions in Exercises (x)—
- No. of Questions Attempted (y)—
- No. of Correct Questions (z)—
(Without referring Explanations)

- Accuracy Level $(z/y \times 100)$—
- Prep Level $(z/x \times 100)$—

In order to expect good rank in JEE, your Accuracy Level should be above 85 & Prep Level should be above 75.

(ix) If $f(x) \le g(x), \forall x$, then $\lim_{x\to a} f(x) \le \lim_{x\to a} g(x)$

(x) $\lim_{x\to a}[f(x)]^{g(x)} = \{\lim_{x\to a} f(x)\}^{\lim_{x\to a} g(x)}$

(xi) $\lim_{x\to a} f\{g(x)\} = f\{\lim_{x\to a} g(x)\} = f(m)$ provided f is continuous at $\lim_{x\to a} g(x) = m$.

(xii) **Sandwich Theorem** If $f(x) \le g(x) \le h(x) \forall x \in (\alpha, \beta) - \{a\}$ and $\lim_{x\to a} f(x) = \lim_{x\to a} h(x) = l$, then $\lim_{x\to a} g(x) = l$ where $a \in (\alpha, \beta)$

Important Results on Limit

Some important results on limits are given below

1. Algebraic Limits

(i) $\lim_{x\to a} \dfrac{x^n - a^n}{x - a} = na^{n-1}, n \in Q, a > 0$

(ii) $\lim_{x\to\infty} \dfrac{1}{x^n} = 0, n \in N$

(iii) If m, n are positive integers and a_0, b_0 are non-zero real numbers, then

$$\lim_{x\to\infty} \frac{a_0x^m + a_1x^{m-1} + \ldots + a_{m-1}x + a_m}{b_0x^n + b_1x^{n-1} + \ldots + b_{n-1}x + b_n} = \begin{cases} \dfrac{a_0}{b_0} & \text{if } m = n \\ 0 & \text{if } m < n \\ \infty & \text{if } m > n, a_0b_0 > 0 \\ -\infty & \text{if } m > n, a_0b_0 < 0 \end{cases}$$

(iv) $\lim_{x\to 0} \dfrac{(1+x)^n - 1}{x} = n$

(v) $\lim_{x\to 0} \dfrac{(1+x)^m - 1}{(1+x)^n - 1} = \dfrac{m}{n}$

2. Trigonometric Limits

(i) $\lim_{x\to 0} \dfrac{\sin x}{x} = 1 = \lim_{x\to 0} \dfrac{x}{\sin x}$

(ii) $\lim_{x\to 0} \dfrac{\tan x}{x} = 1 = \lim_{x\to 0} \dfrac{x}{\tan x}$

(iii) $\lim_{x\to 0} \dfrac{\sin^{-1} x}{x} = 1 = \lim_{x\to 0} \dfrac{x}{\sin^{-1} x}$

(iv) $\lim_{x\to 0} \dfrac{\tan^{-1} x}{x} = 1 = \lim_{x\to 0} \dfrac{x}{\tan^{-1} x}$

(v) $\lim_{x\to 0} \dfrac{\sin x^\circ}{x} = \dfrac{\pi}{180}$

(vi) $\lim_{x\to\infty} \dfrac{\sin x}{x} = \lim_{x\to\infty} \dfrac{\cos x}{x} = 0$

(vii) $\lim_{x\to\infty} \sin x$ or $\lim_{x\to\infty} \cos x$ oscillates between -1 to 1.

(viii) $\lim_{x\to 0} \dfrac{\sin^P mx}{\sin^P nx} = \left(\dfrac{m}{n}\right)^P$

(ix) $\lim_{x\to 0} \dfrac{\tan^P mx}{\tan^P nx} = \left(\dfrac{m}{n}\right)^P$

(x) $\lim_{x\to 0} \dfrac{1 - \cos mx}{1 - \cos nx} = \dfrac{m^2}{n^2}$; $\lim_{x\to 0} \dfrac{\cos ax - \cos bx}{\cos cx - \cos dx} = \dfrac{a^2 - b^2}{c^2 - d^2}$

(xi) $\lim_{x\to 0} \dfrac{\cos mx - \cos nx}{x^2} = \dfrac{n^2 - m^2}{2}$

3. Logarithmic Limits

(i) $\lim_{x\to 0} \dfrac{\log_a(1+x)}{x} = \log_a e;\ a > 0, \ne 1$

(ii) In particular, $\lim_{x\to 0} \dfrac{\log_e(1+x)}{x} = 1$ and $\lim_{x\to 0} \dfrac{\log_e(1-x)}{x} = -1$

4. Expotential Limits

(i) $\lim_{x\to 0} \dfrac{a^x - 1}{x} = \log_e a, a > 0$

(ii) In particular, $\lim_{x\to 0} \dfrac{e^x - 1}{x} = 1$ and $\lim_{x\to 0} \dfrac{e^{\lambda x} - 1}{x} = \lambda$

(iii) $\lim_{x\to\infty} a^x = \begin{cases} 0, & 0 \le a < 1 \\ 1, & a = 1 \\ \infty, & a > 1 \\ \text{Does not exist}, & a < 0 \end{cases}$

5. 1^∞ Form Limits

(i) If $\lim_{x\to a} f(x) = \lim_{x\to a} g(x) = 0$, then $\lim_{x\to a}\{1 + f(x)\}^{1/g(x)} = e^{\lim_{x\to a} \frac{f(x)}{g(x)}}$

(ii) If $\lim_{x\to a} f(x) = 1$ and $\lim_{x\to a} g(x) = \infty$, then $\lim_{x\to a}\{f(x)\}^{g(x)} = e^{\lim_{x\to a}\{f(x) - 1\}g(x)}$

In General Cases

(i) $\lim_{x\to 0}(1 + x)^{1/x} = e$

(ii) $\lim_{x\to\infty}\left(1 + \dfrac{1}{x}\right)^x = e$

(iii) $\lim_{x\to 0}(1 + \lambda x)^{1/x} = e^\lambda$

(iv) $\lim_{x\to\infty}\left(1 + \dfrac{\lambda}{x}\right)^x = e^\lambda$

(v) $\lim_{x\to 0}(1 + ax)^{b/x} = \lim_{x\to\infty}\left(1 + \dfrac{a}{x}\right)^{bx} = e^{ab}$

Methods To Evaluate Limits

To find $\lim_{x\to a} f(x)$, we substitute $x = a$ in the function.

If $f(a)$ is finite, then $\lim_{x\to a} f(x) = f(a)$.

If $f(a)$ leads to one of the following form $\dfrac{0}{0}; \dfrac{\infty}{\infty}; \infty - \infty; 0 \times \infty; 1^\infty$, 0 and ∞^0 (called indeterminate forms), then $\lim_{x\to a} f(x)$ can be evaluated by using following methods

(i) **Factorization Method** This method is particularly used when on substituting the value of x, the expression take the form 0/0.

(ii) **Rationalization Method** This method is particularly used when either the numerator or the denominator or both involved square roots and on substituting the value of x, the expression take the form $\dfrac{0}{0}, \dfrac{\infty}{\infty}$.

NOTE To evaluate $x \to \infty$ type limits write the given expression in the form N/D and then divide both N and D by highest power of x occurring in both N and D to get a meaningful form.

L'Hospital's Rule

If $f(x)$ and $g(x)$ be two functions of x such that

(i) $\lim_{x\to a} f(x) = \lim_{x\to a} g(x) = 0$.

(ii) both are continuous at $x = a$.

(iii) both are differentiable at $x = a$.

(iv) $f'(x)$ and $g'(x)$ are continuous at the point $x = a$, then $\lim_{x\to a} \frac{f(x)}{g(x)} = \lim_{x\to a} \frac{f'(x)}{g'(x)}$ provided that $g(a) \neq 0$.

Above rule is also applicable, if $\lim_{x\to a} f(x) = \infty$ and $\lim_{x\to a} g(x) = \infty$.

If $f'(x), g'(x)$ satisfy all the conditions embeded in L'Hospital's rule, then we can repeat the application of this rule on

$$\frac{f'(x)}{g'(x)} \text{ to get } \lim_{x\to a} \frac{f'(x)}{g'(x)} = \lim_{x\to a} \frac{f''(x)}{g''(x)}.$$

Sometimes, following expansions are useful in evaluating limits.

- $\log(1+x) = x - \frac{x^2}{2} + \frac{x^3}{3} - \frac{x^4}{4} + \frac{x^5}{5} + \ldots + (-1 < x \le 1)$
- $\log(1-x) = -x - \frac{x^2}{2} - \frac{x^3}{3} - \frac{x^4}{4} - \frac{x^5}{5} - \ldots (-1 < x < 1)$
- $e^x = 1 + \frac{x}{1!} + \frac{x^2}{2!} + \frac{x^3}{3!} + \frac{x^4}{4!} + \ldots$
- $a^x = 1 + x(\log_e a) + \frac{x^2}{2!}(\log_e a)^2 + \ldots$
- $\sin x = x - \frac{x^3}{3!} + \frac{x^5}{5!} - \frac{x^7}{7!} + \ldots$
- $\cos x = 1 - \frac{x^2}{2!} + \frac{x^4}{4!} - \frac{x^6}{6!} - \ldots$

Continuity

If the graph of a function has no break (or gap), then it is **continuous**. A function which is not continuous is called a **discontinuous** function. e.g. x^2 and e^x are continuous while $\frac{1}{x}$ and $[x]$, where $[\cdot]$ denotes the greatest integer function, are discontinuous.

Continuity of a Function at a Point

A function $f(x)$ is said to be continuous at a point $x = a$ of its domain if and only if it satisfies the following conditions

(i) $f(a)$ exists, where ('a' lies in the domain of f

(ii) $\lim_{x\to a} f(x)$ exist, i.e. $\lim_{x\to a^-} f(x) = \lim_{x\to a^+} f(x)$

or LHL = RHL

(iii) $\lim_{x\to a} f(x) = f(a)$, $f(x)$ is said to be

left continuous at $x = a$, if $\lim_{x\to a^-} f(x) = f(a)$

right continuous at $x = a$, if $\lim_{x\to a^+} f(x) = f(a)$

Continuity of a Function in an Interval

A function $f(x)$ is said to be continuous in (a,b) if it is continuous at every point of the interval (a,b). A function $f(x)$ is said to be continuous in $[a,b]$, if $f(x)$ is continuous in (a,b). Also, in addition $f(x)$ is continuous at $x = a$ from right and continuous at $x = b$ from left.

Results on Continuous Functions

(i) Sum, difference product and quotient of two continuous functions are always a continuous function. However, $r(x) = \frac{f(x)}{g(x)}$ is continuous at $x = a$ only if $g(a) \neq 0$.

(ii) Every polynomial is continuous at each point of real line.

(iii) Every rational function is continuous at each point where its denominator is different from zero.

(iv) Logarithmic functions, exponential functions, trigonometric functions, inverse circular functions and modulus function are continuous in their domain.

(v) $[x]$ is discontinuous when x is an integer.

(vi) If $g(x)$ is continuous at $x = a$ and f is continuous at $g(a)$, then fog is continuous at $x = a$.

(vii) $f(x)$ is a continuous function defined on $[a, b]$ such that $f(a)$ and $f(b)$ are of opposite signs, then there is atleast one value of x for which $f(x)$ vanishes, i.e. $f(a) > 0$, $f(b) < 0 \Rightarrow \exists\, c \in (a, b)$ such that $f(c) = 0$.

Differentiability

The function $f(x)$ is differentiable at a point P iff there exists a unique tangent at point P. In other words, $f(x)$ is differentiable at a point P iff the curve does not have P as a corner point, i.e. the function is not differentiable at those points where graph of the function has holes or sharp edges. Let us consider the function $f(x) = |x-1|$. It is not differentiable at $x = 1$. Since, $f(x)$ has sharp edge at $x = 1$.

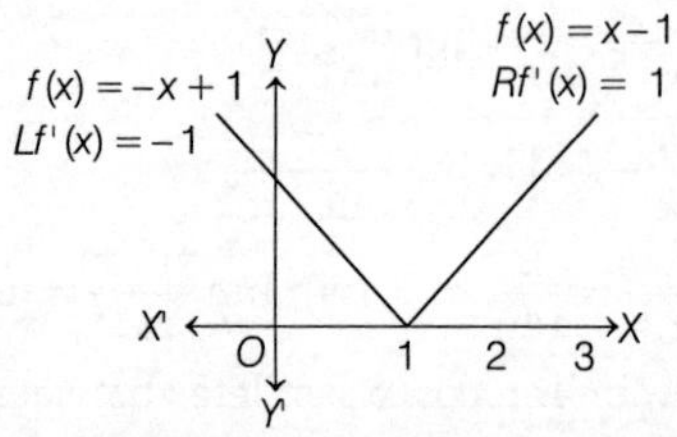

(graph of $f(x)$ describe differentiability)

Differentiability of a Function at a Point

A function f is said to be differentiable at $x = c$, if **left hand** and **right hand** derivatives at c exist and are equal.

- **Right hand derivative** of $f(x)$ at $x = a$ denoted by $f'(a+0)$ or $f'(a^+)$ is $\lim_{h\to 0} \frac{f(a+h)-f(a)}{h}$.

- **Left hand derivative** of $f(x)$ at $x = a$ denoted by $f'(a-0)$ or $f'(a^-)$ is $\lim_{h \to 0} \frac{f(a-h)-f(a)}{-h}$.
- Thus, f is said to be **differentiable** at $x = a$, if $f'(a+0) = f'(a-0) =$ finite.
- The common limit is called the **derivative** of $f(x)$ at $x = a$ denoted by $f'(a)$. i.e. $f'(a) = \lim_{x \to a} \frac{f(x)-f(a)}{x-a}$.

Results on Differentiability

(i) Every polynomial, constant and exponential function is differentiable at each $x \in R$.

(ii) The logarithmic, trigonometric and inverse trigonometric function are differentiable in their domain.

(iii) The sum, difference, product and quotient of two differentiable functions is differentiable.

(iv) Every differentiable function is continuous but converse may or may not be true.

DAY PRACTICE SESSION 1

FOUNDATION QUESTIONS EXERCISE

1 $\lim_{x \to 0} |x|^{[\cos x]}$, where [.] is the greatest integer function, is

(a) 1 (b) 0 (c) Does not exist (d) None of these

2 Let $f : R \to [0, \infty)$ be such that $\lim_{x \to 5} f(x)$ exists and $\lim_{x \to 5} \frac{[f(x)]^2 - 9}{\sqrt{|x-5|}} = 0$. Then, $\lim_{x \to 5} f(x)$ is equal to

(a) 3 (b) 0 (c) 1 (d) 2

3 If $\lim_{x \to \infty} \frac{2}{x}\left[\frac{x}{5}\right] = \frac{m}{n}$ (where [·] denotes greatest integer function), then $m + n$ (where m, n are relatively prime) is

(a) 2 (b) 7 (c) 5 (d) 6

4 The value of the constant α and β such that $\lim_{x \to \infty} \left(\frac{x^2+1}{x+1} - \alpha x - \beta\right) = 0$ are respectively

(a) (1, 1) (b) (−1, 1) (c) (1, −1) (d) (0, 1)

5 $\lim_{n \to \infty} \frac{3 \cdot 2^{n+1} - 4 \cdot 5^{n+1}}{5 \cdot 2^n + 7 \cdot 5^n}$ is equal to

(a) 0 (b) $\frac{3}{5}$ (c) $-\frac{4}{7}$ (d) $-\frac{20}{7}$

6 $\lim_{x \to \infty} \left[\sqrt{x + \sqrt{x + \sqrt{x}}} - \sqrt{x}\right]$ is equal to

(a) 0 (b) $\frac{1}{2}$ (c) log 2 (d) e^4

7 The value of $\lim_{n \to \infty} \cos\left(\frac{x}{2}\right) \cos\left(\frac{x}{4}\right) \cos\left(\frac{x}{8}\right) \ldots \cos\left(\frac{x}{2^n}\right)$ is

(a) 1 (b) $\frac{\sin x}{x}$ (c) $\frac{x}{\sin x}$ (d) None of these

8 If f is periodic with period T and $f(x) > 0 \forall x \in R$, then $\lim_{n \to \infty} n\left(\frac{f(x+T) + 2f(x+2T) + \ldots + nf(x+nT)}{f(x+T) + 4f(x+4T) + \ldots + n^2 f(x+n^2T)}\right)$ is equal to

(a) 2 (b) $\frac{2}{3}$ (c) $\frac{3}{2}$ (d) None of these

9 $\lim_{x \to 0} \frac{(1-\cos 2x)(3+\cos x)}{x \tan 4x}$ is equal to → JEE Mains 2015, 13

(a) 4 (b) 3 (c) 2 (d) $\frac{1}{2}$

10 If $\lim_{x \to 0} \frac{[(a-n)nx - \tan x] \sin nx}{x^2} = 0$, where n is non-zero real number, then a is equal to

(a) 0 (b) $\frac{n+1}{n}$ (c) n (d) $n + \frac{1}{n}$

11 $\lim_{x \to 0} \frac{\sin(\pi \cos^2 x)}{x^2}$ is equal to → JEE Mains 2014

(a) $\frac{\pi}{2}$ (b) 1 (c) $-\pi$ (d) π

12 If $f(x) = \begin{vmatrix} \sin x & \cos x & \tan x \\ x^3 & x^2 & x \\ 2x & 1 & 1 \end{vmatrix}$, then $\lim_{x \to 0} \frac{f(x)}{x^2}$ is equal to

(a) 3 (b) −1 (c) 0 (d) 1

13 The limit of the following is

$$\lim_{x \to 3} \frac{\sqrt{1 - \cos(x^2 - 10x + 21)}}{(x-3)}$$

(a) $-(2)^{3/2}$ (b) $(2)^{1/2}$ (c) $(2)^{-3/2}$ (d) 3

14 The value of $\lim_{x \to 0} \frac{1}{x}\left[\tan^{-1}\left(\frac{x+1}{2x+1}\right) - \frac{\pi}{4}\right]$ is → JEE Mains 2013

(a) 1 (b) $-\frac{1}{2}$ (c) 2 (d) 0

15 $\lim_{x \to \pi/2} \frac{\cot x - \cos x}{(\pi - 2x)^3}$ equals → JEE Mains 2017

(a) $\frac{1}{24}$ (b) $\frac{1}{16}$ (c) $\frac{1}{8}$ (d) $\frac{1}{4}$

16 If m and n are positive integers, then $\lim_{x\to 0}\frac{(\cos x)^{1/m}-(\cos x)^{1/n}}{x^2}$ equals to

(a) $m-n$ (b) $\frac{1}{n}-\frac{1}{m}$ (c) $\frac{m-n}{2mn}$ (d) None of these

17 $\lim_{x\to\infty}\frac{\cot^{-1}(\sqrt{x+1}-\sqrt{x})}{\sec^{-1}\left\{\left(\frac{2x+1}{x-1}\right)^x\right\}}$ is equal to

(a) 1 (b) 0 (c) $\frac{\pi}{2}$ (d) Does not exists

18 Let $p=\lim_{x\to 0}\frac{\ln(\cos 2x)}{3x^2}$, $q=\lim_{x\to 0}\frac{\sin^2 2x}{x(1-e^x)}$ and $r=\lim_{x\to 1}\frac{\sqrt{x}-x}{\ln x}$. Then p, q, r satisfy

(a) $p<q<r$ (b) $q<r<p$ (c) $p<r<q$ (d) $q<p<r$

19 Let $p=\lim_{x\to 0^+}(1+\tan^2\sqrt{x})^{1/2x}$, then $\log p$ is equal to → JEE Mains 2016

(a) 2 (b) 1 (c) $\frac{1}{2}$ (d) $\frac{1}{4}$

20 The value of $\lim_{x\to\infty}\left(\frac{3x-4}{3x+2}\right)^{\frac{x+1}{3}}$ is

(a) $e^{-1/3}$ (b) $e^{-2/3}$ (c) e^{-1} (d) e^{-2}

21 If $\lim_{x\to\infty}\left(1+\frac{a}{x}+\frac{b}{x^2}\right)^{2x}=e^2$, then the values of a and b are

(a) $a\in R, b\in R$ (b) $a=1, b\in R$
(c) $a\in R, b=2$ (d) $a=1, b=2$

22 If $f'(2)=6$ and $f'(1)=4$, then $\lim_{h\to 0}\frac{f(2h+2+h^2)-f(2)}{f(h-h^2+1)-f(1)}$ is equal to

(a) 3 (b) $-3/2$ (c) $3/2$ (d) Does not exist

23 Let $f(a)=g(a)=k$ and their nth derivatives $f^n(a), g^n(a)$ exist and are not equal for some n. Further, if
$$\lim_{x\to a}\frac{f(a)g(x)-f(a)-g(a)f(x)+g(a)}{g(x)-f(x)}=4,$$
then the value of k is

(a) 4 (b) 2 (c) 1 (d) 0

24 If $f:R\to R$ be such that $f(1)=3$ and $f'(1)=6$. Then, $\lim_{x\to 0}\left(\frac{f(1+x)}{f(1)}\right)^{1/x}$ is equal to

(a) 1 (b) $e^{1/2}$ (c) e^2 (d) e^3

25 Let $f(x)=\begin{cases}\sin^2 x, & x \text{ is rational}\\ -\sin^2 x, & x \text{ is irrational}\end{cases}$, then set of points, where $f(x)$ is continuous, is

(a) $\left\{(2n+1)\frac{\pi}{2}, n\in I\right\}$ (b) a null set
(c) $\{n\pi, n\in I\}$ (d) set of all rational numbers

26 Let $f:[a,b]\to R$ be any function which is such that $f(x)$ is rational for irrational x and $f(x)$ is irrational for rational x. Then, in $[a,b]$

(a) f is discontinuous everywhere
(b) f is continuous only at $x=0$
(c) f is continuous for all irrational x and discontinuous for all rational x
(d) f is continuous for all rational x and discontinuous for all irrational x

27 Let $f(x)=1+|x-2|$ and $g(x)=1-|x|$, then the set of all points, where fog is discontinuous, is → JEE Mains 2013

(a) {0, 2} (b) {0, 1, 2} (c) {0} (d) an empty set

28 If $f(x)=\begin{cases}-1, & x<0\\ 0, & x=0\\ 1, & x>0\end{cases}$ and $g(x)=\sin x+\cos x$, then the points of discontinuity of $f\{g(x)\}$ in $(0,2\pi)$ is

(a) $\left\{\frac{\pi}{2},\frac{3\pi}{4}\right\}$ (b) $\left\{\frac{3\pi}{4},\frac{7\pi}{4}\right\}$ (c) $\left\{\frac{2\pi}{3},\frac{5\pi}{3}\right\}$ (d) $\left\{\frac{5\pi}{4},\frac{7\pi}{3}\right\}$

29 If $f(x)$ is differentiable at $x=1$ and $\lim_{h\to 0}\frac{1}{h}f(1+h)=5$, then $f'(1)$ is equal to

(a) 6 (b) 5 (c) 4 (d) 3

30 If $f(x)=3x^{10}-7x^8+5x^6-21x^3+3x^2-7$, then the value of $\lim_{h\to 0}\frac{f(1-h)-f(1)}{h^3+3h}$ is

(a) $\frac{53}{3}$ (b) $\frac{22}{3}$ (c) 13 (d) $\frac{22}{13}$

31 Let $f(2)=4$ and $f'(2)=4$. Then, $\lim_{x\to 2}\frac{xf(2)-2f(x)}{x-2}$ is given by

(a) 2 (b) -2 (c) -4 (d) 3

32 If f is a real-valued differentiable function satisfying $|f(x)-f(y)|\le(x-y)^2; x, y\in R$ and $f(0)=0$, then $f(1)$ is equal to

(a) 1 (b) 2 (c) 0 (d) -1

33 If $\lim_{x\to 0}\frac{\log(a+x)-\log a}{x}+k\lim_{x\to 0}\frac{\log x-1}{x-e}=1$, then

(a) $k=e\left(1-\frac{1}{a}\right)$ (b) $k=e(1+a)$
(c) $k=e(2-a)$ (d) Equality is not possible

34 The left hand derivative of $f(x)=[x]\sin(\pi x)$ at $x=k$, k is an integer, is

(a) $(-1)^k(k-1)\pi$ (b) $(-1)^{k-1}(k-1)\pi$
(c) $(-1)^k k\pi$ (d) $(-1)^{k-1}k\pi$

35 If $f(x)=\begin{cases}e^x, & x\le 0\\ |1-x|, & x>0\end{cases}$, then

(a) $f(x)$ is differentiable at $x=0$
(b) $f(x)$ is continuous at $x=0, 1$
(c) $f(x)$ is differentiable at $x=1$
(d) None of the above

36 If $f(x)=\begin{cases}xe^{-\left(\frac{1}{|x|}+\frac{1}{x}\right)}, & x\ne 0\\ 0, & x=0\end{cases}$, then $f(x)$ is

(a) continuous as well as differentiable for all x
(b) continuous for all x but not differentiable at $x = 0$
(c) neither differentiable nor continuous at $x = 0$
(d) discontinuous everywhere

37 The set of points, where $f(x) = \dfrac{x}{1+|x|}$ is differentiable, is

(a) $(-\infty, -1) \cup (-1, \infty)$ (b) $(-\infty, \infty)$
(c) $(0, \infty)$ (d) $(-\infty, 0) \cup (0, \infty)$

38 Let $f(x) = \cos x$ and
$g(x) = \begin{cases} \min\{f(t): 0 \le t \le x\}, & x \in [0, \pi] \\ (\sin x) - 1, & x > \pi \end{cases}$ then

(a) $g(x)$ is discontinuous at $x = \pi$
(b) $g(x)$ is continuous for $x \in [0, \infty]$
(c) $g(x)$ is differentiable at $x = \pi$
(d) $g(x)$ is differentiable for $x \in [0, \infty]$

39 If the function $g(x) = \begin{cases} k\sqrt{x+1}, & 0 \le x \le 3 \\ mx + 2, & 3 < x \le 5 \end{cases}$ is differentiable, then the value of $k + m$ is

(a) 2 (b) $\dfrac{16}{5}$ (c) $\dfrac{10}{5}$ (d) 4

40 If $f(x) = \begin{cases} \sin(\cos^{-1} x) + \cos(\sin^{-1} x), & x \le 0 \\ \sin(\cos^{-1} x) - \cos(\sin^{-1} x), & x > 0 \end{cases}$

then at $x = 0$

(a) $f(x)$ is continuous and differentiable
(b) $f(x)$ is continuous but not differentiable
(c) f is not continuous but differentiable
(d) f is neither continuous nor differentiable

41 If $f(x) = [\sin x] + [\cos x]$, $x \in [0, 2\pi]$, where $[.]$ denotes the greatest integer function. Then, the total number of points, where $f(x)$ is non-differentiable, is

(a) 2 (b) 3 (c) 5 (d) 4

42 If $f(x) = |\sin x|$, then

(a) f is everywhere differentiable
(b) f is everywhere continuous but not differentiable at $x = n\pi, n \in Z$
(c) f is everywhere continuous but not differentiable at $x = (2n+1)\dfrac{\pi}{2}, n \in Z$
(d) None of the above

43 **Statement I** $f(x) = |\log x|$ is differentiable at $x = 1$.
Statement II Both $\log x$ and $-\log x$ are differentiable at $x = 1$.

(a) Statement I is false, Statement II is true
(b) Statemnt I is true, Statement II is true; Statement II is a correct explanation of Statement I
(c) Statement I is true, Statement II is true; Statement II is not a correct explanation of Statement I
(d) Statement I is true, Statement II is false

DAY PRACTICE SESSION 2

PROGRESSIVE QUESTIONS EXERCISE

1 For each $t \in R$, let $[t]$ be the greatest integer less than or equal to t. Then, $\lim\limits_{x \to 0^+} x\left(\left[\dfrac{1}{x}\right] + \left[\dfrac{2}{x}\right] + \dots + \left[\dfrac{15}{x}\right]\right)$

→ JEE Mains 2018

(a) is equal to 0 (b) is equal to 15
(c) is equal to 120 (d) does not exist (in R)

2 $\lim\limits_{x \to \infty} \left(\dfrac{x^2 + 5x + 3}{x^2 + x + 2}\right)^x$ is equal to

(a) e^4 (b) e^2 (c) e^3 (d) e

3 If α and β are the distinct roots of $ax^2 + bx + c = 0$, then $\lim\limits_{x \to \alpha} \dfrac{1 - \cos(ax^2 + bx + c)}{(x - \alpha)^2}$ is equal to

(a) $\dfrac{1}{2}(\alpha - \beta)^2$ (b) $-\dfrac{a^2}{2}(\alpha - \beta)^2$ (c) 0 (d) $\dfrac{a^2}{2}(\alpha - \beta)^2$

4 $\lim\limits_{n \to \infty} \sin[\pi\sqrt{n^2 + 1}]$ is equal to

(a) ∞ (b) 0
(c) Does not exist (d) None of these

5 If $x > 0$ and g is a bounded function, then $\lim\limits_{n \to \infty} \dfrac{f(x) \cdot e^{nx} + g(x)}{e^{nx} + 1}$ is equal to

(a) 0 (b) $f(x)$
(c) $g(x)$ (d) None of these

6 The value of $\lim\limits_{x \to 0} \left(\dfrac{a^x + b^x + c^x}{3}\right)^{2/x}$ (where $a, b, c > 0$) is

(a) $(abc)^3$ (b) abc
(c) $(abc)^{1/3}$ (d) None of these

7 If $f(x) = \cot^{-1}\left(\dfrac{3x - x^3}{1 - 3x^2}\right)$ and $g(x) = \cos^{-1}\left(\dfrac{1 - x^2}{1 + x^2}\right)$, then $\lim\limits_{x \to a} \dfrac{f(x) - f(a)}{g(x) - g(a)}$, where $0 < a < \dfrac{1}{2}$, is equal to

(a) $\dfrac{3}{2(1 + a^2)}$ (b) $\dfrac{3}{2(1 + x^2)}$ (c) $\dfrac{3}{2}$ (d) $-\dfrac{3}{2}$

8 If $f : R \to R$ is a function defined by $f(x) = [x] \cos\left(\dfrac{2x - 1}{2}\right)\pi$, where $[x]$ denotes the greatest integer function, then f is

(a) continuous for every real x
(b) discontinuous only at $x = 0$
(c) discontinuous only at non-zero integral values of x
(d) continuous only at $x = 0$

9 Let f be a composite function of x defined by $f(u)=\dfrac{1}{u^2+u-2}, u(x)=\dfrac{1}{x-1}$.

Then, the number of points x, where f is discontinuous, is → JEE Mains 2013

(a) 4 (b) 3 (c) 2 (d) 1

10 If $f:R\to R$ be a positive increasing function with $\lim\limits_{x\to\infty}\dfrac{f(3x)}{f(x)}=1$. Then, $\lim\limits_{x\to\infty}\dfrac{f(2x)}{f(x)}$ is equal to

(a) 1 (b) 2/3 (c) 3/2 (d) 3

11 Let $f(x)=\max\{\tan x, \sin x, \cos x\}$, where $x\in\left[-\dfrac{\pi}{2},\dfrac{3\pi}{2}\right]$.

Then, the number of points of non-differentiability is

(a) 1 (b) 3 (c) 0 (d) 2

12 Let $S=\{t\in R: f(x)=|x-\pi|\cdot(e^{|x|}-1)\}\sin|x|$ is not differentiable at $t\}$. Then, the set S is equal to

(a) ϕ (an empty set) (b) $\{0\}$ → JEE Mains 2018

(c) $\{\pi\}$ (d) $\{0,\pi\}$

13 If $f(x)=\lim\limits_{n\to\infty}\dfrac{x^n+\left(\dfrac{\pi}{3}\right)^n}{x^{n-1}+\left(\dfrac{\pi}{3}\right)^{n-1}}$, where n is an even integer,

Then which of the following is incorrect?

(a) If $f:\left[\dfrac{\pi}{3},\infty\right)\to\left[\dfrac{\pi}{3},\infty\right)$, then f is both one-one and onto

(b) $f(x)=f(-x)$ has infinitely many solutions

(c) $f(x)$ is one-one for all $x\in R$ (d) None of these

14 Let $f(x)=\lim\limits_{n\to\infty}\sum\limits_{k=0}^{n-1}\dfrac{x}{(kx+1)\{(k+1)x+1\}}$. Then,

(a) f is continuous but not differentiable at $x=0$

(b) f is differentiable at $x=0$

(c) f is neither continuous nor differentiable at $x=0$

(d) None of the above

15 If $x_1, x_2, x_3, \dots, x_4$ are the roots of $x^n+ax+b=0$, then the value of $(x_1-x_2)(x_1-x_3)\dots(x_1-x_n)$ is equal to

(a) nx_1+b (b) $nx_1^{n-1}+a$

(c) nx_1^{n-1} (d) nx_1^n

16 If $^*\lim\limits_{x\to\infty} x\ln\left(\begin{vmatrix}\dfrac{1}{x} & -b & c\\ 0 & \dfrac{1}{x} & -1\\ 1 & 0 & a/x\end{vmatrix}\right)=-4$, where a, b, c are real numbers, then

(a) $a=1, b\in R, c=-1$ (b) $a\in R, b=2, c=4$

(c) $a=1, b=1, c\in R$ (d) $a\in R, b=1, c=4$

17 If $\lim\limits_{x\to 0}\left[1+x+\dfrac{f(x)}{x}\right]^{1/x}=e^3$, then $\lim\limits_{x\to 0}\left[1+\dfrac{f(x)}{x}\right]^{1/x}$ is equal to

(a) e (b) e^2 (c) e^3 (d) None of these

18 The function $f(x)$ is discontinuous only at $x=0$ such that $f^2(x)=1\ \forall x\in R$. The total number of such function is

(a) 2 (b) 3

(c) 6 (d) None of these

19 Let $f(x)=x|x|$ and $g(x)=\sin x$

Statement I gof is differentiable at $x=0$ and its derivative is continuous at that point.

Statement II gof is twice differentiable at $x=0$.

(a) Statement I is false, Statement II is true

(b) Statement I is true, Statement II is true; Statement II is a correct explanation of Statement I

(c) Statement I is true, Statement II is true; Statement II is not a correct explanation of Statement I

(d) Statement I is true, Statement II is false

20 **Statement I** The function $f(x)=(3x-1)|4x^2-12x+5|\cos\pi x$ is differentiable at $x=\dfrac{1}{2}$ and $\dfrac{5}{2}$.

Statement II $\cos(2n+1)\dfrac{\pi}{2}=0, \forall n\in I$.

(a) Statement I is true, Statement II is true; Statement II is a correct explanation for Statement I.

(b) Statement I is true, Statement II is true; Statement II is not a correct explanation for Statement I.

(c) Statement I is true; Statement II is false.

(d) Statement I is false; Statement II is true.

21 Define $f(x)$ as the product of two real functions $f_1(x)=x$, $x\in IR$

and $f_2(x)=\begin{cases}\sin\dfrac{1}{x}, & \text{if } x\neq 0\\ 0, & \text{if } x=0\end{cases}$ as follows

$$f(x)=\begin{cases} f_1(x)\cdot f_2(x), & \text{if } x\neq 0\\ 0, & \text{if } x=0\end{cases}$$

Statement I $f(x)$ is continuous on IR.

Statement II $f_1(x)$ and $f_2(x)$ are continuous on IR.

(a) Statement I is false, Statement II is true

(b) Statement I is true, Statement II is true; Statement II is correct explanation of Statement I.

(c) Statement I is true, Statement II is true; Statement II is not a correct explanation of Statement I

(d) Statement I is true, Statement II is false

22 Consider the function $f(x)=|x-2|+|x-5|, x\in R$.

Statement I $f'(4)=0$

Statement II f is continuous in $[2,5]$ and differentiable in $(2,5)$ and $f(2)=f(5)$.

(a) Statement I is true, Statement II is true; Statement II is a correct explanation for Statement I

(b) Statement I is true, Statement II is true; Statement II is not correct explanation for Statement I

(c) Statement I is true; Statement II is false

(d) Statement I is false; Statement II is true

ANSWERS

SESSION 1

1 (a)	**2** (a)	**3** (b)	**4** (c)	**5** (d)	**6** (b)	**7** (b)	**8** (c)	**9** (c)	**10** (d)
11 (d)	**12** (d)	**13** (a)	**14** (b)	**15** (b)	**16** (c)	**17** (a)	**18** (d)	**19** (c)	**20** (b)
21 (b)	**22** (a)	**23** (a)	**24** (c)	**25** (c)	**26** (a)	**27** (d)	**28** (b)	**29** (b)	**30** (a)
31 (c)	**32** (c)	**33** (a)	**34** (a)	**35** (b)	**36** (b)	**37** (b)	**38** (b)	**39** (a)	**40** (d)
41 (c)	**42** (b)	**43** (a)							

SESSION 2

1 (c)	**2** (a)	**3** (d)	**4** (b)	**5** (b)	**6** (d)	**7** (d)	**8** (a)	**9** (b)	**10** (a)
11 (b)	**12** (d)	**13** (c)	**14** (c)	**15** (b)	**16** (d)	**17** (b)	**18** (c)	**19** (b)	**20** (a)
21 (d)	**22** (b)								

Hints and Explanations

SESSION 1

1 RHL $= \lim_{x \to 0^+} (x)^0 = 1$

LHL $= \lim_{x \to 0^-} (-x)^0 = \lim_{x \to 0^-} 1 = 1$

$\because$ RHL = LHL

$\therefore \lim_{x \to 0} |x|^{[\cos x]} = 1$

2 Given, $\lim_{x \to 5} f(x)$ exists and

$$\lim_{x \to 5} \frac{[f(x)]^2 - 9}{\sqrt{|x - 5|}} = 0$$

$$\Rightarrow \lim_{x \to 5} [f(x)]^2 - 9 = 0$$

$$\Rightarrow \left(\lim_{x \to 5} [f(x)]\right)^2 = 9$$

$$\therefore \lim_{x \to 5} f(x) = 3, -3$$

But $f : R \to [0, \infty)$

$\therefore$ Range of $f(x) \geq 0$

$\Rightarrow \lim_{x \to 5} f(x) = 3$

3 Clearly, $\lim_{x \to \infty} \frac{2}{x}\left[\frac{x}{5}\right] = \lim_{x \to \infty} \frac{2}{x}\left(\frac{x}{5} - \left\{\frac{x}{5}\right\}\right)$

$$= \frac{2}{5} - 0 = \frac{2}{5}$$

$\therefore m + n = 7$

4 $\lim_{x \to \infty}\left(\frac{x^2 + 1}{x + 1} - \alpha x - \beta\right) = 0$

$$\Rightarrow \lim_{x \to \infty} \frac{x^2(1 - \alpha) - x(\alpha + \beta) + 1 - \beta}{x + 1} = 0$$

$\therefore 1 - \alpha = 0, \alpha + \beta = 0$

$\Rightarrow \alpha = 1, \beta = -1$

5 Clearly,

$$\lim_{n \to \infty} \frac{3 \cdot 2^{n+1} - 4 \cdot 5^{n+1}}{5 \cdot 2^n + 7 \cdot 5^n} = \lim_{n \to \infty} \frac{6 \cdot 2^n - 20 \cdot 5^n}{5 \cdot 2^n + 7 \cdot 5^n}$$

$$= \lim_{n \to \infty} \frac{6 \cdot \left(\frac{2}{5}\right)^n - 20}{5 \cdot \left(\frac{2}{5}\right)^n + 7} = \frac{0 - 20}{0 + 7} = -\frac{20}{7}$$

6 $\lim_{x \to \infty}\left[\sqrt{x + \sqrt{x + \sqrt{x}}} - \sqrt{x}\right]$

$$= \lim_{x \to \infty} \frac{x + \sqrt{x + \sqrt{x}} - x}{\sqrt{x + \sqrt{x + \sqrt{x}}} + \sqrt{x}}$$

$$= \lim_{x \to \infty} \frac{\sqrt{x + \sqrt{x}}}{\sqrt{x + \sqrt{x + \sqrt{x}}} + \sqrt{x}}$$

$$= \lim_{x \to \infty} \frac{\sqrt{1 + x^{-1/2}}}{\sqrt{1 + \sqrt{x^{-1} + x^{-3/2}}} + 1} = \frac{1}{2}$$

7 We know that,

$$\cos A \cdot \cos 2A \cdot \cos 4A \ldots \cos 2^{n-1} A = \frac{\sin 2^n A}{2^n \sin A}$$

Take $A = \frac{x}{2^n}$,

then $\cos\left(\frac{x}{2^n}\right) \cdot \cos\left(\frac{x}{2^{n-1}}\right) \ldots \cos\left(\frac{x}{4}\right) \cos\left(\frac{x}{2}\right)$

$$= \frac{\sin x}{2^n \sin\left(\frac{x}{2^n}\right)}$$

$$\therefore \lim_{n \to \infty} \cos\left(\frac{x}{2}\right) \cos\left(\frac{x}{4}\right) \ldots \cos\left(\frac{x}{2^{n-1}}\right) \cdot \cos\left(\frac{x}{2^n}\right)$$

$$= \lim_{n \to \infty} \frac{\sin x}{2^n \sin\left(\frac{x}{2^n}\right)}$$

$$= \lim_{n \to \infty} \frac{\sin x}{x} \cdot \frac{x / 2^n}{\sin(x / 2^n)}$$

$$= \frac{\sin x}{x}$$

8 Clearly, $f(x + T) = f(x + 2T) = \ldots$

$$= f(x + nT) = f(x)$$

$$\therefore \lim_{n \to \infty} n \left(\frac{f(x + T) + 2f(x + 2T) + \ldots + nf(x + nT)}{f(x + T) + 4f(x + 4T) + \ldots + n^2 f(x + x^2 T)}\right)$$

$$= \lim_{n \to \infty} \frac{nf(x)(1 + 2 + 3 + \ldots + n)}{f(x)(1 + 2^2 + 3^2 + \ldots + n^2)}$$

$$= \lim_{n \to \infty} \frac{n\left(\frac{n(n + 1)}{2}\right)}{\frac{n(n + 1)(2n + 1)}{6}} = \frac{3}{2}$$

9 We have, $\lim_{x \to 0} \frac{(1 - \cos 2x)(3 + \cos x)}{x \tan 4x}$

$$= \lim_{x \to 0} \frac{2\sin^2 x(3 + \cos x)}{x \times \frac{\tan 4x}{4x} \times 4x}$$

$$= \lim_{x \to 0} \frac{2\sin^2 x}{x^2} \times \lim_{x \to 0} \frac{(3 + \cos x)}{4} \times \frac{1}{\lim_{x \to 0} \frac{\tan 4x}{4x}}$$

$$= 2 \times \frac{4}{4} \times 1 = 2$$

$$\left[\because \lim_{x \to 2} \frac{\sin \theta}{\theta} = 1 \text{ and } \lim_{x \to 0} \frac{\tan \theta}{\theta} = 1\right]$$

10 Since,

$$\lim_{x \to 0}\left[(a - n)n - \frac{\tan x}{x}\right] \cdot \frac{\sin nx}{x} = 0$$

$\Rightarrow \quad [(a-n)n-1]\,n = 0$

$\Rightarrow \quad (a-n)n = 1$

$\therefore \quad a = n + \frac{1}{n}$

11 $\lim_{x\to 0}\frac{\sin(\pi\cos^2 x)}{x^2} = \lim_{x\to 0}\frac{\sin(\pi-\pi\sin^2 x)}{x^2}$

$= \lim_{x\to 0}\frac{\sin(\pi\sin^2 x)}{x^2}$ $\quad [\because \sin(\pi-\theta) = \sin\theta]$

$= \lim_{x\to 0}\frac{\sin(\pi\sin^2 x)}{\pi\sin^2 x}\times(\pi)\left(\frac{\sin^2 x}{x^2}\right)$

$= \pi \quad \left[\because \lim_{\theta\to 0}\frac{\sin\theta}{\theta} = 1\right]$

12 $\because f(x) = x(x-1)\sin x - (x^3-2x^2)\cos x - x^3\tan x$

$= x^2\sin x - x^3\cos x - x^3\tan x + 2x^2\cos x - x\sin x$

$\therefore \lim_{x\to 0}\frac{f(x)}{x^2} = \lim_{x\to 0}\left(\sin x - x\cos x - x\tan x + 2\cos x - \frac{\sin x}{x}\right)$

$= 2 - 1 = 1$

13 $\lim_{x\to 3}\frac{\sqrt{1-\cos(x^2-10x+21)}}{(x-3)}$

$= \lim_{x\to 3}\frac{\sqrt{2}\sin\frac{(x-3)(x-7)}{2}}{(x-3)}$

$= \lim_{x\to 3}\frac{\sqrt{2}\sin\frac{(x-3)(x-7)}{2}}{\frac{(x-3)\cdot(x-7)}{2}}\cdot\frac{(x-7)}{2}$

$= \lim_{x\to 3}(x-7)\cdot\lim_{x\to 3}\frac{\sin\frac{(x-3)(x-7)}{2}}{\frac{(x-3)(x-7)}{2}}\times\frac{1}{\sqrt{2}}$

$= -(2)^{3/2}$

14 $\lim_{x\to 0}\frac{1}{x}\left\{\tan^{-1}\left(\frac{x+1}{2x+1}\right)-\frac{\pi}{4}\right\}$

$= \lim_{x\to 0}\left\{\tan^{-1}\left(\frac{x+1}{2x+1}\right)-\tan^{-1}(1)\right\}$

$= \lim_{x\to 0}\frac{1}{x}\cdot\tan^{-1}\left\{\frac{\frac{x+1}{2x+1}-1}{1+\frac{x+1}{2x+1}}\right\}$

$= \lim_{x\to 0}\frac{1}{x}\cdot\tan^{-1}\left(\frac{x+1-2x-1}{2x+1+x+1}\right)$

$= \lim_{x\to 0}\frac{1}{x}\cdot\tan^{-1}\left(\frac{-x}{4x+2}\right)\left[\text{form }\frac{0}{0}\right]$

[using L' Hospital rule]

$= \lim_{x\to 0}\frac{1}{1+\frac{x^2}{(4x+2)^2}}\times\left[-\left(\frac{4x+2-4x}{(4x+2)^2}\right)\right]$

$= \lim_{x\to 0} -\frac{(+2)}{x^2+(4x+2)^2}$

$= -\frac{(+2)}{0+(0+2)^2} = \frac{-2}{4} = \frac{-1}{2}$

15 $\lim_{x\to\pi/2}\frac{\cot x-\cos x}{(\pi-2x)^3}$

$= \lim_{x\to\pi/2}\frac{1}{8}\cdot\frac{\cos x(1-\sin x)}{\sin x\left(\frac{\pi}{2}-x\right)^3}$

$= \lim_{h\to 0}\frac{1}{8}\cdot\frac{\cos\left(\frac{\pi}{2}-h\right)\left[1-\sin\left(\frac{\pi}{2}-h\right)\right]}{\sin\left(\frac{\pi}{2}-h\right)\left(\frac{\pi}{2}-\frac{\pi}{2}+h\right)^3}$

$= \frac{1}{8}\lim_{h\to 0}\frac{\sin h\,(1-\cos h)}{\cos h\cdot h^3}$

$= \frac{1}{8}\lim_{h\to 0}\frac{\sin h\left(2\sin^2\frac{h}{2}\right)}{\cos h\cdot h^3}$

$= \frac{1}{4}\lim_{h\to 0}\frac{\sin h\cdot\sin^2\left(\frac{h}{2}\right)}{h^3\cos h}$

$= \frac{1}{4}\lim_{h\to 0}\left(\frac{\sin h}{h}\right)\left(\frac{\sin\frac{h}{2}}{\frac{h}{2}}\right)^2\cdot\frac{1}{\cos h}\cdot\frac{1}{4}$

$= \frac{1}{4}\times\frac{1}{4} = \frac{1}{16}$

16 $\lim_{x\to 0}\frac{(\cos x)^{1/m}-(\cos x)^{1/n}}{x^2}$

$= \lim_{x\to 0}\frac{(\cos x)^{\frac{1}{m}-\frac{1}{n}}-1}{x^2}\cdot\lim_{x\to 0}\frac{1}{(\cos x)^{1/n}}$

$= \lim_{x\to 0}\frac{\left(1-2\sin^2\frac{x}{2}\right)^{\frac{1}{m}-\frac{1}{n}}-1}{x^2}$

$= \lim_{x\to 0} -2\left(\frac{1}{m}-\frac{1}{n}\right)\frac{\sin^2\frac{x}{2}}{x^2} = \frac{m-n}{2mn}$

17 Clearly, $= \lim_{x\to\infty}(\sqrt{x+1}-\sqrt{x})$

$= \lim_{x\to\infty}\frac{1}{\sqrt{x+1}+\sqrt{x}} = 0$

and

$\lim_{x\to\infty}\left(\frac{2x+1}{x-1}\right)^x = \lim_{x\to\infty}\left(\frac{2+\frac{1}{x}}{1-1/x}\right)^x = \infty$

$\therefore \lim_{x\to\infty}\frac{\cot^{-1}(\sqrt{x+1}-\sqrt{x})}{\sec^{-1}\left(\frac{2x+1}{x-1}\right)^x} = \frac{\cot^{-1}(0)}{\sec^{-1}(\infty)}$

$= \frac{\pi/2}{\pi/2} = 1$

18 Clearly, $p = \lim_{x\to 0}\frac{\ln(1+\cos 2x-1)}{3x^2}$

$= \lim_{x\to 0}\frac{\ln(1+\cos 2x-1)}{(\cos 2x-1)}\cdot\frac{\cos 2x-1}{3x^2}$

$= -\frac{2}{3}$

$q = \lim_{x\to 0}\frac{\sin^2 2x}{4x^2}\cdot\frac{4x^2}{x(1-e^x)} = -4$

and $r = \lim_{x\to 1}\frac{\sqrt{x}-x}{\ln(1+x-1)}$

$= \lim_{x\to 1}\frac{\sqrt{x}(1-\sqrt{x})}{\ln\left(\frac{1+x-1}{x-1}\right)\cdot(x-1)}$

$= \lim_{x\to 1}\frac{\sqrt{x}(1-x)}{\ln\left(\frac{1+(x-1)}{x-1}\right)\cdot(x-1)(1+\sqrt{x})}$

$= -\frac{1}{2}$

Hence, $q < p < r$.

19 Given, $p = \lim_{x\to 0^+}(1+\tan^2\sqrt{x})^{\frac{1}{2x}}$ $\quad (1^\infty \text{ form})$

$= e^{\lim_{x\to 0^+}\frac{\tan^2\sqrt{x}}{2x}} = e^{\frac{1}{2}\lim_{x\to 0^+}\left(\frac{\tan\sqrt{x}}{\sqrt{x}}\right)^2}$

$= e^{\frac{1}{2}}$

$\therefore \log p = \log e^{\frac{1}{2}} = \frac{1}{2}$

20 $\lim_{x\to\infty}\left(\frac{3x-4}{3x+2}\right)^{\frac{x+1}{3}}$

$= \lim_{x\to\infty}\left(\frac{3x+2-6}{3x+2}\right)^{\frac{x+1}{3}}$

$= \lim_{x\to\infty}\left(1-\frac{6}{3x+2}\right)^{\frac{x+1}{3}}$

$= \lim_{x\to\infty}\left[\left(1-\frac{6}{3x+2}\right)^{\frac{3x+2}{-6}}\right]^{\frac{-6}{3x+2}\cdot\frac{x+1}{3}}$

$= \lim_{x\to\infty} e^{\frac{-2(x+1)}{3x+2}} = e^{-2/3}$

$\left[\because \lim_{x\to\infty}\frac{-2(x+1)}{3x+2} = \frac{-2}{3}\right]$

21 Now, $\lim_{x\to\infty}\left(1+\frac{a}{x}+\frac{b}{x^2}\right)^{2x}$

$= \lim_{x\to\infty}\left(1+\frac{a}{x}+\frac{b}{x^2}\right)^{2x\left(\frac{\frac{a}{x}+\frac{b}{x^2}}{\frac{a}{x}+\frac{b}{x^2}}\right)}$

$= e^{\lim_{x\to\infty}2x\left(\frac{a}{x}+\frac{b}{x^2}\right)}$

$\left[\because \lim_{x\to\infty}(1+x)^{1/x} = e\right] = e^{2a}$

But $\lim_{x\to\infty}\left(1+\frac{a}{x}+\frac{b}{x^2}\right)^{2x} = e^2$

$\Rightarrow \quad e^{2a} = e^2$

$\Rightarrow \quad a = 1$

and $\quad b \in R$

22 $\lim_{h\to 0} \dfrac{f(2h+2+h^2)-f(2)}{f(h-h^2+1)-f(1)}$

$= \lim_{h\to 0} \dfrac{f'(2h+2+h^2)(2+2h)}{f'(h-h^2+1)(1-2h)}$

$= \dfrac{f'(2)\times 2}{f'(1)\times 1}$

$= \dfrac{6\times 2}{4\times 1} = 3$

23 $\lim_{x\to a} \dfrac{f(a)g(x)-f(a)-g(a)f(x)+g(a)}{g(x)-f(x)} = 4$

Applying L'Hospital rule, we get

$\lim_{x\to a} \dfrac{f(a)g'(x)-g(a)f'(x)}{g'(x)-f'(x)} = 4$

$\Rightarrow \lim_{x\to a} \dfrac{kg'(x)-kf'(x)}{g'(x)-f'(x)} = 4$

$\therefore \quad k = 4$

24 Let $y = \left(\dfrac{f(1+x)}{f(1)}\right)^{1/x}$

$\Rightarrow \log y = \dfrac{1}{x}[\log f(1+x) - \log f(1)]$

$\Rightarrow \lim_{x\to 0} \log y = \lim_{x\to 0}\left[\dfrac{1}{f(1+x)} f'(1+x)\right]$

$\Rightarrow \lim_{x\to 0} \log y = \dfrac{f'(1)}{f(1)} = \dfrac{6}{3}$

$\therefore \quad \lim_{x\to 0} y = e^2$

25 Clearly, $f(x)$ is continuous only when $\sin^2 x = -\sin^2 x \Rightarrow 2\sin^2 x = 0$

$\Rightarrow \quad x = n\pi$

26 We have,

$f(x) = \begin{cases} \text{rational}, & \text{if } x \notin Q \ \text{ in}[a,b] \\ \text{irrational}, & \text{if } x \in Q \ \text{ in}[a,b] \end{cases}$

Let $C \in [a,b]$ and $c \in Q$. Then,

$f(c) = \text{irrational}$

and $\lim_{x\to c} f(x) = \lim_{h\to 0} f(c+h) =$ rational or irrational

Thus, f is discontinuous everywhere.

27 $g(x) = \begin{cases} 1+x, & x<0 \\ 1-x, & x\ge 0 \end{cases}$

$\therefore f\{g(x)\} = \begin{cases} 1+|x-1|, & x<0 \\ 1+|-x-1|, & x\ge 0 \end{cases}$

$= \begin{cases} 1+1-x, & x<0 \\ 1+x+1, & x\ge 0 \end{cases}$

$= \begin{cases} 2-x, & x<0 \\ 2+x, & x\ge 0 \end{cases}$

It is a polynomial function, so it is continuous in everywhere except at $x = 0$.

Now, LHL $= \lim_{x\to 0} 2 - x = 2$,

RHL $= \lim_{x\to 0} 2 + x = 2$

Also, $f(0) = 2 + 0 = 2$

Hence, it is continuous everywhere.

28 $f\{g(x)\} = \begin{cases} 1, & 0<x<3\pi/4 \\ 0, & x = 3\pi/4, 7\pi/4 \\ -1, & 3\pi/4 < x < 7\pi/4 \\ & \text{or } 7\pi/4 < x < 2 \end{cases}$

Clearly, $[f\{g(x)\}]$ is not continuous at $x = \dfrac{3\pi}{4}, \dfrac{7\pi}{4}$.

29 $f'(1) = \lim_{h\to 0} \dfrac{f(1+h)-f(1)}{h}$

$= \lim_{h\to 0} \dfrac{f(1+h)}{h} - \lim_{h\to 0} \dfrac{f(1)}{h}$

Since, $\lim_{h\to 0} \dfrac{f(1+h)}{h} = 5$, so $\lim_{h\to 0} \dfrac{f(1)}{h}$ must be finite as $f'(1)$ exists and $\lim_{h\to 0} \dfrac{f(1)}{h}$ can be finite only, if $f(1) = 0$ and $\lim_{h\to 0} \dfrac{f(1)}{h} = 0$.

$\therefore \quad f'(1) = \lim_{h\to 0} \dfrac{f(1+h)}{h} = 5$

30 $\lim_{h\to 0} \dfrac{f(1-h)-f(1)}{h^3+3h}$

$= \lim_{h\to 0} \dfrac{f(1-h)-f(1)}{-h} \cdot \dfrac{-1}{h^2+3}$

$= f'(1)\cdot\left(\dfrac{-1}{3}\right) = \dfrac{53}{3}$

31 $\lim_{x\to 2} \dfrac{xf(2)-2f(x)}{x-2}$

$= \lim_{x\to 2} \dfrac{xf(2)-2f(2)+2f(2)-2f(x)}{x-2}$

$= \lim_{x\to 2} \dfrac{f(2)(x-2)-2\{f(x)-f(2)\}}{x-2}$

$= f(2) - 2\lim_{x\to 2} \dfrac{f(x)-f(2)}{x-2}$

$\left[\because f'(x) = \lim_{x\to a} \dfrac{f(x)-f(a)}{x-a}\right]$

$= f(2) - 2f'(2)$

$= 4 - 2\times 4 = -4$

32 $\because \quad |f(x)-f(y)| \le (x-y)^2$

$\therefore \quad \lim_{x\to y} \dfrac{|f(x)-f(y)|}{|x-y|} \le \lim_{x\to y} |x-y|$

$\Rightarrow \quad |f'(y)| \le 0 \Rightarrow f'(y) = 0$

$\Rightarrow \quad f(y) = \text{Constant}$

$\Rightarrow \quad f(y) = 0 \qquad [\because f(0) = 0, \text{ given}]$

$\Rightarrow \quad f(1) = 0$

33 Let $f(x) = \log x$

$\Rightarrow \quad f'(x) = \dfrac{1}{x}$

Therefore, given function

$= f'(a) + kf'(e) = 1$

$\Rightarrow \quad \dfrac{1}{a} + \dfrac{k}{e} = 1$

$\Rightarrow \quad k = e\left(\dfrac{a-1}{a}\right)$

34 If x is just less than k, then $[x] = k-1$

$\therefore \quad f(x) = (k-1)\sin \pi x$

LHD of $f(x) = \lim_{x\to k} \dfrac{(k-1)\sin \pi x - k\sin \pi k}{x-k}$

$= \lim_{x\to k} \dfrac{(k-1)\sin \pi x}{x-k}$,

where $x = k-h$

$= \lim_{h\to 0} \dfrac{(k-1)\sin \pi(k-h)}{-h}$

$= (k-1)(-1)^k \pi$

35 $f(x) = \begin{cases} e^x, & x\le 0 \\ 1-x, & 0<x\le 1 \\ x-1, & x>1 \end{cases}$

$Rf'(0) = \lim_{h\to 0} \dfrac{f(0+h)-f(0)}{h}$

$= \lim_{h\to 0} \dfrac{1-h-1}{h} = -1$

$Lf'(0) = \lim_{h\to 0} \dfrac{f(0-h)-f(0)}{-h}$

$= \lim_{h\to 0} \dfrac{e^{-h}-1}{-h} = 1$

So, it is not differentiable at $x = 0$.

Similarly, it is not differentiable at $x = 1$ but it is continuous at $x = 0$ and 1.

36 RHL $= \lim_{h\to 0} (0+h)e^{-2/h} = \lim_{h\to 0} \dfrac{h}{e^{2/h}} = 0$

LHL $= \lim_{h\to 0} (0-h)e^{-\left(\frac{1}{h}-\frac{1}{h}\right)} = 0$

Hence, $f(x)$ is continuous at $x = 0$.

Now, $Rf'(x) = \lim_{h\to 0} \dfrac{(0+h)e^{-\left(\frac{1}{h}+\frac{1}{h}\right)} - 0}{h}$

$= \lim_{h\to 0} e^{-2/h} = \infty$

and $Lf'(x) = \lim_{h\to 0} \dfrac{(0-h)e^{-\left(\frac{1}{h}-\frac{1}{h}\right)} - 0}{-h}$

$= \lim_{h\to 0} e^{-0} = 1$

$\therefore \quad Lf'(x) \ne Rf'(x)$

Hence, $f(x)$ is not differentiable at $x = 0$.

37 Since, $f(x) = \dfrac{x}{1+|x|} = \dfrac{g(x)}{h(x)}$ [say]

It is clear that $g(x)$ and $h(x)$ are differentiable on $(-\infty, \infty)$ and $(-\infty, 0) \cup (0, \infty)$.

Now,

$\lim_{x\to 0} \dfrac{f(x)-f(0)}{x-0} = \lim_{x\to 0} \dfrac{\dfrac{x}{1+|x|} - 0}{x} = 1$

Hence, $f(x)$ is differentiable on $(-\infty, \infty)$.

38 Clearly, $g(x) = \begin{cases} \cos x, & x \in [0, \pi] \\ \sin x - 1, & x > \pi \end{cases}$

Also, $g(\pi^-) = g(\pi) = g(\pi^+) = -1$

and $g'(\pi^-) \ne g'(\pi^+)$

$\therefore \quad g$ is continuous at $x = \pi$ but not differentiable at $x = \pi$

39 Since, $g(x)$ is differentiable $\Rightarrow g(x)$ must be continuous.

$$g(x)=\begin{cases} k\sqrt{x+1}, & 0\le x\le 3 \\ mx+2, & 3< x\le 5\end{cases}$$

At, $x=3$, RHL $=3m+2$

and at $x=3$, LHL $=2k$

$\therefore \quad 2k=3m+2$

Also, $g'(x)=\begin{cases} \dfrac{k}{2\sqrt{x+1}}, & 0\le x<3 \\ m\ , & 3<x\le 5\end{cases}$

$\therefore \quad L(g'(3))=\dfrac{k}{4}$

and $R\{g'(3)\}=m \Rightarrow \dfrac{k}{4}=m$

i.e. $\quad k=4m$

On solving Eqs (i) and (ii), we get

$$k=\frac{8}{5}, m=\frac{2}{5}$$

$\Rightarrow \quad k+m=2$

40 Clearly, $f(x)=\begin{cases} 2\sqrt{1-x^2}, & x\le 0 \\ 0, & x>0\end{cases}$

$\therefore f(x)$ is discontinuous and hence non-differentiable at $x=0$.

41 $[\sin x]$ is non-differentiable at

$x=\dfrac{\pi}{2}, \pi, 2\pi$ and $[\cos x]$ is

non-differentiable at

$x=0, \dfrac{\pi}{2}, \dfrac{3\pi}{2}$ and 2π.

Thus, $f(x)$ is definitely non-differentiable at $x=\pi, \dfrac{3\pi}{2}, 0$.

Also,

$f\left(\dfrac{\pi}{2}\right)=1, f\left(\dfrac{\pi}{2}-0\right)=0,$

$f(2\pi)=1, \qquad f(2\pi-0)=-1$

Thus, $f(x)$ is also non-differentiable at $x=\dfrac{\pi}{2}$ and 2π.

42 Let $u(x)=\sin x$

$v(x)=|x|$

$\therefore f(x)=vou(x)=v(u(x))=v(\sin x)=|\sin x|$

$\because \ u(x)=\sin x$ is a continuous function and $v(x)=|x|$ is a continuous function.

$\therefore f(x)=vou(x)$ is also continuous everywhere but $v(x)$ is not differentiable at $x=0$

$\Rightarrow f(x)$ is not differentiable where $\sin x=0$

$\Rightarrow \ x=n\pi, n\in Z$

Hence, $f(x)$ is continuous everywhere but not differentiable at

$x=n\pi, n\in Z.$

43 $f(x)=\begin{cases} -\log x, x<1 \\ \log x, x\ge 1\end{cases}$

$f'(x)=\begin{cases} -1/x, & x<1 \\ \dfrac{1}{x}, & x>1\end{cases}$

$\therefore f'(1^-)=-1$ and $f'(1^+)=1$

Hence, $f(x)$ is not differentiable.

SESSION 2

1 We have,

$$\lim_{x\to 0^+} x\left(\left[\frac{1}{x}\right]+\left[\frac{1}{x}\right]+\ldots+\left[\frac{15}{x}\right]\right)$$

We know, $[x]=x-\{x\}$

$\therefore \quad \left[\dfrac{1}{x}\right]=\dfrac{1}{x}-\left\{\dfrac{1}{x}\right\}$

Similarly, $\left[\dfrac{n}{x}\right]=\dfrac{n}{x}-\left\{\dfrac{n}{x}\right\}$

$\therefore$ Given limit

$$=\lim_{x\to 0^+} x\left(\frac{1}{x}-\left\{\frac{1}{x}\right\}+\frac{2}{x}-\left\{\frac{2}{x}\right\}+\ldots\frac{15}{x}-\left\{\frac{15}{x}\right\}\right)$$

$$=\lim_{x\to 0^+}(1+2+3+\ldots+15)-x\left(\left\{\frac{1}{x}\right\}+\left\{\frac{2}{x}\right\}+\ldots+\left\{\frac{15}{x}\right\}\right)$$

$=120-0=120$

$$\left[\begin{array}{l}\because 0\le\left\{\dfrac{n}{x}\right\}<1, \text{therefore} \\ 0\le x\left\{\dfrac{n}{x}\right\}<x\Rightarrow \lim\limits_{x\to 0^+} x\left\{\dfrac{n}{x}\right\}=0\end{array}\right]$$

2 Now, $\lim\limits_{x\to\infty}\left(\dfrac{x^2+5x+3}{x^2+x+2}\right)^x$

$$=\lim_{x\to\infty}\left(1+\frac{4x+1}{x^2+x+2}\right)^x$$

$$=\lim_{x\to\infty}\left[\left(1+\frac{4x+1}{x^2+x+2}\right)^{1/\frac{(4x+1)}{x^2+x+2}}\right]^{\frac{(4x+1)x}{x^2+x+2}}$$

$$=e^{\lim\limits_{x\to\infty}\frac{\left(4+\frac{1}{x}\right)}{1+\frac{1}{x}+\frac{2}{x^2}}}=e^4$$

$$\left[\because \lim_{x\to 0}\left(1+\frac{1}{x}\right)^x=e\right]$$

3 Now, $\lim\limits_{x\to\alpha}\dfrac{1-\cos(ax^2+bx+c)}{(x-\alpha)^2}$

$$=\lim_{x\to\alpha}\frac{2\sin^2\left(\dfrac{ax^2+bx+c}{2}\right)}{(x-\alpha)^2}$$

$$=\lim_{x\to\alpha}\frac{2\sin^2\left(\dfrac{a}{2}(x-\alpha)(x-\beta)\right)}{\left(\dfrac{a}{2}\right)^2(x-\alpha)^2(x-\beta)^2}$$

$$\left(\frac{a}{2}\right)^2(x-\beta)^2$$

$$=\lim_{x\to\alpha}\frac{a^2}{2}(x-\beta)^2 \quad \left[\because \lim_{x\to 0}\frac{\sin x}{x}=1\right]$$

$$=\frac{a^2}{2}(\alpha-\beta)^2$$

4 $\lim\limits_{n\to\infty}\sin\left\{n\pi\left(1+\dfrac{1}{n^2}\right)^{1/2}\right\}$

$$=\lim_{n\to\infty}\sin\left\{n\pi\left(1+\frac{1}{2n^2}-\frac{1}{8n^4}+\ldots\infty\right)\right\}$$

$$=\lim_{n\to\infty}\sin\left\{n\pi+\frac{\pi}{2n}-\frac{\pi}{8n^3}+\ldots\infty\right\}$$

$$=\lim_{n\to\infty}(-1)^n\sin\pi\left(\frac{1}{2n}-\frac{1}{8n^3}+\ldots\infty\right)$$

$=0$

5 Given, $x>0$ and g is a bounded function.

Then, $\lim\limits_{n\to\infty}\dfrac{f(x)\cdot e^{nx}+g(x)}{e^{nx}+1}$

$$=\lim_{n\to\infty}\left[\frac{f(x)}{1+\left(\dfrac{1}{e^{nx}}\right)}+\frac{g(x)}{e^{nx}+1}\right]$$

$$=\frac{f(x)}{1+0}+\frac{\text{Finite}}{\infty}=f(x)$$

6 Let $\quad y=\lim\limits_{x\to 0}\left(\dfrac{a^x+b^x+c^x}{3}\right)^{2/x}$

$\Rightarrow \log y=\lim\limits_{x\to 0}\dfrac{2}{x}\log\left(\dfrac{a^x+b^x+c^x}{3}\right)$

$$=2\lim_{x\to 0}\frac{\log(a^x+b^x+c^x)-\log 3}{x}$$

Apply L'Hospital's rule,

$$=2\lim_{x\to 0}\frac{\dfrac{a^x\log a+b^x\log b+c^x\log c}{a^x+b^x+c^x}}{1}$$

$\log y=\log(abc)^{2/3}$

$\Rightarrow \quad y=(abc)^{2/3}$

7 $f(x)=\cot^{-1}\left(\dfrac{3x-x^3}{1-3x^2}\right)$

and $g(x)=\cos^{-1}\left(\dfrac{1-x^2}{1+x^2}\right)$

On putting $x=\tan\theta$ in both equations, we get

$$f(\theta)=\cot^{-1}\left(\frac{3\tan\theta-\tan^3\theta}{1-3\tan^2\theta}\right)$$

$\Rightarrow f(\theta)=\cot^{-1}(\tan 3\theta)$

$\Rightarrow f(\theta)=\cot^{-1}\left\{\cot\left(\dfrac{\pi}{2}-3\theta\right)\right\}$

$=\dfrac{\pi}{2}-3\theta$

$\therefore \quad f'(\theta)=-3 \quad \ldots$ (i)

and $g(\theta)=\cos^{-1}\left(\dfrac{1-\tan^2\theta}{1+\tan^2\theta}\right)$

$=\cos^{-1}(\cos 2\theta)=2\theta$

$\therefore \quad g'(\theta) = 2$

Now,

$$\lim_{x \to a}\left(\frac{f(x)-f(a)}{g(x)-g(a)}\right) = \lim_{x \to a}\left(\frac{f(x)-f(a)}{x-a}\right) \times \frac{1}{\lim\limits_{x \to a}\left(\frac{g(x)-g(a)}{x-a}\right)}$$

$$= f'(a)\cdot\frac{1}{g'(a)} = -3 \times \frac{1}{2} = -\frac{3}{2}$$

8 Now, $\cos x$ is continuous, $\forall x \in R$

$\Rightarrow \cos \pi\left(x - \frac{1}{2}\right)$ is also continuous,

$\forall\ x \in R$.

Hence, the continuity of f depends upon the continuity of $[x]$, which is discontinuous, $\forall\ x \in I$.

So, we should check the continuity of f at $x = n$, $\forall\ n \in I$

LHL at $x = n$ is given by

$$f(n^-) = \lim_{x \to n^-} f(x) = \lim_{x \to n^-} [x] \cos \pi\left(x - \frac{1}{2}\right) = (n-1)\cos\frac{(2n-1)\pi}{2} = 0$$

RHL at $x = n$ is given by

$$f(n^+) = \lim_{x \to n^+} f(x) = \lim_{x \to n^+} [x] \cos \pi\left(x - \frac{1}{2}\right) = (n)\cos\frac{(2n-1)\pi}{2} = 0$$

Also, value of the function at $x = n$ is

$$f(n) = [n]\cos \pi\left(n - \frac{1}{2}\right) = (n)\cos\frac{(2n-1)\pi}{2} = 0$$

$\therefore \quad f(n^+) = f(n^-) = f(n)$

Hence, f is continuous at $x = n$, $\forall\ n \in I$.

9 The function $u(x) = \dfrac{1}{x-1}$ is discontinuous at the point $x = 1$.

The function $y = f(u)$

$$= \frac{1}{u^2 + u - 2} = \frac{1}{(u+2)(u-1)}$$

is discontinuous at $u = -2$ and $u = 1$.

When $\quad u = -2$

$$\Rightarrow \quad \frac{1}{x-1} = -2$$

$$\Rightarrow \quad x = \frac{1}{2}$$

When $u = 1$

$$\Rightarrow \quad \frac{1}{x-1} = 1$$

$$\Rightarrow \quad x = 2$$

Hence, the composite function $y = f(x)$ is discontinuous at three points, $x = \frac{1}{2}$, $x = 1$ and $x = 2$.

10 Since, $f(x)$ is a positive increasing function.

$\therefore \quad 0 < f(x) < f(2x) < f(3x)$

$$\Rightarrow \quad 0 < 1 < \frac{f(2x)}{f(x)} < \frac{f(3x)}{f(x)}$$

$$\Rightarrow \quad \lim_{x \to \infty} 1 \le \lim_{x \to \infty}\frac{f(2x)}{f(x)} \le \lim_{x \to \infty}\frac{f(3x)}{f(x)}$$

By Sandwich theorem,

$$\lim_{x \to \infty}\frac{f(2x)}{f(x)} = 1$$

11 We have, $f(x) = \max\{\tan x, \sin x, \cos x\}$, where $x \in \left[-\frac{\pi}{2}, \frac{3\pi}{2}\right]$

Let us draw the graph of $y = \tan x$, $y = \sin x$ and $y = \cos x$ in $\left[-\frac{\pi}{2}, \frac{3\pi}{2}\right]$

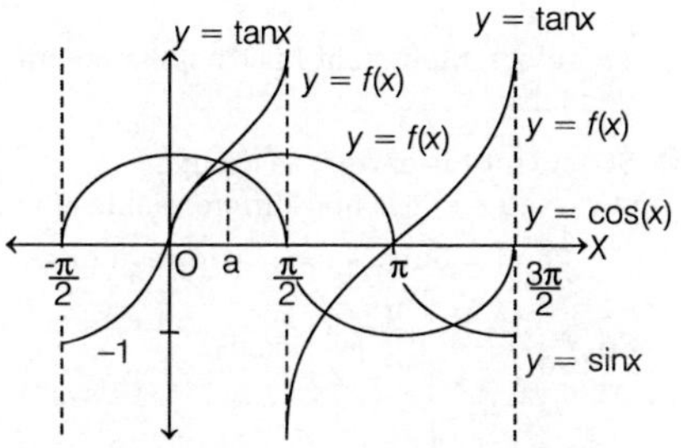

From the graph, it clear that $f(x)$ is non-differentiable at $x = a, \frac{\pi}{2}$ and π.

12 We have,

$$f(x) = |x - \pi|(e^{|x|} - 1)\sin|x|$$

$$f(x) = \begin{cases} (x-\pi)(e^{-x} - 1)\sin x, & x < 0 \\ -(x-\pi)(e^{x} - 1)\sin x, & 0 \le x < \pi \\ (x-\pi)(e^{x} - 1)\sin x, & x \ge \pi \end{cases}$$

We check the differentiability at $x = 0$ and π.

We have,

$$f'(x) = \begin{cases} (x-\pi)(e^{-x} - 1)\cos x + (e^{x} - 1)\sin x \\ \quad + (x-\pi)\sin x e^{-x}(-1), x < 0 \\ -[(x-\pi)(e^{x} - 1)\cos x + (e^{x} - 1)\sin x \\ \quad + (x-\pi)\sin x e^{x}], 0 < x < \pi \\ (x-\pi)(e^{x} - 1)\cos x + (e^{x} - 1)\sin x \\ \quad + (x-\pi)\sin x e^{x}, x > \pi \end{cases}$$

Clearly, $\lim\limits_{x \to 0^-} f'(x) = 0 = \lim\limits_{x \to 0^+} f'(x)$

and $\quad \lim\limits_{x \to \pi^-} f'(x) = 0$

$$= \lim_{x \to \pi^+} f'(x)$$

$\therefore$ f is differentiable at $x = 0$ and $x = \pi$

Hence, f is equal to the set $\{0, \pi\}$.

13 We have, $f(x) = \lim\limits_{n \to \infty}\dfrac{x^n + \left(\frac{\pi}{3}\right)^n}{x^{n-1} + \left(\frac{\pi}{3}\right)^{n-1}}$

$$= \lim_{n \to \infty}\frac{x\left(1 + \left(\frac{\pi}{3x}\right)^n\right)}{1 + \left(\frac{\pi}{3x}\right)^{n-1}} = x, \text{ if } x > \frac{\pi}{3}$$

$$\text{Also, } f(x) = \lim_{n \to \infty}\frac{\pi}{3}\frac{\left(\left(\frac{3x}{\pi}\right)^n + 1\right)}{\left(\left(\frac{3x}{\pi}\right)^{n-1} + 1\right)} = \frac{\pi}{3},$$

if $x < \pi/3$

Note that $f\left(\frac{\pi}{3}\right) = \frac{\pi}{3}$

$$\therefore \quad f(x) = \begin{cases} x, & \text{if } x \ge \frac{\pi}{3} \\ \frac{\pi}{3}, & \text{if } x < \frac{\pi}{3} \end{cases}$$

From the given options, it is clear that option (c) is incorrect.

14 Let $a_{k+1} = \dfrac{x}{(kx+1)\{(k+1)x+1\}}$

$$= \left\{\frac{1}{kx+1} - \frac{1}{(k+1)x+1}\right\}$$

$$\therefore f(x) = \lim_{n \to \infty}\sum_{k=0}^{n-1}\left\{\frac{1}{kx+1} - \frac{1}{(k+1)x+1}\right\}$$

$$= \lim_{n \to \infty}\left[1 - \frac{1}{nx+1}\right]$$

$$= \begin{cases} 1, & \text{if } x \ne 0 \\ 0, & \text{if } x = 0 \end{cases}$$

Clearly, $f(x)$ is neither continuous nor differentiable at $x = 0$.

15 Clearly, $x^n + xa + b = (x - x_1)(x - x_2)\ldots(x - x_n)$

$$\Rightarrow \quad \frac{x^n + xa + b}{x - x_1} = (x - x_2)(x - x_3)\ldots(x - x_n)$$

$$\Rightarrow \quad \lim_{x \to x_1}\frac{x^n + ax + b}{x - x_1} = (x_1 - x_2)(x_1 - x_3)\ldots(x_1 - x_n)$$

$$\Rightarrow \quad \lim_{x \to x_1}\frac{nx^{n-1} + a}{1} = (x_1 - x_2)(x_1 - x_3)\ldots(x_1 - x_n)$$

[using L'Hospital rule]

$$\Rightarrow \quad nx_1^{n-1} + a = (x_1 - x_2)(x_1 - x_3)\ldots(x_1 - x_n)$$

16 Let $L = \lim_{x\to\infty} x \ln\left(\begin{vmatrix} \frac{1}{x} & -b & c \\ 0 & \frac{1}{x} & -1 \\ 1 & 0 & \frac{a}{x} \end{vmatrix}\right)$

$= \lim_{x\to\infty} x \ln\left(\frac{a}{x^3} + b - \frac{c}{x}\right)$

Clearly, for limit to be exist, $b = 1$.

Thus, $L = \lim_{x\to\infty} x \ln\left(1 + \frac{a}{x^3} - \frac{c}{x}\right)$

$= \lim_{x\to\infty} x\left(\frac{a}{x^3} - \frac{c}{x}\right)$

$\left[\because \ln(1+x) = x - \frac{x^2}{2} + \frac{x^3}{3} - \frac{x^4}{4} + \ldots\right]$

$= -c$

$\because \quad L = -4$

$\therefore \quad c = 4$

Hence, $a \in R, b = 1$ and $c = 4$.

17 We have, $\lim_{x\to 0}\left[1 + x + \frac{f(x)}{x}\right]^{1/x} = e^3$

$\Rightarrow e^{\lim_{x\to 0}\left\{1 + x + \frac{f(x)}{x} - 1\right\}\frac{1}{x}} = e^3$

$\Rightarrow e^{\lim_{x\to 0}\left[1 + \frac{f(x)}{x^2}\right]} = e^3$

$\Rightarrow \lim_{x\to 0}\frac{f(x)}{x^2} = 2$

Now,

$\lim_{x\to 0}\left[1 + \frac{f(x)}{x}\right]^{1/x} = e^{\lim_{x\to 0}\left[1 + \frac{f(x)}{x} - 1\right]\frac{1}{x}}$

$= e^{\lim_{x\to 0}\frac{f(x)}{x^2}} = e^2$

18 We have, $f^2(x) = 1 \;\forall x \in R$

$\therefore$ f can take values +1 or −1

Since f is discontinuous only at $x = 0$

$\therefore f$ may be one of the followings

(i) $f(x) = \begin{cases} 1, & x \le 0 \\ -1, & x > 0 \end{cases}$

(ii) $f(x) = \begin{cases} 1, & x < 0 \\ -1, & x \ge 0 \end{cases}$

(iii) $f(x) = \begin{cases} -1, & x \le 0 \\ 1, & x > 0 \end{cases}$

(iv) $f(x) = \begin{cases} -1, & x < 0 \\ 1, & x \ge 0 \end{cases}$

(v) $f(x) = \begin{cases} 1, & x > 0 \\ 1, & x < 0 \\ -1 & x = 0 \end{cases}$

(vi) $f(x) = \begin{cases} -1, & x > 0 \\ -1, & x < 0 \\ 1, & x = 0 \end{cases}$

19 $f(x) = x|x|$ and $g(x) = \sin x$

$gof\,(x) = \sin(x|x|) = \begin{cases} -\sin x^2, & x < 0 \\ \sin x^2, & x \ge 0 \end{cases}$ $\therefore$

$(gof)'(x) = \begin{cases} -2x\cos x^2, & x < 0 \\ 2x\cos x^2, & x \ge 0 \end{cases}$

Clearly, $L\,(gof)'(0) = 0 = R\,(gof)'(0)$

So, gof is differentiable at $x = 0$ and also its derivative is continuous at $x = 0$.

Now,

$(gof)''(x) = \begin{cases} -2\cos x^2 + 4x^2\sin x^2, & x < 0 \\ 2\cos x^2 - 4x^2\sin x^2, & x \ge 0 \end{cases}$

$\therefore L\,(gof)''(0) = -2$ and $R\,(gof)''(0) = 2$

$\therefore \quad L\,(gof)''(0) \ne R\,(gof)''(0)$

Hence, $gof(x)$ is not twice differentiable at $x = 0$.

Therefore, Statement I is true, Statement II is false.

20 Statement I is correct as though $|4x^2 - 12x + 5|$ is non-differentiable at $x = \frac{1}{2}$ and $\frac{5}{2}$ but $\cos \pi x = 0$ those points.

So, $f'\left(\frac{1}{2}\right)$ and $f'\left(\frac{5}{2}\right)$ exists.

21 Here, $f(x) = \begin{cases} x\cdot\sin\left(\frac{1}{x}\right), & x \ne 0 \\ 0, & x = 0 \end{cases}$

To check continuity at $x = 0$,

$\text{LHL} = \lim_{h\to 0}\left\{(-h)\sin\left(-\frac{1}{h}\right)\right\} = 0$

$\text{RHL} = \lim_{h\to 0}\left\{h\sin\left(\frac{1}{h}\right)\right\} = 0$

$f(0) = 0$

So, $f(x)$ is continuous at $x = 0$.

Hence, Statement I is correct.

$f_2(x) = \begin{cases} \sin\left(\frac{1}{x}\right), & x \ne 0 \\ 0, & x = 0 \end{cases}$

Here, $\lim_{x\to 0} f_2(x) = \lim_{x\to 0}\sin\left(\frac{1}{x}\right)$

which does not exist.

So, $f_2(x)$ is not continuous at $x = 0$.

Hence, Statement II is false.

22 $\therefore f(x) = |x-2| + |x-5|$

$= \begin{cases} (2-x) + (5-x), & x < 2 \\ (x-2) + (5-x), & 2 \le x \le 5 \\ (x-2) + (x-5), & x > 5 \end{cases}$

$= \begin{cases} 7 - 2x, & x < 2 \\ 3, & 2 \le x \le 5 \\ 2x - 7, & x > 5 \end{cases}$

Now, we can draw the graph of f very easily.

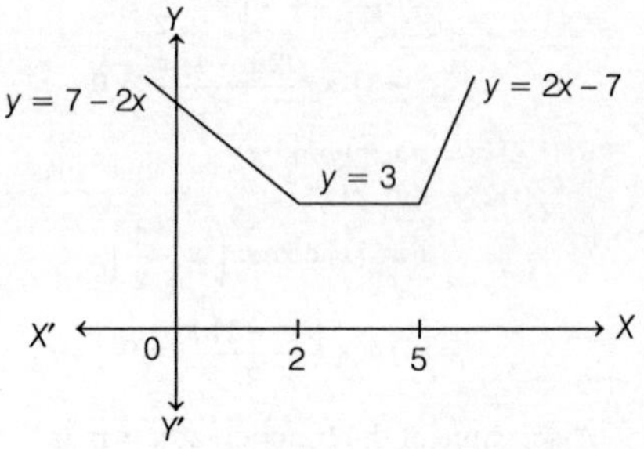

Statement I $f'(4) = 0$

It is obviously clear that, f is constant around $x = 4$, hence $f'(4) = 0$. Hence, Statement I is correct.

Statement II It can be clearly seen that

(i) f is continuous, $\forall\, x \in [2, 5]$

(ii) f is differentiable, $\forall\, x \in (2, 5)$

(iii) $f(2) = f(5) = 3$

Hence, Statement II is also correct but obviously not a correct explanation of Statement I.

DAY TWELVE

Differentiation

Learning & Revision for the Day

- Derivative (Differential Coefficient)
- Geometrical Meaning of Derivative at a point
- Methods of Differentiation
- Second Order Derivative
- Differentiation of a Determinant

Derivative (Differential Coefficient)

The rate of change of a quantity y with respect to another quantity x is called the **derivative or differential coefficient** of y with respect to x. The process of finding derivative of a function called **differentiation.**

Geometrical Meaning of Derivative at a Point

Geometrically derivative of a function at a point $x = c$ is the slope of the tangent to the curve $y = f(x)$ at the point $P\{c, f(c)\}$.

Slope of tangent at $P = \lim_{x \to c} \dfrac{f(x) - f(c)}{x - c} = \left\{\dfrac{df(x)}{dx}\right\}_{x=c}$ or $f'(c)$.

Derivative of Some Standard Functions

- $\dfrac{d}{dx}(\text{constant}) = 0$
- $\dfrac{d}{dx} x^n = nx^{n-1}$
- $\dfrac{d}{dx}(\sqrt{x}) = \dfrac{1}{2\sqrt{x}}$
- $\dfrac{d}{dx}\left(\dfrac{1}{x^n}\right) = -\dfrac{n}{x^{n+1}}$
- $\dfrac{d}{dx}(\sin x) = \cos x$
- $\dfrac{d}{dx}(\cos x) = -\sin x$
- $\dfrac{d}{dx}(\tan x) = \sec^2 x$
- $\dfrac{d}{dx}(\sec x) = \sec x \tan x$
- $\dfrac{d}{dx}(\cot x) = -\operatorname{cosec}^2 x$
- $\dfrac{d}{dx}(\operatorname{cosec} x) = -\operatorname{cosec} x \cot x$
- $\dfrac{d}{dx}(\log x) = \dfrac{1}{x}$, for $x > 0$
- $\dfrac{d}{dx}(e^x) = e^x$
- $\dfrac{d}{dx}(a^x) = a^x \log a$, for $a > 0$
- $\dfrac{d}{dx}(\log_a x) = \dfrac{1}{x \log a}$, for $x > 0, a > 0, a \neq 1$

PREP MIRROR
Your Personal Preparation Indicator

- No. of Questions in Exercises (x)—
- No. of Questions Attempted (y)—
- No. of Correct Questions (z)—
  (Without referring Explanations)

- Accuracy Level ($z/y \times 100$)—
- Prep Level ($z/x \times 100$)—

In order to expect good rank in JEE, your Accuracy Level should be above 85 & Prep Level should be above 75.

- $\frac{d}{dx}(\sin^{-1} x) = \frac{1}{\sqrt{1-x^2}}$, for $-1 < x < 1$
- $\frac{d}{dx}(\cos^{-1} x) = -\frac{1}{\sqrt{1-x^2}}$, for $-1 < x < 1$
- $\frac{d}{dx}(\sec^{-1} x) = \frac{1}{|x|\sqrt{x^2-1}}$, for $|x| > 1$
- $\frac{d}{dx}(\text{cosec}^{-1} x) = -\frac{1}{|x|\sqrt{x^2-1}}$, for $|x| > 1$
- $\frac{d}{dx}(\tan^{-1} x) = \frac{1}{1+x^2}$, for $x \in R$
- $\frac{d}{dx}(\cot^{-1} x) = -\frac{1}{1+x^2}$, for $x \in R$

Methods of Differentiation

(i) If $y = f(x) \pm g(x)$, then $\frac{dy}{dx} = \frac{d}{dx}\{f(x) \pm g(x)\}$

$$= f'(x) \pm g'(x)$$

(ii) If $y = c \cdot f(x)$, where c is any constant, then
$\frac{dy}{dx} = \frac{d}{dx}(c \cdot f(x)) = c \cdot f'(x)$. [Scalar multiple rule]

(iii) If $y = f(x) \cdot g(x)$, then $\frac{dy}{dx} = \frac{d}{dx}\{f(x) \cdot g(x)\}$

$= f(x) \cdot g'(x) + g(x) \cdot f'(x)$ [Product rule]

(iv) If $y = \frac{f(x)}{g(x)}$, then $\frac{dy}{dx} = \frac{d}{dx}\left\{\frac{f(x)}{g(x)}\right\} = \frac{g(x) \cdot f'(x) - f(x) \cdot g'(x)}{\{g(x)\}^2}$

$g(x) \neq 0$

(v) If $y = f(u)$ and $u = g(x)$, then

$\frac{dy}{dx} = \frac{df}{du} \cdot \frac{du}{dx} = f'(u) \cdot g'(x)$ [Chain rule]

This rule can be extended as follows. If $y = f(u), u = g(v)$ and $v = h(x)$, then $\frac{dy}{dx} = \frac{df}{du} \cdot \frac{du}{dv} \cdot \frac{dv}{dx}$,

(vi) $\frac{d}{dx}(f\{g(x)\}) = f'(g(x)) \cdot g'(x)$

(vii) If given function cannot be expressed in the form $y = f(x)$ but can be expressed in the form $f(x, y) = 0$, then to find derivatives of each term of $f(x, y) = 0$ w.r.t x. [differentiation of implicit function]

(viii) If y is the product or the quotient of a number of complicated functions or if it is of the form $(f(x))^{g(x)}$, then the derivative of y can be found by first taking log on both sides and then differentiating it. [logarithmic differentiation rule]

When $y = (f(x))^{g(x)}$, then $\frac{dy}{dx} = (f(x))^{g(x)}$

$$\left[\frac{g(x)}{f(x)} \cdot f'(x) + \log f(x) \cdot g'(x)\right]$$

(ix) If $x = \phi(t)$ and $y = \Psi(t)$, where t is parameter, then

$\frac{dy}{dx} = \frac{dy/dt}{dx/dt}$ [Parametric differentiation rule]

(x) If $u = f(x)$ and $v = g(x)$, then the differentiation of u with respect to v is $\frac{du}{dv} = \frac{\left(\frac{du}{dx}\right)}{\left(\frac{dv}{dx}\right)}$.

[Differentiation of a function w.r.t another function]

Differentiation Using Substitution

In order to find differential coefficients of complicated expressions, some substitution are very helpful, which are listed below

S. No.	Function	Substitution
(i)	$\sqrt{a^2 - x^2}$	$x = a \sin\theta$ or $a \cos\theta$
(ii)	$\sqrt{x^2 - a^2}$	$x = a \sec\theta$ or $a \text{ cosec}\theta$
(iii)	$\sqrt{x^2 + a^2}$	$x = a \tan\theta$ or $a \cot\theta$
(iv)	$\sqrt{\frac{a+x}{a-x}}$ or $\sqrt{\frac{a-x}{a+x}}$	$x = a \cos 2\theta$
(v)	$\sqrt{\frac{a^2+x^2}{a^2-x^2}}$ or $\sqrt{\frac{a^2-x^2}{a^2+x^2}}$	$x^2 = a^2 \cos 2\theta$
(vi)	$\sqrt{\frac{x}{a+x}}$	$x = a \tan^2\theta$
(vii)	$\sqrt{(x-a)(x-b)}$	$x = a \sec^2\theta - b \tan^2\theta$
(viii)	$\sqrt{ax - x^2}$	$x = a \sin^2\theta$
(ix)	$\sqrt{\frac{x}{a-x}}$	$x = a \sin^2\theta$
(x)	$\sqrt{(x-a)(b-x)}$	$x = a \cos^2\theta + b \sin^2\theta$

Usually this is done in case of inverse trigonometric functions.

Second Order Derivative

If $y = f(x)$, then $\frac{d}{dx}\left(\frac{dy}{dx}\right)$ is called the **second order derivative** of y w.r.t. x. It is denoted by $\frac{d^2y}{dx^2}$ or $f''(x)$ or y'' or y_2.

Differentiation of a Determinant

If $y = \begin{vmatrix} p & q & r \\ u & v & w \\ l & m & n \end{vmatrix}$, then $\frac{dy}{dx} = \begin{vmatrix} \frac{dp}{dx} & \frac{dq}{dx} & \frac{dr}{dx} \\ u & v & w \\ l & m & n \end{vmatrix}$

$$+ \begin{vmatrix} p & q & r \\ \frac{du}{dx} & \frac{dv}{dx} & \frac{dw}{dx} \\ l & m & n \end{vmatrix} + \begin{vmatrix} p & q & r \\ u & v & w \\ \frac{dl}{dx} & \frac{dm}{dx} & \frac{dn}{dx} \end{vmatrix}$$

DAY PRACTICE SESSION 1

FOUNDATION QUESTIONS EXERCISE

1 If $f(x) = |\cos x|$, then $f'\left(\frac{3\pi}{4}\right)$ is equal to → NCERT Exemplar

(a) $\frac{1}{\sqrt{2}}$ (b) $\sqrt{2}$ (c) $\frac{1}{2}$ (d) $2\sqrt{2}$

2 If $f(x) = |x-1|$ and $g(x) = f[f\{f(x)\}]$, then for $x > 2$, $g'(x)$ is equal to

(a) -1, if $2 \le x < 3$ (b) 1, if $2 \le x < 3$
(c) 1, if $x > 2$ (d) None of these

3 The derivative of $y = (1-x)(2-x)\ldots(n-x)$ at $x = 1$ is

(a) 0 (b) $(-1)(n-1)!$ (c) $n! - 1$ (d) $(-1)^{n-1}(n-1)!$

4 If $f(x) = x^n$, then the value of

$$f(1) - \frac{f'(1)}{1!} + \frac{f''(1)}{2!} - \frac{f'''(1)}{3!} + \ldots + \frac{(-1)^n f^n(1)}{n!}$$ is

(a) 2^n (b) 0 (c) 2^{n-1} (d) None of these

5 If $f(x) = \frac{x^2 - x}{x^2 + 2x}$, where $x \ne 0, -2$, then $\frac{d}{dx}[f^{-1}(x)]$ (whenever it is defined) is equal to → JEE Mains 2013

(a) $\frac{-1}{(1-x)^2}$ (b) $\frac{3}{(1-x)^2}$ (c) $\frac{1}{(1-x)^2}$ (d) $\frac{-3}{(1-x)^2}$

6 If $f(x) = 2 + |x| - |x-1| - |x+1|$, then $f'\left(-\frac{1}{2}\right) + f'\left(\frac{1}{2}\right) + f'\left(\frac{3}{2}\right) + f'\left(\frac{5}{2}\right)$ is equal to

(a) 1 (b) -1 (c) 2 (d) -2

7 If $f(x) = |\log_e |x||$, then $f'(x)$ equals

(a) $\frac{1}{|x|}$, where $x \ne 0$ (b) $\frac{1}{x}$ for $|x| > 1$
(c) $-\frac{1}{x}$ for $|x| > 1$
(d) $\frac{1}{x}$ for $x > 0$ and $-\frac{1}{x}$ for $x < 0$

8 If $f(x) = |\cos x - \sin x|$, then $f'\left(\frac{\pi}{2}\right)$ is equal to

(a) 1 (b) -1 (c) 0 (d) None of these

9 If $f(x) = \cos x \cdot \cos 2x \cdot \cos 4x \cdot \cos 8x \cdot \cos 16x$, then $f'\left(\frac{\pi}{4}\right)$ is equal to

(a) $\sqrt{2}$ (b) $\frac{1}{\sqrt{2}}$ (c) 1 (d) $\frac{\sqrt{3}}{2}$

10 If $\sin y = x \sin(a+y)$, then $\frac{dy}{dx}$ is equal to

(a) $\frac{\sin a}{\sin^2(a+y)}$ (b) $\frac{\sin^2(a+y)}{\sin a}$
(c) $\sin a \sin^2(a+y)$ (d) $\frac{\sin^2(a-y)}{\sin a}$

11 If $y = (1-x)(1+x^2)(1+x^4)\ldots(1+x^{2n})$, then $\frac{dy}{dx}$ at $x = 0$ is equal to

(a) -1 (b) $\frac{1}{(1+x)^2}$ (c) $\frac{x}{(1+x^2)}$ (d) $\frac{x}{(1-x)^2}$

12 If $f : (-1, 1) \to R$ be a differentiable function with $f(0) = 1$ and $f'(0) = 1$. Let $g(x) = [f(2f(x) + 2)]^2$. The, $g'(0)$ is equal to

(a) 4 (b) -4 (c) 0 (d) -2

13 Let $f(x)$ be a polynomial function of second degree. If $f(1) = f(-1)$ and a, b, c are in AP, then $f'(a), f'(b)$ and $f'(c)$ are in.

(a) AP
(b) GP
(c) Arithmetic-Geometric progression
(d) None of the above

14 If $y = f(x)$ is an odd differentiable function defined on $(-\infty, \infty)$ such that $f'(3) = -2$, then $f'(-3)$ is equal to

(a) 4 (b) 2 (c) -2 (d) 0

15 If f and g are differentiable function satisfying $g'(a) = 2, g(a) = b$ and $fog = I$ (identity function). Then, $f'(b)$ is equal to

(a) $\frac{1}{2}$ (b) 2 (c) $\frac{2}{3}$ (d) None of these

16 If y is an implicit function of x defined by $x^{2x} - 2x^x \cot y - 1 = 0$. Then, $y'(1)$ is equal to

(a) -1 (b) 1 (c) $\log 2$ (d) $-\log 2$

17 If $x^m y^n = (x+y)^{m+n}$, then $\frac{dy}{dx}$ is equal to

(a) $\frac{x+y}{xy}$ (b) xy (c) $\frac{x}{y}$ (d) $\frac{y}{x}$

18 If $y = (x + \sqrt{1+x^2})^n$, then $(1+x^2)\frac{d^2y}{dx^2} + x\frac{dy}{dx}$ is equal to

(a) $n^2 y$ (b) $-n^2 y$ (c) $-y$ (d) $2x^2 y$

19 If $x = \frac{2t}{1+t^2}$ and $y = \frac{1-t^2}{1+t^2}$, then $\frac{dy}{dx}$ is equal to

(a) $\frac{2t}{t^2+1}$ (b) $\frac{2t}{t^2-1}$
(c) $\frac{2t}{1-t^2}$ (d) None of these

20 For $a > 0, t \in \left(0, \frac{\pi}{2}\right)$, let $x = \sqrt{a^{\sin^{-1} t}}$ and $y = \sqrt{a^{\cos^{-1} t}}$. Then, $1 + \left(\frac{dy}{dx}\right)^2$ equals → JEE Mains 2013

(a) $\frac{x^2}{y^2}$ (b) $\frac{y^2}{x^2}$ (c) $\frac{x^2+y^2}{y^2}$ (d) $\frac{x^2+y^2}{x^2}$

21 If $y = \sec^{-1}\left[\frac{\sqrt{x}+1}{\sqrt{x}-1}\right] + \sin^{-1}\left[\frac{\sqrt{x}-1}{\sqrt{x}+1}\right]$, then $\frac{dy}{dx}$ is equal to

(a) 0 (b) $\frac{1}{\sqrt{x}+1}$
(c) 1 (d) None of these

22 For $x \in \left(0, \frac{1}{4}\right)$, if the derivative of $\tan^{-1}\left(\frac{6x\sqrt{x}}{1-9x^3}\right)$ is $\sqrt{x} \cdot g(x)$, then $g(x)$ equals → JEE Mains 2017

(a) $\frac{9}{1+9x^3}$ (b) $\frac{3x\sqrt{x}}{1-9x^3}$ (c) $\frac{3x}{1-9x^3}$ (d) $\frac{3}{1+9x^3}$

23 If $y = \sec(\tan^{-1} x)$, then $\frac{dy}{dx}$ at $x = 1$ is equal to → JEE Mains 2013

(a) $\frac{1}{\sqrt{2}}$ (b) $\frac{1}{2}$ (c) 1 (d) $\sqrt{2}$

24 If $y = \tan^{-1}\left(\frac{\sqrt{1+x^2} - \sqrt{1-x^2}}{\sqrt{1+x^2} + \sqrt{1-x^2}}\right)$, then $\frac{dy}{dx}$ is equal to

(a) $\frac{x}{\sqrt{1-x^4}}$ (b) $-\frac{x}{\sqrt{1-x^2}}$ (c) $\frac{2x}{\sqrt{1-x^4}}$ (d) None of these

25 If $\sqrt{1-x^2} + \sqrt{1-y^2} = a(x-y)$, then $\frac{dy}{dx}$ is equal to

(a) $\sqrt{\frac{1-x^2}{1-y^2}}$ (b) $\sqrt{\frac{1-y^2}{1-x^2}}$ (c) $\sqrt{\frac{x^2-1}{1-y^2}}$ (d) $\sqrt{\frac{y^2-1}{1-x^2}}$

26 If $y = \sin^{-1}(x\sqrt{1-x} + \sqrt{x}\sqrt{1-x^2})$, then $\frac{dy}{dx}$ is equal to

(a) $\frac{-2x}{\sqrt{1-x^2}} + \frac{1}{2\sqrt{x-x^2}}$ (b) $\frac{-1}{\sqrt{1-x^2}} - \frac{1}{2\sqrt{x-x^2}}$
(c) $\frac{1}{\sqrt{1-x^2}} + \frac{1}{2\sqrt{x-x^2}}$ (d) None of these

27 If $y = \frac{a^{\cos^{-1} x}}{1+a^{\cos^{-1} x}}$ and $z = a^{\cos^{-1} x}$, then $\frac{dy}{dz}$ is equal to

(a) $-\frac{1}{1+a^{\cos^{-1} x}}$ (b) $\frac{1}{1+a^{\cos^{-1} x}}$
(c) $\frac{1}{(1+a^{\cos^{-1} x})^2}$ (d) None of these

28 Let $g(x)$ be the inverse of $f(x)$ such that $f'(x) = \frac{1}{1+x^5}$, then $\frac{d^2(g(x))}{dx^2}$ is equal to

(a) $\frac{1}{1+(g(x))^5}$ (b) $\frac{g'(x)}{1+(g(x))^5}$
(c) $5(g(x))^4(1+(g(x))^5)$ (d) $1+(g(x))^5$

29 $\frac{d^2x}{dy^2}$ is equal to → AIEEE 2011

(a) $-\left(\frac{d^2y}{dx^2}\right)^{-1}\left(\frac{dy}{dx}\right)^{-3}$ (b) $\left(\frac{d^2y}{dx^2}\right)\left(\frac{dy}{dx}\right)^{-2}$
(c) $-\left(\frac{d^2y}{dx^2}\right)\left(\frac{dy}{dx}\right)^{-3}$ (d) $\left(\frac{d^2y}{dx^2}\right)^{-1}$

30 If $y = \tan^{-1}\left(\frac{\log(e/x^2)}{\log(ex^2)}\right) + \tan^{-1}\left(\frac{3+2\log x}{1-6\log x}\right)$, then $\frac{d^2y}{dx^2}$ is

(a) 2 (b) 1
(c) 0 (d) –1

31 If graph of $y = f(x)$ is symmetrical about the Y-axis and that of $y = g(x)$ is symmetrical about the origin and if $h(x) = f(x) \cdot g(x)$, then $\frac{d^2h(x)}{dx^2}$ at $x = 0$ is

(a) $f(0)g(0)$
(b) 0
(c) can't be determined
(d) None of the above

32 If $\begin{vmatrix} f'(x) & f(x) \\ f''(x) & f'(x) \end{vmatrix} = 0$, where $f(x)$ is continuously differentiable function with $f'(x) \neq 0$ and satisfies $f(0) = 1$ and $f'(0) = 2$, then $\lim_{x \to 0} \frac{f(x)-1}{x}$ is

(a) 1 (b) 2
(c) $\frac{1}{2}$ (d) 0

33 If $f(x) = \begin{vmatrix} 3x^2 & \cos x & \sin x \\ 6 & -1 & 0 \\ P & P^2 & P^3 \end{vmatrix}$, where P is a constant.

Then, $\frac{d^2}{dx^2}\{f(x)\}$ at $x = 0$ is equal to

(a) P (b) $P + P^2$
(c) $P + P^3$ (d) independent of P

34 Which of the following statements is/are true?

Statement I If $y = (\log x)^{\log x}$, then
$$\frac{dy}{dx} = (\log x)^{\log x}\left[\frac{1}{x} + \frac{\log(\log x)}{x}\right].$$

Statement II If $y = \cos(a \cos x + b \sin x)$ for some constants a and b, then
$$y' = (a \sin x - b \cos x) \sin(a \cos x + b \sin x)$$

(a) Only I is true
(b) Only II is true
(c) Both I and II are true
(d) Neither I nor II is true

35 **Statement I** If $u = f(\tan x), v = g(\sec x)$ and $f'(1) = 2$, $g'(\sqrt{2}) = 4$, then $\left(\frac{du}{dv}\right)_{x=\pi/4} = \frac{1}{\sqrt{2}}$.

Statement II If $u = f(x), v = g(x)$, then the derivative of f with respect to g is $\frac{du}{dv} = \frac{du/dx}{dv/dx}$.

(a) Statement I is true, Statement II is true; Statement II is a correct explanation for Statement I
(b) Statement I is true, Statement II is true; Statement II is not a correct explanation for Statement I
(c) Statement I is true; Statement II is false
(d) Statement I is false; Statement II is true

DAY PRACTICE SESSION 2

PROGRESSIVE QUESTIONS EXERCISE

1 For $x \in R, f(x) = |\log 2 - \sin x|$ and $g(x) = f(f(x))$, then
(a) g is not differentiable at $x = 0$
(b) $g'(0) = \cos(\log 2)$
(c) $g'(0) = -\cos(\log 2)$
(d) g is differentiable at $x = 0$ and $g'(0) = -\sin(\log 2)$

2 If $y = \sin x \cdot \sin 2x \cdot \sin 3x \dots \sin nx$, then y' is
(a) $\sum_{k=1}^{n} k \cdot \tan kx$ (b) $y \cdot \sum_{k=1}^{n} k \cot kx$
(c) $y \cdot \sum_{k=1}^{n} k \cdot \tan kx$ (d) $\sum_{k=1}^{n} \cot kx$

3 If $3f(x) - 2f(1/x) = x$, then $f'(2)$ is equal to
(a) $\frac{2}{7}$ (b) $\frac{1}{2}$ (c) 2 (d) $\frac{7}{2}$

4 If $f(x) = (\cos x + i \sin x) \cdot (\cos 2x + i \sin 2x)$ $(\cos 3x + i \sin 3x) \dots (\cos nx + i \sin nx)$ and $f(1) = 1$, then $f''(1)$ is equal to
(a) $\frac{n(n+1)}{2}$ (b) $\left[\frac{n(n+1)}{2}\right]^2$
(c) $-\left[\frac{n(n+1)}{2}\right]^2$ (d) None of these

5 If $\sqrt{x^2 + y^2} = ae^{\tan^{-1}\left(\frac{y}{x}\right)}$, $a>0$ assuming $y > 0$, then $y''(0)$ is equal to
(a) $\frac{2}{a}e^{-\pi/2}$ (b) $-\frac{2}{a}e^{\pi/2}$
(c) $-\frac{2}{a}e^{-\pi/2}$ (d) None of these

6 If $y = |\sin x|^{|x|}$, then the value of $\frac{dy}{dx}$ at $x = -\frac{\pi}{6}$ is
(a) $\frac{2^{\frac{-\pi}{6}}}{6}[6\log 2 - \sqrt{3}\pi]$ (b) $\frac{2^{\frac{\pi}{6}}}{6}[6\log 2 + \sqrt{3}\pi]$
(c) $\frac{2^{-\frac{\pi}{6}}}{6}[6\log 2 + \sqrt{3}\pi]$ (d) None of these

7 The solution set of $f'(x) > g'(x)$, where $f(x) = \frac{1}{2}(5)^{2x+1}$ and $g(x) = 5^x + 4x \log_e 5$ is
(a) $(1, \infty)$ (b) $(0, 1)$ (c) $(\infty, 0)$ (d) $(0, \infty)$

8 Let $f''(x) = -f(x)$, where $f(x)$ is a continuous double differentiable function and $g(x) = f'(x)$.
If $F(x) = \left[f\left(\frac{x}{2}\right)\right]^2 + \left[g\left(\frac{x}{2}\right)\right]^2$ and $F(5) = 5$, then $F(10)$ is equal to
(a) 0 (b) 5
(c) 10 (d) 25

9 If $f(2) = 4, f'(2) = 3, f''(2) = 1$, then $(f^{-1})''(4)$ is equal to
(a) $\frac{-1}{9}$ (b) $\frac{-1}{81}$
(c) $\frac{-1}{27}$ (d) $\frac{-1}{3}$

10 If $f(x) = \sin(\sin x)$ and $f''(x) + \tan x f'(x) + g(x) = 0$, then $g(x)$ is equal to → JEE Mains 2013
(a) $\cos^2 x \cos(\sin x)$ (b) $\sin^2 x \cos(\cos x)$
(c) $\sin^2 x \sin(\cos x)$ (d) $\cos^2 x \sin(\sin x)$

11 If $x = a \cos t \sqrt{\cos 2t}$ and $y = a \sin t \sqrt{\cos 2t}$ (where, $a > 0$), then $\left|\frac{\left[1 + \left(\frac{dy}{dx}\right)^2\right]^{\frac{3}{2}}}{\frac{d^2y}{dx^2}}\right|$ at $\frac{\pi}{6}$ is given by
(a) $\frac{a}{3}$ (b) $a\sqrt{2}$
(c) $\frac{\sqrt{2}}{3a}$ (d) $\frac{\sqrt{2}a}{3}$

12 Let $f(x) = e^{\ln e^{g(x)}}$ and $g(x+1) = x + g(x) \ \forall \ x \in R$. If $n \in I^+$, then $\frac{f'\left(n + \frac{1}{3}\right)}{f\left(n + \frac{1}{3}\right)} - \frac{f'\left(\frac{1}{3}\right)}{f\left(\frac{1}{3}\right)}$ is equal to
(a) $3\left(1 + \frac{1}{2} + \frac{1}{3} + \dots + \frac{1}{n}\right)$ (b) $3\left(1 + \frac{1}{3} + \frac{1}{5} + \dots + \frac{1}{2n-1}\right)$
(c) n (d) 1

13 If the function $f(x) = -4e^{\frac{1-x}{2}} + 1 + x + \frac{x^2}{2} + \frac{x^3}{3}$ and $g(x) = f^{-1}(x)$, then the value of $g'\left(\frac{-7}{6}\right)$ is equal to
(a) $\frac{1}{5}$ (b) $-\frac{1}{5}$
(c) $\frac{6}{7}$ (d) $-\frac{6}{7}$

14. If $f(x) = (x-1)^{100}(x-2)^{2(99)}(x-3)^{3(98)} \dots (x-100)^{100}$, then the value of $\frac{f'(101)}{f(101)}$ is
(a) 5050 (b) 2575
(c) 3030 (d) 1250

15 The derivative of the function represented parametrically as $x = 2t - |t|$, $y = t^3 + t^2|t|$ at $t = 0$ is
(a) -1 (b) 0
(c) 1 (d) does not exist.

ANSWERS

SESSION 1	**1** (a)	**2** (a)	**3** (b)	**4** (b)	**5** (b)	**6** (d)	**7** (b)	**8** (a)	**9** (a)	**10** (b)
	11 (a)	**12** (b)	**13** (a)	**14** (c)	**15** (a)	**16** (a)	**17** (d)	**18** (a)	**19** (b)	**20** (d)
	21 (a)	**22** (a)	**23** (a)	**24** (a)	**25** (b)	**26** (c)	**27** (c)	**28** (c)	**29** (c)	**30** (c)
	31 (b)	**32** (b)	**33** (d)	**34** (c)	**35** (a)					

SESSION 2	**1** (b)	**2** (b)	**3** (b)	**4** (c)	**5** (c)	**6** (a)	**7** (d)	**8** (b)	**9** (c)	**10** (d)
	11 (d)	**12** (c)	**13** (a)	**14** (a)	**15** (b)					

Hints and Explanations

SESSION 1

1 When $\frac{\pi}{2} < x < \pi$, $\cos x < 0$, so that

$|\cos x| = -\cos x$,

i.e. $f(x) = -\cos x$, $f'(x) = \sin x$

Hence, $f'\left(\frac{3\pi}{4}\right) = \sin\left(\frac{3\pi}{4}\right) = \frac{1}{\sqrt{2}}$

2 We have, $f(x) = |x-1|$ $\quad [\because x > 2]$

$$f[f(x)] = f(x-1) = |(x-1)-1| = |x-2|$$

$$g(x) = f[f\{f(x)\}] = f(x-2) = |(x-2)-1| = |x-3| = \begin{cases} x-3, & \text{if } x \geq 3 \\ -x+3, & \text{if } 2 \leq x < 3 \end{cases}$$

$$\therefore g'(x) = \begin{cases} 1, & \text{if } x \geq 3 \\ -1, & \text{if } 2 \leq x < 3 \end{cases}$$

3 $\frac{dy}{dx} = -[(2-x)(3-x)\ldots(n-x) + (1-x)(3-x)\ldots(n-x) + \ldots + (1-x)(2-x)\ldots(n-1-x)]$

$$\Rightarrow \left(\frac{dy}{dx}\right)_{x=1} = -[(n-1)! + 0 + \ldots + 0] = (-1)(n-1)!$$

4 We have, $f(x) = x^n$

$$\Rightarrow f(1) = 1 = {}^nC_0$$

$$\frac{f'(1)}{1!} = \frac{n}{1!} = {}^nC_1$$

$$\Rightarrow \frac{f''(1)}{2!} = \frac{n(n-1)}{2!} = {}^nC_2$$

$$\frac{f'''(1)}{3!} = \frac{n(n-1)(n-2)}{3!} = {}^nC_3$$

$$\vdots \qquad \vdots$$

$$\frac{f^n(1)}{n!} = \frac{n!}{n!} = {}^nC_n$$

$$\therefore f(1) - \frac{f'(1)}{1!} + \frac{f''(1)}{2!} - \frac{f'''(1)}{3!} + \ldots + \frac{(-1)^n f^n(1)}{n!}$$

$$= {}^nC_0 - {}^nC_1 + {}^nC_2 - {}^nC_3 + \ldots + (-1)^n\, {}^nC_n$$

$$= (1-1)^n = 0$$

5 Let $y = \frac{x^2 - x}{x^2 + 2x}$

$$\Rightarrow x = \frac{2y+1}{-y+1};\ x \neq 0$$

$$\Rightarrow f^{-1}(x) = \frac{2x+1}{-x+1}$$

$$\therefore \frac{d}{dx}[f^{-1}(x)] = \frac{(-x+1)\cdot 2 - (2x+1)(-1)}{(-x+1)^2} = \frac{3}{(-x+1)^2}$$

6 We have, $f(x) = 2 + |x| - |x-1| - |x+1|$

$$\therefore f(x) = \begin{cases} 2 - x + (x-1) + (x+1), & \text{if } x < -1 \\ 2 - x + (x-1) - (x+1), & \text{if } -1 \leq x < 0 \\ 2 + x + (x-1) - (x+1), & \text{if } 0 \leq x < 1 \\ 2 + x - (x-1) - (x+1), & \text{if } x \geq 1 \end{cases}$$

$$= \begin{cases} x+2, & \text{if } x < -1 \\ -x, & \text{if } -1 \leq x < 0 \\ x, & \text{if } 0 \leq x < 1 \\ 2-x, & \text{if } x \geq 1 \end{cases}$$

$$\Rightarrow f'(x) = \begin{cases} 1, & \text{if } x < -1 \\ -1, & \text{if } -1 \leq x < 0 \\ 1, & \text{if } 0 \leq x < 1 \\ -1, & \text{if } x \geq 1 \end{cases}$$

Hence, $f'\left(-\frac{1}{2}\right) + f'\left(\frac{1}{2}\right) + f'\left(\frac{3}{2}\right) + f'\left(\frac{5}{2}\right) = (-1) + 1 + (-1) + (-1) = -2$

7 We have, $f(x) = |\log_e |x||$

$$\therefore f(x) = \begin{cases} \log(-x), & x < -1 \\ -\log(-x), & -1 < x < 0 \\ -\log x, & 0 < x < 1 \\ \log x, & x > 1 \end{cases}$$

$$\Rightarrow f'(x) = \begin{cases} \frac{1}{x}, & x < -1 \\ -\frac{1}{x}, & -1 < x < 0 \\ -\frac{1}{x}, & 0 < x < 1 \\ \frac{1}{x}, & x > 1 \end{cases}$$

Clearly, $f'(x) = \frac{1}{x}$ for $|x| > 1$

8 When $0 < x < \frac{\pi}{4}$, $\cos x > \sin x$

$\therefore$ $\cos x - \sin x > 0$

Also, when $\frac{\pi}{4} < x < \pi$, $\cos x < \sin x$

$\therefore \cos x - \sin x < 0$

$\therefore |\cos x - \sin x| = -(\cos x - \sin x)$, when $\frac{\pi}{4} < x < \pi$

$$\Rightarrow f'(x) = \sin x + \cos x$$

$$\Rightarrow f'\left(\frac{\pi}{2}\right) = \sin\frac{\pi}{2} + \cos\frac{\pi}{2} = 1 + 0 = 1$$

9 $$f(x) = \frac{2\sin x \cdot \cos x \cdot \cos 2x \cdot \cos 4x \cdot \cos 8x \cdot \cos 16x}{2\sin x}$$

$$= \frac{\sin 2x \cos 2x \cos 4x \cdot \cos 8x \cdot \cos 16x}{2\sin x}$$

$$= \frac{\sin 32x}{2^5 \sin x}$$

$\therefore\ f'(x) = \frac{1}{32} \cdot \frac{32\cos 32x \cdot \sin x - \cos x \cdot \sin 32x}{\sin^2 x}$

$\Rightarrow f'\left(\frac{\pi}{4}\right) = \frac{32 \times \frac{1}{\sqrt{2}} - \frac{1}{\sqrt{2}} \times 0}{32\left[\frac{1}{\sqrt{2}}\right]^2} = \sqrt{2}$

10 $\therefore\ \sin y = x \sin(a + y)$

$\Rightarrow\ x = \frac{\sin y}{\sin(a + y)}$

On differentiating w.r.t. y, we get

$\frac{dx}{dy} = \frac{\sin(a + y)\cos y - \sin y \cos(a + y)}{\sin^2(a + y)}$

$\Rightarrow \frac{dx}{dy} = \frac{\sin a}{\sin^2(a + y)}$

$\Rightarrow \frac{dy}{dx} = \frac{\sin^2(a + y)}{\sin a}$

11 Given, $y = (1 - x)(1 + x^2)(1 + x^4)\ldots(1 + x^{2n})$

or $y = \frac{(1 - x^2)(1 + x^2)\ldots(1 + x^{2n})}{(1 + x)}$

$= \frac{1 - (x)^{4n}}{(1 + x)}$

$\therefore \frac{dy}{dx} = \frac{(1 + x)\cdot(0 - 4n \cdot x^{4n-1}) - (1 - x^{4n})\cdot 1}{(1 + x)^2}$

$\therefore \left(\frac{dy}{dx}\right)_{x=0} = -1$

12 We have, $f:(-1,1) \to R$

$f(0) = -1,\ f'(0) = 1$

$g(x) = [f(2f(x) + 2)]^2$

$\Rightarrow g'(x) = 2[f(2f(x) + 2)] \times f'(2f(x) + 2) \times 2f'(x)$

$\Rightarrow g'(0) = 2[f\{2f(0) + 2\}] \times f'\{2f(0) + 2\} \times 2f'(0)$

$= 2[f(0)] \times f'(0) \times 2f'(0)$

$= 2 \times (-1) \times 1 \times 2 \times 1 = -4$

13 Let $f(x) = Ax^2 + Bx + C$

$\therefore\ f(1) = A + B + C$

and $f(-1) = A - B + C$

$\because\ f(1) = f(-1)$ [given]

$\Rightarrow A + B + C = A - B + C$

$\Rightarrow\ 2B = 0 \Rightarrow B = 0$

$\therefore\ f(x) = Ax^2 + C$

$\Rightarrow f'(x) = 2Ax$

$\therefore\ f'(a) = 2Aa$

$f'(b) = 2Ab$ and $f'(c) = 2Ac$

Also, a, b, c are in AP.

So, $2Aa, 2Ab$ and $2Ac$ are in AP.

Hence, $f'(a), f'(b)$ and $f'(c)$ are also in AP.

14 Since, $f(x)$ is odd.

$\therefore\ f(-x) = -f(x)$

$\Rightarrow\ f'(-x)(-1) = -f'(x)$

$\Rightarrow\ f'(-x) = f'(x)$ $\therefore$

$f'(-3) = f'(3) = -2$

15 Since, $fog = I \Rightarrow fog(x) = x$ for all x

$\Rightarrow\ f'(g(x))g'(x) = 1$ for all x

$\Rightarrow\ f'(g(a)) = \frac{1}{g'(a)} = \frac{1}{2}$

$\Rightarrow\ f'(b) = \frac{1}{2}$ $\quad [\because g(a) = b]$

16 $x^{2x} - 2x^x \cot y - 1 = 0$ ...(i)

Now, $x = 1$,

$1 - 2\cot y - 1 = 0$

$\Rightarrow\ \cot y = 0$

$\Rightarrow\ y = \frac{\pi}{2}$

On differentiating Eq. (i) w.r.t. x, we get

$2x^{2x}(1 + \log x) - 2[x^x(-\operatorname{cosec}^2 y)\frac{dy}{dx} + \cot y\ x^x(1 + \log x)] = 0$

At $\left(1, \frac{\pi}{2}\right)$, $2(1 + \log 1)$

$-2\left\{1(-1)\left(\frac{dy}{dx}\right)_{\left(1,\frac{\pi}{2}\right)} + 0\right\} = 0$

$\Rightarrow 2 + 2\left(\frac{dy}{dx}\right)_{\left(1,\frac{\pi}{2}\right)} = 0$

$\therefore\ \left(\frac{dy}{dx}\right)_{\left(1,\frac{\pi}{2}\right)} = -1$

17 Given that, $x^m y^n = (x + y)^{m+n}$

Taking log on both sides, we get

$m\log x + n\log y = (m + n)\log(x + y)$

On differentiating w.r.t. x, we get

$\frac{m}{x} + \frac{n}{y}\frac{dy}{dx} = \frac{(m + n)}{(x + y)}\left(1 + \frac{dy}{dx}\right)$

$\Rightarrow \frac{dy}{dx}\left(\frac{m + n}{x + y} - \frac{n}{y}\right) = \frac{m}{x} - \frac{m + n}{x + y}$

$\Rightarrow \frac{dy}{dx}\left(\frac{my + ny - nx - ny}{y(x + y)}\right) = \frac{mx + my - mx - nx}{x(x + y)}$

$\therefore\ \frac{dy}{dx} = \frac{y}{x}$

18 $\frac{d}{dx}(y) = n(x + \sqrt{1 + x^2})^{n-1}\left(1 + \frac{x}{\sqrt{1 + x^2}}\right)$

$\Rightarrow (\sqrt{1 + x^2})\frac{dy}{dx} = n(x + \sqrt{1 + x^2})^n$

$\Rightarrow \frac{d^2y}{dx^2}(\sqrt{1 + x^2}) + \frac{dy}{dx}\left(\frac{x}{\sqrt{1 + x^2}}\right)$

$= n^2(x + \sqrt{1 + x^2})^{n-1}\left(1 + \frac{x}{\sqrt{1 + x^2}}\right)$

$\Rightarrow \frac{\frac{d^2y}{dx^2}(1 + x^2) + \frac{dy}{dx}\cdot x}{\sqrt{1 + x^2}} = \frac{n^2(x + \sqrt{1 + x^2})^n}{\sqrt{1 + x^2}}$

$\Rightarrow (1 + x^2)\frac{d^2y}{dx^2} + x\frac{dy}{dx} = n^2(x + \sqrt{1 + x^2})^n$

$\Rightarrow (1 + x^2)\frac{d^2y}{dx^2} + x\frac{dy}{dx} = n^2 y$

19 We have, $x = \frac{2t}{1 + t^2}, y = \frac{1 - t^2}{1 + t^2}$

Put $t = \tan\theta$

$\therefore\ x = \frac{2\tan\theta}{1 + \tan^2\theta} = \sin 2\theta$ and

$y = \frac{1 - \tan^2\theta}{1 + \tan^2\theta} = \cos 2\theta$

$\therefore \frac{dy}{dx} = \frac{dy/d\theta}{dx/d\theta} = \frac{-2\sin 2\theta}{2\cos 2\theta} = -\tan 2\theta$

$= \frac{-2\tan\theta}{1 - \tan^2\theta}$

$= \frac{-2t}{1 - t^2} = \frac{2t}{t^2 - 1}$

20 $\because \frac{dx}{dt} = \frac{1}{2\sqrt{a^{\sin^{-1}t}}}\left(a^{\sin^{-1}t} \times \frac{1}{\sqrt{1 - t^2}}\right)$

and $\frac{dy}{dt} = \frac{1}{2\sqrt{a^{\cos^{-1}t}}}\left(a^{\cos^{-1}t} \times \frac{-1}{\sqrt{1 - t^2}}\right)$

$\therefore \frac{dy}{dx} = -\frac{\sqrt{a^{\sin^{-1}t}}}{\sqrt{a^{\cos^{-1}t}}}\left(\frac{a^{\cos^{-1}t}}{a^{\sin^{-1}t}} \times 1\right)$

$= -\frac{a^{\sqrt{\cos^{-1}t}}}{a^{\sqrt{\sin^{-1}t}}}$

$\therefore\ 1 + \left(\frac{dy}{dx}\right)^2 = 1 + \frac{a^{\cos^{-1}t}}{a^{\sin^{-1}t}} = 1 + \frac{y^2}{x^2}$

$= \frac{x^2 + y^2}{x^2}$

21 $\because y = \sec^{-1}\left[\frac{\sqrt{x} + 1}{\sqrt{x} - 1}\right] + \sin^{-1}\left[\frac{\sqrt{x} - 1}{\sqrt{x} + 1}\right]$

$= \cos^{-1}\left[\frac{\sqrt{x} - 1}{\sqrt{x} + 1}\right] + \sin^{-1}\left[\frac{\sqrt{x} - 1}{\sqrt{x} + 1}\right] = \frac{\pi}{2}$

$\Rightarrow \frac{dy}{dx} = 0\left[\because \sin^{-1}x + \cos^{-1}x = \frac{\pi}{2}\right]$

22 Let $y = \tan^{-1}\left(\frac{6x\sqrt{x}}{1 - 9x^1}\right)$

$= \tan^{-1}\left[\frac{2\cdot(3x^{3/2})}{1 - (3x^{3/2})^2}\right]$

$= 2\tan^{-1}(3x^{3/2})$

$\left[\because 2\tan^{-1}x = \tan^{-1}\frac{2x}{1 - x^2}\right]$

$\therefore \frac{dy}{dx} = 2\cdot\frac{1}{1 + (3x^{3/2})^2}\cdot 3 \times \frac{3}{2}(x)^{1/2}$

$= \dfrac{9}{1+9x^3} \cdot \sqrt{x}$

$\therefore g(x) = \dfrac{9}{1+9x^3}$

23 Given, $y = \sec(\tan^{-1} x)$

Let $\tan^{-1} x = \theta$

$\Rightarrow \quad x = \tan\theta$

$\therefore \quad y = \sec\theta = \sqrt{1+x^2}$

On differentiating both sides w.r.t. x, we get

$\dfrac{dy}{dx} = \dfrac{1}{2\sqrt{1+x^2}} \cdot 2x$

At $x = 1$

$\dfrac{dy}{dx} = \dfrac{1}{\sqrt{2}}$

24 Given, $y = \tan^{-1}\left[\dfrac{\sqrt{1+x^2} - \sqrt{1-x^2}}{\sqrt{1+x^2} + \sqrt{1-x^2}}\right]$.

Put $x^2 = \cos 2\theta$

$\therefore \quad y = \tan^{-1}\left[\dfrac{\cos\theta - \sin\theta}{\cos\theta + \sin\theta}\right]$

$= \tan^{-1}\left[\dfrac{1-\tan\theta}{1+\tan\theta}\right]$

$= \tan^{-1}\left[\tan\left(\dfrac{\pi}{4} - \theta\right)\right]$

$= \dfrac{\pi}{4} - \theta = \dfrac{\pi}{4} - \dfrac{1}{2}\cos^{-1} x^2$

$\therefore \quad \dfrac{dy}{dx} = 0 + \dfrac{x}{\sqrt{1-x^4}} = \dfrac{x}{\sqrt{1-x^4}}$

25 On putting $x = \sin\theta$ and $y = \sin\phi$, we get

Given equation becomes

$\cos\theta + \cos\phi = a(\sin\theta - \sin\phi)$

$\Rightarrow 2\cos\left(\dfrac{\theta+\phi}{2}\right)\cos\left(\dfrac{\theta-\phi}{2}\right)$

$= a\left\{2\cos\left(\dfrac{\theta+\phi}{2}\right)\sin\left(\dfrac{\theta-\phi}{2}\right)\right\}$

$\Rightarrow \dfrac{\theta-\phi}{2} = \cot^{-1} a$

$\Rightarrow \theta - \phi = 2\cot^{-1} a$

$\Rightarrow \sin^{-1} x - \sin^{-1} y = 2\cot^{-1} a$

$\Rightarrow \dfrac{1}{\sqrt{1-x^2}} - \dfrac{1}{\sqrt{1-y^2}}\dfrac{dy}{dx} = 0$

$\therefore \quad \dfrac{dy}{dx} = \sqrt{\dfrac{1-y^2}{1-x^2}}$

26 On putting $x = \sin A$ and $\sqrt{x} = \sin B$

$y = \sin^{-1}(\sin A\sqrt{1-\sin^2 B}$

$+ \sin B\sqrt{1-\sin^2 A})$

$= \sin^{-1}(\sin A\cos B + \sin B\cos A)$

$= \sin^{-1}[\sin(A+B)]$

$= A + B = \sin^{-1} x + \sin^{-1}\sqrt{x}$

$\Rightarrow \quad \dfrac{dy}{dx} = \dfrac{1}{\sqrt{1-x^2}} + \dfrac{1}{2\sqrt{x-x^2}}$

27 $y = \dfrac{a^{\cos^{-1} x}}{1+a^{\cos^{-1} x}}, z = a^{\cos^{-1} x}$

$\Rightarrow \quad y = \dfrac{z}{1+z}$

$\Rightarrow \quad \dfrac{dy}{dz} = \dfrac{(1+z)1 - z(1)}{(1+z)^2}$

$= \dfrac{1}{(1+z)^2}$

$= \dfrac{1}{(1+a^{\cos^{-1} x})^2}$

28 Since $g(x)$ is the inverse of $f(x)$

$\therefore \quad f(g(x)) = x$

$\Rightarrow \quad f'(g(x))\cdot g'(x) = 1$

$\Rightarrow \quad g'(x) = \dfrac{1}{f'(g(x))} = 1 + (g(x))^5$

$\Rightarrow \quad g''(x) = 5(g(x))^4 \cdot g'(x)$.

$= 5(g(x))^4(1+(g(x))^5)$

29 Since, $\dfrac{dx}{dy} = \left(\dfrac{dy}{dx}\right)^{-1}$

$\Rightarrow \quad \dfrac{d^2x}{dy^2} = -\left(\dfrac{dy}{dx}\right)^{-2}\dfrac{d^2y}{dx^2}\cdot\dfrac{dx}{dy}$

$= -\left(\dfrac{d^2y}{dx^2}\right)\left(\dfrac{dy}{dx}\right)^{-3}$

30 Given, $y = \tan^{-1}\left(\dfrac{\log(e/x^2)}{\log(ex^2)}\right)$

$+ \tan^{-1}\left(\dfrac{3+2\log x}{1-6\log x}\right)$

$\therefore \quad y = \tan^{-1}\left(\dfrac{\log e - \log x^2}{\log e + \log x^2}\right)$

$+ \tan^{-1}\left(\dfrac{3+2\log x}{1-6\log x}\right)$

$= \tan^{-1}\left(\dfrac{1-2\log x}{1+2\log x}\right)$

$+ \tan^{-1}\left(\dfrac{3+2\log x}{1-6\log x}\right)$

$= \tan^{-1}(1) - \tan^{-1}(2\log x)$

$+ \tan^{-1}(3) + \tan^{-1}(2\log x)$

$= \tan^{-1}(1) + \tan^{-1}(3)$

Now, $\dfrac{dy}{dx} = 0$ and $\dfrac{d^2y}{dx^2} = 0$

31 Since, $y = f(x)$ is symmetrical about the Y-axis

$\therefore \quad f(x)$ is an even function.

Also, as $y = g(x)$ is symmetrical about the origin

$\therefore \quad g(x)$ is an odd function.

Thus, $h(x) = f(x)\cdot g(x)$ is an odd function.

or $\quad h(x) = -h(-x)$

Now, $h'(x) = h'(-x)$

and $\quad h''(x) = -h''(-x)$

$\Rightarrow h''(0) = -h''(0)$

$\Rightarrow h''(0) = 0$

32 Since, $\begin{vmatrix} f'(x) & f(x) \\ f''(x) & f'(x) \end{vmatrix} = 0$

$\therefore \quad (f'(x))^2 - f''(x)\cdot f(x) = 0$

$\Rightarrow \quad \dfrac{(f'(x))^2 - f''(x)\cdot f(x)}{(f'(x))^2} = 0$

$\Rightarrow \quad \dfrac{d}{dx}\left[\dfrac{f(x)}{f'(x)}\right] = 0$

$\Rightarrow \quad \dfrac{f(x)}{f'(x)} = c$, (constant)

On putting $x = 0$, we get

$\dfrac{1}{2} = c$

$\Rightarrow \quad \dfrac{f(x)}{f'(x)} = \dfrac{1}{2}$

$\Rightarrow \quad \dfrac{f'(x)}{f(x)} = 2$

$\Rightarrow \quad \dfrac{d}{dx}(\log f(x)) = 2$

$\Rightarrow \quad \log(f(x)) = 2x + k$

On putting $x = 0$, we get $0 = k$

$\Rightarrow \log(f(x)) = 2x$

$\Rightarrow \quad f(x) = e^{2x}$

Now, $\lim\limits_{x\to 0}\dfrac{f(x)-1}{x} = \lim\limits_{x\to 0}\dfrac{e^{2x}-1}{2x}\cdot 2 = 2$.

33 $f''(x)$

$= \begin{vmatrix} \dfrac{d^2}{dx^2}(3x^2) & \dfrac{d^2}{dx^2}(\cos x) & \dfrac{d^2}{dx^2}(\sin x) \\ 6 & -1 & 0 \\ P & P^2 & P^3 \end{vmatrix}$

$= \begin{vmatrix} 6 & -\cos x & -\sin x \\ 6 & -1 & 0 \\ P & P^2 & P^3 \end{vmatrix}$

$\therefore \quad f''(0) = \begin{vmatrix} 6 & -1 & 0 \\ 6 & -1 & 0 \\ P & P^2 & P^3 \end{vmatrix} = 0$, which is independent of P.

34 I. Let $y = (\log x)^{\log x}$

On taking log both sides, we get

$\log y = \log(\log x)\log x$

$\Rightarrow \quad \log y = \log x \log[\log x]$

$[\because \log m^n = n\log m]$

On differentiating both sides w.r.t. x, we get

$\dfrac{1}{y}\dfrac{dy}{dx} = (\log x)\dfrac{d}{dx}\{\log(\log x)\}$

$+ \log(\log x)\dfrac{d}{dx}\log x$

$= (\log x)\dfrac{1}{\log x}\dfrac{1}{x} + \log(\log x)\dfrac{1}{x}$

$= \dfrac{1}{x}\{1 + \log(\log x)\}$

$\Rightarrow \frac{dy}{dx} = \frac{y}{x}\{1 + \log(\log x)\}$

$= \frac{(\log x)^{\log x}}{x}\{1 + \log(\log x)\}$

$= (\log x)^{\log x}\left[\frac{1}{x} + \frac{\log(\log x)}{x}\right]$

II. Let $y = \cos(a\cos x + b\sin x)$.
On differentiating w.r.t. x, we get

$\frac{d}{dx}\{\cos(a\cos x + b\sin x)\}$

$= -\sin(a\cos x + b\sin x)\frac{d}{dx}(a\cos x + b\sin x)$

$= -\sin(a\cos x + b\sin x)[-a\sin x + b\cos x]$

$= (a\sin x - b\cos x)\sin(a\cos x + b\sin x)$

35 Given, $u = f(\tan x)$

$\Rightarrow \frac{du}{dx} = f'(\tan x)\sec^2 x$

and $v = g(\sec x)$

$\Rightarrow \frac{dv}{dx} = g'(\sec x)\sec x\tan x$

$\therefore \frac{du}{dv} = \frac{(du/dx)}{(dv/dx)} = \frac{f'(\tan x)}{g'(\sec x)}\cdot\frac{1}{\sin x}$

$\therefore \left(\frac{du}{dv}\right)_{x=\pi/4} = \frac{f'(1)}{g'(\sqrt{2})}\cdot\sqrt{2}$

$= \frac{2}{4}\cdot\sqrt{2} = \frac{1}{\sqrt{2}}$

SESSION 2

1 We have, $f(x) = |\log 2 - \sin x|$ and $g(x) = f(f(x)), x \in R$

Note that, for $x \to 0$, $\log 2 > \sin x$

$\therefore f(x) = \log 2 - \sin x$

$\Rightarrow g(x) = \log 2 - \sin(f(x))$

$= \log 2 - \sin(\log 2 - \sin x)$

Clearly, $g(x)$ is differentiable at $x = 0$ as $\sin x$ is differentiable.

Now,

$g'(x) = -\cos(\log 2 - \sin x)(-\cos x)$

$= \cos x.\cos(\log 2 - \sin x)$

$\Rightarrow g'(0) = 1\cdot\cos(\log 2)$

2 We have,

$y = \sin x\cdot\sin 2x\cdot\sin 3x\ldots\sin nx$

$\therefore y' = \cos x\cdot\sin 2x\cdot\sin 3x\ldots\sin nx$
$+ \sin x\cdot(2\cos 2x)\sin 3x\ldots\sin nx$
$+ \sin x\cdot\sin 2x(3\cos 3x)\ldots\sin nx$
$+\ldots+ \sin x\sin 2x\sin 3x\ldots(\cos nx)$

(by product rule)

$\Rightarrow y' = \cot x\cdot y + 2\cdot\cot 2x\cdot y + 3\cdot\cot 3x\cdot y + \ldots + n\cdot\cot nx\cdot y$

$\Rightarrow y' = y[\cot x + 2\cot 2x + 3\cot 3x + \ldots + n\cot nx]$

$\Rightarrow y' = y\cdot\sum_{k=1}^{n} k\cot kx$

3 $3f(x) - 2f(1/x) = x$ ...(i)

Let $1/x = y$, then

$3f(1/y) - 2f(y) = 1/y$

$\Rightarrow -2f(y) + 3f(1/y) = 1/y$

$\Rightarrow -2f(x) + 3f(1/x) = 1/x$ ...(ii)

On multiplying Eq. (i) by 3 and Eq. (ii) by 2 and adding, we get

$5f(x) = 3x + \frac{2}{x}$

$\Rightarrow f(x) = \frac{1}{5}\left(3x + \frac{2}{x}\right)$

$\Rightarrow f'(x) = \frac{1}{5}\left(3 - \frac{2}{x^2}\right)$

$\Rightarrow f'(2) = \frac{1}{5}\left(3 - \frac{2}{4}\right) = \frac{1}{2}$

4 $f(x) = (\cos x + i\sin x)$

$(\cos 2x + i\sin 2x)(\cos 3x + i\sin 3x)\ldots(\cos nx + i\sin nx)$

$= \cos(x + 2x + 3x + \ldots + nx) + i\sin(x + 2x + 3x + \ldots + nx)$

$= \cos\frac{n(n+1)}{2}x + i\sin\frac{n(n+1)}{2}x$

$\Rightarrow f'(x) = \left[\frac{n(n+1)}{2}\right]\left[-\sin\frac{n(n+1)}{2}x + i\cos\frac{n(n+1)}{2}x\right]$

$f''(x) = -\left[\frac{n(n+1)}{2}\right]^2\left[\cos\frac{n(n+1)}{2}x + i\sin\frac{n(n+1)}{2}x\right]$

$= -\left[\frac{n(n+1)}{2}\right]^2\cdot f(x)$

$\therefore f''(1) = -\left[\frac{n(n+1)}{2}\right]^2 f(1)$

$= -\left[\frac{n(n+1)}{2}\right]^2$

5 When $x = 0$, $y > 0 \Rightarrow y = ae^{\pi/2}$

On taking log both sides of the given equation, we get

$\frac{1}{2}\log(x^2 + y^2) = \log a + \tan^{-1}\left(\frac{y}{x}\right)$

On differentiating both sides w.r.t. x, we get

$\frac{1}{2}\times\frac{2x + 2yy'}{x^2 + y^2} = \frac{1}{1 + \left(\frac{y}{x}\right)^2}\times\frac{xy' - y}{x^2}$

$\Rightarrow x + yy' = xy' - y$ ...(i)

Again, on differentiating both sides w.r.t. x, we get

$1 + (y')^2 + yy'' = xy'' + y' - y'$

$\Rightarrow 1 + (y')^2 = (x - y)y''$

$\Rightarrow y'' = \frac{1 + (y')^2}{x - y}$

When $x = 0$, we get from Eq. (i),

$y' = -1$

$\Rightarrow y''(0) = \frac{2}{-ae^{\pi/2}} = \frac{-2}{a}e^{-\pi/2}$

6 Given, $y = |\sin x|^{|x|}$

In the neighbourhood of $-\frac{\pi}{6}$, $|x|$ and $|\sin x|$ both are negative

i.e. $y = (-\sin x)^{(-x)}$

On taking log both sides, we get

$\log y = (-x)\cdot\log(-\sin x)$

On differentiating both sides, we get

$\frac{1}{y}\cdot\frac{dy}{dx} = (-x)\left(\frac{1}{-\sin x}\right)\cdot(-\cos x) + \log(-\sin x)\cdot(-1)$

$= -x\cot x - \log(-\sin x)$

$= -[x\cot x + \log(-\sin x)]$

$\Rightarrow \frac{dy}{dx} = -y[x\cot x + \log(-\sin x)]$

$\therefore \left(\frac{dy}{dx}\right)_{\text{at } x=-\frac{\pi}{6}} = \frac{(2)^{-\frac{\pi}{6}}}{6}[6\log 2 - \sqrt{3}\pi]$

7 Since, $f'(x) > g'(x)$

$\Rightarrow \left(\frac{1}{2}\right)5^{2x+1}\log_e 5\times 2 > 5^x\log_e 5 + 4\log_e 5$

$\Rightarrow 5^{2x}\cdot 5 > 5^x + 4$

$\Rightarrow 5\cdot 5^{2x} - 5^x - 4 > 0$

$\Rightarrow (5^x - 1)(5\cdot 5^x + 4) > 0$

$\therefore 5^x > 1$

$\Rightarrow x > 0$

8 Given, $\frac{d}{dx}\{f'(x)\} = -f(x)$

$\Rightarrow g'(x) = -f(x)$

$[\because g(x) = f'(x),\ \text{given}]$

Also, given $F(x)$

$= \left\{f\left(\frac{x}{2}\right)\right\}^2 + \left\{g\left(\frac{x}{2}\right)\right\}^2$

$\Rightarrow F'(x) = 2\left\{f\left(\frac{x}{2}\right)\right\}f'\left(\frac{x}{2}\right)\frac{1}{2} + 2\left\{g\left(\frac{x}{2}\right)\right\}\cdot g'\left(\frac{x}{2}\right)\frac{1}{2} = 0$

Hence, $f(x)$ is constant. Therefore, $F(10) = 5$.

9 Let $y = f(x)$, then $x = f^{-1}(y)$.

Now, $\frac{d^2x}{dy^2} = (f^{-1})''(y)$

$\because \frac{dx}{dy} = \left(\frac{dy}{dx}\right)^{-1}$

$\therefore \frac{d^2x}{dy^2} = \frac{d}{dy}\left(\frac{dy}{dx}\right)^{-1}$

$= \frac{d}{dx}\left(\frac{dy}{dx}\right)^{-1}\cdot\frac{dx}{dy}$

$$= -\left(\frac{dy}{dx}\right)^{-2} \cdot \frac{d^2 y}{dx^2} \cdot \frac{dx}{dy}$$

$$= \frac{-\dfrac{d^2 y}{dx^2}}{\left(\dfrac{dy}{dx}\right)^3}$$

Since, $y = 4$ when $x = 2$

$$\therefore (f^{-1})''(4) = -\frac{\left.\dfrac{d^2 y}{dx^2}\right|_{x=2}}{\left.\left(\dfrac{dy}{dx}\right)^3\right|_{x=2}} = \frac{-1}{27}$$

10 $f(x) = \sin(\sin x)$

$\Rightarrow f'(x) = \cos x \cdot \cos(\sin x)$

$\Rightarrow f''(x) = -\sin x \cdot \cos(\sin x) - \cos^2 x \cdot \sin(\sin x)$

Now, $g(x) = -[f''(x) + f'(x) \cdot \tan x]$

$= \sin x \cdot \cos(\sin x) + \cos^2 x \cdot \sin(\sin x) - \tan x \cdot \cos x \cdot \cos(\sin x)$

$= \sin x \cdot \cos(\sin x) + \cos^2 x \cdot \sin(\sin x) - \sin x \cdot \cos(\sin x)$

$= \cos^2 x \cdot \sin(\sin x)$

11 We have,

$$\frac{dx}{dt} = a\left[-\sin t\sqrt{\cos 2t} - \frac{\cos t \cdot \sin 2t}{\sqrt{\cos 2t}}\right]$$

$$= \frac{-a\sin 3t}{\sqrt{\cos 2t}}$$

and $\dfrac{dy}{dt} = a\left[\cos t\sqrt{\cos 2t} - \dfrac{\sin t \cdot \sin 2t}{\sqrt{\cos 2t}}\right]$

$$= \frac{a\cos 3t}{\sqrt{\cos 2t}}$$

$$\therefore \frac{dy}{dx} = \frac{dy/dt}{dx/dt} = -\cot 3t$$

$$\Rightarrow \frac{d^2 y}{dx^2} = 3\text{cosec}^2 3t \cdot \frac{dt}{dx}$$

$$= \frac{-3\text{cosec}^2 3t \cdot \sqrt{\cos 2t}}{a\sin 3t}$$

$$= -\left(\frac{3}{a}\right)\text{cosec}^3 3t \cdot \sqrt{\cos 2t}$$

$$\therefore \left[1 + \left(\frac{dy}{dx}\right)^2\right]^{3/2} \Bigg/ \frac{d^2 y}{dx^2}$$

$$= (1 + \cot^2 3t)^{3/2} \Bigg/ \left(\frac{-3}{a}\right)\text{cosec}^3 3t\sqrt{\cos 2t}$$

$\Rightarrow \left|\left[1 + \left(\dfrac{dy}{dx}\right)^2\right]^{3/2} \Bigg/ \dfrac{d^2 y}{dx^2}\right|$ at $t = \dfrac{\pi}{6}$ is

$$\frac{a}{3\sqrt{\cos\dfrac{\pi}{3}}} = \frac{\sqrt{2}a}{3}$$

12 Clearly, $f(x) = e^{g(x)}$

Now, as $g(x+1) = x + g(x)$

$\therefore \quad e^{g(x+1)} = e^{x + g(x)} = e^x \cdot e^{g(x)}$

$\Rightarrow \quad f(x+1) = e^x f(x)$

On taking log both sides, we get

$\ln f(x+1) = \ln(e^x \cdot f(x))$

$$\Rightarrow \frac{1}{f(x+1)} \cdot f'(x+1) = 1 + \frac{1}{f(x)} \cdot f'(x)$$

$$\Rightarrow \frac{f'(x+1)}{f(x+1)} - \frac{f'(x)}{f(x)} = 1$$

$$\Rightarrow \frac{f'\left(1+\frac{1}{3}\right)}{f\left(1+\frac{1}{3}\right)} - \frac{f'\left(\frac{1}{3}\right)}{f\left(\frac{1}{3}\right)} = 1$$

$$\frac{f'\left(2+\frac{1}{3}\right)}{f\left(2+\frac{1}{3}\right)} - \frac{f'\left(1+\frac{1}{3}\right)}{f\left(1+\frac{1}{3}\right)} = 1$$

$$\frac{f'\left(n+\frac{1}{3}\right)}{f\left(n+\frac{1}{3}\right)} - \frac{f'\left((n-1)+\frac{1}{3}\right)}{f\left((n-1)+\frac{1}{3}\right)} = 1$$

on adding columnwise, we get

$$\frac{f'\left(n+\frac{1}{3}\right)}{f\left(n+\frac{1}{3}\right)} - \frac{f'\left(\frac{1}{3}\right)}{f\left(\frac{1}{3}\right)} = n$$

13 Since, $g(x) = f^{-1}(x)$

$\therefore \quad f(g(x)) = x \Rightarrow f'(g(x)) \cdot g'(x) = 1$

$$\Rightarrow g'(x) = \frac{1}{f'(g(x))}$$

$$\Rightarrow g'\left(\frac{-7}{6}\right) = \frac{1}{f'\left(g\left(-\frac{7}{6}\right)\right)}$$

$$= \frac{1}{f'\left(f^{-1}\left(\frac{-7}{6}\right)\right)}$$

$$\left[\because f(1) = -4 + 1 + 1 + \frac{1}{2} + \frac{1}{3} = -\frac{7}{6} \quad \therefore f^{-1}\left(\frac{7}{6}\right) = 1\right]$$

$$= \frac{1}{5}$$

$$\left[\because f'(x) = -4e^{\frac{1-x}{2}}\left(-\frac{1}{2}\right) + 1 + x + x^2\right]$$

14 We have, $f(x) = \prod_{i=1}^{100} (x-i)^{i(101-i)}$

$\Rightarrow \log f(x) = \sum_{i=1}^{100} i(101-i)\log(x-i)$

$$\frac{1}{f(x)} \cdot f'(x) = \sum_{i=1}^{100} i(101-i) \cdot \frac{1}{x-i}$$

$$\Rightarrow \frac{f'(101)}{f(101)} = \sum_{i=1}^{100} i\frac{(101-i)}{(101-i)}$$

$$= \sum_{i=1}^{100} i = \frac{100(101)}{2} = 5050$$

15 Given, $x = 2t - |t|$ and $y = t^3 + t^2|t|$

Clearly, $x = t$, $y = 2t^3$ when $t \ge 0$

and $x = 3t$, $y = 0$ when $t < 0$

On eliminating the parameter t, we get

$$y = \begin{cases} 2x^3, & \text{when } x \ge 0 \\ 0, & \text{when } x < 0 \end{cases}$$

Now, $\dfrac{dy}{dx} = \begin{cases} 6x^2, & \text{when } x > 0 \\ 0, & \text{when } x < 0 \end{cases}$

$\because \quad (\text{LHD})_{\text{at } x=0} = (\text{RHD})_{\text{at } x=0} = 0$

$\therefore$ Its derivative at $x = 0$

(i.e. at $t = 0$ is 0)

DAY THIRTEEN

Applications of Derivatives

Learning & Revision for the Day

- Derivatives as the Rate of Change
- Increasing and Decreasing Function
- Tangent and Normal to a Curve
- Angle of Intersection of Two Curves
- Rolle's Theorem
- Lagrange's Mean Value Theorem

Derivatives as the Rate of Change

$\frac{dy}{dx}$ is nothing but the rate of change of y, relative to x. If a variable quantity y is some function of time t i.e. $y = f(t)$, then small change in time Δt have a corresponding change Δy in y. Thus, the average rate of change $= \frac{\Delta y}{\Delta t}$

When limit $\Delta t \to 0$ is applied, the rate of change becomes instantaneous and we get the rate of change with respect to at the instant t, i.e.

$$\lim_{\Delta t \to 0} \frac{\Delta y}{\Delta t} = \frac{dy}{dt}$$

$\frac{dy}{dt}$ is positive if y increases as t increase and it is negative if y decrease as t increase.

Increasing and Decreasing Function

- A function f is said to be an increasing function in $]a, b[$, if $x_1 < x_2 \Rightarrow f(x_1) \le f(x_2)$, $\forall\, x_1, x_2 \in]a, b[$.
- A function f is said to be a decreasing function in $]a, b[$, if $x_1 < x_2 \Rightarrow f(x_1) \ge f(x_2)$, $\forall\, x_1, x_2 \in]a, b[$.
- $f(x)$ is known as **increasing**, if $f'(x) \ge 0$ and **decreasing**, if $f'(x) \le 0$.
- $f(x)$ is known as strictly increasing, if $f'(x) > 0$ and strictly decreasing, if $f'(x) < 0$.
- Let $f(x)$ be a function that is continuous in $[a, b]$ and differentiable in (a, b). Then,
  (i) $f(x)$ is an increasing function in $[a, b]$, if $f'(x) > 0$ in (a, b).
  (ii) $f(x)$ is strictly increasing function in (a, b), if $f'(x) > 0$ in (a, b).

PREP MIRROR
Your Personal Preparation Indicator

- No. of Questions in Exercises (x)—
- No. of Questions Attempted (y)—
- No. of Correct Questions (z)—
  (Without referring Explanations)

- Accuracy Level ($z/y \times 100$)—
- Prep Level ($z/x \times 100$)—

In order to expect good rank in JEE, your Accuracy Level should be above 85 & Prep Level should be above 75.

(iii) $f(x)$ is a decreasing function in $[a, b]$, if $f'(x) < 0$ in (a, b).
(iv) $f(x)$ is a strictly decreasing function in $[a, b]$, if $f'(x) < 0$ in (a, b).

Monotonic Function

A function f is said to be monotonic in an interval, if it is either increasing or decreasing in that interval.

Results on Monotonic Function

(i) If $f(x)$ is a strictly increasing function on an interval $[a, b]$, then f^{-1} exists and it is also a strictly increasing function.
(ii) If $f(x)$ is strictly increasing function on an interval $[a, b]$ such that it is continuous, then f^{-1} is continuous on $[f(a), f(b)]$.
(iii) If $f(x)$ is continuous on $[a, b]$ such that $f'(c) \ge 0$ $[f'(c) > 0]$ for each $c \in (a, b)$, then $f(x)$ is monotonically increasing on $[a, b]$.
(iv) If $f(x)$ is continuous on $[a, b]$ such that $f'(c) \le 0$ $(f'(c) < 0)$ for each $c \in [a, b]$, then $f(x)$ is monotonically decreasing function on $[a, b]$.
(v) Monotonic function have atmost one root.

Tangent and Normal to a Curve

(i) If a **tangent** is drawn to the curve $y = f(x)$ at a point $P(x_1, y_1)$ and this tangent makes an angle ψ with positive X-direction, then

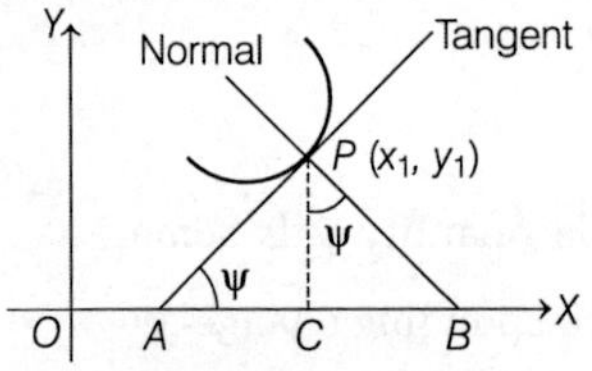

graph desenber tangents and normal

(a) The slope of the tangent is

$$\left(\frac{dy}{dx}\right)_{(x_1, y_1)} = \tan\psi$$

(b) Equation of tangent is $y - y_1 = \left(\frac{dy}{dx}\right)_{(x_1, y_1)} (x - x_1)$

(c) Length of tangent is

$$PA = y_1 \operatorname{cosec}\psi = \left| y_1 \frac{\sqrt{1 + \left(\left(\frac{dy}{dx}\right)_{(x_1, y_1)}\right)^2}}{\left(\frac{dy}{dx}\right)_{(x_1, y_1)}} \right|$$

(d) Length of subtangent $AC = y_1 \cot\psi = \left| \frac{y_1}{(dy/dx)_{(x_1, y_1)}} \right|$

(ii) The **normal** to a curve at a point $P(x_1, y_1)$ is a line perpendicular to tangent at P and passing through P, then

(a) The slope of the normal is $-\frac{1}{(dy/dx)_{(x_1, y_1)}}$

(b) Equation of normal is $y - y_1 = -\frac{1}{\left(\frac{dy}{dx}\right)_{(x_1, y_1)}} (x - x_1)$

(c) Length of normal is $PB = y_1 \sec\psi = \left| y_1 \sqrt{1 + \left(\left(\frac{dy}{dx}\right)_{(x_1, y_1)}\right)^2} \right|$

(d) Length of subnormal is $BC = y_1 \tan\psi = \left| y_1 \left(\frac{dy}{dx}\right)_{(x_1, y_1)} \right|$

Angle of Intersection of Two Curves

The angle of intersection of two curves is defined to be the angle between their tangents, to the two curves at their point of intersection.

The angle between the tangents of the two curves $y = f_1(x)$ and $y = f_2(x)$ is given by

$$\tan\phi = \left| \frac{\left(\frac{dy}{dx}\right)_{\text{I}(x_1, y_1)} - \left(\frac{dy}{dx}\right)_{\text{II}(x_1, y_1)}}{1 + \left(\frac{dy}{dx}\right)_{\text{I}(x_1, y_1)} \left(\frac{dy}{dx}\right)_{\text{II}(x_1, y_1)}} \right| \text{ or } \left| \frac{m_1 - m_2}{1 + m_1 m_2} \right|$$

Orthogonal Curves

If the angle of intersection of two curves is a right angle, the two curves are said to intersect orthogonally and the curves are called orthogonal curves.

If $\phi = \frac{\pi}{2}$, $m_1 m_2 = -1 \Rightarrow \left(\frac{dy}{dx}\right)_{\text{I}} \left(\frac{dy}{dx}\right)_{\text{II}} = -1$

NOTE • Two curves touch each other, if $m_1 = m_2$.

Rolle's Theorem

Let f be a real-valued function defined in the closed interval $[a, b]$, such that

(i) $f(x)$ is continuous in the closed interval $[a, b]$.
(ii) $f(x)$ is differentiable in the open interval (a, b).
(iii) $f(a) = f(b)$, then there is some point c in the open interval (a, b), such that $f'(c) = 0$.

Geometrically,

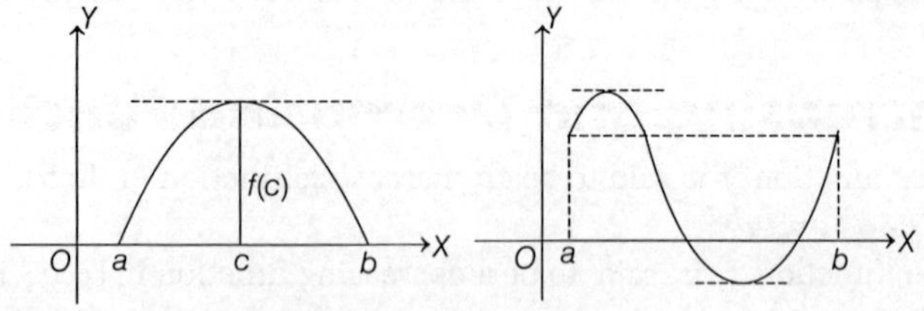

graph of a differentiable function, satisfying the hypothesis of Rolle's theorem.

There is atleast one point c between a and b, such that the tangent to the graph at $(c, f(c))$ is parallel to the X-axis.

Algebraic Interpretation of Rolle's Theorem

Between any two roots of a polynomial $f(x)$, there is always a root of its derivative $f'(x)$.

Lagrange's Mean Value Theorem

Let f be a real function, continuous on the closed interval $[a, b]$ and differentiable in the open interval (a, b). Then, there is atleast one point c in the open interval (a, b), such that

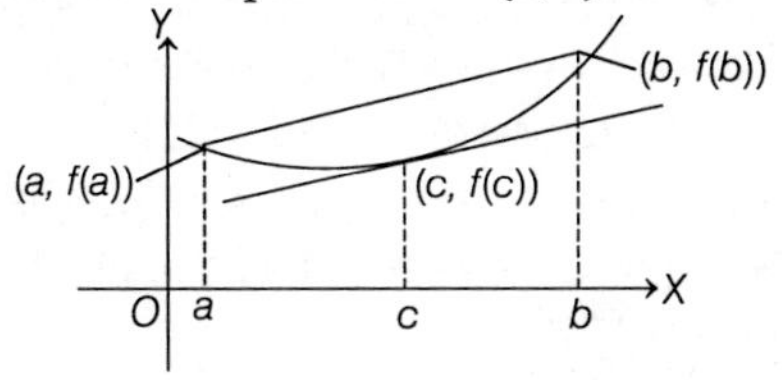

graph of a continuous functions explain Lagrange's mean value theorem.

$$f'(c) = \frac{f(b) - f(a)}{b - a}.$$

Geometrically For any chord of the curve $y = f(x)$, there is a point on the graph, where the tangent is parallel to this chord.

Remarks In the particular case, where $f(a) = f(b)$.

The expression $\dfrac{f(b) - f(a)}{b - a}$ becomes zero.

Thus, when $f(a) = f(b)$, $f'(c) = 0$ for some c in (a, b).

Thus, Rolle's theorem becomes a particular case of the **mean value theorem.**

DAY PRACTICE SESSION 1

FOUNDATION QUESTIONS EXERCISE

1 If the volume of a sphere is increasing at a constant rate, then the rate at which its radius is increasing, is
(a) a constant
(b) proportional to the radius
(c) inversely proportional to the radius
(d) inversely proportional to the surface area

2 Moving along the X-axis there are two points with $x = 10 + 6t$, $x = 3 + t^2$. The speed with which they are reaching from each other at the time of encounter is (x is in centimetre and t is in seconds)
(a) 16 cm/s (b) 20 cm/s (c) 8 cm/s (d) 12 cm/s

3 An object is moving in the clockwise direction around the unit circle $x^2 + y^2 = 1$. As it passes through the point $\left(\frac{1}{2}, \frac{\sqrt{3}}{2}\right)$, its y-coordinate is decreasing at the rate of 3 units per second. The rate at which the x-coordinate changes at this point is (in unit per second)
(a) 2 (b) $3\sqrt{3}$ (c) $\sqrt{3}$ (d) $2\sqrt{3}$

4 The position of a point in time 't' is given by $x = a + bt - ct^2$, $y = at + bt^2$. Its acceleration at time 't' is
(a) $b - c$ (b) $b + c$ (c) $2b - 2c$ (d) $2\sqrt{b^2 + c^2}$

5 Water is dripping out from a conical funnel of semi-vertical angle $\frac{\pi}{4}$ at the uniform rate of 2 cm^2/s in the surface area, through a tiny hole at the vertex of the bottom. When the slant height of cone is 4 cm, the rate of decrease of the slant height of water, is → NCERT Exemplar
(a) $\frac{\sqrt{2}}{4\pi}$ cm/s (b) $\frac{1}{4\pi}$ cm/s
(c) $\frac{1}{\pi\sqrt{2}}$ cm/s (d) None of these

6 A spherical balloon is being inflated at the rate of 35 cc/min. The rate of increase in the surface area (in cm^2/min) of the balloon when its diameter is 14 cm, is → JEE Mains 2013
(a) 10 (b) $\sqrt{10}$ (c) 100 (d) $10\sqrt{10}$

7 Oil is leaking at the rate of 16 cm^3/s from a vertically kept cylindrical drum containing oil. If the radius of the drum is 7 cm and its height is 60 cm. Then, the rate at which the level of the oil is changing when oil level is 18 cm, is
(a) $\frac{-16}{49\pi}$ (b) $\frac{-16}{48\pi}$ (c) $\frac{16}{49\pi}$ (d) $\frac{-16}{47\pi}$

8 Two men A and B start with velocities v at the same time from the junction of two roads inclined at 45° to each other. If they travel by different roads, the rate at which they are being separated. → NCERT Exemplar
(a) $\sqrt{2 - \sqrt{2}} \cdot v$ (b) $\sqrt{2 + \sqrt{2}} \cdot v$
(c) $\sqrt{\sqrt{2} - 1} \cdot v$ (d) $\sqrt{2 + \sqrt{2}} \cdot v$

9 The interval in which the function $f(x) = x^{1/x}$ is increasing, is
(a) $(-\infty, e)$ (b) (e, ∞)
(c) $(-\infty, \infty)$ (d) None of these

10 The function $f(x) = \dfrac{x}{1 + |x|}$ is
(a) strictly increasing
(b) strictly decreasing
(c) neither increasing nor decreasing
(d) not differential at $x = 0$

11 The length of the longest interval, in which the function $3 \sin x - 4 \sin^3 x$ is increasing, is
(a) $\frac{\pi}{3}$ (b) $\frac{\pi}{2}$ (c) $\frac{3\pi}{2}$ (d) π

12 An angle θ, $0 < \theta < \frac{\pi}{2}$, which increases twice as fast as its sine, is → NCERT Exemplar

(a) $\frac{\pi}{2}$ (b) $\frac{3\pi}{2}$ (c) $\frac{\pi}{4}$ (d) $\frac{\pi}{3}$

13 The value of x for which the polynomial $2x^3 - 9x^2 + 12x + 4$ is a decreasing function of x, is

(a) $-1 < x < 1$ (b) $0 < x < 2$
(c) $x > 3$ (d) $1 < x < 2$

14 If $f(x) = \frac{1}{x+1} - \log(1+x)$, $x > 0$, then f is

(a) an increasing function
(b) a decreasing function
(c) both increasing and decreasing function
(d) None of the above

15 If $f(x) = \sin x - \cos x$, the interval in which function is decreasing in $0 \le x \le 2\pi$, is

(a) $\left[\frac{5\pi}{6}, \frac{3\pi}{4}\right]$ (b) $\left[\frac{\pi}{4}, \frac{\pi}{2}\right]$
(c) $\left[\frac{3\pi}{2}, \frac{5\pi}{2}\right]$ (d) None of these

16 If $f(x) = -2x^3 + 21x^2 - 60x + 41$, then

(a) $f(x)$ is decreasing in $(-\infty, 1)$
(b) $f(x)$ is decreasing in $(-\infty, 2)$
(c) $f(x)$ is increasing in $(-\infty, 1)$
(d) $f(x)$ is increasing in $(-\infty, 2)$

17 Function $f(x) = \frac{\lambda \sin x + 6 \cos x}{2 \sin x + 3 \cos x}$ is monotonic increasing, if

(a) $\lambda > 1$ (b) $\lambda < 1$ (c) $\lambda < 4$ (d) $\lambda > 4$

18 The sum of intercepts on coordinate axes made by tangent to the curve $\sqrt{x} + \sqrt{y} = \sqrt{a}$, is

(a) a (b) $2a$
(c) $2\sqrt{a}$ (d) None of these

19 Line joining the points (0, 3) and (5, –2) is a tangent to the curve $y = \frac{ax}{1+x}$, then

(a) $a = 1 \pm \sqrt{3}$ (b) $a = \phi$
(c) $a = -1 \pm \sqrt{3}$ (d) $a = -2 \pm 2\sqrt{3}$

20 The equation of the tangent to the curve $y = x + \frac{4}{x^2}$, that is parallel to the X-axis, is → AIEEE 2010

(a) $y = 0$ (b) $y = 1$ (c) $y = 2$ (d) $y = 3$

21 The slope of the tangent to the curve $x = 3t^2 + 1, y = t^3 - 1$, at $x = 1$ is

(a) 0 (b) $\frac{1}{2}$ (c) ∞ (d) -2

22 Coordinates of a point on the curve $y = x \log x$ at which the normal is parallel to the line $2x - 2y = 3$, are

(a) (0, 0) (b) (e, e)
(c) $(e^2, 2e^2)$ (d) $(e^{-2}, -2e^{-2})$

23 The tangent drawn at the point (0, 1) on the curve $y = e^{2x}$, meets X-axis at the point

(a) $\left(\frac{1}{2}, 0\right)$ (b) $\left(-\frac{1}{2}, 0\right)$ (c) (2, 0) (d) (0, 0)

24 If the normal to the curve $y^2 = 5x - 1$ at the point $(1, -2)$ is of the form $ax - 5y + b = 0$, then a and b are

(a) $4, -14$ (b) $4, 14$ (c) $-4, 14$ (d) $4, 2$

25 The curve $y = ax^3 + bx^2 + cx + 5$ touches the X-axis at $P(-2, 0)$ and cuts the Y-axis at a point Q, where its gradient is 3. Then,

(a) $a = -\frac{1}{2}, b = -\frac{3}{4}$ and $c = 3$
(b) $a = \frac{1}{2}, b = -\frac{3}{4}$ and $c = -3$
(c) $a = \frac{1}{2}, b = -\frac{1}{4}$ and $c = 3$
(d) None of the above

26 The product of the lengths of subtangent and subnormal at any point of a curve is

(a) square of the abscissa (b) square of the ordinate
(c) constant (d) None of these

27 The tangent at $(1, 7)$ to the curve $x^2 = y - 6$ touches the circle $x^2 + y^2 + 16x + 12y + c = 0$ at

(a) (6, 7) (b) (–6, 7) (c) (6, –7) (d) (–6, –7)

28 If the line $ax + by + c = 0$ is normal to curve $xy + 5 = 0$, then

(a) $a + b = 0$ (b) $a > 0$ (c) $a < 0, b < 0$ (d) $a = -2b$

29 The length of subnormal to the curve $y = \frac{x}{1-x^2}$ at the point having abscissa $\sqrt{2}$ is

(a) $5\sqrt{2}$ (b) $3\sqrt{3}$ (c) $\sqrt{3}$ (d) $3\sqrt{2}$

30 If m is the slope of a tangent to the curve $e^y = 1 + x^2$, then

(a) $|m| \le 1$ (b) $m > -1$ (c) $m > 1$ (d) $|m| > 1$

31 If the curves $y = a^x$ and $y = b^x$ intersects at angle α, then $\tan\alpha$ is equal to

(a) $\frac{a-b}{1+ab}$ (b) $\frac{\log a - \log b}{1 + \log a \log b}$
(c) $\frac{a+b}{1-ab}$ (d) $\frac{\log a + \log b}{1 - \log a \log b}$

32 If the curves $y^2 = 6x$, $9x^2 + by^2 = 16$ intersect each other at right angles, then the value of b is

(a) 6 (b) $\frac{7}{2}$
(c) 4 (d) $\frac{9}{2}$

33 Angle between the tangents to the curve $y = x^2 - 5x + 6$ at the points (2, 0) and (3, 0) is

(a) $\frac{\pi}{2}$ (b) $\frac{\pi}{6}$
(c) $\frac{\pi}{4}$ (d) $\frac{\pi}{3}$

34 If the curves $\frac{x^2}{\alpha}+\frac{y^2}{4}=1$ and $y^3=16x$ intersect at right angles, then the value of α is → JEE Mains 2013

(a) 2 (b) $\frac{4}{3}$ (c) $\frac{1}{2}$ (d) $\frac{3}{4}$

35 $f(x)$ satisfies the conditions of Rolle's theorem in [1, 2] and $f(x)$ is continuous in [1, 2], then $\int_1^2 f'(x)\,dx$ is equal to

(a) 3 (b) 0 (c) 1 (d) 2

36 If the function $f(x)=x^3-6x^2+ax+b$ satisfies Rolle's theorem in the interval [1, 3] and $f'\left(\frac{2\sqrt{3}+1}{\sqrt{3}}\right)=0$, then

(a) $a=-11$ (b) $a=-6$ (c) $a=6$ (d) $a=11$

37 If $f(x)$ satisfies the conditions for Rolle's theorem in [3, 5], then $\int_3^5 f(x)\,dx$ is equal to

(a) 2 (b) −1 (c) 0 (d) $-\frac{4}{3}$

38 A value of C for which the conclusion of mean value theorem holds for the function $f(x)=\log_e x$ on the interval [1, 3] is → AIEEE 2007

(a) $2\log_3 e$ (b) $\frac{1}{2}\log_e 3$
(c) $\log_3 e$ (d) $\log_e 3$

39 The abscissa of the points of the curve $y=x^3$ in the interval [−2, 2], where the slope of the tangents can be obtained by mean value theorem for the interval [−2, 2], are

(a) $\pm\frac{2}{\sqrt{3}}$ (b) $+\sqrt{3}$
(c) $\pm\frac{\sqrt{3}}{2}$ (d) 0

40 In the mean value theorem, $f(b)-f(a)=(b-a)f'(c)$, if $a=4$, $b=9$ and $f(x)=\sqrt{x}$, then the value of c is

(a) 8.00 (b) 5.25 (c) 4.00 (d) 6.25

DAY PRACTICE SESSION 2

PROGRESSIVE QUESTIONS EXERCISE

1 A kite is moving horizontally at a height of 151.5 m. If the speed of kite is 10 m/s, how fast is the string being let out, when the kite is 250 m away from the boy who is flying the kite? The height of boy is 1.5 m. → NCERT Exemplar

(a) 8 m/s (b) 12 m/s (c) 16 m/s (d) 19 m/s

2 The normal to the curve $y(x-2)(x-3)=x+6$ at the point, where the curve intersects the Y-axis passes through the point

(a) $\left(-\frac{1}{2},-\frac{1}{2}\right)$ (b) $\left(\frac{1}{2},\frac{1}{2}\right)$
(c) $\left(\frac{1}{2},-\frac{1}{3}\right)$ (d) $\left(\frac{1}{2},\frac{1}{3}\right)$

3 The values of a for which the function $(a+2)x^3-3ax^2+9ax-1=0$ decreases monotonically throughout for all real x, are

(a) $a<-2$ (b) $a>-2$
(c) $-3<a<0$ (d) $-\infty<a\le-3$

4 If $f(x)=\frac{x}{\sin x}$ and $g(x)=\frac{x}{\tan x}$, where $0<x\le1$, then in this interval

(a) both $f(x)$ and $g(x)$ are increasing functions
(b) both $f(x)$ and $g(x)$ are decreasing functions
(c) $f(x)$ is an increasing function
(d) $g(x)$ is an increasing function

5 $f(x)=\int_0^x |\log_2[\log_3\{\log_4(\cos t+a)\}]|\,dt$. If $f(x)$ is increasing for all real values of x, then

(a) $a\in(-1,1)$ (b) $a\in(1,5)$
(c) $a\in(1,\infty)$ (d) $a\in(5,\infty)$

6 If $f(x)$ satisfy all the conditions of mean value theorem in [0, 2]. If $f(0)=0$ and $|f'(x)|\le\frac{1}{2}$ for all x in [0, 2], then

(a) $f(x)<2$ (b) $|f(x)|\le1$
(c) $f(x)=2x$ (d) $f(x)=3$ for atleast one x in [0, 2]

7 If $f'(\sin x)<0$ and $f''(\sin x)>0$, $\forall x\in\left(0,\frac{\pi}{2}\right)$ and $g(x)=f(\sin x)+f(\cos x)$, then $g(x)$ is decreasing in

(a) $\left(\frac{\pi}{4},\frac{\pi}{2}\right)$ (b) $\left(0,\frac{\pi}{4}\right)$ (c) $\left(0,\frac{\pi}{2}\right)$ (d) $\left(\frac{\pi}{6},\frac{\pi}{2}\right)$

8 If $f(x)=(x-p)(x-q)(x-r)$, where $p<q<r$, are real numbers, then application of Rolle's theorem on f leads to

(a) $(p+q+r)(pq+qr+rp)=3$
(b) $(p+q+r)^2=3(pq+qr+rp)$
(c) $(p+q+r)^2>3(pq+qr+rp)$
(d) $(p+q+r)^2<3(pq+qr+rp)$

9 If $f(x)$ is a monotonic polynomial of $2m-1$ degree, where $m\in N$, then the equation $[f(x)+f(3x)+f(5x)+\ldots+f(2m-1)x]=2m-1$ has

(a) atleast one real root (b) $2m$ roots
(c) exactly one real root (d) $(2m+1)$ roots

10 A spherical balloon is filled with 4500 π cu m of helium gas. If a leak in the balloon causes the gas to escape at the rate of 72π cu m/min, then the rate (in m/min) at which the radius of the balloon decreases 49 min after the leakage began is → AIEEE 2012

(a) $\frac{9}{7}$ (b) $\frac{7}{9}$ (c) $\frac{2}{9}$ (d) $\frac{9}{2}$

11 The normal to the curve $x^2 + 2xy - 3y^2 = 0$ at $(1, 1)$ → JEE Mains 2015

(a) does not meet the curve again
(b) meets the curve again in the second quadrant
(c) meets the curve again in the third quadrant
(d) meets the curve again in the fourth quadrant

12 If f and g are differentiable functions in $(0, 1)$ satisfying $f(0) = 2 = g(1), g(0) = 0$ and $f(1) = 6$, then for some $c \in]0, 1[$ → JEE Mains 2014

(a) $2f'(c) = g'(c)$ (b) $2f'(c) = 3g'(c)$
(c) $f'(c) = g'(c)$ (d) $f'(c) = 2g'(c)$

13 If $y = f(x)$ is the equation of a parabola which is touched by the line $y = x$ at the point where $x = 1$, then

(a) $2f'(0) = 3f'(1)$ (b) $f'(1) = 1$
(c) $f(0) + f'(1) + f''(1) = 2$ (d) $2f(0) = 1 + f'(0)$

14 Let $a + b = 4, a < 2$ and $g(x)$ be a monotonically increasing function of x. Then,

$$f(x) = \int_0^a g(x)\,dx + \int_0^b g(x)\,dx$$

(a) increases with increase in $(b - a)$
(b) decreases with increase in $(b - a)$
(c) increases with decreases in $(b - a)$
(d) None of the above

15 The angle of intersection of curves, $y = [|\sin x| + |\cos x|]$ and $x^2 + y^2 = 5$, where $[\cdot]$ denotes greatest integral function is

(a) $\frac{\pi}{4}$ (b) $\tan^{-1}\left(\frac{1}{2}\right)$
(c) $\tan^{-1}(2)$ (d) None of these

16 In [0, 1], Lagrange's mean value theorem is not applicable to

(a) $f(x) = \begin{cases} \frac{1}{2} - x, & x < \frac{1}{2} \\ \left(\frac{1}{2} - x\right)^2, & x \geq \frac{1}{2} \end{cases}$ (b) $f(x) = \begin{cases} \frac{\sin x}{x}, & x \neq 0 \\ 1, & x = 0 \end{cases}$

(c) $f(x) = x|x|$ (d) $f(x) = |x|$

ANSWERS

SESSION 1	**1** (d)	**2** (c)	**3** (b)	**4** (d)	**5** (a)	**6** (a)	**7** (c)	**8** (a)	**9** (a)	**10** (a)
	11 (a)	**12** (d)	**13** (d)	**14** (b)	**15** (d)	**16** (b)	**17** (d)	**18** (a)	**19** (b)	**20** (d)
	21 (a)	**22** (d)	**23** (b)	**24** (a)	**25** (a)	**26** (b)	**27** (d)	**28** (c)	**29** (d)	**30** (a)
	31 (b)	**32** (d)	**33** (a)	**34** (b)	**35** (b)	**36** (d)	**37** (d)	**38** (a)	**39** (a)	**40** (d)
SESSION 2	**1** (a)	**2** (b)	**3** (d)	**4** (c)	**5** (d)	**6** (b)	**7** (b)	**8** (c)	**9** (a)	**10** (c)
	11 (d)	**12** (d)	**13** (b)	**14** (a)	**15** (c)	**16** (a)				

Hints and Explanations

SESSION 1

1 Given that, $\frac{dV}{dt} = k$ (say)

$\because \quad V = \frac{4}{3}\pi R^3 \Rightarrow \frac{dV}{dt} = 4\pi R^2 \frac{dR}{dt}$

$\Rightarrow \frac{dR}{dt} = \frac{k}{4\pi R^2}$

Rate of increasing radius is inversely proportional to its surface area.

2 They will encounter, if

$10 + 6t = 3 + t^2$

$\Rightarrow \quad t^2 - 6t - 7 = 0 \Rightarrow t = 7$

At $t = 7$ s, moving in a first point

$v_1 = \frac{d}{dt}(10 + 6t) = 6$ cm/s

At $t = 7$ s, moving in a second point

$v_2 = \frac{d}{dt}(3 + t^2) = 2t = 2 \times 7 = 14$ cm/s

$\therefore$ Resultant velocity

$= v_2 - v_1 = 14 - 6 = 8$ cm/s

3 The equation of given circle is

$x^2 + y^2 = 1$

On differentiating w.r.t. t, we get

$2x\frac{dx}{dt} + 2y\frac{dy}{dt} = 0$

$\Rightarrow \quad x\frac{dx}{dt} + y\frac{dy}{dt} = 0$

But we have, $x = \frac{1}{2}, y = \frac{\sqrt{3}}{2}$

and $\frac{dy}{dt} = -3$, then $\frac{1}{2}\frac{dx}{dt} + \frac{\sqrt{3}}{2}(-3) = 0$

$\Rightarrow \quad \frac{dx}{dt} = 3\sqrt{3}$

4 Given point is $x = a + bt - ct^2$

Acceleration in x direction and point is $y = at + bt^2 = \frac{d^2x}{dt^2} = -2c$

and point is $y = at + bt^2$ acceleration in y direction

$= \frac{d^2y}{dt^2} = 2b$

$\therefore$ Resultant acceleration

$= \sqrt{\left(\frac{d^2x}{dt^2}\right)^2 + \left(\frac{d^2y}{dt^2}\right)^2}$

$= \sqrt{(-2c)^2 + (2b)^2} = 2\sqrt{b^2 + c^2}$

5 If S represents the surface area, then

$\frac{dS}{dt} = 2 \text{ cm}^2/\text{s}$

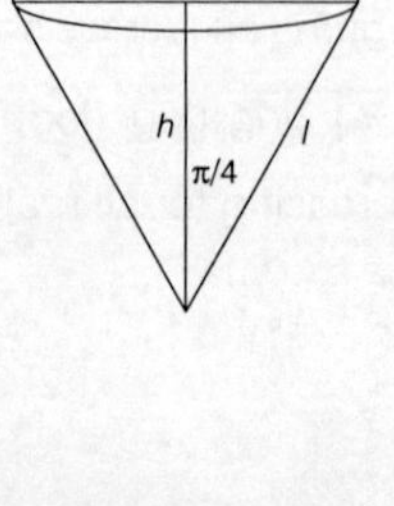

$$S = \pi r l = \pi l \cdot \sin\frac{\pi}{4} l = \frac{\pi}{\sqrt{2}} l^2$$

Therefore,

$$\frac{dS}{dt} = \frac{2\pi}{\sqrt{2}} l \cdot \frac{dl}{dt} = \sqrt{2}\pi l \cdot \frac{dl}{dt}$$

when $l = 4$ cm, $\frac{dl}{dt} = \frac{2}{\sqrt{2}\pi \cdot 4}$

$$= \frac{1}{2\sqrt{2}\pi} = \frac{\sqrt{2}}{4\pi} \text{ cm / s}$$

6 $V = \frac{4}{3}\pi r^3 \Rightarrow \frac{dV}{dt} = \frac{4}{3}\pi 3r^2 \frac{dr}{dt}$

$\Rightarrow 35 = 4\pi(7)^2 \frac{dr}{dt} \Rightarrow \frac{dr}{dt} = \frac{5}{28\pi}$

Surface area of balloon, $S = 4\pi r^2$

$\therefore \frac{dS}{dt} = 8\pi r \frac{dr}{dt}$

$$= 8\pi \times 7 \times \frac{5}{28\pi} = 10 \text{ cm}^2 / \text{min}$$

7 Let h be height of oil level at any instant t and V be the volume of oil in cylindrical drum.

Given, $h = 60$ cm, $r = 7$ cm

and $\frac{dV}{dt} = -16 \text{ cm}^3/\text{s}$

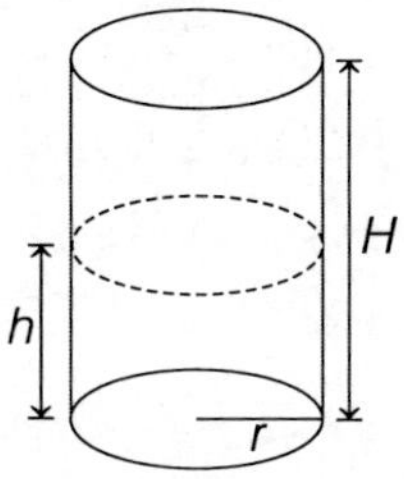

$\because V = \pi r^2 h \Rightarrow \frac{dV}{dt} = \pi r^2 \frac{dh}{dt}$

(since, r is constant all the time)

$\Rightarrow -16 = \pi(7)^2 \frac{dh}{dt} \Rightarrow \frac{dh}{dt} = -\frac{16}{49\pi}$

$\Rightarrow \left(\frac{dh}{dt}\right)_{\text{at } h=18} = -\frac{16}{49\pi}$

So, height of oil is decreasing at the rate of $\frac{16}{49\pi}$ cm/s.

8 Let L and M be the positions of two men A and B at any time t.

Let $OL = x$ and $LM = y$

Then, $OM = x$

Given, $\frac{dx}{dt} = v$ and we have to find $\frac{dy}{dt}$

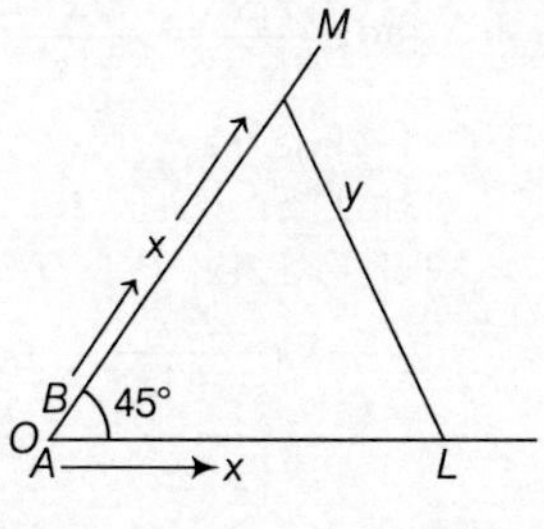

From ΔLOM,

$$\cos 45° = \frac{OL^2 + OM^2 - LM^2}{2 \cdot OL \cdot OM}$$

$\Rightarrow \frac{1}{\sqrt{2}} = \frac{x^2 + x^2 - y^2}{2 \cdot x \cdot x} = \frac{2x^2 - y^2}{2x^2}$

$\Rightarrow \sqrt{2}x^2 = 2x^2 - y^2$

$\Rightarrow (2 - \sqrt{2})x^2 = y^2$

$\therefore y = \sqrt{2 - \sqrt{2}}\, x$

On differentiating w.r.t. t, we get

$$\frac{dy}{dt} = \sqrt{2 - \sqrt{2}} \frac{dx}{dt}$$

$$= \sqrt{2 - \sqrt{2}} v \qquad \left(\because \frac{dx}{dt} = v\right)$$

Hence, they are being separated from each other at the rate $\sqrt{2 - \sqrt{2}} v$.

9 Given, $f(x) = x^{1/x}$

$\Rightarrow f'(x) = \frac{1}{x^2}(1 - \log x) x^{1/x}$

$f'(x) > 0$, if $1 - \log x > 0$

$\Rightarrow \log x < 1 \Rightarrow x < e$

$\therefore f(x)$ is increasing in the interval $(-\infty, e)$.

10 Given, $f(x) = \frac{x}{1 + |x|}$

$$\therefore f'(x) = \frac{(1 + |x|) \cdot 1 - x \cdot \frac{|x|}{x}}{(1 + |x|)^2}$$

$$= \frac{1}{(1 + |x|)^2} > 0, \forall x \in R$$

$\Rightarrow f(x)$ is strictly increasing.

11 Let $f(x) = 3 \sin x - 4 \sin^3 x = \sin 3x$

Since, $\sin x$ is increasing in the interval $\left[-\frac{\pi}{2}, \frac{\pi}{2}\right]$.

$\therefore -\frac{\pi}{2} \le 3x \le \frac{\pi}{2} \Rightarrow -\frac{\pi}{6} \le x \le \frac{\pi}{6}$

Thus, length of interval

$$= \left|\frac{\pi}{6} - \left(-\frac{\pi}{6}\right)\right| = \frac{\pi}{3}$$

12 Given, $2\frac{d}{dt}(\sin\theta) = \frac{d\theta}{dt}$

$\Rightarrow 2 \times \cos\theta \frac{d\theta}{dt} = \frac{d\theta}{dt}$

$\Rightarrow 2\cos\theta = 1 \Rightarrow \cos\theta = \frac{1}{2} \Rightarrow \theta = \frac{\pi}{3}$

13 Let $f(x) = 2x^3 - 9x^2 + 12x + 4$

$\Rightarrow f'(x) = 6x^2 - 18x + 12$

$f'(x) < 0$ for function to be decreasing

$\Rightarrow 6(x^2 - 3x + 2) < 0$

$\Rightarrow (x^2 - 2x - x + 2) < 0$

$\Rightarrow (x - 2)(x - 1) < 0 \Rightarrow 1 < x < 2$

14 Given curve is $f(x) = \frac{1}{x + 1} - \log(1 + x)$

On differentiating w.r.t. x, we get

$$f'(x) = -\frac{1}{(x + 1)^2} - \frac{1}{1 + x}$$

$\Rightarrow f'(x) = -\left[\frac{1}{x + 1} + \frac{1}{(x + 1)^2}\right]$

$\Rightarrow f'(x) = -$ve, when $x > 0$

$\therefore f(x)$ is a decreasing function.

15 $\because f(x) = \sin x - \cos x$

On differentiating w.r.t. x, we get

$f'(x) = \cos x + \sin x$

$$= \sqrt{2}\left(\frac{1}{\sqrt{2}} \cos x + \frac{1}{\sqrt{2}} \sin x\right)$$

$$= \sqrt{2}\left(\cos\frac{\pi}{4} \cos x + \sin\frac{\pi}{4} \sin x\right)$$

$$= \sqrt{2}\left[\cos\left(x - \frac{\pi}{4}\right)\right]$$

For decreasing, $f'(x) < 0$

$$\frac{\pi}{2} < \left(x - \frac{\pi}{4}\right) < \frac{3\pi}{2}$$

(within $0 \le x \le 2\pi$)

$\Rightarrow \frac{\pi}{2} + \frac{\pi}{4} < \left(x - \frac{\pi}{4} + \frac{\pi}{4}\right) < \frac{3\pi}{2} + \frac{\pi}{4}$

$\Rightarrow \frac{3\pi}{4} < x < \frac{7\pi}{4}$

16 Given,

$f(x) = -2x^3 + 21x^2 - 60x + 41$ ...(i)

On differentiating Eq. (i) w.r.t. x, we get

$f'(x) = -6x^2 + 42x - 60$

$= -6(x^2 - 7x + 10)$

$= -6(x - 2)(x - 5)$

If $x < 2$, $f'(x) < 0$ i.e. $f(x)$ is decreasing.

17 $\because f(x) = \frac{\lambda \sin x + 6\cos x}{2\sin x + 3\cos x}$ ...(i)

On differentiating w.r.t. x, we get

$$f'(x) = \frac{\left[\begin{array}{l}(2\sin x + 3\cos x)\\(\lambda\cos x - 6\sin x)\\-(\lambda\sin x + 6\cos x)\\(2\cos x - 3\sin x)\end{array}\right]}{(2\sin x + 3\cos x)^2}$$

The function is monotonic increasing, if $f'(x) > 0$

$\Rightarrow \frac{3\lambda(\sin^2 x + \cos^2 x) - 12}{(\sin^2 x + \cos^2 x)} > 0$

$\Rightarrow 3\lambda - 12 > 0 \quad (\because \sin^2 x + \cos^2 x = 1)$

$\Rightarrow \lambda > 4$

18 $\sqrt{x} + \sqrt{y} = \sqrt{a}$; (i) $\frac{1}{2\sqrt{x}} + \frac{1}{2\sqrt{y}} \frac{dy}{dx} = 0$

$\therefore \frac{dy}{dx} = -\frac{\sqrt{y}}{\sqrt{x}}$

Hence, tangent at (x, y) is

$$Y - y = -\frac{\sqrt{y}}{\sqrt{x}}(X - x)$$

$\Rightarrow X\sqrt{y} + Y\sqrt{x} = \sqrt{xy}(\sqrt{x} + \sqrt{y})$

$\Rightarrow X\sqrt{y} + Y\sqrt{x} = \sqrt{xy}\sqrt{a}$

(using Eq. (i))

$\Rightarrow \quad \dfrac{X}{\sqrt{a}\sqrt{x}} + \dfrac{Y}{\sqrt{a}\sqrt{y}} = 1$

Clearly, its intercepts on the axes are $\sqrt{a}\cdot\sqrt{x}$ and $\sqrt{a}\cdot\sqrt{y}$.

Sum of intercepts $= \sqrt{a}(\sqrt{x}+\sqrt{y})$

$= \sqrt{a}\cdot\sqrt{a} = a$

19 Equation of line joining the points (0, 3) and (5, −2) is $y = 3 - x$. If this line is tangent to $y = \dfrac{ax}{(x+1)}$, then

$(3-x)(x+1) = ax$ should have equal roots.

Thus, $(a-2)^2 + 12 = 0 \Rightarrow$ no value of a

$\Rightarrow a \in \phi$.

20 We have, $y = x + \dfrac{4}{x^2}$

On differentiating w.r.t. x, we get

$\dfrac{dy}{dx} = 1 - \dfrac{8}{x^3}$.

Since, the tangent is parallel to X-axis, therefore

$\dfrac{dy}{dx} = 0 \Rightarrow x^3 = 8$

$\therefore \quad x = 2$ and $y = 3$

21 Given curve is $x = 3t^2 + 1, y = t^3 - 1$

For $x = 1, 3t^2 + 1 = 1 \Rightarrow t = 0$

$\therefore \quad \dfrac{dx}{dt} = 6t, \dfrac{dy}{dt} = 3t^2$

Now, $\quad \dfrac{dy}{dx} = \left(\dfrac{\frac{dy}{dt}}{\frac{dx}{dt}}\right) = \dfrac{3t^2}{6t} = \dfrac{t}{2}$

$\therefore \quad \left(\dfrac{dy}{dx}\right)_{(t=0)} = \dfrac{0}{2} = 0$

22 Given curve is $y = x\log x$

On differentiating w.r.t. x, we get

$\dfrac{dy}{dx} = 1 + \log x$

The slope of the normal

$= -\dfrac{1}{(dy/dx)} = \dfrac{-1}{1+\log x}$

The slope of the given line $2x - 2y = 3$ is 1.

Since, these lines are parallel.

$\therefore \quad \dfrac{-1}{1+\log x} = 1$

$\Rightarrow \quad \log x = -2$

$\Rightarrow \quad x = e^{-2}$

and $\quad y = -2e^{-2}$

$\therefore$ Coordinates of the point are $(e^{-2}, -2e^{-2})$.

23 Given curve is $y = e^{2x}$

On differentiating w.r.t. x, we get

$\dfrac{dy}{dx} = 2e^{2x} \Rightarrow \left(\dfrac{dy}{dx}\right)_{(0,1)} = 2e^0 = 2$

Equation of tangent at (0, 1) with slope 2 is

$y - 1 = 2(x-0) \Rightarrow y = 2x + 1$

This tangent meets X-axis.

$\therefore \quad y = 0 \Rightarrow 0 = 2x + 1 \Rightarrow x = -\dfrac{1}{2}$

$\therefore$ Coordinates of the point on X-axis is $\left(-\dfrac{1}{2}, 0\right)$.

24 We have, $y^2 = 5x - 1$ ...(i)

At $(1,-2)$, $\dfrac{dy}{dx} = \left(\dfrac{5}{2y}\right)_{(1,-2)} = \dfrac{-5}{4}$

$\therefore$ Equation of normal at the point $(1,-2)$ is

$[y-(-2)]\left(\dfrac{-5}{4}\right) + x - 1 = 0$

$\therefore \quad 4x - 5y - 14 = 0$ ...(ii)

As the normal is of the form

$ax - 5y + b = 0$

On comparing this with Eq. (ii), we get

$a = 4$ and $b = -14$

25 Since, we have the curve $y = ax^3 + bx^2 + cx + 5$ touches X-axis at $P(-2, 0)$, then X-axis is the tangent at $(-2, 0)$. The curve meets Y-axis in (0, 5).

$\dfrac{dy}{dx} = 3ax^2 + 2bx + c$

$\Rightarrow \quad \left(\dfrac{dy}{dx}\right)_{0,5} = 0 + 0 + c = 3$ (given)

$\Rightarrow \quad c = 3$ ...(i)

and $\left(\dfrac{dy}{dx}\right)_{(-2,0)} = 0$

$\Rightarrow \quad 12a - 4b + c = 0$ [from Eq. (i)]

$\Rightarrow \quad 12a - 4b + 3 = 0$ ...(ii)

and (−2, 0) lies on the curve, then

$0 = -8a + 4b - 2c + 5$

$\Rightarrow \quad 0 = -8a + 4b - 1$ $(\because c = 3)$

$\Rightarrow \quad 8a - 4b + 1 = 0$ ...(iii)

From Eqs. (ii) and (iii), we get

$a = -\dfrac{1}{2}, b = -\dfrac{3}{4}$

26 Length of subtangent $= y\dfrac{dx}{dy}$

and length of subnormal $= y\dfrac{dy}{dx}$

$\therefore$ Product $= y^2$

$\Rightarrow$ Required product is the square of the ordinate.

27 The tangent to the parabola $x^2 = y - 6$ at (1, 7) is

$x(1) = \dfrac{1}{2}(y+7) - 6 \Rightarrow y = 2x + 5$

which is also a tangent to the given circle.

i.e. $x^2 + (2x+5)^2 + 16x + 12(2x+5) + c = 0$

$\Rightarrow (5x^2 + 60x + 85 + c = 0)$ must have equal roots.

Let the roots be $\alpha = \beta$.

$\therefore \quad \alpha + \beta = -\dfrac{60}{5}$

$\Rightarrow \quad \alpha = -6$

$\therefore \quad x = -6$ and

$y = 2x + 5 = -7$

28 Given curve, is $xy = -5 < 0$

$\Rightarrow \quad x\dfrac{dy}{dx} + y = 0 \Rightarrow \dfrac{dy}{dx} = \dfrac{-y}{x} > 0$

[as $xy = -5 < 0$]

Slope of normal $= \dfrac{-1}{\frac{dy}{dx}} = \dfrac{x}{y} < 0$

[as $xy = -5 < 0$]

Hence, slope of normal will be negative.

The line $ax + by + c = 0$

$\Rightarrow \quad by = -ax - c$

$\Rightarrow \quad y = \dfrac{-a}{b}x - \dfrac{c}{b}$

Slope of normal $\dfrac{-a}{b}$ is negative.

$\Rightarrow \quad \dfrac{-a}{b} < 0 \Rightarrow \dfrac{a}{b} > 0$

$\Rightarrow a > 0, b > 0$ or $a < 0, b < 0$

29 Given, $y = \dfrac{x}{1-x^2}$

At $\quad x = \sqrt{2}, y = -\sqrt{2}$, therefore point is $(\sqrt{2}, -\sqrt{2})$.

$\therefore \quad \dfrac{dy}{dx} = \dfrac{1+x^2}{(1-x^2)^2}$

$\Rightarrow \quad \left(\dfrac{dy}{dx}\right)_{(\sqrt{2},-\sqrt{2})} = \dfrac{1+2}{(1-2)^2} = 3$

$\therefore$ length of subnormal at $(\sqrt{2}, -\sqrt{2})$

$= |(-\sqrt{2})(3)| = 3\sqrt{2}$

30 We have, $e^y = 1 + x^2$

$\Rightarrow \quad e^y\dfrac{dy}{dx} = 2x$

$\Rightarrow \quad \dfrac{dy}{dx} = \dfrac{2x}{1+x^2}$

$\Rightarrow \quad m = \dfrac{2x}{1+x^2}$,

$|m| = \dfrac{2|x|}{|1+x^2|} = \dfrac{2|x|}{1+|x|^2} \le 1$

$\because \quad |x|^2 + 1 - 2|x| \ge 0 \ge$

$\Rightarrow \quad (|x|-1)^2 \ge 0$

$\Rightarrow \quad |x|^2 + 1 \ge 2|x|$

$\Rightarrow \quad 1 \ge \dfrac{2|x|}{1+|x|^2}$

31 Clearly, the point of intersection of curves is (0, 1).

Now, slope of tangent of first curve,

$$m_1 = \frac{dy}{dx} = a^x \log a$$

$$\Rightarrow \left(\frac{dy}{dx}\right)_{(0,1)} = m_1 = \log a$$

Slope of tangent of second curve,

$$m_2 = \frac{dy}{dx} = b^x \log b$$

$$\Rightarrow m_2 = \left(\frac{dy}{dx}\right)_{(0,1)} = \log b$$

$$\therefore \tan\alpha = \frac{m_1 - m_2}{1 + m_1 m_2} = \frac{\log a - \log b}{1 + \log a \log b}$$

32 We have, $y^2 = 6x$

$$\Rightarrow 2y\frac{dy}{dx} = 6$$

$$\Rightarrow \frac{dy}{dx} = \frac{3}{y}$$

Slope of tangent at (x_1, y_1) is $m_1 = \frac{3}{y_1}$

Also, $9x^2 + by^2 = 16$

$$\Rightarrow 18x + 2by\frac{dy}{dx} = 0$$

$$\Rightarrow \frac{dy}{dx} = \frac{-9x}{by}$$

Slope of tangent at (x_1, y_1) is $m_2 = \frac{-9x_1}{by_1}$

Since, these are intersection at right angle.

$$\therefore m_1 m_2 = -1 \Rightarrow \frac{27x_1}{by_1^2} = 1$$

$$\Rightarrow \frac{27x_1}{6bx_1} = 1 \quad [\because y_1^2 = 6x_1]$$

$$\Rightarrow b = \frac{9}{2}$$

33 $\because y = x^2 - 5x + 6$

$$\therefore \frac{dy}{dx} = 2x - 5$$

Now, $m_1 = \left(\frac{dy}{dx}\right)_{(2,0)} = 4 - 5 = -1$

and $m_2 = \left(\frac{dy}{dx}\right)_{(3,0)} = 6 - 5 = 1$

Now, $m_1 m_2 = -1 \times 1 = -1$

Hence, angle between the tangents is $\frac{\pi}{2}$.

34 Slope of the curve, $\frac{x^2}{\alpha} + \frac{y^2}{4} = 1$ is

$$m_1 = \frac{-4x}{\alpha y}.$$

Now, slope of the curve, $y^3 = 16x$ is

$$m_2 = \frac{16}{3y^2}.$$

Now, apply the condition of perpendicularity of two curves,

i.e. $m_1 m_2 = -1$

and get $\alpha = \frac{4}{3}$ with the help of equation of curves.

35 $\int_1^2 f'(x)dx = [f(x)]_1^2 = f(2) - f(1) = 0$

[$\because f(x)$ satisfies the conditions of Rolle's theorem]

$\therefore f(2) = f(1)$

36 $\because f(x) = x^3 - 6x^2 + ax + b$

On differentiating w.r.t. x, we get

$$f'(x) = 3x^2 - 12x + a$$

By the definition of Rolle's theorem

$$f'(c) = 0 \Rightarrow f'\left(2 + \frac{1}{\sqrt{3}}\right) = 0$$

$$\Rightarrow 3\left(2 + \frac{1}{\sqrt{3}}\right)^2 - 12\left(2 + \frac{1}{\sqrt{3}}\right) + a = 0$$

$$\Rightarrow 3\left(4 + \frac{1}{3} + \frac{4}{\sqrt{3}}\right) - 12\left(2 + \frac{1}{\sqrt{3}}\right) + a = 0$$

$$\Rightarrow 12 + 1 + 4\sqrt{3} - 24 - 4\sqrt{3} + a = 0$$

$$\Rightarrow a = 11$$

37 Since, $f(x)$ satisfies all the conditions of Rolle's theorem in [3, 5].

Let $f(x) = (x-3)(x-5) = x^2 - 8x + 15$

Now, $\int_3^5 f(x)dx = \int_3^5 (x^2 - 8x + 15)dx$

$$= \left[\frac{x^3}{3} - \frac{8x^2}{2} + 15x\right]_3^5$$

$$= \left(\frac{125}{3} - 100 + 75\right) - (9 - 36 + 45)$$

$$= \frac{50}{3} - 18 = -\frac{4}{3}$$

38 Using mean value theorem,

$$f'(c) = \frac{f(3) - f(1)}{3 - 1} \quad \left[\because f'(c) = \frac{f(b) - f(a)}{b - a}\right]$$

$$\Rightarrow \frac{1}{c} = \frac{\log e\ 3 - \log e\ 1}{2}$$

$$\therefore c = \frac{2}{\log e\ 3} = 2\log_3 e$$

39 Given that, equation of curve

$y = x^3 = f(x)$

So, $f(2) = 8$ and $f(-2) = -8$

Now, $f'(x) = 3x^2 \Rightarrow f'(x) = \frac{f(2) - f(-2)}{2 - (-2)}$

$$\Rightarrow \frac{8 - (-8)}{4} = 3x^2$$

$$\therefore x = \pm\frac{2}{\sqrt{3}}$$

40 $f(x) = \sqrt{x}$

$\therefore f(a) = \sqrt{4} = 2$

$f(b) = \sqrt{9} = 3;\ f'(x) = \frac{1}{2\sqrt{x}}$

Also, $f'(c) = \frac{f(b) - f(a)}{b - a} = \frac{3 - 2}{9 - 4} = \frac{1}{5}$

$$\therefore \frac{1}{2\sqrt{c}} = \frac{1}{5} \Rightarrow C = \frac{25}{4} = 6.25$$

SESSION 2

1 Let AB be the position of boy who is flying the kite and C be the position of the kite at any time t.

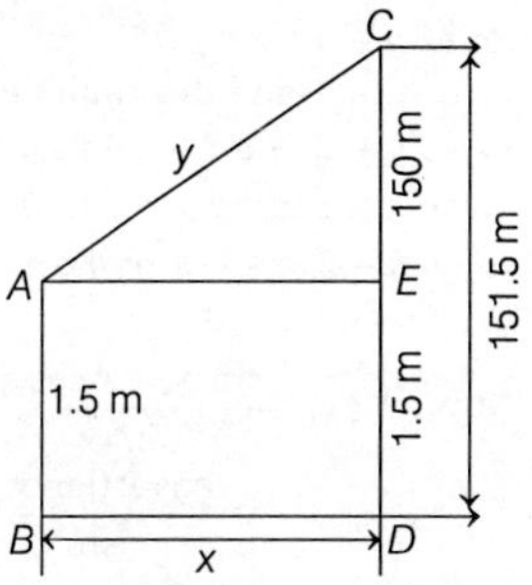

Let $BD = x$ and $AC = y$, then $AE = x$

Given, $AB = 1.5$ m, $CD = 151.5$ m

$\therefore CE = 150$ m

Given, $\frac{dx}{dt} = 10$ m/s

Here, we have to find $\frac{dy}{dt}$ when

$y = 250$ m

Now, from ΔCAE, $y^2 = x^2 + 150^2$

On differentiating, we get

$$2y\frac{dy}{dt} = 2x\frac{dx}{dt}$$

$$\Rightarrow \frac{dy}{dt} = \frac{x}{y}\cdot\frac{dx}{dt} = \frac{x}{y}\cdot 10 \quad \ldots(i)$$

In ΔACE, $x = \sqrt{250^2 - 150^2}$ $(\because y = 250)$

$= 200$ m

$\therefore$ From Eq. (i), we get

$$\frac{dy}{dt} = \frac{200}{250} \times 10 = 8 \text{ m/s}$$

2 Given curve is

$$y(x-2)(x-3) = x + 6 \quad \ldots(i)$$

Put $x = 0$ in Eq. (i), we get

$y(-2)(-3) = 6 \Rightarrow y = 1$

So, point of intersection is (0, 1).

Now, $y = \frac{x + 6}{(x-2)(x-3)}$

$$\Rightarrow \frac{dy}{dx} = \frac{1(x-2)(x-3) - (x+6)(x - 3 + x - 2)}{(x-2)^2(x-3)^2}$$

$\Rightarrow \left(\frac{dy}{dx}\right)_{(0,1)} = \frac{6+30}{4\times 9} = \frac{36}{36} = 1$

$\therefore$ Equation of normal at (0, 1) is given by

$y - 1 = \frac{-1}{1}(x - 0) \Rightarrow x + y - 1 = 0$

which passes through the point $\left(\frac{1}{2}, \frac{1}{2}\right)$.

3 Let $f(x) = (a+2)x^3 - 3ax^2 + 9ax - 1$ decreases monotonically for all $x \in R$, then $f'(x) \le 0$ for all $x \in R$

$\Rightarrow 3(a+2)x^2 - 6ax + 9a \le 0$ for all $x \in R$

$\Rightarrow (a+2)x^2 - 2ax + 3a \le 0$ for all $x \in R$

$\Rightarrow a + 2 < 0$ and discriminant ≤ 0

$\Rightarrow a < -2, -8a^2 - 24a \le 0$

$\Rightarrow a < -2$ and $a(a+3) \ge 0$

$\Rightarrow a < -2, a \le -3$ or $a \ge 0 \Rightarrow a \le -3$

$\therefore -\infty < a \le -3$

4 Now, $f'(x) = \frac{\sin x - x\cos x}{\sin^2 x}$

$= \frac{\cos x(\tan x - x)}{\sin^2 x}$

$\therefore f'(x) > 0$ for $0 < x \le 1$

So, $f(x)$ is an increasing function.

Now, $g'(x) = \frac{\tan x - x\sec^2 x}{\tan^2 x}$

$= \frac{\sin x\cos x - x}{\sin^2 x} = \frac{\sin 2x - 2x}{2\sin^2 x}$

Again, $\frac{d}{dx}(\sin 2x - 2x) = 2\cos 2x - 2$

$= 2(\cos 2x - 1) < 0$

So, $\sin 2x - 2x$ is decreasing.

$\Rightarrow \sin 2x - 2x < 0$

$\therefore g'(x) < 0$

So, $g(x)$ is decreasing.

5 $f'(x) = |\log_2[\log_3\{\log_4(\cos x + a)\}]|$

Clearly, $f(x)$ is increasing for all values of x, if

$\log_2[\log_3\{\log_4(\cos x + a)\}]$ is defined for all values of x.

$\Rightarrow \log_3[\log_4(\cos x + a)] > 0, \forall x \in R$

$\Rightarrow \log_4(\cos x + a) > 1, \forall x \in R$

$\Rightarrow \cos x + a > 4, \forall x \in R$

$\therefore a > 5$

6 Since, $\frac{f(2) - f(0)}{2 - 0} = f'(x)$

$\Rightarrow \frac{f(2) - 0}{2} = f'(x) \Rightarrow \frac{df(x)}{dx} = \frac{f(2)}{2}$

$\Rightarrow f(x) = \frac{f(2)}{2}x + C$

$\because f(0) = 0 \Rightarrow C = 0$

$\therefore f(x) = \frac{f(2)}{2}x$ ...(i)

Also, $|f'(x)| \le \frac{1}{2} \Rightarrow \left|\frac{f(2)}{2}\right| \le \frac{1}{2}$ ...(ii)

From Eq. (i), $|f(x)|$

$= \left|\frac{f(2)}{2}x\right| = \left|\frac{f(2)}{2}\right||x| \le \frac{1}{2}|x|$ [from Eq. (ii)]

In interval [0, 2], for maximum x,

$|f(x)| \le \frac{1}{2}\cdot 2 \Rightarrow |f(x)| \le 1$ [$\because x = 2$]

7 $g'(x) = f'(\sin x)\cdot\cos x - f'(\cos x)\cdot\sin x$

$\Rightarrow g''(x) = -f'(\sin x)\cdot\sin x + \cos^2 x f''(\sin x) + f''(\cos x)\cdot\sin^2 x - f'(\cos x)\cdot\cos x > 0, \forall x \in \left(0, \frac{\pi}{2}\right)$

$\Rightarrow g'(x)$ is increasing in $\left(0, \frac{\pi}{2}\right)$.

Also, $g'\left(\frac{\pi}{4}\right) = 0$

$\Rightarrow g'(x) > 0, \forall x \in \left(\frac{\pi}{4}, \frac{\pi}{2}\right)$

and $g'(x) < 0, \forall x \in \left(0, \frac{\pi}{4}\right)$

Thus, $g(x)$ is decreasing in $\left(0, \frac{\pi}{4}\right)$.

8 We have, $f(x) = (x-p)(x-q)(x-r)$

$\Rightarrow f(p) = 0 = f(q) = f(r)$

$\Rightarrow p, q$ and r are three distinct real roots of $f(x) = 0$

So, by Rolle's theorem, $f'(x)$ has one real root in the interval (p, q) and other in the interval (q, r). Thus, $f'(x)$ has two distinct real roots.

Now, $f(x) = (x-p)(x-q)(x-r)$

$\Rightarrow f(x) = x^3 - x^2(p+q+r) + x(pq+qr+rp) - pqr$

$\Rightarrow f'(x) = 3x^2 - 2(p+q+r)x + (pq+qr+rp)$

As $f'(x)$ has distinct real roots

$\therefore 4(p+q+r)^2 - 12(pq+qr+rp) > 0$

$\Rightarrow (p+q+r)^2 > 3(pq+qr+rp)$

9 Given that, $f(x)$ is monotonic.

$\Rightarrow f'(x) = 0$ or $f'(x) > 0, \forall x \in R$

$\Rightarrow f'(px) < 0$ or $f'(px) > 0, \forall x \in R$

So, $f'(px)$ is also monotonic.

Hence, $f(x) + f(3x) + \ldots + f[(2m-1)x]$ is a monotonic.

Polynomial of odd degree $(2m-1)$, so it will attain all real values only once.

10 Since, the balloon is spherical in shape, hence the volume of the balloon is

$V = \frac{4}{3}\pi r^3$.

On differentiating both the sides w.r.t. t, we get

$\frac{dV}{dt} = \frac{4}{3}\pi\left(3r^2 \times \frac{dr}{dt}\right)$

$\Rightarrow \frac{dr}{dt} = \frac{dV/dt}{4\pi r^2}$ ...(i)

Now, to find $\frac{dr}{dt}$ at the rate $t = 49$ min, we require $\frac{dV}{dt}$ the radius (r) at that stage. $\frac{dV}{dt} = -72\pi m^3/\text{min}$

Also, amount of volume lost in 49 min $= 72\pi \times 49 m^3$

$\therefore$ Final volume at the end of 49 min $= (4500\pi - 3528\pi)m^3$

$= 972\pi m^3$

If r is the radius at the end of 49 min, then $\frac{4}{3}\pi r^3 = 972\pi \Rightarrow r^3 = 729$

$\Rightarrow r = 9$

Radius of the balloon at the end of 49 min = 9 m

From Eq. (i),

$\frac{dr}{dt} = \frac{dV/dt}{4\pi r^2} \Rightarrow \left(\frac{dr}{dt}\right)_{t=49} = \frac{\left(\frac{dV}{dt}\right)_{t=49}}{4\pi(r^2)_{t=49}}$

$\left(\frac{dr}{dt}\right)_{t=49} = \frac{72\pi}{4\pi(9^2)} = \frac{2}{9}$ m/min

11 Given equation of curve is

$x^2 + 2xy - 3y^2 = 0$

On differentiating w.r.t. x, we get

$2x + 2xy' + 2y - 6yy' = 0$

$\Rightarrow y' = \frac{x+y}{3y-x}$

At $x = 1, y = 1, y' = 1$ i.e. $\left(\frac{dy}{dx}\right)_{(1,1)} = 1$

Equation of normal at (1, 1) is

$y - 1 = -\frac{1}{1}(x-1) \Rightarrow y - 1 = -(x-1)$

$\Rightarrow x + y = 2$

On solving Eqs. (i) and (ii) simultaneously, we get

$x^2 + 2x(2-x) - 3(2-x)^2 = 0$

$\Rightarrow x^2 + 4x - 2x^2 - 3(4 + x^2 - 4x) = 0$

$\Rightarrow -x^2 + 4x - 12 - 3x^2 + 12x = 0$

$\Rightarrow -4x^2 + 16x - 12 = 0$

$\Rightarrow 4x^2 - 16x + 12 = 0$

$\Rightarrow x^2 - 4x + 3 = 0$

$\Rightarrow (x-1)(x-3) = 0$

$\Rightarrow x = 1, 3$

Now, when $x = 1$, then $y = 1$

and when $x = 3$, then $y = -1$

$\therefore P = (1, 1)$ and $Q = (3, -1)$

Hence, normal meets the curve again at $(3, -1)$ in fourth quadrant.

Alternate Method

Given, $x^2 + 2xy - 3y^2 = 0$

$\Rightarrow (x-y)(x+3y) = 0$

$\Rightarrow x - y = 0$ or $x + 3y = 0$

Equation of normal at (1, 1)

$y - 1 = -1(x-1) \Rightarrow x + y - 2 = 0$

It intersects $x + 3y = 0$ at $(3, -1)$ and hence normal meets the curve in fourth quadrant.

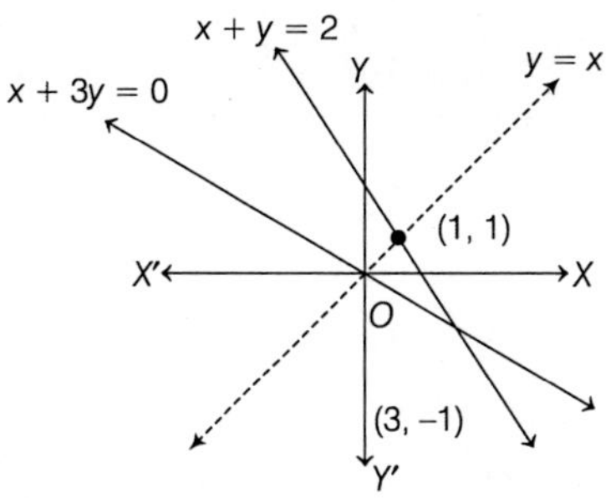

12 Given, $f(0) = 2 = g(1), g(0) = 0$

and $\quad f(1) = 6$

f and g are differentiable in (0,1).

Let $\quad h(x) = f(x) - 2g(x) \quad \ldots$(i)

$\Rightarrow \quad h(0) = f(0) - 2g(0)$

$\Rightarrow \quad h(0) = 2 - 0 \Rightarrow h(0) = 2$

and $\quad h(1) = f(1) - 2g(1) = 6 - 2(2)$

$\Rightarrow \quad h(1) = 2, h(0) = h(1) = 2$

Hence, using Rolle's theorem, we get

$h'(c) = 0$, such that $c \in (0,1)$

On differentiating Eq.(i) at c, we get

$f'(c) - 2g'(c) = 0 \Rightarrow f'(c) = 2g'(c)$

13 Let $y = ax^2 + bx + c$ [equation of parabola]

As it touches $y = x$ at $x = 1$.

$\therefore \quad y = a + b + c$

and $\quad y = 1 \Rightarrow a + b + c = 1$

Now, $\dfrac{dy}{dx} = 2ax + b$

$\Rightarrow \left(\dfrac{dy}{dx}\right)_{\text{at } x=1} = 2a + b \Rightarrow 2a + b = 1$

[from $y = x$, slope $= 1$]

Now, $f(x) = ax^2 + bx + c$

$\Rightarrow f'(x) = 2ax + b \Rightarrow f''(x) = 2a$

$\therefore \quad f(0) = c, f'(0) = b, f''(0) = 2a,$

$f'(1) = 2a + b = 1$

14 $a + b = 4 \Rightarrow b = 4 - a$

and $b - a = 4 - 2a = t$ (say)

Now, $\int_0^a g(x)\,dx + \int_0^b g(x)\,dx = \int_0^a g(x)\,dx + \int_0^{4-a} g(x)\,dx = I(a)$

$\Rightarrow \dfrac{dI(a)}{da} = g(a) - g(4 - a)$

As $a < 2$ and $g(x)$ is increasing.

$\Rightarrow 4 - a > a \Rightarrow g(a) - g(4 - a) < 0$

$\Rightarrow \dfrac{dI(a)}{da} < 0$

Now, $\dfrac{dI(a)}{d(a)} = \dfrac{dI(a)}{dt}\dfrac{dt}{da} = -2 \cdot \dfrac{dI(a)}{dt}$

$\Rightarrow \dfrac{dI(a)}{dt} > 0$

Thus, $I(a)$ is an increasing function of t. Hence, the given expression increasing with $(b - a)$.

15 We know that,

$1 \le |\sin x| + |\cos x| \le \sqrt{2}$

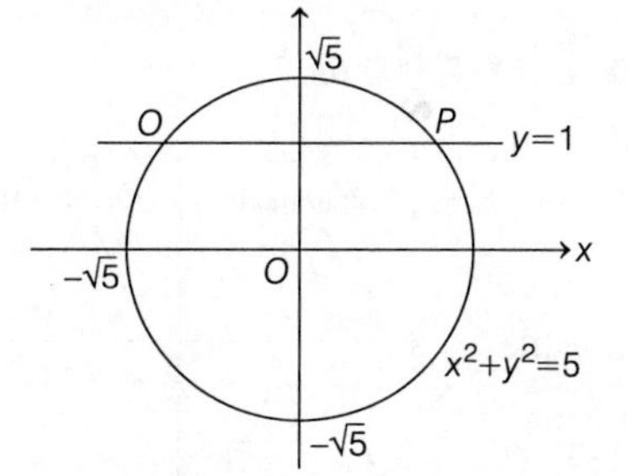

$\Rightarrow y = [|\sin x| + |\cos x|] = 1$

Let P and Q be the points of intersection of given curves.

Clearly, the given curves meet at points where $y = 1$, so we get

$x^2 + 1 = 5 \Rightarrow x = \pm 2$

Now, $\quad P(2, 1)$ and $Q(-2, 1)$

On differentiating $x^2 + y^2 = 5$ w.r.t. x, we get

$2x + 2y\dfrac{dy}{dx} = 0 \Rightarrow \dfrac{dy}{dx} = -\dfrac{x}{y}$

$\left(\dfrac{dy}{dx}\right)_{(2,1)} = -2$

and $\left(\dfrac{dy}{dx}\right)_{(-2,1)} = 2$

Clearly, the slope of line $y = 1$ is zero and the slope of the tangents at P and Q are (-2) and (2), respectively.

Thus, the angle of intersection is $\tan^{-1}(2)$.

16 There is only one function in option (a), whose critical point $\dfrac{1}{2} \in (0,1)$ but in other parts critical point $0 \notin (0,1)$. Then, we can say that functions in options (b), (c) and (d) are continuous on [0, 1] and differentiable in (0, 1).

Now, for $f(x) = \begin{cases} \dfrac{1}{2} - x, & x < \dfrac{1}{2} \\ \left(\dfrac{1}{2} - x\right)^2, & x \ge \dfrac{1}{2} \end{cases}$

Here, $Lf'\left(\dfrac{1}{2}\right) = -1$

and $Rf'\left(\dfrac{1}{2}\right) = 2\left(\dfrac{1}{2} - \dfrac{1}{2}\right)(-1) = 0$

$\therefore \quad Lf'\left(\dfrac{1}{2}\right) \ne Rf'\left(\dfrac{1}{2}\right)$

$\Rightarrow f$ is non-differentiable at $x = \dfrac{1}{2} \in (0,1)$.

$\therefore$ Lagrange mean value theorem is not applicable to $f(x)$ in [0, 1].

DAY FOURTEEN

Maxima and Minima

Learning & Revision for the Day

- Maxima and Minima of a Function
- Concept of Global Maximum/Minimum

Maxima and Minima of a Function

A function $f(x)$ is said to attain a **maximum** at $x = a$, if there exists a neighbourhood $(a - \delta, a + \delta)$, $x \neq a$ i.e. $f(x) < f(a)$, $\forall\, x \in (a - \delta, a + \delta)$, $x \neq a \cdot h > 0$ (very small quantity)

In such a case $f(a)$ is said to be the maximum value of $f(x)$ at $x = a$.

A function $f(x)$ is said to attain a **minimum** at $x = a$, if there exists a neighbourhood $(a - \delta, a + \delta)$ such that $f(x) > f(a)$, $\forall x \in (a - \delta, a + \delta)$, $x \neq a$.

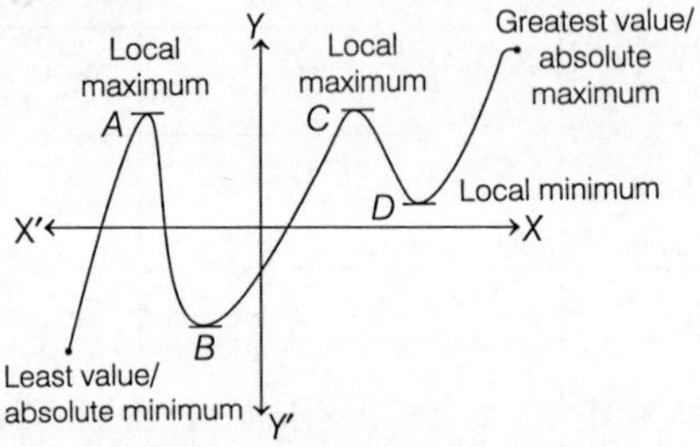

Graph of a continuous function explained local maxima (minima) and absolute maxima (minima). In such a case $f(a)$ is said to be the minimum value of $f(x)$ at $x = a$.

The points at which a function attains either the maximum or the minimum values are known as the **extreme points** or **turning points** and both minimum and maximum values of $f(x)$ are called extreme values. The turning points A and C are called **local maximum** and points B and D are called **local minimum.**

Critical Point

- A point c in the domain of a function f at which either $f'(c) = 0$ or f is not differentiable is called a **critical point** of f. Note that, if f is continuous at point c and $f'(c) = 0$, then there exists $h > 0$ such that f is differentiable in the interval $(c - h, c + h)$.
- The converse of above theorem need not be true, that is a point at which the derivative vanishes need not be a point of local maxima or local minima.

Method to Find Local Maxima or Local Minima

First Derivative Test

Let f be a function defined on an open interval I and f be continuous at a critical point c in I. Then,

(i) If $f'(x)$ changes sign from positive to negative as x increases through c, i.e. if $f'(x) > 0$ at every point sufficiently close to and to the left of c and $f'(x) < 0$ at every point sufficiently close to and to the right of c, then c is a point of **local maxima.**

PREP MIRROR
Your Personal Preparation Indicator

- No. of Questions in Exercises (x)—
- No. of Questions Attempted (y)—
- No. of Correct Questions (z)—
  (Without referring Explanations)

- Accuracy Level $(z/y \times 100)$—
- Prep Level $(z/x \times 100)$—

In order to expect good rank in JEE, your Accuracy Level should be above 85 & Prep Level should be above 75.

(ii) If $f'(x)$ changes sign from negative to positive as x increases through point c, i.e. if $f'(x) < 0$ at every point sufficiently close to and to the left of c and $f'(x) > 0$ at every point sufficiently close to and to the right of c, then c is a point of **local minima**.

(iii) If $f'(x)$ does not change sign as x increases through c, then c is neither a point of local maxima nor a point of local minima. Infact, such a point is called **point of inflection**.

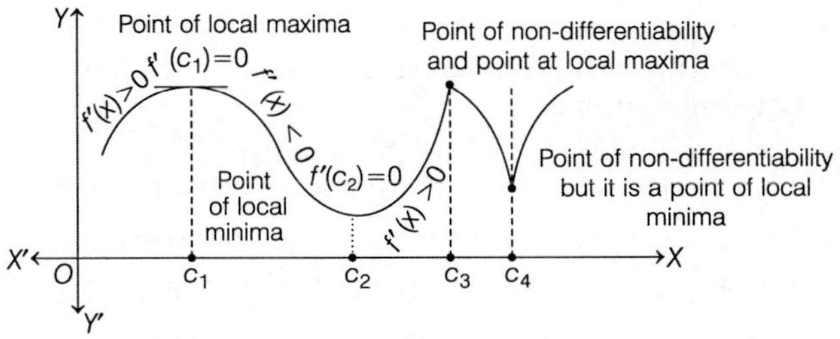

Graph of f around c explained following points.

(iv) If c is a point of local maxima of f, then $f(c)$ is a local maximum value of f. Similarly, if c is a point of local minima of f, then $f(c)$ is a local minimum value of f.

Second or Higher Order Derivative Test

(i) Find $f'(x)$ and equate it to zero. Solve $f'(x) = 0$ let its roots be $x = a_1, a_2, \ldots$

(ii) Find $f''(x)$ and at $x = a_1$,

(a) if $f''(a_1)$ is positive, then $f(x)$ is minimum at $x = a_1$.

(b) if $f''(a_1)$ is negative, then $f(x)$ is maximum at $x = a_1$.

(iii) (a) If at $x = a_1, f''(a_1) = 0$, then find $f'''(x)$. If $f'''(a_1) \neq 0$, then $f(x)$ is neither maximum nor minimum at $x = a$.

(b) If $f'''(a_1) = 0$, then find $f^{iv}(x)$.

(c) If $f^{iv}(x)$ is positive (minimum value) and $f^{iv}(x)$ is negative (maximum value).

(iv) If at $x = a_1, f^{iv}(a_1) = 0$, then find $f^{v}(x)$ and proceed similarly.

Point of Inflection

At point of inflection

(i) It is not necessary that 1st derivative is zero.

(ii) 2nd derivative must be zero or 2nd derivative changes sign in the neighbourhood of point of inflection.

nth Derivative Test

Let f be a differentiable function on an interval I and a be an interior point of I such that

(i) $f'(a) = f''(a) = f'''(a) = \ldots = f^{n-1}(a) = 0$ and

(ii) $f^{n}(a)$ exists and is non-zero.

Important Results

- If n is even and $f^{n}(a) < 0 \Rightarrow x = a$ is a point of local maximum.
- If n is even and $f^{n}(a) > 0 \Rightarrow x = a$ is a point of local minimum.
- If n is odd $\Rightarrow x = a$ is a point of neither local maximum nor a point of local minimum.
- The function $f(x) = \dfrac{ax+b}{cx+d}$ has no local maximum or minimum regardless of values of a, b, c and d.
- The function $f(\theta) = \sin^{m}\theta \cdot \cos^{n}\theta$ attains maximum values at $\theta = \tan^{-1}\left(\sqrt{\dfrac{m}{n}}\right)$.
- If AB is diameter of circle and C is any point on the circumference, then area of the ΔABC will be maximum, if triangle is isosceles.

Concept of Global Maximum/Minimum

- Let $y = f(x)$ be a given function with domain D and $[a, b] \subseteq D$, then global maximum/minimum of $f(x)$ in $[a, b]$ is basically the greatest / least value of $f(x)$ in $[a, b]$.
- Global maxima/minima in $[a, b]$ would always occur at critical points of $f(x)$ within $[a, b]$ or at end points of the interval.

Global Maximum/Minimum in $[a, b]$

In order to find the global maximum and minimum of $f(x)$ in $[a, b]$.

Step I Find out all critical points of $f(x)$ in $[a, b]$ [i.e. all points at which $f'(x) = 0$] and let these points are $c_1, c_2, \ldots, c_n$.

Step II Find the value of $f(c_1), f(c_2), \ldots, f(c_n)$ and also at the end points of domain i.e. $f(a)$ and $f(b)$.

Step III Find $M_1 \rightarrow$ Global maxima or greatest value and $M_2 \rightarrow$ Global minima or least value.

where, $M_1 = \max\{f(a), f(c_1), f(c_2), \ldots, f(c_n), f(b)\}$

and $M_2 = \min\{f(a), f(c_1), f(c_2), \ldots, f(c_n), f(b)\}$

Some Important Results on Maxima and Minima

(i) Maxima and minima occur alternatively i.e. between two maxima there is one minimum and *vice-versa*.

(ii) If $f(x) \rightarrow \infty$ as $x \rightarrow a$ or b and $f'(x) = 0$ only for one value of x (say c) between a and b, then $f(c)$ is necessarily the minimum and the least value.

(iii) If $f(x) \rightarrow -\infty$ as $x \rightarrow a$ or b, then $f(c)$ is necessarily the maximum and greatest value.

(iv) The **stationary points** are the points of the domain, where $f'(x) = 0$.

DAY PRACTICE SESSION 1

FOUNDATION QUESTIONS EXERCISE

1 If f is defined as $f(x) = x + \dfrac{1}{x}$, then which of following is true? → NCERT Exemplar

(a) Local maximum value of $f(x)$ is -2
(b) Local minimum value of $f(x)$ is 2
(c) Local maximum value of $f(x)$ is less than local minimum value of $f(x)$
(d) All the above are true

2 If the sum of two numbers is 3, then the maximum value of the product of the first and the square of second is → NCERT Exemplar

(a) 4 (b) 1 (c) 3 (d) 0

3 If $y = a \log x + bx^2 + x$ has its extremum value at $x = 1$ and $x = 2$, then (a, b) is equal to

(a) $\left(1, \dfrac{1}{2}\right)$ (b) $\left(\dfrac{1}{2}, 2\right)$ (c) $\left(2, \dfrac{-1}{2}\right)$ (d) $\left(\dfrac{-2}{3}, \dfrac{-1}{6}\right)$

4 The function $f(x) = a \cos x + b \tan x + x$ has extreme values at $x = 0$ and $x = \dfrac{\pi}{6}$, then

(a) $a = -\dfrac{2}{3}, b = -1$ (b) $a = \dfrac{2}{3}, b = -1$
(c) $a = -\dfrac{2}{3}, b = 1$ (d) $a = \dfrac{2}{3}, b = 1$

5 The minimum radius vector of the curve $\dfrac{4}{x^2} + \dfrac{9}{y^2} = 1$ is of length

(a) 1 (b) 5 (c) 7 (d) None of these

6 The function $f(x) = 4x^3 - 18x^2 + 27x - 7$ has → NCERT Exemplar

(a) one local maxima
(b) one local minima
(c) one local maxima and two local minima
(d) neither maxima nor minima

7 The function $f(x) = \dfrac{x^2 - 2}{x^2 - 4}$ has

(a) no point of local minima
(b) no point of local maxima
(c) exactly one point of local minima
(d) exactly one point of local maxima

8 Let $f : R \to R$ be defined by $f(x) = \begin{cases} k - 2x, & \text{if } x \le -1 \\ 2x + 3, & \text{if } x > -1 \end{cases}$.

If f has a local minimum at $x = -1$, then a possible value of k is → AIEEE 2010

(a) 1 (b) 0 (c) $-\dfrac{1}{2}$ (d) -1

9 The minimum value of $9x + 4y$, where $xy = 16$ is

(a) 48 (b) 28 (c) 38 (d) 18

10 If the function $f(x) = 2x^3 - 9ax^2 + 12a^2x + 1$, where $a > 0$ attains its maximum and minimum at p and q respectively such that $p^2 = q$, then a is equal to

(a) 3 (b) 1 (c) 2 (d) $\dfrac{1}{2}$

11 If $f(x) = x^2 + 2bx + 2c^2$ and $g(x) = -x^2 - 2cx + b^2$ such that minimum $f(x) >$ maximum $g(x)$, then the relation between b and c is

(a) $0 < c < b\sqrt{2}$ (b) $|c| < |b|\sqrt{2}$
(c) $|c| > |b|\sqrt{2}$ (d) No real values of b and c

12 Let $f(x)$ be a polynomial of degree four having extreme values at $x = 1$ and $x = 2$. If $\lim\limits_{x \to 0}\left[1 + \dfrac{f(x)}{x^2}\right] = 3$, then $f(2)$ is equal to → JEE Mains 2015

(a) -8 (b) -4
(c) 0 (d) 4

13 If a differential function $f(x)$ has a relative minimum at $x = 0$, then the function $\phi(x) = f(x) + ax + b$ has a relative minimum at $x = 0$ for

(a) all a and all b (b) all b, if $a = 0$
(c) all $b > 0$ (d) all $a > 0$

14 The denominator of a fraction is greater than 16 of the square of numerator, then least value of fraction is

(a) $-1/4$ (b) $-1/8$
(c) $1/12$ (d) $1/16$

15 The function $f(x) = ax + \dfrac{b}{x}$, b, $x > 0$ takes the least value at x equal to

(a) b (b) $\sqrt{a}$ (c) $\sqrt{b}$ (d) $\sqrt{\dfrac{b}{a}}$

16 Let f be a function defined by $f(x) = \begin{cases} \dfrac{\tan x}{x}, & x \ne 0 \\ 1, & x = 0 \end{cases}$

Statement I $x = 0$ is point of minima of f.

Statement II $f'(0) = 0$. → AIEEE 2011

(a) Statement I is true, Statement II is true; Statement II is a correct explanation for Statement I
(b) Statement I is true, Statement II is true; Statement II is not a correct explanation for Statement I
(c) Statement I is true; Statement II is false
(d) Statement I is false; Statement II is true

17 The absolute maximum and minimum values of the function f given by $f(x) = \cos^2 x + \sin x$, $x \in [0, \pi]$ → NCERT Exemplar

(a) 2.25 and 2 (b) 1.25 and 1
(c) 1.75 and 1.5 (d) None of these

18 The maximum value of $f(x)=\frac{x}{4+x+x^2}$ on $[-1,1]$ is

(a) $-\frac{1}{4}$ (b) $-\frac{1}{3}$ (c) $\frac{1}{6}$ (d) $\frac{1}{5}$

19 In interval $[1, e]$, the greatest value of $x^2 \log x$ is

(a) e^2 (b) $\frac{1}{e}\log\frac{1}{\sqrt{e}}$ (c) $e^2 \log\sqrt{e}$ (d) None of these

20 If x is real, the maximum value of $\frac{3x^2+9x+17}{3x^2+9x+7}$ is → AIEEE 2007

(a) 41 (b) 1 (c) $\frac{17}{7}$ (d) $\frac{1}{4}$

21 The maximum and minimum values of $f(x)=\sec x+\log\cos^2 x$, $0<x<2\pi$ are respectively → NCERT Exemplar

(a) $(1,-1)$ and $\{2(1-\log 2), 2(1+\log 2)\}$
(b) $(1,-1)$ and $\{2(1-\log 2), 2(1-\log 2)\}$
(c) $(1,-1)$ and $(2,-3)$
(d) None of the above

22 The difference between greatest and least values of the function $f(x)=\sin 2x-x$, on $\left[-\frac{\pi}{2},\frac{\pi}{2}\right]$ is → NCERT Exemplar

(a) π (b) 2π (c) 3π (d) $\frac{\pi}{2}$

23 The point of inflection for the curve $y=x^{5/2}$ is

(a) (1, 1) (b) (0, 0) (c) (1, 0) (d) (0, 1)

24 The maximum area of a right angled triangle with hypotenuse h is → JEE Main 2013

(a) $\frac{h^3}{2\sqrt{2}}$ (b) $\frac{h^2}{2}$ (c) $\frac{h^2}{\sqrt{2}}$ (d) $\frac{h^2}{4}$

25 A straight line is drawn through the point $P(3, 4)$ meeting the positive direction of coordinate axes at the points A and B. If O is the origin, then minimum area of ΔOAB is equal to

(a) 12 sq units (b) 6 sq units
(c) 24 sq units (d) 48 sq units

26 Suppose the cubic x^3-px+q has three distinct real roots, where $p>0$ and $q>0$. Then, which one of the following holds?

(a) The cubic has maxima at both $\sqrt{\frac{p}{3}}$ and $-\sqrt{\frac{p}{3}}$
(b) The cubic has minima at $\sqrt{\frac{p}{3}}$ and maxima at $-\sqrt{\frac{p}{3}}$
(c) The cubic has minima at $-\sqrt{\frac{p}{3}}$ and maxima at $\sqrt{\frac{p}{3}}$
(d) The cubic has minima at both $\sqrt{\frac{p}{3}}$ and $-\sqrt{\frac{p}{3}}$

27 If $A(x_1, y_1)$, $B(x_2, y_2)$ and $C(x_3, y_3)$ are the vertices of a ΔABC. A parallelogram $AFDE$ is drawn with D, E and F on the line segment BC, CA and AB, respectively. Then, maximum area of such parallelogram is

(a) $\frac{1}{2}$ (area of ΔABC) (b) $\frac{1}{4}$ (area of ΔABC)
(c) $\frac{1}{6}$ (area of ΔABC) (d) $\frac{1}{8}$ (area of ΔABC)

28 If $y=f(x)$ is a parametrically defined expression such that $x=3t^2-18t+7$ and $y=2t^3-15t^2+24t+10$, $\forall x\in[0,6]$.
Then, the maximum and minimum values of $y=f(x)$ are

(a) 36, 3 (b) 46, 6 (c) 40, –6 (d) 46, –6

29 The value of a, so that the sum of the squares of the roots of the equation $x^2-(a-2)x-a+1=0$ assume the least value is

(a) 2 (b) 1 (c) 3 (d) 0

30 The minimum intercepts made by the axes on the tangent to the ellipse $\frac{x^2}{16}+\frac{y^2}{9}=1$ is

(a) 25 (b) 7 (c) 1 (d) None of these

31 The curved surface of the cone inscribed in a given sphere is maximum, if

(a) $h=\frac{4R}{3}$ (b) $h=\frac{R}{3}$ (c) $h=\frac{2R}{3}$ (d) None of these

32 The volume of the largest cone that can be inscribed in a sphere of radius R is → NCERT

(a) $\frac{3}{8}$ of the volume of the sphere
(b) $\frac{8}{27}$ of the volume of the sphere
(c) $\frac{2}{7}$ of the volume of the sphere
(d) None of the above

33 Area of the greatest rectangle that can be inscribed in the ellipse $\frac{x^2}{a^2}+\frac{y^2}{b^2}=1$ is

(a) $\sqrt{ab}$ (b) $\frac{a}{b}$ (c) $2ab$ (d) ab

34 The real number x when added to its inverse gives the minimum value of the sum at x equal to → AIEEE 2003

(a) 2 (b) 1 (c) –1 (d) –2

35 The greatest value of $f(x)=(x+1)^{1/3}-(x-1)^{1/3}$ on [0, 1] is → AIEEE 2002

(a) 1 (b) 2 (c) 3 (d) 1/3

36 The coordinate of a point on the parabola $y^2=8x$ whose distance from the circle $x^2+(y+6)^2=1$ is minimum, is

(a) (2,–4) (b) (2,4) (c) (18,–12) (d) (8, 8)

37 The volume of the largest cylinder that can be inscribed in a sphere of radius r cm is

(a) $\frac{4\pi r^3}{\sqrt{3}}$ (b) $\frac{4\pi r^3}{3\sqrt{3}}$ (c) $\frac{4\pi r^3}{2\sqrt{3}}$ (d) $\frac{4\pi r^3}{5\sqrt{2}}$

38 Maximum slope of the curve $y=-x^3+3x^2+9x-27$ is

(a) 0 (b) 12 (c) 16 (d) 32

39 If $ab = 2a + 3b$, $a > 0$, $b > 0$, then the minimum value of ab is

(a) 12 (b) 24

(c) $\frac{1}{4}$ (d) None of these

40 The perimeter of a sector is p. The area of the sector is maximum, when its radius is

(a) $\sqrt{p}$ (b) $\frac{1}{\sqrt{p}}$ (c) $\frac{p}{2}$ (d) $\frac{p}{4}$

DAY PRACTICE SESSION 2

PROGRESSIVE QUESTIONS EXERCISE

1 The minimum radius vector of the curve $\frac{a^2}{x^2} + \frac{b^2}{y^2} = 1$ is of length

(a) $a - b$ (b) $a + b$

(c) $2a + b$ (d) None of these

2 $f(x) = x^2 - 4|x|$ and

$$g(x) = \begin{cases} \min\{f(t): -6 \le t \le x\}, & x \in [-6, 0] \\ \max\{f(t): 0 < t \le x\}, & x \in (0, 6] \end{cases}$$, then $g(x)$ has

(a) exactly one point of local minima

(b) exactly one point of local maxima

(c) no point to local maxima but exactly one point of local minima

(d) neither a point of local maxima nor minima

3 $f(x) = \begin{cases} 4x - x^3 + \log(a^2 - 3a + 3), & 0 \le x < 3 \\ x - 18, & x \ge 3 \end{cases}$

Complete the set of values of a such that $f(x)$ has a local maxima at $x = 3$, is

(a) $[-1, 2]$ (b) $(-\infty, 1) \cup (2, \infty)$

(c) $[1, 2]$ (d) $(-\infty, -1) \cup (2, \infty)$

4 The point in the interval $[0, 2\pi]$, where $f(x) = e^x \sin x$ has maximum slope is

(a) $\frac{\pi}{4}$ (b) $\frac{\pi}{2}$

(c) π (d) None of these

5 The total number of local maxima and local minima of the function $f(x) = \begin{cases} (2 + x)^3, & -3 < x \le -1 \\ x^{2/3}, & -1 < x < 2 \end{cases}$ is

(a) 0 (b) 1 (c) 2 (d) 3

6 If 20 m of wire is available for fencing off a flower-bed in the form of a circular sector, then the maximum area (in sq m) of the flower-bed is → JEE Mains 2017

(a) 12.5 (b) 10 (c) 25 (d) 30

7 The cost of running a bus from A to B, is ₹ $\left(av + \frac{b}{v}\right)$, where v km/h is the average speed of the bus. When the bus travels at 30 km/h, the cost comes out to be ₹ 75 while at 40 km/h, it is ₹ 65. Then, the most economical speed (in km/h) of the bus is → JEE Mains 2013

(a) 45 (b) 50 (c) 60 (d) 40

8 If $f(x) = \begin{cases} |x^2 - 2|, & -1 \le x < \sqrt{3} \\ \frac{x}{\sqrt{3}}, & \sqrt{3} \le x < 2\sqrt{3} \\ 3 - x, & 2\sqrt{3} \le x \le 4 \end{cases}$, then the points, where $f(x)$ takes maximum and minimum values, are

(a) 1, 4 (b) 0, 4

(c) 2, 4 (d) None of these

9 Let $f(x) = \begin{cases} \left|x^3 + x^2 + 3x + \sin x\right| \left(3 + \sin\frac{1}{x}\right), & x \ne 0 \\ 0, & x = 0 \end{cases}$, then number of points [where, $f(x)$ attains its minimum value] is

(a) 1 (b) 2

(c) 3 (d) infinite many

10 A wire of length 2 units is cut into two parts which are bent respectively to form a square of side = x units and a circle of radius = r units. If the sum of the areas of the square and the circle so formed is minimum, then

(a) $2x = (\pi + 4)r$ (b) $(4 - \pi)x = \pi r$

(c) $x = 2r$ (d) $2x = r$

11 Let $f(x) = x^2 + \frac{1}{x^2}$ and $g(x) = x - \frac{1}{x}$, $x \in R - \{-1, 0, 1\}$. If $h(x) = \frac{f(x)}{g(x)}$, then the local minimum value of $h(x)$ is → JEE Mains 2018

(a) 3 (b) -3 (c) $-2\sqrt{2}$ (d) $2\sqrt{2}$

12 The largest term in the sequence $a_n = \frac{n^2}{n^3 + 200}$ is given by

(a) $\frac{529}{49}$ (b) $\frac{8}{89}$

(c) $\frac{49}{543}$ (d) None of these

13 All possible values of the parameter a so that the function $f(x) = x^3 - 3(7 - a)x^2 - 3(9 - a^2)x + 2$ has a negative point of local minimum are

(a) all real values (b) no real values

(c) $(0, \infty)$ (d) $(-\infty, 0)$

14 The circle $x^2 + y^2 = 1$ cuts the X-axis at P and Q. Another circle with centre at Q and variable radius intersects the first circle at R above the X-axis and the line segment PQ at S. Then, the maximum area of the ΔQSR is

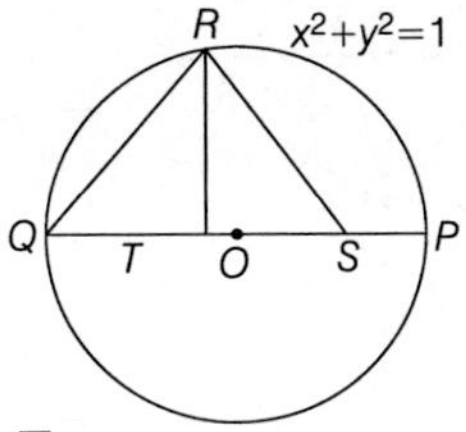

(a) $4\sqrt{3}$ sq units (b) $14\sqrt{3}$ sq units
(c) $\frac{4\sqrt{3}}{9}$ sq units (d) $15\sqrt{3}$ sq units

15 Given, $P(x) = x^4 + ax^3 + bx^2 + cx + d$ such that $x = 0$ is the only real root of $P'(x) = 0$. If $P(-1) < P(1)$, then in the interval $[-1, 1]$. → AIEEE 2009

(a) $P(-1)$ is the minimum and $P(1)$ is the maximum of P
(b) $P(-1)$ is not minimum but $P(1)$ is the maximum of P
(c) $P(-1)$ is the minimum and $P(1)$ is not the maximum of P
(d) Neither $P(-1)$ is the minimum nor $P(1)$ is the maximum of P

ANSWERS

SESSION 1	1 (d)	2 (a)	3 (d)	4 (a)	5 (b)	6 (d)	7 (d)	8 (d)	9 (a)	10 (c)
	11 (c)	12 (c)	13 (b)	14 (b)	15 (d)	16 (b)	17 (b)	18 (c)	19 (a)	20 (a)
	21 (b)	22 (a)	23 (b)	24 (d)	25 (c)	26 (b)	27 (a)	28 (d)	29 (b)	30 (b)
	31 (a)	32 (b)	33 (c)	34 (b)	35 (b)	36 (a)	37 (b)	38 (b)	39 (b)	40 (d)
SESSION 2	1 (b)	2 (d)	3 (c)	4 (b)	5 (c)	6 (c)	7 (c)	8 (b)	9 (a)	10 (c)
	11 (d)	12 (c)	13 (b)	14 (c)	15 (b)					

Hints and Explanations

SESSION 1

1 Let $y = x + \frac{1}{x} \Rightarrow \frac{dy}{dx} = 1 - \frac{1}{x^2}$

Now, $\frac{dy}{dx} = 0 \Rightarrow x^2 = 1$

$\Rightarrow x = \pm 1$

$\Rightarrow \frac{d^2y}{dx^2} = \frac{2}{x^3}$, therefore

$\frac{d^2y}{dx^2}$ (at $x = 1$) > 0

and $\frac{d^2y}{dx^2}$ (at $x = -1$) < 0

Hence, local maximum value of y is at $x = -1$ and the local maximum value $= -2$.

Local minimum value of y is at $x = 1$ and local minimum value $= 2$.

Therefore, local maximum value -2 is less than local minimum value 2.

2 Let two numbers be x and $(3 - x)$.

Then, product $P = x(3 - x)^2$

$\frac{dP}{dx} = -2x(3 - x) + (3 - x)^2$

$\frac{dP}{dx} = (3 - x)(3 - 3x)$ and $\frac{d^2P}{dx^2} = 6x - 12$

For maxima or minima, put $\frac{dP}{dx} = 0$

$\Rightarrow (3 - x)(3 - 3x) = 0 \Rightarrow x = 3, 1$

At $x = 3$,

$\frac{d^2P}{dx^2} = 18 - 12 = 6 > 0$ [minima]

At $x = 1$,

$\frac{d^2P}{dx^2} = -6 < 0$

So, P is maximum at $x = 1$.

$\therefore$ Maximum value of $P = 1(3 - 1)^2 = 4$

3 $\because \frac{dy}{dx} = \frac{a}{x} + 2bx + 1$

$\Rightarrow \left(\frac{dy}{dx}\right)_{x=1} = a + 2b + 1 = 0$

$\Rightarrow a = -2b - 1$

and $\left(\frac{dy}{dx}\right)_{x=2} = \frac{a}{2} + 4b + 1 = 0$

$\Rightarrow \frac{-2b - 1}{2} + 4b + 1 = 0$

$\Rightarrow -b + 4b + \frac{1}{2} = 0 \Rightarrow 3b = \frac{-1}{2}$

$\Rightarrow b = \frac{-1}{6}$ and $a = \frac{1}{3} - 1 = \frac{-2}{3}$

4 $f'(x) = -a \sin x + b \sec^2 x + 1$

Now, $f'(0) = 0$ and $f'\left(\frac{\pi}{6}\right) = 0$

$\Rightarrow b + 1 = 0$ and $-\frac{a}{2} + \frac{4b}{3} + 1 = 0$

$\Rightarrow b = -1, a = -\frac{2}{3}$

5 The given curve is $\frac{4}{x^2} + \frac{9}{y^2} = 1$

Put $x = r\cos\theta$, $y = r\sin\theta$, we get

$r^2 = (2\sec\theta)^2 + (3\,\text{cosec}\,\theta)^2$

So, r^2 will have minimum value $(2 + 3)^2$.

or r have minimum value equal to 5.

6 $f(x) = 4x^3 - 18x^2 + 27x - 7$

$f'(x) = 12x^2 - 36x + 27$

$= 3(4x^2 - 12x + 9) = 3(2x - 3)^2$

$f'(x) = 0 \Rightarrow x = \frac{3}{2}$ (critical point)

Since, $f'(x) > 0$ for all $x < \frac{3}{2}$ and for all $x > \frac{3}{2}$

Hence, $x = \frac{3}{2}$ is a point of inflection *i.e.*, neither a point of maxima nor a point of minima.

$x = \frac{3}{2}$ is the only critical point and f has neither maxima nor minima.

7 For $y = \frac{x^2 - 2}{x^2 - 4} \Rightarrow \frac{dy}{dx} = \frac{-4x}{(x^2 - 4)^2}$

$\Rightarrow \quad \frac{dy}{dx} > 0$, for $x < 0$

and $\quad \frac{dy}{dx} < 0$, for $x > 0$

Thus, $x = 0$ is the point of local maxima for y. Now, $(y)_{x=0} = \frac{1}{2}$ (positive). Thus, $x = 0$ is also the point of local maximum for $y = \left|\frac{x^2 - 2}{x^2 - 4}\right|$.

8 If $f(x)$ has a local minimum at $x = -1$, then

$$\lim_{x \to -1^+} f(x) = \lim_{x \to -1^-} f(x)$$

$\Rightarrow \quad \lim_{x \to -1^+} 2x + 3 = \lim_{x \to -1^-} 1 < -2x$

$\Rightarrow \quad -2 + 3 = k + 2 \quad \Rightarrow \quad k = -1$

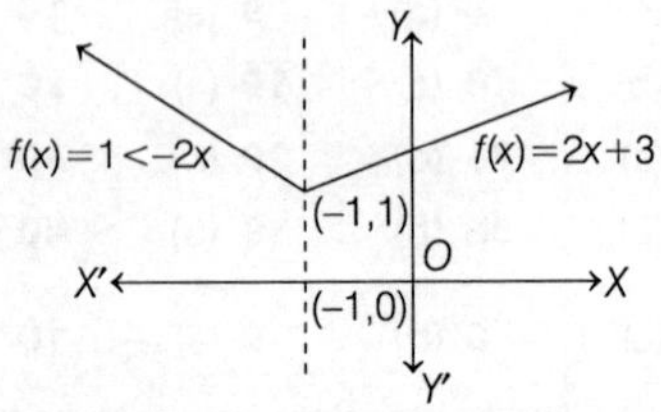

9 Let $S = 9x + 4y$

Since, $xy = 16$ is given.

$\therefore \quad y = \frac{16}{x}$ or $S = 9x + \frac{64}{x}$

On differentiating both sides, we get

$$\frac{dS}{dx} = 9 - \frac{64}{x^2} \quad \text{...(i)}$$

$\because \quad \frac{dS}{dx} = 0 \Rightarrow \frac{64}{x^2} = 9 \Rightarrow x = \pm \frac{8}{3}$

Again, on differentiating Eq. (i) w.r.t. x, we get $\frac{d^2S}{dx^2} = \frac{128}{x^3}$

Hence, it is minimum at $x = \frac{8}{3}$ and minimum value of S is

$$S_{\min} = 9\left(\frac{8}{3}\right) + 4(6) = 48$$

10 We have,

$$f(x) = 2x^3 - 9ax^2 + 12a^2 x + 1$$
$$f'(x) = 6x^2 - 18ax + 12a^2$$
$$f''(x) = 12x - 18a$$

For maximum and minimum,

$6x^2 - 18ax + 12a^2 = 0$

$\Rightarrow \quad x^2 - 3ax + 2a^2 = 0$

$\Rightarrow \quad x = a$ or $x = 2a$

At $x = a$ maximum and at $x = 2a$ minimum.

$\because \quad p^2 = q$

$\therefore \quad a^2 = 2a \Rightarrow a = 2$ or $a = 0$

But $a > 0$, therefore $a = 2$

11 Minimum of $f(x) = -\frac{D}{4a}$

$$= \frac{-(4b^2 - 8c^2)}{4}$$
$$= 2c^2 - b^2$$

and maximum of $g(x) = -\frac{(4c^2 + 4b^2)}{4(-1)}$

$= b^2 + c^2$

Since, min $f(x) >$ max $g(x)$

$\Rightarrow \quad 2c^2 - b^2 > b^2 + c^2$

$\Rightarrow \quad c^2 > 2b^2$

$\Rightarrow \quad |c| > \sqrt{2}\,|b|$

12 **Central Idea** Any function have extreme values (maximum or minimum) at its critical points, where $f'(x) = 0$.

Since, the function have extreme values at $x = 1$ and $x = 2$.

$\therefore \quad f'(x) = 0$ at $x = 1$ and $x = 2$

$\Rightarrow \quad f'(1) = 0$ and $f'(2) = 0$

Also it is given that

$$\lim_{x \to 0}\left[1 + \frac{f(x)}{x^2}\right] = 3 \Rightarrow 1 + \lim_{x \to 0}\frac{f(x)}{x^2} = 3$$

$\Rightarrow \quad \lim_{x \to 0}\frac{f(x)}{x^2} = 2$

$\Rightarrow \quad f(x)$ will be of the form

$$ax^4 + bx^3 + 2x^2$$

[$\because f(x)$ is of four degree polynomial]

Let $f(x) = ax^4 + bx^3 + 2x^2 \Rightarrow f'(x)$

$$= 4ax^3 + 3bx^2 + 4x$$

$\Rightarrow \quad f'(1) = 4a + 3b + 4 = 0 \quad \text{...(i)}$

and $f'(2) = 32a + 12b + 8 = 0$

$\Rightarrow \quad 8a + 3b + 2 = 0 \quad \text{...(ii)}$

On solving Eqs. (i) and (ii), we get

$a = \frac{1}{2}, b = -2$

$\therefore \quad f(x) = \frac{x^4}{2} - 2x^3 + 2x^2$

$\Rightarrow \quad f(2) = 8 - 16 + 8 = 0$

13 $\phi'(x) = f'(x) + a$

$\because \quad \phi'(0) = 0 \Rightarrow f'(0) + a = 0$

$\Rightarrow \quad a = 0 \quad [\because f'(0) = 0]$

Also, $\phi'(0) > 0 \; [\because f''(0) > 0]$

$\Rightarrow \phi(x)$ has relative minimum at $x = 0$ for all b, if $a = 0$

14 Let the number be x, then

$$f(x) = \frac{x}{x^2 + 16}$$

On differentiating w.r.t. x, we get

$$f'(x) = \frac{(x^2 + 16) \cdot 1 - x(2x)}{(x^2 + 16)^2}$$

$$= \frac{x^2 + 16 - 2x^2}{(x^2 + 16)^2} = \frac{16 - x^2}{(x^2 + 16)^2} \quad \text{...(i)}$$

Put $f'(x) = 0$ for maxima or minima

$f'(x) = 0 \quad \Rightarrow \quad 16 - x^2 = 0$

$\Rightarrow \quad x = 4, -4$

Again, on differentiating w.r.t. x, we get

$$f''(x) = \frac{(x^2 + 16)^2(-2x) - (16 - x^2)\,2(x^2 + 16)\,2x}{(x^2 + 16)^4}$$

At $x = 4$, $f''(x) < 0$

$\therefore f(x)$ is maximum at $x = 4$.

and at $x = -4$, $f''(x) > 0$, $f(x)$ is minimum.

$\therefore$ Least value of

$$f(x) = \frac{-4}{16 + 16} = -\frac{1}{8}$$

15 Given, $f(x) = ax + \frac{b}{x}$

On differentiating w.r.t. x, we get

$$f'(x) = a - \frac{b}{x^2}$$

For maxima or minima, put $f'(x) = 0$

$\Rightarrow \quad x = \sqrt{\frac{b}{a}}$

Again, differentiating w.r.t. x, we get

$$f''(x) = \frac{2b}{x^3}$$

At $x = \sqrt{\frac{b}{a}}$, $f''(x) =$ positive

$\Rightarrow f(x)$ is minimum at $x = \sqrt{\frac{b}{a}}$.

$\therefore f(x)$ has the least value at $x = \sqrt{\frac{b}{a}}$.

16 $f(x) = \begin{cases} \frac{\tan x}{x}, & x \neq 0 \\ 1, & x = 0 \end{cases}$

As $\quad \frac{\tan x}{x} > 1, \forall x \neq 0$

$\therefore f(0 + h) > f(0)$ and $f(0 - h) > f(0)$

At $x = 0$, $f(x)$ attains minima.

Now, $f'(0) = \lim_{h \to 0}\frac{f(h) - f(0)}{h}$

$$= \lim_{h \to 0}\frac{\frac{\tan h}{h} - 1}{h} = \lim_{h \to 0}\frac{\tan h - h}{h^2}$$

[using L' Hospital's rule]

$$= \lim_{h \to 0}\frac{\sec^2 h - 1}{2h} \; [\because \tan^2\theta = \sec^2\theta - 1]$$

$$= \lim_{h \to 0}\frac{\tan^2 h}{2h^2} \cdot h = \frac{1}{2} \cdot 0 = 0$$

Therefore, Statement II is true.

Hence, both statements are true but Statement II is not the correct explanation of Statement I.

17 Given, $f(x) = \cos^2 x + \sin x, x \in [0, \pi]$

Now,

$f'(x) = 2\cos x(-\sin x) + \cos x$

$= -2\sin x\cos x + \cos x$

For maximum or minimum put $f'(x) = 0$

$\Rightarrow \quad -2\sin x\cos x + \cos x = 0$

$\Rightarrow \quad \cos x(-2\sin x + 1) = 0$

$\Rightarrow \quad \cos x = 0$ or $\sin x = \frac{1}{2}$

$\Rightarrow \quad x = \frac{\pi}{6}, \frac{\pi}{2}$

For absolute maximum and absolute minimum, we have to evaluate

$$f(0), f\left(\frac{\pi}{6}\right), f\left(\frac{\pi}{2}\right), f(\pi)$$

At $x = 0$,

$f(0) = \cos^2 0 + \sin 0 = 1^2 + 0 = 1$

At $x = \frac{\pi}{6}$, $f\left(\frac{\pi}{6}\right) = \cos^2\left(\frac{\pi}{6}\right) + \sin\frac{\pi}{6}$

$$= \left(\frac{\sqrt{3}}{2}\right)^2 + \frac{1}{2} = \frac{5}{4} = 1.25$$

At $x = \frac{\pi}{2}$,

$$f\left(\frac{\pi}{2}\right) = \cos^2\left(\frac{\pi}{2}\right) + \sin\frac{\pi}{2} = 0^2 + 1 = 1$$

At $x = \pi$,

$f(\pi) = \cos^2\pi + \sin\pi = (-1)^2 + 0 = 1$

Hence, the absolute maximum value of f is 1.25 occurring at $x = \frac{\pi}{6}$ and the absolute minimum value of f is 1 occurring at $x = 0, \frac{\pi}{2}$ and π.

Note If close interval is given, to determine global maximum (minimum), check the value at all critical points as well as end points of a given interval.

18 $\because \quad f(x) = \frac{x}{4 + x + x^2}$

On differentiating w.r.t. x, we get

$$f'(x) = \frac{4 + x + x^2 - x(1 + 2x)}{(4 + x + x^2)^2}$$

For maximum, put $f'(x) = 0$

$$\Rightarrow \frac{4 - x^2}{(4 + x + x^2)^2} = 0 \Rightarrow x = 2, -2$$

Both the values of x are not in the interval $[-1, 1]$.

$$\therefore \quad f(-1) = \frac{-1}{4 - 1 + 1} = \frac{-1}{4}$$

$$f(1) = \frac{1}{4 + 1 + 1} = \frac{1}{6} \text{ (maximum)}$$

19 Given, $f(x) = x^2 \log x$

On differentiating w.r.t. x, we get

$f'(x) = (2\log x + 1)\, x$

For a maximum, put $f'(x) = 0$

$\Rightarrow \quad (2\log x + 1)\, x = 0$

$\Rightarrow \quad x = e^{-1/2}, 0$

$\because \quad 0 < e^{-1/2} < 1$

None of these critical points lies in the interval $[1, e]$.

So, we only compute the value of $f(x)$ at the end points 1 and e.

We have, $f(1) = 0, \quad f(e) = e^2$

Hence, greatest value of $f(x) = e^2$

20 Let $\quad f(x) = 1 + \frac{10}{3\left(x^2 + 3x + \frac{7}{3}\right)}$

$$= 1 + \frac{10}{3\left[\left(x + \frac{3}{2}\right)^2 + \frac{1}{12}\right]}$$

So, the maximum value of $f(x)$ at $x = -\frac{3}{2}$ is

$$f\left(-\frac{3}{2}\right) = 1 + \frac{10}{3\left(\frac{1}{12}\right)} = 1 + 40 = 41$$

21 Given, $f(x) = \sec x + \log\cos^2 x$

$\Rightarrow \quad f(x) = \sec x + 2\log(\cos x)$

Therefore,

$$f'(x) = \sec x \tan x - 2\tan x$$
$$= \tan x(\sec x - 2)$$
$$f'(x) = 0$$

$\Rightarrow \quad \tan x = 0$ or $\sec x = 2 \Rightarrow \cos x = \frac{1}{2}$

Therefore, possible values of x are $x = 0$, $x = \pi$ and $x = \frac{\pi}{3}$ or $x = \frac{5\pi}{3}$.

Again, $f''(x) = \sec^2 x(\sec x - 2) + \tan x(\sec x \tan x)$

$= \sec^3 x + \sec x \tan^2 x - 2\sec^2 x$

$= \sec x(\sec^2 x + \tan^2 x - 2\sec x)$

$\Rightarrow f''(0) = 1(1 + 0 - 2) = -1 < 0$

Therefore, $x = 0$ is a point of maxima.

$f''(\pi) = -1(1 + 0 + 2) = -3 < 0$

Therefore, $x = \pi$ is a point of maxima.

$$f''\left(\frac{\pi}{3}\right) = 2(4 + 3 - 4) = 6 > 0$$

Therefore, $x = \frac{\pi}{3}$ is a point of minima.

$$f''\left(\frac{5\pi}{3}\right) = 2(4 + 3 - 4) = 6 > 0$$

Therefore, $x = \frac{5\pi}{3}$ is a point of minima.

Maximum value of y at $x = 0$ is $1 + 0 = 1$.

Maximum value of y at $x = \pi$ is $-1 + 0 = -1$.

Minimum value of y at $x = \frac{\pi}{3}$ is

$2 + 2\log\frac{1}{2} = 2(1 - \log 2)$.

Minimum value of y at $x = \frac{5\pi}{3}$ is

$2 + 2\log\frac{1}{2} = 2(1 - \log 2)$.

22 Given, $f(x) = \sin 2x - x$

$\Rightarrow f'(x) = 2\cos 2x - 1$

Put $f'(x) = 0 \Rightarrow \cos 2x = \frac{1}{2}$

$$\Rightarrow 2x = -\frac{\pi}{3} \text{ or } \frac{\pi}{3} \Rightarrow x = -\frac{\pi}{6} \text{ or } \frac{\pi}{6}$$

$$\text{Now, } f\left(-\frac{\pi}{2}\right) = \sin(-\pi) + \frac{\pi}{2} = \frac{\pi}{2}$$

$$f\left(-\frac{\pi}{6}\right) = \sin\left(-\frac{2\pi}{6}\right) + \frac{\pi}{6} = -\frac{\sqrt{3}}{2} + \frac{\pi}{6}$$

$$f\left(\frac{\pi}{6}\right) = \sin\left(\frac{2\pi}{6}\right) - \frac{\pi}{6} = \frac{\sqrt{3}}{2} - \frac{\pi}{6}$$

and $f\left(\frac{\pi}{2}\right) = \sin(\pi) - \frac{\pi}{2} = -\frac{\pi}{2}$

Clearly, $\frac{\pi}{2}$ is the greatest value and $-\frac{\pi}{2}$ is the least.

Therefore, difference $= \frac{\pi}{2} + \frac{\pi}{2} = \pi$

23 Given, $y = x^{5/2}$

$$\therefore \quad \frac{dy}{dx} = \frac{5}{2}x^{3/2}, \frac{d^2y}{dx^2} = \frac{15}{4}x^{1/2}$$

At $x = 0$, $\frac{dy}{dx} = 0$, $\frac{d^2y}{dx^2} = 0$

and $\frac{d^3y}{dx^3}$ is not defined,

when $x = 0$, $y = 0$

$\therefore (0, 0)$ is a point of inflection.

24 Area of triangle, $\Delta = \frac{1}{2}x\sqrt{h^2 - x^2}$

$$\frac{d\Delta}{dx} = \frac{1}{2}\left[\sqrt{h^2 - x^2} + \frac{x(-2x)}{2\sqrt{h^2 - x^2}}\right] = 0$$

$$\Rightarrow \quad x = \frac{h}{\sqrt{2}}$$

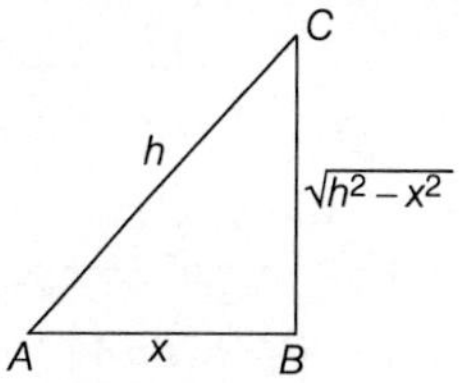

$$\Rightarrow \quad \frac{d^2\Delta}{dx^2} < 0 \text{ at } x = \frac{h}{\sqrt{2}}$$

$$\therefore \quad \Delta = \frac{1}{2} \times \frac{h}{\sqrt{2}}\sqrt{h^2 - \frac{h^2}{2}} = \frac{h^2}{4}$$

25 Let the equation of drawn line be $\frac{x}{a} + \frac{y}{b} = 1$, where $a > 3$, $b > 4$, as the line passes through (3, 4) and meets the positive direction of coordinate axes.

We have, $\frac{3}{a} + \frac{4}{b} = 1 \Rightarrow b = \frac{4a}{(a - 3)}$

Now, area of ΔAOB,

$$\Delta = \frac{1}{2}ab = \frac{2a^2}{(a - 3)}$$

$$\frac{d\Delta}{da} = \frac{2a(a - 6)}{(a - 3)^2}$$

Clearly, $a = 6$ is the point of minima for Δ.

Thus, $\Delta_{\min} = \frac{2 \times 36}{3} = 24$ sq units

26 Let $f(x) = x^3 - px + q$

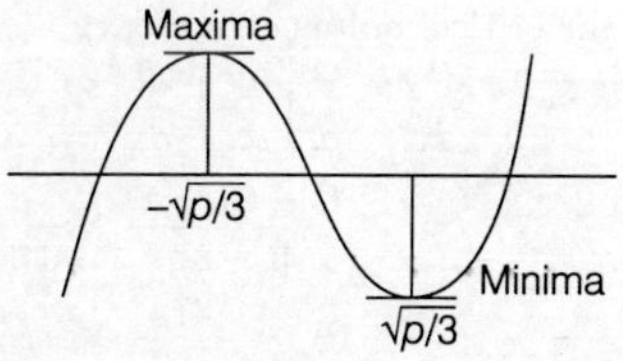

Then, $f'(x) = 3x^2 - p$

Put $f'(x) = 0$

$\Rightarrow \quad x = \sqrt{\frac{p}{3}}, -\sqrt{\frac{p}{3}}$

Now, $f''(x) = 6x$

At $x = \sqrt{\frac{p}{3}}$, $f''(x) = 6\sqrt{\frac{p}{3}} > 0$ [minima]

and at $x = -\sqrt{\frac{p}{3}}$ $f''(x) < 0$ [maxima]

27 We have, $AF \parallel DE$ and $AE \parallel FD$

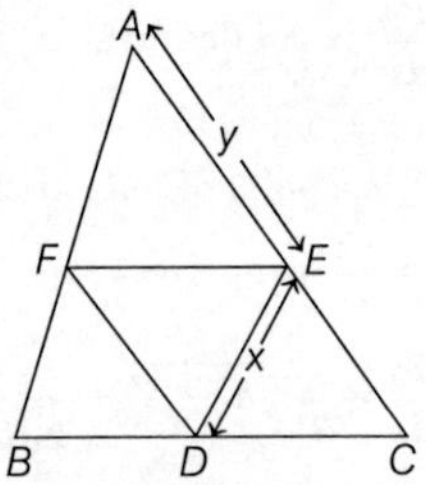

Now, in ΔABC and ΔEDC,

$\angle DEC = \angle BAC$, $\angle ACB$ is common.

$\Rightarrow \quad \Delta ABC \cong \Delta EDC$

Now, $\frac{b-y}{b} = \frac{x}{c} \Rightarrow x = \frac{c}{b}(b-y)$

Now, S = Area of parallelogram

$AFDE = 2$ (area of ΔAEF)

$\Rightarrow \quad S = 2\left(\frac{1}{2}xy \sin A\right)$

$= \frac{c}{b}(b-y)y \sin A$

$\frac{dS}{dy} = \left(\frac{c}{b}\sin A\right)(b-2y)$

Sign scheme of $\frac{dS}{dy}$,

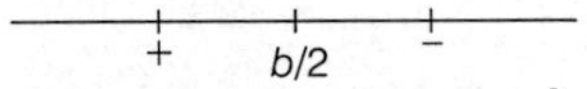

Hence, S is maximum when $y = \frac{b}{2}$.

$\therefore S_{\max} = \frac{c}{b}\left(\frac{b}{2}\right) \times \frac{b}{2}\sin A$

$= \frac{1}{2}\left(\frac{1}{2}bc\sin A\right) = \frac{1}{2}$(area of ΔABC)

28 We have,

$\frac{dy}{dt} = 6t^2 - 30t + 24 = 6(t-1)(t-4)$

and $\frac{dx}{dt} = 6t - 18 = 6(t-3)$

Thus, $\frac{dy}{dx} = \frac{(t-1)(t-4)}{(t-3)}$

which indicates that $t = 1, 3$ and 4 are the critical points of $y = f(x)$.

Now, $\frac{d^2y}{dx^2} = \frac{d}{dt}\left(\frac{dy}{dx}\right) \cdot \frac{dt}{dx}$

$= \frac{t^2 - 6t + 11}{(t-3)^2} \times \frac{1}{6(t-3)}$

At $(t = 1)$, $\frac{d^2y}{dx^2} < 0$

$\Rightarrow t = 1$ is a point of local maxima.

At $(t = 4)$, $\frac{d^2y}{dx^2} > 0$

$\Rightarrow t = 4$ is a point of local minima.

At $(t = 3)$, $\frac{dy}{dx}$ and $\frac{d^2y}{dx^2}$ are not defined and change its sign.

$\frac{d^2y}{dx^2}$ is unknown in the vicinity of $t = 3$, thus $t = 3$ is a point of neither maxima nor minima.

Finally, maximum and minimum values of expression $y = f(x)$ are 46 and -6, respectively.

29 Let α and β be the roots of the equation

$$x^2 - (a-2)x - a + 1 = 0$$

Then, $\alpha + \beta = a - 2, \alpha\beta = -a + 1$

Let $z = \alpha^2 + \beta^2$

$= (\alpha + \beta)^2 - 2\alpha\beta$

$= (a-2)^2 + 2(a-1)$

$= a^2 - 2a + 2$

$\Rightarrow \quad \frac{dz}{da} = 2a - 2$

Put $\frac{dz}{da} = 0$, then

$\Rightarrow \quad a = 1$

$\therefore \quad \frac{d^2z}{da^2} = 2 > 0$

So, z has minima at $a = 1$.

So, $\alpha^2 + \beta^2$ has least value for $a = 1$.

This is because we have only one stationary value at which we have minima.

Hence, $a = 1$.

30 Any tangent to the ellipse is

$\frac{x}{4}\cos t + \frac{y}{3}\sin t = 1$, where the point of contact is $(4\cos t, 3\sin t)$

or $\frac{x}{4\sec t} + \frac{y}{3\operatorname{cosec} t} = 1$,

It means the axes $Q(4\sec t, 0)$ and $R(0, 3\operatorname{cosec} t)$.

$\therefore$ The distance of the line segment QR is

$$QR^2 = D = 16\sec^2 t + 9\operatorname{cosec}^2 t$$

So, the minimum value of D is $(4+3)^2$ or $QR = 7$.

31 Let S be the curved surface area of a cone.

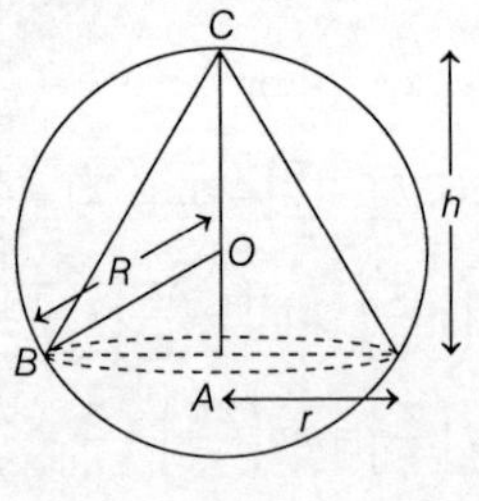

$OA = AC - OC = h - R$

In ΔOAB, $R^2 = r^2 + (h-R)^2$

$\Rightarrow \quad r = \sqrt{2Rh - h^2}$

$\therefore \quad S = \pi rl = \pi(\sqrt{2Rh - h^2})(\sqrt{h^2 + r^2})$

$= (\pi\sqrt{2Rh - h^2})(\sqrt{2Rh})$

Let $S^2 = P$

$\therefore \quad P = \pi^2 2R(2Rh^2 - h^3)$

Since, S is maximum, if P is maximum, then

$\frac{dP}{dh} = 2\pi^2 R(4Rh - 3h^2) = 0$

$\therefore \quad h = 0, \frac{4R}{3}$

Again, on differentiating $\frac{dP}{dh}$, we get

$\frac{d^2P}{dh^2} = 2\pi^2 R(4R - 6h)$

$\frac{d^2P}{dh^2} < 0$ at $h = \frac{4R}{3}$

32 Let $OC = x$, $CQ = r$

Now, $OA = R$ [given]

Height of the cone $= h = x + R$

$\therefore$ Volume of the cone

$= V = \frac{1}{3}\pi r^2 h$ ...(i)

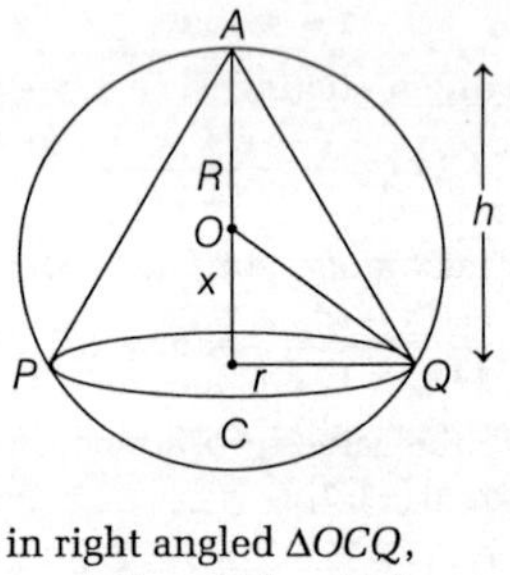

Also, in right angled ΔOCQ,

$OC^2 + CQ^2 = OQ^2$

$\Rightarrow \quad x^2 + r^2 = R^2$

$\Rightarrow \quad r^2 = R^2 - x^2$ ...(ii)

From Eqs. (i) and (ii),

$V = \frac{1}{3}\pi(R^2 - x^2)(x + R)$ ...(iii)

[$\because h = x + R$]

On differentiating Eq. (iii) w.r.t. x, we get

$\frac{dV}{dx} = \frac{1}{3}\pi[(R^2 - x^2) - 2x(x + R)]$

$\Rightarrow \quad \frac{dV}{dx} = \frac{\pi}{3}(R^2 - x^2 - 2x^2 - 2xR)$

$\Rightarrow \quad \frac{dV}{dx} = \frac{\pi}{3}(R^2 - 2xR - 3x^2)$

$\Rightarrow \quad \frac{dV}{dx} = \frac{\pi}{3}(R - 3x)(R + x)$ ...(iv)

For maxima, put $\frac{dV}{dx} = 0$

$\Rightarrow \quad \frac{\pi}{3}(R - 3x)(R + x) = 0$

$\Rightarrow \quad x = \frac{R}{3}$ or $x = -R \Rightarrow x = \frac{R}{3}$

[since, x cannot be negative]

On differentiating Eq. (iv) w.r.t. x, we get

$$\frac{d^2V}{dx^2} = \frac{\pi}{3}[(-3)(R+x)+(R-3x)]$$
$$= \frac{\pi}{3}(-2R-6x) = -\frac{\pi}{3}(2R+6x)$$

At $x = \frac{R}{3}$, $\frac{d^2V}{dx^2} = \frac{-\pi}{3}\left(2R + \frac{6R}{3}\right)$
$$= -\frac{4\pi}{3}R < 0$$

So, V has a local maxima at $x = R/3$.
Now, on substituting the value of x in Eq. (iii), we get

$$V = \frac{\pi}{3}\left(R^2 - \frac{R^2}{9}\right)\left(R + \frac{R}{3}\right)$$
$$= \frac{\pi}{3}\cdot\frac{8R^2}{9}\cdot\frac{4R}{3} = \frac{8}{27}\left(\frac{4}{3}\pi R^3\right)$$

$\Rightarrow$ $V = \frac{8}{27} \times$ Volume of sphere

33

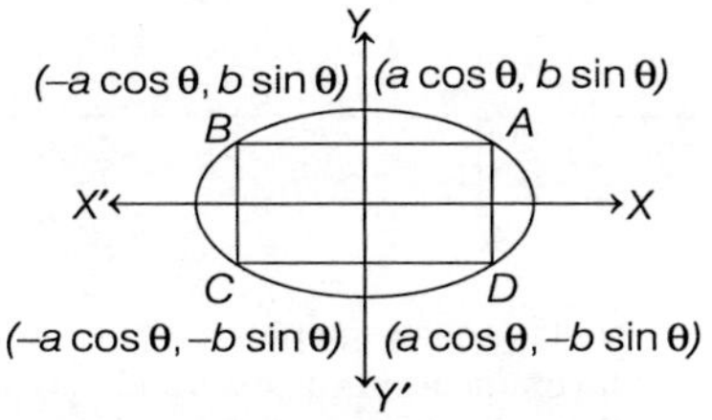

Area of rectangle $ABCD$
$= (2a\cos\theta)(2b\sin\theta) = 2ab\sin 2\theta$
Hence, area of greatest rectangle is equal to $2ab$ when $\sin 2\theta = 1$.

34 Let $f(x) = x + \frac{1}{x}$

$$f'(x) = 1 - \frac{1}{x^2}$$

For maxima and minima, put $f'(x) = 0$

$\Rightarrow 1 - \frac{1}{x^2} = 0 \Rightarrow x = \pm 1$

Now, $f''(x) = \frac{2}{x^3}$

At $x = 1$, $f''(x) = +\text{ve}$ [minima]
and at $x = -1$, $f''(x) = -\text{ve}$ [maxima]
Thus, $f(x)$ attains minimum value at $x = 1$.

35 Given that, $f(x) = (x+1)^{1/3} - (x-1)^{1/3}$
On differentiating w.r.t. x, we get

$$f'(x) = \frac{1}{3}\left[\frac{1}{(x+1)^{2/3}} - \frac{1}{(x-1)^{2/3}}\right]$$
$$= \frac{(x-1)^{2/3} - (x+1)^{2/3}}{3(x^2-1)^{2/3}}$$

Clearly, $f'(x)$ does not exist at $x = \pm 1$.
Now, put $f'(x) = 0$, then
$$(x-1)^{2/3} = (x+1)^{2/3} \Rightarrow x = 0$$
At $x = 0$
$f(x) = (0+1)^{1/3} - (0-1)^{1/3} = 2$
Hence, the greatest value of $f(x)$ is 2.

36 $\because$ $y^2 = 8x$. But $y^2 = 4ax$
$\Rightarrow 4a = 8 \quad \Rightarrow a = 2$

Any point on parabola is $(at^2, 2at)$, i.e., $(2t^2, 4t)$.
For its minimum distance from the circle means its distance from the centre $(0, -6)$ of the circle.

Let $z = (2t^2)^2 + (4t+6)^2$
$\quad = 4(t^4 + 4t^2 + 12t + 9)$

$\therefore$ $\frac{dz}{dt} = 4(4t^3 + 8t + 12)$

$\Rightarrow$ $16(t^3 + 2t + 3) = 0$

$\Rightarrow$ $(t+1)(t^2 - t + 3) = 0$

$\Rightarrow$ $t = -1$

$\Rightarrow$ $\frac{d^2z}{dt^2} = 16(3t^2 + 2) > 0$, hence minimum.

So, point is $(2, -4)$.

37 We know that, volume of cylinder, $V = \pi R^2 h$

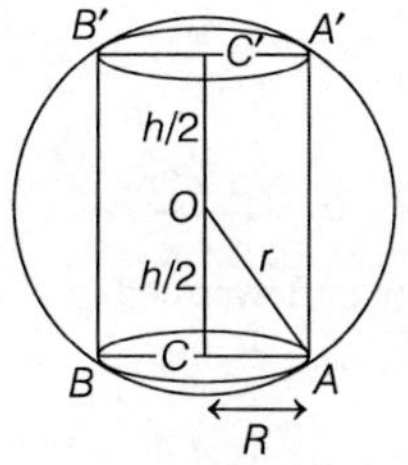

In ΔOCA, $r^2 = \left(\frac{h}{2}\right)^2 + R^2$

$\Rightarrow$ $R^2 = r^2 - \frac{h^2}{4}$

$\therefore$ $V = \pi\left(r^2 - \frac{h^2}{4}\right)h$

$\Rightarrow$ $V = \pi r^2 h - \frac{\pi}{4}h^3$ ...(i)

On differentiating Eq. (i) both sides w.r.t. h, we get

$$\frac{dV}{dh} = \pi r^2 - \frac{3\pi h^2}{4}$$

$\Rightarrow$ $\frac{d^2V}{dh^2} = \frac{-3\pi h}{2}$

For maximum or minimum value of V,

$$\frac{dV}{dh} = 0 \quad \Rightarrow \pi r^2 - \frac{3\pi h^2}{4} = 0$$

$\Rightarrow$ $h^2 = \frac{4r^2}{3} \Rightarrow h = \frac{2}{\sqrt{3}}r$

Now, $\left(\frac{d^2V}{dh^2}\right)_{h = \frac{2r}{\sqrt{3}}} = -\sqrt{3}\pi r < 0$

Thus, V is maximum when $h = \frac{2r}{\sqrt{3}}$, then

$$R^2 = r^2 - \frac{h^2}{4} = r^2 - \frac{1}{4}\left(\frac{2r}{\sqrt{3}}\right)^2 = \frac{2}{3}r^2$$

$$\text{Max } V = \pi R^2 h = \frac{4\pi r^3}{3\sqrt{3}}$$

38 Let $f(x) = -x^3 + 3x^2 + 9x - 27$
The slope of this curve
$$f'(x) = -3x^2 + 6x + 9$$

Let $g(x) = f'(x) = -3x^2 + 6x + 9$
On differentiating w.r.t. x, we get
$$g'(x) = -6x + 6$$
For maxima or minima put $g'(x) = 0$
$\Rightarrow$ $x = 1$
Now, $g''(x) = -6 < 0$ and hence, at $x = 1$, $g(x)$ (slope) will have maximum value.
$\therefore [g(1)]_{max} = -3 \times 1 + 6(1) + 9 = 12$

39 Given,
$ab = 2a + 3b \Rightarrow (a-3)b = 2a$

$\Rightarrow$ $b = \frac{2a}{a-3}$

Now, let $z = ab = \frac{2a^2}{a-3}$

On differentiating w.r.t. x, we get

$$\frac{dz}{da} = \frac{2[(a-3)2a - a^2]}{(a-3)^2} = \frac{2[a^2 - 6a]}{(a-3)^2}$$

For a minimum, put $\frac{dz}{da} = 0$

$\Rightarrow$ $a^2 - 6a = 0$
$\Rightarrow$ $a = 0, 6$

At $a = 6$, $\frac{d^2z}{da^2}$ = positive

When $a = 6$, $b = 4$
$\therefore$ $(ab)_{min} = 6 \times 4 = 24$

40 $\because$ Perimeter of a sector $= p$

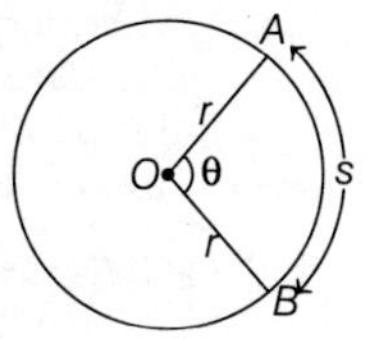

Let AOB be the sector with radius r.
If angle of the sector be θ radians, then area of sector,

$$A = \frac{1}{2}r^2\theta \qquad \text{...(i)}$$

and length of arc, $s = r\theta \Rightarrow \theta = \frac{s}{r}$

$\therefore$ Perimeter of the sector

$$p = r + s + r = 2r + s \qquad \text{...(ii)}$$

On substituting $\theta = \frac{s}{r}$ in Eq. (i), we get

$$A = \left(\frac{1}{2}r^2\right)\left(\frac{s}{r}\right) = \frac{1}{2}rs \Rightarrow s = \frac{2A}{r}$$

Now, on substituting the value of s in Eq. (ii), we get

$$p = 2r + \left(\frac{2A}{r}\right) \Rightarrow 2A = pr - 2r^2$$

On differentiating w.r.t. r, we get

$$2\frac{dA}{dr} = p - 4r$$

For the maximum area, put

$$\frac{dA}{dr} = 0$$

$\Rightarrow$ $p - 4r = 0$

$\Rightarrow$ $r = \frac{p}{4}$

SESSION 2

1 Let radius vector is r.

$\therefore\ r^2 = x^2 + y^2$

$\Rightarrow r^2 = \dfrac{a^2 y^2}{y^2 - b^2} + y^2 \quad \left(\because \dfrac{a^2}{x^2} + \dfrac{b^2}{y^2} = 1\right)$

For minimum value of r,

$\dfrac{d(r^2)}{dy} = 0 \Rightarrow \dfrac{-2yb^2a^2}{(y^2 - b^2)^2} + 2y = 0$

$\Rightarrow\ y^2 = b(a + b)$

$\therefore\ x^2 = a(a + b)$

$\Rightarrow\ r^2 = (a + b)^2 \Rightarrow r = a + b$

2 Bold line represents the graph of $y = g(x)$, clearly $g(x)$ has neither a point of local maxima nor a point of local minima.

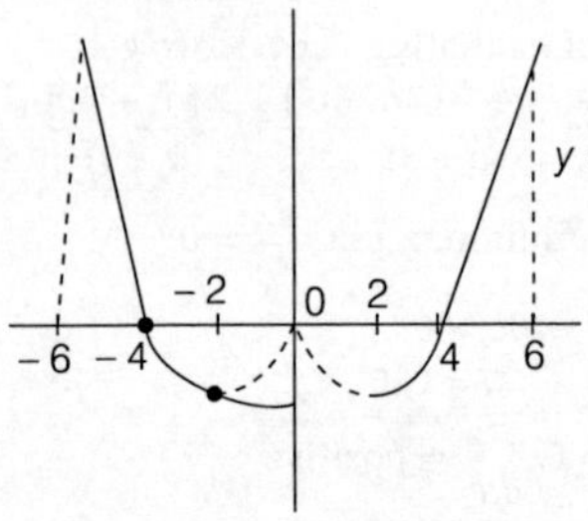

3 Clearly, $f(x)$ in increasing just before $x = 3$ and decreasing after $x = 3$. For $x = 3$ to be the point of local maxima.

$f(3) \ge f(3 - 0)$

$\Rightarrow -15 \ge 12 - 27 + \log(a^2 - 3a + 3)$

$\Rightarrow\ 0 < a^2 - 3a + 3 \le 1 \Rightarrow 1 \le a \le 2$

4 (Slope) $f'(x) = e^x \cos x + \sin x e^x$

$= e^x\sqrt{2}\sin(x + \pi/4)$

$f''(x) = \sqrt{2}e^x\{\sin(x + \pi/4) + \cos(x + \pi/4)\}$

$= 2e^x \cdot \sin(x + \pi/2)$

For maximum slope, put $f''(x) = 0$

$\Rightarrow\ \sin(x + \pi/2) = 0$

$\Rightarrow\ \cos x = 0$

$\therefore\ x = \pi/2, 3\pi/2$

$f'''(x) = 2e^x \cos(x + \pi/2)$

$f'''(\pi/2) = 2e^x \cdot \cos\pi = -\text{ve}$

Maximum slope is at $x = \pi/2$.

5 $f'(x) = \begin{cases} 3(2 + x)^2, & -3 < x \le -1 \\ \dfrac{2}{3}x^{-1/3}, & -1 < x < 2 \end{cases}$

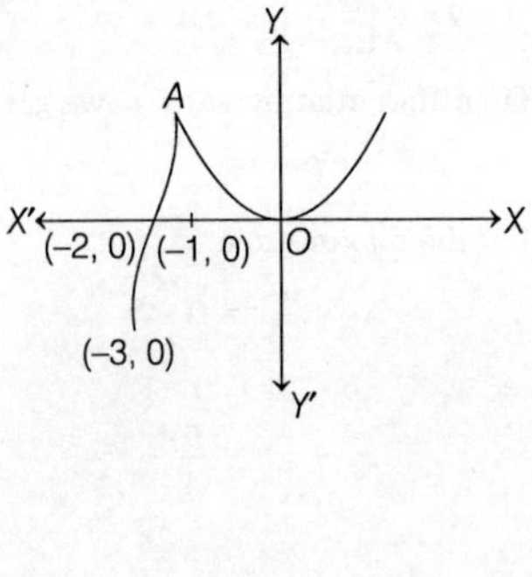

Clearly, $f'(x)$ changes its sign at $x = -1$ from positive to negative and so $f(x)$ has local maxima at $x = -1$.

Also, $f'(0)$ does not exist but $f'(0^-) < 0$ and $f'(0^+) > 0$. It can only be inferred that $f(x)$ has a possibility of a minimum at $x = 0$. Hence, it has one local maxima at $x = -1$ and one local minima at $x = 0$

So, total number of local maxima and local minima is 2.

6 Total length $= 2r + r\theta = 20$

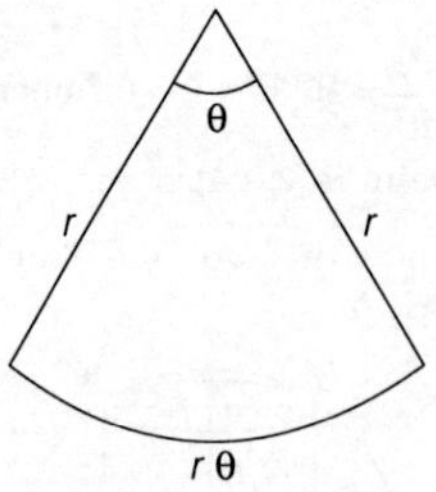

$\Rightarrow\ \theta = \dfrac{20 - 2r}{r}$

Now, area of flower-bed,

$A = \dfrac{1}{2}r^2\theta$

$\Rightarrow\ A = \dfrac{1}{2}r^2\left(\dfrac{20 - 2r}{r}\right)$

$\Rightarrow\ A = 10r - r^2$

$\therefore\ \dfrac{dA}{dr} = 10 - 2r$

For maxima or minima, put $\dfrac{dA}{dr} = 0$.

$\Rightarrow\ 10 - 2r = 0 \Rightarrow r = 5$

$\therefore\ A_{\max} = \dfrac{1}{2}(5)^2\left[\dfrac{20 - 2(5)}{5}\right]$

$= \dfrac{1}{2} \times 25 \times 2 = 25\ \text{sq m}$

7 Let $c = av + \dfrac{b}{v}$ ...(i)

When $v = 30$ km/h, then $c =$ ₹ 75

$\therefore\ 75 = 30a + \dfrac{b}{30}$ ...(ii)

When $v = 40$ km/h, then $c =$ ₹ 65

$\therefore\ 65 = 40a + \dfrac{b}{40}$ ...(iii)

On solving Eqs. (ii) and (iii), we get

$a = \dfrac{1}{2}$ and $b = 1800$

On differentiating w.r.t. v in Eq. (i),

$\dfrac{dc}{dv} = a - \dfrac{b}{v^2}$

For maximum or minimum c,3

$\dfrac{dc}{dv} = 0 \Rightarrow v = \pm\sqrt{\dfrac{b}{a}}$

$\Rightarrow\ \dfrac{d^2c}{dv^2} = \dfrac{2b}{v^3}$ at $v = \sqrt{\dfrac{b}{a}}, \dfrac{dx}{dv^2} > 0$

So, at $v = \sqrt{\dfrac{b}{a}}$ the speed is most economical.

$\therefore$ Most economical speed is

$c = a\sqrt{\dfrac{b}{a}} + b\sqrt{\dfrac{a}{b}} = 2\sqrt{ab}$

$c = 2\sqrt{\dfrac{1}{2} \times 1800} = 2 \times 30$

$\Rightarrow\ c = 60$

8 $f(x) = \begin{cases} |x^2 - 2|, & -1 \le x < \sqrt{3} \\ \dfrac{x}{\sqrt{3}}, & \sqrt{3} \le x < 2\sqrt{3} \\ 3 - x, & 2\sqrt{3} \le x \le 4 \end{cases}$

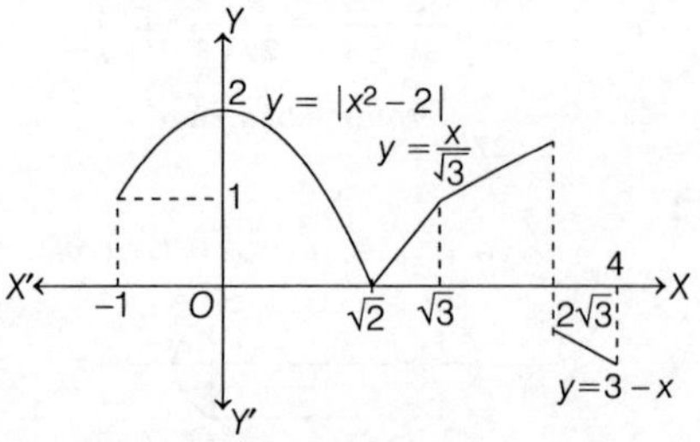

From the above graph,

Maximum occurs at $x = 0$ and minimum at $x = 4$.

9 $f(x) = \begin{cases} |x^3 + x^2 + 3x + \sin x|\left(3 + \sin\left(\dfrac{1}{x}\right)\right), & x \ne 0 \\ 0, & x = 0 \end{cases}$

Let $g(x) = x^3 + x^2 + 3x + \sin x$

$g'(x) = 3x^2 + 2x + 3 + \cos x$

$= 3\left(x^2 + \dfrac{2x}{3} + 1\right) + \cos x$

$= 3\left\{\left(x + \dfrac{1}{3}\right)^2 + \dfrac{8}{9}\right\} + \cos x > 0$

and $2 < 3 + \sin\left(\dfrac{1}{x}\right) < 4$

Hence, minimum value of $f(x)$ is 0 at $x = 0$.

Hence, number of points = 1

10 According to given information, we have

Perimeter of square + Perimeter of circle = 2 units

$\Rightarrow\ 4x + 2\pi r = 2$

$\Rightarrow\ r = \dfrac{1 - 2x}{\pi}$ ...(i)

Now, let A be the sum of the areas of the square and the circle. Then,

$A = x^2 + \pi r^2$

$= x^2 + \pi\dfrac{(1 - 2x)^2}{\pi^2}$

$\Rightarrow\ A(x) = x^2 + \dfrac{(1 - 2x)^2}{\pi}$

Now, for minimum value of $A(x)$,

$$\frac{dA}{dx} = 0$$

$$\Rightarrow \quad 2x + \frac{2(1-2x)}{\pi} \cdot (-2) = 0$$

$$\Rightarrow \quad x = \frac{2-4x}{\pi}$$

$$\Rightarrow \quad \pi x + 4x = 2$$

$$\Rightarrow \quad x = \frac{2}{\pi + 4} \quad \text{...(ii)}$$

Now, from Eq. (i), we get

$$r = \frac{1 - 2 \cdot \frac{2}{\pi+4}}{\pi}$$

$$= \frac{\pi + 4 - 4}{\pi(\pi+4)} = \frac{1}{\pi+4} \quad \text{...(iii)}$$

From Eqs. (ii) and (iii), we get

$$x = 2r$$

11 We have,

$$f(x) = x^2 + \frac{1}{x^2} \text{ and } g(x) = x - \frac{1}{x}$$

$$\Rightarrow \quad h(x) = \frac{f(x)}{g(x)}$$

$$\therefore \quad h(x) = \frac{x^2 + \frac{1}{x^2}}{x - \frac{1}{x}} = \frac{\left(x - \frac{1}{x}\right)^2 + 2}{x - \frac{1}{x}}$$

$$\Rightarrow h(x) = \left(x - \frac{1}{x}\right) + \frac{2}{x - \frac{1}{x}}$$

$$x - \frac{1}{x} > 0, \left(x - \frac{1}{x}\right) + \frac{2}{x - \frac{1}{x}} \in [2\sqrt{2}, \infty)$$

$$x - \frac{1}{x} < 0,$$

$$\left(x - \frac{1}{x}\right) + \frac{2}{x - \frac{1}{x}} \in (-\infty, 2\sqrt{2}]$$

$\therefore$ Local minimum value is $2\sqrt{2}$.

12 Consider the function

$$f(x) = \frac{x^2}{(x^3 + 200)}$$

$$f'(x) = x\frac{(400 - x^3)}{(x^3 + 200)^2} = 0$$

when $x = (400)^{1/3}$, ($\because x \neq 0$)

$$x = (400)^{1/3} - h \Rightarrow f'(x) > 0$$

$$x = (400)^{1/3} + h \Rightarrow f'(x) < 0$$

$\therefore f(x)$ has maxima at $x = (400)^{1/3}$

Since, $7 < (400)^{1/3} < 8$, either a_7 or a_8 is the greatest term of the sequence.

$$\because \quad a_7 = \frac{49}{543}$$

and $$a_8 = \frac{8}{89}$$

and $$\frac{49}{543} > \frac{8}{89}$$

$\Rightarrow \quad a_7 = \frac{49}{543}$ is the greatest term.

13 $f(x) = x^3 - 3(7-a)x^2 - 3(9 - a^2)x + 2$

$f'(x) = 3x^2 - 6(7-a)x - 3(9-a^2)$

For real root $D \geq 0$,

$\Rightarrow \quad 49 + a^2 - 14a + 9 - a^2 \geq 0$

$\Rightarrow a \leq \frac{58}{14}$

For local minimum

$f''(x) = 6x - 6(7-a) > 0$

$\Rightarrow \quad 7 - x$

has x must be negative

$\Rightarrow \quad 7 - a < 0$

$\Rightarrow \quad a > 7$

Thus constradictory, *i.e.*, for real roots $a \leq \frac{58}{14}$ and for negative point of local minimum $a > 7$.

No possible values of a.

14 From the given figure coordinate of Q is $(-1, 0)$.

The equation of circle centre at Q with variable radius r is

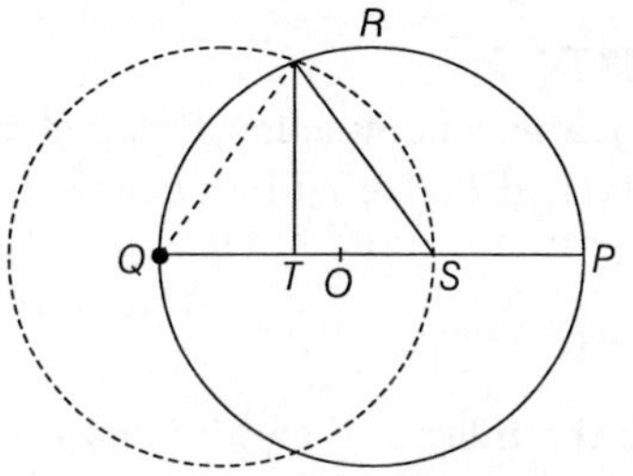

$$(x+1)^2 + y^2 = r^2 \quad \text{...(i)}$$

This circle meets the line segment QP at S, where $QS = r$

It meets the circle $x^2 + y^2 = 1$ at ...(ii)

$$R\left(\frac{r^2 - 2}{2}, \frac{r}{2}\sqrt{4 - r^2}\right)$$

[on solving Eqs. (i) and (ii)]

A = Area of ΔQSR

$$= \frac{1}{2} \times QS \times RT$$

$$= \frac{1}{2} r \left(\frac{r}{2} \cdot \sqrt{4 - r^2}\right)$$

[since, RT is the y-coordinate of R]

$$= \frac{1}{4}[r^2\sqrt{4 - r^2}]$$

$$\therefore \quad \frac{dA}{dr} = \frac{1}{4}\left\{2r\sqrt{4 - r^2} + \frac{r^2(-r)}{\sqrt{4 - r^2}}\right\}$$

$$= \frac{\{2r(4 - r^2) - r^3\}}{4\sqrt{4 - r^2}}$$

$$= \frac{8r - 3r^3}{4\sqrt{4 - r^2}}$$

$\frac{dA}{dr} = 0$, when $r(8 - 3r^2) = 0$ giving

$$r = \sqrt{\frac{8}{3}}$$

$$\Rightarrow \frac{d^2A}{dr^2} = \frac{4\sqrt{4 - r^2}(8 - 9r^2) - (8r - 3r^3)\frac{(-4r)}{\sqrt{4 - r^2}}}{16(4 - r^2)}$$

When $r = \sqrt{\frac{8}{3}}$, then $\frac{d^2A}{dr^2} < 0$

Hence, A is maximum when $r = \sqrt{\frac{8}{3}}$.

Then, maximum area

$$= \frac{8}{4 \times 3}\sqrt{4 - \frac{8}{3}} = \frac{4\sqrt{3}}{9} \text{ sq unit}$$

15 Given,

$P(x) = x^4 + ax^3 + bx^2 + cx + d$

$\Rightarrow P'(x) = 4x^3 + 3ax^2 + 2bx + c$

Since, $x = 0$ is a solution for $P'(x) = 0$, then

$$c = 0$$

$\therefore \quad P(x) = x^4 + ax^3 + bx^2 + d \quad \text{...(i)}$

Also, we have $P(-1) < P(1)$

$\Rightarrow 1 - a + b + d < 1 + a + b + d$

$\Rightarrow \quad a > 0$

Since, $P'(x) = 0$, only when $x = 0$ and $P(x)$ is differentiable in $(-1, 1)$, we should have the maximum and minimum at the points $x = -1$, 0 and 1.

Also, we have $P(-1) < P(1)$

$\therefore$ Maximum of $P(x) =$ Max $\{P(0), P(1)\}$ and minimum of $P(x) =$ Min $\{P(-1), P(0)\}$

In the interval [0, 1],

$$P'(x) = 4x^3 + 3ax^2 + 2bx$$

$$= x(4x^2 + 3ax + 2b)$$

Since, $P'(x)$ has only one root $x = 0$, then $4x^2 + 3ax + 2b = 0$ has no real roots.

$\therefore \quad (3a)^2 - 32b < 0$

$\Rightarrow \quad \frac{9a^2}{32} < b$

$\therefore \quad b > 0$

Thus, we have $a > 0$ and $b > 0$.

$\therefore P'(x) = 4x^3 + 3ax^2 + 2bx > 0$, $\forall \ x \in (0, 1)$

Hence, $P(x)$ is increasing in [0, 1].

$\therefore$ Maximum of $P(x) = P(1)$

Similarly, $P(x)$ is decreasing in $[-1, 0]$. Therefore, minimum $P(x)$ does not occur at $x = -1$.

DAY FIFTEEN

Indefinite Integrals

Learning & Revision for the Day

- Integral as an Anti-derivative
- Fundamental Integration Formulae
- Methods of Integration

Integral as an Anti-derivative

A function $\phi(x)$ is called a **primitive** or **anti-derivative** of a function $f(x)$, if $\phi'(x) = f(x)$. If $f_1(x)$ and $f_2(x)$ are two anti-derivatives of $f(x)$, then $f_1(x)$ and $f_2(x)$ differ by a constant. The collection of all its anti-derivatives is called **indefinite integral** of $f(x)$ and is denoted by $\int f(x)\,dx$.

Thus, $\frac{d}{dx}\{\phi(x) + C\} = f(x) \quad \Rightarrow \quad \int f(x)\,dx = \phi(x) + C$

where, $\phi(x)$ is an anti-derivative of $f(x)$, $f(x)$ is the **integrand** and C is an arbitrary constant known as the **constant of integration**. Anti-derivative of odd function is always even and of even function is always odd.

Properties of Indefinite Integrals

- $\int \{f(x) \pm g(x)\}\,dx = \int f(x)\,dx \pm \int g(x)\,dx$
- $\int k \cdot f(x)\,dx = k \cdot \int f(x)\,dx,$ where k is any non-zero real number.
- $\int [k_1 f_1(x) + k_2 f_2(x) + \ldots + k_n f_n(x)]\,dx = k_1 \int f_1(x)\,dx + k_2 \int f_2(x)\,dx + \ldots + k_n \int f_n(x)\,dx,$ where $k_1, k_2, \ldots k_n$ are non-zero real numbers.

Fundamental Integration Formulae

There are some important fundamental formulae, which are given below

1. Algebraic Formulae

(i) $\int x^n dx = \frac{x^{n+1}}{n+1} + C,\ n \neq -1$

(ii) $\int (ax+b)^n dx = \frac{1}{a} \cdot \frac{(ax+b)^{n+1}}{n+1} + C,\ n \neq -1$

PREP MIRROR
Your Personal Preparation Indicator

- No. of Questions in Exercises (x)—
- No. of Questions Attempted (y)—
- No. of Correct Questions (z)—
  (Without referring Explanations)

- Accuracy Level $(z/y \times 100)$—
- Prep Level $(z/x \times 100)$—

In order to expect good rank in JEE, your Accuracy Level should be above 85 & Prep Level should be above 75.

(iii) $\int \frac{1}{x} dx = \log|x| + C$

(iv) $\int \frac{1}{ax+b} dx = \frac{1}{a}(\log|ax+b|) + C$

(v) $\int \frac{1}{a^2 - x^2} dx = \frac{1}{2a} \log\left|\frac{a+x}{a-x}\right| + C$

(vi) $\int \frac{1}{x^2 - a^2} dx = \frac{1}{2a} \log\left|\frac{x-a}{x+a}\right| + C$

(vii) $\int \frac{1}{a^2 + x^2} dx = \frac{1}{a} \tan^{-1}\left(\frac{x}{a}\right) + C$

(viii) $\int \frac{-1}{a^2 + x^2} dx = \frac{1}{a} \cot^{-1}\left(\frac{x}{a}\right) + C$

(ix) $\int \frac{1}{\sqrt{x^2 - a^2}} dx = \log|x + \sqrt{x^2 - a^2}| + C$

(x) $\int \frac{1}{\sqrt{x^2 + a^2}} dx = \log|x + \sqrt{x^2 + a^2}| + C$

(xi) $\int \frac{1}{\sqrt{a^2 - x^2}} dx = \sin^{-1}\left(\frac{x}{a}\right) + C$

(xii) $\int \frac{-1}{\sqrt{a^2 - x^2}} dx = \cos^{-1}\left(\frac{x}{a}\right) + C$

(xiii) $\int \frac{1}{x\sqrt{x^2 - a^2}} dx = \frac{1}{a} \sec^{-1}\left(\frac{x}{a}\right) + C$

(xiv) $\int \frac{-1}{x\sqrt{x^2 - a^2}} dx = \frac{1}{a} \operatorname{cosec}^{-1}\left(\frac{x}{a}\right) + C$

(xv) $\int \sqrt{a^2 - x^2}\, dx = \frac{1}{2} x\sqrt{a^2 - x^2} + \frac{1}{2} a^2 \sin^{-1}\left(\frac{x}{a}\right) + C$

(xvi) $\int \sqrt{x^2 - a^2}\, dx = \frac{1}{2} x\sqrt{x^2 - a^2} - \frac{1}{2} a^2 \log|x + \sqrt{x^2 - a^2}| + C$

(xvii) $\int \sqrt{x^2 + a^2}\, dx = \frac{1}{2} x\sqrt{x^2 + a^2} + \frac{1}{2} a^2 \log|x + \sqrt{x^2 + a^2}| + C$

2. Trigonometric Formulae

(i) $\int \sin x\, dx = -\cos x + C$

(ii) $\int \cos x\, dx = \sin x + C$

(iii) $\int \tan x\, dx = -\log|\cos x| + C = \log|\sec x| + C$

(iv) $\int \cot x\, dx = \log|\sin x| + C = -\log|\operatorname{cosec} x| + C$

(v) $\int \sec x\, dx = \log|\sec x + \tan x| + C = \log\left|\tan\left(\frac{\pi}{4} + \frac{x}{2}\right)\right| + C$

(vi) $\int \operatorname{cosec} x\, dx = \log|\operatorname{cosec} x - \cot x| + C = \log\left|\tan\frac{x}{2}\right| + C$

(vii) $\int \sec^2 x\, dx = \tan x + C$

(viii) $\int \operatorname{cosec}^2 x\, dx = -\cot x + C$

(ix) $\int \sec x \cdot \tan x\, dx = \sec x + C$

(x) $\int \operatorname{cosec} x \cdot \cot x\, dx = -\operatorname{cosec} x + C$

3. Exponential Formulae

(i) $\int e^x dx = e^x + C$

(ii) $\int e^{(ax+b)}\, dx = \frac{1}{a} \cdot e^{(ax+b)} + C$

(iii) $\int a^x\, dx = \frac{a^x}{\log_e a} + C,\ a > 0$ and $a \neq 1$

(iv) $\int a^{(bx+c)} dx = \frac{1}{b} \cdot \frac{a^{(bx+c)}}{\log_e a} + C,\ a > 0$ and $a \neq 1$

Methods of Integration

Following methods are used for integration

1. Integration by Substitutions

The method of reducing a given integral into one of the standard integrals, by a proper substitution, is called **method of substitution.**

To evaluate an integral of the form $\int f\{g(x)\} \cdot g'(x) dx$, we substitute $g(x) = t$, so that $g'(x) dx = dt$ and given integral reduces to $\int f(t) dt$.

NOTE
- $\int [f(x)]^n \cdot f'(x) = \frac{[f(x)]^{n+1}}{n+1} + C$
- If $\int f(x)\, dx = \phi(x)$, then $\int f(ax+b)\, dx = \frac{1}{a}\phi(ax+b) + C$

(i) To evaluate integrals of the form

$\int \frac{dx}{ax^2 + bx + c}$ or $\int \frac{dx}{\sqrt{ax^2 + bx + c}}$ or $\int \sqrt{ax^2 + bx + c}\, dx$

We write, $ax^2 + bx + c = a\left(x^2 + \frac{b}{a}x + \frac{c}{a}\right)$

$$= a\left(x + \frac{b}{2a}\right)^2 + \frac{c}{a} - \frac{b^2}{4c}$$

This process reduces the integral to one of following forms

$$= \int \frac{dX}{X^2 - A^2}, \int \frac{dX}{X^2 + A^2} \text{ or } \int \frac{dX}{A^2 - X^2},$$

$$\int \frac{dX}{\sqrt{A^2 - X^2}}, \int \frac{dX}{\sqrt{X^2 - A^2}}, \int \frac{dX}{\sqrt{X^2 + A^2}}$$

or $\int \sqrt{A^2 - X^2} dX, \int \sqrt{X^2 - A^2}\, dX, \int \sqrt{A^2 + X^2}\, dX$

(ii) To evaluate integrals of the form

$\int \frac{(px+q)}{ax^2 + bx + c} dx$ or $\int \frac{(px+q)}{\sqrt{ax^2 + bx + c}} dx$

or $\int (px+q)\sqrt{ax^2 + bx + c}\, dx$

We put $px + q = A$ {differentiation of $(ax^2 + bx + c)$} + B, where A and B can be found by comparing the coefficients of like powers of x on the two sides.

2. Integration using Trigonometric Identities

In this method, we have to evaluate integrals of the form

- $\int \sin mx \cdot \cos nx\, dx$ or $\int \sin mx \cdot \sin nx\, dx$ or $\int \cos mx \cdot \cos nx\, dx$ or $\int \cos mx \cdot \sin nx\, dx$

In this method, we use the following trigonometrical identities

(i) $2 \sin A \cdot \cos B = \sin (A + B) + \sin (A - B)$
(ii) $2 \cos A \cdot \sin B = \sin (A + B) - \sin (A - B)$
(iii) $2 \cos A \cdot \cos B = \cos (A + B) + \cos (A - B)$
(iv) $2 \sin A \cdot \sin B = \cos (A - B) - \cos (A + B)$
(v) $2 \sin A \cdot \cos A = \sin 2A$
(vi) $\cos^2 A = \left(\frac{1 + \cos 2A}{2}\right)$
(vii) $\sin^2 A = \left(\frac{1 - \cos 2A}{2}\right)$
(viii) $\cos^2 A - \sin^2 A = \cos 2A$
(ix) $\sin^2 A + \cos^2 A = 1$

3. Integration of Different Types of Functions

- To evaluate integrals of the form $\int \sin^p x \cos^q x\, dx$

Where $p, q \in Q$, we use the following rules :

(i) If p is odd, then put $\cos x = t$
(ii) If q is odd, then put $\sin x = t$
(iii) If both p, q are odd, then put either $\sin x = t$ or $\cos x = t$
(iv) If both p, q are even, then use trigonometric identities only.
(v) If p, q are rational numbers and $\left(\frac{p+q-2}{2}\right)$ is a negative integer, then put $\cot x = t$ or $\tan x = t$ as required.

- To evaluate integrals of the form $\int \frac{dx}{a + b\cos^2 x}$ or $\int \frac{dx}{a + b\sin^2 x}$ or $\int \frac{dx}{a\sin^2 x + b\cos^2 x}$, $\int \frac{dx}{(a\sin x + b\cos x)^2}$ or $\int \frac{dx}{a + b\sin^2 x + c\cos^2 x}$

(i) Divide both the numerator and denominator by $\cos^2 x$.
(ii) Replace $\sec^2 x$ by $1 + \tan^2 x$ in the denominator, if any.
(iii) Put $\tan x = t$, so that $\sec^2 x\, dx = dt$

- To evaluate integrals of the form $\int \frac{1}{a \sin x + b \cos x} dx$ or $\int \frac{1}{a + b \sin x} dx$ or $\int \frac{1}{a + b \cos x} dx$ or $\int \frac{1}{a \sin x + b \cos x + c} dx$

(i) Put $\sin x = \frac{2 \tan \frac{x}{2}}{1 + \tan^2 \frac{x}{2}}$ and $\cos x = \frac{1 - \tan^2 \frac{x}{2}}{1 + \tan^2 \frac{x}{2}}$
(ii) Replace $1 + \tan^2 \frac{x}{2}$ by $\sec^2 \frac{x}{2}$ and put $\tan \frac{x}{2} = t$.

- To evaluate integral of form $\int \frac{a\sin x + b\cos x}{c\sin x + d\cos x} dx$,

we write $a\sin x + b\cos x = A\frac{d}{dx}(c\sin x + d\cos x) + B(c\sin x + d\cos x)$

Where A and B can be found by equating the coefficient of $\sin x$ and $\cos x$ on both sides.

To evaluate integral of the form $\int \frac{a\sin x + b\cos x + c}{p\sin x + q\cos x + r} dx$.

We write $a\sin x + b\cos x + c = A\frac{d}{dx}(p\sin x + q\cos x + r) + B(p\sin x + q\cos x + r) + C$

Where A, B and C can be found by equating the coefficient of $\sin x$, $\cos x$ and the constant term.

- To evaluate integrals of the form $\int \frac{x^2 + 1}{x^4 + kx^2 + 1} dx$ or $\int \frac{x^2 - 1}{x^4 + kx^2 + 1} dx$

We divide the numerator and denominator by x^2 and make perfect square in denominator as $\left(x \pm \frac{1}{x}\right)^2$ and then put $x + \frac{1}{x} = t$ or $x - \frac{1}{x} = t$ as required.

- Substitution for Some Irrational Integrand

(i) $\sqrt{\frac{a-x}{a+x}}, \sqrt{\frac{a+x}{a-x}}$, $x = a \cos 2\theta$
(ii) $\sqrt{\frac{x}{a+x}}, \sqrt{\frac{a+x}{x}}, \sqrt{x(a+x)}, \frac{1}{\sqrt{x(a+x)}}$, $x = a \tan^2 \theta$ or $x = a \cot^2 \theta$
(iii) $\sqrt{\frac{x}{a-x}}, \sqrt{\frac{a-x}{x}}, \sqrt{x(a-x)}, \frac{1}{\sqrt{x(a-x)}}$, $x = a \sin^2 \theta$ or $x = a \cos^2 \theta$

(iv) $\sqrt{\dfrac{x}{x-a}}, \sqrt{\dfrac{x-a}{x}}, \sqrt{x(x-a)}, \dfrac{1}{\sqrt{x(x-a)}}$, $x = a\sec^2\theta$

(v) $\int \dfrac{dx}{(x-\alpha)(\beta-x)}, \int \sqrt{\left(\dfrac{x-\alpha}{\beta-x}\right)}dx$

$\int \sqrt{(x-\alpha)(\beta-x)}\,dx$, put $x = \alpha\cos^2\theta + \beta\sin^2\theta$

(vi) $\int \dfrac{dx}{(px+q)\sqrt{ax+b}}$, put $ax+b=t^2$

(vii) $\int \dfrac{dx}{(ax^2+bx+c)\sqrt{px+q}}$, put $px+q=t^2$

(viii) $\int \dfrac{dx}{(px+q)\sqrt{ax^2+bx+c}}$, put $px+q=\dfrac{1}{t}$

(ix) $\int \dfrac{dx}{(px^2+q)\sqrt{(ax^2+b)}}$ first put $x=\dfrac{1}{t}$

and then $a+bt^2=z^2$

4. Integration by Parts

(i) If u and v are two functions of x, then

$$\int \underset{\text{I}}{u}\,\underset{\text{II}}{v}\,dx = u\int v\,dx - \int\left(\frac{du}{dx}\cdot\int v\,dx\right)dx$$

We use the following preference in order to select the first function

I $\to$ Inverse function
L $\to$ Logarithmic function
A $\to$ Algebraic function
T $\to$ Trigonometric function
E $\to$ Exponential function

(ii) If one of the function is not directly integrable, then we take it as the first function.

(iii) If both the functions are directly integrable, then the first function is chosen in such a way that its derivative vanishes easily or the function obtained in integral sign is easily integrable.

(iv) If only one which is not directly integrable, function is there e.g. $\int \log x\,dx$, then 1 (unity) is taken as second function.

Some more Special Integrals Based on Integration by Parts

(i) $\int e^x\{f(x)+f'(x)\}dx = f(x)e^x + C$

(ii) $\int e^{ax}\sin(bx+c)\,dx = \dfrac{e^{ax}}{a^2+b^2}\{a\sin(bx+c) - b\cos(bx+c)\} + k$

(iii) $\int e^{ax}\cos(bx+c)dx = \dfrac{e^{ax}}{a^2+b^2}\{a\cos(bx+c) + b\sin(bx+c)\} + k$

Here, c and k are integration constant.

5. Integration by Partial Fractions

To evaluate the integral of the form $\int \dfrac{P(x)}{Q(x)}dx$, where $P(x), Q(x)$ are polynomial in x with degree of $P(x) <$ degree of $Q(x)$ and $Q(x) \neq 0$, we use the method of partial fraction.

The partial fractions depend on the nature of the factors of $Q(x)$.

(i) According to nature of factors of $Q(x)$, corresponding form of partial fraction is given below:
If $Q(x) = (x-a_1)(x-a_2)(x-a_3)\ldots(x-a_n)$, then we assume that

$$\frac{P(x)}{Q(x)} = \frac{A_1}{(x-a_1)} + \frac{A_2}{(x-a_2)} + \frac{A_3}{(x-a_3)} + \ldots + \frac{A_n}{(x-a_n)},$$

where the constants $A_1, A_2, \ldots, A_n$ can be determined by equating the coefficients of like power of x or by substituting $x = a_1, a_2, \ldots, a_n$.

(ii) If $Q(x) = (x-a)^k(x-a_1)(x-a_2)\ldots(x-a_r)$, then we assume that

$$\frac{P(x)}{Q(x)} = \frac{A_1}{(x-a)} + \frac{A_2}{(x-a)^2} + \ldots + \frac{A_k}{(x-a)^k} + \frac{B_1}{(x-a_1)} + \frac{B_2}{(x-a_2)} + \ldots + \frac{B_r}{(x-a_r)}$$

where the constants $A_1, A_2, \ldots, A_k, B_1, B_2 \ldots, B_r$ can be obtained by equating the coefficients of like power of x.

(iii) If some of the factors in $Q(x)$ are quadratic and non-repeating, corresponding to each quadratic factor ax^2+bx+c (non-factorisable), we assume the partial fraction of the type $\dfrac{Ax+B}{ax^2+bx+c}$, where A and B are constants to be determined by comparing coefficients of like powers of x.

(iv) If some of the factors in $Q(x)$ are quadratic and repeating, for every quadratic repeating factor of the type $(ax^2+bx+c)^k$ where ax^2+bx+c cannot be further factorise, we assume

$$\frac{A_1x+A_2}{ax^2+bx+c} + \frac{A_3x+A_4}{(ax^2+bx+c)^2} + \ldots + \frac{A_{2k-1}x+A_{2k}}{(ax^2+bx+c)^k}$$

If degree of $P(x) >$ degree of $Q(x)$, then we first divide $P(x)$ by $Q(x)$ so that $\dfrac{P(x)}{Q(x)}$ is expressed in the form of $T(x) + \dfrac{P_1(x)}{Q(x)}$, where $T(x)$ is a polynomial in x and $\dfrac{P_1(x)}{Q(x)}$ is a proper rational function (i.e. degree of $P_1(x) <$ degree of $Q(x)$)

DAY PRACTICE SESSION 1

FOUNDATION QUESTIONS EXERCISE

1 If $\int \frac{dx}{x+x^7} = p(x)$, then $\int \frac{x^6}{x+x^7} dx$ is equal to → JEE Mains 2013

(a) $\log|x| - p(x) + C$ (b) $\log|x| + p(x) + C$

(c) $x - p(x) + C$ (d) $x + p(x) + C$

2 $\int \frac{x^3 - 1}{(x^4+1)(x+1)} dx$ is equal to

(a) $\frac{1}{4}\log(1+x^4) + \frac{1}{3}\log(1+x^3) + C$

(b) $\frac{1}{4}\log(1+x^4) - \frac{1}{3}\log(1+x^3) + C$

(c) $\frac{1}{4}\log(1+x^4) - \log(1+x) + C$

(d) $\frac{1}{4}\log(1+x^4) + \log(1+x) + C$

3 $\int (x+1)(x+2)^7(x+3)\, dx$ is equal to

(a) $\frac{(x+2)^{10}}{10} - \frac{(x+2)^8}{8} + C$

(b) $\frac{(x+1)^2}{2} - \frac{(x+2)^8}{8} - \frac{(x+3)^2}{2} + C$

(c) $\frac{(x+2)^{10}}{10} + C$

(d) $\frac{(x+1)^2}{2} + \frac{(x+2)^8}{8} + \frac{(x+3)^2}{2} + C$

4 The integral $\int \frac{dx}{x^2(x^4+1)^{\frac{3}{4}}}$ equals → JEE Mains 2015

(a) $\left(\frac{x^4+1}{x^4}\right)^{\frac{1}{4}} + C$ (b) $(x^4+1)^{\frac{1}{4}} + C$

(c) $-(x^4+1)^{\frac{1}{4}} + C$ (d) $-\left(\frac{x^4+1}{x^4}\right)^{\frac{1}{4}} + C$

5 If $\int \frac{\sin x}{\sin(x-\alpha)} dx = Ax + B\log\sin(x-\alpha) + C$, then the value of (A, B) is

(a) $(\sin\alpha, \cos\alpha)$ (b) $(\cos\alpha, \sin\alpha)$

(c) $(-\sin\alpha, \cos\alpha)$ (d) $(-\cos\alpha, \sin\alpha)$

6 If $\int \frac{f(x)}{\log\sin x} dx = \log\log\sin x + C$, then $f(x)$ is equal to

(a) $\sin x$ (b) $\cos x$

(c) $\log\sin x$ (d) $\cot x$

7 $\int \left\{\frac{(\log x - 1)}{1+(\log x)^2}\right\}^2 dx$ is eqal to

(a) $\frac{x}{(\log x)^2+1} + C$ (b) $\frac{xe^x}{1+x^2} + C$

(c) $\frac{x}{x^2+1} + C$ (d) $\frac{\log x}{(\log x)^2+1} + C$

8 If $\int \sqrt{x+\sqrt{x^2+5}}\, dx = P\{x+\sqrt{x^2+5}\}^{3/2} + \frac{Q}{\sqrt{x\sqrt{x^2+5}}} + C$, then the value of $3PQ$ is

(a) -1 (b) -4 (c) -3 (d) -5

9 $\int \frac{dx}{\cos x - \sin x}$ is equal to

(a) $\frac{1}{\sqrt{2}}\log\left|\tan\left(\frac{x}{2} - \frac{\pi}{8}\right)\right| + C$ (b) $\frac{1}{\sqrt{2}}\log\left|\cot\left(\frac{x}{2}\right)\right| + C$

(c) $\frac{1}{\sqrt{2}}\log\left|\tan\left(\frac{x}{2} - \frac{3\pi}{8}\right)\right| + C$ (d) $\frac{1}{\sqrt{2}}\log\left|\tan\left(\frac{x}{2} + \frac{3\pi}{8}\right)\right| + C$

10 $\int \frac{\sin^8 x - \cos^8 x}{1 - 2\sin^2 x \cos^2 x} dx$ is equal to

(a) $\sin 2x + C$ (b) $-\frac{1}{2}\sin 2x + C$

(c) $\frac{1}{2}\sin 2x + C$ (d) $-\sin 2x + C$

11 $\int \frac{(\sqrt[3]{x+\sqrt{2-x^2}})(\sqrt[6]{1-x\sqrt{2-x^2}})}{\sqrt[3]{1-x^2}} dx; x \in (0,1)$ equals

(a) $2^{1/6}x + C$ (b) $2^{1/12}x + C$ (c) $2^{1/3}x + C$ (d) None of these

12 $\int \left(\frac{\cos 6x + 6\cos 4x + 15\cos 2x + 10}{10\cos^2 x + 5\cos x\cos 3x + \cos x\cos 5x}\right) dx = f(x) + C$, then $f(10)$ is equal to

(a) 20 (b) 10 (c) $2\sin 10$ (d) $2\cos 10$

13 The integral $\int \frac{2x^{12} + 5x^9}{(x^5 + x^3 + 1)^3} dx$ is equal to → JEE Mains 2016

(a) $\frac{-x^5}{(x^5+x^3+1)^2} + C$ (b) $\frac{x^{10}}{2(x^5+x^3+1)^2} + C$

(c) $\frac{x^5}{2(x^5+x^3+1)^2} + C$ (d) $\frac{-x^{10}}{2(x^5+x^3+1)^2} + C$

14 $\int \frac{x^2 - 1}{x^3\sqrt{2x^4 - 2x^2 + 1}} dx$ is equal to

(a) $\frac{\sqrt{2x^4-2x^2+1}}{x^2} + C$ (b) $\frac{\sqrt{2x^4-2x^2+1}}{x} + C$

(c) $\frac{\sqrt{2x^4-2x^2+1}}{2x} + C$ (d) $\frac{\sqrt{2x^4-2x^2+1}}{2x^2} + C$

15 $\int \frac{dx}{(1+x^2)\sqrt{p^2 + q^2(\tan^{-1}x)^2}}$ is equal to

(a) $\frac{1}{q}\log[q\tan^{-1}x + \sqrt{p^2 + q^2(\tan^{-1}x)^2}] + C$

(b) $\log[q\tan^{-1}x + \sqrt{p^2 + q^2(\tan^{-1}x)^2}] + C$

(c) $\frac{2}{3q}(p^2 + q^2\tan^{-1}x)^{3/2} + C$

(d) None of the above

16 In the integral $\int \frac{\cos 8x + 1}{\cot 2x - \tan 2x} dx = A \cos 8x + k$, where k is an arbitrary constant, then A is equal to → JEE Mains 2013

(a) $-\frac{1}{16}$ (b) $\frac{1}{16}$ (c) $\frac{1}{8}$ (d) $-\frac{1}{8}$

17 $\int \frac{(\sin\theta + \cos\theta)}{\sqrt{\sin 2\theta}} d\theta$ is equal to → JEE Mains 2017

(a) $\log|\cos\theta - \sin\theta + \sqrt{\sin 2\theta}|$
(b) $\log|\sin\theta - \cos\theta + \sqrt{\sin 2\theta}|$
(c) $\sin^{-1}(\sin\theta - \cos\theta) + C$
(d) $\sin^{-1}(\sin\theta + \cos\theta) + C$

18 $\int \left(\frac{f(x)\cdot g'(x) - f'(x)\cdot g(x)}{f(x)\cdot g(x)}\right)((\log g(x) - \log f(x)) dx$ is equal to

(a) $\log\left(\frac{g(x)}{f(x)}\right) + C$ (b) $\frac{1}{2}\left(\frac{g(x)}{f(x)}\right)^2$

(c) $\frac{1}{2}\left(\log\left(\frac{g(x)}{f(x)}\right)\right)^2 + C$ (d) $\log\left(\frac{g(x)}{f(x)}\right)^2 + C$

19 Let $I_n = \int \tan^n x\, dx\ (n > 1)$. If $I_4 + I_6 = a\tan^5 x + bx^5 + C$, where C is a constant of integration, then the ordered pair (a, b) is equal to

(a) $\left(-\frac{1}{5}, 1\right)$ (b) $\left(\frac{1}{5}, 0\right)$ (c) $\left(\frac{1}{5}, -1\right)$ (d) $\left(-\frac{1}{5}, 0\right)$

20 If $\int \frac{\tan x}{1 + \tan x + \tan^2 x} dx = x - \frac{2}{\sqrt{A}} \tan^{-1}\left(\frac{2\tan x + 1}{\sqrt{A}}\right) + C$, then the value of A is

(a) 1 (b) 2
(c) 3 (d) None of these

21 $\int \cos^{-\frac{3}{7}} x \sin^{-\frac{11}{7}} x\, dx$ is equal to

(a) $\log|\sin^{\frac{4}{7}} x| + C$ (b) $\frac{4}{7}\tan^{\frac{4}{7}} x + C$

(c) $\frac{-7}{4}\tan^{-\frac{4}{7}} x + C$ (d) $\log|\cos^{\frac{3}{7}} x| + C$

22 $\int \frac{dx}{2 + \sin x + \cos x}$ is equal to → NCERT Exemplar

(a) $\sqrt{2}\tan^{-1}\left(\frac{\tan(x/2) + 1}{\sqrt{2}}\right) + C$ (b) $\tan^{-1}\left(\frac{\tan(x/2) + 1}{\sqrt{2}}\right) + C$

(c) $\sqrt{2}\tan^{-1}\left(\frac{\tan(x/2)}{\sqrt{2}}\right) + C$ (d) None of these

23 If the integral $\int \frac{5\tan x}{\tan x - 2} dx = x + a\ln|\sin x - 2\cos x| + k$, then a is equal to

(a) -1 (b) -2
(c) 1 (d) 2

24 $\int \frac{x^2 - 1}{x^4 + x^2 + 1} dx$ is equal to

(a) $\frac{1}{2}\log\left(\frac{x^2 + x + 1}{x^2 - x + 1}\right) + C$ (b) $\frac{1}{2}\log\left(\frac{x^2 - x - 1}{x^2 + x + 1}\right) + C$

(c) $\log\left(\frac{x^2 - x + 1}{x^2 + x + 1}\right) + C$ (d) $\frac{1}{2}\log\left(\frac{x^2 - x + 1}{x^2 + x + 1}\right) + C$

25 $\int \sqrt{\frac{x}{a^3 - x^3}} dx$ is equal to

(a) $\sin^{-1}\left(\frac{x}{a}\right)^{3/2} + C$ (b) $\frac{2}{3}\sin^{-1}\left(\frac{x}{a}\right)^{3/2} + C$

(c) $\frac{3}{2}\sin^{-1}\left(\frac{x}{a}\right)^{3/2} + C$ (d) $\frac{3}{2}\sin^{-1}\left(\frac{x}{a}\right)^{2/3} + C$

26 If an anti-derivative of $f(x)$ is e^x and that of $g(x)$ is $\cos x$, then $\int f(x)\cos x\, dx + \int g(x) e^x dx$ is equal to

(a) $f(x)\cdot g(x) + C$ (b) $f(x) + g(x) + C$
(c) $e^x \cos x + C$ (d) $f(x) - g(x) + C$

27. If $\int f(x) dx = \Psi(x)$, then $\int x^5 f(x^3) dx$ is equal to → JEE Mains 2013

(a) $\frac{1}{3}[x^3\Psi(x^3)] - \int x^2\Psi(x^3) dx + C$

(b) $\frac{1}{3}[x^3\Psi(x^3)] - 3\int x^3\Psi(x^3) dx + C$

(c) $\frac{1}{3}[x^3\Psi(x^3) - \int x^2\Psi(x^3) dx] + C$

(d) $\frac{1}{3}[x^3\Psi(x^3)] - \int x^3\Psi(x^3) dx + C$

28 If $\int \frac{1 - 6\cos^2 x}{\sin^6 x \cos^2 x} dx = \frac{f(x)}{(\sin x)^6} + C$, then $f(x)$ is equal to

(a) $\sin x$ (b) $\cos x$ (c) $\tan x$ (d) $\cot x$

29 $\int \tan^{-1}\sqrt{x}\, dx$ is equal to → NCERT Exemplar

(a) $(x + 1)\tan^{-1}\sqrt{x} - \sqrt{x} + C$ (b) $x\tan^{-1}\sqrt{x} - \sqrt{x} + C$
(c) $\sqrt{x} - x\tan^{-1}\sqrt{x} + C$ (d) $\sqrt{x} - (x + 1)\tan^{-1}\sqrt{x} + C$

30 If $I_n = \int (\log x)^n dx$, then $I_n + n I_{n-1}$ is equal to

(a) $x(\log x)^n$ (b) $(x\log x)^n$ (c) $(\log x)^{n-1}$ (d) $n(\log x)^n$

31 If $\int f(x) dx = g(x)$, then $\int f^{-1}(x) dx$ is equal to

(a) $g^{-1}(x)$ (b) $xf^{-1}(x) - g(f^{-1}(x))$
(c) $xf^{-1}(x) - g^{-1}(x)$ (d) $f^{-1}(x)$

32 $\int \frac{(x + 3)e^x}{(x + 4)^2} dx$ is equal to

(a) $\frac{1}{(x + 4)^2} + C$ (b) $\frac{e^x}{(x + 4)^2} + C$

(c) $\frac{e^x}{x + 4} + C$ (d) $\frac{e^x}{x + 3} + C$

33 If $\int \frac{x^2 - x + 1}{x^2 + 1} e^{\cot^{-1} x} dx = A(x) e^{\cot^{-1} x} + C$, then $A(x)$ is equal to → JEE Mains 2013

(a) $-x$ (b) x (c) $\sqrt{1 - x}$ (d) $\sqrt{1 + x}$

34 If $g(x)$ is a differentiable function satisfying $\frac{d}{dx}\{g(x)\} = g(x)$ and $g(0) = 1$, then $\int g(x)\left(\frac{2 - \sin 2x}{1 - \cos 2x}\right) dx$ is equal to

(a) $g(x)\cot x + C$ (b) $-g(x)\cot x + C$

(c) $\frac{g(x)}{1-\cos 2x} + C$ (d) None of these

35 $\int \frac{dx^3}{x^3(x^n + 1)}$ is equal to

(a) $\frac{3}{n}\ln\left(\frac{x^n}{x^n+1}\right) + C$ (b) $\frac{1}{n}\ln\left(\frac{x^n}{x^n+1}\right) + C$

(c) $\frac{3}{n}\ln\left(\frac{x^n+1}{x^n}\right) + C$ (d) $3n\ln\left(\frac{x^{n+1}}{x^n}\right) + C$

DAY PRACTICE SESSION 2

PROGRESSIVE QUESTIONS EXERCISE

1 If $\int f(x)\, dx = f(x)$, then $\int \{f(x)\}^2\, dx$ is equal to

(a) $\frac{1}{2}\{f(x)\}^2$ (b) $\{f(x)\}^3$ (c) $\frac{\{f(x)\}^3}{3}$ (d) $\{f(x)\}^2$

2 If $f(x) = \begin{vmatrix} 0 & x^2 - \sin x & \cos x - 2 \\ \sin x - x^2 & 0 & 1 - 2x \\ 2 - \cos x & 2x - 1 & 0 \end{vmatrix}$,

then $\int f(x)\, dx$ is equal to

(a) $\frac{x^3}{3} - x^2 \sin x + \sin 2x + C$

(b) $\frac{x^3}{3} - x^2 \sin x - \cos 2x + C$

(c) $\frac{x^3}{3} - x^2 \cos x - \cos 2x + C$

(d) None of the above

3 $\int e^{2ax} \frac{1 - \cos 2a x}{1 + \sin 2a x} dx$ is equal to

(a) $-\frac{1}{a} e^{2ax} \cos\left(\frac{\pi}{4} + ax\right) + C$

(b) $-\frac{1}{2a} e^{2ax} \cot\left(\frac{\pi}{4} + ax\right) + C$

(c) $-\frac{1}{2a} e^{2ax} \cos\left(\frac{\pi}{4} + ax\right) + C$

(d) $-\frac{1}{a} e^{2ax} \operatorname{cosec}\left(\frac{\pi}{4} + ax\right) + C$

4 If $x^2 \neq n\pi - 1$, $n \in N$. Then, the value of $\int x \sqrt{\frac{2\sin(x^2+1) - \sin 2(x^2+1)}{2\sin(x^2+1) + \sin 2(x^2+1)}}\, dx$ is equal to

(a) $\log\left|\frac{1}{2}\sec(x^2+1)\right| + C$ (b) $\log\left|\sec\left(\frac{x^2+1}{2}\right)\right| + C$

(c) $\frac{1}{2}\log|\sec(x^2+1)| + C$ (d) None of these

5 The integral

$$\int \frac{\sin^2 x \cos^2 x}{(\sin^5 x + \cos^3 x \sin^2 x + \sin^3 x \cos^2 x + \cos^5 x)^2} dx$$

is equal to → JEE Mains 2018

(a) $\frac{1}{3(1+\tan^3 x)} + C$ (b) $\frac{-1}{3(1+\tan^3 x)} + C$

(c) $\frac{1}{1+\cot^3 x} + C$ (d) $\frac{-1}{1+\cot^3 x} + C$

(where C is a constant of integration)

6 $\int \cos 2\theta \log\left(\frac{\cos\theta + \sin\theta}{\cos\theta - \sin\theta}\right) d\theta$ is equal to

(a) $(\cos\theta - \sin\theta)^2 \log\left(\frac{\cos\theta + \sin\theta}{\cos\theta - \sin\theta}\right) + C$

(b) $(\cos\theta + \sin\theta)^2 \log\left(\frac{\cos\theta + \sin\theta}{\cos\theta - \sin\theta}\right) + C$

(c) $\frac{(\cos\theta - \sin\theta)^2}{2} \log\left(\frac{\cos\theta - \sin\theta}{\cos\theta + \sin\theta}\right) + C$

(d) $\frac{1}{2}\sin 2\theta \log\tan\left(\frac{\pi}{4} + \theta\right) - \frac{1}{2}\log\sec 2\theta + C$

7 $\int \frac{x^2}{(x\sin x + \cos x)^2} dx$ is equal to

(a) $\frac{\sin x + \cos x}{x\sin x + \cos x} + C$

(b) $\frac{x\sin x - \cos x}{x\sin x + \cos x} + C$

(c) $\frac{\sin x - x\cos x}{x\sin x + \cos x} + C$

(d) None of the above

8 If $f(x) = \int \frac{x^2\, dx}{(1+x^2)(1+\sqrt{1+x^2})}$ and $f(0) = 0$, then the value of $f(1)$ is

(a) $\log(1+\sqrt{2})$ (b) $\log(1+\sqrt{2}) - \frac{\pi}{4}$

(c) $\log(1+\sqrt{2}) + \frac{\pi}{2}$ (d) None of these

9 If $I = \int \frac{dx}{\sqrt[3]{(x+1)^2(x-1)^4}} = k\sqrt[3]{\frac{1+x}{1-x}} + C$, then k is equal to

(a) 2/3 (b) 3/2

(c) 1/3 (d) 1/2

10 $\int \frac{dx}{(\sin x + 2)(\sin x - 1)}$ is equal to

(a) $\frac{2}{3\left(\tan\frac{x}{2}-1\right)} - \frac{2}{3\sqrt{3}}\tan^{-1}\left[\frac{2\left(\tan\frac{x}{2}+\frac{1}{2}\right)}{\sqrt{3}}\right] + C$

(b) $\frac{2}{\left(\tan\frac{x}{2}+1\right)} + C$

(c) $-\frac{2}{3}\frac{1}{\left(\tan\frac{x}{2}-1\right)} + \frac{2}{3\sqrt{3}}\tan^{-1}\left[\frac{2\tan\frac{x}{2}-1}{\sqrt{3}}\right] + C$

(d) $-\frac{2}{3}\frac{2}{\left(\tan\frac{x}{2}-1\right)} + \frac{2}{3\sqrt{3}}\tan^{-1}\left[\frac{2\tan\frac{x}{2}-1}{\sqrt{3}}\right] + C$

11 $\int \frac{x^2}{(2+3x^2)^{5/2}}\,dx$ is equal to

(a) $\frac{1}{5}\left[\frac{x^2}{2+3x^2}\right]^{3/2} + C$

(b) $\frac{1}{6}\left[\frac{x^2}{2+3x^2}\right]^{3/2} + C$

(c) $\frac{1}{6}\left[\frac{x^2}{2+3x^2}\right]^{7/2} + C$

(d) None of the above

12 The integral $\int \left(1 + x - \frac{1}{x}\right) e^{x+\frac{1}{x}}\,dx$ is equal to → JEE Mains 2014

(a) $(x-1)\,e^{x+\frac{1}{x}} + C$ (b) $xe^{x+\frac{1}{x}} + C$

(c) $(x+1)\,e^{x+\frac{1}{x}} + C$ (d) $-xe^{x+\frac{1}{x}} + C$

13 $\int (\sin(101\,x)\cdot \sin^{99} x)\,dx$ is equal to

(a) $\frac{\sin(100x)(\sin x)^{100}}{100} + C$ (b) $\frac{\cos(100x)(\sin x)^{100}}{100} + C$

(c) $\frac{\cos(100x)(\cos x)^{100}}{100} + C$ (d) $\frac{\cos(100x)(\cos x)^{100}}{100} + C$

14 $\int \sqrt{\frac{(2018)^{2x}}{1-(2018)^{2x}}}\,(2018)^{\sin^{-1}(2018)^x}\,dx$ is equal to

(a) $(\log_{2018} e)^2 (2018)^{\sin^{-1}(2018)^x} + C$

(b) $(\log_{2018} e)^2 (2018)^{x+\sin^{-1}(2018)^x} + C$

(c) $(\log_{2018} e)^2 (2018)^{x-\sin^{-1}(2018)^x} + C$

(d) $\frac{(2018)^{\sin^{-1}(2018)^x}}{(\log_{2018} e)^2} + C$

15 $\int (\int e^x \left(\log x + \frac{2}{x} - \frac{1}{x^2}\right))\,dx$ is equal to

(a) $e^x \log x + C_1 x + C_2$ (b) $\log x + \frac{1}{x} + C_1 x + C_2$

(c) $\frac{\log x}{x} + C_1 x + C_2$ (d) None of these

ANSWERS

SESSION 1	**1** (a)	**2** (c)	**3** (a)	**4** (d)	**5** (b)	**6** (d)	**7** (a)	**8** (d)	**9** (d)	**10** (b)
	11 (a)	**12** (a)	**13** (b)	**14** (d)	**15** (a)	**16** (a)	**17** (c)	**18** (c)	**19** (b)	**20** (c)
	21 (c)	**22** (a)	**23** (d)	**24** (d)	**25** (b)	**26** (c)	**27** (a)	**28** (c)	**29** (a)	**30** (a)
	31 (b)	**32** (c)	**33** (b)	**34** (b)	**35** (a)					
SESSION 2	**1** (a)	**2** (d)	**3** (b)	**4** (b)	**5** (b)	**6** (d)	**7** (c)	**8** (b)	**9** (b)	**10** (a)
	11 (b)	**12** (b)	**13** (a)	**14** (a)	**15** (a)					

Hints and Explanations

1 Let $I = \int \frac{x^6}{x + x^7} dx = \int \frac{x^6}{x(1 + x^6)} dx$

$$= \int \frac{(1 + x^6) - 1}{x(1 + x^6)} dx$$

$$\Rightarrow I = \int \frac{dx}{x} - \int \frac{dx}{x + x^7}$$

$$= \log|x| - p(x) + C$$

2 Let $I = \int \frac{x^3 - 1}{(x^4 + 1)(x + 1)} dx$

$$= \int \frac{x^3 + x^4 - x^4 - 1}{(x^4 + 1)(x + 1)} dx$$

$$= \int \frac{x^3(x + 1) - (x^4 + 1)}{(x^4 + 1)(x + 1)} dx$$

$$= \int \left[\frac{x^3}{x^4 + 1} - \frac{1}{x + 1}\right] dx$$

$$= \frac{1}{4}\log(x^4 + 1) - \log(x + 1) + C$$

3 Let $I = \int (x + 1)(x + 2)^7 (x + 3) dx$

Put $x + 2 = t$

$\Rightarrow \quad x = t - 2$ and $dx = dt$

$\therefore \quad I = \int (t - 1)t^7(t + 1) dx$

$$= \int (t^2 - 1) \cdot t^7 \, dx = \int (t^9 - t^7) dx$$

$$= \frac{t^{10}}{10} - \frac{t^8}{8} + C = \frac{(x + 2)^{10}}{10} - \frac{(x + 2)^8}{8} + C$$

4 $\int \frac{dx}{x^2(x^4 + 1)^{\frac{3}{4}}} = \int \frac{dx}{x^5\left(1 + \frac{1}{x^4}\right)^{\frac{3}{4}}}$

Put $1 + \frac{1}{x^4} = t^4$

$\Rightarrow \quad -\frac{4}{x^5} dx = 4t^3 dt$

$\Rightarrow \quad \frac{dx}{x^5} = -t^3 dt$

Hence, the integral becomes

$$\int \frac{-t^3 dt}{t^3} = -\int dt = -t + C$$

$$= -\left(1 + \frac{1}{x^4}\right)^{1/4} + C = -\left(\frac{x^4 + 1}{x^4}\right)^{1/4} + C$$

5 Let $I = \int \frac{\sin x}{\sin(x - \alpha)} dx$

Put $x - \alpha = t \Rightarrow dx = dt$

$\therefore I = \int \frac{\sin(t + \alpha)}{\sin t} dt$

$$= \int \cos\alpha \, dt + \int \sin\alpha \cdot \frac{\cos t}{\sin t} dt$$

$$= \cos\alpha \cdot t + \sin\alpha \log \sin t + C$$

$$= x\cos\alpha + \sin\alpha \log\{\sin(x - \alpha)\} + C$$

$\therefore \quad A = \cos\alpha, B = \sin\alpha$

6 We have, $\int \frac{f(x)}{\log \sin x} = \log(\log \sin x) + C$

$\therefore \; f(x) = \frac{d}{dx}(\log \sin x)$

$$\left[\because \int \frac{f'(x)}{f(x)} dx = \log(f(x)) + C\right]$$

$$= \cot x$$

7 $\int \left\{\frac{(\log x - 1)}{1 + (\log x)^2}\right\}^2 dx$

$$= \int \frac{(\log x)^2 + 1 - 2\log x}{[(\log x)^2 + 1]^2} dx$$

$$= \int \frac{(\log x)^2 + 1 - 2x\left(\log x \cdot \frac{1}{x}\right)}{[(\log x)^2 + 1]^2} dx$$

$$= \int \frac{d}{dx}\left[\frac{x}{(\log x)^2 + 1}\right] dx$$

$$= \frac{x}{(\log x)^2 + 1} + C$$

8 Let $I = \int \sqrt{x + \sqrt{x^2 + 5}} \, dx$

Put $x + \sqrt{x^2 + 5} = t$

$\Rightarrow \quad \sqrt{x^2 + 5} = t - x$

$\Rightarrow \quad x^2 + 5 = t^2 + x^2 - 2xt$

$\Rightarrow \quad 5 = t^2 - 2xt$

$\Rightarrow \quad 2xt = t^2 - 5$

$\Rightarrow \quad x = \frac{1}{2}\left(t - \frac{5}{t}\right)$

and $\quad dx = \frac{1}{2}\left(1 + \frac{5}{t^2}\right)$

Now, $\quad I = \int t^{1/2} \cdot \frac{1}{2}\left(1 + \frac{5}{t^2}\right) dt$

$$= \frac{1}{2}\int (t^{1/2} + 5t^{-3/2}) dt$$

$$= \frac{1}{2}\left(\frac{2}{3}t^{3/2} - \frac{10}{\sqrt{t}}\right) + C = \frac{1}{3}t^{3/2} - \frac{5}{\sqrt{t}} + C$$

Clearly, $3PQ = -5$

9 Let $I = \frac{1}{\sqrt{2}} \int \frac{dx}{\left(\frac{1}{\sqrt{2}}\cos x - \frac{1}{\sqrt{2}}\sin x\right)}$

$$= \frac{1}{\sqrt{2}} \int \sec\left(x + \frac{\pi}{4}\right) dx$$

$$= \frac{1}{\sqrt{2}} \log\left|\tan\left(\frac{\pi}{4} + \frac{x}{2} + \frac{\pi}{8}\right)\right| + C$$

$$= \frac{1}{\sqrt{2}} \log\left|\tan\left(\frac{x}{2} + \frac{3\pi}{8}\right)\right| + C$$

10 Let $\quad I = \int \frac{\sin^8 x - \cos^8 x}{1 - 2\sin^2 x \cos^2 x} dx$

$$I = \int \frac{(\sin^4 x + \cos^4 x)(\sin^4 x - \cos^4 x)}{(\sin^2 x + \cos^2 x)^2 - 2\sin^2 x\cos^2 x} dx$$

$$= \int (\sin^4 x - \cos^4 x) dx$$

$$= \int (\sin^2 x - \cos^2 x) dx$$

$$= \int -\cos 2x \, dx$$

$$= -\frac{\sin 2x}{2} + C$$

11 Let $I = \int \frac{(\sqrt[3]{x + \sqrt{2 - x^2}})(\sqrt[6]{1 - x\sqrt{2 - x^2}})}{\sqrt[3]{1 - x^2}} dx$

$$= \int \frac{\sqrt[3]{x + \sqrt{2 - x^2}}\left(\sqrt[6]{\frac{1}{2}(2 - 2x\sqrt{2 - x^2})}\right)}{\sqrt[3]{1 - x^2}} dx$$

$$= \int \frac{\sqrt[3]{x + \sqrt{2 - x^2}}\left(\sqrt[6]{\frac{[x^2 + (\sqrt{2 - x^2})^2 - 2x\sqrt{2 - x^2}]}{2}}\right)}{\sqrt[3]{1 - x^2}} dx$$

$$= \int \frac{\sqrt[3]{x + \sqrt{2 - x^2}}\sqrt[3]{\sqrt{2 - x^2} - x}}{2^{1/6}\sqrt[3]{1 - x^2}} dx$$

$$= \int \frac{\sqrt[3]{(2 - x^2) - x^2}}{2^{1/6}\sqrt[3]{1 - x^2}} dx$$

$$= 2^{1/6}\int dx = 2^{1/6} x + C$$

12 Let $\quad I = \int \left(\frac{\cos 6x + 6\cos 4x + 15\cos 2x + 10}{10\cos^2 x + 5\cos x\cos 3x + \cos x\cos 5x}\right) dx$

$$= \int \frac{(\cos 6x + \cos 4x) + 5(\cos 4x + \cos 2x) + 10(\cos 2x + 1)}{10\cos^2 x + 5\cos x\cos 3x + \cos x\cos 5x} dx$$

$$= \int \frac{2\cos 5x \cdot \cos x + 10 \cdot \cos 3x \cdot \cos x + 10 \cdot (2\cos^2 x)}{10\cos^2 x + 5\cos x \cdot \cos 3x + \cos x\cos 5x} dx = 2\int dx$$

$$= 2x + C$$

Clearly, $f(10) = 20$

13 Let $I = \int \frac{2x^{12} + 5x^9}{(x^5 + x^3 + 1)^3} dx$

$$= \int \frac{2x^{12} + 5x^9}{x^{15}(1 + x^{-2} + x^{-5})^3} dx$$

$$= \int \frac{2x^{-3} + 5x^{-6}}{(1 + x^{-2} + x^{-5})^3} dx$$

Now, put $\quad 1 + x^{-2} + x^{-5} = t$

$\Rightarrow \quad (-2x^{-3} - 5x^{-6}) dx = dt$

$\Rightarrow \quad (2x^{-3} + 5x^{-6}) dx = -dt$

$\therefore \quad I = -\int \frac{dt}{t^3} = -\int t^{-3} dt$

$$= -\frac{t^{-3+1}}{-3+1} + C = \frac{1}{2t^2} + C$$

$$= \frac{x^{10}}{2(x^5 + x^3 + 1)^2} + C$$

14 Let $I = \int \frac{x^2 - 1}{x^3\sqrt{2x^4 - 2x^2 + 1}}dx$

$$= \int \frac{x^2 - 1}{x^5\sqrt{2 - \frac{2}{x^2} + \frac{1}{x^4}}}dx$$

$$= \int \frac{\frac{1}{x^3} - \frac{1}{x^5}}{\sqrt{2 - \frac{2}{x^2} + \frac{1}{x^4}}}dx$$

Now, putting $2 - \frac{2}{x^2} + \frac{1}{x^4} = t$, we get

$$4\left(\frac{1}{x^3} - \frac{1}{x^5}\right)dx = dt$$

$$\therefore I = \frac{1}{4}\int \frac{dt}{\sqrt{t}}$$

$$= \frac{1}{4}\cdot 2\sqrt{t} + C = \frac{\sqrt{2x^4 - 2x^2 + 1}}{2x^2} + C$$

15 Put $q \tan^{-1} x = t$

$$\Rightarrow \frac{q}{1 + x^2}dx = dt \Rightarrow \frac{1}{1 + x^2}dx = \frac{dt}{q}$$

$$\therefore \int \frac{dt}{q\sqrt{p^2 + t^2}} = \frac{1}{q}\log\,[t + \sqrt{p^2 + t^2}]$$

$$= \frac{1}{q}\log\,[q\tan^{-1} x + \sqrt{p^2 + q^2(\tan^{-1} x)^2}] + C$$

16 LHS $= \int \frac{\frac{2\cos^2 4x}{\cos^2 2x - \sin^2 2x}}{\cos 2x \sin 2x}dx$

$$= \int \frac{2\cos^2 4x \times \cos 2x \sin 2x}{\cos 4x}dx$$

$$= \int \cos 4x \times \sin 4x dx = \frac{1}{2}\int \sin 8x dx$$

$$= \frac{-1}{2}\frac{\cos 8x}{8} + k$$

Hence, we get $A = \frac{-1}{16}$

17 Let $I = \int \frac{\sin\theta + \cos\theta}{\sqrt{1 - (1 - 2\sin\theta\cos\theta)}}d\theta$

$$= \int \frac{\sin\theta + \cos\theta}{\sqrt{1 - (\sin\theta - \cos\theta)^2}}d\theta$$

Put $\sin\theta - \cos\theta = t$

$\Rightarrow (\cos\theta + \sin\theta)\,d\theta = dt$

$$\therefore \quad I = \int \frac{dt}{\sqrt{1 - t^2}} = \sin^{-1}(t) + C$$

$$= \sin^{-1}(\sin\theta - \cos\theta) + C$$

18 Let $I = \int \left(\frac{f(x)\cdot g'(x) - f'(x)\cdot g(x)}{f(x)\cdot g(x)}\right)\log\left(\frac{g(x)}{f(x)}\right)dx$

Put $\frac{g(x)}{f(x)} = t$

$$\Rightarrow \frac{f(x)\cdot g'(x) - g(x)\cdot f'(x)}{(f(x))^2}dx = dt$$

$$\Rightarrow \frac{f(x)\cdot g'(x) - g(x)\cdot f'(x)}{f(x)\cdot g(x)}\cdot\frac{g(x)}{f(x)}dx = dt$$

$$\Rightarrow \left(\frac{f(x)\cdot g'(x) - g(x)\cdot f'(x)}{f(x)\cdot g(x)}\right)dx = \frac{dt}{t}$$

Now, $I = \int \frac{1}{t}\cdot\log t\,dt = \frac{(\log t)^2}{2} + C$

$$= \frac{1}{2}\left(\log\left(\frac{g(x)}{f(x)}\right)\right)^2 + C$$

19 We have, $I_n = \int \tan^n x\,dx$

$$\therefore \quad I_n + I_{n+2} = \int \tan^n x\,dx + \int \tan^{n+2} x\,dx$$

$$= \int \tan^n x(1 + \tan^2 x)\,dx$$

$$= \int \tan^n x \sec^2 x\,dx$$

$$= \frac{\tan^{n+1} x}{n+1} + C$$

Put $n = 4$, we get $I_4 + I_6 = \frac{\tan^5 x}{5} + C$

$\therefore \quad a = \frac{1}{5}$ and $b = 0$

20 Let $I = \int \frac{\tan x}{1 + \tan x + \tan^2 x}$

$$dx = \int \frac{\frac{\sin x}{\cos x}}{1 + \frac{\sin^2 x}{\cos^2 x} + \frac{\sin x}{\cos x}}dx$$

$$= \int \frac{\sin 2x}{2 + \sin 2x}dx$$

$$= \int dx - 2\int \frac{dx}{2 + \sin 2x}$$

$$= x - 2\int \frac{\sec^2 x}{2\sec^2 x + 2\tan x}dx$$

Let $\tan x = t$

$\Rightarrow \sec^2 x\,dx = dt$

$$= x - \frac{2}{2}\int \frac{dt}{t^2 + t + 1}$$

$$= x - \int \frac{dt}{\left(t + \frac{1}{2}\right)^2 + \left(\frac{\sqrt{3}}{2}\right)^2}$$

$$\Rightarrow I = x - \frac{2}{\sqrt{3}}\tan^{-1}\left(\frac{2\tan x + 1}{\sqrt{3}}\right) + C$$

Hence, we get $A = 3$.

21 Here, $m + n = \frac{-3}{7} + \left(\frac{-11}{7}\right) = -2$

$$I = \int \cos^{-3/7} x(\sin^{(-2+3/7)} x)dx$$

$$= \int \cos^{-3/7} x \sin^{-2} x \sin^{3/7} x dx$$

$$= \int \frac{\operatorname{cosec}^2 x}{\left(\frac{\cos^{3/7} x}{\sin^{3/7} x}\right)}dx = \int \frac{\operatorname{cosec}^2 x}{\cot^{3/7} x}dx$$

Put $\cot x = t \Rightarrow -\operatorname{cosec}^2 x\,dx = dt$

$$\therefore \quad I = -\int \frac{dt}{t^{3/7}} = \frac{-7}{4}t^{4/7} + C$$

$$= -\frac{7}{4}\tan^{-4/7} x + C$$

22 Let $I = \int \frac{dx}{2 + \sin x + \cos x}$

$$\Rightarrow I = \int \frac{dx}{2 + \frac{2\tan\frac{x}{2}}{1 + \tan^2\frac{x}{2}} + \frac{1 - \tan^2\frac{x}{2}}{1 + \tan^2\frac{x}{2}}}$$

$$= \int \frac{\sec^2\frac{x}{2}dx}{2 + 2\tan^2\frac{x}{2} + 2\tan\frac{x}{2} + 1 - \tan^2\frac{x}{2}}$$

$$I = \int \frac{\sec^2\frac{x}{2}dx}{\tan^2\frac{x}{2} + 2\tan\frac{x}{2} + 3}$$

Put $\tan\frac{x}{2} = t \Rightarrow \frac{1}{2}\sec^2\frac{x}{2}dx = dt$

$$\therefore I = \int \frac{2dt}{t^2 + 2t + 3}$$

$$= 2\int \frac{dt}{t^2 + 2t + 1 + 2} = 2\int \frac{2dt}{(t + 1)^2 + (\sqrt{2})^2}$$

$$= 2\cdot\frac{1}{\sqrt{2}}\tan^{-1}\left(\frac{t + 1}{\sqrt{2}}\right) + C$$

$$\Rightarrow I = \sqrt{2}\tan^{-1}\left(\frac{\tan\frac{x}{2} + 1}{\sqrt{2}}\right) + C$$

23 Given, $\int \frac{5\tan x}{\tan x - 2}dx$

$$= x + a\ln|\sin x - 2\cos x| + k \quad \ldots\text{(i)}$$

Now, let us assume that

$$I = \int \frac{5\tan x}{\tan x - 2}dx$$

On multiplying by $\cos x$ in numerator and denominator, we get

$$I = \int \frac{5\sin x}{\sin x - 2\cos x}dx$$

Let $5\sin x = A(\sin x - 2\cos x) + B(\cos x + 2\sin x)$

$\Rightarrow 0\cdot\cos x + 5\sin x = (A + 2B)\sin x + (B - 2A)\cos x$

On comparing the coefficients of $\sin x$ and $\cos x$, we get

$$A + 2B = 5 \text{ and } B - 2A = 0$$

$\Rightarrow \quad A = 1$ and $B = 2$

$\Rightarrow \quad 5\sin x = (\sin x - 2\cos x) + 2(\cos x + 2\sin x)$

$\Rightarrow I = \int \frac{5\sin x}{\sin x - 2\cos x}dx$

$= \int \frac{(\sin x - 2\cos x) + 2(\cos x + 2\sin x)}{(\sin x - 2\cos x)}dx$

$\Rightarrow \quad I = \int 1\,dx + 2\int \frac{d(\sin x - 2\cos x)}{(\sin x - 2\cos x)}$

$I = x + 2\log|(\sin x - 2\cos x)| + k$ ...(ii)

where, k is the constant of integration.
On comparing the value of I in Eqs. (i) and (ii), we get $a = 2$

24 $\int \frac{\left(1 - \frac{1}{x^2}\right)}{\left(x + \frac{1}{x}\right)^2 - 1}dx$

Put $x + \frac{1}{x} = t \Rightarrow \left(1 - \frac{1}{x^2}\right)dx = dt$

$\therefore \int \frac{dt}{t^2 - 1} = \frac{1}{2}\log\left(\frac{t-1}{t+1}\right) + C$

$= \frac{1}{2}\log\left(\frac{x^2 - x + 1}{x^2 + x + 1}\right) + C$

25 Let $I = \int \sqrt{\frac{x}{a^3 - x^3}}dx$

Put $x = a(\sin\theta)^{2/3}$

$\Rightarrow \ dx = \frac{2}{3}a(\sin\theta)^{-1/3}\cos\theta d\theta$

$a^{1/2}(\sin\theta)^{1/3}\cdot\frac{2}{3}$

$\therefore \quad I = \int \frac{a(\sin\theta)^{-1/3}\cdot\cos\theta}{\sqrt{a^3 - a^3\sin^2\theta}}d\theta$

$= \frac{2}{3}\int \frac{a^{3/2}\cdot\cos\theta}{a^{3/2}\cos\theta}d\theta = \frac{2}{3}\int d\theta$

$= \frac{2}{3}\theta + C = \frac{2}{3}\sin^{-1}\left(\frac{x}{a}\right)^{3/2} + C$

26 $\int f(x)\cos x\,dx + \int g(x)e^x dx$

$= \frac{e^x}{2}(\cos x + \sin x)$

$- \frac{e^x}{2}(\sin x - \cos x) + C$

$= \frac{e^x}{2}(2\cos x) + C$

$= e^x\cos x + C$

27 Given, $\int f(x)\,dx = \Psi(x)$

Let $\quad I = \int x^5 f(x^3)\,dx$

Put $\quad x^3 = t$

$\Rightarrow \ x^2 dx = \frac{dt}{3}$ ...(i)

$\therefore I = \frac{1}{3}\int t\ f(t)\,dt = \frac{1}{3}[t\,\Psi(t) - \int \Psi(t)\,dt]$

$= \frac{1}{3}[x^3\Psi(x^3) - 3\int x^2\Psi(x^3)\,dx] + C$

[from Eq. (i)]

$= \frac{1}{3}x^3\Psi(x^3) - \int x^2\Psi(x^3)\,dx + C$

28 Let $I = \int \frac{1 - 6\cos^2 x}{\sin^6 x\cos^2 x}dx$

$= \int \frac{1}{\sin^6 x\cos^2 x}dx - 6\int \frac{dx}{\sin^6 x}$

$= I_1 - I_2$ (say)

Here, $I_1 = \int \frac{\sec^2 x}{\sin^6 x}dx$

$= \int \frac{1}{\sin^6 x}\cdot\sec^2 x\,dx = \frac{1}{\sin^6 x}\cdot\tan x$

$-\int \frac{(-6)}{\sin^7 x}\cdot\cos x\tan x\,dx$

$= \frac{\tan x}{\sin^6 x} + I_2 \Rightarrow I_1 - I_2 = \frac{\tan x}{\sin^6 x} + C$

Thus, $\quad I = \frac{\tan x}{\sin^6 x} + C$

Hence, $\quad f(x) = \tan x$

29 Let $I = \int \tan^{-1}\sqrt{x}(1)\,dx$

$= \tan^{-1}\sqrt{x}\int 1dx$

$-\left[\int \frac{d}{dx}(\tan^{-1}\sqrt{x})\int(1)\,dx\right]dx$

$= \tan^{-1}\sqrt{x}\cdot x - \int \frac{1}{1+x}\times\frac{1}{2\sqrt{x}}.x dx$

$= x\tan^{-1}\sqrt{x} - \frac{1}{2}\int \frac{x}{(1+x)\sqrt{x}}dx$

Put $x = t^2 \Rightarrow dx = 2t\cdot dt$

$\therefore I = x\tan^{-1}\sqrt{x} - \frac{1}{2}\int \frac{t^2}{(1+t^2)\cdot t}2t\ dt$

$= x\tan^{-1}\sqrt{x} - \int \frac{t^2}{1+t^2}dt$

$= x\tan^{-1}\sqrt{x} - t + \tan^{-1}t + C$

$= x\tan^{-1}\sqrt{x} - \sqrt{x} + \tan^{-1}\sqrt{x} + C$

$= (x+1)\tan^{-1}\sqrt{x} - \sqrt{x} + C$

30 $I_n = \int(\log x)^n dx = x(\log x)^n$

$-n\int(\log x)^{n-1}\cdot\frac{1}{x}\cdot x\,dx$

$\therefore \ I_n + nI_{n-1} = x(\log x)^n$

31 Consider, $\int f^{-1}(x)dx = \int f^{-1}(x)\cdot 1dx$

$= f^{-1}(x)\cdot x - \int \frac{d}{dx}(f^{-1}(x))\cdot x dx$

Now, let $f^{-1}(x) = t$, then

$\frac{d}{dx}(f^{-1}(x)) = \frac{dt}{dx}$

$\Rightarrow \quad \frac{d}{dx}(f^{-1}(x))\cdot dx = dt$

$\therefore \quad \int f^{-1}(x)dx = x\cdot f^{-1}(x) - \int f(t)dt$

$[\because f^{-1}(x) = t \Rightarrow x = f(t)]$

$= x\cdot f^{-1}(x) - g(t)$

$= x\cdot f^{-1}(x) - g(f^{-1}(x))$

32 Let $I = \int e^x\left(\frac{x+3}{(x+4)^2}\right)dx$

$= \int e^x\left(\frac{x+4-1}{(x+4)^2}\right)dx$

$= \int e^x\left(\frac{1}{x+4} - \frac{1}{(x+4)^2}\right)dx$

$= e^x\cdot\frac{1}{x+4} + C$

$[\because \int e^x(f(x) + f'(x)dx = e^x\cdot f(x) + C]$

33 LHS $= \int\left[\frac{x^2+1}{x^2+1} - \frac{x}{x^2+1}\right]e^{\cot^{-1}x}dx$

$= \int 1\cdot e^{\cot^{-1}x}dx - \int \frac{x}{x^2+1}e^{\cot^{-1}x}dx$

$= xe^{\cot^{-1}x} - \int x\cdot e^{\cot^{-1}x}\left(-\frac{1}{1+x^2}\right)dx$

$-\int \frac{x}{1+x^2}e^{\cot^{-1}x}dx + C = xe^{\cot^{-1}x} + C$

$\therefore \ A(x) = x$

34 We have, $\frac{d}{dx}\{g(x)\} = g(x)$

$\Rightarrow \quad g'(x) = g(x)$

$\Rightarrow \quad \int \frac{g'(x)}{g(x)}dx = \int 1\,dx$

$\Rightarrow \quad \log_e\{g(x)\} = x + \log C_1$

$\Rightarrow \quad g(x) = C_1 e^x$

Now, $\quad g(0) = 1 \Rightarrow C_1 = 1$

$\therefore \quad g(x) = e^x$

$\therefore \ \int g(x)\left(\frac{2 - \sin 2x}{1 - \cos 2x}\right)dx$

$= \int e^x(\text{cosec}^2 x - \cot x)\,dx$

$= -e^x\cot x + C$

$= -g(x)\cot x + C$

35 Let $I = \int \frac{dx^3}{x^3(x^n+1)}dx = \int \frac{3x^2\,dx}{x^3(x^n+1)}dx$

$= 3\int \frac{dx}{x(x^n+1)} = 3\int \frac{x^{n-1}\,dx}{x^n(x^n+1)}$

On putting $x^n = t$, we get

$I = \frac{3}{n}\int \frac{dt}{t(t+1)} = \frac{3}{n}\int\left[\frac{1}{t} - \frac{1}{t+1}\right]dt$

$= \frac{3}{n}[\log t - \log(t+1)] + C$

$= \frac{3}{n}\log\left(\frac{t}{t+1}\right) + C$

$= \frac{3}{n}\log\left(\frac{x^n}{x^n+1}\right) + C$

SESSION 2

1 We have, $\int f(x)\,dx = f(x)$

$\Rightarrow \quad \frac{d}{dx}\{f(x)\} = f(x)$

$\Rightarrow \quad \frac{1}{f(x)}d[f(x)] = dx$

$\Rightarrow \quad \log\{f(x)\} = x + \log C$

$\Rightarrow \quad f(x) = Ce^x$

$\Rightarrow \quad \{f(x)\}^2 = C^2 e^{2x}$

$\therefore \quad \int\{f(x)\}^2\,dx = \int C^2 e^{2x}dx$

$= \frac{C^2 e^{2x}}{2} = \frac{1}{2}\{f(x)\}^2$

2 We have,

$$f(x)=\begin{vmatrix} 0 & x^2-\sin x & \cos x-2 \\ \sin x-x^2 & 0 & 1-2x \\ 2-\cos x & 2x-1 & 0 \end{vmatrix}$$

$$\Rightarrow f(x)=\begin{vmatrix} 0 & \sin x-x^2 & 2-\cos x \\ x^2-\sin x & 0 & 2x-1 \\ \cos x-2 & 1-2x & 0 \end{vmatrix}$$

[interchanging rows and columns]

$$\Rightarrow f(x)=(-1)^3\begin{vmatrix} 0 & x^2-\sin x & \cos x-2 \\ \sin x-x^2 & 0 & 1-2x \\ 2-\cos x & 2x-1 & 0 \end{vmatrix}$$

[taking (−1) common from each column]

$\Rightarrow \quad f(x)=-f(x) \Rightarrow f(x)=0$

$\therefore \quad \int f(x)\,dx=\int 0\,dx=C$

3 Let $I=\int e^{2ax}\dfrac{1-\cos 2ax}{1+\sin 2ax}dx$

$\Rightarrow I=\dfrac{1}{a}\int e^{2t}\dfrac{1-\cos 2t}{1+\sin 2t}dt$, [where, $ax=t$]

$$\Rightarrow I=\frac{1}{a}\int e^{2t}\frac{1-2\sin\left(\frac{\pi}{4}+t\right)\cdot\cos\left(\frac{\pi}{4}+t\right)}{2\sin^2\left(\frac{\pi}{4}+t\right)}dt$$

$$\Rightarrow I=\frac{1}{a}\int e^{2t}\left\{\frac{1}{2}\text{cosec}^2\left(\frac{\pi}{4}+t\right)-\cot\left(\frac{\pi}{4}+t\right)\right\}dt$$

$$\Rightarrow I=\frac{1}{2a}\int \underset{\text{I}}{e^{2t}}\,\underset{\text{II}}{\text{cosec}^2}\left(\frac{\pi}{4}+t\right)dt-\frac{1}{a}\int e^{2t}\cot\left(\frac{\pi}{4}+t\right)dt$$

$$\Rightarrow I=-\frac{1}{2a}e^{2t}\cot\left(\frac{\pi}{4}+t\right)+\frac{1}{a}\int e^{2t}\cot\left(\frac{\pi}{4}+t\right)dt-\frac{1}{a}\int e^{2t}\cot\left(\frac{\pi}{4}+t\right)dt+C$$

$$\Rightarrow I=-\frac{1}{2a}e^{2t}\cot\left(\frac{\pi}{4}+t\right)+C$$

$$\therefore \quad I=-\frac{1}{2a}e^{2ax}\cot\left(\frac{\pi}{4}+ax\right)+C$$

4 We have,

$$\int x\sqrt{\frac{2\sin(x^2+1)-\sin 2(x^2+1)}{2\sin(x^2+1)+\sin 2(x^2+1)}}\,dx$$

$$=\int x\sqrt{\frac{2\sin(x^2+1)-2\sin(x^2+1)\cdot\cos(x^2+1)}{2\sin(x^2+1)+2\sin(x^2+1)\cdot\cos(x^2+1)}}\,dx$$

$$=\int x\sqrt{\frac{1-\cos(x^2+1)}{1+\cos(x^2+1)}}\,dx$$

$$=\int x\tan\left(\frac{x^2+1}{2}\right)dx$$

$$\therefore \int \tan\left(\frac{x^2+1}{2}\right)d\left(\frac{x^2+1}{2}\right)$$

$$=\log\left|\sec\left(\frac{x^2+1}{2}\right)\right|+C$$

5 We have,

$$I=\int\frac{\sin^2 x\cdot\cos^2 x}{(\sin^5 x+\cos^3 x\cdot\sin^2 x+\sin^3 x\cdot\cos^2 x+\cos^5 x)^2}dx$$

$$=\int\frac{\sin^2 x\cos^2 x}{\{\sin^3 x(\sin^2 x+\cos^2 x)+\cos^3 x(\sin^2 x+\cos^2 x)\}^2}dx$$

$$=\int\frac{\sin^2 x\cos^2 x}{(\sin^3 x+\cos^3 x)^2}dx$$

$$=\int\frac{\sin^2 x\cos^2 x}{\cos^6 x(1+\tan^3 x)^2}dx$$

$$=\int\frac{\tan^2 x\sec^2 x}{(1+\tan^3 x)^2}dx$$

Put $\tan^3 x=t$

$\Rightarrow \; 3\tan^2 x\sec^2 x\,dx=dt$

$\therefore I=\dfrac{1}{3}\int\dfrac{dt}{(1+t)^2} \Rightarrow I=\dfrac{-1}{3(1+t)}+C$

$\Rightarrow \quad I=\dfrac{-1}{3(1+\tan^3 x)}+C$

6 Since,

$\log\left(\dfrac{\cos\theta+\sin\theta}{\cos\theta-\sin\theta}\right)=\log\tan\left(\dfrac{\pi}{4}+\theta\right)$

and $\int\sec\theta\,d\theta=\log\tan\left(\dfrac{\pi}{4}+\dfrac{\theta}{2}\right)$

$\Rightarrow \quad \int\sec 2\theta\,d\theta=\dfrac{1}{2}\log\tan\left(\dfrac{\pi}{4}+\theta\right)$

$2\sec 2\theta=\dfrac{d}{d\theta}\log\tan\left(\dfrac{\pi}{4}+\theta\right)$

$\therefore I=\dfrac{1}{2}\sin 2\theta\log\tan\left(\dfrac{\pi}{4}+\theta\right)-\int\tan 2\theta\,d\theta$

$=\dfrac{1}{2}\sin 2\theta\log\tan\left(\dfrac{\pi}{4}+\theta\right)-\dfrac{1}{2}\log\sec 2\theta+C$

7 Let $I=\int\dfrac{x^2}{(x\sin x+\cos x)^2}dx$

$=\int\dfrac{x\cos x}{(x\sin x+\cos x)^2}\cdot\dfrac{x}{\cos x}dx$

$\left[\because \dfrac{d}{dx}(x\sin x+\cos x)=x\cos x\right]$

$$\therefore \quad I=\frac{-1}{(x\sin x+\cos x)}\cdot\frac{x}{\cos x}+\int\frac{1}{(x\sin x+\cos x)}\cdot\frac{\cos x-x(-\sin x)}{\cos^2 x}dx$$

$$=\frac{-x}{(x\sin x+\cos x)\cos x}+\tan x+C$$

$$=\frac{-x+x\sin^2 x+\sin x\cdot\cos x}{(x\sin x+\cos x)\cdot\cos x}+C$$

$$=\frac{\sin x-x\cos x}{x\sin x+\cos x}+C$$

8 $f(x)=\int\dfrac{x^2\,dx}{(1+x^2)(1+\sqrt{1+x^2})}$

Put $x=\tan\theta \Rightarrow dx=\sec^2\theta\,d\theta$
$=(1+x^2)\,d\theta$

$\therefore f(x)=\int\dfrac{\tan^2\theta\sec^2\theta}{\sec^2\theta(1+\sec\theta)}d\theta$

$=\int\dfrac{1-\cos^2\theta}{\cos\theta(1+\cos\theta)}d\theta$

$=\int\sec\theta\,d\theta-\int d\theta$

$=\log(\sec\theta+\tan\theta)-\theta+C$

$=\log(x+\sqrt{1+x^2})-\tan^{-1}x+C$

$\Rightarrow \quad f(0)=\log(0+\sqrt{1+0})-\tan^{-1}(0)+C$

$\Rightarrow \quad C=0$

$\therefore \quad f(1)=\log(1+\sqrt{2})-\dfrac{\pi}{4}+0$

9 Let $I=\int\dfrac{dx}{(1-x)^2\sqrt[3]{\left(\dfrac{x+1}{1-x}\right)^2}}$

Put $\dfrac{1+x}{1-x}=t \Rightarrow \dfrac{2}{(1-x)^2}dx=dt$

$\therefore \quad I=\dfrac{1}{2}\int\dfrac{dt}{t^{2/3}}=\dfrac{3}{2}[t^{1/3}]+C$

$=\dfrac{3}{2}\left[\sqrt{\dfrac{1+x}{1-x}}+C\right]$

$\therefore \quad k=\dfrac{3}{2}$

10 $\int\dfrac{dx}{(\sin x+2)(\sin x-1)}$

$=\dfrac{1}{3}\int\dfrac{dx}{(\sin x-1)}-\dfrac{1}{3}\int\dfrac{dx}{(\sin x+2)}$

$$=\frac{1}{3}\int\frac{dx}{\left(\dfrac{2\tan\frac{x}{2}}{1+\tan^2\frac{x}{2}}-1\right)}-\frac{1}{3}\int\frac{dx}{\left(\dfrac{2\tan\frac{x}{2}}{1+\tan^2\frac{x}{2}}+2\right)}$$

Put $\tan\dfrac{x}{2}=t$

$\Rightarrow \quad \dfrac{1}{2}\sec^2\dfrac{x}{2}dx=dt$

$\therefore \quad \dfrac{1}{3}\int\dfrac{2dt}{2t-1-t^2}-\dfrac{1}{3}\int\dfrac{2dt}{2t+2t^2+2}$

$$= -\frac{2}{3}\int \frac{dt}{(t-1)^2} - \frac{1}{3}\int \frac{dt}{\left(t+\frac{1}{2}\right)^2 + \left(\frac{\sqrt{3}}{2}\right)^2}$$

$$= \frac{2}{3}\frac{1}{(t-1)} - \frac{1}{3}\frac{2}{\sqrt{3}}\tan^{-1}\frac{\left(t+\frac{1}{2}\right)}{\frac{\sqrt{3}}{2}} + C$$

$$= \frac{2}{3}\frac{1}{\left(\tan\frac{x}{2}-1\right)} - \frac{2}{3\sqrt{3}}\tan^{-1}\frac{\left(\tan\frac{x}{2}+\frac{1}{2}\right)}{\frac{\sqrt{3}}{2}} + C$$

$$= \frac{2}{3\left(\tan\frac{x}{2}-1\right)} - \frac{2}{3\sqrt{3}}\tan^{-1}\left[\frac{2\left(\tan\frac{x}{2}+\frac{1}{2}\right)}{\sqrt{3}}\right] + C$$

11 $\int \frac{x^2}{(2+3x^2)^{5/2}}dx$

On substituting $2+3x^2 = t^2x^2$

$\Rightarrow \quad x^2 = \frac{2}{(t^2-3)}$

$\therefore \quad dx = -\frac{2t}{x(t^2-3)^2}dt$

$\therefore \int \frac{x^2}{(tx)^5}\cdot\left(\frac{-2t}{x(t^2-3)^2}\right)dt = -2\int\frac{dt}{4t^4}$

$$= -\frac{1}{2}\int\frac{dt}{t^4} = \frac{1}{6t^3} + C = \frac{1}{6}\left(\frac{x^2}{2+3x^2}\right)^{3/2} + C$$

12 $\int\left(1+x-\frac{1}{x}\right)e^{x+\frac{1}{x}}dx$

$= \int e^{x+\frac{1}{x}}dx + \int x\left(1-\frac{1}{x^2}\right)e^{x+\frac{1}{x}}dx$

$= \int e^{x+\frac{1}{x}}dx + x e^{x+\frac{1}{x}} - \int\frac{d}{dx}(x)e^{x+\frac{1}{x}}dx$

$= \int e^{x+\frac{1}{x}}dx + xe^{x+\frac{1}{x}} - \int e^{x+\frac{1}{x}}dx$

$\left[\because \int\left(1-\frac{1}{x^2}\right)e^{x+\frac{1}{x}}dx = e^{x+\frac{1}{x}}\right]$

$= \int e^{x+1/x}dx + xe^{x+1/x} - \int e^{x+1/x}dx$

$= xe^{x+\frac{1}{x}} + C$

13 Let $I = \int(\sin(101x)\cdot\sin^{99}x)dx$

$= \int\sin(100x+x)\sin^{99}x\,dx$

$= \int(\sin(100x)\cdot\cos x + \cos(100x)\cdot\sin x)\sin^{99}x\,dx$

$= \int\sin 100x\cdot(\cos x\cdot\sin^{99}x)dx + \int\cos(100x)\cdot\sin^{100}x\,dx$

$= \left[\sin(100x)\cdot\frac{\sin^{100}x}{100} - \int\cos(100x)\cdot 100\cdot\frac{\sin^{100}x}{100}dx\right] + \int\cos(100x)\cdot\sin^{100}x\,dx$

$= \frac{\sin(100x)\cdot\sin^{100}x}{100} - \int\sin^{100}x\cdot\cos(100x)dx + \int\cos(100x)\cdot\sin^{100}x\,dx$

$= \frac{\sin(100x)\cdot\sin^{100}x}{100} + C$

14 Let $I = \int\sqrt{\frac{(2018)^{2x}}{1-(2018)^{2x}}}\cdot(2018)^{\sin^{-1}(2018)^x}dx$

$= \int\frac{(2018)^x}{\sqrt{1-(2018)^{2x}}}\cdot(2018)^{\sin^{-1}(2018)^x}$

Put $\sin^{-1}(2018)^x = t$

$\Rightarrow \frac{1}{\sqrt{1-(2018^x)^2}}\cdot(2018)^x l_n(2018)dx = dt$

$\therefore I = \frac{1}{\ln(2018)}\int(2018)^t dt$

$= \frac{1}{\ln(2018)}\cdot\frac{(2018)^t}{\ln 2018} + C$

$= \frac{(2018)^t}{\ln^2(2018)} + C$

$= (\log_{2018}e)^2\cdot(2018)^{\sin^{-1}(2018)^x} + C$

15 Let $I = \int(\int e^x(\log x + \frac{2}{x} - \frac{1}{x^2})dx)dx$

$= \int\left(\int e^x\left(\log x + \frac{1}{x} + \frac{1}{x} - \frac{1}{x^2}\right)dx\right)dx$

$= \int\left[\int e^x\left(\log x + \frac{1}{x}\right)dx + \int e^x\left(\frac{1}{x} - \frac{1}{x^2}\right)\right]dx$

$= \int\left(e^x\log x + e^x\frac{1}{x} + C_1\right)dx$

$[\because \int e^x(f(x) + f'(x))dx = e^x\cdot f(x) + C]$

$= \int e^x\left(\log x + \frac{1}{x}\right)dx + \int C_1\,dx$

$= e^x\log x + C_1x + C_2$

DAY SIXTEEN

Definite Integrals

Learning & Revision for the Day

- Concept of Definite Integrals
- Leibnitz Theorem
- Walli's Formula
- Inequalities in Definite Integrals
- Definite Integration as the Limit of a sum

Concept of Definite Integrals

Let $\phi(x)$ be an anti-derivative of a function $f(x)$ defined on $[a,b]$ i.e. $\frac{d}{dx}[\phi(x)] = f(x)$. Then, definite integral of $f(x)$ over $[a, b]$ is denoted by $\int_a^b f(x)\,dx$ and is defined as $[\phi(b) - \phi(a)]$ i.e. $\int_a^b f(x)\,dx = \phi(b) - \phi(a)$. The numbers a and b are called the limits of integration, where a is called **lower limit** and b is **upper limit**.

NOTE
- Every definite integral has a unique value.
- The above definition is nothing but the statement of second fundamental theorem of integral calculus.

Geometrical Interpretation of Definite Integral

In general, $\int_a^b f(x)dx$ represents an algebraic sum of areas of the region bounded by the curve $y = f(x)$, the X-axis, and the ordinates $x = a$ and $x = b$ as show in the following figure.

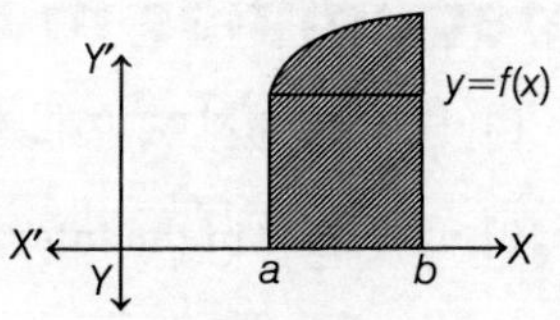

PREP MIRROR
Your Personal Preparation Indicator

- No. of Questions in Exercises (x)—
- No. of Questions Attempted (y)—
- No. of Correct Questions (z)—
  (Without referring Explanations)

- Accuracy Level ($z/y \times 100$)—
- Prep Level ($z/x \times 100$)—

In order to expect good rank in JEE, your Accuracy Level should be above 85 & Prep Level should be above 75.

Evaluation of Definite Integrals by Substitution

When the variable in a definite integral is changed due to substitution, then the limits of the integral will accordingly be changed.

For example, to evaluate definite integral of the form $\int_a^b f[g(x)] \cdot g'(x)\,dx$, we use the following steps

***Step* I** Substitute $g(x) = t$ so that $g'(x)\,dx = dt$

***Step* II** Find the limits of integration in new system of variable. Here, the lower limit is $g(a)$, the upper limit is $g(b)$ and the integral is now $\int_{g(a)}^{g(b)} f(t)\,dt$.

***Step*III** Evaluate the integral, so obtained by usual method.

Properties of Definite Integrals

(i) $\int_\alpha^\alpha f(x)dx = 0$

(ii) $\int_\alpha^\beta f(x)\,dx = \int_\alpha^\beta f(t)\,dt$

(iii) $\int_\alpha^\beta f(x)\,dx = -\int_\beta^\alpha f(x)\,dx$

(iv) $\int_\alpha^\beta f(x)\,dx = \int_\alpha^{c_1} f(x)dx + \int_{c_1}^{c_2} f(x)dx + \ldots + \int_{c_n}^\beta f(x)\,dx$

where, $\alpha < c_1 < c_2 < \ldots < c_n < \beta$

(v) (a) $\int_\alpha^\beta f(x)\,dx = \int_\alpha^\beta f(\alpha + \beta - x)\,dx$

(b) $\int_0^\alpha f(x)\,dx = \int_0^\alpha f(\alpha - x)\,dx$

(vi) $\int_{-\alpha}^\alpha f(x)\,dx$

$$= \begin{cases} 2\int_0^\alpha f(x)\,dx, \text{ if } f(-x) = f(x) \\ \quad \text{i.e. } f(x) \text{ is an even function} \\ 0, \quad \text{if } f(-x) = -f(x) \\ \quad \text{i.e. } f(x) \text{ is an odd function} \end{cases}$$

(vii) $\int_0^{2\alpha} f(x)dx = \int_0^\alpha f(x)dx + \int_0^\alpha f(2\alpha - x)dx$

(viii) $\int_0^{2\alpha} f(x)\,dx = \begin{cases} 2\int_0^\alpha f(x)dx, & \text{if } f(2\alpha - x) = f(x) \\ 0, & \text{if } f(2\alpha - x) = -f(x) \end{cases}$

(ix) $\int_\alpha^\beta f(x)\,dx = (\beta - \alpha)\int_0^1 f[(\beta - \alpha)\,x + \alpha]dx$

(x) If $f(x)$ is a periodic function with period T, then

(a) $\int_\alpha^{\alpha + nT} f(x)\,dx = n\int_0^T f(x)\,dx, n \in I$

(b) $\int_{\alpha T}^{\beta T} f(x)\,dx = (\beta - \alpha)\int_0^T f(x)\,dx, \quad \alpha, \beta \in I$

(c) $\int_{\alpha + nT}^{\beta + nT} f(x)\,dx = \int_\alpha^\beta f(x)\,dx, n \in I$

(xi) Some important integrals, which can be obtained with the help of above properties.

(a) $\int_0^{\pi/2} \log \sin x dx = \int_0^{\pi/2} \log \cos x dx = \frac{\pi}{2} \log\left(\frac{1}{2}\right).3$

(b) $\int_0^{\pi/4} \log(1 + \tan x)dx = \frac{\pi}{8} \log 2.$

(xii) If a function $f(x)$ is discontinuous at points $x_1, x_2, \ldots, x_n$ in (a, b), then we can define sub-intervals $(a, x_1), (x_1, x_2), \ldots, (x_{n-1}, x_n), (x_n, b)$ such that $f(x)$ is continuous in each of these sub-intervals and

$$\int_a^b f(x)dx = \int_a^{x_1} f(x)dx + \int_{x_1}^{x_2} f(x)dx + \ldots + \int_{x_{n-1}}^{x_n} f(x)dx + \int_{x_n}^b f(x)dx.$$

Leibnitz Theorem

If function $\phi(x)$ and $\psi(x)$ are defined on $[\alpha, \beta]$ and differentiable on $[\alpha, \beta]$ and $f(t)$ is continuous on $[\psi(\alpha), \phi(\beta)]$, then

$$\frac{d}{dx}\left[\int_{\phi(x)}^{\psi(x)} f(t)\,dt\right] = \left\{\frac{d}{dt}\,\psi(x)\right\} f(\psi(x)) - \left\{\frac{d}{dx}\{\phi(x)\}\right\}\{f(\phi(x)\}$$

Walli's Formula

This is a special type of integral formula whose limits from 0 to $\pi/2$ and integral is either integral power of $\cos x$ or $\sin x$ or $\cos x \sin x$.

$$\int_0^{\pi/2} \sin^n x\,dx = \int_0^{\pi/2} \cos^n x\,dx$$

$$= \begin{cases} \dfrac{(n-1)(n-3)(n-5)\ldots 5\cdot 3\cdot 1}{n(n-2)(n-4)\ldots 6\cdot 4\cdot 2} \times \dfrac{\pi}{2}, & \text{if } n = 2m \text{ (even)} \\ \dfrac{(n-1)(n-3)(n-5)\ldots 6\cdot 4\cdot 2}{n(n-2)(n-4)\ldots 5\cdot 3\cdot 1}, & \text{if } n = 2m + 1 \text{ (odd)} \end{cases}$$

where, n is positive integer.

$$\int_0^{\pi/2} \sin^m x \cdot \cos^n x\,dx$$

$$= \begin{cases} \dfrac{(m-1)(m-3)\ldots(2 \text{ or } 1).(n-1)(n-3)\ldots(2 \text{ or } 1)}{(m+n)(m+n-2)\ldots(2 \text{ or } 1)} \dfrac{\pi}{2}, \\ \text{when both } m \text{ and } n \text{ are even positive integers} \\ \dfrac{(m-1)(m-3)\ldots(2 \text{ or } 1)\cdot(n-1)(n-3)\ldots(2 \text{ or } 1)}{(m+n)(m+n-2)\ldots(2 \text{ or } 1)}, \\ \text{when either } m \text{ or } n \text{ or both are odd} \\ \text{positive integers} \end{cases}$$

Inequalities in Definite Integrals

(i) If $f(x) \ge g(x)$ on $[\alpha, \beta]$, then $\int_\alpha^\beta f(x)\,dx \ge \int_\alpha^\beta g(x)\,dx$

(ii) If $f(x) \ge 0$ in the interval $[\alpha, \beta]$, then $\int_\alpha^\beta f(x)\,dx \ge 0$

(iii) If $f(x), g(x)$ and $h(x)$ are continuous on $[a, b]$ such that $g(x) \le f(x) \le h(x)$, then $\int_a^b g(x)dx \le \int_a^b f(x)dx \le \int_a^b h(x)dx$

(iv) If f is continuous on $[\alpha, \beta]$ and $l \le f(x) \le M, \forall$ $x \in [\alpha, \beta]$, then $l(\beta - \alpha) \le \int_\alpha^\beta f(x)\,dx \le M(\beta - \alpha)$

(v) If f is continuous on $[\alpha, \beta]$, then $\left|\int_\alpha^\beta f(x)\,dx\right| \le \int_\alpha^\beta |f(x)|\,dx$

(vi) If f is continuous on $[\alpha, \beta]$ and $|f(x)| \le k, \forall x \in [\alpha, \beta]$, then $\left|\int_\alpha^\beta f(x)\,dx\right| \le k(\beta - \alpha)$

Definite Integration as the Limit of a Sum

Let $f(x)$ be a continuous function defined on the closed interval $[a, b]$, then $\int_a^b f(x)\,dx = \lim_{n\to\infty} h \sum_{r=0}^{n-1} f(a + rh)$

where, $h = \dfrac{b-a}{n} \to 0$ as $n \to \infty$

The converse is also true, i.e. if we have an infinite series of the above form, it can be expressed as definite integral.

Some Particular Cases

(i) $\lim_{n\to\infty} \sum_{r=1}^{n} \frac{1}{n} f\left(\frac{r}{n}\right)$ or $\lim_{n\to\infty} \sum_{r=1}^{n-1} \frac{1}{n} f\left(\frac{r}{n}\right) = \int_0^1 f(x)\,dx$

(ii) $\lim_{n\to\infty} \sum_{r=1}^{pn} \frac{1}{n} f\left(\frac{r}{n}\right) = \int_\alpha^\beta f(x)\,dx$

where, $\alpha = \lim_{n\to\infty} \dfrac{r}{h} = 0$ $\quad (\because r = 1)$

and $\beta = \lim_{n\to\infty} \dfrac{r}{h} = p$ $\quad (\because r = pn)$

DAY PRACTICE SESSION 1

FOUNDATION QUESTIONS EXERCISE

1 $\int_{\pi/4}^{3\pi/4} \dfrac{dx}{1 + \cos x}$ is equal to → JEE Mains 2017

(a) -2 (b) 2 (c) 4 (d) -1

2 If $f(x)$ is continuous function, then

(a) $\int_{-2}^{2} f(x)\,dx = \int_0^2 [f(x) - f(-x)]\,dx$

(b) $\int_{-3}^{5} 2f(x)\,dx = \int_6^{10} f(x-1)\,dx$

(c) $\int_{-3}^{5} f(x)\,dx = \int_{-4}^{4} f(x-1)\,dx$

(d) $\int_{-3}^{5} f(x)\,dx = \int_{-2}^{6} f(x-1)\,dx$

3 $\int_0^{\pi/4} [\sqrt{\tan x} + \sqrt{\cot x}]\,dx$ is equal to

(a) $\sqrt{2}\pi$ (b) $\dfrac{\pi}{2}$ (c) $\dfrac{\pi}{\sqrt{2}}$ (d) 2π

4 $\int_0^1 \sin\left(2\tan^{-1}\sqrt{\dfrac{1+x}{1-x}}\right) dx$ is equal to

(a) $\pi/6$ (b) $\pi/4$ (c) $\pi/2$ (d) π

5 If $I_{1(n)} = \int_0^{\frac{\pi}{2}} \dfrac{\sin(2n-1)}{\sin x}\,dx$ and $I_{2(n)} = \int_0^{\frac{\pi}{2}} \dfrac{\sin^2 nx}{\sin^2 x}\,dx$, $n \in N$, then

(a) $I_{2(n+1)} - I_{2(n)} = I_{1(n)}$ (b) $I_{2(n+1)} - I_{2(n)} = I_{1(n+1)}$

(c) $I_{2(n+1)} + I_{1(n)} = I_{2(n)}$ (d) $I_{2(n+1)} + I_{1(n+1)} = I_{2(n)}$

6 $\int_{-1}^{1} \{x^2 + x - 3\}\,dx$, where $\{x\}$ denotes the fractional part of x, is equal to

(a) $\dfrac{1}{3}(1 + 3\sqrt{5})$ (b) $\dfrac{1}{6}(1 + 3\sqrt{5})$

(c) $\dfrac{1}{3}(3\sqrt{5} - 1)$ (d) $\dfrac{1}{6}(3\sqrt{5} - 1)$

7 $\int_0^2 [x^2]\,dx$ is equal to

(a) $2 - \sqrt{2}$ (b) $2 + \sqrt{2}$

(c) $\sqrt{2} - 1$ (d) $-\sqrt{2} - \sqrt{3} + 5$

8 If $\int_{-2}^{x} |2t|\,dt = f(x)$, then for any $x \ge 0, f(x)$ is equal to

(a) $4 + x^2$ (b) $4 - x^2$

(c) $\dfrac{1}{2}(4 + x^2)$ (d) $\dfrac{1}{4}(4 - x^2)$

9 $\int_{-\pi/2}^{\pi/2} \dfrac{\cos x}{1 + e^x}\,dx$ is equal to

(a) 1 (b) 0

(c) -1 (d) None of these

10 Let a, b and c be non-zero real numbers such that $\int_0^3 (3ax^2 + 2bx + c)\,dx = \int_1^3 (3ax^2 + 2bx + c)\,dx$, then

(a) $a + b + c = 3$ (b) $a + b + c = 1$

(c) $a + b + c = 0$ (d) $a + b + c = 2$

11 The value of $\int_1^a [x] f'(x)\,dx, a > 1$, where $[x]$ denotes the greatest integer not exceeding x, is

(a) $[a]f(a) - \{f(1) + f(2) + \ldots + f([a])\}$

(b) $[a]f([a]) - \{f(1) + f(2) + \ldots + f(a)\}$

(c) $af([a]) - \{f(1) + f(2) + \ldots + f(a)\}$

(d) $af(a) - \{f(1) + f(2) + \ldots + f([a])\}$

12 The correct evaluation of $\int_0^{\pi/2} \left|\sin\left(x - \dfrac{\pi}{4}\right)\right| dx$ is

(a) $2 + \sqrt{2}$ (b) $2 - \sqrt{2}$

(c) $-2 + \sqrt{2}$ (d) 0

13 $\int_0^{3\pi/2} \sin\left(\left[\frac{2x}{\pi}\right]\right) dx$, where $[\cdot]$ denotes the greatest integer function is equal to

(a) $\frac{\pi}{2}(\sin 1 + \cos 1)$ (b) $\frac{\pi}{2}(\sin 1 - \sin 2)$
(c) $\frac{\pi}{2}(\sin 1 - \cos 1)$ (d) $\frac{\pi}{2}(\sin 1 + \sin 2)$

14 If [.] denotes the greatest integer function, then the value of $\int_0^{1.5} x[x^2]\,dx$ is

(a) $\frac{5}{4}$ (b) 0 (c) $\frac{3}{2}$ (d) $\frac{3}{4}$

15 The value of $\int_0^1 \frac{8\log(1+x)}{1+x^2} dx$ is

(a) $\frac{\pi}{8}\log 2$ (b) $\frac{\pi}{2}\log 2$ (c) $\log 2$ (d) $\pi \log 2$

16 $\int_0^{\pi/2} \sqrt{1-\sin 2x}$ is equal to → NCERT Exemplar

(a) $2\sqrt{2}$ (b) $2(\sqrt{2}+1)$ (c) 2 (d) $2(\sqrt{2}-1)$

17 The value of $\int_{e^{-1}}^{e^2} \left|\frac{\log_e x}{x}\right| dx$ is

(a) 3/2 (b) 5/2 (c) 3 (d) 5

18 The value of $\sum_{k=1}^{n} \int_0^1 f(k-1+x)\,dx$ is

(a) $\int_0^1 f(x)\,dx$ (b) $\int_0^2 f(x)\,dx$ (c) $\int_0^n f(x)\,dx$ (d) $n\int_0^1 f(x)\,dx$

19 $f(x)$ is a continuous function for all real values of x and satisfies $\int_n^{n+1} f(x)\,dx = \frac{n^2}{2}, \forall n \in I$, then $\int_{-3}^{5} f(|x|)\,dx$ is equal to

(a) $\frac{19}{2}$ (b) $\frac{35}{2}$
(c) $\frac{17}{2}$ (d) None of these

20 If $I_1 = \int_0^1 2^{x^2} dx, I_2 = \int_0^1 2^{x^3} dx, I_3 = \int_1^2 2^{x^2} dx$ and $I_4 = \int_1^2 2^{x^3} dx$, then

(a) $I_3 > I_4$ (b) $I_3 = I_4$
(c) $I_1 > I_2$ (d) $I_2 > I_1$

21 The value of $\int_{7\pi/4}^{7\pi/3} \sqrt{\tan^2 x}\,dx$ is → JEE Mains 2013

(a) $\log 2\sqrt{2}$ (b) $\frac{3}{2}\log 2$
(c) $2\log 2$ (d) $\log\sqrt{2}$

22 $\int_0^{\pi/2} \frac{\sin^2 x}{\sin x + \cos x} dx$ is equal to → NCERT Exemplar

(a) $\frac{1}{\sqrt{2}}\log(\sqrt{2}+1)$ (b) $\frac{1}{2}\log(\sqrt{2}+1)$
(c) $-\log(\sqrt{2}+1)$ (d) None of these

23 The value of $\lim_{n\to\infty} \frac{1}{n}\left\{\frac{1}{n+1} + \frac{2}{n+2} + \ldots + \frac{3n}{4n}\right\}$ is

(a) $5 - 2\ln 2$ (b) $4 - 2\ln 2$
(c) $3 - 2\ln 2$ (d) $2 - 2\ln 2$

24 If f and g are continuous functions in [0, 1] satisfying $f(x) = f(a-x)$ and $g(x) + g(a-x) = a$, then $\int_0^a f(x)\cdot g(x)\,dx$ is equal to → NCERT Exemplar

(a) $\frac{a}{2}$ (b) $\frac{a}{2}\int_0^a f(x)\,dx$
(c) $\int_0^a f(x)\,dx$ (d) $a\int_0^a f(x)\,dx$

25 The value of $\int_{-\pi/2}^{\pi/2} \frac{\sin^2 x}{1+2^x} dx$ is → JEE Mains 2018, 13

(a) π (b) $\frac{\pi}{2}$ (c) 4π (d) $\frac{\pi}{4}$

26 $\int_0^{\pi} [\cot x]\,dx$, where [] denotes the greatest integer function, is equal to

(a) $\frac{\pi}{2}$ (b) 1 (c) -1 (d) $-\frac{\pi}{2}$

27 The integral $\int_2^4 \frac{\log x^2}{\log x^2 + \log(36-12x+x^2)} dx$ is equal to → JEE Mains 2015

(a) 2 (b) 4 (c) 1 (d) 6

28 $\int_{-3\pi/2}^{-\pi/2} [(x+\pi)^3 + \cos^2(x+3\pi)]\,dx$ is equal to

(a) $\left(\frac{\pi^4}{32}\right) + \frac{\pi}{2}$ (b) $\frac{\pi}{2}$ (c) $\left(\frac{\pi}{4}\right) - 1$ (d) $\frac{\pi^4}{32}$

29 If $f: R \to R$ and $g: R \to R$ are one to one, real valued function, then the value of the integral $\int_{-\pi}^{\pi} [f(x) + f(-x)][g(x) - g(-x)]\,dx$ is

(a) 0 (b) π (c) 1 (d) None of these

30 The value of $\lim_{n\to\infty}\left(\frac{n}{(n+1)\sqrt{2n+1}} + \frac{n}{(n+2)\sqrt{2(2n+2)}} + \frac{n}{(n+3)\sqrt{3(2n+3)}} + \ldots + \frac{1}{2n\sqrt{3}}\right)$ is

(a) $\frac{\pi}{3}$ (b) $\frac{\pi}{2}$
(c) $\frac{\pi}{4}$ (d) None of these

31 If $P = \int_0^{3\pi} f(\cos^2 x)\,dx$ and $Q = \int_0^{\pi} f(\cos^2 x)\,dx$, then

(a) $P - Q = 0$ (b) $P - 2Q = 0$
(c) $P - 3Q = 0$ (d) $P - 5Q = 0$

32 $\int_{-\pi}^{\pi} \frac{2x(1+\sin x)}{1+\cos^2 x} dx$ is equal to

(a) $\frac{\pi^2}{4}$ (b) π^2 (c) 0 (d) $\frac{\pi}{2}$

33 If $f(x)$ is differentiable and $\int_0^{t^2} x\,f(x)\,dx = \frac{2}{5}t^5$, then $f\left(\frac{4}{25}\right)$ is equal to

(a) $\frac{2}{5}$ (b) $-\frac{5}{2}$ (c) 1 (d) $\frac{5}{2}$

34 If $f(x) = \frac{1}{x^2}\int_4^x [4t^2 - 2f'(t)]\,dt$, then $f'(4)$ is equal to

(a) 32 (b) $\frac{32}{3}$
(c) $\frac{32}{9}$ (d) None of these

35 The value of $\lim_{x \to 0} \frac{\int_0^{x^2} \sin\sqrt{t}\,dt}{x^3}$ is

(a) 0 (b) 2/9
(c) 1/3 (d) 2/3

36 If $I_1 = \int_1^2 \frac{dx}{\sqrt{1+x^2}}$ and $I_2 = \int_1^2 \frac{dx}{x}$, then

(a) $I_1 > I_2$ (b) $I_2 > I_1$ (c) $I_1 = I_2$ (d) $I_1 > 2I_2$

37 If $x = \int_0^y \frac{dt}{\sqrt{1+t^2}}$, then $\frac{d^2y}{dx^2}$ is equal to

→ JEE Mains 2013

(a) y (b) $\sqrt{1+y^2}$
(c) $\frac{y}{\sqrt{1+y^2}}$ (d) y^2

38 Let $f : R \to R$ be a differentiable function having $f(2) = 6$, $f'(2) = \left(\frac{1}{48}\right)$. Then, $\lim_{x \to 2} \int_6^{f(x)} \frac{4t^3}{x-2}\,dt$ is equal to

(a) 18 (b) 12 (c) 36 (d) 24

39 $\lim_{n \to \infty} \frac{[1+2^4+3^4+\ldots+n^4]}{n^5} - \frac{[1+2^3+3^3+\ldots+n^3]}{n^5}$ is equal to

(a) $\frac{1}{30}$ (b) 0 (c) $\frac{1}{4}$ (d) $\frac{1}{5}$

40 If $I = \int_0^{\pi/2} \frac{dx}{\sqrt{1+\sin^3 x}}\,dx$, then

(a) $0 < I < 1$ (b) $I > \frac{\pi}{2\sqrt{2}}$
(c) $I < \sqrt{2}\,\pi$ (d) $I < 2\pi$

41 If $I = \int_0^1 \frac{\sin x}{\sqrt{x}}\,dx$ and $J = \int_0^1 \frac{\cos x}{\sqrt{x}}\,dx$. Then, which one of the following is true?

(a) $I > \frac{2}{3}$ and $J < 2$ (b) $I > \frac{2}{3}$ and $J > 2$
(c) $I < \frac{2}{3}$ and $J < 2$ (d) $I < \frac{2}{3}$ and $J > 2$

42 **Statement I** $\int_0^2 f(x)\,dx = \frac{4(\sqrt{2}-1)}{3}$,

where, $f(x) = \begin{cases} x^2, & \text{for } 0 \le x < 1 \\ \sqrt{x}, & \text{for } 1 \le x \le 2 \end{cases}$

Statement II $f(x)$ is continuous in [0, 2].

(a) Statement I is true, Statement II is true; Statement II is a correct explanation for statement I
(b) Statement I is true, Statement II is true; Statement II is not a correct explanation for statement I
(c) Statement I is true; Statement II is false
(d) Statement I is false; Statement II is true

43 **Statement I** The value of the integral $\int_{\pi/6}^{\pi/3} \frac{dx}{1+\sqrt{\tan x}}$ is $\frac{\pi}{6}$.

Statement II $\int_a^b f(x)\,dx = \int_a^b f(a+b-x)\,dx$

→ JEE Mains 2013

(a) Statement I is true; Statement II is true; Statement II is a correct explanation for Statement I
(b) Statement I is true; Statement II is true; Statement II is not a correct explanation for Statement I
(c) Statement I is true; Statement II is false
(d) Statement I is false; Statement II is true

44 **Statement I** If $\int_0^1 e^{\sin x}dx = \lambda$, then $\int_0^{200} e^{\sin x}dx = 200\lambda$

Statement II If $\int_0^{na} f(x)\,dx = n\int_0^a f(x)\,dx$, $n \in I$ and $f(a+x) = f(x)$

(a) Statement I is true; Statement II is true; Statement II is a correct explanation for Statement I
(b) Statement I is true; Statement II is true, Statement II is not a correct explanation for Statement I
(c) Statement I is true; Statement II is false
(d) Statement I is false; Statement II is true

45 **Statement I** $\int_0^1 e^{-x}\cos^2 x\,dx < \int_0^1 e^{-x^2}\cos^2 x\,dx$

Statement II $\int_a^b f(x)\,dx < \int_a^b g(x)\,dx, \forall f(x) < g(x)$

(a) Statement I is true; Statement II is true; Statement II is a correct explanation for Statement I
(b) Statement I is true; Statement II is true, Statement II is not a correct explanation for Statement I
(c) Statement I is true; Statement II is false
(d) Statement I is false; Statement II is true

DAY PRACTICE SESSION 2

PROGRESSIVE QUESTIONS EXERCISE

1 If $f(x)$ is a function satisfying $f\left(\frac{1}{x}\right) + x^2 f(x) = 0$ for all non-zero x, then $\int_{\sin\theta}^{\operatorname{cosec}\theta} f(x)\,dx$ is equal to

(a) $\sin\theta + \operatorname{cosec}\theta$ (b) $\sin^2\theta$
(c) $\operatorname{cosec}^2\theta$ (d) None of these

2 If $\frac{d}{dx} f(x) = \frac{e^{\sin x}}{x}$; $x > 0$. If $\int_1^4 \frac{3}{x} e^{\sin x^3}\,dx = f(k) - f(1)$, then the possible value of k, is

(a) 15 (b) 16 (c) 63 (d) 64

3 If $g(1) = g(2)$, then $\int_1^2 [f\{g(x)\}]^{-1} f'\{g(x)\}\, g'(x)\,dx$ is equal to

(a) 1 (b) 2 (c) 0 (d) None of these

4 For $0 \le x \le \frac{\pi}{2}$, the value of $\int_0^{\sin^2 x} \sin^{-1}(\sqrt{t})\,dt + \int_0^{\cos^2 x} \cos^{-1}(\sqrt{t})\,dt$ is → JEE Mains 2013

(a) $\frac{\pi}{4}$ (b) 0 (c) 1 (d) $-\frac{\pi}{4}$

5 If $F(x) = f(x) + f\left(\frac{1}{x}\right)$, where $f(x) = \int_1^x \frac{\log t}{1+t}\,dt$. Then, $F(e)$ is equal to

(a) 1/2 (b) 0 (c) 1 (d) 2

6 The expression $\frac{\int_0^n [x]\,dx}{\int_0^n \{x\}\,dx}$, where $[x]$ and $\{x\}$ are integral and fractional part of x and $n \in N$, is equal to

(a) $\frac{1}{n-1}$ (b) $\frac{1}{n}$ (c) n (d) $n-1$

7 If $f(x) = \frac{e^x}{1+e^x}$, $I_1 = \int_{f(-a)}^{f(a)} x g[x(1-x)]\,dx$ and $I_2 = \int_{f(-a)}^{f(a)} g[x(1-x)]\,dx$, then the value of $\frac{I_2}{I_1}$ is

(a) 2 (b) -3 (c) -1 (d) 1

8 The value of $\lim_{x\to\infty} \frac{\{\int_0^x e^{t^2}\,dt\}^2}{\int_0^x e^{2t^2}\,dt}$ is

(a) 1 / 3 (b) 2 / 3 (c) 1 (d) None of these

9 The integral $\int_0^{\pi} \sqrt{1 + 4\sin^2\frac{x}{2} - 4\sin\frac{x}{2}}\,dx$ is equal to → JEE Mains 2014

(a) $\pi - 4$ (b) $\frac{2\pi}{3} - 4 - 4\sqrt{3}$
(c) $4\sqrt{3} - 4$ (d) $4\sqrt{3} - 4 - \frac{\pi}{3}$

10 $\int_0^{\pi} x f(\sin x)\,dx$ is equal to

(a) $\pi \int_0^{\pi} f(\sin x)\,dx$ (b) $\frac{\pi}{2}\int_0^{\pi/2} f(\sin x)\,dx$
(c) $\pi \int_0^{\pi/2} f(\cos x)\,dx$ (d) $\pi \int_0^{\pi} f(\cos x)\,dx$

11 If $f(x) = \int_{x^2}^{x^2+1} e^{-t^2}\,dt$, then $f(x)$ increases in

(a) (2, 2) (b) no value of x
(c) $(0, \infty)$ (d) $(-\infty, 0)$

12 $\lim_{n\to\infty} \left(\frac{(n+1)(n+2)\dots 3n}{n^{2n}}\right)^{1/n}$ is equal to → JEE Mains 2016

(a) $\frac{18}{e^4}$ (b) $\frac{27}{e^2}$
(c) $\frac{9}{e^2}$ (d) $3\log 3 - 2$

13 The least value of the function $f(x) = \int_{5\pi/4}^{x} (3\sin u + 4\cos u)\,du$ on the interval $\left[\frac{5\pi}{4}, \frac{4\pi}{3}\right]$ is

(a) $\sqrt{3} + \frac{3}{2}$ (b) $-2\sqrt{3} + \frac{3}{2} + \frac{1}{\sqrt{2}}$
(c) $\frac{3}{2} + \frac{1}{\sqrt{2}}$ (d) None of these

14 If $g(x) = \int_0^x f(t)\,dt$, where f is such that, $\frac{1}{2} \le f(t) \le 1$, for $t \in [0, 1]$ and $0 \le f(t) \le \frac{1}{2}$, for $t \in [1, 2]$. Then, $g(2)$ satisfies the inequality

(a) $-\frac{3}{2} \le g(z) < \frac{1}{2}$ (b) $\frac{1}{2} \le g(2) \le \frac{3}{2}$
(c) $\frac{3}{2} < g(2) \le \frac{5}{2}$ (d) $2 < g(2) < 4$

15 Let $n \ge 1, n \in z$. The real number $a \in (0, 1)$ that minimizes the integral $\int_0^1 |x^n - a^n|\,dx$ is

(a) $\frac{1}{2}$ (b) 2
(c) 1 (d) $\frac{1}{3}$

16 The value of

$$\lim_{n\to\infty} \left\{\tan\left(\frac{\pi}{2n}\right)\tan\left(\frac{2\pi}{2n}\right)\tan\left(\frac{3\pi}{2n}\right)\dots\tan\left(\frac{n\pi}{2n}\right)\right\}^{1/n}$$ is

(a) 1 (b) 2
(c) 3 (d) Not defined

17 If $f(x) = \frac{x-1}{x+1}, f^2(x) = f(x), \ldots, f^{k+1}(x) = f\{f^k(x)\}$, $k = 1, 2, 3, \ldots$ and $g(x) = f^{1998}(x)$, then $\int_{1/e}^{1} g(x)\,dx$ is equal to

(a) 0 (b) 1 (c) −1 (d) e

18 If $f(x)$ is a function satisfying $f'(x) = f(x)$ with $f(0) = 1$ and $g(x)$ is a function that satisfies $f(x) + g(x) = x^2$. Then, the value of $\int_0^1 f(x)g(x)\,dx$, is

(a) $e - \frac{e^2}{2} - \frac{5}{2}$ (b) $e + \frac{e^2}{2} - \frac{3}{2}$
(c) $e - \frac{e^2}{2} - \frac{3}{2}$ (d) $e + \frac{e^2}{2} + \frac{5}{2}$

19 If $n > 1$, then

Statement I $\int_0^{\infty} \frac{dx}{1+x^n} = \int_0^1 \frac{dx}{(1-x^n)^{1/n}}$

Statement II $\int_a^b f(x)\,dx = \int_a^b f(a+b-x)\,dx$

(a) Statement I is true, Statement II is true; Statement II is a correct explanation for Statement I
(b) Statement I is true, Statement II is true; Statement II is not a correct explanation for Statement I
(c) Statement I is true; Statement II is false
(d)Statement I is false; Statement II is true

20 Consider $\sin^6 x$ and $\cos^6 x$ is a periodic function with π.

Statement I $\int_0^{\pi/2} (\sin^6 x + \cos^6 x)\,dx$ lie in the interval $\left(\frac{\pi}{8}, \frac{\pi}{2}\right)$.

Statement II $\sin^6 x + \cos^6 x$ is periodic with period $\pi/2$.

(a) Statement I is true, Statement II is true; Statement II is a correct explanation for Statement I
(b) Statement I is true, Statement II is true; Statement II is not a correct explanation for Statement I
(c) Statement I is true; Statement II is false
(d)Statement I is false; Statement II is true

ANSWERS

SESSION 1

1 (b)	**2** (d)	**3** (c)	**4** (b)	**5** (b)	**6** (b)	**7** (d)	**8** (a)	**9** (a)	**10** (c)
11 (a)	**12** (b)	**13** (d)	**14** (d)	**15** (d)	**16** (d)	**17** (b)	**18** (c)	**19** (b)	**20** (c)
21 (b)	**22** (a)	**23** (c)	**24** (b)	**25** (d)	**26** (d)	**27** (c)	**28** (b)	**29** (a)	**30** (a)
31 (c)	**32** (b)	**33** (a)	**34** (c)	**35** (d)	**36** (b)	**37** (a)	**38** (a)	**39** (d)	**40** (b)
41 (c)	**42** (d)	**43** (d)	**44** (d)	**45** (a)					

SESSION 2

1 (d)	**2** (d)	**3** (c)	**4** (a)	**5** (a)	**6** (d)	**7** (a)	**8** (d)	**9** (d)	**10** (c)
11 (d)	**12** (b)	**13** (b)	**14** (b)	**15** (a)	**16** (a)	**17** (c)	**18** (c)	**19** (b)	**20** (b)

Hints and Explanations

SESSION 1

1 Let $I = \int_{\pi/4}^{3\pi/4} \frac{dx}{1+\cos x}$

$= \int_{\pi/4}^{3\pi/4} \frac{1-\cos x}{1-\cos^2 x} dx$

$= \int_{\pi/4}^{3\pi/4} \frac{1-\cos x}{\sin^2 x} dx$

$= \int_{\pi/4}^{3\pi/4} (\text{cosec}^2 x - \text{cosec}\, x \cot x) dx$

$= [-\cot x + \text{cosec}\, x]_{\pi/4}^{3\pi/4}$

$= [(1+\sqrt{2}) - (-1+\sqrt{2})] = 2$

2 Since, f is continuous function.

Let $x = t - 1$

$\therefore dx = dt$

When x tends to -3 and 5, then t tends to -2, 6.

Therefore, $\int_{-3}^{5} f(x)\,dx = \int_{-2}^{6} f(t-1)\,dt$

$= \int_{-2}^{6} f(x-1)\,dx$

3 Let $I = \int_0^{\pi/4} [\sqrt{\tan x} + \sqrt{\cot x}]\, dx$

$= \int_0^{\pi/4} \frac{\sin x + \cos x}{\sqrt{\sin x \cos x}} dx$

$= \sqrt{2} \int_0^{\pi/4} \frac{\sin x + \cos x}{\sqrt{1 - (\sin x - \cos x)^2}} dx$

Put $\sin x - \cos x = t$

$\Rightarrow (\cos x + \sin x)\, dx = dt$

$\therefore I = \sqrt{2} \int_{-1}^{0} \frac{dt}{\sqrt{1-t^2}} \Rightarrow I = \sqrt{2}\, [\sin^{-1} t]_{-1}^{0}$

$= \sqrt{2}[0 - (-\pi/2)] = \frac{\pi}{\sqrt{2}}$

4 $\int_0^1 \sin\left[2\tan^{-1}\sqrt{\frac{1+x}{1-x}}\right] dx$

Put $x = \cos\theta$, then

$\sin\left[2\tan^{-1}\sqrt{\frac{1+\cos\theta}{1-\cos\theta}}\right]$

$= \sin\left[2\tan^{-1}\left(\cot\frac{\theta}{2}\right)\right]$

$= \sin\left[2\tan^{-1}\left\{\tan\left(\frac{\pi}{2} - \frac{\theta}{2}\right)\right\}\right]$

$= \sin\left[2\left(\frac{\pi}{2} - \frac{\theta}{2}\right)\right]$

$= \sin(\pi - \theta) = \sin\theta$

$= \sqrt{1 - \cos^2\theta} = \sqrt{1 - x^2}$

$\therefore \int_0^1 \sin\left[2\tan^{-1}\sqrt{\frac{1+x}{1-x}}\right] dx$

$= \int_0^1 \sqrt{1-x^2}\, dx$

$= \left[\frac{1}{2} x\sqrt{1-x^2}\right]_0^1 + \frac{1}{2}[\sin^{-1} x]_0^1 = \frac{\pi}{4}$

5 $I_{2(n)} - I_{2(n-1)}$

$= \int_0^{\pi/2} \frac{[\sin^2 nx - \sin^2 (n-1)x]}{\sin^2 x} dx$

$= \int_0^{\pi/2} \frac{\sin(2n-1)x \cdot \sin x}{\sin^2 x} dx$

$= \int_0^{\pi/2} \frac{\sin(2n-1)x}{\sin x} dx = I_{1(n)}$

$\Rightarrow I_{2(n+1)} - I_{2(n)} = I_{1(n+1)}$

6 $\int_{-1}^{1} \{x^2 + x - 3\}\, dx = \int_{-1}^{1} \{x^2 + x\}\, dx$

$= \int_{-1}^{1} (x^2 + x - [x^2 + x])\, dx$

$= \left(\frac{x^3}{3} + \frac{x^2}{2}\right)_{-1}^{1} - \int_{-1}^{0} [x^2 + x]\, dx$

$- \int_0^{\frac{\sqrt{5}-1}{2}} [x^2 + x]\, dx$

$- \int_{(\sqrt{5}-1)/2}^{1} [x^2 + x]\, dx$

$= \frac{2}{3} + 1 - 0 - 1\left(1 - \frac{\sqrt{5}-1}{2}\right)$

$= \frac{2}{3} + \frac{\sqrt{5}-1}{2} = \frac{1}{6}(1 + 3\sqrt{5})$

7 $\int_0^2 [x^2]\, dx = \int_0^1 [x^2]\, dx$

$+ \int_1^{\sqrt{2}} [x^2]\, dx + \int_{\sqrt{2}}^{\sqrt{3}} [x^2]\, dx$

$+ \int_{\sqrt{3}}^{2} [x^2]\, dx$

$= \int_0^1 0\, dx + \int_1^{\sqrt{2}} 1\, dx$

$+ \int_{\sqrt{2}}^{\sqrt{3}} 2\, dx + \int_{\sqrt{3}}^{2} 3\, dx$

$= [x]_1^{\sqrt{2}} + [2x]_{\sqrt{2}}^{\sqrt{3}} + [3x]_{\sqrt{3}}^{2}$

$= \sqrt{2} - 1 + 2\sqrt{3} - 2\sqrt{2} + 6 - 3\sqrt{3}$

$= 5 - \sqrt{3} - \sqrt{2}$

8 $\int_{-2}^{x} |2t|\, dt = f(x)$

$= \int_{-2}^{0} |2t|dt + \int_0^x 2t\, dt$

$= -2\left[\frac{t^2}{2}\right]_{-2}^{0} + 2\left[\frac{t^2}{2}\right]_0^x$

$= -2[0 - 2] + 2\left[\frac{x^2}{2} - 0\right] = 4 + x^2$

9 Let $I = \int_{-\pi/2}^{\pi/2} \frac{\cos x}{1+e^x} dx$

$= \int_{-\pi/2}^{0} \frac{\cos x}{1+e^x} dx$

$+ \int_0^{\pi/2} \frac{\cos x}{1+e^x} dx$ ...(i)

On putting $x = -x$ in 1st integral, we get

$\int_{-\pi/2}^{0} \frac{\cos x}{1+e^x} dx = \int_0^{\pi/2} \frac{e^x \cos x}{1+e^x} dx$

$\therefore I = \int_0^{\pi/2} \frac{e^x \cos x}{1+e^x} dx + \int_0^{\pi/2} \frac{\cos x}{1+e^x} dx$

$= \int_0^{\pi/2} \frac{(1+e^x)\cos x}{(1+e^x)} dx$

$= \int_0^{\pi/2} \cos x\, dx = [\sin x]_0^{\pi/2} = 1$

10 $\int_0^3 (3ax^2 + 2bx + c)\, dx$

$= \int_1^3 (3ax^2 + 2bx + c)\, dx$

$\Rightarrow \int_0^1 (3ax^2 + 2bx + c)\, dx$

$+ \int_1^3 (3ax^2 + 2bx + c)\, dx$

$= \int_1^3 (3ax^2 + 2bx + c)\, dx$

$\Rightarrow \int_0^1 (3ax^2 + 2bx + c)\, dx = 0$

$\Rightarrow \left[\frac{3ax^3}{3} + \frac{2bx^2}{2} + cx\right]_0^1 = 0$

$\therefore a + b + c = 0$

11 Since, $\int_1^a [x] f'(x)\, dx = \int_1^2 f'(x)\, dx$

$+ \int_2^3 2f'(x)\, dx + \ldots + \int_{[a]}^{a} [a] f'(x)\, dx$

$= [f(x)]_1^2 + 2[f(x)]_2^3 + \ldots + [a][f(x)]_{[a]}^{a}$

$= f(2) - f(1) + 2f(3) - 2f(2) + \ldots$

$+ [a] f(a) - [a] f([a])$

$= [a] f(a) - \{f(1) + f(2) + \ldots + f([a])\}$

12 Let $I = \int_0^{\pi/2} \left|\sin\left(x - \frac{\pi}{4}\right)\right| dx$

$= -\int_0^{\pi/4} \sin\left(x - \frac{\pi}{4}\right) dx$

$+ \int_{\pi/4}^{\pi/2} \sin\left(x - \frac{\pi}{4}\right) dx$

$= \left[\cos\left(x - \frac{\pi}{4}\right)\right]_0^{\pi/4} - \left[\cos\left(x - \frac{\pi}{4}\right)\right]_{\pi/4}^{\pi/2}$

$= 1 - \frac{1}{\sqrt{2}} - \left(\frac{1}{\sqrt{2}} - 1\right) = 2 - \sqrt{2}$

13 $\int_0^{3\pi/2} \sin\left[\frac{2x}{\pi}\right] dx = \int_0^{\pi/2} \sin\left[\frac{2x}{\pi}\right] dx$

$+ \int_{\pi/2}^{\pi} \sin\left[\frac{2x}{\pi}\right] dx + \int_{\pi}^{3\pi/2} \sin\left[\frac{2x}{\pi}\right] dx$

$= 0 + \sin 1 \int_{\pi/2}^{\pi} dx + \sin 2 \int_{\pi}^{3\pi/2} dx$

$= \frac{\pi}{2}(\sin 1 + \sin 2)$

14 Here, $\int_0^{1.5} x[x^2]\, dx$

$I = \int_0^1 x \cdot 0\, dx + \int_1^{\sqrt{2}} x \cdot 1\, dx + \int_{\sqrt{2}}^{1.5} x \cdot 2dx$

$$= 0 + \left[\frac{x^2}{2}\right]_1^{\sqrt{2}} + [x^2]_{\sqrt{2}}^{1.5}$$

$$= \frac{1}{2}\{2-1\} + \{(1.5)^2 - 2\}$$

$$= \frac{1}{2} + 2.25 - 2 = \frac{1}{2} + \frac{1}{4} = \frac{3}{4}$$

15 Let $I = \int_0^1 \frac{8 \log(1+x)}{(1+x^2)} dx$

Put $x = \tan\theta \Rightarrow dx = \sec^2\theta d\theta$

When $x = 0 \Rightarrow \tan\theta = 0$

$\therefore \quad \theta = 0$

When $x = 1 = \tan\theta \Rightarrow \theta = \frac{\pi}{4}$

$$\therefore I = \int_0^{\pi/4} \frac{8\log(1+\tan\theta)}{1+\tan^2\theta} \cdot \sec^2\theta d\theta$$

$$= 8\int_0^{\pi/4} \log(1+\tan\theta)\, d\theta$$

$$I = 8\int_0^{\pi/4} \log(1+\tan\theta)\, d\theta \qquad \ldots\text{(i)}$$

$$= 8 \cdot \left(\frac{\pi}{2} \cdot \log 2\right)$$

$$\left[\because \int_0^{\pi/4} \log(1+\tan\theta) d\theta = \frac{\pi}{8}\log 2\right]$$

$$= \pi \log 2$$

16 Let $I = \int_0^{\pi/2} \sqrt{1-\sin 2x}\, dx$

$$= \int_0^{\pi/4} \sqrt{(\cos x - \sin x)^2}\, dx + \int_{\pi/4}^{\pi/2} \sqrt{(\sin x - \cos x)^2} dx$$

$\left[\because \cos x - \sin x > 0 \text{ when } x \in \left(0, \frac{\pi}{4}\right) \text{ and}\right.$

$\left.\cos x - \sin x < 0 \text{ when } x \in (\pi/4, \pi/2)\right]$

$$= \int_0^{\pi/4} (\cos x - \sin x)\, dx + \int_{\pi/4}^{\pi/2} (\sin x - \cos x)\, dx$$

$$= [\sin x + \cos x]_0^{\pi/4} + [-\cos x - \sin x]_{\pi/4}^{\pi/2}$$

$$= \left[\left(\sin\frac{\pi}{4} + \cos\frac{\pi}{4}\right) - (\sin 0 + \cos 0)\right] + \left[\left(-\cos\frac{\pi}{2} - \sin\frac{\pi}{2}\right) - \left(-\cos\frac{\pi}{4} - \sin\frac{\pi}{4}\right)\right]$$

$$= \left[\left(\frac{1}{\sqrt{2}} + \frac{1}{\sqrt{2}} - 1\right)\right] + \left[-1 - \left(-\frac{1}{\sqrt{2}} - \frac{1}{\sqrt{2}}\right)\right]$$

$$= \left[\frac{2}{\sqrt{2}} - 1\right] + [-1 + \sqrt{2}]$$

$$= (\sqrt{2} - 1) + (-1 + \sqrt{2}) = 2(\sqrt{2} - 1)$$

17 $\int_{e^{-1}}^{e^2} \left|\frac{\log_e x}{x}\right| dx = \int_{e^{-1}}^{1} \left|\frac{\log_e x}{x}\right| dx + \int_1^{e^2} \left|\frac{\log_e x}{x}\right| dx$

$$= \int_{e^{-1}}^{1} \frac{-\log_e x}{x} dx + \int_1^{e^2} \frac{\log_e x}{x} dx$$

$$= \int_{-1}^{0} -z\, dz + \int_0^2 z\, dz$$

[put $\log_e x = z \Rightarrow (1/x)dx = dz$]

$$\therefore I = \left[-\frac{z^2}{2}\right]_{-1}^{0} + \left[\frac{z^2}{2}\right]_0^2 = \frac{1}{2} + 2 = \frac{5}{2}$$

18 Let $I = \int_0^1 f(k-1+x)\, dx$

$\Rightarrow \quad I = \int_{k-1}^{k} f(t)\, dt$, where

$t = k - 1 + x$

$\Rightarrow \quad dt = dx$

$\Rightarrow \quad I = \int_{k-1}^{k} f(x)\, dx$

$$\therefore \sum_{k=1}^{n} \int_{k-1}^{k} f(x)\, dx$$

$$= \int_0^1 f(x)\,dx + \int_1^2 f(x)\,dx + \ldots + \int_{n-1}^{n} f(x)\, dx$$

$$= \int_0^n f(x)\, dx$$

19 $\int_{-3}^{5} f(|x|)dx = \int_{-3}^{3} f(|x|)\, dx + \int_3^5 f(|x|)dx$

$$= 2\int_0^3 f(x)\, dx + \int_3^5 f(x)\, dx$$

$$= 2\left(\int_0^1 f(x)dx + \int_1^2 f(x)dx + \int_2^3 f(x)\, dx\right) + \left(\int_3^4 f(x)\, dx + \int_4^5 f(x)\, dx\right)$$

$$= 2\left(0 + \frac{1}{2} + \frac{2^2}{2}\right) + \left(\frac{9}{2} + \frac{16}{2}\right) = \frac{35}{2}$$

20 Given that, $I_1 = \int_0^1 2^{x^2} dx$

$I_2 = \int_0^1 2^{x^3} dx,$

$I_3 = \int_1^2 2^{x^2} dx$

and $I_4 = \int_1^2 2^{x^3} dx$

Since, $2^{x^3} < 2^{x^2}$ for $0 < x < 1$

and $\quad 2^{x^3} > 2^{x^2}$ for $x > 1$

$\therefore \quad \int_0^1 2^{x^3}\, dx < \int_0^1 2^{x^2}\, dx$

and $\quad \int_1^2 2^{x^3}\, dx > \int_1^2 2^{x^2}\, dx$

$\Rightarrow \quad I_2 < I_1$ and $I_4 > I_3$

21 $I = \int_{7\pi/4}^{7\pi/3} \sqrt{\tan^2 x}\, dx$

$$= \int_{7\pi/4}^{2\pi} |\tan x|\, dx + \int_{2\pi}^{7\pi/3} |\tan x|\, dx$$

$$= -\int_{7\pi/4}^{2\pi} \tan x\, dx + \int_{2\pi}^{7\pi/3} \tan x\, dx$$

$$= -[\log \sec x]_{7\pi/4}^{2\pi} + [\log \sec x]_{2\pi}^{7\pi/3}$$

$$= -\left[\log \sec 2\pi - \log \sec\frac{7\pi}{4}\right] + \left[\log \sec\frac{7\pi}{3} - \log \sec 2\pi\right]$$

$$= -\left[\log 1 - \log \sec\frac{\pi}{4}\right] + \left[\log \sec\frac{\pi}{3} - \log 1\right]$$

$$= \log\sqrt{2} + \log 2 = \frac{1}{2}\log 2 + \log 2$$

$$= \frac{3}{2}\log 2$$

22 We have, $I = \int_0^{\pi/2} \frac{\sin^2 x}{\sin x + \cos x} dx$

$$= \int_0^{\pi/2} \frac{\sin^2\left(\frac{\pi}{2} - x\right)}{\sin\left(\frac{\pi}{2} - x\right) + \cos\left(\frac{\pi}{2} - x\right)} dx$$

$$I = \int_0^{\pi/2} \frac{\cos^2 x}{\sin x + \cos x} dx$$

Thus, $2I = \frac{1}{\sqrt{2}} \int_0^{\pi/2} \frac{dx}{\cos\left(x - \frac{\pi}{4}\right)}$

$$= \frac{1}{\sqrt{2}} \int_0^{\pi/2} \sec\left(x - \frac{\pi}{4}\right) dx$$

$$= \frac{1}{\sqrt{2}}\left[\log\left\{\sec\left(x - \frac{\pi}{4}\right) + \tan\left(x - \frac{\pi}{4}\right)\right\}\right]_0^{\pi/2}$$

$$= \frac{1}{\sqrt{2}}\left[\log\left(\sec\frac{\pi}{4} + \tan\frac{\pi}{4}\right) - \log\left\{\sec\left(-\frac{\pi}{4}\right) + \tan\left(-\frac{\pi}{4}\right)\right\}\right]$$

$$= \frac{1}{\sqrt{2}}[\log(\sqrt{2}+1) - \log(\sqrt{2}-1)]$$

$$= \frac{1}{\sqrt{2}} \log\left|\frac{\sqrt{2}+1}{\sqrt{2}-1}\right| = \frac{1}{\sqrt{2}} \log\left\{\frac{(\sqrt{2}+1)^2}{1}\right\}$$

$$= \frac{2}{\sqrt{2}} \log(\sqrt{2}+1)]$$

$$\therefore I = \frac{1}{\sqrt{2}} \log(\sqrt{2}+1)$$

23 $\lim_{n\to\infty} \frac{1}{n}\left\{\frac{1}{n+1} + \frac{2}{n+2} + \ldots + \frac{3n}{n+3n}\right\}$

$$= \lim_{n\to\infty} \sum_{r=1}^{3n} \frac{1}{n}\left(\frac{r}{n+r}\right)$$

$$= \int_0^3 \frac{x}{1+x} dx$$

$$= \int_0^3 \left(1 - \frac{1}{1+x}\right) dx$$

$$= [x - \ln(1+x)]_0^3 = 3 - \ln 4$$
$$= 3 - 2\ln 2$$

24 $\because I = \int_0^a f(x) \cdot g(x)\, dx$

$$= \int_0^a f(a-x)\, g(a-x) dx$$

$$= \int_0^a f(x)\{a - g(x)\} dx$$

$$= a\int_0^a f(x)dx - \int_0^a f(x) \cdot g(x) dx$$

$$= a\int_0^a f(x)\, dx - I$$

$$\therefore I = \frac{a}{2}\int_0^a f(x)\, dx$$

25 Let $I = \int_{-\pi/2}^{\pi/2} \frac{\sin^2 x}{1+2^x} dx$

$$\Rightarrow I = \int_{-\pi/2}^{\pi/2} \frac{\sin^2\left(\frac{-\pi}{2} + \frac{\pi}{2} - x\right) dx}{1 + 2^{\frac{-\pi}{2} + \frac{\pi}{2} - x}} dx$$

$$\left[\because \int_a^b f(x)\, dx = \int_a^b f(a+b-x)\, dx\right]$$

$\Rightarrow I = \int_{-\pi/2}^{\pi/2} \frac{\sin^2 x}{1+2^{-x}} dx$

$\Rightarrow \quad I = \int_{-\pi/2}^{\pi/2} \frac{2^x \cdot \sin^2 x}{1+2^x} dx$

$I = \int_{-\pi/2}^{\pi/2} \frac{2^x \cdot \sin^2 x}{1+2^x} dx$

$\Rightarrow 2I = \int_{-\pi/2}^{\pi/2} \sin^2 x\, dx = \int_0^{\pi/2} 2\sin^2 x\, dx$

$= \int_0^{\pi/2} (1 - \cos 2x)\, dx$

$\Rightarrow 2I = \left[x - \frac{\sin 2x}{2}\right]_0^{\pi/2} = \frac{\pi}{2}$

$\therefore \quad I = \frac{\pi}{4}$

26 Let $I = \int_0^{\pi} [\cot x]\, dx \quad \ldots(i)$

$\Rightarrow \quad I = \int_0^{\pi} [\cot(\pi - x)]\, dx$

$= \int_0^{\pi} [-\cot x]\, dx \quad \ldots(ii)$

On adding Eqs. (i) and (ii), we get

$2I = \int_0^{\pi} [\cot x]\, dx + \int_0^{\pi} [-\cot x]\, dx$

$= \int_0^{\pi} (-1)\, dx$

$\left[\because [x] + [-x] = \begin{cases} -1 \text{ if } x \notin Z \\ 0, \text{if } x \in Z \end{cases}\right]$

$= [-x]_0^{\pi} = -\pi$

$\therefore I = -\frac{\pi}{2}$

27 $I = \int_2^4 \frac{\log x^2}{\log x^2 + \log(36 - 12x + x^2)} dx$

$= \int_2^4 \frac{2\log x}{2\log x + \log(6-x)^2} dx$

$= \int_2^4 \frac{2\log x\, dx}{2[\log x + \log(6-x)]}$

$\Rightarrow I = \int_2^4 \frac{\log x dx}{[\log x + \log(6-x)]} \quad \ldots(i)$

$\Rightarrow I = \int_2^4 \frac{\log(6-x)}{\log(6-x) + \log x} dx \quad \ldots(ii)$

$\left[\because \int_a^b f(x)\, dx = \int_a^b f(a+b-x)\, dx\right]$

On adding Eqs.(i) and (ii),we get

$2I = \int_2^4 \frac{\log x + \log(6-x)}{\log x + \log(6-x)} dx$

$\Rightarrow \quad 2I = \int_2^4 dx = [x]_2^4$

$\Rightarrow \quad 2I = 2 \Rightarrow I = 1$

28 Let $I = \int_{-3\pi/2}^{-\pi/2} [(x+\pi)^3 + \cos^2 x]\, dx$

$\Rightarrow I = \int_{-3\pi/2}^{-\pi/2} \left[\left(-\frac{\pi}{2} - \frac{3\pi}{2} - x + \pi\right)^3 + \cos^2\left(-\frac{\pi}{2} - \frac{3\pi}{2} - x\right)\right] dx$

$\left(\because \int_a^b f(x)\, dx = \int_a^b f(a+b-x)\, dx\right)$

$\Rightarrow I = \int_{-3\pi/2}^{-\pi/2} [-(x+\pi)^3 + \cos^2 x]\, dx$

$\Rightarrow 2I = \int_{-3\pi/2}^{-\pi/2} 2\cos^2 x\, dx$

$= \int_{-3\pi/2}^{-\pi/2} (1 + \cos 2x)\, dx$

$= \left[x + \frac{\sin 2x}{2}\right]_{-3\pi/2}^{-\pi/2}$

$= \left[-\frac{\pi}{2} + \frac{\sin(-\pi)}{2} - \left(-\frac{3\pi}{2} + \frac{\sin(-3\pi)}{2}\right)\right] = \pi$

$\therefore \quad I = \frac{\pi}{2}$

29 Let $\phi(x) = [f(x) + f(-x)]\,[g(x) - g(-x)]$

$\therefore \phi(-x) = [f(-x) + f(x)]\,[g(-x) - g(x)] = -\phi(x)$

$\therefore \quad \int_{-\pi}^{\pi} \phi(x)\, dx = 0$

($\because \phi(x)$ is an odd function)

30 $S = \lim_{n \to \infty} \sum_{r=1}^{n} \frac{n}{(n+r)\sqrt{r(2n+r)}}$

$= \lim_{n \to \infty} \sum_{r=1}^{n} \frac{1}{n\left(1 + \frac{r}{n}\right)\sqrt{\frac{r}{n}\left(2 + \frac{r}{n}\right)}}$

$S = \int_0^1 \frac{dx}{(1+x)\sqrt{2x + x^2}}$

$= \int_0^1 \frac{dx}{(1+x)\sqrt{(1+x)^2 - 1}}$

$S = [\sec^{-1}(1+x)]_0^1$

$= \sec^{-1} 2 - \sec^{-1} 1 = \frac{\pi}{3}$

31 $P = \int_0^{3\pi} f(\cos^2 x)\, dx$ and

$Q = \int_0^{\pi} f(\cos^2 x)\, dx$

Also, $P = 3\int_0^{\pi} f(\cos^2 x) dx = 3Q$

$\therefore \quad P - 3Q = 0$

32 Let $I = \int_{-\pi}^{\pi} \frac{2x(1 + \sin x)}{1 + \cos^2 x} dx$

$= \int_{-\pi}^{\pi} \frac{2x}{1+\cos^2 x} dx + \int_{-\pi}^{\pi} \frac{2x \sin x}{1 + \cos^2 x} dx$

$\Rightarrow I = 0 + 4\int_0^{\pi} \frac{x \sin x}{1 + \cos^2 x} dx \quad \ldots(i)$

$\left[\because \frac{2x}{1+\cos^2 x} \text{ is an odd function and } \frac{2x \sin x}{1+\cos^2 x} \text{ is an even function}\right]$

$\Rightarrow I = 4\int_0^{\pi} \frac{(\pi - x)\sin(\pi - x)}{1 + \cos^2(\pi - x)} dx$

$\Rightarrow I = 4\int_0^{\pi} \frac{\pi \sin x}{1+\cos^2 x} dx - 4\int_0^{\pi} \frac{x \sin x}{1 + \cos^2 x} dx$

$\Rightarrow I = 4\pi \int_0^{\pi} \frac{\sin x}{1+\cos^2 x} dx - I$

[from Eq. (i)]

$\Rightarrow I = 2\pi \int_0^{\pi} \frac{\sin x}{1 + \cos^2 x} dx$

Put $\cos x = t$

$\Rightarrow -\sin x\, dx = dt$

$\therefore I = -2\pi \int_1^{-1} \frac{1}{1+t^2} dt$

$= 2\pi[\tan^{-1} t]_{-1}^{1} = 2\pi\left[\frac{\pi}{4} + \frac{\pi}{4}\right]$

$= \pi^2$

33 Using Newton-Leibnitz's formula, we get

$t^2 \{f(t^2)\} \left\{\frac{d}{dt}(t)^2\right\} - 0 \cdot f(0)\left\{\frac{d}{dt}(0)\right\} = 2t^4$

$\Rightarrow \quad t^2 \{f(t^2)\}\, 2t = 2t^4$

$\Rightarrow \quad f(t^2) = t$

$\Rightarrow \quad f\left(\frac{4}{25}\right) = \pm \frac{2}{5} \quad \left[\text{put } t = \pm\frac{2}{5}\right]$

$\therefore \quad f\left(\frac{4}{25}\right) = \frac{2}{5}$

[neglecting negative sign]

34 We have,

$f(x) = \frac{1}{x^2}\int_4^x [4t^2 - 2f'(t)]\, dt$

$\therefore \quad f'(x) = \frac{1}{x^2}[4x^2 - 2f'(x)] - \frac{2}{x^3}\int_4^x [4t^2 - 2f'(t)]\, dt$

$\Rightarrow \quad f'(4) = \frac{1}{16}[64 - 2f'(4)] - 0$

$\therefore \quad f'(4) = \frac{32}{9}$

35 $\lim_{x \to 0} \frac{\int_0^{x^2} \sin\sqrt{t}\, dt}{x^3} \quad \left[\text{form } \frac{0}{0}\right]$

$= \lim_{x \to 0} \frac{\sin x \cdot 2x}{3x^2}$

$= \frac{2}{3}\lim_{x \to 0} \frac{\sin x}{x}$

$= \frac{2}{3} \cdot 1 = \frac{2}{3}$

36 We have, $(1 + x^2) > x^2, \forall\, x$

$\Rightarrow \quad \sqrt{1+x^2} > x, \forall\, x \in (1, 2)$

$\Rightarrow \quad \frac{1}{\sqrt{1+x^2}} < \frac{1}{x}, \forall\, x \in (1, 2)$

$\therefore \quad \int_1^2 \frac{dx}{\sqrt{1+x^2}} < \int_1^2 \frac{dx}{x} \Rightarrow I_1 < I_2$

$\Rightarrow \quad I_2 > I_1$

37 We have, $x = \int_0^y \frac{dt}{\sqrt{1+t^2}}$

By Leibnitz rule, we get

$$1 = \frac{1}{\sqrt{1+y^2}} \cdot \frac{dy}{dx}$$

$$\Rightarrow \frac{dy}{dx} = \sqrt{1+y^2}$$

$$\Rightarrow \frac{d^2y}{dx^2} = \frac{y}{\sqrt{1+y^2}} \cdot \frac{dy}{dx}$$

$$\therefore \frac{d^2y}{dx^2} = \frac{y}{\sqrt{1+y^2}} \cdot \sqrt{1+y^2} = y$$

38 $\lim_{x\to 2} \int_6^{f(x)} \frac{4t^3}{x-2} dt = \lim_{x \to 2} \frac{\int_6^{f(x)} 4t^3\, dt}{(x-2)}$ $\left[\text{form} \frac{0}{0}\right]$

[by Leibnitz's rule]

$$= \lim_{x\to 2} \frac{4\{f(x)\}^3}{1} f'(x) = 4\{f(2)\}^3 f'(2)$$

$$= 4 \times (6)^3 \times \frac{1}{48}$$

$$\left[\because f(2) = 6 \text{ and } f'(2) = \frac{1}{48}, \text{ given}\right]$$

$= 18$

39 $\lim_{n\to\infty} \frac{\left[\begin{matrix}1 + 2^4 + 3^4 \\ + \ldots + n^4\end{matrix}\right]}{n^5} - \lim_{n\to\infty} \frac{\left[\begin{matrix}1 + 2^3 + 3^3 \\ + \ldots + n^3\end{matrix}\right]}{n^5}$

$$= \lim_{n\to\infty} \frac{1}{n} \sum_{r=1}^{n} \left(\frac{r}{n}\right)^4 - \lim_{n\to\infty} \frac{1}{n} \times \lim_{n\to\infty} \frac{1}{n} \sum_{r=1}^{n} \left(\frac{r}{n}\right)^3$$

$$= \int_0^1 x^4\, dx - \lim_{n\to\infty} \frac{1}{n} \times \int_0^1 x^3\, dx$$

$$= \left[\frac{x^5}{5}\right]_0^1 - 0 = \frac{1}{5}$$

40 Since, $x \in \left[0, \frac{\pi}{2}\right] \Rightarrow 1 \le 1 + \sin^3 x \le 2$

$$\Rightarrow \frac{1}{\sqrt{2}} \le \frac{1}{\sqrt{1+\sin^3 x}} \le 1$$

$$\Rightarrow \int_0^{\pi/2} \frac{1}{\sqrt{2}} dx \le \int_0^{\pi/2} \frac{dx}{\sqrt{1+\sin^3 x}}$$

$$\le \int_0^{\pi/2} dx$$

$$\therefore \frac{\pi}{2\sqrt{2}} \le I \le \frac{\pi}{2}$$

41 Since, $I = \int_0^1 \frac{\sin x}{\sqrt{x}} dx < \int_0^1 \frac{x}{\sqrt{x}} dx,$

because in $x \in (0, 1)$, $x > \sin x$

$$\Rightarrow I < \int_0^1 \sqrt{x}\, dx = \frac{2}{3}[x^{3/2}]_0^1$$

$$\Rightarrow I < \frac{2}{3}$$

and $J = \int_0^1 \frac{\cos x}{\sqrt{x}} dx < \int_0^1 \frac{1}{\sqrt{x}}$

$$= \int_0^1 x^{-\frac{1}{2}} dx = 2[x^{1/2}]_0^1 = 2$$

$$\therefore J < 2$$

42 Since, $f(x)$ is continuous in [0, 2].

$$\therefore \int_0^2 f(x)\,dx = \int_0^1 f(x)\,dx + \int_1^2 f(x)\,dx$$

$$= \int_0^1 x^2\, dx + \int_1^2 \sqrt{x}\, dx$$

$$= \left[\frac{x^3}{3}\right]_0^1 + \left[\frac{x^{3/2}}{3/2}\right]_1^2$$

$$= \frac{1}{3} + \frac{2}{3}(2^{3/2} - 1)$$

$$= \frac{1}{3} + \frac{4\sqrt{2}}{3} - \frac{2}{3} = \left(\frac{4\sqrt{2}-1}{3}\right)$$

43 Let $I = \int_{\pi/6}^{\pi/3} \frac{dx}{1+\sqrt{\tan x}}$ ...(i)

$$\therefore I = \int_{\pi/6}^{\pi/3} \frac{dx}{1+\sqrt{\tan\left(\frac{\pi}{2} - x\right)}}$$

$$\Rightarrow I = \int_{\pi/6}^{\pi/3} \frac{\sqrt{\tan x}\,dx}{1+\sqrt{\tan x}} \quad \text{...(ii)}$$

On adding Eqs. (i) and (ii), we get

$$2I = \int_{\pi/6}^{\pi/3} dx$$

$$\Rightarrow 2I = [x]_{\pi/6}^{\pi/3}$$

$$\Rightarrow I = \frac{1}{2}\left(\frac{\pi}{3} - \frac{\pi}{6}\right) = \frac{\pi}{12}$$

Hence, Statement I is false but $\int_a^b f(x)dx = \int_a^b f(a+b-x)\,dx$ is a true statement by property of definite integrals.

44 Since, period of $e^{\sin x}$ is 2π.

$$\therefore \int_0^{200} e^{\sin x}\, dx \ne 200\lambda$$

45 For $0 < x < 1, x > x^2$

$$\Rightarrow -x < -x^2 \Rightarrow e^{-x} < e^{-x^2}$$

$$\Rightarrow \int_0^1 e^{-x} \cos^2 x\, dx < \int_0^1 e^{-x^2} \cos^2 x\, dx$$

If $f(x) \ge g(x)$, then $\int_a^b f(x)\,dx \ge \int_a^b g(x)\,dx$

SESSION 2

1 We have, $f\left(\frac{1}{x}\right) + x^2 f(x) = 0$

$$\Rightarrow f(x) = -\frac{1}{x^2} f\left(\frac{1}{x}\right)$$

$$I = \int_{\sin\theta}^{\operatorname{cosec}\theta} f(x)\,dx$$

$$= \int_{\sin\theta}^{\operatorname{cosec}\theta} \left\{-\frac{1}{x^2} f\left(\frac{1}{x}\right)\right\} dx$$

$$= \int_{\operatorname{cosec}\theta}^{\sin\theta} f(t)\,dt, \text{ where } t = \frac{1}{x}$$

$$\Rightarrow I = -\int_{\sin\theta}^{\operatorname{cosec}\theta} f(t)\,dt = -I$$

$$\Rightarrow 2I = 0 \Rightarrow I = 0$$

2 $\frac{d}{dx} f(x) = \frac{e^{\sin x}}{x}$

$$\therefore \int_1^4 \frac{3}{x} e^{\sin x^3} dx = \int_1^4 \frac{3x^2}{x^3} e^{\sin x^3} dx$$

Put $x^3 = t$

$$\Rightarrow 3x^2\, dx = dt$$

$$\therefore f(t) = \int_1^{64} \frac{e^{\sin t}}{t} dt$$

$$= [f(t)]_1^{64}$$

$$= f(64) - f(1)$$

On comparing, we get

$$k = 64$$

3 Let $I = \int_1^2 [f\{g(x)\}]^{-1} f'\{g(x)\}\{g'(x)\}\,dx$

Put $f\{g(x)\} = z$

$\Rightarrow f'\{g(x)\}\, g'(x)\,dx = dz$

When $x = 1$, then $z = f\{g(1)\}$

When $x = 2$, then $z = f\{g(2)\}$

$$\therefore I = \int_{f\{g(1)\}}^{f\{g(2)\}} \frac{1}{z} dz = [\log z]_{f\{g(1)\}}^{f\{g(2)\}}$$

$$\Rightarrow I = \log f\{g(2)\} - \log f\{g(1)\} = 0$$

$[\because g(2) = g(1)]$

4 Put $t = \sin^2 z$ in 1st integral and $t = \cos^2 u$ in 2nd integral, we get

$dt = 2\sin z \cos z$ and

$dt = -2\cos u \sin u\,du$

$$\therefore I = \int_0^x z(2\sin z \cos z)\,dz + \int_{\pi/2}^x -u(2\cos u \sin u\,du)$$

$$= \int_0^x \underset{\mathrm{I}}{z} \underset{\mathrm{II}}{\sin 2z}\, dz - \int_{\pi/2}^x u \sin 2u\, du$$

$$= \left[-z \cdot \frac{\cos 2z}{2} + \frac{\sin 2z}{4}\right]_0^x - \left[\frac{-u\cos 2u}{2} + \frac{\sin 2u}{4}\right]_{\pi/2}^x$$

$$= \left[-x \cdot \frac{\cos 2x}{2} + \frac{\sin 2x}{4} - \{0+0\}\right] - \left[-x \cdot \frac{\cos 2x}{2} + \frac{\sin 2x}{4} - \left(\frac{\pi}{4} + 0\right)\right]$$

$$= \frac{\pi}{4}$$

5 Since, $f(x) = \int_1^x \frac{\log t}{1+t} dt$

and $F(x) = f(x) + f\left(\frac{1}{x}\right)$

$$\therefore F(e) = f(e) + f\left(\frac{1}{e}\right)$$

$$\Rightarrow F(e) = \int_1^e \frac{\log t}{1+t} dt + \int_1^{1/e} \frac{\log t}{1+t} dt$$

Put $t = \frac{1}{t}$ in second integration

$$\therefore F(e) = \int_1^e \frac{\log t}{1+t} dt + \int_1^e \frac{-\log t}{1+\frac{1}{t}} d\left(\frac{1}{t}\right)$$

$$= \int_1^e \frac{\log t}{1+t}\,dt - \int_1^e \frac{t \log t}{(1+t)} \times \left(\frac{-dt}{t^2}\right)$$

$$= \int_1^e \frac{\log t}{1+t}\,dt + \int_1^e \frac{\log t}{t(1+t)}\,dt$$

$$= \int_1^e \frac{\log t}{1+t}\,dt + \int_1^e \frac{\log t}{t}\,dt - \int_1^e \frac{\log t}{(1+t)}\,dt$$

$$= \int_1^e \frac{\log t}{t}\,dt \qquad \left[\because \frac{1}{t(1+t)} = \frac{1}{t} - \frac{1}{t+1}\right]$$

$$= \left[\frac{(\log t)^2}{2}\right]_1^e$$

$$= \frac{1}{2}[(\log e)^2 - (\log 1)^2]$$

$$= \frac{1}{2}$$

6 We have, $\int_0^n [x]dx$

$$= \int_0^1 0\,dx + \int_1^2 1\,dx + \int_2^3 2\,dx + \ldots + \int_{n-1}^n (n-1)\,dx$$

$$= 1(2-1) + 2(3-2) + 3(4-3) + \ldots + (n-1)\{n-(n-1)\}$$

$$= 1 + 2 + 3 + \ldots + (n-1)$$

$$= \frac{n(n-1)}{2}$$

and $\int_0^n \{x\}\,dx = \int_0^n (x - [x])\,dx = \frac{n}{2}$

$$\therefore \quad \frac{\int_0^n [x]\,dx}{\int_0^n \{x\}\,dx} = n - 1$$

7 Given that, $f(x) = \dfrac{e^x}{1+e^x}$

$$\therefore \quad f(a) = \frac{e^a}{1+e^a} \qquad \ldots\text{(i)}$$

and $f(-a) = \dfrac{e^{-a}}{1+e^{-a}} = \dfrac{1}{1+e^a}$ ...(ii)

On adding Eqs. (i) and (ii), we get

$$f(a) + f(-a) = 1$$

$\Rightarrow \quad f(a) = 1 - f(-a)$

Let $\quad f(-a) = t$

$\Rightarrow \quad f(a) = 1 - t$

Now, $I_1 = \int_t^{1-t} xg\,[x(1-x)]\,dx$ ...(iii)

$$\left[\because I = \int_a^b f(x)\,dx = \int_a^b f(a+b-x)\,dx\right]$$

$\Rightarrow \quad I_1 = \int_t^{1-t} (1-x)g[x(1-x)]\,dx$ ...(iv)

On adding Eqs. (iii) and (iv), we get

$2I_1 = \int_t^{1-t} g[x(1-x)]\,dx = I_2$ [given]

$$\therefore \quad \frac{I_2}{I_1} = 2$$

8 $\displaystyle \lim_{x\to\infty} \frac{\left(\int_0^x e^{t^2}dt\right)^2}{\int_0^x e^{2t^2}dt}$

$$= \lim_{x\to\infty} \frac{\left(2\int_0^x e^{t^2}\,dt\right)(e^{x^2})}{e^{2x^2}} \qquad \left[\text{form } \frac{0}{0}\right]$$

$$= \lim_{x\to\infty} \frac{2\int_0^x e^{t^2}\,dt}{e^{x^2}} \left[\text{form } \frac{0}{0}\right]$$

$$= \lim_{x\to\infty} \frac{2e^{x^2}}{2xe^{x^2}} = \lim_{x\to\infty} \frac{1}{x} = 0$$

9 Use the formula,

$|x-a| = \begin{cases} x-a, & x \ge a \\ -(x-a) & x < a \end{cases}$ to break given integral in two parts and then integrate separately.

$$\int_0^\pi \sqrt{\left(1 - 2\sin\frac{x}{2}\right)^2}\,dx = \int_0^\pi \left|1 - 2\sin\frac{x}{2}\right|\,dx$$

$$= \int_0^{\frac{\pi}{3}} \left(1 - 2\sin\frac{x}{2}\right)dx - \int_{\frac{\pi}{3}}^{\pi} \left(1 - 2\sin\frac{x}{2}\right)dx$$

$$= \left(x + 4\cos\frac{x}{2}\right)_0^{\pi/3} - \left(x + 4\cos\frac{x}{2}\right)_{\frac{\pi}{3}}^{\pi}$$

$$= 4\sqrt{3} - 4 - \frac{\pi}{3}$$

10 Let $\quad I = \int_0^\pi x f(\sin x)\,dx$ ...(i)

$\Rightarrow \quad I = \int_0^\pi (\pi - x) f[\sin(\pi - x)]\,dx$

$\Rightarrow \quad I = \int_0^\pi (\pi - x) f(\sin x)\,dx$ ...(ii)

On adding Eqs. (i) and (ii), we get

$$2I = \int_0^\pi \pi f(\sin x)\,dx$$

$\Rightarrow I = \dfrac{\pi}{2}\int_0^\pi f(\sin x)\,dx$ ...(iii)

$$\left[\because \int_0^{2a} f(x)\,dx = \begin{cases} 2\int_0^a f(x)\,dx, & \text{if } f(2a-x) = f(x) \\ 0 & \text{, if } f(2a-x) = -f(x) \end{cases}\right]$$

$$\Rightarrow I = \pi \int_0^{\pi/2} f(\sin x)\,dx$$

Put $\dfrac{\pi}{2} - x = t \Rightarrow x = \dfrac{\pi}{2} - t$

Put $dx = -dt$ in Eq. (iii), we get

$$I = \frac{\pi}{2}\int_{-\pi/2}^{\pi/2} f(\cos t)\,dt$$

$$= \frac{\pi}{2}\int_{-\pi/2}^{\pi/2} f(\cos x)\,dx$$

$$= \pi\int_0^{\pi/2} f(\cos x)\,dx$$

[$\because f(\cos x)$ is an even function]

11 On differentiate the given interval by using Newton-Leibnitz formula,

we get $\quad f'(x) = e^{-(x^2+1)^2} \cdot \left\{\dfrac{d}{dx}(x^2+1)\right\}$

$$- e^{-(x^2)^2} \cdot \left\{\frac{d}{dx}(x)^2\right\}$$

$$= e^{-(x^2+1)^2} \cdot 2x - e^{-(x^2)^2} \cdot 2x$$

$$= 2xe^{-(x^4+2x^2+1)}(1 - e^{2x^2+1})$$

For $f'(x) > 0$,

then $2x(1 - e^{2x^2+1}) > 0$

$\Rightarrow \quad 2x < 0$

$\Rightarrow \quad x < 0$

12 Let $I = \lim_{n\to\infty} \left(\dfrac{(n+1)\cdot(n+2)\ldots(3n)}{n^{2n}}\right)^{\frac{1}{n}}$

$$= \lim_{n\to\infty} \left(\frac{(n+1)\cdot(n+2)\ldots(n+2n)}{n^{2n}}\right)^{\frac{1}{n}}$$

$$= \lim_{n\to\infty} \left(\left(\frac{n+1}{n}\right)\left(\frac{n+2}{n}\right)\ldots\left(\frac{n+2n}{n}\right)\right)^{\frac{1}{n}}$$

On taking log on both sides, we get

$$\log I = \lim_{n\to\infty} \frac{1}{n}\left[\log\left\{\left(1+\frac{1}{n}\right)\left(1+\frac{2}{n}\right)\ldots\left(1+\frac{2n}{n}\right)\right\}\right]$$

$$\Rightarrow \quad \log I = \lim_{n\to\infty} \frac{1}{n}\left[\log\left(1+\frac{1}{n}\right) + \log\left(1+\frac{2}{n}\right) + \ldots + \log\left(1+\frac{2n}{n}\right)\right]$$

$$\Rightarrow \quad \log I = \lim_{n\to\infty} \frac{1}{n}\sum_{r=1}^{2n} \log\left(1+\frac{r}{n}\right)$$

$$\Rightarrow \quad \log I = \int_0^2 \log(1+x)\,dx$$

$$\Rightarrow \quad \log I = \left[\log(1+x) \cdot x - \int \frac{1}{1+x} \cdot x\,dx\right]_0^2$$

$$\Rightarrow \log I = [\log(1+x) \cdot x]_0^2 - \int_0^2 \frac{x+1-1}{1+x}\,dx$$

$$\Rightarrow \log I = 2 \cdot \log 3 - \int_0^2 \left(1 - \frac{1}{1+x}\right)dx$$

$$\Rightarrow \quad \log I = 2 \cdot \log 3 - [x - \log|1+x|]_0^2$$

$$\Rightarrow \quad \log I = 2 \cdot \log 3 - [2 - \log 3]$$

$$\Rightarrow \quad \log I = 3 \cdot \log 3 - 2$$

$$\Rightarrow \quad \log I = \log 27 - 2$$

$$\Rightarrow \quad I = e^{\log 27 - 2} = 27 \cdot e^{-2} = \frac{27}{e^2}$$

13 We have, $f'(x) = 3\sin x + 4\cos x$

Since, in $\left[\dfrac{5\pi}{4}, \dfrac{4\pi}{3}\right]$, $f'(x) < 0$, so assume the least value at the point $x = \dfrac{4\pi}{3}$.

Thus,

$$f\left[\frac{4\pi}{3}\right] = \int_{5\pi/4}^{4\pi/3} (3\sin u + 4\cos u)\,du$$

$= [-3\cos u + 4\sin u]_{5\pi/4}^{4\pi/3}$

$= \dfrac{3}{2} - 2\sqrt{3} + \dfrac{1}{\sqrt{2}}$

14 We have, $g(x) = \int_0^x f(t)\,dt$

$\Rightarrow \quad g(2) = \int_0^2 f(t)\,dt$

$\Rightarrow \; g(2) = \int_0^1 f(t)dt + \int_1^2 f(t)\,dt$

We know that, $m \le f(x) \le M$ for $x \in [a, b]$

$\Rightarrow m(b-a) \le \int_a^b f(x)\,dx \le M(b-a)$

$\therefore \quad \dfrac{1}{2} \le f(t) \le 1$, for $t \in [0, 1]$

and $0 \le f(t) \le \dfrac{1}{2}$, for $t \in [1, 2]$

$\Rightarrow \; \dfrac{1}{2}(1-0) \le \int_0^1 f(t)\,dt \le 1\,(1-0)$

and $0\,(2-1) \le \int_1^2 f(t)\,dt \le \dfrac{1}{2}(2-1)$

$\Rightarrow \dfrac{1}{2} \le \int_0^1 f(t)\,dt \le 1$ and $0 \le \int_1^2 f(t)\,dt \le \dfrac{1}{2}$

$\Rightarrow \; \dfrac{1}{2} \le \int_0^1 f(t)\,dt + \int_1^2 f(t)\,dt \le \dfrac{3}{2}$

$\therefore \quad \dfrac{1}{2} \le g(2) \le \dfrac{3}{2}$

15 Let $f(a) = \int_0^1 |x^n - a^n|\,dx$

$= \int_0^a (a^n - x^n)\,dx + \int_a^1 (x^n - a^n)\,dx$

$= \left[a^n x - \dfrac{x^{n+1}}{n+1}\right]_0^a + \left[\dfrac{x^{n+1}}{n+1} - a^n \cdot x\right]_a^1$

$= \left[a^{n+1} - \dfrac{a^{n+1}}{n+1}\right] + \left[\dfrac{1}{n+1} - a^n - \dfrac{a^{n+1}}{n+1} + a^{n+1}\right]$

$= 2a^{n+1} - \dfrac{2a^{n+1}}{n+1} - a^n + \dfrac{1}{n+1}$

$= 2a^{n+1}\left[\dfrac{n}{n+1}\right] - a^n + \dfrac{1}{n+1}$

$\Rightarrow \; f'(a) = n(2a-1)a^{n-1}$

Thus, only critical point in $(0, 1)$ is $a = 1/2$

Also, $f'(a) < 0$ for $a \in \left(0, \dfrac{1}{2}\right)$

and $f'(a) > 0$ for $a \in \left(\dfrac{1}{2}, 1\right)$.

$\therefore \quad f(a)$ is minimum for $a = \dfrac{1}{2}$.

16 Let $P = \lim_{n\to\infty} \left\{\prod_{r=1}^{n} \tan\left(\dfrac{r\pi}{2n}\right)\right\}^{1/n}$

$\therefore \; \ln P = \lim_{n\to\infty} \sum_{r=1}^{n} \dfrac{1}{n} \ln\tan\left(\dfrac{r\pi}{2n}\right)$

$= \int_0^1 \ln\tan\left(\dfrac{\pi x}{2}\right) dx$

$\Rightarrow \; \ln P = \dfrac{2}{\pi}\int_0^{\pi/2} \ln\tan x\,dx \quad \ldots(\text{i})$

and $\ln P = \dfrac{2}{\pi}\int_0^{\pi/2} \ln\cot x\,dx \quad \ldots(\text{ii})$

On adding Eqs. (i) and (ii), we get

$2\ln P = \dfrac{2}{\pi}\int_0^{\pi/2} (\ln\tan x + \ln\cot x)dx$

$= \dfrac{2}{\pi}\int_0^{\pi/2} \ln 1\,dx = 0$

$\Rightarrow \quad \ln P = 0$

$\therefore \quad P = 1$

17 We have, $f(x) = \dfrac{x-1}{x+1}$

$\Rightarrow \; f^2(x) = f[f(x)] = f\left(\dfrac{x-1}{x+1}\right)$

$= \dfrac{\dfrac{x-1}{x+1} - 1}{\dfrac{x-1}{x+1} + 1} = -\dfrac{1}{x}$

$\Rightarrow \; f^4(x) = f^2[f^2(x)] = f^2\left(-\dfrac{1}{x}\right)$

$= \dfrac{-1}{-\dfrac{1}{x}} = x$

$\therefore \quad g(x) = f^{1998}(x) = f^2 o f^{1996}(x)$

$\Rightarrow \quad g(x) = f^2[f^{1996}(x)]$

$\Rightarrow \quad g(x) = f^2(x) \quad [\because f^{1996}(x)$

$= \{\underbrace{(f^4 o f^4 o f^4 o \ldots f^4)}_{499 \text{ times}}\}(x) = x]$

$\Rightarrow \quad g(x) = -\dfrac{1}{x}$

$\therefore \int_{1/e}^1 g(x)\,dx = \int_{1/e}^1 \left(-\dfrac{1}{x}\right)dx$

$= -[\log_e x]_{1/e}^1$

$\Rightarrow \quad \int_{1/e}^1 g(x)\,dx = -\left[\log_e 1 - \log_e \dfrac{1}{e}\right]$

$= -[0 + 1] = -1$

18 Given, $f'(x) = f(x)$

and $\quad f(0) = 1$

Let $\quad f(x) = e^x \quad \ldots(\text{i})$

Also, $\; f(x) + g(x) = x^2$

$\Rightarrow \quad g(x) = x^2 - e^x \quad \ldots(\text{ii})$

Now, $\int_0^1 f(x)g(x)\,dx$

$= \int_0^1 e^x(x^2 - e^x)\,dx$

[from Eqs. (i) and (ii)]

$= \int_0^1 x^2 e^x\,dx - \int_0^1 e^{2x}\,dx$

$= [x^2 e^x - \int 2xe^x\,dx]_0^1 - \dfrac{1}{2}[e^{2x}]_0^1$

$= [x^2 e^x - 2xe^x + 2e^x]_0^1 - \dfrac{1}{2}(e^2 - 1)$

$= [(x^2 - 2x + 2)e^x]_0^1 - \dfrac{1}{2}e^2 + \dfrac{1}{2}$

$= [(1 - 2 + 2)e^1 - (0 - 0 + 2)e^0]$

$- \dfrac{1}{2}e^2 + \dfrac{1}{2}$

$= e - 2 - \dfrac{e^2}{2} + \dfrac{1}{2} = e - \dfrac{e^2}{2} - \dfrac{3}{2}$

19 In LHS, put $x^n = \tan^2\theta$

$\Rightarrow \quad nx^{n-1}dx = 2\tan\theta\sec^2\theta\,d\theta$

$\therefore \int_0^\infty \dfrac{dx}{1 + x^n} = \dfrac{2}{n}\int_0^{\pi/2} \tan^{1-2+2/n}\theta\,d\theta$

$= \dfrac{2}{n}\int_0^{\pi/2} \tan^{(2/n)-1}\theta\,d\theta$

In RHS, put $x^n = \sin^2\theta$

$\Rightarrow \quad nx^{n-1}dx = 2\sin\theta\cos\theta\,d\theta$

$\therefore \int_0^1 \dfrac{dx}{(1 - x^n)^{1/n}} = \dfrac{2}{n}\int_0^{\pi/2} \dfrac{1}{\cos^{2/n}\theta}$

$\sin^{\frac{2}{n}-1}\theta\cos\theta\,d\theta = \dfrac{2}{n}\int_0^{\pi/2} \tan^{(2/n)-1}\theta\,d\theta$

20 $\sin^6 x + \cos^6 x = (\sin^2 x)^3 + (\cos^2 x)^3$

$= (\sin^2 x + \cos^2 x)^3 - 3\sin^2 x\cos^2 x(\sin^2 x + \cos^2 x)$

$= 1 - 3\sin^2 x\cos^2 x$

$= 1 - \dfrac{3}{4}\sin^2 2x \quad \left[\because \text{period} = \dfrac{\pi}{2}\right]$

So, the least and greatest value of $\sin^6 x + \cos^6 x$ are $\dfrac{1}{4}$ and 1.

Hence, $\left(\dfrac{\pi}{2} - 0\right) \times \dfrac{1}{4}$

$< \int_0^{\pi/2} (\sin^6 x + \cos^6 x)dx < \left(\dfrac{\pi}{2} - 0\right) \times 1$

$\therefore \; \dfrac{\pi}{8} < \int_0^{\pi/2} (\sin^6 x + \cos^6 x)\,dx < \dfrac{\pi}{2}$

DAY SEVENTEEN

Area Bounded by the Curves

Learning & Revision for the Day

- Curve Area
- Area between a Curve and Lines
- Area between Two Curves

Curve Area

The space occupied by a continuous curve, which is bounded under the certain conditions, is called **curve area** or the area of bounded by the curve.

Area between a Curve and Lines

1. The area of region shown in the following figure, bounded by the curve $y = f(x)$ defined on $[\alpha, \beta]$, X-axis and the lines $x = \alpha$ and $x = \beta$ is given by $\int_{\alpha}^{\beta} y\,dx$ or $\int_{\alpha}^{\beta} f(x)\,dx$.

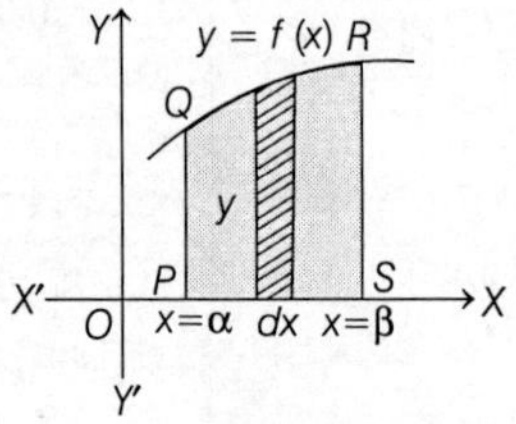

2. If the curve $y = f(x)$ lies below X-axis, then area of region bounded by the curve $y = f(x)$, X-axis and the lines $x = \alpha$ and $x = \beta$ will be negative as shown in the following figure. So, we consider the area as $\left|\int_{\alpha}^{\beta} y\,dx\right|$ or $\left|\int_{\alpha}^{\beta} f(x)\,dx\right|$

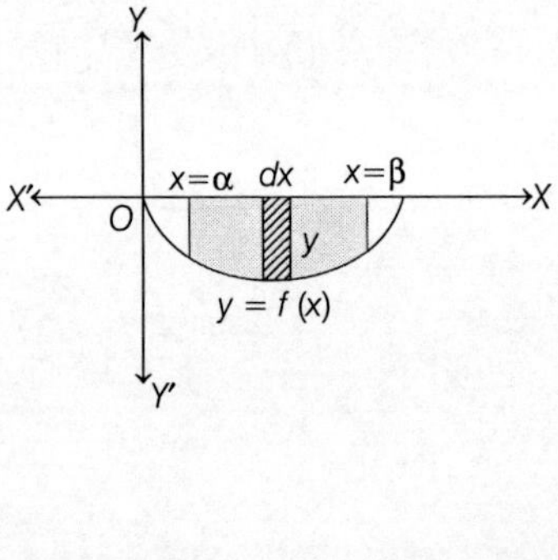

PREP MIRROR
Your Personal Preparation Indicator

- No. of Questions in Exercises (x)—
- No. of Questions Attempted (y)—
- No. of Correct Questions (z)—
  (Without referring Explanations)

- Accuracy Level ($z/y \times 100$)—
- Prep Level ($z/x \times 100$)—

In order to expect good rank in JEE, your Accuracy Level should be above 85 & Prep Level should be above 75.

3. Area of region shown in the following figure, bounded by the curve $x = f(y)$, Y-axis and the lines $y = c$ and $y = d$ is given by $\int_c^d f(y)\,dy$ or $\int_c^d x\,dy$

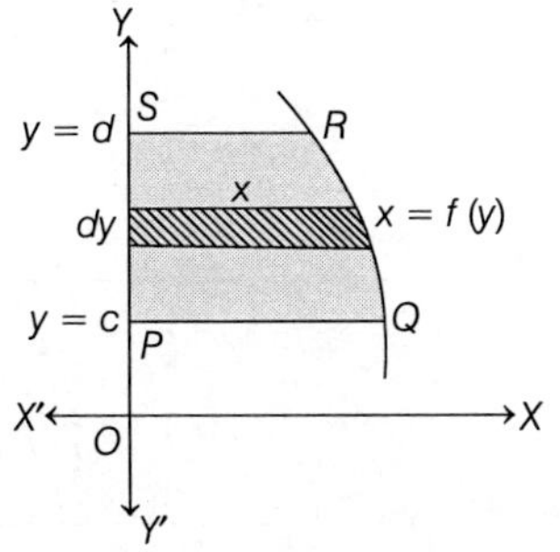

If the position of the curve under consideration is on the left side of Y-axis, then area is given by $\left|\int_c^d f(y)\,dy\right|$.

4. If the curve crosses the X-axis number of times, the area of region shown in the following figure, enclosed between the curve $y = f(x)$, X-axis and the lines $x = \alpha$ and $x = \beta$ is given by

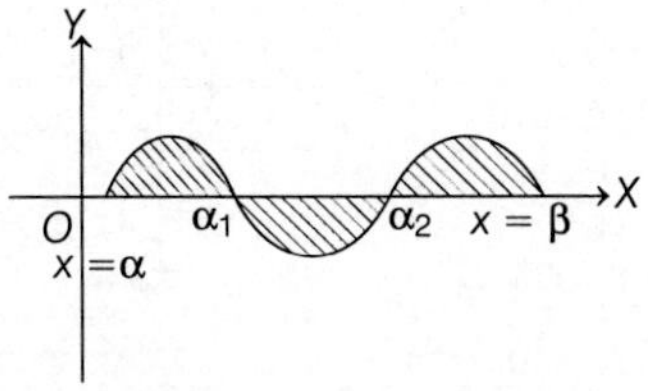

$$\int_\alpha^{\alpha_1} f(x)\,dx + \left|\int_{\alpha_1}^{\alpha_2} f(x)\,dx\right| + \int_{\alpha_2}^{\beta} f(x)\,dx$$

Area between Two Curves

1. (i) Area of region shown in the following figure, bounded between the curves $y = f(x)$, $y = g(x)$, where $f(x) \le g(x)$, and the lines $x = a$, $x = b$ $(a < b)$ is given by

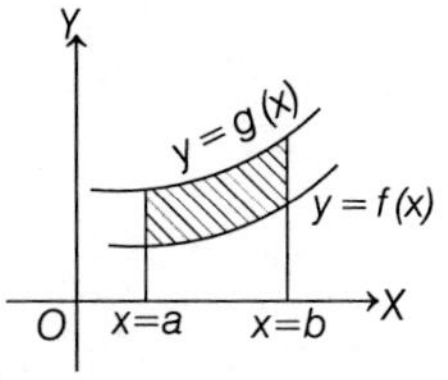

$$\text{Area} = \int_a^b [g(x) - f(x)]\,dx$$

(ii) If $f(x) \ge g(x)$ in $[a, c]$ and $f(x) \le g(x)$ in $[c, b]$ where $a < c < b$, then Area

$$= \int_a^c [f(x) - g(x)]\,dx + \int_c^b [g(x) - f(x)]\,dx$$

2. Area of region shown in the following figure, bounded by the curves $y = f(x)$, $y = g(x)$, X-axis and lines $x = a$, $x = b$ is given by

$$\text{Area} = \int_a^c f(x)\,dx + \int_c^b g(x)\,dx$$

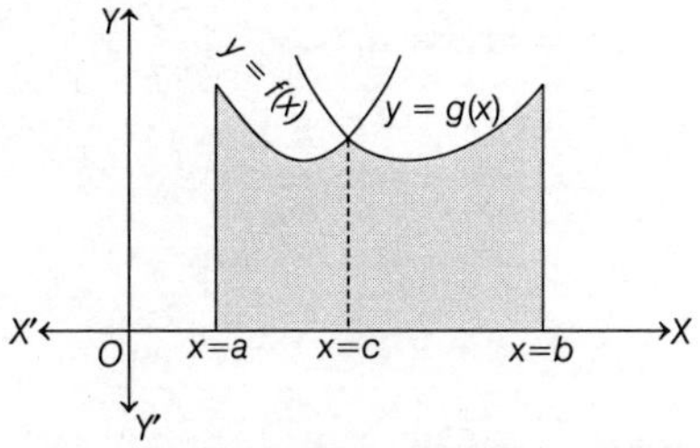

Area and the shape of some important curves

S.No.	Curves	Point of intersection	Area of shaded region	
(i)	$f(x, y): \frac{x^2}{a^2} + \frac{y^2}{b^2} \le 1, \frac{x}{a} + \frac{y}{b} \ge 1$ $\Rightarrow$ or $\frac{x^2}{a^2} + \frac{y^2}{b^2} \le 1 \le \frac{x}{a} + \frac{y}{b}$	$A(a, 0), B(0, b)$	Area $= ab\frac{(\pi - 2)}{4}$ sq units	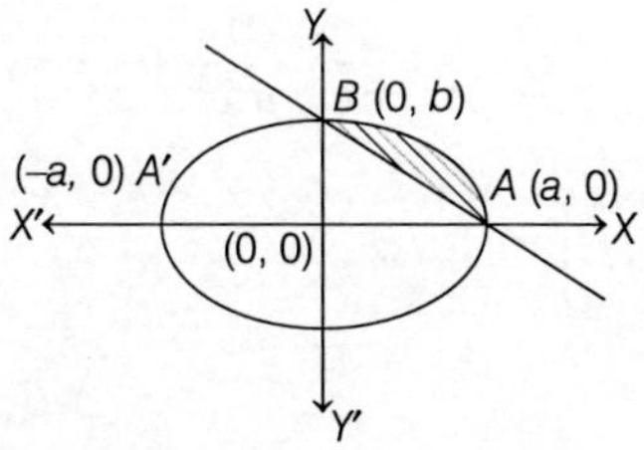
(ii)	Parabola $y^2 = 4ax$ and its latusrectum $x = a$	$A(a, 2a), B(a, -2a)$	Area $= \frac{8}{3}a^2$ sq units	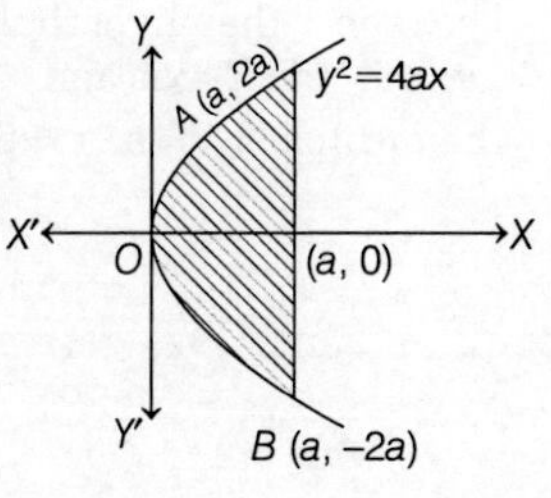

(iii)	$f(x, y): y^2 = 4ax$ and $y = \|mx\|$	$O(0, 0)$, $A\left(\dfrac{4a}{m^2}, \dfrac{4a}{m}\right)$	Area $= \dfrac{8a^2}{3m^3}$ sq units	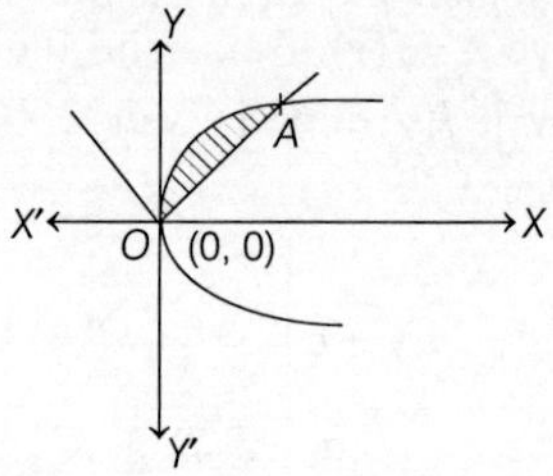
(iv)	$f(x, y): x^2 = 4ay$ $y^2 = 4bx$	$O(0, 0)$, $A\,(4a^{2/3}\, b^{1/3}, 4a^{1/3} b^{2/3})$	Area $= \dfrac{16}{3}(ab)$ sq units	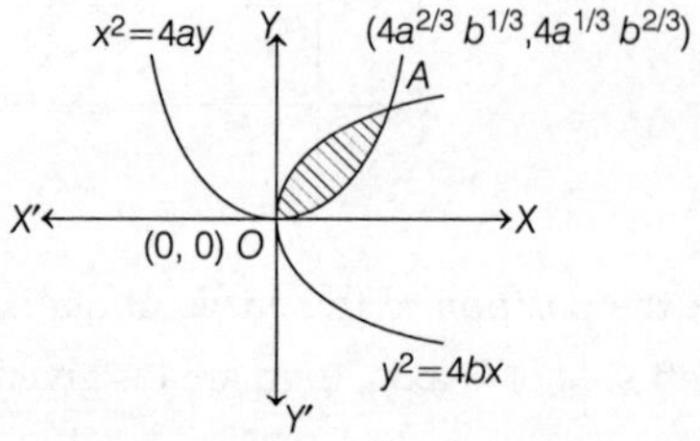
(v)	Area bounded by $y^2 = 4a\,(x + a)$ and $y^2 = 4b\,(b - x)$	$A(b - a, 2\sqrt{ab})$ $B(b - a, -2\sqrt{ab})$	Area $= \dfrac{8}{3}\sqrt{ab}\,(a + b)$ sq units	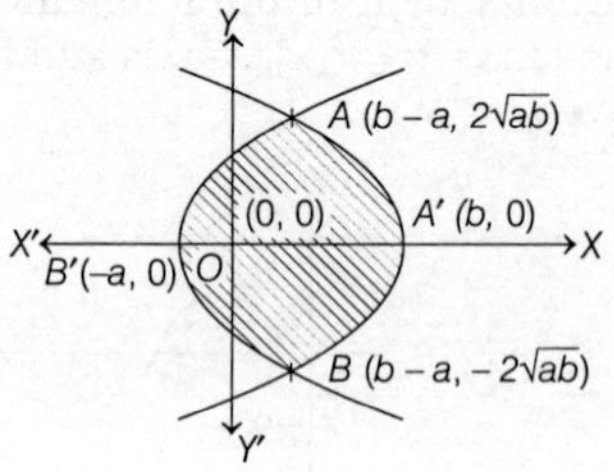
(vi)	Common area bounded by the ellipses $\dfrac{x^2}{b^2} + \dfrac{y^2}{a^2} = \dfrac{1}{a^2b^2}$ and $\dfrac{x^2}{a^2} + \dfrac{y^2}{b^2} = \dfrac{1}{a^2b^2}, 0 < a < b$	$x = y = \dfrac{1}{\sqrt{a^2 + b^2}}$ $x = y = -\dfrac{1}{\sqrt{a^2 + b^2}}$	Area = Area of region $PQRS$ $= 4 \times$ Area of $OLQM$ $= \dfrac{4}{ab}\tan^{-1}\left(\dfrac{a}{b}\right)$ sq units	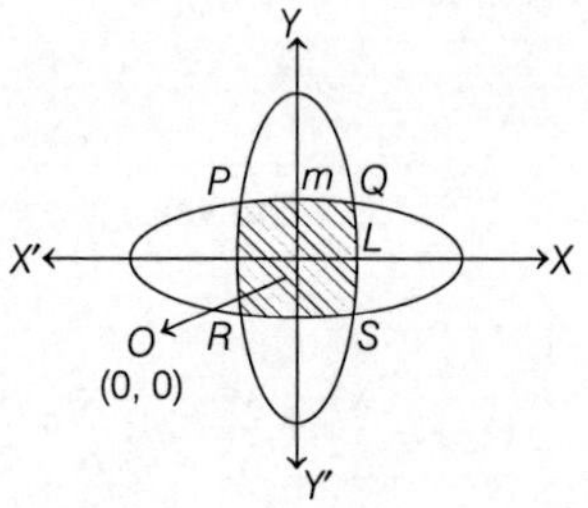
(vii)	If $\alpha, \beta > 0$, $\alpha > \beta$ the area between the hyperbola $xy = p^2$, the X-axis and the ordinates $x = \alpha$, $x = \beta$	—	Area $= p^2 \log\left(\dfrac{\alpha}{\beta}\right)$	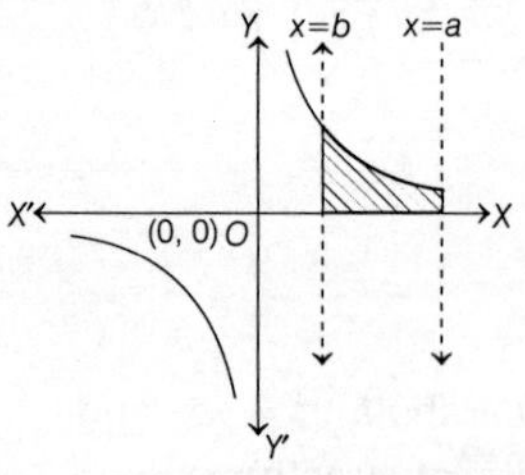

DAY PRACTICE SESSION 1

FOUNDATION QUESTIONS EXERCISE

1 Let $f:[-1,2]\rightarrow[0,\infty)$ be a continuous function such that $f(x)=f(1-x)$ for all $x\in[-1,2]$. Let $R_1=\int_{-1}^{2} x\,f(x)\,dx$, and R_2 be the area of region bounded by $y=f(x), x=-1$, $x=2$ and the X-axis. Then

(a) $R_1=2R_2$ (b) $R_1=3R_2$ (c) $2R_1=R_2$ (d) $3R_1=R_2$

2 Let A_1 be the area of the parabola $y^2=4ax$ lying between the vertex and the latus rectum and A_2 be the area between the latus rectum and the double ordinate $x=2a$. Then A_1/A_2 is

(a) $(2\sqrt{2}-1)/7$ (b) $(2\sqrt{2}+1)/7$
(c) $(2\sqrt{2}+1)$ (d) $(2\sqrt{2}-1)$

3 The area of smaller segment cut off from the circle $x^2+y^2=9$ by $x=1$ is

(a) $\frac{1}{2}(9\sec^{-1}3-\sqrt{8})$ (b) $9\sec^{-1}(3)-\sqrt{8}$
(c) $\sqrt{8}-9\sec^{-1}(3)$ (d) None of these

4 The ratio of the area bounded by the curves $y=\cos x$ and $y=\cos 2x$ between $x=0, \pi/3$ and X-axis, is

(a) $\sqrt{2}:1$ (b) $1:1$ (c) $1:2$ (d) $2:1$

5 The area bounded by the curve $y=f(x)$, X-axis and the lines $x=1$, $x=b$ is $(\sqrt{b^2+1}-\sqrt{2})$ for all $b>1$, then $f(x)$ equals to

(a) $\sqrt{x^2+1}$ (b) $\sqrt{x+1}$ (c) $x/\sqrt{x^2+1}$ (d) None of these

6 If the ordinate $x=a$ divides the area bounded by the curve $y=\left(1+\frac{8}{x^2}\right)$, X-axis and the ordinates $x=2, 4$ into two equal parts, then a is equal to

(a) 8 (b) $2\sqrt{2}$ (c) 2 (d) $\sqrt{2}$

7 Let the straight line $x=b$ divide the area enclosed by $y=(1-x)^2$, $y=0$ and $x=0$ into two parts $R_1(0\le x\le b)$ and $R_2(b\le x\le 1)$ such that $R_1-R_2=1/4$. Then b equals

(a) 3/4 (b) 1/2 (c) 1/3 (d) 1/4

8 The curve $y=a\sqrt{x}+bx$ passes through the point (1, 2) and the area enclosed by the curve, the X-axis and the line $x=4$ is 8 square units. Then $a-b$ is equal to

(a) 2 (b) -1 (c) -2 (d) 4

9 If $y=f(x)$ makes positive intercepts of 2 and 1 unit on x and y-coordinates axes and encloses an area of $\frac{3}{4}$ sq unit with the axes, then $\int_0^2 xf'(x)\,dx$, is

(a) $\frac{3}{2}$ (b) 1 (c) $\frac{5}{4}$ (d) $-\frac{3}{4}$

10 The area of the region (in sq units), in the first quadrant, bounded by the parabola $y=9x^2$ and the lines $x=0$, $y=1$ and $y=4$, is → JEE Mains 2013

(a) $\frac{12}{9}$ (b) $\frac{14}{3}$ (c) $\frac{7}{3}$ (d) $\frac{14}{9}$

11 The area bounded by the curve $y=\ln(x)$ and the lines $y=0$, $y=\ln(3)$ and $x=0$ is equal to → JEE Mains 2013

(a) 3 (b) $3\ln(3)-2$ (c) $3\ln(3)+2$ (d) 2

12 The area of the region bounded by the curve $ay^2=x^3$, the Y-axis and the lines $y=a$ and $y=2a$, is → NCERT Exemplar

(a) $\frac{3}{5}a^2(2\cdot 2^{2/3}-1)$ sq unit (b) $\frac{2}{5}a\ (2^{2/3}-1)$ sq unit
(c) $\frac{3}{5}a^2(2^{2/3}+1)$ sq unit (d) None of these

13 The area of the region bounded by
$y=|x-3|$ and $y=2$, is

(a) 4.5 sq units (b) 6.3 sq units
(c) 3.5 sq units (d) None of these

14 The area between the curve $y=4-|x|$ and X-axis is

(a) 16 sq units (b) 20 sq units
(c) 12 sq units (d) 18 sq units

15 The area under the curve $y=|\cos x-\sin x|, 0\le x\le\frac{\pi}{2}$ and above X-axis is → JEE Mains 2013

(a) $2\sqrt{2}$ (b) $2\sqrt{2}-2$ (c) $2\sqrt{2}+2$ (d) 0

16 The area of the region enclosed by the curves $y=x$, $x=e$, $y=\frac{1}{x}$ and the positive X-axis is

(a) 1 sq unit (b) $\frac{3}{2}$ sq units (c) $\frac{5}{2}$ sq units (d) $\frac{1}{2}$ sq unit

17 The area bounded by $y=|\sin x|$, X-axis and the lines $|x|=\pi$ is

(a) 2 sq units (b) 3 sq units
(c) 4 sq units (d) None of these

18 For $0\le x\le\pi$, the area bounded by $y=x$ and $y=x+\sin x$, is

(a) 2 (b) 4
(c) 2π (d) 4π

19 The area (in sq unit) of the region described by $\{(x,y): y^2\le 2x \text{ and } y\ge 4x-1\}$ is → JEE Mains 2015

(a) $\frac{7}{32}$ (b) $\frac{5}{34}$
(c) $\frac{15}{64}$ (d) $\frac{9}{32}$

20 The area bounded by the curves $y^2 = 4x$ and $x^2 = 4y$, is

(a) 0 (b) $\frac{32}{3}$ (c) $\frac{16}{3}$ (d) $\frac{8}{3}$

21 If the area bounded by $y = ax^2$ and $x = ay^2$, $a > 0$ is 1, then a is equal to

(a) 1 (b) $\frac{1}{\sqrt{3}}$

(c) $\frac{1}{3}$ (d) None of these

22 The area bounded between the parabolas $x^2 = \frac{y}{4}$ and $x^2 = 9y$ and the straight line $y = 2$, is

(a) $20\sqrt{2}$ (b) $\frac{10\sqrt{2}}{3}$ (c) $\frac{20\sqrt{2}}{3}$ (d) $10\sqrt{2}$

23 The area enclosed the curves $y = x^3$ and $y = \sqrt{x}$ is

(a) $\frac{5}{3}$ sq units (b) $\frac{5}{4}$ sq units

(c) $\frac{5}{12}$ sq unit (d) $\frac{12}{5}$ sq units

24 The area (in sq units) of the region $\{(x, y): y^2 \ge 2x$ and $x^2 + y^2 \le 4x, x \ge 0, y \ge 0\}$ is → JEE Mains 2016

(a) $\pi - \frac{4}{3}$ (b) $\pi - \frac{8}{3}$

(c) $\pi - \frac{4\sqrt{2}}{3}$ (d) $\frac{\pi}{2} - \frac{2\sqrt{2}}{3}$

25 The area of the region described by $A = \{(x, y): x^2 + y^2 \le 1$ and $y^2 \le 1 - x\}$ is → JEE Mains 2014

(a) $\frac{\pi}{2} + \frac{4}{3}$ (b) $\frac{\pi}{2} - \frac{4}{3}$ (c) $\frac{\pi}{2} - \frac{2}{3}$ (d) $\frac{\pi}{2} + \frac{2}{3}$

26 The parabola $y^2 = 2x$ divides the circle $x^2 + y^2 = 8$ in two parts. Then, the ratio of the areas of these parts is

(a) $\frac{3\pi - 2}{10\pi + 2}$ (b) $\frac{3\pi + 2}{9\pi - 2}$

(c) $\frac{6\pi - 3}{11\pi - 5}$ (d) $\frac{2\pi - 9}{9\pi + 2}$

27 The area bounded by the curves $y = 2 - |2 - x|$ and $|x| y = 3$ is

(a) $(5 - 4 \ln 2)/3$ (b) $(2 - \ln 3)/2$

(c) $(4 - 3 \ln 3)/2$ (d) None of these

28 The area bounded by the curves $y = |x| - 1$ and $y = -|x| + 1$, is

(a) 1 sq unit (b) 2 sq units

(c) $2\sqrt{2}$ sq units (d) 4 sq units

29 The area of the smaller region bounded by the circle $x^2 + y^2 = 1$ and the lines $|y| = x + 1$ is

(a) $(\pi - 2)/4$ (b) $(\pi - 2)/2$

(c) $(\pi + 2)/2$ (d) None of these

30 The area (in sq units) bounded by the curves $y = \sqrt{x}$, $2y - x + 3 = 0$, X-axis and lying in the first quadrant is → JEE Mains 2013

(a) 9 (b) 36

(c) 18 (d) $\frac{27}{4}$

31 The area bounded by the curves $y = x^2$ and $y = 2/(1 + x^2)$ is

(a) $(3\pi + 2)/3$ (b) $(3\pi - 2)/3$

(c) $(3\pi - 2)/6$ (d) None of these

32 The area bounded by $y = \tan x$, $y = \cot x$, X-axis in $0 \le x \le \frac{\pi}{2}$ is

(a) $3\log 2$ (b) $\log 2$

(c) $2\log 2$ (d) None of these

33 Area of the region bounded by curves $x = 1/2$, $x = 2$, $y = \log_e x$ and $y = 2^x$ is equal to $(4 - \sqrt{2})/\log 2 + b - c\log 2$. Then, $b + c$ equals

(a) 2 (b) -1

(c) 4 (d) None of these

34 The area bounded by the curve $y = (x + 1)^2$, $y = (x - 1)^2$ and the line $y = \frac{1}{4}$ is

(a) $\frac{1}{6}$ sq unit (b) $\frac{2}{3}$ sq unit

(c) $\frac{1}{4}$ sq unit (d) $\frac{1}{3}$ sq unit

35 The area of bounded region by the curve $y = \log_e x$ and $y = (\log_e x)^2$, is

(a) $3 - e$ (b) $e - 3$

(c) $\frac{1}{2}(3 - e)$ (d) $\frac{1}{2}(e - 3)$

DAY PRACTICE SESSION 2

PROGRESSIVE QUESTIONS EXERCISE

1 Let $f(x) = x^2$, $g(x) = \cos x$ and $h(x) = f(g(x))$. Then area bounded by the curve $y = h(x)$, and X-axis between $x = x_1$ and $x = x_2$, where x_1, x_2 are roots of the equation $18x^2 - 9\pi x + \pi^2 = 0$, is

(a) $\pi/12$ (b) $\pi/6$ (c) $\pi/3$ (d) None of these

2 The area bounded by $f(x) = \min(|x|, |x-1|, |x+1|)$ in $[-1, 1]$ and X-axis, is (in sq unit)

(a) $\frac{1}{5}$ (b) $\frac{1}{4}$ (c) $\frac{1}{3}$ (d) $\frac{1}{2}$

3 A point $P(x, y)$ moves in such a way that $[|x|] + [|y|] = 1$, $[\cdot]$ = G.I.F. Area of the region representing all possible positions of the point P is equal to

(a) 8 (b) 4 (c) 16 (d) None of these

4 The area bounded by the curve $xy^2 = 4(2 - x)$ and Y-axis is

(a) 2π (b) 4π (c) 12π (d) 6π

5 The area of the region $R = \{(x, y) : |x| \le |y|$ and $x^2 + y^2 \le 1\}$ is

(a) $\frac{3\pi}{8}$ sq units (b) $\frac{5\pi}{8}$ sq units (c) $\frac{\pi}{2}$ sq units (d) $\frac{\pi}{8}$ sq unit

6 Area of the region bounded by the parabola $(y-2)^2 = x - 1$, the tangent to it at the point with the ordinate 3, and the X-axis is given by

(a) 9/2 sq units (b) 9 sq units
(c) 18 sq units (d) None of these

7 Let the circle $x^2 + y^2 = 4$ divide the area bounded by tangent and normal at $(1, \sqrt{3})$ and X-axis in A_1 and A_2. Then $\frac{A_1}{A_2}$ equals to

(a) $\pi/(3\sqrt{3} - \pi)$ (b) $\pi/(3\sqrt{3} + \pi)$
(c) $\pi/(3 - \pi\sqrt{3})$ (d) None of these

8 The area (in sq units) of the region $\{(x, y) : x \ge 0, x + y \le 3, x^2 \le 4y$ and $y \le 1 + \sqrt{x}\}$ is

→ JEE Mains 2017

(a) $\frac{7}{3}$ (b) $\frac{5}{2}$ (c) $\frac{59}{12}$ (d) $\frac{3}{2}$

9 The area enclosed between the curves $y = \log_e(x + e)$, $x = \log(1/y)$ and the X-axis is equal to

(a) $2e$ (b) 2 (c) $2/e$ (d) None of these

10 Let $f(x)$ be a real valued function satisfying the relation $f(x/y) = f(x) - f(y)$ and $\lim_{x \to 0} \frac{f(1+x)}{x} = 3$. The area bounded by the curve $y = f(x)$, y-axis and the line $y = 3$ is equal to

(a) e (b) $2e$ (c) $3e$ (d) None of these

11 The area bounded by the lines $y = 2$, $x = 1$, $x = a$ and the curve $y = f(x)$, which cuts the last two lines above the first line for all $a \ge 1$, is equal to $\frac{2}{3}[(2a)^{3/2} - 3a + 3 - 2\sqrt{2}]$.

Then $f(x) =$

(a) $2\sqrt{2x}, x \ge 1$ (b) $\sqrt{2x}, x \ge 1$
(c) $2\sqrt{x}, x \ge 1$ (d) None of these

12 The area bounded by the curve $y = \cos^{-1}(\sin x) - \sin^{-1}(\cos x)$ and the lines $y = 0$, $x = \frac{3\pi}{2}$, $x = 2\pi$ is

(a) $\frac{\pi^2}{4}$ (b) $\frac{\pi^2}{2}$ (c) π^2 (d) $2\pi^2$

13 Let $f(x)$ be continuous function such that $f(0) = 1$, $f(x) - f(x/7) = x/7 \ \forall x \in R$. The area bounded by the curve $y = f(x)$ and the coordinate axes is

(a) 2 (b) 3 (c) 6 (d) 9

14 If area bounded by the curves $y = x - bx^2$ and $by = x^2$ is maximum, then b is equal to

(a) 1 (b) -1 (c) $b = \pm 1$ (d) None of these

15 Let A_n be the area bounded by the curve $y = (\tan x)^n$ and the lines $x = 0$ and $x = \pi/4$. Then, for $n > 2$

(a) $\frac{1}{2n} < A_n < \frac{1}{2n-2}$ (b) $\frac{1}{2n+1} < A_n < \frac{1}{2n-1}$
(c) $\frac{1}{2n+2} < A_n < \frac{1}{2n-2}$ (d) $\frac{1}{2n+2} < A_n < \frac{1}{2n}$

ANSWERS

SESSION 1

1 (c)	**2** (b)	**3** (b)	**4** (d)	**5** (c)	**6** (b)	**7** (b)	**8** (d)	**9** (d)	**10** (d)
11 (d)	**12** (a)	**13** (d)	**14** (a)	**15** (b)	**16** (b)	**17** (c)	**18** (a)	**19** (d)	**20** (c)
21 (b)	**22** (c)	**23** (c)	**24** (b)	**25** (a)	**26** (b)	**27** (c)	**28** (b)	**29** (b)	**30** (a)
31 (b)	**32** (b)	**33** (c)	**34** (d)	**35** (a)					

SESSION 2

1 (a)	**2** (d)	**3** (a)	**4** (b)	**5** (c)	**6** (b)	**7** (a)	**8** (b)	**9** (b)	**10** (c)
11 (a)	**12** (a)	**13** (b)	**14** (c)	**15** (c)					

Hints and Explanations

SESSION 1

1 Given, $R_1 = \int_{-1}^{2} x\, f(x)\, dx$,

and $R_2 = \int_{-1}^{2} f(x)\, dx$ and $f(1-x) = f(x)$

Consider, $R_1 = \int_{-1}^{2} x\, f(x)\, dx$

$= \int_{-1}^{2} (1-x)\, f(1-x)\, dx$

$= \int_{-1}^{2} (1-x)\, f(x)\, dx$

$= R_2 - R_1$

$\therefore\ R_2 = 2R_1$

2 $A_1 = 2\int_0^a 2\sqrt{a}\sqrt{x}\, dx$

$= 4\sqrt{a} \times \frac{2}{3}[x^{3/2}]_0^a = 8a^2/3$

$A_2 = 2\int_a^{2a} 2\sqrt{a}\sqrt{x}\, dx$

$= 4\sqrt{a} \times \frac{2}{3}[x^{3/2}]_a^{2a} = \frac{8a^2}{3}(2\sqrt{2}-1)$

$\therefore\ A_1/A_2 = \frac{1}{2\sqrt{2}-1} = \frac{(2\sqrt{2}+1)}{7}$

3 Required area, $A = 2\int_1^3 \sqrt{9-x^2}\, dx$

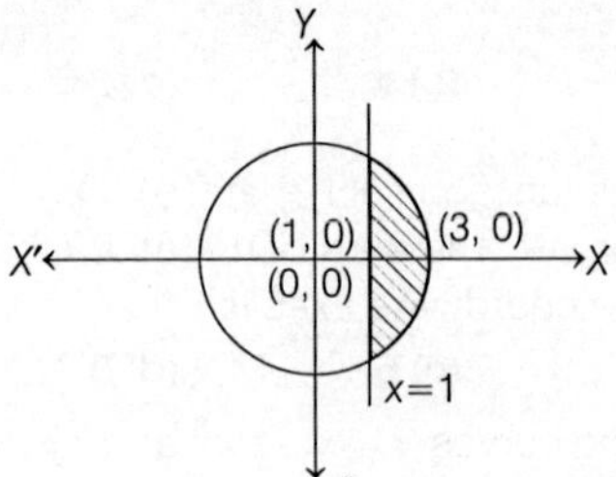

$= 2\cdot\frac{1}{2}\left[x\sqrt{9-x^2} + 9\sin^{-1}\frac{x}{3}\right]_1^3$

$= \left[9\,\frac{\pi}{2} - \sqrt{8} - 9\sin^{-1}\left(\frac{1}{3}\right)\right]$

$= \left[9\left(\frac{\pi}{2} - \sin^{-1}\left(\frac{1}{3}\right)\right) - \sqrt{8}\right]$

$= \left[9\cos^{-1}\left(\frac{1}{3}\right) - \sqrt{8}\right]$

$= [9\sec^{-1}(3) - \sqrt{8}]$

4 Here, $A_1 = \int_0^{\pi/3} \cos x\, dx$

$= [\sin x]_0^{\pi/3} = \frac{\sqrt{3}}{2}$

and $A_2 = \int_0^{\pi/3} \cos 2x\, dx$

$= \left[\frac{\sin 2x}{2}\right]_0^{\pi/3} = \frac{\sqrt{3}}{4}$

$\therefore\ A_1 : A_2 = 2 : 1$

5 $\int_1^b f(x)\, dx = \sqrt{b^2+1} - \sqrt{2}$

Now, on differentiating both sides w.r.t. b, we get

$\Rightarrow\quad f(b) = \frac{2b}{2\sqrt{b^2+1}}\ \forall\, b > 1$

$\Rightarrow\quad f(x) = x/\sqrt{x^2+1}.$

6 Area of region $AMNB$

$= \int_2^4 \left(1 + \frac{8}{x^2}\right) dx = \left[x - \frac{8}{x}\right]_2^4 = 4$

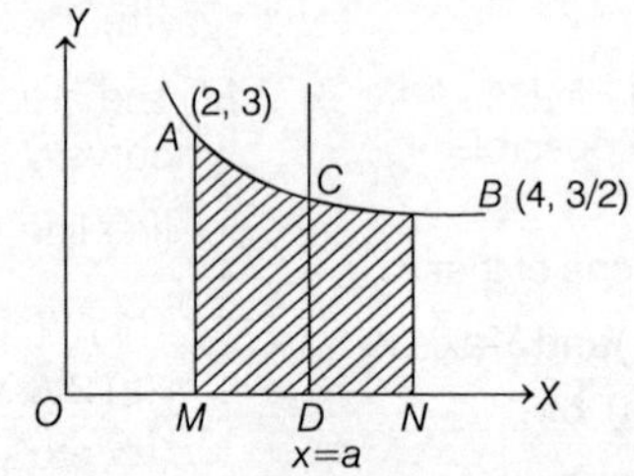

Area of region, $ACDM$

$= \int_2^a \left(1 + \frac{8}{x^2}\right) dx \Rightarrow \left[x - \frac{8}{x}\right]_2^a = 2$

$\Rightarrow\ a = \pm 2\sqrt{2} \Rightarrow a = 2\sqrt{2}$ $\quad[\because a > 0]$

7 We have, $R_1 - R_2 = \int_0^b (1-x)^2\, dx - \int_b^1 (1-x)^2\, dx = \frac{1}{4}$

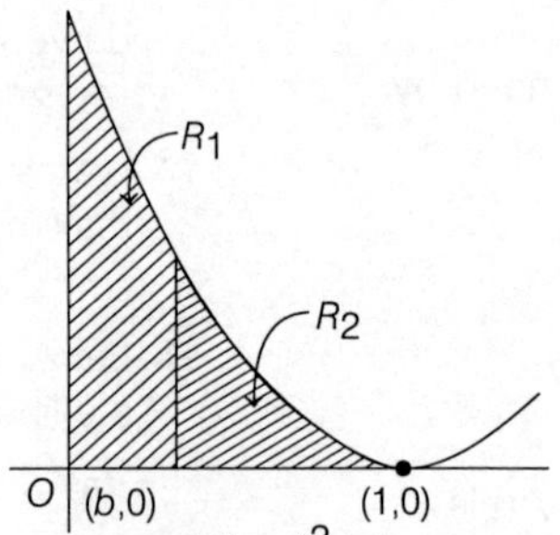

$\Rightarrow\ \frac{1}{3} - \frac{2(1-b)^3}{3} = \frac{1}{4} \Rightarrow (1-b)^3 = \frac{1}{8}$

$\therefore\ b = \frac{1}{2}$

8 $y = a\sqrt{x} + bx\ (x \ge 0)$.

At $x = 1$, $y = 2$, we get $2 = a + b$ $\quad$ ...(i)

$\int_0^4 (a\sqrt{x} + bx)\, dx = 8$

$\Rightarrow\quad \frac{16a}{3} + 8b = 8$ $\quad$ ...(ii)

On solving Eqs. (i) and (ii), we get $a = 3$ and $b = -1$

$\therefore\quad a - b = 4$

9 Clearly, $y = f(x)$ passes through (2, 0) and (0, 1).

$\therefore\ 0 = f(2)$ and $1 = f(0)$

Also, $\int_0^2 f(x)\, dx = \frac{3}{4}$ $\quad$ [given]

Now, $\int_0^2 x f'(x)\, dx = [xf(x)]_0^2 - \int_0^2 f(x)\, dx$

$\Rightarrow\ \int_0^2 x f'(x)\, dx = [2f(2) - 0f(0)] - \frac{3}{4}$

$\Rightarrow\ \int_0^2 x f'(x)\, dx = 2 \times 0 - 0 \times 1 - \frac{3}{4}$

$= -\frac{3}{4}$

10 Required area $= \int_1^4 \frac{\sqrt{y}}{3}\, dy = \frac{1}{3}\left[\frac{y^{3/2}}{3/2}\right]_1^4$

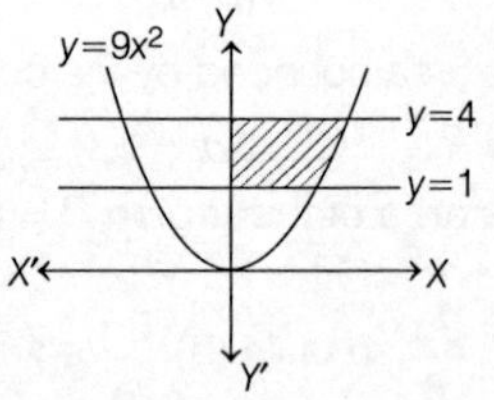

$= \frac{2}{9}(4^{3/2} - 1^{3/2}) = \frac{2}{9} \times 7 = \frac{14}{9}$

11 Required area

$= \int_0^{\log 3} x\, dy = \int_0^{\log 3} e^y\, dy$

$= [e^y]_0^{\log 3} = [e^{\log 3} - e^0] = 3 - 1 = 2$

12 We have,

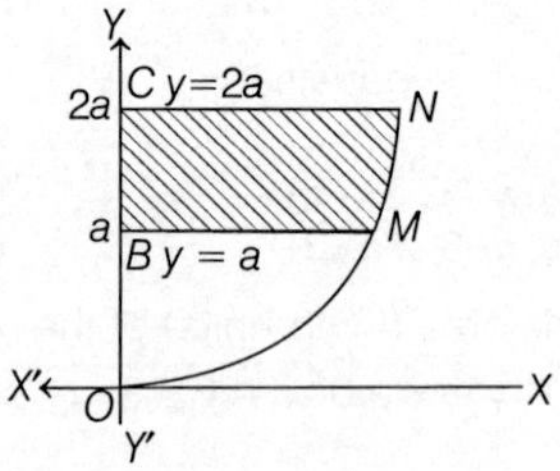

Required area =

Area $BMNC = \int_a^{2a} x\, dy$

$= \int_a^{2a} a^{1/3} y^{2/3}\, dy = \frac{3a^{\frac{1}{3}}}{5}[y^{5/3}]_a^{2a}$

$= \frac{3a^{\frac{1}{3}}}{5}\left((2a)^{\frac{5}{3}} - a^{\frac{5}{3}}\right)$

$= \frac{3}{5}a^{\frac{1}{3}}a^{\frac{5}{3}}\left((2)^{\frac{5}{3}} - 1\right)$

$= \frac{3}{5}a^2\left(2\cdot 2^{\frac{2}{3}} - 1\right)$ sq unit

13 The curve $y = |x-3|$ meets the line $y = 2$, when $x = 1$ and $x = 5$

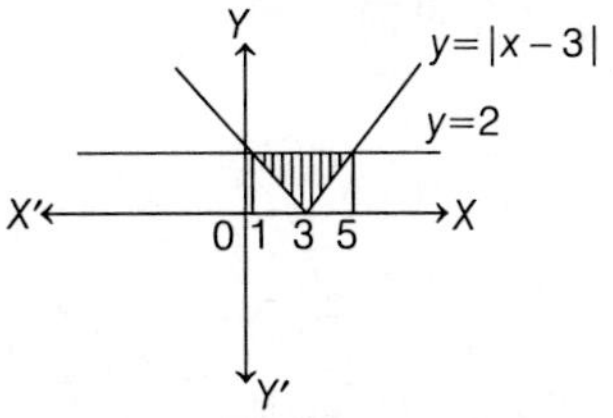

The area of the shaded region

$$= \frac{1}{2} \times 2 \times 4 = 4 \text{ sq units}$$

14 $y = 4 - |x|$ represents two curves as,

$$y = \begin{cases} 4 + x, & x < 0 \\ 4 - x, & x > 0 \end{cases}$$

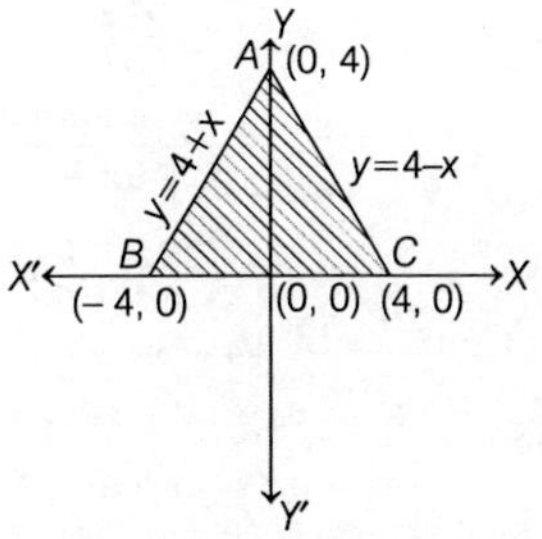

The area of the shaded portion

$$= \frac{1}{2} \times 8 \times 4 = 16 \text{ sq units}$$

15 Required area

$$= \int_0^{\pi/2} |\cos x - \sin x| dx$$

$$= \int_0^{\pi/4} (\cos x - \sin x) dx + \int_{\pi/4}^{\pi/2} (\sin x - \cos x) dx$$

$$= [\sin x + \cos x]_0^{\pi/4} + [-\cos x - \sin x]_{\pi/4}^{\pi/2}$$

$$= (\sqrt{2} - 1) - (1 - \sqrt{2}) = 2\sqrt{2} - 2$$

16 Given, $y = x$, $x = e$ and $y = \frac{1}{x}$, $x \geq 0$

Since, $y = x$ and $x \geq 0 \Rightarrow y \geq 0$

$\therefore$ Area to be calculated in 1st quadrant shown in figure

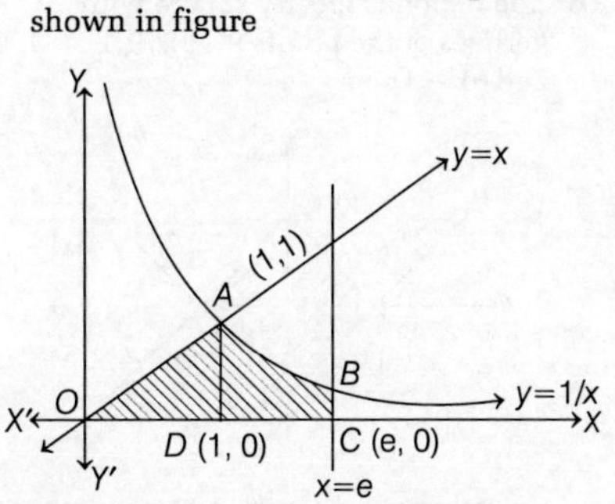

$\therefore$ Required area = Area of ΔODA + Area of $DABCD$

$$= \frac{1}{2}(1 \times 1) + \int_1^e \frac{1}{x} dx = \frac{1}{2} + [\log |x|]_1^e$$

$$= \frac{1}{2} + [\log|x|]_1^e = \frac{1}{2} + 1 = \frac{3}{2} \text{ sq units}$$

17 We have,

$$y = |\sin x| = \begin{cases} \sin x, & \text{if } x \geq 0 \\ -\sin x, & \text{if } x < 0 \end{cases}$$

and $|x| = \pi \Rightarrow x = \pm \pi$

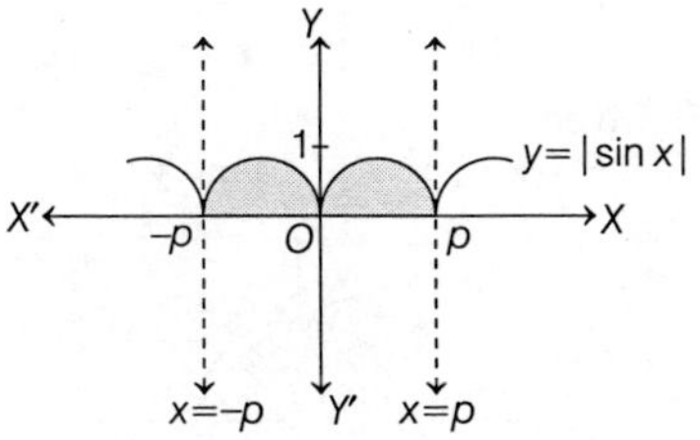

Now, required area $= 2\int_2^{\pi} \sin x dx$

$$= 2[-\cos x]_0^{\pi}$$

$$= -2[\cos \pi - \cos 0]$$

$$= -2[-1 - 1] = 4 \text{ sq units}$$

18 Given, curves $y = x$ and $y = x + \sin x$, which intersect at $(0, 0)$ and (π, π).

$$\therefore \text{ Area, } A = \int_0^{\pi} (x + \sin x) dx - \int_0^{\pi} x dx$$

$$= \int_0^{\pi} \sin x \, dx = [-\cos x]_0^{\pi}$$

$$= -\cos \pi + \cos 0$$

$$= -(-1) + 1 = 2$$

19 Given region is

$$\{(x, y): y^2 \leq 2x \text{ and } y \geq 4x - 1\}$$

$y^2 \leq 2x$ represents a region inside the parabola $y^2 = 2x$ ...(i)

and $y \geq 4x - 1$ represents a region to the left of the line $y = 4x - 1$ ...(ii)

The point of intersection of the curves (i) and (ii) is given by

$$(4x - 1)^2 = 2x \Rightarrow 16x^2 + 1 - 8x = 2x$$

$$\Rightarrow 16x^2 - 10x + 1 = 0$$

$$\Rightarrow \quad x = \frac{1}{2}, \frac{1}{8}$$

So, the points where these curves intersect are $\left(\frac{1}{2}, 1\right)$ and $\left(\frac{1}{8}, -\frac{1}{2}\right)$

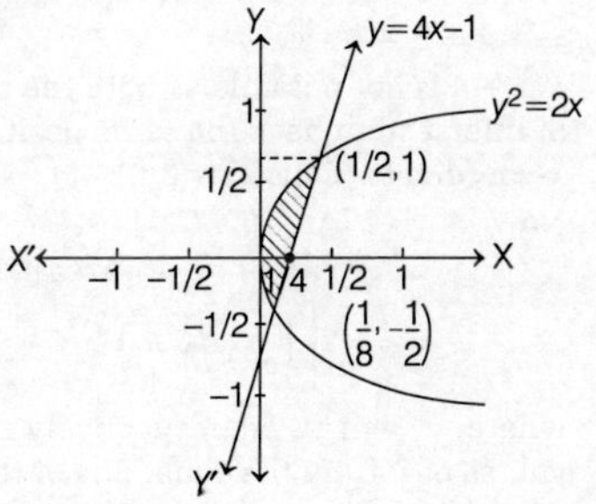

Hence, required area

$$= \int_{-\frac{1}{2}}^{1} \left(\frac{y+1}{4} - \frac{y^2}{2} \right) dy$$

$$= \frac{1}{4}\left(\frac{y^2}{2} + y \right)_{-1/2}^{1} - \frac{1}{6}(y^3)_{-1/2}^{1}$$

$$= \frac{1}{4}\left\{\left(\frac{1}{2} + 1\right) - \left(\frac{1}{8} - \frac{1}{2}\right)\right\} - \frac{1}{6}\left\{1 + \frac{1}{8}\right\}$$

$$= \frac{1}{4}\left\{\frac{3}{2} + \frac{3}{8}\right\} - \frac{1}{6}\left\{\frac{9}{8}\right\}$$

$$= \frac{1}{4} \times \frac{15}{8} - \frac{3}{16} = \frac{9}{32}$$

20 For the point of intersection of

$$y^2 = 4x \text{ and } x^2 = 4y$$

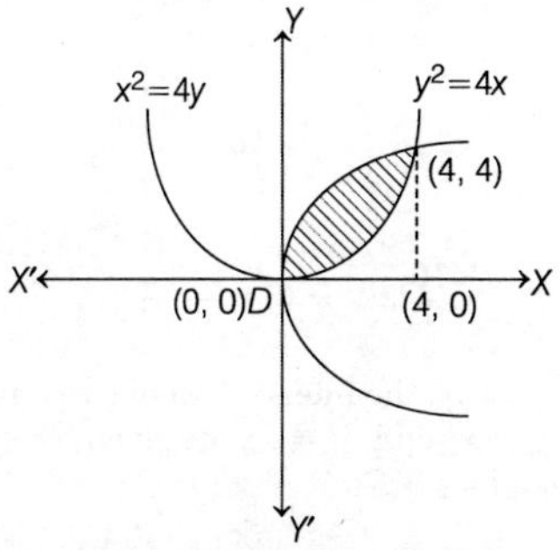

Substitute $y = \frac{x^2}{4}$ in $y^2 = 4x$

$$\Rightarrow \left(\frac{x^2}{4}\right)^2 = 4x \Rightarrow x^4 = 4^3 x \Rightarrow x = 0, 4$$

$\therefore$ Area bounded between curves

$$= \int_0^4 \left(\sqrt{4x} - \frac{x^2}{4} \right) dx$$

$$= \left[2 \cdot \frac{x^{3/2}}{\frac{3}{2}} - \frac{x^3}{12} \right]_0^4$$

$$= \frac{4}{3} \cdot (4)^{3/2} - \frac{(4)^3}{12}$$

$$= \frac{32}{3} - \frac{16}{3} = \frac{16}{3}$$

21 The intersection point of two curves is $\left(\frac{1}{a}, \frac{1}{a}\right)$.

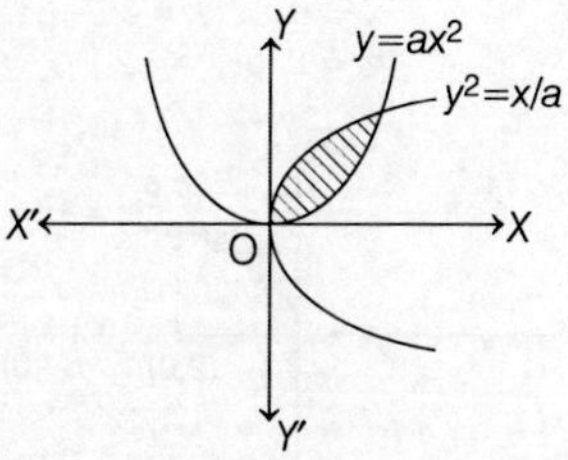

$$\therefore \text{ Area, } A = \int_0^{1/a} \left(\sqrt{\frac{x}{a}} - ax^2 \right) dx$$

$$\Rightarrow \quad 1 = \frac{1}{\sqrt{a}} \cdot \frac{2}{3}[x^{3/2}]_0^{1/a} - \frac{a}{3}[x^3]_0^{1/a}$$

$$\Rightarrow \quad a^2 = \frac{1}{3} \Rightarrow a = \frac{1}{\sqrt{3}}$$

22

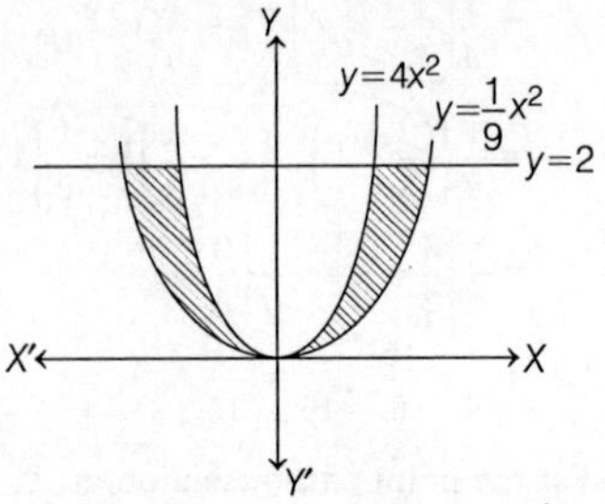

$$\text{Required area} = 2\int_0^2 \left(3\sqrt{y} - \frac{\sqrt{y}}{2}\right) dy$$

$$= 2\int_0^2 \left(\frac{5}{2}\sqrt{y}\right) dy = 5\left[\frac{y^{3/2}}{\frac{3}{2}}\right]_{y=0}^{y=2}$$

$$= \frac{10}{3}(2^{3/2} - 0) = \frac{20\sqrt{2}}{3}$$

23 Clearly, the intersection of two curves $y = x^3$ and $y = \sqrt{x}$ are given by $x = 0$ and $x = 1$.

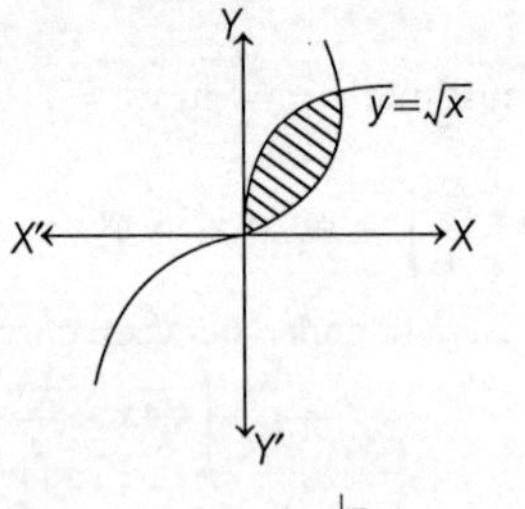

$$\therefore A = \left|\int_0^1 (x^3 - \sqrt{x})\, dx\right| = \left|\left[\frac{x^4}{4} - \frac{2x^{3/2}}{3}\right]_0^1\right|$$

$$= \left|\left[\frac{1}{4} - \frac{2}{3}\right]\right| = \frac{5}{12} \text{ sq unit}$$

24 We have, $x^2 + y^2 \le 4x$ and $y^2 \ge 2x$

To find point of intersection, substitute $y^2 = 2x$ in $x^2 + y^2 = 4x$

$$x^2 + y^2 = 4x \Rightarrow x^2 + 2x = 4x$$

$$\Rightarrow \quad x^2 - 2x = 0 \quad \Rightarrow x(x-2) = 0$$

$$\Rightarrow \quad x = 0 \text{ or } x = 2$$

$$\Rightarrow \quad y = 0 \text{ or } y = 2$$

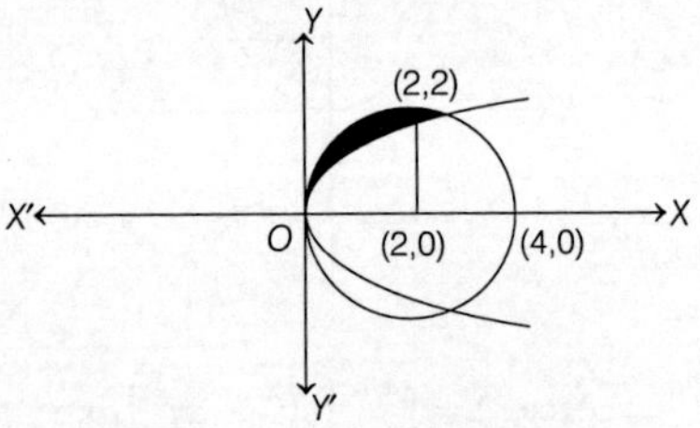

$$\text{Required area} = \int_0^2 (y_1 - y_2)\, dx$$

$$= \int_0^2 Y_{\text{circle}} - \int_0^2 Y_{\text{parabola}}\, dx$$

$$= \frac{\pi r^2}{4} - \int_0^2 \sqrt{2}(x)^{1/2}\, dx$$

$$= \frac{\pi \times 4}{4} - \sqrt{2}\frac{2}{3}\left[x^{\frac{3}{2}}\right]_0^2$$

$$= \pi - \frac{2\sqrt{2}}{3}(2^{3/2} - 0) = \pi - \frac{8}{3}$$

25 Given,

$$A = \{(x, y): x^2 + y^2 \le 1 \text{ and } y^2 \le 1 - x\}$$

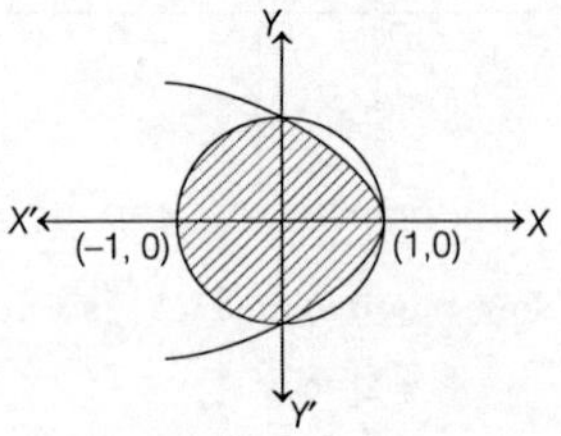

$$\text{Required area} = \frac{1}{2}\pi r^2 + 2\int_0^1 (1 - y^2)dy$$

$$= \frac{1}{2}\pi(1)^2 + 2\left(y - \frac{y^3}{3}\right)_0^1 = \frac{\pi}{2} + \frac{4}{3}$$

26

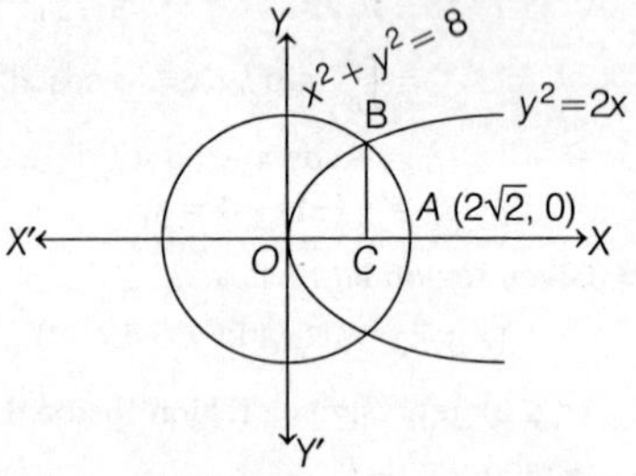

Let the area of the smaller part be A_1 and that of the bigger part be A_2. We have to find $\frac{A_1}{A_2}$.

The point B is a point of intersection (lying in the first quadrant) of the given parabola and the circle, whose coordinates can be obtained by solving the two equations $y^2 = 2x$ and $x^2 + y^2 = 8$.

$$\Rightarrow x^2 + 2x = 8 \Rightarrow (x-2)(x+4) = 0$$

$$\Rightarrow x = 2, -4$$

$x = -4$ is not possible as both the points of intersection have the same positive x-coordinate. Thus, $C \equiv (2, 0)$.

Now, $A_1 = 2$ [Area $(OBCO)$ + Area $(CBAC)$]

$$= 2\left[\int_0^2 y_1\, dx + \int_2^{2\sqrt{2}} y_2\, dx\right],$$

where, y_1 and y_2 are respectively the values of y from the equations of the parabola and that of the circle.

$$\Rightarrow A_1 = 2\left[\int_0^2 \sqrt{2x}\,dx + \int_2^{2\sqrt{2}} \sqrt{8 - x^2}\, dx\right]$$

$$\Rightarrow A_1 = 2\left[\sqrt{2} \cdot \frac{2}{3} \cdot x^{3/2}\right]_0^2 + 2\left[\frac{x}{2}\sqrt{8 - x^2} + \frac{8}{2}\sin^{-1}\frac{x}{2\sqrt{2}}\right]_2^{2\sqrt{2}}$$

$$= \frac{16}{3} + 2\left[2\pi - \left(2 + 4 \times \frac{\pi}{4}\right)\right]$$

$$= \left(\frac{4}{3} + 2\pi\right) \text{ sq units}$$

Clearly, area of the circle $= \pi(2\sqrt{2})^2$

$$= 8\pi \text{ sq units}$$

Now, $A_2 = 8\pi - A_1 = 6\pi - \dfrac{4}{3}$

and the required ratio, $\dfrac{A_1}{A_2}$ is

$$= \frac{\frac{4}{3} + 2\pi}{6\pi - \frac{4}{3}} = \frac{2 + 3\pi}{9\pi - 2}$$

27 $y = 2 - |2 - x| = \begin{cases} x & : x < 2 \\ 4 - x : x \ge 2 \end{cases}$

and $y = \dfrac{3}{|x|} \Rightarrow y = \begin{cases} -3/x : x < 0 \\ 3/x \quad : x > 0 \end{cases}$

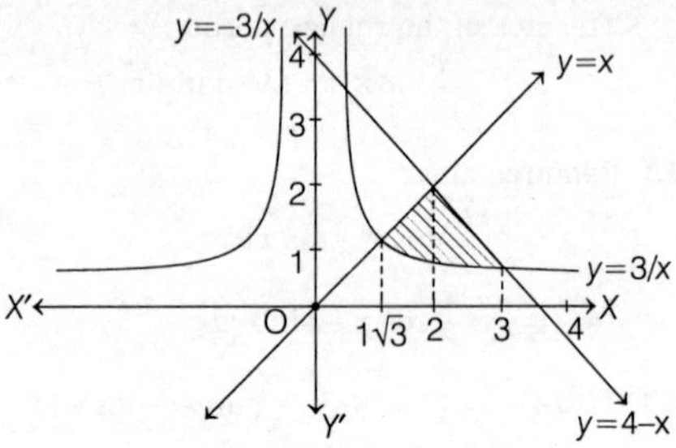

$$\Rightarrow \quad x = 3/x \Rightarrow x = \sqrt{3}$$

and $4 - x = 3/x \Rightarrow x = 3\ (x > 2)$.

$\therefore$ Required area

$$= \int_{\sqrt{3}}^2 \left(x - \frac{3}{x}\right) dx + \int_2^3 \left(4 - x - \frac{3}{x}\right) dx$$

$$= (4 - 3\log 3)/2$$

28 The region is clearly square with vertices at the points $(1, 0), (0, 1), (-1, 0)$ and $(0, -1)$.

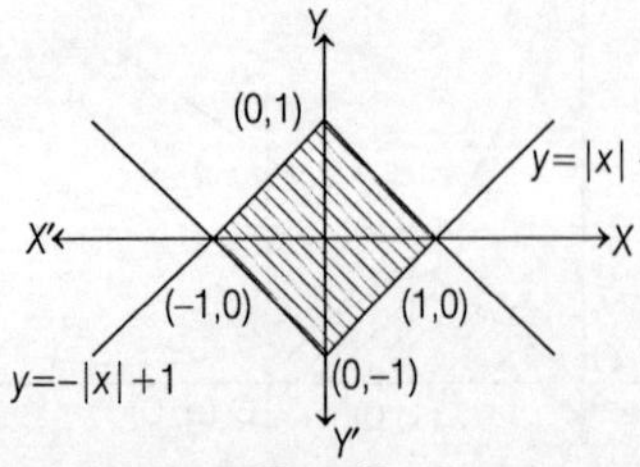

So, its area $= \sqrt{2} \times \sqrt{2} = 2$ sq units

29 Due to symmetry, required area

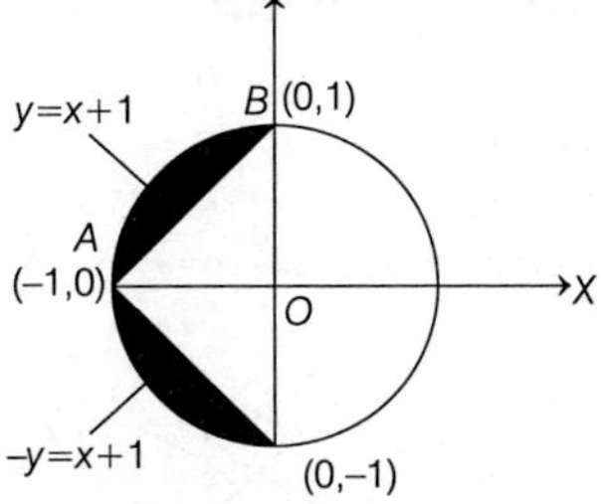

$$= 2\int_{-1}^{0} [\sqrt{1-x^2} - (x+1)]\, dx$$

or Area = 2 [area of sector $AOBA - \Delta AOB$]

$$= 2\left[\frac{\pi}{4} - \frac{1}{2}\right] = \frac{(\pi - 2)}{2}$$

30 Given curves are

$$y = \sqrt{x} \qquad \text{...(i)}$$

and $2y - x + 3 = 0$ ...(ii)

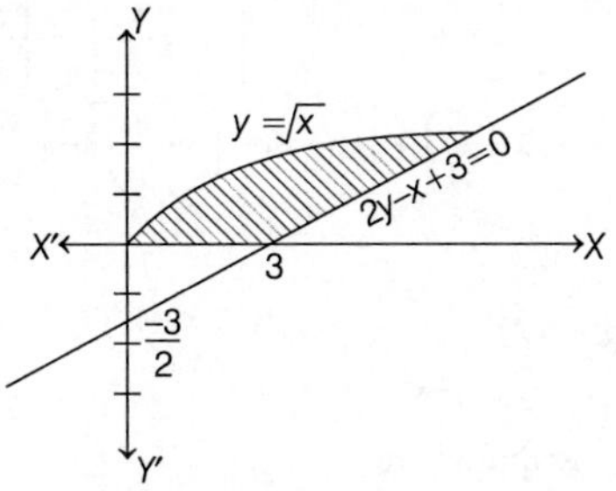

On solving Eqs. (i) and (ii) , we get

$$2\sqrt{x} - (\sqrt{x})^2 + 3 = 0$$

$$\Rightarrow \quad (\sqrt{x})^2 - 2\sqrt{x} - 3 = 0$$

$$\Rightarrow \quad (\sqrt{x} - 3)(\sqrt{x} + 1) = 0$$

$$\Rightarrow \quad \sqrt{x} = 3,$$

$[\because \sqrt{x} = -1 \text{ is not possible}]$

$\therefore\ y = 3$

$\therefore$ Required area $= \int_0^3 (x_2 - x_1)\, dy$

$$= \int_0^3 \{(2y+3) - y^2\}\, dy$$

$$= \left[y^2 + 3y - \frac{y^3}{3}\right]_0^3 = 9 + 9 - 9 = 9$$

31 Curves $x^2 = y$ and $y = 2/(1+x^2)$ are symmetrical about Y-axis.

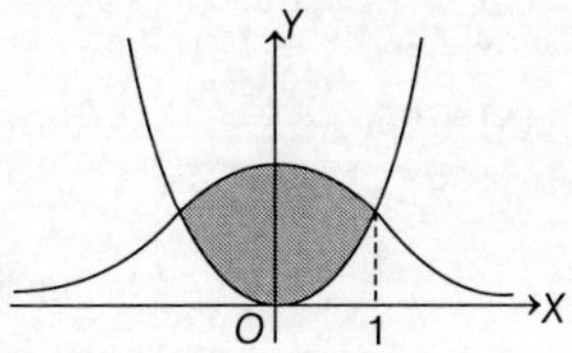

$$\text{Area} = 2\int_0^1 \left(\frac{2}{1+x^2} - x^2\right) dx$$

$$= \left[4\tan^{-1} x - \frac{2x^3}{3}\right]_0^1 = \pi - 2/3 = \frac{3\pi - 2}{3}$$

32 Clearly, the two curves will intersect at $\left(\frac{\pi}{4}, 1\right)$.

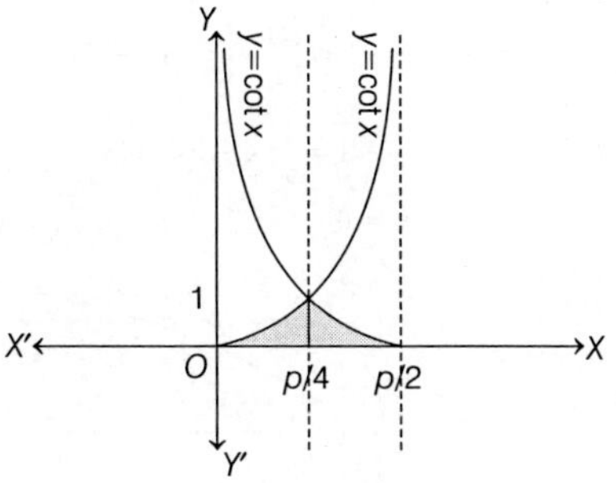

Now, required area

$$= \int_0^{\pi/4} \tan x\, dx + \int_{\pi/4}^{\pi/2} \cot x\, dx$$

$$= [\log \sec x]_0^{\pi/4} + [\log \sin x]_{\pi/4}^{\pi/2}$$

$$= [\log \sqrt{2} - 0] + [0 + \log \sqrt{2}]$$

$$= 2\log \sqrt{2} = \log 2 \text{ sq units}$$

33 Required area

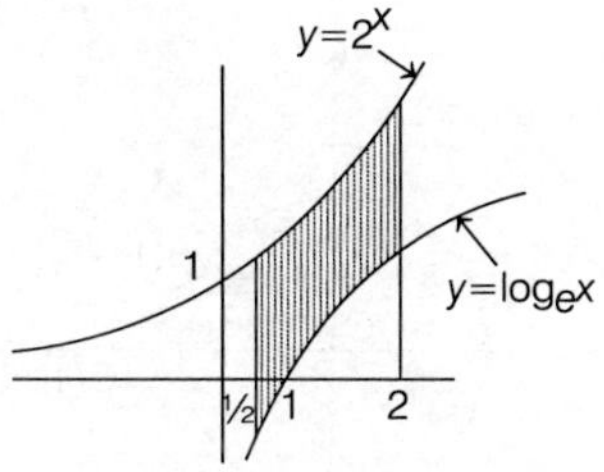

$$= \int_{1/2}^{2} (2^x - \log_e x)\, dx$$

$$= \left[\frac{2^x}{\log 2} - x\log x + x\right]_{1/2}^{2}$$

$$= \frac{4 - \sqrt{2}}{\log 2} - \frac{5}{2}\log 2 + \frac{3}{2}$$

$$= \frac{4 - \sqrt{2}}{\log 2} + b - c\log 2$$

$$\Rightarrow \quad b = 3/2, c = 5/2 \Rightarrow b + c = 4$$

34

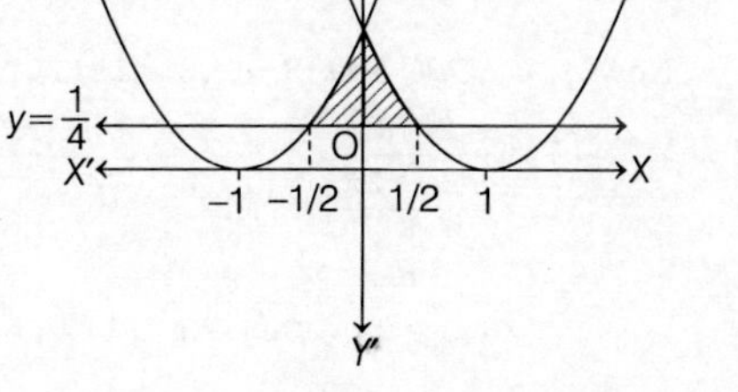

$\therefore$ Required area

$$= 2\int_0^{1/2} \left[(x-1)^2 - \frac{1}{4}\right] dx$$

$$= 2\left[\frac{(x-1)^3}{3} - \frac{x}{4}\right]_0^{1/2} = 2\left[-\frac{1}{24} - \frac{1}{8} + \frac{1}{3}\right]$$

$$= \frac{1}{3} \text{ sq unit}$$

35 Required area, A

$$= \int_1^e [\log x - (\log x)^2]\, dx$$

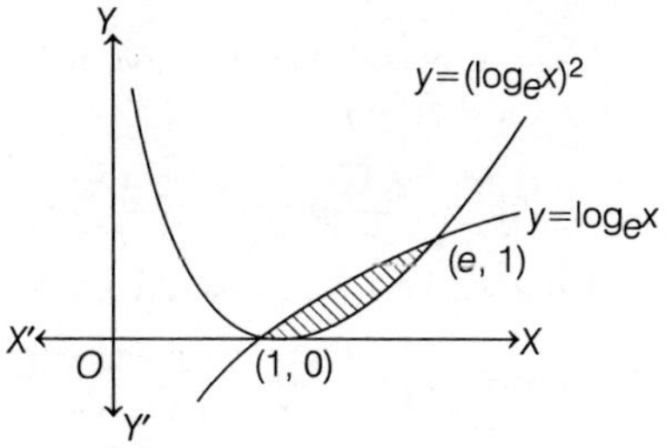

$$A = \int_1^e \log x\, dx - \int_1^e (\log x)^2 dx$$

$$= [x\log x - x]_1^e - [x(\log x)^2 - 2(x\log x - x)]_1^e$$

$$= [e - e - (-1)] - [e(1)^2 - 2e + 2e - (2)]$$

$$= 1 - (e - 2) = 3 - e$$

SESSION 2

1 $h(x) = f(\cos x) = \cos^2 x$

$$18x^2 - 9\pi x + \pi^2 = 0$$

$$\Rightarrow (6x - \pi)(3x - \pi) = 0$$

$$\Rightarrow x_1 = \pi/6,\ x_2 = \pi/3$$

$\therefore$ Required area $= \int_{\pi/6}^{\pi/3} \cos^2 x\, dx = \pi/12$

2 Graph of

$f(x) = \min(|x|, |x-1|, |x+1|)$ is

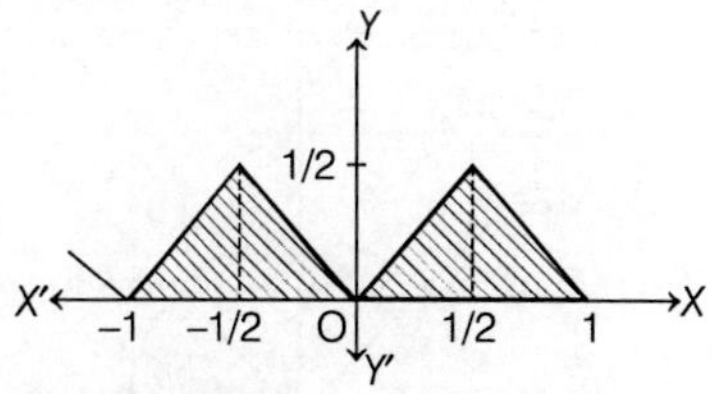

Required area

$$= 2 \times \left[\frac{1}{2} \times 1 \times \frac{1}{2}\right] = \frac{1}{2} \text{ sq unit}$$

3

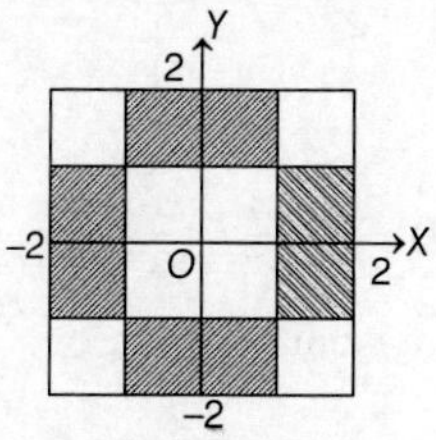

$$[|x|] + [|y|] = 1$$

$\Rightarrow \quad [|x|] = 1, [|y|] = 0$

or $\quad [|x|] = 0, [|y|] = 1$

$\Rightarrow \quad 1 \le |x| < 2,\ 0 \le |y| < 1$

or $0 \le |x| < 1$ or $1 \le |y| < 2$

$\Rightarrow x \in (-2, -1] \cup [1, 2),\ y \in (-1, 1)$

or $x \in (-1, 1),\ y \in (-2, -1] \cup [1, 2)$

$\therefore$ Required area $= 8 \times 1 \times 1 = 8$

4 In the equation of curve $xy^2 = 4(2 - x)$, the degree of y is even. Therefore, the curve is symmetrical about X-axis and lies in $0 < x \le 2$.

The area bounded by the curve and the Y-axis is $2\int_0^2 y\,dx$

$$= 2\int_0^2 2\sqrt{\frac{2-x}{x}}\,dx = 4\int_0^2 \sqrt{\frac{2-x}{x}}\,dx$$

Put $x = 2\sin^2\theta \Rightarrow dx = 4\sin\theta \cdot \cos\theta\, d\theta$

$$\therefore \text{Required area} = 4\int_0^{\pi/2} \sqrt{\frac{2 - 2\sin^2\theta}{2\sin^2\theta}} \cdot 4\sin\theta \cdot \cos\theta\, d\theta$$

$$= 8\int_0^{\pi/2} 2\cos^2\theta\, d\theta$$

$$= 8\int_0^{\pi/2} (1 + \cos 2\theta)\, d\theta$$

$$= 8\left[\theta + \sin\frac{2\theta}{2}\right]_0^{\frac{\pi}{2}}$$

$$= 8\left[\frac{\pi}{2} + 0 - 0\right] = 4\pi$$

5 Required area = Area of the shaded region

= 4 (Area of the shaded region in first quadrant)

$$= 4\int_0^{1/\sqrt{2}} (y_1 - y_2)\, dx$$

$$= 4\int_0^{1/\sqrt{2}} (\sqrt{1 - x^2} - x)\, dx$$

$$= 4\left[\frac{1}{2}x\sqrt{1 - x^2} + \frac{1}{2}\sin^{-1} x - \frac{x^2}{2}\right]_0^{1/\sqrt{2}}$$

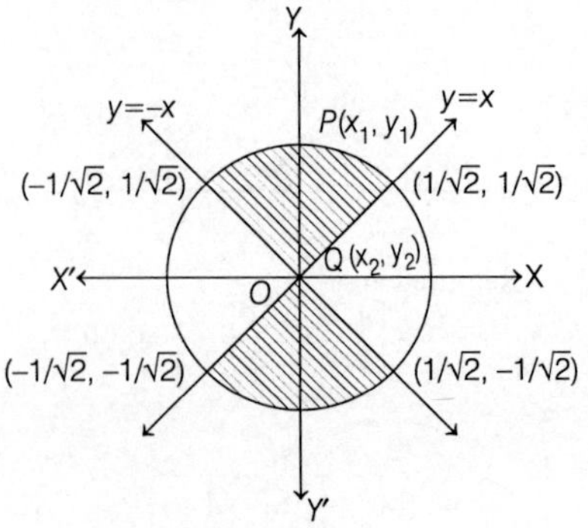

$$= 4\left[\frac{1}{2\sqrt{2}} \times \frac{1}{\sqrt{2}} + \frac{1}{2} \times \frac{\pi}{4} - \frac{1}{4}\right]$$

$$= \frac{\pi}{2} \text{ sq units}$$

6 $(y - 2)^2 = x - 1 \quad \ldots(\text{i})$

Curve (i) is a parabola with vertex at the point $A(1, 2)$, axis $y - 2 = 0$ i.e. $y = 2$ and concavity towards positive X-axis.

For the point, say P, at which ordinate $y = 3$ and $x = 2$

Equation of tangent at $P(2, 3)$ is

$$y - 3 = \left[\frac{dy}{dx}\right]_{(2,3)} (x - 2)$$

or $\quad y - 3 = \frac{1}{2}(x - 2)$ i.e.

$x - 2y + 4 = 0 \quad \ldots(\text{ii})$

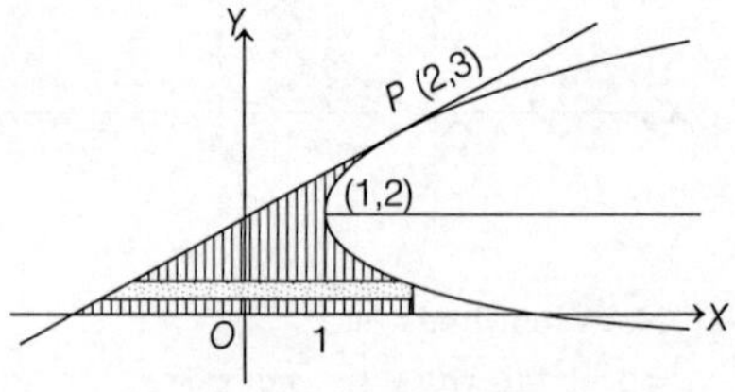

$$\therefore \text{Area} = \int_0^3 [y^2 - 4y + 5 - (2y - 4)]\, dy$$

$$= \left[\frac{y^3}{3} - 3y^2 + 9y\right]_0^3$$

$$= 9 - 27 + 27 - 0 = 9.$$

7

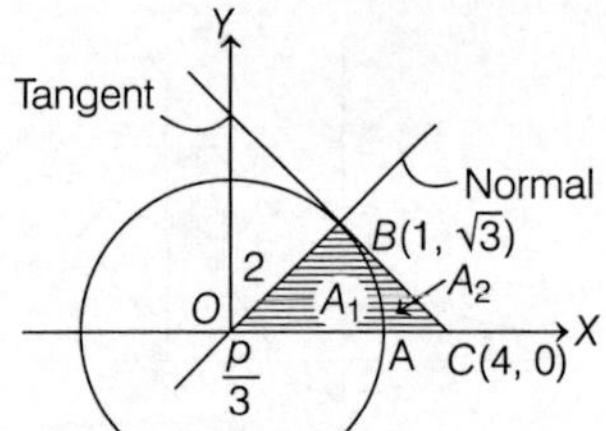

Let area of portion $OAB = A_1$

and area of portion $ABC = A_2$

The equation of tangent at $(1, \sqrt{3})$ is

$x + \sqrt{3}y = 4$

[$\because xx_1 + yy_1 = a^2$ is the tangent for the circle $x^2 + y^2 = a^2$ at (x_1, y_1)]

Now, the area of the

$$\Delta OBC = \frac{1}{2} \times OB \times BC$$

$$= \frac{1}{2} \times 2 \times 2\sqrt{3} = 2\sqrt{3}$$

The area of portion OAB i.e. $A_1 = \dfrac{r^2\theta}{2}$

$$= \frac{4 \cdot \pi/3}{2} = \frac{2\pi}{3}.$$

Now, $A_2 = \Delta OBC - OAB = 2\sqrt{3} - 2\pi/3$

$$\frac{A_1}{A_2} = \frac{2\pi/3}{\dfrac{6\sqrt{3} - 2\pi}{3}}$$

$$= \frac{2\pi}{6\sqrt{3} - 2\pi} = \frac{\pi}{3\sqrt{3} - \pi}.$$

8 On solving $x^2 = 4y$ and $x + y = 3$

we get, $\dfrac{x^2}{4} + x = 3$

$\Rightarrow \quad x^2 + 4x - 12 = 0$

$\Rightarrow \quad (x + 6)(x - 2) = 0 \Rightarrow x = 2, y = 1$

Solving $y = 1 + \sqrt{x}$ and $y = 3 - x$,

we get $1 + \sqrt{x} = 3 - x \Rightarrow x = 1, y = 2$

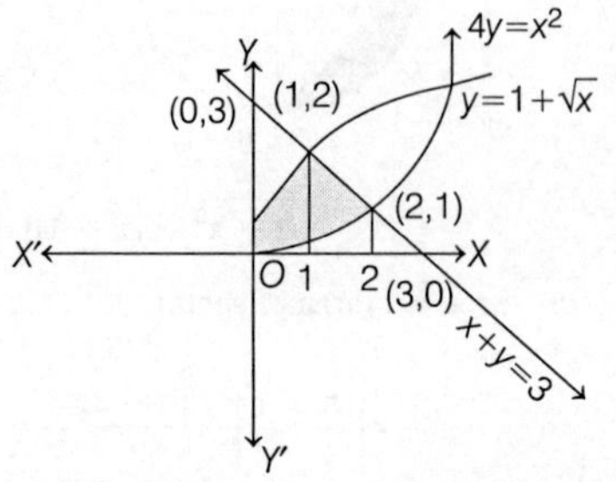

Required Area

$$\int_0^1 (1 + \sqrt{x})\, dx + \int_1^2 (3 - x)dx - \int_0^2 \frac{x^2}{4} dx$$

$$\left[x + \frac{2}{3}x^{3/2}\right]_0^1 + \left[3x - \frac{x^2}{2}\right]_1^2 - \left[\frac{x^3}{12}\right]_0^2$$

$$= \left(1 + \frac{2}{3}\right) + \left[(6 - 2) - \left(3 - \frac{1}{2}\right)\right] - \left[\frac{8}{12}\right]$$

$$= \frac{5}{2}$$

9 $\quad y = \log_e (x + e) \quad \ldots(\text{i})$

$\quad y = e^{-x} \quad \ldots(\text{ii})$

Required area

$$= \int_{1-e}^{0} \log(x + e)\, dx + \int_0^{\infty} e^{-x}\, dx$$

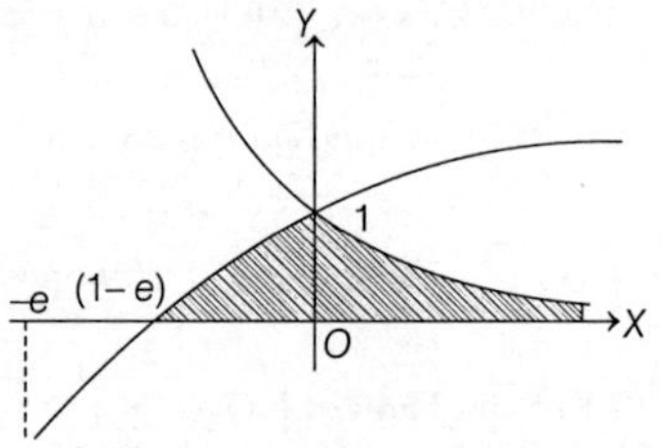

$$= [x \log(x + e)]_{1-e}^{0}$$

$$- \int_{1-e}^{0} \frac{x}{x + e}\, dx + [-e^{-x}]_0^{\infty}$$

$$= 0 + 1 - [x - e\log(x + e)]_{1-e}^{0}$$

$$= 1 + 1 = 2$$

10 $f(x/y) = f(x) - f(y) \quad \ldots(\text{i})$

$x = y = 1 \Rightarrow f(1) = 0$

$$f'(x) = \lim_{h \to 0} \frac{f(x + h) - f(x)}{h}$$

$$= \lim_{h \to 0} \frac{f\left(\dfrac{x + h}{x}\right)}{h} \quad \text{[using Eq. (i)]}$$

$$= \lim_{h \to 0} \frac{f(1 + h/x)}{h/x} \cdot \frac{1}{x} = \frac{3}{x}$$

$\Rightarrow f(x) = 3\ln x + c$

$f(1) = 0 \Rightarrow c = 0.$

$\therefore \quad y = f(x) = 3\log_e x$

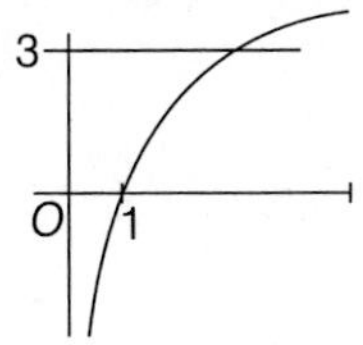

Required area $= \int_{-\infty}^{3} e^{y/3}\, dy$

$= 3[e^{y/3}]_{-\infty}^{3} = 3e$

11 According to given condition, we have

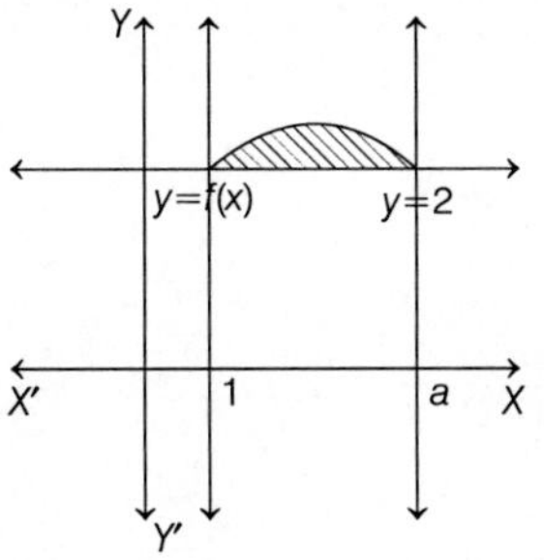

$\int_1^a [f(x)-2]dx = \frac{2}{3}[(2a)^{3/2} - 3a + 3 - 2\sqrt{2}]$

On differentiating both sides w.r.t. a, we get

$f(a) - 2 = \frac{2}{3}\left[\frac{3}{2}(2a)^{1/2}\cdot 2 - 3\right]$

$\Rightarrow \quad f(a) - 2 = 2\sqrt{2a} - 2$

$\Rightarrow \quad f(a) = 2\sqrt{2a}$

$\Rightarrow \quad f(x) = 2\sqrt{2}x,\ x \geq 1$

12 $\frac{3\pi}{2} \leq x \leq 2\pi \Rightarrow -2\pi \leq -x \leq -\frac{3\pi}{2}$

$\cos^{-1}(\sin x) = \cos^{-1}\cos\left(2\pi + \frac{\pi}{2} - x\right)$

$= 2\pi + \frac{\pi}{2} - x$

and $\sin^{-1}(-\cos x)$

$= \sin^{-1}\sin\left(\frac{3\pi}{2} - x\right) = \frac{3\pi}{2} - x$

$\therefore \quad y = 2\pi + \pi/2 - x + 3\pi/2 - x$

$= 4\pi - 2x$

Required area $= \int_{3\pi/2}^{2\pi}(4\pi - 2x)dx = \frac{\pi^2}{4}$

13 $f(x) - f(x/7) = x/7$

$f(x/7) - f(x/7^2) = x/7^2$

$f(x/7^2) - f(x/7^3) = x/7^3$

$\vdots \quad \vdots \quad \vdots \quad \vdots \quad \vdots \quad \vdots \quad \vdots$

$f(x/7^{n-1}) - f(x/7^n) = x/7^n.$

Adding, we get

$f(x) - f(x/7^n) = \frac{x}{7}\left(1 + \frac{1}{7} + \ldots + \frac{1}{7^{n-1}}\right)$

$= \frac{x}{6}\left(1 - \frac{1}{7^n}\right)$

Taking limit as $n \to \infty$, we get

$f(x) - f(0) = \frac{x}{6} \Rightarrow f(x) = 1 + \frac{x}{6}$

$[\because f(0) = 1]$

Required area $= \int_{-6}^{0}\left(1 + \frac{x}{6}\right) = 3$

14 Given curves are

$y = x - bx^2$ and $by = x^2$

Solving these, we get $x = 0, b/(1 + b^2)$

$\therefore \quad \Delta(b) = \left|\int_0^{b/b+1}\left(\frac{x^2}{b} - x + bx^2\right)dx\right|$

$= \left|\left[\left(\frac{b^2+1}{b}\right)\frac{x^3}{3} - \frac{x^2}{2}\right]_0^{b/(b^2+1)}\right|$

$= \frac{1}{6}\frac{b^2}{(b^2+1)^2}$

$\Delta'(b) = \frac{1}{6}\cdot\frac{2b(b^2+1)^2 - 2b^2(b^2+1)\times 2b}{(b^2+1)^4}$

$= \frac{2b(1-b)(1+b)}{(b^2+1)^3}$

$\Rightarrow \quad \Delta(b)$ is max. for $b = 1, -1.$

$\left[\because \Delta'(b) = \frac{\oplus\ \ominus\ \oplus\ \ominus}{-1\ \ 0\ \ 1}\right]$

15 In $(0, \pi/4)$, $\tan x > 0$

$\therefore A_n = \int_0^{\pi/4}\tan^n x\, dx$

$= \int_0^{\pi/4}\tan^{n-2} x(\sec^2 x - 1)\, dx$

$= \left[\frac{\tan^{n-1} x}{n-1}\right]_0^{\pi/4} - \int_0^{\pi/4}\tan^{n-2} x\, dx$

$\Rightarrow \quad A_n = \frac{1}{n-1} - A_{n-2}$

$\Rightarrow \quad A_n + A_{n-2} = \frac{1}{n-1}$

For $n > 2$, $0 < x < \pi/4$

$\Rightarrow 0 < \tan x < 1 \Rightarrow \tan^{n-2} x > \tan^n x$

$\Rightarrow \quad A_{n-2} > A_n$

$\Rightarrow \quad 2A_n < \frac{1}{n-1}$

$\Rightarrow \quad A_n < \frac{1}{2n-2}$

Also, $A_{n+2} + A_n = \frac{1}{n+1}$

$\Rightarrow \quad 2A_n > \frac{1}{n+1}$

$\therefore \quad \frac{1}{2n+2} < A_n < \frac{1}{2n-2}$

DAY EIGHTEEN

Differential Equations

Learning & Revision for the Day

- Solution of Differential Equation
- Solutions of Differential Equations of First Order and First Degree
- Linear Differential Equation
- Inspection Method

An equation involving independent variables, a dependent variable and the derivatives of dependent variable w.r.t. independent variables, is called **differential equation**.

A differential equation which contains only one independent variable, is called an ordinary differential equation.

Order and Degree of a Differential Equation

- The **order** of a differential equation is the order of the highest derivative occurring in the differential equation.
- The **degree** of differential equation is the degree of the highest order derivative, when differential coefficients are made free from radicals and fractions.

Solution of Differential Equation

A solution of the differential equation is a relation between the dependent and independent variables of the equation not containing the derivatives, but satisfying the given differential equation.

General Solution

A general solution of a differential equation is a relation between the variables (not involving the derivatives) which contains the same number of the arbitrary constants as the order of the differential equation.

Particular Solution

Particular solution of the differential equation is obtained from the general solution by assigning particular values to the arbitrary constant in the general solution.

PREP MIRROR
Your Personal Preparation Indicator

- No. of Questions in Exercises (x)—
- No. of Questions Attempted (y)—
- No. of Correct Questions (z)—
  (Without referring Explanations)

- Accuracy Level $(z/y \times 100)$—
- Prep Level $(z/x \times 100)$—

In order to expect good rank in JEE, your Accuracy Level should be above 85 & Prep Level should be above 75.

Solutions of Differential Equations of the First Order and First Degree

A differential equation of first order and first degree may be of the following types:

Differential Equation with Variable Separable

'Variable separable method' is used to solve such an equation in which variable can be separated completely, e.g. the term $f(x)$ with dx and the term $g(y)$ with dy

i.e. $g(y)dy = f(x)dx$

Reducible to Variable Separable Form

Differential equations of the form $\frac{dy}{dx} = f(ax + by + c)$ can be reduced to variable separable form by the substitution $ax + by + c = z$.

$$\Rightarrow \quad a + b\frac{dy}{dx} = \frac{dz}{dx} \quad \Rightarrow \quad \left(\frac{dz}{dx} - a\right)\frac{1}{b} = f(z) \quad \Rightarrow \quad \frac{dz}{dx} = a + bf(z)$$

Homogeneous Differential Equation

A function $f(x, y)$ is said to be a homogeneous function of degree n if $f(\lambda x, \lambda y) = \lambda^n f(x, y)$ for some non-zero constant λ.

- Let $f(x, y)$ and $g(x, y)$ be two homogeneous functions of same degree, then a differential equation expressible in the form
$$\frac{dy}{dx} = \frac{f(x, y)}{g(x, y)}$$
is called a **homogeneous differential equation.**
- To solve a homogeneous differential equation of the type $\frac{dy}{dx} = \frac{f(x, y)}{g(x, y)} = h\left(\frac{y}{x}\right)$, we put $y = vx$ and to solve a homogeneous differential equation of type $\frac{dx}{dy} = \frac{f(x, y)}{g(x, y)} = h\left(\frac{x}{y}\right)$, we put $x = vy$.

Linear Differential Equation

A differential equation of the form $\frac{dy}{dx} + Py = Q$, where, P and Q are the functions of x (or constants), is called a linear differential equation.

To solve such an equation, first find integrating Factor, $\text{IF} = e^{\int P dx}$. Then, the solution of the differential equation is given by $y(\text{IF}) = \int Q(\text{IF})\, dx + C$.

NOTE
- Sometimes given equation becomes a linear differential equation of the form $\frac{dx}{dy} + Rx = S$, where R and S are function of y (or constants).
- The integrating factor in this case is $\text{IF} = e^{\int R\, dy}$ and the solution is given by $x \cdot (\text{IF}) = \int (S \times \text{IF})dy + C$.

Bernoulli's Differential Equation (Reducible to Linear Form)

Let the differential equation be of the form

$$\frac{dy}{dx} + Py = Qy^n \quad \Rightarrow \quad y^{-n}\frac{dy}{dx} + Py^{-n+1} = Q$$

Here, put $y^{-n+1} = z \Rightarrow (-n + 1)y^{-n}\frac{dy}{dx} = \frac{dz}{dx}$

$$\Rightarrow \quad \frac{dz}{dx} + (1 - n)\, Pz = (1 - n)\, Q$$

which is a linear differential equation in z.

The solution of this equation is given by

$$ze^{\int (1-n)P dx} = \int \left[(1-n)\cdot Q \cdot e^{\int (1-n)P dx}\right] dx + C$$

Inspection Method

If we can write the differential equation in the form $f\{f_1(x, y)\}\, d\{f_1(x, y)\} + \phi\{f_2(x, y)\}\, d\{f_2(x, y)\} + \ldots = 0$, then each term can be easily integrated separately.

For this use the following results.

(i) $d(x \pm y) = dx \pm dy$ (ii) $d(xy) = x\,dy + y\,dx$

(iii) $d\left(\frac{x}{y}\right) = \frac{y\,dx - x\,dy}{y^2}$

(iv) $d\left(\tan^{-1}\frac{x}{y}\right) = \frac{y\,dx - x\,dy}{x^2 + y^2}$

(v) $d[\log(xy)] = \frac{x\,dy + y\,dx}{xy}$

(vi) $d\left[\log\left(\frac{x}{y}\right)\right] = \frac{y\,dx - x\,dy}{xy}$

(vii) $d\left[\frac{1}{2}\log(x^2 + y^2)\right] = \frac{x\,dx + y\,dy}{x^2 + y^2}$

(viii) $d\left(-\frac{1}{xy}\right) = \frac{x\,dy + y\,dx}{x^2y^2}$

(ix) $d\left(\frac{e^x}{y}\right) = \frac{ye^x dx - e^x\,dy}{y^2}$

(x) $d(\sqrt{x^2 + y^2}) = \frac{x\,dx + y\,dy}{\sqrt{x^2 + y^2}}$

(xi) $d\left(\frac{1}{2}\log\frac{x + y}{x - y}\right) = \frac{x\,dy - y\,dx}{x^2 - y^2}$

DAY PRACTICE SESSION 1

FOUNDATION QUESTIONS EXERCISE

1 The order of the differential equation whose general solution is given by

$y = c_1 e^{2x + c_2} + c_3 e^x + c_4 \sin(x + c_5)$, is

(a) 5 (b) 4 (c) 3 (d) 2

2 The degree of the differential equation

$\frac{d^2y}{dx^2} + 3\left[\frac{dy}{dx}\right]^2 = x^2 \log\left[\frac{d^2y}{dx^2}\right]$, is

(a) 1 (b) 2
(c) 3 (d) None of these

3 Order and degree of the differential equation, representing the family of curves $y^2 = 2c(x + \sqrt{c})$, where c is a positive parameter, are respectively equal to

(a) 1, 3 (b) 2, 3
(c) 2, 4 (d) 1, 2

4 The differential equation of all circles in the first quadrant which touch the coordinate axes is of order

(a) 1 (b) 2
(c) 3 (d) None of these

5 The differential equation of the family of curves $v = \frac{A}{r} + B$, where A and B are arbitrary constants, is

(a) $\frac{d^2v}{dr^2} + \frac{1}{r}\frac{dv}{dr} = 0$ (b) $\frac{d^2v}{dr^2} - \frac{2}{r}\frac{dv}{dr} = 0$
(c) $\frac{d^2v}{dr^2} + \frac{2}{r}\frac{dv}{dr} = 0$ (d) None of these

6 The solution of the differential equation

$x\,dx + y\,dy + \frac{x\,dy - y\,dx}{x^2 + y^2} = 0$, is

(a) $y = x \tan\left(\frac{x^2 + y^2 + C}{2}\right)$

(b) $x = y \tan\left(\frac{x^2 + y^2 + C}{2}\right)$

(c) $y = x \tan\left(\frac{C - x^2 - y^2}{2}\right)$

(d) None of the above

7 The differential equation of the family of parabolas with focus at the origin and the X-axis as axis, is

(a) $y\left(\frac{dy}{dx}\right)^2 + 4x\frac{dy}{dx} = 4y$ (b) $y\left(\frac{dy}{dx}\right)^2 = 2x\frac{dy}{dx} - y$
(c) $y\left(\frac{dy}{dx}\right)^2 + y = 2xy\frac{dy}{dx}$ (d) $y\left(\frac{dy}{dx}\right)^2 + 2xy\frac{dy}{dx} + y = 0$

8 The differential equation whose solution is $x^2 + y^2 + 2ax + 2by + c = 0$, where a, b, c are arbitrary constants, is

(a) $3y_1y_2 - (1 + y_1^2)y_3 = 0$ (b) $3y_1^2y_2 - (1 + y_1^2)y_3 = 0$
(c) $3y_1y_2^2 + (1 + y_1^2)y_3 = 0$ (d) $3y_1y_2^2 - (1 + y_1^2)y_3 = 0$

9 If $x\,dy = y(dx + y\,dy)$, $y(1) = 1$ and $y(x) > 0$. Then, $y(-3)$ is equal to

(a) 3 (b) 2
(c) 1 (d) 0

10 If $\frac{dy}{dx} = y + 3 > 0$ and $y(0) = 2$, then $y(\log 2)$ is equal to

(a) 5 (b) 13 (c) −2 (d) 7

11 The differential equation $\frac{dy}{dx} = \frac{\sqrt{1 - y^2}}{y}$ determines family of circles with,

(a) variable radii and a fixed centre at (0, 1)
(b) variable radii and a fixed centre at (0, −1)
(c) fixed radius 1 and variable centre along the X-axis
(d) fixed radius 1 and variables centre along the Y-axis

12 If the solution of the differential equation $\frac{dy}{dx} = \frac{ax + 3}{2y + f}$ represents a circle, then the value of a is

(a) 2 (b) −2 (c) 3 (d) −4

13 Solution of the differential equation $y - x\frac{dy}{dx} = y^2 + \frac{dy}{dx}$ is

(a) $Cy = (1 - x)(1 - y)$ (b) $Cx = (1 + x)(1 - y)$
(c) $Cy = (1 + x)(1 - y)$ (d) $Cx = (1 - x)(1 + y)$

14 Solution of the equation $ln\left(\frac{dy}{dx}\right) = ax + by$, is

(a) $ae^{-by} + be^{ax} + C = 0$
(b) $ae^{by} + be^{ax} + C = 0$
(c) $ae^{by} + be^{-ax} + C = 0$
(d) None of the above

15 The solution of the differential equation

$\frac{dy}{dx} = e^{x-y} + x^2e^{-y}$ is

(a) $e^y = e^x + \frac{x^3}{3} + C$ (b) $e^y = e^x + 2x + C$
(c) $e^y = e^x + x^3 + C$ (d) $y = e^x + C$

16 If $(2 + \sin x)\frac{dy}{dx} + (y + 1)\cos x = 0$ and $y(0) = 1$, then $y\left(\frac{\pi}{2}\right)$ is equal to → JEE Mains 2017

(a) $-\frac{1}{3}$ (b) $\frac{4}{3}$ (c) $\frac{1}{3}$ (d) $-2/3$

17 If a curve passes through the point $\left(2, \frac{7}{2}\right)$ and has slope $\left(1 - \frac{1}{x^2}\right)$ at any point (x, y) on it, then the ordinate of the point on the curve, whose abscissa is – 2, is → JEE Mains 2013

(a) $-\frac{3}{2}$ (b) $\frac{3}{2}$ (c) $\frac{5}{2}$ (d) $-\frac{5}{2}$

18 Solution of the equation $(x+y)^2\frac{dy}{dx}=4, y(0)=0$ is

(a) $y=2\tan^{-1}\left(\frac{x+y}{2}\right)$ (b) $y=4\tan^{-1}\left(\frac{x+y}{4}\right)$

(c) $y=4\tan^{-1}\left(\frac{x+y}{2}\right)$ (d) None of these

19 The solution of $\frac{dy}{dx}+1=e^{x+y}$ is

(a) $e^{-(x+y)}+x+C=0$ (b) $e^{-(x+y)}-x+C=0$

(c) $e^{x+y}+x+C=0$ (d) $e^{x+y}-x+C=0$

20 Solution of the equation $xdy=(y+xf(y/x)/f'(y/x))\,dx$ is

(a) $|f(y/x)|=C|x|,\ C\in R$ (b) $|f(y/x)|=|x|+C, C>0$

(c) $|f(y/x)|=C|x|, C>0$ (d) None of these

21 If $x\frac{dy}{dx}=y(\log y-\log x+1)$, then the solution of the equation is

(a) $\log\left(\frac{x}{y}\right)=Cy$ (b) $\log\left(\frac{y}{x}\right)=Cx$

(c) $x\log\left(\frac{y}{x}\right)=Cy$ (d) $y\log\left(\frac{x}{y}\right)=Cx$

22 The general solution of $y^2\,dx+(x^2-xy+y^2)\,dy=0$, is

(a) $\tan^{-1}\left(\frac{x}{y}\right)+\log y+C=0$

(b) $2\tan^{-1}\left(\frac{x}{y}\right)+\log x+C=0$

(c) $\log(y+\sqrt{x^2+y^2})+\log y+C=0$

(d) $\sin^{-1}\left(\frac{x}{y}\right)+\log y+C=0$

23 Solution of the differential equation $x^2dy+y(x+y)dx=0$ is

(a) $y+2x=c^2x^2y$ (b) $y-2x=c^2x^2/y$

(c) $y+2x=c^2x^2/y$ (d) None of these

24 If $(x^2+y^2)dy=xydx$ and $y(x_0)=e, y(1)=1$, then x_0 is

(a) $e\sqrt{3}$ (b) $\sqrt{2e^2-1}/\sqrt{2}$

(c) $\sqrt{e^2-1}/\sqrt{2}$ (d) $\sqrt{e^2+1}/\sqrt{2}$

25 If a curve $y=f(x)$ passes through the point (1,–1) and satisfies the differential equation $y(1+xy)dx=xdy$, then $f\left(-\frac{1}{2}\right)$ is equal to → JEE Mains 2016

(a) $-\frac{2}{5}$ (b) $-\frac{4}{5}$ (c) $\frac{2}{5}$ (d) $\frac{4}{5}$

26 The equation of the curve passing through the origin and satisfying the differential equation

$$(1+x^2)\frac{dy}{dx}+2xy=4x^2 \text{ is}$$

→ JEE Mains 2013

(a) $(1+x^2)y=x^3$ (b) $3(1+x^2)y=2x^3$

(c) $(1+x^2)y=3x^3$ (d) $3(1+x^2)\,y=4x^3$

27 The solution of the differential equation $\frac{dy}{dx}+\frac{3x^2}{1+x^3}y=\frac{\sin^2 x}{1+x^3}$, is

(a) $y(1+x^3)=x+\frac{1}{2}\sin 2x+C$

(b) $y(1+x^3)=Cx+\frac{1}{2}\sin 2x$

(c) $y(1+x^3)=Cx-\frac{1}{2}\sin 2x$

(d) $y(1+x^3)=\frac{x}{2}-\frac{1}{4}\sin 2x+C$

28 If $y(x)$ satisfies the differential equation $y'-y\tan x=2x\sec x$ and $y(0)=0$, then → IIT 2012

(a) $y\left(\frac{\pi}{4}\right)=\frac{\pi^2}{8\sqrt{2}}$ (b) $y'\left(\frac{\pi}{4}\right)=\frac{\pi^2}{18}$

(c) $y\left(\frac{\pi}{3}\right)=\frac{\pi^2}{9}$ (d) $y'\left(\frac{\pi}{3}\right)=\frac{4\pi}{3}+\frac{\pi^2}{3\sqrt{3}}$

29 The solution of the equation $\frac{dy}{dx}+y\tan x=x^m\cos x$, is

(a) $(m+1)\,y=x^{m+1}\cos x+C(m+1)\cos x$

(b) $my=(x^m+C)\cos x$

(c) $y=(x^{m+1}+C)\cos x$

(d) None of the above

30 Let the population of rabbits surviving at a time t be governed by the differential equation $\frac{dP(t)}{dt}=\frac{1}{2}P(t)-200$. If $P(0)=100$, then $P(t)$ is equal to → JEE Mains 2014

(a) $400-300\,e^{\frac{t}{2}}$ (b) $300-200\,e^{-\frac{t}{2}}$

(c) $600-500\,e^{\frac{t}{2}}$ (d) $400-300\,e^{-\frac{t}{2}}$

31 The solution of differential equation

$$(xy^5+2y)dx-xdy=0, \text{ is}$$

(a) $9x^8+4x^9y^4=9y^4C$ (b) $9x^8-4x^9y^4-9y^4C=0$

(c) $x^8(9+4y^4)=10y^4C$ (d) None of these

32 Let $Y=y(x)$ be the solution of the differential equation $\sin x\frac{dy}{dx}+y\cos x=4x, x\in(0,\pi)$. If $y\left(\frac{\pi}{2}\right)=0$, then $y(\pi/6)$ is equal to → JEE Mains 2018

(a) $-\frac{8}{9\sqrt{3}}\pi^2$ (b) $-\frac{8}{9}\pi^2$ (c) $-\frac{4}{9}\pi^2$ (d) $\frac{4}{9\sqrt{3}}\pi^2$

33 **Statement I** The slope of the tangent at any point P on a parabola, whose axis is the axis of x and vertex is at the origin, is inversely proportional to the ordinate of the point P.

Statement II The system of parabolas $y^2=4ax$ satisfies a differential equation of degree 1 and order 1.

→ JEE Mains 2013

(a) Statement I is true, Statement II is true; Statement II is a correct explanation for Statement I

(b) Statement I is true, Statement II is true; Statement II is not a correct explanation for Statement I
(c) Statement I is true; Statement II is false
(d) Statement I is false; Statement II is true

34 **Statement I** The elimination of four arbitrary constants in $y = (c_1 + c_2 + c_3 e^{c_4})x$ results into a differential equation of the first order $x\dfrac{dy}{dx} = y$.

Statement II Elimination of n independent arbitrary constants results in a differential equation of the nth order.

(a) Statement I is true, Statement II is true; Statement II is a correct explanation for Statement I
(b) Statement I is true, Statement II is true; Statement II is not a correct explanation for Statement I
(c) Statement I is true; Statement II is false
(d) Statement I is false; Statement II is true

DAY PRACTICE SESSION 2

PROGRESSIVE QUESTIONS EXERCISE

1 The differential equation which represents the family of curves $y = c_1 e^{c_2 x}$ where c_1 and c_2 are arbitrary constant, is

(a) $y'' = y'y$
(b) $yy'' = y'$
(c) $yy'' = (y')^2$
(d) $y' = y^2$

2 The population $p(t)$ at time t of a certain mouse species satisfies the differential equation $\dfrac{dp(t)}{dt} = 0.5(t) - 450$. If $p(0) = 850$, then the time at which the population becomes zero is

(a) $2\log 18$
(b) $\log 9$
(c) $\dfrac{1}{2}\log 18$
(d) $\log 18$

3 A curve passes through the point $\left(1, \dfrac{\pi}{6}\right)$. Let the slope of the curve at each point (x, y) be $\dfrac{y}{x} + \sec\left(\dfrac{y}{x}\right), x > 0$.

Then the equation of the curve is → JEE Advanced 2013

(a) $\sin\left(\dfrac{y}{x}\right) = \log x + \dfrac{1}{2}$
(b) $\text{cosec}\left(\dfrac{y}{x}\right) = \log x + 2$
(c) $\sec\left(\dfrac{2y}{x}\right) = \log x + 2$
(d) $\cos\left(\dfrac{2y}{x}\right) = \log x + \dfrac{1}{2}$

4 Let $y'(x) + y(x)g'(x) = g(x)g'(x), y(0) = 0, x \in R$, where $f'(x)$ denotes $\dfrac{df(x)}{dx}$ and $g(x)$ is a given non-constant differentiable function on R with $g(0) = g(2) = 0$. Then the value of $y(2)$ is

(a) 2
(b) 4
(c) 5
(d) 0

5 Consider the differential equation $y^2 dx + \left(x - \dfrac{1}{y}\right) dy = 0$. If $y(1) = 1$, then x is given by

(a) $1 - \dfrac{1}{y} + \dfrac{e^{\frac{1}{y}}}{e}$
(b) $4 - \dfrac{2}{y} - \dfrac{e^{\frac{1}{y}}}{e}$
(c) $3 - \dfrac{1}{y} + \dfrac{e^{\frac{1}{y}}}{e}$
(d) $1 + \dfrac{1}{y} - \dfrac{e^{\frac{1}{y}}}{e}$

6 The equation of the curve $y = f(x)$ passing through the origin which satisfies the differential equation $\dfrac{dy}{dx} = \sin(10x + 6y)$ is

(a) $y = \dfrac{1}{3}\left\{\tan^{-1}\dfrac{5\tan 4x}{4 - 3\tan 4x} - 5x\right\}$
(b) $y = \dfrac{1}{3}\left\{\tan^{-1}\dfrac{5\tan 4x}{4 - 3\tan 4x} + 5x\right\}$
(c) $y = \dfrac{1}{3}\left\{\tan^{-1}\dfrac{5\tan 4x}{4 + 3\tan 4x} - 5x\right\}$
(d) None of the above

7 Let I be the purchase value of an equipment and $V(t)$ be the value after it has been used for t yr. The value $V(t)$ depreciates at a rate given by differential equation $\dfrac{dV(t)}{dt} = -k(T - t)$, where $k > 0$ is a constant and T is the total life in years of the equipment. Then, the scrap value $V(T)$ of the equipment is

(a) $I - \dfrac{kT^2}{2}$
(b) $I - \dfrac{k(T - t)^2}{2}$
(c) e^{-kT}
(d) $T^2 - \dfrac{1}{k}$

8 The degree of the differential equation satisfying the relation $\sqrt{1 + x^2} + \sqrt{1 + y^2} = \lambda(x\sqrt{1 + y^2} - y\sqrt{1 + x^2})$, is

(a) 1
(b) 2
(c) 3
(d) None of these

9 If length of tangent at any point on the curve $y = f(x)$ intercepted between the point of contact and X-axis is of length 1, the equation of the curve is

(a) $\sqrt{1-y^2}+\ln\left|(1-\sqrt{1-y^2})/y\right|=\pm x+C$
(b) $\sqrt{1-y^2}-\ln\left|(1-\sqrt{1-y^2})/y\right|=\pm x+C$
(c) $\sqrt{1-y^2}+\ln\left|(1+\sqrt{1-y^2})/y\right|=\pm x+C$
(d) None of the above

10 The solution of differential equation $\cos x\,dy = y(\sin x - y)\,dx$, where, $0 < x < \frac{\pi}{2}$, is

(a) $\sec x = (\tan x + C)y$ (b) $y\sec x = \tan x + C$
(c) $y\tan x = \sec x + C$ (d) $\tan x = (\sec x + C)y$

11 The curves satisfying the differential equation $(1-x^2)y' + xy = ax$ are

(a) ellipse and parabola (b) ellipse and circles
(c) ellipse and hyperbola (d) None of these

12 A curve passes through (2, 0) and the slope of tangent at a point $P(x, y)$ is equal to $((x+1)^2 + y - 3)/(x+1)$. Then equation of the curve is → IIT 2004

(a) $y = x^2 + 2x$ (b) $y = x^2 - 2x$
(c) $y = 2x^2 - x$ (d) None of these

13 The solution of the differential equation

$$\frac{x+\frac{x^3}{3!}+\frac{x^5}{5!}+\dots}{1+\frac{x^2}{2!}+\frac{x^4}{4!}+\dots}=\frac{dx-dy}{dx+dy} \text{ is}$$

(a) $2ye^{2x} = Ce^{2x} + 1$ (b) $2ye^{2x} = Ce^{2x} - 1$
(c) $ye^{2x} = Ce^{2x} + 2$ (d) None of these

14 Let f be a real-valued differentiable function on R (the set of all real numbers) such that $f(1) = 1$. If the y-intercept of the tangent at any point $P(x, y)$ on the curve $y = f(x)$ is equal to the cube of the abscissa of P, then the value of $f(-3)$ is equal to

(a) 3 (b) 6 (c) 9 (d) 1

15 Let $f(x)$ be differentiable in the interval $(0, \pi)$ such that $f(1) = 1$ and $\lim_{t\to x}\frac{t^2 f(x) - x^2 f(t)}{t - x} = 1$ for each $x > 0$. Then $f(x)$ is

(a) $\frac{1}{3x}+\frac{2x^2}{3}$ (b) $-\frac{1}{3x}+\frac{4x^2}{3}$
(c) $-\frac{1}{2x}+\frac{2}{x^2}$ (d) $\frac{1}{x}$

16 Let a solution $y = y(x)$ of the differential equation $x\sqrt{x^2-1}\,dy - y\sqrt{y^2-1}\,dx = 0$ satisfy $y(2) = \frac{2}{\sqrt{3}}$.

Statement I $y(x) = \sec\left(\sec^{-1} x - \frac{\pi}{6}\right)$.

Statement II $y(x)$ is given by $\frac{1}{y} = \frac{2\sqrt{3}}{x} - \sqrt{1-\frac{1}{x^2}}$.

(a) Statement I is true, Statement II is true; Statement II is a correct explanation for Statement I
(b) Statement I is true, Statement II is true; Statement II is not a correct explanation for Statement I
(c) Statement I is true; Statement II is false
(d) Statement I is false; Statement II is true

ANSWERS

SESSION 1	1 (b)	2 (d)	3 (a)	4 (a)	5 (c)	6 (c)	7 (b)	8 (d)	9 (a)	10 (d)
	11 (c)	12 (b)	13 (c)	14 (a)	15 (a)	16 (c)	17 (a)	18 (a)	19 (a)	20 (c)
	21 (b)	22 (a)	23 (a)	24 (a)	25 (d)	26 (d)	27 (d)	28 (a)	29 (a)	30 (a)
	31 (a)	32 (b)	33 (b)	34 (a)						
SESSION 2	1 (c)	2 (a)	3 (a)	4 (d)	5 (d)	6 (a)	7 (a)	8 (a)	9 (a)	10 (a)
	11 (c)	12 (b)	13 (b)	14 (c)	15 (a)	16 (c)				

Hints and Explanations

SESSION 1

1 Given,
$y = c_1 \cdot e^{2x+c_2} + c_3 e^x + c_4 \sin(x + c_5)$
$= c_1 \cdot e^{c_2} e^{2x} + c_3 e^x + c_4(\sin x \cos c_5 + \cos x \sin c_5)$
$= Ae^{2x} + c_3 e^x + B\sin x + D\cos x$
Here, $A = c_1 e^{c_2}$, $B = c_4 \cos c_5$ and $D = c_4 \sin c_5$
Since, equation consists four arbitrary constants.
So, the order of differential equation is 4.

2 Since, the equation is not a polynomial in all differential coefficients, so its degree is not defined.

3 Given, $y^2 = 2c(x + \sqrt{c})$ ...(i)
On differentiating both side w.r.t. x, we get
$$2y\frac{dy}{dx} = 2c \Rightarrow c = y\frac{dy}{dx}$$
On putting in Eq. (i),
$$y^2 = 2xy\frac{dy}{dx} + 2y^{3/2}\left(\frac{dy}{dx}\right)^{3/2}$$
$$\Rightarrow 8y^3\left(\frac{dy}{dx}\right)^3 = \left(y^2 - 2xy\frac{dy}{dx}\right)^2$$
Which is the differential equation of order one and degree 3.

4 Clearly, the equation of family of circle which touch both the axes is $(x-a)^2 + (y-a)^2 = a^2$, where a is a parameter.
Since, there is only one parameter, therefore order of differential equation representing this family is 1.

5 We have, $v = \frac{A}{r} + B$, where A and B are parameters
On differentiating twice w.r.t. r, we get
$$\Rightarrow \frac{dv}{dr} = \frac{-A}{r^2} \quad \text{... (i)}$$
$$\text{and} \quad \frac{d^2v}{dr^2} = \frac{2A}{r^3} \quad \text{... (ii)}$$
Now, on substituting the value of A from Eq. (i) in Eq. (ii), we get
$$\frac{d^2v}{dr^2} = \frac{2}{r^3}\left(-r^2\frac{dv}{dr}\right) = \frac{-2}{r}\frac{dv}{dr}$$
$$\Rightarrow \frac{d^2v}{dr^2} + \frac{2}{r}\frac{dv}{dr} = 0,$$ which is the required differential equation.

6 We have, $x\,dx + y\,dy + \frac{x dy - y dx}{x^2 + y^2} = 0$
$$\Rightarrow \frac{1}{2}d(x^2 + y^2) + d\left(\tan^{-1}\frac{y}{x}\right) = 0$$
On integrating, we get
$$\frac{1}{2}(x^2 + y^2) + \tan^{-1}\left(\frac{y}{x}\right) = \frac{C}{2}$$
$$\Rightarrow \frac{C - x^2 - y^2}{2} = \tan^{-1}\left(\frac{y}{x}\right)$$
$$\therefore \quad y = x\tan\left(\frac{C - x^2 - y^2}{2}\right)$$

7 Equation of family of parabolas with focus at (0, 0) and axis as X-axis is
$$y^2 = 4a(x - a) \quad \text{...(i)}$$
On differentiating Eq. (i) w.r.t. x, we get
$$2yy_1 = 4a$$
$$\therefore \quad y^2 = 2yy_1\left(x - \frac{yy_1}{2}\right)$$
$$\Rightarrow \quad y = 2xy_1 - yy_1^2$$
$$\Rightarrow \quad yy_1^2 = 2xy_1 - y$$

8 $x^2 + y^2 + 2ax + 2by + c = 0$ ...(i)
On differentiating Eq. (i) three times, we get
$$2x + 2yy_1 + 2a + 2by_1 = 0$$
$$\Rightarrow \quad x + yy_1 + a + by_1 = 0 \quad \text{...(ii)}$$
$$1 + y_1^2 + yy_2 + by_2 = 0 \quad \text{...(iii)}$$
$$3y_1y_2 + yy_3 + by_3 = 0 \quad \text{...(iv)}$$
On eliminating b from Eqs. (iii) and (iv), we get
$$3y_1y_2^2 - y_3 - y_1^2y_3 = 0$$

9 We have, $x\,dy = y\,(dx + y\,dy)$, $y > 0$
$$\therefore \quad \frac{x\,dy - y\,dx}{y^2} = dy$$
$$\Rightarrow \quad \frac{x}{y} = -y + C \quad \text{[integrating]}$$
Now, $y(1) = 1 \Rightarrow C = 2$
$$\therefore \quad \frac{x}{y} + y = 2$$
For $x = -3$, $-3 + y^2 = 2y$
$\Rightarrow y^2 - 2y - 3 = 0$
$\Rightarrow (y + 1)(y - 3) = 0$
$\therefore \quad y = 3$ $[\because y > 0]$

10 Here, $\frac{dy}{dx} = y + 3 > 0$ and $y(0) = 2$
$$\Rightarrow \quad \int\frac{dy}{y + 3} = \int dx$$
$$\Rightarrow \quad \log_e|y + 3| = x + C$$
But $y(0) = 2$
$\therefore \quad \log_e|2 + 3| = 0 + C$
$\Rightarrow \quad C = \log_e 5$
$\Rightarrow \quad \log_e|y + 3| = x + \log_e 5$
When $x = \log_e 2$, then
$\Rightarrow \log_e|y + 3| = \log_e 2 + \log_e 5 = \log_e 10$
$\therefore \quad y + 3 = 10 \Rightarrow y = 7$

11 Clearly, $\int\frac{y}{\sqrt{1 - y^2}}dy = \int dx$
$$\Rightarrow \quad -\sqrt{1 - y^2} = x + C$$
$$\Rightarrow (x + C)^2 + y^2 = 1$$
Hence, the centre is $(-C, 0)$ and radius is 1.

12 We have, $\frac{dy}{dx} = \frac{ax + 3}{2y + f}$
$$\Rightarrow \quad (ax + 3)\,dx = (2y + f)\,dy$$
On integrating, we obtain
$$a\cdot\frac{x^2}{2} + 3x = y^2 + fy + C$$
$$\Rightarrow -\frac{a}{2}x^2 + y^2 - 3x + fy + C = 0$$
This will represent a circle, if $-\frac{a}{2} = 1$
[$\because$ coefficient of x^2 = coefficient of y^2]
$\Rightarrow \quad a = -2$

13 $(x + 1)\frac{dy}{dx} = y - y^2$
$$\Rightarrow \quad \frac{dy}{y(1 - y)} = \frac{dx}{x + 1}$$
$$\Rightarrow \quad \left(\frac{1}{y} + \frac{1}{1 - y}\right)dy = \frac{dx}{x + 1}$$
$$\Rightarrow \quad \log y - \log(1 - y) = \log(x + 1) + \log C$$
$$\Rightarrow \quad \frac{y}{1 - y} = C(x + 1)$$
$$\Rightarrow \quad (x + 1)(1 - y) = Cy.$$

14 We have, $\frac{dy}{dx} = e^{ax}.e^{by}$
$$\Rightarrow \quad e^{-by}dy = e^{ax}dx$$
$$\Rightarrow \quad -\frac{1}{b}e^{-by} = \frac{1}{a}e^{ax} + C$$
$$\Rightarrow \quad be^{ax} + ae^{-by} + C' = 0 \quad [C' = abC]$$

15 We have,
$$\frac{dy}{dx} = e^{x-y} + x^2e^{-y} = e^{-y}(x^2 + e^x)$$
On separating the variables, we get
$$e^y dy = (x^2 + e^x)dx$$
On integrating both sides, we get
$$e^y = \frac{x^3}{3} + e^x + C$$

16 We have,
$$(2 + \sin x)\frac{dy}{dx} + (y + 1)\cos x = 0$$
$$\Rightarrow \quad \frac{dy}{y + 1} = -\frac{\cos x}{2 + \sin x}dx$$
$$\Rightarrow \quad \int\frac{dy}{y + 1} = \int -\frac{\cos x}{2 + \sin x}dx$$
$$\Rightarrow \quad \ln(y + 1) = -\ln(2 + \sin x) + \ln C$$
$$\Rightarrow \quad (y + 1)(2 + \sin x) = C$$
Now, $\quad y(0) = 1$

$\Rightarrow \quad (2)(2+0) = C$

$\Rightarrow \quad C = 4$

Thus, $(y+1)(2+\sin x) = 4$

Now, at $x = \frac{\pi}{2}$, $\quad (y+1)\left(2+\sin\frac{\pi}{2}\right) = 4$

$$(y+1)(2+1) = 4$$

$$y = \frac{4}{3} - 1 = \frac{1}{3}$$

17 Given, $\frac{dy}{dx} = 1 - \frac{1}{x^2}$

$\Rightarrow dy = \left(1 - \frac{1}{x^2}\right)dx$

$\Rightarrow y = x + \frac{1}{x} + C \quad$ [integrating]

Since, the curve passing through the point $\left(2, \frac{7}{2}\right)$.

i.e. $\frac{7}{2} = 2 + \frac{1}{2} + C \Rightarrow C = 1$

$\therefore \quad y = x + \frac{1}{x} + 1 \quad \ldots(i)$

Now, at $x = -2$

Ordinate, $y = -2 - \frac{1}{2} + 1 = -3/2$

18 We have, $(x+y)^2 \frac{dy}{dx} = 4$

On putting $x + y = v$ and $1 + \frac{dy}{dx} = \frac{dv}{dx}$,

we get

$\Rightarrow \quad v^2\left(\frac{dv}{dx} - 1\right) = 4$

$\Rightarrow \quad v^2 \frac{dv}{dx} = v^2 + 4.$

$\therefore \quad \frac{v^2 + 4 - 4}{v^2 + 4} dv = dx$

$\Rightarrow v - 2\tan^{-1}(v/2) = x + C$

$\Rightarrow x + y - 2\tan^{-1}\left(\frac{x+y}{2}\right) = x + C$

Now, $\quad y(0) = 0 \Rightarrow C = 0.$

$\therefore \quad y = 2\tan^{-1}\left(\frac{x+y}{2}\right)$

19 We have, $\frac{dy}{dx} + 1 = e^{x+y}$

On putting $x + y = z$ and $1 + \frac{dy}{dx} = \frac{dz}{dx}$,

we get

$\therefore \quad \frac{dz}{dx} = e^z$

$\Rightarrow \quad e^{-z}dz = dx$

On integrating both sides, we get

$$\frac{e^{-z}}{-1} = x + C$$

$\Rightarrow \quad x + e^{-(x+y)} + C = 0$

20 $\frac{dy}{dx} = \frac{y}{x} + \frac{f(y/x)}{f'(y/x)}$

On putting $y/x = v$ and $\frac{dy}{dx} = v + x\frac{dv}{dx}$,

we get

$$\frac{f'(v)}{f(v)}dv = \frac{dx}{x}$$

$\Rightarrow \log|f(v)| = \log|x| + \log C, C > 0$

$\Rightarrow |f(y/x)| = C|x|, C > 0.$

21 Given, $\frac{dy}{dx} = \left(\frac{y}{x}\right)\left[\log\left(\frac{y}{x}\right) + 1\right]$

Put $y = tx$,

and $\quad \frac{dy}{dx} = t + x\frac{dt}{dx}$

Then, we get $t + x\frac{dt}{dx} = t\log t + t$

$\Rightarrow \quad \frac{dt}{t\log t} = \frac{dx}{x}$

$\Rightarrow \quad \log\log t = \log x + \log C$

$\therefore \quad \log\left(\frac{y}{x}\right) = Cx$

22 Given, $\frac{dx}{dy} + \frac{x^2 - xy + y^2}{y^2} = 0$

$\Rightarrow \quad \frac{dx}{dy} + \left(\frac{x}{y}\right)^2 - \left(\frac{x}{y}\right) + 1 = 0$

Put $\quad v = \frac{x}{y}$ or $x = vy$

and $\quad \frac{dx}{dy} = v + y\frac{dv}{dy}$

Then, we get $v + y\frac{dv}{dy} + v^2 - v + 1 = 0$

$\Rightarrow \quad \frac{dv}{v^2+1} + \frac{dy}{y} = 0$

$\Rightarrow \tan^{-1}(v) + \log y + C = 0 \quad$ [integrating]

$\therefore \quad \tan^{-1}\left(\frac{x}{y}\right) + \log y + C = 0$

23 We have $\frac{dy}{dx} + \frac{y(x+y)}{x^2} = 0$

On putting $\quad y = vx$, we get

$$v + x\frac{dv}{dx} + v(1+v) = 0$$

or $\quad \frac{dx}{x} + \frac{dv}{v(v+2)} = 0$

or $\frac{dx}{x} + \frac{1}{2}\left(\frac{1}{v} - \frac{1}{v+2}\right)dv = 0$

On integrating, we get

$\log x + \frac{1}{2}[\log v - \log(v+2)] + \log C = 0$

or $\quad \log(v+2) = \log x^2 v C^2$

or $\quad y + 2x = C^2 x^2 y.$

24 Given, $\frac{dy}{dx} = \frac{xy}{x^2 + y^2}$

Put $y = vx \Rightarrow \frac{dy}{dx} = v + x\frac{dv}{dx}$, we get

$\Rightarrow \quad v + \frac{xdv}{dx} = \frac{v}{1+v^2}$

$\Rightarrow \quad x\frac{dv}{dx} = \frac{v}{1+v^2} - v = -\frac{v^3}{1+v^2}$

$\Rightarrow \quad \frac{1+v^2}{v^3}dv = -\frac{dx}{x}$

$\Rightarrow -\frac{1}{2v^2} + \log v = -\log x + C$

$\Rightarrow \frac{x^2}{2y^2} + C = \log\frac{y}{x} + \log x = \log y.$

Now, $\quad x = 1, y = 1$

$\Rightarrow \quad C = -1/2$

and $\quad x = x_0, y = e$

$\Rightarrow \quad \frac{x_0^2}{2e^2} - \frac{1}{2} = 1$

$\Rightarrow \quad x_0^2 = 3e^2.$

25 We have,

$$y(1+xy)dx = xdy$$

$\Rightarrow \quad \frac{dy}{dx} = \frac{y(1+xy)}{x} = \frac{y}{x} + y^2$

On putting $y = vx$ and $\frac{dy}{dx} = v + x\frac{dv}{dx}$,

we get

$$v + x\frac{dv}{dx} = v + v^2x^2$$

$\Rightarrow \quad \frac{dv}{v^2} = xdx$

$\Rightarrow \quad \int\frac{dv}{v^2} = \int xdx$

$\Rightarrow \quad -\frac{1}{v} = \frac{x^2}{2} + C$

$\Rightarrow \quad -\frac{x}{y} = \frac{x^2}{2} + C$

Put $(1, -1)$, then

$$C = \frac{1}{2}$$

$\therefore \quad -\frac{x}{y} = \frac{x^2}{2} + \frac{1}{2}$

Now, put $x = -\frac{1}{2}, y = \frac{4}{5}$

$\therefore \quad f\left(-\frac{1}{2}\right) = \frac{4}{5}$

26 Given, $(1+x^2)\frac{dy}{dx} + 2xy = 4x^2$

$\Rightarrow \quad \frac{dy}{dx} + \frac{2x}{1+x^2}y = \frac{4x^2}{1+x^2}$

$\therefore \quad \text{IF} = e^{\int\frac{2x}{1+x^2}dx} = e^{\log(1+x^2)} = 1 + x^2$

and the solution is

$$y\cdot(1+x^2) = \int 4x^2\,dx + C$$

$$y(1+x^2) = \frac{4x^3}{3} + C$$

Now, $x = 0, y = 0$

$\Rightarrow \quad C = 0$

$\therefore \quad y(1+x^2) = \frac{4x^3}{3}$

$\Rightarrow 3y(1+x^2) = 4x^3$

27 We have, $\frac{dy}{dx} + \frac{3x^2}{1+x^3}y = \frac{\sin^2 x}{1+x^3}$

Since, it is a linear equation with

$$P = \frac{3x^2}{1+x^3}$$

$\therefore \text{IF} = e^{\int P\,dx} = e^{\log(1+x^3)} = 1 + x^3$

and the solution is

$$y(1+x^3)=\int \frac{\sin^2 x}{1+x^3}(1+x^3)\,dx$$

$$=\int \frac{1-\cos 2x}{2}\,dx$$

$$\therefore\ y(1+x^3)=\frac{1}{2}x-\frac{\sin 2x}{4}+C$$

28 Given, $\frac{dy}{dx}-y\tan x=2x\sec x, y(0)=0$

$$\text{IF}=e^{-\int \tan x dx}=e^{-\log \sec x}$$

$$\text{IF}=\cos x$$

$\therefore\ \cos x\cdot y=\int 2x\sec x\cdot\cos x dx$

$\Rightarrow\ \cos x\cdot y=x^2+C$

$\Rightarrow\ y(0)=0\Rightarrow C=0$

$\therefore\ y=x^2\sec x$

and $y'=2x\cdot\sec x+x^2\sec x\cdot\tan x$

$$y\left(\frac{\pi}{4}\right)=\frac{\pi^2}{16}\cdot\sqrt{2}=\frac{\pi^2}{8\sqrt{2}}$$

$$y'\left(\frac{\pi}{4}\right)=\frac{\pi}{2}\cdot\sqrt{2}+\frac{\pi^2}{16}\cdot\sqrt{2}$$

$$y\left(\frac{\pi}{3}\right)=\frac{\pi^2}{9}\cdot 2=\frac{2\pi^2}{9}$$

$$y'\left(\frac{\pi}{3}\right)=2\frac{\pi}{3}\cdot 2+\frac{\pi^2}{9}\cdot 2\cdot\sqrt{3}$$

$$=\frac{4\pi}{3}+\frac{2\pi^2\sqrt{3}}{9}$$

29 This is the linear equation of the form $\frac{dy}{dx}+Py=Q$.

where, $P=\tan x$ and $Q=x^m\cos x$

Now, integrating factor (IF)

$$=e^{\int P dx}=e^{\int \tan x dx}$$

$$=e^{\log\sec x}=\sec x$$

and the solution is given by,

$$y\cdot e^{\int P dx}=\int Q\cdot e^{\int P dx}\,dx+C$$

$$\Rightarrow y\cdot\sec x=\int x^m\cdot\cos x\cdot\sec x dx+C$$

$$\Rightarrow\ y\sec x=\frac{x^{m+1}}{m+1}+C$$

$$\therefore (m+1)y=x^{m+1}\cos x + C(m+1)\cos x$$

30 Given differential equation is $\frac{dP}{dt}-\frac{1}{2}P(t)=-200$, which is a linear differential equation.

Here, $P(t)=\frac{-1}{2}$ and $Q(t)=-200$

Now, $\text{IF}=e^{\int -\left(\frac{1}{2}\right)dt}=e^{-\frac{t}{2}}$

and the solution is

$P(t).\text{IF}=\int Q(t)\,\text{IF}\,dt+K$

$$P(t)\cdot e^{-\frac{t}{2}}=-\int 200\,e^{-\frac{t}{2}}\,dt+K$$

$$P(t)\cdot e^{-\frac{t}{2}}=400\,e^{-\frac{t}{2}}+K$$

$$\Rightarrow P(t)=400+Ke^{\frac{t}{2}}$$

If $P(0)=100$, then $K=-300$

$$\Rightarrow P(t)=400-300\,e^{\frac{t}{2}}$$

31 We have, $(xy^5+2y)dx=xdy$

$$\Rightarrow\ x\frac{dy}{dx}-2y=xy^5$$

$$\Rightarrow\ \frac{dy}{dx}-\frac{2y}{x}=y^5$$

$$\Rightarrow y^{-5}\frac{dy}{dx}-\frac{2y^{-4}}{x}=1 \quad \ldots\text{(i)}$$

Put, $y^{-4}=t$

$$\Rightarrow\ -4y^{-5}\frac{dy}{dx}=\frac{dt}{dx}$$

$$\Rightarrow\ y^{-5}\frac{dy}{dx}=\frac{-1}{4}\frac{dt}{dx} \quad \ldots\text{(ii)}$$

From Eqs. (i) and (ii), we get

$$-\frac{1}{4}\frac{dt}{dx}-\frac{2t}{x}=1$$

$$\Rightarrow\ \frac{dt}{dx}+\frac{8t}{x}=-4$$

Now, $\text{IF}=e^{\int\frac{8}{x}dx}=e^{8\log x}=x^8$

and the solution is

$$t\cdot x^8=\int(-4)x^8dx+C$$

$$\Rightarrow\ \frac{x^8}{y^4}=-\frac{4\cdot x^9}{9}+C$$

$$\Rightarrow 9x^8+4x^9\cdot y^4=9y^4C$$

32 We have, $\sin x\frac{dy}{dx}+y\cos x=4x$

$\Rightarrow\ \sin x dy+y\cos x dx=4x dx$

$\Rightarrow\ d(y\sin x)=4x dx$

On integrating both sides, we get

$$y\sin x=2x^2+C$$

Since, it is passes through $\left(\frac{\pi}{2},0\right)$

$$\therefore\ 0=\frac{\pi^2}{2}+C\Rightarrow C=-\pi^2/2$$

$$\Rightarrow\ y\sin x=2x^2-\pi^2/2$$

$$\Rightarrow\ y=2x^2\operatorname{cosec}x-\frac{\pi^2}{2}\operatorname{cosec}x$$

$$\Rightarrow y(\pi/6)=2\left(\frac{\pi^2}{36}\right)\operatorname{cosec}\frac{\pi}{6}-\frac{\pi^2}{2}\operatorname{cosec}\pi/6$$

$$=2\left(\frac{\pi^2}{36}\right)2-\frac{\pi^2}{2}\cdot 2$$

$$=-\frac{8\pi^2}{9}$$

33 **Statement I** Let the equation of parabola whose axis is the axis of x and vertex at the origin is

$$y^2=4ax$$

$$\Rightarrow\ 2y\frac{dy}{dx}=4a$$

$$\Rightarrow\ \frac{dy}{dx}=\frac{2a}{y}$$

$$\Rightarrow\ \frac{dy}{dx}\propto\frac{1}{y}$$

[where, $a\to$ parameter]

Statement II $y^2=4ax$ ...(i)

$$\Rightarrow\ 2y\frac{dy}{dx}=4a$$

$$\Rightarrow\ a=\frac{y}{2}\cdot\frac{dy}{dx} \quad \text{[from Eq. (i)]}$$

$$\Rightarrow\ y^2=4x\cdot\frac{y}{2}\cdot\frac{dy}{dx}$$

$$\Rightarrow\ y^2=2xy\frac{dy}{dx}$$

$$\Rightarrow\ y=2x\cdot\frac{dy}{dx},$$

which has order = 1 and degree = 1

34 Let $c_1+c_2+c_3e^{c_4}=A$ [constant]

Then, $y=Ax$

$$\Rightarrow\ \frac{dy}{dx}=A$$

$$\Rightarrow\ y=x\frac{dy}{dx}$$

$$\therefore\ x\frac{dy}{dx}=y$$

SESSION 2

1 Given, $y=c_1e^{c_2x}$

$$\Rightarrow\ y'=c_1c_2e^{c_2x}\Rightarrow c_2=\frac{y'}{y}$$

and $y''=c_1c_2^2e^{c_2x}\Rightarrow y''=y.\left(\frac{y'}{y}\right)^2$

$\Rightarrow\ yy''=(y')^2$

2 Given,

$$p'(t)=\frac{dp(t)}{dt}=0.5\,p(t)-450$$

$$\Rightarrow\ \frac{2dp(t)}{p(t)-900}=dt$$

$$\Rightarrow\ \int\frac{2dp(t)}{p(t)-900}=\int dt$$

$\Rightarrow\ 2\log|p(t)-900|=t+C$

To find the value of C, let's substitute $t=0$ and $p(0)=850$

$\Rightarrow\ 2\log|p(0)-900|=0+C$

$\Rightarrow\ C=2\log|850-900|$

$\Rightarrow\ C=2\log 50$

Now, $2\log|p(t)-900|=t+2\log 50$

Now, put $p(t)=0$, then

$$2\log|0-900|=t+2\log 50$$

$$\Rightarrow\ t=2\log\left|\frac{900}{50}\right|=2\log 18$$

3 Given slope at (x,y) is

$$\frac{dy}{dx}=\frac{y}{x}+\sec(y/x)$$

Let $\frac{y}{x}=t\Rightarrow y=xt$

and $\frac{dy}{dx}=t+x\frac{dt}{dx}$

Now, $t+x\frac{dt}{dx}=t+\sec(t)$

$$\int \cos t \, dt = \int \frac{1}{x} dx$$
$$\sin t = \ln x + C$$
$$\sin(y/x) = \ln x + C$$
$\because$ This curve passes through $(1, \pi/6)$
$$\sin(\pi/6) = \ln(1) + C \Rightarrow C = \frac{1}{2}$$
Thus, $\sin\left(\frac{y}{x}\right) = \ln x + \frac{1}{2}$

4 We have, $y'(x) + y(x)g'(x) = g(x)g'(x)$

Linear differential equation with integrating factor $e^{g(x)}$

$\Rightarrow \; y(x)\cdot e^{g(x)} = \int g(x)\cdot g'(x)\cdot e^{g(x)} dx$

$\Rightarrow \; y(x)\cdot e^{g(x)} = e^{g(x)}(g(x) - 1) + C$

Since, $y(0) = 0$ and $g(0) = 0$, therefore $C = 1$

$\Rightarrow \; y(x) = (g(x) - 1) + e^{-g(x)}$

$\Rightarrow \; y(2) = (g(2) - 1) + e^{-g(2)} = 0$, as $g(2) = 0$.

5 Here, $\frac{dx}{dy} + \frac{1}{y^2}\cdot x = \frac{1}{y^3}$

[linear differential equation in x]

Clearly, $\quad \text{IF} = e^{\int \frac{1}{y^2} dy} = e^{-\frac{1}{y}}$

Now, complete solution is,
$$x\cdot e^{-\frac{1}{y}} = \int \frac{1}{y^3}\cdot e^{-\frac{1}{y}} dy$$
$$\Rightarrow \; x\cdot e^{-\frac{1}{y}} = \int \frac{1}{y}\cdot\frac{1}{y^2}\cdot e^{-\frac{1}{y}} dy$$
Put $\; -\frac{1}{y} = t \Rightarrow \frac{1}{y^2} dy = dt$
$$\Rightarrow \; xe^{-\frac{1}{y}} = \int -t\cdot e^t dt$$
$$\Rightarrow \; xe^{-\frac{1}{y}} = -\{t\cdot e^t - \int 1\cdot e^t dt\} + C$$
$$\Rightarrow \; xe^{-\frac{1}{y}} = -te^t + e^t + C$$
$$\Rightarrow \; xe^{-\frac{1}{y}} = \frac{1}{y}\cdot e^{-\frac{1}{y}} + e^{-\frac{1}{y}} + C$$
$$\Rightarrow \; e^{-1} = e^{-1} + e^{-1} + C \quad [\because y(1) = 1]$$
$$\Rightarrow \; C = -\frac{1}{e}$$
$$\Rightarrow \; xe^{-\frac{1}{y}} = \frac{1}{y}\cdot e^{-\frac{1}{y}} + e^{-\frac{1}{y}} - \frac{1}{e}$$
$$\therefore \; x = \frac{1}{y} + 1 - \frac{1}{e}\cdot e^{\frac{1}{y}}$$

6 On putting $10x + 6y = t$ and $10 + 6\frac{dy}{dx} = \frac{dt}{dx}$, we get
$$\frac{dt}{dx} = 6\sin t + 10 = 6\left(\frac{2\tan t/2}{1 + \tan^2 t/2}\right) + 10$$
$$\Rightarrow \frac{\sec^2 t/2}{10\tan^2 t/2 + 12\tan t/2 + 10} dt = dx$$
$$\Rightarrow \; \frac{dz}{5z^2 + 6z + 5} = dx \qquad [z = \tan(t/2)]$$
$$\Rightarrow \; \frac{1}{5}\int \frac{dz}{\left(z + \frac{3}{5}\right)^2 + (4/5)^2} = \int dx$$
$$\Rightarrow \; \frac{1}{4}\tan^{-1}\frac{5z + 3}{4} = x + C$$
$$\Rightarrow \; 5\tan t/2 + 3 = 4\tan 4(x + C)$$
$$\Rightarrow \; 5\tan(5x + 3y) + 3 = 4\tan 4(x + C)$$
Now, $x = 0, y = 0 \Rightarrow C = \frac{1}{4}\tan^{-1}(3/4)$

Hence, equation of the curve is
$$5\tan(5x + 3y) = \frac{25\tan 4x}{4 - 3\tan 4x}$$
$$\Rightarrow \; y = \frac{1}{3}\left\{\tan^{-1}\frac{5\tan 4x}{4 - 3\tan 4x} - 5x\right\}.$$

7 Given, $\frac{d\{V(t)\}}{dt} = -k(T - t)$
$$\therefore \; d\{V(t)\} = -k(T - t)\,dt \qquad \ldots\text{(i)}$$
$$\Rightarrow \; \int_0^T d\{V(t)\} = \int_0^T -k(T - t)dt$$
$$\Rightarrow \; V(T) - V(0) = k\left[\frac{(t - T)^2}{2}\right]_0^T$$
$$\Rightarrow V(T) - I = \frac{k}{2}[(T - T)^2 - (0 - T)^2]$$
[$\because$ when $t = 0$, then $V(t) = I$]
$$\therefore \; V(T) = I - \frac{k}{2}T^2$$

8 On putting $x = \tan A$, and $y = \tan B$ in the given relation, we get

$\cos A + \cos B = \lambda(\sin A - \sin B)$
$$\Rightarrow \; \tan\left(\frac{A - B}{2}\right) = \frac{1}{\lambda}$$
$$\Rightarrow \tan^{-1} x - \tan^{-1} y = 2\tan^{-1}\left(\frac{1}{\lambda}\right)$$
On differentiating w.r.t. x, we get
$$\frac{1}{1 + x^2} - \frac{1}{1 + y^2}\cdot\frac{dy}{dx} = 0$$
$$\Rightarrow \; \frac{dy}{dx} = \frac{1 + y^2}{1 + x^2}$$
Clearly, it is a differential equation of degree 1.

9 $|y|\sqrt{1 + (dx/dy)^2} = 1$
$$\Rightarrow \; y^2\left(1 + \left(\frac{dx}{dy}\right)^2\right) = 1$$
$$\therefore \; \frac{dy}{dx} = \pm\frac{y}{\sqrt{1 - y^2}}$$
$$\Rightarrow \int \frac{\sqrt{1 - y^2}}{y} dy = \pm x + C.$$
On putting $y = \sin\theta$ and $dy = \cos\theta d\theta$, we get
$$C \pm x = \int \frac{\cos\theta\cdot\cos\theta d\theta}{\sin\theta}$$
$$= \int(\text{cosec}\theta - \sin\theta)d\theta$$
$$\Rightarrow \; C \pm x = \log|\text{cosec}\theta - \cot\theta| + \cos\theta$$
$$\Rightarrow \; C \pm x = \log\left|\frac{1 - \sqrt{1 - y^2}}{y}\right| + \sqrt{1 - y^2}.$$

10 Given, $\quad \cos x\,dy = y\sin x\,dx - y^2 dx$
$$\Rightarrow \; \frac{1}{y^2}\frac{dy}{dx} - \frac{1}{y}\tan x = -\sec x$$
Put $\quad -\frac{1}{y} = z \Rightarrow \frac{1}{y^2}\frac{dy}{dx} = \frac{dz}{dx}$
$$\therefore \; \frac{dz}{dx} + (\tan x)z = -\sec x$$
This is a linear differential equation.

Now, $\text{IF} = e^{\int \tan x\,dx} = e^{\log\sec x} = \sec x$

Hence, the solution is

$z\cdot(\sec x) = \int -\sec x\cdot\sec x\,dx + C_1$
$$\Rightarrow \; -\frac{1}{y}\sec x = -\tan x + C_1$$
$$\therefore \; \sec x = y(\tan x + C)$$
where $C = -C_1$.

11 We have, $(1 - x^2)\frac{dy}{dx} + xy = ax$
$$\Rightarrow \; \frac{dy}{dx} + \frac{x}{1 - x^2}y = \frac{ax}{1 - x^2} \qquad \text{[L. D. E.]}$$
$$\text{IF} = e^{\int(x/1 - x^2)dx} = e^{-\frac{1}{2}\log|1 - x^2|}$$
$$= \begin{cases} \frac{1}{\sqrt{1 - x^2}}, & \text{if } -1 < x < 1 \\ \frac{1}{\sqrt{x^2 - 1}}, & \text{if } x < -1 \text{ or } x > 1. \end{cases}$$
If $-1 < x < 1$, then solution is
$$y\cdot\frac{1}{\sqrt{1 - x^2}} = a\int \frac{x}{(1 - x^2)^{3/2}} dx + C$$
$$y\cdot\frac{1}{\sqrt{1 - x^2}} = \frac{a}{\sqrt{1 - x^2}} + C$$
$$\Rightarrow \; y = a + C\sqrt{1 - x^2}$$
$$\Rightarrow \; (y - a)^2 = C^2(1 - x^2)$$
$$\Rightarrow \; C^2x^2 + (y - a)^2 = C^2$$
which is an ellipse.

If $x < -1$ or $x > 1$, solution is
$$y\cdot\frac{1}{\sqrt{x^2 - 1}} = C + a\int \frac{x}{(1 - x^2)\sqrt{x^2 - 1}} dx$$
$$= C - a\int \frac{x}{(x^2 - 1)^{3/2}} dx = C + \frac{a}{\sqrt{x^2 - 1}}$$
$$\Rightarrow \; y = a + C\sqrt{x^2 - 1}$$
$$\Rightarrow \; (y - a)^2 = C^2x^2 - c^2$$
$$\Rightarrow \; C^2x^2 - (y - a)^2 = C^2$$
which represents a hyperbola.

12 Given,
$$\frac{dy}{dx} = \frac{(x + 1)^2 + y - 3}{x + 1} = (x + 1) + \frac{y - 3}{x + 1}$$
Putting $x + 1 = X, y - 3 = Y, \frac{dy}{dx} = \frac{dY}{dX}$,

the equation becomes
$$\frac{dY}{dX} = X + \frac{Y}{X}$$

or $\dfrac{dY}{dX} - \dfrac{1}{X} \cdot Y = X$ [L. D. E.]

$\text{IF} = e^{\int (-1/X) dX} = e^{-\log X} = X^{-1} = \dfrac{1}{X}$

$\therefore$ The solution is

$$Y \cdot \left(\frac{1}{X}\right) = C + \int X \cdot \left(\frac{1}{X}\right) dx = C + X$$

or $\dfrac{(y-3)}{(x+1)} = C + x + 1$

Now, $x = 2, y = 0$

$\Rightarrow \dfrac{0-3}{2+1} = C + 2 + 1$

$\Rightarrow C = -4$

$\therefore$ The equation of the curve is

$$\frac{y-3}{x+1} = x - 3 \text{ or } y = x^2 - 2x.$$

13 We have, $\dfrac{x + \dfrac{x^3}{3!} + \dfrac{x^5}{5!} + \dots}{1 + \dfrac{x^2}{2!} + \dfrac{x^4}{4!} + \dots} = \dfrac{dx - dy}{dx + dy}$

On applying componendo and dividendo, we get

$$\frac{\left(x + \frac{x^3}{3!} + \frac{x^5}{5!} + \dots\right) + \left(1 + \frac{x^2}{2!} + \frac{x^4}{4!} + \dots\right)}{\left(x + \frac{x^3}{3!} + \frac{x^4}{4!} + \dots\right) - \left(1 + \frac{x^2}{2!} + \frac{x^4}{4!} + \dots\right)}$$

$$= \frac{(dx - dy) + (dx + dy)}{(dx - dy) - (dx - dy)}$$

$$\Rightarrow \frac{\left(1 + x + \frac{x^2}{2!} + \frac{x^3}{3!} + \dots\right)}{-\left(1 - x + \frac{x^2}{2!} - \frac{x^3}{3!} + \dots\right)} = \frac{2dx}{-2dy} = -\frac{dx}{dy}$$

$\Rightarrow \dfrac{e^x}{-e^{-x}} = -\dfrac{dx}{dy}$

$\Rightarrow dy = e^{-2x} dx$

On integrating both sides, we get

$$y = \frac{e^{-2x}}{-2} + C_1$$

$\Rightarrow 2ye^{2x} = -1 + 2C_1 e^{2x}$

$\Rightarrow 2ye^{2x} = Ce^{2X} - 1$, where $C = 2C_1$

14 Equation of tangent is $Y - y = \dfrac{dy}{dx}(X - x)$

Y-intercept of tangent is $y - x\dfrac{dy}{dx}$

From given condition, $y - x\dfrac{dy}{dx} = x^3$,

$\Rightarrow \dfrac{dy}{dx} - \dfrac{1}{x} \cdot y = -x^2$ ...(i)

Now, $\text{IF} = e^{-\int \frac{1}{x} dx} = e^{-\log_e x} = \dfrac{1}{x}$

and solution is

$$\frac{1}{x} \cdot y = -\int x dx = -\frac{x^2}{2} + C$$

$\Rightarrow f(x) = \dfrac{-x^3}{2} + Cx$

Now, $f(1) = 1 \Rightarrow C = 1 + \dfrac{1}{2} = \dfrac{3}{2}$

$\therefore f(x) = \dfrac{-x^3}{2} + \dfrac{3}{2}x$

$\Rightarrow f(-3) = \dfrac{27}{2} - \dfrac{9}{2} = 9$

15 $\lim\limits_{t \to x} \dfrac{t^2 f(x) - x^2 f(x) + x^2 f(x) - x^2 f(t)}{t - x} = 1$

$\lim\limits_{t \to x} \left[(t + x) f(x) + x^2 \left(\dfrac{f(x) - f(t)}{t - x}\right)\right] = 1$

$2xf(x) + x^2(-f'(x)) = 1$

$\Rightarrow \dfrac{x^2 \dfrac{dy}{dx} - 2xy}{x^4} = -\dfrac{1}{x^4}$,

where $y = f(x)$

$\dfrac{d}{dx}\left(\dfrac{y}{x^2}\right) = -\dfrac{1}{x^4}$

$\Rightarrow x^{-2} y = \dfrac{1}{3x^3} + C$

$\Rightarrow x^{-2} y = \dfrac{1}{3x^3} + \dfrac{2}{3}$ [$\because f(1) = 1$]

$\Rightarrow y = \dfrac{1}{3x} + \dfrac{2x^2}{3}$.

16 Given, $\dfrac{dy}{dx} = \dfrac{y\sqrt{y^2 - 1}}{x\sqrt{x^2 - 1}}$

$\Rightarrow \displaystyle\int \frac{dy}{y\sqrt{y^2 - 1}} = \int \frac{dx}{x\sqrt{x^2 - 1}}$

$\Rightarrow \sec^{-1} y = \sec^{-1} x + C$

Now, $x = 2$, $y = \dfrac{2}{\sqrt{3}}$

$\Rightarrow \dfrac{\pi}{6} = \dfrac{\pi}{3} + C \Rightarrow C = -\dfrac{\pi}{6}$

$\therefore y = \sec\left(\sec^{-1} x - \dfrac{\pi}{6}\right)$

$\Rightarrow \dfrac{1}{y} = \cos\left(\cos^{-1} \dfrac{1}{x} - \cos^{-1} \dfrac{\sqrt{3}}{2}\right)$

$$\Rightarrow \frac{1}{y} = \cos\left[\cos^{-1}\left(\frac{\sqrt{3}}{2x} + \sqrt{1 - \frac{1}{x^2}} \cdot \sqrt{1 - \frac{3}{4}}\right)\right]$$

$\therefore \dfrac{1}{y} = \dfrac{\sqrt{3}}{2x} + \dfrac{1}{2}\sqrt{1 - \dfrac{1}{x^2}}$

DAY NINTEEN

Unit Test 2
(Calculus)

1 Let $f:(2,3)\to(0,1)$ be defined by $f(x)=x-[x]$, then $f^{-1}(x)$ is equal to

(a) $x-2$ (b) $x+1$ (c) $x-1$ (d) $x+2$

2 $\int \frac{dx}{\sin x - \cos x + \sqrt{2}}$ is equal to

(a) $-\frac{1}{\sqrt{2}}\tan\left(\frac{x}{2}+\frac{\pi}{8}\right)+C$ (b) $\frac{1}{\sqrt{2}}\tan\left(\frac{x}{2}+\frac{\pi}{8}\right)+C$

(c) $\frac{1}{\sqrt{2}}\cot\left(\frac{x}{2}+\frac{\pi}{8}\right)+C$ (d) $-\frac{1}{\sqrt{2}}\cot\left(\frac{x}{2}+\frac{\pi}{8}\right)+C$

3 $\lim_{n\to\infty}\left(\frac{n^2-n+1}{n^2-n-1}\right)^{n(n-1)}$ is equal to

(a) e (b) e^2 (c) e^{-1} (d) 1

4 If the normal to the curve $y=f(x)$ at the point $(3, 4)$ makes an angle $3\pi/4$ with the positive X-axis, then $f'(3)$ is equal to

(a) -1 (b) $-\frac{3}{4}$ (c) $\frac{4}{3}$ (d) 1

5 The area bounded by $y=\sin^{-1}x$, $x=1/\sqrt{2}$ and X-axis is

(a) $\left(\frac{1}{\sqrt{2}}+1\right)$ sq units (b) $\left(1-\frac{1}{\sqrt{2}}\right)$ sq units

(c) $\frac{\pi}{4\sqrt{2}}$ sq units (d) $\left(\frac{\pi}{4\sqrt{2}}+\frac{1}{\sqrt{2}}-1\right)$ sq units

6 The value of $\int_{-\pi/2}^{\pi/2}\log\left(\frac{2-\sin\theta}{2+\sin\theta}\right)d\theta$ is

(a) 0 (b) 1 (c) 2 (d) None of these

7 The general solution of the differential equation $(2x-y+1)\,dx+(2y-x+1)\,dy=0$ is

(a) $x^2+y^2+xy-x+y=C$
(b) $x^2+y^2-xy+x+y=C$
(c) $x^2-y^2+2xy-x+y=C$
(d) $x^2-y^2-2xy+x-y=C$

8 The function $f(x)=x^4-\frac{x^3}{3}$ is

(a) increasing for $x>\frac{1}{4}$ and decreasing for $x<\frac{1}{4}$
(b) increasing for every value of x
(c) decreasing for every value of x
(d) None of the above

9 If $y=1+x+\frac{x^2}{2!}+\frac{x^3}{3!}+\ldots+\frac{x^n}{n!}$, then $\frac{dy}{dx}$ is equal to

(a) y (b) $y+\frac{x^n}{n!}$ (c) $y-\frac{x^n}{n!}$ (d) $y-1-\frac{x^n}{n!}$

10 The value of $\lim_{x\to1}\frac{x^{1/4}-x^{1/5}}{x^3-1}$ is

(a) $\frac{1}{20}$ (b) $\frac{1}{40}$ (c) $\frac{1}{60}$ (d) $\frac{3}{20}$

11 The differential coefficient of the function $|x-1|+|x-3|$ at the point $x=2$ is

(a) -2 (b) 0
(c) 2 (d) not defined

12 The difference between the greatest and least values of the function $f(x)=\cos x+\frac{1}{2}\cos 2x-\frac{1}{3}\cos 3x$ is

(a) $\frac{2}{3}$ (b) $\frac{8}{7}$ (c) $\frac{9}{4}$ (d) $\frac{3}{8}$

13 In the mean value theorem $\frac{f(b)-f(a)}{b-a}=f'(c)$, if $a=0, b=\frac{1}{2}$ and $f(x)=x(x-1)(x-2)$, then value of c is

(a) $1-\frac{\sqrt{15}}{6}$ (b) $1+\sqrt{15}$ (c) $1-\frac{\sqrt{21}}{6}$ (d) $1+\sqrt{21}$

14 If the sides and angles of a plane triangle vary in such a way that its circumradius remains constant. Then, $\frac{da}{\cos A}+\frac{db}{\cos B}+\frac{dc}{\cos C}$ is equal to (where, da, db and dc are small increments in the sides a, b and c, respectively).

(a) 1 (b) 2 (c) 0 (d) 3

15 $\int_0^{1.5}[x^2]\,dx$, where [.] denotes the greatest integer function, is equal to

(a) $2+\sqrt{2}$ (b) $2-\sqrt{2}$ (c) $-2+\sqrt{2}$ (d) $-2-\sqrt{2}$

16 The integrating factor of the differential equation $\frac{dy}{dx} = y \tan x - y^2 \sec x$, is

(a) $\tan x$ (b) $\sec x$ (c) $-\sec x$ (d) $\cot x$

17 The area bounded by the straight lines $x = 0, x = 2$ and the curve $y = 2^x, y = 2x - x^2$ is

(a) $\frac{4}{3} - \frac{1}{\log 2}$ (b) $\frac{3}{\log 2} + \frac{4}{3}$
(c) $\frac{4}{\log 2} - 1$ (d) $\frac{3}{\log 2} - \frac{4}{3}$

18 The area of the region bounded by $y = |x - 1|$ and $y = 1$ is

(a) 2 sq units (b) 1 sq unit
(c) 1/2 sq unit (d) None of these

19 If $I_1 = \int_{1-k}^{k} x \sin\{x(1-x)\}\, dx$ and

$I_2 = \int_{1-k}^{k} \sin\{x(1-x)\}\, dx$, then

(a) $I_1 = 2I_2$ (b) $2I_1 = I_2$
(c) $I_1 = I_2$ (d) None of these

20 If $I_1 = \int_{\sec^2 z}^{2-\tan^2 z} x f\{x(3-x)\}\, dx$

and $I_2 = \int_{\sec^2 z}^{2-\tan^2 z} f\{x(3-x)\}\, dx$, where f is a continuous function and z is any real number, then $\frac{I_1}{I_2}$ is equal to

(a) $\frac{3}{2}$ (b) $\frac{1}{2}$
(c) 1 (d) None of these

21 If f and g are continuous functions on $[0, \pi]$ satisfying $f(x) + f(\pi - x) = g(x) + g(\pi - x) = 1$, then $\int_0^{\pi} [f(x) + g(x)]\, dx$ is equal to

(a) π (b) 2π (c) $\frac{\pi}{2}$ (d) $\frac{3\pi}{2}$

22 The function $f(x) = \begin{cases} |2x-3| \cdot [x], & x \ge 1 \\ \sin\left(\frac{\pi x}{2}\right), & x < 1 \end{cases}$

(where, $[x]$ denotes the greatest integer $\le x$) is

(a) continuous at $x = 2$
(b) differentiable at $x = 1$
(c) continuous but not differentiable at $x = 1$
(d) None of the above

23 If $y = \sqrt{f(x) + \sqrt{f(x) + \sqrt{f(x) + \dots \infty}}}$, then $\frac{dy}{dx}$ is equal to

(a) $\frac{f'(x)}{2y-1}$ (b) $\frac{f'(x)}{2y+1}$ (c) $\frac{f'(x)}{1-2y}$ (d) $\frac{f'(x)}{4+2y}$

24 If $h(x) = f(x) - (f(x))^2 + (f(x))^3$ for every real numbers x, then

(a) h is increasing whenever f is increasing
(b) h is increasing whenever f is decreasing
(c) h is decreasing whenever f is increasing
(d) Nothing can be said in general

25 If $u = \int e^{ax} \cos bx\, dx$ and $v = \int e^{ax} \sin bx\, dx$, then $(a^2 + b^2)(u^2 + v^2)$ is equal to

(a) $2e^{ax}$ (b) $(a^2 + b^2)e^{2ax}$
(c) e^{2ax} (d) $(a^2 - b^2)e^{2ax}$

26 If $\frac{d[f(x)]}{dx} = g(x)$ for $a \le x \le b$, then $\int_a^b f(x) \cdot g(x)\, dx$ is equal to

(a) $f(b) - f(a)$ (b) $g(b) - g(a)$
(c) $\frac{[f(b)]^2 - [f(a)]^2}{2}$ (d) $\frac{[g(b)]^2 - [g(a)]^2}{2}$

27 If y be the function which passes through $(1, 2)$ having slope $(2x + 1)$. Then, the area bounded between the curve and X-axis is

(a) 6 sq units (b) $\frac{5}{6}$ sq unit
(c) $\frac{1}{6}$ sq unit (d) None of these

28 If $h(x) = \min\{x, x^2\}$, for every real number of x. Then,

(a) h is continuous for all x (b) h is differentiable for all x
(c) $h'(x) = 2$ for all $x > 1$ (d) None of these

29 The function $f(x) = \begin{cases} e^{2x} - 1, & x < 0 \\ ax + \frac{bx^2}{2} - 1, & x \ge 0 \end{cases}$ is continuous and differentiable for

(a) $a = 1, b = 2$ (b) $a = 2, b = 4$
(c) $a = 2$, any b (d) any a, $b = 4$

30 The value of $\lim_{x \to \frac{\pi}{2}} \frac{\int_{\frac{\pi}{2}}^{x} t\, dt}{\sin(2x - \pi)}$ is

(a) ∞ (b) $\frac{\pi}{2}$ (c) $\frac{\pi}{4}$ (d) $\frac{\pi}{8}$

31 Let $f(x) = \left[\sqrt{2} \cos\left(x + \frac{\pi}{4}\right)\right]$, $0 < x \le 2\pi$ (where, $[.]$ denotes the greatest integer $\le x$). The number of points of discontinuity of $f(x)$ are

(a) 3 (b) 4 (c) 5 (d) 6

32 If $f(x) = \tan^{-1}\left[\frac{\sin x}{1 + \cos x}\right]$, then $f'\left[\frac{\pi}{3}\right]$ is equal to

(a) $\frac{1}{2(1 + \cos x)}$ (b) $\frac{1}{2}$
(c) $\frac{1}{4}$ (d) None of these

33 Let $f(x)$ be a polynomial function of the second degree. If $f(1) = f(-1)$ and a_1, a_2, a_3 are in AP, then $f'(a_1), f'(a_2)$ and $f'(a_3)$ are in

(a) AP (b) GP
(c) HP (d) None of these

34 If $y = \cos^{-1}(\cos x)$, then $y'(x)$ is equal to

(a) 1 for all x
(b) -1 for all x
(c) 1 in 2nd and 3rd quadrants
(d) -1 in 3rd and 4th quadrants

35 In $(-4, 4)$ the function $f(x)=\int_{-10}^{x}(t^4-4)\,e^{-4t}dt$ has

(a) no extreme (b) one extreme
(c) two extreme (d) four extreme

36 If $f(x)=kx^3-9x^2+9x+3$ is monotonically increasing in each interval, then

(a) $k<3$ (b) $k\le 3$
(c) $k>3$ (d) None of these

37 The integral

$\int_{-1}^{3}\left(\tan^{-1}\frac{x}{x^2+1}+\tan^{-1}\frac{x^2+1}{x}\right)dx$ is equal to

(a) π (b) 2π
(c) 3π (d) None of these

38 $\lim_{n\to\infty}\frac{1^{99}+2^{99}+3^{99}+\ldots+n^{99}}{n^{100}}$ is equal to

(a) $\frac{9}{100}$ (b) $\frac{1}{100}$ (c) $\frac{1}{99}$ (d) $\frac{1}{101}$

39 If $\int(\sin 2x+\cos 2x)\,dx=\frac{1}{\sqrt{2}}\sin(2x-c)+a$, then the value of a and c is

(a) $c=\frac{\pi}{4}$ and $a=k$

(b) $c=-\frac{\pi}{4}$ and $a=\frac{\pi}{2}$

(c) $c=\frac{\pi}{2}$ and a is an arbitrary constant

(d) None of the above

40 $\int\frac{\cos 2x}{(\cos x+\sin x)^2}dx$ is equal to

(a) $\log\sqrt{\cos x+\sin x}+C$ (b) $\log(\cos x-\sin x)+C$
(c) $\log(\cos x+\sin x)+C$ (d) $-\frac{1}{\cos x+\sin x}+C$

41 If $\int\frac{\cos 4x+1}{\cot x-\tan x}dx=k\cos 4x+C$, then

(a) $k=-\frac{1}{2}$ (b) $k=-\frac{1}{8}$
(c) $k=-\frac{1}{4}$ (d) None of these

42 $\int\frac{dx}{\sin(x-a)\,\sin(x-b)}$ is equal to

(a) $\frac{1}{\sin(a-b)}\log\left|\frac{\sin(x-a)}{\sin(x-b)}\right|+C$

(b) $\frac{-1}{\sin(a-b)}\log\left|\frac{\sin(x-a)}{\sin(x-b)}\right|+C$

(c) $\log\sin(x-a)\cdot\sin(x-b)+C$

(d) $\log\left|\frac{\sin(x-a)}{\sin(x-b)}\right|+C$

43 $\int e^x\frac{(x^2+1)}{(x+1)^2}dx$ is equal to

(a) $\left(\frac{x-1}{x+1}\right)e^x+C$ (b) $e^x\left(\frac{x+1}{x-1}\right)+C$
(c) $e^x(x+1)(x-1)+C$ (d) None of these

44 $I_1=\int\sin^{-1}x\,dx$ and $I_2=\int\sin^{-1}\sqrt{1-x^2}\,dx$, then

(a) $I_1=I_2$ (b) $I_2=\frac{\pi}{2I_1}$
(c) $I_1+I_2=\frac{\pi x}{2}$ (d) $I_1+I_2=\frac{\pi}{2}$

45 $\int_0^{\pi/4}\frac{dx}{\cos^4x-\cos^2x\,\sin^2x+\sin^4x}$ is equal to

(a) $\frac{\pi}{2}$ (b) $\frac{\pi}{4}$
(c) $\frac{\pi}{3}$ (d) None of these

46 If $f(x)=|x-1|+|x-3|+|5-x|,\forall x\in R$. If $f(x)$ is increases, then x belongs to

(a) $(1,\infty)$ (b) $(3,\infty)$ (c) $(5,\infty)$ (d) $(1,3)$

47 The value of $\int\frac{dx}{(1+x)^{1/2}-(1+x)^{1/3}}$ is

(a) $2\lambda^{1/2}+3\lambda^{1/3}+6\lambda^{1/6}+6\ln|\lambda^{1/6}-1|+C$
(b) $2\lambda^{1/2}-3\lambda^{1/3}+6\lambda^{1/6}+6\ln|\lambda^{1/6}-1|+C$
(c) $2\lambda^{1/2}+3\lambda^{1/3}-6\lambda^{1/6}+6\ln|\lambda^{1/6}-1|+C$
(d) $2\lambda^{1/2}+3\lambda^{1/3}+6\lambda^{1/6}-6\ln|\lambda^{1/6}-1|+C$
(where, $\lambda=1+x$)

48 The value of $\int\frac{\sqrt[4]{x}}{\sqrt{x}-1}dx$ is

(a) $-\frac{4}{3}x^{3/4}+4x^{1/4}+2\ln\left|\frac{x^{1/4}-1}{x^{1/4}+1}\right|+C$

(b) $\frac{4}{3}x^{3/4}+4x^{1/4}+2\ln\left|\frac{x^{1/4}-1}{x^{1/4}+1}\right|+C$

(c) $-\frac{4}{3}x^{3/4}-4x^{1/4}+2\ln\left|\frac{x^{1/4}-1}{x^{1/4}+1}\right|+C$

(d) $\frac{4}{3}x^{3/4}-4x^{1/4}+2\ln\left|\frac{x^{1/4}-1}{x^{1/4}+1}\right|+C$

49 The value of n for which the function

$$f(x)=\begin{cases}\dfrac{((5)^x-1)^3}{\sin\left(\dfrac{x}{n}\right)\cdot\log\left(1+\dfrac{x^2}{3}\right)}, & x\ne 0\\ 15(\log 5)^3, & x=0\end{cases}$$

may be continuous at $x=0$, is

(a) 5 (b) 3 (c) 4 (d) 2

50 The longest distance of the point $(4, 0)$ from the curve $2x(1-x)=y^2$ is equal to

(a) 3 units (b) 4.5 units
(c) 5 units (d) None of these

51 The normal to the curve $5x^5-10x^3+x+2y+4=0$ at $P(0,-2)$ meets the curve again at two points at which equation of tangents to the curve is equal to

(a) $y=3x+2$ (b) $y=2(x-1)$
(c) $3y+2x+7=0$ (d) None of these

52 $\int_{-1}^{2} \{|x-1|+[x]\}\, dx$, where $[x]$ is greatest integer is equal to

(a) 8 (b) 9 (c) $\frac{5}{2}$ (d) 4

53 The equation of the curve passing through the point $\left(\frac{1}{2}, \frac{\pi}{8}\right)$ and having slope of tangent at any point (x, y) as $(y/x) - \cos^2(y/x)$ is equal to

(a) $y = x \tan^{-1}\left(\log \frac{e}{2x}\right)$ (b) $x = y \tan^{-1}\left(\frac{e}{x^2}\right)$

(c) $y = x^2 \tan^{-1}\left(\frac{2x}{e}\right)$ (d) None of these

54 If $x\, dy = y\,(dx + y\, dy)$, $y(1) = 1$ and $y(x) > 0$, then $y(-3)$ is equal to

(a) 3 (b) 2 (c) 1 (d) 0

55 $\int_0^{\pi/4} \frac{\sin x + \cos x}{9 + 16 \sin 2x}\, dx$ is equal to

(a) $\frac{1}{20} \log 3$ (b) $\log 3$

(c) $\frac{1}{20} \log 5$ (d) None of these

56 If $f(x) = \begin{cases} e^{\cos x} \sin x, & |x| \le 2 \\ 2, & \text{otherwise} \end{cases}$, then $\int_{-2}^{3} f(x)\, dx$ is equal to

(a) 0 (b) 1 (c) 2 (d) 3

57 The area bounded by curves $y = \cos x$ and $y = \sin x$ and ordinates $x = 0$ and $x = \frac{\pi}{4}$ is

(a) $\sqrt{2}$ (b) $\sqrt{2} + 1$ (c) $\sqrt{2} - 1$ (d) $\sqrt{2}(\sqrt{2} - 1)$

Direction (Q. Nos. 58-66) *Each of these questions contains two statements : Statement* I *(Assertion) and Statement* II *(Reason). Each of these questions also has four alternative choices, only one of which is the correct answer. You have to select one of the codes (a), (b), (c) and (d) given below.*

(a) Statement I is true, Statement II is true; Statement II is a correct explanation for Statement I

(b) Statement I is true, Statement II is true; Statement II is not a correct explanation for Statement I

(c) Statement I is true; Statement II is false

(d) Statement I is false; Statement II is true

58 **Statement I** The tangents to curve $y = 7x^3 + 11$ at the points, where $x = 2$ and $x = -2$ are parallel.

Statement II The slope of the tangents at the points, where $x = 2$ and $x = -2$, are equal.

59 **Statement I** The derivative of

$$f(x) = \int_{1/x}^{\sqrt{x}} \cos t^2 dt, (x > 0) \text{ at } x = 1 \text{ is } \left(\frac{3}{2}\right) \cos 1.$$

Statement II $\frac{d}{dx} \int_{\psi(x)}^{\phi(x)} f(t)\, dt = f(\phi(x)) - f(\psi(x))$

60 **Statement I** The solution of the equation

$$x \frac{dy}{dx} + 6y = 3xy^{4/3} \text{ is } y(x) = \frac{1}{(x + cx^2)^3}.$$

Statement II The solution of a linear equation is obtained by multiplying with its integrating factor.

61 **Statement I** $\int \frac{x^2 - 1}{(x^2 + 1)\sqrt{x^4 + 1}}\, dx = \sec^{-1}\left|\frac{x^2 + 1}{x\sqrt{2}}\right| + C$

Statement II $\int \frac{dt}{t\sqrt{t^2 - a}} = \frac{1}{\sqrt{a}} \sec^{-1}\left|\frac{t}{\sqrt{a}}\right| + C$

62 **Statement I** The absolute minimum value of $|x-1| + |x-2| + |x-3|$ is 2.

Statement II The function $|x-1| + |x-2| + |x-3|$ is differentiable on $R \sim \{1, 2\}$.

63 **Statement I** If $f(x) = \max\{x^2 - 2x + 2, |x-1|\}$, then the greatest value of $f(x)$ on the interval $[0, 3]$ is 5.

Statement II Greatest value, $f(3) = \max\{5, 2\} = 5$

64 **Statement I** The point of contact of the vertical tangents to $x = 2 - 3\sin\theta$, $y = 3 + 2\cos\theta$ are $(-1, 3)$ and $(5, 3)$.

Statement II For vertical tangent, $dx/d\theta = 0$

65 Let $f: R \to R$ be differentiable and strictly increasing function throughout its domain.

Statement I If $|f(x)|$ is also strictly increasing function, then $f(x) = 0$ has no real roots.

Statement II At ∞ or $-\infty$ $f(x)$ may approach to 0, but cannot be equal to zero.

66 **Statement I** The area by region $|x + y| + |x - y| \le 2$ is 4 sq units.

Statement II Area enclosed by region $|x + y| + |x - y| \le 2$ is symmetric about X-axis.

ANSWERS

1 (d)	**2** (d)	**3** (b)	**4** (d)	**5** (d)	**6** (a)	**7** (b)	**8** (a)	**9** (c)	**10** (c)
11 (b)	**12** (c)	**13** (c)	**14** (c)	**15** (b)	**16** (b)	**17** (d)	**18** (b)	**19** (b)	**20** (a)
21 (a)	**22** (c)	**23** (a)	**24** (a)	**25** (c)	**26** (c)	**27** (c)	**28** (a)	**29** (c)	**30** (c)
31 (d)	**32** (b)	**33** (a)	**34** (d)	**35** (c)	**36** (c)	**37** (b)	**38** (b)	**39** (a)	**40** (c)
41 (b)	**42** (a)	**43** (a)	**44** (c)	**45** (a)	**46** (b)	**47** (a)	**48** (b)	**49** (a)	**50** (c)
51 (b)	**52** (c)	**53** (a)	**54** (a)	**55** (a)	**56** (c)	**57** (c)	**58** (a)	**59** (c)	**60** (b)
61 (d)	**62** (c)	**63** (b)	**64** (a)	**65** (b)	**66** (b)				

Hints and Explanations

1 Given, $f:(2,3)\to(0,1)$
and $f(x)=x-[x]$

$\therefore \quad f(x)=y=x-2$

$\Rightarrow \quad x=y+2=f^{-1}(y)$

$\Rightarrow \quad f^{-1}(x)=x+2$

2 Let

$$I=\int \frac{dx}{\sqrt{2}\left(\sin x\cdot\sin\frac{\pi}{4}-\cos x\cdot\cos\frac{\pi}{4}+1\right)}$$

$$=\frac{1}{\sqrt{2}}\int \frac{dx}{1-\cos 2\left(\frac{x}{2}+\frac{\pi}{8}\right)}$$

$$=\frac{1}{2\sqrt{2}}\int \operatorname{cosec}^2\left(\frac{x}{2}+\frac{\pi}{8}\right)dx$$

$$=\frac{1}{2\sqrt{2}}\frac{-\cot\left(\frac{x}{2}+\frac{\pi}{8}\right)}{1/2}+C$$

$$=-\frac{1}{\sqrt{2}}\cot\left(\frac{x}{2}+\frac{\pi}{8}\right)+C$$

3 $$\lim_{n\to\infty}\left(\frac{n^2-n+1}{n^2-n-1}\right)^{n(n-1)}$$

$$=\lim_{n\to\infty}\left(\frac{n(n-1)+1}{n(n-1)-1}\right)^{n(n-1)}$$

$$=\lim_{n\to\infty}\frac{\left\{1+\frac{1}{n(n-1)}\right\}^{n(n-1)}}{\left\{1-\frac{1}{n(n-1)}\right\}^{n(n-1)}}=\frac{e}{e^{-1}}=e^2$$

4 Slope of the normal $=\dfrac{-1}{\frac{dy}{dx}}$

$$\Rightarrow \tan\frac{3\pi}{4}=\frac{-1}{\left(\frac{dy}{dx}\right)_{(3,4)}}\Rightarrow\left(\frac{dy}{dx}\right)_{(3,4)}=1$$

$\therefore \quad f'(3)=1$

5 Required area

= Area of rectangle $OABC$ – Area of curve $OBCO$

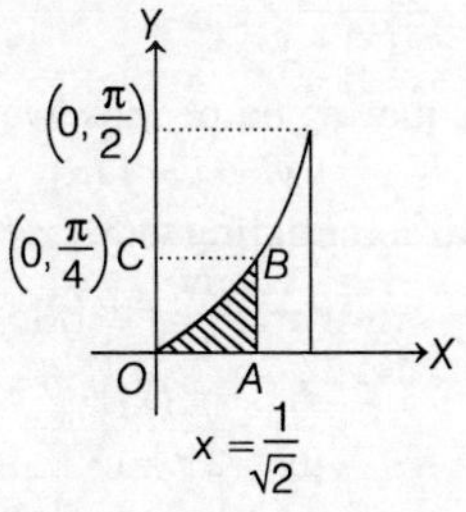

$$=\frac{\pi}{4\sqrt{2}}-\int_0^{\pi/4}\sin y\,dy$$

$$=\frac{\pi}{4\sqrt{2}}+[\cos y]_0^{\pi/4}$$

$$=\left(\frac{\pi}{4\sqrt{2}}+\frac{1}{\sqrt{2}}-1\right)\text{ sq units}$$

6 $$f(-\theta)=\log\left(\frac{2-\sin\theta}{2+\sin\theta}\right)^{-1}$$

$$=-\log\left(\frac{2-\sin\theta}{2+\sin\theta}\right)=-f(\theta)$$

$f(\theta)$ is an odd function of θ.

$$\therefore \quad I=\int_{-\pi/2}^{\pi/2}\log\left(\frac{2-\sin\theta}{2+\sin\theta}\right)d\theta=0$$

7. Given,

$(2x-y+1)\,dx+(2y-x+1)\,dy=0$

$$\Rightarrow \quad \frac{dy}{dx}=\frac{2x-y+1}{x-2y-1},$$

Put $\quad x=X+h,\ y=Y+k$

$$\Rightarrow \quad \frac{dY}{dX}=\frac{2X-Y+2h-k+1}{X-2Y+h-2k-1}$$

Again, put $2h-k+1=0$
and $\quad h-2k-1=0$

On solving,

$h=-1, k=-1$

$$\therefore \quad \frac{dY}{dX}=\frac{2X-Y}{X-2Y}$$

On putting $Y=vX$, we get

$$\Rightarrow \quad v+X\frac{dv}{dX}=\frac{2X-vX}{X-2vX}=\frac{2-v}{1-2v}$$

$$\Rightarrow X\frac{dv}{dX}=\frac{2-2v+2v^2}{1-2v}=\frac{2(v^2-v+1)}{1-2v}$$

$$\therefore \quad \frac{dX}{X}=\frac{(1-2v)}{2(v^2-v+1)}dv$$

Put $\quad v^2-v+1=t$

$\Rightarrow \quad (2v-1)\,dv=dt$

$$\therefore \quad \frac{dX}{X}=-\frac{dt}{2t}$$

On integrating,

$\log X=\log t^{-1/2}+\log C^{1/2}$

$\therefore \quad X=t^{-1/2}C^{1/2}$

$\Rightarrow \quad X=(v^2-v+1)^{-1/2}C^{1/2}$

$\Rightarrow \quad X^2(v^2-v+1)=C$

$$\Rightarrow (x+1)^2\left[\frac{(y+1)^2}{(x+1)^2}-\frac{(y+1)}{(x+1)}+1\right]=C$$

$$\Rightarrow (y+1)^2-(y+1)(x+1)+(x+1)^2=C$$

$\Rightarrow \quad y^2+x^2-xy+x+y=C-1$

$\Rightarrow \quad x^2+y^2-xy+x+y=C$

8 $f(x)=x^4-\dfrac{x^3}{3}$

$\Rightarrow f'(x)=4x^3-x^2$

For increasing,

$4x^3-x^2>0\Rightarrow x^2(4x-1)>0$

Therefore, the function is increasing for $x>1/4$.

Similarly, decreasing for $x<\dfrac{1}{4}$.

9 Given,

$$y=1+x+\frac{x^2}{2!}+\frac{x^3}{3!}+\ldots+\frac{x^n}{n!}$$

$$\Rightarrow \frac{dy}{dx}=0+1+x+\frac{x^2}{2!}+\ldots+\frac{x^{n-1}}{(n-1)!}$$

$$\Rightarrow \frac{dy}{dx}+\frac{x^n}{n!}=1+x+\frac{x^2}{2!}+\ldots+\frac{x^{n-1}}{(n-1)!}+\frac{x^n}{n!}$$

$$\Rightarrow \quad \frac{dy}{dx}=y-\frac{x^n}{n!}$$

10 $$\lim_{x\to1}\frac{x^{1/4}-x^{1/5}}{x^3-1}=\lim_{x\to1}\frac{\frac{1}{4}x^{-3/4}-\frac{1}{5}x^{-4/5}}{3x^2}$$

$$=\frac{\frac{1}{4}-\frac{1}{5}}{3}=\frac{1}{60}$$

11 Given, $f(x)=|x-1|+|x-3|$

$$f(x)=\begin{cases}-(x-1)-(x-3), & x<1\\ (x-1)-(x-3), & 1\le x<3\\ (x-1)+(x-3), & x>3\end{cases}$$

$$=\begin{cases}4-2x, & x<1\\ 2, & 1<x<3\\ 2x-4, & x\ge 3\end{cases}$$

In the neighbourhood of $x=2$, $f(x)=2$

Hence, $f'(x)=0$

12 The given function is periodic with period 2π. So, the difference between the greatest and least values of the function is the difference between these values on the interval $[0,2\pi]$.

Now,

$f'(x)=-(\sin x+\sin 2x-\sin 3x)$

$=-4\sin x\sin(3x/2)\sin(x/2)$

Hence, $x = 0, 2\pi/3, \pi$ and 2π are the critical points.

Also, $f(0) = 1 + \frac{1}{2} - \frac{1}{3} = \frac{7}{6}$

$$f\left(\frac{2\pi}{3}\right) = -\frac{13}{12}, f(\pi) = -\frac{1}{6}$$

and $f(2\pi) = \frac{7}{6}$

∴ Required difference

$$= \frac{7}{6} - \left(-\frac{13}{12}\right) = \frac{27}{12} = \frac{9}{4}$$

13 From mean value theorem,

$$f'(c) = \frac{f(b) - f(a)}{b - a}$$

Given, $a = 0 \Rightarrow f(a) = 0$

and $b = \frac{1}{2} \Rightarrow f(b) = \frac{3}{8}$

$$f'(x) = (x-1)(x-2) + x(x-2) + x(x-1)$$

$$f'(c) = (c-1)(c-2) + c(c-2) + c(c-1)$$

$$= c^2 - 3c + 2 + c^2 - 2c + c^2 - c$$

$\Rightarrow f'(c) = 3c^2 - 6c + 2$

According to mean value theorem,

$$f'(c) = \frac{f(b) - f(a)}{b - a}$$

$$\Rightarrow 3c^2 - 6c + 2 = \frac{(3/8) - 0}{(1/2) - 0} = \frac{3}{4}$$

$$\Rightarrow 3c^2 - 6c + \frac{5}{4} = 0$$

$$\Rightarrow c = \frac{6 \pm \sqrt{36 - 15}}{2 \times 3} = \frac{6 \pm \sqrt{21}}{6}$$

$$= 1 \pm \frac{\sqrt{21}}{6} = 1 - \frac{\sqrt{21}}{6}$$

$$\left[\because \left(1 + \frac{\sqrt{21}}{6}\right) \notin \left(0, \frac{1}{2}\right)\right]$$

14 We know that, in a triangle

$$\angle A + \angle B + \angle C = \pi$$

$\therefore \quad dA + dB + dC = 0$

If R is circumradius, then

$$\frac{a}{\sin A} = \frac{b}{\sin B} = \frac{c}{\sin C} = 2R$$

$$a = 2R \sin A$$

On differentiating, we get

$$da = 2R \cos A \, dA$$

$$\frac{da}{\cos A} = 2R \, dA$$

Similarly, $\frac{db}{\cos B} = 2R \, dB$

and $\frac{dc}{\cos C} = 2R \, dC$

$$\therefore \frac{da}{\cos A} + \frac{db}{\cos B} + \frac{dc}{\cos C} = 2R(dA + dB + dC) = 0$$

15 $\therefore \int_0^{1.5} [x^2] \, dx$

$$= \int_0^1 [x^2] \, dx + \int_1^{\sqrt{2}} [x^2] \, dx + \int_{\sqrt{2}}^{1.5} [x^2] \, dx$$

$$= \int_0^1 0 \, dx + \int_1^{\sqrt{2}} 1 \, dx + \int_{\sqrt{2}}^{1.5} 2 \, dx$$

$$= 0 + (x)_1^{\sqrt{2}} + 2(x)_{\sqrt{2}}^{1.5}$$

$$= \sqrt{2} - 1 + 3 - 2\sqrt{2} = 2 - \sqrt{2}$$

16 The differential equation is

$$\frac{dy}{dx} - y \tan x = -y^2 \sec x$$

This is Bernoulli's equation, which can be reducible to linear equation. On dividing the equation by y^2, we get

$$\frac{1}{y^2}\frac{dy}{dx} - \frac{1}{y}\tan x = -\sec x \quad \ldots \text{(i)}$$

Put $\quad \frac{1}{y} = Y \Rightarrow -\frac{1}{y^2}\frac{dy}{dx} = \frac{dY}{dx}$

Eq. (i) reduces to

$$-\frac{dY}{dx} - Y \tan x = -\sec x$$

$\Rightarrow \frac{dY}{dx} + Y \tan x = \sec x$, which is a linear equation.

Hence, $\text{IF} = e^{\int \tan x \, dx} = \sec x$

17 $A = \int_0^2 [2^x - (2x - x^2)] \, dx$

$$= \left[\frac{2^x}{\log 2} - x^2 + \frac{x^3}{3}\right]_0^2 = \frac{3}{\log 2} - \frac{4}{3}$$

18 $y = x - 1$, if $x > 1$ and $y = -(x - 1)$; if $x < 1$

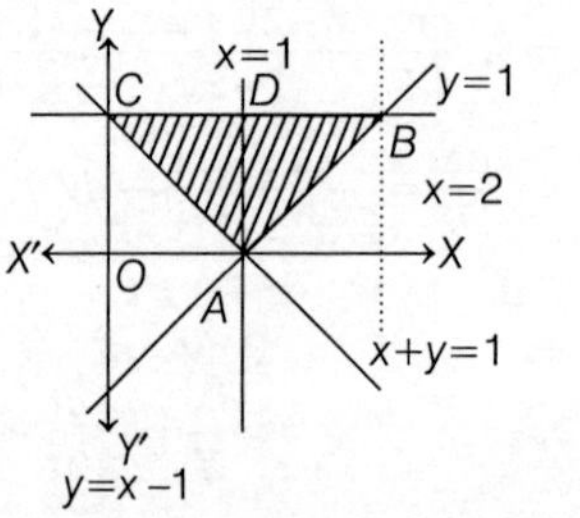

Area of bounded region,

A = Area of ΔABC

$$= \frac{1}{2} \times BC \times AD = \frac{1}{2} \times 2 \times 1$$

$= 1$ sq unit

19 $I_1 = \int_{1-k}^{k} (1 - x) \sin [x(1 - x)] \, dx$

[by property]

$$= \int_{1-k}^{k} \sin x (1 - x) \, dx - \int_{1-k}^{k} x \sin x (1 - x) \, dx$$

$$= I_2 - I_1 \text{ or } 2I_1 = I_2$$

20 $I_1 = \int_{\sec^2 z}^{2 - \tan^2 z} (3 - x) f(x(3 - x)) \, dx$

[by property]

$$I_1 = 3\int_{\sec^2 z}^{2 - \tan^2 z} f(x(3 - x)) \, dx - \int_{\sec^2 z}^{2 - \tan^2 z} x f(x(3 - x)) \, dx$$

$$I_1 = 3I_2 - I_1 \Rightarrow \frac{I_1}{I_2} = \frac{3}{2}$$

21 Let $I = \int_0^{\pi} [f(x) + g(x)] \, dx$

$$I = \int_0^{\pi} [f(\pi - x) + g(\pi - x)] \, dx$$

$$= \int_0^{\pi} [1 - f(x) + 1 - g(x)] \, dx$$

$$= 2\int_0^{\pi} dx - \int_0^{\pi} [f(x) + g(x)] \, dx$$

$\Rightarrow 2I = 2\pi$

$\therefore \quad I = \pi$

22 At $x = 1, f(x) = 1$,

$$\lim_{x \to 1^+} f(x) = \lim_{x \to 1^+} |2x - 3| [x] = 1$$

and $\lim_{x \to 1^-} f(x) = \lim_{x \to 1^-} \sin \frac{\pi x}{2} = 1$

Hence, $f(x)$ is continuous at $x = 1$.

Now,

$$\lim_{h \to 0^+} \frac{f(1 + h) - 1}{h} = \lim_{h \to 0^+} \frac{|2h - 1| - 1}{h}$$

$$= \lim_{h \to 0^+} \frac{1 - 2h - 1}{h} = -2$$

and $\lim_{h \to 0^-} \frac{f(1 + h) - 1}{h}$

$$= \lim_{h \to 0^-} \frac{\sin\left(\frac{\pi}{2} + \frac{\pi h}{2}\right) - 1}{h}$$

$$= \lim_{h \to 0^-} \frac{\cos \frac{\pi h}{2} - 1}{h} = 0$$

Hence, $f(x)$ is not differentiable at $x = 1$.

23 Given that,

$$y = \sqrt{f(x) + \sqrt{f(x) + \sqrt{f(x) + \ldots \infty}}}$$

$$y = \sqrt{f(x) + y}$$

On squaring both sides, we get

$$y^2 - y = f(x)$$

On differentiating w.r.t. x, we get

$$(2y - 1)\frac{dy}{dx} = f'(x)$$

$$\therefore \quad \frac{dy}{dx} = \frac{f'(x)}{2y - 1}$$

24 We have,

$h(x) = f(x) - [f(x)]^2 + [f(x)]^3$

$h'(x) = f'(x) - 2f(x)\,f'(x) + 3\,[f(x)]^2 \cdot f'(x)$

$= f'(x)\,[1 - 2f(x) + 3\,[f(x)]^2]$

$= 3f'(x)\left\{\left(f(x) - \frac{1}{3}\right)^2 + \frac{2}{9}\right\}$

Hence, $h'(x)$ and $f'(x)$ have same sign.

25 We have,

$u = \int e^{ax} \cos bx\, dx = e^{ax} \cdot \frac{\sin bx}{b} - \frac{a}{b}\int e^{ax} \cdot \sin bx\, dx$

$= \frac{e^{ax} \cdot \sin bx}{b} - \frac{a}{b} v$

$\Rightarrow bu + av = e^{ax} \cdot \sin bx$ ...(i)

Similarly, $bv - au = -e^{ax} \cdot \cos bx$ ...(ii)

On squaring and adding Eqs. (i) and (ii), we get

$(a^2 + b^2)(u^2 + v^2) = e^{2ax}$

26 Let $I = \int_a^b f(x) \cdot g(x)\, dx$

Put $f(x) = t \Rightarrow f'(x)\,dx = dt$

$\Rightarrow g(x)\,dx = dt$

$I = \int_{f(a)}^{f(b)} t\, dt = \left|\frac{t^2}{2}\right|_{f(a)}^{f(b)} = \frac{[f(b)]^2 - [f(a)]^2}{2}$

27 Given, $\frac{dy}{dx} = 2x + 1 \Rightarrow y = x^2 + x + C$

$\Rightarrow y = x^2 + x,$

[$\because C = 0$ by putting $x = 1, y = 2$]

$\Rightarrow \left(x + \frac{1}{2}\right)^2 = y + \frac{1}{4}$, which is an equation of parabola whose vertex is $V\left(\frac{-1}{2}, \frac{-1}{4}\right)$.

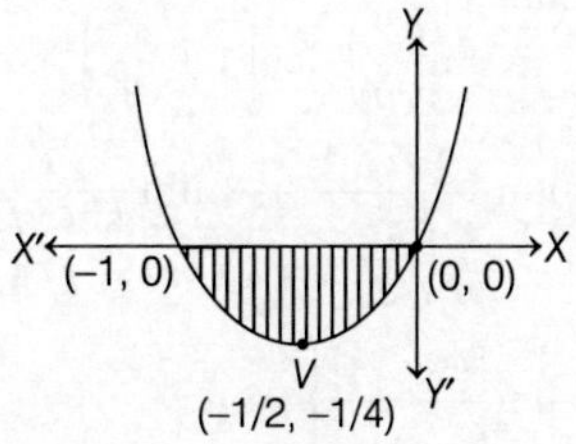

$\therefore$ Required area

$= \left|\int_{-1}^{0} (x^2 + x)\, dx\right| = \left(\frac{x^3}{3} + \frac{x^2}{2}\right)_{-1}^{0}$

$= \left|\frac{-1}{3} + \frac{1}{2}\right| = \frac{1}{6}$ sq unit

28 If $x \in R - (0, 1)$, then

$x \le x^2 \Rightarrow x(1 - x) \le 0$

$\Rightarrow x(x - 1) \ge 0 \Rightarrow x \le 0$ or $x \ge 1$,

$\therefore h(x) = \begin{cases} x, & x \le 0 \\ x^2, & 0 < x < 1 \\ x, & x \ge 1 \end{cases}$

$h(x)$ is continuous for everywhere but not differentiable at $x = 0$ and 1.

i.e.

$h'(x) = \begin{bmatrix} 1, & x < 0 \\ 2x, & 0 < x < 1 \\ \text{not exist}, & x = 0 \\ 1, & x > 1 \\ \text{not exist}, & x = 1 \end{bmatrix}$

$\therefore h'(x) = 1, \forall\, x > 1$

29 Since, f is continuous at $x = 0$.

$\therefore f(0^-) = f(0^+) = f(0) = -1$

Also, f is differentiable at $x = 0$, therefore $Lf'(0) = Rf'(0)$

$\Rightarrow \lim_{h \to 0} \frac{f(0 - h) - f(0)}{-h} = \lim_{h \to 0} \frac{f(0 + h) - f(0)}{h}$

$\Rightarrow \lim_{h \to 0}\left(\frac{e^{-2h} - 1 + 1}{-h}\right) = \lim_{h \to 0} \frac{\left(ah + \frac{bh^2}{2} - 1 + 1\right)}{h}$

$\Rightarrow \lim_{h \to 0}\left(\frac{-2e^{-2h}}{-1}\right) = \lim_{h \to 0}(a + bh)$

[L' Hospital's rule]

$\Rightarrow 2 = a + 0 \Rightarrow a = 2, b$ any number

30 $y = \lim_{x \to \pi/2} \frac{\int_{\pi/2}^{x} t\, dt}{\sin(2x - \pi)}$

$\Rightarrow y = \lim_{x \to \pi/2} \frac{\left[\frac{t^2}{2}\right]_{\pi/2}^{x}}{\sin(2x - \pi)}$

$\Rightarrow y = \lim_{x \to \pi/2} \frac{\left(\frac{x^2}{2} - \frac{\pi^2}{8}\right)}{\sin(2x - \pi)}$

$\Rightarrow y = \lim_{x \to \pi/2} \frac{1}{8} \frac{(4x^2 - \pi^2)}{\sin(2x - \pi)}$

$\Rightarrow y = \lim_{x \to \pi/2} \frac{1}{8} \frac{(2x - \pi)(2x + \pi)}{\sin(2x - \pi)}$

$\Rightarrow y = \frac{1}{8} \frac{\lim_{x \to \pi/2}(2x + \pi)}{\lim_{x \to \pi/2} \frac{\sin(2x - \pi)}{2x - \pi}}$

$\left[\because \lim_{\theta \to 0} \frac{\sin\theta}{\theta} = 1\right]$

$\Rightarrow y = \frac{1}{8} \times 2\pi = \frac{1}{4}\pi$

31 Using the fact that $[x]$ is discontinuous at all integer numbers.

$\therefore f(x) = \sqrt{2}\cos\left(x + \frac{\pi}{4}\right)$ is an integer for

$x + \frac{\pi}{4} = \frac{\pi}{2}, \frac{3\pi}{4}, \frac{5\pi}{4}, \frac{3\pi}{2}, \frac{7\pi}{4}, \frac{9\pi}{4}$

$\Rightarrow x = \frac{\pi}{4}, \frac{\pi}{2}, \pi, \frac{5\pi}{4}, \frac{3\pi}{2}, 2\pi$

32 $f(x) = \tan^{-1}\left[\frac{\sin x}{1 + \cos x}\right]$

$= \tan^{-1}\left[\tan\frac{x}{2}\right] = \frac{x}{2} \Rightarrow f'(x) = \frac{1}{2}$

Hence, $f'\left(\frac{\pi}{3}\right) = \frac{1}{2}$

33 Let $f(x) = ax^2 + bx + c$

Then, $f'(x) = 2ax + b$

Also, $f(1) = f(-1)$

$\Rightarrow a + b + c = a - b + c$

$\Rightarrow b = 0$

$\therefore f'(x) = 2ax$;

$\Rightarrow f'(a_1) = 2aa_1,\ f'(a_2) = 2aa_2,\ f'(a_3) = 2aa_3$

As a_1, a_2, a_3 are in AP,

$f'(a_1), f'(a_2), f'(a_3)$ are in AP.

34 Given, $y = \cos^{-1}(\cos x)$

$\Rightarrow y'(x) = \frac{1}{\sqrt{1 - \cos^2 x}} \sin x = \frac{\sin x}{|\sin x|}$

$= \begin{cases} 1, & \text{1st and 2nd quadrants} \\ -1, & \text{3rd and 4th quadrants} \end{cases}$

35 $f(x) = \int_{-10}^{x} (t^4 - 4)e^{-4t} dt$

$\Rightarrow f'(x) = (x^4 - 4)e^{-4x}$

Now, $f'(x) = 0 \Rightarrow x = \pm\sqrt{2}$

Now,

$f''(x) = -4(x^4 - 4)e^{-4x} + 4x^3e^{-4x}$

At $x = \sqrt{2}$ and $x = -\sqrt{2}$, the given function has extreme value.

36 $f'(x) = 3kx^2 - 18x + 9$

$= 3[kx^2 - 6x + 3] > 0, \forall x \in R$

$\therefore \Delta = b^2 - 4ac < 0, k > 0$

i.e. $36 - 12k < 0$ or $k > 3$.

37 Let $I = \int_{-1}^{3}\left\{\tan^{-1}\left(\frac{x}{x^2 + 1}\right) + \tan^{-1}\left(\frac{x^2 + 1}{x}\right)\right\}dx$

$$= \int_{-1}^{3} \left\{\tan^{-1}\left(\frac{x}{x^2+1}\right) + \cot^{-1}\left(\frac{x}{x^2+1}\right)\right\}dx$$

$$= \int_{-1}^{3} \frac{\pi}{2}\, dx = 2\pi$$

$$\left[\because \tan^{-1} x + \cot^{-1} x = \frac{\pi}{2}, \forall\, x \in R\right]$$

38 $\lim_{n\to\infty} \sum_{r=1}^{n} \left(\frac{r^{99}}{n^{100}}\right) = \int_0^1 x^{99} dx$

$$= \left[\frac{x^{100}}{100}\right]_0^1 = \frac{1}{100}$$

39 $I = \frac{-\cos 2x}{2} + \frac{\sin 2x}{2} + k$

$$= \frac{1}{\sqrt{2}}\left(\sin 2x \cos\frac{\pi}{4} - \cos 2x \sin\frac{\pi}{4}\right) + k$$

$$= \frac{1}{\sqrt{2}} \sin\left(2x - \frac{\pi}{4}\right) + k$$

$\therefore \quad c = \frac{\pi}{4};\ a = k$

40 $\int \frac{(\cos x - \sin x)(\cos x + \sin x)}{(\cos x + \sin x)^2} dx$

$$= \int \frac{\cos x - \sin x}{\cos x + \sin x} dx$$

Put $\sin x + \cos x = t$

$\Rightarrow (\cos x - \sin x)\, dx = dt$

$\therefore \quad \int \frac{1}{t}\, dt = \log t + C$

$$= \log(\sin x + \cos x) + C$$

41 $\int \frac{2\cos^2 2x}{\cos^2 x - \sin^2 x} \cdot \sin x \cos x\, dx$

$$= \int \cos 2x \cdot \sin 2x\, dx$$

$$= \frac{1}{2}\int \sin 4x\, dx = -\frac{1}{8}\cos 4x + C$$

$\therefore \quad k = -\frac{1}{8}$

42 $\int \frac{dx}{\sin(x-a)\sin(x-b)}$

$$= \frac{1}{\sin(a-b)} \int \frac{\sin\{(x-b)-(x-a)\}}{\sin(x-a)\sin(x-b)} dx$$

$$= \frac{1}{\sin(a-b)} \left[\int \cot(x-a)\, dx - \int \cot(x-b)\, dx\right]$$

$$= \frac{1}{\sin(a-b)} \log\left|\frac{\sin(x-a)}{\sin(x-b)}\right| + C$$

43 $\int \frac{e^x(x^2 - 1 + 2)}{(x+1)^2} dx$

$$= \int e^x \left[\frac{x-1}{x+1} + \frac{2}{(x+1)^2}\right] dx$$

$$= \int e^x\, [f(x) + f'(x)]\, dx$$

$$\left[\therefore f(x) = \frac{x-1}{x+1}, f'(x) = \frac{2}{(x+1)^2}\right]$$

$$= e^x \left(\frac{x-1}{x+1}\right) + C$$

44 $I_1 = \int \sin^{-1} x\, dx.$

Let $\sin^{-1} x = \theta \Rightarrow x = \sin\theta$

$\Rightarrow \quad dx = \cos\theta\, d\theta$

$\therefore \quad I_1 = \int \theta \cos\theta\, d\theta$

$= \theta\sin\theta - \int \sin\theta\, d\theta = \theta\sin\theta + \cos\theta$

$= x\sin^{-1} x + \sqrt{1 - x^2}$

and $\quad I_2 = \int \sin^{-1}\sqrt{1-x^2}\, dx$

$= \int \cos^{-1} x\, dx$

Let $\cos\phi = x$

$\Rightarrow -\sin\phi\, d\phi = dx$

$\therefore\ I_2 = -\int \phi\sin\phi\, d\phi = \phi\cos\phi + \int -\cos\phi\, d\phi$

$= \phi\cos\phi - \sin\phi$

$= x\cos^{-1} x - \sqrt{1 - x^2}$

$\therefore \quad I_1 + I_2 = x(\cos^{-1} x + \sin^{-1} x) = \frac{\pi x}{2}$

45 Divide numerator and denominator by $\cos^4 x$,

$\therefore \quad I = \int_0^{\pi/4} \frac{\sec^2 x \sec^2 x\, dx}{1 - \tan^2 x + \tan^4 x}$

Put $\tan x = t \Rightarrow \sec^2 x\, dx = dt$

$\therefore \quad I = \int_0^1 \frac{1 + t^2}{t^4 - t^2 + 1} dt$

$$= \int_0^1 \frac{1 + \frac{1}{t^2}}{t^2 - 1 + \frac{1}{t^2}} dt = \int_0^1 \frac{\left(1 + \frac{1}{t^2}\right)}{\left(t - \frac{1}{t}\right)^2 + 1} dt$$

Put $\quad z = t - \frac{1}{t}$

and $\quad dz = \left(1 + \frac{1}{t^2}\right) dt$

$$I = \int_{-\infty}^{0} \frac{dz}{1 + z^2} dz = [\tan^{-1} z]_{-\infty}^{0}$$

$$= \tan^{-1}(0) - \tan^{-1}(-\infty) = \frac{\pi}{2}$$

46 Given,

$f(x) = |x - 1| + |x - 3| + |5 - x|$, $\forall\, x \in R$

$$\therefore \quad f(x) = \begin{cases} 9 - 3x, & x < 1 \\ 7 - x, & 1 \le x < 3 \\ x + 1, & 3 \le x < 5 \\ 3x - 9, & x \ge 5 \end{cases}$$

$$\Rightarrow \quad f'(x) = \begin{cases} -3, & x < 1 \\ -1, & 1 < x < 3 \\ 1, & 3 < x < 5 \\ 3, & x > 5 \end{cases}$$

It is clear that $f'(x) > 0$, when $x \in (3, \infty)$.

47 Let $I = \int \frac{dx}{(1+x)^{1/2} - (1+x)^{1/3}}$

Put $1 + x = t^6 \quad \Rightarrow \quad dx = 6t^5\, dt$

Then, $\quad I = \int \frac{6t^5 dt}{(t^3 - t^2)} = 6\int \frac{t^3}{(t-1)} dt$

$$= 6\int \frac{(t^3 - 1) + 1}{(t-1)} dt$$

$$= 6\int \left(t^2 + t + 1 + \frac{1}{t-1}\right) dt$$

$$= 6\left(\frac{t^3}{3} + \frac{t^2}{2} + t + \ln|t - 1|\right) + C$$

$= 2(1+x)^{1/2} + 3(1+x)^{1/3} + 6(1+x)^{1/6} + 6\ln|(1+x)^{1/6} - 1| + C$

$= 2\lambda^{1/2} + 3\lambda^{1/3} + 6\lambda^{1/6} + 6\ln|\lambda^{1/6} - 1| + C$ [where, $\lambda = 1 + x$]

48 Let $\quad I = \int \frac{x^{1/4}}{x^{1/2} - 1} dt$

Put $\quad x = t^4 \Rightarrow dx = 4t^3 dt$

$$\therefore I = \int \frac{t \cdot 4t^3 dt}{(t^2 - 1)} = 4\int \left(\frac{t^4 - 1 + 1}{t^2 - 1}\right) dt$$

$$= 4\int \left(t^2 + 1 + \frac{1}{t^2 - 1}\right) dt$$

$$= 4\left\{\frac{t^3}{3} + t + \frac{1}{2}\ln\left|\frac{t-1}{t+1}\right|\right\} + C$$

$$= \frac{4}{3}x^{3/4} + 4x^{1/4} + 2\ln\left|\frac{x^{1/4} - 1}{x^{1/4} + 1}\right| + C$$

49 Clearly,

$$= \lim_{x\to 0} \frac{(5^x - 1)^3}{\sin\left(\frac{x}{n}\right)\cdot \log\left(1 + \frac{x^2}{3}\right)}$$

$$= \lim_{x\to 0} \frac{(5^x - 1)^3}{\frac{x}{n}\left(\frac{x^2}{3} + \dots\right)} \cdot \lim_{x\to 0} \frac{1}{\frac{\sin x/n}{x/n}}$$

$$= \lim_{x\to 0} \left\{\frac{(5^x - 1)}{x}\right\}^3 \cdot 3n$$

$= 3n(\log 5)^3 \quad \dots(i)$

Since, the value of the function at $x = 0$ is $15(\log 5)^3$.

$\therefore \quad 3n(\log 5)^3 = 15(\log 5)^3$

$\Rightarrow \quad n = 5$

50 Let the distance of point (4, 0) from the point (x, y) lying on the curve be

$D^2 = (x-4)^2 + y^2$

$\Rightarrow \quad D^2 = (x-4)^2 + 2x - 2x^2$

$= x^2 + 16 - 8x + 2x - 2x^2$

$= -x^2 - 6x + 16 \quad \ldots(i)$

On differentiating Eq. (i), we get

$2D\dfrac{dD}{dx} = -2x - 6 \quad \ldots(ii)$

$= -2(x+3)$

For maximum or minimum value,

put $\dfrac{dD}{dx} = 0$

$\therefore \quad x = -3$

Again, on differentiating Eq. (ii), we get

$\dfrac{d^2D}{dx^2}$ = negative on putting $x = -3$

$\therefore$ The longest distance is

$D^2 = -9 - 6(-3) + 16$

$= -9 + 18 + 16 = 25$

$\therefore \quad D = 5$ units

51 The given curve is

$5x^5 - 10x^3 + x + 2y + 4 = 0 \quad \ldots(i)$

On differentiating Eq. (i), we get

$25x^4 - 30x^2 + 1 + 2\dfrac{dy}{dx} = 0$

$\Rightarrow \dfrac{dy}{dx} = \dfrac{-25x^4 + 30x^2 - 1}{2}$

$\Rightarrow \quad \dfrac{dy}{dx} = -\dfrac{1}{2}$ at P

$\therefore$ Slope of normal is 2.

Therefore, its equation is

$(y+2) = 2(x-0)$

$\Rightarrow \quad y = 2x - 2 \quad \ldots(ii)$

On solving Eqs. (i) and (ii), we get

$5x^5 - 10x^3 + x + 4x - 4 + 4 = 0$

$\Rightarrow \quad 5x[x^4 - 2x^2 + 1] = 0$

$\Rightarrow \quad 5x(x^2-1)^2 = 0$

$\Rightarrow \quad x = 0$

or $x^2 = 1$ or $x = 0, 1, -1$

$\therefore \quad y = -2, 0, -4$

Since, the other two points are $(1, 0), (-1, -4)$.

The tangents at these points are

$(y - 0) = 2(x-1)$

and $(y+4) = 2(x+1)$ or $y = 2(x-1)$

52 Let $I = \int_{-1}^{2} \{|x-1| + [x]\}\,dx$

$= \int_{-1}^{2} (|x-1|)\,dx + \int_{-1}^{2} [x]\,dx$

$= \int_{-1}^{1} -(x-1)\,dx + \int_{1}^{2} (x-1)\,dx$

$+ \int_{-1}^{2} [x]\,dx$

$I = I_1 + I_2$

where,

$I_1 = -\dfrac{1}{2}[(x-1)^2]_{-1}^{1} + \dfrac{1}{2}[(x-1)^2]_{1}^{2}$

$= \dfrac{1}{2}\{[(x-1)^2]_1^2 - [(x-1)^2]_{-1}^{1}\}$

$= \dfrac{1}{2}\{1+4\} = \dfrac{5}{2} \quad \ldots(i)$

and $I_2 = \int_{-1}^{0} -dx + \int_{0}^{1} 0\cdot dx + \int_{1}^{2} dx$

$= -1 + 0 + 1 = 0 \quad \ldots(ii)$

From Eqs. (i) and (ii), we get

$I = I_1 + I_2 = \dfrac{5}{2} + 0 = \dfrac{5}{2}$

53 According to the question,

$\dfrac{dy}{dx} = \dfrac{y}{x} - \cos^2\left(\dfrac{y}{x}\right)$

$\Rightarrow \quad \dfrac{xdy - ydx}{x} = -\left[\cos^2 \dfrac{y}{x}\right]dx$

$\Rightarrow \sec^2 \dfrac{y}{x}\dfrac{(xdx - ydx)}{x^2} = -\dfrac{dx}{x}$

$\sec^2 \dfrac{y}{x} d\left(\dfrac{y}{x}\right) = -\dfrac{dx}{x} \quad \ldots(i)$

On integrating both sides of Eq. (i), we get

$\tan\left(\dfrac{y}{x}\right) = -\log x + C$

When $x = 1/2$ and $y = \pi/8$, then

$1 = -\log\dfrac{1}{2} + C = -[-\log 2] + C$

$1 - \log 2 = C$

$\therefore \tan\dfrac{y}{x} = -\log x + 1 - \log 2$

$= -\log 2x + \log e = \log\left(\dfrac{e}{2x}\right)$

$\Rightarrow \quad \dfrac{y}{x} = \tan^{-1}\left(\log\dfrac{e}{2x}\right)$

$\Rightarrow \quad y = x\tan^{-1}\left(\log\dfrac{e}{2x}\right)$

54. Given, $\dfrac{xdy - y\,dx}{y^2} = dy$

$\Rightarrow d\left(\dfrac{x}{y}\right) = -dy$

$\Rightarrow \quad \dfrac{x}{y} = -y + C \quad$ [integrating]

As $y(1) = 1 \Rightarrow C = 2$

$\therefore \quad \dfrac{x}{y} + y = 2$

Again, for $x = -3$,

$-3 + y^2 = 2y$

$\Rightarrow (y+1)(y-3) = 0$

Also, $y > 0$

$\Rightarrow \quad y = 3 \quad$ [neglecting $y = -1$]

55 Let $I = \int_0^{\pi/4} \dfrac{\sin x + \cos x}{9 + 16\sin 2x}\,dx$

Put $\sin x - \cos x = t$

Then, $(\sin x + \cos x)\,dx = dt$

$\therefore I = \int_{-1}^{0} \dfrac{dt}{9 + 16(1-t^2)}$

$= \int_{-1}^{0} \dfrac{dt}{25 - 16t^2}$

$= \dfrac{1}{10}\int_{-1}^{0}\left(\dfrac{1}{5-4t} + \dfrac{1}{5+4t}\right)dt$

$= \left[\dfrac{1}{10}\cdot\dfrac{1}{4}[\log(5+4t) - \log(5-4t)]\right]_{-1}^{0}$

$= \dfrac{1}{40}(\log 9 - \log 1) = \dfrac{1}{20}\log 3$

56. $\int_{-2}^{3} f(x)\,dx = \int_{-2}^{2} f(x)\,dx + \int_{2}^{3} f(x)\,dx$

Since, $e^{\cos x}\sin x$ is an odd function.

$\therefore \quad \int_{-2}^{3} f(x)\,dx = 0 + 2(3-2) = 2$

57. Required area, $A = \int_{x_1}^{x_2} y\,dx$

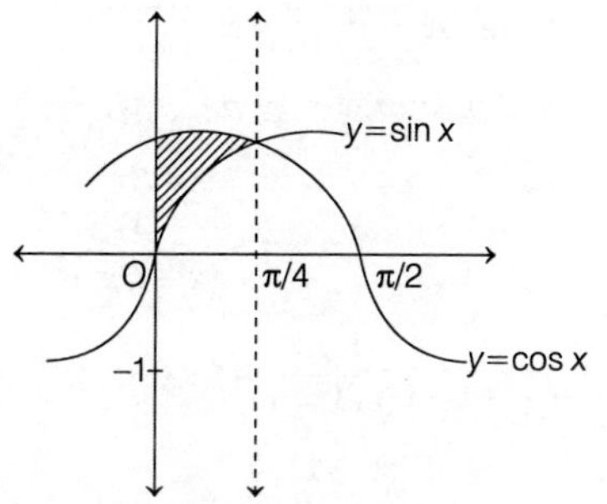

$= \int_0^{\pi/4} \cos x\,dx - \int_0^{\pi/4} \sin x\,dx$

$= [\sin x]_0^{\pi/4} - [-\cos x]_0^{\pi/4}$

$= \left(\dfrac{1}{\sqrt{2}} - 0\right) + \left(\dfrac{1}{\sqrt{2}} - 1\right) = \sqrt{2} - 1$

58 The equation of the given curve is

$y = 7x^3 + 11 \quad \ldots(i)$

$\Rightarrow \quad \dfrac{dy}{dx} = 7 \times 3x^2 = 21x^2$

[differentiating w.r.t. x]

$\therefore$ Slope of tangent at $x = 2$ is

$\left(\dfrac{dy}{dx}\right)_{x=2} = 21(2)^2 = 84$

Slope of tangent at $x = -2$ is

$\left(\dfrac{dy}{dx}\right)_{x=-2} = 21(-2)^2 = 84$

It is observed that the slopes of the tangents at the points where, $x = 2$ and $x = -2$ are equal. Hence, the two tangents are parallel.

Hence, both the statements are true and Statement II is correct explanation of Statement I.

59 $f'(x) = \cos(\sqrt{x})^2 \dfrac{d}{dx}(\sqrt{x})$

$$- \cos\left(\frac{1}{x}\right)^2 \frac{d}{dx}\left(\frac{1}{x}\right)$$

$$= \frac{1}{2}\frac{\cos x}{\sqrt{x}} + \cos\left(\frac{1}{x^2}\right)\frac{1}{x^2}$$

$$\Rightarrow f'(1) = \frac{1}{2}\cos 1 + \cos 1$$

$$= \frac{3}{2}\cos 1$$

60 Given equation can be rewritten as

$$\frac{x}{y^{4/3}}\frac{dy}{dx} + 6y^{-1/3} = 3x$$

Put $y^{-1/3} = v$

$$\Rightarrow \quad y^{-4/3}\frac{dy}{dx} = -3\frac{dv}{dx}$$

$$\therefore \quad \frac{dv}{dx} - \frac{2}{x}v = -1$$

Here, IF $= e^{\int -\frac{2}{x}dx} = x^{-2}$

$\therefore$ Solution is $x^{-2}v = \dfrac{1}{x} + C$

$$\Rightarrow \quad v = x + Cx^2$$

$$\Rightarrow \quad y(x) = \frac{1}{(x + Cx^2)^3}$$

61 $\displaystyle\int \frac{x^2 - 1}{(x^2 + 1)\sqrt{x^4 + 1}}\,dx$

$$= \int \frac{1 - \frac{1}{x^2}}{\left(x + \frac{1}{x}\right)\sqrt{x^2 + \frac{1}{x^2}}}\,dx$$

$$= \int \frac{dt}{t\sqrt{t^2 - 2}} \left[\text{put } t = x + \frac{1}{x}\right]$$

$$= \frac{1}{\sqrt{2}}\sec^{-1}\frac{x^2 + 1}{x\sqrt{2}} + C$$

62 Let $f(x) = |x - 1| + |x - 2| + |x - 3|$

$$= \begin{cases} 6 - 3x, & x \le 1 \\ 4 - x, & 1 < x \le 2 \\ x, & 2 < x \le 3 \\ 3x - 6, & x > 3 \end{cases}$$

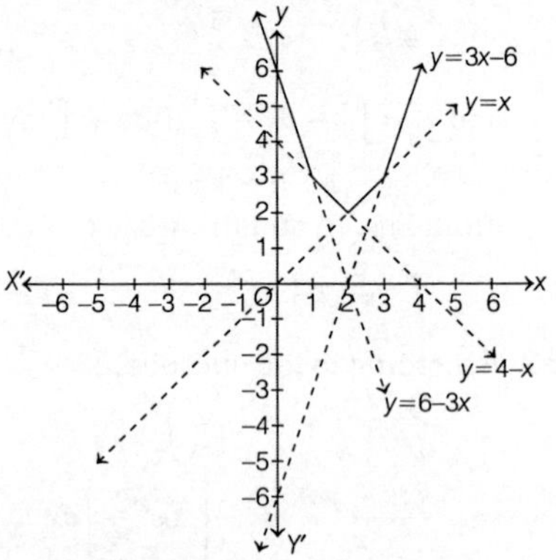

Clearly, the function has absolute minimum at $x = 2$.

So, the absolute minimum is equal to 2.

Also, the curve is taking sharp turn at $x = 1$, 2 and 3.

$\therefore f$ is not differentiable at $x = 1$, 2 and 3.

63 Given,

$f(x) = \max\{(x - 1)^2 + 1, |x - 1|\}$

$= (x - 1)^2 + 1$

$\therefore \quad f'(x) = 2(x - 1) = 0$ [say]

$\Rightarrow \quad x = 1 \in [0, 3]$

Now, $f(0) = 2, f(1) = 1, f(3) = 5$

$\therefore$ Greatest value of

$f(x) = \max\{f(0), f(1), f(3)\} = 5$

64 For vertical tangent, $\dfrac{dx}{d\theta} = 0$

$\therefore -3\cos\theta = 0 \Rightarrow \cos\theta = 0$

$$\Rightarrow \quad \theta = \frac{\pi}{2}, \frac{3\pi}{2}$$

At $\theta = \dfrac{\pi}{2}$, $x = 2 - 3$

$= -1$, $y = 3 + 0 = 3$

i.e. $(-1, 3)$ and

At $\theta = \dfrac{3\pi}{2}$, $x = 2 + 3 = 5$

and $y = 3 + 0 = 3$

i.e. $(5, 3)$.

65 Suppose $f(x) = 0$ has a real root say $x = a$, then $f(x) < 0$ for all $x < a$.

Thus, $|f(x)|$ becomes strictly decreasing on $(-\infty, a)$.

So, Statement I is true.

66 As the area enclosed by $|x| + |y| \le a$ is the area of square i.e. $2a^2$.

$\therefore$ Area enclosed by

$|x + y| + |x - y| \le 2$

is area of square shown as

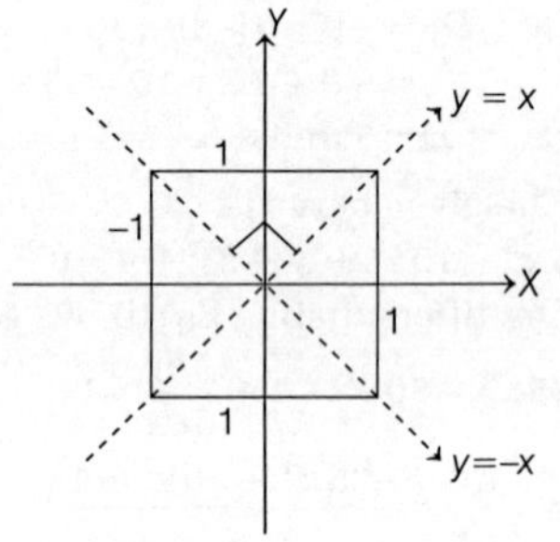

$\therefore$ Required area

$$= 4\left(\frac{1}{2} \times 2 \times 1\right) = 4 \text{ sq units}$$

Also, the area enclosed by

$|x + y| + |x - y| \le 2$

is symmetric about X-axis, Y-axis, $y = x$ and $y = -x$.

Hence, both the statements are true but Statement II is not the correct explanation of Statement I.

DAY TWENTY

Trigonometric Functions and Equations

Learning & Revision for the Day

- Angle on Circular System
- Trigonometric Functions
- Trigonometric Identities
- Trigonometric Ratios/Functions of Acute Angles
- Trigonometric Ratios of Compound and Multiple Angles
- Transformation Formulae
- Conditional Identities
- Maximum and Minimum Values
- Trigonometric Equations
- Summation of Some Trigonometric Series

Angle on Circular System

If the angle subtended by an arc of length l at the centre of a circle of radius r is θ, then $\theta = \dfrac{l}{r}$

If the length of arc is equal to the radius of the circle, then the angle subtended at the centre of the circle will be one radian. One radian is denoted by 1^c and $1^c = 57°16'22''$ approximately.

(Figure shows the angle whose measure are one radian.)

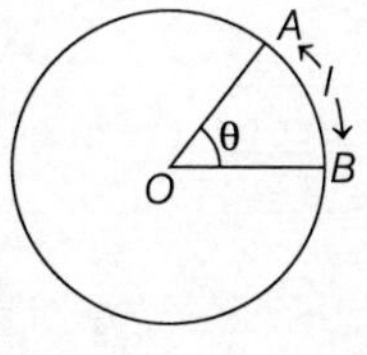

$$2\pi^C = 360°, 1^C = \frac{180°}{\pi}, 1° = \left(\frac{\pi}{180}\right)^C$$

Trigonometric Functions

Let $X'OX$ and YOY' be the coordinate axes. Taking O as the centre and a unit radius, draw a circle, cutting the coordinate axes at A, B, A' and B' as shown in the figure

Also, let $P(x, y)$ be any on the circle with $\angle AOP = \theta$ radian, i.e. length of arc $AP = \theta$

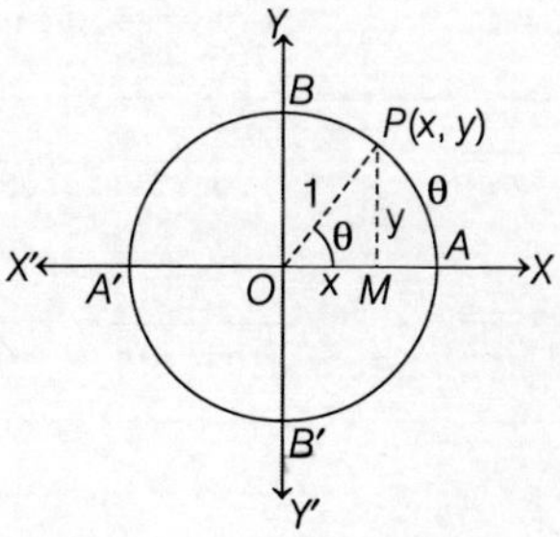

PREP MIRROR
Your Personal Preparation Indicator

- No. of Questions in Exercises (x)—
- No. of Questions Attempted (y)—
- No. of Correct Questions (z)—
  (Without referring Explanations)

- Accuracy Level ($z/y \times 100$)—
- Prep Level ($z/x \times 100$)—

In order to expect good rank in JEE, your Accuracy Level should be above 85 & Prep Level should be above 75.

Then, the six trigonometric functions, can be defined as

(i) $\cos\theta = \dfrac{OM}{OP} = x$

(ii) $\sin\theta = \dfrac{PM}{OP} = y$

(iii) $\sec\theta = \dfrac{OP}{OM} = \dfrac{1}{x}, x \neq 0$

(iv) $\text{cosec}\,\theta = \dfrac{OP}{PM} = \dfrac{1}{y}, y \neq 0$

(v) $\tan\theta = \dfrac{PM}{OM} = \dfrac{y}{x}, x \neq 0$

(vi) $\cot\theta = \dfrac{OM}{PM} = \dfrac{x}{y}, y \neq 0$

Trigonometric Identities

An equation involving trigonometric functions which is true for all those angles for which the functions which is true for all those angles for which the functions are defined is called trigonometric identity.

Some identities are given below

(i) $\sin^2\theta + \cos^2\theta = 1$

(ii) $\sec^2\theta - \tan^2\theta = 1$

(iii) $\text{cosec}^2\theta - \cot^2\theta = 1$

Trigonometric Ratios/Functions of Acute Angles

The ratios of the sides of a triangles with respect to its acute angles are called trigonometric ratios or T-ratios.

In a right angled ΔABC, if $\angle CAB = \theta$, then

1. $\sin\theta = \dfrac{BC}{AC} = \dfrac{\text{Perpendicular}}{\text{Hypotenuse}}$
2. $\cos\theta = \dfrac{AB}{AC} = \dfrac{\text{Base}}{\text{Hypotenuse}}$
3. $\tan\theta = \dfrac{BC}{AB} = \dfrac{\text{Perpendicular}}{\text{Base}}$
4. $\text{cosec}\,\theta = \dfrac{1}{\sin\theta} = \dfrac{AC}{BC} = \dfrac{\text{Hypotenuse}}{\text{Perpendicular}}$
5. $\sec\theta = \dfrac{1}{\cos\theta} = \dfrac{AC}{AB} = \dfrac{\text{Hypotenuse}}{\text{Base}}$
6. $\cot\theta = \dfrac{1}{\tan\theta} = \dfrac{AB}{BC} = \dfrac{\text{Base}}{\text{Perpendicular}}$

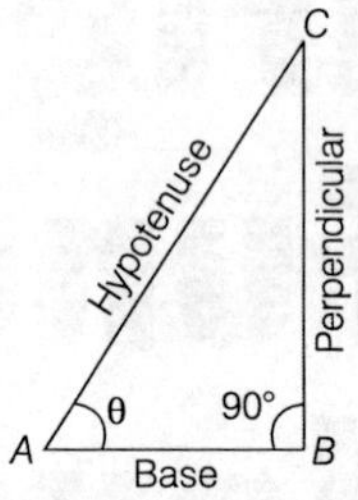

Sign for Trigonometric Ratios in four Quadrants

Quadrant	$\sin\theta$	$\cos\theta$	$\tan\theta$	$\cot\theta$	$\sec\theta$	$\text{cosec}\,\theta$
I. $(0, 90°)$	+	+	+	+	+	+
II. $(90°, 180°)$	+	–	–	–	–	+
III. $(180°, 270°)$	–	–	+	+	–	–
IV. $(270°, 360°)$	–	+	–	–	+	–

Trigonometric ratios of some useful angles between 0°and 90°

Angle	0°/0	$15°/\frac{\pi}{12}$	$18°/\frac{\pi}{10}$	$22.5°/\frac{\pi}{8}$	$30°/\frac{\pi}{6}$	$36°/\frac{\pi}{5}$	$45°/\frac{\pi}{4}$	$54°/\frac{3\pi}{10}$	$60°/\frac{\pi}{3}$	$67.5°/\frac{3\pi}{8}$	$72°/\frac{2\pi}{5}$	$75°/\frac{5\pi}{12}$	$90°/\frac{\pi}{2}$
$\sin\theta$	0	$\frac{\sqrt{3}-1}{2\sqrt{2}}$	$\frac{\sqrt{5}-1}{4}$	$\frac{\sqrt{2-\sqrt{2}}}{2}$	$\frac{1}{2}$	$\frac{\sqrt{10-2\sqrt{5}}}{4}$	$\frac{1}{\sqrt{2}}$	$\frac{\sqrt{5}+1}{4}$	$\frac{\sqrt{3}}{2}$	$\frac{\sqrt{2+\sqrt{2}}}{2}$	$\frac{\sqrt{10+2\sqrt{5}}}{4}$	$\frac{\sqrt{3}+1}{2\sqrt{2}}$	1
$\cos\theta$	1	$\frac{\sqrt{3}+1}{2\sqrt{2}}$	$\frac{\sqrt{10+2\sqrt{5}}}{4}$	$\frac{\sqrt{2+\sqrt{2}}}{2}$	$\frac{\sqrt{3}}{2}$	$\frac{\sqrt{5}+1}{4}$	$\frac{1}{\sqrt{2}}$	$\frac{\sqrt{10-2\sqrt{5}}}{4}$	$\frac{1}{2}$	$\frac{\sqrt{2-\sqrt{2}}}{2}$	$\frac{\sqrt{5}-1}{4}$	$\frac{\sqrt{3}-1}{2\sqrt{2}}$	0
$\tan\theta$	0	$\frac{\sqrt{3}-1}{\sqrt{3}+1}$	$\frac{\sqrt{5}-1}{\sqrt{10+2\sqrt{5}}}$	$\sqrt{2}-1$	$\frac{1}{\sqrt{3}}$	$\frac{\sqrt{10-2\sqrt{5}}}{\sqrt{5}+1}$	1	$\frac{\sqrt{5}+1}{\sqrt{10-2\sqrt{5}}}$	$\sqrt{3}$	$\sqrt{2}+1$	$\frac{\sqrt{10+2\sqrt{5}}}{\sqrt{5}-1}$	$\frac{\sqrt{3}+1}{\sqrt{3}-1}$	∞
$\cot\theta$	∞	$\frac{\sqrt{3}+1}{\sqrt{3}-1}$	$\frac{\sqrt{10+2\sqrt{5}}}{\sqrt{5}-1}$	$\sqrt{2}+1$	$\sqrt{3}$	$\frac{\sqrt{5}+1}{\sqrt{10-2\sqrt{5}}}$	1	$\frac{\sqrt{10-2\sqrt{5}}}{\sqrt{5}+1}$	$\frac{1}{\sqrt{3}}$	$\sqrt{2}-1$	$\frac{\sqrt{5}-1}{\sqrt{10+2\sqrt{5}}}$	$\frac{\sqrt{3}-1}{\sqrt{3}+1}$	0
$\sec\theta$	1	$\frac{2\sqrt{2}}{\sqrt{3}+1}$	$\frac{4}{\sqrt{10+2\sqrt{5}}}$	$\sqrt{4-2\sqrt{2}}$	$\frac{2}{\sqrt{3}}$	$\sqrt{5}-1$	$\sqrt{2}$	$\frac{4}{\sqrt{10-2\sqrt{5}}}$	2	$\sqrt{4+2\sqrt{2}}$	$\sqrt{5}+1$	$\frac{2\sqrt{2}}{\sqrt{3}-1}$	∞
$\text{cosec}\,\theta$	∞	$\frac{2\sqrt{2}}{\sqrt{3}-1}$	$\sqrt{5}+1$	$\sqrt{4+2\sqrt{2}}$	2	$\frac{4}{\sqrt{10-2\sqrt{5}}}$	$\sqrt{2}$	$\sqrt{5}-1$	$\frac{2}{\sqrt{3}}$	$\sqrt{4-2\sqrt{2}}$	$\frac{4}{\sqrt{10+2\sqrt{5}}}$	$\frac{2\sqrt{2}}{\sqrt{3}+1}$	1

Trigonometric ratios of allied angles

θ	$\sin\theta$	$\text{cosec}\,\theta$	$\cos\theta$	$\sec\theta$	$\tan\theta$	$\cot\theta$
$-\theta$	$-\sin\theta$	$-\text{cosec}\theta$	$\cos\theta$	$\sec\theta$	$-\tan\theta$	$-\cot\theta$
$90°-\theta$	$\cos\theta$	$\sec\theta$	$\sin\theta$	$\text{cosec}\,\theta$	$\cot\theta$	$\tan\theta$
$90°+\theta$	$\cos\theta$	$\sec\theta$	$-\sin\theta$	$-\text{cosec}\,\theta$	$-\cot\theta$	$-\tan\theta$
$180°-\theta$	$\sin\theta$	$\text{cosec}\,\theta$	$-\cos\theta$	$-\sec\theta$	$-\tan\theta$	$-\cot\theta$
$180°+\theta$	$-\sin\theta$	$-\text{cosec}\,\theta$	$-\cos\theta$	$-\sec\theta$	$\tan\theta$	$\cot\theta$
$270°-\theta$	$-\cos\theta$	$-\sec\theta$	$-\sin\theta$	$-\text{cosec}\,\theta$	$\cot\theta$	$\tan\theta$
$270°+\theta$	$-\cos\theta$	$-\sec\theta$	$\sin\theta$	$\text{cosec}\,\theta$	$-\cot\theta$	$-\tan\theta$
$360°-\theta$	$-\sin\theta$	$-\text{cosec}\,\theta$	$\cos\theta$	$\sec\theta$	$-\tan\theta$	$-\cot\theta$

Trigonometric Ratios of Compound and Multiple Angles

Compound Angles

(i) $\sin(A+B)=\sin A\cos B+\cos A\sin B$

(ii) $\sin(A-B)=\sin A\cos B-\cos A\sin B$

(iii) $\cos(A+B)=\cos A\cos B-\sin A\sin B$

(iv) $\cos(A-B)=\cos A\cos B+\sin A\sin B$

(v) $\tan(A+B)=\dfrac{\tan A+\tan B}{1-\tan A\tan B}$

(vi) $\tan(A-B)=\dfrac{\tan A-\tan B}{1+\tan A\tan B}$

(vii) $\cot(A+B)=\dfrac{\cot A\cot B-1}{\cot A+\cot B}$

(viii) $\cot(A-B)=\dfrac{\cot A\cot B+1}{\cot B-\cot A}$

(ix) (a) $\dfrac{1+\tan A}{1-\tan A}=\tan\left(\dfrac{\pi}{4}+A\right)$

(b) $\dfrac{1-\tan A}{1+\tan A}=\tan\left(\dfrac{\pi}{4}-A\right)$

Multiple Angles

(i) $\sin 2A=2\sin A\cos A=\dfrac{2\tan A}{1+\tan^2 A}$

(ii) $\cos 2A=\cos^2 A-\sin^2 A=\dfrac{1-\tan^2 A}{1+\tan^2 A}$

$=2\cos^2 A-1=1-2\sin^2 A$

(iii) $\tan 2A=\dfrac{2\tan A}{1-\tan^2 A}$

(iv) $\sin 3A=3\sin A-4\sin^3 A$

(v) $\cos 3A=4\cos^3 A-3\cos A$

(vi) $\tan 3A=\dfrac{3\tan A-\tan^3 A}{1-3\tan^2 A}$

Transformation Formulae

(i) $2\sin A\cos B=\sin(A+B)+\sin(A-B)$

(ii) $2\cos A\sin B=\sin(A+B)-\sin(A-B)$

(iii) $2\cos A\cos B=\cos(A+B)+\cos(A-B)$

(iv) $2\sin A\sin B=\cos(A-B)-\cos(A+B)$

(v) $\sin C+\sin D=2\sin\dfrac{C+D}{2}\cos\dfrac{C-D}{2}$

(vi) $\sin C-\sin D=2\cos\dfrac{C+D}{2}\sin\dfrac{C-D}{2}$

(vii) $\cos C+\cos D=2\cos\dfrac{C+D}{2}\cos\dfrac{C-D}{2}$

(viii) $\cos C-\cos D=-2\sin\dfrac{C+D}{2}\sin\dfrac{C-D}{2}$

(ix) $\sin(A+B)\sin(A-B)=\sin^2 A-\sin^2 B$

(x) $\cos(A+B)\cos(A-B)=\cos^2 A-\sin^2 B$

(xi) $\cos A\cos 2A\cos 4A\cos 8A\ldots\cos 2^{n-1}A$

$=\dfrac{1}{2^n\sin A}\sin(2^n A)$

Conditional Identities

If $A+B+C=180°$, then

(i) $\sin 2A+\sin 2B+\sin 2C=4\sin A\sin B\sin C$

(ii) $\cos 2A+\cos 2B+\cos 2C=-1-4\cos A\cos B\cos C$

(iii) $\sin A+\sin B+\sin C=4\cos\dfrac{A}{2}\cos\dfrac{B}{2}\cos\dfrac{C}{2}$

(iv) $\cos A+\cos B+\cos C=1+4\sin\dfrac{A}{2}\sin\dfrac{B}{2}\sin\dfrac{C}{2}$

(v) $\tan A+\tan B+\tan C=\tan A\tan B\tan C$

(vi) $\tan\dfrac{A}{2}\tan\dfrac{B}{2}+\tan\dfrac{B}{2}\tan\dfrac{C}{2}+\tan\dfrac{C}{2}\tan\dfrac{A}{2}=1$

Maximum and Minimum Values

1. $-1\le\sin x\le 1, |\sin x|\le 1$
2. $-1\le\cos x\le 1, |\cos x|\le 1$
3. $|\sec x|\ge 1, |\text{cosec}\,x|\ge 1$
4. $\tan x, \cot x$ take all real values

NOTE
- Maximum value of $a\cos\theta \pm b\sin\theta = \sqrt{a^2+b^2}$
- Minimum value of $a\cos\theta \pm b\sin\theta = -\sqrt{a^2+b^2}$
- Maximum value of $a\cos\theta \pm b\sin\theta + c = c + \sqrt{a^2+b^2}$
- Minimum value of $a\cos\theta \pm b\sin\theta + c = c - \sqrt{a^2+b^2}$

Trigonometric Equations

An equation involving one or more trigonometrical ratios of unknown angle is called a **trigonometrical equation**.

e.g. $\sin\theta + \cos^2\theta = 0$

Principal Solution

The value of the unknown angle (say θ) which satisfies the trigonometric equation is known as principal solution, if $0 \le \theta < 2\pi$.

General Solution

Since, trigonometrical functions are periodic function, solution of trigonometric equation can be generalised with the help of the periodicity of the trigonometrical functions.

The solution consisting of all possible solutions of a trigonometric equation is called its general solution.

Some general trigonometric equations and their solutions

Equations	Solutions	Equations	Solutions
$\sin x = \sin\alpha \left(-\frac{\pi}{2} \le \alpha \le \frac{\pi}{2}\right)$	$x = n\pi + (-1)^n\alpha$, $n \in I$	$\sin^2 x = \sin^2\alpha$	$x = n\pi \pm \alpha, n \in I$
$\cos x = \cos\alpha\ (0 \le \alpha \le \pi)$	$x = 2n\pi \pm \alpha$, $n \in I$	$\cos^2 x = \cos^2\alpha$	$x = n\pi \pm \alpha, n \in I$
$\tan x = \tan\alpha \left(-\frac{\pi}{2} < \alpha < \frac{\pi}{2}\right)$	$x = n\pi + \alpha, n \in I$	$\tan^2 x = \tan^2\alpha$	$x = n\pi \pm \alpha, n \in I$
$\sin x = 0$	$x = n\pi, n \in I$	$\sin x = 1$	$x = (4n+1)\pi/2, n \in I$
$\cos x = 0$	$x = (2n+1)\pi/2, n \in I$	$\cos x = 1$	$x = 2n\pi, n \in I$
$\tan x = 0$	$x = n\pi, n \in I$	$\cos x = -1$	$x = (2n+1)\pi, n \in I$
		$\sin x = \sin\alpha$ and $\cos x = \cos\alpha$	$x = 2n\pi + \alpha, n \in I$

Summation of Some Trigonometric Series

(i) $\sin\alpha + \sin(\alpha+\beta) + \sin(\alpha+2\beta) + \ldots$ to n terms

$$= \frac{\sin\left(\frac{n\beta}{2}\right)}{\sin\left(\frac{\beta}{2}\right)} \cdot \sin\left\{\alpha + (n-1)\left(\frac{\beta}{2}\right)\right\}$$

(ii) $\cos\alpha + \cos(\alpha+\beta) + \cos(\alpha+2\beta) + \ldots$ to n terms

$$= \frac{\sin\left(\frac{n\beta}{2}\right)}{\sin\left(\frac{\beta}{2}\right)} \cdot \cos\left\{\alpha + (n-1)\left(\frac{\beta}{2}\right)\right\}$$

DAY PRACTICE SESSION 1

FOUNDATION QUESTIONS EXERCISE

1 If $\frac{2\sin\theta}{1+\cos\theta+\sin\theta}=x$, then $\frac{1-\cos\theta+\sin\theta}{1+\sin\theta}$ is equal to

(a) x (b) $\frac{1}{x}$ (c) $1-x$ (d) $1+x$

2 If $\frac{3\pi}{4}<\alpha<\pi$, then $\sqrt{\text{cosec}^2\alpha+2\cot\alpha}$ is equal to

(a) $1+\cot\alpha$ (b) $1-\cot\alpha$
(c) $-1-\cot\alpha$ (d) $-1+\cot\alpha$

3 The least value of $\text{cosec}^2 x+25\sec^2 x$ is

(a) 0 (b) 26 (c) 28 (d) 36

4 If $\tan A+\sin A=m$ and $\tan A-\sin A=n$, then $\frac{(m^2-n^2)^2}{mn}$ is equal to

(a) 4 (b) 3 (c) 16 (d) 9

5 If $a\cos^3\alpha+3a\cos\alpha\sin^2\alpha=m$ and $a\sin^3\alpha+3a\cos^2\alpha\sin\alpha=n$, then $(m+n)^{2/3}+(m-n)^{2/3}$ is equal to

(a) $2a^2$ (b) $2a^{1/3}$ (c) $2a^{2/3}$ (d) $2a^3$

6 $\left(1+\cos\frac{\pi}{8}\right)\left(1+\cos\frac{3\pi}{8}\right)\left(1+\cos\frac{5\pi}{8}\right)\left(1+\cos\frac{7\pi}{8}\right)$ is equal to

(a) 1 (b) $\cos\frac{\pi}{8}$ (c) $\frac{1}{8}$ (d) $\frac{1+\sqrt{2}}{2\sqrt{2}}$

7 If $\sin\theta=-\frac{4}{5}$ and θ lies in the third quadrant, then $\cos\frac{\theta}{2}$ is equal to → NCERT Exemplar

(a) $\frac{1}{\sqrt{5}}$ (b) $-\frac{1}{\sqrt{5}}$ (c) $\sqrt{\frac{2}{5}}$ (d) $-\sqrt{\frac{2}{5}}$

8 The expression $\frac{\tan A}{1-\cot A}+\frac{\cot A}{1-\tan A}$ can be written as → JEE Mains 2013

(a) $\sin A\cos A+1$ (b) $\sec A\,\text{cosec}\,A+1$
(c) $\tan A+\cot A$ (d) $\sec A+\text{cosec}\,A$

9 If $\tan\theta$, $2\tan\theta+2$ and $3\tan\theta+3$ are in GP, then the value of $\frac{7-5\cot\theta}{9-4\sqrt{\sec^2\theta-1}}$ is

(a) $\frac{12}{5}$ (b) $-\frac{33}{28}$ (c) $\frac{33}{100}$ (d) $\frac{12}{13}$

10 If $\sin(\alpha+\beta)=1$ and $\sin(\alpha-\beta)=\frac{1}{2}$, then $\tan(\alpha+2\beta)\tan(2\alpha+\beta)$ is equal to

(a) 1 (b) -1
(c) zero (d) None of these

11 In an acute angled triangle, the least value of $\sec A+\sec B+\sec C$ is

(a) 3 (b) 4 (c) 5 (d) 6

12 If $0<A<B<\pi$, $\sin A+\sin B=\sqrt{\frac{3}{2}}$ and $\cos A+\cos B=\frac{1}{\sqrt{2}}$, then A is equal to

(a) 15° (b) 30° (c) 45° (d) $22\frac{1}{2}°$

13 In a ΔPQR, $\angle R=\frac{\pi}{2}$. If $\tan\left(\frac{P}{2}\right)$ and $\tan\left(\frac{Q}{2}\right)$ are the roots of $ax^2+bx+c=0$, where $a\neq 0$, then

(a) $b=a+c$ (b) $b=c$
(c) $c=a+b$ (d) $a=b+c$

14 If $\cos(\alpha+\beta)=\frac{4}{5}$ and $\sin(\alpha-\beta)=\frac{5}{13}$, where $0\le\alpha,\ \beta\le\frac{\pi}{4}$. Then, $\tan 2\alpha$ is equal to → AIEEE 2010

(a) $\frac{25}{16}$ (b) $\frac{56}{33}$ (c) $\frac{19}{12}$ (d) $\frac{20}{7}$

15 If $\sin\alpha=x$, $\sin\beta=y$, $\sin(\alpha+\beta)=z$, then $\cos(\alpha+\beta)$ as a rational function is

(a) $\frac{z^2-x^2-y^2}{xy}$ (b) $\frac{z^2-x^2-y^2}{2xy}$
(c) $\frac{z^2+x^2+y^2}{xy}$ (d) $\frac{z^2+x^2+y^2}{2xy}$

16 Let α and β be such that $\pi<\alpha-\beta<3\pi$. If $\sin\alpha+\sin\beta=-\frac{21}{65}$ and $\cos\alpha+\cos\beta=-\frac{27}{65}$, then the value of $\cos\left(\frac{\alpha-\beta}{2}\right)$ is → AIEEE 2004

(a) $-\frac{3}{\sqrt{130}}$ (b) $\frac{3}{\sqrt{130}}$ (c) $\frac{6}{65}$ (d) $-\frac{6}{65}$

17 If $A+B+C=\pi$ and $\cos A=\cos B\cos C$, then $\tan B\tan C$ is equal to

(a) $\frac{1}{2}$ (b) 2 (c) 1 (d) $-\frac{1}{2}$

18 If $\tan\alpha=(1+2^{-x})^{-1}$, $\tan\beta=(1+2^{x+1})^{-1}$, then $\alpha+\beta$ equals

(a) $\frac{\pi}{6}$ (b) $\frac{\pi}{4}$ (c) $\frac{\pi}{3}$ (d) $\frac{\pi}{2}$

19 If $0<x<\pi$ and $\cos x+\sin x=\frac{1}{2}$, then $\tan x$ is equal to

(a) $\frac{-(4+\sqrt{7})}{3}$ (b) $\frac{1+\sqrt{7}}{4}$ (c) $\frac{1-\sqrt{7}}{4}$ (d) $\frac{4-\sqrt{7}}{3}$

20 $\text{cosec}\,10°-\sqrt{3}\sec 10°$ is equal to

(a) 1 (b) 1/2 (c) 2 (d) 4

21 The value of $\sin 12°\sin 48°\sin 54°$ is equal to

(a) $\frac{1}{16}$ (b) $\frac{1}{32}$ (c) $\frac{1}{8}$ (d) $\frac{1}{4}$

22 The value of $1 + \cos 56° + \cos 58° - \cos 66°$ is equal to
(a) $2 \cos 28° \cos 29° \cos 33°$ (b) $4 \cos 28° \cos 29° \sin 33°$
(c) $4 \cos 28° \cos 29° \cos 33°$ (d) $2 \cos 28° \cos 29° \sin 33°$

23 Let n be a positive integer such that
$\sin \frac{\pi}{2n} + \cos \frac{\pi}{2n} = \frac{\sqrt{n}}{2}$. Then,
(a) $n = 6$ (b) $n = 1, 2, 3, \ldots, 8$
(c) $n = 5$ (d) None of these

24 The value of $\tan A + 2 \tan 2A + 4 \tan 4A + 8 \cot 8A$ is
(a) $\cot A$ (b) $\tan A$ (c) $\cos A$ (d) $\sin A$

25 The maximum value of $\sin\left(x + \frac{\pi}{6}\right) + \cos\left(x + \frac{\pi}{6}\right)$ in the interval $\left(0, \frac{\pi}{2}\right)$ is attained at
(a) $x = \frac{\pi}{12}$ (b) $x = \frac{\pi}{6}$ (c) $x = \frac{\pi}{3}$ (d) $x = \frac{\pi}{2}$

26 If $\cos A = m \cos B$ and $\cot \frac{A + B}{2} = \lambda \tan \frac{B - A}{2}$, then λ is
(a) $\frac{m}{m - 1}$ (b) $\frac{m + 1}{m}$
(c) $\frac{m + 1}{m - 1}$ (d) None of these

27 If $x \cos \theta = y \cos\left(\theta + \frac{2\pi}{3}\right) = z \cos\left(\theta + \frac{4\pi}{3}\right)$, then the value of $\frac{1}{x} + \frac{1}{y} + \frac{1}{z}$ is equal to
(a) 1 (b) 2 (c) 0 (d) $3\cos\theta$

28 The maximum value of $\cos^2\left(\frac{\pi}{3} - x\right) - \cos^2\left(\frac{\pi}{3} + x\right)$ is
(a) $-\frac{\sqrt{3}}{2}$ (b) $\frac{1}{2}$ (c) $\frac{\sqrt{3}}{2}$ (d) $\frac{3}{2}$

29 If sum of all the solution of the equation
$$8 \cos x \cdot \left(\cos\left(\frac{\pi}{6} + x\right) \cdot \cos\left(\frac{\pi}{6} - x\right) - \frac{1}{2}\right)$$
$= 1$ in $[0, \pi]$ is $k\pi$, then k is equal to
(a) $\frac{2}{3}$ (b) $\frac{13}{9}$ (c) $\frac{8}{9}$ (d) $\frac{20}{9}$

30 $\sin^2 \theta = \frac{4xy}{(x + y)^2}$ is true, if and only if → AIEEE 2002
(a) $x - y \neq 0$ (b) $x = -y$
(c) $x + y \neq 0$ (d) $x \neq 0, y \neq 0$

31 The smallest positive integral value of p for which the equation $\cos(p \sin x) = \sin(p \cos x)$ in x has a solution in $[0, 2\pi]$ is
(a) 2 (b) 1 (c) 3 (d) 5

32 If $A = \sin^2 x + \cos^4 x$, then for all real x
(a) $\frac{13}{16} \leq A \leq 1$ (b) $1 \leq A \leq 2$
(c) $\frac{3}{4} \leq A \leq \frac{13}{16}$ (d) $\frac{3}{4} \leq A \leq 1$

33 $\cos 2\theta + 2 \cos \theta$ is always
(a) greater than $-\frac{3}{2}$ (b) less than or equal to $\frac{3}{2}$
(c) greater than or equal to $-\frac{3}{2}$ and less than or equal to 3
(d) None of the above

34 If $f : R \to S$, defined by $f(x) = \sin x - \sqrt{3} \cos x + 1$, is onto, then the interval of S is
(a) $[0, 3]$ (b) $[-1, 1]$ (c) $[0, 1]$ (d) $[-1, 3]$

35 If $x \in \left[-\frac{5\pi}{2}, \frac{5\pi}{2}\right]$, then the greatest positive solution of $1 + \sin^4 x = \cos^2 3x$ is
(a) π (b) 2π (c) $\frac{5\pi}{2}$ (d) None of these

36 The number of solution of $\cos x = |1 + \sin x|$, $0 \leq x \leq 3\pi$ is
(a) 2 (b) 3 (c) 4 (d) 5

37 If $\sin 2x + \cos x = 0$, then which among the following is/are true?
I. $\cos x = 0$ II. $\sin x = -\frac{1}{2}$
III. $x = (2n + 1)\frac{\pi}{2}, n \in Z$ IV. $x = n\pi + (-1)^n \frac{7\pi}{6}, n \in Z$
(a) I is true (b) I and II are true
(c) I, II and III are true (d) All are true

38 The number of solutions of the equation $\sin 2x - 2 \cos x + 4 \sin x = 4$ in the interval $[0, 5\pi]$ is → JEE Mains 2013
(a) 3 (b) 5 (c) 4 (d) 6

39 The possible values of $\theta \in (0, \pi)$ such that $\sin(\theta) + \sin(4\theta) + \sin(7\theta) = 0$ are
(a) $\frac{2\pi}{9}, \frac{\pi}{4}, \frac{4\pi}{9}, \frac{\pi}{2}, \frac{3\pi}{4}, \frac{8\pi}{9}$ (b) $\frac{\pi}{4}, \frac{5\pi}{12}, \frac{\pi}{2}, \frac{2\pi}{3}, \frac{3\pi}{4}, \frac{8\pi}{9}$
(c) $\frac{2\pi}{9}, \frac{\pi}{4}, \frac{\pi}{2}, \frac{2\pi}{3}, \frac{3\pi}{4}, \frac{35\pi}{36}$ (d) $\frac{2\pi}{9}, \frac{\pi}{4}, \frac{\pi}{2}, \frac{2\pi}{3}, \frac{3\pi}{4}, \frac{8\pi}{9}$

40 The number of values of x in the interval $[0, 3\pi]$ satisfying the equation $2 \sin^2 x + 5 \sin x - 3 = 0$ is
(a) 6 (b) 1 (c) 2 (d) 4

41 If α is a root of $25 \cos^2 \theta + 5 \cos \theta - 12 = 0$, $\frac{\pi}{2} < \alpha < \pi$, then $\sin 2\alpha$ is equal to
(a) $\frac{24}{25}$ (b) $-\frac{24}{25}$ (c) $\frac{13}{18}$ (d) $-\frac{13}{18}$

42 **Statement I** $\cos \frac{\pi}{7} \cdot \cos \frac{2\pi}{7} \cdot \cos \frac{4\pi}{7} = -1/8$

Statement II $\cos\theta \cos 2\theta \cos 4\theta \ldots$
$$\cos 2^{n-1}\theta = -\frac{1}{2^n} \text{ for } \theta = \frac{\pi}{2^n - 1}$$
(a) Statement I is true, Statement II is true; Statement II is the correct explanation for Statement I
(b) Statement I is true, Statement II is true; Statement II is not the correct explanation for Statement I
(c) Statement I is true, Statement II is false
(d) Statement I is false, Statement II is true

DAY PRACTICE SESSION 2

PROGRESSIVE QUESTIONS EXERCISE

1 If $f_k(x)=\frac{1}{k}(\sin^k x+\cos^k x)$, where $x\in R, k\geq 1$, then $f_4(x)-f_6(x)$ is equal to → JEE Mains 2014

(a) $\frac{1}{6}$ (b) $\frac{1}{3}$ (c) $\frac{1}{4}$ (d) $\frac{1}{12}$

2 If $\tan\alpha$, $\tan\beta$ and $\tan\gamma$ are the roots of the equation $x^3-px^2-r=0$, then the value of $(1+\tan^2\alpha)(1+\tan^2\beta)(1+\tan^2\gamma)$ is equal to

(a) $(p-r)^2$ (b) $1+(p-r)^2$
(c) $1-(p-r)^2$ (d) None of these

3 For $0<\phi<\frac{\pi}{2}$, if $x=\sum_{n=0}^{\infty}\cos^{2n}\phi$, $y=\sum_{n=0}^{\infty}\sin^{2n}\phi$ and $z=\sum_{n=0}^{\infty}\cos^{2n}\phi\sin^{2n}\phi$, then

(a) $xyz=xz+y$ (b) $xyz=xy-z$
(c) $xyz=x+y+z$ (d) $xyz=yz+x$

4 If $\sin^4 x+\cos^4 y+2=4\sin x\cos y$ and $0\leq x, y\leq\frac{\pi}{2}$, then $\sin x+\cos y$ is equal to

(a) -2 (b) 0 (c) 2 (d) 3/2

5 If $0\leq x<2\pi$, then the number of real values of x, which satisfy the equation

$$\cos x+\cos 2x+\cos 3x+\cos 4x=0, \text{ is}$$

(a) 3 (b) 5
(c) 7 (d) 9

6 If $\sin(\theta+\alpha)=a$ and $\sin(\theta+\beta)=b$, then $\cos 2(\alpha-\beta)-4ab\cos(\alpha-\beta)$ is equal to

(a) $1-a^2-b^2$ (b) $1-2a^2-2b^2$
(c) $2+a^2-b^2$ (d) $2-a^2-b^2$

7 If $\alpha=\sin^8\theta+\cos^{14}\theta$, then which of the following is true?

(a) $\alpha>1$ (b) $\alpha\leq 1$ (c) $\alpha=0$ (d) $\alpha<0$

8 Let n be an odd integer. If $\sin n\theta=\sum_{r=0}^{n}b_r\sin^r\theta$ for all real θ, then

(a) $b_0=1, b_1=3$ (b) $b_0=0, b_1=n$
(c) $b_0=-1, b_1=n$ (d) $b_0=0, b_1=n^2-3n-3$

9 If $x\sin a+y\sin 2a+z\sin 3a=\sin 4a$
$x\sin b+y\sin 2b+z\sin 3b=\sin 4b$
$x\sin c+y\sin 2c+z\sin 3c=\sin 4c$

Then, the roots of the equation $t^3-\left(\frac{z}{2}\right)t^2-\left(\frac{y+2}{4}\right)t+\left(\frac{z-x}{8}\right)=0$, $a, b, c\neq n\pi$, are

(a) $\sin a, \sin b, \sin c$ (b) $\cos a, \cos b, \cos c$
(c) $\sin 2a, \sin 2b, \sin 2c$ (d) $\cos 2a, \cos 2b, \cos 2c$

10 The number of solutions of the equation $|\sin\theta\cdot\cos\theta|+\sqrt{2+\tan^2\theta+\cot^2\theta}=\sqrt{3}, \theta\in[0, 4\pi]$ is/are

(a) 0 (b) 1 (c) 2 (d) 3

11 The number of values of θ in the interval $\left(-\frac{\pi}{2},\frac{\pi}{2}\right)$ satisfying the equation $(\sqrt{3})^{\sec^2\theta}=\tan^4\theta+2\tan^2\theta$ is

(a) 1 (b) 2 (c) 3 (d) None of these

12 The equation $e^{\sin x}-e^{-\sin x}-4=0$ has AIEEE 2012

(a) infinite number of real roots
(b) no real roots
(c) exactly one real root
(d) exactly four real roots

13 Find the general solution of the equation $(\sqrt{3}-1)\cos\theta+(\sqrt{3}+1)\sin\theta=2$.

(a) $2n\pi\pm\frac{\pi}{4}-\frac{5\pi}{12}$ (b) $2n\pi\pm\frac{\pi}{4}+\frac{5\pi}{12}$
(c) $2n\pi\pm\pi-\frac{3\pi}{12}$ (d) None of these

14 At how many points the curve $y=81^{\sin^2 x}+81^{\cos^2 x}-30$ will intersect X-axis in the region $-\pi\leq x\leq\pi$?

(a) 4 (b) 6 (c) 8 (d) None of these

15 In a ΔPQR, if $3\sin P+4\cos Q=6$ and $4\sin Q+3\cos P=1$, then the angle R is equal to → AIEEE 2012

(a) $\frac{5\pi}{6}$ (b) $\frac{\pi}{6}$ (c) $\frac{\pi}{4}$ (d) $\frac{3\pi}{4}$

ANSWERS

SESSION 1

1. (a)	**2.** (c)	**3.** (d)	**4.** (c)	**5.** (c)	**6.** (c)	**7.** (b)	**8.** (b)	**9.** (c)	**10.** (a)
11. (d)	**12.** (a)	**13.** (c)	**14.** (b)	**15.** (b)	**16.** (a)	**17.** (b)	**18.** (b)	**19.** (a)	**20.** (d)
21. (c)	**22.** (b)	**23.** (a)	**24.** (a)	**25.** (a)	**26.** (c)	**27.** (c)	**28.** (c)	**29.** (b)	**30.** (c)
31. (a)	**32.** (d)	**33.** (c)	**34.** (d)	**35.** (b)	**36.** (b)	**37.** (d)	**38.** (a)	**39.** (a)	**40.** (d)
41. (b)	**42.** (a)								

SESSION 2

1. (d)	**2.** (b)	**3.** (c)	**4.** (c)	**5.** (c)	**6.** (b)	**7.** (b)	**8.** (b)	**9.** (b)	**10.** (a)
11. (b)	**12.** (b)	**13.** (b)	**14.** (c)	**15.** (b)					

Hints and Explanations

SESSION 1

1 Given, $\dfrac{2\sin\theta}{1+\cos\theta+\sin\theta}=x$

$\therefore\ x=\dfrac{2\sin\theta\,(1+\sin\theta-\cos\theta)}{(1+\sin\theta)^2-\cos^2\theta}$

$=\dfrac{2\sin\theta\,(1+\sin\theta-\cos\theta)}{1^2+\sin^2\theta+2\sin\theta-(1-\sin^2\theta)}$

$=\dfrac{1+\sin\theta-\cos\theta}{1+\sin\theta}$

2 We have, $\sqrt{\operatorname{cosec}^2\alpha+2\cot\alpha}$

$=\sqrt{1+\cot^2\alpha+2\cot\alpha}$

$=|1+\cot\alpha|$

But $\dfrac{3\pi}{4}<\alpha<\pi$

$\Rightarrow\quad \cot\alpha<-1$

$\Rightarrow\quad 1+\cot\alpha<0$

Hence, $|1+\cot\alpha|=-(1+\cot\alpha)$

3 $\operatorname{cosec}^2x+25\sec^2x$

$=1+\cot^2x+25(1+\tan^2x)$

$=26+\cot^2x+25\tan^2x$

$=26+10+(\cot x-5\tan x)^2\geq 36$

4 Since, $\tan A+\sin A=m$

and $\tan A-\sin A=n$

$\therefore\quad m+n=2\tan A$

and $\quad m-n=2\sin A$

Also, $mn=(\tan A+\sin A)(\tan A-\sin A)$

$=\tan^2A-\sin^2A$

Now, $\dfrac{(m^2-n^2)^2}{mn}=\dfrac{(m+n)^2(m-n)^2}{mn}$

$=\dfrac{(2\tan A)^2(2\sin A)^2}{\tan^2A-\sin^2A}$

$=\dfrac{16\tan^2A\sin^2A}{\sin^2A\tan^2A}=16$

5 Given,

$a\cos^3\alpha+3a\cos\alpha\sin^2\alpha=m$

and $a\sin^3\alpha+3a\cos^2\alpha\sin\alpha=n$

$\therefore (m+n)=a\cos^3\alpha+3a\cos\alpha\sin^2\alpha+3a\cos^2\alpha\sin\alpha+a\sin^3\alpha$

$=a(\cos\alpha+\sin\alpha)^3$

and similarly,

$(m-n)=a(\cos\alpha-\sin\alpha)^3$

$\therefore\quad (m+n)^{2/3}+(m-n)^{2/3}$

$=a^{2/3}\{(\cos\alpha+\sin\alpha)^2+(\cos\alpha-\sin\alpha)^2\}$

$=a^{2/3}\{2(\cos^2\alpha+\sin^2\alpha)\}$

$=2a^{2/3}$

6 Given expression

$=\left(1+\cos\dfrac{\pi}{8}\right)\left(1+\cos\dfrac{3\pi}{8}\right)\left(1-\cos\dfrac{3\pi}{8}\right)\left(1-\cos\dfrac{\pi}{8}\right)$

$=\left(1-\cos^2\dfrac{\pi}{8}\right)\left(1-\cos^2\dfrac{3\pi}{8}\right)$

$=\sin^2\dfrac{\pi}{8}\sin^2\dfrac{3\pi}{8}=\sin^2\dfrac{\pi}{8}\cos^2\dfrac{\pi}{8}$

$=\dfrac{1}{4}\sin^2\dfrac{\pi}{4}=\dfrac{1}{8}$

7 Given that, $\sin\theta=-\dfrac{4}{5}$ and θ lies in the 3rd quadrant.

$\therefore\quad \cos\theta=-\sqrt{1-\dfrac{16}{25}}=-\dfrac{3}{5}$

Now, $\cos\dfrac{\theta}{2}=\pm\sqrt{\dfrac{1+\cos\theta}{2}}$

$=\pm\sqrt{\dfrac{1-\dfrac{3}{5}}{2}}=\pm\sqrt{\dfrac{1}{5}}$

But we take $\cos\dfrac{\theta}{2}=-\dfrac{1}{\sqrt{5}}$. Since, if θ lies in 3rd quadrant, then $\dfrac{\theta}{2}$ will be in 2nd quadrant.

Hence, $\cos\dfrac{\theta}{2}=-\dfrac{1}{\sqrt{5}}$

8 Given expression is

$\dfrac{\tan A}{1-\cot A}+\dfrac{\cot A}{1-\tan A}=\dfrac{\sin A}{\cos A}\times\dfrac{\sin A}{\sin A-\cos A}+\dfrac{\cos A}{\sin A}\times\dfrac{\cos A}{\cos A-\sin A}$

$=\dfrac{1}{\sin A-\cos A}\left\{\dfrac{\sin^3A-\cos^3A}{\cos A\sin A}\right\}$

$=\dfrac{\sin^2A+\sin A\cos A+\cos^2A}{\sin A\cos A}$

$=\dfrac{1+\sin A\cos A}{\sin A\cos A}$

$=1+\sec A\operatorname{cosec}A$

9 We have,

$(2\tan\theta+2)^2=\tan\theta(3\tan\theta+3)$

$\Rightarrow\quad 4\tan^2\theta+8\tan\theta+4=3\tan^2\theta+3\tan\theta$

$\Rightarrow\quad \tan^2\theta+5\tan\theta+4=0$

$\Rightarrow\quad (\tan\theta+4)(\tan\theta+1)=0$

$\Rightarrow\quad \tan\theta=-4\qquad(\because \tan\theta\neq-1)$

$$\therefore \quad \frac{7-5\cot\theta}{9-4\sqrt{\tan^2\theta}} = \frac{7+\frac{5}{4}}{9-4(-4)} = \frac{33}{100}$$

10 Since, $\sin(\alpha+\beta)=1$

$\therefore \quad \alpha+\beta = \frac{\pi}{2}$ ...(i)

and $\sin(\alpha-\beta) = \frac{1}{2}$

$\Rightarrow \quad \alpha-\beta = \frac{\pi}{6}$ ...(ii)

On solving Eqs. (i) and (ii), we get

$\alpha = \frac{\pi}{3}$ and $\beta = \frac{\pi}{6}$

Now, $\tan(\alpha+2\beta)\tan(2\alpha+\beta)$

$= \tan\left(\frac{2\pi}{3}\right)\tan\left(\frac{5\pi}{6}\right)$

$= \tan\left(\pi-\frac{\pi}{3}\right)\tan\left(\pi-\frac{\pi}{6}\right)$

$= \left(-\cot\frac{\pi}{3}\right)\left(-\cot\frac{\pi}{6}\right)$

$= \frac{1}{\sqrt{3}}\times\sqrt{3} = 1$

11 $\sec A$, $\sec B$ and $\sec C$ are positive in an acute angled triangle.

Also, arithmetic mean $\geq$ harmonic mean

$\Rightarrow \quad \frac{\sec A+\sec B+\sec C}{3} \geq \frac{3}{\cos A+\cos B+\cos C}$

We have, in ΔABC,

$\cos A+\cos B+\cos C \leq \frac{3}{2}$

$\Rightarrow \quad \frac{1}{\cos A+\cos B+\cos C} \geq \frac{2}{3}$

$\therefore \quad \frac{\sec A+\sec B+\sec C}{3} \geq 2$

$\Rightarrow \quad \sec A+\sec B+\sec C \geq 6$

12 Clearly, $(\sin A+\sin B)^2 + (\cos A+\cos B)^2 = 2$

$\therefore \quad 2+2(\sin A\sin B + \cos A\cos B) = 2$

$\Rightarrow \cos(B-A)=0 \Rightarrow B = A+90°$

Second equation gives,

$\cos A - \sin A = \frac{1}{\sqrt{2}}$

$\Rightarrow \cos(A+45°) = \cos 60°$

$\therefore \quad A = 15°$

13 Since, $\tan\frac{P}{2}+\tan\frac{Q}{2} = -\frac{b}{a}$

and $\tan\frac{P}{2}\tan\frac{Q}{2} = \frac{c}{a}$

Also, $\frac{P}{2}+\frac{Q}{2}+\frac{R}{2} = \frac{\pi}{2}$

$\Rightarrow \quad \frac{P+Q}{2} = \frac{\pi}{4}$ $\quad\left[\because \angle R = \frac{\pi}{2}\right]$

$\Rightarrow \quad \frac{\tan\frac{P}{2}+\tan\frac{Q}{2}}{1-\tan\frac{P}{2}\tan\frac{Q}{2}} = 1 \Rightarrow \frac{-\frac{b}{a}}{1-\frac{c}{a}} = 1$

$\Rightarrow \quad c = a+b$

14 $\cos(\alpha+\beta) = \frac{4}{5}$

$\Rightarrow \quad \alpha+\beta \in$ 1st quadrant

$\sin(\alpha-\beta) = \frac{5}{13}$

$\Rightarrow \quad (\alpha-\beta) \in$ 1st quadrant

Now, as $\quad 2\alpha = (\alpha+\beta)+(\alpha-\beta)$

$\therefore \tan 2\alpha = \frac{\tan(\alpha+\beta)+\tan(\alpha-\beta)}{1-\tan(\alpha+\beta)\tan(\alpha-\beta)}$

$= \frac{\frac{3}{4}+\frac{5}{12}}{1-\frac{3}{4}\cdot\frac{5}{12}} = \frac{56}{33}$

15 We have,

$z = \sin(\alpha+\beta) = x\sqrt{1-y^2}+y\sqrt{1-x^2}$

$\Rightarrow \quad z^2 = x^2+y^2-2x^2y^2 + 2xy\sqrt{1-x^2}\sqrt{1-y^2}$

Now,

$\cos(\alpha+\beta) = \sqrt{1-x^2}\sqrt{1-y^2}-xy$

$= \frac{z^2-x^2-y^2+2x^2y^2}{2xy} - xy$

$= \frac{z^2-x^2-y^2}{2xy}$

16 Given that, $\sin\alpha+\sin\beta = -\frac{21}{65}$ ...(i)

and $\cos\alpha+\cos\beta = -\frac{27}{65}$ ...(ii)

On squaring and adding Eqs. (i) and (ii), we get

$\sin^2\alpha+\sin^2\beta+2\sin\alpha\sin\beta + \cos^2\alpha+\cos^2\beta+2\cos\alpha\cos\beta$

$= \left(-\frac{21}{65}\right)^2+\left(-\frac{27}{65}\right)^2$

$\Rightarrow \quad 2+2(\cos\alpha\cos\beta+\sin\alpha\sin\beta) = \frac{441}{4225}+\frac{729}{4225}$

$\Rightarrow \quad 2[1+\cos(\alpha-\beta)] = \frac{1170}{4225}$

$\Rightarrow \quad \cos^2\left(\frac{\alpha-\beta}{2}\right) = \frac{1170}{4\times 4225}$

$\Rightarrow \quad \cos^2\left(\frac{\alpha-\beta}{2}\right) = \frac{9}{130}$

$\therefore \quad \cos\left(\frac{\alpha-\beta}{2}\right) = -\frac{3}{\sqrt{130}}$

$\left[\because \pi < \alpha-\beta < 3\pi \Rightarrow \frac{\pi}{2} < \frac{\alpha-\beta}{2} < \frac{3\pi}{2}\right]$

17 Since, $A+B+C=\pi$

$\therefore \quad A = \pi-(B+C)$

We have, $\cos A = \cos B\cos C$

$\Rightarrow \quad \cos[\pi-(B+C)] = \cos B\cos C$

$\Rightarrow \quad -\cos(B+C) = \cos B\cos C$

$\Rightarrow \quad -[\cos B\cos C - \sin B\sin C] = \cos B\cos C$

$\Rightarrow \quad \sin B\sin C = 2\cos B\cos C$

$\Rightarrow \quad \tan B\tan C = 2$

18 We know,

$\tan(\alpha+\beta) = \frac{\tan\alpha+\tan\beta}{1-\tan\alpha\tan\beta}$

$\because \quad \tan\alpha = \frac{1}{1+2^{-x}}$

and $\quad \tan\beta = \frac{1}{1+2^{x+1}}$

$\therefore \quad \tan(\alpha+\beta) = \frac{\frac{1}{1+\frac{1}{2^x}}+\frac{1}{1+2^{x+1}}}{1-\frac{1}{1+\frac{1}{2^x}}\cdot\frac{1}{1+2^{x+1}}}$

$\Rightarrow \tan(\alpha+\beta)$

$= \frac{2^x+2\cdot 2^{2x}+2^x+1}{1+2^x+2\cdot 2^x+2\cdot 2^{2x}-2^x}$

$\Rightarrow \tan(\alpha+\beta) = 1$

$\Rightarrow \quad \alpha+\beta = \frac{\pi}{4}$

19 We have, $\cos\left(x-\frac{\pi}{4}\right) = \frac{1}{2\sqrt{2}}$

$\Rightarrow \quad \tan\left(x-\frac{\pi}{4}\right) = \sqrt{7}$

$\Rightarrow \quad \frac{\tan x-1}{\tan x+1} = \sqrt{7}$

$\therefore \quad \tan x = \frac{\sqrt{7}+1}{1-\sqrt{7}} = \frac{-(4+\sqrt{7})}{3}$

20 $\operatorname{cosec} 10° - \sqrt{3}\sec 10°$

$= \frac{1}{\sin 10°} - \frac{\sqrt{3}}{\cos 10°}$

$= \left(\frac{\cos 10°}{2} - \frac{\sqrt{3}}{2}\sin 10°\right)\frac{4}{\sin 20°}$

$= \frac{4\sin(30°-10°)}{\sin 20°} = 4$

21 Now, $\sin 12°\sin 48°\sin 54°$

$= \frac{1}{2}(\cos 36° - \cos 60°)\cos 36°$

$$=\frac{1}{2}\left[\frac{\sqrt{5}+1}{4}-\frac{1}{2}\right]\left[\frac{\sqrt{5}+1}{4}\right]$$
$$=\frac{1}{2}\left[\frac{\sqrt{5}-1}{4}\right]\left[\frac{\sqrt{5}+1}{4}\right]$$
$$=\frac{5-1}{32}=\frac{4}{32}=\frac{1}{8}$$

22 $1+\cos 56^\circ+\cos 58^\circ-\cos 66^\circ$
$$=2\cos^2 28^\circ+2\sin 62^\circ\sin 4^\circ$$
$$=2\cos^2 28^\circ+2\cos 28^\circ\cos 86^\circ$$
$$=2\cos 28^\circ(\cos 28^\circ+\cos 86^\circ)$$
$$=2\cos 28^\circ(2\cos 57^\circ\cos 29^\circ)$$
$$=4\cos 28^\circ\cos 29^\circ\sin 33^\circ$$

23 We have,
$$\sin\left(\frac{\pi}{2n}\right)+\cos\left(\frac{\pi}{2n}\right)=\frac{\sqrt{n}}{2}$$
On squaring both sides, we get
$$\sin^2\left(\frac{\pi}{2n}\right)+\cos^2\left(\frac{\pi}{2n}\right)+\sin\left(\frac{\pi}{n}\right)=\frac{n}{4}$$
$$\Rightarrow \quad \sin\left(\frac{\pi}{n}\right)=\frac{n}{4}-1$$
$$\Rightarrow \quad \sin\left(\frac{\pi}{n}\right)=\frac{n-4}{4}$$
$$\Rightarrow \quad n=6 \text{ only}$$

24 $\tan A+2\tan 2A+4\tan 4A+8\left(\frac{1-\tan^2 4A}{2\tan 4A}\right)$
$$=\tan A+2\tan 2A+\left(\frac{4\tan^2 4A+4-4\tan^2 4A}{\tan 4A}\right)$$
$$=\tan A+2\tan 2A+4\cot 4A$$
$$=\tan A+2\tan 2A+4\left(\frac{1-\tan^2 2A}{2\tan 2A}\right)$$
$$=\tan A+\left[\frac{2\tan^2 2A+2-2\tan^2 2A}{\tan 2A}\right]$$
$$=\tan A+2\cot 2A$$
$$=\tan A+2\left(\frac{1-\tan^2 A}{2\tan A}\right)$$
$$=\frac{\tan^2 A+1-\tan^2 A}{\tan A}=\cot A$$

25 We have, $\sin\left(x+\frac{\pi}{6}\right)+\cos\left(x+\frac{\pi}{6}\right)$
$$=\sqrt{2}\left[\frac{1}{\sqrt{2}}\sin\left(x+\frac{\pi}{6}\right)+\frac{1}{\sqrt{2}}\cos\left(x+\frac{\pi}{6}\right)\right]$$
$$=\sqrt{2}\cos\left[x+\frac{\pi}{6}-\frac{\pi}{4}\right]$$
$$=\sqrt{2}\cos\left(x-\frac{\pi}{12}\right)$$
Hence, maximum value will be at $x=\frac{\pi}{12}$.

26 We have, $\cos A=m\cos B$
$$\Rightarrow \frac{\cos A}{\cos B}=\frac{m}{1}$$
$$\Rightarrow \frac{\cos A+\cos B}{\cos A-\cos B}=\frac{m+1}{m-1}$$
$$\Rightarrow \frac{2\cos\frac{A+B}{2}\cos\frac{B-A}{2}}{2\sin\frac{A+B}{2}\sin\frac{B-A}{2}}=\frac{m+1}{m-1}$$
$$\Rightarrow \cot\frac{A+B}{2}=\left(\frac{m+1}{m-1}\right)\tan\frac{B-A}{2}$$
But $\cot\frac{A+B}{2}=\lambda\tan\frac{B-A}{2}$
$$\therefore \quad \lambda=\frac{m+1}{m-1}$$

27 We have, $x\cos\theta=y\cos\left(\theta+\frac{2\pi}{3}\right)$
$$=z\cos\left(\theta+\frac{4\pi}{3}\right)=k \text{ (say)}$$
$$\Rightarrow \cos\theta=\frac{k}{x},\quad \cos\left(\theta+\frac{2\pi}{3}\right)=\frac{k}{y}$$
and $\cos\left(\theta+\frac{4\pi}{3}\right)=\frac{k}{z}$
$$\therefore \frac{k}{x}+\frac{k}{y}+\frac{k}{z}=\cos\theta+\cos\left(\theta+\frac{2\pi}{3}\right)+\cos\left(\theta+\frac{4\pi}{3}\right)$$
$$=\cos\theta-\cos\left(\frac{\pi}{3}-\theta\right)-\cos\left(\frac{\pi}{3}+\theta\right)$$
$$=\cos\theta-2\cos\frac{\pi}{3}\cos\theta$$
$$\Rightarrow \cos\theta-2\times\frac{1}{2}\cos\theta=0$$
$$\Rightarrow \frac{1}{x}+\frac{1}{y}+\frac{1}{z}=0$$

28 $\cos^2\left(\frac{\pi}{3}-x\right)-\cos^2\left(\frac{\pi}{3}+x\right)$
$$=\left[\cos\left(\frac{\pi}{3}-x\right)+\cos\left(\frac{\pi}{3}+x\right)\right]\left[\cos\left(\frac{\pi}{3}-x\right)-\cos\left(\frac{\pi}{3}+x\right)\right]$$
$$=\left(2\cos\frac{\pi}{3}\cos x\right)\left(2\sin\frac{\pi}{3}\sin x\right)$$
$$=\sin\frac{2\pi}{3}\sin 2x=\frac{\sqrt{3}}{2}\sin 2x$$
Hence, maximum value of given expression is $\frac{\sqrt{3}}{2}$.

29 **Key idea** Apply the identify $\cos(x+y)\cos(x-y)=\cos^2 x-\sin^2 y$ and $\cos 3x=4\cos^3 x-3\cos x$

We have,
$$8\cos x\left(\cos\left(\frac{\pi}{6}+x\right)\cos\left(\frac{\pi}{6}-x\right)-\frac{1}{2}\right)=1$$
$$\Rightarrow 8\cos x\left(\cos^2\frac{\pi}{6}-\sin^2 x-\frac{1}{2}\right)=1$$
$$\Rightarrow 8\cos x\left(\frac{3}{4}-\sin^2 x-\frac{1}{2}\right)=1$$
$$\Rightarrow 8\cos x\left(\frac{3}{4}-\frac{1}{2}-1+\cos^2 x\right)=1$$
$$\Rightarrow 8\cos x\left(\frac{-3+4\cos^2 x}{4}\right)=1$$
$$\Rightarrow 2(4\cos^3 x-3\cos x)=1$$
$$\Rightarrow 2\cos 3x=1 \Rightarrow \cos 3x=\frac{1}{2}$$
$$\Rightarrow 3x=\frac{\pi}{3},\frac{5\pi}{3},\frac{7\pi}{3} \quad [0\le 3x\le 3\pi]$$
$$\Rightarrow x=\frac{\pi}{9},\frac{5\pi}{9},\frac{7\pi}{9}$$
Now, $\text{Sum}=\frac{\pi}{9}+\frac{5\pi}{9}+\frac{7\pi}{9}=\frac{13\pi}{9}$
$$\Rightarrow k\pi=\frac{13\pi}{9}$$
Hence, $k=\frac{13}{9}$

30 $\because \quad \sin^2\theta\le 1$
$$\therefore \frac{4xy}{(x+y)^2}\le 1$$
$$\Rightarrow x^2+y^2+2xy-4xy\ge 0$$
$$\Rightarrow (x-y)^2\ge 0$$
which is true for all real values of x and y provided $x+y\ne 0$, otherwise $\frac{4xy}{(x+y)^2}$ will be meaningless.

31 Clearly, $p\sin x=\frac{\pi}{2}\pm p\cos x$
$$\Rightarrow \sin x\pm\cos x=\frac{\pi}{2p}$$
$$\Rightarrow \sin\left(x\pm\frac{\pi}{4}\right)=\frac{\pi}{2\sqrt{2}p}\Rightarrow\left|\frac{\pi}{2\sqrt{2}p}\right|\le 1$$
For positive p, $p\ge\frac{\pi}{2\sqrt{2}}$ but $1<\frac{\pi}{2\sqrt{2}}<2$

Hence, the smallest positive integral value of p for which equation has solution is $p=2$.

32 $A=\sin^2 x+\cos^4 x$
$$\Rightarrow \quad A=1-\cos^2 x+\cos^4 x$$

$= \cos^4 x - \cos^2 x + \frac{1}{4} + \frac{3}{4}$

$= \left(\cos^2 x - \frac{1}{2}\right)^2 + \frac{3}{4}$ ... (i)

where, $0 \le \left(\cos^2 x - \frac{1}{2}\right)^2 \le \frac{1}{4}$ ... (ii)

$\therefore \quad \frac{3}{4} \le A \le 1$

33 $\cos 2\theta + 2\cos\theta = 2\cos^2\theta - 1 + 2\cos\theta$

$= 2\left(\cos\theta + \frac{1}{2}\right)^2 - \frac{3}{2} \ge -\frac{3}{2}$

$\left[\because 2\left(\cos\theta + \frac{1}{2}\right)^2 \ge 0, \forall \theta\right]$

and the maximum value of $\cos 2\theta + 2\cos\theta$ is 3.

34 Given that, $f(x) = \sin x - \sqrt{3}\cos x + 1$

$\because \quad -2 \le \sin x - \sqrt{3}\cos x \le 2$

$[\because \sqrt{a^2 + b^2} \le a\sin x + b\cos x \le \sqrt{a^2 + b^2}]$

$\Rightarrow -1 \le \sin x - \sqrt{3}\cos x + 1 \le 3$

$\therefore$ Range of $f(x) = [-1, 3]$

35 We have, $\sin^2 3x + \sin^4 x = 0$

$\Rightarrow \sin^2 x\{(3 - 4\sin^2 x)^2 + \sin^2 x\} = 0$

$\therefore \quad \sin x = 0 \Rightarrow x = n\pi$

Hence, greatest positive solution is 2π.

36 Clearly, $1 + \sin x \ge 0$

$\therefore$ The given equation becomes

$\cos x - \sin x = 1 \Rightarrow \cos\left(x + \frac{\pi}{4}\right) = \frac{1}{\sqrt{2}}$

$\Rightarrow \quad x + \frac{\pi}{4} = \frac{\pi}{4}, \frac{7\pi}{4}, \frac{9\pi}{4}, \frac{15\pi}{4}, \ldots$

$\Rightarrow \quad x = 0, \frac{3\pi}{2}, 2\pi, \frac{7\pi}{2}, \ldots$

$\because \quad 0 \le x \le 3\pi$

$\therefore \quad x = 0, \frac{3\pi}{2}, 2\pi$

37 $\sin 2x + \cos x = 0$

$\Rightarrow 2\sin x\cos x + \cos x = 0$ $[\because \sin 2x = 2\sin x\cos x]$

$\Rightarrow \cos x(2\sin x + 1) = 0$

$\Rightarrow \cos x = 0$ or $\sin x = -\frac{1}{2}$

When $\cos x = 0$, then $x = (2n + 1)\frac{\pi}{2}$

When $\quad \sin x = -\frac{1}{2}$,

then $\quad \sin x = -\sin\frac{\pi}{6}$

$\sin x = \sin\left(\pi + \frac{\pi}{6}\right)$

$[\because \sin(\pi + \theta) = -\sin\theta]$

$\sin x = \sin\frac{7\pi}{6}$

$\Rightarrow \quad x = n\pi + (-1)^n\frac{7\pi}{6} \quad [\because n \in Z]$

38 $\sin 2x - 2\cos x + 4\sin x = 4$

$\Rightarrow (\sin x - 1)(2\cos x + 4) = 0$

$\Rightarrow \sin x = 1, \cos x \ne -2$

$\therefore \quad x = n\pi + (-1)^n\frac{\pi}{2}$,

where $n \in z$

Hence, the required value of $x = \frac{\pi}{2}, \frac{5\pi}{2}, \frac{9\pi}{2}$ in interval $[0, 5\pi]$.

39 We have,

$\sin\theta + \sin 4\theta + \sin 7\theta = 0$

$\Rightarrow \sin 4\theta + (\sin\theta + \sin 7\theta) = 0$

$\Rightarrow \sin 4\theta + 2\sin 4\theta \cdot \cos 3\theta = 0$

$\Rightarrow \quad \sin 4\theta\{1 + 2\cos 3\theta\} = 0$

$\Rightarrow \quad \sin 4\theta = 0, \cos 3\theta = -\frac{1}{2}$

As, $0 < \theta < \pi$

$\therefore \quad 0 < 4\theta < 4\pi$

$\therefore \quad 4\theta = \pi, 2\pi, 3\pi$

Also, $\quad 0 < 3\theta < 3\pi$

$\Rightarrow \quad 3\theta = \frac{2\pi}{3}, \frac{4\pi}{3}, \frac{8\pi}{3}$

$\Rightarrow \quad \theta = \frac{\pi}{4}, \frac{\pi}{2}, \frac{3\pi}{4}, \frac{2\pi}{9}, \frac{4\pi}{9}, \frac{8\pi}{9}$

40 Given equation is

$2\sin^2 x + 5\sin x - 3 = 0.$

$\Rightarrow (2\sin x - 1)(\sin x + 3) = 0$

$\Rightarrow \quad \sin x = \frac{1}{2} \quad [\because \sin x \ne -3]$

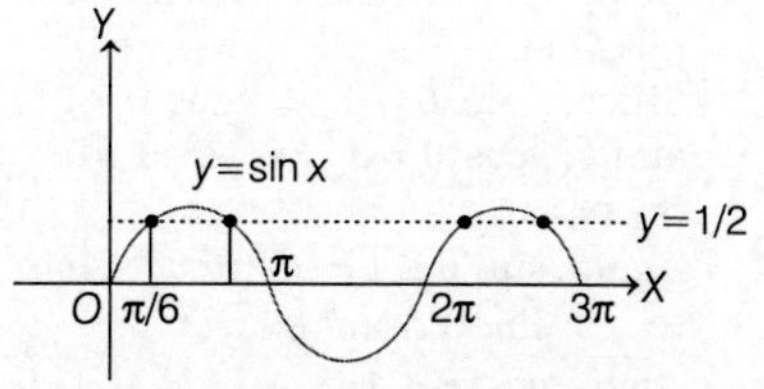

It is clear from figure that the curve intersect the line at four points in the given interval.

Hence, number of solutions are 4.

41 Since, α is a root of

$25\cos^2\theta + 5\cos\theta - 12 = 0$

$\therefore \quad 25\cos^2\alpha + 5\cos\alpha - 12 = 0$

$\Rightarrow \quad (5\cos\alpha - 3)(5\cos\alpha + 4) = 0$

$\Rightarrow \quad \cos\alpha = -\frac{4}{5}$ and $\frac{3}{5}$

But $\frac{\pi}{2} < \alpha < \pi$ i.e. in second quadrant.

$\therefore \quad \cos\alpha = -\frac{4}{5}$

$\Rightarrow \quad \sin\alpha = \frac{3}{5}$

Now, $\sin 2\alpha = 2\sin\alpha\cos\alpha$

$= 2 \times \frac{3}{5} \times \left(-\frac{4}{5}\right) = -\frac{24}{25}$

42 $\cos\frac{\pi}{7} \cdot \cos\frac{2\pi}{7} \cdot \cos\frac{4\pi}{7}$

$= \frac{\left(2\sin\frac{\pi}{7} \cdot \cos\frac{\pi}{7} \cdot \cos\frac{2\pi}{7} \cdot \cos\frac{4\pi}{7}\right)}{2\sin\frac{\pi}{7}}$

$= \frac{1}{8}\frac{\sin\frac{8\pi}{7}}{\sin\frac{\pi}{7}} = \frac{1}{8}\frac{\sin\left(\pi + \frac{\pi}{7}\right)}{\sin\frac{\pi}{7}} = -\frac{1}{8}$

$(\because 2\sin x\cos x = \sin 2x)$

Again, $\theta = \frac{\pi}{2^n - 1} \Rightarrow 2^n\theta = \pi + \theta$

$\therefore \quad \sin(2^n\theta) = -\sin\theta$

We know,

$\cos\theta \cdot \cos 2\theta \cdot \ldots \cdot \cos 2^{n-1}\theta$

$= \frac{1}{2^n}\frac{\sin(2^n\theta)}{\sin\theta}$

$= -\frac{1}{2^n}\frac{\sin\theta}{\sin\theta} = -\frac{1}{2^n}$

So, Statements I and II both are true and Statement II is a correct explanation for Statement I.

SESSION 2

1 Given, $f_k(x) = \frac{1}{k}(\sin^k x + \cos^k x)$,

where $x \in R$ and $k \ge 1$

Now, $f_4(x) - f_6(x)$

$= \frac{1}{4}(\sin^4 x + \cos^4 x) - \frac{1}{6}(\sin^6 x + \cos^6 x)$

$= \frac{1}{4}(1 - 2\sin^2 x \cdot \cos^2 x) - \frac{1}{6}(1 - 3\sin^2 x \cdot \cos^2 x)$

$= \frac{1}{4} - \frac{1}{6} = \frac{1}{12}$

2 From the given equations, we have

$\Sigma\tan\alpha = p$

$\Sigma\tan\alpha\tan\beta = 0$

and $\quad \tan\alpha\tan\beta\tan\gamma = r$

Now,

$(1 + \tan^2\alpha)(1 + \tan^2\beta)(1 + \tan^2\gamma)$

$= 1 + \Sigma \tan^2\alpha + \Sigma \tan^2\alpha \tan^2\beta + \tan^2\alpha \tan^2\beta \tan^2\gamma$

$= 1 + (\Sigma \tan\alpha)^2 - 2\Sigma \tan\alpha \tan\beta + (\Sigma \tan\alpha \tan\beta)^2 - 2\tan\alpha \tan\beta \tan\gamma \Sigma \tan\alpha + \tan^2\alpha \tan^2\beta \tan^2\gamma$

$= 1 + p^2 - 2pr + r^2 = 1 + (p - r)^2$

3 $x = \dfrac{1}{1-\cos^2\phi} = \dfrac{1}{\sin^2\phi},\ y = \dfrac{1}{\cos^2\phi}$

and $z = \dfrac{1}{1-\sin^2\phi\cdot\cos^2\phi}$

Clearly, $\dfrac{1}{x} + \dfrac{1}{y} = 1$

$\Rightarrow xy = x + y$ and $\dfrac{1}{z} = 1 - \dfrac{1}{xy}$

$\Rightarrow xy = x + y$

and $xy = xyz - z$

$\therefore xyz = xy + z = x + y + z$

4 The given equation is

$\sin^4 x + \cos^4 y + 2 = 4\sin x \cos y$

$\Rightarrow (\sin^2 x - 1)^2 + (\cos^2 y - 1)^2 + 2\sin^2 x + 2\cos^2 y - 4\sin x \cos y = 0$

$\Rightarrow (\sin^2 x - 1)^2 + (\cos^2 y - 1)^2 + 2(\sin x - \cos y)^2 = 0$

which is possible only when

$\sin^2 x - 1 = 0,\ \cos^2 y - 1 = 0,$

$\sin x - \cos y = 0$

$\Rightarrow \sin^2 x = 1, \cos^2 y = 1, \sin x = \cos y$

As $0 \le x, y \le \dfrac{\pi}{2}$

$\therefore$ We get $\sin x = \cos y = 1$ and so

$\sin x + \cos y = 1 + 1 = 2$

5 Given equation is

$\cos x + \cos 2x + \cos 3x + \cos 4x = 0$

$\Rightarrow (\cos x + \cos 3x) + (\cos 2x + \cos 4x) = 0$

$\Rightarrow 2\cos 2x \cos x + 2\cos 3x \cos x = 0$

$\Rightarrow 2\cos x(\cos 2x + \cos 3x) = 0$

$\Rightarrow 2\cos x\left(2\cos\dfrac{5x}{2}\cos\dfrac{x}{2}\right) = 0$

$\Rightarrow \cos x\cdot\cos\dfrac{5x}{2}\cdot\cos\dfrac{x}{2} = 0$

$\Rightarrow \cos x = 0$ or $\cos\dfrac{5x}{2} = 0$ or $\cos\dfrac{x}{2} = 0$

Now, $\cos x = 0$

$\Rightarrow x = \dfrac{\pi}{2}, \dfrac{3\pi}{2}$ $[\because 0 \le x < 2\pi]$

$\cos\dfrac{5x}{2} = 0$

$\Rightarrow \dfrac{5x}{2} = \dfrac{\pi}{2}, \dfrac{3\pi}{2}, \dfrac{5\pi}{2}, \dfrac{7\pi}{2}, \dfrac{9\pi}{2}, \dfrac{11\pi}{2}, \ldots,$

$\Rightarrow x = \dfrac{\pi}{5}, \dfrac{3\pi}{5}, \pi, \dfrac{7\pi}{5}, \dfrac{9\pi}{5}$ $[\because 0 \le x < 2x]$

and $\cos\dfrac{x}{2} = 0$

$\Rightarrow \dfrac{x}{2} = \dfrac{\pi}{2}, \dfrac{3\pi}{2}, \dfrac{5\pi}{2}, \ldots$

$\Rightarrow x = \pi$ $[\because 0 \le x < 2\pi]$

Hence, $x = \dfrac{\pi}{2}, \dfrac{3\pi}{2}, \pi, \dfrac{\pi}{5}, \dfrac{3\pi}{5}, \dfrac{7\pi}{5}, \dfrac{9\pi}{5}$

6 We have, $\sin(\theta + \alpha) = a$

and $\sin(\theta + \beta) = b$

$\Rightarrow \theta + \alpha = \sin^{-1} a$ and $\theta + \beta = \sin^{-1} b$

$\therefore \alpha - \beta = \sin^{-1} a - \sin^{-1} b$

$= \dfrac{\pi}{2} - \cos^{-1} a - \dfrac{\pi}{2} + \cos^{-1} b$

$= \cos^{-1} b - \cos^{-1} a$

$= \cos^{-1}\{ab + \sqrt{1-b^2}\sqrt{1-a^2}\}$

$\Rightarrow \cos(\alpha - \beta) = ab + \sqrt{(1 - a^2 - b^2 + a^2b^2)}$

$\Rightarrow \cos^2(\alpha - \beta) = a^2b^2 + 1 - a^2 - b^2 + a^2b^2 + 2ab\sqrt{(1 - a^2 - b^2 + a^2b^2)}$

Now, $\cos 2(\alpha - \beta) - 4ab\cos(\alpha - \beta)$

$= 2\cos^2(\alpha - \beta) - 1 - 4ab\cos(\alpha - \beta)$

$= 4a^2b^2 + 2 - 2a^2 - 2b^2 + 4ab\sqrt{1 - a^2 - b^2 + a^2b^2} - 1 - 4a^2b^2 - 4ab\sqrt{1 - a^2 - b^2 + a^2b^2}$

$= 1 - 2a^2 - 2b^2$

7 Given, $\alpha = \sin^8\theta + \cos^{14}\theta$

Since, $\sin^8\theta \ge 0$ and $\cos^{14}\theta \ge 0$.

So, $\alpha \ge 0$

Also, $\sin^8\theta + \cos^{14}\theta = 0$ is not possible.

Since, $\sin\theta = 0 \Rightarrow \cos\theta \ne 0$

and $\cos\theta = 0 \Rightarrow \sin\theta \ne 0$

So, $\alpha > 0$

Again, $\sin^2\theta \le 1 \Rightarrow (\sin^2\theta)^4 \le \sin^2\theta$

$\Rightarrow \sin^8\theta \le \sin^2\theta$

Also, $\cos^2\theta \le 1$

$\Rightarrow (\cos^2\theta)^7 \le \cos^2\theta$

$\Rightarrow \cos^{14}\theta \le \cos^2\theta$

So, $\alpha = \sin^8\theta + \cos^{14}\theta \le \sin^2\theta + \cos^2\theta = 1$

$\alpha \le 1$ and $\alpha > 0$

8 Given, $\sin n\theta = \sum_{r=0}^{n} b_r \sin^r\theta = b_0 + b_1\sin\theta + b_2\sin^2\theta + \ldots + b_n\sin^n\theta$ ...(i)

Putting $\theta = 0$ in Eq. (i), we get $0 = b_0$

Again, Eq. (i) can be written as

$\sin n\theta = \sum_{r=1}^{n} b_r \sin^r\theta$

$\dfrac{\sin n\theta}{\sin\theta} = \sum_{r=1}^{n} b_r \sin^{r-1}\theta$

On taking limit as $\theta \to 0$, we get

$\lim_{\theta\to 0}\dfrac{\sin n\theta}{\sin\theta} = b_1$

$\Rightarrow \lim_{\theta\to 0} n\left(\dfrac{\sin n\theta}{n\theta}\right)\left(\dfrac{\theta}{\sin\theta}\right) = b_1$

$\Rightarrow n = b_1$

Hence, $b_0 = 0; b_1 = n$

9 Equation first can be written as

$x\sin a + y \times 2\sin a\cos a + z \times \sin a(3 - 4\sin^2 a)$

$= 2\times 2\sin a\cos a\cos 2a$

$\Rightarrow x + 2y\cos a + z(3 + 4\cos^2 a - 4)$

$= 4\cos a(2\cos^2 a - 1)$ as $\sin a \ne 0$

$\Rightarrow 8\cos^3 a - 4z\cos^2 a - (2y + 4)\cos a + (z - x) = 0$

$\Rightarrow \cos^3 a - \left(\dfrac{z}{2}\right)\cos^2 a - \left(\dfrac{y+2}{4}\right)\cos a + \left(\dfrac{z-x}{8}\right) = 0$

which shows that $\cos a$ is a root of the equation

$t^3 - \left(\dfrac{z}{2}\right)t^2 - \left(\dfrac{y+2}{4}\right)t + \left(\dfrac{z-x}{8}\right) = 0$

Similarly, from second and third equation we can verify that $\cos b$ and $\cos c$ are the roots of the given equation.

10 Given, $|\sin\theta.\cos\theta| + \sqrt{2 + \tan^2\theta + \cot^2\theta} = \sqrt{3}$

$\Rightarrow |\sin\theta\cdot\cos\theta| + \sqrt{(\tan\theta + \cot\theta)^2} = \sqrt{3}$

$\Rightarrow |\sin\theta\cdot\cos\theta| + |\tan\theta + \cot\theta| = \sqrt{3}$

$\Rightarrow |\sin\theta\cdot\cos\theta| + \left|\dfrac{\sin^2\theta + \cos^2\theta}{\sin\theta\cdot\cos\theta}\right| = \sqrt{3}$

$\Rightarrow |\sin\theta\cdot\cos\theta| + \dfrac{1}{|\sin\theta\cdot\cos\theta|} = \sqrt{3}$

We know that,

$|\sin\theta\cdot\cos\theta| + \dfrac{1}{|\sin\theta\cdot\cos\theta|} \ge 2$

Hence, there is no solution of this equation.

11 $(\sqrt{3})^{\sec^2\theta} = (\tan^2\theta + 1)^2 - 1$

$= (\sec^2\theta)^2 - 1$

Put $\sec^2\theta = x \quad (x \ge 1)$

Then, $(\sqrt{3})^x = x^2 - 1$

Let $y = (\sqrt{3})^x = (x^2 - 1) \quad (x > 1)$

Now, graphs of $y = (\sqrt{3})^x$ and $y = x^2 - 1$ intersect at one point

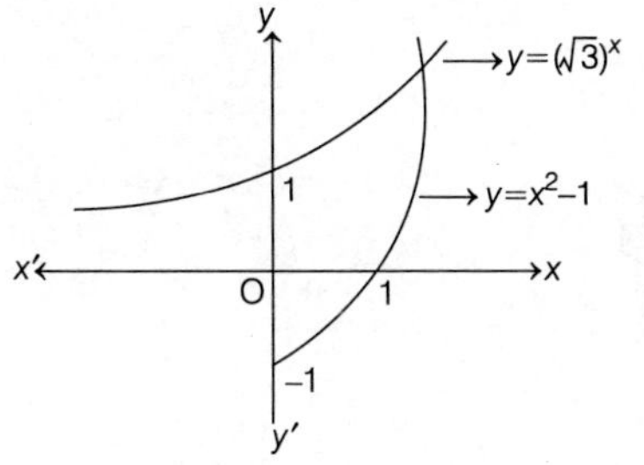

i.e. $x = 2$, then $y = 3$

Thus, $\sec^2\theta = 2 \Rightarrow \sec\theta = \pm\sqrt{2}$

Therefore, there are two values of θ in $\left(-\frac{\pi}{2}, \frac{\pi}{2}\right)$.

12 Given equation is

$$e^{\sin x} - e^{-\sin x} = 4 \Rightarrow e^{\sin x} - \frac{1}{e^{\sin x}} = 4$$

Now, let $y = e^{\sin x}$

Then, we get

$$y - \frac{1}{y} = 4 \Rightarrow y^2 - 4y - 1 = 0$$

$$\therefore \quad y = \frac{4 \pm \sqrt{16 + 4}}{2} \Rightarrow y = 2 \pm \sqrt{5}$$

$$\Rightarrow e^{\sin x} = 2 \pm \sqrt{5}$$

Since, sine is a bounded function i.e. $-1 \le \sin x \le 1$. Therefore, we get

$$e^{-1} \le e^{\sin x} \le e$$

$$\Rightarrow \quad e^{\sin x} \in \left[\frac{1}{e}, e\right]$$

Also, it is obvious that $2 + \sqrt{5} > e$

and $2 - \sqrt{5} < \frac{1}{e} \Rightarrow 2 \pm \sqrt{5} \notin \left[\frac{1}{e}, e\right]$

So, $e^{\sin x} = 2 + \sqrt{5}$ is not possible for any $x \in R$

and $e^{\sin x} = 2 - \sqrt{5}$ is also not possible for any $x \in R$

Hence, we can say that the given equation has no solution.

13 Given, $(\sqrt{3} - 1)\cos\theta + (\sqrt{3} + 1)\sin\theta = 2$ ... (i)

Let $(\sqrt{3} - 1) = r\cos\alpha$

and $(\sqrt{3} + 1) = r\sin\alpha$

Then, $r^2(\cos^2\alpha + \sin^2\alpha) = (\sqrt{3} - 1)^2 + (\sqrt{3} + 1)^2$

$\Rightarrow r^2 = 3 + 1 - 2\sqrt{3} + 3 + 1 + 2\sqrt{3}$

$\Rightarrow r^2 = 8 \Rightarrow r = 2\sqrt{2}$

and $\dfrac{r\sin\alpha}{r\cos\alpha} = \dfrac{\sqrt{3} + 1}{\sqrt{3} - 1} = -\dfrac{1 + 1/\sqrt{3}}{1 - 1/\sqrt{3}}$

$= \tan\left(\frac{\pi}{4} + \frac{\pi}{6}\right)$

$\Rightarrow \tan\alpha = \tan\left(\frac{5\pi}{12}\right) \Rightarrow \alpha = \frac{5\pi}{12}$

Also, Eq. (i) we have,

$r(\cos\alpha\cos\theta + \sin\alpha\sin\theta) = 2$

$\Rightarrow 2\sqrt{2}\cos(\alpha - \theta) = 2$

$\Rightarrow \cos(\theta - \alpha) = \frac{1}{\sqrt{2}}$

$\Rightarrow \theta - \alpha = 2n\pi \pm \frac{\pi}{4}$

$\therefore \theta = 2n\pi \pm \frac{\pi}{4} + \frac{5\pi}{12}$

14 Given, $y = 81^{\sin^2 x} + 81^{\cos^2 x} - 30$

[For intersection X –axis, put $y = 0$]

$\Rightarrow 81^{\sin^2 x} + 81^{1 - \sin^2 x} - 30 = 0$

$\Rightarrow 81^{2\sin^2 x} + 81 - 30 \cdot 81^{\sin^2 x} = 0$

[multiplying by $81\sin^2 x$]

$\Rightarrow 81^{2(\sin^2 x)} - 3 \cdot 81^{\sin^2 x} - 27 \cdot 81^{\sin^2 x} + 81 = 0$

$\Rightarrow (81^{\sin^2 x} - 3)(81^{\sin^2 x} - 27) = 0$

$\Rightarrow [(3)^4]^{\sin^2 x} = (3)^1$ or $[(3^4)]^{\sin^2 x} = (3)^3$

$\Rightarrow \sin x = \pm\frac{1}{2}$ or $\sin x = \pm\frac{\sqrt{3}}{2}$

$\Rightarrow x = \pm\frac{\pi}{6}, \pm\frac{5\pi}{6}$ or $x = \pm\frac{\pi}{3}, \pm\frac{2\pi}{3}$

Clearly, the graph of $y = 81^{\sin^2 x} + 81^{\cos^2 x} - 30$ intersects the X-axis at eight points in $-\pi \le x \le \pi$.

15 Given A ΔPQR such that

$3\sin P + 4\cos Q = 6$ ... (i)

$4\sin Q + 3\cos P = 1$ ... (ii)

On squaring and adding the Eqs. (i) and (ii), we get

$(3\sin P + 4\cos Q)^2 + (4\sin Q + 3\cos P)^2 = 36 + 1$

$\Rightarrow 9(\sin^2 P + \cos^2 P) + 16(\sin^2 Q + \cos^2 Q) + 2 \times 3 \times 4(\sin P\cos Q + \sin Q\cos P) = 37$

$\Rightarrow 24[\sin(P + Q)] = 37 - 25$

$\Rightarrow \sin(P + Q) = \frac{1}{2}$

Since, P and Q are angles of ΔPQR, therefore,

$0° < P, Q < 180°$

$\Rightarrow P + Q = 30°$ or $150°$

$\Rightarrow R = 150°$ or $30°$

Hence, two cases aries here.

Case I When, $R = 150°$

$R = 150° \Rightarrow P + Q = 30°$

$\Rightarrow 0 < P, Q < 30°$

$\Rightarrow \sin P < \frac{1}{2}, \cos Q < 1$

$\Rightarrow 3\sin P + 4\cos Q < \frac{3}{2} + 4$

$\Rightarrow 3\sin P + 4\cos Q < \frac{11}{2} < 6$

$\Rightarrow 3\sin P + 4\cos Q < 6$ not possible

Case II $R = 30°$

Hence, $R = 30°$ is the only possibility.

DAY TWENTY ONE

Properties of Triangle, Height and Distances

Learning & Revision for the Day

- Properties Related to Triangle
- Some Important Theorems
- Circles Connected with Triangle
- Angle of Elevation and Depression

Properties Related to Triangle

In any ΔABC,

(i) perimeter, $2s = a + b + c$

(ii) sum of all angles of a triangle is 180°, i.e. $\angle A + \angle B + \angle C = 180°$

(iii) $a+b > c, b+c > a, c+a > b$

(iv) $|a-b| < c, |b-c| < a, |c-a| < b$

(v) $a > 0, b > 0, c > 0$

Relations between the Sides and Angles of Triangle

For a triangle ΔABC with sides a, b, c and opposite angles are respectively A, B and C, then

(i) **Sine Rule** In any ΔABC, $\dfrac{a}{\sin A} = \dfrac{b}{\sin B} = \dfrac{c}{\sin C}$

(ii) **Cosine Rule**

(a) $\cos A = \dfrac{b^2 + c^2 - a^2}{2bc}$ (b) $\cos B = \dfrac{c^2 + a^2 - b^2}{2ca}$

(c) $\cos C = \dfrac{a^2 + b^2 - c^2}{2ab}$

PREP MIRROR
Your Personal Preparation Indicator

- No. of Questions in Exercises (x)—
- No. of Questions Attempted (y)—
- No. of Correct Questions (z)—
  (Without referring Explanations)

- Accuracy Level ($z/y \times 100$)—
- Prep Level ($z/x \times 100$)—

In order to expect good rank in JEE, your Accuracy Level should be above 85 & Prep Level should be above 75.

(iii) **Projection Rule**

(a) $a = b\cos C + c\cos B$ (b) $b = c\cos A + a\cos C$

(c) $c = a\cos B + b\cos A$

(iv) **Napier's Analogy**

(a) $\tan\frac{B-C}{2} = \frac{b-c}{b+c}\cot\frac{A}{2}$ (b) $\tan\frac{C-A}{2} = \frac{c-a}{c+a}\cot\frac{B}{2}$

(c) $\tan\frac{A-B}{2} = \frac{a-b}{a+b}\cot\frac{C}{2}$

(v) **Half Angle of Triangle**

(a) $\sin\frac{A}{2} = \sqrt{\frac{(s-b)(s-c)}{bc}}$ (b) $\sin\frac{B}{2} = \sqrt{\frac{(s-c)(s-a)}{ca}}$

(c) $\sin\frac{C}{2} = \sqrt{\frac{(s-a)(s-b)}{ab}}$ (d) $\cos\frac{A}{2} = \sqrt{\frac{s(s-a)}{bc}}$

(e) $\cos\frac{B}{2} = \sqrt{\frac{s(s-b)}{ca}}$ (f) $\cos\frac{C}{2} = \sqrt{\frac{s(s-c)}{ab}}$

(g) $\tan\frac{A}{2} = \sqrt{\frac{(s-b)(s-c)}{s(s-a)}}$ (h) $\tan\frac{B}{2} = \sqrt{\frac{(s-a)(s-c)}{s(s-b)}}$

(i) $\tan\frac{C}{2} = \sqrt{\frac{(s-a)(s-b)}{s(s-c)}}$

(vi) Area of a Triangle $\Delta = \frac{1}{2}bc\sin A = \frac{1}{2}ca\sin B = \frac{1}{2}ab\sin C$
$= \sqrt{s(s-a)(s-b)(s-c)}$

Some Important Theorems

1. m-n Theorem (Trigonometric Theorem)

If in a ΔABC, D divides AB in the ratio $m:n$, then (shown as in given figure)

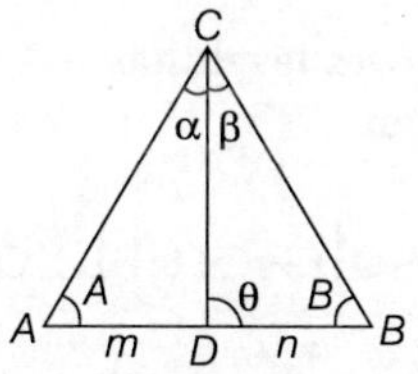

(i) $(m+n)\cot\theta = n\cot A - m\cot B$

(ii) $(m+n)\cot\theta = m\cot\alpha - n\cot\beta$

2. Appolonius Theorem

If in ΔABC, AD is median, then

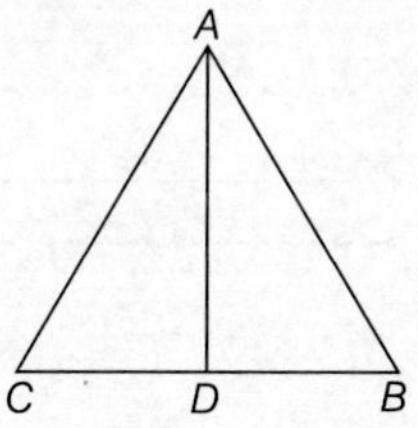

$$AB^2 + AC^2 = 2(AD^2 + BD^2)$$

The length of medians AD, BE and CF of a ΔABC are (shown as in given figure)

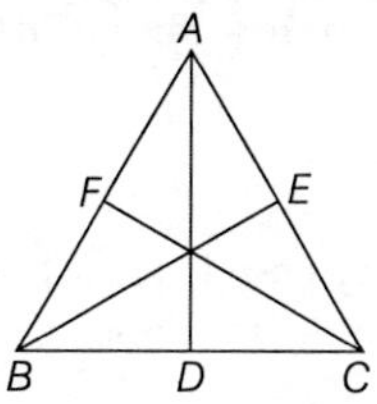

$AD = \frac{1}{2}\sqrt{2b^2 + 2c^2 - a^2}$, $BE = \frac{1}{2}\sqrt{2c^2 + 2a^2 - b^2}$

and $CF = \frac{1}{2}\sqrt{2a^2 + 2b^2 - c^2}$

Circles Connected with Triangle

1. Circumcircle

The circle passing through the vertices of the ΔABC is called the circumcircle. (shown as in given figure)

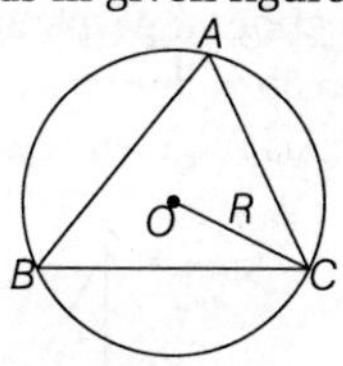

Its radius R is called the circumradius,

and $R = \frac{a}{2\sin A} = \frac{b}{2\sin B} = \frac{c}{2\sin C} = \frac{abc}{4\Delta}$

NOTE
- The mid-point of the hypotenuse of a right angled triangle is equidistant from the three vertices of the triangle.
- The mid-point of the hypotenuse of a right angled triangle is the circumcentre of the triangle.
- Distance of circumcentre from the side AC is $R\cos B$.
- Radius of circumcircle of a n-sided regular polygon with each side a is $R = \frac{a}{2}\text{cosec}\frac{\pi}{n}$.

2. Incircle

The circle touching the three sides of the triangle internally is called the inscribed circle or the incircle of the triangle. Its

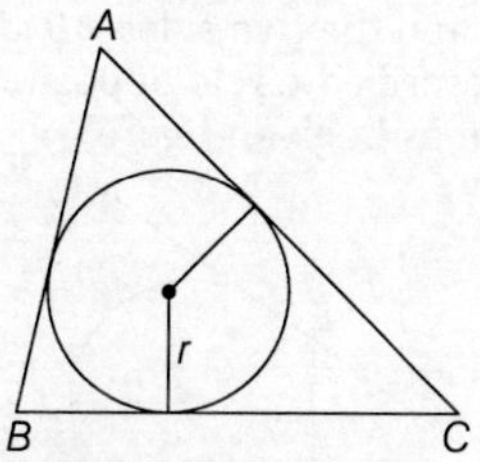

radius r is called inradius of the circle.

(i) $r = \frac{\Delta}{s}$

(ii) $r = (s-a)\tan\frac{A}{2} = (s-b)\tan\frac{B}{2} = (s-c)\tan\frac{C}{2}$

(iii) $r = 4R\sin\frac{A}{2}\sin\frac{B}{2}\sin\frac{C}{2}$

(iv) $r = \dfrac{a\sin\frac{B}{2}\sin\frac{C}{2}}{\cos\frac{A}{2}} = \dfrac{b\sin\frac{C}{2}\sin\frac{A}{2}}{\cos\frac{B}{2}} = \dfrac{c\sin\frac{A}{2}\sin\frac{B}{2}}{\cos\frac{C}{2}}$

NOTE Radius of incircle of a n-sided regular polygon with each side a is $r = \frac{a}{2}\cot\frac{\pi}{n}$. (shown as in given figure)

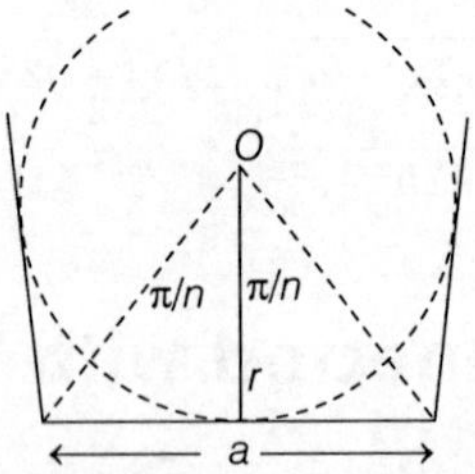

3. Orthocentre and Pedal Triangle

- The point of intersection of perpendiculars drawn from the vertices on the opposite sides of a triangle is called orthocentre. (shown as in given figure)

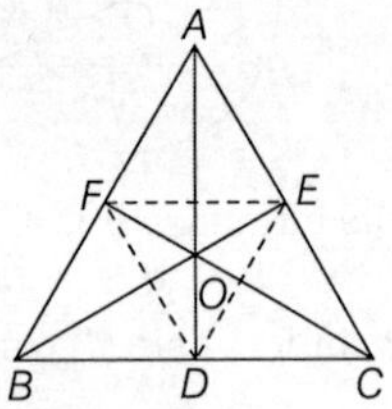

- The ΔDEF formed by joining the feet of the altitudes is called the pedal triangle.
- Orthocentre of the triangle is the incentre of the pedal triangle.
- Distance of the orthocentre of the triangle from the angular points are $2R\cos A$, $2R\cos B$, $2R\cos C$ and its distances from the sides are $2R\cos B\cos C$, $2R\cos C\cos A$, $2R\cos A\cos B$.

4. Escribed Circle

The circle touching BC and the two sides AB and AC produced of ΔABC, is called the escribed circle opposite to A. Its radius is denoted by r_1. (shown as in given figure)

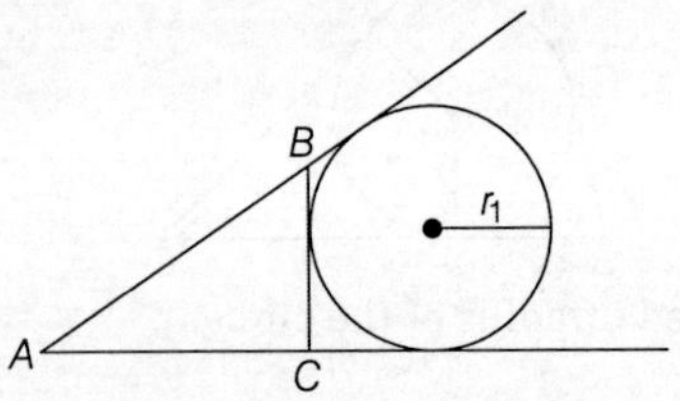

Similarly, r_2 and r_3 denote the radii of the escribed circles opposite to angles B and C, respectively.

r_1, r_2 and r_3 are called the exradius of ΔABC.

(i) $r_1 = \dfrac{\Delta}{s-a} = s\tan\dfrac{A}{2} = 4R\sin\dfrac{A}{2}\cos\dfrac{B}{2}\cos\dfrac{C}{2} = \dfrac{a\cos\frac{B}{2}\cos\frac{C}{2}}{\cos\frac{A}{2}}$

(ii) $r_2 = \dfrac{\Delta}{s-b} = s\tan\dfrac{B}{2} = 4R\sin\dfrac{B}{2}\cos\dfrac{C}{2}\cos\dfrac{A}{2} = \dfrac{b\cos\frac{C}{2}\cos\frac{A}{2}}{\cos\frac{B}{2}}$

(iii) $r_3 = \dfrac{\Delta}{s-c} = s\tan\dfrac{C}{2} = 4R\sin\dfrac{C}{2}\cos\dfrac{A}{2}\cos\dfrac{B}{2} = \dfrac{c\cos\frac{A}{2}\cos\frac{B}{2}}{\cos\frac{C}{2}}$

NOTE
- $r_1 + r_2 + r_3 = 4R + r$
- $r_1r_2 + r_2r_3 + r_3r_1 = \dfrac{r_1r_2r_3}{r}$
- $(r_1 - r)(r_2 - r)(r_3 - r) = 4Rr^2$
- $\dfrac{1}{r_1} + \dfrac{1}{r_2} + \dfrac{1}{r_3} = \dfrac{1}{r}$

Angle of Elevation and Depression

Let O be the observer's eye and OX be the horizontal line through O. (shown as in following figures)

If a object P is at a higher level than eye, then $\angle POX$ is called the **angle of elevation**.

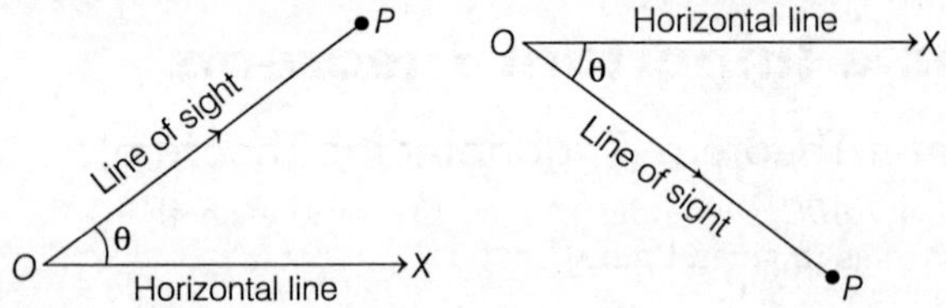

If a object P is at a lower level than eye, then $\angle POX$ is called the **angle of depression**.

Important Results on Heights and Distances

Results shown by the following figures.

(i) $a = h(\cot\alpha - \cot\beta)$

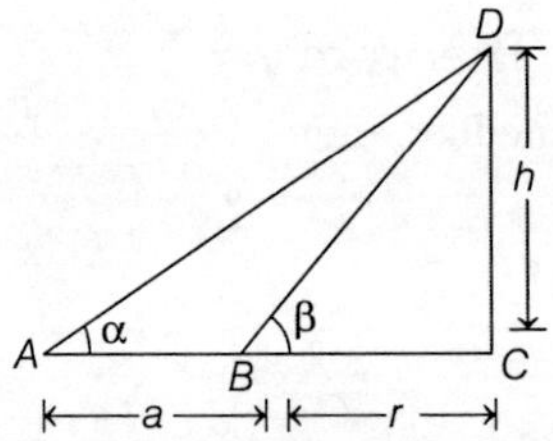

(ii) If $AB = CD$, then $x = y \tan\left(\frac{\alpha+\beta}{2}\right)$

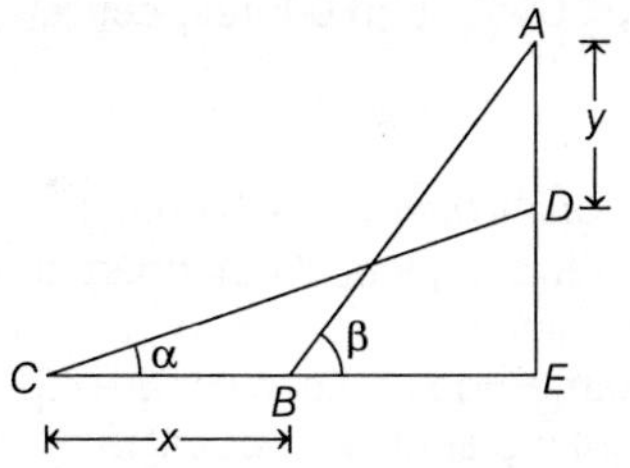

(iv) $H = \frac{h \cot\beta}{\cot\alpha}$

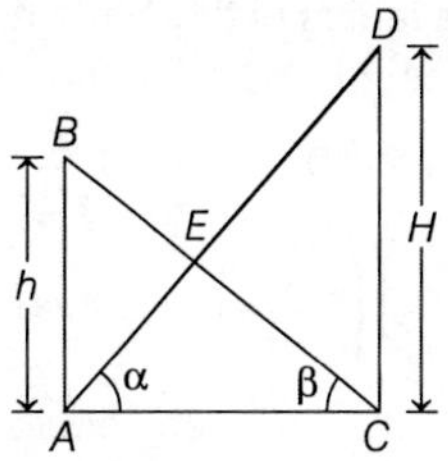

(iii) $h = \frac{H \sin(\beta-\alpha)}{\cos\alpha \sin\beta}$ and $H = \frac{h \cot\alpha}{\cot\alpha - \cot\beta}$

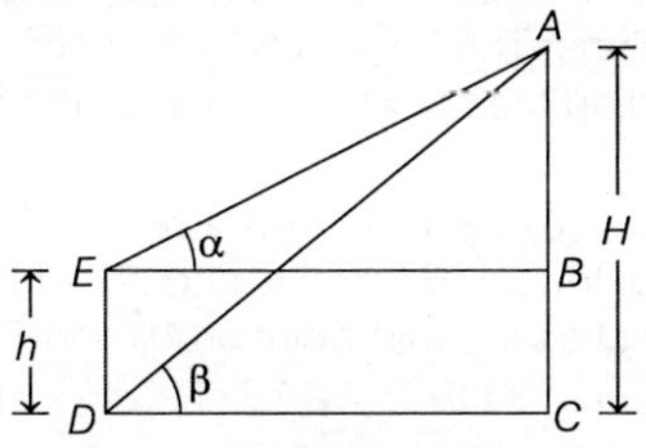

(v) $a = h(\cot\alpha + \cot\beta)$, $h = a \sin\alpha \sin\beta \operatorname{cosec}(\alpha+\beta)$ and $d = h\cot\beta = a\sin\alpha\cos\beta\operatorname{cosec}(\alpha+\beta)$

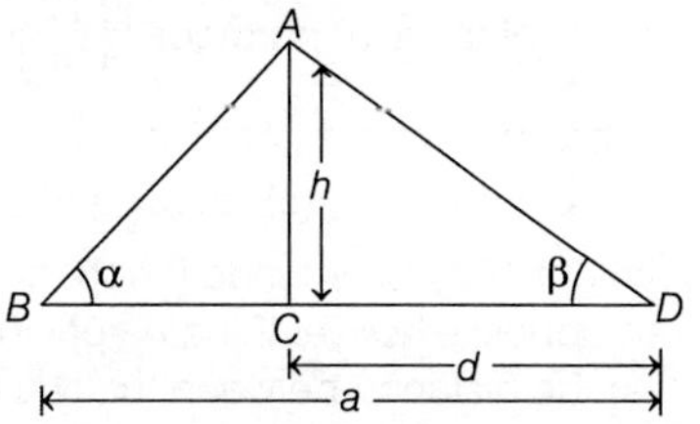

DAY PRACTICE SESSION 1

FOUNDATION QUESTIONS EXERCISE

1 If a, b and c are sides of a triangle, then
(a) $\sqrt{a} + \sqrt{b} > \sqrt{c}$
(b) $|\sqrt{a} - \sqrt{b}| > \sqrt{c}$ (if c is smallest)
(c) $\sqrt{a} + \sqrt{b} < \sqrt{c}$
(d) None of the above

2 If in a ΔABC, $A = 30°$, $B = 45°$ and $a = 1$, then the values of b and c are respectively
(a) $\sqrt{2}, \frac{\sqrt{3}+1}{\sqrt{2}}$ (b) $\sqrt{2}, \frac{\sqrt{3}-1}{\sqrt{2}}$
(c) $\sqrt{3}, \frac{\sqrt{3}-1}{\sqrt{2}}$ (d) $\sqrt{2}, \frac{\sqrt{3}+2}{\sqrt{2}}$

3 If $A = 75°$, $B = 45°$, then $b + c\sqrt{2}$ is equal to
(a) $2a$ (b) $2a + 1$ (c) $3a$ (d) $2a - 1$

4 If in a ΔABC, $2b^2 = a^2 + c^2$, then $\frac{\sin 3B}{\sin B}$ is equal to
(a) $\frac{c^2 - a^2}{2ca}$ (b) $\frac{c^2 - a^2}{ca}$
(c) $\left(\frac{c^2 - a^2}{ca}\right)^2$ (d) $\left(\frac{c^2 - a^2}{2ca}\right)^2$

5 The sides of a triangle are $\sin\alpha$, $\cos\alpha$ and $\sqrt{1 + \sin\alpha\cos\alpha}$ for some $0 < \alpha < \frac{\pi}{2}$. Then, the greatest angle of the triangle is
(a) 60° (b) 90° (c) 120° (d) 150°

6 In a ΔABC, $(c + a + b)(a + b - c) = ab$. The measure of $\angle C$ is
(a) $\frac{\pi}{3}$ (b) $\frac{\pi}{6}$
(c) $\frac{2\pi}{3}$ (d) None of these

7 In ΔABC, if $\tan\frac{A}{2}\tan\frac{C}{2} = \frac{1}{2}$, then a, b and c are in
(a) AP (b) GP
(c) HP (d) None of these

8 In a ΔABC, $a : b : c = 4 : 5 : 6$. The ratio of the radius of the circumcircle to that of the incircle is
(a) $\frac{16}{9}$ (b) $\frac{16}{7}$ (c) $\frac{11}{7}$ (d) $\frac{7}{16}$

9 In any ΔABC, $4\left(\frac{s}{a}-1\right)\left(\frac{s}{b}-1\right)\left(\frac{s}{c}-1\right)$ is equal to
(a) $\frac{r}{R}$ (b) $\frac{2r}{R}$
(c) $\frac{3r}{R}$ (d) None of these

10 In a ΔABC, R = circumradius and r = inradius. The value of $\frac{a\cos A + b\cos B + c\cos C}{a + b + c}$ is
(a) $\frac{R}{r}$ (b) $\frac{R}{2r}$ (c) $\frac{r}{R}$ (d) $\frac{2r}{R}$

11 Let the orthocentre and centroid of a triangle be $A(-3,5)$ and $B(3,3)$, respectively. If C is the circumcentre of this triangle, then the radius of the circle having line segment AC as diameter, is

(a) $\sqrt{10}$ (b) $2\sqrt{10}$ (c) $3\sqrt{\frac{5}{2}}$ (d) $\frac{3\sqrt{5}}{2}$

12 In a triangle $\left(1-\frac{r_1}{r_2}\right)\left(1-\frac{r_1}{r_3}\right)=2$, then the triangle is

(a) right angled (b) equilateral
(c) isosceles (d) None of these

13 There are two stations A and B due North, due South of a tower of height 15 m. The angle of depression of A and B as seen from top of the tower are $\cot^{-1}\left(\frac{12}{5}\right)$, $\sin^{-1}\left(\frac{3}{5}\right)$, then the distance between A and B is

(a) 48 m (b) 56 m (c) 25 m (d) None of these

14 A house of height 100 m subtends a right angle at the window of an opposite house. If the height of the window be 64 m, then the distance between the two houses is

(a) 48 m (b) 36 m (c) 54 m (d) 72 m

15 The angle of elevation of the top of a tower from the top and bottom of a building of height 'a' are 30° and 45°, respectively. If the tower and the building stand at the same level, then the height of the tower is

(a) $\frac{a(3+\sqrt{3})}{2}$ (b) $a(\sqrt{3}+1)$ (c) $a\sqrt{3}$ (d) $a(\sqrt{3}-1)$

16 A person walking along a straight road observes that at two points 1 km apart, the angles of elevation of a pole in front of line are 30° and 75°. The height of the pole is

(a) $250(\sqrt{3}+1)$ m (b) $250(\sqrt{3}-1)$ m
(c) $225(\sqrt{2}-1)$ m (d) $225(\sqrt{2}+1)$ m

17 A man is walking towards a vertical pillar in a straight path, at a uniform speed. At a certain point A on the path, he observes that the angle of elevation of the top of the pillar is 30°. After walking for 10 min from A in the same direction, at a point B, he observes that the angle of elevation of the top of the pillar is 60°. Then, the time taken (in minutes) by him, from B to reach the pillar, is
→ JEE Mains 2016

(a) 6 (b) 10 (c) 20 (d) 5

18 A bird is sitting on the top of a vertical pole 20 m high and its elevation from a point O on the ground is 45°. It flies off horizontally straight away from the point O. After 1s, the elevation of the bird from O is reduced to 30°. Then, the speed (in m/s) of the bird is → JEE Mains 2014

(a) $40(\sqrt{2}-1)$ (b) $40(\sqrt{3}-\sqrt{2})$ (c) $20\sqrt{2}$ (d) $20(\sqrt{3}-1)$

19 The shadow of a pole of height $(\sqrt{3}+1)$ m standing on the ground is found to be 2 m longer, when the elevation is 30° than when elevation was α, then α is equal to

(a) 15° (b) 30° (c) 45° (d) 75°

20 If the angles of elevation of the top of a tower from three collinear points A, B and C on a line leading to the foot of the tower are 30°, 45° and 60° respectively, then the ratio $AB:BC$ is → JEE Mains 2015

(a) $\sqrt{3}:1$ (b) $\sqrt{3}:\sqrt{2}$ (c) $1:\sqrt{3}$ (d) $2:3$

21 From the top of a tower 100 m height, the angles of depression of two objects 200 m apart on the horizontal plane and in a line passing through the foot of the tower and on the same side of the tower are $45°-A$ and $45°+A$. Then, the angle A is equal to

(a) 25° (b) 30° (c) $22\frac{1°}{2}$ (d) 45°

22 Two vertical poles 20 m and 80 m stands apart on a horizontal plane. The height of the point of intersection of the lines joining the top of each pole to the foot of the other is

(a) 15 m (b) 16 m (c) 18 m (d) 50 m

23 Let a vertical tower AB have its end A on the level ground. Let C be the mid-point of AB and P be a point on the ground such that $AP=2AB$. If $\angle BPC=\beta$, then $\tan\beta$ is equal to → JEE Mains 2017

(a) $\frac{6}{7}$ (b) $\frac{1}{4}$ (c) $\frac{2}{9}$ (d) $\frac{4}{9}$

24 From the tower 60 m high angles of depression of the top and bottom of a house are α and β, respectively. If the height of the house is $\frac{60\sin(\beta-\alpha)}{x}$, then x is equal to

(a) $\sin\alpha\sin\beta$ (b) $\cos\alpha\cos\beta$
(c) $\sin\alpha\cos\beta$ (d) $\cos\alpha\sin\beta$

25 A vertical tower stands on a declivity which is inclined at 15° to the horizon. From the foot of the tower a man ascends the declivity for 80 ft and then, finds that the tower subtends an angle of 30°. The height of tower is

(a) $20(\sqrt{6}-\sqrt{2})$ ft (b) $40(\sqrt{6}-\sqrt{2})$ ft
(c) $40(\sqrt{6}+\sqrt{2})$ ft (d) None of these

26 A ladder rest against a wall at an $\angle\alpha$ to the horizontal. Its foot is pulled away through a distance a_1, so that it slides a distance b_1 down the wall and rests inclined at $\angle\beta$ with the horizontal. It foot is further pulled aways through a_2, so that it slides a further distance b_2 down the wall and is now, inclined at an $\angle\gamma$. If $a_1a_2=b_1b_2$, then

(a) $\alpha+\beta+\gamma$ is greater than π
(b) $\alpha+\beta+\gamma$ is equal to π
(c) $\alpha+\beta+\gamma$ is less than π
(d) nothing can be said about $\alpha+\beta+\gamma$

27 $ABCD$ is a trapezium such that AB and CD are parallel and $BC\perp CD$. If $\angle ADB=\theta$, $BC=p$ and $CD=q$, then AB is equal to → JEE Mains 2013

(a) $\frac{(p^2+q^2)\sin\theta}{p\cos\theta+q\sin\theta}$ (b) $\frac{p^2+q^2\cos\theta}{p\cos\theta+q\sin\theta}$
(c) $\frac{p^2+q^2}{p^2\cos\theta+q^2\sin\theta}$ (d) $\frac{(p^2+q^2)\sin\theta}{(p\cos\theta+q\sin\theta)^2}$

28 The angle of elevation of the top of the tower observed from each of the three points A, B, C on the ground forming a triangle is the same $\angle \alpha$. If R is the circumradius of the ΔABC, then the height of the tower is

(a) $R \sin \alpha$ (b) $R \cos \alpha$
(c) $R \cot \alpha$ (d) $R \tan \alpha$

29 A tower stands at the centre of a circular park. A and B are two points on the boundary of the park such that $AB(= a)$ subtends an angle of 60° at the foot of the tower and the angle of elevation of the top of the tower from A or B is 30°. The height of the tower is

(a) $\frac{2a}{\sqrt{3}}$ (b) $2a\sqrt{3}$ (c) $\frac{a}{\sqrt{3}}$ (d) $\sqrt{3}$

30 $ABCD$ is a square plot. The angle of elevation of the top of a pole standing at D from A or C is 30° and that from B is θ, then $\tan \theta$ is equal to

(a) $\sqrt{6}$ (b) $1/\sqrt{6}$ (c) $\sqrt{3}/2$ (d) $\sqrt{2/3}$

DAY PRACTICE SESSION 2

PROGRESSIVE QUESTIONS EXERCISE

1 For a regular polygon, let r and R be the radii of the inscribed and the circumscribed circles. A false statement among the following is

(a) there is a regular polygon with $\frac{r}{R} = \frac{1}{2}$

(b) there is a regular polygon with $\frac{r}{R} = \frac{1}{\sqrt{2}}$

(c) there is a regular polygon with $\frac{r}{R} = \frac{2}{3}$

(d) there is a regular polygon with $\frac{r}{R} = \frac{\sqrt{3}}{2}$

2 If $\cos A + \cos B + 2 \cos C = 2$, then the sides of the ΔABC are in

(a) AP (b) GP
(c) HP (d) None of these

3 If A_0, A_1, A_2, A_3, A_4 and A_5 are the consecutive vertices of a regular hexagon inscribed in a unit circle. Then, the product of length of $A_0A_1 \times A_0A_2$ is

(a) $\sqrt{3}$ (b) $2\sqrt{3}$ (c) 2 (d) $3\sqrt{3}$

4 Two ships leave a port at the same time.One goes 24 km/h in the direction North 45° East and other travels 32 km/h in the direction South 75° East. The distance between the ships at the end of 3 h is approximately

(a) 81.4 km (b) 82 km (c) 85 km (d) 86.4 km

5 A round balloon of radius r subtends an angle α at the observer, while the angle of elevation of its centre is β. The height of the centre of balloon is

(a) $r \operatorname{cosec} \alpha \sin\frac{\beta}{2}$ (b) $r \sin\beta \operatorname{cosec} \frac{\alpha}{2}$

(c) $r \sin\frac{\alpha}{2} \operatorname{cosec} \beta$ (d) $r \operatorname{cosec} \frac{\alpha}{2} \sin\beta$

6 PQR is a triangular park with $PQ = PR = 200$ m. A TV tower stands at the mid-point of QR. If the angles of elevation of the top of the tower at P, Q and R are respectively 45°, 30° and 30°, then the height of the tower (in m) is → JEE Mains 2018

(a) 100 (b) 50
(c) $100\sqrt{3}$ (d) $50\sqrt{2}$

7 At the foot of a mountain the elevation of its summit is 45°. After ascending 2 km towards the mountain up an incline of 30°, the elevation changes to 60°. The height of mountain is

(a) 2.732 km (b) 1.732 km
(c) 2.03 km (d) 1.045 km

8 In a cubical hall $ABCDPQRS$ with each side 10 m, G is the centre of the wall $BCRQ$ and T is the mid-point of the side AB. The angle of elevation of G at the point T is

(a) $\sin^{-1} (1/\sqrt{3})$ (b) $\cos^{-1} (1/\sqrt{3})$
(c) $\cot^{-1} (1/\sqrt{3})$ (d) None of these

9 A tree stands vertically on a hill side which makes an angle of 15° with the horizontal.From a point on the ground 35 m down the hill from the base of the tree, the angle of elevation of the top of the tree is 60°.The height of the tree is

(a) $31\sqrt{2}$m (b) $33\sqrt{2}$m
(c) $35\sqrt{2}$m (d) $34\sqrt{2}$m

10 A flag staff stands in the centre of a rectangular field whose diagonal is 1200 m, and subtends angles 15° and 45° at the mid-points of the sides of the field. The height of the flag staff is

(a) 200 m (b) $300(\sqrt{2 + \sqrt{3}})$ m
(c) $300(\sqrt{2 - \sqrt{3}})$ m (d) 400 m

ANSWERS

SESSION 1	1. (a)	2. (a)	3. (a)	4. (d)	5. (c)	6. (c)	7. (d)	8. (b)	9. (a)	10. (c)
	11. (c)	12. (a)	13. (b)	14. (a)	15. (a)	16. (a)	17. (d)	18. (d)	19. (c)	20. (a)
	21. (c)	22. (b)	23. (c)	24. (d)	25. (b)	26. (c)	27. (a)	28. (d)	29. (c)	30. (b)
SESSION 2	1. (c)	2. (a)	3. (a)	4. (d)	5. (b)	6. (a)	7. (a)	8. (a)	9. (c)	10. (c)

Hints and Explanations

SESSION 1

1 Clearly, $(\sqrt{a}+\sqrt{b}+\sqrt{c})(\sqrt{a}+\sqrt{b}-\sqrt{c})$

$$= (\sqrt{a}+\sqrt{b})^2 - c$$
$$= a + b - c + 2\sqrt{ab} > 0$$

$\therefore \sqrt{a}+\sqrt{b} > \sqrt{c}$

[$\because a + b > c$, as sides of triangle]

2 Given, $A = 30°, B = 45°, a = 1$

$\therefore \quad C = 105°$

Using sine rule,

$\dfrac{a}{\sin A} = \dfrac{b}{\sin B} = \dfrac{c}{\sin C}$, we get

$$\frac{1}{\left(\frac{1}{2}\right)} = \frac{b}{\left(\frac{1}{\sqrt{2}}\right)} = \frac{c}{\sin 105°}$$

$\Rightarrow \quad b = \sqrt{2}$

and

$c = 2\sin 105° = 2\cos 15°$

$\therefore \quad c = 2\left(\dfrac{\sqrt{3}+1}{2\sqrt{2}}\right) = \dfrac{\sqrt{3}+1}{\sqrt{2}}$

3 Given, $A = 75°$ and $B = 45°$

$\Rightarrow \quad C = 60°$

By sine rule, $\dfrac{a}{\sin A} = \dfrac{b}{\sin B} = \dfrac{c}{\sin C}$, we get $\dfrac{a}{\sin 75°} = \dfrac{b}{\sin 45°} = \dfrac{c}{\sin 60°}$

Now, $b + c\sqrt{2} = \dfrac{\sin 45°}{\sin 75°}a + \sqrt{2}\,\dfrac{\sin 60°}{\sin 75°}a$

$$= \frac{\frac{1}{\sqrt{2}}}{\frac{\sqrt{3}+1}{2\sqrt{2}}}a + \sqrt{2}\,\frac{\frac{\sqrt{3}}{2}}{\frac{\sqrt{3}+1}{2\sqrt{2}}}a$$
$$= \frac{2}{\sqrt{3}+1}a + \frac{2\sqrt{3}a}{\sqrt{3}+1} = 2a$$

4 $\dfrac{\sin 3B}{\sin B} = \dfrac{3\sin B - 4\sin^3 B}{\sin B}$

$$= 3 - 4\sin^2 B$$
$$= 3 - 4(1 - \cos^2 B)$$
$$= -1 + \frac{4(a^2+c^2-b^2)^2}{4(ac)^2}$$
$$= -1 + \frac{\left(\frac{a^2+c^2}{2}\right)^2}{(ac)^2}$$

[$\because 2b^2 = a^2 + c^2$ (given)]

$$= -1 + \frac{(a^2+c^2)^2}{4(ac)^2}$$
$$= \frac{(a^2+c^2)^2 - 4a^2c^2}{4(ac)^2} = \left(\frac{c^2-a^2}{2ac}\right)^2$$

5 Let $a = \sin\alpha, b = \cos\alpha$

and $c = \sqrt{1 + \sin\alpha\cos\alpha}$

Here, we can see that the greatest side is c.

$\therefore \quad \cos C = \dfrac{a^2+b^2-c^2}{2ab}$

$\Rightarrow \cos C = \dfrac{\sin^2\alpha + \cos^2\alpha - 1 - \sin\alpha\cos\alpha}{2ab}$

$\Rightarrow \quad \cos C = -\dfrac{\sin\alpha\cos\alpha}{2\sin\alpha\cos\alpha}$

$\Rightarrow \quad \cos C = -\dfrac{1}{2} = \cos 120°$

$\Rightarrow \quad \angle C = 120°$

6 We have,

$2s(2s - 2c) = ab \Rightarrow \dfrac{s(s-c)}{ab} = \dfrac{1}{4}$

$\Rightarrow \quad \cos^2\dfrac{C}{2} = \dfrac{1}{4}$

we get, $\cos\dfrac{C}{2} = \dfrac{1}{2}$

$\left[\text{since, } \dfrac{C}{2} \text{ must be acute}\right]$

$\Rightarrow \quad \dfrac{C}{2} = 60° \Rightarrow C = \dfrac{2\pi}{3}$

7 We have, $\tan\dfrac{A}{2}\tan\dfrac{C}{2} = \dfrac{1}{2}$

$\Rightarrow \quad \sqrt{\dfrac{(s-b)(s-c)}{s(s-a)}}\sqrt{\dfrac{(s-a)(s-b)}{s(s-c)}} = \dfrac{1}{2}$

$\Rightarrow \quad \dfrac{s-b}{s} = \dfrac{1}{2} \Rightarrow 2s - 2b - s = 0$

$\Rightarrow \quad a + c - 3b = 0$

8 We have, $R = \dfrac{abc}{4\Delta}$ and $r = \dfrac{\Delta}{s}$

$\therefore \quad \dfrac{R}{r} = \dfrac{abc}{4\Delta}\cdot\dfrac{s}{\Delta} = \dfrac{abc}{4(s-a)(s-b)(s-c)}$

Since, $a : b : c = 4 : 5 : 6$

$\Rightarrow \quad \dfrac{a}{4} = \dfrac{b}{5} = \dfrac{c}{6} = k$ (say)

Thus, $\dfrac{R}{r} = \dfrac{(4k)(5k)(6k)}{4\left(\frac{15k}{2} - 4k\right)\left(\frac{15k}{2} - 5k\right)\left(\frac{15k}{2} - 6k\right)}$

$$= \frac{120k^3\cdot 2}{k^3\cdot 7\cdot 5\cdot 3} = \frac{16}{7}$$

9 We have, $4\left(\dfrac{s}{a} - 1\right)\left(\dfrac{s}{b} - 1\right)\left(\dfrac{s}{c} - 1\right)$

$$= 4\cdot\frac{(s-a)(s-b)(s-c)}{abc}$$
$$= 4\cdot\frac{s(s-a)(s-b)(s-c)}{s\cdot abc}$$
$$= 4\cdot\frac{\Delta^2}{s\cdot abc} = \frac{4\Delta}{abc}\cdot\frac{\Delta}{s} = \frac{r}{R}$$

10 We have, $\dfrac{a\cos A + b\cos B + c\cos C}{a+b+c}$

$$= \frac{2R\sin A\cos A + 2R\sin B\cos B + 2R\sin C\cos C}{2s}$$
$$= \frac{R}{2s}\cdot(\sin 2A + \sin 2B + \sin 2C)$$
$$= \frac{R}{2s}\cdot 4\sin A\sin B\sin C$$
$$= \frac{2R}{s}\cdot\frac{abc}{8R^3} = \frac{abc}{4sR^2}$$
$$= \frac{4\Delta R}{4\cdot\frac{\Delta}{r}\cdot R^2} = \frac{r}{R} \quad \left[\because R = \frac{abc}{4\Delta}, r = \frac{\Delta}{s}\right]$$

11 **Key Idea** Orthocentre, centroid and circumcentre are collinear and centroid divide orthocentre and circumcentre in 2:1 ratio.

We have orthocentre $A(-3, 5)$ and centroid $B(3, 3)$. Let C be the circumcentre.

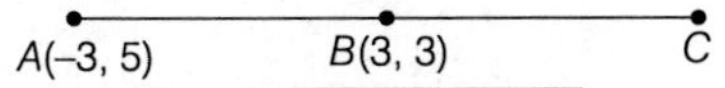

Clearly, $AB = \sqrt{(3+3)^2 + (3-5)^2}$

$= \sqrt{36+4} = 2\sqrt{10}$

We know that, $AB : BC = 2:1 \Rightarrow BC = \sqrt{10}$

Now, $AC = AB + BC = 2\sqrt{10} + \sqrt{10}$

$= 3\sqrt{10}$

Since, AC is a diameter of circle.

$\therefore \quad r = \frac{AC}{2} \Rightarrow r = \frac{3\sqrt{10}}{2} = 3\sqrt{\frac{5}{2}}$

12 Since, $\left(1 - \frac{r_1}{r_2}\right)\left(1 - \frac{r_1}{r_3}\right) = 2$

$\therefore \quad \left(1 - \frac{s-b}{s-a}\right)\left(1 - \frac{s-c}{s-a}\right) = 2$

$\Rightarrow \quad \frac{(b-a)(c-a)}{(s-a)^2} = 2$

$\Rightarrow \quad \frac{bc - ab - ac + a^2}{\left(\frac{a+b+c}{2} - a\right)^2} = 2$

$\Rightarrow \quad \frac{2(bc - ab - ac + a^2)}{(b+c-a)^2} = 1$

$\Rightarrow \quad 2bc - 2ab - 2ac + 2a^2$
$= b^2 + c^2 + a^2 + 2bc - 2ab - 2ac$

$\Rightarrow a^2 = b^2 + c^2$

So, triangle is right angled.

13 Given, $\cot\alpha = \frac{12}{5}$ and $\sin\beta = \frac{3}{5}$

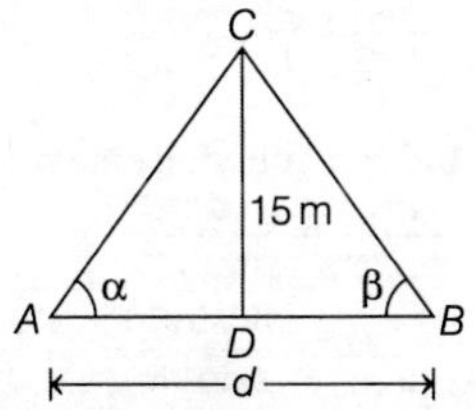

In ΔDAC and ΔDBC,

$AD = 15\cot\alpha, \; BD = 15\cot\beta$

$\Rightarrow \quad d = 15(\cot\alpha + \cot\beta)$

$= 15\left(\frac{12}{5} + \frac{4}{3}\right) = 56 \text{ m}$

14 In ΔDAB, $\tan\theta = \frac{64}{d}$

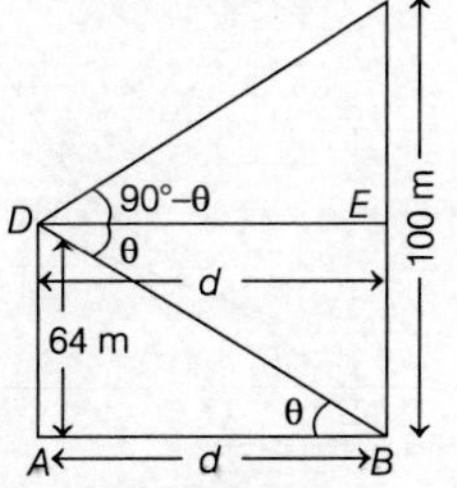

$\Rightarrow \quad d = 64\cot\theta \quad \ldots\text{(i)}$

In ΔCDE, $\tan(90° - \theta) = \frac{(100-64)}{d}$

$\Rightarrow \quad d = 36\tan\theta \quad \ldots\text{(ii)}$

From Eqs. (i) and (ii), we get

$d^2 = 36 \times 64 \Rightarrow d = 48 \text{ m}$

15 Let CD be a tower of height h and AB is building of height a.

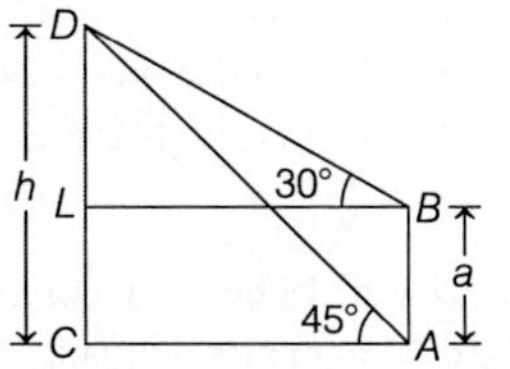

In ΔBLD, $\tan 30° = \frac{h-a}{LB}$

$\therefore \quad LB = \frac{(h-a)}{\tan 30°} = \sqrt{3}(h-a) \quad \ldots\text{(i)}$

In ΔACD, $\tan 45° = \frac{h}{CA} \Rightarrow h = CA \Rightarrow$

$h = LB \quad [\because LB = CA] \ldots\text{(ii)}$

From Eqs. (i) and (ii), we get

$h(\sqrt{3} - 1) = \sqrt{3}a$

$\therefore \quad h = \frac{\sqrt{3}a}{\sqrt{3}-1} = \frac{\sqrt{3}(\sqrt{3}+1)a}{2}$

$= \left(\frac{3+\sqrt{3}}{2}\right)a$

16 Let OP be the pole of height x m.

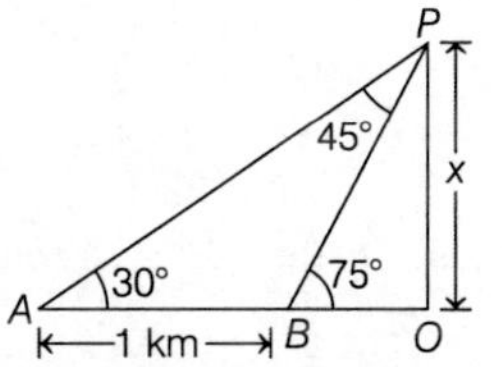

Using sine rule in APB, $\frac{\sin 30°}{\sin 45°} = \frac{PB}{AB}$

$\Rightarrow PB = 1000 \times \frac{1/2}{1/\sqrt{2}} = 500\sqrt{2} \text{ m}$

In ΔPBO, $x = 500\sqrt{2} \cdot \sin 75°$

$= 500\sqrt{2} \times \frac{\sqrt{3}+1}{2\sqrt{2}} = 250(\sqrt{3}+1) \text{ m}$

17 According to given information, we have the following figure

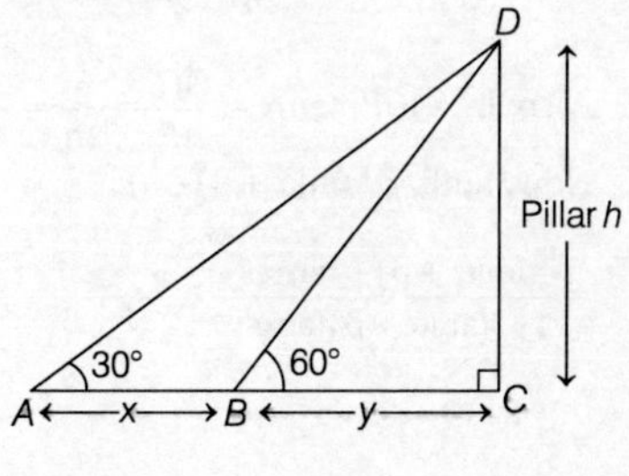

Now, from ΔACD and ΔBCD, we have

$\tan 30° = \frac{h}{x+y}$

and $\quad \tan 60° = \frac{h}{y}$

$\Rightarrow \quad h = \frac{x+y}{\sqrt{3}} \quad \ldots\text{(i)}$

and $\quad h = \sqrt{3}y \quad \ldots\text{(ii)}$

From Eqs. (i) and (ii)

$\frac{x+y}{\sqrt{3}} = \sqrt{3}y$

$\Rightarrow \quad x + y = 3y$

$\Rightarrow \quad x - 2y = 0 \Rightarrow y = \frac{x}{2}$

$\because$ Speed is uniform

$\therefore$ Distance y will be cover in 5 min.

($\because$ Distance x covered in 10 min.)

$\Rightarrow$ Distance $\frac{x}{2}$ will be cover in 5 min.

18 In ΔOA_1B_1, $\tan 45° = \frac{A_1B_1}{OB_1}$

$\Rightarrow \quad \frac{20}{OB_1} = 1 \Rightarrow OB_1 = 20$

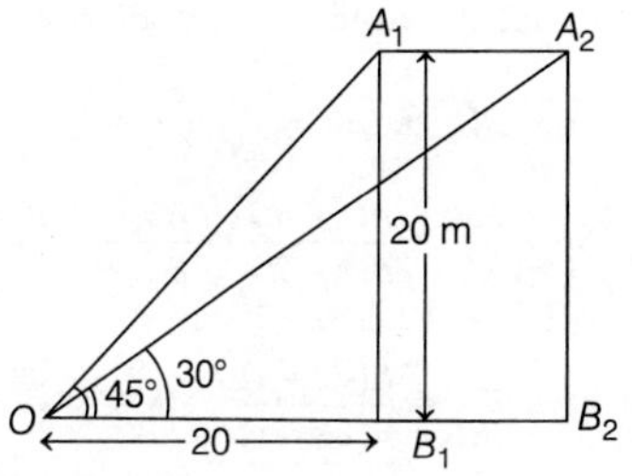

In ΔOA_2B_2,

$\tan 30° = \frac{20}{OB_2} \Rightarrow OB_2 = 20\sqrt{3}$

$\Rightarrow \quad B_1B_2 + OB_1 = 20\sqrt{3}$

$\Rightarrow \quad B_1B_2 = 20\sqrt{3} - 20$

$\Rightarrow \quad B_1B_2 = 20(\sqrt{3} - 1) \text{m}$

Now, Speed $= \frac{\text{Distance}}{\text{Time}} = \frac{20(\sqrt{3}-1)}{1}$

$= 20(\sqrt{3} - 1) \text{ m/s}$

19 Let OP be a tower with height $(\sqrt{3}+1)$ m and $AB = 2$ m.

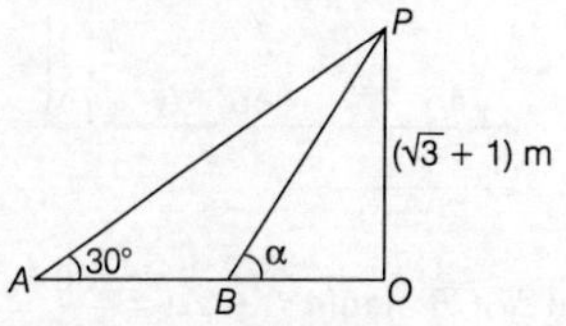

In ΔAOP,

$\tan 30° = \frac{\sqrt{3}+1}{OA}$

$\Rightarrow \quad OA = (\sqrt{3}+1)\sqrt{3}$

and in ΔBOP, $\tan\alpha = \dfrac{\sqrt{3}+1}{OB}$

$\Rightarrow \quad OB = (\sqrt{3}+1)\cot\alpha$

Now,
$OA - OB = (3+\sqrt{3}) - (\sqrt{3}+1)\cot\alpha$

$\Rightarrow \quad 2 = 3 + \sqrt{3} - (\sqrt{3}+1)\cot\alpha$

$\Rightarrow \quad \cot\alpha = 1$

$\therefore \quad \alpha = 45^\circ$

20 According to the given information the figure should be as follows.

Let the height of tower $= h$

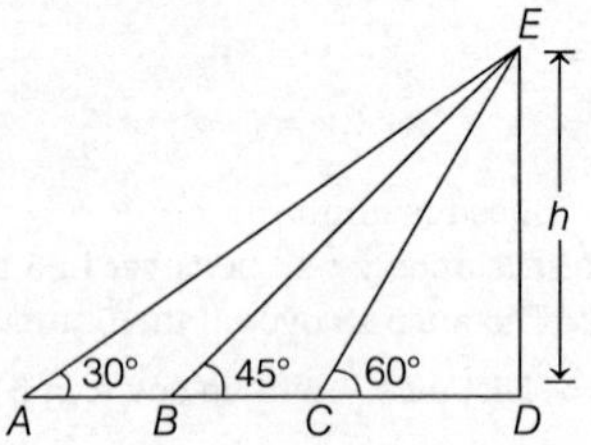

In ΔEDA, $\tan 30^\circ = \dfrac{ED}{AD}$

$\Rightarrow \quad \dfrac{1}{\sqrt{3}} = \dfrac{ED}{AD} = \dfrac{h}{AD} \Rightarrow AD = h\sqrt{3}$

In ΔEDB, $\tan 45^\circ = \dfrac{h}{BD} \Rightarrow BD = h$

In ΔEDC, $\tan 60^\circ = \dfrac{h}{CD} \Rightarrow CD = \dfrac{h}{\sqrt{3}}$

Now, $\dfrac{AB}{BC} = \dfrac{AD - BD}{BD - CD} \Rightarrow \dfrac{AB}{BC} = \dfrac{h\sqrt{3} - h}{h - \dfrac{h}{\sqrt{3}}}$

$\Rightarrow \quad \dfrac{AB}{BC} = \dfrac{h(\sqrt{3}-1)}{\dfrac{h(\sqrt{3}-1)}{\sqrt{3}}} \Rightarrow \dfrac{AB}{BC} = \dfrac{\sqrt{3}}{1}$

$\therefore \quad AB : BC = \sqrt{3} : 1$

21 Let AB be the tower with height 100 m.

Distance between the objects,

$CD = 200$ m

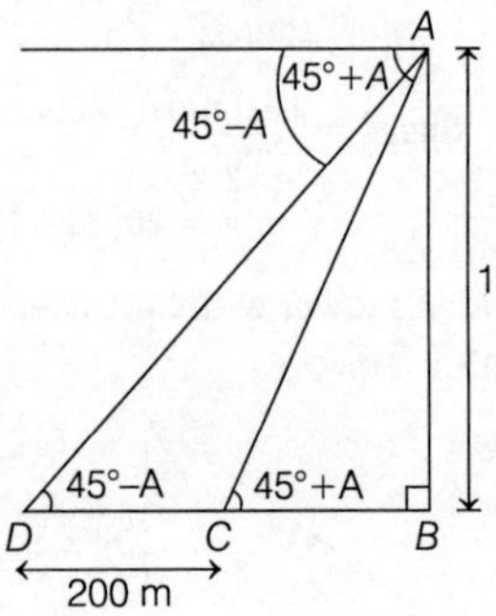

In ΔACB, $\tan(45^\circ + A) = \dfrac{100}{BC}$

$\Rightarrow \quad BC = 100\cot(45^\circ + A) \quad \ldots$(i)

and in ΔADB,

$\tan(45^\circ - A) = \dfrac{100}{BD}$

$\Rightarrow \quad BD = 100\cot(45^\circ - A)$

$\Rightarrow \quad BC + CD = 100\cot(45^\circ - A) \quad \ldots$(ii)

$\therefore$ From Eqs. (i) and (ii), we get

$CD = 100[\cot(45^\circ - A) - \cot(45^\circ + A)]$

$= 100\left[\dfrac{1+\tan A}{1-\tan A} - \dfrac{1-\tan A}{1+\tan A}\right]$

$= 100\left[\dfrac{4\tan A}{1-\tan^2 A}\right] = 200\tan 2A$

$\Rightarrow \quad 200 = 200\tan 2A \Rightarrow \tan 2A = 1$

$\Rightarrow \quad \tan 2A = \tan 45^\circ \Rightarrow 2A = 45^\circ$

$\therefore \quad A = 22\dfrac{1^\circ}{2}$

22 Let PQ and RS be the poles of height 20 m and 80 m subtending angles α and β at R and P, respectively. Let h be the height of the point T, the intersection of QR and PS.

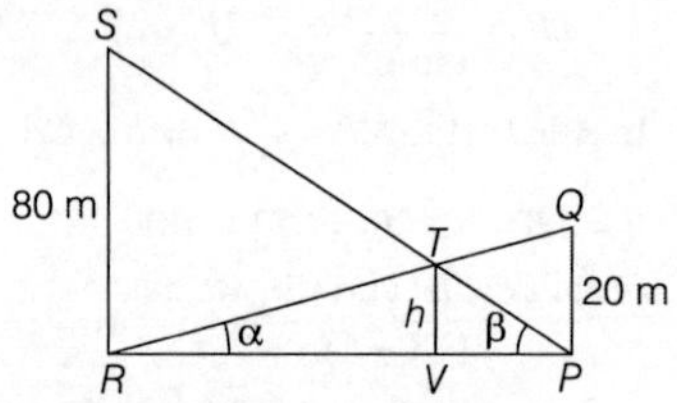

Then, $PR = h\cot\alpha + h\cot\beta$

$= 20\cot\alpha = 80\cot\beta$

$\Rightarrow \quad \cot\alpha = 4\cot\beta \Rightarrow \dfrac{\cot\alpha}{\cot\beta} = 4$

Again, $h\cot\alpha + h\cot\beta = 20\cot\alpha$

$\Rightarrow \quad (h - 20)\cot\alpha = -h\cot\beta$

$\Rightarrow \quad \dfrac{\cot\alpha}{\cot\beta} = \dfrac{h}{20-h} = 4$

$\Rightarrow \quad h = 80 - 4h$

$\therefore \quad h = 16$ m

23 Let $AB = h$, then $AP = 2h$ and $AC = BC = \dfrac{h}{2}$

Again, let $\angle CPA = \alpha$

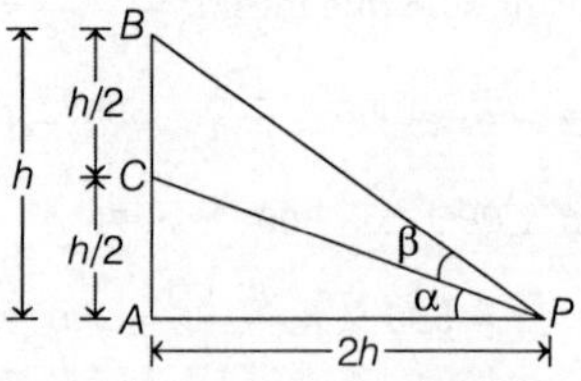

Now, in ΔABP,

$\tan(\alpha + \beta) = \dfrac{AB}{AP} = \dfrac{h}{2h} = \dfrac{1}{2}$

Also, in ΔACP, $\tan\alpha = \dfrac{AC}{AP} = \dfrac{\frac{h}{2}}{2h} = \dfrac{1}{4}$

Now, $\tan\beta = \tan[(\alpha + \beta) - \alpha]$

$= \dfrac{\tan(\alpha+\beta) - \tan\alpha}{1 + \tan(\alpha+\beta)\tan\alpha} = \dfrac{\frac{1}{2} - \frac{1}{4}}{1 + \frac{1}{2}\times\frac{1}{4}} = \dfrac{\frac{1}{4}}{\frac{9}{8}}$

$= \dfrac{2}{9}$

24 In ΔABD, $\tan\beta = \dfrac{60}{d}$

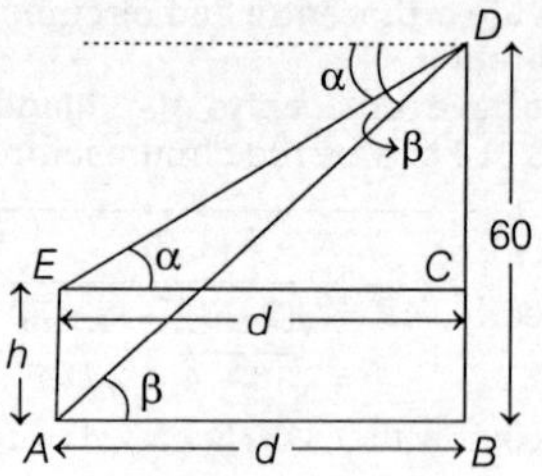

$\Rightarrow \quad d = 60\cot\beta \quad \ldots$(i)

In ΔDEC, $\tan\alpha = \dfrac{DC}{EC}$

$\Rightarrow \quad DC = d\tan\alpha$

$\Rightarrow \quad 60 - h = d\tan\alpha \quad (\because BC = EA = h)$

$\Rightarrow \quad 60 - h = 60\cot\beta\tan\alpha \quad$ [from Eq. (i)]

$\Rightarrow \quad h = 60\left(1 - \dfrac{\cos\beta}{\sin\beta}\cdot\dfrac{\sin\alpha}{\cos\alpha}\right)$

$\Rightarrow \quad h = \dfrac{60\sin(\beta - \alpha)}{\cos\alpha\sin\beta}$

$\Rightarrow \dfrac{60\sin(\beta-\alpha)}{x} = \dfrac{60\sin(\beta-\alpha)}{\cos\alpha\sin\beta}$ (given)

$\Rightarrow \quad x = \cos\alpha\sin\beta$

25 Let BC be the declivity and BA be the tower.

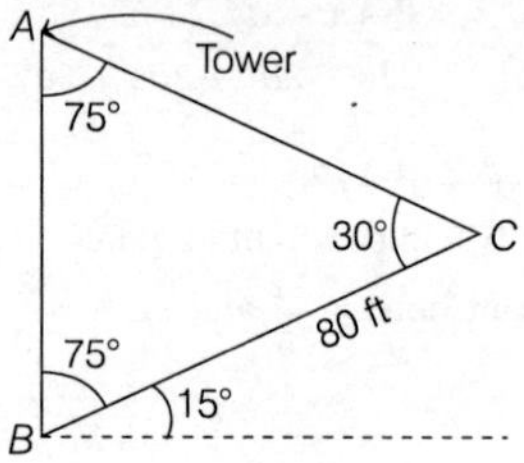

In ΔABC, on applying sine rule

$\dfrac{BC}{\sin 75^\circ} = \dfrac{AB}{\sin 30^\circ}$

$\Rightarrow \quad AB = \dfrac{80\sin 30^\circ}{\sin 75^\circ} = \dfrac{40\times 2\sqrt{2}}{\sqrt{3}+1}$

$= 40(\sqrt{6} - \sqrt{2})$ ft

26 Clearly, $\dfrac{a_1}{b_1} = \tan\left(\dfrac{\alpha+\beta}{2}\right) \quad \ldots$(i)

and $\quad \dfrac{a_2}{b_2} = \tan\left(\dfrac{\beta+\gamma}{2}\right) \quad \ldots$(ii)

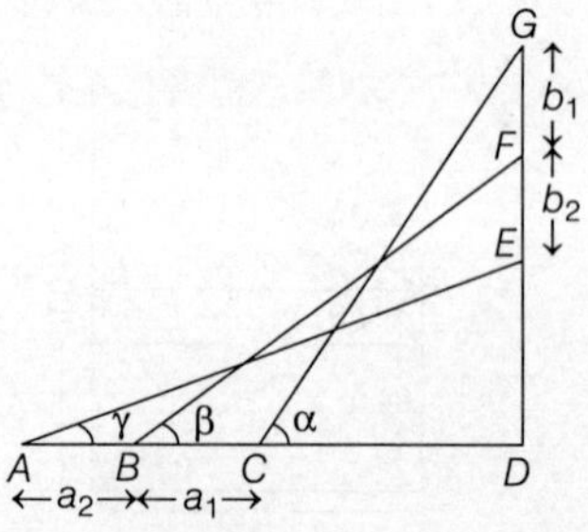

Since, $a_1 a_2 = b_1 b_2$

$\therefore \frac{a_1}{b_1} = \frac{b_2}{a_2} \Rightarrow \tan\left(\frac{\alpha+\beta}{2}\right) = \frac{1}{\tan\left(\frac{\beta+\gamma}{2}\right)}$

$\Rightarrow \tan\left(\frac{\alpha+\beta}{2}\right)\tan\left(\frac{\beta+\gamma}{2}\right) = 1$

$\therefore \frac{\alpha+\beta}{2} + \frac{\beta+\gamma}{2} = \frac{\pi}{2}$

$\Rightarrow \alpha+\beta+\gamma = \pi - \beta < \pi$

27 Let $AB = x$

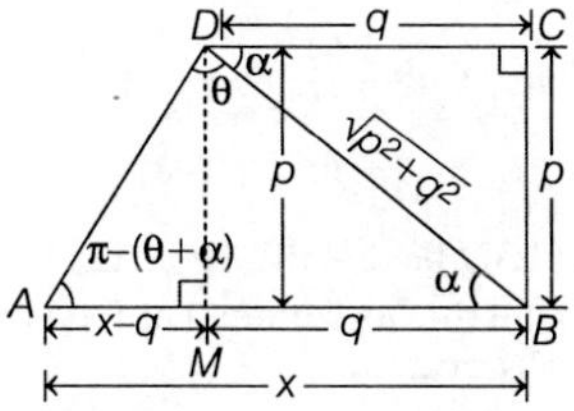

In ΔDAM, $\tan(\pi - \theta - \alpha) = \frac{p}{x-q}$

$\Rightarrow \tan(\theta+\alpha) = \frac{p}{q-x}$

$\Rightarrow q - x = p\cot(\theta+\alpha)$

$\Rightarrow x = q - p\cot(\theta+\alpha)$

$= q - p\left(\frac{\cot\theta\cot\alpha - 1}{\cot\alpha + \cot\theta}\right)$

$= q - p\left(\frac{\frac{q}{p}\cot\theta - 1}{\frac{q}{p} + \cot\theta}\right)$

$\left[\because \text{in } \Delta BDC, \cot\alpha = \frac{q}{p}\right]$

$= q - p\left(\frac{q\cot\theta - p}{q + p\cot\theta}\right)$

$= q - p\left(\frac{q\cos\theta - p\sin\theta}{q\sin\theta + p\cos\theta}\right)$

$\Rightarrow x = \frac{q^2\sin\theta + pq\cos\theta - pq\cos\theta + p^2\sin\theta}{p\cos\theta + q\sin\theta}$

$\Rightarrow AB = \frac{(p^2+q^2)\sin\theta}{p\cos\theta + q\sin\theta}$

Alternate Solution

Applying sine rule in ΔABD,

$\left(\frac{a}{\sin A} = \frac{b}{\sin B} = \frac{c}{\sin C}\right)$

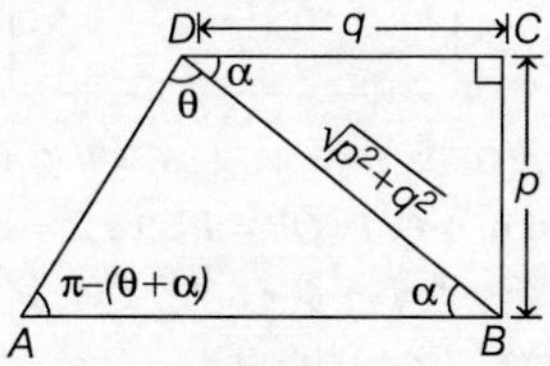

$\frac{AB}{\sin\theta} = \frac{\sqrt{p^2+q^2}}{\sin\{\pi - (\theta+\alpha)\}}$

$\Rightarrow \frac{AB}{\sin\theta} = \frac{\sqrt{p^2+q^2}}{\sin(\theta+\alpha)}$

$\Rightarrow AB = \frac{\sqrt{p^2+q^2}\sin\theta}{\sin\theta\cos\alpha + \cos\theta\sin\alpha}$

$= \frac{(p^2+q^2)\sin\theta}{q\sin\theta + p\cos\theta}$

$\left[\because \cos\alpha = \frac{q}{\sqrt{p^2+q^2}}, \sin\alpha = \frac{p}{\sqrt{p^2+q^2}}\right]$

28 Let OP be the tower. Since, the tower make equal angles at the vertices of the triangle, therefore foot of tower is at the circumcentre.

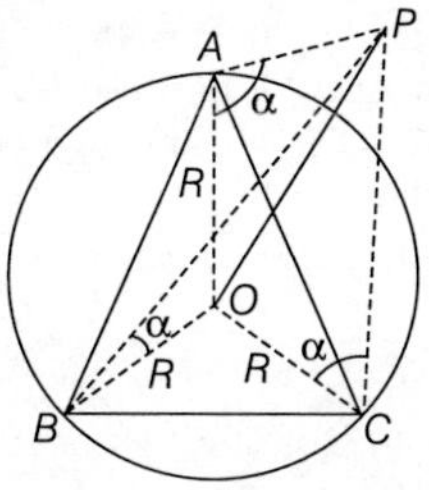

In ΔOAP, $\tan\alpha = \frac{OP}{OA} \Rightarrow OP = OA\tan\alpha$

$\Rightarrow OP = R\tan\alpha \quad (\because OA = R, \text{given})$

29 Let h be the height of a tower

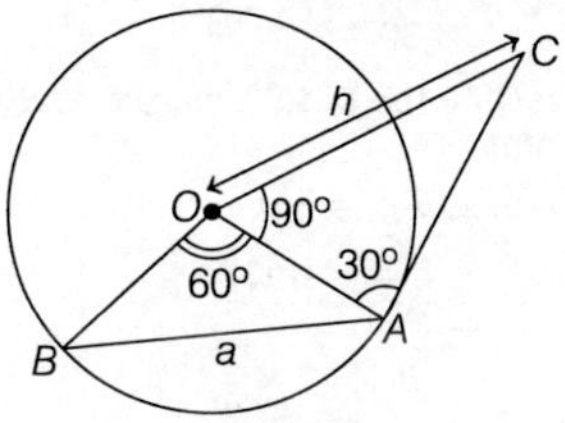

Since, $\angle AOB = 60°$

Also, $OB = OA$ = radii

$\therefore \angle OBA = \angle OAB = 60°$

So, ΔOAB is an equilateral

$\therefore OA = OB = AB = a$

In ΔOAC, $\tan 30° = \frac{h}{a}$

$\Rightarrow \frac{1}{\sqrt{3}} = \frac{h}{a} \Rightarrow h = \frac{a}{\sqrt{3}}$

30 Let PD be a pole.

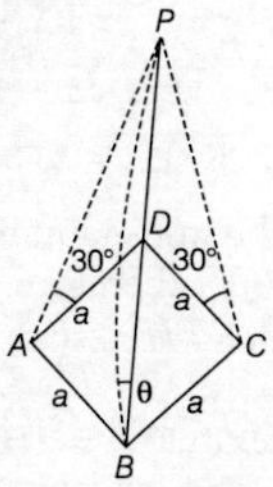

In ΔDAP, $\tan 30° = \frac{DP}{AD}$

$\Rightarrow DP = \frac{a}{\sqrt{3}}$

In ΔPDB, $\tan\theta = \frac{DP}{BD}$

$\Rightarrow \tan\theta = \frac{a/\sqrt{3}}{\sqrt{2}a} = \frac{1}{\sqrt{6}}$

SESSION 2

1

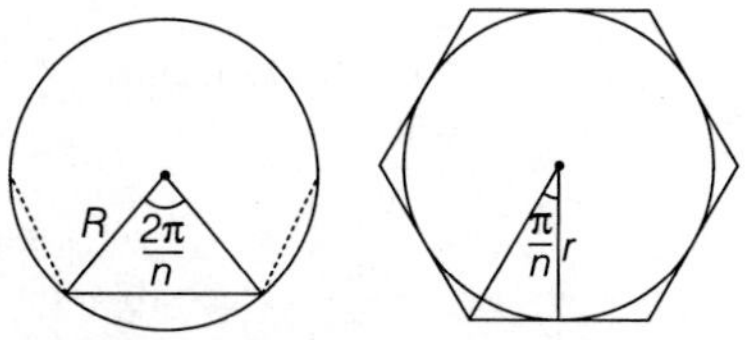

$\therefore \frac{a}{2R} = \sin\frac{\pi}{n}$ and $\frac{a}{2r} = \tan\frac{\pi}{n}$

$\therefore \frac{r}{R} = \cos\frac{\pi}{n}$

$n = 3$ gives, $\frac{r}{R} = \frac{1}{2}$

$n = 4$ gives, $\frac{r}{R} = \frac{1}{\sqrt{2}}$

$n = 6$ gives, $\frac{r}{R} = \frac{\sqrt{3}}{2}$

2 $2\cos\frac{A+B}{2}\cos\frac{A-B}{2} = 4\sin^2\frac{C}{2}$

$\Rightarrow \cos\frac{A-B}{2} = 2\cos\frac{A+B}{2}$

$\left[\because \cos\frac{A+B}{2} \neq 0\right]$

$\Rightarrow \cos\frac{A}{2}\cos\frac{B}{2} + \sin\frac{A}{2}\cdot\sin\frac{B}{2} = 2\left(\cos\frac{A}{2}\cos\frac{B}{2} - \sin\frac{A}{2}\cdot\sin\frac{B}{2}\right)$

$\Rightarrow 3\sin\frac{A}{2}\cdot\sin\frac{B}{2} = \cos\frac{A}{2}\cos\frac{B}{2}$

$\Rightarrow \tan\frac{A}{2}\cdot\tan\frac{B}{2} = \frac{1}{3}$

$\Rightarrow \sqrt{\frac{(s-b)(s-c)}{s(s-a)}}\cdot\sqrt{\frac{(s-c)(s-a)}{s(s-b)}} = \frac{1}{3}$

$\Rightarrow \frac{s-c}{s} = \frac{1}{3} \Rightarrow \frac{a+b-c}{a+b+c} = \frac{1}{3}$

$\Rightarrow 3a + 3b - 3c = a + b + c$

$\Rightarrow a + b = 2c$

Hence, a, c and b are in AP.

3 Clearly,

$A_0A_1 = 2 \times 1\cos 60° = 1 = A_1A_2$

[using cosine rule]]

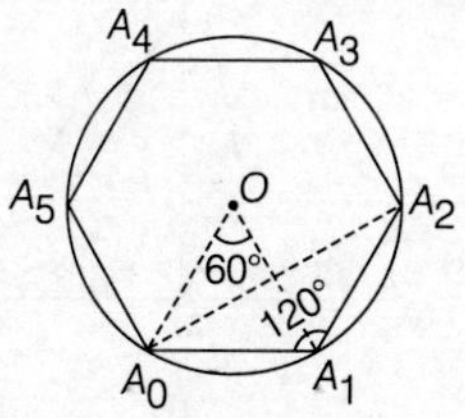

and

$$\cos 120° = \frac{(A_0A_1)^2 + (A_1A_2)^2 - (A_0A_2)^2}{2 \cdot A_0A_1 \cdot A_1A_2}$$

$$= \frac{1 + 1 - (A_0A_2)^2}{2 \cdot 1 \cdot 1}$$

$\Rightarrow A_0A_2 = \sqrt{3}$

$\therefore \; A_0A_1 \times A_0A_2 = 1 \times \sqrt{3} = \sqrt{3}$

4 Let A and B be the positions of the two ships at the end of 3 h.
Then, $OA = (24 \times 3) = 72$ km and $OB = (32 \times 3) = 96$ km

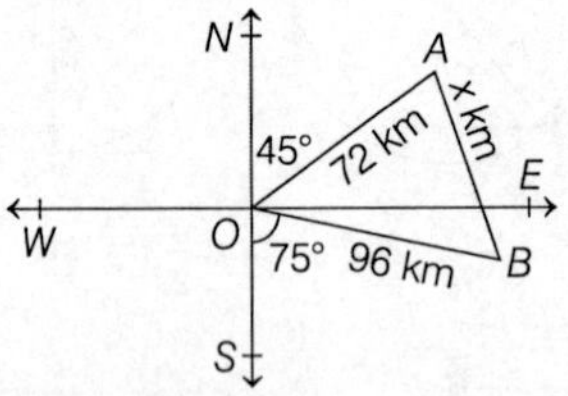

Let $AB = x$ km
We have, $\angle NOA = 45°$ and $\angle SOB = 75°$
$\therefore \; \angle AOB = 180° - (45° + 75°) = 60°$
Using cosine formula on ΔAOB, we get

$$\cos 60° = \frac{(72)^2 + (96)^2 - x^2}{2 \times 72 \times 96}$$

$\Rightarrow \quad x^2 = 14400 - 6912 = 7488$

$\Rightarrow \quad x = \sqrt{7488} = 86.53$ km

Hence, the distance between the ships at the end of 3 h is
86.53 km $\approx$ 86.4 km (approx.).

5 In ΔAPC,

$$\sin \frac{\alpha}{2} = \frac{r}{AC} \Rightarrow AC = r \operatorname{cosec} \frac{\alpha}{2}$$

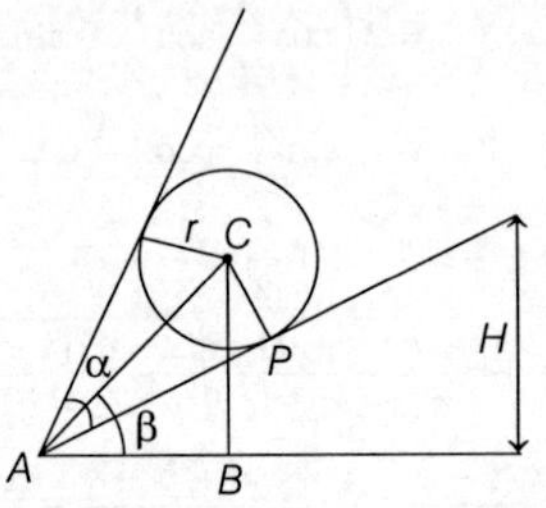

In $\Delta\, ABC$, $\sin \beta = \dfrac{BC}{AC}$

$\Rightarrow \; H = AC \sin \beta \Rightarrow H = r \sin \beta \operatorname{cosec} \dfrac{\alpha}{2}$

6

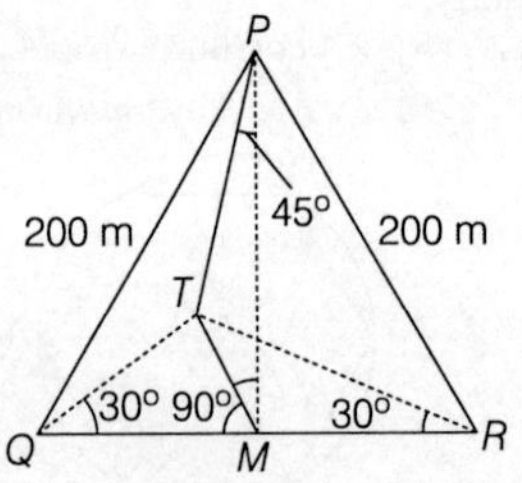

Let height of tower TM be h.

In ΔPMT, $\tan 45° = \dfrac{TM}{PM}$

$\Rightarrow \quad 1 = \dfrac{h}{PM}$

$\Rightarrow \quad PM = h$

In ΔTQM, $\tan 30° = \dfrac{h}{QM}$; $QM = \sqrt{3}h$

In ΔPMQ, $PM^2 + QM^2 = PQ^2$

$$h^2 + (\sqrt{3}h)^2 = (200)^2$$

$\Rightarrow \quad 4h^2 = (200)^2 \Rightarrow h = 100$ m

7 Let P be the summit and A be the foot of the mountain.
Then, $\quad \angle OAP = 45°$
$\therefore \quad \angle OPA = 45°$
Hence, ΔOAP is an isosceles.
Let $AO = OP = h$ km height of mountain. [say]

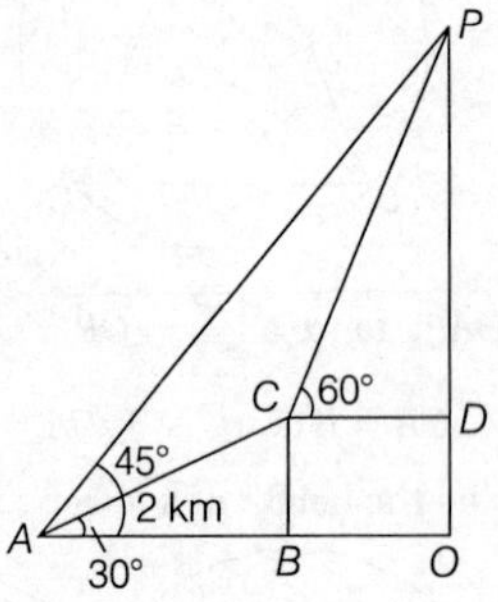

Let C be the point where elevation is 60°.
Then, $\angle BAC = 30°$ and $AC = 2$ km

$\therefore \quad \dfrac{BC}{AC} = \sin 30°$

$\Rightarrow \quad BC = 2 \times \dfrac{1}{2} = 1$

i.e. $\quad BC = 1$ km and $\dfrac{AB}{AC} = \cos 30°$

$\Rightarrow \quad AB = \dfrac{2\sqrt{3}}{2} = \sqrt{3}$

i.e. $\quad AB = \sqrt{3}$ km

Now, $\quad PD = OP - OD$
$\quad = OP - BC = h - 1$

and $\quad CD = BO = AO - AB = h - \sqrt{3}$

In ΔDCP, $\dfrac{PD}{CD} = \tan 60° \Rightarrow \dfrac{h-1}{h-\sqrt{3}} = \sqrt{3}$

$\Rightarrow \quad \sqrt{3}h - 3 = h - 1 \Rightarrow (\sqrt{3} - 1)h = 2$

$\therefore \quad h = \dfrac{2}{\sqrt{3} - 1} \times \dfrac{\sqrt{3} + 1}{\sqrt{3} + 1} = \sqrt{3} + 1$

$\quad = 1.732 + 1 = 2.732$ km

8 Let H be the mid-point of BC. Since, $\angle TBH = 90°$, therefore,
$TH^2 = BT^2 + BH^2 = 5^2 + 5^2 = 50$
Also since,
$\angle THG = 90°$, $TG^2 = TH^2 + GH^2$
$= 50 + 25 = 75$

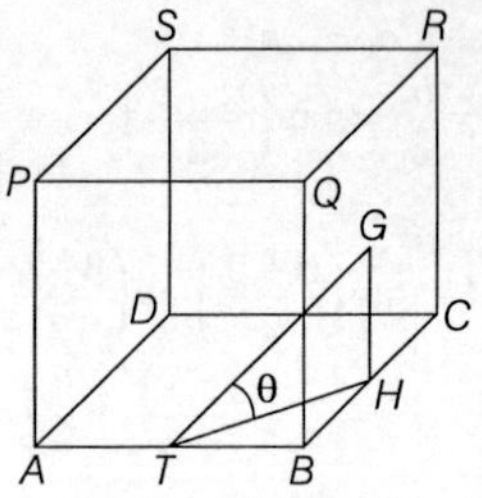

Let θ be the required angle of elevation of G at T.

Then, $\quad \sin \theta = \dfrac{GH}{TG} = \dfrac{5}{5\sqrt{3}} = \dfrac{1}{\sqrt{3}}$

$\Rightarrow \quad \theta = \sin^{-1}(1/\sqrt{3})$

9 Let PAQ be the hill, AB be the tree and PD be the horizontal. Let P be the point of observation.

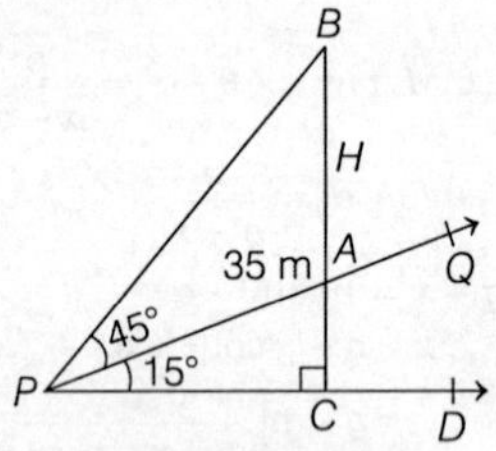

Produce BA to meet PD at C.
Let $AB = H$ m.
Then, $\angle DPA = 15°$, $PA = 35$m
$\angle CPB = 60°$ and $\angle PCA = 90°$.
$\therefore \quad \angle APB = (60° - 15°) = 45°$
In ΔPAC, $\angle PAC = 180° - (15° + 90°) = 75°$
$\therefore \quad \angle PAB = (180° - 75°) = 105°$
and $\angle PBA = 180° - (45° + 105°) = 30°$
Applying sine rule on ΔPAB, we get

$$\frac{PA}{\sin \angle PBA} = \frac{AB}{\sin \angle APB} \Rightarrow \frac{35}{\sin 30°} = \frac{H}{\sin 45°}$$

$\therefore \quad 35 \times 2 = H \times \sqrt{2} \Rightarrow H = 35\sqrt{2}$ m

Hence, the height of the tree is $35\sqrt{2}$ m.

10 Let OP be the flag staff of height h standing at the centre O of the field.

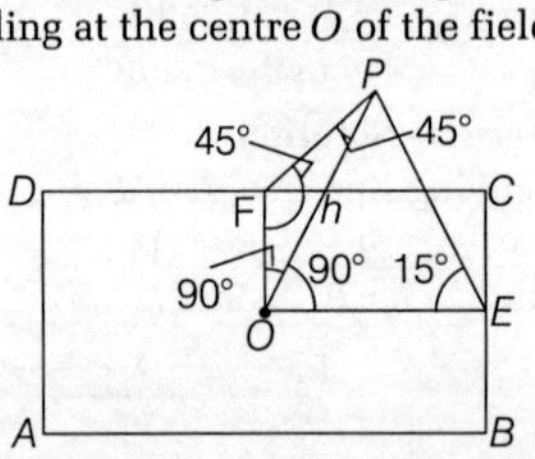

In $\Delta\, OEP$, $OE = h \cot 15° = h\,(2 + \sqrt{3})$

and in $\Delta\, OFP$, $OF = h \cot 45° = h$

$\therefore \; EF = h\sqrt{1 + (2 + \sqrt{3})^2} = 2h\sqrt{2 + \sqrt{3}}$

Since, $BD = 1200$ m

$\Rightarrow \quad 2EF = 4h\sqrt{(2 + \sqrt{3})} = 1200$

$\therefore \; h = \dfrac{300}{\sqrt{2 + \sqrt{3}}} = (300\sqrt{2 - \sqrt{3}})$ m

DAY TWENTY TWO

Inverse Trigonometric Function

Learning & Revision for the Day

- Inverse Trigonometric Function
- Properties of Inverse Trigonometric Function

Inverse Trigonometric Function

Trigonometric functions are not one-one and onto on their natural domains and ranges, so their inverse do not exists in the whole domain. If we restrict their domain and range, then their inverse may exists.

$y = f(x) = \sin x$. Then, its inverse is $x = \sin^{-1} y$.

NOTE
- $\sin^{-1} y \neq (\sin y)^{-1}$
- $\sin^{-1} y \neq \sin\left(\frac{1}{y}\right)$

The value of an inverse trigonometric functions which lies in its principal value branch is called the principal value of that inverse trigonometric function.

Domain and range of inverse trigonometric functions

Function	Domain	Range (Principal Value Branch)
$\sin^{-1} x$	$[-1, 1]$	$\left[-\frac{\pi}{2}, \frac{\pi}{2}\right]$
$\cos^{-1} x$	$[-1, 1]$	$[0, \pi]$
$\tan^{-1} x$	R	$\left(-\frac{\pi}{2}, \frac{\pi}{2}\right)$
$\cot^{-1} x$	R	$(0, \pi)$
$\sec^{-1} x$	$R-(-1, 1)$	$[0, \pi] - \left\{\frac{\pi}{2}\right\}$
$\text{cosec}^{-1} x$	$R-(-1, 1)$	$\left[-\frac{\pi}{2}, \frac{\pi}{2}\right] - \{0\}$

PREP MIRROR
Your Personal Preparation Indicator

- No. of Questions in Exercises (x)—
- No. of Questions Attempted (y)—
- No. of Correct Questions (z)—
  (Without referring Explanations)

- Accuracy Level $(z/y \times 100)$—
- Prep Level $(z/x \times 100)$—

In order to expect good rank in JEE, your Accuracy Level should be above 85 & Prep Level should be above 75.

Properties of Inverse Trigonometric Functions

1. (i) $\sin^{-1} x + \cos^{-1} x = \frac{\pi}{2}$; $(-1 \le x \le 1)$

   (ii) $\tan^{-1} x + \cot^{-1} x = \frac{\pi}{2}$; $x \in R$

   (iii) $\sec^{-1} x + \text{cosec}^{-1} x = \frac{\pi}{2}$; $(x \le -1 \text{ or } x \ge 1)$

2. (i) $\sin^{-1}(-x) = -\sin^{-1} x$; $(-1 \le x \le 1)$

   (ii) $\cos^{-1}(-x) = \pi - \cos^{-1} x$; $(-1 \le x \le 1)$

   (iii) $\tan^{-1}(-x) = -\tan^{-1}(x)$; $(-\infty < x < \infty)$

   (iv) $\cot^{-1}(-x) = \pi - \cot^{-1} x$; $(-\infty < x < \infty)$

   (v) $\sec^{-1}(-x) = \pi - \sec^{-1} x$; $x \le -1$ or $x \ge 1)$

   (vi) $\text{cosec}^{-1}(-x) = -\text{cosec}^{-1} x$; $(x \le -1 \text{ or } x \ge 1)$

3. (i) $\sin^{-1}(\sin x)$ is a periodic function with period 2π.

$$\sin^{-1}(\sin x) = \begin{cases} x, & x \in \left[-\frac{\pi}{2}, \frac{\pi}{2}\right] \\ \pi - x, & x \in \left[\frac{\pi}{2}, \frac{3\pi}{2}\right] \\ x - 2\pi, & x \in \left[\frac{3\pi}{2}, \frac{5\pi}{2}\right] \\ 3\pi - x, & x \in \left[\frac{5\pi}{2}, \frac{7\pi}{2}\right] \end{cases}$$

   (ii) $\cos^{-1}(\cos x)$ is a periodic function with period 2π.

$$\cos^{-1}(\cos x) = \begin{cases} x, & x \in [0, \pi] \\ 2\pi - x, & x \in [\pi, 2\pi] \\ x - 2\pi, & x \in [2\pi, 3\pi] \\ 4\pi - x, & x \in [3\pi, 4\pi] \end{cases}$$

   (iii) $\tan^{-1}(\tan x)$ is a periodic function with period π.

$$\tan^{-1}(\tan x) = \begin{cases} x, & x \in \left[-\frac{\pi}{2}, \frac{\pi}{2}\right] \\ x - \pi, & x \in \left[\frac{\pi}{2}, \frac{3\pi}{2}\right] \\ x - 2\pi, & x \in \left[\frac{3\pi}{2}, \frac{5\pi}{2}\right] \\ x - 3\pi, & x \in \left[\frac{5\pi}{2}, \frac{7\pi}{2}\right] \end{cases}$$

   (iv) $\cot^{-1}(\cot x)$ is a periodic function with period π.
   $\cot^{-1}(\cot x) = x$; $0 < x < \pi$

   (v) $\sec^{-1}(\sec x)$ is a periodic function with period 2π.
   $\sec^{-1}(\sec x) = x$; $0 \le x < \frac{\pi}{2}$ or $\frac{\pi}{2} < x \le \pi$

   (vi) $\text{cosec}^{-1}(\text{cosec } x)$ is a periodic function with period 2π.
   $\text{cosec}^{-1}(\text{cosec } x) = x$; $-\frac{\pi}{2} \le x < 0$ or $0 < x \le \frac{\pi}{2}$

4. (i) $\sin^{-1}\left(\frac{1}{x}\right) = \text{cosec}^{-1} x$, if $x \in (-\infty, -1] \cup [1, \infty)$

   (ii) $\cos^{-1}\left(\frac{1}{x}\right) = \sec^{-1} x$, if $x \in (-\infty, -1] \cup [1, \infty)$

   (iii) $\tan^{-1}\left(\frac{1}{x}\right) = \begin{cases} \cot^{-1} x, & \text{if } x > 0 \\ -\pi + \cot^{-1} x, & \text{if } x < 0 \end{cases}$

5. (i) $\sin^{-1} x = \cos^{-1}\sqrt{1 - x^2}$
   $= \tan^{-1}\left(\frac{x}{\sqrt{1 - x^2}}\right) = \cot^{-1}\left(\frac{\sqrt{1 - x^2}}{x}\right)$
   $= \sec^{-1}\left(\frac{1}{\sqrt{1 - x^2}}\right) = \text{cosex}^{-1}\left(\frac{1}{x}\right)$, if $x \in (0, 1)$

   (ii) $\cos^{-1} x = \sin^{-1}\sqrt{1 - x^2}$
   $= \tan^{-1}\left(\frac{\sqrt{1 - x^2}}{x}\right) = \cot^{-1}\left(\frac{x}{\sqrt{1 - x^2}}\right) = \sec^{-1}\left(\frac{1}{x}\right)$
   $= \text{cosec}^{-1}\left(\frac{1}{\sqrt{1 - x^2}}\right)$, if $x \in (0, 1)$

   (iii) $\tan^{-1} x = \sin^{-1}\left(\frac{x}{\sqrt{1 + x^2}}\right)$
   $= \cos^{-1}\left(\frac{1}{\sqrt{1 + x^2}}\right) = \cot^{-1}\left(\frac{1}{x}\right)$
   $= \text{cosec}^{-1}\left(\frac{\sqrt{1 + x^2}}{x}\right) = \sec^{-1}(\sqrt{1 + x^2})$, if $x \in (0, \infty)$

6. (i) $\sin^{-1} x + \sin^{-1} y$

$$= \begin{cases} \sin^{-1}(x\sqrt{1 - y^2} + y\sqrt{1 - x^2}); \ |x|, |y| \le 1 \text{ and} \\ \quad x^2 + y^2 \le 1 \text{ or } (xy < 0 \text{ and } x^2 + y^2 > 1) \\ \pi - \sin^{-1}(x\sqrt{1 - y^2} + y\sqrt{1 - x^2}); \ 0 < x, y \le 1 \\ \quad \text{and } x^2 + y^2 > 1 \\ -\pi - \sin^{-1}(x\sqrt{1 - y^2} + y\sqrt{1 - x^2}); \\ \quad -1 \le x, y < 0 \text{ and } x^2 + y^2 > 1 \end{cases}$$

   (ii) $\sin^{-1} x - \sin^{-1} y$

$$= \begin{cases} \sin^{-1}(x\sqrt{1 - y^2} - y\sqrt{1 - x^2}); \ |x|, |y| \le 1 \\ \quad \text{and } x^2 + y^2 \le 1 \text{ or } (xy > 0 \text{ and } x^2 + y^2 > 1) \\ \pi - \sin^{-1}(x\sqrt{1 - y^2} - y\sqrt{1 - x^2}); \\ \quad 0 < x \le 1, -1 \le y < 0 \text{ and } x^2 + y^2 > 1 \\ -\pi - \sin^{-1}(x\sqrt{1 - y^2} - y\sqrt{1 - x^2}); \\ \quad -1 \le x < 0, 0 < y \le 1 \text{ and } x^2 + y^2 > 1 \end{cases}$$

   (iii) $\cos^{-1} x + \cos^{-1} y$

$$= \begin{cases} \cos^{-1}\{xy - \sqrt{(1 - x^2)}\sqrt{(1 - y^2)}\}; \ |x|, |y| \le 1 \\ \quad \text{and } x + y \ge 0 \\ 2\pi - \cos^{-1}\{xy - \sqrt{(1 - x^2)}\sqrt{(1 - y^2)}\}; \\ \quad |x|, |y| \le 1 \text{ and } x + y \le 0 \end{cases}$$

   (iv) $\cos^{-1} x - \cos^{-1} y$

$$= \begin{cases} \cos^{-1}\{xy + \sqrt{(1 - x^2)}\sqrt{(1 - y^2)}\}; \ |x|, |y| \le 1 \\ \quad \text{and } x \le y \\ -\cos^{-1}\{xy + \sqrt{(1 - x^2)}\sqrt{(1 - y^2)}\}; \\ \quad -1 \le y \le 0, 0 < x \le 1 \text{ and } x \ge y \end{cases}$$

(v) $\tan^{-1} x + \tan^{-1} y$

$$= \begin{cases} \tan^{-1}\left(\dfrac{x+y}{1-xy}\right); & xy < 1 \\ \pi + \tan^{-1}\left(\dfrac{x+y}{1-xy}\right); & x > 0, y > 0, xy > 1 \\ -\pi + \tan^{-1}\left(\dfrac{x+y}{1-xy}\right); & x < 0, y < 0, xy > 1 \end{cases}$$

(vi) $\tan^{-1} x - \tan^{-1} y$

$$= \begin{cases} \tan^{-1}\left(\dfrac{x-y}{1+xy}\right); & xy > -1 \\ \pi + \tan^{-1}\left(\dfrac{x-y}{1+xy}\right); & xy < -1, x > 0, y < 0 \\ -\pi + \tan^{-1}\left(\dfrac{x-y}{1+xy}\right); & xy < -1, x < 0, y > 0 \end{cases}$$

7. (i) $2 \sin^{-1} x$

$$= \begin{cases} \sin^{-1}\{2x\sqrt{(1-x^2)}\}; & -\dfrac{1}{\sqrt{2}} \le x \le \dfrac{1}{\sqrt{2}} \\ \pi - \sin^{-1}(2x\sqrt{1-x^2}); & \dfrac{1}{\sqrt{2}} \le x \le 1 \\ -\pi - \sin^{-1}(2x\sqrt{1-x^2}); & -1 \le x < -\dfrac{1}{\sqrt{2}} \end{cases}$$

(ii) $2 \cos^{-1} x = \begin{cases} \cos^{-1}(2x^2 - 1); & 0 \le x \le 1 \\ 2\pi - \cos^{-1}(2x^2-1); & -1 \le x < 0 \end{cases}$

(iii) $2 \tan^{-1} x = \begin{cases} \tan^{-1}\left(\dfrac{2x}{1-x^2}\right); & -1 < x < 1 \\ \pi + \tan^{-1}\left(\dfrac{2x}{1-x^2}\right); & x > 1 \\ -\pi + \tan^{-1}\left(\dfrac{2x}{1-x^2}\right); & x < -1 \end{cases}$

(iv) $2 \tan^{-1} x = \begin{cases} \sin^{-1}\left(\dfrac{2x}{1+x^2}\right); & -1 \le x \le 1 \\ \pi - \sin^{-1}\left(\dfrac{2x}{1+x^2}\right); & x > 1 \\ -\pi - \sin^{-1}\left(\dfrac{2x}{1+x^2}\right); & x < -1 \end{cases}$

(v) $2 \tan^{-1} x = \begin{cases} \cos^{-1}\left(\dfrac{1-x^2}{1+x^2}\right); & 0 \le x < \infty \\ -\cos^{-1}\left(\dfrac{1-x^2}{1+x^2}\right); & -\infty < x \le 0 \end{cases}$

NOTE
- If $\sin^{-1} x + \sin^{-1} y = \theta$, then $\cos^{-1} x + \cos^{-1} y = \pi - \theta$
- If $\cos^{-1} x + \cos^{-1} y = \theta$, then $\sin^{-1} x + \sin^{-1} y = \pi - \theta$

8. (i) $3 \sin^{-1} x = \begin{cases} \sin^{-1}(3x - 4x^3), & \text{if } \dfrac{-1}{2} \le x \le \dfrac{1}{2} \\ \pi - \sin^{-1}(3x-4x^3), & \text{if } \dfrac{1}{2} < x \le 1 \\ -\pi - \sin^{-1}(3x-4x^3), & \text{if } -1 \le x < \dfrac{-1}{2} \end{cases}$

(ii) $3 \cos^{-1} x = \begin{cases} \cos^{-1}(4x^3 - 3x), & \text{if } \dfrac{1}{2} \le x \le 1 \\ 2\pi - \cos^{-1}(4x^3-3x), & \text{if } \dfrac{-1}{2} \le x \le \dfrac{1}{2} \\ 2\pi + \cos^{-1}(4x^3-3x), & \text{if } -1 \le x \le \dfrac{1}{2} \end{cases}$

(iii) $3 \tan^{-1} x = \begin{cases} \tan^{-1}\left(\dfrac{3x-x^3}{1-3x^2}\right), & \text{if } \dfrac{-1}{\sqrt{3}} < x < \dfrac{1}{\sqrt{3}} \\ \pi + \tan^{-1}\left(\dfrac{3x-x^3}{1-3x^2}\right), & \text{if } x > \dfrac{1}{\sqrt{3}} \\ -\pi + \tan^{-1}\left(\dfrac{3x-x^3}{1-3x^2}\right), & \text{if } x < \dfrac{-1}{\sqrt{3}} \end{cases}$

DAY PRACTICE SESSION 1

FOUNDATION QUESTIONS EXERCISE

1 The principal value of $\sin^{-1}\left(\cos\dfrac{33\pi}{5}\right)$ is → NCERT Exemplar

(a) $\dfrac{3\pi}{5}$ (b) $\dfrac{7\pi}{5}$ (c) $\dfrac{\pi}{10}$ (d) $-\dfrac{\pi}{10}$

2 If $\sum_{i=1}^{20} \sin^{-1} x_i = 10\pi$, then $\sum_{i=1}^{20} x_i$ is equal to

(a) 20 (b) 10 (c) 0 (d) None of these

3 The domain of the function defined by $f(x) = \sin^{-1}\sqrt{x-1}$ is → NCERT Exemplar

(a) [1, 2] (b) [– 1, 1]
(c) [0, 1] (d) None of these

4 The value of $\cos(2\cos^{-1} x + \sin^{-1} x)$ at $x = \dfrac{1}{5}$ is

(a) 1 (b) 3 (c) 0 (d) $-\dfrac{2\sqrt{6}}{5}$

5 The value of $\cos[\tan^{-1}\{\sin(\cot^{-1} x)\}]$ is

(a) $\dfrac{1}{\sqrt{x^2+2}}$ (b) $\sqrt{\dfrac{x^2+2}{x^2+1}}$ (c) $\sqrt{\dfrac{x^2+1}{x^2+2}}$ (d) $\dfrac{1}{\sqrt{x^2+1}}$

6 The equation $\tan^{-1} x - \cot^{-1} x = \tan^{-1}\left(\dfrac{1}{\sqrt{3}}\right)$ has

(a) no solution → NCERT Exemplar
(b) unique solution
(c) infinite number of solutions
(d) two solutions

7 Let $\tan^{-1} y = \tan^{-1} x + \tan^{-1}\left(\frac{2x}{1-x^2}\right)$, where $|x| < \frac{1}{\sqrt{3}}$. Then, a value of y is → JEE Mains 2015

(a) $\frac{3x - x^3}{1 - 3x^2}$ (b) $\frac{3x + x^3}{1 - 3x^2}$
(c) $\frac{3x - x^3}{1 + 3x^2}$ (d) $\frac{3x + x^3}{1 + 3x^2}$

8 If $\theta = \tan^{-1} a$, $\phi = \tan^{-1} b$ and $ab = -1$, then $(\theta - \phi)$ is equal to

(a) 0 (b) $\frac{\pi}{4}$ (c) $\frac{\pi}{2}$ (d) None of these

9 The range of

$f(x) = |3 \tan^{-1} x - \cos^{-1}(0)| - \cos^{-1}(-1)$ is

(a) $[-\pi, \pi)$ (b) $(-\pi, \pi)$ (c) $[-\pi, \pi]$ (d) $\left[-\frac{\pi}{2}, \frac{\pi}{2}\right)$

10 The number of solutions of the equation $\cos(\cos^{-1} x) = \text{cosec}(\text{cosec}^{-1} x)$ is

(a) 2 (b) 3 (c) 4 (d) 1

11 The value of $\cot\left(\text{cosec}^{-1}\frac{5}{3} + \tan^{-1}\frac{2}{3}\right)$ is → AIEEE 2008

(a) $\frac{5}{17}$ (b) $\frac{6}{17}$ (c) $\frac{3}{17}$ (d) $\frac{4}{17}$

12 If $\tan(\cos^{-1} x) = \sin\left(\cot^{-1}\frac{1}{2}\right)$, then x is equal to

(a) $\pm\frac{5}{3}$ (b) $\pm\frac{\sqrt{5}}{3}$ (c) $\pm\frac{5}{\sqrt{3}}$ (d) None of these

13 If $\cos^{-1} x + \cos^{-1} y = \frac{\pi}{2}$ and $\tan^{-1} x - \tan^{-1} y = 0$, then $x^2 + xy + y^2$ is equal to

(a) 0 (b) $\frac{1}{\sqrt{2}}$ (c) $\frac{3}{2}$ (d) $\frac{1}{8}$

14 If $\sin^{-1}\left(\frac{x}{5}\right) + \text{cosec}^{-1}\left(\frac{5}{4}\right) = \frac{\pi}{2}$, then the value of x is → AIEEE 2007

(a) 1 (b) 3 (c) 4 (d) 5

15 If $\sin^{-1} x + \sin^{-1} y + \sin^{-1} z = \frac{3\pi}{2}$, then the value of

$\sum \frac{(x^{101} + y^{101})(x^{202} + y^{202})}{(x^{303} + y^{303})(x^{404} + y^{404})}$ is

(a) 0 (b) 1 (c) 2 (d) 3

16 The root of the equation

$\tan^{-1}\left(\frac{x-1}{x+1}\right) + \tan^{-1}\left(\frac{2x-1}{2x+1}\right) = \tan^{-1}\left(\frac{23}{36}\right)$ is

(a) $-\frac{3}{8}$ (b) $-\frac{1}{2}$ (c) $\frac{3}{4}$ (d) $\frac{4}{3}$

17 The number of solutions of

$\tan^{-1}\sqrt{x(x+1)} + \sin^{-1}\sqrt{x^2 + x + 1} = \frac{\pi}{2}$ is

→ NCERT Exemplar

(a) 0 (b) 1 (c) 2 (d) infinite

18 The maximum value of $(\sec^{-1} x)^2 + (\text{cosec}^{-1} x)^2$ is

(a) $\frac{\pi^2}{2}$ (b) $\frac{5\pi^2}{4}$
(c) π^2 (d) None of these

19 The trigonometric equation $\sin^{-1} x = 2\sin^{-1} a$, has a solution for → AIEEE 2003

(a) $\frac{1}{2} < |a| < \frac{1}{\sqrt{2}}$ (b) all real values of a
(c) $|a| \le \frac{1}{\sqrt{2}}$ (d) $|a| \ge \frac{1}{\sqrt{2}}$

20 If $x \in \left(-\frac{\pi}{2}, \frac{\pi}{2}\right)$, then the value of

$\tan^{-1}\left(\frac{\tan x}{4}\right) + \tan^{-1}\left(\frac{3\sin 2x}{5 + 3\cos 2x}\right)$ is

(a) $\frac{x}{2}$ (b) $2x$ (c) $3x$ (d) x

21 If $\tan\theta + \tan\left(\frac{\pi}{3} + \theta\right) + \tan\left(-\frac{\pi}{3} + \theta\right) = a\tan 3\theta$, then a is equal to

(a) 1/3 (b) 1
(c) 3 (d) None of these

22 If $\cos^{-1} x = \tan^{-1} x$, then $\sin(\cos^{-1} x)$ is equal to

(a) $-x$ (b) x^2 (c) x^3 (d) $-\frac{1}{x^2}$

23 The real solution of

$\tan^{-1}\sqrt{x(x+1)} + \sin^{-1}\sqrt{x^2 + x + 1} = \frac{\pi}{2}$ is

(a) 2, 3 (b) 1, 0 (c) −1, 0 (d) 3, 1

24 If $\sin^{-1}\left(\frac{x}{5}\right) + \text{cosec}^{-1}\left(\frac{5}{4}\right) = \frac{\pi}{2}$, then the value of x is → AIEEE 2007

(a) 1 (b) 3 (c) 4 (d) 5

25 If $\text{cosec}^{-1} x + \cos^{-1} y + \sec^{-1} z \ge \alpha^2 - \sqrt{2\pi}\,\alpha + 3\pi$, where α is a real number, then

(a) $x = 1, y = -1$ (b) $x = -1, z = -1$
(c) $x = 2, y = 1$ (d) $x = 1, y = -2$

26 The solution of $\sin^{-1} x \le \cos^{-1} x$ is

(a) $\left(-1, \frac{1}{\sqrt{2}}\right)$ (b) $\left[-1, \frac{1}{\sqrt{2}}\right]$
(c) $\left[1, \frac{1}{\sqrt{2}}\right]$ (d) $\left(1, \frac{1}{\sqrt{2}}\right)$

27 If m and M are the least and the greatest value of $(\cos^{-1} x)^2 + (\sin^{-1} x)^2$, then $\frac{M}{m}$ is equal to

(a) 10 (b) 5 (c) 4 (d) 2

28 The number of real solutions of the equation $\sqrt{1 + \cos 2x} = \sqrt{2}\cos^{-1}(\cos x)$ in $\left[\frac{\pi}{2}, \pi\right]$ is

→ NCERT Exemplar

(a) 0 (b) 1 (c) 2 (d) infinite

29 The sum of the infinite series

$$\sin^{-1}\left(\frac{1}{\sqrt{2}}\right)+\sin^{-1}\left(\frac{\sqrt{2}-1}{\sqrt{6}}\right)+\sin^{-1}\left(\frac{\sqrt{3}-\sqrt{2}}{\sqrt{12}}\right)+\ldots+\sin^{-1}\left(\frac{\sqrt{n}-\sqrt{(n-1)}}{\sqrt{n(n+1)}}\right)$$ is

(a) $\frac{\pi}{8}$ (b) $\frac{\pi}{4}$ (c) $\frac{\pi}{2}$ (d) π

30 A root of the equation

$$17x^2+17x\tan\left[2\tan^{-1}\left(\frac{1}{5}\right)-\frac{\pi}{4}\right]-10=0 \text{ is}$$

(a) $\frac{10}{17}$ (b) -1 (c) $-\frac{7}{17}$ (d) 1

31 The value of x for which $\sin[\cot^{-1}(1+x)]=\cos(\tan^{-1}x)$, is → JEE Mains 2013

(a) $-\frac{1}{2}$ (b) 1 (c) 0 (d) $\frac{1}{2}$

32 If $0<x<1$, then $\sqrt{1+x^2}[\{x\cos(\cot^{-1}x)+\sin(\cot^{-1}x)\}^2-1]^{1/2}$ is equal to

(a) $\frac{x}{\sqrt{1+x^2}}$ (b) x (c) $x\sqrt{1+x^2}$ (d) $\sqrt{1+x^2}$

33 If x, y and z are in AP and $\tan^{-1}x$, $\tan^{-1}y$ and $\tan^{-1}z$ are also in AP, then → AIEEE 2012

(a) $x=y=z$ (b) $2x=y=6z$
(c) $6x=3y=2z$ (d) $6x=4y=3z$

DAY PRACTICE SESSION 2

PROGRESSIVE QUESTIONS EXERCISE

1 If $\sin^{-1}\left(x-\frac{x^2}{2}+\frac{x^3}{4}-\ldots\right)+\cos^{-1}\left(x^2-\frac{x^4}{2}+\frac{x^6}{4}-\ldots\right)=\frac{\pi}{2}$, where $0<|x|<\sqrt{2}$, then x is equal to

(a) $\frac{1}{2}$ (b) 1 (c) $-\frac{1}{2}$ (d) -1

2 If the mapping $f(x)=ax+b, a>0$ maps $[-1,1]$ onto $[0,2]$, then $\cot[\cot^{-1}7+\cot^{-1}8+\cot^{-1}18]$ is equal to

(a) $f(-1)$ (b) $f(0)$ (c) $f(1)$ (d) $f(2)$

3 If $S=\tan^{-1}\left(\frac{1}{n^2+n+1}\right)+\tan^{-1}\left(\frac{1}{n^2+3n+3}\right)+\ldots+\tan^{-1}\left\{\frac{1}{1+(n+19)(n+20)}\right\}$, then $\tan S$ is equal to → JEE Mains 2013

(a) $\frac{20}{401+20n}$ (b) $\frac{n}{n^2+20n+1}$
(c) $\frac{20}{n^2+20n+1}$ (d) $\frac{n}{401+20n}$

4 If $f(x)=e^{\cos^{-1}\sin\left(x+\frac{\pi}{3}\right)}$, then

(a) $f\left(-\frac{7\pi}{4}\right)=e^{\frac{\pi}{11}}$ (b) $f\left(\frac{8\pi}{9}\right)=e^{\frac{13\pi}{18}}$
(c) $f\left(-\frac{7\pi}{4}\right)=e^{\frac{3\pi}{12}}$ (d) $f\left(-\frac{7\pi}{4}\right)=e^{\frac{11\pi}{13}}$

5 If $\sin^{-1}x+\sin^{-1}y+\sin^{-1}z=\frac{3\pi}{2}$ and $f(1)=2$, $f(p+q)=f(p)\cdot f(q), \forall\, p,q\in R$, then $x^{f(1)}+y^{f(2)}+z^{f(3)}-\frac{(x+y+z)}{x^{f(1)}+y^{f(2)}+z^{f(3)}}$ is equal to

(a) 0 (b) 1
(c) 2 (d) 3

6 If $[\cot^{-1}x]+[\cos^{-1}x]=0$, where x is a non-negative real number and [.] denotes the greatest integer function, then complete set of values of x is

(a) (cos 1, 1] (b) (cot 1, 1)
(c) (cos 1, cot 1) (d) None of these

7 $\cot^{-1}(\sqrt{\cos\alpha})-\tan^{-1}(\sqrt{\cos\alpha})=x$, then $\sin x$ is equal to → AIEEE 2002

(a) $\tan^2\left(\frac{\alpha}{2}\right)$ (b) $\cot^2\left(\frac{\alpha}{2}\right)$ (c) $\tan\alpha$ (d) $\cot\left(\frac{\alpha}{2}\right)$

8 If $(\tan^{-1}x)^2+(\cot^{-1}x)^2=\frac{5\pi^2}{8}$, then the value of x is

(a) -2 (b) -3 (c) -1 (d) 2

9 The solution set of $\tan^2(\sin^{-1}x)>1$ is

(a) $\left(-1,-\frac{1}{\sqrt{2}}\right)\cup\left(\frac{1}{\sqrt{2}},1\right)$ (b) $\left(-\frac{1}{\sqrt{2}},\frac{1}{\sqrt{2}}\right)\sim\{0\}$
(c) $(-1,1)\sim\{0\}$ (d) None of these

10 If θ and ϕ are the roots of the equation $8x^2+22x+5=0$, then

(a) both $\sin^{-1}\theta$ and $\sin^{-1}\phi$ are equal
(b) both $\sec^{-1}\theta$ and $\sec^{-1}\phi$ are real
(c) both $\tan^{-1}\theta$ and $\tan^{-1}\phi$ are real
(d) None of the above

11 $2\tan^{-1}(-2)$ is equal to

(a) $\cos^{-1}\left(\frac{-3}{5}\right)$ (b) $\pi+\cos^{-1}\frac{3}{5}$
(c) $-\frac{\pi}{2}+\tan^{-1}\left(-\frac{3}{4}\right)$ (d) $-\pi+\cot^{-1}\left(-\frac{3}{4}\right)$

12 Let $x\in(0,1)$. The set of all x such that $\sin^{-1}x>\cos^{-1}x$, is the interval → JEE Mains 2013

(a) $\left(\frac{1}{2},\frac{1}{\sqrt{2}}\right)$ (b) $\left(\frac{1}{\sqrt{2}},1\right)$
(c) (0, 1) (d) $\left(0,\frac{\sqrt{3}}{2}\right)$

ANSWERS

SESSION 1

1 (d)	**2** (a)	**3** (a)	**4** (d)	**5** (c)	**6** (b)	**7** (a)	**8** (c)	**9** (a)	**10** (a)
11 (b)	**12** (b)	**13** (c)	**14** (b)	**15** (d)	**16** (d)	**17** (b)	**18** (b)	**19** (c)	**20** (d)
21 (c)	**22** (b)	**23** (c)	**24** (b)	**25** (a)	**26** (b)	**27** (a)	**28** (a)	**29** (c)	**30** (d)
31 (a)	**32** (c)	**33** (a)							

SESSION 2

1 (b)	**2** (d)	**3** (c)	**4** (b)	**5** (c)	**6** (b)	**7** (a)	**8** (c)	**9** (a)	**10** (c)
11 (c)	**12** (b)								

Hints and Explanations

SESSION 1

1 $\cos\left(\frac{33\pi}{5}\right) = \cos\left(6\pi + \frac{3\pi}{5}\right) = \cos\frac{3\pi}{5}$

$= \sin\left(\frac{\pi}{2} - \frac{3\pi}{5}\right) = \sin\left(-\frac{\pi}{10}\right)$

$\therefore \quad \sin^{-1}\left(\cos\frac{33\pi}{5}\right)$

$= \sin^{-1}\left(\sin\left(-\frac{\pi}{10}\right)\right) = -\frac{\pi}{10}$

2 Since, $-\frac{\pi}{2} \le \sin^{-1} x \le \frac{\pi}{2}$

$\therefore \quad \sin^{-1} x_i = \frac{\pi}{2}, \quad 1 \le i \le 20$

$\Rightarrow \quad x_i = 1, 1 \le i \le 20$

Thus, $\sum_{i=1}^{20} x_i = 20$

3 Given, $f(x) = \sin^{-1}\sqrt{x-1}$

For domain of $f(x)$ $\quad -1 \le \sqrt{x-1} \le 1$

$\Rightarrow \quad 0 \le (x-1) \le 1 \Rightarrow 1 \le x \le 2$

$\therefore \quad x \in [1, 2]$

4 $\cos(2\cos^{-1} x + \sin^{-1} x)$

$= \cos[2(\cos^{-1} x + \sin^{-1} x) - \sin^{-1} x]$

$= \cos(\pi - \sin^{-1} x) = -\cos(\sin^{-1} x)$

$= -\cos\left[\sin^{-1}\left(\frac{1}{5}\right)\right] \qquad \left[\because x = \frac{1}{5}\right]$

$= -\cos\left[\cos^{-1}\sqrt{1 - \left(\frac{1}{5}\right)^2}\right]$

$= -\cos\left(\cos^{-1}\frac{2\sqrt{6}}{5}\right) = -\frac{2\sqrt{6}}{5}$

5 We have, $\cos[\tan^{-1}\{\sin(\cot^{-1} x)\}]$

Let $\cot^{-1} x = \alpha$

$\Rightarrow \quad \cot\alpha = x$

$\Rightarrow \quad \operatorname{cosec}\alpha = \sqrt{1+x^2}$

$\Rightarrow \quad \sin\alpha = \frac{1}{\sqrt{1+x^2}}$

$\Rightarrow \quad \alpha = \sin^{-1}\frac{1}{\sqrt{1+x^2}}$

Hence, $\cos[\tan^{-1}\{\sin(\cot^{-1} x)\}]$

$= \cos\left[\tan^{-1}\left\{\sin\left(\sin^{-1}\frac{1}{\sqrt{1+x^2}}\right)\right\}\right]$

$= \cos\left[\tan^{-1}\frac{1}{\sqrt{1+x^2}}\right]$

$= \cos\left[\cos^{-1}\sqrt{\frac{x^2+1}{x^2+2}}\right]$

$\left[\begin{array}{l} \because \text{let } \tan^{-1}\left(\frac{1}{\sqrt{1+x^2}}\right) = \beta \\ \tan\beta = \frac{1}{\sqrt{1+x^2}} \\ \sec\beta = \sqrt{1 + \frac{1}{1+x^2}} = \sqrt{\frac{x^2+2}{x^2+1}} \\ \cos\beta = \sqrt{\frac{x^2+1}{x^2+2}} \end{array}\right]$

$= \sqrt{\frac{x^2+1}{x^2+2}}$

6 Given, $\tan^{-1} x - \cot^{-1} x = \tan^{-1}\left(\frac{1}{\sqrt{3}}\right)$

$\Rightarrow \tan^{-1} x - \cot^{-1} x = \frac{\pi}{6} \qquad \ldots\text{(i)}$

But $\tan^{-1} x + \cot^{-1} x = \frac{\pi}{2} \qquad \ldots\text{(ii)}$

On adding Eqs. (i) and (ii), we get

$2\tan^{-1} x = \frac{\pi}{6} + \frac{\pi}{2} = \frac{2\pi}{3}$

$\Rightarrow \tan^{-1} x = \frac{\pi}{3}$

$\Rightarrow \quad x = \tan\frac{\pi}{3} \Rightarrow x = \sqrt{3}$

It has unique solution.

7 Given,

$\tan^{-1} y = \tan^{-1} x + \tan^{-1}\left(\frac{2x}{1-x^2}\right)$,

where $|x| < \frac{1}{\sqrt{3}}$

$\Rightarrow \tan^{-1} y = \tan^{-1}\left\{\frac{x + \frac{2x}{1-x^2}}{1 - x\left(\frac{2x}{1-x^2}\right)}\right\}$

$\left[\because \tan^{-1} x + \tan^{-1} y = \tan^{-1}\left(\frac{x+y}{1-xy}\right), \quad xy < 1\right]$

$= \tan^{-1}\left(\frac{x - x^3 + 2x}{1 - x^2 - 2x^2}\right)$

$\tan^{-1} y = \tan^{-1}\left(\frac{3x - x^3}{1 - 3x^2}\right)$

$\Rightarrow \quad y = \frac{3x - x^3}{1 - 3x^2}$

Alternate Method

$|x| < \frac{1}{\sqrt{3}} \Rightarrow -\frac{1}{\sqrt{3}} < x < \frac{1}{\sqrt{3}}$

Let $x = \tan\theta \Rightarrow -\frac{\pi}{6} < \theta < \frac{\pi}{6}$

$\therefore \quad \tan^{-1} y = \theta + \tan^{-1}(\tan 2\theta)$

$= \theta + 2\theta = 3\theta$

$\Rightarrow \quad y = \tan 3\theta$

$\Rightarrow \quad y = \frac{3\tan\theta - \tan^3\theta}{1 - 3\tan\theta}$

$\Rightarrow \quad y = \frac{3x - x^3}{1 - 3x^2}$

8 Given that, $\theta = \tan^{-1} a$

and $\quad \phi = \tan^{-1} b$

and $\quad ab = -1$

$\therefore \tan\theta \tan\phi = ab = -1$

$\Rightarrow \quad \tan\theta = -\cot\phi$

$\Rightarrow \quad \tan\theta = \tan\left(\frac{\pi}{2}+\phi\right)$

$\Rightarrow \quad \theta - \phi = \frac{\pi}{2}$

9 $f(x) = |3\tan^{-1}x - \cos^{-1}(0)| - \cos^{-1}(-1)$

$= \left|3\tan^{-1}x - \left(\frac{\pi}{2}\right)\right| - \pi$

We know that, $-\frac{\pi}{2} < \tan^{-1}x < \frac{\pi}{2}$

$\Rightarrow \frac{-3\pi}{2} < 3\tan^{-1}x < \frac{3\pi}{2}$

$\Rightarrow -2\pi < 3\tan^{-1}x - \frac{\pi}{2} < \pi$

$\Rightarrow 0 \le \left|3\tan^{-1}x - \frac{\pi}{2}\right| < 2\pi$

$\Rightarrow -\pi \le \left|3\tan^{-1}x - \frac{\pi}{2}\right| - \pi < \pi$

10 $\cos(\cos^{-1}x) = x, \forall x \in [-1, 1]$ and $\text{cosec}(\text{cosec}^{-1}x) = x, \forall x \in (-\infty, -1] \cup [1, \infty)$

$\Rightarrow \cos(\cos^{-1}x) = \text{cosec}(\text{cosec}^{-1}x)$ for $x = \pm 1$ only.

Hence, there are two roots.

11 Since, $\text{cosec}^{-1}\left(\frac{5}{3}\right) = \tan^{-1}\left(\frac{3}{4}\right)$

$\therefore \cot\left(\tan^{-1}\frac{3}{4} + \tan^{-1}\frac{2}{3}\right)$

$= \cot\left(\tan^{-1}\left[\frac{\frac{3}{4}+\frac{2}{3}}{1-\frac{1}{2}}\right]\right)$

$= \cot\left(\tan^{-1}\left[\frac{\left(\frac{17}{12}\right)}{\left(\frac{1}{2}\right)}\right]\right)$

$= \cot\left[\tan^{-1}\left(\frac{17}{6}\right)\right] = \frac{6}{17}$

12 Let $\cot^{-1}\frac{1}{2} = \phi \Rightarrow \frac{1}{2} = \cot\phi$

$\Rightarrow \sin\phi = \frac{1}{\sqrt{1+\cot^2\phi}} = \frac{2}{\sqrt{5}}$

Let $\cos^{-1}x = \theta \Rightarrow \sec\theta = \frac{1}{x}$

$\Rightarrow \tan\theta = \sqrt{\sec^2\theta - 1}$

$\Rightarrow \tan\theta = \sqrt{\frac{1}{x^2} - 1} \Rightarrow \tan\theta = \frac{\sqrt{1-x^2}}{x}$

Now, $\tan(\cos^{-1}x) = \sin\left(\cot^{-1}\frac{1}{2}\right)$

$\Rightarrow \tan\left(\tan^{-1}\frac{\sqrt{1-x^2}}{x}\right) = \sin\left(\sin^{-1}\frac{2}{\sqrt{5}}\right)$

$\Rightarrow \frac{\sqrt{1-x^2}}{x} = \frac{2}{\sqrt{5}} \Rightarrow \sqrt{(1-x^2)5} = 2x$

On squaring both sides, we get

$(1-x^2)5 = 4x^2$

$\Rightarrow 9x^2 = 5 \Rightarrow x = \pm\frac{\sqrt{5}}{3}$

13 $\because \tan^{-1}x - \tan^{-1}y = 0 \Rightarrow x = y$

Also, $\cos^{-1}x + \cos^{-1}y = \frac{\pi}{2}$

$\Rightarrow 2\cos^{-1}x = \frac{\pi}{2} \Rightarrow \cos^{-1}x = \frac{\pi}{4}$

$\Rightarrow x = \frac{1}{\sqrt{2}} \Rightarrow x^2 = \frac{1}{2}$

Hence, $x^2 + xy + y^2 = 3x^2 = \frac{3}{2}$

14 $\because \sin^{-1}\left(\frac{x}{5}\right) + \text{cosec}^{-1}\left(\frac{5}{4}\right) = \frac{\pi}{2}$

$\Rightarrow \sin^{-1}\left(\frac{x}{5}\right) + \sin^{-1}\left(\frac{4}{5}\right) = \frac{\pi}{2}$

$\Rightarrow \sin^{-1}\left(\frac{x}{5}\right) = \frac{\pi}{2} - \sin^{-1}\left(\frac{4}{5}\right)$

$\Rightarrow \sin^{-1}\left(\frac{x}{5}\right) = \cos^{-1}\left(\frac{4}{5}\right)$

$\Rightarrow \sin^{-1}\left(\frac{x}{5}\right) = \sin^{-1}\left(\frac{3}{5}\right)$

$\therefore \quad x = 3$

15 Given, $\sin^{-1}x + \sin^{-1}y + \sin^{-1}z = \frac{3\pi}{2}$

$\Rightarrow \sin^{-1}x = \sin^{-1}y = \sin^{-1}z = \frac{\pi}{2}$

$\Rightarrow x = y = z = 1$

$\therefore \Sigma\frac{(x^{101}+y^{101})(x^{202}+y^{202})}{(x^{303}+y^{303})(x^{404}+y^{404})}$

$= \Sigma\frac{(1+1)(1+1)}{(1+1)(1+1)} = \Sigma 1 = 3$

16 $\tan^{-1}\left[\frac{\frac{x-1}{x+1}+\frac{2x-1}{2x+1}}{1-\left(\frac{x-1}{x+1}\right)\left(\frac{2x-1}{2x+1}\right)}\right] = \tan^{-1}\left(\frac{23}{36}\right)$

$\Rightarrow \frac{2x^2-1}{3x} = \frac{23}{36}$

$\Rightarrow 24x^2 - 12 - 23x = 0$

$\Rightarrow x = \frac{4}{3}, -\frac{3}{8}$

But x cannot be negative.

$\therefore \quad x = \frac{4}{3}$

17 For existence, $x(x+1) \ge 0$ ...(i)

and $x^2 + x + 1 \le 1$

$\Rightarrow x^2 + x \le 0$

$\Rightarrow x(x+1) \le 0$ ...(ii)

From Eqs. (i) and (ii), we get

$x(x+1) = 0 \Rightarrow x = 0, -1$

But $x = -1$ is not satisfied the given equation.

18 Let $I = (\sec^{-1}x)^2 + (\text{cosec}^{-1}x)^2$

$= (\sec^{-1}x + \text{cosec}^{-1}x)^2 - 2\sec^{-1}x\,\text{cosec}^{-1}x$

$= \frac{\pi^2}{4} - 2\sec^{-1}x\left(\frac{\pi}{2} - \sec^{-1}x\right)$

$= \frac{\pi^2}{4} + 2\left[(\sec^{-1}x)^2 - \frac{\pi}{2}(\sec^{-1}x) + \frac{\pi^2}{16} - \frac{\pi^2}{16}\right]$

$= \frac{\pi^2}{8} + 2\left[\left(\sec^{-1}x - \frac{\pi}{4}\right)^2\right]$

$\therefore I_{max} = \frac{\pi^2}{8} + 2\left[\frac{9\pi^2}{16}\right] = \frac{5\pi^2}{4}$

19 Given that, $\sin^{-1}x = 2\sin^{-1}a$

$\because -\frac{\pi}{2} \le \sin^{-1}x \le \frac{\pi}{2}$

$\Rightarrow -\frac{\pi}{2} \le 2\sin^{-1}a \le \frac{\pi}{2}$

$\Rightarrow -\frac{\pi}{4} \le \sin^{-1}a \le \frac{\pi}{4}$

$\Rightarrow \sin\left(-\frac{\pi}{4}\right) \le a \le \sin\frac{\pi}{4}$

$\Rightarrow -\frac{1}{\sqrt{2}} \le a \le \frac{1}{\sqrt{2}}$

$\Rightarrow |a| \le \frac{1}{\sqrt{2}}$

20 $\tan^{-1}\left(\frac{\tan x}{4}\right) + \tan^{-1}\left(\frac{3\sin 2x}{5+3\cos 2x}\right)$

$= \tan^{-1}\left(\frac{\tan x}{4}\right) + \tan^{-1}\left(\frac{\frac{6\tan x}{1+\tan^2 x}}{5+\frac{3(1-\tan^2 x)}{1+\tan^2 x}}\right)$

$= \tan^{-1}\left(\frac{\tan x}{4}\right) + \tan^{-1}\left(\frac{6\tan x}{8+2\tan^2 x}\right)$

$= \tan^{-1}\left(\frac{\tan x}{4}\right) + \tan^{-1}\left(\frac{3\tan x}{4+\tan^2 x}\right)$

$= \tan^{-1}\left(\frac{\frac{\tan x}{4}+\frac{3\tan x}{4+\tan^2 x}}{1-\frac{3\tan^2 x}{4(4+\tan^2 x)}}\right)$

$\left(\text{as}\left|\frac{\tan x}{4}\cdot\frac{3\tan x}{4\tan^2 x}\right| < 1\right)$

$= \tan^{-1}\left(\frac{16\tan x + \tan^3 x}{16+\tan^2 x}\right)$

$= \tan^{-1}(\tan x) = x$

21 $\tan\theta + \tan\left(\frac{\pi}{3}+\theta\right) + \tan\left(-\frac{\pi}{3}+\theta\right) = a\tan 3\theta$

$\Rightarrow \tan\theta + \frac{\sqrt{3}+\tan\theta}{1-\sqrt{3}\tan\theta} + \frac{\tan\theta-\sqrt{3}}{1+\sqrt{3}\tan\theta} = a\tan 3\theta$

$\Rightarrow \tan\theta + \frac{8\tan\theta}{1-3\tan^2\theta} = a\tan 3\theta$

$\Rightarrow \frac{3(3\tan\theta-\tan^3\theta)}{1-3\tan^2\theta} = a\tan 3\theta$

$\Rightarrow 3\tan 3\theta = a\tan 3\theta$

$\Rightarrow a = 3$

22 Let $\cos^{-1}x = \tan^{-1}x = \theta$

$\Rightarrow x = \cos\theta = \tan\theta$

$\Rightarrow \cos\theta = \tan\theta \Rightarrow \cos\theta = \frac{\sin\theta}{\cos\theta}$

$\Rightarrow \cos^2\theta = \sin\theta$

$\Rightarrow \sin^2\theta + \sin\theta - 1 = 0$

$\Rightarrow \sin\theta = \frac{-1\pm\sqrt{1+4}}{2}$

$\Rightarrow \sin\theta = \frac{\sqrt{5}-1}{2}$

$\therefore x^2 = \cos^2\theta = \sin\theta = \frac{\sqrt{5}-1}{2}$

and $\sin(\cos^{-1}x) = \sin\theta = \frac{\sqrt{5}-1}{2} = x^2$

23 Given, $\tan^{-1}\sqrt{x(x+1)} + \sin^{-1}\sqrt{x^2+x+1} = \frac{\pi}{2}$

$\Rightarrow \cos^{-1}\frac{1}{\sqrt{1+(x^2+x)}} + \sin^{-1}\sqrt{x^2+x+1} = \frac{\pi}{2}$

$\Rightarrow \cos^{-1}\frac{1}{\sqrt{1+(x^2+x)}} = \frac{\pi}{2} - \sin^{-1}\sqrt{x^2+x+1}$

$\Rightarrow \cos^{-1}\frac{1}{\sqrt{x^2+x+1}} = \cos^{-1}\sqrt{x^2+x+1}$

$\Rightarrow \frac{1}{\sqrt{x^2+x+1}} = \sqrt{x^2+x+1}$

$\Rightarrow x^2+x+1 = 1 \Rightarrow x = -1, 0$

24 Since, $\sin^{-1}\left(\frac{x}{5}\right) + \operatorname{cosec}^{-1}\left(\frac{5}{4}\right) = \frac{\pi}{2}$

$\Rightarrow \sin^{-1}\left(\frac{x}{5}\right) + \sin^{-1}\left(\frac{4}{5}\right) = \frac{\pi}{2}$

$\Rightarrow \sin^{-1}\left(\frac{x}{5}\right) = \frac{\pi}{2} - \sin^{-1}\left(\frac{4}{5}\right)$

$\Rightarrow \sin^{-1}\left(\frac{x}{5}\right) = \cos^{-1}\left(\frac{4}{5}\right)$

$\Rightarrow \sin^{-1}\left(\frac{x}{5}\right) = \sin^{-1}\left(\frac{3}{5}\right)$

$\therefore x = 3$

25 Given,

$\operatorname{cosec}^{-1}x + \cos^{-1}y + \sec^{-1}z \geq \alpha^2 - \sqrt{2\pi}\,\alpha + 3\pi$

RHS $= \alpha^2 - \sqrt{2\pi}\,\alpha + 3\pi$

$= \alpha^2 - 2\sqrt{\frac{\pi}{2}}\,\alpha + \frac{\pi}{2} + 3\pi - \frac{\pi}{2}$

$= \left(\alpha - \sqrt{\frac{\pi}{2}}\right)^2 + \frac{5\pi}{2} \geq \frac{5\pi}{2}$ ...(i)

$\because$ LHS, $\operatorname{cosec}^{-1}x \in \left[-\frac{\pi}{2}, \frac{\pi}{2}\right] - \{0\}$

$\cos^{-1}y \in [0, \pi]$

and $\sec^{-1}z \in [0, \pi] - \left\{\frac{\pi}{2}\right\}$

$\therefore$ LHS $\leq \frac{5\pi}{2}$ ...(ii)

From Eqs. (i) and (ii), we get only possibility is sign of equality

$x = 1, y = -1, z = -1$

26 Given, $\cos^{-1}x \geq \sin^{-1}x$

$\Rightarrow \frac{\pi}{2} \geq 2\sin^{-1}x$

$\left[\because \sin^{-1}x + \cos^{-1}x = \frac{\pi}{2}, \forall x \in [-1,1]\right]$

$\Rightarrow \sin^{-1}x \leq \frac{\pi}{4} \Rightarrow -\frac{\pi}{2} \leq \sin^{-1}x \leq \frac{\pi}{2}$

$\left(\because \text{range of } \sin^{-1}x \text{ is } \left[-\frac{\pi}{2}, \frac{\pi}{2}\right]\right)$

$\Rightarrow -1 \leq x \leq \sin\left(\frac{\pi}{4}\right)$

$\Rightarrow x \in \left[-1, \frac{1}{\sqrt{2}}\right]$

27 $\left(\frac{\pi}{2} - \sin^{-1}x\right)^2 + (\sin^{-1}x)^2$

$= \frac{\pi^2}{4} + 2(\sin^{-1}x)^2 - \pi\sin^{-1}x$

$= \frac{\pi^2}{8} + 2\left[\sin^{-1}x - \frac{\pi}{4}\right]^2$

Here, $m = \frac{\pi^2}{8}, M = \frac{5\pi^2}{4}$

$\therefore \frac{M}{m} = 10$

28 Given, $\sqrt{1+\cos 2x} = \sqrt{2}\cos^{-1}(\cos x)$

$\therefore \sqrt{2\cos^2 x} = \sqrt{2}\,x$

$\Rightarrow \sqrt{2}\,|\cos x| = \sqrt{2}\,x$

For $x \in \left[\frac{\pi}{2}, \pi\right]$, $|\cos x| = -\cos x$

$-\sqrt{2}\cos x = \sqrt{2}\,x$

$\Rightarrow -\cos x = x$

$\therefore \cos x = -x$

Hence, no solution exist.

29 $\because T_r = \sin^{-1}\left\{\frac{\sqrt{r}-\sqrt{(r-1)}}{\sqrt{r(r+1)}}\right\}$

$= \tan^{-1}\left\{\frac{\sqrt{r}-\sqrt{(r-1)}}{1+\sqrt{r}(\sqrt{r-1})}\right\}$

$\therefore S_n = \sum_{r=1}^{n}\tan^{-1}\left(\frac{\sqrt{r}-\sqrt{r-1}}{1+\sqrt{r}\sqrt{r-1}}\right)$

$= \sum_{r=1}^{n}[\tan^{-1}\sqrt{r} - \tan^{-1}\sqrt{(r-1)}]$

$= \tan^{-1}\sqrt{n} - \tan^{-1}\sqrt{0}$

$= \tan^{-1}\sqrt{n} - 0$

$\therefore S_\infty = \tan^{-1}\infty = \frac{\pi}{2}$

30 Now, $\tan\left\{2\tan^{-1}\left(\frac{1}{5}\right)\right\} = \frac{2\times\frac{1}{5}}{1-\frac{1}{25}} = \frac{5}{12}$

Given equation can be rewritten as

$17x^2 - 17x\tan\left\{\frac{\pi}{4} - 2\tan^{-1}\left(\frac{1}{5}\right)\right\} - 10 = 0$

$\Rightarrow 17x^2 - 17x\cdot\frac{1-\frac{5}{12}}{1+\frac{5}{12}} - 10 = 0$

$\Rightarrow 17x^2 - 7x - 10 = 0$

$\Rightarrow (x-1)(17x+10) = 0$

Hence, $x = 1$ is a root of the given equation.

31 $\sin\left(\sin^{-1}\frac{1}{\sqrt{(1+x)^2+1}}\right) = \cos\left(\cos^{-1}\frac{1}{\sqrt{1+x^2}}\right)$

$\frac{1}{\sqrt{(1+x)^2+1}} = \frac{1}{\sqrt{1+x^2}}$

$\Rightarrow (1+x)^2 + 1 = 1 + x^2$

$\Rightarrow 2x + 1 = 0$

$\Rightarrow x = -\frac{1}{2}$

32 We have, $0 < x < 1$

Now, $\sqrt{1+x^2}[\{x\cos(\cot^{-1}x) + \sin(\cot^{-1}x)\}^2 - 1]^{1/2}$

$= \sqrt{1+x^2}\left[\left\{x\frac{x}{\sqrt{1+x^2}} + \frac{1}{\sqrt{1+x^2}}\right\}^2 - 1\right]^{1/2}$

$= \sqrt{1+x^2}\left[\left(\frac{1+x^2}{\sqrt{1+x^2}}\right)^2 - 1\right]^{1/2}$

$= \sqrt{1+x^2}[1+x^2-1]^{1/2}$

$= x\sqrt{1+x^2}$

33 Since, x, y and z are in AP.

$\therefore \quad 2y = x + z$

Also, $\tan^{-1} x$, $\tan^{-1} y$ and $\tan^{-1} z$ are in AP.

$\therefore 2\tan^{-1} y = \tan^{-1} x + \tan^{-1} z$

$\Rightarrow \quad \tan^{-1}\left(\dfrac{2y}{1-y^2}\right) = \tan^{-1}\left(\dfrac{x+z}{1-xz}\right)$

$\Rightarrow \quad \dfrac{x+z}{1-y^2} = \dfrac{x+z}{1-xz}$

$\Rightarrow \quad y^2 = xz \quad [\because 2y = x + z]$

Since x, y and z are in AP as well as in GP.

$\therefore \quad x = y = z$

SESSION 2

1 Now, $x - \dfrac{x^2}{2} + \dfrac{x^3}{4} - \ldots$

$= \dfrac{x}{1+\dfrac{x}{2}} = \dfrac{2x}{2+x}$

and $x^2 - \dfrac{x^4}{2} + \dfrac{x^6}{4} - \ldots$

$= \dfrac{x^2}{1+\dfrac{x^2}{2}} = \dfrac{2x^2}{2+x^2}$

$\because \sin^{-1}\alpha + \cos^{-1}\alpha = \dfrac{\pi}{2}$

$\therefore \quad \dfrac{2x}{2+x} = \dfrac{2x^2}{2+x^2}, x \neq 0$

$\Rightarrow \quad 2 + x^2 = 2x + x^2$

$\therefore \quad x = 1$

2 $\because \quad f(x) = ax + b$

$\therefore f'(x) = a > 0$

So, $f(x)$ is an increasing function.

$\Rightarrow \quad f(-1) = 0$ and $f(1) = 2$

$\Rightarrow \quad -a + b = 0$

and $\quad a + b = 2$

Then, $\quad a = b = 1$

$\therefore \quad f(x) = x + 1$

Now, $\cot[\cot^{-1} 7 + \cot^{-1} 8 + \cot^{-1} 18]$

$= \cot\left[\tan^{-1}\left(\dfrac{1}{7}\right) + \tan^{-1}\left(\dfrac{1}{8}\right) + \tan^{-1}\left(\dfrac{1}{18}\right)\right]$

$= \cot\left[\tan^{-1}\left(\dfrac{\dfrac{1}{7}+\dfrac{1}{8}}{1-\dfrac{1}{7}\cdot\dfrac{1}{8}}\right) + \tan^{-1}\left(\dfrac{1}{18}\right)\right]$

$= \cot\left[\tan^{-1}\left(\dfrac{15}{55}\right) + \tan^{-1}\left(\dfrac{1}{18}\right)\right]$

$= \cot\left[\tan^{-1}\left(\dfrac{3}{11}\right) + \tan^{-1}\left(\dfrac{1}{18}\right)\right]$

$= \cot\left[\tan^{-1}\left(\dfrac{\dfrac{3}{11}+\dfrac{1}{18}}{1-\dfrac{3}{11}\cdot\dfrac{1}{18}}\right)\right]$

$= \cot\left[\tan^{-1}\left(\dfrac{65}{195}\right)\right]$

$= \cot\left[\tan^{-1}\left(\dfrac{1}{3}\right)\right] = \cot(\cot^{-1} 3) = 3$

$= 1 + 2 = f(2) \quad [\because f(x) = x + 1]$

3 $S = \tan^{-1}\left\{\dfrac{(n+1)-(n+0)}{1+(n+0)(n+1)}\right\}$

$+ \tan^{-1}\left\{\dfrac{(n+2)-(n+1)}{1+(n+1)(n+2)}\right\}$

$+ \ldots + \tan^{-1}\left\{\dfrac{(n+20)-(n+19)}{1+(n+19)(n+20)}\right\}$

$= \tan^{-1}(n+1) - \tan^{-1} n$

$+ \tan^{-1}(n+2) - \tan^{-1}(n+1)$

$+ \ldots + \tan^{-1}(n+20) - \tan^{-1}(n+19)$

$= \tan^{-1}(n+20) - \tan^{-1} n$

$= \tan^{-1}\left\{\dfrac{n+20-n}{1+n(n+20)}\right\}$

$= \tan^{-1}\left(\dfrac{20}{n^2+20n+1}\right)$

$\Rightarrow \tan S = \tan\left\{\tan^{-1}\left(\dfrac{20}{n^2+20n+1}\right)\right\}$

$\Rightarrow \tan S = \dfrac{20}{n^2+20n+1}$

4 Given, $f(x) = e^{\cos^{-1}\sin\left(x+\frac{\pi}{3}\right)}$

$\Rightarrow \quad f\left(\dfrac{8\pi}{9}\right) = e^{\cos^{-1}\sin\left(\frac{8\pi}{9}+\frac{\pi}{3}\right)}$

$= e^{\cos^{-1}\sin\left(\frac{11\pi}{9}\right)}$

$\Rightarrow \quad f\left(\dfrac{8\pi}{9}\right) = e^{\cos^{-1}\cos\left(\frac{13\pi}{18}\right)} = e^{\frac{13\pi}{18}}$

Also, $f\left(-\dfrac{7\pi}{4}\right) = e^{\cos^{-1}\sin\left(-\frac{7\pi}{4}+\frac{\pi}{3}\right)}$

$= e^{\cos^{-1}\sin\left(-\frac{17\pi}{12}\right)} = e^{\cos^{-1}\cos\frac{\pi}{12}} = e^{\frac{\pi}{12}}$

5 $\because -\dfrac{\pi}{2} \leq \sin^{-1} x \leq \dfrac{\pi}{2}, -\dfrac{\pi}{2} \leq \sin^{-1} y \leq \dfrac{\pi}{2}$

and $-\dfrac{\pi}{2} \leq \sin^{-1} z \leq \dfrac{\pi}{2}$

Given that,

$\sin^{-1} x + \sin^{-1} y + \sin^{-1} z = \dfrac{3\pi}{2}$

which is possible only when

$\sin^{-1} x = \sin^{-1} y = \sin^{-1} z = \dfrac{\pi}{2}$

$\Rightarrow \quad x = y = z = 1$

Put $p = q = 1$

Then, $f(2) = f(1)f(1) = 2 \cdot 2 = 4$

and put $p = 1, q = 2$

Then, $f(3) = f(1)f(2) = 2 \cdot 2^2 = 8$

$\therefore x^{f(1)} + y^{f(2)} + z^{f(3)} - \dfrac{x+y+z}{x^{f(1)}+y^{f(2)}+z^{f(3)}}$

$= 1 + 1 + 1 - \dfrac{3}{1+1+1} = 3 - 1 = 2$

6 $\because \quad 0 \leq \cos^{-1} x \leq \pi$

and $0 < \cot^{-1} x < \pi$

Given, $[\cot^{-1} x] + [\cos^{-1} x] = 0$

$\Rightarrow [\cot^{-1} x] = 0$ and $[\cos^{-1} x] = 0$

$\Rightarrow 0 < \cot^{-1} x < 1$ and $0 \leq \cos^{-1} x < 1$

$\therefore \quad x \in (\cot 1, \infty)$

and $\quad x \in (\cos 1, 1) \Rightarrow x \in (\cot 1, 1)$

7 Given that,

$\cot^{-1}(\sqrt{\cos\alpha}) - \tan^{-1}(\sqrt{\cos\alpha}) = x \quad \ldots(i)$

We know that,

$\cot^{-1}(\sqrt{\cos\alpha}) + \tan^{-1}(\sqrt{\cos\alpha}) = \dfrac{\pi}{2}$ 3 $\quad \ldots(ii)$

$\left[\because \cot^{-1} x + \tan^{-1} x = \dfrac{\pi}{2}\right]$

On adding Eqs. (i) and (ii), we get

$2\cot^{-1}(\sqrt{\cos\alpha}) = \dfrac{\pi}{2} + x$

$\Rightarrow \quad \sqrt{\cos\alpha} = \cot\left(\dfrac{\pi}{4} + \dfrac{x}{2}\right)$

$\Rightarrow \quad \sqrt{\cos\alpha} = \dfrac{\cot\dfrac{x}{2} - 1}{1 + \cot\dfrac{x}{2}}$

$\Rightarrow \quad \sqrt{\cos\alpha} = \dfrac{\cos\dfrac{x}{2} - \sin\dfrac{x}{2}}{\cos\dfrac{x}{2} + \sin\dfrac{x}{2}}$

On squaring both sides, we get

$\Rightarrow \quad \cos\alpha = \dfrac{\cos^2\dfrac{x}{2} + \sin^2\dfrac{x}{2} - 2\sin\dfrac{x}{2}\cos\dfrac{x}{2}}{\cos^2\dfrac{x}{2} + \sin^2\dfrac{x}{2} + 2\sin\dfrac{x}{2}\cos\dfrac{x}{2}}$

$\Rightarrow \quad \cos\alpha = \dfrac{1 - \sin x}{1 + \sin x}$

$\Rightarrow \quad \dfrac{1 - \tan^2\dfrac{\alpha}{2}}{1 + \tan^2\dfrac{\alpha}{2}} = \dfrac{1 - \sin x}{1 + \sin x}$

On applying componendo and dividendo rule, we get

$\sin x = \tan^2\left(\dfrac{\alpha}{2}\right)$

8 Given,

$(\tan^{-1} x)^2 + (\cot^{-1} x)^2 = \frac{5\pi^2}{8}$

$\Rightarrow (\tan^{-1} x + \cot^{-1} x)^2 - 2\tan^{-1} x \cdot \cot^{-1} x = \frac{5\pi^2}{8}$

$\Rightarrow \left(\frac{\pi}{2}\right)^2 - 2\tan^{-1} x\left(\frac{\pi}{2} - \tan^{-1} x\right) = \frac{5\pi^2}{8}$ $\left[\because \tan^{-1} x + \cot^{-1} x = \frac{\pi}{2}\right]$

$\Rightarrow \frac{\pi^2}{4} - 2\cdot\frac{\pi}{2}\tan^{-1} x + 2(\tan^{-1} x)^2 = \frac{5\pi^2}{8}$

$\Rightarrow 2(\tan^{-1} x)^2 - \pi\tan^{-1} x - \frac{3\pi^2}{8} = 0$

$\Rightarrow \tan^{-1} x = -\frac{\pi}{4}, \frac{3\pi}{4}$

$\Rightarrow \tan^{-1} x = -\frac{\pi}{4}$

$\Rightarrow x = -1$ $\left\{\text{neglecting } \tan^{-1} x = \frac{3\pi}{4} \text{ as principal value of } \tan^{-1} x \in \left(-\frac{\pi}{2}, \frac{\pi}{2}\right)\right\}$

9 $\tan^2(\sin^{-1} x) > 1$

$\Rightarrow \frac{\pi}{4} < \sin^{-1} x < \frac{\pi}{2}$

or $-\frac{\pi}{2} < \sin^{-1} x < -\frac{\pi}{4}$

$\Rightarrow x \in \left(\frac{1}{\sqrt{2}}, 1\right)$ or $x \in \left(-1, -\frac{1}{\sqrt{2}}\right)$

$\Rightarrow x \in \left(-1, -\frac{1}{\sqrt{2}}\right) \cup \left(\frac{1}{\sqrt{2}}, 1\right)$

10 $8x^2 + 22x + 5 = 0$

$\Rightarrow x = -\frac{1}{4}, -\frac{5}{2}$

$\because -1 < -\frac{1}{4} < 1$ and $-\frac{5}{2} < -1$

$\therefore \sin^{-1}\left(-\frac{1}{4}\right)$ exists but $\sin^{-1}\left(-\frac{5}{2}\right)$ does not exist.

$\sec^{-1}\left(-\frac{5}{2}\right)$ exists but $\sec^{-1}\left(-\frac{1}{4}\right)$ does not exist.

So, $\tan^{-1}\left(-\frac{1}{4}\right)$

and $\tan^{-1}\left(-\frac{5}{2}\right)$ both exist.

11 Let $\tan^{-1}(-2) = \theta$

$\Rightarrow \tan\theta = -2$

$\Rightarrow \theta \in \left(-\frac{\pi}{2}, 0\right)$

$\Rightarrow 2\theta \in (-\pi, 0)$

Now, $\cos(-2\theta) = \cos 2\theta$

$= \frac{1 - \tan^2\theta}{1 + \tan^2\theta}$

$= \frac{-3}{5}$

$\Rightarrow -2\theta = \cos^{-1}\left(\frac{-3}{5}\right)$

$= \pi - \cos^{-1}\frac{3}{5}$

$\Rightarrow 2\theta = -\pi + \cos^{-1}\frac{3}{5}$

$\Rightarrow 2\theta = -\pi + \tan^{-1}\frac{4}{3}$

$= -\pi + \cot^{-1}\frac{3}{4}$

$= -\pi + \frac{\pi}{2} - \tan^{-1}\frac{3}{4}$

$= -\frac{\pi}{2} - \tan^{-1}\frac{3}{4}$

$= -\frac{\pi}{2} + \tan^{-1}\left(-\frac{3}{4}\right)$

12

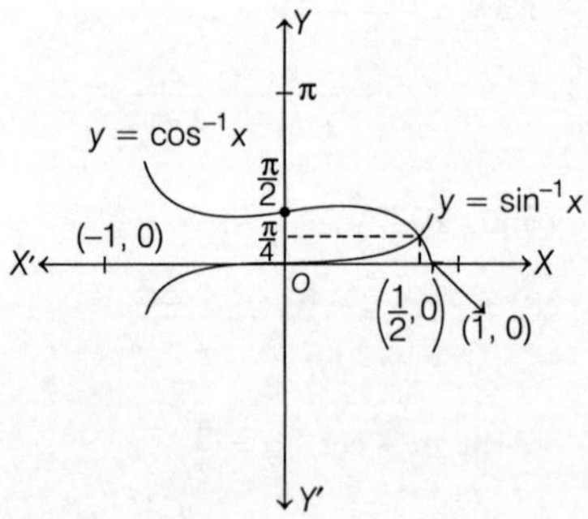

$\therefore \sin^{-1} x > \cos^{-1} x, \forall x \in \left(\frac{1}{\sqrt{2}}, 1\right)$

DAY TWENTY THREE

Unit Test 3
(Trigonometry)

1 The range of $f(x) = \sin^{-1} x + \tan^{-1} x + \sec^{-1}x$, is

(a) $\left(\frac{\pi}{4}, \frac{3\pi}{4}\right)$ (b) $\left[\frac{\pi}{4}, \frac{3\pi}{4}\right]$

(c) $\left\{\frac{\pi}{4}, \frac{3\pi}{4}\right\}$ (d) None of these

2 From a point a m above a lake the angle of elevation of a cloud is α and the angle of depression of its reflection is β. The height of the cloud is

(a) $\frac{a \sin(\alpha+\beta)}{\sin(\alpha-\beta)}$ m (b) $\frac{a \sin(\alpha+\beta)}{\sin(\beta-\alpha)}$ m

(c) $\frac{a \sin(\beta-\alpha)}{\sin(\alpha+\beta)}$ m (d) None of these

3 If $\sum_{i=1}^{2n} \cos^{-1} x_i = 0$, then $\sum_{i=1}^{2n} x_i$ is equal to

(a) n (b) $2n$

(c) $\frac{n(n+1)}{2}$ (d) None of these

4 The value of x for which $\cos^{-1}(\cos 4) > 3x^2 - 4x$, is

(a) $\left(0, \frac{2+\sqrt{6\pi-8}}{3}\right)$

(b) $\left(\frac{2-\sqrt{6\pi-8}}{3}, 0\right)$

(c) $(-2, 2)$

(d) $\left(\frac{2-\sqrt{6\pi-8}}{3}, \frac{2+\sqrt{6\pi-8}}{3}\right)$

5 If $\alpha \le \sin^{-1} x + \cos^{-1}x + \tan^{-1}x \le \beta$, then

(a) $\alpha = 0$ (b) $\beta = \frac{\pi}{2}$ (c) $\alpha = \frac{\pi}{4}$ (d) $\beta = \frac{\pi}{3}$

6 The angles of elevation of the top of a tower at the top and the foot of a pole of height 10 m are 30° and 60°, respectively. The height of the tower is

(a) 10 m (b) 15 m

(c) 20 m (d) None of these

7 The maximum value of $12 \sin\theta - 9 \sin^2 \theta$ is

(a) 3 (b) 4

(c) 5 (d) None of these

8 The minimum value of the expression $\sin\alpha + \sin\beta + \sin\gamma$, when α, β and γ are real numbers satisfying $\alpha + \beta + \gamma = \pi$, is

(a) -3 (b) negative (c) positive (d) zero

9 If $\cos^{-1}\sqrt{x} + \cos^{-1}\sqrt{1-x} + \cos^{-1}\sqrt{1-y} = \pi$, then the value of y is

(a) $\frac{1}{2}$ (b) $\frac{3}{2}$ (c) 1 (d) $\frac{1}{4}$

10 If α and β are the roots of the equation $5\cos\theta + 4\sin\theta = 3$, then $\cos(\alpha+\beta)$ is equal to

(a) $\frac{9}{40}$ (b) $\frac{9}{41}$ (c) $\frac{3}{10}$ (d) $\frac{21}{31}$

11 The angular depressions of the top and the foot of a chimney as seen from the top of a second chimney, which is 150 m high and standing on the same level as the first are θ and ϕ, respectively. If $\tan\theta = \frac{4}{3}$ and $\tan\phi = \frac{5}{2}$, then the distance between their tops is

(a) 120 m (b) 110 m

(c) 100 m (d) None of these

12 The elevation of the hill from a place P due West of it is 60° and at a place Q due South of it is 30°. If the distance PQ be 200 m, then the height of the hill is

(a) 109.54 m (b) 108.70 m

(c) 110.6 m (d) None of these

13 $\cos^{-1}\left\{\frac{1}{\sqrt{2}}\left(\cos\frac{7\pi}{5} - \sin\frac{2\pi}{5}\right)\right\}$ is equal to

(a) $\frac{13\pi}{20}$ (b) $\frac{21\pi}{20}$

(c) $\frac{33\pi}{20}$ (d) None of these

14 $\sec^2\theta = \dfrac{4ab}{(a+b)^2}$, where $a, b \in R$ is true if and only if

(a) $a+b \neq 0$ (b) $a=b, a \neq 0$
(c) $a=b$ (d) $a \neq 0, b \neq 0$

15 If $x = \cos^{-1}(\cos 4)$ and $y = \sin^{-1}(\sin 3)$, then which of the following conditions holds?

(a) $x - y = 1$ (b) $x + y + 1 = 0$
(c) $x + 2y = 2$ (d) $\tan(x+y) = -\tan 7$

16 If $\log_2 x \geq 0$, then $\log_{1/\pi}\left\{\sin^{-1}\dfrac{2x}{1+x^2} + 2\tan^{-1}x\right\}$ is equal to

(a) $\log_{1/\pi}(4\tan^{-1}x)$ (b) 0
(c) -1 (d) None of these

17 If A lies in the third quadrant and $3\tan A - 4 = 0$, then $5\sin 2A + 3\sin A + 4\cos A$ is equal to

(a) 0 (b) $-\dfrac{24}{5}$
(c) $\dfrac{24}{5}$ (d) $\dfrac{48}{5}$

18 The minimum value of $27^{\cos x} + 81^{\sin x}$ is

(a) $\dfrac{2}{3\sqrt{3}}$ (b) $\dfrac{1}{3\sqrt{3}}$
(c) $\dfrac{2}{9\sqrt{3}}$ (d) None of these

19 If $\sin^2 x + a\sin x + 1 = 0$ has no real number solution, then

(a) $|a| \geq 2$ (b) $|a| \geq 1$
(c) $|a| < 2$ (d) None of these

20 If $\sin\theta = n\sin(\theta + 2\alpha)$, then $\tan(\theta + \alpha)$ is equal to

(a) $\dfrac{n+1}{n-1}\tan\alpha$ (b) $\dfrac{1+n}{1-n}\tan\alpha$
(c) $\dfrac{n}{1+n}\tan\alpha$ (d) None of these

21 $\sin^6 x + \cos^6 x$ lies between

(a) $\dfrac{1}{4}$ and 1 (b) $\dfrac{1}{4}$ and 2
(c) 0 and 1 (d) None of these

22 If $n = \dfrac{\pi}{4\alpha}$, then $\tan\alpha \cdot \tan 2\alpha \cdot \tan 3\alpha \ldots \tan(2n-1)\alpha$ is equal to

(a) 1 (b) -1
(c) ∞ (d) None of these

23 The ratio of the greatest value of $2 - \cos x + \sin^2 x$ to its least value is

(a) $\dfrac{1}{4}$ (b) $\dfrac{9}{4}$ (c) $\dfrac{13}{4}$ (d) $\dfrac{17}{4}$

24 If $m = a\cos^3\theta + 3a\cos\theta\sin^2\theta, n = a\sin^3\theta + 3a\cos^2\theta\sin\theta$, then the value of $(m+n)^{2/3} + (m-n)^{2/3}$ is

(a) $2a^{2/3}$ (b) $a^{2/3}$ (c) $a^{3/2}$ (d) $4a^{2/3}$

25 If $a\sin^2\alpha - \dfrac{1}{a}\text{cosec}^2\alpha = 0, 0 < \alpha < \dfrac{\pi}{2}$, then $\cos^2\alpha + 5\sin\alpha\cos\alpha + 6\sin^2\alpha$ is equal to

(a) 5 (b) $\dfrac{a^2+5a+6}{a^2}$
(c) $\dfrac{a^2-5a+6}{a^2}$ (d) None of these

26 If $\tan\dfrac{x}{2} = \text{cosec}\, x - \sin x$, then $\tan^2\dfrac{x}{2}$ is equal to

(a) $2-\sqrt{5}$ (b) $\sqrt{5}-2$
(c) $(9-4\sqrt{5})(2+\sqrt{5})$ (d) $(9+4\sqrt{5})(2-\sqrt{5})$

27 If $0° < \theta < 180°$, then $\sqrt{2+\sqrt{2+\sqrt{2+\ldots+\sqrt{2(1+\cos\theta)}}}}$ (where, number of 2's is n) is equal to

(a) $2\cos\left(\dfrac{\theta}{2^n}\right)$ (b) $2\cos\left(\dfrac{\theta}{2^{n-1}}\right)$
(c) $2\cos\left(\dfrac{\theta}{2^{n+1}}\right)$ (d) None of these

28 In ΔABC, $\tan A + \tan B + \tan C = 6$, $\tan B\tan C = 2$, then $\sin^2 A : \sin^2 B : \sin^2 C$ is equal to

(a) $\dfrac{9}{10}:\dfrac{5}{10}:\dfrac{8}{10}$ (b) $\dfrac{9}{10}:\dfrac{7}{10}:\dfrac{8}{10}$
(c) $\dfrac{9}{10}:\dfrac{8}{10}:\dfrac{7}{10}$ (d) None of these

29 If $0 \leq x \leq 3\pi$, $0 \leq y \leq 3\pi$ and $\cos x \cdot \sin y = 1$, then the possible number of values of the ordered pair (x, y) is

(a) 6 (b) 12 (c) 8 (d) 15

30 The general solution of the equation $1 + \sin^4 2x = \cos^2 6x$ is

(a) $\dfrac{n\pi}{3}$ (b) $\dfrac{n\pi}{2}$ (c) $3n\pi$ (d) None of these

31 The equation $2\cos^2\dfrac{x}{2}\sin^2 x = x^2 + x^{-2}, 0 < x \leq \dfrac{\pi}{2}$ has

(a) no real solution
(b) a unique real solution
(c) finitely many real solutions
(d) infinitely many real solutions

32 The n poles standing at equal distance on a straight road subtend the same angle α at a point O on the road. If the height of the largest pole is h and the distance of the foot of the smallest pole from O is a, the distance between two consecutive poles is

(a) $\dfrac{h\cos\alpha - a\sin\alpha}{(n-1)\sin\alpha}$ (b) $\dfrac{h\cos\alpha + a\sin\alpha}{(n-1)\sin\alpha}$
(c) $\dfrac{h\cos\alpha - a\sin\alpha}{(n+1)\sin\alpha}$ (d) $\dfrac{a\cos\alpha - h\sin\alpha}{(n-1)\sin\alpha}$

33 The solution of the equation

$\tan^{-1}(x+1) + \tan^{-1}(x-1) = \tan^{-1}\left(\dfrac{8}{31}\right)$ is

(a) $\dfrac{1}{4}$ (b) $\dfrac{1}{4}$ or -8
(c) -8 (d) None of these

34 The solution of $\cos^{-1} x\sqrt{3} + \cos^{-1} x = \frac{\pi}{2}$, is

(a) $-\frac{1}{2}$ (b) $-\frac{1}{2}$ or $\frac{1}{2}$
(c) $\frac{1}{2}$ (d) None of these

35 The number of solutions of the equation $\tan x + \sec x = 2\cos x$ lying in the interval $[0, 2\pi]$, is

(a) 0 (b) 1 (c) 2 (d) 3

36 If $\cot^2 x + \operatorname{cosec} x - a = 0$ has atleast one solution, then complete set of value of a is

(a) $[-1, \infty)$ (b) $(-1, 1)$ (c) $(-\infty, 1)$ (d) $(-\infty, 1]$

37 From the top of the cliff 400 m high the top of a tower was observed at an angle of depression 45° and from the foot of the tower the top of the cliff was observed at an angle of elevation 60°. The height of the tower is

(a) 169.06 m (b) 179.2 m
(c) 180.0 m (d) None of these

38 If $\cos 5\theta = P\cos\theta - 20\cos^3\theta + Q\cos^5\theta + R$, then $P + Q + R$ is equal to

(a) 21 (b) 18 (c) 15 (d) 31

39 In ΔABC, if $\frac{\cos A}{a} = \frac{\cos B}{b} = \frac{\cos C}{c}$, then the triangle is

(a) right angled (b) obtuse angled
(c) equilateral (d) isosceles

40 If the area of a ΔABC be λ, then $a^2 \sin 2B + b^2 \sin 2A$ is equal to

(a) 2λ (b) λ
(c) 4λ (d) None of these

41 If $\cos 2x + 2\cos x = 1$, then the value of $\sin^2 x(2 - \cos^2 x)$ is

(a) 0 (b) 1 (c) 2 (d) -1

42 If $\sin x(1 + \sin x) + \cos x(1 + \cos x) = A$ and $\sin x(1 - \sin x) + \cos x(1 - \cos x) = B$, then the value of $A^2 - 2A - \sin 2x = B^2 + 2B - \sin 2x$ is

(a) 1 (b) 2 (c) 3 (d) 0

43 If the sides a, b and c of a ΔABC are roots of the equation $x^3 - 15x^2 + 47x - 82 = 0$, then the value of $\frac{\cos A}{a} + \frac{\cos B}{b} + \frac{\cos C}{c}$ is

(a) $\frac{225}{164}$ (b) $\frac{131}{164}$ (c) $\frac{131}{82}$ (d) $\frac{169}{82}$

44 If in a ΔABC, $a^4 + b^4 + c^4 - 2b^2c^2 - 2c^2a^2 = 0$. Then, the angle c is

(a) 30° or 150° (b) 45° or 135°
(c) 60° or 120° (d) None of these

45 Let $f(\theta, \alpha) = 2\sin^2\theta + 4\cos(\theta + \alpha)$ $\sin\theta \sin\alpha + \cos 2(\theta + \alpha)$. Then, the value of $f\left(\frac{\pi}{3}, \frac{\pi}{4}\right)$ is

(a) 0 (b) 1 (c) 2 (d) 3

46 Which among the following is/are correct statement/statements?

I. The general value of θ satisfying the equations $\sin^2\theta = \sin^2\alpha$, $\cos^2\theta = \cos^2\alpha$ and $\tan^2\theta = \tan^2\alpha$ is given by $\theta = n\pi \pm \alpha$

II. The general value of θ satisfying equations $\sin\theta = \sin\alpha$ and $\cos\theta = \cos\alpha$ simultaneously is given by $\theta = 2n\pi \pm \alpha$, $n \in Z$.

(a) I is correct (b) II is correct
(c) Both I and II are correct (d) Both I and II are incorrect

Directions (Q. Nos. 47-50) *Each of these questions contains two statements : Statement* I *and Statement* II. *Each of these questions also has four alternative choices, only one of which is the correct answer. You have to select one of the codes (a), (b), (c) and (d) given below.*

(a) Statement I is true, Statement II is true; Statement II is a correct explanation for Statement I
(b) Statement I is true, Statement II is true; Statement II is not a correct explanation for Statement I
(c) Statement I is true; Statement II is false
(d) Statement I is false; Statement II is true

47 **Statement I** If $\frac{1}{2} \le x \le 1$, then

$\cos^{-1} x - \sin^{-1}\left[\frac{x}{2} + \frac{\sqrt{3 - 3x^2}}{2}\right]$ is equal to $-\frac{\pi}{3}$.

Statement II $\sin^{-1}(2x\sqrt{1 - x^2}) = 2\sin^{-1} x$, if $x \in \left[-\frac{1}{\sqrt{2}}, \frac{1}{\sqrt{2}}\right]$

48 Suppose any angle is satisfied both equations, then it is said to be solution of an equation.

Statement I The number of solutions of the pair of equations

$$2\sin^2\theta - \cos 2\theta = 0$$
and
$$2\cos^2\theta - 3\sin\theta = 0$$
in the interval $[0, 2\pi]$ is two.

Statement II If $2\cos^2\theta - 3\sin\theta = 0$, then θ cannot lie in III or IV quadrant.

49 **Statement I** If $2\sin\left(\frac{\theta}{2}\right) = \sqrt{1 + \sin\theta} + \sqrt{1 - \sin\theta}$, then $\frac{\theta}{2}$ lies between $2n\pi + \frac{\pi}{4}$ and $2n\pi + \frac{3\pi}{4}$.

Statement II If $\frac{\pi}{4} \le \theta \le \frac{3\pi}{4}$, then $\sin\frac{\theta}{2} > 0$.

50 Suppose a person is standing on a ground and see the tower, then they make an angle of elevation.

Statement I A tower subtends angles $\alpha, 2\alpha$ and 3α respectively at points A, B and C all lying on a horizontal line through the foot of AB tower, then $AB : BC = \sin 3\alpha : \sin\alpha$.

Statement II In ΔABC, if $\frac{a}{\sin A} = \frac{b}{\sin B} = \frac{c}{\sin C}$, then ΔABC is an equilateral triangle.

ANSWERS

1 (c)	2 (b)	3 (b)	4 (d)	5 (a)	6 (b)	7 (b)	8 (c)	9 (c)	10 (b)
11 (c)	12 (a)	13 (d)	14 (b)	15 (d)	16 (c)	17 (a)	18 (c)	19 (c)	20 (b)
21 (a)	22 (a)	23 (c)	24 (a)	25 (d)	26 (c)	27 (a)	28 (a)	29 (a)	30 (b)
31 (a)	32 (a)	33 (a)	34 (c)	35 (c)	36 (a)	37 (a)	38 (a)	39 (c)	40 (c)
41 (b)	42 (d)	43 (b)	44 (b)	45 (a)	46 (a)	47 (b)	48 (a)	49 (b)	50 (c)

Hints and Explanations

1 $f(x) = \sin^{-1} x + \tan^{-1} x + \sec^{-1} x$,
Hence, domain of $f(x)$ is ± 1. So, the range is $\{f(1), f(-1)\}$ i.e. $\left\{\frac{\pi}{4}, \frac{3\pi}{4}\right\}$.

2 In ΔPMC, $\tan\alpha = \dfrac{h-a}{PM}$

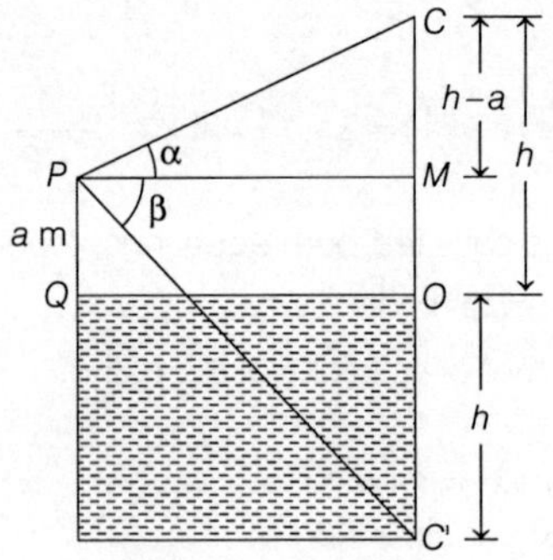

$\Rightarrow \quad PM = (h-a)\cot\alpha \quad \ldots(i)$

and in $\Delta PMC'$, $\tan\beta = \dfrac{h+a}{PM}$

$\Rightarrow h + a = PM \tan\beta$

$\therefore \quad h = (h-a)\cot\alpha \tan\beta - a$ [from Eq. (i)]

$\Rightarrow h(1 - \cot\alpha\tan\beta) = -a(\cot\alpha\tan\beta + 1)$

$\Rightarrow \quad h = \dfrac{a(\sin\alpha\cos\beta + \cos\alpha\sin\beta)}{\sin\beta\cos\alpha - \sin\alpha\cos\beta}$

$\Rightarrow \quad h = \dfrac{a\sin(\alpha+\beta)}{\sin(\beta-\alpha)}$ m

3 Since, $0 \le \cos^{-1} x_i \le \pi$

$\therefore \quad \cos^{-1} x_i = 0 \ \forall\, i$

$\therefore \quad x_i = 1, \forall\, i$

$\therefore \quad \sum_{i=1}^{2n} x_i = 2n$

4 Now, $\cos^{-1}(\cos 4) = \cos^{-1}[\cos(2\pi - 4)]$

$= 2\pi - 4$

$\therefore \quad \cos^{-1}(\cos 4) > 3x^2 - 4x$

$\therefore \quad 2\pi - 4 > 3x^2 - 4x$

$\Rightarrow \quad 3x^2 - 4x - (2\pi - 4) < 0$

$\Rightarrow \quad \dfrac{2 - \sqrt{6\pi - 8}}{3} < x < \dfrac{2 + \sqrt{6\pi - 8}}{3}$

5 Since, $\sin^{-1} x + \cos^{-1} x + \tan^{-1} x = \dfrac{\pi}{2} + \tan^{-1} x$

Now, $-\dfrac{\pi}{2} \le \tan^{-1} x \le \dfrac{\pi}{2}$

$\Rightarrow \quad 0 \le \dfrac{\pi}{2} + \tan^{-1} x \le \pi$

$\therefore \quad \alpha = 0, \beta = \pi$

6 Let AB and CD be the pole and tower, respectively.

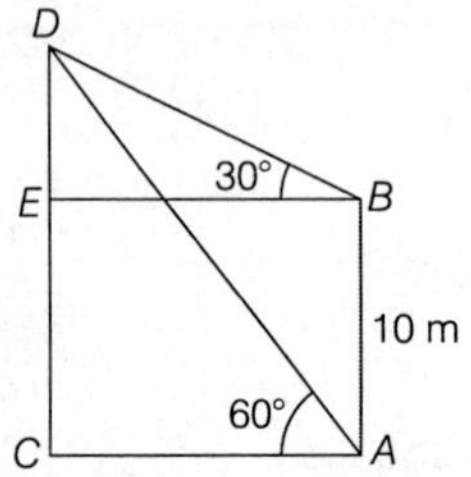

In ΔACD, $\tan 60° = \dfrac{CD}{AC}$

$\Rightarrow \quad AC = \dfrac{CD}{\sqrt{3}} \quad \ldots(i)$

In ΔDBE, $\tan 30° = \dfrac{DE}{BE} = \dfrac{DE}{CA}$

$\Rightarrow \quad \dfrac{1}{\sqrt{3}} = \dfrac{DE}{AC}$

$\Rightarrow \quad \dfrac{1}{\sqrt{3}} = \dfrac{DE}{CD/\sqrt{3}}$ [from Eq. (i)]

$\Rightarrow \quad \dfrac{CD}{DE} = 3 \Rightarrow \dfrac{DE + EC}{DE} = 3$

$\Rightarrow \quad DE = \dfrac{EC}{2} = \dfrac{10}{2} = 5$ m

$\therefore \quad CD = DE + EC = 10 + 5 = 15$ m

7 $12\sin\theta - 9\sin^2\theta$

$= -(3\sin\theta - 2)^2 + 4 \le 4$

Hence, maximum value is 4.

8 Since, $\sin\alpha + \sin\beta + \sin\gamma$

$= 4\cos\dfrac{\alpha}{2}\cos\dfrac{\beta}{2}\cos\dfrac{\gamma}{2}$

Also, since each of $\dfrac{\alpha}{2}, \dfrac{\beta}{2}$ and $\dfrac{\gamma}{2}$ is less than $\dfrac{\pi}{2}$. So, $\cos\dfrac{\alpha}{2}$, $\cos\dfrac{\beta}{2}$ and $\cos\dfrac{\gamma}{2}$ are all positive.

Hence, minimum value of $4\cos\dfrac{\alpha}{2}\cos\dfrac{\beta}{2}\cos\dfrac{\gamma}{2}$ is positive.

9 $\therefore \quad \cos^{-1}\sqrt{x} + \cos^{-1}\sqrt{1-x}$

$= \cos^{-1}[\sqrt{x}\sqrt{1-x} - \sqrt{1-x}\sqrt{x}]$

$= \cos^{-1}(0) = \dfrac{\pi}{2}$

$\therefore \quad \pi = \dfrac{\pi}{2} + \cos^{-1}\sqrt{1-y}$

$\Rightarrow \dfrac{\pi}{2} = \cos^{-1}\sqrt{1-y} \Rightarrow \sqrt{1-y} = 0$

$\therefore \quad y = 1$

10 Since, α and β are the roots of the equation $5\cos\theta + 4\sin\theta = 3$.

$\therefore \quad 5\cos\alpha + 4\sin\alpha = 3 \quad \ldots(i)$

and $\quad 5\cos\beta + 4\sin\beta = 3 \quad \ldots(ii)$

On subtracting Eq. (ii) from Eq. (i), we get

$5(\cos\alpha - \cos\beta) + 4(\sin\alpha - \sin\beta) = 0$

$\Rightarrow \quad -5 \times 2 \cdot \sin\dfrac{\alpha+\beta}{2} \cdot \sin\dfrac{\alpha-\beta}{2}$

$+ 4 \cdot 2 \cdot \cos\dfrac{\alpha+\beta}{2} \cdot \sin\dfrac{\alpha-\beta}{2} = 0$

$\Rightarrow \quad \left(4\cos\dfrac{\alpha+\beta}{2} - 5\sin\dfrac{\alpha+\beta}{2}\right) = 0$

$\Rightarrow \quad \tan\frac{\alpha+\beta}{2} = \frac{4}{5}$

$\therefore \quad \cos(\alpha+\beta) = \dfrac{1-\tan^2\left(\frac{\alpha+\beta}{2}\right)}{1+\tan^2\left(\frac{\alpha+\beta}{2}\right)}$

$= \dfrac{1-\frac{16}{25}}{1+\frac{16}{25}} = \dfrac{\frac{9}{25}}{\frac{41}{25}} = \dfrac{9}{41}$

11 Let $AR = x$ and the height of the chimney, $AB = h$.

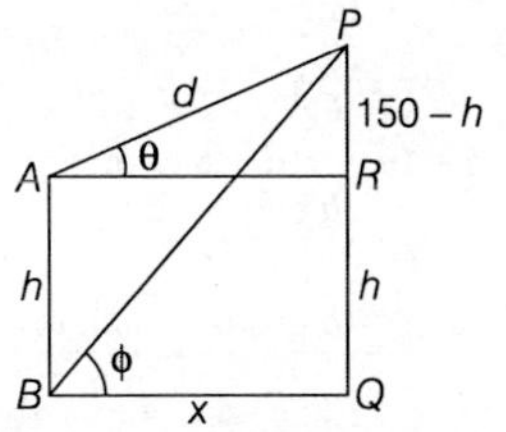

Now, $PR = PQ - RQ = 150 - h$

In ΔPAR, $\tan\theta = \dfrac{PR}{AR}$

$\Rightarrow \quad \dfrac{4}{3} = \dfrac{150-h}{x}$ ...(i)

and in ΔPBQ, $\tan\phi = \dfrac{PQ}{BQ}$

$\Rightarrow \quad \dfrac{5}{2} = \dfrac{150}{x}$ ...(ii)

$\therefore \quad \dfrac{150-h}{150} = \dfrac{4}{3}\times\dfrac{2}{5} = \dfrac{8}{15}$

[dividing Eq. (i) by Eq. (ii)]

$\Rightarrow 1 - \dfrac{h}{150} = \dfrac{8}{15} \Rightarrow h = 70\text{ m}$

From Eq. (i), $\dfrac{150-70}{x} = \dfrac{4}{3}$

$\Rightarrow x = 60$ m and

$PR = 150 - 70 = 80$ m

In ΔPAR, $AP^2 = AR^2 + PR^2$

$\Rightarrow \quad d^2 = 60^2 + 80^2$

$\therefore \quad d = 100$ m

12 Let the height of the hill be h and let A be its top.

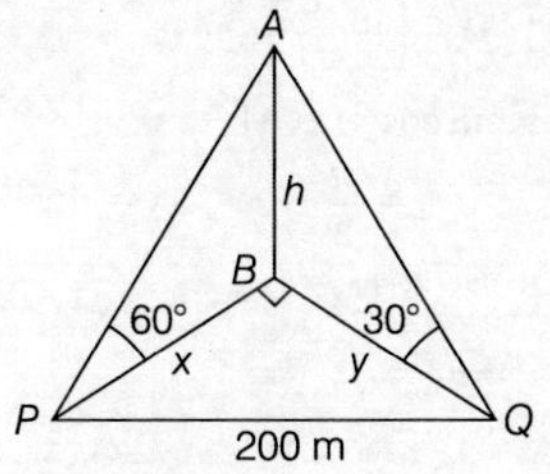

Since, BQ and BP represents South and West, respectively.

In ΔAPB, $\tan 60° = \dfrac{AB}{PB}$

$\Rightarrow \sqrt{3} = \dfrac{h}{x} \Rightarrow x = \dfrac{h}{\sqrt{3}}$

Again, in ΔAQB, $\tan 30° = \dfrac{AB}{BQ}$

$\Rightarrow \quad \dfrac{1}{\sqrt{3}} = \dfrac{h}{y} \Rightarrow y = h\sqrt{3}$

In right angled ΔPBQ,

$PQ^2 = PB^2 + BQ^2 = x^2 + y^2$

$\Rightarrow (200)^2 = \dfrac{h^2}{3} + 3h^2 = h^2\left(\dfrac{10}{3}\right)$

$\Rightarrow \quad h^2 = 200^2 \times \dfrac{3}{10}$

$\therefore \quad AB = 200\sqrt{\dfrac{3}{10}} = 109.54$ m

13 Now, $\dfrac{1}{\sqrt{2}}\left(\cos\dfrac{7\pi}{5} - \sin\dfrac{2\pi}{5}\right)$

$= \cos\dfrac{\pi}{4}\cos\dfrac{7\pi}{5} - \sin\dfrac{\pi}{4}\sin\dfrac{2\pi}{5}$

$= \cos\dfrac{\pi}{4}\cos\dfrac{7\pi}{5} + \sin\dfrac{\pi}{4}\sin\dfrac{7\pi}{5}$

$= \cos\left(\dfrac{7\pi}{5} - \dfrac{\pi}{4}\right) = \cos\left(\dfrac{23\pi}{20}\right)$

$= \cos\left(2\pi - \dfrac{17\pi}{20}\right) = \cos\left(\dfrac{17\pi}{20}\right)$

$\therefore \cos^{-1}\left\{\dfrac{1}{\sqrt{2}}\left(\cos\dfrac{7\pi}{5} - \sin\dfrac{2\pi}{5}\right)\right\} = \dfrac{17\pi}{20}$

14 Since, $\sec^2\theta \ge 1 \Rightarrow \dfrac{4ab}{(a+b)^2} \ge 1$ and

$a, b \ne 0$

$\Rightarrow \quad a, b \ne 0$ and $-\dfrac{(a-b)^2}{(a+b)^2} \ge 0$

$\Rightarrow \quad a, b \ne 0$ and $-(a-b)^2 \ge 0$

$\Rightarrow \quad a = b$ and $a \ne 0$

15 Given, $\quad x = \cos^{-1}(\cos 4)$

$\Rightarrow \; x = \cos^{-1}\cos(2\pi - 4)$

$\Rightarrow \; x = 2\pi - 4$ and $y = \sin^{-1}(\sin 3)$

$\Rightarrow \; y = \sin^{-1}\sin(\pi - 3) \Rightarrow y = \pi - 3$

$\therefore \quad x + y = 3\pi - 7$

$\therefore \quad \tan(x+y) = -\tan 7$

16 Since, $\log_2 x \ge 0 \Rightarrow x \ge 1$

For $x \ge 1$, we have

$\sin^{-1}\left(\dfrac{2x}{1+x^2}\right) = \pi - 2\tan^{-1}x$

$\therefore \quad \log_{1/\pi}\left\{\sin^{-1}\dfrac{2x}{1+x^2} + 2\tan^{-1}x\right\}$

$= \log_{1/\pi}\{\pi - 2\tan^{-1}x + 2\tan^{-1}x\}$

$= \log_{1/\pi}\pi = -1$

17 $\because \; 3\tan A - 4 = 0 \Rightarrow \tan A = 4/3$

$\Rightarrow \quad \cos A = -\dfrac{3}{5}$ and $\sin A = -\dfrac{4}{5}$

[since, A lies in III quadrant]

$\therefore \; \sin 2A = \dfrac{2\tan A}{1+\tan^2 A} = \dfrac{24}{25}$

$\therefore \; 5\sin 2A + 3\sin A + 4\cos A$

$= \dfrac{24}{5} - \dfrac{12}{5} - \dfrac{12}{5} = 0$

18 $27^{\cos x} + 81^{\sin x} = 3^{3\cos x} + 3^{4\sin x}$

$\ge 2\cdot\sqrt{3^{3\cos x}\cdot 3^{4\sin x}}$ [$\because$ AM$\ge$ GM]

$= 2\cdot 3^{(3\cos x + 4\sin x)/2} \ge 2\cdot 3^{\frac{1}{2}(-5)}$

[$\because -5 \le 3\cos x + 4\sin x \le 5$]

$= 2\cdot 3^{-\frac{5}{2}} = 2\cdot 3^{-2}\cdot 3^{-\frac{1}{2}}$

$= \dfrac{2}{9\sqrt{3}}$

19 Let $\sin x = t$

$\therefore \quad t^2 + at + 1 = 0 \Rightarrow t + \dfrac{1}{t} = -a$

$\Rightarrow \quad |a| = \left|t + \dfrac{1}{t}\right| \ge 2$

$\Rightarrow \quad |a| \ge 2$ [$\because$ AM$\ge$ GM]

Hence, for no real solution $|a| < 2$.

20 Given, $\quad \dfrac{1}{n} = \dfrac{\sin(\theta+2\alpha)}{\sin\theta}$

On applying componendo and dividendo, we get

$\Rightarrow \quad \dfrac{1+n}{1-n} = \dfrac{\sin(\theta+2\alpha)+\sin\theta}{\sin(\theta+2\alpha)-\sin\theta}$

$\Rightarrow \quad \dfrac{1+n}{1-n} = \dfrac{2\sin(\theta+\alpha)\cos\alpha}{2\cos(\theta+\alpha)\sin\alpha}$

$= \tan(\theta+\alpha)\cot\alpha$

$\Rightarrow \quad \dfrac{1+n}{1-n}\tan\alpha = \tan(\theta+\alpha)$

21 $(\sin^2 x)^3 + (\cos^2 x)^3$

$= (\sin^2 x + \cos^2 x)^3 - 3\sin^2 x\cos^2 x\,(\sin^2 x + \cos^2 x)$

$= 1 - 3\sin^2 x\cos^2 x = 1 - \dfrac{3}{4}(\sin 2x)^2$

$\therefore$ Maximum value $= 1 - \dfrac{3}{4}\times 0 = 1$

and minimum value $= 1 - \dfrac{3}{4}\times 1 = \dfrac{1}{4}$

22 Now, $\tan\alpha\cdot\tan(2n-1)\alpha$

$= \tan\alpha\,\tan\left(\dfrac{\pi}{2\alpha} - 1\right)\alpha = \tan\alpha\,\cot\alpha = 1$

Hence, the value of given expression is 1.

23 $2 - \cos x + \sin^2 x = 2 - \cos x + 1 - \cos^2 x$

$= 3 - (\cos^2 x + \cos x)$

$= 3 - \left[\left(\cos x + \dfrac{1}{2}\right)^2 - \dfrac{1}{4}\right]$

Hence, the maximum value occurs at $\cos x = -\dfrac{1}{2}$ and its value

$$= 2 - \left(-\frac{1}{2}\right) + \left(1 - \frac{1}{4}\right) = \frac{13}{4}$$

and minimum value occurs at $\cos x = 1$ and its value $= 2 - 1 + (1 - 1) = 1$.

$\therefore$ Required ratio $= \dfrac{13}{4}$

24 Given, $\dfrac{m}{a} = \cos^3\theta + 3\cos\theta\sin^2\theta$

and $\dfrac{n}{a} = \sin^3\theta + 3\cos^2\theta\sin\theta$

$$\therefore \left(\frac{m}{a} + \frac{n}{a}\right) = (\sin\theta + \cos\theta)^3$$

$$\Rightarrow \left(\frac{m+n}{a}\right)^{1/3} = \sin\theta + \cos\theta$$

$$\text{and } \left(\frac{m-n}{a}\right)^{1/3} = \cos\theta - \sin\theta$$

$$\therefore \left(\frac{m+n}{a}\right)^{2/3} + \left(\frac{m-n}{a}\right)^{2/3} = 2(\sin^2\theta + \cos^2\theta)$$

$$(m+n)^{2/3} + (m-n)^{2/3} = 2a^{2/3}$$

25 Given, $a\sin^2\alpha - \dfrac{1}{a}\operatorname{cosec}^2\alpha = 0$

$$\Rightarrow \sin^2\alpha = \frac{1}{a}$$

$$\therefore \cos^2\alpha + 5\sin\alpha\cos\alpha + 6\sin^2\alpha$$

$$= 1 - \frac{1}{a} + \frac{5}{\sqrt{a}}\sqrt{1 - \frac{1}{a}} + \frac{6}{a}$$

$$= 1 + \frac{5\sqrt{a-1} - 1}{a} + \frac{6}{a}$$

$$= \frac{a + 5\sqrt{a-1} + 5}{a}$$

26 $$\tan\frac{x}{2} = \frac{1 + \tan^2\frac{x}{2}}{2\tan\frac{x}{2}} - \frac{2\tan\frac{x}{2}}{1 + \tan^2\frac{x}{2}}$$

$$\Rightarrow 2\tan^2\frac{x}{2}\left(1 + \tan^2\frac{x}{2}\right) = \left(1 + \tan^2\frac{x}{2}\right)^2 - 4\tan^2\frac{x}{2}$$

$$\Rightarrow 2y(1+y) = (1+y)^2 - 4y \quad \left[\text{put } \tan^2\frac{x}{2} = y\right]$$

$$\Rightarrow y^2 + 4y - 1 = 0$$

$$\therefore y = \frac{-4 \pm \sqrt{16+4}}{2} = -2 \pm \sqrt{5}$$

Since, $y \geq 0$, we get

$$y = \sqrt{5} - 2 = \frac{(\sqrt{5}-2)^2}{\sqrt{5}-2} \cdot \frac{2+\sqrt{5}}{2+\sqrt{5}}$$

$$= (9 - 4\sqrt{5})(2 + \sqrt{5})$$

27 Now, $\sqrt{2 + \sqrt{2 + \sqrt{2 + \ldots + \sqrt{2(1+\cos\theta)}}}}$

$$= \sqrt{2 + \sqrt{2 + \ldots + \sqrt{2 + 2\cos\frac{\theta}{2}}}}$$

....

....

$$= \sqrt{2 + 2\cos\left(\frac{\theta}{2^{n-1}}\right)}$$

$$= \sqrt{2\left(1 + \cos\frac{\theta}{2^{n-1}}\right)}$$

$$= \sqrt{2 \cdot 2\cos^2\left(\frac{\theta}{2 \cdot 2^{n-1}}\right)} = 2\cos\left(\frac{\theta}{2^n}\right)$$

28 In ΔABC, $A + B + C = \pi$

$$\therefore \tan A + \tan B + \tan C = \tan A \tan B \tan C$$

$$\Rightarrow 6 = 2\tan A \Rightarrow \tan A = 3$$

$$\therefore \tan B + \tan C = 3$$

$$\text{and } \tan B \tan C = 2$$

$$\Rightarrow \tan B = 1 \text{ or } 2 \text{ and } \tan C = 2 \text{ or } 1$$

Now, $\sin^2 A = \dfrac{\tan^2 A}{1 + \tan^2 A} = \dfrac{9}{10}$

$$\sin^2 B = \frac{\tan^2 B}{1 + \tan^2 B} = \frac{1}{1+1}, \frac{4}{1+4}$$

$$= \frac{1}{2}, \frac{4}{5} = \frac{5}{10}, \frac{8}{10}$$

and $\sin^2 C = \dfrac{\tan^2 C}{1 + \tan^2 C} = \dfrac{8}{10}, \dfrac{5}{10}$

$\therefore \sin^2 A : \sin^2 B : \sin^2 C$

$$= \frac{9}{10} : \frac{5}{10} : \frac{8}{10} \text{ or } \frac{9}{10} : \frac{8}{10} : \frac{5}{10}$$

29 Maximum value of $\sin\theta$ and $\cos\theta$ is 1.

$$\because \cos x \cdot \sin y = 1$$

$$\Rightarrow \cos x = 1, \sin y = 1$$

$$\text{or} \quad \cos x = -1, \sin y = -1$$

$$\Rightarrow x = 0, 2\pi, y = \frac{\pi}{2}, \frac{5\pi}{2}$$

$$\text{or} \quad x = \pi, 3\pi, y = \frac{3\pi}{2}$$

$\therefore$ Required number of ordered pair $= 2 \times 2 + 2 \times 1 = 6$

30 Given, $(1 - \cos^2 6x) + \sin^4 2x = 0$

$$\Rightarrow \sin^2 6x + \sin^4 2x = 0$$

$$\Rightarrow \sin^2 6x = 0 \text{ and } \sin^4 2x = 0$$

$$\Rightarrow 6x = n\pi \text{ and } 2x = n\pi$$

$$\Rightarrow x = \frac{n\pi}{6}$$

$$\text{and} \quad x = \frac{n\pi}{2}$$

31 Since, $x^2 + x^{-2} \geq 2$ $\quad[\because \text{AM} \geq \text{GM}]$

therefore the equation is valid only if

$$2\cos^2\frac{x}{2}\sin^2 x = 2$$

$$\Leftrightarrow \cos\frac{x}{2} = \operatorname{cosec} x$$

i.e. iff $\operatorname{cosec} x = \cos\dfrac{x}{2} = 1$

which cannot be true.

32 Consider $A_1, A_2, \ldots, A_n$ be the foot of the n poles subtending angle α to O, such that $OA_1 = a$, if d be the distance between two consecutive poles

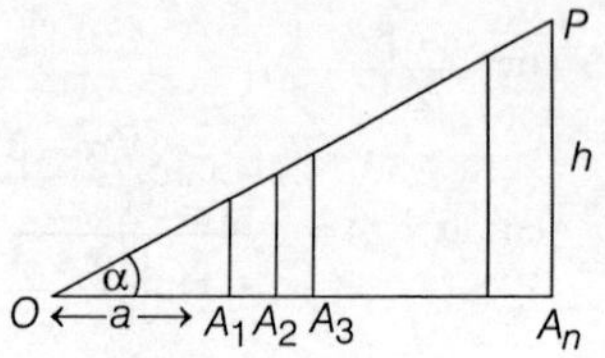

$$OA_2 = a + d$$

$$OA_3 = a + 2d$$

$$\vdots \quad \vdots \quad \vdots$$

$$OA_n = a + (n-1)d$$

Now, in ΔPOA_n,

$$\tan\alpha = \frac{h}{OA_n}$$

$$OA_n = h\cot\alpha$$

$$\Rightarrow a + (n-1)d = h\cot\alpha$$

$$d = \frac{h\cot\alpha - a}{n-1}$$

$$= \frac{h\cos\alpha - a\sin\alpha}{(n-1)\sin\alpha}$$

33 $$\tan^{-1}\left(\frac{(x+1)+(x-1)}{1-(x+1)(x-1)}\right) = \tan^{-1}\left(\frac{8}{31}\right)$$

Provided $(x+1)(x-1) < 0$

i.e. $\quad x^2 < 1 \quad \ldots$(i)

$$\Rightarrow \tan^{-1}\frac{2x}{1-(x^2-1)} = \tan^{-1}\left(\frac{8}{31}\right)$$

$$\Rightarrow \frac{2x}{2-x^2} = \frac{8}{31}$$

$$\Rightarrow 4x^2 + 31x - 8 = 0$$

$$\Rightarrow (4x-1)(x+8) = 0$$

$$\Rightarrow x = \frac{1}{4} \text{ or } -8 \Rightarrow x = \frac{1}{4}$$

Since, $x = -8$ is not satisfied the Eq. (i).

34 Obviously $x > 0$ and $x\sqrt{3} < 1$

i.e. $x < \dfrac{1}{\sqrt{3}}$

If $x > \dfrac{1}{\sqrt{3}}$, then $\cos^{-1}(x\sqrt{3})$ will be undefined. If $x < 0$, then $x\sqrt{3} < 0$. Hence,

$\cos^{-1} x > \dfrac{\pi}{2}$ and $\cos^{-1} x\sqrt{3} > \dfrac{\pi}{2}$

which is not satisfied the equation.

$$\therefore x \in \left(0, \frac{1}{\sqrt{3}}\right)$$

Given, $\cos^{-1}(x\sqrt{3})$

$$= \frac{\pi}{2} - \cos^{-1} x = \sin^{-1} x$$

$$\Rightarrow \cos^{-1}(x\sqrt{3}) = \cos^{-1}\sqrt{1-x^2}$$

$$\Rightarrow x\sqrt{3} = \sqrt{1-x^2}$$

$$\Rightarrow 3x^2 = 1 - x^2 \Rightarrow x = \pm\frac{1}{2}$$

$$x = \frac{1}{2} \quad \left[\because x = -\frac{1}{2} \notin \left(0, \frac{1}{\sqrt{3}}\right)\right]$$

35 $\because \dfrac{1+\sin x}{\cos x} = 2\cos x, \cos x \neq 0$

$\therefore \quad 1 + \sin x = 2(1 - \sin^2 x)$

$\Rightarrow 1 + \sin x = 2(1 + \sin x)(1 - \sin x)$

$\Rightarrow (1 + \sin x)[1 - 2(1 - \sin x)] = 0$

$\Rightarrow \sin x = \dfrac{1}{2}, -1$

$\therefore \quad x = \dfrac{\pi}{6}, \pi - \dfrac{\pi}{6}, \dfrac{3\pi}{2}$

$x = \dfrac{\pi}{6}, \dfrac{5\pi}{6}, x \neq \dfrac{3\pi}{2}$ $[\because \cos x \neq 0]$

36 Given, $\cot^2 x + \operatorname{cosec} x - a = 0$

$\Rightarrow \operatorname{cosec}^2 x + \operatorname{cosec} x - 1 - a = 0$

$\Rightarrow \left(\operatorname{cosec} x + \dfrac{1}{2}\right)^2 = 1 + a + \dfrac{1}{4} = \dfrac{5}{4} + a$

$\therefore \quad \operatorname{cosec} x \geq 1 \text{ or} \leq -1$

$\Rightarrow \quad \operatorname{cosec} x + \dfrac{1}{2} \geq \dfrac{3}{2} \text{ or} \leq -\dfrac{1}{2}$

$\Rightarrow \left(\operatorname{cosec} x + \dfrac{1}{2}\right)^2 \geq \dfrac{1}{4} \Rightarrow \dfrac{5}{4} + a \geq \dfrac{1}{4}$

$\therefore \quad a \geq -1$

37 Let h be the height of the tower QR.

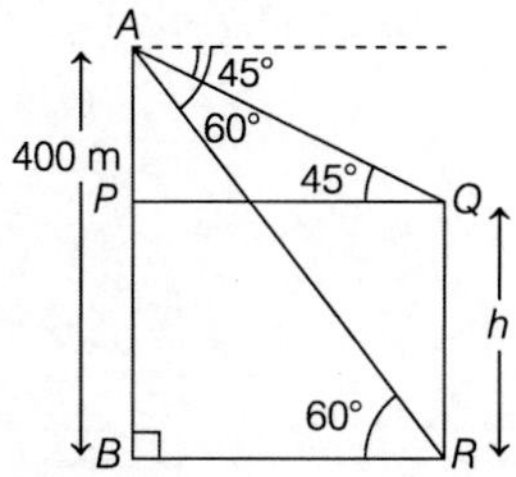

Then, $\quad PA = 400 - h$

In ΔAPQ, $\dfrac{AP}{PQ} = 1 \Rightarrow AP = PQ$

$\Rightarrow \quad 400 - h = PQ$

Again in ΔABR, $\tan 60° = \dfrac{400}{BR}$

$[\because BR = PQ]$

$\therefore \quad \sqrt{3} = \dfrac{400}{400 - h}$

$\Rightarrow 400\sqrt{3} - h\sqrt{3} = 400$

$\Rightarrow (400\sqrt{3} - 400) = h\sqrt{3}$

$\Rightarrow \dfrac{400(\sqrt{3} - 1)}{\sqrt{3}} = h$

$= \dfrac{400(3 - \sqrt{3})}{3}$ m

$= 169.06$ m

38 $\cos 5\theta = \cos(3\theta + 2\theta)$

$= \cos 3\theta \cdot \cos 2\theta - \sin 3\theta \cdot \sin 2\theta$

$= (4\cos^3\theta - 3\cos\theta)(2\cos^2\theta - 1)$
$\quad - (3\sin\theta - 4\sin^3\theta) \times (2\sin\theta \cdot \cos\theta)$

$= 8\cos^5\theta - 10\cos^3\theta + 3\cos\theta$
$\quad - 2\sin^2\theta(3 - 4\sin^2\theta) \cdot \cos\theta$

$= 8\cos^5\theta - 10\cos^3\theta + 3\cos\theta$
$\quad - 2\cos\theta(4\cos^2\theta - 1)(1 - \cos^2\theta)$

$= 8\cos^5\theta - 10\cos^3\theta + 3\cos\theta$
$\quad - 2\cos\theta[4\cos^2\theta - 4\cos^4\theta - 1 + \cos^2\theta]$

$= 16\cos^5\theta - 20\cos^3\theta + 5\cos\theta$

On comparing the coefficient of $\cos^5\theta$, $\cos\theta$ and constant term, we get $P = 5, Q = 16$ and $R = 0$

$\therefore \quad P + Q + R = 21$

39 $\because \dfrac{\cos A}{a} = \dfrac{\cos B}{b} = \dfrac{\cos C}{c}$

$\therefore \quad \dfrac{\cos A}{k \sin A} = \dfrac{\cos B}{k \sin B} = \dfrac{\cos C}{k \sin C}$

$\Rightarrow \quad \cot A = \cot B = \cot C$

$\Rightarrow \quad A = B = C = 60°$

40 $a^2 \sin 2B + b^2 \sin 2A$

$= 2a^2 \sin B \cdot \cos B + 2b^2 \sin A \cdot \cos A$

$= \dfrac{a^2 b}{R} \cos B + \dfrac{b^2 a}{R} \cos A$

$= \dfrac{ab}{R}(a\cos B + b\cos A) = \dfrac{abc}{R}$

$= 2bc \sin A = 4\left(\dfrac{1}{2} bc \sin A\right)$

$= 4\lambda$

41 Given, $\cos 2x + 2\cos x = 1$

$\Rightarrow \quad 2\cos^2 x - 1 + 2\cos x - 1 = 0$

$\Rightarrow \quad \cos^2 x + \cos x - 1 = 0$

$\Rightarrow \quad \cos x = \dfrac{-1 + \sqrt{5}}{2}$

$\left[\text{neglecting } \dfrac{-1 - \sqrt{5}}{2}, \text{As } -1 \leq \cos x \leq 1 \text{ and } \left(\dfrac{-1 - \sqrt{5}}{2}\right) < -1\right]$

$\therefore \cos^2 x = \left(\dfrac{\sqrt{5} - 1}{2}\right)^2$

$= \dfrac{6 - 2\sqrt{5}}{4} = \dfrac{3 - \sqrt{5}}{2}$

$\therefore \quad \sin^2 x(2 - \cos^2 x)$

$= \left(1 - \dfrac{3 - \sqrt{5}}{2}\right)\left(2 - \dfrac{3 - \sqrt{5}}{2}\right)$

$= \left(\dfrac{\sqrt{5} - 1}{2}\right)\left(\dfrac{\sqrt{5} + 1}{2}\right) = 1$

42 From the given parts of question, we get

$\cos x + \sin x = A - 1 = B + 1$

$\Rightarrow \quad A = B + 2 \quad \ldots \text{(i)}$

and $\quad A \cdot B = (\sin x + \cos x + 1)(\sin x + \cos x - 1)$

$= (\sin x + \cos x)^2 - 1$

$1 + \sin 2x - 1 = \sin 2x$

$\Rightarrow \quad (B + 2) \cdot B = \sin 2x \quad$ [from Eq. (i)]

$\Rightarrow \quad B^2 + 2B - \sin 2x = 0$

$\Rightarrow \quad (A - 2)^2 + 2(A - 2) - \sin 2x = 0$

$\Rightarrow \quad A^2 - 2A - \sin 2x = 0$

43 Given equation is

$x^3 - 15x^2 + 47x - 82 = 0$

$\Sigma a = 15$

$\Sigma ab = 47$

$abc = 82$

Now, $\dfrac{\cos A}{a} + \dfrac{\cos B}{b} + \dfrac{\cos C}{c}$

$\dfrac{b^2 + c^2 - a^2}{2abc} + \dfrac{c^2 + a^2 - b^2}{2abc} + \dfrac{a^2 + b^2 - c^2}{2abc}$

(by cosine rule)

$= \dfrac{a^2 + b^2 + c^2}{2abc}$

$= \dfrac{(\Sigma a)^2 - 2\Sigma ab}{2abc}$

$= \dfrac{225 - 94}{2 \cdot 82} = \dfrac{131}{164}$

44 Given, $a^4 + b^4 + c^4 - 2b^2c^2 - 2c^2a^2 = 0$

$\Rightarrow \quad (a^2 + b^2 - c^2)^2 = 2a^2b^2$

$\Rightarrow \quad \dfrac{(a^2 + b^2 - c^2)^2}{4a^2b^2} = \dfrac{1}{2}$

$\Rightarrow \quad \dfrac{a^2 + b^2 - c^2}{2ab} = \pm \dfrac{1}{\sqrt{2}}$

$\Rightarrow \quad \cos C = \pm \dfrac{1}{\sqrt{2}}$

$\Rightarrow \quad C = \dfrac{\pi}{4} \text{ or } \dfrac{3\pi}{4}$

So, the angle is 45° or 135°.

45 $f(\theta, \alpha) = 2\sin^2\theta + 4\cos(\theta + \alpha)$
$\quad \sin\theta \sin\alpha + 2\cos^2(\theta + \alpha) - 1$

$= 2\sin^2\theta + 2\cos(\theta + \alpha)$
$\quad [2\sin\theta \sin\alpha + \cos(\theta + \alpha)] - 1$

$= 2\sin^2\theta + 2\cos(\theta + \alpha)$
$\quad [\sin\theta \sin\alpha + \cos\theta \cos\alpha] - 1$

$= 2\sin^2\theta + 2\cos(\theta + \alpha)\cos(\theta - \alpha) - 1$

$= 2\sin^2\theta + 2\cos^2\theta - 2\sin^2\alpha - 1$

$= 1 - 2\sin^2\alpha = \cos 2\alpha$

$\therefore \quad f\left(\dfrac{\pi}{3}, \dfrac{\pi}{4}\right) = \cos\left(2 \times \dfrac{\pi}{4}\right) = 0$

46 I. The general value of θ satisfying any of the equations $\sin^2\theta = \sin^2\alpha, \cos^2\theta = \cos^2\alpha$ and $\tan^2\theta = \tan^2\alpha$ is given by $\theta = n\pi \pm \alpha$.

II. The general value of θ satisfying equations $\sin\theta = \sin\alpha$ and $\cos\theta = \cos\alpha$ simultaneously is given by $\theta = 2n\pi + \alpha, n \in Z$.

So, Statement I is correct and Statement II is incorrect.

47 **Statement I** Put $x = \cos\theta$, then $0 \le \theta \le \frac{\pi}{3}$

$$\text{LHS} = \cos^{-1}(\cos\theta) - \sin^{-1}\left[\frac{1}{2}\cos\theta + \frac{\sqrt{3}}{2}\sin\theta\right]$$

$$= \theta - \sin^{-1}\left[\sin\left(\theta + \frac{\pi}{3}\right)\right]$$

$$= \theta - \theta - \frac{\pi}{3} = -\frac{\pi}{3}$$

Statement II Put $x = \sin\theta$,

then $-\frac{\pi}{4} \le \theta \le \frac{\pi}{4}$

$$\text{LHS} = \sin^{-1}(2\sin\theta\cos\theta)$$
$$= \sin^{-1}(\sin 2\theta)$$
$$= 2\theta = 2\sin^{-1} x$$

48 Given, $2\sin^2\theta - \cos 2\theta = 0$

$\Rightarrow 4\sin^2\theta = 1 \Rightarrow \sin\theta = \pm\left(\frac{1}{2}\right)$

So, $\sin\theta = \frac{1}{2}$

[$\because \sin\theta = -\frac{1}{2}$ does not satisfy the second equation]

$\therefore \theta = \frac{\pi}{6}$ or $\frac{5\pi}{6}$

which also satisfy $2\cos^2\theta - 3\sin\theta = 0$.
Hence, the number of solutions are two.

49 Given, $2\sin\left(\frac{\theta}{2}\right) = \sqrt{\left[\cos\left(\frac{\theta}{2}\right) + \sin\left(\frac{\theta}{2}\right)\right]^2} + \sqrt{\left[\cos\left(\frac{\theta}{2}\right) - \sin\left(\frac{\theta}{2}\right)\right]^2}$

$$= \left|\cos\left(\frac{\theta}{2}\right) + \sin\left(\frac{\theta}{2}\right)\right| + \left|\cos\left(\frac{\theta}{2}\right) - \sin\left(\frac{\theta}{2}\right)\right|$$

$\Rightarrow \cos\left(\frac{\theta}{2}\right) + \sin\left(\frac{\theta}{2}\right) > 0$

$\Rightarrow \sin\left(\frac{\theta}{2} + \frac{\pi}{4}\right) > 0$

and $\cos\left(\frac{\theta}{2}\right) - \sin\left(\frac{\theta}{2}\right) < 0$

and $\cos\left(\frac{\theta}{2} + \frac{\pi}{4}\right) < 0$

$\Rightarrow 2n\pi + \frac{\pi}{2} < \frac{\theta}{2} + \frac{\pi}{4} < 2n\pi + \pi$

$\therefore 2n\pi + \frac{\pi}{4} < \frac{\theta}{2} < 2n\pi + \frac{3\pi}{4}$

50 **Statement I** $\angle APB = 2\alpha - \alpha = \alpha$

and $\angle BPC = 3\alpha - 2\alpha = \alpha$

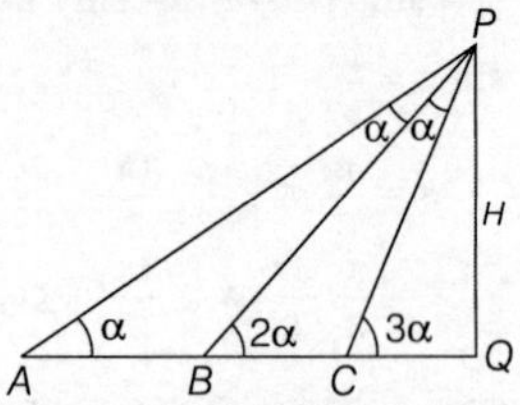

Hence, PB is an angle bisector of $\angle APC$.

Then, $\frac{AB}{BC} = \frac{AP}{CP}$

$$= \frac{H \operatorname{cosec}\alpha}{H \operatorname{cosec} 3\alpha}$$

$$= \frac{\sin 3\alpha}{\sin\alpha}$$

Statement II But Statement II is not always true.

DAY TWENTY FOUR

Cartesian System of Rectangular Coordinates

Learning & Revision for the Day

- Rectangular Coordinates
- Distance Formula
- Section Formulae
- Area of a Triangle
- Area of Some Geometric Figures
- Coordinates of Different Points of a Triangle
- Translation of Axes
- Slope of a Line
- Locus and its Equation

Rectangular Coordinates

Let XOX' and YOY' be two perpendicular axes in the plane intersecting at O (as shown in the figure). Let P be any point in the plane. Draw PM perpendicular to OX.

The ordered pair (x, y) is called the rectangular or cartesian **coordinates of point** P.

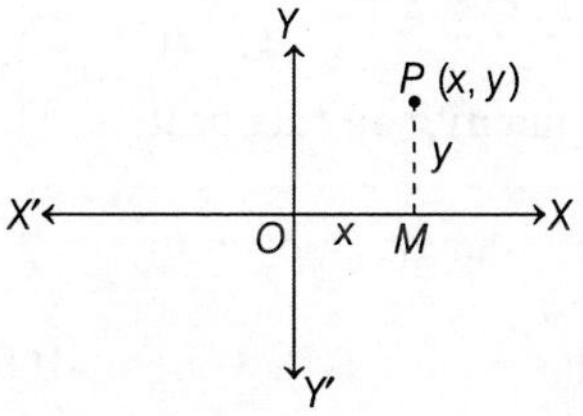

Distance Formula

Let $P(x_1, y_1)$ and $Q(x_2, y_2)$ be the two points.

Then, $PQ = d = \sqrt{(x_2 - x_1)^2 + (y_2 - y_1)^2}$ or $\sqrt{(x_1 - x_2)^2 + (y_1 - y_2)^2}$

Distance between the points $(0, 0)$ and (x, y) is $\sqrt{x^2 + y^2}$.

Section Formulae

The coordinates of a point which divide the line segment joining two points $P(x_1, y_1)$ and $Q(x_2, y_2)$ in the ratio $m_1 : m_2$ are

(i) $\left(\dfrac{m_1 x_2 + m_2 x_1}{m_1 + m_2}, \dfrac{m_1 y_2 + m_2 y_1}{m_1 + m_2}\right)$ [internal division]

(ii) $\left(\dfrac{m_1 x_2 - m_2 x_1}{m_1 - m_2}, \dfrac{m_1 y_2 - m_2 y_1}{m_1 - m_2}\right)$ [external division]

When m_1 and m_2 are of opposite signs, then division is **external**.

PREP MIRROR
Your Personal Preparation Indicator

- No. of Questions in Exercises (x)—
- No. of Questions Attempted (y)—
- No. of Correct Questions (z)—
  (Without referring Explanations)

- Accuracy Level ($z/y \times 100$)—
- Prep Level ($z/x \times 100$)—

In order to expect good rank in JEE, your Accuracy Level should be above 85 & Prep Level should be above 75.

- Mid-point of the line joining (x_1, y_1) and (x_2, y_2) is $\left(\frac{x_1 + x_2}{2}, \frac{y_1 + y_2}{2}\right)$.
- Coordinates of any point on one line segment which divide the line segment joining two points $P(x_1, y_1)$ and $Q(x_2, y_2)$ in the ratio $\lambda : 1$ are given by

$$\left(\frac{x_1 + \lambda x_2}{\lambda + 1}, \frac{y_1 + \lambda y_2}{\lambda + 1}\right), (\lambda \neq -1)$$

- X-axis and Y-axis divide the line segment joining the points (x_1, y_1) and (x_2, y_2) in the ratio of $-\frac{y_1}{y_2}$ and $-\frac{x_1}{x_2}$ respectively.

If the ratio is positive, then the axis divides it internally and if ratio is negative, then the axis divides externally.

Area of a Triangle

Area of a triangle whose three vertices has coordinates $(x_1, y_1), (x_2, y_2)$ and (x_3, y_3) as shown in the figure below is given by

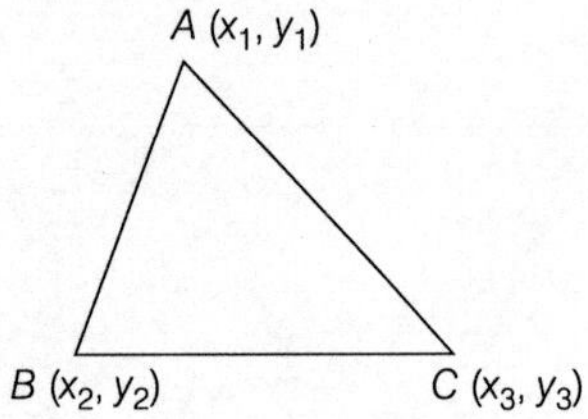

$$\text{Area of a } \Delta ABC = \frac{1}{2}\left|\begin{matrix}(x_1 y_2 + x_2 y_3 + x_3 y_1) \\ -(y_1 x_2 + y_2 x_3 + y_3 x_1)\end{matrix}\right|$$

$$= \frac{1}{2}|x_1(y_2 - y_3) + x_2(y_3 - y_1) + x_3(y_1 - y_2)| = \frac{1}{2}\left|\begin{vmatrix} x_1 & y_1 & 1 \\ x_2 & y_2 & 1 \\ x_3 & y_3 & 1 \end{vmatrix}\right|$$

It should be noted that area is a positive quantity and its unit is square of unit of length.

In the inverse problems, i.e. when area of a triangle is given to be a square units, then we have

$$\Delta ABC = \frac{1}{2}\left|\begin{vmatrix} x_1 & y_1 & 1 \\ x_2 & y_2 & 1 \\ x_3 & y_3 & 1 \end{vmatrix}\right| = a \Rightarrow \frac{1}{2}\begin{vmatrix} x_1 & y_1 & 1 \\ x_2 & y_2 & 1 \\ x_3 & y_3 & 1 \end{vmatrix} = \pm a$$

NOTE If area of ΔABC is zero. It mean points are collinear.

Area of Some Geometrical Figures

(i) Suppose a and b are the adjacent sides of a parallelogram and θ be the angle between them as shown in the figure below, then area of parallelogram $ABCD = ab \sin \theta$.

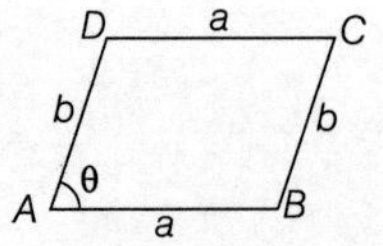

(ii) Area of convex quadrilateral with vertices $(x_1, y_1), (x_2, y_2), (x_3, y_3), (x_4, y_4)$ in that order is

$$\frac{1}{2}\begin{vmatrix} x_1 - x_3 & y_1 - y_3 \\ x_2 - x_4 & y_2 - y_4 \end{vmatrix}.$$

(iii) A triangle having vertices $(at_1^2, 2at_1), (at_2^2, 2at_2)$ and $(at_3^2, 2at_3)$, then area of triangle $= a^2[(t_1 - t_2)(t_2 - t_3)(t_3 - t_1)]$

(iv) Area of triangle formed by coordinate axes and the lines $ax + by + c$ is $= \frac{c^2}{2ab}$.

Coordinates of Different Points of a Triangle

1. Centroid

The centroid of a triangle is the point of intersection of its medians. It divides the medians in the ratio 2 : 1. If $A(x_1, y_1)$, $B(x_2, y_2)$ and $C(x_3, y_3)$ are the vertices of ΔABC, then the coordinates of its centroid G are $\left(\frac{x_1 + x_2 + x_3}{3}, \frac{y_1 + y_2 + y_3}{3}\right)$

2. Orthocentre

The orthocentre of a triangle is the point of intersection of its altitudes. If $A(x_1, y_1)$, $B(x_2, y_2)$ and $C(x_3, y_3)$ are the vertices of a ΔABC, then the coordinates of its orthocentre O are

$$\left(\frac{x_1 \tan A + x_2 \tan B + x_3 \tan C}{\tan A + \tan B + \tan C}, \frac{y_1 \tan A + y_2 \tan B + y_3 \tan C}{\tan A + \tan B + \tan C}\right)$$

3. Circumcentre

The circumcentre of a triangle is the point of intersection of the perpendicular bisectors of its sides. It is the centre of the circle passing through the vertices of a triangle and so it is equidistant from the vertices of the triangle.

Here, $OA = OB = OC$, where O is the centre of circle and A, B and C are the vertices of a triangle. The coordinates of the circumcentre are also given by

$$S\left(\frac{x_1 \sin 2A + x_2 \sin 2B + x_3 \sin 2C}{\sin 2A + \sin 2B + \sin 2C}, \frac{y_1 \sin 2A + y_2 \sin 2B + y_3 \sin 2C}{\sin 2A + \sin 2B + \sin 2C}\right)$$

Incentre

The point of intersection of the internal bisectors of the angles of a triangle is called its incentre.

If $A(x_1, y_1)$, $B(x_2, y_2)$ and $C(x_3, y_3)$ are the vertices of a ΔABC such that $BC = a, CA = b$ and $AB = c$, then the coordinates of the incentre are $I\left(\frac{ax_1 + bx_2 + cx_3}{a + b + c}, \frac{ay_1 + by_2 + cy_3}{a + b + c}\right)$.

Excentre

Coordinate of excentre opposite of $\angle A$ is given by

$$I_1 \equiv \left(\frac{-ax_1 + bx_2 + cx_3}{-a + b + c}, \frac{-ay_1 + by_2 + cy_3}{-a + b + c}\right) \text{ and similarly for}$$

excentres (I_2 and I_3) opposite to $\angle B$ and $\angle C$ are given by

$$I_2 \equiv \left(\frac{ax_1 - bx_2 + cx_3}{a - b + c}, \frac{ay_1 - by_2 + cy_3}{a - b + c}\right)$$

and $I_3 \equiv \left(\frac{ax_1 + bx_2 - cx_3}{a+b-c}, \frac{ay_1 + by_2 - cy_3}{a+b-c}\right)$.

In an equilateral triangle, orthocentre, centroid, circumcentre, incentre, coincide.

Important Results

- Circumcentre of the right angled ΔABC, right angled at A is $\frac{B+C}{2}$.
- Orthocentre of the right angled ΔABC, right angled at A is A.
- Orthocentre, centroid, circumcentre of a triangle are collinear.
- Centroid divides the line joining the orthocentre and circumcentre in the ratio 2 : 1.
- The circumcentre of right angled triangle is the mid-point of the hypotenuse.
- A triangle is isosceles, if any two of its medians are equal.

Translation of Axes

1. To Alter the Origin of Coordinates Without Altering the Direction of the Axes

Let origin $O(0,0)$ be shifted to a point (a, b) by moving the X and Y-axes parallel to themselves. If the coordinates of point P with reference to old axis are (x_1, y_1), then coordinates of this point with respect to new axis will be $(x_1 - a, y_1 - b)$.

2. To Change the Direction of the Axes of Coordinates without Changing Origin

Let OX and OY be the old axes and OX' and OY' be the new axes obtained by rotating the old OX and OY through an angle θ, then the coordinates of $P(x, y)$ with respect to new coordinate axes will be given by

	$x\downarrow$	$y\downarrow$
$x' \rightarrow$	$\cos\theta$	$\sin\theta$
$y' \rightarrow$	$-\sin\theta$	$\cos\theta$

(i) x and y are old coordinates, x', y' are new coordinates.

(ii) The axes rotation in anti-clockwise is positive and clockwise rotation of axes is negative.

3. To Change the Direction of the Axes of Coordinates by Changing the Origin

If $P(x, y)$ and the axes are shifted parallel to the original axis, so that new origin is (α, β) and then the axes are rotated about the new origin (α, β) by angle ϕ in the anti-clockwise (x', y'), then the coordinates of P will be given by

$$x = \alpha + x'\cos\phi - y'\sin\phi$$
$$y = \beta + x'\sin\phi + y'\cos\phi$$

Slope of a Line

The tangent of the angle that a line makes with the positive direction of the X-axis is called the **slope** or **gradient** of the line. The slope of a line is generally denoted by m.

Thus, $m = \tan\theta$.

Slope of a Line in Terms of Coordinates of any Two Points on it

Let $P(x_1, y_1)$ and $Q(x_2, y_2)$ be two points on a line making an angle θ with the positive direction of X-axis. Then, its slope m is given by

$$m = \frac{y_2 - y_1}{x_2 - x_1} = \frac{\text{Difference of ordinates}}{\text{Difference of abscissa}}$$

Parallel and Perpendicular Lines on the Coordinate Axes

A line parallel to X-axis makes an angle of 0° with X-axis. Therefore, its slope is $\tan 0° = 0$. A line parallel to Y-axis i.e. perpendicular to X-axis makes an angle of 90° with X-axis, so its slope is $\tan\frac{\pi}{2} = \infty$. Also, the slope of a line equally inclined with axes is 1 or −1 as it makes an angle of 45° or 135° with X-axis.

Locus and its Equation

It is the path or curve traced by a moving point satisfying the given condition.

Equation to the Locus of a Point

The equation to the locus of a point is the algebraic relation which is satisfied by the coordinates of every point on the locus of the point.

Steps to Find the Locus of a Point

The following steps are used to find the locus of a point

Step I Assume the coordinates of the point say (h, k) whose locus is to be find.

Step II Write the given condition involving (h, k).

Step III Eliminate the variable(s), if any.

Step IV Replace $h \rightarrow x$ and $k \rightarrow y$. The equation, so obtained is the locus of the point which moves under some definite conditions.

DAY PRACTICE SESSION 1

FOUNDATION QUESTIONS EXERCISE

1 Length of the median from B to AC where $A(-1, 3)$, $B(1, -1)$, $c(5, 1)$ is

(a) $\sqrt{18}$ (b) $\sqrt{10}$ (c) $2\sqrt{3}$ (d) 4

2 Three points $(p+1, 1)$, $(2p+1, 3)$ and $(2p+2, 2p)$ are collinear if p is equal to

(a) -1 (b) 1 (c) 2 (d) 0

3 The point A divides the join of $P \equiv (-5, 1)$ and $Q \equiv (3, 5)$ in the ratio $k:1$. The two values of k for which the area of ΔABC, where $B \equiv (1, 5)$, $C \equiv (7, -2)$ is equal to 2 sq units are

(a) $\left(7, \frac{30}{9}\right)$ (b) $\left(7, \frac{31}{9}\right)$ (c) $\left(4, \frac{31}{9}\right)$ (d) $\left(7, \frac{31}{3}\right)$

4 If Δ_1 is the area of the triangle with vertices (0, 0), $(a \tan \alpha, b \cot \alpha)$, $(a \sin \alpha, b \cos \alpha)$, Δ_2 is the area of the triangle with vertices (a, b), $(a \sec^2 \alpha, b \operatorname{cosec}^2 \alpha)$, $(a + a \sin^2 \alpha, b + b \cos^2 \alpha)$ and Δ_3 is the area of the triangle with vertices $(0, 0)$, $(a \tan \alpha, -b \cot \alpha)$, $(a \sin \alpha, b \cos \alpha)$. Then,

(a) $\Delta_1, \Delta_2, \Delta_3$ are in GP
(b) $\Delta_1, \Delta_2, \Delta_3$ are not in GP
(c) Cannot be discussed
(d) None of these

5 If the area of the triangle formed by the points $O(0, 0)$, $A(a^{x^2}, 0)$ and $B(0, a^{6x})$ is $\frac{1}{2a^5}$ sq units, then $x =$

(a) 1, 5 (b) -1, 5 (c) 1, -5 (d) -1, -5

6 The value of k for which the distinct points $(k, 2-2k)$, $(1-k, 2k)$ and $(-4-k, 6-2k)$ are collinear is (are)

(a) -1 or $1/2$
(b) Only $1/2$
(c) Only -1
(d) can not be found

7 If the line $2x + y = k$ passes through the point which divides the line segment joining the points (1, 1) and (2, 4) in the ratio 3 : 2, then k is equal to

(a) $\frac{29}{5}$ (b) 5 (c) 6 (d) $\frac{11}{5}$

8 A line L intersects the three sides BC, CA and AB of a ΔABC at P, Q and R, respectively. Then, $\frac{BP}{PC} \cdot \frac{CQ}{QA} \cdot \frac{AR}{RB}$ is equal to

(a) 1
(b) 0
(c) -1
(d) None of these

9 If the coordinates of the vertices of a triangle are integers, then the triangle cannot be

(a) equilateral
(b) isosceles
(c) scalene
(d) None of these

10 Let $O(0, 0)$, $P(3, 4)$ and $Q(6, 0)$ be the vertices of the ΔOPQ. The point R inside the ΔOPQ is such that ΔOPR, ΔPQR and ΔOQR are of equal area. Then, R is equal to

(a) $\left(\frac{4}{3}, 3\right)$
(b) $\left(3, \frac{2}{3}\right)$
(c) $\left(3, \frac{4}{3}\right)$
(d) $\frac{4}{3}$

11 The number of points having both coordinates as integers that lie in the interior of the triangle with vertices (0, 0), (0, 41) and (41, 0) is → JEE Mains 2015

(a) 901 (b) 861 (c) 820 (d) 780

12 If (0, 0), (1, 1) and (1, 0) be the middle points of the sides of a triangle, its incentre is

(a) $(2 + \sqrt{2}, 2 + \sqrt{2})$
(b) $[(2 + \sqrt{2}, -(2 + \sqrt{2})]$
(c) $(2 - \sqrt{2}, 2 - \sqrt{2})$
(d) $[(2 - \sqrt{2}, -(2 - \sqrt{2})]$

13 Vertices of a triangle are (1,2), (2,3) and (3,1) Its circumcentre is

(a) (11/6, 13/6)
(b) (11/6, 2)
(c) (13/6, 11/6)
(d) None of these

14 If a vertex of a triangle be (1,1) and the middle points of two sides through it be $(-2, 3)$ and (5, 2), then the centroid of the triangle is

(a) (3, 5/3)
(b) (3, 5)
(c) (5/3, 3)
(d) None of these

15 The centroid of the triangle is (3,3) and the orthocentre is $(-3, 5)$ then its circumcentre is

(a) (0, 4) (b) (0, 8) (c) (6, 2) (d) $(6, -2)$

16 Let the orthocentre and centroid of a triangle be $A(-3, 5)$ and $B(3, 3)$ respectively. If C is the circumcentre of this triangle, then the radius of the circle having line segment AC as diameter, is → JEE Mains 2018

(a) $\sqrt{10}$ (b) $2\sqrt{10}$ (c) $3\sqrt{\frac{5}{2}}$ (d) $\frac{3\sqrt{5}}{2}$

17 If G is the centroid of ΔABC with vertices $A(a, 0)$, $B(-a, 0)$ and $C(b, c)$, then $\frac{(AB^2 + BC^2 + CA^2)}{(GA^2 + GB^2 + GC^2)}$ is equal to

(a) 1 (b) 2 (c) 3 (d) 4

18 Let a, b, c and d be non-zero numbers. If the point of intersection of the lines $4ax + 2ay + c = 0$ and $5bx + 2by + d = 0$ lies in the fourth quadrant and is equidistant from the two axes, then → JEE Mains 2014

(a) $2bc - 3ad = 0$
(b) $2bc + 3ad = 0$
(c) $2ad - 3bc = 0$
(d) $3bc - 2ad = 0$

19 The origin is shifted to (1,2). The equation $y^2 - 8x - 4y + 12 = 0$ changes to $y^2 = 4ax$, then a is equal to

(a) 1 (b) 2 (c) –2 (d) –1

20 If the axes are rotated through an angle of 60°, the coordinates of a point in the new system are $(2, -\sqrt{3})$, then its original coordinates are

(a) $\left(\frac{5}{3}, -\frac{\sqrt{2}}{3}\right)$ (b) $\left(-\frac{5}{3}, \frac{\sqrt{2}}{3}\right)$

(c) $\left(\frac{5}{2}, \frac{\sqrt{3}}{2}\right)$ (d) $\left(-\frac{5}{2}, -\frac{\sqrt{3}}{2}\right)$

21 By rotating the axes through 180°, the equation $x - 2y + 3 = 0$ changes to

(a) $x + 2y - 3 = 0$ (b) $x - 2y + 3 = 0$

(c) $x - 2y - 3 = 0$ (d) None of these

22 Let $0 < \alpha < \frac{\pi}{2}$ be a fixed angle. If $P = (\cos\theta, \sin\theta)$ and $Q = \{\cos(\alpha - \theta), \sin(\alpha - \theta)\}$, then Q is obtained from P by

(a) clockwise rotation around the origin through angles α

(b) anti-clockwise rotation around origin through angle α

(c) reflection in the line through the origin with slope $\tan\alpha$

(d) reflection in the line through the origin with slope $\tan\alpha/2$

23 The point (4, 1) undergoes the following transformations

(i) Reflection in the line $x - y = 0$

(ii) Translation through a distance of 2 units along positive direction of X-axis.

(iii) Projection on X-axis.

The coordinate of the point in its final position is

(a) (3, 4) (b) (3, 0) (c) (1, 0) (d) (4, 3)

24 If the points are $A(0, 4)$ and $B(0, -4)$, then find the locus of $P(x, y)$ such that $|AP - BP| = 6$.

(a) $9x^2 - 7y^2 + 63 = 0$ (b) $9x^2 + 7y^2 - 63 = 0$

(c) $9x^2 + 7y^2 + 63 = 0$ (d) None of these

25 ABC is a variable triangle with the fixed vertex $C(1, 2)$ and A, B having the coordinates $(\cos t, \sin t)$, $(\sin t, -\cos t)$ respectively, where t is a parameter. The locus of the centroid of the ΔABC is

(a) $3(x^2 + y^2) - 2x - 4y - 1 = 0$

(b) $3(x^2 + y^2) - 2x - 4y + 1 = 0$

(c) $3(x^2 + y^2) + 2x + 4y - 1 = 0$

(d) $3(x^2 + y^2) + 2x + 4y + 1 = 0$

26 If $A(2, -3)$ and $B(-2, 1)$ are two vertices of a triangle and third vertex moves on the line $2x + 3y = 9$, then the locus of the centroid of the triangle is

(a) $2x - 3y = 1$ (b) $x - y = 1$

(c) $2x + 3y = 1$ (d) $2x + 3y = 3$

27 Let $A(-3, 2)$ and $B(-2, 1)$ be the vertices of a ΔABC. If the centroid of this triangle lies on the line $3x + 4y + 2 = 0$, then the vertex C lies on the line

→ JEE Mains 2013

(a) $4x + 3y + 5 = 0$ (b) $3x + 4y + 3 = 0$

(c) $4x + 3y + 3 = 0$ (d) $3x + 4y + 5 = 0$

28 The coordinates of points A and B are $(ak, 0)$ and $\left(\frac{a}{k}, 0\right)$, where $(k \neq \pm 1)$ if p moves in such a way that $PA = kPB$, the locus of P is

(a) $k^2(x^2 + y^2) = a^2$ (b) $x^2 + y^2 = k^2a^2$

(c) $x^2 + y^2 + a^2 = 0$ (d) $x^2 + y^2 = a^2$

29 If $A(-a, 0)$ and $B(a, 0)$ are two fixed points, then the locus of the point at which AB subtends a right angle is

(a) $x^2 + y^2 = 2a^2$ (b) $x^2 - y^2 = a^2$

(c) $x^2 + y^2 + a^2 = 0$ (d) $x^2 + y^2 = a^2$

30 A point moves in such a way that the sum of its distances from two fixed points $(ae, 0)$ and $(-ae, 0)$ is $2a$. Then the locus of the points is

(a) $\frac{x^2}{a^2} + \frac{y^2}{a^2(1-e^2)} = 1$

(b) $\frac{x^2}{a^2} - \frac{y^2}{a^2(1-e^2)} = 1$

(c) $\frac{x^2}{a^2(1-e^2)} + \frac{y^2}{a^2} = 1$

(d) None of the above

Directions (Q. Nos. 31-35) *Each of these questions contains two statements : Statement* I *and Statement* II. *Each of these questions also has four alternative choices, only one of which is the correct answer. You have to select one of the codes (a), (b), (c) and (d) given below.*

(a) Statement I is true, Statement II is true; Statement II is a correct explanation for Statement I

(b) Statement I is true, Statement II is true; Statement II is not a correct explanation for Statement I

(c) Statement I is true; Statement II is false

(d) Statement I is false; Statement II is true

31 **Statement I** If $A(2a, 4a)$ and $B(2a, 6a)$ are two vertices of a equilateral ΔABC and the vertex C is given by $(2a + a\sqrt{3}, 5a)$.

Statement II An equilateral triangle all the coordinates of three vertices can be rational.

32 **Statement I** If the circumcentre of a triangle lies at the origin and centroid is the middle point of the line joining the points (2, 3) and (4, 7), then its orthocentre lies on the line $5x - 3y = 0$.

Statement II The circumcentre, centroid and the orthocentre of a triangle lie on the same line.

33 **Statement I** If the origin is shifted to the centroid of the triangle with vertices (0, 0), (3, 3) and (3, 6) without rotation of axes, then the vertices of the triangle in the new system of coordinates are (–2, 0), (1, 3) and (1, – 3).

Statement II If the origin is shifted to the point (2, 3) without rotation of the axes, then the coordinates of the point $P(\alpha - 1, \alpha + 1)$ in the new system of coordinates are $(\alpha - 3, \alpha - 2)$.

34 Let the equation of the line $ax + by + c = 0$.

Statement I If a, b and c are in AP, then $ax + by + c = 0$ pass through a fixed point $(1, -2)$.

Statement II Any family of lines always pass through a fixed point.

35 The lines $L_1 : y - x = 0$ and $L_2 : 2x + y = 0$ intersect the line $L_3 : y + 2 = 0$ at P and Q, respectively. The bisector of the acute angle between L_1 and L_2 intersects L_3 at R.

Statement I The ratio $PR : RQ$ equals $2\sqrt{2} : \sqrt{5}$.

Statement II In any triangle, bisector of an angle divides the triangle into two similar triangles.

DAY PRACTICE SESSION 2

PROGRESSIVE QUESTIONS EXERCISE

1 If the coordinates of two points A and B are $(3, 4)$ and $(5, -2)$, respectively. Then, the coordinates of any point P, if $PA = PB$ and area of $\Delta PAB = 10$, are

(a) (7,5), (1, 0) (b) (7, 2), (1, 0)
(c) (7, 2), (−1, 0), (d) None of these

2 Lat $A(a, b)$ be a fixed point and O be the origin an coordinates. If A_1 is the mid-point at OA, A_2 is the mid-point at AA_1, A_3 is the mid-point at AA_2 and so on. Then, the coordinates of A_n are

(a) $(a(1-2^{-n}), b(1-2^{-n}))$ (b) $(a(2^{-n}-1), b(2^{-n}-1))$
(c) $(a(1-2^{-(n-1)}), b(1-2^{-(n-1)}))$ (d) None of these

3 The coordinates of points A, B, C are (x_1, y_1), (x_2, y_2) and (x_3, y_3) and point D divides AB in the ratio $l : k$. If P divides line DC in the ratio $m : (k + l)$, coordinates of P are

(a) $\left(\frac{kx_1 + lx_2 + mx_3}{k + l + m}, \frac{ky_1 + ly_2 + my_3}{k + l + m}\right)$

(b) $\left(\frac{lx_1 + mx_2 + kx_3}{l + m + k}, \frac{ly_1 + my_2 + ky_3}{l + m + k}\right)$

(c) $\left(\frac{mx_1 + kx_2 + lx_3}{m + k + l}, \frac{my_1 + ky_2 + ly_3}{m + K + l}\right)$

(d) None of the above

4 The locus of a point P which moves such that $2PA = 3PB$, where $A(0, 0)$ and $B(4, -3)$ are points, is

(a) $5x^2 - 5y^2 - 72x + 54y + 225 = 0$
(b) $5x^2 + 5y^2 - 72x + 54y + 225 = 0$
(c) $5x^2 + 5y^2 + 72x - 54y + 225 = 0$
(d) $5x^2 + 5y^2 - 72x - 54y - 225 = 0$

5 Two points $P(a, 0)$ and $Q(-a, 0)$ are given, R is a variable point on one side of the line PQ such that $\angle RPQ - \angle RQP$ is 2α, then

(a) locus of R is $x^2 - y^2 + 2xy \cot 2\alpha - a^2 = 0$
(b) locus of R is $x^2 + y^2 + 2xy \cot \alpha - a^2 = 0$
(c) locus of R is a hyperbola, if $\alpha = \pi/4$
(d) locus of R is a circle, if $\alpha = \pi/4$

6 If the axis be turned through an angle $\tan^{-1} 2$, then what does the equation $4xy - 3x^2 = a^2$ become?

(a) $X^2 - 4Y^2 = a^2$ (b) $X^2 + 4Y^2 = a^2$
(c) $X^2 + 4Y^2 = -a^2$ (d) None of these

7 The orthocentre of the triangle whose vertices are $\{at_1 t_2, a(t_1 + t_2)\}$, $\{at_2 t_3, a(t_2 + t_3)\}$, $\{at_3 t_1, a(t_3 + t_1)\}$ is

(a) $\{-a, a(t_1 + t_2 + t_3 + t_1 + t_2 t_3)\}$
(b) $\{-a, a(t_1 + t_2 + t_3 + t_1 t_2 t_3)\}$
(c) $\{-a, a(t_1 - t_2 - t_3 - t_1 t_2 t_3)\}$
(d) $\{-a, a(t_1 + t_2 - t_3 - t_1 t_2 t_3)\}$

8 ABC is an isosceles triangle of area $\frac{25}{6}$ sq unit if the coordinates of base are $B(1, 3)$ and $C(-2, 7)$, the coordinates of A are

(a) $(1, 6), \left(-\frac{11}{6}, \frac{5}{6}\right)$ (b) $\left(-\frac{1}{2}, 5\right), \left(4, \frac{5}{6}\right)$
(c) $\left(\frac{5}{6}, 6\right), \left(-\frac{11}{6}, 4\right)$ (d) $\left(5, \frac{5}{6}\right), \left(\frac{11}{6}, 4\right)$

9 If $A(6, -3)$, $B(-3, 5)$, $C(4, -2)$, $P(\alpha, \beta)$, then the ratio of the areas of the triangles PBC and ABC is

(a) $|\alpha + \beta|$ (b) $|\alpha - \beta|$
(c) $|\alpha + \beta + 2|$ (d) $|\alpha + \beta - 2|$

10 If O be the origin and if $P_1(x_1, y_1)$ and $P_2(x_2, y_2)$ be two points, then $|OP_1| \cdot |OP_2| \cos(\angle P_1OP_2)$ is equal to

(a) $x_1y_2 + x_2y_1$ (b) $(x_1^2 + y_1^2)(x_2^2 + y_2^2)$
(c) $(x_1 - x_2)^2 + (y_1 - y_2)^2$ (d) $x_1x_2 + y_1y_2$

11 If points $(0, 0)$, $(2, 2\sqrt{3})$ and (a, b) are vertices of an equilateral triangle, then (a, b) is equal to

(a) (0, − 4) (b) (0, 4) (c) (4, 0) (d) (− 4, 0)

12 If the equation of the locus of a point equidistant from the points (a_1, b_1) and (a_2, b_2) is $(a_1 - a_2)x + (b_1 - b_2)y + c = 0$, then the value of c is

(a) $a_1^2 - a_2^2 + b_1^2 - b_2^2$ (b) $\sqrt{a_1^2 + b_1^2 - a_2^2 - b_2^2}$
(c) $\frac{1}{2}(a_1^2 + a_2^2 + b_1^2 + b_2^2)$ (d) $\frac{1}{2}(a_2^2 + b_2^2 - a_1^2 - b_1^2)$

13 If $(2, 1)$, $(5, 2)$ and $(3, 4)$ are vertices of a triangle, its circumcentre is

(a) $\left(\frac{13}{2}, \frac{9}{2}\right)$ (b) $\left(\frac{13}{4}, \frac{9}{4}\right)$
(c) $\left(\frac{9}{4}, \frac{13}{4}\right)$ (d) $\left(\frac{9}{2}, \frac{13}{2}\right)$

14 A point moves is such a way that the sum of squares of its distances from $A(2, 0)$ and $B(-2, 0)$ is always equal to the square of the distance between A and B, then the locus of point P is

(a) $x^2 + y^2 - 2 = 0$ (b) $x^2 + y^2 + 2 = 0$
(c) $x^2 + y^2 + 4 = 0$ (d) $x^2 + y^2 - 4 = 0$

15 The area of a triangle is 5 and its two vertices are $A(2, 1)$ and $B(3, -2)$. The third vertex lies on $y = x + 3$. Then, third vertex is

(a) $\left(\frac{7}{2}, \frac{13}{2}\right)$ (b) $\left(\frac{5}{2}, \frac{5}{2}\right)$
(c) $\left(-\frac{3}{2}, -\frac{3}{2}\right)$ (d) (0, 0)

ANSWERS

SESSION 1	**1** (b)	**2** (c)	**3** (b)	**4** (b)	**5** (d)	**6** (c)	**7** (c)	**8** (c)	**9** (a)	**10** (c)
	11 (d)	**12** (c)	**13** (c)	**14** (c)	**15** (c)	**16** (c)	**17** (c)	**18** (c)	**19** (b)	**20** (c)
	21 (c)	**22** (d)	**23** (b)	**24** (a)	**25** (b)	**26** (c)	**27** (b)	**28** (d)	**29** (d)	**30** (a)
	31 (c)	**32** (a)	**33** (a)	**34** (a)	**35** (c)					
SESSION 2	**1** (b)	**2** (a)	**3** (a)	**4** (b)	**5** (a)	**6** (a)	**7** (b)	**8** (c)	**9** (d)	**10** (d)
	11 (c)	**12** (d)	**13** (b)	**14** (d)	**15** (a)					

Hints and Explanations

SESSION 1

1 Let BD be the median from B to AC, where D is the mid-point of AC.

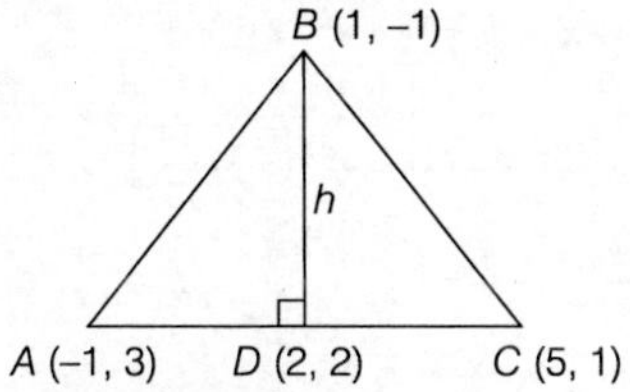

According to mid-point formula coordinates of D are

$$\left(\frac{-1+5}{2}, \frac{3+1}{2}\right) = (2, 2)$$

$\therefore$ Length of median BD

$$= \sqrt{(2-1)^2 + (2+1)^2} = \sqrt{1+9} = \sqrt{10}$$

2 Condition of collinearity

$$\Delta = \begin{vmatrix} p+1 & 1 & 1 \\ 2p+1 & 3 & 1 \\ 2p+2 & 2p & 1 \end{vmatrix} = 0$$

$$\Rightarrow (p+1)(3-2p) - 1(2p+1-2p-2) + 1(4p^2 + 2p - 6p - 6) = 0$$

$$\Rightarrow -2p^2 + p + 3 + 1 + 4p^2 - 4p - 6 = 0$$

$$\Rightarrow \quad 2p^2 - 3p - 2 = 0$$

$$\Rightarrow 2p^2 - 4p + p - 2 = 0$$

$$\therefore \quad p = 2, -\frac{1}{2}$$

3 Coordinates of A, dividing the join of $P \equiv (-5, 1)$ and $Q \equiv (3, 5)$ in the ratio $k:1$ are given by $A\left(\frac{3k-5}{k+1}, \frac{5k+1}{k+1}\right)$.

Also, area of ΔABC is given by

$$\Delta = \left|\frac{1}{2}\Sigma x_1 (y_2 - y_3)\right|$$

$$= \frac{1}{2}|[x_1(y_2 - y_3) + x_2(y_3 - y_1) + x_3(y_1 - y_2)]|$$

$$\Rightarrow \left|\frac{1}{2}\left\{\frac{3k-5}{k+1}(7) + 1\left(-2 - \frac{5k+1}{k+1}\right) + 7\left(\frac{5k+1}{k+1} - 5\right)\right\}\right| = 2$$

$$\Rightarrow \frac{1}{2}\left\{\frac{3k-5}{k+1}(7) + \left(-2 - \frac{5k+1}{k+1}\right) + 7\left(\frac{5k+1}{k+1} - 5\right)\right\} = \pm 2$$

$$\Rightarrow \quad 14k - 66 = 4k + 4$$

$$\Rightarrow \quad 10k = 70 \Rightarrow k = 7$$

or $\quad 14k - 66 = -4k - 4$

$$\Rightarrow \quad 18k = 62$$

$$\Rightarrow \quad k = \left(\frac{31}{9}\right)$$

Therefore, the values of k are 7 and $\frac{31}{9}$.

4 We have, $\Delta_1 = \frac{1}{2}\left\|\begin{matrix} 0 & 0 & 1 \\ a\tan\alpha & b\cot\alpha & 1 \\ a\sin\alpha & b\cos\alpha & 1 \end{matrix}\right\|$

$$= \frac{1}{2}ab\,|\sin\alpha - \cos\alpha|$$

and $\Delta_2 = \frac{1}{2}\left\|\begin{matrix} a & b & 1 \\ a\sec^2\alpha & b\,\mathrm{cosec}^2\alpha & 1 \\ a + a\sin^2\alpha & b + b\cos^2\alpha & 1 \end{matrix}\right\|$

On applying $C_1 \to C_1 - aC_3$ and $C_2 \to C_2 - bC_3$, we get

$$\Delta_2 = \frac{1}{2}ab\left\|\begin{matrix} 0 & 0 & 1 \\ \tan^2\alpha & \cot^2\alpha & 1 \\ \sin^2\alpha & \cos^2\alpha & 1 \end{matrix}\right\|$$

$$= \frac{1}{2}ab\,|\sin^2\alpha - \cos^2\alpha|$$

and $\Delta_3 = \frac{1}{2}\left\|\begin{matrix} 0 & 0 & 1 \\ a\tan\alpha & -b\cot\alpha & 1 \\ a\sin\alpha & b\cos\alpha & 1 \end{matrix}\right\|$

$$= \frac{1}{2}ab\,|\sin\alpha + \cos\alpha|$$

So that, $\Delta_1 \Delta_3 = \frac{1}{2}ab\Delta_2$

Suppose, Δ_1, Δ_2 and Δ_3 are in GP.

Then, $\Delta_1\Delta_3 = \Delta_2^2 \Rightarrow \frac{1}{2}ab\Delta_2 = \Delta_2^2$

$$\Rightarrow \quad \Delta_2 = \frac{1}{2}ab$$

$\Rightarrow \frac{1}{2}ab(\sin^2\alpha - \cos^2\alpha) = \frac{1}{2}ab$

$\Rightarrow \quad \sin^2\alpha - \cos^2\alpha = 1$

i.e. $\alpha = (2m+1)\frac{\pi}{2}, m \in I.$

But for this value of α, the vertices of the given triangles are not defined.

Hence, Δ_1, Δ_2 and Δ_3 cannot be in GP for any value of α.

5 We have,

area of $\Delta OAB = \frac{1}{2a^5}$ sq units

$\Rightarrow \frac{1}{2} \times a^{x^2} \times a^{6x} = \frac{1}{2}a^{-5}$

$\Rightarrow a^{x^2 + 6x} = a^{-5}$

$\Rightarrow x^2 + 6x + 5 = 0$

$\Rightarrow x = -1, -5$

6 Points are collinear so

$$\begin{vmatrix} k & 2-2k & 1 \\ 1-k & 2k & 1 \\ -4-k & 6-2k & 1 \end{vmatrix} = 0$$

$$\Rightarrow \begin{vmatrix} k & 2-2k & 1 \\ 1-2k & 4k-2 & 0 \\ -4-2k & 4 & 0 \end{vmatrix} = 0$$

[applying $C_2 \to C_2 - C_1$ and $C_3 \to C_3 - C_2$]

$\Rightarrow 4 - 8k + (4k-2)(4+2k) = 0$

$\Rightarrow \quad 2k^2 + k - 1 = 0$

so $k = -1$ and $1/2$

But for $k = \frac{1}{2}$, points are $(1/2, 1)$,

$(1/2, 1)$ and $\left(-\frac{9}{5}, 5\right)$

Which is a contradiction as given points are distinct.

7 Using section formula, the coordi- nates of the point P, which divides AB internally in the ratio 3 : 2 are

$$P\left(\frac{3\times 2 + 2\times 1}{3+2}, \frac{3\times 4 + 2\times 1}{3+2}\right)$$

$$\equiv P\left(\frac{8}{5}, \frac{14}{5}\right)$$

Also, since the line L passes through P, hence substituting the coordinates of $P\left(\frac{8}{5}, \frac{14}{5}\right)$ in the equation of $L: 2x + y = k$, we get

$$2\left(\frac{8}{5}\right) + \left(\frac{14}{5}\right) = k \Rightarrow k = 6$$

8 Let $A(x_1, y_1), B(x_2, y_2)$ and $C(x_3, y_3)$ be the vertices of ΔABC and $lx + my + n = 0$ be the equation of the line. If P divides BC in the ratio λ:1, then the coordinates of P are $\left(\frac{\lambda x_3 + x_2}{\lambda + 1}, \frac{\lambda y_3 + y_2}{\lambda + 1}\right)$.

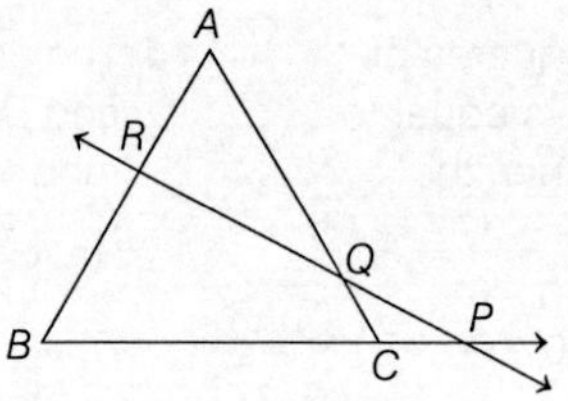

Also, as P lies on L, we have

$$l\left(\frac{\lambda x_3 + x_2}{\lambda + 1}\right) + m\left(\frac{\lambda y_3 + y_2}{\lambda + 1}\right) + n = 0$$

$$\Rightarrow \quad -\frac{lx_2 + my_2 + n}{lx_3 + my_3 + n} = \frac{BP}{PC} = \lambda \quad \text{...(i)}$$

Similarly, we obtain

$$\frac{CQ}{QA} = -\frac{lx_3 + my_3 + n}{lx_1 + my_1 + n} \quad \text{...(ii)}$$

and $$\frac{AR}{RB} = -\frac{lx_1 + my_1 + n}{lx_2 + my_2 + n} \quad \text{...(iii)}$$

On multiplying Eqs. (i), (ii) and (iii), we get

$$\frac{BP}{PC} \cdot \frac{CQ}{QA} \cdot \frac{AR}{PB} = -1$$

9 Let $A = (x_1, y_1), B = (x_2, y_2), C = (x_3, y_3)$ be the vertices of a triangle and x_1, x_2, x_3 and y_1, y_2, y_3 be integers.

So, $BC^2 = (x_2 - x_3)^2 + (y_2 - y_3)^2$ is a positive integers.

If the triangle is equilateral, then

$AB = BC = CA = a$ [say]

and $\angle A = \angle B = \angle C = 60°$.

$\therefore$ Area of the triangle $= \left(\frac{1}{2}\right)\sin A \cdot bc$

$$= \left(\frac{1}{2}\right)a^2 \sin 60°$$

$$= \left(\frac{a^2}{2}\right)\left(\frac{\sqrt{3}}{2}\right)$$

$$= \frac{\sqrt{3}}{4}a^2$$

which is irrational.

[since, a^2 is a positive integer]

Now, the area of the triangle in terms of the coordinates

$$= \frac{1}{2}[x_1(y_2 - y_3) + x_2(y_3 - y_1) + x_3(y_1 - y_2)]$$

which is a rational number.

This contradicts that the area is an irrational number, if the triangle is equilateral.

10 If the centroid is joined to the vertices, we get three triangles of equal area.

$\therefore \quad R = G = \left(3, \frac{4}{3}\right)$

11 Required points (x, y) are such that, it satisfy $x + y < 41$ and $x > 0, y > 0$.

Number of positive integral solution of the equation $x + y + k = 41$ will be number of integral coordinates in the bounded region.

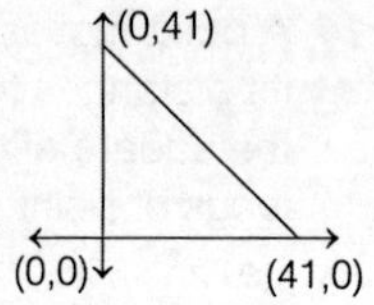

$\therefore$ Total number of integral coordinates

$$= {}^{41-1}C_{3-1} = {}^{40}C_2$$

$$= \frac{40!}{2!\,38!} = 780$$

12 If $A(x_1, y_1), B(x_2, y_2), C(x_3, y_3)$ be the vertices of the triangle and if (0, 0), (1, 1) and (1,0) are the middle points of AB, BC and CA respectively, then

$x_1 + x_2 = 0, x_2 + x_3 = 2, x_3 + x_1 = 2$

$y_1 + y_2 = 2, y_2 + y_3 = 2, y_3 + y_1 = 0$

So, $A(0, 0)$, $B(0, 2)$ and $C(2, 0)$ are the vertices of the ΔABC.

Now, $a = BC = 2\sqrt{2}$, $b = CA = 2$, $c = AB = 2$

The coordinates (α, β) of the in-centre are given by

$$\alpha = \frac{ax_1 + bx_2 + cx_3}{a + b + c} = 2 - \sqrt{2},$$

$$\beta = \frac{ay_1 + by_2 + cy_3}{a + b + c} = 2 - \sqrt{2}$$

i.e. The in-centre is $(2 - \sqrt{2}, 2 - \sqrt{2})$.

13 $(x-1)^2 + (y-2)^2$

$= (x-2)^2 + (y-3)^2$

$= (x-3)^2 + (y-1)^2$

$\Rightarrow x + y = 4, 4x - 2y = 5$

$\Rightarrow \quad x = 13/6, y = 11/6$

$\therefore$ Circumcentre $= \left(\frac{13}{6}, \frac{11}{6}\right)$

14 $\frac{x_1 + 1}{2} = -2$

$\Rightarrow \quad x_1 = -5$

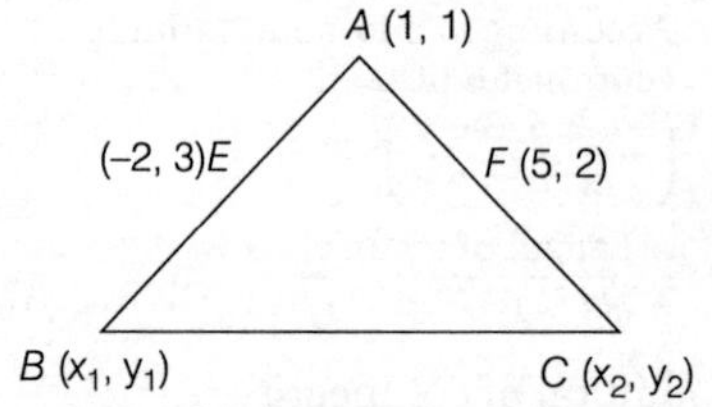

$\frac{y_1 + 1}{2} = 3 \Rightarrow y_1 = 5$

$B = (-5, 5)$

$\frac{x_2 + 1}{2} = 5 \Rightarrow x_2 = 9$

$\frac{y_2 + 1}{2} = 2 \Rightarrow y_2 = 3$

$\therefore \quad C = (9, 3)$

$$G = \left(\frac{1 + 9 - 5}{3}, \frac{1 + 5 + 3}{3}\right)$$

$= (5/3, 3)$

15 Since, we know that Centroid divides the join of orthocenter and circumcenter in the ratio of 2 : 1

Let the circumcenter of Δ is (α, β)

$$\Rightarrow G\ (3, 3) = G\left(\frac{2\alpha - 3}{3}, \frac{2\beta + 5}{3}\right)$$

$$\therefore \frac{2\alpha - 3}{3} = 3 \text{ and } \frac{2\beta + 5}{3} = 3$$

or $\alpha = 6, \beta = 2$

$\therefore$ Circumcentre of Δ is $C(6, 2)$.

16 We have, orthocentre and centroid of a triangle be $A(-3, 5)$ and $B(3, 3)$ respectively and C circumcentre

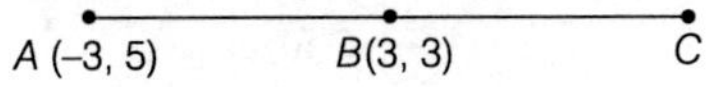

We know that,

$$AB : BC = 2 : 1$$

$$AB = \sqrt{(3 + 3)^2 + (3 - 5)^2}$$

$$= \sqrt{36 + 4} = 2\sqrt{10}$$

$$\therefore \quad BC = \sqrt{10}$$

$$AC = AB + BC$$

$$= 2\sqrt{10} + \sqrt{10} = 3\sqrt{10}$$

Since, AC is a diametre of circle

$$\therefore \quad r = \frac{AC}{2} = \frac{3\sqrt{10}}{2} = 3\sqrt{\frac{5}{2}}$$

17 Coordinates of point G is $G\left(\frac{b}{3}, \frac{c}{3}\right)$

$$\text{Let } E = \frac{(AB)^2 + (BC)^2 + (CA)^2}{(GA)^2 + (GB)^2 + (GC)^2}$$

$$\Rightarrow E = \frac{4a^2 + (a+b)^2 + c^2 + (a-b)^2 + c^2}{\left(\frac{b}{3} - a\right)^2 + \left(\frac{c}{3}\right)^2 + \left(\frac{b}{3} + a\right)^2 + \left(\frac{c}{3}\right)^2 + \left(\frac{2b}{3}\right)^2 + \left(\frac{2c}{3}\right)^2}$$

$$\Rightarrow E = \frac{4a^2 + 2c^2 + 2a^2 + 2b^2}{\frac{2b^2}{9} + 2a^2 + \frac{6c^2}{9} + \frac{4b^2}{9}}$$

$$\Rightarrow E = \frac{6a^2 + 2b^2 + 2c^2}{\frac{1}{9}(6b^2 + 18a^2 + 6c^2)}$$

$$\Rightarrow E = \frac{2(3a^2 + b^2 + c^2)}{\frac{1}{9}6(3a^2 + b^2 + c^2)} = 3$$

18 Let coordinate of the intersection point in fourth quadrant be $(\alpha, -\alpha)$.

Since, $(\alpha, -\alpha)$ lies on both lines $4ax + 2ay + c = 0$ and $5bx + 2by + d = 0$.

$$\therefore \quad 4a\alpha - 2a\alpha + c = 0$$

$$\Rightarrow \quad \alpha = \frac{-c}{2a} \quad \text{...(i)}$$

and $5b\alpha - 2b\alpha + d = 0$

$$\Rightarrow \quad \alpha = \frac{-d}{3b} \quad \text{...(ii)}$$

From Eqs. (i) and (ii), we get

$$\frac{-c}{2a} = \frac{-d}{3b} \Rightarrow 3bc = 2ad$$

$$\Rightarrow \quad 2ad - 3bc = 0 \quad \text{...(iii)}$$

19 Let $P(x, y)$ be the original position of the point w.r.t the original axes. Let us move the origin at new position to (h, k). Hence, the position of the same point P in the new system is

$$x' = x - h$$

$$y' = y - k$$

Here, $(h, k) = (1, 2)$

$$\therefore \quad x' = (x - 1)$$

$$y' = (y - 2)$$

As per the given situation

$$y^2 - 8x - 4y + 12 = (y - 2)^2 - 4a(x - 1)$$

$$\Rightarrow y^2 - 8x - 4y + 12 = y^2 - 4y + 4 - 4ax + 4a$$

Comparing respective coefficients, we have

$$4a = 8$$

$$\therefore \quad a = 2$$

20 Let $P(a', y')$ be the coordinates of the point obtained by rotating the axes through an angle of 60°.

$\therefore$ The transformation matrix can be written as

$$\begin{bmatrix} x' \\ y' \end{bmatrix} = \begin{bmatrix} \cos\theta & \sin\theta \\ -\sin\theta & \cos\theta \end{bmatrix} \begin{bmatrix} x \\ y \end{bmatrix}$$

$$\Rightarrow \begin{bmatrix} 2 \\ -\sqrt{3} \end{bmatrix} = \begin{bmatrix} \frac{1}{2} & \frac{\sqrt{3}}{2} \\ -\frac{\sqrt{3}}{2} & \frac{1}{2} \end{bmatrix} \begin{bmatrix} x \\ y \end{bmatrix}$$

$$\Rightarrow \begin{bmatrix} 2 \\ -\sqrt{3} \end{bmatrix} = \begin{bmatrix} \frac{x}{2} + \frac{\sqrt{3}y}{2} \\ -\frac{\sqrt{3}x}{2} + \frac{y}{2} \end{bmatrix}$$

$$\Rightarrow \quad x + \sqrt{3}y = 4 \text{ and } \sqrt{3}x - y = 2\sqrt{3}$$

Solving the above equations,

we have $(x, y) = \left(\frac{5}{2}, \frac{\sqrt{3}}{2}\right)$

21 Let $P(x', y')$ be the coordinates of the point $P(x, y)$ after rotation of axes at an angle of 180°

$$\Rightarrow \begin{bmatrix} x' \\ y' \end{bmatrix} = \begin{bmatrix} \cos\theta & \sin\theta \\ -\sin\theta & \cos\theta \end{bmatrix} \begin{bmatrix} x \\ y \end{bmatrix}$$

Since, here $\theta = 180°$

$$\Rightarrow \begin{bmatrix} x' \\ y' \end{bmatrix} = \begin{bmatrix} -1 & 0 \\ 0 & -1 \end{bmatrix} \begin{bmatrix} x \\ y \end{bmatrix}$$

$$\Rightarrow \begin{bmatrix} x' \\ y' \end{bmatrix} = \begin{bmatrix} -x \\ -y \end{bmatrix}$$

$$\therefore \quad x = -x' \text{ and } y = -y'$$

Hence, the new equation of curve,

$x - 2y + 3 = 0$ is $(-x') - 2(-y') + 3 = 0$

$$\Rightarrow -x' + 2y' + 3 = 0$$

$$\Rightarrow \quad x' - 2y' - 3 = 0$$

or $\quad x - 2y - 3 = 0$ in general

22 OP is inclined at angle θ with X-axis OQ is inclined at angle $\alpha - 2\theta$ with X-axis. The bisector of angle POQ is inclined at angle

$$\frac{\alpha - 2\theta}{2} + \theta = \frac{\alpha}{2} \text{ with } X\text{-axis.}$$

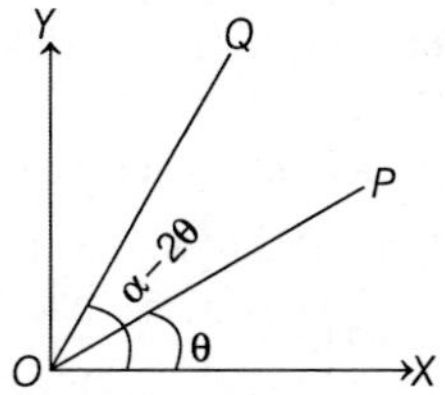

23 Image of (4, 1) in the line $x = y$ is (1, 4) on translating this point along positive direction of X-axis by 2 units, this point is transformed into (3, 4) and projection of the point (3, 4) on X-axis is (3, 0).

24 $BP - AP = \pm 6$ or $BP = AP \pm 6$

$$\Rightarrow \sqrt{x^2 + (y + 4)^2} = \sqrt{x^2 + (y - 4)^2} \pm 6$$

On squaring and simplifying, we get

$$4y - 9 = \pm 3\sqrt{x^2 + (y - 4)^2}$$

Again on squaring, we get

$$9x^2 - 7y^2 + 63 = 0$$

25 Let $G(\alpha, \beta)$ be the centroid in any position. Then,

$$(\alpha, \beta) = \left(\frac{1 + \cos t + \sin t}{3}, \frac{2 + \sin t - \cos t}{3}\right)$$

$$\therefore \quad \alpha = \frac{1 + \cos t + \sin t}{3}$$

$$\text{and } \beta = \frac{2 + \sin t - \cos t}{3}$$

$$\Rightarrow \quad 3\alpha - 1 = \cos t + \sin t \quad \text{...(i)}$$

$$\text{and } 3\beta - 2 = \sin t - \cos t \quad \text{...(ii)}$$

On squaring and adding Eqs. (i) and (ii), we get

$$(3\alpha - 1)^2 + (3\beta - 2)^2 = (\cos t + \sin t)^2 + (\sin t - \cos t)^2$$

$$= 2(\cos^2 t + \sin^2 t) = 2$$

$\therefore$ The equation of the locus of the centroid is $(3x - 1)^2 + (3y - 2)^2 = 2$

$$\Rightarrow \quad 9(x^2 + y^2) - 6x - 12y + 3 = 0$$

$$\Rightarrow \quad 3(x^2 + y^2) - 2x - 4y + 1 = 0$$

26 The third vertex lies on $2x + 3y = 9$

i.e. $\left(x, \frac{9 - 2x}{3}\right)$

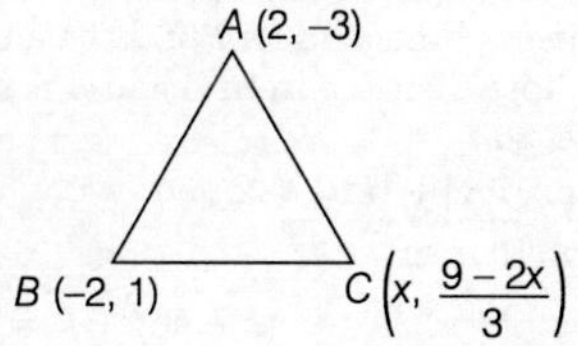

$\therefore$ Locus of centroid is

$$\left(\frac{2-2+x}{3}, \frac{-3+\dfrac{9-2x}{3}+1}{3}\right) = (h,k)$$

$$\therefore \quad h = \frac{x}{3} \text{ and } k = \frac{3-2x}{9}$$

$\Rightarrow \; 9k = 3 - 2(3h) \Rightarrow \; 9k = 3 - 6h$

$\Rightarrow \; 2h + 3k = 1$

Hence, locus of a point is $2x + 3y = 1$.

27 Let third vertex be $C(x_1, y_1)$.

$\therefore$ Centroid $\left(\frac{-3-2+x_1}{3}, \frac{2+1+y_1}{3}\right)$ lies on line

$$3x + 4y + 3 = 0$$

28 Let $P(\alpha, \beta)$ be any point such that

$$(PA) = k(PB)$$

$\Rightarrow \; (PA)^2 = k^2(PB)^2$

$\Rightarrow \; (\alpha - ak)^2 + \beta^2 = k^2\left\{\left(\alpha - \frac{a}{k}\right)^2 + \beta^2\right\}$

$\Rightarrow \alpha^2 + \beta^2 - 2ak\alpha + a^2k^2 = k^2\alpha^2 + k^2\beta^2 - \frac{2ak^2}{k}\alpha + a^2$

$\Rightarrow (1-k^2)\alpha^2 + (1-k^2)\beta^2 = (1-k^2)a^2$

$\Rightarrow (1-k^2)\{\alpha^2 + \beta^2\} = (1-k^2)a^2$

$\{\because k \neq \pm 1\}$

$\therefore \quad \alpha^2 + \beta^2 = a^2$

Replace α by x and β by y, we have

$$x^2 + y^2 = a^2$$

29 Let $P(h, k)$ represents all those points subtending a right angle at A and B

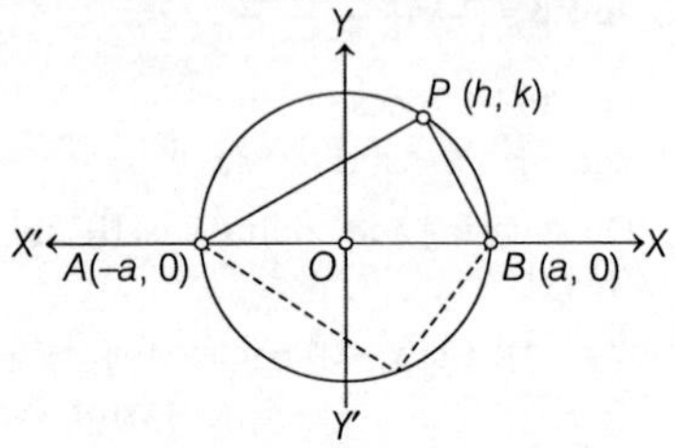

$\therefore \quad m_{AP} \cdot m_{PB} = -1$

$\Rightarrow \left(\frac{k-0}{h+a}\right)\left(\frac{k-0}{h-a}\right) = -1$

$\Rightarrow \quad k^2 = -(h^2 - a^2)$

$\Rightarrow \quad k^2 + h^2 = a^2$

Replace $k \to y$ and $h \to x$, we get

$$x^2 + y^2 = a^2$$

30 Since, $A(\text{ae}, 0)$ and $B(-\text{ae}, 0)$ be the given points and let $P(h, k)$ be any point whose distance from A and B is constant i.e. $2a$.

i.e. $|PA| + |PB| = 2a$

$\Rightarrow \sqrt{(h-\text{ae})^2 + k^2} + \sqrt{(h+\text{ae})^2 + k^2} = 2a$ ...(i)

Let us assume

$\{(h-\text{ae})^2 + k^2\} - \{(h+\text{ae})^2 + k^2\} = -4aeh$ ...(ii)

On dividing Eqs. (ii) by (i), we have

$$\frac{\{(h-\text{ae})^2 + k^2\} - \{(h+\text{ae})^2 + k^2\}}{\sqrt{(h-\text{ae})^2 + k^2} + \sqrt{(h+\text{ae})^2 + k^2}} = \frac{-4aeh}{2a}$$

$\Rightarrow \sqrt{(h-\text{ae})^2 + k^2} - \sqrt{(h+\text{ae})^2 + k^2} = -2eh$ ...(iii)

$\{\because a - b = (\sqrt{a} + \sqrt{b})(\sqrt{a} - \sqrt{b})\}$

Adding Eqs. (i) and (iii), we have

$$2\sqrt{(h-\text{ae})^2 + k^2} = (2a - 2eh)$$

$\Rightarrow \; 2\sqrt{(h-ae)^2 + k^2} = 2(a - eh)$

Squaring both sides, we have

$$(h - ae)^2 + k^2 = (a - eh)^2$$

$\Rightarrow h^2 + a^2e^2 - 2aeh + k^2 = a^2 + e^2h^2 - 2aeh$

$\Rightarrow \quad h^2 - e^2h^2 + k^2 = a^2 - a^2e^2$

$\Rightarrow \; h^2(1-e^2) + k^2 = a^2(1-e^2)$

Replacing h by x and k by y, we get the locus of point $P(h, k)$ which is the locus of an ellipse.

$$\frac{x^2}{a^2} + \frac{y^2}{a^2(1-e^2)} = 1$$

31 **Statement I** : $AB = BC = CA$

$\therefore A, B, C$ are the vertices of triangle ABC.

Statement II : Let $A(x_1, y_1), B(x_2, y_2)$ and $C(x_3, y_3)$ are all rational coordinates.

$$\therefore \text{ Area } (\Delta ABC) = \frac{1}{2}\begin{vmatrix} x_1 & y_1 & 1 \\ x_2 & y_2 & 1 \\ x_3 & y_3 & 1 \end{vmatrix} = \frac{\sqrt{3}}{4}[(x_1 - x_2)^2 + (y_1 - y_2)^2]$$

LHS = rational, RHS = irrational

Hence, $(x_1, y_1), (x_2, y_2)$ and (x_3, y_3) cannot be all rational.

32 The orthocentre lies on the line joining the points (0, 0) and (3, 5) i.e.

$5x - 3y = 0$.

Also, Statement II is true.

33 Statement II is true as the coordinates of the point P in new system are $(\alpha - 1 - 2, \alpha + 1 - 3)$.

In Statement I, the centroid is (2, 3), so the coordinates of the vertices in the new system of coordinates are $(-2, -3)$, (1, 0), (1, 3).

34 Statement II is false as $L_1 + \lambda L_2 = 0$

$\Rightarrow$ Family of concurrent lines, if L_1 and L_2 are intersect.

$\Rightarrow$ Family of parallel lines, if L_1 and L_2 are parallel.

$\Rightarrow$ Family of coincident lines, if L_1 and L_2 are coincident.

As a, b and c are in AP.

$\Rightarrow \; 2b = a + c \Rightarrow a - 2b + c = 0$

On comparing with $ax + by + c = 0$, it passes through fixed points $(1, -2)$.

35 Here, $L_1 : y - x = 0$ and $L_2 : 2x + y = 0$ and $L_3 : y + 2 = 0$ as shown below,

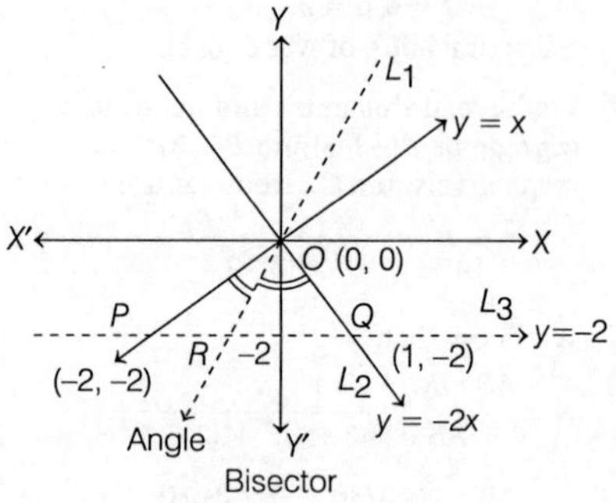

$|PO| = \sqrt{4+4}$

$= 2\sqrt{2}; |OQ| = \sqrt{1+4} = \sqrt{5}$

Since, OR is angle bisector

$$\frac{OP}{OQ} = \frac{PR}{RQ}$$

$$\Rightarrow \quad \frac{PR}{RQ} = \frac{2\sqrt{2}}{\sqrt{5}}$$

Hence, Statement I is true.

But, it does not divide the triangle in two similar triangles.

Hence, Statement II is false.

SESSION 2

1 Let the coordinates of P be (x, y).

Then, $PA = PB \Rightarrow PA^2 = PB^2$

$\Rightarrow (x-3)^2 + (y-4)^2 = (x-5)^2 + (y+2)^2$

$\Rightarrow \quad x - 3y - 1 = 0$ ...(i)

Now, area of $\Delta PAB = 10$

$$\Rightarrow \frac{1}{2}\begin{vmatrix} x & y & 1 \\ 3 & 4 & 1 \\ 5 & -2 & 1 \end{vmatrix} = \pm 10$$

$\Rightarrow \quad 6x + 2y - 26 = \pm 20$

$\Rightarrow \quad 6x + 2y - 46 = 0$

or $\quad 6x + 2y - 6 = 0$

$\Rightarrow \quad 3x + y - 23 = 0$

or $\quad 3x + y - 3 = 0$ ...(ii)

On solving, $x - 3y - 1 = 0$ and $3x + y - 23 = 0$, we get

$$x = 7, y = 2$$

On solving $x - 3y - 1 = 0$ and $3x + y - 3 = 0$, we get

$$x = 1, y = 0$$

Thus, the coordinates of P are (7,2) or (1, 0).

2 The coordinates of A_1 are $\left(\frac{a}{2},\frac{b}{2}\right)$

The coordinates of A_2 are $\left(\frac{a+\frac{a}{2}}{2},\frac{b+\frac{b}{2}}{2}\right)$

$$=\left(\frac{a}{2}+\frac{a}{2^2},\frac{b}{2}+\frac{b}{2^2}\right)$$

The coordinates of A_3 are

$$=\left(\frac{a+\frac{a}{2}+\frac{a}{2^2}}{2},\frac{b+\frac{b}{b^2}+\frac{b}{2^2}}{2}\right)$$

$$=\left(\frac{a}{2}+\frac{a}{2^2}+\frac{a}{2^3},\frac{b}{2}+\frac{b}{2^2}+\frac{b}{2^3}\right)$$

Continuing in this manner we observe that the coordinates of A_n are

$$\left(\frac{a}{2}+\frac{a}{2^2}+\frac{a}{2^3}+\ldots+\frac{a}{2^n},\ \frac{b}{2}+\frac{b}{2^2}+\frac{b}{2^3}+\ldots+\frac{b}{2^n}\right)$$

$$=\left(a\left(1-\frac{1}{2^n}\right),b\left(1-\frac{1}{2^n}\right)\right)$$

$$=(a(1-2^{-n}),b(1-2^{-n}))$$

3

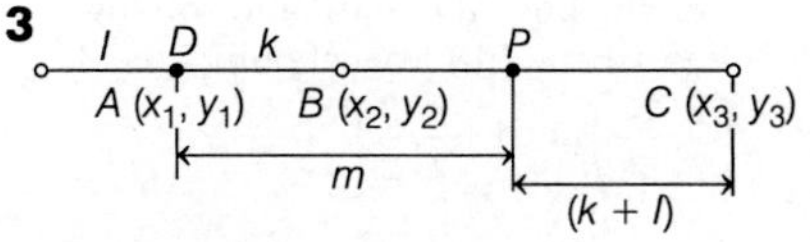

Coordinates of point D are

$$D\left(\frac{lx_2+kx_1}{k+l},\frac{ly_2+ky_1}{k+l}\right)$$

Coordinates of point P are

$$P\left(\frac{lx_2+kx_1+mx_3}{k+l+m},\frac{ly_2+ky_1+my_3}{k+l+m}\right)$$

4 Let $P(h,k)$ be any point such that

$$2(PA)=3(PB)$$

$$\Rightarrow\quad 4(PA)^2=9(PB)^2$$

$$\Rightarrow 4(h^2+k^2)=9\{(h-4)^2+(k+3)^2\}$$

$$\Rightarrow\ 4(h^2+k^2)=9(h^2+k^2-8h+6k+25)$$

$$\Rightarrow\ 5h^2+5k^2-72h+54k+225=0$$

$\therefore$ Required locus is

$$5x^2+5y^2-72x+54y+225=0$$

5 Let $\angle RPQ=\theta$ and $\angle RQP=\phi$

$$\therefore\qquad \theta-\phi=2\alpha$$

Let $RM\perp PQ$, so that $RM=k$,
$MP=a-h$

and $\quad MQ=a+h$

Then, $\tan\theta=\dfrac{RM}{MP}=\dfrac{k}{a-h}$

and $\quad\tan\phi=\dfrac{RM}{MQ}=\dfrac{k}{a+h}$

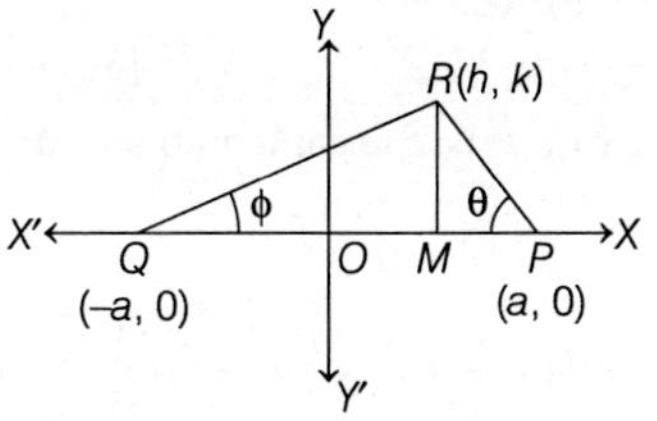

Again, now $2\alpha=\theta-\phi$

$$\therefore\qquad \tan 2\alpha=\tan(\theta-\phi)$$

$$=\frac{\tan\theta-\tan\phi}{1+\tan\theta\tan\phi}$$

$$=\frac{k(a+h)-k(a-h)}{a^2-h^2+k^2}$$

$\Rightarrow a^2-h^2+k^2=2hk\cot 2\alpha$

Hence, the locus is

$$x^2-y^2+2xy\cot 2\alpha-a^2=0$$

6 Here, $\tan\theta=2$

So, $\quad\cos\theta=\dfrac{1}{\sqrt5},\sin\theta=\dfrac{2}{\sqrt5}$

For x and y, we have

$$x=X\cos\theta-Y\sin\theta=\frac{X-2Y}{\sqrt5}$$

and $\ y=X\sin\theta+Y\cos\theta=\dfrac{2X+Y}{\sqrt5}$

The equation $4xy-3x^2=a^2$ reduces to

$$\frac{4(X-2Y)}{\sqrt5}\cdot\frac{(2X+Y)}{\sqrt5}-3\left(\frac{X-2Y}{\sqrt5}\right)^2=a^2$$

$$\Rightarrow\quad 4(2X^2-2Y^2-3XY)-3(X^2-4XY+4Y^2)=5a^2$$

$$\Rightarrow\qquad 5X^2-20Y^2=5a^2$$

$$\therefore\qquad X^2-4Y^2=a^2$$

7 Let the vertices be C,A and B, respectively. The altitude from A is

$$\frac{y-a(t_2+t_3)}{x-at_2t_3}=-t_1$$

$$\Rightarrow xt_1+y=at_1t_2t_3+a(t_2+t_3)\qquad\ldots\text{(i)}$$

The altitude from B is

$$xt_2+y=at_1t_2t_3+a(t_3+t_1)\quad\ldots\text{(ii)}$$

On subtracting Eq. (ii) from Eq. (i), we get $x=-a$

Hence, $y=a(t_1+t_2+t_3+t_1t_2t_3)$

So, the orthocentre is
$\{-a,a(t_1+t_2+t_3+t_1t_2t_3)\}$.

8 Given that, the triangle ABC is isosceles

$\therefore\quad |AB|=|AC|$

Let the coordinate of A are $A(h,k)$

$$\therefore\ \sqrt{(h-1)^2+(k-3)^2}=\sqrt{(h+2)^2+(k-7)^2}$$

$$\Rightarrow\quad (h-1)^2+(k-3)^2=(h+2)^2+(k-7)^2$$

$$\Rightarrow -2h+1-6k+9=4h+4-14k+49$$

$$\Rightarrow\quad 6h-8k+43=0\qquad\ldots\text{(i)}$$

Since, the area of triangle is 10 sq unit {given}

$$\text{ar}(\Delta ABC)=\frac12|BC||AC|$$

$$\Rightarrow\ \left|\frac12(5)\sqrt{\left(h+\frac12\right)^2+(k-5)^2}\right|=\frac{25}{6}$$

On squaring, we get

$$\Rightarrow\ \left(h+\frac12\right)^2+(k-5)^2=\frac{25}{9}$$

Using Eq. (i), we have

$$\Rightarrow\left(\frac{8k-43}{6}+\frac12\right)^2+(k-5)^2=\frac{25}{9}$$

$$\Rightarrow(4k-20)^2+9(k-5)^2=25$$

$$\Rightarrow\qquad 25\cdot(k-5)^2=25$$

$$\Rightarrow\qquad (k-5)^2=1$$

$$\Rightarrow\qquad |k-5|=1$$

$$\Rightarrow\qquad k-5=\pm1$$

$\therefore\ k=1+5$ or $k=-1+5$

$\Rightarrow k=6$ or $k=4$

Using Eq. (i), we have $h=\dfrac56$

Using Eq. (i), we have

$$h=-\frac{11}{6}$$

Therefore, the vertex A of the isosceles ΔABC is $A\left(\frac56,6\right)$ or $A\left(-\frac{11}{6},4\right)$.

9 $\text{ar}(\Delta PBC)=\dfrac12\left\|\begin{matrix}\alpha&\beta&1\\-3&5&1\\4&-2&1\end{matrix}\right\|$

$$\Rightarrow\text{ar}(\Delta PBC)=\frac12|7\alpha+7\beta-14|$$

$$=\frac72|\alpha+\beta-2|$$

Also, $\text{ar}(\Delta ABC)=\dfrac12\left\|\begin{matrix}6&-3&1\\-3&5&1\\4&-2&1\end{matrix}\right\|$

$$\Rightarrow\text{ar}(\Delta ABC)=\frac12|42-21-14|=\frac72$$

$$\frac{\text{ar}(\Delta PBC)}{\text{ar}(\Delta\ ABC)}=|\alpha+\beta-2|$$

10 By PROJECTION FORMULA, we have

$$\cos\angle P_1OP_2=\frac{|OP_1|^2+|OP_2|^2-|P_1P_2|^2}{2|OP_1||OP_2|}$$

Let $E=|OP_1||OP_2|\cos\angle P_1OP_2$

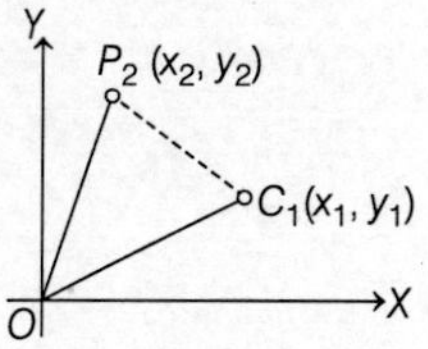

$$\Rightarrow E = \frac{(x_1^2 + y_1^2) + (x_2^2 + y_2^2) - \{(x_2 - x_1)^2 + (y_2 - y_1)^2\}}{2}$$

$$\Rightarrow E = \frac{[x_1^2 + x_2^2 + y_1^2 + y_2^2 - (x_2 - x_1)^2 - (y_2 - y_1)^2]}{2}$$

$$\Rightarrow E = \frac{2x_1x_2 + 2y_1y_2}{2}$$

$\therefore |OP_1||OP_2| \cos \angle P_1OP_2 = x_1x_2 + y_1y_2$

11 The points $A(0, 0)$, $B(2, 2\sqrt{3})$ and $C(a, b)$ are the vertices of an equilateral triangle if

$|AB| = |BC| = |CA|$

$\Rightarrow |AB|^2 = |BC|^2 = |CA|^2$

$\Rightarrow 4 + 12 = (a - 2)^2 + (b - 2\sqrt{3})^2$

$= a^2 + b^2$

Now, $(a - 2)^2 + (b - 2\sqrt{3})^2 = a^2 + b^2$

$a^2 + b^2 - 4a - 4\sqrt{3}b + 16 = a^2 + b^2$

$a + \sqrt{3}b = 4$

$a = 4 - \sqrt{3}b$ ...(i)

Also, $a^2 + b^2 = 16$

$(4 - \sqrt{3}b)^2 + b^2 = 16$ [using Eq. (i)]

$\Rightarrow 4b^2 - 8\sqrt{3}b + 16 = 16$

$\Rightarrow 4b(b - 2\sqrt{3}) = 0$

$\Rightarrow b = 0$ or $b = 2\sqrt{3}$

If $b = 0$

$\Rightarrow a = 4$

or if $b = 2\sqrt{3}$

$\Rightarrow a = -2$ [using Eq. (i)]

12 Let (h, k) be the point on the locus. Then by the given conditions

$(h - a_1)^2 + (k - b_1)^2 = (h - a_2)^2 + (k - b_2)^2$

$\Rightarrow 2h(a_1 - a_2) + 2k(b_1 - b_2) + a_2^2 + b_2^2 - a_1^2 - b_1^2 = 0$

$\Rightarrow h(a_1 - a_2) + k(b_1 - b_2) + \frac{1}{2}(a_2^2 + b_2^2 - a_1^2 - b_1^2) = 0$ ...(i)

Since, the locus of (h, k) is the line

$(a_1 - a_2)h + (b_1 - b_2)k + c = 0$ ...(ii)

$\therefore$ Comparing Eqs. (i) and (ii), we get

$c = \frac{1}{2}(a_2^2 + b_2^2 - a_1^2 - b_1^2)$

13 Circumcentre of a triangle is the point which is equidistant from the vertices of a triangle.

Let the circumcentre of triangle be $C(x, y)$ and the three vertices of the triangle are represented by $P(2, 1)$, $Q(5, 2)$, $R(3, 4)$

$\therefore$ According to given condition, we have

$|PC| = |QC| = |RC|$

Case I $|PC| = |QC|$

$(x - 2)^2 + (y - 1)^2 = (x - 5)^2 + (y - 2)^2$

$\Rightarrow 6x + 2y = 24$ ...(i)

Case II $|PC| = |RC|$

$(x - 2)^2 + (y - 1)^2 = (x - 3)^2 + (y - 4)^2$

$\Rightarrow 2x + 6y = 20$ ...(ii)

Solving Eqs. (i) and (ii) for x and y, we have

$\therefore$ Co-ordinates of circumcentre are

$C(x, y) = C\left(\frac{13}{4}, \frac{9}{4}\right)$

14 Let $P(h, k)$ be the point such that

$|PA|^2 + |PB|^2 = |AB|^2$

$\Rightarrow (h - 2)^2 + k^2 + (h + 2)^2 + k^2 = 4^2 + 0$

$\Rightarrow 2h^2 + 8 + 2k^2 = 16$

$\Rightarrow h^2 + k^2 = 4$

$\therefore$ Locus of P is $x^2 + y^2 = 4$

15. Let the third vertex be (p, q).

$\Rightarrow q = p + 3$ ...(i)

Now, $\Delta = |5|$

$\Delta = \pm 5$

$$\frac{1}{2}\begin{vmatrix} p & q & 1 \\ 2 & 1 & 1 \\ 3 & -2 & 1 \end{vmatrix} = \pm 5$$

$\Rightarrow q + 3p - 7 = \pm 10$

$\Rightarrow 3p + q = 17$ ...(ii)

and $3p + q = -3$ ...(iii)

Solving Eqs. (i) and (ii) and solving Eqs. (i) and (iii), we get points $\left(\frac{7}{2}, \frac{13}{2}\right)$ and $\left(-\frac{3}{2}, \frac{3}{2}\right)$

DAY TWENTY FIVE

Straight Line

Learning & Revision for the Day

- Concept of Straight Line
- Angle between Two Lines
- Conditions for Concurrence of Three Lines
- Distance of a Point from a Line

Concept of Straight Line

Any curve is said to be a **straight line,** if for any two points taken on the curve, each and every point on the line segment joining these two points lies on the curve.

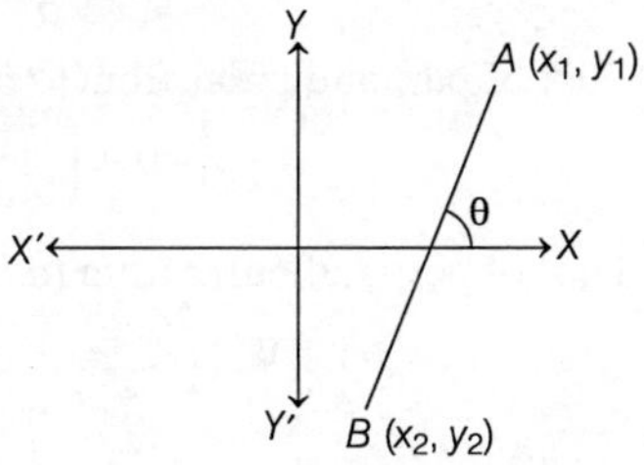

The slope of a line AB is $m = \tan\theta = \dfrac{y_2 - y_1}{x_2 - x_1}$

Various Forms of Equations of a Line

The equation of a line in the general form can be written as $ax + by + c = 0$

1. **Slope Intercept Form** The equation of a line with slope m and making an intercept c on Y-axis is $y = mx + c$
2. **Point Slope Form** The equation of a line which passes through the point (x_1, y_1) and has the slope m is $y - y_1 = m(x - x_1)$.
3. **Two Points Form** The equation of a line passing through two points (x_1, y_1) and (x_2, y_2) is
$$(y - y_1) = \left(\frac{y_2 - y_1}{x_2 - x_1}\right)(x - x_1).$$
4. **Intercept Form of a Line** The equation of a line which cuts off intercepts a and b respectively from the X and Y-axes is $\dfrac{x}{a} + \dfrac{y}{b} = 1$.

PREP MIRROR

Your Personal Preparation Indicator

- No. of Questions in Exercises (x)—
- No. of Questions Attempted (y)—
- No. of Correct Questions (z)—
  (*Without referring Explanations*)

- Accuracy Level $(z/y \times 100)$—
- Prep Level $(z/x \times 100)$—

In order to expect good rank in JEE, your Accuracy Level should be above 85 & Prep Level should be above 75.

5. **Normal or Perpendicular Form** The equation of the straight line upon which the length of the perpendicular from the origin is p and this perpendicular makes an angle α with positive direction of X-axis in anti-clockwise sense is

$$x\cos\alpha + y\sin\alpha = p, \text{ where } 0 \le \alpha \le 2\pi.$$

6. **General Equation of a Line to the Normal Form** The general equation of a line is

$$Ax + By + C = 0$$

Now, to reduce the general equation of a line to normal form, we first shift the constant term on the RHS and make it positive, if it is not so and then divide both sides by $\sqrt{(\text{coefficient of } x)^2 + (\text{coefficient of } y)^2}$

$$\Rightarrow \left(\frac{A}{\sqrt{A^2+B^2}}\right)x + \left(\frac{B}{\sqrt{A^2+B^2}}\right)y = \left(\frac{-C}{\sqrt{A^2+B^2}}\right)$$

Now, take $\cos\alpha = \frac{A}{\sqrt{A^2+B^2}}$, $\sin\alpha = \frac{B}{\sqrt{A^2+B^2}}$ and $p = \frac{-C}{\sqrt{A^2+B^2}}$, which gives the required normal form.

7. **Intersection of lines** Let the equation of lines be $a_1x + b_1y + c_1 = 0$ and $a_2x + b_2y + c_2 = 0$, then their point of intersection is $\left(\frac{b_1c_2 - b_2c_1}{a_1b_2 - a_2b_1}, \frac{c_1a_2 - c_2a_1}{a_1b_2 - a_2b_1}\right)$

8. **Distance Form or Parametric form** The equation of the straight line passing through (x_1, y_1) and making an angle θ with the positive direction of X-axis is $\frac{x - x_1}{\cos\theta} = \frac{y - y_1}{\sin\theta} = r$,

where r is the distance of any point (x, y) on the line from the point (x_1, y_1).

Angle between Two Lines

The acute angle θ between the lines having slopes m_1 and m_2 is given by $\tan\theta = \left|\frac{m_2 - m_1}{1 + m_1m_2}\right|$.

Condition of Parallel Lines

Let m_1, m_2 be slope of two lines, then lines are parallel, if $m_1 = m_2$.

Equation of any line parallel to $ax + by + c = 0$ can be taken as $ax + by + \lambda = 0$

Condition of Perpendicular Lines

Let m_1, m_2 be slope of two lines, then the lines are perpendicular, if $m_1m_2 = -1$

If one line is parallel to X-axis, then its perpendicular line is parallel to Y-axis

Equation of the line perpendicular to $ax + by + c = 0$ is taken as $bx - ay + \lambda = 0$

Straight line $ax + by + c = 0$ and $a'x + b'y + c' = 0$ are right angle if $aa' + bb' = 0$

Conditions for Concurrence of Three Lines

1. Three lines are said to be concurrent, if they pass through a common point i.e. they meet at a point.
2. If three lines are concurrent, the point of intersection of two lines lies on the third line.
3. The lines $a_1x + b_1y + c_1 = 0$, $a_2x + b_2y + c_2 = 0$ and $a_3x + b_3y + c_3 = 0$, are concurrent iff $\begin{vmatrix} a_1 & b_1 & c_1 \\ a_2 & b_2 & c_2 \\ a_3 & b_3 & c_3 \end{vmatrix} = 0$

This is the required condition of concurrence of three lines.

Distance of a Point from a Line

(i) The length of the perpendicular from a point (x_1, y_1) to a line $ax + by + c = 0$ is $\left|\frac{ax_1 + by_1 + c}{\sqrt{a^2 + b^2}}\right|$.

(ii) Distance between two parallel lines $ax + by + c_1 = 0$ and $ax + by + c_2 = 0$ is $\frac{|c_2 - c_1|}{\sqrt{a^2 + b^2}}$.

Important Results

- The foot of the perpendicular (h, k) from (x_1, y_1) to the line $ax + by + c = 0$ is given by $\frac{h - x_1}{a} = \frac{k - y_1}{b} = -\frac{(ax_1 + by_1 + c)}{a^2 + b^2}$
- Foot of perpendicular from (a, b) on $x - y = 0$ is $\left(\frac{a+b}{2}, \frac{a+b}{2}\right)$.
- Foot of perpendicular from (a, b) on $x + y = 0$ is $\left(\frac{a-b}{2}, \frac{b-a}{2}\right)$.
- Image (h, k) from (x_1, y_1) w.r.t. the line mirror $ax + by + c = 0$ is given by $\frac{h - x_1}{a} = \frac{k - y_1}{b} = \frac{-2(ax_1 + by_1 + c)}{a^2 + b^2}$
- Area of the parallelogram formed by the lines $a_1x + b_1y + c_1 = 0$; $a_2x + b_2y + c_2 = 0$ $a_1x + b_1y + d_1 = 0$; $a_2x + b_2y + d_2 = 0$ is $\left|\frac{(d_1 - c_1)(d_2 - c_2)}{a_1b_2 - a_2b_1}\right|$.

Equation of Internal and External Bisectors of Angles between Two Lines

The bisectors of the angles between two straight lines are the locus of a point which is equidistant from the two lines. The equation of the bisector of the angles between the lines

$$a_1x + b_1y + c_1 = 0 \quad \ldots\text{(i)}$$

and

$$a_2x + b_2y + c_2 = 0 \quad \ldots\text{(ii)}$$

are given by, $\frac{a_1x + b_1y + c_1}{\sqrt{a_1^2 + b_1^2}} = \pm\frac{a_2x + b_2y + c_2}{\sqrt{a_2^2 + b_2^2}}$

where,

(i) if $a_1a_2 + b_1b_2 > 0$, the positive sign for obtuse and negative sign for acute.

(ii) $a_1a_2 + b_1b_2 < 0$, negative sign for obtuse and positive sign for acute.

Equation of Family of Lines Through the Intersection of Two given Lines

The equation of the family of lines passing through the intersection of the lines

$a_1x + b_1y + c_1 = 0$ and $a_2x + b_2y + c_2 = 0$ is

$(a_1x + b_1y + c_1) + \lambda(a_2x + b_2y + c_2) = 0$.

where λ is a parameter.

Important Properties

(i) The **position** of a **point** (x_1, y_1) and (x_2, y_2) relative to the line $ax + by + c = 0$

(a) If $\dfrac{(ax_1 + by_1 + c)}{ax_2 + by_2 + c} > 0$, then points lie on the same side.

(b) If $\dfrac{ax_1 + by_1 + c}{ax_2 + by_2 + c} < 0$, then the points lie on opposite side.

(ii) The equations of the straight lines which pass through a given point (x_1, y_1) and make a given angle α with the given straight line $y = mx + c$ are

$(y - y_1) = \dfrac{m \pm \tan\alpha}{1 \mp \tan\alpha}(x - x_1)$.

DAY PRACTICE SESSION 1

FOUNDATION QUESTIONS EXERCISE

1 The equation of the line, the reciprocals of whose intercepts on the axes are a and b, is given by:

(a) $\dfrac{x}{a} + \dfrac{y}{b} = 1$ (b) $ax + by = 1$

(c) $ax + by = ab$ (d) $ax - by = 1$

2 The equation of the straight line passing through the point (4, 3) and making intercepts on the coordinate axes whose sum is -1, is

(a) $\dfrac{x}{2} + \dfrac{y}{3} = -1, \dfrac{x}{-2} + \dfrac{y}{1} = -1$

(b) $\dfrac{x}{2} - \dfrac{y}{3} = -1, \dfrac{x}{-2} + \dfrac{y}{1} = -1$

(c) $\dfrac{x}{2} + \dfrac{y}{3} = 1, \dfrac{x}{2} + \dfrac{y}{1} = 1$

(d) $\dfrac{x}{2} - \dfrac{y}{3} = 1, \dfrac{x}{-2} + \dfrac{y}{1} = 1$

3 The equation of a line passing through $(-4, 3)$ and this point divided the portion of line between axes in the ratio 5:3 internally, is

(a) $9x + 20y + 96 = 0$ (b) $20x + 9y + 96 = 0$

(c) $9x - 20y + 96 = 0$ (d) $20x - 9y - 96 = 0$

4 A straight line through a fixed point (2,3) intersects the coordinate axes at distinct points P and Q. If O is the origin and the rectangle $OPRQ$ is completed, then the locus of R is

(a) $3x + 2y = 6$ (b) $2x + 3y = xy$

(c) $3x + 2y = xy$ (d) $3x + 2y = 6xy$

5 If the x-intercept of some line L is double as that of the line, $3x + 4y = 12$ and the y-intercept of L is half as that of the same line, then the slope of L is ➜ JEE Mains 2013

(a) -3 (b) $-\dfrac{3}{8}$ (c) $-\dfrac{3}{2}$ (d) $-\dfrac{3}{16}$

6 For which values of a and b, intercepts on axes by line $ax + by + 8 = 0$ are equal and opposite in sign of intercepts on axis by line $2x - 3y + 6 = 0$

(a) $a = \dfrac{8}{3}, b = -4$ (b) $a = \dfrac{-8}{3}, b = -4$

(c) $a = \dfrac{8}{3}, b = 4$ (d) $a = \dfrac{-8}{3}, b = 4$

7 Equation of the line passing through the points of intersection of the parabola $x^2 = 8y$ and the ellipse $\dfrac{x^2}{3} + y^2 = 1$ is ➜ JEE Mains 2013

(a) $y - 3 = 0$ (b) $y + 3 = 0$ (c) $3y + 1 = 0$ (d) $3y - 1 = 0$

8 A ray of light along $x + \sqrt{3}y = \sqrt{3}$ gets reflected upon reaching X-axis, the equation of the reflected ray is

(a) $y = x + \sqrt{3}$ (b) $\sqrt{3}y = x - \sqrt{3}$

(c) $y = \sqrt{3}x - \sqrt{3}$ (d) $\sqrt{3}y = x - 1$

9 The range of values of α such that $(0, \alpha)$ lie on or inside the triangle formed by the lines $3x + y + 2 = 0$, $2x - 3y + 5 = 0$ and $x + 4y - 14 = 0$ is

(a) $1/2 \le \alpha \le 1$ (b) $5/3 \le \alpha \le 7/2$

(c) $5 \le \alpha \le 7$ (d) None of these

10 The lines $x + y = |a|$ and $ax - y = 1$ intersect each other in the first quadrant. Then, the set of all possible values of a is the interval

(a) $(-1, 1]$ (b) $(0, \infty)$

(c) $[1, \infty)$ (d) $(-1, \infty)$

11 Area of the parallelogram formed by the lines $y = mx, y = mx + 1, y = nx, y = nx + 1$ is equal to

(a) $\dfrac{|m + n|}{(m - n)^2}$ (b) $\dfrac{2}{|m + n|}$

(c) $\dfrac{1}{|m + n|}$ (d) $\dfrac{1}{|m - n|}$

12 If PS is the median of the triangle with vertices $P(2,2)$, $Q(6,-1)$ and $R(7,3)$, then equation of the line passing through $(1,-1)$ and parallel to PS is → JEE Mains 2014

(a) $4x-7y-11=0$ (b) $2x+9y+7=0$
(c) $4x+7y+3=0$ (d) $2x-9y-11=0$

13 If $A(2,-1)$ and $B(6,5)$ are two points, then the ratio in which the foot of the perpendicular from $(4,1)$ to AB divides it, is

(a) 8 : 15 (b) 5 : 8 (c) –5 : 8 (d) –8 : 5

14 The line L given by $\frac{x}{5}+\frac{y}{b}=1$ passes through the point (13, 32). The line K is parallel to L and has the equation $\frac{x}{c}+\frac{y}{3}=1$. Then, the distance between L and K is

(a) $\frac{23}{\sqrt{15}}$ (b) $\sqrt{17}$ (c) $\frac{17}{\sqrt{15}}$ (d) $\frac{23}{\sqrt{17}}$

15 The nearest point on the line $3x-4y=25$ from the origin is

(a) $(-4,5)$ (b) $(3,-4)$ (c) $(3,4)$ (d) $(3,5)$

16 Two sides of rhombus are along the lines, $x-y+1=0$ and $7x-y-5=0$ if its diagonals intersect at $(-1,-2)$, then which one of the following is a vertex of this rhombus → JEE Mains 2016

(a) $(-3,-9)$ (b) $(-3,-8)$
(c) $\left(\frac{1}{3},\frac{-8}{3}\right)$ (d) $\left(\frac{-10}{3},\frac{-7}{3}\right)$

17 If the lines $ax+2y+1=0$, $bx+3y+1=0$, $cx+4y+1=0$ are concurrent, then a, b, c are in

(a) AP (b) GP
(c) HP (d) None of these

18 For all real values of a and b lines $(2a+b)x+(a+3b)y+(b-3a)=0$ and $mx+2y+6=0$ are concurrent, then m is equal to

(a) -2 (b) -3 (c) -4 (d) -5

19 If p is the length of perpendicular from origin to the line which intercepts a and b on axes, then

(a) $a^2+b^2=p^2$ (b) $a^2+b^2=\frac{1}{p^2}$
(c) $\frac{1}{a^2}+\frac{1}{b^2}=\frac{2}{p^2}$ (d) $\frac{1}{a^2}+\frac{1}{b^2}=\frac{1}{p^2}$

20 A straight line through the origin O meets the parallel lines $4x+2y=9$ and $2x+y=-6$ at points P and Q, respectively. Then, the point O divides the segment PQ in the ratio

(a) 1 : 2 (b) 3 : 4 (c) 2 : 1 (d) 4 : 3

21 If p is the length the perpendicular from the origin on the line $\frac{x}{a}+\frac{y}{b}=1$ and a^2, p^2, b^2 are in AP then $a^4+b^4=$

(a) 0 (b) 1
(c) data is inconsistent (d) None of these

22 If p and p' be perpendiculars from the origin upon the straight lines $x\sec\theta+y\,\text{cosec}\,\theta=a$ and $x\cos\theta-y\sin\theta=a\cos 2\theta$ respectively, then the value of the expression $4p^2+p'^2$ is

(a) a^2 (b) $3a^2$ (c) $2a^2$ (d) $4a^2$

23 If p_1, p_2, p_3, are lengths of perpendiculars from points $(m^2, 2m)$, $(mm', m+m')$ and $(m'^2, 2m')$ to the line $x\cos\alpha+y\sin\alpha+\frac{\sin^2\alpha}{\cos\alpha}=0$, then p_1, p_2, p_3 are in

(a) AP (b) GP
(c) HP (d) None

24 The equation of bisector of acute angle between lines $3x-4y+7=0$ and $12x+5y-2=0$ is

(a) $21x+77y-101=0$ (b) $11x-3y+9=0$
(c) $31x+77y+101=0$ (d) $11x-3x-9=0$

25 A ray of light coming along the line $3x+4y-5=0$ gets reflected from the line $ax+by-1=0$ and goes along the line $5x-12y-10=0$, then

(a) $a=\frac{64}{115}, b=\frac{112}{15}$ (b) $a=-\frac{64}{115}, b=\frac{8}{115}$
(c) $a=\frac{64}{115}, b=-\frac{8}{115}$ (d) $a=-\frac{64}{115}, b=-\frac{8}{115}$

26 The sides BC, CA, AB of ΔABC are respectively $x+2y=1$, $3x+y+5=0$, $x-y+2=0$. The altitude through B is

(a) $x-3y+1=0$ (b) $x-3y+4=0$
(c) $3x-y+4=0$ (d) $x-y+2=0$

27 A variable straight line drawn through the point of intersection of the lines $\frac{x}{a}+\frac{y}{b}=1$ and $\frac{x}{b}+\frac{y}{a}=1$ meets the coordinates axes at A and B, the locus of the mid-point of AB is

(a) $2xy(a+b)=ab(x+y)$
(b) $2xy(a-b)=ab(x-y)$
(c) $2xy(a+b)=ab(x-y)$
(d) None of the above

28 The base BC of ΔABC is bisected at the point (p,q) and equations of AB and AC are $px+qy=1$ and $qx+py=1$ respectively, then equation of the median passing through A is

(a) $(2pq-1)(px+qy-1)=(p^2+q^2-1)(qx+py-1)$
(b) $(2pq+1)(px+qy-1)=(p^2+q^2-1)(qx+py-1)$
(c) $(2pq+1)(px+qy-1)=(p^2+q^2+1)(qx+py-1)$
(d) None of the above

29 If P is a point (x, y) on the line $y=-3x$ such that P and the point $(3, 4)$ are on the opposite sides of the line $3x-4y=8$, then

(a) $x>\frac{8}{15}, y<-\frac{8}{5}$ (b) $x>\frac{8}{5}, y<\frac{8}{15}$
(c) $x=\frac{8}{15}, y=-\frac{8}{5}$ (d) None of these

30 If $P\left(1+\frac{\alpha}{\sqrt{2}}, 2+\frac{\alpha}{\sqrt{2}}\right)$ be any point on a line, then the range of values of α for which the point P lies between the parallel lines $x+2y=1$ and $2x+4y=15$ is

(a) $-\frac{4\sqrt{2}}{3}<\alpha<\frac{5\sqrt{2}}{6}$ (b) $0<\alpha<\frac{5\sqrt{2}}{6}$
(c) $\frac{-4\sqrt{2}}{3}<\alpha<0$ (d) None of these

31 If (a, a^2) falls inside the angle made by the lines $x-2y=0, x>0$ and $y=3x(x>0)$, then a belongs to :

(a) $(0,1/2)$ (b) $(3,\infty)$
(c) $(1/2,3)$ (d) $(-3,-1/2)$

32 The lines passing through $(3,-2)$ and inclined at angle $60°$ with $\sqrt{3}x+y=1$ is

(a) $y+2=0$ (b) $x+2=0$
(c) $x+y=2$ (d) $x-y=\sqrt{3}$

Directions (Q. Nos. 33-36) *Each of these questions contains two statements : Statement I (Assertion) and Statement II (Reason). Each of these questions also has four alternative choices, only one of which is the correct answer. You have to select one of the codes (a), (b), (c) and (d) given below.*

(a) Statement I is true, Statement II is true; Statement II is a correct explanation for Statement I
(b) Statement I is true, Statement II is true; Statement II is not a correct explanation for Statement I
(c) Statement I is true; Statement II is false
(d) Statement I is false; Statement II is true

33 **Statement I** Consider the points $A(0,1)$ and $B(2,0)$ and P be a point on the line $4x+3y+9=0$, then coordinates of P such that $|PA-PB|$ is maximum, is $\left(\frac{-12}{5}, \frac{17}{5}\right)$.

Statement II $|PA-PB| \le |AB|$

34 **Statement I** If point of intersection of the lines $4x+3y=\lambda$ and $3x-4y=\mu, \forall \lambda, \mu \in R$ is (x_1, y_1), then the locus of (λ, μ) is $x+7y=0$, $\forall x_1 = y_1$.

Statement II If $4\lambda+3\mu>0$ and $3\lambda-4\mu>0$, then (x_1, y_1) is in first quadrant.

35 Let θ_1 be the angle between two lines $2x+3y+c_1=0$ and $-x+5y+c_2=0$ and θ_2 be the angle between two lines $2x+3y+c_1=0$ and $-x+5y+c_3=0$, where c_1, c_2, c_3 are any real numbers.

Statement I If c_2 and c_3 are proportional, then $\theta_1=\theta_2$.

Statement II $\theta_1=\theta_2$ for all c_2 and c_3. → JEE Mains 2013

36 **Statement I** Each point on the line $y-x+12=0$ is equidistant from the lines $4y+3x-12=0$, $3y+4x-24=0$.

Statement II The locus of a point which is equidistant from two given lines is the angular bisector of the two lines.

DAY PRACTICE SESSION 2

PROGRESSIVE QUESTIONS EXERCISE

1 A line $4x+y=1$ through the point $A(2,-7)$ meets the line BC whose equation is $3x-4y+1=0$. The equation to the line AC, so that $AB=AC$ is

(a) $52x-89y-519=0$ (b) $52x+89y-519=0$
(c) $52x-89y+519=0$ (d) $52x+89y+519=0$

2 If the lines $y=m_r x, r=1,2,3$ cut off equal intercepts on the transversal $x+y=1$, then $1+m_1, 1+m_2, 1+m_3$ are in:

(a) AP (b) GP
(c) HP (d) None of these

3 In triangle ABC, equation of the right bisectors of the sides AB and AC are $x+y=0$ and $x-y=0$ respectively. If $A\equiv(5,7)$ then equation of side BC is

(a) $7y=5x$ (b) $5x=y$ (c) $5y=7x$ (d) $5y=x$

4 Let k be an integer such that the triangle with vertices $(k,-3k)$, $(5,k)$ and $(-k,2)$ has area 28 sq units. Then, the orthocentre of this triangle is at the point

(a) $\left(2,-\frac{1}{2}\right)$ (b) $\left(1,\frac{3}{4}\right)$ (c) $\left(1,-\frac{3}{4}\right)$ (d) $\left(2,\frac{1}{2}\right)$

5 A variable line through the point (p,q) cuts the x and y axes at A and B respectively. The lines through A and B parallel to Y-axis and the X-axis respectively meet at P. If the locus of P is $3x+2y-xy=0$, then

(a) $p=2, q=3$ (b) $p=3, q=2$
(c) $p=-2, q=-3$ (d) $p=-3, q=-2$

6 If the three lines $x-3y=p$, $ax+2y=q$ and $ax+y=r$ from a right angled triangle, then → JEE Mains 2013

(a) $a^2-9a+18=0$
(b) $a^2-6a-12=0$
(c) $a^2-6a-18=0$
(d) $a^2-9a+12=0$

7 A light ray emerging from the point source placed at $P(1,3)$ is reflected at a point Q in the axis of x. If the reflected ray passes through the point $R(6,7)$, then the abscissa of Q is → JEE Mains 2013

(a) 1 (b) 3
(c) $\frac{7}{2}$ (d) $\frac{5}{2}$

8 Locus of the image of the point $(2, 3)$ in the line $(2x-3y+4)+k(x-2y+3)=0$, $k \in R$, is a → JEE Mains 2015
(a) straight line parallel to X-axis
(b) straight line parallel to Y-axis
(c) circle of radius $\sqrt{2}$
(d) circle of radius $\sqrt{3}$

9 A square of side a lies above the X-axis and has one vertex at the origin. The side passing through the origin makes an angle $\alpha \left(\text{where, } 0<\alpha<\frac{\pi}{4}\right)$ with the positive direction X-axis. The equation of its diagonal not passing through the origin is
(a) $y(\cos\alpha - \sin\alpha) - x(\sin\alpha - \cos\alpha) = a$
(b) $y(\cos\alpha + \sin\alpha) - x(\sin\alpha - \cos\alpha) = a$
(c) $y(\cos\alpha + \sin\alpha) + x(\sin\alpha + \cos\alpha) = a$
(d) $y(\cos\alpha + \sin\alpha) + x(\sin\alpha - \cos\alpha) = a$

10 A variable line 'L' is drawn through $O(0,0)$ to meet the line $L_1: y - x - 10 = 0$ and $L_2: y - x - 20 = 0$ at the point A and B respectively. A point P is taken on 'L' such that $\frac{2}{OP} = \frac{1}{OA} + \frac{1}{OB}$. Locus of '$P$' is
(a) $3x + 3y = 40$ (b) $3x + 3y + 40 = 0$
(c) $3x - 3y = 40$ (d) $3y - 3x = 40$

11 Consider the family of lines $5x + 3y - 2 + \lambda_1(3x - y - 4) = 0$ and $x - y + 1 + \lambda_2(2x - y - 2) = 0$. Equation of a straight line that belongs to both families is
(a) $25x - 62y + 86 = 0$ (b) $62x - 25y + 86 = 0$
(c) $25x - 62y = 86$ (d) $5x - 2y - 7 = 0$

12 Equation of the straight line which belongs to the system of straight lines $a(2x + y - 3) + b(3x + 2y - 5) = 0$ and is farthest from the point $(4, -3)$ is
(a) $4x + 11y - 15 = 0$ (b) $3x - 4y + 1 = 0$
(c) $7x + y - 8 = 0$ (d) None of these

13 One diagonal of a square is along the line $8x - 15y = 0$ and its one vertex $(1,2)$, then equations of a side passing through this vertex are
(a) $7x + 23y - 53 = 0$, $23x - 7y - 9 = 0$
(b) $7x - 23y - 53 = 0$, $23x + 7y - 9 = 0$
(c) $7x + 23y + 53 = 0$, $23x - 7y + 9 = 0$
(d) $7x + 23y + 53 = 0$, $23x + 7y + 9 = 0$

14 The equations of the straight lines through $(-2, -7)$ and having intercept of length 3 between the lines $4x + 3y = 12$ and $4x + 3y = 3$ is
(a) $7x - 24y - 182 = 0$ (b) $7x + 24y + 182 = 0$
(c) $7x + 24y - 182 = 0$ (d) None of these

15 If angle between lines $ax + by + p = 0$ and $x\cos\alpha + y\sin\alpha = p$ is $\frac{\pi}{4}$ and these lines with other line $x\sin\alpha - y\cos\alpha = 0$ are concurrent, then $a^2 + b^2 =$
(a) 1 (b) 2 (c) 3 (d) 4

16 Straight lines $y = mx + c_1$ and $y = mx + c_2$, where $m \in R^+$ meet the X – axis at A_1 and A_2 respectively and Y – axis at B_1 and B_2 respectively. It is given that point A_1, A_2, B_1 and B_2 are concyclic. Locus of intersection of lines A_1B_2 and A_2B_1 is
(a) $y = x$ (b) $y + x = 0$
(c) $xy = 1$ (d) $xy + 1 = 0$

ANSWERS

SESSION 1	1 (b)	2 (d)	3 (c)	4 (c)	5 (d)	6 (d)	7 (d)	8 (b)	9 (b)	10 (c)
	11 (d)	12 (b)	13 (b)	14 (d)	15 (b)	16 (c)	17 (a)	18 (a)	19 (d)	20 (b)
	21 (c)	22 (a)	23 (b)	24 (b)	25 (c)	26 (b)	27 (a)	28 (a)	29 (a)	30 (a)
	31 (c)	32 (a)	33 (d)	34 (b)	35 (a)	36 (a)				
SESSION 2	1 (d)	2 (c)	3 (a)	4 (d)	5 (a)	6 (a)	7 (d)	8 (c)	9 (b)	10 (d)
	11 (d)	12 (b)	13 (a)	14 (b)	15 (b)	16 (b)				

Hints and Explanations

SESSION 1

1 If a_1, b_1 are intercepts of the line on the axes, then

$$1/a_1 = a, 1/b_1 = b$$

$$\Rightarrow \quad a_1 = 1/a, b_1 = 1/b$$

$\therefore$ Equation of the line is

$$x/a_1 + y/b_1 = 1 \text{ or } ax + by = 1$$

2 Let x-intercept $= a$

and y-intercept $= b$

Since, $a + b = -1 \Rightarrow b = -(a+1)$

$\therefore$ Equation of line is $\dfrac{x}{a} - \dfrac{y}{a+1} = 1$

Clearly,

$$\frac{4}{a} - \frac{3}{a+1} = 1 \Rightarrow \frac{4a + 4 - 3a}{a(a+1)} = 1$$

$$\Rightarrow a + 4 = a^2 + a \Rightarrow a = \pm 2$$

Hence, equation of line is

$$\frac{x}{2} - \frac{y}{3} = 1 \text{ or } \frac{x}{-2} + \frac{y}{1} = 1.$$

3 Let the equation of line be $\dfrac{x}{a} + \dfrac{y}{b} = 1$

$\therefore$ According to given condition, we have

$$C(-4, 3) \equiv C\left(\frac{3a}{8}, \frac{5b}{8}\right)$$

$$\Rightarrow \quad a = -\frac{32}{3} \text{ and } b = \frac{24}{5}$$

$\therefore$ Equation of line is

$$-\frac{3x}{32} + \frac{5y}{24} = 1$$

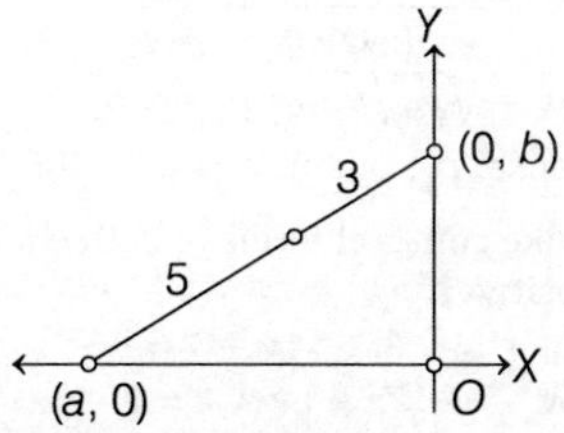

$$\Rightarrow -9x + 20y - 96 = 0$$

$$\Rightarrow 9x - 20y + 96 = 0$$

4 Equation of PQ is

$$\frac{x}{h} + \frac{y}{k} = 1$$

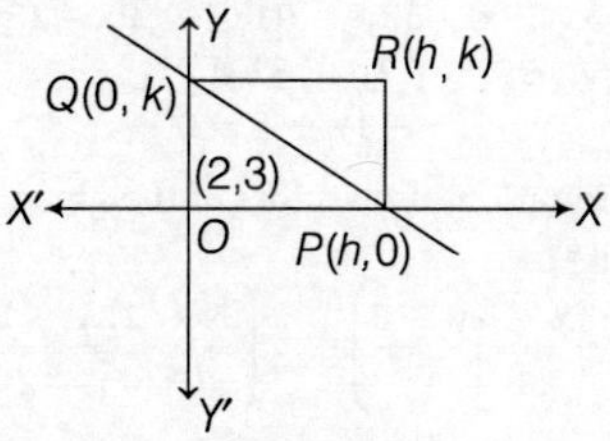

Since, it is passes through the points (2,3)

$$\therefore \quad \frac{2}{h} + \frac{3}{k} = 1 \Rightarrow 2k + 3h = hk$$

So, locus is $3x + 2y = xy$

5 We have, $\dfrac{x}{4} + \dfrac{y}{3} = 1$

For line L, x-intercept $= 2 \times 4 = 8$

$$y\text{-intercept} = \frac{1}{2} \times 3 = \frac{3}{2}$$

$\therefore$ Line L is $\dfrac{x}{8} + \dfrac{y}{3/2} = 1$, Slope, $m = -\dfrac{3}{16}$

6 $ax + by + 8 = 0$

$$\Rightarrow \quad ax + by = -8$$

$$\Rightarrow \quad \frac{x}{-\frac{8}{a}} + \frac{y}{-\frac{8}{b}} = 1 \text{ (intercept form)}$$

Also, $\quad 2x - 3y = -6 \Rightarrow -\dfrac{x}{3} + \dfrac{y}{2} = 1$

According to given condition, we have

$$-\frac{8}{a} = -(-3) \text{ and } -\frac{8}{b} = -2$$

$$\Rightarrow \quad a = -\frac{8}{3} \text{ and } b = 4$$

7 On solving both the equations, we get

$$\frac{8y}{3} + y^2 = 1$$

$$\Rightarrow \quad 3y^2 + 8y - 3 = 0$$

$$\Rightarrow \quad (3y - 1)(y + 3) = 0$$

$$\Rightarrow \quad y = -3, \frac{1}{3} \text{ here } y \neq -3$$

At $y = \dfrac{1}{3}$, $x = \pm 2\sqrt{\dfrac{2}{3}}$

So, the points of intersection are

$$\left(2\sqrt{\frac{2}{3}}, \frac{1}{3}\right) \text{ and } \left(-2\sqrt{\frac{2}{3}}, \frac{1}{3}\right).$$

From option (d); $3y - 1 = 0$ is the required equation which satisfied the intersection points.

8 Take any point $B(0,1)$ on given line.

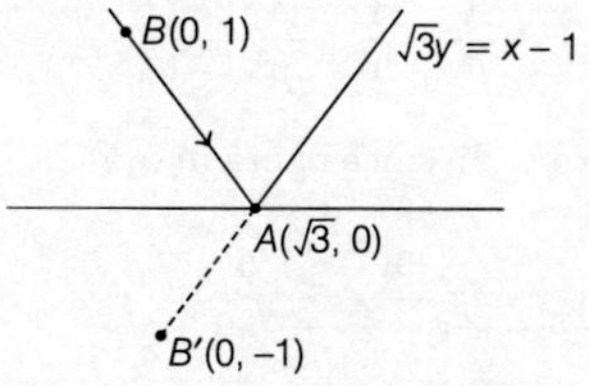

Equation of AB' is

$$y - 0 = \frac{-1-0}{0-\sqrt{3}}(x - \sqrt{3})$$

$$\Rightarrow \quad -\sqrt{3}y = -x + \sqrt{3}$$

$$\Rightarrow \quad x - \sqrt{3}y = \sqrt{3}$$

$$\Rightarrow \quad \sqrt{3}y = x - \sqrt{3}$$

9 From figure, for $(0, \alpha)$ to be inside or on the triangle,

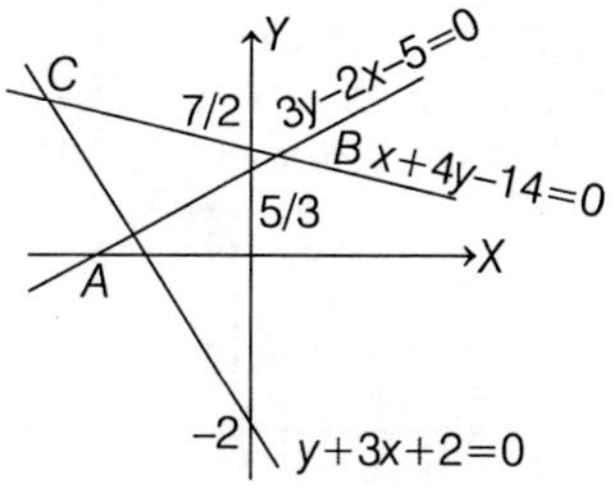

$$\frac{5}{3} \le \alpha \le \frac{7}{2}$$

10 As $x + y = |a|$ and $ax - y = 1$.

Intersect in first quadrant.

So, x and y-coordinates are positive.

$$\therefore x = \frac{1+|a|}{1+a} \ge 0 \text{ and } y = \frac{a|a|-1}{a+1} \ge 0$$

$$\Rightarrow \quad 1 + a \ge 0 \text{ and } a|a| - 1 \ge 0$$

$$\Rightarrow \quad a \ge -1 \text{ and } a|a| \ge 1 \quad \text{...(i)}$$

If $-1 \le a < 0 \Rightarrow -a^2 \ge 1$ [not possible]

If $a \ge 0 \Rightarrow a^2 \ge 1 \Rightarrow a \ge 1 \Rightarrow a \in [1, \infty)$

11 Let lines $OB: y = mx$, $CA: y = mx + 1$

$BA: y = nx + 1$ and $OC: y = nx$

So, the point of intersection B of OB and AB has x-coordinate $\dfrac{1}{m-n}$.

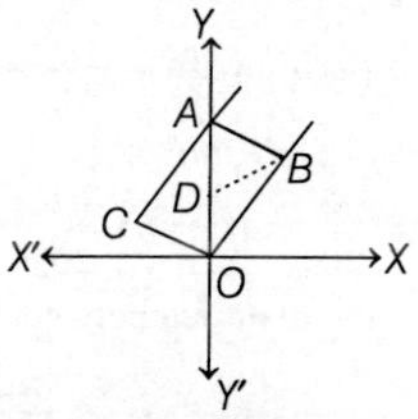

Now, area of a parallelogram $OBAC = 2 \times$ Area of ΔOBA

$$= 2 \times \frac{1}{2} \times OA \times DB = 2 \times \frac{1}{2} \times \frac{1}{m-n}$$

$$= \frac{1}{m-n} = \frac{1}{|m-n|}$$

depending upon whether $m > n$ or $m < n$.

12 Coordinate of

$$S = \left(\frac{7+6}{2}, \frac{3-1}{2}\right) = \left(\frac{13}{2}, 1\right)$$

[since, S is mid-point of line QR]

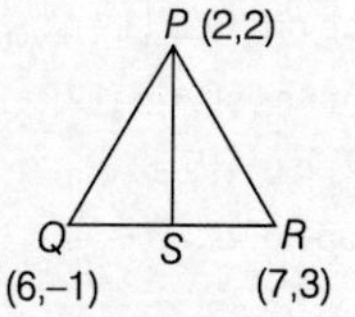

Slope of the line PS is $\frac{-2}{9}$.

Required equation of line passes through $(1, -1)$ and parallel to PS is

$$y + 1 = \frac{-2}{9}(x - 1)$$

$$\Rightarrow \quad 2x + 9y + 7 = 0$$

13 Let $P(4,1)$ and $PD \perp AB$.

Equation of AB is $3x - 2y - 8 = 0$

$\therefore$ Equation of PD is $2x + 3y - 11 = 0$

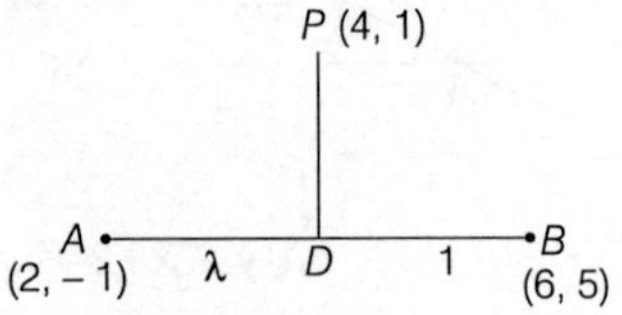

Let line AB is divided by PD in the ratio $\lambda:1$, then intersecting point $D\left(\frac{6\lambda+2}{\lambda+1}, \frac{5\lambda-1}{\lambda+1}\right)$ lies on $2x + 3y - 11 = 0$.

$$\Rightarrow \quad 2\left(\frac{6\lambda+2}{\lambda+1}\right) + 3\left(\frac{5\lambda-1}{\lambda+1}\right) - 11 = 0$$

$$\Rightarrow \quad 16\lambda - 10 = 0 \Rightarrow \lambda:1 = 5:8$$

14 Since, the line L is passing through the point $(13, 32)$.

Therefore, $\frac{13}{5} + \frac{32}{b} = 1 \Rightarrow \frac{32}{b} = -\frac{8}{5}$

$\Rightarrow b = -20$

The line K is parallel to the line L, its equation must be

$$\frac{x}{5} - \frac{y}{20} = a \text{ or } \frac{x}{5a} - \frac{y}{20a} = 1$$

On comparing with $\frac{x}{c} + \frac{y}{3} = 1$, we get

$$20a = -3, c = 5a$$

$$a = \frac{-3}{20} \text{ and } c = 5 \times \frac{-3}{20} = \frac{-3}{4}$$

Hence, the distance between lines

$$= \frac{|a-1|}{\sqrt{\frac{1}{25} + \frac{1}{400}}} = \frac{\left|\frac{-3}{20} - 1\right|}{\sqrt{\frac{17}{400}}} = \frac{23}{\sqrt{17}}$$

15 The desired point is the foot of the perpendicular from the origin on the line $3x - 4y = 25$.

The equation of a line passing through the origin and perpendicular to $3x - 4y = 25$ is $4x + 3y = 0$.

Solving these two equations we get $x = 3$, $y = -4$.

Hence, the nearest point on the line from the origin is $(3, -4)$.

16 Coordinates of $A \equiv (1,2)$

$\therefore$ Slope of $AE = 2$

$\Rightarrow$ Slope of $BD = -\frac{1}{2}$

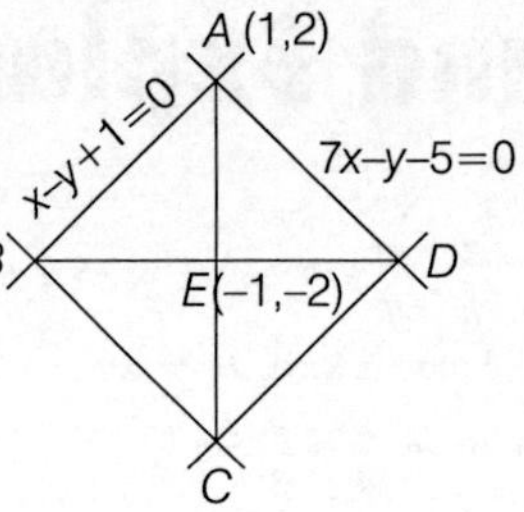

$\Rightarrow$ Equation of BD is $\frac{y+2}{x+1} = \frac{-1}{2}$

$\Rightarrow \quad x + 2y + 5 = 0$

$\therefore$ Coordinates of $D = \left(\frac{1}{3}, \frac{-8}{3}\right)$

17 It is given that the lines $ax + 2y + 1 = 0$, $bx + 3y + 1 = 0$, $cx + 4y + 1 = 0$ are concurrent

$$\therefore \quad \begin{vmatrix} a & 2 & 1 \\ b & 3 & 1 \\ c & 4 & 1 \end{vmatrix} = 0$$

$$\Rightarrow \quad -a + 2b - c = 0 \Rightarrow 2b = a + c$$

$\therefore$ a, b, c are in AP.

18 Given equations, $(2a + b)x + (a + 3b)y + (b - 3a) = 0$ and $mx + 2y + 6 = 0$ are concurrent for all real values of a and b, so they must represent the same line for some values of a and b. Therefore, we get

$$\frac{2a+b}{m} = \frac{(a+3b)}{2} = \frac{b-3a}{6}$$

On taking last two ratios,

$$\frac{a+3b}{2} = \frac{-3a+b}{6} \Rightarrow b = -\frac{3}{4}a$$

On taking first two ratios,

$$m = \frac{2(2a+b)}{a+3b} = \frac{2\{2a - (3/4)a\}}{a + 3(-3/4)a} = -\frac{10}{5} = -2$$

19 The length of perpendicular from $(0,0)$ to line $\frac{x}{a} + \frac{y}{b} = 1$, is $p = \frac{1}{\sqrt{\frac{1}{a^2} + \frac{1}{b^2}}}$

$$\therefore \quad \frac{1}{a^2} + \frac{1}{b^2} = \frac{1}{p^2}$$

20 Now, distance of origin from $4x + 2y - 9 = 0$ is

$$\frac{|-9|}{\sqrt{4^2 + 2^2}} = \frac{9}{\sqrt{20}}$$

and distance of origin from $2x + y + 6 = 0$ is

$$\frac{|6|}{\sqrt{2^2 + 1^2}} = \frac{6}{\sqrt{5}}$$

$$\therefore \text{ Required ratio} = \frac{9/\sqrt{20}}{6/\sqrt{5}} = \frac{3}{4}$$

21 $p = \frac{1}{\sqrt{(1/a^2) + (1/b^2)}} = \frac{ab}{\sqrt{a^2 + b^2}}$

a^2, p^2, b^2 are in AP.

$$\Rightarrow \quad \frac{2a^2b^2}{a^2 + b^2} = a^2 + b^2$$

$\Rightarrow \quad a^4 + b^4 = 0$ i.e. $a = b = 0$

This is impossible, therefore given information is inconsistent.

22 Since, p = length of the perpendicular from $(0,0)$ on $x\sec\theta + y\,\text{cosec}\,\theta = a$

$$\therefore \quad p = \frac{a}{\sqrt{\sec^2\theta + \text{cosec}^2\theta}} = \frac{a\sin 2\theta}{2}$$

$\Rightarrow \quad 2p = a\sin 2\theta$ ...(i)

Also, p' = length of perpendicular from $(0,0)$ on $x\cos\theta - y\sin\theta = a\cos 2\theta$

$$\therefore \quad p' = \frac{a\cos 2\theta}{\sqrt{\cos^2\theta + \sin^2\theta}}$$

$= a\cos 2\theta$ ...(ii)

On squaring and adding Eqs. (i), (ii), we get

$$4p^2 + p'^2 = a^2$$

23 $p_1 = \left|m^2\cos\alpha + 2m\sin\alpha + \frac{\sin^2\alpha}{\cos\alpha}\right|$

$$p_2 = \left|mm'\cos\alpha + (m + m')\sin\alpha + \frac{\sin^2\alpha}{\cos\alpha}\right|$$

$$p_3 = \left|m'^2\cos\alpha + 2m'\sin\alpha + \frac{\sin^2\alpha}{\cos\alpha}\right|$$

$$p_2 = \frac{mm'\cos^2\alpha + (m + m')\sin\alpha\cos\alpha + \sin^2\alpha}{\cos\alpha}$$

$$p_1 = \frac{(m\cos\alpha + \sin\alpha)^2}{\cos\alpha}$$

$$p_3 = \frac{(m'\cos\alpha + \sin\alpha)^2}{\cos\alpha}$$

$\therefore p_2 = \sqrt{p_1}\sqrt{p_3} \Rightarrow p_2^2 = p_1 p_3$

Hence, p_1, p_2 and p_3 are in GP.

24 Make constant terms of both equation positive.

$$3x - 4y + 7 = 0$$

and $\quad -12x - 5y + 2 = 0$

Since, $a_1a_2 + b_1b_2 = -36 + 20 < 0$

$\therefore$ Bisector of acute angle is given by with positive sign

$$\frac{3x - 4y + 7}{\sqrt{9 + 16}} = +\left(\frac{-12x - 5y + 2}{\sqrt{144 + 25}}\right)$$

$\Rightarrow \quad 39x - 52y + 91 = -60x - 25y + 10$

$\Rightarrow \quad 99x - 27y + 81 = 0$

$\therefore \quad 11x - 3y + 9 = 0$

25 Equation of bisectors of the given lines are

$$\left(\frac{3x + 4y - 5}{\sqrt{3^2 + 4^2}}\right) = \pm\left(\frac{5x - 12y - 10}{\sqrt{5^2 + (-12)^2}}\right)$$

$\therefore \quad (39x + 52y - 65) = \pm(25x - 60y - 50)$

$\Rightarrow \quad 14x + 112y - 15 = 0$

or $\quad 64x - 8y - 115 = 0$

$\Rightarrow \quad \frac{14}{15}x + \frac{112}{15}y - 1 = 0$

or $\quad \frac{64}{115}x - \frac{8}{115}y - 1 = 0$

$\therefore \quad a = \frac{14}{15}, b = \frac{112}{15}$

or $\quad a = \frac{64}{115}, b = -\frac{8}{115}$

26 The required line is given by

$x + 2y - 1 + \lambda(x - y + 2) = 0 \quad ...(i)$

It is perpendicular to $3x + y + 5 = 0$

$\therefore\ 3(1 + \lambda) + 2 - \lambda = 0 \Rightarrow \lambda = -\frac{5}{2}$

From Eq. (i), we get

$x - 3y + 4 = 0$

27 The intersection of given lines is

$$\frac{x}{a} + \frac{y}{b} - 1 + \lambda\left(\frac{x}{b} + \frac{y}{a} - 1\right) = 0$$

meets the coordinate axes at

$$A\left[\frac{1+\lambda}{\frac{1}{a}+\frac{\lambda}{b}}, 0\right] \text{ and } B\left[0, \frac{1+\lambda}{\frac{1}{b}+\frac{\lambda}{a}}\right]$$

The mid-point of AB is given by

$$2x = \frac{1+\lambda}{\frac{1}{a}+\frac{\lambda}{b}}, 2y = \frac{1+\lambda}{\frac{1}{b}+\frac{\lambda}{a}}$$

$$\Rightarrow (1+\lambda)\left[\frac{1}{x} + \frac{1}{y}\right]$$

$$= 2\left[\frac{1}{a} + \frac{\lambda}{b}\right] + 2\left[\frac{1}{b} + \frac{\lambda}{a}\right]$$

$$= 2(1+\lambda)\left[\frac{1}{a} + \frac{1}{b}\right]$$

$\therefore\ (x + y)ab = 2xy(a + b)$

28

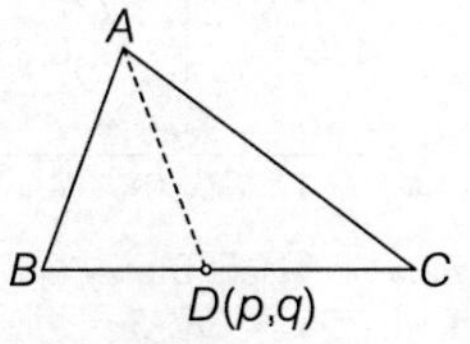

Equation of line AB

$px + qy = 1$

Equation of line AC

$qx + py = 1$

The equation of line passing through the intersection point of above lines is

$px + qy - 1 + \lambda(qx + py - 1) = 0$

which passes through (p, q)

$\therefore\ p^2 + q^2 - 1 + \lambda(pq + pq - 1) = 0 \quad ...(i)$

$\Rightarrow \quad \lambda = -\frac{p^2 + q^2 - 1}{2pq - 1}$

$\therefore$ Substituting the value of λ in Eq. (i), of the line through A is $(px + qy - 1)$

$$-\frac{p^2 + q^2 - 1}{2pq - 1}(qx + py - 1) = 0$$

$\Rightarrow \quad (2pq - 1)(px + qy - 1)$

$= (p^2 + q^2 - 1)(qx + py - 1)$

29 Let $L_1 = 3x - 4y - 8$

At (3, 4), $L_1 = 9 - 16 - 8 = -15 < 0$

For the point $P(x, y)$, we should have $L_1 > 0$.

$\Rightarrow \quad 3x - 4y - 8 > 0 \quad [\because y = -3x]$

$\Rightarrow \quad 3x - 4(-3x) - 8 > 0$

$[\because P(x, y) \text{ lies on } y = -3x]$

$\Rightarrow \quad x > 8/15 \text{ and } -y - 4y - 8 > 0$

$\Rightarrow \quad y < -8/5$

30 Since, $P\left(1 + \frac{\alpha}{\sqrt{2}}, 2 + \frac{\alpha}{\sqrt{2}}\right)$ lies between the parallel lines $x + 2y = 1$ and $2x + 4y = 15$, therefore

$$\frac{\left(1 + \frac{\alpha}{\sqrt{2}}\right) + 2\left(2 + \frac{\alpha}{\sqrt{2}}\right) - 1}{2\left(1 + \frac{\alpha}{\sqrt{2}}\right) + 4\left(2 + \frac{\alpha}{\sqrt{2}}\right) - 15} < 0$$

$$\Rightarrow \frac{4 + \frac{3\alpha}{\sqrt{2}}}{-5 + \frac{6\alpha}{\sqrt{2}}} < 0 \Rightarrow \frac{\left(\alpha + \frac{4\sqrt{2}}{3}\right)}{\left(\alpha - \frac{5\sqrt{2}}{6}\right)} < 0$$

$$\therefore \quad \frac{-4\sqrt{2}}{3} < \alpha < \frac{5\sqrt{2}}{6}$$

31 Clearly, $a^2 - \frac{a}{2} > 0, a^2 - 3a < 0$

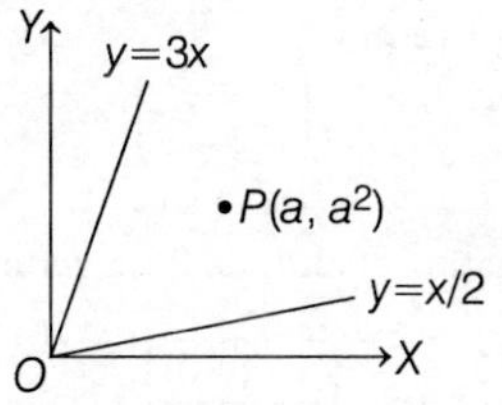

$\Rightarrow \quad \frac{1}{2} < a < 3.$

32

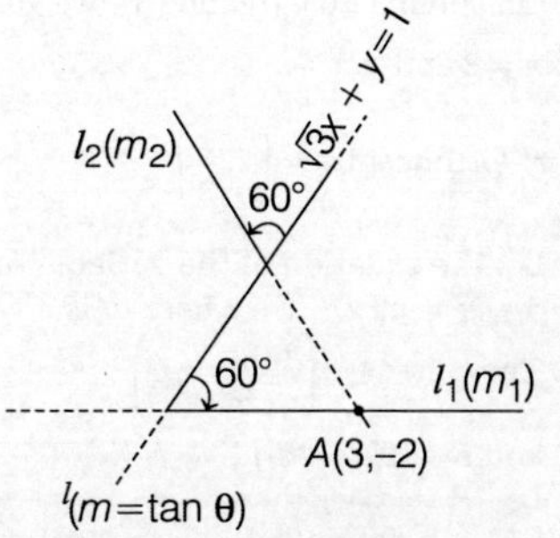

Let l_1 and l_2 are the equations of the lines inclined at an angle of 60° with the line l.

$\therefore$ Slopes of lines are $\tan(\theta \pm 60°)$

Now, equation of lines l_1 and l_2 are

$y + 2 = \tan(\theta \pm 60°)(x - 3)$

$\Rightarrow \quad y + 2 = \frac{\tan\theta \pm \tan 60°}{1 \mp \tan\theta \tan 60°}(x - 3)$

$\Rightarrow \quad y + 2 = \frac{-\sqrt{3} \pm \sqrt{3}}{1 \mp (-\sqrt{3})\sqrt{3}}(x - 3)$

$\Rightarrow \quad y + 2 = 0$

or $\quad y + 2 = \sqrt{3}(x - 3)$

33 Equation of line AB is

$y - 1 = \frac{0 - 1}{2 - 0}(x - 0)$

$\Rightarrow \quad x + 2y - 2 = 0$

Here, $\quad |PA - PB| \le |AB|$

Thus, for $|PA - PB|$ to be maximum, A, B and P must be collinear.

34 The point of intersection of lines $4x + 3y = \lambda$ and $3x - 4y = \mu$ is

$x_1 = \frac{4\lambda + 3\mu}{25}$

and $y_1 = \frac{3\lambda - 4\mu}{25}$

$\because \quad x_1 = y_1 \Rightarrow \frac{4\lambda + 3\mu}{25} = \frac{3\lambda - 4\mu}{25}$

$\Rightarrow \quad \lambda + 7\mu = 0$

Hence, locus of a point (λ, μ) is $x + 7y = 0$.

35 Here, angle between the lines

$2x + 3y + c_1 = 0$

and $-x + 5 + c_2 = 0$ is θ_1.

$\therefore \quad \tan\theta_1 = \left|\frac{1/5 + 2/3}{1 - 2/15}\right| = \left|\frac{13/15}{13/15}\right|$

$= 1 = \tan 45°$

$\Rightarrow \quad \theta_1 = 45°$

Also, the angle between the lines

$2x + 3y + c_1 = 0$ and

$-x + 5y + c_3 = 0$ is θ_2.

$\therefore \quad \tan\theta_2 = \left|\frac{1/5 + 2/3}{1 - 2/15}\right| = \left|\frac{13/15}{13/15}\right|$

$= 1 = \tan 45°$

$\Rightarrow \quad \theta_2 = 45°$

Here, we observe that the value of c_1, c_2 and c_3 is not depend on measuring the angle between the lines.

So, c_2 and c_3 are proportional or for all c_2 and c_3 $\theta_1 = \theta_2$

36 Equation of bisector of

$4y + 3x - 12 = 0$

and $3y + 4x - 24 = 0$ is

$$\frac{4y + 3x - 12}{\sqrt{16 + 9}} = \pm \frac{3y + 4x - 24}{\sqrt{9 + 16}}$$

$\Rightarrow \quad y - x + 12 = 0$

and $7y + 7x - 36 = 0$

So, the line $y - x + 12 = 0$ is the angular bisector.

SESSION 2

1 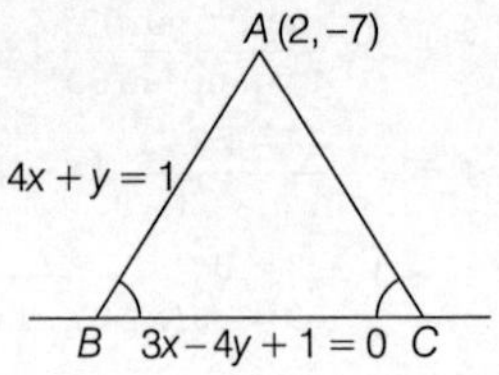

Let m be the slope of AC, then

$$\tan B = \tan C \Rightarrow \frac{\frac{3}{4}+4}{1-3} = \frac{m-\frac{3}{4}}{1+\frac{3m}{4}}$$

$$\Rightarrow -\frac{19}{8} = \frac{4m-3}{4+3m} \Rightarrow m = -\frac{52}{89}$$

$\therefore$ Equation of AC is

$$y+7 = -\frac{52}{89}(x-2)$$

$$\Rightarrow 52x + 89y + 519 = 0$$

2 $AB = BC \Rightarrow B$ is mid-point of AC.

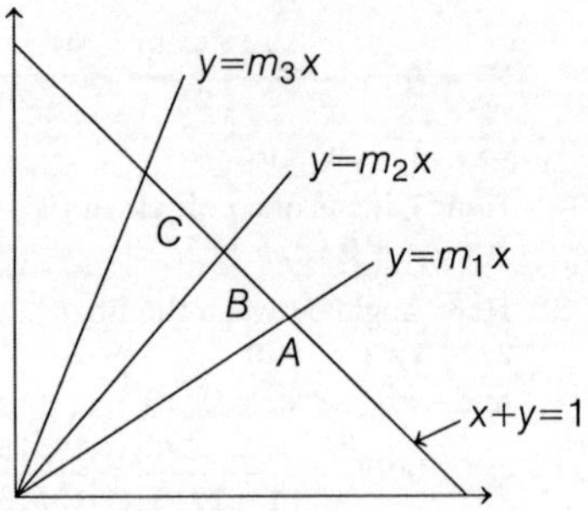

Clearly, $A = \left(\frac{1}{m_1+1}, \frac{m_1}{m_1+1}\right)$

$B = \left(\frac{1}{m_2+1}, \frac{m_2}{m_2+1}\right)$

$C = \left(\frac{1}{m_3+1}, \frac{m_3}{m_3+1}\right)$

Now, $\frac{2}{m_2+1} = \frac{1}{m_1+1} + \frac{1}{m_3+1}$

$\therefore$ m_1+1, m_2+1, m_3+1 are in HP.

3 Clearly, $B =$ reflection of $A(5,7)$ in the line $x+y=0$

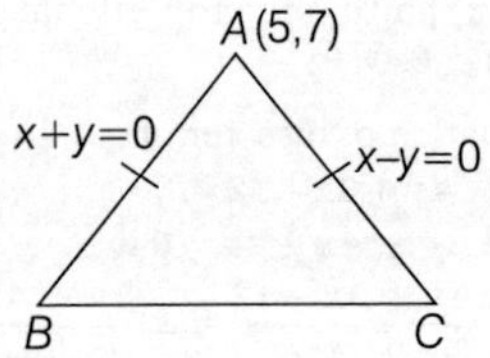

$\Rightarrow \quad B \equiv (-7,-5)$

$C =$ reflection of $A(5,7)$ in the line $x-y=0$

$\Rightarrow \quad C = (7,5)$

Equation of BC is $7y = 5x$.

4 Given, vertices of triangle are $(k, -3k)$, $(5, k)$ and $(-k, 2)$.

$$\therefore \quad \frac{1}{2}\begin{vmatrix} k & -3k & 1 \\ 5 & k & 1 \\ -k & 2 & 1 \end{vmatrix} = \pm 28$$

$$\Rightarrow \quad \begin{vmatrix} k & -3k & 1 \\ 5 & k & 1 \\ -k & 2 & 1 \end{vmatrix} = \pm 56$$

$$\Rightarrow k(k-2) + 3k(5+k) + 1(10+k^2) = \pm 56$$

$$\Rightarrow k^2 - 2k + 15k + 3k^2 + 10 + k^2 = \pm 56$$

$$\Rightarrow \quad 5k^2 + 13k + 10 = \pm 56$$

$$\Rightarrow \quad 5k^2 + 13k - 66 = 0$$

or $\quad 5k^2 + 13k - 46 = 0$

$$\Rightarrow \quad k = 2 \quad [\because k \in I]$$

Thus, the coordinates of vertices of triangle are $A(2, -6)$, $B(5, 2)$ and $C(-2, 2)$.

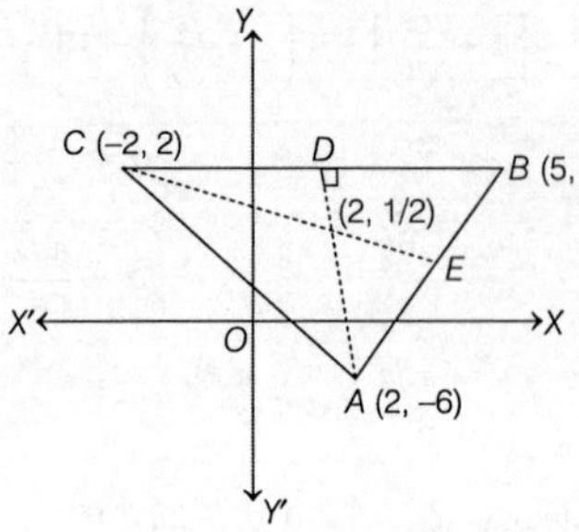

Now, equation of altitude from vertex A is $y - (-6) = \frac{-1}{\left(\frac{2-2}{-2-5}\right)}(x-2)$

$$\Rightarrow \quad x = 2 \quad \ldots\text{(i)}$$

Equation of altitude from vertex C is

$$y - 2 = \frac{-1}{\left[\frac{2-(-6)}{5-2}\right]}[x-(-2)]$$

$$\Rightarrow \quad 3x + 8y - 10 = 0 \quad \ldots\text{(ii)}$$

On solving Eqs. (i) and (ii), we get $x = 2$ and $y = \frac{1}{2}$.

$\therefore$ Orthocentre $= \left(2, \frac{1}{2}\right)$

5 Let the equation of the variable line be $y - q = m(x-p)$, where m is a variable

Then, $\quad A \equiv \left(\frac{mp-q}{m}, 0\right)$

and $B \equiv (0, q - mp)$

Let $P \equiv (x, y)$, then

$$x = \frac{mp-q}{m} \text{ or } m(p-x) = q \quad \ldots\text{(i)}$$

and $y = q - mp$ or $mp = q - y \quad \ldots$(ii)

On eliminating m from Eqs. (i) and (ii), we get

$$\frac{p-x}{q} = \frac{p}{q-y}$$

$$\Rightarrow \quad pq - qx - py + xy = pq$$

$$\Rightarrow \quad py + qx = xy$$

or $\quad \frac{p}{x} + \frac{q}{y} = 1$

This is the locus of P.

But locus of P is $3x + 2y = xy$ (given)

or $\quad \frac{2}{x} + \frac{3}{y} = 1$

$\therefore \quad p = 2$

and $\quad q = 3$

6 ***Case* I** Let line $l_1 \equiv x - 3y = p$ and $l_2 \equiv ax + 2y = p$ are perpendicular, then

$$\frac{1}{3} \times -\frac{a}{2} = -1$$

$$\Rightarrow \quad a = 6$$

***Case* II** Let line $l_2 \equiv ax + 2y = p$ and $l_3 \equiv ax + y = r$ are perpendicular, then

$$\frac{-a}{2} \times -a = -1$$

$$\Rightarrow \quad a^2 = -2 \quad \text{[not possible]}$$

***Case* III** Let line $l_3 \equiv ax + y = r$ and $l_1 \equiv x - 3y = p$ are perpendicular, then $-a \times \frac{1}{3} = -1 \Rightarrow a = 3$. So, formation of quadratic equation in a, whose roots are 3 and 6, is

$$a^2 - (6+3)a + (6 \cdot 3) = 0$$

$$\Rightarrow \quad a^2 - 9a + 18 = 0$$

7 Here, $QS \perp OX$

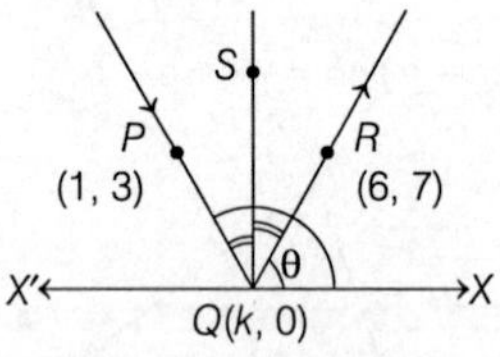

It means QS bisect the $\angle PQR$.

Then, $\quad \angle PQS = \angle RQS$

$$\Rightarrow \quad \angle RQX = \angle PQO = \theta \quad \text{[let]}$$

$$\Rightarrow \quad \angle XQP = 180° - \theta$$

Slope of $QR = \tan\theta = \frac{y_2 - y_1}{x_2 - x_1}$

$$= \frac{7-0}{6-k} \quad \ldots\text{(i)}$$

Slope of $QP = \tan(180° - \theta) = -\tan\theta$

$$= \frac{y_2 - y_1}{x_2 - x_1} = \frac{3-0}{1-k} \quad \ldots\text{(ii)}$$

$\therefore$ From Eqs. (i) and (ii),

$$\frac{7}{6-k} = -\frac{3}{1-k}$$

$$\Rightarrow \quad 7 - 7k = -18 + 3k$$

$$10k = 25$$

$$\Rightarrow \quad k = \frac{5}{2}$$

Hence, the coordinate of Q is $\left(\frac{5}{2}, 0\right)$.

8 **Key Idea** First of all find the point of inter- section of the lines $2x - 3y + 4 = 0$ and $x - 2y + 3 = 0$ (say A). Now, the line $(2x - 3y + 4) + k(x - 2y + 3) = 0$ is the perpendicular bisector of the line joining points $P(2, 3)$ and image $P'(h, k)$. Now, $AP = AP'$ and simplify.

Given line is

$$(2x - 3y + 4) + k(x - 2y + 3) = 0, \quad k \in R \quad \ldots(i)$$

This line will pass through the point of intersection of the lines

$$2x - 3y + 4 = 0 \quad \ldots(ii)$$

and $\quad x - 2y + 3 = 0 \quad \ldots(iii)$

On solving Eqs. (ii) and (iii), we get

$$x = 1, y = 2$$

Thus, point of intersection of lines (ii) and (iii) is (1, 2).

Let M be the mid-point of PP', then AM is perpendicular bisector of PP' (where, A is the point of intersection of given lines).

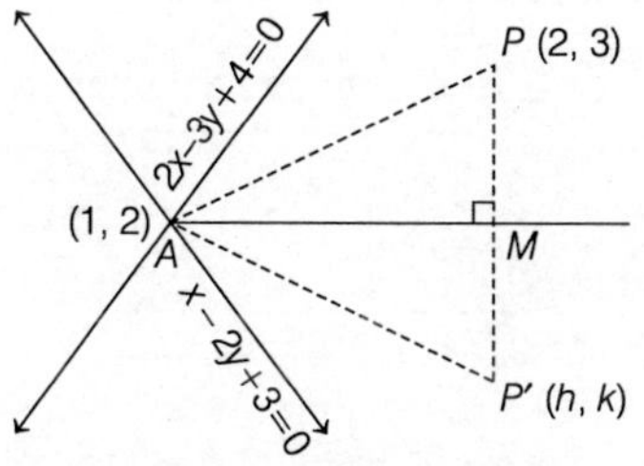

Clearly, $\quad AP = AP'$

$$\Rightarrow \sqrt{(2-1)^2 + (3-2)^2} = \sqrt{(h-1)^2 + (k-2)^2}$$

$$\Rightarrow \sqrt{2} = \sqrt{h^2 + k^2 - 2h - 4k + 1 + 4}$$

$$\Rightarrow \sqrt{2} = \sqrt{h^2 + k^2 - 2h - 4k + 5}$$

$$\Rightarrow h^2 + k^2 - 2h - 4k + 5 = 2$$

$$\Rightarrow h^2 + k^2 - 2h - 4k + 3 = 0$$

Thus, the required locus is

$$x^2 + y^2 - 2x - 4y + 3 = 0$$

which is an equation of circle with radius $= \sqrt{1 + 4 - 3} = \sqrt{2}$

9 Slope of the diagonal $= \tan\left(\frac{3\pi}{4} + \alpha\right)$

$$= \frac{-1 + \tan\alpha}{1 + \tan\alpha}$$

$$= \frac{\sin\alpha - \cos\alpha}{\sin\alpha + \cos\alpha}$$

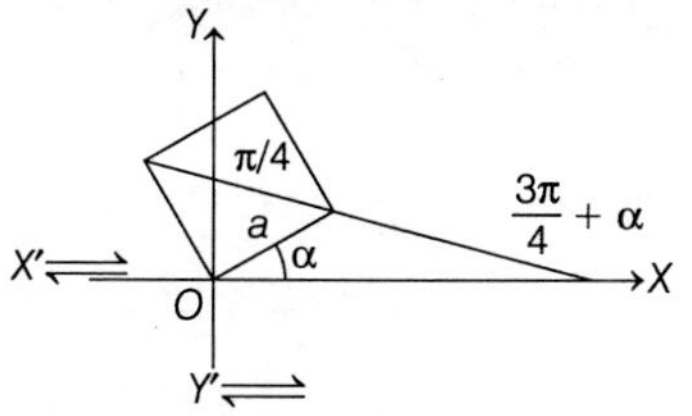

The equation is

$$\frac{y - a\sin\alpha}{x - a\cos\alpha} = \frac{\sin\alpha - \cos\alpha}{\sin\alpha + \cos\alpha}$$

$$\Rightarrow y(\cos\alpha + \sin\alpha) - x(\sin\alpha - \cos\alpha) = a$$

10 Let equation of the line OAB be

$$\frac{x}{\cos\theta} = \frac{y}{\sin\theta} = r$$

$$\Rightarrow \quad x = r\cos\theta, y = r\sin\theta.$$

For A, let $OA = r_1$ then $A(r_1\cos\theta, r_1\sin\theta)$ lies on L_1.

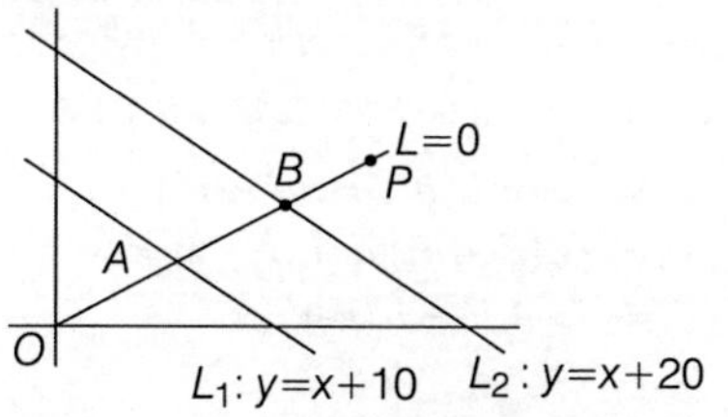

$$\Rightarrow \quad \frac{1}{OA} = \frac{1}{r_1} = \frac{\sin\theta - \cos\theta}{10}$$

Similarly, $\frac{1}{OB} = \frac{1}{r_2} = \frac{\sin\theta - \cos\theta}{20}$

Let $\quad P = (h, k) = (r\cos\theta, r\sin\theta)$

Then, $\frac{2}{OP} = \frac{2}{r} = \frac{1}{r_1} + \frac{1}{r_2}$

$$\Rightarrow \quad \frac{2}{r} = \frac{\sin\theta - \cos\theta}{10} + \frac{\sin\theta - \cos\theta}{20}$$

$$\Rightarrow \quad 40 = 3r\sin\theta - 3r\cos\theta$$

$$= 3k - 3h$$

$\Rightarrow$ Locus of P is $3x - 3y + 40 = 0$

$$\Rightarrow \quad 3y - 3x = 40$$

11 Lines $5x + 3y - 2 + \lambda_1(3x - y - 4) = 0$ are concurrent at the point of intersection of the lines $5x + 3y - 2 = 0$ and $3x - y - 4 = 0$ i.e. at $A(1, -1)$.

Similarly, lines $x - y + 1 + \lambda_2(2x - y - 2) = 0$ are concurrent at $B(3, 4)$.

The line, that belongs to both families is AB, whose equation is

$$y - 4 = \frac{-1-4}{1-3}(x-3)$$

i.e. $\quad 5x - 2y - 7 = 0.$

12 $a(2x + y - 3) + b(3x + 2y - 5) = 0$ passes through the point of intersection of the lines $2x + y - 3 = 0$ and $3x + 2y - 5 = 0$.

i.e. Through (1,1).

The line of this family which is farthest from $(4, -3)$ is the line through (1,1) and perpendicular to the line joining (1,1) and $(4, -3)$.

$\therefore$ Required line is $y - 1 = 3/4(x - 1)$

i.e. $3x - 4y + 1 = 0$

13

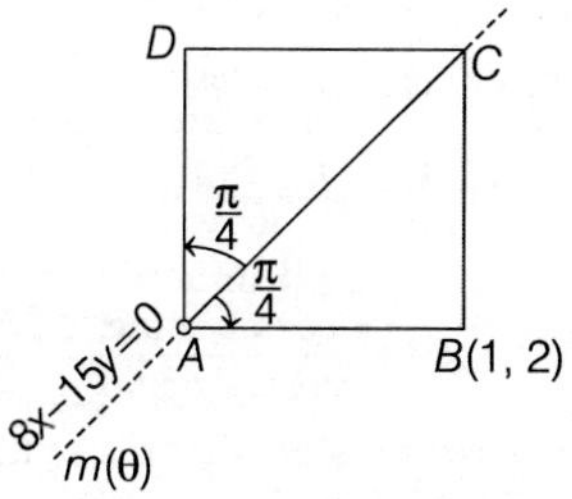

As can be seen from the figure, AB and AD are the line segments inclined at an angle of 45° with the diagonal line

$$AC(8x - 15y = 0)$$

Now, slope of line $AB = m_{AB}$

$$\Rightarrow \quad m_{AB} = \tan\left(\theta - \frac{\pi}{4}\right)$$

where, θ is the angle of inclination of diagonal with the positive X-axis.

$$\therefore \quad m_{AB} = \frac{\tan\theta - 1}{1 + \tan\theta} = \frac{8/15 - 1}{1 + 8/15}$$

$$\Rightarrow \quad m_{AB} = -\frac{7}{23}$$

$\therefore$ Equation of line AB is

$$y - 2 = -\frac{7}{23}(x - 1)$$

$$\Rightarrow \quad 23y - 46 = -7x + 7$$

$$\therefore \quad 7x + 23y = 53$$

Also, slope of line

$$AD(\perp AB) = m_{BC} = \frac{23}{7}$$

$$\therefore \quad y - 2 = \frac{23}{7}(x - 1)$$

$$\Rightarrow \quad 23x - 7y = 23 - 14$$

$$\therefore \quad 23x - 7y = 9$$

14 Let m be the slope of the line and angle θ it makes with the parallel line.

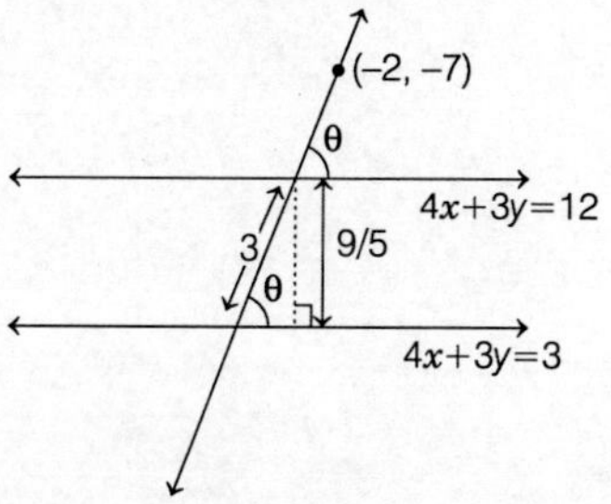

$$\therefore \quad \sin\theta = \frac{3}{5} \text{ or } \tan\theta = \frac{3}{4}$$

Hence, slope of the parallel lines is $-\frac{4}{3}$.

$$\therefore \left|\frac{m+\frac{4}{3}}{1-\frac{4m}{3}}\right| = \tan\theta = \frac{3}{4}$$

$$\Rightarrow \frac{3m+4}{3-4m} = \pm\frac{3}{4} \Rightarrow m = -\frac{7}{24}$$

The line is $y + 7 = -\frac{7}{24}(x+2)$ or

$7x + 24y + 182 = 0.$

15 According to the question,

$$\tan\frac{\pi}{4} = \frac{-\frac{a}{b}+\frac{\cos\alpha}{\sin\alpha}}{1+\frac{a\cos\alpha}{b\sin\alpha}}$$

$$\Rightarrow 1 = \frac{b\cos\alpha - a\sin\alpha}{a\cos\alpha + b\sin\alpha}$$

$\Rightarrow a\cos\alpha + b\sin\alpha = b\cos\alpha - a\sin\alpha$

$\Rightarrow (a-b)\cos\alpha = -(b+a)\sin\alpha$

$$\Rightarrow \tan\alpha = \frac{b-a}{b+a} \quad \text{...(i)}$$

Intersection point of $ax+by+p=0$ and $y = x\tan\alpha$ given by is $ax + bx\tan\alpha = p$

$$\Rightarrow x = \frac{p}{a+b\tan\alpha}$$

and $$y = \frac{p\tan\alpha}{a+b\tan\alpha}$$

Intersection point of $x\cos\alpha + y\sin\alpha = p$ and $y = x\tan\alpha$ is given by

$x\cos\alpha + x\tan\alpha\sin\alpha = p$

$\Rightarrow x = p\cos\alpha,\ y = p\cos\alpha\tan\alpha$

$y = p\sin\alpha$

According to the question,

$$x = \frac{p}{a+b\tan\alpha} = p\cos\alpha \quad \text{... (ii)}$$

and $$y = \frac{p\tan\alpha}{a+b\tan\alpha} = p\sin\alpha \quad \text{...(iii)}$$

$$\therefore \frac{p}{a+b\left\{\frac{b-a}{b+a}\right\}} = \frac{p}{\sec\alpha}$$

$$\Rightarrow \frac{p}{a+b\left(\frac{b-a}{b+a}\right)} = \frac{p}{\sqrt{\sec^2\alpha}}$$

$$= \frac{p}{\sqrt{1+\tan^2\alpha}}$$

$$\Rightarrow \frac{b+a}{ab+a^2+b^2-ab} = \frac{1}{\sqrt{1+\left(\frac{b-a}{b+a}\right)^2}} \quad \text{[using Eq. (i)]}$$

$$= \frac{b+a}{\sqrt{(b+a)^2+(b-a)^2}}$$

$\Rightarrow a^2+b^2 = \sqrt{2(a^2+b^2)}$

$\Rightarrow (a^2+b^2)^2 = 2(a^2+b^2)$

$\Rightarrow (a^2+b^2)(a^2+b^2-2) = 0$

$\Rightarrow a^2+b^2 \neq 0$

$\therefore a^2+b^2 = 2$

16 $A_1B_1 \equiv y = mx + c_1$

$A_2B_2 \equiv y = mx + c_2$

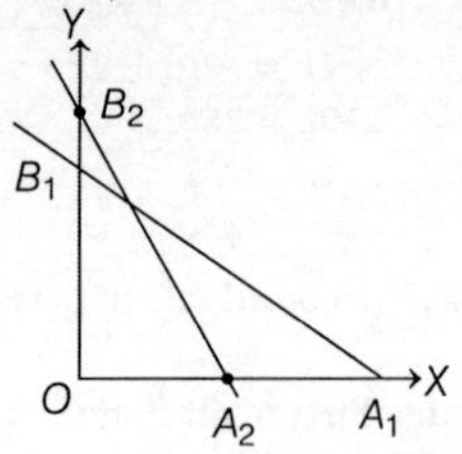

$$\therefore A_1 = \left(-\frac{c_1}{m}, 0\right), B_1 = (0, c_1),$$

$$A_2 = \left(-\frac{c_2}{m}, 0\right), B_2 = (0, c_2)$$

Since A_1, A_2, B_1, B_2 are concyclic,

$$OA_1\,OA_2 = OB_1\,OB_2 \Rightarrow \frac{c_1c_2}{m_2} = c_1c_2$$

$\therefore m^2 = 1 \Rightarrow m = 1\ (m>0)$

$\therefore A_1 = (-c_1, 0), A_2 = (-c_2, 0),$

$B_1 = (0, c_1), B_2 = (0, c_2)$

Now, $A_1B_2 \equiv -\frac{x}{c_1} + \frac{y}{c_2} = 1$

and $A_2B_1 \equiv -\frac{x}{c_2} + \frac{y}{c_1} = 1$

For point of intersection, consider

$$-\frac{x}{c_1} + \frac{y}{c_2} = -\frac{x}{c_2} + \frac{y}{c_1}$$

$(c_1 - c_2)x + (c_1 - c_2)y = 0 \Rightarrow x + y = 0$

DAY TWENTY SIX

The Circle

Learning & Revision for the Day

- Concept of Circle
- Line and Circle
- Equation of Tangents
- Equation of Normal
- Pair of Tangents
- Common Tangents of Two Circles
- Director Circle
- Chord of Contact
- Pole and Polar
- Angle of Intersection of Two Circles
- Family of Circles
- Radical Axis
- Coaxial System of Circles

Concept of Circle

Circle is the locus of a point which moves in a plane, such that its distance from a fixed point in the plane is a constant. The fixed point is the **centre** and the constant distance is the **radius**.

Standard Form of Equation of a Circle

The equation of a circle whose centre is at (h, k) and radius r is given, is $(x-h)^2 + (y-k)^2 = r^2$. It is also known as the central form of the equation of a circle.

The equation $x^2 + y^2 + 2gx + 2fy + c = 0$ always represents a circle, whose centre is $(-g, -f)$ and radius is $\sqrt{g^2 + f^2 - c}$. This is known as the general equation of a circle.

Equation of Circle when the End Points of a Diameter are Given

If A and B are end points of a diameter of a circle whose coordinates are (x_1, y_1) and (x_2, y_2), respectively.

Then, the equation of circle is $(x-x_1)(x-x_2)+(y-y_1)(y-y_2)=0$

NOTE Parametric equations of a circle is $(x-h)^2 + (y-k)^2 = r^2$, where $x = h + r\cos\theta$, $y = k + r\sin\theta$, $0 \le \theta \le 2\pi$.

Intercept on Axes

The length of intercepts made by the circle $x^2 + y^2 + 2gx + 2fy + c = 0$ with X and Y-axes are $2\sqrt{g^2 - c}$ and $2\sqrt{f^2 - c}$, respectively.

PREP MIRROR
Your Personal Preparation Indicator

- No. of Questions in Exercises (x)—
- No. of Questions Attempted (y)—
- No. of Correct Questions (z)—
  (Without referring Explanations)

- Accuracy Level $(z/y \times 100)$—
- Prep Level $(z/x \times 100)$—

In order to expect good rank in JEE, your Accuracy Level should be above 85 & Prep Level should be above 75.

Position of a Point w.r.t. a Circle

A point (x_1, y_1) lies outside, on or inside a circle

$$S = x^2 + y^2 + 2gx + 2fy + c = 0$$

according as $S_1 >, =,$ or < 0,

where, $S_1 = x_1^2 + y_1^2 + 2gx_1 + 2fy_1 + c$

Greatest and **least distance** of a point $A(x_1, y_1)$ from a circle with centre C and radius r as shown in the figure below, is $|AC + r|$ and $|AC - r|$.

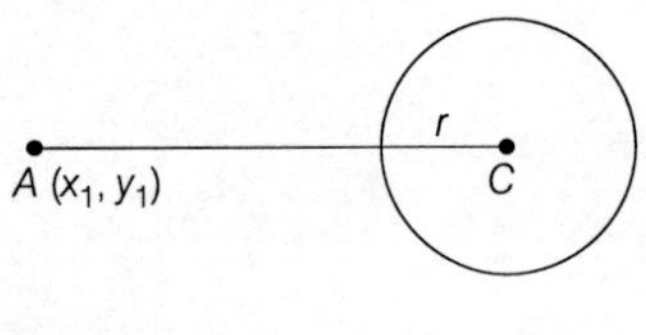

Line and Circle

Let $y = mx + c$ be a line and $x^2 + y^2 = r^2$ be a circle. If r is the radius of circle and $p = \left|\frac{c}{\sqrt{1 + m^2}}\right|$ is the length of the perpendicular from the centre on the line, then

(i) $p > r \Leftrightarrow$ the line passes outside the circle.

(ii) $p = r \Leftrightarrow$ the line touches the circle or the line is a tangent to the circle.

(iii) $p < r \Leftrightarrow$ the line intersect the circle at two points or the line is secant of the circle.

(iv) $p = 0 \Leftrightarrow$ the line is a diameter of the circle.

- The length of the intercept cut-off from the line $y = mx + c$ by the circle $x^2 + y^2 = r^2$ is

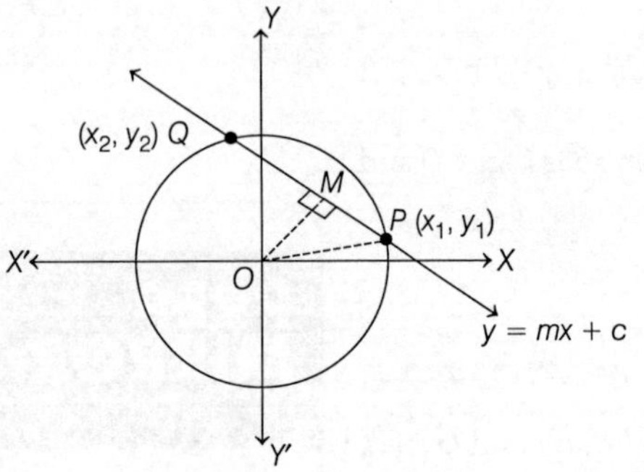

$$PQ = 2\sqrt{\frac{r^2(1 + m^2) - c^2}{1 + m^2}}$$

Equation of Tangents

A line which touch only one point of a circle is called its tangent as shown in the following figure. This tangent may be in slope or point form as given below.

1. Slope Form

(i) The equation of tangents of slope m to the circle $(x - a)^2 + (y - b)^2 = r^2$ are given by

$$y - b = m(x - a) \pm r\sqrt{1 + m^2}$$

and the coordinates of the points of contact are

$$\left(a \pm \frac{mr}{\sqrt{1 + m^2}}, b \mp \frac{r}{\sqrt{1 + m^2}}\right).$$

(ii) Point of intersection of the tangent drawn to the circle $x^2 + y^2 = r^2$ at the point $P(\alpha)$ and $Q(\beta)$ is

$$h = \frac{r\cos\frac{\alpha + \beta}{2}}{\cos\frac{\alpha - \beta}{2}} \text{ and } k = \frac{r\sin\frac{\alpha + \beta}{2}}{\cos\frac{\alpha - \beta}{2}}.$$

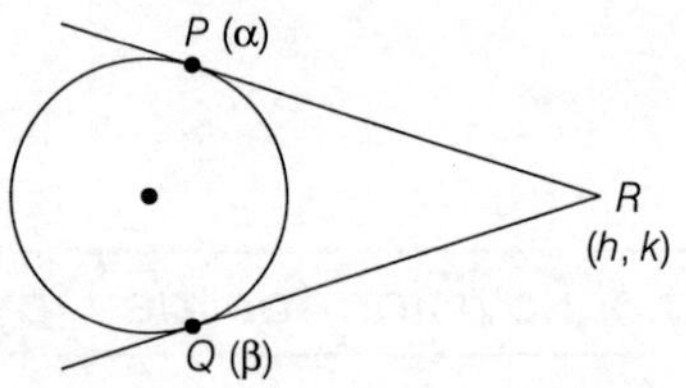

2. Point Form

(i) Equation of tangent for $x^2 + y^2 = r^2$ at (x_1, y_1) is $xx_1 + yy_1 = r^2$.

(ii) The equation of the tangent at the point $P(x_1, y_1)$ to a circle $x^2 + y^2 + 2gx + 2fy + c = 0$ is $xx_1 + yy_1 + g(x + x_1) + f(y + y_1) + c = 0$.

3. Parametric form

(i) Parametric coordinates of circle $x^2 + y^2 = r^2$ is $(r\cos\theta, r\sin\theta)$, then equation of tangent at $(r\cos\theta, r\sin\theta)$ is $x\cos\theta + y\sin\theta = r$.

(ii) Equation of the tangent to the circle $(x - h)^2 + (y - k)^2 = r^2$ at $(h + r\cos\theta, k + r\sin\theta) = r^2$ is $(x - h)\cos\theta + (y - k)\sin\theta = r$.

Length of the Tangents

The length of the tangent from the point $P(x_1, y_1)$ to the circle $x^2 + y^2 + 2gx + 2fy + c = 0$ is equal to

$$\sqrt{x_1^2 + y_1^2 + 2gx_1 + 2fy_1 + c} = \sqrt{S_1}$$

Equation of Normal

The normal at any point on a curve is a straight line which is perpendicular to the tangent to the curve at that point.

1. Point Form

The equation of normal to the circle

$$x^2 + y^2 + 2gx + 2fy + c = 0 \text{ or } x^2 + y^2 = a^2$$

at any point (x_1, y_1) is $\dfrac{x - x_1}{x_1 + g} = \dfrac{y - y_1}{y_1 + f}$

or $\dfrac{x}{x_1} = \dfrac{y}{y_1}$

2. Parametric Form

The equation of normal to the circle $x^2 + y^2 = a^2$ at point $(a\cos\theta, a\sin\theta)$ is $\frac{x}{\cos\theta} = \frac{y}{\sin\theta}$ or $y = x\tan\theta$.

Pair of Tangents

From a given point, two tangents can be drawn to a circle which are real and distinct, coincident or imaginary according as the given point lies outside, on or inside the circle.

The combined equation of the pair of tangents drawn from a point $P(x_1, y_1)$ to the circle $x^2 + y^2 = a^2$ is $SS_1 = T^2$.

where, $S = x^2 + y^2 - a^2, S_1 = x_1^2 + y_1^2 - a^2$ and $T = xx_1 + yy_1 - a^2$

Common Tangents of Two Circles

Let the centres and radii of two circles be C_1, C_2 and r_1, r_2, respectively.

(i) When one circle contains other as shown in the figure below, no common tangent is possible. Condition $C_1C_2 < r_1 - r_2$.

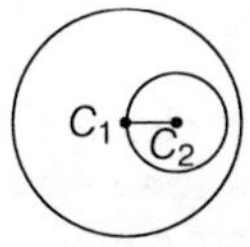

(ii) When two circles touch internally as shown in the figure, one common tangent is possible.

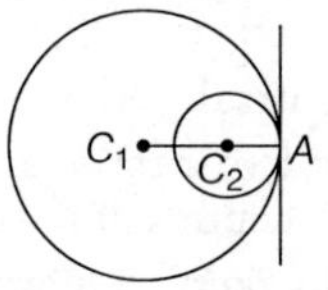

Condition, $C_1C_2 = r_1 - r_2$

(iii) When two circles intersect as shown in the figure below, two common tangents are possible.

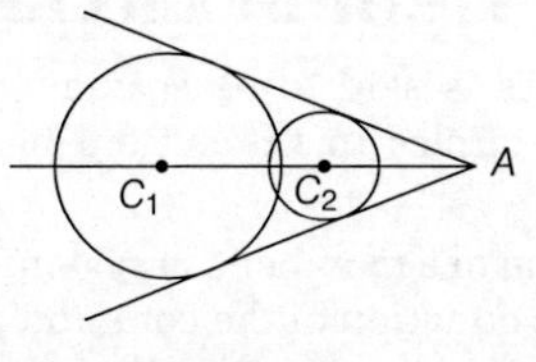

Condition, $|r_1 - r_2| < C_1C_2 < r_1 + r_2$

(iv) When two circles touch externally as shown in the figure below, three common tangents are possible.

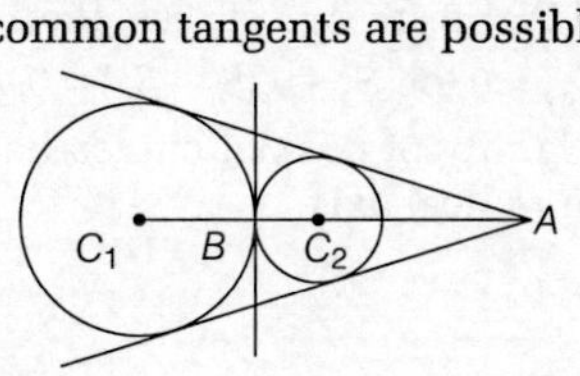

Condition, $C_1C_2 = r_1 + r_2$

A divides C_1C_2 externally in the ratio $r_1 : r_2$. ***B*** divides C_1C_2 internally in the ratio $r_1 : r_2$.

(v) When two circles are separately as shown in the figure below, four common tangents are possible.

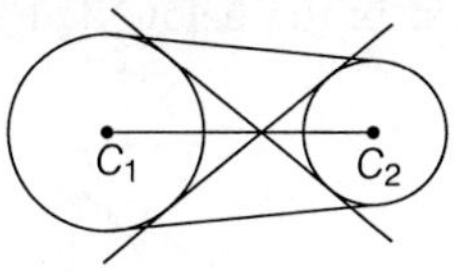

Condition, $C_1C_2 > r_1 + r_2$

Director Circle

The locus of the point of intersection of two perpendicular tangents to a given circle is known as its director circle. The equation of the director circle of the circle

$$x^2 + y^2 = a^2 \text{ is } x^2 + y^2 = 2a^2.$$

Chord of Contact

- The chord joining the points of contact of the two tangents from a point, which is outside is called the chord of contact of tangents.
- The equation of the chord of contact of tangents drawn from a point (x_1, y_1) to the circle $x^2 + y^2 = a^2$ is $xx_1 + yy_1 = a^2$ or $T = 0$.
- If AB is a chord of contact of tangents from C to the circle $x^2 + y^2 = r^2$ and M is the mid-point of AB as shown in figure. Then,

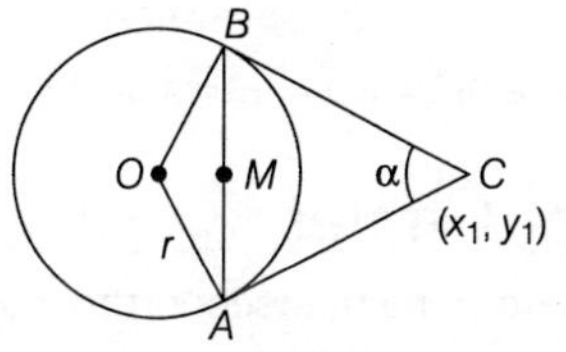

Angle between two tangents $\angle ACB$ is $2\tan^{-1}\frac{r}{\sqrt{S_1}}$.

Chord Bisected at a Given Point

The equation of the chord of the circle

$$x^2 + y^2 = a^2$$

bisected at the point (x_1, y_1) is given by

$$xx_1 + yy_1 - a^2 = x_1^2 + y_1^2 - a^2$$

or $$T = S_1$$

Of all the chords which passes through a given point $M(a,b)$ inside the circle the shortest chord is one whose middle point is (a,b).

Pole and Polar

If through a point $P(x_1, y_1)$ (inside or outside a circle) there be drawn any straight line to meet the given circle at Q and R as shown in the following figure, the locus of the point of intersection of the tangents at Q and R is called the **polar** of P and P is called the **pole** of the polar.

The polar of a point $P(x_1, y_1)$ with respect to the circle $x^2 + y^2 = a^2$ as shown in the below figure is $xx_1 + yy_1 = a^2$ or $T = 0$.

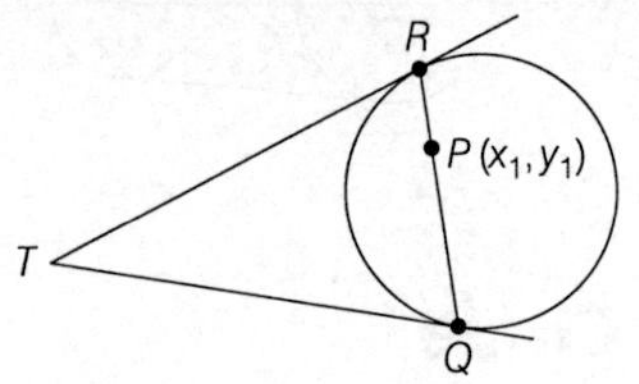

1. Conjugate Points

Two points A and B are conjugate points with respect to a given circle, if each lies on the polar of the other with respect to the circle.

2. Conjugate Lines

If two lines be such that the pole of one line lies on the other, then they are called conjugate lines with respect to the given circle.

Angle of Intersection of Two Circles

The angle of intersection of two circles is defined as the angle between the tangents to the two circles at their point of intersection is given by $\cos\theta = \dfrac{r_1^2 + r_2^2 - d^2}{2r_1 r_2}$

where, d is distance between centres of the circles.

Orthogonal Circles

Two circles are said to be intersect orthogonally, if their angle of intersection is a right angle.

(Radius of Ist circle)^2 + (Radius of IInd circle)^2

= (Distance between centres)^2

$\Rightarrow \quad 2(g_1g_2 + f_1f_2) = c_1 + c_2$

which is the condition of orthogonality of two circles. The circles having radii r_1 and r_2 intersect orthogonally. Then, length of their common chord is $\dfrac{2r_1r_2}{\sqrt{r_1^2 + r_2^2}}$.

Family of Circles

1. The equation of a family of circles passing through the intersection of a circle $x^2 + y^2 + 2gx + 2fy + c = 0$ and line $L \equiv lx + my + n = 0$ is
$x^2 + y^2 + 2gx + 2fy + c + \lambda\,(lx + my + n) = 0$ or $S + \lambda L = 0$
where, λ is any real number.
2. The equation of the family of circles passing through the point $A(x_1, y_1)$ and $B(x_2, y_2)$ is
$$(x - x_1)(x - x_2) + (y - y_1)(y - y_2) + \lambda \begin{vmatrix} x & y & 1 \\ x_1 & y_1 & 1 \\ x_2 & y_2 & 1 \end{vmatrix} = 0$$
$$\Rightarrow \quad (x - x_1)(x - x_2) + (y - y_1)(y - y_2) + \lambda L = 0$$
where, $L = 0$ represents the line passing through $A(x_1, y_1)$ and $B(x_2, y_2)$ and $\lambda \in R$.
3. The equation of the family of circles touching the circle $S = x^2 + y^2 + 2gx + 2fy + c = 0$ at point $P(x_1, y_1)$ is
$$x^2 + y^2 + 2gx + 2fy + c + \lambda\,\{xx_1 + yy_1 + g(x + x_1) + f(y + y_1) + c\} = 0$$
or $\quad S + \lambda L = 0$
where, $L = 0$ is the equation of the tangent to $S = 0$ at (x_1, y_1) and $\lambda \in R$.
4. The equation of a family of circles passing through the intersection of the circles
$S_1 = x^2 + y^2 + 2g_1x + 2f_1y + c_1 = 0$
and $S_2 = x^2 + y^2 + 2g_2x + 2f_2y + c_2 = 0$ is
$S_1 + \lambda S_2 = 0$, where ($\lambda \neq -1$) is an arbitrary real number.

Radical Axis

The radical axis of two circles is the locus of a point which moves in such a way that the lengths of the tangents drawn from it to the two circles are equal.

The radical axis of two circles $S_1 = 0$ and $S_2 = 0$ is given by, $S_1 - S_2 = 0$.

(i) The equations of radical axis and the common chord of two circles are identical.

(ii) The radical axis of two circles is always perpendicular to the line joining the centres of the circles.

(iii) **Radical centre** The point of intersection of radical axis of three circles whose centres are non-collinear, taken in pairs is called their radical centre.

Coaxial System of Circles

A system of circles is said to be coaxial system of circles, if every pair of the circles in the system has the same radical axis.

1. If the equation of a member of a system of coaxial circles is $S = 0$ and the equation of the common radical axis is $L = 0$, then the equation representing the coaxial system of circle is $S + \lambda L = 0$, where $\lambda \in R$.
2. If $S_1 = 0$ and $S_2 = 0$ are two circles, then $S_1 + \lambda S_2 = 0$
$S_1 + \lambda(S_1 - S_2) = 0$ or $S_2 + \lambda(S_1 - S_2) = 0, \lambda \in R$
represent a family of coaxial circles having $S_1 - S_2 = 0$ as the common radical axis.

DAY PRACTICE SESSION 1

FOUNDATION QUESTIONS EXERCISE

1 Equation of a circle whose two diameters are along the lines $2x-3y+4=0$ and $3x+4y-5=0$ and passes through the origin is

(a) $x^2+y^2+2x-44y=0$
(b) $17x^2+17y^2-2x+44y=0$
(c) $17x^2+17y^2+2x-44y=0$
(d) None of the above

2 Points (2, 0), (0, 1), (4, 5), and (0, a) are concyclic. Then a is equal to

(a) $\frac{14}{3}$ or 1 (b) 14 or $\frac{1}{3}$
(c) $-\frac{14}{3}$ or -1 (d) None of these

3 The abscissae of two points A and B are the roots of the equation $x^2+2ax-4=0$ and their ordinates are the roots of the equation $x^2+2bx-9=0$. Then equation of the circle with AB as diameter is

(a) $x^2+y^2-ax-bx+13=0$
(b) $x^2+y^2+ax+by-13=0$
(c) $x^2+y^2+2ax+2by-13=0$
(d) None of the above

4 If the lines $a_1x+b_1y+c_1=0$ and $a_2x+b_2y+c_2=0$ cut the coordinate axes in concyclic points, then

(a) $a_1a_2=b_1b_2$ (b) $a_1b_1=a_2b_2$
(c) $a_1b_1=a_2b_1$ (d) None of these

5 The equation of circle which passes through the points (2, 0) and whose centre is the limit of the point of intersection of the lines $3x+5y=1$ and $(2+c)x+5c^2y=1$ as $c\to 1$ is

(a) $25(x^2+y^2)-20x+2y+60=0$
(b) $25(x^2+y^2)-20x+2y-60=0$
(c) $25(x^2-y^2)-20x-2y-60=0$
(d) None of the above

6 The lines $3x-y+3=0$ and $x-3y-6=0$ cut the coordinate axes at concyclic points. The equation of the circle through these points is

(a) $x^2+y^2-5x-y-6=0$ (b) $x^2+y^2+5x+y+6=0$
(c) $x^2+y^2+2xy=0$ (d) None of these

7 The circle passing through $(1,-2)$ and touching the X-axis to at (3, 0) also passes through the point

→ JEE Mains 2013

(a) $(-5, 2)$ (b) $(2, -5)$
(c) $(5, -2)$ (d) $(-2, 5)$

8 AB is chord of the circle $x^2+y^2=25$. The tangents of A and B intersect at C. If (2, 3) is the mid-point of AB, then area of the quadrilateral $OACB$ is

(a) $50\sqrt{\frac{13}{3}}$ (b) $50\sqrt{\frac{3}{13}}$ (c) $50\sqrt{3}$ (d) $\frac{50}{\sqrt{3}}$

9 Two vertices of an equilateral triangle are $(-1, 0)$ and (1, 0) and the third vertex lies above the X-axis. Find the equation of its circumcircle.

(a) $x^2-y^2+\frac{2y}{\sqrt{3}}+1=0$ (b) $x^2+y^2-\frac{2y}{\sqrt{3}}-1=0$
(c) $x^2-y^2-\frac{y}{\sqrt{3}}=0$ (d) None of these

10 Circles are drawn through the point (2, 0) to cut intercept of length 5 units on the X-axis. If their centres lie in the first quadrant, then their equation for $k>0$ is

(a) $x^2+y^2-9x+2ky+14=0$
(b) $3x^2+3y^2+27x-2ky+42=0$
(c) $x^2+y^2-9x-2ky+14=0$
(d) $x^2+y^2-2kx-9y+14=0$

11 Circles are drawn through the points (a,b) and $(b,-a)$ such that the chord joining the two points subtends an angle of 45° at any point of the circumference. Then, the distance between the centres is

(a) $\sqrt{3}$ times the radius of either circle
(b) 2 times the radius of either circle
(c) $\frac{1}{\sqrt{2}}$ times the radius of either circle
(d) $\sqrt{2}$ times the radius of either circle

12 Let A be the centre of the circle $x^2+y^2-2x-4y-20=0$. If the tangents at the points $B(1,7)$ and $D(4,-2)$ on the circle meet at C, then find the area of the quadrilateral $ABCD$.

(a) 78 (b) 75 (c) 79 (d) 85

13 Find the equation of a circle concentric with the circle $x^2+y^2-6x+12y+15=0$ and has double of its area.

(a) $x^2+y^2-6x+12y-15=0$
(b) $x^2+y^2-6x-12y+15=0$
(c) $x^2+y^2-6x+12y+15=0$
(d) None of the above

14 The equation of the locus of the mid-points of the chords of the circle $4x^2+4y^2-12x+4y+1=0$ that subtend an angle of $2\pi/3$ at its centre is

(a) $x^2+y^2+3x-y+31/16=0$
(b) $x^2+y^2-3x+y+31/16=0$
(c) $x^2+y^2-3x+y-31/16=0$
(d) None of the above

15 Let PQ and RS be tangents at the extremities of the diameter PR of the circle of radius r. If PS and RQ intersect at a point X on the circumference of the circle, then $2r$ equals

(a) $\sqrt{(PQ\cdot RS)}$ (b) $(PQ+RS)/2$
(c) $2\,PQ\cdot RS/(PQ+RS)$ (d) $\sqrt{\left[\frac{(PQ^2+RS^2)}{2}\right]}$

16 If the line $ax + by = 0$ touches the circle $x^2 + y^2 + 2x + 4y = 0$ and is normal to the circle $x^2 + y^2 - 4x + 2y - 3 = 0$, (a, b) is given by

(a) (2, 1) (b) (1,–2) (c) (1, 2) (d) (–1, 2)

17 A circle touches the hypotenuse of a right angle triangle at its middle point and passes through the mid-point of the shorter side. If a and $b\,(a < b)$ be the length of the sides, then the radius is

(a) $\frac{b}{a}\sqrt{a^2 + b^2}$ (b) $\frac{b}{2a}\sqrt{a^2 - b^2}$

(c) $\frac{b}{4a}\sqrt{a^2 + b^2}$ (d) None of these

18 The number of common tangents to the circles $x^2 + y^2 - 4x - 6y - 12 = 0$ and $x^2 + y^2 + 6x + 18y + 26 = 0$ is → JEE Mains 2015

(a) 1 (b) 2 (c) 3 (d) 4

19 If the two circles $(x-1)^2 + (y-3)^2 = r^2$ and $x^2 + y^2 - 8x + 2y + 8 = 0$ intersect in two distinct points, then

(a) $2 < r < 8$ (b) $r < 2$
(c) $r = 2$ (d) $r > 2$

20 Let C be the circle with centre at (1, 1) and radius 1. If T is the circle centred at (0, k) passing through origin and touching the circle C externally, then the radius of T is equal to → JEE Mains 2014

(a) $\frac{\sqrt{3}}{\sqrt{2}}$ (b) $\frac{\sqrt{3}}{2}$

(c) $\frac{1}{2}$ (d) $\frac{1}{4}$

21 The tangent to the circle $x^2 + y^2 = 5$ at the point (1, – 2), also touches the circle $x^2 + y^2 - 8x + 6y + 20 = 0$, find its point of contact.

(a) $x = 2, y = 1$ (b) $x = 3, y = -1$
(c) $x = 5, y = 7$ (d) None of these

22 If the tangent at (1, 7) to the curve $x^2 = y - 6$ touches the circle $x^2 + y^2 + 16x + 12y + c = 0$, then the value of c is → JEE Mains 2018

(a) 195 (b) 185 (c) 85 (d) 95

23 For the circle $x^2 + y^2 = r^2$, find the value of r for which the area enclosed by the tangents from the point $P(6, 8)$ to the circle and the chord of contact is maximum.

(a) $r = 4$ (b) $r = 5$ (c) $r = 3$ (d) $r = 1$

24 From any point on the circle $x^2 + y^2 = a^2$ tangents are drawn to the circle $x^2 + y^2 = a^2 \sin^2 \alpha$. The angle between them is

(a) $\alpha/2$ (b) α (c) 2α (d) None of these

25 If one of the diameters of the circle, given by the equation $x^2 + y^2 - 4x + 6y - 12 = 0$, is a chord of a circle S, whose centre is at $(-3, 2)$, then the radius of S is → JEE Mains 2016

(a) $5\sqrt{2}$ (b) $5\sqrt{3}$ (c) 5 (d) 10

26 The length of the common chord of two circles $(x-a)^2 + (y-b)^2 = c^2$ and $(x-b)^2 + (y-a)^2 = c^2$ is

(a) $\sqrt{4c^2 + 2(a-b)^2}$ (b) $\sqrt{4c^2 - (a-b)^2}$
(c) $\sqrt{4c^2 - 2(a-b)^2}$ (d) $\sqrt{2c^2 - 2(a-b)^2}$

27 If P, Q and R are the centres and r_1, r_2 and r_3 are the corresponding radii of the three circles form a system of coaxial circle, then $r_1^2 \cdot QR + r_2^2 \cdot RP + r_3^2 \cdot PQ$ is equal to

(a) $PQ \cdot QR \cdot RP$ (b) $-PQ \cdot QR \cdot RP$
(c) $PQ + QR + RP$ (d) $\frac{PQ}{QR} \times RP$

28 The condition that the chord $x \cos\alpha + y \sin\alpha - p = 0$ of $x^2 + y^2 - a^2 = 0$ may subtend a right angle at the centre of circle, is

(a) $a^2 = 2p^2$ (b) $p^2 = 2a^2$ (c) $a = 2p$ (d) $p = 2a$

29 The equation of the circle of minimum radius which contains the three circles $x^2 + y^2 - 4y - 5 = 0$, $x^2 + y^2 + 12x + 4y + 31 = 0$ and $x^2 + y^2 + 6x + 12y + 36 = 0$ is

(a) $\left(x - \frac{31}{18}\right)^2 + \left(y - \frac{23}{12}\right)^2 = \left(3 - \frac{5}{36}\sqrt{949}\right)^2$

(b) $\left(x + \frac{23}{12}\right)^2 + \left(y - \frac{31}{18}\right)^2 = \left(3 + \frac{5}{36}\sqrt{949}\right)^2$

(c) $\left(x + \frac{31}{18}\right)^2 + \left(y + \frac{23}{12}\right)^2 = \left(3 + \frac{5}{36}\sqrt{949}\right)^2$

(d) None of the above

30 The centres of two circles C_1 and C_2 each of unit radius are at a distance of 6 units from each other. Let P be the mid-point of the line segment joining the centres of C_1 and C_2 and C be a circle touching circles C_1 and C_2 externally. If a common tangent to C_1 and C passing through P is also a common tangent to C_2 and C, then the radius of the circle C is

(a) 16 (b) 4 (c) 8 (d) 2

Direction (Q. Nos. 31-35) *Each of these questions contains two statements : Statement* I *(Assertion) and Statement* II *(Reason). Each of these questions also has four alternative choices, only one of which is the correct answer. You have to select one of the codes (a), (b), (c) and (d) given below.*

(a) Statement I is true, Statement II is true; Statement II is a correct explanation for Statement I
(b) Statement I is true, Statement II is true; Statement II is not a correct explanation for Statement I
(c) Statement I is true; Statement II is false
(d) Statement I is false; Statement II is true

31 Tangents are drawn from the point (17, 7) to the circle $x^2 + y^2 = 169$.

Statement I The tangents are mutually perpendicular.

Statement II The locus of the points from which mutually perpendicular tangents can be drawn to the given circle is $x^2 + y^2 = 338$.

32 Consider the radius should be zero in limiting points.

Statement I Equation of a circle through the origin and belonging to the coaxial system, of which the limiting points are (1, 1) and (3, 3) is
$2x^2 + 2y^2 - 3x - 3y = 0$.

Statement II Equation of a circle passing through the point, (1, 1) and (3, 3) is $x^2+y^2 - 2x - 6y + 6 = 0$.

33 **Statement I** The circle of smallest radius passing through two given points A and B must be of radius $\frac{1}{2}AB$.

Statement II A straight line is a shortest distance between two points.

34 Consider $L_1 \equiv 2x + 3y + p - 3 = 0$,
$L_2 \equiv 2x + 3y + p + 3 = 0$, where p is a real number and $C \equiv x^2 + y^2 + 6x - 10y + 30 = 0$.

Statement I If line L_1 is a chord of circle C, then L_2 is not always a diameter of circle C.

Statement II If line L_1 is a diameter of circle C, then L_2 is not a chord of circle C.

35 **Statement I** The only circle having radius $\sqrt{10}$ and a diameter along line $2x + y = 5$ is
$x^2 + y^2 - 6x + 2y = 0$.

Statement II $2x + y = 5$ is a normal to the circle
$x^2 + y^2 - 6x + 2y = 0$ → JEE Mains 2013

DAY PRACTICE SESSION 2

PROGRESSIVE QUESTIONS EXERCISE

1 If the line $y = x + 3$ meets the circle $x^2 + y^2 = a^2$ in A and B, then equation of the circle on AB as diameter is

(a) $x^2 + y^2 + 3x - 3y - a^2 + 9 = 0$
(b) $x^2 + y^2 - 3x + 3y - a^2 + 9 = 0$
(c) $x^2 + y^2 + 3x + 3y - a^2 + 9 = 0$
(d) None of the above

2 A circle C_1 of radius 2 units lies in the first quadrant and touches both the axes. Equation of the circle having centre at (6, 5) and touching the circle C_1 externally is

(a) $x^2 + y^2 - 12x - 10y + 52 = 0$
(b) $x^2 + y^2 - 12x - 10y + 12 = 0$
(c) $x^2 + y^2 - 12x - 10y - 52 = 0$
(d) None of the above

3 The line $(x - 2)\cos\theta + (y - 2)\sin\theta = 1$ touches the circle $x^2 + y^2 + 2gx + 2fy + c = 0$ for all values of θ, then $(g^2 + f^2 + c)/(g + f + c)$ is equal to

(a) 5 (b) 1
(c) −15 (d) None of these

4 Tangents PA and PB are drawn to $x^2 + y^2 = a^2$ from the point $P(x_1, y_1)$. Equation of the circumcircle of triangle PAB is

(a) $x^2 + y^2 - xx_1 - yy = 0$ (b) $x^2 + y^2 + xx_1 - yy_1 = 0$
(c) $x^2 + y^2 - xx_1 + yy_1 = 0$ (d) $x^2 + y^2 + xx_1 + yy_1 = 0$

5 If $a > 2b > 0$ then the positive value of m for which $y = mx - b\sqrt{(1 + m^2)}$ is a common tangent to $x^2 + y^2 = b^2$ and $(x - a)^2 + y^2 = b^2$ is

(a) $\frac{2b}{\sqrt{a^2 - 4b^2}}$ (b) $\frac{\sqrt{a^2 - 4b^2}}{2b}$ (c) $\frac{2b}{1 - 2b}$ (d) $\frac{b}{a - 2b}$

6 The straight line $2x - 3y = 1$ divides the circular region $x^2 + y^2 \le 6$ into two parts.

If $S = \left\{\left(2, \frac{3}{4}\right), \left(\frac{5}{2}, \frac{3}{4}\right), \left(\frac{1}{4}, -\frac{1}{4}\right), \left(\frac{1}{8}, \frac{1}{4}\right)\right\}$, then the number of point(s) in S lying inside the smaller part is

(a) 0 (b) 1 (c) 2 (d) 4

7 Let $ABCD$ be a quadrilateral with area 18, with side AB parallel to the side CD and $AB = 2CD$. Let AD be perpendicular to AB and CD. If a circle is drawn inside the quadrilateral $ABCD$ touching all the sides, then its radius is

(a) 3 (b) 2 (c) 3 / 2 (d) 1

8 If one of the diameters of the circle $S = x^2 + y^2 - 2x - 6y + 6 = 0$ is the common chord to the circle C with centre (2, 1), then the radius of the circle is

(a) 1 (b) 3 (c) $\sqrt{3}$ (d) 2

9 A rational point is a point both of whose coordinates are rational numbers. Let C be any circle with centre $(0, \sqrt{2})$. Then, the maximum number of rational points on the circle is

(a) 0 (b) 2
(c) Infinitely many (d) None of these

10 Let L_1 be a line passing through the origin and L_2 be the line $x + y = 1$. If the intercepts made by the circle $x^2 + y^2 - x + 3y = 0$ on L_1 and L_2 are equal, then L_1 is

(a) $x + y = 0$ (b) $x + y = 2$
(c) $x + 7y = 0$ (d) $x - 7y = 0$

11 If a_n, $n = 1, 2, 3, 4$ represent four distinct positive real numbers other than unit such that each pair of the logarithm of a_n and the reciprocal of logarithm denotes a point on a circle, whose centre lies on Y-axis. Then, the product of these four members is

(a) 0 (b) 1 (c) 2 (d) 3

12 The set of values of a for which the point $(2a, a+1)$ is an interior point of the larger segment of the circle $x^2 + y^2 - 2x - 2y - 8 = 0$ made by the chord $x - y + 1 = 0$, is

(a) $\left(\frac{5}{9}, \frac{9}{5}\right)$ (b) $\left(0, \frac{5}{9}\right)$
(c) $\left(0, \frac{9}{5}\right)$ (d) $\left(1, \frac{9}{5}\right)$

13 Three concentric circles of which biggest circle is $x^2 + y^2 = 1$, have their radii in AP. If the line $y = x + 1$ cuts all the circles in real and distinct points, then the interval in which the common difference of AP will lie, is

(a) $\left[0, \left(1 - \frac{1}{\sqrt{2}}\right)\right]$ (b) $\left(0, \frac{1}{2}\right)$
(c) $(1, 1)$ (d) $\left[0, \frac{1}{2}\left(1 - \frac{1}{\sqrt{2}}\right)\right]$

14 If the distance from the origin of the centres of the three circles $x^2 + y^2 - 2\lambda_i\ x = c^2$, $(i = 1, 2, 3)$ are in G.P., then the lengths of the tangents drawn to them from any point on the circle $x^2 + y^2 = c^2$ are in

(a) AP (b) GP
(c) HP (d) None of these

15 The limiting points of the coaxial system of circles given by $x^2 + y^2 + 2gx + c + \lambda(x^2 + y^2 + 2fy + k) = 0$ Subtend a tight angle at the origin, if

(a) $-\frac{c}{g^2} - \frac{k}{f^2} = 2$ (b) $\frac{c}{g^2} + \frac{k}{f^2} = -2$
(c) $\frac{c}{g^2} - \frac{k}{f^2} = 2$ (d) $\frac{c}{g^2} + \frac{k}{f^2} = 2$

16 A ray of light incident at the point (3, 1) gets reflected from the tangent at $(0, 1)$ to the circle $x^2 + y^2 = 1$. The reflected ray touches the circle. The equation of the line along which the incident ray moves is

(a) $3x + 4y - 13 = 0$ (b) $4x - 3y - 10 = 0$
(c) $4x + 3y - 13 = 0$ (d) $3x - 4y - 5 = 0$

17 The number of integral values of λ for which $x^2 + y^2 + \lambda x + (1 - \lambda)y + 5 = 0$ is the equation of a circle whose radius cannot exceed 5, is

(a) 14 (b) 18 (c) 16 (d) 10

18 The point $([P+1], [P])$ (where $[x]$ is the greatest integer less than or equal to x), lying inside the region bounded by the circles $x^2 + y^2 - 2x - 15 = 0$ and $x^2 + y^2 - 2x - 7 = 0$, then

(a) $P \in [-1, 2) - \{0, 1\}$ (b) $P \in [-1, 0) \cup [0, 1] \cup [1, 2)$
(c) $P \in (-1, 2)$ (d) None of these

19 The line $3x - 4y - k = 0$ touches the circle $x^2 + y^2 - 4x - 8y - 5 = 0$ at (a, b). Then $k, (a, b)$ is

(a) 15, (5, 0) (b) −35, (−1, 8)
(c) Both (a) and (b) (d) None of these

20 If the equation of the circle obtained by reflecting the circle $x^2 + y^2 - a^2 = 0$ in the line $y = mx + c$ is $x^2 + y^2 + 2gx + 2fy + c = 0$, then

(a) $g = \frac{2cm}{1+m^2}, a^2 + c = \frac{4c^2}{1+m^2}$
(b) $g = -\frac{2cm}{1+m^2}, a^2 + c = \frac{4c^2}{1+m^2}$
(c) $f = \frac{4c}{1+m^2}, a^2 + c = \frac{4c^2}{1+m^2}$
(d) None of the above

ANSWERS

SESSION 1

1 (c)	**2** (a)	**3** (c)	**4** (a)	**5** (b)	**6** (a)	**7** (c)	**8** (b)	**9** (b)	**10** (c)
11 (d)	**12** (b)	**13** (a)	**14** (b)	**15** (a)	**16** (c)	**17** (c)	**18** (c)	**19** (a)	**20** (d)
21 (b)	**22** (d)	**23** (b)	**24** (c)	**25** (b)	**26** (c)	**27** (b)	**28** (a)	**29** (c)	**30** (c)
31 (a)	**32** (b)	**33** (b)	**34** (c)	**35** (c)					

SESSION 2

1 (a)	**2** (a)	**3** (a)	**4** (a)	**5** (a)	**6** (c)	**7** (b)	**8** (b)	**9** (b)	**10** (c)
11 (b)	**12** (c)	**13** (d)	**14** (b)	**15** (d)	**16** (a)	**17** (d)	**18** (d)	**19** (c)	**20** (a)

Hints and Explanations

SESSION I

1 Equation of two diameters are
$2x-3y+4=0$ and $3x+4y-5=0$
$\therefore$ Centre is $(-1/17, 22/17)$
Circle passes through origin
$\therefore$ Equation of the circle is

$$\left(x+\frac{1}{17}\right)^2+\left(y-\frac{22}{17}\right)^2=\left(\frac{1}{17}\right)^2+\left(\frac{22}{17}\right)^2$$

$$\Rightarrow x^2+y^2+\frac{2}{17}x-\frac{44}{17}y=0$$

$$\Rightarrow 17x^2+17y^2+2x-44y=0$$

2 Let circle be
$x^2+y^2+2gx+2fy+c=0$
This passes through (2, 0), (0, 1), (4, 5)
So, $4+4g+c=0, 1+2f+c=0,$
$41+8g+10f+c=0$
Solving these equations, we get
$g=-13/6, f=-17/6, c=14/3.$
So, circle is
$3(x^2+y^2)-13x-17y+14=0$
Since $(0, a)$ also lies on it, we get
$\therefore \quad 3a^2-17a+14=0$
So, $\quad a=1$ or $14/3$

3 Let A and B be $(x_1, y_1), (x_2, y_2)$.
x_1 and x_2 are roots of $x^2+2ax-4=0$
Then, $x_1+x_2=-2a, x_1x_2=-4.$
y_1, y_2 are roots of $x^2+2bx-9=0.$
Then, $y_1+y_2=-2b, y_1y_2=-9.$
Equations of circle on AB as diameter is
$(x-x_1)(x-x_2)+(y-y_1)(y-y_2)=0$
$\Rightarrow x^2+y^2-(x_1+x_2)x-(y_1+y_2)y$
$+x_1x_2+y_1y_2=0$
$\Rightarrow \quad x^2+y^2+2ax+2by-13=0$

4 The line $a_1x+b_1y+c_1=0$ cut X and Y-axis in $A(-c_1/a_1, 0)$ and $B(0, -c_1/b_1)$ and the line
$a_2x+b_2y+c_2=0$ cut axes in $C(-c_2/a_2, 0)$ and $D(0, -c_2/b_2)$,
So, AC and BD are chords along X and Y-axes intersecting at origin O.
Since A, B, C, D are concyclic so
$OA\cdot OC=OB\cdot OD$

or $\left(-\frac{c_1}{a_1}\right)\left(-\frac{c_2}{a_2}\right)=\left(-\frac{c_1}{b_1}\right)\left(\frac{c_2}{b_2}\right)$

$\Rightarrow \quad a_1a_2=b_1b_2.$

5 Given lines are $3x+5y=1$ ...(i)
and $(2+c)x+5c^2y=1$ ...(ii)
From Eqs. (i) and (ii), we get
$(1-c)x+5(1-c^2)y=0$
$c=1, x+10y=0$

$\therefore$ Centre $=\left(\frac{2}{5}, -\frac{1}{25}\right)$

$\therefore$ Equation of circle is

$$\left(x-\frac{2}{5}\right)^2+\left(y+\frac{1}{25}\right)^2=\left(2-\frac{2}{5}\right)^2+\frac{1}{25^2}$$

$\Rightarrow 25(x^2+y^2)-20x+2y-60=0$

6 Here, $3\times 1=(-1)(-3)=3$
Hence, the points are concyclic.
$\therefore \quad L_1L_2+\lambda xy=0$
$\Rightarrow (3x-y+3)(x-3y-6)+\lambda xy=0$
$3(x^2+y^2)+(\lambda-10)xy-15x-3y-18=0$
Now for circle coefficient of $xy=0$
$\lambda-10=0\Rightarrow\lambda=10$
$\therefore \quad 3(x^2+y^2)-15x-3y-18=0$
$\Rightarrow \quad x^2+y^2-5x-y-6=0$

7 Let the equation of circle be
$(x-3)^2+(y-0)^2+\lambda y=0$

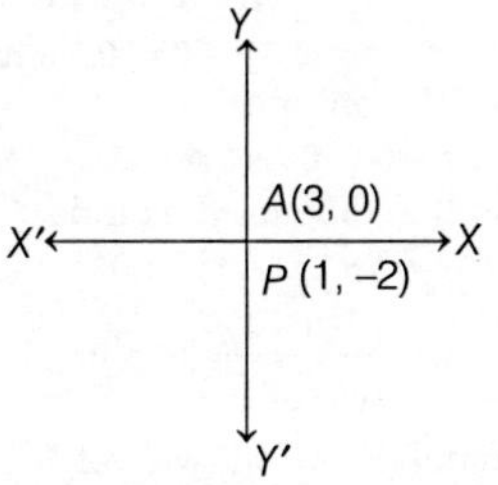

As it passes through $(1, -2)$.
$\therefore \quad (1-3)^2+(-2)^2+\lambda(-2)=0$
$\Rightarrow \quad 4+4-2\lambda=0$
$\Rightarrow \quad \lambda=4$
$\therefore$ Equation of circle is
$(x-3)^2+y^2+4y=0$
Now, by hit and trial method, we see that point $(5, -2)$ satisfies equation of circle.

8 Area of quadrilateral $OACB$,

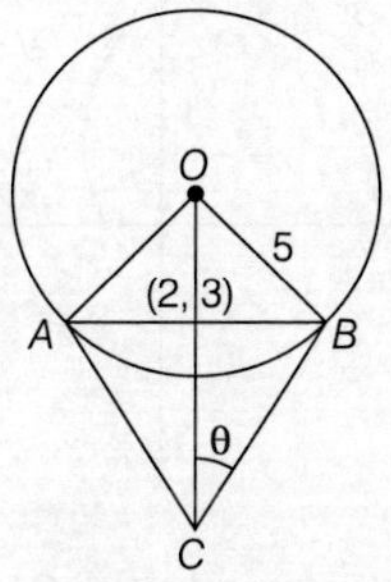

$A=OB\cdot BC=5^2\cot\theta$

$=50\sqrt{\frac{3}{13}} \quad \left[\because \sin\theta=\frac{\sqrt{13}}{5}\right]$

9 Length of side of triangle is 2.
The equation of circle passing through $(-1, 0)$ and $(1, 0)$ is
$(x+1)(x-1)+y^2+y\lambda=0.$
The coordinates of third vertex will be $(0, \sqrt{3})$, which is passing by the circle.

$\therefore \quad \lambda=-\frac{2}{\sqrt{3}}$

So, the equation of circle is

$x^2+y^2-\frac{2y}{\sqrt{3}}-1=0.$

10 Here, clearly, x-coordinate of centre is x- coordinate of the mid-point of AB or AB' i.e. $9/2$ or $-1/2$.

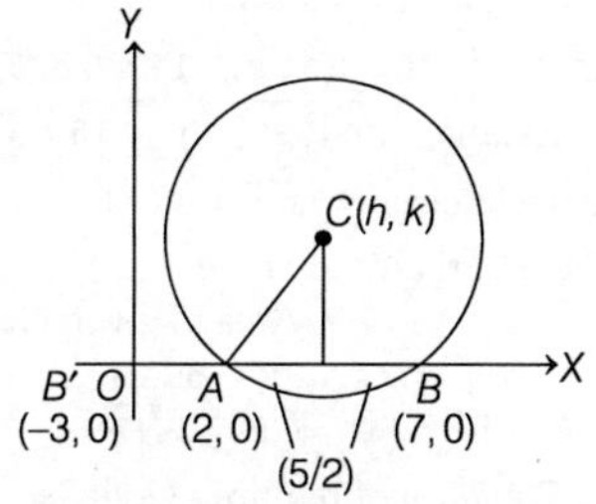

Since centre lies in the first quadrant,
$h=9/2$
$\therefore$ Equation of circle is
$(x-9/2)^2+(y-k)^2=(h-2)^2$
$+k^2=25/4+k^2$
or $\quad x^2+y^2-9x-2ky+14=0$

11 Let $P(x, y)$ be any point on the circumference of the circle.

Then, $m_1=$ Slope of $PA=\frac{b-y}{a-x}$

and $\quad m_2=$ Slope of $PB=\frac{-a-y}{b-x}$

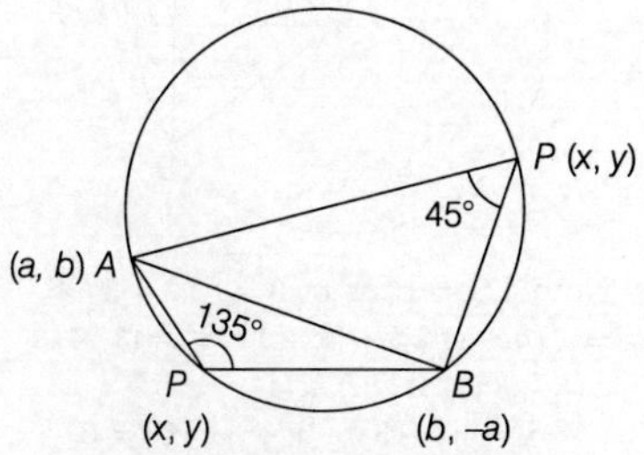

We have, $\angle APB=45°$ or $135°$

$$\Rightarrow \quad \frac{m_1-m_2}{1+m_1m_2}=\tan 45° \text{ or } \tan 135°$$

$$\Rightarrow \quad \frac{\frac{b-y}{a-x}-\frac{-a-y}{b-x}}{1+\frac{b-y}{a-x}\times\frac{-a-y}{b-x}}=\pm 1$$

$$\Rightarrow \frac{(b-y)(b-x)+(a+y)(a-x)}{(a-x)(b-x)-(b-y)(a+y)} = \pm 1$$

$$\Rightarrow \quad x^2 + y^2 = a^2 + b^2$$

$$\Rightarrow \{x-(a+b)\}^2 + \{y-(b-a)\}^2 = a^2 + b^2$$

The centres of these circles are $O(0,0)$ and $C(a+b, b-a)$.

$\therefore$ Distance between the centre

$$= \sqrt{(a+b)^2 + (a-b)^2} = \sqrt{2}\sqrt{a^2+b^2}$$

$= \sqrt{2}$ (Radius of either circle)

12 The tangent at $B(1,7)$ is $y = 7$ and $D(4,-2)$ is $3x - 4y - 20 = 0$.

Then, meet at $C(16, 7)$.

Now, $AB = 5$, $BC = 15$

Area of quadrilateral $ABCD$

$$= AB \cdot BC = 75$$

13 Centre of given circle

$x^2 + y^2 - 6x + 12y + 15 = 0$ is $(3, -6)$.

$\therefore$ Radius $= \sqrt{(3)^2 + (-6)^2 - 15} = \sqrt{30}$

Area of circle $= \pi r^2 = \pi(\sqrt{30})^2 = 30\pi$

Area of required circle

= 2 (Area of given circle)

$$\therefore \quad \pi R^2 = 2 \times 30\pi = 60\pi$$

$$\Rightarrow \quad R^2 = 60 \Rightarrow R = 2\sqrt{15}$$

$\therefore$ Equation of required circle is

$$(x-3)^2 + (y+6)^2 = (2\sqrt{15})^2$$

$$\Rightarrow x^2 + 9 - 6x + y^2 + 36 + 12y = 60$$

$$\Rightarrow \quad x^2 + y^2 - 6x + 12y - 15 = 0$$

14 Let $Q(h, k)$ be the mid-point of chord AB.

$\therefore \angle ACQ = \angle BCQ = \pi/3$

Coordinate of centre are $(3/2, -1/2)$ and radius $= 3/2$

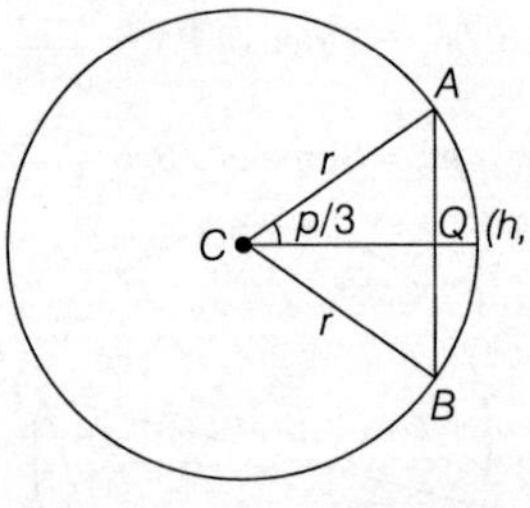

Now, $CQ = r\cos\pi/3 = r/2 = 3/4$.

$$\Rightarrow (h - 3/2)^2 + (k + 1/2)^2 = (3/4)^2$$

$\therefore$ Locus of $Q(h, k)$ is

$$x^2 + y^2 - 3x + y + 31/16 = 0.$$

15 Taking diameter PR as X-axis and centre O as origin, tangents at P and R are given by

$$x = -r \quad \ldots(i)$$

and $$x = r \quad \ldots(ii)$$

Let coordinates of X on circle be $(r\cos\alpha, r\sin\alpha)$

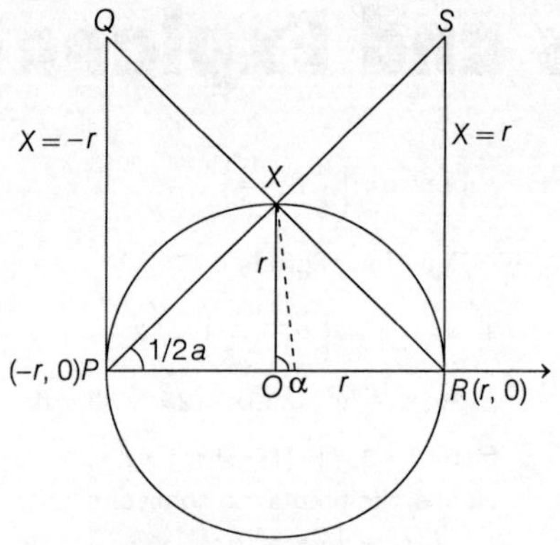

$$\angle XPR = \frac{1}{2}\angle XOR = \frac{1}{2}\alpha$$

$$\therefore \quad RS = 2r\tan\frac{1}{2}\alpha;$$

$$\angle XRP = 90^\circ - \alpha/2$$

$$\therefore \quad PQ = 2r\tan(90^\circ - \alpha/2) = 2r\cot\alpha/2$$

$$\therefore \quad PQ.RS = 4r^2 \Rightarrow 2r = \sqrt{(PQ.RS)}$$

16 Line $ax + by = 0$ is normal to the circle $x^2 + y^2 - 4x + 2y - 3 = 0$, so its centre $(2, -1)$ lie on line.

$\therefore \quad 2a - b = 0$ *i.e.*, $b = 2a$ $\quad\ldots(i)$

Also line touches the circle

$$x^2 + y^2 + 2x + 4y = 0$$

$$\therefore \quad \left|\frac{-a-2b}{\sqrt{(a^2+b^2)}}\right| = \sqrt{5} \quad \ldots(ii)$$

Solving (i) and (ii), we get $4a^2 = b^2$.

From the choices, only solution is $(1, 2)$

17 The equation of the circle touching $\frac{x}{a} + \frac{y}{b} = 1$ at $P\left(\frac{a}{2}, \frac{b}{2}\right)$ is $\left(x - \frac{a}{2}\right)^2$

$$+\left(y - \frac{b}{2}\right)^2 + \lambda\left(\frac{x}{a} + \frac{y}{b} - 1\right) = 0 \quad \ldots(i)$$

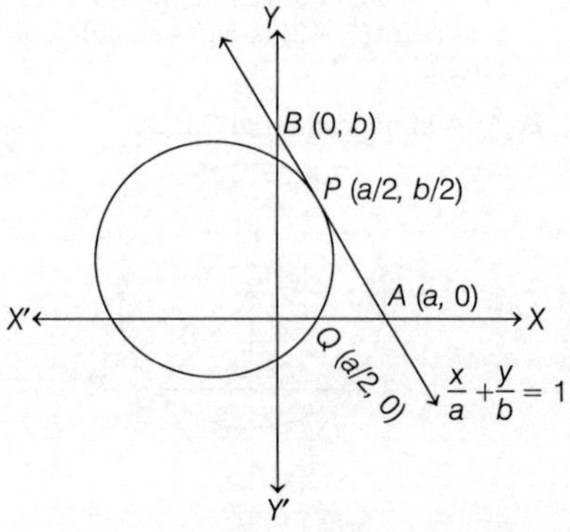

Since, it passes through $Q\left(\frac{a}{2}, 0\right)$.

Therefore, $\lambda = \frac{b^2}{2}$ [from Eq. (i)]

On putting the value of λ in Eq. (i), we get

$$\left(x - \frac{a}{2}\right)^2 + \left(y - \frac{b}{2}\right)^2 + \frac{b^2}{2}\left(\frac{x}{a} + \frac{y}{b} - 1\right) = 0$$

$$\Rightarrow \quad x^2 + y^2 - \left(a - \frac{b^2}{2a}\right)x - \frac{by}{2} + \frac{a^2 - b^2}{4} = 0$$

Let r be the radius of this circle. Then,

$$r^2 = \frac{1}{4}\left(a - \frac{b^2}{2a}\right)^2 + \frac{b^2}{16} - \left(\frac{a^2 - b^2}{4}\right)$$

$$= \frac{b^2(a^2+b^2)}{16a^2} \Rightarrow r = \frac{b}{4a}\sqrt{a^2+b^2}$$

18 Given equations of circles are

$$x^2 + y^2 - 4x - 6y - 12 = 0 \quad \ldots(i)$$

$$x^2 + y^2 + 6x + 18y + 26 = 0 \quad \ldots(ii)$$

Centre of circle (i) is $C_1(2, 3)$ and radius

$$= \sqrt{4 + 9 + 12} = 5(r_1) \quad \text{[say]}$$

Centre of circle (ii) is $C_2(-3, -9)$ and radius

$$= \sqrt{9 + 81 - 26} = 8(r_2) \quad \text{[say]}$$

Now, $C_1C_2 = \sqrt{(2+3)^2 + (3+9)^2}$

$$\Rightarrow \quad C_1C_2 = \sqrt{5^2 + 12^2}$$

$$\Rightarrow \quad C_1C_2 = \sqrt{25 + 144} = 13$$

Also, $r_1 + r_2 = 5 + 8 = 13$

$$\therefore \quad C_1C_2 = r_1 + r_2$$

Thus, both circles touch each other externally. Hence, there are three common tangents.

19 Centres and radii of given circles are $C_1(1, 3)$, $r_1 = r$, $C_2(4, -1)$ and $r_2 = \sqrt{4^2 + 1^2 - 8} = 3$

Since, $r_1 - r_2 < C_1C_2 < r_1 + r_2$

$$\Rightarrow r - 3 < \sqrt{(4-1)^2 + (-1-3)^2} < r + 3$$

$$\Rightarrow \quad r - 3 < 5 < r + 3$$

$$\Rightarrow \quad r - 3 < 5 \text{ and } 5 < r + 3$$

$$\Rightarrow \quad r < 8 \text{ and } 2 < r \quad \Rightarrow \quad 2 < r < 8$$

20 Use the property, when two circles touch each other externally, then distance between the centre is equal to sum of their radii, to get required radius.

Let the coordinate of the centre of T be $(0, k)$.

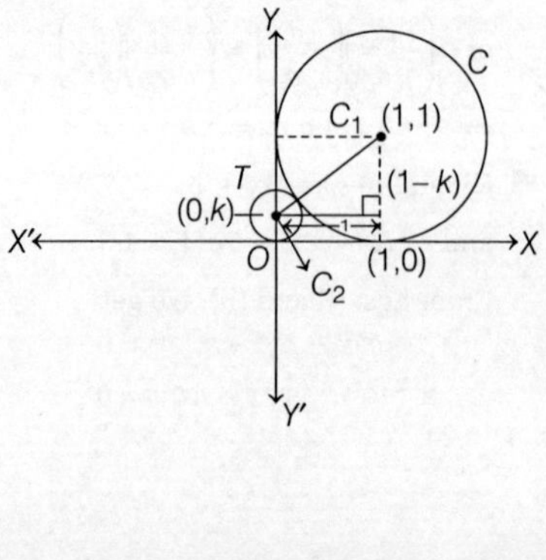

Distance between their centre

$$k+1=\sqrt{1+(k-1)^2} \quad [\because C_1C_2=k+1]$$

$$\Rightarrow k+1=\sqrt{1+k^2+1-2k}$$

$$\Rightarrow k+1=\sqrt{k^2+2-2k}$$

$$\Rightarrow k^2+1+2k=k^2+2-2k$$

$$\Rightarrow \quad k=\frac{1}{4}$$

So, the radius of circle T is k i.e. $\frac{1}{4}$.

21 Equation of tangent at $(1,-2)$ is $x-2y-5=0$. For IInd circle centre $(4,-3)$ and $r=\sqrt{5}$.

Point of contact is $\frac{x-4}{1}=\frac{y+3}{-2}$

$$=-\frac{(4+6-5)}{1^2+2^2}=-1$$

$$\Rightarrow \frac{x-4}{1}=-1, \frac{y+3}{-2}=-1$$

$$\therefore \quad x=3, y=-1$$

22 Equation of tangent at $(1, 7)$ to the curve $x^2=y-6$ is

$$x=\left(\frac{y+7}{2}\right)-6$$

$$\Rightarrow \quad 2x-y+5=0$$

Since, $2x-y+5=0$ touches the circle

$$x^2+y^2+16x+12y=C=0$$

$$\therefore \sqrt{(8)^2+(6)^2-c}=\left|\frac{2(-8)-(-6)+5}{\sqrt{(2)^2+(1)^2}}\right|$$

$$\Rightarrow \quad \sqrt{100-c}=\sqrt{5}$$

$$\Rightarrow \quad c=95$$

23 Now, $OP=\sqrt{6^2+8^2}=10$

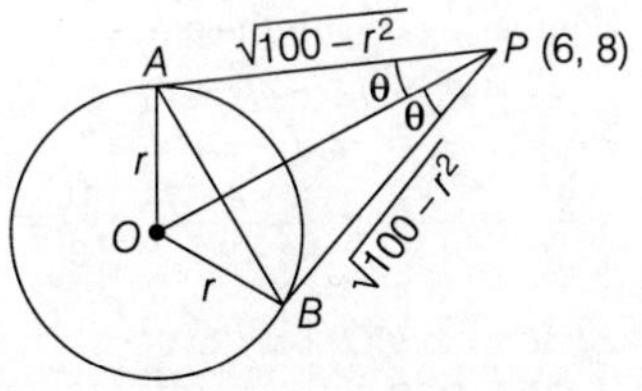

$$PA=\sqrt{S_1}=\sqrt{100-r^2}$$

Let $f(r)=\Delta PAB=\frac{1}{2}PA\cdot PB\cdot\sin 2\theta$

$$=(100-r^2)\sin\theta\cdot\cos\theta$$

$$=\frac{r}{100}(100-r^2)^{3/2}$$

Put $f'(r)=0$

$$\Rightarrow \frac{3}{2}(100-r^2)^{1/2}(-2r^2)+(100-r^2)^{3/2}$$

$$\Rightarrow \sqrt{100-r^2}\,(-3r^2+100-r^2)=0$$

$$\Rightarrow \quad r=\pm 10 \text{ or } r=5$$

Hence, $r=5$

24 Let angles between the tangents $=2\theta$, then

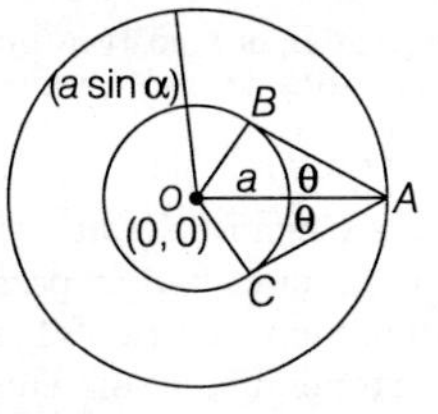

$\sin\theta=OB/OA=(a\sin\alpha/a)=\sin\alpha$

So, $\theta=\alpha$.

Required angle $=2\theta=2\alpha$

25 We have, $x^2+y^2-4x+6y-12=0$

Centre $(2,-3)$

Radius $=\sqrt{(2)^2+(-3)^2+12}$

$$=\sqrt{4+9+12}=5$$

Distance between two centres $c_1(2,-3)$ and $c_2(-3, 2)$

$$d=\sqrt{(2+3)^2+(-3-2)^2}$$

$$=\sqrt{25+25}=\sqrt{50}$$

Radius of circle $S=\sqrt{5^2+(\sqrt{50})^2}$

$$=\sqrt{25+50}=5\sqrt{3}$$

26 The equations of two circles are

$$S_1\equiv(x-a)^2+(y-b)^2=c^2 \quad \ldots\text{(i)}$$

and $S_2\equiv(x-b)^2+(y-a)^2=c^2 \quad \ldots\text{(ii)}$

The equation of the common chord of these circles is

$$S_1-S_2=0$$

$$\Rightarrow (x-a)^2-(x-b)^2+(y-b)^2-(y-a)^2=0$$

$$\Rightarrow (2x-a-b)(b-a)+(2y-b-a)(a-b)=0$$

$$\Rightarrow 2x-a-b-2y+b+a=0$$

$$\Rightarrow x-y=0$$

The centre coordinates of circles S_1 and S_2 are $C_1(a,b)$ and $C_2(b,a)$, respectively.

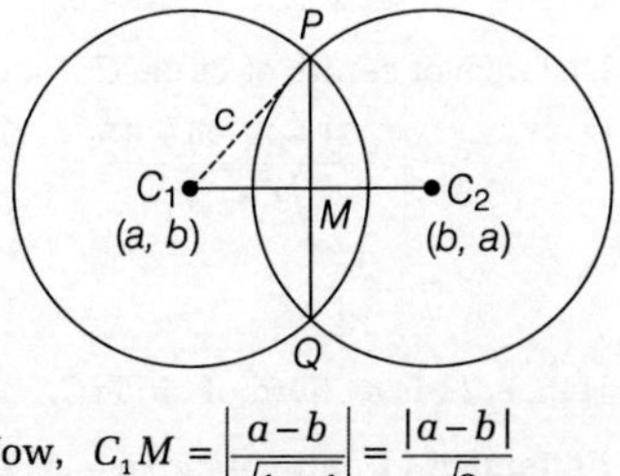

Now, $C_1M=\left|\frac{a-b}{\sqrt{1+1}}\right|=\frac{|a-b|}{\sqrt{2}}$

In right ΔC_1PM,

$$PM=\sqrt{C_1P^2-C_1M^2}=\sqrt{c^2-\frac{(a-b)^2}{2}}$$

$$\therefore PQ=2PM=2\sqrt{c^2-\frac{(a-b)^2}{2}}$$

$$=\sqrt{4c^2-2(a-b)^2}$$

27 Let equation of circle be

$$x^2+y^2+2gx+c=0.$$

where, g is a variable and c is a constant, be a coaxial system of circle having common radical axis as X-axis.

Let $x^2+y^2+2g_ix+c=0; i=1,2,3$ be three members of the given coaxial system of circles.

Then, the coordinates of their centres and radii are

$$P(-g_1, 0), Q(-g_2, 0), R(-g_3, 0)$$

and $r_1^2=g_1^2-c, r_2^2=g_2^2-c, r_3^2=g_3^2-c$

Now, $r_1^2\cdot QR+r_2^2\cdot RP+r_3^2\cdot PQ$

$$=(g_1^2-c)(g_2-g_3)+(g_2^2-c)(g_3-g_1)+(g_3^2-c)(g_1-g_2)$$

$$=g_1^2(g_2-g_3)+g_2^2(g_3-g_1)+g_3^2(g_1-g_2)-c\{(g_2-g_3)+(g_3-g_1)+(g_1-g_2)\}$$

$$=-(g_1-g_2)\cdot(g_2-g_3)\cdot(g_3-g_1)$$

$$=-PQ\cdot QR\cdot RP$$

28 The combined equation of the lines joining the origin to the points of intersection of $x\cos\alpha+y\sin\alpha=p$ and $x^2+y^2-a^2=0$ is a homogeneous equation of second degree in given by

$$x^2+y^2-a^2\left(\frac{x\cos\alpha+y\sin\alpha}{p}\right)^2=0$$

$$\Rightarrow x^2(p^2-a^2\cos^2\alpha)+y^2(p^2-a^2\sin^2\alpha)-2xya^2\sin\alpha\cos\alpha=0$$

The lines given by this equation are at right angle if

$$(p^2-a^2\cos^2\alpha)+(p^2-a^2\sin^2\alpha)=0$$

$$\Rightarrow 2p^2=a^2(\sin^2\alpha+\cos^2\alpha)$$

$$\Rightarrow a^2=2p^2$$

29 The coordinates of the centres and radii of three given circles are as given

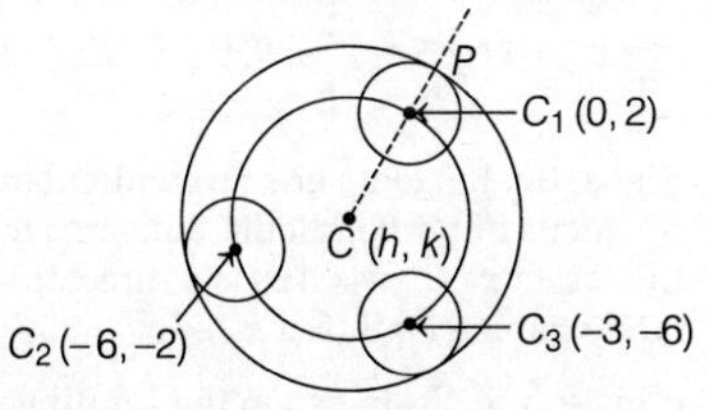

Circle	Centre	Radius
Ist Circle	$C_1(0,2)$	$r_1=3$
IInd Circle	$C_2(-6,-2)$	$r_2=3$
IIIrd Circle	$C_3(-3,-6)$	$r_3=3$

Let $C(h,k)$ be the centre of the circle passing through the centres of the circles Ist, IInd and IIIrd.

Then, $CC_1=CC_2=CC_3$

$$\Rightarrow CC_1^2=CC_2^2=CC_3^2$$

$$\Rightarrow (h-0)^2+(k-2)^2=(h+6)^2+(k+2)^2=(h+3)^2+(k+6)^2$$

$\Rightarrow \quad -4k + 4 = 12h + 4k + 40$
$\qquad = 6h + 12k + 45$
$\Rightarrow \quad 12h + 8k + 36 = 0$
and $\quad 6h - 8k - 5 = 0$
$\Rightarrow \quad 3h + 2k + 9 = 0$
and $\quad 6h - 8k - 5 = 0$
$\Rightarrow \quad h = \dfrac{-31}{18}, k = \dfrac{-23}{12}$

$\therefore \quad CC_1 = \sqrt{\left(0 + \dfrac{31}{18}\right)^2 + \left(2 + \dfrac{23}{12}\right)^2}$

$= \dfrac{5}{36}\sqrt{949}$

Now, $CP = CC_1 + C_1P$

$\Rightarrow \quad CP = \left(\dfrac{5}{36}\sqrt{949} + 3\right)$

Thus, required circle has its centre at $\left(\dfrac{-31}{18}, \dfrac{-23}{12}\right)$ and radius

$= CP = \left(\dfrac{5}{36}\sqrt{949} + 3\right).$

Hence, its equation is

$$\left(x + \frac{31}{18}\right)^2 + \left(y + \frac{23}{12}\right)^2 = \left(3 + \frac{5}{36}\sqrt{949}\right)^2$$

30 $(r + 1)^2 = \alpha^2 + 9; r^2 + 8 = \alpha^2$

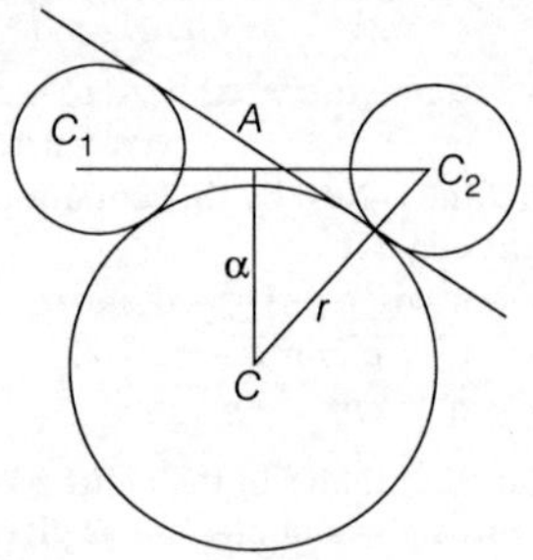

$\Rightarrow \quad r^2 + 2r + 1 = r^2 + 8 + 9$
$\qquad 2r = 16 \Rightarrow r = 8$

31 Since, the tangents are perpendicular. So, locus of perpendicular tangents to the circle $x^2 + y^2 = 169$ is a director circle having equation $x^2 + y^2 = 338$.

32 Equation of circle, when the limiting points are $(1, 1)$ and $(3, 3)$ is

$(x - 1)^2 + (y - 1)^2 = 0$
and $\quad (x - 3)^2 + (y - 3)^2 = 0$
$\Rightarrow \quad x^2 + y^2 - 2x - 2y + 2 = 0$
and $x^2 + y^2 - 6x - 6y + 18 = 0$

Equation of the coaxial system of circle is

$$x^2 + y^2 - 6x - 6y + 18 + \lambda(x^2 + y^2 - 2x - 2y + 2) = 0$$

It passes through origin, therefore

$\lambda = -9$

Hence, required circle is

$2x^2 + 2y^2 - 3x - 3y = 0.$

Statement II is also true but it is not a correct explanation for Statement I.

33 Let C_1 be a circle which passes through A, B and C whose diameter is AB and C_2 be another circle which passes through A and B, then centres of C_1 and C_2 must lie on perpendicular bisector of AB. Indeed centre of C_1 is mid-point M of AB and centre of any other circle lies somewhere else on bisector.

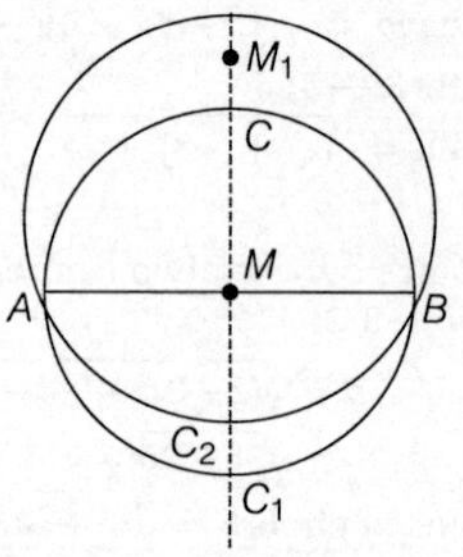

Then, $\quad M_1A > AM$
[hypotenuse of right angled ΔAMM_1]

$\Rightarrow$ Radius of $C_2 > \dfrac{1}{2}AB$

So, C_1 is the circle whose radius is least.

Thus, Statement I is true but does not actually follow from Statement II which is certainly true.

34 Equation of circle C is

$(x + 3)^2 + (y - 5)^2 = 9 + 25 - 30$
i.e. $(x + 3)^2 + (y - 5)^2 = 4$
$\therefore$ Centre is $(3, -5)$.

If L_1 is a diameter of a circle, then

$2(3) + 3(-5) + p - 3 = 0$
$\Rightarrow \quad p = 12$
$\therefore \quad L_1$ is $2x + 3y + 9 = 0$
and $\quad L_2$ is $2x + 3y + 15 = 0.$

Distance of centre of circle C from L_2

$= \left|\dfrac{2(3) + 3(-5) + 15}{\sqrt{2^2 + 3^2}}\right|$

$= \dfrac{6}{\sqrt{13}} < 2$

Hence, L_2 is a chord of circle C.

35 **Statement I** Centre of circle $= (3, -1)$

Now, $2(3) + (-1) = 5 = 5$, [true]

Statement II Centre $= (3, -1)$, which lies on given line.

Simplify it and get the result.

SESSION 2

1 Equation of circle through points of intersections of circle $x^2 + y^2 - a^2 = 0$ and line $x - y + 3 = 0$, (AB) is

$(x^2 + y^2 - a^2) + \lambda(x - y + 3) = 0$

Since, AB is diameter, centre $\left(-\dfrac{\lambda}{2}, \dfrac{\lambda}{2}\right)$ lies on it.

$\therefore \quad -\dfrac{\lambda}{2} - \dfrac{\lambda}{2} + 3 = 0$ i.e. $\lambda = 3$

Hence, equation of required circle is

$x^2 + y^2 + 3x - 3y + 9 - a^2 = 0$

2 $AB = \sqrt{(6 - 2)^2 + (5 - 2)^2} = 5$

$AC = 2$
$\Rightarrow \quad BC = 3$

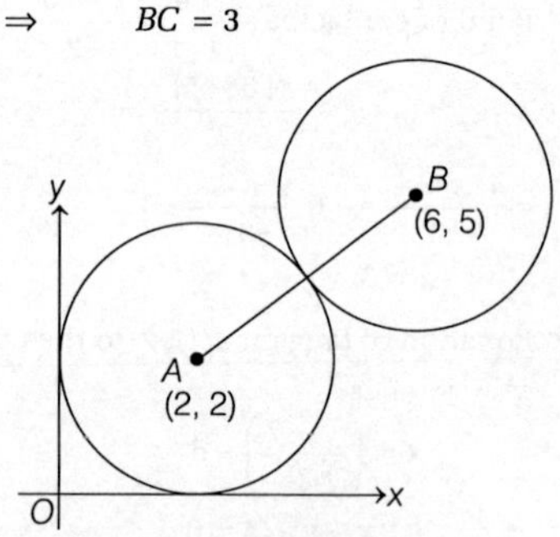

Equation of required circle is

$(x - 6)^2 + (y - 5)^2 = 9.$
$\Rightarrow \quad x^2 + 36 - 12x + y^2 + 25 - 10y = 9$
$\Rightarrow \quad x^2 + y^2 - 12x - 10y + 52 = 0$

3 The line $(x - 2)\cos\theta + (y - 2)\sin\theta = 1$ touches the circle

$(x - 2)^2 + (y - 2)^2 = 1$
i.e. $x^2 + y^2 - 4x - 4y + 7 = 0$

for all values of θ.

Comparing it with the circle

$x^2 + y^2 + 2gx + 2fy + c = 0$

we get, $\quad g = -2, f = -2, c = 7$

$\therefore \quad \dfrac{g^2 + f^2 + c}{g + f + c} = \dfrac{4 + 4 + 7}{-2 - 2 + 7} = \dfrac{15}{3} = 5$

4 $\angle PAO = \angle PBO = \pi/2.$

$\therefore$ P, A, O, B are concyclic.

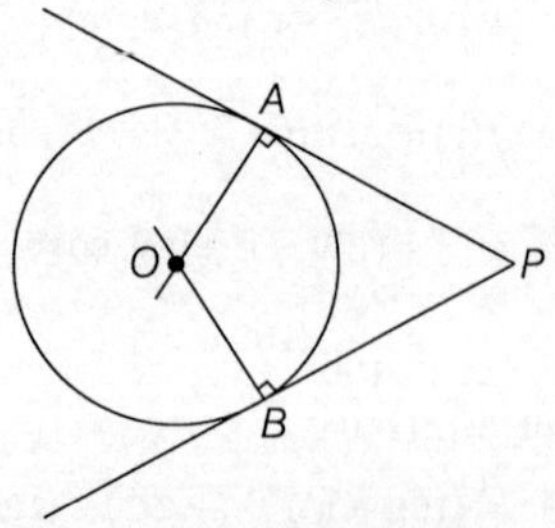

$\therefore$ Equation of circumcircle of ΔABP is

$x(x - x_1) + y(y - y_1) = 0$
i.e. $\quad x^2 + y^2 - xx_1 - yy_1 = 0$

5 The line $y = mx - b\sqrt{1+m^2}$ ...(i)

is tangent to the circle $x^2 + y^2 = b^2$

It is also tangent to the circle

$$(x-a)^2 + y^2 = b^2 \quad ...(ii)$$

It is length of $\perp$ on (i) from centre $(a, 0)$

= radius of circle (ii)

$$\Rightarrow \left|\frac{ma - b\sqrt{(1+m^2)}}{\sqrt{1+m^2}}\right| = b$$

$$\Rightarrow ma - b\sqrt{(1+m^2)} = \pm b\sqrt{(1+m^2)} \quad ...(iii)$$

– ve sign givess $m = 0$, which is none of the given options.

$\therefore$ Taking + ve sign on RHS of (iii), we get

$$ma = 2b\sqrt{(1+m^2)}$$

$$\Rightarrow m = 2b/\sqrt{(a^2 - 4b^2)}.$$

6 $L: 2x - 3y - 1; S: x^2 + y^2 - 6$

If $L_1 > 0$ and $S_1 < 0$

Then, point lies in the smaller part are as

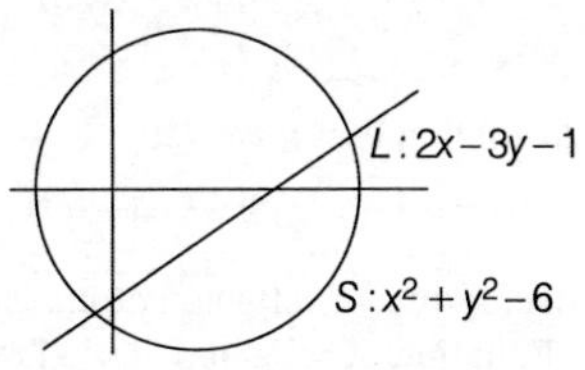

$\therefore \left(2, \frac{3}{4}\right)$ and $\left(\frac{1}{4}, -\frac{1}{4}\right)$ lie inside.

7 $\frac{1}{2}(x + 2x)2r = 18$

$(3x - 2r)^2 = 4r^2 + x^2$

On solving for r, we get $r = 2$

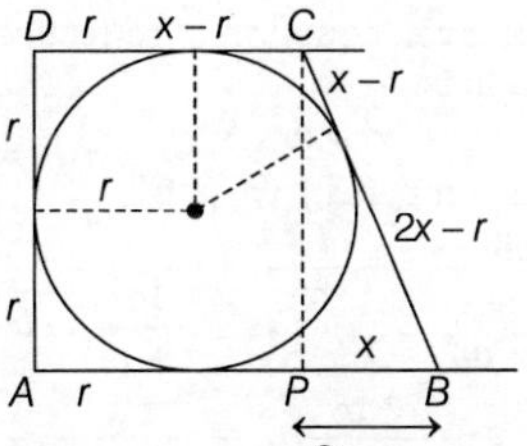

8 $S = x^2 + y^2 - 2x - 6y + 6 = 0,$

Centre = (1, 3)

Let radius of circle $C = r$.

Then, $C = (x-2)^2 + (y-1)^2 = r^2$

$= x^2 + y^2 - 4x - 2y + 5 - r^2 = 0$

Common chord of circles S and C is

$$2x - 4y + 1 - r^2 = 0$$

It is a diameter of circle S.

$\therefore 2 - 12 + 1 - r^2 = 0 \Rightarrow r = 3$

9 $S = x^2 + y^2 - 2\sqrt{2}y + c = 0$

Let us assume that there are more than two rational points on the circle. Let $(x_1, y_1), (x_2, y_2), (x_3, y_3)$

$x_i, y_i \in Q, i = 1, 2, 3$ be three rational points.

Then, $x_1^2 + y_1^2 - 2\sqrt{2}y_1 = x_2^2 + y_2^2 - 2\sqrt{2}y_2 = x_3^2 + y_3^2 - 2\sqrt{2}y_3$

$\Rightarrow x_1^2 + y_1^2 = x_2^2 + y_2^2 = x_3^2 + y_3^2$

and $y_1 = y_2 = y_3 \Rightarrow x_1^2 = x_2^2 = x_3^3$

$\Rightarrow$ There exists two rational points of the form (a, b) and $(-a, b)$, $a, b \in Q$.

10 The chords are equal length, then the distances of the centre from the lines are equal. Let L_1 be $y - mx = 0$ centre is $\left(\frac{1}{2}, -\frac{3}{2}\right)$.

$$\therefore \frac{\left|-\frac{3}{2} - \frac{m}{2}\right|}{\sqrt{m^2+1}} = \frac{\left|\frac{1}{2} - \frac{3}{2} - 1\right|}{\sqrt{2}}$$

$$\Rightarrow 7m^2 - 6m - 1 = 0$$

$$\Rightarrow m = 1, -\frac{1}{7}$$

Hence, L_1 be $y + \frac{1}{7}x = 0 \Rightarrow x + 7y = 0$

11 Let $(0, b)$ be the centre and r be the radius of the given circle, then its equation is

$$(x-0)^2 + (y-b)^2 = r^2$$

$$\Rightarrow x^2 + y^2 - 2yb + b^2 - r^2 = 0 \quad ...(i)$$

It is given that the point

$P_n\left(\log a_n, \frac{1}{\log a_n}\right)$; $n = 1, 2, 3, 4$ lie on the circle given by Eq. (i).

Therefore, $(\log a_n)^2 + \frac{1}{(\log a_n)^2} - \frac{2b}{(\log a_n)}, + b^2 - r^2 = 0$

$n = 1, 2, 3, 4$

Since, $\log a_1$, $\log a_2$, $\log a_3$ and $\log a_4$ are roots of the equation.

Then, $\lambda^4 + (b^2 - r^2)\lambda^2 - 2b\lambda + 1 = 0$

$\therefore$ Sum of the roots = 0

$\Rightarrow \log a_1 + \log a_2 + \log a_3 + \log a_4 = 0$

$\Rightarrow \log(a_1 a_2 a_3 a_4) = 0 \Rightarrow a_1 a_2 a_3 a_4 = 1$

12 The point $(2a, a+1)$ will be an interior point of the larger segment of the circle $x^2 + y^2 - 2x - 2y - 8 = 0$

(i) The point $(2a, a+1)$ is an interior point.

(ii) The point $(2a, a+1)$ and the centre $(1, 1)$ are on the same side of the chord $x - y + 1 = 0$.

$\therefore (2a)^2 + (a+1)^2 - 2(2a) - 2(a+1) - 8 < 0$

and $(2a - a - 1 + 1)(1 - 1 + 1) > 0$

$\Rightarrow 5a^2 - 4a - 9 < 0$ and $a > 0$

$\Rightarrow (5a - 9)(a + 1) < 0$ and $a > 0$

$\Rightarrow -1 < a < \frac{9}{5}$ and $a > 0$

$\Rightarrow a \in \left(0, \frac{9}{5}\right)$

13 The equation of the biggest circle is $x^2 + y^2 = 1^2$

Clearly, it is centred at $O(0, 0)$ and has radius 1. Let the radii of the other two circles be $1 - r, 1 - 2r$, where $r > 0$.

Thus, the equations of the concentric circles are

$$x^2 + y^2 = 1 \quad ...(i)$$

$$x^2 + y^2 = (1-r)^2 \quad ...(ii)$$

$$x^2 + y^2 = (1-2r)^2 \quad ...(iii)$$

Clearly, $y = x + 1$ cuts the circle (i) at (1, 0) and (0, 1). This line will cut circles (ii) and (iii) in real and distinct points, if

$$\left|\frac{1}{\sqrt{2}}\right| < 1 - r \text{ and } \left|\frac{1}{\sqrt{2}}\right| < 1 - 2r$$

$$\Rightarrow \frac{1}{\sqrt{2}} < 1 - r \text{ and } \frac{1}{\sqrt{2}} < 1 - 2r$$

$$\Rightarrow r < 1 - \frac{1}{\sqrt{2}} \text{ and } r < \frac{1}{2}\left(1 - \frac{1}{\sqrt{2}}\right)$$

$$\Rightarrow r < \frac{1}{2}\left(1 - \frac{1}{\sqrt{2}}\right)$$

$$\Rightarrow r \in \left[0, \frac{1}{2}\left(1 - \frac{1}{\sqrt{2}}\right)\right] \quad [\because r > 0]$$

14 Given circles are

$$x^2 + y^2 - 2\lambda_i x - c^2 = 0, i = 1, 2, 3$$

$\therefore$ Centres are $(\lambda_1, 0), (\lambda_2, 0), (\lambda_3, 0)$

Distances of the centres from origin are in G.P.,

$\therefore \lambda_2^2 = \lambda_1 \lambda_3$

Let $(x_1 y_1)$ be any point on the circle $x^2 + y^2 = c^2$.

$\therefore x_1^2 + y_1^2 - c^2 = 0$

Lengths of tangents from (x_1, y_1) to the three given circles are

$$l_i = \sqrt{x_1^2 + y_1^2 - 2\lambda_i x_1 - c^2} = \sqrt{-2\lambda_i x_1}$$

$$\therefore l_1^2 l_3^2 = (-2\lambda_1 x_1)(-2\lambda_3 x_1)$$

$$= 4\lambda_1 \lambda_3 x_1^2 = 4\lambda_2^2 x_1^2 = \lambda_2^4$$

$\therefore l_1, l_2$ and l_3 are in G.P.

15 The equation representing the coaxial system of circle is

$$x^2 + y^2 + 2gx + c + \lambda(x^2 + y^2 + 2fy + k) = 0$$

$$\Rightarrow x^2 + y^2 + \frac{2g}{1+\lambda}x + \frac{2f\lambda}{1+\lambda}y + \frac{c + k\lambda}{1+\lambda} = 0 \quad ...(i)$$

The coordinates of the centre of this circle are

$$\left(-\frac{g}{1+\lambda}, -\frac{f\lambda}{1+\lambda}\right) \quad ...(ii)$$

and radius

$$= \sqrt{\frac{g^2 + f^2\lambda^2 - (c + k\lambda)(1+\lambda)}{(1+\lambda)^2}}$$

For the limiting points, we must have

Radius = 0

$$\Rightarrow g^2 + f^2\lambda^2 - (c + k\lambda)(1+\lambda) = 0$$

$$\Rightarrow \lambda^2(f^2 - k) - \lambda(c + k) + (g^2 - c) = 0 \quad \text{...(iii)}$$

Let λ_1 and λ_2 be the roots of this equation.

Then, $\lambda_1 + \lambda_2 = \dfrac{c + k}{f^2 - k}$

and $\lambda_1\lambda_2 = \dfrac{g^2 - c}{f^2 - k}$ ...(iv)

Thus, the coordinates of limiting points L_1 and L_2 are,

$L_1\left(\dfrac{-g}{1+\lambda_1}, \dfrac{-f\lambda_1}{1+\lambda_1}\right)$ and

$L_2\left(\dfrac{-g}{1+\lambda_2}, \dfrac{-f\lambda_2}{1+\lambda_2}\right)$ [from Eq. (iv)]

Now, L_1L_2 will subtend a right angle at the origin.

If slope of OL_1 × slope of $OL_2 = -1$

$$\Rightarrow \frac{f\lambda_1}{g} \times \frac{f\lambda_2}{g} = -1$$

$$\Rightarrow f^2\lambda_1\lambda_2 = -g^2$$

$$\Rightarrow f^2\left(\frac{g^2 - c}{f^2 - k}\right) = -g^2$$

$$\Rightarrow f^2(g^2 - c) + g^2(f^2 - k) = 0$$

$$\Rightarrow \frac{c}{g^2} + \frac{k}{f^2} = 2$$

16 Tangent at (0, 1) to the circle $x^2 + y^2 = 1$ is $y = 1$. Incident ray, incident at (3, 1) is $y - 1 = m(x - 3)$

Incident and reflected rays are equally and reflected rays are equally inclined to the line is $y = 1$, so slope of reflected ray is $-m$.

∴ Equation of reflected ray is

$$y - 1 = -m(x - 3)$$

i.e. $mx + y - 3m - 1 = 0$

It touches the circle so $(p = r)$

$$\left|\frac{3m + 1}{\sqrt{1 + m^2}}\right| = 1$$

$$\Rightarrow 9m^2 + 6m + 1 = 1 + m^2$$

$$\Rightarrow m = 0, -3/4.$$

$m = 0$ gives the slope for the tangent $y = 1$, so equation of reflected ray is

$$y - 1 = -\frac{3}{4}(x - 3)$$

i.e., $3x + 4y - 13 = 0$

17 $S = x^2 + \lambda x + (1 - \lambda)y + 5 = 0$

is a circle of radius not exceeding 5.

$$\therefore \sqrt{\frac{\lambda^2}{4} + \frac{(1-\lambda)^2}{4} - 5} \le 5$$

and $\dfrac{\lambda^2}{4} + \dfrac{(1-\lambda)^2}{4} - 5 \ge 0$

$$\therefore \lambda^2 + (1-\lambda)^2 - 20 \le 100$$

and $\lambda^2 + (1-\lambda)^2 - 20 > 0$

$$\Rightarrow \frac{1 - \sqrt{239}}{2} \le \lambda \le \frac{1 + \sqrt{239}}{2}$$

i.e. $-7.2 \le \lambda \le 8.2$

and $\lambda < \dfrac{1 - \sqrt{39}}{2}$ or $\lambda > \dfrac{1 + \sqrt{39}}{2}$

i.e. $\lambda - 2.62$ or $\lambda > 3.62$

∴ Possible integral values of λ are

$-7, -6, -5, -4, -3, 4, 5, 6, 7, 8$

∴ In all, there are 10 possible integral values of λ.

18 $[P + 1] = [P] + 1$. Let $[P] = n$, then n is integer.

$([P + 1], [P]) = (n + 1, n)$ lies inside the region of circles.

$S_1 = x^2 + y^2 - 2x - 15 = 0,$

$C_1 = (1, 0), r_1 = 4.$

and $S_2 = x^2 + y^2 - 2x - 7 = 0,$

$C_2 = (1, 0), r_2 = 2\sqrt{2}$

Both circles are concentric.

$\therefore (n + 1)^2 + n^2 - 2(n + 1) - 7 > 0$

and $(n + 1)^2 + n^2 - 2(n + 1) - 15 < 0$

$\Rightarrow 4 < n^2 < 8$

which is not possible for any integer.

19 $L = 3x - 4y - k = 0$ touches the circle $S = x^2 + y^2 - 4x - 8y - 5 = 0$ at (a, b)

Tangent at (a, b) is

$$ax + by - 2(x + a) - 4(y + b) - 5 = 0$$

$$T = x(a - 2) + y(b - 4) - 2a - 4b - 5 = 0$$

Comparing $L = 0, T = 0$, we get

$$\frac{a - 2}{3} = \frac{b - 4}{-4} = \frac{2a + 4b + 5}{k}$$

Also, using $p = r$, we get

$$\frac{6 - 10 - k}{5} = \pm 5 \Rightarrow k = 15 \text{ or } -35$$

Using $k = 15$, we get

$4a + 3b = 20$ and $9a - 12b = 45$

Solving $a = 5, b = 0$ and using

$k = -35$, we get $a = -1, b = 8$

We can verify that (5, 0) and (−1, 8) both satisfy equation of circle. Hence

$k, (a, b) = 15\,(5, 0)$ or $-35, (-1, 8)$

20 Centre of the circle $x^2 + y^2 = a^2$ is (0, 0).

Let its reflection about the line $y = mx + c$ be (h, k).

Then, $(h/2, k/2)$ lies on this line.

$$m\frac{h}{2} + c = k \text{ and } -\frac{k}{h} \times m = -1$$

$$\Rightarrow mk = -h$$

On solving these, we get

$$h = \frac{2cm}{1 + m^2}, k = \frac{2c}{1 + m^2}$$

Also, radius of reflected circle is a.

∴ Equation of relfected circle is

$$\left(x + \frac{2cm}{1 + m^2}\right)^2 + \left(y - \frac{2c}{1 + m^2}\right)^2 = a^2$$

$$\Rightarrow x^2 + y^2 + \frac{4cm}{1 + m^2}x - \frac{4c}{1 + m^2} + \frac{4c^2(1 + m^2)}{(1 + m^2)^2} - a^2 = 0$$

But given equation of reflected circle is

$$x^2 + y^2 + 2gx + 2fy + c = 0$$

On comparing, we get

$$\frac{g}{\frac{2cm}{1 + m^2}} = \frac{f}{\frac{-2c}{1 + m^2}} = \frac{c}{\frac{4c^2}{1 + m^2} - a^2} = 1$$

$$\therefore g = \frac{2cm}{1 + m^2}, f = -\frac{2c}{1 + m^2}, \frac{4c^2}{1 + m^2} - a^2 = c$$

DAY TWENTY SEVEN

Parabola

Learning & Revision for the Day

- Conic Section
- Concept of Parabola
- Line and a Parabola
- Equation of Tangent
- Equation of Normal
- Equation of a Pair of Tangents
- Equations of Chord of Contact
- Director Circle
- Conormal Points
- Diameter

Conic Section

A conic is the locus of a point whose distance from a fixed point bears a constant ratio to its distance from a fixed line. The fixed point is the **focus** S and the fixed line is the **directrix,** l.

The constant ratio is the **eccentricity** denoted by e.

- If $0 < e < 1$, then conic is an ellipse.
- If $e = 1$, then conic is a parabola.
- If $e > 1$, then conic is a hyperbola.

General Equation of Conic Section

A second degree equation $ax^2 + 2hxy + by^2 + 2gx + 2fy + c = 0$ represents

***Case* I** When the focus lies on the directrix

(i) Pair of straight lines, if $\Delta = \begin{vmatrix} a & h & g \\ h & b & f \\ g & f & c \end{vmatrix} = 0$

(ii) If $e > 1$, then the lines will be real and distinct intersecting at fixed point.

(iii) If $e = 1$, then the lines will coincident passing through a fixed point.

(iv) If $e < 1$, then the lines will be imaginary.

***Case* II** When the focus does not lie on the directrix

(i) Circle : $a = b, h = 0, e = 0$ and $\Delta \neq 0$

(ii) Parabola : $h^2 = ab, \Delta \neq 0, e = 1$

(iii) Ellipse : $h^2 < ab, \Delta \neq 0, 0 < e < 1$

(iv) Hyperbola : $h^2 > ab, 0\ \Delta \neq 0, e > 1$

(v) Rectangular hyperbola : $a + b = 0, \Delta \neq 0, e > 1, h^2 > ab$

PREP MIRROR

Your Personal Preparation Indicator

- No. of Questions in Exercises (x)—
- No. of Questions Attempted (y)—
- No. of Correct Questions (z)—
  (Without referring Explanations)

- Accuracy Level ($z/y \times 100$)—
- Prep Level ($z/x \times 100$)—

In order to expect good rank in JEE, your Accuracy Level should be above 85 & Prep Level should be above 75.

Concept of Parabola

Parabola is the locus of a point which moves in a plane such that its distance from a fixed point (focus, S) is equal to its distance from a fixed straight line (directrix, L).

Let $S \equiv (x_1, y_1)$ and $L \equiv ax + by + c = 0$.
Then, equation of parabola is

$(a^2 + b^2)[(x - x_1)^2 + (y - y_1)^2] = (ax + by + c)^2$.

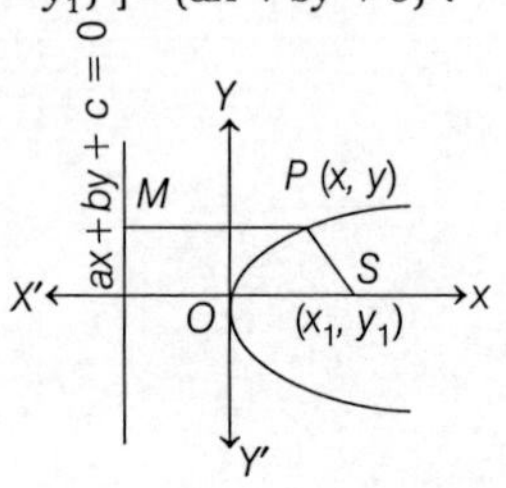

If S lies on L, parabola reduces to a straight line through S and perpendicular to L.

Definitions Related to Parabola

1. **Vertex** The intersection point of parabola and axis.
2. **Centre** The point which bisects every chord of the conic passing through it.
3. **Focal chord** Any chord passing through the focus.
4. **Double ordinate** A chord perpendicular to the axis of a conic.
5. **Latusrectum** A double ordinate passing through the focus of the parabola.
6. **Focal distance** The distance of a point $P(x, y)$ from the focus S is called the focal distance of the point P.

Some related terms of parabolas (in standard form)

S. No.	Related Terms	$y^2 = 4ax$	$y^2 = -4ax$	$x^2 = 4ay$	$x^2 = -4ay$
		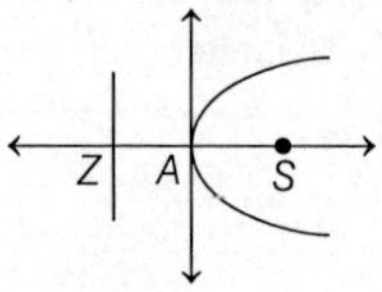	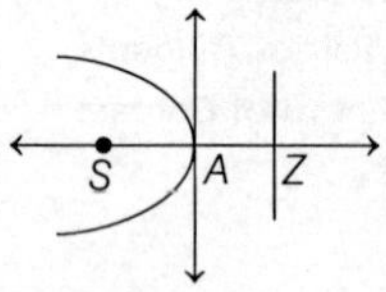	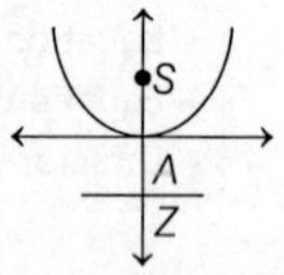	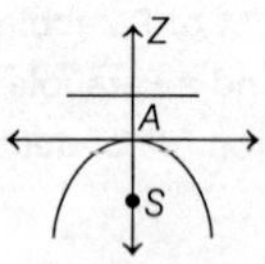
1.	Vertex	$A(0, 0)$	$A(0, 0)$	$A(0, 0)$	$A(0, 0)$
2.	Focus	$S(a, 0)$	$S(-a, 0)$	$S(0, a)$	$S(0, -a)$
3.	Equation of axis	$y = 0$	$y = 0$	$x = 0$	$x = 0$
4.	Equation of directrix	$x + a = 0$	$x - a = 0$	$y + a = 0$	$y - a = 0$
5.	Eccentricity	$e = 1$	$e = 1$	$e = 1$	$e = 1$
6.	Extremities of latusrectum	$(a, \pm 2a)$	$(-a, \pm 2a)$	$(\pm 2a, a)$	$(\pm 2a, -a)$
7.	Length of latusrectum	$4a$	$4a$	$4a$	$4a$
8.	Equation of tangent at vertex	$x = 0$	$x = 0$	$y = 0$	$y = 0$
9.	Parametric equation	$\begin{cases} x = at^2 \\ y = 2at \end{cases}$	$\begin{cases} x = -at^2 \\ y = 2at \end{cases}$	$\begin{cases} x = 2at \\ y = at^2 \end{cases}$	$\begin{cases} x = 2at \\ y = -at^2 \end{cases}$
10.	Focal distance of any point $P(h, k)$ on the parabola	$h + a$	$h - a$	$k + a$	$k - a$
11.	Equation of latusrectum	$x = a$	$x + a = 0$	$y = a$	$y + a = 0$

Results on Parabola $y^2 = 4ax$

(i) Length of latusrectum = 2 (Harmonic mean of focal segment)

(ii) If y_1, y_2 and y_3 are the ordinates of the vertices of triangle inscribed in the parabola $y^2 = 4ax$,
then its area $= \dfrac{1}{8a}(y_1 - y_2)(y_2 - y_3)(y_3 - y_1)$

(iii) For the ends of latusrectum of the parabola $y^2 = 4ax$, the values of the parameter are ± 1.

Position of a Point

A point (h, k) with respect to the parabola S lies inside, on or outside the parabola, if $S_1 < 0, S_1 = 0$ or $S_1 > 0$.

Line and a Parabola

(i) The line $y = mx + c$ meets the parabola $y^2 = 4ax$ in two points real, coincident or imaginary according to $c > \frac{a}{m}$, $c = \frac{a}{m}$ or $c < \frac{a}{m}$ respectively.

(ii) Length of the chord intercepted by the parabola on the line $y = mx + c$ is $= \frac{4\sqrt{a(1+m^2)(a-mc)}}{m^2}$

(iii) Length of the focal chord making an angle α with the X-axis is $4a\,\text{cosec}^2\alpha$.

(iv) If t_1 and t_2 are the end points of a focal chord of the parabola $y^2 = 4ax$, then $t_1t_2 = -1$

Equation of Tangent

A line which intersects the parabola at only one point is called the tangent to the parabola.

Equation of tangent to parabola in different cases are given below;

- In point (x_1, y_1) form, $yy_1 = 2a(x + x_1)$
- In slope (m) form, $y = mx + \frac{a}{m}$
- In parametric (t) form, $ty = x + at^2$
- The line $y = mx + c$ touches a parabola iff $c = \frac{a}{m}$ and the coordinates of the point of contact are $\left(\frac{a}{m^2}, \frac{2a}{m}\right)$.

Results on Tangent

(i) Points of intersection of tangents at two points $P(at_1^2, 2at_1)$, $Q(at_2^2, 2at_2)$ on the parabola $y^2 = 4ax$ is $R\{at_1t_2, a(t_1 + t_2)\}$ (where, R is GM of x-coordinates of P, Q and AM of y-coordinates of P, Q).

(ii) Angle θ between tangents at two points $P(at_1^2, 2at_1)$, $Q(at_2^2, 2at_2)$ on the parabola $y^2 = 4ax$ is given by $\tan\theta = \left|\frac{t_2 - t_1}{1 + t_1t_2}\right|$.

(iii) Locus of the point of intersection of perpendicular tangents to the parabola is its directrix.

(iv) If the tangents at the points P and Q on a parabola meet T, then ST is the GM between SP and SQ i.e. $ST^2 = SP \cdot SQ$

(v) If the tangent and normal at any point P of the parabola intersect the axis at T and G, then $ST = SG = SP$, where S is the focus.

(vi) Any tangent to a parabola and the perpendicular on it from the focus meet on the tangent at the vertex.

(vii) The orthocentre of any triangle formed by three tangents to a parabola lies on the directrix.

(viii) The length of the subtangent at any point on a parabola is equal to twice the abscissae of the point.

(ix) Two tangents can be drawn from a point to a parabola. Two tangents are real and distinct or coincident or imaginary according as given point lies outside, on or inside the parabola.

Equation of Normal

A line which is perpendicular to the tangent of the parabola is called the normal to the parabola.

Equation of normal to parabola in different cases are given below;

- In point (x_1, y_1) form, $(y - y_1) = -\frac{y_1}{2a}(x - x_1)$.
- In slope m form, $y = mx - 2am - am^3$.
- In parametric t form, $y + tx = 2at + at^3$.

NOTE Point of intersection of normals of t_1 and t_2 are $[a(t_1^2 + t_2^2 + t_1t_2 + 2), -at_1t_2(t_1 + t_2)]$.

Results on Normal

(i) If the normals at two points P and Q of a parabola $y^2 = 4ax$ intersects at a third point R on the curve, then the product of the ordinates of P and Q is $8a^2$.

(ii) Normal at the ends of latusrectum of the parabola $y^2 = 4ax$ meet at right angles on the axis of the parabola.

(iii) Tangents and normals at the extremities of the latusrectum of a parabola $y^2 = 4ax$ constitute a square, their points of intersection being $(-a, 0)$ and $(3a, 0)$.

(iv) The normal at any point of a parabola is equally inclined to the focal distance of the point and the axis of the parabola.

(v) The normal drawn at a point $P(at_1^2, 2at_1)$ to the parabola $y^2 = 4ax$ meets again the parabola at $Q(at_2^2, 2at_2)$, then $t_2 = -t_1 - \frac{2}{t_1}$.

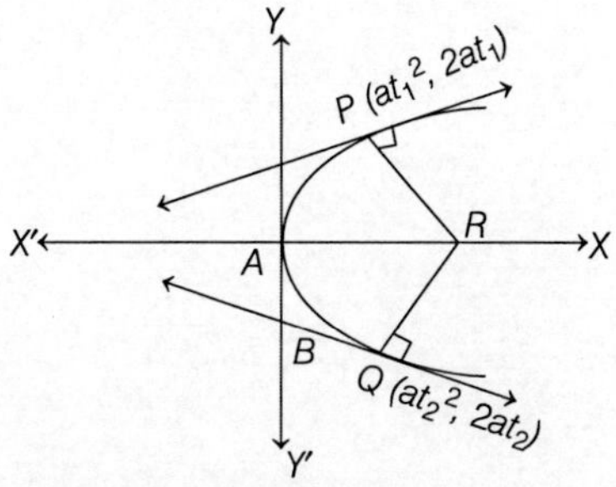

(vi) The normal chord of a parabola at a point whose ordinate is equal to the abscissae, subtends a right angle at the focus.

(vii) Three normals can be drawn from a point to a parabola.

Equation of a Pair of Tangents

The equation of pair of tangents drawn from an external point $P(x_1, y_1)$ to the parabola is $SS_1 = T^2$.

where, $S = y^2 - 4ax$, $S_1 = y_1^2 - 4ax_1$ and $T = yy_1 - 2a(x + x_1)$

Equations of Chord of Contact

1. The equation of chord of contact is
$$yy_1 - 2a(x + x_1) = 0$$
2. The equation of chord of parabola, whose mid-point (x_1, y_1) is $T = S_1$, i.e. $yy_1 - 2a(x + x_1) = y_1^2 - 4ax_1$
3. Length of the chord of contact is
$$l = \frac{\sqrt{(y_1^2 - 4ax_1)(y_1^2 + 4a^2)}}{a}.$$
4. Area of the ΔPAB formed by the pair of tangents and their chord of contact is
$$A = \frac{(y_1^2 - 4ax_1)^{3/2}}{2a}.$$

NOTE
- Equation of the chord joining points $P(at_1^2, 2at_1)$, $Q(at_2^2, 2at_2)$ is $(t_1 + t_2)y = 2x + 2at_1t_2$.
- For PQ to be focal chord, $t_1t_2 = -1$.
- Length of the focal chord having t_1, t_2 as end points is $a(t_2 - t_1)^2$.

Director Circle

The locus of the point of intersection of perpendicular tangents to a conic is known as director circle. The director circle of a parabola is its directrix.

Conormal Points

The points on the parabola at which the normals pass through a common point are called conormal points. The conormal points are called the feet of the normals.

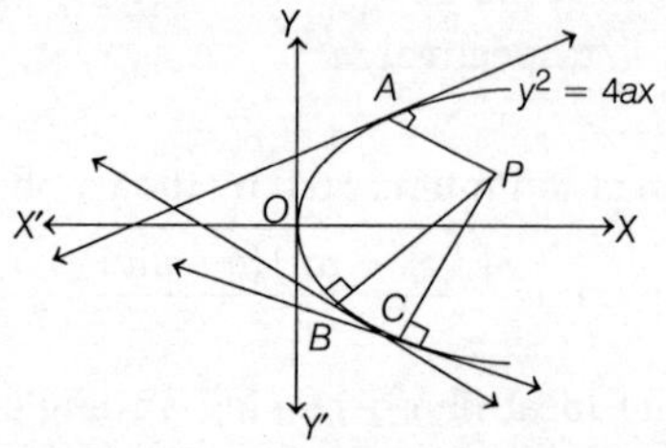

Points A, B and C are called conormal points with respect to point P.

1. The algebraic sum of the slopes of the normals at conormals point is 0.
2. The sum of the ordinates of the conormal points is 0.
3. The centroid of the triangle formed by the conormal points on a parabola lies on its axis.

Diameter

Diameter is the locus of mid-points of a system of parallel chords of parabola.

1. The tangent at the extremities of a focal chord intersect at right angles on the directrix and hence a circle on any focal chord as diameter touches the directrix.
2. A circle on any focal radii of a point $P(at^2, 2at)$ as diameter touches the tangent at the vertex and intercepts a chord of length $a\sqrt{1+t^2}$ on a normal at the point P.
3. The diameter bisecting chords of slope m to the parabola $y^2 = 4ax$ is $y = \frac{2a}{m}$.

DAY PRACTICE SESSION 1

FOUNDATION QUESTIONS EXERCISE

1 Equation of the parabola whose vertex is $(-1, -2)$, axis is vertical and which passes through the point $(3, 6)$ is

(a) $x^2 + 4x + 28y - 136 = 0$
(b) $x^2 + 2x - 2y - 3 = 0$
(c) $y^2 + 4y - 16x - 12 = 0$
(d) None of the above

2 A focal chord of the parabola $y^2 = 8x$ in inclined to X-axis at an angle $\tan^{-1} 3$. Then its length is equal to

(a) 80/3 (b) 80/9 (c) 40/3 (d) 40/9

3 Latus rectum of the parabola whose axis is parallel to the Y-axis and which passes through the points $(0, 4)$, $(1, 9)$, and $(-2, 6)$ is equal to

(a) 1/2 (b) 1 (c) 2 (d) None of these

4 If the line $x - 1 = 0$ is the directrix of the parabola $y^2 - kx + 8 = 0$, then one of the value of k is

(a) 1/8 (b) 8 (c) 4 (d) 1/4

5 The locus of trisection point of any double ordinate of the parabola $y^2 = 4ax$ is

(a) $y^2 = 9ax$ (b) $y^2 = ax$
(c) $9y^2 = 4ax$ (d) None of these

6 Let O be the vertex and Q be any point on the parabola $x^2 = 8y$. If the point P divides the line segment OQ internally in the ratio 1:3, then the locus of P is → JEE Mains 2015

(a) $x^2 = y$ (b) $y^2 = x$
(c) $y^2 = 2x$ (d) $x^2 = 2y$

7 At any points P on the parabola $y^2 - 2y - 4x + 5 = 0$, a tangent is drawn which meets the directrix at Q the locus of the points R which divides QP externally in the ratio $\frac{1}{2} : 1$, is

(a) $(x + 1)(1 - y)^2 + 4 = 0$ (b) $x + 1 = 0$
(c) $(1 - y)^2 - 4 = 0$ (d) None of these

8 The line $x - b + \lambda y = 0$ cuts the parabola $y^2 = 4ax$ at $P(at_1^2, 2at_1)$ and $Q(at_2^2, 2at_2)$. If $b \in [2a, 4a]$ and $\lambda \in R$, then $t_1 t_2$ belongs to

(a) $[-4, -2]$ (b) $[4, -3]$
(c) $[-3, -2]$ (d) None of these

9 The centre of the circle passing through the point $(0, 1)$ and touching the curve $y = x^2$ at $(2, 4)$ is

(a) $\left(\frac{-16}{5}, \frac{27}{10}\right)$ (b) $\left(\frac{-16}{7}, \frac{53}{10}\right)$
(c) $\left(\frac{-16}{5}, \frac{53}{10}\right)$ (d) None of these

10 Equation of common tangents to parabolas $y = x^2$ and $y = -x^2 + 4x - 4$ is/are

(a) $y = 4(x - 1); y = 0$ (b) $y = 0, y = -4(x - 1)$
(c) $y = 0, y = -10(x + 5)$ (d) None of these

11 Angle between the tangents drawn from the point $(1, 4)$ to the parabola $y^2 = 4x$ is

(a) $\pi/6$ (b) $\pi/4$
(c) $\pi/3$ (d) $\pi/2$

12 If the lines $y - b = m_1 (x + a)$ and $y - b = m_2 (x + a)$ are the tangents of the parabola $y^2 = 4ax$, then

(a) $m_1 + m_2 = 0$ (b) $m_1 m_2 = 1$
(c) $m_1 m_2 = -1$ (d) $m_1 + m_2 = 1$

13 Set of values of h for which the number of distinct common normals of $(x - 2)^2 = 4(y - 3)$ and $x^2 + y^2 - 2x - hy - c = 0$ where, $(c > 0)$ is 3, is

(a) $(2, \infty)$ (b) $(4, \infty)$ (c) $(2, 4)$ (d) $(10, \infty)$

14 Tangent and normal are drawn at $P(16, 16)$ on the parabola $y^2 = 16x$, which intersect the axis of the parabola at A and B respectively. If C is the centre of the circle through the points P, A and B and $\angle CPB = \theta$, then a value of $\tan\theta$ is → JEE Mains 2018

(a) $\frac{1}{2}$ (b) 2 (c) 3 (d) $\frac{4}{3}$

15 P is a point on the parabola $y^2 = 4x$ and Q is a point on the line $2x + y + 4 = 0$. If the line $x - y + 1 = 0$ is the perpendicular bisector of PQ, then the coordinates of P is

(a) $(8, 9), (10, 11)$ (b) $(1, -2), (9, -6)$
(c) $(7, 8), (9, 8)$ (d) None of these

16 The locus of the vertices of the family of parabolas $y = \frac{a^3 x^2}{3} + \frac{a^2 x}{2} - 2a$ is

(a) $xy = 105/64$ (b) $xy = 3/4$
(c) $xy = 35/16$ (d) $xy = 64/105$

17 The parabola $y^2 = \lambda x$ and $25[(x - 3)^2 + (y + 2)^2] = (3x - 4y - 2)^2$ are equal, if λ is equal to

(a) 1 (b) 2 (c) 3 (d) 6

18 A line is drawn from $A(-2, 0)$ to intersect the curve $y^2 = 4x$ in P and Q in the first quadrant such the $\frac{1}{AP} + \frac{1}{AQ} < \frac{1}{4}$, then slope of the line is always

(a) $< \sqrt{3}$ (b) $> \sqrt{3}$
(c) $\geq \sqrt{3}$ (d) None of these

19 Vertex A of a parabola $y^2 = 4ax$ is joined to any point P on it and line PQ is drawn at right angle to AP to meet the axis at Q. Then, the projection of PQ on the axis is always equal to

(a) $3a$ (b) $2a$
(c) $\sqrt{3}a$ (d) $4a$

20. The slope of the line touching both the parabolas $y^2 = 4x$ and $x^2 = -32y$ is → JEE Mains 2014

(a) $\frac{1}{2}$ (b) $\frac{3}{2}$ (c) $\frac{1}{8}$ (d) $\frac{2}{3}$

21 If the normals at the end points of variable chord PQ of the parabola $y^2 - 4y - 2x = 0$ are perpendicular, then the tangents at P and Q will intersect on the line

(a) $x + y = 3$ (b) $3x - 7 = 0$ (c) $y + 3 = 0$ (d) $2x + 5 = 0$

22 Find the length of the normal drawn from the point on the axis of the parabola $y^2 = 8x$ whose distance from the focus is 8.

(a) 10 (b) 8 (c) 9 (d) None of these

23 If $x + y = k$ is a normal to the parabola $y^2 = 12x$, p is the length of the perpendicular from the focus of the parabola on this normal, then $3k^3 + 2p^2$ is equal to

(a) 2223 (b) 2224
(c) 2222 (d) None of these

24 If $a \neq 0$ and the line $2bx + 3cy + 4d = 0$ passes through the points of intersection of the parabolas $y^2 = 4ax$ and $x^2 = 4ay$, then

(a) $d^2 + (2b + 3c)^2 = 0$ (b) $d^2 + (3b + 2c)^2 = 0$
(c) $d^2 + (2b - 3c)^2 = 0$ (d) $d^2 + (3b - 2c)^2 = 0$

25 Slopes of the normals to the parabola $y^2 = 4ax$ intersecting at a point on the axis of the parabola at a distance $4a$ from its vertex are in

(a) HP (b) GP
(c) AP (d) None of these

26 The area of the triangle formed by the tangent and the normal to the parabola $y^2 = 4ax$, both drawn at the same end of the latusrectum and the axis of the parabola is

(a) $2\sqrt{2}\,a^2$ (b) $2a^2$
(c) $4a^2$ (d) None of these

27 If the tangent at the point $P(2, 4)$ to the parabola $y^2 = 8x$ meets the parabola $y^2 = 8x + 5$ at Q and R, then mid-point of QR is

(a) (2, 4) (b) (4, 2)
(c) (7, 9) (d) None of these

28 The equation of the common tangent touching the circle $(x - 3)^2 + y^2 = 9$ and the parabola $y^2 = 4x$ above the X-axis is

(a) $\sqrt{3}y = 3x + 1$ (b) $\sqrt{3}y = -(x + 3)$
(c) $\sqrt{3}y = x + 3$ (d) $\sqrt{3}y = -(3x + 1)$

29 If tangents drawn from point P to the parabola $y^2 = 4x$ are inclined to X-axis at angles θ_1 and θ_2 such that $\cot\theta_1 + \cot\theta_2 = 2$, then locus of the point P is

(a) $y = 2$ (b) $y = 8$ (c) $y = 1$ (d) None of these

30 Tangents to the parabola $y^2 = 4x$ are drawn from the point (1, 3). The length of chord of contact is

(a) 5 (b) 13 (c) $\sqrt{65}$ (d) None of these

Directions (Q. Nos. 31-35) *Each of these questions contains two statements : Statement I (Assertion) and Statement II (Reason). Each of these questions also has four alternative choices, only one of which is the correct answer. You have to select one of the codes (a), (b), (c) and (d) given below.*

(a) Statement I is true, Statement II is true; Statement II is a correct explanation for Statement I
(b) Statement I is true, Statement II is true; Statement II is not a correct explanation for Statement I
(c) Statement I is true; Statement II is false
(d) Statement I is false; Statement II is true

31 **Statement I** The perpendicular bisector of the line segment joining the points $(-a, 2at)$ and $(a, 0)$ is tangent to the parabola $y^2 = -4ax$, where $t \in R$.

Statement II Number of parabolas with a given point as vertex and length of latusrectum equal to 4 is 2.

32 Consider the equation of the parabola is $y^2 = 4ax$.

Statement I Length of focal chord of a parabola having focus (2, 0) making an angle of 60° with X-axis is 32.

Statement II Length of focal chord of a parabola $y^2 = 4ax$ making an angle α with X-axis is $4a\,\text{cosec}^2\alpha$.

33 Consider the equation of the parabola is $y^2 = 4ax$.

Statement I Area of triangle formed by pair of tangents drawn from a point (12, 8) to the parabola having focus (1,0) and their corresponding chord of contact is 32 sq units.

Statement II If from a point $P(x_1, y_1)$ tangents are drawn to a parabola, then area of triangle formed by these tangents and their corresponding chord of contact is $\frac{(y_1^2 - 4ax_1)^{3/2}}{4\,|a|}$ sq units.

34 **Statement I** The latusrectum of a parabola is 4 units, axis is the line $3x + 4y - 4 = 0$ and the tangent at the vertex is the line $4x - 3y + 7 = 0$, then the equation of directrix of the parabola is $4x - 3y + 8 = 0$.

Statement II If P is any point on the parabola and PM and PN are perpendiculars from P on the axis and tangent at the vertex respectively, then $(PM)^2 = (\text{latusrectum})\,(PN)$.

35 A circle, $2x^2 + 2y^2 = 5$ and a parabola, $y^2 = 4\sqrt{5}x$.

Statement I An equation of a common tangent to these curves is $y = x + \sqrt{5}$.

Statement II If the line, $y = mx + \frac{\sqrt{5}}{m}$ where, $m \neq 0$ is the common tangent, then m satisfies $m^4 - 3m^2 + 2 = 0$.

→ JEE Mains 2013

DAY PRACTICE SESSION 2

PROGRESSIVE QUESTIONS EXERCISE

1 A ray of light moving parallel to the X-axis gets reflected from a parabolic mirror whose equation is $(y-2)^2 = 4(x+1)$. After reflection, the ray must pass through the point

(a) (0, 2) (b) (2, 0)
(c) (0, −2) (d) (−1, 2)

2 Mutually perpendicular tangents TA and TB are drawn to the parabola $y^2 = 8x$. The minimum length of AB is

(a) 16 (b) 4
(c) 8 (d) None of these

3 If a line $x + y = 1$ cuts the parabola $y^2 = 4x$ at points A and B and normals at A and B meet on C. The normals to the parabola from C, other than above two, meet the parabola in D, the coordinates of D are

(a) (2, 1) (b) (−4, 4)
(c) (4, 4) (d) None of these

4 A chord PP' of a parabola cuts the axis of the parabola at A. The feet of the perpendiculars from P and P' on the axis are M and M' respectively. If V is the vertex, then VM, VA, VM' are in

(a) AP (b) GP
(c) HP (d) None of these

5 The set of points on the axis of the parabola $y^2 = 4x + 8$ from which the 3 normals to the parabola are all real and different, is

(a) $\{(k, 0) | k \le -2\}$ (b) $\{(k, 0) | k > -2\}$
(c) $\{(k, 0) | k > 0\}$ (d) None of these

6 Normals drawn to $y^2 = 4ax$ at the points where it is intersected by the line $y = mx + c$, intersect at the point P. Foot of another normal drawn to the parabola from the point P may be

(a) $(a/m^2, -2a/m)$
(b) $(9a/m^2, -6a/m)$
(c) $(4a/m^2, -4a/m)$
(d) None of the above

7 Sides of an equilateral triangle ABC touch the parabola $y^2 = 4ax$, then points A, B, C lie on

(a) $y^2 = 3(x+a)^2 + 4ax$
(b) $y^2 = (x+a)^2 + ax$
(c) $y^2 = 3(x+a)^2 + ax$
(d) None of the above

8 Minimum distance between the curves $y^2 = 4x$ and $x^2 + y^2 - 12x + 31 = 0$ is

(a) $\sqrt{21}$ (b) $\sqrt{5}$
(c) $2\sqrt{7} - \sqrt{5}$ (d) None of these

9 The triangle formed by the tangent to the parabola $y = x^2$ at the point whose abscissa is $x_0, 1 \le x_0 \le 2$, the Y-axis and the straight line $y = x_0^2$ has the greatest area if $x_0 =$

(a) 1 (b) 2
(c) 3/2 (d) None of these

10 The equation of the curve obtained by reflecting the parabola $y^2 = 4x$ about the line $x - y + 13 = 0$ is

(a) $(2y - x - 13)^2 = 4(y + 13)$
(b) $(2y + x - 13)^2 = 4(y - 13)$
(c) $(2y - x - 13)^2 = 4(y - 13)$
(d) None of the above

11 Let P be the point on the parabola, $y^2 = 8x$ which is at a minimum distance from the centre C of the circle, $x^2 + (y+6)^2 = 1$. Then the equation of the circle, passing through C and having its centre at P is → JEE Mains 2016

(a) $x^2 + y^2 - 4x + 8y + 12 = 0$
(b) $x^2 + y^2 - x + 4y - 12 = 0$
(c) $x^2 + y^2 - \frac{x}{4} + 2y - 24 = 0$
(d) $x^2 + y^2 - 4x + 9y + 18 = 0$

12 The radius of a circle, having minimum area, which touches the curve $y = 4 - x^2$ and the line $y = |x|$ is → JEE Mains 2017

(a) $4(\sqrt{2} - 1)$ (b) $4(\sqrt{2} + 1)$
(c) $2(\sqrt{2} + 1)$ (d) $2(\sqrt{2} - 1)$

13 If y_1, y_2 are the ordinates of two points P and Q on the parabola and y_3 is the ordinate of the point of intersection of tangents at P and Q, then

(a) y_1, y_2, y_3 are in AP
(b) y_1, y_3, y_2 are in AP
(c) y_1, y_2, y_3 are in GP
(d) y_1, y_3, y_2 are in GP

14 The number of points with integral coordinates that lie in the interior of the region common to the circle $x^2 + y^2 = 16$ and the parabola $y^2 = 4x$ is

(a) 8 (b) 10
(c) 16 (d) None of these

15 The tangent and normal at the point $P = (16, 16)$ to the parabola $y^2 = 16x$ intersect the X-axis at the points Q and R respectively. The equation to the circum circle of ΔPQR is

(a) $x^2 + y^2 - 8x - 384 = 0$
(b) $x^2 + y^2 - 2x - 8y - 352 = 0$
(c) $x^2 + y^2 + 2y - 544 = 0$
(d) None of the above

ANSWERS

SESSION 1									
1 (b)	2 (b)	3 (a)	4 (c)	5 (c)	6 (d)	7 (a)	8 (a)	9 (c)	10 (a)
11 (c)	12 (c)	13 (d)	14 (b)	15 (b)	16 (a)	17 (d)	18 (b)	19 (d)	20 (a)
21 (d)	22 (b)	23 (a)	24 (a)	25 (c)	26 (c)	27 (a)	28 (c)	29 (a)	30 (c)
31 (c)	32 (d)	33 (c)	34 (d)	35 (b)					

SESSION 2									
1 (a)	2 (c)	3 (c)	4 (b)	5 (c)	6 (c)	7 (a)	8 (b)	9 (b)	10 (d)
11 (a)	12 (a)	13 (b)	14 (d)	15 (a)					

Hints and Explanations

SESSION 1

1 Axis is vertical i.e. parallel to Y-axis so its equation should be
$(x+1)^2 = 4a(y+2)$
It passes through $(3, 6)$ so $4a = 2$.
Hence the equation of the required parabola is $x^2 + 2x - 2y - 3 = 0$,

2 Length of focal chord $= 4a \operatorname{cosec}^2\alpha$.
Here $a = 2$, $\alpha = \tan^{-1} 3$ i.e. $\tan\alpha = 3$.
$\therefore$ Length of focal chord
$= 4 \times 2 \times (1 + 1/9) = 80/9$.

3 Let vertex be (b, c). Then equation of parabola is $(x-b)^2 = 4a(y-c)$. It passes through the points $(0, 4)$, $(1, 9)$ and $(-2, 6)$.

$\therefore \quad b^2 = 4a(4-c)$
$(1-b^2) = 4a(9-c)$
and $(-2-b)^2 = 4a(6-c)$.
Solving these equations, latus rectum $4a = 1/2$.

4 $y^2 - kx + 8 = 0 \Rightarrow y^2 = k(x - 8/k)$.
$\therefore$ Directrix is $x - 8/k = -k/4$
or $x = 8/k - k/4 = 1$
$\Rightarrow k^2 + 4k - 32 = 0 \Rightarrow k = -8$ or 4.
$\therefore$ One value of k is 4.

5 Let $P(\alpha, \beta)$ be the trisection point.

$\therefore \alpha = at^2, \beta = \dfrac{2(2at) + 1(-2at)}{3}$

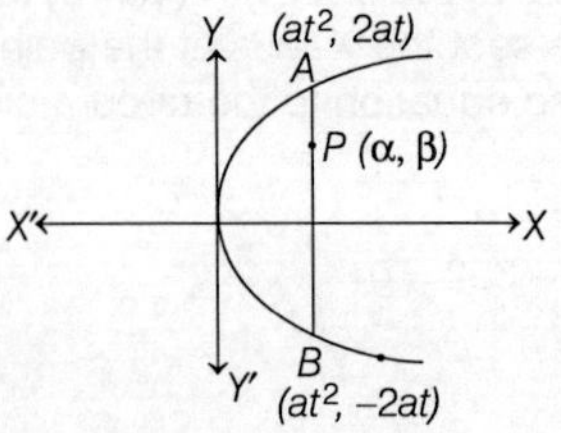

$\Rightarrow \quad \beta = \dfrac{2}{3}at \quad \Rightarrow t = \dfrac{3\beta}{2a}$

$\therefore \quad \alpha = a\left(\dfrac{3\beta}{2a}\right)^2 \Rightarrow 9\beta^2 = 4a\alpha$

Hence, the locus of P is $9y^2 = 4ax$.

6 Any point on the parabola $x^2 = 8y$ is $(4t, 2t^2)$. Point P divides the line segment joining of $O(0,0)$ and $Q(4t, 2t^2)$ in the ratio 1: 3. Apply the section formula for internal division.

Equation of parabola is $x^2 = 8y$
Let any point Q on the parabola (i) is $(4t, 2t^2)$.
Let $P(h,k)$ be the point which divides the line segment joining $(0,0)$ and $(4t, 2t^2)$ in the ratio 1:3.

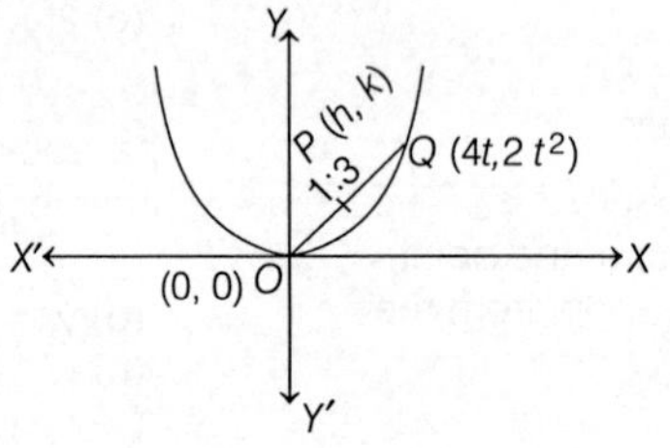

$\therefore \quad h = \dfrac{1 \times 4t + 3 \times 0}{4} \Rightarrow h = t$

and $k = \dfrac{1 \times 2t^2 + 3 \times 0}{4} \Rightarrow k = \dfrac{t^2}{2}$

$\Rightarrow \quad k = \dfrac{1}{2}h^2 \qquad [\because t = h]$

$\Rightarrow \quad 2k = h^2 \Rightarrow 2y = x^2$, which is required locus.

7 Given, $(y-1)^2 = 4(x-1)$. P has coordinates $x = 1 + t^2$, $y = 1 + 2t$.
Tangent at P is
$(x-1) - (y-1)t + t^2 = 0$.
So, the directrix is $x = 0$.

$\therefore \quad Q = \left[0, t + 1 - \dfrac{1}{t}\right]$

If $R(x, y)$ divides QP externally in the ratio $1:2$.

$\therefore x = -(1+t^2)$ and $y = 1 - \dfrac{2}{t} \Rightarrow t = \dfrac{2}{1-y}$

$\therefore \quad x + 1 + \dfrac{4}{(1-y)^2} = 0$

$\Rightarrow (x+1)(1-y)^2 + 4 = 0$

8 Line $x - b + \lambda y = 0$ always passes through $(b, 0)$.

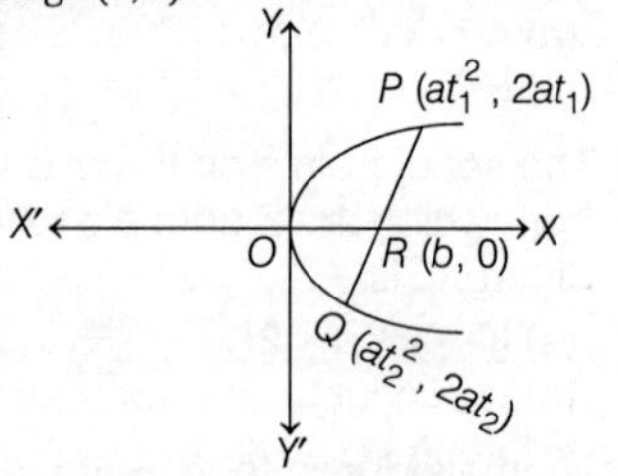

Slope of PR = Slope of RQ

$\Rightarrow \quad t_1 t_2 = -\dfrac{b}{a}$

$\therefore$ Minimum value of $t_1 t_2 = -4$
and maximum value of $t_1 t_2 = -2$

9 The slope of the tangent to $y = x^2$ at $(2, 4)$ is 4 and the equation of the tangent is $4x - y - 4 = 0$
Equation of the circle is
$(x-2)^2 + (y-4)^2 + \lambda(4x - y - 4) = 0$ ...(i)
Since, it passes through $(0, 1)$.

Hence, $\lambda = \dfrac{13}{5}$

On putting the value of λ in Eq. (i), we get
$5(x-2)^2 + 5(y-4)^2 + 13(4x - y - 4) = 0$
$\Rightarrow 5(x^2 + 4 - 4x) + 5(y^2 + 16 - 8y)$
$+ 52x - 13y - 52 = 0$
$\Rightarrow 5x^2 + 5y^2 + 32x - 53y + 48 = 0$
$\Rightarrow x^2 + y^2 + \dfrac{32}{5}x - \dfrac{53}{5}y + \dfrac{48}{5}$

So, the centre of the circle is $\left(\dfrac{-16}{5}, \dfrac{53}{10}\right)$.

10 Tangent to parabola is $y = mx - am^2$.

$\therefore$ Tangents to two given parabolas are

$y = mx - (m^2/4)$ and

$y = m(x-2) + (m^2/4)$

These are identical $\Rightarrow m = 0$ or 4.

$\therefore$ Common tangents are $y = 0$ and $y = 4x - 4$.

11 $y = mx + 1/m$ passes through (1, 4).

$\Rightarrow m^2 - 4m + 1 = 0$.

$$\therefore \tan\theta = \frac{m_1 - m_2}{1 + m_1 m_2} = \frac{\sqrt{(m_1+m_2)^2 - 4m_1m_2}}{1+m_1m_2} = \frac{\sqrt{16-4}}{1+1} = \frac{2\sqrt{3}}{2} = \sqrt{3}$$

$\Rightarrow \theta = \pi/3$.

12 Both lines pass through $(-a, b)$ which is a point on the directrix $x = -a$. Therefore, tangents drawn from $(-a, b)$ are perpendicular, so $m_1 m_2 = -1$.

13 The equation of any normal of $(x-2)^2 = 4(y-3)$ is

$$(x-2) = m(y-3) - 2m - m^3.$$

If it passes through $\left(1, \frac{h}{2}\right)$, then

$$1 - 2 = m\left(\frac{h}{2} - 3\right) - 2m - m^3$$

$\Rightarrow 2m^3 + m(10-h) - 2 = 0 = f(m)$ [say]

This equation will given three distinct values of m.

If $f'(m) = 0$ has two distinct roots, where

$f(m) = 2m^3 + m(10-h) - 2$

Now, $f'(m) = 6m^2 + (10-h)$

Put $f'(m) = 0 \Rightarrow m = \pm\sqrt{\frac{h-10}{6}}$

So, the values of m are real and distinct, if $h > 10$ i.e. $h \in (10, \infty)$.

14 Equation of tangent at $P(16, 16)$ is

$x - 2y + 16 = 0$

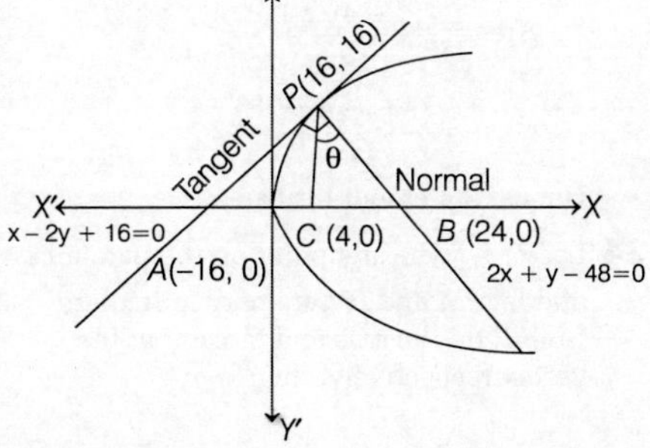

Slope of $PC = \frac{4}{3}$

Slope of $PB = -2$

$$\text{Hence, } \tan\theta = \left|\frac{\frac{4}{3} + 2}{1 - \frac{4}{3} \times 2}\right| = 2$$

15 Any point on the parabola is $P = (t^2, 2t)$.

Q is its image of the line $x - y + 1 = 0$.

$$\therefore \frac{x - t^2}{1} = \frac{y - 2t}{-1} = -(t^2 - 2t + 1)$$

$\Rightarrow Q = (2t - 1, t^2 + 1)$

Since, it lies on the line

$2x + y + 4 = 0$

$\therefore 4t - 2 + t^2 + 1 + 4 = 0$

$\Rightarrow t^2 + 4t + 3 = 0 \Rightarrow t = -1, -3$

So, the possible positions of P are $(1, -2)$ and $(9, -6)$.

16 $y = \frac{1}{3}a^3x^2 + \frac{a^2}{2}x - 2a$

$$\Rightarrow \left(x + \frac{3}{4a}\right)^2 = \frac{3}{a^3}\left(y + \frac{35}{16}a\right)$$

$$\text{Vertex } P(h, k) = \left(-\frac{3}{4a}, -\frac{35}{16}a\right)$$

$$\Rightarrow a = -\frac{3}{4h},\ a = -\frac{16k}{35}$$

$\Rightarrow$ Locus of vertex P is $xy = 105/64$.

17 Let us recall that two parabolas are equal, if the length of their latusrectum are equal.

Length of the latusrectum of $y^2 = \lambda x$ is λ.

The equation of the second parabola is

$25\{(x-3)^2 + (y+2)^2\} = (3x - 4y - 2)^2$

$$\Rightarrow \sqrt{(x-3)^2 + (y+2)^2} = \frac{|3x - 4y - 2|}{\sqrt{3^2 + 4^2}}$$

Clearly, it represents a parabola having focus at $(3, -2)$ and equation of the directrix as $3x - 4y - 2 = 0$.

$\therefore$ Length of the latusrectum

= 2 (Distance between focus and directrix)

$$= 2\left|\frac{3 \times 3 - 4 \times (-2) - 2}{\sqrt{3^2 + (-4)^2}}\right| = 6$$

Thus, the two parabolas are equal, if $\lambda = 6$.

18 Let $P(-2 + r\cos\theta, r\sin\theta)$ and P lies on parabola.

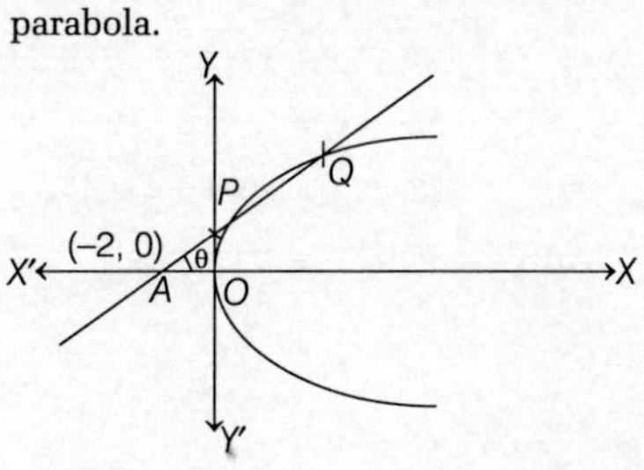

$\Rightarrow r^2\sin^2\theta - 4(-2 + r\cos\theta) = 0$

$$\Rightarrow r_1 + r_2 = \frac{4\cos\theta}{\sin^2\theta} \Rightarrow r_1r_2 = \frac{8}{\sin^2\theta}$$

$$\therefore \frac{r_1 + r_2}{r_1r_2} = \frac{1}{AP} + \frac{1}{AQ}$$

$$\Rightarrow \cos\theta < \frac{1}{2} \Rightarrow \tan\theta > \sqrt{3}$$

[because $\cos\theta$ is decreasing and $\tan\theta$ is increasing in $\left(0, \frac{\pi}{2}\right)$]

$\Rightarrow m > \sqrt{3}$

19 Let $P \equiv (at^2, 2at)$

Equation of the line PQ is

$$y - 2at = -\frac{t}{2}(x - at^2).$$

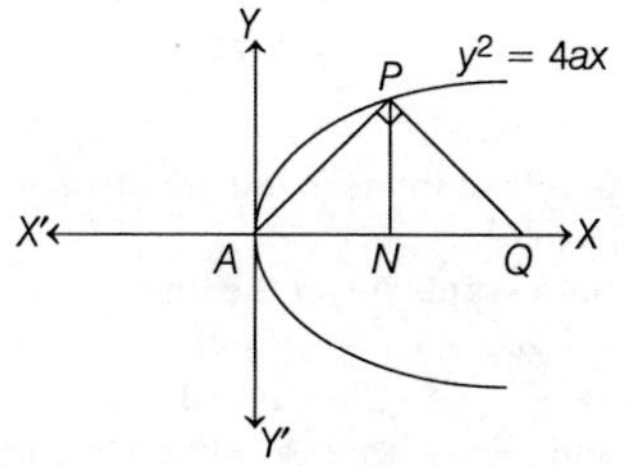

On putting $y = 0$, we get $x = 4a + at^2$

So, the coordinates of Q and N are $(4a + at^2, 0)$ and $(at^2, 0)$, respectively.

So, length of projection

$= 4a + at^2 - at^2 = 4a$

20 Let the tangent to parabola be $y = mx + a/m$, if it touches the other curve, then $D = 0$, to get the value of m.

For parabola, $y^2 = 4x$

Let $y = mx + \frac{1}{m}$ be tangent line and it touches the parabola $x^2 = -32y$.

$$\therefore x^2 = -32\left(mx + \frac{1}{m}\right)$$

$$\Rightarrow x^2 + 32mx + \frac{32}{m} = 0$$

$\because D = 0$

$$\therefore (32m)^2 - 4\left(\frac{32}{m}\right) = 0 \Rightarrow m^3 = \frac{1}{8}$$

$$\therefore m = \frac{1}{2}$$

21 The tangents and normals form a rectangle.

Hence, tangents meet on the directrix.

Now, $(y-2)^2 = 2(x+2)$

Vertex $= (-2, 2)$ and directrix, $x = -\frac{5}{2}$

$\Rightarrow 2x + 5 = 0$

22 Here, $a = 2$ normal at t is

$xt + y = 2t^3 + 4t$. Focus $= (2, 0)$

So, the point on the axis is (10, 0).
Normal passes through (10, 0).

$\therefore \quad 10 = 2t^2 + 4 \Rightarrow t^2 = 3$

So, the normal is at the point $(6, 4\sqrt{3})$.

So, the required length is

$\sqrt{(10-6)^2 + (4\sqrt{3})^2} = \sqrt{16+48} = 8$

23 The equation of normal to the parabola $y^2 = 12x$ with slope -1 is

$$y = -x - 2(3)(-1) - 3(-1)^3$$

$\Rightarrow \quad y = -x + 9 \Rightarrow x + y = 9$

$\therefore \quad k = 9$

Since, the focus of the parabola is (3, 0).

$\therefore \quad p = \left|\dfrac{3-9}{\sqrt{2}}\right|$

$\Rightarrow \quad 2p^2 = 36$

$\therefore \quad 3k^3 + 2p^2 = 3(9)^3 + 36$

$= 2223$

24 Solving $y^2 = 4ax$ and $x^2 = 4ay (a \neq 0)$, points of intersection are (0, 0) and $(4a, 4a)$.

Both points lie on the line

$2bx + 3cy + 4d = 0$

$\Rightarrow \quad d = 0$

and $\quad 2b + 3c = 0 \quad (\because a \neq 0)$

$\therefore \quad d^2 + (2b + 3c)^2 = 0$

25 .The normal $y = mx - 2am - am^3$ passes through $(4a, 0)$.

$\therefore \quad m^3 - 2m = 0 \Rightarrow m = 0, \pm\sqrt{2}$

$\therefore \quad$ Slopes of normals are $-\sqrt{2}, 0, \sqrt{2}$ which are in AP.

26 The coordinate of end of the latusrectum is $(a, 2a)$. The equation of the tangent at $(a, 2a)$ is $y \cdot 2a = 2a(x + a)$, i.e. $y = x + a$. The normal at $(a, 2a)$ is $y + x = 2a + a$, i.e. $y + x = 3a$.

On solving $y = 0$ and $y = x + a$, we get

$$x = -a, y = 0$$

On solving $y = 0$ and $y + x = 3a$, we get

$$x = 3a, y = 0$$

The area of the triangle with vertices $(a, 2a), (-a, 0), (3a, 0)$

$$= \frac{1}{2} \times 4a \times 2a = 4a^2$$

27 Equation of tangent to $y^2 = 8x$ at (2, 4) is

$4y = 4(x + 2)$ i.e. $x - y + 2 = 0 \quad$ ...(i)

Let mid-point of QR be (x_1, y_1). Then, equation of $QR (T = S_1)$ is

$yy_1 - 4(x + x_1) - 5 = y_1^2 - 8x_1 - 5$

$\Rightarrow 4x - yy_1 - 4x_1 + y_1^2 = 0 \quad$ ...(ii)

On comparing Eqs. (i) and (ii), we get

$$\frac{4}{1} = \frac{y_1}{1} = \frac{y_1^2 - 4x_1}{2}$$

$\Rightarrow \; y_1 = 4$ and $8 = -4x_1 + y_1^2$

$\Rightarrow \; y_1 = 4$ and $x_1 = 2$

Hence, required mid-point is (2, 4).

28 As common tangent is above X-axis, its slope is positive.

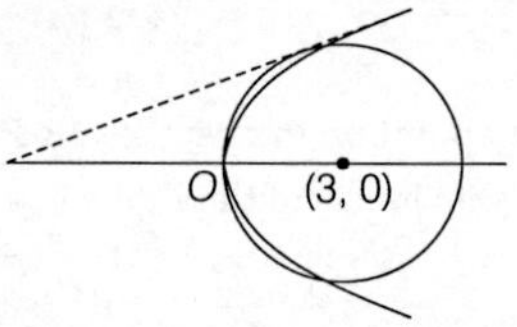

$y = mx + 1/m$ is a tangent to the parabola.

It touches the circle $(x-3)^2 + y^2 = 9$ if

$$\left|\frac{3m - 0 + 1/m}{\sqrt{1+m^2}}\right| = 3$$

$\Rightarrow \; (3m + 1/m)^2 = 9(1 + m^2)$

$\Rightarrow \; 6 + 1/m^2 = 9$ i.e. $m^2 = 1/3$.

As $m > 0$, $m = 1/\sqrt{3}$.

$\therefore$ Equation of common tangent above X-axis is

$$y = \frac{1}{\sqrt{3}}x + \sqrt{3}$$

$\Rightarrow \quad \sqrt{3}y = x + 3.$

29 Let $P(x_1, y_1)$. Equation of any tangent making angle θ with X-axis (slope = $\tan\theta$) is

$y = x\tan\theta + \dfrac{1}{\tan\theta} \quad (\because y = mx + a/m)$

It passes through $P(x_1, y_1)$

$\therefore \quad x_1 \tan\theta + \dfrac{1}{\tan\theta} = y_1$

$\Rightarrow \; x_1 \tan^2\theta - y_1 \tan\theta + 1 = 0$

$\tan\theta_1 + \tan\theta_2 = y_1 / x_1,$

$\tan\theta_1 \tan\theta_2 = 1/x_1$

Given that, $\cot\theta_1 + \cot\theta_2 = 2$

$\Rightarrow \tan\theta_1 + \tan\theta_2 = 2\tan\theta_1 \tan\theta_2$

$\Rightarrow \; y_1 / x_1 = 2/x_1$

$\Rightarrow$ Locus of $P(x_1, y_1)$ is $y = 2$.

30 Equation of chord of contact of (1,3) to the parabola $y^2 = 4x$ is

$3y = 2(x + 1) \quad$ ...(i)

Solving Eq. (i) and parabola, we get

$$\frac{4}{9}(x+1)^2 = 4x$$

$\Rightarrow \quad x^2 - 7x + 1 = 0$

$\therefore \quad x_1 + x_2 = 7, x_1 x_2 = 1$

$\Rightarrow \quad (x_1 - x_2)^2 = 49 - 4 = 45$

Also, $\quad y_1 - y_2 = \dfrac{2}{3}(x_1 - x_2)$

$\Rightarrow \quad (y_1 - y_2)^2 = 20$

Hence, length of chord of contact

$= \sqrt{45 + 20} = \sqrt{65}$

31 Image of $(a, 0)$ with respect to tangent $yt = x + at^2$ is $(-a, 2at)$.

So, perpendicular bisector of $(a, 0)$ and $(-a, 2at)$ is the tangent line $yt = x + at^2$ to the parabola.

Hence, Statement I is true.

Statement II Infinitely many parabolas are possible.

Hence, Statement II is false.

32 Let AB be a focal chord slope of $AB = \dfrac{2t}{t^2 - 1} = \tan\alpha$

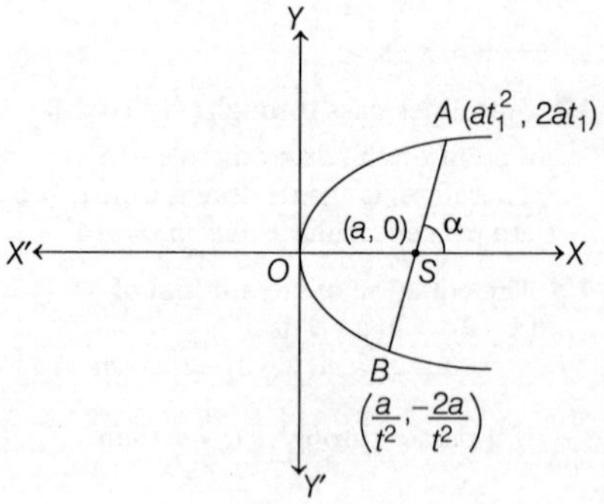

$\Rightarrow \quad \tan\dfrac{\alpha}{2} = \dfrac{1}{t}$

$\Rightarrow \quad t = \cot\dfrac{\alpha}{2}$

Length of $AB = a\left(t + \dfrac{1}{t}\right)^2$

$= 4a\,\text{cosec}^2\alpha$

When $a = 2, \alpha = 60°$

$\therefore$ Length of $AB = 4(2)\,\text{cosec}^2(60°)$

$= \dfrac{32}{3}$

33 **Statement II** Area of triangle formed by these tangents and their corresponding chord of contact is $\dfrac{(y_1^2 - 4ax_1)^{3/2}}{2|a|}$.

Hence, Statement II is false.

Statement I $x_1 = 12, y_1 = 8$

$\therefore \quad$ Area $= \dfrac{(y_1^2 - 4ax_1)^{3/2}}{2}$

$= \dfrac{(64-48)^{3/2}}{2} = 32$

Hence, Statement I is true.

34 Let $P(x, y)$ be any point on the parabola and let PM and PN are perpendiculars from P on the axis and tangent at the vertex respectively, then

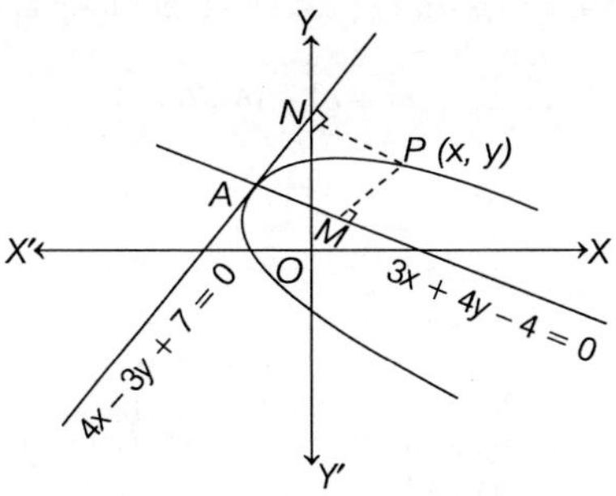

$(PM)^2 = (\text{Latusrectum})(PN)$

$$\Rightarrow \left(\frac{3x+4y-4}{\sqrt{3^2+4^2}}\right)^2 = 4\left(\frac{4x+3y+7}{\sqrt{4^2+(-3)^2}}\right)$$

$\Rightarrow Y^2 = 4AX$

$$\therefore \quad A = 1, Y = \frac{3x+4y-4}{5},$$

$$X = \frac{4x-3y+7}{5}$$

So, the directrix is $X + A = 0$.

$$\Rightarrow \quad \frac{4x-3y+7}{5} + 1 = 0$$

$$\Rightarrow \quad 4x - 3y + 12 = 0$$

35 Equation of circle can be rewritten as

$$x^2 + y^2 = \frac{5}{2}$$

Let common tangent be $y = mx + \frac{\sqrt{5}}{m}$

So, the perpendicular from centre to the tangent is equal to radius.

$$\therefore \quad \frac{\frac{\sqrt{5}}{m}}{\sqrt{1+m^2}} = \sqrt{\frac{5}{2}}$$

$$\Rightarrow \quad m\sqrt{1+m^2} = \sqrt{2}$$

$$\Rightarrow \quad m^2(1+m^2) = 2$$

$$\Rightarrow \quad m^4 + m^2 - 2 = 0$$

$$\Rightarrow \quad (m^2+2)(m^2-1) = 0$$

$$\Rightarrow \quad m = \pm 1$$

$[\because m^2 + 2 \neq 0, \text{ as } m \in R]$

$$\therefore \quad y = \pm x \pm \sqrt{5}$$

Both statements are correct as

$m = \pm 1$

satisfies the given equation of Statement II.

But, Statement II is not a correct explanation of Statement I.

SESSION 2

1 Equation of axis is $y = 2$ which is parallel to X-axis.

Therefore, reflected ray will pass through the focus, which is (0, 2)

2 Tangents TA, TB are perpendicular

$\Rightarrow$ AB is focal chord

$\Rightarrow$ AB is latusrectum.

$\therefore$ $AB = 8$

3 Here A, B and D are co-normal points,

Let A, B and D be (x_1, y_1), (x_2, y_2) and (x_3, y_3) respectively. AB is the chord $x + y = 1$.

Solving $x + y = 1$ and $y^2 = 4x$, we get

$$\Rightarrow \quad y^2 = 4(1-y)$$

$$\Rightarrow \quad y^2 + 4y - 4 = 0$$

$$\therefore \quad y_1 + y_2 = -4$$

$$y_1 + y_2 + y_3 = 0$$

$$\Rightarrow \quad y_3 = 4$$

$\therefore 16 = 4x_3 \Rightarrow x_3 = 4.$

Hence, D is (4, 4).

4 $VM = at_1^2$, $VM' = at_2^2$ and $VA = k$, then

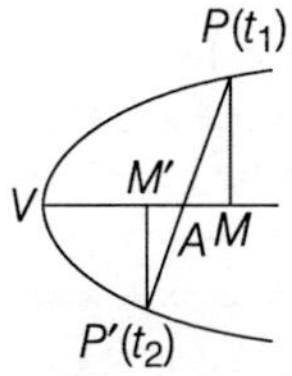

$$\begin{vmatrix} at_1^2 & 2at_1 & 1 \\ at_2^2 & 2at_2 & 1 \\ k & 0 & 1 \end{vmatrix}$$

$[\because P, A, P' \text{ are collinear}]$

$\Rightarrow k + at_1t_2 = 0$

$\Rightarrow VM \cdot VM' = (at_1t_2)^2 = k^2 = VA^2$

$\Rightarrow$ VM, VA and VM' are in GP.

5 Let $P(k, 0)$ be a point on the axis on the parabola $y^2 = 4(x+2)$

Equation of normal at $(-2+t^2, 2t)$ is

$$t(x+2) + y = 2t + t^3$$

$$\Rightarrow \quad y + tx = t^3.$$

This passes through $(k, 0)$

$\therefore t^3 - kt = 0$ or $t = 0, t^2 = k$

For three real and distinct normals $k > 0$.

$\therefore$ Set of all such point

$$= \{(k, 0) \mid k > 0\}.$$

6 Let $y = mx + c$ intersect $y^2 = 4ax$ at $A(t_1)$ and $B(t_2)$. Then

$$m = \frac{2}{t_1 + t_2}$$

$\Rightarrow t_1 + t_2 = 2/m$

Normals at A and B meet at P. Let another normal from P meet the parabola at $C(t_3)$.

Then A, B and C are co-normal points.

$$\therefore \quad t_1 + t_2 + t_3 = 0$$

$$\Rightarrow \quad t_3 = -2/m$$

$$\therefore \quad C \text{ may be } \left(\frac{4a}{m^2}, -\frac{4a}{m}\right).$$

7 Let the sides of the triangle touch the parabola $y^2 = 4ax$ at t_1, t_2 and t_3.

Tangent at t_1, t_2, t_3 meets in

$A(at_1t_2, a(t_1+t_2))$, $B(at_1t_3, a(t_1+t_3))$ and $C(at_1t_3, a(t_2+t_3))$.

Triangle ABC is equilateral.

$$m_{AB} = \frac{a(t_3 - t_2)}{at_1(t_3 - t_2)} = \frac{1}{t_1}, \text{ and}$$

$m_{AC} = 1/t_2$.

$$\therefore \quad \sqrt{3} = \left|\frac{1/t_1 - 1/t_2}{1 + 1/t_1t_2}\right| = \frac{|t_2 - t_1|}{|1 + t_1t_2|}$$

$\Rightarrow (t_2 - t_1)^2 = 3(1 + t_1t_2)^2$

$\Rightarrow (t_1 + t_2)^2 - 4t_1t_2 = 3 + 6t_1t_2 + 3(t_1t_2)^2$

Let A be (x, y). Then

$$\frac{y^2}{a^2} = 3 + 10\frac{x}{a} + \frac{3x^2}{a^2}$$

$$\Rightarrow \quad y^2 = 3a^2 + 10ax + 3x^2$$

$$= 3(x+a)^2 + 4ax$$

8 Circle is $x^2 + y^2 - 12x + 31 = 0$, $C(6, 0)$, $r = \sqrt{5}$.

Shortest distance will take place along the common normal. Normal to $y^2 = 4x$ at $A(t^2, 2t)$ is

$y = -tx + 2t + t^3$. It must passes through (6, 0).

$\therefore \quad t^3 - 4t = 0 \Rightarrow t = 0$ or $\pm\sqrt{2}$.

$\therefore$ Distances between the curves along common normal are $6 - \sqrt{5}$, and $\sqrt{5}$.

Hence, minimum distance between the curves $= \sqrt{5}$.

9 Area $A = 2x_0^2 \times x_0 \times \frac{1}{2} = x_0^3$

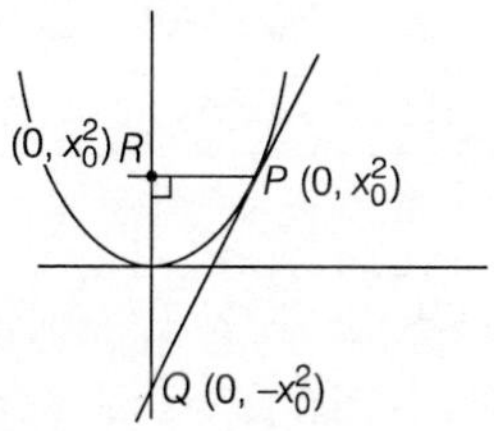

Since $1 \le x_0 \le 2$, then area is max. at $x_0 = 2$.

10 Let $A(x_1, y_1)$ be any point on the parabola

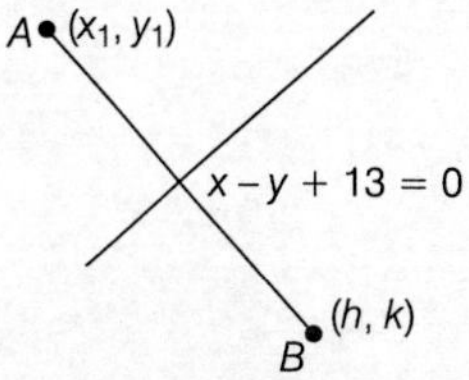

$y^2 = 4x$ and $B(h, k)$ be the reflection of A with respect to the line $x - y + 13 = 0$. Then,

$$\frac{h + x_1}{2} - \frac{k + y_1}{2} + 13 = 0$$

$$\text{and} \quad \frac{k - y_1}{h - x_1} \cdot (1) = -1$$

Then, $x_1 = k - 13$, $y_1 = h + 13$

$\therefore \quad (h + 13)^2 = 4(k - 13)$

$\therefore$ Locus of the point B is

$$(x + 13)^2 = 4(y - 13).$$

11 Normal at $P(at^2, 2at)$ is

$y + tx = 2at + at^3$

Given it passes $(0, -6)$

$\Rightarrow \quad -6 = 2at + at^3 \quad [\because a = 2]$

$\Rightarrow \quad -6 = 4t + 2t^3$

$\Rightarrow \quad t^3 + 2t + 3 = 0$

$\Rightarrow \quad t = -1$

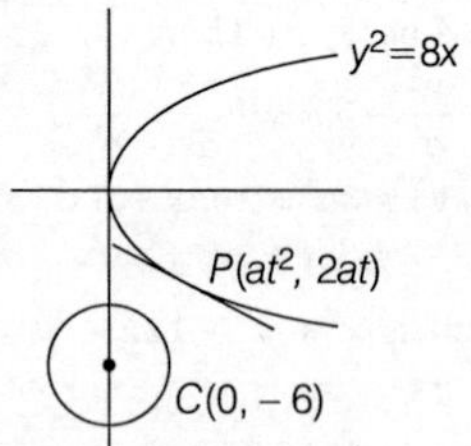

So, $P(a, -2a) = (2, -4) \quad [\because a = 1]$

Radius of circle

$$= CP = \sqrt{2^2 + (-4 + 6)^2} = 2\sqrt{2}$$

Equation of circle is

$$(x - 2)^2 + (y + 4)^2 = (2\sqrt{2})^2$$

$$x^2 + y^2 - 4x + 8y + 12 = 0$$

12 $x^2 + x - 4 = 0$

$$x = \frac{-1 \pm \sqrt{1 + 16}}{2}$$

$$x = \frac{-1 + \sqrt{17}}{2}$$

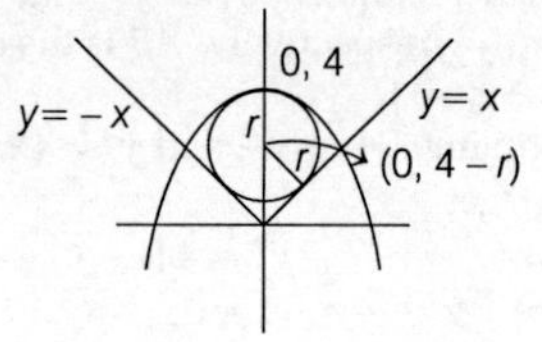

$$\frac{4 - r - 0}{\sqrt{2}} = r$$

$$4 - r = \pm \sqrt{2}r$$

$$r = \frac{4}{\sqrt{2} + 1} \quad \left(\because \frac{4}{1 - \sqrt{2}} < 0\right)$$

$$r = 4(\sqrt{2} - 1)$$

13 Let $P(x_1, y_1)$ and $Q(x_2, y_2)$.

Tangents at P and Q to the parabola $y^2 = 4ax$ are

$$yy_1 = 2a(x + x_1)$$

and $yy_2 = 2a(x + x_2)$.

$\therefore \quad y(y_1 - y_2) = 2a(x_1 - x_2)$

$$= \frac{2a(y_1^2 - y_2^2)}{4a}$$

$\Rightarrow \quad y_3 = \dfrac{y_1 + y_2}{2}$

$\Rightarrow \quad y_1, y_3, y_2$ are in AP.

14 Let (p, q), $p, q \in Z$ be an interior point of both the curves.

Then, $\quad p^2 + q^2 - 16 < 0$

and $q^2 - 4p < 0$, $p \geq 0$.

$\Rightarrow \quad p > (q/2)^2$ and $p^2 < 16 - q^2$.

$q = 0 \Rightarrow p = 1, 2, 3$

$q = 1 \Rightarrow p = 1, 2, 3$

$q = 2 \Rightarrow p = 2, 3$

$q = 3 \Rightarrow p =$ has no value.

$\therefore$ There only 8 points $(1, 0)$, $(2, 0)$, $(3, 0)$, $(1, 1)$, $(2, 1)$, $(2, 2)$ $(3, 1)$, $(3, 2)$ in upper half.

Due to symmetry about X-axis.

$(1, -1)$, $(2, -1)$, $(2, -2)$, $(3, -1)$, $(3, -2)$ are also interior points. Hence in all, three are 13 interior integral points.

15 Clearly, QR is the diameter of the required circle.

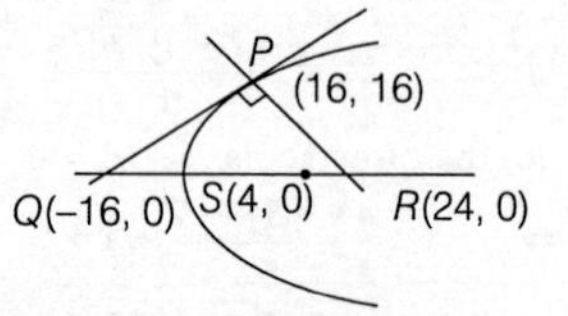

$16y = 8(x + 16) \Rightarrow Q = (-16, 0)$

$y - 16 = -2(x - 16) \Rightarrow R = (24, 0)$

$\therefore$ Equation of required circle is

$(x + 16)(x - 24) + y^2 = 0$

$\Rightarrow \quad x^2 + y^2 - 8x - 384 = 0$

DAY TWENTY EIGHT

Ellipse

Learning & Revision for the Day

- Concept of Ellipse
- Tangent to an Ellipse
- Normal to an Ellipse
- Auxiliary Circle
- Eccentric Angle of a Point
- Diameter and Conjugate Diameters

Concept of Ellipse

Ellipse is the locus of a point in a plane which moves in such a way that the ratio of its distances from a fixed point (focus) in the same plane to its distance from a fixed straight line (directrix) is always constant, which is always less than unity.

Equations of Ellipse in Standard Form

Different forms of an ellipse and their equations are given below

1. Ellipse of the Form $\frac{x^2}{a^2}+\frac{y^2}{b^2}=1,\ 0<b<a$

If the coefficient of x^2 has the larger denominator, then its major axis lies along the X-axis and it is said to be horizontal ellipse as shown below.

Various elements of horizontal ellipse are as follows

(i) Centre, $O(0, 0)$

(ii) Coordinates of the vertices : $A(a, 0)$ and $A_1(-a, 0)$

(iii) Equation of the major axis, $y=0$

(iv) Equation of the minor axis, $x=0$

(v) Focal distance of a point (x, y) is $a \pm ex$.

(vi) Major axis, $AA_1 = 2a$, Minor axis, $BB_1 = 2b$

(vii) Foci are $S(ae, 0)$ and $S_1(-ae, 0)$.

(viii) Equations of directrices are $l : x=\frac{a}{e}$, $l' : x=-\frac{a}{e}$

(ix) Length of latusrectum, $LL_1 = L'L_1' = \frac{2b^2}{a}$

(x) Eccentricity, $e=\sqrt{1-\frac{b^2}{a^2}}$

(xi) Sum of focal distances of a point (x, y) is $2a$.

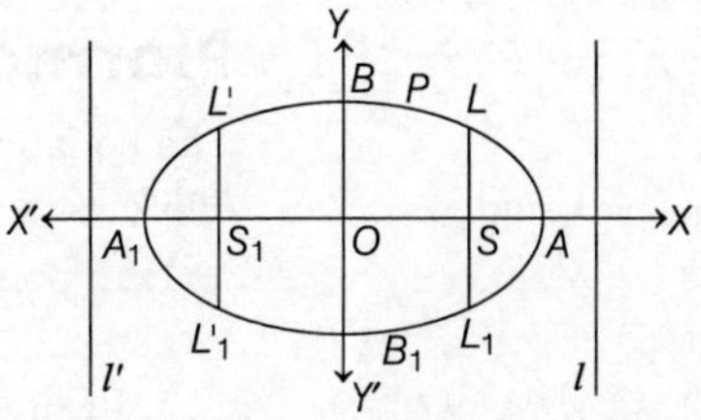

PREP MIRROR
Your Personal Preparation Indicator

- No. of Questions in Exercises (x)—
- No. of Questions Attempted (y)—
- No. of Correct Questions (z)—
  (Without referring Explanations)

- Accuracy Level $(z/y \times 100)$—
- Prep Level $(z/x \times 100)$—

In order to expect good rank in JEE, your Accuracy Level should be above 85 & Prep Level should be above 75.

2. Ellipse of the Form $\frac{x^2}{a^2}+\frac{y^2}{b^2}=1, 0<a<b$

If the coefficient of x^2 has the smaller denominator, then its major axis lies along the Y-axis and it is said to be vertical ellipse as shown below.

(i) Centre $O(0, 0)$
(ii) Coordinates of the vertices $B(0, b)$ and $B_1(0, -b)$.
(iii) Equation of the major axis, $x = 0$
(iv) Equation of the minor axis, $y = 0$
(v) Focal distance of a point (x, y) is $b \pm ey$.
(vi) Major axis, $BB_1 = 2b$, Minor axis, $AA_1 = 2a$
(vii) Foci are $S(0, be)$ and $S_1(0, -be)$.
(viii) Equation of directrices are

$$l: y = \frac{b}{e}, l': y = -\frac{b}{e}.$$

(ix) Length of latusrectum,

$$LL_1 = L'L_1' = \frac{2a^2}{b}$$

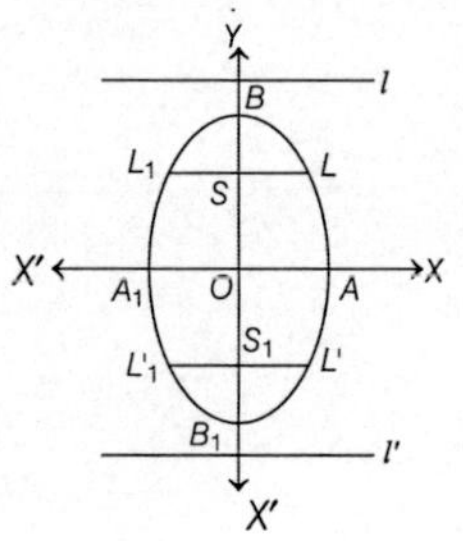

(x) Eccentricity, $e = \sqrt{1-\frac{a^2}{b^2}}$
(xi) Sum of focal distances of a point (x, y) is $2b$.

Results on Ellipse $\left(\frac{x^2}{a^2}+\frac{y^2}{b^2}=1\right)$

(i) The equations $x = a\cos\theta$, $y = b\sin\theta$ taken together are called the parametric equations of the ellipse $\frac{x^2}{a^2}+\frac{y^2}{b^2}=1$, where θ is the parameter.
(ii) A point (x_1, y_1) with respect to ellipse 'S' lie inside, on or outside the ellipse, if $S_1 < 0, S_1 = 0$ or $S_1 > 0$

where, $S_1 \equiv \frac{x_1^2}{a^2}+\frac{y_1^2}{b^2}-1$

(iii) Locus of mid-points of focal chords of an ellipse $\frac{x^2}{a^2}+\frac{y^2}{b^2}=1, (a>b)$ is $\frac{x^2}{a^2}+\frac{y^2}{b^2}=\frac{ex}{a}$.
(iv) Let $P(a\cos\theta, b\sin\theta)$ and $Q(a\cos\phi, b\sin\phi)$ be any two points of the ellipse $\frac{x^2}{a^2}+\frac{y^2}{b^2}=1$ $(a>b)$. Then, equation of the chord joining these two points is

$$\frac{x}{a}\cos\left(\frac{\theta+\phi}{2}\right)+\frac{y}{b}\sin\left(\frac{\theta+\phi}{2}\right)=\cos\left(\frac{\theta-\phi}{2}\right)$$

Tangent to an Ellipse

The equation of tangent to an ellipse for different forms are given below.

(i) In point (x_1, y_1) form, $\frac{xx_1}{a^2}+\frac{yy_1}{b^2}=1$.
(ii) In slope 'm' form, $y = mx \pm \sqrt{a^2m^2+b^2}$. and the point of contact is $\left(\pm\frac{a^2m}{\sqrt{a^2m^2+b^2}}, \mp\frac{b^2}{\sqrt{a^2m^2+b^2}}\right)$
(iii) In parametric $(a\cos\theta, b\sin\theta)$ form, $\frac{x}{a}\cos\theta+\frac{y}{b}\sin\theta=1$.

Results on Tangent to an Ellipse

(i) The line $lx+my+n=0$ is a tangent to the ellipse $\frac{x^2}{a^2}+\frac{y^2}{b^2}=1$, then $n^2 = a^2l^2+b^2m^2$
(ii) The line $y = mx + c$ touches an ellipse, iff $c^2 = a^2m^2+b^2$ and the point of contact is $\left(\pm\frac{a^2m}{c}, \mp\frac{b^2}{c}\right)$.
(iii) The equation of pair of tangents drawn from an external point $P(x_1, y_1)$ to the ellipse is $SS_1 = T^2$,

where $S \equiv \frac{x^2}{a^2}+\frac{y^2}{b^2}-1$, $S_1 \equiv \frac{x_1^2}{a^2}+\frac{y_1^2}{b^2}-1$

and $T \equiv \frac{xx_1}{a^2}+\frac{yy_1}{b^2}-1$

(iv) The equation of chord of contact of tangents is $\frac{xx_1}{a^2}+\frac{yy_1}{b^2}=1$ or $T = 0$.
(v) The equation of chord of an ellipse, whose mid-point is (x_1, y_1), is $T = S_1$, i.e. $\frac{xx_1}{a^2}+\frac{yy_1}{b^2}=\frac{x_1^2}{a^2}+\frac{y_1^2}{b^2}$
(vi) The locus of the point of intersection of perpendicular tangents to the ellipse is a director circle, i.e. $x^2+y^2=a^2+b^2$.
(vii) The point of intersection of the tangents at α and β is

$$\left(\frac{a\cos\frac{\alpha+\beta}{2}}{\cos\frac{\alpha-\beta}{2}}, \frac{b\sin\frac{\alpha+\beta}{2}}{\cos\frac{\alpha-\beta}{2}}\right).$$

Normal to an Ellipse

The equations of normal in different form to an ellipse are given below

1. In point (x_1, y_1) form, $\frac{a^2x}{x_1}-\frac{b^2y}{y_1}=a^2-b^2$
2. In slope 'm' form, $y = mx \pm \frac{m(a^2-b^2)}{\sqrt{a^2+b^2m^2}}$ at the points $\left(\pm\frac{a^2}{\sqrt{a^2+b^2m^2}}, \pm\frac{b^2m}{\sqrt{a^2+b^2m^2}}\right)$.
3. In parametric $(a\cos\theta, b\sin\theta)$ form,

$$ax\sec\theta - by\,\mathrm{cosec}\,\theta = a^2-b^2.$$

4. The point of intersection of normals to the ellipse at two points $(a\cos\theta_1, b\sin\theta_1)$ and $(a\cos\theta_2, b\sin\theta_2)$ are (λ, μ),

where $\lambda = \dfrac{(a^2-b^2)}{(a)} \cdot \cos\theta_1 \cdot \cos\theta_2 \cdot \dfrac{\cos\left(\dfrac{\theta_1+\theta_2}{2}\right)}{\cos\left(\dfrac{\theta_1-\theta_2}{2}\right)}$

and $\mu = -\dfrac{(a^2-b^2)}{(b)} \cdot \sin\theta_1 \cdot \sin\theta_2 \cdot \dfrac{\sin\left(\dfrac{\theta_1+\theta_2}{2}\right)}{\cos\left(\dfrac{\theta_1-\theta_2}{2}\right)}$

Results on Normal to an Ellipse

(i) The line $lx + my + n = 0$ is a normal to the ellipse $\dfrac{x^2}{a^2}+\dfrac{y^2}{b^2}=1$, then $\dfrac{a^2}{l^2}+\dfrac{b^2}{m^2}=\dfrac{(a^2-b^2)^2}{n^2}$

(ii) Four normals can be drawn from a point to an ellipse.

(iii) If the line $y = mx + c$ is a normal to the ellipse $\dfrac{x^2}{a^2}+\dfrac{y^2}{b^2}=1$, then $c^2 = \dfrac{m^2(a^2-b^2)^2}{a^2+b^2m^2}$ is the condition of normality of the line to the ellipse.

(iv) The points on the ellipse, the normals at which to the ellipse pass through a given point are called **conormal** points.

(v) Tangent at an end of a latusrectum (Ist quadrant) is $\dfrac{ex}{a}+\dfrac{\sqrt{1-e^2}}{b}y=1$, or $ex + y = a$ and normal is $\dfrac{ax}{e}-\dfrac{by}{\sqrt{1-e^2}}=a^2-b^2$ or $x - ye = ae^3$

Auxiliary Circle

The circle described on the major axis of an ellipse as diameter is called an auxiliary circle.

If $\dfrac{x^2}{a^2}+\dfrac{y^2}{b^2}=1$ is an ellipse, then its auxiliary circle is $x^2+y^2=a^2$

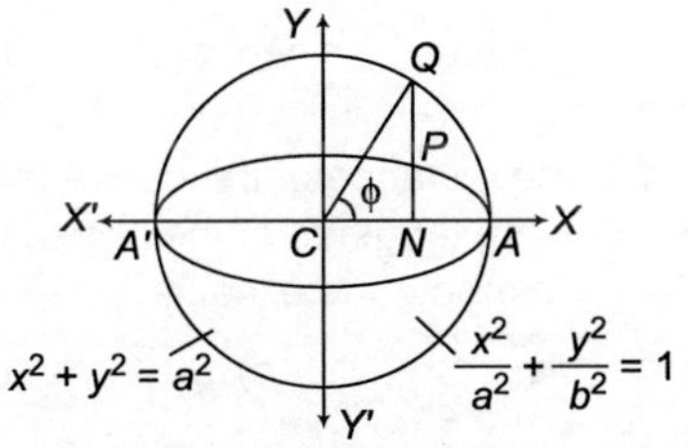

Eccentric Angle of a Point

Let P be any point on the ellipse $\dfrac{x^2}{a^2}+\dfrac{y^2}{b^2}=1$. Draw PN perpendicular from P on the major axis of the ellipse and produce NP to meet the auxiliary circle in Q. Then, $\angle XCQ = \phi$ is called the eccentric angle of the point P on the ellipse.

So, the coordinates of Q and P are $(a\cos\phi, a\sin\phi)$ and $(a\cos\phi, b\sin\phi)$, where ϕ is an eccentric angle.

Results on Eccentric Angles

(i) Eccentric angles of the extremities of latusrectum of the ellipse $\dfrac{x^2}{a^2}+\dfrac{y^2}{b^2}=1$ are $\tan^{-1}\left(\pm\dfrac{b}{ae}\right)$.

(ii) A circle cut an ellipse in four points real or imaginary. The sum of the eccentric angles of these four concyclic points on the ellipse is an even multiple of π.

(iii) The sum of the eccentric angles of the conormal points on the ellipse $\dfrac{x^2}{a^2}+\dfrac{y^2}{b^2}=1$ is an odd multiple of π.

Diameter and Conjugate Diameters

The locus of the mid-points of a system of parallel chords is called a **diameter.** If $y = mx + c$ represents a system of parallel chords of an ellipse $\dfrac{x^2}{a^2}+\dfrac{y^2}{b^2}=1$, then the line $y = -\dfrac{b^2}{a^2m}x$ is the equation of the diameter.

The two diameters are said to be **conjugate diameters,** when each bisects all chords parallel to the other.

If $y = mx$ and $y = m_1x$ be two conjugate diameters of an ellipse, then $m\,m_1 = -\dfrac{b^2}{a^2}$

Results on Conjugate Diameters

(i) The area of a parallelogram formed by the tangents at the ends of conjugate diameters of an ellipse is constant and is equal to the product of the axes.

(ii) The sum of the squares of any two conjugate semi-diameters of an ellipse is constant and equal to the sum of the squares of the semi-axes of the ellipse, i.e. $CP^2 + CD^2 = a^2 + b^2$.

(iii) The product of the focal distance of a point on an ellipse is equal to the square of the semi-diameter which is conjugate to the diameter through the point.

Important Points

(i) The eccentric angles of the ends of a pair of conjugate diameter of an ellipse differ by a right angle.

(ii) The tangents at the ends of a pair of conjugate diameters of an ellipse form a parallelogram.

Conjugate Points Two points P and Q are conjugate points with respect to an ellipse, if the polar of P passes through Q and the polar of Q passes through P.

Conjugate Lines Two lines are said to be conjugate lines with respect to an ellipse, if each passes through the pole of the polar.

Pole and Polar Let P be a point inside or outside an ellipse. Then, the locus of the point of intersection of tangents to the ellipse at the point, where secants drawn through 'P' intersect the ellipse is called the **polar** of point P with respect to the ellipse and the point P is called the **pole** of the polar. The polar of a point (x_1, y_1) with respect to the ellipse $\dfrac{x^2}{a^2}+\dfrac{y^2}{b^2}=1$ is $\dfrac{xx_1}{a^2}+\dfrac{yy_1}{b^2}=1$

DAY PRACTICE SESSION 1

FOUNDATION QUESTIONS EXERCISE

1 The equation $\frac{x^2}{8-a}+\frac{y^2}{a-2}=1$ will represent an ellipse if

(a) $a \in (1, 4)$ (b) $a \in (-\infty, 2) \cup (8, \infty)$
(c) $a \in (2, 8)$ (d) None of these

2 Equation of ellipse whose minor axis is equal to the distance between the foci and whose latusrectum is 10, is given by (take origin as centre and major axis along X-axis)

(a) $2x^2 + y^2 = 100$ (b) $x^2 + 2y^2 = 100$
(c) $2x^2 + y^2 = 50$ (d) None of these

3 Let AB be a rod of length 4 units with A on x and B on Y-axis. Rod AB slides on axes. If point P divides AB in the ratio $1:2$, locus of P is an ellipse. The eccentricity of this ellipse is

(a) $\frac{3}{4}$ (b) $\frac{\sqrt{3}}{2}$ (c) $\frac{2}{3}$ (d) None

4 Equation of the ellipse whose axes are the axes of coordinates and which passes through the point $(-3, 1)$ and has eccentricity $\sqrt{\frac{2}{5}}$ is

(a) $5x^2 + 3y^2 - 48 = 0$ (b) $3x^2 + 5y^2 - 15 = 0$
(c) $5x^2 + 3y^2 - 32 = 0$ (d) $3x^2 + 5y^2 - 32 = 0$

5 If the angle between the straight lines joining foci and the end of minor axis of the ellipse $\frac{x^2}{a^2}+\frac{y^2}{b^2}=1$ is 90°, then its eccentricity is

(a) $\frac{1}{2}$ (b) $\frac{1}{4}$ (c) $\frac{1}{\sqrt{2}}$ (d) None

6 The ellipse $4x^2 + 9y^2 = 36$ and the straight line $y = mx + c$ intersect in real points only if

(a) $9m^2 \le c^2 - 4$ (b) $9m^2 > c^2 - 4$
(c) $9m^2 \ge c^2 - 4$ (d) None of these

7 If the line $3x + 4y = \sqrt{7}$ touches the ellipse $3x^2 + 4y^2 = 1$ then the point of contact is

(a) $\left(\frac{1}{\sqrt{7}}, \frac{1}{\sqrt{7}}\right)$ (b) $\left(\frac{1}{\sqrt{3}}, \frac{-1}{\sqrt{3}}\right)$
(c) $\left(\frac{1}{\sqrt{7}}, \frac{-1}{\sqrt{7}}\right)$ (d) None of these

8 The equation of common tangent of the curves $x^2 + 4y^2 = 8$ and $y^2 = 4x$ are

(a) $x - 2y + 4 = 0,\ x + 2y + 4 = 0$
(b) $2x - y + 4 = 0,\ 2x + y + 4 = 0$
(c) $2x - y + 2 = 0,\ 2x + y + 2 = 0$
(d) None of the above

9 Tangents are drawn to the ellipse $x^2 + 2y^2 = 4$ from any arbitrary point on the line $x + y = 6$. The corresponding chord of contact will always pass through the fixed point

(a) $\left(\frac{1}{3}, \frac{2}{3}\right)$ (b) $\left(\frac{2}{3}, \frac{1}{3}\right)$
(c) $\left(\frac{1}{3}, \frac{1}{3}\right)$ (d) no such fixed point exist.

10 The equation of the circle passing through the foci of the ellipse $\frac{x^2}{16}+\frac{y^2}{9}=1$ and having centre at (0, 3) is

→ JEE Mains 2013

(a) $x^2 + y^2 - 6y - 7 = 0$ (b) $x^2 + y^2 - 6y + 7 = 0$
(c) $x^2 + y^2 - 6y - 5 = 0$ (d) $x^2 + y^2 - 6y + 5 = 0$

11 If tangent at any point P on the ellipse $7x^2 + 16y^2 = 12$ cuts the tangent at the end points of the major axis at the points A and B, then the circle with AB as diameter passes through a fixed point whose coordinates are

(a) $(\pm \sqrt{a^2 - b^2}, 0)$ (b) $(\pm \sqrt{a^2 + b^2}, 0)$
(c) $(0, \pm \sqrt{a^2 - b^2})$ (d) $(0, \pm \sqrt{a^2 + b^2})$

12 The length of the common tangent to the ellipse $\frac{x^2}{25}+\frac{y^2}{4}=1$ and the circle $x^2 + y^2 = 16$ intercepted by the coordinate axes is

(a) 5 (b) $2\sqrt{7}$ (c) $\frac{7}{\sqrt{3}}$ (d) $\frac{14}{\sqrt{3}}$

13 The distance of the centre of ellipse $x^2 + 2y^2 - 2 = 0$ to those tangents of the ellipse which are equally inclined from both the axes, is

(a) $\frac{3}{\sqrt{2}}$ (b) $\sqrt{3/2}$ (c) $\sqrt{2}/3$ (d) $\frac{\sqrt{3}}{2}$

14 PQ is a chord of the ellipse through the centre. If the square of its length is the HM of the squares of major and minor axes, find its inclination with X-axis.

(a) $\frac{\pi}{4}$ (b) $\frac{\pi}{2}$
(c) $\frac{2\pi}{3}$ (d) None of these

15 If straight line $ax + by = c$ is a normal to the ellipse $4x^2 + 9y^2 = 36$, then $4a^2 + 9b^2$ is equal to

(a) $\frac{169a^2b^2}{c^2}$ (b) $\frac{25a^2b^2}{c^2}$
(c) $\frac{13a^2b^2}{c^2}$ (d) None of these

16 If the normal at the point $P(\theta)$ to the ellipse $5x^2+14y^2=70$ intersects it again at the point $Q(2\theta)$, then $\cos\theta$ is equal to

(a) $\frac{2}{3}$ (b) $-\frac{2}{3}$ (c) $\frac{1}{3}$ (d) $-\frac{1}{3}$

17 An ellipse is drawn by taking a diameter of the circle $(x-1)^2+y^2=1$ as its semi-minor axis and a diameter of the circle $x^2+(y-2)^2=4$ as its semi-major axis. If the centre of the ellipse is at the origin and its axes are the coordinate axes, then the equation of the ellipse is

(a) $4x^2+y^2=4$ (b) $x^2+4y^2=8$
(c) $4x^2+y^2=8$ (d) $x^2+4y^2=16$

18 The area (in sq units) of the quadrilateral formed by the tangents at the end points of the latusrectum to the ellipse $\frac{x^2}{9}+\frac{y^2}{5}=1$ is → JEE Mains 2015

(a) $\frac{27}{4}$ (b) 18 (c) $\frac{27}{2}$ (d) 27

19 The locus of the foot of perpendicular drawn from the centre of the ellipse $x^2+3y^2=6$ on any tangent to it is → JEE Mains 2014

(a) $(x^2-y^2)^2=6x^2+2y^2$ (b) $(x^2-y^2)^2=6x^2-2y^2$
(c) $(x^2+y^2)^2=6x^2+2y^2$ (d) $(x^2+y^2)^2=6x^2-2y^2$

20 Tangent is drawn to the ellipse $\frac{x^2}{27}+y^2=1$ at $(3\sqrt{3}\cos\theta,\sin\theta)$ (where, $\theta\in(0,\pi/2)$). Then, the value of θ such that the sum of intercepts on axes made by this tangent is minimum, is

(a) $\frac{\pi}{3}$ (b) $\frac{\pi}{6}$ (c) $\frac{\pi}{8}$ (d) $\frac{\pi}{4}$

21 If the tangent at a point $(a\cos\theta, b\sin\theta)$ on the ellipse $\frac{x^2}{a^2}+\frac{y^2}{b^2}=1$ meets the auxiliary circle in two points, the chord joining them subtends a right angle at the centre, then the eccentricity of the ellipse is given by

(a) $(1+\cos^2\theta)^{-1/2}$ (b) $(1+\sin^2\theta)$
(c) $(1+\sin^2\theta)^{-1/2}$ (d) $(1+\cos^2\theta)$

22 Chord of contact of tangents drawn from the point $P(h,k)$ to the ellipse $x^2+4y^2=4$ subtends a right angle at the centre. Locus of the point P is

(a) $x^2+16y^2=20$ (b) $x^2+8y^2=10$
(c) $16x^2+y^2=20$ (d) None of these

23 Tangent at a point P on the ellipse $x^2+4y^2=4$ meets the X-axis at B and AP is ordinate of P. If Q is a point on AP produced such that $AQ=AB$, then locus of Q is

(a) $x^2+xy-4=0$ (b) $x^2-xy+4=0$
(c) $x^2+xy-1=0$ (d) $x^2-xy+1=0$

24 A parabola is drawn whose focus is one of the foci of the ellipse $\frac{x^2}{a^2}+\frac{y^2}{b^2}=1$ (where, $a>b$) and whose directrix passes through the other focus and perpendicular to the major axis of the ellipse. Then, the eccentricity of the ellipse for which the latusrectum of the ellipse and the parabola are same, is

(a) $\sqrt{2}-1$ (b) $2\sqrt{2}+1$ (c) $\sqrt{2}+1$ (d) $2\sqrt{2}-1$

25 At a point P on the ellipse $\frac{x^2}{a^2}+\frac{y^2}{b^2}=1$ tangent PQ is drawn. If the point Q be at a distance $1/p$ from the point P, where p is distance of the tangent from the origin, then the locus of the point Q is

(a) $\frac{x^2}{a^2}+\frac{y^2}{b^2}=1+\frac{1}{a^2b^2}$ (b) $\frac{x^2}{a^2}-\frac{y^2}{b^2}=1-\frac{1}{a^2b^2}$
(c) $\frac{x^2}{a^2}+\frac{y^2}{b^2}=\frac{1}{a^2b^2}$ (d) $\frac{x^2}{a^2}-\frac{y^2}{b^2}=\frac{1}{a^2b^2}$

26 If CP and CD are semi-conjugate diameters of an ellipse $\frac{x^2}{14}+\frac{y^2}{8}=1$, then CP^2+CD^2 is equal to

(a) 20 (b) 22 (c) 24 (d) 26

27 If θ and ϕ are eccentric angles of the ends of a pair of conjugate diameters of the ellipse $\frac{x^2}{a^2}+\frac{y^2}{b^2}=1$, then $\theta-\phi$ is equal to

(a) $\pm\frac{\pi}{2}$ (b) $\pm\pi$ (c) 0 (d) None of these

28 The coordinates of all the points P on the ellipse $\frac{x^2}{a^2}+\frac{y^2}{b^2}=1$, for which the area of the ΔPON is maximum, where O denotes the origin and N, the foot of the perpendicular from O to the tangent at P, is

(a) $\left(\pm\frac{a^2}{\sqrt{a^2+b^2}},\pm\frac{b^2}{\sqrt{a^2+b^2}}\right)$
(b) $\left(\pm\frac{a^2}{\sqrt{a^2-b^2}},\pm\frac{b^2}{\sqrt{a^2-b^2}}\right)$
(c) $\left(\pm\frac{2a^2}{\sqrt{a^2+b^2}},\pm\frac{2b^2}{\sqrt{a^2+b^2}}\right)$
(d) $\left(\frac{2a^2}{\sqrt{a^2-b^2}},\frac{2b^2}{\sqrt{a^2-b^2}}\right)$

29 If α, β, γ are the eccentric angles of three points on the ellipse $4x^2+9y^2=36$, the normals at which are concurrent, then $\sin(\alpha+\beta)+\sin(\beta+\gamma)+\sin(\gamma+\alpha)$ is equal to

(a) $\frac{2}{3}$ (b) $\frac{4}{9}$ (c) 0 (d) None

30 A triangle is drawn such that it is right angled at the centre of the ellipse $\frac{x^2}{a^2}+\frac{y^2}{b^2}=1$ (where, $a>b$) and its other two vertices lie on the ellipse with eccentric angles α and β. Then, $\dfrac{1-e^2\cos^2\left(\frac{\alpha+\beta}{2}\right)}{\cos^2\left(\frac{\alpha-\beta}{2}\right)}$ is equal to

(a) $\frac{a^2}{a^2+b^2}$ (b) $\frac{a^2+b^2}{a^2}$ (c) $\frac{a^2}{a^2-b^2}$ (d) $\frac{a^2-b^2}{a^2}$

Directions (Q. Nos. 31-35) *Each of these questions contains two statements : Statement I (Assertion) and Statement II (Reason). Each of these questions also has four alternative choices, only one of which is the correct answer. You have to select one of the codes* (a), (b), (c) and (d) *given below.*

(a) Statement I is true, Statement II is true; Statement II is a correct explanation for Statement I
(b) Statement I is true, Statement II is true; Statement II is not a correct explanation for Statement I
(c) Statement I is true; Statement II is false
(d) Statement I is false; Statement II is true

31 Let $S_1 \equiv (x-1)^2+(y-2)^2=0$ and $S_2 \equiv (x+2)^2+(y-1)^2=0$ be the equations of two circles.

Statement I Locus of centre of a variable circle touching two circles S_1 and S_2 is an ellipse.

Statement II If a circle $S_1=0$ lies completely inside the circle $S_2=0$, then locus of centre of variable circle $S=0$ which touches both the circles is an ellipse.

32 **Statement I** If $P\left(\frac{3\sqrt{3}}{2}, 1\right)$ is a point on the ellipse $4x^2+9y^2=36$. Circle drawn AP as diameter touches another circle $x^2+y^2=9$, where $A \equiv (-\sqrt{5}, 0)$.

Statement II Circle drawn with focal radius as diameter touches the auxiliary circle.

33 **Statement I** The product of the focal distances of a point on an ellipse is equal to the square of the semi-diameter which is conjugate to the diameter through the point.

Statement II If $y=mx$ and $y=m_1x$ are two conjugate diameters of an ellipse, then $mm_1=-\frac{b^2}{a^2}$.

34 **Statement I** The condition on a and b for which two distinct chords of the ellipse $\frac{x^2}{2a^2}+\frac{y^2}{2b^2}=1$ passing through $(a,-b)$ are bisected by the line $x+y=b$ is $a^2+6ab-7b^2 \geq 0$.

Statement II Equation of chord of the ellipse $\frac{x^2}{a^2}+\frac{y^2}{b^2}=1$ whose mid-point is (x_1, y_1), is $T=S_1$.

35 **Statement I** An equation of a common tangent to the parabola $y^2=16\sqrt{3}x$ and the ellipse $2x^2+y^2=4$ is $y=2x+2\sqrt{3}$.

Statement II If the line $y=mx+\frac{4\sqrt{3}}{m}$, $(m \neq 0)$ is a common tangent to the parabola $y^2=16\sqrt{3}x$ and the ellipse $2x^2+y^2=4$, then m satisfies $m^4+2m^2=24$.

DAY PRACTICE SESSION 2

PROGRESSIVE QUESTIONS EXERCISE

1 The eccentricity of an ellipse whose centre is at origin is $\frac{1}{2}$. If one of its directrices is $x=-4$, then the equation of the normal to it at $\left(1, \frac{3}{2}\right)$ is

→ JEE Mains 2017

(a) $4x+2y=7$ (b) $x+2y=4$
(c) $2y-x=2$ (d) $4x-2y=1$

2 The locus of the mid-points of a focal chord of the ellipse $\frac{x^2}{a^2}+\frac{y^2}{b^2}=1$ is

(a) $\frac{x^2}{a^2}+\frac{y^2}{b^2}=\frac{ex}{a}$ (b) $\frac{x^2}{a^2}-\frac{y^2}{b^2}=\frac{ex}{a}$
(c) $x^2+y^2=a^2+b^2$ (d) None of these

3 Foot of normal to the ellipse $4x^2+9y^2=36$ having slope 2 may be

(a) $\left(\frac{9}{5}, \frac{-8}{5}\right)$ (b) $\left(\frac{9}{5}, \frac{8}{5}\right)$
(c) $\left(\frac{-9}{5}, \frac{8}{5}\right)$ (d) None of these

4 On the ellipse $4x^2+9y^2=1$, the points at which the tangents are parallel to the line $8x=9y$ are

(a) $\left(\frac{2}{5}, \frac{1}{5}\right), \left(\frac{-2}{5}, \frac{1}{5}\right)$ (b) $\left(\frac{-2}{5}, \frac{1}{5}\right), \left(\frac{2}{5}, \frac{-1}{5}\right)$
(c) $\left(\frac{-3}{5}, \frac{-1}{5}\right), \left(\frac{3}{5}, \frac{1}{5}\right)$ (d) None of these

5 The normal at a point P on the ellipse $x^2+4y^2=16$ meets the X-axis at Q. If M is the mid-point of the line segment PQ, then the locus of M intersects the latusrectums of the given ellipse at the points

(a) $\left(\pm\frac{3\sqrt{5}}{2}, \pm\frac{2}{7}\right)$ (b) $\left(\pm\frac{3\sqrt{5}}{2}, \pm\frac{\sqrt{19}}{4}\right)$
(c) $\left(\pm 2\sqrt{3}, \pm\frac{1}{7}\right)$ (d) $\left(\pm 2\sqrt{3}, \pm\frac{4\sqrt{3}}{7}\right)$

6 The line passing through the extremity A of the major axis and extremity B of the minor axis of ellipse $x^2+9y^2=9$ meets its auxiliary circle at the point M. Then the area of the triangle with vertices at AM and the origin O is

(a) $\frac{31}{10}$ (b) $\frac{29}{10}$ (c) $\frac{21}{10}$ (d) $\frac{27}{10}$

7 The ellipse $x^2 + 4y^2 = 4$ is inscribed in a rectangle alingent with the coordinate axes, which in turn is inscribed in another ellipse that passes through the point (4, 0). Then the equation of the ellipse is

(a) $x^2 + 12y^2 = 16$ (b) $4x^2 + 48y^2 = 48$
(c) $4x^2 + 64y^2 = 48$ (d) $x^2 + 16y^2 = 16$

8 Let $P(x_1, y_1)$ and $Q(x_2, y_2)$, $y_1 < 0$, $y_2 < 0$, be the end points of the latusrectum of the ellipse $x^2 + 4y^2 = 4$. The equations of parabolas with latusrectum PQ are

(a) $x^2 - 2\sqrt{3}y = 3 + \sqrt{3}$ (b) $x^2 + 2\sqrt{3}y = 3 - \sqrt{3}$
(c) both (a) and (b) (d) None

9 The ellipse $E_1 : \frac{x^2}{9} + \frac{y^2}{4} = 1$ is inscribed in a rectangle R whose sides are parallel to the co-ordinate axes. Another ellipse E_2 passing through the point (0, 4) circumscribes the rectangle R. The eccentricity of the ellipse E_2 is

(a) $\frac{\sqrt{2}}{2}$ (b) $\frac{\sqrt{3}}{2}$ (c) $\frac{1}{2}$ (d) $\frac{3}{4}$

10 If the line $px + qy = r$ intersects the ellipse $x^2 + 4y^2 = 4$ in points whose eccentric angles differ by $\frac{\pi}{3}$, then r^2 is equal to

(a) $\frac{3}{4}(4p^2 + q^2)$ (b) $\frac{4}{3}(4p^2 + q^2)$
(c) $\frac{2}{3}(4p^2 + q^2)$ (d) None of these

11 The sum of the square of the reciprocals of two perpendicular diameters of an ellipse $\frac{x^2}{a^2} + \frac{y^2}{b^2} = 1$, is

(a) $\frac{1}{4}\left[\frac{1}{a^2} + \frac{1}{b^2}\right]$ (b) $\frac{1}{2}\left[\frac{1}{a^2} + \frac{1}{b^2}\right]$
(c) $\frac{1}{a^2} + \frac{1}{b^2}$ (d) None of these

12 The area of a triangle inscribed in an ellipse bears a constant ratio of the area of triangle formed by joining the corresponding points on the auxiliary circle of the vertices of the first triangle. This ratio is

(a) $\frac{a}{b}$ (b) $\frac{b}{a}$ (c) $\frac{a^2}{b^2}$ (d) $\frac{b^2}{a^2}$

13 If α, β are the eccentric angles of the extremities of a focal chord of the ellipse $16x^2 + 25y^2 = 400$, then $\tan\left(\frac{\alpha}{2}\right)\tan\left(\frac{\beta}{2}\right)$ is equal to

(a) -4 (b) $\frac{-1}{4}$ (c) $\frac{-3}{8}$ (d) None

14 The tangent and normal drawn to the ellipse $x^2 + 4y^2 = 4$ at the point $P(\theta)$ meets the X-axis at the points A and B. If $AB = 2$, then $\cos^2\theta$ is equal to

(a) $\frac{4}{9}$ (b) $\frac{8}{9}$ (c) $\frac{2}{9}$ (d) None

15 Given an ellipse $\frac{x^2}{a^2} + \frac{y^2}{b^2} = 1 (a > b)$ with foci at S and S' and vertices at A and A'. A tangent is drawn at any point P on the ellipse and let R, R', B, B' respectively be the feet of the perpendiculars drawn from S, S', A, A' on the tangent at P. Then, the ratio of the areas of the quadrilaterals $S'R'RS$ and $A'B'BA$ is

(a) $e:2$ (b) $e:3$
(c) $e:1$ (d) $e:4$

ANSWERS

SESSION 1	1 (d)	2 (b)	3 (b)	4 (d)	5 (c)	6 (c)	7 (a)	8 (a)	9 (b)	10 (a)
	11 (a)	12 (d)	13 (b)	14 (a)	15 (b)	16 (b)	17 (d)	18 (d)	19 (c)	20 (b)
	21 (c)	22 (a)	23 (a)	24 (a)	25 (a)	26 (b)	27 (a)	28 (a)	29 (c)	30 (b)
	31 (d)	32 (a)	33 (b)	34 (a)	35 (b)					
SESSION 2	1 (d)	2 (a)	3 (b)	4 (b)	5 (c)	6 (d)	7 (a)	8 (c)	9 (c)	10 (a)
	11 (a)	12 (b)	13 (b)	14 (a)	15 (c)					

Hints and Explanations

SESSION 1

1 $\frac{x^2}{8-a} + \frac{y^2}{a-2} = 1$ will represent an ellipse if

$$8 - a > 0,\ a - 2 > 0,\ 8 - a \neq a - 2$$

$\Rightarrow\ 2 < a < 8,\ a \neq 5$

$\Rightarrow\ a \in (2, 8) - \{5\}$

2 $2b = 2ae$

$\Rightarrow\quad a^2e^2 = b^2 = a^2(1 - e^2)$

$\Rightarrow\quad e^2 = \frac{1}{2}$

$\frac{2b^2}{a} = 10 \Rightarrow b^2 = 5a$

$\Rightarrow\ a^2(1 - 1/2) = 5a$

$\Rightarrow\quad a = 10$

$\therefore\quad b^2 = 50,\ a^2 = 100$

Hence, equation of ellipse is

$$\frac{x^2}{100} + \frac{y^2}{50} = 1$$

$\Rightarrow\quad x^2 + 2y^2 = 100$

3 Let $A = (a, 0)$, $B(0, b)$.

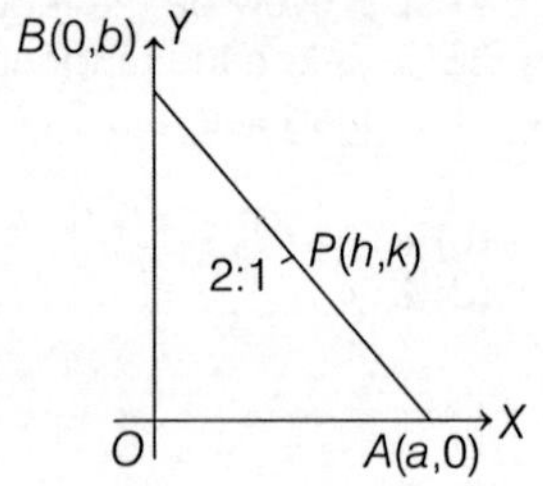

P divides AB in the ratio 1:2.

$\therefore\quad h = \frac{2a}{3},\ k = \frac{b}{3}$

$\therefore\ a^2 + b^2 = \frac{9h^2}{4} + 9k^2 = 16$ $\quad[\because AB = 4]$

$\Rightarrow$ Locus of P is $\frac{x^2}{\left(\frac{64}{9}\right)} + \frac{y^2}{\left(\frac{16}{9}\right)} = 1$

$\therefore\quad \frac{16}{9} = \frac{64}{9}(1 - e^2) \Rightarrow e^2 = \frac{3}{4}$ i.e. $e = \frac{\sqrt{3}}{2}$

4 Let the equation of ellipse is

$\frac{x^2}{a^2} + \frac{y^2}{b^2} = 1,\ (a > b)$

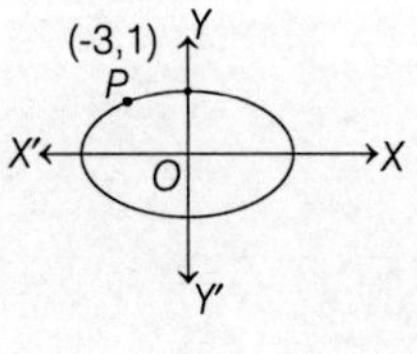

It passes through $p(-3,1)$ and $e = \sqrt{\frac{2}{5}}$.

$\therefore\quad \frac{9}{a^2} + \frac{1}{b^2} = 1 \quad$ ...(i)

and $\quad e^2 = 1 - \frac{b^2}{a^2}$

$\Rightarrow\quad \frac{2}{5} = 1 - \frac{b^2}{a^2} \Rightarrow \frac{b^2}{a^2} = \frac{3}{5}$

From Eq. (i), $\frac{9}{a^2} + \frac{5}{3a^2} = 1$

$\Rightarrow\quad \frac{27 + 5}{3a^2} = 1$

$\Rightarrow\quad a^2 = \frac{32}{3}$ and then $b^2 = \frac{32}{5}$

$\therefore$ Equation of ellipse is

$$\frac{3x^2}{32} + \frac{5y^2}{32} = 1 \Rightarrow 3x^2 + 5y^2 = 32$$

5 Let equation of ellipse be

$$\frac{x^2}{a^2} + \frac{y^2}{b^2} = 1$$

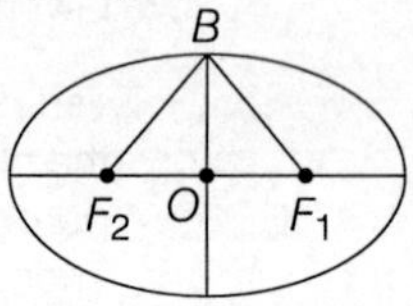

Then, $F_1 = (ae, 0)$, $F_2 = (-ae, 0)$, $B = (0, b)$.

$\angle F_1BF_2 = \frac{\pi}{2}$

$\Rightarrow\quad \left(-\frac{b}{ae}\right)\left(\frac{-b}{-ae}\right) = -1$

$\Rightarrow\quad a^2e^2 = b^2 = a^2(1 - e^2)$

$\Rightarrow\quad e^2 = \frac{1}{2} \Rightarrow e = \frac{1}{\sqrt{2}}$

6 Solving $4x^2 + 9y^2 = 36$ and $y = mx + c$, we get

$$4x^2 + 9(mx + c)^2 - 36 = 0$$

$\Rightarrow\quad (9m^2 + 4)x^2 + 18cmx + 9c^2 - 36 = 0$

Roots are real, so

$$18 \times 18c^2m^2 - 4(9m^2 + 4)(9c^2 - 36) \geq 0$$

$\Rightarrow\quad 9m^2 - c^2 + 4 \geq 0$

$\Rightarrow\quad 9m^2 \geq c^2 - 4$

7 Equation of ellipse is $3x^2 + 4y^2 = 1$

Tangent at $P(x_1, y_1)$ is

$3xx_1 + 4yy_1 = 1 \quad$ ...(i)

On comparing Eq. (i) with the given tangent $3x + 4y = \sqrt{7}$, $\frac{3x_1}{3} = \frac{4y_1}{4} = \frac{1}{\sqrt{7}} \Rightarrow$

$P(x_1, y_1) = P\left(\frac{1}{\sqrt{7}}, \frac{1}{\sqrt{7}}\right)$

Clearly, $P\left(\frac{1}{\sqrt{7}}, \frac{1}{\sqrt{7}}\right)$ lies on the ellipse, therefore it is a point of contact.

8 Any tangent to parabola $y^2 = 4x$ is

$$y = mx + \frac{1}{m}$$

It touches the ellipse $\frac{x^2}{8} + \frac{y^2}{2} = 1$

$\Rightarrow \frac{1}{m^2} = 8m^2 + 2\ [\because c^2 = a^2m^2 + b^2]$

$\Rightarrow 8m^4 + 2m^2 - 1 = 0$

$\Rightarrow m^2 = \frac{1}{4}$ i.e. $m = \pm\frac{1}{2}$

Hence, required tangent are

$$x - 2y + 4 = 0,\ x + 2y + 4 = 0$$

9 Any point on the line can be taken as $P(h, 6 - h)$

Equation of chord of contact of P w.r.t. $x^2 + 2y^2 = 4$ is $hx + 2(6 - h)y = 4$ $[T = 0]$

$\Rightarrow\ h(x - 2y) + 4(3y - 1) = 0$

which passes through the fixed point $y = \frac{1}{3}$, $x = \frac{2}{3}$ i.e., through $\left(\frac{2}{3}, \frac{1}{3}\right)$.

10 Given equation of ellipse is

$$\frac{x^2}{16} + \frac{y^2}{9} = 1.$$

Here, $a = 4$, $b = 3$, $e = \sqrt{1 - \frac{9}{16}} = \frac{\sqrt{7}}{4}$

$\therefore$ Foci is $(\pm\ ae, 0)$

$= \left(\pm 4 \times \frac{\sqrt{7}}{4}, 0\right) = (\pm\sqrt{7}, 0)$

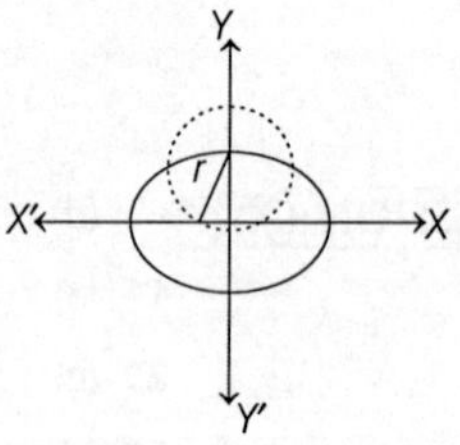

$\therefore\quad$ Radius of the circle,

$$r = \sqrt{(ae)^2 + b^2} = \sqrt{7 + 9}$$
$$= \sqrt{16} = 4$$

Now, the equation of circle is

$$(x - 0)^2 + (y - 3)^2 = 16$$

$\Rightarrow\quad x^2 + y^2 - 6y - 7 = 0$

11 Equation of tangent at any point $P(a\cos\theta, b\sin\theta)$ is $\frac{x}{a}\cos\theta + \frac{y}{b}\sin\theta = 1$.

The equation of tangents at the end

points of the major axis are $x = a, x = -a$.

$\therefore$ The intersection point of these tangents are

$$A = \left(a, b\tan\frac{\theta}{2}\right), B = \left(-a, b\cot\frac{\theta}{2}\right).$$

Equation of circle with AB as diameter

$$(x-a)(x+a) + \left(y - b\tan\frac{\theta}{2}\right)\left(y - b\cot\frac{\theta}{2}\right) = 0$$

$$\Rightarrow x^2 - a^2 + y^2 - by\left(\tan\frac{\theta}{2} + \cot\frac{\theta}{2}\right) + b^2 = 0$$

$$\Rightarrow x^2 + y^2 - a^2 + b^2 - 2by\operatorname{cosec}\theta = 0$$

which is the equation of family of circles passing through the point of intersection of the circle

$$x^2 + y^2 - a^2 + b^2 = 0 \text{ and } y = 0$$

So, the fixed point is $(\pm\sqrt{a^2 - b^2}, 0)$.

12 The tangent to $\dfrac{x^2}{25} + \dfrac{y^2}{4} = 1$ is $\dfrac{x}{5}\cos\theta + \dfrac{y}{2}\sin\theta = 1$. If it is also tangent to the circle, then

$$16 = \frac{1}{\dfrac{\cos^2\theta}{25} + \dfrac{\sin^2\theta}{4}} = \frac{100}{4 + 21\sin^2\theta}$$

$$\Rightarrow \sin^2\theta = \frac{3}{28}, \cos^2\theta = \frac{25}{28}$$

If the tangents meets the axes at A and B, then

$$A = \left(\frac{5}{\cos\theta}, 0\right) \text{ and } B = \left(0, \frac{2}{\sin\theta}\right)$$

$$\therefore AB^2 = \frac{25}{\cos^2\theta} + \frac{4}{\sin^2\theta}$$

$$= 28 + \frac{4}{3}\cdot 28 = \frac{196}{3} \Rightarrow AB = \frac{14}{\sqrt{3}}$$

13 Given, equation of ellipse is $\dfrac{x^2}{2} + \dfrac{y^2}{1} = 1$. General equation of tangent to the ellipse of slope m is

$$y = mx \pm \sqrt{2m^2 + 1}$$

Since, this is equally inclined to axes, so $m = \pm 1$.

Then, tangents are

$$y = \pm x \pm \sqrt{2+1} = \pm x \pm \sqrt{3}$$

Distance of any tangent from origin

$$= \frac{|0 + 0 \pm \sqrt{3}|}{\sqrt{1^2 + 1^2}} = \sqrt{3/2}$$

14 The straight line $x = r\cos\theta, y = r\sin\theta$ meets the ellipse $\dfrac{x^2}{a^2} + \dfrac{y^2}{b^2} = 1$, where r is given.

$$\therefore \frac{1}{r^2} = \frac{\cos^2\theta}{a^2} + \frac{\sin^2\theta}{b^2} \quad \text{...(i)}$$

But $PQ^2 = 4r^2 = \text{HM of } 4a^2, 4b^2$

$$\therefore \frac{1}{r^2} = \frac{1}{2a^2} + \frac{1}{2b^2} \quad \text{...(ii)}$$

From Eqs. (i) and (ii), we get

$$\frac{\cos^2\theta}{a^2} + \frac{\sin^2\theta}{b^2} = \frac{1}{2a^2} + \frac{1}{2b^2}$$

It is possible, when $\theta = \dfrac{\pi}{4}$

15 Given ellipse is $\dfrac{x^2}{9} + \dfrac{y^2}{4} = 0$, $(a = 3, b = 2)$

Equation of normal is

$3x\sec\theta - 2y\operatorname{cosec}\theta = 9 - 4 = 5$

Comparing it with given normal $ax + by = c$,

$$\frac{3\sec\theta}{a} = \frac{-2\operatorname{cosec}\theta}{b} = \frac{5}{c}$$

$$\Rightarrow \cos\theta = \frac{3c}{5a}, \sin\theta = -\frac{2c}{5b}$$

$$\Rightarrow \frac{9c^2}{25a^2} + \frac{4c^2}{25b^2} = 1$$

$$\Rightarrow 9b^2 + 4a^2 = \frac{25a^2b^2}{c^2}$$

16 Given ellipse is $\dfrac{x^2}{14} + \dfrac{y^2}{5} = 1$

$\therefore$ $P(\theta)$ is $(\sqrt{14}\cos\theta, \sqrt{5}\sin\theta)$

Equation of normal at $P(\theta)$ is

$$\frac{\sqrt{14}x}{\cos\theta} - \frac{\sqrt{5}y}{\sin\theta} = 14 - 5 = 9$$

$Q(2\theta)$ lies on this normal, therefore

$$\frac{14\cos 2\theta}{\cos\theta} - \frac{5\sin 2\theta}{\sin\theta} = 9$$

$$\Rightarrow 14\cos 2\theta\sin\theta - 10\sin\theta\cos^2\theta = 9\cos\theta\sin\theta$$

$$\Rightarrow 18\cos^2\theta - 9\cos\theta - 14 = 0$$

$$\Rightarrow (3\cos\theta + 2)(6\cos\theta - 7) = 0$$

$$\Rightarrow \cos\theta = \frac{-2}{3} \text{ or } \frac{7}{6}$$

$$\Rightarrow \cos\theta = \frac{-2}{3} \text{ as } \cos\theta \neq \frac{7}{6}$$

17 Given,

(i) An ellipse whose semi-minor axis coincides with one of the diameters of the circle $(x-1)^2 + y^2 = 1$.

(ii) The semi-major axis of the ellipse coincides with one of the diameters of circle $x^2 + (y-2)^2 = 4$.

(iii) The centre of the ellipse is at origin.

(iv) The axes of the ellipse are coordinate axes.

Now, diameter of circle $(x-1)^2 + y^2 = 1$ is 2 units and that of circle $x^2 + (y-2)^2 = 4$ is 4 units.

Semi-minor axis of ellipse, $b = 2$ units and semi-major axis of ellipse, $a = 4$ units.

Hence, the equation of the ellipse is

$$\frac{x^2}{a^2} + \frac{y^2}{b^2} = 1 \Rightarrow \frac{x^2}{16} + \frac{y^2}{4} = 1$$

$$\Rightarrow x^2 + 4y^2 = 16$$

18 Area of quadrilateral formed by tangents at the ends of latusrectum $= \dfrac{2a^2}{e}$.

Here, $a^2 = 9, b^2 = 5, e = \dfrac{2}{3}$

$$\therefore \text{Area} = 2 \times 9 \times \frac{3}{2} = 27$$

19 $x^2 + 3y^2 = 6$

$$\Rightarrow \frac{x^2}{6} + \frac{y^2}{2} = 1 \; (a^2 = 6, b^2 = 2) \quad \text{...(i)}$$

Equation of any tangent to Eqs. (i) is

$$y = mx \pm \sqrt{6m^2 + 2} \quad \text{...(ii)}$$

Equation of perpendicular line drawn from centre $(0, 0)$ to Eqs. (ii) is

$$y = -\frac{1}{m}x \quad \text{...(iii)}$$

Eliminating m from Eqs. (ii) and (iii), required locus of foot of perpendicular is

$$y = \left(-\frac{x}{y}\right)x \pm \sqrt{\left(6\frac{x^2}{y^2} + 2\right)}$$

$$\Rightarrow (y^2 + x^2)^2 = 6x^2 + 2y^2$$

20 The equation of tangent from the point $(3\sqrt{3}\cos\theta, \sin\theta)$ to the curve is

$$\frac{x\cos\theta}{3\sqrt{3}} + \frac{y\sin\theta}{1} = 1.$$

Thus, sum of intercepts

$= 3\sqrt{3}\sec\theta + \operatorname{cosec}\theta = f(\theta)$ [say]

$$\Rightarrow f'(\theta) = \frac{3\sqrt{3}\sin^3\theta - \cos^3\theta}{\sin^2\theta\cos^2\theta}$$

Put $f'(\theta) = 0$

$$\Rightarrow \sin^3\theta = \frac{1}{3^{3/2}}\cos^3\theta$$

$$\Rightarrow \tan\theta = \frac{1}{\sqrt{3}} \Rightarrow \theta = \frac{\pi}{6}$$

Now, $f''(\theta) > 0$, for $\theta = \pi/6$ [i.e. minimum]

Hence, value of θ is $\dfrac{\pi}{6}$.

21 Equation of tangent at $(a\cos\theta, b\sin\theta)$ to the ellipse is $\dfrac{x}{a}\cos\theta + \dfrac{y}{b}\sin\theta = 1$ ...(i)

The joint equation of the lines joining the points of intersection of Eq. (i) and the auxiliary circle $x^2 + y^2 = a^2$ to the origin, which is the centre of the circle,

is $x^2 + y^2 = a^2\left[\dfrac{x}{a}\cos\theta + \dfrac{y}{b}\sin\theta\right]^2$

Since, these lines are at right angles.

$$\therefore 1 - a^2\left(\frac{\cos^2\theta}{a^2}\right) + 1 - a^2\left(\frac{\sin^2\theta}{b^2}\right) = 0$$

[$\because$ coefficient of x^2 + coefficient of $y^2 = 0$]

$\Rightarrow \sin^2\theta\left(1-\dfrac{a^2}{b^2}\right)+1=0$

$\Rightarrow \sin^2\theta\,(b^2-a^2)+b^2=0$

$\Rightarrow \quad (1+\sin^2\theta)(a^2e^2)=a^2$

$\Rightarrow e=(1+\sin^2\theta)^{-1/2}$

22 Chord of contact of $P(h,k)$ to the ellipse $x^2+4y^2=4$ is

$hx+4ky-4=0 \qquad \text{...(i)}$

Making $x^2+4y^2-4=0$ homogeneous with Eq. (i), we get

$x^2+4y^2-4\left(\dfrac{hx+4ky}{4}\right)^2=0$

Chord of contact subtends right angles at the centre, so

coefficient of x^2 + coefficient of $y^2=0$

$\therefore \quad 1+4-\dfrac{1}{4}h^2-4k^2=0$

$\Rightarrow$ locus of $P(h,k)$ is $x^2+16y^2=20$

23 Ellipse is $x^2+4y^2=4$. $P=(2\cos\theta,\sin\theta)$

Equation of tangent at P is

$2\cos\theta x+4y\sin\theta=4$

It meets X-axis at B.

$\therefore B=(2\sec\theta,0)$, $A=(2\cos\theta,0)$

$AQ=AB=2\,|\sec\theta-\cos\theta|=\dfrac{2\sin^2\theta}{\cos\theta}$

Let $Q=(h,k)$. Then,

$h=2\cos\theta,\ k=\dfrac{2\sin^2\theta}{\cos\theta}=\dfrac{2(1-\cos^2\theta)}{\cos\theta}$

$\Rightarrow hk=4\left(\dfrac{1-h^2}{4}\right)$

$\therefore$ Locus of Q is $xy=4-x^2$.

$\Rightarrow \quad x^2+xy-4=0$

24 Equation of the ellipse is $\dfrac{x^2}{a^2}+\dfrac{y^2}{b^2}=1$.

Equation of the parabola with focus $S(ae,0)$ and directrix $x+ae=0$ is $y^2=4aex$.

Now, length of latusrectum of the ellipse is $\dfrac{2b^2}{a}$ and that of the parabola is $4ae$.

For the two latusrectum to be equal,

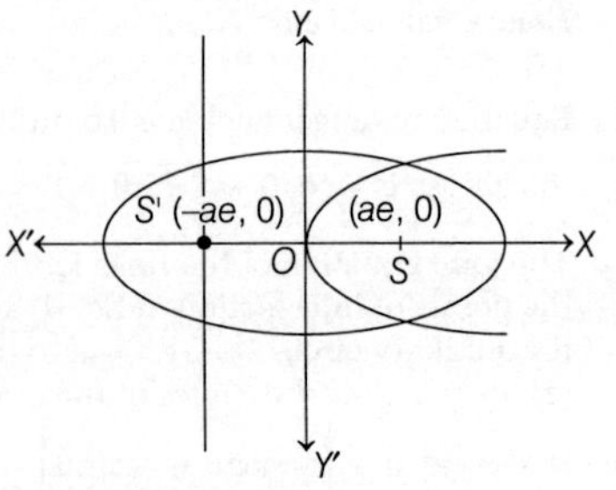

$\dfrac{2b^2}{a}=4ae \Rightarrow \dfrac{2a^2(1-e^2)}{a}=4ae$

$\Rightarrow 1-e^2=2e \Rightarrow e^2+2e-1=0$

Therefore, $e=\dfrac{-2\pm\sqrt{8}}{2}=-1\pm\sqrt{2}$

Hence, $e=\sqrt{2}-1$ as $0<e<1$ for ellipse.

25 Equation of the tangent at P is

$\dfrac{x-a\cos\theta}{a\sin\theta}=\dfrac{y-b\sin\theta}{-b\cos\theta}$

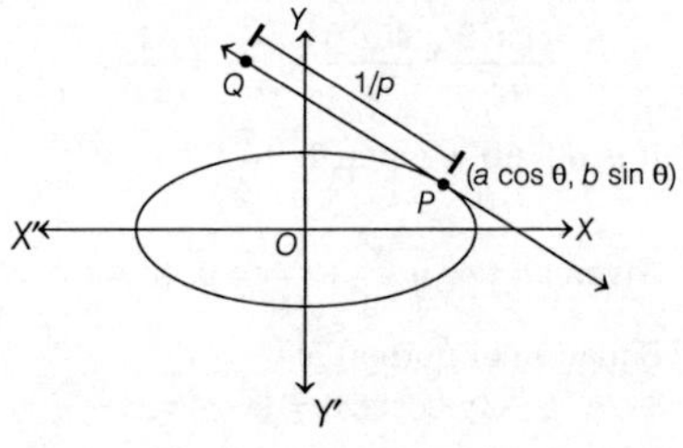

The distance of the tangent from the origin is

$p=\left|\dfrac{ab}{\sqrt{b^2\cos^2\theta+a^2\sin^2\theta}}\right|$

$\Rightarrow \dfrac{1}{p}=\dfrac{\sqrt{b^2\cos^2\theta+a^2\sin^2\theta}}{ab}$

Now, the coordinates of the point Q are given as follows,

$\dfrac{\dfrac{x-a\cos\theta}{-a\sin\theta}}{\sqrt{b^2\cos^2\theta+a^2\sin^2\theta}}=\dfrac{\dfrac{y-b\sin\theta}{b\cos\theta}}{\sqrt{b^2\cos^2\theta+a^2\sin^2\theta}}=\dfrac{1}{p}=\dfrac{\sqrt{b^2\cos^2\theta+a^2\sin^2\theta}}{ab}$

$\Rightarrow \quad x=a\cos\theta-\dfrac{a\sin\theta}{ab}$

and $\quad y=b\sin\theta+\dfrac{b\cos\theta}{ab}$

$\Rightarrow \quad \left(\dfrac{x}{a}\right)^2+\left(\dfrac{y}{b}\right)^2=1+\dfrac{1}{a^2b^2}$

is the required locus.

26 CP and CD are semi-conjugate diameters of an ellipse $\dfrac{x^2}{14}+\dfrac{y^2}{8}=1$ and let eccentric angle of P is ϕ, then eccentric angle of D is $\dfrac{\pi}{2}+\phi$, therefore the coordinates of P and D are $(a\cos\phi, b\sin\phi)$ and

$\left[\sqrt{14}\cos\left(\dfrac{\pi}{2}+\phi\right),\ \sqrt{8}\sin\left(\dfrac{\pi}{2}+\phi\right)\right]$

i.e. $CP^2+CD^2=(a^2\cos^2\phi+b^2\sin^2\phi)+(a^2\sin^2\phi+b^2\cos^2\phi)$

$=a^2+b^2=14+8=22$

27 Let $y=m_1x$ and $y=m_2x$ be a pair of conjugate diameters of an ellipse $\dfrac{x^2}{a^2}+\dfrac{y^2}{b^2}=1$ and let $P(a\cos\theta, b\sin\theta)$ and $Q(a\cos\phi, b\sin\phi)$ be ends of these two diameters.

Then, $m_1m_2=-\dfrac{b^2}{a^2}$

$\Rightarrow \dfrac{b\sin\theta-0}{a\cos\theta-0}\times\dfrac{b\sin\phi-0}{a\cos\phi-0}=-\dfrac{b^2}{a^2}$

$\Rightarrow \quad \sin\theta\sin\phi=-\cos\theta\cos\phi$

$\Rightarrow \quad \cos(\theta-\phi)=0$

$\Rightarrow \quad \theta-\phi=\pm\dfrac{\pi}{2}$

28 Equation of tangent at P is

$\dfrac{x\cos\theta}{a}+\dfrac{y\sin\theta}{b}=1.$

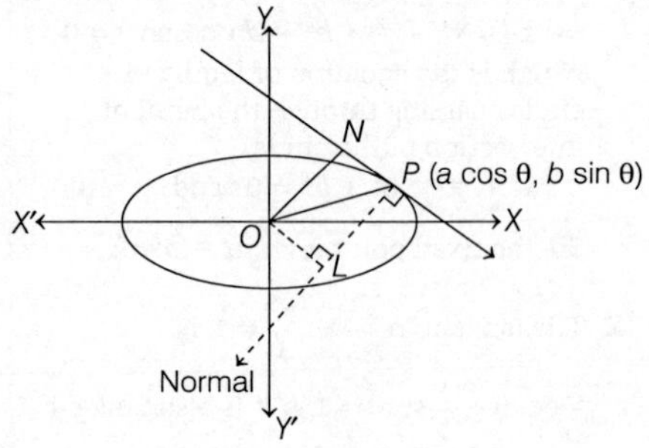

$ON=\dfrac{1}{\sqrt{\dfrac{\cos^2\theta}{a^2}+\dfrac{\sin^2\theta}{b^2}}}$

$=\dfrac{ab}{\sqrt{b^2\cos^2\theta+a^2\sin^2\theta}}$

Equation of the normal at P is

$ax\sec\theta-by\,\text{cosec}\,\theta=a^2-b^2$.

$OL=\dfrac{a^2-b^2}{\sqrt{a^2\sec^2\theta+b^2\text{cosec}^2\theta}}$

$=\dfrac{(a^2-b^2)\sin\theta\cos\theta}{\sqrt{a^2\sin^2\theta+b^2\cos^2\theta}}$

where, L is the foot of perpendicular from O on the normal.

Area of $\Delta PON=\dfrac{1}{2}\times ON\times OL$

$[\because NP=OL]$

$=\dfrac{(a^2-b^2)ab\tan\theta}{a^2\tan^2\theta+b^2}=\dfrac{(a^2-b^2)ab}{a^2\tan\theta+b^2\cot\theta}$

which is minimum when $a^2\tan\theta+b^2\cot\theta$ is maximum.

Thus, for area to be minimum, $\tan\theta=\dfrac{b}{a}$

$\therefore \cos\theta=\pm\dfrac{a}{\sqrt{a^2+b^2}},\ \sin\theta=\pm\dfrac{b}{\sqrt{a^2+b^2}}$

So, the required coordinates is

$\left(\pm\dfrac{a^2}{\sqrt{a^2+b^2}},\ \pm\dfrac{b^2}{\sqrt{a^2+b^2}}\right).$

29 If $\alpha,\beta,\gamma,\delta$ are eccentric angles of four co-normal points.

Then, $\alpha+\beta+\gamma+\delta=(2n+1)\pi$

and $\Sigma \sin(\alpha+\beta)=0$

Now, $\sin(\alpha+\beta)$

$= \sin[2n\pi + \pi - (\gamma+\delta)] = \sin(\gamma+\delta)$

Similarly, $\sin(\beta+\gamma)=\sin(\alpha+\delta)$

and $\sin(\gamma+\alpha)=\sin(\beta+\delta)$

$\therefore 2[\sin(\alpha+\beta)+\sin(\beta+\gamma)+\sin(\gamma+\alpha)]=0$

30. Equation of the chord with eccentric angles of the extremities as α and β is

$$\frac{x}{a}\cos\left(\frac{\alpha+\beta}{2}\right)+\frac{y}{b}\sin\left(\frac{\alpha+\beta}{2}\right) = \cos\left(\frac{\alpha-\beta}{2}\right)$$

Let $\frac{\alpha+\beta}{2}=\Delta_1$ and $\frac{\alpha-\beta}{2}=\Delta_2$,

so that $\frac{x}{a}\cos\Delta_1+\frac{y}{b}\sin\Delta_1=\cos\Delta_2$

As the triangle is right angled, homogenising the equation of the curve, we get

$$\frac{x^2}{a^2}+\frac{y^2}{b^2}-\left(\frac{x\cos\Delta_1}{a\cos\Delta_2}+\frac{y\sin\Delta_1}{b\cos\Delta_2}\right)^2=0$$

$$\Rightarrow \left(\frac{1}{a^2}-\frac{\cos^2\Delta_1}{a^2\cos^2\Delta_2}\right)+\left(\frac{1}{b^2}-\frac{\sin^2\Delta_1}{b^2\cos^2\Delta_2}\right)=0$$

[as coefficient of x^2 + coefficient of $y^2=0$]

$\Rightarrow b^2(\cos^2\Delta_2-\cos^2\Delta_1)+a^2(\cos^2\Delta_2-\sin^2\Delta_1)=0$

$\Rightarrow \cos^2\Delta_2(a^2+b^2)-b^2\cos^2\Delta_1-a^2+a^2\cos^2\Delta_1=0$

$\Rightarrow \cos^2\Delta_2(a^2+b^2)=a^2(1-e^2\cos^2\Delta_1)$

$$\Rightarrow \frac{1-e^2\cos^2\left(\frac{\alpha+\beta}{2}\right)}{\cos^2\left(\frac{\alpha-\beta}{2}\right)}=\frac{a^2+b^2}{a^2}$$

31 Let C_1 and C_2 be the centres and R_1 and R_2 be the radii of the two circles. Let $S_1=0$ lie completely inside in the circle $S_2=0$.

Let 'C' and 'r' be the centre and radius of the variable centre.

Then, $CC_2=R_2-r$

and $C_1C=R_1+r$

$\therefore C_1C+C_2C=R_1+R_2$ [constant]

So, the locus of C is an ellipse.

Therefore, Statement II is true.

Hence, Statement I is false (two circles are intersecting).

32 The ellipse is $\frac{x^2}{9}+\frac{y^2}{4}=1$.

So, auxiliary circle is $x^2+y^2=9$ and $(-\sqrt{5},0)$ and $(\sqrt{5},0)$ are foci.

Hence, Statement I is true, then Statement II is also true.

33 Let PCP' and DCD' be the conjugate diameters of an ellipse and let the eccentric angle of P is ϕ, then coordinate of P is $(a\cos\phi, b\sin\phi)$.

So, coordinate of D is $(-a\sin\phi, b\cos\phi)$.

Let S and S' be two foci of the ellipse.

Then, $SP\cdot S'P$

$=(a-ae\cos\phi)\cdot(a+ae\cos\phi)$

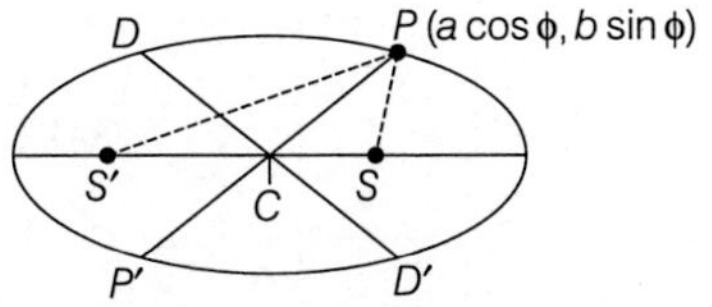

$=a^2-a^2e^2\cos^2\phi$ $\left[\begin{array}{l}\Rightarrow b^2=a^2(1-e^2)\\ \Rightarrow a^2-b^2=a^2e^2\end{array}\right]$

$=a^2-(a^2-b^2)\cos^2\phi$

$=a^2\sin^2\phi+b^2\cos^2\phi=CD^2$

34 Let $(t, b-t)$ be a point on the line $x+y=b$, then equation of chord whose mid-point is $(t, b-t)$, is

$$\frac{tx}{2a^2}+\frac{(b-t)y}{2b^2}-1=\frac{t^2}{2a^2}+\frac{(b-t)^2}{2b^2}-1 \quad \ldots(i)$$

Since, point $(a,-b)$ lies on Eq. (i), then

$$\frac{ta}{2a^2}-\frac{b(b-t)}{2b^2}=\frac{t^2}{2a^2}+\frac{(b-t)^2}{2b^2}$$

$\Rightarrow t^2(a^2+b^2)-ab(3a+b)t+2a^2b^2=0$

Since, t is real.

$\therefore B^2-4AC\ge 0$

$\Rightarrow a^2b^2(3a+b)^2-4(a^2+b^2)2a^2b^2\ge 0$

$\Rightarrow 9a^2+6ab+b^2-8a^2-8b^2\ge 0$

$\therefore a^2+6ab-7b^2\ge 0$

35. **Statement I** Given, a parabola $y^2=16\sqrt{3}x$ and an ellipse $2x^2+y^2=4$.

To find the equation of common tangent to the given parabola and the ellipse. This can be very easily done by comparing the standard equation of tangents. Standard equation of tangent with slope 'm' to the parabola $y^2=16\sqrt{3}x$ is

$$y=mx+\frac{4\sqrt{3}}{m} \quad \ldots(i)$$

Standard equation of tangent with slope 'm' to the ellipse $\frac{x^2}{2}+\frac{y^2}{4}=1$ is

$$y=mx\pm\sqrt{2m^2+4} \quad \ldots(ii)$$

If a line L is a common tangent to both parabola and ellipse, then L should be tangent to parabola i.e. its equation should be like Eq. (i) and L should be tangent to ellipse i.e. its equation should be like Eq. (ii) i.e. L must be like both of the Eqs. (i) and (ii).

Hence, comparing Eqs. (i) and (ii), we get

$$\frac{4\sqrt{3}}{m}=\pm\sqrt{2m^2+4}$$

On squaring both sides, we get

$m^2(2m^2+4)=48$

$\Rightarrow m^4+2m^2-24=0$

$\Rightarrow (m^2+6)(m^2-4)=0$

$\Rightarrow m^2=4$ $[\because m^2\ne -6]$

$\Rightarrow m=\pm 2$

On substituting $m=\pm 2$ in Eq. (i), we get the required equation of the common tangent as

$y=2x+2\sqrt{3}$ and $y=-2x-2\sqrt{3}$

Hence, Statement I is correct.

Statement II We have already seen that, if the line $y=mx+\frac{4\sqrt{3}}{m}$ is a common tangent to the parabola $y^2=16\sqrt{3}x$ and the ellipse $\frac{x^2}{2}+\frac{y^2}{4}=1$, then it satisfies the equation $m^4+2m^2-24=0$.

Hence, Statement II is also correct but is not able to explain the Statement I. It is an intermediate step in the final answer.

SESSION 2

1 Let equation of ellipse be

$$\frac{x^2}{a^2}+\frac{y^2}{b^2}=1$$

Equation of directrix is $x=\pm\frac{a}{e}$

$\therefore \frac{a}{e}=4\Rightarrow a=4e\Rightarrow a=4\times\frac{1}{2}=2$ $\left[\because e=\frac{1}{2}\right]$

$e^2=1-\frac{b^2}{a^2}\Rightarrow b^2=3$

$\therefore$ Equation of ellipse $=\frac{x^2}{4}+\frac{y^2}{3}=1$

Equation of normal

$$\frac{a^2x}{x_1}-\frac{b^2y}{y_1}=a^2-b^2,\ \frac{4x}{1}-\frac{3y}{\left(\frac{3}{2}\right)}=1$$

$\left[\because (x_1,y_1)=\left(1,\frac{3}{2}\right)\right]$

$\Rightarrow 4x-2y=1$

2 Let $P(h,k)$ be the mid-point of the focal chord. Then its equation is

$$\frac{hx}{a^2}+\frac{ky}{b^2}=\frac{h^2}{a^2}+\frac{k^2}{b^2} \quad \ldots(i)$$

Eq. (i) is focal chord so it passes through $(ae, 0)$

$\therefore \frac{h\,ea}{a^2}=\frac{h^2}{a^2}+\frac{k^2}{b^2}$

$\therefore$ Locus of P is $\frac{x^2}{a^2}+\frac{y^2}{b^2}=\frac{ex}{a}$

3 Let foot of normal be $P(x_1, y_1)$. Then

slope of normal $=\frac{a^2}{b^2}\frac{y_1}{x_1}=2\Rightarrow 9y_1=8x_1$

P lies on the ellipse $4x^2+9y^2=36$

$\Rightarrow \quad 4x_1^2 + 9\left(\frac{64x_1^2}{81}\right) = 36 \Rightarrow x_1^2 = \frac{81}{25}$

On taking $x_1 = \frac{9}{5}$, we get $y_1 = \frac{8}{5}$

$\therefore$ One possible foot of normal is $\left(\frac{9}{5}, \frac{8}{5}\right)$

4 Slope of tangent $m = \frac{8}{9}$

Let $P(x_1, y_1)$ be the point of contact.
Then, equation of tangent is

$$4xx_1 + 9yy_1 = 1$$

$$\therefore \quad -\frac{4x_1}{9y_1} = \frac{8}{9} \Rightarrow x_1 = -2y_1$$

$$4x_1^2 + 9y_1^2 = 1 \Rightarrow 16y_1^2 + 9y_1^2 = 1$$

$$\Rightarrow \quad y_1 = \pm\frac{1}{5}$$

$\therefore$ Required points of contact are $\left(\frac{-2}{5}, \frac{1}{5}\right)$ and $\left(\frac{2}{5}, \frac{-1}{5}\right)$

5 Normal is $4x\sec\phi - 2y\,\text{cosec}\,\phi = 12$

$Q \equiv (3\cos\phi, 0), M = (\alpha, \beta)$

$$\alpha = \frac{3\cos\phi + 4\cos\phi}{2} = \frac{7}{2}\cos\phi$$

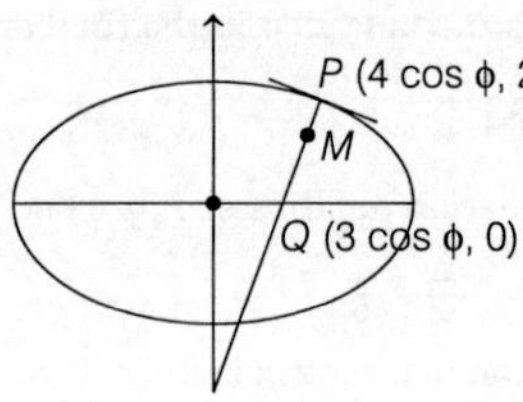

$\Rightarrow \quad \cos\phi = \frac{2}{7}\alpha$ and $\beta = \sin\phi$

$$\cos^2\phi + \sin^2\phi = 1$$

$\Rightarrow \quad \frac{4}{49}\alpha^2 + \beta^2 = 1 \Rightarrow \frac{4}{49}x^2 + y^2 = 1$

$\Rightarrow$ Latusrectum $x = \pm 2\sqrt{3}$

$$\frac{48}{49} + y^2 = 1 \Rightarrow y = \pm\frac{1}{7}\left(\pm 2\sqrt{3}, \pm\frac{1}{7}\right)$$

6 Equation of auxiliary circle is

$$x^2 + y^2 = 9 \quad \text{...(i)}$$

Equation of AM is $\frac{x}{3} + \frac{y}{1} = 1$ ...(ii)

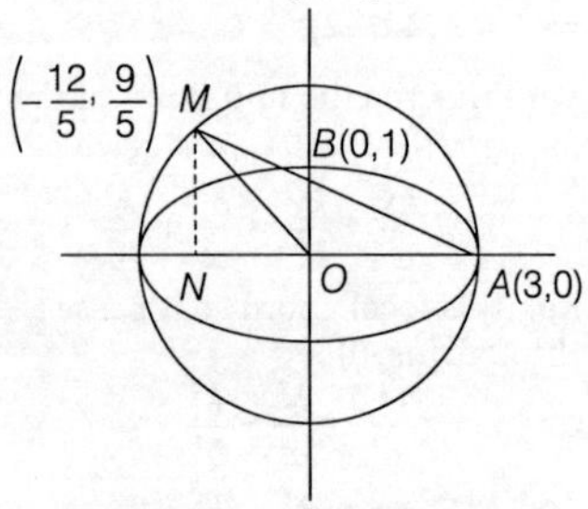

On solving Eqs. (i) and (ii), we get

$M\left(\frac{-12}{5}, \frac{9}{5}\right)$

Now, area of ΔAOM

$$= \frac{1}{2}\cdot OA\cdot MN = \frac{27}{10} \text{ sq unit.}$$

7 $x^2 + 4y^2 = 4 \Rightarrow \frac{x^2}{4} + \frac{y^2}{1} = 1$

$\Rightarrow \quad a = 2, b = 1 \Rightarrow P = (2, 1)$

Required Ellipse is $\frac{x^2}{a^2} + \frac{y^2}{b^2} = 1$

$\Rightarrow \quad \frac{x^2}{4^2} + \frac{y^2}{b^2} = 1$ (2, 1) lies on it

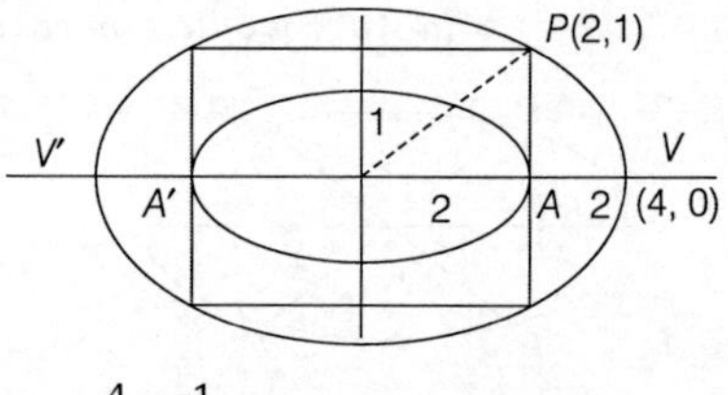

$$\Rightarrow \quad \frac{4}{16} + \frac{1}{b^2} = 1$$

$$\Rightarrow \quad \frac{1}{b^2} = 1 - \frac{1}{4} = \frac{3}{4} \Rightarrow b^2 = \frac{4}{3}$$

$$\therefore \quad \frac{x^2}{16} + \frac{y^2}{\left(\frac{4}{3}\right)} = 1$$

$$\Rightarrow \quad x^2 + 12y^2 = 16$$

8 $\frac{x^2}{4} + \frac{y^2}{1} = 1$ and $b^2 = a^2(1 - e^2)$

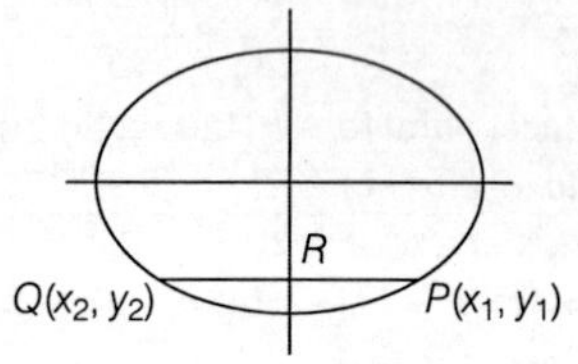

$$\Rightarrow \quad e = \frac{\sqrt{3}}{2} \Rightarrow P\left(\sqrt{3}, -\frac{1}{2}\right)$$

and $Q = \left(-\sqrt{3}, -\frac{1}{2}\right)$ (given y_1 and y_2 less than 0)

Co-ordinates of mid-point of PQ are

$R \equiv \left(0, -\frac{1}{2}\right)$

$PQ = 2\sqrt{3}$ = length of latusrectum.

$\Rightarrow$ Two parabolas are possible whose vertices are $\left(0, -\frac{\sqrt{3}}{2} - \frac{1}{2}\right)$ and $\left(0, \frac{\sqrt{3}}{2} - \frac{1}{2}\right)$

Hence, the equations of the parabolas are $x^2 - 2\sqrt{3}y = 3 + \sqrt{3}$ and $x^2 + 2\sqrt{3}y = 3 - \sqrt{3}$

9 Let required ellipse is

$$E_2 : \frac{x^2}{a^2} + \frac{y^2}{b^2} = 1$$

It passes through (0, 4)

$$0 + \frac{16}{b^2} = 1 \Rightarrow b^2 = 16$$

It also passes through $(\pm 3, \pm 2)$

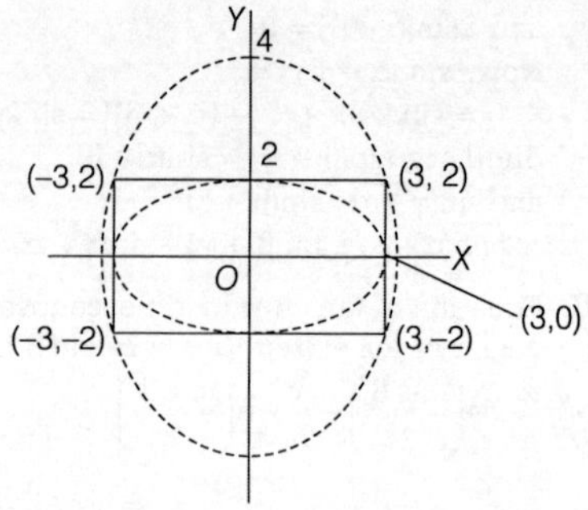

$$\frac{9}{a^2} + \frac{4}{b^2} = 1, \frac{9}{a^2} + \frac{1}{4} = 1 \Rightarrow a^2 = 12$$

$$a^2 = b^2(1 - e^2) \Rightarrow \frac{12}{16} = 1 - e^2$$

$$\Rightarrow e^2 = 1 - \frac{12}{16} = \frac{14}{16} = \frac{1}{4} \Rightarrow e = \frac{1}{2}$$

10 Ellipse is $\frac{x^2}{4} + \frac{y^2}{1} = 1$

Let $A(\theta) = (2\cos\theta, \sin\theta)$

Then, $B = \left(2\cos\left(\theta + \frac{\pi}{3}\right), \sin\left(\theta + \frac{\pi}{3}\right)\right)$

Equation of chord AB is

$$\frac{x}{2}\cos\left(\theta + \frac{\pi}{6}\right) + \frac{y}{1}\sin\left(\theta + \frac{\pi}{6}\right) = \cos\frac{\pi}{6} = \frac{\sqrt{3}}{2}$$

or given chord is $px + qy = r$
Comparing the coefficient, we get

$$\frac{\cos\left(\theta + \frac{\pi}{6}\right)}{2p} = \frac{\sin\left(\theta + \frac{\pi}{6}\right)}{q} = \frac{\sqrt{3}}{2r}$$

$$\frac{3p^2}{r^2} + \frac{3q^2}{4r^2} = 1$$

$$\Rightarrow \quad 3(4p^2 + q^2) = 4r^2$$

$$\Rightarrow \quad r^2 = \frac{3}{4}(4p^2 + q^2)$$

11 Let POP', QOQ' be two perpendicular diameters.

Let $\angle AOP = \theta$, then $\angle AOQ = \frac{\pi}{2} + \theta$

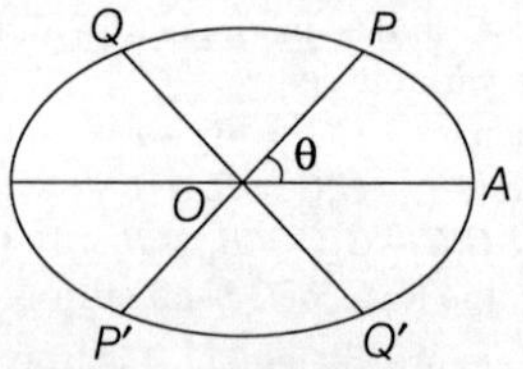

Then, $P = (OP\cos\theta, OP\sin\theta)$

$$\therefore \quad \frac{OP^2\cos^2\theta}{a^2} + \frac{OP^2\sin^2\theta}{b^2} = 1$$

$$\Rightarrow \quad \frac{1}{OP^2} = \frac{\cos^2\theta}{a^2} + \frac{\sin^2\theta}{b^2} \quad \text{...(i)}$$

Also $\quad Q = (OQ\cos(90° + \theta),$
$OQ\sin(90° + \theta))$

$\therefore \quad \frac{1}{OQ^2} = \frac{\sin^2\theta}{a^2} + \frac{\cos^2\theta}{b^2}$

$\Rightarrow \quad \frac{1}{OP^2} + \frac{1}{OQ^2} = \frac{1}{a^2} + \frac{1}{b^2}$

$\therefore \frac{1}{(POP')^2} + \frac{1}{(QOQ')^2} = \frac{1}{4}\left(\frac{1}{a^2} + \frac{1}{b^2}\right)$

$[\because POP' = 2OP, QOQ' = 2OQ]$

12 Let $P(\theta_1), Q(\theta_2), R(\theta_3)$ be the vertices of ΔPQR, inscribed in the ellipse

$$\frac{x^2}{a^2} + \frac{y^2}{b^2} = 1$$

Then, P', Q', R', corresponding points on auxiliary circle $x^2 + y^2 = a^2$ are $P'(a\cos\theta_1, a\sin\theta_1)$, $Q'(a\cos\theta_2, a\sin\theta_2)$ and $R'(a\cos\theta_3, a\sin\theta_3)$

Δ_1 = Area of ΔPQR

$$= \frac{1}{2}\begin{vmatrix} a\cos\theta_1 & b\sin\theta_1 & 1 \\ a\cos\theta_2 & b\sin\theta_2 & 1 \\ a\cos\theta_3 & b\sin\theta_3 & 1 \end{vmatrix}$$

$$= \frac{1}{2}ab\begin{vmatrix} \cos\theta_1 & \sin\theta_1 & 1 \\ \cos\theta_2 & \sin\theta_2 & 1 \\ \cos\theta_3 & \sin\theta_3 & 1 \end{vmatrix}$$

Δ_2 = area of $P'Q'R'$

$$= \frac{1}{2}a^2\begin{vmatrix} \cos\theta_1 & \sin\theta_1 & 1 \\ \cos\theta_2 & \sin\theta_2 & 1 \\ \cos\theta_3 & \sin\theta_3 & 1 \end{vmatrix}$$

$\therefore \quad \frac{\Delta_1}{\Delta_2} = \frac{b}{a}$ = Constant

13 Equation of chord $P(\alpha), Q(\beta)$ is

$$\frac{x}{a}\cos\left(\frac{\alpha+\beta}{2}\right) + \frac{y}{b}\sin\left(\frac{\alpha+\beta}{2}\right) = \cos\left(\frac{\alpha-\beta}{2}\right) \quad \text{...(i)}$$

Here, $a = 5, b = 4$. PQ is a focal chord.

$$e^2 = \frac{a^2 - b^2}{a^2} = \frac{9}{25} \Rightarrow e = \frac{3}{5}$$

One of the foci $= (ae, 0) = (3, 0)$

Eq. (i) passes through (3, 0)

$\therefore \quad \frac{3}{5}\cos\left(\frac{\alpha+\beta}{2}\right) = \cos\left(\frac{\alpha-\beta}{2}\right)$

$\Rightarrow \quad 5\left[\cos\frac{\alpha}{2}\cos\frac{\beta}{2} + \sin\frac{\alpha}{2}\sin\frac{\beta}{2}\right]$

$= 3\left[\cos\frac{\alpha}{2}\cos\frac{\beta}{2} - \sin\frac{\alpha}{2}\sin\frac{\beta}{2}\right]$

$\Rightarrow \quad 8\sin\frac{\alpha}{2}\sin\frac{\beta}{2} = -2\cos\frac{\alpha}{2}\cos\frac{\beta}{2}$

$\Rightarrow \quad \tan\frac{\alpha}{2}\tan\frac{\beta}{2} = -\frac{1}{4}.$

14 Any point $P(\theta)$ on the ellipse $\frac{x^2}{4} + \frac{y^2}{1} = 1$ is $P(2\cos\theta, \sin\theta)$

Equation of tangent at $P(\theta)$ is

$$2x\cos\theta + 4y\sin\theta = 4$$

$\therefore \quad A = (2\sec\theta, 0)$

Equation of normal of $P(\theta)$ is

$$\frac{2x}{\cos\theta} - \frac{y}{\sin\theta} = 4 - 1 = 3$$

$\therefore \quad B = \left(\frac{3}{2}\cos\theta, 0\right)$

$$AB = 2 \Rightarrow \left|2\sec\theta - \frac{3}{2}\cos\theta\right| = 2$$

$\Rightarrow \quad \left(2 - \frac{3}{2}\cos^2\theta\right)^2 = 4\cos^2\theta$

$\Rightarrow \quad 9\cos^4\theta - 40\cos^2\theta + 16 = 0$

$\Rightarrow \quad (9\cos^2\theta - 4)(\cos^2\theta - 4) = 0$

$\Rightarrow \quad \cos^2\theta = \frac{4}{9}$ as $\cos^2\theta \neq 4$

15 Equation of tangent at P is

$$y = mx + \sqrt{a^2m^2 + b^2} \quad \text{...(i)}$$

$\therefore$ $S'R'RS$ is a trapezium and its area

$$\Delta_1 = \frac{1}{2}(SR + S'R') \times SD$$

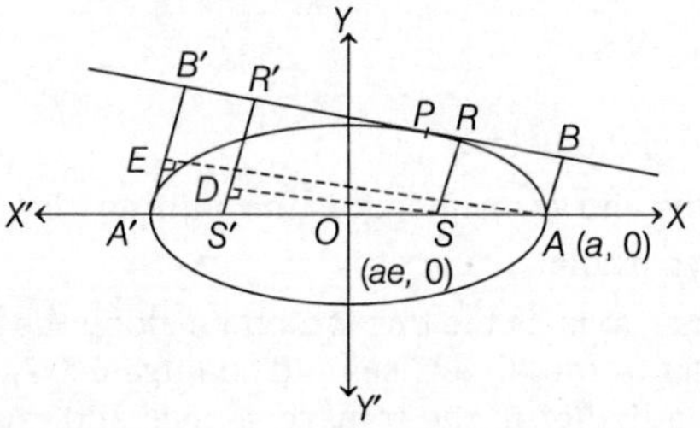

Equation of the line $S'R'$ is

$$y = -\frac{1}{m}(x + ae)$$

$\Rightarrow \quad x + my + ae = 0 \quad \text{...(ii)}$

Therefore, $SD = \left|\frac{ae + ae}{\sqrt{1+m^2}}\right|$

$$SR = \left|\frac{aem + \sqrt{a^2m^2 + b^2}}{\sqrt{1+m^2}}\right|$$

and $S'R' = \left|\frac{-aem + \sqrt{a^2m^2 + b^2}}{\sqrt{1+m^2}}\right|$

Then, $\Delta_1 = \frac{1}{2}(S'R' + SR) \times SD$

$\Rightarrow \quad \Delta_1 = 2ae\left(\frac{\sqrt{a^2m^2 + b^2}}{1+m^2}\right)$

Area of $A'B'BA$ is

$$\Delta_2 = \frac{1}{2}(A'B' + AB) \times AE$$

Equation of $A'B'$ is

$$y = -\frac{1}{m}(x + a)$$

$\Rightarrow \quad x + my + a = 0 \quad \text{...(iii)}$

Therefore, $AE = \left|\frac{a + a}{\sqrt{1+m^2}}\right|$

and $AB = \left|\frac{ma + \sqrt{a^2m^2 + b^2}}{\sqrt{1+m^2}}\right|$

and $A'B' = \left|\frac{-ma + \sqrt{a^2m^2 + b^2}}{\sqrt{1+m^2}}\right|$

Then, $\Delta_2 = \frac{1}{2}(A'B' + AB) \times AE$

$\Rightarrow \quad \Delta_2 = 2a\left(\frac{\sqrt{a^2m^2 + b^2}}{1+m^2}\right)$

Hence, $\Delta_1 : \Delta_2 = e : 1$

DAY TWENTY NINE

Hyperbola

Learning & Revision for the Day

- Concept of Hyperbola
- Equations of Hyperbola in Standard Form
- Tangent to a Hyperbola
- Normal to a Hyperbola
- Pole and Polar
- Director Circle
- Asymptotes
- Rectangular or Equilateral Hyperbola

Concept of Hyperbola

Hyperbola is the locus of a point in a plane which moves in such a way that the ratio of its distance from a fixed point (focus) in the same plane to its distance from a fixed line (directrix) is always constant which is always greater than unity.

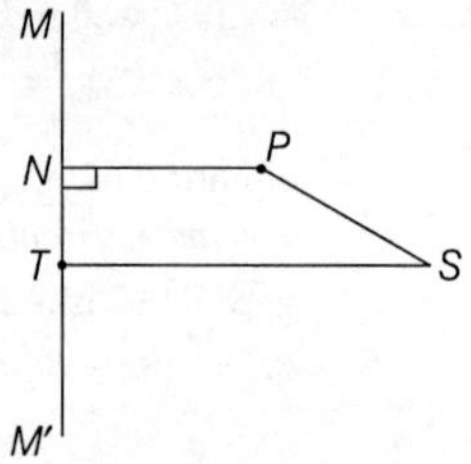

Mathematically, $\frac{SP}{PN} = e$, where $e > 1$.

Terms Related to Hyperbola

Some important terms related to hyperbola are given below.

1. **Vertices** The points A and A', where the curve meets the line joining the foci S and S', are called the vertices of the hyperbola.
2. **Transverse and conjugate axes** Transverse axis is the one which lie along the line passing through the foci and perpendicular to the directrices and conjugate axis and conjugate axis is the one which is perpendicular to the transverse axis and passes through the mid-point of the foci i.e. centre.
3. **Centre** The mid-point C of AA' bisects every chord of the hyperbola passing through it and is called the centre of the hyperbola.
4. **Focal chord** A chord of a hyperbola which is passing through the focus is called a focal chord of the hyperbola.
5. **Directrix** A line which is perpendicular to the axis and it lies between centre and vertex. The equation of directrix is $x = \pm \frac{a}{e}$.
6. **Double ordinates** If Q be a point on the hyperbola draw QN perpendicular to the axis of the hyperbola and produced to meet the curve again at Q'. Then, QQ' is called a double ordinate of Q.

PREP MIRROR
Your Personal Preparation Indicator

- No. of Questions in Exercises (x)—
- No. of Questions Attempted (y)—
- No. of Correct Questions (z)—
  (Without referring Explanations)

- Accuracy Level $(z/y \times 100)$—
- Prep Level $(z/x \times 100)$—

In order to expect good rank in JEE, your Accuracy Level should be above 85 & Prep Level should be above 75.

7. **Latusrectum** The double ordinate passing through focus is called latusrectum.

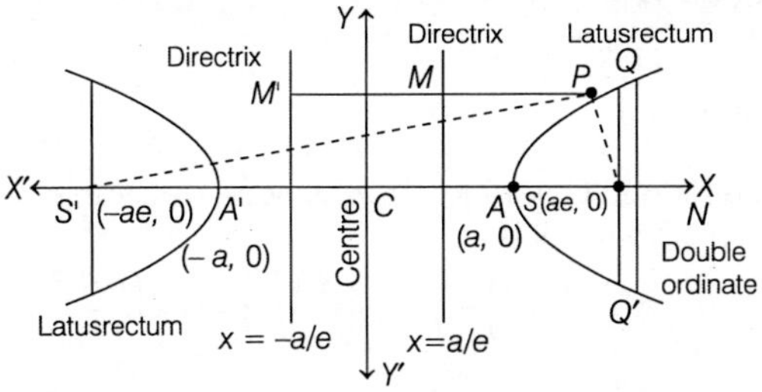

NOTE
- The vertex divides the join of focus and the point of intersection of directrix with axis internally and externally in the ratio e : 1.
- Domain and range of a hyperbola $\frac{x^2}{a^2}-\frac{y^2}{b^2}=1$ are $x \le -a$ or $x \ge a$ and $y \in R$, respectively.
- The line through the foci of the hyperbola is called its transverse axis.
- The line through the centre and perpendicular to the transverse axis of the hyperbola is called its conjugate axis.

Equations of Hyperbola in Standard Form

If the centre of the hyperbola is at the origin and foci are on the X-axis or Y-axis, then that types of equation are called standard equation of an ellipse.

1. Hyperbola of the Form $\frac{x^2}{a^2}-\frac{y^2}{b^2}=1$

When the hyperbola is in the given form, then it is also called the equation of auxiliary circle.

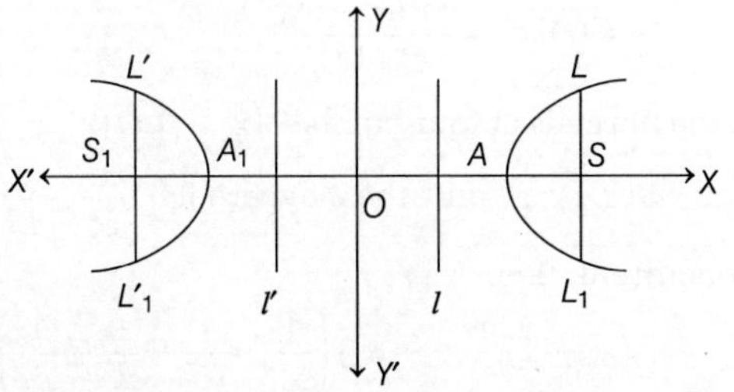

(i) Centre, $O(0,0)$
(ii) Foci : $S(ae,0), S_1(-ae,0)$
(iii) Vertices : $A(a,0), A_1(-a,0)$
(iv) Directrices $l : x=\frac{a}{e}, l' : x=-\frac{a}{e}$
(v) Length of latusrectum, $LL_1 = L'L'_1 = \frac{2b^2}{a}$
(vi) Eccentricity, $e=\sqrt{1+\left(\frac{b}{a}\right)^2}$ or $b^2=a^2(e^2-1)$
(vii) Length of transverse axis, $2a$
(viii) Length of conjugate axis, $2b$
(ix) Equation of transverse axis, $y=0$
(x) Equation of conjugate axis, $x=0$
(xi) Focal distances of a point on the hyperbola is $ex \pm a$.
(xii) Difference of the focal distances of a point on the hyperbola is $2a$.

2. Conjugate Hyperbola $-\frac{x^2}{a^2}+\frac{y^2}{b^2}=1$

The hyperbola whose transverse and conjugate axes are respectively the conjugate and transverse axes of a given hyperbola is called the conjugate hyperbola of the given hyperbola. The conjugate hyperbola of the hyperbola $\frac{x^2}{a^2}-\frac{y^2}{b^2}=1$ is $-\frac{x^2}{a^2}+\frac{y^2}{b^2}=1$.

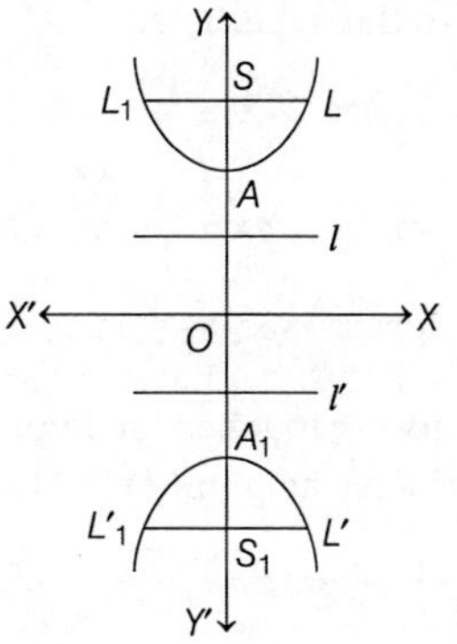

(i) Centre, $O(0,0)$
(ii) Foci, $S(0,be), S_1(0,-be)$
(iii) Vertices, $A(0,b), A_1(0,-b)$
(iv) Directrices $l : y=\frac{b}{e}, l' : y=-\frac{b}{e}$
(v) Length of latusrectum $LL_1 = L'L_1' = \frac{2a^2}{b}$
(vi) Eccentricity, $e=\sqrt{1+\left(\frac{a}{b}\right)^2}$ as $a^2=b^2(e^2-1)$
(vii) Length of transverse axis, $2b$
(viii) Length of conjugate axis, $2a$
(ix) Equation of transverse axis, $x=0$
(x) Equation of conjugate axis, $y=0$
(xi) Focal distances of a point on the hyperbola is $ey \pm b$.
(xii) Difference of the focal distances of a point on the hyperbola is $2b$.

NOTE If the centre of the hyperbola is (h,k) and axes are parallel to the coordinate axes, then its equation is $\frac{(x-h)^2}{a^2}-\frac{(y-k)^2}{b^2}=1$.

Results on Hyperbola $\frac{x^2}{a^2}-\frac{y^2}{b^2}=1$

1. The foci of a hyperbola and its conjugate are concyclic and form the vertices of a square.
2. The points of intersection of the directrix with the transverse axis are known as foot of the directrix.
3. Latusrectum $(l) = 2e$ (distance between the focus and the foot of the corresponding directrix).
4. The parametric equation of a hyperbola is $x = a\sec\theta$ and $y = b\tan\theta$, where $\theta \in [0, 2\pi)$.
5. The position of a point (h, k) with respect to the hyperbola S lie inside, on or outside the hyperbola, if

$$S_1 > 0,\ S_1 = 0 \text{ or } S_1 < 0 \text{ where, } S_1 \equiv \frac{x_1^2}{a^2}-\frac{y_1^2}{b^2}-1$$

Tangent to a Hyperbola

A line which intersects the hyperbola at only one point is called the tangent to the hyperbola.

(i) In point (x_1, y_1) form, $\frac{xx_1}{a^2}-\frac{yy_1}{b^2}=1$

(ii) In slope 'm' form, $y = mx \pm \sqrt{a^2m^2-b^2}$

(iii) In parametric form, $\frac{x}{a}\sec\theta-\frac{y}{b}\tan\theta=1$ at $(a\sec\theta,\ b\tan\theta)$.

(iv) The line $y = mx + c$ touches the hyperbola, iff $c^2 = a^2m^2 - b^2$ and the point of contact is $\left(\pm\frac{a^2m}{c}, \pm\frac{b^2}{c}\right)$, where $c=\sqrt{a^2m^2-b^2}$.

Results on Tangent

1. Two tangents can be drawn from a point to a hyperbola.
2. The point of intersection of tangents at t_1 and t_2 to the curve $xy = c^2$ is $\left(\frac{2ct_1t_2}{t_1+t_2}, \frac{2c}{t_1+t_2}\right)$.
3. The tangent at the point $P(a\sec\theta_1, b\tan\theta_1)$ and $Q(a\sec\theta_2, b\tan\theta_2)$ intersect at the point

$$R\left(\frac{a\cos\left(\frac{\theta_1-\theta_2}{2}\right)}{\cos\left(\frac{\theta_1+\theta_2}{2}\right)}, \frac{b\sin\left(\frac{\theta_1+\theta_2}{2}\right)}{\cos\left(\frac{\theta_1+\theta_2}{2}\right)}\right).$$

4. The equation of pair of tangents drawn from an external point $P(x_1, y_1)$ to the hyperbola is $SS_1 = T^2$

where $S=\frac{x^2}{a^2}-\frac{y^2}{b^2}-1$, $S_1=\frac{x_1^2}{a^2}-\frac{y_1^2}{b^2}-1$

and $T=\frac{xx_1}{a^2}-\frac{yy_1}{b^2}-1$

5. The equation of chord of contact is $\frac{xx_1}{a^2}-\frac{yy_1}{b^2}=1$ or $T = 0$.
6. The equation of chord of the hyperbola, whose mid-point is (x_1, y_1) is $T = S_1$ i.e. $\frac{xx_1}{a^2}-\frac{yy_1}{b^2}=\frac{x_1^2}{a^2}-\frac{y_1^2}{b^2}$.
7. Equation of chord joining the points $(a\sec\alpha, b\tan\alpha)$ and $(a\sec\beta, b\tan\beta)$ is

$$\frac{x}{a}\cos\left(\frac{\alpha-\beta}{2}\right)-\frac{y}{b}\sin\left(\frac{\alpha+\beta}{2}\right)=\cos\left(\frac{\alpha+\beta}{2}\right).$$

Normal to a Hyperbola

A line which is perpendicular to the tangent of the hyperbola is called the normal to the hyperbola.

(i) In point (x_1, y_1) form $\frac{a^2x}{x_1}+\frac{b^2y}{y_1}=a^2+b^2$.

(ii) In slope 'm' form $y = mx \mp \frac{m(a^2+b^2)}{\sqrt{a^2-m^2b^2}}$ and the point of intersection is $\left(\pm\frac{a^2}{\sqrt{a^2-b^2m^2}}, \mp\frac{b^2m}{\sqrt{a^2-b^2m^2}}\right)$

(iii) In parametric form, $ax\cos\theta + by\cot\theta = a^2 + b^2$ at $(a\sec\theta, b\tan\theta)$

Results on Normals

1. If the straight line $lx + my + n = 0$ is a normal to the hyperbola $\frac{x^2}{a^2}-\frac{y^2}{b^2}=1$, then $\frac{a^2}{l^2}-\frac{b^2}{m^2}=\frac{(a^2+b^2)^2}{n^2}$
2. Four normals can be drawn from any point to a hyperbola.
3. The line $y = mx + c$ will be a normal to the hyperbola $\frac{x^2}{a^2}-\frac{y^2}{b^2}=1$, if $c^2=\frac{m^2(a^2+b^2)^2}{a^2-b^2m^2}$.
4. If the normals at four points $P(x_1, y_1), Q(x_2, y_2), R(x_3, y_3)$ and $S(x_4, y_4)$ on the hyperbola $\frac{x^2}{a^2}-\frac{y^2}{b^2}=1$ are concurrent, then

$$(x_1+x_2+x_3+x_4)\left(\frac{1}{x_1}+\frac{1}{x_2}+\frac{1}{x_3}+\frac{1}{x_4}\right)=4.$$

Pole and Polar

Let P be a point inside or outside a hyperbola. Then, the locus of the point of intersection of two tangents to the hyperbola at the point where secants drawn through P intersect the hyperbola, is called the **polar** of point P with respect to the hyperbola and the point P is called the **pole** of the polar.

(i) Polars of two points $P(x_1 y_1)$ and $Q(x_2, y_2)$ of hyperbola $\frac{x^2}{a^2}-\frac{y^2}{b^2}=1$ are $\frac{xx_1}{a^2}-\frac{yy_1}{b^2}=1$ and $\frac{xx_2}{a^2}-\frac{yy_2}{b^2}=1$

(ii) If the polar of $P(x_1, y_1)$ passes through $Q(x_2, y_2)$, then $\frac{x_1x_2}{a^2}-\frac{y_1y_2}{b^2}=1$ and the polar of $Q(x_2, y_2)$ passes through $P(x_1, y_1)$, then $\frac{x_1x_2}{a^2}-\frac{y_1y_2}{b^2}=1$

Conjugate Points and Conjugate Lines

(i) Two points are said to be conjugate points with respect to a hyperbola, if each lies on the polar of the other.

(ii) Two lines are said to be conjugate lines with respect to a hyperbola $\frac{x^2}{a^2}-\frac{y^2}{b^2}=1$, if each passes through the pole of the other.

(iii) Two lines $l_1x+m_1y+n_1=0$ and $l_2x+m_2y+n_2=0$ are conjugate lines with respect to the hyperbola $\frac{x^2}{a^2}-\frac{y^2}{b^2}=1$, if $a^2l_1l_2-b^2m_1m_2=n_1n_2$.

Director Circle

The locus of the point of intersection of the tangents to the hyperbola $\frac{x^2}{a^2}-\frac{y^2}{b^2}=1$, which are perpendicular to each other, is called a director circle.

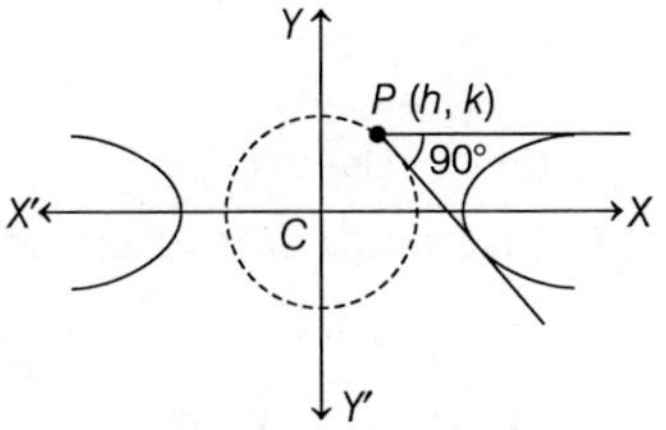

The equation of director circle is $x^2+y^2=a^2-b^2$.

The circle $x^2+y^2=a^2$ is known the auxiliary circle of both hyperbola.

Asymptotes

An asymptote to a curve is a straight line, which touches on it two points at infinity but which itself does not lie entirely at infinity.

Results on Asymptotes

1. A hyperbola and its conjugate hyperbola have the same asymptotes.
2. The angle between the asymptotes of hyperbola $\frac{x^2}{a^2}-\frac{y^2}{b^2}=1$ is $2\tan^{-1}\left(\frac{b}{a}\right)$.
3. Asymptotes always passes through the centre of the hyperbola.
4. The bisectors of the angle between the asymptotes are the coordinate axes.
5. The asymptotes of a hyperbola are the diagonals of the rectangle formed by the lines drawn through the extrimities of each axis parallel to the other axis.

Rectangular or Equilateral Hyperbola

A hyperbola whose asymptotes include a right angle is said to be rectangular hyperbola or if the lengths of transverse and conjugate axes of any hyperbola is equal, then it is said to be a rectangular hyperbola.

Rectangular Hyperbola of the Form $x^2-y^2=a^2$

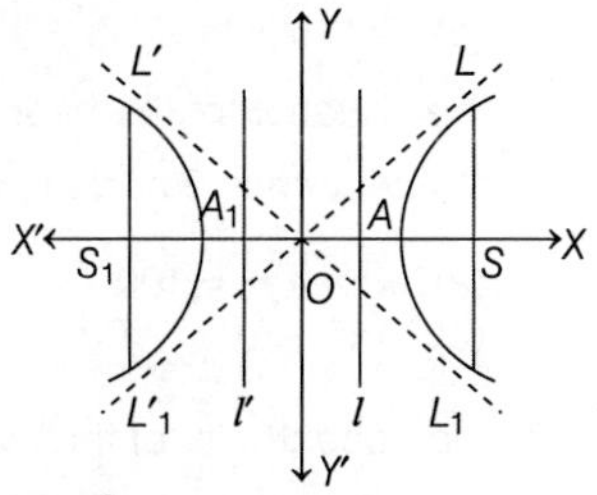

1. Asymptotes are perpendicular lines i.e. $x \pm y = 0$
2. Eccentricity, $e=\sqrt{2}$,
3. Centre, $O(0,0)$
4. Foci, S and S_1 $(\pm\sqrt{2}\,a, 0)$
5. Directrices, $x=\pm\frac{a}{\sqrt{2}}$
6. Latusrectum $=2a$
7. Point form, $x=x_1, y=y_1$
   Equation of tangent, $xx_1-yy_1=a^2$
   Equation of normal, $\frac{x_1}{x}+\frac{y_1}{y}=2$
8. Parametric form, $x=a\sec\theta, y=a\tan\theta$
   Equation of tangent, $x\sec\theta-y\tan\theta=a$
   Equation of normal, $\frac{x}{\sec\theta}+\frac{y}{\tan\theta}=2a$

Rectangular Hyperbola of the Form $xy=c^2$

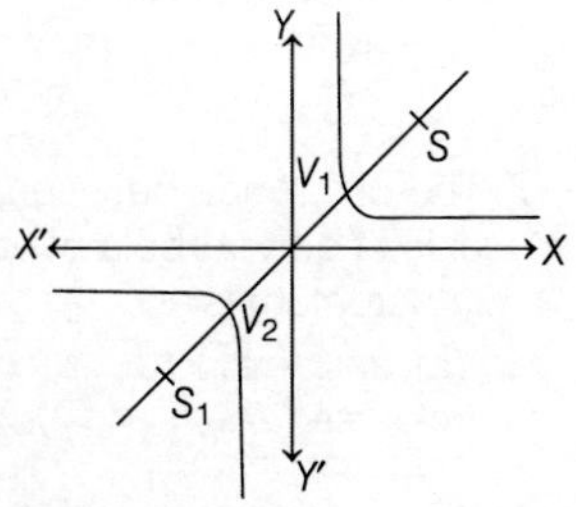

1. Asymptotes are perpendicular lines i.e. $x=0$ and $y=0$
2. Eccentricity, $e=\sqrt{2}$
3. Centre, $(0,0)$
4. Foci
   $S(\sqrt{2}c, \sqrt{2}c), S_1(-\sqrt{2}c, -\sqrt{2}c)$
5. Vertices, $V_1(c, c), V_2(-c, -c)$
6. Directrices, $x+y=\pm\sqrt{2}\,c$
7. Latusrectum $=2\sqrt{2}\,c$
8. Point form, $x=x_1, y=y_1$
   Equation of tangent, $xy_1+yx_1=2c^2$
   $\Rightarrow \frac{x}{x_1}+\frac{y}{y_1}=2$
   Equation of Normal, $xx_1-yy_1=x_1^2-y_1^2$
9. Parametric form : $x=ct, y=\frac{c}{t}$
   Equation of tangent, $x+yt^2=2ct$
   Equation of normal, $t^2x-y=c\left(t^3-\frac{1}{t}\right)$

DAY PRACTICE SESSION 1

FOUNDATION QUESTIONS EXERCISE

1 If the vertices of hyperbola are $(0, \pm 6)$ and its eccentricity is $\frac{5}{3}$, then → NCERT Exemplar

I. The equation of hyperbola is $\frac{y^2}{36} - \frac{x^2}{64} = 1$.

II. The foci of hyperbola are $(0, \pm 10)$.

(a) Both I and II are true (b) Only I is true
(c) Only II is true (d) Both I and II are false

2 The equation of the hyperbola, the lenght of whose latusrectum is 8 and eccentricity is $\frac{3}{\sqrt{5}}$, is

(a) $5x^2 - 4y^2 = 100$ (b) $4x^2 - 5y^2 = 100$
(c) $-4x^2 + 5y^2 = 100$ (d) $-5x^2 + 4y^2 = 100$

3 The eccentricity of the hyperbola $\frac{x^2}{a^2} - \frac{y^2}{b^2} = 1$ which passes through the points (3, 0) and $(3\sqrt{2}, 2)$, is → NCERT Exemplar

(a) $\frac{13}{3}$ (b) $\sqrt{\frac{13}{3}}$ (c) $\frac{\sqrt{13}}{9}$ (d) $\frac{\sqrt{13}}{3}$

4 Eccentricity of hyperbola $\frac{x^2}{k} + \frac{y^2}{k^2} = 1 (k < 0)$ is

(a) $\sqrt{1+k}$ (b) $\sqrt{1-k}$ (c) $\sqrt{1+\frac{1}{k}}$ (d) $\sqrt{1-\frac{1}{k}}$

5 The equation of the hyperbola whose foci are $(-2, 0)$ and $(2, 0)$ and eccentricity is 2, is given by → AIEEE 2011

(a) $-3x^2 + y^2 = 3$ (b) $x^2 - 3y^2 = 3$
(c) $3x^2 - y^2 = 3$ (d) $-x^2 + 3y^2 = 3$

6 The length of transverse axis of the hyperbola $3x^2 - 4y^2 = 32$ is

(a) $\frac{8\sqrt{2}}{\sqrt{3}}$ (b) $\frac{16\sqrt{2}}{\sqrt{3}}$ (c) $\frac{3}{32}$ (d) $\frac{64}{3}$

7 The difference between the length $2a$ of the transverse axis of a hyperbola of eccentricity e and the length of its latusrectum is

(a) $2a(3 - e^2)$ (b) $2a|2 - e^2|$
(c) $2a(e^2 - 1)$ (d) $a(2e^2 - 1)$

8 If eccentricity of hyperbola $\frac{x^2}{a^2} - \frac{y^2}{b^2} = 1$ is e and e' is the eccentricity of its conjugate hyperbola, then

(a) $e = e'$ (b) $ee' = 1$
(c) $\frac{1}{e^2} + \frac{1}{(e')^2}$ (d) None of these

9 A hyperbola, having the transverse axis of length $2\sin\theta$, is confocal with the ellipse $3x^2 + 4y^2 = 12$. Then, its equation is

(a) $x^2 \operatorname{cosec}^2\theta - y^2 \sec^2\theta = 1$ (b) $x^2 \sec^2\theta - y^2 \operatorname{cosec}^2\theta = 1$
(c) $x^2 \sin^2\theta - y^2 \cos^2\theta = 1$ (d) $x^2 \cos^2\theta - y^2 \sin^2\theta = 1$

10 Length of the latusrectum of the hyperbola $xy - 3x - 4y + 8 = 0$ is

(a) 4 (b) $4\sqrt{2}$ (c) 8 (d) None of these

11 The eccentricity of the hyperbola whose latusrectum is 8 and conjugate axis is equal to half of the distance between the foci is → NCERT Exemplar

(a) $\frac{4}{3}$ (b) $\frac{4}{\sqrt{3}}$ (c) $\frac{2}{\sqrt{3}}$ (d) None of these

12 A general point on the hyperbola $\frac{x^2}{a^2} - \frac{y^2}{b^2} = 1$ is

(a) $(a\sin\theta, b\cos\theta)$ (where, θ is parameter)
(b) $(a\tan\theta, b\sec\theta)$ (where, θ is parameter)
(c) $\left(a\frac{e^t + e^{-t}}{2}, b\frac{e^t - e^{-t}}{2}\right)$ (where, t is parameter)
(d) None of the above

13 For the hyperbola $\frac{x^2}{\cos^2\alpha} - \frac{y^2}{\sin^2\alpha} = 1$. Which of the following remains constant when α varies?

(a) Directrix (b) Abscissae of vertices
(c) Abscissae of foci (d) Eccentricities

14 The product of the perpendicular from two foci on any tangent to the hyperbola $\frac{x^2}{a^2} - \frac{y^2}{b^2} = 1$, is

(a) a^2 (b) b^2 (c) $-a^2$ (d) $-b^2$

15 If a line $21x + 5y = 116$ is tangent to the hyperbola $7x^2 - 5y^2 = 232$, then point of contact is

(a) $(-6, 3)$ (b) $(6, -2)$ (c) $(8, 2)$ (d) None of these

16 If e_1 is the eccentricity of the ellipse $\frac{x^2}{16} + \frac{y^2}{25} = 1$ and e_2 is the eccentricity of the hyperbola passing through the foci of the ellipse and $e_1 e_2 = 1$, then equation of the hyperbola is

(a) $\frac{x^2}{9} - \frac{y^2}{16} = 1$ (b) $\frac{x^2}{16} - \frac{y^2}{9} = -1$
(c) $\frac{x^2}{9} - \frac{y^2}{25} = 1$ (d) None of these

17 If the line $2x + \sqrt{6}y = 2$ touches the hyperbola $x^2 - 2y^2 = 4$, then the point of contact is

(a) $(-2, \sqrt{6})$ (b) $(-5, 2\sqrt{6})$ (c) $\left(\frac{1}{2}, \frac{1}{\sqrt{6}}\right)$ (d) $(4, -\sqrt{6})$

18 If a hyperbola passes through the point $P(\sqrt{2}, \sqrt{3})$ and has foci at $(\pm 2, 0)$, then the tangent to this hyperbola at P also passes through the point → JEE Mains 2017

(a) $(3\sqrt{2}, 2\sqrt{3})$ (b) $(2\sqrt{2}, 3\sqrt{3})$ (c) $(\sqrt{3}, \sqrt{2})$ (d) $(-\sqrt{2}, -\sqrt{3})$

19 The common tangent to $9x^2 - 4y^2 = 36$ and $x^2 + y^2 = 3$ is

(a) $y - 2\sqrt{3}x - \sqrt{39} = 0$ (b) $y + 2\sqrt{3}x + \sqrt{39} = 0$
(c) $y - 2\sqrt{3}x + \sqrt{39} = 0$ (d) None of the above

20 The locus of a point $P(\alpha, \beta)$ moving under the condition that the line $y = \alpha x + \beta$ is a tangent to the hyperbola $\frac{x^2}{a^2} - \frac{y^2}{b^2} = 1$, is → AIEEE 2005

(a) a hyperbola (b) a parabola (c) a circle (d) an ellipse

21 The straight line $x + y = \sqrt{2}p$ will touch the hyperbola $4x^2 - 9y^2 = 36$, if

(a) $p^2 = 2$ (b) $p^2 = 5$ (c) $5p^2 = 2$ (d) $2p^2 = 5$

22 The locus of the point of intersection of perpendicular tangents to the hyperbola $\frac{x^2}{3} - \frac{y^2}{1} = 1$ is

(a) $x^2 + y^2 = 2$ (b) $x^2 + y^2 = 3$
(c) $x^2 - y^2 = 3$ (d) $x^2 + y^2 = 4$

23 If P is a point on the hyperbola $\frac{x^2}{9} - \frac{y^2}{4} = 1$ and N is foot of the perpendicular from P on the transverse axis. The tangent to the hyperbola at P meets the transverse axis at T. If O is the centre of the hyperbola, then $OT \cdot ON$ is equal to

(a) 9 (b) 4 (c) e^2 (d) None of these

24 The locus of the points of intersection of perpendicular tangents to $\frac{x^2}{16} - \frac{y^2}{9} = 1$ is

(a) $x^2 - y^2 = 7$ (b) $x^2 - y^2 = 25$
(c) $x^2 + y^2 = 25$ (d) $x^2 + y^2 = 7$

25 Tangents are drawn from points on the hyperbola $\frac{x^2}{9} - \frac{y^2}{4} = 1$ to the circle $x^2 + y^2 = 9$. The locus of the mid-point of the chord of contact is

(a) $x^2 + y^2 = \frac{x^2}{9} - \frac{y^2}{4}$ (b) $(x^2 + y^2)^2 = \frac{x^2}{9} - \frac{y^2}{4}$
(c) $(x^2 + y^2)^2 = 81\left(\frac{x^2}{9} - \frac{y^2}{4}\right)$ (d) $(x^2 + y^2)^2 = 9\left(\frac{x^2}{9} - \frac{y^2}{4}\right)$

26 A tangent to the hyperbola $\frac{x^2}{4} - \frac{y^2}{2} = 1$ meets X-axis at P and Y-axis at Q. Lines PR and QR are drawn such that $OPRQ$ is a rectangle (where, O is the origin). Then, R lies on → JEE Mains 2013

(a) $\frac{4}{x^2} + \frac{2}{y^2} = 1$ (b) $\frac{2}{x^2} - \frac{4}{y^2} = 1$
(c) $\frac{2}{x^2} + \frac{4}{y^2} = 1$ (d) $\frac{4}{x^2} - \frac{2}{y^2} = 1$

27 Consider a branch of the hyperbola $x^2 - 2y^2 - 2\sqrt{2}x - 4\sqrt{2}y - 6 = 0$ with vertex at the point A. Let B be one of the end points of its latusrectum. If C is the focus of the hyperbola nearest to the point A, then the area of the ΔABC is

(a) $\left(1 - \sqrt{\frac{2}{3}}\right)$ sq unit (b) $\left(\sqrt{\frac{3}{2}} - 1\right)$ sq unit
(c) $\left(1 + \sqrt{\frac{2}{3}}\right)$ sq units (d) $\left(\sqrt{\frac{3}{2}} + 1\right)$ sq units

28 Given the base BC of ΔABC and if $\angle B - \angle C = K$, a constant, then locus of the vertex A is a hyperbola.

(a) No (b) Yes (c) Both (d) None

29 Let $P(a \sec \theta, b \tan \theta)$ and $Q(a \sec \phi, b \tan \phi)$, where $\theta + \phi = \frac{\pi}{2}$, be two points on $\frac{x^2}{a^2} - \frac{y^2}{b^2} = 13$. If (h, k) is the point of intersection of normals at P and Q, then k is

(a) $\frac{a^2 + b^2}{a}$ (b) $-\frac{(a^2 + b^2)}{a}$
(c) $\frac{a^2 + b^2}{b}$ (d) $-\frac{(a^2 + b^2)}{b}$

30 If the chords of contact of tangents from two points (x_1, y_1) and (x_2, y_2) to the hyperbola $4x^2 - 9y^2 - 36 = 0$ are at right angles, then $\frac{x_1 x_2}{y_1 y_2}$ is equal to

(a) $\frac{9}{4}$ (b) $-\frac{9}{4}$ (c) $\frac{81}{16}$ (d) $-\frac{81}{16}$

31 If $x = 9$ is the chord of contact of the hyperbola $x^2 - y^2 = 9$, then the equation of the corresponding pair of tangents is

(a) $9x^2 - 8y^2 + 18x - 9 = 0$ (b) $9x^2 - 8y^2 - 18x + 9 = 0$
(c) $9x^2 - 8y^2 - 18x - 9 = 0$ (d) $9x^2 - 8y^2 + 18x + 9 = 0$

32 If chords of the hyperbola $x^2 - y^2 = a^2$ touch the parabola $y^2 = 4ax$. Then, the locus of the middle points of these chords is

(a) $y^2 = (x - a)x^3$ (b) $y^2(x - a) = x^3$
(c) $x^2(x - a) = x^3$ (d) None of these

33 The locus of middle points of chords of hyperbola $3x^2 - 2y^2 + 4x - 6y = 0$ parallel to $y = 2x$ is

(a) $3x - 4y = 4$ (b) $3y - 4x + 4 = 0$
(c) $4x - 3y = 3$ (d) $3x - 4y = 2$

34 The equation of the chord joining two points (x_1, y_1) and (x_2, y_2) on the rectangular hyperbola $xy = c^2$ is → AIEEE 2002

(a) $\frac{x}{x_1 + x_2} + \frac{y}{y_1 + y_2} = 1$ (b) $\frac{x}{x_1 - x_2} + \frac{y}{y_1 - y_2} = 1$
(c) $\frac{x}{y_1 + y_2} + \frac{y}{x_1 + x_2} = 1$ (d) $\frac{x}{y_1 - y_2} + \frac{y}{x_1 - x_2} = 1$

35 Match the vertices (v) and foci (f) of hyperbola given in Column I with their corresponding equation given in Column II and choose the correct option from the codes given below.

Column I	Column II
A. $v(\pm 2, 0)$, $f(\pm 3, 0)$	1. $\frac{y^2}{25} - \frac{x^2}{39} = 1$
B. $v(0, \pm 5)$, $f(0, \pm 8)$	2. $\frac{y^2}{9} - \frac{x^2}{16} = 1$
C. $v(0, \pm 3)$, $f(0 \pm 5)$	3. $\frac{x^2}{4} - \frac{y^2}{5} = 1$

Codes

	A	B	C
(a)	3	1	2
(b)	1	3	2
(c)	3	2	1
(d)	2	1	3

DAY PRACTICE SESSION 2

PROGRESSIVE QUESTIONS EXERCISE

1 If chords of the hyperbola $x^2 - y^2 = a^2$ touch the parabola $y^2 = 4ax$. Then, the locus of their middle point is

(a) $y^2(x-a) = 2x^2$ (b) $y^2(x-a) = x^3$
(c) $y^2(x-a) = x^4$ (d) $y^2(x+a) = x^3$

2 The equation of a tangent to the hyperbola $3x^2 - y^2 = 3$ parallel to the line $y = 2x + 4$ is

(a) $y = 3x + 4$ (b) $y = 2x + 1$
(c) $y = 2x - 2$ (d) $y = 3x + 5$

3 Let $P(a\sec\theta, b\tan\theta)$ and $Q(a\sec\phi, b\tan\phi)$, where $\theta+\phi = \frac{\pi}{2}$ be two points on the hyperbola $\frac{x^2}{a^2} - \frac{y^2}{b^2} = 1$. If (h,k) is the point of intersection of normals at P and Q, then k is equal to

(a) $\frac{a^2+b^2}{a}$ (b) $-\left[\frac{a^2+b^2}{a}\right]$ (c) $\frac{a^2+b^2}{b}$ (d) $-\left[\frac{a^2+b^2}{b}\right]$

4 The equation of the asymptotes of the hyperbola $3x^2 + 4y^2 + 8xy - 8x - 4y - 6 = 0$ is

(a) $3x^2 + 4y^2 + 8xy - 8x - 4y - 3 = 0$
(b) $3x^2 + 4y^2 + 8xy - 8x - 4y + 3 = 0$
(c) $3x^2 + 4y^2 + 8xy - 8x - 4y + 6 = 0$
(d) $4x^2 + 3y^2 + 2xy - x + y + 3 = 0$

5 The angle between the rectangular hyperbolas $(y-mx)(my+x) = a^2$ and $(m^2-1)(y^2-x^2) + 4mxy = b^2$ is

(a) $\frac{\pi}{2}$ (b) $\frac{\pi}{3}$ (c) $\frac{\pi}{4}$ (d) $\frac{\pi}{6}$

6 If PQ is a double ordinate of the hyperbola $\frac{x^2}{a^2} - \frac{y^2}{b^2} = 1$ such that OPQ is an equilateral triangle, O being the centre of the hyperbola. Then, the eccentricity e of the hyperbola satisfies

(a) $1 < e < \frac{2}{\sqrt{3}}$ (b) $e = \frac{2}{\sqrt{3}}$
(c) $e = \frac{\sqrt{3}}{2}$ (d) $e > \frac{2}{\sqrt{3}}$

7 The normal at P to a hyperbola of eccentricity e, intersects its transverse and conjugate axes at L and M respectively. If locus of the mid-point of LM is hyperbola, then eccentricity of the hyperbola is

(a) $\left(\frac{e+1}{e-1}\right)$ (b) $\frac{e}{\sqrt{(e^2-1)}}$ (c) e (d) None of these

8 Tangents are drawn to the hyperbola $4x^2 - y^2 = 36$ at the points P and Q. If these tangents intersect at the point $T(0, 3)$, then the area (in sq units) of ΔPTQ is

→ JEE Mains 2018

(a) $45\sqrt{5}$ (b) $54\sqrt{3}$
(c) $60\sqrt{3}$ (d) $36\sqrt{5}$

9 If tangents PQ and PR are drawn from variable point P to the hyperbola $\frac{x^2}{a^2} - \frac{y^2}{b^2} = 1 (a > b)$ so that the fourth vertex S of parallelogram $PQSR$ lies on circumcircle of ΔPQR, then locus of P is

(a) $x^2 + y^2 = b^2$ (b) $x^2 + y^2 = a^2$
(c) $x^2 + y^2 = (a^2 - b^2)$ (d) None of these

10 The equation of the common tangent to the curves $y^2 = 8x$ and $xy = -1$ is

(a) $3y = 9x + 2$ (b) $y = 2x + 1$
(c) $2y = y + 8$ (d) $y = x + 2$

11 A series of hyperbolas is drawn having a common transverse axis of length $2a$. Then, the locus of a point P on each hyperbola, such that its distance from the transverse axis is equal to its distance from an asymptote, is

(a) $(y^2 - x^2)^2 = 4x^2(x^2 - a^2)$ (b) $(x^2 - y^2)^2 = x^2(x^2 - a^2)$
(c) $(x^2 - y^2) = 4x^2(x^2 - a^2)$ (d) None of these

12 The exhaustive set of values of α^2 such that there exists a tangent to the ellipse $x^2 + \alpha^2 y^2 = \alpha^2$ and the portion of the tangent intercepted by the hyperbola $\alpha^2 x^2 - y^2 = 1$ subtends a right angle at the center of the curves is

(a) $\left[\frac{\sqrt{5}+1}{2}, 2\right]$ (b) $(1, 2]$
(c) $\left[\frac{\sqrt{5}-1}{2}, 1\right)$ (d) $\left[\frac{\sqrt{5}-1}{2}, 1\right) \cup \left(1, \frac{\sqrt{5}+1}{2}\right]$

13 A rectangular hyperbola whose centre is C is cut by any circle of radius r in four points P, Q, R and S. Then, $CP^2 + CQ^2 + CR^2 + CS^2$ is equal to

(a) r^2 (b) $2r^2$
(c) $3r^2$ (d) $4r^2$

14 A point P moves such that sum of the slopes of the normals drawn from it to the hyperbola $xy = 4$ is equal to the sum of ordinates of feet of normals. Then, the locus of P is

(a) a parabola (b) a hyperbola
(c) an ellipse (d) a circle

15 A triangle is inscribed in the rectangular hyperbola $xy = c^2$, such that two of its sides are parallel to the lines $y = m_1 x$ and $y = m_2 x$. Then, the third side of the triangle touches the hyperbola

(a) $xy = \left\{\frac{c^2(m_1+m_2)^2}{4m_1 m_2}\right\}$ (b) $xy = \left\{\frac{c^2(m_1-m_2)^2}{4m_1 m_2}\right\}$
(c) $xy = \left\{\frac{c^2(m_1-m_2)^2}{m_1 m_2}\right\}$ (d) $xy = \left\{\frac{c^2(m_1+m_2)^2}{m_1 m_2}\right\}$

ANSWERS

SESSION 1	**1.** (a)	**2.** (b)	**3.** (d)	**4.** (d)	**5.** (c)	**6.** (a)	**7.** (b)	**8.** (c)	**9.** (a)	**10.** (b)
	11. (c)	**12.** (c)	**13.** (c)	**14.** (b)	**15.** (b)	**16.** (b)	**17.** (d)	**18.** (b)	**19.** (a)	**20.** (a)
	21. (d)	**22.** (a)	**23.** (a)	**24.** (d)	**25.** (c)	**26.** (d)	**27.** (b)	**28.** (b)	**29.** (d)	**30.** (d)
	31. (b)	**32.** (b)	**33.** (a)	**34.** (a)	**35.** (a)					
SESSION 2	**1.** (b)	**2.** (b)	**3.** (d)	**4.** (a)	**5.** (a)	**6.** (d)	**7.** (b)	**8.** (a)	**9.** (c)	**10.** (d)
	11. (a)	**12.** (a)	**13.** (d)	**14.** (a)	**15.** (a)					

Hints and Explanations

SESSION 1

1 Since, the vertices are on the Y-axis (with origin as the mid-point), the equation is of the form $\frac{y^2}{a^2} - \frac{x^2}{b^2} = 1$

As vertices are $(0, \pm 6)$

$\therefore \quad a = 6, b^2 = a^2(e^2 - 1) = 36\left(\frac{25}{9} - 1\right)$
$= 64$

So, the required equation of the hyperbola is $\frac{y^2}{36} - \frac{x^2}{64} = 1$ and the foci are $(0, \pm ae) = (0, \pm 10)$.

2 Let the equation of the hyperbola be

$$\frac{x^2}{a^2} - \frac{y^2}{b^2} = 1 \quad \ldots\text{(i)}$$

we have, length of the latusrectum = 8

$\Rightarrow \quad \frac{2b^2}{a} = 8 \Rightarrow b^2 = 4a$

$\Rightarrow \quad a^2(e^2 - 1) = 4a \Rightarrow a(e^2 - 1) = 4$

$\Rightarrow \quad a\left(\frac{9}{5} - 1\right) = 4 \Rightarrow a = 5$

Putting $a = 5$ in $b^2 = 4a$, we get $b^2 = 20$

Hence, the equation of the required hyperbola is $\frac{x^2}{25} - \frac{y^2}{20} = 1$.

3 Given that the hyperbola $\frac{x^2}{a^2} - \frac{y^2}{b^2} = 1$ is passing through the points $(3, 0)$ and $(3\sqrt{2}, 2)$, so we get $a^2 = 9$ and $b^2 = 4$.

Again, we know that $b^2 = a^2(e^2 - 1)$.

This gives $4 = 9(e^2 - 1)$

$\Rightarrow \quad e^2 = \frac{13}{9} \Rightarrow e = \frac{\sqrt{13}}{3}$

4 Given equation can be rewritten as

$$\frac{y^2}{k^2} - \frac{x^2}{(-k)} = 1 (-k > 0)$$

$$e^2 = 1 + \frac{(-k)}{k^2} = 1 - \frac{k}{k^2}$$

$\Rightarrow \quad e = \sqrt{1 - \frac{1}{k}}$

5

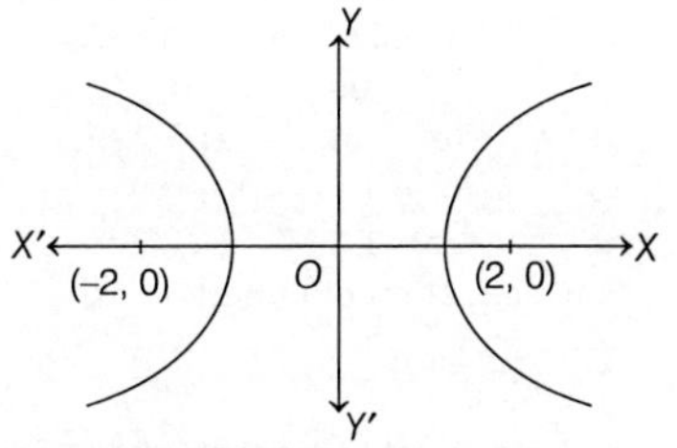

Let equation of hyperbola be

$$\frac{x^2}{a^2} - \frac{y^2}{b^2} = 1$$

where, $2ae = 4$ and $e = 2$

$\Rightarrow \quad a = 1$

$\because \quad a^2e^2 = a^2 + b^2 \Rightarrow 4 = 1 + b^2$

$\therefore \quad b^2 = 3$

Thus, equation of hyperbola is

$$\frac{x^2}{1} - \frac{y^2}{3} = 1 \text{ or } 3x^2 - y^2 = 3$$

6 The given equation may be written as

$$\frac{x^2}{\frac{32}{3}} - \frac{y^2}{8} = 1 \Rightarrow \frac{x^2}{\left(\frac{4\sqrt{2}}{\sqrt{3}}\right)^2} - \frac{y^2}{(2\sqrt{2})^2} = 1$$

On comparing the given equation with the standard equation

$\frac{x^2}{a^2} - \frac{y^2}{b^2} = 1$, we get

$$a^2 = \left(\frac{4\sqrt{2}}{\sqrt{3}}\right)^2 \text{ and } b^2 = (2\sqrt{2})^2$$

$\therefore$ Length of transverse axis of a hyperbola $= 2a = 2 \times \frac{4\sqrt{2}}{\sqrt{3}} = \frac{8\sqrt{2}}{\sqrt{3}}$

7 Let equation of hyperbola be

$\frac{x^2}{a^2} - \frac{y^2}{b^2} = 1$

Length of transverse axis is $2a$ and

Length of latusrectum is $\frac{2b^2}{a}$.

Now, difference

$$E = \left|2a - \frac{2b^2}{a}\right| = \frac{2}{|a|}|2a^2 - a^2e^2|$$

$\therefore$ Difference $= 2a|2 - e^2|$

8 Given equation of hyperbola is $\frac{x^2}{a^2} - \frac{y^2}{b^2} = 1$ and equation of conjugate hyperbola is $\frac{x^2}{b^2} - \frac{y^2}{a^2} = 1$

Since, e and e' are the eccentricities of the respective hyperbola, then

$$e^2 = 1 + \frac{b^2}{a^2}, (e')^2 = 1 + \frac{a^2}{b^2}$$

$\therefore \quad \frac{1}{e^2} + \frac{1}{e'^2} = \frac{a^2}{a^2 + b^2} + \frac{b^2}{a^2 + b^2} = 1$

9 Here, $a = \sin\theta$

Since, foci of the ellipse are $(\pm 1, 0)$.

$\therefore \pm 1 = \sqrt{a^2 + b^2} \Rightarrow b^2 = \cos^2\theta$

Then, equation is $\frac{x^2}{\sin^2\theta} - \frac{y^2}{\cos^2\theta} = 1$

$\Rightarrow \quad x^2 \operatorname{cosec}^2\theta - y^2 \sec^2\theta = 1$

10 Given equation can be rewritten as $(x - 4)(y - 3) = 4$ which is a rectangular hyperbola of the type $xy = c^2$.

$\therefore \quad c = 2$

Then, $\quad a = b = c\sqrt{2} = 2\sqrt{2}$

$\therefore$ Length of latusrectum

$$= \frac{2b^2}{a} = 2a = 4\sqrt{2}$$

11 Let equation of hyperbola be

$$\frac{x^2}{a^2} - \frac{y^2}{b^2} = 1.$$

Given, $\frac{2b^2}{a} = 8 \Rightarrow \frac{b^2}{a} = 4$

and $2b = \frac{1}{2}(2ae) \Rightarrow 2b = ae$

$\Rightarrow \quad 4b^2 = a^2e^2 \Rightarrow 4\left(\frac{b^2}{a^2}\right) = e^2$

$\Rightarrow \quad 4(e^2 - 1) = e^2 \quad [\because b^2 = a^2(e^2 - 1)]$

$\Rightarrow \quad 3e^2 = 4 \Rightarrow e^2 = \frac{4}{3} \Rightarrow e = \frac{2}{\sqrt{3}}$

12 Now, taking option (c).

Let $x = a\dfrac{e^t + e^{-t}}{2} \Rightarrow \dfrac{2x}{a} = e^t + e^{-t}$ ...(i)

and $\dfrac{2y}{b} = e^t - e^{-t}$ ...(ii)

On squaring and subtracting Eq. (ii) from Eq. (i), we get

$$\frac{4x^2}{a^2} - \frac{4y^2}{b^2} = 4 \Rightarrow \frac{x^2}{a^2} - \frac{y^2}{b^2} = 1$$

13 The given equation of hyperbola is

$$\frac{x^2}{\cos^2\alpha} - \frac{y^2}{\sin^2\alpha} = 1$$

Here, $a^2 = \cos^2\alpha$ and $b^2 = \sin^2\alpha$

So, the coordinates of foci are $(\pm\, ae, 0)$.

$$\therefore \quad e = \sqrt{1 + \frac{b^2}{a^2}} \Rightarrow e = \sqrt{1 + \frac{\sin^2\alpha}{\cos^2\alpha}}$$

$$\Rightarrow \quad e = \sqrt{1 + \tan^2\alpha} = \sec\alpha$$

Hence, abscissae of foci remain constant when α varies.

14 Let equation of tangent to hyperbola

$\dfrac{x^2}{a^2} - \dfrac{y^2}{b^2} = 1$ is $y = mx + \sqrt{a^2m^2 - b^2}$

i.e. $\quad mx - y + \sqrt{a^2m^2 - b^2} = 0$

$\therefore$ Required product

$$= \left|\frac{mae + \sqrt{a^2m^2 - b^2}}{\sqrt{m^2 + 1}}\right| \left|\frac{-\,mae + \sqrt{a^2m^2 - b^2}}{\sqrt{m^2 + 1}}\right|$$

$$= \left|\frac{a^2m^2 - b^2 - m^2a^2e^2}{m^2 + 1}\right|$$

$$= \left|\frac{m^2a^2\,(1 - e^2) - b^2}{m^2 + 1}\right| = \left|\frac{-\,m^2b^2 - b^2}{m^2 + 1}\right|$$

$[\because b^2 = a^2(e^2 - 1)]$

$= b^2$

15. Here, $a^2 = \dfrac{232}{7}, b^2 = \dfrac{232}{5}$

and $y = -\dfrac{21}{5}x + \dfrac{116}{5}$ with slope $-\dfrac{21}{5}$.

Now, $a^2m^2 - b^2 = \left(\dfrac{116}{5}\right)^2$

[since, line is tangent]

If (x_1, y_1) is the point of contact, then tangent is $S_1 = 0$

$\therefore\ 7x_1x - 5y_1y = 232$

On comparing it with

$21x + 5y = 116$, we get $x_1 = 6, y_1 = -2$

So, the point of contact is $(6, -2)$.

16 Here, $e_1 = \sqrt{1 - \dfrac{16}{25}} = \dfrac{3}{5}$

$\because\ e_1e_2 = 1 \Rightarrow e_2 = \dfrac{5}{3}$

Since, foci of ellipse are $(0, \pm\, 3)$.

Hence, equation of hyperbola is

$$\frac{x^2}{16} - \frac{y^2}{9} = -1.$$

17 As we know, equation of tangent at (x_1, y_1) is $xx_1 - 2yy_1 = 4$, which is same as $2x + \sqrt{6}y = 2$

$$\therefore \quad \frac{x_1}{2} = -\frac{2y_1}{\sqrt{6}} = \frac{4}{2} \Rightarrow x_1 = 4$$

and $y_1 = -\sqrt{6}$

18 Let the equation of hyperbola be

$$\frac{x^2}{a^2} - \frac{y^2}{b^2} = 1.$$

$\therefore \quad ae = 2 \Rightarrow a^2e^2 = 4$

$\Rightarrow \quad a^2 + b^2 = 4 \Rightarrow b^2 = 4 - a^2$

$\therefore \quad \dfrac{x^2}{a^2} - \dfrac{y^2}{4 - a^2} = 1$

Since, $(\sqrt{2}, \sqrt{3})$ lie on hyperbola.

$\therefore \quad \dfrac{2}{a^2} - \dfrac{3}{4 - a^2} = 1$

$\Rightarrow \quad 8 - 2a^2 - 3a^2 = a^2(4 - a^2)$

$\Rightarrow \quad 8 - 5a^2 = 4a^2 - a^4$

$\Rightarrow \quad a^4 - 9a^2 + 8 = 0$

$\Rightarrow \quad (a^2 - 8)(a^2 - 1) = 0$

$\Rightarrow \quad a^2 = 8, a^2 = 1$

$\therefore \quad a = 1$

Now, equation of hyperbola is

$$\frac{x^2}{1} - \frac{y^2}{3} = 1.$$

$\therefore$ Equation of tangent at $(\sqrt{2}, \sqrt{3})$ is given by

$$\sqrt{2}x - \frac{\sqrt{3}y}{3} = 1 \Rightarrow \sqrt{2}x - \frac{y}{\sqrt{3}} = 1$$

which passes through point $(2\sqrt{2}, 3\sqrt{3})$.

19 Suppose the common tangent is $y = mx + c$ to

$9x^2 - 4y^2 = 36$ and $x^2 + y^2 = 3$

$\therefore \quad c^2 = a^2m^2 - b^2 = 4m^2 - 9$ ...(i)

and $c^2 = 3 + 3m^2$ ...(ii)

From Eqs. (i) and (ii), we get

$4m^2 - 9 = 3m^2 + 3 \Rightarrow m^2 = 12$

$\Rightarrow\ m = 2\sqrt{3}$

$\therefore \quad c = \sqrt{3 + 3 \times 12} = \sqrt{39}$

Hence, the common tangent is

$y = 2\sqrt{3}x + \sqrt{39}$

20 Since, the line $y = \alpha x + \beta$ is tangent to the hyperbola $\dfrac{x^2}{a^2} - \dfrac{y^2}{b^2} = 1$.

$\therefore \quad \beta^2 = a^2\alpha^2 - b^2$.

So, locus of (α, β) is $y^2 = a^2x^2 - b^2$

$\Rightarrow a^2x^2 - y^2 - b^2 = 0$

Since, this equation represents a hyperbola, so locus of a point $P(\alpha, \beta)$ is a hyperbola.

21 Given equation of hyperbola is

$$\frac{x^2}{9} - \frac{y^2}{4} = 1.$$

Here, $\quad a^2 = 9, b^2 = 4$

and equation of line is

$$y = -x + \sqrt{2}\,p \qquad \text{...(i)}$$

If the line $y = mx + c$ touches the hyperbola

$\dfrac{x^2}{a^2} - \dfrac{y^2}{b^2} = 1$, then $c^2 = a^2m^2 - b^2$ ...(ii)

From Eq. (i), we get $m = -1, c = \sqrt{2}p$

On putting these values in Eq. (ii), we get

$$(\sqrt{2}p)^2 = 9(1) - 4 \Rightarrow 2p^2 = 5$$

22 We know that the locus of the point of intersection of perpendicular tangents to the hyperbola

$\dfrac{x^2}{a^2} - \dfrac{y^2}{b^2} = 1$ is a circle $x^2 + y^2 = a^2 - b^2$

Thus, locus of the point of intersection of perpendicular tangents to the hyperbola

$\dfrac{x^2}{3} - \dfrac{y^2}{1} = 1$ is a circle

$$x^2 + y^2 = 3 - 1 \Rightarrow x^2 + y^2 = 2$$

23 The point on the hyperbola is $P(x_1, y_1)$, then N is $(x, 0)$.

$\therefore$ Tangent at (x_1, y_1) is $\dfrac{xx_1}{9} - \dfrac{yy_1}{4} = 1$

This meets X-axis at $T\left(\dfrac{9}{x_1}, 0\right)$

$\therefore\ OT \cdot ON = \dfrac{9}{x_1} \cdot x_1 = 9$

24 The equation of tangent in slope form to the hyperbola $\dfrac{x^2}{16} - \dfrac{y^2}{9} = 1$ is

$y = mx + \sqrt{16\,m^2 - 9}$

Since, it passes through (h, k).

$\therefore \quad k = mh + \sqrt{16\,m^2 - 9^2}$

$\Rightarrow\ (k - mh)^2 = (16\,m^2 - 9^2)$

$\Rightarrow k^2 + m^2h^2 - 2m\,kh - 16\,m^2 + 9^2 = 0$

It is quadratic in m and let the slope of two tangents be m_1 and m_2, then

$$m_1m_2 = \frac{k^2 + 9}{h^2 - 16}$$

$$\Rightarrow \quad -1 = \frac{k^2 + 9}{h^2 - 16} \Rightarrow h^2 + k^2 = 7$$

The required locus is $x^2 + y^2 = 7$

25 Any point on the hyperbola is $(3\sec\theta, 2\tan\theta)$. The chord of contact to the circle is $3x\sec\theta + 2y\tan\theta = 9$ ...(i)

If (x_1, y_1) is the mid-point of the chord, then its equation is

$$xx_1 + yy_1 = x_1^2 + y_1^2 \qquad \text{...(ii)}$$

From Eqs. (i) and (ii), we get

$$\frac{3\sec\theta}{x_1} = \frac{2\tan\theta}{y_1} = \frac{9}{x_1^2 + y_1^2}$$

Eliminating θ, $\dfrac{1}{81}(x_1^2 + y_1^2)^2 = \dfrac{x_1^2}{9} - \dfrac{y_1^2}{4}$

Hence, the locus is

$$(x^2 + y^2)^2 = 81\left(\frac{x^2}{9} - \frac{y^2}{4}\right)$$

26 Given hyperbola is $\frac{x^2}{4} - \frac{y^2}{2} = 1$.

Here, $a^2 = 4, b^2 = 2 \Rightarrow a = 2, b = \sqrt{2}$

The equation of tangent is

$$\frac{x}{a}\sec\theta - \frac{y}{b}\tan\theta = 1$$

$$\Rightarrow \frac{x}{2}\sec\theta - \frac{y}{\sqrt{2}}\tan\theta = 1$$

So, the coordinates of P and Q are $P(2\cos\theta, 0)$ and $Q(0, -\sqrt{2}\cot\theta)$, respectively.

Let coordinates of R are (h, k).

$\therefore \quad h = 2\cos\theta, k = -\sqrt{2}\cot\theta$

$$\Rightarrow \frac{k}{h} = \frac{-\sqrt{2}}{\sin\theta} \Rightarrow \sin\theta = \frac{-\sqrt{2}h}{2k}$$

On squaring both sides, we get

$$\sin^2\theta = \frac{2h^2}{4k^2} \Rightarrow 1 - \cos^2\theta = \frac{2h^2}{4k^2}$$

$$\Rightarrow 1 - \frac{h^2}{4} = \frac{2h^2}{4k^2} \Rightarrow \frac{2h^2}{4k^2} + \frac{h^2}{4} = 1$$

$$\Rightarrow \frac{h^2}{4}\left(\frac{2}{k^2} + 1\right) = 1 \Rightarrow \frac{2}{k^2} + 1 = \frac{4}{h^2}$$

$$\Rightarrow \frac{4}{h^2} - \frac{2}{k^2} = 1$$

Hence, R lies on $\frac{4}{x^2} - \frac{2}{y^2} = 1$.

27 The given equation can be rewritten as

$$\frac{(x - \sqrt{2})^2}{4} - \frac{(y + \sqrt{2})^2}{2} = 1$$

For $A(x, y)$,

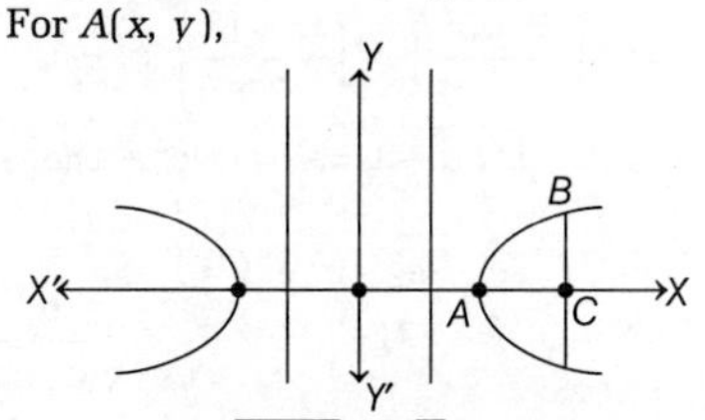

$$e = \sqrt{1 + \frac{2}{4}} = \sqrt{\frac{3}{2}}$$

$\therefore \quad x - \sqrt{2} = 2 \Rightarrow x = 2 + \sqrt{2}$

For $C(x, y)$, $x - \sqrt{2} = ae = \sqrt{6}$

$\therefore \quad x = \sqrt{6} + \sqrt{2}$

Now, $AC = \sqrt{6} + \sqrt{2} - 2 - \sqrt{2} = \sqrt{6} - 2$

and $\quad BC = \frac{b^2}{a} = \frac{2}{2} = 1$

$\therefore$ Area of $\Delta ABC = \frac{1}{2} \times AC \times BC$

$$= \frac{1}{2} \times (\sqrt{6} - 2) \times 1 = \left(\sqrt{\frac{3}{2}} - 1\right) \text{ sq unit}$$

28 Let $B(-\lambda, 0)$, $C(\lambda, 0)$ and $A(x, y)$.

Given, $K = \angle B - \angle C$

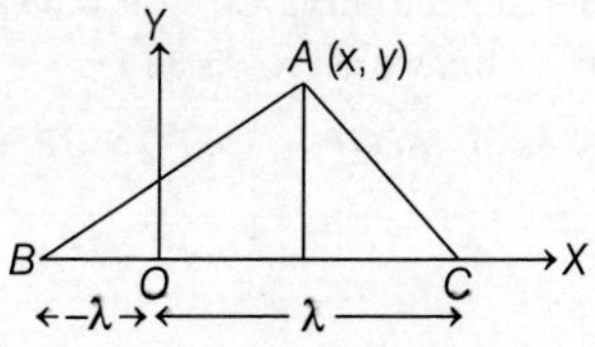

$$\therefore \tan K = \frac{\tan B - \tan C}{1 + \tan B \cdot \tan C}$$

$$= \frac{\frac{y}{\lambda + x} - \frac{y}{\lambda - x}}{1 + \frac{y^2}{\lambda^2 - x^2}}$$

$$\Rightarrow \quad \lambda^2 - x^2 + y^2 = -2xy\cot K$$

$$\Rightarrow x^2 - 2\cot K \cdot xy - y^2 = \lambda^2$$

which is a hyperbola.

29 Equation of normal at θ is $\frac{ax}{\sec\theta} + \frac{by}{\tan\theta} = a^2 + b^2$ and normal at $\phi = \frac{\pi}{2} - \theta$ is $\frac{ax}{\operatorname{cosec}\theta} + \frac{by}{\cot\theta} = a^2 + b^2$

Eliminating x, we get

$$by\left(\frac{1}{\sin\theta} - \frac{1}{\cos\theta}\right) = (a^2 + b^2)\left(\frac{1}{\cos\theta} - \frac{1}{\sin\theta}\right)$$

$$\Rightarrow by = -(a^2 + b^2) \text{ or } k = -\frac{(a^2 + b^2)}{b}$$

30 The equation of hyperbola is

$$4x^2 - 9y^2 = 36 \Rightarrow \frac{x^2}{9} - \frac{y^2}{4} = 1 \quad \text{...(i)}$$

The equation of the chords of contact of tangents from (x_1, y_1) and (x_2, y_2) to the given hyperbola are

$$\frac{xx_1}{9} - \frac{yy_1}{4} = 1 \quad \text{...(ii)}$$

and $$\frac{xx_2}{9} - \frac{yy_2}{4} = 1 \quad \text{...(iii)}$$

Lines (ii) and (iii) are at right angles.

$$\frac{9}{4}\cdot\frac{x_1}{y_1} \times \frac{4}{9}\cdot\frac{x_2}{y_2} = -1 \Rightarrow \frac{x_1x_2}{y_1y_2} = -\left(\frac{9}{4}\right)^2 = -\frac{81}{16}$$

31 Let (h, k) be the point whose chord of contact w.r.t. hyperbola $x^2 - y^2 = 9$ is $x = 9$. We know that chord of (h, k) w.r.t. hyperbola $x^2 - y^2 = 9$ is $T = 0$

$\Rightarrow \quad hx - ky - 9 = 0$

But it is the equation of line $x = 9$. This is possible only when $h = 1, k = 0$. Again, equation of pair of tangents is $T^2 = SS_1$

$\Rightarrow \quad (x - 9)^2 = (x^2 - y^2 - 9)(1 - 9)$

$\Rightarrow \quad x^2 - 18x + 81 = (x^2 - y^2 - 9)(-8)$

$\Rightarrow \quad 9x^2 - 8y^2 - 18x + 9 = 0$

32 Equation of chord of hyperbola $x^2 - y^2 = a^2$ with mid-point as (h, k) is given by

$$xh - yk = h^2 - k^2 \Rightarrow y = \frac{h}{k}x - \frac{(h^2 - k^2)}{k}$$

This will touch the parabola $y^2 = 4ax$, if

$$-\left(\frac{h^2 - k^2}{k}\right) = \frac{a}{h/k} \Rightarrow ak^2 = -h^3 + k^2h$$

$\therefore$ Locus of the mid-point is $x^3 = y^2(x - a)$

33 Let (h, k) is mid-point of chord. Then, its equation is

$$3hx - 2ky + 2(x + h) - 3(y + k) = 3h^2 - 2k^2 + 4h - 6k$$

$$\Rightarrow x(3h + 2) + y(-2k - 3) = 3h^2 - 2k^2 + 2h - 3k$$

Since, this line is parallel to $y = 2x$.

$$\therefore \quad \frac{3h + 2}{2k + 3} = 2 \Rightarrow 3h + 2 = 4k + 6$$

$\Rightarrow \quad 3h - 4k = 4$

Thus, locus of mid-point is $3x - 4y = 4$

34 The mid-point of the chord is $\left(\frac{x_1 + x_2}{2}, \frac{y_1 + y_2}{2}\right)$.

The equation of the chord $T = S_1$.

$$\therefore \quad x\left(\frac{y_1 + y_2}{2}\right) + y\left(\frac{x_1 + x_2}{2}\right) = 2\left(\frac{x_1 + x_2}{2}\right)\left(\frac{y_1 + y_2}{2}\right)$$

$$\Rightarrow x(y_1 + y_2) + y(x_1 + x_2) = (x_1 + x_2)(y_1 + y_2)$$

$$\Rightarrow \frac{x}{x_1 + x_2} + \frac{y}{y_1 + y_2} = 1$$

35 A. Since, vertices $(\pm 2, 0)$ and foci $(\pm 3, 0)$ lie on X-axis, as coefficient of Y-axis is zero.

Hence, equation of hyperbola will be of the form

$$\frac{x^2}{a^2} - \frac{y^2}{b^2} = 1 \quad \text{...(i)}$$

Here, it is given that $a = 2$ and $c = 3$

$\because \quad c^2 = a^2 + b^2$

$\Rightarrow \quad 9 = 4 + b^2 \Rightarrow b^2 = 5$

Put the values of $a^2 = 4$ and $b^2 = 5$ in Eq. (i), we get

$$\frac{x^2}{4} - \frac{y^2}{5} = 1 \quad \text{...(ii)}$$

B. Since, vertices $(0, \pm 5)$ and foci $(0, \pm 8)$ lie on Y-axis as coefficient of X-axis is zero. Hence, equation of hyperbola will be of the form

$$\frac{y^2}{a^2} - \frac{x^2}{b^2} = 1 \quad \text{...(i)}$$

Here, it is given that $(0, \pm a) = (0, \pm 5)$ and foci $(0, \pm c) = (0, \pm 8)$

$\Rightarrow \quad a = 5$ and $c = 8$

$\because \quad c^2 = a^2 + b^2 \Rightarrow b^2 = 39$

Put $a^2 = 25$ and $b^2 = 39$ in Eq. (i), we get

$$\frac{y^2}{25} - \frac{x^2}{39} = 1$$

C. Since, vertices $(0, \pm 3)$ and foci $(0, \pm 5)$ lie on Y- axis as coordinate of x is zero.

Hence, equation of hyperbola will be of the form

$$\frac{y^2}{a^2} - \frac{x^2}{b^2} = 1 \quad \text{...(i)}$$

Here, it is given that vertices $(0, \pm 3) = (0, \pm a)$ and foci

$$(0, \pm 5) = (0, \pm c)$$

$\Rightarrow \quad a = 3$ and $c = 5$

$\because \quad c^2 = a^2 + b^2$

$\Rightarrow \quad 25 = 9 + b^2 \Rightarrow b^2 = 16$

Put $a^2 = 9$ and $b^2 = 16$ in Eq. (i), we get

$$\frac{y^2}{9} - \frac{x^2}{16} = 1$$

SESSION 2

1 If (h, k) is the mid-point of the chord, then its equation by $T = S_1$ is

$$hx - ky = h^2 - k^2$$

$$\Rightarrow \quad y = \frac{h}{k}x + \frac{k^2 - h^2}{k}$$

If it touches the parabola $y^2 = 4ax$, then we get

$$\frac{k^2 - h^2}{k} = \frac{a \cdot k}{h} \Rightarrow ak^2 = hk^2 - h^3$$

$$\Rightarrow \quad ay^2 = xy^2 - x^3$$

So, required locus is $y^2 (x - a) = x^3$.

2 Equation of the hyperbola is

$$3x^2 - y^2 = 3 \Rightarrow \frac{x^2}{1} - \frac{y^2}{3} = 1$$

Equation of tangent in terms of slope

$$y = mx \pm \sqrt{(m^2 - 3)}$$

Given, $\quad m = 2$

$\therefore \quad y = 2x \pm 1$

3 Equation of the tangents at $P(a\sec\theta, b\tan\theta)$ is $\frac{x}{a}\sec\theta - \frac{y}{b}\tan\theta = 1$

$\therefore$ Equation of the normal at P is

$$ax + b\,\text{cosec}\theta\; y = (a^2 + b^2)\sec\theta \quad \text{...(i)}$$

Similarly, the equation of normal at $Q(a\sec\phi, b\tan\phi)$ is

$$ax + b\,\text{cosec}\phi\, y = (a^2 + b^2)\sec\phi \quad \text{...(ii)}$$

On subtracting Eq. (ii) from Eq. (i), we get

$$y = \frac{a^2 + b^2}{b} \cdot \frac{\sec\theta - \sec\phi}{\text{cosec}\theta - \text{cosec}\phi}$$

$$\text{So,} \quad k = y = \frac{a^2 + b^2}{b} \cdot \frac{\sec\theta - \sec\left(\frac{\pi}{2} - \theta\right)}{\text{cosec}\theta - \text{cosec}\left(\frac{\pi}{2} - \theta\right)}$$

$$= \frac{a^2 + b^2}{b} \cdot \frac{\sec\theta - \text{cosec}\theta}{\text{cosec}\theta - \sec\theta} = -\left[\frac{a^2 + b^2}{b}\right]$$

4 The equation of the asymptotes of the hyperbola

$$3x^2 + 4y^2 + 8xy - 8x - 4y - 6 = 0$$

is $3x^2 + 4y^2 + 8xy - 8x - 4y + \lambda = 0$

It should represent a pair of straight lines.

$$\therefore \quad abc + 2fgh - af^2 - bg^2 - ch^2 = 0$$

$$3 \cdot 4 \cdot \lambda + 2 \cdot (-2)(-4)\,4 - 3(-2)^2$$

$$- 4(-4)^2 - \lambda(4)^2 = 0$$

$$\Rightarrow \quad 12\lambda + 64 - 12 - 64 - 16\lambda = 0$$

$$\Rightarrow \quad -4\lambda - 12 = 0 \Rightarrow \lambda = -3$$

$\therefore$ Required equation is

$$3x^2 + 4y^2 + 8xy - 8x - 4y - 3 = 0$$

5 Given equations are

$$(y - mx)(my + x) = a^2 \quad \text{...(i)}$$

and $(m^2 - 1)(y^2 - x^2) + 4mxy = b^2$ ...(ii)

On differentiating Eq.(i), we get

$$(y - mx)\left(m\frac{dy}{dx} + 1\right) + (my + x)\left(\frac{dy}{dx} - m\right) = 0$$

$$\Rightarrow \quad \frac{dy}{dx}(my + x + my - m^2x) + y - mx - m^2y - mx = 0$$

$$\Rightarrow \quad \frac{dy}{dx} = \frac{-y + m^2y + 2mx}{2my + x - m^2x} = m_1 \quad \text{[say]}$$

On differentiating Eq.(ii), we get

$$(m^2 - 1)\left(2y\frac{dy}{dx} - 2x\right) + 4m\left(x\frac{dy}{dx} + y\right) = 0$$

$$\Rightarrow \quad \frac{dy}{dx}[2y(m^2 - 1) + 4mx] = -4my + 2x(m^2 - 1)$$

$$\Rightarrow \quad \frac{dy}{dx} = \frac{-2my + m^2x - x}{m^2y - y + 2mx} = m_2 \quad \text{[say]}$$

$\because \quad m_1 m_2 = -1$

So, angle between the hyperbola $= \frac{\pi}{2}$.

6 $\because$ PQ is the double ordinate. Let $MP = MQ = l$.

Given that ΔOPQ is an equilateral, then $OP = OQ = PQ$

$$\Rightarrow \quad (OP)^2 = (OQ)^2 = (PQ)^2$$

$$\Rightarrow \frac{a^2}{b^2}(b^2 + l^2) + l^2 = \frac{a^2}{b^2}(b^2 + l^2) + l^2 = 4l^2$$

$$\Rightarrow \quad \frac{a^2}{b^2}(b^2 + l^2) = 3l^2$$

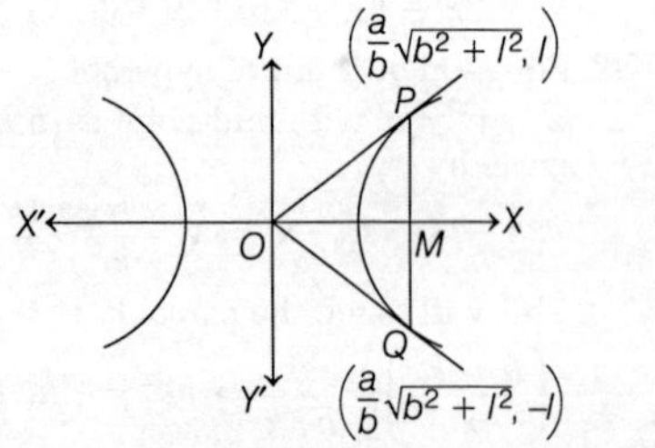

$$\Rightarrow \quad a^2 = l^2\left(3 - \frac{a^2}{b^2}\right)$$

$$\Rightarrow \quad l^2 = \frac{a^2b^2}{(3b^2 - a^2)} > 0$$

$$\therefore \quad 3b^2 - a^2 > 0$$

$$\Rightarrow \quad 3b^2 > a^2$$

$$\Rightarrow \quad 3a^2(e^2 - 1) > a^2$$

$$\Rightarrow \quad e^2 > 4/3 \;\therefore\; e > \frac{2}{\sqrt{3}}$$

7 Equation of normal at $P(a\sec\phi, b\tan\phi)$ is $ax\cos\phi + by\cot\phi = a^2 + b^2$.

Then, coordinates of L and M are

$$\left(\frac{a^2 + b^2}{a} \cdot \sec\phi, 0\right) \text{ and } \left(0, \frac{a^2 + b^2}{b}\tan\phi\right)$$

respectively.

Let mid-point of ML is $Q(h, k)$,

$$\text{then} \quad h = \frac{(a^2 + b^2)}{2a}\sec\phi$$

$$\therefore \quad \sec\phi = \frac{2ah}{(a^2 + b^2)} \quad \text{...(i)}$$

$$\text{and} \quad k = \frac{(a^2 + b^2)}{2b}\tan\phi$$

$$\therefore \quad \tan\phi = \frac{2bk}{(a^2 + b^2)} \quad \text{...(ii)}$$

From Eqs. (i) and (ii), we get

$$\sec^2\phi - \tan^2\phi = \frac{4a^2h^2}{(a^2 + b^2)^2} - \frac{4b^2k^2}{(a^2 + b^2)^2}$$

Hence, required locus is

$$\frac{x^2}{\left(\frac{a^2 + b^2}{2a}\right)^2} - \frac{y^2}{\left(\frac{a^2 + b^2}{2b}\right)^2} = 1$$

Let eccentricity of this curve is e_1.

$$\Rightarrow \quad \left(\frac{a^2 + b^2}{2b}\right)^2 = \left(\frac{a^2 + b^2}{2a}\right)^2 (e_1^2 - 1)$$

$$\Rightarrow a^2 = b^2(e_1^2 - 1) \Rightarrow a^2 = a^2(e^2 - 1)(e_1^2 - 1)$$

$$[\because b^2 = a^2(e^2 - 1)]$$

$$\Rightarrow \quad e^2e_1^2 - e^2 - e_1^2 + 1 = 1$$

$$\Rightarrow \quad e_1^2(e^2 - 1) = e^2 \Rightarrow e_1 = \frac{e}{\sqrt{(e^2 - 1)}}$$

8 Tangents are drawn to the hyperbola $4x^2 - y^2 = 36$ at the point P and Q. Tangent intersects at point $T(0, 3)$

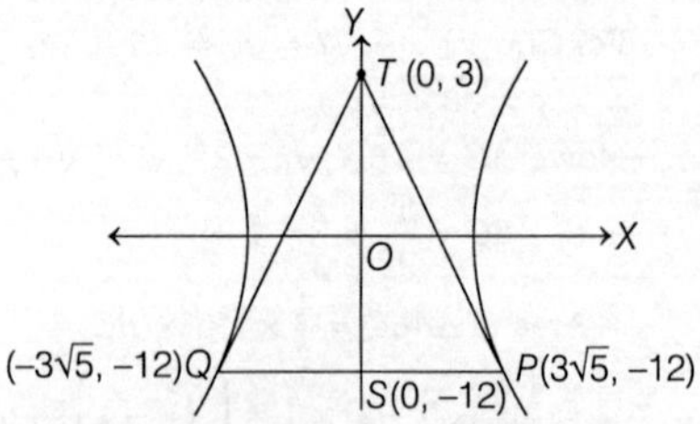

Clearly, PQ is chord of contact.

$\therefore$ Equation of PQ is $-3y = 36$

$\Rightarrow \quad y = -12$

Solving the curve $4x^2 - y^2 = 36$ and $y = -12$, we get $x = \pm 3\sqrt{5}$

$$\text{Area of} \quad \Delta PQT = \frac{1}{2} \times PQ \times ST$$

$$= \frac{1}{2}(6\sqrt{5} \times 15) = 45\sqrt{5}$$

9 Fourth vertex of parallelogram lies on circumcircle

$\Rightarrow$ Parallelogram is cyclic.

$\Rightarrow$ Parallelogram is a rectangle.

$\Rightarrow$ Tangents are perpendicular

$\Rightarrow$ Locus of P is the director circle

i.e., $x^2 + y^2 = a^2 - b^2$

10 Any point on parabola $y^2 = 8x$ is $(2t^2, 4t)$. The equation of tangent at that point is

$$yt = x + 2t^2 \quad \text{...(i)}$$

Given that, $xy = -1$ ...(ii)

On solving Eqs. (i) and (ii), we get

$y(yt - 2t^2) = -1 \Rightarrow ty^2 - 2t^2y + 1 = 0$

$\because$ It is common tangent. It means they are intersect only at one point and the value of discriminant is equal to zero.

i.e. $4t^4 - 4t = 0 \Rightarrow t = 0, 1$

$\therefore$ The common tangent is $y = x + 2$, (when $t = 0$, it is $x = 0$ which can touch $xy = -1$ at infinity only)

11 The equation of a hyperbola of the series is $\frac{x^2}{a^2} - \frac{y^2}{\lambda^2} = 1$ where, λ is a parameter. The asymptotes of this hyperbola $\frac{x}{a} = \pm\frac{y}{\lambda}$. Suppose (x', y') is a point P on the hyperbola which is equidistant from the transverse axis and asymptote.

Then, $\frac{x'^2}{a^2} - \frac{y'^2}{\lambda^2} = 1$ ...(i)

and $y' = \frac{x'/a - y'/\lambda}{\sqrt{\frac{1}{a^2} + \frac{1}{\lambda^2}}}$ ...(ii)

i.e. $\frac{y'^2}{\lambda^2} = \frac{x'^2}{a^2} - 1$ [from Eq. (i)] ...(iii)

and $y'^2\left(\frac{1}{a^2} + \frac{1}{\lambda^2}\right) = \frac{x'^2}{a^2} + \frac{y'^2}{\lambda^2} - \frac{2x'y'}{a\lambda}$

[from Eq. (ii)] ...(iv)

On simplification the second relation gives

$$(y'^2 - x'^2)^2 = \frac{4x'^2 y'^2 a^2}{\lambda^2}$$

$$= 4x'^2(x'^2 - a^2) \text{ [using Eq. (iii)]}$$

So, the locus of P is

$$(y^2 - x^2)^2 = 4x^2(x^2 - a^2).$$

12 The equation of tangent at point $P(\alpha\cos\theta, \sin\theta)$

$$\frac{x}{\alpha}\cos\theta + \frac{y}{1}\sin\theta = 1 \quad \text{...(i)}$$

Let it cut the hyperbola at points P and Q. Homogenising hyperbola $\alpha^2x^2 - y^2 = 1$ with the help of Eq. (i), we get

$$\alpha^2x^2 - y^2 = \left(\frac{x}{\alpha}\cos\theta + y\sin\theta\right)^2$$

This is a pair of straight lines OP OQ.

Given $\angle POQ = \pi/2$.

Coefficient of x^2 + Coefficient of $y^2 = 0$

or $\alpha^2 - \frac{\cos^2\theta}{\alpha^2} - 1 - \sin^2\theta = 0$

or $\alpha^2 - \frac{\cos^2\theta}{\alpha^2} - 1 - 1 + \cos^2\theta = 0$

or $\cos^2\theta = \frac{\alpha^2(2-\alpha^2)}{\alpha^2 - 1}$

Now, $0 \le \cos^2\theta \le 1$

or $0 \le \frac{\alpha^2(2-\alpha^2)}{\alpha^2-1} \le 1$

After solving, we get $\alpha^2 \in \left[\frac{\sqrt{5}+1}{2}, 2\right]$

13 Let equation of the rectangular hyperbola be $xy = c^2$ ...(i)

and equation of circle be

$$x^2 + y^2 = r^2. \quad \text{...(ii)}$$

From Eq. (i) and Eq.(ii) eliminating y, we get

$$x^4 - r^2x^2 + c^4 = 0 \quad \text{...(iii)}$$

Let x_1, x_2, x_3 and x_4 are the roots of Eq. (iii).

$\therefore$ Sum of roots $= \sum_{i=1}^{4} x_i = 0$

Sum of products of the roots taken two at a time $= \Sigma x_1x_2 = -r^2$

From Eq. (i) and Eq. (ii) eliminating x, we get

$$y^4 - r^2y^2 + c^4 = 0 \quad \text{...(iv)}$$

Let y_1, y_2, y_3 and y_4 are the roots of Eq. (iv).

$\therefore \sum_{i=1}^{4} y_i = 0$ and $\Sigma y_1y_2 = -r^2$

Now, $CP^2 + CQ^2 + CR^2 + CS^2$

$= x_1^2 + y_1^2 + x_2^2 + y_2^2 + x_3^2 + y_3^2 + x_4^2 + y_4^2$

$= (x_1^2 + x_2^2 + x_3^2 + x_4^2) + (y_1^2 + y_2^2 + y_3^2 + y_4^2)$

$= \left[(\sum_{i=1}^{4} x_i)^2 - 2\Sigma x_1x_2\right] + \left[(\sum_{i=1}^{4} y_i)^2 - 2\Sigma y_1y_2\right]$

$= (0 + 2r^2) + (0 + 2r^2)$

$\left[\because (\sum_{i=1}^{4} x_i) = 0, \Sigma x_1x_2 = -r^2, \sum_{i=1}^{4} y_i = 0, \Sigma y_1y_2 = -r^2\right]$

$= 4r^2$

14 Any point on the hyperbola $xy = 4$ is $\left(2t, \frac{2}{t}\right)$. Now, normal at $\left(2t, \frac{2}{t}\right)$ is

$y - \frac{2}{t} = t^2(x - 2t)$. [its slope is t^2]

If the normal passes through $P(h, k)$, then

$$k - \frac{2}{t} = t^2(h - 2t)$$

$\Rightarrow 2t^4 - ht^3 + tk - 2 = 0$ ...(i)

Roots of Eq. (i) give parameters of feet of normals passing through (h, k). Let roots be t_1, t_2, t_3 and t_4, then

$t_1 + t_2 + t_3 + t_4 = \frac{h}{2}$ ...(ii)

$t_1t_2 + t_2t_3 + t_1t_3 + t_1t_4 + t_2t_4 + t_3t_4 = 0$ ...(iii)

$t_1t_2t_3 + t_1t_2t_4 + t_1t_3t_4 + t_2t_3t_4 = -\frac{k}{2}$ ...(iv)

and $t_1t_2t_3t_4 = -1$ ...(v)

On dividing Eq. (iv) by Eq. (v), we get

$$\frac{1}{t_1} + \frac{1}{t_2} + \frac{1}{t_3} + \frac{1}{t_4} = \frac{k}{2}$$

$\Rightarrow y_1 + y_2 + y_3 + y_4 = k \left[\because y = \frac{2}{t}\right]$

From Eqs. (ii) and (iii), we get

$$t_1^2 + t_2^2 + t_3^2 + t_4^2 = \frac{h^2}{4}$$

Given that, $\frac{h^2}{4} = k$

Hence, locus of (h, k) is $x^2 = 4ay$, which is a parabola.

15 Let $P\left(ct_1, \frac{c}{t_1}\right), Q\left(ct_2, \frac{c}{t_2}\right), R\left(ct_3, \frac{c}{t_3}\right)$ be the vertices of a ΔPQR inscribed in the rectangular hyperbola $xy = c^2$ such that the sides PQ and QR are parallel to $y = m_1x$ and $y = m_2x$, respectively.

$\therefore m_1$ = Slope of PQ

and m_2 = Slope of QR

$\Rightarrow m_1 = -\frac{1}{t_1t_2}$

and $m_2 = -\frac{1}{t_2t_3}$

$\therefore \frac{m_1}{m_2} = \frac{t_3}{t_1} \Rightarrow t_3 = \left(\frac{m_1}{m_2}\right)t_1$

The equation of PR is

$$x + yt_1t_3 = c(t_1 + t_3)$$

$\Rightarrow x + y\left(\frac{m_1}{m_2}\right)t_1^2 = c\left(t_1 + \frac{m_1}{m_2}t_1\right)$

$\Rightarrow x + y\left(\frac{m_1}{m_2}\right)t_1^2 = c\left(1 + \frac{m_1}{m_2}\right)t_1$

$\Rightarrow x + y\left(\sqrt{\frac{m_1}{m_2}}t_1\right)^2 = 2\left\{\frac{c(m_1 + m_2)}{2\sqrt{m_1m_2}}\right\}\sqrt{\frac{m_1}{m_2}} \cdot t_1$

$\Rightarrow x + yt^2 = 2\lambda t$, where,

$t = \sqrt{\frac{m_1}{m_2}} \cdot t_1$ and $\lambda = \frac{c(m_1 + m_2)}{2\sqrt{m_1m_2}}$

Clearly, it touches the hyperbola,

$$xy = \left\{\frac{c(m_1 + m_2)}{2\sqrt{m_1m_2}}\right\}^2$$

or $xy = \left\{\frac{c^2(m_1 + m_2)^2}{4m_1m_2}\right\}$

DAY THIRTY

Unit Test 4
(Coordinate Geometry)

1 The parabola $y^2 = 4x$ and the circle $(x-6)^2 + y^2 = r^2$ will have no common tangent, then
(a) $r > \sqrt{20}$ (b) $r < \sqrt{20}$
(c) $r > \sqrt{18}$ (d) $r \in (\sqrt{20}, \sqrt{28})$

2 The lines $lx + my + n = 0$, $mx + ny + l = 0$ and $nx + ly + m = 0$ are concurrent, if
(a) $l + m + n = 0$ (b) $l + m - n = 0$
(c) $l - m + n = 0$ (d) None of these

3 If the latusrectum of a hyperbola through one focus substends 60° angle at the other focus, then its eccentricity e is
(a) $\sqrt{2}$ (b) $\sqrt{3}$ (c) $\sqrt{5}$ (d) $\sqrt{6}$

4 Set of values of m for which a chord of slope m of the circle $x^2 + y^2 = 4$ touches the parabola $y^2 = 4x$, is
(a) $\left(-\infty, -\sqrt{\frac{\sqrt{2}-1}{2}}\right) \cup \left(\sqrt{\frac{\sqrt{2}-1}{2}}, \infty\right)$
(b) $(-\infty, 1) \cup (1, \infty)$
(c) $(-1, 1)$
(d) R

5 If ω is one of the angles between the normals to the ellipse $\frac{x^2}{a^2} + \frac{y^2}{b^2} = 1$ at the points whose eccentric angles are θ and $\frac{\pi}{2} + \theta$, then $\frac{2\cot\omega}{\sin 2\theta}$ is equal to
(a) $\frac{e^2}{\sqrt{1-e^2}}$ (b) $\frac{e^2}{\sqrt{1+e^2}}$ (c) $\frac{e^2}{1-e^2}$ (d) $\frac{e^2}{1+e^2}$

6 If in a ΔABC (whose circumcentre is origin), $a \le \sin A$, then for any point (x, y) inside the circumcircle of ΔABC
(a) $|xy| < \frac{1}{8}$ (b) $|xy| > \frac{1}{8}$
(c) $\frac{1}{8} < xy < \frac{1}{2}$ (d) None of these

7 If $A(n, n^2)$ (where, $n \in N$) is any point in the interior of the quadrilateral formed by the lines $x = 0$, $y = 0$, $3x + y - 4 = 0$ and $4x + y - 21 = 0$, then the possible number of positions of the point A is
(a) 0 (b) 1
(c) 2 (d) 3

8 The range of values of r for which the point $\left(-5 + \frac{r}{\sqrt{2}}, -3 + \frac{r}{\sqrt{2}}\right)$ is an interior point of the major segment of the circle $x^2 + y^2 = 16$, cut off by the line $x + y = 2$, is
(a) $(-\infty, 5\sqrt{2})$
(b) $(4\sqrt{2} - \sqrt{14}, 5\sqrt{2})$
(c) $(4\sqrt{2} - \sqrt{14}, 4\sqrt{2} + \sqrt{14})$
(d) None of the above

9 AB is a double ordinate of the parabola $y^2 = 4ax$. Tangents drawn to parabola at A and B meet Y-axis at A_1 and B_1, respectively. If the area of trapezium AA_1B_1B is equal to $24a^2$, then angle subtended by A_1B_1 at the focus of the parabola is equal to
(a) $2\tan^{-1}(3)$ (b) $\tan^{-1}(3)$
(c) $2\tan^{-1}(2)$ (d) $\tan^{-1}(2)$

10 If two tangents can be drawn to the different branches of hyperbola $\frac{x^2}{1} - \frac{y^2}{4} = 1$ from the point (α, α^2), then
(a) $\alpha \in (-2, 0)$ (b) $\alpha \in (-3, 0)$
(c) $\alpha \in (-\infty, -2)$ (d) $\alpha \in (-\infty, -3)$

11 A hyperbola has the asymptotes $x + 2y = 3$ and $x - y = 0$ and passes through $(2, 1)$. Its centre is
(a) (1, 2) (b) (2, 2)
(c) (1, 1) (d) (2, 1)

12 The equation of the ellipse having vertices at $(\pm 5, 0)$ and foci at $(\pm 4, 0)$ is

(a) $\frac{x^2}{25}+\frac{y^2}{16}=1$ (b) $4x^2+5y^2=20$
(c) $9x^2+25y^2=225$ (d) None of these

13 The foci of an ellipse are $(0, \pm 1)$ and minor axis is of unit length. The equation of the ellipse is

(a) $2x^2+y^2=2$ (b) $x^2+2y^2=2$
(c) $4x^2+20y^2=5$ (d) $20x^2+4y^2=5$

14 The radius of the circle passing through the foci of the ellipse $\frac{x^2}{16}+\frac{y^2}{9}=1$ and having its centre at (0, 3) is

(a) 3 (b) 4
(c) $\sqrt{12}$ (d) $\frac{7}{2}$

15 If the straight lines $2x+3y-1=0$, $x+2y-1=0$ and $ax+by-1=0$ form a triangle with origin as orthocentre, then (a, b) is given by

(a) $(-3, 3)$ (b) (6, 4)
(c) $(-8, 8)$ (d) (0, 7)

16 If $P(1, 0)$, $Q(-1, 0)$ and $R(2, 0)$ are three given points, then the locus of S satisfying the relation $SQ^2+SR^2=2SP^2$ is

(a) a straight line parallel to X-axis
(b) a circle through the origin
(c) a circle with centre at the origin
(d) a straight line parallel to Y-axis

17 The point $(a^2, a+1)$ lies in the angle between the lines $3x-y+1=0$ and $x+2y-5=0$ containing the origin, if

(a) $a \in (-3, 0) \cup \left(\frac{1}{3}, 1\right)$ (b) $a \in (-\infty, -3) \cup \left(\frac{1}{3}, 1\right)$
(c) $a \in \left(-3, \frac{1}{3}\right)$ (d) $a \in \left(\frac{1}{3}, \infty\right)$

18 The diameter of $16x^2-9y^2=144$ which is conjugate to $x=2y$ is

(a) $y=\frac{16x}{9}$ (b) $y=\frac{32x}{9}$
(c) $x=\frac{16y}{9}$ (d) $x=\frac{32y}{9}$

19 The locus of poles with respect to the parabola $y^2=12x$ of tangent to the hyperbola $x^2-y^2=9$ is

(a) $4x^2+y^2=36$ (b) $x^2+4y^2=9$
(c) $x^2+4y^2=36$ (d) $4x^2+y^2=81$

20 A point moves such that the area of the triangle formed by it with the points (1, 5) and $(3, -7)$ is 21 sq units. Then, locus of the point is

(a) $6x+y-32=0$ (b) $6x-y+32=0$
(c) $6x-y-32=0$ (d) $x+6y-32=0$

21 The length of the latusrectum of the parabola $169[(x-1)^2+(y-3)^2]=(5x-12y+17)^2$ is

(a) $\frac{14}{13}$ (b) $\frac{28}{13}$ (c) $\frac{12}{13}$ (d) $\frac{16}{13}$

22 If two vertices of an equilateral triangle are (0, 0) and $(3, 3\sqrt{3})$, then the third vertex is

(a) $(3, -3)$ (b) $(-3, 3)$
(c) $(-3, 3\sqrt{3})$ (d) None of these

23 Let ABC is a triangle with vertices $A(-1, 4)$, $B(6, -2)$ and $C(-2, 4)$. If D, E and F are the points which divide each AB, BC and CA respectively, in the ratio 3 : 1 internally. Then, the centroid of the ΔDEF is

(a) (3, 6) (b) (1, 2)
(c) (4, 8) (d) None of these

24 A variable circle through the fixed point $A(p, q)$ touches the X-axis. The locus of the outer end of the diameter through A is

(a) $(x-p)^2=4qy$ (b) $(x-q)^2=4py$
(c) $(x-p)^2=4qx$ (d) $(x-q)^2=4px$

25 The exhaustive range of values of a such that the angle between the pair of tangents drawn from (a, a) to the circle $x^2+y^2-2x-2y-6=0$ lies in the range $\left(\frac{\pi}{3}, \pi\right)$ is

(a) $(0, \infty)$ (b) $(-3, -1) \cup (3, 5)$
(c) $(-2, -1) \cup (2, 3)$ (d) $-(3, 0) \cup (1, 2)$

26 The four distinct points (0, 0), (2, 0), $(0, -2)$ and $(k, -2)$ are concyclic, if k is equal to

(a) 0 (b) -2 (c) 2 (d) 1

27 If a point $P(4, 3)$ is rotated through an angle 45° in anti-clockwise direction about origin, then coordinates of P in new position are

(a) $\left(\frac{1}{\sqrt{2}}, \frac{7}{\sqrt{2}}\right)$ (b) $\left(-\frac{7}{\sqrt{2}}, -\frac{1}{\sqrt{2}}\right)$
(c) $\left(-\frac{1}{\sqrt{2}}, \frac{7}{\sqrt{2}}\right)$ (d) $\left(\frac{1}{\sqrt{2}}, -\frac{7}{\sqrt{2}}\right)$

28 The number of integral values of λ for which the equation $x^2+y^2-2\lambda x+2\lambda y+14=0$ represents a circle whose radius cannot exceed 6, is

(a) 9 (b) 10 (c) 11 (d) 12

29 The slopes of tangents to the circle $(x-6)^2+y^2=2$ which passes through the focus of the parabola $y^2=16x$ are

(a) ± 2 (b) 1/2, -2
(c) $-1/2$, 2 (d) ± 1

30 The range of values of n for which $(n, -1)$ is exterior to both the parabolas $y^2=|x|$ is

(a) (0, 1) (b) $(-1, 1)$
(c) $(-1, 0)$ (d) None of these

31 The parameters t and t' of two points on the parabola $y^2=4ax$, are connected by the relation $t=k^2t'$. The tangents at these points intersect on the curve

(a) $y^2=ax$ (b) $y^2=k^2x$
(c) $y^2=ax\left(k+\frac{1}{k}\right)^2$ (d) None of these

32 Triangle ABC is right angled at A. The circle with centre A and radius AB cuts BC and AC internally at D and E respettvely if $BD = 20$ and $DC = 16$ then the length of AC equals.

(a) $6\sqrt{21}$ (b) $6\sqrt{26}$ (c) 30 (d) 32

33 If the line $y - \sqrt{3}x + 3 = 0$ cuts the parabola $y^2 = x + 2$ at A and B, and if $P \equiv (\sqrt{3}, 0)$, then $PA \cdot PB$ is equal to

(a) $2\left(\frac{2+\sqrt{3}}{1}\right)$ (b) $\frac{4(2-\sqrt{3})}{3}$

(c) $\frac{4(2+\sqrt{3})}{3}$ (d) $\frac{2(2-\sqrt{3})}{3}$

34 If S and S' are the foci of the ellipse $\frac{x^2}{25} + \frac{y^2}{16} = 1$ and P is any point on it, then difference of maximum and minimum of $SP \cdot S'P$ is equal to

(a) 16 (b) 9 (c) 15 (d) 25

35 The locus of a point which moves, such that the chord of contact of the tangent from the point to two fixed given circles are perpendicular to each other is

(a) circle (b) parabola

(c) ellipse (d) None of these

36 Tangent is drawn to the ellipse $\frac{x^2}{27} + y^2 = 1$ at $(3\sqrt{3}\cos\theta, \sin\theta)\left[\text{where}, \theta \in \left(0, \frac{\pi}{2}\right)\right]$. Then, the value of θ such that sum of intercept on axes made by this tangent is minimum, is

(a) $\frac{\pi}{3}$ (b) $\frac{\pi}{6}$ (c) $\frac{\pi}{8}$ (d) $\frac{\pi}{4}$

37 The condition for the line $px + qy + r = 0$ to be tangent to the rectangular hyperbola $x = ct$, $y = \frac{c}{t}$ is

(a) $p < 0, q > 0$ (b) $p > 0, q > 0$

(c) $p > 0, q < 0$ (d) None of these

38 If the line $x + 3y + 2 = 0$ and its perpendicular line are conjugate w.r.t. $3x^2 - 5y^2 = 15$, then equation of conjugate line is

(a) $3x - y = 15$ (b) $3x - y + 12 = 0$

(c) $3x - y + 10 = 0$ (d) $3x - y = 4$

39 The product of the lengths of perpendicular drawn from any point on the hyperbola $x^2 - 2y^2 - 2 = 0$ to its asymptotes, is

(a) 1/2 (b) 2/3 (c) 3/2 (d) 2

40 Tangents at any point on the hyperbola $\frac{x^2}{a^2} - \frac{y^2}{b^2} = 1$ cut the axes at A and B respectively, if the rectangle $OAPB$, where O is the origin is completed, then locus of the point P is given by

(a) $\frac{a^2}{x^2} - \frac{b^2}{y^2} = 1$ (b) $\frac{a^2}{k^2} + \frac{b^2}{y^2} = 1$

(c) $\frac{a^2}{y^2} - \frac{b^2}{x^2} = 1$ (d) None of these

41 In a triangle ABC, if $A(2,-1)$ and $7x - 10y + 1 = 0$ and $3x - 2y + 5 = 0$ are equation of on altitude and on angle bisector respectively drawn from B, the equation of BC is.

(a) $x + y + 1 = 0$

(b) $5x + y + 17 = 0$

(c) $4x + 9y + 30 = 0$

(d) $x - 5y - 7 = 0$

42 If $\frac{a}{\sqrt{bc}} - 2 = \sqrt{\frac{b}{c}} + \sqrt{\frac{c}{b}}$ where $a, b, c > 0$ the family of lines $\sqrt{a}x + \sqrt{b}y + \sqrt{c} = 0$ passes through the point.

(a) (1, 1) (b) (1, −2) (c) (−1, 2) (d) (−1, 1)

43 A series of ellipses $E_1, E_2, E_3, \ldots, E_n$ are drawn such that E_n touches E_{n-1} at the extremeties of the major axis of E_{n-1} and the foci of E_n coincide with the extremeties of minor axis of E_{n-1}. If eccentricity of the ellipse B independent of x then the value of eccentricity is.

(a) $\frac{\sqrt{5}-1}{2}$ (b) $\frac{\sqrt{5}+1}{3}$

(c) $\frac{\sqrt{5}+1}{4}$ (d) $\frac{\sqrt{5}-1}{4}$

44 If one of the diagonal of a square is along the line $x = 2y$ and one of its vertices is (3, 0), then its sides through this vertex are given by the equations

(a) $y - 3x + 9 = 0, 3y + x - 3 = 0$

(b) $y + 3x + 9 = 0, 3y + x - 3 = 0$

(c) $y - 3x + 9 = 0, 3y - x + 3 = 0$

(d) $y - 3x + 3 = 0, 3y + x + 9 = 0$

45 Given $A(0,0)$ and $B(x, y)$ with $x \in (0,1)$ and $y > 0$. Let the slope of line AB equals m_1. Point C lies on the line $x = 1$ such that the slope of BC equals m_2 where $0 < m_2 < m_1$. If the area of the triangle ABC can be expressed as $(m_1 - m_2)f(x)$, then the largest possible value of $f(x)$ is.

(a) 1/8 (b) 1/2 (c) 1/4 (d) 1

46 A circle is inscribed into a rhombus $ABCD$ with one angle 60°. The distance from the centre of the circle to the nearest vertex is equal to 1. If P is any point of the circle, then

$$|PA|^2 + |PB|^2 + |PC|^2 + |PD|^2 \text{ is equal to.}$$

(a) 8 (b) 9 (c) 10 (d) 11

47 The equation of common tangent touching the circle $(x-3)^2 + y^2 = 9$ and the parabola $y^2 = 4x$ above the X-axis is

(a) $\sqrt{3}y = 3x + 1$ (b) $\sqrt{3}y = -(x + 3)$

(c) $\sqrt{3}y = x + 3$ (d) $\sqrt{3}y = -(3x + 1)$

48 If the tangent at the point ϕ on the ellipse $\frac{x^2}{16} + \frac{11y^2}{256} = 1$ touches the circle $x^2 + y^2 - 2x - 15 = 0$, then ϕ is equal to

(a) $\pm\frac{\pi}{2}$ (b) $\pm\frac{\pi}{4}$

(c) $\pm\frac{\pi}{3}$ (d) $\pm\frac{\pi}{6}$

Directions (Q.Nos. 49-55) *Each of these questions contains two statements : Statement I (Assertion) and Statement II (Reason). Each of these questions also has four alternative choices, only one of which is the correct answer. You have to select one of the codes (a), (b), (c) and (d) given below.*

(a) Statement I is true, Statement II is true; Statement II is a correct explanation for Statement I
(b) Statement I is true, Statement II is true; Statement II is not a correct explanation for Statement I
(c) Statement I is true; Statement II is false
(d) Statement I is false; Statement II is true

49 If p, x_1, x_2, x_3 and q, y_1, y_2, y_3 form two arithmetic progression with common differences a and b.

Statement I The centroid of triangle formed by points $(x_1, y_1), (x_2, y_2)$ and (x_3, y_3), lies on a straight line.

Statement II The point (h, k) given by $h = \frac{x_1 + x_2 + \dots + x_n}{n}$ and $k = \frac{y_1 + y_2 + \dots + y_n}{n}$ always lies on the line $b(x - p) = a(y - q)$ for all values of n.

50 **Statement I** A is a point on the parabola $y^2 = 4ax$. The normal at A cuts the parabola again at point B.

If AB subtends a right angle at the vertex of the parabola, then slope of AB is $\frac{1}{\sqrt{2}}$.

Statement II If normal at $(at_1^2, 2at_1)$ cuts again the parabola at $(at_2^2, 2at_2)$, then $t_2 = -t_1 - \frac{2}{t_1}$.

51 Suppose $ABCD$ is a cyclic quadrilateral inscribed in a circle.

Statement I If radius is one unit and $AB \cdot BC \cdot CD \cdot DA \geq 4$, then $ABCD$ is a square.

Statement II A cyclic quadrilateral is a square, if its diagonals are the diameters of the circle.

52 If a circle $S = 0$ intersects a hyperbola $xy = c^2$ at four points.

Statement I If $c = 2$ and three of the intersection points are (2, 2), (4, 1) and $\left(6, \frac{2}{3}\right)$, then coordinates of the fourth point are $\left(\frac{1}{4}, 16\right)$.

Statement II If a circle intersects a hyperbola at t_1, t_2, t_3, t_4, then $t_1 \cdot t_2 \cdot t_3 \cdot t_4 = 1$.

53 The auxiliary circle of an ellipse is described on the major axis of an ellipse.

Statement I The circle $x^2 + y^2 = 4$ is auxiliary circle of an ellipse $\frac{x^2}{4} + \frac{y^2}{b^2} = 1$ (where, $b < 2$).

Statement II A given circle is auxiliary circle of exactly one ellipse.

54 The tangent at a point P on the ellipse $\frac{x^2}{a^2} + \frac{y^2}{b^2} = 1$, which is not an extremity of major axis, meets a directrix at T.

Statement I The circle on PT as diameter passes through the focus of the ellipse corresponding to the directrix on which T lies.

Statement II PT subtends a right angle at the focus of the ellipse corresponding to the directrix on which T lies.

55 **Statement I** If the perpendicular bisector of the line segment joining $P(1, 4)$ and $Q(k, 3)$ has y-intercept -4, then $k^2 - 16 = 0$.

Statement II Locus of a point equidistant from two given points is the perpendicular bisector of the line joining the given points.

ANSWERS

1 (b)	2 (a)	3 (b)	4 (a)	5 (a)	6 (a)	7 (b)	8 (b)	9 (d)	10 (c)
11 (c)	12 (c)	13 (d)	14 (b)	15 (c)	16 (d)	17 (a)	18 (b)	19 (a)	20 (a)
21 (b)	22 (c)	23 (b)	24 (a)	25 (b)	26 (c)	27 (a)	28 (c)	29 (d)	30 (b)
31 (c)	32 (b)	33 (c)	34 (b)	35 (a)	36 (b)	37 (b)	38 (b)	39 (b)	40 (a)
41 (b)	42 (d)	43 (a)	44 (a)	45 (a)	46 (d)	47 (c)	48 (c)	49 (a)	50 (d)
51 (c)	52 (d)	53 (c)	54 (a)	55 (a)					

Hints and Explanations

1 Any normal of parabola is $y = -tx + 2t + t^3$.

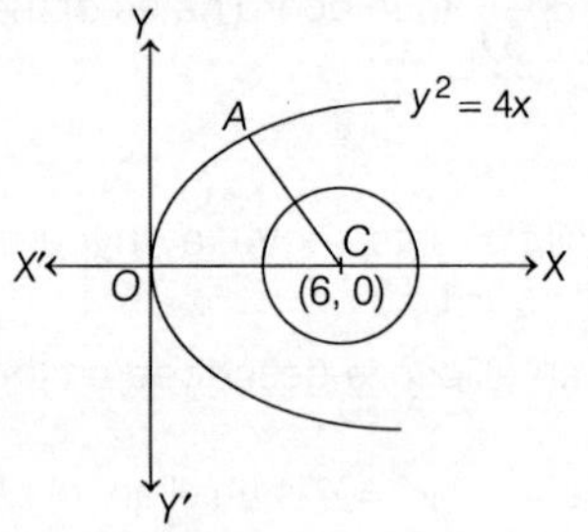

If it passes through (6,0), then

$$-6t + 2t + t^3 = 0$$

$$\Rightarrow \quad t = 0, t^2 = 4$$

Thus, $\quad A \equiv (4, 4)$

Thus, for no common tangent,

$$AC = \sqrt{4 + 16} > r$$

$$\Rightarrow \quad r < \sqrt{20}$$

2 Since, lines are concurrent.

$$\therefore 1\,(lx + my + n) + 1\,(mx + ny + l) + 1\,(nx + ly + m) = 0$$

$$\Rightarrow x\,(l + m + n) + y\,(l + m + n) + (l + m + n) = 0$$

$$\therefore \quad l + m + n = 0$$

3 Let LSL' be a latusrectum through the focus $S\,(ae, 0)$ of the hyperbola

$$\frac{x^2}{a^2} - \frac{y^2}{b^2} = 1.$$

It substends angle 60° at the otherfocus $S'(-ae, 0)$.

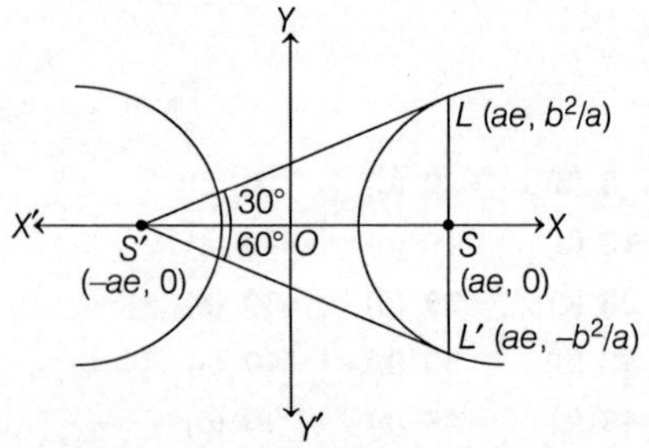

We have, $\angle LS'L' = 60°$

$$\therefore \quad \angle LS'S = 30°$$

In $\Delta LS'S$, $\tan 30° = \dfrac{LS}{S'S}$

$$\Rightarrow \quad \frac{1}{\sqrt{3}} = \frac{b^2/a}{2ae}$$

$$\Rightarrow \quad \frac{1}{\sqrt{3}} = \frac{b^2}{2a^2e}$$

$$= \frac{e^2 - 1}{2e}$$

$$\Rightarrow \quad \sqrt{3}\,e^2 - 2e - \sqrt{3} = 0$$

$$\Rightarrow \quad (e - \sqrt{3})(\sqrt{3}e + 1) = 0$$

$$\therefore \quad e = \sqrt{3}$$

4 The equation of tangent of slope m to the parabola $y^2 = 4x$ is $y = mx + \dfrac{1}{m}$.

This will be a chord of the circle $x^2 + y^2 = 4$, if length of the perpendicular from the centre (0, 0) is less than the radius.

$$\text{i.e.} \quad \left|\frac{1}{m\sqrt{m^2 + 1}}\right| < 2$$

$$\Rightarrow 4m^4 + 4m^2 - 1 > 0$$

$$\Rightarrow \left(m^2 - \frac{\sqrt{2} - 1}{2}\right)\left(m^2 + \frac{1 + \sqrt{2}}{2}\right) > 0$$

$$\Rightarrow \left(m^2 - \frac{\sqrt{2} - 1}{2}\right) > 0$$

$$\Rightarrow \left(m - \sqrt{\frac{\sqrt{2} - 1}{2}}\right)\left(m + \sqrt{\frac{\sqrt{2} - 1}{2}}\right) > 0$$

$$\Rightarrow m \in \left(-\infty, -\sqrt{\frac{\sqrt{2} - 1}{2}}\right) \cup \left(\sqrt{\frac{\sqrt{2} - 1}{2}}, \infty\right)$$

5 The equations of the normals to the ellipse $\dfrac{x^2}{a^2} + \dfrac{y^2}{b^2} = 1$ at the points whose eccentric angles are θ and $\dfrac{\pi}{2} + \theta$ are

$ax \sec\theta - by \operatorname{cosec}\theta = a^2 - b^2$

and $-ax \operatorname{cosec}\theta - by \sec\theta = a^2 - b^2$, respectively.

Since, ω is the angle between these two normals.

Therefore,

$$\tan\omega = \left|\frac{\dfrac{a}{b}\tan\theta + \dfrac{a}{b}\cot\theta}{1 - \dfrac{a^2}{b^2}}\right|$$

$$= \left|\frac{ab\,(\tan\theta + \cot\theta)}{b^2 - a^2}\right|$$

$$\Rightarrow \tan\omega = \left|\frac{2\,ab}{\sin 2\theta\,(b^2 - a^2)}\right|$$

$$= \frac{2ab}{(a^2 - b^2)\sin 2\theta} = \frac{2a^2\sqrt{1 - e^2}}{a^2e^2 \sin 2\theta}$$

$$\therefore \frac{2\cot\omega}{\sin 2\theta} = \frac{e^2}{\sqrt{1 - e^2}}$$

6 Given, $a \le \sin A \Rightarrow \dfrac{a}{\sin A} \le 1$

$$\Rightarrow 2R \le 1 \Rightarrow R \le \frac{1}{2}$$

So, for any point (x, y) inside the circumcircle,

$$x^2 + y^2 < \frac{1}{4}$$

Using AM≥ GM, $\left(\dfrac{x^2 + y^2}{2}\right) \ge |\,xy\,|$

$$\Rightarrow \quad |\,xy\,| < \frac{1}{8}$$

7 Origin is on the left of PS, as

$$0 + 0 - 4 < 0$$

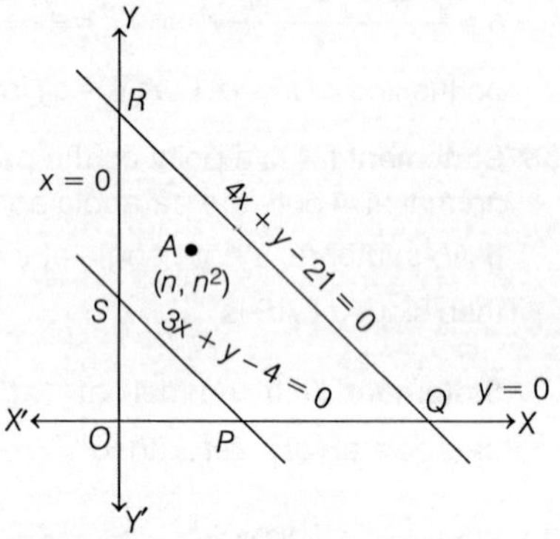

$\therefore$ At point $A\,(n, n^2)$,

$$3n + n^2 - 4 > 0$$

$$\Rightarrow \quad n^2 + 3n - 4 > 0$$

$$\Rightarrow \quad (n + 4)(n - 1) > 0$$

$$\Rightarrow \quad n - 1 > 0 \qquad \ldots\text{(i)}$$

Now, as A and O lies on the same sides of QR.

and $\quad 4x + y - 21 = 0 + 0 - 21 < 0$

$\therefore$ At point $A\,(n, n^2)$,

$$4n + n^2 - 21 < 0$$

$$\Rightarrow \quad n^2 + 4n - 21 < 0$$

$$\Rightarrow \quad (n + 7)(n - 3) < 0$$

$$\Rightarrow \quad 0 < n < 3 \qquad [\because n \in N\,] \ldots\text{(ii)}$$

From Eqs. (i) and (ii),

$$1 < n < 3 \Rightarrow n = 2$$

Hence, A (2, 4) is only one point.

8 The given point is an interior point, if

$$\left(-5 + \frac{r}{\sqrt{2}}\right)^2 + \left(-3 + \frac{r}{\sqrt{2}}\right)^2 - 16 < 0$$

$$\Rightarrow r^2 - 8\sqrt{2}\,r + 18 < 0$$

$$\Rightarrow 4\sqrt{2} - \sqrt{14} < r < 4\sqrt{2} + \sqrt{14}$$

Since, the point is on the major segment, the centre and the point are on the same side of the line $x + y = 2$.

$\Rightarrow \quad -5+\frac{r}{\sqrt{2}}-3+\frac{r}{\sqrt{2}}-2<0$

$\Rightarrow \quad r<5\sqrt{2}$

Hence,

$4\sqrt{2}-\sqrt{14}<r<5\sqrt{2}$

9 Let $A \equiv (at_1^2, 2at_1)$, $B \equiv (at_1^2, -2at_1)$.
Equation of tangents at A and B are $yt_1 = x + at_1^2$ and $yt_1 = x - at_1^2$, respectively.
Now, $A_1 \equiv (0, at_1)$, $B_1 = (0, -at_1)$
Area of trapezium

$AA_1B_1B = \frac{1}{2}(AB + A_1B_1)\cdot OC$

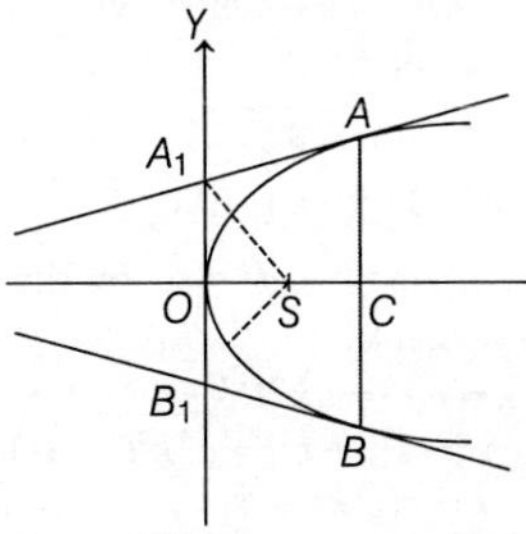

$\Rightarrow \quad 24a^2 = \frac{1}{2}\cdot(4at_1 + 2at_1)(at_1^2)$

$\Rightarrow t_1^3 = 8 \Rightarrow t_1 = 2 \Rightarrow A_1 = (0, 2a)$

If $\angle OSA_1 = \theta$, then

$\tan\theta = \frac{2a}{a} = 2 \Rightarrow \theta = \tan^{-1}(2)$

10 Given that, $\frac{x^2}{1}-\frac{y^2}{4}=1$

Since, (α, α^2) lies on the parabola $y = x^2$, then (α, α^2) must lie between the asymptotes of hyperbola $\frac{x^2}{1}-\frac{y^2}{4}=1$ in 1st and 2nd quadrant.

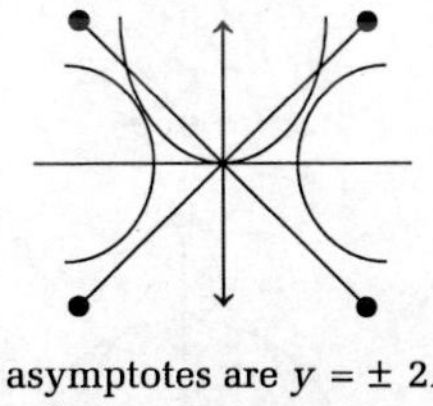

So, the asymptotes are $y = \pm 2x$.

$\therefore \quad 2\alpha < \alpha^2 \Rightarrow \alpha < 0$ or $\alpha > 2$

and $-2\alpha < \alpha^2$

$\alpha < -2$ or $\alpha > 0$

$\therefore \quad \alpha \in (-\infty, -2)$ or $(2, \infty)$

11 Given equations of asymptotes are

$x + 2y = 3 \quad$ ...(i)

and $\quad x - y = 0 \quad$ ...(ii)

On solving Eqs. (i) and (ii), we get

$x = 1, y = 1$

So, the centre of hyperbola is (1, 1).

12 The line joining foci and vertices is X-axis and the centre is (0, 0). So, axes of the ellipse coincide with coordinate axes.

Here, $a = 5$ and $ae = 4$

$\Rightarrow \quad e = \frac{4}{5}$

Now, $b^2 = a^2(1-e^2)$

$= 25\left[1-\left(\frac{4}{5}\right)^2\right] = 9$

Hence, the equation of the ellipse is

$\frac{x^2}{5^2}+\frac{y^2}{3^2}=1$

$\Rightarrow \quad 9x^2 + 25y^2 = 225$

13 Given, $2b = 1 \Rightarrow b = \frac{1}{2}$ and $a\cdot e = 1$

Since, $a^2(1-e^2) = \frac{1}{4}$

$\Rightarrow a^2 - 1^2 = \frac{1}{4} \quad \Rightarrow \quad a^2 = \frac{5}{4}$

Hence, the equation of the ellipse $\frac{x^2}{b^2}+\frac{y^2}{a^2}=1$ is $\frac{x^2}{1/4}+\frac{y^2}{5/4}=1$ or

$20x^2 + 4y^2 = 5$

14 Since, $a^2(1-e^2) = 9$

$\Rightarrow 16 - a^2e^2 = 9 \Rightarrow ae = \sqrt{7}$

So, foci are at $(\sqrt{7}, 0)$ and $(-\sqrt{7}, 0)$.

$\therefore$ Required radius

$= \sqrt{(\sqrt{7}-0)^2 + (3-0)^2} = 4$

15 Equation of AO is

$2x + 3y - 1 + \lambda(x + 2y - 1) = 0$.

Since, it passes through (0, 0), then $\lambda = -1$.

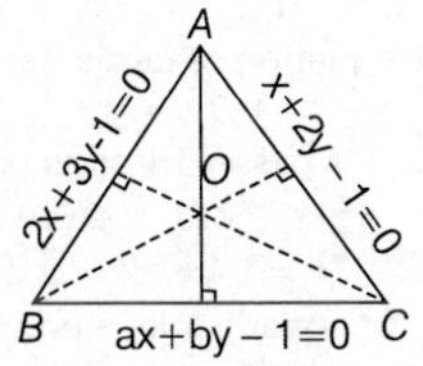

$\therefore \quad x + y = 0$

Since, AO is perpendicular to BC.

$\therefore \quad (-1)\left(-\frac{a}{b}\right) = -1 \Rightarrow a = -b$

Similarly,

$(2x + 3y - 1) + \mu(ax - ay - 1) = 0$ will be equation of BO for $\mu = -1$.

Thus, BO is perpendicular to AC.

$\Rightarrow \quad -\frac{2-a}{3+a}\cdot\left(\frac{-1}{2}\right) = -1$

$\Rightarrow \quad 2 - a = -6 - 2a$

$\Rightarrow \quad a = -8$ and $b = 8$

16 Let the coordinates of a point S be (x, y).

Since, $SQ^2 + SR^2 = 2SP^2$

$\Rightarrow (x+1)^2 + y^2 + (x-2)^2 + y^2$
$= 2[(x-1)^2 + y^2]$

$\Rightarrow \quad 2x + 3 = 0$

Hence, it is a straight line parallel to Y-axis.

17 Since, origin and the point $(a^2, a+1)$ lie on the same side of both the lines.

$\therefore \quad 3a^2 - (a+1) + 1 > 0$

and $\quad a^2 + 2(a+1) - 5 < 0$

i.e. $a(3a-1) > 0$ and $a^2 + 2a - 3 < 0$

i.e. $a(3a-1) > 0$ and $(a-1)(a+3) < 0$

$\Rightarrow \quad a \in (-\infty, 0) \cup \left(\frac{1}{3}, \infty\right)$

and $\quad a \in (-3, 1)$

$\therefore \quad a \in (-3, 0) \cup \left(\frac{1}{3}, 1\right)$

18 Diameters $y = m_1x$ and $y = m_2x$ are conjugate diameters of the hyperbola $\frac{x^2}{a^2}-\frac{y^2}{b^2}=1$, if $m_1m_2 = \frac{b^2}{a^2}$.

Here, $a^2 = 9$, $b^2 = 16$ and $m_1 = \frac{1}{2}$

$\because \quad m_1m_2 = \frac{b^2}{a^2}$

$\Rightarrow \quad \frac{1}{2}(m_2) = \frac{16}{9}$

$\Rightarrow \quad m_2 = \frac{32}{9}$

Thus, the required diameter is $y = \frac{32x}{9}$.

19 Let the pole be (h, k), so that polar is

$ky = 6(x + h)$

$\Rightarrow \quad y = \frac{6x}{k} + \frac{6h}{k}$

Since, it is tangent to the hyperbola,

$x^2 - y^2 = 9$

$\therefore \quad c^2 = 9m^2 - 9$

$\Rightarrow \quad \frac{36h^2}{k^2} = \frac{324}{k^2} - 9$

$\left[\because c = \frac{6h}{k}, m = \frac{6}{k}\right]$

$\Rightarrow \quad 4h^2 + k^2 = 36$

Hence, the locus is $4x^2 + y^2 = 36$.

20 Let (x, y) be the required point.

Then, $\frac{1}{2}\begin{vmatrix} x & y & 1 \\ 1 & 5 & 1 \\ 3 & -7 & 1 \end{vmatrix} = 21$

$\Rightarrow \quad (5+7)x - (1-3)y + (-7-15) = 42$

$\Rightarrow \quad 12x + 2y - 22 = 42$

$\Rightarrow \quad 6x + y - 32 = 0$

21 Given parabola is

$(x-1)^2 + (y-3)^2 = \left(\frac{5x - 12y + 17}{13}\right)^2$

Focus = (1, 3), directrix is

$5x - 12y + 17 = 0$

$\therefore$ Length of latusrectum

$= 2\left|\frac{5 - 36 + 17}{13}\right| = \frac{28}{13}$

22 Since, $\angle AOM$ is 30°.

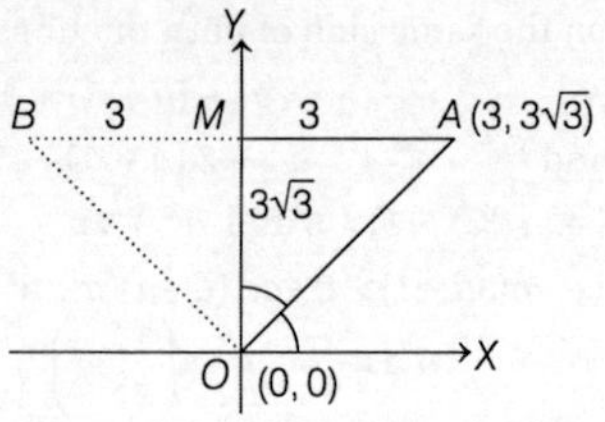

Hence, the required point B is $(-3, 3\sqrt{3})$.

23 Let (x_1, y_1), (x_2, y_2) and (x_3, y_3) are coordinates of the points D, E and F which divide each AB, BC and CA respectively in the ratio 3 : 1 (internally).

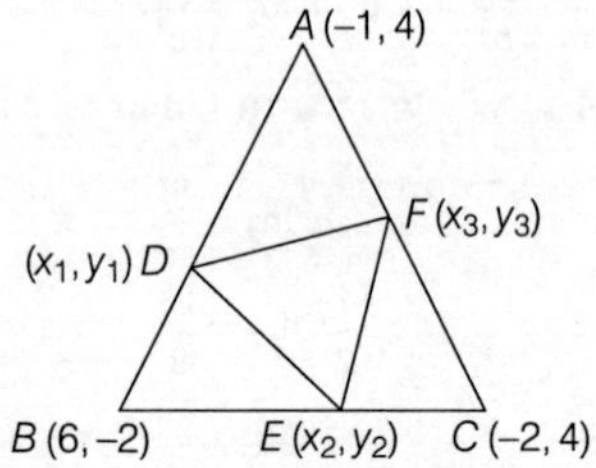

$\therefore\ x_1 = \dfrac{3\times 6 - 1\times 1}{4} = \dfrac{17}{4}$

$y_1 = \dfrac{-2\times 3 + 4\times 1}{4} = -\dfrac{2}{4} = -\dfrac{1}{2}$

Similarly, $x_2 = 0,\ y_2 = \dfrac{5}{2}$

and $\quad x_3 = -\dfrac{5}{4},\ y_3 = 4$

Let (x, y) be the coordinates of centroid of ΔDEF.

$\therefore\quad x = \dfrac{1}{3}\left(\dfrac{17}{4} + 0 - \dfrac{5}{4}\right) = 1$

and $y = \dfrac{1}{3}\left(-\dfrac{1}{2} + \dfrac{5}{2} + 4\right) = 2$

So, the coordinates of centroid are $(1, 2)$.

24 The circle touching the X-axis is

$$x^2 + y^2 + 2gx + 2fy + g^2 = 0.$$

Since, it passes through (p, q).

$\therefore\quad p^2 + q^2 + 2gp + 2fq + g^2 = 0 \quad$ ...(i)

If (x, y) is the other end of the diameter, then

$$p + x = -2g,\ q + y = -2f$$

Now, Eq. (i) gives

$$p^2 + q^2 - p(p + x) - q(q + y) + \frac{(p + x)^2}{4} = 0$$

$\Rightarrow\quad (x + p)^2 = 4px + 4qy$

$\Rightarrow\quad (x - p)^2 = 4qy$

25 Given that,

$$x^2 + y^2 - 2x - 2y - 6 = 0$$

Centre $= C(1, 1)$, radius $= 2\sqrt{2}$

Since, point (a, a) must lie outside the circle.

So, $\quad 2a^2 - 4a - 6 > 0$

$\Rightarrow\quad a < -1$ or $a > 3$

Now, in ΔPAC, $\tan\dfrac{\theta}{2} = \dfrac{2\sqrt{2}}{\sqrt{2a^2 - 4a - 6}}$

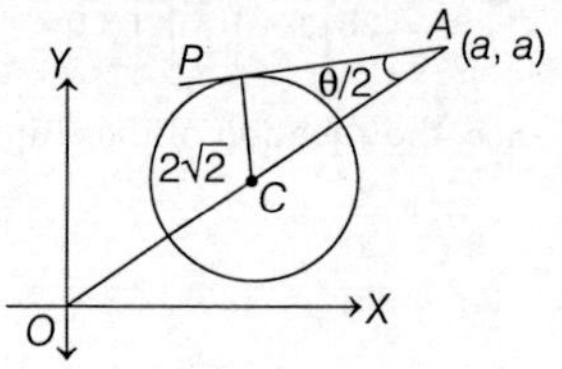

As given that, $\dfrac{\pi}{3} < \theta < \pi$

$\Rightarrow\quad \dfrac{\pi}{6} < \dfrac{\theta}{2} < \dfrac{\pi}{2}$

$\therefore\quad \dfrac{2\sqrt{2}}{\sqrt{2a^2 - 4a - 6}} > \dfrac{1}{\sqrt{3}}$

$\Rightarrow\ \sqrt{a^2 - 2a - 3} < 2\sqrt{3}$

$\therefore\quad a^2 - 2a - 3 < 12$

$\Rightarrow\quad a^2 - 2a - 15 < 0$

$\Rightarrow\quad -3 < a < 5$

$\therefore\quad a \in (-3, -1) \cup (3, 5)$

26 Let the general equation of the circle be

$$x^2 + y^2 + 2gx + 2fy + c = 0.$$

$\because$ The equation of circle passing through $(0, 0)$, $(2, 0)$ and $(0, -2)$.

$\therefore\quad c = 0 \quad$ ...(i)

$4 + 4g + c = 0 \quad$ ...(ii)

and $\quad 4 - 4f + c = 0 \quad$ ...(iii)

On solving Eqs. (i), (ii) and (iii), we get $c = 0,\ g = -1,\ f = 1$

$\therefore$ The equation of circle becomes $x^2 + y^2 - 2x + 2y = 0$

Since, it is passes through $(k, -2)$,

$$k^2 + 4 - 2k - 4 = 0$$

$\Rightarrow\ k^2 - 2k = 0 \Rightarrow k = 0, 2$

We have already take a point $(0, -2)$, so we take only $k = 2$.

27 Slope of line $OP = \dfrac{3}{4}$, let new position is $Q\ (x, y)$.

Slope of $OQ = \dfrac{y}{x}$,

also $x^2 + y^2 = OQ^2 = 25 = (OP^2)$

$\therefore\ \tan 45^\circ = \left|\dfrac{\dfrac{y}{x} - \dfrac{3}{4}}{1 + \dfrac{3y}{4x}}\right|$

$\Rightarrow\quad \pm 1 = \dfrac{4y - 3x}{4x + 3y}$

$\Rightarrow\quad 4x + 3y = 4y - 3x$

or $\quad -4x - 3y = 4y - 3x$

$\Rightarrow\quad x = \dfrac{1}{7}y \quad$ ...(i)

or $\quad -x = 7y \quad$ ...(ii)

Correct relation is $x = \dfrac{1}{7}y$ as new point must lies in Ist quadrant.

$\therefore\quad x^2 + 49x^2 = 25$

$\Rightarrow\quad x = +\dfrac{1}{\sqrt{2}},\ y = \dfrac{7}{\sqrt{2}}$

28 Since, $(\text{radius})^2 \le 36$

$\Rightarrow \lambda^2 + \lambda^2 - 14 \le 36$

$\Rightarrow \lambda^2 \le 25 \Rightarrow -5 \le \lambda \le 5 \Rightarrow \lambda$
$= 0, \pm 1, \pm 2, \pm 3, \pm 4, \pm 5$

Hence, number of integer values of λ is 11.

29 Tangent to the circle with slope m is

$$y = m(x - 6) \pm \sqrt{2(1 + m^2)}.$$

Since, it passes through $(4, 0)$.

$\therefore\quad 4m^2 = 2 + 2m^2$

$\Rightarrow\quad m = \pm 1$

30 Since, $1 - |n| > 0$

$\Rightarrow\ |n| < 1$ or $n \in (-1, 1)$

31 Tangents at t and t' meet on the point (x, y) given by

$x = att' = ak^2t'^2 \quad$ ...(i)

and $\ y = a(t + t') = a(k^2t' + t')$
$= at'(k^2 + 1) \quad$ ...(ii)

From Eqs. (i) and (ii),

$$x = \frac{ak^2 \cdot y^2}{a^2(k^2 + 1)^2} = \frac{k^2y^2}{a(k^2 + 1)^2}$$

$$\Rightarrow y^2 = \frac{ax(k^2 + 1)^2}{k^2} = ax\left(k + \frac{1}{k}\right)^2$$

32. In $\Delta\ ABC$

$AC^2 + AB^2 = BC^2$

$AC^2 + r^2 = 36^2 \quad$ ...(i)

and $\quad CF \times CE = BC \times CD$

$\Rightarrow\ (AC + r)(AC - r) = 36 \times 16$

$\Rightarrow\quad AC^2 - r^2 = 36 \times 16 \quad$ ...(ii)

From Eqs. (i) and (ii), we get

$2AC^2 = 36(36 + 16)$

$\Rightarrow\quad AC^2 = 18 \times 52$

$\Rightarrow\quad AC = 6\sqrt{26}$

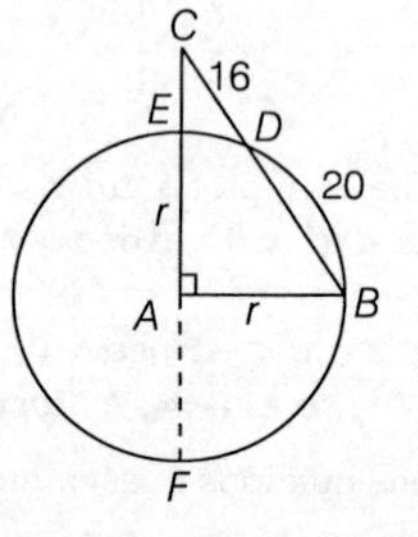

33 Let $PA = r_1,\ PB = -r_2$

Put $(\sqrt{3} + r\cos\theta, r\sin\theta)$ in $y^2 = x + 2$

$\Rightarrow\ r^2\sin^2\theta = (\sqrt{3} + r\cos\theta) + 2$

$\Rightarrow\ r^2\sin^2\theta - r\cos\theta - (\sqrt{3} + 2) = 0$

$PA \cdot PB = -r_1 . r_2 = \dfrac{\sqrt{3} + 2}{\sin^2\theta}$

$= (\sqrt{3} + 2)(1 + \cot^2\theta)$

$= (\sqrt{3}+2)\left(1+\frac{1}{3}\right)$ $\quad [\because \tan\theta = \sqrt{3}]$

$PA \cdot PB = \frac{4(2+\sqrt{3})}{3}$

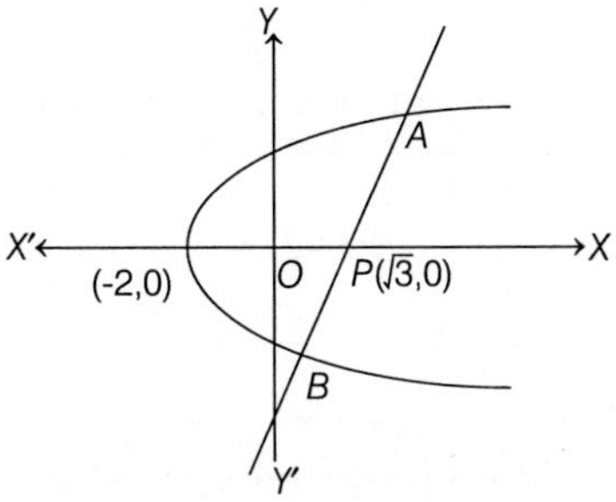

34 $(SP)(S'P) = a(1 - e\cos\theta)\,a(1+e\cos\theta)$

$= a^2(1-e^2\cos^2\theta)$

$= a^2 - a^2e^2\cos^2\theta$

$= 25 - 9\cos^2\theta$

Maximum $= 25 - 9(0) = 25$ $\quad [\theta = 90°]$

Minimum $= 25 - 9(1) = 16$ $\quad [\theta = 0°]$

Maximum–Minimum $= 25 - 16 = 9$

35 Let the equations of two given circles are

$x^2 + y^2 + 2g_1x + 2f_1y + c_1 = 0$ ...(i)

and $x^2 + y^2 + 2g_2x + 2f_2y + c_2 = 0$ ...(ii)

Now, the equations of the chords of contacts from $P(h, k)$ to Eqs. (i) and (ii) are

$x(h+g_1) + y(k+f_1) + g_1h + f_1k + c_1 = 0$

and $x(h+g_2) + y(k+f_2) + g_2h + f_2k + c_2 = 0$

According to the given condition,

$$\frac{(h+g_1)}{(k+f_1)} \times \frac{(h+g_2)}{(k+f_2)} = -1$$

$\Rightarrow h^2 + (g_1+g_2)h + g_1g_2 + k^2 + k(f_1+f_2) + f_1f_2 = 0$

Hence, the locus of point is

$x^2 + y^2 + (g_1+g_2)x + (f_1+f_2)y + g_1g_2 + f_1f_2 = 0$

which is the equation of a circle.

36 Equation of tangent at $(3\sqrt{3}\cos\theta, \sin\theta)$ to the ellipse $\frac{x^2}{27} + y^2 = 1$ is

$$\frac{x\cos\theta}{3\sqrt{3}} + y\sin\theta = 1.$$

This intersect on the coordinate axes at $(3\sqrt{3}\sec\theta, 0)$ and $(0, \operatorname{cosec}\theta)$

$\therefore$ Sum of intercepts on axes is

$3\sqrt{3}\sec\theta + \operatorname{cosec}\theta = f(\theta)$ [say]

On differentiating w.r.t. θ, we get

$f'(\theta) = 3\sqrt{3}\sec\theta\tan\theta - \operatorname{cosec}\theta\cot\theta$

$$= \frac{3\sqrt{3}\sin^3\theta - \cos^3\theta}{\sin^2\theta\cos^2\theta}$$

For maxima and minima, put $f'(\theta) = 0$

$3\sqrt{3}\sin^3\theta - \cos^3\theta = 0$

$\Rightarrow \tan\theta = \frac{1}{\sqrt{3}}$

$\Rightarrow \theta = \frac{\pi}{6}$

At $\theta = \frac{\pi}{3}$, $f''(\theta) > 0$. So, $f(\theta)$ is minimum at $\theta = \frac{\pi}{6}$.

37 Given, $x = ct$, $y = c/t$

Then, $\frac{dy}{dt} = \frac{-c}{t^2}$ and $\frac{dx}{dt} = c$

$\therefore \frac{dy}{dx} = \frac{-1}{t^2}$

But equation of tangent is $px + qy + r = 0$.

$\therefore -\frac{p}{q} = -\frac{1}{t^2} \Rightarrow \frac{p}{q} = \frac{1}{t^2} > 0$

$\Rightarrow \frac{p}{q} > 0$

$\Rightarrow p > 0, q > 0$ or $p < 0, q < 0$

38 Since, the lines $x + 3y + 2 = 0$ and $3x - y + k = 0$ are conjugate w.r.t. $\frac{x^2}{5} - \frac{y^2}{3} = 1$.

$\therefore 5(1)(3) - 3(3)(-1) = 2k$

$\Rightarrow k = 12$

Hence, equation of conjugate line is $3x - y + 12 = 0$.

39 Given, equation can be rewritten as

$\frac{x^2}{2} - \frac{y^2}{1} = 1$

Here, $a^2 = 2, b^2 = 1$

The product of length of perpendicular drawn from any point on the hyperbola to the asymptotes is

$$\frac{a^2b^2}{a^2+b^2} = \frac{2(1)}{2+1} = \frac{2}{3}.$$

40 The equation of tangent is

$\frac{x}{a}\sec\theta - \frac{y}{b}\tan\theta = 1.$

So, the coordinates of A and B are $(a\cos\theta, 0)$ and $(0, -b\cot\theta)$, respectively.

Let coordinates of P are (h, k).

$\therefore h = a\cos\theta$, $k = -b\cot\theta$

$\Rightarrow \frac{k}{h} = -\frac{b}{a\sin\theta}$

$\Rightarrow \frac{b^2h^2}{a^2k^2} = \sin^2\theta$

$\Rightarrow \frac{b^2h^2}{a^2k^2} + \frac{h^2}{a^2} = 1$

$[\because \sin^2\theta + \cos^2\theta = 1]$

$\Rightarrow \frac{a^2}{h^2} - \frac{b^2}{k^2} = 1$

Hence, the locus of P is $\frac{a^2}{x^2} - \frac{b^2}{y^2} = 1$.

41 Image of $A(2,-1)$ with respect to line $3x - 2y + 5 = 0$. A' is given by

$$\frac{x-2}{3} = \frac{y+1}{-2} = \frac{-2(6+2+5)}{13} = -2$$

$A' \in (-4, 3)$

Coordinate of B is intersection point of $7x - 10y + 1 = 0$ and $3x - 2y + 5 = 0$

i.e. $(-3,-2)$

$\therefore$ Equation of BC is

$y - 3 = \frac{3+2}{-4+3}(x+4)$

$\Rightarrow 5x + y + 17 = 0$

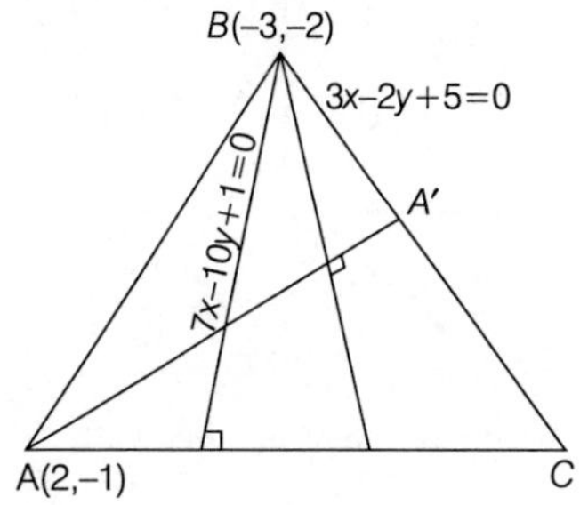

42 We have, $\frac{a}{\sqrt{bc}} - 2 = \sqrt{\frac{b}{c}} + \sqrt{\frac{c}{b}}$

$\Rightarrow a - 2\sqrt{bc} = b + c$

$\Rightarrow a = b + c + 2\sqrt{bc}$

$\Rightarrow (\sqrt{a})^2 = (\sqrt{b} + \sqrt{c})^2$

$\Rightarrow (\sqrt{b}+\sqrt{c})^2 - (\sqrt{a})^2 = 0$

$\Rightarrow \sqrt{b} + \sqrt{c} - \sqrt{a} = 0$

or $\sqrt{b} + \sqrt{c} + \sqrt{a} = 0$ (rejected)

$\Rightarrow \sqrt{ax} + \sqrt{by} + \sqrt{c} = 0$

passes through fixed point $(-1,1)$

43 Here, $b' = a$, $a'e = b$

$(b')^2 = (a')^2 - (a'e)^2$

$\Rightarrow a^2 = \frac{b^2}{e^2} - b^2$

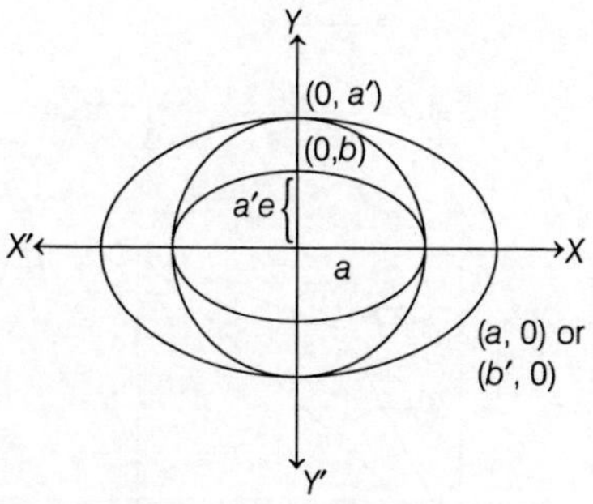

$\Rightarrow e^2 = \frac{b^2}{a^2}(1-e^2)$ $\quad \left[\because 1 - e^2 = \frac{b^2}{a^2}\right]$

$\Rightarrow e^2 = (1-e^2)(1-e^2)$

$\Rightarrow 1 - e^2 = \pm e$

$\Rightarrow \quad e^2 - e - 1 = 0$ or $e^2 + e - 1 = 0$

$\Rightarrow \quad e = \dfrac{-1 \pm \sqrt{5}}{2}$

or $\quad \dfrac{1 \pm \sqrt{5}}{2}$

$\Rightarrow \quad e = \dfrac{\sqrt{5} - 1}{2} \qquad [0 < e < 1]$

44 Equation of diagonal

AC is $y - 0 = -2(x - 3)$

$\Rightarrow \quad 2x + y = 6$

On solving $2x + y = 6$ and $x = 2y$, we get $y = \dfrac{6}{5}$ and $x = \dfrac{12}{5}$

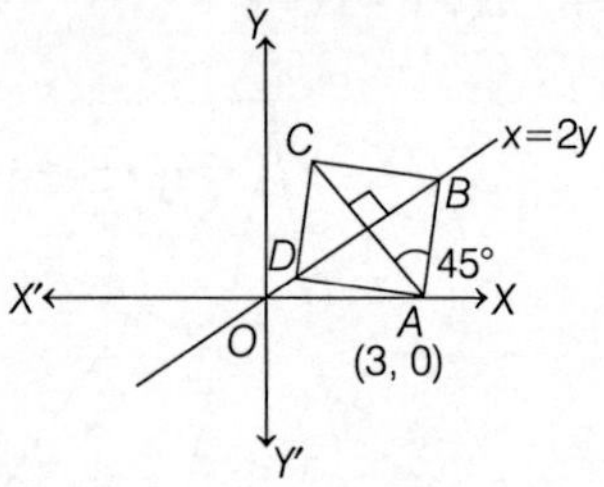

So, the centre of square is $\left(\dfrac{12}{5}, \dfrac{6}{5}\right)$.

Let slope of side AB or AD is m, then

$$\left|\frac{m - (-2)}{1 + m(-2)}\right| = 1$$

$\Rightarrow \quad (m + 2) = \pm (1 - 2m)$

$\Rightarrow \quad m = -\dfrac{1}{3}$ and $m = 3$

Hence, slopes of AB and AD are 3 and $-\dfrac{1}{3}$, respectively.

$\therefore$ Equations of sides AB and AD are

$y - 0 = 3(x - 3)$

and $\quad y - 0 = -\dfrac{1}{3}(x - 3)$

or $y - 3x + 9 = 0$ and $3y + x - 3 = 0$, respectively.

45 Let the coordinate of C be (1,c)

$$m_2 = \frac{c - y}{1 - x}$$

$\Rightarrow \quad m_2 = \dfrac{c - m_1 x}{1 - x} \qquad \left(\because m_1 = \dfrac{y}{x}\right)$

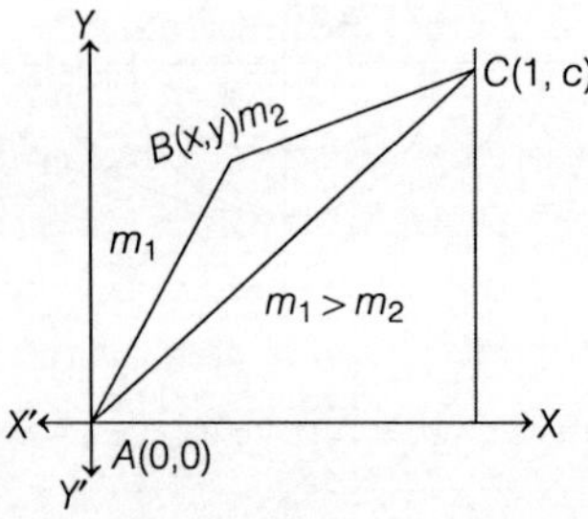

$\Rightarrow \quad (m_1 - m_2)x = c - m_2$

$\Rightarrow \quad c = (m_1 - m_2)x + m_2$

Now area of

$$\Delta ABC = \frac{1}{2}\begin{vmatrix} 0 & 0 & 1 \\ x & m_1 x & 1 \\ 1 & c & 1 \end{vmatrix} = \frac{1}{2}|cx - m_1 x|$$

$$= \frac{1}{2}|((m_1 - m_2)x + m_2)x - m_1 x|$$

$$= \frac{1}{2}(m_1 - m_2)(x - x^2)$$

$[\because x > x^2 \text{ in}(0, 1)]$

Hence $f(x) = \dfrac{1}{2}(x - x^2)$

$f(x)_{\max} = \dfrac{1}{8}$,

when $x = \dfrac{1}{2}$

46 $r = \sqrt{3}\sin 30° = \dfrac{\sqrt{3}}{2}$

$P(x, y)$ is any point on circle

$(PA)^2 + (PB)^2 + (PC)^2 + (PD)^2$

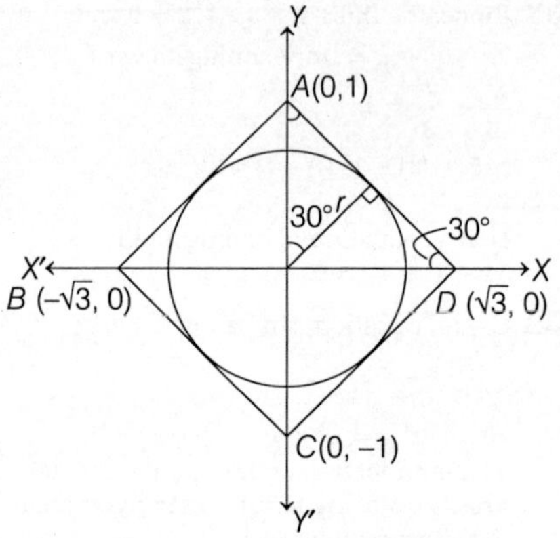

$$= x^2 + (y - 1)^2 + (x + \sqrt{3})^2 + y^2 + x^2 + (y + 1)^2 + (x - \sqrt{3})^2 + y^2$$

$= 4(x^2 + y^2 + 2)$

$= 4(r^2 + 2) \qquad [\because x^2 + y^2 = r^2]$

$= 4\left(\dfrac{3}{4} + 2\right) = 11$

47 Any tangent to $y^2 = 4x$ is of the form

$y = mx + \dfrac{1}{m}$, $(\because a = 1)$,

This touches the circle $(x - 3)^2 + y^2 = 9$, whose centre is (3, 0) and radius is 3.

So, $\left|\dfrac{m(3) + \dfrac{1}{m} - 0}{\sqrt{m^2 + 1}}\right| = 3$

$3m^2 + 1 = \pm\, 3m\sqrt{m^2 + 1}$

$\Rightarrow \quad 9m^4 + 1 + 6m^2 = 9m^2(m^2 + 1)$

[on squaring both sides]

$\Rightarrow \quad 3m^2 = 1$

$\Rightarrow \quad m = \pm \dfrac{1}{\sqrt{3}}$

If the tangent touches the parabola and circle above X-axis, then slope m should be positive.

$\therefore \quad m = \dfrac{1}{\sqrt{3}}$ and the equation is

$$y = \frac{x}{\sqrt{3}} + \sqrt{3}$$

$\Rightarrow \quad \sqrt{3}y = x + 3$

which is the required equation of tangent.

48 Given equation is

$$\frac{x^2}{16} + \frac{y^2}{(16/\sqrt{11})^2} = 1.$$

Thus, the parametric coordinates are $\left(4\cos\phi, \dfrac{16}{\sqrt{11}}\sin\phi\right)$. The equation of tangent at this point is

$$\frac{x\cos\phi}{4} + \frac{\sqrt{11}y\sin\phi}{16} = 1.$$

This touches the circle

$x^2 + y^2 - 2x - 15 = 0$

$$\therefore \quad \frac{\left|\dfrac{\cos\phi}{4} - 1\right|}{\sqrt{\dfrac{\cos^2\phi}{16} + \dfrac{11\sin^2\phi}{256}}} = 4$$

$\Rightarrow \cos^2\phi + 16 - 8\cos\phi = 256\left(\dfrac{\cos^2\phi}{16} + \dfrac{11\sin^2\phi}{256}\right)$

$\Rightarrow \quad 15\cos^2\phi + 11(1 - \cos^2\phi) + 8\cos\phi - 16 = 0$

$\Rightarrow \quad 4\cos^2\phi + 8\cos\phi - 5 = 0$

$\Rightarrow \quad \cos\phi = \dfrac{1}{2} \qquad \left[\because \cos\phi \neq \dfrac{5}{2}\right]$

$\Rightarrow \quad \phi = \dfrac{\pi}{3}$ or $-\dfrac{\pi}{3}$

49 Since, $p, x_1, x_2, \ldots$ and $q, y_1, y_2, \ldots$ are in AP with common differences a and b, respectively.

$\Rightarrow \quad x_i = p + ai$ and $y_i = q + ib$

$\therefore \quad h = \dfrac{x_1 + x_2 + \ldots + x_n}{n}$

and $k = \dfrac{y_1 + y_2 + \ldots + y_n}{n}$

$\Rightarrow \quad nh = \sum_{i=1}^{n} x_i$ and $nk = \sum_{i=1}^{n} y_i$

$\Rightarrow \quad nh = \sum_{i=1}^{n} (p + ia)$

and $nh = \sum_{i=1}^{n} (q + ib)$

$\Rightarrow nh = np + \frac{n(n+1)}{2}a$

and $nk = nq + \frac{n(n+1)}{2}b$

$\Rightarrow \frac{h-p}{a} = \frac{n+1}{2}$ and $\frac{k-q}{b} = \frac{n+1}{2}$

$\therefore \frac{h-p}{a} = \frac{k-q}{b}$

Hence, locus of (h, k) is

$b(x-p) = a(k-q)$.

Hence, Statement II is true and since for Statement I, $n = 3$

So, Statement I is true and Statement II is a correct explanation of Statement I.

50 If t_1 and t_2 are parameters of a and b, then

$t_1 t_2 = -4$ ...(i)

Also, $t_1\left(-t_1 - \frac{2}{t_1}\right) = -4$ ...(ii)

On solving Eqs. (i) and (ii), we get

$t_1 = \sqrt{2}$

Also, $m_{AB} = \frac{2}{t_2 + t_1} = -t_1 = -\sqrt{2}$

51 Clearly, $ac + bd = AC \cdot BD \le 4$

[using Ptolemy's theorem]

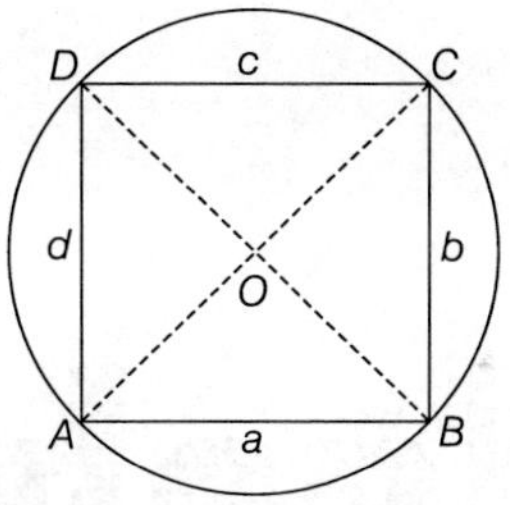

$\Rightarrow ac + bd = 4$ and $AC \cdot BD = 4$

but $ac + bd \ge 4$ [$\because$ AM$\ge$ GM]

$\Rightarrow AC = BD = 2$ and $ac = bd = 2$

[$\because abcd \ge 4$]

$\Rightarrow a = b = c = d = \sqrt{2}$

52 Statement II is true.

For the point (2, 2), $t_1 = 1$

For the point (4, 1), $t_2 = 2$

For the point (6, 2/3), $t_3 = 3$

For the point $\left(\frac{1}{4}, 16\right)$, $t_4 = \frac{1}{8}$

Now, $t_1 \cdot t_2 \cdot t_3 \cdot t_4 = \frac{3}{4} \ne 1$

Hence, Statement I is false.

53 The auxiliary circle of an ellipse $\frac{x^2}{4} + \frac{y^2}{b^2} = 1, b < 2$ is $x^2 + y^2 = 4$

54 The equation of tangent to the ellipse is $\frac{x}{a}\cos\theta + \frac{y}{b}\sin\theta = 1$ and it meets the directrix $x = \frac{a}{e}$ at

$T\left[\frac{a}{e}, \frac{b(e-\cos\theta)}{e\sin\theta}\right]$.

Since, focus is $S(ae, 0)$.

$\therefore$ Slope of $SP = \frac{b\sin\theta}{a(\cos\theta - e)}$

and slope of $ST = \frac{b(e-\cos\theta)}{a\sin\theta(1-e^2)}$

Now, as product of slopes $= -\frac{b^2}{a^2(1-e^2)}$

$= -1$, therefore PT substends a right angle at the focus.

Hence, circle with PT as diameter passes through the focus.

55 Statement II is true, using in Statement I, $(x-1)^2 + (y-4)^2$

$= (x-k)^2 + (y-3)^2$

$\Rightarrow 2(k-1)x - 2y = k^2 - 8$

y-intercept $= -\frac{k^2-8}{2} = -4$ [given]

$\Rightarrow k^2 = 16 \Rightarrow k^2 - 16 = 0$

DAY THIRTY ONE

Vector Algebra

Learning & Revision for the Day

- Vectors and Scalars
- Types of Vectors
- Addition, Subtraction and Scalar Multiplication of Vectors
- Angular Bisectors
- Position Vector (PV)
- Components of a Vector in 2D and 3D
- Scalar (Dot) Product
- Vector (Cross) Product
- Scalar Triple Product
- Vector Triple Product
- Linear Combination, Linear Independence and Dependence

Vectors and Scalars

Physical quantities are divided into two categories **Scalar Quantities** and **Vector Quantities.** Those quantities which have only magnitude and not related to any fixed direction in space are called **scalar quantities** or simply scalars. Examples of scalars are mass, volume, density, work, temperature etc.,

Vectors are those quantities which have both magnitude as well as direction.

Displacement, velocity, acceleration, momentum, weight, force etc., are examples of vector quantities. A directed line-segment is a vector, denoted as $\overrightarrow{AB}$ (or **AB**) or simply $\vec{\mathbf{a}}$ (or **a**).

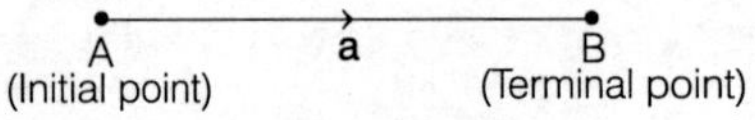

- Magnitude (or length) of a vector **a** is denoted by $|\mathbf{a}|$ and it is always a **non-negative scalar.**

Types of Vectors

(i) A vector whose initial and terminal points coincide is called the **zero** or **null** vector and it is denoted as **0**.

(ii) A vector whose magnitude is 1, is called a **unit vector.** The unit vector in the direction of **a** is given by $\frac{\mathbf{a}}{|\mathbf{a}|}$ and is denoted by $\hat{\mathbf{a}}$. Unit vectors parallel to X-axis, Y-axis and Z-axis are denoted by **i**, **j** and **k**, respectively.

(iii) Vectors are said to be **like** when they have the same sense of direction and **unlike** when they have opposite directions.

(iv) Two vectors **a** and **b** are said to be **equal,** written as $\mathbf{a} = \mathbf{b}$, if they have same length and same direction.

(v) Vectors which are parallel to the same line are called **collinear vectors** or **parallel vector,** otherwise they are called **non-collinear vector**. If **a** and **b** are two collinear vectors, then $\mathbf{a} = \lambda \mathbf{b}$ for same $\lambda \in R$.

(vi) Vectors having the same initial point are called **coinitial vectors.**

PREP MIRROR
Your Personal Preparation Indicator

- No. of Questions in Exercises (x)—
- No. of Questions Attempted (y)—
- No. of Correct Questions (z)—
(Without referring Explanations)

- Accuracy Level ($z/y \times 100$)—
- Prep Level ($z/x \times 100$)—

In order to expect good rank in JEE, your Accuracy Level should be above 85 & Prep Level should be above 75.

(vii) Vectors having the same terminal point are called **coterminous vectors.**

(viii) A system of vectors is said to be **coplanar**, if they are parallel to the same plane or lie in the same plane otherwise they are called **non-coplanar vectors.**

(ix) The vector which has the same magnitude as that of a given vector **a** but opposite direction, is called the **negative** of **a** and is denoted by $-\mathbf{a}$.

(x) A vector having the same direction as that of a given vector **a** but magnitude equal to the reciprocal of the given vector is known as the **reciprocal** of **a** and is denoted by $\mathbf{a}^{-1}$.

(xi) A vector which is drawn parallel to a given vector through a specified point in space is called a **localised vector.**

(xii) Vector whose initial points are not specified are called **free vectors.**

(xiii) When a particle is displaced from point A to other point B, then the displacement AB is a vector, called **displacement vector** of the particle.

(xiv) Two vectors are called **orthogonal,** if angle between the two is a right angle.

Addition, Subtraction and Scalar Multiplication of Vectors

There are three laws for vector addition, which are given below.

(i) **Triangle law** If the vectors **a** and **b** lie along the two sides of a triangle in consecutive order (as shown in the adjoining figure), then their sum (resultant) $\mathbf{a}+\mathbf{b}$ is represented by the third side, but in opposite direction.

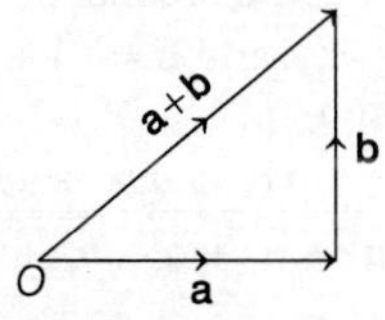

(ii) **Parallelogram law** If the vectors lie along two adjacent sides of a parallelogram (as shown in the adjoining figure), then diagonal of the parallelogram through the common vertex represents their sum.

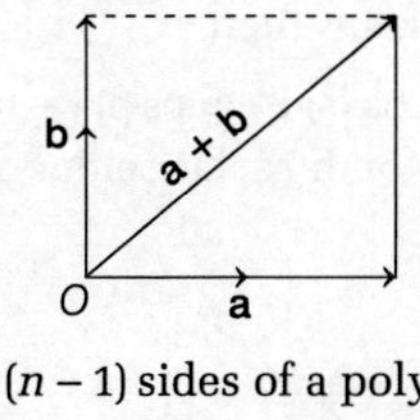

(iii) **Polygon law** If $(n-1)$ sides of a polygon represents vector

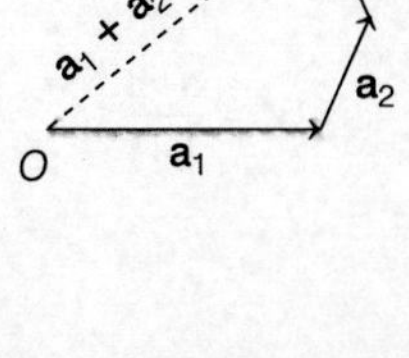

$\mathbf{a}_1, \mathbf{a}_2, \mathbf{a}_3, \ldots$ in consecutive order, then their sum is represented by the nth side, but in opposite direction (as shown in the adjoining figure).

(iv) If $\vec{a}$ and $\vec{b}$ are two vectors, then the **subtraction** of $\vec{b}$ from $\vec{a}$ is defined as the vector sum of $\vec{a}$ and $-\vec{b}$ and it is denoted by $\vec{a}-\vec{b}$, i.e., $\vec{a}-\vec{b} = \vec{a}+(-\vec{b})$.

(v) To subtract **b** from **a**, we reverse the direction of **b** and add it to vector **a**.

(vi) Let $PQRS$ be a parallelogram such that $\mathbf{PQ} = \mathbf{a} = \mathbf{SR}$ and $\mathbf{PS} = \mathbf{b} = \mathbf{QR}$. Then, diagonal $\mathbf{PR} = \mathbf{a}+\mathbf{b}$ (addition of vectors) and diagonal $\mathbf{SQ} = \mathbf{a}-\mathbf{b}$ (subtraction of vectors).

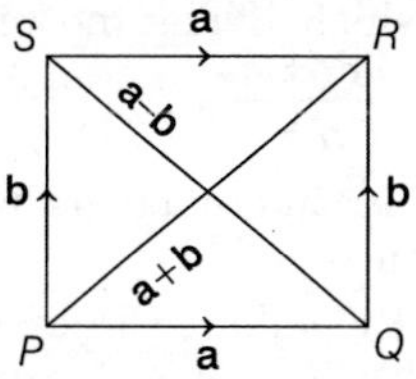

(ii) If **a** is a vector and λ is a scalar (i.e. a real number), then $\lambda\mathbf{a}$ is a vector whose magnitude is λ times that of **a** and whose direction is the same as that of **a** if $\lambda > 0$ and opposite of **a** if $\lambda < 0$.
Thus, $|\lambda\mathbf{a}| = |\lambda||\mathbf{a}|$

Important Results

(i) $\mathbf{a}+\mathbf{b} = \mathbf{b}+\mathbf{a}$

(ii) $\mathbf{a}+(\mathbf{b}+\mathbf{c}) = (\mathbf{a}+\mathbf{b})+\mathbf{c}$

(iii) $\mathbf{a}-\mathbf{b} \neq \mathbf{b}-\mathbf{a}$

(iv) $|\mathbf{a}\pm\mathbf{b}| \le |\mathbf{a}|+|\mathbf{b}|$

(v) $|\mathbf{a}\pm\mathbf{b}| \ge |\mathbf{a}|-|\mathbf{b}|$

(vi) $\lambda_1(\lambda_2\mathbf{a}) = (\lambda_1\lambda_2)\mathbf{a} = \lambda_2(\lambda_1\mathbf{a})$

(vii) $(\lambda_1+\lambda_2)\mathbf{a} = \lambda_1\mathbf{a}+\lambda_2\mathbf{a}$

(viii) $\lambda(\mathbf{a}+\mathbf{b}) = \lambda\mathbf{a}+\lambda\mathbf{b}$

NOTE When the sides of a triangle are taken in order, it leads to zero resultant, e.g., In ABC, $\mathbf{AB}+\mathbf{BC}+\mathbf{CA} = 0$.

Angular Bisectors

Let **a** and **b** are unit vectors, the internal bisector of angle between **a** and **b** is along $\mathbf{a}+\mathbf{b}$ and external bisector of angle is along $\mathbf{a}-\mathbf{b}$.

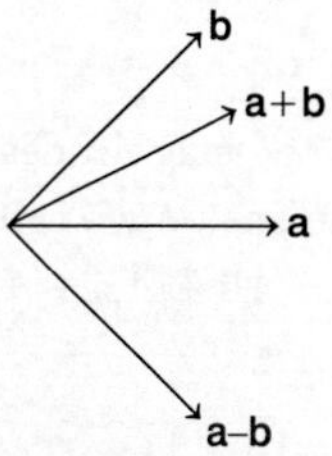

If **a** and **b** are not unit vectors, then above angle bisectors are along $\dfrac{\mathbf{a}}{|\mathbf{a}|}+\dfrac{\mathbf{b}}{|\mathbf{b}|}$ and $\dfrac{\mathbf{a}}{|\mathbf{a}|}-\dfrac{\mathbf{b}}{|\mathbf{b}|}$, respectively.

These bisectors are perpendicular to each other.

Position Vector (PV)

Every point $P(x, y, z)$ in space is associated with a vector whose initial point is O(origin) and terminal point is P. This vector is called **position vector** and it is given by

$$\mathbf{OP} \text{ (or } \mathbf{r}) = x\,\mathbf{i} + y\,\mathbf{j} + z\,\mathbf{k}.$$

Using distance formula, the magnitude of **OP** (or **r**) is $|\mathbf{OP}| = \sqrt{x^2 + y^2 + z^2}$

- If $A(a_1, a_2, a_3)$ and $B(b_1, b_2, b_3)$ have position vectors then **a** and **b** respectively, then $\mathbf{AB} = \mathbf{b} - \mathbf{a}$
  and $AB = |\mathbf{b} - \mathbf{a}| = \sqrt{(a_1 - b_1)^2 + (a_2 - b_2)^2 + (a_3 - b_3)^2}$
- If **a** and **b** are the PV of A and B respectively and **r** be the PV of the point P which divides the join of A and B in the ratio $m : n$, then $\mathbf{r} = \dfrac{m\mathbf{b} \pm n\mathbf{a}}{m \pm n}$.
  Here, '+' sign takes for internal division and '–' sign takes for external division.
- If **a**, **b** and **c** be the PV of three vertices of ΔABC and **r** be the PV of the centroid of ΔABC, then $\mathbf{r} = \dfrac{\mathbf{a} + \mathbf{b} + \mathbf{c}}{3}$

Components of a Vector in 2D and 3D

The process of splitting a vector is called resolution of a vector. The parts of the vector obtained after splitting the vectors are known as the components of the vector.

Let $\mathbf{A}_x$ is the resolved part of **A** along X-axis i.e. the projection of **A** on X-axis. Similarly, $\mathbf{A}_y$ is the resolved part of **A** along Y-axis, i.e. the projection of **A** on Y-axis.

Then, by the parallelogram law,

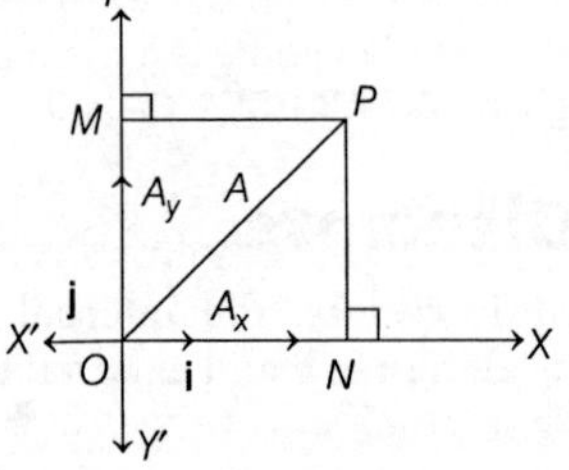

$$\mathbf{OP} = \mathbf{ON} + \mathbf{NP} \quad \text{or} \quad \mathbf{A} = \mathbf{A}_x + \mathbf{A}_y = A_x\,\hat{\mathbf{i}} + A_y\,\hat{\mathbf{j}}$$

Where A_x and A_y are the magnitudes of $\mathbf{A}_x$ and $\mathbf{A}_y$.

Similarly, in three dimension we can represents vector A.

as $\mathbf{A} = \mathbf{A}_x + \mathbf{A}_y + \mathbf{A}_z = A_x\mathbf{i} + A_y\mathbf{j} + A_z\mathbf{k}$

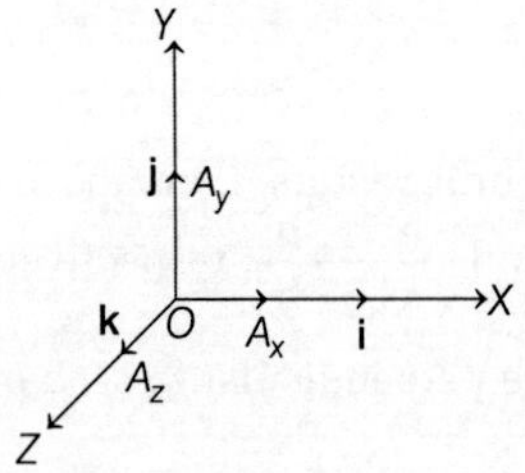

where A_x, A_y and A_z are the magnitudes of $\mathbf{A}_x$, $\mathbf{A}_y$ and $\mathbf{A}_z$, respectively.

Here A_x, A_y and A_z are called **scalar components** of A and $\mathbf{A}_x$, $\mathbf{A}_y$ and $\mathbf{A}_z$ are called **Vector components** of A.

If two vectors $\mathbf{a} = (a_1, a_2, a_3)$ and $\mathbf{b} = (b_1, b_2, b_3)$ are equal, then their resolved parts will also equal i.e. $a_1 = b_1, a_2 = b_2$ and $a_3 = b_3$. If $\mathbf{a} = a_1\mathbf{i} + a_2\mathbf{j} + a_3\mathbf{k}$ and $\mathbf{b} = b_1\mathbf{i} + b_2\mathbf{j} + b_3\mathbf{k}$, then

(i) $\mathbf{a} + \mathbf{b} = (a_1 + b_1)\mathbf{i} + (a_2 + b_2)\mathbf{j} + (a_3 + b_3)\mathbf{k}$

(ii) $\mathbf{a} - \mathbf{b} = (a_1 - b_1)\mathbf{i} + (a_2 - b_2)\mathbf{j} + (a_3 - b_3)\mathbf{k}$

(iii) $\lambda\mathbf{a} = \lambda a_1\mathbf{i} + \lambda a_2\mathbf{j} + \lambda a_3\mathbf{k}$

Scalar (Dot) Product

The scalar product of two vectors **a** and **b** is given by $\mathbf{a} \cdot \mathbf{b} = |\mathbf{a}||\mathbf{b}|\cos\theta, 0 \le \theta \le \pi$.

Properties of Scalar Product is listed below:

1. $\mathbf{a} \cdot \mathbf{b} = \mathbf{b} \cdot \mathbf{a}$ [commutative law]
2. $\mathbf{a} \cdot \mathbf{b} = 0$, if $\mathbf{a} \perp \mathbf{b}$
3. $\mathbf{a} \cdot \mathbf{a} = |\mathbf{a}|^2$
4. $\mathbf{i} \cdot \mathbf{i} = \mathbf{j} \cdot \mathbf{j} = \mathbf{k} \cdot \mathbf{k} = 1$
   and $\mathbf{i} \cdot \mathbf{j} = \mathbf{i} \cdot \mathbf{k} = \mathbf{j} \cdot \mathbf{k} = \mathbf{j} \cdot \mathbf{i} = \mathbf{k} \cdot \mathbf{i} = \mathbf{k} \cdot \mathbf{j} = 0$
5. If $\mathbf{a} = a_1\mathbf{i} + a_2\mathbf{j} + a_3\mathbf{k}$ and $\mathbf{b} = b_1\mathbf{i} + b_2\mathbf{j} + b_3\mathbf{k}$,
   then $\mathbf{a} \cdot \mathbf{b} = a_1 b_1 + a_2 b_2 + a_3 b_3$
6. $\mathbf{a} \cdot (\alpha \cdot \mathbf{b}) = \alpha\,(\mathbf{a} \cdot \mathbf{b})$
7. $\mathbf{a} \cdot (\mathbf{b} \pm \mathbf{c}) = \mathbf{a} \cdot \mathbf{b} \pm \mathbf{a} \cdot \mathbf{c}$ [distributive law]
8. For any two vectors **a** and **b**, we have
   (i) $|\mathbf{a} + \mathbf{b}|^2 = |\mathbf{a}|^2 + |\mathbf{b}|^2 + 2\,(\mathbf{a} \cdot \mathbf{b})$
   (ii) $|\mathbf{a} - \mathbf{b}|^2 = |\mathbf{a}|^2 + |\mathbf{b}|^2 - 2\,(\mathbf{a} \cdot \mathbf{b})$
   (iii) $(\mathbf{a} + \mathbf{b}) \cdot (\mathbf{a} - \mathbf{b}) = |\mathbf{a}|^2 - |\mathbf{b}|^2$
   (iv) $\mathbf{a} \cdot \mathbf{b} < 0$ iff **a** and **b** are inclined at an obtuse angle.
   (v) $\mathbf{a} \cdot \mathbf{b} > 0$ iff **a** and **b** are inclined at an acute angle.
9. If $\mathbf{a} = a_1\mathbf{i} + a_2\mathbf{j} + a_3\mathbf{k}$ and $\mathbf{b} = b_1\mathbf{i} + b_2\mathbf{j} + b_3\mathbf{k}$ are inclined at an angle θ, then
   $$\cos\theta = \frac{a_1 b_1 + a_2 b_2 + a_3 b_3}{\sqrt{a_1^2 + a_2^2 + a_3^2}\,\sqrt{b_1^2 + b_2^2 + b_3^2}}$$
10. If **r** is a vector making angles α, β and γ with OX, OY and OZ respectively, then
    $$\cos\alpha = \mathbf{r} \cdot \mathbf{i}, \cos\beta = \mathbf{r} \cdot \mathbf{j}, \cos\gamma = \mathbf{r} \cdot \mathbf{k}$$
    $$\mathbf{r} = |\mathbf{r}|\cos\alpha\,\mathbf{i} + |\mathbf{r}|\cos\beta\,\mathbf{j} + |\mathbf{r}|\cos\gamma\,\mathbf{k}$$
    If **r** is a unit vector, then
    $$\mathbf{r} = (\cos\alpha)\,\mathbf{i} + (\cos\beta)\,\mathbf{j} + (\cos\gamma)\,\mathbf{k}$$
11. Projection of **a** on **b**(scalar component of **a** along **b**) is

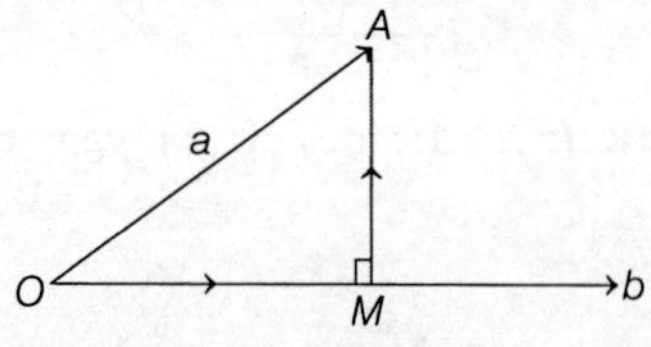

$$OM = \frac{\mathbf{a} \cdot \mathbf{b}}{|\mathbf{b}|}$$

12. Components of **a** along and perpendicular to **b** are $OM = \frac{\mathbf{a}\cdot\mathbf{b}}{|\mathbf{b}|}\cdot\hat{\mathbf{b}}$ and $MA = \mathbf{a} - \left(\frac{\mathbf{a}\cdot\mathbf{b}}{|\mathbf{b}|}\right)\cdot\hat{\mathbf{b}}$, respectively.
13. **Work done** If a particle acted on by a force **F** has displacement **d**, then work done $= \mathbf{F}\cdot\mathbf{d}$

Vector (Cross) Product

The vector product of two vectors **a** *and* **b** is given by $\mathbf{a}\times\mathbf{b} = |\mathbf{a}||\mathbf{b}|\sin\theta\cdot\hat{\mathbf{n}}$, where $\hat{\mathbf{n}}$ is a unit vector perpendicular to **a** and **b** such that **a**, **b** and $\hat{\mathbf{n}}$ form a right handed system and θ $(0\le\theta\le\pi)$ is the angle between **a** *and* **b**.

Properties of vector product are listed below:

1. $\mathbf{a}\times\mathbf{b} = -(\mathbf{b}\times\mathbf{a})$
2. $(\mathbf{a}\times\mathbf{b})^2 = |\mathbf{a}|^2|\mathbf{b}|^2 - (\mathbf{a}\cdot\mathbf{b})^2$
3. $m\mathbf{a}\times\mathbf{b} = m(\mathbf{a}\times\mathbf{b}) = \mathbf{a}\times m\mathbf{b}$
4. $\mathbf{a}\times(\mathbf{b}\pm\mathbf{c}) = \mathbf{a}\times\mathbf{b}\pm\mathbf{a}\times\mathbf{c}$ and $(\mathbf{b}\pm\mathbf{c})\times\mathbf{a} = \mathbf{b}\times\mathbf{a}\pm\mathbf{c}\times\mathbf{a}$
5. $\mathbf{a}\times\mathbf{b} = \mathbf{0} \Leftrightarrow \mathbf{a}\,||\,\mathbf{b}$, where, **a** and **b** are non-zero vectors.
6. If $\mathbf{a} = a_1\mathbf{i} + a_2\mathbf{j} + a_3\mathbf{k}$ and $\mathbf{b} = b_1\mathbf{i} + b_2\mathbf{j} + b_3\mathbf{k}$ then, $\mathbf{a}\times\mathbf{b} = \begin{vmatrix}\mathbf{i} & \mathbf{j} & \mathbf{k}\\ a_1 & a_2 & a_3\\ b_1 & b_2 & b_3\end{vmatrix}$
7. The vector perpendicular to both **a** and **b** is given by $\mathbf{a}\times\mathbf{b}$.
8. The unit vectors perpendicular to the plane of **a** and **b** are $\pm\frac{\mathbf{a}\times\mathbf{b}}{|\mathbf{a}\times\mathbf{b}|}$ and a vector of magnitude λ perpendicular to the plane of (**a** and **b** or **b** and **a**) is $\frac{\lambda(\mathbf{a}\times\mathbf{b})}{|\mathbf{a}\times\mathbf{b}|}$.
9. If **i**, **j** and **k** are three unit vectors along three mutually perpendicular lines, then $\mathbf{i}\times\mathbf{i} = \mathbf{j}\times\mathbf{j} = \mathbf{k}\times\mathbf{k} = 0$, $\mathbf{i}\times\mathbf{j} = \mathbf{k}$, $\mathbf{j}\times\mathbf{i} = -\mathbf{k}$, $\mathbf{j}\times\mathbf{k} = \mathbf{i}$, $\mathbf{k}\times\mathbf{j} = -\mathbf{i}$, $\mathbf{k}\times\mathbf{i} = \mathbf{j}$, $\mathbf{i}\times\mathbf{k} = -\mathbf{j}$
10. (i) The area of parallelogram with adjacent sides **a** and **b** is $|\mathbf{a}\times\mathbf{b}|$.
    (ii) The area of quadrilateral with diagonals $\mathbf{d}_1$ and $\mathbf{d}_2$ is $\frac{1}{2}|\mathbf{d}_1\times\mathbf{d}_2|$.
    (iii) The area of triangle with adjacent sides **a** and **b**, is $\frac{1}{2}|\mathbf{a}\times\mathbf{b}|$.
    (iv) If **a**, **b**, **c** are position vectors of a ΔABC, then the area $= \frac{1}{2}|(\mathbf{a}\times\mathbf{b}) + (\mathbf{b}\times\mathbf{c}) + (\mathbf{c}\times\mathbf{a})|$.
11. Three points with position vectors **a**, **b**, **c** are collinear if $\mathbf{a}\times\mathbf{b} + \mathbf{b}\times\mathbf{c} + \mathbf{c}\times\mathbf{a} = \mathbf{0}$

Scalar Triple Product

If **a**, **b**, **c** are three vectors, then their scalar triple product is defined as the dot product of **a** and $\mathbf{b}\times\mathbf{c}$. It is denoted by $[\mathbf{a}\ \mathbf{b}\ \mathbf{c}]$. Thus, $[\mathbf{a}\ \mathbf{b}\ \mathbf{c}] = \mathbf{a}\cdot(\mathbf{b}\times\mathbf{c})$.

Properties of Scalar Triple Product are listed below:

1. $[\mathbf{a}\ \mathbf{b}\ \mathbf{c}] = [\mathbf{b}\ \mathbf{c}\ \mathbf{a}] = [\mathbf{c}\ \mathbf{a}\ \mathbf{b}]$
2. $[\mathbf{a}\ \mathbf{b}\ \mathbf{c}] = -[\mathbf{b}\ \mathbf{a}\ \mathbf{c}] = -[\mathbf{c}\ \mathbf{b}\ \mathbf{a}] = -[\mathbf{a}\ \mathbf{c}\ \mathbf{b}]$
3. If λ is a scalar, then $[\lambda\mathbf{a}\ \mathbf{b}\ \mathbf{c}] = \lambda[\mathbf{a}\ \mathbf{b}\ \mathbf{c}]$
4. $[\mathbf{a}\ \mathbf{b}\ \mathbf{c}_1+\mathbf{c}_2] = [\mathbf{a}\ \mathbf{b}\ \mathbf{c}_1] + [\mathbf{a}\ \mathbf{b}\ \mathbf{c}_2]$
5. $(\mathbf{a}\times\mathbf{b})\cdot\mathbf{c} = \mathbf{a}\cdot(\mathbf{b}\times\mathbf{c})$
6. The scalar triple product of three vectors is zero, if any two of them are equal or parallel or collinear.
7. If **a**, **b** and **c** are coplanar, then $[\mathbf{a}\ \mathbf{b}\ \mathbf{c}] = 0$
8. If $[\mathbf{a}\ \mathbf{b}\ \mathbf{c}] = 0$, then any two of the vectors are parallel or **a**, **b** and **c** are coplanar or $\mathbf{c} = \alpha\mathbf{a} + \beta\mathbf{b}$.
9. Four points with position vectors **a**, **b**, **c** and **d** will be coplanar, if $[\mathbf{b}-\mathbf{a}, \mathbf{c}-\mathbf{a}, \mathbf{d}-\mathbf{a}] = 0$.
10. Volume of parallelopiped, whose coterminous edges are **a**, **b** and **c** is $|[\mathbf{a}\ \mathbf{b}\ \mathbf{c}]|$.
11. If $\mathbf{a} = a_1\mathbf{i} + a_2\mathbf{j} + a_3\mathbf{k}$, $\mathbf{b} = b_1\mathbf{i} + b_2\mathbf{j} + b_3\mathbf{k}$ and $\mathbf{c} = c_1\mathbf{i} + c_2\mathbf{j} + c_3\mathbf{k}$, then $[\mathbf{a}\ \mathbf{b}\ \mathbf{c}] = \begin{vmatrix} a_1 & a_2 & a_3\\ b_1 & b_2 & b_3\\ c_1 & c_2 & c_3\end{vmatrix}$
12. $[\mathbf{a}\times\mathbf{b}\ \ \mathbf{b}\times\mathbf{c}\ \ \mathbf{c}\times\mathbf{a}] = [\mathbf{a}\ \mathbf{b}\ \mathbf{c}]^2 = \begin{vmatrix}\mathbf{a}\cdot\mathbf{a} & \mathbf{a}\cdot\mathbf{b} & \mathbf{a}\cdot\mathbf{c}\\ \mathbf{b}\cdot\mathbf{a} & \mathbf{b}\cdot\mathbf{b} & \mathbf{b}\cdot\mathbf{c}\\ \mathbf{c}\cdot\mathbf{a} & \mathbf{c}\cdot\mathbf{b} & \mathbf{c}\cdot\mathbf{c}\end{vmatrix}$
13. $[\mathbf{a}\ \mathbf{b}\ \mathbf{c}]\cdot[\mathbf{u}\ \mathbf{v}\ \mathbf{w}] = \begin{vmatrix}\mathbf{a}\cdot\mathbf{u} & \mathbf{a}\cdot\mathbf{v} & \mathbf{a}\cdot\mathbf{w}\\ \mathbf{b}\cdot\mathbf{u} & \mathbf{b}\cdot\mathbf{v} & \mathbf{b}\cdot\mathbf{w}\\ \mathbf{c}\cdot\mathbf{u} & \mathbf{c}\cdot\mathbf{v} & \mathbf{c}\cdot\mathbf{w}\end{vmatrix}$
14. If $\mathbf{a} = a_1\mathbf{l}+a_2\mathbf{m}+a_3\mathbf{n}$; $\mathbf{b} = b_1\mathbf{l}+b_2\mathbf{m}+b_3\mathbf{n}$ and $\mathbf{c} = c_1\mathbf{l}+c_2\mathbf{m}+c_3\mathbf{n}$, then $[\mathbf{a}\ \mathbf{b}\ \mathbf{c}] = \begin{vmatrix} a_1 & a_2 & a_3\\ b_1 & b_2 & b_3\\ c_1 & c_2 & c_3\end{vmatrix}[\mathbf{l}, \mathbf{m}, \mathbf{n}]$

Tetrahedron and Its Volume

A tetrahedron is a three dimensional figure formed by four triangles, as shown in figure

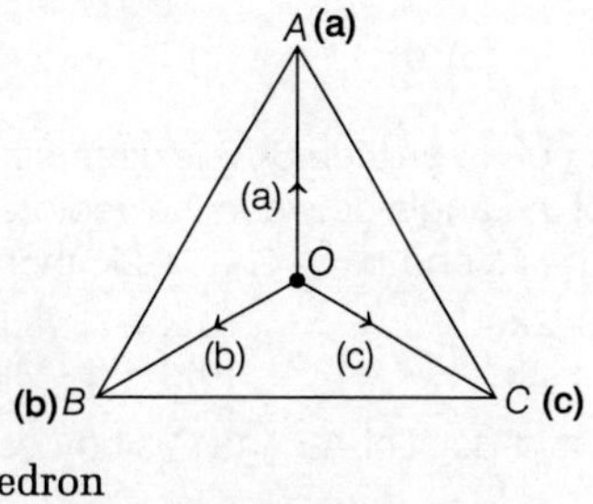

Volume of tetrahedron

$$OABC = \frac{1}{6}[\mathbf{a}\ \mathbf{b}\ \mathbf{c}]$$

If **a**, **b**, **c** and **d** are position vectors of vertices A, B, C and D of a tetrahedron $ABCD$, then its volume

$$= \frac{1}{6}[\mathbf{a}-\mathbf{d}\ \ \mathbf{b}-\mathbf{d}\ \ \mathbf{c}-\mathbf{d}].$$

Vector Triple Product

If **a**, **b** and **c** are three vector quantities, then the vectors $\mathbf{a}\times(\mathbf{b}\times\mathbf{c})$ and $(\mathbf{a}\times\mathbf{b})\times\mathbf{c}$ represents the vector triple product and is given by

$$\mathbf{a}\times(\mathbf{b}\times\mathbf{c})=(\mathbf{a}\cdot\mathbf{c})\,\mathbf{b}-(\mathbf{a}\cdot\mathbf{b})\,\mathbf{c}$$

$$(\mathbf{a}\times\mathbf{b})\times\mathbf{c}=(\mathbf{a}\cdot\mathbf{c})\,\mathbf{b}-(\mathbf{b}\cdot\mathbf{c})\,\mathbf{a}$$

Properties of Vector Triple Product are listed below:

1. $(\mathbf{a}\times\mathbf{b})\times\mathbf{c}=\mathbf{a}\times(\mathbf{b}\times\mathbf{c})$ if some or all **a**, **b** and **c** are zero vectors or **a** and **c** are collinear.
2. $\mathbf{a}\times\mathbf{b}\times\mathbf{a}=(\mathbf{a}\times\mathbf{b})\times\mathbf{a}=\mathbf{a}\times(\mathbf{b}\times\mathbf{a})$
3. Vector $\mathbf{a}\times(\mathbf{b}\times\mathbf{c})$ is perpendicular to **a** and lies in the plane of **b** and **c**.
4. $(\mathbf{a}\times\mathbf{b})\cdot(\mathbf{c}\times\mathbf{d})=\begin{vmatrix}\mathbf{a}\cdot\mathbf{c} & \mathbf{a}\cdot\mathbf{d}\\ \mathbf{b}\cdot\mathbf{c} & \mathbf{b}\cdot\mathbf{d}\end{vmatrix}$.

Linear Combination, Linear Independence and Dependence

A vector **r** is said to be a **linear combination** *of vectors* **a**, **b**, **c**, ... etc., if there exist scalars $x, y, z, \ldots$ etc., such that $\mathbf{r}=x\mathbf{a}+y\mathbf{b}+z\mathbf{c}+\ldots$

- If two non-zero vectors **a** and **b** are **linearly dependent**, then it means
  (i) there are non-zero scalar α and β such that $\alpha\,\mathbf{a}+\beta\mathbf{b}=\mathbf{0}$.
  (ii) **a** and **b** are parallel.
  (iii) **a** and **b** are collinear.
  (iv) $\mathbf{a}\times\mathbf{b}=\mathbf{0}$
  Otherwise, **a** and **b** are **linearly independent.**
- If three non-zero vectors **a**, **b** and **c** are **linearly dependent**, then it means
  (i) there are non-zero scalars α, β and γ such that
  $$\alpha\,\mathbf{a}+\beta\,\mathbf{b}+\gamma\mathbf{c}=\mathbf{0}$$
  (ii) **a**, **b** and **c** are parallel to same plane.
  (iii) **a**, **b** and **c** are coplanar.
  (iv) $\mathbf{a}=\alpha_1\mathbf{b}+\alpha_2\mathbf{c}$ etc.
  (v) $[\mathbf{a}\ \mathbf{b}\ \mathbf{c}]=0$
  Otherwise **a**, **b** and **c** are **linearly independent.**
- A set of non-zero vectors $\mathbf{a}_1, \mathbf{a}_2, \mathbf{a}_3, \ldots\ldots, \mathbf{a}_n$ is said to be **linearly independent**, if $x_1\mathbf{a}_1+x_2\mathbf{a}_2+\ldots+x_n\mathbf{a}_n=0$
  $\Rightarrow x_1=x_2=\ldots=x_n=0$, where $x_1, x_2, \ldots x_n$ are scalars.

DAY PRACTICE SESSION 1

FOUNDATION QUESTIONS EXERCISE

1 Let us define the length of a vector $a\mathbf{i}+b\mathbf{j}+c\mathbf{k}$ as $|a|+|b|+|c|$. This definition coincides with the usual definition of length of a vector $a\mathbf{i}+b\mathbf{j}+c\mathbf{k}$ iff
(a) $a=b=c=0$
(b) any two of a, b and c are zero
(c) any one of a, b and c are zero
(d) $a+b+c=0$

2 The non-zero vectors **a**, **b** and **c** are related by $\mathbf{a}=8\mathbf{b}$ and $\mathbf{c}=-7\mathbf{b}$. Then, the angle between **a** and **c** is → AIEEE 2008
(a) π (b) 0 (c) $\frac{\pi}{4}$ (d) $\frac{\pi}{2}$

3 If a vector **r** of magnitude $3\sqrt{6}$ is directed along the bisector of the angle between the vectors $\mathbf{a}=7\mathbf{i}-4\mathbf{j}-4\mathbf{k}$ and $\mathbf{b}=-2\mathbf{i}-\mathbf{j}+2\mathbf{k}$, then **r** is equal to
(a) $\mathbf{i}-7\mathbf{j}+2\mathbf{k}$ (b) $\mathbf{i}+7\mathbf{j}-2\mathbf{k}$
(c) $\mathbf{i}+7\mathbf{j}+2\mathbf{k}$ (d) $\mathbf{i}-7\mathbf{j}-2\mathbf{k}$

4 If C is the mid-point of AB and P is any point out side AB then → AIEEE 2005
(a) $\mathbf{PA}+\mathbf{PB}+\mathbf{PC}=\mathbf{0}$ (b) $\mathbf{PA}+\mathbf{PB}+2\mathbf{PC}=\mathbf{0}$
(c) $\mathbf{PA}+\mathbf{PB}=\mathbf{PC}$ (d) $\mathbf{PA}+\mathbf{PB}=2\mathbf{PC}$

5 If the vectors $\mathbf{AB}=3\mathbf{i}+4\mathbf{k}$ and $\mathbf{AC}=5\mathbf{i}-2\mathbf{j}+4\mathbf{k}$ are the sides of a ΔABC, then the length of the median through A is → JEE Mains 2013
(a) $\sqrt{18}$ (b) $\sqrt{72}$ (c) $\sqrt{33}$ (d) $\sqrt{45}$

6 Let $|\mathbf{a}|=2\sqrt{2}$, $|\mathbf{b}|=3$ and the angle between **a** and **b** is $\frac{\pi}{4}$. If a parallelogram is constructed with adjacent sides $2\mathbf{a}-3\mathbf{b}$ and $\mathbf{a}+\mathbf{b}$, then its longer diagonal is of length
(a) 10 (b) 8
(c) $2\sqrt{26}$ (d) 6

7 If **a**, **b** and **c** are unit vectors, then $|\mathbf{a}-\mathbf{b}|^2+|\mathbf{b}-\mathbf{c}|^2+|\mathbf{c}-\mathbf{a}|^2$ does not exceed to
(a) 4 (b) 9 (c) 8 (d) 6

8 **a** and **c** are unit collinear vectors and $|\mathbf{b}|=6$, then $\mathbf{b}-3\mathbf{c}=\lambda\mathbf{a}$, if λ is
(a) $-9, 3$ (b) $9, 3$
(c) $3, -3$ (d) None of these

9 If **a** and **b** are unit vectors, then what is the angle between **a** and **b** for $\sqrt{3}\,\mathbf{a}-\mathbf{b}$ to be a unit vector? → NCERT Exemplar
(a) 30° (b) 45°
(c) 60° (d) 90°

10 If **a** and **b** are unit vectors inclined at an angle α, $\alpha\in[0,\pi]$ to each other and $|\mathbf{a}+\mathbf{b}|<1$. Then, α belong to
(a) $\left(\frac{\pi}{3},\frac{2\pi}{3}\right)$ (b) $\left(\frac{2\pi}{3},\pi\right)$
(c) $\left(0,\frac{\pi}{3}\right)$ (d) $\left(\frac{\pi}{4},\frac{3\pi}{4}\right)$

11 If **a**, **b** and **c** are unit vectors satisfying $\mathbf{a} - \sqrt{3}\mathbf{b} + \mathbf{c} = 0$, then the angle between the vectors **a** and **c** is

(a) $\frac{\pi}{4}$ (b) $\frac{\pi}{3}$ (c) $\frac{\pi}{6}$ (d) $\frac{\pi}{2}$

12 The value of a, for which the points, A, B, C with position vectors $2\mathbf{i} - \mathbf{j} + \mathbf{k}, \mathbf{i} - 3\mathbf{j} - 5\mathbf{k}$ and $a\mathbf{i} - 3\mathbf{j} + \mathbf{k}$ respectively are the vertices of a right angled triangle with $C = \frac{\pi}{2}$ are → AIEEE 2006

(a) –2 and –1 (b) –2 and 1
(c) 2 and –1 (d) 2 and 1

13 If the vectors $\mathbf{a} = \mathbf{i} - \mathbf{j} + 2\mathbf{k}$, $\mathbf{b} = 2\mathbf{i} + 4\mathbf{j} + \mathbf{k}$ and $\mathbf{c} = \lambda\mathbf{i} + \mathbf{j} + \mu\mathbf{k}$ are mutually orthogonal, then (λ, μ) is equal to → AIEEE 2010

(a) (–3, 2) (b) (2, –3)
(c) (–2, 3) (d) (3, –2)

14 Let **a** and **b** be two unit vectors. If the vectors $\mathbf{c} = \mathbf{a} + 2\mathbf{b}$ and $\mathbf{d} = 5\mathbf{a} - 4\mathbf{b}$ are perpendicular to each other, then the angle between **a** and **b** is → AIEEE 2012

(a) $\frac{\pi}{6}$ (b) $\frac{\pi}{2}$ (c) $\frac{\pi}{3}$ (d) $\frac{\pi}{4}$

15 If **p**, **q** and **r** are perpendicular to $\mathbf{q} + \mathbf{r}, \mathbf{r} + \mathbf{p}$ and $\mathbf{p} + \mathbf{q}$ respectively and if $|\mathbf{p} + \mathbf{q}| = 6, |\mathbf{q} + \mathbf{r}| = 4\sqrt{3}$ and $|\mathbf{r} + \mathbf{p}| = 4$, then $|\mathbf{p} + \mathbf{q} + \mathbf{r}|$ is

(a) $5\sqrt{2}$ (b) 10
(c) 5 (d) 15

16 Let **a** and **b** be the position vectors of points A and B with respect to origin and $|\mathbf{a}| = a, |\mathbf{b}| = b$. The points C and D divide AB internally and externally in the ratio 2 : 3. If **OC** and **OD** are perpendicular, then

(a) $9a^2 = 4b^2$ (b) $4a^2 = 9b^2$
(c) $9a = 4b$ (d) $4a = 9b$

17 A vector of magnitude $\sqrt{2}$ coplanar with $\mathbf{i} + \mathbf{j} + 2\mathbf{k}$ and $\mathbf{i} + 2\mathbf{j} + \mathbf{k}$ and perpendicular to $\mathbf{i} + \mathbf{j} + \mathbf{k}$ is

(a) $-\mathbf{j} + \mathbf{k}$ (b) $\mathbf{i} - \mathbf{k}$ (c) $\mathbf{i} - \mathbf{j}$ (d) $\mathbf{i} - 2\mathbf{j} + \mathbf{k}$

18 If the positive numbers a, b and c are the pth, qth and rth terms of GP, then the vectors $\log a \cdot \mathbf{i} + \log b \cdot \mathbf{j} + \log c \cdot \mathbf{k}$ and $(q - r)\mathbf{i} + (r - p)\mathbf{j} + (p - q)\mathbf{k}$ are

(a) equal (b) parallel
(c) perpendicular (d) None of these

19 The distance of the point B with position vector $\mathbf{i} + 2\mathbf{j} + 3\mathbf{k}$ from the line passing through the point A, whose position vector is $4\mathbf{i} + 2\mathbf{j} + 2\mathbf{k}$ and parallel to the vector $2\mathbf{i} + 3\mathbf{j} + 6\mathbf{k}$ is

(a) $\sqrt{10}$ (b) $\sqrt{5}$ (c) $\sqrt{6}$ (d) $\sqrt{8}$

20 Let $\mathbf{a} = 2\mathbf{i} - \mathbf{j} + \mathbf{k}, \mathbf{b} = \mathbf{i} + 2\mathbf{j} - \mathbf{k}$ and $\mathbf{c} = \mathbf{i} + \mathbf{j} - 2\mathbf{k}$ be three vectors. A vector of the type $\mathbf{b} + \lambda\mathbf{c}$ for some scalar λ, whose projection on **a** is of magnitude $\sqrt{\frac{2}{3}}$, is

(a) $2\mathbf{i} + \mathbf{j} + 5\mathbf{k}$ (b) $2\mathbf{i} + 3\mathbf{j} - 3\mathbf{k}$
(c) $2\mathbf{i} - \mathbf{j} + 5\mathbf{k}$ (d) $2\mathbf{i} + 3\mathbf{j} + 3\mathbf{k}$

21 Let **u**, **v** and **w** be three vectors such that

$$|\mathbf{u}| = 1, |\mathbf{v}| = 2, |\mathbf{w}| = 3$$

If the projection of **v** along **u** is equal to that of **w** along **u** and **v** and **w** are perpendicular to each other, then $|\mathbf{u} - \mathbf{v} + \mathbf{w}|$ is equal to

(a) 4 (b) $\sqrt{7}$ (c) $\sqrt{14}$ (d) 2

22 If **a**, **b** and **c** are three mutually perpendicular vectors, then the projection of the vector $l\frac{\mathbf{a}}{|\mathbf{a}|} + m\frac{\mathbf{b}}{|\mathbf{b}|} + n\frac{(\mathbf{a} \times \mathbf{b})}{|\mathbf{a} \times \mathbf{b}|}$ along the angle bisector of the vector **a** and **b** is

(a) $\frac{l^2 + m^2}{\sqrt{l^2 + m^2 + n^2}}$ (b) $\sqrt{l^2 + m^2 + n^2}$

(c) $\frac{\sqrt{l^2 + m^2}}{\sqrt{l^2 + m^2 + n^2}}$ (d) $\frac{l + m}{\sqrt{2}}$

23 Resolved part of vector **a** along the vector **b** is $\mathbf{a}_1$ and that perpendicular to **b** is $\mathbf{a}_2$, then $\mathbf{a}_1 \times \mathbf{a}_2$ is equal to

(a) $\frac{(\mathbf{a} \times \mathbf{b})\,\mathbf{b}}{|\mathbf{b}|^2}$ (b) $\frac{(\mathbf{a} \times \mathbf{b})\,\mathbf{a}}{|\mathbf{a}|^2}$

(c) $\frac{(\mathbf{a} \cdot \mathbf{b})\,(\mathbf{b} \times \mathbf{a})}{|\mathbf{b}|^2}$ (d) $\frac{(\mathbf{a} \cdot \mathbf{b})\,(\mathbf{b} \times \mathbf{a})}{|\mathbf{b} \times \mathbf{a}|}$

24 A particle is acted upon by constant forces $4\mathbf{i} + \mathbf{j} - 3\mathbf{k}$ and $3\mathbf{i} + \mathbf{j} - \mathbf{k}$ which displace it from a point $\mathbf{i} + 2\mathbf{j} + 3\mathbf{k}$ to the point $5\mathbf{i} + 4\mathbf{j} + \mathbf{k}$ The work done in standard units by the forces is given by → AIEEE 2004

(a) 40 units (b) 30 units
(c) 25 units (d) 15 units

25 If **u** and **v** are unit vectors and θ is the acute angle between them, then $2\mathbf{u} \times 3\mathbf{v}$ is a unit vector for → AIEEE 2007

(a) exactly two values of θ
(b) more than two values of θ
(c) no value of θ
(d) exactly one value of θ

26 If the vectors $\mathbf{c}, \mathbf{a} = x\mathbf{i} + y\mathbf{j} + z\mathbf{k}$ and $\mathbf{b} = \mathbf{j}$ are such that **a**, **c** and **b** form a right handed system, then **c** is → AIEEE 2002

(a) $z\mathbf{i} - x\mathbf{k}$ (b) **0**
(c) $y\mathbf{j}$ (d) $-z\mathbf{i} + x\mathbf{k}$

27 If $\mathbf{a} = \mathbf{i} + \mathbf{j} + \mathbf{k}$ and $\mathbf{b} = \mathbf{j} - \mathbf{k}$, then a vector c such that $\mathbf{a} \times \mathbf{c} = \mathbf{b}$ and $\mathbf{a} \cdot \mathbf{c} = 3$ is → NCERT Exemplar

(a) $\frac{5}{3}\mathbf{i} + \frac{2}{3}\mathbf{j} + \frac{2}{3}\mathbf{k}$ (b) $\frac{2}{3}\mathbf{i} + \frac{5}{3}\mathbf{j} + \frac{5}{3}\mathbf{k}$

(c) $\frac{5}{3}\mathbf{i} + \frac{2}{3}\mathbf{j} + \frac{1}{3}\mathbf{k}$ (d) None of these

28 Vectors **a** and **b** are not perpendicular and **c** and **d** are two vectors satisfying $\mathbf{b} \times \mathbf{c} = \mathbf{b} \times \mathbf{d}$ and $\mathbf{a} \cdot \mathbf{d} = 0$. Then, the vector **d** is equal to → AIEEE 2011

(a) $\mathbf{c} + \left(\frac{\mathbf{a} \cdot \mathbf{c}}{\mathbf{a} \cdot \mathbf{b}}\right)\mathbf{b}$ (b) $\mathbf{b} + \left(\frac{\mathbf{b} \cdot \mathbf{c}}{\mathbf{a} \cdot \mathbf{b}}\right)\mathbf{c}$

(c) $\mathbf{c} - \left(\frac{\mathbf{a} \cdot \mathbf{c}}{\mathbf{a} \cdot \mathbf{b}}\right)\mathbf{b}$ (d) $\mathbf{b} - \left(\frac{\mathbf{b} \cdot \mathbf{c}}{\mathbf{a} \cdot \mathbf{b}}\right)\mathbf{c}$

29 If **u**, **v** and **w** are non-coplanar vectors and p, q are real numbers, then the equality
$[3\mathbf{u}\ p\mathbf{v}\ p\mathbf{w}]-[p\mathbf{v}\ \mathbf{w}\ q\mathbf{u}]-[2\mathbf{w}\ q\mathbf{v}\ q\mathbf{u}]=0$
holds for → AIEEE 2009
(a) exactly two values of (p, q)
(b) more than two but not all values of (p, q)
(c) all values of (p, q)
(d) exactly one value of (p, q)

30 Let $\mathbf{v}=2\mathbf{i}+\mathbf{j}-\mathbf{k}$ and $\mathbf{w}=\mathbf{i}+3\mathbf{k}$. If **u** is a unit vector, then the maximum value of $[\mathbf{u}\ \mathbf{v}\ \mathbf{w}]$ is
(a) -1 (b) $\sqrt{10}+\sqrt{6}$
(c) $\sqrt{59}$ (d) $\sqrt{60}$

31 If $\mathbf{a}=\mathbf{i}-\mathbf{j}$, $\mathbf{b}=\mathbf{j}-\mathbf{k}$, $\mathbf{c}=\mathbf{k}-\mathbf{i}$ and **d** is a unit vector such that $\mathbf{a}\cdot\mathbf{d}=0=[\mathbf{b}\ \mathbf{c}\ \mathbf{d}]$, then **d** is/are
(a) $\pm\dfrac{\mathbf{i}+\mathbf{j}-\mathbf{k}}{\sqrt{3}}$ (b) $\pm\dfrac{\mathbf{i}+\mathbf{j}-2\mathbf{k}}{\sqrt{6}}$
(c) $\pm\dfrac{\mathbf{i}+\mathbf{j}+\mathbf{k}}{\sqrt{3}}$ (d) $\pm\mathbf{k}$

32 The vector $(\mathbf{i}\times\mathbf{a}\cdot\mathbf{b})\mathbf{i}+(\mathbf{j}\times\mathbf{a}\cdot\mathbf{b})\mathbf{j}+(\mathbf{k}\times\mathbf{a}\cdot\mathbf{b})\mathbf{k}$ is equal to
(a) $\mathbf{b}\times\mathbf{a}$ (b) **a** (c) $\mathbf{a}\times\mathbf{b}$ (d) **b**

33 The points with position vectors $\alpha\mathbf{i}+\mathbf{j}+\mathbf{k}$, $\mathbf{i}-\mathbf{j}-\mathbf{k}$, $\mathbf{i}+2\mathbf{j}-\mathbf{k}$, $\mathbf{i}+\mathbf{j}+\beta\mathbf{k}$ are coplanar if
(a) $(1-\alpha)(1+\beta)=0$ (b) $(1-\alpha)(1-\beta)=0$
(c) $(1+\alpha)(1+\beta)=0$ (d) $(1+\alpha)(1-\beta)=0$

34 Let $\mathbf{a}=\mathbf{i}+\mathbf{j}+\mathbf{k}$, $\mathbf{b}=\mathbf{i}-\mathbf{j}+2\mathbf{k}$ and $\mathbf{c}=x\mathbf{i}+(x-2)\mathbf{j}-\mathbf{k}$ If the vector **c** lies in the plane of **a** and **b**, then x equal to → AIEEE 2007
(a) 0 (b) 1
(c) -4 (d) -2

35 If the vectors $p\mathbf{i}+\mathbf{j}+\mathbf{k}$, $\mathbf{i}+q\mathbf{j}+\mathbf{k}$ and $\mathbf{i}+\mathbf{j}+r\mathbf{k}$ $(p\neq q\neq r\neq 1)$ are coplanar, then the value of $pqr-(p+q+r)$ is → AIEEE 2011
(a) -2 (b) 2
(c) 0 (d) -1

36 Let $\alpha=a\mathbf{i}+b\mathbf{j}+c\mathbf{k}$, $\beta=b\mathbf{i}+c\mathbf{j}+a\mathbf{k}$ and $\gamma=c\mathbf{i}+a\mathbf{j}+b\mathbf{k}$ be three coplanar vectors with $a\neq b$ and $\mathbf{v}=\mathbf{i}+\mathbf{j}+\mathbf{k}$. Then, **v** is perpendicular to
(a) α (b) β (c) γ (d) All of these

37 If **a**, **b**, **c** are non-coplanar vectors and λ is a real number, then $[\lambda(\mathbf{a}+\mathbf{b})\ \ \lambda^2\mathbf{b}\ \ \lambda\mathbf{c}]=[\mathbf{a}\ \ \mathbf{b}+\mathbf{c}\ \ \mathbf{b}]$ for
(a) exactly two values of λ → AIEEE 2005
(b) exactly three values of λ
(c) no value of λ
(d) exactly one value of λ

38 Let **a**, **b** and **c** be three non-zero vectors which are pairwise non-collinear. If $\mathbf{a}+3\mathbf{b}$ is collinear with **c** and $\mathbf{b}+2\mathbf{c}$ is collinear with **a**, then $\mathbf{a}+3\mathbf{b}+6\mathbf{c}$ is equal to
(a) a + c (b) a (c) c (d) 0

39 If $[\mathbf{a}\times\mathbf{b}\ \ \mathbf{b}\times\mathbf{c}\ \ \mathbf{c}\times\mathbf{a}]=\lambda[\mathbf{a}\,\mathbf{b}\,\mathbf{c}]^2$, then λ is equal to → JEE Mains 2014
(a) 0 (b) 1
(c) 2 (d) 3

40 If V is the volume of the parallelopiped having three coterminous edges, as a, b and c, then the volume of the parallelopiped having three coterminous edges as
$$\alpha=(\mathbf{a}\cdot\mathbf{a})\mathbf{a}+(\mathbf{a}\cdot\mathbf{b})\mathbf{b}+(\mathbf{a}\cdot\mathbf{c})\mathbf{c}$$
$$\beta=(\mathbf{a}\cdot\mathbf{b})\mathbf{a}+(\mathbf{b}\cdot\mathbf{b})\mathbf{b}+(\mathbf{b}\cdot\mathbf{c})\mathbf{c}$$
$$\gamma=(\mathbf{a}\cdot\mathbf{c})\mathbf{a}+(\mathbf{b}\cdot\mathbf{c})\mathbf{b}+(\mathbf{c}\cdot\mathbf{c})\mathbf{c}$$
(a) V^3 (b) $3V$
(c) V^2 (d) $2V$

41 If **a**, **b** and **c** are non-coplanar vectors and λ is a real number, then the vectors $\mathbf{a}+2\mathbf{b}+3\mathbf{c}$, $\lambda\mathbf{b}+4\mathbf{c}$ and $(2\lambda-1)\mathbf{c}$ are non-coplanar for
(a) no value of λ
(b) all except one value of λ
(c) all except two values of λ
(d) all values of λ

42 If $\mathbf{a}=\dfrac{1}{\sqrt{10}}(3\mathbf{i}+\mathbf{k})$ and $\mathbf{b}=\dfrac{1}{7}(2\mathbf{i}+3\mathbf{j}-6\mathbf{k})$, then the value of $(2\mathbf{a}-\mathbf{b})\cdot[(\mathbf{a}\times\mathbf{b})\times(\mathbf{a}+2\mathbf{b})]$ is → AIEEE 2011
(a) -3 (b) 5
(c) 3 (d) -5

43 Let $\hat{\mathbf{a}}$, $\hat{\mathbf{b}}$ and $\hat{\mathbf{c}}$ be three unit vectors such that $\hat{\mathbf{a}}\times(\hat{\mathbf{b}}\times\hat{\mathbf{c}})=\dfrac{\sqrt{3}}{2}(\hat{\mathbf{b}}+\hat{\mathbf{c}})$. If $\hat{\mathbf{b}}$ is not parallel to $\hat{\mathbf{c}}$, then the angle between $\hat{\mathbf{a}}$ and $\hat{\mathbf{b}}$ is → JEE Mains 2016
(a) $\dfrac{3\pi}{4}$ (b) $\dfrac{\pi}{2}$ (c) $\dfrac{2\pi}{3}$ (d) $\dfrac{5\pi}{6}$

44 If $\mathbf{a}=\mathbf{j}-\mathbf{k}$ and $\mathbf{c}=\mathbf{i}-\mathbf{j}-\mathbf{k}$. Then, the vector **b** satisfying $\mathbf{a}\times\mathbf{b}+\mathbf{c}=0$ and $\mathbf{a}\cdot\mathbf{b}=3$, is → AIEEE 2010
(a) $-\mathbf{i}+\mathbf{j}-2\mathbf{k}$ (b) $2\mathbf{i}-\mathbf{j}+2\mathbf{k}$
(c) $\mathbf{i}-\mathbf{j}-2\mathbf{k}$ (d) $\mathbf{i}+\mathbf{j}-2\mathbf{k}$

45 Let **u** be a vector coplanar with the vectors $\mathbf{a}=2\hat{\mathbf{i}}+3\hat{\mathbf{j}}-\hat{\mathbf{k}}$ and $\mathbf{b}=\hat{\mathbf{j}}+\hat{\mathbf{k}}$. If **u** is perpendicular to **a** and $\mathbf{u}\cdot\mathbf{b}=24$, then $|\mathbf{u}|^2$ is equal to → JEE Mains 2018
(a) 336 (b) 315 (c) 256 (d) 84

46 Let **a**, **b** and **c** be three unit vectors such that **a** is perpendicular to the plane of **b** and **c**. If the angle between **b** and **c** is $\dfrac{\pi}{3}$, then $|\mathbf{a}\times\mathbf{b}-\mathbf{a}\times\mathbf{c}|^2$ is equal to
(a) $\dfrac{1}{3}$ (b) $\dfrac{1}{2}$
(c) 1 (d) 2

47 The vectors **a** and **b** are non-collinear. The value of x, for which the vectors, $\mathbf{c}=(2x+3)\mathbf{a}+\mathbf{b}$ and $\mathbf{d}=(2x+3)\mathbf{a}-\mathbf{b}$ are collinear is
(a) $\dfrac{1}{2}$ (b) $\dfrac{1}{3}$
(c) $-\dfrac{3}{2}$ (d) None of these

48 It is given that **a**, **b**, **c** are mutually perpendicular vectors of equal magnitudes.
Statement I Vector $(\mathbf{a}+\mathbf{b}+\mathbf{c})$ is equally inclined to **a**, **b** and **c**.

Statement II If α, β and γ are the angles at which $(\mathbf{a}+\mathbf{b}+\mathbf{c})$ is inclined to **a**, **b** and **c**, then $\alpha=\beta=\gamma$.

(a) Statement I is true, Statement II is true; Statement II is a correct explanation for Statement I
(b) Statement I is true, Statement II is true; Statement II is not a correct explanation for Statement I
(c) Statement I is true; Statement II is false
(d) Statement I is false; Statement II is true

49 **Statement I** For $a=-\frac{1}{\sqrt{3}}$ the volume of the parallelopiped formed by vectors $\mathbf{i}+a\,\mathbf{j}$, $a\,\mathbf{i}+\mathbf{j}+\mathbf{k}$ and $\mathbf{j}+a\,\mathbf{k}$ is maximum.

Statement II The volume of the parallelopiped having three coterminous edges a, b and c is $|[\mathrm{a\ b\ c}]|$.

(a) Statement I is true, Statement II is true; Statement II is not a correct explanation for Statement I
(b) Statement I is true; Statement II is false
(c) Statement I is false; Statement II is true
(d) Statement I is true, Statement II is true; Statement II is a correct explanation for Statement I

50 **Statement I** A relation between the vectors **r**, **a** and **b** is $\mathbf{r}\times\mathbf{a}=\mathbf{b}\Rightarrow\mathbf{r}=\frac{\mathbf{a}\times\mathbf{b}}{\mathbf{a}\cdot\mathbf{a}}$.

Statement II $\mathbf{r}\cdot\mathbf{a}=0$.

(a) Statement I is true, Statement II is true; Statement II is a correct explanation for Statement I
(b) Statement I is true, Statement II is true; Statement II is not a correct explanation for Statement I
(c) Statement I is true; Statement II is false
(d) Statement I is false; Statement II is true

DAY PRACTICE SESSION 2

PROGRESSIVE QUESTIONS EXERCISE

1 Let $\mathbf{a}=2\hat{\mathbf{i}}+\hat{\mathbf{j}}-2\hat{\mathbf{k}}$, $\mathbf{b}=\hat{\mathbf{i}}+\hat{\mathbf{j}}$ and **c** be a vector such that $|\mathbf{c}-\mathbf{a}|=3$, $|(\mathbf{a}\times\mathbf{b})\times\mathbf{c}|=3$ and the angle between **a** and $\mathbf{a}\times\mathbf{b}$ is 30°. Then, $\mathbf{a}\cdot\mathbf{c}$ is equal to ➜ JEE Mains 2017

(a) $\frac{25}{8}$ (b) 2 (c) 5 (d) $\frac{1}{8}$

2 Given, two vectors are $\mathbf{i}-\mathbf{j}$ and $\mathbf{i}+2\mathbf{j}$, the unit vector coplanar with the two vectors and perpendicular to first is

(a) $\frac{1}{\sqrt{2}}(\mathbf{i}+\mathbf{j})$ (b) $\frac{1}{\sqrt{5}}(2\mathbf{i}+\mathbf{j})$
(c) $\pm\frac{1}{\sqrt{2}}(\mathbf{i}+\mathbf{j})$ (d) None of these

3 Vectors **a** and **b** are such that $|\mathbf{a}|=1$, $|\mathbf{b}|=4$ and $\mathbf{a}\cdot\mathbf{b}=2$. If $\mathbf{c}=2\mathbf{a}\times\mathbf{b}-3\mathbf{b}$, then the angle between **b** and **c** is

(a) $\frac{\pi}{6}$ (b) $\frac{5\pi}{6}$ (c) $\frac{\pi}{3}$ (d) $\frac{2\pi}{3}$

4 A unit vector **d** is equally inclined at an angle α with the vectors $\mathbf{a}=\cos\theta\cdot\mathbf{i}+\sin\theta\cdot\mathbf{j}$, $\mathbf{b}=-\sin\theta\cdot\mathbf{i}+\cos\theta\cdot\mathbf{j}$ and $\mathbf{c}=\mathbf{k}$. Then, α is equal to

(a) $\cos^{-1}\left(\frac{1}{\sqrt{2}}\right)$ (b) $\cos^{-1}\left(\frac{1}{\sqrt{3}}\right)$
(c) $\cos^{-1}\frac{1}{3}$ (d) $\frac{\pi}{2}$

5 Let $\mathbf{p}=3ax^2\,\mathbf{i}-2(x-1)\,\mathbf{j}$, $\mathbf{q}=b(x-1)\,\mathbf{i}+x\,\mathbf{j}$ and $ab<0$. Then, **p** and **q** are parallel for atleast one x in

(a) (0, 1) (b) (1, 0)
(c) (1, 2) (d) None of these

6 If $|\mathbf{a}|=|\mathbf{b}|=|\mathbf{c}|=1$ and $\mathbf{a}\cdot\mathbf{b}=\mathbf{b}\cdot\mathbf{c}=\mathbf{c}\cdot\mathbf{a}=\cos\theta$, then the maximum value of θ is

(a) $\frac{\pi}{3}$ (b) $\frac{\pi}{2}$ (c) $\frac{2\pi}{3}$ (d) $\frac{2\pi}{5}$

7 If $\mathbf{b}=4\mathbf{i}+3\mathbf{j}$ and **c** be two vectors perpendicular to each other in the xy-plane. Then, a vector in the same plane having projections 1 and 2 along **b** and **c** respectively, is

(a) $\mathbf{i}+2\mathbf{j}$ (b) $2\mathbf{i}-\mathbf{j}$
(c) $2\mathbf{i}+\mathbf{j}$ (d) None of these

8 The unit vector which is orthogonal to the vector $3\mathbf{i}+2\mathbf{j}+6\mathbf{k}$ and is coplanar with the vectors $2\mathbf{i}+\mathbf{j}+\mathbf{k}$ and $\mathbf{i}-\mathbf{j}+\mathbf{k}$, is

(a) $\frac{2\mathbf{i}-6\mathbf{j}+\mathbf{k}}{\sqrt{41}}$ (b) $\frac{2\mathbf{i}-3\mathbf{j}}{\sqrt{13}}$
(c) $\frac{3\mathbf{j}-\mathbf{k}}{\sqrt{10}}$ (d) $\frac{4\mathbf{i}+3\mathbf{j}-3\mathbf{k}}{\sqrt{34}}$

9 The values of x for which the angle between the vectors $2x^2\,\mathbf{i}+4x\,\mathbf{j}+\mathbf{k}$ and $7\mathbf{i}-2\mathbf{j}+x\,\mathbf{k}$ is obtuse and the angle between the Z-axis and $7\mathbf{i}-2\mathbf{j}+x\,\mathbf{k}$ is acute and less than $\frac{\pi}{6}$ is given by

(a) $0<x<\frac{1}{2}$ (b) $x>\frac{1}{2}$ or $x<0$
(c) $\frac{1}{2}<x<15$ (d) No such value for x

10 If **a** and **b** are unit vectors, then the greatest value of $|\mathbf{a}+\mathbf{b}|+|\mathbf{a}-\mathbf{b}|$ is

(a) 2 (b) 4
(c) $2\sqrt{2}$ (d) $\sqrt{2}$

11 Let $\mathbf{u}=\mathbf{i}+\mathbf{j}$, $\mathbf{v}=\mathbf{i}-\mathbf{j}$ and $\mathbf{w}=\mathbf{i}+2\mathbf{j}+3\mathbf{j}$. If **n** is a unit vector such that $\mathbf{u}\cdot\mathbf{n}=0$ and $\mathbf{v}\cdot\mathbf{n}=0$, then $|\mathbf{w}\cdot\mathbf{n}|$ is equal to

(a) 3 (b) 0
(c) 1 (d) 2

12 Let **b** and **c** be non-collinear vectors. If **a** is a vector such that $\mathbf{a}\cdot(\mathbf{b}+\mathbf{c})=4$ and $\mathbf{a}\times(\mathbf{b}\times\mathbf{c})=(x^2-2x+6)\,\mathbf{b}+\sin y\cdot\mathbf{c}$ then (x, y) lies on the line

(a) $x+y=0$ (b) $x-y=0$
(c) $x=1$ (d) $y=\pi$

13 If $\mathbf{a}=a_1\mathbf{i}+a_2\mathbf{j}+a_3\mathbf{k}$, $\mathbf{b}=b_1\mathbf{i}+b_2\mathbf{j}+b_3\mathbf{k}$, $\mathbf{c}=c_1\mathbf{i}+c_2\mathbf{j}+c_3\mathbf{k}$, $|\mathbf{c}|=1$ and $(\mathbf{a}\times\mathbf{b})\times\mathbf{c}=0$, then

$\begin{vmatrix} a_1 & a_2 & a_3 \\ b_1 & b_2 & b_3 \\ c_1 & c_2 & c_3 \end{vmatrix}^2$ is equal to

(a) 0 (b) 1 (c) $|\mathbf{a}|^2\,|\mathbf{b}|^2$ (d) $|\mathbf{a}\times\mathbf{b}|^2$

14 If **a** is a unit vector and projection of x along **a** is 2 and $\mathbf{a}\times\mathbf{r}+\mathbf{b}=\mathbf{r}$, then **r** is equal to

(a) $\frac{1}{2}[\mathbf{a}-\mathbf{b}+\mathbf{a}\times\mathbf{b}]$ (b) $\frac{1}{2}[2\mathbf{a}+\mathbf{b}+\mathbf{a}\times\mathbf{b}]$
(c) $\mathbf{a}+\mathbf{a}\times\mathbf{b}$ (d) $\mathbf{a}-\mathbf{a}\times\mathbf{b}$

15 Let **a**, **b** and **c** be unit vectors such that $\mathbf{a}+\mathbf{b}+\mathbf{c}=0$. Which one of the following is correct?

(a) $\mathbf{a}\times\mathbf{b}=\mathbf{b}\times\mathbf{c}=\mathbf{c}\times\mathbf{a}=0$
(b) $\mathbf{a}\times\mathbf{b}=\mathbf{b}\times\mathbf{c}=\mathbf{c}\times\mathbf{a}\neq 0$
(c) $\mathbf{a}\times\mathbf{b}=\mathbf{b}\times\mathbf{c}=\mathbf{a}\times\mathbf{c}=0$
(d) $\mathbf{a}\times\mathbf{b}$, $\mathbf{b}\times\mathbf{c}$ and $\mathbf{c}\times\mathbf{a}$ are mutually perpendicular

16 Let G_1, G_2 and G_3 be the centroids of the triangular faces *OBC*, *OCA* and *OAB* of a tetrahedron *OABC*. If V_1 denote the volume of the tetrahedron *OABC* and V_2 that of the parallelopiped with OG_1, OG_2 and OG_3 as three concurrent edges, then

(a) $4V_1=9V_2$ (b) $9V_1=4V_2$
(c) $3V_1=2V_2$ (d) $3V_2=2V_1$

17 *ABC* is triangle, right angled at *A*. The resultant of the forces acting along **AB** and **AC** with magnitudes $\frac{1}{AB}$ and $\frac{1}{AC}$ respectively is the force along **AD**, where *D* is the foot of the perpendicular from *A* onto *BC*. The magnitude of the resultant is

(a) $\frac{(AB)\,(AC)}{AB+AC}$ (b) $\frac{1}{AB}+\frac{1}{AC}$
(c) $\frac{1}{AD}$ (d) $\frac{AB^2+AC^2}{(AB)^2\,(AC)^2}$

18 Let **a**, **b** and **c** be three non-zero vectors such that no two of them are collinear and $(\mathbf{a}\times\mathbf{b})\times\mathbf{c}=\frac{1}{3}|\mathbf{b}||\mathbf{c}|\mathbf{a}$. If θ is the angle between vectors **b** and **c**, then a value of $\sin\theta$ is → JEE Mains 2015

(a) $\frac{2\sqrt{2}}{3}$ (b) $\frac{-\sqrt{2}}{3}$ (c) $\frac{2}{3}$ (d) $\frac{-2\sqrt{3}}{3}$

19. Let *ABCD* be a parallelogram such that $\mathbf{AB}=\mathbf{q}$, $\mathbf{AD}=\mathbf{p}$, and $\angle BAD$ be an acute angle. If **r** is the vector that coincides with the altitude directed from the vertex *B* to the side *AD*, then **r** is given by

(a) $\mathbf{r}=3\mathbf{q}-\frac{3(\mathbf{p}\cdot\mathbf{q})}{(\mathbf{p}\cdot\mathbf{q})}\mathbf{p}$ (b) $\mathbf{r}=-\mathbf{q}+\left(\frac{\mathbf{q}\cdot\mathbf{p}}{\mathbf{p}\cdot\mathbf{p}}\right)\mathbf{p}$
(c) $\mathbf{r}=\mathbf{q}-\left(\frac{\mathbf{p}\cdot\mathbf{q}}{\mathbf{p}\cdot\mathbf{q}}\right)\mathbf{p}$ (d) $\mathbf{r}=-3\mathbf{q}+\frac{3(\mathbf{p}\cdot\mathbf{q})}{(\mathbf{p}\cdot\mathbf{q})}\mathbf{p}$

20 **Statement I** If **u** and **v** are unit vectors inclined at an angle α and *x* is a unit vector bisecting the angle between them, then $x=\frac{\mathbf{u}+\mathbf{v}}{2\cos\frac{\alpha}{2}}$.

Statement II If ΔABC is an isosceles triangle with $AB=AC=1$, then vector representing bisector of angle *A* is given by $\mathbf{AD}=\frac{\mathbf{AB}+\mathbf{AC}}{2}$.

(a) Statement I is true, Statement II is true; Statement II is a correct explanation for Statement I
(b) Statement I is true, Statement II is true; Statement II is not a correct explanation for Statement I
(c) Statement I is true; Statement II is false
(d) Statement I is false; Statement II is true

ANSWERS

SESSION 1

1 (b)	**2** (a)	**3** (a)	**4** (d)	**5** (c)	**6** (c)	**7** (b)	**8** (a)	**9** (a)	**10** (b)
11 (b)	**12** (d)	**13** (a)	**14** (c)	**15** (a)	**16** (a)	**17** (a)	**18** (c)	**19** (a)	**20** (b)
21 (c)	**22** (d)	**23** (c)	**24** (a)	**25** (d)	**26** (a)	**27** (a)	**28** (c)	**29** (d)	**30** (c)
31 (b)	**32** (c)	**33** (a)	**34** (d)	**35** (a)	**36** (d)	**37** (c)	**38** (d)	**39** (b)	**40** (a)
41 (c)	**42** (d)	**43** (d)	**44** (a)	**45** (a)	**46** (c)	**47** (c)	**48** (a)	**49** (c)	**50** (b)

SESSION 2

1 (b)	**2** (a)	**3** (b)	**4** (b)	**5** (a)	**6** (c)	**7** (b)	**8** (c)	**9** (d)	**10** (c)
11 (a)	**12** (c)	**13** (d)	**14** (b)	**15** (b)	**16** (a)	**17** (c)	**18** (a)	**19** (b)	**20** (a)

Hints and Explanations

SESSION 1

1 We have,

$|a\mathbf{i} + b\mathbf{j} + c\mathbf{k}| = |a| + |b| + |c|$

$\Rightarrow \sqrt{a^2 + b^2 + c^2} = |a| + |b| + |c|$

$\Rightarrow a^2 + b^2 + c^2 = a^2 + b^2 + c^2 + 2[|a||b| + |b||c| + |c||a|]$

$\Rightarrow |a||b| + |b||c| + |c||a| = 0$

$\Rightarrow ab = bc = ca = 0$

Hence, any two of a, b and c are zero.

2 Since, $\mathbf{a} = 8\mathbf{b}$ and $\mathbf{c} = -7\mathbf{b}$

So, $\mathbf{a}$ is parallel to $\mathbf{b}$ and $\mathbf{c}$ is anti-parallel to $\mathbf{b}$.

$\Rightarrow$ $\mathbf{a}$ and $\mathbf{c}$ are anti-parallel.

So, the angle between $\mathbf{a}$ and $\mathbf{c}$ is π.

3 The required vector

$\mathbf{r} = \lambda\left(\frac{\mathbf{a}}{|\mathbf{a}|} + \frac{\mathbf{b}}{|\mathbf{b}|}\right)$, where λ is a scalar.

$\Rightarrow \mathbf{r} = \lambda\left(\frac{1}{9}(7\mathbf{i} - 4\mathbf{j} - 4\mathbf{k}) + \frac{1}{3}(-2\mathbf{i} - \mathbf{j} + 2\mathbf{k})\right)$

$\Rightarrow \mathbf{r} = \frac{\lambda}{9}(\mathbf{i} - 7\mathbf{j} + 2\mathbf{k})$

Given, $|\mathbf{r}|^2 = 54$

$\Rightarrow \frac{\lambda^2}{81}(1 + 49 + 4) = 54 \Rightarrow \lambda = \pm 9$

Thus, the required vector is

$\mathbf{r} = \pm(\mathbf{i} - 7\mathbf{j} + 2\mathbf{k})$.

4 Let P be the origin outside of AB and C is mid-point of AB, then

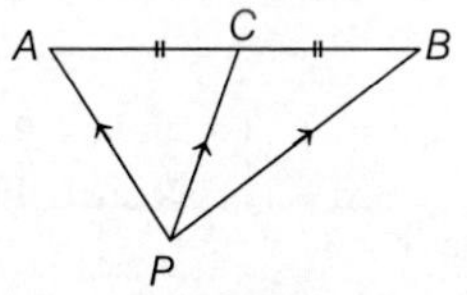

$\mathbf{PC} = \frac{\mathbf{PA} + \mathbf{PB}}{2}$

$\Rightarrow 2\mathbf{PC} = \mathbf{PA} + \mathbf{PB}$

5 We know that, the sum of three vectors of a triangle is zero.

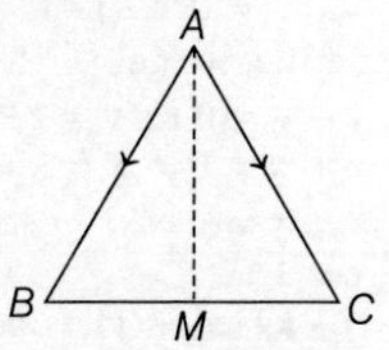

$\therefore \quad \mathbf{AB} + \mathbf{BC} + \mathbf{CA} = 0$

$\Rightarrow \quad \mathbf{BC} = \mathbf{AC} - \mathbf{AB}$

$\Rightarrow \quad \mathbf{BM} = \frac{\mathbf{AC} - \mathbf{AB}}{2}$

[since, M is a mid-point of BC]

Also, $\mathbf{AB} + \mathbf{BM} + \mathbf{MA} = 0$

[by properties of a triangle]

$\Rightarrow \quad \mathbf{AB} + \frac{\mathbf{AC} - \mathbf{AB}}{2} = \mathbf{AM}$

$\Rightarrow \quad \mathbf{AM} = \frac{\mathbf{AB} + \mathbf{AC}}{2}$

$= \frac{3\mathbf{i} + 4\mathbf{k} + 5\mathbf{i} - 2\mathbf{j} + 4\mathbf{k}}{2}$

$= 4\mathbf{i} - \mathbf{j} + 4\mathbf{k}$

$\therefore \quad |\mathbf{AM}| = \sqrt{4^2 + 1^2 + 4^2} = \sqrt{33}$

6 We have, $\mathbf{a} \cdot \mathbf{b} = 2\sqrt{2} \cdot 3 \cdot \frac{1}{\sqrt{2}} = 6$

The diagonals are $(2\mathbf{a} - 3\mathbf{b}) \pm (\mathbf{a} + \mathbf{b})$

i.e. $3\mathbf{a} - 2\mathbf{b}$ and $\mathbf{a} - 4\mathbf{b}$

$\therefore$ Length of diagonals are

$|3\mathbf{a} - 2\mathbf{b}|^2 = 9|\mathbf{a}| + 4|\mathbf{b}| - 12\mathbf{a} \cdot \mathbf{b}$

$= 9 \cdot 8 + 4 \cdot 9 - 12 \cdot 6 = 36$

and $|\mathbf{a} - 4\mathbf{b}|^2 = |\mathbf{a}| + 16|\mathbf{b}| - 8\mathbf{a} \cdot \mathbf{b}$

$= 8 + 16 \cdot 9 - 8 \cdot 6 = 104$

So, the length of the longer diagonal is $\sqrt{104}$ i.e. $2\sqrt{26}$.

7 Since, $(\mathbf{a} + \mathbf{b} + \mathbf{c})^2 \geq 0 \Rightarrow 3 + 2\sum \mathbf{a} \cdot \mathbf{b} \geq 0$

or $\quad -2\sum \mathbf{a} \cdot \mathbf{b} \leq 3$

$\therefore \quad |\mathbf{a} - \mathbf{b}|^2 + |\mathbf{b} - \mathbf{c}|^2 + |\mathbf{c} - \mathbf{a}|^2$

$= 6 - 2\sum \mathbf{a} \cdot \mathbf{b} \leq 9$

8 We have, $\mathbf{b} - 3\mathbf{c} = \lambda\mathbf{a}$

Taking scalar product with $\mathbf{c}$, we have

$(\mathbf{b} - 3\mathbf{c}) \cdot \mathbf{c} = \lambda(\mathbf{a} \cdot \mathbf{c})$

$\Rightarrow \quad \mathbf{b} \cdot \mathbf{c} - 3(\mathbf{c} \cdot \mathbf{c}) = \lambda(\mathbf{a} \cdot \mathbf{a})$

[$\because |\mathbf{a}| = |\mathbf{c}| = 1$ and $\mathbf{a}$ and $\mathbf{c}$ are collinear vectors]

$\Rightarrow \quad \mathbf{b} \cdot \mathbf{c} - 3 = \lambda$

$\Rightarrow \quad \mathbf{b} \cdot \mathbf{c} = 3 + \lambda \quad$ ...(i)

Again, $\quad \mathbf{b} - 3\mathbf{c} = \lambda\mathbf{a}$

$\Rightarrow \quad |\mathbf{b} - 3\mathbf{c}| = |\lambda\mathbf{a}|$

$\Rightarrow \quad |\mathbf{b} - 3\mathbf{c}|^2 = \lambda^2|\mathbf{a}|^2$

$\Rightarrow \quad |\mathbf{b}|^2 + 9|\mathbf{c}|^2 - 6(\mathbf{b} \cdot \mathbf{c}) = \lambda^2|\mathbf{a}|^2$

$\Rightarrow \quad 36 + 9 - 6(3 + \lambda) = \lambda^2$

[from Eq. (i)]

$\Rightarrow \quad 27 - 6\lambda = \lambda^2 \Rightarrow \lambda^2 + 6\lambda - 27 = 0$

$\therefore \quad \lambda = -9, 3$

9 We have, $(\sqrt{3}\mathbf{a} - \mathbf{b})^2$

$= 3\mathbf{a}^2 + \mathbf{b}^2 - 2\sqrt{3}\mathbf{a} \cdot \mathbf{b} = 1$

$3 + 1 - 2\sqrt{3}\mathbf{a} \cdot \mathbf{b} = 1$

[since, $\mathbf{a}$ and $\mathbf{b}$ are unit vectors]

Thus, $3 = 2\sqrt{3}\mathbf{a} \cdot \mathbf{b}$

$\mathbf{a} \cdot \mathbf{b} = \frac{\sqrt{3}}{2} \Rightarrow |\mathbf{a}||\mathbf{b}|\cos\theta = \frac{\sqrt{3}}{2}$

$\Rightarrow \quad \cos\theta = \frac{\sqrt{3}}{2}$

$\Rightarrow \quad \theta = 30° \quad$ [$\because |\mathbf{a}| = |\mathbf{b}| = 1$]

10 Since, $|\mathbf{a} + \mathbf{b}|^2 < 1$

$\Rightarrow \quad 2 + 2\cos\alpha < 1$

$\Rightarrow \quad 4\cos^2\frac{\alpha}{2} < 1$

$\Rightarrow \quad \cos\frac{\alpha}{2} < \frac{1}{2} \Rightarrow \alpha \in \left(\frac{2\pi}{3}, \pi\right)$

11 $\sqrt{3}\mathbf{b} = (\mathbf{a} + \mathbf{c})$

$\Rightarrow \quad 3|\mathbf{b}|^2 = |\mathbf{a}|^2 + |\mathbf{c}|^2 + 2\mathbf{a} \cdot \mathbf{c}$

$\Rightarrow \quad 3(1) = 1 + 1 + 2\mathbf{a} \cdot \mathbf{c}$

$\Rightarrow \quad 2\mathbf{a} \cdot \mathbf{c} = 1$

$\Rightarrow \quad |\mathbf{a}||\mathbf{c}|\cos\theta = \frac{1}{2}$

$\Rightarrow \quad \cos\theta = \frac{1}{2}$

$\therefore \quad \theta = \frac{\pi}{3}$

12 Since, position vectors of A, B, C are $2\mathbf{i} - \mathbf{j} + \mathbf{k}$, $\mathbf{i} - 3\mathbf{j} - 5\mathbf{k}$ and $a\mathbf{i} - 3\mathbf{j} + \mathbf{k}$, respectively.

Now,

$\mathbf{AC} = (a\mathbf{i} - 3\mathbf{j} + \mathbf{k}) - (2\mathbf{i} - \mathbf{j} + \mathbf{k})$

$= (a - 2)\mathbf{i} - 2\mathbf{j}$

and $\mathbf{BC} = (a\mathbf{i} - 3\mathbf{j} + \mathbf{k}) - (\mathbf{i} - 3\mathbf{j} - 5\mathbf{k})$

$= (a - 1)\mathbf{i} + 6\mathbf{k}$

Since, the ΔABC is right angled at C, then

$\mathbf{BC} \cdot \mathbf{BC} = 0$

$\Rightarrow \quad \{(a - 2)\mathbf{i} - 2\mathbf{j}\} \cdot \{(a - 1)\mathbf{i} + 6\mathbf{k}\} = 0$

$\Rightarrow \quad (a - 2)(a - 1) = 0$

$\therefore \quad a = 1$ and $a = 2$

13 Since, the given vectors are mutually orthogonal, therefore

$\mathbf{a} \cdot \mathbf{b} = 2 - 4 + 2 = 0$

$\mathbf{a} \cdot \mathbf{c} = \lambda - 1 + 2\mu = 0 \quad$ ...(i)

$\mathbf{b} \cdot \mathbf{c} = 2\lambda + 4 + \mu = 0 \quad$ ...(ii)

On solving Eqs. (i) and (ii), we get

$\mu = 2$ and $\lambda = -3$

Hence, $\quad (\lambda, \mu) = (-3, 2)$

14 Given that,

(i) $\mathbf{a}$ and $\mathbf{b}$ are unit vectors, i.e. $|\mathbf{a}| = |\mathbf{b}| = 1$

(ii) $\mathbf{c} = \mathbf{a} + 2\mathbf{b}$ and $\mathbf{d} = 5\mathbf{a} - 4\mathbf{b}$

(iii) $\mathbf{c}$ and $\mathbf{d}$ are perpendicular to each other, i.e. $\mathbf{c} \cdot \mathbf{d} = 0$

Now, $\mathbf{c} \cdot \mathbf{d} = 0$

$\Rightarrow \quad (\mathbf{a} + 2\mathbf{b}) \cdot (5\mathbf{a} - 4\mathbf{b}) = 0$

$\Rightarrow 5\mathbf{a} \cdot \mathbf{a} - 4\mathbf{a} \cdot \mathbf{b} + 10\mathbf{b} \cdot \mathbf{a} - 8\mathbf{b} \cdot \mathbf{b} = 0$

$\Rightarrow \quad 6\mathbf{a} \cdot \mathbf{b} = 3$

$\Rightarrow \quad \mathbf{a} \cdot \mathbf{b} = \frac{1}{2}$

So, the angle between $\mathbf{a}$ and $\mathbf{b}$ is $\frac{\pi}{3}$.

15 $\mathbf{p} \perp \mathbf{q}+\mathbf{r}, \mathbf{q} \perp \mathbf{r}+\mathbf{p}$ and $\mathbf{r} \perp \mathbf{p}+\mathbf{q}$

$\therefore \quad \mathbf{p} \cdot (\mathbf{q}+\mathbf{r}) = 0$

$\mathbf{q} \cdot (\mathbf{r}+\mathbf{p}) = 0$

$\mathbf{r} \cdot (\mathbf{p}+\mathbf{q}) = 0$

$\Rightarrow \quad \mathbf{p}\cdot\mathbf{q}+\mathbf{p}\cdot\mathbf{r} = 0$

$\mathbf{q}\cdot\mathbf{r}+\mathbf{q}\cdot\mathbf{p} = 0$

$\mathbf{r}\cdot\mathbf{p}+\mathbf{r}\cdot\mathbf{q} = 0$

On adding, we get

$$2(\mathbf{p}\cdot\mathbf{q}+\mathbf{q}\cdot\mathbf{r}+\mathbf{r}\cdot\mathbf{p}) = 0$$

Also, $\quad |\mathbf{p}+\mathbf{q}| = 6$

$\Rightarrow \quad |\mathbf{p}+\mathbf{q}|^2 = 36$

$\Rightarrow \quad \mathbf{p}^2+\mathbf{q}^2+2\mathbf{p}\cdot\mathbf{q} = 36$

Similarly, $\mathbf{q}^2+\mathbf{r}^2+2\mathbf{q}\cdot\mathbf{r} = 48$

and $\quad \mathbf{r}^2+\mathbf{p}^2+2\,\mathbf{r}\cdot\mathbf{p} = 16$

Adding all, we get

$2(\mathbf{p}^2+\mathbf{q}^2+\mathbf{r}^2+\mathbf{p}\cdot\mathbf{q}+\mathbf{q}\cdot\mathbf{r}+\mathbf{r}\cdot\mathbf{p}) = 100$

$\Rightarrow \quad 2(\mathbf{p}^2+\mathbf{q}^2+\mathbf{r}^2) = 100$

$[\because \mathbf{p}\cdot\mathbf{q}+\mathbf{q}\cdot\mathbf{r}+\mathbf{r}\cdot\mathbf{p} = 0]$

$\Rightarrow \quad \mathbf{p}^2+\mathbf{q}^2+\mathbf{r}^2 = 50$

$\Rightarrow \quad |\mathbf{p}+\mathbf{q}+\mathbf{r}|^2 = 50$

$|\mathbf{p}+\mathbf{q}+\mathbf{r}| = 5\sqrt{2}$

16. $\left(\dfrac{3\mathbf{a}+2\mathbf{b}}{5}\right)\cdot(3\mathbf{a}-2\mathbf{b}) = 0$

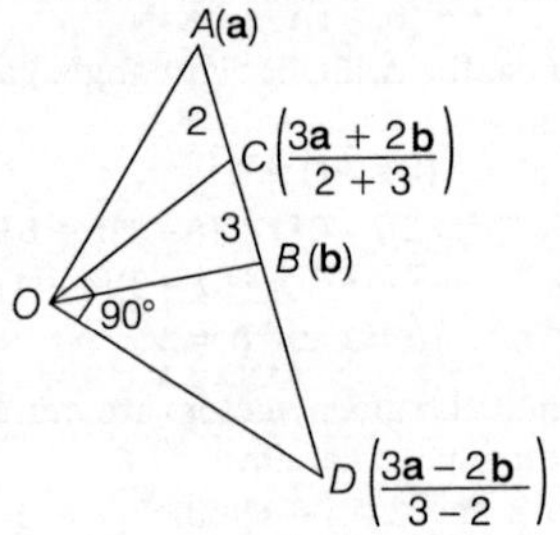

$\Rightarrow \quad 9|\mathbf{a}|^2 - 4|\mathbf{b}|^2 = 0$

$\therefore \quad 9\mathbf{a}^2 = 4\mathbf{b}^2$

17 A vector coplanar with $\mathbf{i}+\mathbf{j}+2\mathbf{k}$ and $\mathbf{i}+2\mathbf{j}+\mathbf{k}$ is $\mathbf{i}+\mathbf{j}+2\mathbf{k}+\lambda(\mathbf{i}+2\mathbf{j}+\mathbf{k})$

$= (1+\lambda)\mathbf{i}+(1+2\lambda)\mathbf{j}+(2+\lambda)\mathbf{k}.$

It is perpendicular to $\mathbf{i}+\mathbf{j}+\mathbf{k}$.

$\therefore 1+\lambda+1+2\lambda+2+\lambda = 0 \Rightarrow \lambda = -1$

So, the required vector is $-\mathbf{j}+\mathbf{k}$.

18 Let first term and common ratio of a GP be α and β. Then,

$a = \alpha\cdot\beta^{p-1}, b = \alpha\cdot\beta^{q-1}, c = \alpha\cdot\beta^{r-1}$

$\therefore \log a = \log\alpha + (p-1)\log\beta$, etc.

The dot product of the given two vectors is

$\Sigma\{\log\alpha + (p-1)\log\beta\}(q-r)$

$= (\log\alpha - \log\beta)\Sigma(q-r)$

$+ \log\beta\,\Sigma\, p(q-r) = 0$

Hence, given vectors are perpendicular.

19 Here, $\mathbf{AB} = -3\mathbf{i}+\mathbf{k}$

Now, $\mathbf{AB}\cdot(2\mathbf{i}+3\mathbf{j}+6\mathbf{k}) = -6+6 = 0$

Hence, **AB** is perpendicular to the given line.

Thus, the required distance

$= |\mathbf{AB}| = \sqrt{9+1} = \sqrt{10}$

20 Given, $\mathbf{a} = 2\mathbf{i}-\mathbf{j}+\mathbf{k}, \mathbf{b} = \mathbf{i}+2\mathbf{j}-\mathbf{k}$ and $\mathbf{c} = \mathbf{i}+\mathbf{j}-2\mathbf{k}$

Now, we have;

$\mathbf{b}+\lambda\mathbf{c} = (1+\lambda)\mathbf{i}+(2+\lambda)\mathbf{j} + (-1-2\lambda)\mathbf{k}$

$\therefore$ Projection of $(\mathbf{b}+\lambda\mathbf{c})$ on

$\mathbf{a} = \left|\dfrac{(\mathbf{b}+\lambda\mathbf{c})\cdot\mathbf{a}}{|\mathbf{a}|}\right| = \sqrt{\dfrac{2}{3}}$ [given]

$\Rightarrow \left|\dfrac{2(1+\lambda)-(2+\lambda)+(-1-2\lambda)}{\sqrt{4+1+1}}\right| = \sqrt{\dfrac{2}{3}}$

$\Rightarrow \quad \left|\dfrac{-\lambda-1}{\sqrt{6}}\right| = \dfrac{\sqrt{2}}{\sqrt{3}}$

$\Rightarrow \lambda+1 = 2 \Rightarrow \lambda = 1$

$\therefore \quad \mathbf{b}+\lambda\mathbf{c} = 2\mathbf{i}+3\mathbf{j}-3\mathbf{k}$

21 We have, projection of **v** along **u** = Projection of **w** along **u**

$\Rightarrow \quad \dfrac{\mathbf{v}\cdot\mathbf{u}}{|\mathbf{u}|} = \dfrac{\mathbf{w}\cdot\mathbf{u}}{|\mathbf{u}|}$

$\Rightarrow \quad \mathbf{v}\cdot\mathbf{u} = \mathbf{w}\cdot\mathbf{u}$ ...(i)

Also, **v** and **w** are perpendicular to each other.

$\therefore \quad \mathbf{v}\cdot\mathbf{w} = 0$ ...(ii)

Now,

$|\mathbf{u}-\mathbf{v}+\mathbf{w}| = |\mathbf{u}|^2+|\mathbf{v}|^2+|\mathbf{w}|^2 - 2(\mathbf{u}\cdot\mathbf{v}) - 2(\mathbf{v}\cdot\mathbf{w}) + 2(\mathbf{u}\cdot\mathbf{w})$

$\Rightarrow \quad |\mathbf{u}-\mathbf{v}+\mathbf{w}|^2 = 1+4+9$

[from Eqs. (i) and (ii)]

$\Rightarrow \quad |\mathbf{u}-\mathbf{v}+\mathbf{w}| = \sqrt{14}$

22 A vector parallel to the bisector of the angle between the vectors **a** and b is

$\dfrac{\mathbf{a}}{|\mathbf{a}|}+\dfrac{\mathbf{b}}{|\mathbf{b}|} = \mathbf{a}+\mathbf{b}.$

$\therefore$ Unit vector along the bisector

$= \dfrac{\mathbf{a}+\mathbf{b}}{|\mathbf{a}+\mathbf{b}|}$

$= \dfrac{1}{\sqrt{2}}(\mathbf{a}+\mathbf{b})$

$\left[\begin{array}{l}\because |\mathbf{a}+\mathbf{b}|^2 = |\mathbf{a}|^2+|\mathbf{b}|^2+2\mathbf{a}\cdot\mathbf{b}\\ \Rightarrow |\mathbf{a}+\mathbf{b}|^2 = 1+1+0 = 2\end{array}\right]$

$\therefore$ Required projection

$= \left\{l\cdot\dfrac{\mathbf{a}}{|\mathbf{a}|} + m\dfrac{\mathbf{b}}{|\mathbf{a}|} + n\dfrac{\mathbf{a}\times\mathbf{b}}{|\mathbf{a}\times\mathbf{b}|}\right\}\cdot\dfrac{1}{\sqrt{2}}(\mathbf{a}+\mathbf{b})$

$= \dfrac{1}{\sqrt{2}}(l+m)$

$[\because |\mathbf{a}| = |\mathbf{b}| = 1$ and $\mathbf{a}\cdot\mathbf{b} = \mathbf{a}\cdot(\mathbf{a}\times\mathbf{b}) = \mathbf{b}\cdot(\mathbf{a}\times\mathbf{b}) = 0]$

23 $\mathbf{a}_1 = \dfrac{(\mathbf{a}\cdot\mathbf{b})\mathbf{b}}{|\mathbf{b}|^2}$

$\Rightarrow \quad \mathbf{a}_2 = \mathbf{a}-\mathbf{a}_1 = \mathbf{a} - \dfrac{(\mathbf{a}\cdot\mathbf{b})\mathbf{b}}{|\mathbf{b}|^2}$

Thus, $\mathbf{a}_1\times\mathbf{a}_2 = \dfrac{(\mathbf{a}\cdot\mathbf{b})\mathbf{b}}{|\mathbf{b}|^2}\times\left(\mathbf{a}-\dfrac{(\mathbf{a}\cdot\mathbf{b})\mathbf{b}}{|\mathbf{b}|^2}\right)$

$= \dfrac{(\mathbf{a}\cdot\mathbf{b})(\mathbf{b}\times\mathbf{a})}{|\mathbf{b}|^2}$

24 Total force,

$\mathbf{F} = (4\mathbf{i}+\mathbf{j}-3\mathbf{k})+(3\mathbf{i}+\mathbf{j}-\mathbf{k})$

$\therefore \quad \mathbf{F} = 7\mathbf{i}+2\mathbf{j}-4\mathbf{k}$

The particle is displaced from

$A(\mathbf{i}+2\mathbf{j}+3\mathbf{k})$

to $\quad B(5\mathbf{i}+4\mathbf{j}+\mathbf{k})$

Now, displacement,

$\mathbf{AB} = (5\mathbf{i}+4\mathbf{j}+\mathbf{k})-(\mathbf{i}+2\mathbf{j}+3\mathbf{k})$

$= 4\mathbf{i}+2\mathbf{j}-2\mathbf{k}$

$\therefore$ Work done $= \mathbf{F}\cdot\mathbf{AB}$

$= (7\mathbf{i}+2\mathbf{j}-4\mathbf{k})\cdot(4\mathbf{i}+2\mathbf{j}-2\mathbf{k})$

$= 28+4+8$

$= 40$ units

25 Since, $(2\mathbf{u}\times 3\mathbf{v})$ is a unit vector.

$\Rightarrow \quad |2\mathbf{u}\times 3\mathbf{v}| = 1$

$\Rightarrow \quad 6|\mathbf{u}||\mathbf{v}|\sin\theta| = 1$

$\Rightarrow \quad \sin\theta = \dfrac{1}{6}$

$[\because |\mathbf{u}| = |\mathbf{v}| = 1]$

Since, θ is an acute angle, then there is exactly one value of θ for which $(2\mathbf{u}\times 3\mathbf{v})$ is a unit vector.

26 Since, the vectors $\mathbf{a} = x\mathbf{i}+y\mathbf{j}+z\mathbf{k}$ and $\mathbf{b} = \mathbf{j}$ are such that a, c and b form a right handed system.

$\therefore \quad \mathbf{c} = \mathbf{b}\times\mathbf{a} = \begin{vmatrix}\mathbf{i} & \mathbf{j} & \mathbf{k}\\ 0 & 1 & 0\\ x & y & z\end{vmatrix}$

$= z\mathbf{i}-x\mathbf{k}$

27 Given, $\mathbf{a} = \mathbf{i}+\mathbf{j}+\mathbf{k}$ and $\mathbf{b} = 0\,\mathbf{i}+\mathbf{j}-\mathbf{k}$

Let $\quad \mathbf{c} = x\mathbf{i}+y\mathbf{j}+z\mathbf{k}$ such that $\mathbf{a}\times\mathbf{c} = \mathbf{b}$ and $\mathbf{a}\cdot\mathbf{c} = 3$

Now, $\mathbf{a}\times\mathbf{c} = \mathbf{b}$

$\Rightarrow \begin{vmatrix}\mathbf{i} & \mathbf{j} & \mathbf{k}\\ 1 & 1 & 1\\ x & y & z\end{vmatrix} = 0\,\mathbf{i}+\mathbf{j}-\mathbf{k}$

$\Rightarrow (z-y)\mathbf{i}-\mathbf{j}(z-x)+\mathbf{k}(y-x)$

$= 0\,\mathbf{i}+\mathbf{j}-\mathbf{k}$

On comparing, we get

$z-y = 0 \Rightarrow y = z$ ...(i)

$-z+x = 1 \Rightarrow x = 1+z$ ...(ii)

and $\quad y-x = -1$ ...(iii)

Also, $\mathbf{a}\cdot\mathbf{c} = 3$

$\Rightarrow \quad (\mathbf{i}+\mathbf{j}+\mathbf{k})\cdot(x\mathbf{i}+y\mathbf{j}+z\mathbf{k}) = 3$

$\Rightarrow \quad x+y+z = 3$ ...(iv)

On putting the values of x and y from Eqs. (i) and (ii) in Eq. (iv), we get

$(1+z)+z+z = 3$

$\Rightarrow \quad 3z = 2 \Rightarrow z = \frac{2}{3}$

On putting the value of z in Eqs. (i) and (ii), we get

$y = \frac{2}{3}$ and $x = \frac{5}{3}$

These values of x and y also satisfy Eq. (iii), we get

$x = \frac{5}{3}, y = \frac{2}{3}, z = \frac{2}{3}$

Hence, $\mathbf{c} = \frac{5}{3}\mathbf{i} + \frac{2}{3}\mathbf{j} + \frac{2}{3}\mathbf{k}$, which is the required vector.

28 Given, $\mathbf{a} \cdot \mathbf{b} \neq 0, \mathbf{a} \cdot \mathbf{d} = 0$ ...(i)

and $\mathbf{b} \times \mathbf{c} = \mathbf{b} \times \mathbf{d}$

$\Rightarrow \quad \mathbf{b} \times (\mathbf{c} - \mathbf{d}) = 0$

$\therefore \quad \mathbf{b} \parallel (\mathbf{c} - \mathbf{d})$

$\Rightarrow \quad \mathbf{c} - \mathbf{d} = \lambda \mathbf{b}$

$\Rightarrow \quad \mathbf{d} = \mathbf{c} - \lambda \mathbf{d}$ ...(ii)

On taking dot product with $\mathbf{a}$, we get

$\mathbf{a} \cdot \mathbf{d} = \mathbf{a} \cdot \mathbf{c} - \lambda \mathbf{a} \cdot \mathbf{b}$

$\Rightarrow \quad 0 = \mathbf{a} \cdot \mathbf{c} - \lambda (\mathbf{a} \cdot \mathbf{b})$ [from Eq. (i)]

$\Rightarrow \quad \lambda = \frac{\mathbf{a} \cdot \mathbf{c}}{\mathbf{a} \cdot \mathbf{b}}$ ...(iii)

$\therefore \quad \mathbf{d} = \mathbf{c} - \frac{(\mathbf{a} \cdot \mathbf{c})}{(\mathbf{a} \cdot \mathbf{b})}\mathbf{b}$

29 Since, $[3\mathbf{u}\ p\mathbf{v}\ p\mathbf{w}] - [p\mathbf{v}\ \mathbf{w}\ q\mathbf{u}] - [2\mathbf{w}\ q\mathbf{v}\ q\mathbf{u}] = 0$

$\therefore 3p^2[\mathbf{u} \cdot (\mathbf{v} \times \mathbf{w})] - pq\,[\mathbf{v} \cdot (\mathbf{w} \times \mathbf{u})] - 2q^2\,[\mathbf{w} \cdot (\mathbf{v} \times \mathbf{u})] = 0$

$\Rightarrow (3p^2 - pq + 2q^2)[\mathbf{u} \cdot (\mathbf{v} \times \mathbf{w})] = 0$

But $[\mathbf{u}\ \mathbf{v}\ \mathbf{w}] \neq 0$

$\Rightarrow \quad 3p^2 - pq + 2q^2 = 0$

$\therefore \quad p = q = 0$

30 Here, $|\mathbf{u}| = 1$ and $\mathbf{v} \times \mathbf{w} = \begin{vmatrix} \mathbf{i} & \mathbf{j} & \mathbf{k} \\ 2 & 1 & -1 \\ 1 & 0 & 3 \end{vmatrix}$

$= 3\mathbf{i} - 7\mathbf{j} - \mathbf{k}$

$|\mathbf{v} \times \mathbf{w}| = \sqrt{9 + 49 + 1} = \sqrt{59}$

$[\mathbf{u}\ \mathbf{v}\ \mathbf{w}] = \mathbf{u} \cdot (\mathbf{v} \times \mathbf{w}) \leq |\mathbf{u}||\mathbf{v} \times \mathbf{w}| \leq \sqrt{59}$

31 Let $\mathbf{d} = d_1\mathbf{i} + d_2\mathbf{j} + d_3\mathbf{k}$

$\mathbf{a} \cdot \mathbf{d} = (\mathbf{i} - \mathbf{j}) \cdot (d_1\mathbf{i} + d_2\mathbf{j} + d_3\mathbf{k})$

$\Rightarrow \quad d_1 - d_2 = 0$ [$\because \mathbf{a} \cdot \mathbf{d} = 0$]

$\Rightarrow \quad d_1 = d_2$ ...(i)

Also, $\mathbf{d}$ is a unit vector.

$\Rightarrow \quad d_1^2 + d_2^2 + d_3^2 = 1$ ...(ii)

Also, $[\mathbf{b}\,\mathbf{c}\,\mathbf{d}] = 0$

$\Rightarrow \begin{vmatrix} 0 & 1 & -1 \\ -1 & 0 & 1 \\ d_1 & d_2 & d_3 \end{vmatrix} = 0$

$\Rightarrow -1(-d_3 - d_1) - 1(-d_2) = 0$

$\Rightarrow \quad d_1 + d_2 + d_3 = 0$

$\Rightarrow \quad 2d_1 + d_3 = 0$ [from Eq. (i)]

$\Rightarrow \quad d_3 = -2d_1$ ...(iii)

Using Eqs. (iii) and (i) in Eq. (ii), we get

$d_1^2 + d_1^2 + 4d_1^2 = 1 \Rightarrow d_1 = \pm \frac{1}{\sqrt{6}}$

$\therefore \quad d_2 = \pm \frac{1}{\sqrt{6}}$

and $\quad d_3 = \mp \frac{2}{\sqrt{6}}$

Hence, required vector is

$\pm \frac{1}{\sqrt{6}}(\mathbf{i} + \mathbf{j} - 2\mathbf{k})$.

32 $(\mathbf{i} \times \mathbf{a} \cdot \mathbf{b})\mathbf{i} + (\mathbf{j} \times \mathbf{a} \cdot \mathbf{b})\mathbf{j} + (\mathbf{k} \times \mathbf{a} \cdot \mathbf{b})\mathbf{k}$

$= [\mathbf{i}\,\mathbf{a}\,\mathbf{b}]\mathbf{i} + [\mathbf{j}\,\mathbf{a}\,\mathbf{b}]\mathbf{j} + [\mathbf{k}\,\mathbf{a}\,\mathbf{b}]\mathbf{k}$

Let $\mathbf{a} = a_1\mathbf{i} + a_2\mathbf{j} + a_3\mathbf{k}$

$\mathbf{b} = b_1\mathbf{i} + b_2\mathbf{j} + b_3\mathbf{k}$

$\therefore \ [\mathbf{iab}] = \begin{vmatrix} 1 & 0 & 0 \\ a_1 & a_2 & a_3 \\ b_1 & b_2 & b_3 \end{vmatrix} = (a_2b_3 - b_2a_3)$

$[\mathbf{jab}] = \begin{vmatrix} 0 & 1 & 0 \\ a_1 & a_2 & a_3 \\ b_1 & b_2 & b_3 \end{vmatrix} = (b_1a_3 - a_1b_3)$

and $[\mathbf{kab}] = \begin{vmatrix} 0 & 0 & 1 \\ a_1 & a_2 & a_3 \\ b_1 & b_2 & b_3 \end{vmatrix}$

$= (a_1b_2 - a_2b_1)$

$\therefore [\mathbf{i}\,\mathbf{a}\,\mathbf{b}]\mathbf{i} + [\mathbf{j}\,\mathbf{a}\,\mathbf{b}]\mathbf{j} + [\mathbf{k}\,\mathbf{a}\,\mathbf{b}]\mathbf{k}$

$= (a_2b_3 - b_2a_3)\mathbf{i} + (b_1a_3 - a_1b_3)\mathbf{j} + (a_1b_2 - a_2b_1)\mathbf{k} = \mathbf{a} \times \mathbf{b}$

33 Let P, Q, R and S be the given points with position vectors $\alpha\mathbf{i} + \mathbf{j} + \mathbf{k}$, $\mathbf{i} - \mathbf{j} - \mathbf{k}$, $\mathbf{i} + 2\mathbf{j} - \mathbf{k}$ and $\mathbf{i} + \mathbf{j} + \beta\mathbf{k}$ respectively. Then,

$\mathbf{QP} = (\alpha - 1)\mathbf{i} + 2\mathbf{j} + 2\mathbf{k}$,

$\mathbf{QR} = 0\mathbf{i} + 3\mathbf{j} + 0\mathbf{k}$

and $\mathbf{QS} = 0\mathbf{i} + 2\mathbf{j} + (\beta + 1)\mathbf{k}$ are coplanar

$\therefore \quad [\mathbf{QP}\,\mathbf{QR}\,\mathbf{QS}] = 0$

$\Rightarrow \begin{vmatrix} \alpha - 1 & 2 & 2 \\ 0 & 3 & 0 \\ 0 & 2 & \beta + 1 \end{vmatrix} = 0$

$\Rightarrow \quad (\alpha - 1)(\beta + 1) = 0$

$\Rightarrow \quad (1 - \alpha)(1 + \beta) = 0$

34 Since, given vectors $\mathbf{a}$, $\mathbf{b}$ and $\mathbf{c}$ are coplanar.

$\therefore \begin{vmatrix} 1 & 1 & 1 \\ 1 & -1 & 2 \\ x & x - 2 & -1 \end{vmatrix} = 0$

$\Rightarrow 1\{1 - 2(x - 2)\} - 1(-1 - 2x) + 1(x - 2 + x) = 0$

$\Rightarrow 1 - 2x + 4 + 1 + 2x + 2x - 2 = 0$

$\Rightarrow 2x = -4$

$\Rightarrow \quad x = -2$

35 Given, $\mathbf{a} = p\mathbf{i} + \mathbf{j} + \mathbf{k}$, $\mathbf{b} = \mathbf{i} + q\mathbf{j} + \mathbf{k}$ and $\mathbf{c} = \mathbf{i} + \mathbf{j} + r\mathbf{k}$ are coplanar and $p \neq q \neq r \neq 1$.

Since, $\mathbf{a}, \mathbf{b}$ and $\mathbf{c}$ are coplanar.

$\Rightarrow \quad [\mathbf{a}\,\mathbf{b}\,\mathbf{c}] = 0$

$\Rightarrow \begin{vmatrix} p & 1 & 1 \\ 1 & q & 1 \\ 1 & 1 & r \end{vmatrix} = 0$

$\Rightarrow p(qr - 1) - 1(r - 1) + 1(1 - q) = 0$

$\Rightarrow \quad pqr - p - r + 1 + 1 - q = 0$

$\therefore \quad pqr - (p + q + r) = -2$

36 It is given that α, β and γ are coplanar vectors.

$\therefore \quad [\alpha\ \beta\ \gamma] = 0$

$\Rightarrow \begin{vmatrix} a & b & c \\ b & c & a \\ c & a & b \end{vmatrix} = 0$

$\Rightarrow \quad 3abc - a^3 - b^3 - c^3 = 0$

$\Rightarrow \quad a^3 + b^3 + c^3 - 3abc = 0$

$\Rightarrow \ (a + b + c)(a^2 + b^2 + c^2 - ab - bc - ca) = 0$

$\Rightarrow \quad a + b + c = 0$

$[\because a^2 + b^2 + c^2 - ab - bc - ca \neq 0]$

$\Rightarrow \quad \mathbf{v} \cdot \alpha = \mathbf{v} \cdot \beta = \mathbf{v} \cdot \gamma = 0$

Hence, $\mathbf{v}$ is perpendicular to α, β and γ.

37 Given that,

$[\lambda(\mathbf{a} + \mathbf{b})\ \ \lambda^2\mathbf{b}\ \ \lambda\mathbf{c}] = [\mathbf{a}\ \ \mathbf{b} + \mathbf{c}\ \ \mathbf{b}]$

$\therefore \begin{vmatrix} \lambda(a_1 + b_1) & \lambda(a_2 + b_2) & \lambda(a_3 + b_3) \\ \lambda^2 b_1 & \lambda^2 b_2 & \lambda^2 b_3 \\ \lambda c_1 & \lambda c_2 & \lambda c_3 \end{vmatrix}$

$= \begin{vmatrix} a_1 & a_2 & a_3 \\ b_1 + c_1 & b_2 + c_2 & b_3 + c_3 \\ b_1 & b_2 & b_3 \end{vmatrix}$

$\Rightarrow \quad \lambda^4 \begin{vmatrix} a_1 & a_2 & a_3 \\ b_1 & b_2 & b_3 \\ c_1 & c_2 & c_3 \end{vmatrix} = -\begin{vmatrix} a_1 & a_2 & a_3 \\ b_1 & b_2 & b_3 \\ c_1 & c_2 & c_3 \end{vmatrix}$

$\Rightarrow \quad \lambda^4 = -1$

So, no real value of λ exists.

38 As $\mathbf{a} + 3\mathbf{b}$ is collinear with $\mathbf{c}$.

$\Rightarrow \quad \mathbf{a} + 3\mathbf{b} = \lambda\mathbf{c}$ ...(i)

Also, $\mathbf{b} + 2\mathbf{c}$ is collinear with $\mathbf{a}$.

$\Rightarrow \quad \mathbf{b} + 2\mathbf{c} = \mu\mathbf{a}$ ...(ii)

From Eq. (i),

$\mathbf{a} + 3\mathbf{b} + 6\mathbf{c} = (\lambda + 6)\mathbf{c}$ ...(iii)

From Eq. (ii),

$\mathbf{a} + 3\mathbf{b} + 6\mathbf{c} = (1 + 3\mu)\mathbf{a}$ ...(iv)

From Eqs. (iii) and (iv), we get

$(\lambda + 6)\mathbf{c} = (1 + 3\mu)\mathbf{a}$

Since, $\mathbf{a}$ is not collinear with $\mathbf{c}$.

$\Rightarrow \quad \lambda + 6 = 1 + 3\mu = 0$

From Eq. (iv), $\mathbf{a} + 3\mathbf{b} + 6\mathbf{c} = 0$

39 We know, $[\mathbf{a} \times \mathbf{b}\ \ \mathbf{b} \times \mathbf{c}\ \ \mathbf{c} \times \mathbf{a}] = [\mathbf{a}\ \mathbf{b}\ \mathbf{c}]^2$

$\therefore \quad \lambda = 1$

40. We have, $|[\mathbf{a}\ \mathbf{b}\ \mathbf{c}]| = V$

Let V_1 be the volume of the parallelopiped formed by the vectors α, β and γ.

Then, $\quad V_1 = |[\alpha\ \beta\ \gamma]|$

Now, $[\alpha\ \beta\ \gamma] = \begin{vmatrix} \mathbf{a}\cdot\mathbf{a} & \mathbf{a}\cdot\mathbf{b} & \mathbf{a}\cdot\mathbf{c} \\ \mathbf{a}\cdot\mathbf{b} & \mathbf{b}\cdot\mathbf{b} & \mathbf{b}\cdot\mathbf{c} \\ \mathbf{a}\cdot\mathbf{c} & \mathbf{c}\cdot\mathbf{b} & \mathbf{c}\cdot\mathbf{c} \end{vmatrix} [\mathbf{a\ b c}]$

$\Rightarrow [\alpha\ \beta\ \gamma] = [\mathbf{abc}]^2\ [\mathbf{abc}]$

$\Rightarrow [\alpha\ \beta\ \gamma] = [\mathbf{abc}]^3$

$\therefore V_1 = |[\alpha\ \beta\ \gamma]|$

$= |[\mathbf{a\ bc}]^3| = V^3$

41 Let $\alpha = \mathbf{a} + 2\mathbf{b} + 3\mathbf{c}, \beta = \lambda\mathbf{b} + 4\mathbf{c}$ and $\gamma = (2\lambda - 1)\mathbf{c}$

Then, $[\alpha\ \beta\ \gamma] = \begin{vmatrix} 1 & 2 & 3 \\ 0 & \lambda & 4 \\ 0 & 0 & (2\lambda-1) \end{vmatrix} [\mathbf{a\ bc}]$

$\Rightarrow [\alpha\ \beta\ \gamma] = \lambda(2\lambda - 1)[\mathbf{abc}]$

$\Rightarrow [\alpha\ \beta\ \gamma] = 0$, if $\lambda = 0, \frac{1}{2}$ $[\because [\mathbf{a\ bc}] \neq 0]$

Hence, α, β and γ are non-coplanar for all values of λ except two values 0 and $\frac{1}{2}$.

42 Given, $\mathbf{a} = \frac{1}{\sqrt{10}}(3\mathbf{i} + \mathbf{k})$

and $\mathbf{b} = \frac{1}{7}(2\mathbf{i} + 3\mathbf{j} - 6\mathbf{k})$

$\therefore (2\mathbf{a} - \mathbf{b})\cdot\{(\mathbf{a}\times\mathbf{b})\times(\mathbf{a}+2\mathbf{b})\}$

$= (2\mathbf{a}-\mathbf{b})\cdot\{(\mathbf{a}\times\mathbf{b})\times\mathbf{a} + (\mathbf{a}\times\mathbf{b})\times 2\mathbf{b}\}$

$= (2\mathbf{a}-\mathbf{b})\cdot\{(\mathbf{a}\cdot\mathbf{a})\mathbf{b} - (\mathbf{b}\cdot\mathbf{a})\mathbf{a} + 2(\mathbf{a}\cdot\mathbf{b})\mathbf{b} - 2(\mathbf{b}\cdot\mathbf{b})\mathbf{a}\}$

$= (2\mathbf{a}-\mathbf{b})\cdot\{1(\mathbf{b}) - (0)\mathbf{a} + 2(0)\mathbf{b} - 2(1)\mathbf{a}\}$

$[\mathbf{a}\cdot\mathbf{b} = 0$ and $\mathbf{a}\cdot\mathbf{a} = \mathbf{b}\cdot\mathbf{b} = 1]$

$= (2\mathbf{a}-\mathbf{b})(\mathbf{b}-2\mathbf{a})$

$= -(4|\mathbf{a}|^2 - 4\mathbf{a}\cdot\mathbf{b} + |\mathbf{b}|^2)$

$= -\{4 - 0 + 1\} = -5$

43 Given $|\hat{\mathbf{a}}| = |\hat{\mathbf{b}}| = |\hat{\mathbf{c}}| = 1$

and $\hat{\mathbf{a}}\times(\hat{\mathbf{b}}\times\hat{\mathbf{c}}) = \frac{\sqrt{3}}{2}(\hat{\mathbf{b}}+\hat{\mathbf{c}})$

Now, consider

$\hat{\mathbf{a}}\times(\hat{\mathbf{b}}\times\hat{\mathbf{c}}) = \frac{\sqrt{3}}{2}(\hat{\mathbf{b}}+\hat{\mathbf{c}})$

$\Rightarrow (\hat{\mathbf{a}}\cdot\hat{\mathbf{c}})\hat{\mathbf{b}} - (\hat{\mathbf{a}}\cdot\hat{\mathbf{b}})\hat{\mathbf{c}} = \frac{\sqrt{3}}{2}\hat{\mathbf{b}} + \frac{\sqrt{3}}{2}\hat{\mathbf{c}}$

On comparing, we get

$\hat{\mathbf{a}}\cdot\hat{\mathbf{b}} = -\frac{\sqrt{3}}{2} \Rightarrow |\hat{\mathbf{a}}||\hat{\mathbf{b}}|\cos\theta = -\frac{\sqrt{3}}{2}$

$\Rightarrow \cos\theta = -\frac{\sqrt{3}}{2}$ $[\because |\hat{\mathbf{a}}| = |\hat{\mathbf{b}}| = 1]$

$\Rightarrow \cos\theta = \cos\left(\pi - \frac{\pi}{6}\right) \Rightarrow \theta = \frac{5\pi}{6}$

44 We have, $\mathbf{a}\times\mathbf{b} + \mathbf{c} = 0$

$\Rightarrow \mathbf{a}\times(\mathbf{a}\times\mathbf{b}) + \mathbf{a}\times\mathbf{c} = 0$

$\Rightarrow (\mathbf{a}\cdot\mathbf{b})\mathbf{a} - (\mathbf{a}\cdot\mathbf{a})\mathbf{b} + \mathbf{a}\times\mathbf{c} = 0$

$\Rightarrow 3\mathbf{a} - 2\mathbf{b} + \mathbf{a}\times\mathbf{c} = 0$

$\Rightarrow 2\mathbf{b} = 3\mathbf{a} + \mathbf{a}\times\mathbf{c}$

$\Rightarrow 2\mathbf{b} = 3\mathbf{j} - 3\mathbf{k} - 2\mathbf{i} - \mathbf{j} - \mathbf{k}$

$= -2\mathbf{i} + 2\mathbf{j} - 4\mathbf{k}$

$\therefore \mathbf{b} = -\mathbf{i} + \mathbf{j} - 2\mathbf{k}$

45 **Key Idea** If any vector **x** is coplanar with the vector **y** and **z**, then

$\mathbf{x} = \lambda\mathbf{y} + \mu\mathbf{b}$

Here, **u** is coplanar with **a** and **b**

$\therefore \mathbf{u} = \lambda\mathbf{a} + \mu\mathbf{b}$

Dot product with **a**, we get ... (i)

$[\because \mathbf{a} = 2\hat{\mathbf{i}} + 3\hat{\mathbf{j}} - \hat{\mathbf{k}}, \mathbf{b} = \mathbf{j} + \mathbf{k}, \mathbf{u}\cdot\mathbf{a} = 0]$

Dot product with **b**, we get

$\mathbf{u}\cdot\mathbf{b} = \lambda(\mathbf{a}\cdot\mathbf{b}) + \mu(\mathbf{b}\cdot\mathbf{b})$

$24 = 2\lambda + 2\mu$ ... (ii)

$[\because \mathbf{u}\cdot\mathbf{b} = 24]$

Solving Eqs. (i) and (ii), we get

$\lambda = -2, \mu = 14$

Dot product with **u**, we get

$|\mathbf{u}|^2 = \lambda(\mathbf{u}\cdot\mathbf{a}) + \mu(\mathbf{u}\cdot\mathbf{b})$

$|\mathbf{u}|^2 = -2(0) + 14(24)$

$\Rightarrow |\mathbf{u}|^2 = 336$

46 Since, $\mathbf{a}\cdot\mathbf{b} = \mathbf{a}\cdot\mathbf{c} = 0$ and $\mathbf{b}\cdot\mathbf{c} = \frac{1}{2}$

$\therefore |\mathbf{a}\times\mathbf{b}| = |\mathbf{a}\times\mathbf{c}| = 1$

Now, $|\mathbf{a}\times\mathbf{b} - \mathbf{a}\times\mathbf{c}|^2$

$= |\mathbf{a}\times\mathbf{b}|^2 + |\mathbf{a}\times\mathbf{c}|^2 - 2(\mathbf{a}\times\mathbf{b})\cdot(\mathbf{a}\times\mathbf{c})$

$= 1 + 1 - 2\begin{vmatrix} 1 & 0 \\ 0 & \frac{1}{2} \end{vmatrix} = 1$

47 Since, **d** is collinear to vector **c**.

$\therefore \mathbf{c} = \lambda\mathbf{d}$

$\Rightarrow (2x+3)\mathbf{a} + \mathbf{b} = \lambda[(2x+3)\mathbf{a} - \mathbf{b}]$

$\Rightarrow 2x\mathbf{a} + 3\mathbf{a} + \mathbf{b} = 2\lambda x\mathbf{a} + 3\lambda\mathbf{a} - \mathbf{b}\lambda$

$\Rightarrow (2x - 2\lambda x\mathbf{a} + 3\mathbf{a} - 3\lambda\mathbf{a}) + (\mathbf{b}\lambda + \mathbf{b}) = 0$

$\Rightarrow (2x - 2\lambda x + 3 - 3\lambda)\mathbf{a} + (\lambda+1)\mathbf{b} = 0$

$\therefore (2x - 2\lambda x + 3 - 3\lambda) = 0$ and $(\lambda + 1) = 0$

$\Rightarrow (2x+3) - \lambda(2x+3) = 0$ and $\lambda = -1$

$\Rightarrow (2x+3)(1-\lambda) = 0$

$\therefore x = -\frac{3}{2}$ and $\lambda = 1$

48. We have, $\mathbf{a}\cdot\mathbf{b} = \mathbf{b}\cdot\mathbf{c} = \mathbf{c}\cdot\mathbf{a} = 0$

and $|\mathbf{a}| = |\mathbf{b}| = |\mathbf{c}|$

Let vector $(\mathbf{a}+\mathbf{b}+\mathbf{c})$ be inclined to **a**, **b** and **c** at angles α, β and γ, respectively. Then,

$\cos\alpha = \frac{(\mathbf{a}+\mathbf{b}+\mathbf{c})\cdot\mathbf{a}}{|\mathbf{a}+\mathbf{b}+\mathbf{c}||\mathbf{a}|}$

$= \frac{\mathbf{a}\cdot\mathbf{a} + \mathbf{b}\cdot\mathbf{a} + \mathbf{c}\cdot\mathbf{a}}{|\mathbf{a}+\mathbf{b}+\mathbf{c}||\mathbf{a}|}$

$= \frac{|\mathbf{a}|}{|\mathbf{a}+\mathbf{b}+\mathbf{c}|}$

Similarly, $\cos\beta = \frac{|\mathbf{b}|}{|\mathbf{a}+\mathbf{b}+\mathbf{c}|}$

and $\cos\gamma = \frac{|\mathbf{c}|}{|\mathbf{a}+\mathbf{b}+\mathbf{c}|}$

Now, as $|\mathbf{a}| = |\mathbf{b}| = |\mathbf{c}|$, therefore $\cos\alpha = \cos\beta = \cos\gamma$

$\therefore \alpha = \beta = \gamma$

Hence, the vector $(\mathbf{a}+\mathbf{b}+\mathbf{c})$ is equally inclined to **a**, **b** and **c**.

49 $V = \begin{vmatrix} 1 & a & 0 \\ a & 1 & 1 \\ 0 & 1 & a \end{vmatrix} = a - 1 - a^3$

$\therefore \frac{dV}{da} = 1 - 3a^2 = 0$ (say)

Now, $a = \pm\frac{1}{\sqrt{3}}$ and $\frac{d^2V}{da^2} = -6a$

$\Rightarrow \left(\frac{d^2V}{da^2}\right)_{\left(a=\frac{1}{\sqrt{3}}\right)} = -\frac{6}{\sqrt{3}}$ (– ve)

Hence, V is maximum at $a = \frac{1}{\sqrt{3}}$.

50 Since, $\mathbf{b} = \mathbf{r}\times\mathbf{a}$

We have, $\mathbf{a}\times\mathbf{b} = \mathbf{a}\times(\mathbf{r}\times\mathbf{a})$

$= (\mathbf{a}\cdot\mathbf{a})\mathbf{r} - (\mathbf{a}\cdot\mathbf{r})\mathbf{a}$

$= (\mathbf{a}\cdot\mathbf{a})\mathbf{r}$

$\because \mathbf{a}\cdot\mathbf{r} = 0$

$\therefore \mathbf{r} = \frac{\mathbf{a}\times\mathbf{b}}{\mathbf{a}\cdot\mathbf{a}}$

SESSION 2

1 We have, $\mathbf{a} = 2\hat{\mathbf{i}} + \hat{\mathbf{j}} - 2\hat{\mathbf{k}}$

$\Rightarrow |\mathbf{a}| = \sqrt{4+1+4} = 3$

and $\mathbf{b} = \hat{\mathbf{i}} + \hat{\mathbf{j}} \Rightarrow |\mathbf{b}| \sqrt{1+1} = \sqrt{2}$

Now, $|\mathbf{c}-\mathbf{a}| = 3 \Rightarrow |\mathbf{c}-\mathbf{a}|^2 = 9$

$\Rightarrow (\mathbf{c}-\mathbf{a})\cdot(\mathbf{c}-\mathbf{a}) = 9$

$\Rightarrow |\mathbf{c}|^2 + |\mathbf{a}|^2 - 2\mathbf{c}\cdot\mathbf{a} = 9$ ... (i)

Again, $|(\mathbf{a}\times\mathbf{b})\times\mathbf{c}| = 3$

$\Rightarrow |\mathbf{a}\times\mathbf{b}||\mathbf{c}|\sin 30° = 3$

$\Rightarrow |\mathbf{c}| = \frac{6}{|\mathbf{a}\times\mathbf{b}|}$

But $\mathbf{a}\times\mathbf{b} = \begin{vmatrix} \hat{\mathbf{i}} & \hat{\mathbf{j}} & \hat{\mathbf{k}} \\ 2 & 1 & -2 \\ 1 & 1 & 0 \end{vmatrix} = 2\hat{\mathbf{i}} - 2\hat{\mathbf{j}} + \hat{\mathbf{k}}$

$\therefore |\mathbf{c}| = \frac{6}{\sqrt{4+4+1}} = 2$ ... (ii)

From Eqs. (i) and (ii), we get

$(2)^2 + (3)^2 - 2\mathbf{c}\cdot\mathbf{a} = 9$

$\Rightarrow 4 + 9 - 2\mathbf{c}\cdot\mathbf{a} = 9 \Rightarrow \mathbf{c}\cdot\mathbf{a} = 2$

2 Given two vectors lie in xy-plane. So, a vector coplanar with them is

$\mathbf{a} = x\mathbf{i} + y\mathbf{j}$

Since, $\mathbf{a} \perp (\mathbf{i} - \mathbf{j})$

$\Rightarrow (x\mathbf{i} + y\mathbf{j})\cdot(\mathbf{i} - \mathbf{j}) = 0$

$\Rightarrow x - y = 0$

$\Rightarrow x = y$

$\therefore \mathbf{a} = x\mathbf{i} + x\mathbf{j}$

and $|\mathbf{a}| = \sqrt{x^2 + x^2} = x\sqrt{2}$

∴ Required unit vector

$$= \frac{\mathbf{a}}{|\mathbf{a}|} = \frac{x(\mathbf{i} + \mathbf{j})}{x\sqrt{2}} = \frac{1}{\sqrt{2}}(\mathbf{i} + \mathbf{j})$$

3 Now, $|\mathbf{a} \times \mathbf{b}|^2 = |\mathbf{a}|^2 |\mathbf{b}|^2 - (\mathbf{a} \cdot \mathbf{b})^2$

$$= 16 - 4 = 12$$

and $|\mathbf{c}|^2 = (2\mathbf{a} \times \mathbf{b} - 3\mathbf{b}) \cdot (2\mathbf{a} \times \mathbf{b} - 3\mathbf{b})$

$$= 4|\mathbf{a} \times \mathbf{b}|^2 + 9|\mathbf{b}|^2$$

$$= 4 \cdot 12 + 9 \cdot 16$$

$$= 192$$

$\Rightarrow \quad |\mathbf{c}| = 8\sqrt{3}$

Now, $\mathbf{b} \cdot \mathbf{c} = \mathbf{b} \cdot (2\mathbf{a} \times \mathbf{b} - 3\mathbf{b})$

$$= -3|\mathbf{b}|^2 = -48$$

$\because \quad \cos\theta = \dfrac{\mathbf{b} \cdot \mathbf{c}}{|\mathbf{b}||\mathbf{c}|}$

$$= -\frac{48}{4 \cdot 8\sqrt{3}} = -\frac{\sqrt{3}}{2}$$

$\therefore \quad \theta = \dfrac{5\pi}{6}$

4 Let $\mathbf{d} \cdot \mathbf{a} = \mathbf{d} \cdot \mathbf{b} = \mathbf{b} \cdot \mathbf{c} = \cos\alpha$

$\Rightarrow \mathbf{d} \cdot (\mathbf{a} - \mathbf{k}) = 0$ and $\mathbf{d} \cdot (\mathbf{b} - \mathbf{k}) = 0$

$\mathbf{d}$ is parallel to $(\mathbf{a} - \mathbf{k}) \times (\mathbf{b} - \mathbf{k})$

$$\therefore \quad \mathbf{d} = \begin{vmatrix} \mathbf{i} & \mathbf{j} & \mathbf{k} \\ \cos\theta & \sin\theta & -1 \\ -\sin\theta & \cos\theta & -1 \end{vmatrix}$$

$$\Rightarrow \quad \mathbf{d} = \frac{(\cos\theta - \sin\theta)\,\mathbf{i} + (\cos\theta + \sin\theta)\,\mathbf{j} + \mathbf{k}}{\sqrt{3}}$$

$$\cos\alpha = \mathbf{d} \cdot \mathbf{k} = \frac{1}{\sqrt{3}}$$

$$\Rightarrow \quad \alpha = \cos^{-1}\left(\frac{1}{\sqrt{3}}\right)$$

5 Hence, $\mathbf{p} \times \mathbf{q} = \{3\,ax^3 + 2b\,(x-1)^2\}\mathbf{k} = f(x)\,\mathbf{k}$,

where, $f(0)\,f(1) = 6\,ab < 0$

∴ By intermediate value theorem there exists, x in $(0, 1)$ such that $f(x) = 0$.

6
$$[\mathbf{a}\ \mathbf{b}\ \mathbf{c}] = \begin{vmatrix} \mathbf{a} \cdot \mathbf{a} & \mathbf{a} \cdot \mathbf{b} & \mathbf{a} \cdot \mathbf{c} \\ \mathbf{b} \cdot \mathbf{a} & \mathbf{b} \cdot \mathbf{b} & \mathbf{b} \cdot \mathbf{c} \\ \mathbf{c} \cdot \mathbf{a} & \mathbf{c} \cdot \mathbf{b} & \mathbf{c} \cdot \mathbf{c} \end{vmatrix}$$

$$= \begin{vmatrix} 1 & \cos\theta & \cos\theta \\ \cos\theta & 1 & \cos\theta \\ \cos\theta & \cos\theta & 1 \end{vmatrix}$$

$$= 1 - 3\cos^2\theta + 2\cos^3\theta$$

$$= (1 - \cos\theta)^2\,(1 + 2\cos\theta)$$

$\Rightarrow \quad 1 + 2\cos\theta \geq 0 \Rightarrow \theta \leq \dfrac{2\pi}{3}$

7 Let $\mathbf{c} = x\mathbf{i} + y\mathbf{j}$

Then, $\mathbf{b} \perp \mathbf{c} \Rightarrow \mathbf{b} \cdot \mathbf{c} = 0 \Rightarrow 4x + 3y = 0$

$\Rightarrow \quad \dfrac{x}{3} = \dfrac{y}{-4} = \lambda$ [say]

$\Rightarrow \quad x = 3\lambda,\ y = -4\lambda$

$\therefore \quad \mathbf{c} = \lambda\,(3\mathbf{i} - 4\mathbf{j})$

Let the required vector be $\alpha = p\,\mathbf{i} + q\,\mathbf{j}$.

Then, the projections of α on $\mathbf{b}$ and $\mathbf{c}$ are $\dfrac{\alpha \cdot \mathbf{b}}{|\mathbf{b}|}$ and $\dfrac{\alpha \cdot \mathbf{c}}{|\mathbf{c}|}$, respectively.

$\therefore \quad \alpha \cdot \dfrac{\mathbf{b}}{|\mathbf{b}|} = 1$ and $\dfrac{\alpha \cdot \mathbf{c}}{|\mathbf{c}|} = 2$ [given]

$\Rightarrow \quad 4p + 3q = 5$ and $3p - 4q = 10$

$\Rightarrow \quad p = 2, q = -1$

$$\alpha = 2\mathbf{i} - \mathbf{j}$$

8 A vector coplanar to $(2\,\mathbf{i} + \mathbf{j} + \mathbf{k}), (\mathbf{i} - \mathbf{j} + \mathbf{k})$ and orthogonal to $(3\,\mathbf{i} + 2\,\mathbf{j} + 6\,\mathbf{k})$

$$= \lambda\,[\{(2\,\mathbf{i} + \mathbf{j} + \mathbf{k}) \times (\mathbf{i} - \mathbf{j} + \mathbf{k})\} \times (3\,\mathbf{i} + 2\,\mathbf{j} + 6\,\mathbf{k})]$$

$$= \lambda\,(21\,\mathbf{j} - 7\,\mathbf{k})$$

∴ Required unit vector is

$$\pm \frac{(21\,\mathbf{j} - 7\,\mathbf{k})}{\sqrt{(21)^2 + (7)^2}} = \pm \frac{(3\,\mathbf{j} - \mathbf{k})}{\sqrt{10}}$$

9 Let $\mathbf{a} = 2x^2\,\mathbf{i} + 4x\,\mathbf{j} + \mathbf{k}$

and $\mathbf{b} = 7\,\mathbf{i} - 2\,\mathbf{j} + x\,\mathbf{k}$.

The angle between a and b is obtuse.

$\Rightarrow \quad \mathbf{a} \cdot \mathbf{b} < 0 \Rightarrow 14x^2 - 8x + x < 0$

$\Rightarrow \quad 7x(2x - 1) < 0$

$\therefore \quad x \in \left(0, \dfrac{1}{2}\right)$ ...(i)

Also, it is given, $\mathbf{b} \cdot \mathbf{k} = x$

and $\dfrac{\mathbf{b} \cdot \mathbf{k}}{|\mathbf{b}|} < \cos\dfrac{\pi}{6} = \dfrac{\sqrt{3}}{2}$

$\Rightarrow \quad 2x > \sqrt{3}\sqrt{53 + x^2}$

$\therefore \quad x^2 > 159$ ...(ii)

Hence, there is no common value for Eqs. (i) and (ii).

10 Let θ be an angle between unit vectors $\mathbf{a}$ and $\mathbf{b}$.

Then, $\mathbf{a} \cdot \mathbf{b} = \cos\theta$

Now, $|\mathbf{a} + \mathbf{b}|^2 = |\mathbf{a}|^2 + |\mathbf{b}|^2 + 2\mathbf{a} \cdot \mathbf{b}$

$$= 2 + 2\cos\theta = 4\cos^2\frac{\theta}{2}$$

$$\Rightarrow |\mathbf{a} + \mathbf{b}| = 2\cos\frac{\theta}{2}$$

Similarly, $|\mathbf{a} - \mathbf{b}| = 2\sin\dfrac{\theta}{2}$

$$\Rightarrow |\mathbf{a} + \mathbf{b}| + |\mathbf{a} - \mathbf{b}| = 2\left(\cos\frac{\theta}{2} + \sin\frac{\theta}{2}\right) \leq 2\sqrt{2}$$

11 We have, $\mathbf{u} \cdot \mathbf{n} = 0$ and $\mathbf{v} \cdot \mathbf{n} = 0$

$\Rightarrow \quad \mathbf{n} \perp \mathbf{u}$ and $\mathbf{n} \perp \mathbf{v}$

$\Rightarrow \quad \mathbf{n} = \pm \dfrac{\mathbf{u} \times \mathbf{v}}{|\mathbf{u} \times \mathbf{v}|}$

Now, $\mathbf{u} \times \mathbf{v} = (\mathbf{i} + \mathbf{j}) \times (\mathbf{i} - \mathbf{j}) = -2\,\mathbf{k}$

$\therefore \quad \mathbf{n} = \pm\,\mathbf{k}$

Hence,

$|\mathbf{w} \cdot \mathbf{n}| = |(\mathbf{i} + 2\mathbf{j} + 3\,\mathbf{k}) \cdot (\pm\,\mathbf{k})| = 3$

12 By expanding $\mathbf{a} \times (\mathbf{b} \times \mathbf{c})$, we get

$\mathbf{a} \cdot \mathbf{c} = x^2 - 2x + 6,\ \mathbf{a} \cdot \mathbf{b} = -\sin y$

Given, $\quad \mathbf{a} \cdot (\mathbf{b} + \mathbf{c}) = 4$

$\Rightarrow \quad x^2 - 2x + 2 = \sin y$

$\Rightarrow \quad \sin y = x^2 - 2x + 2$

$$= (x - 1)^2 + 1 \geq 1$$

But $\sin y \leq 1$

So, both sides are equal only for $x = 1$.

13 If $(\mathbf{a} \times \mathbf{b}) \times \mathbf{c} = \mathbf{0}$, then $\mathbf{c} = \dfrac{\mathbf{a} \times \mathbf{b}}{|\mathbf{a} \times \mathbf{b}|}$,

$\therefore \quad (\mathbf{a} \times \mathbf{b}) \cdot \mathbf{c} = |\mathbf{a} \times \mathbf{b}|$

$$\Rightarrow \quad \begin{vmatrix} a_1 & a_2 & a_3 \\ b_1 & b_2 & b_3 \\ c_1 & c_2 & c_3 \end{vmatrix}^2 = |\mathbf{a} \times \mathbf{b}|^2$$

14 Here, $\mathbf{a} \cdot \mathbf{x} = 2$ and $\mathbf{a} \times \mathbf{r} + \mathbf{b} = \mathbf{r}$ ...(i)

Dot product of Eq. (i) with $\mathbf{a}$ gives,

$\mathbf{a} \cdot \mathbf{b} = \mathbf{a} \cdot \mathbf{r} = 2$

Cross product of Eq. (i) with a gives

$$\mathbf{a} \times (\mathbf{a} \times \mathbf{r}) + \mathbf{a} \times \mathbf{b}$$

$$= \mathbf{a} \times \mathbf{r} = \mathbf{r} - \mathbf{b} \quad \text{[from Eq. (i)]}$$

$\Rightarrow \quad 2\mathbf{a} - \mathbf{r} + \mathbf{a} \times \mathbf{b} = \mathbf{r} - \mathbf{b}$

$\therefore \quad \mathbf{r} = \dfrac{1}{2}\,[2\mathbf{a} + \mathbf{b} + \mathbf{a} \times \mathbf{b}]$

15 Since, $\mathbf{a} + \mathbf{b} + \mathbf{c} = \mathbf{0}$

Taking cross product with $\mathbf{a}$, we get

$\mathbf{a} \times \mathbf{b} + \mathbf{a} \times \mathbf{c} = 0$ or $\mathbf{a} \times \mathbf{b} = \mathbf{c} \times \mathbf{a}$

Similarly, $\quad \mathbf{b} \times \mathbf{c} = \mathbf{a} \times \mathbf{b}$

Thus $\quad \mathbf{a} \times \mathbf{b} = \mathbf{b} \times \mathbf{c} = \mathbf{c} \times \mathbf{a}$

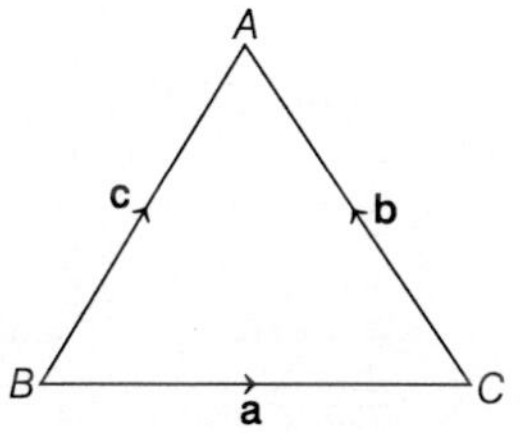

Now, as $\mathbf{a}$, $\mathbf{b}$ and $\mathbf{c}$ are unit vectors and $\mathbf{a} + \mathbf{b} + \mathbf{c} = 0$, therefore $\mathbf{a}$, $\mathbf{b}$ and $\mathbf{c}$ represents an equilateral triangle.

Hence, $\quad \mathbf{a} \times \mathbf{b} = \mathbf{b} \times \mathbf{c} = \mathbf{c} \times \mathbf{a} \neq 0$

16 Taking O as the origin, let the position vectors of A, B and C be $\mathbf{a}$, $\mathbf{b}$ and $\mathbf{c}$, respectively.

Then, the position vectors G_1, G_2 and G_3 are $\dfrac{\mathbf{b} + \mathbf{c}}{3}, \dfrac{\mathbf{c} + \mathbf{a}}{3}$

and $\dfrac{\mathbf{a} + \mathbf{b}}{3}$, respectively.

$\therefore \quad V_1 = \dfrac{1}{6}\,[\mathbf{a}\ \mathbf{b}\ \mathbf{c}]$

and $\quad V_2 = [\mathbf{OG}_1\ \mathbf{OG}_2\ \mathbf{OG}_3]$

Now, $\quad V_2 = [\mathbf{OG}_1\ \mathbf{OG}_2\ \mathbf{OG}_3]$

$$= \left[\frac{\mathbf{b} + \mathbf{c}}{3}\ \frac{\mathbf{c} + \mathbf{a}}{3}\ \frac{\mathbf{a} + \mathbf{b}}{3}\right]$$

$$= \frac{1}{27}[\mathbf{b}+\mathbf{c}\ \mathbf{c}+\mathbf{a}\ \mathbf{a}+\mathbf{b}]$$

$$= \frac{2}{27}[\mathbf{a}\ \mathbf{b}\ \mathbf{c}] = \frac{2}{27} \times 6V_1$$

$$\Rightarrow 9V_2 = 4V_1$$

17 Let $|\mathbf{BC}| = l$

In ΔABC, $l = \sqrt{AB^2 + AC^2}$

$$\therefore \tan\theta = \frac{AB}{AC}$$

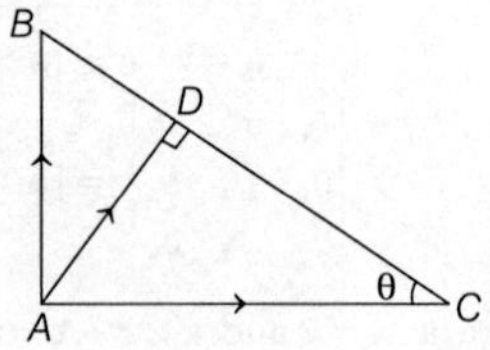

$$\Rightarrow \sin\theta = \frac{AB}{l} \text{ and } \cos\theta = \frac{AC}{l}$$

$$\therefore \text{Resultant vector} = \frac{1}{AB}\mathbf{i} + \frac{1}{AC}\mathbf{j}$$

$$= \left(\frac{1}{l\sin\theta}\mathbf{i} + \frac{1}{l\cos\theta}\mathbf{j}\right)$$

In ΔADC,

$$AD = AC\sin\theta = l\sin\theta\cos\theta = \frac{AB \cdot AC}{l}$$

$\therefore$ Magnitude of resultant vector

$$= \sqrt{\frac{1}{l^2}\left(\frac{1}{\sin^2\theta} + \frac{1}{\cos^2\theta}\right)}$$

$$= \frac{l}{(AB)(AC)} = \frac{1}{AD}$$

18 Given, $(\mathbf{a} \times \mathbf{b}) \times \mathbf{c} = \frac{1}{3}|\mathbf{b}||\mathbf{c}|\mathbf{a}$

$$\Rightarrow -\mathbf{c} \times (\mathbf{a} \times \mathbf{b}) = \frac{1}{3}|\mathbf{b}||\mathbf{c}|\mathbf{a}$$

$$\Rightarrow -(\mathbf{c}\cdot\mathbf{b})\cdot\mathbf{a} + (\mathbf{c}\cdot\mathbf{a})\mathbf{b} = \frac{1}{3}|\mathbf{b}||\mathbf{c}|\mathbf{a}$$

$$\left[\frac{1}{3}|\mathbf{b}||\mathbf{c}| + (\mathbf{c}\cdot\mathbf{b})\right]\mathbf{a} = (\mathbf{c}\cdot\mathbf{a})\,\mathbf{b}$$

Since, **a** and **b** are not collinear.

$$\mathbf{c}\cdot\mathbf{b} + \frac{1}{3}|\mathbf{b}||\mathbf{c}| = 0 \text{ and } \mathbf{c}\cdot\mathbf{a} = 0$$

$$\Rightarrow |\mathbf{c}||\mathbf{b}|\cos\theta + \frac{1}{3}|\mathbf{b}||\mathbf{c}| = 0$$

$$\Rightarrow |\mathbf{b}||\mathbf{c}|\left(\cos\theta + \frac{1}{3}\right) = 0$$

$$\Rightarrow \cos\theta + \frac{1}{3} = 0$$

$$[\because |\mathbf{b}| \neq 0, |\mathbf{c}| \neq 0]$$

$$\Rightarrow \cos\theta = -\frac{1}{3}$$

$$\Rightarrow \sin\theta = \frac{\sqrt{8}}{3} = \frac{2\sqrt{2}}{3}$$

19 Given,

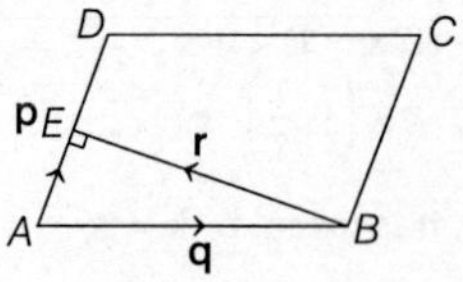

(i) A parallelogram $ABCD$ such that $\mathbf{AB} = \mathbf{q}$ and $\mathbf{AD} = \mathbf{p}$.

(ii) The altitude from vertex B to side AD coincides with a vector **r**.

To find The vector **r** in terms of **p** and **q**.

Let E be the foot of perpendicular from B to side AD.

AE = Projection of vector **q**

AE = Vector along AE of length AE

$$= |\mathbf{AE}|\,\mathbf{AE} = \frac{(\mathbf{q}\cdot\mathbf{p})\,\mathbf{p}}{|\mathbf{p}|^2}$$

Now, applying triangles law in ΔABE, we get

$$\mathbf{AB} + \mathbf{BE} = \mathbf{AE}$$

$$\Rightarrow \mathbf{q} + \mathbf{r} = \frac{(\mathbf{q}\cdot\mathbf{p})\,\mathbf{p}}{|\mathbf{p}|^2}$$

$$\Rightarrow \mathbf{r} = \frac{(\mathbf{q}\cdot\mathbf{p})\,\mathbf{p}}{|\mathbf{p}|^2} - \mathbf{q}$$

$$= -\mathbf{q} + \left(\frac{\mathbf{q}\cdot\mathbf{p}}{\mathbf{p}\cdot\mathbf{p}}\right)\mathbf{p}$$

20 In an isosceles ΔABC in which $AB = AC$, the median and bisector from A must be same line. Statement II is true.

Now, $\mathbf{AD} = \dfrac{\mathbf{u}+\mathbf{v}}{2}$

and $|\mathbf{AD}|^2 = \frac{1}{2}\cdot 2\cos^2\frac{\alpha}{2}$

So, $|\mathbf{AD}| = \cos\frac{\alpha}{2}$

Unit vector along AD, i.e. x is given by

$$\mathbf{x} = \frac{\mathbf{AD}}{|\mathbf{AD}|} = \frac{\mathbf{u}+\mathbf{v}}{2\cos\frac{\alpha}{2}}.$$

DAY THIRTY TWO

Three Dimensional Geometry

Learning & Revision for the Day

- Coordinates of a Point in a Space
- Section Formula
- Direction Cosines and Ratios
- Equation of Line in Space
- Skew-Lines
- Coplanar Lines
- Plane
- Angle between a Line and a Plane

Coordinates of a Point in a Space

From the adjoining figure, we have

- The three mutually perpendicular lines in a space which divides the space into eight parts are called coordinates axes.
- The coordinates of a point are the distances from the origin to the feet of the perpendiculars from the point on the respective coordinate axes.
- The coordinates of any point on the X, Y and Z-axes will be as $(x,0,0)$ $(0,y,0)$ and $(0,0,z)$ respectively and the coordinates of any point P in space will be as (x,y,z).

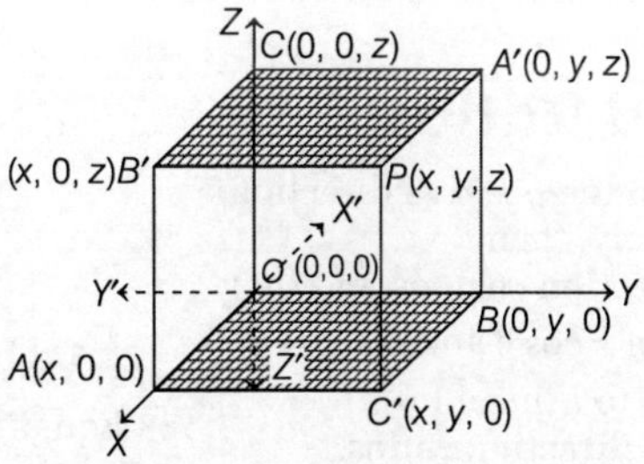

PREP MIRROR
Your Personal Preparation Indicator

- No. of Questions in Exercises (x)—
- No. of Questions Attempted (y)—
- No. of Correct Questions (z)—
(Without referring Explanations)

- Accuracy Level $(z/y \times 100)$—
- Prep Level $(z/x \times 100)$—

In order to expect good rank in JEE, your Accuracy Level should be above 85 & Prep Level should be above 75.

Distance between Two Points

The distance between two points $A(x_1, y_1, z_1)$ and $B(x_2, y_2, z_2)$ is

$$|AB| = \sqrt{(x_2 - x_1)^2 + (y_2 - y_1)^2 + (z_2 - z_1)^2}$$

Section Formula

If $M(x, y, z)$ divides the line joining of points $P(x_1, y_1, z_1)$ and $Q(x_2, y_2, z_2)$ in the ratio $m:n$, then

For Internal Division

$$x = \frac{mx_2 + nx_1}{m+n}, y = \frac{my_2 + ny_1}{m+n} \text{ and } z = \frac{mz_2 + nz_1}{m+n}$$

For External Division

$$x = \frac{mx_2 - nx_1}{m-n}, y = \frac{my_2 - ny_1}{m-n} \text{ and } z = \frac{mz_2 - nz_1}{m-n}$$

The coordinates of the mid-point of the line joining $P(x_1, y_1, z_1)$ and $Q(x_2, y_2, z_2)$ are $\left(\frac{x_1 + x_2}{2}, \frac{y_1 + y_2}{2}, \frac{z_1 + z_2}{2}\right)$

Some Important Results

1. If $A(x_1, y_1, z_1), B(x_2, y_2, z_2)$ and $C(x_3, y_3, z_3)$ are the vertices of a ΔABC, then

   (i) Centroid of triangle $= \left(\frac{x_1 + x_2 + x_3}{3}, \frac{y_1 + y_2 + y_3}{3}\right)$

   (ii) Area of $\Delta ABC = \frac{1}{2}\begin{vmatrix} \mathbf{i} & \mathbf{j} & \mathbf{k} \\ x_2 - x_1 & y_2 - y_1 & z_2 - z_1 \\ x_3 - x_1 & y_3 - y_1 & z_3 - z_1 \end{vmatrix}$

   (iii) If area of $\Delta ABC = 0$, then these points are collinear.

2. Four non-coplanar points $A(x_1, y_1, z_1)$, $B(x_2, y_2, z_2)$, $C(x_3, y_3, z_3)$ and $D(x_4, y_4, z_4)$ form a tetrahedron with vertices A, B, C and D, edges AB, AC, AD, BC, BD and CD, faces ABC, ABD, ACD and BCD, then

   (i) Centroid $\left(\frac{x_1 + x_2 + x_3 + x_4}{4}, \frac{y_1 + y_2 + y_3 + y_4}{4}, \frac{z_1 + z_2 + z_3 + z_4}{4}\right)$

   (ii) Volume $= \frac{1}{6}\begin{vmatrix} x_2 - x_1 & y_2 - y_1 & z_2 - z_1 \\ x_3 - x_1 & y_3 - y_1 & z_3 - z_1 \\ x_4 - x_1 & y_4 - y_1 & z_4 - z_1 \end{vmatrix}$

Direction Cosines and Ratios

If a vector makes angles α, β and γ with the positive directions of X-axis, Y-axis and Z-axis respectively, then $\cos\alpha, \cos\beta$ and $\cos\gamma$ are called its **direction cosines** and they are denoted by l, m, n, i.e. $l = \cos\alpha$, $m = \cos\beta$ and $n = \cos\gamma$.

If numbers a, b and c are proportional to l, m and n respectively, then a, b and c are called **direction ratios.**

Thus, a, b and c are the direction ratios of a vector, provided $\frac{l}{a} = \frac{m}{b} = \frac{n}{c}$

Important Results

- A line in space can be extended in two opposite directions and so it has two sets of direction cosines.
- In order to get unique set of direction cosines, we must take the given line as a directed line.
- Let L is a directed line which makes α, β and γ with positive direction of X, Y and Z-axis, respectively. If we reverse the direction of L, then direction angles are replaced by their supplements, i.e. $\pi - \alpha, \pi - \beta, \pi - \gamma$.
- If the line does not pass through origin, then draw a line through origin and parallel to given line and then find its direction cosines as two parallel lines have same set of direction cosines.

Some Important Deductions

(i) Direction ratios of the line joining two points $P(x_1, y_1, z_1)$ and $Q(x_2, y_2, z_2)$ are $x_2 - x_1, y_2 - y_1, z_2 - z_1$ and its direction cosines are

$$\frac{x_2 - x_1}{|\mathbf{PQ}|}, \frac{y_2 - y_1}{|\mathbf{PQ}|}, \frac{z_2 - z_1}{|\mathbf{PQ}|}.$$

(ii) If $P(x, y, z)$ is a point in space and $OP = \mathbf{r}$ then

(a) $x = l|\mathbf{r}|, y = m|\mathbf{r}|, z = n|\mathbf{r}|$

(b) $l|\mathbf{r}|, m|\mathbf{r}|$ and $n|\mathbf{r}|$ are projections of $\mathbf{r}$ on OX, OY and OZ, respectively.

(c) $\mathbf{r} = |\mathbf{r}|(l\mathbf{i} + m\mathbf{j} + n\mathbf{k})$ and $\hat{\mathbf{r}} = l\mathbf{i} + m\mathbf{j} + n\mathbf{k}$

(d) If $\mathbf{r} = a\mathbf{i} + b\mathbf{j} + c\mathbf{k}$, then a, b, c are DR's of vector and DC's are given by $l = \frac{a}{|\mathbf{r}|}, m = \frac{b}{|\mathbf{r}|}, n = \frac{c}{|\mathbf{r}|}$

(iii) The sum of squares of direction cosines is always unity, i.e. $l^2 + m^2 + n^2 = 1$

(iv) Direction cosines are unique but direction ratio are not unique and it can be infinite.

(v) If a, b, c are DR's of a line and l, m, n are DC's of a line, then $l = \pm\frac{a}{\sqrt{a^2 + b^2 + c^2}}$, $m = \pm\frac{b}{\sqrt{a^2 + b^2 + c^2}}$ and $n = \pm\frac{c}{\sqrt{a^2 + b^2 + c^2}}$

(vi) The DC's of a line which is equally inclined to the coordinate axes are $\left(\pm\frac{1}{\sqrt{3}}, \pm\frac{1}{\sqrt{3}}, \pm\frac{1}{\sqrt{3}}\right)$.

(vii) If l, m and n are the DC's of a line, then the maximum value of $lmn = \frac{1}{3\sqrt{3}}$.

Equations of a Line in Space

Equation of line passing through point $A(\mathbf{a})$ and parallel to vector $(\mathbf{b})$ is $\mathbf{r} = \mathbf{a} + \lambda\,\mathbf{b}$.

If coordinates of A be (x_1, y_1, z_1) and the direction ratios of line be a, b and c, then equation of line is $\frac{x - x_1}{a} = \frac{y - y_1}{b} = \frac{z - z_1}{c}$.

NOTE
- Equation of X-axis is $\frac{x-0}{1} = \frac{y-0}{0} = \frac{z-0}{0}$ or $y = 0, z = 0$
- Equation of Y-axis is $\frac{x-0}{0} = \frac{y-0}{1} = \frac{z-0}{0}$ or $x = 0, z = 0$
- Equation of Z-axis is $\frac{x-0}{0} = \frac{y-0}{0} = \frac{z-0}{1}$ or $x = 0, y = 0$

Equation of a line passing through two given points $A(x_1, y_1, z_1)$ and $B(x_2, y_2, z_2)$ is $\frac{x - x_1}{x_2 - x_1} = \frac{y - y_1}{y_2 - y_1} = \frac{z - z_1}{z_2 - z_1}$

Its vector form is $\mathbf{r} = \mathbf{a} + \lambda(\mathbf{b} - \mathbf{a})$.

The parametric equations of a line through (a_1, a_2, a_3) with DC's l, m and n are $x = a_1 + lr$, $y = a_2 + mr$ and $z = a_3 + nr$.

Angle between Two Intersecting Lines

1. If DR's of two lines are a_1, b_1, c_1 and a_2, b_2, c_2, then

$$\cos\theta = \frac{a_1a_2 + b_1b_2 + c_1c_2}{\sqrt{(a_1^2 + b_1^2 + c_1^2)(a_2^2 + b_2^2 + c_2^2)}}$$

(i) Condition for parallel lines, $\frac{a_1}{a_2} = \frac{b_1}{b_2} = \frac{c_1}{c_2}$

(ii) Condition for perpendicular lines,

$$a_1a_2 + b_1b_2 + c_1c_2 = 0$$

2. Angle between two lines with DC's l_1, m_1, n_1 and l_2, m_2, n_2 is $\cos^{-1}(l_1l_2 + m_1m_2 + n_1n_2)$

or $\sin^{-1}\left(\sqrt{\Sigma(m_1n_2 - m_2n_1)^2}\right)$.

(i) Condition for parallel lines, $\frac{l_1}{l_2} = \frac{m_1}{m_2} = \frac{n_1}{n_2}$

(ii) Condition for perpendicular lines,

$$l_1l_2 + m_1m_2 + n_1n_2 = 0$$

NOTE
- The angle between any two diagonals of a cube is $\cos^{-1}\left(\frac{1}{3}\right)$.
- The angle between a diagonal of a cube and a face is $\cos^{-1}\left(\sqrt{\frac{2}{3}}\right)$
- The angle between the diagonal of a cube and edge of cube is $\cos^{-1}\left(\frac{1}{\sqrt{3}}\right)$
- If a straight line makes angles α, β, γ and δ with the diagonals of a cube, then

$$\cos^2\alpha + \cos^2\beta + \cos^2\gamma + \cos^2\delta = \frac{4}{3}$$

Skew-Lines

Two straight lines in a space which are neither parallel nor intersecting are called skew-lines. Thus, skew-lines are those lines which do not lie in the same plane.

Shortest Distance between Two Skew-Lines

- If l_1 and l_2 are two skew-lines, then there is one and only one line perpendicular to each of the line l_1 and l_2, which is known as the line of shortest distance.
- The shortest distance between two lines l_1 and l_2 is the distance PQ between the points P and Q, where the line of shortest distance intersects the two given lines.
- The shortest distance between two skew-lines $\mathbf{r} = \mathbf{a} + \lambda\mathbf{b}$ and $\mathbf{r} = \mathbf{c} + \mu\,\mathbf{d}$ is given by SD $= \left|\frac{(\mathbf{c} - \mathbf{a})\cdot(\mathbf{b} \times \mathbf{d})}{|\mathbf{b} \times \mathbf{d}|}\right|$
- The shortest distance between the lines $\frac{x - x_1}{a_1} = \frac{y - y_1}{b_1} = \frac{z - z_1}{c_1}$ and $\frac{x - x_2}{a_2} = \frac{y - y_2}{b_2} = \frac{z - z_2}{c_2}$ is

$$\text{SD} = \left|\frac{\begin{vmatrix} x_2 - x_1 & y_2 - y_1 & z_2 - z_1 \\ a_1 & b_1 & c_1 \\ a_2 & b_2 & c_2 \end{vmatrix}}{\sqrt{\sum (b_1c_2 - b_2c_1)^2}}\right|$$

- Two lines $\mathbf{r} = \mathbf{a} + \lambda\mathbf{b}$ and $\mathbf{r} = \mathbf{c} + \mu\mathbf{d}$ are intersecting if shortest distance between them is zero.

i.e. $\left|\frac{(\mathbf{c} - \mathbf{a})\cdot(\mathbf{b} \times \mathbf{d})}{|\mathbf{b} \times \mathbf{d}|}\right| = 0 \Rightarrow (\mathbf{c} - \mathbf{a})\cdot(\mathbf{b} \times \mathbf{d}) = 0$

or $\begin{vmatrix} x_2 - x_1 & y_2 - y_1 & z_2 - z_1 \\ a_1 & b_1 & c_1 \\ a_2 & b_2 & c_2 \end{vmatrix} = 0$

Distance or shortest distance between two parallel lines

- Shortest distance between parallel lines will be the perpendicular distance.
- If the parallel lines are given by $\mathbf{r} = \mathbf{a} + \lambda\mathbf{b}$ and $\mathbf{r} = \mathbf{c} + \mu\mathbf{b}$ then distance between them is $d = \left|\frac{|(\mathbf{c} - \mathbf{a}) \times \mathbf{b}|}{|\mathbf{b}|}\right|$

Coplanar Lines

Lines which lie in the same plane are called coplanar lines. Any two coplanar lines are either parallel or intersecting.

Condition for Coplanarity of Two non-parallel Lines

Two lines $\mathbf{r} = \mathbf{a} + \lambda\,\mathbf{b}$ and $\mathbf{r} = \mathbf{c} + \mu\,\mathbf{d}$ are coplanar or intersecting, if $(\mathbf{c} - \mathbf{a})\cdot(\mathbf{b} \times \mathbf{d}) = 0 \Rightarrow [\mathbf{a}\ \mathbf{b}\ \mathbf{d}] = [\mathbf{c}\ \mathbf{b}\ \mathbf{d}]$

The lines $\frac{x - x_1}{l_1} = \frac{y - y_1}{m_1} = \frac{z - z_1}{n_1}$

and $\frac{x - x_2}{l_2} = \frac{y - y_2}{m_2} = \frac{z - z_2}{n_2}$ are coplanar,

if $\begin{vmatrix} x_2 - x_1 & y_2 - y_1 & z_2 - z_1 \\ l_1 & m_1 & n_1 \\ l_2 & m_2 & n_2 \end{vmatrix} = 0.$

Plane

A plane is a surface such that line joining any two points of the plane totally lies in it.

Equation of a Plane in Different Forms

1. The general equation of a plane is $ax + by + cz + d = 0$ and $a^2 + b^2 + c^2 \neq 0$, where, a, b and c are the DR's of the normal to the plane.

(i) Plane through the origin is $ax + by + cz = 0$.

(ii) Planes parallel to the coordinate planes (perpendicular to coordinate axes) $x = k$ parallel to YOZ plane, $y = k$ parallel to ZOX plane and $z = k$ parallel to XOY plane.

(iii) Planes parallel to coordinate axes
$by + cz + d = 0$ parallel to X-axis
$ax + cz + d = 0$ parallel to Y-axis
$ax + by + d = 0$ parallel to Z-axis

2. If a, b and c are the intercepts of plane with the coordinate axes, then equation of plane is $\frac{x}{a} + \frac{y}{b} + \frac{z}{c} = 1$. It meets the coordinate axes at $A(a, 0, 0), B(0, b, 0)$ and $C(0, 0, c)$.

3. (i) If l, m and n are DC's of normal to the plane, p is the distance of the origin from the plane, then equation of plane is $lx + my + nz = p$.

(ii) Coordinates of foot of perpendicular, drawn from the origin to the plane, is (lp, mp, np).

(iii) If $\mathbf{ON}$ is the normal from the origin to the plane and $\hat{\mathbf{n}}$ is the unit vector along $\mathbf{ON}$. Then $\mathbf{ON} = p\hat{\mathbf{n}}$ and equation of plane is $\hat{\mathbf{r}} \cdot \hat{\mathbf{n}} = p$, where $\mathbf{r} = x\mathbf{i} + y\mathbf{j} + z\mathbf{k}$.

4. Plane through a point (x_1, y_1, z_1) is
$a(x - x_1) + b(y - y_1) + c(z - z_1) = 0$.
where a, b, c are DR's of normal to the plane.

5. Plane through three non-collinear points $(x_1, y_1, z_1), (x_2, y_2, z_2)$ and (x_3, y_3, z_3) is

$$\begin{vmatrix} x - x_1 & y - y_1 & z - z_1 \\ x_2 - x_1 & y_2 - y_1 & z_2 - z_1 \\ x_3 - x_1 & y_3 - y_1 & z_3 - z_1 \end{vmatrix} = 0$$

6. (i) Equation of plane, passing through a point A with position vector $\mathbf{a}$ and is parallel to given vectors $\mathbf{b}$ and $\mathbf{c}$, is $(\mathbf{r} - \mathbf{a}) \cdot (\mathbf{b} \times \mathbf{c}) = 0$ or $[\mathbf{r} - \mathbf{a}\ \mathbf{b}\ \mathbf{c}] = 0$

(ii) Its cartesian equation is $\begin{vmatrix} x - a_1 & y - a_2 & z - a_3 \\ b_1 & b_2 & b_3 \\ c_1 & c_2 & c_3 \end{vmatrix} = 0$

7. Plane parallel to the given plane $ax + by + cz + d = 0$ is $ax + by + cz + k = 0$, where k is a constant determined by the given condition.

8. (i) Any plane passing through the line of intersection of the planes $ax + by + cz + d = 0$ and $a_1x + b_1y + c_1z + d_1 = 0$ is $(ax + by + cz + d) + \lambda(a_1x + b_1y + c_1z + d_1) = 0$

(ii) If $\mathbf{r} \cdot \mathbf{n}_1 = d_1$ and $\mathbf{r} \cdot \mathbf{n}_2 = d_2$ are two planes, then their line of intersection is perpendicular to both $\mathbf{n}_1$ and $\mathbf{n}_2$, i.e. line is parallel to the vectors $\mathbf{n}_1 \times \mathbf{n}_2$.

Some Important Results on plane

- If $ax + by + cz + d_1 = 0$ and $a_1x + b_1y + c_1z + d_2 = 0$ are the equations of any two planes, then $ax + bx + cz + d_1 = 0$ $= a_1x + b_1y + c_1z + d_2$ gives the equation of straight line.
- Plane $ax + by + cz + d = 0$ intersecting a line segment joining $A(x_1, y_1, z_1)$ and $B(x_2, y_2, z_2)$ divides it in the ratio $-\frac{ax_1 + by_1 + cz_1 + d}{ax_2 + by_2 + cz_2 + d}$
  (i) If this ratio is positive, then A and B are on opposite sides of the plane.
  (ii) If this ratio is negative, then A and B are on the same side of the plane.
- If θ be the angle between the planes $a_1x + b_1y + c_1z + d_1 = 0$ and $a_2x + b_2y + c_2z + d_2 = 0$, then
$$\theta = \cos^{-1}\left(\left|\frac{a_1a_2 + b_1b_2 + c_1c_2}{\sqrt{a_1^2 + b_1^2 + c_1^2}\sqrt{a_2^2 + b_2^2 + c_2^2}}\right|\right)$$
- Two planes are parallel if their normals are parallel and the planes are perpendicular if their normals are perpendicular.
- If $\mathbf{r} \cdot \mathbf{n}_1 = d_1$ (or $a_1x + b_1y + c_1z = d_1$) and $\mathbf{r} \cdot n_2 = d_2$ (or $a_2x + b_2y + c_2z = d_2$) are two planes, then they are
  (i) parallel if $\mathbf{n}_1 = \lambda\mathbf{n}_2$ or $\frac{a_1}{a_2} = \frac{b_1}{b_2} = \frac{c_1}{c_2}$.
  (ii) perpendicular if $\mathbf{n}_1 \cdot \mathbf{n}_2 = 0$ or $a_1a_2 + b_1b_2 + c_1c_2 = 0$.
- Distance of a point (x_1, y_1, z_1) from the plane $ax + by + cz + d = 0$ is $\frac{|ax_1 + by_1 + cz_1 + d|}{\sqrt{a^2 + b^2 + c^2}}$.
  Distance of the origin is $\frac{|d|}{\sqrt{a^2 + b^2 + c^2}}$.
- The distance between two parallel planes $ax + by + cz + d_1 = 0$ and $ax + by + cz + d_2 = 0$ is $\frac{|d_1 - d_2|}{\sqrt{a^2 + b^2 + c^2}}$.

Angle between a Line and a Plane

If angle between the line $\frac{x - x_1}{a} = \frac{y - y_1}{b} = \frac{z - z_1}{c}$ and the plane $a_1x + b_1y + c_1z + d = 0$ is θ, then $(90° - \theta)$ is the angle between normal and the line, therefore

$$\cos(90° - \theta) = \left|\frac{aa_1 + bb_1 + cc_1}{\sqrt{a^2 + b^2 + c^2}\sqrt{a_1^2 + b_1^2 + c_1^2}}\right|$$

- If $\frac{a}{a_1} = \frac{b}{b_1} = \frac{c}{c_1}$, then line is perpendicular to plane.
- If $a \cdot a_1 + b \cdot b_1 + c \cdot c_1 = 0$, then line is parallel to plane.
- If $a \cdot a_1 + b \cdot b_1 + c \cdot c_1 = 0$, and $a_1x_1 + b_1y_1 + c_1z_1 + d = 0$, then line lies in the plane.

Important Points Related to Line and Plane

- Projection of a line segment joining the points (x_1, y_1, z_1) and (x_2, y_2, z_2) on a line with direction cosine l, m, n is $|(x_2 - x_1)l + (y_2 - y_1)m + (z_2 - z_1)n|$

- Foot of the perpendicular from a point (x_1, y_1, z_1) on the plane $ax + by + cz + d = 0$ is (x, y, z), where $\frac{x - x_1}{a} = \frac{y - y_1}{b} = \frac{z - z_1}{c} = -\frac{(ax_1 + by_1 + cz_1 + d)}{a^2 + b^2 + c^2}$.
- Image of the point (x_1, y_1, z_1) in the plane $ax + by + cz + d = 0$ is (x, y, z), where $\frac{x - x_1}{a} = \frac{y - y_1}{b} = \frac{z - z_1}{c} = -\frac{2(ax_1 + by_1 + cz_1 + d)}{a^2 + b^2 + c^2}$.
- Four points (x_i, y_i, z_i), where $i = 1, 2, 3$ and 4 are coplanar, if $\begin{vmatrix} x_2 - x_1 & y_2 - y_1 & z_2 - z_1 \\ x_3 - x_1 & y_3 - y_1 & z_3 - z_1 \\ x_4 - x_1 & y_4 - y_1 & z_4 - z_1 \end{vmatrix} = 0$.
- Planes bisecting the angle between two intersecting planes $a_1x + b_1y + c_1z + d_1 = 0$ and $a_2x + b_2y + c_2z + d_2 = 0$ are given by $\frac{a_1x + b_1y + c_1z + d_1}{\sqrt{a_1^2 + b_1^2 + c_1^2}} = \pm \frac{(a_2x + b_2y + c_2z + d_2)}{\sqrt{a_2^2 + b_2^2 + c_2^2}}$

  (i) If $a_1a_2 + b_1b_2 + c_1c_2 < 0$, then origin is in acute angle and the acute angle bisector is obtained by taking positive sign in the above equation. The obtuse angle bisector is obtained by taking negative sign in the above equation.

  (ii) If $a_1a_2 + b_1b_2 + c_1c_2 > 0$, then origin lies in obtuse angle and the obtuse angle bisector is obtained by taking positive sign in above equation. Acute angle bisector is obtained by taking negative sign.

DAY PRACTICE SESSION 1

FOUNDATION QUESTIONS EXERCISE

1 If the orthocentre and centroid of a triangle are $(-3, 5, 1)$ and $(3, 3, -1)$ respectively, then its circumcentre is

(a) $(6, 2, -2)$ (b) $(1, 2, 0)$ (c) $(6, 2, 2)$ (d) $(6, -2, 2)$

2 A line makes the same angle θ with each of the X and Z-axes. If the angle β, which it makes with Y-axis, is such that $\sin^2 \beta = 3 \sin^2 \theta$, then $\cos^2 \theta$ is equal to → AIEEE 2004

(a) $\frac{2}{3}$ (b) $\frac{1}{5}$ (c) $\frac{3}{5}$ (d) $\frac{2}{5}$

3 A line makes an angle θ with X and Y-axes both. A possible value of θ is in

(a) $\left[0, \frac{\pi}{4}\right]$ (b) $\left[0, \frac{\pi}{2}\right]$ (c) $\left[\frac{\pi}{4}, \frac{\pi}{2}\right]$ (d) $\left[\frac{\pi}{3}, \frac{\pi}{6}\right]$

4 The projections of a vector on the three coordinate axes are 6, – 3 and 2, respectively. The direction cosines of the vector are → AIEEE 2009

(a) $6, -3, 2$ (b) $\frac{6}{5}, -\frac{3}{5}, \frac{2}{5}$ (c) $\frac{6}{7}, -\frac{3}{7}, \frac{2}{7}$ (d) $-\frac{6}{7}, -\frac{3}{7}, \frac{2}{7}$

5 If the projections of a line segment on the X, Y and Z-axes in 3-dimensional space are 2, 3 and 6 respectively, then the length of the line segment is → JEE Mains 2013

(a) 12 (b) 7 (c) 9 (d) 6

6 A vector **r** is inclined at equal angles to OX, OY and OZ. If the magnitude of **r** is 6 units, then **r** is equal to

(a) $\sqrt{3}\,(\mathbf{i} + \mathbf{j} + \mathbf{k})$ (b) $-\sqrt{3}\,(\mathbf{i} + \mathbf{j} + \mathbf{k})$
(c) $-2\sqrt{3}\,(\mathbf{i} + \mathbf{j} + \mathbf{k})$ (d) None of these

7 A line L_1 passes through the point $3\mathbf{i}$ and is parallel to the vector $-\mathbf{i} + \mathbf{j} + \mathbf{k}$ and another line L_2 passes through the point $\mathbf{i} + \mathbf{j}$ and is parallel to the vector $\mathbf{i} + \mathbf{k}$, then point of intersection of the lines is

(a) $\mathbf{i} + 2\mathbf{j} + \mathbf{k}$ (b) $2\mathbf{i} + \mathbf{j} + \mathbf{k}$ (c) $\mathbf{i} - 2\mathbf{j} - \mathbf{k}$ (d) $\mathbf{i} - 2\mathbf{j} + \mathbf{k}$

8 The line passing through the points $(5, 1, a)$ and $(3, b, 1)$ crosses the YZ-plane at the point $\left(0, \frac{17}{2}, -\frac{13}{2}\right)$. Then, → AIEEE 2008

(a) $a = 8, b = 2$ (b) $a = 2, b = 8$
(c) $a = 4, b = 6$ (d) $a = 6, b = 4$

9 The angle between the lines $2x = 3y = -z$ and $6x = -y = -4z$ is → AIEEE 2005

(a) 30° (b) 45° (c) 90° (d) 0°

10 The angle between a diagonal of a cube and an edge of the cube intersecting the diagonal is

(a) $\cos^{-1} \frac{1}{3}$ (b) $\cos^{-1} \sqrt{\frac{2}{3}}$
(c) $\tan^{-1} \sqrt{2}$ (d) None of these

11 The angle between the lines whose direction cosines satisfy the equations $l + m + n = 0$ and $l^2 = m^2 + n^2$ is → JEE Mains 2014

(a) $\frac{\pi}{3}$ (b) $\frac{\pi}{4}$ (c) $\frac{\pi}{6}$ (d) $\frac{\pi}{2}$

12 If the line, $\frac{x - 3}{2} = \frac{y + 2}{-1} = \frac{z + 4}{3}$ lies in the plane, $lx + my - z = 9$, then $l^2 + m^2$ is equal to → JEE Mains 2016

(a) 26 (b) 18
(c) 5 (d) 2

13 The direction cosines of two lines at right angles are $(1, 2, 3)$ and $\left(-2, \frac{1}{2}, \frac{1}{3}\right)$, then the direction cosine perpendicular to both the given lines are

(a) $\sqrt{\frac{25}{2198}}, \frac{19}{\sqrt{2198}}, \sqrt{\frac{729}{2198}}$ (b) $\sqrt{\frac{24}{2198}}, \sqrt{\frac{38}{2198}}, \sqrt{\frac{730}{2198}}$
(c) $\frac{1}{3}, -2, \frac{-7}{2}$ (d) None of these

14 The foot of perpendicular from (0, 2, 3) to the line $\frac{x+3}{5}=\frac{y-1}{2}=\frac{z+4}{3}$, is

(a) (−2, 3, 4) (b) (2, −1, 3) (c) (2, 3, −1) (d) (3, 2, −1)

15 The projection of the line segment joining (2, 5, 6) and (3, 2, 7) on the line with direction ratios 2, 1, −2, is

(a) $\frac{1}{2}$ (b) $\frac{1}{3}$ (c) 2 (d) 1

16 The shortest distance between the lines $\frac{x+1}{7}=\frac{y+1}{-6}=\frac{z+1}{1}$ and $\frac{x-3}{1}=\frac{y-5}{-2}=\frac{z-7}{1}$, is → NCERT

(a) $\sqrt{29}$ units (b) 29 units (c) $\frac{29}{2}$ units (d) $2\sqrt{29}$ units

17 The shortest distance between the diagonals of a rectangular parallelopiped whose sides are a, b, c and the edges not meeting it, are

(a) $\frac{bc}{\sqrt{b^2-c^2}}, \frac{ca}{\sqrt{c^2-a^2}}, \frac{ab}{\sqrt{a^2-b^2}}$

(b) $\frac{bc}{\sqrt{b^2+c^2}}, \frac{ca}{\sqrt{c^2+a^2}}, \frac{ab}{\sqrt{a^2+b^2}}$

(c) $\frac{2bc}{\sqrt{b^2-c^2}}, \frac{2ca}{\sqrt{c^2-a^2}}, \frac{2ab}{\sqrt{a^2-b^2}}$

(d) None of the above

18 If the line $\frac{x-1}{2}=\frac{y+1}{3}=\frac{z-1}{4}$ and $\frac{x-3}{1}=\frac{y-k}{2}=\frac{z}{1}$ intersect, then k is equal to → AIEEE 2012

(a) −1 (b) $\frac{2}{9}$ (c) $\frac{9}{2}$ (d) 0

19 If the straight lines $x=1+s, y=-3-\lambda s, z=1+\lambda s$ and $x=\frac{t}{2}, y=1+t, z=2-t$, with parameters s and t respectively are coplanar, then λ is equal to → AIEEE 2004

(a) −2 (b) −1 (c) $-\frac{1}{2}$ (d) 0

20 The distance of the point (1, 0, 2) from the point of intersection of the line $\frac{x-2}{3}=\frac{y+1}{4}=\frac{z-2}{12}$ and the plane $x-y+z=16$ is → JEE Mains 2015

(a) $2\sqrt{14}$ (b) 8 (c) $3\sqrt{21}$ (d) 13

21 A vector **n** is inclined to X-axis at 45°, to Y-axis at 60° and at an acute angle to Z-axis. If **n** is a normal to a plane passing through the point $(\sqrt{2}, -1, 1)$, then the equation of the plane is → JEE Mains 2013

(a) $4\sqrt{2}x+7y+z=2$ (b) $\sqrt{2}x+y+z=2$
(c) $3\sqrt{2}x-4y-3z=7$ (d) $\sqrt{2}x-y-z=2$

22 Let Q be the foot of perpendicular from the origin to the plane $4x-3y+z+13=0$ and R be a point $(-1, 1, -6)$ on the plane. Then, length QR is → JEE Mains 2013

(a) $\sqrt{14}$ (b) $\sqrt{\frac{19}{2}}$ (c) $3\sqrt{\frac{7}{2}}$ (d) $\frac{3}{\sqrt{2}}$

23 The plane passing through the point (−2, −2, 2) and containing the line joining the points (1, −1, 2) and (1, 1, 1) makes intercepts on the coordinate axes and the sum of whose length is

(a) 3 (b) 6 (c) 12 (d) 20

24 The coordinates of the point where the line through (3, −4, −5) and (2, −3, 1) crosses the plane passing through three points (2, 2, 1), (3, 0, 1) and (4, −1, 0), is → NCERT Exemplar

(a) (1, 2, 7) (b) (−1, 2, −7)
(c) (1, −2, 7) (d) None of these

25 The volume of the tetrahedron formed by coordinate planes and $2x+3y+z=6$, is

(a) 5 (b) 4 (c) 6 (d) 0

26 The equation of the plane passing through (2, 1, 5) and parallel to the plane $3x-4y+5z=4$ is

(a) $3x-4y+5z-27=0$ (b) $3x-4y+5z+21=0$
(c) $3x-4y+5z+26=0$ (d) $3x-4y+5z+17=0$

27 If Q is the image of the point $P(2, 3, 4)$ under the reflection in the plane $x-2y+5z=6$, then the equation of the line PQ is

(a) $\frac{x-2}{-1}=\frac{y-3}{2}=\frac{z-4}{5}$ (b) $\frac{x-2}{1}=\frac{y-3}{-2}=\frac{z-4}{5}$
(c) $\frac{x-2}{-1}=\frac{y-3}{-2}=\frac{z-4}{5}$ (d) $\frac{x-2}{1}=\frac{y-3}{2}=\frac{z-4}{5}$

28 If the points (1, 2, 3) and (2, −1, 0) lie on the opposite sides of the plane $2x+3y-2z=k$, then

(a) $k<1$ (b) $k>2$
(c) $k<1$ or $k>2$ (d) $1<k<2$

29 The equation of the plane containing the lines $2x-5y+z=3$, $x+y+4z=5$ and parallel to the plane $x+3y+6z=1$ is → JEE Mains 2015

(a) $2x+6y+12z=13$ (b) $x+3y+6z=-7$
(c) $x+3y+6z=7$ (d) $2x+6y+12z=-13$

30 The equation of a plane through the line of intersection of the planes $x+2y=3$, $y-2z+1=0$ and perpendicular to the first plane is → JEE Mains 2013

(a) $2x-y-10z=9$ (b) $2x-y+7z=11$
(c) $2x-y+10z=11$ (d) $2x-y-9z=10$

31 An equation of a plane parallel to the plane $x-2y+2z-5=0$ and at a unit distance from the origin is → AIEEE 2012

(a) $x-2y+2z\pm 3=0$ (b) $x-2y+2z+1=0$
(c) $x-2y+2z-1=0$ (d) $x-2y+2z+5=0$

32 Two systems of rectangular axes have the same origin. If a plane cuts them at distances a, b, c and a', b', c' from the origin, then → AIEEE 2003

(a) $\frac{1}{a^2}+\frac{1}{b^2}+\frac{1}{c^2}+\frac{1}{a'^2}+\frac{1}{b'^2}+\frac{1}{c'^2}=0$

(b) $\frac{1}{a^2}+\frac{1}{b^2}-\frac{1}{c^2}+\frac{1}{a'^2}+\frac{1}{b'^2}-\frac{1}{c'^2}=0$

(c) $\frac{1}{a^2}-\frac{1}{b^2}-\frac{1}{c^2}+\frac{1}{a'^2}-\frac{1}{b'^2}-\frac{1}{c'^2}=0$

(d) $\frac{1}{a^2}+\frac{1}{b^2}+\frac{1}{c^2}-\frac{1}{a'^2}-\frac{1}{b'^2}-\frac{1}{c'^2}=0$

33 The distance of the point (1, – 5, 9) from the plane $x - y + z = 5$ measured along a straight line $x = y = z$ is → JEE Mains 2016

(a) $3\sqrt{10}$ (b) $10\sqrt{3}$ (c) $\frac{10}{\sqrt{3}}$ (d) $\frac{20}{3}$

34 Distance between two parallel planes $2x + y + 2z = 8$ and $4x + 2y + 4z + 5 = 0$ is → JEE Mains 2013

(a) $\frac{3}{2}$ (b) $\frac{5}{2}$ (c) $\frac{7}{2}$ (d) $\frac{9}{2}$

35 Find the planes bisecting the acute angle between the planes $x - y + 2z + 1 = 0$ and $2x + y + z + 2 = 0$.

(a) $x + z - 1 = 0$ (b) $x + z + 1 = 0$
(c) $x - z - 1 = 0$ (d) None of these

36 The angle between the lines $\frac{x-1}{2} = \frac{y-2}{1} = \frac{z+3}{-2}$ and the plane $x + y + 4 = 0$ is

(a) 0° (b) 30° (c) 45° (d) 90°

37 If the angle between the line $x = \frac{y-1}{2} = \frac{z-3}{\lambda}$ and the plane $x + 2y + 3z = 4$ is $\cos^{-1}\left(\sqrt{\frac{5}{14}}\right)$, then λ is equal to → AIEEE 2011

(a) $\frac{3}{2}$ (b) $\frac{2}{5}$ (c) $\frac{5}{3}$ (d) $\frac{2}{3}$

38 The distance between the line $\mathbf{r} = 2\mathbf{i} - 2\mathbf{j} + 3\mathbf{k} + \lambda(\mathbf{i} - \mathbf{j} + 4\mathbf{k})$ and the plane $\mathbf{r} \cdot (\mathbf{i} + 5\mathbf{j} + \mathbf{k}) = 5$ is

(a) $\frac{10}{3\sqrt{3}}$ (b) $\frac{10}{9}$ (c) $\frac{10}{3}$ (d) $\frac{3}{10}$

39 Consider the following statements.

Statement I If the coordinates of the points A, B, C, D are $(1, 2, 3), (4, 5, 7), (-4, 3, -6)$ and $(2, 9, 2)$ respectively, then the angle between the lines AB and CD is $\frac{\pi}{6}$.

Statement II The straight lines $\frac{x-1}{1} = \frac{y-2}{2} = \frac{z-3}{3}$ and $\frac{x-1}{2} = \frac{y-2}{2} = \frac{z-3}{-2}$ are parallel.

Choose the correct option.

(a) Statement I is true (b) Statement II is true
(c) Both statements are true (d) Both statements are false

40 Consider a line is perpendicular to the plane, then DR's of plane is proportional to the line.

Statement I The lines $\frac{x-1}{1} = \frac{y}{-1} = \frac{z+1}{1}$ and $\frac{x-2}{1} = \frac{y+1}{2} = \frac{z}{3}$ are coplanar and equation of the plane containing them is $5x + 2y - 3z - 8 = 0$

Statement II The line $\frac{x-2}{1} = \frac{y+1}{2} = \frac{z}{3}$ is perpendicular to the plane $3x + 6y + 9z - 8 = 0$ and parallel to the plane $x + y - z = 0$

(a) Statement I is true, Statement II is true; Statement II is a correct explanation of Statement I
(b) Statement I is true; Statement II is true; Statement II is not a correct explanation of Statement I
(c) Statement I is true; Statement II is false
(d) Statement I is false; Statement II is true

41 **Statement I** The point A (1, 0, 7) is the mirror image of the point B (1, 6, 3) in the line $\frac{x}{1} = \frac{y-1}{2} = \frac{z-2}{3}$.

Statement II The line $\frac{x}{1} = \frac{y-1}{2} = \frac{z-2}{3}$ bisects the line segment joining A (1, 0, 7) and B (1, 6, 3). → AIEEE 2011

(a) Statement I is true, Statement II is true; Statement II is a correct explanation for Statement I
(b) Statement I is true, Statement II is true; Statement II is not a correct explanation for Statement I
(c) Statement I is true; Statement II is false
(d) Statement I is false; Statement II is true

42 Consider the following statements

Statement I If the line drawn from the point $(-2, -1, -3)$ meets a plane at right angle at the point $(1, -3, 3)$, then the equation of plane is $3x - 2y + 6z - 27 = 0$.

Statement II The equation of the plane through the points $(2, 1, 0), (3, -2, -2)$ and $(3, 1, 7)$ is $7x + 3y - z = 17$.

Choose the correct option.

(a) Statement I is true (b) Statement II is true
(c) Both statements are true (d) Both statements are false

43 **Statement I** The point A (3, 1, 6) is the mirror image of the point B (1, 3, 4) in the plane $x - y + z = 5$.

Statement II The plane $x - y + z = 5$ bisects the line segment joining A (3, 1, 6) and B (1, 3, 4). → AIEEE 2010

(a) Statement I is true, Statement II is true; Statement II is a correct explanation for Statement I
(b) Statement I is true, Statement II is true; Statement II is not a correct explanation for Statement I
(c) Statement I is true; Statement II is false
(d) Statement I is false; Statement II is true

44 **Statement I** A point on the straight line $2x + 3y - 4z = 5$ and $3x - 2y + 4z = 7$ can be determined by taking $x = k$ and then solving the two equations for y and z, where k is any real number.

Statement II If $c' \neq kc$, then the straight line $ax + by + cz + d = 0, kax + kby + c'z + d' = 0$, does not intersect the plane $z = \alpha$, where α is any real number.

(a) Statement I is true, Statement II is true; Statement II is a correct explanation of Statement I
(b) Statement I is true; Statement II is true; Statement II is not a correct explanation of Statement I
(c) Statement I is true; Statement II is false
(d) Statement I is false; Statement II is true

45 Consider the lines

$$L_1 : \frac{x+1}{3} = \frac{y+2}{1} = \frac{z+1}{2},\ L_2 : \frac{x-2}{1} = \frac{y+2}{2} = \frac{z-3}{3}.$$

Statement I The distance of the point (1, 1, 1) from the plane passing through the point (−1, −2, −1) and whose normal is perpendicular to both the lines L_1 and L_2 is $\frac{13}{5\sqrt{3}}$.

Statement II The unit vector perpendicular to both the lines L_1 and L_2 is $\frac{-\mathbf{i}-7\mathbf{j}+5\mathbf{k}}{5\sqrt{3}}$.

(a) Statement I is true, Statement II is true; Statement II is a correct explanation of Statement I
(b) Statement I is true; Statement II is true; Statement II is not a correct explanation of Statement I
(c) Statement I is true; Statement II is false
(d) Statement I is false; Statement II is true

DAY PRACTICE SESSION 2

PROGRESSIVE QUESTIONS EXERCISE

1 The direction ratios of normal to the plane through (1, 0, 0), (0, 1, 0) which makes an angle $\frac{\pi}{4}$ with the plane $x + y = 3$ are

(a) $1, \sqrt{2}, 1$ (b) $1, 1, \sqrt{2}$ (c) 1, 1, 2 (d) $\sqrt{2}, 1, 1$

2 The angle between the lines whose direction cosines are given by $2l - m + 2n = 0, lm + mn + nl = 0$, is

(a) $\frac{\pi}{6}$ (b) $\frac{\pi}{4}$ (c) $\frac{\pi}{3}$ (d) $\frac{\pi}{2}$

3 If L_1 is the line of intersection of the planes $2x - 2y + 3z - 2 = 0$, $x - y + z + 1 = 0$ and L_2 is the line of intersection of the planes $x + 2y - z - 3 = 0$, $3x - y + 2z - 1 = 0$, then the distance of the origin from the plane, containing the lines L_1 and L_2 is → **JEE Mains 2018**

(a) $\frac{1}{4\sqrt{2}}$ (b) $\frac{1}{3\sqrt{2}}$ (c) $\frac{1}{2\sqrt{2}}$ (d) $\frac{1}{\sqrt{2}}$

4 If the plane $x + y + z = 1$ is rotated through an angle 90° about its line of intersection with the plane $x - 2y + 3z = 0$, the new position of the plane is

(a) $x - 5y + 4z = 1$ (b) $x - 5y + 4z = -1$
(c) $x - 8y + 7z = 2$ (d) $x - 8y + 7z = -2$

5 A variable plane at a distance of 1 unit from the origin cut the coordinate axes at A, B and C. If the centroid $D(x, y, z)$ of ΔABC satisfies the relation $\frac{1}{x^2} + \frac{1}{y^2} + \frac{1}{z^2} = k$, then k is equal to

(a) 3 (b) 1 (c) $\frac{1}{3}$ (d) 9

6 The lines $x = py + q, z = ry + s$ and $x = p'y + q'$, $z = r'y + s'$ are perpendicular, if → **NCERT Exemplar**

(a) $pr + p'r' + 1 = 0$ (b) $pp' + rr' + 1 = 0$
(c) $pr + p'r' = 0$ (d) $pp' + rr' = 0$

7 A line with direction cosines proportional to 2, 1, 2 meets each of the lines $x = y + a = z$ and $x + a = 2y = 2z$. The coordinates of each of the points of intersection are given by

(a) $(3a, 3a, 3a), (a, a, a)$ (b) $(3a, 2a, 3a), (a, a, a)$
(c) $(3a, 2a, 3a), (a, a, 2a)$ (d) $(2a, 3a, 3a), (2a, a, a)$

8 A parallelopiped is formed by planes drawn through the points (2, 3, 5) and (5, 9, 7), parallel to the coordinate planes. The length of a diagonal of the parallelopiped is

(a) 7 units (b) $\sqrt{38}$ units
(c) $\sqrt{155}$ units (d) None of these

9 The distance of the point (1, 3, −7) from the plane passing through the point (1, −1, −1) having normal perpendicular to both the lines $\frac{x-1}{1} = \frac{y+2}{-2} = \frac{z-4}{3}$ and $\frac{x-2}{2} = \frac{y+1}{-1} = \frac{z+7}{-1}$, is → **JEE Mains 2017**

(a) $\frac{20}{\sqrt{74}}$ units (b) $\frac{10}{\sqrt{83}}$ units (c) $\frac{5}{\sqrt{83}}$ units (d) $\frac{10}{\sqrt{74}}$ units

10 Find the distance of the plane $x + 2y - z = 2$ from the point (2, −1, 3) as measured in the direction with DR's (2, 2, 1).

(a) 2 (b) −3 (c) −2 (d) 3

11 ΔABC is such that the mid-points of the sides BC, CA and AB are $(l, 0, 0), (0, m, 0), (0, 0, n)$, respectively. Then, $\frac{AB^2 + BC^2 + CA^2}{l^2 + m^2 + n^2}$ is equal to

(a) 2 (b) 4 (c) 8 (d) 16

12 If α, β, γ and δ are the angles between a straight line with the diagonals of a cube, then $\sin^2\alpha + \sin^2\beta + \sin^2\gamma + \sin^2\delta$ is equal to

(a) $\frac{5}{3}$ (b) $\frac{8}{3}$ (c) $\frac{7}{4}$ (d) None of these

13 The equation of the line passing through the points (3, 0, 1) and parallel to the planes $x + 2y = 0$ and $3y - z = 0$, is → **NCERT Exemplar**

(a) $\frac{x-3}{-2} = \frac{y-0}{1} = \frac{z-1}{3}$ (b) $\frac{x-3}{1} = \frac{y-0}{-2} = \frac{z-1}{3}$
(c) $\frac{x-3}{3} = \frac{y-0}{1} = \frac{z-1}{-2}$ (d) None of these

14 The length of the projection of the line segment joining the points (5, −1, 4) and (4, −1, 3) on the plane, $x + y + z = 7$ is → JEE Mains 2018

(a) $\frac{2}{\sqrt{3}}$ (b) $\frac{2}{3}$ (c) $\frac{1}{3}$ (d) $\sqrt{\frac{2}{3}}$

15 Let ABC be a triangle with vertices at points $A(2, 3, 5)$, $B(-1, 3, 2)$ and $C(\lambda, 5, \mu)$ in three dimensional space. If the median through A is equally inclined with the axes, then (λ, μ) is equal to → JEE Mains 2013

(a) (10, 7) (b) (7, 5) (c) (7, 10) (d) (5, 7)

16 A plane passes through the point $(1, -2, 3)$ and is parallel to the plane $2x - 2y + z = 0$. The distance of the point $(-1, 2, 0)$ from the plane, is

(a) 2 (b) 3 (c) 4 (d) 5

17 The equation of the plane through the line intersection of the planes $x + y + z = 1$ and $2x + 3y + 4z = 5$, which is perpendicular to plane $x - y + z = 0$, is

(a) $x + 2y + 3z - 4 = 0$
(b) $5x + 6y + 7z - 8 = 0$
(c) $120x + 144y + 168z - 5 = 0$
(d) $x - z + 2 = 0$

18 Let L be the line of intersection of the planes $2x + 3y + z = 1$ and $x + 3y + 2z = 2$. If L makes an angle α with the positive X-axis, then $\cos \alpha$ is equal to

(a) $\frac{1}{2}$ (b) 1 (c) $\frac{1}{\sqrt{2}}$ (d) $\frac{1}{\sqrt{3}}$

19 If the image of the point $P(1, -2, 3)$ in the plane $2x + 3y - 4z + 22 = 0$ measured parallel to the line $\frac{x}{1} = \frac{y}{4} = \frac{z}{5}$ is Q, then PQ is equal to → JEE Mains 2017

(a) $3\sqrt{5}$ (b) $2\sqrt{42}$ (c) $\sqrt{42}$ (d) $6\sqrt{5}$

20 The image of the line $\frac{x-1}{3} = \frac{y-3}{1} = \frac{z-4}{-5}$ in the plane $2x - y + z + 3 = 0$ is the line → JEE Mains 2014

(a) $\frac{x+3}{3} = \frac{y-5}{1} = \frac{z-2}{-5}$
(b) $\frac{x+3}{-3} = \frac{y-5}{-1} = \frac{z+2}{5}$
(c) $\frac{x-3}{3} = \frac{y+5}{1} = \frac{z-2}{-5}$.
(d) $\frac{x-3}{-3} = \frac{y+5}{-1} = \frac{z-2}{5}$

ANSWERS

SESSION 1	**1.** (a)	**2.** (c)	**3.** (c)	**4.** (c)	**5.** (b)	**6.** (c)	**7.** (b)	**8.** (d)	**9.** (c)	**10.** (c)
	11. (a)	**12.** (d)	**13.** (a)	**14.** (c)	**15.** (d)	**16.** (d)	**17.** (b)	**18.** (c)	**19.** (a)	**20.** (d)
	21. (b)	**22.** (c)	**23.** (c)	**24.** (c)	**25.** (c)	**26.** (a)	**27.** (b)	**28.** (d)	**29.** (c)	**30.** (c)
	31. (a)	**32.** (d)	**33.** (b)	**34.** (c)	**35.** (b)	**36.** (c)	**37.** (d)	**38.** (a)	**39.** (d)	**40.** (b)
	41. (b)	**42.** (c)	**43.** (a)	**44.** (b)	**45.** (a)					
SESSION 2	**1.** (b)	**2.** (d)	**3.** (b)	**4.** (d)	**5.** (d)	**6.** (d)	**7.** (b)	**8.** (a)	**9.** (b)	**10.** (d)
	11. (c)	**12.** (b)	**13.** (a)	**14.** (d)	**15.** (c)	**16.** (d)	**17.** (d)	**18.** (d)	**19.** (b)	**20.** (a)

Hints and Explanations

SESSION 1

1 Since, S divides OG in the ratio $3 : -1$.

Then, $S = \left(\frac{9+3}{2}, \frac{-5+9}{2}, \frac{-3-1}{2}\right)$

$= (6, 2, -2)$

2 A line makes angle θ with X-axis and Z-axis and β with Y-axis.

$\therefore \quad l = \cos\theta,\ m = \cos\beta,\ n = \cos\theta$

$\because \quad l^2 + m^2 + n^2 = 1$

$\therefore \quad \cos^2\theta + \cos^2\beta + \cos^2\theta = 1$

$\Rightarrow \quad 2\cos^2\theta = 1 - \cos^2\beta$

$\Rightarrow \quad 2\cos^2\theta = \sin^2\beta \qquad \ldots(i)$

But it is given that,

$\sin^2\beta = 3\sin^2\theta \qquad \ldots(ii)$

From Eqs. (i) and (ii), we get

$3\sin^2\theta = 2\cos^2\theta$

$\Rightarrow \quad 3(1 - \cos^2\theta) = 2\cos^2\theta$

$\Rightarrow \quad 3 = 5\cos^2\theta$

$\therefore \quad \cos^2\theta = \frac{3}{5}$

3 We know that,

$\cos^2\theta + \cos^2\theta + \cos^2\gamma = 1$

$\Rightarrow \quad \cos^2\gamma = -\cos 2\theta$

$\Rightarrow \quad \cos 2\theta \le 0$

$\therefore \quad \theta \in \left[\frac{\pi}{4}, \frac{\pi}{2}\right]$

4 Projection of a vector on coordinate axes are

$x_2 - x_1,\ y_2 - y_1,\ z_2 - z_1$

$\Rightarrow \quad x_2 - x_1 = 6,\ y_2 - y_1 = -3,$

$z_2 - z_1 = 2$

Now, $\sqrt{(x_2 - x_1)^2 + (y_2 - y_1)^2 + (z_2 - z_1)^2}$

$= \sqrt{36 + 9 + 4} = 7$

So, the DC's of the vector are $\frac{6}{7}, -\frac{3}{7}$ and $\frac{2}{7}$.

5 Given that, the projections of a line segment on the X, Y and Z-axes in $3D$-space are, $lr = 2$, $mr = 3$ and $nr = 6$

$\because (lr)^2 + (mr)^2 + (nr)^2 = (2)^2 + (3)^2 + (6)^2$

$\Rightarrow \quad (l^2 + m^2 + n^2)r^2 = 4 + 9 + 36$

$\Rightarrow \quad r^2 = 49 \Rightarrow r = 7$

6 Let $\mathbf{r}$ be inclined at an angle α to each axis, then $l = m = n = \cos\alpha$

Since, $\quad l^2 + m^2 + n^2 = 1$

$\Rightarrow \quad 3\cos^2\alpha = 1$

If α is acute, then $l = m = n = \frac{1}{\sqrt{3}}$ and

$|\mathbf{r}| = 6$

$\therefore \quad \mathbf{r} = |\mathbf{r}|(l\mathbf{i} + m\mathbf{j} + n\mathbf{k})$

$= 6\left(\frac{1}{\sqrt{3}}\mathbf{i} + \frac{1}{\sqrt{3}}\mathbf{j} + \frac{1}{\sqrt{3}}\mathbf{k}\right)$

$= 2\sqrt{3}(\mathbf{i} + \mathbf{j} + \mathbf{k})$

If α is obtuse, then

$l = m = n = -\frac{1}{\sqrt{3}}$ and $|\mathbf{r}| = 6$

$\therefore \quad \mathbf{r} = |\mathbf{r}|(l\mathbf{i} + m\mathbf{j} + n\mathbf{k})$

$= 6\left(-\frac{1}{\sqrt{3}}\mathbf{i} - \frac{1}{\sqrt{3}}\mathbf{j} - \frac{1}{\sqrt{3}}\mathbf{k}\right)$

$= -2\sqrt{3}(\mathbf{i} + \mathbf{j} + \mathbf{k})$

7 Equation of L_1 is

$\mathbf{r} = 3\mathbf{i} + \lambda(-\mathbf{i} + \mathbf{j} + \mathbf{k})$

$= (3 - \lambda)\mathbf{i} + \lambda\mathbf{j} + \lambda\mathbf{k}$

Equation of L_2 is

$\mathbf{r} = (\mathbf{i} + \mathbf{j}) + \mu(\mathbf{i} + \mathbf{k})$

$= (1 + \mu)\mathbf{i} + \mathbf{j} + \mu\mathbf{k}$

For point of intersection, we get

$\lambda = \mu = 1 \Rightarrow \mathbf{r} = 2\mathbf{i} + \mathbf{j} + \mathbf{k}$

8 Equation of line passing through $(5, 1, a)$ and $(3, b, 1)$ is

$\frac{x-3}{5-3} = \frac{y-b}{1-b} = \frac{z-1}{a-1} \qquad \ldots(i)$

$\left[\because \frac{x - x_1}{x_2 - x_1} = \frac{y - y_1}{y_2 - y_1} = \frac{z - z_1}{z_2 - z_1}\right]$

Point $\left(0, \frac{17}{2}, -\frac{13}{2}\right)$ satisfies Eq. (i), we get

$-\frac{3}{2} = \frac{\frac{17}{2} - b}{1 - b} = \frac{-\frac{13}{2} - 1}{a - 1}$

$\Rightarrow \quad a - 1 = \frac{\left(-\frac{15}{2}\right)}{\left(-\frac{3}{2}\right)} = 5 \Rightarrow a = 6$

Also, $-3(1 - b) = 2\left(\frac{17}{2} - b\right)$

$\Rightarrow 3b - 3 = 17 - 2b \Rightarrow 5b = 20 \Rightarrow b = 4$

9 The given equations of lines can be rewritten as

$\frac{x}{3} = \frac{y}{2} = \frac{z}{-6}$ and $\frac{x}{2} = \frac{y}{-12} = \frac{z}{-3}$

$\therefore$ Angle between the lines is

$\theta = \cos^{-1}\left(\frac{3 \times 2 + 2(-12) - 6(-3)}{\sqrt{3^2 + 2^2 + (-6)^2}\sqrt{(2)^2 + (-12)^2 + (-3)^2}}\right)$

$\left[\because \cos\theta = \frac{a_1 \cdot a_2 + b_2 \cdot b_2 + c_1 \cdot c_2}{\sqrt{a_1^2 + b_1^2 + c_1^2}\sqrt{a_2^2 + b_2^2 + c_2^2}}\right]$

$= \cos^{-1}(0) = 90°$

10 If three edges of the cube are along x, y and z, then diagonal has DR's 1, 1, 1 and edge along X-axis has DR's 1, 0, 0. The angle between them is

$\cos^{-1}\frac{1}{\sqrt{3}} = \tan^{-1}\sqrt{2}$

11 We know that, angle between two lines is

$\cos\theta = \frac{a_1a_2 + b_1b_2 + c_1c_2}{\sqrt{a_1^2 + b_1^2 + c_1^2}\sqrt{a_2^2 + b_2^2 + c_2^2}}$

$l + m + n = 0$

$\Rightarrow \quad l = -(m + n)$

$\Rightarrow \quad (m + n)^2 = l^2$

$\Rightarrow \quad m^2 + n^2 + 2mn = m^2 + n^2$

$[\because l^2 = m^2 + n^2, \text{ given}]$

$\Rightarrow \quad 2mn = 0$

When, $m = 0 \Rightarrow l = -n$

Hence, (l, m, n) is $(1, 0, -1)$.

When $n = 0$, then $l = -m$

Hence, (l, m, n) is $(1, 0, -1)$.

$\therefore \quad \cos\theta = \frac{1 + 0 + 0}{\sqrt{2} \times \sqrt{2}} = \frac{1}{2} \Rightarrow \theta = \frac{\pi}{3}$

12 Since, the line $\frac{x-3}{2} = \frac{y+2}{-1} = \frac{z+4}{3}$ lies in the plane $lx + my - z = 9$, therefore we have $2l - m - 3 = 0$

[$\because$ normal will be perpendicular to the line]

$\Rightarrow \quad 2l - m = 3 \qquad \ldots(i)$

and $\quad 3l - 2m + 4 = 9$

[$\because$ point $(3, -2, -4)$ lies on the plane]

$\Rightarrow \quad 3l - 2m = 5 \qquad \ldots(ii)$

On solving Eqs. (i) and (ii), we get

$l = 1$ and $m = -1$

$\therefore \quad l^2 + m^2 = 2$

13 Let the direction cosine of the line perpendicular to two given lines is (l, m, n), then $l + 2m + 3n = 0$ and $-2l + \frac{m}{2} + \frac{n}{3} = 0$

From the above equation,

$\frac{l}{2 \times \frac{1}{3} - \frac{1}{2} \times 3} = -\frac{m}{-3 \times (-2) - 1 \times \frac{1}{3}}$

$= \frac{n}{1 \times \frac{1}{2} - 2(-2)}$

$\Rightarrow \quad \frac{l^2}{\frac{25}{36}} = \frac{m^2}{\frac{361}{9}} = \frac{n^2}{\frac{81}{4}} = \frac{1}{\frac{25}{36} + \frac{361}{9} + \frac{81}{4}}$

$\therefore \quad l = \sqrt{\frac{25}{2198}},\ m = \frac{19}{\sqrt{2198}},\ n = \sqrt{\frac{729}{2198}}$

14 Let L be foot of perpendicular from $P(0, 2, 3)$ on the line

$$\frac{x-(-3)}{5} = \frac{y-1}{2} = \frac{z-(-4)}{3} = t \quad \ldots(i)$$

Any point on Eq. (i) is
$L(-3+5t, 1+2t, -4+3t)$.
Then, DR's of PL are
$(-3+5t-0, 1+2t-2, -4+3t-3)$
or $(5t-3, 2t-1, 3t-7)$.
Since, PL is perpendicular to Eq. (i), therefore $5(5t-3)+2(2t-1)$
$+3(3t-7) = 0 \Rightarrow t = 1$
So, the coordinate of L is $(2, 3, -1)$.

15 The vector joining the points is $\mathbf{i} - 3\mathbf{j} + \mathbf{k}$. Its projection along the vector $2\mathbf{i} + \mathbf{j} - 2\mathbf{k}$

$$= \frac{|(\mathbf{i}-3\mathbf{j}+\mathbf{k})\cdot(2\mathbf{i}+\mathbf{j}-2\mathbf{k})|}{\sqrt{2^2+1^2+2^2}}$$

$$= \frac{|2-3-2|}{3} = 1$$

16 The given lines are

$$\frac{x+1}{7} = \frac{y+1}{-6} = \frac{z+1}{1}$$

and $\frac{x-3}{1} = \frac{y-5}{-2} = \frac{z-7}{1}$

For line Ist DR's = $(7, -6, 1)$ and it passes through $(-1, -1, -1)$, then equation of given lines (in vector form) is
$\mathbf{r}_1 = -\mathbf{i}-\mathbf{j}-\mathbf{k}+\lambda(7\mathbf{i}-6\mathbf{j}+\mathbf{k})$
Similarly, $\mathbf{r}_2 = 3\mathbf{i}+5\mathbf{j}+7\mathbf{k}+\mu(\mathbf{i}-2\mathbf{j}+\mathbf{k})$
which are of the form $\mathbf{r}_1 = \mathbf{a}_1 + \lambda\mathbf{b}_1$ and $\mathbf{r}_2 = \mathbf{a}_2 + \mu\mathbf{b}_2$ where,
$\mathbf{a}_1 = -\mathbf{i}-\mathbf{j}-\mathbf{k}, \mathbf{b}_1 = 7\mathbf{i}-6\mathbf{j}+\mathbf{k}$
and $\mathbf{a}_2 = 3\mathbf{i}+5\mathbf{j}+7\mathbf{k}, \mathbf{b}_2 = \mathbf{i}-2\mathbf{j}+\mathbf{k}$
Now, $\mathbf{a}_2 - \mathbf{a}_1 = (3\mathbf{i}+5\mathbf{j}+7\mathbf{k})$
$-(-\mathbf{i}-\mathbf{j}-\mathbf{k}) = 4\mathbf{i}+6\mathbf{j}+8\mathbf{k}$

and $\mathbf{b}_1 \times \mathbf{b}_2 = \begin{vmatrix} \mathbf{i} & \mathbf{j} & \mathbf{k} \\ 7 & -6 & 1 \\ 1 & -2 & 1 \end{vmatrix}$

$= \mathbf{i}(-6+2) - \mathbf{j}(7-1) + \mathbf{k}(-14+6)$
$= -4\mathbf{i} - 6\mathbf{j} - 8\mathbf{k}$
$|\mathbf{b}_1 \times \mathbf{b}_2| = \sqrt{(-4)^2+(-6)^2+(-8)^2}$
$= \sqrt{16+36+64}$
$= \sqrt{116} = 2\sqrt{29}$

So, the shortest distance between the given lines

$$d = \left|\frac{(\mathbf{b}_1 \times \mathbf{b}_2)\cdot(\mathbf{a}_2 - \mathbf{a}_1)}{|\mathbf{b}_1 \times \mathbf{b}_2|}\right|$$

$$= \frac{|(-4\mathbf{i}-6\mathbf{j}-8\mathbf{k})\cdot(4\mathbf{i}+6\mathbf{j}+8\mathbf{k})|}{2\sqrt{29}}$$

$$= \frac{|(-4)\times 4+(-6)\times 6+(-8)\times 8|}{2\sqrt{29}}$$

$$= \frac{|-16-36-64|}{2\sqrt{29}} = \frac{116}{2\sqrt{29}}$$

$$= \frac{58}{\sqrt{29}} = 2\sqrt{29} \text{ units}$$

17 Let one vertex of the parallelopiped be at the origin O and three coterminous edges OA, OB and OC be along OX, OY and OZ, respectively. The coordinates of the vertices of the parallelopiped are marked in figure.
The edges which do not meet the diagonal OF are AH, AD and BD and their parallels are BE, CE and CH, respectively.
The vector equation of the diagonal OF is

$$\mathbf{r} = 0 + \lambda(a\mathbf{i} + b\mathbf{j} + c\mathbf{k}) \quad \ldots(i)$$

The vector equation of the edge BD is

$$\mathbf{r} = b\,\mathbf{j} + \mu\, a\mathbf{i} \quad \ldots(ii)$$

We have,
$(a\mathbf{i}+b\mathbf{j}+c\mathbf{k}) \times a\mathbf{i} = ba(\mathbf{j}\times\mathbf{i}) + ca(\mathbf{k}\times\mathbf{i})$
$= -ba\mathbf{k} + ca\mathbf{j}$
$\therefore |(a\mathbf{i}\times b\mathbf{j}\times c\mathbf{k})\times a\mathbf{i}| = \sqrt{b^2a^2+c^2a^2}$
and $\{(a\mathbf{i}+b\mathbf{j}+c\mathbf{k})\times a\mathbf{i}\}\cdot(b\mathbf{j}-\mathbf{0})$
$= (-ba\mathbf{k}+ca\mathbf{j})\cdot b\mathbf{j}$
$= abc$

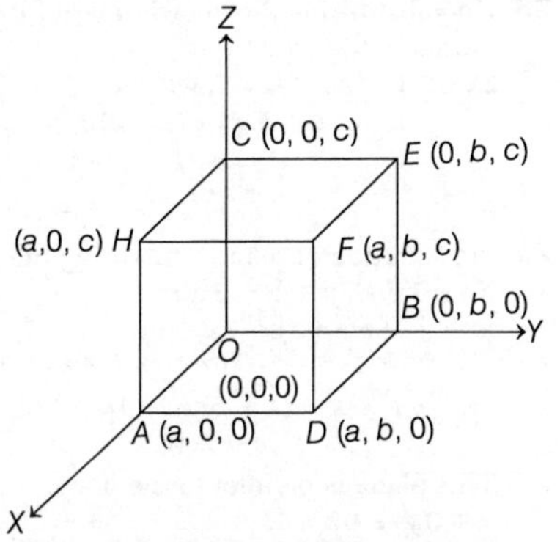

Thus, the shortest distance between Eqs. (i) and (ii) is given by

$$\text{SD} = \frac{|\{(a\mathbf{i}+b\mathbf{j}+c\mathbf{k})\times a\mathbf{i}\}\cdot(b\mathbf{j}-\mathbf{0})|}{|(a\mathbf{i}+b\mathbf{j}+c\mathbf{k})\times a\mathbf{i}|}$$

$$= \frac{abc}{\sqrt{b^2a^2+c^2a^2}} = \frac{bc}{\sqrt{b^2+c^2}}$$

Similarly, it can be shown that the shortest distance between OF and AD is $\frac{ca}{\sqrt{a^2+c^2}}$ and that between OF and AH is $\frac{ab}{\sqrt{a^2+b^2}}$.

18 Let $L_1 : \frac{x-1}{2} = \frac{y+1}{3} = \frac{z-1}{4} = p$

and $L_2 : \frac{x-3}{1} = \frac{y-k}{2} = \frac{z-0}{1} = q$

$\Rightarrow$ Any point P on line L_1 is of type $P(2p+1, 3p-1, 4p+1)$ and any point Q on line L_2 is of type $Q(q+3, 2q+k, q)$
Since, L_1 and L_2 are intersecting each other, hence both point P and Q should coincide at the point of intersection, i.e. corresponding coordinates of P and Q should be same.

$$2p+1 = q+3,\ 4p+1 = q$$

and $3p - 1 = 2q + k$
On solving $2p+1 = q+3$ and $4p+1 = q$, we get the values of p and q as

$$p = \frac{-3}{2} \text{ and } q = -5$$

On substituting the values of p and q in the third equation $3p - 1 = 2q + k$, we get

$$\therefore 3\left(\frac{-3}{2}\right) - 1 = 2(-5) + k \Rightarrow k = \frac{9}{2}$$

19 The given straight line can be rewritten as

$$\frac{x-1}{1} = \frac{y+3}{-\lambda} = \frac{z-1}{\lambda} = s$$

and $\frac{x-0}{1} = \frac{y-1}{2} = \frac{z-2}{-2} = t$

These two lines are coplanar, if

$$\begin{vmatrix} x_1-x_2 & y_1-y_2 & z_1-z_2 \\ l_1 & m_1 & n_1 \\ l_2 & m_2 & n_2 \end{vmatrix} = 0$$

$$\Rightarrow \begin{vmatrix} 1-0 & -3-1 & 1-2 \\ 1 & -\lambda & \lambda \\ 1 & 2 & -2 \end{vmatrix} = 0$$

$$\Rightarrow \begin{vmatrix} 1 & -4 & -1 \\ 1 & -\lambda & \lambda \\ 1 & 2 & -2 \end{vmatrix} = 0$$

$\Rightarrow 1(2\lambda - 2\lambda) + 4(-2-\lambda) - 1(2+\lambda) = 0$
$\Rightarrow -8 - 4\lambda - 2 - \lambda = 0$
$\Rightarrow \quad -10 = 5\lambda \Rightarrow \lambda = -2$

20 Given equation of line is

$$\frac{x-2}{3} = \frac{y+1}{4} = \frac{z-2}{12} = \lambda \quad [\text{say}] \ldots(i)$$

and equation of plane is

$$x - y + z = 16 \quad \ldots(ii)$$

Any point on the line (i) is
$(3\lambda+2, 4\lambda-1, 12\lambda+2)$
Let this point be point of intersection of the line and plane.
$\therefore (3\lambda+2) - (4\lambda-1) + (12\lambda+2) = 16$
$\Rightarrow \quad 11\lambda + 5 = 16$
$\Rightarrow \quad 11\lambda = 11 \Rightarrow \lambda = 1$
$\therefore$ Point of intersection is $(5, 3, 14)$.
Now, distance between the points $(1, 0, 2)$ and $(5, 3, 14)$
$= \sqrt{(5-1)^2+(3-0)^2+(14-2)^2}$
$= \sqrt{16+9+144} = \sqrt{169} = 13$

21 $\because \cos^2\alpha + \cos^2\beta + \cos^2\gamma = 1$
$\Rightarrow \cos^2 45^\circ + \cos^2 60^\circ + \cos^2\gamma = 1$
$\Rightarrow \cos^2\gamma = 1 - \frac{1}{2} - \frac{1}{4} = 1 - \frac{3}{4} = \frac{1}{4}$
$\Rightarrow \cos\gamma = \frac{1}{2}$

$\therefore$ Direction Ratio's of normal to the plane is $< \cos 45°; \cos 60°, \frac{1}{2} >$

$$= < \frac{1}{\sqrt{2}}, \frac{1}{2}, \frac{1}{2} >$$

$\therefore$ Equation of plane passing through $(\sqrt{2}, -1, 1)$ is

$$\frac{1}{\sqrt{2}}(x - \sqrt{2}) + \frac{1}{2}(y + 1) + \frac{1}{2}(z - 1) = 0$$

$$\Rightarrow 2(x - \sqrt{2}) + \sqrt{2}(y + 1) + \sqrt{2}(z - 1) = 0$$

$$\Rightarrow \sqrt{2}(x - \sqrt{2}) + (y + 1) + (z - 1) = 0$$

$$\Rightarrow \sqrt{2}x - 2 + y + 1 + z - 1 = 0$$

$$\Rightarrow \sqrt{2}x + y + z = 2$$

22 Let foot of perpendicular $Q(x, y, z)$ from $O(0, 0, 0)$

$$\frac{x-0}{4} = \frac{y-0}{-3} = \frac{z-0}{1} = -\frac{\{4(0) - 3(0) + 1(0) + 13\}}{4^2 + 3^2 + 1^2}$$

$$\Rightarrow \frac{x}{4} = \frac{y}{-3} = \frac{z}{1} = \frac{-13}{26} = -\frac{1}{2}$$

$$x = -2, y = \frac{3}{2}, z = -\frac{1}{2}$$

$$\therefore Q\left(-2, \frac{3}{2}, -\frac{1}{2}\right)$$

$$\therefore PQ = \sqrt{(-1+2)^2 + \left(1 - \frac{3}{2}\right)^2 + \left(-6 + \frac{1}{2}\right)^2}$$

$$= \sqrt{1 + \frac{1}{4} + \frac{121}{4}} = \frac{\sqrt{126}}{2} = 3\sqrt{\frac{7}{2}}$$

23 Required equation of plane is

$$\begin{vmatrix} x-1 & y-1 & z-1 \\ -3 & -3 & 1 \\ 0 & -2 & 1 \end{vmatrix} = 0$$

$$\Rightarrow x - 3y - 6z + 8 = 0$$

Since, the intercepts are $8, \frac{8}{3}, \frac{8}{6}$.

So, their sum is 12.

24 Equation of plane through three points $(2, 2, 1)$, $(3, 0, 1)$ and $(4, -1, 0)$ is

$[(\mathbf{r} - \mathbf{i} + 2\mathbf{j} + \mathbf{k})] \cdot [(\mathbf{i} - 2\mathbf{j}) \times (\mathbf{i} - \mathbf{j} - \mathbf{k})] = 0$

i.e. $\mathbf{r} \cdot (2\mathbf{i} + \mathbf{j} + \mathbf{k}) = 7$

or $2x + y + z - 7 = 0$ ...(i)

Equation of line through $(3, -4, -5)$ and $(2, -3, 1)$ is

$$\frac{x-3}{-1} = \frac{y+4}{1} = \frac{z+5}{6} \quad \text{...(ii)}$$

Any point on line (ii) is $(-\lambda + 3, \lambda - 4, 6\lambda - 5)$. This point lies on plane (i).

Therefore,

$$2(-\lambda + 3) + (\lambda - 4) + (6\lambda - 5) - 7 = 0$$

$$\Rightarrow \lambda = 2$$

Hence, the required point is $(1, -2, 7)$.

25 Since, the vertices of the tetrahedron are $(0, 0, 0)$, $(3, 0, 0)$, $(0, 2, 0)$ and $(0, 0, 6)$.

$\therefore$ Volume of tetrahedron

$$= \frac{1}{6}\begin{vmatrix} 3 & 0 & 0 \\ 0 & 2 & 0 \\ 0 & 0 & 6 \end{vmatrix} = 6$$

26 The equation of the plane parallel to the plane $3x - 4y + 5z = 4$ is

$3x - 4y + 5z + k = 0$

Since, this plane passes through $(2, 1, 5)$. On substituting coordinates $(2, 1, 5)$, we get

$3 \times 2 - 4 \times 1 + 5 \times 5 + k = 0 \Rightarrow k = -27$

So, the equation of plane is

$3x - 4y + 5z - 27 = 0$.

27 Since Q is the image of P, therefore PQ is perpendicular to the plane $x - 2y + 5z = 6$.

$\therefore$ Required equation of line is

$$\frac{x-2}{1} = \frac{y-3}{-2} = \frac{z-4}{5}$$

28 On substituting the coordinates of the points in the equation $2x + 3y - 2z - k = 0$, we get

$$(2 + 6 - 6 - k)(4 - 3 - k) < 0$$

$$\Rightarrow (k - 1)(k - 2) < 0$$

$$\therefore 1 < k < 2$$

29 Let equation of plane containing the lines $2x - 5y + z = 3$ and $x + y + 4z = 5$ be $(2x - 5y + z - 3) + \lambda(x + y + 4z - 5) = 0$

$$\Rightarrow (2 + \lambda)x + (\lambda - 5)y + (4\lambda + 1)z - 3 - 5\lambda = 0 \quad \text{...(i)}$$

This plane is parallel to the plane $x + 3y + 6z = 1$.

$$\therefore \frac{2+\lambda}{1} = \frac{\lambda - 5}{3} = \frac{4\lambda + 1}{6}$$

On taking first two equalities, we get

$$6 + 3\lambda = \lambda - 5 \Rightarrow 2\lambda = -11$$

$$\Rightarrow \lambda = -\frac{11}{2}$$

On taking last two equalities, we get

$$6\lambda - 30 = 3 + 12\lambda$$

$$\Rightarrow -6\lambda = 33 \Rightarrow \lambda = -\frac{11}{2}$$

So, the equation of required plane is

$$\left(2 - \frac{11}{2}\right)x + \left(\frac{-11}{2} - 5\right)y + \left(-\frac{44}{2} + 1\right)z - 3 + 5 \times \frac{11}{2} = 0$$

$$\Rightarrow -\frac{7}{2}x - \frac{21}{2}y - \frac{42}{2}z + \frac{49}{2} = 0$$

$$\Rightarrow x + 3y + 6z - 7 = 0$$

30 Intersection of two planes is

$(x + 2y - 3) + \lambda(y - 2z + 1)$

$\Rightarrow x + (2 + \lambda)y - 2\lambda z + \lambda - 3 = 0$

$\therefore 1(1) + 2(2 + \lambda) + 0(-2\lambda) = 0 \Rightarrow \lambda = -\frac{5}{2}$

$$\therefore (x + 2y - 3) - \frac{5}{2}(y - 2z + 1) = 0$$

$$\Rightarrow 2x + 4y - 6 - 5y + 10z - 5 = 0$$

$$\Rightarrow 2x - y + 10z - 11 = 0$$

$$\Rightarrow 2x - y + 10z = 11$$

31 Given, a plane $P: x - 2y + 2z - 5 = 0$

Equation of family of planes parallel to the given plane P is

$$Q: x - 2y + 2z + d = 0$$

Also, perpendicular distance of Q from origin is 1 unit.

$$\Rightarrow \left|\frac{0 - 2(0) + 2(0) + d}{\sqrt{1^2 + 2^2 + 2^2}}\right| = 1$$

$$\Rightarrow \left|\frac{d}{3}\right| = 1 \Rightarrow d = \pm 3$$

Hence, the required equation of the plane parallel to P and at unit distance from origin is $x - 2y + 2z \pm 3 = 0$.

32 Consider OX, OY, OZ and Ox, Oy, Oz are two systems of rectangular axes.

Let their corresponding equations of plane be

$$\frac{x}{a} + \frac{y}{b} + \frac{z}{c} = 1 \quad \text{...(i)}$$

and $$\frac{x}{a'} + \frac{y}{b'} + \frac{z}{c'} = 1 \quad \text{...(ii)}$$

Length of perpendicular from origin to Eqs. (i) and (ii) must be same.

$$\therefore \frac{1}{\sqrt{\frac{1}{a^2} + \frac{1}{b^2} + \frac{1}{c^2}}} = \frac{1}{\sqrt{\frac{1}{a'^2} + \frac{1}{b'^2} + \frac{1}{c'^2}}}$$

$$\Rightarrow \frac{1}{a^2} + \frac{1}{b^2} + \frac{1}{c^2} = \frac{1}{a'^2} + \frac{1}{b'^2} + \frac{1}{c'^2}$$

$$\Rightarrow \frac{1}{a^2} + \frac{1}{b^2} + \frac{1}{c^2} - \frac{1}{a'^2} - \frac{1}{b'^2} - \frac{1}{c'^2} = 0$$

33 Equation of PQ is

$$\frac{x-1}{1} = \frac{y+5}{1} = \frac{z-9}{1} = \lambda$$

So, $x = \lambda + 1$, $y = \lambda - 5$ and $z = \lambda + 9$ lies on the plane $x - y + z = 5$.

$$\Rightarrow \lambda + 1 - \lambda + 5 + \lambda + 9 = 5$$

$$\therefore \lambda = -10$$

So, the coordinate of Q is $(-9, -15, -1)$ and coordinate of P is $(1, -5, 9)$.

$$\therefore |PQ| = \sqrt{(10)^2 + (10)^2 + (10)^2} = 10\sqrt{3}$$

34 Given planes are,

$$2x + y + 2z - 8 = 0$$

and $$2x + y + 2z + \frac{5}{2} = 0$$

Distance between two planes

$$= \frac{|c_1 - c_2|}{\sqrt{a^2 + b^2 + c^2}} = \left|\frac{-8 - \frac{5}{2}}{\sqrt{2^2 + 1^2 + 2^2}}\right|$$

$$= \frac{\frac{21}{2}}{3} = \frac{7}{2}$$

35 Now, $a_1 a_2 + b_1 b_2 + c_1 c_2 = 2 - 1 + 2 > 0$.
The acute angle bisecting plane is
$x - y + 2z + 1 = -(2x + y + z + 2)$
i.e. $x + z + 1 = 0$

36 DR's of line are 2, 1, −2 and DR's of normal to the plane are 1, 1, 0.
$\therefore$ Therefore, their DC's are $\frac{2}{3},\frac{1}{3},-\frac{2}{3}$ and $\frac{1}{\sqrt{2}},\frac{1}{\sqrt{2}},0$ respectively.
Now, let θ be the angle b/w line and the plane, then
$$\cos(90^\circ - \theta) = \frac{2}{3}\cdot\frac{1}{\sqrt{2}} + \frac{1}{3}\cdot\frac{1}{\sqrt{2}} + \left(-\frac{2}{3}\right)\cdot 0$$
$$\Rightarrow \quad \sin\theta = \frac{1}{\sqrt{2}} \Rightarrow \theta = 45^\circ$$

37 Angle between straight line $\mathbf{r} = \mathbf{a} + \lambda\,\mathbf{b}$ and plane $\mathbf{r}\cdot\mathbf{n} = d$,
$$\sin\theta = \frac{\mathbf{b}\cdot\mathbf{n}}{|\mathbf{b}|\,|\mathbf{n}|}$$
$$\therefore \quad \sin\theta = \frac{(\mathbf{i} + 2\,\mathbf{j} + \lambda\,\mathbf{k})\cdot(\mathbf{i} + 2\,\mathbf{j} + 3\,\mathbf{k})}{\sqrt{1 + 4 + \lambda^2}\,\sqrt{1 + 4 + 9}}$$
$$\Rightarrow \sin\theta = \frac{5 + 3\lambda}{\sqrt{\lambda^2 + 5}\cdot\sqrt{14}}$$
Given, $\cos\theta = \sqrt{\frac{5}{14}}$
$$\therefore \quad \sin\theta = \frac{3}{\sqrt{14}}$$
$$\Rightarrow \quad \frac{3}{\sqrt{14}} = \frac{5 + 3\lambda}{\sqrt{\lambda^2 + 5}\cdot\sqrt{14}}$$
$$\Rightarrow 9(\lambda^2 + 5) = 9\lambda^2 + 30\lambda + 25$$
$$\Rightarrow 9\lambda^2 + 45 = 9\lambda^2 + 30\lambda + 25$$
$$\Rightarrow \quad 30\lambda = 20$$
$$\therefore \quad \lambda = \frac{2}{3}$$

38 Clearly, given line is parallel to the plane.
Given point on the line is $A(2, -2, 3)$ and a point on the plane is $B(0, 0, 5)$
$\therefore$ $\mathbf{AB} = (2 - 0)\mathbf{i} + (-2 - 0)\mathbf{j} + (3 - 5)\mathbf{k}$
$= 2\mathbf{i} - 2\mathbf{j} - 2\mathbf{k}$
Now, required distance = Projection of $\mathbf{AB}$ or $\mathbf{i} + 5\mathbf{j} + \mathbf{k}$
$$= \left|\frac{(2\mathbf{i} - 2\mathbf{j} - 2\mathbf{k})\cdot(\mathbf{i} + 5\mathbf{j} + \mathbf{k})}{\sqrt{1 + 25 + 1}}\right|$$
$$= \left|\frac{2 - 10 - 2}{\sqrt{27}}\right|$$
$$= \frac{10}{3\sqrt{3}}$$

39 I. $\mathbf{AB} = \mathbf{OB} - \mathbf{OA}$
$= (4\mathbf{i}+5\mathbf{j}+7\mathbf{k}) - (\mathbf{i}+2\mathbf{j}+3\mathbf{k})$
$= 3\mathbf{i}+3\mathbf{j}+4\mathbf{k}$
$\mathbf{CD} = \mathbf{OD} - \mathbf{OC}$
$= (2\mathbf{i}+9\mathbf{j}+2\mathbf{k}) - (-4\mathbf{i}+3\mathbf{j} - 6\mathbf{k})$
$= 6\mathbf{i}+6\mathbf{j}+8\mathbf{k}$
$$\because \cos\theta = \left|\frac{\mathbf{AB}\cdot\mathbf{CD}}{|\mathbf{AB}||\mathbf{CD}|}\right|$$
$$\cos\theta = \left|\frac{(3\mathbf{i}+3\mathbf{j}+4\mathbf{k})\cdot(6\mathbf{i}+6\mathbf{j}+8\mathbf{k})}{\sqrt{3^2 + 3^2 + 4^2}\sqrt{6^2 + 6^2 + 8^2}}\right|$$
$$= \left|\frac{18 + 18 + 32}{\sqrt{34}\sqrt{136}}\right| = \left|\frac{68}{2\times 34}\right| = 1$$
$\therefore$ $\theta = \cos^{-1} 1 = 0$
Hence, Statement I is false.
II. Given, $a_1 = 1, b_1 = 2, c_1 = 3$
$a_2 = 2, b_2 = 2, c_2 = -2$
$$\frac{a_1}{a_2} = \frac{1}{2}, \frac{b_1}{b_2} = \frac{2}{2}$$
$$\because \quad \frac{a_1}{a_2} \neq \frac{b_1}{b_2}$$
Hence, given lines are not parallel.
Therefore, Statement II is false.

40 **Statement I** The equation of the plane containing them is
$$\begin{vmatrix} x - 1 & y & z + 1 \\ 1 & -1 & 1 \\ 1 & 2 & 3 \end{vmatrix} = 0$$
$$\Rightarrow \quad -(5x + 2y - 3z - 8) = 0$$
Statement II Here, $\frac{1}{3} = \frac{2}{6} = \frac{3}{9}$
$$\Rightarrow \quad \frac{1}{3} = \frac{1}{3} = \frac{1}{3}$$
and $1(1) + 2(1) + 3(-1) = 0$
$\therefore$ Statement II is true.

41 Since, mid-point on AB is $M(1, 3, 5)$.
which lies on $\frac{x}{1} = \frac{y - 1}{2} = \frac{z - 2}{3}$.

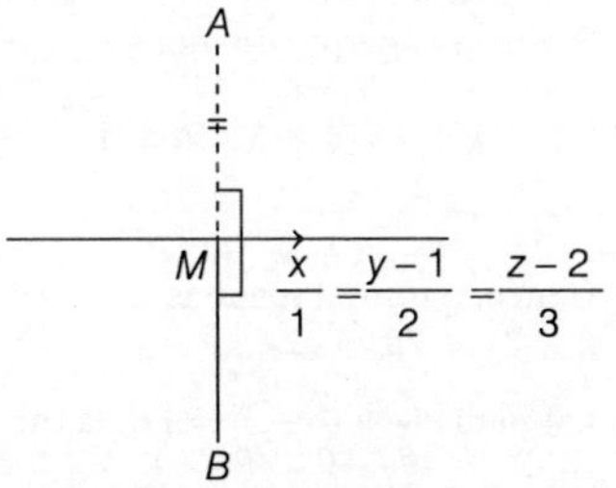

$$\therefore \quad \frac{1}{1} = \frac{3 - 1}{2} = \frac{5 - 2}{3}$$
$$\Rightarrow \quad 1 = 1 = 1$$
Hence, Statement II is true.
Also, direction ratio of AB is
$(1 - 1, 6 - 0, 3 - 7) = (0, 6, -4)$ ...(i)
and direction ratio of straight line is
$(1, 2, 3)$ ... (ii)
These two lines are perpendicular, if
$0(1) + 6(2) - 4(3) = 12 - 12 = 0$
Hence, Statement I is true.

42 I. $\mathbf{N} = \mathbf{OB} - \mathbf{OA}$
$= (\mathbf{i} - 3\mathbf{j} + 3\mathbf{k}) - (-2\mathbf{i} - \mathbf{j} - 3\mathbf{k})$
$= 3\mathbf{i} - 2\mathbf{j} + 6\mathbf{k}$

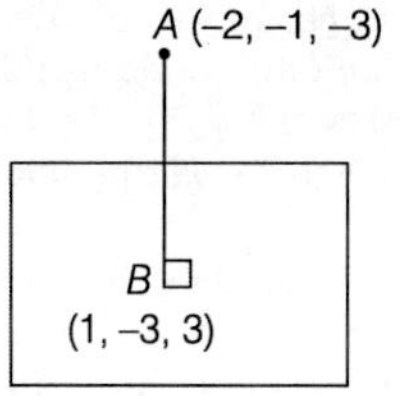

Equation of a plane is given by
$$A(x - x_1) + B(y - y_1) + C(z - z_1) = 0$$
$$\Rightarrow 3(x - 1) + (-2)(y + 3) + 6(z - 3) = 0$$
$$\Rightarrow \quad 3x - 3 - 2y - 6 + 6z - 18 = 0$$
$$\Rightarrow \quad 3x - 2y + 6z - 27 = 0$$
II. Equation of any plane through $(2, 1, 0)$ is
$a(x - 2) + b(y - 1) + c(z - 0) = 0$ ...(i)
Since, it passes through the points $(3,-2,-2)$ and $(3,1,7)$. Then, we get
$a - 3b - 2c = 0$ ...(ii)
and $a + 0b + 7c = 0$ ...(iii)
On solving Eqs.(ii) and (iii) by cross-multiplication, we get
$a = 7\lambda, b = 3\lambda, c = -\lambda$
On substituting the value of a, b, c in Eq. (i), we get
$$7\lambda(x - 2) + 3\lambda(y - 1) - \lambda z = 0$$
$$\Rightarrow \quad 7x + 3y - z = 17$$
This is the required equation of the plane.

43 The image of the point (3, 1, 6) with respect to the plane $x - y + z = 5$ is
$$\frac{x - 3}{1} = \frac{y - 1}{-1} = \frac{z - 6}{1}$$
$$= \frac{-2(3 - 1 + 6 - 5)}{1 + 1 + 1}$$
$$\Rightarrow \quad \frac{x - 3}{1} = \frac{y - 1}{-1} = \frac{z - 6}{1} = -2$$
$$\Rightarrow \quad x = 3 - 2 = 1$$
$$y = 1 + 2 = 3$$
$$z = 6 - 2 = 4$$
which shows that Statement I is true.
We observe that the line segment joining the points $A(3, 1, 6)$ and $B(1, 3, 4)$ has direction ratios $2, -2, 2$ which are proportional to $1, -1, 1$ the direction ratios of the normal to the plane. Hence, Statement II is true.
Thus, the Statements I and II are true and Statement II is correct explanation of Statement I.

44 **Statement I** $3y - 4z = 5 - 2k$
$-2y + 4z = 7 - 3k$
$\therefore$ $x = k, y = 12 - 5k$ and
$z = \frac{31 - 13k}{4}$ is a point on the line for all real values of k.
Statement I is true.
Statement II Direction ratios of the straight line are

$< bc' - kbc, kac - ac', 0 >$ and direction ratios of normal the plane are $< 0, 0, 1 >$.

Now, $0 \times (bc' - kbc) + 0 \times (kac - ac') + 1 \times 0 = 0$

Hence, the straight line is parallel to the plane.

45 **Statement II** Lines L_1 and L_2 are parallel to the vectors $\mathbf{a} = 3\mathbf{i} + \mathbf{j} + 2\mathbf{k}$ and $\mathbf{b} = \mathbf{i} + 2\mathbf{j} + 3\mathbf{k}$, respectively. The unit vector perpendicular to both L_1 and L_2 is

$$\frac{\mathbf{a} \times \mathbf{b}}{|\mathbf{a} \times \mathbf{b}|} = \frac{-\mathbf{i} - 7\mathbf{j} + 5\mathbf{k}}{\sqrt{1 + 49 + 25}} = \frac{-\mathbf{i} - 7\mathbf{j} + 5\mathbf{k}}{5\sqrt{3}}$$

Hence, Statement II is true.

Statement I Plane is

$-(x + 1) - 7(y + 2) + 5(z + 1) = 0$, whose distance from $(1, 1, 1)$ is $\frac{13}{5\sqrt{3}}$.

Hence, Statement I is true.

Thus, statement I is true, statement II is true; Statement II is a correct explanation of Statement I.

SESSION 2

1 Let the equation of plane in intercept form is $\frac{x}{a} + \frac{y}{b} + \frac{z}{c} = 1$.

Since, it passes through the point (1, 0, 0) and (0,1,0).

$\therefore$ Equation of plane is $\frac{x}{1} + \frac{y}{1} + \frac{z}{c} = 1$

DR's of normal are $1, 1, \frac{1}{c}$ and of given plane are 1,1,0.

$$\text{Now, } \cos\frac{\pi}{4} = \frac{1 \cdot 1 + 1 \cdot 1 + \frac{1}{c} \cdot 0}{\left(\sqrt{\frac{1}{c^2} + 2}\right)\sqrt{2}}$$

$$\Rightarrow \frac{1}{\sqrt{2}} = \frac{2}{\left(\sqrt{\frac{1}{c^2} + 2}\right)\sqrt{2}}$$

$$\Rightarrow \frac{1}{c^2} + 2 = 4 \Rightarrow c^2 = \frac{1}{2}$$

$$\therefore \quad c = \frac{1}{\sqrt{2}}$$

So, the DR's of normal are $1, 1, \sqrt{2}$.

2 On eliminating m from given equations, we get

$$2(l + n)^2 + nl = 0$$

[$\because$ put $m = 2l + 2n$]

$\Rightarrow \quad (2l + n)(l + 2n) = 0$

$\Rightarrow \quad n = -2l \Rightarrow m = -2l$

or $\quad l = -2n$

$\Rightarrow \quad m = -2n$

The DR's are $1, -2, -2$ and $-2, -2, 1$.

Now, $1(-2) - 2(-2) - 2(1) = 0$

Hence, lines are perpendicular.

So, angle between them is $\pi/2$.

3 L_1 is the line of intersection of the plane $2x - 2y + 3z - 2 = 0$ and $x - y + z + 1 = 0$ and L_2 is the line of intersection of the plane $x + 2y - z - 3 = 0$ and $3x - y + 2z - 1 = 0$

Since L_1 is parallel to

$$\begin{vmatrix} \hat{i} & \hat{j} & \hat{k} \\ 2 & -2 & 3 \\ 1 & -1 & 1 \end{vmatrix} = \hat{i} + \hat{j}$$

L_2 is parallel to

$$\begin{vmatrix} \hat{i} & \hat{j} & \hat{k} \\ 1 & 2 & -1 \\ 3 & -1 & 2 \end{vmatrix} = 3\hat{i} - 5\hat{j} - 7\hat{k}$$

Also, L_2 passes through $\left(\frac{5}{7}, \frac{8}{7}, 0\right)$.

[put $z = 0$ in last two planes]

So, equation of plane is

$$\begin{vmatrix} x - \frac{5}{7} & y - \frac{8}{7} & z \\ 1 & 1 & 0 \\ 3 & -5 & -7 \end{vmatrix} = 0$$

$\Rightarrow \quad 7x - 7y + 8z + 3 = 0$

Now, perpendicular distance from origin is

$$\left|\frac{3}{\sqrt{7^2 + 7^2 + 8^2}}\right| = \frac{3}{\sqrt{162}} = \frac{1}{3\sqrt{2}}$$

4 The new position of plane is

$$x - 2y + 3z + \lambda(x + y + z - 1) = 0$$

$\Rightarrow (1 + \lambda)x + (\lambda - 2)y + (\lambda + 3)z - \lambda = 0$

Since, it is perpendicular to $x + y + z - 1 = 0$.

$\therefore 1 + \lambda + \lambda - 2 + \lambda + 3 = 0$

$\Rightarrow \lambda = -\frac{2}{3}$

Hence, required plane is $x - 8y + 7z = -2$.

5 Let plane $\frac{x}{a} + \frac{y}{b} + \frac{z}{c} = 1$ cuts the axes at $A(a, 0, 0)$, $B(0, b, 0)$ and $C(0, 0, c)$.

Centroid of plane ABC is $D\left(\frac{a}{3}, \frac{b}{3}, \frac{c}{3}\right)$.

Distance of the plane from the origin

$$d = \frac{1}{\sqrt{\frac{1}{a^2} + \frac{1}{b^2} + \frac{1}{c^2}}} = 1 \quad \text{[given]}$$

$\Rightarrow \quad 1 = \frac{1}{a^2} + \frac{1}{b^2} + \frac{1}{c^2}$

$\therefore \quad D(x, y, z)$

$\Rightarrow \quad a = 3x, b = 3y, c = 3z$

$\Rightarrow \quad \frac{1}{x^2} + \frac{1}{y^2} + \frac{1}{z^2} = 9$

Hence, the value of k is 9.

6 Given lines are

$$x = py + q, z = ry + s$$

$\Rightarrow \quad \frac{x - q}{p} = y = \frac{z - s}{r}$

and $\quad x = p'y + q', z = r'y + s'$

$\Rightarrow \quad \frac{x - q'}{p'} = y = \frac{z - s'}{r'}$

Two lines $\frac{x - x_1}{a_1} = \frac{y - y_1}{b_1} = \frac{z - z_1}{c_1}$

and $\frac{x - x_2}{a_2} = \frac{y - y_2}{b_2} = \frac{z - z_2}{c_2}$ are perpendicular, if $a_1 a_2 + b_1 b_2 + c_1 c_2 = 0$

$\therefore$ Given lines are perpendicular, if

$$pp' + rr' = 0$$

7 Let the equation of line AB be

$$\frac{x - 0}{1} = \frac{y + a}{1} = \frac{z - 0}{1} = k \quad \text{[say]}$$

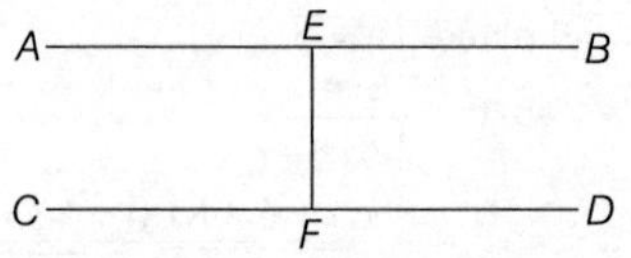

Any point on the line is $F(k, k - a, k)$.

Also, the equation of other line CD is

$$\frac{x + a}{2} = \frac{y - 0}{1} = \frac{z - 0}{1} = \lambda \quad \text{[say]}$$

Any point on the line is $E(2\lambda - a, \lambda, \lambda)$.

Direction ratios of EF are $[(k - 2\lambda + a), (k - a - \lambda), (k - \lambda)]$.

Since, it is given that direction ratios of EF are proportional to 2, 1, 2.

$$\therefore \frac{k - 2\lambda + a}{2} = \frac{k - \lambda - a}{1} = \frac{k - \lambda}{2}$$

On solving first and second fractions, we get

$$k - 2\lambda + a = 2k - 2\lambda - 2a$$

$\Rightarrow \quad k = 3a \quad$ ...(i)

On solving second and third fractions, we get

$2k - 2\lambda - 2a = k - \lambda$

$\Rightarrow \quad k - \lambda = 2a$

$\Rightarrow \quad \lambda = 3a - 2a$

[from Eq. (i)]

$\therefore \lambda = a$

Hence, coordinates of E are $(3a, 2a, 3a)$ and coordinates of F are (a, a, a).

8 A parallelopiped is formed by planes drawn through the points (2, 3, 5) and (5, 9, 7), parallel to the coordinate planes.

Let a, b and c be the lengths of edges, then

$$a = 5 - 2 = 3, b = 9 - 3 = 6$$

and $c = 7 - 5 = 2$

So, the length of diagonal of a parallelopiped

$= \sqrt{a^2 + b^2 + c^2}$
$= \sqrt{9 + 36 + 4}$
$= \sqrt{49} = 7$ units

9 Given, equations of lines are

$$\frac{x-1}{1} = \frac{y+2}{-2} = \frac{z-4}{3}$$

and $\frac{x-2}{2} = \frac{y+1}{-1} = \frac{z+7}{-1}$

Let $\mathbf{n}_1 = \hat{\mathbf{i}} - 2\hat{\mathbf{j}} + 3\hat{\mathbf{k}}$

and $\mathbf{n}_2 = 2\hat{\mathbf{i}} - \hat{\mathbf{j}} - \hat{\mathbf{k}}$

$\therefore$ Any vector $\mathbf{n}$ perpendicular to both $\mathbf{n}_1, \mathbf{n}_2$ is given by

$$\mathbf{n} = \mathbf{n}_1 \times \mathbf{n}_2$$

$$\Rightarrow \mathbf{n} = \begin{vmatrix} \hat{\mathbf{i}} & \hat{\mathbf{j}} & \hat{\mathbf{k}} \\ 1 & -2 & 3 \\ 2 & -1 & -1 \end{vmatrix}$$

$$= 5\hat{\mathbf{i}} + 7\hat{\mathbf{j}} + 3\hat{\mathbf{k}}$$

$\therefore$ Equation of a plane passing through $(1, -1, -1)$ and perpendicular to $\mathbf{n}$ is given by

$$5(x-1) + 7(y+1) + 3(z+1) = 0$$

$$\Rightarrow 5x + 7y + 3z + 5 = 0$$

$\therefore$ Required distance

$$= \left| \frac{5 + 21 - 21 + 5}{\sqrt{5^2 + 7^2 + 3^2}} \right|$$

$$= \frac{10}{\sqrt{83}} \text{ units}$$

10 Consider the line through $(2, -1, 3)$ with DC's $\frac{2}{3}, \frac{2}{3}, \frac{1}{3}$ is

$$\frac{x-2}{2/3} = \frac{y+1}{2/3} = \frac{z-3}{1/3} = r \quad \text{[say]}$$

$\therefore x = 2 + \frac{2r}{3}, y = -1 + \frac{2r}{3}, z = 3 + \frac{r}{3}$

Since, it lies on the plane $x + 2y - z = 2$.

$\therefore 2 + \frac{2r}{3} - 2 + \frac{4r}{3} - 3 - \frac{r}{3} = 2$

$\Rightarrow r = 3$

11 Given, mid-points of sides are $D\ (l, 0, 0)$, $E\ (0, m, 0)$ and $F\ (0, 0, n)$

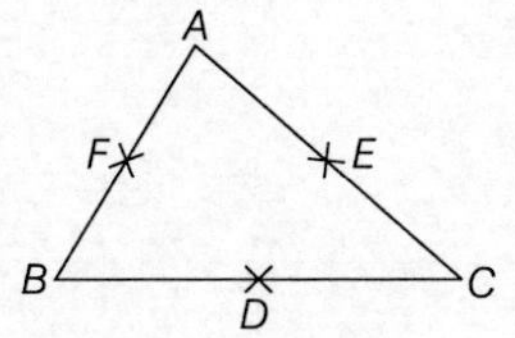

Also, $EF^2 = \frac{BC^2}{4}$ [by mid-point theorem]

$\Rightarrow BC^2 = 4(m^2 + n^2)$

Similarly, $AB^2 = 4(l^2 + m^2)$

and $CA^2 = 4(l^2 + n^2)$

$\therefore \frac{AB^2 + BC^2 + CA^2}{l^2 + m^2 + n^2} = 8$

12 $\because \cos^2\alpha + \cos^2\beta + \cos^2\gamma + \cos^2\delta = \frac{4}{3}$

and $\cos^2\alpha = 1 - \sin^2\alpha$, similarly for all other angles.

$\therefore \cos^2\alpha + \cos^2\beta + \cos^2\gamma + \cos^2\delta$
$= 4 - (\sin^2\alpha + \sin^2\beta + \sin^2\gamma + \sin^2\delta)$

$\Rightarrow \sin^2\alpha + \sin^2\beta + \sin^2\gamma + \sin^2\delta = \left(4 - \frac{4}{3}\right) = \frac{8}{3}$

13 Let a, b and c be the direction ratios of the required line.

Then, its equation is

$$\frac{x-3}{a} = \frac{y-0}{b} = \frac{z-1}{c} \quad \text{...(i)}$$

Since, Eq. (i) is parallel to the planes $x + 2y + 0z = 0$ and $0x + 3y - z = 0$. Therefore, normal to the plane is perpendicular to the line.

$\therefore a(1) + b(2) + c(0) = 0$

and $a(0) + b(3) + c(-1) = 0$

On solving these two equations by cross-multiplication, we get

$$\frac{a}{(2)(-1) - (0)(3)} = \frac{b}{(0)(0) - (1)(-1)} = \frac{c}{(1)(3) - (0)(2)}$$

$$\Rightarrow \frac{a}{-2} = \frac{b}{1} = \frac{c}{3} = \lambda \quad \text{[say]}$$

$\Rightarrow a = -2\lambda, b = \lambda$ and $c = 3\lambda$

On substituting the values of a, b and c in Eq. (i), we get the equation of the required line as

$$\frac{x-3}{-2} = \frac{y-0}{1} = \frac{z-1}{3}$$

14 Key idea *length of projection of the line segment joining* $\mathbf{a}_1$ *and* $\mathbf{a}_2$ *on the plane* $\mathbf{r} \cdot \mathbf{n} = d$ *is* $\left| \frac{(\mathbf{a}_2 - \mathbf{a}_1) \times \mathbf{n}}{|\mathbf{n}|} \right|$

Length of projection the line segment joining the points $(5, -1, 4)$ and $(4, -1, 3)$ on the plane $x + y + z = 7$ is

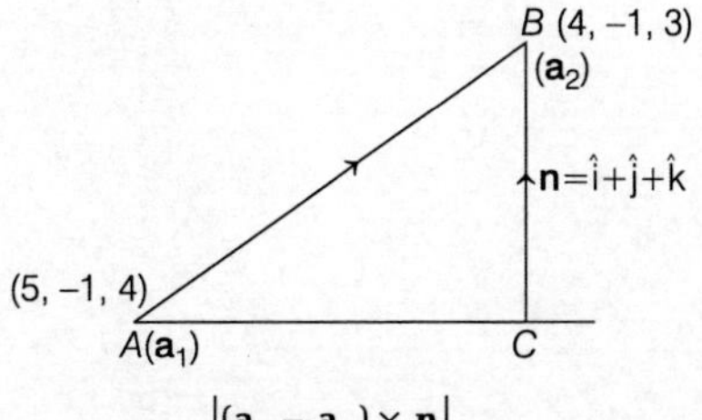

$$AC = \left| \frac{(\mathbf{a}_2 - \mathbf{a}_1) \times \mathbf{n}}{|\mathbf{n}|} \right|$$

$$= \frac{|(-\hat{i} - \hat{k}) \times (\hat{i} + \hat{j} + \hat{k})|}{|i + j + k|}$$

$$AC = \frac{|\hat{i} - \hat{k}|}{\sqrt{3}}$$

$$\Rightarrow AC = \frac{\sqrt{2}}{\sqrt{3}} = \sqrt{\frac{2}{3}}$$

Alternative Method

Clearly, DR's of AB are $4 - 5, -1 + 1, 3 - 4$, i.e. $-1, 0, -1$ and DR's of normal to plane are $1, 1, 1$.

Now, let θ be the angle between the line and plane, then θ is given by

$$\sin\theta = \frac{|-1 + 0 - 1|}{\sqrt{(-1)^2 + (-1)^2}\sqrt{1^2 + 1^2 + 1^2}}$$

$$= \frac{2}{\sqrt{2}\sqrt{3}} = \sqrt{\frac{2}{3}}$$

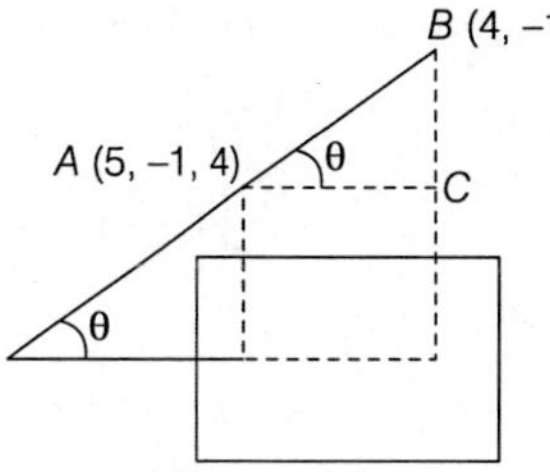

Plane : $x + y + z = 7$

$\Rightarrow \sin\theta = \sqrt{\frac{2}{3}} \Rightarrow \cos\theta = \sqrt{1 - \sin^2\theta}$

$= \sqrt{1 - \frac{2}{3}} = \frac{1}{\sqrt{3}}$

Clearly, length of projection

$= AB\cos\theta = \sqrt{2}\frac{1}{\sqrt{3}} \quad [\because AB = \sqrt{2}]$

$= \sqrt{\frac{2}{3}}$

15 Centroid of ΔABC,

$$G = \left(\frac{2 - 1 + \lambda}{3}, \frac{3 + 3 + 5}{3}, \frac{5 + 2 + \mu}{3} \right)$$

$$= \left(\frac{1 + \lambda}{3}, \frac{11}{3}, \frac{7 + \mu}{3} \right)$$

Since, median is always passes through centroid and they are equally inclined.

$\therefore \frac{1 + \lambda}{3} - 2 = \frac{11}{3} - 3 = \frac{7 + \mu}{3} - 5$

$\Rightarrow \frac{\lambda - 5}{3} = \frac{2}{3} = \frac{\mu - 8}{3}$

$\Rightarrow \lambda = 7, \mu = 10$

16 Let parallel plane be $2x - 2y + z + \lambda = 0$. It passes through $(1, -2, 3)$.

$\therefore \lambda = -9$

The distance of $(-1, 2, 0)$ from the plane $2x - 2y + z - 9 = 0$ is $\left| \frac{-2 - 4 - 9}{3} \right| = 5$

17 Equation of any plane through the intersection of planes can be written as

$x + y + z - 1 + \lambda(2x + 3y + 4z - 5) = 0$

$\Rightarrow (1 + 2\lambda)x + (1 + 3\lambda)y + (1 + 4\lambda)z - 1 - 5\lambda = 0$ ...(i)

The direction ratios, a_1, b_1, c_1, of the plane are $(1+2\lambda), (3\lambda+1)$ and $(4\lambda+1)$.
The plane in Eq. (i) is perpendicular to $x - y + z$ = direction ratios, a_2, b_2, c_2 are 1,–1 and 1.
Since, the planes are perpendicular.

$$\therefore \quad a_1a_2 + b_1b_2 + c_1c_2 = 0$$

$$\Rightarrow 1(1+2\lambda) - 1(1+3\lambda) + 1(1+4\lambda) = 0$$

$$\Rightarrow \quad 1 + 2\lambda - 1 - 3\lambda + 1 + 4\lambda = 0$$

$$\Rightarrow \quad 3\lambda = -1 \Rightarrow \lambda = -\frac{1}{3}$$

On substituting this value of λ in Eq. (i), we obtain the required plane as

$$\left(1-\frac{2}{3}\right)x + \left(1-\frac{3}{3}\right)y + \left(1-\frac{4}{3}\right)z - 1 + \frac{5}{3} = 0$$

$$\Rightarrow \quad \frac{1}{3}x - \frac{1}{3}z + \frac{2}{3} = 0$$

$$\Rightarrow \quad x - z + 2 = 0$$

This is the required equation of the plane.

18 The two normal vectors are
$\mathbf{m} = 2\mathbf{i} + 3\mathbf{j} + \mathbf{k}$ and $\mathbf{n} = \mathbf{i} + 3\mathbf{j} + 2\mathbf{k}$

The line L is along, $\mathbf{m} \times \mathbf{n} = \begin{vmatrix} \mathbf{i} & \mathbf{j} & \mathbf{k} \\ 2 & 3 & 1 \\ 1 & 3 & 2 \end{vmatrix}$

$= 3(\mathbf{i} - \mathbf{j} + \mathbf{k})$
and DC's of X-axis are $(1, 0, 0)$.

$$\therefore \cos\alpha = \frac{3(\mathbf{i} - \mathbf{j} + \mathbf{k}) \cdot (\mathbf{i})}{\sqrt{3^2(1+1+1)}\sqrt{1}} = \frac{1}{\sqrt{3}}$$

19 Any line parallel to $\frac{x}{1} = \frac{y}{4} = \frac{z}{5}$ and passing through $P(1, -2, 3)$ is

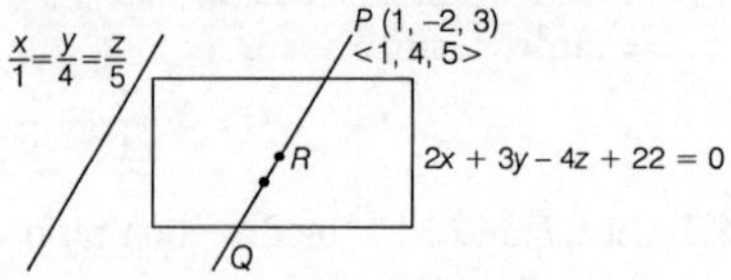

$$\frac{x-1}{1} = \frac{y+2}{4} = \frac{z-3}{5} = \lambda \quad \text{(say)}$$

Any point on above line can be written as $(\lambda + 1, 4\lambda - 2, 5\lambda + 3)$.

$\therefore$ Coordinates of R are $(\lambda + 1, 4\lambda - 2, 5\lambda + 3)$.

Since, point R lies on the above plane.

$$\therefore \quad 2(\lambda+1) + 3(4\lambda - 2) - 4(5\lambda + 3) + 22 = 0$$

$$\Rightarrow \quad \lambda = 1$$

So, point R is $(2, 2, 8)$.
Now,

$$PR = \sqrt{(2-1)^2 + (2+2)^2 + (8-3)^2} = \sqrt{42}$$

$$\therefore \quad PQ = 2PR = 2\sqrt{42}$$

20 Here, plane, line and its image are parallel to each other. So, find any point on the normal to the plane from which the image line will be passed and then find equation of image line.
Here, plane and line are parallel to each other. Equation of normal to the plane through the point $(1, 3, 4)$ is

$$\frac{x-1}{2} = \frac{y-3}{-1} = \frac{z-4}{1} = k \quad \text{[say]}$$

Any point on this normal is $(2k+1, -k+3, 4+k)$.
Then,

$$\left(\frac{2k+1+1}{2}, \frac{3-k+3}{2}, \frac{4+k+4}{2}\right) \text{ lies}$$

on plane.

$$\Rightarrow \quad 2(k+1) - \left(\frac{6-k}{2}\right) + \left(\frac{8+k}{2}\right) + 3 = 0$$

$$\Rightarrow \quad k = -2$$

Hence, point through which this image pass is

$$(2k+1, 3-k, 4+k)$$

i.e. $[2(-2)+1, 3+2, 4-2] = (-3, 5, 2)$

Hence, equation of image line is

$$\frac{x+3}{3} = \frac{y-5}{1} = \frac{z-2}{-5}$$

DAY THIRTY THREE

Unit Test 5

(Vectors and 3D Geometry)

1 The equation of plane perpendicular to $2x + 6y + 6z = 1$ and passing through the points (2, 2,1) and (9, 3, 6), is

(a) $3x + 4y + 5z - 9 = 0$
(b) $3x + 4y - 5z + 9 = 0$
(c) $3x + 4y - 5z - 9 = 0$
(d) $3x + 4y + 5z + 9 = 0$

2 Let **a**, **b** and **c** be the unit vectors such that **a** and **b** are mutually perpendicular and **c** is equally inclined to **a** and **b** at an angle θ. If $\mathbf{c} = x\mathbf{a} + y\mathbf{b} + z(\mathbf{a} \times \mathbf{b})$, then

(a) $z^2 = 1 - 2x^2$
(b) $z^2 = 1 - x^2 + y^2$
(c) $z^2 = 1 + 2y^2$
(d) None of the above

3 If **a**, **b** and **c** are three non-coplanar vectors such that $\mathbf{a} + \mathbf{b} + \mathbf{c} = \alpha\mathbf{d}$ and $\mathbf{b} + \mathbf{c} + \mathbf{d} = \beta\mathbf{a}$, then $\mathbf{a} + \mathbf{b} + \mathbf{c} + \mathbf{d}$ is equal to

(a) $\mathbf{0}$
(b) $\alpha\mathbf{a}$
(c) β
(d) $(\alpha + \beta)\mathbf{c}$

4 The plane $ax + by = 0$ is rotated through an angle α about its line of intersection with the plane $z = 0$. Then, the equation to the plane in new position is

(a) $ax - by \pm z\sqrt{a^2 + b^2}\cot\alpha = 0$
(b) $ax + by \pm z\sqrt{a^2 + b^2}\cot\alpha = 0$
(c) $ax - by \pm z\sqrt{a^2 + b^2}\tan\alpha = 0$
(d) $ax + by \pm z\sqrt{a^2 + b^2}\tan\alpha = 0$

5 If the axes are rectangular, the distance from the point (3, 4, 5) to the point, where the line $\frac{x-3}{1} = \frac{y-4}{2} = \frac{z-5}{2}$ meets the plane $x + y + z = 17$ is

(a) 1
(b) 2
(c) 3
(d) None of these

6 Let $A(3, 2, 0)$, $B(5, 3, 2)$, $C(-9, 6, -3)$ are three points forming a triangle. If AD, the bisector of $\angle BAC$ meets BC in D, then coordinates of D are

(a) $\left(-\frac{19}{8}, \frac{57}{16}, \frac{17}{16}\right)$
(b) $\left(\frac{19}{8}, -\frac{57}{16}, \frac{17}{16}\right)$
(c) $\left(\frac{19}{8}, \frac{57}{16}, -\frac{17}{16}\right)$
(d) $\left(\frac{19}{8}, \frac{57}{16}, \frac{17}{16}\right)$

7 Vectors **a** and **b** are inclined at an angle $\theta = 120°$. If $|\mathbf{a}| = 1$, $|\mathbf{b}| = 2$, then $[(\mathbf{a} + 3\mathbf{b}) \times (3\mathbf{a} - \mathbf{b})]^2$ is equal to

(a) 300 (b) 325 (c) 275 (d) 225

8 A point moves so that the sum of the squares of its distances from the six faces of a cube given by $x = \pm 1$, $y = \pm 1$, $z = \pm 1$ is 10 units. The locus of the point is

(a) $x^2 + y^2 + z^2 = 1$
(b) $x^2 + y^2 + z^2 = 2$
(c) $x + y + z = 1$
(d) $x + y + z = 2$

9 If $(\mathbf{a} \times \mathbf{b})^2 + (\mathbf{a} \cdot \mathbf{b})^2 = 144$ and $|\mathbf{a}| = 4$, then $|\mathbf{b}|$ is equal to

(a) 3 (b) 8 (c) 12 (d) 16

10 The values of x for which the angle between $\mathbf{a} = 2x^2\mathbf{i} + 4x\mathbf{j} + \mathbf{k}$, $\mathbf{b} = 7\mathbf{i} - 2\mathbf{j} + x\mathbf{k}$ is obtuse, is

(a) $x > 1/2$ or $x < 0$
(b) $0 < x < 1/2$
(c) $1/2 < x < 15$
(d) None of these

11 Let $\mathbf{a} = 3\mathbf{i} + 2\mathbf{k}$ and $\mathbf{b} = 2\mathbf{j} + \mathbf{k}$. If **c** is a unit vector, then the maximum value of the vector triple product $[\mathbf{a}\ \mathbf{b}\ \mathbf{c}]$, is

(a) $\sqrt{61}$
(b) $\sqrt{59}$
(c) $\sqrt{3} \cdot \sqrt{36}$
(d) None of these

12 The ratio of lengths of diagonals of the parallelogram constructed on the vectors $\mathbf{a} = 3\mathbf{p} - \mathbf{q}$, $\mathbf{b} = \mathbf{p} + 3\mathbf{q}$ is (given that $|\mathbf{p}| = |\mathbf{q}| = 2$ and the angle between **p** and **q** is $\frac{\pi}{3}$)

(a) $\sqrt{7} : \sqrt{3}$ (b) $\sqrt{3} : \sqrt{2}$ (c) $\sqrt{5} : \sqrt{7}$ (d) $\sqrt{5} : \sqrt{3}$

13 Let **p**, **q** and **r** be three mutually perpendicular vectors of the same magnitude. If a vector x satisfies equation $\mathbf{p} \times \{(\mathbf{x}-\mathbf{q}) \times \mathbf{p}\} + \mathbf{q} \times \{(\mathbf{x}-\mathbf{r}) \times \mathbf{q}\} + \mathbf{r} \times \{(\mathbf{x}-\mathbf{p}) \times \mathbf{r}\} = \mathbf{0}$ Then, **x** is given by

(a) $\frac{1}{2}(\mathbf{p}+\mathbf{q}-2\mathbf{r})$ (b) $\frac{1}{2}(\mathbf{p}+\mathbf{q}+\mathbf{r})$
(c) $\frac{1}{3}(2\mathbf{p}+\mathbf{q}-\mathbf{r})$ (d) None of these

14 A line with direction cosines proportional to 2, 1, 2 meets each of the lines given by the equation $x = y + 2 = z$; $x + 2 = 2y = 2z$.

The coordinates of the point of intersection are

(a) (6, 4, 6), (2, 4, 2) (b) (6, 6, 6), (2, 6, 2)
(c) (6, 4, 6), (2, 2, 0) (d) None of these

15 The vector **B** satisfying the vector equation $\mathbf{A}+\mathbf{B}=\mathbf{a}$, $\mathbf{A}\times\mathbf{B}=\mathbf{b}$ and $\mathbf{A}\cdot\mathbf{a}=1$, where **a** and **b** are given vectors is

(a) $\frac{(\mathbf{b}\times\mathbf{a})+\mathbf{a}\,(a^2-1)}{a^2}$ (b) $\frac{(\mathbf{a}\times\mathbf{b})+\mathbf{a}}{\mathbf{a}}$
(c) $\frac{\mathbf{a}\,(a^2-1)+\mathbf{b}(b^2-1)}{a^2}$ (d) None of these

16 The vector **c**, directed along the bisectors of the angle between the vectors $\mathbf{a} = 7\mathbf{i} - 4\mathbf{j} - 4\mathbf{k}$, $\mathbf{b} = -2\mathbf{i} - \mathbf{j} + 2\mathbf{k}$ and $|\mathbf{c}| = 5\sqrt{6}$ is

(a) $\pm\frac{5}{3}(\mathbf{i}-7\mathbf{j}+2\mathbf{k})$ (b) $\pm\frac{5}{3}(5\mathbf{i}+5\mathbf{j}+2\mathbf{k})$
(c) $\pm\frac{5}{3}(\mathbf{i}+7\mathbf{j}+2\mathbf{k})$ (d) $\pm\frac{5}{3}(-5\mathbf{i}+5\mathbf{j}+2\mathbf{k})$

17 Let **a**, **b** and **c** be three non-zero and non-coplanar vectors and **p**, **q** and **r** be three vectors given by $\mathbf{p}=\mathbf{a}+\mathbf{b}-2\mathbf{c}$, $\mathbf{q}=3\mathbf{a}-2\mathbf{b}+\mathbf{c}$ and $\mathbf{r}=\mathbf{a}-4\mathbf{b}+2\mathbf{c}$. If the volume of the parallelopiped determined by **a**, **b** and **c** is V_1 and the volume of the parallelopiped determined by **p**, **q** and **r** is V_2, then $V_2 : V_1$ is equal to

(a) 7 : 1 (b) 3 : 1
(c) 11 : 1 (d) None of these

18 A non-zero vector **a** is parallel to the line of intersection of the plane determined by the vectors $\mathbf{i}, \mathbf{i}+\mathbf{j}$ and the plane determined by the vectors $\mathbf{i}-\mathbf{j}, \mathbf{i}+\mathbf{k}$. Then, the angle between **a** and the vector $\mathbf{i}-2\mathbf{j}+2\mathbf{k}$ is

(a) $\frac{\pi}{6}$ (b) $\frac{\pi}{4}$
(c) $\frac{\pi}{3}$ (d) $\frac{\pi}{2}$

19 Let **a**, **b** and **c** be three vectors having magnitudes 1, 1 and 2, respectively. If $\mathbf{a}\times(\mathbf{a}\times\mathbf{c})+\mathbf{b}=\mathbf{0}$, then the acute angle between **a** and **c** is

(a) $\frac{\pi}{6}$ (b) $\frac{\pi}{4}$
(c) $\frac{\pi}{3}$ (d) None of these

20 If **a**, **b** and **c** are unit coplanar vectors, then the scalar triple product $[2\mathbf{a}-\mathbf{b}\;\; 2\mathbf{b}-\mathbf{c}\;\; 2\mathbf{c}-\mathbf{a}]$ is equal to

(a) 0 (b) 1 (c) $-\sqrt{3}$ (d) $\sqrt{3}$

21 Point (α, β, γ) lies on the plane $x + y + z = 2$. Let $\mathbf{a} = \alpha\mathbf{i} + \beta\mathbf{j} + \gamma\mathbf{k}$, $\mathbf{k}\times(\mathbf{k}\times\mathbf{a}) = \mathbf{0}$. Then, γ is equal to

(a) 0 (b) 1
(c) 2 (d) $\frac{1}{2}$

22 If the four plane faces of a tetrahedron are represented by the equation $\mathbf{r}\cdot(l\mathbf{i}+m\mathbf{j})=0$, $\mathbf{r}\cdot(m\mathbf{j}+n\mathbf{k})=0$, $\mathbf{r}\cdot(n\mathbf{k}+p\mathbf{i})=0$ and $\mathbf{r}\cdot(l\mathbf{i}+m\mathbf{j}+n\mathbf{k})=p$, then the volume of the tetrahedron is

(a) $\frac{p^3}{6\,lmn}$ (b) $\frac{2p^3}{3\,lmn}$
(c) $\frac{3\,p^3}{lmn}$ (d) $\frac{6p^3}{lmn}$

23 If a variable plane forms a tetrahedron of constant volume $64\,k^3$ with the coordinate planes, then the locus of the centroid of the tetrahedron is

(a) $xyz = k^3$ (b) $xyz = 2k^3$
(c) $xyz = 12k^3$ (d) $xyz = 6k^3$

24 The line through $\mathbf{i}+3\mathbf{j}+2\mathbf{k}$ and perpendicular to the line $\mathbf{r}=(\mathbf{i}+2\mathbf{j}-\mathbf{k})+\lambda\,(2\mathbf{i}+\mathbf{j}+\mathbf{k})$ and $\mathbf{r}=(2\mathbf{i}+6\mathbf{j}+\mathbf{k})+\mu\,(\mathbf{i}+2\mathbf{j}+3\mathbf{k})$ is

(a) $\mathbf{r}=(\mathbf{i}+2\mathbf{j}-\mathbf{k})+\lambda\,(-\mathbf{i}+5\mathbf{j}-3\mathbf{k})$
(b) $\mathbf{r}=(\mathbf{i}+3\mathbf{j}+2\mathbf{k})+\lambda\,(\mathbf{i}-5\mathbf{j}+3\mathbf{k})$
(c) $\mathbf{r}=(\mathbf{i}+3\mathbf{j}+2\mathbf{k})+\lambda\,(\mathbf{i}+5\mathbf{j}+3\mathbf{k})$
(d) None of the above

25 The orthogonal projection A' of the point A with position vector (1, 2, 3) on the plane $3x - y + 4z = 0$, is

(a) $(-1, 3, -1)$ (b) $\left(-\frac{1}{2}, \frac{5}{2}, 1\right)$
(c) $\left(\frac{1}{2}, -\frac{5}{2}, -1\right)$ (d) $(6, -7, -5)$

26 The equation of the plane containing the points $A(1, 0, 1)$ and $B(3, 1, 2)$ and parallel to the line joining the origin to the point $C(1, -1, 2)$ is

(a) $x + y - z = 0$ (b) $x + y + z = 0$
(c) $x - y + z = 0$ (d) $x - y - z = 0$

27 The planes $3x - y + z + 1 = 0$ and $5x + y + 3z = 0$ intersect in the line PQ. The equation of the plane through the point (2, 1, 4) and perpendicular to PQ is

(a) $x + y - 2z = 5$ (b) $x + y - 2z = -5$
(c) $x + y + 2z = 5$ (d) $x + y + 2z = -5$

28 The line of intersection of the planes $\mathbf{r}\cdot(3\mathbf{i}-\mathbf{j}+\mathbf{k})=1$ and $\mathbf{r}\cdot(\mathbf{i}+4\mathbf{j}-2\mathbf{k})=2$ is parallel to the vector

(a) $-2\mathbf{i}+7\mathbf{j}+13\mathbf{k}$ (b) $-2\mathbf{i}-7\mathbf{j}+13\mathbf{k}$
(c) $2\mathbf{i}+7\mathbf{j}-13\mathbf{k}$ (d) None of these

29 A line with direction cosines proportional to 2, 1, 2 meets each of the lines $x = y + a = z$ and $x + a = 2y = 2z$. The coordinates of each of the points of intersection are given by

(a) $(3a, 2a, 3a), (a, a, 2a)$ (b) $(3a, 3a, 3a), (a, a, a)$
(c) $(3a, 2a, 3a), (a, a, a)$ (d) None of these

30 The sides of a parallelogram are $3\mathbf{i}+4\mathbf{j}-6\mathbf{k}$ and $2\mathbf{i}-3\mathbf{j}+5\mathbf{k}$. The unit vector parallel to one of the diagonals is

(a) $\frac{5\mathbf{i}+\mathbf{j}-\mathbf{k}}{\sqrt{27}}$ (b) $\frac{4\mathbf{i}+3\mathbf{j}-2\mathbf{k}}{\sqrt{29}}$
(c) $\frac{\mathbf{i}-2\mathbf{j}+4\mathbf{k}}{\sqrt{21}}$ (d) None of these

31 Let $\mathbf{p}=8\mathbf{i}+6\mathbf{j}$ and $\mathbf{q}$ be two vectors perpendicular to each other in the xy-plane. Then, the vector in the same plane having projections 2 and 4 along $\mathbf{p}$ and $\mathbf{q}$ respectively is

(a) $\pm 3(\mathbf{i}-2\mathbf{j})$ (b) $\pm(\mathbf{i}+2\mathbf{j})$
(c) $\pm 2(2\mathbf{i}-\mathbf{j})$ (d) None of these

32 The equation of the plane containing the line $2x-y+z-3=0$ and $3x+y+z=5$ at a distance of $\frac{1}{\sqrt{6}}$ from the point $(2,1,-1)$ is

(a) $x+y+z-3=0$
(b) $2x-y-z-3=0$
(c) $2x-y+z+3=0$
(d) $62x+29y+19z-105=0$

33 The equation of the plane through the point $(2,-1,-3)$ and parallel to the lines $\frac{x-1}{3}=\frac{y+2}{2}=\frac{z}{-4}$ and $\frac{x}{2}=\frac{y-1}{-3}=\frac{z-2}{2}$ is

(a) $8x+14y+13z+37=0$ (b) $8x-14y+13y+37=0$
(c) $8x+14y-13z+37=0$ (d) None of these

34 The shortest distance between the lines
$\mathbf{r}=-(\mathbf{i}+\mathbf{j}+\mathbf{k})+\lambda(2\mathbf{i}+3\mathbf{j}+4\mathbf{k})$
and $\mathbf{r}=-\mathbf{i}+\mu(3\mathbf{i}+4\mathbf{j}+5\mathbf{k})$ is

(a) 1 (b) $\frac{1}{\sqrt{2}}$ (c) $\frac{1}{\sqrt{3}}$ (d) $\frac{1}{\sqrt{6}}$

35 A tetrahedron has vertices at $O(0,0,0)$, $A(1,2,1)$, $B(2,1,3)$ and $C(-1,1,2)$. The angle between the faces OAB and ABC is

(a) $\cos^{-1}\left(\frac{19}{35}\right)$ (b) $\cos^{-1}\left(\frac{17}{31}\right)$
(c) 30° (d) 90°

36 If $\mathbf{p}=2\mathbf{i}-3\mathbf{j}+3\mathbf{k}$ and $\mathbf{q}=4\mathbf{i}-2\mathbf{j}+\mathbf{k}$ be two vectors and $\mathbf{r}$ is a vector perpendicular to $\mathbf{p}$ and $\mathbf{q}$ and satisfying the condition. $\mathbf{r}(2\mathbf{i}-4\mathbf{j}+2\mathbf{k})=-12$, then $\mathbf{r}$ is equal to

(a) $2\mathbf{i}-\frac{20}{3}\mathbf{j}+16\mathbf{k}$ (b) $\frac{2}{3}(3\mathbf{i}+10\mathbf{j}+8\mathbf{k})$
(c) $\frac{1}{3}(3\mathbf{i}-10\mathbf{j}+8\mathbf{k})$ (d) None of these

37 The direction ratios of a normal to the plane through (2, 0, 0) and (0, 2, 0) that makes an angle $\frac{\pi}{3}$ with the plane $2x+3y=5$ is

(a) $1:1:2$ (b) $1:1:\sqrt{3}$
(c) $\sqrt{2}:1:3$ (d) $1:1:\sqrt{5.7}$

38 The shortest distance between the lines $\frac{x-2}{3}=\frac{y-1}{2}=\frac{z-3}{2}$ and $\frac{x-1}{2}=\frac{y-2}{3}=\frac{z-4}{4}$ is equal to

(a) 1.14 units (b) 2.01 units
(c) 3.16 units (d) None of these

39 The intersecting point of lines, $L_1=\frac{x+1}{-3}=\frac{y-3}{2}=\frac{y+2}{1}$ and $L_2=\frac{x}{1}=\frac{y-7}{-3}=\frac{z+7}{2}$ is

(a) $(-3,2,1)$ (b) $(2,1,-3)$
(c) $(1,-3,2)$ (d) None of these

40 If $\mathbf{a}$ and $\mathbf{b}$ are unit vectos, then the greatest value of $|\mathbf{a}+\mathbf{b}|+|\mathbf{a}-\mathbf{b}|$ is

(a) 2 (b) 4 (c) $2\sqrt{2}$ (d) $\sqrt{2}$

41 If $\mathbf{a}$, $\mathbf{b}$ and $\mathbf{c}$ are non-coplanar vectors and r is a real number, then the vectors $\mathbf{a}+2\mathbf{b}+3\mathbf{c}$, $\lambda\mathbf{b}+4\mathbf{c}$ and $(2\lambda-1)\mathbf{c}$ are non-coplanar for

(a) no value of λ (b) all except one value of λ
(c) all except two values of λ (d) all values of λ

42 Let $\mathbf{a}$, $\mathbf{b}$ and $\mathbf{c}$ be three unit vectors such that $\mathbf{a}$ is perpendicular to the plane of $\mathbf{b}$ and $\mathbf{c}$. If the angle between $\mathbf{b}$ and $\mathbf{c}$ is $\frac{\pi}{3}$, then $|\mathbf{a}\times\mathbf{b}-\mathbf{a}\times\mathbf{c}|$ is equal to

(a) 1/3 (b) 1/2 (c) 1 (d) 2

43 The distance of the point $A(-2,3,1)$ from the line PQ through $P(-3,5,2)$ which make equal angles with the axes is

(a) $\frac{2}{\sqrt{3}}$ (b) $\sqrt{\frac{14}{3}}$ (c) $\frac{16}{\sqrt{3}}$ (d) $\frac{5}{\sqrt{3}}$

44 The plane passing through the point (5, 1, 2) perpendicular to the line $2(x-2)=y-4=z-5$ will meet the line in the point

(a) (1, 2, 3) (b) (2, 3, 1)
(c) (1, 3, 2) (d) (3, 2, 1)

Direction (Q. Nos. 45-48) *Each of these questions contains two statements : Statement* I *(Assertion) and Statement* II *(Reason). Each of these questions also has four alternative choices, only one of which is the correct answer. You have to select one of the codes (a), (b), (c) and (d) given below.*

(a) Statement I is true, Statement II is true; Statement II is a correct explanation for Statement I
(b) Statement I is true, Statement II is true; Statement II is not a correct explanation for Statement I
(c) Statement I is true; Statement II is false
(d) Statement I is false; Statement II is true

45 Consider vectors $\mathbf{a}$ and $\mathbf{c}$ are non-collinear, then

Statement I The lines $\mathbf{r}=6\mathbf{a}-\mathbf{c}+\lambda(2\mathbf{c}-\mathbf{a})$ and $\mathbf{r}=\mathbf{a}-\mathbf{c}+\mu(\mathbf{a}+3\mathbf{c})$ are coplanar.

Statement II There exist λ and μ such that the two values of r become same.

46 Consider **u** and **v** are unit vectors inclined at an angle α and **a** is a unit vector bisecting the angle between them,

Statement I Then, $\mathbf{a} = \dfrac{\mathbf{u}+\mathbf{v}}{2\cos(\alpha/2)}$.

Statement II If ABC is an isosceles triangle with $AB=AC=1$, then vector representing bisector of $\angle A$ is $\dfrac{\mathbf{AB}+\mathbf{AC}}{2}$.

47 Suppose $\pi : x + y - 2z = 3, P:(2,1,6), Q:(6,5,-2)$

Statement I The line joining PQ is perpendicular to the normal to the plane π.

Statement II Q is the image of P in the plane π.

48 Consider the equation of planes $P_1 = x + y + z - 6 = 0$ and $P_2 = 2x + 3y + 4z + 5 = 0$.

Statement I The equation of the plane through the intersection of the planes P_1 and P_2 and the point (4, 4, 4) is $29x + 23y + 17y = 276$.

Statement II Equation of the plane through the line of intersection of the planes $P_1 = 0$ and $P_2 = 0$ is $P_1 + \lambda P_2 = 0$, $\lambda \neq 0$.

49 The two adjacent sides of a parallelogram are $2\mathbf{i} - 4\mathbf{j} + 5\mathbf{k}$ and $\mathbf{i} - 2\mathbf{j} - 3\mathbf{k}$.

Statement I The unit vector parallel to its diagonal is $\frac{3}{5}\mathbf{i} - \frac{6}{5}\mathbf{j} + \frac{2}{5}\mathbf{k}$.

Statement II Area of parallelogram is $11\sqrt{5}$ sq units.

(a) Only Statement I is true (b) Only Statement II is true
(c) Both statements are ture (d) Both statements are false

ANSWERS

1. (c)	**2.** (a)	**3.** (a)	**4.** (d)	**5.** (c)	**6.** (d)	**7.** (a)	**8.** (b)	**9.** (a)	**10.** (b)
11. (a)	**12.** (a)	**13.** (b)	**14.** (d)	**15.** (a)	**16.** (a)	**17.** (d)	**18.** (b)	**19.** (a)	**20.** (a)
21. (c)	**22.** (b)	**23.** (d)	**24.** (b)	**25.** (b)	**26.** (d)	**27.** (b)	**28.** (a)	**29.** (c)	**30.** (a)
31. (c)	**32.** (d)	**33.** (a)	**34.** (d)	**35.** (a)	**36.** (d)	**37.** (d)	**38.** (d)	**39.** (b)	**40.** (c)
41. (c)	**42.** (c)	**43.** (b)	**44.** (a)	**45.** (a)	**46.** (a)	**47.** (d)	**48.** (a)	**49.** (b)	

Hints and Explanations

1 The plane passing through (2, 2, 1), is

$a(x-2) + b(y-2) + c(z-1) = 0$

Since, it passes through (9, 3, 6).

$\therefore \quad 7a + b + 5c = 0 \quad \ldots(i)$

Since, it is perpendicular to $2x + 6y + 6z - 1 = 0$.

$\therefore \quad 2a + 6b + 6c = 0 \quad \ldots(ii)$

$\Rightarrow \quad \frac{a}{-24} = \frac{b}{-32} = \frac{c}{40}$ [from Eqs. (i) and (ii)]

$\Rightarrow \quad \frac{a}{3} = \frac{b}{4} = \frac{c}{-5}$

The required plane is

$3(x-2) + 4(y-2) - 5(z-1) = 0$

$\Rightarrow \quad 3x + 4y - 5z - 9 = 0$

2 Now, $\mathbf{a}\cdot\mathbf{c} = x(\mathbf{a}\cdot\mathbf{a}) + y(\mathbf{a}\cdot\mathbf{b}) + z\mathbf{a}\cdot(\mathbf{a}\times\mathbf{b})$

$\Rightarrow \quad x = \cos\theta$

Similarly, $y = \cos\theta$

Now, $|\mathbf{c}|^2 = x^2|\mathbf{a}|^2 + y^2|\mathbf{b}|^2 + z^2|\mathbf{a}\times\mathbf{b}|^2$

$\Rightarrow 1 - 2\cos^2\theta = z^2 \Rightarrow 1 - 2x^2 = z^2$

where, $z = |\mathbf{a}\ \mathbf{b}\ \mathbf{c}|$

3 Given, $\mathbf{a} + \mathbf{b} + \mathbf{c} = \alpha\mathbf{d}$ and $\mathbf{b} + \mathbf{c} + \mathbf{d} = \beta\mathbf{a}$

$\therefore \quad \mathbf{a} + \mathbf{b} + \mathbf{c} + \mathbf{d} = (\alpha + 1)\mathbf{d}$

and $\mathbf{a} + \mathbf{b} + \mathbf{c} + \mathbf{d} = (\beta + 1)\mathbf{a}$

$\Rightarrow \quad (\alpha + 1)\mathbf{d} = (\beta + 1)\mathbf{a}$

If $\alpha \neq -1$, then $(\alpha + 1)\mathbf{d} = (\beta + 1)\mathbf{a}$

$\Rightarrow \quad \mathbf{d} = \frac{\beta + 1}{\alpha + 1}\mathbf{a}$

$\therefore \quad \mathbf{a} + \mathbf{b} + \mathbf{c} = \alpha\left(\frac{\beta + 1}{\alpha + 1}\right)\mathbf{a}$

$\Rightarrow \quad \left(1 - \frac{\alpha(\beta + 1)}{\alpha + 1}\right)\mathbf{a} + \mathbf{b} + \mathbf{c} = 0$

Hence, **a**, **b** and **c** are coplanar, which is a contradiction to the given condition.

$\therefore \quad \alpha = -1$

$\Rightarrow \quad \mathbf{a} + \mathbf{b} + \mathbf{c} + \mathbf{d} = 0$

4 Equation of any plane passing through the line of intersection of given plane, is

$ax + by + kz = 0 \quad \ldots(i)$

$\therefore$ DC's of Eq. (i) are

$\frac{a}{\sqrt{a^2 + b^2 + k^2}}, \frac{b}{\sqrt{a^2 + b^2 + k^2}}, \frac{k}{\sqrt{a^2 + b^2 + k^2}}$

The DC's of a normal to the given plane is

$\frac{a}{\sqrt{a^2 + b^2}}, \frac{b}{\sqrt{a^2 + b^2}}, 0.$

$\therefore \quad \cos\alpha = \frac{a\cdot a + b\cdot b + k\cdot 0}{\sqrt{a^2 + b^2 + k^2}\sqrt{a^2 + b^2}}$

$= \sqrt{\frac{a^2 + b^2}{a^2 + b^2 + k^2}}$

$\Rightarrow \quad k^2\cos^2\alpha = a^2(1 - \cos^2\alpha) + b^2(1 - \cos^2\alpha)$

$\Rightarrow \quad k^2 = \frac{(a^2 + b^2)\sin^2\alpha}{\cos^2\alpha}$

$\therefore \quad k = \pm\sqrt{a^2 + b^2}\tan\alpha$

From Eq. (i),

$ax + by \pm z\sqrt{a^2 + b^2}\tan\alpha = 0$

5 Any point on the line is $(r + 3, 2r + 4, 2r + 5)$ which lies on the plane $x + y + z = 17$.

$\because \quad (r + 3) + (2r + 4) + (2r + 5) = 17$

$\therefore \quad r = 1$

Thus, the point of intersection is (4, 6, 7).

So, the required distance

$= \sqrt{(4-3)^2 + (6-4)^2 + (7-5)^2}$
$= \sqrt{1+4+4} = 3$

6 Here, $AB = \sqrt{(5-3)^2 + (3-2)^2 + (2-0)^2} = 3$

and $AC = \sqrt{(-9-3)^2 + (6-2)^2 + (-3-0)^2} = 13$

Since, AD is the bisector of $\angle BAC$.

$\therefore \frac{BD}{DC} = \frac{AB}{AC} = \frac{3}{13}$

Since, D divides BC in the ratio 3 : 13.

$\therefore$ The coordinates of D are

$\left[\frac{3(-9)+13(5)}{3+13}, \frac{3(6)+13(3)}{3+13}, \frac{3(-3)+13(2)}{3+13}\right]$

$= \left(\frac{19}{8}, \frac{57}{16}, \frac{17}{16}\right)$

7 $[(\mathbf{a}+3\mathbf{b})\times(3\mathbf{a}-\mathbf{b})]^2 = [10(\mathbf{b}\times\mathbf{a})]^2$
$= 100[|\mathbf{a}|^2|\mathbf{b}|^2 - (\mathbf{a}\cdot\mathbf{b})^2]$
$= 100[1\times 4 - (1\times 2\times \cos 120°)^2]$
$= 100(4-1) = 300$

8 Let $P(x, y, z)$ be any point on the locus, then the distances from the six faces are $|x+1|$, $|x-1|$, $|y+1|$, $|y-1|$, $|z+1|$, $|z-1|$

According to the given condition,

$|x+1|^2 + |x-1|^2 + |y+1|^2 + |y-1|^2 + |z+1|^2 + |z-1|^2 = 10$

$\Rightarrow 2(x^2+y^2+z^2) = 10 - 6 = 4$

$\therefore x^2+y^2+z^2 = 2$

9 $(\mathbf{a}\times\mathbf{b})^2 + (\mathbf{a}\cdot\mathbf{b})^2 = |\mathbf{a}|^2|\mathbf{b}|^2\sin^2\theta + |\mathbf{a}|^2|\mathbf{b}|^2\cos^2\theta$
$= |\mathbf{a}|^2|\mathbf{b}|^2$

$\Rightarrow 144 = (4)^2|\mathbf{b}|^2 \Rightarrow |\mathbf{b}| = 3$

10 Since, $\mathbf{a}\cdot\mathbf{b} < 0$

$\Rightarrow 14x^2 - 8x + x < 0$

$\Rightarrow 14x^2 - 7x < 0$

$\Rightarrow 7x(2x-1) < 0$

$\therefore 0 < x < \frac{1}{2}$

11 Here, $\mathbf{a}\times\mathbf{b} = -4\mathbf{i} - 3\mathbf{j} + 6\mathbf{k}$

Now, $[\mathbf{a}\ \mathbf{b}\ \mathbf{c}] = (\mathbf{a}\times\mathbf{b})\cdot\mathbf{c}$
$= (-4\mathbf{i} - 3\mathbf{j} + 6\mathbf{k})\cdot\mathbf{c}$

$\therefore$ The maximum value of $[\mathbf{a}\ \mathbf{b}\ \mathbf{c}]$
$= |-4\mathbf{i} - 3\mathbf{j} + 6\mathbf{k}||\mathbf{c}| = \sqrt{61}$

12 Now, $\mathbf{d}_1 = \mathbf{a}+\mathbf{b} = 4\mathbf{p} + 2\mathbf{q}$

and $\mathbf{d}_2 = \mathbf{a}-\mathbf{b} = 2\mathbf{p} - 4\mathbf{q}$

$\Rightarrow \mathbf{d}_1^2 = 16\mathbf{p}^2 + 4\mathbf{q}^2 + 16\,\mathbf{p}\cdot\mathbf{q}$
$= 16(4) + 4(4) + 16\left(2\times 2\times \cos\frac{\pi}{3}\right)$
$= 112$

$\Rightarrow |\mathbf{d}_1| = 4\sqrt{7}$

Similarly, $|\mathbf{d}_2| = 4\sqrt{3}$

$\therefore \mathbf{d}_1 : \mathbf{d}_2 = \sqrt{7} : \sqrt{3}$

13 $|\mathbf{p}| = |\mathbf{q}| = |\mathbf{r}| = c$ [say]

and $\mathbf{p}\cdot\mathbf{q} = 0 = \mathbf{p}\cdot\mathbf{r} = \mathbf{q}\cdot\mathbf{r}$

Given that,

$\mathbf{p}\times\{(\mathbf{x}-\mathbf{q})\times\mathbf{p}\} + \mathbf{q}\times\{(\mathbf{x}-\mathbf{r})\times\mathbf{q}\} + \mathbf{r}\times\{(\mathbf{x}-\mathbf{p})\times\mathbf{r}\} = \mathbf{0}$

$\Rightarrow (\mathbf{p}\cdot\mathbf{p})(\mathbf{x}-\mathbf{q}) - \{\mathbf{p}\cdot(\mathbf{x}-\mathbf{q})\}\mathbf{p} + \ldots = 0$

$\Rightarrow c^2(\mathbf{x}-\mathbf{q}+\mathbf{x}-\mathbf{r}+\mathbf{x}-\mathbf{p}) - (\mathbf{p}\cdot\mathbf{x})\mathbf{p} - (\mathbf{q}\cdot\mathbf{x})\mathbf{q} - (\mathbf{r}\cdot\mathbf{x})\mathbf{r} = 0$

$\Rightarrow c^2\{3\mathbf{x} - (\mathbf{p}+\mathbf{q}+\mathbf{r})\} - \{(\mathbf{p}\cdot\mathbf{x})\mathbf{p} + (\mathbf{q}\cdot\mathbf{x})\mathbf{q} + (\mathbf{r}\cdot\mathbf{x})\mathbf{r}\} = 0$

which is satisfied by $\mathbf{x} = \frac{1}{2}(\mathbf{p}+\mathbf{q}+\mathbf{r})$

14 Let $P(r, r-2, r)$ and $Q(2k-2, k, k)$ are the general coordinates of points on the two given lines.

$\therefore$ DR's of PQ are

$(r-2k+2, r-k-2, r-k)$

$\therefore \frac{r-2k+2}{2} = \frac{r-k-2}{1} = \frac{r-k}{2}$

$\Rightarrow r = 6, k = 2$

So, the points of intersection are (6, 4, 6) and (2, 2, 2).

15 Given, $\mathbf{A}+\mathbf{B} = \mathbf{a}$ ...(i)

$\Rightarrow \mathbf{A}\cdot\mathbf{a} + \mathbf{B}\cdot\mathbf{a} = \mathbf{a}\cdot\mathbf{a}$

$\Rightarrow 1 + \mathbf{B}\cdot\mathbf{a} = a^2$

$\Rightarrow \mathbf{B}\cdot\mathbf{a} = a^2 - 1$ ...(ii)

Also, $\mathbf{A}\times\mathbf{B} = \mathbf{b}$

$\Rightarrow \mathbf{a}\times(\mathbf{A}\times\mathbf{B}) = \mathbf{a}\times\mathbf{b}$

$\Rightarrow (\mathbf{a}\cdot\mathbf{B})\mathbf{A} - (\mathbf{a}\cdot\mathbf{A})\mathbf{B} = \mathbf{a}\times\mathbf{b}$

$\Rightarrow (a^2-1)\mathbf{A} - \mathbf{B} = \mathbf{a}\times\mathbf{b}$

[from Eq. (ii)] ...(iii)

From Eqs. (i) and (iii), we get

$\mathbf{A} = \frac{(\mathbf{a}\times\mathbf{b}) + \mathbf{a}}{a^2}$

and $\mathbf{B} = \mathbf{a} - \left\{\frac{(\mathbf{a}\times\mathbf{b}) + \mathbf{a}}{a^2}\right\}$

$\therefore \mathbf{B} = \frac{(\mathbf{b}\times\mathbf{a}) + \mathbf{a}(a^2-1)}{a^2}$

16 The required vector $\mathbf{c}$ is given by

$\mathbf{c} = \pm\lambda\left(\frac{\mathbf{a}}{|\mathbf{a}|} + \frac{\mathbf{b}}{|\mathbf{b}|}\right)$

$= \pm\lambda\left\{\frac{1}{9}(7\mathbf{i} - 4\mathbf{j} - 4\mathbf{k}) + \frac{1}{3}(-2\mathbf{i} - \mathbf{j} + 2\mathbf{k})\right\}$

$\Rightarrow \mathbf{c} = \pm\frac{\lambda}{9}(\mathbf{i} - 7\mathbf{j} + 2\mathbf{k})$

$\Rightarrow 5\sqrt{6} = \frac{\lambda}{9}\sqrt{1+49+4}$

$= \frac{\lambda}{9}\sqrt{54}$ $[\because |\mathbf{c}| = 5\sqrt{6}]$

$\Rightarrow \lambda = 15$

$\therefore \mathbf{c} = \pm\frac{15}{9}(\mathbf{i} - 7\mathbf{j} + 2\mathbf{k})$

$= \pm\frac{5}{3}(\mathbf{i} - 7\mathbf{j} + 2\mathbf{k})$

17 Given, $V_1 = [\mathbf{a}\ \mathbf{b}\ \mathbf{c}]$ and $V_2 = [\mathbf{p}\ \mathbf{q}\ \mathbf{r}]$

Then, $V_2 = \begin{vmatrix} 1 & 1 & -2 \\ 3 & -2 & 1 \\ 1 & -4 & 2 \end{vmatrix} [\mathbf{a}\ \mathbf{b}\ \mathbf{c}]$

$= 15[\mathbf{a}\ \mathbf{b}\ \mathbf{c}]$

$\therefore V_2 : V_1 = 15 : 1$

18 The normal to the first plane is along $\mathbf{i}\times(\mathbf{i}+\mathbf{j}) = \mathbf{k}$ and the normal to the second plane is along

$(\mathbf{i}-\mathbf{j})\times(\mathbf{i}+\mathbf{k}) = -\mathbf{i} - \mathbf{j} + \mathbf{k}$

Since, $\mathbf{a}$ is perpendicular to the two normals.

So, $\mathbf{a}$ is along $\mathbf{k}\times(-\mathbf{i}-\mathbf{j}+\mathbf{k}) = \mathbf{i} - \mathbf{j}$

Hence, the angle between $\mathbf{a}$ and the vector $\mathbf{i} - 2\mathbf{j} + 2\mathbf{k}$ is

$\cos^{-1}\frac{\mathbf{a}\cdot(\mathbf{i}-2\mathbf{j}+2\mathbf{k})}{|\mathbf{a}||\mathbf{i}-2\mathbf{j}+2\mathbf{k}|}$

$= \cos^{-1}\frac{(\mathbf{i}-\mathbf{j})}{\sqrt{2}}\cdot\frac{(\mathbf{i}-2\mathbf{j}+2\mathbf{k})}{3}$

$= \cos^{-1}\frac{1}{\sqrt{2}} = \frac{\pi}{4}$

19 Since, $\mathbf{b} = (\mathbf{a}\times\mathbf{c})\times\mathbf{a}$

$\Rightarrow |\mathbf{b}| = |\mathbf{a}\times\mathbf{c}||\mathbf{a}| \Rightarrow 1 = 2\sin\theta$

$\therefore \theta = \frac{\pi}{6}$

20 If $\mathbf{a}$, $\mathbf{b}$ and $\mathbf{c}$ lie in a plane, then $2\mathbf{a}-\mathbf{b}$, $2\mathbf{b}-\mathbf{c}$ and $2\mathbf{c}-\mathbf{a}$, also lie in the same plane. So, their scalar triple product is '0'.

21 Since, $\mathbf{k}\times(\mathbf{k}\times\mathbf{a}) = 0$

$\Rightarrow (\alpha\mathbf{i} + \beta\mathbf{j}) = 0$

$\Rightarrow \alpha = \beta = 0$

$\because \alpha + \beta + \gamma = 2$

$\therefore \gamma = 2$

22 The first three planes meet at the point whose position vector is (0, 0, 0). The first two and the fourth planes meet at the point whose position vector is $\left(\frac{p}{l}, \frac{-p}{m}, \frac{p}{n}\right)$. Similarly, the other two vertices of the tetrahedron have position vectors $\left(\frac{-p}{l}, \frac{p}{m}, \frac{p}{n}\right)$ and $\left(\frac{p}{l}, \frac{p}{m}, \frac{-p}{n}\right)$.

$\therefore$ Volume of the tetrahedron

$= \frac{1}{6}\left\|\begin{matrix} p/l & -p/m & p/n \\ -p/l & p/m & p/n \\ p/l & p/m & -p/n \end{matrix}\right\|$

$= \frac{p^3}{6\,lmn}\left\|\begin{matrix} 1 & -1 & 1 \\ -1 & 1 & 1 \\ 1 & 1 & -1 \end{matrix}\right\|$

$= \frac{p^3}{6\,lmn}\left\|\begin{matrix} 0 & 0 & 2 \\ 0 & 2 & 0 \\ 1 & 1 & -1 \end{matrix}\right\|$ $\left[\begin{matrix} R_1 \to R_1 + R_2, \\ R_2 \to R_2 + R_3 \end{matrix}\right]$

$= \frac{|-4|p^3}{6\,lmn} = \frac{2p^3}{3\,lmn}$

23 Let the variable plane intersects the coordinate axes at $A(a, 0, 0)$, $B(0, b, 0)$ and $C(0, 0, c)$.
Then, the equation of the plane will be

$$\frac{x}{a}+\frac{y}{b}+\frac{z}{c}=1 \quad \ldots(i)$$

Let $P(\alpha, \beta, \gamma)$ be the centroid of tetrahedron $OABC$, then

$$\alpha=\frac{a}{4}, \beta=\frac{b}{4} \text{ and } \gamma=\frac{c}{4}$$

$\therefore \quad a=4\alpha, b=4\beta, c=4\gamma$

Now, volume of tetrahedron

$$=(\text{Area of } \Delta AOB)\, OC$$

$$\Rightarrow \quad 64k^3=\frac{1}{3}\left(\frac{1}{2}ab\right)c=\frac{abc}{6}$$

$$\Rightarrow \quad 64k^3=\frac{(4\alpha)(4\beta)\times(4\gamma)}{3\times 2}$$

$$\therefore \quad k^3=\frac{\alpha\beta\gamma}{6}$$

Hence, locus of $P(\alpha, \beta, \gamma)$ is $xyz=6k^3$.

24 The required line passes through the point $\mathbf{i}+3\mathbf{j}+2\mathbf{k}$ and is perpendicular to the lines

$$\mathbf{r}=(\mathbf{i}+2\mathbf{j}-\mathbf{k})+\lambda(2\mathbf{i}+\mathbf{j}+\mathbf{k})$$

and $\mathbf{r}=(2\mathbf{i}+6\mathbf{j}+\mathbf{k})+\mu(\mathbf{i}+2\mathbf{j}+3\mathbf{k})$

So, it is parallel to the vector.

$$\therefore \mathbf{b}=(2\mathbf{i}+\mathbf{j}+\mathbf{k})\times(\mathbf{i}+2\mathbf{j}+3\mathbf{k})=(\mathbf{i}-5\mathbf{j}+3\mathbf{k})$$

The required equation is

$$\mathbf{r}=(\mathbf{i}+3\mathbf{j}+2\mathbf{k})+\lambda(\mathbf{i}-5\mathbf{j}+3\mathbf{k})$$

25 Let $\mathbf{n}=3\mathbf{i}-\mathbf{j}+4\mathbf{k}$

Line through A and parallel to $\mathbf{n}$ is

$$\mathbf{r}=\mathbf{i}+2\mathbf{j}+3\mathbf{k}+\lambda(3\mathbf{i}-\mathbf{j}+4\mathbf{k})=(3\lambda+1)\mathbf{i}+(2-\lambda)\mathbf{j}+(3+4\lambda)\mathbf{k} \quad \ldots(i)$$

Eq. (i) must satisfy the plane $3x-y+4z=0$.

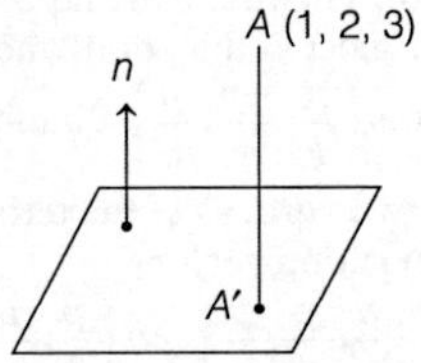

$\therefore \quad 3(3\lambda+1)-(2-\lambda)+4(3+4\lambda)=0$

$\Rightarrow \quad 26\lambda+13=0$

$\Rightarrow \quad \lambda=-\frac{1}{2}$

Hence, A' is $\left(-\frac{1}{2}, \frac{5}{2}, 1\right)$ which is the foot of the perpendicular from A on the given plane.

26 DR's of OC are $(1, -1, 2)$.

Let the equation of plane passing through $(1, 0, 1)$ is

$$a(x-1)+b(y-0)+c(z-1)=0 \quad \ldots(i)$$

Since, its normal is perpendicular to OC

$\therefore \quad 1\cdot a+(-1)b+2c=0$

$\Rightarrow \quad a-b+2c=0 \quad \ldots(ii)$

As Eq. (i) passes through $(3, 1, 2)$.

$\therefore \quad 2a+b+c=0 \quad \ldots(iii)$

$$\Rightarrow \quad \frac{a}{-1}=\frac{b}{1}=\frac{c}{1}$$

[from Eqs. (ii) and (iii)]

Hence, required equation of plane be $x-y-z=0$.

27 Let DC's of PQ be l, m and n.

$\therefore 3l-m+n=0$ and $5l+m+3n=0$

$$\therefore \quad \frac{l}{-3-1}=\frac{m}{5-9}=\frac{n}{3+5}$$

$$\Rightarrow \quad \frac{l}{1}=\frac{m}{1}=\frac{n}{-2}$$

Thus, the equation of plane perpendicular to PQ will have $x+y-2z=\lambda$.

It passes through $(2, 1, 4)$, therefore $\lambda=-5$.

Hence, the required equation of plane be $x+y-2z=-5$

28 The line of intersection of the planes $\mathbf{r}\cdot(3\mathbf{i}-\mathbf{j}+\mathbf{k})=1$ and $\mathbf{r}\cdot(\mathbf{i}+4\mathbf{j}-2\mathbf{k})=2$ is perpendicular to each of the normal vectors.

Here, $\mathbf{n}_1=3\mathbf{i}-\mathbf{j}+\mathbf{k}$ and $\mathbf{n}_2=\mathbf{i}+4\mathbf{j}-2\mathbf{k}$

$\therefore$ It is parallel to the vector $\mathbf{n}_1\times\mathbf{n}_2$

$$=(3\mathbf{i}-\mathbf{j}+\mathbf{k})\times(\mathbf{i}+4\mathbf{j}-2\mathbf{k})=-2\mathbf{i}+7\mathbf{j}+13\mathbf{k}$$

29 Since, lines are $\frac{x}{1}=\frac{y+a}{1}=\frac{z}{1}$

and $\frac{x+a}{2}=\frac{y}{1}=\frac{z}{1}$

Let $P\equiv(r, r-a, r)$ and $Q\equiv(2\lambda-a, \lambda, \lambda)$ be the points of I and II lines.

So, DR's of PQ are $r-2\lambda+a, r-\lambda-a, r-\lambda$.

According to the given question,

$$\frac{r-2\lambda+a}{2}=\frac{r-\lambda-a}{1}=\frac{r-\lambda}{2}$$

From I and II terms, $r-a=2a \Rightarrow r=3a$

From II and III terms, $\lambda=a$

$\therefore P\equiv(3a, 2a, 3a)$ and $Q\equiv(a, a, a)$

30 Let $\mathbf{a}=3\mathbf{i}+4\mathbf{j}-6\mathbf{k}$ and $\mathbf{b}=2\mathbf{i}-3\mathbf{j}+5\mathbf{k}$

$\because$ Diagonals of a parallelogram in terms of its sides are

$$\mathbf{p}=\mathbf{a}+\mathbf{b} \text{ and } \mathbf{q}=\mathbf{b}-\mathbf{a}$$

$\Rightarrow \mathbf{p}=5\mathbf{i}+\mathbf{j}-\mathbf{k}$ and $\mathbf{q}=-\mathbf{i}-7\mathbf{j}+11\mathbf{k}$

The unit vectors along the diagonals are

$$\frac{5\mathbf{i}+\mathbf{j}-\mathbf{k}}{\sqrt{25+1+1}} \text{ and } \frac{-\mathbf{i}-7\mathbf{j}+11\mathbf{k}}{\sqrt{(-1)^2+49+121}}$$

$$\Rightarrow \frac{5\mathbf{i}+\mathbf{j}-\mathbf{k}}{\sqrt{27}} \text{ and } \frac{-\mathbf{i}-7\mathbf{j}+11\mathbf{k}}{\sqrt{171}}$$

31 Since, $\mathbf{p}$ and $\mathbf{q}$ are perpendicular.

$\therefore \quad \mathbf{p}\cdot\mathbf{q}=0$

Let $\mathbf{q}=x\mathbf{i}+y\mathbf{j}$, then

$(8\mathbf{i}+6\mathbf{j})(x\mathbf{i}+y\mathbf{j})=0 \Rightarrow 8x+6y=0$

$$\therefore \quad y=-\frac{8x}{6}=-\frac{4x}{3}$$

$$\mathbf{q}=x\mathbf{i}+\left(\frac{-4x}{3}\right)\mathbf{j}=\frac{3x\mathbf{i}-4x\mathbf{j}}{3}=\frac{x}{3}(3\mathbf{i}-4\mathbf{j})$$

Again, the projection of vector $\mathbf{r}=\pm(x_1\mathbf{i}+x_2\mathbf{j})$ on vector $\mathbf{p}$ is 2 and on $\mathbf{q}$ is 4.

$$\therefore 2=\left|\frac{8x_1+6x_2}{10}\right| \text{ and } 4=\left|\frac{6x_1-8x_2}{10}\right|$$

$\Rightarrow \quad 8x_1+6x_2=20$

and $\quad 6x_1-8x_2=40$

$\Rightarrow \quad 4x_1+3x_2=10$

and $\quad 3x_1-4x_2=20$

$\Rightarrow \quad x_1=4$ and $x_2=-2$

$\therefore \quad \mathbf{r}=\pm(4\mathbf{i}-2\mathbf{j})=\pm 2(2\mathbf{i}-\mathbf{j})$

32 The plane is

$$(2+3\lambda)x+(\lambda-1)y+(\lambda+1)z-5\lambda-3=0$$

Its distance from $(2, 1, -1)$ is $\frac{1}{\sqrt{6}}$.

$$\therefore \frac{(4+6\lambda+\lambda-1-\lambda-1-5\lambda-3)^2}{(2+3\lambda)^2+(\lambda-1)^2+(\lambda+1)^2}=\frac{1}{6}$$

$$\Rightarrow (5\lambda+24)\lambda=0 \Rightarrow \lambda=\frac{-24}{5} \text{ or } 0$$

The planes are $2x-y+z-3=0$ and $62x+29y+19z-105=0$

33 Equation of a plane passing through the point $(2, -1, -3)$ and parallel to the given line is

$$\begin{vmatrix} x-2 & y+1 & z+3 \\ 3 & 2 & -4 \\ 2 & -3 & 2 \end{vmatrix}=0$$

$$\Rightarrow (x-2)(4-12)-(y+1)(6+8)+(z+3)(-9-4)=0$$

$$\Rightarrow 8x+14y+13z+37=0$$

34 The common normal is $\begin{vmatrix} \mathbf{i} & \mathbf{j} & \mathbf{k} \\ 2 & 3 & 4 \\ 3 & 4 & 5 \end{vmatrix}$

$$\mathbf{r}=-\mathbf{i}+2\mathbf{j}-\mathbf{k}$$

$$\therefore \text{ Shortest distance}=(\mathbf{j}+\mathbf{k})\cdot\frac{\mathbf{r}}{|\mathbf{r}|}=\frac{1}{\sqrt{6}}$$

35 Normal to OAB is $\mathbf{OA}\times\mathbf{OB}$

$$=\begin{vmatrix} \mathbf{i} & \mathbf{j} & \mathbf{k} \\ 1 & 2 & 1 \\ 2 & 1 & 3 \end{vmatrix}=5\mathbf{i}-\mathbf{j}-3\mathbf{k}$$

Normal to ABC is

$$\mathbf{AB}\times\mathbf{AC}=\begin{vmatrix} \mathbf{i} & \mathbf{j} & \mathbf{k} \\ 1 & -1 & 2 \\ -2 & -1 & 1 \end{vmatrix}=\mathbf{i}-5\mathbf{j}-3\mathbf{k}$$

If θ is angle between the planes, then

$$\cos\theta = \frac{5+5+9}{\sqrt{35}\cdot\sqrt{35}} = \frac{19}{35}$$

36 Let $\mathbf{r} = a_1\mathbf{i} + a_2\mathbf{j} + a_3\mathbf{k}$

Since, $\mathbf{r}\perp\mathbf{p}$ and $\mathbf{r}\perp\mathbf{q}$

$\therefore\quad \mathbf{r}\cdot\mathbf{p} = 0$ and $\mathbf{r}\cdot\mathbf{q} = 0$

$\Rightarrow\quad 2a_1 - 3a_2 + 3a_3 = 0$

and $\quad 4a_1 - 2a_2 + a_3 = 0$

$\Rightarrow\quad 2a_1 - 3a_2 + 3a_3 = 0$

and $\quad 4a_1 - 2a_2 + a_3 = 0$

$$\therefore\ \frac{a_1}{-3-(-6)} = \frac{-a_2}{12-2} = \frac{a_3}{-4-(-12)} = k \quad \text{[say]}$$

$$\Rightarrow\ \frac{a_1}{3} = \frac{a_2}{10} = \frac{a_3}{8} = k$$

$\therefore\quad a_1 = 3k, a_2 = 10k, a_3 = 8k \quad$ ...(i)

Again, $(a_1\mathbf{i} + a_2\mathbf{j} + a_3\mathbf{k})$

$(2\mathbf{i} - 4\mathbf{j} + 2\mathbf{k}) = -12$

$\Rightarrow 2a_1 - 4a_2 + 2a_3 = -12 \quad$ ...(ii)

From Eqs. (i) and (ii), we get

$6k - 4(10k) + 2(8k) = -12$

$6k - 40k + 16k = -12$

$$-18k = -12 \Rightarrow k = \frac{12}{18} = \frac{2}{3}$$

$$\therefore\quad a_1 = 2, a_2 = \frac{20}{3}, a_3 = \frac{16}{3}$$

$$\therefore\quad \mathbf{r} = 2\mathbf{i} + \frac{20}{3}\mathbf{j} + \frac{16}{3}\mathbf{k}$$

37 Plane passing through a point (x_1, y_1, z_1) is

$A(x - x_1) + B(y - y_1) + C(z - z_1) = 0$

$\therefore$ Plane through $(2, 0, 0)$ is

$a(x-2) + b(y-0) + c(z-0) = 0 \quad$ ...(i)

contains $(0, 2, 0)$, if

$-2a + 2b = 0 \Rightarrow -a + b = 0 \quad$ ...(ii)

Since, plane

$a(x-2) + b(y-0) + c(z-0) = 0$

makes an angle $\frac{\pi}{3}$ with the plane

$2x + 3y = 5$

$$\therefore\quad \cos\frac{\pi}{3} = \frac{2a+3b}{\sqrt{a^2+b^2+c^2}\sqrt{4+9}}$$

$$\Rightarrow\quad \frac{1}{2} = \frac{2a+3b}{\sqrt{(a^2+b^2+c^2)(13)}}$$

$$\Rightarrow\quad \frac{1}{4} = \frac{(2a+3b)^2}{[\sqrt{(a^2+b^2+c^2)(13)}]^2}$$

$\Rightarrow\quad 13(a^2+b^2+c^2) = 100a^2$

$\because\quad a = b$

$\therefore\quad 13(2a^2+c^2) = 100a^2$

$\Rightarrow\quad 26a^2 + 13c^2 = 100a^2$

$$\Rightarrow\quad 13c^2 = 74a^2 \Rightarrow c = \sqrt{\frac{74}{13}}\,a$$

$$\therefore\quad a:b:c = a:a:\sqrt{\frac{74}{13}}\,a$$

$= a : a : \sqrt{5.7}\,a$

$= 1 : 1 : \sqrt{5.7}$

38 Shortest distance

$$= \frac{\left\|\begin{matrix} x_2 - x_1 & y_2 - y_1 & z_2 - z_1 \\ a_1 & b_1 & c_1 \\ a_2 & b_2 & c_2 \end{matrix}\right\|}{\left\|\begin{matrix} \mathbf{i} & \mathbf{j} & \mathbf{k} \\ a_1 & b_1 & c_1 \\ a_2 & b_2 & c_2 \end{matrix}\right\|}$$

$$= \frac{\left\|\begin{matrix} -1 & 1 & 1 \\ 3 & 2 & 2 \\ 2 & 3 & 4 \end{matrix}\right\|}{\left\|\begin{matrix} \mathbf{i} & \mathbf{j} & \mathbf{k} \\ 3 & 2 & 2 \\ 2 & 3 & 4 \end{matrix}\right\|}$$

$$= \frac{|-1(8-6) + 1(4-12) + 1(9-4)|}{|\mathbf{i}(8-6) + \mathbf{j}(4-12) + \mathbf{k}(9-4)|}$$

$$= \left|\frac{-2-8+5}{\sqrt{93}}\right|$$

$$= \frac{5}{\sqrt{93}} = 0.52 \text{ unit}$$

39 Any point on the lines L_1 and L_2 are $(-3r_1 - 1, 2r_1 + 3, r_1 - 2)$ and $(r_2, -3r_2 + 7, 2r_2 - 7)$

Since, they intersect each other, therefore

$-3r_1 - 1 = r_2, 2r_1 + 3 = -3r_2 + 7$

and $\quad r_1 - 2 = 2r_2 - 7$

On solving, we get

$r_2 = 2$ and $r_1 = -1$

Hence, the required point is $(2, 1, -3)$.

40 Let θ be an angle between unit vectors $\mathbf{a}$ and $\mathbf{b}$. Then, $\mathbf{a}\cdot\mathbf{b} = \cos\theta$

Now, $|\mathbf{a} + \mathbf{b}|^2 = |\mathbf{a}|^2 + |\mathbf{b}|^2 + 2\mathbf{a}\cdot\mathbf{b}$

$$= 2 + 2\cos\theta = 4\cos^2\frac{\theta}{2}$$

$$\Rightarrow\ |\mathbf{a}+\mathbf{b}| = 2\cos\frac{\theta}{2}\ |\mathbf{a}-\mathbf{b}| = 2\sin\frac{\theta}{2}$$

$\Rightarrow\ |\mathbf{a}+\mathbf{b}| + |\mathbf{a}-\mathbf{b}|$

$$= 2\left(\cos\frac{\theta}{2} + \sin\frac{\theta}{2}\right) \le 2\sqrt{2}$$

41 Let $\alpha = \mathbf{a} + 2\mathbf{b} + 3\mathbf{c}, \beta = \lambda\mathbf{b} + 4\mathbf{c}$ and $\gamma = (2\lambda - 1)\mathbf{c}$

$$\text{Then, } [\alpha\ \beta\ \gamma] = \begin{vmatrix} 1 & 2 & 3 \\ 0 & \lambda & 4 \\ 0 & 0 & (2\lambda-1) \end{vmatrix} [\mathbf{a}\ \mathbf{b}\ \mathbf{c}]$$

$[\alpha\ \beta\ \gamma] = \lambda(2\lambda - 1)[\mathbf{abc}]$

$\Rightarrow [\alpha\ \beta\ \gamma] = 0,$

If $\lambda = 0, \frac{1}{2} \quad [\because [\mathbf{a}\ \mathbf{b}\ \mathbf{c}] \ne 0]$

Hence, α, β and γ are non-coplanar for all value of λ except two values 0 and $\frac{1}{2}$.

42 Since, $\mathbf{a}\cdot\mathbf{b} = \mathbf{a}\cdot\mathbf{c} = 0, \mathbf{b}\cdot\mathbf{c} = \frac{1}{2}$

$\therefore\quad |\mathbf{a}\times\mathbf{b}| = |\mathbf{a}\times\mathbf{c}| = 1$

Now, $|\mathbf{a}\times\mathbf{b} - \mathbf{a}\times\mathbf{c}|^2 = |\mathbf{a}\times\mathbf{b}|^2 + |\mathbf{a}\times\mathbf{c}|^2 - 2(\mathbf{a}\times\mathbf{b})\cdot(\mathbf{a}\times\mathbf{c})$

$$= 1 + 1 - 2\begin{vmatrix} 1 & 0 \\ 0 & 1/2 \end{vmatrix} = 1$$

43 Here, $\alpha = \beta = \gamma$

$\because\quad \cos^2\alpha + \cos^2\beta + \cos^2\gamma = 1$

$$\therefore\quad \cos\alpha = \frac{1}{\sqrt{3}}$$

DC's of PQ are $\left(\frac{1}{\sqrt{3}}, \frac{1}{\sqrt{3}}, \frac{1}{\sqrt{3}}\right)$

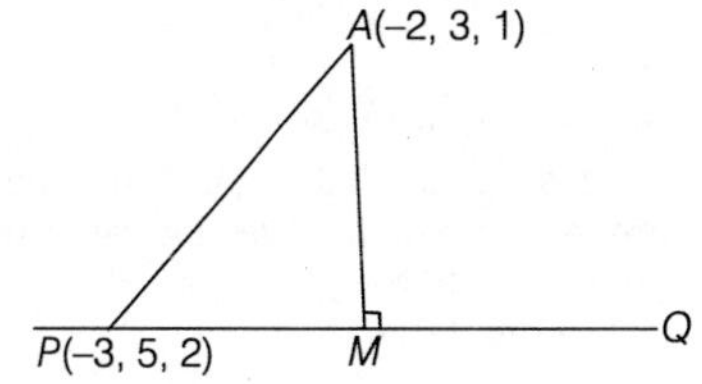

PM = Projection of AP on PQ

$$= \left|(-2+3)\frac{1}{\sqrt{3}} + (3-5)\cdot\frac{1}{\sqrt{3}} + (1-2)\cdot\frac{1}{\sqrt{3}}\right| = \frac{2}{\sqrt{3}}$$

and

$AP = \sqrt{(-2+3)^2 + (3-5)^2 + (1-2)^2}$

$= \sqrt{6}$

$$AM = \sqrt{(AP)^2 - (PM)^2} = \sqrt{6 - \frac{4}{3}} = \sqrt{\frac{14}{3}}$$

44 Equation of the plane through $(5, 1, 2)$ is

$a(x-5) + b(y-1) + c(z-2) = 0 \quad$ ...(i)

Given plane (i) is perpendicular to the line

$$\frac{x-2}{1/2} = \frac{y-4}{1} = \frac{z-5}{1} \quad \text{...(ii)}$$

$\therefore$ Equation of normal of Eq. (i) and straight line (ii) are parallel

$$\text{i.e.}\quad \frac{a}{1/2} = \frac{b}{1} = \frac{c}{1} = k \text{ (say)}$$

$$\therefore\quad a = \frac{k}{2}, b = k, c = k$$

From Eq. (i),

$$\frac{k}{2}(x-5) + k(y-1) + k(z-2) = 0$$

$\Rightarrow\ x + 2y + 2z = 11$

Any point on Eq. (ii) is

$$\left(2 + \frac{\lambda}{2}, 4 + \lambda, 5 + \lambda\right)$$

which lies on Eq. (iii), then $\lambda = -2$.

$\therefore$ Required point is $(1, 2, 3)$.

45 If the lines have a common point, then there exists λ and μ such that

$6 - \lambda = 1 + \mu$

and $\quad -1 + 2\lambda = -1 + 3\mu$

$\Rightarrow\quad \lambda = 3, \mu = 2$

$\therefore\quad \mathbf{r} = 3\mathbf{a} + 5\mathbf{c}$

46 In an isosceles ΔABC in which $AB = AC$ the median and bisector from A must be same line, so Statement II is true.

Now, $\mathbf{AD} = \dfrac{\mathbf{u} + \mathbf{v}}{2}$

$\Rightarrow \quad |\mathbf{AD}|^2 = \dfrac{1}{4}[|\mathbf{u}|^2 + |\mathbf{v}|^2 + 2\mathbf{u} \cdot \mathbf{v}]$

$= \dfrac{1}{4}(2 + 2\cos\alpha) = \cos^2\dfrac{\alpha}{2}$

$\therefore$ Unit vector along

$\mathbf{AD} = \dfrac{1}{2\cos\dfrac{\alpha}{2}}(\mathbf{u} + \mathbf{v})$

47 Direction ratios of PQ are

$6 - 2, 5 - 1, -2 - 6$ i.e. $4, 4, -8$ which are proportional to the direction ratios of the normal to the plane π, so PQ is perpendicular to π.

Hence, Statement I is false and Statement II is true.

48 The equation of plane through the line of intersection of the planes $x + y + z = 6$ and $2x + 3y + 4z + 5 = 0$ is $(x + y + z - 6) + \lambda(2x + 3y + 4z + 5) = 0$ ...(i)

Since, it passes through (4, 4, 4), then

$(4 + 4 + 4 - 6) + \lambda(8 + 12 + 16 + 5) = 0$

$\Rightarrow \quad 6 + 41\lambda = 0 \Rightarrow \lambda = -\dfrac{6}{41}$

From Eq. (i), we get

$41(x + y + z - 6) - 6(2x + 3y + 4z + 5) = 0$

$\therefore \ 29x + 23y + 17z = 276$

49 Adjacent sides of a parallelogram are given as $\mathbf{a} = 2\mathbf{i} - 4\mathbf{j} + 5\mathbf{k}$ and $\mathbf{b} = \mathbf{i} - 2\mathbf{j} - 3\mathbf{k}$.

Then, the diagonal of a parallelogram is given by $\mathbf{v} = \mathbf{a} + \mathbf{b}$.

[since, from the figure, it is clear that resultant of adjacent sides of a parallelogram is given by the diagonal]

$\therefore \ \mathbf{v} = 2\mathbf{i} - 4\mathbf{j} + 5\mathbf{k} + \mathbf{i} - 2\mathbf{j} - 3\mathbf{k}$

$= (2 + 1)\mathbf{i} + (-4 - 2)\mathbf{j} + (5 - 3)\mathbf{k}$

$= 3\mathbf{i} - 6\mathbf{j} + 2\mathbf{k}$

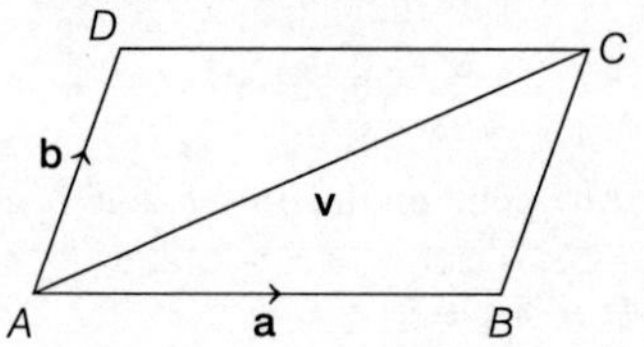

$\therefore \ |\mathbf{v}| = \sqrt{x^2 + y^2 + z^2}$

$= \sqrt{(3)^2 + (-6)^2 + (2)^2}$

$= \sqrt{9 + 36 + 4}$

$= \sqrt{49} = 7$

Thus, the unit vector parallel to the diagonal is

$\dfrac{\mathbf{v}}{|\mathbf{v}|} = \dfrac{3\mathbf{i} - 6\mathbf{j} + 2\mathbf{k}}{7}$

$= \dfrac{3}{7}\mathbf{i} - \dfrac{6}{7}\mathbf{j} + \dfrac{2}{7}\mathbf{k}$

Also, area of parallelogram

$ABCD = |\mathbf{a} \times \mathbf{b}| = \begin{vmatrix} \mathbf{i} & \mathbf{j} & \mathbf{k} \\ 2 & -4 & 5 \\ 1 & -2 & -3 \end{vmatrix}$

$= |\mathbf{i}(12 + 10) - \mathbf{j}(-6 - 5) + \mathbf{k}(-4 + 4)|$

$= |22\mathbf{i} + 11\mathbf{j} + 0\mathbf{k}|$

$= \sqrt{(22)^2 + (11)^2 + 0^2}$

$= \sqrt{(11)^2 + (2^2 + 1^2)}$

$= 11\sqrt{5}$ sq units

DAY THIRTY FOUR

Statistics

Learning & Revision for the Day

- Measures of Central Tendency
- Measure of Dispersion

Measures of Central Tendency

A value which describes the characteristics of entire data, is called an **average** or a **central value**. Generally, an average lies in the central part of the data and therefore such values are called the **measure of central tendency**.

There are five measures of central tendency, which are given below :

1. Mean (Arithmetic Mean)

The sum of all the observations divided by the number of observations, is called **mean** and it is denoted by $\bar{x}$. The most stable measure of central tendency is mean.

- If $x_1, x_2, \dots, x_n$ be the n observations, then mean is given by

$$\bar{x} = \frac{x_1 + x_2 + \dots + x_n}{n} = \frac{\Sigma x_i}{n}$$

or $\bar{x} = A + \dfrac{\sum_{i=1}^{n} d_i}{n}$, where $d_i = x_i - A$ and A is assumed mean.

- If corresponding frequencies of n observations are $f_1, f_2, \dots, f_n$, then

$$\bar{x} = \frac{f_1x_1 + f_2x_2 + \dots + f_nx_n}{f_1 + f_2 + \dots + f_n} = \frac{\sum_{i=1}^{n} f_i x_i}{\sum_{i=1}^{n} f_i} \text{ or } \bar{x} = A + \frac{\sum_{i=1}^{n} f_i d_i}{N},$$

where $d_i = x_i - A$ and $N = \sum_{i=1}^{n} f_i$

- If corresponding weights are $w_1, w_2, \dots, w_n$, then **weighted mean**, $\bar{x}_w = \dfrac{\Sigma w_i x_i}{\Sigma w_i}$.

- If $\bar{x}_1, \bar{x}_2, \dots, \bar{x}_k$ be the means of k sets of observations of size $n_1, n_2, \dots, n_k$ respectively, then their **combined mean**,

$$\bar{x}_{1k} = \frac{n_1\bar{x}_1 + n_2\bar{x}_2 + \dots + n_k\bar{x}_k}{n_1 + n_2 + \dots + n_k}$$

Here, $\bar{x}_1$ = Mean of first set of observations

n_1 = Number of observations in first set

PREP MIRROR
Your Personal Preparation Indicator

- No. of Questions in Exercises (x)—
- No. of Questions Attempted (y)—
- No. of Correct Questions (z)— *(Without referring Explanations)*

- Accuracy Level ($z/y \times 100$)—
- Prep Level ($z/x \times 100$)—

In order to expect good rank in JEE, your Accuracy Level should be above 85 & Prep Level should be above 75.

$\bar{x}_2$ = Mean of second set of observations
n_2 = Number of observations in second set and so on.

NOTE
- The sum of the deviations of the individual values from AM is always zero, i.e. $\Sigma(x_i - \bar{x}) = 0$.
- The sum of squares of deviations of the individual values is least when taken from AM i.e. $\Sigma(x_i - \bar{x})^2$ is least.

2. Geometric Mean (GM)

If $x_1, x_2, \dots x_n$ are n observations, then nth root of the product of all observations is called **geometric mean.**

- If $x_1, x_2, \dots, x_n$ are n non-zero positive observations, then
$$GM = (x_1 \cdot x_2 \dots x_n)^{1/n} = \text{antilog}\left(\frac{1}{n}\sum_{i=1}^{n} \log x_i\right)$$
- If corresponding frequencies of each observation are $f_1, f_2, \dots, f_n$, then
$$GM = [x_1^{f_1} \cdot x_2^{f_2} \dots x_n^{f_n}]^{\frac{1}{N}} = \text{antilog}\left[\frac{1}{N}\sum_{i=1}^{n} f_i \log x_i\right]$$
where, $N = \sum_{i=1}^{n} f_i$

3. Harmonic Mean (HM)

The harmonic mean of any set of non-zero observations, is the reciprocal of the arithmetic mean of the reciprocals of the observations.

- The harmonic mean of n items $x_1, x_2, \dots, x_n$ is defined as
$$HM = \frac{n}{\frac{1}{x_1} + \frac{1}{x_2} + \dots + \frac{1}{x_n}} = \frac{1}{\frac{1}{n}\sum_{i=1}^{n}\frac{1}{x_i}} = \left[\frac{1}{n}\sum_{i=1}^{n}\frac{1}{x_i}\right]^{-1}$$
- If corresponding frequencies of each observation are $f_1, f_2, \dots, f_n$, then $HM = \left[\frac{1}{N}\sum_{i=1}^{n}\frac{f_i}{x_i}\right]^{-1}$
- Relation between AM, GM and HM $(GM)^2 = (AM)\cdot(HM)$

4. Median

If the observations are arranged in ascending or descending order, then the value of the middle observation is defined as the **median.**

- Let $x_1, x_2, \dots, x_n$ be n observations, arranged in ascending or descending order, then

If n is odd, then Median $= \frac{n+1}{2}$ th observation

If n is even, then Median $= \frac{\left[\left(\frac{n}{2}\right)\text{th} + \left(\frac{n}{2}+1\right)\text{th}\right]\text{observation}}{2}$

- If in a continuous distribution the total frequency is N, then the class whose cumulative frequency is either equal to $N/2$ or just greater than $N/2$ is called **median class** and in that case,
$$\text{Median} = l + \frac{\frac{N}{2} - c}{f} \times h,$$
where, l = Lower limit of median class
f = Frequency of median class
h = Size of median class
c = Cumulative frequency of class preceding the median class

5. Mode

Mode is the observation which has maximum frequency.

- If $x_1, x_2, \dots, x_n$ are the n observations and corresponding frequencies are $f_1, f_2, \dots, f_n$, then the observation of maximum frequency is a modal value.
- In a continuous distribution the interval which has maximum frequency called **modal class** and in that case,
$$\text{mode} = l + \frac{f_1 - f_0}{2f_1 - f_0 - f_2} \times h,$$
where, l = Lower limit of modal class
f_1 = Frequency of modal class
f_0 = Frequency of the class preceding the modal class
f_2 = Frequency of the class succeeding the modal class
h = Class size
- Relation between Mean, Median and Mode
Mode = 3 Median − 2 Mean (Emperical formula)
- It is not necessary that a distribution has unique mode.

Measure of Dispersion

A measure of dispersion is designed to state the extent to which the individual observations vary from their average. The commonly used measures of dispersion are :

Range

The difference between the maximum and the minimum observations is called **range.**

i.e. Range $= L - S$.
where, L = Maximum observation
and S = Minimum observation

Coefficient of range $= \frac{L-S}{L+S}$

Mean Deviation (MD)

The mean, of the absolute deviations of the values of the variable from a measure of their average is called **Mean Deviation** (MD).

- If $x_1, x_2, \dots, x_n$ are n observations, then $MD = \frac{\sum_{i=1}^{n}|x_i - z|}{n}$
where, z = mean or mode or median

- If corresponding frequencies of each observation are $f_1, f_2, \ldots, f_n$, then

$$\text{MD} = \frac{\sum_{i=1}^{n} f_i |x_i - z|}{N}, \text{ where } N = \sum_{i=1}^{n} f_i$$

- Coefficient of MD $= \dfrac{\text{Mean deviation}}{\text{Corresponding average}}$
- Mean deviation is least when deviations are taken from median.

Standard Deviation (SD)

The square root of the arithmetic mean of the squares of deviations of the observations from their arithmetic mean is called **standard deviation** and it is denoted by σ.

- If $x_1, x_2, \ldots, x_n$ are n observations, then

$$\sigma = \sqrt{\frac{\sum_{i=1}^{n}(x_i - \bar{x})^2}{n}} = \sqrt{\frac{1}{n}\sum_{i=1}^{n} x_i^2 - \left(\frac{1}{n}\sum_{i=1}^{n} x_i\right)^2}$$

- If corresponding frequencies of each observation are $f_1, f_2, \ldots, f_n$, then

$$\sigma = \sqrt{\frac{\sum_{i=1}^{n} f_i(x_i - \bar{x})^2}{N}} = \sqrt{\frac{1}{N}\left(\sum_{i=1}^{n} f_i x_i^2\right) - \left(\frac{1}{N}\sum_{i=1}^{n} f_i x_i\right)^2},$$

where $N = \sum_{i=1}^{n} f_i$

Variance

The square of SD is called **variance** and it is denoted by σ^2.

- Coefficient of variation $= \dfrac{\sigma}{\bar{x}} \times 100\%$

- Two different series having n_1 and n_2 observations and whose corresponding means and variances are $\bar{x}_1, \bar{x}_2$ and σ_1^2, σ_2^2. Then, their **combined variance,**

$$\sigma^2 = \frac{n_1\sigma_1^2 + n_2\sigma_2^2 + n_1 d_1^2 + n_2 d_2^2}{n_1 + n_2},$$

where $d_1 = (\bar{x}_1 - \bar{x}_{12}),\ d_2 = (\bar{x}_2 - \bar{x}_{12})$

and $\bar{x}_{12} = \dfrac{n_1\bar{x}_1 + n_2\bar{x}_2}{n_1 + n_2}$

NOTE
- Standard deviation is always less than range.
- Standard deviation of n natural numbers is $\sigma = \left[\frac{1}{12}(n^2 - 1)\right]^{1/2}$.
- Mean deviation $= \frac{4}{5}\sigma$.

Effect of average and dispersion on change of origin and scale

	Change of origin	Change of scale
Mean	Dependent	Dependent
Median	Dependent	Dependent
Mode	Dependent	Dependent
Standard deviation	Independent	Dependent
Variance	Independent	Dependent

- If $\bar{x}$, m_e, m_o and σ represent the mean, median, mode and standard deviation, respectively, of $x_1, x_2, \ldots, x_n$. Then,
  (i) Mean of $ax_1 + b, ax_2 + b, \ldots, ax_n + b$, is $a\bar{x} + b$
  (ii) Median of $ax_1 + b, ax_2 + b, \ldots, ax_n + b$, is $am_e + b$
  (iii) Mode of $ax_1 + b, ax_2 + b, \ldots, ax_n + b$, is $am_o + b$
  (iv) Standard deviation of $ax_1 + b, ax_2 + b, \ldots, ax_n + b$, is $|a| \cdot \sigma$.

DAY PRACTICE SESSION 1

FOUNDATION QUESTIONS EXERCISE

1 The mean of a data set consisting of 20 observations is 40. If one observation 53 was wrongly recorded as 33, then the correct mean will be → JEE Mains 2013

(a) 41 (b) 49 (c) 40.5 (d) 42.5

2 If the variance of 1, 2, 3, 4, 5, ..., 10 is $\frac{99}{12}$, then the standard deviation of 3, 6, 9, 12, ..., 30 is

(a) $\frac{297}{4}$ (b) $\frac{3}{2}\sqrt{33}$ (c) $\frac{3}{2}\sqrt{99}$ (d) $\sqrt{\frac{99}{12}}$

3 The mean of n terms is $\bar{x}$. If the first term is increased by 1, second by 2 and so on, then the new mean is

(a) $\bar{x} + n$ (b) $\bar{x} + \frac{n}{2}$
(c) $\bar{x} + \frac{n+1}{2}$ (d) None of these

4 The mean of the data set comprising of 16 observations is 16. If one of the observation valued 16 is deleted and three new observations valued 3, 4 and 5 are added to the data, then the mean of the resultant data is → JEE Mains 2015

(a) 16.8 (b) 16.0 (c) 15.8 (d) 14.0

5 If the sum of deviation of a set of values $x_1, x_2, \ldots x_n$ measure from 59 is 20 and from 54 is 70, then sample size (n) and the sample mean is

(a) 10, 61 (b) 10, 55.67 (c) 6, 55.67 (d) 6, 44

6 The average marks of boys in a class is 52 and that of girls is 42. The average marks of boys and girls combined is 50. The percentage of boys in the class is

(a) 40% (b) 20% (c) 80% (d) 60%

7 The weighted mean of first n natural numbers whose weights are equal to the squares of corresponding numbers is

(a) $\frac{n+1}{2}$ (b) $\frac{3n(n+1)}{2(2n+1)}$ (c) $\frac{(n+1)(2n+1)}{6}$ (d) $\frac{n(n+1)}{2}$

8 A distribution consists of three components with frequencies 20, 25 and 30 having means 25, 10 and 15 respectively. The mean of the combined distribution is

(a) 14 (b) 16 (c) 17.5 (d) 20

9 A car completes the first half of its journey with a velocity v_1 and the rest half with a velocity v_2. Then the average velocity of the car for the whole journey is

(a) $\frac{v_1+v_2}{2}$ (b) $\sqrt{v_1v_2}$ (c) $\frac{2v_1v_2}{v_1+v_2}$ (d) None of these

10 The median of a set of 9 distinct observations is 20.5. If each of the largest 4 observations of the set is increased by 2, then the median of the new set is

(a) increased by 2
(b) decreased by 2
(c) two times the original median
(d) remains the same as that of original set

11 The median of 19 observations of a group is 30. If two observations with values 8 and 32 are further included, then the median of the new group of 21 observations will be

(a) 28 (b) 30 (c) 32 (d) 34

12 If a variable takes the discrete values

$\alpha+4, \alpha-\frac{7}{2}, \alpha-\frac{5}{2}, \alpha-3, \alpha-2, \alpha+\frac{1}{2}, \alpha-\frac{1}{2}, \alpha+5$

(where, $\alpha > 0$), then the median is

(a) $\alpha-\frac{5}{4}$ (b) $\alpha-\frac{1}{2}$ (c) $\alpha-2$ (d) $\alpha+\frac{5}{4}$

13 Median of ${}^{2n}C_0, {}^{2n}C_1, {}^{2n}C_2, {}^{2n}C_3, \dots, {}^{2n}C_n$ (where, n is even) is

(a) ${}^{2n}C_{\frac{n}{2}}$ (b) ${}^{2n}C_{\frac{n+1}{2}}$
(c) ${}^{2n}C_{\frac{n-1}{2}}$ (d) None of these

14 If the median and the range of four numbers $\{x, y, 2x+y, x-y\}$, where $0 < y < x < 2y$, are 10 and 28 respectively, then the mean of the numbers is → JEE Mains 2013

(a) 18 (b) 10 (c) 5 (d) 14

15 Find the mean deviation from the median of the following data. → NCERT Exemplar

Class interval	0-6	6-12	12-18	18-24	24-30
Frequency	4	5	3	6	2

(a) 7.08 (b) 7 (c) 7.1 (d) 7.05

16 If the mean deviations about the median of the numbers $a, 2a, \dots, 5a$ is 50, then $|a|$ is equal to

(a) 3 (b) 4 (c) 5 (d) 2

17 If the mean deviation of number $1, 1+d, 1+2d, \dots, 1+100d$ from their mean is 255, then d is equal to

(a) 10.0 (b) 20.0 (c) 10.1 (d) 20.2

18 Consider any set of 201 observations $x_1, x_2, \dots, x_{200}, x_{201}$. It is given that $x_1 < x_2 < \dots < x_{200} < x_{201}$. Then, the mean deviation of this set of observations about a point k is minimum when k is equal to

(a) x_{110} (b) x_1 (c) x_{101} (d) x_{201}

19 If the standard deviation of the numbers 2, 3, a and 11 is 3.5, then which of the following is true? → JEE Mains 2016

(a) $3a^2-26a+55=0$ (b) $3a^2-32a+84=0$
(c) $3a^2-34a+91=0$ (d) $3a^2-23a+44=0$

20 Mean and standard deviation of 100 observations were found to be 40 and 10, respectively. If at the time of calculation two observations were wrongly taken as 30 and 70 in place of 3 and 27 respectively, find the correct standard deviation. → NCERT Exemplar

(a) 10.20 (b) 10.24 (c) 10.29 (d) 10.27

21 Coefficient of variation of two distributions are 50 and 60 and their arithmetic means are 30 and 25, respectively. Difference of their standard deviation is → NCERT Exemplar

(a) 0 (b) 1 (c) 1.5 (d) 2.5

22 If MD is 12, the value of SD will be

(a) 15 (b) 12 (c) 24 (d) None of these

23 If SD of variate x is σ_x, then the SD of $\frac{ax+b}{p}, \forall a, b, p \in R$ is

(a) $\left|\frac{a}{p}\right|\sigma_x$ (b) $\left|\frac{p}{a}\right|\sigma_x$ (c) $\frac{p}{a}\sigma_x$ (d) $\frac{a}{p}\sigma_x$

24 If the standard deviation of the observations $-5, -4, -3, -2, -1, 0, 1, 2, 3, 4, 5$ is $\sqrt{10}$. The standard deviation of the observations 15, 16, 17, 18, 19, 20, 21, 22, 23, 24, 25 will be

(a) $\sqrt{10}+20$ (b) $\sqrt{10}+10$
(c) $\sqrt{10}$ (d) None of these

25 A scientist is weighing each of 30 fishes. Their mean weight worked out is 30 g and a standard deviation of 2 g. Later, it was found that the measuring scale was misaligned and always under reported every fish weight by 2 g. The correct mean and standard deviation (in gram) of fishes are respectively → AIEEE 2011

(a) 28, 4 (b) 32, 2 (c) 32, 4 (d) 28, 2

26 The variance of first 50 even natural numbers is → JEE Mains 2014

(a) $\frac{833}{4}$ (b) 833 (c) 437 (d) $\frac{437}{4}$

27 Mean of 5 observations is 7. If four of these observations are 6, 7, 8, 10 and one is missing, then the variance of all the five observations is → JEE Mains 2013

(a) 4 (b) 6 (c) 8 (d) 2

28 The mean of the numbers a, b, 8, 5, 10 is 6 and the variance is 6.80. Then, which one of the following gives possible values of a and b?

(a) $a = 3, b = 4$ (b) $a = 0, b = 7$
(c) $a = 5, b = 2$ (d) $a = 1, b = 6$

29 In an experiment with 15 observations on x, the following results were available $\Sigma x^2 = 2830$, $\Sigma x = 170$. One observation that was 20 was found to be wrong and was replaced by the correct value 30. Then, the corrected variance is

(a) 78.0 (b) 188.66 (c) 177.33 (d) 8.33

30 For two data sets, each of size 5, the variances are given to be 4 and 5 and the corresponding means are given to be 2 and 4, respectively. The variance of the combined data set is

(a) $\frac{5}{2}$ (b) $\frac{11}{2}$ (c) 6 (d) $\frac{13}{2}$

31 If a variable x takes values x_i such that $a \le x_i \le b$, for $i = 1, 2, \ldots, n$, then

(a) $a^2 \le \text{Var}(x) \le b^2$ (b) $a \le \text{Var}(x) \le b$
(c) $\frac{a^2}{4} \le \text{Var}(x)$ (d) $(b - a)^2 \ge \text{Var}(x)$

32 All the students of a class performed poorly in Mathematics. The teacher decided to give grace marks of 10 to each of the students. Which of the following statistical measures will not change even after the grace marks were given? → JEE Mains 2013

(a) Mean (b) Median (c) Mode (d) Variance

33 Let $x_1, x_2, \ldots, x_n$ be n observations. Let $w_i = l \cdot x_i + k$ for $i = 1, 2, \ldots, n$, where l and k are constants. If the mean of x_i's is 48 and their standard deviation is 12, the mean of w_i's is 55 and standard deviation of w_i's is 15. The values of l and k should be → NCERT Exemplar

(a) $l = 1.25, k = -5$ (b) $l = -1.25, k = 5$
(c) $l = 2.5, k = -5$ (d) $l = 2.5, k = 5$

34 If n is a natural number, then

Statement I The mean of the squares of first n natural number is $\frac{(n+1)(2n+1)}{6}$.

Statement II $\Sigma n = \frac{n(n+1)}{2}$

(a) Statement I is true, Statement II is true; Statement II is a correct explanation for Statement I
(b) Statement I is true, Statement II is true; Statement II is not a correct explanation for Statement I
(c) Statement I is true; Statement II is false
(d) Statement I is false; Statement II is true

35 Let $x_1, x_2, \ldots, x_n$ be n observations and let $\bar{x}$ be their arithmetic mean and σ^2 be the variance.

Statement I Variance of $2x_1, 2x_2, \ldots, 2x_n$ is $4\sigma^2$.

Statement II Arithmetic mean $2x_1, 2x_2, \ldots, 2x_n$ is $4\bar{x}$.

→ AIEEE 2012

(a) Statement I is true, Statement II is true; Statement II is a correct explanation for Statement I
(b) Statement I is true, Statement II is true; Statement II is not a correct explanation for Statement I
(c) Statement I is true; Statement II is false
(d) Statement I is false; Statement II is true

DAY PRACTICE SESSION 2

PROGRESSIVE QUESTIONS EXERCISE

1 The mean age of a combined group of men and women is 25 yr. If the mean age of the group of men is 26 and that of the group of women is 21, then the percentage of men and women in the group is

(a) 40, 60 (b) 80, 20 (c) 20, 80 (d) 60, 40

2 The average of the four-digit numbers that can be formed using each of the digits 3, 5, 7 and 9 exactly once in each number, is

(a) 4444 (b) 5555 (c) 6666 (d) 7777

3 Suppose a population A has 100 observations 101, 102, ..., 200 and another population B has 100 observations 151, 152, ..., 250. If V_A and V_B represent the variances of the two populations respectively, then $\frac{V_A}{V_B}$ is equal to

(a) $\frac{9}{4}$ (b) $\frac{4}{9}$ (c) $\frac{2}{3}$ (d) 1

4 If the value observed are 1, 2, 3, ..., n each with frequency 1 and n is even, then the mean deviation from mean equals to

(a) n (b) $\frac{n}{2}$ (c) $\frac{n}{4}$ (d) None of these

5 In a class of 19 students, seven boys failed in a test. Those who passed scored 12, 15, 17, 15, 16, 15, 19, 19, 17, 18, 18 and 19 marks. The median score of the 19 students in the class is

(a) 15 (b) 16 (c) 17 (d) 18

6 If G is the geometric mean of the product of r sets of observations with geometric means $G_1, G_2, G_3, \ldots, G_r$ respectively, then G is equal to

(a) $\log G_1 + \log G_2 + \ldots + \log G_r$
(b) $G_1 \cdot G_2 \ldots G_r$
(c) $\log G_1 \cdot \log G_2 \cdot \log G_3 \ldots \log G_r$
(d) None of the above

7 In a class of 100 students, there are 70 boys whose average marks in a subject are 75. If the average marks of the complete class is 72, then what is the average marks of the girls?

(a) 73 (b) 65 (c) 68 (d) 74

8 An aeroplane flies around a squares the sides of which measure 100 mile each. The aeroplane covers at a speed of 100 m/h the first side, at 200 m/h the second side, at 300 m/h the third side and 400 m/h the fourth side. The average speed of the aeroplane around the square is

(a) 190 m/h (b) 195 m/h (c) 192 m/h (d) 200 m/h

9 The first of two samples has 100 items with mean 15 and SD = 3. If the whole group has 250 items with mean 15.6 and SD = $\sqrt{13.44}$, the SD of the second group is

(a) 4 (b) 5 (c) 6 (d) 3.52

10 If a variable takes the values 0, 1, 2, ... , n with frequencies proportional to the binomial coefficients ${}^nC_0, {}^nC_1, \ldots, {}^nC_n$, then mean of the distribution is

(a) $\frac{n}{2}$ (b) $\frac{n(n+1)}{2}$ (c) $\frac{n(n-1)}{2}$ (d) $\frac{2}{n}$

11 The marks of some students were listed out of 75. The SD of marks was found to be 9. Subsequently the marks were raised to a maximum of 100 and variance of new marks was calculated. The new variance is

(a) 81 (b) 122 (c) 144 (d) 125

12 If $x_1, x_2, \ldots, x_n$ are n observations such that $\Sigma x_i^2 = 400$ and $\Sigma x_i = 80$. Then, a possible value of n among the following is

(a) 12 (b) 9 (c) 14 (d) 16

13 In a set of $2n$ observations, half of them are equal to a and the remaining half are equal to $-a$. If the standard deviation of all the observations is 2, then the value of $|a|$ is → JEE Mains 2013

(a) 2 (b) $\sqrt{2}$ (c) 4 (d) $2\sqrt{2}$

14 If $\sum_{i=1}^{9}(x_i - 5) = 9$ and $\sum_{i=1}^{9}(x_i - 5)^2 = 45$, then the standard deviation of the 9 items $x_1, x_2, \ldots, x_9$ is → JEE Mains 2018

(a) 9 (b) 4 (c) 2 (d) 3

15 If $\bar{x}_1$ and $\bar{x}_2$ are the means of two distributions such that $\bar{x}_1 < \bar{x}_2$ and $\bar{x}$ is the mean of the combined distribution, then

(a) $\bar{x} < \bar{x}_1$ (b) $\bar{x} > \bar{x}_2$

(c) $\bar{x} = \frac{\bar{x}_1 + \bar{x}_2}{2}$ (d) $\bar{x}_1 < \bar{x} < \bar{x}_2$

ANSWERS

SESSION 1

1. (a)	**2.** (b)	**3.** (c)	**4.** (d)	**5.** (a)	**6.** (c)	**7.** (b)	**8.** (b)	**9.** (c)	**10.** (d)
11. (b)	**12.** (a)	**13.** (a)	**14.** (d)	**15.** (b)	**16.** (b)	**17.** (c)	**18.** (c)	**19.** (b)	**20.** (b)
21. (a)	**22.** (a)	**23.** (a)	**24.** (c)	**25.** (b)	**26.** (b)	**27.** (a)	**28.** (a)	**29.** (a)	**30.** (b)
31. (d)	**32.** (d)	**33.** (a)	**34.** (b)	**35.** (c)					

SESSION 2

1. (b)	**2.** (c)	**3.** (d)	**4.** (c)	**5.** (a)	**6.** (b)	**7.** (b)	**8.** (c)	**9.** (a)	**10.** (a)
11. (c)	**12.** (d)	**13.** (a)	**14.** (c)	**15.** (d)					

Hints and Explanations

SESSION 1

1 Given; $\frac{\Sigma x_{\text{incorrect}}}{20} = 40$

$\Rightarrow \Sigma x_{\text{incorrect}} = 20 \times 40 = 800$

$\therefore \Sigma x_{\text{correct}} = 800 - 33 + 53 = 820$

$\Rightarrow \frac{\Sigma x_{\text{correct}}}{20} = \frac{820}{20}$

$\therefore$ Correct mean = 41

2 Given, $\sigma^2 = \frac{99}{12} = \frac{33}{4} \Rightarrow \sigma = \frac{\sqrt{33}}{2}$

Clearly, SD of required series

$= 3\sigma = \frac{3\sqrt{33}}{2}$

3 Let the observations be $x_1, x_2, \ldots, x_n$.

Then, $\bar{x} = \frac{x_1 + x_2 + \ldots + x_n}{n}$

Now, when the first term is increased by 1, second term by 2 and so on, then the observations will be $(x_1 + 1), (x_2 + 2), \ldots, (x_n + n)$.

$\therefore$ New mean

$= \frac{(x_1 + 1) + (x_2 + 2) + \ldots + (x_n + n)}{n}$

$= \frac{(x_1 + x_2 + \ldots + x_n)}{n} + \frac{(1 + 2 + \ldots + n)}{n}$

$= \bar{x} + \frac{n(n+1)}{2n} = \bar{x} + \frac{n+1}{2}$

4 Given, $\frac{x_1 + x_2 + x_3 + \ldots + x_{16}}{16} = 16$

$\Rightarrow \sum_{i=1}^{16} x_i = 16 \times 16$

Sum of new observations

$= \sum_{i=1}^{18} y_i = (16\times16 - 16) + (3 + 4 + 5) = 252$

Number of observations = 18

$\therefore$ New mean $= \frac{\sum_{i=1}^{18} y_i}{18} = \frac{252}{18} = 14$

5 We have, $\sum_{i=1}^{n}(x_i - 59) = 20$ ...(i)

and $\sum_{i=1}^{n}(x_i - 54) = 70$ ...(ii)

From Eq. (i), we get

$\sum_{i=1}^{n} x_i - 59n = 20$

$\Rightarrow \sum_{i=1}^{n} x_i = 20 + 59n$ ...(iii)

and similarly, from Eq. (ii), we get

$\sum_{i=1}^{n} x_i = 70 + 54n$ ...(iv)

From Eqs. (iii) and (iv), we get

$70 + 54n = 20 + 59n$

$\Rightarrow 5n = 50 \Rightarrow n = 10$

Now, from Eq. (iii), we get

$\sum_{i=1}^{n} x_i = 610$

$\therefore$ Sample mean = 61

6 Let the number of boys and girls be x and y.

$\therefore 52x + 42y = 50(x + y)$

$\Rightarrow 52x + 42y = 50x + 50y$

$\Rightarrow 2x = 8y$

$\Rightarrow x = 4y$

$\therefore$ Total number of students in the class

$= x + y = 4y + y = 5y$

$\therefore$ Required percentage of boys

$= \left(\frac{4y}{5y} \times 100\right)\% = 80\%$

7 Required mean

$= \frac{1\cdot 1^2 + 2\cdot 2^2 + \dots + n\cdot n^2}{1^2 + 2^2 + \dots + n^2}$

$= \frac{1^3 + 2^3 + \dots + n^3}{1^2 + 2^2 + \dots + n^2} = \frac{\frac{n^2(n+1)^2}{4}}{\frac{n(n+1)(2n+1)}{6}}$

$= \frac{n(n+1)}{4} \times \frac{6}{2n+1} = \frac{3n(n+1)}{2(2n+1)}$

8 Here, $\bar{x}_1 = 25, \bar{x}_2 = 10, \bar{x}_3 = 15$

and $n_1 = 20, n_2 = 25, n_3 = 30$

Now, combined mean

$x_{13} = \frac{n_1\bar{x}_1 + n_2\bar{x}_2 + n_3\bar{x}_3}{n_1 + n_2 + n_3}$

$= \frac{500 + 250 + 450}{75} = \frac{1200}{75} = 16$

9 Clearly, average velocity

$= \text{HM of } (v_1, v_2) = \frac{2v_1 v_2}{v_1 + v_2}$

10 After arranging the terms in ascending order median is the $\left(\frac{n+1}{2}\right)$th term,

i.e. 5th term.

Here, we increase largest four observations of the set which will come after 5th term.

Hence, median remains the same as that of original set.

11 Since, there are 19 observations. So, the middle term is 10th.

After including 8 and 32, i.e. 8 will come before 30 and 32 will come after 30.

Here, new median will remain 30.

12 Firstly arrange the given data in ascending order.

$\alpha - \frac{7}{2}, \alpha - 3, \alpha - \frac{5}{2}, \alpha - 2, \alpha - \frac{1}{2},$

$\alpha + \frac{1}{2}, \alpha + 4, \alpha + 5$

$\therefore$ Median $= \frac{1}{2}$ [Value of 4th item + Value of 5th item]

$= \frac{\alpha - 2 + \alpha - \frac{1}{2}}{2} = \frac{2\alpha - \frac{5}{2}}{2} = \alpha - \frac{5}{4}$

13 Total number of terms $= n + 1$ (which is odd)

$\therefore$ Median $= \left(\frac{n+1+1}{2}\right)$th term

$= \left(\frac{n}{2} + 1\right)$th term $= {}^{2n}C_{n/2}$

14 First we arrange four numbers according to the condition

$0 < y < x < 2y$ i.e. $x - y, y, x, 2x + y$

Median $= \frac{\text{2nd term + 3rd term}}{2} = 10$

$\Rightarrow y + x = 20$ ...(i)

Range $= (2x + y) - (x - y) = 28$

$\Rightarrow x + 2y = 28$ ...(ii)

On solving Eqs. (i) and (ii), we get

$x = 12, y = 8$

So, four numbers are 4, 8, 12, 32.

$\therefore$ Mean $= \frac{4 + 8 + 12 + 32}{4} = \frac{56}{4} = 14$

15

Class interval	Mid value x_i	f	cf	$\lvert x - M_d\rvert$	$f\lvert x - M_d\rvert$
0-6	3	4	4	11	44
6-12	9	5	9	5	25
12-18	15	3	12	1	3
18-24	21	6	18	7	42
24-30	27	2	20	13	26
Total		20			140

Now, $\frac{N}{2} = \frac{20}{2} = 10$, which lies in the interval 12-18.

$l = 12, cf = 9, f = 3$

$\therefore M_d = 12 + \frac{10 - 9}{3} \times 6$

$\left[\because M_d = l + \frac{\frac{N}{2} - cf}{2} \times h\right]$

$= 12 + 2 = 14$

$\therefore$ Mean deviation $= \frac{\Sigma f_i\lvert x_i - M_d\rvert}{N} = \frac{140}{20} = 7$

16 Median of $a, 2a, 3a, 4a, \dots, 50a$ is

$\frac{25a + 26a}{2} = (25.5)a$

Mean deviation about median

$= \frac{\sum_{i=1}^{50}\lvert x_i - \text{median}\rvert}{n}$

$\Rightarrow 50 = \frac{1}{50}\{2\lvert a\rvert(0.5 + 1.5 + 2.5 + \dots + 24.5)\}$

$\Rightarrow 2500 = 2\lvert a\rvert\frac{25}{2}(2 \times 0.5 + 24 \times 1)$

$= 2\lvert a\rvert\cdot\frac{25}{2}(25)$

$\therefore \lvert a\rvert = 4$

17 Clearly,

$(\bar{x}) = \frac{\text{Sum of quantities}}{n} = \frac{\frac{n}{2}(a + l)}{n}$

$= \frac{1}{2}[1 + 1 + 100d] = 1 + 50d$

Now, MD $= \frac{1}{n}\Sigma\lvert x_i - \bar{x}\rvert \Rightarrow 255$

$= \frac{1}{101}[50d + 49d + 48d + \dots + d + 0 + d + \dots + 50d]$

$= \frac{2d}{101}\left(\frac{50 \times 51}{2}\right)$

$\therefore d = \frac{255 \times 101}{50 \times 51} = 10.1$

18 Given that, $x_1 < x_2 < x_3 < \dots < x_{201}$

Hence, median of the given observation

$= \left(\frac{201 + 1}{2}\right)$th item $= x_{101}$

Now, deviation will be minimum of taken from the median.

Hence, mean deviation will be minimum, if $k = x_{101}$.

19 We know that, if $x_1, x_2, \dots, x_n$ are n observations, then their standard deviation is given by $\sqrt{\frac{1}{n}\Sigma x_i^2 - \left(\frac{\Sigma x_i}{n}\right)^2}$

We have, $(3.5)^2 = \frac{(2^2 + 3^2 + a^2 + 11^2)}{4} - \left(\frac{2 + 3 + a + 11}{4}\right)^2$

$\Rightarrow \frac{49}{4} = \frac{4 + 9 + a^2 + 121}{4} - \left(\frac{16 + a}{4}\right)^2$

$\Rightarrow \frac{49}{4} = \frac{134 + a^2}{4} - \frac{256 + a^2 + 32a}{16}$

$\Rightarrow \frac{49}{4} = \frac{4a^2 + 536 - 256 - a^2 - 32a}{16}$

$\Rightarrow 49 \times 4 = 3a^2 - 32a + 280$

$\Rightarrow 3a^2 - 32a + 84 = 0$

20 $\therefore$ New mean,

$\bar{x} = \frac{100 \times 40 + 3 + 27 - 30 - 70}{100}$

$= \frac{4000-70}{100} = \frac{3930}{100} = 39.3$

$\because \Sigma x^2 = N(\sigma^2 + \bar{x}^2)$

$\therefore$ SD $= 100(100+1600) = 170000$

New $\Sigma x^2 = 170000-(30)^2-(70)^2+(3)^2+(27)^2$
$= 170000-900-4900+9+729 = 164938$

$\therefore$ New SD $= \sqrt{\frac{\text{New } \Sigma x^2}{N} - (\text{New } \bar{x})^2}$

$= \sqrt{\frac{164938}{100} - (39.3)^2}$

$= \sqrt{1649.38 - 1544.49} = \sqrt{104.89}$

$= 10.24$

21 Given, coefficient of variation, $C_1 = 50$ and coefficient of variation, $C_2 = 60$

We have, $\bar{x}_1 = 30$ and $\bar{x}_2 = 25$

$\because \quad C = \frac{\sigma}{\bar{x}} \times 100 \Rightarrow 50 = \frac{\sigma_1}{30} \times 100$

$\Rightarrow \sigma_1 = 15$ and $60 = \frac{\sigma_2}{25} \times 100$

$\Rightarrow \sigma_2 = 15$

$\therefore$ Required difference,

$\sigma_1 - \sigma_2 = 15 - 15 = 0$

22 We know that MD $= \frac{4}{5}$ SD

$\therefore$ SD $= \frac{5}{4}$ MD $= \frac{5}{4} \times 12 = 15$

23 Let $u = \frac{ax+b}{p}$, then $\bar{u} = \frac{a\bar{x}+b}{p}$

$\therefore$ SD $= \sqrt{\frac{\Sigma(u-\bar{u})^2}{\Sigma f}} = \sqrt{\frac{a^2}{p^2}\frac{\Sigma(x-\bar{x})^2}{\Sigma f}}$

$= \sqrt{\frac{a^2}{p^2}\sigma_x^2} = \left|\frac{a}{p}\right|\sigma_x$

24 The new observations are obtained by adding 20 to each previous observation. Hence, the standard deviation of new observations will be same i.e. $\sqrt{10}$.

25 Correct mean = Old mean +2

$= 30 + 2 = 32$

As standard deviation is independent of change of origin.

$\therefore$ It remains same.

$\Rightarrow$ Standard deviation = 2

26 Here, $\bar{x} = \frac{\Sigma x_i}{n}$

$= \frac{2+4+6+8+\ldots+100}{50}$

$= \frac{50 \times 51}{50} = 51$

$[\because \Sigma 2n = n(n+1)$, here $n = 50]$

Variance, $\sigma^2 = \frac{1}{n}\Sigma x_i^2 - (\bar{x})^{-2}$

$\sigma^2 = \frac{1}{50}(2^2+4^2+\ldots+100^2) - (51)^2 = 833$

27 Mean, $7 = \frac{6+7+8+10+x}{5}$

$\Rightarrow \quad x = 4$

Variance

$= \frac{(6-7)^2+(7-7)^2+(8-7)^2+(10-7)^2+(4-7)^2}{5}$

$= \frac{1^2+0+1^2+3^2+3^2}{5} = \frac{20}{5} = 4$

28 According to the given condition,

$6.80 = \frac{\left[\begin{array}{c}(6-a)^2+(6-b)^2+(6-8)^2 \\ +(6-5)^2+(6-10)^2\end{array}\right]}{5}$

$\Rightarrow 34 = (6-a)^2 + (6-b)^2 + 4 + 1 + 16$

$\Rightarrow \quad (6-a)^2 + (6-b)^2 = 13 = 9 + 4$

$\Rightarrow \quad (6-a)^2 + (6-b)^2 = 3^2 + 2^2$

$\therefore \quad a = 3, b = 4$

29 Given, $n = 15$, $\Sigma x^2 = 2830$ and $\Sigma x = 170$

Since, one observation 20 was replaced by 30, then

$\Sigma x^2 = 2830 - 400 + 900 = 3330$

and $\Sigma x = 170 - 20 + 30 = 180$

$\therefore$ Variance, $\sigma^2 = \frac{\Sigma x^2}{n} - \left(\frac{\Sigma x}{n}\right)^2$

$= \frac{3330}{15} - \left(\frac{180}{15}\right)^2$

$= \frac{3330-2160}{15} = 78.0$

30 Here, $\sigma_1^2 = 4, \sigma_2^2 = 5, \bar{x}_1 = 2, \bar{x}_2 = 4$, $n_1 = n_2 = 5$

Clearly, combined mean

$x_{12} = \frac{n_1\bar{x}_1 + n_2\bar{x}_2}{n_1+n_2}$

$= \frac{1}{2}(2+4) = 3$

Combined variance,

$\sigma^2 = \frac{n_1\sigma_1^2 + n_2\sigma_2^2 + n_1d_1^2 + n_2d_2^2}{n_1+n_2}$

$= \frac{1}{2}[4+5+1+1] = \frac{11}{2}$

31 Since, SD < Range $\Rightarrow \sigma \le (b-a)$

$\sigma^2 \le (b-a)^2 \Rightarrow (b-a)^2 \ge \text{Var}(x)$

32 Since, variance is independent of change of origin. So, variance will not change whereas mean, median and mode will increase by 10.

33 Given, $w_i = lx_i + k$

$M(x_i) = \bar{x} = 48, \sigma(x_i) = 12$

$M(\bar{w}) = 55$ and $\sigma(w) = 15$

$M(w_i) = lM(x_i) + M(k)$

$55 = l \times 48 + k \quad \ldots$(i)

and $\sigma(w_i) = l\sigma(x_i) + \sigma(k)$

$\Rightarrow \quad 15 = l(12) + 0$

$\Rightarrow \quad l = \frac{15}{12} = 1.25$

From Eq. (i),

$55 = 1.25 \times 48 + k$

$\Rightarrow \quad k = 55 - 60$

$\therefore \quad k = -5$

34 Required mean $= \frac{\Sigma n^2}{n}$

$= \frac{n(n+1)(2n+1)}{6n} = \frac{(n+1)(2n+1)}{6}$

and Statement II is also a true Statement.

35 Clearly, variance of $2x_1, 2x_2, \ldots, 2x_n$ is $2^2 \cdot \sigma^2 = 4\sigma^2$.

and AM $= \frac{2x_1 + 2x_2 + 2x_3 + \ldots + 2x_n}{n}$

$= 2\left(\frac{x_1 + x_2 + x_3 + \ldots + x_n}{n}\right) = 2\bar{x}$

SESSION 2

1 Let x be the number of men and y be the number of women.

Then, we have

$25 = \frac{x \cdot 26 + y \cdot 21}{x+y}$

$\left[\because \text{Combined mean } = \frac{n_1\bar{x}_1 + n_2\bar{x}_2}{n_1+n_2}\right]$

$\Rightarrow 25x + 25y = 26x + 21y$

$\Rightarrow x = 4y$

Now, percentage of men

$= \frac{x}{x+y} \times 100$

$= \frac{4y}{5y} \times 100 = 80\%$

and hence percentage of women

$= 20\%$

2 Number of required four-digit numbers

$= 4 \times 3 \times 2 \times 1 = 24$

and the sum of all the required four-digit numbers

$= (3+5+7+9) \times (4-1)! \times \left(\frac{10^4-1}{10-1}\right)$

$= 24 \times 6 \times \frac{9999}{9}$

Now, required average

$= \frac{24 \times 6 \times 9999}{9 \times 24} = 6666$

3 Since, variance is independent of change of origin. So, variance of observations 101, 102, 200 is same as variance of 151, 152, ..., 250.

$\therefore \quad V_A = V_B$

$\Rightarrow \quad \frac{V_A}{V_B} = 1$

4 Clearly, mean

$$(\bar{x}) = \frac{1+2+3+\ldots+n}{n} = \frac{n(n+1)}{2n} = \frac{n+1}{2}$$

Now, mean deviation from mean

$$= \frac{1}{n}\Sigma|x_i - \bar{x}|$$

$$= \frac{1}{n}\left[\left|1-\frac{n+1}{2}\right| + \left|2-\frac{n+1}{2}\right| + \ldots + \left|n-\frac{n+1}{2}\right|\right]$$

$$= \frac{1}{n}\left[\left|\frac{1-n}{2}\right| + \left|\frac{3-n}{2}\right| + \ldots + \left|\frac{n-1}{2}\right|\right]$$

$$= \frac{1}{n}\left[\frac{n-1}{2} + \frac{n-3}{2} + \ldots + \frac{n-3}{2} + \frac{n-1}{2}\right]$$

$$= \frac{1}{n}[(n-1)+(n-3)+\ldots+1]\frac{1}{2}$$

$$\frac{1}{2n}\left[\frac{n}{2}[n-1+1]\right] = \frac{n}{4}$$

5 Clearly, median score = score of $\left(\frac{19+1}{2}\right)$ = 10th student.

Since, seven boys are failed out of 19, therefore according to given information there score will be less than 12.

Now, on arranging these scores in ascending order, we get score of 10th student is 15.

6 Taking X as the product of variates $X_1, X_2, \ldots, X_r$, we get

$$\log X = \log X_1 + \log X_2 + \log X_3 + \ldots + \log X_r$$

$$\Rightarrow \Sigma\log X = \Sigma\log X_1 + \Sigma\log X_2 + \Sigma\log X_3 + \ldots + \Sigma\log X_r$$

$$\Rightarrow \frac{1}{n}\Sigma\log X = \frac{1}{n}\Sigma\log X_1 + \frac{1}{n}\Sigma\log X_2 + \frac{1}{n}\Sigma\log X_3 + \ldots + \frac{1}{n}\Sigma\log X_r$$

$$\Rightarrow \log G = \log G_1 + \log G_2 + \ldots + \log G_r$$

$$\Rightarrow G = G_1 \cdot G_2 \cdot \ldots G_r$$

7 Since, total number of students = 100 and number of boys = 70

$\therefore$ Number of girls = (100 − 70) = 30

Now, the total marks of 100 students

$$= 100 \times 72 = 7200$$

And total marks of 70 boys

$$= 70 \times 75 = 5250$$

Total marks of 30 girls

$$= 7200 - 5250 = 1950$$

$\therefore$ Average marks of 30 girls

$$= \frac{1950}{30} = 65$$

8 Using the harmonic mean formula,

$$\frac{1}{H} = \frac{1}{N}\sum_{i=1}^{n}\frac{f_i}{x_i} \Rightarrow H = \frac{1}{\frac{1}{N}\sum_{i=1}^{n}\frac{f_i}{x_i}}$$

$\therefore$ Average speed

$$= \frac{400}{100\left(\frac{1}{100}+\frac{1}{200}+\frac{1}{300}+\frac{1}{400}\right)}$$

$$= 192 \text{ m/h}$$

9 We know,

$$\sigma^2 = \frac{n_1(\sigma_1^2 + d_1^2) + n_2(\sigma_2^2 + d_2^2)}{n_1 + n_2},$$

where $d_1 = m_1 - a, d_2 = m_2 - a$, a being the mean of the whole group.

$$\therefore \quad 15.6 = \frac{100\times 15 + 150\times m_2}{250}$$

$$\Rightarrow \quad m_2 = 16$$

Thus,

$$13.44 = \frac{\left[\begin{array}{c}(100\times 9 + 150\times\sigma^2) + 100\\ \times(0.6)^2 + 150\times(0.4)^2\end{array}\right]}{250}$$

$$\Rightarrow \quad \sigma = 4$$

10 Here,

$$N = \sum_{i=1}^{n} f_i = k({}^nC_0 + {}^nC_1 + \ldots + {}^nC_n)$$

$$= k(1+1)^n = k2^n$$ where, k is a constant of proportionality.

and $\Sigma f_i x_i = k(1\cdot{}^nC_1 + 2\cdot{}^nC_2 + \ldots + n\cdot{}^nC_n)$

$$= kn\left[1 + (n-1) + \frac{(n-1)(n-2)}{2!} + \ldots + 1\right]$$

$$= kn2^{n-1}$$

$$\therefore \text{Mean, } \bar{x} = \frac{1}{2^n}(n\cdot 2^{n-1}) = \frac{n}{2}$$

11 Given, $\sigma = 9$

Let a student obtains x marks out of 75. Then, his marks out of 100 are $\frac{4x}{3}$. Each observation is multiply by $\frac{4}{3}$.

$$\therefore \text{New SD, } \sigma = \frac{4}{3}\times 9 = 12$$

Hence, variance is $\sigma^2 = 144$.

12 Given that, $\Sigma x_i^2 = 400$ and $\Sigma x_i = 80$, since $\sigma^2 \geq 0$

$$\Rightarrow \quad \frac{\Sigma x_i^2}{n} - \left(\frac{\Sigma x_i}{n}\right)^2 \geq 0$$

$$\Rightarrow \quad \frac{400}{n} - \frac{6400}{n^2} \geq 0$$

$$\therefore \quad n \geq 16$$

13

$$\text{Mean} = \frac{\underbrace{(a + a + \ldots + n \text{ times}) + (-a - a - \ldots - n \text{ times})}}{2n}$$

$$= 0$$

$$\therefore \quad SD = \sqrt{\frac{(x_i - \bar{x})^2}{2n}}$$

$$= \sqrt{\frac{a^2 + a^2 + \ldots + 2n \text{ times}}{2n}}$$

$$2 = \sqrt{\frac{(a)^2\, 2n}{2n}} = |a|$$

14 Since, standard deviation is remain unchanged, if observations are added or subtracted by a fixed number

We have, $\sum_{i=1}^{9}(x_1 - 5) = 9$

and $\sum_{i=1}^{9}(x_1 - 5)^2 = 45$

$$SD = \sqrt{\frac{\sum_{i=1}^{9}(x_1-5)^2}{9} - \left(\frac{\sum_{i=1}^{9}(x_1-5)}{9}\right)^2}$$

$$\Rightarrow \quad SD = \sqrt{\frac{45}{9} - \left(\frac{9}{9}\right)^2}$$

$$\Rightarrow \quad SD = \sqrt{5-1} = \sqrt{4} = 2$$

15 Let n_1 and n_2 be the number of observations in two distributions having means $\bar{x}_1$ and $\bar{x}_2$ respectively. Then,

$$\bar{x} = \frac{n_1\bar{x}_1 + n_2\bar{x}_2}{n_1 + n_2}$$

Now consider,

$$\bar{x} - \bar{x}_1 = \frac{n_1\bar{x}_1 + n_2\bar{x}_2}{n_1 + n_2} - \bar{x}_1$$

$$= \frac{n_2(\bar{x}_2 - \bar{x}_1)}{n_1 + n_2} > 0 \quad [\because \bar{x}_2 > \bar{x}_1]$$

$$\Rightarrow \quad \bar{x} > \bar{x}_1 \quad \ldots\text{(i)}$$

Similarly, $\bar{x} - \bar{x}_2 = \frac{n_1(\bar{x}_1 - \bar{x}_2)}{n_1 + n_2} < 0$

$$\Rightarrow \quad \bar{x} < \bar{x}_2 \quad \ldots\text{(ii)}$$

From Eqs. (i) and (ii), we get

$$\bar{x}_1 < \bar{x} < \bar{x}_2.$$

DAY THIRTY FIVE

Probability

Learning & Revision for the Day

- Sample Space and Event
- Probability of an Event
- Conditional Probability
- Multiplication Theorem on Probability
- Theorem of Total Probability
- Baye's Theorem
- Probability Distribution of a Random Variable
- Bernoulli Trials and Binomial Distribution

Sample Space and Event

The **sample space** is the set of all possible outcomes of an random experiment and it is generally denoted by S(e.g. tossing a coin, rolling a die, drawing a card from a pack of playing cards etc.).

An event is a subset of S. If a die is rolled, then $S = \{1, 2, 3, 4, 5, 6\}$ is the sample space and getting an odd number $A = \{1, 3, 5\}$ is an event.

Types of Events

Equally Likely Event

The given events are said to be **equally likely,** if none of them is expected to occur in preference to the other.

Mutually Exclusive/Disjoint

A set of events is said to be **mutually exclusive,** if occurrence of one of them prevents or denies the occurrence of any of the remaining events.

If a set of events $E_1, E_2, \ldots, E_n$ are mutually exclusive events, then $E_1 \cap E_2 \cap \ldots \cap E_n = \phi$.

Exhaustive Events

A set of events is said to be **exhaustive**, if atleast one of the events compulsorily occurs. If a set of events $E_1, E_2, \ldots, E_n$ are exhaustive events, then $E_1 \cup E_2 \cup \ldots \cup E_n = S$

NOTE If the set of events $E_1, E_2, \ldots E_n$ are mutually exclusive and exhaustive events then $E_i \cap E_j = \phi$, $i \neq j$ and $\bigcup_{i=1}^{n} E_i = S$

Complementary Events

In a random experiment, let S be the sample space and E be an event. If $E \subseteq S$, then $E^c = S - E$, S is called the complement of E.

PREP MIRROR
Your Personal Preparation Indicator

- No. of Questions in Exercises (x)—
- No. of Questions Attempted (y)—
- No. of Correct Questions (z)—
  (Without referring Explanations)
- Accuracy Level $(z/y \times 100)$—
- Prep Level $(z/x \times 100)$—

In order to expect good rank in JEE, your Accuracy Level should be above 85 & Prep Level should be above 75.

Probability of an Event

If the sample space has n points (all possible cases) and an event A has m points (all favourable cases), then the probability of A is $P(A) = \frac{m}{n}$.

(i) $0 \le P(A) \le 1$

(ii) $P(S) = 1, P(\phi) = 0$

(iii) Probability of Odds in favour of $A = \frac{P(A)}{P(\bar{A})} = \frac{m}{n-m}$

(iv) Probability of Odds in against $A = \frac{P(\bar{A})}{P(A)} = \frac{n-m}{m}$

Important Results

If n letters corresponding to n envelopes are placed in the envelopes at random, then

(i) probability that letters are in right envelopes $= \frac{1}{n!}$

(ii) probability that letters are not in right envelopes $= 1 - \frac{1}{n!}$

(iii) probability that no letter is in right envelope $= \frac{1}{2!} - \frac{1}{3!} + \frac{1}{4!} - \ldots + (-1)^n \frac{1}{n!}$

(iv) probability that exactly r letters are in right envelopes $= \frac{1}{r!}\left(\frac{1}{2!} - \frac{1}{3!} + \frac{1}{4!} - \ldots + (-1)^{n-r} \frac{1}{(n-r)!}\right)$

Addition Theorem of Probability

If A and B are two events associated with a random experiment, then $P(A \cup B) = P(A) + P(B) - P(A \cap B)$

If A, B and C are three events associated with a random experiment, then

$$P(A \cup B \cup C) = P(A) + P(B) + P(C) - P(A \cap B) - P(B \cap C) - P(C \cap A) + P(A \cap B \cap C).$$

For any two events A and B,

$$P(A \cap B) \le P(A) \le P(A \cup B) \le P(A) + P(B).$$

Booley's Inequalities

(i) $P(A \cap B) \ge P(A) + P(B) - 1$

(ii) $P(A \cap B \cap C) \ge P(A) + P(B) + P(C) - 2$

NOTE
- $P(\bar{A} \cap \bar{B}) = 1 - P(A \cup B)$
- $P(\bar{A} \cup \bar{B}) = 1 - P(A \cap B)$
- $P(A) = P(A \cap B) + P(A \cap \bar{B})$
- $P(B) = P(B \cap A) + P(B \cap \bar{A})$
- P (exactly one of E_1, E_2 occurs)
  $= P(E_1 \cap E_2') + P(E_1' \cap E_2)$
  $= P(E_1) - P(E_1 \cap E_2) + P(E_2) - P(E_1 \cap E_2)$
  $= P(E_1) + P(E_2) - 2P(E_1 \cap E_2)$
- P (neither E_1 nor E_2) $= P(E_1' \cap E_2') = 1 - P(E_1 \cup E_2)$

Conditional Probability

If A and B are two events associated with the sample space of a random experiment, then conditional probability of the event A, given that B has occurred

i.e. $$P\left(\frac{A}{B}\right) = \frac{P(A \cap B)}{P(B)}, \ P(B) \ne 0$$

Properties of Conditional Probability

Let A and B be two events of a sample space S of an experiment, then

(i) $P\left(\frac{S}{A}\right) = P\left(\frac{A}{A}\right) = 1$ (ii) $P\left(\frac{A'}{B}\right) = 1 - P\left(\frac{A}{B}\right)$,

where A' is complement of A.

Multiplication Theorem of Probability

If A and B are two events associated with a random experiment, then

$$P(A \cap B) = P(A) \cdot P\left(\frac{B}{A}\right), = P(B) \cdot P\left(\frac{A}{B}\right),$$

$P(A) \ne 0$ and $P(B) \ne 0$

Independent Events

Two events are said to be independent, if the occurrence of one does not depend upon the other.

- If $E_1, E_2, \ldots, E_n$ are independent events, then
  $P(E_1 \cap E_2 \cap E_3 \cap \ldots \cap E_n) = P(E_1) \cdot P(E_2) \ldots P(E_n)$.
- If E and F are independent events, then the pairs E and $\bar{F}$, $\bar{E}$ and F, $\bar{E}$ and $\bar{F}$ are also independent.

Theorem of Total Probability

If an event A can occur with one of the n mutually exclusive and exhaustive events $B_1, B_2, \ldots, B_n$ and the probabilities $P(A/B_1), P(A/B_2), \ldots P(A/B_n)$ are known, then

$$P(A) = \sum_{i=1}^{n} P(B_i) \cdot P(A/B_i)$$

Baye's Theorem

Let the sample space S be the union of n non-empty disjoint subsets (mutually exclusive and exhaustive events).

i.e. $S = A_1 \cup A_2 \cup \ldots \cup A_n$ and $A_i \cap A_j = \phi, i \ne j$

For any event B such that

$$B = (A_1 \cap B) \cup (A_2 \cap B) \cup \ldots \cup (A_n \cap B),$$

$$P(A_i/B) = \frac{P(A_i \cap B)}{P(B)} = \frac{P(A_i) P(B/A_i)}{\sum_{i=1}^{n} P(A_i) P(B/A_i)}$$

Probability Distribution of a Random Variable

- A random variable is a real valued function whose domain is the sample space of a random experiment.
- A random variable is usually denoted by the capital letters $X, Y, Z, \ldots$ and so on.
- If a random variable X takes values, $x_1, x_2, x_3, \ldots, x_n$ with respective probabilities $p_1, p_2, p_3, \ldots, p_n$, then

X	x_1	x_2	x_3	...	x_n
$P(X)$	p_1	p_2	p_3	...	p_n

  is known as the probability distribution of X.
- The probability distribution of random variable X is defined only when the various values of the random variable, e.g. $x_1, x_2, x_3, \ldots, x_n$ together with respective probabilities $p_1, p_2, p_3, \ldots, p_n$ satisfy $p_i > 0$ and $\sum_{i=1}^{n} p_i = 1$, where $i = 1, 2, n$

Mean

If a random variable X assumes values $x_1, x_2, x_3, \ldots, x_n$ with respective probabilities $p_1, p_2, p_3, \ldots, p_n$, then the mean $\overline{X}$ of X is defined as

$$\overline{X} = p_1 x_1 + p_2 x_2 + \ldots + p_i x_i \quad \text{or} \quad \overline{X} = \sum_{i=1}^{n} p_i x_i$$

The mean of a random variable X is also known as its mathematical expectation and it is denoted by E (X).

Variance

If a random variable X assumes values $x_1, x_2, x_3, \ldots, x_n$ with the respective probabilities $p_1, p_2, \ldots, p_n$, then variance of X is given by $\text{Var}(X) = \sum_{i=1}^{n} p_i x_i^2 - \left(\sum_{i=1}^{n} p_i x_i\right)^2$.

Bernoulli Trials and Binomial Distribution

Bernoulli Trials

Trials of a random experiment are called Bernoulli trials, if they satisfy the following conditions:

(i) There should be a finite number of trials
(ii) The trials should be independent
(iii) Each trial has exactly two outcomes; success or failure
(iv) The probability of success (or failure) remains the same in each trial.

Binomial Distribution

The probability of r successes in n independent Bernoulli trials is denoted by $P(X = r)$ and is given by

$$P(X = r) = {}^nC_r p^r q^{n-r}$$

where, $p =$ Probability of success
and $q =$ Probability of failure and $p + q = 1$

(i) Mean $= np$ (ii) Variance $= npq$
(iii) Mean is always greater than variance.

DAY PRACTICE SESSION 1

FOUNDATION QUESTIONS EXERCISE

1 If three distinct numbers are chosen randomly from the first 100 natural numbers, then the probability that all three of them are divisible by both 2 and 3 is

(a) $\frac{4}{55}$ (b) $\frac{4}{35}$ (c) $\frac{4}{33}$ (d) $\frac{4}{1155}$

2 If the letters of the word 'MATHEMATICS' are arranged arbitrarily, the probability that C comes before E, E before H, H before I and I before S, is

(a) $\frac{1}{75}$ (b) $\frac{1}{24}$ (c) $\frac{1}{120}$ (d) $\frac{1}{720}$

3 If A and B are two events, then the probability that exactly one of them occurs is given by

(a) $P(A) + P(B) - 2P(A \cap B)$
(b) $P(A \cap B') - P(A' \cap B)$
(c) $P(A \cup B) + P(A \cap B)$
(d) $P(A') + P(B') + 2P(A' \cap B')$

4 The probability that atleast one of the events A and B occur is 3/5. If A and B occur simultaneously with probability 1/5, then $P(A') + P(B')$ is equal to

(a) $\frac{2}{5}$ (b) $\frac{4}{5}$ (c) $\frac{6}{5}$ (d) $\frac{7}{5}$

5 A die is thrown. Let A be the event that the number obtained is greater than 3. Let B be the event that the number obtained is less than 5. Then, $P(A \cup B)$ is
→ AIEEE 2008

(a) $\frac{2}{5}$ (b) $\frac{3}{5}$
(c) 0 (d) 1

6 For three events, A, B and C, if P (exactly one of A or B occurs) $= P$ (exactly one of B or C occurs) $= P$ (exactly one of C or A occurs) $= \frac{1}{4}$ and P (all the three events occur simultaneously) $= \frac{1}{16}$, then the probability that atleast one of the events occurs, is → JEE Mains 2017

(a) $\frac{7}{32}$ (b) $\frac{7}{16}$ (c) $\frac{7}{64}$ (d) $\frac{3}{16}$

7 In a leap year, the probability of having 53 Sunday or 53 Monday is → NCERT Exemplar

(a) $\frac{2}{7}$ (b) $\frac{3}{7}$ (c) $\frac{4}{7}$ (d) $\frac{5}{7}$

8 A number is chosen at random among the first 120 natural numbers. The probability of the number chosen being a multiple of 5 or 15 is

(a) $\frac{1}{8}$ (b) $\frac{1}{5}$ (c) $\frac{1}{24}$ (d) $\frac{1}{6}$

9 Consider two events A and B. If odds against A are as 2 : 1 and those in favour of $A \cup B$ are as 3 : 1, then

(a) $\frac{1}{2} \le P(B) \le \frac{3}{4}$ (b) $\frac{5}{12} \le P(B) \le \frac{3}{4}$

(c) $\frac{1}{4} \le P(B) \le \frac{3}{5}$ (d) None of these

10 A and B are events such that $P(A) = 0.4$, $P(B) = 0.3$ and $P(A \cup B) = 0.5$. Then, $P(B/A)$ is equal to → NCERT Exemplar

(a) $\frac{2}{3}$ (b) $\frac{1}{2}$ (c) $\frac{3}{10}$ (d) $\frac{1}{5}$

11 If A and B are two events such that $P(A) = \frac{1}{2}$, $P(B) = \frac{1}{3}$, $P\left(\frac{A}{B}\right) = \frac{1}{4}$, then $P(A' \cap B')$ is equal to → NCERT Exemplar

(a) $\frac{1}{12}$ (b) $\frac{3}{4}$ (c) $\frac{1}{4}$ (d) $\frac{3}{16}$

12 It is given that the events A and B are such that $P(A) = \frac{1}{4}$, $P(A/B) = \frac{1}{2}$ and $P(B/A) = \frac{2}{3}$. Then, $P(B)$ is equal to → AIEEE 2008

(a) $\frac{1}{2}$ (b) $\frac{1}{6}$ (c) $\frac{1}{3}$ (d) $\frac{2}{3}$

13 Let A and B be two events such that $P(A) = 0.6$, $P(B) = 0.2$ and $P(A/B) = 0.5$. Then, $P(A'/B')$ equals → NCERT Exemplar

(a) $\frac{1}{10}$ (b) $\frac{3}{10}$ (c) $\frac{3}{8}$ (d) $\frac{6}{7}$

14 If two events A and B are such that $P(A') = 0.3$, $P(B) = 0.4$ and $(A \cap B') = 0.5$, then $P\left(\frac{B}{A \cup B'}\right)$ is equal to

(a) $\frac{1}{4}$ (b) $\frac{1}{5}$ (c) $\frac{3}{5}$ (d) $\frac{2}{5}$

15 If C and D are two events such that $C \subset D$ and $P(D) \ne 0$, then the correct statement among the following is

(a) $P(C/D) \ge P(C)$ (b) $P(C/D) < P(C)$

(c) $P(C/D) = \frac{P(D)}{P(C)}$ (d) $P(C/D) = P(C)$

16 Let E and F be two independent events such that $P(E) > P(F)$. The probability that both E and F happen is $\frac{1}{12}$ and the probability that neither E nor F happens is $\frac{1}{2}$, then

(a) $P(E) = \frac{1}{3}, P(F) = \frac{1}{4}$ (b) $P(E) = \frac{1}{2}, P(F) = \frac{1}{6}$

(c) $P(E) = 1, P(F) = \frac{1}{12}$ (d) $P(E) = \frac{1}{3}, P(F) = \frac{1}{2}$

17 An unbiased coin is tossed. If the result is a head, a pair of unbiased dice is rolled and the sum of the numbers obtained is noted. If the result is a tail, a card from a well-shuffled pack of eleven cards numbered 2, 3, 4, ..., 12 is picked and the number on the card is noted. The probability that the noted number is either 7 or 8 is

(a) $\frac{192}{401}$ (b) $\frac{193}{401}$ (c) $\frac{193}{792}$ (d) $\frac{17}{75}$

18 A bag contains 3 red and 3 white balls. Two balls are drawn one-by-one. The probability that they are of different colours, is

(a) $\frac{3}{10}$ (b) $\frac{2}{5}$

(c) $\frac{3}{5}$ (d) None of these

19 For two events A and B, if $P(A) = P\left(\frac{A}{B}\right) = \frac{1}{4}$ and $P\left(\frac{B}{A}\right) = \frac{1}{2}$, then

(a) A and B are independent (b) $P\left(\frac{A'}{B}\right) = \frac{3}{4}$

(c) $P\left(\frac{B'}{A'}\right) = \frac{1}{2}$ (d) All of these

20 Let A and B be two events such that $P(\overline{A \cup B}) = \frac{1}{6}$, $P(A \cap B) = \frac{1}{4}$ and $P(\overline{A}) = \frac{1}{4}$, where $\overline{A}$ stands for the complement of the event A. Then, the events A and B are → JEE Mains 2014

(a) independent but not equally likely
(b) independent and equally likely
(c) mutually exclusive and independent
(d) equally likely but not independent

21 For independent events $A_1, \ldots, A_n \cdot P(A_i) = \frac{1}{i+1}$, $i = 1, 2, \ldots, n$. Then, the probability that none of the events will occur is

(a) $\frac{n}{(n+1)}$ (b) $\frac{(n-1)}{(n+1)}$ (c) $\frac{1}{(n+1)}$ (d) $n + \left(\frac{1}{(n+1)}\right)$

22 Let $0 < P(A) < 1$, $0 < P(B) < 1$ and $P(A \cup B) = P(A) + P(B) - P(A) \cdot P(B)$. Then,

(a) $P\left(\frac{B}{A}\right) = P(B) - P(A)$ (b) $P(A^c \cup B^c) = P(A^c) + P(B^c)$

(c) $P(A \cup B)^c = P(A^c) \cdot P(B^c)$ (d) $P\left(\frac{A}{B}\right) = P(B/A)$

23 Let two fair six-faced dice A and B be thrown simultaneously. If E_1 is the event that die A shows up four, E_2 is the event that die B shows up two and E_3 is the event that the sum of numbers on both dice is odd, then which of the following statements is not true?

(a) E_1 and E_2 are independent
(b) E_2 and E_3 are independent
(c) E_1 and E_3 are independent
(d) E_1, E_2 and E_3 are not independent

24 Let A, B, C be three mutually independent events. Consider the two Statements S_1 and S_2

S_1 : A and $B \cup C$ are independent
S_2 : A and $B \cap C$ are independent

Then,

(a) Both S_1 and S_2 are true (b) Only S_1 is true
(c) Only S_2 is true (d) Neither S_1 nor S_2 is true

25 Two independent events namely A and B and the probability that both A and B occurs is $\frac{1}{10}$ and the probability that neither of them occurs is $\frac{3}{10}$. Then, the probability of occurrence of event B is

(a) $\frac{4-\sqrt{7}}{3}$ (b) $\frac{4+\sqrt{7}}{3}$ (c) $\frac{4+\sqrt{6}}{2}$ (d) $\frac{4-\sqrt{6}}{10}$

26 A committee of 4 students is selected at random from a group consisting 8 boys and 4 girls. Given that there is atleast one girl on the committee, the probability that there are exactly 2 girls on the committee, is

→ NCERT Exemplar

(a) $\frac{7}{99}$ (b) $\frac{13}{99}$ (c) $\frac{14}{99}$ (d) None of these

27 Two aeroplanes I and II bomb a target in succession. The probabilities of I and II scoring a hit correctly are 0.3 and 0.2, respectively. The second plane will bomb only, if the first misses the target. The probability that the target is hit by the second plane, is

(a) 0.06 (b) 0.14 (c) 0.32 (d) 0.7

28 Three machines E_1, E_2 and E_3 in a certain factory produce 50%, 25% and 25%, respectively, of the total daily output of electric tubes. It is known that 4% of the tubes produced on each of machines E_1 and E_2 are defective and that 5% of those produced on E_3 are defective. If on tube is picked up at random from a day's production, the probability that it is defective, is → NCERT Exemplar

(a) 0.025 (b) 0.125 (c) 0.325 (d) 0.0425

29 A bag contains 4 red and 6 black balls. A ball is drawn at random from the bag, its colour is observed and this ball along with two additional balls of the same colour are returned to the bag. If now a ball is drawn at random from the bag, then the probability that this drawn ball is red, is → JEE Mains 2018

(a) $\frac{3}{10}$ (b) $\frac{2}{5}$ (c) $\frac{1}{5}$ (d) $\frac{3}{4}$

30 A person goes to office either by car, scooter, bus or train the probabilities of which being $\frac{1}{7}, \frac{3}{7}, \frac{2}{7}$ and $\frac{1}{7}$, respectively. The probability that he reaches office late, if he takes car, scooter, bus or train is $\frac{2}{9}, \frac{1}{9}, \frac{4}{9}$ and $\frac{1}{9}$, respectively. If he reaches office in time, the probability that he travelled by car is

(a) $\frac{1}{5}$ (b) $\frac{1}{9}$ (c) $\frac{2}{11}$ (d) $\frac{1}{7}$

31 A pack of playing cards was found to contain only 51 cards. If the first 13 cards which are examined are all red, then the probability that the missing card is black, is

(a) $\frac{2}{3}$ (b) $\frac{15}{26}$ (c) $\frac{16}{39}$ (d) $\frac{37}{52}$

32 A discrete random variable X has the following probability distribution.

X	1	2	3	4	5	6	7
$P(X)$	C	$2C$	$2C$	$3C$	C^2	$2C^2$	$7C^2 + C$

The value of C and the mean of the distribution are

→ NCERT Exemplar

(a) $\frac{1}{10}$ and 3.66 (b) $\frac{1}{20}$ and 2.66
(c) $\frac{1}{15}$ and 1.33 (d) None of these

33 For a random variable X, $E(X) = 3$ and $E(X^2) = 11$. Then, variable of X is

(a) 8 (b) 5 (c) 2 (d) 1

34 A random variable X has the probability distribution

X	1	2	3	4	5	6	7	8
$P(X)$	0.15	0.23	0.12	0.10	0.20	0.08	0.07	0.05

For the events $E = \{X \text{ is a prime number}\}$ and $F = \{X < 4\}$, then the probability $P(E \cup F)$ is

(a) 0.87 (b) 0.77 (c) 0.35 (d) 0.50

35 Two cards are drawn successively with replacement from a well-shuffled deck of 52 cards, then the mean of the number of aces is

(a) $\frac{1}{13}$ (b) $\frac{3}{13}$ (c) $\frac{2}{13}$ (d) None of these

36 Consider 5 independent Bernoulli's trials each with probability of success p. If the probability of atleast one failure is greater than or equal to 31/32, then p lies in the interval

(a) $\left(\frac{3}{4}, \frac{11}{12}\right]$ (b) $\left[0, \frac{1}{2}\right]$ (c) $\left(\frac{11}{12}, 1\right]$ (d) $\left(\frac{1}{2}, \frac{3}{4}\right]$

37 A fair coin is tossed a fixed number of times. If the probability of getting 7 heads is equal to getting 9 heads, then the probability of getting 2 heads is

(a) $\frac{15}{2^8}$ (b) $\frac{2}{15}$ (c) $\frac{15}{2^{13}}$ (d) None of these

38 A fair coin is tossed n times. If X is the number of times heads occur and $P(X = 4), P(X = 5)$ and $P(X = 6)$ are in AP, then n is equal to

(a) 13 (b) 7 (c) 11 (d) None of these

39 A box contains 15 green and 10 yellow balls. If 10 balls are randomly drawn one-by-one with replacement, then the variance of the number of green balls drawn is → JEE Mains 2017

(a) $\frac{12}{5}$ (b) 6 (c) 4 (d) $\frac{6}{25}$

40 A multiple choice examination has 5 questions. Each question has three alternative answers of which exactly one is correct. The probability that a student will get 4 or more correct answers just by guessing is → JEE Mains 2013

(a) $\frac{17}{3^5}$ (b) $\frac{13}{3^5}$ (c) $\frac{11}{3^5}$ (d) $\frac{10}{3^5}$

DAY PRACTICE SESSION 2

PROGRESSIVE QUESTIONS EXERCISE

1 7 white balls and 3 black balls are placed in a row at random. The probability that no two black balls are adjacent is

(a) $\frac{1}{2}$ (b) $\frac{7}{15}$ (c) $\frac{2}{15}$ (d) $\frac{1}{3}$

2 Events A, B and C are mutually exclusive events such that $P(A) = \frac{3x+1}{3}, P(B) = \frac{1-x}{4}$ and $P(C) = \frac{1-2x}{2}$. Then, the set of possible values of x are in the interval

(a) $\left[\frac{1}{3}, \frac{1}{2}\right]$ (b) $\left[\frac{1}{3}, \frac{2}{3}\right]$ (c) $\left[\frac{1}{3}, \frac{13}{3}\right]$ (d) $[0, 1]$

3 If 12 balls are to be placed in 3 identical boxes, then the probability that one of the boxes contains exactly 3 balls, is → JEE Mains 2015

(a) $\frac{55}{3}\left(\frac{2}{3}\right)^{11}$ (b) $55\left(\frac{2}{3}\right)^{10}$ (c) $220\left(\frac{1}{3}\right)^{12}$ (d) $22\left(\frac{1}{3}\right)^{11}$

4 If 6 objects are distributed at random among 6 persons, the probability that atleast one person does not get any object is

(a) $\frac{313}{324}$ (b) $\frac{315}{322}$ (c) $\frac{317}{324}$ (d) $\frac{319}{324}$

5 10 apples are distributed at random among 6 persons. The probability that atleast one of them will receive none is

(a) $\frac{6}{143}$ (b) $\frac{^{14}C_4}{^{15}C_5}$ (c) $\frac{137}{143}$ (d) $\frac{135}{143}$

6 A draws two cards at random from a pack of 52 cards. After returning them to the pack and shuffling it, B draws two cards at random. The probability that their draws contain exactly one common card is

(a) $\frac{25}{546}$ (b) $\frac{50}{663}$ (c) $\frac{25}{663}$ (d) None of these

7 In a dice game, a player pays a stake of ₹ 1 for each throw of a die. She receives ₹ 5, if the die shows a 3, ₹ 2, if the die shows a 1 or 6 and nothing otherwise. What is the player's expected profit per throw over a long series of throws? → NCERT Exemplar

(a) 0.50 (b) 0.20 (c) 0.70 (d) 0.90

8 In an examination, 20 questions of true-false type are asked. Suppose a students tosses a fair coin to determine his answer to each question. If the coin falls heads, he answers true, if it falls tails, he answers false. The probability that he answers atleast 12 questions correctly is

(a) $\left(\frac{1}{2}\right)^{20} ({}^{20}C_{12} + {}^{20}C_{13} + ... + {}^{20}C_{20})$

(b) $\left(\frac{1}{2}\right)^{10} ({}^{20}C_{11} + {}^{20}C_{12} + ... + {}^{20}C_{20})$

(c) $\left(\frac{1}{2}\right)^{20} ({}^{20}C_{11} + {}^{20}C_{12} + ... + {}^{20}C_{20})$

(d) None of the above

9 In a multiple choice question there are four alternative answers of which one or more than one is correct. A candidate will get marks on the question only if he ticks the correct answer. The candidate decides to tick answers at random. If he is allowed up to three chances of answer the question, then the probability that he will get marks on it is

(a) $\frac{1}{3}$ (b) $\frac{2}{3}$ (c) $\frac{1}{5}$ (d) $\frac{2}{15}$

10 Three natural numbers are taken at random from the set $A = \{x \mid 1 \le x \le 100, x \in N\}$. The probability that the AM of the numbers taken is 25, is

(a) $\frac{^{77}C_2}{^{100}C_3}$ (b) $\frac{^{25}C_2}{^{100}C_3}$ (c) $\frac{^{74}C_{72}}{^{100}C_{97}}$ (d) $\frac{^{75}C_2}{^{100}C_3}$

11 Given that $x \in [0, 1]$ and $y \in [0, 1]$. Let A be the event of (x, y) satisfying $y^2 \le x$ and B be the event of (x, y) satisfying $x^2 \le y$. Then,

(a) $P(A \cap B) = \frac{1}{3}$ (b) A, B are exhaustive

(c) A, B are mutually exclusive (d) A, B are independent

12 If two different numbers are taken from the set $\{0, 1, 2, 3, \dots, 10\}$, then the probability that their sum as well as absolute difference are both multiple of 4, is

→ JEE Mains 2017

(a) $\frac{6}{55}$ (b) $\frac{12}{55}$ (c) $\frac{14}{45}$ (d) $\frac{7}{55}$

13 One Indian and four American men and their wives are to be seated randomly around a circular table. Then, the conditional probability that the Indian man is seated adjacent to his wife given that each American man is seated adjacent to his wife, is

(a) $\frac{1}{2}$ (b) $\frac{1}{3}$ (c) $\frac{2}{5}$ (d) $\frac{1}{5}$

14 If the integers m and n are chosen at random from 1 to 100, then the probability that a number of the form $7^n + 7^m$ is divisible by 5 equals

(a) $\frac{1}{4}$ (b) $\frac{1}{2}$ (c) $\frac{1}{8}$ (d) $\frac{1}{3}$

ANSWERS

SESSION 1

1 (d)	**2** (c)	**3** (a)	**4** (c)	**5** (d)	**6** (b)	**7** (b)	**8** (b)	**9** (b)	**10** (b)
11 (c)	**12** (c)	**13** (c)	**14** (a)	**15** (a)	**16** (a)	**17** (c)	**18** (c)	**19** (d)	**20** (a)
21 (c)	**22** (c)	**23** (d)	**24** (a)	**25** (d)	**26** (d)	**27** (c)	**28** (d)	**29** (b)	**30** (d)
31 (a)	**32** (a)	**33** (c)	**34** (b)	**35** (c)	**36** (b)	**37** (c)	**38** (b)	**39** (a)	**40** (c)

SESSION 2

1 (b)	**2** (a)	**3** (a)	**4** (d)	**5** (c)	**6** (b)	**7** (a)	**8** (a)	**9** (c)	**10** (c)
11 (a)	**12** (a)	**13** (c)	**14** (a)						

Hints and Explanations

SESSION 1

1 Here, $n(S) = {}^{100}C_3$

Let E = All three of them are divisible by both 2 and 3.

$\Rightarrow$ Divisible by 6 i.e. $\{6, 12, 18, \dots, 96\}$

Thus, out of 16 we have to select 3.

$\therefore \quad n(E) = {}^{16}C_3$

$\therefore$ Required probability $= \dfrac{{}^{16}C_3}{{}^{100}C_3} = \dfrac{4}{1155}$

2 Total number of arrangement is

$$\frac{11!}{2!2!2!} = \frac{11!}{8}$$

Number of arrangement in which C, E, H, I and S appear in that order

$$= ({}^{11}C_5)\frac{6!}{2!2!2!} = \frac{11!}{8.5!}$$

$\therefore$ Required probability

$$= \frac{11!}{8.5!} \div \frac{11!}{8} = \frac{1}{5!} = \frac{1}{120}$$

3 P(Exactly one of the events occurs)

$= P(A \cap B') \cup (B \cap A')$

$= P(A \cap B') + P(B \cap A')$

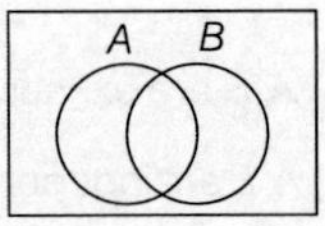

$= P(A) + P(B) - 2P(A \cap B)$

4 Here, $P(A \cup B) = \frac{3}{5}$ and $P(A \cap B) = \frac{1}{5}$

From addition theorem, we get

$$\frac{3}{5} = P(A) + P(B) - \frac{1}{5}$$

$$\Rightarrow \quad \frac{4}{5} = 1 - P(A') + 1 - P(B')$$

$$\therefore \quad P(A') + P(B') = 2 - \frac{4}{5} = \frac{6}{5}$$

5 Clearly, $A = \{4, 5, 6\}$ and $B = \{1, 2, 3, 4\}$

$\therefore A \cap B = \{4\}$

Now, by addition theorem of probability

$\therefore \quad P(A \cup B) = P(A) + P(B) - P(A \cap B)$

$\Rightarrow P(A \cup B) = \frac{3}{6} + \frac{4}{6} - \frac{1}{6} = 1$

6 We have, P (exactly one of A or B occurs) $= P(A \cup B) - P(A \cap B)$

$= P(A) + P(B) - 2P(A \cap B)$

According to the question,

$$P(A) + P(B) - 2P(A \cap B) = \frac{1}{4} \quad \dots \text{(i)}$$

$$P(B) + P(C) - 2P(B \cap C) = \frac{1}{4} \quad \dots \text{(ii)}$$

and $P(C) + P(A) - 2P(C \cap A) = \frac{1}{4} \dots$(iii)

On adding Eqs. (i), (ii) and (iii), we get

$$2\,[P(A) + P(B) + P(C) - P(A \cap B) - P(B \cap C) - P(C \cap A)] = \frac{3}{4}$$

$$\Rightarrow P(A) + P(B) + P(C) - P(A \cap B) - P(B \cap C) - P(C \cap A) = \frac{3}{8}$$

$\therefore P$ (atleast one event occurs)

$= P(A \cup B \cup C)$

$= P(A) + P(B) + P(C) - P(A \cap B) - P(B \cap C) - P(C \cap A) + P(A \cap B \cap C)$

$$= \frac{3}{8} + \frac{1}{16} = \frac{7}{16} \left[\because P(A \cap B \cap C) = \frac{1}{16}\right]$$

7 Since, a leap year has 366 days and hence 52 weeks and 2 days. The 2 days can be (Sun, Mon), (Mon, Tue), (Tue, Wed), (Wed, Thu), (Thu, Fri), (Fri, Sat), (Sat, Sun). Therefore,

$$P\text{(53 Sunday or 53 Monday)} = \frac{3}{7}$$

8 In first 120 natural numbers, total number of multiples of 5, $n(A) = 24$ and total number of multiples of 15, $n(B) = 8$ and $n(A \cap B) = 8$.

$\therefore \quad n(A \cup B) = n(A) + n(B) - n(A \cap B)$

$= 24 + 8 - 8 = 24$

$\therefore$ Required probability $= \dfrac{24}{120} = \dfrac{1}{5}$

9 Here, $P(A) = \frac{1}{3}$ and $P(A \cup B) = \frac{3}{4}$

$\therefore \; P(A \cup B) = P(A) + P(B) - P(A \cap B)$

$\leq P(A) + P(B)$

$\Rightarrow \frac{3}{4} \le \frac{1}{3} + P(B) \Rightarrow P(B) \ge \frac{5}{12}$

Also, $B \subseteq A \cup B$

$\Rightarrow P(B) \le P(A \cup B) = \frac{3}{4}$

$\therefore \quad \frac{5}{12} \le P(B) \le \frac{3}{4}$

10 Given, $P(A) = 0.4$, $P(B) = 0.3$, $P(A \cup B) = 0.5$

$$\therefore P\left(\frac{B}{A}\right) = \frac{P(A \cap B)}{P(A)} = \frac{P(A) + P(B) - P(A \cup B)}{P(A)} = \frac{0.4 + 0.3 - 0.5}{0.4} = \frac{1}{2}$$

11 Given, $P(A) = \frac{1}{2}$, $P(B) = \frac{1}{3}$

and $P\left(\frac{A}{B}\right) = \frac{1}{4}$

$\therefore P(A' \cap B') = 1 - P(A \cup B) = 1 - P(A) - P(B) + P(A \cap B)$

$$= 1 - P(A) - P(B) + P\left(\frac{A}{B}\right) \cdot P(B)$$

$$= 1 - \frac{1}{2} - \frac{1}{3} + \frac{1}{4} \cdot \frac{1}{3} = 1 - \frac{1}{2} - \frac{1}{3} + \frac{1}{12}$$

$$= \frac{12 - 6 - 4 + 1}{12} = \frac{3}{12} = \frac{1}{4}$$

12 Given that, $P(A) = \frac{1}{4}$, $P\left(\frac{A}{B}\right) = \frac{1}{2}$

and $P\left(\frac{B}{A}\right) = \frac{2}{3}$

We know that,

$$P\left(\frac{A}{B}\right) = \frac{P(A \cap B)}{P(B)}$$

and $P\left(\frac{B}{A}\right) = \frac{P(B \cap A)}{P(A)}$

$$\therefore P(B) = \frac{P\left(\frac{B}{A}\right) \cdot P(A)}{P\left(\frac{A}{B}\right)} = \frac{\left(\frac{2}{3}\right)\left(\frac{1}{4}\right)}{\left(\frac{1}{2}\right)} = \frac{1}{3}$$

13 Given that, $P(A) = 0.6, P(B) = 0.2$

$P(A / B) = 0.5$

$P(A \cap B) = P(A / B) \cdot P(B) = (0.5)(0.2) = 0.1$

$$P(A'/B') = \frac{P(A' \cap B')}{P(B')} = \frac{P[(A \cup B)']}{P(B')} = \frac{1 - P(A \cup B)}{1 - P(B)} = \frac{1 - P(A) - P(B) + P(A \cap B)}{1 - 0.2} = \frac{3}{8}$$

14 $P(B / A \cup B') = \frac{P\{B \cap (A \cup B')\}}{P(A \cup B')}$

$$= \frac{P(A \cap B)}{P(A) + P(B') - P(A \cap B')} = \frac{P(A) - P(A \cap B')}{0.7 + 0.6 - 0.5} = \frac{0.7 - 0.5}{0.8} = \frac{1}{4}$$

15 As, $P\left(\frac{C}{D}\right) = \frac{P(C \cap D)}{P(D)}$ $\left[\because C \subset D \quad \therefore P(C \cap D) = P(C)\right]$

$$= \frac{P(C)}{P(D)} \quad \text{...(i)}$$

Also, as $P(D) \le 1$

$\therefore \frac{1}{P(D)} \ge 1$ and $\frac{P(C)}{P(D)} \ge P(C)$ ...(ii)

From Eqs. (i) and (ii), we get

$$P\left(\frac{C}{D}\right) = \frac{P(C)}{P(D)} \ge P(C)$$

16 $P(E \cap F) = P(E)P(F) = \frac{1}{12}$ ...(i)

$P(E^C \cap F^C) = P(E^C) \cdot P(F^C) = \frac{1}{2}$

$\Rightarrow (1 - P(E))(1 - P(F)) = \frac{1}{2}$ ...(ii)

On solving Eqs. (i) and (ii), we get

$P(E) = \frac{1}{3}$ and $P(F) = \frac{1}{4}$,

as $P(E) > P(F)$

17 The probability of getting the sum 7 or 8 from two dice is $\frac{6}{36} + \frac{5}{36} = \frac{11}{36}$.

The probability of getting the card with number 7 or 8 is $\frac{2}{11}$.

$\therefore$ Required probability

$$= \frac{1}{2} \cdot \frac{11}{36} + \frac{1}{2} \cdot \frac{2}{11} = \frac{11}{72} + \frac{2}{22} = \frac{193}{792}$$

18 Let A be event that drawn ball is red and B be event that drawn ball is white. Then, AB and BA are two disjoint cases of the given event.

$\therefore P(AB + BA) = P(AB) + P(BA)$

$$= P(A)P\left(\frac{B}{A}\right) + P(B) \cdot P\left(\frac{A}{B}\right) = \frac{3}{6} \cdot \frac{3}{5} + \frac{3}{6} \cdot \frac{3}{5} = \frac{3}{5}$$

19 Given that,

$$P\left(\frac{B}{A}\right) = \frac{1}{2} \Rightarrow \frac{P(B \cap A)}{P(A)} = \frac{1}{2}$$

$\Rightarrow P(B \cap A) = \frac{1}{2} \times \frac{1}{4} = \frac{1}{8}$

$$P\left(\frac{A}{B}\right) = \frac{1}{4} \Rightarrow \frac{P(A \cap B)}{P(B)} = \frac{1}{4}$$

$\Rightarrow P(B) = 4P(A \cap B) \Rightarrow P(B) = \frac{1}{2}$

$\because P(A \cap B) = \frac{1}{8} = \frac{1}{2} \cdot \frac{1}{4} = P(A) \cdot P(B)$

$\therefore$ Events A and B are independent.

Now, $P\left(\frac{A'}{B}\right) = \frac{P(A' \cap B)}{P(B)} = \frac{P(A')P(B)}{P(B)} = \frac{3}{4}$

and $P\left(\frac{B'}{A'}\right) = \frac{P(B' \cap A')}{P(A')} = \frac{P(B')P(A')}{P(A')} = \frac{1}{2}$

20 Given, $P(\overline{A \cup B}) = \frac{1}{6}$, $P(A \cap B) = \frac{1}{4}$

and $P(\bar{A}) = \frac{1}{4}$

$\therefore P(A \cup B) = 1 - P(\overline{A \cup B}) = 1 - \frac{1}{6} = \frac{5}{6}$

and $P(A) = 1 - P(\bar{A}) = 1 - \frac{1}{4} = \frac{3}{4}$

$P(A \cup B) = P(A) + P(B) - P(A \cap B)$

$\Rightarrow \frac{5}{6} = \frac{3}{4} + P(B) - \frac{1}{4} \Rightarrow P(B) = \frac{1}{3}$

$\Rightarrow A$ and B are not equally likely.

Also, $P(A \cap B) = P(A) \cdot P(B) = \frac{1}{4}$

So, events are independent.

21 P (non-occurrence of (A_i))

$$= 1 - \frac{1}{(i+1)} = \frac{i}{(i+1)}$$

$\therefore P$ (non-occurrence of any of events)

$$= \left(\frac{1}{2}\right) \cdot \left(\frac{2}{3}\right) \cdots \left\{\frac{n}{(n+1)}\right\} = \frac{1}{(n+1)}$$

22 $P(A \cap B) = P(A) \cdot P(B)$

It means A and B are independent events, so A^C and B^C will also be independent.

Hence, $P(A \cup B)^C = P(A^C \cap B^C)$ [De-Morgan's law]

$= P(A^C)P(B^C)$

As A is independent of B,

$$P\left(\frac{A}{B}\right) = P(A) \quad \left[\because P(A \cap B) = P(B) \cdot P\left(\frac{A}{B}\right)\right]$$

23 Clearly, $E_1 = \{(4, 1), (4, 2), (4, 3), (4, 4), (4, 5), (4, 6)\}$

$E_2 = \{(1, 2), (2, 2), (3, 2), (4, 2), (5, 2), (6, 2)\}$

and $E_3 = \{(1, 2), (1, 4), (1, 6), (2, 1), (2, 3), (2, 5), (3, 2), (3, 4), (3, 6), (4, 1), (4, 3), (4, 5), (5, 2), (5, 4), (5, 6), (6, 1), (6, 3), (6, 5)\}$

$\Rightarrow P(E_1) = \frac{6}{36} = \frac{1}{6}$, $P(E_2) = \frac{6}{36} = \frac{1}{6}$

and $P(E_3) = \frac{18}{36} = \frac{1}{2}$

Now, $P(E_1 \cap E_2) = P$ [getting 4 on die A and 2 on die B]

$= \frac{1}{36} = P(E_1) \cdot P(E_2)$

$P(E_2 \cap E_3) = P$ [getting 2 on die B and sum of numbers on both dice is odd]

$= \frac{3}{36} = P(E_2) \cdot P(E_3)$

$P(E_1 \cap E_3) = P$
[getting 4 on die A and sum of numbers on both dice is odd]

$$= \frac{3}{36} = P(E_1)\cdot P(E_3)$$

and $P(E_1 \cap E_2 \cap E_3) = P$
[getting 4 on die A, 2 on die B and sum of numbers is odd]

$= P$ (impossible event) $= 0$

Hence, E_1, E_2 and E_3 are not independent.

24 Consider,

$$P\{A \cap (B \cap C)\} = P(A \cap B \cap C)$$
$$= P(A)\cdot P(B)\cdot P(C) = P(A)\cdot (B \cap C)$$

Now, consider

$$P[A \cap (B \cup C)] = P[(A \cap B) \cup (A \cap C)]$$
$$= P(A \cap B) + P(A \cap C) - P[(A \cap B) \cap (A \cap C)]$$
$$= P(A \cap B) + P(A \cap C) - P(A \cap B \cap C)$$
$$= P(A)\cdot P(B) + P(A)\,P(C) - P(A)\,P(B)\,P(C)$$
$$= P(A)[P(B) + P(C) - P(B)\,P(C)]$$
$$= P(A)\cdot P(B \cup C)$$

$B \cup C$ is independent to A, so S_1 is true.
$B \cap C$ is also independent to A, so S_2 is true.

25 We have,

$$P(A \cap B) = \frac{1}{10} \text{ and } P(\overline{A \cup B}) = \frac{3}{10}.$$

Then, $P(A \cup B) = 1 - P(\overline{A \cup B}) = \frac{7}{10}$

$\because P(A \cup B) = P(A) + P(B) - P(A \cap B)$

$\therefore \quad P(A) + P(B) = P(A \cup B) + P(A \cap B)$

$\Rightarrow \quad P(A) + P(B) = \frac{7}{10} + \frac{1}{10} = \frac{4}{5}$ $\quad$ ...(i)

$\because \quad P(A)\cdot P(B) = \frac{1}{10}$

$\Rightarrow \quad P(A) = \frac{1}{10P(B)}$

From Eq. (i), $P(B) + \frac{1}{10\{P(B)\}} = \frac{4}{5}$

$\Rightarrow \quad 10\{P(B)\}^2 + 1 = 8P(B)$

Let $P(B) = t$, then $10t^2 + 1 = 8t$

$\Rightarrow \quad 10t^2 - 8t + 1 = 0$

$\therefore \quad t = \frac{8 \pm \sqrt{64 - 4 \times 10 \times 1}}{2 \times 10} = \frac{8 \pm 2\sqrt{6}}{20}$

So, $P(B) = \frac{4 - \sqrt{6}}{10}$ is possible.

26 Let A denote the event that atleast one girl will be chosen and B the event that exactly 2 girls will be chosen. We require $P(B|A)$.

Since, A denotes the event that atleast one girl will be chosen. A denotes that no girl is chosen i.e.

4 boys are chosen. Then,

$$P(A') = \frac{{}^8C_4}{{}^{12}C_4} = \frac{70}{495} = \frac{14}{99}$$

$$P(A) = 1 - \frac{14}{99} = \frac{85}{99}$$

Now, $P(A \cap B) = P$ (2 boys and 2 girls)

$$= \frac{{}^8C_2 \cdot {}^4C_2}{{}^{12}C_4} = \frac{6 \times 28}{495} = \frac{56}{165}$$

Thus, $\quad P(B/A) = \frac{P(A \cap B)}{P(A)}$

$$= \frac{56}{165} \times \frac{99}{85} = \frac{168}{425}$$

27 Let the events be A = Ist aeroplane hit the target

B = IInd aeroplane hit the target

and their corresponding probabilities are

$P(A) = 0.3$ and $P(B) = 0.2$

$\Rightarrow \quad P(\overline{A}) = 0.7$ and $P(\overline{B}) = 0.8$

$\therefore$ Required probability

$$= P(\overline{A})P(B) + P(\overline{A})P(\overline{B})P(\overline{A})P(B) + \ldots$$
$$= (0.7)(0.2) + (0.7)(0.8)(0.7)(0.2) + (0.7)(0.8)(0.7)(0.8)(0.7)(0.2) + \ldots$$
$$= 0.14\,[1 + (0.56) + (0.56)^2 + \ldots]$$
$$= 0.14\left(\frac{1}{1 - 0.56}\right) = \frac{0.14}{0.44} = 0.32$$

28 Let D be the event that the picked up tube is defective.

Let A_1, A_2 and A_3 be the events that the tube is produced on machines E_1, E_2 and E_2, respectively.

$$P(D) = P(A_1)\,P(D \mid A_1) + P(A_2)\,P(D \mid A_2) + P(A_3)\,P(D \mid A_3) \ldots \text{(i)}$$

$$P(A_1) = \frac{50}{100} = \frac{1}{2},\ P(A_2) = \frac{1}{4},\ P(A_3) = \frac{1}{4}$$

Also, $P(D \mid A_1) = P(D \mid A_2) = \frac{4}{100} = \frac{1}{25}$

$$P(D \mid A_3) = \frac{5}{100} = \frac{1}{20}.$$

On putting these values in Eq. (i), we get

$$P(D) = \frac{1}{2} \times \frac{1}{25} + \frac{1}{4} \times \frac{1}{25} + \frac{1}{4} \times \frac{1}{20}$$
$$= \frac{1}{50} + \frac{1}{100} + \frac{1}{80} = \frac{17}{400} = 0.0425$$

29 **Key idea** Use the theorem of total probability

Let E_1 = Event that first ball drawn is red

E_2 = Event that first ball drawn is black

A = Event that second ball drawn is red

$$P(E_1) = \frac{4}{10},\ P\left(\frac{A}{E_1}\right) = \frac{6}{12}$$

$$\Rightarrow \quad P(E_2) = \frac{6}{10},\ P\left(\frac{A}{E_2}\right) = \frac{4}{12}$$

By law of total probability

$$P(A) = P(E_1) \times P\left(\frac{A}{E_1}\right) + P(E_2) \times P\left(\frac{A}{E_2}\right)$$
$$= \frac{4}{10} \times \frac{6}{12} + \frac{6}{10} \times \frac{4}{12} = \frac{24 + 24}{120}$$
$$= \frac{48}{120} = \frac{2}{5}$$

30 Let C, S, B and T be the events of the person going by car, scooter, bus and train, respectively.

$$P(C) = \frac{1}{7},\ P(S) = \frac{3}{7},\ P(B) = \frac{2}{7},\ P(T) = \frac{1}{7}$$

Let L be the event that the person reaching the office in time. Then, $\overline{L}$ be the event that the person reaching the office in late.

$$P(L/C) = \frac{7}{9},\ P(L/S) = \frac{8}{9},$$
$$P(L/B) = \frac{5}{9},\ P(L/T) = \frac{8}{9}$$

$\therefore$ Required probability =

$$P(C/L) = \frac{P(C)\cdot P(L/C)}{\left[\begin{array}{l} P(C)\cdot P(L/C) \\ + P(S)\cdot P(L/S) \\ + P(B)\cdot P(L/B) \\ + P(T)\cdot P(L/T) \end{array}\right]}$$

$$= \frac{\frac{1}{7}\cdot\frac{7}{9}}{\frac{1}{7}\cdot\frac{7}{9} + \frac{3}{7}\cdot\frac{8}{9} + \frac{2}{7}\cdot\frac{5}{9} + \frac{1}{7}\cdot\frac{8}{9}} = \frac{7}{49} = \frac{1}{7}$$

31 Let B stand for the event that black card is missing, then $P(B) = P(\overline{B}) = \frac{1}{2}$.

Let E be the event that all the first 13 cards are red.

$$\therefore P(E/B) = \frac{26}{51}\cdot\frac{25}{50}\ldots\frac{14}{39},$$
$$P(E/\overline{B}) = \frac{25}{51}\cdot\frac{24}{50}\ldots\frac{13}{39}$$
$$P(B/E) = \frac{P(B)\cdot P(E/B)}{P(B)\cdot P(E/B) + P(\overline{B})\cdot P(E/\overline{B})}$$
$$= \frac{26\cdot 25\ldots 14}{26\cdot 25\ldots 14 + 25\cdot 24\ldots 13}$$
$$= \frac{26}{26 + 13} = \frac{26}{39} = \frac{2}{3}$$

32 Since, $\Sigma P_i = 1$, we have

$$C + 2C + 2C + 3C + C^2 + 2C^2 + 7C^2 + C = 1$$

i.e. $\quad 10C^2 + 9C - 1 = 0$

i.e. $\quad (10C - 1)(C + 1) = 0$

$\Rightarrow \quad C = \frac{1}{10},\ C = -1$

Therefore, the permissible value of $C = \frac{1}{10}$

$$\text{Mean} = \sum_{i=1}^{n} x_i p_i = \sum_{i=1}^{7} x_i p_i$$
$$= 1 \times \frac{1}{10} + 2 \times \frac{2}{10} + 3 \times \frac{2}{10} + 4 \times \frac{3}{10} + 5\left(\frac{1}{10}\right)^2$$

$$+ 6 \times 2\left(\frac{1}{10}\right)^2 + 7\left(7\left(\frac{1}{10}\right)^2 + \frac{1}{10}\right)$$

$$= \frac{1}{10} + \frac{4}{10} + \frac{6}{10} + \frac{12}{10} + \frac{5}{100} + \frac{12}{100} + \frac{49}{100} + \frac{7}{10}$$

$= 3.66$

33 Given that, $E(X) = 3$ and $(E(X^2)) = 11$

Variance of $X = E(X^2) - [E(X)]^2$

$= 11 - (3)^2 = 11 - 9 = 2$

34 Given, $E = \{X \text{ is a prime number}\}$

$= \{2, 3, 5, 7\}$

$\therefore \quad P(E) = P(X = 2) + P(X = 3) + P(X = 5) + P(X = 7)$

$\Rightarrow P(E) = 0.23 + 0.12 + 0.20 + 0.07 = 0.62$

and $F = \{X < 4\} = \{1, 2, 3\}$

$\Rightarrow P(F) = P(X = 1) + P(X = 2) + P(X = 3)$

$\Rightarrow P(F) = 0.15 + 0.23 + 0.12 = 0.5$

and $E \cap F = \{X$ is prime number as well as $< 4\} = \{2, 3\}$

$P(E \cap F) = P(X = 2) + P(X = 3)$

$= 0.23 + 0.12 = 0.35$

$\therefore$ Required probability

$P(E \cup F) = P(E) + P(F) - P(E \cap F)$

$= 0.62 + 0.5 - 0.35 = 0.77$

35 Let X denote the number of aces.

Probability of selecting a ace,

$$p = \frac{4}{52} = \frac{1}{13}$$

And probability of not selecting ace,

$$q = 1 - \frac{1}{13} = \frac{12}{13}$$

$$P(X = 0) = \left(\frac{12}{13}\right)^2,$$

$$P(X = 1) = 2 \cdot \left(\frac{1}{13}\right) \cdot \left(\frac{12}{13}\right) = \frac{24}{169}$$

$$P(X = 2) = \left(\frac{1}{13}\right)^2 \cdot \left(\frac{12}{13}\right)^0 = \frac{1}{169}$$

$$\text{Mean} = \Sigma P_i X_i = 0 + 1 \times \frac{24}{169} + 2 \times \frac{1}{169} = \frac{24}{169} + \frac{2}{169} = \frac{2}{13}$$

36 We have, $n = 5$ and $r \geq 1$

$\therefore \quad P(X = r) = {}^nC_r p^{n-r} q^r,$

$\therefore \quad P(X \geq 1) = 1 - P(X = 0)$

$$= 1 - {}^5C_0 . p^5 . q^0 \geq \frac{31}{32}$$

$$\Rightarrow \quad p^5 \leq 1 - \frac{31}{32} = \frac{1}{32}$$

$$\therefore \ p \leq \frac{1}{2} \text{ and } p \geq 0 \Rightarrow p \in \left[0, \frac{1}{2}\right]$$

37 Let the coin was tossed n times and X be the random variable representing the number of head appearing in n trials.

According to the given condition,

$\therefore \quad P(X = 7) = P(X = 9)$

$$\Rightarrow {}^nC_7 \cdot \left(\frac{1}{2}\right)^7 \cdot \left(\frac{1}{2}\right)^{n-7} = {}^nC_9 \cdot \left(\frac{1}{2}\right)^9 \cdot \left(\frac{1}{2}\right)^{n-9}$$

$\Rightarrow \quad {}^nC_7 = {}^nC_9 \Rightarrow n = 16$

$[\because {}^nC_x = {}^nC_y \Rightarrow x + y = n\ ]$

$$\therefore \quad P(X = 2) = {}^{16}C_2 \cdot \left(\frac{1}{2}\right)^2 \cdot \left(\frac{1}{2}\right)^{14}$$

$$= \frac{{}^{16}C_2}{2^{16}} = \frac{16 \cdot 15}{2^{17}} = \frac{15}{2^{13}}$$

38 Since, ${}^nC_4 \dfrac{1}{2^n}, {}^nC_5 \dfrac{1}{2^n}$ and ${}^nC_6 \dfrac{1}{2^n}$ are in AP.

$\therefore \quad {}^nC_4, {}^nC_5$ and ${}^nC_6$ are also in AP.

$\therefore \quad 2 \cdot {}^nC_5 = {}^nC_4 + {}^nC_6$

$\therefore$ On dividing by ${}^nC_5$ both sides, we get

$$2 = \frac{{}^nC_4}{{}^nC_5} + \frac{{}^nC_6}{{}^nC_5} = \frac{5}{n-4} + \frac{n-5}{6}$$

$$= \frac{n^2 - 9n + 50}{6(n-4)}$$

$\Rightarrow \quad n^2 - 21n + 98 = 0$

$\Rightarrow \quad n = 7, 14$

39 Given box contains 15 green and 10 yellow balls.

$\therefore$ Total number of balls $= 15 + 10 = 25$

$$P(\text{green balls}) = \frac{15}{25} = \frac{3}{5} = p$$

= Probability of success

$$P(\text{yellow balls}) = \frac{10}{25} = \frac{2}{5} = q$$

= Probability of failure

and $n = 10$ = Number of trials.

$$\therefore \text{ Variance} = npq = 10 \times \frac{3}{5} \times \frac{2}{5} = \frac{12}{5}$$

40 Clearly, probability of guessing correct answer, $p = \dfrac{1}{3}$ and probability of guessing a wrong answer, $q = \dfrac{2}{3}$

$\therefore$ The probability of guessing a 4 or more correct answer

$$= {}^5C_4 \left(\frac{1}{3}\right)^4 \cdot \frac{2}{3} + {}^5C_5 \left(\frac{1}{3}\right)^5$$

$$= 5 \cdot \frac{2}{3^5} + \frac{1}{3^5} = \frac{11}{3^5}$$

SESSION 2

1 $\overset{B}{-}W\overset{B}{-}W\overset{B}{-}W\overset{B}{-}W\overset{B}{-}W\overset{B}{-}W\overset{B}{-}W\overset{B}{-}$

$$n(S) = \frac{10!}{(7!)(3!)} \quad n(E) = {}^8C_3 = \frac{8!}{(3!)(5!)}$$

[because there are 8 places for 3 black balls]

$$\therefore \quad P(E) = \frac{\dfrac{8!}{(3!)(5!)}}{\dfrac{10!}{(7!)(3!)}} = \frac{(8!)(7!)}{(10!)(5!)} = \frac{7 \cdot 6}{10 \cdot 9} = \frac{7}{15}$$

2 Since, $0 \leq P(A) \leq 1$, $0 \leq P(B) \leq 1$, $0 \leq P(C) \leq 1$

and $0 \leq P(A) + P(B) + P(C) \leq 1$

$$\therefore \ 0 \leq \frac{3x+1}{3} \leq 1 \Rightarrow -\frac{1}{3} \leq x \leq \frac{2}{3} \quad \text{...(i)}$$

$$0 \leq \frac{1-x}{4} \leq 1 \Rightarrow -3 \leq x \leq 1 \quad \text{...(ii)}$$

$$0 \leq \frac{1-2x}{2} \leq 1 \Rightarrow -\frac{1}{2} \leq x \leq \frac{1}{2} \quad \text{...(iii)}$$

$$\text{and } 0 \leq \frac{3x+1}{3} + \frac{1-x}{4} + \frac{1-2x}{2} \leq 1$$

$\Rightarrow \quad 0 \leq 13 - 3x \leq 12$

$$\Rightarrow \quad \frac{1}{3} \leq x \leq \frac{13}{3} \quad \text{...(iv)}$$

From Eqs. (i), (ii), (iii) and (iv), we get

$$\frac{1}{3} \leq x \leq \frac{1}{2}$$

3 There seems to be ambiguity in this question. It should be mentioned that boxes are different and one particular box has 3 balls. Then, the required probability

$$= \frac{{}^{12}C_3 \times 2^9}{3^{12}} = \frac{55}{3}\left(\frac{2}{3}\right)^{11}$$

4 Number of ways of distributing 6 objects to 6 persons $= 6^6$

Number of ways of distributing 1 object to each person $= 6!$

$\therefore$ Required probability

$$= 1 - \frac{6!}{6^6} = 1 - \frac{5!}{6^5} = \frac{319}{324}$$

5 The required probability

$= 1 -$ probability of each receiving atleast one $= 1 - \dfrac{n(E)}{n(S)}$

Now, the number of integral solutions of

$x_1 + x_2 + x_3 + x_4 + x_5 + x_6 = 10$

such that $x_1 \geq 1, x_2 \geq 1, \ldots, x_6 \geq 1$ gives $n(E)$ and the number of integral solutions of $x_1 + x_2 + \ldots + x_5 + x_6 = 10$ such that $x_1 \geq 0, x_2 \geq 0, \ldots, x_6 \geq 0$ gives $n(S)$.

$\therefore$ The required probability

$$= 1 - \frac{{}^{10-1}C_{6-1}}{{}^{10+6-1}C_{6-1}} = 1 - \frac{{}^9C_5}{{}^{15}C_5} = \frac{137}{143}$$

6 The probability of both drawing the common card x, $P(X)$ = (Probability of A drawing the card x and any other card y) $\times$ (Probability of B drawing the card x and a card other than y)

$$= \frac{{}^{51}C_1}{{}^{52}C_2} \times \frac{{}^{50}C_1}{{}^{52}C_2} \ \forall \ x,$$

where x has 52 values.

$\therefore$ Required probability $= \Sigma\, P(X)$

$$= 52 \times \frac{51 \times 50 \times 4}{52 \times 51 \times 52 \times 51} = \frac{50}{663}$$

7 Let X be the money won in one throw.
Money lost in 1 throw = ₹ 1
Also, probability of getting 3 = $\frac{1}{6}$
Probability of getting 1 or 6
$\Rightarrow \quad \frac{1}{6}+\frac{1}{6}=\frac{2}{6}$
Probability of getting any other number i.e. 2 or 4 or 5
$=\frac{1}{6}+\frac{1}{6}+\frac{1}{6}=\frac{3}{6}$
Then, probability distribution is

X	5	2	0
$P(X)$	$\frac{1}{6}$	$\frac{2}{6}$	$\frac{3}{6}$

Then, expected money that player can be won
$$E(X)=\frac{5}{6}+\frac{4}{6}+0=\frac{9}{6}=₹\,1.5$$
Thus, player's expected profit
$= ₹\,1.5 - ₹\,1 = 0.50$

8 Let X denote the number of correct answer given by the student. The repeated tosses of a coin are Bernoulli trials. Since, head on a coin represent the true answer and tail represents the false answer, the correctly answered of the question are Bernoulli trials.
$\therefore\ p = P(\text{a success}) = P(\text{coin show up a head}) = \frac{1}{2}$ and $q = 1 - p = 1 - \frac{1}{2} = \frac{1}{2}$
So, X has a binomial distribution with $n = 20,\ p = \frac{1}{2}$ and $q = \frac{1}{2}$
$$\therefore P(X = r) = {}^{20}C_r \cdot \left(\frac{1}{2}\right)^r \left(\frac{1}{2}\right)^{20-r}$$
Hence, P (atleast 12 questions are answered as true)
$$= P(X \ge 12) = P(12) + P(13) + P(14) + P(15) + P(16) + P(17) + P(18) + P(19) + P(20)$$
$$= {}^{20}C_{12}\, p^{12}q^{8} + {}^{20}C_{13}\, p^{13}q^{7} + {}^{20}C_{14}\, p^{14}q^{6} + {}^{20}C_{15}\, p^{15}q^{5} + {}^{20}C_{16}\, p^{16}q^{4} + {}^{20}C_{17}\, p^{17}q^{3} + {}^{20}C_{18}\, p^{18}q^{2} + {}^{20}C_{19}\, p^{19}q^{1} + {}^{20}C_{20}\, p^{20}$$
$$= ({}^{20}C_{12} + {}^{20}C_{13} + {}^{20}C_{14} + {}^{20}C_{15} + {}^{20}C_{16} + {}^{20}C_{17} + {}^{20}C_{18} + {}^{20}C_{19} + {}^{20}C_{20}) \cdot \frac{1}{2^{20}}$$
$$= \left(\frac{1}{2}\right)^{20} ({}^{20}C_{12} + {}^{20}C_{13} + \ldots + {}^{20}C_{20})$$

9 The total number of ways of ticking one or more alternatives out of 4 is ${}^4C_1 + {}^4C_2 + {}^4C_3 + {}^4C_4 = 15$. Out of these 15 combinations only one combination is correct. The probability of ticking the alternative correctly at the first trial is $\frac{1}{15}$ that at the second trial is $\left(\frac{14}{15}\right)\left(\frac{1}{14}\right) = \frac{1}{15}$ and that at the third trial is
$$\left(\frac{14}{15}\right)\left(\frac{13}{14}\right)\left(\frac{1}{13}\right) = \frac{1}{15}$$
Thus, the probability that the candidate will get marks on the question, if he is allowed up to three trials is
$$\frac{1}{15}+\frac{1}{15}+\frac{1}{15}=\frac{1}{5}$$

10 Here, $n(S) = {}^{100}C_3$
As the AM of three numbers is 25, their sum = 75
$\therefore n(E)$ = the number of integral solutions of $x_1 + x_2 + x_3 = 75$,
where, $x_1 \ge 1,\ x_2 \ge 1,\ x_3 \ge 1$
$${}^{75-1}C_{3-1} = {}^{74}C_2 = {}^{74}C_{72}$$
$$\therefore P(E) = \frac{{}^{74}C_{72}}{{}^{100}C_3} = \frac{{}^{74}C_{72}}{{}^{100}C_{97}}$$

11 A = the event of (x, y) belonging to the area $OTQPO$
B = the event of (x, y) belonging to the area $OSQRO$

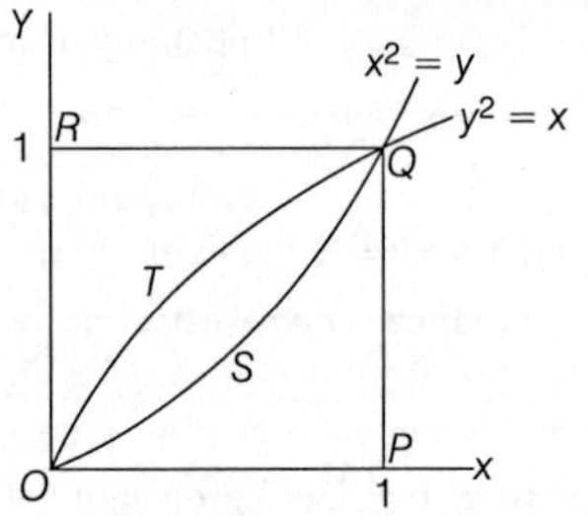

$$P(A) = \frac{\text{ar}(OTQPO)}{\text{ar}(OPQRO)} = \frac{\int_0^1 \sqrt{x}\,dx}{1 \times 1} = \left[\frac{2}{3}x^{3/2}\right]_0^1 = \frac{2}{3}$$
$$P(B) = \frac{\text{ar}(OSQRO)}{\text{ar}(OPQRO)} = \frac{\int_0^1 \sqrt{y}\,dy}{1 \times 1} = \frac{2}{3}$$
$$P(A \cap B) = \frac{\text{ar}(OTQS)}{\text{ar}(OPQRO)} = \frac{\int_0^1 \sqrt{x}\,dx - \int_0^1 x^2\,dx}{1 \times 1} = \frac{2}{3} - \frac{1}{3} = \frac{1}{3}$$
$$\because P(A) + P(B) = \frac{2}{3} + \frac{2}{3} \ne 1$$
So, A and B are not exhaustive.
$$P(A) \cdot P(B) = \frac{2}{3} \cdot \frac{2}{3} = \frac{4}{9} \ne P(A \cap B)$$
So, A and B are not independent.
$$P(A \cup B) = 1,\ P(A) + P(B) = \frac{2}{3} + \frac{2}{3} \ne P(A \cup B)$$
So, A and B are not mutually exclusive.

12 Total number of ways of selecting 2 different numbers from $\{0, 1, 2, \ldots, 10\} = {}^{11}C_2 = 55$
Let two numbers selected be x and y.
Then, $x + y = 4m$ ...(i)
and $x - y = 4n$ ...(ii)
$\Rightarrow 2x = 4(m + n)$ and $2y = 4(m - n)$
$\Rightarrow x = 2(m + n)$ and $y = 2(m - n)$
Thus, x and y both are even numbers.

x	y
0	4, 8
2	6, 10
4	0, 8
6	2, 10
8	0, 4
10	2, 6

$\therefore$ Required probability $= \frac{6}{55}$

13 Let E = event when each American man is seated adjacent to his wife and A = event when Indian man is seated adjacent of his wife.
Now, $n(A \cap E) = (4!) \times (2!)^5$
$\because$ While grouping each couple, we get 5 groups which can be arranged in $(5-1)!$ ways, and each of the group can be arranged in 2! ways.
and $n(E) = (5!) \times (2!)^4$
$\because$ While grouping each American man with his wife, we get 4 groups. These 4 groups together with indian man and his wife (total 6) can be arranged in $(6-1)!$ ways and each of the group can be arranged in 2! ways.
$$\therefore P\left(\frac{A}{E}\right) = \frac{n(A \cap E)}{n(E)} = \frac{(4!) \times (2!)^5}{(5!) \times (2!)^4} = \frac{2}{5}$$

14 We observe that $7^1, 7^2, 7^3$ and 7^4 ends in 7, 9, 3 and 1 respectively. Thus, $7l$ ends in 7, 9, 3 or 1 according as l is of the form $4k + 1$, $4k + 2$, $4k - 1$ or $4k$, respectively. If S is the sample space, then $n(S) = (100)^2$. $7^m + 7^n$ is divisible by 5 if (i) m is of the form $4k + 1$ and n is of the form $4k - 1$ or (ii) m is of the form $4k + 2$ and n is of the form $4k$ or (iii) m is of the form $4k - 1$ and n is of the form $4k + 1$ or (iv) m is of the form $4k$ and n is of the form $4k + 1$.
Thus, number of favourable ordered pairs $(m, n) = 4 \times 25 \times 25$.
Hence, required probability is $\frac{1}{4}$.

DAY THIRTY SIX

Mathematical Reasoning

Learning & Revision for the Day

- Statement (Proposition)
- Truth Value and Truth Table
- Elementary Operations of Logic
- Converse, Inverse and Contrapositive of an Implication
- Tautology
- Contradiction (Fallacy)
- Algebra of Statement

Sentence

A sentence is a relatively independent grammatical unit. It can stand alone or it can be combined with other sentences to form a text, a story etc. Sentences can be divided into different types as declarative, interrogative, imperative, exclamatory and operative sentences.

Statement (Proposition)

A statement is a sentence which is either true or false but not both simultaneously i.e. ambiguous sentence are not considered as statements.

The working nature of statement in logic is same as nature of switch in circuit.

i.e. Switch ⟨ ON (1) / OFF (0) and Statement ⟨ True (T) / False (F)

Types of statements

1. **Simple statement** A statement, which cannot be broken into two or more sentences, is a simple statement.

   Generally, small letters $p, q, r, \dots$ denote simple statements.
2. **Compound statement** A statement formed by two or more simple statements using the words such as "and", "or", "not", "if then", "if and only if", is called a compound statement.

PREP MIRROR
Your Personal Preparation Indicator

- No. of Questions in Exercises (x)—
- No. of Questions Attempted (y)—
- No. of Correct Questions (z)—
  (Without referring Explanations)

- Accuracy Level $(z/y \times 100)$—
- Prep Level $(z/x \times 100)$—

In order to expect good rank in JEE, your Accuracy Level should be above 85 & Prep Level should be above 75.

3. **Substatements** Simple statements which when combined to form a compound statement are called substatements or components.

NOTE
- A true statement is known as a valid statement.
- A false statement is known as an invalid statement.
- Imperative, exclamatory, interrogative, optative sentences are not statements.
- Mathematical identities are considered to be statements because they can either be true or false but not both.

4. **Open statement** A sentence which contains one or more variables such that when certain values are given to the variable it becomes a statement, is called an open statement.

Truth Value and Truth Table

- A statement can be either 'true' or 'false', which are called truth values of a statement and it is represented by the symbols 'T' and 'F', respectively.
- A table that shows the relationship between the truth values of compound statement, $S(p, q, r, \dots)$ and the truth values of its substatements $p, q, r, \dots$ etc., is called the truth table of statement S.
- If a compound statement has simply n substatements, then there are 2^n rows representing logical possibilities.

Logical Connectives or Sentencial Connectives

Two or more statements are combined to form a compound statement by using symbols. These symbols are called logical connectives.

Logical connectives are given below

Words	Symbols
and	$\wedge$
or	$\vee$
implies that (if ..., then)	$\Rightarrow$
If and only if (implies and is implied by)	$\Leftrightarrow$

Elementary Operations of Logic

Formation of compound sentences from simple sentence using logical connectives are termed as elementary operation of logic.

There are five such operations, which are discussed below.

1. Negation (Inversion) of Statement

(i) A statement which is formed by changing the truth value of a given statement by using word like 'no' or 'not' is called negation of a given statement. It represents the symbol '$\sim$'.

(ii) If p is statement, then negation of p is denoted by '$\sim p$'.

(iii) The truth table for NOT is given by

p	$\sim p$
T	F
F	T

2. Conjunction (AND)

A compound sentence formed by two simple sentences p and q using connective "AND" is called the conjunction of p and q and is represented by $p \wedge q$.

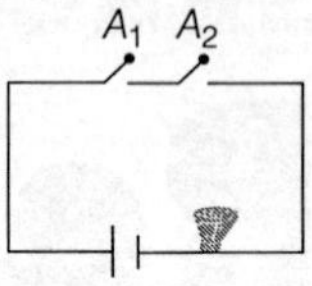

The truth table for operation 'AND' is given by

p	q	$p \wedge q$
T	T	T
T	F	F
F	T	F
F	F	F

NOTE
- The statement $p \wedge q$ is true, if both p and q are true.
- The statement $p \wedge q$ is false, if atleast one of p and q or both are false.

3. Disjunction (OR)

A compound sentence formed by two simple sentences p and q using connective "OR" is called the disjunction of p and q and is represented by $p \vee q$.

The truth table for operation 'OR' is given by

p	q	$p \vee q$
T	T	T
T	F	T
F	T	T
F	F	F

NOTE
- The statement $p \vee q$ is true, if atleast one of p and q or both are true.
- The statement $p \vee q$ is false, if both p and q are false.

4. Implication (Conditional)

A compound sentence formed by two simple sentences p and q using connective "if ... then ..." is called the implication of p and q and represented by $p \Rightarrow q$ which is read as "p implies q".

Here, p is called antecedent or hypothesis and q is called consequent or conclusion.

The truth table for if ... then is given by

p	q	$p \Rightarrow q$	$\sim p$	$\sim p \vee q$
T	T	T	F	T
T	F	F	F	F
F	T	T	T	T
F	F	T	T	T

It is clear from the truth table that column III is equal to column V. i.e. statement $p \Rightarrow q$ is equivalent to $\sim p \vee q$.

5. Biconditional Statement

Two simple sentences connected by the phrase "if and only if," form a biconditional statement. It is represented by the symbol '$\Leftrightarrow$'.

The truth table for if and only if is given by

p	q	$p \Leftrightarrow q$	$\sim p$	$\sim q$	$\sim p \vee q$	$p \vee \sim q$	$(\sim p \vee q) \wedge (p \vee \sim q)$
T	T	T	F	F	T	T	T
T	F	F	F	T	F	T	F
F	T	F	T	F	T	F	F
F	F	T	T	T	T	T	T

NOTE
- It is clear from the truth table that column III is equal to column VIII. i.e. statement $p \Leftrightarrow q$ is equivalent to $(\sim p \vee q) \wedge (p \vee \sim q)$.
- The statement $p \Leftrightarrow q$ is true, if either both are true or both are false.
- The statement $p \Leftrightarrow q$ is false, if exactly one of them is false.

Converse, Inverse and Contrapositive of an Implication

For any two statements p and q,

(i) Converse of the implication 'if p, then q' is 'if q, then p' i.e. $q \Rightarrow p$

(ii) Inverse of the implication 'if p, then q' is 'if $\sim p$, then $\sim q$' i.e. $\sim p \to \sim q$.

(iii) Contrapositive of the implication 'if p, then q' is 'if $\sim q$, then $\sim p$' i.e. $\sim q \to \sim p$.

NOTE
- $\sim (p \Rightarrow q) \equiv \sim (\sim p \vee q) = \{p \wedge (\sim q)\}$ $\therefore$ $\sim (p \Leftrightarrow q) \equiv \{p \wedge (\sim q)\}$
- $\sim (p \Rightarrow q) \equiv (p \wedge \sim q) \vee (q \wedge \sim p)$
- $p \Rightarrow q = \sim p \vee q$
- $(p \Leftrightarrow q) \Leftrightarrow r = p \Leftrightarrow (q \Leftrightarrow r)$
- $p \Leftrightarrow q = (p \Rightarrow q) \wedge (q \Rightarrow p)$

Tautology

A compound statement is called a tautology, if it has truth value T whatever may be the truth value of its compounds.

Statement $(p \Rightarrow q) \wedge p \Rightarrow q$ is a tautology.

The truth table of above statement is prepared as follows.

p	q	$p \Rightarrow q$	$p \Rightarrow q \wedge p$	$(p \Rightarrow q) \wedge p \Rightarrow q$
T	T	T	T	T
T	F	F	F	T
F	T	T	F	T
F	F	T	F	T

Contradiction (Fallacy)

A compound statement is called contradiction, if its truth value is F whatever may be the truth value of its components.

Statement $\sim p \wedge p$ is a contradiction.

The truth table of above statement is prepared as follows

p	$\sim p$	$\sim p \wedge p$
T	F	F
F	T	F

A statement which is neither a tautology nor a contradiction is a contigency.

Algebra of Statement

Some of the important laws considered under the category of algebra of statements are given as

1. **Idempotent Laws**
   For any statement p, we have
   (a) $p \wedge p \equiv p$ (b) $p \vee p \equiv p$
2. **Commutative Laws**
   For any two statements p and q, we have
   (a) $p \wedge q \equiv q \wedge p$ (b) $p \vee q \equiv q \vee p$
3. **Associative Laws**
   For any three statements p, q and r, we have
   (a) $(p \wedge q) \wedge r \equiv p \wedge (q \wedge r)$
   (b) $(p \vee q) \vee r \equiv p \vee (q \vee r)$
4. **Distributive Laws**
   (a) $p \vee (q \wedge r) \equiv (p \vee q) \wedge (p \vee r)$
   (b) $p \wedge (q \vee r) \equiv (p \wedge q) \vee (p \wedge r)$
5. **Involution Laws**
   For any statement p, we have $\sim (\sim p) \equiv p$
6. **De-morgan's Laws**
   (a) $\sim (p \wedge q) = \sim p \vee \sim q$
   (b) $\sim (p \vee q) = \sim p \wedge \sim q$

7. **Complement Laws**
For any statement p, we have
(a) $p \vee \sim p \equiv T$ (b) $p \wedge \sim p \equiv F$
(c) $\sim T \equiv F$ (d) $\sim F \equiv T$

8. **Identity Laws**
For any statement p, we have
(a) $p \wedge T \equiv p$ (b) $p \vee F \equiv p$
(c) $p \vee T \equiv T$ (d) $p \wedge F \equiv F$
where, T and F are the true and false statement.

9. **Duality**
- Two compound statements S_1 and S_2 are said to be duals of each other, if one can be obtained from the other by replacing $\wedge$ by $\vee$ and $\vee$ by $\wedge$.
- The connectives $\wedge$ and $\vee$ are also called duals of each other.
- Symbolically, it can be written as, if $S(p,q) = p \wedge q$, then its dual is $S^*(p,q) = p \vee q$.

DAY PRACTICE SESSION 1

FOUNDATION QUESTIONS EXERCISE

1 Which of the following is not a statement? → NCERT Exemplar
(a) Smoking is injurious to health
(b) $2 + 2 = 4$
(c) 2 is the only even prime number
(d) Come here

2 The negation of the statement "72 is divisible by 2 and 3" is → NCERT Exemplar
(a) 72 is not divisible by 2 or 72 is not divisible by 3
(b) 72 is not divisible by 2 and 72 is not divisible by 3
(c) 72 is divisible by 2 and 72 is not divisible by 3
(d) 72 is not divisible by 2 and 72 is divisible by 3

3 The dual of the statement $(p \vee q) \vee r$ is
(a) $(p \wedge q) \vee r$ (b) $(p \wedge q) \wedge r$
(c) $(p \vee q) \wedge r$ (d) None of these

4 Choose disjunction among the following sentences
(a) It is raining and the Sun is shining
(b) Ram and Shyam are good friends
(c) 2 or 3 is a prime number
(d) Everyone who lies in India is an Indian

5 Which among the following is not a conjunction?
(a) Gautam and Rahul are good friends
(b) The Earth is round and the Sun is hot
(c) $9 > 4$ and $12 > 15$
(d) None of the above

6 If p : a natural number n is odd and q : natural number n is not divisible by 2, then the biconditional statement $p \Leftrightarrow q$ is → NCERT
(a) A natural number n is odd if and only if it is divisible by 2
(b) A natural number n is odd if and only if it is not divisible by 2
(c) If a natural number n is odd, then it is not divisible by 2
(d) None of the above

7 The logically equivalent proposition of $p \Leftrightarrow q$ is
(a) $(p \wedge q) \vee (p \wedge q)$ (b) $(p \Rightarrow q) \wedge (q \Rightarrow p)$
(c) $(p \wedge q) \vee (q \Rightarrow p)$ (d) $(p \wedge q) \Rightarrow (p \vee q)$

8 If both p and q are false, then
(a) $p \wedge q$ is true (b) $p \vee q$ is true
(c) $p \Rightarrow q$ is true (d) $p \Leftrightarrow q$ is false

9 The negation of '$12 > 4$' is
(a) $12 \le 4$ (b) $13 > 5$ (c) $12 > 3$ (d) $12 \ge 4$

10 Which among the following is not a negation of 'The Sun is a star'?
(a) The Sun is not a star.
(b) It is not the case that Sun is a star.
(c) It is false that the Sun is a star.
(d) It is true that the Sun is a star.

11 Which of the following is not equivalent to $p \leftrightarrow q$?
(a) p if and only if q
(b) p is necessary and sufficient for q
(c) q if and only if p
(d) None of the above

12 For integers m and n, both greater than 1, consider the following three statements → JEE Mains 2013
$P : m$ divides n $Q : m$ divides n^2 $R : m$ is prime, then
(a) $Q \wedge R \to P$ (b) $P \wedge Q \to R$
(c) $Q \to R$ (d) $Q \to P$

13 If p and q are two statements such that
p : the questions paper is easy
q : we shall pass,
then the symbolic statement $\sim p \to \sim q$ means
(a) If the question paper is easy, then we shall pass
(b) If the question paper is not easy, then we shall not pass
(c) The question paper is easy and we shall pass
(d) The question paper is easy or we shall pass

14 For the following three statements
p : 2 is an even number
q : 2 is a prime number.
r : Sum of two prime numbers is always even, then the symbolic statement $(p \wedge q) \to \sim r$ means
(a) 2 is an even and prime number and the sum of two prime numbers is always even
(b) 2 is an even and prime number and the sum of two prime numbers is not always even
(c) 2 is an even and prime number, then the sum of two prime numbers is not always even
(d) 2 is an even and prime number, then the sum of two prime numbers is always even

15 The converse of the statement
"If $x > y$, then $x + a > y + a$" is → NCERT Exemplar
(a) If $x < y$, then $x + a < y + a$ (b) If $x + a > y + a$, then $x > y$
(c) If $x < y$, then $x + a > y + a$ (d) If $x > y$, then $x + a < y + a$

16 Consider the following statements
p : If a number is divisible by 10, then it is divisible by 5.
q : If a number is divisible by 5, then it is divisible by 10.
Then, the correct option is
(a) q is converse of p (b) p is converse of q
(c) p is not converse of q (d) Both (a) and (b)

17 The negation of the statement. "If I become a teacher, then I will open a school", is → AIEEE 2012
(a) I will become a teacher and I will not open a school
(b) Either I will not become a teacher or I will not open a school
(c) Neither I will become a teacher nor I will open a school
(d) I will not become a teacher or I will open a school

18 Find the contrapositive of "If two triangles are identical, then these are similar".
(a) If two triangles are not similar, then these are not identical
(b) If two triangles are not identical, then these are not similar
(c) If two triangles are not identical, then these are similar
(d) If two triangles are not similar, then these are identical

19 The contrapositive of the statement
"If 7 is greater than 5, then 8 is greater than 6" is → NCERT Exemplar
(a) If 8 is greater than 6, then 7 is greater than 5
(b) If 8 is not greater than 6, then 7 is greater than 5
(c) If 8 is not greater than 6, then 7 is not greater than 5
(d) If 8 is greater than 6,then 7 is not greater than 5

20 The conditional statement of
"You will get a sweet dish after the dinner" is → NCERT Exemplar
(a) If you take the dinner, then you will get a sweet dish
(b) If you take the dinner, you will get a sweet dish
(c) You get a sweet dish if and only if you take the dinner
(d) None of the above

21 Let S be a non-empty subset of R. Consider the following statement
P : There is a rational number $x \in S$ such that $x > 0$.
Which of the following statements is the negation of the statement P? → AIEEE 2010
(a) There is a rational number $x \in S$ such that $x \le 0$
(b) There is no rational number $x \in S$ such that $x \le 0$
(c) Every rational number $x \in S$ satisfies $x \le 0$
(d) $x \in S$ and $x \le 0 \Rightarrow x$ is not rational

22 The contra positive of $p \to (\sim q \to \sim r)$ is
(a) $(\sim q \wedge r) \to \sim p$ (b) $(q \wedge \sim r) \to \sim p$
(c) $p \to (\sim r \vee q)$ (d) $p \wedge (q \wedge r)$

23 $p \vee \sim (p \wedge q)$ is a
(a) contradiction (b) contingency
(c) tautology (d) None of these

24 If p, q and r are simple propositions with truth values T, F, T respectively, then the truth value of $(\sim p \vee q) \wedge \sim q \to p$ is
(a) true (b) false
(c) true, if r is false (d) None of these

25 The statement $p \to (q \to p)$ is equivalent to → JEE Mains 2013
(a) $p \to q$ (b) $p \to (p \vee q)$
(c) $p \to (p \to q)$ (d) $p \to (p \wedge q)$

26 The statement $(p \to q) \to [(\sim p \to q) \to q]$ is → JEE Mains 2017
(a) a tautology
(b) equivalent to $\sim p \to q$
(c) equivalent to $p \to \sim q$
(d) a fallacy

27 Let p and q be two statements. Then, $(\sim p \vee q) \wedge (\sim p \wedge \sim q)$ is a
(a) tautology
(b) contradiction
(c) neither tautology nor contradiction
(d) both tautology and contradiction

28 The proposition $(p \Rightarrow \sim p) \wedge (\sim p \Rightarrow p)$ is
(a) contigency
(b) neither tautology nor contradiction
(c) contradiction
(d) tautology

29 If p and q are two statements, then
$(p \Rightarrow q) \Leftrightarrow (\sim q \Rightarrow \sim p)$ is a
(a) contradiction (b) tautology
(c) neither (a) nor (b) (d) None of these

30 The proposition $S : (p \Rightarrow q) \Leftrightarrow (\sim p \vee q)$ is
(a) a tautology (b) a contradiction
(c) either (a) or (b) (d) neither (a) nor (b)

31 The statement $\sim(p \leftrightarrow \sim q)$ is
(a) equivalent to $p \leftrightarrow q$ (b) equivalent to $\sim p \leftrightarrow q$
(c) a tautology (d) a fallacy

32 The false statement in the following is
(a) $p \wedge (\sim p)$ is a contradiction
(b) $(p \to q) \leftrightarrow (\sim q \to \sim p)$ is a contradiction
(c) $\sim(\sim p) \leftrightarrow p$ is a tautology
(d) $p \vee (\sim p)$ is a tautology

33 Among the following statements, which is a tautology?
(a) $p \wedge (p \vee q)$ (b) $p \vee (p \wedge q)$
(c) $[p \wedge (p \to q)] \to q$ (d) $q \to [p \wedge (p \to q)]$

34 Consider the following statements
P : Suman is brilliant.
Q : Suman is rich.
R : Suman is honest.
The negative of the statement.'Suman is brilliant and dishonest if and only if Suman is rich.' can be expressed as → AIEEE 2011
(a) $\sim(Q \leftrightarrow (P \wedge \sim R))$ (b) $\sim(Q \leftrightarrow P \wedge R)$
(c) $\sim(P \wedge \sim R) \leftrightarrow Q$ (d) $\sim P \wedge (Q \leftrightarrow \sim R)$

35 The statement $(p \Rightarrow q) \Leftrightarrow (\sim p \wedge q)$ is a
(a) tautology (b) contradiction
(c) neither (a) nor (b) (d) None of these

36 The proposition $\sim(p \Rightarrow q) \Rightarrow (\sim p \vee \sim q)$ is
(a) a tautology (b) a contradiction
(c) either (a) or (b) (d) neither (a) nor (b)

37 The negation of the compound proposition is $p \vee (\sim p \vee q)$
(a) $(p \wedge \sim q) \wedge \sim p$ (b) $(p \vee \sim q) \vee \sim p$
(c) $(p \wedge \sim q) \vee \sim p$ (d) None of these

38 The negation of $\sim s \vee (\sim r \wedge s)$ is equivalent to → JEE Mains 2015
(a) $s \wedge \sim r$ (b) $s \wedge (r \wedge \sim s)$
(c) $s \vee (r \vee \sim s)$ (d) $s \wedge r$

39 $\sim S(p, q)$ is equivalent to
(a) $S^*(\sim p, \sim q)$ (b) $S^*(p, \sim q)$
(c) $S^*(\sim p, q)$ (d) None of these

40 Let p : 25 is a multiple of 5.
q : 25 is a multiple of 8.
Statement I The compound statement "p and q" is false.
Statement II The compound statement "p or q" is false.
Choose the correct option
(a) Only Statement I is correct
(b) Only Statement II is correct
(c) Both statements are correct
(d) Both statements are incorrect

DAY PRACTICE SESSION 2

PROGRESSIVE QUESTIONS EXERCISE

1 If p : 4 is an even prime number, q : 6 is divisor of 12 and r : the HCF of 4 and 6 is 2, then which one of the following is true?
(a) $(p \wedge q)$ (b) $(p \vee q) \wedge \sim r$
(c) $\sim(q \wedge r) \vee p$ (d) $\sim p \vee (q \wedge r)$

2 An equivalent expression for $(p \Rightarrow q \wedge r) \vee (r \Leftrightarrow s)$ which contains neither the biconditional nor the conditional is
(a) $(\sim p \vee q \wedge r) \vee ((\sim r \vee s) \wedge (r \vee \sim s))$
(b) $(\sim p \wedge q \wedge r) \vee ((\sim r \vee s) \wedge (r \vee \sim s))$
(c) $(\sim p \vee q \wedge r) \wedge ((\sim r \vee s) \vee (r \vee \sim s))$
(d) None of the above

3 If $(p \wedge \sim r) \Rightarrow (\sim p \vee q)$ is false, then the truth values of p, q and r respectively
(a) T, F and F (b) F, F and T
(c) F, T and T (d) T, F and T

4 The Boolean expression $(p \wedge \sim q) \vee q \vee (\sim p \wedge q)$ is equivalent to → JEE Mains 2016
(a) $\sim p \wedge q$ (b) $p \wedge q$
(c) $p \vee q$ (d) $p \vee \sim q$

5 Which of the following is a tautology?
(a) $(p \to q) \wedge (p \to q)$ (b) $(p \to q) \vee (p \to q)$
(c) $(p \to q) \vee (q \to p)$ (d) None of these

6 Let p and q stand for the statements '$2 \times 4 = 8$' and '4 divides 7', respectively. Then, what are the truth values for following biconditional statements?
(i) $p \leftrightarrow q$ (ii) $\sim p \leftrightarrow q$
(iii) $\sim q \leftrightarrow p$ (iv) $\sim p \leftrightarrow \sim q$
(a) T T T T (b) F T T T
(c) F T F F (d) F T T F

7 The only statement among the followings that is a tautology is → AIEEE 2011
(a) $B \to [A \wedge (A \to B)]$ (b) $A \wedge (A \vee B)$
(c) $A \vee (A \wedge B)$ (d) $[A \wedge (A \to B)] \to B$

8 Which of the following is not correct?
(a) $\sim(p \wedge q) = (\sim p) \vee (\sim q)$
(b) Truth value of $p \wedge q$ = truth value of $q \wedge p$
(c) $\sim(\sim p) = p$
(d) $p \Leftrightarrow q \equiv (p \Rightarrow q) \vee (q \Rightarrow p)$

9 $\sim (p \Rightarrow q) \Leftrightarrow \sim p \vee \sim q$ is

(a) a tautology (b) a contradiction
(c) neither a tautology nor a contradiction
(d) cannot come to any conclusion

10 Which of the following is wrong statement?

(a) $p \to q$ is logically equivalent to $\sim p \vee q$
(b) If the truth values of p, q, r are T, F, T respectively, then the truth value of $(p \vee q) \wedge (q \vee r)$ is T
(c) $\sim (\vee q \vee q \vee r) \equiv \sim p \wedge \sim q \wedge \sim r$
(d) The truth value of $p \wedge \sim (p \vee q)$ is always T

11 If p, q and r are simple propositions, then $(p \wedge q) \wedge (q \wedge r)$ is true, then

(a) p, q and r are true (b) p, q are true and r is false
(c) p is true and q, r are false (d) p, q and r are false

12 $\sim p \wedge q$ is logically equivalent to

(a) $p \Rightarrow q$ (b) $q \Rightarrow p$
(c) $\sim (p \Rightarrow q)$ (d) $\sim (q \Rightarrow p)$

13 Which of the following is true for any two statements p and q?

(a) $\sim [p \vee \sim q] \equiv \sim p \wedge q$ (b) $\sim p \wedge q$ is a fallacy
(c) $p \vee \sim q$ is a tautology (d) $p \vee \sim p$ is a contradiction

14 Consider

Statement I $(p \wedge \sim q) \wedge (\sim p \wedge q)$ is a fallacy.
Statement II $(p \to q) \leftrightarrow (\sim q \to p)$ is a tautology.

➜ **JEE Mains 2013**

(a) Statement I is true; Statement II is true; Statement II is a correct explanation for Statement I
(b) Statement I is true; Statement II is true; Statement II is not a correct explanation for Statement I
(c) Statement I is true; Statement II is false
(d) Statement I is false; Statement II is true

15 **Statement I** The statement $A \to (B \to A)$ is equivalent to $A \to (A \vee B)$.

Statement II The statement $\sim \{(A \wedge B) \to (\sim A \vee B)\}$ is tautology.

➜ **JEE Mains 2013**

(a) Statement I is true, Statement II is false
(b) Statement I is true, Statement II is true; Statement II is not a correct explanation for Statement I
(c) Statement I is true, Statement II is true; Statement II is a correct explanation for Statement I
(d) Statement I is false, Statement II is true

ANSWERS

SESSION 1	**1** (d)	**2** (a)	**3** (b)	**4** (c)	**5** (a)	**6** (b)	**7** (b)	**8** (c)	**9** (a)	**10** (d)
	11 (d)	**12** (a)	**13** (b)	**14** (c)	**15** (b)	**16** (d)	**17** (a)	**18** (a)	**19** (c)	**20** (a)
	21 (c)	**22** (a)	**23** (c)	**24** (a)	**25** (b)	**26** (a)	**27** (c)	**28** (c)	**29** (b)	**30** (a)
	31 (a)	**32** (b)	**33** (c)	**34** (a)	**35** (c)	**36** (a)	**37** (a)	**38** (d)	**39** (a)	**40** (a)
SESSION 2	**1** (d)	**2** (a)	**3** (a)	**4** (c)	**5** (c)	**6** (d)	**7** (d)	**8** (d)	**9** (c)	**10** (d)
	11 (a)	**12** (d)	**13** (a)	**14** (b)	**15** (c)					

Hints and Explanations

SESSION 1

1 In option (a), (b) and (c).
It is a declarative sentence, which is clearly true, therefore it is a true statement.
In option (d), it is an imperative sentence, therefore it is not a statement.

2 The negation of the given statement is "72 is not divisible by 2 or 72 is not divisible by 3".

3 According to definition let $s(p,q) = p \wedge q$ be a compound statement. Then, $s^*(p,q) = p \vee q$,
Therefore, the dual of $(p \vee q) \vee r$ is $(p \wedge q) \wedge r$.

4 In disjunction, two sentences are connected with 'OR'.

5 In the sentence, 'Gautam and Rahul are good friends', the word 'and' is not a connective. So, this sentence is not a conjunction.

6 Given, p : A natural number n is odd and q : natural number n is not divisible by 2. The biconditional statement $p \Leftrightarrow q$ i.e. "A natural number n is odd if and only if it is not divisible by 2".

7 $(p \Rightarrow q) \wedge (q \Rightarrow p)$ means $p \Leftrightarrow q$. Hence, option (b) is correct.

8 If both p and q are false, then $p \Rightarrow q$ is true.

9 Negation is a denial of a statement.
So, $12 \leq 4$ is correct option.

10 The denial of any statement is called its negation.
'It is true that the Sun is a star' is an assertion of the given statement.

11 Options (a), (b) and (c) are equivalent to $p \leftrightarrow q$.

12 $P : \frac{n}{m}; Q : \frac{n^2}{m}; R : m$ is prime.

Let $m = 5, p = 10, n^2 = 100, m = 3, n = 12, n^2 = 144$

$m = 7, n = 14, n^2 = 196$

So, P,Q and R are true statements.

$\therefore \quad Q \wedge R = T \wedge T \rightarrow T = P$

13 Given, p : The question paper is easy and q : We shall pass given.
The symbolic representation of given option are
(a) : $p \rightarrow q$ (b) : $\sim p \rightarrow \sim q$ (c) : $p \wedge q$ (d) : $p \vee q$

14 Given, p : 2 is an even number
q : 2 is a prime number
r : sum of two prime numbers is always even.
The symbolic representation of given option (s) are
(a) : $p \wedge q \wedge r$ (b) : $p \wedge q \wedge \sim r$
(c) : $p \wedge q \Rightarrow \sim r$ (d) : $p \wedge q \Rightarrow r$

15 Converse statement is "If $x + a > y + a$, then $x > y$".

16 The converse of the statement "if p then q" is "if q then p".
Given statements are
p : if a number is divisible by 10, it is divisible by 5.
q : if a number is divisible by 5, then it is divisible by 10.
It is clear that q is converse of p and p is converse of q.

17 Let us assume that p : 'I become a teacher' and q : I will open a school. Then, we can easily ascertain that
Negation of $(p \rightarrow q)$ is $\sim(p \rightarrow q) = p \wedge \sim q$, which means that 'I will become a teacher and I will not open a school.'

18 Consider the following statements
p : Two triangles are identical.
q : Two triangles are similar.
Clearly, the given statement in symbolic form is $p \Rightarrow q$.
So, its contrapositive is given by $\sim q \Rightarrow \sim p$.

19 The contrapositive of the given statement is "If 8 is not greater than 6, then 7 is not greater than 5.

20 The conditional statement of given statement is "If you take the dinner, then you will get a sweet dish".

21 P : There is rational number $x \in S$ such that $x > 0$.
$\sim P$: Every rational number $x \in S$ satisfies $x \leq 0$.

22 The contrapositive of $p \rightarrow q$ is $\sim q \rightarrow \sim p$
$\therefore$ Contrapositive of $p \rightarrow (\sim q \rightarrow \sim r)$ is $\sim(\sim q \rightarrow \sim r) \rightarrow \sim p$

$$\equiv (\sim q \wedge r) \rightarrow \sim p \quad \begin{bmatrix} \because \sim(p \rightarrow q) \equiv p \wedge \sim q \\ \therefore \sim(\sim q \rightarrow \sim r) \equiv \sim q \wedge r \end{bmatrix}$$

23

p	q	$p \wedge q$	$\sim(p \wedge q)$	$p \vee \sim(p \wedge q)$
T	T	T	F	T
T	F	F	T	T
F	T	F	T	T
F	F	F	T	T

24

p	q	r	$\sim p$	$\sim q$	$\sim p \vee q$	$(\sim p \vee q) \wedge \sim q$	$(\sim p \vee q) \wedge \sim q \rightarrow p$
T	F	T	F	T	F	F	T

25

p	q	$q \rightarrow p$	$p \rightarrow (q \rightarrow p)$	$p \vee q$	$p \rightarrow (p \vee q)$
T	T	T	T	T	T
T	F	T	T	T	T
F	T	F	T	T	T
F	F	T	T	F	T

So, statement $p \rightarrow (q \rightarrow p)$ is logically equivalent to $p \rightarrow (p \vee q)$.

26 The truth table of the given expression is given below :

p	q	$x \equiv p \rightarrow q$	$\sim p$	$\sim p \rightarrow q$	$y \equiv (\sim p \rightarrow q) \rightarrow q$	$x \rightarrow y$
T	T	T	F	T	T	T
T	F	F	F-	T	F	T
F	T	T	T	T	T	T
F	F	T	T	F	T	T

Hence, it is a tautology.

27

p	q	$\sim p$	$\sim q$	$(\sim p \vee q)$	$(\sim p \wedge \sim q)$	$(\sim p \vee q) \wedge (\sim p \wedge \sim q)$
T	T	F	F	T	F	F
T	F	F	T	F	F	F
F	T	T	F	T	F	F
F	F	T	T	T	T	T

Hence, it is neither tautology nor contradiction.

28

p	$\sim p$	$p \Rightarrow \sim p$	$\sim p \Rightarrow p$	$(p \Rightarrow \sim p) \wedge (\sim p \Rightarrow p)$
F	T	T	F	F
T	F	F	T	F

$\therefore$ Statement is contradiction.

29

p	q	$\sim q$	$\sim p$	$\sim q \Rightarrow \sim p$	$p \Leftrightarrow q$	$(p \Rightarrow q) \Leftrightarrow (\sim q \Rightarrow \sim p)$
T	T	F	F	T	T	T
T	F	T	F	F	F	T
F	T	F	T	T	T	T
F	F	T	T	T	T	T

Hence, it is tautology.

30

p	q	$p \Rightarrow q$	$\sim p$	$(\sim p \vee q)$	$(p \Rightarrow q) \Leftrightarrow (\sim p \vee q)$
T	T	T	F	T	T
T	F	F	F	F	T
F	T	T	T	T	T
F	F	T	T	T	T

Since, all values of given proposition is true, hence it is a tautology.

31

p	q	$\sim p$	$\sim q$	$q \leftrightarrow q$	$p \leftrightarrow \sim q$	$\sim p \leftrightarrow q$	$\sim(p \leftrightarrow \sim q)$
T	F	F	T	F	T	T	F
F	T	T	F	F	T	T	F
T	T	F	F	T	F	F	T
F	F	T	T	T	F	F	T

$\sim(p \leftrightarrow \sim q)$ is equivalent to $(p \leftrightarrow q)$.

32 $p \rightarrow q$ is equivalent to $\sim q \rightarrow \sim p$

$\therefore \quad (p \rightarrow q) \leftrightarrow (\sim q \rightarrow \sim p)$

is a tautology but not a contradiction.

33 $p \wedge (p \vee q)$ is F, when $p \equiv$ F

$p \vee (p \wedge q)$ is F, when $p \equiv$ F, $q \equiv$ F

and $q \rightarrow [p \wedge (p \rightarrow q)]$ is F, when $p \equiv$ F, $q \equiv$ T

So, for $[p \wedge (p \rightarrow q)] \rightarrow q \equiv [p \wedge (\sim p \vee q)] \rightarrow q$

$\equiv [\{p \wedge (\sim p)\} \vee (p \wedge q)] \rightarrow q$

34 Suman is brilliant and dishonest, if and only if Suman is rich, is expressed as, $Q \leftrightarrow (P \wedge \sim R)$

So, negation of it will be $\sim(Q \leftrightarrow (P \wedge \sim R))$.

35

p	q	$\sim p$	$p \Rightarrow q$	$\sim p \wedge q$	$(p \Rightarrow q) \Leftrightarrow (\sim p \wedge q)$
T	T	F	T	F	F
T	F	F	F	F	T
F	T	T	T	T	T
F	F	T	T	F	F

36 By truth table

p	q	$\sim p$	$\sim q$	$p \Rightarrow q$	$\sim(p \Rightarrow q)$	$\sim p \vee \sim q$	$\sim(p \Rightarrow q) \Rightarrow (\sim p \vee \sim q)$
T	T	F	F	T	F	F	T
T	F	F	T	F	T	T	T
F	T	T	F	T	F	T	T
F	F	T	T	T	F	T	T

Hence, given proposition is a tautology.

37 Since, $S : \sim(p \vee (\sim p \vee q))$

$\Rightarrow \quad S : \sim p \wedge \sim(\sim p \vee q)$ (De-Morgan's Law)

$\Rightarrow \quad S : \sim p \wedge (p \wedge \sim q)$ (De-Morgan's law)

38 $\sim(\sim s \vee (\sim r \wedge s)) \equiv s \wedge (\sim(\sim r \wedge s)) \equiv s \wedge (r \vee \sim s)$

$\equiv (s \wedge r) \vee (s \wedge \sim s) \equiv (s \wedge r) \vee F \qquad [\because s \wedge \sim s \text{ is false}]$

$\equiv s \wedge r$

39 $\because \quad \sim S(p, q) = \sim(p \wedge q) = (\sim p) \vee (\sim q) = S^*(\sim p, \sim q)$

40 I. Compound statement with 'AND' is 25 is a multiple of 5 and 8.

This is a false statement. Since, p is true but q is false.

[since, 25 is divisible by 5 but not divisible by 8]

II. Compound statement with 'OR' is

25 is a multiple of 5 or it is a multiple of 8.

This is a true statement. Since, p is true and q is false.

SESSION 2

1 Given that, p : 4 is an even prime number.

q : 6 is a divisor of 12 and r : the HCF of 4 and 6 is 2.

So, the truth value of the statements p,q and r are F, T and T, respectively.

Hence, $\sim p \vee (q \wedge r)$ is true.

2 $(p \Rightarrow q \wedge r) \vee (r \Leftrightarrow s) \equiv (p \Rightarrow q \wedge r) \vee [(\sim r \vee s) \wedge (r \vee \sim s)]$

$\equiv (\sim p \vee q \wedge r) \vee [(\sim r \vee s) \wedge (r \vee \sim s)]$

$[\because p \Rightarrow q \wedge r \equiv \sim p \vee (q \wedge r)]$

3 Truth table

p	q	r	$\sim p$	$\sim r$	$p \wedge \sim r$	$\sim p \vee q$	$(p \wedge \sim r) \Rightarrow (\sim p \vee q)$
T	T	T	F	F	F	T	T
T	T	F	F	T	T	T	T
T	F	T	F	F	F	F	T
T	F	F	F	T	T	F	F
F	T	T	T	F	F	T	T
F	T	F	T	T	F	T	T
F	F	T	T	F	F	T	T
F	F	F	T	T	F	T	T

Since, $(p \wedge \sim r) \Rightarrow (\sim p \vee q)$ is F.
Then, $p = $ T, $q = $ F, $r = $ F

4 Consider, $(p \wedge \sim q) \vee q \vee (\sim p \wedge q)$

$\equiv [(p \wedge \sim q) \vee q] \vee (\sim p \wedge q)$
$\equiv [(p \vee q) \wedge (\sim q \vee q)] \vee (\sim p \wedge q)$
$\equiv [(p \vee q) \wedge t] \vee (\sim p \wedge q)$
$\equiv (p \vee q) \vee (\sim p \wedge q)$
$\equiv (p \vee q \vee \sim p) \wedge (p \vee q \vee q)$
$\equiv (q \vee t) \wedge (p \vee q) \equiv t \wedge (p \vee q) \equiv p \vee q$

5 Truth Table

p	q	$p \to q$	$q \to p$	$(p \to q) \wedge (p \to q)$	$(p \to q) \vee (p \to q)$	$(p \to q) \vee (q \to p)$
T	T	T	T	T	T	T
T	F	F	T	F	F	T
F	T	T	F	T	T	T
F	F	T	T	T	T	T

So, only $(p \to q) \vee (q \to p)$ is a tautology.

6 Since, p is true and q is false $\Rightarrow \sim p$ is false and $\sim q$ is true.

$p \leftrightarrow q$ is F [since, p is true, q is false]
$\sim p \leftrightarrow q$ is T [since, $\sim p$ is false, q is false]
$\sim q \leftrightarrow p$ is T [since, $\sim q$ is true, p is true]
$\sim p \leftrightarrow \sim q$ is F [since, $\sim p$ is false, $\sim q$ is true]

7

A	B	$A \vee B$	$A \wedge B$	$A \vee (A \vee B)$	$A \vee (A \wedge B)$
T	T	T	T	T	T
T	F	T	F	T	T
F	T	T	F	T	F
F	F	F	F	F	F

$A \to B$	$A \wedge (A \to B)$	$A \wedge (A \to B) \to B$	$B \to (A \wedge (A \to B))$
T	T	T	T
F	F	T	T
T	F	T	F
T	F	T	T

Since, the truth value of all the elements in the column $A \wedge (A \to B) \to B$
So, $A \wedge (A \to B) \to B$ is tautology.

8

p	q	$\sim p$	$\sim q$	$(p \wedge q)$	$\sim(p \wedge q)$
T	T	F	F	T	F
T	F	F	T	F	T
F	T	T	F	F	T
F	F	T	T	F	T

$\sim p \vee \sim q$	$q \wedge p$	$p \Leftrightarrow q$	$p \Rightarrow q$	$q \Rightarrow p$	$(p \Leftrightarrow q) \vee (q \Rightarrow p)$
F	T	T	T	T	T
T	F	F	F	T	T
T	F	F	T	F	T
T	F	T	T	F	T

It is clear from the table that false statement is $p \Leftrightarrow q \equiv (p \Rightarrow q) \vee (q \Rightarrow p)$

Hence, it is clear from the table that $p \Leftrightarrow q$ and $(p \Rightarrow q) \vee (q \Rightarrow p)$ is not logically equilibrium.

9

p	q	$p \Leftrightarrow q$	$\sim(p \Leftrightarrow q)$
T	T	T	F
T	F	F	T
F	T	F	T
F	F	T	F

$\sim p$	$\sim q$	$\sim p \vee \sim q$	$\sim(p \Leftrightarrow q) \Leftrightarrow \sim p \vee \sim q$
F	F	F	T
F	T	T	T
T	F	T	T
T	T	T	F

Last column shows that result is neither a tautology nor a contradiction.

10 The truth tables of $p \to q$ and $\sim p \vee q$ are given below

p	q	$\sim p$	$p \to q$	$\sim p \vee q$
T	T	F	T	T
T	F	F	F	F
F	T	T	T	T
F	F	T	T	T

Clearly, truth tables of $p \to q$ and $\sim p \vee q$ are same.
So, $p \to q$ is logically equivalent to $\sim p \vee q$.
Hence, option (a) is correct.
If the truth value of p, q, r are T, F, T respectively, then the truth values of $p \vee q$ and $q \vee r$ are each equal to T. Therefore, the truth value of $(p \vee q) \wedge (q \vee r)$ is T.
Hence, option (b) is correct.
We know that, $\sim(p \vee q \vee r) \equiv (\sim p \wedge \sim q \wedge \sim r)$
So, option (c) is correct.
If p is true and q is false, then $p \vee q$ is true. Consequently $\sim(p \vee q)$ is false and hence $p \wedge \sim(p \vee q)$ is false.
Hence, option (d) is wrong.

11

p	q	r	$p \wedge q$	$p \wedge r$	$(p \wedge q) \wedge (q \wedge r)$
T	F	F	F	F	F
T	F	T	F	T	F
T	T	F	T	F	F
T	T	T	T	T	T
F	F	F	F	F	F
F	F	T	F	F	F
F	T	F	F	F	F
F	T	T	F	F	F

12.

p	q	$\sim p$	$\sim p \wedge q$	$p \Rightarrow q$	$q \Rightarrow p$	$\sim (p \Rightarrow q)$	$\sim (q \Rightarrow p)$
T	T	F	F	T	T	F	F
T	F	F	F	F	T	T	F
F	T	T	T	T	F	F	T
F	F	T	F	T	T	F	F

It is clear from the above table that columns 4 and 8 are equal. Hence, $\sim p \wedge q$ is equivalent to $\sim (q \Rightarrow p)$.

13

p	q	$\sim p$	$\sim q$	$\sim p \wedge q$	$p \vee \sim q$	$p \vee \sim p$	$\sim [p \vee \sim q]$
T	T	F	F	F	T	T	F
T	F	F	T	F	T	T	F
F	T	T	F	T	F	T	T
F	F	T	T	F	T	T	F

So, $\sim p \wedge q \equiv \sim [p \vee \sim q]$ and $p \vee \sim p$ is a tautology.

14 **Statement II** $(p \to q) \leftrightarrow (\sim q \to \sim p)$

$$\equiv (p \to q) \leftrightarrow (p \to q)$$

which is always true, so Statement II is true.

Statement I $(p \wedge \sim q) \wedge (\sim p \wedge q)$

$$\equiv p \wedge \sim q \wedge \sim p \wedge q \equiv p \wedge \sim p \wedge \sim q \wedge q \equiv f \wedge f \equiv f$$

Hence, it is a fallacy statement. So, Statement I is true.

Alternate Method

Statement II $(p \to q) \leftrightarrow (\sim q \to \sim p)$

$\sim q \to \sim p$ is contrapositive of $p \to q$

Hence, $(p \to q) \leftrightarrow (p \to q)$ will be a tautology.

Statement I $(p \wedge \sim q) \wedge (\sim p \wedge q)$

p	q	$\sim p$	$\sim q$	$p \wedge \sim q$	$\sim p \wedge q$	$(p \wedge \sim q) \wedge (\sim p \wedge q)$
T	T	F	F	F	F	F
T	F	F	T	T	F	F
F	T	T	F	F	T	F
F	F	T	T	F	F	F

Hence, it is a fallacy.

15

A	B	$A \vee B$	$B \to A$	$A \wedge B$	$\sim A$	$\sim A \vee B$	$A \to (A \vee B)$
T	T	T	T	T	F	T	T
T	F	T	T	F	F	F	T
F	T	T	F	F	T	T	T
F	F	F	T	F	T	T	T

DAY THIRTY SEVEN

Unit Test 6

(Statistics, Probability & Mathematical Reasoning)

1 While shuffling a pack of playing cards, four are accidently dropped. The probability that the cards are dropped one from each suit is

(a) $\frac{1}{256}$ (b) $\frac{2197}{20825}$ (c) $\frac{3}{20825}$ (d) None of these

2 If p : Ajay works hard, q : Ajay gets good marks, then proposition $\sim p \Rightarrow \sim q$ is equivalent to

(a) Ajay does not work hard and yet he gets good marks
(b) Ajay work hard if and only if he gets good marks
(c) If Ajay does not work hard, then he does not get good marks
(d) None of the above

3 $\sim [(p \vee q) \wedge \sim (p \wedge q)]$ is equivalent to

(a) $p \Leftrightarrow q$ (b) $\sim p \wedge q$
(c) $\sim (p \Leftrightarrow q)$ (d) None of these

4 Five-digit numbers are formed using the digits 0, 2, 4, 6, 8 without repeating the digits. If a number so formed is chosen at random, probability that it is divisible by 20 is

(a) 1/2 (b) 1/3 (c) 1/4 (d) 2/5

5 For a frequency distribution consisting of 18 observations, the mean and the standard deviation were found to be 7 and 4, respectively. But on comparison with the original data, it was found that a figure 12 was miscopied as 21 in calculations. The correct mean and standard deviation are

(a) 6.7, 2.7 (b) 6.5, 2.5
(c) 6.34, 2.34 (d) None of these

6 A boy is throwing stones at a target. The probability of hitting the target at any trial is 1/2. The probability of hitting the target 5th time at the 10th throw is

(a) $\frac{5}{2^{10}}$ (b) $\frac{63}{2^9}$ (c) $\frac{{}^{10}C_5}{2^{10}}$ (d) $\frac{{}^{10}C_4}{2^{10}}$

7 An automobile driver travels from plane to a hill station, 120 km distant at an average speed of 30 km/h. Then, he makes the return trip at an average speed of 25 km/h. He covers another 120 km distance on plane at an average speed of 50 km/h. His average speed over the entire distance of 360 km will be

(a) $\frac{30+25+50}{3}$ km/h (b) $\frac{25+35+15}{3}$ km/h
(c) $\frac{3}{\frac{1}{30}+\frac{1}{25}+\frac{1}{50}}$ km/h (d) None of these

8 The marks obtained by 60 students in a certain test are given below

Marks	10-20	20-30	30-40	40-50	50-60	60-70	70-80	80-90	90-100
No. of Students	2	3	4	5	6	12	14	10	4

Find the median of the above data.

(a) 68.33 (b) 70
(c) 71.11 (d) None of these

9 Assuming $(p \vee q)$ is true and $(p \wedge q)$ is false, state which of the following proposition have true values?

(a) $\sim p \wedge q$ (b) $\sim p \vee \sim q$ (c) $p \Leftrightarrow q$ (d) None of these

10 Let S be the universal set and $n(X) = k$. The probability of selecting two subsets A and B of the set X such that $B = \overline{A}$, is

(a) $\frac{1}{2}$ (b) $\frac{1}{2^k - 1}$ (c) $\frac{1}{2^k}$ (d) $\frac{1}{3^k}$

11 The probability that when 12 balls are distributed among three boxes, the first box will contain three balls, is

(a) $\frac{2^9}{3^{12}}$ (b) $\frac{{}^{12}C_3 \cdot 2^9}{3^{12}}$ (c) $\frac{{}^{12}C_3 \cdot 2^{12}}{3^{12}}$ (d) None

12 A fair coin is tossed 100 times. The probability of getting tails an odd number of times is

(a) $\frac{1}{2}$ (b) $\frac{1}{8}$ (c) $\frac{3}{8}$ (d) None of these

13 One mapping is selected at random from all the mappings of the set $A = \{1, 2, 3, \ldots, n\}$ into itself. The probability that the mapping selected is one to one is given by

(a) $\frac{1}{n^n}$ (b) $\frac{1}{n!}$
(c) $\frac{(n-1)!}{n^{n-1}}$ (d) None of these

14 A natural number is selected at random from the set $X = \{x : 1 \le x \le 100\}$. The probability that the number satisfies the inequation $x^2 - 13x \le 30$ is

(a) $\frac{9}{50}$ (b) $\frac{3}{20}$ (c) $\frac{2}{11}$ (d) None of these

15 If $0 < P(A) < 1$, $0 < P(B) < 1$ and $P(A \cup B) = P(A) + P(B) - P(A)P(B)$, then

(a) $P\left(\frac{A}{B}\right) = 0$ (b) $P\left(\frac{B}{A}\right) = 0$
(c) $P(A' \cap B') = P(A')P(B')$ (d) $P(A/B) + P(B/A) = 1$

16 A man is known to speak the truth 3 out of 4 times. He throws a die and reports that it is a six. The probability that it is actually a six is

(a) $\frac{3}{8}$ (b) $\frac{1}{5}$ (c) $\frac{3}{4}$ (d) None of these

17 A letter is taken at random from the letters of the word 'STATISTICS' and another letter is taken at random from the letters of the word 'ASSISTANT'. The probability that they are the same letters is

(a) $\frac{1}{45}$ (b) $\frac{13}{90}$ (c) $\frac{19}{90}$ (d) $\frac{5}{18}$

18 An ellipse of eccentricity $\frac{2\sqrt{2}}{3}$ is inscribed in a circle and a point within the circle is chosen at random. The probability that this point lies outside the ellipse is

(a) $\frac{1}{3}$ (b) $\frac{2}{3}$ (c) $\frac{1}{9}$ (d) $\frac{2}{9}$

19 Four-digit numbers are formed using each of the digit 1, 2, ..., 8 only once. One number from them is picked up at random. The probability that the selected number contains unity is

(a) $\frac{1}{2}$ (b) $\frac{1}{4}$
(c) $\frac{1}{8}$ (d) None of these

20 The mean of 10 numbers is 12.5, the mean of the first six is 15 and the last five is 10. The sixth number is

(a) 12 (b) 15
(c) 18 (d) None of these

21 The geometric mean of numbers $7, 7^2, 7^3, \ldots, 7^n$ is

(a) $7^{7/n}$ (b) $7^{n/7}$
(c) $7^{(n-1)/2}$ (d) $7^{(n+1)/2}$

22 The probability that the 14th day of a randomly chosen month is a Saturday, is

(a) $\frac{1}{12}$ (b) $\frac{1}{7}$
(c) $\frac{1}{84}$ (d) None of these

23 The mean of the numbers

$$\frac{{}^{30}C_0}{1}, \frac{{}^{30}C_2}{3}, \frac{{}^{30}C_4}{5}, \ldots\ldots, \frac{{}^{30}C_{30}}{31}$$ equals to

(a) $\frac{2^{30}}{31}$ (b) $\frac{2^{26}}{31}$
(c) $\frac{2^{26}}{31 \times 15}$ (d) None of these

24 If the mean of a binomial distribution is 25, then its standard deviation lies in the interval

(a) $[0, 5)$ (b) $(0, 5]$
(c) $[0, 25)$ (d) $(0, 25)$

25 Let p : She is intelligent and q : She is studious. The symbolic form of "it is not true that she is not intelligent or she is not studious" is

(a) $p \wedge \sim q$ (b) $\sim p \wedge q$
(c) $p \wedge q$ (d) None of these

26 The negation of $p \rightarrow (\sim p \vee q)$ is

(a) $p \vee (p \vee \sim q)$ (b) $p \rightarrow \sim (p \vee q)$
(c) $p \rightarrow q$ (d) $p \wedge \sim q$

27 The proposition of $(p \vee r) \wedge (q \vee r)$ is equivalent to

(a) $(p \wedge q) \vee r$ (b) $(p \vee q) \wedge r$
(c) $p \wedge (q \vee r)$ (d) $p \vee (q \wedge r)$

28 Which of the following statement has the truth value 'F'?

(a) A quadratic equation has always a real root.
(b) The number of ways of seating 2 persons in two chairs out of n persons is $P(n, 2)$.
(c) The cube roots of unity are in GP.
(d) None of the above

29 The variates x and u are related by $hu = x - a$, then correct relation between σ_x and σ_u is

(a) $\sigma_x = h\sigma_u$ (b) $\sigma_u = h\sigma_x$
(c) $\sigma_x = a + h\sigma_u$ (d) $\sigma_u = a + h\sigma_x$

30 Given that, $x \in [0, 1]$ and $y \in [0, 1]$. If A is the event of (x, y) satisfying $y^2 \le x$ and B is the event of (x, y) satisfying $x^2 \le y$. Then,

(a) $P(A \cap B) = \frac{1}{3}$
(b) A, B are exhaustive
(c) A, B are mutually exclusive
(d) A, B are independent

31 There are two independent events A and B. The probability that both A and B occurs is $\frac{1}{8}$ and the probability that neither of them occurs is $\frac{1}{4}$. Then, the probability of the two events are, respectively

(a) $\frac{7 \pm \sqrt{17}}{16}, \frac{2}{7 \pm \sqrt{17}}$ (b) $\frac{5 \pm \sqrt{14}}{16}, \frac{3}{5 \pm \sqrt{14}}$
(c) $\left(\frac{1}{3}, \frac{1}{2}\right)\left(\frac{1}{2}, \frac{1}{3}\right)$ (d) None of these

32 If $p \Rightarrow (q \vee r)$ is false, then the truth values of p, q, r are respectively.
(a) T, F, F (b) F, F, F (c) F, T, T (d) T, T, F

33 In a factory, workers work in three shifts say shift A, shift B and shift C and they get wages in the ratio 4 : 5 : 6 depending on the shift A, B and C, respectively. Number of workers in the shifts are in the ratio 3: 2 :1. If total number of workers is 1500 and wages per worker in shift A is ₹ 400. The mean wage of a worker is
(a) ₹467 (b) ₹500 (c) ₹600 (d) ₹400

34 The odd in favour of standing first of three students appearing in an examination are 1:2, 2:5 and 1:7, respectively. The probability that either of them will stand first, is
(a) $\frac{125}{168}$ (b) $\frac{75}{168}$ (c) $\frac{32}{168}$ (d) $\frac{4}{168}$

35 Median of ${}^{2n}C_0, {}^{2n}C_1, {}^{2n}C_2, \ldots\ldots, {}^{2n}C_n$ (when n is odd) is
(a) $\frac{1}{2}\left({}^{2n}C_{\frac{(n-1)}{2}} + {}^{2n}C_{\frac{(n+1)}{2}}\right)$
(b) ${}^{2n}C_{n/2}$
(c) ${}^{2n}C_n$
(d) None of the above

36 The contrapositive of $(p \vee q) \rightarrow r$ is
(a) $\sim r \rightarrow (p \vee q)$ (b) $r \rightarrow (p \vee q)$
(c) $\sim r \rightarrow (\sim p \wedge \sim q)$ (d) $p \rightarrow (q \vee r)$

Direction (Q. Nos. 37-40) *Each of these questions contains two statements : Statement I (Assertion) and Statement II (Reason). Each of these questions also has four alternative choices, only one of which is the correct answer. You have to select one of the codes (a), (b), (c) and (d) given below.*
(a) Statement I is true, Statement II is true; Statement II is a correct explanation for Statement I
(b) Statement I is true, Statement II is true; Statement II is not a correct explanation for Statement I
(c) Statement I is true; Statement II is false
(d) Statement I is false; Statement II is true

37 Suppose two groups of scores A and B are such that $A = (x, x + 2, x + 4)$ and $B = (x - 2, x + 2, x + 6)$
Statement I Group B has more variability than group A.
Statement II The value of mean for group B is more than that of group A.

38 Let p : He is poor. q : He is happy.
Statement I The symbolic form of the statement "It is not true that if he is poor, then he is happy" is $\sim (p \Rightarrow q)$.
Statement II The negation of the above statement is $(p \Rightarrow \sim q)$.

39 Let A and B be two independent events.
Statement I If $P(A) = 0.3$ and $P(A \cup \overline{B}) = 0.8$, then $P(B)$ is $\frac{2}{7}$.
Statement II $P(\overline{E}) = 1 - P(E)$, where E is any event.

40 **Statement I** The statements $(p \vee q) \wedge \sim p$ and $\sim p \wedge q$ are logically equivalent.
Statement II The end columns of the truth table of both statements are identical.

ANSWERS

1 (b)	2 (c)	3 (a)	4 (c)	5 (b)	6 (b)	7 (c)	8 (a)	9 (b)	10 (b)
11 (b)	12 (a)	13 (c)	14 (b)	15 (c)	16 (a)	17 (c)	18 (b)	19 (a)	20 (b)
21 (d)	22 (c)	23 (b)	24 (a)	25 (c)	26 (d)	27 (a)	28 (a)	29 (a)	30 (a)
31 (a)	32 (a)	33 (a)	34 (a)	35 (a)	36 (c)	37 (c)	38 (c)	39 (b)	40 (a)

Hints and Explanations

1 Required probability $= \frac{13^4}{^{52}C_4} = \frac{2197}{20825}$

2 Given, proposition is equivalent to "If Ajay does not work hard, then he does not get good marks".

3 It is clear from the table that column IIIrd and IVth are identical.

p	q	$p \Leftrightarrow q$	$\sim[(p \vee q) \wedge \sim (p \wedge q)]$
T	T	T	T
T	F	F	F
F	T	F	F
F	F	T	T

4 The total number of numbers formed is $5! - 4! = 96$.
If a number is divisible by 20, then the last digit is 0 and tens digit must be even. The number of such numbers with zero at the end is $4! = 24$.
$\therefore$ Required probability $= \frac{24}{96} = \frac{1}{4}$

5 Given, $\frac{\Sigma x_i}{18} = 7$

$\Rightarrow \Sigma x_i = 126$

Now, correct $\Sigma x_i = 126 - 21 + 12 = 117$

$\therefore$ True mean $= \frac{\Sigma x_i}{18} = \frac{117}{18} = 6.5$

Since, $\frac{\Sigma x_i^2}{18} - (\text{Mean})^2 = 4^2$

$\therefore \frac{\Sigma x_i^2}{18} = 4^2 + (7)^2$

$\Rightarrow \Sigma x_i^2 = 1170$

Now, correct $\Sigma x_i^2 = 1170 - 21^2 + 12^2 = 873$

$\therefore$ True variance

$= \frac{\Sigma x_i^2}{18} - (\text{Mean})^2$

$= \frac{873}{18} - (6.5)^2$

$= 48.5 - 42.25 = 6.25$

$\therefore$ True standard deviation

$= \sqrt{\text{True variance}}$

$= \sqrt{6.25} = 2.5$

6 Here, $p = \frac{1}{2}$ and $q = \frac{1}{2}$

$\therefore$ Required probability

$= {^9C_4}\left(\frac{1}{2}\right)^4\left(\frac{1}{2}\right)^5 \times \frac{1}{2}$

$= {^9C_4}\left(\frac{1}{2}\right)^{10}$

$= \frac{9 \times 8 \times 7 \times 6}{(1 \times 2 \times 3 \times 4) \times 2^{10}} = \frac{63}{2^9}$

7 Average speed $= \dfrac{120 + 120 + 120}{\frac{120}{30} + \frac{120}{25} + \frac{120}{50}}$

$= \dfrac{3}{\frac{1}{30} + \frac{1}{25} + \frac{1}{50}}$ km/h

8

Class	f_i	cf
10-20	2	2
20-30	3	5
30-40	4	9
40-50	5	14
50-60	6	20
60-70	12	32
70-80	14	46
80-90	10	56
90-100	4	60

Here, $N = 60 \Rightarrow \frac{N}{2} = 30$

$\Rightarrow$ median class is 60-70.

$\therefore$ Median $= l + \dfrac{\frac{N}{2} - c}{f} \times h$

$= 60 + \frac{30 - 20}{12} \times 10$

$= 60 + \frac{100}{12}$

$= 68.33$

9

p	q	$p \vee q$	$p \wedge q$	$\sim p \wedge q$	$\sim p \vee \sim q$	$p \Leftrightarrow q$
T	F	T	F	F	T	F
F	T	T	F	T	T	F

10 Total number of subsets of X is 2^k.
$\therefore$ Total number of possible out comes $= {^{2^k}C_2}$

Let $n(E)$ = The number of selections of two non-intersecting subsets whose union is X.

$= \frac{1}{2}({^kC_0} + {^kC_1} + {^kC_2} + \dots)$

$= \frac{1}{2} \times 2^k$

$\therefore$ Required probability

$= \dfrac{\frac{1}{2} \times 2^k}{^{2^k}C_2} = \dfrac{2^{k-1}}{2^k\left(\frac{2^k - 1}{2}\right)}$

$= \frac{1}{2^k - 1}$

11 Since, each ball can be put into any one of the three boxes, so that total number of ways in which 12 balls can be put into three boxes is 3^{12}.
The three balls can be chosen in $^{12}C_3$ ways and remaining 9 balls can be put in the remaining 2 boxes in 2^9 ways.
$\therefore$ Required probability

$= \frac{^{12}C_3 \cdot 2^9}{3^{12}}$

12 $P(X = \text{odd number})$

$= P(X = 1) + P(X = 3) + \dots + P(X = 99)$

$= {^{100}C_1}\left(\frac{1}{2}\right)^{100} + {^{100}C_3}\left(\frac{1}{2}\right)^{100} + \dots + {^{100}C_{99}}\left(\frac{1}{2}\right)^{100}$

$= ({^{100}C_1} + {^{100}C_3} + \dots + {^{100}C_{99}})\left(\frac{1}{2}\right)^{100}$

$= 2^{99} \cdot \left(\frac{1}{2}\right)^{100} = \frac{1}{2}$

13 Total number of cases $= n^n$
$\therefore$ The number of favourable cases
$= n(n-1)\dots 2 \cdot 1 = n!$
$\therefore$ Required probability
$= \frac{n!}{n^n} = \frac{(n-1)!}{n^{n-1}}$

14 Total number of ways = 100

Given, $x^2 - 13x \le 30$

$\Rightarrow \left(x - \frac{13}{2}\right)^2 \le \frac{289}{4}$

$\Rightarrow -\frac{17}{2} \le x - \frac{13}{2} \le \frac{17}{2}$

$\Rightarrow -2 \le x \le 15$

$\because x \in N$

$\therefore x \in \{1, 2, 3, \dots, 15\}$

$\therefore$ Required probability

$= \frac{15}{100} = \frac{3}{20}$

15 Given,

$P(A \cup B) = P(A) + P(B) - P(A)P(B)$

$\therefore P(A \cap B) = P(A)P(B)$

$\because$ A and B are independent events.

$\Rightarrow P(A' \cap B') = P(A')P(B')$

16 Let E = The event that six occurs and A = The event that the man reports that it is a six

$\therefore P\left(\frac{E}{A}\right) = \frac{P(E)P(A/E)}{P(E)P(A/E) + P(E')P(A/E')}$

$= \frac{(1/6)(3/4)}{(1/6)(3/4) + (5/6)(1/4)} = \frac{3}{8}$

17 Letters of the word STATISTICS are A, C, I, I, S, S, S, T, T, T. Letters of the word ASSISTANT are A, A, I, N, S, S, S, T, T. Common letters are A, I, S and T.

Probability of choosing A is

$$\frac{1}{10} \times \frac{2}{9} = \frac{2}{90}.$$

Probability of choosing I is

$$\frac{2}{10} \times \frac{1}{9} = \frac{2}{90}.$$

Probability of choosing S is

$$\frac{3}{10} \times \frac{3}{9} = \frac{9}{90}.$$

Probability of choosing T is

$$\frac{3}{10} \times \frac{2}{9} = \frac{6}{90}.$$

$\therefore$ Probability of required event

$$= \frac{2}{90} + \frac{2}{90} + \frac{9}{90} + \frac{6}{90} = \frac{19}{90}$$

18 Let equation of ellipse be

$$\frac{x^2}{a^2} + \frac{y^2}{b^2} = 1 \text{ (where, } a > b)$$

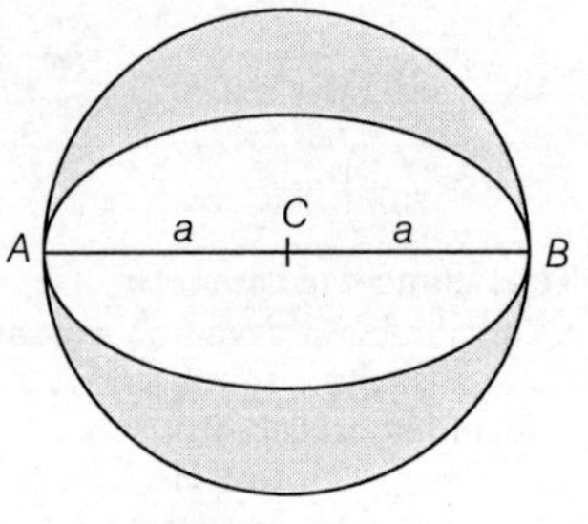

$\therefore$ Area of an ellipse $= \pi ab$

$$= \pi a \times a\sqrt{1 - e^2}$$

$$= \pi a^2 \sqrt{1 - \frac{8}{9}} = \frac{\pi a^2}{3}$$

and area of circle $= \pi a^2$

Now, required probability

$$= \frac{\pi a^2 - \frac{\pi a^2}{3}}{\pi a^2} = \frac{2}{3}$$

19 The total number of four-digit numbers formed with the digits $1, 2, \ldots, 8$ is ${}^8C_4 \times 4!$.

Now, the total number of four-digits numbers formed with the digits $1, 2, \ldots, 8$ and containing unity as one of the digits is ${}^7C_3 \times 4!$.

$\therefore$ Required probability

$$= \frac{{}^7C_3 \times 4!}{{}^8C_4 \times 4!} = \frac{1}{2}$$

20 Let the mean of the last four digit be A_2.

Then, $\quad 12.5 = \dfrac{6 \times 15 + 4 \times A_2}{6 + 4}$

$\Rightarrow \quad 125 = 90 + 4A_2$

$\Rightarrow \quad A_2 = \dfrac{35}{4}$

Let the sixth number be x, then

$$10 = \frac{1 \times x + 4 \times \frac{35}{4}}{1 + 4}$$

$\Rightarrow \quad 50 = x + 35$

$\Rightarrow \quad x = 15$

21 $\text{GM} = (7 \cdot 7^2 \ldots 7^n)^{1/n}$

$$= \left(7^{\frac{n(n+1)}{2}}\right)^{1/n} = 7^{\frac{n+1}{2}}$$

22 Clearly any month, out of 12 months, can be chosen with probability $= \dfrac{1}{12}$

Now, as there are 7 possible ways in which the month can start and it will be a Saturday on 14th day if the first day of the month is Sunday.

$\therefore$ Its probability $= \dfrac{1}{7}$

Hence, the required probability

$$= \frac{1}{12} \times \frac{1}{7} = \frac{1}{84}.$$

23 Consider

$$(1 + x)^{30} = {}^{30}C_0 + {}^{30}C_1 x + {}^{30}C_2 x^2 + \ldots + {}^{30}C_{30} x^{30} \quad \ldots \text{(i)}$$

and $(1 - x)^{30} = {}^{30}C_0 - {}^{30}C_1 x + {}^{30}C_2 x^2 - \ldots + {}^{30}C_{30} x^{30} \quad \ldots$(ii)

On adding Eqs. (i) and (ii), we get

$${}^{30}C_0 + {}^{30}C_2 x^2 + \ldots + {}^{30}C_{30} x^{30}$$

$$= \frac{1}{2}[(1 + x)^{30} + (1 - x)^{30}]$$

Now, on integrating over 0 to 1, we get

$$\left[{}^{30}C_0 x + \frac{{}^{30}C_2 x^3}{3} + \ldots + \frac{{}^{30}C_{30} x^{31}}{31}\right]_0^1$$

$$= \frac{1}{2}\left[\frac{(1 + x)^{31}}{31} - \frac{(1 - x)^{31}}{31}\right]_0^1$$

$$\Rightarrow \quad \frac{{}^{30}C_0}{1} + \frac{{}^{30}C_2}{3} + \ldots + \frac{{}^{30}C_{30}}{31}$$

$$= \frac{1}{2 \times 31}[2^{31}] = \frac{2^{30}}{31}$$

Now, as there are 16 terms, therefore required mean

$$= \frac{\frac{{}^{30}C_0}{1} + \frac{{}^{30}C_2}{3} + \ldots + \frac{{}^{30}C_{30}}{31}}{16}$$

$$= \frac{\frac{2^{30}}{31}}{16} = \frac{2^{30}}{31 \times 16} = \frac{2^{26}}{31}$$

24 Since, $\quad 0 \le \sqrt{npq} < \sqrt{np}$

$\Rightarrow \quad 0 \le \text{SD} < 5 \quad [\because p \ne 0]$

$\Rightarrow \quad \text{SD} \in [0, 5)$

25 Here, $\sim p$: She is not intelligent.

$\sim q$: She is not studious.

So, the required symbolic form is $\sim(\sim p \vee \sim q)$ or $p \wedge q$.

26 Clearly,

$$p \to (\sim p \vee q) \equiv (\sim p) \vee ((\sim p) \vee q)$$

$$\equiv ((\sim p) \vee (\sim p)) \vee q \quad \text{[by associative law]}$$

$$\equiv (\sim p) \vee q \text{ [by idempotent law]}$$

Now, $\sim(p \to (\sim p \vee q)) \equiv \sim((\sim p) \vee q)$

$$\equiv \sim(\sim p) \wedge (\sim q)$$

$$\equiv p \wedge \sim q$$

27 Using Distributive law,

$$(p \vee r) \wedge (q \vee r) \equiv (p \wedge q) \vee r$$

28 Clearly, the roots of a quadratic equation can be imaginary. So, this statement has truth value F.

29 Here, $\quad u = \dfrac{x}{h} - \dfrac{a}{h}$

We know that standard deviation is not dependent on change of origin.

$\therefore \quad \sigma_u = \dfrac{\sigma_x}{h} \Rightarrow \sigma_x = h\sigma_u$

30 A = The event of (x, y) belonging to the area $OTQPO$

and B = The event of (x, y) belonging to the area $OSQRO$

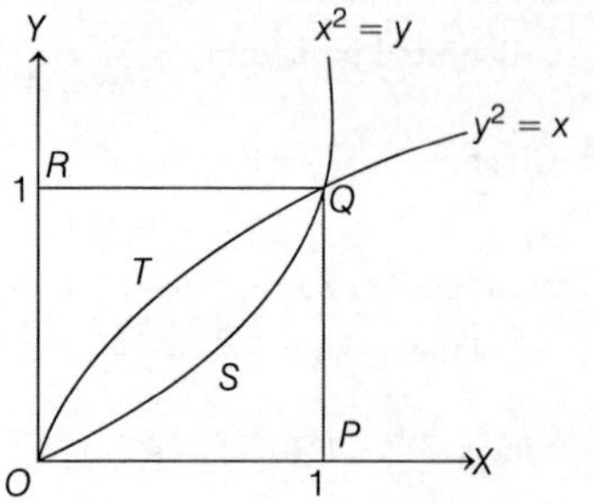

$$P(A) = \frac{\text{ar}(OTQPO)}{\text{ar}(OPQRO)} = \frac{\int_0^1 \sqrt{x}\, dx}{1 \times 1}$$

$$= \left[\frac{2}{3} x^{3/2}\right]_0^1 = \frac{2}{3}$$

$$P(B) = \frac{\text{ar}(OSQRO)}{\text{ar}(OPQRO)} = \frac{\int_0^1 \sqrt{y}\, dy}{1 \times 1} = \frac{2}{3}$$

$$P(A \cap B) = \frac{\text{ar}(OTQS)}{\text{ar}(OPQRO)}$$

$$= \frac{\int_0^1 \sqrt{x}\, dx - \int_0^1 x^2 dx}{1 \times 1} = \frac{2}{3} - \frac{1}{3} = \frac{1}{3}$$

Note (i) $P(A) + P(B) = \dfrac{2}{3} + \dfrac{2}{3} \ne 1$

So, A and B are not exhaustive.

(ii) $P(A) \cdot P(B) = \dfrac{2}{3} \cdot \dfrac{2}{3} = \dfrac{4}{9} \ne P(A \cap B)$

So, A and B are not independent.

(iii) $P(A \cap B) \ne 0$, so A and B are not mutually exclusive.

31 Given, $P(A \cap B) = \frac{1}{8}$

and $P(\overline{A} \cap \overline{B}) = \frac{1}{4}$

$\therefore \quad P(A) \cdot P(B) = \frac{1}{8}$

and $P(\overline{A \cup B}) = \frac{1}{4}$

$\Rightarrow \quad P(A) \cdot P(B) = \frac{1}{8}$

and $P(A \cup B) = 1 - \frac{1}{4} = \frac{3}{4}$

$\Rightarrow \quad \frac{3}{4} = P(A) + P(B) - \frac{1}{8}$

$[\because P(A \cup B) = P(A) + P(B) - P(A \cap B)]$

$\Rightarrow \quad P(A) + P(B) = \frac{7}{8}$

Let $P(A) = x$, then $P(B) = \frac{1}{8x}$

$\therefore \quad x + \frac{1}{8x} = \frac{7}{8}$

$\Rightarrow \quad \frac{8x^2 + 1}{8x} = \frac{7}{8}$

$\Rightarrow \quad 8x^2 + 1 = 7x$

or $\quad 8x^2 - 7x + 1 = 0$

$\therefore \quad x = \frac{7 \pm \sqrt{49 - 32}}{16}$

$P(A) = \frac{7 \pm \sqrt{17}}{16}$

$\Rightarrow \quad P(B) = \frac{2}{7 \pm \sqrt{17}}$

32 We know, $p \Rightarrow q$ is false when p is true and q is false.

$\therefore p \Rightarrow (q \vee r)$ is false when p is true and $(q \vee r)$ is false, and we know $q \vee r$ is false only when both q and r are false.

Hence, truth values of p, q, and r are respectively T, F, F.

33 Clearly, number of workers in shift A

$= \frac{3}{6} \times 1500 = 750$

Number of workers in shift B

$= \frac{2}{6} \times 1500 = 500$

and number of workers in shift $C = 250$

Now, let the sum of wages per person in three shifts $= x$, then

Wages in shift $A = \frac{4}{15} \times x$

$\Rightarrow \quad \frac{4}{15} \times x = 400$

$\Rightarrow \quad x = 1500$

Now, wages in shift B

$= \frac{5}{15} \times 1500 = 500$ per person

and wages in shift C

$= \frac{6}{15} \times 1500 = 600$ per person

$\therefore$ Mean wage

$\frac{750 \times 400 + 500 \times 500 + 250 \times 600}{1500}$

$=$ ₹ 467 per worker

34 Let the three students be A, B and C. Also, let E, F and G denote the events of standing first of three students A, B and C respectively. Then, we have

$P(E) = \frac{1}{1+2} = \frac{1}{3}; P(F) = \frac{2}{2+5} = \frac{2}{7}$

and $P(G) = \frac{1}{1+7} = \frac{1}{8}$.

Since, the events E, F and G are mutually exclusive.

$\therefore P(E \cup F \cup G) = P(E) + P(F) + P(G)$

$= \frac{1}{3} + \frac{2}{7} + \frac{1}{8}$

$= \frac{56 + 48 + 21}{168} = \frac{125}{168}$

35 Since, n is odd, therefore ${}^{2n}C_0, {}^{2n}C_1, {}^{2n}C_2, \ldots, {}^{2n}C_n$ are even in number.

Now,

$$\text{median} = \frac{1}{2}\left[\left(\frac{n+1}{2}\right)^{\text{th}} \text{observation} + \left(\frac{n+1}{2} + 1\right)^{\text{th}} \text{observation}\right]$$

$$= \frac{1}{2}\left[{}^{2n}C_{\frac{n+1}{2}-1} + {}^{2n}C_{\frac{n+1}{2}+1-1}\right]$$

$$= \frac{1}{2}\left[{}^{2n}C_{\frac{n-1}{2}} + {}^{2n}C_{\frac{n+1}{2}}\right]$$

36 Contrapositive of $(p \vee q) \rightarrow r$ is

$\sim r \rightarrow \sim(p \vee q) \equiv \sim r \rightarrow (\sim p \wedge \sim q)$

37 Since, $A = (x, x+2, x+4)$

and $B = (x-2, x+2, x+6)$

$\therefore$ Mean of A

$= \frac{x + x + 2 + x + 4}{3} = x + 2$

and mean of

$B = \frac{x - 2 + x + 2 + x + 6}{3} = x + 2$

Hence, group B has more variability than group A.

[$\because$ From the given data difference in scores of group A is 2 but difference in scores of group B is 4.]

38 Clearly, Statement I is true but Statement II is false, as negation of given statement is $p \Rightarrow q$.

39 $P(A \cup \overline{B}) = 1 - P(\overline{A \cup \overline{B}})$

$\Rightarrow \quad 0.8 = 1 - P(\overline{A} \cap B)$

$\Rightarrow \quad 0.8 = 1 - P(\overline{A}) P(B)$

$\Rightarrow \quad 0.8 = 1 - 0.7 \times P(B)$

$\therefore \quad P(B) = \frac{2}{7}$

40

p	q	$\sim p$	$p \vee q$	$(p \vee q) \wedge \sim p$	$\sim p \wedge q$
T	T	F	T	F	F
T	F	F	T	F	F
F	T	T	T	T	T
F	F	T	F	F	F

It is clear from the table both statements are true and Statement II is a correct explanation for Statement I.

DAY THIRTY EIGHT

Mock Test 1

(Based on Complete Syllabus)

Instructions

1. This question paper contains of 30 Questions of Mathematic, divided into two Sections :
 Section A Objective Type Questions and **Section B** Numerical Type Questions.
2. Section A contains 20 Objective questions and all Questions are compulsory (**Marking Scheme** : Correct +4, Incorrect −1).
3. Section B contains 10 Numerical value questions out of which only 5 questions are to be attempted (**Marking Scheme** : Correct +4, Incorrect 0).

Section A : Objective Type Questions

1 Let $f(x)$ satisfies the requirements of Lagrange's mean value theorem in [0, 2]. If $f(0)=0$ and $|f'(x)|\le\frac{1}{2}$ for all x in [0, 2], then

(a) $f(x)\le 2$ (b) $|f(x)|\le 1$
(c) $f(x)=2x$ (d) $f(x)=3$ for atleast one x in [0, 2]

2 A woman purchases 1 kg of onions from each of the 4 places at the rate of 1kg, 2kg, 3kg, 4kg per rupee respectively. On the average she has purchased x kg of onions per rupee, then the value of x is

(a) 2 (b) 2.5
(c) 1.92 (d) None of these

3 The statement $\sim(p\vee q)\vee(\sim p\wedge q)$ is logically equivalent to

(a) p (b) $\sim p$ (c) q (d) $\sim q$

4 If $A=\{\theta: 2\cos^2\theta+\sin\theta\le 2\}$ and $B=\left\{\theta:\frac{\pi}{2}\le\theta\le\frac{3\pi}{2}\right\}$, then $A\cap B$ is equal to

(a) $\left\{\theta:\pi\le\theta\le\frac{3\pi}{2}\right\}$ (b) $\left\{\theta:\frac{\pi}{2}\le\theta\le\frac{5\pi}{2}\right\}$
(c) $\left\{\theta:\frac{\pi}{2}\le\theta\le\frac{5\pi}{6}\text{ or }\pi\le\theta\le\frac{3\pi}{2}\right\}$
(d) None of the above

5 A square $OABC$ is formed by line pairs $xy=0$ and $xy+1=x+y$, where O is the origin. A circle with centre C_1 inside the square is drawn to touch the line pair $xy=0$ and another circle with centre C_2 and radius twice that of C_1, is drawn to touch the circle C_1 and the other line pair. The radius of the circle with centre C_1 is

(a) $\frac{\sqrt{2}}{\sqrt{3}(\sqrt{2}+1)}$ (b) $\frac{2\sqrt{2}}{\sqrt{2}(\sqrt{2}+1)}$
(c) $\frac{\sqrt{2}}{3(\sqrt{2}+1)}$ (d) $\frac{\sqrt{2}+1}{3\sqrt{2}}$

6 A function $y=f(x)$ has a second order derivative $f''(x)=6(x-1)$. If its graph passes through the point (2, 1) and at that point the tangent to the graph is $y=3x-5$, then the function is

(a) $(x-1)^3$ (b) $(x+1)^3$
(c) $(x+1)^2$ (d) $(x-1)^2$

7 If z is non-real and $i=\sqrt{-1}$, then $\sin^{-1}\left(\frac{1}{i}(z-1)\right)$ be the angle of a triangle, if

(a) $\text{Re}(z)=1, \text{Im}(z)=2$
(b) $\text{Re}(z)=1, -1\le\text{Im}(z)\le 1$
(c) $\text{Re}(z)+\text{Im}(z)=0$
(d) None of these

8 The ends A and B of a rod of length $\sqrt{5}$ are sliding along the curve $y = 2x^2$. Let x_A and x_B be the x-coordinate of the ends. At the moment when A is at (0, 0) and B is at (1, 2) the derivative $\frac{dx_B}{dx_A}$ has the value equal to

(a) $\frac{1}{3}$ (b) $\frac{1}{5}$ (c) $\frac{1}{8}$ (d) $\frac{1}{9}$

9 Suppose the function $g_n(x) = x^{2n+1} + a_n x + b_n$; $(n \in N)$ satisfies the equation $\int_{-1}^{1} (px + q) g_n(x)\, dx = 0$ for all linear functions $(px+q)$, then

(a) $a_n = b_n = 0$ (b) $b_n = 0, a_n = -\frac{3}{2n+3}$

(c) $a_n = 0, b_n = -\frac{3}{2n+3}$ (d) $a_n = \frac{3}{2n+3}, b_n = -\frac{3}{2n+3}$

10 The function $f(x) = \max.\{(1-x), (1+x), 2\}$, $x \in (-\infty, \infty)$, is

(a) continuous at all points except at $x = 1$ and $x = -1$
(b) differentiable at all points except at $x = 1$ and $x = -1$
(c) differentiable at all points
(d) None of the above

11 The two vertices of a triangle are $(4, -3)$ and $(-2, 5)$. If the orthocentre of the triangle is at (1, 2), then the third vertex is

(a) (– 33, – 26) (b) (33, 26)
(c) (26, 33) (d) None of these

12 The unit vectors **a** and **b** are perpendicular and the unit vector **c** is inclined at an angle θ to both **a** and **b**. If $c = \alpha a + \beta b + \gamma(a \times b)$, then which is not true ?

(a) $\gamma^2 = 1 - 2\alpha^2$ (b) $\alpha = 2\beta$

(c) $\gamma^2 = -\cos 2\theta$ (d) $\beta^2 = \frac{1+\cos 2\theta}{2}$

13 If $0° \le \theta \le 180°$ and $81^{\sin^2\theta} + 81^{\cos^2\theta} = 30$, then θ is

(a) 30° (b) 45° (c) 120° (d) 150°

14 The sum of n terms of the following series

$1 + (1 + x) + (1 + x + x^2) + \dots$ will be

(a) $\frac{1-x^n}{1-x}$ (b) $\frac{x(1-x^n)}{1-x}$

(c) $\frac{n(1-x) - x(1-x^n)}{(1-x)^2}$ (d) None of these

15 Let function $f : R \to R$ be defined by $f(x) = 2x + \sin x$ for $x \in R$, then f is

(a) one-one and onto (b) one-one but not onto
(c) onto but not one-one (d) Neither one-one nor onto

16 For the arithmetic progression $a, a + d, a + 2d, a + 3d, \dots, a + 2nd$, the mean deviation from mean is

(a) $\frac{n(n+1)d}{2n-1}$ (b) $\frac{n(n+1)d}{2n+1}$

(c) $\frac{n(n-1)d}{2n+1}$ (d) None of these

17 At a point P on the ellipse $\frac{x^2}{a^2} + \frac{y^2}{b^2} = 1$, tangent PQ is drawn. If the point Q be at a distance $\frac{1}{p}$ from the point P, where p is distance of the tangent from the origin, then the locus of the point Q is

(a) $\frac{x^2}{a^2} + \frac{y^2}{b^2} = 1 + \frac{1}{a^2b^2}$ (b) $\frac{x^2}{a^2} - \frac{y^2}{b^2} = 1 - \frac{1}{a^2b^2}$

(c) $\frac{x^2}{a^2} + \frac{y^2}{b^2} = \frac{1}{a^2b^2}$ (d) $\frac{x^2}{a^2} - \frac{y^2}{b^2} = \frac{1}{a^2b^2}$

18 The lines $\mathbf{r} = \mathbf{a} + \lambda(\mathbf{b} \times \mathbf{c})$ and $\mathbf{r} = \mathbf{b} + \mu(\mathbf{c} \times \mathbf{a})$ will intersect if

(a) $\mathbf{a} \times \mathbf{c} = \mathbf{b} \times \mathbf{c}$ (b) $\mathbf{a} \cdot \mathbf{c} = \mathbf{b} \cdot \mathbf{c}$
(c) $\mathbf{b} \times \mathbf{a} = \mathbf{c} \times \mathbf{a}$ (d) None of these

19 The volume of the parallelopiped formed by vectors $\hat{\mathbf{i}} + a\hat{\mathbf{j}}$, $a\hat{\mathbf{i}} + \hat{\mathbf{j}} + \hat{\mathbf{k}}$ and $\hat{\mathbf{j}} + a\hat{\mathbf{k}}$ is maximum, when a equals to

(a) $\sqrt{3}$ (b) $-\sqrt{3}$ (c) $\frac{-1}{\sqrt{3}}$ (d) $\frac{1}{\sqrt{3}}$

20 Solution of the differential equation $(3xy^2 + x\sin(xy))dy + (y^3 + y\sin(xy))dx = 0$ is

(a) $xy^3 - \cos xy = C$ (b) $xy^3 + \cos xy = C$
(c) $xy^2 - \cos xy = C$ (d) $xy^2 + \sin xy = C$

Section B : Numerical Type Questions

21 The number of ordered pairs (α, β), where $\alpha, \beta \in (-\pi, \pi)$ satisfying $\cos(\alpha - \beta) = 1$ and $\cos(\alpha + \beta) = \frac{1}{e}$ is

22 The coefficient of x^{50} in $(1 + x)^{41}(1 - x + x^2)^{40}$ is

23 The value of $\log_7 \log_7 \sqrt{7\sqrt{7\sqrt{7}}}$ is $k + m\log_7 n$, then value of $|k| + |m| + |n|$ is equal to

24 The integral $\int_{-1/2}^{1/2} \left[[x] + \log\left(\frac{1+x}{1-x}\right) \right] dx$ is equal to $\frac{p}{q}$, then value of $p^2 + q^2$ is equal to

25 A natural number x is chosen at random from the first one hundred natural numbers. The probability that $\frac{(x-20)(x-40)}{(x-30)} < 0$ is $7/a$, then a is equal to

26 If $A_r = \begin{bmatrix} r & r-1 \\ r-1 & r \end{bmatrix}$, where r is a natural number, then $|A_1| + |A_2| + \dots + |A_{2018}|$ must be equal to k^2, then value of k is equal to

27 $\lim_{\theta \to 0} \frac{4\theta(\tan\theta - 2\theta\tan\theta)}{(1 - \cos 2\theta)}$ is equal to

28 Normals AO, AA_1, AA_2 are drawn to parabola $y^2 = 8x$ from the point $A(h, 0)$. If ΔOA_1A_2 is equilateral, then possible values of h is

29 A dictionary is printed consisting of 7 lettered words only that can be made with a letter of the word CRICKET. If the words are printed in alphabetical order, as in an ordinary dictionary, then the number of words before the word CRICKET is

30 If $a, b, c > 0$, $a^2 = bc$ and $a + b + c = abc$, then the least value of $a^4 + a^2 + 7$ must be equal to

Hints and Explanations

1 *(b)* Since, $f(x)$ satisfies the Lagrange's mean value theorem.

$\therefore \quad f'(c) = \dfrac{f(x) - f(0)}{x - 0}$

where, $0 < c < x < 2$ i.e. $0 < c < 2$

$\Rightarrow \quad f(x) = x f'(c)$

$\Rightarrow \quad |f(x)| = |x f'(c)|$

$= |x||f'(c)| \le 2 \cdot \dfrac{1}{2} = 1$

$\Rightarrow \quad |f(x)| \le 1$

2 *(c)* Cost of 1kg onion, purchased from place 1 = ₹ 1

Cost of 1kg onion, purchased from place 2 = ₹ $\dfrac{1}{2}$

Cost of 1kg onion, purchased from place 3 = ₹ $\dfrac{1}{3}$

Cost of 1kg onion, purchased from place 4 = ₹ $\dfrac{1}{4}$

Now, average rate of 1kg onion

$= ₹\left(\dfrac{1 + \frac{1}{2} + \frac{1}{3} + \frac{1}{4}}{4}\right) = ₹\dfrac{25}{48}$

Thus, in ₹ $\dfrac{25}{48}$, we get 1 kg onoin.

$\therefore$ In ₹1, we get

$\dfrac{48}{25}$ kg onion = 1.92 kg

Alternate Method

Harmonic mean will give the correct answer, here

$HM = \dfrac{4}{1 + \frac{1}{2} + \frac{1}{3} + \frac{1}{4}} = \dfrac{4 \times 12}{12 + 6 + 4 + 3}$

$= \dfrac{48}{25} = 1.92$ kg

3 *(b)* $\sim(p \vee q) \vee (\sim p \wedge q)$

$\equiv (\sim p \wedge \sim q) \vee (\sim p \wedge q)$

$\equiv \sim p \wedge (\sim q \vee q)$

$\equiv \sim p \wedge t \equiv \sim p$

4 *(c)* Consider, $2\cos^2\theta + \sin\theta \le 2$ and $\dfrac{\pi}{2} \le \theta \le \dfrac{3\pi}{2} \Rightarrow 2 - 2\sin^2\theta + \sin\theta \le 2$

$\Rightarrow \sin\theta\,(2\sin\theta - 1) \ge 0$

Case I $\quad \sin\theta \ge 0$ and $2\sin\theta - 1 \ge 0$

$\therefore \quad \sin\theta \ge 0$ and $\sin\theta \ge \dfrac{1}{2}$

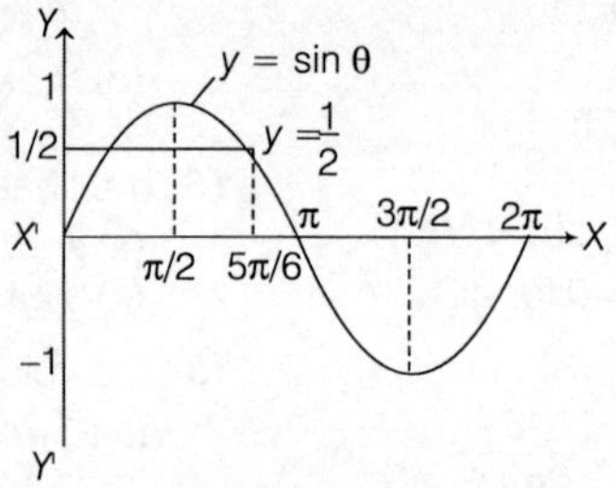

$\Rightarrow \quad \sin\theta \ge \dfrac{1}{2} \Rightarrow \dfrac{\pi}{2} \le \theta \le \dfrac{5\pi}{6} \quad$ ...(i)

Case II $\quad \sin\theta \le 0$ and $2\sin\theta - 1 \le 0$

$\therefore \quad \sin\theta \le 0$ and $\sin\theta \le \dfrac{1}{2}$

$\Rightarrow \quad \sin\theta \le 0$

$\Rightarrow \quad \pi \le \theta \le \dfrac{3\pi}{2} \quad$ ...(ii)

From Eqs. (i) and (ii), we get

$A \cap B = \left\{\theta : \dfrac{\pi}{2} \le \theta \le \dfrac{5\pi}{6} \text{ or } \pi \le \theta \le \dfrac{3\pi}{2}\right\}$

5 *(a)* Diagonal of the square = $\sqrt{2}$

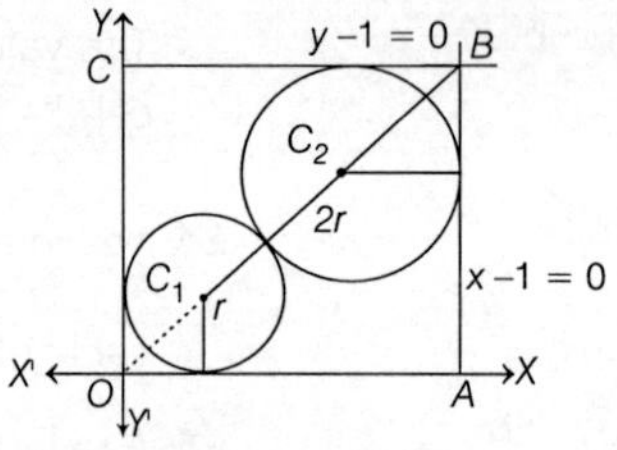

Also, $\quad d = r\sqrt{2} + 3r + 2\sqrt{2}r$

$\Rightarrow \quad \sqrt{2} = 3\sqrt{2}r + 3r \Rightarrow r = \dfrac{\sqrt{2}}{3(\sqrt{2} + 1)}$

6 *(a)* Given that, $f''(x) = 6(x - 1)$

$f'(x) = 3(x - 1)^2 + C_1 \quad$ ...(i)

But at point (2, 1) the line $y = 3x - 5$ is tangent to the graph $y = f(x)$.

$\therefore \quad \left(\dfrac{dy}{dx}\right)_{(x=2)} = 3$ or $f'(2) = 3$

From Eq. (i), $f'(2) = 3(2 - 1)^2 + C_1$

$\Rightarrow \quad 3 = 3 + C_1 \Rightarrow C_1 = 0$

$\therefore \quad f'(x) = 3(x - 1)^2$

$\Rightarrow \quad f(x) = (x - 1)^3 + C_2$

$\because \quad f(2) = 1$

$\therefore \quad 1 + C_2 = 1 \Rightarrow C_2 = 0$

Hence, $\quad f(x) = (x - 1)^3$

7 *(b)* By the properties of inverse trigonometric function $\dfrac{z - 1}{i}$ = real

$\Rightarrow \dfrac{x - 1 + iy}{i}$ = real $\Rightarrow \dfrac{x - 1}{i} + y$ = real

$\Rightarrow \quad x - 1 = 0 \Rightarrow x = 1$

$\therefore \quad \sin^{-1}\left(\dfrac{z - 1}{i}\right) = \sin^{-1}(y)$

So, $\; -1 \le y \le 1$

$\therefore \mathrm{Re}(z) = x = 1, \; -1 \le \mathrm{Im}(z) \le 1$

8 *(d)* We have, $y = 2x^2$

$(AB)^2 = (x_B - x_A)^2 + (2x_B^2 - 2x_A^2)^2 = 5$

$\Rightarrow \quad (x_B - x_A)^2 + 4(x_B^2 - x_A^2)^2 = 5$

On differentiating w.r.t. x_A and denoting $\dfrac{dx_B}{dx_A} = D$, we get

$2(x_B - x_A)(D - 1) + 8(x_B^2 - x_A^2)(2x_B D - 2x_A) = 0$

On putting $x_A = 0$; $x_B = 1$, then

$2(1 - 0)(D - 1) + 8(1 - 0)(2D - 0) = 0$

$\Rightarrow \quad 2D - 2 + 16D = 0$

$\Rightarrow \quad D = \dfrac{1}{9}$

9 *(b)* We have,

$\int_{-1}^{1} (px + q)(x^{2n+1} + a_n x + b_n)\,dx = 0$

Equating the odd component to be zero and integrating, we get

$\dfrac{2p}{2n + 3} + \dfrac{2a_n p}{3} + 2b_n q = 0$ for all p, q

Hence, $\quad b_n = 0$

and $\quad a_n = -\dfrac{3}{2n + 3}$

10 *(c)* We have, $f(x) = \max \cdot \{1 - x, x + 1, 2\}$

Let us draw the graph of $y = f(x)$, as shown below

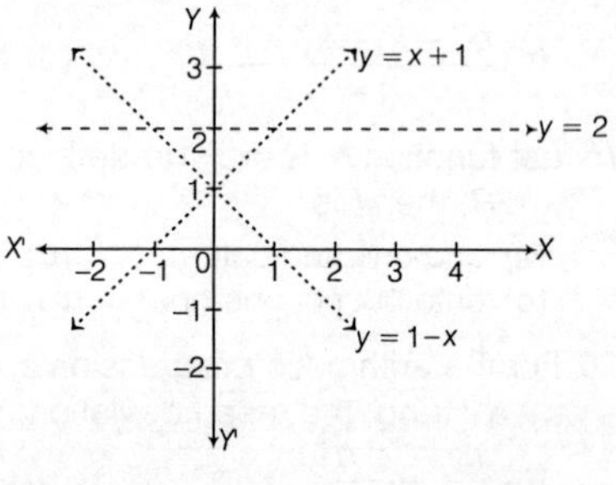

From the graph it is clear that, $f(x)$ is continuous everywhere but not differentiable at $x = -1, 1$.

11 *(b)* Let the third vertex be (h, k).

Now, the slope of AO or AD is $\dfrac{k - 2}{h - 1}$.

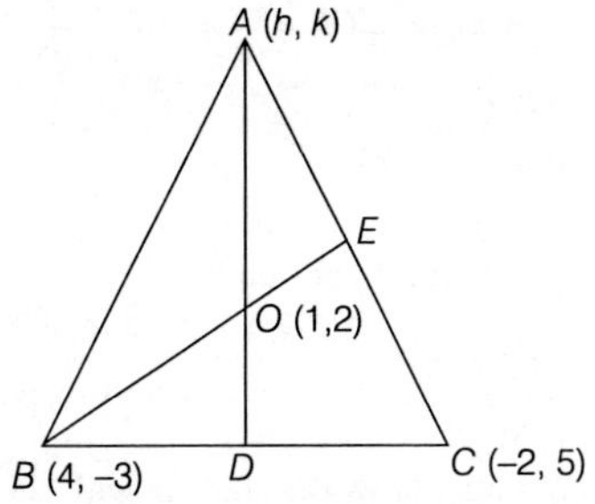

Slope of BC is $\frac{5+3}{-2-4} = -\frac{4}{3}$

Slope of BE is $\frac{-3-2}{4-1} = -\frac{5}{3}$

and slope of AC is $\frac{k-5}{h+2}$.

Since, $AD \perp BC$, $\frac{k-2}{h-1} \times \left(-\frac{4}{3}\right) = -1$

$\Rightarrow \quad 3h = 4k - 5 \quad \ldots(i)$

Again, $BE \perp AC$,

$-\frac{5}{3} \times \frac{k-5}{h+2} = -1$

$\Rightarrow \quad 3h = 5k - 31 \quad \ldots(ii)$

On solving Eqs. (i) and (ii), we get

$h = 33$ and $k = 26$

12 *(b)* Here, $|\mathbf{a}| = |\mathbf{b}| = |\mathbf{c}| = 1, \mathbf{a} \cdot \mathbf{b} = 0$

and $\cos\theta = \mathbf{a} \cdot \mathbf{c} = \mathbf{b} \cdot \mathbf{c}$

Now, $\mathbf{c} = \alpha\,\mathbf{a} + \beta\,\mathbf{b} + \gamma\,(\mathbf{a} \times \mathbf{b}) \quad \ldots(i)$

$\Rightarrow \mathbf{a} \cdot \mathbf{c} = \alpha(\mathbf{a} \cdot \mathbf{a}) + \beta\,(\mathbf{a} \cdot \mathbf{b}) + \gamma\,\{\mathbf{a} \cdot (\mathbf{a} \times \mathbf{b})\}$

$\Rightarrow \cos\theta = \alpha\,|\mathbf{a}|^2 \Rightarrow \cos\theta = \alpha$

Similarly, by taking dot product on both sides of Eq. (i) by $\mathbf{b}$, we get

$\beta = \cos\theta$

$\therefore \quad \alpha = \beta$

$\therefore$ From Eq. (i), we get

$|\mathbf{c}|^2 = |\alpha\,\mathbf{a} + \beta\mathbf{b} + \gamma(\mathbf{a} \times \mathbf{b})|^2$

$= \alpha^2\,|\mathbf{a}|^2 + \beta^2\,|\mathbf{b}|^2 + \gamma^2\,|\mathbf{a} \times \mathbf{b}|^2 + 2\alpha\beta\,(\mathbf{a} \cdot \mathbf{b}) + 2\alpha\gamma\,\{\mathbf{a} \cdot (\mathbf{a} \times \mathbf{b})\} + 2\beta\gamma\,\{\mathbf{b} \cdot (\mathbf{a} \times \mathbf{b})\}$

$\Rightarrow \quad 1 = \alpha^2 + \beta^2 + \gamma^2\,|\mathbf{a} \times \mathbf{b}|^2$

$\Rightarrow \quad 1 = 2\alpha^2 + \gamma^2 \left\{|\mathbf{a}|^2 \cdot |\mathbf{b}|^2 \sin^2\frac{\pi}{2}\right\}$

$\Rightarrow \quad 1 = 2\alpha^2 + \gamma^2$

$\Rightarrow \quad \gamma^2 = 1 - 2\cos^2\theta = -\cos 2\theta$

$\Rightarrow \quad \alpha^2 = \beta^2 = \frac{1-\gamma^2}{2} = \frac{1+\cos 2\theta}{2}$

13 *(a)* Let $81^{\sin^2\theta} = t$

Given, $81^{\sin^2\theta} + 81^{\cos^2\theta} = 30$

$\therefore \quad t + \frac{81}{t} = 30 \Rightarrow t^2 - 30t + 81 = 0$

$\Rightarrow (t-27)(t-3) = 0 \Rightarrow t = 27, 3$

$\Rightarrow 81^{\sin^2\theta} = 3^{4\sin^2\theta} = 3^3, 3^1$

$\Rightarrow 4\sin^2\theta = 3, 4\sin^2\theta = 1$

$\Rightarrow \sin\theta = \pm\frac{\sqrt{3}}{2}, \pm\frac{1}{2} \Rightarrow \theta = \frac{\pi}{6}, \frac{\pi}{3}$

14 *(c)* Let S_n

$= 1 + (1+x) + (1+x+x^2) + \ldots + (1 + x + x^2 + x^3 + \ldots + x^{n-1})$

Then, $S_n = \frac{1}{(1-x)}\{(1-x) + (1-x^2) + (1-x^3) + (1-x^4) + \ldots + \text{upto } n \text{ terms}\}$

$= \frac{1}{(1-x)}[n - (x + x^2 + x^3 + \ldots + \text{upto } n \text{ terms})]$

$= \frac{1}{(1-x)}\left[n - \frac{x(1-x^n)}{1-x}\right]$

$= \frac{n(1-x) - x(1-x^n)}{(1-x)^2}$

15 *(a)* $\because \; f(x) = 2x + \sin x$

$\therefore \; f'(x) = 2 + \cos x > 0$ for all x

Since, $f(x)$ is strictly increasing. So, f is one-one.

Here, $\lim_{x \to -\infty} f(x) = -\infty$ and $\lim_{x \to \infty} f(x) = \infty$

Hence, f is onto.

16 *(b)* Clearly, mean

$\bar{x} = \frac{1}{(2n+1)}[a + (a+d) + (a+2d) + \ldots + (a+2nd)]$

$= \frac{1}{(2n+1)}\left[\frac{2n+1}{2}(a + a + 2nd)\right]$

$= a + nd$

Now, mean deviation from mean

$= \frac{1}{(2n+1)}\sum_{r=0}^{2n}|(a+rd) - (a+nd)|$

$= \frac{1}{(2n+1)}\sum_{r=0}^{2n}|(r-n)d|$

$= \frac{1}{(2n+1)} \times 2d\,(1 + 2 + \ldots + n)$

$= \frac{n(n+1)}{2n+1}d$

17 *(a)* Equation of the tangent at P is

$\frac{x - a\cos\theta}{a\sin\theta} = \frac{y - b\sin\theta}{-b\cos\theta}$

$\Rightarrow \; xb\cos\theta + ay\sin\theta = ab$

The distance of the tangent from the origin is $p = \left|\frac{ab}{\sqrt{b^2\cos\theta + a^2\sin^2\theta}}\right|$

$\Rightarrow \; \frac{1}{p} = \frac{\sqrt{b^2\cos^2\theta + a^2\sin^2\theta}}{ab}$

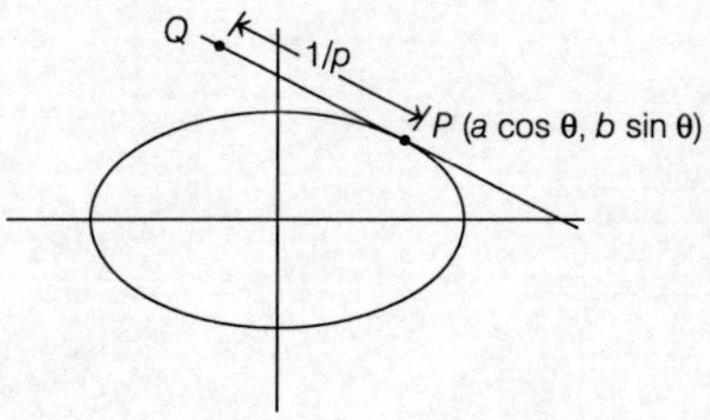

Now, the coordinates of the point Q are given as follows

$\frac{x - a\cos\theta}{\frac{-a\sin\theta}{\sqrt{b^2\cos^2\theta + a^2\sin^2\theta}}} = \frac{y - b\sin\theta}{\frac{b\cos\theta}{\sqrt{b^2\cos^2\theta + a^2\sin^2\theta}}} = \frac{1}{p} = \frac{\sqrt{b^2\cos^2\theta + a^2\sin^2\theta}}{ab}$

$x = a\cos\theta - \frac{a\sin\theta}{ab}$

and $y = b\sin\theta + \frac{b\cos\theta}{ab}$

$\Rightarrow \left(\frac{x}{a}\right)^2 + \left(\frac{y}{b}\right)^2 = 1 + \frac{1}{a^2b^2}$ is the required locus.

18 *(b)* Clearly, the lines $\mathbf{r} = \mathbf{a} + \lambda(\mathbf{b} \times \mathbf{c})$ and $\mathbf{r} = \mathbf{b} + \mu(\mathbf{c} \times \mathbf{a})$ will intersect, if the shortest distance between them is zero.

i.e. $(\mathbf{a} - \mathbf{b}) \cdot \{(\mathbf{b} \times \mathbf{c}) \times (\mathbf{c} \times \mathbf{a})\} = 0$

$\Rightarrow \; (\mathbf{a} - \mathbf{b}) \cdot \{[\mathbf{b}\,\mathbf{c}\,\mathbf{a}]\mathbf{c} - [\mathbf{b}\,\mathbf{c}\,\mathbf{c}]\mathbf{a}\} = 0$

$\Rightarrow \; ((\mathbf{a} - \mathbf{b}) \cdot \mathbf{c})[\mathbf{b}\,\mathbf{c}\,\mathbf{a}] = 0$

$\Rightarrow \quad \mathbf{a} \cdot \mathbf{c} - \mathbf{b} \cdot \mathbf{c} = 0$

$\Rightarrow \quad \mathbf{a} \cdot \mathbf{c} = \mathbf{b} \cdot \mathbf{c}$

19 *(d)* Let $V = \begin{vmatrix} 1 & a & 0 \\ a & 1 & 1 \\ 0 & 1 & a \end{vmatrix} = a - 1 - a^3$

$\therefore \quad \frac{dV}{da} = 1 - 3a^2 = 0$

$\Rightarrow \quad a = \pm\frac{1}{\sqrt{3}}$

Now, $\quad \frac{d^2V}{da^2} = -6a$

$\Rightarrow \left(\frac{d^2V}{da^2}\right)_{\left(a = \frac{1}{\sqrt{3}}\right)} = -\frac{6}{\sqrt{3}}$

Hence, it is maximum at

$a = \frac{1}{\sqrt{3}}$.

20 *(a)* We have, $(3xy^2 + x\sin(xy))dy + (y^3 + y\sin(xy))dx = 0$

$\Rightarrow \; (3xy^2dy + y^3dx) + \sin(xy)(xdy + ydx) = 0$

$\Rightarrow \; d(xy^3) + \sin(xy)d(xy) = 0$

On integrating, we get

$xy^3 - \cos(xy) = C$

21 *(4)* Since, $\cos(\alpha - \beta) = 1, \; \alpha - \beta = 2n\pi$

But $\quad -2\pi < \alpha - \beta < 2\pi$

$[\because \alpha, \beta \in (-\pi, \pi)]$

$\therefore \quad \alpha - \beta = 0$

Now, $\cos(\alpha+\beta)=\frac{1}{e}$

$\Rightarrow \cos 2\alpha = \frac{1}{e} < 1$, which is true for four values of α, as $-2\pi < 2\alpha < 2\pi$.

22. *(0)* Coefficient of x^{50} in

$(1+x)^{41}(1-x+x^2)^{40}$

$=$ Coefficient of x^{50} in $(1+x)(1+x^3)^{40}$

$=$ Coefficient of x^{50} in

$(1+x)(1+{}^{40}C_1 x^3+\ldots+{}^{40}C_{16}(x^3)^{16}+{}^{40}C_{17}(x^3)^{17}+\ldots)$

$=0$

23 *(6)* $\log_7 \log_7 7^{\frac{1}{2}+\frac{1}{4}+\frac{1}{8}}$

$=\log_7\left(\frac{1}{2}+\frac{1}{4}+\frac{1}{8}\right)$

$=\log_7\left(\frac{7}{8}\right)=1-\log_7 2^3 = 1-3\log_7 2$

$\Rightarrow k=1, m=-3$ and $n=2$

$\therefore |k|+|m|+|n|$

$=1+3+2=6$

24 *(5)* Let $I=\int_{-1/2}^{1/2}\left[[x]+\log\left(\frac{1+x}{1-x}\right)\right]dx$

$=\int_{-1/2}^{0}[x]\,dx+\int_{0}^{1/2}[x]\,dx+0$

$\left[\because \log\left(\frac{1+x}{1-x}\right) \text{ is an odd function}\right]$

$=\int_{-1/2}^{0}-1\,dx+\int_{0}^{1/2}0\,dx$

$=-(x)_{-1/2}^{0}+0=-\frac{1}{2}$

$\therefore p^2+q^2=1+4=5$

25 *(25)* Since, $\frac{(x-20)(x-40)}{(x-30)}<0$

$\Rightarrow x\in(-\infty,20)\cup(30,40)$

Let $E=\{1,2,3,\ldots,19,31,32,\ldots,39\}$, then $n(E)=28$

Now, required probability

$P(E)=\frac{28}{100}=\frac{7}{25}$

$\therefore a=25$

26 *(2018)* $|A_r|=r^2-(r-1)^2$

$\therefore |A_1|+|A_2|+\ldots+|A_{2018}|$

$=\sum_{r=1}^{2018}\{r^2-(r-1)^2\}$

$=(2018)^2-(0)^2=(2018)^2$

$\therefore k=2018$

27 *(2)* $\lim_{\theta\to0}\frac{4(\theta\tan\theta-2\theta^2\tan\theta)}{1-\cos2\theta}$

$=\frac{4(\theta\sec^2\theta+\tan\theta-4\theta\tan\theta-2\theta^2\sec^2\theta)}{2\sin2\theta}$

[using L'Hospital's rule]

$=\lim_{\theta\to0}\frac{\begin{bmatrix}4(\sec^2\theta+2\theta\sec^2\theta\tan\theta\\+\sec^2\theta-4\tan\theta-4\theta\sec^2\theta\\-4\theta\sec^2\theta-4\theta^2\sec^2\theta\tan\theta)\end{bmatrix}}{4\cos2\theta}$

[using L'Hospital's rule]

$=\frac{4(1+0+1)}{4}=2$

28 *(28)* Let $A_1=(2t_1^2,4t_1)$, $A_2=(2t_1^2,-4t_1)$

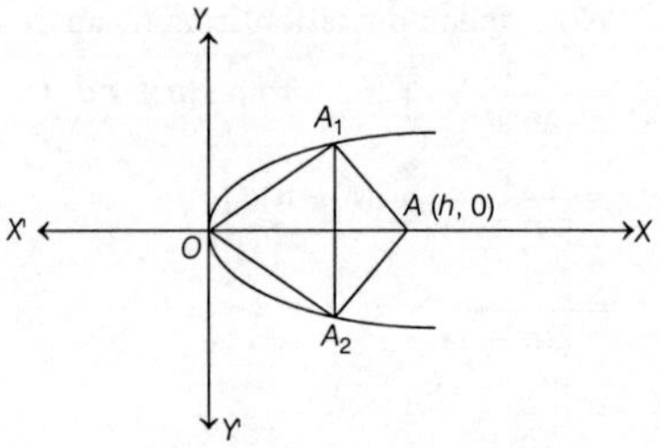

Clearly, $\angle A_1OA=\frac{\pi}{6}\Rightarrow\frac{2}{t_1}=\frac{1}{\sqrt3}$

$\Rightarrow t_1=2\sqrt3$

Equation of normal at A_1 is

$y=-t_1x+4t_1+2t_1^3$

Since, $A(h,0)$ lies on it,

$\Rightarrow h=4+2t_1^2=4+2\cdot12=28$

29 *(530)* Number of words starting with CC is 5!

Number of words starting with CE is 5!

Number of words starting with CI is 5!

Number of words starting with CK is 5!

Number of words starting with CRC is 4!

Number of words starting with CRE is 4!

Now, the first word starting with CRI is CRICEKT and next of it is CRICETK and next of it is CRICKET.

Hence, number of words before the word CRICKET

$=4\times5!+2\times4!+2$

$=480+48+2=530$

30 *(19)* $\because bc=a^2$

and $b+c=abc-a$

$=a(a^2)-a$

$=a^3-a$

$\therefore b$ and c are the roots of the equation

$x^2-(b+c)x+bc=0$

ie, $x^2-(a^3-a)x+a^2=0$

$\because$ Roots (b,c) are real.

$\therefore (a^3-a)^2-4\cdot1\cdot a^2\ge0$

$\Rightarrow (a^3-a+2a)(a^3-a-2a)\ge0$

or $a^2(a^2+1)(a^2-3)\ge0$

$\Rightarrow a^2\ge3$

or $a^4+a^2+7\ge3^2+3+7=19$

or $a^4+a^2+7\ge19$

DAY THIRTY NINE

Mock Test 2

(Based on Complete Syllabus)

Instructions

1. This question paper contains of 30 Questions of Mathematic, divided into two Sections :
 Section A Objective Type Questions and **Section B** Numerical Type Questions.
2. Section A contains 20 Objective questions and all Questions are compulsory (**Marking Scheme** : Correct +4, Incorrect −1).
3. Section B contains 10 Numerical value questions out of which only 5 questions are to be attempted (**Marking Scheme** : Correct +4, Incorrect 0).

Section A : Objective Type Questions

1 The average weight of students in a class of 35 students is 40 kg. If the weight of the teacher be included, then average rises by $\frac{1}{2}$ kg; the weight of the teacher is

(a) 40.5 kg (b) 50 kg (c) 41 kg (d) 58 kg

2 Let $f(x)=\frac{1}{[\sin x]}$, $[\cdot]$ being the greatest integer function, then

(a) $f(x)$ is not continuous, where $x \in (2n\pi, 2n\pi+\pi), n \in I$
(b) $f(x)$ is differentiable at $x=\frac{\pi}{4}$
(c) $f(x)$ is differentiable at $x=\frac{\pi}{2}$
(d) None of the above

3 If $\int f(x)\sin x\cos x\,dx=\frac{1}{2(b^2-a^2)}\log[f(x)]+C$, then $f(x)$ is equal to

(a) $\frac{1}{a^2\sin^2 x+b^2\cos^2 x}$ (b) $\frac{1}{a^2\sin^2 x-b^2\cos^2 x}$
(c) $\frac{1}{a^2\cos^2 x-b^2\sin^2 x}$ (d) $\frac{1}{a^2\cos^2 x+b^2\sin^2 x}$

4 The solution of the differential equation $\frac{dy}{dx}=\frac{y}{x}+\frac{\phi\left(\frac{y}{x}\right)}{\phi'\left(\frac{y}{x}\right)}$ is

(a) $x\phi\left(\frac{y}{x}\right)=k$ (b) $\phi\left(\frac{y}{x}\right)=kx$
(c) $y\phi\left(\frac{y}{x}\right)=k$ (d) $\phi\left(\frac{y}{x}\right)=ky$

5 If $a \le \tan^{-1}x+\cot^{-1}x+\sin^{-1}x \le b$, then $a+b$ is equal to

(a) 0 (b) $\frac{\pi}{2}$ (c) $\frac{\pi}{4}$ (d) π

6 The two consecutive terms in the expansion of $(3+2x)^{74}$ whose coefficients are equal, are

(a) 11, 12 (b) 7, 8
(c) 30, 31 (d) None of these

7 A parabola is drawn with its focus at (3, 4) and vertex at the focus of the parabola $y^2-12x-4y+4=0$. The equation of the parabola is

(a) $y^2-8x-6y+25=0$ (b) $y^2-6x+8y-25=0$
(c) $x^2-6x-8y+25=0$ (d) $x^2+6x-8y-25=0$

8 If p, p' denote the lengths of the perpendiculars from the focus and the centre of an ellipse with semi-major axis of length a respectively on a tangent to the ellipse and r denotes the focal distance of the point, then

(a) $ap' = rp + 1$ (b) $rp = ap'$
(c) $ap = rp' + \frac{1}{\sqrt{3}}$ (d) $ap = rp'$

9 The equation of the locus of the pole with respect to the ellipse $\frac{x^2}{a^2} + \frac{y^2}{b^2} = 1$, of any tangent line to the auxiliary circle, is the curve $\frac{x^2}{a^4} + \frac{y^2}{b^4} = \lambda^2$, where

(a) $\lambda^2 = a^2$ (b) $\lambda^2 = \frac{1}{a^2}$
(c) $\lambda^2 = b^2$ (d) $\lambda^2 = \frac{1}{b^2}$

10 The solution set for $\frac{(2x+3)(4-3x)^3(x-4)}{(x-2)^2 x^5} \le 0$

(a) $\left(-\infty, -\frac{3}{2}\right] \cup \left(0, \frac{4}{3}\right] \cup (4, \infty)$
(b) $\left[-\frac{3}{2}, 0\right) \cup \left[\frac{4}{3}, 4\right]$
(c) $(-\infty, 0) \cup (2, \infty)$
(d) None of the above

11 $f(x) = x - \cot^{-1} x - \log(x + \sqrt{1+x^2})$ is increasing in

(a) $(-\infty, \infty)$ (b) $(-\infty, 2)$
(c) $(2, 5)$ (d) $(-\infty, 10)$

12 The contrapositive of the statement, "if x is a prime number and x divides ab, then x divides a or x divides b", can be symbolically represented using logical connectives on appropriately defined statements p, q, r and s as

(a) $(\sim r \vee \sim s) \to (\sim p \wedge \sim q)$
(b) $(r \wedge s) \to (\sim p \wedge \sim q)$
(c) $(\sim r \wedge \sim s) \to (\sim p \vee \sim q)$
(d) $(r \vee s) \to (\sim p \vee \sim q)$

13 The area bounded by the lines $y = 2$, $x = 1$, $x = a$ and the curve $y = f(x)$, which cuts the last two lines above the first line for all $a > 1$, is equal to $\frac{2}{3}[(2a)^{3/2} - 3a + 3 - 2\sqrt{2}]$..

Then, $f(x)$ equals

(a) $2\sqrt{2x}, x > 1$ (b) $\sqrt{2x}, x > 1$
(c) $2\sqrt{x}, x > 1$ (d) None of these

14 Area of a triangle with vertices (a, b), (x_1, y_1) and (x_2, y_2), where a, x_1 and x_2 are in GP with common ratio r and b, y_1 and y_2 are in GP with common ratio s, is given by

(a) $\frac{1}{2} ab(r-1)(s-1)(s-r)$
(b) $\frac{1}{2} ab(r+1)(s+1)(s-r)$
(c) $ab(r-1)(s-1)(s-r)$
(d) None of the above

15 The equation of perpendicular bisectors of sides AB and AC of a ΔABC are $x - y + 5 = 0$ and $x + 2y = 0$, respectively. If the coordinates of vertex A are $(1, -2)$, then equation of BC is

(a) $14x + 23y - 40 = 0$ (b) $14x - 23y + 40 = 0$
(c) $23x + 14y - 40 = 0$ (d) $23x - 23y + 40 = 0$

16 If $|z - 25i| \le 15$, then $|\max \text{amp}(z) - \min \text{amp}(z)|$ is equal to

(a) $\cos^{-1}\left(\frac{3}{5}\right)$ (b) $\pi - 2\cos^{-1}\left(\frac{3}{5}\right)$
(c) $\frac{\pi}{2} + \cos^{-1}\left(\frac{3}{5}\right)$ (d) $\sin^{-1}\left(\frac{3}{5}\right) - \cos^{-1}\left(\frac{3}{5}\right)$

17 If $ABCD$ is a cyclic quadrilateral such that $12\tan A - 5 = 0$ and $5\cos B + 3 = 0$, then the quadratic equation whose roots are $\cos C$ and $\tan D$ is

(a) $39x^2 + 88x + 48 = 0$
(b) $39x^2 - 16x - 48 = 0$
(c) $39x^2 - 88x + 48 = 0$
(d) None of the above

18 A mirror and a source of light are situated at the origin O and a point on OX, respectively. A ray of light from the source strikes the mirror and is reflected. If the DR's of the normal to the plane of mirror are $1, -1, 1$, then DC's for the reflected ray are

(a) $\frac{1}{3}, \frac{2}{3}, \frac{2}{3}$ (b) $-\frac{1}{3}, \frac{2}{3}, \frac{2}{3}$
(c) $-\frac{1}{3}, -\frac{2}{3}, -\frac{2}{3}$ (d) $-\frac{1}{3}, -\frac{2}{3}, \frac{2}{3}$

19 If both roots of the equation $4x^2 - 2x + a = 0$, $a \in R$, lies in the interval $(-1, 1)$, then

(a) $-2 < a \le \frac{1}{4}$ (b) $-6 \le a \le \frac{1}{4}$
(c) $a > \frac{1}{4}$ (d) $a < -2$

20 Let λ and θ be real numbers. Then, the set of all values of λ for which the system of linear equations

$$\lambda x + (\sin\theta)y + (\cos\theta)z = 0$$
$$x + (\cos\theta)y + (\sin\theta)z = 0$$
$$-x + (\sin\theta)y - (\cos\theta)z = 0$$

has a non-trivial solution, is

(a) $[0, \sqrt{2}]$ (b) $[-\sqrt{2}, 0]$
(c) $[-\sqrt{2}, \sqrt{2}]$ (d) None of these

Section B : Numerical Type Questions

21 If the value of integral $\left|\int_0^{\pi} [2\sin x]\,dx\right|$, where $[x]$ denotes greatest integer function, is $\frac{2\pi}{T}$, then value of T is equal to

22 Suppose $A_1, A_2, \ldots, A_{30}$ are thirty sets, each having 5 elements and $B_1, B_2, \ldots, B_n$ are n sets each with 3 elements, let $\bigcup_{i=1}^{30} A_i = \bigcup_{j=1}^{n} B_j = S$ and each element of

S belongs to exactly 10 of the A_i's and exactly 9 of the B_j's Then, n is equal to

23 An experiment succeeds twice as often as it fails. Then, the probability that in the next 4 trials there will be atleast 2 successes, is p/q, then value of pq is equal to

24 A variable plane at a constant distance P, from origin cuts axes at A, B and C, the locus of centroid of tetrahedron $OABC$ is $\frac{1}{x^2}+\frac{1}{y^2}+\frac{1}{z^2}=\frac{k^2}{p^2}$, then value of k^2 is equal to

25 A parallelogram is constructed on the vectors $\mathbf{a}=3\alpha-\beta$, $\mathbf{b}=\alpha+3\beta$, if $|\alpha|=|\beta|=2$ and angle between α and β is $\frac{\pi}{3}$, length of a diagonal of the parallelogram is $p\sqrt{3}$, then value of p is equal to ....

26 Let $A=\begin{bmatrix}1 & \sin\theta & 1\\ -\sin\theta & 1 & \sin\theta\\ -1 & -\sin\theta & 1\end{bmatrix}$, where $0\le\theta\le 2\pi$.

Then, largest value of $|A|$ is

27 The value of the expression $\frac{\sin^3 x}{1+\cos x}+\frac{\cos^3 x}{1-\sin x}$ is $\frac{1}{\sqrt{a}}\cos\left(\frac{\pi}{4}-x\right)$, then value of a is equal to ....

28 Number of ways in which 5 boys and 4 girls can be arranged on a circular table such that no two girls sit together and two particular boys are always together is ....

29 If $S_n=1+\frac{1}{2}+\frac{1}{2^2}+\ldots+\frac{1}{2^{n-1}}$ and $2-S_n<\frac{1}{100}$, then the least value of n must be ....

30 If X and Y are independent binomial variates $B\left(5,\frac{1}{2}\right)$ and $B\left(7,\frac{1}{2}\right)$. If the value of $P(X+Y=3)$ is λ, then the value of 4096λ must be ....

Hints and Explanations

1 *(d)* Let the weight of teacher be x kg, then

$$40+\frac{1}{2}=\frac{35\times 40+x}{35+1}$$

$$\Rightarrow \quad \frac{81}{2}\times 36=35\times 40+x$$

$$\Rightarrow \quad 81\times 18=1400+x$$

$$\Rightarrow \quad 1458=1400+x\Rightarrow x=58$$

Hence, the weight of teacher is 58 kg.

2 *(a)* We have, $f(x)=\frac{1}{[\sin x]}$

Clearly, $\sin x\notin[0,1)$

$[\because \text{if } 0\le \sin x<1, [\sin x]=0]$

$$\Rightarrow x\notin[2n\pi,(2n+1)\pi]-(4n+1)\frac{\pi}{2}, n\in I$$

Thus, $f(x)$ is not continuous if $x\in(2n\pi,2n\pi+\pi), n\in I$.

3 *(a)* Given that, $\int f(x)\sin x\cos x\,dx$

$$=\frac{1}{2(b^2-a^2)}\log[f(x)]+C$$

On differentiating both sides, we get

$$f(x)\sin x\cos x=\frac{d}{dx}\left\{\frac{\log[f(x)]}{2(b^2-a^2)}+C\right\}$$

$$\Rightarrow f(x)\sin x\cos x=\frac{1}{2(b^2-a^2)}\cdot\frac{1}{f(x)}f'(x)$$

$$\Rightarrow \quad 2(b^2-a^2)\sin x\cos x=\frac{f'(x)}{[f(x)]^2}$$

On integrating both sides, we get

$$-b^2\cos^2 x-a^2\sin^2 x=-\frac{1}{f(x)}$$

$$\therefore \quad f(x)=\frac{1}{a^2\sin^2 x+b^2\cos^2 x}$$

4 *(b)* Given, $\frac{dy}{dx}-\frac{y}{x}=\frac{\phi\left(\frac{y}{x}\right)}{\phi'\left(\frac{y}{x}\right)}$

$$\Rightarrow \quad \frac{\phi'\left(\frac{y}{x}\right)\left(\frac{x\,dy-y\,dx}{x^2}\right)}{\phi\left(\frac{y}{x}\right)}=\frac{1}{x}dx$$

$$\Rightarrow \int\frac{\phi'\left(\frac{y}{x}\right)d\left(\frac{y}{x}\right)}{\phi\left(\frac{y}{x}\right)}=\int\frac{1}{x}dx+\log k$$

$$\Rightarrow \quad \log\phi\left(\frac{y}{x}\right)=\log x+\log k$$

$$\Rightarrow \quad \phi\left(\frac{y}{x}\right)=kx$$

$\therefore$ Sum of two

5 *(d)* We know, $-\frac{\pi}{2}\le\sin^{-1}x\le\frac{\pi}{2}$

$$\therefore \quad 0\le\frac{\pi}{2}+\sin^{-1}x\le\pi$$

$$\Rightarrow 0\le\tan^{-1}x+\cot^{-1}x+\sin^{-1}x\le\pi$$

$$\Rightarrow a=0 \text{ and } b=\pi$$

Hence, $a+b=\pi$

6 *(c)* General term of $(3+2x)^{74}$ is

$$T_{r+1}={}^{74}C_r(3)^{74-r}2^r x^r$$

Let two consecutive terms be T_{r+1} th and T_{r+2} th terms.

According to the question

Coefficient of T_{r+1} = Coefficient of T_{r+2}

$$\Rightarrow \quad {}^{74}C_r 3^{74-r}2^r={}^{74}C_{r+1}3^{74-(r+1)}2^{r+1}$$

$$\Rightarrow \quad \frac{{}^{74}C_{r+1}}{{}^{74}C_r}=\frac{3}{2}$$

$$\Rightarrow \quad \frac{74-r}{r+1}=\frac{3}{2}$$

$$\Rightarrow \quad 148-2r=3r+3$$

$$\therefore \quad r=29$$

Hence, two consecutive terms are 30 and 31.

7 *(c)* Given, equation can be rewritten as $(y-2)^2=12x$

Here, vertex and focus are (0, 2) and (3, 2).

$\therefore$ Vertex of the required parabola is (3, 2) and focus is (3, 4).

The axis of symmetry is $x=3$ and latusrectum $=4\cdot 2=8$

Hence, required equation is

$$(x-3)^2=8(y-2)$$

$$\Rightarrow \quad x^2-6x-8y+25=0$$

8 *(d)* Tangent to the ellipse $\frac{x^2}{a^2}+\frac{y^2}{b^2}=1$ at $P(a\cos\theta, b\sin\theta)$ is

$$\frac{x}{a}\cos\theta+\frac{y}{b}\sin\theta=1 \qquad \ldots\text{(i)}$$

Now, p = perpendicular distance from focus $S(ae, 0)$ to the line (i)

$$=\frac{\left|\frac{ae}{a}\cos\theta+0-1\right|}{\sqrt{\frac{\cos^2\theta}{a^2}+\frac{\sin^2\theta}{b^2}}}$$

$$=\frac{1-e\cos\theta}{\sqrt{\frac{\cos^2\theta}{a^2}+\frac{\sin^2\theta}{b^2}}} \qquad \ldots\text{(ii)}$$

Also, p' = perpendicular distance from centre (0, 0) to the line (i) .

$$=\frac{1}{\sqrt{\frac{\cos^2\theta}{a^2}+\frac{\sin^2\theta}{b^2}}} \qquad \ldots\text{(iii)}$$

Again, $r=SP=a(1-e\cos\theta)$

$$\therefore \quad ap=\frac{a-ae\cos\theta}{\sqrt{\frac{\cos^2\theta}{a^2}+\frac{\sin^2\theta}{b^2}}}=rp'$$

[using Eqs. (ii) and (iii)]

9 *(b)* The equation of the auxiliary circle is $x^2+y^2=a^2$.

Let (h, k) be the pole, then equation of the polar of (h, k) with respect to the given ellipse is

$$\frac{hx}{a^2}+\frac{ky}{b^2}=1$$

Since, this is tangent to the circle.

$$\Rightarrow \quad \frac{|0+0-1|}{\sqrt{\left(\frac{h}{a^2}\right)^2+\left(\frac{k}{b^2}\right)^2}}=\pm a$$

$$\Rightarrow \quad \frac{h^2}{a^4}+\frac{y^2}{b^4}=\frac{1}{a^2}$$

Hence, locus of (h, k) is

$$\frac{x^2}{a^4}+\frac{y^2}{b^4}=\frac{1}{a^2}$$

10 *(d)* Given inequality can be rewritten as

$$\frac{(2x+3)(3x-4)^3(x-4)}{(x-2)^2x^5}\ge 0$$

$$\Rightarrow (2x+3)(3x-4)^3(x-4)(x-2)^2 x^5\ge 0, x\ne 0, 2$$

$$\Rightarrow (2x+3)(3x-4)^3(x-4)x^5\ge 0;$$

$x\ne 0, 2$ $[\because (x-2)^2>0]$

$$x=-\frac{3}{2}, \frac{4}{3}, 4, 0$$

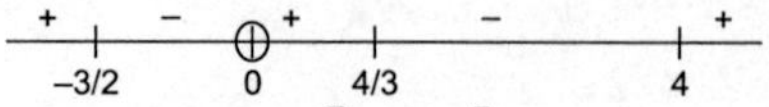

$$\Rightarrow \quad x\in\left(-\infty,-\frac{3}{2}\right]\cup\left(0,\frac{4}{3}\right]\cup[4,\infty)$$

11 *(a)* Given that,

$$f(x)=x-\cot^{-1}x-\log(x+\sqrt{1+x^2})$$

On differentiating w.r.t. x, we get

$$f'(x)=1+\frac{1}{1+x^2}-\frac{1}{(x+\sqrt{1+x^2})}\left(1+\frac{x}{\sqrt{1+x^2}}\right)$$

$$=1+\frac{1}{1+x^2}-\frac{1}{\sqrt{1+x^2}}$$

$$= \frac{1 + x^2 + 1 - \sqrt{1 + x^2}}{1 + x^2}$$

$$= \frac{2 + x^2 - \sqrt{1 + x^2}}{1 + x^2}$$

So, $f(x)$ is an increasing function in $(-\infty, \infty)$.

12 *(c)* Let $p = x$ is a prime number

$q = x$ divides ab

$r = x$ divides a

and $s = x$ divides b

The given statement becomes in logical form is $p \wedge q \rightarrow r \vee s$

Its contrapositive is

$$\sim(r \vee s) \rightarrow \sim(p \wedge q)$$

$$\Rightarrow \quad (\sim r \wedge \sim s) \rightarrow (\sim p \vee \sim q)$$

13 *(a)* We have,

$$\int_1^a [f(x) - 2]dx = \frac{2}{3}[(2a)^{3/2} - 3a + 3 - 2\sqrt{2}]$$

On differentiating both sides w.r.t. a, we get

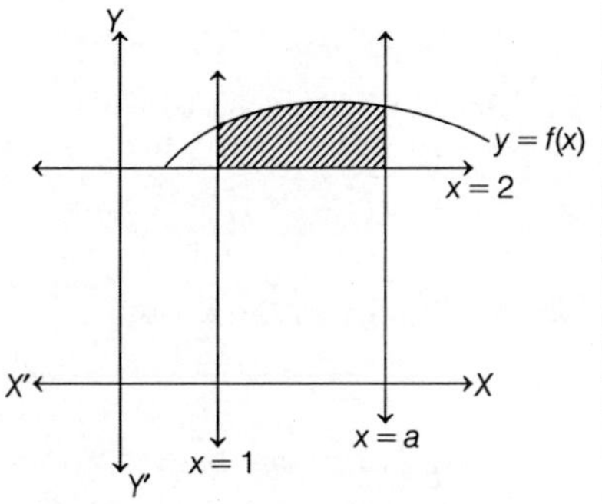

$$f(a) - 2 = \frac{2}{3}\left[\frac{3}{2}\sqrt{2a} \cdot 2 - 3\right]$$

$$\Rightarrow \quad f(a) - 2 = 2\sqrt{2a} - 2$$

$$\Rightarrow \quad f(a) = 2\sqrt{2a}, a > 1$$

$$\Rightarrow \quad f(x) = 2\sqrt{2x}, x > 1$$

14 *(a)* Since, a, x_1 and x_2 are in GP with common ratio r.

$\therefore \quad x_1 = ar, x_2 = ar^2$

Also, b, y_1 and y_2 are in GP with common ratio s.

$\therefore \quad y_1 = bs, y_2 = bs^2$

The area of triangle is given by

$$\Delta = \frac{1}{2}\begin{vmatrix} a & b & 1 \\ x_1 & y_1 & 1 \\ x_2 & y_2 & 1 \end{vmatrix} = \frac{1}{2}\begin{vmatrix} a & b & 1 \\ ar & bs & 1 \\ ar^2 & bs^2 & 1 \end{vmatrix}$$

$$= \frac{1}{2}ab\begin{vmatrix} 1 & 1 & 1 \\ r & s & 1 \\ r^2 & s^2 & 1 \end{vmatrix}$$

$$= \frac{ab}{2}\begin{vmatrix} 1 & 0 & 0 \\ r & s - r & 1 - r \\ r^2 & s^2 - r^2 & 1 - r^2 \end{vmatrix}$$

[applying $C_2 \rightarrow C_2 - C_1$ and $C_3 \rightarrow C_3 - C_1$]

$$= \frac{ab}{2}\{(s - r)(1 - r^2) - (1 - r)(s^2 - r^2)\}$$

$$= \frac{ab}{2}(s - r)(1 - r)\{1 + r - (s + r)\}$$

$$= \frac{ab}{2}(s - r)(1 - r)(1 - s)$$

$$= \frac{ab}{2}(s - r)(r - 1)(s - 1)$$

15 *(a)* Let the coordinates of B and C are (x_1, y_1) and (x_2, y_2) respectively.

Then, $P\left(\frac{x_1 + 1}{2}, \frac{y_1 - 2}{2}\right)$ lies on perpendicular bisector $x - y + 5 = 0$.

$$\therefore \quad \frac{x_1 + 1}{2} - \frac{y_1 - 2}{2} = -5$$

$$\Rightarrow \quad x_1 - y_1 = -13 \quad \ldots\text{(i)}$$

Also, PN is perpendicular to AB.

$$\therefore \quad \frac{y_1 + 2}{x_1 - 1} \times 1 = -1$$

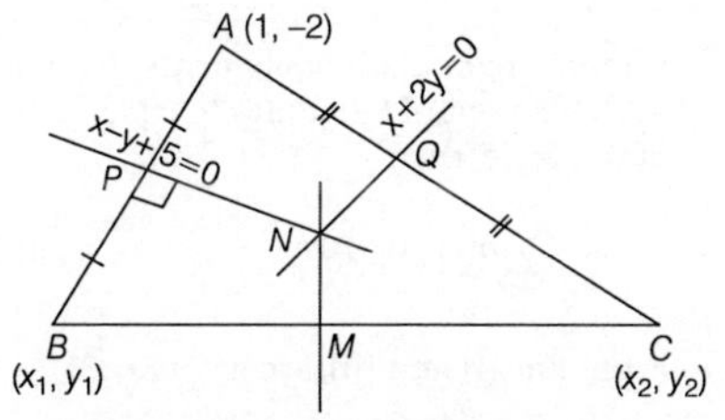

$$\Rightarrow \quad y_1 + 2 = -x_1 + 1$$

$$\Rightarrow \quad x_1 + y_1 = -1 \quad \ldots\text{(ii)}$$

On solving Eqs. (i) and (ii), we get

$$x_1 = -7, y_1 = 6$$

So, the coordinates of B are $(-7, 6)$.

Similarly, the coordinates of C are $\left(\frac{11}{5}, \frac{2}{5}\right)$.

Hence, equation of BC is

$$y - 6 = \frac{\frac{2}{5} - 6}{\frac{11}{5} + 7}(x + 7)$$

$$\Rightarrow \quad y - 6 = -\frac{14}{23}(x + 7)$$

$$\Rightarrow \quad 14x + 23y - 40 = 0$$

16 *(b)* We have,

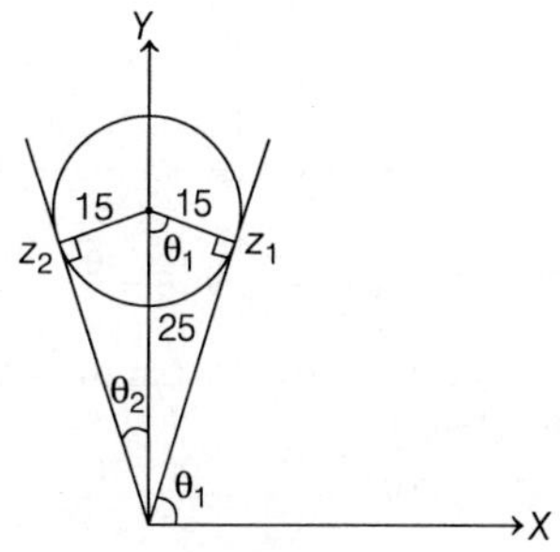

Clearly, max amp (z) = amp (z_2) and min amp(z) = amp(z_1)

Now,

$$\text{amp}(z_1) = \theta_1 = \cos^{-1}\left(\frac{15}{25}\right) = \cos^{-1}\left(\frac{3}{5}\right)$$

and
$$\text{amp}(z_2) = \frac{\pi}{2} + \theta_2 = \frac{\pi}{2} + \sin^{-1}\left(\frac{15}{25}\right)$$

$$= \frac{\pi}{2} + \sin^{-1}\left(\frac{3}{5}\right)$$

$\therefore$ $|\text{max amp}(z) - \text{min amp}(z)|$

$$= \left|\frac{\pi}{2} + \sin^{-1}\frac{3}{5} - \cos^{-1}\frac{3}{5}\right|$$

$$= \left|\frac{\pi}{2} + \frac{\pi}{2} - \cos^{-1}\frac{3}{5} - \cos^{-1}\frac{3}{5}\right|$$

$$= \pi - 2\cos^{-1}\left(\frac{3}{5}\right)$$

17 *(b)* Clearly, $\tan A = \frac{5}{12} = -\tan C$,

$$\cos B = -\frac{3}{5} = -\cos D$$

[$\because$ in cyclic quadrilateral, $A + C = \pi$ and $B + D = \pi$]

Now, $\tan C = -\frac{5}{12}$

$$\Rightarrow \quad \cos C = -\frac{12}{13} = \alpha \quad \text{(say)}$$

and
$$\cos D = \frac{3}{5} \Rightarrow \tan D = \frac{4}{3} = \beta \quad \text{(say)}$$

$\therefore$ Required equation is

$$x^2 - (\alpha + \beta)x + \alpha\beta = 0$$

$$x^2 - \left(-\frac{12}{13} + \frac{4}{3}\right)x + \left(-\frac{12}{13} \times \frac{4}{3}\right) = 0$$

$$\Rightarrow 39x^2 - 16x - 48 = 0$$

18 *(d)* Let the source of light be situated at $A(a, 0, 0)$, where $a \neq 0$.

Let AO be the incident ray and OB be the reflected ray, ON is the normal to the mirror at O.

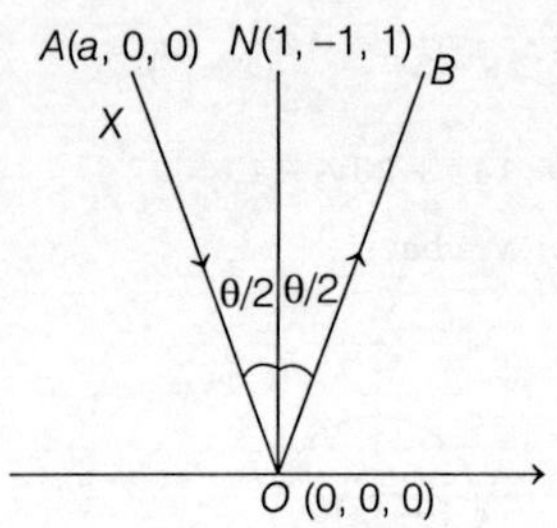

Then, $\angle AON = \angle NOB = \dfrac{\theta}{2}$ (say)

DR's of OA are $(a, 0, 0)$ and so its DC's are $(1, 0, 0)$.

DC's of ON are $\dfrac{1}{\sqrt{3}}, -\dfrac{1}{\sqrt{3}}, \dfrac{1}{\sqrt{3}}$

$\therefore \quad \cos\dfrac{\theta}{2} = \dfrac{1}{\sqrt{3}}$

Let l, m and n be the DC's of the reflected ray OB.

Then, $\dfrac{l+1}{2\cos\dfrac{\theta}{2}} = \dfrac{1}{\sqrt{3}}, \quad \dfrac{m+0}{2\cos\dfrac{\theta}{2}} = -\dfrac{1}{\sqrt{3}}$

and $\quad \dfrac{n+0}{2\cos\dfrac{\theta}{2}} = \dfrac{1}{\sqrt{3}}$

$\Rightarrow l = \dfrac{2}{3} - 1,$

$m = -\dfrac{2}{3}, n = \dfrac{2}{3}$

$\Rightarrow \quad l = -\dfrac{1}{3}, m = -\dfrac{2}{3}, n = \dfrac{2}{3}$

Hence, DC's of the reflected ray are $-\dfrac{1}{3}, -\dfrac{2}{3}, \dfrac{2}{3}$.

19 *(a)* Let $f(x) = 4x^2 - 2x + a$

Since, both roots of $f(x) = 0$ lie in the interval $(-1, 1)$, we can take

$D \ge 0, f(-1) > 0$ and $f(1) > 0$

1. Consider $D \ge 0$,

$\Rightarrow \quad (-2)^2 - 4 \cdot 4 \cdot a \ge 0$

$\Rightarrow \quad a \le \dfrac{1}{4}$ ...(i)

2. Consider $f(-1) > 0$,

$4(-1)^2 - 2(-1) + a > 0$

$\Rightarrow \quad a > -6$ ...(ii)

3. Consider $f(1) > 0$,

$4(1)^2 - 2(1) + a > 0$

$\Rightarrow \quad a > -2$ ...(iii)

From Eqs. (i), (ii) and (iii), we get

$-2 < a \le \dfrac{1}{4}$

20 *(c)* Since the system of given equation has a non-trivial solution, therefore

$$\begin{vmatrix} \lambda & \sin\theta & \cos\theta \\ 1 & \cos\theta & \sin\theta \\ -1 & \sin\theta & -\cos\theta \end{vmatrix} = 0$$

$\Rightarrow \lambda[-\cos^2\theta - \sin^2\theta] - \sin\theta[-\cos\theta + \sin\theta] + \cos\theta\,[\sin\theta + \cos\theta] = 0$

$\Rightarrow \quad -\lambda + \sin\theta\cos\theta - \sin^2\theta + \sin\theta\cos\theta + \cos^2\theta = 0$

$\Rightarrow \quad -\lambda + 2\sin\theta\cos\theta + \cos^2\theta - \sin^2\theta = 0$

$\Rightarrow \quad -\lambda + \sin 2\theta + \cos 2\theta = 0$

$\Rightarrow \quad \lambda = \sqrt{2}\cos\left(2\theta - \dfrac{\pi}{4}\right)$ ...(i)

$\because \quad -1 \le \cos\left(2\theta - \dfrac{\pi}{4}\right) \le 1 \;\forall \theta \in R$

$\therefore -\sqrt{2} \le \sqrt{2}\cos\left(2\theta - \dfrac{\pi}{4}\right) \le \sqrt{2} \;\forall \theta \in R$

$\Rightarrow -\sqrt{2} \le \lambda \le \sqrt{2}$ [using Eq. (i)]

Hence, $\lambda \in [-\sqrt{2}, \sqrt{2}]$.

21 *(3)* Let $I = \int_0^{\pi} [2\sin x]\, dx$

$= \int_0^{\pi/6} [2\sin x]\, dx + \int_{\pi/6}^{\pi/2} [2\sin x]\, dx + \int_{\pi/2}^{5\pi/6} [2\sin x]\, dx + \int_{5\pi/6}^{\pi} [2\sin x]\, dx$

$= 0 + \int_{\pi/6}^{\pi/2} 1\, dx + \int_{\pi/2}^{5\pi/6} 1\, dx + 0$

$= [x]_{\pi/6}^{\pi/2} + [x]_{\pi/2}^{5\pi/6}$

$= \dfrac{\pi}{2} - \dfrac{\pi}{6} + \dfrac{5\pi}{6} - \dfrac{\pi}{2} = \dfrac{2\pi}{3} \quad \therefore \quad T = 3$

22 *(45)* Given, A's are 30 sets with five elements each, so

$\sum_{i=1}^{30} n(A_i) = 5 \times 30 = 150$ ...(i)

If there are m distinct elements in S and each element of S belongs to exactly 10 of the A_i's, then

$\sum_{i=1}^{30} n(A_i) = 10m$ ...(ii)

From Eqs.(i) and (ii), we get $m = 15$

Similarly, $\sum_{j=1}^{n} n(B_j) = 3n$

and $\quad \sum_{j=1}^{n} n(B_j) = 9m$

$\therefore \quad 3n = 9m$

$\Rightarrow \quad n = \dfrac{9m}{3} = 3 \times 15 = 45$

23 *(72)* Given, $p = 2q$

$\because \quad p + q = 1 \Rightarrow p = \dfrac{2}{3}$ and $q = \dfrac{1}{3}$

Now, required probability

$= {}^4C_2\left(\dfrac{2}{3}\right)^2\left(\dfrac{1}{3}\right)^2 + {}^4C_3\left(\dfrac{2}{3}\right)^3\left(\dfrac{1}{3}\right)^1 + {}^4C_4\left(\dfrac{2}{3}\right)^4$

$= 6 \times \dfrac{4}{81} + \dfrac{4 \times 8}{81} + \dfrac{1 \times 16}{81}$

$= \dfrac{72}{81} = \dfrac{8}{9}$

$\therefore p = 8, q = 9$

Hence, $pq = 8 \times 9 = 72$

24 *(16)* Let the equation of plane be $\dfrac{x}{a} + \dfrac{y}{b} + \dfrac{z}{c} = 1.$

Let the centroid of tetrahedron $OABC$ is (α, β, γ), then

$\alpha = \dfrac{0 + a + 0 + 0}{4}, \; \beta = \dfrac{0 + 0 + b + 0}{4},$

$\gamma = \dfrac{0 + 0 + 0 + c}{4}$

$\Rightarrow \quad a = 4\alpha, b = 4\beta, c = 4\gamma$

Since, distance of plane from origin is P,

$$P = \dfrac{1}{\sqrt{\dfrac{1}{a^2} + \dfrac{1}{b^2} + \dfrac{1}{c^2}}}$$

$\Rightarrow \quad \dfrac{1}{a^2} + \dfrac{1}{b^2} + \dfrac{1}{c^2} = \dfrac{1}{P^2}$

On putting the values of a, b and c, then

$\dfrac{1}{16\alpha^2} + \dfrac{1}{16\beta^2} + \dfrac{1}{16\gamma^2} = \dfrac{1}{P^2}$

Hence, the locus of centroid is

$\dfrac{1}{x^2} + \dfrac{1}{y^2} + \dfrac{1}{z^2} = \dfrac{16}{P^2}$

$\therefore \quad k^2 = 16$

25 *(4)* Clearly, $\quad \mathbf{AC} = \mathbf{a} + \mathbf{b}$

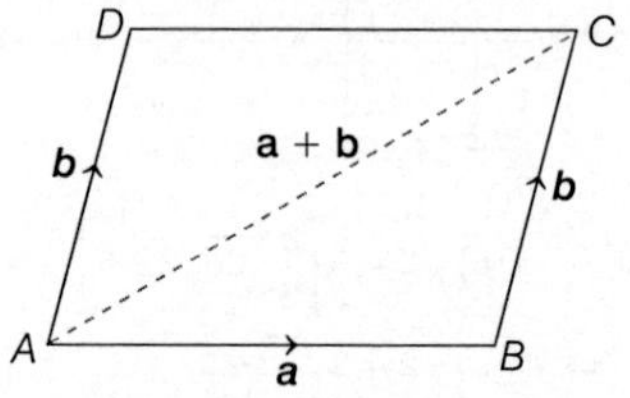

$\Rightarrow |\mathbf{AC}|^2 = |\mathbf{a}|^2 + |\mathbf{b}|^2 + 2\mathbf{a} \cdot \mathbf{b}$

$\Rightarrow |\mathbf{AC}|^2 = \{|3\alpha - \beta|^2 + |\alpha + 3\beta|^2 + 2(3\alpha - \beta)(\alpha + 3\beta)\}$

$= 9\alpha^2 + \beta^2 - 6\alpha\beta + \alpha^2 + 9\beta^2 + 6\alpha \cdot \beta + 6\alpha^2 - 6\beta^2 + 16\alpha \cdot \beta$

$\Rightarrow |\mathbf{AC}|^2 = 16\alpha^2 + 4\beta^2 + 16\alpha.\beta$

$\Rightarrow |\mathbf{AC}|^2 = 64 + 16 + 16\,|\alpha|\,|\beta|\cos\dfrac{\pi}{3}$

$\Rightarrow |\mathbf{AC}|^2 = 64 + 16 + 16 \times 2 \times 2 \times \dfrac{1}{2}$

$\Rightarrow |\mathbf{AC}| = 4\sqrt{7}$

Similarly, $|\mathbf{BD}| = |\mathbf{a} - \mathbf{b}| = 4\sqrt{3}$

$\therefore \quad p = 4$

26 *(4)* Clearly,

$$|A| = \begin{vmatrix} 1 & \sin\theta & 1 \\ -\sin\theta & 1 & \sin\theta \\ -1 & -\sin\theta & 1 \end{vmatrix}$$

$\Rightarrow \quad |A| = 2(1 + \sin^2\theta)$

Now, $0 \le \sin^2\theta \le 1$, for all $\theta \in [0, 2\pi]$.

$\Rightarrow \quad 2 \le 2 + 2\sin^2\theta \le 4$, for all $\theta \in [0, 2\pi]$

$\therefore$ Largest value of $|A|$ is 4.

27 *(2)* Let

$$A = \frac{(\sin^3 x + \cos^3 x) + (\cos^4 x - \sin^4 x)}{(1 + \cos x)(1 - \sin x)}$$

$$= \frac{(\sin^3 x + \cos^3 x) + (\cos x + \sin x)(\cos^2 x + \sin^2 x)(\cos x - \sin x)}{(1 + \cos x)(1 - \sin x)}$$

$$= \frac{(\sin x + \cos x)\{(1 - \sin x \cos x) + (\cos x - \sin x)\}}{1 + \cos x - \sin x - \sin x \cos x}$$

$\Rightarrow \quad A = \sin x + \cos x \quad$ ...(i)

$$= \frac{1}{\sqrt{2}} \sin\left(\frac{\pi}{4} + x\right)$$

Again from Eq. (i), we get

$$A = \frac{1}{\sqrt{2}} \cos\left(\frac{\pi}{4} - x\right)$$

$\therefore \quad a = 2$

28 *(288)* Let two particular boys as one boy, we have only four boys.

$\therefore$ 5 boys can be seated at a round table when two particular boys are always together = 3!2!

$\Rightarrow$ 4 girls have are 4 places.

$\therefore$ 4 girls can be arranged in 4! ways.

$\therefore$ Required number = $3!2!4! = 6 \cdot 2 \cdot 24$

$= 288$

29 *(8)* $\therefore S_n = \dfrac{1\left(1 - \left(\frac{1}{2}\right)^n\right)}{\left(1 - \frac{1}{2}\right)} = 2 - \dfrac{1}{2^{n-1}}$

or $\quad 2 - S_n = \dfrac{1}{2^{n-1}} < \dfrac{1}{100} \quad \left[\because 2 - S_n < \dfrac{1}{100}\right]$

or $\quad 2^{n-1} > 100 > 2^6$

$\Rightarrow \quad 2^{n-1} > 2^6$

$\Rightarrow \quad n - 1 > 6$

$\therefore \quad n > 7$

Hence, least value of n is 8.

30 *(220)* We have,

$$P(X + Y = 3) = P(X = 0, Y = 3) + P(X = 1, Y = 2) + P(X = 2, Y = 1) + P(X = 3, Y = 0)$$

$$P(X = 0)P(Y = 3) + P(X = 1)P(Y = 2) + P(X = 2)P(Y = 1) + P(X = 3)P(Y = 0)$$

[$\therefore X$ and Y are independent]

$$= {}^5C_0\left(\frac{1}{2}\right)^5 \cdot {}^7C_3\left(\frac{1}{2}\right)^7 + {}^5C_1\left(\frac{1}{2}\right)^5 {}^7C_2\left(\frac{1}{2}\right)^7 + {}^5C_2\left(\frac{1}{2}\right)^5 {}^7C_2\left(\frac{1}{2}\right)^7 + {}^5C_3\left(\frac{1}{2}\right)^5 {}^7C_0\left(\frac{1}{2}\right)^7$$

$$= \left(\frac{1}{2}\right)^{12} [(1)(35) + (5)(21) + (10)(7) + (10)(1)]$$

$$= \frac{220}{2^{12}} = \frac{55}{1024} = \lambda \quad \text{(given)}$$

$\therefore \; 4096\lambda = 4096 \times \dfrac{55}{1024} = 4 \times 55 = 220$

DAY FOURTY

Mock Test 3

(Based on Complete Syllabus)

Instructions

1. This question paper contains of 30 Questions of Mathematic, divided into two Sections : **Section A** Objective Type Questions and **Section B** Numerical Type Questions.
2. Section A contains 20 Objective questions and all Questions are compulsory (**Marking Scheme** : Correct +4, Incorrect –1).
3. Section B contains 10 Numerical value questions out of which only 5 questions are to be attempted (**Marking Scheme** : Correct +4, Incorrect 0).

Section A : Objective Type Questions

1 Two circles in complex plane are

$C_1 : |z - i| = 2$

$C_2 : |z - 1 - 2i| = 4$. Then,

(a) C_1 and C_2 touch each other
(b) C_1 and C_2 intersect at two distinct points
(c) C_1 lies within C_2
(d) C_2 lies within C_1

2 If $\int \frac{\sqrt{\cos 2x}}{\sin x}\, dx = -\log|\cot x + \sqrt{\cot^2 x - 1}| + A + C$,

then A is equal to

(a) $\frac{1}{\sqrt{2}} \log\left|\frac{\sqrt{2} + \sqrt{1 - \tan^2 x}}{\sqrt{2} - \sqrt{1 - \tan^2 x}}\right|$

(b) $\frac{1}{\sqrt{2}} \log\left|\frac{\sqrt{2} + \sqrt{1 - \sec^2 x}}{\sqrt{2} - \sqrt{1 - \sec^2 x}}\right|$

(c) $\frac{1}{\sqrt{2}} \log\left|\frac{\sqrt{2} + \sqrt{1 - \sin^2 x}}{\sqrt{2} - \sqrt{1 - \sin^2 x}}\right|$

(d) None of the above

3 If m_1 and m_2 are the roots of the equation $x^2 + (\sqrt{3} + 2)x + \sqrt{3} - 1 = 0$, then the area of the triangle formed by the lines $y = m_1 x$, $y = -m_2 x$ and $y = 1$ is

(a) $\frac{1}{2}\left(\frac{\sqrt{3} + 2}{\sqrt{3} - 1}\right)$
(b) $\frac{1}{2}\left(\frac{\sqrt{3} + 2}{\sqrt{3} + 1}\right)$
(c) $\frac{1}{2}\left(\frac{-\sqrt{3} + 2}{\sqrt{3} - 1}\right)$
(d) None of these

4 The equation of the ellipse whose axes are coincident with the coordinate axes and which touches the straight lines $3x - 2y - 20 = 0$ and $x + 6y - 20 = 0$, is

(a) $\frac{x^2}{5} + \frac{y^2}{8} = 1$
(b) $\frac{x^2}{40} + \frac{y^2}{10} = 10$
(c) $\frac{x^2}{40} + \frac{y^2}{10} = 1$
(d) $\frac{x^2}{10} + \frac{y^2}{40} = 1$

5 A variable straight line through the point of intersection of the lines $\frac{x}{a} + \frac{y}{b} = 1$ and $\frac{x}{b} + \frac{y}{a} = 1$ meets the coordinate axes in A and B.

I. The locus of mid-point of AB is the curve $2xy(a + b) = ab(x + y)$.

II. The locus of mid-point of AB is the curve $2xy(x + y) = ab(a + b)$.

(a) Both I and II are true
(b) Only I is true
(c) Only II is true
(d) Both I and II are false

6 If the function $f(x)=\dfrac{2x-\sin^{-1}x}{2x+\tan^{-1}x}$, $x\neq 0$ is continuous at each point of its domain, then the value of $f(0)$ is

(a) 2 (b) $\frac{1}{3}$ (c) $\frac{2}{3}$ (d) $-\frac{1}{3}$

7 If two distinct chords, drawn from the point (p, q) on the circle $x^2+y^2=px+qy$, (where $pq\neq 0$) are bisected by the X-axis, then

(a) $p^2=q^2$ (b) $p^2=8q^2$
(c) $p^2<8q^2$ (d) $p^2>8q^2$

8 If A and B are different matrices satisfying $A^3=B^3$ and $A^2B=B^2A$, then

(a) det (A^2+B^2) must be zero
(b) det $(A-B)$ must be zero
(c) det (A^2+B^2) as well as det $(A-B)$ must be zero
(d) Atleast one of det (A^2+B^2) or det $(A-B)$ must be zero

9 A fair coin is tossed 100 times. The probability of getting tails 1, 3, ..., 49 times is

(a) $\frac{1}{2}$ (b) $\frac{1}{4}$ (c) $\frac{1}{8}$ (d) $\frac{1}{16}$

10 The distance between the line

$$\mathbf{r}=2\hat{\mathbf{i}}-2\hat{\mathbf{j}}+3\hat{\mathbf{k}}+\lambda(\hat{\mathbf{i}}-\hat{\mathbf{j}}+4\hat{\mathbf{k}})$$

and the plane $\mathbf{r}\cdot(\hat{\mathbf{i}}+5\hat{\mathbf{j}}+\hat{\mathbf{k}})=5$ is

(a) $\frac{10}{3}$ (b) $\frac{3}{10}$ (c) $\frac{10}{3\sqrt{3}}$ (d) $\frac{10}{9}$

11 The negation of the compound preposition $p\vee(\sim p\vee q)$ is

(a) $(p\wedge\sim q)\wedge\sim p$ (b) $(p\wedge\sim q)\vee\sim p$
(c) $(p\wedge\sim q)\vee\sim p$ (d) $(p\wedge q)\wedge q$

12 If $f(x)=\begin{cases}\dfrac{x-1}{2x^2-7x+5}, & \text{for } x\neq 1\\ -1/2, & \text{for } x=1\end{cases}$, then $f'(1)$ is equal to

(a) $-\frac{1}{9}$ (b) $-\frac{2}{9}$ (c) $-\frac{1}{3}$ (d) $\frac{1}{3}$

13 The value of $\tan\left\{\cos^{-1}\left(\frac{-2}{7}\right)-\frac{\pi}{2}\right\}$ is

(a) $\frac{2}{3\sqrt{5}}$ (b) $\frac{2}{3}$ (c) $\frac{1}{\sqrt{5}}$ (d) $\frac{4}{\sqrt{5}}$

14 A boat is being rowed away from a cliff of 150 m height. At the top of the cliff the angle of depression of boat changes from 60° to 45° in 2 min. Then, the speed of the boat (in m/h) is

(a) $\frac{4500}{\sqrt{3}}$ (b) $\frac{4500}{\sqrt{3}}(\sqrt{3}-1)$
(c) $\frac{4300}{\sqrt{3}}$ (d) $\frac{4500}{\sqrt{3}}(\sqrt{3}+1)$

15 The equation of a line of intersection of planes $4x+4y-5z=12$ and $8x+12y-13z=32$ can be written as

(a) $\frac{x-1}{2}=\frac{y+2}{-3}=\frac{z}{4}$ (b) $\frac{x-1}{2}=\frac{y-2}{3}=\frac{z}{4}$
(c) $\frac{x}{2}=\frac{y+1}{3}=\frac{z-2}{4}$ (d) $\frac{x}{2}=\frac{y}{3}=\frac{z-2}{4}$

16 If the value of the determinant $\begin{vmatrix}a & 1 & 1\\ 1 & b & 1\\ 1 & 1 & c\end{vmatrix}$ is positive, then

(a) $abc>1$ (b) $abc>-8$
(c) $abc<-8$ (d) $abc>-2$

17 A bag contains a white and b black balls. Two players A and B alternately draw a ball from the bag, replacing the ball each time after the draw. A begins the game.

If the probability of A winning (that is drawing a white ball) is twice the probability of B winning (that is drawing a black ball), then the ratio $a:b$ is equal to

(a) 1 : 2 (b) 2 : 1
(c) 1 : 1 (d) None of these

18 If $(5+2\sqrt{6})^n=I+f;\ n, I\in N$ and $0\leq f<1$, then I equals

(a) $\frac{1}{f}-f$ (b) $\frac{1}{1+f}-f$ (c) $\frac{1}{1+f}+f$ (d) $\frac{1}{1-f}-f$

19 The median of a set of 11 distinct observations is 20.5. If each of the last 5 observations of the set is increased by 4, then the median of the new set

(a) is increased by 2
(b) is decreased by 2
(c) is two times the original a median
(d) remains the same as that of the original set

20 Solution of the differential equation $\dfrac{\sqrt{x}\,dx+\sqrt{y}\,dy}{\sqrt{x}\,dx-\sqrt{y}\,dy}=\sqrt{\dfrac{y^3}{x^3}}$ is given by

(a) $\frac{3}{2}\log\left(\frac{y}{x}\right)+\log\left|\frac{x^{3/2}+y^{3/2}}{x^{3/2}}\right|+\tan^{-1}\left(\frac{y}{x}\right)^{3/2}+c=0$

(b) $\frac{2}{3}\log\left(\frac{y}{x}\right)+\log\left|\frac{x^{3/2}+y^{3/2}}{x^{3/2}}\right|+\tan^{-1}\left(\frac{y}{x}\right)+c=0$

(c) $\frac{2}{3}\log\left(\frac{y}{x}\right)+\log\left(\frac{x+y}{x}\right)+\tan^{-1}\left(\frac{y^{3/2}}{x^{3/2}}\right)+c=0$

(d) None of the above

Section B : Numerical Type Questions

21 Sixteen metre of wire is available to fence off a flower bed in the form of a sector. If the flower bed has the maximum surface then radius is ....

22 A circle is drawn to pass through the extremities of the latusrectum of the parabola $y^2=8x$. It is given that, this circle also touches the directrix of the parabola. The radius of this circle is equal to ....

23 The number of permutations of the letters a, b, c, d such that b does not follow a, c does not follow b and d does not follow c, is ....

24 If $f'(x)=\sin(\log x)$ and $y=f\left(\dfrac{2x+3}{3-2x}\right)$ and $\dfrac{dy}{dx}$ at $x=1$ is equal to $p\sin\log(q)$, then value of $p+q$ is equal to ....

25 If $x+y+z=1$ and x, y, z are positive numbers such that $(1-x)(1-y)(1-z) \geq kxyz$, then k^2 equals ....

26 If $f(x)$ is a function satisfying $f(x+y)=f(x)f(y)$ for all $x, y \in N$ such that $f(1)=3$ and $\sum_{x=1}^{n} f(x)=120$, then the value of n^4 is ....

27 Number of solutions of the equation $\sin x=[x]$, where [.] denotes the largest integer function, is

28 If the value of $\int_0^1 \frac{x^4(1-x)^4}{1+x^2}dx$ is $\frac{m}{n}-\pi$, then value of mn is equal to

29 If the area inside the parabola $5x^2-y=0$ but outside the parabola $2x^2-y+9=0$ is $p\sqrt{3}$, then value of p is equal to

30 If $\hat{i}-3\hat{j}+5\hat{k}$ bisects the angle between $\hat{a}$ and $-\hat{i}+2\hat{j}+2\hat{k}$ when $\hat{a}$ is a unit vector, then $\hat{a}=\frac{\lambda\hat{i}+\mu\hat{j}+\gamma\hat{k}}{105}$, then the value of $\lambda^2-2\mu+v$ must be

Hints and Explanations

1 *(c)* Given equations can be rewritten as

$$x^2+(y-1)^2=2^2$$

and $\quad (x-1)^2+(y-2)^2=4^2$

Here, centres are C_1 (0, 1) and C_2 (1, 2) and radii are

$$r_1=2, r_2=4.$$

Now, $\quad C_1C_2=\sqrt{(1-0)^2+(2-1)^2}$

$$=\sqrt{1^2+1^2}=\sqrt{2}$$

$\because \quad |r_1-r_2|=|2-4|=2$

$$C_1C_2<|r_1-r_2|$$

2 *(a)* Let $I=\int\sqrt{\frac{\cos 2x}{\sin^2 x}}dx$

$$=\int\sqrt{\frac{\cos^2 x-\sin^2 x}{\sin^2 x}}dx$$

$$=\int\sqrt{\cot^2 x-1}\,dx$$

On putting $\cot x=\sec\theta$

$\Rightarrow \ -\text{cosec}^2 x\,dx=\sec\theta\tan\theta\,d\theta$,

we get

$$I=\int\sqrt{\sec^2\theta-1}\times\frac{\sec\theta\tan\theta}{-\text{cosec}^2 x}d\theta$$

$$=-\int\frac{\sec\theta\tan^2\theta}{1+\sec^2\theta}d\theta$$

$$=-\int\frac{\sin^2\theta}{\cos\theta+\cos^3\theta}d\theta$$

$$=-\int\frac{1-\cos^2\theta}{\cos\theta+\cos^3\theta}d\theta$$

$$=-\int\frac{(1+\cos^2\theta)-2\cos^2\theta}{\cos\theta(1+\cos^2\theta)}d\theta$$

$$=-\int\sec\theta\,d\theta+2\int\frac{d(\sin\theta)}{1+\cos^2\theta}$$

$$=-\log|\sec\theta+\sqrt{\sec^2\theta-1}|$$

$$+\frac{1}{\sqrt{2}}\log\left|\frac{\sqrt{2}+\sqrt{1-\cos^2\theta}}{\sqrt{2}-\sqrt{1-\cos^2\theta}}\right|+C$$

$$=-\log|\cot x+\sqrt{\cot^2 x-1}|$$

$$+\frac{1}{\sqrt{2}}\log\left|\frac{\sqrt{2}+\sqrt{1-\tan^2 x}}{\sqrt{2}-\sqrt{1-\tan^2 x}}\right|+C$$

But $I=-\log|\cot x+\sqrt{\cot^2 x-1}|$

$$+A+C\ \text{[given]}$$

$$\therefore\ A=\frac{1}{\sqrt{2}}\log\left|\frac{\sqrt{2}+\sqrt{1-\tan^2 x}}{\sqrt{2}-\sqrt{1-\tan^2 x}}\right|$$

3 *(a)* Since, m_1 and m_2 are the roots of the equation

$$x^2+(\sqrt{3}+2)x+(\sqrt{3}-1)=0.$$

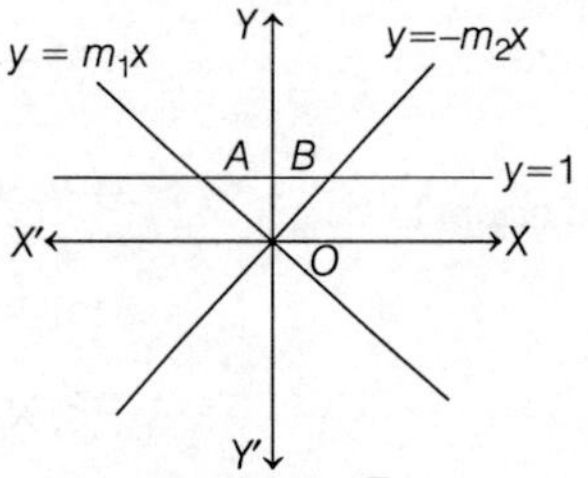

$\therefore \quad m_1+m_2=-(2+\sqrt{3})$

and $\quad m_1m_2=\sqrt{3}-1$

$\Rightarrow \quad m_1<0, m_2<0$

So, the point of intersections are

$$A\left(\frac{1}{m_1}, 1\right)\text{ and } B\left(-\frac{1}{m_2}, 1\right)$$

and $O(0, 0)$.

$$\therefore\ \text{Area of }\Delta OAB=-\frac{1}{2}\left\|\begin{matrix}\frac{1}{m_1} & 1 & 1\\ \frac{-1}{m_2} & 1 & 1\\ 0 & 0 & 1\end{matrix}\right\|$$

$$=\frac{1}{2}\left|\frac{1}{m_1}+\frac{1}{m_2}\right|=\frac{1}{2}\left|\frac{m_1+m_2}{m_1\,m_2}\right|$$

$$=\frac{1}{2}\left(\frac{2+\sqrt{3}}{\sqrt{3}-1}\right)$$

4 *(c)* The general equation of the tangent to the ellipse is

$$y=mx\pm\sqrt{a^2m^2+b^2} \quad \text{...(i)}$$

Since, $y=\frac{3}{2}x-10$ is a tangent to the ellipse, therefore its comparing with Eq. (i), we get

$$m=\frac{3}{2}\text{ and } a^2m^2+b^2=100$$

$$\Rightarrow\ 9a^2+4b^2=400 \quad \text{...(ii)}$$

Similarly, line

$$y=-\frac{1}{6}x+\frac{10}{3}$$

is a tangent to the ellipse, therefore its comparing with Eq. (i), we get

$$m=-\frac{1}{6}$$

and $\quad a^2m^2+b^2=\frac{100}{9}$

$$\Rightarrow\quad a^2+36b^2=400 \quad \text{...(iii)}$$

On solving Eqs. (ii) and (iii), we get

$$a^2=40$$

and $\quad b^2=10$

Hence, required equation of the ellipse is

$$\frac{x^2}{40}+\frac{y^2}{10}=1.$$

5 *(b)* Any line through the point of intersection of given lines is

$$\left(\frac{x}{a}+\frac{y}{b}-1\right)+\lambda\left(\frac{x}{b}+\frac{y}{a}-1\right)=0$$

$$x\left(\frac{1}{a}+\frac{\lambda}{b}\right)+y\left(\frac{1}{b}+\frac{\lambda}{a}\right)=(1+\lambda)$$

$$\Rightarrow x\left(\frac{b+a\lambda}{ab}\right)+y\left(\frac{a+b\lambda}{ab}\right)=(1+\lambda)$$

$$\Rightarrow\ \frac{x}{\left[\frac{ab(1+\lambda)}{b+a\lambda}\right]}+\frac{y}{\left[\frac{ab(1+\lambda)}{a+b\lambda}\right]}=1$$

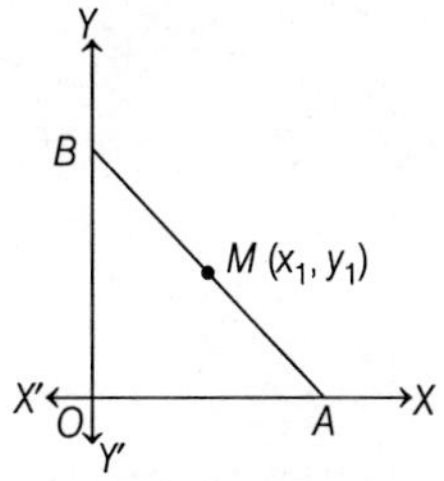

This meets the X-axis at

$$A \equiv \left(\frac{ab(1+\lambda)}{b+a\lambda}, 0\right)$$

and meets the Y-axis at

$$B \equiv \left(0, \frac{ab(1+\lambda)}{a+b\lambda}\right)$$

Let the mid-point of AB is $M(x_1, y_1)$.
Then,

$$x_1 = \frac{ab(1+\lambda)}{2(b+a\lambda)}$$

and $$y_1 = \frac{ab(1+\lambda)}{2(a+b\lambda)}$$

$$\therefore \frac{1}{x_1}+\frac{1}{y_1} = \frac{2(b+a\lambda)}{ab(1+\lambda)}+\frac{2(a+b\lambda)}{ab(1+\lambda)}$$

$$= \frac{2}{ab(1+\lambda)}(b+a\lambda+a+b\lambda)$$

$$= \frac{2}{ab(1+\lambda)}(b+a)(1+\lambda)$$

$$\Rightarrow \frac{(x_1+y_1)}{x_1 y_1} = \frac{2(a+b)}{ab}$$

$$\Rightarrow 2x_1 y_1 (a+b) = ab(x_1+y_1)$$

Hence, the locus of mid-point of AB is $2xy(a+b) = ab(x+y)$

6 *(b)* Since, $f(x)$ is continuous

$$\therefore \lim_{x\to 0}\left(\frac{2x-\sin^{-1}x}{2x+\tan^{-1}x}\right) = f(0)$$

$$\therefore f(0) = \lim_{x\to 0}\frac{\left(2-\frac{1}{\sqrt{1-x^2}}\right)}{2+\frac{1}{1+x^2}}$$

$$= \frac{2-\frac{1}{\sqrt{1}}}{2+\frac{1}{1}} = \frac{1}{3}$$

[$\because$ apply L'Hospital's Rule]

7 *(d)* Let AB be a chord of the circle through $A(p,q)$ and $M(h, 0)$ be the mid-point of AB. Therefore, the coordinates of B are $(-p+2h, -q)$.
Since, B lies on the circle $x^2+y^2 = px+qy$, then

$$(-p+2h)^2+(-q)^2 = p(-p+2h)+q(-q)$$

$$\Rightarrow 2p^2+2q^2-6ph+4h^2 = 0$$

$$\Rightarrow 2h^2-3ph+(p^2+q^2) = 0 \quad \text{...(i)}$$

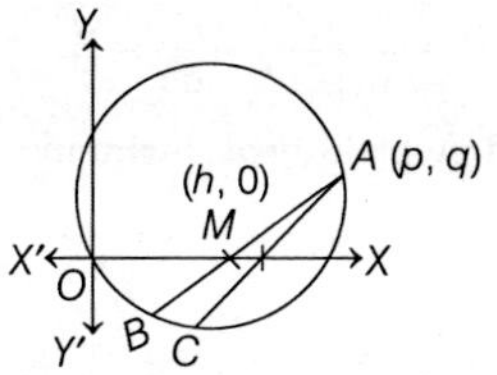

As, there are two distinct chords AB and AC from $A(p,q)$ which are bisected on X-axis there must be two distinct values of h satisfying Eq. (i), then $D = (b^2-4ac) > 0$, we have

$$(-3p)^2-4(2)(p^2+q^2) > 0$$

$$\Rightarrow p^2 > 8q^2$$

8 *(c)* Since, $A^3 = B^3$ and $A^2B = B^2A$

$$\therefore A^3 - A^2B = B^3 - B^2A$$

$$\Rightarrow (A^2+B^2)(A-B) = 0$$

$$\Rightarrow \det(A^2+B^2)\det(A-B) = 0$$

9 *(b)* Here, $p = \frac{1}{2}$ and $q = \frac{1}{2}$

$$\therefore P(X=1)+P(X=3)+\ldots+P(X=49)$$

$$= {}^{100}C_1\left(\frac{1}{2}\right)^{100} + {}^{100}C_3\left(\frac{1}{2}\right)^{100} + \ldots + {}^{100}C_{49}\left(\frac{1}{2}\right)^{100}$$

$$= \left(\frac{1}{2}\right)^{100}({}^{100}C_1+{}^{100}C_3+\ldots+{}^{100}C_{49})$$

$$\left[\begin{array}{l}\because ({}^{100}C_1+{}^{100}C_3+\ldots+{}^{100}C_{99}) = 2^{99}\\ \text{but } {}^{100}C_{99} = {}^{100}C_1\\ {}^{100}C_{97} = {}^{100}C_3, \ldots, {}^{100}C_{51} = {}^{100}C_{49}\\ \therefore 2({}^{100}C_1+{}^{100}C_3+\ldots+{}^{100}C_{49}) = 2^{99}\end{array}\right]$$

$$= \left(\frac{1}{2}\right)^{100}\times 2^{98} = \frac{1}{4}$$

10 *(c)* Line is parallel to plane as $(\hat{\mathbf{i}}-\hat{\mathbf{j}}+4\hat{\mathbf{k}})\cdot(\hat{\mathbf{i}}+5\hat{\mathbf{j}}+\hat{\mathbf{k}}) = 0$
General point on the line is $(\lambda+2, -\lambda-2, 4\lambda+3)$. For $\lambda = 0$ point on this line is $(2, -2, 3)$ and distance from $\mathbf{r}\cdot(\hat{\mathbf{i}}+5\hat{\mathbf{j}}+\hat{\mathbf{k}}) = 5$ or $x+5y+z = 5$, is

$$d = \frac{|2+5(-2)+3-5|}{\sqrt{1+25+1}}$$

$$\Rightarrow d = \frac{|-10|}{3\sqrt{3}} = \frac{10}{3\sqrt{3}}$$

11 *(a)* Negation of $p\vee(\sim p\vee q)$

$$\Rightarrow \sim[p\vee(\sim p\vee q)] \equiv \sim p\wedge\sim(\sim p\vee q)$$

$$\equiv \sim p\wedge(\sim(\sim p)\wedge\sim q)$$

$$\equiv \sim p\wedge(p\wedge\sim q)$$

$$\equiv (p\wedge\sim q)\wedge\sim p$$

12 *(b)* Given,

$$f(x) = \begin{cases}\frac{x-1}{2x^2-7x+5}, & x\neq 1\\ -\frac{1}{3}, & x = 1\end{cases}$$

$$f(x) = \begin{cases}\frac{1}{2x-5}, & x\neq 1\\ -\frac{1}{3}, & x = 1\end{cases}$$

$$f'(1) = \lim_{h\to 0}\frac{f(1+h)-f(1)}{h}$$

$$= \lim_{h\to 0}\frac{\frac{1}{2(1+h)-5}-\left(-\frac{1}{3}\right)}{h}$$

$$= \lim_{h\to 0}\frac{\frac{1}{2h-3}+\frac{1}{3}}{h}$$

$$= \lim_{h\to 0}\frac{3+2h-3}{3h(2h-3)} = -\frac{2}{9}$$

$$Lf'(1) = \lim_{h\to 0}\frac{f(1-h)-f(1)}{-h}$$

$$= \lim_{h\to 0}\frac{\frac{1}{2(1-h)-5}-\left(-\frac{1}{3}\right)}{-h}$$

$$= \lim_{h\to 0} -\frac{2}{3(2h+3)} = -\frac{2}{9}$$

$$\therefore f'(1) = -\frac{2}{9}$$

13 *(a)* $\tan\left\{\cos^{-1}\left(-\frac{2}{7}\right)-\frac{\pi}{2}\right\}$

$$= \tan\left\{\pi-\cos^{-1}\left(\frac{2}{7}\right)-\frac{\pi}{2}\right\}$$

$$= \tan\left\{\sin^{-1}\left(\frac{2}{7}\right)\right\}$$

$$= \tan\left\{\tan^{-1}\left(\frac{2}{3\sqrt{5}}\right)\right\} = \frac{2}{3\sqrt{5}}$$

14 *(b)* Let $PQ = 150$ m

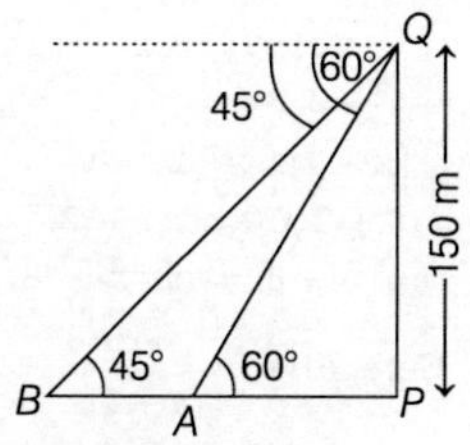

In ΔAPQ, $\tan 60° = \frac{PQ}{AP}$

$$\Rightarrow AP = \frac{150}{\sqrt{3}} \quad \text{...(i)}$$

and in ΔBPQ,

$$\tan 45° = \frac{PQ}{AB+AP}$$

$$\Rightarrow AB+\frac{150}{\sqrt{3}} = 150$$

$\Rightarrow \quad AB = \frac{150}{\sqrt{3}}(\sqrt{3}-1)$

$\therefore$ Speed of boat $= \frac{AB}{2}$

$= \frac{1}{2} \times \frac{150}{\sqrt{3}}(\sqrt{3}-1) \times 60$

$= \frac{4500}{\sqrt{3}}(\sqrt{3}-1)$ m/h

15 *(b)* Given equation of planes are

$4x + 4y - 5z = 12 \quad \ldots(i)$

and $8x + 12y - 13z = 32 \quad \ldots(ii)$

Let DR's of required line be (l, m, n).

From Eqs. (i) and (ii), we get

$4l + 4m - 5n = 0$

and $8l + 12m - 13n = 0$

$\Rightarrow \frac{l}{8} = \frac{m}{12} = \frac{n}{16}$

$\Rightarrow \frac{l}{2} = \frac{m}{3} = \frac{n}{4}$

Now, we take intersection point with $z = 0$ is given by

$4x + 4y = 12 \quad \ldots(iii)$

and $8x + 12y = 32 \quad \ldots(iv)$

On solving Eqs. (i) and (ii), we get the point (1, 2, 0).

$\therefore$ Required line is $\frac{x-1}{2} = \frac{y-2}{3} = \frac{z}{4}$.

16 *(b)* Let $\Delta' = \begin{vmatrix} a & 1 & 1 \\ 1 & b & 1 \\ 1 & 1 & c \end{vmatrix}$

$= abc + 2 - a - b - c > 0$

or $abc + 2 > a + b + c \quad \ldots(i)$

$\because$ AM > GM $\Rightarrow \frac{a+b+c}{3} > (abc)^{1/3}$

$a + b + c > 3(abc)^{1/3} \quad \ldots(ii)$

From Eqs. (i) and (ii),

$abc + 2 > 3(abc)^{1/3}$

Let $(abc)^{1/3} = x$

Then, $x^3 + 2 > 3x$

$\Rightarrow (x-1)^2(x+2) > 0$

$\therefore x + 2 > 0 \Rightarrow x > -2$

$\Rightarrow x^3 > -8 \Rightarrow abc > -8$

17 *(c)* Here, $P(W) = \frac{a}{a+b}$

and $P(B) = \frac{b}{a+b}$

$\therefore$ Probability of A winning

$= P(W) + P(\overline{W})P(\overline{B})P(W) + \ldots$

$= \frac{P(W)}{1 - P(\overline{W})P(\overline{B})}$

$= \dfrac{\frac{a}{a+b}}{1 - \frac{b}{a+b}\cdot\frac{a}{a+b}}$

$= \frac{a(a+b)}{a^2+b^2+ab} = P_1 \quad$ [say]

and probability of B winning

$= 1 - P_1 = 1 - \frac{a^2+ab}{a^2+b^2+ab}$

$= \frac{b^2}{a^2+b^2+ab} = P_2 \quad$ [say]

Given, $P_1 = 2P_2$

$\Rightarrow \frac{a^2+ab}{a^2+b^2+ab} = \frac{2b^2}{a^2+b^2+ab}$

$\Rightarrow a^2 + ab - 2b^2 = 0$

$\Rightarrow (a-b)(a+2b) = 0$

$\Rightarrow a - b = 0 \quad [\because a + 2b \neq 0]$

$\Rightarrow a = b$

$\therefore a : b = 1 : 1$

18 *(d)* $I + f + f'$

$= (5 + 2\sqrt{6})^n + (5 - 2\sqrt{6})^n$

$= 2k \quad$ [even integer]

$\therefore f + f' = 1$

Now, $(I + f)f' = (5 + 2\sqrt{6})^n (5 - 2\sqrt{6})^n$

$= (1)^n = 1$

$\Rightarrow (I + f)(1 - f) = 1$

$\Rightarrow I = \frac{1}{1-f} - f$

19 *(d)* Since, $n = 11$, then median term

$= \left(\frac{11+1}{2}\right)$th term = 6 th term

As, last five observations are increased by 4. Hence, the median of the 6th observations will remain same.

20 *(d)* We have, $\frac{\sqrt{x}\,dx + \sqrt{y}\,dy}{\sqrt{x}\,dx - \sqrt{y}\,dy} = \sqrt{\frac{y^3}{x^3}}$

$\Rightarrow \frac{d(x^{3/2}) + d(y^{3/2})}{d(x^{3/2}) - d(y^{3/2})} = \frac{y^{3/2}}{x^{3/2}}$

$\Rightarrow \frac{du + dv}{du - dv} = \frac{v}{u}$

where $u = x^{3/2}$ and $v = y^{3/2}$

$\Rightarrow u\,du + u\,dv = v\,du - v\,dv$

$\Rightarrow u\,du + v\,dv = v\,du - u\,dv$

$\Rightarrow \frac{u\,du + v\,dv}{u^2 + v^2} = \frac{v\,du - u\,dv}{u^2 + v^2}$

$\Rightarrow \frac{d(u^2 + v^2)}{u^2 + v^2} = -2d\tan^{-1}\left(\frac{v}{u}\right) + c$

On integrating, we get

$\log(u^2 + v^2) = -2\tan^{-1}\left(\frac{v}{u}\right) + c$

$\Rightarrow \frac{1}{2}\log(x^3 + y^3) + \tan^{-1}\left(\frac{y}{x}\right)^{3/2} = \frac{c}{2}$

21 *(4)* Let r be the radius of the sector and angle subtended at the centre be θ.

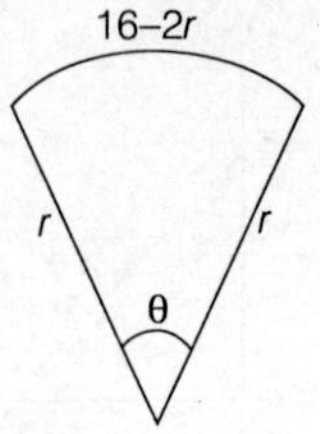

Then, S = surface area of sector

$= \frac{\theta}{360} \times \pi r^2$

We know, $\theta = \frac{\text{length of arc}}{\text{radius}}$

$= \frac{16-2r}{r}$

$\therefore \quad S = \frac{\theta}{2\pi}\pi r^2 = \frac{16-2r}{2r}\cdot r^2$

$\Rightarrow \quad S = (8 - r)\cdot r = 8r - r^2$

$\Rightarrow \quad \frac{dS}{dr} = 8 - 2r$

Now, for area to be maximum,

$\frac{dS}{dr} = 0$

$\Rightarrow \quad r = 4$

22 *(4)* Extremities of the latusrectum of the parabola are (2, 4) and (2, – 4).

Since, any circle drawn with any focal chord at its diameter touches the directrix, thus equation of required circle is

$(x-2)^2 + (y-4)(y+4) = 0$

$\Rightarrow x^2 + y^2 - 4x - 12 = 0$

$\therefore$ Radius $= \sqrt{(2)^2 + 12} = 4$

23 *(11)* We have the following cases.

Case I

a			

In this case we have only two possibilities, namely, *acbd* and *adcb*

Case II

b			

In this case we have only three possibilities, namely, *badc*, *bdac* and *bdca*

Case III

c			

In this case we have only three possibilities, namely, *cadb*, *cbda* and *cbad*

Case IV

d			

In this case we have only three possibilities, namely, $d\,c\,b\,a$, $d\,a\,c\,b$ and $d\,b\,a\,c$

Hence, the total number of ways

$$= 2 + 3 + 3 + 3 = 11$$

24 *(17)* Given that, $y = f\left(\frac{2x+3}{3-2x}\right)$

$$\Rightarrow \frac{dy}{dx} = f'\left(\frac{2x+3}{3-2x}\right)\frac{d}{dx}\left(\frac{2x+3}{3-2x}\right)$$

$$= \sin\left[\log\left(\frac{2x+3}{3-2x}\right)\right]$$

$$\left[\frac{(3-2x)(2)-(2x+3)(-2)}{(3-2x)^2}\right]$$

$$= \frac{12}{(3-2x)^2}\sin\left[\log\left(\frac{2x+3}{3-2x}\right)\right]$$

$$\Rightarrow \left(\frac{dy}{dx}\right)_{(x=1)} = \frac{12}{(3-2)^2}\sin\log(5)$$

$$= 12\sin\log(5)$$

Here, $p = 12, q = 5$

$\therefore \quad p + q = 17$

25 *(64)* Clearly, $\frac{x+y}{2} \geq \sqrt{xy}; \frac{y+z}{2} \geq \sqrt{yz}$

and $\frac{x+z}{2} \geq \sqrt{xz}$

$$\therefore \frac{(x+y)}{2}\cdot\frac{(y+z)}{2}\cdot\frac{(x+z)}{2} \geq \sqrt{xy}\cdot\sqrt{yz}\cdot\sqrt{xz}$$

$\Rightarrow (1-z)(1-x)(1-y) \geq 8xyz$

$[\because x + y + z = 1]$

$\therefore \quad k = 8$

Hence, $k^2 = 64$

26 *(256)* $f(x) = f(1+1+1+\ldots+x$ times)

$= f(1)f(1)f(1)\ldots\ldots\ldots x$ times

$= [f(1)]^x = 3^x$

$$\therefore \sum_{x=1}^{n} f(x) = \sum_{x=1}^{n} 3^x = 3^1 + 3^2 + \ldots + 3^n$$

$$= \frac{3^1 - 3^n\cdot 3}{1-3}$$

$$= \frac{3^{n+1}-3}{2} \quad \left[\because \text{sum} = \frac{a - lr}{1-r}\right]$$

$$\therefore \frac{3^{n+1}-3}{2} = 120 \Rightarrow 3^{n+1} = 243 = 3^5$$

$\Rightarrow \quad n + 1 = 5$

$\Rightarrow \quad n = 4$

$\therefore \quad n^4 = 256$

27 *(2)* $y = \sin x = [x]$

Graphs of $y = \sin x$ and $y = [x]$ are as shown.

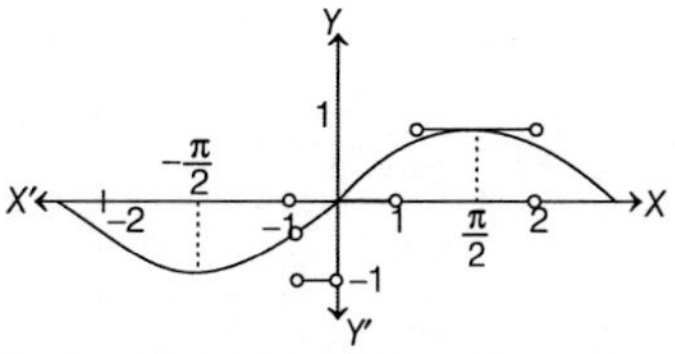

Hence, two solutions are $x = 0$ and $x = \frac{\pi}{2}$.

28 *(154)* Let $I = \int_0^1 \frac{x^4(1-x)^4}{1+x^2}dx$

$$= \int_0^1 x^4\left(x^2 - 4x + 5 - \frac{4}{1+x^2}\right)dx$$

$$= \int_0^1 (x^6 - 4x^5 + 5x^4)\,dx - 4\int_0^1 \frac{x^4}{(1+x^2)}dx$$

$$= \int_0^1 (x^6 - 4x^5 + 5x^4)\,dx - 4\int_0^1\left(x^2 - 1 + \frac{1}{1+x^2}\right)dx$$

$$= \left(\frac{x^7}{7} - \frac{4x^6}{6} + x^5\right)_0^1 - 4\left(\frac{x^3}{3} - x + \tan^{-1}x\right)_0^1$$

$$= \left(\frac{1}{7} - \frac{4}{6} + 1\right) - 4\left(\frac{1}{3} - 1 + \frac{\pi}{4}\right)$$

$$= \frac{22}{7} - \pi$$

$\therefore m = 22, n = 7$

Hence, $m \times n = 22 \times 7 = 154$

29 *(12)* Given parabolas are

$$5x^2 - y = 0$$

and $\quad 2x^2 - y + 9 = 0$

Now, eliminating y from above equations, we get

$$5x^2 - (2x^2 + 9) = 0$$

$$\Rightarrow 3x^2 = 9 \Rightarrow x = \pm\sqrt{3}$$

Given parabolas intersect at $(\sqrt{3}, 15)$ and $(-\sqrt{3}, 15)$.

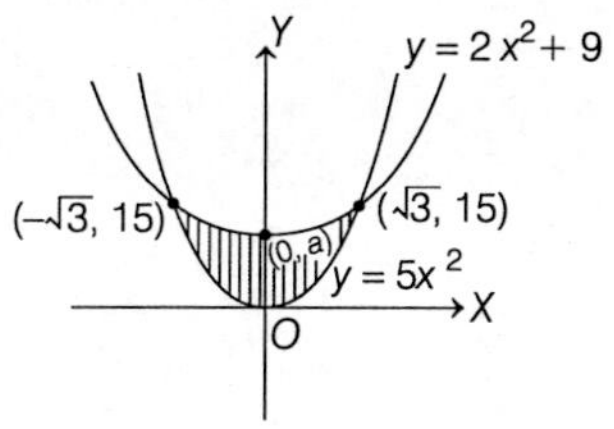

The two parabolas are

$$x^2 = \frac{1}{5}y, x^2 = \frac{1}{2}(y-9)$$

$$\therefore \text{Area} = 2\int_0^{\sqrt{3}}(y_1 - y_2)\,dx$$

$$= 2\int_0^{\sqrt{3}}[(2x^2+9) - 5x^2)]\,dx$$

$$= 2\int_0^{\sqrt{3}}(9 - 3x^2)\,dx$$

$$= 2[9x - x^3]_0^{\sqrt{3}}$$

$\therefore$ Area $= 12\sqrt{3}$

$\therefore \quad p = 12$

30 *(1817)* We must have

$$\alpha(\hat{i} - 3\hat{j} + 5\hat{k}) = \hat{a} + \frac{2\hat{k} + 2\hat{j} - \hat{i}}{3}$$

$$\Rightarrow 3\hat{a} = \alpha(3\hat{i} - 9\hat{j} + 15\hat{k}) - 2\hat{k} - 2\hat{j} + \hat{i}$$

$$\Rightarrow 3\hat{a} = (3\alpha+1)\hat{i} - (9\alpha+2)\hat{j} + (15\alpha - 2)\hat{k}$$

$$\therefore 3|\hat{a}| = |(3\alpha+1)\hat{i} - (9\alpha+2)\hat{j} + (15\alpha-2)\hat{k}|$$

$$\Rightarrow 3 = \sqrt{\{(3\alpha+1)^2 + (9\alpha+2)^2 + (15\alpha-2)^2\}}$$

$$\Rightarrow 9 = 135\alpha^2 - 18\alpha + 9$$

$$\therefore \quad \alpha = 0, \frac{2}{35}$$

For $\alpha = 0, \hat{a} = \frac{\hat{i} - 2\hat{j} - 2\hat{k}}{3}$

(not acceptable)

For $\alpha = \frac{2}{35}, \hat{a} = \frac{2}{35}(\hat{i} - 3\hat{j} + 5\hat{k}) - \frac{(2\hat{k} + 2\hat{j} - \hat{i})}{3}$

$$= \frac{1}{105}(41\hat{i} - 88\hat{j} - 40\hat{k})$$

On comparing, we get

$\lambda = 41, \mu = -88, v = -40$

$\therefore \lambda^2 - 2\mu + v = (41)^2 + 2(88) - 40$

$= 1681 + 176 - 40 = 1817$

JEE Main 2021

(February, March, July & August Attempt)

Instructions

- This question paper contains of 30 Questions of Mathematics, divided into two Sections : **Section A** Objective Type Questions and **Section B** Numerical Type Questions.
- Section A contains 20 questions and all Questions are compulsory (**Marking Scheme :** Correct + 4, Incorrect –1) .
- Section B contains 10 questions out of which only 5 questions are to be attempted (**Marking Scheme :** Correct + 4, Incorrect 0) .

24 FEBRUARY SHIFT I

Section A : Objective Type Questions

1. Let $f : R \to R$ be defined as $f(x) = 2x - 1$ and $g : R - \{1\} \to R$ be defined as $g(x) = \dfrac{x - (1/2)}{x - 1}$

Then, the composition function $f(g(x))$ is

a. one-one but not onto **b.** onto but not one-one

c. Neither one-one nor onto **d.** Both one-one and onto

2. Let p and q be two positive numbers, such that $p + q = 2$ and $p^4 + q^4 = 272$. Then, p and q are roots of the equation

a. $x^2 - 2x + 136 = 0$ **b.** $x^2 - 2x + 16 = 0$

c. $x^2 - 2x + 8 = 0$ **d.** $x^2 - 2x + 2 = 0$

3. The system of linear equations

$3x - 2y - kz = 10$

$2x - 4y - 2z = 6$

$x + 2y - z = 5m$

is inconsistent, if

a. $k \neq 3, m \neq \dfrac{4}{5}$ **b.** $k = 3, m = \dfrac{4}{5}$

c. $k = 3, m \neq \dfrac{4}{5}$ **d.** $k \neq 3, m \in R$

4. The value of $-{}^{15}C_1 + 2 \cdot {}^{15}C_2 - 3 \cdot {}^{15}C_3 + \ldots - 15 \cdot {}^{15}C_{15} + {}^{14}C_1 + {}^{14}C_3 + {}^{14}C_5 + \ldots + {}^{14}C_{11}$ is

a. $2^{16} - 1$ **b.** $2^{13} - 14$

c. $2^{13} - 13$ **d.** 2^{14}

5. If $e^{(\cos^2 x + \cos^4 x + \cos^6 x + \ldots \infty) \log_e 2}$ satisfies the equation $t^2 - 9t + 8 = 0$, then the value of $\dfrac{2 \sin x}{\sin x + \sqrt{3} \cos x}, \left(0 < x < \dfrac{\pi}{2}\right)$ is

a. $\dfrac{1}{2}$ **b.** $\sqrt{3}$

c. $\dfrac{3}{2}$ **d.** $2\sqrt{3}$

6. $\lim\limits_{x \to 0} \dfrac{\int_0^{x^2} (\sin \sqrt{t}) dt}{x^3}$ is equal to

a. $\dfrac{2}{3}$ **b.** $\dfrac{3}{2}$

c. $\dfrac{1}{15}$ **d.** 0

7. The function

$f(x) = \dfrac{4x^3 - 3x^2}{6} - 2 \sin x + (2x - 1) \cos x$

a. increases in $\left[\dfrac{1}{2}, \infty\right)$ **b.** decreases in $\left[\dfrac{1}{2}, \infty\right)$

c. increases in $\left(-\infty, \dfrac{1}{2}\right]$ **d.** decreases in $\left(-\infty, \dfrac{1}{2}\right]$

8. A scientific committee is to be formed from 6 Indians and 8 foreigners, which includes at least 2 Indians and double the number of foreigners as Indians. Then, the number of ways, the committee can be formed, is

a. 1050 **b.** 1625

c. 560 **d.** 575

9. If $f : R \to R$ is a function defined by $f(x) = [x - 1] \cos\left(\dfrac{2x - 1}{2}\right)\pi$, where [.] denotes the greatest integer function, then f is

a. discontinuous only at $x = 1$

b. discontinuous at all integral values of x except at $x = 1$

c. continuous only at $x = 1$

d. continuous for every real x

10. If $\int \frac{\cos x - \sin x}{\sqrt{8 - \sin 2x}} dx = a \sin^{-1}\left(\frac{\sin x + \cos x}{b}\right) + c$, where c is a constant of integration, then the ordered pair (a, b) is equal to

a. (3, 1) **b.** (1, 3) **c.** (– 1, 3) **d.** (1, – 3)

11. The area (in sq. units) of the part of the circle $x^2 + y^2 = 36$, which is outside the parabola $y^2 = 9x$, is

a. $24\pi + 3\sqrt{3}$ **b.** $24\pi - 3\sqrt{3}$

c. $12\pi + 3\sqrt{3}$ $12\pi - 3\sqrt{3}$

12. The population $P = P(t)$ at time t of a certain species follows the differential equation $\frac{dP}{dt} = 0.5P - 450$. If $P(0) = 850$, then the time at which population becomes zero is

a. $\log_e 9$ **b.** $\frac{1}{2}\log_e 18$

c. $\log_e 18$ **d.** $2\log_e 18$

13. A man is walking on a straight line. The arithmetic mean of the reciprocals of the intercepts of this line on the coordinate axes is $\frac{1}{4}$. Three stones *A*, *B* and *C* are placed at the points (1,1), (2, 2) and (4, 4), respectively. Then, which of these stones is / are on the path of the man ?

a. *A* only **b.** *B* only

d. All the three **c.** *C* only

14. The locus of the mid-point of the line segment joining the focus of the parabola $y^2 = 4ax$ to a moving point of the parabola, is another parabola whose directrix is

a. $x = a$ **b.** $x = -\frac{a}{2}$ **c.** $x = 0$ **d.** $x = \frac{a}{2}$

15. If the tangent to the curve $y = x^3$ at the point $P(t, t^3)$ meets the curve again at Q, then the ordinate of the point which divides PQ internally in the ratio 1 : 2 is

a. 0 **b.** $2t^3$ **c.** $-t^3$ **d.** $-2t^3$

16. The equation of the plane passing through the point (1, 2, –3) and perpendicular to the planes $3x + y - 2z = 5$ and $2x - 5y - z = 7$, is

a. $6x - 5y + 2z + 10 = 0$ **b.** $11x + y + 17z + 38 = 0$

c. $6x - 5y - 2z - 2 = 0$ **d.** $3x - 10y + 2z + 11 = 0$

17. The distance of the point (1, 1, 9) from the point of intersection of the line $\frac{x-3}{1} = \frac{y-4}{2} = \frac{z-5}{2}$ and the plane $x + y + z = 17$ is

a. $2\sqrt{19}$ **b.** $19\sqrt{2}$ **c.** $\sqrt{38}$ **d.** 38

18. An ordinary dice is rolled for a certain number of times. If the probability of getting an odd number 2 times is equal to the probability of getting an even number 3 times, then the probability of getting an odd number for odd number of times is

a. $\frac{1}{32}$ **b.** $\frac{3}{16}$

c. $\frac{5}{16}$ **d.** $\frac{1}{2}$

19. Two vertical poles are 150 m apart and the height of one is three times that of the other. If from the middle point of the line joining their feet, an observer finds the angles of elevation of their tops to be complementary, then the height of the shorter pole (in metres) is

a. 25 **b.** 30

c. $20\sqrt{3}$ **d.** $25\sqrt{3}$

20. The statement among the following that is a tautology is

a. $A \wedge (A \vee B)$ **b.** $A \vee (A \wedge B)$

c. $[A \wedge (A \to B)] \to B$ **d.** $B \to [A \wedge (A \to B)]$

Section B : Numerical Type Questions

21. If the least and the largest real values of α, for which the equation $z + \alpha|z - 1| + 2i = 0$ ($z \in C$ and $i = \sqrt{-1}$) has a solution, are p and q respectively, then $4(p^2 + q^2)$ is equal to ………… .

22. Let $B_i (i = 1, 2, 3)$ be three independent events in a sample space. The probability that only B_1 occur is α, only B_2 occurs is β and only B_3 occurs is γ. Let P be the probability that none of the events B_i occurs and these 4 probabilities satisfy the equations $(\alpha - 2\beta)P = \alpha\beta$ and $(\beta - 3\gamma)P = 2\beta\gamma$ (All the probabilities are assumed to lie in the interval (0,1)). Then, $\frac{P(B_1)}{P(B_3)}$ is equal to ……… .

23. Let $P = \begin{bmatrix} 3 & -1 & -2 \\ 2 & 0 & \alpha \\ 3 & -5 & 0 \end{bmatrix}$, where $\alpha \in R$. Suppose $Q = [q_{ij}]$ is a matrix satisfying $PQ = kI_3$ for some non-zero $k \in R$. If $q_{23} = -\frac{k}{8}$ and $|Q| = \frac{k^2}{2}$, then $\alpha^2 + k^2$ is equal to ……… .

24. Let M be any 3×3 matrix with entries from the set {0, 1, 2}. The maximum number of such matrices, for which the sum of diagonal elements of $M^T M$ is seven, is ……… .

25. Let $A = \{n \in N : n$ is a 3-digit number$\}$

$B = \{9k + 2 : k \in N\}$

and $C = \{9k + l : k \in N\}$ for some $l (0 < l < 9)$

If the sum of all the elements of the set $A \cap (B \cup C)$ is 274 $\times$ 400, then l is equal to ……….. .

26. The minimum value of α for which the equation $\frac{4}{\sin x} + \frac{1}{1 - \sin x} = \alpha$ has at least one solution in $\left(0, \frac{\pi}{2}\right)$ is …… .

27. If $\int_{-a}^{a} (|x| + |x - 2|) dx = 22, (a > 2)$ and $[x]$ denotes the greatest integer $\leq x$, then $\int_{a}^{-a} (x + [x]) dx$ is equal to …….. .

28. If one of the diameters of the circle $x^2 + y^2 - 2x - 6y + 6 = 0$ is a chord of another circle 'C', whose centre is at (2, 1), then its radius is ....

29. Let three vectors **a**, **b** and **c** be such that **c** is coplanar with **a** and **b**, $\mathbf{a} \cdot \mathbf{c} = 7$ and **b** is perpendicular to **c**, where $\mathbf{a} = -\hat{\mathbf{i}} + \hat{\mathbf{j}} + \hat{\mathbf{k}}$ and $\mathbf{b} = 2\hat{\mathbf{i}} + \hat{\mathbf{k}}$, then the value of $2|\mathbf{a}+\mathbf{b}+\mathbf{c}|^2$ is ....

30. $\lim_{n \to \infty} \tan\left\{\sum_{r=1}^{n} \tan^{-1}\left(\frac{1}{1+r+r^2}\right)\right\}$ is equal to ....

Answers

For solution scan the QR code

1. (a)	2. (b)	3. (c)	4. (b)	5. (a)	6. (a)	7. (a)	8. (b)	9. (d)	10. (b)
11. (b)	12. (d)	13. (b)	14. (c)	15. (d)	16. (b)	17. (c)	18. (d)	19. (d)	20. (c)
21. (10)	22. (6)	23. (17)	24. (540)	25. (5)	26. (9)	27. (3)	28. (3)	29. (75)	30. (1)

24 FEBRUARY SHIFT II

Section A : Objective Type Questions

1. For the statements p and q, consider the following compound statements

A. $[\sim q \wedge (p \to q)] \to \sim p$

B. $[(p \vee q) \wedge \sim p] \to q$

Then, which of the following statement(s) is/are correct?

a. (A) and (B) both are not tautologies.
b. (A) and (B) both are tautologies.
c. (A) is a tautology but not (B).
d. (B) is a tautology but not (A).

2. Let $a, b \in R$. If the mirror image of the point $P(a, 6, 9)$ with respect to the line $\frac{x-3}{7} = \frac{y-2}{5} = \frac{z-1}{-9}$ is $(20, b, -a-9)$, then $|a+b|$ is equal to

a. 88 **b.** 86
c. 84 **d.** 90

3. The vector equation of the plane passing through the intersection of the planes $\mathbf{r} \cdot (\hat{\mathbf{i}} + \hat{\mathbf{j}} + \hat{\mathbf{k}}) = 1$ and $\mathbf{r} \cdot (\hat{\mathbf{i}} - 2\hat{\mathbf{j}}) = -2$ and the point (1, 0, 2) is

a. $\mathbf{r} \cdot (\hat{\mathbf{i}} + 7\hat{\mathbf{j}} + 3\hat{\mathbf{k}}) = \frac{7}{3}$ **b.** $\mathbf{r} \cdot (3\hat{\mathbf{i}} + 7\hat{\mathbf{j}} + 3\hat{\mathbf{k}}) = 7$
c. $\mathbf{r} \cdot (\hat{\mathbf{i}} + 7\hat{\mathbf{j}} + 3\hat{\mathbf{k}}) = 7$ **d.** $\mathbf{r} \cdot (\hat{\mathbf{i}} - 7\hat{\mathbf{j}} + 3\hat{\mathbf{k}}) = \frac{7}{3}$

4. If P is a point on the parabola $y = x^2 + 4$ which is closest to the straight line $y = 4x - 1$, then the coordinates of P are

a. (3, 13) **b.** (1, 5)
c. (− 2, 8) **d.** (2, 8)

5. The angle of elevation of a jet plane from a point A on the ground is 60°. After a flight of 20 s at the speed of 432 km/h, the angle of elevation changes to 30°. If the jet plane is flying at a constant height, then its height is

a. $1800\sqrt{3}$ m **b.** $3600\sqrt{3}$ m
c. $2400\sqrt{3}$ m **d.** $1200\sqrt{3}$ m

6. If $n \geq 2$ is a positive integer, then the sum of the series ${}^{n+1}C_2 + 2({}^2C_2 + {}^3C_2 + {}^4C_2 + \ldots + {}^nC_2)$ is

a. $\frac{n(n-1)(2n+1)}{6}$ **b.** $\frac{n(n+1)(2n+1)}{6}$
c. $\frac{n(2n+1)(3n+1)}{6}$ **d.** $\frac{n(n+1)^2(n+2)}{12}$

7. Let $f: R \to R$ be defined as

$$f(x) = \begin{cases} -55x, & \text{if } x < -5 \\ 2x^3 - 3x^2 - 120x, & \text{if } -5 \leq x < 4 \\ 2x^3 - 3x^2 - 36x - 336, & \text{if } x \geq 4 \end{cases}$$

Let $A = \{x \in R : f \text{ is increasing}\}$. Then, A is equal to

a. $(-\infty, -5) \cup (4, \infty)$ **b.** $(-5, \infty)$
c. $(-\infty, -5) \cup (-4, \infty)$ **d.** $(-5, -4) \cup (4, \infty)$

8. Let f be a twice differentiable function defined on R, such that $f(0) = 1$, $f'(0) = 2$ and $f''(x) \neq 0$ for all $x \in R$. If $\begin{vmatrix} f(x) & f'(x) \\ f'(x) & f''(x) \end{vmatrix} = 0$, for all $x \in R$, then the value of $f(1)$ lies in the interval

a. (9, 12) **b.** (6, 9) **c.** (0, 3) **d.** (3, 6)

9. For which of the following curves, the line $x + \sqrt{3}y = 2\sqrt{3}$ is the tangent at the point $\left(\frac{3\sqrt{3}}{2}, \frac{1}{2}\right)$?

a. $x^2 + y^2 = 7$ **b.** $y^2 = \frac{1}{6\sqrt{3}}x$
c. $2x^2 - 18y^2 = 9$ **d.** $x^2 + 9y^2 = 9$

10. The value of the integral $\int_1^3 [x^2 - 2x - 2]\,dx$, where $[x]$ denotes the greatest integer less than or equal to x, is

a. $-\sqrt{2} - \sqrt{3} + 1$ **b.** $-\sqrt{2} - \sqrt{3} - 1$
c. − 5 **d.** − 4

11. A possible value of $\tan\left(\frac{1}{4}\sin^{-1}\frac{\sqrt{63}}{8}\right)$ is

a. $1/\sqrt{7}$ **b.** $2\sqrt{2} - 1$ **c.** $\sqrt{7} - 1$ **d.** $1/2\sqrt{2}$

12. The negative of the statement $\sim p \wedge (p \vee q)$ is

a. $\sim p \vee q$ **b.** $p \vee \sim q$ **c.** $\sim p \wedge q$ **d.** $p \wedge \sim q$

13. If the curve $y = ax^2 + bx + c, x \in R$, passes through the point (1,2) and the tangent line to this curve at origin is $y = x$, then the possible values of a, b, c are

a. $a = \frac{1}{2}, b = \frac{1}{2}, c = 1$ **b.** $a = 1, b = 0, c = 1$
c. $a = 1, b = 1, c = 0$ **d.** $a = -1, b = 1, c = 1$

14. The area of the region $R = \{(x, y): 5x^2 \le y \le 2x^2 + 9\}$ is

a. $11\sqrt{3}$ square units **b.** $12\sqrt{3}$ square units
c. $9\sqrt{3}$ square units **d.** $6\sqrt{3}$ square units

15. If a curve $y = f(x)$ passes through the point (1, 2) and satisfies $x\frac{dy}{dx} + y = bx^4$, then for what value of b, $\int_1^2 f(x)\,dx = \frac{62}{5}$?

a. 5 **b.** 10
c. 62.5 **d.** 31/5

16. Let $f(x)$ be a differentiable function defined on [0, 2], such that $f'(x) = f'(2 - x)$, for all $x \in (0, 2)$, $f(0) = 1$ and $f(2) = e^2$. Then, the value of $\int_0^2 f(x)\,dx$ is

a. $1 - e^2$ **b.** $1 + e^2$
c. $2(1 - e^2)$ **d.** $2(1 + e^2)$

17. Let A and B be 3×3 real matrices, such that A is symmetric matrix and B is skew-symmetric matrix. Then, the system of linear equations $(A^2B^2 - B^2A^2)X = O$, where X is a 3×1 column matrix of unknown variables and O is a 3×1 null matrix, has

a. no solution **b.** exactly two solutions
c. infinitely many solutions **d.** a unique solution

18. Let a, b, c be in an arithmetic progression. Let the centroid of the triangle with vertices (a, c), $(2, b)$ and (a, b) be $\left(\frac{10}{3}, \frac{7}{3}\right)$. If α, β are the roots of the equation $ax^2 + bx + 1 = 0$, then the value of $\alpha^2 + \beta^2 - \alpha\beta$ is

a. $\frac{71}{256}$ **b.** $\frac{69}{256}$
c. $-\frac{69}{256}$ **d.** $-\frac{71}{256}$

19. For the system of linear equations $x - 2y = 1, x - y + kz = -2, ky + 4z = 6, k \in R$, consider the following statements

(A) The system has unique solution, if $k \ne 2, k \ne -2$.
(B) The system has unique solution, if $k = -2$.
(C) The system has unique solution, if $k = 2$.
(D) The system has no solution, if $k = 2$.
(E) The system has infinite number of solutions, if $k \ne -2$.

Which of the following statements are correct?

a. (C) and (D) **b.** (B) and (E)
c. (A) and (E) **d.** (A) and (D)

20. The probability that two randomly selected subsets of the set {1, 2, 3, 4, 5} have exactly two elements in their intersection, is

a. $65/2^7$ **b.** $65/2^8$ **c.** $135/2^9$ **d.** $35/2^7$

Section B : Numerical Type Questions

21. For integers n and r, let $\binom{n}{r} = \begin{cases} {}^nC_r, & \text{if } n \ge r \ge 0 \\ 0, & \text{otherwise} \end{cases}$

The maximum value of k for which the sum, $\sum_{i=0}^{k} \binom{10}{i}\binom{15}{k-i} + \sum_{i=0}^{k+1} \binom{12}{i}\binom{13}{k+1-i}$ exists, is equal to ....

22. Let λ be an integer. If the shortest distance between the lines $x - \lambda = 2y - 1 = -2z$ and $x = y + 2\lambda = z - \lambda$ is $\frac{\sqrt{7}}{2\sqrt{2}}$, then the value of $|\lambda|$ is ....

23. If $a + \alpha = 1, b + \beta = 2$ and $af(x) + \alpha f\left(\frac{1}{x}\right) = bx + \frac{\beta}{x}, x \ne 0$, then the value of expression $\dfrac{f(x) + f\left(\frac{1}{x}\right)}{x + \frac{1}{x}}$ is ....

24. Let a point P be such that its distance from the point (5, 0) is thrice the distance of P from the point (− 5, 0). If the locus of the point P is a circle of radius r, then $4r^2$ is equal to ....

25. If the area of the triangle formed by the positive X-axis, the normal and the tangent to the circle $(x - 2)^2 + (y - 3)^2 = 25$ at the point (5, 7) is A, then $24A$ is equal to ....

26. If the variance of 10 natural numbers 1, 1, 1,...., 1, k is less than 10, then the maximum possible value of k is ....

27. The sum of first four terms of a geometric progression (G.P.) is $\frac{65}{12}$ and the sum of their respective reciprocals is $\frac{65}{18}$. If the product of first three terms of the G.P. is 1 and the third term is α, then 2α is ....

28. The students $S_1, S_2, \ldots, S_{10}$ are to be divided into 3 groups A, B and C such that each group has at least one student and the group C has at most 3 students. Then, the total number of possibilities of forming such groups is ....

29. Let $i = \sqrt{-1}$. If $\frac{(-1 + i\sqrt{3})^{21}}{(1 - i)^{24}} + \frac{(1 + i\sqrt{3})^{21}}{(1 + i)^{24}} = k$ and $n = [|k|]$ be the greatest integral part of $|k|$.

Then, $\sum_{j=0}^{n+5} (j + 5)^2 - \sum_{j=0}^{n+5} (j + 5)$ is equal to ....

30. The number of the real roots of the equation $(x + 1)^2 + |x - 5| = \frac{27}{4}$ is ....

Answers

For solution scan the QR code

1. *(b)*	2. *(a)*	3. *(c)*	4. *(d)*	5. *(d)*	6. *(b)*	7. *(d)*	8. *(b)*	9. *(d)*	10. *(b)*
11. *(a)*	12. *(b)*	13. *(c)*	14. *(b)*	15. *(b)*	16. *(b)*	17. *(c)*	18. *(d)*	19. *(d)*	20. *(c)*
21. *(*)*	22. *(1)*	23. *(2)*	24. *(56.25)*	25. *(1225)*	26. *(11)*	27. *(3)*	28. *(31650)*	29. *(310)*	30. *(2)*

***Note** (*) None of the option is correct.*

25 FEBRUARY SHIFT I

Section A : Objective Type Questions

1. When a missile is fired from a ship, the probability that it is intercepted is 1/3 and the probability that the missile hits the target, given that it is not intercepted, is 3/4. If three missiles are fired independently from the ship, then the probability that all three hit the target, is

a. 1/27 **b.** 3/4 **c.** 1/8 **d.** 3/8

2. If $0 < \theta, \phi < \frac{\pi}{2}$, $x = \sum_{n=0}^{\infty} \cos^{2n}\theta$, $y = \sum_{n=0}^{\infty} \sin^{2n}\phi$ and $z = \sum_{n=0}^{\infty} \cos^{2n}\theta \cdot \sin^{2n}\phi$, then

a. $xy - z = (x + y)z$ **b.** $xy + yz + zx = z$
c. $xyz = 4$ **d.** $xy + z = (x + y)z$

3. Let $f, g : N \to N$, such that $f(n + 1) = f(n) + f(1)\ \forall\ n \in N$ and g be any arbitrary function. Which of the following statements is not true?

a. if *fog* is one-one, then g is one-one
b. if f is onto, then $f(n) = n\ \forall\ n \in N$
c. f is one-one
d. if g is onto, then *fog* is one-one

4. The equation of the line through the point (0,1,2) and perpendicular to the line $\frac{x-1}{2} = \frac{y+1}{3} = \frac{z-1}{-2}$ is

a. $\frac{x}{3} = \frac{y-1}{4} = \frac{z-2}{3}$ **b.** $\frac{x}{3} = \frac{y-1}{-4} = \frac{z-2}{3}$
c. $\frac{x}{3} = \frac{y-1}{4} = \frac{z-2}{-3}$ **d.** $\frac{x}{-3} = \frac{y-1}{4} = \frac{z-2}{3}$

5. Let α be the angle between the lines whose direction cosines satisfy the equations $l + m - n = 0$ and $l^2 + m^2 - n^2 = 0$. Then, the value of $\sin^4\alpha + \cos^4\alpha$ is

a. 3/4 **b.** 3/8 **c.** 5/8 **d.** 1/2

6. The value of the integral

$$\int \left[\frac{\sin\theta \cdot \sin 2\theta\,(\sin^6\theta + \sin^4\theta + \sin^2\theta)\sqrt{2\sin^4\theta + 3\sin^2\theta + 6}}{1 - \cos 2\theta} \right] d\theta$$

is (where, c is a constant of integration)

a. $\frac{1}{18}[11 - 18\sin^2\theta + 9\sin^4\theta - 2\sin^6\theta]^{\frac{3}{2}} + c$

b. $\frac{1}{18}[9 - 2\cos^6\theta\ 3\cos^4\theta - 6\cos^2\theta]^{\frac{3}{2}} + c$

c. $\frac{1}{18}[9 - 2\sin^6\theta - 3\sin^4\theta - 6\sin^2\theta]^{\frac{3}{2}} + c$

d. $\frac{1}{18}[11 - 18\cos^2\theta + 9\cos^4\theta - 2\cos^6\theta]^{\frac{3}{2}} + c$

7. The value of $\int_{-1}^{1} x^2 e^{[x^3]}\,dx$, where $[t]$ denotes the greatest integer $\le t$, is

a. $\frac{e-1}{3e}$ **b.** $\frac{e+1}{3}$ **c.** $\frac{e+1}{3e}$ **d.** $\frac{1}{3e}$

8. A man is observing, from the top of a tower, a boat speeding towards the tower from a certain point A, with uniform speed. At that point, angle of depression of the boat with the man's eye is 30° (ignore man's height). After sailing for 20 s, towards the base of the tower (which is at the level of water), the boat has reached a point B, where the angle of depression is 45°. Then, the time taken (in seconds) by the boat from B to reach the base of the tower is

a. 10 **b.** $10\sqrt{3}$ **c.** $10(\sqrt{3} + 1)$ **d.** $10(\sqrt{3} - 1)$

9. A tangent is drawn to the parabola $y^2 = 6x$, which is perpendicular to the line $2x + y = 1$. Which of the following points does not lie on it?

a. (–6, 0) **b.** (4, 5) **c.** (5, 4) **d.** (0, 3)

10. All possible values of $\theta \in [0, 2\pi]$ for which $\sin 2\theta + \tan 2\theta > 0$ lie in

a. $\left(0, \frac{\pi}{2}\right) \cup \left(\pi, \frac{3\pi}{2}\right)$

b. $\left(0, \frac{\pi}{2}\right) \cup \left(\frac{\pi}{2}, \frac{3\pi}{4}\right) \cup \left(\pi, \frac{7\pi}{6}\right)$

c. $\left(0, \frac{\pi}{4}\right) \cup \left(\frac{\pi}{2}, \frac{3\pi}{4}\right) \cup \left(\frac{3\pi}{2}, \frac{11\pi}{6}\right)$

d. $\left(0, \frac{\pi}{4}\right) \cup \left(\frac{\pi}{2}, \frac{3\pi}{4}\right) \cup \left(\pi, \frac{5\pi}{4}\right) \cup \left(\frac{3\pi}{2}, \frac{7\pi}{4}\right)$

11. Let the lines $(2 - i)z = (2 + i)\bar{z}$ and $(2 + i)z + (i - 2)\bar{z} - 4i = 0$, (here $i^2 = -1$) be normal to a circle C. If the line $iz + \bar{z} + 1 + i = 0$ is tangent to this circle C, then its radius is

a. $\frac{3}{\sqrt{2}}$ **b.** $\frac{1}{2\sqrt{2}}$ **c.** $3\sqrt{2}$ **d.** $\frac{3}{2\sqrt{2}}$

12. The image of the point (3, 5) in the line $x - y + 1 = 0$, lies on

a. $(x-2)^2 + (y-2)^2 = 12$
b. $(x-4)^2 + (y+2)^2 = 16$
c. $(x-4)^2 + (y-4)^2 = 8$
d. $(x-2)^2 + (y-4)^2 = 4$

13. If the curves, $\frac{x^2}{a} + \frac{y^2}{b} = 1$ and $\frac{x^2}{c} + \frac{y^2}{d} = 1$, intersect each other at an angle of 90°, then which of the following relations is true?

a. $a + b = c + d$ b. $a - b = c - d$
c. $a - c = b + d$ d. $ab = \frac{c+d}{a+b}$

14. $\lim_{n\to\infty} \left(1 + \frac{1 + \frac{1}{2} + \ldots\ldots + \frac{1}{n}}{n^2}\right)^n$ is equal to

a. $\frac{1}{2}$ b. 0 c. $\frac{1}{e}$ d. 1

15. The coefficients a, b and c of the quadratic equation, $ax^2 + bx + c = 0$ are obtained by throwing a dice three times. The probability that this equation has equal roots is

a. $\frac{1}{72}$ b. $\frac{5}{216}$ c. $\frac{1}{36}$ d. $\frac{1}{54}$

16. The total number of positive integral solutions (x, y, z), such that $xyz = 24$ is

a. 36 b. 24
c. 45 d. 30

17. The integer 'k', for which the inequality $x^2 - 2(3k-1)x + 8k^2 - 7 > 0$ is valid for every x in R, is

a. 3 b. 2 c. 0 d. 4

18. If a curve passes through the origin and the slope of the tangent to it at any point (x, y) is $\frac{x^2 - 4x + y + 8}{x - 2}$, then this curve also passes through the point

a. (5, 4) b. (4, 5) c. (4, 4) d. (5, 5)

19. The statement $A \to (B \to A)$ is equivalent to

a. $A \to (A \wedge B)$ b. $A \to (A \to B)$
c. $A \to (A \leftrightarrow B)$ d. $A \to (A \vee B)$

20. If Rolle's theorem holds for the function $f(x) = x^3 - ax^2 + bx + 4$, $x \in [1, 2]$ with $f'\left(\frac{4}{3}\right) = 0$, then ordered pair (a, b) is equal to

a. (5, 8) b. (–5, 8)
c. (5, –8) d. (–5, –8)

Section B : Numerical Type Questions

21. Let $f(x)$ be a polynomial of degree 6 in x, in which the coefficient of x^6 is unity and it has extrema at $x = -1$ and $x = 1$. If $\lim_{x\to 0} \frac{f(x)}{x^3} = 1$, then $5f(2)$ is equal to …………… .

22. The number of points at which the function $f(x) = |2x+1| - 3|x+2| + |x^2 + x - 2|$, $x \in R$ is not differentiable, is ………. .

23. The graph of sine and cosine functions, intersect each other at a number of points and between two consecutive points of intersection, the two graphs enclose the same area A. Then A^4 is equal to ………… .

24. Let $A_1, A_2, A_3, \ldots\ldots$ be squares, such that for each $n \geq 1$, the length of the side of A_n equals the length of diagonal of A_{n+1}. If the length of A_1 is 12 cm, then the smallest value of n for which area of A_n is less than one, is ……… .

25. Let $A = \begin{bmatrix} x & y & z \\ y & z & x \\ z & x & y \end{bmatrix}$, where x, y and z are real numbers, such that $x + y + z > 0$ and $xyz = 2$. If $A^2 = I_3$, then the value of $x^3 + y^3 + z^3$ is ……… .

26. If $A = \begin{bmatrix} 0 & -\tan\left(\frac{\theta}{2}\right) \\ \tan\left(\frac{\theta}{2}\right) & 0 \end{bmatrix}$ and $(I_2 + A)(I_2 - A)^{-1} = \begin{bmatrix} a & -b \\ b & a \end{bmatrix}$, then $13(a^2 + b^2)$ is equal to ………… .

27. The total number of numbers, lying between 100 and 1000 that can be formed with the digits 1, 2, 3, 4, 5, if the repetition of digits is not allowed and numbers are divisible by either 3 or 5, is ………… .

28. Let $\mathbf{a} = \hat{\mathbf{i}} + 2\hat{\mathbf{j}} - \hat{\mathbf{k}}$, $\mathbf{b} = \hat{\mathbf{i}} - \hat{\mathbf{j}}$ and $\mathbf{c} = \hat{\mathbf{i}} - \hat{\mathbf{j}} - \hat{\mathbf{k}}$ be three given vectors. If $\mathbf{r}$ is a vector such that $\mathbf{r} \times \mathbf{a} = \mathbf{c} \times \mathbf{a}$ and $\mathbf{r} \cdot \mathbf{b} = 0$, then $\mathbf{r} \cdot \mathbf{a}$ is equal to ……… .

29. If the system of equations

$kx + y + 2z = 1$ $3x - y - 2z = 2$

$-2x - 2y - 4z = 3$

has infinitely many solutions, then k is equal to ……… .

30. The locus of the point of intersection of the lines $(\sqrt{3})kx + ky - 4\sqrt{3} = 0$ and $\sqrt{3}x - y - 4(\sqrt{3})k = 0$ is a conic, whose eccentricity is

Answers

For solution scan the QR code

1. *(c)*	2. *(d)*	3. *(d)*	4. *(d)*	5. *(c)*	6. *(d)*	7. *(c)*	8. *(c)*	9. *(c)*	10. *(d)*
11. *(d)*	12. *(d)*	13. *(b)*	14. *(d)*	15. *(b)*	16. *(d)*	17. *(a)*	18. *(d)*	19. *(d)*	20. *(a)*
21. *(144)*	22. *(2)*	23. *(64)*	24. *(9)*	25. *(7)*	26. *(13)*	27. *(32)*	28. *(12)*	29. *(21)*	30. *(2)*

25 FEBRUARY SHIFT II

Section A : Objective Type Questions

1. If for the matrix, $A = \begin{bmatrix} 1 & -\alpha \\ \alpha & \beta \end{bmatrix}$, $AA^T = I_2$, then the value of $\alpha^4 + \beta^4$ is

a. 4 **b.** 1 **c.** 2 **d.** 3

2. Let A be a 3×3 matrix with det $(A) = 4$. Let R_i denote the ith row of A. If a matrix B is obtained by performing the operation $R_2 \to 2R_2 + 5R_3$ on $2A$, then det (B) is equal to

a. 16 **b.** 80 **c.** 64 **d.** 128

3. The following system of linear equations

$2x + 3y + 2z = 9$ $3x + 2y + 2z = 9$

$x - y + 4z = 8$

a. does not have any solution

b. has a unique solution

c. has infinitely many solutions

d. has a solution (α, β, γ) satisfying $\alpha + \beta^2 + \gamma^3 = 12$

4. If $I_n = \int_{\pi/4}^{\pi/2} \cot^n x \, dx$, then

a. $\frac{1}{I_2 + I_4}, \frac{1}{I_3 + I_5}, \frac{1}{I_4 + I_6}$ are in AP

b. $I_2 + I_4, I_3 + I_5, I_4 + I_6$ are in AP

c. $\frac{1}{I_2 + I_4}, \frac{1}{I_3 + I_5}, \frac{1}{I_4 + I_6}$ are in GP

d. $I_2 + I_4, (I_3 + I_5)^2, I_4 + I_6$ are in GP

5. A function $f(x)$ is given by $f(x) = \frac{5^x}{5^x + 5}$, then the sum of the series

$f\left(\frac{1}{20}\right) + f\left(\frac{2}{20}\right) + f\left(\frac{3}{20}\right) + \ldots + f\left(\frac{39}{20}\right)$ is equal to

a. $\frac{29}{2}$ **b.** $\frac{49}{2}$

c. $\frac{39}{2}$ **d.** $\frac{19}{2}$

6. Let α and β be the roots of $x^2 - 6x - 2 = 0$. If $a_n = \alpha^n - \beta^n$ for $n \geq 1$, then the value of $\frac{a_{10} - 2a_8}{3a_9}$ is

a. 4 **b.** 3 **c.** 2 **d.** 1

7. The minimum value of $f(x) = a^{a^x} + a^{1-a^x}$, where $a, x \in R$ and $a > 0$, is equal to

a. $a + 1$ **b.** $a + \frac{1}{a}$ **c.** $2\sqrt{a}$ **d.** $2a$

8. The integral $\int \frac{e^{3\log_e 2x} + 5e^{2\log_e 2x}}{e^{4\log_e x} + 5e^{3\log_e x} - 7e^{2\log_e x}} dx, x > 0$, is equal to

(where, c is a constant of integration)

a. $\log_e |x^2 + 5x - 7| + c$ **b.** $4\log_e |x^2 + 5x - 7| + c$

c. $\frac{1}{4}\log_e |x^2 + 5x - 7| + c$ **d.** $\log_e \sqrt{x^2 + 5x - 7} + c$

9. If $\alpha, \beta \in R$ are such that $1 - 2i$ (here $i^2 = -1$) is a root of $z^2 + \alpha z + \beta = 0$, then $(\alpha - \beta)$ is equal to

a. 3 **b.** -3 **c.** 7 **d.** -7

10. If the curve $x^2 + 2y^2 = 2$ intersects the line $x + y = 1$ at two points P and Q, then the angle subtended by the line segment PQ at the origin is

a. $\frac{\pi}{2} + \tan^{-1}\left(\frac{1}{4}\right)$ **b.** $\frac{\pi}{2} - \tan^{-1}\left(\frac{1}{4}\right)$

c. $\frac{\pi}{2} + \tan^{-1}\left(\frac{1}{3}\right)$ **d.** $\frac{\pi}{2} - \tan^{-1}\left(\frac{1}{3}\right)$

11. The shortest distance between the line $x - y = 1$ and the curve $x^2 = 2y$ is

a. $\frac{1}{\sqrt{2}}$ **b.** $\frac{1}{2\sqrt{2}}$ **c.** 0 **d.** $\frac{1}{2}$

12. A hyperbola passes through the foci of the ellipse $\frac{x^2}{25} + \frac{y^2}{16} = 1$ and its transverse and conjugate axes coincide with major and minor axes of the ellipse, respectively. If the product of their eccentricities is one, then the equation of the hyperbola is

a. $\frac{x^2}{9} - \frac{y^2}{16} = 1$ **b.** $\frac{x^2}{9} - \frac{y^2}{4} = 1$

c. $\frac{x^2}{9} - \frac{y^2}{25} = 1$ **d.** $x^2 - y^2 = 9$

13. A plane passes through the points $A(1, 2, 3)$, $B(2, 3, 1)$ and $C(2, 4, 2)$. If O is the origin and P is $(2, -1, 1)$, then the projection of **OP** on this plane is of length

a. $\sqrt{\frac{2}{3}}$ **b.** $\sqrt{\frac{2}{11}}$ **c.** $\sqrt{\frac{2}{7}}$ **d.** $\sqrt{\frac{2}{5}}$

14. $\lim_{n \to \infty} \left[\frac{1}{n} + \frac{n}{(n+1)^2} + \frac{n}{(n+2)^2} + \ldots + \frac{n}{(2n-1)^2}\right]$ is equal to

a. 1 **b.** $\frac{1}{2}$ **c.** $\frac{1}{3}$ **d.** $\frac{1}{4}$

15. In a group of 400 people, 160 are smokers and non-vegetarian; 100 are smokers and vegetarian and the remaining 140 are non-smokers and vegetarian. Their chances of getting a particular chest disorder are 35%, 20% and 10%, respectively. A person is chosen from the group at random and is found to be suffering from the chest disorder. The probability that the selected person is a smoker and non-vegetarian is

a. $\frac{7}{45}$ **b.** $\frac{8}{45}$ **c.** $\frac{28}{45}$ **d.** $\frac{14}{45}$

16. Let A be a set of all 4-digit natural numbers whose exactly one digit is 7. Then, the probability that a randomly chosen element of A leaves remainder 2 when divided by 5 is

a. $\frac{1}{5}$ **b.** $\frac{2}{9}$ **c.** $\frac{97}{297}$ **d.** $\frac{122}{297}$

17. If $0 < x, y < \pi$ and $\cos x + \cos y - \cos(x + y) = \frac{3}{2}$, then $\sin x + \cos y$ is equal to

a. $\frac{1}{2}$ b. $\frac{\sqrt{3}}{2}$

c. $\frac{1-\sqrt{3}}{2}$ d. $\frac{1+\sqrt{3}}{2}$

18. Let x denote the total number of one-one functions from a set A with 3 elements to a set B with 5 elements and y denote the total number of one-one functions from the set A to the set $A \times B$. Then,

a. $2y = 91x$ b. $2y = 273x$

c. $y = 91x$ d. $y = 273x$

19. $\operatorname{cosec}\left[2\cot^{-1}(5) + \cos^{-1}\left(\frac{4}{5}\right)\right]$ is equal to

a. $\frac{56}{33}$ b. $\frac{65}{33}$ c. $\frac{65}{56}$ d. $\frac{75}{56}$

20. The contrapositive of the statement; "If you will work, you will earn money" is

a. to earn money, you need to work

b. you will earn money, if you will not work

c. if you will not earn money, you will not work

d. if you will earn money, you will work

Section B : Numerical Type Questions

21. A function f is defined on $[-3, 3]$ as

$$f(x) = \begin{cases} \min\{|x|, 2 - x^2\}, & -2 \le x \le 2 \\ [|x|], & 2 < |x| \le 3 \end{cases}$$

where, $[x]$ denotes the greatest integer $\le x$. The number of points, where f is not differentiable in $(-3, 3)$ is ....

22. If the curve $y = y(x)$ represented by the solution of the differential equation $(2xy^2 - y)\,dx + x\,dy = 0$, passes through the intersection of the lines $2x - 3y = 1$ and $3x + 2y = 8$, then $|y(1)|$ is equal to ....

23. The total number of two digit numbers 'n', such that $3^n + 7^n$ is a multiple of 10, is ....

24. If $\lim_{x \to 0} \frac{ax - (e^{4x} - 1)}{ax(e^{4x} - 1)}$ exists and is equal to b, then the value of $a - 2b$ is ....

25. If the curves $x = y^4$ and $xy = k$ cut at right angles, then $(4k)^6$ is equal to ....

26. The value of $\int_{-2}^{2} |3x^2 - 3x - 6|\, dx$ is ....

27. If the remainder when x is divided by 4 is 3, then the remainder when $(2020 + x)^{2022}$ is divided by 8 is ....

28. A line 'l' passing through origin is perpendicular to the lines

$l_1 : \mathbf{r} = (3 + t)\hat{\mathbf{i}} + (-1 + 2t)\hat{\mathbf{j}} + (4 + 2t)\hat{\mathbf{k}}$

$l_2 : \mathbf{r}\,(3 + 2s)\hat{\mathbf{i}} + (3 + 2s)\hat{\mathbf{j}} + (2 + s)\hat{\mathbf{k}}$

If the coordinates of the point in the first octant on 'l_2' at a distance of $\sqrt{17}$ from the point of intersection of 'l' and 'l_1' are (a, b, c), then $18(a + b + c)$ is equal to ....

29. A line is a common tangent to the circle $(x - 3)^2 + y^2 = 9$ and the parabola $y^2 = 4x$. If the two points of contact (a, b) and (c, d) are distinct and lie in the first quadrant, then $2(a + c)$ is equal to ....

30. Let $\mathbf{a} = \hat{\mathbf{i}} + \alpha\hat{\mathbf{j}} + 3\hat{\mathbf{k}}$ and $\mathbf{b} = 3\hat{\mathbf{i}} - \alpha\hat{\mathbf{j}} + \hat{\mathbf{k}}$. If the area of the parallelogram whose adjacent sides are represented by the vectors $\mathbf{a}$ and $\mathbf{b}$ is $8\sqrt{3}$ square units, then $\mathbf{a} \cdot \mathbf{b}$ is equal to ....

Answers

For solution scan the QR code

1. (b)	2. (c)	3. (b)	4. (a)	5. (c)	6. (c)	7. (c)	8. (b)	9. (d)	10. (a)
11. (b)	12. (a)	13. (b)	14. (b)	15. (c)	16. (c)	17. (d)	18. (a)	19. (c)	20. (c)
21. (5)	22. (1)	23. (45)	24. (5)	25. (4)	26. (19)	27. (1)	28. (44)	29. (9)	30. (2)

26 FEBRUARY SHIFT I

Section A : Objective Type Questions

1. If $\mathbf{a}$ and $\mathbf{b}$ are perpendicular, then $\mathbf{a} \times (\mathbf{a} \times (\mathbf{a} \times (\mathbf{a} \times \mathbf{b})))$ is equal to

a. 0 c. $\frac{1}{2}|\mathbf{a}|^4\,\mathbf{b}$ c. $\mathbf{a} \times \mathbf{b}$ d. $|\mathbf{a}|^4\,\mathbf{b}$

2. A fair coin is tossed a fixed number of times. If the probability of getting 7 heads is equal to probability of getting 9 heads, then the probability of getting 2 heads is

a. $\frac{15}{2^{13}}$ b. $\frac{15}{2^{12}}$

c. $\frac{15}{2^8}$ d. $\frac{15}{2^{14}}$

3. Let A be a symmetric matrix of order 2 with integer entries. If the sum of the diagonal elements of A^2 is 1, then the possible number of such matrices is

a. 4 b. 1 c. 6 d. 12

4. In an increasing geometric series, the sum of the second and the sixth term is $\frac{25}{2}$ and the product of the third and fifth term is 25. Then, the sum of 4th, 6th and 8th terms is equal to

a. 30 **b.** 26
c. 35 **d.** 32

5. The value of $\sum_{n=1}^{100} \int_{n-1}^{n} e^{x-[x]} dx$, where $[x]$ is the greatest integer $\le x$, is

a. $100(e-1)$ **b.** $100(1-e)$
c. $100e$ **d.** $100(1+e)$

6. In the circle given below, let $OA = 1$ unit, $OB = 13$ unit and $PQ \perp OB$. Then, the area of the triangle PQB (in square units) is

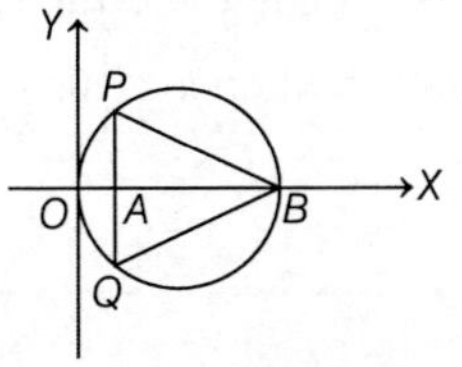

a. $24\sqrt{2}$ **b.** $24\sqrt{3}$
c. $26\sqrt{3}$ **d.** $26\sqrt{2}$

7. The sum of the infinite series $1+\frac{2}{3}+\frac{7}{3^2}+\frac{12}{3^3}+\frac{17}{3^4}+\frac{22}{3^5}+\ldots$ is equal to

a. $\frac{13}{4}$ **b.** $\frac{9}{4}$ **c.** $\frac{15}{4}$ **d.** $\frac{11}{4}$

8. The value of $\lim_{h\to 0} 2\left\{\frac{\sqrt{3}\sin\left(\frac{\pi}{6}+h\right)-\cos\left(\frac{\pi}{6}+h\right)}{\sqrt{3}h(\sqrt{3}\cos h-\sin h)}\right\}$ is

a. $\frac{4}{3}$ **b.** $\frac{2}{\sqrt{3}}$ **c.** $\frac{3}{4}$ **d.** $\frac{2}{3}$

9. The maximum value of the term independent of 't' in the expansion of $\left(tx^{1/5}+\frac{(1-x)^{1/10}}{t}\right)^{10}$, where $x\in(0,1)$ is

a. $\frac{10!}{\sqrt{3}(5!)^2}$ **b.** $\frac{2\cdot 10!}{3\sqrt{3}(5!)^2}$
c. $\frac{2\cdot 10!}{3(5!)^2}$ **d.** $\frac{10!}{3(5!)^2}$

10. The rate of growth of bacteria in a culture is proportional to the number of bacteria present and the bacteria count is 1000 at initial time $t=0$. The number of bacteria is increased by 20% in 2 h. If the population of bacteria is 2000 after $\frac{k}{\log_e(6/5)}$ h, then $\left(\frac{k}{\log_e 2}\right)^2$ is equal to

a. 4 **b.** 8 **c.** 2 **d.** 16

11. If $(1, 5, 35)$, $(7, 5, 5)$, $(1, \lambda, 7)$ and $(2\lambda, 1, 2)$ are coplanar, then the sum of all possible values of λ is

a. $\frac{39}{5}$ **b.** $-\frac{39}{5}$ **c.** $\frac{44}{5}$ **d.** $-\frac{44}{5}$

12. If $\frac{\sin^{-1}(x)}{a}=\frac{\cos^{-1}x}{b}=\frac{\tan^{-1}y}{c}$, $0<x<1$, then the value of $\cos\left(\frac{\pi c}{a+b}\right)$ is

a. $\frac{1-y^2}{y\sqrt{y}}$ **b.** $1-y^2$ **c.** $\frac{1-y^2}{1+y^2}$ **d.** $\frac{1-y^2}{2y}$

13. The number of seven digit integers with sum of the digits equal to 10 and formed by using the digits 1, 2 and 3 only is

a. 42 **b.** 82 **c.** 77 **d.** 35

14. Let f be any function defined on R and let it satisfy the condition
$|f(x)-f(y)|\le|(x-y)^2|, \forall (x,y)\in R$
If $f(0)=1$, then

a. $f(x)$ can take any value in R
b. $f(x)<0, \forall x\in R$
c. $f(x)=0, \forall x\in R$
d. $f(x)>0, \forall x\in R$

15. The maximum slope of the curve $y=\frac{1}{2}x^4-5x^3+18x^2-19x$ occurs at the point

a. (2, 2) **b.** (0, 0)
c. (2, 9) **d.** $\left(3, \frac{21}{2}\right)$

16. The intersection of three lines $x-y=0$, $x+2y=3$ and $2x+y=6$ is a

a. right angled triangle **b.** equilateral triangle
c. isosceles triangle **d.** None of these

17. Consider the three planes
$P_1: 3x+15y+21z=9$,
$P_2: x-3y-z=5$ and
$P_3: 2x+10y+14z=5$

Then, which one of the following is true ?

a. P_1 and P_2 are parallel
b. P_1 and P_3 are parallel
c. P_2 and P_3 are parallel
d. P_1, P_2 and P_3 all are parallel

18. The value of $\begin{vmatrix}(a+1)(a+2) & a+2 & 1\\ (a+2)(a+3) & a+3 & 1\\ (a+3)(a+4) & a+4 & 1\end{vmatrix}$ is

a. $(a+2)(a+3)(a+4)$ **b.** -2
c. $(a+1)(a+2)(a+3)$ **d.** 0

19. The value of $\int_{-\pi/2}^{\pi/2}\frac{\cos^2 x}{1+3^x}dx$ is

a. $\frac{\pi}{4}$ **b.** 4π **c.** $\frac{\pi}{2}$ **d.** 2π

20. Let $R=\{(P,Q)\mid P$ and Q are at the same distance from the origin$\}$ be a relation, then the equivalence class of $(1,-1)$ is the set

a. $S=\{(x,y)\mid x^2+y^2=4\}$ **b.** $S=\{(x,y)\mid x^2+y^2=1\}$
c. $S=\{(x,y)\mid x^2+y^2=\sqrt{2}\}$ **d.** $S=\{(x,y)\mid x^2+y^2=2\}$

Section B : Numerical Type Questions

21. The difference between degree and order of a differential equation that represents the family of curves given by $y^2 = a\left(x + \frac{\sqrt{a}}{2}\right), a > 0$ is ....

22. The number of integral values of k for which the equation $3 \sin x + 4 \cos x = k + 1$ has a solution, $k \in R$ is ....

23. The number of solutions of the equation $\log_4 (x-1) = \log_2 (x-3)$ is ....

24. The sum of 162th power of the roots of the equation $x^3 - 2x^2 + 2x - 1 = 0$ is

25. Let $m, n \in N$ and $gcd(2, n) = 1$. If
$30\binom{30}{0} + 29\binom{30}{1} + \dots + 2\binom{30}{28} + 1\binom{30}{29} = n \cdot 2^m$, then $n + m$ is .... (Here, $\binom{n}{k} = {}^nC_k$)

26. If $y = y(x)$ is the solution of the equation
$e^{\sin y} \cos y \frac{dy}{dx} + e^{\sin y} \cos x = \cos x$, $y(0) = 0$, then
$1 + y\left(\frac{\pi}{6}\right) + \frac{\sqrt{3}}{2} y\left(\frac{\pi}{3}\right) + \frac{1}{\sqrt{2}} y\left(\frac{\pi}{4}\right)$ is ....

27. Let $(\lambda, 2, 1)$ be a point on the plane which passes through the point $(4, -2, 2)$. If the plane is perpendicular to the line joining the points $(-2, -21, 29)$ and $(-1, -16, 23)$, then $\left(\frac{\lambda}{11}\right)^2 - \frac{4\lambda}{11} - 4$ is ....

28. The area bounded by the lines $y = ||x-1| - 2|$ is ....

29. The value of the integral $\int_0^{\pi} |\sin 2x| \, dx$ is ....

30. If $\sqrt{3} (\cos^2 x) = (\sqrt{3} - 1) \cos x + 1$, the number of solutions of the given equation when $x \in \left[0, \frac{\pi}{2}\right]$ is ....

Answers

For solution scan the QR code

1. (d)	2. (a)	3. (a)	4. (c)	5. (a)	6. (b)	7. (a)	8. (a)	9. (b)	10. (a)
11. (c)	12. (c)	13. (c)	14. (d)	15. (a)	16. (c)	17. (b)	18. (b)	19. (a)	20. (d)
21. (2)	22. (11)	23. (1)	24. (3)	25. (45)	26. (1)	27. (8)	28. (4)	29. (2)	30. (1)

26 FEBRUARY SHIFT II

Section A : Objective Type Questions

1. If vectors $\mathbf{a}_1 = x\hat{\mathbf{i}} - \hat{\mathbf{j}} + \hat{\mathbf{k}}$ and $\mathbf{a}_2 = \hat{\mathbf{i}} + y\hat{\mathbf{j}} + z\hat{\mathbf{k}}$ are collinear, then a possible unit vector parallel to the vector $x\hat{\mathbf{i}} + y\hat{\mathbf{j}} + z\hat{\mathbf{k}}$ is

a. $\frac{1}{\sqrt{2}}(-\hat{\mathbf{j}} + \hat{\mathbf{k}})$ **b.** $\frac{1}{\sqrt{2}}(\hat{\mathbf{i}} - \hat{\mathbf{j}})$

c. $\frac{1}{\sqrt{3}}(\hat{\mathbf{i}} + \hat{\mathbf{j}} - \hat{\mathbf{k}})$ **d.** $\frac{1}{\sqrt{3}}(\hat{\mathbf{i}} - \hat{\mathbf{j}} + \hat{\mathbf{k}})$

2. Let $A = \{1, 2, 3, \dots, 10\}$ and $f : A \to A$ be defined as

$$f(k) = \begin{cases} k + 1, & \text{if } k \text{ is odd} \\ k, & \text{if } k \text{ is even} \end{cases}$$

Then, the number of possible functions $g : A \to A$, such that $gof = f$ is

a. 10^5 **b.** ${}^{10}C_5$ **c.** 5^5 **d.** $5!$

3. Let $f : R \to R$ be defined as

$$f(x) = \begin{cases} 2\sin\left(-\frac{\pi x}{2}\right), & \text{if } x < -1 \\ |ax^2 + x + b|, & \text{if } -1 \le x \le 1 \\ \sin(\pi x), & \text{if } x > 1 \end{cases}$$

If $f(x)$ is continuous on R, then $a + b$ equals

a. -3 **b.** -1

c. 3 **d.** 1

4. For $x > 0$, if $f(x) = \int_1^x \frac{\log_e t}{(1+t)} dt$, then $f(e) + f\left(\frac{1}{e}\right)$ is equal to

a. 1 **b.** -1 **c.** $\frac{1}{2}$ **d.** 0

5. A natural number has prime factorisation given by $n = 2^x 3^y 5^z$, where y and z are such that $y + z = 5$ and $y^{-1} + z^{-1} = \frac{5}{6}, y > z$. Then, the number of odd divisors of n, including 1, is

a. 11 **b.** 6

c. $6x$ **d.** 12

6. Let $f(x) = \sin^{-1} x$ and $g(x) = \frac{x^2 - x - 2}{2x^2 - x - 6}$.

If $g(2) = \lim_{x \to 2} g(x)$, then the domain of the function fog is

a. $(-\infty, -2] \cup \left[-\frac{3}{2}, \infty\right)$ **b.** $(-\infty, -2] \cup [-1, \infty)$

c. $(-\infty, -2] \cup \left[-\frac{4}{3}, \infty\right)$ **d.** $(-\infty, -1] \cup [2, \infty)$

7. The triangle of maximum area that can be inscribed in a given circle of radius r is

a. an isosceles triangle with base equal to $2r$
b. an equilateral triangle of height $\frac{2r}{3}$
c. an equilateral triangle having each of its side of length $\sqrt{3}r$
d. a right angle triangle having two of its sides of length $2r$ and r

8. Let L be a line obtained from the intersection of two planes $x + 2y + z = 6$ and $y + 2z = 4$. If point $P(\alpha, \beta, \gamma)$ is the foot of perpendicular from (3, 2, 1) on L, then the value of $21(\alpha + \beta + \gamma)$ equals

a. 142 b. 68
c. 136 d. 102

9. Let $F_1(A, B, C) = (A \wedge \sim B) \vee [\sim C \wedge (A \vee B)] \vee \sim A$ and $F_2(A, B) = (A \vee B) \vee (B \to \sim A)$ be two logical expressions. Then,

a. F_1 and F_2 both are tautologies
b. F_1 is a tautology but F_2 is not a tautology
c. F_1 is not tautology but F_2 is a tautology
d. Both F_1 and F_2 are not tautologies

10. Let slope of the tangent line to a curve at any point $P(x, y)$ be given by $\frac{xy^2 + y}{x}$. If the curve intersects the line $x + 2y = 4$ at $x = -2$, then the value of y, for which the point $(3, y)$ lies on the curve, is

a. $\frac{18}{35}$ b. $-\frac{4}{3}$
c. $-\frac{18}{19}$ d. $-\frac{18}{11}$

11. If the locus of the mid-point of the line segment from the point (3, 2) to a point on the circle, $x^2 + y^2 = 1$ is a circle of radius r, then r is equal to

a. 1 b. 1/2
c. 1/3 d. 1/4

12. Consider the following system of equations

$x + 2y - 3z = a$
$2x + 6y - 11z = b$
$x - 2y + 7z = c$

where, a, b and c are real constants. Then, the system of equations

a. has a unique solution, when $5a = 2b + c$
b. has infinite number of solutions when $5a = 2b + c$
c. has no solution for all a, b and c
d. has a unique solution for all a, b and c

13. If $0 < a, b < 1$ and $\tan^{-1} a + \tan^{-1} b = \frac{\pi}{4}$, then the value of

$$(a + b) - \left(\frac{a^2 + b^2}{2}\right) + \left(\frac{a^3 + b^3}{3}\right) - \left(\frac{a^4 + b^4}{4}\right) + \ldots \text{is}$$

a. $\log_e 2$ b. $e^2 - 1$
c. e d. $\log_e \left(\frac{e}{2}\right)$

14. The sum of the series $\sum_{n=1}^{\infty} \frac{n^2 + 6n + 10}{(2n + 1)!}$ is equal to

a. $\frac{41}{8}e + \frac{19}{8}e^{-1} - 10$ b. $\frac{41}{8}e - \frac{19}{8}e^{-1} - 10$
c. $\frac{41}{8}e + \frac{19}{8}e^{-1} + 10$ d. $-\frac{41}{8}e + \frac{19}{8}e^{-1} - 10$

15. Let $f(x)$ be a differentiable function at $x = a$ with $f'(a) = 2$ and $f(a) = 4$.
Then, $\lim_{x \to a} \frac{xf(a) - af(x)}{x - a}$ equals

a. $2a + 4$ b. $4 - 2a$ c. $2a - 4$ d. $a + 4$

16. Let $A(1, 4)$ and $B(1, -5)$ be two points. Let P be a point on the circle $(x - 1)^2 + (y - 1)^2 = 1$, such that $(PA)^2 + (PB)^2$ have maximum value, then the points P, A and B lie on

a. a straight line b. a hyperbola
c. an ellipse d. a parabola

17. If the mirror image of the point (1, 3, 5) with respect to the plane $4x - 5y + 2z = 8$ is (α, β, γ), then $5(\alpha + \beta + \gamma)$ equals

a. 47 b. 43 c. 39 d. 41

18. Let $f(x) = \int_0^x e^t f(t)\, dt + e^x$ be a differentiable function for all $x \in R$. Then, $f(x)$ equals

a. $2e^{(e^x - 1)} - 1$ b. $e^{e^x} - 1$ c. $2e^{e^x} - 1$ d. $e^{(e^x - 1)}$

19. Let A_1 be the area of the region bounded by the curves $y = \sin x$, $y = \cos x$ and y-axis in the first quadrant. Also, let A_2 be the area of the region bounded by the curves $y = \sin x$, $y = \cos x$, x-axis and $x = \frac{\pi}{2}$ in the first quadrant.
Then,

a. $A_1 : A_2 = 1 : \sqrt{2}$ and $A_1 + A_2 = 1$
b. $A_1 = A_2$ and $A_1 + A_2 = \sqrt{2}$
c. $2A_1 = A_2$ and $A_1 + A_2 = 1 + \sqrt{2}$
d. $A_1 : A_2 = 1 : 2$ and $A_1 + A_2 = 1$

20. A seven digit number is formed using digits 3, 3, 4, 4, 4, 5, 5. The probability, that number so formed is divisible by 2, is

a. 6/7 b. 1/7 c. 3/7 d. 4/7

Section B : Numerical Type Questions

21. Let z be those complex numbers which satisfy $|z + 5| \leq 4$ and $z(1 + i) + \bar{z}(1 - i) \geq -10$, $i = \sqrt{-1}$. If the maximum value of $|z + 1|^2$ is $\alpha + \beta\sqrt{2}$, then the value of $(\alpha + \beta)$ is ....

22. Let the normals at all the points on a given curve pass through a fixed point (a, b). If the curve passes through $(3, -3)$ and $(4, -2\sqrt{2})$ and given that $a - 2\sqrt{2}b = 3$, then $(a^2 + b^2 + ab)$ is equal to ______.

23. Let α and β be two real numbers, such that $\alpha + \beta = 1$ and $\alpha\beta = -1$. Let $p_n = (\alpha)^n + (\beta)^n$, $p_{n-1} = 11$ and $p_{n+1} = 29$, for some integer $n \geq 1$. Then, the value of p_n^2 is ______.

24. If $I_{m,n} = \int_0^1 x^{m-1}(1-x)^{n-1}\,dx$, for $m, n \geq 1$ and $\int_0^1 \frac{x^{m-1} + x^{n-1}}{(1+x)^{m+n}}\,dx = \alpha I_{m,n}, \alpha \in R$, then α equals ______.

25. If the arithmetic mean and geometric mean of the pth and qth terms of the sequence $-16, 8, -4, 2, \ldots$ satisfy the equation $4x^2 - 9x + 5 = 0$, then $p + q$ is equal to ______.

26. The total number of 4-digit numbers whose greatest common divisor with 18 is 3, is ______.

27. Let L be a common tangent line to the curves $4x^2 + 9y^2 = 36$ and $(2x)^2 + (2y)^2 = 31$. Then, the square of the slope of the line L is ______.

28. Let a be an integer, such that all the real roots of the polynomial $2x^5 + 5x^4 + 10x^3 + 10x^2 + 10x + 10$ lie in the interval $(a, a+1)$. Then, $|a|$ is equal to ______.

29. Let $X_1, X_2, \ldots, X_{18}$ be eighteen observations, such that $\sum_{i=1}^{18}(X_i - \alpha) = 36$ and $\sum_{i=1}^{18}(X_i - \beta)^2 = 90$, where α and β are distinct real numbers. If the standard deviation of these observations is 1, then the value of $|\alpha - \beta|$ is ______.

30. If the matrix $A = \begin{bmatrix} 1 & 0 & 0 \\ 0 & 2 & 0 \\ 3 & 0 & -1 \end{bmatrix}$ satisfies the equation $A^{20} + \alpha A^{19} + \beta A = \begin{bmatrix} 1 & 0 & 0 \\ 0 & 4 & 0 \\ 0 & 0 & 1 \end{bmatrix}$ for some real numbers α and β, then $\beta - \alpha$ is equal to ______.

Answers

For solution scan the QR code

1. *(d)*	2. *(a)*	3. *(b)*	4. *(c)*	5. *(d)*	6. *(c)*	7. *(c)*	8. *(d)*	9. *(c)*	10. *(c)*
11. *(b)*	12. *(b)*	13. *(a)*	14. *(b)*	15. *(b)*	16. *(a)*	17. *(a)*	18. *(a)*	19. *(a)*	20. *(c)*
21. *(48)*	22. *(9)*	23. *(324)*	24. *(1)*	25. *(10)*	26. *(1000)*	27. *(3)*	28. *(2)*	29. *(4)*	30. *(4)*

16 MARCH SHIFT I

Section A : Objective Type Questions

1. The number of elements in the set $\{x \in R: (|x| - 3)|x + 4| = 6\}$ is equal to

a. 3 **b.** 2
c. 4 **d.** 1

2. Let a vector $\alpha\hat{\mathbf{i}} + \beta\hat{\mathbf{j}}$ be obtained by rotating the vector $\sqrt{3}\hat{\mathbf{i}} + \hat{\mathbf{j}}$ by an angle 45° about the origin in counterclockwise direction in the first quadrant. Then the area of triangle having vertices (α, β), $(0, \beta)$ and $(0,0)$ is equal to

a. $\frac{1}{2}$ **b.** 1 **c.** $\frac{1}{\sqrt{2}}$ **d.** $2\sqrt{2}$

3. If for $a > 0$, the feet of perpendiculars from the points $A(a, -2a, 3)$ and $B(0, 4, 5)$ on the plane $lx + my + nz = 0$ are points $C(0, -a, -1)$ and D respectively, then, the length of line segment CD is equal to

a. $\sqrt{31}$ **b.** $\sqrt{41}$
c. $\sqrt{55}$ **d.** $\sqrt{66}$

4. Consider three observations a, b and c, such that $b = a + c$. If the standard deviation of $a + 2, b + 2, c + 2$ is d, then which of the following is true?

a. $b^2 = 3(a^2 + c^2) + 9d^2$ **b.** $b^2 = a^2 + c^2 + 3d^2$
c. $b^2 = 3(a^2 + c^2 + d^2)$ **d.** $b^2 = 3(a^2 + c^2) - 9d^2$

5. If for $x \in \left(0, \frac{\pi}{2}\right)$, $\log_{10}\sin x + \log_{10}\cos x = -1$ and $\log_{10}(\sin x + \cos x) = \frac{1}{2}(\log_{10} n - 1)$, $n > 0$, then the value of n is equal to

a. 20 **b.** 12 **c.** 9 **d.** 16

6. Let $A = \begin{bmatrix} i & -i \\ -i & i \end{bmatrix}$, $i = \sqrt{-1}$. Then, the system of linear equations $A^8\begin{bmatrix} x \\ y \end{bmatrix} = \begin{bmatrix} 8 \\ 64 \end{bmatrix}$ has

a. a unique solution **b.** infinitely many solutions
c. no solution **d.** exactly two solutions

7. If the three normals drawn to the parabola, $y^2 = 2x$ pass through the point $(a, 0)$, $a \neq 0$, then a must be greater than

a. $\frac{1}{2}$ **b.** $-\frac{1}{2}$ **c.** -1 **d.** 1

8. Let the position vectors of two points P and Q be $3\hat{\mathbf{i}} - \hat{\mathbf{j}} + 2\hat{\mathbf{k}}$ and $\hat{\mathbf{i}} + 2\hat{\mathbf{j}} - 4\hat{\mathbf{k}}$, respectively. Let R and S be two points such that the direction ratios of lines PR and QS are $(4, -1, 2)$ and $(-2, 1, -2)$, respectively. Let lines PR and QS intersect at T. If the vector **TA** is perpendicular to both **PR** and **QS** and the length of vector **TA** is $\sqrt{5}$ units, then the modulus of a position vector of A is

a. $\sqrt{482}$ **b.** $\sqrt{171}$
c. $\sqrt{5}$ **d.** $\sqrt{227}$

9. Let the functions $f: R \to R$ and $g: R \to R$ be defined as

$$f(x) = \begin{cases} x + 2, & x < 0 \\ x^2, & x \geq 0 \end{cases} \text{ and } g(x) = \begin{cases} x^3, & x < 1 \\ 3x - 2, & x \geq 1 \end{cases}$$

Then, the number of points in R, where $(fog)(x)$ is NOT differentiable is equal to

a. 3 **b.** 1 **c.** 0 **d.** 2

10. Which of the following Boolean expression is a tautology?

a. $(p \wedge q) \vee (p \vee q)$ **b.** $(p \wedge q) \vee (p \to q)$
c. $(p \wedge q) \wedge (p \to q)$ **d.** $(p \wedge q) \to (p \to q)$

11. Let a complex number z, $|z| \neq 1$, satisfy $\log_{\frac{1}{\sqrt{2}}}\left(\frac{|z| + 11}{(|z| - 1)^2}\right) \leq 2$. Then, the largest value of $|z|$ is equal to

a. 8 **b.** 7
c. 6 **d.** 5

12. If n is the number of irrational terms in the expansion of $(3^{1/4} + 5^{1/8})^{60}$, then $(n - 1)$ is divisible by

a. 26 **b.** 30
c. 8 **d.** 7

13. Let P be a plane $lx + my + nz = 0$ containing the line, $\frac{1 - x}{1} = \frac{y + 4}{2} = \frac{z + 2}{3}$. If plane P divides the line segment AB joining points $A(-3, -6, 1)$ and $B(2, 4, -3)$ in ratio $k : 1$, then the value of k is equal to

a. 1.5 **b.** 3
c. 2 **d.** 4

14. The range of $a \in R$ for which the function $f(x) = (4a - 3)(x + \log_e 5) + 2(a - 7)\cot\left(\frac{x}{2}\right)\sin^2\left(\frac{x}{2}\right)$, $x \neq 2n\pi$, $n \in N$, has critical points, is

a. $(-3, 1)$ **b.** $\left[-\frac{4}{3}, 2\right]$
c. $[1, \infty)$ **d.** $(-\infty, -1]$

15. A pack of cards has one card missing. Two cards are drawn randomly and are found to be spades. The probability that the missing card is not a spade, is

a. $\frac{3}{4}$ **b.** $\frac{52}{867}$
c. $\frac{39}{50}$ **d.** $\frac{22}{425}$

16. Let $[x]$ denote greatest integer less than or equal to x. If for $n \in N$, $(1 - x + x^3)^n = \sum_{j=0}^{3n} a_j x^j$, then

$$\sum_{j=0}^{\left[\frac{3n}{2}\right]} a_{2j} + 4\sum_{j=0}^{\left[\frac{3n-1}{2}\right]} a_{2j} + 1 \text{ is equal to}$$

a. 2 **b.** 2^{n-1}
c. 1 **d.** n

17. If $y = y(x)$ is the solution of the differential equation, $\frac{dy}{dx} + 2y\tan x = \sin x$, $y\left(\frac{\pi}{3}\right) = 0$, then the maximum value of the function $y(x)$ over R is equal to

a. 8 b. $\frac{1}{2}$ c. $-\frac{15}{4}$ d. $\frac{1}{8}$

18. The locus of the mid-points of the chord of the circle, $x^2 + y^2 = 25$ which is tangent to the hyperbola, $\frac{x^2}{9} - \frac{y^2}{16} = 1$ is

a. $(x^2 + y^2)^2 - 16x^2 + 9y^2 = 0$
b. $(x^2 + y^2)^2 - 9x^2 + 144y^2 = 0$
c. $(x^2 + y^2)^2 - 9x^2 - 16y^2 = 0$
d. $(x^2 + y^2)^2 - 9x^2 + 16y^2 = 0$

19. The number of roots of the equation, $(81)^{\sin^2 x} + (81)^{\cos^2 x} = 30$ in the interval $[0, \pi]$ is equal to

a. 3 b. 4
c. 8 d. 2

20. Let $S_k = \sum_{r=1}^{k} \tan^{-1}\left(\frac{6^r}{2^{2r+1} + 3^{2r+1}}\right)$. Then, $\lim_{k\to\infty} S_k$ is equal to

a. $\tan^{-1}\left(\frac{3}{2}\right)$ b. $\frac{\pi}{2}$

c. $\cot^{-1}\left(\frac{3}{2}\right)$ d. $\tan^{-1}(3)$

Section B : Numerical Type Questions

21. Consider an arithmetic series and a geometric series having four initial terms from the set {11, 8, 21, 16, 26, 32, 4}. If the last terms of these series are the maximum possible four digit numbers, then the number of common terms in these two series is equal to......... .

22. Let $f: (0, 2) \to R$ be defined as

$$f(x) = \log_2\left[1 + \tan\left(\frac{\pi x}{4}\right)\right].$$

Then, $\lim_{n\to\infty} \frac{2}{n}\left[f\left(\frac{1}{n}\right) + f\left(\frac{2}{n}\right) + \ldots + f(1)\right]$ is equal to......... .

23. Let $ABCD$ be a square of side of unit length. Let a circle C_1 centred at A with unit radius is drawn. Another circle C_2, which touches C_1 and the lines AD and AB are tangent to it, is also drawn. Let a tangent line from the point C to the circle C_2 meet the side AB at E. If the length of EB is $\alpha + \sqrt{3}\beta$, where α, β are integers, then $\alpha + \beta$ is equal to ....

24. If $\lim_{x\to 0} \frac{ae^x - b\cos x + ce^{-x}}{x\sin x} = 2$, then $a + b + c$ is equal to ....

25. The total number of 3×3 matrices A having entries from the set (0, 1, 2, 3), such that the sum of all the diagonal entries of AA^T is 9, is equal to ....

26. Let

$$P = \begin{bmatrix} -30 & 20 & 56 \\ 90 & 140 & 112 \\ 120 & 60 & 14 \end{bmatrix} \text{ and } A = \begin{bmatrix} 2 & 7 & \omega^2 \\ -1 & -\omega & 1 \\ 0 & -\omega & -\omega + 1 \end{bmatrix}$$

where, $\omega = \frac{-1 + i\sqrt{3}}{2}$, and I_3 be the identity matrix of order 3. If the determinant of the matrix $(P^{-1}AP - I_3)^2$ is $\alpha\omega^2$, then the value of α is equal to......... .

27. If the normal to the curve $y(x) = \int_0^x (2t^2 - 15t + 10)dt$ at a point (a, b) is parallel to the line $x + 3y = -5$, $a > 1$, then the value of $|a + 6b|$ is equal to......... .

28. Let the curve $y = y(x)$ be the solution of the differential equation $\frac{dy}{dx} = 2(x + 1)$.

If the numerical value of area bounded by the curve $y = y(x)$ and x-axis is $\frac{4\sqrt{8}}{3}$, then the value of $y(1)$ is equal to......... .

29. Let $f: R \to R$ be a continuous function such that $f(x) + f(x + 1) = 2$, for all $x \in R$. If $I_1 = \int_0^8 f(x)dx$ and $I_2 = \int_{-1}^3 f(x)dx$, then the value of $I_1 + 2I_2$ is equal to......... .

30. Let z and w be two complex numbers, such that $w = z\bar{z} - 2z + 2$, $\left|\frac{z + i}{z - 3i}\right| = 1$ and $Re(w)$ has minimum value.

Then, the minimum value of $n \in N$ for which w^n is real, is equal to ....

Answers

For solution scan the QR code

1. (b)	2. (a)	3. (d)	4. (d)	5. (b)	6. (c)	7. (d)	8. (b)	9. (b)	10. (d)
11. (b)	12. (a)	13. (c)	14. (b)	15. (c)	16. (c)	17. (d)	18. (d)	19. (b)	20. (c)
21. (3)	22. (1)	23. (1)	24. (4)	25. (766)	26. (36)	27. (406)	28. (2)	29. (16)	30. (4)

16 MARCH SHIFT II

Section A : Objective Type Questions

1. The maximum value of

$$f(x)=\begin{vmatrix}\sin^2 x & 1+\cos^2 x & \cos 2x\\ 1+\sin^2 x & \cos^2 x & \cos 2x\\ \sin^2 x & \cos^2 x & \sin 2x\end{vmatrix}, x \in R \text{ is}$$

a. $\sqrt{7}$ **b.** 3/4 **c.** $\sqrt{5}$ **d.** 5

2. Let A denote the event that a 6-digit integer formed by 0, 1, 2, 3, 4, 5, 6 without repetitions, be divisible by 3. Then, probability of event A is equal to

a. 9/56 **b.** 4/9 **c.** 3/7 **d.** 11/27

3. Let $\alpha \in R$ be such that the function

$$f(x)=\begin{cases}\dfrac{\cos^{-1}(1-\{x\}^2)\sin^{-1}(1-\{x\})}{\{x\}-\{x\}^3}, & x \neq 0\\ \alpha, & x=0\end{cases}$$

is continuous at $x=0$, where $\{x\}=x-[x]$, $[x]$ is the greatest integer less than or equal to x. Then,

a. $\alpha=\dfrac{\pi}{\sqrt{2}}$ **b.** $\alpha=0$

c. no such α exists **d.** $\alpha=\dfrac{\pi}{4}$

4. If (x, y, z) be an arbitrary point lying on a plane P, which passes through the points (42, 0, 0), (0, 42, 0) and (0, 0, 42), then the value of expression

$$3+\frac{x-11}{(y-19)^2(z-12)^2}+\frac{y-19}{(x-11)^2(z-12)^2}+\frac{z-12}{(x-11)^2(y-19)^2}-\frac{x+y+z}{14(x-11)(y-19)(z-12)}$$ is equal to

a. 0 **b.** 3 **c.** 39 **d.** −45

5. Consider the integral

$$I=\int_0^{10}\frac{[x]e^{[x]}}{e^{x-1}}dx,$$

where $[x]$ denotes the greatest integer less than or equal to x. Then, the value of I is equal to

a. $9(e-1)$ **b.** $45(e+1)$

c. $45(e-1)$ **d.** $9(e+1)$

6. Let C be the locus of the mirror image of a point on the parabola $y^2=4x$ with respect to the line $y=x$. Then, the equation of tangent to C at $P(2, 1)$ is

a. $x-y=1$ **b.** $2x+y=5$

c. $x+3y=5$ **d.** $x+2y=5$

7. If $y=y(x)$ is the solution of the differential equation $\dfrac{dy}{dx}+(\tan x)y=\sin x, 0\le x\le\dfrac{\pi}{3}$, with $y(0)=0$, then $y\left(\dfrac{\pi}{4}\right)$ is equal to

a. $\dfrac{1}{4}\log_e 2$ **b.** $\left(\dfrac{1}{2\sqrt{2}}\right)\log_e 2$

c. $\log_e 2$ **d.** $\dfrac{1}{2}\log_e 2$

8. Let $A=\{2, 3, 4, 5, \ldots, 30\}$ and '$\simeq$' be an equivalence relation on $A\times A$, defined by $(a, b)\simeq(c, d)$, if and only if $ad=bc$. Then, the number of ordered pairs, which satisfy this equivalence relation with ordered pair (4, 3) is equal to

a. 5 **b.** 6 **c.** 8 **d.** 7

9. Let the lengths of intercepts on X-axis and Y-axis made by the circle $x^2+y^2+ax+2ay+c=0, (a<0)$ be $2\sqrt{2}$ and $2\sqrt{5}$, respectively. Then, the shortest distance from origin to a tangent to this circle which is perpendicular to the line $x+2y=0$, is equal to

a. $\sqrt{11}$ **b.** $\sqrt{7}$ **c.** $\sqrt{6}$ **d.** $\sqrt{10}$

10. The least value of $|z|$, where z is a complex number which satisfies the inequality

$$\exp\left(\frac{(|z|+3)(|z|-1)}{(|z|+1)}\log_e 2\right)\ge\log_{\sqrt{2}}|5\sqrt{7}+9i|.$$

$i=\sqrt{-1}$, is equal to

a. 3 **b.** $\sqrt{5}$ **c.** 2 **d.** 8

11. Consider a rectangle $ABCD$ having 5, 7, 6, 9 points in the interior of the line segments AB, CD, BC, DA, respectively. Let α be the number of triangles having these points from different sides as vertices and β be the number of quadrilaterals having these points from different sides as vertices. Then, $(\beta-\alpha)$ is equal to

a. 795 **b.** 1173

c. 1890 **d.** 717

12. If the point of intersections of the ellipse $\dfrac{x^2}{16}+\dfrac{y^2}{b^2}=1$ and the circle $x^2+y^2=4b, b>4$, lie on the curve $y^2=3x^2$, then b is equal to

a. 12 **b.** 5 **c.** 6 **d.** 10

13. Given that the inverse trigonometric functions take principal values only. Then, the number of real values of x which satisfy $\sin^{-1}\left(\dfrac{3x}{2}\right)+\sin^{-1}\left(\dfrac{4x}{5}\right)=\sin^{-1}x$ is equal to

a. 2 **b.** 1 **c.** 3 **d.** 0

14. Let $A(-1, 1)$, $B(3, 4)$ and $C(2, 0)$ be given three points. A line $y=mx, m>0$ intersects lines AC and BC at point P and Q, respectively. Let A_1 and A_2 be the areas of ΔABC and ΔPQC, respectively, such that $A_1=3A_2$, then the value of m is equal to

a. $\dfrac{4}{15}$ **b.** 1 **c.** 2 **d.** 3

15. Let f be a real valued function, defined on $R-\{-1, 1\}$ and given by

$$f(x)=3\log_e\left|\frac{x-1}{x+1}\right|-\frac{2}{x-1}.$$

Then, in which of the following intervals, function $f(x)$ is increasing?

a. $(-\infty, -1) \cup \left(\left[\frac{1}{2}, \infty\right) - \{1\}\right)$ b. $(-\infty, \infty) - \{-1, 1\}$

c. $\left(-1, \frac{1}{2}\right]$ d. $\left(-\infty, \frac{1}{2}\right] - [-1]$

16. Let $f: S \to S$, where $S = (0, \infty)$ be a twice differentiable function, such that $f(x+1) = x\, f(x)$. If $g: S \to R$ be defined as $g(x) = \log_e f(x)$, then the value of $|g''(5) - g''(1)|$ is equal to

a. $\frac{205}{144}$ b. $\frac{197}{144}$ c. $\frac{187}{144}$ d. 1

17. Let $P(x) = x^2 + bx + c$ be a quadratic polynomial with real coefficients, such that $\int_0^1 P(x)dx = 1$ and $P(x)$ leaves remainder 5 when it is divided by $(x-2)$. Then, the value of $9(b+c)$ is equal to

a. 9 b. 15 c. 7 d. 11

18. If the foot of the perpendicular from point $(4, 3, 8)$ on the line $L_1: \frac{x-a}{l} = \frac{y-2}{3} = \frac{z-b}{4}$, $l \neq 0$ is $(3, 5, 7)$, then the shortest distance between the line L_1 and line $L_2: \frac{x-2}{3} = \frac{y-4}{4} = \frac{z-5}{5}$ is equal to

a. 1/2 b. $1/\sqrt{6}$

c. $\sqrt{2/3}$ d. $1/\sqrt{3}$

19. Let C_1 be the curve obtained by the solution of differential equation $2xy\frac{dy}{dx} = y^2 - x^2$, $x > 0$. Let the curve C_2 be the solution of $\frac{2xy}{x^2 - y^2} = \frac{dy}{dx}$. If both the curves pass through (1, 1), then the area enclosed by the curves C_1 and C_2 is equal to

a. $\pi - 1$ b. $\frac{\pi}{2} - 1$

c. $\pi + 1$ d. $\frac{\pi}{4} + 1$

20. Let $\mathbf{a} = \hat{\mathbf{i}} + 2\hat{\mathbf{j}} - 3\hat{\mathbf{k}}$ and $\mathbf{b} = 2\hat{\mathbf{i}} - 3\hat{\mathbf{j}} + 5\hat{\mathbf{k}}$. If $\mathbf{r} \times \mathbf{a} = \mathbf{b} \times \mathbf{r}$, $\mathbf{r}.(\alpha\hat{\mathbf{i}} + 2\hat{\mathbf{j}} + \hat{\mathbf{k}}) = 3$ and $\mathbf{r}.(2\hat{\mathbf{i}} + 5\hat{\mathbf{j}} - \alpha\hat{\mathbf{k}}) = -1$, $\alpha \in R$, then the value of $\alpha + |\mathbf{r}|^2$ is equal to

a. 9 b. 15 c. 13 d. 11

Section B : Numerical Type Questions

21. If the distance of the point $(1, -2, 3)$ from the plane $x + 2y - 3z + 10 = 0$ measured parallel to the line, $\frac{x-1}{3} = \frac{2-y}{m} = \frac{z+3}{1}$ is $\sqrt{\frac{7}{2}}$, then the value of $|m|$ is equal to.........

22. Consider the statistics of two sets of observations as follows

	Size	Mean	Variance
Observation I	10	2	2
Observation II	n	3	1

If the variance of the combined set of these two observations is 17/9, then the value of n is equal to ....

23. Let $A = \begin{bmatrix} a_1 \\ a_2 \end{bmatrix}$ and $B = \begin{bmatrix} b_1 \\ b_2 \end{bmatrix}$ be two 2×1 matrices with real entries such that $A = XB$, where $X = \frac{1}{\sqrt{3}}\begin{bmatrix} 1 & -1 \\ 1 & k \end{bmatrix}$ and $k \in R$. If $a_1^2 + a_2^2 = \frac{2}{3}(b_1^2 + b_2^2)$ and $(k^2 + 1)b_2^2 \neq -2b_1b_2$, then the value of k is ....

24. For real numbers α, β, γ and δ, if

$$\int \frac{(x^2 - 1) + \tan^{-1}\left(\frac{x^2+1}{x}\right)}{(x^4 + 3x^2 + 1)\tan^{-1}\left(\frac{x^2+1}{x}\right)} dx$$

$$= \alpha \log_e\left[\tan^{-1}\left(\frac{x^2+1}{x}\right)\right]$$

$$+ \beta \tan^{-1}\left(\frac{\gamma(x^2-1)}{x}\right) + \delta \tan^{-1}\left(\frac{x^2+1}{x}\right) + C$$

where C is an arbitrary constant, then the value of $10(\alpha + \beta\gamma + \delta)$ is equal to......... .

25. Let $f: R \to R$ and $g: R \to R$ be defined as

$$f(x) = \begin{cases} x + a, & x < 0 \\ |x - 1|, & x \geq 0 \end{cases} \text{ and } g(x) = \begin{cases} x + 1, & x < 0 \\ (x-1)^2 + b, & x \geq 0 \end{cases}$$

where a, b are non-negative real numbers. If $(gof)(x)$ is continuous for all $x \in R$, then $a + b$ is equal to......... .

26. Let $\frac{1}{16}$, a and b be in G. P. and $\frac{1}{a}, \frac{1}{b}, 6$ be in A. P., where $a, b > 0$. Then, $72(a + b)$ is equal to ....

27. In ΔABC, the lengths of sides AC and AB are 12 cm and 5 cm, respectively. If the area of ΔABC is 30 cm^2 and R and r are respectively the radii of circumcircle and incircle of ΔABC, then the value of $2R + r$ (in cm) is equal to ....

28. Let n be a positive integer. Let

$$A = \sum_{k}^{n} (-1)^k\, {}^nC_k \left[\left(\frac{1}{2}\right)^k + \left(\frac{3}{4}\right)^k + \left(\frac{7}{8}\right)^k + \left(\frac{15}{16}\right)^k + \left(\frac{31}{32}\right)^k\right]$$

If $63A = 1 - \frac{1}{2^{30}}$, then n is equal to ....

29. Let $\mathbf{c}$ be a vector perpendicular to the vectors $\mathbf{a} = \hat{\mathbf{i}} + \hat{\mathbf{j}} - \hat{\mathbf{k}}$ and $\mathbf{b} = \hat{\mathbf{i}} + 2\hat{\mathbf{j}} + \hat{\mathbf{k}}$

If $\mathbf{c} \cdot (\hat{\mathbf{i}} + \hat{\mathbf{j}} + 3\hat{\mathbf{k}}) = 8$ then the value of $\mathbf{c} \cdot (\mathbf{a} \times \mathbf{b})$ is equal to ....

30. Let $S_n(x) = \log_{a^{1/2}} x + \log_{a^{1/3}} x + \log_{a^{1/6}} x + \log_{a^{1/11}} x + \log_{a^{1/18}} x + \log_{a^{1/27}} x + \ldots$ up to n-terms, where $a > 1$. If $S_{24}(x) = 1093$ and $S_{12}(2x) = 265$, then value of a is equal to ....

Answers

For solution scan the QR code

1. (c)	2. (b)	3. (c)	4. (b)	5. (c)	6. (a)	7. (b)	8. (d)	9. (c)	10. (a)
11. (d)	12. (a)	13. (c)	14. (b)	15. (a)	16. (a)	17. (c)	18. (b)	19. (b)	20. (b)
21. (2)	22. (5)	23. (1)	24. (6)	25. (1)	26. (14)	27. (15)	28. (6)	29. (28)	30. (16)

17 MARCH SHIFT I

Section A : Objective Type Questions

1. The inverse of $y = 5^{\log x}$ is

a. $x = 5^{\log y}$ **b.** $x = y^{\log 5}$

c. $x = y^{\frac{1}{\log 5}}$ **d.** $x = 5^{\frac{1}{\log y}}$

2. Let $\mathbf{a} = 2\hat{\mathbf{i}} - 3\hat{\mathbf{j}} + 4\hat{\mathbf{k}}$ and $\mathbf{b} = 7\hat{\mathbf{i}} + \hat{\mathbf{j}} - 6\hat{\mathbf{k}}$. If $\mathbf{r} \times \mathbf{a} = \mathbf{r} \times \mathbf{b}$, $\mathbf{r} \cdot (\hat{\mathbf{i}} + 2\hat{\mathbf{j}} + \hat{\mathbf{k}}) = -3$, then, $\mathbf{r} \cdot (2\hat{\mathbf{i}} - 3\hat{\mathbf{j}} + \hat{\mathbf{k}})$ is equal to

a. 12 **b.** 8 **c.** 13 **d.** 10

3. In a ΔPQR, the coordinates of the points P and Q are $(-2, 4)$ and $(4, -2)$, respectively. If the equation of the perpendicular bisector of PR is $2x - y + 2 = 0$, then the centre of the circumcircle of the ΔPQR is

a. $(-1, 0)$ **b.** $(-2, -2)$ **c.** $(0, 2)$ **d.** $(1, 4)$

4. The system of equations $kx + y + z = 1$, $x + ky + z = k$ and $x + y + zk = k^2$ has no solution, if k is equal to

a. 0 **b.** 1 **c.** -1 **d.** -2

5. If $\cot^{-1}(\alpha) = \cot^{-1}2 + \cot^{-1}8 + \cot^{-1}18 + \cot^{-1}32 + \ldots$ upto 100 terms, then α is

a. 1.01 **b.** 1.00 **c.** 1.02 **d.** 1.03

6. The equation of the plane which contains the Y-axis and passes through the point $(1, 2, 3)$ is

a. $x + 3z = 10$ **b.** $x + 3z = 0$

c. $3x + z = 6$ **d.** $3x - z = 0$

7. If $A = \begin{pmatrix} 0 & \sin\alpha \\ \sin\alpha & 0 \end{pmatrix}$ and $\det\left(A^2 - \frac{1}{2}I\right) = 0$, then a possible value of α is

a. $\frac{\pi}{2}$ **b.** $\frac{\pi}{3}$ **c.** $\frac{\pi}{4}$ **d.** $\frac{\pi}{6}$

8. If the Boolean expression $(p \Rightarrow q) \Leftrightarrow (q * (\sim p))$ is a tautology, then the Boolean expression $(p * (\sim q))$ is equivalent to

a. $q \Rightarrow p$ **b.** $\sim q \Rightarrow p$

c. $p \Rightarrow \sim q$ **d.** $p \Rightarrow q$

9. Two dices are rolled. If both dices have six faces numbered 1, 2, 3, 5, 7 and 11, then the probability that the sum of the numbers on the top faces is less than or equal to 8 is

a. $\frac{4}{9}$ **b.** $\frac{17}{36}$ **c.** $\frac{5}{12}$ **d.** $\frac{1}{2}$

10. If the fourth term in the expansion of $(x + x^{\log_2 x})^7$ is 4480, then the value of x, where $x \in N$ is equal to

a. 2 **b.** 4

c. 3 **d.** 1

11. In a school, there are three types of games to be played. Some of the students play two types of games, but none play all the three games. Which Venn diagram can justify the above statement?

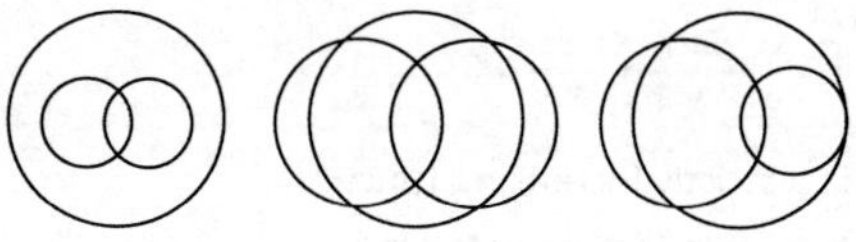

(P) (Q) (R)

a. P and Q **b.** P and R

c. None of these **d.** Q and R

12. The sum of possible values of x for $\tan^{-1}(x + 1) + \cot^{-1}\left(\frac{1}{x - 1}\right) = \tan^{-1}\left(\frac{8}{31}\right)$ is

a. $\frac{-32}{4}$ **b.** $-\frac{31}{4}$

c. $-\frac{30}{4}$ **d.** $-\frac{33}{4}$

13. The area of the triangle with vertices $A(z)$, $B(iz)$ and $C(z + iz)$ is

a. 1 **b.** $\frac{1}{2}|z|^2$

c. $\frac{1}{2}$ **d.** $\frac{1}{2}|z + iz|^2$

14. The line $2x - y + 1 = 0$ is a tangent to the circle at the point $(2, 5)$ and the centre of the circle lies on $x - 2y = 4$. Then, the radius of the circle is

a. $3\sqrt{5}$ **b.** $5\sqrt{3}$

c. $5\sqrt{4}$ **d.** $4\sqrt{5}$

15. Team A consists of 7 boys and n girls and Team B has 4 boys and 6 girls. If a total of 52 single matches can be arranged between these two teams, when a boy plays against a boy and a girl plays against a girl, then n is equal to

a. 5 **b.** 2

c. 4 **d.** 6

16. The value of $4+\cfrac{1}{5+\cfrac{1}{4+\cfrac{1}{5+\cfrac{1}{4+\ldots\infty}}}}$ is

a. $2+\frac{2}{5}\sqrt{30}$ **b.** $2+\frac{4}{\sqrt{5}}\sqrt{30}$

c. $4+\frac{4}{\sqrt{5}}\sqrt{30}$ **d.** $5+\frac{2}{5}\sqrt{30}$

17. Choose the incorrect statement about the two circles whose equations are given below
$x^2+y^2-10x-10y+41=0$
and $x^2+y^2-16x-10y+80=0$

a. Distance between two centres is the average of radii of both the circles.
b. Both circles' centres lie inside region of one another.
c. Both circles pass through the centre of each other.
d. Circles have two intersection points.

18. Which of the following statements is correct for the function $g(\alpha)$ for $\alpha \in R$, such that

$$g(\alpha)=\int_{\pi/6}^{\pi/3}\frac{\sin^\alpha x}{\cos^\alpha x+\sin^\alpha x}dx$$

a. $g(\alpha)$ is a strictly increasing function
b. $g(\alpha)$ has an inflection point at $\alpha=-\frac{1}{2}$
c. $g(\alpha)$ is a strictly decreasing function
d. $g(\alpha)$ is an even function

19. Which of the following is true for $y(x)$ that satisfies the differential equation
$\frac{dy}{dx}=xy-1+x-y;\ y(0)=0$

a. $y(1)=e^{-\frac{1}{2}}-1$ **b.** $y(1)=e^{\frac{1}{2}}-e^{-\frac{1}{2}}$

c. $y(1)=1$ **d.** $y(1)=e^{\frac{1}{2}}-1$

20. The value of

$$\lim_{x\to 0^+}\frac{\cos^{-1}(x-[x]^2).\sin^{-1}(x-[x]^2)}{x-x^3},$$ where $[x]$ denotes the greatest integer $\le x$ is

a. π **b.** 0 **c.** $\pi/4$ **d.** $\pi/2$

Section B : Numerical Type Questions

21. The maximum value of z in the following equation $z=6xy+y^2$, where $3x+4y\le 100$ and $4x+3y\le 75$ for $x\ge 0$ and $y\ge 0$ is ....

22. If the function $f(x)=\frac{\cos(\sin x)-\cos x}{x^4}$ is continuous at each point in its domain and $f(0)=\frac{1}{k}$, then k is ....

23. If $f(x)=\sin\left(\cos^{-1}\left(\frac{1-2^{2x}}{1+2^{2x}}\right)\right)$ and its first derivative with respect to x is $-\frac{b}{a}\log_e 2$ when $x=1$, where a and b are integers, then the minimum value of $|a^2-b^2|$ is ....

24. Let there be three independent events E_1, E_2 and E_3. The probability that only E_1 occurs is α, only E_2 occurs is β and only E_3 occurs is γ. Let p denote the probability of none of events occur that satisfies the equations $(\alpha-2\beta)p=\alpha\beta$ and $(\beta-3\gamma)p=2\beta\gamma$. All the given probabilities are assumed to lie in the interval (0, 1).

Then, $\frac{\text{probability of occurrence of } E_1}{\text{probability of occurrence of } E_3}$ is equal to ....

25. If $\mathbf{a}=\alpha\hat{\mathbf{j}}+\beta\hat{\mathbf{j}}+3\hat{\mathbf{k}}, \mathbf{b}=-\beta\hat{\mathbf{i}}-\alpha\hat{\mathbf{j}}-\hat{\mathbf{k}}$ and $\mathbf{c}=\hat{\mathbf{i}}-2\hat{\mathbf{j}}-\hat{\mathbf{k}}$, such that $\mathbf{a}.\mathbf{b}=1$ and $\mathbf{b}.\mathbf{c}=-3$, then $\frac{1}{3}[(\mathbf{a}\times\mathbf{b})\cdot\mathbf{c}]$ is equal to ....

26. If $A=\begin{bmatrix}2 & 3\\ 0 & -1\end{bmatrix}$, then the value of $\det(A^4)+\det[A^{10}-\text{Adj}(2A)^{10}]$ is equal to ....

27. If [·] represents the greatest integer function, then the value of $\left|\int_0^{\sqrt{\pi/2}}[x^2-\cos x]dx\right|$ is ....

28. The minimum distance between any two points P_1 and P_2 while considering point P_1 on one circle and point P_2 on the other circle for the given circles equations
$x^2+y^2-10x-10y+41=0$
$x^2+y^2-24x-10y+160=0$ is ....

29. If the equation of the plane passing through the line of intersection of the planes $2x-7y+4z-3=0$, $3x-5y+4z+11=0$ and the point $(-2, 1, 3)$ is $ax+by+cz-7=0$, then the value of $2a+b+c-7$ is ....

30. If $(2021)^{3762}$ is divided by 17, then the remainder is ....

Answers

For solution scan the QR code

1. *(c)*	2. *(a)*	3. *(b)*	4. *(d)*	5. *(a)*	6. *(d)*	7. *(c)*	8. *(a)*	9. *(b)*	10. *(a)*
11. *(c)*	12. *(a)*	13. *(b)*	14. *(a)*	15. *(c)*	16. *(a)*	17. *(b)*	18. *(d)*	19. *(a)*	20. *(d)*
21. *(904)*	22. *(6)*	23. *(481)*	24. *(6)*	25. *(2)*	26. *(16)*	27. *(1)*	28. *(1)*	29. *(4)*	30. *(4)*

17 MARCH SHIFT II

Section A : Objective Type Questions

1. Let $f : R \to R$ be defined as $f(x) = e^{-x}\sin x$. If $F : [0, 1] \to R$ is a differentiable function, such that $F(x) = \int_0^x f(t)\,dt$, then the value of $\int_0^1 [F'(x) + f(x)]\,e^x\,dx$ lies in the interval

a. $\left[\frac{327}{360}, \frac{329}{360}\right]$ **b.** $\left[\frac{330}{360}, \frac{331}{360}\right]$ **c.** $\left[\frac{331}{360}, \frac{334}{360}\right]$ **d.** $\left[\frac{335}{360}, \frac{336}{360}\right]$

2. If the integral $\int_0^{10} \frac{[\sin 2\pi x]}{e^{x-[x]}}dx = \alpha e^{-1} + \beta e^{-\frac{1}{2}} + \gamma$, where α, β, γ are integers and $[x]$ denotes the greatest integer less than or equal to x, then the value of $\alpha + \beta + \gamma$ is equal to

a. 0 **b.** 20 **c.** 25 **d.** 10

3. Let $y = y(x)$ be the solution of the differential equation $\cos x\,(3\sin x + \cos x + 3)dy =$ $[1 + y\sin x\,(3\sin x + \cos x + 3)]dx, 0 \le x \le \frac{\pi}{2}, y(0) = 0$. Then, $y\left(\frac{\pi}{3}\right)$ is equal to

a. $2\log_e\left(\frac{2\sqrt{3}+9}{6}\right)$ **b.** $2\log_e\left(\frac{2\sqrt{3}+10}{11}\right)$

c. $2\log_e\left(\frac{\sqrt{3}+7}{2}\right)$ **d.** $2\log_e\left(\frac{3\sqrt{3}-8}{4}\right)$

4. The value of $\sum_{r=0}^{6} ({}^6C_r \cdot {}^6C_{6-r})$ is equal to

a. 1124 **b.** 1324 **c.** 1024 **d.** 924

5. The value of $\lim_{n\to\infty} \frac{[r] + [2r] + \ldots\ldots + [nr]}{n^2}$, where r is non-zero real number and $[r]$ denotes the greatest integer less than or equal to r, is equal to

a. $r/2$ **b.** r **c.** $2r$ **d.** 0

6. The number of solutions of the equation $\sin^{-1}\left[x^2 + \frac{1}{3}\right] + \cos^{-1}\left[x^2 - \frac{2}{3}\right] = x^2$, for $x \in [-1, 1]$, and $[x]$ denotes the greatest integer less than or equal to x, is

a. 2 **b.** 0 **c.** 4 **d.** infinite

7. Let a computer program generate only the digits 0 and 1 to form a string of binary numbers with probability of occurrence of 0 at even places be $\frac{1}{2}$ and probability of occurrence of 0 at the odd place be $\frac{1}{3}$. Then, the probability that '10' is followed by '01' is equal to

a. $\frac{1}{18}$ **b.** $\frac{1}{3}$ **c.** $\frac{1}{6}$ **d.** $\frac{1}{9}$

8. The number of solutions of the equation $x + 2\tan x = \pi/2$ in the interval $[0, 2\pi]$ is

a. 3 **b.** 4 **c.** 2 **d.** 5

9. Let S_1, S_2 and S_3 be three sets defined as

$S_1 = \{z \in C : |z - 1| \le \sqrt{2}\}$

$S_2 = \{z \in C : \text{Re}[(1 - i)z] \ge 1\}$

$S_3 = \{z \in C : \text{Im}(z) \le 1\}$

Then, the set $S_1 \cap S_2 \cap S_3$

a. is a singleton

b. has exactly two elements

c. has infinitely many elements

d. has exactly three elements

10. If the curve $y = y(x)$ is the solution of the differential equation $2(x^2 + x^{5/4})dy - y(x + x^{1/4})\,dx = 2x^{9/4}\,dx, x > 0$ which passes through the point $\left(1, 1 - \frac{4}{3}\log_e 2\right)$, then the value of $y(16)$ is equal to

a. $4\left(\frac{31}{3} + \frac{8}{3}\log_e 3\right)$ **b.** $\left(\frac{31}{3} + \frac{8}{3}\log_e 3\right)$

c. $4\left(\frac{31}{3} - \frac{8}{3}\log_e 3\right)$ **d.** $\left(\frac{31}{3} - \frac{8}{3}\log_e 3\right)$

11. If the sides AB, BC and CA of a triangle ΔABC have 3, 5 and 6 interior points respectively, then the total number of triangles that can be constructed using these points as vertices, is equal to

a. 364 **b.** 240 **c.** 333 **d.** 360

12. If x, y, z are in arithmetic progression with common difference d, $x \ne 3d$, and the determinant of the matrix $\begin{bmatrix} 3 & 4\sqrt{2} & x \\ 4 & 5\sqrt{2} & y \\ 5 & k & z \end{bmatrix}$ is zero, then the value of k^2 is

a. 72 **b.** 12 **c.** 36 **d.** 6

13. Let O be the origin. Let $\mathbf{OP} = x\hat{\mathbf{i}} + y\hat{\mathbf{j}} - \hat{\mathbf{k}}$ and $\mathbf{OQ} = -\hat{\mathbf{i}} + 2\hat{\mathbf{j}} + 3x\hat{\mathbf{k}}$, $x, y \in R, x > 0$, be such that $|\mathbf{PQ}| = \sqrt{20}$ and the vector $\mathbf{OP}$ is perpendicular to $\mathbf{OQ}$. If $\mathbf{OR} = 3\hat{\mathbf{i}} + z\hat{\mathbf{j}} - 7\hat{\mathbf{k}}$, $z \in R$, is coplanar with $\mathbf{OP}$ and $\mathbf{OQ}$, then the value of $x^2 + y^2 + z^2$ is equal to

a. 7 **b.** 9 **c.** 2 **d.** 1

14. Two tangents are drawn from a point P to the circle $x^2 + y^2 - 2x - 4y + 4 = 0$, such that the angle between these tangents is $\tan^{-1}\left(\frac{12}{5}\right)$, where $\tan^{-1}\left(\frac{12}{5}\right) \in (0, \pi)$. If the centre of the circle is denoted by C and these tangents touch the circle at points A and B, then the ratio of the areas of ΔPAB and ΔCAB is

a. 11 : 4 **b.** 9 : 4 **c.** 3 : 1 **d.** 2 : 1

15. Consider the function $f : R \to R$ defined by

$$f(x) = \begin{cases} \left(2 - \sin\left(\frac{1}{x}\right)\right)|x|, & x \ne 0 \\ 0, & x = 0 \end{cases}$$

Then, f is

a. monotonic on $(-\infty, 0) \cup (0, \infty)$
b. not monotonic on $(-\infty, 0)$ and $(0, \infty)$
c. monotonic on $(0, \infty)$ only
d. monotonic on $(-\infty, 0)$ only

16. Let L be a tangent line to the parabola $y^2 = 4x - 20$ at $(6, 2)$. If L is also a tangent to the ellipse $\frac{x^2}{2} + \frac{y}{b} = 1$, then the value of b is equal to

a. 11 **b.** 14 **c.** 16 **d.** 20

17. The value of the limit $\lim_{\theta \to 0} \frac{\tan(\pi \cos^2 \theta)}{\sin(2\pi \sin^2 \theta)}$ is equal to

a. $-\frac{1}{2}$ **b.** $-\frac{1}{4}$ **c.** 0 **d.** $\frac{1}{4}$

18. Let the tangent to the circle $x^2 + y^2 = 25$ at the point $R(3, 4)$ meet X-axis and Y-axis at point P and Q, respectively. If r is the radius of the circle passing through the origin O and having centre at the incentre of the triangle OPQ, then r^2 is equal to

a. $\frac{529}{64}$ **b.** $\frac{125}{72}$ **c.** $\frac{625}{72}$ **d.** $\frac{585}{66}$

19. If the Boolean expression $(p \wedge q) * (p \otimes q)$ is a tautology, then $*$ and $\otimes$ are respectively, given by

a. $\rightarrow, \rightarrow$ **b.** $\wedge, \vee$ **c.** $\vee, \rightarrow$ **d.** $\wedge, \rightarrow$

20. If the equation of plane passing through the mirror image of a point $(2, 3, 1)$ with respect to line $\frac{x+1}{2} = \frac{y-3}{1} = \frac{z+2}{-1}$ and containing the line $\frac{x-2}{3} = \frac{1-y}{2} = \frac{z+1}{1}$ is $\alpha x + \beta y + \gamma z = 24$, then $\alpha + \beta + \gamma$ is equal to

a. 20 **b.** 19 **c.** 18 **d.** 21

Section B : Numerical Type Questions

21. If 1, $\log_{10}(4^x - 2)$ and $\log_{10}\left(4^x + \frac{18}{5}\right)$ are in arithmetic progression for a real number x, then the value of the determinant

$$\begin{vmatrix} 2\left(x - \frac{1}{2}\right) & x-1 & x^2 \\ 1 & 0 & x \\ x & 1 & 0 \end{vmatrix}$$ is equal to ……… .

22. Let $f : [-1, 1] \to R$ be defined as $f(x) = ax^2 + bx + c$ for all $x \in [-1, 1]$, where $a, b, c \in R$, such that $f(-1) = 2$, $f'(-1) = 1$ and for $x \in (-1, 1)$ the maximum value of $f''(x)$ is $\frac{1}{2}$. If $f(x) \le \alpha$, $x \in [-1, 1]$, then the least value of α is equal to …………… .

23. Let $f : [-3, 1] \to R$ be given as

$$f(x) = \begin{cases} \min\{(x+6), x^2\}, & -3 \le x \le 0 \\ \max\{\sqrt{x}, x^2\}, & 0 \le x \le 1 \end{cases}.$$

If the area bounded by $y = f(x)$ and x-axis is A, then the value of $6A$ is equal to ………… .

24. Let $\tan\alpha$, $\tan\beta$ and $\tan\gamma$; $\alpha, \beta, \gamma \ne \frac{(2n-1)\pi}{2}$, $n \in N$ be the slopes of three line segments OA, OB and OC, respectively, where O is origin. If circumcentre of ΔABC coincides with origin and its orthocentre lies on y-axis, then the value of $\left(\frac{\cos 3\alpha + \cos 3\beta + \cos 3\gamma}{\cos\alpha \cos\beta \cos\gamma}\right)^2$ is equal to ……… .

25. Consider a set of $3n$ numbers having variance 4. In this set, the mean of first $2n$ numbers is 6 and the mean of the remaining n numbers is 3. A new set is constructed by adding 1 into each of first $2n$ numbers and subtracting 1 from each of the remaining n numbers. If the variance of the new set is k, then $9k$ is equal to ……… .

26. Let the coefficients of third, fourth and fifth terms in the expansion of $\left(x + \frac{a}{x^2}\right)^n$, $x \ne 0$, be in the ratio 12 : 8 : 3. Then, the term independent of x in the expansion, is equal to …………… .

27. Let $A = \begin{bmatrix} a & b \\ c & d \end{bmatrix}$ and $B = \begin{bmatrix} \alpha \\ \beta \end{bmatrix} \ne \begin{bmatrix} 0 \\ 0 \end{bmatrix}$, such that $AB = B$ and $a + d = 2021$, then the value of $ad - bc$ is equal to …………… .

28. Let $\mathbf{x}$ be a vector in the plane containing vectors $\mathbf{a} = 2\hat{\mathbf{i}} - \hat{\mathbf{j}} + \hat{\mathbf{k}}$ and $\mathbf{b} = \hat{\mathbf{i}} + 2\hat{\mathbf{j}} - \hat{\mathbf{k}}$. If the vector $\mathbf{x}$ is perpendicular to $(3\hat{\mathbf{i}} + 2\hat{\mathbf{j}} - \hat{\mathbf{k}})$ and its projection on $\mathbf{a}$ is $\frac{17\sqrt{6}}{2}$, then the value of $|\vec{x}|^2$ is equal to ………… .

29. Let $I_n = \int_1^e x^{19}(\log|x|)^n \, dx$, where $n \in N$. If $(20) I_{10} = \alpha I_9 + \beta I_8$, for natural numbers α and β, then $\alpha - \beta$ is equal to …………… .

30. Let P be an arbitrary point having sum of the squares of the distance from the planes $x + y + z = 0$, $lx - nz = 0$ and $x - 2y + z = 0$, equal to 9. If the locus of the point P is $x^2 + y^2 + z^2 = 9$, then the value of $l - n$ is equal to ……………… .

Answers

For solution scan the QR code

1. (b)	2. (a)	3. (b)	4. (d)	5. (a)	6. (b)	7. (d)	8. (a)	9. (c)	10. (c)
11. (c)	12. (a)	13. (b)	14. (b)	15. (b)	16. (b)	17. (a)	18. (c)	19. (a)	20. (b)
21. (2)	22. (5)	23. (41)	24. (144)	25. (68)	26. (4)	27. (2020)	28. (486)	29. (1)	30. (0)

18 MARCH SHIFT I

Section A : Objective Type Questions

1. The differential equation satisfied by the system of parabolas $y^2 = 4a(x + a)$ is

a. $y\left(\frac{dy}{dx}\right)^2 - 2x\left(\frac{dy}{dx}\right) - y = 0$ **b.** $y\left(\frac{dy}{dx}\right)^2 - 2x\left(\frac{dy}{dx}\right) + y = 0$

c. $y\left(\frac{dy}{dx}\right)^2 + 2x\left(\frac{dy}{dx}\right) - y = 0$ **d.** $y\left(\frac{dy}{dx}\right) + 2x\left(\frac{dy}{dx}\right) - y = 0$

2. The number of integral values of m, so that the abscissa of point of intersection of lines $3x + 4y = 9$ and $y = mx + 1$ is also an integer, is

a. 1 **b.** 2

c. 3 **d.** 0

3. Let $(1 + x + 2x^2)^{20} = a_0 + a_1x + a_2x^2 + \dots + a_{40}x^{40}$, then $a_1 + a_3 + a_5 + \dots + a_{37}$ is equal to

a. $2^{20}(2^{20} - 21)$ **b.** $2^{19}(2^{20} - 21)$

c. $2^{19}(2^{20} + 21)$ **d.** $2^{20}(2^{20} + 21)$

4. The solutions of the equation

$$\begin{vmatrix} 1+\sin^2 x & \sin^2 x & \sin^2 x \\ \cos^2 x & 1+\cos^2 x & \cos^2 x \\ 4\sin 2x & 4\sin 2x & 1+4\sin 2x \end{vmatrix} = 0, (0 < x < \pi), \text{ are}$$

a. $\frac{\pi}{12}, \frac{\pi}{6}$ **b.** $\frac{\pi}{6}, \frac{5\pi}{6}$

c. $\frac{5\pi}{12}, \frac{7\pi}{12}$ **d.** $\frac{7\pi}{12}, \frac{11\pi}{12}$

5. Choose the correct statement about two circles whose equations are given below.

$x^2 + y^2 - 10x - 10y + 41 = 0$

$x^2 + y^2 - 22x - 10y + 137 = 0$

a. circles have same centre

b. circles have no meeting point

c. circles have only one meeting point

d. circles have two meeting points

6. Let α, β, γ be the real roots of the equation, $x^3 + ax^2 + bx + c = 0$, $(a, b, c \in R$ and $a, b \neq 0)$. If the system of equations (in u,v,w) given by $\alpha u + \beta v + \gamma w = 0$, $\beta u + \gamma v + \alpha w = 0$; $\gamma u + \alpha v + \beta w = 0$ has non-trivial solution, then the value of $\frac{a^2}{b}$ is

a. 5 **b.** 3 **c.** 1 **d.** 0

7. The integral $\int \frac{(2x-1)\cos\sqrt{(2x-1)^2+5}}{\sqrt{4x^2-4x+6}}\,dx$ is equal to

(where, c is a constant of integration)

a. $\frac{1}{2}\sin\sqrt{(2x-1)^2+5}+c$ **b.** $\frac{1}{2}\cos\sqrt{(2x+1)^2+5}+c$

c. $\frac{1}{2}\cos\sqrt{(2x-1)^2+5}+c$ **d.** $\frac{1}{2}\sin\sqrt{(2x+1)^2+5}+c$

8. The equation of one of the straight lines which passes through the point (1,3) and makes an angle $\tan^{-1}(\sqrt{2})$ with the straight line, $y + 1 = 3\sqrt{2}x$ is

a. $4\sqrt{2}x + 5y - (15 + 4\sqrt{2}) = 0$ **b.** $5\sqrt{2}x + 4y - (15 + 4\sqrt{2}) = 0$

c. $4\sqrt{2}x + 5y - 4\sqrt{2} = 0$ **d.** $4\sqrt{2}x - 5y - (5 + 4\sqrt{2}) = 0$

9. If $\lim_{x\to 0} \frac{\sin^{-1}x - \tan^{-1}x}{3x^3}$ is equal to L, then the value of $(6L + 1)$ is

a. 1/6 **b.** 1/2 **c.** 6 **d.** 2

10. A vector **a** has components $3p$ and 1 with respect to rectangular cartesian system. This system is rotated through a certain angle about the origin in the counter clockwise sense. If, with respect to new system, **a** has components $p + 1$ and $\sqrt{10}$, then a value of p is equal to

a. 1 **b.** $-\frac{5}{4}$ **c.** $\frac{4}{5}$ **d.** -1

11. If the equation $a|z|^2 + \overline{\alpha z} + \alpha\bar{z} + d = 0$ represents a circle, where a,d are real constants, then which of the following condition is correct ?

a. $|\alpha|^2 - ad \neq 0$ **b.** $|\alpha|^2 - ad > 0$ and $a \in R - \{0\}$

c. $|\alpha|^2 - ad \geq 0$ and $a \in R$ **d.** $\alpha = 0, a, d \in R^+$

12. For the four circles M, N, O and P, following four equations are given

Circle M : $x^2 + y^2 = 1$

Circle N : $x^2 + y^2 - 2x = 0$

Circle O : $x^2 + y^2 - 2x - 2y + 1 = 0$

Circle P : $x^2 + y^2 - 2y = 0$

If the centre of circle M is joined with centre of the circle N, further centre of circle N is joined with centre of the circle O, centre of circle O is joined with the centre of circle P and lastly, centre of circle P is joined with centre of circle M, then these lines form the sides of a

a. rhombus **b.** square

c. rectangle **d.** parallelogram

13. If α, β are natural numbers, such that $100^\alpha - 199\beta = (100)(100) + (99)(101) + (98)(102) + \dots + (1)(199)$, then the slope of the line passing through (α, β) and origin is

a. 540 **b.** 550

c. 530 **d.** 510

14. The real valued function $f(x) = \frac{\text{cosec}^{-1}x}{\sqrt{x - [x]}}$, where $[x]$ denotes the greatest integer less than or equal to x, is defined for all x belonging to

a. all reals except integers

b. all non-integers except the interval $[-1,1]$

c. all integers except $0, -1, 1$

d. all reals except the interval $[-1, 1]$

15. $\frac{1}{3^2-1}+\frac{1}{5^2-1}+\frac{1}{7^2-1}+\ldots+\frac{1}{(201)^2-1}$ is equal to

a. $\frac{101}{404}$ b. $\frac{25}{101}$

c. $\frac{101}{408}$ d. $\frac{99}{400}$

16. If the functions are defined as $f(x)=\sqrt{x}$ and $g(x)=\sqrt{1-x}$, then what is the common domain of the following functions?

$f+g, f-g, f/g, g/f, g-f$, where $(f \pm g)(x) = f(x) \pm g(x)$, $(f/g)(x) = \frac{f(x)}{g(x)}$

a. $0 \le x \le 1$ b. $0 \le x < 1$

c. $0 < x < 1$ d. $0 < x \le 1$

17. If $f(x) = \begin{cases} \frac{1}{|x|} & ; \ |x| \ge 1 \\ ax^2+b & ; \ |x| < 1 \end{cases}$

is differentiable at every point of the domain, then the values of a and b are respectively

a. $\frac{1}{2}, \frac{1}{2}$ b. $\frac{1}{2}, -\frac{3}{2}$

c. $\frac{5}{2}, -\frac{3}{2}$ d. $-\frac{1}{2}, \frac{3}{2}$

18. Let $A+2B = \begin{bmatrix} 1 & 2 & 0 \\ 6 & -3 & 3 \\ -5 & 3 & 1 \end{bmatrix}$ and $2A - B = \begin{bmatrix} 2 & -1 & 5 \\ 2 & -1 & 6 \\ 0 & 1 & 2 \end{bmatrix}$. If $Tr(A)$ denotes the sum of all diagonal elements of the matrix A, then $Tr(A) - Tr(B)$ has value equal to

a. 1 b. 2 c. 0 d. 3

19. The sum of all the 4-digit distinct numbers that can be formed with the digits 1, 2, 2 and 3 is

a. 26664 b. 122664

c. 122234 d. 22264

20. The value of $3+\cfrac{1}{4+\cfrac{1}{3+\cfrac{1}{4+\cfrac{1}{3+\ldots\infty}}}}$ is equal to

a. $1.5+\sqrt{3}$ b. $2+\sqrt{3}$

c. $3+2\sqrt{3}$ d. $4+\sqrt{3}$

Section B : Numerical Type Questions

21. The number of times the digit 3 will be written when listing the integers from 1 to 1000 is ....

22. Let the plane, $ax+by+cz+d=0$ bisect the line joining the points $(4,-3,1)$ and $(2,3,-5)$ at the right angles. If a, b, c, d are integers, then the minimum value of $(a^2+b^2+c^2+d^2)$ is ....

23. Let $f(x)$ and $g(x)$ be two functions satisfying $f(x^2)+g(4-x)=4x^3$ and $g(4-x)+g(x)=0$, then the value of $\int_{-4}^{4} f(x)^2 dx$ is ....

24. The missing value in the following figure is

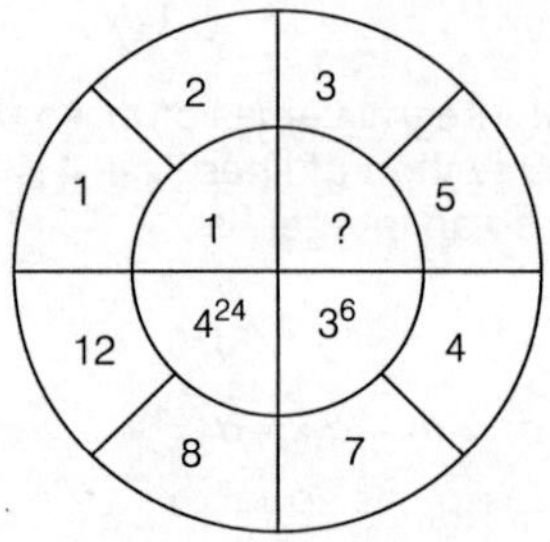

25. Let z_1, z_2 be the roots of the equation $z^2+az+12=0$ and z_1, z_2 form an equilateral triangle with origin. Then, the value of $|a|$ is ....

26. The equation of the planes parallel to the plane $x-2y+2z-3=0$ which are at unit distance from the point $(1, 2, 3)$ is $ax+by+cz+d=0$. If $(b-d)=K(c-a)$, then the positive value of K is ....

27. The mean age of 25 teachers in a school is 40 yr. A teacher retires at the age of 60 yr and a new teacher is appointed in his place. If the mean age of the teachers in this school now is 39 yr, then the age (in years) of the newly appointed teacher is ....

28. If $f(x)=\int \frac{5x^8+7x^6}{(x^2+1+2x^7)^2}dx, (x \ge 0), f(0)=0$ and $f(1)=\frac{1}{K}$, then the value of K is ....

29. A square $ABCD$ has all its vertices on the curve $x^2y^2=1$. The mid-points of its sides also lie on the same curve. Then, the square of area of $ABCD$ is ....

30. The number of solutions of the equation $|\cot x| = \cot x + \frac{1}{\sin x}$ in the interval $[0, 2\pi]$ is ....

Answers

For solution scan the QR code

1. *(c)*	2. *(b)*	3. *(b)*	4. *(d)*	5. *(c)*	6. *(b)*	7. *(a)*	8. *(a)*	9. *(d)*	10. *(d)*
11. *(b)*	12. *(b)*	13. *(b)*	14. *(b)*	15. *(b)*	16. *(c)*	17. *(d)*	18. *(b)*	19. *(a)*	20. *(a)*
21. *(300)*	22. *(28)*	23. *(512)*	24. *(4)*	25. *(6)*	26. *(4)*	27. *(35)*	28. *(4)*	29. *(80)*	30. *(1)*

18 MARCH SHIFT II

Section A : Objective Type Questions

1. Let $y = y(x)$ be the solution of the differential equation $\frac{dy}{dx} = (y+1)[(y+1)e^{x^2/2} - x], 0 < x < 2.1$, with $y(2) = 0$. Then the value of $\frac{dy}{dx}$ at $x = 1$ is equal to

a. $\frac{-e^{3/2}}{(e^2+1)^2}$ **b.** $-\frac{2e^2}{(1+e^2)^2}$

c. $\frac{e^{5/2}}{(1+e^2)^2}$ **d.** $\frac{5e^{1/2}}{(e^2+1)^2}$

2. In a triangle ABC, if $|\mathbf{BC}| = 8$, $|\mathbf{CA}| = 7$, $|\mathbf{AB}| = 10$, then the projection of the vector **AB** on **AC** is equal to

a. $\frac{25}{4}$ **b.** $\frac{85}{14}$

c. $\frac{127}{20}$ **d.** $\frac{115}{16}$

3. Let the system of linear equations

$4x + \lambda y + 2z = 0$

$2x - y + z = 0$

$\mu x + 2y + 3z = 0, \lambda, \mu \in R.$

has a non-trivial solution. Then which of the following is true ?

a. $\mu = 6, \lambda \in R$ **b.** $\lambda = 2, \mu \in R$

c. $\lambda = 3, \mu \in R$ **d.** $\mu = -6, \lambda \in R$

4. Let $f : R - \{3\} \to R - \{1\}$ be defined by $f(x) = \frac{x-2}{x-3}$. Let $g : R \to R$ be given as $g(x) = 2x - 3$. Then, the sum of all the values of x for which $f^{-1}(x) + g^{-1}(x) = \frac{13}{2}$ is equal to

a. 7 **b.** 2 **c.** 5 **d.** 3

5. Let the centroid of an equilateral triangle ΔABC be at the origin. Let one of the sides of the equilateral triangle be along the straight line $x + y = 3$. If R and r be the radius of circumcircle and incircle, respectively of ΔABC, then $(R + r)$ is equal to

a. $\frac{9}{\sqrt{2}}$ **b.** $7\sqrt{2}$

c. $2\sqrt{2}$ **d.** $3\sqrt{2}$

6. Consider a hyperbola $H : x^2 - 2y^2 = 4$. Let the tangent at a point $P(4, \sqrt{6})$ meet the x-axis at Q and latus rectum at $R(x_1, y_1), x_1 > 0$. If F is a focus of H which is nearer to the point P, then the area of ΔQFR is equal to

a. $4\sqrt{6}$ **b.** $\sqrt{6} - 1$ **c.** $\frac{7}{\sqrt{6}} - 2$ **d.** $4\sqrt{6} - 2$

7. If P and Q are two statements, then which of the following compound statement is a tautology ?

a. $(P \Rightarrow Q) \wedge \sim Q] \Rightarrow Q$ **b.** $(P \Rightarrow Q) \wedge \sim Q] \Rightarrow \sim P$

c. $(P \Rightarrow Q) \wedge \sim Q] \Rightarrow P$ **d.** $(P \Rightarrow Q) \wedge \sim Q] \Rightarrow (P \wedge Q)$

8. Let $g(x) = \int_0^x f(t)\, dt$, where f is continuous function in $[0, 3]$ such that $\frac{1}{3} \le f(t) \le 1$ for all $t \in [0, 1]$ and $0 \le f(t) \le \frac{1}{2}$ for all $t \in (1, 3]$. The largest possible interval in which g (3) lies is

a. $\left[-1, -\frac{1}{2}\right]$ **b.** $\left[-\frac{3}{2}, -1\right]$

c. $\left[\frac{1}{3}, 2\right]$ **d.** $[1, 3]$

9. Let S_1 be the sum of first $2n$ terms of an arithmetic progression. Let S_2 be the sum of first $4n$ terms of the same arithmetic progression. If $(S_2 - S_1)$ is 1000, then the sum of the first $6n$ terms of the arithmetic progression is equal to

a. 1000 **b.** 7000 **c.** 5000 **d.** 3000

10. Let a complex number be $w = 1 - \sqrt{3}i$. Let another complex number z be such that $|zw| = 1$ and $\arg(z) - \arg(w) = \frac{\pi}{2}$. Then the area of the triangle with vertices origin, z and w, is equal to

a. 4 **b.** $\frac{1}{2}$ **c.** $\frac{1}{4}$ **d.** 2

11. Let in a series of $2n$ observations, half of them are equal to a and remaining half are equal to $-a$. Also, by adding a constant b in each of these observations, the mean and standard deviation of new set become 5 and 20, respectively. Then, the value of $a^2 + b^2$ is equal to

a. 425 **b.** 650

c. 250 **d.** 925

12. Let $S_1 \Rightarrow x^2 + y^2 = 9$ and $S_2 \Rightarrow (x-2)^2 + y^2 = 1$. Then the locus of center of a variable circle S which touches S_1 internally and S_2 externally always passes through the points

a. $(0, \pm\sqrt{3})$ **b.** $\left(\frac{1}{2}, \pm\frac{\sqrt{5}}{2}\right)$

c. $\left(2, \pm\frac{3}{2}\right)$ **d.** $(1, \pm 2)$

13. Let $\mathbf{a}$ and $\mathbf{b}$ be two non-zero vectors perpendicular to each other and $|\mathbf{a}| = |\mathbf{b}|$. If $|\mathbf{a} \times \mathbf{b}| = |\mathbf{a}|$, then the angle between the vectors $[\mathbf{a} + \mathbf{b} + (\mathbf{a} \times \mathbf{b})]$ and $\mathbf{a}$ is equal to

a. $\sin^{-1}\left(\frac{1}{\sqrt{3}}\right)$ **b.** $\cos^{-1}\left(\frac{1}{\sqrt{3}}\right)$

c. $\cos^{-1}\left(\frac{1}{\sqrt{2}}\right)$ **d.** $\sin^{-1}\left(\frac{1}{\sqrt{6}}\right)$

14. Let in a Binomial distribution, consisting of 5 independent trials, probabilities of exactly 1 and 2 successes be 0.4096 and 0.2048, respectively. Then the probability of getting exactly 3 successes is equal to

a. $\frac{32}{625}$ **b.** $\frac{80}{243}$ **c.** $\frac{40}{243}$ **d.** $\frac{128}{625}$

15. Let a tangent be drawn to the ellipse $\frac{x^2}{27}+y^2=1$ at $(3\sqrt{3}\cos\theta, \sin\theta)$, where $\theta \in \left(0, \frac{\pi}{2}\right)$. Then the value of θ, such that the sum of intercepts on axes made by this tangent is minimum is equal to

a. $\frac{\pi}{8}$ b. $\frac{\pi}{4}$ c. $\frac{\pi}{6}$ d. $\frac{\pi}{3}$

16. Define a relation R over a class of $n \times n$ real matrices A and B as "ARB, if there exists a non-singular matrix P such that $PAP^{-1}=B$". Then which of the following is true ?

a. R is symmetric, transitive but not reflexive
b. R is reflexive, symmetric but not transitive
c. R is an equivalence relation
d. R is reflexive, transitive but not symmetric

17. A pole stands vertically inside a triangular park ABC. Let the angle of elevation of the top of the pole from each corner of the park be $\frac{\pi}{3}$. If the radius of the circumcircle of ΔABC is 2, then the height of the pole is equal to

a. $\frac{2\sqrt{3}}{3}$ b. $2\sqrt{3}$ c. $\sqrt{3}$ d. $\frac{1}{\sqrt{3}}$

18. If $15\sin^4\alpha + 10\cos^4\alpha = 6$, for some $\alpha \in R$, then the value of $27\sec^6\alpha + 8\operatorname{cosec}^6\alpha$ is equal to

a. 350 b. 500 c. 400 d. 250

19. The area bounded by the curve $4y^2 = x^2(4-x)(x-2)$ is equal to

a. $\frac{\pi}{8}$ b. $\frac{3\pi}{8}$ c. $\frac{3\pi}{2}$ d. $\frac{\pi}{16}$

20. Let $f: R \to R$ be a function defined as

$$f(x)=\begin{cases}\frac{\sin(a+1)x+\sin 2x}{2x}, & \text{if } x<0\\ b & \text{if } x<0\\ \frac{\sqrt{x+bx^3}-\sqrt{x}}{bx^{5/2}}, & \text{if } x>0\end{cases}$$

If f is continuous at $x=0$, then the value of $a+b$ is equal to

a. $-\frac{5}{2}$ b. -2 c. -3 d. $-\frac{3}{2}$

Section B : Numerical Type Questions

21. If $f(x)$ and $g(x)$ are two polynomials such that the polynomial $P(x)=f(x^3)+xg(x^3)$ is divisible by x^2+x+1, then $P(1)$ is equal to ....

22. Let I be an identity matrix of order 2×2 and $P=\begin{bmatrix}2 & -1\\ 5 & -3\end{bmatrix}$. Then the value of $n \in N$ for which $P^n = 5I - 8P$ is equal to ....

23. If $\sum_{r=1}^{10} r!(r^3+6r^2+2r+5)=\alpha(11!)$, then the value of α is equal to ....

24. The term independent of x in the expansion of $\left[\frac{x+1}{x^{2/3}-x^{1/3}+1}-\frac{x-1}{x-x^{1/2}}\right]^{10}$, $x \neq 1$, is equal to ....

25. Let $P(x)$ be a real polynomial of degree 3 which vanishes at $x=-3$. Let $P(x)$ have local minima at $x=1$, local maxima at $x=-1$ and $\int_{-1}^{1} P(x)\,dx = 18$, then the sum of all the coefficients of the polynomial $P(x)$ is equal to ....

26. Let the mirror image of the point $(1, 3, a)$ with respect to the plane $\mathbf{r}\cdot(2\hat{i}-\hat{j}+\hat{k})-b=0$ be $(-3, 5, 2)$. Then the value of $|a+b|$ is equal to ....

27. Let $f: R \to R$ satisfy the equation $f(x+y)=f(x)\cdot f(y)$ for all $x, y \in R$ and $f(x)\neq 0$ for any $x \in R$. If the function f is differentiable at $x=0$ and $f'(0)=3$, then $\lim_{h\to 0}\frac{1}{h}(f(h)-1)$ is equal to............ .

28. Let ${}^nC_r$ denote the binomial coefficient of x^r in the expansion of $(1+x)^n$.
If $\sum_{k=0}^{10}(2^2+3k)\,{}^nC_k=\alpha\cdot 3^{10}+\beta\cdot 2^{10}, \alpha, \beta \in R$, then $\alpha+\beta$ is equal to ....

29. Let P be a plane containing the line $\frac{x-1}{3}=\frac{y+6}{4}=\frac{z+5}{2}$ and parallel to the line $\frac{x-3}{4}=\frac{y-2}{-3}=\frac{z+5}{7}$. If the point $(1,-1,\alpha)$ lies on the plane P, then the value of $|5\alpha|$ is equal to ....

30. Let $y=y(x)$ be the solution of the differential equation $x\,dy - y\,dx=\sqrt{(x^2-y^2)}\,dx$, $x \geq 1$, with $y(1)=0$. If the area bounded by the line $x=1$, $x=e^{\pi}$, $y=0$ and $y=y(x)$ is $\alpha e^{2\pi}+\beta$, then the value of $10(\alpha+\beta)$ is equal to ....

Answers

For solution scan the QR code

1. (a)	2. (b)	3. (a)	4. (c)	5. (a)	6. (c)	7. (b)	8. (c)	9. (d)	10. (b)
11. (a)	12. (c)	13. (b)	14. (a)	15. (c)	16. (c)	17. (b)	18. (d)	19. (c)	20. (d)
21. (0)	22. (6)	23. (160)	24. (210)	25. (8)	26. (1)	27. (3)	28. (19)	29. (38)	30. (4)

20 JULY SHIFT I

Section A : Objective Type Questions

1. The Boolean expression $(p \wedge \sim q) \Rightarrow (q \vee \sim p)$ is equivalent to

a. $q \Rightarrow p$ **b.** $p \Rightarrow q$ **c.** $\sim q \Rightarrow p$ **d.** $p \Rightarrow \sim q$

2. Let a be a positive real number such that $\int_0^a e^{x-[x]}dx = 10e - 9$, where $[x]$ is the greatest integer less than or equal to x. Then, a is equal to

a. $10 - \log_e(1+e)$ **b.** $10 + \log_e 2$ **c.** $10 + \log_e 3$ **d.** $10 + \log_e(1+e)$

3. The mean of 6 distinct observations is 6.5 and their variance is 10.25. If 4 out of 6 observations are 2, 4, 5 and 7, then the remaining two observations are

a. 10, 11 **b.** 3, 18 **c.** 8, 13 **d.** 1, 20

4. The value of the integral $\int_{-1}^{1} \log_e(\sqrt{1-x} + \sqrt{1+x})dx$ is equal to

a. $\frac{1}{2}\log_e 2 + \frac{\pi}{4} - \frac{3}{2}$ **b.** $2\log_e 2 + \frac{\pi}{4} - 1$ **c.** $\log_e 2 + \frac{\pi}{2} - 1$ **d.** $2\log_e 2 + \frac{\pi}{2} - \frac{1}{2}$

5. If α and β are the distinct roots of the equation $x^2 + (3)^{1/4}x + 3^{1/2} = 0$, then the value of $\alpha^{96}(\alpha^{12} - 1) + \beta^{96}(\beta^{12} - 1)$ is equal to

a. 56×3^{25} **b.** 56×3^{24} **c.** 52×3^{24} **d.** 28×3^{25}

6. Let $A = \begin{bmatrix} 2 & 3 \\ a & 0 \end{bmatrix}$, $a \in R$ be written as $P + Q$, where P is a symmetric matrix and Q is skew symmetric matrix. If det $(Q) = 9$, then the modulus of the sum of all possible values of determinant of P is equal to

a. 36 **b.** 24 **c.** 45 **d.** 18

7. If z and ω are two complex numbers such that $|z\omega| = 1$ and $\arg(z) - \arg(\omega) = \frac{3\pi}{2}$, then $\arg\left(\frac{1 - 2\bar{z}\omega}{1 + 3\bar{z}\omega}\right)$ is

(Here, arg (z) denotes the principal argument of complex number z)

a. $\frac{\pi}{4}$ **b.** $-\frac{3\pi}{4}$ **c.** $-\frac{\pi}{4}$ **d.** $\frac{3\pi}{4}$

8. If in a triangle ABC, $AB = 5$ units, $\angle B = \cos^{-1}\left(\frac{3}{5}\right)$ and radius of circumcircle of ΔABC is 5 units, then the area (in square units) of ΔABC is

a. $10 + 6\sqrt{2}$ **b.** $8 + 2\sqrt{2}$ **c.** $6 + 8\sqrt{3}$ **d.** $4 + 2\sqrt{3}$

9. Let $[x]$ denote the greatest integer $\le x$, where $x \in R$. If the domain of the real valued function

$$f(x) = \sqrt{\frac{|[x]| - 2}{|[x]| - 3}} \text{ is } (-\infty, a) \cup [b, c) \cup [u, \infty),$$

$a < b < c$, then the value of $a + b + c$ is

a. 8 **b.** 1 **c.** –2 **d.** –3

10. Let $y = y(x)$ be the solution of the differential equation $x\tan\left(\frac{y}{x}\right)dy = \left(y\tan\left(\frac{y}{x}\right) - x\right)dx$, $-1 \le x \le 1$, $y\left(\frac{1}{2}\right) = \frac{\pi}{6}$. Then, the area of the region bounded by the curves $x = 0$, $x = \frac{1}{\sqrt{2}}$ and $y = y(x)$ in the upper half plane is

a. $\frac{1}{8}(\pi - 1)$ **b.** $\frac{1}{12}(\pi - 3)$ **c.** $\frac{1}{4}(\pi - 2)$ **d.** $\frac{1}{6}(\pi - 1)$

11. The coefficient of x^{256} in the expansion of $(1-x)^{101}(x^2 + x + 1)^{100}$ is

a. ${}^{100}C_{16}$ **b.** ${}^{100}C_{15}$ **c.** $-{}^{100}C_{16}$ **d.** $-{}^{100}C_{15}$

12. Let $A = [a_{ij}]$ be a 3×3 matrix, where

$$a_{ij} = \begin{cases} 1, & \text{if } i = j \\ -x, & \text{if } |i - j| = 1 \\ 2x + 1, & \text{otherwise} \end{cases}$$

Let a function $f : R \to R$ be defined as $f(x) = \det(A)$. Then, the sum of maximum and minimum values of f on R is equal to

a. $-\frac{20}{27}$ **b.** $\frac{88}{27}$ **c.** $\frac{20}{27}$ **d.** $-\frac{88}{27}$

13. Let $\mathbf{a} = 2\hat{\mathbf{i}} + \hat{\mathbf{j}} - 2\hat{\mathbf{k}}$ and $\mathbf{b} = \hat{\mathbf{i}} + \hat{\mathbf{j}}$. If $\mathbf{c}$ is a vector such that $\mathbf{a} \cdot \mathbf{c} = |\mathbf{c}|$, $|\mathbf{c} - \mathbf{a}| = 2\sqrt{2}$ and the angle between $(\mathbf{a} \times \mathbf{b})$ and $\mathbf{c}$ is $\frac{\pi}{6}$, then the value of $|(\mathbf{a} \times \mathbf{b}) \times \mathbf{c}|$ is

a. $\frac{2}{3}$ **b.** 4 **c.** 3 **d.** 3/2

14. The number of real roots of the equation $\tan^{-1}\sqrt{x(x+1)} + \sin^{-1}\sqrt{x^2 + x + 1} = \frac{\pi}{4}$ is

a. 1 **b.** 4 **c.** 3 **d.** 0

15. Let $y = y(x)$ be the solution of the differential equation $e^x\sqrt{1 - y^2}dx + \left(\frac{y}{x}\right)dy = 0$, $y(1) = -1$. Then, the value of $[y(3)]^2$ is equal to

a. $1 - 4e^3$ **b.** $1 - 4e^6$ **c.** $1 + 4e^3$ **d.** $1 + 4e^6$

16. Let a be a real number such that the function $f(x) = ax^2 + 6x - 15, x \in R$ is increasing in $\left(-\infty, \frac{3}{4}\right)$ and decreasing in $\left(\frac{3}{4}, \infty\right)$. Then, the function $g(x) = ax^2 - 6x + 15, x \in R$ has a

a. local maximum at $x = -\frac{3}{4}$

b. local minimum at $x = -\frac{3}{4}$

c. local maximum at $x = \frac{3}{4}$

d. local minimum at $x = \frac{3}{4}$

17. Let a function $f : R \to R$ be defined as

$$f(x) = \begin{cases} \sin x - e^x, & \text{if} \quad x \le 0 \\ a + [-x], & \text{if} \quad 0 < x < 1 \\ 2x - b, & \text{if} \quad x \ge 1 \end{cases}$$

where, $[x]$ is the greatest integer less than or equal to x. If f is continuous on R, then $(a + b)$ is equal to

a. 4 **b.** 3

c. 2 **d.** 5

18. Words with or without meaning are to be formed using all the letters of the word EXAMINATION. The probability that the letter M appears at the fourth position in any such word is

a. 1/66 **b.** 1/11

c. 1/9 **d.** 2/11

19. The probability of selecting integers $a \in [-5, 30]$ such that $x^2 + 2(a + 4)x - 5a + 64 > 0$, for all $x \in R$, is

a. $\frac{7}{36}$ **b.** $\frac{2}{9}$

c. $\frac{1}{6}$ **d.** $\frac{1}{4}$

20. Let the tangent to the parabola $S: y^2 = 2x$ at the point $P(2, 2)$ meet the X-axis at Q and normal at it meet the parabola S at the point R. Then, the area (in square units) of ΔPQR is equal to

a. $\frac{25}{2}$ **b.** $\frac{35}{2}$

c. $\frac{15}{2}$ **d.** 25

Section B : Numerical Type Questions

21. Let **a**, **b**, **c** be three mutually perpendicular vectors of the same magnitude and equally inclined at an angle θ, with the vector $\mathbf{a} + \mathbf{b} + \mathbf{c}$. Then, $36\cos^2 2\theta$ is equal to ....

22. Let $A = \begin{pmatrix} 1 & -1 & 0 \\ 0 & 1 & -1 \\ 0 & 0 & 1 \end{pmatrix}$ and $B = 7A^{20} - 20A^7 + 2I$, where I is an identity matrix of order 3×3. If $B = [b_{ij}]$, then b_{13} is equal to ....

23. Let P be a plane passing through the points (1, 0, 1), (1, −2, 1) and (0, 1, −2). Let a vector $\mathbf{a} = \alpha\hat{\mathbf{i}} + \beta\hat{\mathbf{j}} + \gamma\hat{\mathbf{k}}$ be such that **a** is parallel to the plane P, perpendicular to $(\hat{\mathbf{i}} + 2\hat{\mathbf{j}} + 3\hat{\mathbf{k}})$ and $\mathbf{a} \cdot (\hat{\mathbf{i}} + \hat{\mathbf{j}} + 2\hat{\mathbf{k}}) = 2$, then $(\alpha - \beta + \gamma)^2$ equals ....

24. The number of rational terms in the binomial expansion of $(4^{1/4} + 5^{1/6})^{120}$ is ....

25. If the shortest distance between the lines $\mathbf{r}_1 = \alpha\hat{\mathbf{i}} + 2\hat{\mathbf{j}} + 2\hat{\mathbf{k}} + \lambda(\hat{\mathbf{i}} - 2\hat{\mathbf{j}} + 2\hat{\mathbf{k}}), \lambda \in R, \alpha > 0$ and $\mathbf{r}_2 = -4\hat{\mathbf{i}} - \hat{\mathbf{k}} + \mu(3\hat{\mathbf{i}} - 2\hat{\mathbf{j}} - 2\hat{\mathbf{k}}), \mu \in R$ is 9, then α is equal to ....

26. Let T be the tangent to the ellipse $E : x^2 + 4y^2 = 5$ at the point $P(1, 1)$. If the area of the region bounded by the tangent T, ellipse E, lines $x = 1$ and $x = \sqrt{5}$ is $\alpha\sqrt{5} + \beta + \gamma\cos^{-1}\left(\frac{1}{\sqrt{5}}\right)$, then $|\alpha + \beta + \gamma|$ is equal to ....

27. Let a, b, c, d be in arithmetic progression with common difference λ.

If $\begin{vmatrix} x + a - c & x + b & x + a \\ x - 1 & x + c & x + b \\ x - b + d & x + d & x + c \end{vmatrix} = 2$,

then value of λ^2 is equal to ....

28. There are 15 players in a cricket team, out of which 6 are bowlers, 7 are batsmen and 2 are wicketkeepers. The number of ways, a team of 11 players be selected from them so as to include atleast 4 bowlers, 5 batsman and 1 wicketkeeper, is ....

29. Let $y = mx + c, m > 0$ be the focal chord of $y^2 = -64x$, which is tangent to $(x + 10)^2 + y^2 = 4$. Then, the value of $4\sqrt{2}(m + c)$ is equal to ....

30. If the value of $\lim_{x \to 0} (2 - \cos x\sqrt{\cos 2x})^{\left(\frac{x+2}{x^2}\right)}$ is equal to e^a, then a is equal to ....

Answers

For solution scan the QR code

1. (b)	2. (b)	3. (a)	4. (c)	5. (c)	6. (a)	7. (b)	8. (c)	9. (c)	10. (a)
11. (b)	12. (d)	13. (d)	14. (d)	15. (b)	16. (a)	17. (b)	18. (b)	19. (b)	20. (a)
21. (4)	22. (910)	23. (81)	24. (21)	25. (6)	26. (1.25)	27. (1)	28. (777)	29. (34)	30. (3)

20 JULY SHIFT II

Section A : Objective Type Questions

1. For the natural numbers m, n, if $(1-y)^m(1+y)^n = 1 + a_1y + a_2y^2 + \ldots\ldots + a_{m+n}y^{m+n}$ and $a_1 = a_2 = 10$, then the value of $(m+n)$ is equal to

a. 88 **b.** 64 **c.** 100 **d.** 80

2. The value of $\tan\left(2\tan^{-1}\left(\frac{3}{5}\right) + \sin^{-1}\left(\frac{5}{13}\right)\right)$ is equal to

a. $\frac{-181}{69}$ **b.** $\frac{220}{21}$ **c.** $\frac{-291}{76}$ **d.** $\frac{151}{63}$

3. Let r_1 and r_2 be the radii of the largest and smallest circles, respectively, which pass through the point $(-4, 1)$ and having their centres on the circumference of the circle $x^2 + y^2 + 2x + 4y - 4 = 0$. If $\frac{r_1}{r_2} = a + b\sqrt{2}$, then $a + b$ is equal to

a. 3 **b.** 11
c. 5 **d.** 7

4. Consider the following three statements,

(A) If $3 + 3 = 7$, then $4 + 3 = 8$

(B) If $5 + 3 = 8$, then earth is flat

(C) If both (A) and (B) are true, then $5 + 6 = 17$

Then, which of the following statements is correct

a. (A) is false, but (B) and (C) are true
b. (A) and (C) are true while (B) is false
c. (A) is true while (B) and (C) are false
d. (A) and (B) are false while (C) is true

5. The lines $x = ay - 1 = z - 2$ and $x = 3y - 2 = bz - 2$, $(ab \neq 0)$ are coplanar, if

a. $b = 1, a \in R - \{0\}$ **b.** $a = 1, b \in R - \{0\}$
c. $a = 2, b = 2$ **d.** $a = 2, b = 3$

6. If $[x]$ denotes the greatest integer less than or equal to x then the value of the integral $\int_{-\frac{\pi}{2}}^{\frac{\pi}{2}} [[x] - \sin x]dx$ is equal to

a. $-\pi$ **b.** π **c.** 0 **d.** 1

7. If the real part of the complex number $(1 - \cos\theta + 2i\sin\theta)^{-1}$ is $\frac{1}{5}$ for $\theta \in (0, \pi)$, then the value of the integral $\int_0^{\theta} \sin x\,dx$ is equal to

a. 1 **b.** 2 **c.** -1 **d.** 0

8. Let $f : R - \left\{\frac{\alpha}{6}\right\} \longrightarrow R$ be defined by $f(x) = \frac{5x+3}{6x-\alpha}$.

Then the value of α for which $(fof)(x) = x$, for all $x \in R - \left\{\frac{\alpha}{6}\right\}$ is

a. No such α exists **b.** 5
c. 8 **d.** 6

9. If $f : R \to R$ is given by $f(x) = x + 1$, then the value of

$$\lim_{n\to\infty} \frac{1}{n}\left[f(0) + f\left(\frac{5}{n}\right) + f\left(\frac{10}{n}\right) + \ldots + f\left(\frac{5(n-1)}{n}\right)\right]$$

is

a. 3/2 **b.** 5/2
c. 1/2 **d.** 7/2

10. Let A, B and C be three events such that the probability that exactly one of A and B occurs is $(1-k)$, the probability that exactly one of B and C occurs is $(1-2k)$, the probability that exactly one of C and A occurs is $(1-k)$ and the probability of all A, B and C occur simultaneously is k^2, where $0 < k < 1$. Then the probability that at least one of A, B and C occur is

a. greater than 1/8 but less than 1/4
b. greater than 1/2
c. greater than 1/4 but less than 2/2
d. exactly equal to 1/2

11. The sum of all the local minimum values of the twice diffrentiable function $f : R \to R$ defined by $f(x) = x^3 - 3x^2 - \frac{3f''(2)}{2}x + f''(1)$ is

a. -22 **b.** 5
c. -27 **d.** 0

12. Let in a right angled triangle, the smallest angle be θ. If a triangle formed by taking the reciprocal of its sides is also a right angled triangle, then $\sin\theta$ is equal to

a. $\frac{\sqrt{5}+1}{4}$ **b.** $\frac{\sqrt{5}-1}{2}$
c. $\frac{\sqrt{2}-1}{2}$ **d.** $\frac{\sqrt{5}-1}{4}$

13. Let $y = y(x)$ satisfies the equation $\frac{dy}{dx} - |A| = 0$, for all $x > 0$, where $A = \begin{bmatrix} y & \sin x & 1 \\ 0 & -1 & 1 \\ 2 & 0 & \frac{1}{x} \end{bmatrix}$. If $y(\pi) = \pi + 2$, then the value of $y\left(\frac{\pi}{2}\right)$ is

a. $\frac{\pi}{2} + \frac{4}{\pi}$ **b.** $\frac{\pi}{2} - \frac{1}{\pi}$ **c.** $\frac{3\pi}{2} - \frac{1}{\pi}$ **d.** $\frac{\pi}{2} - \frac{4}{\pi}$

14. Consider the line L given by the equation $\frac{x-3}{2} = \frac{y-1}{1} = \frac{z-2}{1}$.

Let Q be the mirror image of the point $(2, 3, -1)$ with respect to L. Let a plane P be such that it passes through Q, and the line L is perpendicular to P. Then which of the following points is on the plane P?

a. $(-1, 1, 2)$ **b.** $(1, 1, 1)$
c. $(1, 1, 2)$ **d.** $(1, 2, 2)$

15. If the mean and variance of six observations 7, 10, 11, 15, a, b are 10 and $\frac{20}{3}$, respectively, then the value of $|a-b|$ is equal to

a. 9 **b.** 11 **c.** 7 **d.** 1

16. Let $g(t)=\int_{-\frac{\pi}{2}}^{\frac{\pi}{2}}\cos\left(\frac{\pi}{4}t+f(x)\right)dx$, where $f(x)=\log_e(x+\sqrt{x^2+1}), x\in R$. Then, which one of the following is correct ?

a. $g(1)=g(0)$
b. $\sqrt{2}g(1)=g(0)$
c. $g(1)=\sqrt{2}g(0)$
d. $g(1)+g(0)=0$

17. Let P be a variable point on the parabola $y=4x^2+1$. Then, the locus of the mid point of the point P and the foot of the perpendicular drawn from the point P to the line $y=x$ is

a. $(3x-y)^2+(x-3y)+2=0$
b. $2(3x-y)^2+(x-3y)+2=0$
c. $(3x-y)^2+2(x-3y)+2=0$
d. $2(x-3y)^2+(3x-y)+2=0$

18. The value of $k\in R$, for which the following system of liner equations

$$3x-y+4z=3$$
$$x+2y-3z=-2$$
$$6x+5y+kz=-3$$

has infinitely many solutions, is

a. 3 **b.** −5 **c.** 5 **d.** −3

19. If sum of the first 21 terms of the series $\log_{9^{1/2}}x+\log_{9^{1/3}}x+\log_{9^{1/4}}x+\ldots$, where $x>0$ is 504, then x is equal to

a. 243 **b.** 9 **c.** 7 **d.** 81

20. In ΔABC, if $|\mathbf{BC}|=3$, $|\mathbf{CA}|=5$ and $|\mathbf{BA}|=7$, then the projection of the vector $\mathbf{BA}$ on $\mathbf{BC}$ is equal to

a. $\frac{19}{2}$
b. $\frac{13}{2}$
c. $\frac{11}{2}$
d. $\frac{15}{2}$

Section B : Numerical Type Questions

21. Let $A=\{a_{ij}\}$ be a 3×3 matrix, where

$$a_{ij}=\begin{cases}(-1)^{j-i} & \text{if } i<j\\ 2 & \text{if } i=j\\ (-1)^{i+j} & \text{if } i>j\end{cases}$$

then $\det[3\text{Adj}(2A^{-1})]$ is equal to

22. The number of solutions of the equation $\log_{(x+1)}(2x^2+7x+5)+\log_{(2x+5)}(x+1)^2-4=0$, $x>0$, is

23. Let a curve $y=y(x)$ be given by the solution of the differential equation

$$\cos\left(\frac{1}{2}\cos^{-1}(e^{-x})\right)dx=\sqrt{e^{2x}-1}dy$$

If it intersects Y-axis at $y=-1$ and the intersection point of the curve with X-axis is $(\alpha,0)$, then e^{α} is equal to

24. For $p>0$, a vector $\mathbf{V}_2=2\hat{\mathbf{i}}+(p+1)\hat{\mathbf{j}}$ is obtained by rotating the vector $\mathbf{v}_1=\sqrt{3}p\,\hat{\mathbf{i}}+\hat{\mathbf{j}}$ by an angle θ about origin in counter clockwise direction. If $\tan\theta=\frac{(\alpha\sqrt{3}-2)}{(4\sqrt{3}+3)}$, then the value of α is equal to

25. Consider a triangle having vertices $A(-2,3)$, $B(1,9)$ and $C(3,8)$. If a line L passing through the circumcentre of ΔABC, bisects line BC, and intersects Y-axis at point $\left(0,\frac{\alpha}{2}\right)$, then the value of real number α is

26. If the point on the curve $y^2=6x$, nearest to the point $\left(3,\frac{3}{2}\right)$ is (α,β), then $2(\alpha+\beta)$ is equal to

27. Let a function $g:[0,4]\to R$ be defined as

$$g(x)=\begin{cases}\max\limits_{0\le t\le x}(t^3-6t^2+9t-3), & 0\le x\le 3\\ 4-x, & 3<x\le 4\end{cases}$$

then the number of points in the interval (0, 4) where $g(x)$ is not differentiable, is ... .

28. For $k\in N$, let

$$\frac{1}{\alpha(\alpha+1)(\alpha+2)\ldots(\alpha+20)}=\sum_{k=0}^{20}\frac{A_k}{\alpha+k},$$

where $\alpha>0$. Then the value of $100\left(\frac{A_{14}+A_{15}}{A_{13}}\right)^2$ is equal to ....

29. Let $\{a_n\}_{n=1}^{\infty}$ be a sequence such that $a_1=1, a_2=1$ and $a_{n+2}=2a_{n+1}+a_n$ for all $n\ge1$. Then the value of $47\sum_{n=1}^{\infty}\frac{a_n}{2^{3n}}$ is equal to ... .

30. If $\lim\limits_{x\to0}\frac{\alpha xe^x-\beta\log_e(1+x)+\gamma x^2e^{-x}}{x\sin^2x}=10, \alpha,\beta,\gamma\in R$, then the value of $\alpha+\beta+\gamma$ is ... .

Answers

For solution scan the QR code

1. *(d)*	2. *(b)*	3. *(c)*	4. *(b)*	5. *(a)*	6. *(a)*	7. *(a)*	8. *(b)*	9. *(d)*	10. *(b)*
11. *(c)*	12. *(b)*	13. *(a)*	14. *(d)*	15. *(d)*	16. *(b)*	17. *(b)*	18. *(b)*	19. *(d)*	20. *(c)*
21. *(108)*	22. *(1)*	23. *(2)*	24. *(6)*	25. *(9)*	26. *(9)*	27. *(1)*	28. *(9)*	29. *(7)*	30. *(3)*

22 JULY SHIFT II

Section A : Objective Type Questions

1. Let L be the line of intersection of planes $\mathbf{r}\cdot(\hat{\mathbf{i}}-\hat{\mathbf{j}}+2\hat{\mathbf{k}})=2$ and $\mathbf{r}\cdot(2\hat{\mathbf{i}}+\hat{\mathbf{j}}-\hat{\mathbf{k}})=2$. If $P(\alpha, \beta, \gamma)$ is the foot of perpendicular on L from the point (1, 2, 0), then the value of $35(\alpha+\beta+\gamma)$ is equal to

a. 101 **b.** 119 **c.** 143 **d.** 134

2. Let S_n denote the sum of first n terms of an arithmetic progression. If $S_{10}=530$, $S_5=140$, then $S_{20}-S_6$ is equal to

a. 1862 **b.** 1842 **c.** 1852 **d.** 1872

3. Let $f: R \to R$ be defined as

$$f(x)=\begin{cases}-\frac{4}{3}x^3+2x^2+3x & , x>0\\ 3xe^x & , x\le 0\end{cases}$$

. Then, f is increasing function in the interval

a. $\left(-\frac{1}{2}, 2\right)$ **b.** (0, 2)

c. $\left(-1, \frac{3}{2}\right)$ **d.** (−3, −1)

4. Let $y=y(x)$ be the solution of the differential equation $\operatorname{cosec}^2 x\,dy+2dx=(1+y\cos 2x)\operatorname{cosec}^2 x\,dx$, with $y\left(\frac{\pi}{4}\right)=0$. Then, the value of $(y(0)+1)^2$ is equal to

a. $e^{1/2}$ **b.** $e^{-1/2}$

c. e^{-1} **d.** e

5. Four dice are thrown simultaneously and the numbers shown on these dice are recorded in 2×2 matrices. The probability that such formed matrices have all different entries and are non singular, is

a. $\frac{45}{162}$ **b.** $\frac{23}{81}$ **c.** $\frac{22}{81}$ **d.** $\frac{43}{162}$

6. Let a vector **a** be coplanar with vectors $\mathbf{b}=2\hat{\mathbf{i}}+\hat{\mathbf{j}}+\hat{\mathbf{k}}$ and $\mathbf{c}=\hat{\mathbf{i}}-\hat{\mathbf{j}}+\hat{\mathbf{k}}$. If **a** is perpendicular to $\mathbf{d}=3\hat{\mathbf{i}}+2\hat{\mathbf{j}}+6\hat{\mathbf{k}}$ and $|\mathbf{a}|=\sqrt{10}$. Then a possible value of $[\mathbf{a}\ \mathbf{b}\ \mathbf{c}]+[\mathbf{a}\ \mathbf{b}\ \mathbf{d}]+[\mathbf{a}\ \mathbf{c}\ \mathbf{d}]$ is equal to

a. −42 **b.** −40

c. −29 **d.** −38

7. If $\int_0^{100\pi}\frac{\sin^2 x}{e^{\left(\frac{x}{\pi}-\left[\frac{x}{\pi}\right]\right)}}dx=\frac{\alpha\pi^3}{1+4\pi^2}$, $\alpha\in R$, where $[x]$ is the greatest integer less than or equal to x, then the value of α is

a. $200(1-e^{-1})$ **b.** $100(1-e)$ **c.** $50(e-1)$ **d.** $150(e^{-1}-1)$

8. Let three vectors **a**, **b** and **c** be such that $\mathbf{a}\times\mathbf{b}=\mathbf{c}, \mathbf{b}\times\mathbf{c}=\mathbf{a}$ and $|\mathbf{a}|=2$. Then, which one of the following is not true ?

a. $\mathbf{a}\times((\mathbf{b}+\mathbf{c})\times(\mathbf{b}-\mathbf{c}))=\mathbf{0}$

b. Projection of **a** on $(\mathbf{b}\times\mathbf{c})$ is 2

c. $[\mathbf{a}\ \mathbf{b}\ \mathbf{c}]+[\mathbf{c}\ \mathbf{a}\ \mathbf{b}]=8$

d. $|3\mathbf{a}+\mathbf{b}-2\mathbf{c}|^2=51$

9. The values of λ and μ such that the system of equations $x+y+z=6, 3x+5y+5z=26, x+2y+\lambda z=\mu$ has no solution, are

a. $\lambda=3, \mu=5$ **b.** $\lambda=3, \mu\neq 10$

c. $\lambda\neq 2, \mu=10$ **d.** $\lambda=2, \mu\neq 10$

10. If the shortest distance between the straight lines $3(x-1)=6(y-2)=2(z-1)$ and $4(x-2)=2(y-\lambda)=(z-3)$, $\lambda\in R$ is $\frac{1}{\sqrt{38}}$, then the integral value of λ is equal to

a. 3 **b.** 2

c. 5 **d.** −1

11. Which of the following Boolean expression is not a tautology ?

a. $(p\Rightarrow q)\vee(\sim q\Rightarrow p)$ **b.** $(q\Rightarrow p)\vee(\sim q\Rightarrow p)$

c. $(p\Rightarrow\sim q)\vee(\sim q\Rightarrow p)$ **d.** $(\sim p\Rightarrow q)\vee(\sim q\Rightarrow p)$

12. Let $A=[a_{ij}]$ be a real matrix of order 3×3, such that $a_{i1}+a_{i2}+a_{i3}=1$, for $i=1, 2, 3$. Then, the sum of all the entries of the matrix A^3 is equal to

a. 2 **b.** 1

c. 3 **d.** 9

13. Let $[x]$ denote the greatest integer less than or equal to x. Then, the values of $x\in R$ satisfying the equation $[e^x]^2+[e^x+1]-3=0$ lie in the interval

a. $\left[0, \frac{1}{e}\right)$ **b.** $[\log_e 2, \log_e 3)$

c. $[1, e)$ **d.** $[0, \log_e 2)$

14. Let the circle $S: 36x^2+36y^2-108x+120y+C=0$ be such that it neither intersects nor touches the coordinate axes. If the point of intersection of the lines $x-2y=4$ and $2x-y=5$ lies inside the circle S, then

a. $\frac{25}{9}<C<\frac{13}{3}$ **b.** $100<C<165$

c. $81<C<156$ **d.** $100<C<156$

15. Let n denote the number of solutions of the equation $z^2+3\bar{z}=0$, where z is a complex number. Then, the value of $\sum_{k=0}^{\infty}\frac{1}{n^k}$ is equal to

a. 1 **b.** $\frac{4}{3}$ **c.** $\frac{3}{2}$ **d.** 2

16. The number of solutions of $\sin^7 x+\cos^7 x=1, x\in[0, 4\pi]$ is equal to

a. 11 **b.** 7 **c.** 5 **d.** 9

17. If the domain of the function $f(x)=\frac{\cos^{-1}\sqrt{x^2-x+1}}{\sqrt{\sin^{-1}\left(\frac{2x-1}{2}\right)}}$ is the interval (α, β), then $\alpha+\beta$ is equal to

a. $\frac{3}{2}$ **b.** 2 **c.** $\frac{1}{2}$ **d.** 1

18. Let $f : R \to R$ be defined as

$$f(x) = \begin{cases} \dfrac{x^3}{(1-\cos 2x)^2} \log_e \left(\dfrac{1+2xe^{-2x}}{(1-xe^{-x})^2} \right), & x \neq 0 \\ \alpha, & x = 0 \end{cases}$$

If f is continuous at $x = 0$, then α is equal to

a. 1 **b.** 3
c. 0 **d.** 2

19. Let a line $L : 2x + y = k, k > 0$ be a tangent to the hyperbola $x^2 - y^2 = 3$. If L is also a tangent to the parabola $y^2 = \alpha x$, then α is equal to

a. 12 **b.** −12 **c.** 24 **d.** −24

20. Let $E_1 : \dfrac{x^2}{a^2} + \dfrac{y^2}{b^2} = 1, a > b$. Let E_2 be another ellipse such that it touches the end points of major axis of E_1 and the foci of E_2 are the end points of minor axis of E_1. If E_1 and E_2 have same eccentricities, then its value is

a. $\dfrac{-1+\sqrt{5}}{2}$ **b.** $\dfrac{-1+\sqrt{8}}{2}$

c. $\dfrac{-1+\sqrt{3}}{2}$ **d.** $\dfrac{-1+\sqrt{6}}{2}$

Section B : Numerical Type Questions

21. Let $A = \{0, 1, 2, 3, 4, 5, 6, 7\}$. Then, the number of bijective functions $F : A \to A$ such that $f(1) + f(2) = 3 - f(3)$ is equal to ....

22. If the digits are not allowed to repeat in any number formed by using the digits 0, 2, 4, 6, 8, then the number of all numbers greater than 10000 is equal to ....

23. Let $A = \begin{bmatrix} 0 & 1 & 0 \\ 1 & 0 & 0 \\ 0 & 0 & 1 \end{bmatrix}$. Then, the number of 3×3 matrices B with entries from the set $\{1, 2, 3, 4, 5\}$ and satisfying $AB = BA$ is ....

24. Consider the following frequency distribution

Class	0-6	6-12	12-18	18-24	24-30
Frequency	a	b	12	9	5

If mean $= \dfrac{309}{22}$ and median = 14, then the value $(a-b)^2$ is equal to ....

25. The sum of all the elements in the set $\{n \in \{1, 2, ...100\}$: HCF of n and 2040 is 1} is equal to ....

26. The area (in square units) of the region bounded by the curves $x^2 + 2y - 1 = 0$, $y^2 + 4x - 4 = 0$ and $y^2 - 4x - 4 = 0$, in the upper half plane is ....

27. Let $f : R \to R$ be a function defined as

$$f(x) = \begin{cases} 3\left(1 - \dfrac{|x|}{2}\right), & \text{if } |x| \le 2 \\ 0, & \text{if } |x| > 2 \end{cases}$$

Let $g : R \to R$ be given by $g(x) = f(x+2) - f(x-2)$. If n and m denote the number of points in R, where g is not continuous and not differentiable respectively, then $n + m$ is equal to ....

28. If the constant term, in Binomial expansion of $\left(2x^r + \dfrac{1}{x^2}\right)^{10}$ is 180, then r is equal to ....

29. Let $y = y(x)$ be the solution of the differential equation

$$\left((x+2)e^{\left(\frac{y+1}{x+2}\right)} + (y+1) \right) dx = (x+2)dy,\ y(1) = 1.$$

If the domain of $y = y(x)$ is an open interval α, β, then $|\alpha + \beta|$ is equal to ....

30. The number of elements in the set $\{n \in \{1, 2, 3, ...,100\} : (11)^n > (10)^n + (9)^n\}$ is ....

Answers

For solution scan the QR code

1. *(b)*	2. *(a)*	3. *(c)*	4. *(c)*	5. *(d)*	6. *(a)*	7. *(a)*	8. *(d)*	9. *(d)*	10. *(a)*
11. *(d)*	12. *(c)*	13. *(d)*	14. *(d)*	15. *(b)*	16. *(c)*	17. *(a)*	18. *(a)*	19. *(d)*	20. *(a)*
21. *(720)*	22. *(96)*	23. *(3125)*	24. *(4)*	25. *(1251)*	26. *(2)*	27. *(4)*	28. *(8)*	29. *(4)*	30. *(96)*

25 JULY SHIFT I

Section A : Objective Type Questions

1. A spherical gas balloon of radius 16 m subtends an angle 60° at the eye of the observer *A* while the angle of elevation of its center from the eye of *A* is 75°. Then the height (in metre) of the top most point of the balloon from the level of the observer's eye is

a. $8(2+2\sqrt{3}+\sqrt{2})$ **b.** $8(\sqrt{6}+\sqrt{2}+2)$
c. $8(\sqrt{2}+2+\sqrt{3})$ **d.** $8(\sqrt{6}-\sqrt{2}+2)$

2. Let $f(x)=3\sin^4 x+10\sin^3 x+6\sin^2 x-3$

$x\in\left[-\frac{\pi}{6},\frac{\pi}{2}\right]$. Then, *f* is

a. increasing in $\left(-\frac{\pi}{6},\frac{\pi}{2}\right)$ **b.** decreasing in $\left(0,\frac{\pi}{2}\right)$
c. increasing in $\left(-\frac{\pi}{6},0\right)$ **d.** decreasing in $\left(-\frac{\pi}{6},0\right)$

3. Let S_n be the sum of the first *n* terms of an arithmetic progression, If $S_{3n}=3S_{2n}$, then the value of $\frac{S_{4n}}{S_{2n}}$ is

a. 6 **b.** 4 **c.** 2 **d.** 8

4. The locus of the centroid of the triangle formed by any point *P* on the hyperbola
$16x^2-9y^2+32x+36y-164=0$, and its foci is

a. $16x^2-9y^2+32x-36y-36=0$
b. $9x^2-16y^2+36x-32y-144=0$
c. $16x^2-9y^2+32x-36y-144=0$
d. $9x^2-16y^2+36x-32y-36=0$

5. Let the vectors
$(2+a+b)\hat{i}+(a+2b+c)\hat{j}-(b+c)\hat{k}$, $(1+b)\hat{i}+2b\hat{j}-b\hat{k}$ and $(2+b)\hat{i}+2b\hat{j}+(1-b)\hat{k}$, $a,b,c\in R$ be co-planar.

Then, which of the following is true?

a. $2b=a+c$ **b.** $3c=a+b$
c. $a=b+2c$ **d.** $2a=b+c$

6. Let $f:R\to R$ be defined as

$$f(x)=\begin{cases}\dfrac{\lambda|x^2-5x+6|}{\mu(5x-x^2-6)}, & x<2\\ e^{\frac{\tan(x-2)}{x-[x]}}, & x>2\\ \mu, & x=2\end{cases}$$

where, $[x]$ is the greatest integer less than or equal to *x*. If *f* is continuous at $x=2$, then $\lambda+\mu$ is equal to

a. $e(-e+1)$ **b.** $e(e-2)$ **c.** 1 **d.** $2e-1$

7. The value of the definite integral $\int_{\frac{\pi}{24}}^{\frac{5\pi}{24}}\frac{dx}{1+\sqrt[3]{\tan 2x}}$ is

a. $\frac{\pi}{3}$ **b.** $\frac{\pi}{6}$ **c.** $\frac{\pi}{12}$ **d.** $\frac{\pi}{18}$

8. If *b* is very small as compared to the value of *a*, so that the cube and other higher powers of $\frac{b}{a}$ can be neglected in the identity $\frac{1}{a-b}+\frac{1}{a-2b}+\frac{1}{a-3b}+\ldots+\frac{1}{a-nb}$ $=\alpha n+\beta n^2+\gamma n^3$, then the value of γ is

a. $\frac{a^2+b}{3a^3}$ **b.** $\frac{a+b}{3a^2}$ **c.** $\frac{b^2}{3a^3}$ **d.** $\frac{a+b^2}{3a^3}$

9. Let $y=y(x)$ be the solution of the differential equation $\frac{dy}{dx}=1+xe^{y-x}$, $-\sqrt{2}<x<\sqrt{2}$, $y(0)=0$, then the minimum value of $y(x)$, $x\in(-\sqrt{2},\sqrt{2})$ is equal to

a. $(2-\sqrt{3})-\log_e 2$ **b.** $(2+\sqrt{3})-\log_e 2$
c. $(1+\sqrt{3})-\log_e(\sqrt{3}-1)$ **d.** $(1-\sqrt{3})-\log_e(\sqrt{3}-1)$

10. The Boolean expression $(p\Rightarrow q)\wedge(q\Rightarrow\sim p)$ is equivalent to

a. $\sim q$ **b.** q **c.** p **d.** $\sim p$

11. The area (in sq. units) of the region, given by the set $\{(x,y)\in R\times R\mid x\ge 0, 2x^2\le y\le 4-2x\}$ is

a. $\frac{8}{3}$ **b.** $\frac{17}{3}$ **c.** $\frac{13}{3}$ **d.** $\frac{7}{3}$

12. The sum of all values of *x* in $[0,2\pi]$, for which $\sin x+\sin 2x+\sin 3x+\sin 4x=0$, is equal to

a. 8π **b.** 11π
c. 12π **d.** 9π

13. Let $g:N\to N$ be defined as
$g(3n+1)=3n+2$,
$g(3n+2)=3n+3$,
$g(3n+3)=3n+1$, for all $n\ge 0$.

Then which of the following statements is true ?

a. There exists an onto function $f:N\to N$ such that $fog=f$
b. There exists a one-one function $f:N\to N$ such that $fog=f$
c. $gogog=g$
d. There exists a function $f:N\to N$ such that $gof=f$

14. Let $f:[0,\infty)\to[0,\infty)$ be defined as $f(x)=\int_0^x[y]dy$

where, $[x]$ is the greatest integer less than or equal to *x*. Which of the following is true ?

a. *f* is continuous at every point in $[0,\infty)$ and differentiable except at the integer points
b. *f* is both continuous and differentiable except at the integer points in $[0,\infty)$
c. *f* is continuous everywhere except at the integer points in $[0,\infty)$
d. *f* is differentiable at every point in $[0,\infty)$

15. The values of *a* and *b*, for which the system of equations
$2x+3y+6z=8$, $x+2y+az=5$
$3x+5y+9z=b$
has no solution, are

a. $a=3, b\ne 13$ **b.** $a\ne 3, b\ne 13$
c. $a\ne 3, b=3$ **d.** $a=3, b=13$

16. Let 9 distinct balls be distributed among 4 boxes, B_1, B_2, B_3 and B_4. If the probability that B_3 contains exactly 3 balls is $k\left(\frac{3}{4}\right)^9$, then k lies in the set

a. $\{x \in R : |x-3| < 1\}$
b. $\{x \in R : |x-2| \le 1\}$
c. $\{x \in R : |x-13| < 1\}$
d. $\{x \in R : |x-5| \le 1\}$

17. Let a parabola P be such that its vertex and focus lie on the positive x-axis at a distance 2 and 4 units from the origin, respectively. If tangents are drawn from $O(0, 0)$ to the parabola P which meet P at S and R, then the area (in sq. units) of ΔSOR is equal to

a. $16\sqrt{2}$ b. 16
c. 32 d. $8\sqrt{2}$

18. The number of real roots of the equation $e^{6x} - e^{4x} - 2e^{3x} - 12e^{2x} + e^x + 1 = 0$ is

a. 2 b. 4
c. 6 d. 1

19. Let an ellipse $E \Rightarrow \frac{x^2}{a^2} + \frac{y^2}{b^2} = 1, a^2 > b^2$, passes through $\left(\sqrt{\frac{3}{2}}, 1\right)$ and has eccentricity $\frac{1}{\sqrt{3}}$. If a circle, centered at focus $F(\alpha, 0), \alpha > 0$, of E and radius $\frac{2}{\sqrt{3}}$, intersects E at two points P and Q, then PQ^2 is equal to

a. $\frac{8}{3}$ b. $\frac{4}{3}$
c. $\frac{16}{3}$ d. 3

20. Let the foot of perpendicular from a point $P(1, 2, -1)$ to the straight line $L \Rightarrow \frac{x}{1} = \frac{y}{0} = \frac{z}{-1}$ be N. Let a line be drawn from P parallel to the plane $x + y + 2z = 0$ which meets L at point Q. If α is the acute angle between the lines PN and PQ, then $\cos\alpha$ is equal to

a. $\frac{1}{\sqrt{5}}$ b. $\frac{\sqrt{3}}{2}$
c. $\frac{1}{\sqrt{3}}$ d. $\frac{1}{2\sqrt{3}}$

Section B : Numerical Type Questions

21. Let $y = y(x)$ be solution of the following differential equation $e^y \frac{dy}{dx} - 2e^y \sin x + \sin x \cos^2 x = 0, y\left(\frac{\pi}{2}\right) = 0$.
If $y(0) = \log_e(\alpha + \beta e^{-2})$, then $4(\alpha + \beta)$ is equal to ....

22. If the value of $\left(1 + \frac{2}{3} + \frac{6}{3^2} + \frac{10}{3^3} + \ldots \text{upto } \infty\right)^{\log_{(0.25)}\left(\frac{1}{3} + \frac{1}{3^2} + \frac{1}{3^3} + \ldots \text{upto } \infty\right)}$ is l, then l^2 is equal to ....

23. Consider the following frequency distribution

Class	10-20	20-30	30-40	40-50	50-60
Frequency	α	110	54	30	β

If the sum of all frequencies is 584 and median is 45, then $|\alpha - \beta|$ is equal to ....

24. Let $\mathbf{p} = 2\hat{i} + 3\hat{j} + \hat{k}$ and $\mathbf{q} = \hat{i} + 2\hat{j} + \hat{k}$ be two vectors. If a vector $\mathbf{r} = (\alpha\hat{j} + \beta\hat{j} + \gamma\hat{k})$ is perpendicular to each of the vectors $(\mathbf{p} + \mathbf{q})$ and $(\mathbf{p} - \mathbf{q})$, and $|\mathbf{r}| = \sqrt{3}$, then $|\alpha| + |\beta| + |\gamma|$ is equal to ....

25. The ratio of the coefficient of the middle term in the expansion of $(1 + x)^{20}$ and the sum of the coefficients of two middle terms in expansion of $(1 + x)^{19}$ is ....

26. Let $M = \left[A = \begin{pmatrix} a & b \\ c & d \end{pmatrix} : a, b, c, d \in \{\pm 3, \pm 2, \pm 1, 0\}\right]$
Define $f : M \to Z$, as $f(A) = \det(A)$, for all $A \in M$, where Z is set of all integers. Then the number of $A \in M$ such that $f(A) = 15$ is equal to ......

27. There are 5 students in class 10, 6 students in class 11 and 8 students in class 12. If the number of ways in which 10 students can be selected from them so as to include at least 2 students from each class and at most 5 students from the total 11 students of class 10 and 11 is 100 k, then k is equal to ....

28. If α, β are roots of the equation.
$x^2 + 5(\sqrt{2})x + 10 = 0, \alpha > \beta$ and $P_n = \alpha^n - \beta^n$ for each positive integer n, then the value of $\left(\frac{P_{17}P_{20} + 5\sqrt{2}P_{17}P_{19}}{P_{18}P_{19} + 5\sqrt{2}P_{18}^2}\right)$ is equal to ....

29. The term independent of x in the expansion of
$\left(\frac{x+1}{x^{23} - x^{1/3} + 1} - \frac{x-1}{x - x^{1/2}}\right)^{10}$, where $x \ne 0, 1$ is equal to ..... .

30. Let $S = \left\{n \in N \,\middle|\, \begin{pmatrix} 0 & i \\ 1 & 0 \end{pmatrix}^n \begin{pmatrix} a & b \\ c & d \end{pmatrix} = \begin{pmatrix} a & b \\ c & d \end{pmatrix}, \forall a, b, c, d \in R\right\}$,
where $i = \sqrt{-1}$. Then the number of 2-digit numbers in the set S is ....

Answers

For solution scan the QR code

1. *(b)*	2. *(d)*	3. *(a)*	4. *(a)*	5. *(a)*	6. *(a)*	7. *(c)*	8. *(c)*	9. *(d)*	10. *(d)*
11. *(d)*	12. *(d)*	13. *(a)*	14. *(a)*	15. *(a)*	16. *(a)*	17. *(b)*	18. *(a)*	19. *(c)*	20. *(c)*
21. *(4)*	22. *(3)*	23. *(164)*	24. *(3)*	25. *(1)*	26. *(16)*	27. *(238)*	28. *(1)*	29. *(210)*	30. *(11)*

25 JULY SHIFT II

Section A : Objective Type Questions

1. The sum of all those terms which are rational numbers in the expansion of $(2^{1/3} + 3^{1/4})^{12}$ is

a. 89 **b.** 27
c. 35 **d.** 43

2. The first of the two samples in a group has 100 items with mean 15 and standard deviation 3. If the whole group has 250 items with mean 15.6 and standard deviation $\sqrt{13.44}$, then the standard deviation of the second sample is

a. 8 **b.** 6
c. 4 **d.** 5

3. If $f(x) = \begin{cases} \int_0^x (5+|1-t|)dt, & x > 2 \\ 5x+1, & x \le 2 \end{cases}$, then

a. $f(x)$ is not continuous at $x = 2$
b. $f(x)$ is everywhere differentiable
c. $f(x)$ is continuous but not differentiable at $x = 2$
d. $f(x)$ is not differentiable at $x = 1$

4. If the greatest value of the term independent of x in the expansion of $\left(x\sin\alpha + a\dfrac{\cos\alpha}{x}\right)^{10}$ is $\dfrac{10!}{(5!)^2}$, then the value of a is equal to

a. −1 **b.** 1 **c.** −2 **d.** 2

5. Consider the statement "The match will be played only if the weather is good and ground is not wet".
Select the correct negation from the following

a. The match will not be played and weather is not good and ground is wet.
b. If the match will not be played, then either weather is not good or ground is wet.
c. The match will be played and weather is not good or ground is wet.
d. The match will not be played or weather is good and ground is not wet.

6. The value of $\cot\dfrac{\pi}{24}$ is

a. $\sqrt{2}+\sqrt{3}+2-\sqrt{6}$ **b.** $\sqrt{2}+\sqrt{3}+2+\sqrt{6}$
c. $\sqrt{2}-\sqrt{3}-2+\sqrt{6}$ **d.** $3\sqrt{2}-\sqrt{3}-\sqrt{6}$

7. The lowest integer which is greater than $\left(1+\dfrac{1}{10^{100}}\right)^{10^{100}}$ is

a. 3 **b.** 4
c. 2 **d.** 1

8. The value of the integral $\int_{-1}^{1} \log(x+\sqrt{x^2+1})dx$ is

a. 2 **b.** 0
c. −1 **d.** 1

9. Let a, b and c be distinct positive numbers. If the vectors $a\hat{\mathbf{i}} + a\hat{\mathbf{j}} + c\hat{\mathbf{k}}$, $\hat{\mathbf{i}} + \hat{\mathbf{k}}$ and $c\hat{\mathbf{i}} + c\hat{\mathbf{j}} + b\hat{\mathbf{k}}$ are coplanar, then c is equal to

a. $\dfrac{2}{\dfrac{1}{a}+\dfrac{1}{b}}$ **b.** $\dfrac{a+b}{2}$
c. $\dfrac{1}{a}+\dfrac{1}{b}$ **d.** $\sqrt{ab}$

10. If $[x]$ be the greatest integer less than or equal to x, then $\sum_{n=8}^{100}\left[\dfrac{(-1)^n n}{2}\right]$ is equal to

a. 0 **b.** 4 **c.** −2 **d.** 2

11. The number of distinct real roots of
$\begin{vmatrix} \sin x & \cos x & \cos x \\ \cos x & \sin x & \cos x \\ \cos x & \cos x & \sin x \end{vmatrix} = 0$ in the interval $-\dfrac{\pi}{4} \le x \le \dfrac{\pi}{4}$ is

a. 4 **b.** 1 **c.** 2 **d.** 3

12. If $|\mathbf{a}| = 2, |\mathbf{b}| = 5$ and $|\mathbf{a}\times\mathbf{b}| = 8$, then $|\mathbf{a}.\,\mathbf{b}|$ is equal to

a. 6 **b.** 4 **c.** 3 **d.** 5

13. The number of real solutions of the equation $x^2 - |x| - 12 = 0$ is

a. 2 **b.** 3 **c.** 1 **d.** 4

14. Consider function $f : A \to B$ and $g : B \longrightarrow C(A, B, C \subseteq R)$ such that $(gof)^{-1}$ exists, then

a. f and g both are one-one
b. f and g both are onto
c. f is one-one and g is 0 into
d. f is onto and g is one-one

15. If $P = \begin{bmatrix} 1 & 0 \\ 1/2 & 1 \end{bmatrix}$, then P^{50} is

a. $\begin{bmatrix} 1 & 0 \\ 25 & 1 \end{bmatrix}$ **b.** $\begin{bmatrix} 1 & 50 \\ 0 & 1 \end{bmatrix}$
c. $\begin{bmatrix} 1 & 25 \\ 0 & 1 \end{bmatrix}$ **d.** $\begin{bmatrix} 1 & 0 \\ 50 & 1 \end{bmatrix}$

16. Let x be a random variable such that the probability function of a distribution is given by $P(X=0) = \dfrac{1}{2}$, $P(X = j) = \dfrac{1}{3^j}(j = 1, 2, 3, \ldots \infty)$. Then, the mean of the distribution and P (X is positive and even) respectively are

a. $\dfrac{3}{8}$ and $\dfrac{1}{8}$ **b.** $\dfrac{3}{4}$ and $\dfrac{1}{8}$ **c.** $\dfrac{3}{4}$ and $\dfrac{1}{9}$ **d.** $\dfrac{3}{4}$ and $\dfrac{1}{16}$

17. If a tangent to the ellipse $x^2 + 4y^2 = 4$ meets the tangents at the extremities of its major axis at B and C, then the circle with BC as diameter passes through the point

a. $(\sqrt{3}, 0)$ **b.** $(\sqrt{2}, 0)$ **c.** $(1, 1)$ **d.** $(-1, 1)$

18. Let the equation of the pair of lines, $y = px$ and $y = qx$ can be written as $(y - px)(y - qx) = 0$
Then, the equation of the pair of the angle bisectors of the line $x^2 - 4xy - 5y^2 = 0$ is

a. $x^2 - 3xy + y^2 = 0$
b. $x^2 + 4xy - y^2 = 0$
c. $x^2 + 3xy - y^2 = 0$
d. $x^2 - 3xy - y^2 = 0$

19. If ${}^nP_r = {}^nP_{r+1}$ and ${}^nC_r = {}^nC_{r-1}$, then the value of r is equal to

a. 1 **b.** 4
c. 2 **d.** 3

20. Let $y = y(x)$ be the solution of the differential equation $xdy = (y + x^3 \cos x)dx$ with $y(\pi) = 0$, then $y\left(\frac{\pi}{2}\right)$ is equal to

a. $\frac{\pi^2}{4} + \frac{\pi}{2}$ **b.** $\frac{\pi^2}{2} + \frac{\pi}{4}$
c. $\frac{\pi^2}{2} - \frac{\pi}{4}$ **d.** $\frac{\pi^2}{4} - \frac{\pi}{2}$

Section B : Numerical Type Questions

21. Let $n \in N$ and $[x]$ denote the greatest integer less than or equal to x. If the sum of $(n + 1)$ terms ${}^nC_0, 3.{}^nC_1, 5.{}^nC_2, 7.{}^nC_3$ ...... is equal to 2^{100}. 101, then $2\left[\frac{n-1}{2}\right]$ is equal to ....

22. Consider the function

$$f(x) = \begin{cases} \dfrac{P(x)}{\sin(x-2)} & , x \neq 2 \\ 7 & , x = 2 \end{cases}$$

where, $P(x)$ is a polynomial such that $P''(x)$ is always a constant and $P(3) = 9$. If $f(x)$ is continuous at $x = 2$, then $P(5)$ is equal to ....

23. The equation of a circle is $\text{Re}(z^2) + 2[\text{Im}(z)]^2 + 2\text{Re}(z) = 0$. where, $z = x + iy$. A line which passes through the center of the given circle and the vertex of the parabola, $x^2 - 6x - y + 13 = 0$, has y-intercept equal to ....

24. If a rectangle is inscribed in an equilateral triangle of side length $2\sqrt{2}$ as shown in the figure, then the square of the largest area of such a rectangle is ....

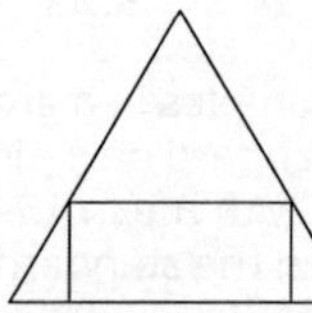

25. If $(\mathbf{a} + 3\mathbf{b})$ is perpendicular to $(7\mathbf{a} - 5\mathbf{b})$ and $(\mathbf{a} - 4\mathbf{b})$ is perpendicular to $(7\mathbf{a} - 2\mathbf{b})$, then the angle between $\mathbf{a}$ and $\mathbf{b}$ (in degrees) is ....

26. Let a curve $y = f(x)$ pass through the point $[2, (\log_e 2)^2]$ and have slope $\frac{2y}{x \log_e x}$ for all positive real value of x. Then the value of $f(e)$ is equal to ....

27. If $a + b + c = 1$, $ab + bc + ca = 2$ and $abc = 3$, then the value of $a^4 + b^4 + c^4$ is equal to ....

28. A fair coin is tossed n times such that the probability of getting at least one head is at least 0.9. Then the minimum value of n is ....

29. If the co-efficient of x^7 and x^8 in the expansion of $\left(2 + \frac{x}{3}\right)^n$ are equal, then the value of n is equal to ....

30. If the lines $\frac{x-k}{1} = \frac{y-2}{2} = \frac{z-3}{3}$ and $\frac{x+1}{3} = \frac{y+2}{2} = \frac{z+3}{1}$ are co-planar, then the value of k is ....

Answers

For solution scan the QR code

1. *(d)*	2. *(c)*	3. *(c)*	4. *(d)*	5. *(c)*	6. *(b)*	7. *(a)*	8. *(b)*	9. *(d)*	10. *(b)*
11. *(b)*	12. *(a)*	13. *(a)*	14. *(c)*	15. *(a)*	16. *(b)*	17. *(a)*	18. *(c)*	19. *(c)*	20. *(a)*
21. *(98)*	22. *(39)*	23. *(1)*	24. *(3)*	25. *(60)*	26. *(1)*	27. *(13)*	28. *(4)*	29. *(55)*	30. *(1)*

27 JULY SHIFT I

Section A : Objective Type Questions

1. If the mean and variance of the following data : 6, 10, 7, 13, a, 12, b, 12 are 9 and $\frac{37}{4}$ respectively, then $(a - b)^2$ is equal to

a. 24 **b.** 12 **c.** 32 **d.** 16

2. The value of $\lim_{n \to \infty} \frac{1}{n} \sum_{j=1}^{n} \frac{(2j-1) + 8n}{(2j-1) + 4n}$ is equal to

a. $5 + \log_e\left(\frac{3}{2}\right)$ **b.** $2 - \log_e\left(\frac{2}{3}\right)$
c. $3 + 2\log_e\left(\frac{2}{3}\right)$ **d.** $1 + 2\log_e\left(\frac{3}{2}\right)$

3. Let $\mathbf{a} = \hat{\mathbf{i}} + \hat{\mathbf{j}} + 2\hat{\mathbf{k}}$ and $\mathbf{b} = -\hat{\mathbf{i}} + 2\hat{\mathbf{j}} + 3\hat{\mathbf{k}}$. Then the vector product $(\mathbf{a} + \mathbf{b}) \times [(\mathbf{a} \times \{(\mathbf{a} - \mathbf{b}) \times \mathbf{b})\} \times \mathbf{b}]$ is equal to

a. $5(34\hat{\mathbf{i}} - 5\hat{\mathbf{j}} + 3\hat{\mathbf{k}})$ **b.** $7(34\hat{\mathbf{i}} - 5\hat{\mathbf{j}} + 3\hat{\mathbf{k}})$
c. $7(30\hat{\mathbf{i}} - 5\hat{\mathbf{j}} + 7\hat{\mathbf{k}})$ **d.** $5(30\hat{\mathbf{i}} - 5\hat{\mathbf{j}} + 7\hat{\mathbf{k}})$

4. The value of the definite integral

$$\int_{-\frac{\pi}{4}}^{\frac{\pi}{4}} \frac{dx}{(1+e^{x\cos x})(\sin^4 x + \cos^4 x)}$$ is equal to

a. $-\frac{\pi}{2}$ **b.** $\frac{\pi}{2\sqrt{2}}$
c. $-\frac{\pi}{4}$ **d.** $\frac{\pi}{\sqrt{2}}$

5. Let C be the set of all complex numbers. Let
$S_1 = \{z \in C \mid |z - 3 - 2i|^2 = 8\}$,
$S_2 = \{z \in C \mid \text{Re}(z) \geq 5\}$ and
$S_3 = \{z \in C \mid |z - \bar{z}| \geq 8\}$.
Then, the number of elements in $S_1 \cap S_2 \cap S_3$ is equal to

a. 1 **b.** 0
c. 2 **d.** Infinite

6. If the area of the bounded region
$R = \left\{(x, y) : \max\{0, \log_e x\} \leq y \leq 2^x, \frac{1}{2} \leq x \leq 2\right\}$
is, $\alpha(\log_e 2)^{-1} + \beta(\log_e 2) + \gamma$, then the value of $(\alpha + \beta - 2\gamma)^2$ is equal to

a. 8 **b.** 2
c. 4 **d.** 1

7. A ray of light through (2, 1) is reflected at a point P on the Y-axis and then passes through the point (5, 3). If this reflected ray is the directrix of an ellipse with eccentricity $\frac{1}{3}$ and the distance of the nearer focus from this directrix is $\frac{8}{\sqrt{53}}$, then the equation of the other directrix can be

a. $11x + 7y + 8 = 0$ or $11x + 7y - 15 = 0$
b. $11x - 7y - 8 = 0$ or $11x + 7y + 15 = 0$
c. $2x - 7y + 29 = 0$ or $2x - 7y - 7 = 0$
d. $2x - 7y - 39 = 0$ or $2x - 7y - 7 = 0$

8. If the coefficients of x^7 in $\left(x^2 + \frac{1}{bx}\right)^{11}$ and x^{-7} in $\left(x - \frac{1}{bx^2}\right)^{11}$, $b \neq 0$, are equal, then the value of b is equal to

a. 2 **b.** -1
c. 1 **d.** -2

9. The compound statement $(P \vee Q) \wedge (\sim P) \Rightarrow Q$ is equivalent to

a. $P \vee Q$ **b.** $P \wedge \sim Q$
c. $\sim(P \Rightarrow Q)$ **d.** $\sim(P \Rightarrow Q) \Leftrightarrow P \wedge \sim Q$

10. If $\sin\theta + \cos\theta = \frac{1}{2}$, then $16[\sin(2\theta) + \cos(4\theta) + \sin(6\theta)]$ is equal to

a. 23 **b.** -27
c. -23 **d.** 27

11. Let $A = \begin{bmatrix} 1 & 2 \\ -1 & 4 \end{bmatrix}$. If $A^{-1} = \alpha I + \beta A$, $\alpha, \beta \in R$, I is a 2×2 identity matrix, then $4(\alpha - \beta)$ is equal to

a. 5 **b.** 8/3
c. 2 **d.** 4

12. Let $f : \left(-\frac{\pi}{4}, \frac{\pi}{4}\right) \rightarrow R$ be defined as

$$f(x) = \begin{cases} (1 + |\sin x|)^{\frac{3a}{|\sin x|}}, & \frac{-\pi}{4} < x < 0 \\ b, & x = 0 \\ e^{\frac{\cot 4x}{\cot 2x}}, & 0 < x < \frac{\pi}{4} \end{cases}$$

If f is continuous at $x = 0$, then the value of $6a + b^2$ is equal to

a. $1 - e$ **b.** $e - 1$
c. $1 + e$ **d.** e

13. Let $y = y(x)$ be solution of the differential equation $\log_e\left(\frac{dy}{dx}\right) = 3x + 4y$, with $y(0) = 0$.
If $y\left(-\frac{2}{3}\log_e 2\right) = \alpha \log_e 2$, then the value of α is equal to

a. $-\frac{1}{4}$ **b.** $\frac{1}{4}$
c. 2 **d.** $-\frac{1}{2}$

14. Let the plane passing through the point $(-1, 0, -2)$ and perpendicular to each of the planes $2x + y - z = 2$ and $x - y - z = 3$ be $ax + by + cz + 8 = 0$. Then the value of $a + b + c$ is equal to

a. 3 **b.** 8
c. 5 **d.** 4

15. Two tangents are drawn from the point $P(-1, 1)$ to the circle $x^2 + y^2 - 2x - 6y + 6 = 0$. If these tangents touch the circle at points A and B, and if D is a point on the circle such that length of the segments AB and AD are equal, then the area of the ΔABD is equal to

a. 2 **b.** $(3\sqrt{2} + 2)$
c. 4 **d.** $3(\sqrt{2} - 1)$

16. Let $f : R \rightarrow R$ be a function such that $f(2) = 4$ and $f'(2) = 1$. Then, the value of $\lim_{x \to 2} \frac{x^2 f(2) - 4f(x)}{x - 2}$ is equal to

a. 4 **b.** 8
c. 16 **d.** 12

17. Let P and Q be two distinct points on a circle which has center at $C(2,3)$ and which passes through origin O. If OC is perpendicular to both the line segments CP and CQ, then the set $\{P, Q\}$ is equal to

a. $\{(4, 0), (0, 6)\}$
b. $\{(2 + 2\sqrt{2}, 3 - \sqrt{3}), (2 - 2\sqrt{2}, 3 + \sqrt{5})\}$
c. $\{(2 + 2\sqrt{2}, 3 + \sqrt{5}), (2 - 2\sqrt{2}, 3 - \sqrt{5})\}$
d. $\{(-1, 5), (5, 1)\}$

18. Let α, β be two roots of the equation $x^2 + (20)^{1/4}x + (5)^{1/2} = 0$. Then, $\alpha^8 + \beta^8$ is equal to

a. 10 **b.** 100
c. 50 **d.** 160

19. The probability that a randomly selected 2-digit number belongs to the set $\{n \in N : (2^n - 2)$ is a multiple of $3\}$ is equal to

a. $\frac{1}{6}$ b. $\frac{2}{3}$ c. $\frac{1}{2}$ d. $\frac{1}{3}$

20. Let $A = \{(x, y) \in R \times R \mid 2x^2 + 2y^2 - 2x - 2y = 1\}$,
$B = \{(x, y) \in R \times R \mid 4x^2 + 4y^2 - 16y + 7 = 0\}$
$C = \{(x, y) \in R \times R \mid x^2 + y^2 - 4x - 2y + 5 \le r^2\}$
Then the minimum value of $|r|$ such that $A \cup B \subseteq C$ is equal to

a. $\frac{3+\sqrt{10}}{2}$ b. $\frac{2+\sqrt{10}}{2}$

c. $\frac{3+2\sqrt{5}}{2}$ d. $1+\sqrt{5}$

Section B : Numerical Type Questions

21. For real numbers α and β, consider the following system of linear equations
$x + y - z = 2, x + 2y + \alpha z = 1,$
$2x - y + z = \beta.$
If the system has infinite solutions, then $\alpha + \beta$ is equal to ……………… .

22. Let $\mathbf{a} = \hat{\mathbf{i}} + \hat{\mathbf{j}} + \hat{\mathbf{k}}$, $\mathbf{b}$ and $\mathbf{c} = \hat{\mathbf{j}} - \hat{\mathbf{k}}$ be three vectors such that $\mathbf{a} \times \mathbf{b} = \mathbf{c}$ and $\mathbf{a} \cdot \mathbf{b} = 1$. If the length of projection vector of the vector $\mathbf{b}$ on the vector $\mathbf{a} \times \mathbf{c}$ is l, then the value of $3l^2$ is equal to ………..

23. If $\log_3 2, \log_3(2^x - 5), \log_3\left(2^x - \frac{7}{2}\right)$ are in an arithmetic progression, then the value of x is equal to ……….. .

24. Let the domain of the function
$f(x) = \log_4[\log_5(\log_3(18x - x^2 - 77))]$ be (a, b).
Then the value of the integral
$$\int_a^b \frac{\sin^3 x}{[\sin^3 x + \sin^3(a + b - x)]} dx$$ is equal to ………..

25. Let $f(x) = \begin{vmatrix} \sin^2 x & -2 + \cos^2 x & \cos 2x \\ 2 + \sin^2 x & \cos^2 x & \cos 2x \\ \sin^2 x & \cos^2 x & 1 + \cos 2x \end{vmatrix}, x \in [0, \pi]$

Then the maximum value of $f(x)$ is equal to ……….. .

26. Let $f : [3, 5] \to R$ be a twice differentiable function on (3, 5) such that
$$f(x) = e^{-x} \int_3^x [3t^2 + 2t + 4f'(t)]\, dt.$$
If $f'(4) = \frac{\alpha e^\beta - 224}{(e^\beta - 4)^2}$, then $\alpha + \beta$ is equal to ……… .

27. Let a plane P pass through the point $(3, 7, -7)$ and contain the line, $\frac{x-2}{-3} = \frac{y-3}{2} = \frac{z+2}{1}$. If distance of the plane P from the origin is d, then d^2 is equal to …………… .

28. Let $\{S = 1, 2, 3, 4, 5, 6, 7\}$. Then the number of possible functions $f : S \to S$ such that $f(m \cdot n) = f(m) \cdot f(n)$ for every $m, n \in S$ and $m \cdot n \in S$ is equal to ……….. .

29. If $y = y(x)$, $y \in \left[0, \frac{\pi}{2}\right)$ is the solution of the differential equation $\sec y \frac{dy}{dx} - \sin(x + y) - \sin(x - y) = 0$, with $y(0) = 0$, then $5y'\left(\frac{\pi}{2}\right)$ is equal to ……….. .

30. Let $f : [0, 3] \to R$ be defined by
$f(x) = \min\{x - [x], 1 + [x] - x\}$
where $[x]$ is the greatest integer less than or equal to x. Let P denote the set containing all $x \in (0, 3)$, where f is discontinuous and Q denote the set containing all $x \in (0, 3)$, where f is not differentiable.

Then the sum of number of elements in P and Q is equal to ……….. .

Answers

For solution scan the QR code

1. (d)	2. (d)	3. (b)	4. (b)	5. (a)	6. (b)	7. (c)	8. (c)	9. (d)	10. (c)
11. (d)	12. (c)	13. (a)	14. (d)	15. (c)	16. (d)	17. (d)	18. (c)	19. (c)	20. (c)
21. (5)	22. (2)	23. (3)	24. (1)	25. (6)	26. (16)	27. (3)	28. (490)	29. (2)	30. (5)

27 JULY SHIFT II

Section A : Objective Type Questions

1. The point $P(a, b)$ undergoes the following three transformations successively

(A) Reflection about the line $y = x$.

(B) Translation through 2 units along the positive direction of X-axis.

(C) Rotation through angle $\frac{\pi}{4}$ about the origin in the anti-clockwise direction.

If the co-ordinates of the final position of the point P are $\left(-\frac{1}{\sqrt{2}}, \frac{7}{\sqrt{2}}\right)$, then the value of $2a + b$ is equal to

a. 13 **b.** 9 **c.** 5 **d.** 7

2. A possible value of 'x', for which the ninth term in the expansion of $\left\{3^{\log_3 \sqrt{25^{x-1}+7}} + 3^{\left(-\frac{1}{8}\right)\log_3 (5^{x-1}+1)}\right\}^{10}$ is equal to 180, is

a. 0 **b.** –1 **c.** 2 **d.** 1

3. For real numbers α and $\beta \neq 0$, if the point of intersection of the straight lines

$$\frac{x-\alpha}{1} = \frac{y-1}{2} = \frac{z-1}{3}$$

and $$\frac{x-4}{\beta} = \frac{y-6}{3} = \frac{z-7}{3}$$

lies on the plane $x + 2y - z = 8$, then $\alpha - \beta$ is equal to

a. 5 **b.** 9

c. 3 **d.** 7

4. Let $f : R \to R$ be defined as

$f(x + y) + f(x - y) = 2f(x)\, f(y)$, $f\left(\frac{1}{2}\right) = -1$. Then, the value of $\sum_{k=1}^{20} \frac{1}{\sin(k)\sin[k + f(k)]}$ is equal to

a. $\text{cosec}^2(21)\cos(20)\cos(2)$

b. $\sec^2(1)\sec(21)\cos(20)$

c. $\text{cosec}^2(1)\,\text{cosec}(21)\sin(20)$

d. $\sec^2(21)\sin(20)\sin(2)$

5. Let C be the set of all complex numbers.
Let $S_1 = \{z \in C : |z - 2| \leq 1\}$ and
$S_2 = \{z \in C : z(1 + i) + \bar{z}(1 - i) \geq 4\}$.

Then, the maximum value of $\left|z - \frac{5}{2}\right|^2$ for $z \in S_1 \cap S_2$ is equal to

a. $\frac{3+2\sqrt{2}}{4}$ **b.** $\frac{5+2\sqrt{2}}{2}$

c. $\frac{3+2\sqrt{2}}{2}$ **d.** $\frac{5+2\sqrt{2}}{4}$

6. A student appeared in an examination consisting of 8 true-false type questions. The student guesses the answers with equal probability. The smallest value of n, so that the probability of guessing at least 'n' correct answers is less than $\frac{1}{2}$, is

a. 5 **b.** 6

c. 3 **d.** 4

7. If $\tan\left(\frac{\pi}{9}\right)$, x, $\tan\left(\frac{7\pi}{18}\right)$ are in arithmetic progression and $\tan\left(\frac{\pi}{9}\right)$, y, $\tan\left(\frac{5\pi}{18}\right)$ are also in arithmetic progression, then $|x - 2y|$ is equal to

a. 4 **b.** 3

c. 0 **d.** 1

8. Let the mean and variance of the frequency distribution

x	$x_1 = 2$	$x_2 = 6$	$x_3 = 8$	$x_4 = 9$
f	4	4	α	β

be 6 and 6.8, respectively. If x_3 is changed from 8 to 7, then the mean for the new data will be

a. 4 **b.** 5 **c.** $\frac{17}{3}$ **d.** $\frac{16}{3}$

9. The area of the region bounded by $y - x = 2$ and $x^2 = y$ is equal to

a. $\frac{16}{3}$ **b.** $\frac{2}{3}$ **c.** $\frac{9}{2}$ **d.** $\frac{4}{3}$

10. Let $y = y(x)$ be the solution of the differential equation $(x - x^3)\,dy = (y + yx^2 - 3x^4)\,dx$, $x > 2$. If $y(3) = 3$, then $y(4)$ is equal to

a. 4 **b.** 12

c. 8 **d.** 16

11. The value of $\lim_{x \to 0}\left(\frac{x}{\sqrt[8]{1-\sin x} - \sqrt[8]{1+\sin x}}\right)$ is equal to

a. 0 **b.** 4

c. – 4 **d.** –1

12. Two sides of a parallelogram are along the lines $4x + 5y = 0$ and $7x + 2y = 0$. If the equation of one of the diagonals of the parallelogram is $11x + 7y = 9$, then other diagonal passes through the point

a. (1, 2) **b.** (2, 2)

c. (2, 1) **d.** (1, 3)

13. Let $\alpha = \max_{x \in R} \{8^{2\sin 3x} \cdot 4^{4\cos 3x}\}$ and $\beta = \max_{x \in R} \{8^{2\sin 3x} \cdot 4^{4\cos 3x}\}$. If $8x^2 + bx + c = 0$ is a quadratic equation whose roots are $\alpha^{1/5}$ and $\beta^{1/5}$, then the value of $c - b$ is equal to

a. 42 **b.** 47

c. 43 **d.** 50

14. Let $f:[0,\infty)\rightarrow[0,3]$ be a function defined by

$$f(x)=\begin{cases}\max\{\sin t:0\le t\le x\}, & 0\le x<\pi\\ 2+\cos x, & x>\pi\end{cases}$$

Then which of the following is true ?

a. f is continuous everywhere but not differentiable exactly at one point in $(0,\infty)$

b. f is differentiable everywhere in $(0,\infty)$

c. f is not continuous exactly at two points in $(0,\infty)$

d. f is continuous everywhere but not differentiable exactly at two points in $(0,\infty)$

15. Let N be the set of natural numbers and a relation R on N be defined by
$R=\{(x,y)\in N\times N: x^3-3x^2y-xy^2+3y^3=0\}$.
Then the relation R is

a. symmetric but neither reflexive nor transitive

b. reflexive but neither symmetric nor transitive

c. reflexive and symmetric, but not transitive

d. an equivalence relation

16. Which of the following is the negation of the statement "for all $M>0$, there exists $x\in S$ such that $x\ge M$" ?

a. there exists $M>0$, such that $x<M$ for all $x\in S$

b. there exists $M>0$, there exists $x\in S$ such that $x\ge M$

c. there exists $M>0$, there exists $x\in S$ such that $x<M$

d. there exists $M>0$, such that $x\ge M$ for all $x\in S$

17. Consider a circle C which touches the Y-axis at (0, 6) and cuts off an intercept $6\sqrt{5}$ on the X-axis. Then the radius of the circle C is equal to

a. $\sqrt{53}$ **b.** 9 **c.** 8 **d.** $\sqrt{82}$

18. Let **a**, **b** and **c** be three vectors such that $\mathbf{a}=\mathbf{b}\times(\mathbf{b}\times\mathbf{c})$. If magnitudes of the vectors **a**, **b** and **c** are $\sqrt{2}$, 1 and 2, respectively and the angle between **b** and **c** is $\theta\left(0<\theta<\frac{\pi}{2}\right)$, then the value of $(1+\tan\theta)$ is equal to

a. $\sqrt{3}+1$ **b.** 2 **c.** 1 **d.** $\frac{\sqrt{3}+1}{\sqrt{3}}$

19. Let A and B be two 3×3 real matrices such that (A^2-B^2) is invertible matrix. If $A^5=B^5$ and $A^3B^2=A^2B^3$, then the value of the determinant of the matrix A^3+B^3 is equal to

a. 2 **b.** 4 **c.** 1 **d.** 0

20. Let $f:(a,b)\rightarrow R$ be twice differentiable function such that $f(x)=\int_a^x g(t)\,dt$ for a differentiable function $g(x)$. If $f(x)=0$ has exactly five distinct roots in (a,b), then $g(x)g'(x)=0$ has at least

a. twelve roots in (a,b)

b. five roots in (a,b)

c. seven roots in (a,b)

d. three roots in (a,b)

Section B : Numerical Type Questions

21. Let $\mathbf{a}=\hat{i}-\alpha\hat{j}+\beta\hat{k}$, $\mathbf{b}=3\hat{i}+\beta\hat{j}-\alpha\hat{k}$ and $\mathbf{c}=-\alpha\hat{i}-2\hat{j}+\hat{k}$, where α and β are integers. If $\mathbf{a}\cdot\mathbf{b}=-1$ and $\mathbf{b}\cdot\mathbf{c}=10$, then $(\mathbf{a}\times\mathbf{b})\cdot\mathbf{c}$ is equal to ....

22. The distance of the point $P(3,4,4)$ from the point of intersection of the line joining the points. $Q(3,-4,-5)$ and $R(2,-3,1)$ and the plane $2x+y+z=7$, is equal to ....

23. If the real part of the complex number $z=\frac{3+2i\cos\theta}{1-3i\cos\theta}$, $\theta\in\left(0,\frac{\pi}{2}\right)$ is zero, then the value of $\sin^2 3\theta+\cos^2\theta$ is equal to ....

24. Let E be an ellipse whose axes are parallel to the co-ordinates axes, having its center at $(3,-4)$, one focus at $(4,-4)$ and one vertex at $(5,-4)$. If $mx-y=4$, $m>0$ is a tangent to the ellipse E, then the value of $5m^2$ is equal to ....

25. If $\int_0^\pi(\sin^3 x)e^{-\sin^2 x}dx=\alpha-\frac{\beta}{e}\int_0^1\sqrt{t}\,e^t dt$, then $(\alpha+\beta)$ is equal to ....

26. The number of real roots of the equation $e^{4x}-e^{3x}-4e^{2x}-e^x+1=0$ is equal to ....

27. Let $y=y(x)$ be the solution of the differential equation $dy=e^{\alpha x+y}dx$; $\alpha\in N$. If $y(\log_e 2)=\log_e 2$ and $y(0)=\log_e\left(\frac{1}{2}\right)$, then the value of α is equal to ....

28. Let n be a non-negative integer. Then the number of divisors of the form "$4n+1$" of the number $(10)^{10}.(11)^{11}.(13)^{13}$ is equal to ....

29. Let $A=\{n\in N\mid n^2\le n+10{,}000\}$, $B=\{3k+1\mid k\in N\}$ and $C=\{2k\mid k\in N\}$, then the sum of all the elements of the set $A\cap(B-C)$ is equal to ....

30. If $A=\begin{bmatrix}1&1&1\\0&1&1\\0&0&1\end{bmatrix}$ and $M=A+A^2+A^3+\ldots\ldots+A^{20}$, then the sum of all the elements of the matrix M is equal to ....

Answers

For solution scan the QR code

1. (b)	2. (d)	3. (d)	4. (c)	5. (d)	6. (a)	7. (c)	8. (c)	9. (c)	10. (b)
11. (c)	12. (b)	13. (a)	14. (b)	15. (b)	16. (a)	17. (b)	18. (b)	19. (d)	20. (c)
21. (9)	22. (7)	23. (1)	24. (3)	25. (5)	26. (2)	27. (2)	28. (924)	29. (832)	30. (2020)

26 AUGUST SHIFT I

Section A : Objective Type Questions

1. The sum of solutions of the equation $\frac{\cos x}{1+\sin x}=|\tan 2x|, x\in\left(-\frac{\pi}{2},\frac{\pi}{2}\right)-\left\{\frac{\pi}{4},-\frac{\pi}{4}\right\}$ is

a. $-\frac{11\pi}{30}$ **b.** $\frac{\pi}{10}$
c. $-\frac{7\pi}{30}$ **d.** $-\frac{\pi}{15}$

2. The mean and standard deviation of 20 observations were calculated as 10 and 2.5 respectively. It was found that by mistake one data value was taken as 25 instead of 35. If α and $\sqrt{\beta}$ are the mean and standard deviation respectively for correct data, then (α, β) is

a. (11, 26) **b.** (10.5, 25)
c. (11, 25) **d.** (10.5, 26)

3. On the ellipse $\frac{x^2}{8}+\frac{y^2}{4}=1$. Let P be a point in the second quadrant such that the tangent at P to the ellipse is perpendicular to the line $x+2y=0$. Let S and S' be the foci of the ellipse and e be its eccentricity. If A is the area of the $\Delta SPS'$ then, the value of $(5-e^2)\cdot A$ is

a. 6 **b.** 12 **c.** 14 **d.** 24

4. Let $y=y(x)$ be a solution curve of the differential equation $(y+1)\tan^2 x\,dx+\tan x$ $dy+y\,dx=0, x\in\left(0,\frac{\pi}{2}\right)$. If $\lim_{x\to 0^+} xy(x)=1$, then the value of $y\left(\frac{\pi}{4}\right)$ is

a. $-\frac{\pi}{4}$ **b.** $\frac{\pi}{4}-1$ **c.** $\frac{\pi}{4}+1$ **d.** $\frac{\pi}{4}$

5. Let A and B be independent events such that $P(A)=p$ and $P(B)=2p$. The largest value of p, for which P (exactly one of A, B occurs) $=\frac{5}{9}$, is

a. $\frac{1}{3}$ **b.** $\frac{2}{9}$ **c.** $\frac{4}{9}$ **d.** $\frac{5}{12}$

6. Let $\theta\in\left(0,\frac{\pi}{2}\right)$. If the system of linear equations

$$(1+\cos^2\theta)x+\sin^2\theta y+4\sin 3\theta z=0$$
$$\cos^2\theta x+(1+\sin^2\theta)y+4\sin 3\theta z=0$$
$$\cos^2\theta x+\sin^2\theta y+(1+4\sin 3\theta)z=0$$

has a non-trivial solution, then the value of θ is

a. $\frac{4\pi}{9}$ **b.** $\frac{7\pi}{18}$ **c.** $\frac{\pi}{18}$ **d.** $\frac{5\pi}{18}$

7. Let $f(x)=\cos\left(2\tan^{-1}\sin\left(\cot^{-1}\sqrt{\frac{1-x}{x}}\right)\right), 0<x<1$. Then,

a. $(1-x)^2 f'(x)-2(f(x))^2=0$ **b.** $(1+x)^2 f'(x)+2(f(x))^2=0$
c. $(1-x)^2 f'(x)+2(f(x))^2=0$ **d.** $(1+x)^2 f'(x)-2f(x))^2=0$

8. The sum of the series $\frac{1}{x+1}+\frac{2}{x^2+1}+\frac{2^2}{x^4+1}+\dots+\frac{2^{100}}{x^{2^{100}}+1}$, when $x=2$ is

a. $1+\frac{2^{101}}{4^{101}-1}$ **b.** $1+\frac{2^{100}}{4^{101}-1}$
c. $1-\frac{2^{100}}{4^{100}-1}$ **d.** $1-\frac{2^{101}}{4^{101}-1}$

9. If ${}^{20}C_r$ is the coefficient of x^r in the expansion of $(1+x)^{20}$, then the value of $\sum_{r=0}^{20} r^2\,{}^{20}C_r$ is equal to

a. 420×2^{19} **b.** 380×2^{19}
c. 380×2^{18} **d.** 420×2^{18}

10. Out of all the patients in a hospital 89% are found to be suffering from heart ailment and 98% are suffering from lungs infection. If K% of them are suffering from both ailments, then K can not belong to the set

a. {80, 83, 86, 89}
b. {84, 86, 88, 90}
c. {79, 81, 83, 85}
d. {84, 87, 90, 93}

11. The equation $\arg\left(\frac{z-1}{z+1}\right)=\frac{\pi}{4}$ represents a circle with

a. centre at (0, –1) and radius $\sqrt{2}$
b. centre at (0, 1) and radius $\sqrt{2}$
c. centre at (0, 0) and radius $\sqrt{2}$
d. centre at (0, 1) and radius 2

12. Let $\mathbf{a}=\hat{\mathbf{i}}+\hat{\mathbf{j}}+\hat{\mathbf{k}}$ and $\mathbf{b}=\hat{\mathbf{j}}-\hat{\mathbf{k}}$. If $\mathbf{c}$ is a vector such that $\mathbf{a}\times\mathbf{c}=\mathbf{b}$ and $\mathbf{a}\cdot\mathbf{c}=3$, then $\mathbf{a}\cdot(\mathbf{b}\times\mathbf{c})$ is equal to

a. –2 **b.** –6
c. 6 **d.** 2

13. If a line along a chord of the circle $4x^2+4y^2+120x+675=0$, passes through the point $(-30, 0)$ and is tangent to the parabola $y^2=30x$, then the length of this chord is

a. 5 **b.** 7
c. $5\sqrt{3}$ **d.** $3\sqrt{5}$

14. The value of $\int_{-1/\sqrt{2}}^{1/\sqrt{2}}\left(\left(\frac{x+1}{x-1}\right)^2+\left(\frac{x-1}{x+1}\right)^2-2\right)^{1/2}dx$ is

a. $\log_e 4$ **b.** $\log_e 16$
c. $2\log_e 16$ **d.** $4\log_e(3+2\sqrt{2})$

15. A plane P contains the line $x+2y+3z+1=0=x-y-z-6$ and is perpendicular to the plane $-2x+y+z+8=0$. Then which of the following points lies on P?

a. (–1, 1, 2) **b.** (0, 1, 1)
c. (1, 0, 1) **d.** (2, –1, 1)

16. If $A = \begin{pmatrix} \frac{1}{\sqrt{5}} & \frac{2}{\sqrt{5}} \\ \frac{-2}{\sqrt{5}} & \frac{1}{\sqrt{5}} \end{pmatrix}$, $B = \begin{pmatrix} 1 & 0 \\ i & 1 \end{pmatrix}$, $i = \sqrt{-1}$, and $Q = A^TBA$, then the inverse of the matrix $AQ^{2021}A^T$ is equal to

a. $\begin{pmatrix} \frac{1}{\sqrt{5}} & -2021 \\ 2021 & \frac{1}{\sqrt{5}} \end{pmatrix}$ b. $\begin{pmatrix} 1 & 0 \\ -2021i & 1 \end{pmatrix}$

c. $\begin{pmatrix} 1 & 0 \\ 2021i & 1 \end{pmatrix}$ d. $\begin{pmatrix} 1 & -2021i \\ 0 & 1 \end{pmatrix}$

17. If the sum of an infinite GP $a, ar, ar^2, ar^3, \ldots$ is 15 and the sum of the squares of its each term is 150, then the sum of $ar^2, ar^4, ar^6, \ldots$ is

a. 5/2 b. 1/2 c. 25/2 d. 9/2

18. The value of $\lim_{n \to \infty} \frac{1}{n} \sum_{r=0}^{2n-1} \frac{n^2}{n^2 + 4r^2}$ is

a. $\frac{1}{2}\tan^{-1}(2)$ b. $\frac{1}{2}\tan^{-1}(4)$

c. $\tan^{-1}(4)$ d. $\frac{1}{4}\tan^{-1}(4)$

19. Let ABC be a triangle with $A(-3, 1)$ and $\angle ACB = \theta, 0 < \theta < \frac{\pi}{2}$. If the equation of the median through B is $2x + y - 3 = 0$ and the equation of angle bisector of C is $7x - 4y - 1 = 0$, then $\tan\theta$ is equal to

a. 1/2 b. 3/4 c. 4/3 d. 2

20. If the truth value of the Boolean expression $((p \vee q) \wedge (q \to r) \wedge (\sim r) \to (p \wedge q)$ is false, then the truth values of the statements, p, q and r respectively can be

a. T F T b. F F T c. T F F d. F T F

Section B : Numerical Type Questions

21. Let $z = \frac{1 - i\sqrt{3}}{2}$ and $i = \sqrt{-1}$. Then the value of

$$21 + \left(z + \frac{1}{z}\right)^3 + \left(z^2 + \frac{1}{z^2}\right)^3 + \left(z^3 + \frac{1}{z^3}\right)^3 + \ldots + \left(z^{21} + \frac{1}{z^{21}}\right)^3$$ is

22. The sum of all integral values of k $(k \neq 0)$ for which the equation $\frac{2}{x-1} - \frac{1}{x-2} = \frac{2}{k}$ in x has no real roots, is

23. Let the line L be the projection of the line $\frac{x-1}{2} = \frac{y-3}{1} = \frac{z-4}{2}$ in the plane $x - 2y - z = 3$. If d is the distance of the point $(0, 0, 6)$ from L, then d^2 is equal to

24. If ${}^1P_1 + 2.{}^2p_1 + 3.{}^3P_3 + \ldots + 15.{}^{15}P_{15} = {}^qP_r - s$, $0 \le s \le 1$, then ${}^{q+s}C_{r-s}$ is equals to

25. A wire of length 36 m is cut into two pieces, one of the pieces is bent to form a square and the other is bent to form a circle. If the sum of the areas of the two figures is minimum and the circumference of the circle is k (m), then $\left(\frac{4}{\pi} + 1\right)k$ is equal to

26. The area of the region $S = \{(x, y) : 3x^2 \le 4y \le 6x + 24\}$ is

27. The locus of a point, which moves such that the sum of squares of its distances from the points (0, 0), (1, 0), (0, 1) (1, 1) is 18 units, is a circle of diameter d. Then, d^2 is equal to

28. If $y = y(x)$ is an implicit function of x such that $\log_e(x + y) = 4xy$, then $\frac{d^2y}{dx^2}$ at $x = 0$ is equal to

29. The number of three-digit even numbers, formed by the digits 0, 1, 3, 4, 6, 7, if the repetition of digits is not allowed, is

30. Let $a, b \in R, b \neq 0$. Define a function

$$f(x) = \begin{cases} a\sin\frac{\pi}{2}(x-1), & \text{for } x \le 0 \\ \frac{\tan 2x - \sin 2x}{bx^3}, & \text{for } x > 0 \end{cases}$$

If f is continuous at $x = 0$, then $10 - ab$ is equal to

Answers

For solution scan the QR code

1. (a)	2. (d)	3. (a)	4. (d)	5. (d)	6. (b)	7. (c)	8. (d)	9. (d)	10. (c)
11. (b)	12. (a)	13. (d)	14. (b)	15. (b)	16. (b)	17. (b)	18. (b)	19. (c)	20. (c)
21. (13)	22. (66)	23. (26)	24. (136)	25. (36)	26. (27)	27. (16)	28. (40)	29. (52)	30. (14)

26 AUGUST SHIFT II

Section A : Objective Type Questions

1. Let $[t]$ denote the greatest integer less than or equal to t. Let $f(x) = x - [x], g(x) = 1 - x + [x]$, and $h(x) = \min\{f(x), g(x)\}, x \in [-2, 2]$. Then h is

a. continuous in $[-2, 2]$ but not differentiable at more than four points in $(-2, 2)$

b. not continuous at exactly three points in $[-2, 2]$

c. continuous in $[-2, 2]$ but not differentiable at exactly three points in $(-2, 2)$

d. not continuous at exactly four points in $[-2, 2]$

2. Let $A = \begin{pmatrix} 1 & 0 & 0 \\ 0 & 1 & 1 \\ 1 & 0 & 0 \end{pmatrix}$. Then $A^{2025} - A^{2020}$ is equal to

a. $A^6 - A$ **b.** A^5
c. $A^5 - A$ **d.** A^6

3. The local maximum value of the function $f(x) = (2/x)^{x^2}$, $x > 0$

a. $(2\sqrt{e})^{\frac{1}{e}}$ **b.** $\left(\frac{4}{\sqrt{e}}\right)^{\frac{e}{4}}$ **c.** $(e)^{\frac{2}{e}}$ **d.** 1

4. If the value of the integral $\int_0^5 \frac{x + [x]}{e^{x-[x]}} dx = \alpha e^{-1} + \beta$, where $\alpha, \beta \in R, 5\alpha + 6\beta = 0$ and $[x]$ denotes the greatest integer less than or equal to x, then the value of $(\alpha + \beta)^2$ is equal to :

a. 100 **b.** 25 **c.** 16 **d.** 36

5. The point $P(-2\sqrt{6}, \sqrt{3})$ lies on the hyperbola $\frac{x^2}{a^2} - \frac{y^2}{b^2} = 1$ having eccentricity $\sqrt{5}/2$. If the tangent and normal at P to the hyperbola intersect its conjugate axis at the point Q and R respectively, then QR is equal to

a. $4\sqrt{3}$ **b.** 6 **c.** $6\sqrt{3}$ **d.** $3\sqrt{6}$

6. Let $y(x)$ be the solution of the differential equation $2x^2dy + (e^y - 2x)\,dx = 0, x > 0$. If $y(e) = 1$, then $y(1)$ is equal to

a. 0 **b.** 2 **c.** $\log_e 2$ **d.** $\log_e(2e)$

7. Consider the two statements :

$(S_1) : (p \to q) \vee (\sim q \to p)$ is a tautology

$(S_2) : (p \wedge \sim q) \wedge (\sim p \vee q)$ is a fallacy

Then,

a. only (S_1) is true **b.** both (S_1) and (S_2) are false
c. both (S_1) and (S_2) are true **d.** only (S_2) is true

8. The domain of the function $\text{cosec}^{-1}\left(\frac{1+x}{x}\right)$ is

a. $\left(-1, -\frac{1}{2}\right] \cup (0, \infty)$ **b.** $\left[-\frac{1}{2}, 0\right) \cup [1, \infty)$
c. $\left(-\frac{1}{2}, \infty\right) - \{0\}$ **d.** $\left[-\frac{1}{2}, \infty\right) - \{0\}$

9. A fair die is tossed until six is obtained on it. Let X be the number of required tosses, then the conditional probability $P(X \geq 5 \mid X > 2)$ is

a. 125/216 **b.** 11/36 **c.** 5/6 **d.** 25/36

10. If $\sum_{r=1}^{50} \tan^{-1}\frac{1}{2r^2} = p$, then the value of $\tan p$ is

a. 101/102 **b.** 50/51 **c.** 100 **d.** 51/50

11. Two fair dice are thrown. The numbers on them are taken as λ and μ, and a system of linear equations

$x + y + z = 5, x + 2y + 3z = \mu$ and

$x + 3y + \lambda z = 1$

is constructed. If p is the probability that the system has a unique solution and q is the probability that the system has no solution, then

a. $p = \frac{1}{6}$ and $q = \frac{1}{36}$ **b.** $p = \frac{5}{6}$ and $q = \frac{5}{36}$
c. $p = \frac{5}{6}$ and $q = \frac{1}{36}$ **d.** $p = \frac{1}{6}$ and $q = \frac{5}{36}$

12. The locus of the mid-points of the chords of the hyperbola $x^2 - y^2 = 4$, which touch the parabola $y^2 = 8x$, is

a. $y^3(x-2) = x^2$ **b.** $x^3(x-2) = y^2$
c. $y^2(x-2) = x^3$ **d.** $x^2(x-2) = y^3$

13. The value of $2\sin\left(\frac{\pi}{8}\right)\sin\left(\frac{2\pi}{8}\right)\sin\left(\frac{3\pi}{8}\right)\sin\left(\frac{5\pi}{8}\right)\sin\left(\frac{6\pi}{8}\right)\sin\left(\frac{7\pi}{8}\right)$ is

a. $\frac{1}{4\sqrt{2}}$ **b.** $\frac{1}{4}$ **c.** $\frac{1}{8}$ **d.** $\frac{1}{8\sqrt{2}}$

14. If $(\sqrt{3} + i)^{100} = 2^{99}(p + iq)$, then p and q are roots of the equation

a. $x^2 - (\sqrt{3} - 1)x - \sqrt{3} = 0$ **b.** $x^2 + (\sqrt{3} + 1)x + \sqrt{3} = 0$
c. $x^2 + (\sqrt{3} - 1)x - \sqrt{3} = 0$ **d.** $x^2 - (\sqrt{3} + 1)x + \sqrt{3} = 0$

15. A hall has a square floor of dimension 10m × 10m (see the figure) and vertical walls. If the $\angle GPH$ between the diagonals AG and BH is $\cos^{-1}\frac{1}{5}$, then the height of the hall (in m) is

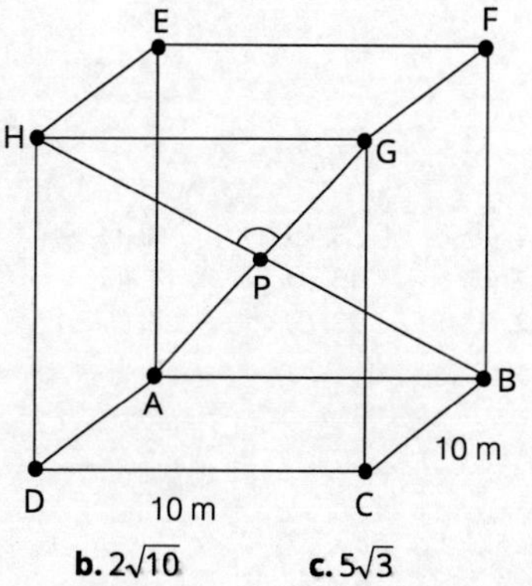

a. 5 **b.** $2\sqrt{10}$ **c.** $5\sqrt{3}$ **d.** $5\sqrt{2}$

16. Let P be the plane passing through the point (1,2,3) and the line of intersection of the planes $\mathbf{r}\cdot(\hat{\mathbf{i}}+\hat{\mathbf{j}}+4\hat{\mathbf{k}})=16$ and $\mathbf{r}\cdot(-\hat{\mathbf{i}}+\hat{\mathbf{j}}+\hat{\mathbf{k}})=6$. Then which of the following points does not lie on P?

a. (3, 3, 2) **b.** (6, – 6, 2)
c. (4, 2, 2) **d.** (– 8, 8, 6)

17. A 10 inches long pencil AB with mid-point C and a small eraser P are placed on the horizontal top of a table such that $PC=\sqrt{5}$ inches and $\angle PCB=\tan^{-1}(2)$. The acute angle through which the pencil must be rotated about C so that the perpendicular distance between eraser and pencil becomes exactly 1 inch is

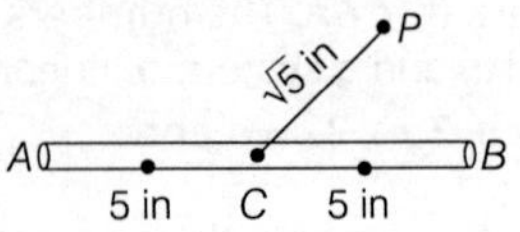

a. $\tan^{-1}\left(\frac{3}{4}\right)$ **b.** $\tan^{-1}(1)$
c. $\tan^{-1}\left(\frac{4}{3}\right)$ **d.** $\tan^{-1}\left(\frac{1}{2}\right)$

18. The value of $\int_{-\frac{\pi}{2}}^{\frac{\pi}{2}}\left(\frac{1+\sin^2 x}{1+\pi^{\sin x}}\right)dx$ is

a. $\frac{\pi}{2}$ **b.** $\frac{5\pi}{4}$ **c.** $\frac{3\pi}{4}$ **d.** $\frac{3\pi}{2}$

19. A circle C touches the line $x=2y$ at the point (2,1) and intersects the circle $C_1: x^2+y^2+2y-5=0$ at two points P and Q such that PQ is a diameter of C_1. Then the diameter of C is

a. $7\sqrt{5}$ **b.** 15 **c.** 285 **d.** $4\sqrt{15}$

20. $\lim_{x\to 2}\left(\sum_{n=1}^{9}\frac{x}{n(n+1)x^2+2(2n+1)x+4}\right)$ is equal to

a. $\frac{9}{44}$ **b.** $\frac{5}{24}$
c. $\frac{1}{5}$ **d.** $\frac{7}{36}$

Section B : Numerical Type Questions

21. The sum of all three-digit numbers less than or equal to 500, that are formed without using the digit 1 and they all are multiple of 11, is

22. Let a be the points of local maximum and local minimum of the function $f(x)=2x^3-3x^2-12x$. If A is the total area of the region bounded by $y=f(x)$, the X-axis and the lines $x=a$ and $x=b$, then 4A is equal to

23. If the projection of the vector $\hat{\mathbf{i}}+2\hat{\mathbf{j}}+\hat{\mathbf{k}}$ on the sum of the two vectors $2\hat{\mathbf{i}}+4\hat{\mathbf{j}}-5\hat{\mathbf{k}}$ and $-\lambda\hat{\mathbf{i}}+2\hat{\mathbf{j}}+3\hat{\mathbf{k}}$ is 1, then λ is equal to

24. Let $a_1, a_2, \ldots\ldots, a_{10}$ be an AP with common difference –3 and $b_1, b_2, \ldots\ldots, b_{10}$ be a GP with common ratio 2. Let $c_k=a_k+b_k$, $k=1,2,\ldots\ldots,10$. If $c_2=12$ and $c_3=13$, then $\sum_{k=1}^{10}c_k$ is equal to

25. Let Q be the foot of the perpendicular from the point P(7, – 2,13) on the plane containing the lines $\frac{x+1}{6}=\frac{y-1}{7}=\frac{z-3}{8}$ and $\frac{x+1}{3}=\frac{y-2}{5}=\frac{z-3}{7}$. Then $(PQ)^2$ is equal to

26. Let $\binom{n}{k}$ denotes ${}^nC_k$ and $\left[\begin{matrix}n\\k\end{matrix}\right]=\begin{cases}\binom{n}{k}, & \text{if } 0\le k\le n\\ 0, & \text{otherwise}\end{cases}$

If $A_k=\sum_{i=0}^{9}\binom{9}{i}\left[\begin{matrix}12\\12-k+i\end{matrix}\right]+\sum_{i=0}^{8}\binom{8}{i}\left[\begin{matrix}13\\13-k+i\end{matrix}\right]$ and $A_4-A_3=190p$, then p is equal to

27. Let $\lambda\neq 0$ be in R. If α and β are the roots of the equation $x^2-x+2\lambda=0$ and α and γ are the roots of equation $3x^2-10x+27\lambda=0$, then $\frac{\beta\gamma}{\lambda}$ is equal to

28. Let the mean and variance of four numbers 3, 7, x and $y (x>y)$ be 5 and 10 respectively. Then, the mean of four numbers $3+2x, 7+2y, x+y$ and $x-y$ is

29. Let A be a 3×3 real matrix.
If $\det(2\text{Adj}(2\text{ Adj}(\text{Adj}(2A))))=2^{41}$, then the value of $\det(A^2)$ equal

30. The least positive integer n such that $\frac{(2i)^n}{(1-i)^{n-2}}$, $i=\sqrt{-1}$, is a positive integer, is

Answers

For solution scan the QR code

1. (a)	2. (a)	3. (c)	4. (b)	5. (c)	6. (c)	7. (c)	8. (d)	9. (d)	10. (b)
11. (b)	12. (c)	13. (c)	14. (a)	15. (d)	16. (c)	17. (a)	18. (c)	19. (a)	20. (a)
21. (7744)	22. (114)	23. (5)	24. (2021)	25. (96)	26. (49)	27. (18)	28. (12)	29. (4)	30. (6)

27 AUGUST SHIFT I

Section A : Objective Type Questions

1. If $0 < x < 1$, then $\frac{3}{2}x^2 + \frac{5}{3}x^3 + \frac{7}{4}x^4 + ...$, is equal to

a. $x\left(\frac{1+x}{1-x}\right) + \log_e(1-x)$

b. $x\left(\frac{1-x}{1+x}\right) + \log_e(1-x)$

c. $\frac{1-x}{1+x} + \log_e(1-x)$

d. $\frac{1+x}{1-x} + \log_e(1-x)$

2. If $x, y \in R, x > 0$

$y = \log_{10}x + \log_{10}x^{1/3} + \log_{10}x^{1/9} + \ldots$ upto ∞ terms and $\frac{2+4+6+\ldots+2y}{3+6+9+\ldots+3y} = \frac{4}{\log_{10}x}$, then the ordered pair (x, y) is equal to

a. $(10^6, 6)$ **b.** $(10^4, 6)$
c. $(10^2, 3)$ **d.** $(10^6, 9)$

3. Let A be a fixed point (0, 6) and B be a moving point $(2t, 0)$. Let M be the mid-point of AB and the perpendicular bisector of AB meets the Y-axis at C.
The locus of the mid-point P of MC is

a. $3x^2 - 2y - 6 = 0$ **b.** $3x^2 + 2y - 6 = 0$
c. $2x^2 + 3y - 9 = 0$ **d.** $2x^2 - 3y + 9 = 0$

4. If $(\sin^{-1}x)^2 - (\cos^{-1}x)^2 = a, 0 < x < 1, a \neq 0$, then the value of $2x^2 - 1$ is

a. $\cos\left(\frac{4a}{\pi}\right)$ **b.** $\sin\left(\frac{2a}{\pi}\right)$
c. $\cos\left(\frac{2a}{\pi}\right)$ **d.** $\sin\left(\frac{4a}{\pi}\right)$

5. If the matrix $A = \begin{pmatrix} 0 & 2 \\ K & -1 \end{pmatrix}$ satisfies $A(A^3 + 3I) = 2I$, then the value of K is

a. $\frac{1}{2}$ **b.** $-\frac{1}{2}$
c. -1 **d.** 1

6. The distance of the point (1, –2, 3) from the plane $x - y + z = 5$ measured parallel to a line, whose direction ratios are 2, 3, –6 is

a. 3 **b.** 5
c. 2 **d.** 1

7. If $S = \left\{z \in C : \frac{z-i}{z+2i} \in R\right\}$, then

a. S contains exactly two elements.
b. S contains only one element.
c. S is a circle in the complex plane.
d. S is a straight line in the complex plane.

8. Let $y = y(x)$ be the solution of the differential equation $\frac{dy}{dx} = 2(y + 2\sin x - 5)\,x - 2\cos x$ such that, $y(0) = 7$. Then $y(\pi)$ is equal to

a. $2e^{\pi^2} + 5$ **b.** $e^{\pi^2} + 5$
c. $3e^{\pi^2} + 5$ **d.** $7e^{\pi^2} + 5$

9. Equation of a plane at a distance $\sqrt{2/21}$ from the origin, which contains the line of intersection of the planes $x - y - z - 1 = 0$ and $2x + y - 3z + 4 = 0$, is

a. $3x - y - 5z + 2 = 0$
b. $3x - 4z + 3 = 0$
c. $-x + 2y + 2z - 3 = 0$
d. $4x - y - 5z + 2 = 0$

10. If $U_n = \left(1 + \frac{1}{n^2}\right)\left(1 + \frac{2^2}{n^2}\right)^2 \ldots \left(1 + \frac{n^2}{n^2}\right)^n$, then $\lim_{n \to \infty} (U_n)^{\frac{-4}{n^2}}$ is equal to

a. $e^2/16$ **b.** $4/e$
c. $16/e^2$ **d.** $4/e^2$

11. The statement $(p \wedge (p \to q) \wedge (q \to r)) \to r$ is

a. a tautology **b.** equivalent to $p \to \sim r$
c. a fallacy **d.** equivalent to $q \to \sim r$

12. Let us consider a curve, $y = f(x)$ passing through the point (– 2, 2) and the slope of the tangent to the curve at any point $(x, f(x))$ is given by $f(x) + xf'(x) = x^2$ Then

a. $x^2 + 2x f(x) - 12 = 0$ **b.** $x^3 + 2x f(x) + 12 = 0$
c. $x^3 - 3x f(x) - 4 = 0$ **d.** $x^2 + 2x f(x) + 4 = 0$

13. $\sum_{k=0}^{20} ({}^{20}C_k)^2$ is equal to

a. ${}^{40}C_{21}$ **b.** ${}^{40}C_{19}$
c. ${}^{40}C_{20}$ **d.** ${}^{41}C_{20}$

14. A tangent and a normal are drawn at the point $P(2, -4)$ on the parabola $y^2 = 8x$, which meet the directrix of the parabola at the points A and B respectively. If $Q(a, b)$ is a point such that $AQBP$ is a square, then $2a + b$ is equal to

a. – 16 **b.** – 18 **c.** – 12 **d.** – 20

15. Let $\frac{\sin A}{\sin B} = \frac{\sin(A-C)}{\sin(C-B)}$, where A, B and C are angles of a ΔABC. If the lengths of the sides opposite these angles are a, b and c respectively, then

a. $b^2 - a^2 = a^2 + c^2$ **b.** b^2, c^2 and a^2 are in AP
c. c^2, a^2 and b^2 are in AP **d.** a^2, b^2 and c^2 are in AP

16. If α, β are the distinct roots of $x^2 + bx + c = 0$, then

$\lim_{x \to \beta} \frac{e^{2(x^2+bx+c)} - 1 - 2(x^2 + bx + c)}{(x-\beta)^2}$ is equal to

a. $b^2 + 4c$ **b.** $2(b^2 + 4c)$
c. $2(b^2 - 4c)$ **d.** $b^2 - 4c$

17. When a certain biased die is rolled, a particular face occurs with probability $\frac{1}{6} - x$ and its opposite face occurs with probability $\frac{1}{6} + x$. All other faces occur with probability 1/6. Note that opposite faces sum to 7 in any die. If $0 < x < \frac{1}{6}$, and the probability of obtaining total sum = 7, when such a die is rolled twice is 13/96, then the value of x is

a. 1/16 **b.** 1/8
c. 1/9 **d.** 1/12

18. If $x^2 + 9y^2 - 4x + 3 = 0$, $x, y \in R$, then x and y respectively lie in the intervals

a. $\left[-\frac{1}{3}, \frac{1}{3}\right]$ and $\left[-\frac{1}{3}, \frac{1}{3}\right]$

b. $\left[-\frac{1}{3}, \frac{1}{3}\right]$ and [1, 3]

c. [1, 3] and [1, 3]

d. [1, 3] and $\left[-\frac{1}{3}, \frac{1}{3}\right]$

19. $\int_6^{16} \frac{\log_e x^2}{\log_e x^2 + \log_e(x^2 - 44x + 484)} dx$ is equal to

a. 6 **b.** 8
c. 5 **d.** 10

20. A wire of length 20 m is to be cut into two pieces. One of the pieces is to be made into a square and the other into a regular hexagon. Then, the length of the side (in m) of the hexagon, so that the combined area of the square and the hexagon is minimum, is

a. $\frac{5}{2+\sqrt{3}}$ **b.** $\frac{10}{2+3\sqrt{3}}$

c. $\frac{5}{3+\sqrt{3}}$ **d.** $\frac{10}{3+2\sqrt{3}}$

Section B : Numerical Type Questions

21. Let $\mathbf{a} = \hat{\mathbf{i}} + 5\hat{\mathbf{j}} + \alpha\hat{\mathbf{k}}$, $\mathbf{b} = \hat{\mathbf{i}} + 3\hat{\mathbf{j}} + \beta\hat{\mathbf{k}}$ and $\mathbf{c} = -\hat{\mathbf{i}} + 2\hat{\mathbf{j}} - 3\hat{\mathbf{k}}$ be three vectors such that, $|\mathbf{b} \times \mathbf{c}| = 5\sqrt{3}$ and $\mathbf{a}$ is perpendicular to $\mathbf{b}$. Then, the greatest amongst the values of $|\mathbf{a}|^2$ is

22. The number of distinct real roots of the equation $3x^4 + 4x^3 - 12x^2 + 4 = 0$ is

23. Let the equation $x^2 + y^2 + px + (1-p)y + 5 = 0$ represent circles of varying radius $r \in (0, 5]$. Then, the number of elements in the set $S = \{q : q = p^2 \text{ and } q \text{ is an integer}\}$ is

24. If $A = \{x \in \mathbf{R} : |x - 2| > 1\}$, $B = \{x \in \mathbf{R} : \sqrt{x^2 - 3} > 1\}$ and $C = \{x \in \mathbf{R} : |x - 4| \geq 2\}$ and $\mathbf{Z}$ is the set of all integers, then the number of subsets of the set $(A \cap B \cap C)^c \cap \mathbf{Z}$ is

25. If $\int \frac{dx}{(x^2 + x + 1)^2} = a \tan^{-1}\left(\frac{2x+1}{\sqrt{3}}\right) + b\left(\frac{2x+1}{x^2+x+1}\right) + C$, $x > 0$ where C is the constant of integration, then the value of $9(\sqrt{3}a + b)$ is equal to

26. If the system of linear equations

$$2x + y - z = 3$$
$$x - y - z = \alpha$$
$$3x + 3y + \beta z = 3$$

has infinitely many solution, then $\alpha + \beta - \alpha\beta$ is equal to

27. Let n be an odd natural number such that the variance of 1, 2, 3, 4, ..., n is 14. Then n is equal to

28. If the minimum area of the triangle formed by a tangent to the ellipse $\frac{x^2}{b^2} + \frac{y^2}{4a^2} = 1$ and the coordinate axis is kab, then k is equal to

29. A number is called a palindrome if it reads the same backward as well as forward. For example 285582 is a six digit palindrome. The number of six digit palindromes, which are divisible by 55, is

30. If $y^{1/4} + y^{-1/4} = 2x$, and $(x^2 - 1)\frac{d^2y}{dx^2} + \alpha x\frac{dy}{dx} + \beta y = 0$ then $|\alpha - \beta|$ is equal to

Answers

For solution scan the QR code

1. (a)	2. (d)	3. (c)	4. (b)	5. (a)	6. (d)	7. (d)	8. (a)	9. (d)	10. (a)
11. (a)	12. (c)	13. (c)	14. (a)	15. (b)	16. (c)	17. (b)	18. (c)	19. (c)	20. (d)
21. (90)	22. (4)	23. (61)	24. (256)	25. (15)	26. (5)	27. (13)	28. (2)	29. (100)	30. (17)

27 AUGUST SHIFT II

Section A : Objective Type Questions

1. The angle between the straight lines, whose direction cosines are given by the equations $2l + 2m - n = 0$ and $mn + nl + lm = 0$, is

a. $\frac{\pi}{2}$ **b.** $\pi - \cos^{-1}\left(\frac{4}{9}\right)$

c. $\cos^{-1}\left(\frac{8}{9}\right)$ **d.** $\frac{\pi}{3}$

2. Let $A = \begin{pmatrix} [x+1] & [x+2] & [x+3] \\ [x] & [x+3] & [x+3] \\ [x] & [x+2] & [x+4] \end{pmatrix}$, where $[t]$ denotes the greatest integer less than or equal to t. If $\det(A) = 192$, then the set of values of x is the interval

a. [68, 69) **b.** [62, 63)

c. [65, 66) **d.** [60, 61)

3. Let M and m respectively be the maximum and minimum values of the function

$f(x) = \tan^{-1}(\sin x + \cos x)$ in $\left[0, \frac{\pi}{2}\right]$, then the value of $\tan(M - m)$ is equal to

a. $2 + \sqrt{3}$ **b.** $2 - \sqrt{3}$

c. $3 + 2\sqrt{2}$ **d.** $3 - 2\sqrt{2}$

4. Each of the persons A and B independently tosses three fair coins. The probability that both of them get the same number of heads is

a. $\frac{1}{8}$ **b.** $\frac{5}{8}$

c. $\frac{5}{16}$ **d.** 1

5. A differential equation representing the family of parabolas with axis parallel to Y-axis and whose length of latus rectum is the distance of the point (2, – 3) from the line $3x + 4y = 5$, is given by

a. $10\frac{d^2y}{dx^2} = 11$ **b.** $11\frac{d^2x}{dy^2} + 10$

c. $10\frac{d^2x}{dy^2} = 11$ **d.** $11\frac{d^2y}{dx^2} = 10$

6. If two tangents drawn from a point P to the parabola $y^2 = 16(x - 3)$ are at right angles, then the locus of point P is

a. $x + 3 = 0$ **b.** $x + 1 = 0$

c. $x + 2 = 0$ **d.** $x + 4 = 0$

7. The equation of the plane passing through the line of intersection of the planes

$\mathbf{r}\cdot(\hat{i} + \hat{j} + \hat{k}) = 1$ and $\mathbf{r}\cdot(2\hat{i} + 3\hat{j} - \hat{k}) + 4 = 0$ and parallel to the X-axis is

a. $\mathbf{r}\cdot(\hat{j} - 3\hat{k}) + 6 = 0$ **b.** $\mathbf{r}\cdot(\hat{i} + 3\hat{k}) + 6 = 0$

c. $\mathbf{r}\cdot(\hat{i} - 3\hat{k}) + 6 = 0$ **d.** $\mathbf{r}\cdot(\hat{j} - 3\hat{k}) - 6 = 0$

8. If the solution curve of the differential equation $(2x - 10y^3)dy + ydx = 0$, passes through the points (0, 1) and $(2, \beta)$, then β is a root of the equation

a. $y^5 - 2y - 2 = 0$ **b.** $2y^5 - 2y - 1 = 0$

c. $2y^5 - y^2 - 2 = 0$ **d.** $y^5 - y^2 - 1 = 0$

9. Let $A(a, 0)$, $B(b, 2b + 1)$ and $C(0, b)$, $b \neq 0, |b| \neq 1$, be points such that the area of ΔABC is 1 sq. unit, then the sum of all possible values of a is

a. $\frac{-2b}{b+1}$ **b.** $\frac{2b}{b+1}$

c. $\frac{2b^2}{b+1}$ **d.** $\frac{-2b^2}{b+1}$

10. Let $|\lambda|$ be the greatest integer less than or equal to λ. The set of all values of λ for which the system of linear equations $x + y + z = 4$, $3x + 2y + 5z = 3$, $9x + 4y + (28 + |\lambda|)z = |\lambda|$ has a solution is

a. R **b.** $(-\infty, -9) \cup (-9, \infty)$

c. $[-9, -8)$ **d.** $(-\infty, -9) \cup [-8, \infty)$

11. The set of all values of $k > -1$, for which the equation $(3x^2 + 4x + 3)^2 - (k + 1)(3x^2 + 4x + 3)$ $(3x^2 + 4x + 2) + k(3x^2 + 4x + 2)^2 = 0$ has real roots, is

a. $\left(1, \frac{5}{2}\right]$ **b.** [2, 3)

c. $\left[-\frac{1}{2}, 1\right)$ **d.** $\left(\frac{1}{2}, \frac{3}{2}\right] - \{1\}$

12. A box open from top is made from a rectangular sheet of dimension $a \times b$ by cutting squares each of side x from each of the four corners and folding up the flaps. If the volume of the box is maximum, then x is equal to

a. $\frac{a + b - \sqrt{a^2 + b^2 - ab}}{12}$ **b.** $\frac{a + b - \sqrt{a^2 + b^2 + ab}}{6}$

c. $\frac{a + b - \sqrt{a^2 + b^2 - ab}}{6}$ **d.** $\frac{a + b + \sqrt{a^2 + b^2 - ab}}{6}$

13. The Boolean expression $(p \wedge q) \Rightarrow ((r \wedge q) \wedge p)$ is equivalent to

a. $(p \wedge q) \Rightarrow (r \wedge q)$ **b.** $(q \wedge r) \Rightarrow (p \wedge q)$

c. $(p \wedge q) \Rightarrow (tr \vee q)$ **d.** $(p \wedge r) \Rightarrow (p \wedge q)$

14. Let Z be the set of all integers,
$A = \{(x, y) \in Z \times Z : (x - 2)^2 + y^2 \leq 4\}$,
$B = \{(x, y) \in Z \times Z : x^2 + y^2 \leq 4\}$ and
$C = \{(x, y) \in Z \times Z : (x - 2)^2 + (y - 2)^2 \leq 4\}$
If the total number of relation from $A \cap B$ to $A \cap B$ is 2^p, then the value of p is

a. 16 **b.** 25 **c.** 49 **d.** 9

15. The area of the region bounded by the parabola $(y - 2)^2 = (x - 1)$, the tangent to it at the point whose ordinate is 3 and the X-axis is

a. 9 **b.** 10 **c.** 4 **d.** 6

16. If $y(x) = \cot^{-1}\left(\frac{\sqrt{1+\sin x} + \sqrt{1-\sin x}}{\sqrt{1+\sin x} - \sqrt{1-\sin x}}\right), \cdot\, x \in \left(\frac{\pi}{2}, \pi\right)$, then $\frac{dy}{dx}$ at $x = \frac{5\pi}{6}$ is

a. $-\frac{1}{2}$ **b.** -1 **c.** $\frac{1}{2}$ **d.** 0

17. Two poles, AB of length a m and CD of length $a + b\ (b \neq a)$ metres are erected at the same horizontal level with bases at B and D. If $BD = x$ and $\tan \angle ACB! = \frac{1}{2}$, then

a. $x^2 + 2(a + 2b)x - b(a + b) = 0$
b. $x^2 + 2(a + 2b)x + a(a + b) = 0$
c. $x^2 - 2ax + b(a + b) = 0$
d. $x^2 - 2ax + a(a + b) = 0$

18. If $0 < x < 1$ and $y = \frac{1}{2}x^2 + \frac{2}{3}x^3 + \frac{3}{4}x^{4+\cdots}$ then the value of e^{1+y} at $x = \frac{1}{2}$ is

a. $\frac{1}{2}e^2$ **b.** $2e$
c. $\frac{1}{2}\sqrt{e}$ **d.** $2e^2$

19. The value of the integral $\int_0^1 \frac{\sqrt{x}\,dx}{(1+x)(1+3x)(3+x)}$ is

a. $\frac{\pi}{8}\left(1-\frac{\sqrt{3}}{2}\right)$ **b.** $\frac{\pi}{4}\left(1-\frac{\sqrt{3}}{6}\right)$
c. $\frac{\pi}{8}\left(1-\frac{\sqrt{3}}{6}\right)$ **d.** $\frac{\pi}{4}\left(1-\frac{\sqrt{3}}{2}\right)$

20. If $\lim_{x \to \infty} (\sqrt{x^2 - x + 1} - ax) = b$, then the ordered pair (a, b) is

a. $\left(1, \frac{1}{2}\right)$ **b.** $\left(1, -\frac{1}{2}\right)$
c. $\left(-1, \frac{1}{2}\right)$ **d.** $\left(-1, -\frac{1}{2}\right)$

Section B : Numerical Type Questions

21. Let S be the sum of all solutions (in radians) of the equation $\sin^4\theta + \cos^4\theta - \sin\theta\cos\theta = 0$ in $[0, 4\pi]$. Then, $8S/\pi$ is equal to

22. Let S be the mirror image of the point Q $(1, 3, 4)$ with respect to the plane $2x - y + z + 3 = 0$ and let $R(3, 5, \gamma)$ be a point of this plane. Then the square of the length of the line segment SR is

23. The probability distribution of random variable X is given by

X	1	2	3	4	5
$P(X)$	K	$2K$	$2K$	$3K$	K

Let $p = P(1 < X < 4 \mid X < 3)$. If $5p = \lambda K$ then, λ equal to

24. Let Z_1 and Z_2 be two complex numbers such that $\arg(Z_1 - Z_2) = \frac{\pi}{4}$ and Z_1, Z_2 satisfy the equation $|Z - 3| = \text{Re}(Z)$. Then, the imaginary part of $Z_1 + Z_2$ is equal to

25. Let $S = \{1, 2, 3, 4, 5, 6, 9\}$. Then, the number of elements in the set $T = \{A \subseteq S : A \neq \phi$ and the sum of all the elements of A is not a multiple of 3$\}$ is

26. Let $A(\sec\theta, 2\tan\theta)$ and $B(\sec\phi, 2\tan\phi)$, where $\theta + \phi = \pi/2$, be two points on the hyperbola $2x^2 - y^2 = 2$. If (α, β) is the point of the intersection of the normals to the hyperbola at A and B, then $(2\beta)^2$ is equal to

27. Two circles each of radius 5 units touch each other at the point $(1, 2)$. If the equation of their common tangent is $4x + 3y = 10$ and $C_1(\alpha, \beta)$ and $C_2(\gamma, \delta)$, $C_1 \neq C_2$ are their centres, then $|(\alpha + \beta)(\gamma + \delta)|$ is equal to

28. $3 \times 7^{22} + 2 \times 10^{22} - 44$ when divided by 18 leaves the remainder

29. An online exam is attempted by 50 candidates out of which 20 are boys. The average marks obtained by boys is 12 with a variance 2. The variance of marks obtained by 30 girls is also 2. The average marks of all 50 candidates is 15. If μ is the average marks of girls and σ^2 is the variance of marks of 50 candidates, then $\mu + \sigma^2$ is equal to

30. If $\int \frac{2e^x + 3e^{-x}}{4e^x + 7e^{-x}}\,dx = \frac{1}{14}(ux + v\log_e(4e^x + 7e^{-x})) + C$, where C is a constant of integration, then $u + v$ is equal to

Answers

For solution scan the QR code

1. (a)	2. (b)	3. (d)	4. (c)	5. (d)	6. (b)	7. (a)	8. (d)	9. (d)	10. (a)
11. (a)	12. (c)	13. (a)	14. (b)	15. (a)	16. (a)	17. (c)	18. (a)	19. (a)	20. (b)
21. (56)	22. (72)	23. (30)	24. (6)	25. (80)	26. (*)	27. (40)	28. (15)	29. (25)	30. (7)

***Note** (*) None of the option is correct.*

31 AUGUST SHIFT I

Section A : Objective Type Questions

1. Let $*, \square \in \{\wedge, \vee\}$ be such that the Boolean expression $(p * \sim q) \Rightarrow (p \square q)$ is a tautology. Then

a. $* = \vee, \square = \vee$ **b.** $* = \wedge, \square = \wedge$
c. $* = \wedge, \square = \vee$ **d.** $* = \vee, \square = \wedge$

2. The number of real roots of the equation $e^{4x} + 2e^{3x} - e^x - 6 = 0$ is

a. 2 **b.** 4 **c.** 1 **d.** 0

3. The sum of 10 terms of the series $\frac{3}{1^2 \times 2^2} + \frac{5}{2^2 \times 3^2} + \frac{7}{3^2 \times 4^2} + \ldots$ is

a. 1 **b.** 120/121 **c.** 99/100 **d.** 143/144

4. Let the equation of the plane, that passes through the point (1, 4, –3) and contains the line of intersection of the planes $3x - 2y + 4z - 7 = 0$ and $x + 5y - 2z + 9 = 0$ be $\alpha x + \beta y + \gamma z + 3 = 0$, then $\alpha + \beta + \gamma$ is equal to

a. –23 **b.** –15 **c.** 23 **d.** 15

5. Let f be a non–negative function in [0, 1] and twice differentiable in (0, 1). If $\int_0^x \sqrt{1-(f'(t))^2}\, dt = \int_0^x f(t)dt, 0 \le x \le 1$ and $f(0) = 0$, then $\lim_{x \to 0} \frac{1}{x^2}\int_0^x f(t)dt$

a. equals 0 **b.** equals 1
c. does not exist **d.** equals 1/2

6. Let **a** and **b** be two vectors such that $|2\mathbf{a} + 3\mathbf{b}| = |3\mathbf{a} + \mathbf{b}|$ and the angle between **a** and **b** is 60°. If $\frac{1}{8}\mathbf{a}$ is a unit vector, then $|\mathbf{b}|$ is equal to

a. 4 **b.** 6 **c.** 5 **d.** 8

7. The function $f(x) = |x^2 - 2x - 3| \cdot e^{|9x^2 - 12x + 4|}$ is not differentiable at exactly

a. four points **b.** three points
c. two points **d.** one point

8. Three numbers are in an increasing geometric progression with common ratio r. If the middle number is doubled, then the new numbers are in an arithmetic progression with common difference d. If the fourth term of GP is $3r^2$, then $r^2 - d$ is equal to

a. $7 - 7\sqrt{3}$ **b.** $7 + \sqrt{3}$ **c.** $7 - \sqrt{3}$ **d.** $7 + 3\sqrt{3}$

9. Which of the following is not correct for relation R on the set of real numbers ?

a. $(x, y) \in R \Leftrightarrow 0 < |x| - |y| \le 1$ is neither transitive nor symmetric.
b. $(x, y) \in R \Leftrightarrow 0 < |x - y| \le 1$ is symmetric and transitive.
c. $(x, y) \in R \Leftrightarrow |x| - |y| \le 1$ is reflexive but not symmetric.
d. $(x, y) \in R \Leftrightarrow |x - y| \le 1$ is reflexive and symmetric.

10. The integral $\int \frac{1}{\sqrt[4]{(x-1)^3 (x+2)^5}} dx$ is equal to (where C is a constant of integration)

a. $\frac{3}{4}\left(\frac{x+2}{x-1}\right)^{1/4} + C$ **b.** $\frac{3}{4}\left(\frac{x+2}{x-1}\right)^{5/4} + C$
c. $\frac{4}{3}\left(\frac{x-1}{x+2}\right)^{1/4} + C$ **d.** $\frac{4}{3}\left(\frac{x-1}{x+2}\right)^{5/4} + C$

11. If p and q are the lengths of the perpendiculars from the origin on the lines,
$x \operatorname{cosec} \alpha - y \sec \alpha = k \cot 2\alpha$ and
$x \sin \alpha + y \cos \alpha = k \sin 2\alpha$
respectively, then k^2 is equal to

a. $4p^2 + q^2$ **b.** $2p^2 + q^2$
c. $p^2 + 2q^2$ **d.** $p^2 + 4q^2$

12. cosec 18° is a root of the equation

a. $x^2 + 2x - 4 = 0$ **b.** $4x^2 + 2x - 1 = 0$
c. $x^2 - 2x + 4 = 0$ **d.** $x^2 - 2x - 4 = 0$

13. If the following system of linear equations
$2x + y + z = 5, x - y + z = 3$ and $x + y + az = b$
has no solution, then

a. $a = -\frac{1}{3}$ and $b \ne \frac{7}{3}$ **b.** $a \ne \frac{1}{3}$ and $b = \frac{7}{3}$
c. $a \ne -\frac{1}{3}$ and $b = \frac{7}{3}$ **d.** $a = \frac{1}{3}$ and $b \ne \frac{7}{3}$

14. The length of the latus rectum of a parabola, whose vertex and focus are on the positive X-axis at a distance R and $S(> R)$ respectively from the origin, is

a. $4(S + R)$ **b.** $2(S - R)$
c. $4(S - R)$ **d.** $2(S + R)$

15. If the function $f(x) = \begin{cases} \frac{1}{x}\log_e\left(\frac{1+\frac{x}{a}}{1-\frac{x}{b}}\right) & , x < 0 \\ k & , x = 0 \\ \frac{\cos^2 x - \sin^2 x - 1}{\sqrt{x^2+1} - 1} & , x > 0 \end{cases}$

is continuous at $x = 0$, then $\frac{1}{a} + \frac{1}{b} + \frac{4}{k}$ is equal to

a. –5 **b.** 5
c. – 4 **d.** 4

16. If $\frac{dy}{dx} = \frac{2^{x+y} - 2^x}{2^y}$, $y(0) = 1$, then $y(1)$ is equal to

a. $\log_2(2 + e)$ **b.** $\log_2(1 + e)$
c. $\log_2(2e)$ **d.** $\log_2(1 + e^2)$

17. $\lim_{x \to 0} \frac{\sin^2(\pi \cos^4 x)}{x^4}$ is equal to

a. π^2 **b.** $2\pi^2$ **c.** $4\pi^2$ **d.** 4π

18. A vertical pole fixed to the horizontal ground is divided in the ratio 3 : 7 by a mark on it with lower part shorter than the upper part. If the two parts subtend equal angles at a point on the ground 18 m away from the base of the pole, then the height of the pole (in m) is

a. $12\sqrt{15}$ **b.** $12\sqrt{10}$ **c.** $8\sqrt{10}$ **d.** $6\sqrt{10}$

19. If $a_r = \cos\frac{2r\pi}{9} + i\sin\frac{2r\pi}{9}, r = 1, 2, 3, \ldots, i = \sqrt{-1}$, then the determinant $\begin{vmatrix} a_1 & a_2 & a_3 \\ a_4 & a_5 & a_6 \\ a_7 & a_8 & a_9 \end{vmatrix}$ is equal to

a. $a_2a_6 - a_4a_8$ **b.** a_9 **c.** $a_1a_9 - a_3a_7$ **d.** a_5

20. The line $12x\cos\theta + 5y\sin\theta = 60$ is tangent to which of the following curves?

a. $x^2 + y^2 = 169$ **b.** $144x^2 + 25y^2 = 3600$
c. $25x^2 + 12y^2 = 3600$ **d.** $x^2 + y^2 = 60$

Section B : Numerical Type Questions

21. Let $[t]$ denote the greatest integer $\le t$. Then the value of $8.\int_{-1/2}^{1}([2x] + |x|)\,dx$ is

22. A point z moves in the complex plane such that $\arg\left(\frac{z-2}{z+2}\right) = \frac{\pi}{4}$, then the minimum value of $|z - 9\sqrt{2} - 2i|^2$ equal to

23. The square of the distance of the point of intersection of the line $\frac{x-1}{2} = \frac{y-2}{3} = \frac{z+1}{6}$ and the plane $2x - y + z = 6$ from the point $(-1, -1, 2)$ is

24. If R is the least value of a such that the function $f(x) = x^2 + ax + 1$ is increasing on [1, 2] and S is the greatest value of a such that the function $f(x) = x^2 + ax + 1$ is decreasing on [1, 2], then the value of $|R - S|$ is

25. The mean of 10 numbers $7\times 8, 10\times 10, 13\times 12, 16\times 14, \ldots$ is

26. If the variable line $3x + 4y = \alpha$ lies between the two circles $(x-1)^2 + (y-1)^2 = 1$ and $(x-9)^2 + (y-1)^2 = 4$, without intercepting a chord on either circle, then the sum of all the integral values of α is

27. The number of six letter words (with or without meaning), formed using all the letters of the word 'VOWELS', so that all the consonants never come together, is

28. If $x\,\phi(x) = \int_5^x (3t^2 - 2\phi'(t))\,dt, x > -2$, and $\phi(0) = 4$, then $\phi(2)$ is

29. If $\left(\frac{3^6}{4^4}\right)k$ is the term, independent of x, in the binomial expansion of $\left(\frac{x}{4} - \frac{12}{x^2}\right)^{12}$, then k is equal to

30. An electric instrument consists of two units. Each unit must function independently for the instrument to operate. The probability that the first unit functions is 0.9 and that of the second unit is 0.8. The instrument is switched on and it fails to operate. If the probability that only the first unit failed and second unit is functioning is P, then $98P$ is equal to

Answers

For solution scan the QR code

1. (c)	2. (c)	3. (b)	4. (a)	5. (d)	6. (c)	7. (c)	8. (b)	9. (b)	10. (c)
11. (a)	12. (d)	13. (d)	14. (c)	15. (a)	16. (b)	17. (c)	18. (b)	19. (c)	20. (b)
21. (5)	22. (98)	23. (61)	24. (2)	25. (398)	26. (105)	27. (576)	28. (4)	29. (55)	30. (28)

31 AUGUST SHIFT II

Section A : Objective Type Questions

1. If $\alpha + \beta + \gamma = 2\pi$, then the system of equations

$x + (\cos\gamma)y + (\cos\beta)z = 0$
$(\cos\gamma)x + y + (\cos\alpha)z = 0$
$(\cos\beta)x + (\cos\alpha)y + z = 0$

has:

a. no solution **b.** infinitely many solution
c. exactly two solutions **d.** a unique solution

2. Let **a**, **b** and **c** be three vectors mutually perpendicular to each other and have same magnitude. If a vector **r** satisfies.

$\mathbf{a} \times \{(\mathbf{r} - \mathbf{b}) \times \mathbf{a}\} + \mathbf{b} \times \{(\mathbf{r} - \mathbf{c}) \times b\} + \mathbf{c} \times \{(\mathbf{r} - \mathbf{a}) \times \mathbf{c}\} = \mathbf{0}$,

then **r** is equal to

a. $\frac{1}{3}(\mathbf{a} + \mathbf{b} + \mathbf{c})$ **b.** $\frac{1}{3}(2\mathbf{a} + \mathbf{b} - \mathbf{c})$
c. $\frac{1}{2}(\mathbf{a} + \mathbf{b} + \mathbf{c})$ **d.** $\frac{1}{2}(\mathbf{a} + \mathbf{b} + 2\mathbf{c})$

3. The domain of the function

$f(x) = \sin^{-1}\left(\frac{3x^2 + x - 1}{(x-1)^2}\right) + \cos^{-1}\left(\frac{x-1}{x+1}\right)$ is

a. $\left[0, \frac{1}{4}\right]$ **b.** $[-2, 0] \cup \left[\frac{1}{4}, \frac{1}{2}\right]$

c. $\left[\frac{1}{4}, \frac{1}{2}\right] \cup \{0\}$ **d.** $\left[0, \frac{1}{2}\right]$

4. Let $S = \{1, 2, 3, 4, 5, 6\}$. Then, the probability that a randomly chosen onto function g from S to S satisfies $g(3) = 2g(1)$ is

a. $\frac{1}{10}$ **b.** $\frac{1}{15}$ **c.** $\frac{1}{5}$ **d.** $\frac{1}{30}$

5. Let $f: \mathbf{N} \to \mathbf{N}$ be a function such that $f(m+n) = f(m) + f(n)$ for every $m, n \in \mathbf{N}$. If $f(6) = 18$, then $f(2) \cdot f(3)$ is equal to

a. 6 **b.** 54 **c.** 18 **c.** 36

6. The distance of the point (–1, 2, –2) from the line of intersection of the planes $2x + 3y + 2z = 0$ and $x - 2y + z = 0$ is

a. $\frac{1}{\sqrt{2}}$ **b.** $\frac{5}{2}$ **c.** $\frac{\sqrt{42}}{2}$ **d.** $\frac{\sqrt{34}}{2}$

7. Negation of the statement $(p \vee q) \Rightarrow (q \vee r)$ is

a. $p \wedge \sim q \wedge \sim r$ **b.** $\sim p \wedge q \wedge \sim r$ **c.** $\sim p \wedge q \wedge r$ **d.** $p \wedge q \wedge r$

8. If $\alpha = \lim_{x \to \frac{\pi}{4}} \frac{\tan^3 x - \tan x}{\cos\left(x + \frac{\pi}{4}\right)}$ and $\beta = \lim_{x \to 0} (\cos x)^{\cot x}$ are the roots of the equation, $ax^2 + bx - 4 = 0$, then the ordered pair (a, b) is

a. (1, – 3) **b.** (– 1, 3) **c.** (–1, – 3) **d.** (1, 3)

9. The locus of mid–points of the line segments joining (– 3, –5) and the points on the ellipse

$\frac{x^2}{4} + \frac{y^2}{9} = 1$ is

a. $9x^2 + 4y^2 + 18x + 8y + 145 = 0$

b. $36x^2 + 16y^2 + 90x + 56y + 145 = 0$

c. $36x^2 + 16y^2 + 108x + 80y + 145 = 0$

d. $36x^2 + 16y^2 + 72x + 32y + 145 = 0$

10. If $\frac{dy}{dx} = \frac{2^x y + 2^y \cdot 2^x}{2^x + 2^{x+y} \log_e 2}$, $y(0) = 0$, then for $y = 1$, the value of x lies in the interval

a. (1,2) **b.** $\left(\frac{1}{2}, 1\right]$

c. (2, 3) **d.** $\left(0, \frac{1}{2}\right]$

11. An angle of intersection of the curves, $\frac{x^2}{a^2} + \frac{y^2}{b^2} = 1$ and $x^2 + y^2 = ab, a > b$, is

a. $\tan^{-1}\left(\frac{a+b}{\sqrt{ab}}\right)$ **b.** $\tan^{-1}\left(\frac{a-b}{2\sqrt{ab}}\right)$

c. $\tan^{-1}\left(\frac{a-b}{\sqrt{ab}}\right)$ **d.** $\tan^{-1}(2\sqrt{ab})$

12. If $y\frac{dy}{dx} = x\left[\frac{y^2}{x^2} + \frac{\phi\left(\frac{y^2}{x^2}\right)}{\phi'\left(\frac{y^2}{x^2}\right)}\right]$, $x > 0$, $\phi > 0$, and $y(1) = -1$, then $\phi\left(\frac{y^2}{4}\right)$ is equal to

a. $4\phi(2)$ **b.** $4\phi(1)$

c. $2\phi(1)$ **d.** $\phi(1)$

13. The sum of the roots of the equation $x + 1 - 2\log_2(3 + 2^x) + 2\log_4(10 - 2^{-x}) = 0$ is

a. $\log_2 14$ **b.** $\log_2 11$ **c.** $\log_2 12$ **d.** $\log_2 13$

14. If z is a complex number such that $\frac{z-i}{z-1}$ is purely imaginary, then the minimum value of $|z - (3 + 3i)|$ is

a. $2\sqrt{2} - 1$ **b.** $3\sqrt{2}$ **c.** $6\sqrt{2}$ **d.** $2\sqrt{2}$

15. Let a_1, a_2, a_3 .... be an A.P. If $\frac{a_1 + a_2 + \ldots + a_{10}}{a_1 + a_2 + \ldots + a_p} = \frac{100}{p^2}$, $p \neq 10$, then $\frac{a_{11}}{a_{10}}$ is equal to

a. $\frac{19}{21}$ **b.** $\frac{100}{121}$ **c.** $\frac{21}{19}$ **d.** $\frac{121}{100}$

16. Let A be the set of all points (α, β) such that the area of triangle formed by the points (5, 6), (3, 2) and (α, β) is 12 sq units. Then, the least possible length of a line segment joining the origin to a point in A, is

a. $\frac{4}{\sqrt{5}}$ **b.** $\frac{16}{\sqrt{5}}$

c. $\frac{8}{\sqrt{5}}$ **d.** $\frac{12}{\sqrt{5}}$

17. The number of solutions of the equation $32^{\tan^2 x} + 32^{\sec^2 x} = 81, 0 \le x \le \frac{\pi}{4}$ is

a. 3 **b.** 1 **c.** 0 **d.** 2

18. Let f be any continuous function on [0, 2] and twice differentiable on (0, 2). If $f(0) = 0$, $f(1) = 1$ and $f(2) = 2$, then

a. $f''(x) = 0$ for all $x \in (0, 2)$

b. $f''(x) = 0$ for some $x \in (0, 2)$

c. $f'(x) = 0$ for some $x \in [0, 2]$

d. $f''(x) > 0$ for all $x \in (0, 2)$

19. If $[x]$ is the greatest integer $\le x$, then $\pi^2 \int_0^2 \left(\sin \frac{\pi x}{2}\right)(x - [x])^{[x]} dx$ is equal to

a. $2(\pi - 1)$ **b.** $4(\pi - 1)$

c. $4(\pi + 1)$ **d.** $2(\pi + 1)$

20. The mean and variance of 7 observations are 8 and 16 respectively. If two observations are 6 and 8, then the variance of the remaining 5 observations is

a. $\frac{92}{5}$ **b.** $\frac{134}{5}$

c. $\frac{536}{25}$ **d.** $\frac{112}{5}$

Section B : Numerical Type Questions

21. If the coefficient of a^7b^8 in the expansion of $(a + 2b + 4ab)^{10}$ is $k \cdot 2^{16}$, then k is equal to

22. Suppose the line $\frac{x-2}{\alpha} + \frac{y-2}{-5} = \frac{z+2}{2}$ lies on the plane $x + 3y - 2z + \beta = 0$. Then, $(\alpha + \beta)$ is equal to

23. The number of 4-digit numbers which are neither multiple of 7 nor multiple of 3 is

24. If $\int \frac{\sin x}{\sin^3 x + \cos^3 x} dx =$
$\alpha \log_e |1 + \tan x| + \beta \log_e |1 - \tan x + \tan^2 x|$
$+ \gamma \tan^{-1}\left(\frac{2\tan x - 1}{\sqrt{3}}\right) + C$, when C is constant of integration, then the value of $18(\alpha + \beta + \gamma^2)$ is

25. A tangent line L is drawn at the point $(2, -4)$ on the parabola $y^2 = 8x$. If the line L is also tangent to the circle $x^2 + y^2 = a$, then a is equal to

26. If $S = \frac{7}{5} + \frac{9}{5^2} + \frac{13}{5^3} + \frac{19}{5^4} + \ldots$, then $160\,S$ is equal to

27. The number of elements in the set
$$\left\{A = \begin{pmatrix} a & b \\ 0 & d \end{pmatrix} : a, b \text{ and } d \in \{-1, 0, 1\} \text{ and } (I - A)^3 = I - A^3\right\},$$
where I is 2×2 identity matrix, is

28. If the line $y = mx$ bisects the area enclosed by the lines $x = 0$ and $y = 0$, $x = 3/2$ and the curve $y = 1 + 4x - x^2$, then 12 m is equal to

29. Let B be the centre of the circle $x^2 + y^2 - 2x + 4y + 1 = 0$. Let the tangents at two points P and Q on the circle intersect at the point $A(3, 1)$. Then $8\left(\frac{\text{area } \Delta APQ}{\text{area } \Delta BPQ}\right)$ is equal to

30. Let $f(x)$ be a cubic polynomial with $f(1) = -10$, $f(-1) = 6$, and has a local minima at $x = 1$, and $f'(x)$ has a local minima at $x = -1$. Then $f(3)$ is equal to.

Answers

For solution scan the QR code

1. (b)	2. (c)	3. (c)	4. (a)	5. (b)	6. (d)	7. (a)	8. (d)	9. (c)	10. (a)
11. (c)	12. (b)	13. (b)	14. (d)	15. (c)	16. (c)	17. (b)	18. (b)	19. (b)	20. (c)
21. (315)	22. (7)	23. (5143)	24. (3)	25. (2)	26. (305)	27. (8)	28. (26)	29. (18)	30. (22)

1 SEPTEMBER SHIFT II

Section A : Objective Type Questions

1. Let $f : R \to R$ be a continuous function. Then,

$$\lim_{x \to \frac{\pi}{4}} \frac{\frac{\pi}{4}\int_{2}^{\sec^2 x} f(x)dx}{x^2 - \frac{\pi^2}{16}}$$ is equal to

a. $f(2)$ **b.** $2f(2)$ **c.** $2f(\sqrt{2})$ **d.** $4f(2)$

2. $\cos^{-1}(\cos(-5)) + \sin^{-1}(\sin(6)) - \tan^{-1}(\tan(12))$ is equal to

(The inverse trigonometric functions take the principal values)

a. $3\pi - 11$ **b.** $4\pi - 9$
c. $4\pi - 11$ **d.** $3\pi + 1$

3. Consider the system of linear equations

$$-x + y + 2z = 0$$
$$3x - ay + 5z = 1$$
$$2x - 2y - az = 7$$

Let S_1 be the set of all $a \in R$ for which the system is inconsistent and S_2 be the set of all $a \in R$ for which the system has infinitely many solutions. If $n(S_1)$ and $n(S_2)$ denote the number of elements in S_1 and S_2 respectively, then

a. $n(S_1) = 2$ and $n(S_2) = 2$ **b.** $n(S_1) = 1$ and $n(S_2) = 0$
c. $n(S_1) = 2$ and $n(S_2) = 0$ **d.** $n(S_1) = 0$ and $n(S_2) = 2$

4. Let the acute angle bisector of the two planes $x - 2y - 2z + 1 = 0$ and $2x - 3y - 6z + 1 = 0$ be the plane P. Then, which of the following points lies on P ?

a. $\left(3, 1, -\frac{1}{2}\right)$ **b.** $\left(-2, 0, -\frac{1}{2}\right)$
c. $(0, 2, -4)$ **d.** $(4, 0, -2)$

5. Which of the following is equivalent to the Boolean expression $p \wedge \sim q$?

a. $\sim (q \to p)$ **b.** $\sim p \to \sim q$
c. $\sim (p \to \sim q)$ **d.** $\sim (p \to q)$

6. Two squares are chosen at random on a chessboard (see figure). The probability that they have a side in common is

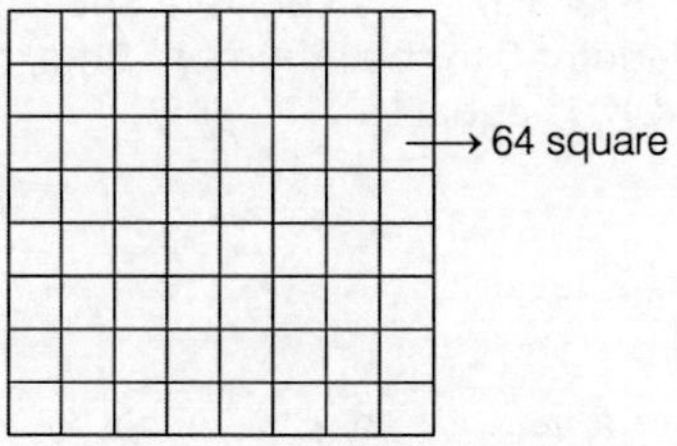

a. $\frac{2}{7}$ **b.** $\frac{1}{18}$ **c.** $\frac{1}{7}$ **d.** $\frac{1}{9}$

7. If $y = y(x)$ is the solution curve of the differential equation $x^2dy + \left(y - \frac{1}{x}\right)dx = 0$; $x > 0$ and $y(1) = 1$, then $y\left(\frac{1}{2}\right)$ is equal to

a. $\frac{3}{2} - \frac{1}{\sqrt{e}}$ **b.** $3 + \frac{1}{\sqrt{e}}$
c. $3 + e$ **d.** $3 - e$

8. If n is the number of solutions of the equation $2\cos x\left(4\sin\left(\frac{\pi}{4} + x\right)\sin\left(\frac{\pi}{4} - x\right) - 1\right) = 1, x \in [0, \pi]$ and S is the sum of all these solutions, then the ordered pair (n, S) is

a. $(3, 13\pi/9)$ **b.** $(2, 2\pi/3)$ **c.** $(2, 8\pi/9)$ **d.** $(3, 5\pi/3)$

9. The function $f(x) = x^3 - 6x^2 + ax + b$ is such that $f(2) = f(4) = 0$. Consider two statements.
(S_1) there exists $x_1, x_2 \in (2, 4), x_1 < x_2$, such that $f'(x_1) = -1$ and $f'(x_2) = 0$. (S_2) there exists $x_3, x_4 \in (2, 4), x_3 < x_4$, such that f is decreasing in $(2, x_4)$, increasing in $(x_4, 4)$ and $2f'(x_3) = \sqrt{3}f(x_4)$. Then,

a. both (S_1) and (S_2) are true **b.** (S_1) is false and (S_2) is true
c. both (S_1) and (S_2) are false **d.** (S_1) is true and (S_2) is false

10. Let $J_{n,m} = \int_0^{1/2} \frac{x^n}{x^m - 1}dx$, $\forall\, n > m$ and $n, m \in N$. Consider a matrix $A = [a_{ij}]_{3 \times 3}$ where

$$a_{ij} = \begin{cases} J_{6+i,3} - J_{i+3,3}, & i \le j \\ 0, & i > j \end{cases}$$ Then, $|\text{adj } A^{-1}|$ is

a. $(15)^2 \times 2^{42}$ **b.** $(15)^2 \times 2^{34}$
c. $(105)^2 \times 2^{38}$ **d.** $(105)^2 \times 2^{36}$

11. The area, enclosed by the curves $y = \sin x + \cos x$ and $y = |\cos x - \sin x|$ and the lines $x = 0, x = \frac{\pi}{2}$, is

a. $2\sqrt{2}(\sqrt{2} - 1)$ **b.** $2(\sqrt{2} + 1)$
c. $4(\sqrt{2} - 1)$ **d.** $2\sqrt{2}(\sqrt{2} + 1)$

12. The distance of line $3y - 2z - 1 = 0 = 3x - z + 4$ from the piont $(2, -1, 6)$ is

a. $\sqrt{26}$ **b.** $2\sqrt{5}$ **c.** $2\sqrt{6}$ **d.** $4\sqrt{2}$

13. Consider the parabola with vertex $\left(\frac{1}{2}, \frac{3}{4}\right)$ and the directrix $y = \frac{1}{2}$. Let P be the point where the parabola meets the line $x = -\frac{1}{2}$. If the normal to the parabola at P intersects the parabola again at the point Q, then $(PQ)^2$ is equal to

a. $\frac{75}{8}$ **b.** $\frac{125}{16}$
c. $\frac{25}{2}$ **d.** $\frac{15}{2}$

14. The numbers of pairs (a, b) of real numbers, such that whenever α is a root of the equation $x^2 + ax + b = 0$, $\alpha^2 - 2$ is also a root of this equation, is

a. 6 b. 2 c. 4 d. 8

15. Let $S_n = 1 \cdot (n-1) + 2 \cdot (n-2) + 3 \cdot (n-3) + \ldots + (n-1) \cdot 1$, $n \geq 4$.

The sum $\sum_{n=4}^{\infty} \left(\frac{2S_n}{n!} - \frac{1}{(n-2)!} \right)$ is equal to

a. $\frac{e-1}{3}$ b. $\frac{e-2}{6}$

c. $\frac{e}{3}$ d. $\frac{e}{6}$

16. Let $P_1, P_2, \ldots\ldots, P_{15}$ be 15 points on a circle. The number of distinct triangles formed by points $P_{\hat{i}}, P_{\hat{j}}$ and $P_{\hat{k}}$, such that $\hat{i} + \hat{j} + \hat{k} \neq 15$ is

a. 12 b. 419 c. 443 d. 455

17. The range of the function,

$$f(x) = \log_{\sqrt{5}}\left(3 + \cos\left(\frac{3\pi}{4} + x\right) + \cos\left(\frac{\pi}{4} + x\right) + \cos\left(\frac{\pi}{4} - x\right) - \cos\left(\frac{3\pi}{4} - x\right)\right)$$ is

a. $(0, \sqrt{5})$ b. $[-2, 2]$

c. $\left[\frac{1}{\sqrt{5}}, \sqrt{5}\right]$ d. $[0, 2]$

18. Let $a_1, a_2, \ldots\ldots, a_{21}$ be an AP such that $\sum_{n=1}^{20} \frac{1}{a_n a_{n+1}} = \frac{4}{9}$. If the sum of this AP is 189, then $a_6 a_{16}$ is equal to

a. 57 b. 72

c. 48 d. 36

19. The function $f(x)$, that satisfies the condition

$$f(x) = x + \int_0^{\pi/2} \sin x \cdot \cos y \; f(y) \, dy,$$ is

a. $x + \frac{2}{3}(\pi - 2)\sin x$ b. $x + (\pi + 2)\sin x$

c. $x + \frac{\pi}{2}\sin x$ d. $x + (\pi - 2)\sin x$

20. Let θ be the acute angle between the tangents to the ellipse $\frac{x^2}{9} + \frac{y^2}{1} = 1$ and the circle $x^2 + y^2 = 3$ at their point of intersection in the first quadrant. Then, $\tan\theta$ is equal to

a. $\frac{5}{2\sqrt{3}}$ b. $\frac{2}{\sqrt{3}}$

c. $\frac{4}{\sqrt{3}}$ d. 2

Section B : Numerical Type Questions

21. Let X be a random variable with distribution.

x	-2	-1	3	4	6
$P(X = x)$	$\frac{1}{5}$	a	$\frac{1}{3}$	$\frac{1}{5}$	b

If the mean of X is 2.3 and variance of X is σ^2, then $100\,\sigma^2$ is equal to

22. Let $f(x) = x^6 + 2x^4 + x^3 + 2x + 3$, $x \in R$. Then, the natural number n for which $\lim_{x \to 1} \frac{x^n f(1) - f(x)}{x - 1} = 44$ is

23. If for the complex numbers z satisfying $|z - 2 - 2i| \leq 1$, the maximum value of $|3iz + 6|$ is attained at $a + ib$, then $a + b$ is equal to

24. Let the points of intersections of the lines $x - y + 1 = 0$, $x - 2y + 3 = 0$ and $2x - 5y + 11 = 0$ are the mid-points of the sides of a ΔABC. Then, the area of the ΔABC is

25. Let $f(x)$ be polynomial of degree 3 such that $f(k) = -\frac{2}{k}$ for $k = 2, 3, 4, 5$. Then, the value of $52 - 10f(10)$ is equal to

26. All the arrangements, with or without meaning, of the word FARMER are written excluding any word that has two R appearing together. The arrangements are listed serially in the alphabetic order as in the English dictionary. Then, the serial number of the word FARMER in this list is

27. If the sum of the coefficients in the expansion of $(x + y)^n$ is 4096, then the greatest coefficient in the expansion is

28. Let $\mathbf{a} = 2\hat{i} - \hat{j} + 2\hat{k}$ and $\mathbf{b} = \hat{i} + 2\hat{j} - \hat{k}$. Let a vector $\mathbf{v}$ be in the plane containing $\mathbf{a}$ and $\mathbf{b}$. If $\mathbf{v}$ is perpendicular to the vector $3\hat{i} + 2\hat{j} - \hat{k}$ and its projection on a is 19 units, then $|2\mathbf{v}|^2$ is equal to

29. Let $[t]$ denote the greatest integer $\leq t$. The number of points where the function

$$f(x) = [x]\,|x^2 - 1| + \sin\left(\frac{\pi}{[x] + 3}\right) - [x + 1], \; x \in (-2, 2)$$ is not continuous is

30. A man starts walking from the point $P(-3, 4)$, touches the X-axis at R, and then turns to reach at the point $Q(0, 2)$. The man is walking at a constant speed. If the man reaches the point Q in the minimum time, then $50((PR)^2 + (RQ)^2)$ is equal to

Answers

For solution scan the QR code

1. (b)	2. (c)	3. (c)	4. (b)	5. (d)	6. (b)	7. (d)	8. (a)	9. (a)	10. (c)
11. (a)	12. (c)	13. (b)	14. (a)	15. (a)	16. (c)	17. (d)	18. (b)	19. (d)	20. (b)
21. (781)	22. (7)	23. (5)	24. (6)	25. (26)	26. (77)	27. (924)	28. (1494)	29. (2)	30. (1250)

JEE Main

Solved Papers
2022

ONLINE SOLVED PAPERS

JEE Main 2022

(June Attempt)

24 JUNE 2022, SHIFT-I

Section A : Objective Type Questions

1 Let $A = \{z \in C : 1 \le |z-(1+i)| \le 2\}$ and $B = \{z \in A : |z-(1-i)| = 1\}$. Then, B

(a) is an empty set
(b) contains exactly two elements
(c) contains exactly three elements
(d) is an infinite set

2 The remainder when 3^{2022} is divided by 5 is

(a) 1 (b) 2 (c) 3 (d) 4

3 The surface area of a balloon of spherical shape being inflated, increases at a constant rate. If initially, the radius of balloon is 3 units and after 5 s, it becomes 7 units, then its radius after 9 s is

(a) 9 (b) 10 (c) 11 (d) 12

4 Bag A contains 2 white, 1 black and 3 red balls and bag B contains 3 black, 2 red and n white balls. One bag is chosen at random and 2 balls drawn from it at random, are found to be 1 red and 1 black. If the probability that both balls come from bag A is $\frac{6}{11}$, then n is equal to ....

(a) 13 (b) 6 (c) 4 (d) 3

5 Let $x^2 + y^2 + Ax + By + C = 0$ be a circle passing through (0, 6) and touching the parabola $y = x^2$ at (2, 4). Then, $A + C$ is equal to ....

(a) 16 (b) 88/5 (c) 72 (d) −8

6 The number of values of α for which the system of equations

$$x + y + z = \alpha,$$
$$\alpha x + 2\alpha y + 3z = -1$$

and $$x + 3\alpha y + 5z = 4$$

is inconsistent, is

(a) 0 (b) 1 (c) 2 (d) 3

7 If the sum of the squares of the reciprocals of the roots α and β of the equation $3x^2 + \lambda x - 1 = 0$ is 15, then $6(\alpha^3 + \beta^3)^2$ is equal to

(a) 18 (b) 24 (c) 36 (d) 96

8 The set of all values of k for which $(\tan^{-1} x)^3 + (\cot^{-1} x)^3 = k\pi^3$, $x \in R$, is the interval

(a) $\left[\frac{1}{32}, \frac{7}{8}\right)$ (b) $\left(\frac{1}{24}, \frac{13}{16}\right)$

(c) $\left[\frac{1}{48}, \frac{13}{16}\right]$ (d) $\left[\frac{1}{32}, \frac{9}{8}\right)$

9 Let $S = \{\sqrt{n} : 1 \le n \le 50 \text{ and } n \text{ is odd}\}$

Let $a \in S$ and $A = \begin{bmatrix} 1 & 0 & a \\ -1 & 1 & 0 \\ -a & 0 & 1 \end{bmatrix}$

If $\sum_{a \in S} \det(\operatorname{adj} A) = 100\,\lambda$, then λ is equal to

(a) 218 (b) 221
(c) 663 (d) 1717

10 For the function $f(x) = 4\log_e(x-1) - 2x^2 + 4x + 5$, $x > 1$, which one of the following is NOT correct?

(a) f is increasing in (1, 2) and decreasing in $(2, \infty)$
(b) $f(x) = -1$ has exactly two solutions
(c) $f'(e) - f''(2) < 0$
(d) $f(x) = 0$ has a root in the interval $(e, e+1)$

11 If the tangent at the point (x_1, y_1) on the curve $y = x^3 + 3x^2 + 5$ passes through the origin, then (x_1, y_1) does NOT lie on the curve

(a) $x^2 + \frac{y^2}{81} = 2$ (b) $\frac{y^2}{9} - x^2 = 8$

(c) $y = 4x^2 + 5$ (d) $\frac{x}{3} - y^2 = 2$

12 The sum of absolute maximum and absolute minimum values of the function

$f(x) = |2x^2 + 3x - 2| + \sin x \cos x$ in the interval [0, 1] is

(a) $3 + \frac{\sin(1)\cos^2(1/2)}{2}$ (b) $3 + \frac{1}{2}(1 + 2\cos(1)\sin(1))$

(c) $5 + \frac{1}{2}(\sin(1) + \sin(2))$ (d) $2 + \sin\left(\frac{1}{2}\right)\cos\left(\frac{1}{2}\right)$

13 If $\{a_i\}_{i=1}^{n}$ where n is an even integer, is an arithmetic progression with common difference 1, and $\sum_{i=1}^{n} a_i = 192, \sum_{i=1}^{n/2} a_{2i} = 120$, then n is equal to

(a) 48 (b) 96 (c) 92 (d) 104

14 If $x = x(y)$ is the solution of the differential equation $y\frac{dx}{dy} = 2x + y^3(y+1)e^y$, $x(1) = 0$; then $x(e)$ is equal to

(a) $e^3(e^e - 1)$ (b) $e^e(e^3 - 1)$
(c) $e^2(e^e + 1)$ (d) $e^e(e^2 - 1)$

15 Let $\lambda x - 2y = \mu$ be a tangent to the hyperbola $a^2x^2 - y^2 = b^2$. Then, $\left(\frac{\lambda}{a}\right)^2 - \left(\frac{\mu}{b}\right)^2$ is equal to

(a) −2 (b) −4
(c) 2 (d) 4

16 Let $\hat{\mathbf{a}}$ and $\hat{\mathbf{b}}$ be unit vectors. If $\mathbf{c}$ is a vector such that the angle between $\hat{\mathbf{a}}$ and $\mathbf{c}$ is $\frac{\pi}{12}$ and $\hat{\mathbf{b}} = \mathbf{c} + 2(\mathbf{c} \times \hat{\mathbf{a}})$, then $|6\mathbf{c}|^2$ is equal to

(a) $6(3-\sqrt{3})$ (b) $3+\sqrt{3}$
(c) $6(3+\sqrt{3})$ (d) $6(\sqrt{3}+1)$

17 If a random variable X follows the Binomial distribution $B(33, p)$ such that $3P(X = 0) = P(X = 1)$, then the value of $\frac{P(X=15)}{P(X=18)} - \frac{P(X=16)}{P(X=17)}$ is equal to

(a) 1320 (b) 1088 (c) $\frac{120}{1331}$ (d) $\frac{1088}{1089}$

18 The domain of the function

$$f(x) = \frac{\cos^{-1}\left(\frac{x^2-5x+6}{x^2-9}\right)}{\log_e(x^2-3x+2)}$$ is

(a) $(-\infty, 1) \cup (2, \infty)$
(b) $(2, \infty)$
(c) $\left[-\frac{1}{2}, 1\right) \cup (2, \infty)$
(d) $\left[-\frac{1}{2}, 1\right) \cup (2, \infty) - \left\{\frac{3+\sqrt{5}}{2}, \frac{3-\sqrt{5}}{2}\right\}$

19 Let $S = \left\{\theta \in [-\pi, \pi] - \left\{\pm\frac{\pi}{2}\right\} : \sin\theta \tan\theta + \tan\theta = \sin 2\theta\right\}$

If $T = \sum_{\theta \in S} \cos 2\theta$, then $T + n(S)$ is equal to

(a) $7+\sqrt{3}$ (b) 9
(c) $8+\sqrt{3}$ (d) 10

20 The number of choices of $\Delta \in \{\wedge, \vee, \Rightarrow, \Leftrightarrow\}$, such that $(p\Delta q) \Rightarrow ((p\Delta \sim q) \vee ((\sim p)\Delta q))$ is a tautology, is

(a) 1 (b) 2 (c) 3 (d) 4

Section B : Numerical Type Questions

21 The number of one-one functions $f : \{a, b, c, d\} \to \{0, 1, 2, \ldots, 10\}$ such that $2f(a) - f(b) + 3f(c) + f(d) = 0$ is ....

22 In an examination, there are 5 multiple choice questions with 3 choices, out of which exactly one is correct There are 3 marks for each correct answer, −2 marks for each wrong answer and 0 mark if the question is not attempted. Then, the number of ways a student appearing in the examination gets 5 marks is ....

23 Let $A\left(\frac{3}{\sqrt{a}}, \sqrt{a}\right)$, $a > 0$, be a fixed point in the xy-plane. The image of A in Y-axis be B and the image of B in X-axis be C. If $D(3\cos\theta, a\sin\theta)$ is a point in the fourth quadrant such that the maximum area of ΔACD is 12 sq units, then a is equal to ....

24 Let a line having direction ratios $1, -4, 2$ intersect the lines $\frac{x-7}{3} = \frac{y-1}{-1} = \frac{z+2}{1}$ and $\frac{x}{2} = \frac{y-7}{3} = \frac{z}{1}$ at the points A and B. Then, $(AB)^2$ is equal to ....

25 The number of points where the function

$$f(x) = \begin{cases} |2x^2 - 3x - 7| & \text{if } x \le -1 \\ [4x^2 - 1] & \text{if } -1 < x < 1 \\ |x+1| + |x-2| & \text{if } x \ge 1 \end{cases}$$

$[t]$ denotes the greatest integer $\le t$, is discontinuous is ....

26 Let $f(\theta) = \sin\theta + \int_{-\pi/2}^{\pi/2} (\sin\theta + t\cos\theta) f(t)\,dt$. Then, the value of $\left|\int_0^{\pi/2} f(\theta)\,d\theta\right|$ is ....

27 Let $\max_{0 \le x \le 2}\left\{\frac{9-x^2}{5-x}\right\} = \alpha$ and $\min_{0 \le x \le 2}\left\{\frac{9-x^2}{5-x}\right\} = \beta$.

If $\int_{\beta-8/3}^{2\alpha-1} \max\left\{\frac{9-x^2}{5-x}, x\right\} dx = \alpha_1 + \alpha_2 \log_e\left(\frac{8}{15}\right)$, then $\alpha_1 + \alpha_2$ is equal to ....

28 If two tangents drawn from a point (α, β) lying on the ellipse $25x^2 + 4y^2 = 1$ to the parabola $y^2 = 4x$ are such that the slope of one tangent is four times the other, then the value of $(10\alpha + 5)^2 + (16\beta^2 + 50)^2$ equals ....

29 Let S be the region bounded by the curves $y = x^3$ and $y^2 = x$. The curve $y = 2|x|$ divides S into two regions of areas R_1 and R_2. If $\max\{R_1, R_2\} = R_2$, then $\frac{R_2}{R_1}$ is equal to ....

30 If the shortest distance between the lines

$$\mathbf{r} = (-\hat{\mathbf{i}} + 3\hat{\mathbf{k}}) + \lambda(\hat{\mathbf{i}} - a\hat{\mathbf{j}})$$

and $\mathbf{r} = (-\hat{\mathbf{j}} + 2\hat{\mathbf{k}}) + \mu(\hat{\mathbf{i}} - \hat{\mathbf{j}} + \hat{\mathbf{k}})$ is $\sqrt{\frac{2}{3}}$, then the integral value of a is equal to ....

24 JUNE 2022, SHIFT-II

Section A : Objective Type Questions

1 Let $x * y = x^2 + y^3$ and $(x * 1) * 1 = x * (1 * 1)$. Then, a value of $2\sin^{-1}\left(\frac{x^4+x^2-2}{x^4+x^2+2}\right)$ is

(a) $\frac{\pi}{4}$ (b) $\frac{\pi}{3}$ (c) $\frac{\pi}{2}$ (d) $\frac{\pi}{6}$

2 The sum of all the real roots of the equation $(e^{2x}-4)(6e^{2x}-5e^x+1)=0$ is

(a) $\log_e 3$ (b) $-\log_e 3$
(c) $\log_e 6$ (d) $-\log_e 6$

3 Let the system of linear equations

$$x + y + \alpha z = 2,$$
$$3x + y + z = 4$$
and $$x + 2z = 1$$

have a unique solution (x^*, y^*, z^*). If (α, x^*), (y^*, α) and $(x^*, -y^*)$ are collinear points, then the sum of absolute values of all possible values of α is

(a) 4 (b) 3 (c) 2 (d) 1

4 Let $x, y > 0$. If $x^3 y^2 = 2^{15}$, then the least value of $3x + 2y$ is

(a) 30 (b) 32
(c) 36 (d) 40

5 Let $f(x) = \begin{cases} \frac{\sin(x-[x])}{x-[x]}, & x \in (-2,-1) \\ \max\{2x, 3[|x|]\}, & |x| < 1 \\ 1, & \text{otherwise} \end{cases}$

where $[t]$ denotes greatest integer $\le t$. If m is the number of points where f is not continuous and n is the number of points where f is not differentiable, then the ordered pair (m, n) is

(a) (3, 3) (b) (2, 4) (c) (2, 3) (d) (3, 4)

6 The value of the integral $\int_{-\pi/2}^{\pi/2} \frac{dx}{(1+e^x)(\sin^6 x + \cos^6 x)}$ is equal to

(a) 2π (b) 0
(c) π (d) $\frac{\pi}{2}$

7 $\lim_{n\to\infty}\left(\frac{n^2}{(n^2+1)(n+1)} + \frac{n^2}{(n^2+4)(n+2)} + \frac{n^2}{(n^2+9)(n+3)} + \ldots + \frac{n^2}{(n^2+n^2)(n+n)}\right)$

is equal to

(a) $\frac{\pi}{8} + \frac{1}{4}\log_e 2$ (b) $\frac{\pi}{4} + \frac{1}{8}\log_e 2$
(c) $\frac{\pi}{4} - \frac{1}{8}\log_e 2$ (d) $\frac{\pi}{8} + \frac{1}{8}\log_e \sqrt{2}$

8 A particle is moving in the xy-plane along a curve C passing through the point (3, 3). The tangent to the curve C at the point P meets the X-axis at Q. If the Y-axis bisects the segment PQ, then C is a parabola with

(a) length of latus rectum 3 (b) length of latus rectum 6
(c) focus $\left(\frac{4}{3}, 0\right)$ (d) focus $\left(0, \frac{3}{4}\right)$

9 Let the maximum area of the triangle that can be inscribed in the ellipse $\frac{x^2}{a^2} + \frac{y^2}{4} = 1, a > 2$, having one of its vertices at one end of the major axis of the ellipse and one of its sides parallel to the Y-axis, be $6\sqrt{3}$. Then, the eccentricity of the ellipse is

(a) $\frac{\sqrt{3}}{2}$ (b) $\frac{1}{2}$
(c) $\frac{1}{\sqrt{2}}$ (d) $\frac{\sqrt{3}}{4}$

10 Let the area of the triangle with vertices $A(1, \alpha)$, $B(\alpha, 0)$ and $C(0, \alpha)$ be 4 sq units. If the points $(\alpha, -\alpha)$, $(-\alpha, \alpha)$ and (α^2, β) are collinear, then β is equal to

(a) 64 (b) –8
(c) –64 (d) 512

11 The number of distinct real roots of the equation $x^7 - 7x - 2 = 0$ is

(a) 5 (b) 7 (c) 1 (d) 3

12 A random variable x has the following probability distribution

x	0	1	2	3	4
$P(x)$	k	$2k$	$4k$	$6k$	$8k$

The value of $P(1 < x < 4 \mid x \le 2)$ is equal to

(a) $\frac{4}{7}$ (b) $\frac{2}{3}$
(c) $\frac{3}{7}$ (d) $\frac{4}{5}$

13 The number of solutions of the equation $\cos\left(x+\frac{\pi}{3}\right)\cos\left(\frac{\pi}{3}-x\right) = \frac{1}{4}\cos^2 2x, x \in [-3\pi, 3\pi]$ is

(a) 8 (b) 5
(c) 6 (d) 7

14 If the shortest distance between the lines $\frac{x-1}{2} = \frac{y-2}{3} = \frac{z-3}{\lambda}$ and $\frac{x-2}{1} = \frac{y-4}{4} = \frac{z-5}{5}$ is $\frac{1}{\sqrt{3}}$, then the sum of all possible values of λ is

(a) 16 (b) 6 (c) 12 (d) 15

15 Let the points on the plane P be equidistant from the points $(-4, 2, 1)$ and $(2, -2, 3)$. Then, the acute angle between the plane P and the plane $2x + y + 3z = 1$ is

(a) $\frac{\pi}{6}$ (b) $\frac{\pi}{4}$
(c) $\frac{\pi}{3}$ (d) $\frac{5\pi}{12}$

16 Let $\mathbf{a}$ and $\mathbf{b}$ be two unit vectors such that $|(\mathbf{a}+\mathbf{b}) + 2(\mathbf{a}\times\mathbf{b}| = 2$. If $\theta \in (0, \pi)$ is the angle between $\mathbf{a}$ and $\mathbf{b}$, then among the statements

Statement I : $2|\mathbf{a}\times\mathbf{b}| = |\mathbf{a}-\mathbf{b}|$

Statement II : The projection of $\mathbf{a}$ on $(\mathbf{a}+\mathbf{a})$ is $\frac{1}{2}$

(a) Only Statement I is true
(b) Only Statement II is true
(c) Both Statement I and Statement II are true
(d) Both Statement I and Statement II are false

17 If $y = \tan^{-1}(\sec x^3 - \tan x^3), \frac{\pi}{2} < x^3 < \frac{3\pi}{2}$, then

(a) $xy''+2y'=0$ (b) $x^2y''-6y+\frac{3\pi}{2}=0$
(c) $x^2y''-6y+3\pi=0$ (d) $xy''-4y'=0$

18 Consider the following statements :
A : Rishi is a judge.
B : Rishi is honest.
C : Rishi is not arrogant.
The negation of the statement "if Rishi is a judge and he is not arrogant, then he is honest" is
(a) $B \to (A \vee C)$ (b) $(\sim B) \wedge (A \wedge C)$
(c) $B \to ((\sim A) \vee (\sim C))$ (d) $B \to (A \wedge C)$

19 The slope of normal at any point (x, y), $x > 0$, $y > 0$ on the curve $y = y(x)$ is given by $\frac{x^2}{xy - x^2y^2 - 1}$. If the curve passes through the point $(1, 1)$, then $e \cdot y(e)$ is equal to

(a) $\frac{1-\tan(1)}{1+\tan(1)}$ (b) $\tan(1)$
(c) 1 (d) $\frac{1+\tan(1)}{1-\tan(1)}$

20 Let λ^* be the largest value of λ for which the function $f_\lambda(x) = 4\lambda x^3 - 36\lambda x^2 + 36x + 48$ is increasing for all $x \in R$. Then, $f_{\lambda}^*(1) + f_{\lambda}^*(-1)$ is equal to

(a) 36 (b) 48
(c) 64 (d) 72

Section B : Numerical Type Questions

21 Let $S = \{z \in C : |z - 3| \le 1 \text{ and } z(4+3i) + \bar{z}(4-3i) \le 24\}$. If $\alpha + i\beta$ is the point in S which is closest to $4i$, then $25(\alpha+\beta)$ is equal to ....

22 Let $S = \left\{\begin{pmatrix} -1 & a \\ 0 & b \end{pmatrix}, a, b \in \{1, 2, 3, \dots 100\}\right\}$ and let $T_n = \{A \in S : A^{n(n+1)} = I\}$. Then, the number of elements in $\bigcap_{n=1}^{100} T_n$ is ....

23 The number of 7-digit numbers which are multiples of 11 and are formed using all the digits 1, 2, 3, 4, 5, 7 and 9 is ....

24 The sum of all the elements of the set $\{\alpha \in \{1, 2, \dots, 100\} : \text{HCF}(\alpha, 24) = 1\}$ is ....

25 The remainder on dividing $1 + 3 + 3^2 + 3^3 + \dots + 3^{2021}$ by 50 is ....

26 The area (in sq units) of the region enclosed between the parabola $y^2 = 2x$ and the line $x + y = 4$ is ....

27 Let a circle $C : (x-h)^2 + (y-k)^2 = r^2, k > 0$, touch the X-axis at $(1, 0)$. If the line $x + y = 0$ intersects the circle C at P and Q such that the length of the chord PQ is 2, then the value of $h + k + r$ is equal to ....

28 In an examination, there are 10 true-false type questions. Out of 10, a student can guess the answer of 4 questions correctly with probability $\frac{3}{4}$ and the remaining 6 questions correctly with probability $\frac{1}{4}$. If the probability that the student guesses the answers of exactly 8 questions correctly out of 10 is $\frac{27k}{4^{10}}$, then k is equal to ....

29 Let the hyperbola $H : \frac{x^2}{a^2} - y^2 = 1$ and the ellipse $E : 3x^2 + 4y^2 = 12$ be such that the length of latus rectum of H is equal to the length of latus rectum of E. If e_H and e_E are the eccentricities of H and E respectively, then the value of $12(e_H^2 + e_E^2)$ is equal to ....

30 Let P_1 be a parabola with vertex $(3, 2)$ and focus $(4, 4)$ and P_2 be its mirror image with respect to the line $x + 2y = 6$. Then, the directrix of P_2 is $x + 2y = \dots\dots$.

25 JUNE 2022, SHIFT-I

Section A : Objective Type Questions

1 Let a circle C touch the lines $L_1 : 4x - 3y + k_1 = 0$ and $L_2 : 4x - 3y + k_2 = 0$, $k_1, k_2 \in R$. If a line passing through the centre of the circle C intersects L_1 at $(-1, 2)$ and L_2 at $(3, -6)$, then the equation of the circle C is

(a) $(x-1)^2 + (y-2)^2 = 4$
(b) $(x+1)^2 + (y-2)^2 = 4$
(c) $(x-1)^2 + (y+2)^2 = 16$
(d) $(x-1)^2 + (y-2)^2 = 16$

2 The value of $\int_0^{\pi} \frac{e^{\cos x} \sin x}{(1 + \cos^2 x)(e^{\cos x} + e^{-\cos x})} dx$ is equal to

(a) $\frac{\pi^2}{4}$ (b) $\frac{\pi^2}{2}$
(c) $\frac{\pi}{4}$ (d) $\frac{\pi}{2}$

3 Let a, b and c be the length of sides of a ΔABC such that $\frac{a+b}{7} = \frac{b+c}{8} = \frac{c+a}{9}$. If r and R are the radius of incircle and radius of circumcircle of the ΔABC, respectively, then the value of R/r is equal to

(a) $\frac{5}{2}$ (b) 2 (c) $\frac{3}{2}$ (d) 1

4 Let $f : N \to R$ be a function such that $f(x + y) = 2f(x) f(y)$ for natural numbers x and y. If $f(1) = 2$, then the value of α for which $\sum_{K=1}^{10} f(\alpha + K) = \frac{512}{3}(2^{20} - 1)$ holds, is

(a) 2 (b) 3
(c) 4 (d) 6

5 Let A be a 3×3 real matrix such that

$$A\begin{bmatrix}1\\1\\0\end{bmatrix} = \begin{bmatrix}1\\1\\0\end{bmatrix}, A\begin{bmatrix}1\\0\\1\end{bmatrix} = \begin{bmatrix}-1\\0\\1\end{bmatrix} \text{ and } A\begin{bmatrix}0\\0\\1\end{bmatrix} = \begin{bmatrix}1\\1\\2\end{bmatrix}$$

If $X = (x_1, x_2, x_3)^T$ and I is an identity matrix of order 3, then the system $(A - 2I) X = \begin{bmatrix}4\\1\\1\end{bmatrix}$ has

(a) no solution (b) infinitely many solutions
(c) unique solution (d) exactly two solutions

6 Let $f : R \to R$ be defined as $f(x) = x^3 + x - 5$. If $g(x)$ is a function such that $f[g(x)] = x, \forall x \in R$, then $g'(63)$ is equal to ...

(a) $\frac{1}{49}$ (b) $\frac{3}{49}$
(c) $\frac{43}{49}$ (d) $\frac{91}{49}$

7 Consider the following two propositions

$P1 : \sim(p \to \sim q)$
$P2 : (p \wedge \sim q) \wedge ((\sim p) \vee q)$

If the proposition $p \to ((\sim p) \vee q)$ is evaluated as FALSE, then

(a) $P1$ is TRUE and $P2$ is FALSE
(b) $P1$ is FALSE and $P2$ is TRUE
(c) Both $P1$ and $P2$ are FALSE
(d) Both $P1$ and $P2$ are TRUE

8 If $\frac{1}{2 \cdot 3^{10}} + \frac{1}{2^2 \cdot 3^9} + \dots \frac{1}{2^{10} \cdot 3} = \frac{K}{2^{10} \cdot 3^{10}}$, then the remainder when K is divided by 6 is

(a) 1 (b) 2 (c) 3 (d) 5

9 Let $f(x)$ be a polynomial function such that $f(x) + f'(X) + f''(x) = x^5 + 64$. Then, the value of $\lim_{x \to 1} \frac{f(x)}{x - 1} =$

(a) -15 (b) -60 (c) 60 (d) 15

10 Let E_1 and E_2 be two events such that the conditional probabilities $P(E_1 \mid E_2) = \frac{1}{2}$, $P(E_2 \mid E_1) = \frac{3}{4}$ and $P(E_1 \cap E_2) = \frac{1}{8}$. Then,

(a) $P(E_1 \cap E_2) = P(E_1) \cdot P(E_2)$
(b) $P(E'_1 \cap E'_2) = P(E'_1) \cdot P(E_2)$
(c) $P(E_1 \cap E'_2) = P(E_1) \cdot P(E_2)$
(d) $P(E'_1 \cap E_2) = P(E_1) \cdot P(E_2)$

11 Let $A = \begin{bmatrix}0 & -2\\2 & 0\end{bmatrix}$. If M and N are two matrices given by $M = \sum_{k=1}^{10} A^{2k}$ and $N = \sum_{k=1}^{10} A^{2k-1}$, then MN is

(a) a non-identity symmetric matrix
(b) a skew-symmetric matrix
(c) neither symmetric nor skew-symmetric matrix
(d) an identity matrix

12 Let $g : (0, \infty) \to R$ be a differentiable function such that $\int \left(\frac{x(\cos x - \sin x)}{e^x + 1} + \frac{g(x)(e^x + 1 - xe^x)}{(e^x + 1)^2} \right) dx = \frac{xg(x)}{e^x + 1} + c$ for all $x > 0$, where c is an arbitrary constant. Then,

(a) g is decreasing in $\left(0, \frac{\pi}{4}\right)$
(b) g' is increasing in $\left(0, \frac{\pi}{4}\right)$
(c) $g + g'$ is increasing in $\left(0, \frac{\pi}{2}\right)$
(d) $g - g'$ is increasing in $\left(0, \frac{\pi}{2}\right)$

13 Let $f : R \to R$ and $g : R \to R$ be two functions defined by $f(x) = \log_e (x^2 + 1) - e^{-x} + 1$ and $g(x) = \dfrac{1 - 2e^{2x}}{e^x}$. Then, for which of the following range of α, the inequality $f\left(g\left(\dfrac{(\alpha - 1)^2}{3}\right)\right) > f\left(g\left(\alpha - \dfrac{5}{3}\right)\right)$ holds?

(a) (2, 3) (b) (– 2, – 1)
(c) (1, 2) (d) (– 1, 1)

14 Let $\mathbf{a} = a_1\hat{\mathbf{i}} + a_2\hat{\mathbf{j}} + a_3\hat{\mathbf{k}}$, $a_i > 0, i = 1, 2, 3$ be a vector which makes equal angles with the coordinates axes OX, OY and OZ. Also, let the projection of $\mathbf{a}$ on the vector $3\hat{\mathbf{i}} + 4\hat{\mathbf{j}}$ be 7. Let $\mathbf{b}$ be a vector obtained by rotating $\mathbf{a}$ with 90°. If $\mathbf{a} \cdot \mathbf{b}$ and X-axis are coplanar, then projection of a vector $\mathbf{b}$ on $3\hat{\mathbf{i}} + 4\hat{\mathbf{j}}$ is equal to

(a) $\sqrt{7}$ (b) $\sqrt{2}$
(c) 2 (d) 7

15 Let $y = y(x)$ be the solution of the differential equation $(x + 1)y' - y = e^{3x}(x + 1)^2$, with $y(0) = \dfrac{1}{3}$. Then, the point $x = -\dfrac{4}{3}$ for the curve $y = y(x)$ is

(a) not a critical point (b) a point of local minima
(c) a point of local maxima (d) a point of inflection

16 If $y = m_1 x + c_1$ and $y = m_2 x + c_2$, $m_1 \neq m_2$ are two common tangents of circle $x^2 + y^2 = 2$ and parabola $y^2 = x$, then the value of $8|m_1 m_2|$ is equal to

(a) $3 + 4\sqrt{2}$ (b) $-5 + 6\sqrt{2}$
(c) $-4 + 3\sqrt{2}$ (d) $7 + 6\sqrt{2}$

17 Let Q be the mirror image of the point $P(1, 0, 1)$ with respect to the plane $S : x + y + z = 5$. If a line L passing through $(1, -1, -1)$, parallel to the line PQ meets the plane S at R, then QR^2 is equal to

(a) 2 (b) 5 (c) 7 (d) 11

18 If the solution curve $y = y(x)$ of the differential equation $y^2 dx + (x^2 - xy + y^2)\,dy = 0$, which passes through the point (1, 1) and intersects the line $y = \sqrt{3}\,x$ at the point $(\alpha, \sqrt{3}\,\alpha)$, then value of $\log_e(\sqrt{3}\,\alpha)$ is equal to

(a) $\dfrac{\pi}{3}$ (b) $\dfrac{\pi}{2}$ (c) $\dfrac{\pi}{12}$ (d) $\dfrac{\pi}{6}$

19 Let $x = 2t$, $y = \dfrac{t^2}{3}$ be a conic. Let S be the focus and B be the point on the axis of the conic such that $SA \perp BA$, where A is any point on the conic. If k is the ordinate of the centroid of ΔSAB, then $\lim\limits_{t \to 1} k$ is equal to

(a) $\dfrac{17}{18}$ (b) $\dfrac{19}{18}$ (c) $\dfrac{11}{18}$ (d) $\dfrac{13}{18}$

20 Let a circle C in complex plane pass through the points $z_1 = 3 + 4i$, $z_2 = 4 + 3i$ and $z_3 = 5i$. If $z\ (\neq z_1)$ is a point on C such that the line through z and z_1 is perpendicular to the line through z_2 and z_3, then $\arg(z)$ is equal to

(a) $\tan^{-1}\left(\dfrac{2}{\sqrt{5}}\right) - \pi$ (b) $\tan^{-1}\left(\dfrac{24}{7}\right) - \pi$
(c) $\tan^{-1}(3) - \pi$ (d) $\tan^{-1}\left(\dfrac{3}{4}\right) - \pi$

Section B : Numerical Type Questions

21 Let C_r denote the binomial coefficient of x^r in the expansion of $(1 + x)^{10}$. If $\alpha, \beta \in R$. $C_1 + 3 \cdot 2\, C_2 + 5 \cdot 3\, C_3 + \dots$ upto 10 terms
$= \dfrac{\alpha \times 2^{11}}{2^\beta - 1}\left(C_0 + \dfrac{C_1}{2} + \dfrac{C_2}{3} + \dots \text{upto 10 terms}\right)$,
then the value of $\alpha + \beta$ is equal to ....

22 The number of 3-digit odd numbers, whose sum of digits is a multiple of 7, is ....

23 Let θ be the angle between the vectors $\mathbf{a}$ and $\mathbf{b}$ where $|\mathbf{a}| = 4$, $|\mathbf{b}| = 3$, $\theta \in \left(\dfrac{\pi}{4}, \dfrac{\pi}{3}\right)$. Then, $|(\mathbf{a} - \mathbf{b}) \times (\mathbf{a} + \mathbf{b})|^2 + 4(\mathbf{a} \cdot \mathbf{b})^2$ is equal to ....

24 Let the abscissae of the two points P and Q be the roots of $2x^2 - rx + p = 0$ and the ordinates of P and Q be the roots of $x^2 - sx - q = 0$. If the equation of the circle described on PQ as diameter is $2(x^2 + y^2) - 11x - 14y - 22 = 0$, then $2r + s - 2q + p$ is equal to ....

25 The number of value of x in the interval $\left(\dfrac{\pi}{4}, \dfrac{7\pi}{4}\right)$ for which $14\,\mathrm{cosec}^2 x - 2\sin^2 x = 21 - 4\cos^2 x$ holds, is ....

26 For a natural number n, let $a_n = 19^n - 12^n$. Then, the value of $\dfrac{31\alpha_9 - \alpha_{10}}{57\alpha_8}$ is ....

27 Let $f : R \to R$ be a function defined by $f(x) = \left(2\left(1 - \dfrac{x^{25}}{2}\right)(2 + x^{25})\right)^{\frac{1}{50}}$. If the function $g(x) = f(f(f(x))) + f(f(x))$, then the greatest integer less than or equal to $g(1)$ is ....

28 Let the lines
$L_1 : \mathbf{r} = \lambda(\hat{\mathbf{i}} + 2\hat{\mathbf{j}} + \hat{\mathbf{k}}), \lambda \in R$
$L_2 : \mathbf{r} = (\hat{\mathbf{i}} + 3\hat{\mathbf{j}} + \hat{\mathbf{k}}) + \mu(\hat{\mathbf{i}} + \hat{\mathbf{j}} + 5\hat{\mathbf{k}}); \mu \in R$
Intersect at the point S. If a plane $ax + by - z + d = 0$ passes through S and is parallel to both the lines L_1 and L_2, then the value of $a + b + d$ is equal to ....

29 Let A be a 3×3 matrix having entries from the set $\{-1, 0, 1\}$. The number of all such matrices A having sum of all the entries equal to 5, is ....

30 The greatest integer less than or equal to the sum of first 100 terms of the sequence $\frac{1}{3}, \frac{5}{9}, \frac{19}{27}, \frac{65}{81}, \ldots$ is equal to ....

25 JUNE 2022, SHIFT-II

Section A : Objective Type Questions

1 Let $A = \{x \in R : |x+1| < 2\}$ and $B = \{x \in R : |x-1| \geq 2\}$. Then, which one of the following statements is NOT true?

(a) $A - B = (-1, 1)$
(b) $B - A = R - (-3, 1)$
(c) $A \cap B = (-3, -1]$
(d) $A \cup B = R - [1, 3)$

2 Let $a, b \in R$ be such that the equation $ax^2 - 2bx + 15 = 0$ has a repeated root α. If α and β are the roots of the equation $x^2 - 2bx + 21 = 0$, then $\alpha^2 + \beta^2$ is equal to

(a) 37
(b) 58
(c) 68
(d) 92

3 Let z_1 and z_2 be two complex numbers such that $\overline{z_1} = i\,\overline{z_2}$ and $\arg\left(\frac{z_1}{\overline{z_2}}\right) = \pi$. Then,

(a) $\arg z_2 = \left(\frac{\pi}{4}\right)$
(b) $\arg z_2 = -\frac{3\pi}{4}$
(c) $\arg z_1 = \frac{\pi}{4}$
(d) $\arg z_1 = -\frac{3\pi}{4}$

4 The system of equations

$$-kx + 3y - 14z = 25,$$
$$-15x + 4y - kz = 3$$
and $$-4x + y + 3z = 4$$

is consistent for all k in the set

(a) R
(b) $R - \{-11, 13\}$
(c) $R - \{13\}$
(d) $R - \{-11, 11\}$

5 $\lim_{x \to \frac{\pi}{2}} \tan^2 x \left((2\sin^2 x + 3\sin x + 4)^{\frac{1}{2}} - (\sin^2 x + 6\sin x + 2)^{\frac{1}{2}} \right)$ is equal to

(a) $\frac{1}{12}$
(b) $-\frac{1}{18}$
(c) $-\frac{1}{12}$
(d) $\frac{1}{6}$

6 The area of the region enclosed between the parabolas $y^2 = 2x - 1$ and $y^2 = 4x - 3$ is

(a) $\frac{1}{3}$
(b) $\frac{1}{6}$
(c) $\frac{2}{3}$
(d) $\frac{3}{4}$

7 The coefficient of x^{101} in the expression $(5+x)^{500} + x(5+x)^{499} + x^2(5+x)^{498} + \ldots + x^{500}$, $x > 0$, is

(a) ${}^{501}C_{101}\,(5)^{399}$
(b) ${}^{501}C_{101}\,(5)^{400}$
(c) ${}^{501}C_{100}\,(5)^{400}$
(d) ${}^{500}C_{101}\,(5)^{399}$

8. The sum $1 + 2\cdot 3 + 3\cdot 3^2 + \ldots + 10\cdot 3^9$ is equal to

(a) $\frac{2\cdot 3^{12} + 10}{4}$
(b) $\frac{19\cdot 3^{10} + 1}{4}$
(c) $5\cdot 3^{10} - 2$
(d) $\frac{9\cdot 3^{10} + 1}{2}$

9 Let P be the plane passing through the intersection of the planes $\mathbf{r}\cdot(\hat{\mathbf{i}} + 3\hat{\mathbf{j}} - \hat{\mathbf{k}}) = 5$ and $\mathbf{r}\cdot(2\hat{\mathbf{i}} - \hat{\mathbf{j}} + \hat{\mathbf{k}}) = 3$ and the point $(2, 1, -2)$. Let the position vectors of the points X and Y be $\hat{\mathbf{i}} - 2\hat{\mathbf{j}} + 4\hat{\mathbf{k}}$ and $5\hat{\mathbf{i}} - \hat{\mathbf{j}} + 2\hat{\mathbf{k}}$ respectively. Then, the points

(a) X and $X + Y$ are on the same side of P
(b) Y and $Y - X$ are on the opposite sides of P
(c) X and Y are on the opposite sides of P
(d) $X + Y$ and $X - Y$ are on the same side of P

10 A circle touches both the Y-axis and the line $x + y = 0$. Then, the locus of its center is

(a) $y = \sqrt{2}\cdot x$
(b) $x = \sqrt{2}\cdot y$
(c) $y^2 - x^2 = 2xy$
(d) $x^2 - y^2 = 2xy$

11 Water is being filled at the rate of 1 cm^3/s in a right circular conical vessel (vertex downwards) of height 35 cm and diameter 14 cm. When the height of the water level is 10 cm, the rate (in cm^2/ s) at which the wet conical surface area of the vessel increases is

(a) 5
(b) $\frac{\sqrt{21}}{5}$
(c) $\frac{\sqrt{26}}{5}$
(d) $\frac{\sqrt{26}}{10}$

12 If $b_n = \int_0^{\frac{\pi}{2}} \frac{\cos^2 nx}{\sin x}\, dx$, $n \in N$, then

(a) $b_3 - b_2$, $b_4 - b_3$, $b_5 - b_4$ are in an AP with common difference -2
(b) $\frac{1}{b_3 - b_2}, \frac{1}{b_4 - b_3}, \frac{1}{b_5 - b_4}$ are in an AP with common difference 2
(c) $b_3 - b_2$, $b_4 - b_3$, $b_5 - b_4$ are in a GP
(d) $\frac{1}{b_3 - b_2}, \frac{1}{b_4 - b_3}, \frac{1}{b_5 - b_4}$ are in an AP with common difference -2

13 If $y = y(x)$ is the solution of the differential equation $2x^2 \frac{dy}{dx} - 2xy + 3y^2 = 0$ such that $y(e) = \frac{e}{3}$, then $y(1)$ is equal to

(a) $\frac{1}{3}$
(b) $\frac{2}{3}$
(c) $\frac{3}{2}$
(d) 3

14 If the angle made by the tangent at the point (x_0, y_0) on the curve $x = 12(t + \sin t \cos t)$, $y = 12(1 + \sin t)^2$, $0 < t < \frac{\pi}{2}$, with the positive X-axis is $\frac{\pi}{3}$, then y_0 is equal to

(a) $6(3 + 2\sqrt{2})$ (b) $3(7 + 4\sqrt{3})$
(c) 27 (d) 48

15 The value of $2\sin(12°) - \sin(72°)$ is

(a) $\frac{\sqrt{5}(1-\sqrt{3})}{4}$ (b) $\frac{1-\sqrt{5}}{8}$
(c) $\frac{\sqrt{3}(1-\sqrt{5})}{2}$ (d) $\frac{\sqrt{3}(1-\sqrt{5})}{4}$

16 A biased die is marked with numbers 2,4, 8, 16, 32, 32 on its faces and the probability of getting a face with mark n is $\frac{1}{n}$. If the die is thrown thrice, then the probability, that the sum of the numbers obtained is 48, is

(a) $\frac{7}{2^{11}}$ (b) $\frac{7}{2^{12}}$
(c) $\frac{3}{2^{10}}$ (d) $\frac{13}{2^{12}}$

17 The negation of the Boolean expression $((\sim q) \wedge p) \Rightarrow ((\sim p) \vee q)$ is logically equivalent to

(a) $p \Rightarrow q$ (b) $q \Rightarrow p$
(c) $\sim(p \Rightarrow q)$ d. $\sim(q \Rightarrow p)$

18. If the line $y = 4 + kx$, $k > 0$, is the tangent to the parabola $y = x - x^2$ at the point P and V is the vertex of the parabola, then the slope of the line through P and V is

(a) $\frac{3}{2}$ (b) $\frac{26}{9}$ (c) $\frac{5}{2}$ d. $\frac{23}{6}$

19 The value of $\tan^{-1}\left(\frac{\cos\left(\frac{15\pi}{4}\right) - 1}{\sin\left(\frac{\pi}{4}\right)}\right)$ is equal to

(a) $-\frac{\pi}{4}$ (b) $-\frac{\pi}{8}$
(c) $-\frac{5\pi}{12}$ d. $-\frac{4\pi}{9}$

20 The line $y = x + 1$ meets the ellipse $\frac{x^2}{4} + \frac{y^2}{2} = 1$ at two points P and Q. If r is the radius of the circle with PQ as diameter, then $(3r)^2$ is equal to

(a) 20 (b) 12
(c) 11 (d) 8

Section B : Numerical Type Questions

21 Let $A = \begin{bmatrix} 2 & -2 \\ 1 & -1 \end{bmatrix}$ and $B = \begin{bmatrix} -1 & 2 \\ -1 & 2 \end{bmatrix}$. Then, the number of elements in the set $\{(n, m): n, m \in \{1, 2, \ldots, 10\}$ and $nA^n + mB^m = I\}$ is ......

22 Let $f(x) = [2x^2 + 1]$ and $g(x) = \begin{cases} 2x - 3, & x < 0 \\ 2x + 3, & x \geq 0 \end{cases}$, where $[t]$ is the greatest integer $\leq t$. Then, in the open interval $(-1, 1)$, the number of points where fog is discontinuous is equal to ....

23 The value of $b > 3$ for which

$$12\int_3^b \frac{1}{(x^2-1)(x^2-4)}\,dx = \log_e\left(\frac{49}{40}\right)$$ is equal to

24 If the sum of the coefficients of all the positive even powers of x in the binomial expansion of $\left(2x^3 + \frac{3}{x}\right)^{10}$ is $5^{10} - \beta \cdot 3^9$, then β is equal to ....

25 If the mean deviation about the mean of the numbers 1, 2, 3, ..., n, where n is odd, is $\frac{5(n+1)}{n}$, then n is equal to ....

26 Let $\mathbf{b} = \hat{\mathbf{i}} + \hat{\mathbf{j}} + \lambda\hat{\mathbf{k}}$, $\lambda \in R$. If $\mathbf{a}$ is a vector such that $\mathbf{a} \times \mathbf{b} = 13\hat{\mathbf{i}} - \hat{\mathbf{j}} - 4\hat{\mathbf{k}}$ and $\mathbf{a} \cdot \mathbf{b} + 21 = 0$, then $(\mathbf{b} - \mathbf{a}) \cdot (\hat{\mathbf{k}} - \hat{\mathbf{j}}) + (\mathbf{b} + \mathbf{a}) \cdot (\hat{\mathbf{i}} - \hat{\mathbf{k}})$ is equal to ....

27 The total number of three-digit numbers, with one digit repeated exactly two times, is

28 Let $f(x) = |(x-1)(x^2 - 2x - 3)| + x - 3$, $x \in R$. If m and M are respectively the number of points of local minimum and local maximum of f in the interval $(0, 4)$, then $m + M$ is equal to ....

29 Let the eccentricity of the hyperbola $\frac{x^2}{a^2} - \frac{y^2}{b^2} = 1$ be $\frac{5}{4}$. If the equation of the normal at the point $\left(\frac{8}{\sqrt{5}}, \frac{12}{5}\right)$ on the hyperbola is $8\sqrt{5}x + \beta y = \lambda$, then $\lambda - \beta$ is equal to ....

30 Let l_1 be the line in xy-plane with x and y intercepts $\frac{1}{8}$ and $\frac{1}{4\sqrt{2}}$ respectively and l_2 be the line in zx-plane with x and z intercepts $-\frac{1}{8}$ and $-\frac{1}{6\sqrt{3}}$ respectively. If d is the shortest distance between the line l_1 and l_2, then d^{-2} is equal to ....

26 JUNE 2022, SHIFT-I

Section A : Objective Type Questions

1 Let $f(x) = \frac{x-1}{x+1}$, $x \in R - \{0, -1, 1\}$. If $f^{n+1}(x) = f(f^n(x))$ for all $n \in N$, then $f^6(6) + f^7(7)$ is equal to

(a) $\frac{7}{6}$ (b) $-\frac{3}{2}$ (c) $\frac{7}{12}$ (d) $-\frac{11}{12}$

2 Let $A = \left\{z \in C : \left|\frac{z+1}{z-1}\right| < 1\right\}$ and

$B = \left\{z \in C : \arg\left(\frac{z-1}{z+1}\right) = \frac{2\pi}{3}\right\}$

Then, $A \cap B$ is

(a) a portion of a circle centred at $\left(0, -\frac{1}{\sqrt{3}}\right)$ that lies in the second and third quadrants only

(b) a portion of a circle centred at $\left(0, -\frac{1}{\sqrt{3}}\right)$ that lies in the second quadrant only

(c) an empty set

(d) a portion of a circle of radius $\frac{2}{\sqrt{3}}$ that lies in the third quadrant only

3 Let A be a 3×3 invertible matrix. If $|\text{adj}(24A)| = |\text{adj}(3\text{adj}(2A))|$, then $|A|^2$ is equal to

(a) 6^6 (b) 2^{12} (c) 2^6 (d) 1

4 The ordered pair (a, b), for which the system of linear equations

$$3x - 2y + z = b$$
$$5x - 8y + 9z = 3$$
$$2x + y + az = -1$$

has no solution, is

(a) $\left(3, \frac{1}{3}\right)$ (b) $\left(-3, \frac{1}{3}\right)$

(c) $\left(-3, -\frac{1}{3}\right)$ (d) $\left(3, -\frac{1}{3}\right)$

5 The remainder when $(2021)^{2023}$ is divided by 7 is

(a) 1 (b) 2 (c) 5 (d) 6

6 $\lim_{x \to \frac{1}{\sqrt{2}}} \frac{\sin(\cos^{-1} x) - x}{1 - \tan(\cos^{-1} x)}$ is equal to

(a) $\sqrt{2}$ (b) $-\sqrt{2}$ (c) $\frac{1}{\sqrt{2}}$ (d) $-\frac{1}{\sqrt{2}}$

7 $f, g : R \to R$ be two real valued functions defined as

$f(x) = \begin{cases} -|x+3|, & x < 0 \\ e^x, & x \geq 0 \end{cases}$ and

$g(x) = \begin{cases} x^2 + k_1 x, & x < 0 \\ 4x + k_2, & x \geq 0 \end{cases}$, where k_1 and k_2 are real constants. If (gof) is differentiable at $x = 0$, then $(gof)(-4) + (gof)(4)$ is equal to

(a) $4(e^4 + 1)$ (b) $2(2e^4 + 1)$

(c) $4e^4$ (d) $2(2e^4 - 1)$

8 The sum of the absolute minimum and the absolute maximum values of the function $f(x) = |3x - x^2 + 2| - x$ in the interval $[-1, 2]$ is

(a) $\frac{\sqrt{17} + 3}{2}$ (b) $\frac{\sqrt{17} + 5}{2}$ (c) 5 (d) $\frac{9 - \sqrt{17}}{2}$

9 Let S be the set of all the natural numbers, for which the line $\frac{x}{a} + \frac{y}{b} = 2$ is a tangent to the curve $\left(\frac{x}{a}\right)^n + \left(\frac{y}{b}\right)^n = 2$ at the point (a, b), $ab \neq 0$. Then,

(a) $S = \phi$ (b) $n(S) = 1$

(c) $S = \{2k : k \in N\}$ (d) $S = N$

10 The area bounded by the curve $y = |x^2 - 9|$ and the line $y = 3$ is

(a) $4(2\sqrt{3} + \sqrt{6} - 4)$ (b) $4(4\sqrt{3} + \sqrt{6} - 4)$

(c) $8(4\sqrt{3} + 3\sqrt{6} - 9)$ (d) $8(4\sqrt{3} + \sqrt{6} - 9)$

11 Let R be the point $(3, 7)$ and let P and Q be two points on the line $x + y = 5$ such that PQR is an equilateral triangle. Then, the area of ΔPQR is

(a) $\frac{25}{4\sqrt{3}}$ (b) $\frac{25\sqrt{3}}{2}$ (c) $\frac{25}{\sqrt{3}}$ (d) $\frac{25}{2\sqrt{3}}$

12 Let C be a circle passing through the points $A(2, -1)$ and $B(3, 4)$. The line segment AB is not a diameter of C. If r is the radius of C and its centre lies on the circle $(x-5)^2 + (y-1)^2 = \frac{13}{2}$, then r^2 is equal to

(a) 32 (b) $\frac{65}{2}$ (c) $\frac{61}{2}$ (d) 30

13 Let the normal at the point P on the parabola $y^2 = 6x$ pass through the point $(5, -8)$. If the tangent at P to the parabola intersects its directrix at the point Q, then the ordinate of the point Q is

(a) -3 (b) $-\frac{9}{4}$

(c) $-\frac{5}{2}$ (d) -2

14 If the two lines $l_1 : \frac{x-2}{3} = \frac{y+1}{-2}$, $z = 2$ and $l_2 : \frac{x-1}{1} = \frac{2y+3}{\alpha} = \frac{z+5}{2}$ are perpendicular, then an angle between the lines l_2 and $l_3 : \frac{1-x}{3} = \frac{2y-1}{-4} = \frac{z}{4}$ is

(a) $\cos^{-1}\left(\frac{29}{4}\right)$ (b) $\sec^{-1}\left(\frac{29}{4}\right)$

(c) $\cos^{-1}\left(\frac{2}{29}\right)$ (d) $\cos^{-1}\left(\frac{2}{\sqrt{29}}\right)$

15 Let the plane $2x+3y+z+20=0$ be rotated through a right angle about its line of intersection with the plane $x-3y+5z=8$. If the mirror image of the point $\left(2,-\frac{1}{2},2\right)$ in the rotated plane is $B(a,b,c)$, then

(a) $\frac{a}{8}=\frac{b}{5}=\frac{c}{-4}$ (b) $\frac{a}{4}=\frac{b}{5}=\frac{c}{-2}$

(c) $\frac{a}{8}=\frac{b}{-5}=\frac{c}{4}$ (d) $\frac{a}{4}=\frac{b}{5}=\frac{c}{2}$

16 If $\mathbf{a}\cdot\mathbf{b}=1$, $\mathbf{b}\cdot\mathbf{c}=2$ and $\mathbf{c}\cdot\mathbf{a}=3$, then the value of $[\mathbf{a}\times(\mathbf{b}\times\mathbf{c}), \mathbf{b}\times(\mathbf{c}\times\mathbf{a}), \mathbf{c}\times(\mathbf{b}\times\mathbf{a})]$ is

(a) 0 (b) $-6\mathbf{a}\cdot(\mathbf{b}\times\mathbf{c})$

(c) $12\mathbf{c}\cdot(\mathbf{a}\times\mathbf{b})$ (d) $-12\mathbf{b}\cdot(\mathbf{c}\times\mathbf{a})$

17 Let a biased coin be tossed 5 times. If the probability of getting 4 heads is equal to the probability of getting 5 heads, then the probability of getting atmost two heads is

(a) $\frac{275}{6^5}$ (b) $\frac{36}{5^4}$ (c) $\frac{181}{5^5}$ (d) $\frac{46}{6^4}$

18 The mean of the numbers a, b, 8, 5, 10 is 6 and their variance is 6.8. If M is the mean deviation of the numbers about the mean, then $25M$ is equal to

(a) 60 (b) 55 (c) 50 (d) 45

19 Let $f(x)=2\cos^{-1}x+4\cot^{-1}x-3x^2-2x+10$, $x\in[-1,1]$. If $[a,b]$ is the range of the function f, then $4a-b$ is equal to

(a) 11 (b) $11-\pi$ (c) $11+\pi$ (d) $15-\pi$

20 Let $\Delta, \nabla \in \{\wedge, \vee\}$ be such that $p\nabla q \Rightarrow ((p\nabla q)\nabla r)$ is a tautology. Then, $(p\nabla q)\Delta r$ is logically equivalent to

(a) $(p\Delta r)\vee q$ (b) $(p\Delta r)\wedge q$

(c) $(p\wedge r)\Delta q$ (d) $(p\nabla r)\wedge q$

Section B : Numerical Type Questions

21. The sum of the cubes of all the roots of the equation $x^4-3x^3-2x^2+3x+1=10$ is ....

22 There are ten boys $B_1, B_2, \ldots, B_{10}$ and five girls $G_1, G_2, \ldots, G_5$ in a class. Then, the number of ways of forming a group consisting of three boys and three girls, if both B_1 and B_2 together should not be the members of a group, is...... .

23 Let the common tangents to the curves $4(x^2+y^2)=9$ and $y^2=4x$ intersect at the point Q. Let an ellipse, centered at the origin O, has lengths of semi-minor and semi-major axes equal to OQ and 6, respectively. If e and l respectively denote the eccentricity and the length of the latusrectum of this ellipse, then $\frac{l}{e^2}$ is equal to...... .

24 Let $f(x)=\max\{|x+1|,|x+2|,\ldots,|x+5|\}$. Then, $\int_{-6}^{0} f(x)\,dx$ is equal to ....

25 Let the solution curve $y=y(x)$ of the differential equation $(4+x^2)\,dy-2x(x^2+3y+4)\,dx=0$ pass through the origin. Then, $y(2)$ is equal to...... .

26 If $\sin^2(10^\circ)\sin(20^\circ)\sin(40^\circ)\sin(50^\circ)\sin(70^\circ)=\alpha-\frac{1}{16}\sin(10^\circ)$, then $16+\alpha^{-1}$ is equal to ....

27 Let $A=\{n\notin N : \text{HCF}(n,45)=1\}$ and let $B=\{2k : k\in\{1,2,\ldots,100\}\}$. Then, the sum of all the elements of $A\cap B$ is ....

28 The value of the integral $\frac{48}{\pi^4}\int_0^{\pi}\left(\frac{3\pi x^2}{2}-x^3\right)\frac{\sin x}{1+\cos^2 x}\,dx$ is equal to ....

29 Let $A=\sum_{i=1}^{10}\sum_{j=1}^{10}\min\{i,j\}$ and $B=\sum_{i=1}^{10}\sum_{j=1}^{10}\max\{i,j\}$. Then, $A+B$ is equal to ....

30 Let $S=(0,2\pi)-\left\{\frac{\pi}{2},\frac{3\pi}{4},\frac{3\pi}{2},\frac{7\pi}{4}\right\}$.

Let $y=y(x)$, $x\in S$, be the solution curve of the differential equation $\frac{dy}{dx}=\frac{1}{1+\sin 2x}$, $y\left(\frac{\pi}{4}\right)=\frac{1}{2}$. If the sum of abscissa of all the points of intersection of the curve $y=y(x)$ with the curve $y=\sqrt{2}\sin x$ is $\frac{k\pi}{12}$, then k is equal to ....

26 JUNE 2022, SHIFT-II

Section A : Objective Type Questions

1 Let $f: R\to R$ be defined as $f(x)=x-1$ and $g: R-\{1,-1\}\to R$ be defined as $g(x)=\frac{x^2}{x^2-1}$.

Then, the function fog is

(a) one-one but not onto function

(b) onto but not one-one function

(c) both one-one and onto function

(d) neither one-one nor onto function

2 If the system of equations $\alpha x+y+z=5$, $x+2y+3z=4$, $x+3y+5z=\beta$, has infinitely many solutions, then the ordered pair (α,β) is equal to

(a) $(1,-3)$ (b) $(-1,3)$ (c) $(1,3)$ (d) $(-1,-3)$

3 If $A=\sum_{n=1}^{\infty}\frac{1}{(3+(-1)^n)^n}$ and $B=\sum_{n=1}^{\infty}\frac{(-1)^n}{(3+(-1)^n)^n}$, then $\frac{A}{B}$ is equal to

(a) $\frac{11}{9}$ (b) 1 (c) $-\frac{11}{9}$ (d) $-\frac{11}{3}$

4 $\lim_{x \to 0} \frac{\cos(\sin x) - \cos x}{x^4}$ is equal to

(a) $\frac{1}{3}$ (b) $\frac{1}{4}$ (c) $\frac{1}{6}$ (d) $\frac{1}{12}$

5 Let $f(x) = \min\{1, 1 + x\sin x\}$, $0 \le x \le 2\pi$. If m is the number of points, where f is not differentiable and n is the number of points, where f is not continuous, then the ordered pair (m, n) is equal to

(a) (2, 0) (b) (1, 0) (c) (1, 1) (d) (2, 1)

6 Consider a cuboid of sides $2x$, $4x$ and $5x$ and a closed hemisphere of radius r. If the sum of their surface areas is a constant k, then the ratio $x : r$, for which the sum of their volumes is maximum, is

(a) 2 : 5 (b) 19 : 45 (c) 3 : 8 (d) 19 : 15

7 The area of the region bounded by $y^2 = 8x$ and $y^2 = 16(3 - x)$ is equal to

(a) $\frac{32}{3}$ (b) $\frac{40}{3}$ (c) 16 (d) 19

8 If $\int \frac{1}{x}\sqrt{\frac{1-x}{1+x}}\,dx = g(x) + c$, $g(1) = 0$, then $g\left(\frac{1}{2}\right)$ is equal to

(a) $\log_e\left(\frac{\sqrt{3}-1}{\sqrt{3}+1}\right) + \frac{\pi}{3}$ (b) $\log_e\left(\frac{\sqrt{3}+1}{\sqrt{3}-1}\right) + \frac{\pi}{3}$

(c) $\log_e\left(\frac{\sqrt{3}+1}{\sqrt{3}-1}\right) - \frac{\pi}{3}$ (d) $\frac{1}{2}\log_e\left(\frac{\sqrt{3}-1}{\sqrt{3}+1}\right) - \frac{\pi}{6}$

9 If $y = y(x)$ is the solution of the differential equation $x\frac{dy}{dx} + 2y = xe^x$, $y(1) = 0$, then the local maximum value of the function $z(x) = x^2 y(x) - e^x$, $x \in R$ is

(a) $1 - e$ (b) 0

(c) $\frac{1}{2}$ (d) $\frac{4}{e} - e$

10 If the solution of the differential equation $\frac{dy}{dx} + e^x(x^2 - 2)\,y = (x^2 - 2x)(x^2 - 2)\,e^{2x}$ satisfies $y(0) = 0$, then the value of $y(2)$ is ....

(a) −1 (b) 1
(c) 0 (d) e

11 If m is the slope of a common tangent to the curves $\frac{x^2}{16} + \frac{y^2}{9} = 1$ and $x^2 + y^2 = 12$, then $12\,m^2$ is equal to

(a) 6 (b) 9
(c) 10 (d) 12

12 The locus of the mid-point of the line segment joining the point (4, 3) and the points on the ellipse $x^2 + 2y^2 = 4$ is an ellipse with eccentricity

(a) $\frac{\sqrt{3}}{2}$ (b) $\frac{1}{2\sqrt{2}}$

(c) $\frac{1}{\sqrt{2}}$ (d) $\frac{1}{2}$

13 The normal to the hyperbola $\frac{x^2}{a^2} - \frac{y^2}{9} = 1$ at the point $(8, 3\sqrt{3})$ on it passes through the point.

(a) $(15, -2\sqrt{3})$ (b) $(9, 2\sqrt{3})$
(c) $(-1, 9\sqrt{3})$ (d) $(-1, 6\sqrt{3})$

14 If the plane $2x + y - 5z = 0$ is rotated about its line of intersection with the plane $3x - y + 4z - 7 = 0$ by an angle of $\frac{\pi}{2}$, then the plane after the rotation passes through the point

(a) (2, −2, 0) (b) (−2, 2, 0)
(c) (1, 0, 2) (d) (−1, 0, −2)

15 If the lines $\mathbf{r} = (\hat{\mathbf{i}} - \hat{\mathbf{j}} + \hat{\mathbf{k}}) + \lambda(3\hat{\mathbf{j}} - \hat{\mathbf{k}})$ and $\mathbf{r} = (\alpha\hat{\mathbf{i}} - \hat{\mathbf{j}}) + \mu(2\hat{\mathbf{i}} - 3\hat{\mathbf{k}})$ are coplanar, then distance of the plane containing these two lines from the point $(\alpha, 0, 0)$ is

(a) $\frac{2}{9}$ (b) $\frac{2}{11}$
(c) $\frac{4}{11}$ (d) 2

16 Let $\mathbf{a} = \hat{\mathbf{i}} + \hat{\mathbf{j}} + 2\hat{\mathbf{k}}$, $\mathbf{b} = 2\hat{\mathbf{i}} - 3\hat{\mathbf{j}} + \hat{\mathbf{k}}$ and $\mathbf{c} = \hat{\mathbf{i}} - \hat{\mathbf{j}} + \hat{\mathbf{k}}$ be three given vectors. Let $\mathbf{v}$ be a vector in the plane of $\mathbf{a}$ and $\mathbf{b}$ whose projection on $\mathbf{c}$ is $\frac{2}{\sqrt{3}}$. If $\mathbf{v} \cdot \hat{\mathbf{j}} = 7$, then $\mathbf{v} \cdot (\hat{\mathbf{i}} + \hat{\mathbf{k}})$ is equal to

(a) 6 (b) 7
(c) 8 (d) 9

17 The mean and standard deviation of 50 observations are 15 and 2 respectively. It was found that one incorrect observation was taken such that the sum of correct and incorrect observations is 70. If the correct mean is 16, then the correct variance is equal to

(a) 10 (b) 36
(c) 43 (d) 60

18 $16\sin(20^\circ)\sin(40^\circ)\sin(80^\circ)$ is equal to

(a) $\sqrt{3}$ (b) $2\sqrt{3}$
(c) 3 (d) $4\sqrt{3}$

19 If the inverse trigonometric functions take principal values, then

$\cos^{-1}\left(\frac{3}{10}\cos\left(\tan^{-1}\frac{4}{3}\right) + \frac{2}{5}\sin\left(\tan^{-1}\left(\frac{4}{3}\right)\right)\right)$ is equal to :

(a) 0 (b) $\frac{\pi}{4}$
(c) $\frac{\pi}{3}$ (d) $\frac{\pi}{6}$

20. Let $r \in \{p, q, \sim p, \sim q\}$ be such that the logical statement $r \vee (\sim p) \Rightarrow (p \wedge q) \vee r$ is a tautology. Then, r is equal to

(a) p (b) q
(c) $\sim p$ (d) $\sim q$

Section B : Numerical Type Questions

21 Let $f: R \to R$ satisfy $f(x+y) = 2^x f(y) + 4^y f(x), \forall x, y \in R$. If $f(2) = 3$, then $14 \cdot \frac{f'(4)}{f'(2)}$ is equal to ....

22 Let p and q be two real numbers such that $p + q = 3$ and $p^4 + q^4 = 369$. Then, $\left(\frac{1}{p} + \frac{1}{q}\right)^{-2}$ is equal to ....

23 If $z^2 + z + 1 = 0, z \in C$, then $\left|\sum_{n=1}^{15}\left(z^n + (-1)^n \frac{1}{z^n}\right)^2\right|$ is equal to ....

24 Let $X = \begin{bmatrix} 0 & 1 & 0 \\ 0 & 0 & 1 \\ 0 & 0 & 0 \end{bmatrix}$, $y = \alpha I + \beta x + \gamma x^2$ and $Z = \alpha^2 I - \alpha\beta x + (\beta^2 - \alpha\gamma) x^2, \alpha, \beta, \gamma \in R$. If $y^{-1} = \begin{bmatrix} 1/5 & -2/5 & 1/5 \\ 0 & 1/5 & -2/5 \\ 0 & 0 & 1/5 \end{bmatrix}$, then $(\alpha - \beta + \gamma)^2$ is equal to ....

25 The total number of 3-digit numbers, whose greatest common divisor with 36 is 2, is ....

26 If $({}^{40}C_0) + ({}^{41}C_1) + ({}^{42}C_2) + \ldots + ({}^{60}C_{20}) = \frac{m}{n} {}^{60}C_{20}$, m and n are coprime, then $m + n$ is equal to ....

27 If $a_1 (> 0), a_2, a_3, a_4, a_5$ are in a GP, $a_2 + a_4 = 2a_3 + 1$ and $3a_2 + a_3 = 2a_4$, then $a_2 + a_4 + 2a_5$ is equal to ....

28 The integral $\frac{24}{\pi}\int_0^{\sqrt{2}} \frac{(2 - x^2)\, dx}{(2 + x^2)\sqrt{4 + x^4}}$ is equal to ....

29 Let a line L_1 be tangent to the hyperbola $\frac{x^2}{16} - \frac{y^2}{4} = 1$ and let L_2 be the line passing through the origin and perpendicular to L_1.

If the locus of the point of intersection of L_1 and L_2 is $(x^2 + y^2)^2 = \alpha x^2 + \beta y^2$, then $\alpha + \beta$ is equal to ....

30 If the probability that a randomly chosen 6-digit number formed by using digits 1 and 8 only is a multiple of 21 is p, then $96\, p$ is equal to ....

27 JUNE 2022, SHIFT-I

Section A : Objective Type Questions

1 The area of the polygon, whose vertices are the non-real roots of the equation $\bar{z} = iz^2$ is

(a) $\frac{3\sqrt{3}}{4}$ (b) $\frac{3\sqrt{3}}{2}$

(c) $\frac{3}{2}$ (d) $\frac{3}{4}$

2 Let the system of linear equations $x + 2y + z = 2$, $\alpha x + 3y - z = \alpha$, $-\alpha x + y + 2z = -\alpha$ be inconsistent. Then, α is equal to

(a) $\frac{5}{2}$ (b) $-\frac{5}{2}$ (c) $\frac{7}{2}$ (d) $-\frac{7}{2}$

3 If $x = \sum_{n=0}^{\infty} a^n$, $y = \sum_{n=0}^{\infty} b^n$, $z = \sum_{n=0}^{\infty} c^n$, where a, b, c are in AP and $|a| < 1, |b| < 1, |c| < 1, abc \neq 0$, then

(a) x, y, z are in AP (b) x, y, z are in GP

(c) $\frac{1}{x}, \frac{1}{y}, \frac{1}{z}$ are in AP (d) $\frac{1}{x} + \frac{1}{y} + \frac{1}{z} = 1 - (a + b + c)$

4 Let $\frac{dy}{dx} = \frac{ax - by + a}{bx + cy + a}$, where a, b, c are constants, represent a circle passing through the point (2, 5). Then, the shortest distance of the point (11, 6) from this circle is

(a) 10 (b) 8 (c) 7 (d) 5

5 Let a be an integer such that $\lim_{x \to 7} \frac{18 - [1 - x]}{[x - 3a]}$ exists, where $[t]$ is greatest integer $\leq t$. Then, a is equal to

(a) -6 (b) -2

(c) 2 (d) 6

6 The number of distinct real roots of $x^4 - 4x + 1 = 0$ is

(a) 4 (b) 2

(c) 1 (d) 0

7 The lengths of the sides of a triangle are $10 + x^2$, $10 + x^2$ and $20 - 2x^2$. If for $x = k$, the area of the triangle is maximum, then $3k^2$ is equal to

(a) 5 (b) 8

(c) 10 (d) 12

8 If $\cos^{-1}\left(\frac{y}{2}\right) = \log_e\left(\frac{x}{5}\right)^5, |y| < 2$, then

(a) $x^2y'' + xy' - 25y = 0$ (b) $x^2y'' - xy' - 25y = 0$

(c) $x^2y'' - xy' + 25y = 0$ (d) $x^2y'' + xy' + 25y = 0$

9 If $\int \frac{(x^2 + 1)e^x}{(x + 1)^2} dx = f(x)\, e^x + C$, where C is a constant, then $\frac{d^3 f}{dx^3}$ at $x = 1$ is equal to

(a) $-\frac{3}{4}$ (b) $\frac{3}{4}$ (c) $-\frac{3}{2}$ (d) $\frac{3}{2}$

10 The value of the integral $\int_{-2}^{2} \frac{|x^3 + x|}{(e^{x|x|} + 1)} dx$ is equal to

(a) $5e^2$ (b) $3e^{-2}$ (c) 4 (d) 6

11 If $\frac{dy}{dx} + \frac{2^{x-y}(2^y - 1)}{2^x - 1} = 0$, $x, y > 0$, $y(1) = 1$, then $y(2)$ is equal to

(a) $2 + \log_2 3$ (b) $2 + \log_3 2$
(c) $2 - \log_3 2$ (d) $2 - \log_2 3$

12 In an isosceles ΔABC, the vertex A is (6, 1) and the equation of the base BC is $2x + y = 4$. Let the point B lie on the line $x + 3y = 7$. If (α, β) is the centroid ΔABC, then $15(\alpha + \beta)$ is equal to

(a) 39 (b) 41 (c) 51 (d) 63

13 Let the eccentricity of an ellipse $\frac{x^2}{a^2} + \frac{y^2}{b^2} = 1$, $a > b$, be $\frac{1}{4}$. If this ellipse passes through the point $\left(-4\sqrt{\frac{2}{5}}, 3\right)$, then $a^2 + b^2$ is equal to

(a) 29 (b) 31 (c) 32 (d) 34

14 If two straight line whose direction cosines are given by relations $l + m - n = 0, 3l^2 + m^2 + cnl = 0$ are parallel, then the positive value of c is

(a) 6 (b) 4 (c) 3(d) 2

15 Let $\mathbf{a} = \hat{\mathbf{i}} + \hat{\mathbf{j}} - \hat{\mathbf{k}}$ and $\mathbf{c} = 2\hat{\mathbf{i}} - 3\hat{\mathbf{j}} + 2\hat{\mathbf{k}}$. Then, the number of vectors $\mathbf{b}$ such that $\mathbf{b} \times \mathbf{c} = \mathbf{a}$ and $|\mathbf{b}| \in \{1, 2, \ldots, 10\}$ is

(a) 0 (b) 1
(c) 2 (d) 3

16 Five numbers x_1, x_2, x_3, x_4, x_5 are randomly selected from the numbers 1, 2, 3, ..., 18 and are arranged in the increasing order($x_1 < x_2 < x_3 < x_4 < x_5$). The probability that $x_2 = 7$ and $x_4 = 11$ is

(a) $\frac{1}{136}$ (b) $\frac{1}{72}$ (c) $\frac{1}{68}$ (d) $\frac{1}{34}$

17 Let X be a random variable having binomial distribution $B(7, p)$. If $P(X = 3) = 5P(X = 4)$, then the sum of the mean and the variance of X is

(a) $\frac{105}{16}$ (b) $\frac{7}{16}$ (c) $\frac{77}{36}$ (d) $\frac{49}{16}$

18 The value of $\cos\left(\frac{2\pi}{7}\right) + \cos\left(\frac{4\pi}{7}\right) + \cos\left(\frac{6\pi}{7}\right)$ is equal to

(a) -1 (b) $-\frac{1}{2}$ (c) $-\frac{1}{3}$ (d) $-\frac{1}{4}$

19 $\sin^{-1}\left(\sin\frac{2\pi}{3}\right) + \cos^{-1}\left(\cos\frac{7\pi}{6}\right) + \tan^{-1}\left(\tan\frac{3\pi}{4}\right)$ is equal to

(a) $\frac{11\pi}{12}$ (b) $\frac{17\pi}{12}$ (c) $\frac{31\pi}{12}$ (d) $-\frac{3\pi}{4}$

20 The Boolean expression $(\sim(p \wedge q)) \vee q$ is equivalent to

(a) $q \to (p \wedge q)$ (b) $p \to q$
(c) $p \to (p \to q)$ d. $p \to (p \vee q)$]

Section B : Numerical Type Questions

21 Let $f : R \to R$ be a function defined by $f(x) = \frac{2e^{2x}}{e^{2x} + e}$. Then, $f\left(\frac{1}{100}\right) + f\left(\frac{2}{100}\right) + f\left(\frac{3}{100}\right) + \ldots + f\left(\frac{99}{100}\right)$ is equal to ....

22 If the sum of all the roots of the equation $e^{2x} - 11e^x - 45e^{-x} + \frac{81}{2} = 0$ is $\log_e p$, then p is equal to ....

23 The positive value of the determinant of the matrix A, whose $\text{Adj}(\text{adj}(A)) = \begin{bmatrix} 14 & 28 & -14 \\ -14 & 14 & 28 \\ 28 & -14 & 14 \end{bmatrix}$, is ....

24 The number of ways, 16 identical cubes, of which 11 are blue and rest are red, can be placed in a row so that between any two red cubes there should be at least 2 blue cubes, is ....

25 If the coefficient of x^{10} in the binomial expansion of $\left(\frac{\sqrt{x}}{5^{1/4}} + \frac{\sqrt{5}}{x^{1/3}}\right)^{60}$ is $5^k l$, where $l, k \in N$ and l is coprime to 5, then k is equal to ....

26 Let $A_1 = \{(x, y) : |x| \le y^2, |x| + 2y \le 8\}$ and $A_2 = \{(x, y) : |x| + |y| \le k\}$. If 27 (Area A_1) = 5 (Area A_2), then k is equal to ....

27 If the sum of the first ten term of the series $\frac{1}{5} + \frac{2}{65} + \frac{3}{325} + \frac{4}{1025} + \frac{5}{2501} + \ldots$ is $\frac{m}{n}$, where m and n and coprime numbers, then $m + n$ is equal to ....

28 A rectangle R with end points of the one of its sides as (1, 2) and (3, 6) is inscribed in a circle. If the equation of a diameter of the circle is $2x - y + 4 = 0$, then the area of R is ....

29 A circle of radius 2 units passes through the vertex and the focus of the parabola $y^2 = 2x$ and touches the parabola $y = \left(x - \frac{1}{4}\right)^2 + \alpha$, where $\alpha > 0$. Then, $(4\alpha - 8)^2$ is equal to ....

30 Let the mirror image of the point (a, b, c) with respect to the plane $3x - 4y + 12z + 19 = 0$ be $(a - 6, \beta, \gamma)$. If $a + b + c = 5$, then $7\beta - 9\gamma$ is equal to ....

27 JUNE 2022, SHIFT-II

Section A : Objective Type Questions

1 The number of points of intersection of $|z-(4+3i)|=2$ and $|z|+|z-4|=6$, $z \in C$ is

(a) 0 (b) 1 (c) 2 (d) 3

2 Let $f(x)=\begin{vmatrix} a & -1 & 0 \\ ax & a & -1 \\ ax^2 & ax & a \end{vmatrix}$, $a \in R$. Then, the sum of the squares of all the value of a for which $2f'(10)-f'(5)+100=0$ is

(a) 117 (b) 106 (c) 125 (d) 136

3 Let for some real numbers α and β, $a=\alpha-i\beta$. If the system of equations $4ix+(1+i)y=0$ and $8\left(\cos\frac{2\pi}{3}+i\sin\frac{2\pi}{3}\right)x+\bar{a}y=0$ has more than one solution, then $\frac{\alpha}{\beta}$ is equal to

(a) $-2+\sqrt{3}$ (b) $2-\sqrt{3}$ (c) $2+\sqrt{3}$ (d) $-2-\sqrt{3}$

4 Let A and B be two 3×3 matrices such that $AB=I$ and $|A|=\frac{1}{8}$, then $|\text{adj}(B\,\text{adj}(2A))|$ is equal to

(a) 16 (b) 32 (c) 64 (d) 128

5 Let $S=2+\frac{6}{7}+\frac{12}{7^2}+\frac{20}{7^3}+\frac{30}{7^4}+\ldots$, then $4S$ is equal to

(a) $\left(\frac{7}{3}\right)^2$ (b) $\frac{7^3}{3^2}$ (c) $\left(\frac{7}{3}\right)^3$ (d) $\frac{7^2}{3^3}$

6 If $a_1, a_2, a_3, \ldots$ and $b_1, b_2, b_3, \ldots$, are AP and $a_1=2$, $a_{10}=3$, $a_1b_1=1=a_{10}b_{10}$, then a_4b_4 is equal to

(a) $\frac{35}{27}$ (b) 1 (c) $\frac{27}{28}$ (d) $\frac{28}{27}$

7 If m and n respectively are the number of local maximum and local minimum points of the function

$f(x)=\int_0^{x^2}\frac{t^2-5t+4}{2+e^t}\,dt$, then the ordered pair (m, n) is equal to

(a) (3, 2) (b) (2, 3)
(c) (2, 2) (d) (3, 4)

8 Let f be a differentiable function in $\left(0, \frac{\pi}{2}\right)$. If $\int_{\cos x}^{1} t^2 f(t)\,dt=\sin^3 x+\cos x$, then $\frac{1}{\sqrt{3}}f'\left(\frac{1}{\sqrt{3}}\right)$ is equal to

(a) $6-9\sqrt{2}$ (b) $6-\frac{9}{\sqrt{2}}$
(c) $\frac{9}{2}-6\sqrt{2}$ (d) $\frac{9}{\sqrt{2}}-6$

9 The integral $\int_0^1 \frac{1}{7^{\left[\frac{1}{x}\right]}}\,dx$, where [.] denotes the greatest integer function is equal to

(a) $1+6\log_e\left(\frac{6}{7}\right)$ (b) $1-6\log_e\left(\frac{6}{7}\right)$
(c) $\log_e\left(\frac{7}{6}\right)$ (d) $1-7\log_e\left(\frac{6}{7}\right)$

10 If the solution curve of the differential equation $((\tan^{-1} y)-x)dy=(1+y^2)dx$ passes through the point (1, 0), then the abscissa of the point on the curve whose ordinate is tan (1) is

(a) $2e$ (b) $\frac{2}{e}$
(c) 2 (d) $\frac{1}{e}$

11 If the equation of the parabola, whose vertex is at (5, 4) and the directrix is $3x+y-29=0$, is $x^2+ay^2+bxy+cx+dy+e=0$, then $a+b+c+d+e$ is equal to

(a) 575 (b) -575
(c) 576 (d) -576

12 The set of values of k for which the circle $C: 4x^2+4y^2-12x+8y+k=0$ lies inside the fourth quadrant and the point $\left(1, -\frac{1}{3}\right)$ lies on or inside the circle C is

(a) An empty set (b) $\left(6, \frac{65}{9}\right]$
(c) $\left[\frac{80}{9}, 10\right)$ (d) $\left(9, \frac{92}{9}\right]$

13 Let the foot of the perpendicular from the point (1, 2, 4) on the line $\frac{x+2}{4}=\frac{y-1}{2}=\frac{z+1}{3}$ be P. Then, the distance of P from the plane $3x+4y+12z+23=0$

(a) 5 (b) $\frac{50}{13}$
(c) 4 (d) $\frac{63}{13}$

14 The shortest distance between the lines $\frac{x-3}{2}=\frac{y-2}{3}=\frac{z-1}{-1}$ and $\frac{x+3}{2}=\frac{y-6}{1}=\frac{z-5}{3}$ is

(a) $\frac{18}{\sqrt{5}}$ (b) $\frac{22}{3\sqrt{5}}$
(c) $\frac{46}{3\sqrt{5}}$ (d) $6\sqrt{3}$

15 Let **a** and **b** be the vectors along the diagonals of a parallelogram having area $2\sqrt{2}$. Let the angle between **a** and **b** be acute, $|\mathbf{a}| = 1$ and $|\mathbf{a} \cdot \mathbf{b}| = |\mathbf{a} \times \mathbf{b}|$. If $\mathbf{c} = 2\sqrt{2}\,(\mathbf{a} \times \mathbf{b}) - 2\mathbf{b}$, then an angle between **b** and **c** is

(a) $\frac{\pi}{4}$ (b) $-\frac{\pi}{4}$
(c) $\frac{5\pi}{6}$ (d) $\frac{3\pi}{4}$

16 The mean and variance of the data 4, 5, 6, 6, 7, 8, x, y where $x < y$ are 6 and $\frac{9}{4}$ respectively. Then, $x^4 + y^2$ is equal to

(a) 162 (b) 320 (c) 674 (d) 420

17 If a point $A(x, y)$ lies in the region bounded by the Y-axis, straight lines $2y + x = 6$ and $5x - 6y = 30$, then the probability that $y < 1$ is

(a) $\frac{1}{6}$ (b) $\frac{5}{6}$
(c) $\frac{2}{3}$ (d) $\frac{6}{7}$

18 The value of $\cot\left(\sum_{n=1}^{50} \tan^{-1}\left(\frac{1}{1 + n + n^2}\right)\right)$ is

(a) $\frac{26}{25}$ (b) $\frac{25}{26}$
(c) $\frac{50}{51}$ (d) $\frac{52}{51}$

19 $\alpha = \sin 36°$ is a root of which of the following equation

(a) $16x^4 - 10x^2 - 5 = 0$ (b) $16x^4 + 20x^2 - 5 = 0$
(c) $16x^4 - 20x^2 + 5 = 0$ (d) $16x^4 - 10x^2 + 5 = 0$

20 Which of the following statement is a tautology?

(a) $((\sim q) \wedge p) \wedge q$ (b) $((\sim q) \wedge p) \wedge (p \wedge (\sim p))$
(c) $((\sim q) \wedge p) \vee (p \vee (\sim p))$ (d) $(p \wedge q) \wedge (\sim (p \wedge q))$

Section B : Numerical Type Questions

21 Let $S = \{1, 2, 3, 4, 5, 6, 7, 8, 9, 10\}$. Define $f : S \to S$ as

$$f(n) = \begin{cases} 2n, & \text{if } n = 1,2,3,4,5 \\ 2n - 11, & \text{if } n = 6, 7, 8, 9, 10 \end{cases}.$$

Let $g : S \to S$ be a function such that

$$fog(n) = \begin{cases} n + 1, & \text{if } n \text{ is odd} \\ n - 1, & \text{if } n \text{ is even} \end{cases}, \text{ then}$$

$g(10)\,[(g(1) + g(2) + g(3) + g(4) + g(5)]$ is equal to ………… .

22 Let α and β be the roots of the equation $x^2 - 4\lambda x + 5 = 0$ and α, γ be the roots of the equation $x^2 - (3\sqrt{2} + 2\sqrt{3})x + 7 + 3\lambda\sqrt{3} = 0$, $\lambda > 0$. If $\beta + \gamma = 3\sqrt{2}$, then $(\alpha + 2\beta + \gamma)^2$ is equal to ……… .

23 Let A be a matrix of order 2×2, whose entries are from the set $\{0, 1, 3, 4, 5\}$. If the sum of all the entries of A is a prime number p, $2 < p < 8$, then the number of such matrices A is ……… .

24 If the sum of the coefficients of all the positive powers of x, in the binomial expansion of $\left(x^n + \frac{2}{x^5}\right)^7$ is 939, then the sum of all the possible integral values of n is ……… .

25 Let $[t]$ denote the greatest integer $\leq t$ and $\{t\}$ denote the fractional part of t. Then, integral value of α for which the left hand limit of the function

$$f(x) = [1 + x] + \frac{\alpha^{2[x] + \{x\}} + [x] - 1}{2[x] + \{x\}} \text{ at } x = 0 \text{ is equal to } \alpha - \frac{4}{3}$$

is ……… .

26 If $y(x) = (x^x)^x$, $x > 0$, then $\frac{d^2x}{dy^2} + 20$ at $x = 1$ is equal to ………… .

27 If the area of the region $\left\{(x, y) : x^{\frac{2}{3}} + y^{\frac{2}{3}} \leq 1, x + y \geq 0, y \geq 0\right\}$ is A, then $\frac{256\,A}{\pi}$ is equal to ……… .

28 Let $y = y(x)$ be the solution of the differential equation $(1 - x^2)dy = (xy + (x^3 + 2)\sqrt{1 - x^2})\,dx$, $-1 < x < 1$ and $y(0) = 0$ if $\int_{-1/2}^{1/2} \sqrt{1 - x^2}\; y(x)\,dx = k$, then k^{-1} is equal to ……… .

29 Let a circle C of radius 5 lie below the X-axis. The line $L_1 : 4x + 3y + 2 = 0$ passes through the centre P of the circle C and intersects the line $L_2 : 3x - 4y - 11 = 0$ at Q. The line L_2 touches C at the point Q. Then, the distance of P from the line $5x - 12y + 51 = 0$ is ………… .

30 Let $S = \{E_1, E_2 \ldots. E_8\}$ be a sample space of random experiment such that $P(E_n) = \frac{n}{36}$ for every $n = 1, 2 \ldots., 8$. Then, the number of elements in the set $\left\{A \subseteq S : P(A) \geq \frac{4}{5}\right\}$ is ……… .

28 JUNE 2022, SHIFT-I

Section A : Objective Type Questions

1 If $\sum_{k=1}^{31} ({}^{31}C_k)({}^{31}C_{k-1}) - \sum_{k=1}^{30} ({}^{30}C_k)({}^{30}C_{k-1}) = \frac{\alpha(60!)}{(30!)(31!)}$, where $\alpha \in R$, then the value of 16α is equal to

(a) 1411 (b) 1320 (c) 1615 (d) 1855

2 Let a function $f : N \to N$ be defined by

$$f(n) = \begin{bmatrix} 2n, & n = 2, 4, 6, 8, \ldots \\ n-1, & n = 3, 7, 11, 15, \ldots \\ \frac{n+1}{2}, & n = 1, 5, 9, 13, \ldots \end{bmatrix}$$ then, f is

(a) one-one but not onto (b) onto but not one-one
(c) neither one-one nor onto (d) one-one and onto

3 If the system of linear equations

$2x + 3y - z = -2$

$x + y + z = 4$

$x - y + |\lambda| z = 4\lambda - 4$

where $\lambda \in R$, has no solution, then

(a) $\lambda = 7$ (b) $\lambda = -7$ (c) $\lambda = 8$ (d) $\lambda^2 = 1$

4 Let A be a matrix of order 3×3 and det $(A) = 2$. Then, det (det (A) adj (5 adj (A^3))) is equal to ____.

(a) 512×10^6 (b) 256×10^6
(c) 1024×10^6 (d) 256×10^{11}

5 The total number of 5-digit numbers, formed by using the digits 1, 2, 3, 5, 6, 7 without repetition, which are multiple of 6, is

(a) 36 (b) 48 (c) 60 (d) 72

6 Let $A_1, A_2, A_3, \ldots$ be an increasing geometric progression of positive real numbers. If $A_1 A_3 A_5 A_7 = \frac{1}{1296}$ and $A_2 + A_4 = \frac{7}{36}$, then the value of $A_6 + A_8 + A_{10}$ is equal to

(a) 33 (b) 37 (c) 43 (d) 47

7 Let $[t]$ denote the greatest integer less than or equal to t. Then, the value of the integral $\int_0^1 [-8x^2 + 6x - 1]dx$ is equal to

(a) -1 (b) $-\frac{5}{4}$ (c) $\frac{\sqrt{17} - 13}{8}$ (d) $\frac{\sqrt{17} - 16}{8}$

8 Let $f : R \to R$ be defined as

$$f(x) = \begin{bmatrix} [e^x], & x < 0 \\ ae^x + [x-1], & 0 \le x < 1 \\ b + [\sin(\pi x)], & 1 \le x < 2 \\ [e^{-x}] - c, & x \ge 2 \end{bmatrix}$$

where $a, b, c \in R$ and $[t]$ denotes greatest integer less than or equal to t. Then, which of the following statement is true ?

(a) There exists $a, b, c \in R$ such that f is continuous on R.
(b) If f is discontinuous at exactly one point, then $a + b + c = 1$.
(c) If f is discontinuous at exactly one point, then $a + b + c \ne 1$.
(d) f is discontinuous at atleast two points, for any values of a, b and c.

9 The area of the region

$S = \{(x, y) : y^2 \le 8x, y \ge \sqrt{2}x, x \ge 1\}$ is

(a) $\frac{13\sqrt{2}}{6}$ (b) $\frac{11\sqrt{2}}{6}$
(c) $\frac{5\sqrt{2}}{6}$ (d) $\frac{19\sqrt{2}}{6}$

10 Let the solution curve $y = y(x)$ of the differential equation $\left[\frac{x}{\sqrt{x^2 - y^2}} + e^{\frac{y}{x}}\right] x \frac{dy}{dx} = x + \left[\frac{x}{\sqrt{x^2 - y^2}} + e^{\frac{y}{x}}\right] y$ pass through the points (1, 0) and $(2\alpha, \alpha)$, $\alpha > 0$. Then, α is equal to

(a) $\frac{1}{2} \exp\left(\frac{\pi}{6} + \sqrt{e} - 1\right)$ (b) $\frac{1}{2} \exp\left(\frac{\pi}{3} + e - 1\right)$
(c) $\exp\left(\frac{\pi}{6} + \sqrt{e} + 1\right)$ (d) $2 \exp\left(\frac{\pi}{3} + \sqrt{e} - 1\right)$

11 Let $y = y(x)$ be the solution of the differential equation $x(1 - x^2)\frac{dy}{dx} + (3x^2 y - y - 4x^3) = 0$, $x > 1$, with $y(2) = -2$. Then, $y(3)$ is equal to

(a) -18 (b) -12
(c) -6 (d) -3

12 The number of real solutions of $x^7 + 5x^3 + 3x + 1 = 0$ is equal to ....

(a) 0 (b) 1 (c) 3 (d) 5

13 Let the eccentricity of the hyperbola $H : \frac{x^2}{a^2} - \frac{y^2}{b^2} = 1$ be $\sqrt{\frac{5}{2}}$ and length of its latus rectum be $6\sqrt{2}$. If $y = 2x + c$ is a tangent to the hyperbola H, then the value of c^2 is equal to

(a) 18 (b) 20 (c) 24 (d) 32

14 If the tangents drawn at the points $O(0, 0)$ and $P(1 + \sqrt{5}, 2)$ on the circle $x^2 + y^2 - 2x - 4y = 0$ intersect at the point Q, then the area of the ΔOPQ is equal to

(a) $\frac{3 + \sqrt{5}}{2}$ (b) $\frac{4 + 2\sqrt{5}}{2}$
(c) $\frac{5 + 3\sqrt{5}}{2}$ (d) $\frac{7 + 3\sqrt{5}}{2}$

15 If two distinct points Q, R lie on the line of intersection of the planes $-x+2y-z=0$ and $3x-5y+2z=0$ and $PQ=PR=\sqrt{18}$, where the point P is $(1, -2, 3)$, then the area of the ΔPQR is equal to

(a) $\frac{2}{3}\sqrt{38}$
(b) $\frac{4}{3}\sqrt{38}$
(c) $\frac{8}{3}\sqrt{38}$
(d) $\sqrt{\frac{152}{3}}$

16 The acute angle between the planes P_1 and P_2, when P_1 and P_2 are the planes passing through the intersection of the planes $5x+8y+13z-29=0$ and $8x-7y+z-20=0$ and the points (2, 1, 3) and (0, 1, 2), respectively, is

(a) $\frac{\pi}{3}$
(b) $\frac{\pi}{4}$
(c) $\frac{\pi}{6}$
(d) $\frac{\pi}{12}$

17 Let the plane $P:\mathbf{r}\cdot\mathbf{a}=d$ contain the line of intersection of two planes $\mathbf{r}\cdot(\hat{\mathbf{i}}+3\hat{\mathbf{j}}-\hat{\mathbf{k}})=6$ and $\mathbf{r}\cdot(-6\hat{\mathbf{i}}+5\hat{\mathbf{j}}-\hat{\mathbf{k}})=7$. If the plane P passes through the point $\left(2, 3, \frac{1}{2}\right)$, then the value of $\frac{|13\mathbf{a}|^2}{d^2}$ is equal to

(a) 90
(b) 93
(c) 95
(d) 97

18 The probability, that in a randomly selected 3-digit number at least two digits are odd, is

(a) $\frac{19}{36}$
(b) $\frac{15}{36}$
(c) $\frac{13}{36}$
(d) $\frac{23}{36}$

19 Let AB and PQ be two vertical poles, 160 m apart from each other. Let C be the middle point of B and Q, which are feet of these two poles. Let $\frac{\pi}{8}$ and θ be the angles of elevation from C to P and A, respectively. If the height of pole PQ is twice the height of pole AB, then $\tan^2\theta$ is equal to

(a) $\frac{3-2\sqrt{2}}{2}$
(b) $\frac{3+\sqrt{2}}{2}$
(c) $\frac{3-2\sqrt{2}}{4}$
(d) $\frac{3-\sqrt{2}}{4}$

20 Let p, q, r be three logical statements. Consider the compound statements $S_1:((\sim p)\vee q)\vee((\sim p)\vee r)$ and $S_2: p\to(q\vee r)$

Then, which of the following is **NOT** true ?

(a) If S_2 is True, then S_1 is True
(b) If S_2 is False, then S_1 is False
(c) If S_2 is False, then S_1 is True
(d) If S_1 is False, then S_2 is False

Section B : Numerical Type Questions

21 Let R_1 and R_2 be relations on the set {1, 2, 3, 50} such that $R_1=\{(p, p^n): p \text{ is a prime and } n\geq 0 \text{ is an integer}\}$ and $R_2=\{(p, p^n): p \text{ is a prime and } n=0 \text{ or } 1\}$.
Then, the number of elements in R_1-R_2 is ....

22 The number of real solutions of the equation $e^{4x}+4e^{3x}-58e^{2x}+4e^x+1=0$ is ....

23 The mean and standard deviation of 15 observations are found to be 8 and 3 respectively. On rechecking it was found that, in the observations, 20 was misread as 5. Then, the correct variance is equal to ....

24 If $\mathbf{a}=2\hat{\mathbf{i}}+\hat{\mathbf{j}}+3\hat{\mathbf{k}}$, $\mathbf{b}=3\hat{\mathbf{i}}+3\hat{\mathbf{j}}+\hat{\mathbf{k}}$ and $\mathbf{c}=c_1\hat{\mathbf{i}}+c_2\hat{\mathbf{j}}+c_3\hat{\mathbf{k}}$ are coplanar vectors and

$\mathbf{a}\cdot\mathbf{c}=5$, $\mathbf{b}\perp\mathbf{c}$, then $122(c_1+c_2+c_3)$ is equal to ....

25 A ray of light passing through the point $P(2, 3)$ reflects on the X-axis at point A and the reflected ray passes through the point $Q(5, 4)$. Let R be the point that divides the line segment AQ internally into the ratio 2 : 1. Let the coordinates of the foot of the perpendicular M from R on the bisector of the $\angle PAQ$ be (α, β). Then, the value of $7\alpha+3\beta$ is equal to ....

26 Let l be a line which is normal to the curve $y=2x^2+x+2$ at a point P on the curve. If the point $Q(6, 4)$ lies on the line l and O is origin, then the area of the ΔOPQ is equal to ....

27 Let $A=\{1, a_1, a_2, ..., a_{18}, 77\}$ be a set of integers with $1<a_1<a_2<.....<a_{18}<77$. Let the set $A+A=\{x+y: x, y\in A\}$ contain exactly 39 elements. Then, the value of $a_1+a_2+...+a_{18}$ is equal to ....

28 The number of positive integers k such that the constant term in the binomial expansion of $\left(2x^3+\frac{3}{x^k}\right)^{12}$, $x\neq 0$ is $2^8\cdot l$, where l is an odd integer, is ....

29 The number of elements in the set $\{z=a+ib\in C: a, b\in Z \text{ and } 1<|z-3+2i|<4\}$ is ......

30 Let the lines $y+2x=\sqrt{11}+7\sqrt{7}$ and $2y+x=2\sqrt{11}+6\sqrt{7}$ be normal to a circle $C: (x-h)^2+(y-k)^2=r^2$. If the line $\sqrt{11}y-3x=\frac{5\sqrt{77}}{3}+11$ is tangent to the circle C, then the value of $(5h-8k)^2+5r^2$ is equal to ....

28 JUNE 2022, SHIFT-II

Section A : Objective Type Questions

1 Let $R_1 = \{(a, b) \in N \times N : |a - b| \le 13\}$ and $R_2 = \{(a, b) \in N \times N : |a - b| \ne 13\}$. Then, on N

(a) Both R_1 and R_2 are equivalence relations
(b) Neither R_1 nor R_2 is an equivalence relation
(c) R_1 is an equivalence relation but R_2 is not
(d) R_2 is an equivalence relation but R_1 is not

2 Let $f(x)$ be a quadratic polynomial such that $f(-2) + f(3) = 0$. If one of the roots of $f(x) = 0$ is -1, then the sum of the roots of $f(x) = 0$ is equal to

(a) $\frac{11}{3}$ (b) $\frac{7}{3}$
(c) $\frac{13}{3}$ (d) $\frac{14}{3}$

3 The number of ways to distribute 30 identical candies among four children C_1, C_2, C_3 and C_4 so that C_2 receives atleast 4 and atmost 7 candies, C_3 receives atleast 2 and atmost 6 candies, is equal to

(a) 205 (b) 615
(c) 510 (d) 430

4 The term independent of x in the expansion of $(1 - x^2 + 3x^3)\left(\frac{5}{2}x^3 - \frac{1}{5x^2}\right)^{11}, x \ne 0$ is

(a) $\frac{7}{40}$ (b) $\frac{33}{200}$
(c) $\frac{39}{200}$ (d) $\frac{11}{50}$

5 If n arithmetic means are inserted between a and 100 such that the ratio of the first mean to the last mean is $1 : 7$ and $a + n = 33$, then the value of n is

(a) 21 (b) 22 (c) 23 (d) 24

6 Let $f, g : R \to R$ be functions defined by

$$f(x) = \begin{cases} [x], & x < 0 \\ |1 - x|, & x \ge 0 \end{cases} \text{ and } g(x) = \begin{cases} e^x - x, & x < 0 \\ (x-1)^2 - 1, & x \ge 0 \end{cases}$$

where $[x]$ denote the greatest integer less than or equal to x. Then, the function fog is discontinuous at exactly

(a) one point (b) two points
(c) three points (d) four points

7 Let $f : R \to R$ be a differentiable function such that $f\left(\frac{\pi}{4}\right) = \sqrt{2}, f\left(\frac{\pi}{2}\right) = 0$ and $f'\left(\frac{\pi}{2}\right) = 1$ and let

$$g(x) = \int_x^{\pi/4} (f'(t) \sec t + \tan t \sec t \, f(t))\, dt \text{ for } x \in \left[\frac{\pi}{4}, \frac{\pi}{2}\right).$$

Then, $\lim_{x \to \left(\frac{\pi}{2}\right)^-} g(x)$ is equal to

(a) 2 (b) 3
(c) 4 (d) -3

8 Let $f : R \to R$ be continuous function satisfying $f(x) + f(x + k) = n$, for all $x \in R$ where $k > 0$ and n is a positive integer. If $I_1 = \int_0^{4nk} f(x)\, dx$ and $I_2 = \int_{-k}^{3k} f(x)\, dx$, then

(a) $I_1 + 2I_2 = 4nk$ (b) $I_1 + 2I_2 = 2nk$
(c) $I_1 + nI_2 = 4n^2k$ (d) $I_1 + nI_2 = 6n^2k$

9 The area of the bounded region enclosed by the curve $y = 3 - \left|x - \frac{1}{2}\right| - |x + 1|$ and the X-axis is

(a) $\frac{9}{4}$ (b) $\frac{45}{16}$
(c) $\frac{27}{8}$ (d) $\frac{63}{16}$

10 Let $x = x(y)$ be the solution of the differential equation $2ye^{x/y^2} dx + (y^2 - 4xe^{x/y^2})\, dy = 0$ such that $x(1) = 0$. Then, $x(e)$ is equal to

(a) $e \log_e(2)$ (b) $-e\log_e(2)$ (c) $e^2 \log_e(2)$ (d) $-e^2 \log_e(2)$

11 Let the slope of the tangent to a curve $y = f(x)$ at (x, y) be given by $2 \tan x (\cos x - y)$. If the curve passes through the point $\left(\frac{\pi}{4}, 0\right)$, then the value of $\int_0^{\pi/2} y\, dx$ is equal to

(a) $(2 - \sqrt{2}) + \frac{\pi}{\sqrt{2}}$ (b) $2 - \frac{\pi}{\sqrt{2}}$
(c) $(2 + \sqrt{2}) + \frac{\pi}{\sqrt{2}}$ (d) $2 + \frac{\pi}{\sqrt{2}}$

12 Let a triangle be bounded by the lines $L_1 : 2x + 5y = 10$; $L_2 : -4x + 3y = 12$ and the line L_3, which passes through the point $P(2, 3)$, intersects L_2 at A and L_1 at B. If the point P divides the line segment AB, internally in the ratio $1 : 3$, then the area of the triangle is equal to

(a) $\frac{110}{13}$ (b) $\frac{132}{13}$
(c) $\frac{142}{13}$ (d) $\frac{151}{13}$

13 Let $a > 0, b > 0$. Let e and l respectively be the eccentricity and length of the latus rectum of the hyperbola $\frac{x^2}{a^2} - \frac{y^2}{b^2} = 1$. Let e' and l' respectively be the eccentricity and length of the latus rectum of its conjugate hyperbola. If $e^2 = \frac{11}{14} l$ and $(e')^2 = \frac{11}{8} l'$, then the value of $77a + 44b$ is equal to

(a) 100 (b) 110
(c) 120 (d) 130

14 Let $\mathbf{a} = \alpha\hat{\mathbf{i}} + 2\hat{\mathbf{j}} - \hat{\mathbf{k}}$ and $\mathbf{b} = -2\hat{\mathbf{i}} + \alpha\hat{\mathbf{j}} + \hat{\mathbf{k}}$, where $\alpha \in R$. If the area of the parallelogram whose adjacent sides are represented by the vectors **a** and **b** is $\sqrt{15(\alpha^2 + 4)}$, then the value of $2|\mathbf{a}|^2 + (\mathbf{a} \cdot \mathbf{b})|\mathbf{b}|^2$ is equal to

(a) 10 (b) 7 (c) 9 (d) 14

15 If vertex of a parabola is (2, – 1) and the equation of its directrix is $4x - 3y = 21$, then the length of its latus rectum is

(a) 2 (b) 8
(c) 12 (d) 16

16 Let the plane $ax + by + cz = d$ passes through (2, 3, – 5) and is perpendicular to the planes $2x + y - 5z = 10$ and $3x + 5y - 7z = 12$. If a, b, c, d are integers $d > 0$ and $gcd(|a|, |b|, |c|, d) = 1$, then the value of $a + 7b + c + 20d$ is equal to

(a) 18 (b) 20
(c) 24 (d) 22

17 The probability that a randomly chosen one-one function from the set $\{a, b, c, d\}$ to the set $\{1, 2, 3, 4, 5\}$ satisfies $f(a) + 2f(b) - f(c) = f(d)$ is

(a) $\frac{1}{24}$ (b) $\frac{1}{40}$
(c) $\frac{1}{30}$ (d) $\frac{1}{20}$

18 The value of $\lim_{n \to \infty} 6 \tan\left\{\sum_{r=1}^{n} \tan^{-1}\left(\frac{1}{r^2 + 3r + 3}\right)\right\}$ is equal to

(a) 1 (b) 2
(c) 3 (d) 6

19 Let **a** be a vector which is perpendicular to the vector $3\hat{\mathbf{i}} - \frac{1}{2}\hat{\mathbf{j}} + 2\hat{\mathbf{k}}$. If $\mathbf{a} \times (2\hat{\mathbf{i}} + \hat{\mathbf{k}}) = 2\hat{\mathbf{i}} - 13\hat{\mathbf{j}} - 4\hat{\mathbf{k}}$, then the projection of the vector **a** on the vector $2\hat{\mathbf{i}} + 2\hat{\mathbf{j}} + \hat{\mathbf{k}}$ is

(a) $\frac{1}{3}$ (b) 1
(c) $\frac{5}{3}$ (d) $\frac{7}{3}$

20 If $\cot\alpha = 1$ and $\sec\beta = -\frac{5}{3}$, where $\pi < \alpha < \frac{3\pi}{2}$ and $\frac{\pi}{2} < \beta < \pi$, then the value of $\tan(\alpha + \beta)$ and the quadrant in which $\alpha + \beta$ lies, respectively are

(a) $-\frac{1}{7}$ and IVth quadrant
(b) 7 and Ist quadrant
(c) – 7 and IVth quadrant
(d) $\frac{1}{7}$ and Ist quadrant

Section B : Numerical Type Questions

21 Let the image of the point $P(1, 2, 3)$ in the line $L: \frac{x-6}{3} = \frac{y-1}{2} = \frac{z-2}{3}$ be Q. let $R(\alpha, \beta, \gamma)$ be a point that divides internally the line segment PQ in the ratio 1 : 3. Then, the value of $22(\alpha + \beta + \gamma)$ is equal to ....

22 Suppose a class has 7 students. The average marks of these students in the mathematics examination is 62, and their variance is 20. A student fails in the examination if he/she gets less than 50 marks, then in worst case, the number of students can fail is ....

23 If one of the diameters of the circle $x^2 + y^2 - 2\sqrt{2}x - 6\sqrt{2}y + 14 = 0$ is a chord of the circle $(x - 2\sqrt{2})^2 + (y - 2\sqrt{2})^2 = r^2$, then the value of r^2 is equal to......... .

24 If $\lim_{x \to 1} \frac{\sin(3x^2 - 4x + 1) - x^2 + 1}{2x^3 - 7x^2 + ax + b} = -2$, then the value of $(a - b)$ is equal to......... .

25 Let for $n = 1, 2, \ldots, 50$, S_n be the sum of the infinite geometric progression whose first term is n^2 and whose common ratio is $\frac{1}{(n+1)^2}$. Then, the value of $\frac{1}{26} + \sum_{n=1}^{50}\left(S_n + \frac{2}{n+1} - n - 1\right)$ is equal to ....

26 If the system of linear equations

$$2x - 3y = \gamma + 5$$
$$\alpha x + 5y = \beta + 1,$$

where $\alpha, \beta, \gamma \in R$ has infinitely many solutions, then the value of $|9\alpha + 3\beta + 5\gamma|$ is equal to ....

27 Let $A = \begin{bmatrix} 1+i & 1 \\ -i & 0 \end{bmatrix}$ where $i = \sqrt{-1}$.

Then, the number of elements in the set $\{n \in \{1, 2, \ldots, 100\} : A^n = A\}$ is ....

28 Sum of squares of modulus of all the complex numbers z satisfying $\bar{z} = iz^2 + z^2 - z$ is equal to ....

29 Let $S = \{1, 2, 3, 4\}$. Then, the number of elements in the set $\{f : S \times S \to S : f$ is onto and $f(a, b) = f(b, a) \geq a, \forall (a, b) \in S \times S\}$ is ....

30 The maximum number of compound propositions, out of
$p \vee r \vee s, p \vee r \vee \sim s, p \vee \sim q \vee s,$
$\sim p \vee \sim r \vee s, \vee \sim p \vee \sim r \vee \sim s, \sim p \vee q \vee \sim s,$
$q \vee r \vee \sim s, q \vee \sim r \vee \sim s, \sim p \vee \sim q \vee \sim s$
that can be made simultaneously true by an assignment of the truth values to p, q, r and s, is equal to ....

29 JUNE 2022, SHIFT-I

Section A : Objective Type Questions

1 The probability that a randomly chosen 2×2 matrix with all the entries from the set of first 10 primes, is singular, is equal to

(a) $\frac{133}{10^4}$ (b) $\frac{18}{10^3}$

(c) $\frac{19}{10^3}$ (d) $\frac{271}{10^4}$

2 Let the solution curve of the differential equation $x\frac{dy}{dx} - y = \sqrt{y^2 + 16x^2}$, $y(1) = 3$ be $y = y(x)$. Then, $y(2)$ is equal to

(a) 15 (b) 11 (c) 13 (d) 17

3 In the mirror image of the point (2, 4, 7) in the plane $3x - y + 4z = 2$ is (a, b, c), then $2a + b + 2c$ is equal to

(a) 54 (b) 50

(c) – 6 (d) – 42

4 Let $f : R \to R$ be a function defined by

$$f(x) = \begin{cases} \max\{t^3 - 3t\}; t \le x & x \le 2 \\ x^2 + 2x - 6; & 2 < x < 3 \\ [x - 3] + 9; & 3 \le x \le 5 \\ 2x + 1; & x > 5 \end{cases}$$

where $[t]$ is the greatest integer less than or equal to t. Let m be the number of points where f is not differentiable and $I = \int_{-2}^{2} f(x)dx$. Then, the ordered pair (m, I) is equal to

(a) $\left(3, \frac{27}{4}\right)$ (b) $\left(3, \frac{23}{4}\right)$

(c) $\left(4, \frac{27}{4}\right)$ (d) $\left(4, \frac{23}{4}\right)$

5 Let $\mathbf{a} = \alpha\hat{\mathbf{i}} + 3\hat{\mathbf{j}} - \hat{\mathbf{k}}$, $\mathbf{b} = 3\hat{\mathbf{i}} - \beta\hat{\mathbf{j}} + 4\hat{\mathbf{k}}$ and $\mathbf{c} = \hat{\mathbf{i}} + 2\hat{\mathbf{j}} - 2\hat{\mathbf{k}}$ where $\alpha, \beta \in R$, be three vectors. If the projection of $\mathbf{a}$ on $\mathbf{c}$ is $\frac{10}{3}$ and $\mathbf{b} \times \mathbf{c} = -6\hat{\mathbf{i}} + 10\hat{\mathbf{j}} + 7\hat{\mathbf{k}}$, then the value of $\alpha + \beta$ is equal to

(a) 3 (b) 4

(c) 5 (d) 6

6 The area enclosed by $y^2 = 8x$ and $y = \sqrt{2}x$ that lies outside the triangle formed by $y = \sqrt{2}x$, $x = 1$, $y = 2\sqrt{2}$, is equal to

(a) $\frac{16\sqrt{2}}{6}$ (b) $\frac{11\sqrt{2}}{6}$

(c) $\frac{13\sqrt{2}}{6}$ (d) $\frac{5\sqrt{2}}{6}$

7 If the system of linear equations $2x + y - z = 7$, $x - 3y + 2z = 1$ and $x + 4y + \delta z = k$, where $\delta, k \in R$ has infinitely many solutions, then $\delta + k$ is equal to

(a) – 3 (b) 3

(c) 6 (d) 9

8 Let α and β be the roots of the equation $x^2 + (2i - 1) = 0$. Then, the value of $|\alpha^8 + \beta^8|$ is equal to

(a) 50 (b) 250

(c) 1250 (d) 1500

9. Let $\Delta \in \{\wedge, \vee, \Rightarrow, \Leftrightarrow\}$ be such that $(p \wedge q) \Delta ((p \vee q) \Rightarrow q)$ is a tautology. Then, Δ is equal to

(a) $\wedge$ (b) $\vee$

(c) $\Rightarrow$ (d) $\Leftrightarrow$

10 Let $A = [a_{ij}]$ be a square matrix of order 3 such that $a_{ij} = 2^{j-i}$, for all $i, j = 1, 2, 3$. Then, the matrix $A^2 + A^3 + \ldots + A^{10}$ is equal to

(a) $\left(\frac{3^{10} - 3}{2}\right)A$ (b) $\left(\frac{3^{10} - 1}{2}\right)A$

(c) $\left(\frac{3^{10} + 1}{2}\right)A$ (d) $\left(\frac{3^{10} + 3}{2}\right)A$

11 Let a set $A = A_1 \cup A_2 \cup \ldots \cup A_k$, where $A_i \cap A_j = \phi$ for $i \ne j$, $1 \le i, j \le k$. Define the relation R from A to A by $R = \{(x, y) : y \in A_i$ if and only if $x \in A_i, 1 \le i \le k\}$. Then, R is

(a) reflexive, symmetric but not transitive
(b) reflexive, transitive but not symmetric
(c) reflexive but not symmetric and transitive
(d) an equivalence relation

12 Let $\{a_n\}_{n=0}^{\infty}$ be a sequence such that $a_0 = a_1 = 0$ and $a_{n+2} = 2a_{n+1} - a_n + 1$ for all $n \ge 0$. Then, $\sum_{n=2}^{\infty} \frac{a_n}{7^n}$ is equal to

(a) $\frac{6}{343}$ (b) $\frac{7}{216}$

(c) $\frac{8}{343}$ (d) $\frac{49}{216}$

13 The distance between the two points A and A' which lie on $y = 2$ such that both the line segments AB and $A'B$ (where B is the point (2, 3)) subtend angle $\frac{\pi}{4}$ at the origin, is equal to

(a) 10 (b) $\frac{48}{5}$

(c) $\frac{52}{5}$ (d) 3

14 A wire of length 22 m is to be cut into two pieces. One of the pieces is to be made into a square and the other into an equilateral triangle. Then, the length of the side of the equilateral triangle, so that the combined area of the square and the equilateral triangle is minimum, is

(a) $\frac{22}{9+4\sqrt{3}}$ (b) $\frac{66}{9+4\sqrt{3}}$ (c) $\frac{22}{4+9\sqrt{3}}$ (d) $\frac{66}{4+9\sqrt{3}}$

15 The domain of the function

$$\cos^{-1}\left(\frac{2\sin^{-1}\left(\frac{1}{4x^2-1}\right)}{\pi}\right)$$ is

(a) $R-\left\{-\frac{1}{2},\frac{1}{2}\right\}$ (b) $(-\infty,-1]\cup[1,\infty)\cup\{0\}$

(c) $\left(-\infty,\frac{-1}{2}\right)\cup\left(\frac{1}{2},\infty\right)\cup\{0\}$ (d) $\left(-\infty,\frac{-1}{\sqrt{2}}\right]\cup\left[\frac{1}{\sqrt{2}},\infty\right)\cup\{0\}$

16 If the constant term in the expansion of $\left(3x^3-2x^2+\frac{5}{x^5}\right)^{10}$ is $2^k \cdot l$, where l is an odd integer, then the value of k is equal to

(a) 6 (b) 7 (c) 8 (d) 9

17 $\int_0^5 \cos\left(\pi\left(x-\left[\frac{x}{2}\right]\right)\right)dx$,

where $[t]$ denotes greatest integer less than or equal to t, is equal to

(a) -3 (b) -2 (c) 2 (d) 0

18 Let PQ be a focal chord of the parabola $y^2=4x$ such that it subtends an angle of $\pi/2$ at the point (3, 0). Let the line segment PQ be also a focal chord of the ellipse $E:\frac{x^2}{a^2}+\frac{y^2}{b^2}=1$, $a^2>b^2$. If e is the eccentricity of the ellipse E, then the value of $\frac{1}{e^2}$ is equal to

(a) $1+\sqrt{2}$ (b) $3+2\sqrt{2}$

(c) $1+2\sqrt{3}$ (d) $4+5\sqrt{3}$

19. Let the tangent to the circle $C_1: x^2+y^2=2$ at the point $M(-1,1)$ intersect the circle $C_2:(x-3)^2+(y-2)^2=5$, at two distinct points A and B. If the tangents to C_2 at the points A and B intersect at N, then the area of the ΔANB is equal to

(a) $\frac{1}{2}$ (b) $\frac{2}{3}$ (c) $\frac{1}{6}$ (d) $\frac{5}{3}$

20 Let the mean and the variance of 5 observations x_1, x_2, x_3, x_4, x_5 be $\frac{24}{5}$ and $\frac{194}{25}$ respectively. If the mean and variance of the first 4 observation are 7/2 and a respectively, then $(4a+x_5)$ is equal to

(a) 13 (b) 15 (c) 17 (d) 18

Section B : Numerical Type Questions

21 Let $S=\{z\in C: |z-2|\le 1, z(1+i)+\bar{z}(1-i)\le 2\}$. Let $|z-4i|$ attains minimum and maximum values, respectively, at $z_1\in S$ and $z_2\in S$. If $5(|z_1|^2+|z_2|^2)=\alpha+\beta\sqrt{5}$, where α and β are integers, then the value of $\alpha+\beta$ is equal to ....

22 Let $y=y(x)$ be the solution of the differential equation

$$\frac{dy}{dx}+\frac{\sqrt{2}y}{2\cos^4 x-\cos 2x}=xe^{\tan^{-1}(\sqrt{2}\cot 2x)},$$

$0<x<\frac{\pi}{2}$ with $y\left(\frac{\pi}{4}\right)=\frac{\pi^2}{32}$.

If $y\left(\frac{\pi}{3}\right)=\frac{\pi^2}{18}e^{-\tan^{-1}(\alpha)}$, then the value of $3\alpha^2$ is equal to ....

23 Let d be the distance between the foot of perpendiculars of the points $P(1,2-1)$ and $Q(2,-1,3)$ on the plane $-x+y+z=1$. Then, d^2 is equal to ....

24 The number of elements in the set $S=\{\theta\in[-4\pi,4\pi]: 3\cos^2 2\theta+6\cos 2\theta-10\cos^2\theta+5=0\}$ is ....

25 The number of solutions of the equation $2\theta-\cos^2\theta+\sqrt{2}=0$ in R is equal to ....

26 $50\tan\left(3\tan^{-1}\left(\frac{1}{2}\right)+2\cos^{-1}\left(\frac{1}{\sqrt{5}}\right)\right)+4\sqrt{2}\tan\left(\frac{1}{2}\tan^{-1}(2\sqrt{2})\right)$ is equal to ....

27 Let $c,k\in R$. If $f(x)=(c+1)x^2+(1-c^2)x+2k$ and $f(x+y)=f(x)+f(y)-xy$, for all $x,y\in R$, then the value of $|2(f(1)+f(2)+f(3)+....+f(20))|$ is equal to ....

28 Let $H:\frac{x^2}{a^2}-\frac{y^2}{b^2}=1$, $a>0$, $b>0$, be a hyperbola such that the sum of lengths of the transverse and the conjugate axes is $4(2\sqrt{2}+\sqrt{14})$. If the eccentricity H is $\frac{\sqrt{11}}{2}$, then value of a^2+b^2 is equal to ....

29. Let $P_1:\mathbf{r}\cdot(2\hat{\mathbf{i}}+\hat{\mathbf{j}}-3\hat{\mathbf{k}})=4$ be a plane. Let P_2 be another plane which passes through the points $(2,-3,2)$ $(2,-2,-3)$ and $(1,-4,2)$. If the direction ratios of the line of intersection of P_1 and P_2 be 16, α, β, then the value of $\alpha+\beta$ is equal to ______ .

30 Let $b_1b_2b_3b_4$ be a 4-element permutation with $b_i\in\{1,2,3,......,100\}$ for $1\le i\le 4$ and $b_i\ne b_j$ for $i\ne j$, such that either b_1, b_2, b_3 are consecutive integers or b_2, b_3, b_4 are consecutive integers.

Then, the number of such permutations $b_1b_2b_3b_4$ is equal to .....

29 JUNE 2022, SHIFT-II

Section A : Objective Type Questions

1 Let α be a root of the equation $1 + x^2 + x^4 = 0$. Then, the value of $\alpha^{1011} + \alpha^{2022} - \alpha^{3033}$ is equal to

(a) 1 (b) α
(c) $1+\alpha$ (d) $1+2\alpha$

2 Let arg (z) represent the principal argument of the complex number z. Then, $|z| = 3$ and $\arg(z-1) - \arg(z+1) = \frac{\pi}{4}$ intersect

(a) exactly at one point
(b) exactly at two points
(c) nowhere
(d) at infinitely many points

3 Let $A = \begin{bmatrix} 2 & -1 \\ 0 & 2 \end{bmatrix}$. If $B = I - {}^5C_1 (\text{adj } A) + {}^5C_2 (\text{adj } A)^2 - \dots - {}^5C_5 (\text{adj } A)^5$, then the sum of all elements of the matrix B is

(a) -5 (b) -6
(c) -7 (d) -8

4 The sum of the infinite series $1 + \frac{5}{6} + \frac{12}{6^2} + \frac{22}{6^3} + \frac{35}{6^4} + \frac{51}{6^5} + \frac{70}{6^6} + \dots$ is equal to

(a) $\frac{425}{216}$ (b) $\frac{429}{216}$
(c) $\frac{288}{125}$ (d) $\frac{280}{125}$

5 The value of $\lim_{x \to 1} \frac{(x^2-1)\sin^2(\pi x)}{x^4 - 2x^3 + 2x - 1}$ is equal to

(a) $\frac{\pi^2}{6}$ (b) $\frac{\pi^2}{3}$
(c) $\frac{\pi^2}{2}$ (d) π^2

6 Let $f : R \to R$ be a function defined by $f(x) = (x-3)^{n_1}(x-5)^{n_2}$, $n_1, n_2 \in N$. Then, which of the following is not true?

(a) For $n_1 = 3$, $n_2 = 4$, there exists $\alpha \in (3, 5)$ where f attains local maxima
(b) For $n_1 = 4$, $n_2 = 3$, there exists $\alpha \in (3, 5)$ where f attains local minim(a)
(c) For $n_1 = 3$, $n_2 = 5$, there exists $\alpha \in (3, 5)$ where f attains local maxim(a)
(d) For $n_1 = 4$, $n_2 = 6$, there exists $\alpha \in (3, 5)$ where f attains local maxim(a)

7 Let f be a real valued continuous function on $[0, 1]$ and $f(x) = x + \int_0^1 (x-t) f(t)dt$. Then, which of the following points (x, y) lies on the curve $y = f(x)$?

(a) (2, 4) (b) (1, 2) (c) (4, 17) (d) (6, 8)

8 If $\int_0^2 (\sqrt{2x} - \sqrt{2x - x^2})\, dx = \int_0^1 \left(1 - \sqrt{1-y^2} - \frac{y^2}{2}\right) dy + \int_1^2 \left(2 - \frac{y^2}{2}\right) dy + I$, then I is equal to

(a) $\int_0^1 (1 + \sqrt{1-y^2})\, dy$ (b) $\int_0^1 \left(\frac{y^2}{2} - \sqrt{1-y^2} + 1\right) dy$
(c) $\int_0^1 (1 - \sqrt{1-y^2})\, dy$ (d) $\int_0^1 \left(\frac{y^2}{2} + \sqrt{1-y^2} + 1\right) dy$

9 If $y = y(x)$ is the solution of the differential equation $(1 + e^{2x})\frac{dy}{dx} + 2(1 + y^2)e^x = 0$ and $y(0) = 0$, then $6\{y'(0) + (y(\log_e \sqrt{3}))^2\}$ is equal to

(a) 2 (b) -2
(c) -4 (d) -1

10 Let $P : y^2 = 4ax$, $a > 0$ be a parabola with focus S. Let the tangents to the parabola P make an angle of $\frac{\pi}{4}$ with the line $y = 3x + 5$ touch the parabola P at A and *(b)* Then, the value of a for which A, B and S are collinear is

(a) 8 only (b) 2 only
(c) $\frac{1}{4}$ only (d) any $a > 0$

11 Let a ΔABC be inscribed in the circle $x^2 - \sqrt{2}(x + y) + y^2 = 0$ such that $\angle BAC = \frac{\pi}{2}$. If the length of side AB is $\sqrt{2}$, then the area of the ΔABC is equal to

(a) $\frac{(\sqrt{2}+\sqrt{6})}{3}$ (b) $\frac{(\sqrt{6}+\sqrt{3})}{2}$
(c) $\frac{(3+\sqrt{3})}{4}$ (d) $\frac{(\sqrt{6}+2\sqrt{3})}{4}$

12 Let $\frac{x-2}{3} = \frac{y+1}{-2} = \frac{z+3}{-1}$ lie on the plane $px - qy + z = 5$, for some $p, q \in R$. The shortest distance of the plane from the origin is

(a) $\sqrt{\frac{3}{109}}$ (b) $\sqrt{\frac{5}{142}}$
(c) $\sqrt{\frac{5}{71}}$ (d) $\sqrt{\frac{1}{142}}$

13 The distance of the origin from the centroid of the triangle whose two sides have the equations $x - 2y + 1 = 0$ and $2x - y - 1 = 0$ and whose orthocenter is $\left(\frac{7}{3}, \frac{7}{3}\right)$, is

(a) $\sqrt{2}$ (b) 2
(c) $2\sqrt{2}$ (d) 4

14 Let Q be the mirror image of the point $P(1, 2, 1)$ with respect to the plane $x + 2y + 2z = 16$. Let T be a plane passing through the point Q and contains the line $\mathbf{r} = -\mathbf{k} + \lambda(\hat{\mathbf{i}} + \hat{\mathbf{j}} + 2\hat{\mathbf{k}}), \lambda \in R$. Then, which of the following points lies on T?

(a) $(2, 1, 0)$ (b) $(1, 2, 1)$
(c) $(1, 2, 2)$ (d) $(1, 3, 2)$

15 Let A, B, C be three points whose position vectors respectively are

$\mathbf{a} = \hat{\mathbf{i}} + 4\hat{\mathbf{j}} + 3\hat{\mathbf{k}}$

$\mathbf{b} = 2\hat{\mathbf{i}} + \alpha\hat{\mathbf{j}} + 4\hat{\mathbf{k}}, \alpha \in R$

$\mathbf{c} = 3\hat{\mathbf{i}} - 2\hat{\mathbf{j}} + 5\hat{\mathbf{k}}$

If α is the smallest positive integer for which $\mathbf{a}, \mathbf{b}, \mathbf{c}$ are non-collinear, then the length of the median in ΔABC, through A is

(a) $\frac{\sqrt{82}}{2}$ (b) $\frac{\sqrt{62}}{2}$
(c) $\frac{\sqrt{69}}{2}$ (d) $\frac{\sqrt{66}}{2}$

16 The probability that a relation R from $\{x,y\}$ to $\{x,y\}$ is both symmetric and transitive is equal to

(a) $\frac{5}{16}$ (b) $\frac{9}{16}$
(c) $\frac{11}{16}$ (d) $\frac{13}{16}$

17 The number of values of $a \in N$ such that the variance of $3, 7, 12\ a, 43 - a$ is a natural number is

(a) 0 (b) 2
(c) 5 (d) infinite

18 From the base of a pole of height 20 m, the angle of elevation of the top of a tower is 60°. The pole subtends an angle 30° at the top of the tower.
Then, the height of the tower is

(a) $15\sqrt{3}$ (b) $20\sqrt{3}$
(c) $20 + 10\sqrt{3}$ (d) 30

19 Negation of the Boolean statement $(p \vee q) \Rightarrow ((\sim r) \vee p)$ is equivalent to

(a) $p \wedge (\sim q) \wedge r$ (b) $(\sim p) \wedge (\sim q) \wedge r$
(c) $(\sim p) \wedge q \wedge r$ (d) $p \wedge q \wedge (\sim r)$

20 Let $n \geq 5$ be an integer. If $9^n - 8n - 1 = 64\alpha$ and $6^n - 5n - 1 = 25\beta$, then $\alpha - \beta$ is equal to

(a) $1 + {}^nC_2(8-5) + {}^nC_3(8^2 - 5^2) + \dots + {}^nC_n(8^{n-1} - 5^{n-1})$
(b) $1 + {}^nC_3(8-5) + {}^nC_4(8^2 - 5^2) + \dots + {}^nC_n(8^{n-2} - 5^{n-2})$
(c) ${}^nC_3(8-5) + {}^nC_4(8^2 - 5^2) + \dots + {}^nC_n(8^{n-2} - 5^{n-2})$
(d) ${}^nC_4(8-5) + {}^nC_5(8^2 - 5^2) + \dots + {}^nC_n(8^{n-3} - 5^{n-3})$

Section B : Numerical Type Questions

21 Let $\mathbf{a} = \hat{\mathbf{i}} - 2\hat{\mathbf{j}} + 3\hat{\mathbf{k}}$, $\hat{\mathbf{b}} = \hat{\mathbf{i}} + \hat{\mathbf{j}} + \hat{\mathbf{k}}$ and $\mathbf{c}$ be a vector such that $\mathbf{a} + (\mathbf{b} \times \mathbf{c}) = \mathbf{0}$ and $\mathbf{b} \cdot \mathbf{c} = 5$. Then, the value of $3(\mathbf{c} \cdot \mathbf{a})$ is equal to ....

22 Let $y = y(x), x > 1$, be the solution of the differential equation $(x-1)\frac{dy}{dx} + 2xy = \frac{1}{x-1}$ with $y(2) = \frac{1+e^4}{2e^4}$. If $y(3) = \frac{e^\alpha + 1}{\beta e^\alpha}$, then the value of $\alpha + \beta$ is equal to ....

23 Let 3, 6, 9, 12,... upto 78 terms and 5, 9, 13 17.. upto 59 terms be two series. Then, the sum of the terms common to both the series is equal to ....

24 The number of solutions of the equation $\sin x = \cos^2 x$ in the interval $(0,10)$ is ....

25 For real numbers $a, b\ (a > b > 0)$, let

$$\text{Area}\left\{(x, y) : x^2 + y^2 \leq a^2 \text{ and } \frac{x^2}{a^2} + \frac{y^2}{b^2} \geq 1\right\} = 30\pi$$

$$\text{and Area}\left\{(x, y) : x^2 + y^2 \geq b^2 \text{ and } \frac{x^2}{a^2} + \frac{y^2}{b^2} \leq 1\right\} = 18\pi$$

Then, the value of $(a - b)^2$ is equal to

26 Let f and g be twice differentiable even functions on $(-2, 2)$ such that $f\left(\frac{1}{4}\right) = 0$, $f\left(\frac{1}{2}\right) = 0$, $f(1) = 1$ and $g\left(\frac{3}{4}\right) = 0$, $g(1) = 2$. Then, the minimum number of solutions of $f(x)g''(x) + f'(x)g'(x) = 0$ in $(-2, 2)$ is equal to ....

27 Let the coefficients of x^{-1} and x^{-3} in the expansion of $\left(2x^{1/5} - \frac{1}{x^{1/5}}\right)^{15}$, $x > 0$, be m and n respectively. If r is a positive integer such that $mn^2 = {}^{15}C_r \cdot 2^r$, then the value of r is equal to ....

28 The total number of four-digit numbers such that each of the first three digits is divisible by the last digit is equal to ....

29 Let $M = \begin{bmatrix} 0 & -\alpha \\ \alpha & 0 \end{bmatrix}$, where α is a non-zero real number and $N = \sum_{k=1}^{49} M^{2k}$. If $(I - M^2)N = 2I$, then the positive integral value of α is ....

30 Let $f(x)$ and $g(x)$ be two real polynomials of degree 2 and 1 respectively. If $f(g(x)) = 8x^2 - 2x$ and $g(f(x)) = 4x^2 + 6x + 1$, then the value of $f(2) + g(2)$ is ....

Answers

For solutions scan the QR code

24 June 2022, Shift-I

1. (d)	2. (d)	3. (a)	4. (c)	5. (a)	6. (b)	7. (b)	8. (a)	9. (b)	10. (c)
11. (d)	12. (*)	13. (b)	14. (a)	15. (d)	16. (c)	17. (a)	18. (d)	19. (b)	20. (b)
21. 31	22. 40	23. 8	24. 84	25. 7	26. 1	27. 34	28. 2929	29. 19	30. 2

24 June 2022, Shift-II

1. (b)	2. (b)	3. (c)	4. (d)	5. (c)	6. (c)	7. (a)	8. (a)	9. (a)	10. (c)
11. (d)	12. (a)	13. (d)	14. (a)	15. (c)	16. (c)	17. (b)	18. (b)	19. (d)	20. (d)
21. 80	22. 100	23. 576	24. 1633	25. 4	26. 18	27. 7	28. 479	29. 42	30. 10

25 June 2022, Shift-I

1. (c)	2. (c)	3. (a)	4. (c)	5. (b)	6. (a)	7. (c)	8. (d)	9. (a)	10. (c)
11. (a)	12. (d)	13. (a)	14. (b)	15. (b)	16. (c)	17. (b)	18. (c)	19. (d)	20. (b)
21. 286	22. 63	23. 576	24. 7	25. 4	26. 4	27. 2	28. 5	29. 414	30. 98

25 June 2022, Shift-II

1. (d)	2. (b)	3. (c)	4. (d)	5. (a)	6. (a)	7. (a)	8. (b)	9. (c)	10. (d)
11. (c)	12. (d)	13. (b)	14. (c)	15. (d)	16. (d)	17. (c)	18. (c)	19. (b)	20. (a)
21. 1	22. 62	23. 6	24. 83	25. 21	26. 14	27. 243	28. 3	29. 85	30. 51

26 June 2022, Shift-I

1. (b)	2. (b)	3. (c)	4. (c)	5. (c)	6. (d)	7. (d)	8. (a)	9. (d)	10. (*)
11. (d)	12. (b)	13. (b)	14. (b)	15. (a)	16. (a)	17. (d)	18. (a)	19. (b)	20. (a)
21. 36	22. 1120	23. 4	24. 21	25. 12	26. 80	27. 5264	28. 6	29. 1100	30. 42

26 June 2022, Shift-II

1. (d)	2. (c)	3. (c)	4. (c)	5. (b)	6. (b)	7. (c)	8. (a)	9. (d)	10. (c)
11. (b)	12. (c)	13. (c)	14. (c)	15. (b)	16. (d)	17. (c)	18. (b)	19. (c)	20. (c)
21. 248	22. 4	23. 2	24. 100	25. 150	26. 102	27. 40	28. 3	29. 12	30. 33

27 June 2022, Shift-I

1. (a)	2. (d)	3. (c)	4. (b)	5. (a)	6. (b)	7. (c)	8. (d)	9. (b)	10. (d)
11. (d)	12. (c)	13. (b)	14. (a)	15. (a)	16. (c)	17. (c)	18. (b)	19. (a)	20. (d)
21. 99	22. 45	23. 14	24. 56	25. 5	26. 6	27. 276	28. 16	29. 63	30. 137

27 June 2022, Shift-II

1. (c)	2. (c)	3. (b)	4. (c)	5. (c)	6. (d)	7. (b)	8. (b)	9. (a)	10. (b)
11. (d)	12. (d)	13. (a)	14. (a)	15. (d)	16. (b)	17. (b)	18. (a)	19. (c)	20. (c)
21. 190	22. 98	23. 180	24. 57	25. 3	26. 16	27. 36	28. 320	29. 11	30. 19

For solutions scan the QR code

28 June 2022, Shift-I

1. (a)	2. (d)	3. (b)	4. (a)	5. (d)	6. (c)	7. (c)	8. (c)	9. (b)	10. (a)
11. (a)	12. (b)	13. (b)	14. (c)	15. (b)	16. (a)	17. (b)	18. (a)	19. (c)	20. (c)
21. 8	22. 2	23. 17	24. 150	25. 31	26. 13	27. 702	28. 2	29. 40	30. 816

28 June 2022, Shift-II

1. (b)	2. (a)	3. (d)	4. (b)	5. (c)	6. (b)	7. (b)	8. (c)	9. (c)	10. (d)
11. (b)	12. (b)	13. (d)	14. (d)	15. (b)	16. (d)	17. (d)	18. (c)	19. (c)	20. (a)
21. 125	22. 0	23. 10	24. 11	25. 41651	26. 58	27. 25	28. 2	29. 37	30. 9

29 June 2022, Shift-I

1. (c)	2. (a)	3. (c)	4. (c)	5. (a)	6. (c)	7. (b)	8. (a)	9. (c)	10. (a)
11. (d)	12. (b)	13. (c)	14. (b)	15. (d)	16. (d)	17. (d)	18. (b)	19. (c)	20. (b)
21. 26	22. 2	23. 26	24. 32	25. 1	26. 29	27. 3395	28. 88	29. 28	30. 18915

29 June 2022, Shift-II

1. (a)	2. (c)	3. (c)	4. (c)	5. (d)	6. (c)	7. (d)	8. (c)	9. (c)	10. (d)
11. (*)	12. (b)	13. (c)	14. (b)	15. (a)	16. (a)	17. (a)	18. (d)	19. (c)	20. (c)
21. 10	22. 14	23. 2223	24. 4	25. 12	26. 4	27. 5	28. 1086	29. 1	30. 18

ONLINE SOLVED PAPERS

JEE Main 2022

(July Attempt)

25 JULY 2022, SHIFT-I

Section A : Objective Type Questions

1 The total number of functions,
$f:\{1, 2, 3, 4\} \rightarrow \{1, 2, 3, 4, 5, 6\}$
such that $f(1) + f(2) = f(3)$ is equal to
(a) 60 (b) 90 (c) 108 (d) 126

2 If $\alpha, \beta, \gamma, \delta$ are the roots of the equation
$x^4 + x^3 + x^2 + x + 1 = 0$, then
$\alpha^{2021} + \beta^{2021} + \gamma^{2021} + \delta^{2021}$ is equal to
(a) -4 (b) -1 (c) 1 (d) 4

3 For $n \in N$, let $S_n = \left\{z \in C : |z - 3 + 2i| = \frac{n}{4}\right\}$ and
$T_n = \left\{z \in C : |z - 2 + 3i| = \frac{1}{n}\right\}$. Then, the number of
elements in the set $\{n \in N : S_n \cap T_n = \phi\}$ is
(a) 0 (b) 2 (c) 3 (d) 4

4 The number of $q \in (0, 4\pi)$ for which the system of linear equations $3(\sin 3\theta)x - y + z = 2$; $3(\cos 2\theta)x + 4y + 3z = 3$
$6x + 7y + 7z = 9$ has no solution, is
(a) 6 (b) 7 (c) 8 (d) 9

5 If $\lim_{n \to \infty} (\sqrt{n^2 - n - 1} + n\alpha + \beta) = 0$, then $8(\alpha + \beta)$ is equal to
(a) 4 (b) -8 (c) -4 (d) 8

6 If the absolute maximum value of the function
$f(x) = (x^2 - 2x + 7)\, e^{(4x^3 - 12x^2 - 180x + 31)}$ in the interval $[-3, 0]$
is $f(\alpha)$, then
(a) $\alpha = 0$ (b) $\alpha = -3$
(c) $\alpha \in (-1, 0)$ (d) $\alpha \in (-3, -1]$

7 The curve $y(x) = ax^3 + bx^2 + cx + 5$ touches the X-axis at the point $P(-2, 0)$ and cuts the Y-axis at the point Q, where y' is equal to 3. Then, the local maximum value of $y(x)$ is
(a) $\frac{27}{4}$ (b) $\frac{29}{4}$
(c) $\frac{37}{4}$ (d) $\frac{9}{2}$

8 The area of the region given by
$A = \{(x, y); x^2 \le y \le \min\{x + 2, 4 - 3x\}\}$ is
(a) $\frac{31}{8}$ (b) $\frac{17}{6}$ (c) $\frac{19}{6}$ (d) $\frac{27}{8}$

9 For any real number x, let $[x]$ denote the largest integer less than or equal to x. Let f be a real valued function defined on the interval $[-10, 10]$ by

$$f(x) = \begin{cases} x - [x], & \text{if } [x] \text{ is odd} \\ 1 + [x] - x, & \text{if } [x] \text{ is even} \end{cases}$$

Then, the value of $\frac{\pi^2}{10} \int_{-10}^{10} f(x) \cos \pi x \, dx$
(a) 4 (b) 2
(c) 1 (d) 0

10 The slope of the tangent to a curve $C : y = y(x)$ at any point (x, y) on it is $\frac{2e^{2x} - 6e^{-x} + 9}{2 + 9e^{-2x}}$. If C passes through the points $\left(0, \frac{1}{2} + \frac{\pi}{2\sqrt{2}}\right)$ and $\left(\alpha, \frac{1}{2}e^{2\alpha}\right)$, then e^{α} is equal to
(a) $\frac{3 + \sqrt{2}}{3 - \sqrt{2}}$ (b) $\frac{3}{\sqrt{2}}\left(\frac{3 + \sqrt{2}}{3 - \sqrt{2}}\right)$
(c) $\frac{1}{\sqrt{2}}\left(\frac{\sqrt{2} + 1}{\sqrt{2} - 1}\right)$ (d) $\frac{\sqrt{2} + 1}{\sqrt{2} - 1}$

11 The general solution of the differential equation
$(x - y^2)dx + y(5x + y^2)\, dy = 0$ is
(a) $(y^2 + x)^4 = C|(y^2 + 2x)^3|$ (b) $(y^2 + 2x)^4 = C|(y^2 + x)^3|$
(c) $|(y^2 + x)^3| = C(2y^2 + x)^4$ (d) $|(y^2 + 2x)^3| = C(2y^2 + x)^4$

12 A line, with the slope greater than one, passes through the point $A(4, 3)$ and intersects the line $x - y - 2 = 0$ at the point B.
If the length of the line segment AB is $\frac{\sqrt{29}}{3}$, then B also lies on the line
(a) $2x + y = 9$ (b) $3x - 2y = 7$
(c) $x + 2y = 6$ (d) $2x - 3y = 3$

13 Let the locus of the centre $(\alpha, \beta), \beta > 0$, of the circle which touches the circle $x^2 + (y-1)^2 = 1$ externally and also touches the X-axis be L. Then, the area bounded by L and the line $y = 4$ is

(a) $\frac{32\sqrt{2}}{3}$ (b) $\frac{40\sqrt{2}}{3}$

(c) $\frac{64}{3}$ (d) $\frac{32}{3}$

14 Let P be the plane containing the straight line $\frac{x-3}{9} = \frac{y+4}{-1} = \frac{z-7}{-5}$ and perpendicular to the plane containing the straight lines $\frac{x}{2} = \frac{y}{3} = \frac{z}{5}$ and $\frac{x}{3} = \frac{y}{7} = \frac{z}{8}$. If d is the distance P from the point $(2, -5, 11)$, then d^2 is equal to

(a) 147/2 (b) 96

(c) 32/3 (d) 54

15 Let ABC be a triangle such that $\mathbf{BC} = \mathbf{a}, \mathbf{CA} = \mathbf{b}, \mathbf{AB} = \mathbf{c}$, $|\mathbf{a}| = 6\sqrt{2}, |\mathbf{b}| = 2\sqrt{3}$ and $\mathbf{b} \cdot \mathbf{c} = 12$. Consider the statements

Statement I $|(\mathbf{a} \times \mathbf{b}) + (\mathbf{c} \times \mathbf{b})| - |\mathbf{c}| = 6(2\sqrt{2} - 1)$

Statement II $\angle ACB = \cos^{-1}\left(\sqrt{\frac{2}{3}}\right)$

Then,

(a) Both statement I and statement II are true

(b) Only statement I is true

(c) Only statement II is true

(d) Both statement I and statement II are false

16 If the sum and the product of mean and variance of a binomial distribution are 24 and 128, respectively, then the probability of one or two successes is

(a) $\frac{33}{2^{32}}$ (b) $\frac{33}{2^{29}}$

(c) $\frac{33}{2^{28}}$ (d) $\frac{33}{2^{27}}$

17 If the numbers appeared on the two throws of a fair six faced die are α and β, then the probability that $x^2 + \alpha x + \beta > 0, \forall x \in R$, is

(a) $\frac{17}{36}$ (b) $\frac{4}{9}$

(c) $\frac{1}{2}$ (d) $\frac{19}{36}$

18 The number of solutions of $|\cos x| = \sin x$, such that $-4\pi \le x \le 4\pi$ is

(a) 4 (b) 6

(c) 8 (d) 12

19 A tower PQ stands on a horizontal ground with base Q on the ground. The point R divides the tower in two parts such that $QR = 15$ m. If from a point A on the ground the angle of elevation of R is 60° and the part PR of the tower subtends an angle of 15° at A, then the height of the tower is

(a) $5(2\sqrt{3} + 3)$ m

(b) $5(\sqrt{3} + 3)$ m

(c) $10(\sqrt{3} + 1)$ m

(d) $10(2\sqrt{3} + 1)$ m

20 Which of the following statements is a tautology?

(a) $((\sim p) \vee q) \Rightarrow p$

(b) $p \Rightarrow ((\sim p) \vee q)$

(c) $((\sim p) \vee q) \Rightarrow q$

(d) $q \Rightarrow ((\sim p) \vee q)$

Section B : Numerical Type Questions

21 Let $A = \begin{bmatrix} 2 & -1 & -1 \\ 1 & 0 & -1 \\ 1 & -1 & 0 \end{bmatrix}$ and $B = A - I$. If $\omega = \frac{\sqrt{3}i - 1}{2}$ then the number of elements in the set $\{n \in \{1, 2, \ldots, 100\} : A^n + (\omega B)^n = A + B\}$ is equal to ....

22 The letters of the word 'MANKIND' are written in all possible orders and arranged in serial order as in an English dictionary. Then, the serial number of the word 'MANKIND' is ....

23 If the maximum value of the term independent of t in the expansion of $\left(t^2 x^{\frac{1}{5}} + \frac{(1-x)^{\frac{1}{10}}}{t}\right)^{15}$, $x \ge 0$ is k, then $8k$ is equal to ....

24 Let a and b be two non-zero real numbers. If p and r are the roots of the equation $x^2 - 8ax + 2a = 0$ and q and s are the roots of the equation $x^2 + 12bx + 6b = 0$, such that $\frac{1}{p}, \frac{1}{q}, \frac{1}{r}, \frac{1}{s}$ are is AP, then $a^{-1} - b^{-1}$ is equal to ....

25 Let $a_1 = b_1 = 1, a_n = a_{n-1} + 2$ and $b_n = a_a + b_{n-1}$ for every natural number $n \ge 2$. Then, $\sum_{n=1}^{15} a_n \cdot b_n$ is equal to ....

26 Let $f(x) = \begin{cases} |4x^2 - 8x + 5|, \text{if } 8x^2 - 6x + 1 \ge 0 \\ [4x^2 - 8x + 5], \text{if } 8x^2 - 6x + 1 < 0 \end{cases}$

where $[\alpha]$ denotes the greatest integer less than or equal to α. Then, the number of points in R where f is not differentiable is ....

27 If $\lim_{n \to \infty} \frac{(n+1)^{k-1}}{n^{k+1}} [(nk+1) + (nk+2) + \ldots + (nk+n)]$

$= 33 \cdot \lim_{n \to \infty} \frac{1}{n^{k+1}} [1^k + 2^k + 3^k + \ldots + n^k]$,

then the integral value of k is equal to ....

28 Let the equation of two diameters of a circle $x^2 + y^2 - 2x + 2fy + 1 = 0$ be $2px - y = 1$ and $2x + py = 4p$. Then, the slope $m \in (0, \infty)$ of the tangent to the hyperbola $3x^2 - y^2 = 3$ passing through the centre of the circle is equal to ....

29 The sum of diameters of the circles that touch (i) the parabola $75x^2 = 64(5y - 3)$ at the point $\left(\frac{8}{5}, \frac{6}{5}\right)$ and (ii) the Y-axis, is equal to ....

30 The line of shortest distance between the lines $\frac{x-2}{0} = \frac{y-1}{1} = \frac{z}{1}$ and $\frac{x-3}{2} = \frac{y-5}{2} = \frac{z-1}{1}$ makes an angle of $\cos^{-1}\left(\sqrt{\frac{2}{27}}\right)$ with the plane $P: ax - y - z = 0$, $(a > 0)$.

If the image of the point $(1, 1, -5)$ in the plane P is (α, β, γ), then $\alpha + \beta - \gamma$ is equal to ....

25 JULY 2022, SHIFT-II

Section A : Objective Type Questions

1 For $z \in C$ if the minimum value of $(|z - 3\sqrt{2}| + |z - p\sqrt{2}i|)$ is $5\sqrt{2}$, then a value of p is ....

(a) 3 (b) $\frac{7}{2}$ (c) 4 (d) $\frac{9}{2}$

2 The number of real values of λ, such that the system of linear equations

$$2x - 3y + 5z = 9,$$
$$x + 3y - z = -18$$
and $$3x - y + (\lambda^2 - |\lambda|)z = 16$$

has no solutions, is

(a) 0 (b) 1 (c) 2 (d) 4

3 The number of bijective function $f: \{1, 3, 5, 7, ..., 99\} \to \{2, 4, 6, 8, ..., 100\}$ such that $f(3) \geq f(9) \geq f(15) \geq f(21) \geq .... \geq f(99)$, is

(a) ${}^{50}P_{17}$ (b) ${}^{50}P_{33}$

(c) $33! \times 17!$ (d) $\frac{50!}{2}$

4 The remainder when $(11)^{1011} + (1011)^{11}$ is divided by 9 is

(a) 1 (b) 4 (c) 6 (d) 8

5 The sum $\sum_{n=1}^{21} \frac{3}{(4n-1)(4n+3)}$ is equal to

(a) $\frac{7}{87}$ (b) $\frac{7}{29}$ (c) $\frac{14}{87}$ (d) $\frac{21}{29}$

6 $\lim_{x \to \frac{\pi}{4}} \frac{8\sqrt{2} - (\cos x + \sin x)^7}{\sqrt{2} - \sqrt{2}\sin 2x}$ is equal to

(a) 14 (b) 7 (c) $14\sqrt{2}$ (d) $7\sqrt{2}$

7 $\lim_{n \to \infty} \frac{1}{2^n}\left(\frac{1}{\sqrt{1 - \frac{1}{2^n}}} + \frac{1}{\sqrt{1 - \frac{2}{2^n}}} + \frac{1}{\sqrt{1 - \frac{3}{2^n}}} + ... + \frac{1}{\sqrt{1 - \frac{2^n - 1}{2^n}}}\right)$

is equal to

(a) $\frac{1}{2}$ (b) 1 (c) 2 (d) -2

8 If A and B are two events such that $P(A) = \frac{1}{3}$, $P(B) = \frac{1}{5}$ and $P(A \cup B) = \frac{1}{2}$, then $P(A | B') + P(B | A')$ is equal to

(a) $\frac{3}{4}$ (b) $\frac{5}{8}$

(c) $\frac{5}{4}$ (d) $\frac{7}{8}$

9 Let $[t]$ denote the greatest integer less than or equal to t. Then, the value of the integral $\int_{-3}^{101} ([\sin(\pi x)] + e^{[\cos(2\pi x)]})\, dx$ is equal to

(a) $\frac{52(1-e)}{e}$ (b) $\frac{52}{e}$

(c) $\frac{52(2+e)}{e}$ (d) $\frac{104}{e}$

10 Let the point $P(\alpha, \beta)$ be at a unit distance from each of the two lines $L_1: 3x - 4y + 12 = 0$ and $L_2: 8x + 6y + 11 = 0$. If P lies below L_1 and above L_2, then $100(\alpha + \beta)$ is equal to

(a) -14 (b) 42

(c) -22 (d) 14

11 Let a smooth curve $y = f(x)$ be such that the slope of the tangent at any point (x, y) on it is directly proportional to $\left(-\frac{y}{x}\right)$. If the curve passes through the points $(1, 2)$ and $(8, 1)$, then $\left|y\left(\frac{1}{8}\right)\right|$ is equal to

(a) $2\log_e 2$ (b) 4

(c) 1 (d) $4\log_e 2$

12 If the ellipse $\frac{x^2}{a^2} + \frac{y^2}{b^2} = 1$ meets the line $\frac{x}{7} + \frac{y}{2\sqrt{6}} = 1$ on the X-axis and the line $\frac{x}{7} - \frac{y}{2\sqrt{6}} = 1$ on the Y-axis, then the eccentricity of the ellipse is

(a) $\frac{5}{7}$ (b) $\frac{2\sqrt{6}}{7}$

(c) $\frac{3}{7}$ (d) $\frac{2\sqrt{5}}{7}$

13 The tangents at the points $A(1, 3)$ and $B(1, -1)$ on the parabola $y^2 - 2x - 2y = 1$ meet at the point P. Then, the area (in sq. unit) of the ΔPAB is

(a) 4 (b) 6 (c) 7 (d) 8

14 Let the foci of the ellipse $\frac{x^2}{16} + \frac{y^2}{7} = 1$ and the hyperbola $\frac{x^2}{144} - \frac{y^2}{\alpha} = \frac{1}{25}$ coincide. Then, the length of the latus rectum of the hyperbola is

(a) $\frac{32}{9}$ (b) $\frac{18}{5}$ (c) $\frac{27}{4}$ (d) $\frac{27}{10}$

15 A plane E is perpendicular to the two planes $2x - 2y + z = 0$ and $x - y + 2z = 4$ and passes through the point $P(1, -1, 1)$. If the distance of the plane E from the point $Q(a, a, 2)$ is $3\sqrt{2}$, then $(PQ)^2$ is equal to

(a) 9 (b) 12 (c) 21 (d) 33

16 The shortest distance between the lines $\frac{x+7}{-6} = \frac{y-6}{7} = z$ and $\frac{7-x}{2} = y - 2 = z - 6$ is

(a) $2\sqrt{29}$ (b) 1 (c) $\sqrt{\frac{37}{29}}$ (d) $\frac{\sqrt{29}}{2}$

17 Let $\mathbf{a} = \hat{\mathbf{i}} - \hat{\mathbf{j}} + 2\hat{\mathbf{k}}$ and let $\mathbf{b}$ be a vector such that $\mathbf{a} \times \mathbf{b} = 2\hat{\mathbf{i}} - \hat{\mathbf{k}}$ and $\mathbf{a} \cdot \mathbf{b} = 3$. Then, the projection of $\mathbf{b}$ on the vector $\mathbf{a} - \mathbf{b}$ is

(a) $\frac{2}{\sqrt{21}}$ (b) $2\sqrt{\frac{3}{7}}$

(c) $\frac{2}{3}\sqrt{\frac{7}{3}}$ (d) $\frac{2}{3}$

18 If the mean deviation about median for the number 3, 5, 7, $2k$, 12, 16, 21, 24 arranged in the ascending order is 6, then the median is

(a) 11.5 (b) 10.5 (c) 12 (d) 11

19 $2\sin\left(\frac{\pi}{22}\right)\sin\left(\frac{3\pi}{22}\right)\sin\left(\frac{5\pi}{22}\right)\sin\left(\frac{7\pi}{22}\right)\sin\left(\frac{9\pi}{22}\right)$ is equal to

(a) $\frac{3}{16}$ (b) $\frac{1}{16}$

(c) $\frac{1}{32}$ (d) $\frac{9}{32}$

20. Consider the following statements

P: Ramu is intelligent

Q: Ramu is rich

R: Ramu is not honest

The negation of the statement "Ramu is intelligent and honest if and only if Ramu is not rich" can be expressed as

(a) $((P \wedge (\sim R)) \wedge Q) \wedge ((\sim Q) \wedge (\sim P) \vee R))$

(b) $((P \wedge R) \wedge Q) \vee ((\sim Q) \wedge ((\sim P) \vee (\sim R)))$

(c) $((P \wedge R) \wedge Q) \wedge ((\sim Q) \wedge ((\sim P) \vee (\sim R)))$

(d) $((P \wedge (\sim R)) \wedge Q) \vee ((\sim Q) \wedge (\sim P) \wedge R))$

Section B : Numerical Type Questions

21 Let $A = \{1, 2, 3, 4, 5, 6, 7\}$. Define $B = (T \subseteq A$: either $1 \notin T$ or $2 \in T)$ and $C = \{T \subseteq A : T$ the sum of all the elements of T is a prime number$\}$. Then, the number of elements in the set $B \cup C$ is ....

22 Let $f(x)$ be a quadratic polynomial with leading coefficient 1 such that $f(0) = p$, $p \neq 0$, and $f(1) = \frac{1}{3}$. If the equations $f(x) = 0$ and of $fofofof(x) = 0$ have a common real root, then $f(-3)$ is equal to ....

23 Let $A = \begin{bmatrix} 1 & a & a \\ 0 & 1 & b \\ 0 & 0 & 1 \end{bmatrix}$, $a, b \in R$.

If for some $n \in N$, $A^n = \begin{bmatrix} 1 & 48 & 2160 \\ 0 & 1 & 96 \\ 0 & 0 & 1 \end{bmatrix}$, then $n + a + b$ is equal to ....

24 The sum of the maximum and minimum values of the function $f(x) = |5x - 7| + [x^2 + 2x]$ in the interval $\left[\frac{5}{4}, 2\right]$, where $[t]$ is the greatest integer $\leq t$, is ....

25 Let $y = y(x)$ be the solution of the differential equation $\frac{dy}{dx} = \frac{4y^3 + 2yx^2}{3xy^2 + x^3}$, $y(1) = 1$.

If for some $n \in N$, $y(2) \in [n - 1, n]$, then n is equal to ....

26 Let f be a twice differentiable function on R. If $f'(0) = 4$ and $f(x) + \int_0^x (x - t) f'(t)\, dt = (e^{2x} + e^{-2x}) \cos 2x + \frac{2}{a}x$, then $(2a + 1)^5 a^2$ is equal to ....

27 Let $a_n = \int_{-1}^{n} \left(1 + \frac{x}{2} + \frac{x^2}{3} + \ldots + \frac{x^{n-1}}{n}\right) dx$ for every $n \in N$.

Then, the sum of all the elements of the set $\{n \in N : a_n \in (2, 30)\}$ is ....

28 If the circle $x^2 + y^2 + 6x + 8y + 16 = 0$ and $x^2 + y^2 + 2(3 - \sqrt{3})x + 2(4 - \sqrt{6})y = k + 6\sqrt{3} + 8\sqrt{6}$, $k > 0$, touch internally at the point $P(\alpha, \beta)$, then $(\alpha + \sqrt{3})^2 + (\beta + \sqrt{6})^2$ is equal to ....

29 Let the area enclosed by the X-axis and the tangent and normal drawn to the curve $4x^3 - 3xy^2 + 6x^2 - 5xy - 8y^2 + 9x + 14 = 0$ at the point $(-2, 3)$ be A. Then, $8A$ is equal to ....

30 Let $x = \sin(2\tan^{-1}\alpha)$ and $y = \sin\left(\frac{1}{2}\tan^{-1}\frac{4}{3}\right)$. If $S = \{\alpha \in R : y^2 = 1 - x\}$, then $\sum_{\alpha \in S} 16\alpha^3$ is equal to ....

26 JULY 2022, SHIFT-I

Section A : Objective Type Questions

1 Let $f: R \to R$ be a continuous function such that $f(3x) - f(x) = x$. If $f(8) = 7$, then $f(14)$ is equal to

(a) 4 (b) 10
(c) 11 (d) 16

2 Let O be the origin and A be the point $z_1 = 1 + 2i$. If B is the point z_2, Re $(z_2 < 2)$, such that OAB is a right angled isosceles triangle with OB as hypotenuse, then which of the following is not true?

(a) $\arg z_2 = \pi - \tan^{-1} 3$ (b) $\arg (z_1 - 2z_2) = -\tan^{-1} \frac{4}{3}$
(c) $|z_2| = \sqrt{10}$ (d) $|2z_1 - z_2| = 5$

3 If the system of linear equations

$$8x + y + 4z = -2$$
$$x + y + z = 0$$
$$\lambda x - 3y = \mu$$

has infinitely many solutions, then the distance of the point $\left(\lambda, \mu - \frac{1}{2}\right)$ from the plane $8x + y + 4z + 2 = 0$ is

(a) $3\sqrt{5}$ (b) 4
(c) 26/9 (d) 10/3

4 Let A be a 2×2 matrix with det $(A) = -1$ and det $((A + I)(\text{Adj}(A) + I)) = 4$. Then, the sum of the diagonal elements of A can be

(a) -1 (b) 2
(c) 1 (d) $-\sqrt{2}$

5 The odd natural number a, such that the area of the region bounded by $y = 1$, $y = 3$, $x = 0$, $x = y^a$ is $364/3$, is equal to

(a) 3 (b) 5
(c) 7 (d) 9

6 Consider two GP's $2, 2^2, 2^3, \ldots\ldots$ and $4, 4^2, 4^3 \ldots\ldots$ of 60 and n terms, respectively. If the geometric mean of all the $60 + n$ terms is $(2)^{\frac{225}{8}}$, then $\sum_{k=1}^{n} k(n-k)$ is equal to

(a) 560 (b) 1540
(c) 1330 (d) 2600

7 If the function

$$f(x) = \begin{cases} \dfrac{\log_e(1 - x + x^2) + \log_e(1 + x + x^2)}{\sec x - \cos x} & x \in \left(\dfrac{-\pi}{2}, \dfrac{\pi}{2}\right) - (0) \\ k & x = 0 \end{cases}$$ is

continuous at $x = 0$, then k is equal to

(a) 1 (b) -1 (c) e (d) 0

8 If $f(x) = \begin{cases} x + a, & x \le 0 \\ |x - 4|, & x > 0 \end{cases}$ and $g(x) = \begin{cases} x + 1, & x < 0 \\ (x-4)^2 + b, & x \ge 0 \end{cases}$ are continuous on R, then $(gof)(2) + (fog)(-2)$ is equal to

(a) -10 (b) 10
(c) 8 (d) -8

9 Let $f(x) = \begin{cases} x^3 - x^2 + 10x - 7, & x \le 1 \\ -2x + \log_2(b^2 - 4), & x > 1 \end{cases}$. Then, the set of all values of b, for which $f(x)$ has maximum value at $x = 1$, is

(a) $(-6, -2)$ (b) $(2, 6)$
(c) $[-6, -2) \cup (2, 6]$ (d) $[-\sqrt{6}, -2) \cup (2, \sqrt{6}]$

10 If $a = \lim_{n \to \infty} \sum_{n=1}^{n} \frac{2n}{n^2 + k^2}$ and $f(x) = \sqrt{\frac{1 - \cos x}{1 + \cos x}}$, $x \in (0, 1)$, then

(a) $2\sqrt{2} f\left(\frac{a}{2}\right) = f'\left(\frac{a}{2}\right)$ (b) $f\left(\frac{a}{2}\right) f'\left(\frac{a}{2}\right) = \sqrt{2}$
(c) $\sqrt{2} f\left(\frac{a}{2}\right) = f'\left(\frac{a}{2}\right)$ (d) $f\left(\frac{a}{2}\right) = \sqrt{2} f'\left(\frac{a}{2}\right)$

11 If $\frac{dy}{dx} + 2y \tan x = \sin x$, $0 < x < \frac{\pi}{2}$ and $y\left(\frac{\pi}{3}\right) = 0$, then, the maximum value of $y(x)$ is

(a) 1/8 (b) 3/4
(c) 1/4 (d) 3/8

12 A point P moves so that the sum of squares of its distances from the points (1, 2) and (–2, 1) is 14. Let $f(x, y) = 0$ be the locus of P, which intersects the X-axis at the points A and B and the Y-axis at the points C and D. Then, the area of the quadrilateral $ABCD$ is equal to

(a) $\frac{9}{2}$ (b) $\frac{3\sqrt{17}}{2}$
(c) $\frac{3\sqrt{17}}{4}$ (d) 9

13. Let the tangent drawn to the parabola $y^2 = 24x$ at the point (α, β) is perpendicular to the line $2x + 2y = 5$. Then, the normal to the hyperbola $\frac{x^2}{\alpha^2} - \frac{y^2}{\beta^2} = 1$ at the point $(\alpha + 4, \beta + 4)$ does not pass through the point

(a) (25, 10) (b) (20, 12)
(c) (30, 8) (d) (15, 13)

14. The length of the perpendicular from the point (1, –2, 5) on the line passing through (1, 2, 4) and parallel to the line $x + y - z = 0$ and $x - 2y + 3z - 5$ is

(a) $\sqrt{21/2}$ (b) $\sqrt{9/2}$
(c) $\sqrt{73/2}$ (d) 1

15 Let $\mathbf{a} = \alpha\hat{\mathbf{i}} + \hat{\mathbf{j}} - \hat{\mathbf{k}}$ and $\mathbf{b} = 2\hat{\mathbf{i}} + \hat{\mathbf{j}} - \alpha\hat{\mathbf{k}}$, $\alpha > 0$. If the projection of $\mathbf{a} \times \mathbf{b}$ on the vector $-\hat{\mathbf{i}} + 2\hat{\mathbf{j}} - 2\hat{\mathbf{k}}$ is 30, then α is equal to

(a) $\frac{15}{2}$ (b) 8 (c) $\frac{13}{2}$ (d) 7

16 The mean and variance of a binomial distribution are α and $\alpha/3$, respectively. If $P(X = 1) = \frac{4}{243}$, then $P(X = 4 \text{ or } 5)$ is equal to

(a) 5/9 (b) 64/81
(c) 16/27 (d) 145/243

17 Let E_1, E_2 and E_3 be three mutually exclusive events such that $P(E_1) = \frac{2+3p}{6}$, $P(E_2) = \frac{2-p}{8}$ and $P(E_3) = \frac{1-p}{2}$. If the maximum and minimum values of p are p_1 and p_2, then $(p_1 + p_2)$ is equal to

(a) 2/3 (b) 5/3
(c) 5/4 (d) 1

18 Let $S = \{\theta \in [0, 2\pi] : 8^{2\sin^2\theta} + 8^{2\cos^2\theta} = 16\}$. Then, $n(S) + \sum_{\theta \in S}\left(\sec\left(\frac{\pi}{4} + 2\theta\right)\operatorname{cosec}\left(\frac{\pi}{4} + 2\theta\right)\right)$ is equal to

(a) 0 (b) −2 (c) −4 (d) 12

19 $\tan\left(2\tan^{-1}\frac{1}{5} + \sec^{-1}\frac{\sqrt{5}}{2} + 2\tan^{-1}\frac{1}{8}\right)$ is equal to

(a) 1 (b) 2 (c) 1/4 (d) 5/4

20 The statements $(\sim(p \Leftrightarrow \sim q)) \wedge q$ is

(a) a tautology (b) a contradiction
(c) equivalent to $(p \Rightarrow q) \wedge q$ (d) equivalent to $(p \Rightarrow q) \wedge p$

Section B : Numerical Type Questions

21 If for some $p, q, r \in R$, not all have same sign, one of the roots of the equation $(p^2 + q^2)x^2 - 2q(p + r)x + q^2 + r^2 = 0$ is also a root of the equation $x^2 + 2x - 8 = 0$, then $\frac{q^2 + r^2}{p^2}$ is equal to ....

22 The number of 5- digit natural numbers, such that the product of their digits is 36, is ....

23 The series of positive multiples of 3 is divided into sets: {3}, {6, 9, 12}, {15, 18, 21, 24, 27},...... Then, the sum of the elements in the 11th set is equal to ....

24 The number of distinct real roots of the equation $x^5(x^3 - x^2 - x + 1) + x(3x^3 - 4x^2 - 2x + 4) - 1 = 0$ is ....

25 If the coefficients of x and x^2 in the expansion of $(1 + x)^p(1 - x)^q$, $p, q \le 15$, are −3 and −5 respectively, then coefficient of x^3 is equal to ......

26 If $n(2n + 1)\int_0^1 (1 - x^n)^{2n}dx = 1177\int_0^1 (1 - x^n)^{2n+1}dx$, then $n \in N$ is equal to ....

27 Let a curve $y = y(x)$ pass through the point (3, 3) and the area of the region under this curve, above the X-axis and between the abscissae 3 and $x(> 3)$ be $\left(\frac{y}{x}\right)^3$. If this curve also passes through the point $(\alpha, 6\sqrt{10})$ in the first quadrant, then α is equal to ....

28 The equation of the sides AB, BC and CA of a ΔABC are $2x + y = 0$, $x + py = 15a$ and $x - y = 3$, respectively. If its orthocentre is $(2, a)$, $-\frac{1}{2} < a < 2$, then p is equal to ....

29 Let the function $f(x) = 2x^2 - \log_e x$, $x > 0$, be decreasing in $(0, a)$ and increasing in $(a, 4)$. A tangent to the parabola $y^2 = 4ax$ at a point p on it passes through the point $(8a, 8a - 1)$ but does not pass through the point $\left(-\frac{1}{a}, 0\right)$.

If the equation of the normal at P is $\frac{x}{\alpha} + \frac{y}{\beta} = 1$, then $\alpha + \beta$ is equal to ....

30 Let Q and R be two points on the line $\frac{x+1}{2} = \frac{y+2}{3} = \frac{z-1}{2}$ at a distance $\sqrt{26}$ from the point $P(4, 2, 7)$. Then, the square of the area of the ΔPQR is ....

26 JULY 2022, SHIFT-II

Section A : Objective Type Questions

1 The minimum value of the sum of the squares of the roots of $x^2 + (3 - a)x + 1 = 2a$ is

(a) 4 (b) 5 (c) 6 (d) 8

2 If $z = x + iy$ satisfies $|z| - 2 = 0$ and $|z - i| - |z + 5i| = 0$, then

(a) $x + 2y - 4 = 0$ (b) $x^2 + y - 4 = 0$
(c) $x + 2y + 4 = 0$ (d) $x^2 - y + 3 = 0$

3 Let $A = \begin{bmatrix} 1 \\ 1 \\ 1 \end{bmatrix}$ and $B = \begin{bmatrix} 9^2 & -10^2 & 11^2 \\ 12^2 & 13^2 & -14^2 \\ -15^2 & 16^2 & 17^2 \end{bmatrix}$, then the value of $A'BA$ is

(a) 1224
(b) 1042
(c) 540
(d) 539

4 $\sum_{\substack{i,j=0\\ i\neq j}}^{n} {}^nC_i\,{}^nC_j$ is equal to

(a) $2^{2n} - {}^{2n}C_n$ (b) $2^{2n-1} - {}^{2n-1}C_{n-1}$
(c) $2^{2n} - \frac{1}{2}\,{}^{2n}C_n$ (d) $2^{n-1} + 2^{2n-1}C_n$

5 Let P and Q be any points on the curves $(x-1)^2 + (y+1)^2 = 1$ and $y = x^2$, respectively. The distance between P and Q is minimum for some value of the abscissa of P in the interval

(a) $\left(0, \frac{1}{4}\right)$ (b) $\left(\frac{1}{2}, \frac{3}{4}\right)$ (c) $\left(\frac{1}{4}, \frac{1}{2}\right)$ (d) $\left(\frac{3}{4}, 1\right)$

6 If the maximum value of a, for which the function $f_a(x) = \tan^{-1} 2x - 3ax + 7$ is non-decreasing in $\left(-\frac{\pi}{6}, \frac{\pi}{6}\right)$, is $\bar{a}$, then $f_{\bar{a}}\left(\frac{\pi}{8}\right)$ is equal to

(a) $8 - \frac{9\pi}{4(9+\pi^2)}$ (b) $8 - \frac{4\pi}{9(4+\pi^2)}$

(c) $8\left(\frac{1+x^2}{9+\pi^2}\right)$ (d) $8 - \frac{\pi}{4}$

7 Let $\beta = \lim_{x \to 0} \frac{\alpha x - (e^{3x} - 1)}{\alpha x (e^{3x} - 1)}$ for some $\alpha \in R$. Then, the value of $\alpha + \beta$ is

(a) $\frac{14}{5}$ (b) $\frac{3}{2}$ (c) $\frac{5}{2}$ (d) $\frac{7}{2}$

8 The value of $\log_e 2 \frac{d}{dx} (\log_{\cos x} \operatorname{cosec} x)$ at $x = \frac{\pi}{4}$ is

(a) $-2\sqrt{2}$ (b) $2\sqrt{2}$ (c) -4 (d) 4

9 $\int_0^{20\pi} (|\sin x| + |\cos x|)^2\, dx$ is equal to

(a) $10(\pi + 4)$ (b) $10(\pi + 2)$
(c) $20(\pi - 2)$ (d) $20(\pi + 2)$

10 Let the solution curve $y = f(x)$ of the differential equation

$$\frac{dy}{dx} + \frac{xy}{x^2 - 1} = \frac{x^4 + 2x}{\sqrt{1 - x^2}},\ x \in (-1, 1)$$ pass through the origin.

Then, $\int_{-\sqrt{3}/2}^{\sqrt{3}/2} f(x)\, dx$ is

(a) $\frac{\pi}{3} - \frac{1}{4}$ (b) $\frac{\pi}{3} - \frac{\sqrt{3}}{4}$

(c) $\frac{\pi}{6} - \frac{\sqrt{3}}{4}$ (d) $\frac{\pi}{6} - \frac{\sqrt{3}}{2}$

11 The acute angle between the pair of tangents drawn to the ellipse $2x^2 + 3y^2 = 5$ from the point $(1, 3)$ is

(a) $\tan^{-1}\left(\frac{16}{7\sqrt{5}}\right)$ (b) $\tan^{-1}\left(\frac{24}{7\sqrt{5}}\right)$

(c) $\tan^{-1}\left(\frac{32}{7\sqrt{5}}\right)$ (d) $\tan^{-1}\left(\frac{3 + 8\sqrt{5}}{35}\right)$

12 The equation of a common tangent to the parabolas $y = x^2$ and $y = -(x-2)^2$ is

(a) $y = 4(x - 2)$ (b) $y = 4(x - 1)$
(c) $y = 4(x + 1)$ (d) $y = 4(x + 2)$

13 Let the abscissae of the two points P and Q on a circle be the roots $x^2 - 4x - 6 = 0$ and the ordinates of P and Q be the roots of $y^2 + 2y - 7 = 0$. If PQ is a diameter of the circle $x^2 + y^2 + 2ax + 2by + c = 0$, then the value of $(a + b - c)$ is

(a) 12 (b) 13 (c) 14 (d) 16

14 If the line $x - 1 = 0$ is directrix of the hyperbola $kx^2 - y^2 = 6$, then the hyperbola passes through the point

(a) $(-2\sqrt{5}, 6)$ (b) $(-\sqrt{5}, 3)$ (c) $(\sqrt{5}, -2)$ (d) $(2\sqrt{5}, 3\sqrt{6})$

15 A vector $\mathbf{a}$ is parallel to the line of intersection of the plane determined by the vectors $\hat{\mathbf{i}}, \hat{\mathbf{i}} + \hat{\mathbf{j}}$ and the plane determined by the vectors $\hat{\mathbf{i}} - \hat{\mathbf{j}}, \hat{\mathbf{i}} + \hat{\mathbf{k}}$. The obtuse angle between $\mathbf{a}$ and the vector $\mathbf{b} = \hat{\mathbf{i}} - 2\hat{\mathbf{j}} + 2\hat{\mathbf{k}}$ is

(a) $\frac{3\pi}{4}$ (b) $\frac{2\pi}{3}$

(c) $\frac{4\pi}{5}$ (d) $\frac{5\pi}{6}$

16 If $0 < x < \frac{1}{\sqrt{2}}$ and $\frac{\sin^{-1} x}{\alpha} = \frac{\cos^{-1} x}{\beta}$, then the value of $\sin\left(\frac{2\pi\alpha}{\alpha + \beta}\right)$ is

(a) $4\sqrt{(1 - x^2)}(1 - 2x^2)$ (b) $4x\sqrt{(1 - x^2)}(1 - 2x^2)$

(c) $2x\sqrt{(1 - x^2)}(1 - 4x^2)$ (d) $4\sqrt{(1 - x^2)}(1 - 4x^2)$

17 Negation of the Boolean expression $p \Leftrightarrow (q \Rightarrow p)$ is

(a) $(\sim p) \wedge q$ (b) $p \wedge (\sim q)$
(c) $(\sim p) \vee (\sim q)$ (d) $(\sim p) \wedge (\sim q)$

18 Let X be a binomially distributed random variable with mean 4 and variance $\frac{4}{3}$. Then, $54\, P(X \le 2)$ is equal to

(a) $\frac{73}{27}$ (b) $\frac{146}{27}$

(c) $\frac{146}{81}$ (d) $\frac{126}{81}$

19 The integral $\int \frac{\left(1 - \frac{1}{\sqrt{3}}\right)(\cos x - \sin x)}{\left(1 + \frac{2}{\sqrt{3}} \sin 2x\right)} dx$ is equal to

(a) $\frac{1}{2}\log_e \left|\frac{\tan\left(\frac{x}{2} + \frac{\pi}{12}\right)}{\tan\left(\frac{x}{2} + \frac{\pi}{6}\right)}\right| + C$ (b) $\frac{1}{2}\log_e \left|\frac{\tan\left(\frac{x}{2} + \frac{x}{6}\right)}{\tan\left(\frac{x}{2} + \frac{\pi}{3}\right)}\right| + C$

(c) $\log_e \left|\frac{\tan\left(\frac{x}{2} + \frac{\pi}{6}\right)}{\tan\left(\frac{x}{2} + \frac{\pi}{12}\right)}\right| + C$ (d) $\frac{1}{2}\log_e \left|\frac{\tan\left(\frac{x}{2} - \frac{\pi}{12}\right)}{\tan\left(\frac{x}{2} - \frac{\pi}{6}\right)}\right| + C$

20 The area bounded by the curves $y = |x^2 - 1|$ and $y = 1$ is

(a) $\frac{2}{3}(\sqrt{2}+1)$ (b) $\frac{4}{3}(\sqrt{2}-1)$
(c) $2(\sqrt{2}-1)$ (d) $\frac{8}{3}(\sqrt{2}-1)$

Section B : Numerical Type Questions

21 Let $A = \{1, 2, 3, 4, 5, 6, 7\}$ and $B = \{3, 6, 7, 9\}$, Then, the number of elements in the set $\{C \subseteq A : C \cap B \neq \phi\}$is ....

22 The largest value of a, for which the perpendicular distance of the plane containing the lines $\mathbf{r} = (\hat{\mathbf{i}} + \hat{\mathbf{j}}) + \lambda(\hat{\mathbf{i}} + a\hat{\mathbf{j}} - \hat{\mathbf{k}})$ and $\mathbf{r} = (\hat{\mathbf{i}} + \hat{\mathbf{j}}) + \mu(-\hat{\mathbf{i}} + \hat{\mathbf{j}} - a\hat{\mathbf{k}})$ from the point $(2, 1, 4)$ is $\sqrt{3}$, is ....

23 Numbers are to be formed between 1000 and 3000, which are divisible by 4, using the digits 1, 2, 3, 4, 5 and 6 without repetition of digits. Then, the total number of such numbers is ....

24 If $\sum_{k=1}^{10} \frac{k}{k^4 + k^2 + 1} = \frac{m}{n}$, where m and n are co-prime, then $m + n$ is equal to ....

25 If the sum of solutions of the system of equations $2\sin^2\theta - \cos 2\theta = 0$ and $2\cos^2\theta + 3\sin\theta = 0$ in the interval $[0, 2\pi]$ is $k\pi$, then k is equal to ....

26 The mean and standard deviation of 40 observations are 30 and 5 respectively. It was noticed that two of these observations 12 and 10 were wrongly recorded. If σ is the standard deviation of the data after omitting the two wrong observations from the data, then $38\sigma^2$ is equal to ....

27 The plane passing through the line L : $lx - y + 3(1 - l)z = 1, x + 2y - z = 2$ and perpendicular to the plane $3x + 2y + z = 6$ is $3x - 8y + 7z = 4$. If θ is the acute angle between the line L and the Y-axis, then $415\cos^2\theta$ is equal to ....

28 Suppose $y = y(x)$ be the solution curve to the differential equation $\frac{dy}{dx} - y = 2 - e^{-x}$ such that $\lim_{x \to \infty} y(x)$ is finite. If a and b are respectively the x and y-intercepts of the tangent to the curve at $x = 0$, then the value of $a - 4b$ is equal to ....

29 Different AP's are constructed with the first term 100, the last term 199, and integral common differences. The sum of the common differences of all such AP's having atleast 3 terms and at most 33 terms is ....

30 The number of matrices $A = \begin{bmatrix} a & b \\ c & d \end{bmatrix}$, where a, b, c and $d \in \{-1, 0, 1, 2, 3, \ldots, 10\}$, such that $A = A^{-1}$, is ....

27 JULY 2022, SHIFT-I

Section A : Objective Type Questions

1 Let R_1 and R_2 be two relations defined on R by $a R_1 b \Leftrightarrow ab \geq 0$ and $a R_2 b \Leftrightarrow a \geq b$. Then,

(a) R_1 is an equivalence relation but not R_2
(b) R_2 is an equivalence relation but not R_1
(c) Both R_1 and R_2 are equivalence relations
(d) Neither R_1 nor R_2 is an equivalence relation

2 Let $f, g : N - \{1\} \to N$ be functions defined by $f(a) = \alpha$, where α is the maximum of the powers of those primes p such that p^α divides a and $g(a) = a + 1$, for all $a \in N - \{1\}$. Then, the function $f + g$ is

(a) one-one but not onto (b) onto but not one one
(c) both one-one and onto (d) neither one-one nor onto

3 Let the minimum value v_0 of $v = |z|^2 + |z-3|^2 + |z - 6i|^2$, $z \in C$ is attained at $z = z_0$. Then, $|2z_0^2 - \bar{z}_0^3 + 3|^2 + v_0^2$ is equal to

(a) 1000 (b) 1024 (c) 1105 (d) 1196

4 Let $A = \begin{bmatrix} 1 & 2 \\ -2 & -5 \end{bmatrix}$. Let $\alpha, \beta, \in R$ be such that $\alpha A^2 + \beta A = 2I$. Then, $\alpha + \beta$ is equal to

(a) −10 (b) −6 (c) 6 (d) 10

5 The remainder when $(2021)^{2022} + (2022)^{2021}$ is divided by 7 is

(a) 0 (b) 1 (c) 2 (d) 6

6 Suppose $a_1, a_2, \ldots, a_n, \ldots\ldots$ be an arithmetic progression of natural numbers. If the ratio of the sum of first five terms to the sum of first nine terms of the progression is $5 : 17$ and $110 < a_{15} < 120$, then the sum of the first ten terms of the progression is equal to

(a) 290 (b) 380 (c) 460 (d) 510

7 Let $R \to R$ be function defined as $f(x) = a\sin\left(\frac{\pi[x]}{2}\right) + [2 - x]$, $a \in R$, where $[t]$ is the greatest integer less than or equal to t. If $\lim_{x \to -1} f(x)$ exists, then the value of $\int_0^4 f(x)\,dx$ is equal to

(a) −1 (b) −2 (c) 1 (d) 2

8 Let $I = \int_{\pi/4}^{\pi/3} \left(\frac{8\sin x - \sin 2x}{x}\right) dx$. Then,

(a) $\frac{\pi}{2} < I < \frac{3\pi}{4}$ (b) $\frac{\pi}{5} < I < \frac{5\pi}{12}$
(c) $\frac{5\pi}{12} < I < \frac{\sqrt{2}}{3}\pi$ (d) $\frac{3\pi}{4} < I < \pi$

9 The area of the smaller region enclosed by the curves $y^2 = 8x + 4$ and $x^2 + y^2 + 4\sqrt{3}x - 4 = 0$ is equal to

(a) $\frac{1}{3}(2 - 12\sqrt{3} + 8\pi)$ (b) $\frac{1}{3}(2 - 12\sqrt{3} + 6\pi)$
(c) $\frac{1}{3}(4 - 12\sqrt{3} + 8\pi)$ (d) $\frac{1}{3}(4 - 12\sqrt{3} + 6\pi)$

10 Let $y = y_1(x)$ and $y = y_2(x)$ be two distinct solution of the differential equation $\frac{dy}{dx} = x + y$, with $y_1(0) = 0$ and $y_2(0) = 1$, respectively. Then, the number of points of intersection of $y = y_1(x)$ and $y = y_2(x)$ is

(a) 0 (b) 1
(c) 2 (d) 3

11 Let $P(a, b)$ be a point on the parabola $y^2 = 8x$ such that the tangent at P passes through the centre of the circle $x^2 + y^2 - 10x - 14y + 65 = 0$. Let A be the product of all possible values of a and B be the product of all possible values of b. Then, the value of $A + B$ is equal to

(a) 0 (b) 25
(c) 40 (d) 65

12 Let $\mathbf{a} = \alpha\hat{\mathbf{i}} + \hat{\mathbf{j}} + \beta\hat{\mathbf{k}}$ and $\mathbf{b} = 3\hat{\mathbf{i}} - 5\hat{\mathbf{j}} + 4\hat{\mathbf{k}}$ be two vectors such that $\mathbf{a} \times \mathbf{b} = -\hat{\mathbf{i}} + 9\hat{\mathbf{j}} + 12\hat{\mathbf{k}}$. Then, the projection of $\mathbf{b} - 2\mathbf{a}$ on $\mathbf{b} + \mathbf{a}$ is equal to

(a) 2 (b) $\frac{39}{5}$
(c) 9 (d) $\frac{46}{5}$

13 Let $\mathbf{a} = 2\hat{\mathbf{i}} - \hat{\mathbf{j}} + 5\hat{\mathbf{k}}$ and $\mathbf{b} = \alpha\hat{\mathbf{i}} + \beta\hat{\mathbf{j}} + 2\hat{\mathbf{k}}$.
If $((\mathbf{a} \times \mathbf{b}) \times \hat{\mathbf{i}}) \cdot \hat{\mathbf{k}} = \frac{23}{2}$, then $|\mathbf{b} \times 2\hat{\mathbf{j}}|$ is equal to

(a) 4 (b) 5
(c) $\sqrt{21}$ (d) $\sqrt{17}$

14 Let S be the sample space of all five digit numbers. If p is the probability that a randomly selected number from S, is multiple of 7 but not divisible by 5, then $9p$ is equal to

(a) 1.0146 (b) 1.2085
(c) 1.0285 (d) 1.1521

15 Let a vertical tower AB of height $2h$ stands on a horizontal ground. Let from a point P on the ground a man can see upto height h of the tower with an angle of elevation 2α. When from P, he moves a distance d in the direction of $\mathbf{AP}$, he can see the top B of the tower with an angle of elevation α. If $d = \sqrt{7}h$, then $\tan\alpha$ is equal to

(a) $\sqrt{5} - 2$ (b) $\sqrt{3} - 1$
(c) $\sqrt{7} - 2$ (d) $\sqrt{7} - \sqrt{3}$

16 $(p \wedge r) \Leftrightarrow (p \wedge (\sim q))$ is equivalent to $(\sim p)$ when r is

(a) p (b) $\sim p$
(c) q (d) $\sim q$

17 If the plane P passes through the intersection of two mutually perpendicular planes $2x + ky - 5z = 1$ and $3kx - ky + z = 5$, $k < 3$ and intercepts a unit length on positive X-axis, then the intercept made by the plane P on the Y-axis is

(a) $\frac{1}{11}$ (b) $\frac{5}{11}$
(c) 6 (d) 7

18 Let $A(1, 1)$, $B(-4, 3)$, $C(-2, -5)$ be vertices of a ΔABC, P be a point on side BC, and Δ_1 and Δ_2 be the areas of ΔAPB and ΔABC, respectively. If $\Delta_1 : \Delta_2 = 4 : 7$, then the area enclosed by the lines AP, AC and the X-axis is

(a) $\frac{1}{4}$ (b) $\frac{3}{4}$ (c) $\frac{1}{2}$ (d) 1

19 If the circle $x^2 + y^2 - 2gx + 6y - 19c = 0$. $g, c \in R$ passes through the point (6, 1) and its centre lies on the line $x - 2cy = 8$, then the length of intercept made by the circle on X-axis is

(a) $\sqrt{11}$ (b) 4 (c) 3 (d) $2\sqrt{23}$

20 Let a function $f : R \to R$ be defined as

$$f(x) = \begin{cases} \int_0^x (5 - |t - 3|)dt & x > 4 \\ x^2 + bx & x \le 4 \end{cases}$$

where, $b \in R$. If f is continuous at $x = 4$, then which of the following statements is not true?

(a) f is not differentiable at $x = 4$
(b) $f'(3) + f'(5) = \frac{35}{4}$
(c) f is increasing in $\left(-\infty, \frac{1}{8}\right) \cup (8, \infty)$
(d) f has a local minima at $x = \frac{1}{8}$

Section B : Numerical Type Questions

21 For $k \in R$, let the solution of the equation $\cos(\sin^{-1}(x\cot(\tan^{-1}(\cos(\sin^{-1}x))))) = k, 0 < |x| < \frac{1}{\sqrt{2}}$.

Inverse trigonometric functions take only principal values. If the solutions of the equation $x^2 - bx - 5 = 0$ are $\frac{1}{\alpha^2} + \frac{1}{\beta^2}$ and $\frac{\alpha}{\beta}$, then $\frac{b}{k^2}$ is equal to ....

22 The mean and variance of 10 observation were calculated as 15 and 15 respectively by a student who took by mistake 25 instead of 15 for one observation. Then, the correct standard deviation is ....

23 Let the line $\frac{x-3}{7} = \frac{y-2}{-1} = \frac{z-3}{-4}$ intersect the plane containing the lines $\frac{x-4}{1} = \frac{y+1}{-2} = \frac{z}{1}$ and $4ax - y + 5z - 7a = 0 = 2x - 5y - z - 3$, $a \in R$ at the point $P(\alpha, \beta, \gamma)$. Then, the value of $\alpha + \beta + \gamma$ equals ....

24 An ellipse $E : \frac{x^2}{a^2} + \frac{y^2}{b^2} = 1$ passes through the vertices of the hyperbola $H : \frac{x^2}{49} - \frac{y^2}{64} = -1$. Let the major and minor axes of the ellipse E coincide with the transverse and conjugate axes of the hyperbola H, respectively. Let the product of the eccentricities of E and H be $\frac{1}{2}$.

If the length of the latus rectum of the ellipse E, then the value of $113l$ is equal to ....

25 Let $y = y(x)$ be the solution curve of the differential equation $\sin(2x^2) \log_e(\tan x^2)dy + \left(4xy - 4\sqrt{2}x \sin\left(x^2 - \frac{\pi}{4}\right)\right)dx = 0, 0 < x < \sqrt{\frac{\pi}{2}}$, which passes through the point $\left(\sqrt{\frac{\pi}{6}}, 1\right)$.

Then, $\left|y\left(\sqrt{\frac{\pi}{3}}\right)\right|$ is equal to ....

26 Let M and N be the number of points on the curve $y^5 - 9xy + 2x = 0$, where the tangents to the curve are parallel to X-axis and Y-axis, respectively. Then, the value of $M + N$ equals ....

27 Let $f(x) = 2x^2 - x - 1$ and $S = \{n \in Z : |f(n)| \le 800\}$. Then, the value of $\sum_{n \in S} f(n)$ is equal to ....

28 Let S be the set containing all 3×3 matrices with entries from $\{-1, 0, 1\}$. The total number of matrices $A \in S$ such that the sum of all the diagonal elements of $A^T A$ is 6, is ....

29 If the length of the latus rectum of the ellipse $x^2 + 4y^2 + 2x + 8y - \lambda = 0$ is 4 and l is the length of its major axis, then $\lambda + l$ is equal to ....

30 Let $S = \{z \in C : z^2 + \bar{z} = 0\}$. Then, $\sum_{z \in S} (\text{Re}(z) + \text{Im}(z))$ is equal to ....

27 JULY 2022, SHIFT-II

Section A : Objective Type Questions

1 The domain of the function $f(x) = \sin^{-1}[2x^2 - 3] + \log_2\left(\log_{\frac{1}{2}}(x^2 - 5x + 5)\right)$, where $[t]$ is the greatest integer function, is

(a) $\left(-\sqrt{\frac{5}{2}}, \frac{5-\sqrt{5}}{2}\right)$ (b) $\left(\frac{5-\sqrt{5}}{2}, \frac{5+\sqrt{5}}{2}\right)$

(c) $\left(1, \frac{5-\sqrt{5}}{2}\right)$ (d) $\left(1, \frac{5+\sqrt{5}}{2}\right)$

2 Let S be the set of all (α, β), $\pi < \alpha, \beta < 2\pi$, for which the complex number $\frac{1 - i \sin \alpha}{1 + 2i \sin \alpha\beta}$ is purely imaginary and $\frac{1 + i \cos \beta}{1 - 2i \cos \beta}$ is purely real. Let $Z_{\alpha\beta} = \sin 2\alpha + i \cos 2\beta$, $(\alpha, \beta) \in S$.

Then, $\sum_{(\alpha, \beta) \in S} \left(iZ_{\alpha\beta} + \frac{1}{i\bar{Z}_{\alpha\beta}}\right)$ is equal to

(a) 3 (b) $3i$ (c) 1 (d) $2 - i$

3 If α and β are the roots of the equation $x^2 - (5 + 3^{\sqrt{\log_3 5}} - 5^{\sqrt{\log_5 3}})\, x + 3\left(3^{(\log_3 5)^{1/3}} - 5^{(\log_5 3)^{2/3}} - 1\right) = 0$, then the equation, whose roots are $\alpha + \frac{1}{\beta}$ and $\beta + \frac{1}{\alpha}$, is

(a) $3x^2 - 20x - 12 = 0$ (b) $3x^2 - 10x - 4 = 0$

(c) $3x^2 - 10x + 2 = 0$ (d) $3x^2 - 20x + 16 = 0$

4 Let $A = \begin{bmatrix} 4 & -2 \\ \alpha & \beta \end{bmatrix}$

If $A^2 + 4A + 18I = 0$, then det (A) is equal to ....

(a) -18 (b) 18 (c) -50 (d) 50

5 If for $p \ne q \ne 0$, the function $f(x) = \frac{\sqrt[7]{p(729 + x)} - 3}{\sqrt[3]{729 + qx} - 9}$ is continuous at $x = 0$, then

(a) $7pq\, f(0) - 1 = 0$ (b) $63q\, f(0) - p^2 = 0$

(c) $21q\, f(0) - p^2 = 0$ (d) $7pq\, f(0) - 9 = 0$

6 Let $f(x) = 2 + |x| - |x - 1| + |x + 1|$ $x \in R$. Consider

Statement I : $f'\left(-\frac{3}{2}\right) + f'\left(-\frac{1}{2}\right) + f'\left(\frac{1}{2}\right) + f'\left(\frac{3}{2}\right) = 2$

Statement II : $\int_{-2}^{2} f(x)\, dx = 12$

Then,

(a) Both statement I and statement II are correct
(b) Both statement I and statement II are wrong
(c) Only statement I is correct
(d) Only statement II is correct

7 Let the sum of an infinite GP, whose first term is a and the common ratio is r, be 5. Let the sum of its first five terms be $\frac{98}{25}$. Then, the sum of the first 21 terms of an AP, whose first term is $10ar$, nth term is a_n and the common difference is $10ar^2$ is equal to

(a) $21\, a_{11}$ (b) $22\, a_{11}$

(c) $15\, a_{16}$ (d) $14\, a_{16}$

8 The area of the region enclosed by $y \le 4x^2$, $x^2 \le 9y$ and $y \le 4$, is equal to

(a) $\frac{40}{3}$ (b) $\frac{56}{3}$ (c) $\frac{112}{3}$ (d) $\frac{80}{3}$

9 $\int_0^2 \left(|2x^2 - 3x| + \left[x - \frac{1}{2} \right] \right) dx$, where $[t]$ is the greatest integer function is equal to

(a) $\frac{7}{6}$ (b) $\frac{19}{12}$ (c) $\frac{31}{12}$ (d) $\frac{3}{2}$

10 Consider a curve $y = y(x)$ in the first quadrant as shown in the figure. Let the area A_1 is twice the area A_2. Then, the normal to the curve perpendicular to the line $2x - 12y = 15$ does **NOT** pass through the point.

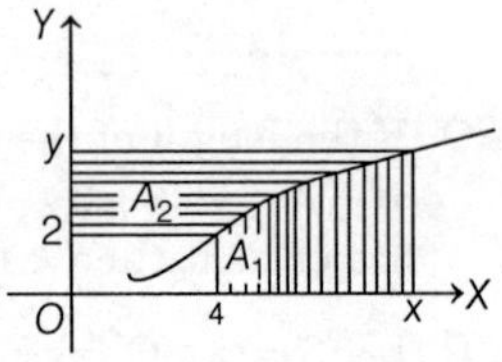

(a) (6, 21) (b) (8, 9)
(c) (10, –4) (d) (12, –15)

11 The equations of the sides AB, BC and CA of a ΔABC are $2x + y = 0$, $x + py = 39$ and $x - y = 3$ respectively and $P(2, 3)$ is its circumcentre. Then, which of the following is **NOT** true?

(a) $(AC)^2 = 9p$ (b) $(AC)^2 + p^2 = 136$
(c) $32 <$ area $(\Delta ABC) < 36$ (d) $34 <$ area $(\Delta ABC) < 38$

12 A circle C_1 passes through the origin O and has diameter 4 on the positive X-axis. The line $y = 2x$ gives a chord OA of circle C_1. Let C_2 be the circle with OA as a diameter. If the tangent to C_2 at the point A meets the X-axis at P and Y-axis at Q, then $QA : AP$ is equal to

(a) 1 : 4 (b) 1 : 5
(c) 2 : 5 (d) 1 : 3

13 If the length of the latus rectum of a parabola, whose focus is (a, a) and the tangent at its vertex is $x + y = a$, is 16, then $|a|$ is equal to

(a) $2\sqrt{2}$ (b) $2\sqrt{3}$ (c) $4\sqrt{2}$ (d) 4

14 If the length of the perpendicular drawn from the point $P(a, 4, 2)$, $a > 0$ on the line $\frac{x+1}{2} = \frac{y-3}{3} = \frac{z-1}{-1}$ is $2\sqrt{6}$ units and $Q(\alpha_1, \alpha_2, \alpha_3)$ is the image of the point P in this line, then $a + \sum_{i=1}^{3} \alpha_i$ is equal to

(a) 7 (b) 8 (c) 12 (d) 14

15 If the line of intersection of the planes $ax + by = 3$ and $ax + by + cz = 0$, $a > 0$ makes an angle 30° with the plane $y - z + 2 = 0$, then the direction cosines of the line are

(a) $\frac{1}{\sqrt{2}}, \frac{1}{\sqrt{2}}, 0$ (b) $\frac{1}{\sqrt{2}}, -\frac{1}{\sqrt{2}}, 0$

(c) $\frac{1}{\sqrt{5}}, -\frac{2}{\sqrt{5}}, 0$ (d) $\frac{1}{2}, -\frac{\sqrt{3}}{2}, 0$

16 Let X have a binomial distribution $B(n, p)$ such that the sum and the product of the mean and variance of X are 24 and 128, respectively. If $P(X > n - 3) = \frac{k}{2^n}$, then k is equal to

(a) 528 (b) 529
(c) 629 (d) 630

17 A six faced die is biased such that $3 \times P$ (a prime number) $= 6 \times P$ (a composite number) $= 2 \times P(1)$. Let X be a random variable that counts the number of times one gets a perfect square on some throws of this die. If the die is thrown twice, then the mean of X is

(a) $\frac{3}{11}$ (b) $\frac{5}{11}$ (c) $\frac{7}{11}$ (d) $\frac{8}{11}$

18 The angle of elevation of the top P of a vertical tower PQ of height 10 from a point A on the horizontal ground is 45°. Let R be a point on AQ and from a point B, vertically above R, the angle of elevation of P is 60°. If $\angle BAQ = 30°$, $AB = d$ and the area of the trapezium $PQRB$ is α, then the ordered pair (d, α) is

(a) $(10(\sqrt{3} - 1), 25)$ (b) $\left(10(\sqrt{3} - 1), \frac{25}{2}\right)$

(c) $(10(\sqrt{3} + 1), 25)$ (d) $\left(10(\sqrt{3} + 1), \frac{25}{2}\right)$

19 Let $S = \left\{ \theta \in \left(0, \frac{\pi}{2}\right) : \sum_{m=1}^{9} \sec\left(\theta + (m-1)\frac{\pi}{6}\right) \sec\left(\theta + \frac{m\pi}{6}\right) = -\frac{8}{\sqrt{3}} \right\}$. Then,

(a) $S = \left\{\frac{\pi}{12}\right\}$ (b) $S = \left\{\frac{2\pi}{3}\right\}$

(c) $\sum_{\theta \in S} \theta = \frac{\pi}{2}$ (d) $\sum_{\theta \in S} \theta = \frac{3\pi}{4}$

20 If the truth value of the statement $(P \wedge (\sim R)) \to ((\sim R) \wedge Q)$ is F, then the truth value of which of the following is F?

(a) $P \vee Q \to \sim R$ (b) $R \vee Q \to \sim P$
(c) $\sim (P \vee Q) \to \sim R$ (d) $\sim (R \vee Q) \to \sim P$

Section B : Numerical Type Questions

21 Consider a matrix $A = \begin{bmatrix} \alpha & \beta & \gamma \\ \alpha^2 & \beta^2 & \gamma^2 \\ \beta+\gamma & \gamma+\alpha & \alpha+\beta \end{bmatrix}$, where α, β, γ are three distinct natural numbers. If $\frac{\det(\text{adj}(\text{adj}(\text{adj}(\text{adj } A))))}{(\alpha-\beta)^{16}(\beta-\gamma)^{16}(\gamma-\alpha)^{16}} = 2^{32} \times 3^{16}$, then the number of such 3-tuples (α, β, γ) is ....

22 The number of functions f, from the set $A = \{x \in N : x^2 - 10x + 9 \le 0\}$ to the set $B = \{n^2 : n \in N\}$ such that $f(x) \le (x-3)^2 + 1$, for every $x \in A$, is ....

23 Let the 9th term in the binomial expansion of $(3+6x)^n$, in the increasing powers of $6x$, to be the greatest for $x = 3/2$, the least value of n is n_0. If k is the ratio of the coefficient of x^6 to the coefficient of x^3, then $k + n_0$ is equal to ....

24 $\frac{2^3-1^3}{1\times 7}+\frac{4^3-3^3+2^2-1^3}{2\times 11}+\frac{6^3-5^3+4^3-3^3+2^3-1^3}{3\times 15}+\ldots+\frac{(30)^3-(29)^3+(28)^3-(27)^3+\ldots+2^3-1^3}{15\times 63}$ is equal to ....

25 A water tank has the shape of a right circular cone with axis vertical and vertex downwards. Its semi-vertical angle is $\tan^{-1}\frac{3}{4}$. Water is poured in it at a constant rate of 6 cubic meter per hour. The rate (in square metre per hour), at which the wet curved surface area of the tank is increasing, when the depth of water in the tank is 4 m, is ....

26 For the curve $C: (x^2+y^2-3)+(x^2-y^2-1)^5=0$, the value of $3y'-y^3y''$, at the point (α, α), $\alpha > 0$, on C is equal to ....

27 Let $f(x) = \min\{[x-1], [x-2], \ldots\ldots, [x-10]\}$ where $[t]$ denotes on greatest integer $\le t$.

Then, $\int_0^{10} f(x)\,dx + \int_0^{10} (f(x))^2\,dx + \int_0^{10} |f(x)|\,dx$ is equal to ....

28 Let f be a differential function satisfying $f(x) = \frac{2}{\sqrt{3}}\int_0^{\sqrt{3}} f\left(\frac{\lambda^2 x}{3}\right) d\lambda$, $x > 0$ and $f(1) = \sqrt{3}$. If $y = f(x)$ passes through the point $(\alpha, 6)$, then α is equal to ....

29 A common tangent T to the curves $C_1 : \frac{x^2}{4}+\frac{y^2}{9}=1$ and $C_2 : \frac{x^2}{42}-\frac{y^2}{143}=1$ does not pass through the fourth quadrant. If T touches C_1 at (x_1, y_1) and C_2 at (x_2, y_2), then $|2x_1 + x_2|$ is equal to ....

30 Let $\mathbf{a}, \mathbf{b}$ and $\mathbf{c}$ be three non-coplanar vectors such that $\mathbf{a}\times\mathbf{b} = 4\mathbf{c}$, $\mathbf{b}\times\mathbf{c} = 9\mathbf{a}$ and $\mathbf{c}\times\mathbf{a} = \alpha\mathbf{b}$, $\alpha > 0$. If $|\mathbf{a}|+|\mathbf{b}|+|\mathbf{c}| = \frac{1}{36}$, then α is equal to ....

28 JULY 2022, SHIFT-I

Section A : Objective Type Questions

1 Let the solution curve of the differential equation $xdy = (\sqrt{x^2+y^2}+y)\,dx$, $x > 0$, intersect the line $x = 1$ at $y = 0$ and the line $x = 2$ at $y = \alpha$. Then, the value of α is

(a) $\frac{1}{2}$ (b) $\frac{3}{2}$ (c) $-\frac{3}{2}$ (d) $\frac{5}{2}$

2 Considering only the principal values of the inverse trigonometric functions, the domain of the function $f(x) = \cos^{-1}\left(\frac{x^2-4x+2}{x^2+3}\right)$ is

(a) $\left(-\infty, \frac{1}{4}\right]$ (b) $\left[-\frac{1}{4}, \infty\right)$

(c) $\left(\frac{-1}{3}, \infty\right)$ (d) $\left(-\infty, \frac{1}{3}\right]$

3 Let the vectors $\mathbf{a} = (1+t)\hat{\mathbf{i}} + (1-t)\hat{\mathbf{j}} + \hat{\mathbf{k}}$, $\mathbf{b} = (1-t)\hat{\mathbf{i}} + (1+t)\hat{\mathbf{j}} + 2\hat{\mathbf{k}}$ and $\mathbf{c} = t\hat{\mathbf{i}} - t\hat{\mathbf{j}} + \hat{\mathbf{k}}$, $t \in R$ be such that for $\alpha, \beta, \gamma \in R$, $\alpha\mathbf{a} + \beta\mathbf{b} + \gamma\mathbf{c} = \mathbf{0} \Rightarrow \alpha = \beta = \gamma = 0$. Then, the set of all values of t is

(a) A non-empty finite set (b) Equal to N

(c) Equal to $R - \{0\}$ (d) Equal to R

4 Considering the principal values of the inverse trigonometric function, the sum of all the solutions of the equation $\cos^{-1}(x) - 2\sin^{-1}(x) = \cos^{-1}(2x)$ is equal to

(a) 0 (b) 1 (c) $\frac{1}{2}$ (d) $-\frac{1}{2}$

5 Let the operations $*, \odot \in \{\wedge, \vee\}$. If $(p * q) \odot (p \odot \sim q)$ is a tautology, then the ordered pair $(*, \odot)$ is

(a) $(\vee, \wedge)$ (b) $(\vee, \vee)$ (c) $(\wedge, \wedge)$ (d) $(\wedge, \vee)$

6 Let a vector $\mathbf{a}$ has magnitude 9. Let a vector $\mathbf{b}$ be such that for every $(x, y) \in R \times R - \{(0, 0)\}$, the vector $(x\mathbf{a} + y\mathbf{b})$ is perpendicular to the vector $(6y\mathbf{a} - 18x\mathbf{b})$. Then, the value of $|\mathbf{a}\times\mathbf{b}|$ is equal to

(a) $9\sqrt{3}$ (b) $27\sqrt{3}$ (c) 9 (d) 81

7 For $t \in (0, 2\pi)$, if ABC is an equilateral triangle with vertices $A(\sin t - \cos t)$, $B(\cos t, \sin t)$ and $C(a, b)$ such that its orthocentre lies on a circle with centre $\left(1, \frac{1}{3}\right)$, then $(a^2 - b^2)$ is equal to

(a) $\frac{8}{3}$ (b) 8 (c) $\frac{77}{9}$ (d) $\frac{80}{9}$

8 For $\alpha \in N$, consider a relation R on N given by $R = \{(x, y) : 3x + \alpha y$ is a multiple of $7\}$. The relation R is an equivalence relation if and only if

(a) $\alpha = 14$

(b) α is a multiple of 4

(c) 4 is the remainder when α is divided by 10

(d) 4 is the remainder with α is divided by 7

9 Out of 60% female and 40% male candidates appearing in an exam, 60% candidates qualify it. The number of females qualifying the exam is twice the number of males qualifying it. A candidate is randomly chosen from the qualified candidates. The probability, that the chosen candidate is a female, is

(a) $\frac{3}{4}$ (b) $\frac{11}{16}$ (c) $\frac{23}{32}$ (d) $\frac{13}{16}$

10 If $y = y(x)$, $x \in \left(0, \frac{\pi}{2}\right)$ be the solution curve of the differential equation $(\sin^2 2x)\frac{dy}{dx} + (8\sin^2 2x + 2\sin 4x)y = 2e^{-4x}(2\sin 2x + \cos 2x)$, with $y\left(\frac{\pi}{4}\right) = e^{-\pi}$, then $y\left(\frac{\pi}{6}\right)$ is equal to

(a) $\frac{2}{\sqrt{3}}e^{-2\pi/3}$ (b) $\frac{2}{\sqrt{3}}e^{2\pi/3}$

(c) $\frac{1}{\sqrt{3}}e^{-2\pi/3}$ (d) $\frac{1}{\sqrt{3}}e^{2\pi/3}$

11 If the tangents drawn at the points P and Q on the parabola $y^2 = 2x - 3$ intersect at the point $R(0, 1)$, then the orthocentre of the ΔPQR is

(a) (0, 1) (b) (2, −1)

(c) (6, 3) (d) (2, 1)

12 Let C be the centre of the circle $x^2 + y^2 - x + 2y = \frac{11}{4}$ and P be a point on the circle. A line passes through the point C, makes an angle of $\frac{\pi}{4}$ with the line CP and intersects the circle at Q and R. Then, area of the ΔPQR (in unit^2) is

(a) 2 (b) $2\sqrt{2}$

(c) $8\sin\left(\frac{\pi}{8}\right)$ (d) $8\cos\left(\frac{\pi}{8}\right)$

13 The remainder when $7^{2022} + 3^{2022}$ is divided by 5 is

(a) 0 (b) 2 (c) 3 (d) 4

13 Let the matrix $A = \begin{vmatrix} 0 & 1 & 0 \\ 0 & 0 & 1 \\ 1 & 0 & 0 \end{vmatrix}$ and the matrix $B_0 = A^{49} + 2A^{98}$. If $B_n = \text{Adj}(B_{n-1})$ for all $n \geq 1$, then det (B_4) is equal to

(a) 3^{28} (b) 3^{30} (c) 3^{32} (d) 3^{36}

15 Let $S_1 = \left\{z_1 \in C : |z_1 - 3| = \frac{1}{2}\right\}$ and $S_2 = \{z_2 \in C : |z_2 - |z_2 + 1|| = |z_2 + |z_2 - 1||\}$. Then, for $z_1 \in S_1$ and $z_2 \in S_2$, the least value of $|z_2 - z_1|$ is

(a) 0 (b) $\frac{1}{2}$ (c) $\frac{3}{2}$ (d) $\frac{5}{2}$

16 The foot of the perpendicular from a point on the circle $x^2 + y^2 = 1$, $z = 0$ to the plane $2x + 3y + z = 6$ lies on which one of the following curves ?

(a) $(6x + 5y - 12)^2 + 4(3x + 7y - 8)^2 = 1, z = 6 - 2x - 3y$

(b) $(5x + 6y - 12)^2 + 4(3x + 5y - 9)^2 = 1, z = 6 - 2x - 3y$

(c) $(6x + 5y - 14)^2 + 9(3x + 5y - 7)^2 = 1, z = 6 - 2x - 3y$

(d) $(5x + 6y - 14)^2 + + 9(3x + 7y - 8)^2 = 1, z = 6 - 2x - 3y$

17 If the minimum value of $f(x) = \frac{5x^2}{2} + \frac{\alpha}{x^5}$, $x > 0$, is 14, then the value of α is equal to

(a) 32 (b) 64 (c) 128 (d) 256

18 Let α, β and γ be three positive real numbers. Let $f(x) = \alpha x^5 + \beta x^3 + \gamma x$, $x \in R$ and $g : R \to R$ be such that $g(f(x)) = x$ for all $x \in R$.
If $a_1, a_2, a_3, \ldots, a_n$ be in arithmetic progression with mean zero, then the value of $f\left(g\left(\frac{1}{n}\sum_{i=1}^{n} f(a_i)\right)\right)$ is equal to

(a) 0 (b) 3 (c) 9 (d) 27

19 Consider the sequence $a_1, a_2, a_3, \ldots$ such that $a_1 = 1$, $a_2 = 2$ and $a_{n+2} = \frac{2}{a_{n+1}} + a_n$ for $n = 1, 2, 3 \ldots$ If

$$\left(\frac{a_1 + \frac{1}{a_2}}{a_3}\right)\cdot\left(\frac{a_2 + \frac{1}{a_3}}{a_4}\right)\left(\frac{a_3 + \frac{1}{a_4}}{a_5}\right)\cdots\cdots\left(\frac{a_{30} + \frac{1}{a_{31}}}{a_{32}}\right)$$

$= 2^{\alpha}({}^{61}C_{31})$, then α is equal to

(a) −30 (b) −31 (c) −60 (d) −61

20 The minimum value of the twice differentiable function $f(x) = \int_0^x e^{x-t} f'(t)dt - (x^2 - x + 1)e^x$, $x \in R$ is

(a) $-\frac{2}{\sqrt{e}}$ (b) $-2\sqrt{e}$ (c) $-\sqrt{e}$ (d) $\frac{2}{\sqrt{e}}$

Section B : Numerical Type Questions

21 Let S be the set of all passwords which are six to eight characters long, where each character is either an alphabet from $\{A, B, C, D, E\}$ or a number from $\{1, 2, 3, 4, 5\}$ with the repetition of characters allowe(d) If the number of passwords in S whose atleast one character is a number from $\{1, 2, 3, 4, 5\}$ is $\alpha \times 5^6$, then α is equal to ....

22 Let $P(-2, -1, 1)$ and $Q\left(\frac{56}{17}, \frac{43}{17}, \frac{111}{17}\right)$ be the vertices of the rhombus $PRQS$. If the direction ratios of the diagonal RS are $\alpha, -1, \beta$ where both α and β are integers of minimum absolute values, then $\alpha^2 + \beta^2$ is equal to ....

23 Let $f : [0, 1] \to R$ be a twice differentiable function in (0, 1) such that $f(0) = 3$ and $f(1) = 5$. If the line $y = 2x + 3$ intersects the graph of f at only two distinct points in (0, 1), then the least number of points $x \in (0, 1)$ at which $f'(x) = 0$ is ....

24 If $\int_0^{\pi/3} \frac{15x^3}{\sqrt{1 + x^2 + \sqrt{(1 + x^2)^3}}} dx = \alpha\sqrt{2} + \beta\sqrt{3}$, where α, β are integers, then $\alpha + \beta$ is equal to ....

25 Let $A = \begin{bmatrix} 1 & -1 \\ 2 & \alpha \end{bmatrix}$ and $B = \begin{bmatrix} \beta & 1 \\ 1 & 0 \end{bmatrix}, \alpha, \beta \in R$. Let α_1 be the value of α which satisfies $(A + B)^2 = A^2 + \begin{bmatrix} 2 & 2 \\ 2 & 2 \end{bmatrix}$ and α_2 be the value of α which satisfies $(A + B)^2 = B^2$. Then, $|\alpha_1 - \alpha_2|$ is equal to ....

26 For p and $q \in R$, consider the real valued function $f(x) = (x - p)^2 - q, x \in R$ and $q > 0$, let a_1, a_2, a_3 and a_4 be in an arithmetic progression with mean p and positive common difference. If $|f(a_i)| = 500$ for all $i = 1, 2, 3$ and 4, then the absolute difference between the roots of $f(x) = 0$ is ....

27 For the hyperbola $H : x^2 - y^2 = 1$ and the ellipse $E : \frac{x^2}{a^2} + \frac{y^2}{b^2} = 1, a > b > 0$, let the

(1) the eccentricity of E be reciprocal of the eccentricity of H, and

(2) the line $y = \sqrt{\frac{5}{2}}x + K$ be a common tangent of E and H.

Then, $4(a^2 + b^2)$ is equal to ....

28 Let $x_1, x_2, x_3, \ldots, x_{20}$ be in geometric progression with $x_1 = 3$ and the common ratio 1/2. A new data is constructed replacing each x_i by $(x_i - i)^2$. If $\bar{x}$ is the mean of new data, then the greatest integer less than or equal to $\bar{x}$ is ....

29 $\lim_{x \to 0} \left(\frac{(x + 2\cos x)^3 + 2(x + 2\cos x)^2 + 3\sin(x + 2\cos x)}{(x + 2)^3 + 2(x + 2)^2 + 3\sin(x + 2)} \right)^{100/x}$ is equal to ....

30 The sum of all real values of x for which $\frac{3x^2 - 9x + 17}{x^2 + 3x + 10} = \frac{5x^2 - 7x + 19}{3x^2 + 5x + 12}$ is equal to ....

28 JULY 2022, SHIFT-II

Section A : Objective Type Questions

1 Let $S = \left\{ x \in [-6, 3] - \{-2, 2\} : \frac{|x + 3| - 1}{|x| - 2} \geq 0 \right\}$ and $T = \{x \in Z : x^2 - 7|x| + 9 \leq 0\}$. Then, the number of elements in $S \cap T$ is

(a) 7 (b) 5
(c) 4 (d) 3

2 Let α and β be the roots of the equation $x^2 - \sqrt{2}x + \sqrt{6} = 0$ and $\frac{1}{\alpha^2} + 1, \frac{1}{\beta^2} + 1$ be the roots of the equation $x^2 + ax + b = 0$. Then, the roots of the equation $x^2 - (a + b - 2)x + (a + b + 2) = 0$ are

(a) non-real complex number
(b) real and both negative
(c) real and both positive
(d) real and exactly one of them is positive

3 Let A and B be any two 3×3 symmetric and skew-symmetric matrices, respectively. Then, which of the following is not true?

(a) $A^4 - B^4$ is a symmetric matrix
(b) $AB - BA$ is a symmetric matrix
(c) $B^5 - A^5$ is a skew-symmetric matrix
(d) $AB + BA$ is a skew-symmetric matrix

4 Let $f(x) = ax^2 + bx + c$ be such that $f(1) = 3, f(-2) = \lambda$ and $f(3) = 4$. If $f(0) + f(1) + f(-2) + f(3) = 14$, then λ is equal to

(a) -4 (b) $\frac{13}{2}$
(c) $\frac{23}{2}$ (d) 4

5 The function $f : R \to R$ defined by $f(x) = \lim_{n \to \infty} \frac{\cos(2\pi x) - x^{2n}\sin(x - 1)}{1 + x^{2n+1} - x^{2n}}$ is continuous for all x in

(a) $R - \{-1\}$ (b) $R - \{-1, 1\}$
(c) $R - \{1\}$ (d) $R - \{0\}$

6 The function $f(x) = xe^{x(1-x)}, x \in R$ is

(a) increasing in $\left(-\frac{1}{2}, 1\right)$ (b) decreasing in $\left(\frac{1}{2}, 2\right)$
(c) increasing in $\left(-1, -\frac{1}{2}\right)$ (d) decreasing in $\left(-\frac{1}{2}, \frac{1}{2}\right)$

7 The sum of the absolute maximum and absolute minimum values of the function $f(x) = \tan^{-1}(\sin x - \cos x)$ in the interval $[0, \pi]$ is

(a) 0 (b) $\tan^{-1}\left(\frac{1}{\sqrt{2}}\right) - \frac{\pi}{4}$
(c) $\cos^{-1}\left(\frac{1}{\sqrt{3}}\right) - \frac{\pi}{4}$ (d) $\frac{-\pi}{12}$

8 Let $x(t) = 2\sqrt{2}\cos t\sqrt{\sin 2t}$ and $y(t) = 2\sqrt{2}\sin t\sqrt{\sin 2t}, t \in \left(0, \frac{\pi}{2}\right)$.

Then, $\frac{1 + \left(\frac{dy}{dx}\right)^2}{\frac{d^2y}{dx^2}}$ at $t = \frac{\pi}{4}$ is equal to

(a) $\frac{-2\sqrt{2}}{3}$ (b) $\frac{2}{3}$ (c) $\frac{1}{3}$ (d) $\frac{-2}{3}$

9 Let $I_n(x) = \int_0^x \frac{1}{(t^2+5)^n} dt, n = 1, 2, 3, \ldots$ Then,

(a) $50I_6 - 9I_5 = xI'(5)$ (b) $50I_6 - 11I_5 = xI'(5)$
(c) $50I_6 - 9I_5 = I'(5)$ (d) $50I_6 - 11I_5 = I'(5)$

10 The area enclosed by the curves $y = \log_e (x + e^2)$, $x = \log_e\left(\frac{2}{y}\right)$ and $x = \log_e 2$, above the line $y = 1$ is

(a) $2 + e - \log_e 2$ (b) $1 + e - \log_e 2$
(c) $e - \log_e 2$ (d) $1 + \log_e 2$

11 Let $y = y(x)$ be the solution curve of the differential equation $\frac{dy}{dx} + \frac{1}{x^2-1} y = \left(\frac{x-1}{x+1}\right)^{1/2}$, $x > 1$ passing through the point $\left(2, \sqrt{\frac{1}{3}}\right)$. Then, $\sqrt{7}y(8)$ is

(a) $11 + 6\log_e 3$ (b) 19
(c) $12 - 2\log_e 3$ (d) $19 - 6\log_e 3$

12 The differential equation of the family of circles passing through the points (0, 2) and (0, −2) is

(a) $2xy\frac{dy}{dx} + (x^2 - y^2 + 4) = 0$
(b) $2xy\frac{dy}{dx} + (x^2 + y^2 - 4) = 0$
(c) $2xy\frac{dy}{dx} + (y^2 - x^2 + 4) = 0$
(d) $2xy\frac{dy}{dx} - (x^2 - y^2 + 4) = 0$

13 Let the tangents at two points A and B on the circle $x^2 + y^2 - 4x + 3 = 0$ meet at origin $O(0, 0)$. Then, the area of the ΔOAB is

(a) $\frac{3\sqrt{3}}{2}$ (b) $\frac{3\sqrt{3}}{4}$ (c) $\frac{3}{2\sqrt{3}}$ (d) $\frac{3}{4\sqrt{3}}$

14 Let the hyperbola $H: \frac{x^2}{a^2} - \frac{y^2}{b^2} = 1$ passes through the point $(2\sqrt{2}, -2\sqrt{2})$. A parabola is drawn whose focus is same as the focus of H with positive abscissa and the directrix of the parabola passes through the other focus of H. If the length of the latusrectum of the parabola is e times the length of the latusrectum of H, where e is the eccentricity of H, then which of the following points lies on the parabola?

(a) $(2\sqrt{3}, 3\sqrt{2})$ (b) $(3\sqrt{3}, -6\sqrt{2})$
(c) $(\sqrt{3}, -\sqrt{6})$ (d) $(3\sqrt{6}, 6\sqrt{2})$

15 Let the lines $\frac{x-1}{\lambda} = \frac{y-2}{1} = \frac{z-3}{2}$ and $\frac{x+26}{-2} = \frac{y+18}{3} = \frac{z+28}{\lambda}$ be coplanar and P be the plane containing these two lines. Then, which of the following points does not lie on P?

(a) (0, −2, −2) (b) (−5, 0, −1)
(c) (3, −1, 0) (d) (0, 4, 5)

16 A plane P is parallel to two lines whose direction ratios are −2, 1, −3 and −1, 2, −2 and it contains the point (2, 2, −2). Let P intersect the coordinate axes at the points A, B and C making the intercepts α, β and γ. If V is the volume of the tetrahedron $OABC$, where O is the origin and $p = \alpha + \beta + \gamma$, then the ordered pair (V, p) is equal to

(a) (48, −13)
(b) (24, −13)
(c) (48, 11)
(d) (24, −5)

17. Let S be the set of all $a \in R$ for which the angle between the vectors $\mathbf{u} = a(\log_e b)\hat{\mathbf{i}} - 6\hat{\mathbf{j}} + 3\hat{\mathbf{k}}$ and $\mathbf{v} = (\log_e b)\hat{\mathbf{i}} + 2\hat{\mathbf{j}} + 2a(\log_e b)\hat{\mathbf{k}}$, $(b > 1)$ is acute. Then, S is equal to

(a) $\left(-\infty, -\frac{4}{3}\right)$ (b) ϕ
(c) $\left(-\frac{4}{3}, 0\right)$ (d) $\left(\frac{12}{7}, \infty\right)$

18 A horizontal park is in the shape of a ΔOAB with $AB = 16$. A vertical lamp post OP is erected at the point O such that $\angle PAO = \angle PBO = 15°$ and $\angle PCO = 45°$, where C is the mid-point of AB. Then, $(OP)^2$ is equal to

(a) $\frac{32}{\sqrt{3}}(\sqrt{3} - 1)$
(b) $\frac{32}{\sqrt{3}}(2 - \sqrt{3})$
(c) $\frac{16}{\sqrt{3}}(\sqrt{3} - 1)$
(d) $\frac{16}{\sqrt{3}}(2 - \sqrt{3})$

19 Let A and B be two events such that $P(B/A) = \frac{2}{5}$, $P(A/B) = \frac{1}{7}$ and $P(A \cap B) = \frac{1}{9}$. Consider

Statement I $P(A' \cup B) = 5/6$,
Statement II $P(A' \cap B') = 1/18$. Then,

(a) Both statement I and statement II are true
(b) Both statement I and statement II are false
(c) Only statement I is true
(d) Only statement II is true

20 Let
p: Ramesh listens to music.
q: Ramesh is out of his village.
r: It is Sunday.
s: It is Saturday.

Then, the statement "Ramesh listens to music only, if he is in his village and it is Sunday or Saturday" can be expressed as

(a) $((\sim q) \wedge (r \vee s)) \Rightarrow p$
(b) $(q \wedge (r \vee s)) \Rightarrow p$
(c) $p \Rightarrow (q \wedge (r \vee s))$
(d) $p \Rightarrow ((\sim q) \wedge (r \vee s))$

Section B : Numerical Type Questions

21 Let the coefficients of the middle terms in the expansion of $\left(\frac{1}{\sqrt{6}}+\beta x\right)^4$, $(1-3\beta x)^2$ and $\left(1-\frac{\beta}{2}x\right)^6$, $\beta>0$, respectively form the first three terms of an AP. If d is the common difference of this AP, then $50-\frac{2d}{\beta^2}$ is equal to ....

22 A class contains b boys and g girls. If the number of ways of selecting 3 boys and 2 girls from the class is 168, then $b+3g$ is equal to ....

23 Let the tangents at the points P and Q on the ellipse $x^2/2+y^2/4=1$ meet at the point $R(\sqrt{2}, 2\sqrt{2}-2)$. If S is the focus of the ellipse on its negative major axis, then SP^2+SQ^2 is equal to ....

24 If $1+(2+{}^{49}C_1+{}^{49}C_2+\ldots\ldots+{}^{49}C_{49})$ $({}^{50}C_2+{}^{50}C_4+\ldots.+{}^{50}C_{50})$ is equal to $2^n\cdot m$, where m is odd, then $n+m$ is equal to ....

25 Two tangent lines l_1 and l_2 are drawn from the point $(2,0)$ to the parabola $2y^2=-x$. If the lines l_1 and l_2 are also tangent to the circle $(x-5)^2+y^2=r$, then $17r$ is equal to ....

26 If $\frac{6}{3^{12}}+\frac{10}{3^{11}}+\frac{20}{3^{10}}+\frac{40}{3^9}+\ldots.+\frac{10240}{3}=2^n\cdot m$ where m is odd, then $m\cdot n$ is equal to ....

27 Let $S=\left[-\pi,\frac{\pi}{2}\right]-\left[-\frac{\pi}{2},-\frac{\pi}{4},-\frac{3\pi}{4},\frac{\pi}{4}\right]$.
Then, the number of elements in the set $A=\{\theta\in S:\tan\theta\,(1+\sqrt{5}\tan(2\theta))=\sqrt{5}-\tan(2\theta)\}$ is ....

28 Let $z=a+ib$, $b\neq 0$ be complex numbers satisfying $z^2=\bar{z}\cdot 2^{1-|Z|}$. Then, the least value of $n\in N$, such that $z^n=(z+1)^n$, is equal to ....

29 A bag contains 4 white and 6 black balls. Three balls are drawn at random from the bag. Let X be the number of white balls, among the drawn balls. If σ^2 is the variance of X, then $100\sigma^2$ is equal to ....

30 The value of the integral $\int_0^{\pi/2} 60\frac{\sin(6x)}{\sin x}\,dx$ is equal to ....

29 JULY 2022, SHIFT-I

Section A : Objective Type Questions

1 Let R be a relation from the set $\{1,2,3,\ldots\ldots,60\}$ to itself such that $R=\{(a,b): b=pq$, where $p,q\geq 3$ are prime numbers$\}$. Then, the number of elements in R is

(a) 600 (b) 660 (c) 540 (d) 720

2 If $z=2+3i$, then $z^5+(\bar{z})^5$ is equal to

(a) 244 (b) 224 (c) 245 (d) 265

3 Let A and B be two 3×3 non-zero real matrices such that AB is a zero matrix. Then,

(a) the system of linear equation $AX=0$ has a unique solution
(b) the system of linear equations $AX=0$ has infinitely many solutions
(c) B is an invertible matrix
(d) Adj (A) is an invertible matrix

4 If $\frac{1}{(20-a)(40-a)}+\frac{1}{(40-a)(60-a)}+\ldots.+\frac{1}{(180-a)(200-a)}=\frac{1}{256}$, then the maximum value of a is

(a) 198 (b) 202 (c) 212 (d) 218

5 If $\lim_{x\to 0}\frac{\alpha e^x+\beta e^{-x}+\gamma\sin x}{x\sin^2 x}=\frac{2}{3}$, where α,β and $\gamma\in R$, then which of the following is not correct ?

(a) $\alpha^2+\beta^2+\gamma^2=6$ (b) $\alpha\beta+\beta\gamma+\gamma\alpha+1=0$
(c) $\alpha\beta^2+\beta\gamma^2+\gamma\alpha^2+3=0$ (d) $\alpha^2-\beta^2+\gamma^2=4$

6 The integral $\int_0^{\pi/2}\frac{1}{3+2\sin x+\cos x}\,dx$ is equal to

(a) $\tan^{-1}(2)$ (b) $\tan^{-1}(2)-\frac{\pi}{4}$
(c) $\frac{1}{2}\tan^{-1}(2)-\frac{\pi}{8}$ (d) $\frac{1}{2}$

7 Let the solution curve $y=y(x)$ of the differential equation $(1+e^{2x})\left(\frac{dy}{dx}+y\right)=1$ passes through the point $\left(0,\frac{\pi}{2}\right)$. Then, $\lim_{x\to\infty}e^x y(x)$ is equal to

(a) $\frac{\pi}{4}$ (b) $\frac{3\pi}{4}$
(c) $\frac{\pi}{2}$ (d) $\frac{3\pi}{2}$

8 Let a line L passes through the point of intersection of the lines $bx+10y-8=0$ and $2x-3y=0$, $b\in R-\left\{\frac{4}{3}\right\}$. If the line L also passes through the point $(1,1)$ and touches the circle $17(x^2+y^2)=16$, then the eccentricity of the ellipse $\frac{x^2}{5}+\frac{y^2}{b^2}=1$ is

(a) $\frac{2}{\sqrt{5}}$ (b) $\sqrt{\frac{3}{5}}$
(c) $\frac{1}{\sqrt{5}}$ (d) $\sqrt{\frac{2}{5}}$

9 If the foot of the perpendicular from the point $A(-1, 4, 3)$ on the plane $P: 2x + my + nz = 4$, is $\left(-2, \frac{7}{2}, \frac{3}{2}\right)$, then the distance of the point A from the plane P, measured parallel to a line with direction ratios $3, -1, -4$ is equal to

(a) 1 (b) $\sqrt{26}$ (c) $2\sqrt{2}$ (d) $\sqrt{14}$

10 Let $\mathbf{a} = 3\hat{\mathbf{i}} + \hat{\mathbf{j}}$ and $\mathbf{b} = \hat{\mathbf{i}} + 2\hat{\mathbf{j}} + \hat{\mathbf{k}}$. Let $\mathbf{c}$ be a vector satisfying $\mathbf{a} \times (\mathbf{b} \times \mathbf{c}) = \mathbf{b} + \lambda\mathbf{c}$. If $\mathbf{b}$ and $\mathbf{c}$ are non-parallel, then the value of λ is

(a) -5 (b) 5 (c) 1 (d) -1

11 The angle of elevation of the top of a tower from a point A due North of it is α and from a point B at a distance of 9 units due West of A is $\cos^{-1}\left(\frac{3}{\sqrt{13}}\right)$. If the distance of the point B from the tower is 15 units, then $\cot\alpha$ is equal to

(a) $\frac{6}{5}$ (b) $\frac{9}{5}$

(c) $\frac{4}{3}$ (d) $\frac{7}{3}$

12 The statement $(p \wedge q) \Rightarrow (p \wedge r)$ is equivalent to

(a) $q \Rightarrow (p \wedge r)$ (b) $p \Rightarrow (p \wedge r)$

(c) $(p \wedge r) \Rightarrow (p \wedge q)$ (d) $(p \wedge q) \Rightarrow r$

13 Let the circumcentre of a triangle with vertices $A(a, 3)$, $B(b, 5)$ and $C(a, b)$, $ab > 0$ be $P(1, 1)$. If the line AP intersects the line BC at the point $Q(k_1, k_2)$, then $k_1 + k_2$ is equal to

(a) 2 (b) $\frac{4}{7}$

(c) $\frac{2}{7}$ (d) 4

14 Let $\mathbf{a}$ and $\mathbf{b}$ be two unit vectors such that the angle between them is $\frac{\pi}{4}$. If θ is the angle between the vectors $(\mathbf{a} + \mathbf{b})$ and $(\mathbf{a} + 2\mathbf{b} + 2(\mathbf{a} \times \mathbf{b}))$. Then, the value of $164\cos^2\theta$ is equal to

(a) $90 + 27\sqrt{2}$ (b) $45 + 18\sqrt{2}$

(c) $90 + 3\sqrt{2}$ (d) $54 + 90\sqrt{2}$

15 If $f(\alpha) = \int_1^{\alpha} \frac{\log_{10} t}{1 + t} dt$, $\alpha > 0$, then $f(e^3) + f(e^{-3})$ is equal to

(a) 9 (b) $\frac{9}{2}$

(c) $\frac{9}{\log_e(10)}$ (d) $\frac{9}{2\log_e(10)}$

16 The area of the region

$\{(x, y): |x - 1| \le y \le \sqrt{5 - x^2}\}$ is equal to

(a) $\frac{5}{2}\sin^{-1}\left(\frac{3}{5}\right) - \frac{1}{2}$ (b) $\frac{5\pi}{4} - \frac{3}{2}$

(c) $\frac{3\pi}{4} + \frac{3}{2}$ (d) $\frac{5\pi}{4} - \frac{1}{2}$

17 Let the focal chord of the parabola $P: y^2 = 4x$ along the line $L: y = mx + c$, $m > 0$ meet the parabola at the points M and N. Let the line L be a tangent to the hyperbola $H: x^2 - y^2 = 4$. If O is the vertex of P and F is the focus of H on the positive X-axis, then the area of the quadrilateral $OMFN$ is

(a) $2\sqrt{6}$ (b) $2\sqrt{14}$

(c) $4\sqrt{6}$ (d) $4\sqrt{14}$

18 The number of points, where the function $f: R \to R$, $f(x) = |x - 1|\cos|x - 2|\sin|x - 1| + (x - 3)|x^2 - 5x + 4|$, is not differentiable, is

(a) 1 (b) 2

(c) 3 (d) 4

19 Let $S = \{1, 2, 3, \ldots, 2022\}$. Then, the probability, that a randomly chosen number n from the set S such that HCF $(n, 2022) = 1$, is

(a) $\frac{128}{1011}$ (b) $\frac{166}{1011}$

(c) $\frac{127}{337}$ (d) $\frac{112}{337}$

20 Let $f(x) = 3^{(x^2 - 2)^3 + 4}$, $x \in R$. Then, which of the following statements are true?

$P: x = 0$ is a point of local minima of f

$Q: x = \sqrt{2}$ is a point of inflection of f

$R: f$ is increasing for $x > \sqrt{2}$

(a) Only P and Q

(b) Only P and R

(c) Only Q and R

(d) All, P, Q and R

Section B : Numerical Type Questions

21 Let $S = \{\theta \in (0, 2\pi): 7\cos^2\theta - 3\sin^2\theta - 2\cos^2 2\theta = 2\}$. Then, the sum of roots of all the equations $x^2 - 2(\tan^2\theta + \cot^2\theta)x + 6\sin^2\theta = 0$, $\theta \in S$ is......

22 Let the mean and the variance of 20 observations $x_1, x_2, \ldots, x_{20}$ be 15 and 9, respectively. For $\alpha \in R$, if the mean of $(x_1 + \alpha)^2, (x_2 + \alpha)^2, \ldots, (x_{20} + \alpha)^2$ is 178, then the square of the maximum value of α is equal to ....

23 Let a line with direction ratios $a, -4a, -7$ be perpendicular to the lines with direction ratios $3, -1, 2b$ and $b, a, -2$. If the point of intersection of the line $\frac{x + 1}{a^2 + b^2} = \frac{y - 2}{a^2 - b^2} = \frac{z}{1}$ and the plane $x - y + z = 0$ is (α, β, γ), then $\alpha + \beta + \gamma$ is equal to ......

24 Let $a_1, a_2, a_3, \ldots$ be an AP. If $\sum_{r=1}^{\infty} \frac{a_r}{2^r} = 4$, then $4a_2$ is equal to ....

25 Let the ratio of the fifth term from the beginning to the fifth term from the end in the binomial expansion of $\left(\sqrt[4]{2}+\frac{1}{\sqrt[4]{3}}\right)^n$, in the increasing powers of $\frac{1}{\sqrt[4]{3}}$ be $\sqrt[4]{6}:1$. If the sixth term from the beginning is $\frac{\alpha}{\sqrt[4]{3}}$, then α is equal to ....

26 The number of matrices of order 3×3, whose entries are either 0 or 1 and the sum of all the entries is a prime number, is ....

27 Let p and $p+2$ be prime numbers and let

$$\Delta=\begin{vmatrix} P! & (P+1)! & (P+2)! \\ (P+1)! & (P+2)! & (P+3)! \\ (P+2)! & (P+3)! & (P+4)! \end{vmatrix}$$

Then, the sum of the maximum values of α and β such that P^{α} and $(P+2)^{\beta}$ divide Δ, is ......

28 If $\frac{1}{2\times3\times4}+\frac{1}{3\times4\times5}+\frac{1}{4\times5\times6}+\ldots+\frac{1}{100\times101\times102}=\frac{K}{101}$, then $34K$ is equal to

29 Let $S=\{4,6,9\}$ and $T=\{9,10,11,\ldots,1000\}$. If $A=\{a_1+a_2+\ldots+a_k : k\in N, a_1,a_2,a_3,\ldots,a_k\in S\}$, then the sum of all the elements in the set $T-A$ is equal to ....

30 Let the mirror image of a circle $C_1: x^2+y^2-2x-6y+\alpha=0$ in line $y=x+1$ be $C_2: 5x^2+5y^2+10gx+10fy+38=0$. If r is the radius of circle C_2, then $\alpha+6r^2$ is equal to

29 JULY 2022, SHIFT-II

Section A : Objective Type Questions

1 If $z\neq0$ be a complex number such that $\left|z-\frac{1}{z}\right|=2$, then the maximum value of $|z|$ is

(a) $\sqrt{2}$ (b) 1 (c) $\sqrt{2}-1$ (d) $\sqrt{2}+1$

2 Which of the following matrices can not be obtained from the matrix $\begin{bmatrix}-1 & 2\\ 1 & -1\end{bmatrix}$ by a single elementary row operation?

(a) $\begin{bmatrix}0 & 1\\ 1 & -1\end{bmatrix}$ (b) $\begin{bmatrix}1 & -1\\ -1 & 2\end{bmatrix}$ (c) $\begin{bmatrix}-1 & 2\\ -2 & 7\end{bmatrix}$ (d) $\begin{bmatrix}-1 & 2\\ -1 & 3\end{bmatrix}$

3 If the system of equations

$$x+y+z=6,$$
$$2x+5y+\alpha z=\beta$$
and $$x+2y+3z=14$$

has infinitely many solutions, then $\alpha+\beta$ is equal to

(a) 8 (b) 36 (c) 44 (d) 48

4 Let the function

$$f(x)=\begin{cases}\dfrac{\log_e(1+5x)-\log_e(1+\alpha x)}{x} & ;\text{ if } x\neq0\\ 10 & ;\text{ if } x=0\end{cases}$$

be continuous at $x=0$. Then, α is equal to

(a) 10 (b) -10 (c) 5 (d) -5

5 If $[t]$ denotes the greatest integer $\leq t$, then the value of $\int_0^1 [2x-|3x^2-5x+2|+1]\,dx$ is

(a) $\frac{\sqrt{37}+\sqrt{13}-4}{6}$ (b) $\frac{\sqrt{37}-\sqrt{13}-4}{6}$

(c) $\frac{-\sqrt{37}-\sqrt{13}+4}{6}$ (d) $\frac{-\sqrt{37}+\sqrt{13}+4}{6}$

6 Let $\{a_n\}_{n=0}^{\infty}$ be a sequence such that $a_0=a_1=0$ and $a_{n+2}=3a_{n+1}-2a_n+1, \forall\, n\geq0$. Then, $a_{25}a_{23}-2a_{25}a_{22}-2a_{23}a_{24}+4a_{22}a_{24}$ is equal to

(a) 483 (b) 528

(c) 575 (d) 624

7 $\sum_{r=1}^{20}(r^2+1)(r!)$ is equal to

(a) $22!-21!$

(b) $22!-2(21!)$

(c) $21!-2(20!)$

(d) $21!-20!$

8 For $I(x)=\int\frac{\sec^2 x-2022}{\sin^{2022}x}dx$, if $I\left(\frac{\pi}{4}\right)=2^{1011}$, then

(a) $3^{1010}I\left(\frac{\pi}{3}\right)-I\left(\frac{\pi}{6}\right)=0$ (b) $3^{1010}I\left(\frac{\pi}{6}\right)-I\left(\frac{\pi}{3}\right)=0$

(c) $3^{1011}I\left(\frac{\pi}{3}\right)-I\left(\frac{\pi}{6}\right)=0$ (d) $3^{1011}I\left(\frac{\pi}{6}\right)-I\left(\frac{\pi}{3}\right)=0$

9 If the solution curve of the differential equation $\frac{dy}{dx}=\frac{x+y-2}{x-y}$ passes through the point (2, 1) and $(k+1, 2)$, $k>0$, then

(a) $2\tan^{-1}\left(\frac{1}{k}\right)=\log_e(k^2+1)$

(b) $\tan^{-1}\left(\frac{1}{k}\right)=\log_e(k^2+1)$

(c) $2\tan^{-1}\left(\frac{1}{k+1}\right)=\log_e(k^2+2k+2)$

(d) $2\tan^{-1}\left(\frac{1}{k}\right)=\log_e\left(\frac{k^2+1}{k^2}\right)$

10 Let $y = y(x)$ be the solution curve of the differential equation $\frac{dy}{dx} + \left(\frac{2x^2 + 11x + 13}{x^3 + 6x^2 + 11x + 6}\right) y = \frac{(x+3)}{x+1}$, $x > -1$, which passes through the point (0, 1). They $y(1)$ is equal to

(a) 1/2 (b) 3/2
(c) 5/2 (d) 7/2

11 Let m_1, m_2 be the slopes of two adjacent sides of a square of side a such that $a^2 + 11a + 3(m_1^2 + m_2^2) = 220$. If one vertex of the square is $(10(\cos\alpha - \sin\alpha),$ $10(\sin\alpha + \cos\alpha))$, where $\alpha \in (0, \pi/2)$ and the equation of one diagonal is $(\cos\alpha - \sin\alpha)x + (\sin\alpha + \cos\alpha)y = 10$, then $72(\sin^4\alpha + \cos^4\alpha) + a^2 - 3a + 13$ is equal to

(a) 119 (b) 128
(c) 145 (d) 155

12 The number of elements in the set $S = \left\{x \in R : 2\cos\left(\frac{x^2 + x}{6}\right) = 4^x + 4^{-x}\right\}$ is

(a) 1 (b) 3
(c) 0 (d) infinite

13 Let $A(\alpha, -2)$, $B(\alpha, 6)$ and $C\left(\frac{\alpha}{4}, -2\right)$ be vertices of a ΔABC. If $\left(5, \frac{\alpha}{4}\right)$ is the circumcentre of ΔABC, then which of the following is NOT correct about ΔABC

(a) area is 24 units (b) perimeter is 25 units
(c) circumradius is 5 units (d) inradius is 2 units

14 Let Q be the foot of perpendicular drawn from the point $P(1, 2, 3)$ to the plane $x + 2y + z = 14$. If R is a point on the plane such that $\angle PRQ = 60°$, then the area of ΔPQR is equal to

(a) $\frac{\sqrt{3}}{2}$ (b) $\sqrt{3}$
(c) $2\sqrt{3}$ (d) 3

15 If (2, 3, 9), (5, 2, 1), (1, λ, 8) and (λ, 2, 3) are coplanar, then the product of all possible values of λ is

(a) $\frac{21}{2}$ (b) $\frac{59}{8}$
(c) $\frac{57}{8}$ (d) $\frac{95}{8}$

16 Bag I contains 3 red, 4 black and 3 white balls and bag II contains 2 red, 5 black and 2 white balls. One ball is transferred from bag I to bag II and then a ball is draw from bag II. The ball so drawn is found to be black in colour. Then, the probability, that the transferred ball is red, is

(a) $\frac{4}{9}$ (b) $\frac{5}{18}$ (c) $\frac{1}{6}$ (d) $\frac{3}{10}$

17 Let $S = \{z = x + iy : |z - 1 + i| \geq |z|, |z| < 2, |z + i| = |z - 1|\}$. Then, the set of all values of x, for which $w = 2x + iy \in S$ for some $y \in R$, is

(a) $\left(-\sqrt{2}, \frac{1}{2\sqrt{2}}\right]$ (b) $\left(-\frac{1}{\sqrt{2}}, \frac{1}{4}\right]$
(c) $\left(-\sqrt{2}, \frac{1}{2}\right]$ (d) $\left(-\frac{1}{\sqrt{2}}, \frac{1}{2\sqrt{2}}\right]$

18 Let **a**, **b** and **c** be three coplanar concurrent vectors such that angles between any two of them is same. If the product of their magnitudes is 14 and $(\mathbf{a} \times \mathbf{b}) \cdot (\mathbf{b} \times \mathbf{c}) + (\mathbf{b} \times \mathbf{c}) \cdot (\mathbf{c} \times \mathbf{a}) + (\mathbf{c} \times \mathbf{a}) \cdot (\mathbf{a} \times \mathbf{b}) = 168$, then $|\mathbf{a}| + |\mathbf{b}| + |\mathbf{c}|$ is equal to

(a) 10 (b) 14 (c) 16 (d) 18

19 The domain of the function $f(x) = \sin^{-1}\left(\frac{x^2 - 3x + 2}{x^2 + 2x + 7}\right)$ is

(a) $[1, \infty)$ (b) $(-1, 2]$ (c) $[-1, \infty)$ (d) $(-\infty, 2]$

20 The statement $(p \Rightarrow q) \vee (p \Rightarrow r)$ is NOT equivalent to

(a) $(p \wedge (\sim r)) \Rightarrow q$ (b) $(\sim q) \Rightarrow ((-r) \vee p)$
(c) $p \Rightarrow (q \vee r)$ (d) $(p \wedge (\sim q)) \Rightarrow r$

Section B : Numerical Type Questions

21 The sum and product of the mean and variance of a binomial distribution are 82.5 and 1350, respectively. Then, the number of trials in the binomial distribution is ....

22 Let α and β (α > β) be the roots of the quadratic equation $x^2 - x - 4 = 0$.

If $P_n = \alpha^n - \beta^n$, $n \in N$, then $\frac{P_{15}P_{16} - P_{14}P_{16} - P_{15}^2 + P_{14}P_{15}}{P_{13}P_{14}}$ is equal to ....

23 Let $X = \begin{bmatrix} 1 \\ 1 \\ 1 \end{bmatrix}$ and $A = \begin{bmatrix} -1 & 2 & 3 \\ 0 & 1 & 6 \\ 0 & 0 & -1 \end{bmatrix}$. For $k \in N$, if $X' A^k X = 33$, then k is equal to ....

24 The number of natural numbers lying between 1012 and 23421 that can be formed using the digits 2, 3, 4, 5, 6 (repetition of digits is not allowed) and divisible by 55 is ....

25 If $\sum_{k=1}^{10} k^2 ({}^{10}C_k)^2 = 22000L$, then L is equal to ....

26 If $[t]$ denotes the greatest integer $\leq t$, then number of points, at which the function $f(x) = 4|2x + 3| + 9\left[x + \frac{1}{2}\right] - 12[x + 20]$ is not differentiable in the open interval (–20, 20), is ....

27 If the tangent to the curve $y = x^3 - x^2 + x$ at the point (a, b) is also tangent to the curve $y = 5x^2 + 2x - 25$ at the point $(2, -1)$, then $|2a + 9b|$ is equal to ....

28 Let AB be a chord of length 12 of the circle $(x-2)^2 + (y+1)^2 = 169/4$.
If tangents drawn to the circle at points A and B intersect at the point P, then five times the distance of point P from chord AB is equal to ....

29 Let $\mathbf{a}$ and $\mathbf{b}$ be two vectors such that $|\mathbf{a} + \mathbf{b}|^2 = |\mathbf{a}|^2 + 2|\mathbf{b}|^2$, $\mathbf{a} \cdot \mathbf{b} = 3$ and $|\mathbf{a} \times \mathbf{b}|^2 = 75$. Then, $|\mathbf{a}|^2$ is equal to ....

30 Let $S = \{(x, y) \in N \times N : 9(x-3)^2 + 16(y-4)^2 \le 144\}$ and $T = \{(x, y) \in R \times R : (x-7)^2 + (y-4)^2 \le 36\}$. Them, $(S \cap T)$ is equal to ....

Answers

For solutions scan the QR code

25 July 2022, Shift-I

1. (b)	2. (b)	3. (*)	4. (b)	5. (c)	6. (b)	7. (a)	8. (b)	9. (a)	10. (b)
11. (a)	12. (c)	13. (c)	14. (c)	15. (c)	16. (c)	17. (a)	18. (c)	19. (a)	20. (d)
21. 17	22. 1492	23. 6006	24. 38	25. 27560	26. 3	27. 5	28. 2	29. 10	30. (*)

25 July 2022, Shift-II

1. (c)	2. (c)	3. (b)	4. (d)	5. (b)	6. (a)	7. (c)	8. (b)	9. (b)	10. (d)
11. (b)	12. (a)	13. (d)	14. (d)	15. (c)	16. (a)	17. (a)	18. (d)	19. (b)	20. (d)
21. 107	22. 25	23. 24	24. 15	25. 3	26. 8	27. 5	28. 25	29. 170	30. 130

26 July 2022, Shift-I

1. (b)	2. (d)	3. (d)	4. (b)	5. (b)	6. (c)	7. (a)	8. (d)	9. (c)	10. (c)
11. (a)	12. (b)	13. (d)	14. (a)	15. (d)	16. (c)	17. (b)	18. (c)	19. (b)	20. (d)
21. 272	22. 180	23. 6993	24. 3	25. 23	26. 24	27. 6	28. 3	29. 45	30. 153

26 July 2022, Shift-II

1. (c)	2. (c)	3. (d)	4. (a)	5. (c)	6. (a)	7. (c)	8. (d)	9. (d)	10. (b)
11. (b)	12. (b)	13. (a)	14. (c)	15. (a)	16. (b)	17. (d)	18. (b)	19. (a)	20. (d)
21. 112	22. 2	23. 30	24. 116	25. 3	26. 238	27. 125	28. 3	29. 53	30. 50

27 July 2022, Shift-I

1. (d)	2. (d)	3. (a)	4. (d)	5. (a)	6. (b)	7. (b)	8. (c)	9. (c)	10. (a)
11. (d)	12. (d)	13. (b)	14. (c)	15. (c)	16. (c)	17. (d)	18. (c)	19. (d)	20. (c)
21. 12	22. 2	23. 12	24. 1552	25. 1	26. 2	27. 10620	28. 5376	29. 75	30. 0

27 July 2022, Shift-II

1. (c)	2. (c)	3. (b)	4. (b)	5. (b)	6. (d)	7. (a)	8. (d)	9. (b)	10. (c)
11. (d)	12. (a)	13. (c)	14. (b)	15. (b)	16. (b)	17. (d)	18. (a)	19. (c)	20. (d)
21. 42	22. 1440	23. 24	24. 120	25. 5	26. 16	27. 385	28. 12	29. 20	30. *

For solutions scan the QR code

28 July 2022, Shift-I

1. *(b)*	2. *(b)*	3. *(c)*	4. *(a)*	5. *(b)*	6. *(b)*	7. *(b)*	8. *(d)*	9. *(*)*	10. *(a)*
11. *(b)*	12. *(b)*	13. *(c)*	14. *(c)*	15. *(c)*	16. *(b)*	17. *(c)*	18. *(a)*	19. *(c)*	20. *(a)*
21. *7073*	22. *450*	23. *2*	24. *10*	25. *2*	26. *50*	27. *3*	28. *142*	29. *1*	30. *6*

28 July 2022, Shift-II

1. *(d)*	2. *(b)*	3. *(c)*	4. *(d)*	5. *(b)*	6. *(a)*	7. *(c)*	8. *(d)*	9. *(a)*	10. *(b)*
11. *(d)*	12. *(a)*	13. *(b)*	14. *(b)*	15. *(d)*	16. *(b)*	17. *(b)*	18. *(b)*	19. *(a)*	20. *(d)*
21. *57*	22. *17*	23. *13*	24. *99*	25. *9*	26. *12*	27. *5*	28. *6*	29. *56*	30. *104*

29 July 2022, Shift-I

1. *(b)*	2. *(a)*	3. *(b)*	4. *(c)*	5. *(c)*	6. *(b)*	7. *(b)*	8. *(b)*	9. *(b)*	10. *(a)*
11. *(a)*	12. *(d)*	13. *(b)*	14. *(a)*	15. *(d)*	16. *(d)*	17. *(b)*	18. *(b)*	19. *(d)*	20. *(d)*
21. *16*	22. *4*	23. *10*	24. *16*	25. *84*	26. *282*	27. *4*	28. *286*	29. *11*	30. *12*

29 July 2022, Shift-II

1. *(d)*	2. *(c)*	3. *(c)*	4. *(d)*	5. *(a)*	6. *(b)*	7. *(b)*	8. *(a)*	9. *(a)*	10. *(b)*
11. *(b)*	12. *(a)*	13. *(b)*	14. *(b)*	15. *(d)*	16. *(b)*	17. *(b)*	18. *(c)*	19. *(c)*	20. *(b)*
21. *96*	22. *16*	23. *10*	24. *6*	25. *221*	26. *79*	27. *195*	28. *72*	29. *14*	30. *27*

ONLINE SOLVED PAPERS

JEE Main 2023

(January Attempt)

24 JANUARY 2023, SHIFT-I

Section A : Objective Type Questions

1 Let Ω be the sample space and $A \subseteq \Omega$ be an event.

Given below are two statements

(S1) : If $P(A) = 0$, then $A = \phi$

(S2) : If $P(A) = 1$, then $A = \Omega$.

Then,

(a) only (S2) is true

(b) Both (S1) and (S2) are true

(c) only (S1) is true

(d) Both (S1) and (S2) are false

2 The distance of the point $(7, -3, -4)$ from the plane passing through the points $(2, -3, 1)$, $(-1, 1, -2)$ and $(3, -4, -2)$ is

(a) 4 (b) $4\sqrt{2}$ (c) 5 (d) $5\sqrt{2}$

3 $\tan^{-1}\left(\dfrac{1+\sqrt{3}}{3+\sqrt{3}}\right) + \sec^{-1}\left(\sqrt{\dfrac{8+4\sqrt{3}}{6+3\sqrt{3}}}\right)$ is equal to

(a) $\frac{\pi}{4}$ (b) $\frac{\pi}{6}$ (c) $\frac{\pi}{2}$ (d) $\frac{\pi}{3}$

4 Let a tangent to the curve $y^2 = 24x$ meet the curve $xy = 2$ at the points A and B. Then, the mid points of such line segment AB lies on a parabola with the

(a) length of latus rectum 2 (b) directrix $4x = -3$

(c) length of latus rectum $\frac{3}{2}$ (d) directrix $4x = 3$

5 The distance of the point $(-1, 9, -16)$ from the plane $2x + 3y - z = 5$ measured parallel to the line $\dfrac{x+4}{3} = \dfrac{2-y}{4} = \dfrac{z-3}{12}$ is

(a) $20\sqrt{2}$ (b) 26 (c) 31 (d) $13\sqrt{2}$

6 Let PQR be a triangle. The points A, B and C are on the sides QR, RP and PQ, respectively such that $\dfrac{QA}{AR} = \dfrac{RB}{BP} = \dfrac{PC}{CQ} = \dfrac{1}{2}$. Then, $\dfrac{\text{Area }(\Delta PQR)}{\text{Area }(\Delta ABC)}$ is equal to

(a) $\frac{5}{2}$ (b) 3 (c) 4 (d) 2

7 The equation $x^2 - 4x + [x] + 3 = x[x]$, where $[x]$ denotes the greatest integer function, has

(a) a unique solution in $(-\infty, \infty)$

(b) no solution

(c) a unique solution in $(-\infty, 1)$

(d) exactly two solutions in $(-\infty, \infty)$

8 The value of $\sum_{r=0}^{22} {}^{22}C_r \, {}^{23}C_r$ is

(a) ${}^{45}C_{23}$ (b) ${}^{44}C_{23}$

(c) ${}^{44}C_{22}$ (d) ${}^{45}C_{24}$

9 Let α be a root of the equation $(a-c)x^2 + (b-a)x + (c-b) = 0$, where a, b, c are distinct real numbers such that the matrix $\begin{bmatrix} \alpha^2 & \alpha & 1 \\ 1 & 1 & 1 \\ a & b & c \end{bmatrix}$ is singular. Then, the value of

$$\frac{(a-c)^2}{(b-a)(c-b)} + \frac{(b-a)^2}{(a-c)(c-b)} + \frac{(c-b)^2}{(a-c)(b-a)} \text{ is}$$

(a) 6 (b) 9 (c) 12 (d) 3

10 $\lim_{t \to 0}\left(1^{\frac{1}{\sin^2 t}} + 2^{\frac{1}{\sin^2 t}} + \ldots + n^{\frac{1}{\sin^2 t}}\right)^{\sin^2 t}$ is equal to

(a) $\frac{n(n+1)}{2}$ (b) n (c) $n^2 + n$ (d) n^2

11 Let $p, q \in R$ and $(1 - \sqrt{3}i)^{200} = 2^{199}(p + iq)$, $i = \sqrt{-1}$. Then, $p + q + q^2$ and $p - q + q^2$ are roots of the equation.

(a) $x^2 - 4x + 1 = 0$ (b) $x^2 - 4x - 1 = 0$

(c) $x^2 + 4x - 1 = 0$ (d) $x^2 + 4x + 1 = 0$

12 For three positive integers p, q, r, $x^{pq^2} = y^{qr} = z^{p^2r}$ and $r = pq + 1$ such that $3, 3\log_y x, 3\log_z y, 7\log_x z$ are in AP with common difference 1/2. Then, r-p-q is equal to

(a) 12 (b) 2 (c) 6 (d) -6

13 The relation $R = \{(a, b) : \gcd(a, b) = 1, 2a \neq b, a, b \in Z\}$ is

(a) Neither symmetric nor transitive
(b) reflexive but not symmetric
(c) symmetric but not transitive
(d) transitive but not reflexive

14 If A and B are two non-zero $n \times n$ matrics such that $A^2 + B = A^2B$, then

(a) $A^2B = BA^2$ (b) $A^2B = I$
(c) $A^2 = I$ or $B = I$ (d) $AB = I$

15 Let $f(x) = \begin{cases} x^2 \sin\left(\frac{1}{x}\right), & x \neq 0 \\ 0, & x = 0 \end{cases}$

Then, at $x = 0$

(a) f and f' both are continuous
(b) f is continuous but f' is not continuous
(c) f is continuous but not differentiable
(d) f' is continuous but not differentiable

16 Let $\mathbf{u} = \hat{\mathbf{i}} - \hat{\mathbf{j}} - 2\hat{\mathbf{k}}$, $\mathbf{v} = 2\hat{\mathbf{i}} + \hat{\mathbf{j}} - \hat{\mathbf{k}}$, $\mathbf{v} \cdot \mathbf{w} = 2$ and $\mathbf{v} \times \mathbf{w} = \mathbf{u} + \lambda\mathbf{v}$. Then, $\mathbf{u} \cdot \mathbf{w}$ is equal to

(a) 1 (b) $-\frac{2}{3}$ (c) $\frac{3}{2}$ (d) 2

17 Let $y = y(x)$ be the solution of the differential equation $x^3 dy + (xy - 1)dx = 0$, $x > 0$, $y\left(\frac{1}{2}\right) = 3 - e$. Then, $y(1)$ is equal to

(a) e (b) $2 - e$ (c) 1 (d) 3

> **Key Idea** For find the solution of given differential equation convert into a linear differential equation as form $\frac{dy}{dx} + py = \theta$, IF $= e^{\int P dx}$ and $y \times \text{IF} = \int \text{IF} \times \theta \, dx$, then solve it using integration.

18 The area enclosed by the curves $y^2 + 4x = 4$ and $y - 2x = 2$ is

(a) $\frac{25}{3}$ (b) $\frac{23}{3}$ (c) 9 (d) $\frac{22}{3}$

19 The compound statement $(\sim P \wedge Q)) \vee ((\sim P) \wedge \sim Q) \Rightarrow ((\sim P) \wedge (\sim Q))$ is equivalent to

(a) $((\sim P) \vee Q) \wedge ((\sim Q) \vee P)$ (b) $(\sim Q) \vee P$
(c) $(\sim P) \vee Q$ (d) $((\sim P) \vee Q) \wedge (\sim Q)$

20 Let N denote the number that turns up when a fair die is rolled. If the probability that the system of equations

$$x + y + z = 1$$
$$2x + Ny + 2z = 2$$
$$3x + 3y + Nz = 3$$

has unique solution is $\frac{k}{6}$, then the sum of value of k and all possible values of N is

(a) 21 (b) 18 (c) 19 (d) 20

Section B : Numerical Value Type Questions

21 Suppose, $\sum_{r=0}^{2023} r^2 \, {}^{2023}C_r = 2023 \times \alpha \times 2^{2022}$. Then, the value of α is ....

22 The value of $\frac{8}{\pi} \int_0^{\pi/2} \frac{(\cos x)^{2023}}{(\sin x)^{2023} + (\cos x)^{2023}} dx$ is ....

23 Let C be the largest circle centred at $(2, 0)$ and inscribed in the ellipse $\frac{x^2}{36} + \frac{y^2}{16} = 1$. If $(1, \alpha)$ lies on C, then $10\alpha^2$ is equal to ....

24 The shortest distance between the lines $\frac{x-2}{3} = \frac{y+1}{2} = \frac{z-6}{2}$ and $\frac{x-6}{3} = \frac{1-y}{2} = \frac{z+8}{0}$ is equal to

> **Key Idea** To find the shortest distance between the lines firstly find the vector form of lines. Then use of formula SD $= \left|\frac{(a_2 - a_1) \cdot (b_1 \times b_2)}{|b_1 \times b_2|}\right|$.

25 The number of 9-digit numbers, that can be formed using all the digits of the number 123412341, so that the even digits occupy only even places, is ....

26 The value of $12\int_0^3 |x^2 - 3x + 2| \, dx$ is ....

27 A boy needs to select five courses from 12 available cources, out of which 5 courses are language courses. If he can choose at most two language courses, then the number of ways he can choose five courses is ....

28 Let $\lambda \in R$ and let the equation E be $|x|^2 - 2|x| + |\lambda - 3| = 0$. Then, the largest element in the set $S = \{x + \lambda : x \text{ is an integer solution of } E\}$ is ....

29 Let a tangent to the curve $9x^2 + 16y^2 = 144$ intersect the co-ordinate axes at the points A and B. Then, the minimum length of the line segment AB is ....

30 The 4th term of GP is 500 and its common ratio is $\frac{1}{m}$, $m \in N$. Let S_n denote the sum of the first n terms of this GP. If $S_6 > S_5 + 1$ and $S_7 < S_6 + \frac{1}{2}$, then the number of possible values of m is ....

24 JANUARY 2023, SHIFT-II

Section A : Objective Type Questions

1 Let p and q be two statements. Then, $\sim (p \wedge (p \Rightarrow \sim q))$ is equivalent to

(a) $p \vee ((\sim p) \wedge q)$ (b) $p \vee (p \wedge q)$
(c) $(\sim p) \vee q$ (d) $p \vee (p \wedge (\sim q))$

2 Let the plane containing the line of intersection of the planes P1 : $x + (\lambda + 4)y + z = 1$ and P2 : $2x + y + z = 2$ pass through the points $(0, 1, 0)$ and $(1, 0, 1)$. Then, the distance of the point $(2\lambda, \lambda, -\lambda)$ from the plane P2 is

(a) $2\sqrt{6}$ (b) $4\sqrt{6}$
(c) $5\sqrt{6}$ (d) $3\sqrt{6}$

3 The set of all values of a for which $\lim_{x \to a} ([x-5] - [2x+2]) = 0$, where $[\alpha]$ denotes the greatest integer less than or equal to α is equal to

(a) $[-7.5, -6.5)$ (b) $(-7.5, -6.5)$
(c) $(-7.5, -6.5]$ (d) $[-7.5, -6.5]$

4 The number of integers, greater than 7000 that can be formed, using the digits 3, 5, 6, 7, 8 without repetition, is

(a) 120 (b) 220
(c) 48 (d) 168

5 If the system of equations

$$x + 2y + 3z = 3$$
$$4x + 3y - 4z = 4$$
$$8x + 4y - \lambda z = 9 + \mu$$

has infinitely many solutions, then the ordered pair (λ, μ) is equal to

(a) $\left(-\frac{72}{5}, -\frac{21}{5}\right)$ (b) $\left(\frac{72}{5}, -\frac{21}{5}\right)$
(c) $\left(-\frac{72}{5}, \frac{21}{5}\right)$ (d) $\left(\frac{72}{5}, \frac{21}{5}\right)$

6 If $f(x) = \frac{2^{2x}}{2^{2x} + 2}$, $x \in R$, then $f\left(\frac{1}{2023}\right) + f\left(\frac{2}{2023}\right) + \ldots + f\left(\frac{2022}{2023}\right)$ is equal to

(a) 2010 (b) 2011 (c) 1011 (d) 1010

7 If $({}^{30}C_1)^2 + 2({}^{30}C_2)^2 + 3({}^{30}C_3)^2 + \ldots + ({}^{30}C_{30})^2 = \frac{\alpha 60!}{(30!)^2}$, then α is equal to

(a) 30 (b) 60 (c) 10 (d) 15

8 If the foot of the perpendicular drawn from $(1, 9, 7)$ to the line passing through the point $(3, 2, 1)$ and parallel to the planes $x + 2y + z = 0$ and $3y - z = 3$ is (α, β, γ), then $\alpha + \beta + \gamma$ is equal to

(a) -1 (b) 5 (c) 3 (d) 1

9 Let $y = y(x)$ be the solution of the differential equation $(x^2 - 3y^2)dx + 3xy\,dy = 0$, $y(1) = 1$. Then, $6y^2(e)$ is equal to

(a) e^2 (b) $2e^2$ (c) $3e^2$ (d) $\frac{3}{2}e^2$

> **Key Idea** This is homogeneous differential equation form $\frac{dy}{dx} = f(x, y)$. This equation is solved by putting $y = vx$, where v is a function of x.

10 The number of square matrices of order 5 with entries from the set $\{0, 1\}$, such that the sum of all the elements in each row is 1 and the sum of all the elements in each column is also 1, is

(a) 120 (b) 125 (c) 225 (d) 150

11 Let $\alpha = 4\hat{i} + 3\hat{j} + 5\hat{k}$ and $\beta = \hat{i} + 2\hat{j} - 4\hat{k}$. Let β_1 be parallel to α and β_2 be perpendicular to α. If $\beta = \beta_1 + \beta_2$, then the value of $5\beta_2 \cdot (\hat{i} + \hat{j} + \hat{k})$ is

(a) 6 (b) 9 (c) 7 (d) 11

12 Let the six numbers $a_1, a_2, a_3, a_4, a_5, a_6$, be in AP and $a_1 + a_3 = 10$. If the mean of these six numbers is $\frac{19}{2}$ and their variance is σ^2, then $8\sigma^2$ is equal to

(a) 105 (b) 200
(c) 210 (d) 220

13 The number of real solutions of the equation $3\left(x^2 + \frac{1}{x^2}\right) - 2\left(x + \frac{1}{x}\right) + 5 = 0$, is

(a) 2 (b) 0 (c) 4 (d) 3

14 Let $f(x)$ be a function such that $f(x+y) = f(x) \cdot f(y)$, $\forall x, y \in N$. If $f(1) = 3$ and $\sum_{k=1}^{n} f(k) = 3279$, then the value of n is

(a) 9 (b) 6 (c) 7 (d) 8

15 The value of $\left(\frac{1 + \sin\frac{2\pi}{9} + i\cos\frac{2\pi}{9}}{1 + \sin\frac{2\pi}{9} - i\cos\frac{2\pi}{9}}\right)^3$ is

(a) $\frac{1}{2}(\sqrt{3} + i)$ (b) $\frac{1}{2}(1 - i\sqrt{3})$
(c) $-\frac{1}{2}(1 - i\sqrt{3})$ (d) $-\frac{1}{2}(\sqrt{3} - i)$

16 If $f(x) = x^3 - x^2 f'(1) + x f''(2) - f'''(3)$, $x \in R$, then

(a) $f(1) + f(2) + f(3) = f(0)$ (b) $3f(1) + f(2) = f(3)$
(c) $f(3) - f(2) = f(1)$ (d) $2f(0) - f(1) + f(3) = f(2)$

17 $\int_{\frac{3\sqrt{2}}{4}}^{\frac{3\sqrt{3}}{4}} \frac{48}{\sqrt{9 - 4x^2}}\,dx$ is equal to

(a) $\frac{\pi}{2}$ (b) 2π (c) $\frac{\pi}{3}$ (d) $\frac{\pi}{6}$

18 Let A be a 3×3 matrix such that $|\text{adj}(\text{adj}(\text{adj } A))| = 12^4$. Then, $|A^{-1} \text{ adj } A|$ is equal to

(a) 12 (b) 1
(c) $2\sqrt{3}$ (d) $\sqrt{6}$

19 The equations of the sides AB and AC of a triangle ABC are $(\lambda + 1)x + \lambda y = 4$ and $\lambda x + (1 - \lambda)y + \lambda = 0$, respectively. Its vertex A is on the Y-axis and its orthocentre is $(1, 2)$. The length of the tangent from the point C to the part of the parabola $y^2 = 6x$ in the first quadrant is

(a) $2\sqrt{2}$ (b) 2
(c) $\sqrt{6}$ (d) 4

20 The locus of the mid points of the chords of the circle $C_1 : (x-4)^2 + (y-5)^2 = 4$ which subtend an angle θ_i at the centre of the circle C_1, is a circle of radius r_i. If $\theta_1 = \frac{\pi}{3}, \theta_3 = \frac{2\pi}{3}$ and $r_1^2 = r_2^2 + r_3^2$, then θ_2 is equal to

(a) $\frac{3\pi}{4}$ (b) $\frac{\pi}{4}$
(c) $\frac{\pi}{6}$ (d) $\frac{\pi}{2}$

Section B : Numerical Value Type Questions

21 Let $S = \{\theta \in [0, 2\pi) : \tan(\pi \cos\theta) + \tan(\pi \sin\theta) = 0\}$. Then, $\sum_{\theta \in S} \sin^2\left(\theta + \frac{\pi}{4}\right)$ is equal to ....

22 If the area of the region bounded by the curves $y^2 - 2y = -x$, $x + y = 0$ is A, then 8A is equal to ....

23 If $\frac{1^3 + 2^3 + 3^3 + \ldots \text{ up to } n \text{ terms}}{1 \cdot 3 + 2 \cdot 5 + 3 \cdot 7 + \ldots \text{ up to } n \text{ terms}} = \frac{9}{5}$, then the value of n is

24 Let $\mathbf{a} = \hat{\mathbf{i}} + 2\hat{\mathbf{j}} + \lambda\hat{\mathbf{k}}$, $\mathbf{b} = 3\hat{\mathbf{i}} - 5\hat{\mathbf{j}} - \lambda\hat{\mathbf{k}}$, $\mathbf{a} \cdot \mathbf{c} = 7$, $2\mathbf{b} \cdot \mathbf{c} + 43 = 0$, $\mathbf{a} \times \mathbf{c} = \mathbf{b} \times \mathbf{c}$. Then, $|\mathbf{a} \cdot \mathbf{b}|$ is equal to

25 If the shortest distance between the lines $\frac{x + \sqrt{6}}{2} = \frac{y - \sqrt{6}}{3} = \frac{z - \sqrt{6}}{4}$ and $\frac{x - \lambda}{3} = \frac{y - 2\sqrt{6}}{4} = \frac{z + 2\sqrt{6}}{5}$ is 6, then the square of sum of all possible values of λ is.

> **Key Idea** If two lines are not parallel, then the shortest distance is, $d = \left|\frac{(b_1 \times b_2) \cdot (a_2 - a_1)}{|b_1 \times b_2|}\right|$.

26 Let the sum of the coefficients of the first three terms in the expansion of $\left(x + \frac{3}{x^2}\right)^n$, $x \neq 0$, $n \in N$, be 376. Then, the coefficient of x^4 is ....

27 let f be a differentiable function defined on $\left[0, \frac{\pi}{2}\right]$ such that $f(x) > 0$ and $f(x) + \int_0^x f(t)\sqrt{1 - (\log_e f(t))^2}\, dt = e$, $\forall x \in \left[0, \frac{\pi}{2}\right]$. Then, $\left(6 \log_e f\left(\frac{\pi}{6}\right)\right)^2$ is equal to

28 The minimum number of elements that must be added to the relation $R = \{(a, b), (b, c), (b, d)\}$ on the set $\{a, b, c, d\}$, so that it is an equivalence relation, is ....

29 The equations of the sides AB, BC and CA of a triangle ABC are: $2x + y = 0$, $x + py = 21a$, $(a \neq 0)$ and $x - y = 3$, respectively. Let $P(2, a)$ be the centroid of ΔABC. Then, $(BC)^2$ is equal to

30 Three urns A, B and C contain 4 red, 6 black, 5 red, 5 black and λ red, 4 black balls, respectively. One of the urns is selected at random and a ball is drawn. If the ball drawn is red and the probability that it is drawn from urn C is 0.4, then the square of the length of the side of the largest equilateral triangle, inscribed in the parabola $y^2 = \lambda x$ with one vertex at the vertex of the parabola, is

25 JANUARY 2023, SHIFT-I

Section A : Objective Type Questions

1 Consider the lines L_1 and L_2 given by

$$L_1 : \frac{x-1}{2} = \frac{y-3}{1} = \frac{z-2}{2}$$

$$L_2 : \frac{x-2}{1} = \frac{y-2}{2} = \frac{z-3}{3}$$

A line L_3 having direction ratios $1, -1, -2$, intersects L_1 and L_2 at the points P and Q, respectively. Then, the length of line segment PQ is

(a) $4\sqrt{3}$ (b) 4
(c) $2\sqrt{6}$ (d) $3\sqrt{2}$

2 Let $f : (0, 1) \to R$ be a function defined by $f(x) = \frac{1}{1 - e^{-x}}$ and $g(x) = (f(-x) - f(x))$. Consider two statements

(I) g is an increasing function in $(0, 1)$.

(II) g is one-one in $(0, 1)$.

Then,

(a) Only (II) is true
(b) Both (I) and (II) are true
(c) Only (I) is true
(d) Neither (I) nor (II) is true

3 The points of intersection of the line $ax + by = 0$, $(a \neq b)$ and the circle $x^2 + y^2 - 2x = 0$ are $A(\alpha, 0)$ and $B(1, \beta)$.

The image of the circle with AB as a diameter in the line $x + y + 2 = 0$ is

(a) $x^2 + y^2 - 5x - 5y + 12 = 0$
(b) $x^2 + y^2 + 3x + 5y + 8 = 0$
(c) $x^2 + y^2 + 5x + 5y + 12 = 0$
(d) $x^2 + y^2 + 3x + 3y + 4 = 0$

4 Let $f(x) = \int \frac{2x}{(x^2+1)(x^2+3)} dx$. If $f(3) = \frac{1}{2}(\log_e 5 - \log_e 6)$, then $f(4)$ is equal to

(a) $\frac{1}{2}(\log_e 19 - \log_e 17)$ (b) $\frac{1}{2}(\log_e 17 - \log_e 19)$
(c) $\log_e 19 - \log_e 20$ (d) $\log_e 17 - \log_e 18$

5 The minimum value of the function $f(x) = \int_0^2 e^{|x-t|} dt$ is

(a) $2(e-1)$ (b) $e(e-1)$ (c) 2 (d) $2e - 1$

6 The distance of the point $P(4, 6, -2)$ from the line passing through the point $(-3, 2, 3)$ and parallel to a line with direction ratios 3, 3, −1 is equal to

(a) $\sqrt{14}$ (b) 3 (c) $2\sqrt{3}$ (d) $\sqrt{6}$

7 If a_r is the coefficient of x^{10-r} in the Binomial expansion of $(1+x)^{10}$, then $\sum_{r=1}^{10} r^3 \left(\frac{a_r}{a_{r-1}}\right)^2$ is equal to

(a) 1210 (b) 4895 (c) 3025 (d) 5445

8 Let S_1 and S_2 be respectively the sets of all $a \in R - \{0\}$ for which the system of linear equations

$ax + 2ay - 3az = 1$
$(2a+1)x + (2a+3)y + (a+1)z = 2$
$(3a+5)x + (a+5)y + (a+2)z = 3$

has unique solution and infinitely many solutions. Then,

(a) $S_1 = \phi$ and $S_2 = R - \{0\}$
(b) S_1 is an infinite set and $n(S_2) = 2$
(c) $n(S_1) = 2$ and S_2 is an infinite set
(d) $S_1 = R - \{0\}$ and $S_2 = \phi$

9 Let $y(x) = (1+x)(1+x^2)(1+x^4)(1+x^8)(1+x^{16})$. Then, $y' - y''$ at $x = -1$ is equal to

(a) 496 (b) 464 (c) 976 (d) 944

10 Let $z_1 = 2 + 3i$ and $z_2 = 3 + 4i$. The set $S = \{z \in C : |z - z_1|^2 - |z - z_2|^2 = |z_1 - z_2|^2\}$ represents a

(a) hyperbola with the length of the transverse axis 7
(b) straight line with the sum of its intercepts on the coordinate axes equals 18
(c) straight line with the sum of its intercepts on the coordinate axes equals 14
(d) hyperbola with eccentricity 2

11 Let M be the maximum value of the product of two positive integers when their sum is 66. Let the sample space $S = \left\{x \in Z : x(66 - x) \geq \frac{5}{9} M\right\}$ and the event $A = \{x \in S : x \text{ is a multiple of } 3\}$. Then, $P(A)$ is equal to

(a) $\frac{1}{5}$ (b) $\frac{15}{44}$ (c) $\frac{1}{3}$ (d) $\frac{7}{22}$

12 The distance of the point $(6, -2\sqrt{2})$ from the common tangent $y = mx + c$, $m > 0$, of the curves $x = 2y^2$ and $x = 1 + y^2$ is

(a) 5 (b) $\frac{1}{3}$ (c) $\frac{14}{3}$ (d) $5\sqrt{3}$

13 Let $y = y(x)$ be the solution curve of the differential equation $\frac{dy}{dx} = \frac{y}{x}(1 + xy^2(1 + \log_e x))$, $x > 0$, $y(1) = 3$. Then, $\frac{y^2(x)}{9}$ is equal to

(a) $\frac{x^2}{3x^3(1 + \log_e x^2) - 2}$ (b) $\frac{x^2}{2x^3(2 + \log_e x^3) - 3}$
(c) $\frac{x^2}{5 - 2x^3(2 + \log_e x^3)}$ (d) $\frac{x^2}{7 - 3x^3(2 + \log_e x^2)}$

Key Idea Convert the given differential equation in the standard form of linear differential equation, $\frac{dy}{dx} + Py = Q$, where P and Q are function of x, then solve the linear differential equation using integration.

14 The statement $(p \wedge (\sim q)) \Rightarrow (p \Rightarrow (\sim q))$ is

(a) a tautology (b) a contradiction
(c) equivalent to $p \vee q$ (d) equivalent to $(\sim p) \vee (\sim q)$

15 Let $x, y, z > 1$ and $A = \begin{bmatrix} 1 & \log_x y & \log_x z \\ \log_y x & 2 & \log_y z \\ \log_z x & \log_z y & 3 \end{bmatrix}$. Then, $|\text{adj}(\text{adj} A^2)|$ is equal to

(a) 6^4 (b) 4^8 (c) 2^8 (d) 2^4

16 Let **a**, **b** and **c** be three non zero vectors such that $\mathbf{b} \cdot \mathbf{c} = 0$ and $\mathbf{a} \times (\mathbf{b} \times \mathbf{c}) = \frac{\mathbf{b} - \mathbf{c}}{2}$. If **d** be a vector such that $\mathbf{b} \cdot \mathbf{d} = \mathbf{a} \cdot \mathbf{b}$, then $(\mathbf{a} \times \mathbf{b}) \cdot (\mathbf{c} \times \mathbf{d})$ is equal to

(a) $\frac{3}{4}$ (b) $-\frac{1}{4}$ (c) $\frac{1}{4}$ (d) $\frac{1}{2}$

17 The vector $\mathbf{a} = -\hat{\mathbf{i}} + 2\hat{\mathbf{j}} + \hat{\mathbf{k}}$ is rotated through a right angle, passing through the Y-axis in its way and the resulting vector is **b**. Then, the projection of $3\mathbf{a} + \sqrt{2}\mathbf{b}$ on $\mathbf{c} = 5\hat{\mathbf{i}} + 4\hat{\mathbf{j}} + 3\hat{\mathbf{k}}$ is

(a) 1 (b) $\sqrt{6}$ (c) $3\sqrt{2}$ (d) $2\sqrt{3}$

18 The mean and variance of the marks obtained by the students in a test are 10 and 4, respectively. Later, the marks of one of the students in increased from 8 to 12. If the new mean of the marks is 10.2, then their new variance is equal to

(a) 4.04 (b) 4.08 (c) 3.92 (d) 3.96

19 Let $x = 2$ be a local minima of the function $f(x) = 2x^4 - 18x^2 + 8x + 12$, $x \in (-4, 4)$. If M is local maximum value of the function f in $(-4, 4)$, then $M =$

(a) $12\sqrt{6} - \frac{33}{2}$ (b) $18\sqrt{6} - \frac{33}{2}$

(c) $18\sqrt{6} - \frac{31}{2}$ (d) $12\sqrt{6} - \frac{31}{2}$

20 The value of

$$\lim_{n \to \infty} \frac{1 + 2 - 3 + 4 + 5 - 6 + \ldots + (3n-2) + (3n-1) - 3n}{\sqrt{2n^4 + 4n + 3} - \sqrt{n^4 + 5n + 4}}$$ is

(a) $\frac{3}{2}(\sqrt{2}+1)$ (b) $\frac{3}{2\sqrt{2}}$ (c) $3(\sqrt{2}+1)$ (d) $\frac{\sqrt{2}+1}{2}$

Section B : Numerical Value Type Questions

21 Let x and y be distinct integers, where $1 \le x \le 25$ and $1 \le y \le 25$. Then, the number of ways of choosing x and y, such that $x + y$ is divisible by 5, is ....

22 Let $S = \{\alpha : \log_2(9^{2\alpha-4} + 13) - \log_2\left(\frac{5}{2} \cdot 3^{2\alpha-4} + 1\right) = 2\}$.

Then, the maximum value of β for which the equation $x^2 - 2\left(\sum_{\alpha \in S} \alpha\right)^2 x + \sum_{\alpha \in S} (\alpha+1)^2 \beta = 0$ has real roots, is

23 The constant term in the expansion of $\left(2x + \frac{1}{x^7} + 3x^2\right)^5$ is ....

24 Let A_1, A_2, A_3 be the three AP with the same common difference d and having their first terms as $A, A+1, A+2$, respectively. Let a, b, c be the 7th, 9th, 17th terms of A_1, A_2, A_3, respectively such that

$$\begin{vmatrix} a & 7 & 1 \\ 2b & 17 & 1 \\ c & 17 & 1 \end{vmatrix} + 70 = 0.$$

If $a = 29$, then the sum of first 20 terms of an AP whose first term is $c - a - b$ and common difference is $\frac{d}{12}$, is equal to .... .

25 If the sum of all the solutions of $\tan^{-1}\left(\frac{2x}{1-x^2}\right) + \cot^{-1}\left(\frac{1-x^2}{2x}\right) = \frac{\pi}{3}$, $-1 < x < 1$, $x \ne 0$, is $\alpha - \frac{4}{\sqrt{3}}$, then α is equal to ....

> **Key Idea** We use property, $\cot^{-1}x = \tan^{-1}\frac{1}{x}$, $x > 0$ and $\cot^{-1}x = \pi + \tan^{-1}\frac{1}{x}$, $x < 0$ to solve $\tan^{-1}\left(\frac{2x}{1-x^2}\right) + \cot^{-1}\left(\frac{1-x^2}{2x}\right) = \frac{\pi}{3}$.

26 For some $a, b, c \in N$, let $f(x) = ax - 3$ and $g(x) = x^b + c$, $x \in R$. If $(fog)^{-1}(x) = \left[\frac{x-7}{2}\right]^{\frac{1}{3}}$, then $(fog)(ac) + (gof)(b)$ is equal to ....

27 Let the equation of the plane passing through the line $x - 2y - z - 5 = 0 = x + y + 3z - 5$ and parallel to the line $x + y + 2z - 7 = 0 = 2x + 3y + z - 2$ be $ax + by + cz = 65$. Then, the distance of the point (a, b, c) from the plane $2x + 2y - z + 16 = 0$ is ....

28 The vertices of a hyperbola H are $(\pm 6, 0)$ and its eccentricity is $\frac{\sqrt{5}}{2}$. Let N be the normal to H at a point in the first quadrant and parallel to the line $\sqrt{2}x + y = 2\sqrt{2}$. If d is the length of the line segment of N between H and the Y-axis, then d^2 equal to ....

29 Let $S = \{1, 2, 3, 5, 7, 10, 11\}$. The number of non-empty subsets of S that have the sum of all elements a multiple of 3, is ....

30 If the area enclosed by the parabolas $P_1 : 2y = 5x^2$ and $P_2 : x^2 - y + 6 = 0$ is equal to the area enclosed by P_1 and $y = \alpha x$, $\alpha > 0$, then α^3 is equal to ....

25 JANUARY 2023, SHIFT-II

Section A : Objective Type Questions

1 Let $y = y(t)$ be a solution of the differential equation $\frac{dy}{dt} + \alpha y = \gamma e^{-\beta t}$,

where $\alpha > 0$, $\beta > 0$ and $\gamma > 0$. Then, $\lim_{t \to \infty} y(t)$

(a) does not exist (b) is 0

(c) is 1 (d) is -1

> **Key Idea** Use the linear differential equation of the form $\frac{dy}{dx} + Py = Q$, where P and Q are the function of x.

2 $\sum_{k=0}^{6} {}^{51-k}C_3$ is equal to

(a) ${}^{52}C_3 - {}^{45}C_3$ (b) ${}^{51}C_4 - {}^{45}C_4$

(c) ${}^{51}C_3 - {}^{45}C_3$ (d) ${}^{52}C_4 - {}^{45}C_4$

3 Let N be the sum of the numbers appeared when two fair dice are rolled and let the probability that $N-2$, $\sqrt{3N}$, $N+2$ are in geometric progression be $\frac{k}{48}$. Then, the value of k is

(a) 2 (b) 8 (c) 16 (d) 4

4 The foot of perpendicular of the point (2, 0, 5) on the line $\frac{x+1}{2}=\frac{y-1}{5}=\frac{z+1}{-1}$ is (α, β, γ). Then, which of the following is not correct?

(a) $\frac{\beta}{\gamma}=-5$ (b) $\frac{\gamma}{\alpha}=\frac{5}{8}$ (c) $\frac{\alpha}{\beta}=-8$ (d) $\frac{\alpha\beta}{\gamma}=\frac{4}{15}$

5 The number of functions $f:\{1,2,3,4\}\to\{a\in Z:|a|\le 8\}$ satisfying $f(n)+\frac{1}{n}f(n+1)=1, \forall\, n\in\{1,2,3\}$ is

(a) 3 (b) 2 (c) 1 (d) 4

6 If the four points, whose position vectors are $3\hat{i}-4\hat{j}+2\hat{k}$, $\hat{i}+2\hat{j}-\hat{k}$, $-2\hat{i}-\hat{j}+3\hat{k}$ and $5\hat{i}-2\alpha\hat{j}+4\hat{k}$ are coplanar, then α is equal to

(a) $\frac{73}{17}$ (b) $-\frac{107}{17}$ (c) $-\frac{73}{17}$ (d) $\frac{107}{17}$

7 Let T and C respectively be the transverse and conjugate axes of the hyperbola $16x^2-y^2+64x+4y+44=0$. Then, the area of the region above the parabola $x^2=y+4$, below the transverse axis T and on the right of the conjugate axis C is

(a) $4\sqrt{6}+\frac{28}{3}$ (b) $4\sqrt{6}-\frac{28}{3}$

(c) $4\sqrt{6}+\frac{44}{3}$ (d) $4\sqrt{6}-\frac{44}{3}$

8 Let $A=\begin{bmatrix}\frac{1}{\sqrt{10}} & \frac{3}{\sqrt{10}}\\ \frac{-3}{\sqrt{10}} & \frac{1}{\sqrt{10}}\end{bmatrix}$ and $B=\begin{bmatrix}1 & -i\\ 0 & 1\end{bmatrix}$, where $i=\sqrt{-1}$.

If $M=A^TBA$, then the inverse of the matrix $AM^{2003}A^T$ is

(a) $\begin{bmatrix}1 & 0\\ 2023i & 1\end{bmatrix}$ (b) $\begin{bmatrix}1 & -2023i\\ 0 & 1\end{bmatrix}$

(c) $\begin{bmatrix}1 & 0\\ -2023i & 1\end{bmatrix}$ (d) $\begin{bmatrix}1 & 2023i\\ 0 & 1\end{bmatrix}$

9 Let A, B, C be 3×3 matrices such that A is symmetric and B and C are skew-symmetric. Consider the statements

(S1) : $A^{13}B^{26}-B^{26}A^{13}$ is symmetric.

(S2) : $A^{26}C^{13}-C^{13}A^{26}$ is symmetric.

Then,

(a) Both S1 and S2 are false (b) only S1 is true

(c) only S2 is true (d) Both S1 and S2 are true

10 Let z be a complex number such that $\left|\frac{z-2i}{z+i}\right|=2$, $z\ne -1$. Then, z lies on the circle of radius 2 and centre

(a) (2, 0) (b) (0, 0) (c) (0, – 2) (d) (0, 2)

11 Let $f(x)=2x^n+\lambda$, $\lambda\in R$, $n\in N$ and $f(4)=133$, $f(5)=255$. Then, the sum of all the positive integer divisors of $(f(3)-f(2))$ is

(a) 60 (b) 59 (c) 58 (d) 61

12 If the function

$$f(x)=\begin{cases}(1+|\cos x|)^{\frac{\lambda}{|\cos x|}}, & 0<x<\frac{\pi}{2}\\ \mu, & x=\frac{\pi}{2}\\ e^{\frac{\cot 6x}{\cot 4x}}, & \frac{\pi}{2}<x<\pi\end{cases}$$ is

continuous at $x=\frac{\pi}{2}$, then $9\lambda+6\log_e\mu+\mu^6-e^{6\lambda}$ is equal to

(a) 10 (b) $2e^4+8$ (c) 8 (d) 11

13 Let $f: R\to R$ be a function defined by $f(x)=\log_{\sqrt{m}}\{\sqrt{2}(\sin x-\cos x)+m-2\}$, for some m, such that the range of f is [0, 2]. Then, the value of m is ....

(a) 3 (b) 4 (c) 2 (d) 5

14 The integral $16\int_1^2\frac{dx}{x^3(x^2+2)^2}$ is equal to

(a) $\frac{11}{6}-\log_e 4$ (b) $\frac{11}{12}-\log_e 4$

(c) $\frac{11}{12}+\log_e 4$ (d) $\frac{11}{6}+\log_e 4$

15 Let $\mathbf{a}=-\hat{i}-\hat{j}+\hat{k}$, $\mathbf{a}\cdot\mathbf{b}=1$ and $\mathbf{a}\times\mathbf{b}=\hat{i}-\hat{j}$. Then, $\mathbf{a}-6\mathbf{b}$ is equal to

(a) $3(\hat{i}-\hat{j}+\hat{k})$ (b) $3(\hat{i}+\hat{j}+\hat{k})$

(c) $3(\hat{i}+\hat{j}-\hat{k})$ (d) $3(\hat{i}-\hat{j}-\hat{k})$

16 Let the function $f(x)=2x^3+(2p-7)x^2+3(2p-9)x-6$ have a maxima, for some value of $x<0$ and a minima for some value of $x>0$. Then, the set of all values of p is

(a) $\left(0,\frac{9}{2}\right)$ (b) $\left(-\infty,\frac{9}{2}\right)$

(c) $\left(\frac{9}{2},\infty\right)$ (d) $\left(-\frac{9}{2},\frac{9}{2}\right)$

17 The shortest distance between the lines $x+1=2y=-12z$ and $x=y+2=6z-6$ is

(a) 3 (b) $\frac{3}{2}$ (c) 2 (d) $\frac{5}{2}$

Key Idea For the calculation of shortest distance between two skew lines, use the formula $d=\left|\frac{(b_1\times b_2)\cdot(a_2-a_1)}{|b_1\times b_2|}\right|$, where b_1 and b_2 are direction vectors and a_1 and a_2 are position vectors.

18 The equations of two sides of a variable triangle are $x=0$ and $y=3$ and its third side is a tangent to the parabola $y^2=6x$. The locus of its circumcentre is

(a) $4y^2-18y-3x+18=0$
(b) $4y^2-18y+3x+18=0$
(c) $4y^2-18y-3x-18=0$
(d) $4y^2+18y+3x+18=0$

19 Let $\Delta, \nabla \in \{\wedge, \vee\}$ be such that $(p\to q)\Delta(p\nabla q)$ is a tautology. Then,

(a) $\Delta=\wedge, \nabla=\wedge$ (b) $\Delta=\vee, \nabla=\wedge$
(c) $\Delta=\vee, \nabla=\vee$ (d) $\Delta=\wedge, \nabla=\vee$

20 The number of numbers, strictly between 5000 and 10000 can be formed using the digits 1,3,5,7,9 without repetition, is

(a) 6 (b) 72 (c) 120 (d) 12

Section B : Numerical Value Type Questions

21 25% of the population are smokers. A smoker has 27 times more chances to develop lung cancer than a non-smoker. A person is diagnosed with lung cancer and the probability that this person is a smoker is $\frac{K}{10}$. Then, the value of K is ....

22 If $\int_{1/3}^{3}|\log_e x|\,dx=\frac{m}{n}\log_e\left(\frac{n^2}{e}\right)$, where m and n are coprime natural numbers, then m^2+n^2-5 is equal to ....

23 The remainder when $(2023)^{2023}$ is divided by 35 is

24 If the shortest distance between the line joining the points (1, 2, 3) and (2, 3, 4) and the line $\frac{x-1}{2}=\frac{y+1}{-1}=\frac{z-2}{0}$ is α, then $28\alpha^2$ is equal to ....

25 If m and n respectively are the numbers of positive and negative values of θ in the interval $[-\pi, \pi]$ that satisfy the equation $\cos 2\theta\cos\frac{\theta}{2}=\cos 3\theta\cos\frac{9\theta}{2}$, then mn is equal to ....

26 Suppose, Anil's mother wants to give 5 whole fruits to Anil from a basket of 7 red apples, 5 white apples and 8 oranges. If in the selected 5 fruits, at least 2 oranges, at least one red apple and at least one white apple must be given, then the number of ways, Anil's mother can offer 5 fruits to Anil is ....

27 Points $P(-3, 2)$, $Q(9, 10)$ and $R(\alpha, 4)$ lie on a circle C with PR as its diameter. The tangents to C at the points Q and R intersect at the point S. If S lies on the line $2x-ky=1$, then k is equal to ....

28 A triangle is formed by X-axis, Y-axis and the line $3x+4y=60$. Then, the number of point $P(a, b)$ which lie strictly inside the triangle, where a is an integer and b is a multiple of a, is ....

29 Let $a\in R$ and let α, β be the roots of the equation $x^2+60^{1/4}x+a=0$. If $\alpha^4+\beta^4=-30$, then the product of all possible values of a is

30 For the two positive numbers a, b, if a, b and $\frac{1}{18}$ are in a geometric progression. while $\frac{1}{a}$, 10 and $\frac{1}{b}$ are in an arithmetic progression, then $16a+12b$ is equal to

29 JANUARY 2023, SHIFT-I

Section A : Objective Type Questions

1 Let $f: R\to R$ be a function such that $f(x)=\frac{x^2+2x+1}{x^2+1}$. Then,

(a) $f(x)$ is one-one in $[1,\infty)$ but not in $(-\infty,\infty)$
(b) $f(x)$ is many-one in $(-\infty,-1)$
(c) $f(x)$ is one-one in $(-\infty,\infty)$
(d) $f(x)$ is many-one in $(1,\infty)$

> **Key Idea** A function $f: A\to B$ is a one-one, if for every $a_1, a_2\in A$ such that $a_1\neq a_2$, then $f(a_1)\neq f(a_2)$.
> And $f: A\to B$ is many-one, if for every $a_1, a_2\in A$ such that $a_1\neq a_2$, then $f(a_1)=f(a_2)$.

2 Let α and β be real numbers. Consider a 3×3 matrix A, such that $A^2=3A+\alpha I$. If $A^4=21A+\beta I$, then

(a) $\alpha=1$ (b) $\alpha=4$
(c) $\beta=8$ (d) $\beta=-8$

3 Three rotten apples are mixed accidently with seven good apples and four apples are drawn one by one without replacement. Let the random variable X denote the number of rotten apples. If μ and σ^2 represent mean and variance of X respectively, then $10(\mu^2+\sigma^2)$ is equal to

(a) 20 (b) 250 (c) 25 (d) 30

4 Let $x=2$ be a root of the equation $x^2+px+q=0$ and

$$f(x)=\begin{cases}\dfrac{1-\cos(x^2-4px+q^2+8q+16)}{(x-2p)^4}, & x\neq 2p\\ 0, & x=2p\end{cases}$$

Then, $\lim_{x\to 2p^+}[f(x)]$, where [.] denotes greatest integer function, is

(a) 1 (b) -1
(c) 2 (d) 0

5 For two non-zero complex numbers z_1 and z_2, if $\text{Re}(z_1 z_2) = 0$ and $\text{Re}(z_1 + z_2) = 0$, then which of the following are possible?

A. $\text{Im}(z_1) > 0$ and $\text{Im}(z_2) > 0$ B. $\text{Im}(z_1) < 0$ and $\text{Im}(z_2) > 0$
C. $\text{Im}(z_1) > 0$ and $\text{Im}(z_2) < 0$ D. $\text{Im}(z_1) < 0$ and $\text{Im}(z_2) < 0$

Choose the correct answer from the options given below

(a) B and D (b) A and B (c) A and C (d) B and C

6 Let the tangents at the points $A(4, -11)$ and $B(8, -5)$ on the circle $x^2 + y^2 - 3x + 10y - 15 = 0$, intersect at the point C. Then, the radius of the circle, whose centre is C and the line joining A and B is its tangent, is equal to

(a) $2\sqrt{13}$ (b) $\frac{2\sqrt{13}}{3}$ (c) $\frac{3\sqrt{3}}{4}$ (d) $\sqrt{13}$

> **Key Idea** The equation of tangent at point $p(x_1, y_1)$ to a circle $x^2 + y^2 + 2gx + 2fy + c = 0$ is
> $xx_1 + yy_1 + g(x + x_1) + f(y + y_1) + c = 0$.

7 Let Δ be the area of the region $\{(x, y) \in R^2 : x^2 + y^2 \le 21, y^2 \le 4x, x \ge 1\}$.

Then, $\frac{1}{2}\left(\Delta - 21 \sin^{-1}\frac{2}{\sqrt{7}}\right)$ is equal to

(a) $2\sqrt{3} - \frac{2}{3}$ (b) $\sqrt{3} - \frac{2}{3}$ (c) $\sqrt{3} - \frac{4}{3}$ (d) $2\sqrt{3} - \frac{1}{3}$

8 Let $\lambda \ne 0$ be a real number. Let α, β be the roots of the equation $14x^2 - 31x + 3\lambda = 0$ and α, γ be the roots of the equation $35x^2 - 53x + 4\lambda = 0$. Then, $\frac{3\alpha}{\beta}$ and $\frac{4\alpha}{\gamma}$ are the roots of the equation

(a) $7x^2 - 245x + 250 = 0$ (b) $7x^2 + 245x - 250 = 0$
(c) $49x^2 - 245x + 250 = 0$ (d) $49x^2 + 245x + 250 = 0$

9 Consider the following system of equations

$\alpha x + 2y + z = 1$, $2\alpha x + 3y + z = 1$,
$3x + \alpha y + 2z = \beta$

for some $\alpha, \beta \in R$. Then, which of the following is not correct.

(a) It has a solution, $\forall \alpha \ne 1$ and $\beta = 2$
(b) It has no solution, if $\alpha = -1$ and $\beta \ne 2$
(c) It has no solution, for $\alpha = 3$ and $\forall \beta \ne 2$
(d) It has no solution, for $\alpha = -1$ and $\forall \beta \in R$

10 Let $y = f(x)$ be the solution of the differential equation $y(x+1)dx - x^2 dy = 0$, $y(1) = e$. Then, $\lim_{x \to 0^+} f(x)$ is equal to

(a) 0 (b) e^2 (c) $\frac{1}{e^2}$ (d) $\frac{1}{e}$

11 A light ray emits from the origin making an angle 30° with the positive X-axis. After getting reflected by the line $x + y = 1$, if this ray intersects X-axis at Q, then the abscissa of Q is

(a) $\frac{2}{3 + \sqrt{3}}$ (b) $\frac{2}{3 - \sqrt{3}}$ (c) $\frac{\sqrt{3}}{2(\sqrt{3} + 1)}$ (d) $\frac{2}{(\sqrt{3} - 1)}$

12 Let $A = \{(x, y) \in R^2 : y \ge 0, 2x \le y \le \sqrt{4 - (x-1)^2}\}$ and $B = \{(x, y) \in R \times R : 0 \le y \le \min\{2x, \sqrt{4 - (x-1)^2}\}\}$.

Then, the ratio of the area of A to the area of B is

(a) $\frac{\pi}{\pi - 1}$ (b) $\frac{\pi + 1}{\pi - 1}$ (c) $\frac{\pi - 1}{\pi + 1}$ (d) $\frac{\pi}{\pi + 1}$

13 If the vectors $\mathbf{a} = \lambda\hat{\mathbf{i}} + \mu\hat{\mathbf{j}} + 4\hat{\mathbf{k}}$, $\mathbf{b} = -2\hat{\mathbf{i}} + 4\hat{\mathbf{j}} - 2\hat{\mathbf{k}}$ and $\mathbf{c} = 2\hat{\mathbf{i}} + 3\hat{\mathbf{j}} + \hat{\mathbf{k}}$ are coplanar and the projection of a on the vector $\mathbf{b}$ is $\sqrt{54}$ units, then the sum of all possible values of $\lambda + \mu$ is equal to

(a) 18 (b) 24 (c) 0 (d) 6

14 Let $f(\theta) = 3\left(\sin^4\left(\frac{3\pi}{2} - \theta\right) + \sin^4(3\pi + \theta)\right) - 2(1 - \sin^2 2\theta)$ and $S = \left\{\theta \in [0, \pi] : f'(\theta) = -\frac{\sqrt{3}}{2}\right\}$. If $4\beta = \sum_{\theta \in S} \theta$, then $f(\beta)$ is equal to

(a) $\frac{5}{4}$ (b) $\frac{3}{2}$ (c) $\frac{11}{8}$ (d) $\frac{9}{8}$

15 If p, q and r are three propositions, then which of the following combination of truth values of p, q and r makes the logical expression $\{(p \vee q) \wedge ((\sim p) \vee r)\} \to ((\sim q) \vee r)$ false?

(a) $p = F, q = T, r = F$ (b) $p = T, q = F, r = F$
(c) $p = T, q = T, r = F$ (d) $p = T, q = F, r = T$

16 Let $f(x) = x + \frac{a}{\pi^2 - 4}\sin x + \frac{b}{\pi^2 - 4}\cos x$, $x \in R$ be a function which satisfies $f(x) = x + \int_0^{\pi/2} \sin(x + y) f(y)\, dy$.

Then, $(a + b)$ is equal to

(a) $-\pi(\pi + 2)$ (b) $-2\pi(\pi - 2)$
(c) $-2\pi(\pi + 2)$ (d) $-\pi(\pi - 2)$

17 The domain of $f(x) = \frac{\log_{(x-1)}(x-2)}{e^{2\log x} - (2x + 3)}$, $x \in R$ is

(a) $R - \{3\}$ (b) $(2, \infty) - \{3\}$
(c) $(-1, \infty) - \{3\}$ (d) $R - \{-1, 3\}$

18 Let $[x]$ denote the greatest integer $\le x$. Consider the function $f(x) = \max\{x^2, 1 + [x]\}$. Then, the value of the integral $\int_0^2 f(x) dx$ is

(a) $\frac{4 + 5\sqrt{2}}{3}$ (b) $\frac{1 + 5\sqrt{2}}{3}$
(c) $\frac{8 + 4\sqrt{2}}{3}$ (d) $\frac{5 + 4\sqrt{2}}{3}$

19 Let B and C be the two points on the line $y + x = 0$ such that B and C are symmetric with respect to the origin. Suppose, A is a point on $y - 2x = 2$ such that ΔABC is an equilateral triangle. Then, the area of the ΔABC is

(a) $\frac{10}{\sqrt{3}}$ (b) $3\sqrt{3}$
(c) $2\sqrt{3}$ (d) $\frac{8}{\sqrt{3}}$

20 Fifteen football players of a club-team are given 15 T-shirts with their names written on the backside. If the players pick up the T-shirts randomly, then the probability that at least 3 players pick the correct T-shirt is

(a) $\frac{5}{36}$ (b) $\frac{5}{24}$

(c) $\frac{1}{6}$ (d) $\frac{2}{15}$

Section B : Numerical Value Type Questions

21 Let **a**, **b** and **c** be three non-zero non-coplanar vectors. Let the position vectors of four points A, B, C and D be $\mathbf{a}-\mathbf{b}+\mathbf{c}$, $\lambda\mathbf{a}-3\mathbf{b}+4\mathbf{c}$, $-\mathbf{a}+2\mathbf{b}-3\mathbf{c}$ and $2\mathbf{a}-4\mathbf{b}+6\mathbf{c}$, respectively. If **AB**, **AC** and **AD** are coplanar, then λ is equal to ....

22 If all the six digit numbers $x_1\,x_2\,x_3\,x_4\,x_5\,x_6$ with $0<x_1<x_2<x_3<x_4<x_5<x_6$ are arranged in the increasing order, then the sum of the digits in the 72th number is ....

23 Five digit numbers are formed using the digits 1, 2, 3, 5, 7 with repetitions and are written in descending order with serial numbers. e.g. the number 77777 has serial number 1. Then, the serial number of 35337 is ....

24 Suppose, f is a function satisfying

$f(x+y)=f(x)+f(y), \forall\, x, y \in N$ and $f(1)=\frac{1}{5}$.

If $\sum_{n=1}^{m} \frac{f(n)}{n(n+1)(n+2)}=\frac{1}{12}$, then m is equal to ....

25 If the coefficient of x^9 in $\left(\alpha x^3+\frac{1}{\beta x}\right)^{11}$ and the coefficient of x^{-9} in $\left(\alpha x-\frac{1}{\beta x^3}\right)^{11}$ are equal, then $(\alpha\beta)^2$ is equal to ....

26 Let the equation of the plane P containing the line $x+10=\frac{8-y}{2}=z$ be $ax+by+3z=2(a+b)$ and the distance of the plane P from the point (1, 27, 7) be c. Then, $a^2+b^2+c^2$ is equal to ....

27 Let the co-ordinates of one vertex of ΔABC be $A(0, 2, \alpha)$ and the other two vertice lie on the line $\frac{x+\alpha}{5}=\frac{y-1}{2}=\frac{z+4}{3}$. For $\alpha \in Z$, if the area of ΔABC is 21 sq. units and the line segment BC has length $2\sqrt{21}$ units, then α^2 is equal to ....

28 Let $a_1, a_2, a_3, \ldots$ be a GP of increasing positive numbers, If the product of fourth and sixth term is 9 and the sum of fifth and seventh terms is 24, then $a_1a_9+a_2\,a_4a_9+a_5+a_7$ is equal to ....

29 Let $f: R \to R$ be a differentiable function that satisfies the relation $f(x+y)=f(x)+f(y)-1, \forall x, y \in R$. If $f'(0)=2$, then $|f(-2)|$ is equal to ....

30 Let the coefficients of three consecutive terms in the binomial expansion of $(1+2x)^n$ be in the ratio 2 : 5 : 8. Then, the coefficient of the term, which is in the middle of these three terms, is ....

29 JANUARY 2023, SHIFT-II

Section A : Objective Type Questions

1 The set of all values of $t \in R$, for which the matrix

$$\begin{bmatrix} e^t & e^{-t}(\sin t-2\cos t) & e^{-t}(-2\sin t-\cos t) \\ e^t & e^{-t}(2\sin t+\cos t) & e^{-t}(\sin t-2\cos t) \\ e^t & e^{-1}\cos t & e^{-t}\sin t \end{bmatrix}$$ is

invertible, is

(a) $\{k\pi, k \in Z\}$ (b) R

(c) $\left\{k\pi+\frac{\pi}{4}, k \in Z\right\}$ (d) $\left\{(2k+1)\frac{\pi}{2}, k \in Z\right\}$

2 The statement $B \Rightarrow ((\sim A) \vee B)$ is equivalent to

(a) $B \Rightarrow (A \Rightarrow B)$ (b) $A \Rightarrow ((\sim A) \Rightarrow B)$

(c) $B \Rightarrow ((\sim A) \Rightarrow B)$ (d) $A \Rightarrow (A \Leftrightarrow B)$

3 If $\mathbf{a}=\hat{\mathbf{i}}+2\hat{\mathbf{k}}$, $\mathbf{b}=\hat{\mathbf{i}}+\hat{\mathbf{j}}+\hat{\mathbf{k}}$, $\mathbf{c}=7\hat{\mathbf{i}}-3\hat{\mathbf{j}}+4\hat{\mathbf{k}}$, $\mathbf{r}\times\mathbf{b}+\mathbf{b}\times\mathbf{c}=\mathbf{0}$ and $\mathbf{r}\cdot\mathbf{a}=0$. Then, $\mathbf{r}\cdot\mathbf{c}$ is equal to

(a) 34 (b) 36

(c) 30 (d) 32

4 Let $\mathbf{a}=4\hat{\mathbf{i}}+3\hat{\mathbf{j}}$ and $\mathbf{b}=3\hat{\mathbf{i}}-4\hat{\mathbf{j}}+5\hat{\mathbf{k}}$. If **c** is a vector such that $\mathbf{c}\cdot(\mathbf{a}\times\mathbf{b})+25=0$, $\mathbf{c}\cdot(\hat{\mathbf{i}}+\hat{\mathbf{j}}+\hat{\mathbf{k}})=4$ and projection of **c** on **a** is 1, then the projection of **c** on **b** equals

(a) $\frac{1}{5}$ (b) $\frac{3}{\sqrt{2}}$ (c) $\frac{5}{\sqrt{2}}$ (d) $\frac{1}{\sqrt{2}}$

5 Let f and g be twice differentiable functions on R such that

$f''(x)=g''(x)+6x$

$f'(1)=4g'(1)-3=9$

$f(2)=3g(2)=12$.

Then, which of the following is not true?

(a) There exists $x_0 \in \left(1, \frac{3}{2}\right)$ such that $f(x_0)=g(x_0)$

(b) $g(-2)-f(-2)=20$

(c) If $-1<x<2$, then $|f(x)-g(x)|<8$

(d) $|f'(x)-g'(x)|<6 \Rightarrow -1<x<1$

6 The plane $2x - y + z = 4$ intersects the line segment joining the points $A(a, -2, 4)$ and $B(2, b, -3)$ at the point C in the ratio 2 : 1 and the distance of the point C from the origin is $\sqrt{5}$. If $ab < 0$ and P is the point $(a - b, b, 2b - a)$, then CP^2 is equal to

(a) $\frac{97}{3}$ (b) $\frac{17}{3}$ (c) $\frac{16}{3}$ (d) $\frac{73}{3}$

7 Let K be the sum of the coefficient of the odd powers of x in the expansion of $(1 + x)^{99}$. Let a be the middle term in the expansion of $\left(2 + \frac{1}{\sqrt{2}}\right)^{200}$. If $\frac{{}^{200}C_{99}K}{a} = \frac{2^l m}{n}$, where m and n are odd numbers, then the ordered pair (l, n) is equal to

(a) (51,99) (b) (50,51) (c) (50,101) (d) (51,101)

> **Key Idea** For the solution firstly find the middle term using formula $T_{r+1} = {}^nC_r x^{n-r} y^r$, then solve it.

8 Let R be a relation defined on N as $a\,R\,b$, if $2a + 3b$ is a multiple of 5, $a, b \in N$. Then, R is

(a) symmetric but not transitive
(b) an equivalence relation
(c) transitive but not symmetric
(d) not reflexive

9 Let $S = \{w_1, w_2, ...\}$ be the sample space associated to a random experiment. Let $P(w_n) = \frac{P(w_{n-1})}{2}, n \geq 2$. Let $A = \{2k + 3l : k, l \in N\}$ and $B = \{w_n : n \in A\}$. Then, $P(B)$ is equal to

(a) $\frac{1}{16}$ (b) $\frac{1}{32}$ (c) $\frac{3}{64}$ (d) $\frac{3}{32}$

10 Consider a function $f : N \to R$, satisfying $f(1) + 2f(2) + 3f(3) + \ldots + xf(x) = x(x + 1) f(x); x \geq 2$ with $f(1) = 1$. Then, $\frac{1}{f(2022)} + \frac{1}{f(2028)}$ is equal to

(a) 8200 (b) 8000 (c) 8400 (d) 8100

11 The area of the region $A = \left\{(x, y) : |\cos x - \sin x| \leq y \leq \sin x, 0 \leq x \leq \frac{\pi}{2}\right\}$ is

(a) $\sqrt{5} + 2\sqrt{2} - 4.5$ (b) $\frac{3}{\sqrt{5}} - \frac{3}{\sqrt{2}} + 1$
(c) $1 - \frac{3}{\sqrt{2}} + \frac{4}{\sqrt{5}}$ (d) $\sqrt{5} - 2\sqrt{2} + 1$

12 The shortest distance between the lines $\frac{x-1}{2} = \frac{y+8}{-7} = \frac{z-4}{5}$ and $\frac{x-1}{2} = \frac{y-2}{1} = \frac{z-6}{-3}$ is

(a) $5\sqrt{3}$ (b) $3\sqrt{3}$
(c) $4\sqrt{3}$ (d) $2\sqrt{3}$

13 Let $y = y(x)$ be the solution of the differential equation $x \log_e x \frac{dy}{dx} + y = x^2 \log_e x, (x > 1)$. If $y(2) = 2$, then $y(e)$ is equal to

(a) $\frac{1 + e^2}{2}$ (b) $\frac{2 + e^2}{2}$ (c) $\frac{4 + e^2}{4}$ (d) $\frac{1 + e^2}{4}$

14 The set of all values of λ for which the equation $\cos^2 2x - 2\sin^4 x - 2\cos^2 x = \lambda$ has a real solution x, is

(a) $\left[-\frac{3}{2}, -1\right]$ (b) $[-2, -1]$
(c) $\left[-2, -\frac{3}{2}\right]$ (d) $\left[-1, -\frac{1}{2}\right]$

15 If the lines $\frac{x-1}{1} = \frac{y-2}{2} = \frac{z+3}{1}$ and $\frac{x-a}{2} = \frac{y+2}{3} = \frac{z-3}{1}$ intersect at the point P, then the distance of the point P from the plane $z = a$ is

(a) 22 (b) 10 (c) 28 (d) 16

16 If the tangent at a point P on the parabola $y^2 = 3x$ is parallel to the line $x + 2y = 1$ and the tangents at the points Q and R on the ellipse $\frac{x^2}{4} + \frac{y^2}{1} = 1$ are perpendicular to the line $x - y = 2$, then the area of the gle ΔPQR is

(a) $5\sqrt{3}$ (b) $\frac{3}{2}\sqrt{5}$ (c) $\frac{9}{\sqrt{5}}$ (d) $3\sqrt{5}$

17 The value of the integral $\int_{1/2}^{2} \frac{\tan^{-1} x}{x} dx$ is equal to

(a) $\frac{1}{2}\log_e 2$ (b) $\frac{\pi}{2}\log_e 2$ (c) $\frac{\pi}{4}\log_e 2$ (d) $\pi \log_e 2$

18 The value of the integral $\int_1^2 \left(\frac{t^4 + 1}{t^6 + 1}\right) dt$ is

(a) $\tan^{-1} 2 - \frac{1}{3}\tan^{-1} 8 + \frac{\pi}{3}$
(b) $\tan^{-1} 2 + \frac{1}{3}\tan^{-1} 8 - \frac{\pi}{3}$
(c) $\tan^{-1}\frac{1}{2} + \frac{1}{3}\tan^{-1} 8 - \frac{\pi}{3}$
(d) $\tan^{-1}\frac{1}{2} - \frac{1}{3}\tan^{-1} 8 + \frac{\pi}{3}$

19 The number of 3-digit numbers that are divisible by either 3 or 4 but not divisible by 48, is

(a) 507 (b) 400 (c) 472 (d) 432

20 The letters of the word 'OUGHT' are written in all possible ways and these words are arranged as in a dictionary in a series. Then, the serial number of the word 'TOUGH' is

(a) 84 (b) 89 (c) 86 (d) 79

Section B : Numerical Value Type Questions

21 Let $X = \{11, 12, 13, \ldots, 40, 41\}$ and $Y = \{61, 62, 63, \ldots, 90, 91\}$ be the two sets of observations. If $\bar{x}$ and $\bar{y}$ are their respective means and σ^2 is the variance of all the observations in $X \cup Y$, then $|\bar{x} + \bar{y} - \sigma^2|$ is equal to ....

> **Key Idea** For find the solution of this type of problems. Lirstly find the mean, then use the value of mean, find the variance of the given data with the help of $\sigma^2 = \frac{\Sigma x_i^2 + y_i^2}{n} - \left(\frac{\Sigma x_i + \Sigma y_i^2}{n}\right)$.

22 A circle with centre (2, 3) and radius 4 intersects the line $x + y = 3$ at the points P and Q. If the tangents at P and Q intersect at the point $S(\alpha, \beta)$, then $4\alpha - 7\beta$ is equal to ....

23 The total number of 4-digit numbers, whose greatest common divisor with 54 is 2, is ....

24 Let $a_1 = b_1 = 1$ and $a_n = a_{n-1} + (n-1)$, $b_n = b_{n-1} + a_{n-1}, \forall n \geq 2$. If $S = \sum_{n=1}^{10} \frac{b_n}{2^n}$ and $T = \sum_{n=1}^{8} \frac{n}{2^{n-1}}$, then $2^7(2S - T)$ is equal to ....

25 Let $\alpha = 8 - 14i$, $A = \left\{z \in c : \frac{\alpha z - \overline{\alpha z}}{z^2 - (\bar{z})^2 - 112i} = 1\right\}$ and $B = \{z \in c : |z + 3i| = 4\}$. Then, $\sum_{z \in A \cap B} (\text{Re } z - \text{Im } z)$ is equal to ....

26 Let $\{a_k\}$ and $\{b_k\}$, $k \in N$ be two GPs with common ratios r_1 and r_2 respectively such that $a_1 = b_1 = 4$ and $r_1 < r_2$. Let $c_k = a_k + b_k$, $k \in N$. If $c_2 = 5$ and $c_3 = \frac{13}{4}$, then $\sum_{k=1}^{\infty} c_k - (12a_6 + 8b_4)$ is equal to ....

27 A triangle is formed by the tangents at the point (2, 2) on the curves $y^2 = 2x$ and $x^2 + y^2 = 4x$ and the line $x + y + 2 = 0$. If r is the radius of its circumcircle, then r^2 is equal to ....

28 If the equation of the normal to the curve $y = \frac{x - a}{(x + b)(x - 2)}$ at the point $(1, -3)$ is $x - 4y = 13$, then the value of $a + b$ is equal to ....

29 Let A be a symmetric matrix such that $|A| = 2$ and $\begin{bmatrix} 2 & 1 \\ 3 & \frac{3}{2} \end{bmatrix} A = \begin{bmatrix} 1 & 2 \\ \alpha & \beta \end{bmatrix}$. If the sum of the diagonal elements of A is s, then $\frac{\beta s}{\alpha^2}$ is equal to ....

30 Let $\alpha_1, \alpha_2, \ldots, \alpha_7$ be the roots of the equation $x^7 + 3x^5 - 13x^3 - 15x = 0$ and $|\alpha_1| \geq |\alpha_2| \geq \ldots \geq |\alpha_7|$. Then, $\alpha_1\alpha_2 - \alpha_3\alpha_4 + \alpha_5\alpha_6$ is equal to ....

30 JANUARY 2023, SHIFT-I

Section A : Objective Type Questions

1 Among the statements

$(S_1) : ((p \vee q) \Rightarrow r) \Leftrightarrow (p \Rightarrow r)$

$(S_2) : ((p \vee q) \Rightarrow r) \Leftrightarrow ((p \Rightarrow r) \vee (q \Rightarrow r)$

(a) Both (S_1) and (S_2) are tautologies
(b) Neither (S_1) nor (S_2) is a tautology
(c) Only (S_2) is a tautology
(d) Only (S_1) is a tautology

2 Let the system of linear equations

$$x + y + kz = 2$$
$$2x + 3y - z = 1$$
$$3x + 4y + 2z = k$$

have infinitely many solutions. Then, the system

$$(k + 1)x + (2k - 1)y = 7$$
$$(2k + 1)x + (k + 5)y = 10$$

has

(a) No solution
(b) unique solution satisfying $x - y = 1$
(c) unique solution satisfying $x + y = 1$
(d) infinitely many solutions

3 The coefficient of x^{301} in $(1 + x)^{500} + x(1 + x)^{499} + x^2(1 + x)^{498} + \ldots + x^{500}$ is

(a) ${}^{500}C_{300}$ (b) ${}^{501}C_{200}$
(c) ${}^{500}C_{301}$ (d) ${}^{501}C_{302}$

4 Let the solution curve $y = y(x)$ of the differential equation $\frac{dy}{dx} - \frac{3x^5 \tan^{-1}(x^3)}{(1 + x^6)^{3/2}} y = 2x \exp\left\{\frac{x^3 - \tan^{-1} x^3}{\sqrt{(1 + x^6)}}\right\}$ pass through the origin. Then, $y(1)$ is equal to

(a) $\exp\left(\frac{4 + \pi}{4\sqrt{2}}\right)$ (b) $\exp\left(\frac{1 - \pi}{4\sqrt{2}}\right)$
(c) $\exp\left(\frac{4 - \pi}{4\sqrt{2}}\right)$ (d) $\exp\left(\frac{\pi - 4}{4\sqrt{2}}\right)$

> **Key Idea** Reduce the given equation into a linear differential equation as form $\frac{dy}{dx} + Py = Q$
> I.F $= e^{\int P dv}$ and $y \times \text{IF} = \int \text{IF} \times Q\, dv$.
> Then, solve it using Integration method.

5 If an unbiased die, marked with $-2, -1, 0, 1, 2, 3$ on its faces, is thrown five times, then the probability that the product of the outcomes is positive, is

(a) $\frac{27}{288}$ (b) $\frac{521}{2592}$ (c) $\frac{440}{2592}$ (d) $\frac{881}{2592}$

6 If $\mathbf{a}, \mathbf{b}, \mathbf{c}$ are three non-zero vectors and $\hat{\mathbf{n}}$ is a unit vector perpendicular to $\mathbf{c}$ such that $\mathbf{a} = \alpha\mathbf{b} - \hat{\mathbf{n}}$, $(\alpha \neq 0)$ and $\mathbf{b} \cdot \mathbf{c} = 12$, then $|\mathbf{c} \times (\mathbf{a} \times \mathbf{b})|$ is equal to

(a) 6 (b) 15 (c) 12 (d) 9

7 The line l_1 passes through the point (2,6,2) and is perpendicular to the plane $2x + y - 2z = 10$. Then, the shortest distance between the line l_1 and the line $\frac{x+1}{2} = \frac{y+4}{-3} = \frac{z}{2}$ is

(a) $\frac{13}{3}$ (b) 9 (c) 7 (d) $\frac{19}{3}$

8 A straight line cut-off the intercepts $OA = a$ and $OB = b$ on the positive directions of X-axis and Y-axis, respectively. If the perpendicular from origin O to this line makes an angle of $\frac{\pi}{6}$ with positive direction of Y-axis and the area of ΔOAB is $\frac{98}{3}\sqrt{3}$, then $a^2 - b^2$ is equal to

(a) $\frac{392}{3}$ (b) $\frac{196}{3}$ (c) 196 (d) 98

9 The number of points on the curve $y = 54x^5 - 135x^4 - 70x^3 + 180x^2 + 210x$ at which the normal lines are parallel to $x + 90y + 2 = 0$ is

(a) 0 (b) 3
(c) 4 (d) 2

> **Key Idea** Find the $\frac{dy}{dx}$ from the given curve, then normal line to the curve is parallel of given line.
>
> $\left.\frac{-1}{(dv/dy)}\right|_{\text{at}(v_1, y_1)}$ = slope of line.

10 If $P(h, k)$ is a point on the parabola $x = 4y^2$, which is nearest to the point $Q(0, 33)$, then the distance of P from the directrix of the parabola $y^2 = 4(x + y)$ is equal to

(a) 2 (b) 8 (c) 4 (d) 6

11 Suppose, $f : R \to (0, \infty)$ be a differentiable function such that $5f(x + y) = f(x) \cdot f(y), \forall x, y \in R$. If $f(3) = 320$, then $\sum_{n=0}^{5} f(n)$ is equal to

(a) 6525 (b) 6825 (c) 6575 (d) 6875

12 The minimum number of elements that must be added to the relation $R = \{(a, b), (b, c)\}$ on the set $\{a, b, c\}$, so that it becomes symmetric and transitive is

(a) 4 (b) 5 (c) 3 (d) 7

13 If $[t]$ denotes the greatest integer $\le t$, then the value of $\frac{3(e-1)}{e}\int_1^2 x^2 e^{[x]+[x^3]}\,dx$ is

(a) $e^7 - 1$ (b) $e^8 - 1$
(c) $e^9 - e$ (d) $e^8 - e$

14 Let $A = \begin{pmatrix} m & n \\ p & q \end{pmatrix}$, $d = |A| \ne 0$ and $|A - d(\text{Adj } A)| = 0$. Then,

(a) $1 + d^2 = (m + q)^2$ (b) $1 + d^2 = m^2 + q^2$
(c) $(1 + d)^2 = m^2 + q^2$ (d) $(1 + d)^2 = (m + q)^2$

15 If the solution of the equation $\log_{\cos x} \cot x + 4\log_{\sin x} \tan x = 1, x \in \left(0, \frac{\pi}{2}\right)$, is $\sin^{-1}\left(\frac{\alpha + \sqrt{\beta}}{2}\right)$, where α, β are integers, then $\alpha + \beta$ is equal to

(a) 3 (b) 5 (c) 6 (d) 4

16 Let a unit vector **OP** make angles α, β, γ with the positive directions of the coordinate axes OX, OY, OZ respectively, where $\beta \in \left(0, \frac{\pi}{2}\right)$. If **OP** is perpendicular to the plane through points (1, 2, 3), (2, 3, 4) and (1, 5, 7), then which one of the following is true?

(a) $\alpha \in \left(0, \frac{\pi}{2}\right)$ and $\gamma \in \left(0, \frac{\pi}{2}\right)$

(b) $\alpha \in \left(\frac{\pi}{2}, \pi\right)$ and $\gamma \in \left(0, \frac{\pi}{2}\right)$

(c) $\alpha \in \left(\frac{\pi}{2}, \pi\right)$ and $\gamma \in \left(\frac{\pi}{2}, \pi\right)$

(d) $\alpha \in \left(0, \frac{\pi}{2}\right)$ and $\gamma \in \left(\frac{\pi}{2}, \pi\right)$

17 Let $y = x + 2$, $4y = 3x + 6$ and $3y = 4x + 1$ be three tangent lines to the circle $(x - h)^2 + (y - k)^2 = r^2$. Then, $h + k$ is equal to

(a) $5(1 + \sqrt{2})$ (b) $5\sqrt{2}$ (c) 5 (d) 6

18 If the coefficient of x^{15} in the expansion of $\left(ax^3 + \frac{1}{bx^{\frac{1}{3}}}\right)^{15}$ is equal to the coefficient of x^{-15} in the expansion of $\left(ax^{\frac{1}{3}} - \frac{1}{bx^3}\right)^{15}$, where a and b are positive real numbers, then for each such ordered pair (a, b)

(a) $ab = 1$ (b) $a = b$
(c) $a = 3b$ (d) $ab = 3$

19 If $a_n = \frac{-2}{4n^2 - 16n + 15}$, then $a_1 + a_2 + \ldots + a_{25}$ is equal to

(a) $\frac{50}{141}$ (b) $\frac{52}{147}$ (c) $\frac{49}{138}$ (d) $\frac{51}{144}$

20 If $\tan 15° + \frac{1}{\tan 75°} + \frac{1}{\tan 105°} + \tan 195° = 2a$, then the value of $\left(a + \frac{1}{a}\right)$ is

(a) $5 - \frac{3}{2}\sqrt{3}$ (b) $4 - 2\sqrt{3}$ (c) 4 (d) 2

Section B : Numerical Value Type Questions

21 Let $z = 1 + i$ and $z_1 = \frac{1 + i\bar{z}}{\bar{z}(1 - z) + \frac{1}{z}}$. Then, $\frac{12}{\pi}\arg(z_1)$ is equal to ....

22 Let α be the area of the larger region bounded by the curve $y^2 = 8x$ and the lines $y = x$ and $x = 2$, which lies in the first quadrant. Then, the value of 3α is equal to ....

23 $\lim_{x \to \infty} \frac{48}{x^4} \int_0^x \frac{t^3}{t^6 + 1} dt$ is equal to ....

24 Let $S = \{1, 2, 3, 4, 5, 6\}$. Then, the number of one-one functions $f : S \to P(S)$, where $P(S)$ denote the power set of S, such that $f(n) \subset f(m)$, where $n < m$ is ....

25 Number of 4-digit numbers (the repetition of digits is allowed) which are made using the digits 1, 2, 3 and 5 are divisible by 15, is equal to ....

26 Let $f^1(x) = \frac{3x + 2}{2x + 3}$, $x \in R - \left\{\frac{-3}{2}\right\}$.

For $n \geq 2$, defines $f^n(x) = f^1 o f^{n-1}(x)$.

If $f^5(x) = \frac{ax + b}{bx + a}$, gcd $(a, b) = 1$, then $a + b$ is equal to ....

27 The mean and variance of 7 observations are 8 and 16, respectively. If one observation 14 is omitted and a and b are respectively mean and variance of remaining 6 observations, then $a + 3b - 5$ is equal to ....

28 If $\lambda_1 < \lambda_2$ are two values of λ such that the angle between the planes $P_1 : \mathbf{r}(3\hat{\mathbf{i}} - 5\hat{\mathbf{j}} + \hat{\mathbf{k}}) = 7$ and $P_2 : \mathbf{r} \cdot (\lambda\hat{\mathbf{i}} + \hat{\mathbf{j}} - 3\hat{\mathbf{k}}) = 9$ is $\sin^{-1}\left(\frac{2\sqrt{6}}{5}\right)$, then the square of the length of perpendicular from the point $(38\lambda_1, 10\lambda_2, 2)$ to the plane P_1 is ....

29 If the equation of the plane passing through the point (1, 1, 2) and perpendicular to the line $x - 3y + 2z - 1 = 0 = 4x - y + z$ is $Ax + By + Cz = 1$, then $140(C - B + A)$ is equal to ....

30 Let $\sum_{n=0}^{\infty} \frac{n^3 (2n)! + (2n - 1)(n!)}{(n!)((2n)!)} = ae + \frac{b}{e} + c$, where $a, b, c \in Z$ and $e = \sum_{n=0}^{\infty} \frac{1}{n!}$. Then, $a^2 - b + c$ is equal to ....

30 JANUARY 2023, SHIFT-II

Section A : Objective Type Questions

1 Let S be the set of all values of a_1 for which the mean deviation about the mean of 100 consecutive positive integers $a_1, a_2, a_3, \ldots, a_{100}$ is 25. Then, S is

(a) N (b) $\{9\}$ (c) ϕ (d) $\{99\}$

2 The solution of the differential equation $\frac{dy}{dx} = -\left(\frac{x^2 + 3y^2}{3x^2 + y^2}\right)$, $y(1) = 0$ is

(a) $\log_e |x + y| - \frac{2xy}{(x + y)^2} = 0$

(b) $\log_e |x + y| - \frac{xy}{(x + y)^2} = 0$

(c) $\log_e |x + y| + \frac{2xy}{(x + y)^2} = 0$

(d) $\log_e |x + y| + \frac{xy}{(x + y)^2} = 0$

> **Key Idea** The given differential equation is a homogenious differential equation. For find the solution of such type of differential equation substitute $y = vx$. Then, solve it using integration.

3 Let q be the maximum integral value of p in [0, 10] for which the roots of the equation $x^2 - px + \frac{5}{4}p = 0$ are rational. Then, the area of the region $\{(x, y) : 0 \leq y \leq (x - q)^2, 0 \leq x \leq q\}$ is

(a) $\frac{125}{3}$ sq units (b) 243 sq units

(c) 25 sq units (d) 164 sq units

> **Key Idea** If q is the maximum integral value of$p$ in [0, 10] for which roots of the equation are rational.
> We know that, if roots of the quadratic equation are rational, then $D > 0$ and perfect square.

4 Consider the following statements :

P: I have fever

Q: I will not take medicine

R: I will take rest

The statement "If I have fever, then I will take medicine and I will take rest" is equivalent to

(a) $((\sim P) \vee \sim Q) \wedge ((\sim P) \vee \sim R)$

(b) $((\sim P) \vee \sim Q) \wedge ((\sim P) \vee R)$

(c) $(P \vee Q) \wedge ((\sim P) \vee R)$

(d) $(P \vee \sim Q) \wedge (P \vee \sim R)$

5 Let $a_1 = 1, a_2, a_3, a_4, \ldots\ldots$ be consecutive natural numbers. Then,

$$\tan^{-1}\left(\frac{1}{1 + a_1a_2}\right) + \tan^{-1}\left(\frac{1}{1 + a_2a_3}\right) + \ldots + \tan^{-1}\left(\frac{1}{1 + a_{2021}a_{2022}}\right)$$ is equal to

(a) $\frac{\pi}{4} - \cot^{-1}(2022)$ (b) $\tan^{-1}(2022) - \frac{\pi}{4}$

(c) $\frac{\pi}{4} - \tan^{-1}(2022)$ (d) $\cot^{-1}(2022) - \frac{\pi}{4}$

6 Let $\mathbf{a}$ and $\mathbf{b}$ be two vectors. Let $|\mathbf{a}| = 1$, $|\mathbf{b}| = 4$ and $\mathbf{a} \cdot \mathbf{b} = 2$. If $\mathbf{c} = (2\mathbf{a} \times \mathbf{b}) - 3\mathbf{b}$, then the value of $\mathbf{b} \cdot \mathbf{c}$ is

(a) – 84 (b) – 60 (c)– 24 (d) – 48

7 Let $a, b, c>1$, a^3, b^3 and c^3 be in AP and $\log_a b$, $\log_c a$ and $\log_b c$ be in GP. If the sum of first 20 terms of an AP, whose first term is $\frac{a+4b+c}{3}$ and the common difference is $\frac{a-8b+c}{10}$ is -444, then abc is equal to

(a) 216 (b) 343 (c) $\frac{343}{8}$ (d) $\frac{125}{8}$

8 The number of ways of selecting two numbers a and b, $a \in \{2, 4, 6, \ldots, 100\}$ and $b \in \{1, 3, 5, \ldots, 99\}$ such that 2 is the remainder when $a+b$ is divided by 23 is

(a) 186 (b) 54 (c) 108 (d) 268

9 Let f, g and h be the real valued functions defined on R as

$$f(x)=\begin{cases}\frac{x}{|x|}, & x\neq 0\\ 1, & x=0\end{cases},$$

$$g(x)=\begin{cases}\frac{\sin(x+1)}{(x+1)}, & x\neq -1\\ 1, & x=-1\end{cases}$$

and $h(x)=2[x]-f(x)$, where $[x]$ is the greatest integer $\leq x$. Then, the value of $\lim_{x\to 1} g(h(x-1))$ is

(a) 1 (b) sin (1) (c) 0 (d) -1

10 A vector $\mathbf{v}$ in the first octant is inclined to the X-axis at 60°, to the Y-axis at 45° and to the Z-axis at an acute angle. If a plane passing through the points $(\sqrt{2}, -1, 1)$ and (a, b, c) is normal to $\mathbf{v}$, then

(a) $a+b+\sqrt{2}c=1$ (b) $a+\sqrt{2}b+c=1$
(c) $\sqrt{2}a+b+c=1$ (d) $\sqrt{2}a-b+c=1$

11 Let A be a point on the X-axis. Common tangents are drawn from A to the curves $x^2+y^2=8$ and $y^2=16x$. If one of these tangents touches the two curves at Q and R, then $(QR)^2$ is equal to

(a) 76 (b) 72 (c) 64 (d) 81

12 If P is a 3×3 real matrix such that $P^T=aP+(a-1)I$, where $a>1$, then

(a) P is a singular matrix (b) $|\text{Adj } P|=1$
(c) $|\text{Adj } P|>1$ (d) $|\text{Adj } P|=\frac{1}{2}$

13 The parabolas $ax^2+2bx+cy=0$ and $dx^2+2ex+fy=0$ intersect on the line $y=1$. If a, b, c, d, e, f are positive real numbers and a, b, c are in GP, then

(a) d, e, f are in AP (b) d, e, f are in GP
(c) $\frac{d}{a}, \frac{e}{b}, \frac{f}{c}$ are in GP (d) $\frac{d}{a}, \frac{e}{b}, \frac{f}{c}$ are in AP

14 Let $\lambda \in R$, $\mathbf{a}=\lambda\hat{\mathbf{i}}+2\hat{\mathbf{j}}-3\hat{\mathbf{k}}$, $\mathbf{b}=\hat{\mathbf{i}}-\lambda\hat{\mathbf{j}}+2\hat{\mathbf{k}}$. If $((\mathbf{a}+\mathbf{b})\times(\mathbf{a}\times\mathbf{b}))\times(\mathbf{a}-\mathbf{b})=8\hat{\mathbf{i}}-40\hat{\mathbf{j}}-24\hat{\mathbf{k}}$, then $|\lambda(\mathbf{a}+\mathbf{b})\times(\mathbf{a}-\mathbf{b})|^2$ is equal to

(a) 144 (b) 136
(c) 132 (d) 140

15 $\lim_{n\to\infty}\frac{3}{n}\left\{4+\left(2+\frac{1}{n}\right)^2+\left(2+\frac{2}{n}\right)^2+\ldots+\left(3-\frac{1}{n}\right)^2\right\}$ is equal to

(a) $\frac{19}{3}$ (b) 0 (c) 12 (d) 19

16 For $\alpha, \beta \in R$, suppose the system of linear equations

$$x-y+z=5$$
$$2x+2y+\alpha z=8$$
$$3x-y+4z=\beta$$

has infinitely many solutions. Then, α and β are the roots of

(a) $x^2+14x+24=0$ (b) $x^2+18x+56=0$
(c) $x^2-18x+56=0$ (d) $x^2-10x+16=0$

17 If a plane passes through the points $(-1, k, 0)$, $(2, k, -1)$, $(1, 1, 2)$ and is parallel to the line $\frac{x-1}{1}=\frac{2y+1}{2}=\frac{z+1}{-1}$, then the value of $\frac{k^2+1}{(k-1)(k-2)}$ is

(a) $\frac{17}{5}$ (b) $\frac{5}{17}$ (c) $\frac{6}{13}$ (d) $\frac{13}{6}$

18 If the functions $f(x)=\frac{x^3}{3}+2bx+\frac{ax^2}{2}$ and $g(x)=\frac{x^3}{3}+ax+bx^2$, $a\neq 2b$ have a common extreme point, then $a+2b+7$ is equal to

(a) 3 (b) $\frac{3}{2}$ (c) 6 (d) 4

19 The range of the function $f(x)=\sqrt{3-x}+\sqrt{2+x}$ is

(a) $[\sqrt{5}, \sqrt{13}]$ (b) $[\sqrt{5}, \sqrt{10}]$
(c) $[\sqrt{2}, \sqrt{7}]$ (d) $[2\sqrt{2}, \sqrt{11}]$

20 Let $x=(8\sqrt{3}+13)^{13}$ and $y=(7\sqrt{2}+9)^9$. If $[t]$ denotes the greatest integer $\leq t$, then

(a) $[x]$ is odd but $[y]$ is even (b) $[x]$ and $[y]$ is even
(c) $[x]$ is even but $[y]$ is odd (d) $[x]$ and $[y]$ are both odd

Section B : Numerical Value Type Questions

21 If $\int\sqrt{\sec 2x-1}\,dx=\alpha\log_e\left|\cos 2x+\beta+\sqrt{\cos 2x\left(1+\cos\frac{1}{\beta}x\right)}\right|+\text{constant}$, then $\beta-\alpha$ is equal to ....

22 Let $P(a_1, b_1)$ and $Q(a_2, b_2)$ be two distinct points on a circle with center $C(\sqrt{2}, \sqrt{3})$. Let O be the origin and OC be perpendicular to both CP and CQ. If the area of ΔOCP is $\frac{\sqrt{35}}{2}$, then $a_1^2+a_2^2+b_1^2+b_2^2$ is equal to ....

23 The 8th common term of the series

$$S_1=3+7+11+15+19+\ldots,$$
$$S_2=1+6+11+16+21+\ldots$$

is ....

24 Let a line L passes through the point $P(2, 3, 1)$ and be parallel to the line $x + 3y - 2z - 2 = 0 = x - y + 2z$. If the distance of L from the point (5,3,8) is α, then $3\alpha^2$ is equal to ....

25 Let A be the area of the region $\{(x, y) : y \geq x^2, y \geq (1-x)^2, y \leq 2x(1-x)\}$. Then, 540 A is equal to ....

26 50th root of a number x is 12 and 50th root of another number y is 18. Then, the remainder obtained on dividing $(x + y)$ by 25 is ....

27 Let $A = \{1, 2, 3, 5, 8, 9\}$. Then, the number of possible functions $f : A \to A$ such that $f(m \cdot n) = f(m) \cdot f(n), \forall\, m, n \in A$ with $m \cdot n \in A$ is equal to ....

28 If the value of real number $a > 0$, for which $x^2 - 5ax + 1 = 0$ and $x^2 - ax - 5 = 0$ have a common real root is $\frac{3}{\sqrt{2\beta}}$, then β is equal to ....

29 The number of seven digits odd numbers, that can be formed using all the seven digits 1, 2, 2, 2, 3, 3, 5 is ....

30 A bag contains six balls of different colours. Two balls are drawn in succession with replacement. The probability that both the balls are of the same colour is p. Next four balls are drawn in succession with replacement and the probability that exactly three balls are of the same colour is q. If $p : q = m : n$, where m and n are coprime, then $m + n$ is equal to ....

31 JANUARY 2023, SHIFT-I

Section A : Objective Type Questions

1 The number of real roots of the equation $\sqrt{x^2 - 4x + 3} + \sqrt{x^2 - 9} = \sqrt{4x^2 - 14x + 6}$, is

(a) 3 (b) 2
(c) 0 (d) 1

2 A wire of length 20 m is to be cut into two pieces. A piece of length l_1 is bent to make a square of area A_1 and the other piece of length l_2 is made into a circle of area A_2. If $2A_1 + 3A_2$ is minimum, then $(\pi l_1) : l_2$ is equal to

(a) 3 : 1 (b) 4 : 1
(c) 1 : 6 (d) 6 : 1

3 Let $\alpha \in (0, 1)$ and $\beta = \log_e (1 - \alpha)$. Let $P_n(x) = x + \frac{x^2}{2} + \frac{x^3}{3} + \dots + \frac{x^n}{n}, x \in (0, 1)$. Then, the integral $\int_0^{\alpha} \frac{t^{50}}{1-t}\, dt$ is equal to

(a) $P_{50}(\alpha) - \beta$ (b) $-(\beta + P_{50}(\alpha))$
(c) $\beta + P_{50}(\alpha)$ (d) $\beta - P_{50}(\alpha)$

4 A bag contains 6 balls. Two balls are drawn from it at random and both are found to be black. The probability that the bag contains at least 5 black balls is

(a) $\frac{5}{7}$ (b) $\frac{2}{7}$ (c) $\frac{5}{6}$ (d) $\frac{3}{7}$

5 For all $z \in C$ on the curve $C_1 : |z| = 4$, let the locus of the point $z + \frac{1}{z}$ be the curve C_2. Then,

(a) the curves C_1 and C_2 intersect at 4 points
(b) the curves C_1 and C_2 intersect at 2 points
(c) the curve C_2 lies inside C_1
(d) the curve C_1 lies inside C_2

6 Let $A = \begin{bmatrix} 1 & 0 & 0 \\ 0 & 4 & -1 \\ 0 & 12 & -3 \end{bmatrix}$. Then, the sum of the diagonal elements of the matrix $(A + I)^{11}$ is equal to

(a) 2050 (b) 4094 (c) 6144 (d) 4097

Key Idea To find the sum of the diagonal elements of the matrix like $(A + I)^{11}$. Expand the given matrix with the help of Binomial expansion.

7 If $\sin^{-1}\frac{\alpha}{17} + \cos^{-1}\frac{4}{5} - \tan^{-1}\frac{77}{36} = 0, 0 < \alpha < 13$, then $\sin^{-1}(\sin \alpha) + \cos^{-1}(\cos \alpha)$ is equal to

(a) 0 (b) π (c) 16 (d) $16 - 5\pi$

8 Let $y = f(x) = \sin^3\left(\frac{\pi}{3}\left(\cos\left(\frac{\pi}{3\sqrt{2}}(-4x^3 + 5x^2 + 1)^{\frac{3}{2}}\right)\right)\right)$.

Then, at $x = 1$,

(a) $2y' + 3\pi^2 y = 0$ (b) $2y' + \sqrt{3}\pi^2 y = 0$
(c) $y' + 3\pi^2 y = 0$ (d) $\sqrt{2}y' - 3\pi^2 y = 0$

9 If the sum and product of four positive consecutive terms of a GP are 126 and 1296 respectively, then the sum of common ratios of all such GPs is

(a) 7 (b) 14 (c) 3 (d) 9/2

10 If the domain of the function $f(x) = \frac{[x]}{1 + x^2}$, where $[x]$ is greatest integer $\leq x$, is $[2, 6)$, then its range is

(a) $\left(\frac{5}{37}, \frac{2}{5}\right] - \left\{\frac{9}{29}, \frac{27}{109}, \frac{18}{89}, \frac{9}{53}\right\}$

(b) $\left(\frac{5}{26}, \frac{2}{5}\right] - \left\{\frac{9}{29}, \frac{27}{109}, \frac{18}{89}, \frac{9}{53}\right\}$

(c) $\left(\frac{5}{26}, \frac{2}{5}\right]$

(d) $\left(\frac{5}{37}, \frac{2}{5}\right]$

11 The value of $\int_{\frac{\pi}{3}}^{\frac{\pi}{2}} \frac{(2+3\sin x)}{\sin x(1+\cos x)}\, dx$ is equal to

(a) $\frac{7}{2} - \sqrt{3} - \log_e \sqrt{3}$ (b) $\frac{10}{3} - \sqrt{3} + \log_e \sqrt{3}$

(c) $\frac{10}{3} - \sqrt{3} - \log_e \sqrt{3}$ (d) $-2 + 3\sqrt{3} + \log_e \sqrt{3}$

12 Let a differentiable function f satisfy $f(x) + \int_3^x \frac{f(t)}{t}\, dt = \sqrt{x+1}$, $x \geq 3$. Then, $12f(8)$ is equal to

(a) 34 (b) 17 (c) 19 (d) 1

13 For the system of linear equations

$$x + y + z = 6$$
$$\alpha x + \beta y + 7z = 3$$
$$x + 2y + 3z = 14,$$

which of the following is not true?

(a) If $\alpha = \beta = 7$, then the system has no solution
(b) There is a unique point (α, β) on the line $x + 2y + 18 = 0$ for which the system has infinitely many solutions
(c) If $\alpha = \beta$ and $\alpha \neq 7$, then the system has a unique solution
(d) For every point $(\alpha, \beta) \neq (7, 7)$ on the line $x - 2y + 7 = 0$, the system has infinitely many solutions

14 If the maximum distance of normal to the ellipse $\frac{x^2}{4} + \frac{y^2}{b^2} = 1$, $b < 2$, from the origin is 1, then the eccentricity of the ellipse is

(a) $\frac{1}{\sqrt{2}}$ (b) $\frac{1}{2}$ (c) $\frac{\sqrt{3}}{4}$ (d) $\frac{\sqrt{3}}{2}$

15 Let the shortest distance between the lines $L: \frac{x-5}{-2} = \frac{y-\lambda}{0} = \frac{z+\lambda}{1}$, $\lambda \geq 0$ and $L_1: x+1 = y-1 = 4-z$ be $2\sqrt{6}$. If (α, β, γ) lies on L, then which of the following is not possible?

(a) $\alpha - 2\gamma = 19$ (b) $\alpha + 2\gamma = 24$

(c) $2\alpha - \gamma = 9$ (d) $2\alpha + \gamma = 7$

16 S1 : $(p \Rightarrow q) \vee (p \wedge (\sim q))$ is a tautology
S2 : $((\sim p) \Rightarrow (\sim q)) \wedge ((\sim p) \vee q)$ is a contradiction.
Then,

(a) only (S2) is correct
(b) only (S1) is correct
(c) both (S1) and (S2) are wrong
(d) both (S1) and (S2) are correct

17 Let $y = f(x)$ represent a parabola with focus $\left(-\frac{1}{2}, 0\right)$ and directrix $y = -\frac{1}{2}$. Then,

$S = \left\{x \in R: \tan^{-1}(\sqrt{f(x)}) + \sin^{-1}(\sqrt{f(x)+1}) = \frac{\pi}{2}\right\}$

(a) contains exactly one element
(b) is an infinite set
(c) is an empty set
(d) contains exactly two elements

18 Let R be a relation of $N \times N$ defined by $(a, b)\, R\, (c, d)$, if and only if $ad(b-c) = bc(a-d)$. Then, R is

(a) transitive but neither reflexive nor symmetric
(b) symmetric and transitive but not reflexive
(c) symmetric but neither reflexive nor transitive
(d) reflexive and symmetric but not transitive

19 Let $\mathbf{a} = 2\hat{\mathbf{i}} + \hat{\mathbf{j}} + \hat{\mathbf{k}}$ and $\mathbf{b}$ and $\mathbf{c}$ be two non-zero vectors such that $|\mathbf{a} + \mathbf{b} + \mathbf{c}| = |\mathbf{a} + \mathbf{b} - \mathbf{c}|$ and $\mathbf{b} \cdot \mathbf{c} = 0$. Consider the following two statements

(A) : $|\mathbf{a} + \lambda \mathbf{c}| \geq |\mathbf{a}|, \forall \lambda \in R$
(B) : $\mathbf{a}$ and $\mathbf{c}$ are always parallel.

Then,

(a) neither (A) nor (B) is correct
(b) both (A) and (B) are correct
(c) only (B) is correct
(d) only (A) is correct

20 Let a circle C_1 be obtained on rolling the circle $x^2 + y^2 - 4x - 6y + 11 = 0$ upwards 4 units on the tangent T to it at the point $(3, 2)$. Let C_2 be the image of C_1 in T. Let A and B be the centres of circles C_1 and C_2 respectively, M and N be respectively the feet of perpendiculars drawn from A and B on the X-axis. Then, the area of the trapezium $AMNB$ is

(a) $3 + 2\sqrt{2}$ (b) $2(2 + \sqrt{2})$

(c) $4(1 + \sqrt{2})$ (d) $2(1 + \sqrt{2})$

Section B : Numerical Value Type Questions

21 Number of 4-digit numbers that are less than or equal to 2800 and either divisible by 3 or 11, is equal to ....

Key Idea For these kind, if problems first write 5^{99} in the form of $11k + P$ and where $k \in I$ and $0 \leq P < 11$. Then, find the solution.

22 The remainder on dividing 5^{99} by 11 is ....

23 Let the line $L: \frac{x-1}{2} = \frac{y+1}{-1} = \frac{z-3}{1}$ intersect the plane $2x + y + 3z = 16$ at the point P. Let the point Q be the foot of perpendicular from the point $R(1, -1, -3)$ on the line L. If α is the area of ΔPQR, then α^2 is equal to ....

24 Let for $x \in R$, $f(x) = \frac{x + |x|}{2}$ and $g(x) = \begin{cases} x, & x < 0 \\ x^2, & x \geq 0 \end{cases}$, then area bounded by the curve $y = (fog)(x)$ and the lines $y = 0$, $2y - x = 15$ is equal to ......

25 If the variance of the frequency distribution

x_i	2	3	4	5	6	7	8
Frequency (f_i)	3	6	16	α	9	5	6

is 3, then α is equal to ....

26 Let $\mathbf{a}$ and $\mathbf{b}$ be two vectors such that $|\mathbf{a}| = \sqrt{14}$, $|\mathbf{b}| = \sqrt{6}$ and $|\mathbf{a} \times \mathbf{b}| = \sqrt{48}$. Then, $(\mathbf{a} \cdot \mathbf{b})^2$ is equal to ....

27 Let 5-digit numbers be constructed using the digits 0, 2, 3, 4, 7, 9 with repetition allowed and are arranged in ascending order with serial numbers. Then, the serial number of the number 42923 is ....

28 Let $a_1, a_2, ..., a_n$ be in AP. If $a_5 = 2a_7$ and $a_{11} = 18$, then

$$12\left(\frac{1}{\sqrt{a_{10}}+\sqrt{a_{11}}}+\frac{1}{\sqrt{a_{11}}+\sqrt{a_{12}}}+\ldots+\frac{1}{\sqrt{a_{17}}+\sqrt{a_{18}}}\right)$$ is equal to ....

29 Let $\alpha > 0$ be the smallest number such that the expansion of $\left(x^{\frac{2}{3}}+\frac{2}{x^3}\right)^{30}$ has a term $\beta x^{-\alpha}, \beta \in N$. Then, α is equal to ....

30 Let θ be the angle between the planes $P_1 : \mathbf{r}\cdot(\hat{\mathbf{i}}+\hat{\mathbf{j}}+2\hat{\mathbf{k}}) = 9$ and $P_2 : \mathbf{r}\cdot(2\hat{\mathbf{i}}-\hat{\mathbf{j}}+\hat{\mathbf{k}}) = 15$. Let L be the line that meets P_2 at the point $(4, -2, 5)$ and makes an angle θ with the normal of P_2. If α is the angle between L and P_2, then $(\tan^2\theta)(\cot^2\alpha)$ is equal to ....

31 JANUARY 2023, SHIFT-II

Section A : Objective Type Questions

1 $\lim_{x\to\infty}\frac{(\sqrt{3x+1}+\sqrt{3x-1})^6+(\sqrt{3x+1}-\sqrt{3x-1})^6}{(x+\sqrt{x^2-1})^6+(x-\sqrt{x^2-1})^6}x^3$

(a) is equal to $\frac{27}{2}$ (b) does not exist

(c) is equal to 27 (d) is equal to 9

2 The foot of perpendicular from the origin O to a plane P which meets the coordinate axes at the points A, B, C is $(2, a, 4)$, $a \in N$. If the volume of the tetrahedron $OABC$ is 144 unit^3, then which of the following points is not on P?

(a) (0, 6, 3) (b) (3, 0, 4)

(c) (0, 4, 4) (d) (2, 2, 4)

3 The number of values of $r \in \{p, q, \sim p, \sim q\}$, for which $((p\wedge q)\Rightarrow(r\vee q))\wedge((p\wedge r)\Rightarrow q)$ is a tautology, is

(a) 2 (b) 3 (c) 4 (d) 1

> **Key Idea** (i) For any two statements p and q
> $p\to q \equiv p \vee q$ (ii) $\sim(p\vee q)\equiv \sim p\wedge\sim q$
> (iii) $p\vee T\equiv T$ (iv) $p\wedge T\equiv p$

4 If $\phi(x) = \frac{1}{\sqrt{x}}\int_{\frac{\pi}{4}}^{x}(4\sqrt{2}\sin t - 3\phi'(t)dt, x > 0$, then $\phi'\left(\frac{\pi}{4}\right)$ is equal to

(a) $\frac{4}{6+\sqrt{\pi}}$ (b) $\frac{8}{6+\sqrt{\pi}}$

(c) $\frac{8}{\sqrt{\pi}}$ (d) $\frac{4}{6-\sqrt{\pi}}$

5 Let $\alpha > 0$. If $\int_0^{\alpha}\frac{x}{\sqrt{x+\alpha}-\sqrt{x}}dx = \frac{16+20\sqrt{2}}{15}$, then α is equal to

(a) 4 (b) 2

(c) $\sqrt{2}$ (d) $2\sqrt{2}$

6 Let P be the plane passing through the point $(1, -1, -5)$ and perpendicular to the line joining the points $(4, 1, -3)$ and $(2, 4, 3)$. Then, the distance of P from the point $(3, -2, 2)$ is

(a) 7 (b) 6 (c) 4 (d) 5

> **Key Idea**
> (i) The cartesian equation of a plane passing through a given point $A(x_1, y_1, z_1)$ and perpendicular to a line having direction ratios (a, b, c) is $a(x-x_1)+b(y-y_1)+c(z-z_1)=0$
> (ii) Let $P(x_1, y_1, z_1)$ be the given point and $ax+by+cz=d$ be the cartesian equation of the given plane.
> Then, perpendicular distance
> $=\left|\frac{ax_1+by_1+cz_1-d}{\sqrt{a^2+b^2+c^2}}\right|$.

7 Let the plane $P: 8x+\alpha_1 y+\alpha_2 z+12=0$ be parallel to the line $L: \frac{x+2}{2}=\frac{y-3}{3}=\frac{z+4}{5}$. If the intercept of P on the Y-axis is 1, then the distance between P and L is

(a) $\sqrt{\frac{7}{2}}$ (b) $\frac{6}{\sqrt{14}}$ (c) $\sqrt{14}$ (d) $\sqrt{\frac{2}{7}}$

8 The equation $e^{4x}+8e^{3x}+13e^{2x}-8e^x+1=0$, $x\in R$ has

(a) two solutions and only one of them is negative

(b) two solutions and both are negative

(c) no solution

(d) four solutions and two of which are negative

9 Let H be the hyperbola, whose foci are $(1\pm\sqrt{2}, 0)$ and eccentricity is $\sqrt{2}$. Then, the length of its latusrectum is ....

(a) 3 (b) 2 (c) $\frac{5}{2}$ (d) $\frac{3}{2}$

10 Let $(a, b)\subset(0, 2\pi)$ be the largest interval for which $\sin^{-1}(\sin\theta)-\cos^{-1}(\sin\theta)>0, \theta\in(0, 2\pi)$, holds. If $\alpha x^2+\beta x+\sin^{-1}(x^2-6x+10)+\cos^{-1}(x^2-6x+10)=0$ and $\alpha-\beta=b-a$, then α is equal to

(a) $\frac{\pi}{12}$ (b) $\frac{\pi}{8}$ (c) $\frac{\pi}{48}$ (d) $\frac{\pi}{16}$

11 Let the mean and standard deviation of marks of class A of 100 students be respectively 40 and α (> 0) and the mean and standard deviation of marks of class B of n students be respectively 55 and $30-\alpha$. If the mean and variance of the marks of the combined class of $100+n$ students are respectively 50 and 350, then the sum of variances of classes A and B is

(a) 650 (b) 450 (c) 500 (d) 900

12 Let $y=y(x)$ be the solution of the differential equation $(3y^2-5x^2)y\,dx+2x(x^2-y^2)\,dy=0$ such that $y(1)=1$. Then, $|(y(2))^3-12y(2)|$ is equal to

(a) $16\sqrt{2}$ (b) 32 (c) $32\sqrt{2}$ (d) 64

13 Among the relations

$S=\{(a,b): a, b\in R-\{0\}, 2+\frac{a}{b}>0\}$ and

$T=\{(a,b): a, b\in R, a^2-b^2\in Z\}$,

(a) both S and T are symmetric
(b) T is symmetric but S is not
(c) S is transitive but T is not
(d) neither S nor T is transitive

14 Let $\mathbf{a}=\hat{\mathbf{i}}+2\hat{\mathbf{j}}+3\hat{\mathbf{k}}$, $\mathbf{b}=\hat{\mathbf{i}}-\hat{\mathbf{j}}+2\hat{\mathbf{k}}$ and $\mathbf{c}=5\hat{\mathbf{i}}-3\hat{\mathbf{j}}+3\hat{\mathbf{k}}$ be three vectors. If $\mathbf{r}$ is a vector such that, $\mathbf{r}\times\mathbf{b}=\mathbf{c}\times\mathbf{b}$ and $\mathbf{r}\cdot\mathbf{a}=0$, then $25|\mathbf{r}|^2$ is equal to

(a) 560 (b) 449 (c) 339 (d) 336

15 Let $a_1, a_2, a_3, \ldots$ be an AP. If $a_7=3$, the product a_1a_4 is minimum and the sum of its first n terms is zero, then $n!-4a_{n(n+2)}$ is equal to

(a) $\frac{33}{4}$ (b) $\frac{381}{4}$ (c) 9 (d) 24

16 If a point $P(\alpha, \beta, \gamma)$ satisfying

$$[\alpha\ \ \beta\ \ \gamma]\begin{bmatrix}2 & 10 & 8\\ 9 & 3 & 8\\ 8 & 4 & 8\end{bmatrix}=[0\ \ 0\ \ 0]$$

lies on the plane $2x+4y+3z=5$, then $6\alpha+9\beta+7\gamma$ is equal to

(a) $\frac{11}{5}$ (b) -1 (c) 11 (d) $\frac{5}{4}$

17 The set of all values of a^2 for which the line $x+y=0$ bisects two distinct chords drawn from a point $P\left(\frac{1+a}{2}, \frac{1-a}{2}\right)$ on the circle $2x^2+2y^2-(1+a)x-(1-a)y=0$, is equal to

(a) (0, 4] (b) $(4, \infty)$ (c) $(8, \infty)$ (d) (2, 12]

18 The absolute minimum value of the function $f(x)=|x^2-x+1|+[x^2-x+1]$, where $[t]$ denotes the greatest integer function, in the interval $[-1, 2]$, is

(a) $\frac{3}{2}$ (b) $\frac{5}{4}$ (c) $\frac{3}{4}$ (d) $\frac{1}{4}$

19 The complex number $z=\dfrac{i-1}{\cos\frac{\pi}{3}+i\sin\frac{\pi}{3}}$ is equal to

(a) $\sqrt{2}\left(\cos\frac{5\pi}{12}+i\sin\frac{5\pi}{12}\right)$ (b) $\cos\frac{\pi}{12}-i\sin\frac{\pi}{12}$

(c) $\sqrt{2}\left(\cos\frac{\pi}{12}+i\sin\frac{\pi}{12}\right)$ (d) $\sqrt{2}i\left(\cos\frac{5\pi}{12}-i\sin\frac{5\pi}{12}\right)$

20 Let $f: R-\{2, 6\}\to R$ be real valued function defined as $f(x)=\dfrac{x^2+2x+1}{x^2-8x+12}$.

Then, range of f is

(a) $\left(-\infty, -\frac{21}{4}\right]\cup[1, \infty)$ (b) $\left(-\infty, -\frac{21}{4}\right]\cup[0, \infty)$

(c) $\left(-\infty, -\frac{21}{4}\right]\cup\left[\frac{21}{4}, \infty\right)$ (d) $\left(-\infty, -\frac{21}{4}\right)\cup(0, \infty)$

Section B : Numerical Value Type Questions

21 Let $A=[a_{ij}]$, $a_{ij}\in Z\cap[0, 4]$, $1\le i, j\le 2$.
The number of matrices A such that the sum of all entries is a prime number $p\in(2, 13)$ is ....

22 If ${}^{2n+1}P_{n-1} : {}^{2n-1}P_n = 11:21$, then n^2+n+15 is equal to ....

23 Let A be the event that the absolute difference between two randomly chosen real numbers in the sample space [0, 60] is less than or equal to a. If $P(A)=\frac{11}{36}$, then a is equal to ....

24 Let A be a $n\times n$ matrix such that $|A|=2$. If the determinant of the matrix Adj $(2\cdot\text{Adj}(2A^{-1}))$ is 2^{84}, then n is equal to ....

25 Let $\mathbf{a}, \mathbf{b}, \mathbf{c}$ be three vectors such that $|\mathbf{a}|=\sqrt{31}$, $4|\mathbf{b}|=|\mathbf{c}|=2$ and $2(\mathbf{a}\times\mathbf{b})=3(\mathbf{c}\times\mathbf{a})$. If the angle between $\mathbf{b}$ and $\mathbf{c}$ is $\frac{2\pi}{3}$, then $\left(\frac{\mathbf{a}\times\mathbf{c}}{\mathbf{a}\cdot\mathbf{b}}\right)$ is equal to ....

26 The sum $1^2-2\cdot3^2+3\cdot5^2-4\cdot7^2+5\cdot9^2-\ldots+15\cdot29^2$ is ....

27 Let the area of the region $\{(x, y): |2x-1|\le y\le|x^2-x|, 0\le x\le 1\}$ be A.
Then, $(6A+11)^2$ is equal to ....

28 If the constant term in the binomial expansion of $\left(\frac{x^{5/2}}{2}-\frac{4}{x^l}\right)^9$ is -84 and the coefficient of x^{-3l} is $2^\alpha\beta$, where $\beta<0$ is an odd number, then $|\alpha l-\beta|$ is equal to ....

29 The coefficient of x^{-6} in the expansion of $\left(\frac{4x}{5}+\frac{5}{2x^2}\right)^9$, is ....

30 Let S be the set of all $a\in N$ such that the area of the triangle formed by the tangent at the point $P(b, c)$, $b, c\in N$, on the parabola $y^2=2ax$ and the lines $x=b$, $y=0$ is 16 unit^2, then $\sum_{a\in S} a$ is equal to ....

1 FEBRUARY 2023, SHIFT-I

Section A : Objective Type Questions

1 The negation of the expression $q \vee ((\sim q) \wedge p)$ is equivalent to

(a) $(\sim p) \vee (\sim q)$ (b) $p \wedge (\sim q)$

(c) $(\sim p) \vee q$ (d) $(\sim p) \wedge (\sim q)$

2 The mean and variance of 5 observations are 5 and 8, respectively. If 3 observations are 1, 3, 5, then the sum of cubes of the remaining two observations is

(a) 1216 (b) 1072

(c) 1792 (d) 1456

3 The combined equation of the two lines $ax + by + c = 0$ and $a'x + b'y + c' = 0$ can be written as $(ax + by + c)(a'x + b'y + c') = 0$.
The equation of the angle bisectors of the lines represented by the equation $2x^2 + xy - 3y^2 = 0$ is

(a) $x^2 - y^2 + 10xy = 0$

(b) $3x^2 + 5xy + 2y^2 = 0$

(c) $3x^2 + xy - 2y^2 = 0$

(d) $x^2 - y^2 - 10xy = 0$

4 Let S denote the set of all real values of λ such that the system of equations

$$\lambda x + y + z = 1$$
$$x + \lambda y + z = 1$$
$$x + y + \lambda z = 1$$

is inconsistent, then $\sum_{\lambda \in S} (|\lambda|^2 |\lambda|)$ is equal to

(a) 12 (b) 6

(c) 4 (d) 2

5 Let S be the set of all solutions of the equation $\cos^{-1}(2x) - 2\cos^{-1}(\sqrt{1-x^2}) = \pi$, $x \in \left[-\frac{1}{2}, \frac{1}{2}\right]$. Then, $\sum_{x \in S} 2\sin^{-1}(x^2 - 1)$ is equal to

(a) $\pi - 2\sin^{-1}\left(\frac{\sqrt{3}}{4}\right)$ (b) $\frac{-2\pi}{3}$

(c) 0 (d) $\pi - \sin^{-1}\left(\frac{\sqrt{3}}{4}\right)$

6 In a binomial distribution $B(n, p)$ the sum and the product of the mean and the variance are 5 and 6 respectively, then $6(n + p - q)$ is equal to

(a) 53 (b) 50

(c) 51 (d) 52

7 The shortest distance between the lines $\frac{x-5}{1} = \frac{y-2}{2} = \frac{z-4}{-3}$ and $\frac{x+3}{1} = \frac{y+5}{4} = \frac{z-1}{-5}$ is

(a) $6\sqrt{3}$ (b) $5\sqrt{3}$

(c) $4\sqrt{3}$ (d) $7\sqrt{3}$

8 The value of $\frac{1}{1!50!} + \frac{1}{3!48!} + \frac{1}{5!46!} + \ldots + \frac{1}{49!2!} + \frac{1}{51!1!}$ is

(a) $\frac{2^{51}}{50!}$ (b) $\frac{2^{51}}{51!}$ (c) $\frac{2^{50}}{51!}$ (d) $\frac{2^{50}}{50!}$

9 Let the image of the point $P(2, -1, 3)$ in the plane $x + 2y - z = 0$ be Q. Then, the distance of the plane $3x + 2y + z + 29 = 0$ from the point Q is

(a) $\frac{24\sqrt{2}}{7}$ (b) $2\sqrt{14}$

(c) $\frac{22\sqrt{2}}{7}$ (d) $3\sqrt{14}$

10 $\lim_{n \to \infty}\left[\frac{1}{1+n} + \frac{1}{2+n} + \frac{1}{3+n} + \ldots + \frac{1}{2n}\right]$ is equl to

(a) $\log_e\left(\frac{3}{2}\right)$ (b) $\log_e 2$

(c) $\log_e\left(\frac{2}{3}\right)$ (d) 0

> **Key Idea** In this type of problem first convert $\lim_{n\to\infty}\left(\frac{1}{1+n} + \frac{1}{2+n} + \ldots \frac{1}{2n}\right)$ to $\lim_{n\to\infty} \frac{1}{n}\sum_{r=1}^{n} \frac{1}{\left(\frac{r}{n}+1\right)}$.
> Then, solve using integration.

11 For a triangle ABC, the value of $\cos 2A + \cos 2B + \cos 2C$ is least. If its inradius is 3 and incentre is M, then which of the following is not correct?

(a) $\sin 2A + \sin 2B + \sin 2C = \sin A + \sin B + \sin C$

(b) perimeter of ΔABC is $18\sqrt{3}$

(c) $\mathbf{MA} \cdot \mathbf{MB} = -18$

(d) area of ΔABC is $\frac{27\sqrt{3}}{2}$

12 The sum to 10 terms of the series $\frac{1}{1+1^2+1^4} + \frac{2}{1+2^2+2^4} + \frac{3}{1+3^2+3^4} + \ldots$ is

(a) $\frac{56}{111}$ (b) $\frac{59}{111}$

(c) $\frac{55}{111}$ (d) $\frac{58}{111}$

13 The area enclosed by the closed curve C given by the differential equation $\frac{dy}{dx} + \frac{x+a}{y-2} = 0$, $y(1) = 0$ is 4π.

Let P and Q be the points of intersection of the curve C and the Y-axis. If normals at P and Q on the curve C intersect X-axis at points R and S respectively, then the length of the line segment RS is

(a) $\frac{4\sqrt{3}}{3}$ (b) $2\sqrt{3}$ (c) 2 (d) $\frac{2\sqrt{3}}{3}$

14 Let $f(x) = 2x + \tan^{-1} x$ and $g(x) = \log_e(\sqrt{1+x^2} + x)$, $x \in [0, 3]$. Then,

(a) there exists $x \in [0, 3]$ such that $f'(\hat{x}) < g'(\hat{x})$

(b) There exist $0 < x_1 < x_2 < 3$ such that $f(x) < g(x)$, $\forall x \in (x_1, x_2)$

(c) $\max f(x) > \max g(x)$

(d) $\min f'(x) = 1 + \max g'(x)$

15 Let R be a relation on R, given by $R = \{(a, b): 3a - 3b + \sqrt{7}$ is an irrational number$\}$. Then, R is

(a) an equivalence relation

(b) reflexive and transitive but not symmetric

(c) reflexive and symmetric but not transitive

(d) reflexive but neither symmetric nor transitive

16 Let $S = \{x : x \in R$ and $(\sqrt{3}+\sqrt{2})^{x^2-4} + (\sqrt{3}-\sqrt{2})^{x^2-4} = 10\}$.

Then, $n(S)$ is equal to

(a) 4 (b) 2

(c) 0 (d) 6

17 If the centre and radius of the circle $\left|\dfrac{z-2}{z-3}\right| = 2$ are respectively, (α, β) and γ, then $3(\alpha + \beta + \gamma)$ is equal to

(a) 9 (b) 10

(c) 11 (d) 12

18 If the orthocentre of the triangle, whose vertices are (1, 2), (2, 3) and (3, 1) is (α, β), then the quadratic equation whose roots are $\alpha + 4\beta$ and $4\alpha + \beta$, is

(a) $x^2 - 20x + 99 = 0$ (b) $x^2 - 19x + 90 = 0$

(c) $x^2 - 18x + 80 = 0$ (d) $x^2 - 22x + 120 = 0$

19 Let $f(x) = \begin{vmatrix} 1+\sin^2 x & \cos^2 x & \sin 2x \\ \sin^2 x & 1+\cos^2 x & \sin 2x \\ \sin^2 x & \cos^2 x & 1+\sin 2x \end{vmatrix}, x \in \left[\dfrac{\pi}{6}, \dfrac{\pi}{3}\right]$.

If α and β respectively are the maximum and the minimum values of f, then

(a) $\beta^2 + 2\sqrt{\alpha} = \dfrac{19}{4}$ (b) $\alpha^2 + \beta^2 = \dfrac{9}{2}$

(c) $\alpha^2 - \beta^2 = 4\sqrt{3}$ (d) $\beta^2 - 2\sqrt{\alpha} = \dfrac{19}{4}$

> **Key Idea** When its given that determinats is a function. Apply elementry operation to reduce it into simple form and then find the solution with the help of given data.

20 If $y = y(x)$ is the solution curve of the differential equation $\dfrac{dy}{dx} + y\tan x = x\sec x$, $0 \le x \le \dfrac{\pi}{3}$, $y(0) = 1$, then $y\left(\dfrac{\pi}{6}\right)$ is equal to

(a) $\dfrac{\pi}{12} + \dfrac{\sqrt{3}}{2}\log_e\left(\dfrac{2}{e\sqrt{3}}\right)$

(b) $\dfrac{\pi}{12} - \dfrac{\sqrt{3}}{2}\log_e\left(\dfrac{2\sqrt{3}}{e}\right)$

(c) $\dfrac{\pi}{12} - \dfrac{\sqrt{3}}{2}\log_e\left(\dfrac{2}{e\sqrt{3}}\right)$

(d) $\dfrac{\pi}{12} + \dfrac{\sqrt{3}}{2}\log_e\left(\dfrac{2\sqrt{3}}{e}\right)$

Section B : Numerical Value Type Questions

21 The number of 3-digit numbers, that are divisible by either 2 or 3 but not divisible by 7, is ....

22 The remainder, when $19^{200} + 23^{200}$ is divided by 49, is ....

23 $A(2, 6, 2)$, $B(-4, 0, \lambda)$, $C(2, 3, -1)$ and $D(4, 5, 0)$, $|\lambda| \le 5$ are the vertices of a quadrilateral $ABCD$. If its area is 18 square units, then $5 - 6\lambda$ is equal to ....

24 let $a_1 = 8, a_2, a_3, \ldots, a_n$ be an AP. If the sum of its first four terms is 50 and the sum of its last four terms is 170, then the product of its middle two terms is ....

25 If $\int_0^1 (x^{21} + x^{14} + x^7)(2x^{14} + 3x^7 + 6)^{1/7} dx = \dfrac{1}{l}(11)^{m/n}$, where $l, m, n \in N$, m and n are coprime, then $l + m + n$ is equal to ....

26 Let A be the area bounded by the curve $y = x|x-3|$, the X-axis and the ordinates $x = -1$ and $x = 2$. Then, $12A$ is equal to ....

27 Let $f : R \to R$ be a differentiable function such that $f'(x) + f(x) = \int_0^2 f(t)dt$. If $f(0) = e^{-2}$, then $2f(0) - f(2)$ is equal to ....

28 Let $\mathbf{v} = \alpha\hat{\mathbf{i}} + 2\hat{\mathbf{j}} - 3\hat{\mathbf{k}}$, $\mathbf{w} = 2\alpha\hat{\mathbf{i}} + \hat{\mathbf{j}} - \hat{\mathbf{k}}$ and $\mathbf{u}$ be a vector such that $|\hat{\mathbf{u}}| = \alpha > 0$. If the minimum value of the scalar triple product $|\mathbf{u}\ \mathbf{v}\ \mathbf{w}|$ is $-\alpha\sqrt{3401}$ and $|\mathbf{u}\cdot\hat{\mathbf{i}}|^2 = \dfrac{m}{n}$, where m and n are coprime natural numbers, then $m + n$ is equal to ....

29 The number of words, with or without meaning that can be formed using all the letters of the word 'ASSASSINATION', so that the vowels occur together, is ....

30 If $f(x) = x^2 + g'(1)x + g''(2)$ and $g(x) = f(1)x^2 + xf'(x) + f''(x)$, then the value of $f(4) - g(4)$ is equal to ....

1 FEBRUARY 2023, SHIFT-II

Section A : Objective Type Questions

1 Let $\alpha x = \exp(x^{\beta} y^{\gamma})$ be the solution of the differential equation $2x^2 y dy - (1 - xy^2)\, dx = 0$, $x > 0$, $y(2) = \sqrt{\log_e 2}$. Then, $\alpha + \beta - \gamma$ equals to

(a) -1 (b) 3
(c) 0 (d) 1

> **Key Idea** To find the solution of given differential equation reduced as form of $\frac{dy}{dx} + py = Q$,
> $IF = e^{\int P dx}$ and $IF \cdot y = \int IF \cdot X\, Q\, dx + C$. Then, solve using integration.

2 The area of the region given by $\{(x, y) : xy \le 8, 1 \le y \le x^2\}$ is

(a) $16 \log_e 2 + \frac{7}{3}$ (b) $16 \log_e 2 - \frac{14}{3}$
(c) $8 \log_e 2 + \frac{7}{6}$ (d) $8 \log_e 2 - \frac{13}{3}$

3 Let $\mathbf{a} = 5\hat{\mathbf{i}} - \hat{\mathbf{j}} - 3\hat{\mathbf{k}}$ and $\mathbf{b} = \hat{\mathbf{i}} + 3\hat{\mathbf{j}} + 5\hat{\mathbf{k}}$ be two vectors. Then, which one of the following is true?

(a) Projection of $\mathbf{a}$ on $\mathbf{b}$ is $\frac{-17}{\sqrt{35}}$ and the direction of the projection vector is opposite to the direction of $\mathbf{b}$.
(b) Projection of $\mathbf{a}$ on $\mathbf{b}$ is $\frac{-17}{\sqrt{35}}$ and the direction of the projection vector is same as of $\mathbf{b}$.
(c) Projection of $\mathbf{a}$ on $\mathbf{b}$ is $\frac{17}{\sqrt{35}}$ and the direction of the projection vector is same as of $\mathbf{b}$.
(d) Projection of $\mathbf{a}$ on $\mathbf{b}$ is $\frac{17}{\sqrt{35}}$ and the direction of the projection vector is opposite to the direction of $\mathbf{b}$.

4 Let
$$S = \left\{ x \in R : 0 < x < 1 \text{ and } 2\tan^{-1}\left(\frac{1-x}{1+x}\right) = \cos^{-1}\left(\frac{1-x^2}{1+x^2}\right) \right\}.$$
If $n(S)$ denotes the number of elements in S, then

(a) $n(S) = 1$ and the element in S is less than $\frac{1}{2}$
(b) $n(S) = 2$ and only one element in S is less than $\frac{1}{2}$
(c) $n(S) = 0$
(d) $n(S) = 1$ and the elements in S is more than $\frac{1}{2}$

5 Let a and b be two real numbers such that $ab < 0$. If the complex number $\frac{1 + ai}{b + i}$ is of unit modulus and $a + ib$ lies on the circle $|z - 1| = |2z|$, then a possible value of $\frac{1 + [a]}{4b}$, where $[t]$ is greatest integer function is

(a) $-\frac{1}{2}$ (b) $\frac{1}{2}$ (c) -1 (d) 1

6 The sum of the absolute maximum and minimum values of the function $f(x) = |x^2 - 5x + 6| - 3x + 2$ in the interval $[-1, 3]$ is equal to

(a) 12 (b) 24 (c) 13 (d) 10

7 Let the plane P passes through the intersection of the planes $2x + 3y - z = 2$ and $x + 2y + 3z = 6$ and be perpendicular to the plane $2x + y - z + 1 = 0$. If d is the distance of P from the point $(-7, 1, 1)$, then d^2 is equal to

(a) $\frac{25}{83}$ (b) $\frac{250}{82}$ (c) $\frac{250}{83}$ (d) $\frac{15}{53}$

> **Key Idea** To find the value of distance P from given point, firstly find the equation of plane. Then, used distance formula to evaluate the value of distance.

8 If $A = \frac{1}{2}\begin{bmatrix} 1 & \sqrt{3} \\ -\sqrt{3} & 1 \end{bmatrix}$, then

(a) $A^{30} - A^{25} = 2I$ (b) $A^{30} + A^{25} - A = I$
(c) $A^{30} + A^{25} + A = I$ (d) $A^{30} = A^{25}$

9 Which of the following statements is a tautology ?

(a) $(p \wedge (p \to q)) \to \sim q$ (b) $(p \wedge q) \to (\sim (p) \to q)$
(c) $p \to (p \wedge (p \to q))$ (d) $p \vee (p \wedge q)$

10 If $y(x) = x^x$, $x > 0$, then $y''(2) - 2y'(2)$ is equal to

(a) $4(\log_e 2)^2 - 2$ (b) $4 \log_e 2 + 2$
(c) $8 \log_e 2 - 2$ (d) $4(\log_e 2)^2 + 2$

11 Two dice are thrown independently. Let A be the event that the number appeared on the 1st die is less than the number appeared on the 2nd die and B be the event that the number appeared on the 1st die is even and that on the second die is odd and C be the event that the number appeared on the 1st die is odd and that on the 2nd is even. Then,

(a) B and C are independent
(b) the number of favourable cases of the event $(A \cup B) \cap C$ is 6
(c) the number of favourable cases of the events A, B and C are 15, 6 and 6, respectively
(d) A and B are mutually exclusive

12 Let $P(x_0, y_0)$ be the point on the hyperbola $3x^2 - 4y^2 = 36$, which is nearest to the line $3x + 2y = 1$. Then, $\sqrt{2}(y_0 - x_0)$ is equal to

(a) 9 (b) -9 (c) -3 (d) 3

13 Let $\mathbf{a} = 2\hat{\mathbf{i}} - 7\hat{\mathbf{j}} - 5\hat{\mathbf{k}}$, $\mathbf{b} = \hat{\mathbf{i}} + \hat{\mathbf{k}}$ and $\mathbf{c} = \hat{\mathbf{i}} + 2\hat{\mathbf{j}} - 3\hat{\mathbf{k}}$ be three given vectors. If $\mathbf{r}$ is a vector such that $\mathbf{r} \times \mathbf{a} = \mathbf{c} \times \mathbf{a}$ and $\mathbf{r} \cdot \mathbf{b} = 0$, then $|\mathbf{r}|$ is equal to

(a) $\frac{11}{7}$ (b) $\frac{11}{5}\sqrt{2}$
(c) $\frac{\sqrt{914}}{7}$ (d) $\frac{11}{7}\sqrt{2}$

14 Let $P(S)$ denote the power set of $S = \{1, 2, 3, \ldots, 10\}$. Define the relations R_1 and R_2 on $P(S)$ as AR_1B, if $(A \cap B^C) \cup (B \cap A^C) = \phi$ and AR_2B, if $A \cup B^C = B \cup A^C$, $\forall A, B \in P(S)$. Then,

(a) only R_1 is an equivalence relation
(b) both R_1 and R_2 are not equivalence relations
(c) only R_2 is an equivalence relation
(d) both R_1 and R_2 are equivalence relations

15 The sum $\sum_{n=1}^{\infty} \frac{2n^2 + 3n + 4}{(2n)!}$ is equal to

(a) $\frac{13e}{4} + \frac{5}{4e} - 4$ (b) $\frac{13e}{4} + \frac{5}{4e}$
(c) $\frac{11e}{2} + \frac{7}{2e} - 4$ (d) $\frac{11e}{2} + \frac{7}{2e}$

16 The number of integral values of k, for which one root of the equation $2x^2 - 8x + k = 0$ lies in the interval (1, 2) and its other root lies in the interval (2, 3), is

(a) 3 (b) 2
(c) 0 (d) 1

17 For the system of linear equations $\alpha x + y + z = 1$, $x + \alpha y + z = 1$, $x + y + \alpha z = \beta$, which one of the following statements is not correct ?

(a) It has no solution, if $\alpha = -2$ and $\beta = 1$
(b) $x + y + z = \frac{3}{4}$, if $\alpha = 2$ and $\beta = 1$
(c) It has infinitely many solutions, if $\alpha = 1$ and $\beta = 1$
(d) It has infinitely many solutions, if $\alpha = 2$ and $\beta = -1$

18 The value of the integral $\int_{-\frac{\pi}{4}}^{\frac{\pi}{4}} \frac{x + \frac{\pi}{4}}{2 - \cos 2x} dx$ is

(a) $\frac{\pi^2}{6}$ (b) $\frac{\pi^2}{12\sqrt{3}}$
(c) $\frac{\pi^2}{3\sqrt{3}}$ (d) $\frac{\pi^2}{6\sqrt{3}}$

19 Let $f : R - \{0, 1\} \to R$ be a function such that $f(x) + f\left(\frac{1}{1-x}\right) = 1 + x$. Then, $f(2)$ is equal to

(a) $\frac{9}{2}$ (b) $\frac{9}{4}$
(c) $\frac{7}{3}$ (d) $\frac{7}{4}$

20 Let $9 = x_1 < x_2 < \ldots < x_7$ be in an AP with common difference d. If the standard deviation of $x_1, x_2 \ldots, x_7$ is 4 and the mean is $\bar{x}$, then $\bar{x} + x_6$ is equal to

(a) $2\left(9 + \frac{8}{\sqrt{7}}\right)$ (b) $18\left(1 + \frac{1}{\sqrt{3}}\right)$
(c) 34 (d) 25

Section B : Numerical Value Type Questions

21 If the term without x in the expansion of $\left(x^{\frac{2}{3}} + \frac{\alpha}{x^3}\right)^{22}$ is 7315, then $|\alpha|$ is equal to ....

22 If the x-intercept of a focal chord of the parabola $y^2 = 8x + 4y + 4$ is 3, then the length of this chord is equal to ....

23 The total number of six digit numbers, formed using the digits 4, 5, 9 only and divisible by 6, is ....

24 The sum of the common terms of the following three Arithmetic progressions,
3, 7, 11, 15, ..., 399,
2, 5, 8, 11, ..., 359
and 2, 7, 12, 17, ..., 197 is equal to ....

25 Let $\alpha x + \beta y + \gamma z = 1$ be the equation of a plane passing through the point $(3, -2, 5)$ and perpendicular to the line joining the points
(1, 2, 3) and $(-2, 3, 5)$. Then, the value of $\alpha\beta\gamma$ is equal to ....

26 The point of intersection C of the plane $8x + y + 2z = 0$ and the line joining the points $A(-3, -6, 1)$ and $B(2, 4, -3)$ divides the line segment AB internally in the ratio $K : 1$. If a, b, c ($|a|, |b|, |c|$ are coprime) are the direction ratios of the perpendicular from the point C on the line $\frac{1-x}{1} = \frac{y+4}{2} = \frac{z+2}{3}$, then $|a + b + c|$ is equal to ....

27 Number of integral solutions to the equation $x + y + z = 21$, where $x \geq 1$, $y \geq 3$, $z \geq 4$, is equal to ....

28 If $\int_0^{\pi} \frac{5^{\cos x}(1 + \cos x \cos 3x + \cos^2 x + \cos^3 x \cos 3x) dx}{1 + 5^{\cos x}} = \frac{k\pi}{16}$, then k is equal to ....

29 The line $x = 8$ is the directirix of the ellipse $E : \frac{x^2}{a^2} + \frac{y^2}{b^2} = 1$ with the corresponding focus (2, 0). If the tangent to E at the point P in the first quadrant passes through the point $(0, 4\sqrt{3})$ and intersects the X-axis at Q, then $(3PQ)^2$ is equal to ....

30 Let the sixth term in the bionomial expansion of $\left(\sqrt{2^{\log_2(10-3^x)}} + \sqrt[5]{2^{(x-2)\log_2 3}}\right)^m$ in the increasing powers of $2^{(x-2)\log_2 3}$ be 21. If the binomial coefficients of the second, third and fourth terms in the expansion are respectively the first, third and fifth terms of an AP, then the sum of the squares of all possible values of x is ....

Answers

For solutions scan the QR code

24 Jan 2023, Shift-I

1. (b)	2. (d)	3. (d)	4. (d)	5. (b)	6. (b)	7. (a)	8. (a)	9. (d)	10. (b)
11. (a)	12. (b)	13. (a)	14. (a)	15. (b)	16. (a)	17. (c)	18. (c)	19. (a)	20. (d)
21. 1012	22. 2	23. 118	24. 14	25. 60	26. 22	27. 546	28. 5	29. 7	30. (12)

24 Jan 2023, Shift-II

1. (c)	2. (d)	3. (a)	4. (d)	5. (b)	6. (c)	7. (d)	8. (b)	9. (b)	10. (a)
11. (c)	12. (c)	13. (b)	14. (c)	15. (d)	16. (d)	17. (b)	18. (c)	19. (a)	20. (d)
21. 2	22. 36	23. 5	24. 8	25. 624	26. 405	27. 27	28. 9	29. 122	30. 432

25 Jan 2023, Shift-I

1. (c)	2. (b)	3. (c)	4. (b)	5. (a)	6. (a)	7. (a)	8. (d)	9. (a)	10. (c)
11. (c)	12. (a)	13. (c)	14. (a)	15. (c)	16. (c)	17. (c)	18. (d)	19. (a)	20. (a)
21. 120	22. 25	23. 1080	24. 495	25. 2	26. 2039	27. 9	28. 216	29. 43	30. 600

25 Jan 2023, Shift-II

1. (b)	2. (d)	3. (d)	4. (a)	5. (b)	6. (a)	7. (a)	8. (d)	9. (c)	10. (c)
11. (a)	12. (a)	13. (d)	14. (a)	15. (b)	16. (b)	17. (c)	18. (b)	19. (c)	20. (b)
21. 9	22. 20	23. 7	24. 18	25. 25	26. 6860	27. 3	28. 31	29. 45	30. 3

29 Jan 2023, Shift-I

1. (a)	2. (d)	3. (a)	4. (d)	5. (d)	6. (b)	7. (c)	8. (c)	9. (a)	10. (a)
11. (a)	12. (c)	13. (b)	14. (a)	15. (a)	16. (c)	17. (b)	18. (d)	19. (d)	20. (c)
21. 2	22. 32	23. 1436	24. 10	25. 1	26. 355	27. 9	28. 60	29. 3	30. 2240

29 Jan 2023, Shift-II

1. (b)	2. (a,b,c)	3. (a)	4. (c)	5. (c)	6. (b)	7. (c)	8. (b)	9. (c)	10. (d)
11. (d)	12. (c)	13. (c)	14. (a)	15. (c)	16. (d)	17. (b)	18. (b)	19. (d)	20. (b)
21. 603	22. 11	23. 3000	24. 461	25. 14	26. 9	27. 10	28. 4	29. 5	30. 9

30 Jan 2023, Shift-I

1. (b)	2. (c)	3. (b)	4. (c)	5. (b)	6. (c)	7. (b)	8. (a)	9. (c)	10. (d)
11. (b)	12. (d)	13. (d)	14. (d)	15. (d)	16. (c)	17. (c)	18. (a)	19. (a)	20. (c)
21. 9	22. 22	23. 12	24. $^{64}P_6$	25. 21	26. 3125	27. 37	28. 315	29. 15	30. 26

30 Jan 2023, Shift-II

1. (a)	2. (c)	3. (b)	4. (b)	5. (b)	6. (d)	7. (a)	8. (c)	9. (a)	10. (b)
11. (b)	12. (b)	13. (d)	14. (d)	15. (d)	16. (c)	17. (d)	18. (c)	19. (b)	20. (b)
21. 1	22. 24	23. 151	24. 158	25. 25	26. 23	27. 48	28. 13	29. 240	30. 14

For solutions scan the QR code

31 Jan 2023, Shift-I

1. *(d)*	2. *(d)*	3. *(b)*	4. *(a)*	5. *(a)*	6. *(d)*	7. *(b)*	8. *(a)*	9. *(c)*	10. *(d)*
11. *(b)*	12. *(b)*	13. *(d)*	14. *(d)*	15. *(b)*	16. *(b)*	17. *(d)*	18. *(c)*	19. *(d)*	20. *(c)*
21. *710*	22. *9*	23. *180*	24. *72*	25. *5*	26. *36*	27. *2997*	28. *8*	29. *2*	30. *9*

31 Jan 2023, Shift-II

1. *(c)*	2. *(b)*	3. *(a)*	4. *(b)*	5. *(b)*	6. *(d)*	7. *(c)*	8. *(b)*	9. *(b)*	10. *(a)*
11. *(c)*	12. *(c)*	13. *(b)*	14. *(c)*	15. *(d)*	16. *(c)*	17. *(c)*	18. *(c)*	19. *(a)*	20. *(b)*
21. *204*	22. *45*	23. *10*	24. *5*	25. *3*	26. *6952*	27. *125*	28. *98*	29. *5040*	30. *146*

1 Feb 2023, Shift-I

1. *(d)*	2. *(b)*	3. *(d)*	4. *(b)*	5. *(b)*	6. *(d)*	7. *(a)*	8. *(c)*	9. *(d)*	10. *(b)*
11. *(d)*	12. *(c)*	13. *(a)*	14. *(c)*	15. *(d)*	16. *(a)*	17. *(d)*	18. *(a)*	19. *(d)*	20. *(c)*
21. *514*	22. *29*	23. *11*	24. *754*	25. *63*	26. *62*	27. *1*	28. *3501*	29. *50400*	30. *14*

1 Feb 2023, Shift-II

1. *(d)*	2. *(b)*	3. *(*)*	4. *(a)*	5. *(*)*	6. *(d)*	7. *(c)*	8. *(b)*	9. *(b)*	10. *(a)*
11. *(b)*	12. *(b)*	13. *(d)*	14. *(d)*	15. *(a)*	16. *(d)*	17. *(d)*	18. *(d)*	19. *(b)*	20. *(c)*
21. *1*	22. *16*	23. *81*	24. *321*	25. *6*	26. *10*	27. *105*	28. *13*	29. *39*	30. *4*

ONLINE SOLVED PAPERS

JEE Main 2023

(April Attempt)

6 APRIL 2023, SHIFT-I

Section A : Objective Type Questions

1 Let $5f(x) + 4f\left(\frac{1}{x}\right) = \frac{1}{x} + 3,\ x > 0$. Then, $18\int_1^2 f(x)\,dx$ is equal to

(a) $10 \log_e 2 - 6$ (b) $10 \log_e 2 + 6$
(c) $5 \log_e 2 - 3$ (d) $5 \log_e 2 + 3$

2 Let $A = [a_{ij}]_{2 \times 2}$, where $a_{ij} \neq 0$ for all i, j and $A^2 = I$. Let a be the sum of all diagonal elements of A and $b = |A|$. Then, $3a^2 + 4b^2$ is equal to

(a) 7 (b) 14 (c) 3 (d) 4

3 Statement $(P \Rightarrow Q) \wedge (R \Rightarrow Q)$ is logically equivalent to

(a) $(P \vee R) \Rightarrow Q$ (b) $(P \Rightarrow R) \wedge (Q \Rightarrow R)$
(c) $(P \Rightarrow R) \vee (Q \Rightarrow R)$ (d) $(P \wedge R) \Rightarrow Q$

4 The sum of all the roots of the equation $|x^2 - 8x + 15| - 2x + 7 = 0$ is

(a) $9 + \sqrt{3}$ (b) $11 - \sqrt{3}$
(c) $11 + \sqrt{3}$ (d) $9 - \sqrt{3}$

5 The straight lines l_1 and l_2 pass through the origin and trisect the line segment of the line $L: 9x + 5y = 45$ between the axes. If m_1 and m_2 are the slopes of the lines l_1 and l_2, then the point of intersection of the line $y = (m_1 + m_2)x$ with L lies on

(a) $6x + y = 10$ (b) $y - x = 5$
(c) $6x - y = 15$ (d) $y - 2x = 5$

6 Let the position vectors of the points A, B, C and D be $5\hat{\mathbf{i}} + 5\hat{\mathbf{j}} + 2\lambda\hat{\mathbf{k}},\ \hat{\mathbf{i}} + 2\hat{\mathbf{j}} + 3\hat{\mathbf{k}},\ -2\hat{\mathbf{i}} + \lambda\hat{\mathbf{j}} + 4\hat{\mathbf{k}}$ and $-\hat{\mathbf{i}} + 5\hat{\mathbf{j}} + 6\hat{\mathbf{k}}$. Let the set $S = \{\lambda \in R :$ the points A, B, C and D are coplanar$\}$. Then, $\sum_{\lambda \in S} (\lambda + 2)^2$ is equal to

(a) 25 (b) 41
(c) $\frac{37}{2}$ (d) 13

7 From the top A of a vertical wall AB of height 30 m, the angles of depression of the top P and bottom Q of a vertical tower PQ are 15° and 60°, respectively, B and Q are on the same horizontal level. If C is a point on AB such that $CB = PQ$, then the area (in m^2) of the quadrilateral $BCPQ$ is equal to

(a) $300(\sqrt{3} - 1)$ (b) $600(\sqrt{3} - 1)$
(c) $200(3 - \sqrt{3})$ (d) $300(3\sqrt{3} + 1)$

8 If $2x^y + 3y^x = 20$, then $\frac{dy}{dx}$ at (2, 2) is equal to :

(a) $-\left(\frac{2 + \log_e 8}{3 + \log_e 4}\right)$ (b) $-\left(\frac{3 + \log_e 4}{2 + \log_e 8}\right)$
(c) $-\left(\frac{3 + \log_e 16}{4 + \log_e 8}\right)$ (d) $-\left(\frac{3 + \log_e 8}{2 + \log_e 4}\right)$

9 If the ratio of the fifth term from the beginning to the fifth term from the end in the expansion of $\left(\sqrt[4]{2} + \frac{1}{\sqrt[4]{3}}\right)^n$ is $\sqrt{6} : 1$, then the third term from the beginning is

(a) $30\sqrt{2}$ (b) $60\sqrt{2}$
(c) $60\sqrt{3}$ (d) $30\sqrt{3}$

10 If the equation of the plane passing through the line of intersection of the planes $2x - y + z = 3$, $4x - 3y + 5z + 9 = 0$ and parallel to the line $\frac{x+1}{-2} = \frac{y+3}{4} = \frac{z-2}{5}$ is $ax + by + cz + 6 = 0$, then $a + b + c$ is equal to

(a) 14 (b) 15
(c) 12 (d) 13

11 The mean and variance of a set of 15 numbers are 12 and 14 respectively. The mean and variance of another set of 15 numbers are 14 and σ^2 respectively. If the variance of all the 30 numbers in the two sets is 13, then σ^2 is equal to

(a) 9 (b) 11 (c) 10 (d) 12

12 If the system of equations

$$x+y+az=b$$
$$2x+5y+2z=6$$
$$x+2y+3z=3$$

has infinitely many solutions, then $2a+3b$ is equal to

(a) 23 (b) 25
(c) 28 (d) 20

13 Let $a_1, a_2, a_3, \ldots, a_n$ be n positive consecutive terms of an arithmetic progression. If $d>0$ is its common difference, then

$$\lim_{n\to\infty}\sqrt{\frac{d}{n}}\left(\frac{1}{\sqrt{a_1}+\sqrt{a_2}}+\frac{1}{\sqrt{a_2}+\sqrt{a_3}}+\ldots+\frac{1}{\sqrt{a_{n-1}}+\sqrt{a_n}}\right) \text{ is}$$

(a) $\sqrt{d}$ (b) 0 (c) 1 (d) $\frac{1}{\sqrt{d}}$

14 Let $A=\{x\in R: [x+3]+[x+4]\le 3\}$,

$B=\left\{x\in R: 3^x\left(\sum_{r=1}^{\infty}\frac{3}{10^r}\right)^{x-3}<3^{-3x}\right\}$, where $[t]$ denotes greatest integer function. Then,

(a) $A=B$ (b) $A\cap B=\phi$
(c) $B\subset C, A\ne B$ (d) $A\subset B, A\ne B$

15 One vertex of a rectangular parallelopiped is at the origin O and the lengths of its edges along X, Y and Z axes are 3, 4 and 5 units respectively. Let P be the vertex $(3,4,5)$. Then, the shortest distance between the diagonal OP and an edge parallel to Z-axis, not passing through O or P is

(a) $\frac{12}{5\sqrt{5}}$ (b) $12\sqrt{5}$ (c) $\frac{12}{\sqrt{5}}$ (d) $\frac{12}{5}$

16 Let $\mathbf{a}=2\hat{\mathbf{i}}+3\hat{\mathbf{j}}+4\hat{\mathbf{k}}$, $\mathbf{b}=\hat{\mathbf{i}}-2\hat{\mathbf{j}}-2\hat{\mathbf{k}}$ and $\mathbf{c}=-\hat{\mathbf{i}}+4\hat{\mathbf{j}}+3\hat{\mathbf{k}}$. If $\mathbf{d}$ is a vector perpendicular to both $\mathbf{b}$ and $\mathbf{c}$, and $\mathbf{a}\cdot\mathbf{d}=18$, then $|\mathbf{a}\times\mathbf{d}|^2$ is equal to

(a) 720 (b) 680 (c) 640 (d) 760

17 Let $I(x)=\int\frac{x^2(x\sec^2 x+\tan x)}{(x\tan x+1)^2}dx$. If $I(0)=0$, then $I\left(\frac{\pi}{4}\right)$ is equal to

(a) $\log_e\frac{(\pi+4)^2}{16}-\frac{\pi^2}{4(\pi+4)}$

(b) $\log_e\frac{(\pi+4)^2}{16}+\frac{\pi^2}{4(\pi+4)}$

(c) $\log_e\frac{(\pi+4)^2}{32}+\frac{\pi^2}{4(\pi+4)}$

(d) $\log_e\frac{(\pi+4)^2}{32}-\frac{\pi^2}{4(\pi+4)}$

18 A pair of dice is thrown 5 times. For each throw, a total of 5 is considered a success. If the probability of at least 4 successes is $\frac{k}{3^{11}}$, then k is equal to

(a) 75 (b) 123
(c) 164 (d) 82

19 If ${}^{2n}C_3 : {}^nC_3 = 10:1$, then the ratio $(n^2+3n):(n^2-3n+4)$ is

(a) 27:11 (b) 2:1
(c) 65:37 (d) 35:16

20 The sum of the first 20 terms of the series $5+11+19+29+41+\ldots$ is

(a) 3450 (b) 3250
(c) 3520 (d) 3420

Section B : Numerical Value Type Questions

21 Let $A=\{1,2,3,4,\ldots,10\}$ and $B=\{0,1,2,3,4\}$. The number of elements in the relation $R=\{(a,b)\in A\times A: 2(a-b)^2+3(a-b)\in B\}$ is

22 If the area of the region $S=\{(x,y): 2y-y^2\le x^2\le 2y, x\ge y\}$ is equal to $\frac{n+2}{n+1}-\frac{\pi}{n-1}$, then the natural number n is equal to ....

23 Let the image of the point $P(1,2,3)$ in the plane $2x-y+z=9$ be Q. If the co-ordinates of the point R are $(6,10,7)$, then the square of the area of the ΔPQR is

24 The coefficient of x^{18} in the expansion of $\left(x^4-\frac{1}{x^3}\right)^{15}$ is

25 Let $a\in Z$ and $[t]$ be the greatest integer $\le t$. Then, the number of points, where the function $f(x)=[a+13\sin x]$, $x\in(0,\pi)$ is not differentiable, is

26 Let $y=y(x)$ be a solution of the differential equation $(x\cos x)dy+(xy\sin x+y\cos x-1)dx=0$, $0<x<\frac{\pi}{2}$. If $\frac{\pi}{3}y\left(\frac{\pi}{3}\right)=\sqrt{3}$, then $\left|\frac{\pi}{6}y''\left(\frac{\pi}{6}\right)+2y'\left(\frac{\pi}{6}\right)\right|$ is equal to ....

27 A circle passing through the point $P(\alpha,\beta)$ in the first quadrant touches the two co-ordinate axes at the points A and B. The point P is above the line AB. The point Q on the line segment AB is the foot of perpendicular from P on AB. If PQ is equal to 11 units, then the value of $\alpha\beta$ is

28 Let the tangent to the curve $x^2+2x-4y+9=0$ at the point $P(1,3)$ on it meet the Y-axis at A. Let the line passing through P and parallel to the line $x-3y=6$ meet the parabola $y^2=4x$ at B. If B lies on the line $2x-3y=8$, then $(AB)^2$ is equal to ....

29 Let the point $(p,p+1)$ lie inside the region $E=\{(x,y): 3-x\le y\le\sqrt{9-x^2}, 0\le x\le 3\}$. If the set of all values of p in the interval (a,b), then b^2+b-a^2 is equal to ......

30 The number of ways of giving 20 distinct oranges to 3 children such that each child gets at least one orange is

6 APRIL 2023, SHIFT-II

Section A : Objective Type Questions

1 Let the sets A and B denote the domain and range respectively of the function $f(x)=\frac{1}{\sqrt{[x]-x}}$, where $[x]$ denotes the smallest integer greater than or equal to x. Then, among the statements

(S1): $A\cap B=(1,\infty)-N$ and

(S2) : $A\cup B=(1,\infty)$

(a) Both (S1) and (S2) are true

(b) Neither (S1) nor (S2) is true

(c) Only (S1) is true

(d) Only (S2) is true

2 All the letters of the word PUBLIC are written in all possible orders and these words are written as in a dictionary with serial numbers. Then, the serial number of the word PUBLIC is

(a) 576 (b) 578 (c) 580 (d) 582

3 Let $f(x)$ be a function satisfying $f(x)+f(\pi-x)=\pi^2$, $\forall\, x\in R$. Then, $\int_0^{\pi} f(x)\sin x\,dx$ is equal to

(a) $\frac{\pi^2}{2}$ (b) $2\pi^2$ (c) π^2 (d) $\frac{\pi^2}{4}$

4 Among the statements:

(S1): $2023^{2022}-1999^{2022}$ is divisible by 8.

(S2): $13(13)^n-12n-13$ is divisible by 144 for infinitely many $n\in N$.

(a) Both (S1) and (S2) are incorrect

(b) Only (S2) is correct

(c) Both (S1) and (S2) are correct

(d) Only (S1) is correct

5 The sum of all values of α, for which the points whose position vectors are $\hat{\mathbf{i}}-2\hat{\mathbf{j}}+3\hat{\mathbf{k}}$, $2\hat{\mathbf{i}}-3\hat{\mathbf{j}}+4\hat{\mathbf{k}}$, $(\alpha+1)\hat{\mathbf{i}}+2\hat{\mathbf{k}}$ and $9\hat{\mathbf{i}}+(\alpha-8)\hat{\mathbf{j}}+6\hat{\mathbf{k}}$ are coplanar is equal to

(a) 2 (b) -2 (c) 4 (d) 6

6 $\lim_{n\to\infty}\left\{\left(2^{\frac{1}{2}}-2^{\frac{1}{3}}\right)\left(2^{\frac{1}{2}}-2^{\frac{1}{5}}\right)\cdots\left(2^{\frac{1}{2}}-2^{\frac{1}{2n-1}}\right)\right\}$ is equal to

(a) $\frac{1}{\sqrt{2}}$ (b) $\sqrt{2}$ (c) 0 (d) 1

7 If the solution curve $f(x,y)=0$ of the differential equation $(1+\log_e x)\frac{dx}{dy}-x\log_e x=e^{-y}$, $x>0$, passes through the points $(1,0)$ and $(\alpha,2)$, then α^{α} is equal to

(a) $e^{2e^{\sqrt{2}}}$ (b) $e^{\sqrt{2}e^2}$

(c) e^{2e^2} (d) e^{e^2}

8 In a group of 100 persons 75 speak English and 40 speak Hindi. Each person speaks at least one of the two languages. If the number of persons, who speak only English is α and the number of persons who speak only Hindi is β, then the eccentricity of the ellipse $25(\beta^2x^2+\alpha^2y^2)=\alpha^2\beta^2$ is

(a) $\frac{3\sqrt{15}}{12}$ (b) $\frac{\sqrt{129}}{12}$ (c) $\frac{\sqrt{117}}{12}$ (d) $\frac{\sqrt{119}}{12}$

9 For the system of equations

$x+y+z=6$

$x+2y+\alpha z=10$

$x+3y+5z=\beta$, which one of the following is not true?

(a) System has infinitely many solutions for $\alpha=3$, $\beta=14$.

(b) System has no solution for $\alpha=3$, $\beta=24$.

(c) System has a unique solution for $\alpha=3$, $\beta\neq 14$.

(d) System has a unique solution for $\alpha=-3$, $\beta=14$.

10 If $\gcd(m,n)=1$ and $1^2-2^2+3^2-4^2+\ldots+(2021)^2-(2022)^2+(2023)^2=1012\,m^2n$, then m^2-n^2 is equal to

(a) 240 (b) 220

(c) 200 (d) 180

11 The area bounded by the curves $y=|x-1|+|x-2|$ and $y=3$ is equal to

(a) 4 (b) 5 (c) 6 (d) 3

12 Let a and b are two non-zero real numbers. Then, the number of elements in the set $X=\{z\in C:\text{Re}(az^2+bz)=a \text{ and } \text{Re}(bz^2+az)=b\}$ is equal to

(a) 2 (b) 1 (c) 3 (d) 0

13 Let the vectors **a**, **b**, **c** represent three coterminous edges of a parallelopiped of volume V. Then, the volume of the parallelopiped, whose coterminous edges are represented by **a**, **b** + **c** and **a** + 2**b** + 3**c** is equal to

(a) V (b) $2V$ (c) $3V$ (d) $6V$

14 A plane P contains the line of intersection of the plane $\mathbf{r}\cdot(\hat{\mathbf{i}}+\hat{\mathbf{j}}+\hat{\mathbf{k}})=6$ and $\mathbf{r}\cdot(2\hat{\mathbf{i}}+3\hat{\mathbf{j}}+4\hat{\mathbf{k}})=-5$. If P passes through the point $(0,2,-2)$, then the square of distance of the point $(12,12,18)$ from the plane P is

(a) 620 (b) 155

(c) 1240 (d) 310

15 If the coefficients of x^7 in $\left(ax^2+\frac{1}{2bx}\right)^{11}$ and x^{-7} in $\left(ax-\frac{1}{3bx^2}\right)^{11}$ are equal, then

(a) $32ab=729$ (b) $243ab=64$

(c) $729ab=32$ (d) $64ab=243$

16 Among the statements

(S1): $(p \Rightarrow q) \vee ((\sim p) \wedge q)$ is a tautology.

(S2): $(q \Rightarrow p) \Rightarrow ((\sim p) \wedge q)$ is a contradiction.

(a) Neither (S1) nor (S2) is true
(b) Only (S1) is true
(c) Both (S1) and (S2) are true
(d) Only (S2) is true

17 Let P be a square matrix such that $P^2 = I - P$. For $\alpha, \beta, \gamma, \delta \in N$. If $P^\alpha + P^\beta = \gamma I - 29P$ and $P^\alpha - P^\beta = \delta I - 13P$, then $\alpha + \beta + \gamma - \delta$ is equal to

(a) 22 (b) 18 (c) 24 (d) 40

18 If the tangents at the points P and Q on the circle $x^2 + y^2 - 2x + y = 5$ meet at the point $R\left(\frac{9}{4}, 2\right)$, then the area of the ΔPQR is

(a) $\frac{5}{4}$ (b) $\frac{13}{4}$ (c) $\frac{13}{8}$ (d) $\frac{5}{8}$

19 Three dice are rolled. If the probability of getting different numbers on the three dice is p/q, where p and q are coprime, then $q - p$ is equal to

(a) 4 (b) 3 (c) 1 (d) 2

20 Let the line L pass through the point (0, 1, 2), intersect the line $\frac{x-1}{2} = \frac{y-2}{3} = \frac{z-3}{1}$ and be parallel to the plane $2x + y - 3z = 4$. Then, the distance of the point $P(1, -9, 2)$ from the line L is

(a) $\sqrt{74}$ (b) $\sqrt{54}$ (c) $\sqrt{69}$ (d) 9

Section B : Numerical Value Type Questions

21 The number of points, where the curve $y = x^5 - 20x^3 + 50x + 2$ crosses the X-axis, is

22 If the mean and variance of the frequency distribution

x_i	2	4	6	8	10	12	14	16
f_i	4	4	α	15	8	β	4	5

are 9 and 15.08 respectively, then the value of $\alpha^2 + \beta^2 - \alpha\beta$ is

23 If the lines $\frac{x-1}{2} = \frac{2-y}{-3} = \frac{z-3}{\alpha}$ and $\frac{x-4}{5} = \frac{y-1}{2} = \frac{z}{\beta}$ intersect, then the magnitude of the minimum value of $8\alpha\beta$ is

24 Let the eccentricity of an ellipse $\frac{x^2}{a^2} + \frac{y^2}{b^2} = 1$ is reciprocal to that of the hyperbola $2x^2 - 2y^2 = 1$. If the ellipse intersects the hyperbola at right angles, then square of length of the latus rectum of the ellipse is

25 The value of $\tan 9° - \tan 27° - \tan 63° + \tan 81°$ is

26 For $\alpha, \beta; z \in C$ and $\lambda > 1$, if $\sqrt{\lambda - 1}$ is the radius of the circle $|z - \alpha|^2 + |z - \beta|^2 = 2\lambda$, then $|\alpha - \beta|$ is equal to ....

27 Let a curve $y = f(x)$, $x \in (0, \infty)$ pass through the points $P\left(1, \frac{3}{2}\right)$ and $Q\left(a, \frac{1}{2}\right)$. If the tangent at any point $R(b, f(b))$ to the given curve cuts the Y-axis at the point $S(0, c)$ such that $bc = 3$, then $(PQ)^2$ is equal to ....

28 Let $f(x) = \frac{x}{(1 + x^n)^{\frac{1}{n}}}$, $x \in R - \{-1\}$, $n \in N$, $n > 2$.

If $f^n(x) = (fofof \ldots \text{ upto } n \text{ times})(x)$, then $\lim_{n \to \infty} \int_0^1 x^{n-2}(f^n(x))dx$ is equal to ....

29 If $(20)^{19} + 2(21)(20)^{18} + 3(21)^2(20)^{17} + \ldots + 20(21)^{19} = k(20)^{19}$, then k is equal to ....

30 The number of 4-letter words, with or without meaning, each consisting of 2 vowels and 2 consonants, which can be formed from the letters of the word UNIVERSE without repetition is

8 APRIL 2023, SHIFT-I

Section A : Objective Type Questions

1 If for $z = \alpha + i\beta$, $|z + 2| = z + 4(1 + i)$, then $\alpha + \beta$ and $\alpha\beta$ are the roots of the equation

(a) $x^2 + 2x - 3 = 0$ (b) $x^2 + 7x + 12 = 0$
(c) $x^2 + 3x - 4 = 0$ (d) $x^2 + x - 12 = 0$

2 Let $f(x) = \frac{\sin x + \cos x - \sqrt{2}}{\sin x - \cos x}$, $x \in [0, \pi] - \left\{\frac{\pi}{4}\right\}$. Then, $f\left(\frac{7\pi}{12}\right) f''\left(\frac{7\pi}{12}\right)$ is equal to

(a) $\frac{2}{9}$ (b) $\frac{-1}{3\sqrt{3}}$
(c) $\frac{-2}{3}$ (d) $\frac{2}{3\sqrt{3}}$

3 Negation of $(p \Rightarrow q) \Rightarrow (q \Rightarrow p)$ is

(a) $q \wedge (\sim p)$ (b) $(\sim q) \wedge p$ (c) $(\sim p) \vee q$ (d) $p \vee (\sim q)$

4 The number of arrangements of the letters of the word "INDEPENDENCE" in which all the vowels always occur together is

(a) 16800 (b) 33600 (c) 18000 (d) 14800

5 If the points with position vectors $\alpha\hat{\mathbf{i}} + 10\hat{\mathbf{j}} + 13\hat{\mathbf{k}}$, $6\hat{\mathbf{i}} + 11\hat{\mathbf{j}} + 11\hat{\mathbf{k}}$, $\frac{9}{2}\hat{\mathbf{i}} + \beta\hat{\mathbf{j}} - 8\hat{\mathbf{k}}$ are collinear, then $(19\alpha - 6\beta)^2$ is equal to

(a) 36 (b) 25
(c) 49 (d) 16

6 Let $S_K = \frac{1+2+\ldots+K}{K}$ and $\sum_{j=1}^{n} S_j^2 = \frac{n}{A}(Bn^2 + Cn + D)$, where $A, B, C, D \in N$ and A has least value. Then,

(a) $A + B$ is divisible by D
(b) $A + B = 5(D - C)$
(c) $A + B + C + D$ is divisible by 5
(d) $A + C + D$ is not divisible by B

7 The area of the region $\{(x, y): x^2 \le y \le 8 - x^2, y \le 7\}$ is

(a) 21 (b) 20 (c) 18 (d) 24

8 Let $P = \begin{bmatrix} \frac{\sqrt{3}}{2} & \frac{1}{2} \\ -\frac{1}{2} & \frac{\sqrt{3}}{2} \end{bmatrix}$, $A = \begin{bmatrix} 1 & 1 \\ 0 & 1 \end{bmatrix}$ and $Q = PAP^T$. If $P^T Q^{2007} P = \begin{bmatrix} a & b \\ c & d \end{bmatrix}$, then $2a + b - 3c - 4d$ equal to

(a) 2007 (b) 2004 (c) 2005 (d) 2006

9 Let $C(\alpha, \beta)$ be the circumcentre of the triangle formed by the lines $4x + 3y = 69$, $4y - 3x = 17$, and $x + 7y = 61$. Then, $(\alpha - \beta)^2 + \alpha + \beta$ is equal to

(a) 16 (b) 17 (c) 18 (d) 15

10 If the equation of the plane containing the line $x + 2y + 3z - 4 = 0 = 2x + y - z + 5$ and perpendicular to the plane $\mathbf{r} = (\hat{\mathbf{i}} - \hat{\mathbf{j}}) + \lambda(\hat{\mathbf{i}} + \hat{\mathbf{j}} + \hat{\mathbf{k}}) + \mu(\hat{\mathbf{i}} - 2\hat{\mathbf{j}} + 3\hat{\mathbf{k}})$ is $ax + by + cz = 4$, then $(a - b + c)$ is equal to

(a) 20 (b) 24 (c) 18 (d) 22

11 Let the number of elements in sets A and B be five and two respectively. Then, the number of subsets of $A \times B$ each having at least 3 and at most 6 elements is

(a) 782 (b) 792 (c) 752 (d) 772

12 Let $A = \begin{bmatrix} 2 & 1 & 0 \\ 1 & 2 & -1 \\ 0 & -1 & 2 \end{bmatrix}$. If $|\text{adj}(\text{adj}(\text{adj}\, 2A))| = (16)^n$, then n is equal to

(a) 8 (b) 10 (c) 12 (d) 9

13 In a bolt factory, machines A, B and C manufacture respectively 20%, 30% and 50% of the total bolts. Of their output 3, 4 and 2 percent are respectively defective bolts. A bolt is drawn at random from the product. If the bolt drawn is found the defective, then the probability that it is manufactured by the machine C is

(a) $\frac{2}{7}$ (b) $\frac{9}{28}$ (c) $\frac{3}{7}$ (d) $\frac{5}{14}$

14 The shortest distance between the lines $\frac{x-4}{4} = \frac{y+2}{5} = \frac{z+3}{3}$ and $\frac{x-1}{3} = \frac{y-3}{4} = \frac{z-4}{2}$ is

(a) $3\sqrt{6}$ (b) $6\sqrt{3}$ (c) $2\sqrt{6}$ (d) $6\sqrt{2}$

15 The number of ways in which 5 girls and 7 boys can be seated at a round table so that no two girls sit together, is

(a) $7(360)^2$ (b) $7(720)^2$
(c) 720 (d) $126(5!)^2$

16 Let $I(x) = \int \frac{(x+1)}{x(1 + xe^x)^2} dx$, $x > 0$. If $\lim_{x \to \infty} I(x) = 0$, then $I(1)$ is equal to

(a) $\frac{e+2}{e+1} - \log_e(e+1)$ (b) $\frac{e+2}{e+1} + \log_e(e+1)$
(c) $\frac{e+1}{e+2} + \log_e(e+1)$ (d) $\frac{e+1}{e+2} - \log_e(e+1)$

17 If the coefficients of three consecutive terms in the expansion of $(1 + x)^n$ are in the ratio 1 : 5 : 20, then the coefficient of the fourth term is

(a) 2436 (b) 5481 (c) 3654 (d) 1827

18 Let α, β, γ be the three roots of the equation $x^3 + bx + c = 0$. If $\beta\gamma = 1 = -\alpha$, then $b^3 + 2c^3 - 3\alpha^3 - 6\beta^3 - 8\gamma^3$ is equal to

(a) 21 (b) 19 (c) $\frac{169}{8}$ (d) $\frac{155}{8}$

19 $\lim_{x \to 0} \left(\left(\frac{1 - \cos^2(3x)}{\cos^3(4x)} \right) \left(\frac{\sin^3(4x)}{(\log_e(2x+1))^5} \right) \right)$ is equal to ....

(a) 24 (b) 9
(c) 18 (d) 15

20 Let R be the focus of the parabola $y^2 = 20x$ and the line $y = mx + c$ intersect the parabola at two points P and Q. Let the point $G(10, 10)$ be the centroid of the ΔPQR. If $c - m = 6$, then $(PQ)^2$ is

(a) 325 (b) 317 (c) 296 (d) 346

Section B : Numerical Value Type Questions

21 Let $[t]$ denote the greatest integer $\le t$. If the constant term in the expansion of $\left(3x^2 - \frac{1}{2x^5}\right)^7$ is α, then $[\alpha]$ is equal to ....

22 Consider a circle $C_1 : x^2 + y^2 - 4x - 2y = \alpha - 5$. Let its mirror image in the line $y = 2x + 1$ be another circle $C_2 : 5x^2 + 5y^2 - 10fx - 10gy + 36 = 0$. Let r be the radius of C_2. Then, $\alpha + r$ is equal to .....

23 If the solution curve of the differential equation $(y - 2\log_e x)dx + (x\log_e x^2)dy = 0$, $x > 1$ passes through the points $\left(e, \frac{4}{3}\right)$ and (e^4, α), then α is equal to ....

24 The largest natural number n such that 3^n divides $66!$ is

25 Let the mean and variance of 8 numbers x, y, 10, 12, 6, 12, 4, 8 be 9 and 9.25, respectively. If $x > y$, then $3x - 2y$ is equal to ....

26 Let $A = \{0, 3, 4, 6, 7, 8, 9, 10\}$ and R be the relation defined on A such that $R = \{(x, y) \in A \times A : x - y$ is odd positive integer or $x - y = 2\}$. The minimum number of elements that must be added to the relation R, so that it is a symmetric relation, is equal to ....

27 Let $\mathbf{a} = 6\hat{\mathbf{i}} + 9\hat{\mathbf{j}} + 12\hat{\mathbf{k}}$, $\mathbf{b} = \alpha\hat{\mathbf{i}} + 11\hat{\mathbf{j}} - 2\hat{\mathbf{k}}$ and $\mathbf{c}$ be vectors such that $\mathbf{a} \times \mathbf{c} = \mathbf{a} \times \mathbf{b}$. If $\mathbf{a} \cdot \mathbf{c} = -12$, $\mathbf{c} \cdot (\hat{\mathbf{i}} - 2\hat{\mathbf{j}} + \hat{\mathbf{k}}) = 5$, then $\mathbf{c} \cdot (\hat{\mathbf{i}} + \hat{\mathbf{j}} + \hat{\mathbf{k}})$ is equal to ....

28 Let λ_1, λ_2 be the values of λ for which the points $\left(\frac{5}{2}, 1, \lambda\right)$ and $(-2, 0, 1)$ are at equal distance from the plane $2x + 3y - 6z + 7 = 0$. If $\lambda_1 > \lambda_2$, then the distance of the point $(\lambda_1 - \lambda_2, \lambda_2, \lambda_1)$ from the line $\frac{x-5}{1} = \frac{y-1}{2} = \frac{z+7}{2}$ is

29 Let $[t]$ denote the greatest integer $\le t$. Then, $\frac{2}{\pi}\int_{\pi/6}^{5\pi/6} (8\,[\text{cosec}\, x] - 5[\cot x])dx$ is equal to ....

30 If a_α is the greatest term in the sequence $a_n = \frac{n^3}{n^4 + 147}$, $n = 1, 2, 3 \ldots$, then α is equal to ....

8 APRIL 2023, SHIFT-II

Section A : Objective Type Questions

1 For $a, b \in Z$ and $|a - b| \le 10$, let the angle between the plane $P: ax + y - z = b$ and the line $l: x - 1 = a - y = z + 1$ be $\cos^{-1}\left(\frac{1}{3}\right)$. If the distance of the point $(6, -6, 4)$ from the plane P is $3\sqrt{6}$, then $a^4 + b^2$ is equal to

(a) 48 (b) 85 (c) 32 (d) 25

2 Let a_n be the nth term of the series $5 + 8 + 14 + 23 + 35 + 50 + \ldots$ and $S_n = \sum_{k=1}^{n} a_k$. Then, $S_{30} - a_{40}$ is equal to

(a) 11260 (b) 11290
(c) 11280 (d) 11310

3 Let the mean and variance of 12 observations be $\frac{9}{2}$ and 4, respectively. Later on, it was observed that two observations were considered as 9 and 10 instead of 7 and 14 respectively. If the correct variance is $\frac{m}{n}$, where m and n are coprime, then $m + n$ is equal to

(a) 316 (b) 315 (c) 314 (d) 317

4 If the probability that the random variable X takes values x is given by $P(X = x) = k(x + 1)3^{-x}$, $X = 0, 1, 2, 3, \ldots$, where k is a constant, then $P(X \ge 2)$ is equal to

(a) $\frac{11}{18}$ (b) $\frac{7}{27}$ (c) $\frac{20}{27}$ (d) $\frac{7}{18}$

5 If $A = \begin{bmatrix} 1 & 5 \\ \lambda & 10 \end{bmatrix}$, $A^{-1} = \alpha A + \beta I$ and $\alpha + \beta = -2$, then $4\alpha^2 + \beta^2 + \lambda^2$ is equal to

(a) 14 (b) 12 (c) 19 (d) 10

6 $25^{190} - 19^{190} - 8^{190} + 2^{190}$ is divisible by

(a) 14 but not by 34 (b) 34 but not by 14
(c) Neither 14 nor 34 (d) Both 14 and 34

7 Let O be the origin and OP and OQ be the tangents to the circle $x^2 + y^2 - 6x + 4y + 8 = 0$ at the points P and Q on it. If the circumcircle of the ΔOPQ passes through the point $\left(\alpha, \frac{1}{2}\right)$, then a value of α is

(a) $\frac{5}{2}$ (b) 1
(c) $-\frac{1}{2}$ (d) $\frac{3}{2}$

8 Let the vectors $\mathbf{u}_1 = \hat{\mathbf{i}} + \hat{\mathbf{j}} + a\hat{\mathbf{k}}$, $\mathbf{u}_2 = \hat{\mathbf{i}} + b\hat{\mathbf{j}} + \hat{\mathbf{k}}$ and $\mathbf{u}_3 = c\hat{\mathbf{i}} + \hat{\mathbf{j}} + \hat{\mathbf{k}}$ be coplanar. If the vectors $\mathbf{v}_1 = (a + b)\hat{\mathbf{i}} + c\hat{\mathbf{j}} + c\hat{\mathbf{k}}$, $\mathbf{v}_2 = a\hat{\mathbf{i}} + (b + c)\hat{\mathbf{j}} + a\hat{\mathbf{k}}$ and $\mathbf{v}_3 = b\hat{\mathbf{i}} + b\hat{\mathbf{j}} + (c + a)\hat{\mathbf{k}}$ are also coplanar, then $6(a + b + c)$ is equal to

(a) 12 (b) 0
(c) 4 (d) 6

9 The absolute difference of the coefficients of x^{10} and x^7 in the expansion of $\left(2x^2 + \frac{1}{2x}\right)^{11}$ is equal to

(a) $13^3 - 13$ (b) $11^3 - 11$
(c) $12^3 - 12$ (d) $10^3 - 10$

10 Let $A = \{1, 2, 3, 4, 5, 6, 7\}$. Then, the relation $R = \{(x, y) \in A \times A : x + y = 7\}$ is

(a) reflexive but neither symmetric nor transitive
(b) an equivalence relation
(c) transitive but neither symmetric nor reflexive
(d) symmetric but neither reflexive nor transitive

11 Let $A = \left\{\theta \in (0, 2\pi) : \frac{1 + 2i \sin\theta}{1 - i \sin\theta} \text{ is purely imaginary}\right\}$. Then, the sum of the elements in A is

(a) 2π (b) π (c) 3π (d) 4π

12 The negation of $(p \wedge (\sim q)) \vee (\sim p)$ is equivalent to

(a) $p \wedge q$ (b) $p \wedge (q \wedge (\sim p))$
(c) $p \wedge (\sim q)$ (d) $p \vee (q \vee (\sim p))$

13 Let $A(0, 1)$, $B(1, 1)$ and $C(1, 0)$ be the mid-points of the sides of a triangle with incentre at the point D. If the focus of the parabola $y^2 = 4ax$ passing through D is $(\alpha + \beta\sqrt{2}, 0)$, where α and β are rational numbers, then $\frac{\alpha}{\beta^2}$ is equal to

(a) 8 (b) $\frac{9}{2}$ (c) 12 (d) 6

14 The integral $\int \left(\left(\frac{x}{2}\right)^x + \left(\frac{2}{x}\right)^x\right) \log_2 x \, dx$ is equal to

(a) $\left(\frac{x}{2}\right)^x \log_2\left(\frac{2}{x}\right) + C$ (b) $\left(\frac{x}{2}\right)^x \log_2\left(\frac{x}{2}\right) + C$
(c) $\left(\frac{x}{2}\right)^x + \left(\frac{2}{x}\right)^x + C$ (d) $\left(\frac{x}{2}\right)^x - \left(\frac{2}{x}\right)^x + C$

15 Let S be the set of all values of $\theta \in [-\pi, \pi]$ for which the system of linear equations

$$x + y + \sqrt{3}z = 0$$
$$-x + (\tan\theta)y + \sqrt{7}z = 0$$
$$x + y + (\tan\theta)z = 0$$

has non-trivial solution. Then, $\frac{120}{\pi} \sum_{\theta \in S} \theta$ is equal to

(a) 40 (b) 10 (c) 30 (d) 20

16 The area of the quadrilateral $ABCD$ with vertices $A(2, 1, 1)$, $B(1, 2, 5)$, $C(-2, -3, 5)$ and $D(1, -6, -7)$ is equal to

(a) 54 (b) $8\sqrt{38}$ (c) 48 (d) $9\sqrt{38}$

17 The value of $36 (4\cos^2 9° - 1)(4\cos^2 27° - 1)(4\cos^2 81° - 1)(4\cos^2 243° - 1)$ is

(a) 27 (b) 36 (c) 18 (d) 54

18 Let P be the plane passing through the line $\frac{x-1}{1} = \frac{y-2}{-3} = \frac{z+5}{7}$ and the point $(2, 4, -3)$. If the image of the point $(-1, 3, 4)$ in the plane P is (α, β, γ), then $\alpha + \beta + \gamma$ is equal to

(a) 9 (b) 11 (c) 12 (d) 10

19 If the number of words, with or without meaning, which can be made using all the letters of the word MATHEMATICS in which C and S do not come together, is $(6!)k$, then k is equal to

(a) 5670 (b) 1890 (c) 945 (d) 2835

20 If $\alpha > \beta > 0$ are the roots of the equation $ax^2 + bx + 1 = 0$, and

$$\lim_{x \to \frac{1}{\alpha}} \left(\frac{1 - \cos(x^2 + bx + a)}{2(1 - \alpha x)^2}\right)^{\frac{1}{2}} = \frac{1}{k}\left(\frac{1}{\beta} - \frac{1}{\alpha}\right),$$ then k is equal to

(a) α (b) 2α (c) 2β (d) β

Section B : Numerical Value Type Questions

21 Let m and n be the numbers of real roots of the quadratic equations $x^2 - 12x + [x] + 31 = 0$ and $x^2 - 5|x + 2| - 4 = 0$ respectively, where $[x]$ denotes the greatest integer $\leq x$. Then, $m^2 + mn + n^2$ is equal to ....

22 Let the solution curve $x = x(y)$, $0 < y < \frac{\pi}{2}$, of the differential equation $(\log_e(\cos y))^2 \cos y \, dx - (1 + 3x \log_e(\cos y)) \sin y \, dy = 0$ satisfy $x\left(\frac{\pi}{3}\right) = \frac{1}{2\log_e 2}$. If $x\left(\frac{\pi}{6}\right) = \frac{1}{\log_e m - \log_e n}$, where m and n are coprime, then mn is equal to......

23 Let $[t]$ denote the greatest integer function. If $\int_0^{2.4} [x^2] dx = \alpha + \beta\sqrt{2} + \gamma\sqrt{3} + \delta\sqrt{5}$, then $\alpha + \beta + \gamma + \delta$ is equal to ....

24 Let k and m be positive real numbers such that the function $f(x) = \begin{cases} 3x^2 + k\sqrt{x+1}, & 0 < x < 1 \\ mx^2 + k^2, & x \geq 1 \end{cases}$ is differentiable for all $x > 0$. Then, $\frac{8f'(8)}{f'\left(\frac{1}{8}\right)}$ is equal to

....

[Given, 64]

25 Let $R = \{a, b, c, d, e\}$ and $S = \{1, 2, 3, 4\}$. Total number of onto functions $f : R \to S$ such that $f(a) \neq 1$, is equal to ....

26 Let P_1 be the plane $3x - y - 7z = 11$ and P_2 be the plane passing through the points $(2, -1, 0)$, $(2, 0, -1)$, and $(5, 1, 1)$. If the foot of the perpendicular drawn from the point $(7, 4, -1)$ on the line of intersection of the planes P_1 and P_2 is (α, β, γ), then $\alpha + \beta + \gamma$ is equal to ....

27 The ordinates of the points P and Q on the parabola with focus $(3, 0)$ and directrix $x = -3$ are in the ratio $3 : 1$. If $R(\alpha, \beta)$ is the point of intersection of the tangents to the parabola at P and Q, then $\frac{\beta^2}{\alpha}$ is equal to ....

28 If domain of the function $\log_e\left(\frac{6x^2+5x+1}{2x-1}\right)+\cos^{-1}\left(\frac{2x^2-3x+4}{3x-5}\right)$ is $(\alpha, \beta) \cup (\gamma, \delta]$, then $18(\alpha^2+\beta^2+\gamma^2+\delta^2)$ is equal to ....

29 Let $0 < z < y < x$ be three real numbers such that $\frac{1}{x}, \frac{1}{y}, \frac{1}{z}$ are in an arithmetic progression and $x, \sqrt{2}y, z$ are in a geometric progression. If $xy+yz+zx=\frac{3}{\sqrt{2}}xyz$, then $3(x+y+z)^2$ is equal to ....

30 Let the area enclosed by the lines $x+y=2$, $y=0$, $x=0$ and the curve $f(x)=\min\left\{x^2+\frac{3}{4}, 1+[x]\right\}$ where $[x]$ denotes the greatest integer $\leq x$, be A. Then, the value of $12A$ is

10 APRIL 2023, SHIFT-I

Section A : Objective Type Questions

1 An arc PQ of a circle subtends a right angle at its centre O. The mid-point of the arc PQ is R. If $\mathbf{OP}=\mathbf{u}$, $\mathbf{OR}=\mathbf{v}$ and $\mathbf{OQ}=\alpha\mathbf{u}+\beta\mathbf{v}$, then α, β^2 are the roots of the equation

(a) $3x^2-2x-1=0$
(b) $3x^2+2x-1=0$
(c) $x^2-x-2=0$
(d) $x^2+x-2=0$

2 Let f be a differentiable function such that $x^2f(x)-x=4\int_0^x tf(t)dt$, $f(1)=\frac{2}{3}$. Then, $18f(3)$ is equal to

(a) 210 (b) 160 (c) 180 (d) 150

3 Let two vertices of a ΔABC be $(2, 4, 6)$ and $(0, -2, -5)$, and its centroid be $(2, 1, -1)$. If the image of the third vertex in the plane $x+2y+4z=11$ is (α, β, γ), then $\alpha\beta+\beta\gamma+\gamma\alpha$ is equal to

(a) 70 (b) 74 (c) 72 (d) 76

4 Let the ellipse $E: x^2+9y^2=9$ intersect the positive X and Y- axes at the points A and B, respectively. Let the major axis of E be a diameter of the circle C. Let the line passing through A and B meet the circle C at the point P. If the area of the triangle with vertices A, P and the origin O is $\frac{m}{n}$, where m and n are coprime, then $m-n$ is equal to

(a) 15 (b) 17 (c) 18 (d) 16

5 A square piece of tin of side 30 cm is to be made into a box without top by cutting a square from each corner and folding up the flaps to form a box. If the volume of the box is maximum, then its surface area (in cm^2) is equal to

(a) 800 (b) 900 (c) 1025 (d) 675

6 If A is a 3×3 matrix and $|A|=2$, then $\left|3\,\text{adj}\,(|3A|A^2)\right|$ is equal to

(a) $3^{12}\cdot6^{10}$ (b) $3^{12}\cdot6^{11}$
(c) $3^{11}\cdot6^{10}$ (d) $3^{10}\cdot6^{11}$

7 The shortest distance between the lines $\frac{x+2}{1}=\frac{y}{-2}=\frac{z-5}{2}$ and $\frac{x-4}{1}=\frac{y-1}{2}=\frac{z+3}{0}$ is

(a) 6 (b) 8 (c) 9 (d) 7

8 If $I(x)=\int e^{\sin^2 x}(\cos x\sin 2x-\sin x)dx$ and $I(0)=1$, then $I\left(\frac{\pi}{3}\right)$ is equal to

(a) $e^{\frac{3}{4}}$ (b) $-e^{\frac{3}{4}}$
(c) $\frac{1}{2}e^{\frac{3}{4}}$ (d) $-\frac{1}{2}e^{\frac{3}{4}}$

9 For the system of linear equations

$2x-y+3z=5$
$3x+2y-z=7$
$4x+5y+\alpha z=\beta$

which of the following is not correct?

(a) The system has infinitely many solutions for $\alpha=-5$ and $\beta=9$
(b) The system has a unique solution for $\alpha\neq-5$ and $\beta=8$
(c) The system has infinitely many solutions for $\alpha=-6$ and $\beta=9$
(d) The system is inconsistent for $\alpha=-5$ and $\beta=8$

10 Let P be the point of intersection of the line $\frac{x+3}{3}=\frac{y+2}{1}=\frac{1-z}{2}$ and the plane $x+y+z=2$. If the distance of the point P from the plane $3x-4y+12z=32$ is q, then q and $2q$ are the roots of the equation

(a) $x^2-18x+72=0$
(b) $x^2+18x+72=0$
(c) $x^2-18x-72=0$
(d) $x^2+18x-72=0$

11 Let the first term a and the common ratio r of a geometric progression be positive integers. If the sum of squares of its first three terms is 33033, then the sum of these three terms is equal to

(a) 210 (b) 220
(c) 231 (d) 241

12 $96 \cos \frac{\pi}{33} \cos \frac{2\pi}{33} \cos \frac{4\pi}{33} \cos \frac{8\pi}{33} \cos \frac{16\pi}{33}$ is equal to

(a) 4 (b) 3
(c) 2 (d) 1

13 Let O be the origin and the position vector of the point P be $-\hat{\mathbf{i}} - 2\hat{\mathbf{j}} + 3\hat{\mathbf{k}}$. If the position vectors of the points A, B and C are $-2\hat{\mathbf{i}} + \hat{\mathbf{j}} - 3\hat{\mathbf{k}}$, $2\hat{\mathbf{i}} + 4\hat{\mathbf{j}} - 2\hat{\mathbf{k}}$ and $-4\hat{\mathbf{i}} + 2\hat{\mathbf{j}} - \hat{\mathbf{k}}$, respectively, then the projection of the vector $\mathbf{OP}$ on a vector perpendicular to the vectors $\mathbf{AB}$ and $\mathbf{AC}$ is

(a) $\frac{7}{3}$ (b) 3
(c) $\frac{8}{3}$ (d) $\frac{10}{3}$

14 A line segment AB of length λ moves such that the points A and B remain on the periphery of a circle of radius λ. Then, the locus of the point, that divides the line segment AB in the ratio 2 : 3, is a circle of radius

(a) $\frac{\sqrt{19}}{7}\lambda$ (b) $\frac{\sqrt{19}}{5}\lambda$
(c) $\frac{3}{5}\lambda$ (d) $\frac{2}{3}\lambda$

15 If the coefficient of x^7 in $\left(ax - \frac{1}{bx^2}\right)^{13}$ and the coefficient of x^{-5} in $\left(ax + \frac{1}{bx^2}\right)^{13}$ are equal, then $a^4 b^4$ is equal to

(a) 11 (b) 33 (c) 22 (d) 44

16 Let N denote the sum of the numbers obtained when two dice are rolled. If the probability that $2^N < N!$ is $\frac{m}{n}$, where m and n are coprime, then $4m - 3n$ is equal to

(a) 10 (b) 12 (c) 6 (d) 8

17 Let the complex number $z = x + iy$ be such that $\frac{2z - 3i}{2z + i}$ is purely imaginary. If $x + y^2 = 0$, then $y^4 + y^2 - y$ is equal to

(a) 2/3 (b) 3/2 (c) 3/4 (d) 4/3

18 The negation of the statement $(p \vee q) \wedge (q \vee (\sim r))$ is

(a) $((\sim p) \vee r)) \wedge (\sim q)$ (b) $(p \vee r) \wedge (\sim q)$
(c) $((\sim p) \vee (\sim q)) \wedge (\sim r)$ (d) $((\sim p) \vee (\sim q)) \vee (\sim r)$

19 If $f(x) = \frac{(\tan 1°)x + \log_e(123)}{x \log_e(1234) - (\tan 1°)}$, $x > 0$, then the least value of $f(f(x)) + f\left(f\left(\frac{4}{x}\right)\right)$ is

(a) 2 (b) 8 (c) 4 (d) 0

20 The slope of tangent at any point (x, y) on a curve $y = y(x)$ is $\frac{x^2 + y^2}{2xy}$, $x > 0$. If $y(2) = 0$, then a value of $y(8)$ is

(a) $-4\sqrt{2}$ (b) $4\sqrt{3}$ (c) $-2\sqrt{3}$ (d) $2\sqrt{3}$

Section B : Numerical Value Type Questions

21 Let a, b, c be three distinct positive real numbers such that $(2a)^{\log_e a} = (bc)^{\log_e b}$ and $b^{\log_e 2} = a^{\log_e c}$.

Then, $6a + 5bc$ is equal to ....

22 Let $f : (-2, 2) \to R$ be defined by

$$f(x) = \begin{cases} x[x], & -2 < x < 0 \\ (x-1)[x], & 0 \le x < 2 \end{cases},$$

where $[x]$ denotes the greatest integer function. If m and n respectively are the number of points in $(-2, 2)$ at which $y = |f(x)|$ is not continuous and not differentiable, then $m + n$ is equal to ......

23 If the mean of the frequency distribution

Class	0-10	10-20	20-30	30-40	40-50
Frequency	2	3	x	5	4

is 28, then its variance is.

24 The number of permutations of the digits 1, 2, 3, ..., 7 without repetition, which neither contain the string 153 nor the string 2467, is

25 Let a common tangent to the curves $y^2 = 4x$ and $(x-4)^2 + y^2 = 16$ touch the curves at the points P and Q. Then, $(PQ)^2$ is equal to

26 Let $y = p(x)$ be the parabola passing through the points $(-1, 0)$, $(0, 1)$ and $(1, 0)$. If the area of the region $\{(x, y) : (x+1)^2 + (y-1)^2 \le 1, y \le p(x)\}$ is A, then $12(\pi - 4A)$ is equal to

27 The coefficient of x^7 in $(1 - x + 2x^3)^{10}$ is

28 The sum of all those terms of the arithmetic progression 3, 8, 13, ..., 373, which are not divisible by 3, is equal to

29 The number of elements in the set $\{n \in \mathbb{Z} : |n^2 - 10n + 19| < 6\}$ is

30 Some couples participated in a mixed doubles badminton tournament. If the number of matches played, so that no couple played in a match, is 840, then the total number of persons, who participated in the tournament, is

10 APRIL 2023, SHIFT-II

Section A : Objective Type Questions

1 The statement $\sim [p \vee (\sim (p \wedge q))]$ is equivalent to

(a) $\sim (p \vee q)$ (b) $\sim (p \wedge q)$
(c) $(\sim (p \wedge q)) \wedge q$ (d) $(p \wedge q) \wedge (\sim p)$

2 If the coefficients of x and x^2 in $(1+x)^p (1-x)^q$ are 4 and -5 respectively, then $2p + 3q$ is equal to

(a) 66 (b) 60 (c) 63 (d) 69

3 Let the number $(22)^{2022} + (2022)^{22}$ leave the remainder α when divided by 3 and β when divided by 7. Then, $(\alpha^2 + \beta^2)$ is equal to

(a) 13 (b) 5 (c) 10 (d) 20

4 Let $S = \left\{x \in \left(-\frac{\pi}{2}, \frac{\pi}{2}\right) : 9^{1-\tan^2 x} + 9^{\tan^2 x} = 10\right\}$ and $\beta = \sum_{x \in S} \tan^2\left(\frac{x}{3}\right)$, then $\frac{1}{6}(\beta - 14)^2$ is equal to

(a) 8 (b) 16 (c) 64 (d) 32

5 For $\alpha, \beta, \gamma, \delta \in N$, if $\int \left(\left(\frac{x}{e}\right)^{2x} + \left(\frac{e}{x}\right)^{2x}\right) \log_e x \, dx$ $= \frac{1}{\alpha}\left(\frac{x}{e}\right)^{\beta x} - \frac{1}{\gamma}\left(\frac{e}{x}\right)^{\delta x} + C$, where $e = \sum_{n=0}^{\infty} \frac{1}{n!}$ and C is constant of integration, then $\alpha + 2\beta + 3\gamma - 4\delta$ is equal to

(a) -4 (b) 1
(c) 4 (d) -8

6 Let the line $\frac{x}{1} = \frac{6-y}{2} = \frac{z+8}{5}$ intersect the lines $\frac{x-5}{4} = \frac{y-7}{3} = \frac{z+2}{1}$ and $\frac{x+3}{6} = \frac{3-y}{3} = \frac{z-6}{1}$ at the points A and B, respectively. Then, the distance of the mid-point of the line segment AB from the plane $2x - 2y + z = 14$ is

(a) 4 (b) 3 (c) $\frac{11}{3}$ (d) $\frac{10}{3}$

7 Let a circle of radius 4 be concentric to the ellipse $15x^2 + 19y^2 = 285$. Then, the common tangents are inclined to the minor axis of the ellipse at the angle

(a) $\frac{\pi}{4}$ (b) $\frac{\pi}{3}$ (c) $\frac{\pi}{12}$ (d) $\frac{\pi}{6}$

8 Let $A = \{2, 3, 4\}$ and $B = \{8, 9, 12\}$. Then, the number of elements in the relation $R = \{((a_1, b_1), (a_2, b_2)) \in (A \times B, A \times B) : a_1$ divides b_2 and a_2 divides $b_1\}$ is

(a) 24 (b) 18 (c) 36 (d) 12

9 Let the image of the point $P(1, 2, 6)$ in the plane passing through the points $A(1, 2, 0)$, $B(1, 4, 1)$ and $C(0, 5, 1)$ be $Q(\alpha, \beta, \gamma)$. Then, $(\alpha^2 + \beta^2 + \gamma^2)$ is equal to

(a) 70 (b) 76 (c) 62 (d) 65

10 Let f be a continuous function satisfying $\int_0^{t^2} (f(x) + x^2) dx = \frac{4}{3} t^3, \forall t > 0$. Then, $f\left(\frac{\pi^2}{4}\right)$ is equal to

(a) $-\pi\left(1 + \frac{\pi^3}{16}\right)$ (b) $\pi\left(1 - \frac{\pi^3}{16}\right)$
(c) $-\pi^2\left(1 + \frac{\pi^2}{16}\right)$ (d) $\pi^2\left(1 - \frac{\pi^2}{16}\right)$

11 Eight persons are to be transported from city A to city B in three cars of different makes. If each car can accommodate at most three persons, then the number of ways, in which they can be transported, is

(a) 1680 (b) 3360 (c) 1120 (d) 560

12 If the points P and Q are respectively the circumcentre and the orthocentre of a ΔABC, then $\mathbf{PA} + \mathbf{PB} + \mathbf{PC}$ is equal to

(a) $\mathbf{PQ}$ (b) $2\mathbf{QP}$ (c) $\mathbf{QP}$ (d) $2\mathbf{PQ}$

13 Let a die be rolled n times. Let the probability of getting odd numbers seven times be equal to the probability of getting odd numbers nine times. If the probability of getting even numbers twice is $\frac{k}{2^{15}}$, then k is equal to

(a) 90 (b) 60 (c) 30 (d) 15

14 If $S_n = 4 + 11 + 21 + 34 + 50 + \ldots$ to n terms, then $\frac{1}{60}(S_{29} - S_9)$ is equal to

(a) 220 (b) 223
(c) 227 (d) 226

15 Let $\mathbf{a} = 2\hat{\mathbf{i}} + 7\hat{\mathbf{j}} - \hat{\mathbf{k}}$, $\mathbf{b} = 3\hat{\mathbf{i}} + 5\hat{\mathbf{k}}$ and $\mathbf{c} = \hat{\mathbf{i}} - \hat{\mathbf{j}} + 2\hat{\mathbf{k}}$. Let $\mathbf{d}$ be a vector which is perpendicular to both $\mathbf{a}$ and $\mathbf{b}$, and $\mathbf{c} \cdot \mathbf{d} = 12$. Then, $(-\hat{\mathbf{i}} + \hat{\mathbf{j}} - \hat{\mathbf{k}}) \cdot (\mathbf{c} \times \mathbf{d})$ is equal to

(a) 24 (b) 44 (c) 48 (d) 42

16 Let $g(x) = f(x) + f(1-x)$ and $f''(x) > 0$, $x \in (0, 1)$. If g is decreasing in the interval $(0, \alpha)$ and increasing in the interval $(\alpha, 1)$, then $\tan^{-1}(2\alpha) + \tan^{-1}\left(\frac{1}{\alpha}\right) + \tan^{-1}\left(\frac{\alpha+1}{\alpha}\right)$ is equal to

(a) $\frac{5\pi}{4}$ (b) $\frac{3\pi}{4}$ (c) $\frac{3\pi}{2}$ (d) π

17 Let A be the point $(1, 2)$ and B be any point on the curve $x^2 + y^2 = 16$. If the centre of the locus of the point P, which divides the line segment AB in the ratio $3:2$ is the point $C(\alpha, \beta)$, then the length of the line segment AC is

(a) $\frac{3\sqrt{5}}{5}$ (b) $\frac{6\sqrt{5}}{5}$ (c) $\frac{2\sqrt{5}}{5}$ (d) $\frac{4\sqrt{5}}{5}$

18 If $A = \frac{1}{5!6!7!}\begin{bmatrix} 5! & 6! & 7! \\ 6! & 7! & 8! \\ 7! & 8! & 9! \end{bmatrix}$, then $|\text{adj}(\text{adj}(2A))|$ is equal to

(a) 2^{12} (b) 2^{20} (c) 2^{16} (d) 2^{8}

19 Let $S = \left\{ z = x + iy : \frac{2z - 3i}{4z + 2i} \text{ is a real number} \right\}$. Then, which of the following is not correct?

(a) $(x, y) = \left(0, -\frac{1}{2}\right)$ (b) $y + x^2 + y^2 \neq -\frac{1}{4}$

(c) $x = 0$ (d) $y \in \left(-\infty, -\frac{1}{2}\right) \cup \left(-\frac{1}{2}, \infty\right)$

20 Let μ be the mean and σ be the standard deviation of the distribution

x_i	0	1	2	3	4	5
f_i	$k + 2$	$2k$	$k^2 - 1$	$k^2 - 1$	$k^2 + 1$	$k - 3$

Where, $\Sigma f_i = 62$. If $[x]$ denotes the greatest integer $\leq x$, then $[\mu^2 + \sigma^2]$ is equal to

(a) 8 (b) 9 (c) 7 (d) 6

Section B : Numerical Value Type Questions

21 In the figure, $\theta_1 + \theta_2 = \frac{\pi}{2}$ and $\sqrt{3}(BE) = 4(AB)$. If the area of ΔCAB is $2\sqrt{3} - 3$ unit^2, when $\frac{\theta_2}{\theta_1}$ is the largest, then the perimeter (in unit) of ΔCED is equal to ....

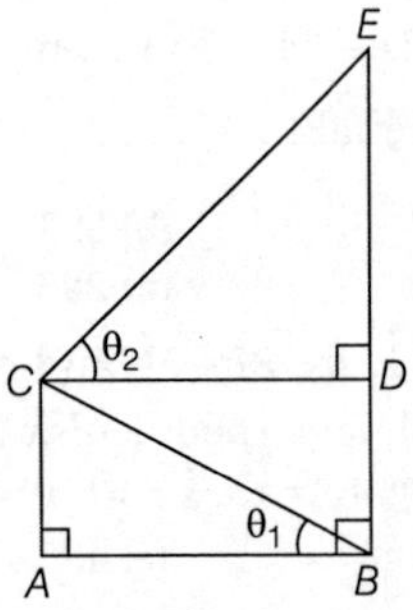

22 If the area of the region $\{(x, y) : |x^2 - 2| \leq y \leq x\}$ is A, then $6A + 16\sqrt{2}$ is equal to ....

23 Let the equations of two adjacent sides of a parallelogram $ABCD$ be $2x - 3y = -23$ and $5x + 4y = 23$. If the equation of its one diagonal AC is $3x + 7y = 23$ and the distance of A from the other diagonal is d, then $50\,d^2$ is equal to ....

[Given 500]

24 Suppose, $a_1, a_2, 2, a_3, a_4$ be in an arithmetico-geometric progression. If the common ratio of the corresponding geometric progression is 2 and the sum of all 5 terms of the arithmetico-geometric progression is 49/2, then a_4 is equal to ....

25 Let the quadratic curve passing through the point $(-1, 0)$ and touching the line $y = x$ at $(1, 1)$ be $y = f(x)$. Then, the x-intercept of the normal to the curve at the point $(\alpha, \alpha + 1)$ in the first quadrant is ....

26 The sum of all the four-digit numbers that can be formed using all the digits 2, 1, 2, 3 is equal to ....

27 Let S be the set of values of λ, for which the system of equations $6\lambda x - 3y + 3z = 4\lambda^2$, $2x + 6\lambda y + 4z = 1$, $3x + 2y + 3\lambda z = \lambda$ has no solution. Then, $12 \sum_{\lambda \in S} |\lambda|$ is equal to ....

28 If the domain of the function $f(x) = \sec^{-1}\left(\frac{2x}{5x + 3}\right)$ is $[\alpha, \beta) \cup (\gamma, \delta]$, then $|3\alpha + 10(\beta + \gamma) + 21\delta|$ is equal to ....

29 Let the foot of perpendicular from the point $A(4, 3, 1)$ on the plane $P : x - y + 2z + 3 = 0$ be N. If $B(5, \alpha, \beta)$, $\alpha, \beta \in Z$ is a point on plane P such that the area of the ΔABN is $3\sqrt{2}$, then $\alpha^2 + \beta^2 + \alpha\beta$ is equal to ....

30 Let the tangent at any point P on a curve passing through the points (1, 1) and $\left(\frac{1}{10}, 100\right)$, intersect positive X-axis and Y-axis at the points A and B respectively. If $PA : PB = 1 : k$ and $y = y(x)$ is the solution of the differential equation $e^{\frac{dy}{dx}} = kx + \frac{k}{2}$, $y(0) = k$, then $4y(1) - 5\log_e 3$ is equal to.........

11 APRIL 2023, SHIFT-I

Section A : Objective Type Questions

1 Let **a** be a non-zero vector parallel to the line of intersection of the two planes described by $\hat{\mathbf{i}} + \hat{\mathbf{j}}$, $\hat{\mathbf{i}} + \hat{\mathbf{k}}$ and $\hat{\mathbf{i}} - \hat{\mathbf{j}}$, $\hat{\mathbf{j}} - \hat{\mathbf{k}}$. If θ is the angle between the vector **a** and the vector $\mathbf{b} = 2\hat{\mathbf{i}} - 2\hat{\mathbf{j}} + \hat{\mathbf{k}}$ and $\mathbf{a} \cdot \mathbf{b} = 6$, then the ordered pair $(\theta, |\mathbf{a} \times \mathbf{b}|)$ is equal to

(a) $\left(\frac{\pi}{3}, 6\right)$ (b) $\left(\frac{\pi}{4}, 3\sqrt{6}\right)$

(c) $\left(\frac{\pi}{3}, 3\sqrt{6}\right)$ (d) $\left(\frac{\pi}{4}, 6\right)$

2 For any vector $\mathbf{a} = a_1\hat{\mathbf{i}} + a_2\hat{\mathbf{j}} + a_3\hat{\mathbf{k}}$, with $10|a_i| < 1$, $i = 1, 2, 3$, consider the following statements

(A) $\max\{|a_1|, |a_2|, |a_3|\} \leq |\mathbf{a}|$

(B) $|\mathbf{a}| \le 3\max\{|a_1|, |a_2|, |a_3|\}$
(a) Neither (A) nor (B) is true
(b) Both (A) and (B) are true
(c) Only (B) is true
(d) Only (A) is true

3 The value of the integral $\int_{-\log_e 2}^{\log_e 2} e^x (\log_e (e^x + \sqrt{1 + e^{2x}}))\, dx$ is equal to

(a) $\log_e\left(\frac{(2+\sqrt{5})^2}{\sqrt{1+\sqrt{5}}}\right) + \frac{\sqrt{5}}{2}$
(b) $\log_e\left(\frac{2(2+\sqrt{5})}{\sqrt{1+\sqrt{5}}}\right) - \frac{\sqrt{5}}{2}$
(c) $\log_e\left(\frac{\sqrt{2}(3-\sqrt{5})^2}{\sqrt{1+\sqrt{5}}}\right) + \frac{\sqrt{5}}{2}$
(d) $\log_e\left(\frac{\sqrt{2}(2+\sqrt{5})^2}{\sqrt{1+\sqrt{5}}}\right) - \frac{\sqrt{5}}{2}$

4 Let R be a rectangle given by the lines $x = 0$, $x = 2$, $y = 0$ and $y = 5$. Let $A(\alpha, 0)$ and $B(0, \beta)$, $\alpha \in [0, 2]$ and $\beta \in [0, 5]$, be such that the line segment AB divides the area of the rectangle R in the ratio 4 : 1. Then, the mid-point of AB lies on a

(a) straight line
(b) hyperbola
(c) circle
(d) parabola

5 Let w_1 be the point obtained by the rotation of $z_1 = 5 + 4i$ about the origin through a right angle in the anti-clockwise direction, and w_2 be the point obtained by the rotation of $z_2 = 3 + 5i$ about the origin through a right angle in the clockwise direction. Then, the principal argument of $w_1 - w_2$ is equal to

(a) $-\pi + \tan^{-1}\frac{8}{9}$
(b) $\pi - \tan^{-1}\frac{8}{9}$
(c) $\pi - \tan^{-1}\frac{33}{5}$
(d) $-\pi + \tan^{-1}\frac{33}{5}$

6 Let sets A and B have 5 elements each. Let the mean of the elements in sets A and B be 5 and 8 respectively and the variance of the elements in sets A and B be 12 and 20 respectively. A new set C of 10 elements is formed by subtracting 3 from each element of A and adding 2 to each element of B. Then, the sum of the mean and variance of the elements of C is

(a) 36
(b) 38
(c) 40
(d) 32

7 Let $f(x) = [x^2 - x] + |-x + [x]|$, where $x \in R$ and $[t]$ denotes the greatest integer less than or equal to t. Then, f is

(a) continuous at $x = 0$ and $x = 1$
(b) continuous at $x = 1$, but not continuous at $x = 0$
(c) not continuous at $x = 0$ and $x = 1$
(d) continuous at $x = 0$, but not continuous at $x = 1$

8 Area of the region $\{(x, y) : x^2 + (y-2)^2 \le 4, x^2 \ge 2y\}$ is

(a) $\pi - \frac{8}{3}$
(b) $2\pi + \frac{16}{3}$
(c) $2\pi - \frac{16}{3}$
(d) $\pi + \frac{8}{3}$

9 Let $x_1, x_2, \ldots, x_{100}$ be in an arithmetic progression, with $x_1 = 2$ and their mean equal to 200. If $y_i = i(x_i - i)$, $1 \le i \le 100$, then the mean of $y_1, y_2, \ldots, y_{100}$ is

(a) 10049.50
(b) 10101.50
(c) 10051.50
(d) 10100

10 The number of elements in the set $S = \{\theta \in [0, 2\pi] : 3\cos^4\theta - 5\cos^2\theta - 2\sin^6\theta + 2 = 0\}$ is

(a) 12
(b) 8
(c) 9
(d) 10

11 Let $S = \{M = [a_{ij}], a_{ij} \in \{0, 1, 2\}, 1 \le i, j \le 2\}$ be a sample space and $A = \{M \in S : M$ is invertible$\}$ be an event. Then, $P(A)$ is equal to

(a) $\frac{47}{81}$
(b) $\frac{16}{27}$
(c) $\frac{50}{81}$
(d) $\frac{49}{81}$

12 The number of triplets (x, y, z), where x, y, z are distinct non negative integers satisfying $x + y + z = 15$, is

(a) 136
(b) 80
(c) 114
(d) 92

13 If equation of the plane that contains the point $(-2, 3, 5)$ and is perpendicular to each of the planes $2x + 4y + 5z = 8$ and $3x - 2y + 3z = 5$ is $\alpha x + \beta y + \gamma z + 97 = 0$, then $\alpha + \beta + \gamma =$

(a) 17
(b) 18
(c) 16
(d) 15

14 An organisation awarded 48 medals in event A, 25 in event B and 18 in event C. If these medals went to total 60 men and only five men got medals in all the three events, then how many received medals in exactly two of three events?

(a) 10
(b) 9
(c) 21
(d) 15

15 Consider ellipses $E_k : kx^2 + k^2y^2 = 1$, $k = 1, 2, \ldots 20$. Let C_k be the circle which touches the four chords joining the end points (one on minor axis and another on major axis) of the ellipse E_k. If r_k is the radius of the circle C_k, then the value of $\sum_{k=1}^{20} \frac{1}{r_k^2}$ is

(a) 3320
(b) 2870
(c) 3080
(d) 3210

16 Let A be a 2×2 matrix with real entries such that $A' = \alpha A + I$, where $\alpha \in R - \{-1, 1\}$. If $\det(A^2 - A) = 4$, then the sum of all possible values of α is equal to

(a) $\frac{5}{2}$
(b) 2
(c) $\frac{3}{2}$
(d) 0

17 Let (α, β, γ) be the image of the point $P(2, 3, 5)$ in the plane $2x + y - 3z = 6$. Then, $\alpha + \beta + \gamma$ is equal to

(a) 10
(b) 12
(c) 5
(d) 9

18 The number of integral solutions x of $\log_{\left(x + \frac{7}{2}\right)}\left(\frac{x-7}{2x-3}\right)^2 \ge 0$ is

(a) 5
(b) 7
(c) 6
(d) 8

19 Let $y = y(x)$ be a solution curve of the differential equation.

$(1 - x^2y^2)dx = ydx + xdy$. If the line $x = 1$ intersects the curve $y = y(x)$ at $y = 2$ and the line $x = 2$ intersects the curve $y = y(x)$ at $y = \alpha$, then a value of α is

(a) $\frac{3e^2}{2(3e^2+1)}$ (b) $\frac{3e^2}{2(3e^2-1)}$

(c) $\frac{1-3e^2}{2(3e^2+1)}$ (d) $\frac{1+3e^2}{2(3e^2-1)}$

20 Let $f : [2, 4] \to R$ be a differentiable function such that $(x \log_e x) f'(x) + (\log_e x) f(x) + f(x) \geq 1$, $x \in [2, 4]$ with $f(2) = \frac{1}{2}$ and $f(4) = \frac{1}{4}$. Consider the following two statements

(A) : $f(x) \leq 1, \forall x \in [2, 4]$

(B) : $f(x) \geq \frac{1}{8}, \forall x \in [2, 4]$. Then,

(a) Both the statements (A) and (B) are true
(b) Neither statement (A) nor statement (B) is true
(c) Only statement (A) is true
(d) Only statement (B) is true

Section B : Numerical Value Type Questions

21 Let $H_n = \frac{x^2}{1+n} - \frac{y^2}{3+n} = 1$, $n \in N$. Let k be the smallest even value of n such that the eccentricity of H_k is a rational number. If l is the length of the latus rectum of H_k, then $21l$ is equal to ....

22 For $m, n > 0$, let $\alpha(m, n) = \int_0^2 t^m (1 + 3t)^n \, dt$. If $11\alpha(10, 6) + 18\alpha(11, 5) = p(14)^6$, then p is equal to ....

23 Let a line l pass through the origin and be perpendicular to the lines

$l_1 : \mathbf{r} = (\hat{i} - 11\hat{j} - 7\hat{k}) + \lambda(\hat{i} + 2\hat{j} + 3\hat{k})$, $\lambda \in R$ and l_2 : $\mathbf{r} = (-\hat{i} + \hat{k}) + \mu(2\hat{i} + 2\hat{j} + \hat{k})$, $\mu \in R$.

If P is the point of intersection of l and l_1, and $Q(\alpha, \beta, \gamma)$ is the foot of perpendicular from P on l_2, then $9(\alpha + \beta + \gamma)$ is equal to ....

24 The number of integral terms in the expansion of $(3^{1/2} + 5^{1/4})^{680}$ is equal to ....

25 In an examination, 5 students have been allotted their seats as per their roll numbers. The number of ways, in which none of the students sits on the allotted seat, is

26 Let $A = \begin{bmatrix} 0 & 1 & 2 \\ a & 0 & 3 \\ 1 & c & 0 \end{bmatrix}$, where $a, c \in R$. If $A^3 = A$ and the positive value of a belongs to the interval $(n-1, n]$, where $n \in N$, then n is equal to ....

27 If a and b are the roots of the equation $x^2 - 7x - 1 = 0$, then the value of $\frac{a^{21} + b^{21} + a^{17} + b^{17}}{a^{19} + b^{19}}$ is equal to ....

28 The mean of the coefficient of $x, x^2, \ldots, x^7$ in the binomial expansion of $(2 + x)^9$ is ....

29 The number of ordered triplets of the truth values of p, q and r such that the truth value of the statement $(p \vee q) \wedge (p \vee r) \Rightarrow (q \vee r)$ is True, is equal to ....

30 Let $S = 109 + \frac{108}{5} + \frac{107}{5^2} + \ldots + \frac{2}{5^{107}} + \frac{1}{5^{108}}$. Then, the value of $(16S - (25)^{-54})$ is equal to ....

11 APRIL 2023, SHIFT-II

Section A : Objective Type Questions

1 If the 1011th term from the end in the binomial expansion of $\left(\frac{4x}{5} - \frac{5}{2x}\right)^{2022}$ is 1024 times 1011th term from the beginning, then $|x|$ is equal to

(a) 10 (b) 12 (c) 8 (d) 15

2 If the radius of the largest circle with centre (2, 0) inscribed in the ellipse $x^2 + 4y^2 = 36$ is r, then $12r^2$ is equal to

(a) 69 (b) 72 (c) 115 (d) 92

3 The angle of elevation of the top P of a tower from the feet of one person standing due South of the tower is 45° and from the feet of another person standing due West of the tower is 30°. If the height of the tower is 5 m, then the distance (in meters) between the two persons is equal to

(a) 10 (b) $5\sqrt{5}$ (c) 5 (d) $\frac{5}{2}\sqrt{5}$

4 Let $A = \{1, 3, 4, 6, 9\}$ and $B = \{2, 4, 5, 8, 10\}$. Let R be a relation defined on $A \times B$ such that $R = \{((a_1, b_1), (a_2, b_2)) : a_1 \leq b_2 \text{ and } b_1 \leq a_2\}$. Then, the number of elements in the set R is

(a) 180 (b) 52 (c) 160 (d) 26

5 Let the mean of 6 observations 1, 2, 4, 5, x and y be 5 and their variance be 10. Then, their mean deviation about the mean is equal to

(a) $\frac{10}{3}$ (b) 3 (c) $\frac{7}{3}$ (d) $\frac{8}{3}$

6 Let f and g be two functions defined by
$$f(x)=\begin{cases} x+1, & x<0 \\ |x-1|, & x\geq 0\end{cases} \text{ and } g(x)=\begin{cases} x+1, & x<0 \\ 1, & x\geq 0\end{cases}$$

Then, $(gof)(x)$ is
(a) continuous everywhere but not differentiable exactly at one point
(b) continuous everywhere but not differentiable at $x=1$
(c) differentiable everywhere
(d) not continuous at $x=-1$

7 If four distinct points with position vectors **a**, **b**, **c** and **d** are coplanar, then [**a b c**] is equal to
(a) [**d c a**] + [**b d a**] + [**c d b**] (b) [**b c d**] + [**d a c**] + [**d b a**]
(c) [**a d b**] + [**d c a**] + [**d b c**]
(d) [**d b a**] + [**a c d**] + [**d b c**]

8 Let P be the plane passing through the points (5, 3, 0), (13, 3, – 2) and (1, 6, 2). For $\alpha \in N$, if the distances of the points $A(3, 4, \alpha)$ and $B(2, \alpha, a)$ from the plane P are 2 and 3 respectively, then the positive value of a is
(a) 3 (b) 6 (c) 5 (d) 4

9 The converse of $((\sim p)\wedge q)\Rightarrow r$ is
(a) $((\sim p)\vee q)\Rightarrow r$ (b) $(\sim r)\Rightarrow((\sim p)\wedge q)$
(c) $(p\vee(\sim q))\Rightarrow(\sim r)$ (d) $(\sim r)\Rightarrow p\wedge q$

10 The sum of the coefficients of three consecutive terms in the binomial expansion of $(1+x)^{n+2}$, which are in the ratio 1 : 3 : 5, is equal to
(a) 25 (b) 63
(c) 92 (d) 41

11 For $a\in C$, let $A=\{z\in C:\text{Re}(a+\bar{z})>\text{Im}(\bar{a}+z)\}$ and $B=\{z\in C:\text{Re}(a+\bar{z})<\text{Im}(\bar{a}+z)\}$. Then, among the two statements :

S1 If Re(a), Im(a) > 0, then the set A contains all the real numbers.

S2 If Re(a), Im(a) < 0, then the set B contains all the real numbers.
(a) Both are true (b) Both are false
(c) Only (S2) is true (d) Only (S1) is true

12 If $f:R\to R$ be a continuous function satisfying
$$\int_0^{\frac{\pi}{2}} f(\sin 2x)\sin x\,dx+\alpha\int_0^{\frac{\pi}{4}} f(\cos 2x)\cos x\,dx=0,$$ then the value of α is
(a) $\sqrt{2}$ (b) $\sqrt{3}$
(c) $-\sqrt{3}$ (d) $-\sqrt{2}$

13 Let the line passing through the points $P(2,-1,2)$ and $Q(5,3,4)$ meet the plane $x-y+z=4$ at the point R. Then, the distance of the point R from the plane $x+2y+3z+2=0$ measured parallel to the line $\frac{x-7}{2}=\frac{y+3}{2}=\frac{z-2}{1}$ is equal to
(a) $\sqrt{189}$ (b) $\sqrt{31}$
(c) 3 (d) $\sqrt{61}$

14 If the letters of the word MATHS are permuted and all possible words formed are arranged as in a dictionary with serial numbers, then the serial number of the word THAMS is
(a) 104 (b) 102 (c) 103 (d) 101

15 Let the function $f:[0,2]\to R$ be defined as
$$f(x)=\begin{cases} e^{\min\{x^2,\,x-[x]\}}, & x\in[0,1) \\ e^{[x-\log_e x]}, & x\in[1,2],\end{cases}$$
where $[t]$ denotes the greatest integer less than or equal to t. Then, the value of the integral $\int_0^2 xf(x)\,dx$ is
(a) $2e-\frac{1}{2}$ (b) $2e-1$
(c) $1+\frac{3e}{2}$ (d) $(e-1)\left(e^2+\frac{1}{2}\right)$

16 If $\begin{vmatrix} x+1 & x & x \\ x & x+\lambda & x \\ x & x & x+\lambda^2\end{vmatrix}=\frac{9}{8}(103x+81)$, then $\lambda, \frac{\lambda}{3}$ are the roots of the equation
(a) $4x^2+24x+27=0$ (b) $4x^2-24x+27=0$
(c) $4x^2+24x-27=0$ (d) $4x^2-24x-27=0$

17 Let a, b, c and d be positive real numbers such that $a+b+c+d=11$. If the maximum value of $a^5b^3c^2d$ is 3750β, then the value of β is
(a) 55 (b) 90 (c) 108 (d) 110

18 If the system of linear equations
$$7x+11y+\alpha z=13$$
$$5x+4y+7z=\beta$$
$$175x+194y+57z=361$$
has infinitely many solutions, then $\alpha+\beta+2$ is equal to
(a) 4 (b) 3 (c) 6 (d) 5

19 Let $y=y(x)$ be the solution of the differential equation $\frac{dy}{dx}+\frac{5}{x(x^5+1)}y=\frac{(x^5+1)^2}{x^7}$, $x>0$. If $y(1)=2$, then $y(2)$ is equal to
(a) $\frac{693}{128}$ (b) $\frac{697}{128}$ (c) $\frac{637}{128}$ (d) $\frac{679}{128}$

20 The domain of the function $f(x)=\frac{1}{\sqrt{[x]^2-3[x]-10}}$ is (where $[x]$ denotes the greatest integer less than or equal to x)
(a) $(-\infty,-2)\cup(5,\infty)$ (b) $(-\infty,-3)\cup[6,\infty)$
(c) $(-\infty,-3)\cup(5,\infty)$ (d) $(-\infty,-2)\cup[6,\infty)$

Section B : Numerical Value Type Questions

21 The number of points, where the curve $f(x)=e^{8x}-e^{6x}-3e^{4x}-e^{2x}+1$, $x\in R$ cuts X-axis, is equal to ....

22 For $k \in N$, if the sum of the series $1+\frac{4}{k}+\frac{8}{k^2}+\frac{13}{k^3}+\frac{19}{k^4}+\ldots$ is 10, then the value of k is ....

23 Let the probability of getting head for a biased coin be $\frac{1}{4}$. It is tossed repeatedly until a head appears. Let N be the number of tosses required. If the probability that the equation $64x^2+5Nx+1=0$ has no real root is $\frac{p}{q}$, where p and q are co-prime, then $q-p$ is equal to ....

24 Let $A=\{1,2,3,4,5\}$ and $B=\{1,2,3,4,5,6\}$. Then, the number of functions $f: A \rightarrow B$ satisfying $f(1)+f(2)=f(4)-1$ is equal to ....

25 If the line $l_1: 3y-2x=3$ is the angular bisector of the lines $l_2: x-y+1=0$ and $l_3: \alpha x+\beta y+17=0$, then $\alpha^2+\beta^2-\alpha-\beta$ is equal to ....

26 Let the line $l: x=\frac{1-y}{-2}=\frac{z-3}{\lambda}, \lambda \in R$ meet the plane $P: x+2y+3z=4$ at the point (α, β, γ). If the angle between the line l and the plane P is $\cos^{-1}\left(\sqrt{\frac{5}{14}}\right)$, then $\alpha+2\beta+6\gamma$ is equal to ....

27 Let the tangent to the parabola $y^2=12x$ at the point $(3, \alpha)$ be perpendicular to the line $2x+2y=3$. Then, the square of distance of the point $(6,-4)$ from the normal to the hyperbola $\alpha^2x^2-9y^2=9\alpha^2$ at its point $(\alpha-1, \alpha+2)$ is equal to ....

28 If A is the area in the first quadrant enclosed by the curve $C: 2x^2-y+1=0$, the tangent to C at the point $(1,3)$ and the line $x+y=1$, then the value of $60A$ is ....

29 Let $\mathbf{a}=\hat{\mathbf{i}}+2\hat{\mathbf{j}}+3\hat{\mathbf{k}}$ and $\mathbf{b}=\hat{\mathbf{i}}+\hat{\mathbf{j}}-\hat{\mathbf{k}}$. If $\mathbf{c}$ is a vector such that $\mathbf{a}\cdot\mathbf{c}=11$, $\mathbf{b}\cdot(\mathbf{a}\times\mathbf{c})=27$ and $\mathbf{b}\cdot\mathbf{c}=-\sqrt{3}|\mathbf{b}|$, then $|\mathbf{a}\times\mathbf{c}|^2$ is equal to ....

30 Let $S=\left\{z \in C-\{i, 2i\}: \frac{z^2+8iz-15}{z^2-3iz-2} \in R\right\}$. If $\alpha-\frac{13}{11}i \in S$, $\alpha \in R-\{0\}$, then $242\alpha^2$ is equal to ....

12 APRIL 2023, SHIFT-I

Section A : Objective Type Questions

1 Let a, b, c be three distinct real numbers, none equal to one. If the vectors $a\hat{\mathbf{i}}+\hat{\mathbf{j}}+\hat{\mathbf{k}}$, $\hat{\mathbf{i}}+b\hat{\mathbf{j}}+\hat{\mathbf{k}}$ and $\hat{\mathbf{i}}+\hat{\mathbf{j}}+c\hat{\mathbf{k}}$ are coplanar, then $\frac{1}{1-a}+\frac{1}{1-b}+\frac{1}{1-c}$ is equal to

(a) -2 (b) 1
(c) 2 (d) -1

2 Let the lines $l_1: \frac{x+5}{3}=\frac{y+4}{1}=\frac{z-\alpha}{-2}$ and $l_2: 3x+2y+z-2=0=x-3y+2z-13$ be coplanar. If the point $P(a, b, c)$ on l_1 is nearest to the point $Q(-4,-3,2)$, then $|a|+|b|+|c|$ is equal to

(a) 14 (b) 12
(c) 10 (d) 8

3 Let $y=y(x)$, $y>0$, be a solution curve of the differential equation $(1+x^2)dy=y(x-y)\,dx$. If $y(0)=1$ and $y(2\sqrt{2})=\beta$, then

(a) $e^{\beta^{-1}}=e^{-2}(5+\sqrt{2})$ (b) $e^{\beta^{-1}}=e^{-2}(3+2\sqrt{2})$
(c) $e^{3\beta^{-1}}=e(3+2\sqrt{2})$ (d) $e^{3\beta^{-1}}=e(5+\sqrt{2})$

4 Among the two statements

S1 : $(p \Rightarrow q) \wedge (p \wedge (\sim q))$ is a contradiction and

S2 : $(p \wedge q) \vee ((\sim p) \wedge q) \vee (p \wedge (\sim q)) \vee ((\sim p) \wedge (\sim q))$ is a tautology

(a) Only S2 is true (b) Only S1 is true
(c) Both are false (d) Both are true

5 The area of the region enclosed by the curve $y=x^3$ and its tangent at the point $(-1,-1)$ is

(a) $\frac{23}{4}$ (b) $\frac{19}{4}$ (c) $\frac{31}{4}$ (d) $\frac{27}{4}$

6 The number of five digit numbers, greater than 40000 and divisible by 5, which can be formed using the digits 0, 1, 3, 5, 7 and 9 without repetition, is equal to

(a) 72 (b) 132 (c) 96 (d) 120

7 Let $P\left(\frac{2\sqrt{3}}{\sqrt{7}}, \frac{6}{\sqrt{7}}\right)$, Q, R and S be four points on the ellipse $9x^2+4y^2=36$. Let PQ and RS be mutually perpendicular and pass through the origin. If $\frac{1}{(PQ)^2}+\frac{1}{(RS)^2}=\frac{p}{q}$, where p and q are coprime, then $p+q$ is equal to

(a) 137 (b) 157 (c) 147 (d) 143

8 Let $<a_n>$ be a sequence such that $a_1+a_2+\ldots+a_n=\frac{n^2+3n}{(n+1)(n+2)}$. If $28\sum_{k=1}^{10}\frac{1}{a_k}=p_1 p_2 p_3 \ldots p_m$, where $p_1, p_2, \ldots, p_m$ are the first m prime numbers, then m is equal to

(a) 5 (b) 8 (c) 6 (d) 7

9 If the point $\left(\alpha, \frac{7\sqrt{3}}{3}\right)$ lies on the curve traced by the mid-points of the line segments of the lines $x\cos\theta + y\sin\theta = 7$, $\theta \in \left(0, \frac{\pi}{2}\right)$ between the co-ordinate axes, then α is equal to

(a) $7\sqrt{3}$ (b) 7 (c) $-7\sqrt{3}$ (d) -7

10 In a ΔABC, if $\cos A + 2\cos B + \cos C = 2$ and the length of the sides opposite to the angles A and C are 3 and 7 respectively, then $\cos A - \cos C$ is equal to

(a) $\frac{5}{7}$ (b) $\frac{3}{7}$ (c) $\frac{10}{7}$ (d) $\frac{9}{7}$

11 Two dice A and B are rolled. Let the numbers obtained on A and B be α and β, respectively. If the variance of $\alpha - \beta$ is $\frac{p}{q}$, where p and q are co-prime, then the sum of the positive divisors of p is equal to

(a) 36 (b) 72 (c) 48 (d) 31

12 Let α, β be the roots of the quadratic equation $x^2 + \sqrt{6}x + 3 = 0$. Then, $\frac{\alpha^{23} + \beta^{23} + \alpha^{14} + \beta^{14}}{\alpha^{15} + \beta^{15} + \alpha^{10} + \beta^{10}}$ is equal to

(a) 81 (b) 72 (c) 9 (d) 729

13 Let D be the domain of the function

$$f(x) = \sin^{-1}\left(\log_{3x}\left(\frac{6 + 2\log_3 x}{-5x}\right)\right).$$

If the range of the function $g: D \to R$ defined by $g(x) = x - [x]$, ($[x]$ is the greatest integer function), is (α, β), then $\alpha^2 + \frac{5}{\beta}$ is equal to

(a) 135 (b) 46
(c) 45 (d) 136

14 Let $\lambda = Z$, $\mathbf{a} = \lambda\hat{\mathbf{i}} + \hat{\mathbf{j}} - \hat{\mathbf{k}}$ and $\mathbf{b} = 3\hat{\mathbf{i}} - \hat{\mathbf{j}} + 2\hat{\mathbf{k}}$. Let $\mathbf{c}$ be a vector such that $(\mathbf{a} + \mathbf{b} + \mathbf{c}) \times \mathbf{c} = \mathbf{0}$, $\mathbf{a} \cdot \mathbf{c} = -17$ and $\mathbf{b} \cdot \mathbf{c} = -20$. Then, $|\mathbf{c} \times (\lambda\hat{\mathbf{i}} + \hat{\mathbf{j}} + \hat{\mathbf{k}})|^2$ is equal to

(a) 53 (b) 46 (c) 62 (d) 49

15 If the local maximum value of the function $f(x) = \left(\frac{\sqrt{3}e}{2\sin x}\right)^{\sin^2 x}$, $x \in \left(0, \frac{\pi}{2}\right)$, is $\frac{k}{e}$, then $\left(\frac{k}{e}\right)^8 + \frac{k^8}{e^5} + k^8$ is equal to

(a) $e^3 + e^6 + e^{10}$ (b) $e^3 + e^6 + e^{11}$
(c) $e^5 + e^6 + e^{11}$ (d) $e^3 + e^5 + e^{11}$

16 If $\frac{1}{n+1}\,{}^nC_n + \frac{1}{n}\,{}^nC_{n-1} + \ldots + \frac{1}{2}\,{}^nC_1 + {}^nC_0 = \frac{1023}{10}$, then n is equal to

(a) 9 (b) 8 (c) 6 (d) 7

17 Let C be the circle in the complex plane with centre $z_0 = \frac{1}{2}(1 + 3i)$ and radius $r = 1$. Let $z_1 = 1 + i$ and the complex number z_2 be outside the circle C such that $|z_1 - z_0||z_2 - z_0| = 1$. If z_0, z_1 and z_2 are collinear, then the smaller value of $|z_2|^2$ is equal to

(a) $\frac{13}{2}$ (b) $\frac{7}{2}$ (c) $\frac{5}{2}$ (d) $\frac{3}{2}$

18 The sum of the coefficient of the first 50 terms in the binomial expansion of $(1 - x)^{100}$, is equal to

(a) $-{}^{101}C_{50}$ (b) ${}^{101}C_{50}$
(c) $-{}^{99}C_{49}$ (d) ${}^{99}C_{49}$

19 Let the plane $P: 4x - y + z = 10$ be rotated by an angle $\frac{\pi}{2}$ about its line of intersection with the plane $x + y - z = 4$. If α is the distance of the point $(2, 3, -4)$ from the new position of the plane P, then 35α is equal to

(a) 126 (b) 90 (c) 85 (d) 105

20 Let $A = \begin{bmatrix} 1 & \frac{1}{51} \\ 0 & 1 \end{bmatrix}$. If $B = \begin{bmatrix} 1 & 2 \\ -1 & -1 \end{bmatrix} A \begin{bmatrix} -1 & -2 \\ 1 & 1 \end{bmatrix}$, then the sum of all the elements of the matrix $\sum_{n=1}^{50} B^n$ is equal to

(a) 100 (b) 75 (c) 125 (d) 50

Section B : Numerical Value Type Questions

21 If $\int_{-0.15}^{0.15} |100x^2 - 1|\, dx = \frac{k}{3000}$, then k is equal to ....

22 Let the plane $x + 3y - 2z + 6 = 0$ meet the co-ordinate axes at the points A, B, C. If the orthocenter of the ΔABC is $\left(\alpha, \beta, \frac{6}{7}\right)$, then $98(\alpha + \beta)^2$ is equal to ....

23 Two circles in the first quadrant of radii r_1 and r_2 touch the co-ordinate axes. Each of them cuts off an intercept of 2 units with the line $x + y = 2$. Then, $r_1^2 + r_2^2 - r_1 r_2$ is equal to ....

24 Let $I(x) = \int \sqrt{\frac{x+7}{x}}\, dx$ and $I(9) = 12 + 7\log_e 7$. If $I(1) = \alpha + 7\log_e(1 + 2\sqrt{2})$, then α^4 is equal to ....

25 The number of relations, on the set {1, 2, 3} containing (1, 2) and (2, 3), which are reflexive and transitive but not symmetric, is ....

26 Let the digits a, b, c be in AP. Nine-digit numbers are to be formed using each of these three digits thrice such that three consecutive digits are in AP at least once. How many such numbers can be formed?

27 Let the positive numbers a_1, a_2, a_3, a_4 and a_5 be in a GP. Let their mean and variance be $\frac{31}{10}$ and $\frac{m}{n}$ respectively, where m and n are co-prime. If the mean of their reciprocals is $\frac{31}{40}$ and $a_3 + a_4 + a_5 = 14$, then $m + n$ is equal to ....

28 A fair $n(n > 1)$ faces die is rolled repeatedly until a number less than n appears. If the mean of the number of tosses required is $\frac{n}{9}$, then n is equal to ....

29 Let $D_k = \begin{vmatrix} 1 & 2k & 2k-1 \\ n & n^2+n+2 & n^2 \\ n & n^2+n & n^2+n+2 \end{vmatrix}$. If $\sum_{k=1}^{n} D_k = 96$, then n is equal to ....

30 Let $[x]$ be the greatest integer $\le x$. Then, the number of points in the interval $(-2, 1)$, where the function $f(x) = |[x]| + \sqrt{x - [x]}$ is discontinuous is ....

13 APRIL 2023, SHIFT-I

Section A : Objective Type Questions

1 The set of all $a \in R$ for which the equation $x|x-1| + |x+2| + a = 0$ has exactly one real root, is

(a) $(-\infty, \infty)$ (b) $(-6, -3)$
(c) $(-6, \infty)$ (d) $(-\infty, -3)$

2 Let the tangent and normal at the point $(3\sqrt{3}, 1)$ on the ellipse $\frac{x^2}{36} + \frac{y^2}{4} = 1$ meet the Y-axis at the points A and B respectively. Let the circle C be drawn taking AB as a diameter and the line $x = 2\sqrt{5}$ intersect C at the points P and Q. If the tangents at the points P and Q on the circle intersect at the point (α, β), then $\alpha^2 - \beta^2$ is equal to

(a) 60 (b) $\frac{314}{5}$ (c) 61 (d) $\frac{304}{5}$

3 Let the equation of plane passing through the line of intersection of the planes $x + 2y + az = 2$ and $x - y + z = 3$ be $5x - 11y + bz = 6a - 1$. For $c \in Z$, if the distance of this plane from the point $(a, -c, c)$ is $\frac{2}{\sqrt{a}}$, then $\frac{a+b}{c}$ is equal to

(a) 2 (b) -4 (c) -2 (d) 4

4 For the system of linear equations

$$2x + 4y + 2az = b$$
$$x + 2y + 3z = 4$$
$$2x - 5y + 2z = 8$$

which of the following is not correct?

(a) It has infinitely many solutions if $a = 3$, $b = 6$
(b) It has unique solution if $a = b = 8$
(c) It has infinitely many solutions if $a = 3$, $b = 8$
(d) It has unique solution if $a = b = 6$

5 Let $y = y_1(x)$ and $y = y_2(x)$ be the solution curves of the differential equation $\frac{dy}{dx} = y + 7$ with initial conditions $y_1(0) = 0$ and $y_2(0) = 1$ respectively. Then, the curves $y = y_1(x)$ and $y = y_2(x)$ intersect at

(a) infinite number of points (b) two points
(c) one point (d) no point

6 $\max_{0 \le x \le \pi} \left\{x - 2\sin x \cos x + \frac{1}{3}\sin 3x\right\}$ is equal to

(a) $\frac{5\pi + 2 + 3\sqrt{3}}{6}$ (b) $\frac{\pi + 2 - 3\sqrt{3}}{6}$
(c) π (d) 0

7 Fractional part of the number $\frac{4^{2022}}{15}$ is equal to

(a) $\frac{8}{15}$ (b) $\frac{1}{15}$ (c) $\frac{14}{15}$ (d) $\frac{4}{15}$

8 A coin is biased so that the head is 3 times as likely to occur as tail. This coin is tossed until a head or three tails occur. If x denotes the number of tosses of the coin, then the mean of x is

(a) $\frac{21}{16}$ (b) $\frac{81}{64}$ (c) $\frac{37}{16}$ (d) $\frac{15}{16}$

9 For $x \in R$, two real valued functions $f(x)$ and $g(x)$ are such that, $g(x) = \sqrt{x} + 1$ and $fog(x) = x + 3 - \sqrt{x}$. Then, $f(0)$ is equal to

(a) 0 (b) 5 (c) 3 (d) 1

10 The negation of the statement $((A \wedge (B \vee C)) \Rightarrow (A \vee B)) \Rightarrow A$ is

(a) equivalent to $\sim A$ (b) a fallacy
(c) equivalent to $B \vee \sim C$ (d) equivalent to $\sim C$

11 $\int_0^\infty \frac{6}{e^{3x} + 6e^{2x} + 11e^x + 6} dx =$

(a) $\log_e\left(\frac{256}{81}\right)$ (b) $\log_e\left(\frac{64}{27}\right)$
(c) $\log_e\left(\frac{32}{27}\right)$ (d) $\log_e\left(\frac{512}{81}\right)$

12 The number of symmetric matrices of order 3, with all the entries from the set {0, 1, 2, 3, 4, 5, 6,7, 8, 9}, is

(a) 9^{10} (b) 10^6 (c) 10^9 (d) 6^{10}

13 Among

S1 $\lim_{n \to \infty} \frac{1}{n^2}(2 + 4 + 6 + \ldots + 2n) = 1$

S2 $\lim_{n \to \infty} \frac{1}{n^{16}}(1^{15} + 2^{15} + 3^{15} + \ldots + n^{15}) = \frac{1}{16}$

(a) Only (S1) is true
(b) Only (S2) is true
(c) Both (S1) and (S2) are true
(d) Both (S1) and (S2) are false

14 Let PQ be a focal chord of the parabola $y^2 = 36x$ of length 100, making an acute angle with the positive X-axis. Let the ordinate of P be positive and M be the point on the line segment PQ such that $PM : MQ = 3:1$. Then, which of the following points does not lie on the line passing through M and perpendicular to the line PQ?

(a) (3, 33) (b) (6, 29) (c) (– 3, 43) (d) (– 6, 45)

15 Let $B = \begin{bmatrix} 1 & 3 & \alpha \\ 1 & 2 & 3 \\ \alpha & \alpha & 4 \end{bmatrix}$, $\alpha > 2$ be the adjoint of a matrix A and $|A| = 2$. Then,

$[\alpha \; -2\alpha \; \alpha]\, B \begin{bmatrix} \alpha \\ -2\alpha \\ \alpha \end{bmatrix}$ is equal to

(a) 32 (b) 16 (c) – 16 (d) 0

16 The area of the region enclosed by the curve $f(x) = \max\{\sin x, \cos x\}$, $-\pi \le x \le \pi$ and the X-axis is

(a) $2\sqrt{2}(\sqrt{2}+1)$ (b) $4(\sqrt{2})$
(c) 4 (d) $2(\sqrt{2}+1)$

17 For the differentiable function

$f : R - \{0\} \to R$, let $3f(x) + 2f\left(\frac{1}{x}\right) = \frac{1}{x} - 10$, then

$\left| f(3) + f'\left(\frac{1}{4}\right) \right|$ is equal to

(a) $\frac{33}{5}$ (b) $\frac{29}{5}$ (c) 7 (d) 13

18 Let $s_1, s_2, s_3, \ldots, s_{10}$ respectively be the sum of 12 terms of 10 AP's whose first terms are 1, 2, 3, ... , 10 and the common differences are 1, 3, 5, ..., 19, respectively. Then, $\sum_{i=1}^{10} s_i$ is equal to

(a) 7380 (b) 7360 (c) 7220 (d) 7260

19 Let $\mathbf{a} = \hat{\mathbf{i}} + 4\hat{\mathbf{j}} + 2\hat{\mathbf{k}}$, $\mathbf{b} = 3\hat{\mathbf{i}} - 2\hat{\mathbf{j}} + 7\hat{\mathbf{k}}$ and $\mathbf{c} = 2\hat{\mathbf{i}} - \hat{\mathbf{j}} + 4\hat{\mathbf{k}}$. If a vector $\mathbf{d}$ satisfies $\mathbf{d} \times \mathbf{b} = \mathbf{c} \times \mathbf{b}$ and $\mathbf{d} \cdot \mathbf{a} = 24$, then $|\mathbf{d}|^2$ is equal to

(a) 323 (b) 313 (c) 423 (d) 413

20 The distance of the point $(-1, 2, 3)$ from the plane $\mathbf{r} \cdot (\hat{\mathbf{i}} - 2\hat{\mathbf{j}} + 3\hat{\mathbf{k}}) = 10$ parallel to the line of the shortest distance between the lines

$\mathbf{r} = (\hat{\mathbf{i}} - \hat{\mathbf{j}}) + \lambda(2\hat{\mathbf{i}} + \hat{\mathbf{k}})$ and $\mathbf{r} = (2\hat{\mathbf{i}} - \hat{\mathbf{j}}) + \mu(\hat{\mathbf{i}} - \hat{\mathbf{j}} + \hat{\mathbf{k}})$ is

(a) $3\sqrt{6}$ (b) $2\sqrt{6}$ (c) $2\sqrt{5}$ (d) $3\sqrt{5}$

Section B : Numerical Value Type Questions

21 The sum to 20 terms of the series

$2 \cdot 2^2 - 3^2 + 2 \cdot 4^2 - 5^2 + 2 \cdot 6^2 - \ldots$ is equal to ....

22 Let m_1 and m_2 be the slopes of the tangents drawn from the point $P(4, 1)$ to the hyperbola $H = \frac{y^2}{25} - \frac{x^2}{16} = 1$. If Q is the point from which the tangents drawn to H have slopes $|m_1|$ and $|m_2|$ and they make positive intercepts α and β on the X-axis, then $\frac{(PQ)^2}{\alpha\beta}$ is equal to ....

23 Let $\mathbf{a} = 3\hat{\mathbf{i}} + \hat{\mathbf{j}} - \hat{\mathbf{k}}$ and $\mathbf{c} = 2\hat{\mathbf{i}} - 3\hat{\mathbf{j}} + 3\hat{\mathbf{k}}$. If $\mathbf{b}$ is a vector such that $\mathbf{a} = \mathbf{b} \times \mathbf{c}$ and $|\mathbf{b}|^2 = 50$, then $|72 - |\mathbf{b} + \mathbf{c}|^2|$ is equal to ....

24 The number of seven digit positive integers formed using the digits 1, 2, 3 and 4 only and sum of the digits equal to 12 is ....

25 Let α be the constant term in the binomial expansion of $\left(\sqrt{x} - \frac{6}{x^{\frac{3}{2}}}\right)^n$, $n \le 15$. If the sum of the coefficients of the remaining terms in the expansion is 649 and the coefficient of x^{-n} is $\lambda\alpha$, then λ is equal to ....

26 Let $w = z\bar{z} + k_1 z + k_2 iz + \lambda(1+i)$, $k_1, k_2 \in R$. Let $\text{Re}(w) = 0$ be the circle C of radius 1 in the first quadrant touching the line $y = 1$ and the Y-axis. If the curve $\text{Im}(w) = 0$ intersects C at A and B, then $30(AB)^2$ is equal to ....

27 Let the image of the point $\left(\frac{5}{3}, \frac{5}{3}, \frac{8}{3}\right)$ in the plane $x - 2y + z - 2 = 0$ be P. If the distance of the point $Q(6, -2, \alpha)$, $\alpha > 0$, from P is 13, then α is equal to ....

28 Let for $x \in R$, $S_0(x) = x$, $S_k(x) = C_k x + k\int_0^x S_{k-1}(t)\,dt$, where $C_0 = 1$, $C_k = 1 - \int_0^1 S_{k-1}(x)\,dx$, $k = 1, 2, 3, \ldots$ Then, $S_2(3) + 6C_3$ is equal to ....

29 Let the mean of the data

x	1	3	5	7	9
Frequency (f)	4	24	28	a	8

be 5. If m and σ^2 are respectively the mean deviation about the mean and the variance of the data, then $\frac{3a}{m + \sigma^2}$ is equal to ....

30 If $S = \left\{ x \in R : \sin^{-1}\left(\frac{x+1}{\sqrt{x^2+2x+2}}\right) - \sin^{-1}\left(\frac{x}{\sqrt{x^2+1}}\right) = \frac{\pi}{4} \right\}$, then

$\sum_{x \in S} \left(\sin\left((x^2 + x + 5)\frac{\pi}{2}\right) - \cos((x^2 + x + 5)\pi) \right)$ is equal to ....

13 APRIL 2023, SHIFT-II

Section A : Objective Type Questions

1 The coefficient of x^5 in the expansion of $\left(2x^3 - \frac{1}{3x^2}\right)^5$ is

(a) $\frac{26}{3}$ (b) 8 (c) $\frac{80}{9}$ (d) 9

2 Let $|\mathbf{a}| = 2$, $|\mathbf{b}| = 3$ and the angle between the vectors $\mathbf{a}$ and $\mathbf{b}$ be $\frac{\pi}{4}$. Then, $|(\mathbf{a} + 2\mathbf{b}) \times (2\mathbf{a} - 3\mathbf{b})|^2$ is equal to

(a) 841 (b) 882 (c) 482 (d) 441

3 The area of the region $\{(x, y) : x^2 \le y \le |x^2 - 4|, y \ge 1\}$ is

(a) $\frac{3}{4}(4\sqrt{2} - 1)$ (b) $\frac{3}{4}(4\sqrt{2} + 1)$

(c) $\frac{4}{3}(4\sqrt{2} - 1)$ (d) $\frac{4}{3}(4\sqrt{2} + 1)$

4 Let for a ΔABC, $\mathbf{AB} = -2\hat{\mathbf{i}} + \hat{\mathbf{j}} + 3\hat{\mathbf{k}}$, $\mathbf{CB} = \alpha\hat{\mathbf{i}} + \beta\hat{\mathbf{j}} + \gamma\hat{\mathbf{k}}$ and $\mathbf{CA} = 4\hat{\mathbf{i}} + 3\hat{\mathbf{j}} + \delta\hat{\mathbf{k}}$

If $\delta > 0$ and the area of the ΔABC is $5\sqrt{6}$, then $\mathbf{CB} \cdot \mathbf{CA}$ is equal to

(a) 60 (b) 54 (c) 108 (d) 120

5 The range of $f(x) = 4\sin^{-1}\left(\frac{x^2}{x^2 + 1}\right)$ is

(a) $[0, \pi)$ (b) $[0, \pi]$ (c) $[0, 2\pi)$ (d) $[0, 2\pi]$

6 Let $S = \{z \in C : \bar{z} = i(z^2 + \text{Re}(\bar{z}))\}$. Then, $\sum_{z \in S} |z|^2$ is equal to

(a) 3 (b) $\frac{7}{2}$ (c) $\frac{5}{2}$ (d) 4

7 The value of $\dfrac{e^{-\pi/4} + \int_0^{\pi/4} e^{-x} \tan^{50} x\,dx}{\int_0^{\pi/4} e^{-x} (\tan^{49} x + \tan^{51} x)\,dx}$ is

(a) 50 (b) 25 (c) 51 (d) 49

8 Let for $A = \begin{bmatrix} 1 & 2 & 3 \\ \alpha & 3 & 1 \\ 1 & 1 & 2 \end{bmatrix}$, $|A| = 2$. If $|2\,\text{adj}\,(2\,\text{adj}\,(2A))| = 32^n$, then $3n + \alpha$ is equal to

(a) 10 (b) 11 (c) 12 (d) 9

9 Let the centre of a circle C be (α, β) and its radius $r < 8$. Let $3x + 4y = 24$ and $3x - 4y = 32$ be two tangents and $4x + 3y = 1$ be a normal to C. Then, $(\alpha - \beta + r)$ is equal to

(a) 6 (b) 9 (c) 7 (d) 5

10 If the system of equations

$$2x + y - z = 5$$
$$2x - 5y + \lambda z = \mu$$
$$x + 2y - 5z = 7$$

has infinitely many solutions, then $(\lambda + \mu)^2 + (\lambda - \mu)^2$ is equal to

(a) 920 (b) 904 (c) 916 (d) 912

11 Let (α, β) be the centroid of the triangle formed by the lines $15x - y = 82$, $6x - 5y = -4$ and $9x + 4y = 17$. Then, $\alpha + 2\beta$ and $2\alpha - \beta$ are the roots of the equation

(a) $x^2 - 10x + 25 = 0$ (b) $x^2 - 7x + 12 = 0$

(c) $x^2 - 14x + 48 = 0$ (d) $x^2 - 13x + 42 = 0$

12 If $\lim_{x \to 0} \dfrac{e^{ax} - \cos(bx) - \dfrac{cxe^{-cx}}{2}}{1 - \cos(2x)} = 17$, then $5a^2 + b^2$ is equal to

(a) 76 (b) 72 (c) 64 (d) 68

13 Let $a_1, a_2, a_3, \ldots$ be a GP of increasing positive numbers. Let the sum of its 6th and 8th terms be 2 and the product of its 3rd and 5th terms be $\frac{1}{9}$. Then, $6(a_2 + a_4)(a_4 + a_6)$ is equal to

(a) $2\sqrt{2}$ (b) 3 (c) $3\sqrt{3}$ (d) 2

14 The statement $(p \wedge (\sim q)) \vee ((\sim p) \wedge q) \vee ((\sim p) \wedge (\sim q))$ is equivalent to

(a) $(\sim p) \vee (\sim q)$ (b) $p \vee q$

(c) $(\sim p) \vee q$ (d) $p \vee (\sim q)$

15 All words, with or without meaning, are made using all the letters of the word MONDAY. These words are written as in a dictionary with serial numbers. The serial number of the word MONDAY is

(a) 327 (b) 328 (c) 324 (d) 326

16 The line that is coplanar to the line $\frac{x + 3}{-3} = \frac{y - 1}{1} = \frac{z - 5}{5}$, is

(a) $\frac{x + 1}{-1} = \frac{y - 2}{2} = \frac{z - 5}{4}$ (b) $\frac{x + 1}{-1} = \frac{y - 2}{2} = \frac{z - 5}{5}$

(c) $\frac{x - 1}{-1} = \frac{y - 2}{2} = \frac{z - 5}{5}$ (d) $\frac{x + 1}{1} = \frac{y - 2}{2} = \frac{z - 5}{5}$

17 Let N be the foot of perpendicular from the point $P(1, -2, 3)$ on the line passing through the points $(4, 5, 8)$ and $(1, -7, 5)$. Then, the distance of N from the plane $2x - 2y + z + 5 = 0$ is

(a) 9 (b) 7 (c) 6 (d) 8

18 Let α, β be the roots of the equation $x^2 - \sqrt{2}x + 2 = 0$. Then, $\alpha^{14} + \beta^{14}$ is equal to

(a) -128 (b) $-128\sqrt{2}$ (c) $-64\sqrt{2}$ (d) -64

19 The random variable X follows binomial distribution $B(n, p)$, for which the difference of the mean and the variance is 1. If $2P(X = 2) = 3P(X = 1)$, then $n^2P(X > 1)$ is equal to

(a) 15 (b) 12 (c) 16 (d) 11

20 The plane, passing through the points $(0, -1, 2)$ and $(-1, 2, 1)$ and parallel to the line passing through $(5, 1, -7)$ and $(1, -1, -1)$, also passes through the point

(a) (2, 0, 1) (b) (– 2, 5, 0) (c) (1, – 2, 1) (d) (0, 5, – 2)

Section B : Numerical Value Type Questions

21 Let $f_n = \int_0^{\frac{\pi}{2}} \left(\sum_{k=1}^{n} \sin^{k-1} x\right)\left(\sum_{k=1}^{n} (2k-1)\sin^{k-1} x\right) \cos x \, dx$, $n \in N$. Then, $f_{21} - f_{20}$ is equal to ....

22 Let $[\alpha]$ denote the greatest integer $\leq \alpha$. Then, $[\sqrt{1}] + [\sqrt{2}] + [\sqrt{3}] + \ldots + [\sqrt{120}]$ is equal to ....

23 Let $f(x) = \sum_{k=1}^{10} kx^k, x \in R$. If $2f(2) + f'(2) = 119(2)^n + 1$, then n is equal to ....

24 If $y = y(x)$ is the solution of the differential equation $\frac{dy}{dx} + \frac{4x}{(x^2-1)} y = \frac{x+2}{(x^2-1)^{5/2}}, x > 1$ such that $y(2) = \frac{2}{9} \log_e (2 + \sqrt{3})$ and $y(\sqrt{2}) = \alpha \log_e (\sqrt{\alpha} + \beta) + \beta - \sqrt{\gamma}, \alpha, \beta, \gamma \in N$, then $\alpha\beta\gamma$ is equal to ....

25 The mean and standard deviation of the marks of 10 students were found to be 50 and 12, respectively. Later, it was observed that two marks 20 and 25 were wrongly read as 45 and 50, respectively. Then, the correct variance is ....

26 Total number of 3-digit numbers that are divisible by 6 and can be formed by using the digits 1, 2, 3, 4, 5 with repetition, is ....

27 The foci of a hyperbola are $(\pm 2, 0)$ and its eccentricity is $\frac{3}{2}$. A tangent, perpendicular to the line $2x + 3y = 6$, is drawn at a point in the first quadrant on the hyperbola. If the intercepts made by the tangent on the X-and Y-axes are a and b respectively, then $|6a| + |5b|$ is equal to ....

28 For $x \in (-1, 1]$, the number of solutions of the equation $\sin^{-1} x = 2 \tan^{-1} x$ is equal to ....

29 Let $A = \{-4, -3, -2, 0, 1, 3, 4\}$ and $R = \{(a, b) \in A \times A : b = |a| \text{ or } b^2 = a + 1\}$ be a relation on A. Then, the minimum number of elements, that must be added to the relation R, so that it becomes reflexive and symmetric, is ....

30 The remainder, when 7^{103} is divided by 17, is ....

15 APRIL 2023, SHIFT-I

Section A : Objective Type Questions

1 The mean and standard deviation of 10 observations are 20 and 8, respectively. Later on, it was observed that one observation was recorded as 50 instead of 40. Then, the correct variance is

(a) 11 (b) 14 (c) 13 (d) 12

2 If the set $\left\{\text{Re}\left(\frac{z - \bar{z} + z\bar{z}}{2 - 3z + 5\bar{z}}\right) : z \in C, \text{Re}(z) = 3\right\}$ is equal to the interval $(\alpha, \beta]$, then $24(\beta - \alpha)$ is equal to

(a) 36 (b) 30 (c) 27 (d) 42

3 Let the system of linear equations

$$-x + 2y - 9z = 7$$
$$-x + 3y + 7z = 9$$
$$-2x + y + 5z = 8$$
$$-3x + y + 13z = \lambda$$

has a unique solution $x = \alpha$, $y = \beta$, $z = \gamma$. Then, the distance of the point (α, β, γ) from the plane $2x - 2y + z = \lambda$ is

(a) 11 (b) 9 (c) 13 (d) 7

4 Let $ABCD$ be a quadrilateral. If E and F are the mid-points of the diagonals AC and BD respectively and $(\mathbf{AB} - \mathbf{BC}) + (\mathbf{AD} - \mathbf{DC}) = k\mathbf{FE}$, then k is equal to

(a) 4 (b) – 4 (c) – 2 (d) 2

5 Let the foot of perpendicular of the point $P(3, -2, -9)$ on the plane passing through the points $(-1, -2, -3)$, $(9, 3, 4)$, $(9, -2, 1)$ be $Q(\alpha, \beta, \gamma)$. Then, distance of Q from the

origin is

(a) $\sqrt{29}$ (b) $\sqrt{42}$ (c) $\sqrt{35}$ (d) $\sqrt{38}$

6 Let $(a + bx + cx^2)^{10} = \sum_{i=0}^{20} p_i x^i$, $a, b, c \in N$. If $p_1 = 20$ and $p_2 = 210$, then $2(a + b + c)$ is equal to

(a) 15 (b) 12 (c) 8 (d) 6

7 Let S be the set of all (λ, μ) for which the vectors $\lambda\hat{\mathbf{i}} - \hat{\mathbf{j}} + \hat{\mathbf{k}}$, $\hat{\mathbf{i}} + 2\hat{\mathbf{j}} + \mu\hat{\mathbf{k}}$ and $3\hat{\mathbf{i}} - 4\hat{\mathbf{j}} + 5\hat{\mathbf{k}}$, where $\lambda - \mu = 5$, are coplanar, then $\sum_{(\lambda, \mu) \in S} 80(\lambda^2 + \mu^2)$ is equal to

(a) 2210 (b) 2130 (c) 2290 (d) 2370

8 The number of common tangents, to the circles $x^2 + y^2 - 18x - 15y + 131 = 0$ and $x^2 + y^2 - 6x - 6y - 7 = 0$, is

(a) 2 (b) 3 (c) 1 (d) 4

9 The total number of three-digit numbers, divisible by 3, which can be formed using the digits 1, 3, 5, 8, if repetition of digits is allowed, is

(a) 22 (b) 18 (c) 20 (d) 21

10 A bag contains 6 white and 4 black balls. A die is rolled once and the number of balls equal to the number obtained on the die are drawn from the bag at random. The probability that all the balls drawn are white is

(a) $\frac{9}{50}$ (b) $\frac{1}{4}$
(c) $\frac{11}{50}$ (d) $\frac{1}{5}$

11 If $\int_0^1 \frac{1}{(5+2x-2x^2)(1+e^{(2-4x)})} dx = \frac{1}{\alpha}\log_e\left(\frac{\alpha+1}{\beta}\right)$, $\alpha, \beta > 0$, then $\alpha^4 - \beta^4$ is equal to

(a) -21 (b) 21 (c) 0 (d) 19

12 Negation of $p \wedge (q \wedge \sim (p \wedge q))$ is

(a) $(\sim (p \wedge q)) \vee p$ (b) $p \vee q$
(c) $\sim (p \vee q)$ (d) $(\sim (p \wedge q)) \wedge q$

13 If the domain of the function $f(x) = \log_e(4x^2 + 11x + 6) + \sin^{-1}(4x+3) + \cos^{-1}\left(\frac{10x+6}{3}\right)$ is $(\alpha, \beta]$, then $36|\alpha + \beta|$ is equal to

(a) 72 (b) 54
(c) 45 (d) 63

14 Let the determinant of a square matrix A of order m be $m - n$, where m and n satisfy $4m + n = 22$ and $17m + 4n = 93$. If $\det(n \operatorname{adj}(\operatorname{adj}(mA))) = 3^a 5^b 6^c$, then $a + b + c$ is equal to

(a) 101 (b) 96
(c) 109 (d) 84

15 If (α, β) is the orthocenter of the ΔABC with vertices $A(3, -7)$, $B(-1, 2)$ and $C(4, 5)$, then $9\alpha - 6\beta + 60$ is equal to

(a) 30 (b) 35
(c) 40 (d) 25

16 Let A_1 and A_2 be two arithmetic means and G_1, G_2, G_3 be three geometric means of two distinct positive numbers. Then, $G_1^4 + G_2^4 + G_3^4 + G_1^2 G_3^2$ is equal to

(a) $(A_1 + A_2)^2 G_1 G_3$ (b) $2(A_1 + A_2) G_1^2 G_3^2$
(c) $(A_1 + A_2) G_1^2 G_3^2$ (d) $2(A_1 + A_2) G_1 G_3$

17 Let $x = x(y)$ be the solution of the differential equation $2(y+2)\log_e(y+2)\, dx + (x + 4 - 2\log_e(y+2))dy = 0$, $y > -1$ with $x(e^4 - 2) = 1$. Then, $x(e^9 - 2)$ is equal to

(a) $\frac{4}{9}$ (b) $\frac{10}{3}$
(c) $\frac{32}{9}$ (d) 3

18 The number of real roots of the equation $x|x| - 5|x+2| + 6 = 0$, is

(a) 4 (b) 3
(c) 6 (d) 5

19 Let S be the set of all values of λ, for which the shortest distance between the lines $\frac{x-\lambda}{0} = \frac{y-3}{4} = \frac{z+6}{1}$ and $\frac{x+\lambda}{3} = \frac{y}{-4} = \frac{z-6}{0}$ is 13. Then, $\left|8\sum_{\lambda \in S} \lambda\right|$ is equal to

(a) 304 (b) 302
(c) 306 (d) 308

20 Let $[x]$ denote the greatest integer function and $f(x) = \max\{1 + x + [x], 2 + x, x + 2[x]\}, 0 \le x \le 2$. Let m be the number of points in $[0, 2]$, where f is not continuous and n be the number of points in $(0, 2)$, where f is not differentiable. Then, $(m+n)^2 + 2$ is equal to

(a) 6 (b) 3
(c) 11 (d) 2

Section B : Numerical Value Type Questions

21 Let the plane P contain the line $2x + y - z - 3 = 0 = 5x - 3y + 4z + 9$ and be parallel to the line $\frac{x+2}{2} = \frac{3-y}{-4} = \frac{z-7}{5}$. Then, the distance of the point $A(8, -1, -19)$ from the plane P measured parallel to the line $\frac{x}{-3} = \frac{y-5}{4} = \frac{2-z}{-12}$ is equal to ....

22 A person forgets his 4-digit ATM pin code. But he remembers that in the code all the digits are different, the greatest digit is 7 and the sum of the first two digits is equal to the sum of the last two digits. Then, the maximum number of trials necessary to obtain the correct code is

23 If the line $x = y = z$ intersects the line $x\sin A + y\sin B + z\sin C - 18 = 0 = x\sin 2A + y\sin 2B + z\sin 2C - 9$, where A, B, C are the angles of a triangle ABC, then $80\left(\sin\frac{A}{2}\sin\frac{B}{2}\sin\frac{C}{2}\right)$ is equal to ....

24 Consider the triangles with vertices $A(2, 1)$, $B(0, 0)$ and $C(t, 4)$, $t \in [0, 4]$. If the maximum and the minimum perimeters of such triangles are obtained at $t = \alpha$ and $t = \beta$, respectively, then $6\alpha + 21\beta$ is equal to ......

25 If the sum of the series $\left(\frac{1}{2} - \frac{1}{3}\right) + \left(\frac{1}{2^2} - \frac{1}{2 \cdot 3} + \frac{1}{3^2}\right) + \left(\frac{1}{2^3} - \frac{1}{2^2 \cdot 3} + \frac{1}{2 \cdot 3^2} - \frac{1}{3^3}\right) + \left(\frac{1}{2^4} - \frac{1}{2^3 \cdot 3} + \frac{1}{2^2 \cdot 3^2} - \frac{1}{2 \cdot 3^3} + \frac{1}{3^4}\right) + \ldots$ is $\frac{\alpha}{\beta}$, where α and β are co-prime, then $\alpha + 3\beta$ is equal to ....

26 If the area bounded by the curve $2y^2 = 3x$, lines $x + y = 3$, $y = 0$ and outside the circle $(x-3)^2 + y^2 = 2$ is A, then $4(\pi + 4A)$ is equal to ....

27 Let an ellipse with centre (1, 0) and latus rectum of length 1/2 have its major axis along X-axis. If its minor axis subtends an angle 60° at the foci, then the square of the sum of the lengths of its minor and major axes is equal to ....

28 Let $A=\{1,2,3,4\}$ and R be a relation on the set $A \times A$ defined by $R=\{((a, b),(c, d)): 2a+3b=4c+5d\}$. Then, the number of elements in R is ....

29 Let $f(x)=\int \frac{dx}{(3+4x^2)\sqrt{4-3x^2}}, |x|<\frac{2}{\sqrt{3}}$. If $f(0)=0$ and $f(1)=\frac{1}{\alpha\beta}\tan^{-1}\left(\frac{\alpha}{\beta}\right), \alpha, \beta>0$, then $\alpha^2+\beta^2$ is equal to ....

30 The number of elements in the set $\{n \in N: 10 \leq n \leq 100$ and 3^n-3 is a multiple of $7\}$ is

Answers

For solutions scan the QR code

6 April, 2023 Shift-I

1. (a)	2. (d)	3. (a)	4. (a)	5. (b)	6. (b)	7. (b)	8. (a)	9. (c)	10. (a)
11. (c)	12. (a)	13. (c)	14. (a)	15. (d)	16. (a)	17. (d)	18. (b)	19. (b)	20. (c)
21. 18	22. 5	23. 594	24. 5005	25. 25	26. 2	27. 121	28. 292	29. 3	30. 3483638676

6 April, 2023 Shift-II

1. (c)	2. (d)	3. (c)	4. (c)	5. (a)	6. (c)	7. (c)	8. (d)	9. (c)	10. (a)
11. (c)	12. (d)	13. (a)	14. (a)	15. (c)	16. (a)	17. (c)	18. (d)	19. (a)	20. (a)
21. 5	22. 25	23. 18	24. 2	25. 4	26. 2	27. 5	28. 0	29. 400	30. 432

8 April, 2023 Shift-I

1. (b)	2. (a)	3. (a)	4. (a)	5. (a)	6. (a)	7. (b)	8. (c)	9. (b)	10. (d)
11. (b)	12. (b)	13. (d)	14. (a)	15. (d)	16. (a)	17. (c)	18. (b)	19. (c)	20. (a)
21. 1275	22. 2	23. 3	24. 31	25. 25	26. 19	27. 11	28. 9	29. 14	30. 5

8 April, 2023 Shift-II

1. (c)	2. (b)	3. (d)	4. (b)	5. (a)	6. (b)	7. (a)	8. (a)	9. (c)	10. (d)
11. (d)	12. (a)	13. (a)	14. (*)	15. (d)	16. (b)	17. (b)	18. (d)	19. (b)	20. (b)
21. 9	22. 12	23. 6	24. 309	25. 180	26. 11	27. 16	28. 20	29. 150	30. 17

10 April, 2023 Shift-I

1. (c)	2. (b)	3. (b)	4. (b)	5. (a)	6. (c)	7. (c)	8. (c)	9. (c)	10. (a)
11. (c)	12. (b)	13. (b)	14. (b)	15. (c)	16. (d)	17. (c)	18. (a)	19. (c)	20. (b)
21. 8	22. 4	23. 151	24. 4898	25. 32	26. 16	27. 960	28. 9525	29. 6	30. 16

10 April, 2023 Shift-II

1. (d)	2. (c)	3. (b)	4. (d)	5. (c)	6. (a)	7. (b)	8. (c)	9. (d)	10. (b)
11. (a)	12. (d)	13. (b)	14. (b)	15. (b)	16. (d)	17. (a)	18. (c)	19. (d)	20. (a)
21. 6	22. 27	23. 529	24. 16	25. 11	26. 26664	27. 24	28. 24	29. 7	30. (log 3 + 4)

11 April, 2023 Shift-I

1. (d)	2. (b)	3. (d)	4. (b)	5. (b)	6. (b)	7. (b)	8. (b)	9. (a)	10. (c)
11. (c)	12. (a)	13. (d)	14. (c)	15. (c)	16. (a)	17. (a)	18. (c)	19. (d)	20. (a)
21. 306	22. 32	23. 5	24. 171	25. 44	26. 2	27. 51	28. 2736	29. 7	30. 2175

For solutions scan the QR code

11 April, 2023 Shift-II

1. (*)	2. (d)	3. (a)	4. (c)	5. (d)	6. (a)	7. (a)	8. (d)	9. (c)	10. (b)
11. (b)	12. (d)	13. (c)	14. (c)	15. (a)	16. (b)	17. (b)	18. (a)	19. (a)	20. (d)
21. 2	22. 2	23. 27	24. 360	25. 348	26. 11	27. 116	28. 16	29. 285	30. 1680

12 April, 2023 Shift-I

1. (b)	2. (c)	3. (c)	4. (d)	5. (d)	6. (d)	7. (b)	8. (c)	9. (b)	10. (c)
11. (c)	12. (a)	13. (a)	14. (b)	15. (b)	16. (a)	17. (c)	18. (c)	19. (a)	20. (a)
21. 575	22. 288	23. 7	24. 64	25. 4	26. 1260	27. 211	28. (10)	29. 6	30. (2)

13 April, 2023 Shift-I

1. (a)	2. (d)	3. (b)	4. (a)	5. (d)	6. (a)	7. (b)	8. (a)	9. (c)	10. (a)
11. (c)	12. (b)	13. (c)	14. (c)	15. (c)	16. (c)	17. (b)	18. (b)	19. (b)	20. (c)
21. 1310	22. 8	23. 66	24. 413	25. 36	26. 24	27. 15	28. 18	29. 8	30. 4

13 April, 2023 Shift-II

1. (c)	2. (b)	3. (c)	4. (a)	5. (c)	6. (d)	7. (a)	8. (b)	9. (c)	10. (c)
11. (d)	12. (d)	13. (b)	14. (a)	15. (a)	16. (b)	17. (b)	18. (a)	19. (d)	20. (b)
21. 41	22. 825	23. 10	24. 6	25. 269	26. 16	27. 12	28. 2	29. 7	30. 12